Abbreviations for Units

A	ampere	h	hour	N	newton
Å	angstrom (10^{-10} m)	Hz	hertz	nm	nanometer (10^{-9} m)
atm	atmosphere	in	inch	Pa	pascal
BTU	British thermal unit	J	joule	rad	radians
Bq	becquerel	K	kelvin	rev	revolution
C	coulomb	kg	kilogram	R	roentgen
°C	degree Celsius	km	kilometer	Sv	sievert
cal	calorie	keV	kilo-electron volt	s	second
Ci	curie	lb	pound	T	tesla
cm	centimeter	L	liter	u	unified mass unit
eV	electron volt	m	meter	V	volt
°F	degree Fahrenheit	MeV	mega-electron volt	W	watt
fm	femtometer, fermi (10^{-15} m)	mi	mile	Wb	weber
ft	foot	min	minute	y	year
G	gauss	mm	millimeter	μm	micrometer (10^{-6} m)
Gy	gray	mmHg	millimeters of mercury	μs	microsecond
g	gram	mol	mole	μC	microcoulomb
H	henry	ms	millisecond	Ω	ohm

Some Conversion Factors

Length

1 m = 39.37 in = 3.281 ft = 1.094 yard
1 m = 10^{15} fm = 10^{10} Å = 10^{9} nm
1 km = 0.6214 mi
1 mi = 5280 ft = 1.609 km
1 light-year = 1 $c \cdot$ y = 9.461×10^{15} m
1 in = 2.540 cm

Volume

1 L = 10^3 cm^3 = 10^{-3} m^3 = 1.057 qt

Time

1 h = 3600 s = 3.6 ks
1 y = 365.24 day = 3.156×10^7 s

Speed

1 km/h = 0.278 m/s = 0.6214 mi/h
1 ft/s = 0.3048 m/s = 0.6818 mi/h

Angle–angular speed

1 rev = 2π rad = 360°
1 rad = 57.30°
1 rev/min (rpm) = 0.1047 rad/s

Force–pressure

1 N = 10^5 dyn = 0.2248 lb
1 lb = 4.448 N
1 atm = 101.3 kPa = 1.013 bar = 760 mmHg = 14.70 lb/in^2

Mass

1 u = $[(10^{-3}\ \text{mol}^{-1})/N_A]$ kg = 1.661×10^{-27} kg
1 tonne = 10^3 kg = 1 Mg
1 kg = 2.205 lb

Energy–power

1 J = 10^7 erg = 0.7376 ft · lb = 9.869×10^{-3} L · atm
1 kW · h = 3.6 MJ
1 cal = 4.186 J
1 L · atm = 101.325 J = 24.22 cal
1 eV = 1.602×10^{-19} J
1 BTU = 778 ft · lb = 252 cal = 1054 J
1 horsepower = 550 ft · lb/s = 746 W

Thermal conductivity

1 W/(m · K) = 6.938 BTU · in/(h · ft^2 · °F)

Magnetic field

1 T = 10^4 G

Viscosity

1 Pa · s = 10 poise

College PHYSICS

Volume I

SECOND EDITION

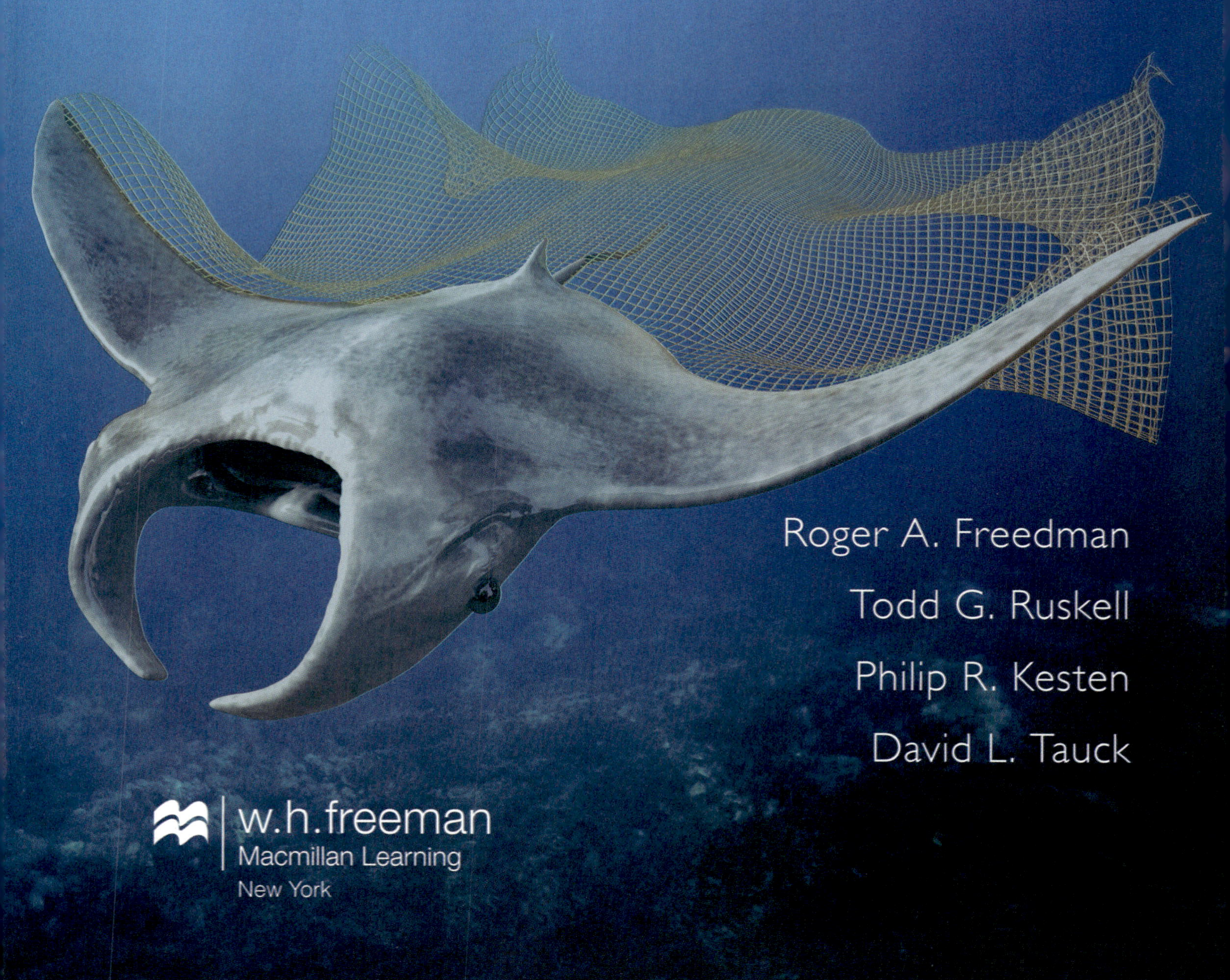

Roger A. Freedman
Todd G. Ruskell
Philip R. Kesten
David L. Tauck

w.h.freeman
Macmillan Learning
New York

Vice President, STEM: *Ben Roberts*
Editorial Program Director: *Brooke Suchomel*
Program Manager: *Lori Stover*
Senior Development Editor: *Blythe Robbins*
Development Editor: *Meg Rosenburg*
Assistant Editor: *Kevin Davidson*
Marketing Manager: *Maureen Rachford*
Marketing Assistant: *Savannah DiMarco*
Content Director: *Natania Mlawer*
Media Editor: *Victoria Garvey*
Media Project Manager: *Daniel Comstock*
Associate Digital Marketing Specialist: *Cate McCaffery*
Physics Content Team Lead: *Josh Hebert*
Director, Content Management Enhancement: *Tracey Kuehn*
Managing Editor: *Lisa Kinne*
Senior Content Project Manager: *Kerry O'Shaughnessy*
Senior Workflow Project Manager: *Paul Rohloff*
Permissions Manager: *Jennifer MacMillan*
Photo Researcher: *Krystyna Borgen*
Director of Design, Content Management: *Diana Blume*
Interior Design: *Vicki Tomaselli*
Cover Design: *Lumina Datamatics, Inc.*
Art Manager: *Matthew McAdams*
Illustrations: *Precision Graphics*
Composition: *Lumina Datamatics, Inc.*
Printing and Binding: *LSC Communications*
Cover Art: *Quade Paul; Frank E. Fish*

Library of Congress Control Number: 2017955775

Volume I:
ISBN-13: 978-1-319-11510-4
ISBN-10: 1-319-11510-1

Printed in the United States of America
First printing

W. H. Freeman and Company
One New York Plaza
Suite 4500
New York, NY 10004-1562
www.macmillanlearning.com

ROGER:
*To the memory of S/Sgt Ann Kazmierczak Freedman,
WAC, and Pvt. Richard Freedman, AUS.*

TODD:
*To Susan and Allison, whose never-ending patience,
love, and support made it possible.
And to my parents, from whom I learned so much—
especially my father, who so effectively demonstrates
what it means to be an effective teacher
both in and out of the classroom.*

PHIL:
*To my parents for instilling in me a love of learning,
to my wife for her unconditional support,
and to my children for letting their kooky dad
infuse so much of their lives with science.*

DAVE:
*To my parents, Bill and Jean, for showing me
how to lead a wonderful life, and to my sister and friends,
teachers and students for helping me do it.*

Preface

MOVE LEARNING FORWARD...

Bridge conceptual understanding and problem solving with resources to support active teaching and learning.

This new integrated learning system brings together a groundbreaking media program with an innovative text presentation of algebra-based physics. An experienced author team provides a unique set of expertise and perspectives to help students master concepts and succeed in developing problem-solving skills necessary for College Physics. Now available for the first time with SaplingPlus—an online learning platform that combines the heavily research-based FlipItPhysics Prelectures (derived from smartPhysics) with the robust Sapling homework system, in which every problem has targeted feedback, hints, and a fully worked and explained solution. This HTML5 platform gives students the ability to actively read with a fully interactive e-Book, watch Prelecture videos, and work or review problems with a mobile-accessible learning experience. Integration is available with Learning Management Systems to provide single sign-on and grade-sync capabilities compatible with the iClicker 2 and other classroom response systems to provide a seamless full course experience for you and your students.

Christine Pang

Roger Freedman This groundbreaking text boasts an exceptionally strong writing team that is uniquely qualified to write a college physics textbook. The *College Physics* author team is led by Roger Freedman, an accomplished textbook author of such best-selling titles as *Universe* (W. H. Freeman), *Investigating Astronomy* (W. H. Freeman), and *University Physics* (Pearson). Dr. Freedman is a lecturer in physics at the University of California, Santa Barbara. He was an undergraduate at the University of California campuses in San Diego and Los Angeles, and did his doctoral research in theoretical nuclear physics at Stanford University. He came to UCSB in 1981 after 3 years of teaching and doing research at the University of Washington. At UCSB, Dr. Freedman has taught in both the Department of physics and the College of Creative Studies, a branch of the university intended for highly gifted and motivated undergraduates. In recent years, he has helped to develop computer-based tools for learning introductory physics and astronomy and has been a pioneer in the use of classroom response systems and the "flipped" classroom model at UCSB. Roger holds a commercial pilot's license and was an early organizer of the San Diego Comic-Con, now the world's largest popular culture convention.

Mark Ramirez

Todd Ruskell As a Teaching Professor of Physics at the Colorado School of Mines, Todd Ruskell focuses on teaching at the introductory level and continually develops more effective ways to help students learn. One method used in large enrollment introductory courses is Studio Physics. This collaborative, hands-on environment helps students develop better intuition about, and conceptual models of, physical phenomena through an active learning approach. Dr. Ruskell brings his experience in improving students' conceptual understanding to the text, as well as a strong liberal arts perspective. Dr. Ruskell's love of physics began with a BA in physics from Lawrence University in Appleton, Wisconsin. He went on to receive an MS and a PhD in optical sciences from the University of Arizona. He has received awards for teaching excellence, including Colorado School of Mines' Alumni Teaching Award. Dr. Ruskell currently serves on the physics panel and advisory board for the NANSLO (North American Network of Science Labs Online) project.

Build a conceptual foundation...before class

We place a high value on learning physics concepts prior to class time instruction and have created resources to make this as effective as possible for student learning and easy to implement for instructors. Groundbreaking Prelecture videos introduce students to physics topics and concepts as well as reinforce understanding with embedded questions ahead of class. In tandem, *College Physics*, Second Edition, provides an exceptional narrative and purposeful pedagogical tools focused on moving both conceptual learning and quantitative skill acquisition prior to class. A unique visual program and seamless blend of biological applications are interwoven throughout the book to provide relevance and interest for students taking algebra-based physics. By providing Prelectures with reading assignments, our goal is to jump-start student learning and allow for more productive class time.

Make classroom engagement meaningful

Instructor-student engagement can become more meaningful in class when students are aware of misconceptions and instructors have better insight into what students know before coming to class. Bridge questions (developed from research-based smartPhysics) provide a vehicle for students to demonstrate and communicate how well they understand the material that they learned before class. This invaluable instructor insight provides you with a way of identifying gaps in understanding and student misconceptions as you develop lectures to make the most efficient and meaningful use of class time. To further support an engaged classroom, Roger Freedman has refined an active classroom approach and shares in-class activities he has written to apply conceptual knowledge and engage students in the process of problem solving in class.

Develop problem-solving skills

Every effort in developing the print and digital materials for *College Physics*, Second Edition has been made to encourage students to develop a deep understanding of physics concepts and foster the reasoning and analytical skills necessary to solve problems. This goal motivated the student-centered pedagogy demonstrated in the worked examples and consistent Set Up—Solve It—Reflect problem-solving strategy found in *College Physics*, as well as the design and development of the Prelecture videos. Text problems incorporate conceptual questions, basic concepts, synthesis of multiple concepts, and life science applications. Paired with the Sapling homework platform, students are provided a tutorial experience with every problem in the system. Sapling adheres to the philosophy that every problem counts—therefore requiring that ALL problems have hints, targeted feedback, and a detailed solution—thus ensuring that students learn how to approach a problem, not just whether they answered correctly or incorrectly.

Philip Kesten

Philip Kesten, Associate Professor of Physics and Associate Vice Provost for Undergraduate Studies at Santa Clara University, holds a BS in physics from the Massachusetts Institute of Technology and received his PhD in high-energy particle physics from the University of Michigan. Since joining the Santa Clara faculty in 1990, Dr. Kesten has also served as Chair of Physics, Faculty Director of the ATOM and da Vinci Residential Learning Communities, and Director of the Ricard Memorial Observatory. He has received awards for teaching excellence and curriculum innovation, was Santa Clara's Faculty Development Professor for 2004–2005, and was named the California Professor of the Year in 2005 by the Carnegie Foundation for the Advancement of Education. Dr. Kesten is co-founder of Docutek (a SirsiDynix Company), an Internet software company, and has served as the Senior Editor for *Modern Dad,* a newsstand magazine.

Philip Kesten

David L. Tauck Unlike any other physics text on the market, *College Physics* includes a physiologist as a primary author. David Tauck, Associate Professor of Biology, holds both a BA in biology and an MA in Spanish from Middlebury College. He earned his PhD in physiology at Duke University and completed postdoctoral fellowships at Stanford University and Harvard University in anesthesia and neuroscience, respectively. Since joining the Santa Clara University faculty in 1987, he has served as Chair of the Biology Department, the College Committee on Rank and Tenure, and the Institutional Animal Care and Use Committee; he has also served as president of the local chapter of Phi Beta Kappa. Dr. Tauck currently serves as the Faculty Director in Residence of the da Vinci Residential Learning Community.

BUILDING A CONCEPTUAL FOUNDATION... BEFORE CLASS

Jump-start student learning with Prelectures, and free up class time for difficult concepts and problem-solving work.

Prelecture videos

Animated, narrated videos introduce core physics topics, laying the groundwork for conceptual understanding before students ever set foot in class. Each video is about 1–3 minutes long and is interspersed with conceptual questions. Each series can either be assigned in its entirety or divided into smaller, more tightly focused assignments. The full Prelecture activity is about 15 minutes long.

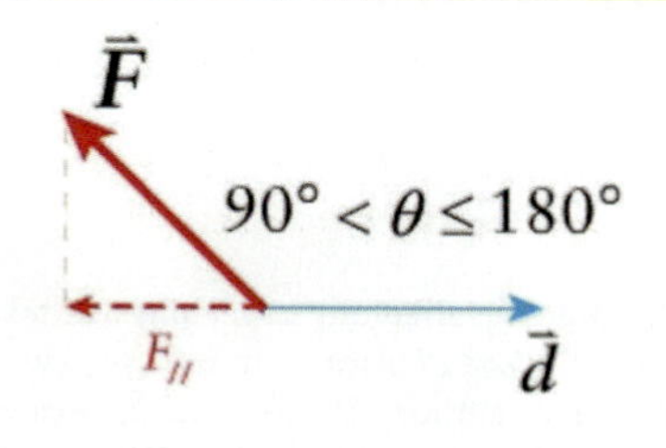

Work $W = F d \cos\theta$ < 0

$W < 0$

Angle between $\vec{F}$ and $\vec{d}$	Value of Work	Effect on Speed
$\theta = 90°$	$W = 0$	
$0° \leq \theta < 90°$	$W > 0$	will increase

Embedded questions reinforce student understanding along the way.

A box is pulled a distance d across the floor by a force F that makes an angle θ with the horizontal, as shown in the figure.

If the magnitude of the force was kept constant but the angle θ was increased toward 90°, the work done by the force in dragging the box would

- ○ either increase or decrease, depending on what the initial angle θ was.
- ○ increase.
- ○ decrease.
- ○ remain the same.
- ○ either increase or decrease, depending on the magnitude of the force F.

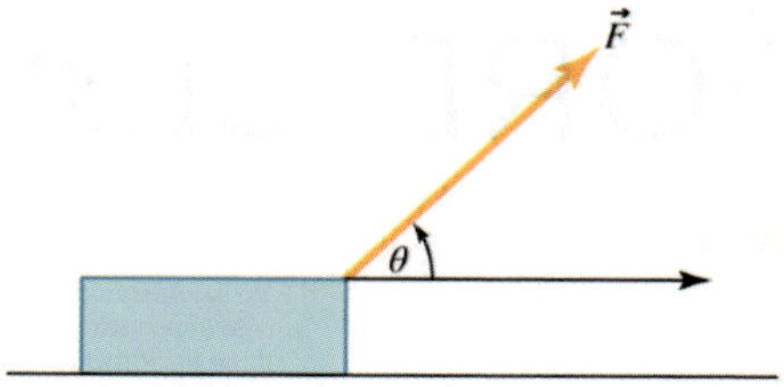

Three objects of the same mass begin their motion at the same height. One object falls straight down, one slides down a low-friction inclined plane, and one swings in a circular arc on the end of a string. All three objects end at the same height.

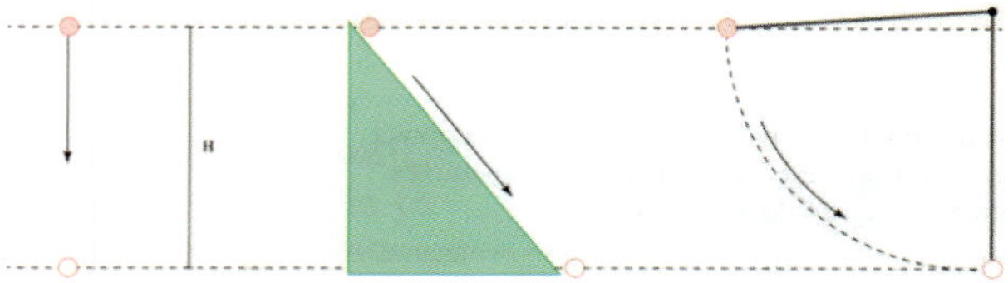

On which object does gravity do the most work?

- ○ Gravity does equal work on all three objects.
- ○ the object traversing the circular arc
- ○ the object sliding down the low-friction incline
- ○ the object in free fall

Briefly explain your choice.

Bridge questions

Multiple-choice and free-response questions review the content covered in the Prelectures and serve as a unique way for students to both demonstrate what they have learned and communicate misunderstandings or questions to an instructor prior to lecture.

A box sits on the horizontal bed of a truck that is accelerating to the left, as shown in the diagram. Static friction between the box and the truck keeps the box from sliding around as the truck accelerates.

The work done on the box by the static friction force as the accelerating truck moves a distance D to the left is

- ○ positive.
- ○ negative.
- ○ zero.
- ○ dependent upon the speed of the truck.

Briefly explain your answer choice.

MAKE CLASSROOM ENGAGEMENT MEANINGFUL

Refine conceptual understanding in class with meaningful instructor–student engagement.

Bridge questions connect the Prelecture activity to the classroom experience, providing instructors with valuable insight into student understanding. With a better-prepared student audience, instructors can devote time to topics needing further explanation, or they can build on the knowledge students acquire before coming to class.

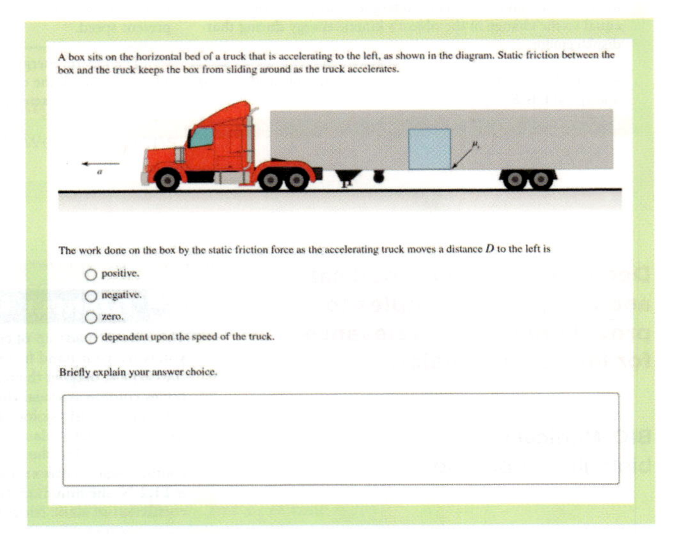
A box sits on the horizontal bed of a truck that is accelerating to the left, as shown in the diagram. Static friction between the box and the truck keeps the box from sliding around as the truck accelerates.

The work done on the box by the static friction force as the accelerating truck moves a distance D to the left is

- positive.
- negative.
- zero.
- dependent upon the speed of the truck.

Briefly explain your answer choice.

Activity 6-15. Energy changes II [Accompanies Sections 6-7 and 6-8]

A block is initially at point A at the bottom of an incline and is moving upward. *There is kinetic friction* between the block and the incline.

The block reaches a maximum height H at point C. Point B is halfway between points A and C. At point A, the gravitational potential energy is zero.

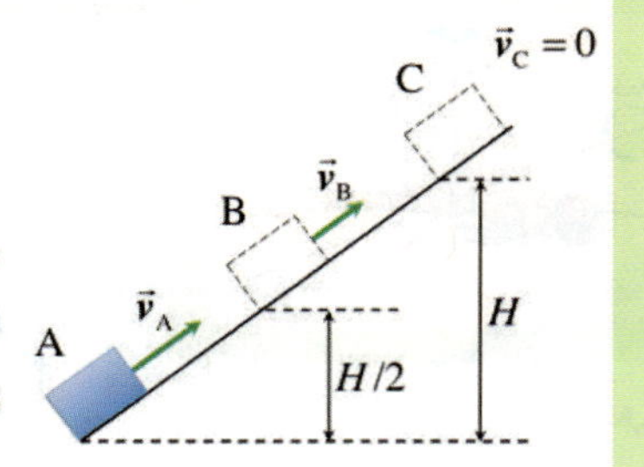

(a) Is mechanical energy conserved in this motion? *Explain* your reasoning.

In-class activities from Roger Freedman provide tough problems to tackle in class with annotations for instructors on learning objectives and student misconceptions. The activities can be used with iClicker Reef.

Integrated with iClicker
iClicker active learning simplified

In-class activities with iClicker

iClicker offers simple, flexible tools to help you give students a voice and facilitate active learning in the classroom. Students can participate with the devices they already bring to class using our iClicker Reef mobile app (which works with smartphones, tablets, or laptops) or iClicker remotes. We've now integrated iClicker with Macmillan's Sapling, and Roger Freedman has authored in-class activities that can be used with iClicker Reef. With Freedman, *College Physics*, Second Edition, and iClicker, it's easier than ever to promote engagement and synchronize student grades both in the classroom and at home.

Work 4

A box sits on the horizontal bed of a moving truck. Static friction between the box and the truck keeps the box from sliding around as the truck drives.

The work done on the box by the static frictional force as the truck moves a distance D is:

A. Positive
B. Negative
C. Zero

Work 4 Explanation

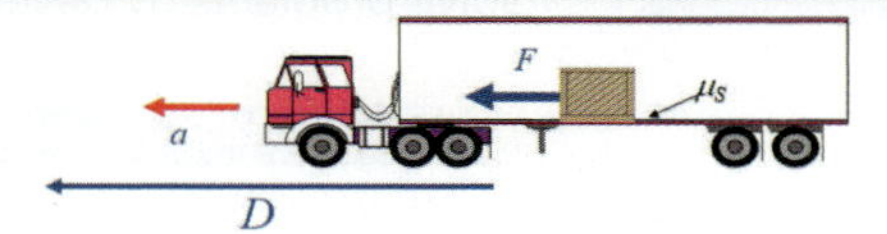

The work done on the box by the static frictional force as the truck moves a distance D is:

A) Positive B) Zero C) Negative

A. In the one example with the box it showed that if the change in x and force were in the same direction, the work would be positive.

B. It is zero because there is no movement.

: is negative because it is working in a negative direction.

iClicker Reef access cards can also be packaged with Sapling or your Macmillan textbook at a significant savings for your students. To learn more, visit **iclicker.com** or talk to your Macmillan Learning representative.

DEVELOP **PROBLEM-SOLVING SKILLS**

Encourage students to develop a deep understanding of physics concepts and foster the reasoning and analytical skills necessary to apply a consistent problem-solving strategy and a student-centered pedagogical framework.

Set Up. The first step in each problem is to determine an overall approach and to gather the necessary pieces of information needed to solve it. These might include sketches, equations related to the physics, and concepts.

Solve. Rather than simply summarizing the mathematical manipulations required to move from first principles to the final answer, the authors show many intermediate steps in working out solutions to the sample problems, highlighting a crucial part of the problem-solving process.

Reflect. An important part of the process of solving a problem is to reflect on the meaning, implications, and validity of the answer. Is it physically reasonable? Do the units make sense? Is there a deeper or wider understanding that can be drawn from the result? The authors address these and related questions when appropriate.

Encourage strategic thinking

Set Up—Solve—Reflect problem-solving strategy

Worked examples mirror the approach scientists take to solve problems by developing reasoning and analysis skills.

BIO-Medical **EXAMPLE 5-7 Terminal Speed**

When diving straight down toward its prey, a peregrine falcon is acted on by two forces: a downward gravitational force, and a drag force of magnitude F_{drag} given by Equation 5-9 directed vertically upward (opposite to the direction of the falcon's motion through the air). As the falcon falls and its speed v increases, the value of F_{drag} also increases. When the drag force becomes equal in magnitude to the gravitational force, the net force on the falcon is zero and the falcon ceases to accelerate. It has reached its *terminal speed*, so it no longer speeds up nor does it slow down. Find the terminal speed of a female peregrine falcon of mass 1.2 kg, for which the value of c is $1.6 \times 10^{-3}\ \text{N}\cdot\text{s}^2/\text{m}^2$.

Set Up

The free-body diagram shows the two forces acting on the falcon. We use Equation 5-9 to find the value of the speed v at which the sum of these forces is zero, so that the acceleration is zero and the downward velocity is constant.

$$\sum \vec{F}_{\text{ext on falcon}} = \vec{F}_{\text{drag on falcon}} + \vec{w}_{\text{falcon}} = m\vec{a}_{\text{falcon}} = 0$$

Drag force for larger objects at faster speeds:

$$F_{drag} = cv^2 \qquad (5\text{-}9)$$

$\vec{F}_{\text{drag on falcon}}$

y x

v = terminal speed

$\vec{w}_{\text{falcon}}$

Illustration based on photo by Ted Swem

Solve

Write Newton's second law in component form and solve for the terminal speed v_{term}.

Newton's second law in component form applied to the falcon:

$$y\text{: } F_{\text{drag on falcon}} + (-w_{\text{falcon}}) = 0$$

At the terminal speed v_{term},

$$F_{\text{drag on falcon}} = cv_{term}^2 \text{ so}$$

$$cv_{term}^2 - w_{falcon} = 0$$

$$cv_{term}^2 = w_{falcon} = mg$$

$$v_{term}^2 = \frac{mg}{c}$$

$$v_{term} = \sqrt{\frac{mg}{c}}$$

Substitute the numerical values of m and c for the falcon.

Using $m = 1.2$ kg and $c = 1.6 \times 10^{-3}\ \text{N}\cdot\text{s}^2/\text{m}^2$,

$$v_{term} = \sqrt{\frac{(1.2\ \text{kg})(9.80\ \text{m/s}^2)}{1.6 \times 10^{-3}\ \text{N}\cdot\text{s}^2/\text{m}^2}} = 86\ \text{m/s} = 310\ \text{km/h} = 190\ \text{mi/h}$$

Reflect

The high diving speeds attained by a peregrine falcon make it the fastest member of the animal kingdom.

The relationship $v_{term} = \sqrt{\frac{mg}{c}}$ explains the common notion that "heavier objects fall faster." A baseball and an iron ball of the same radius falling side by side in air have the same value of c (because they have the same shape and size), but the iron ball will have a greater terminal speed because its mass m is greater. So a heavier object *does* fall faster if we take the drag force into account. If the baseball and iron ball were dropped side by side in a vacuum, however, they would have the same acceleration of magnitude g and so would always have the same speed.

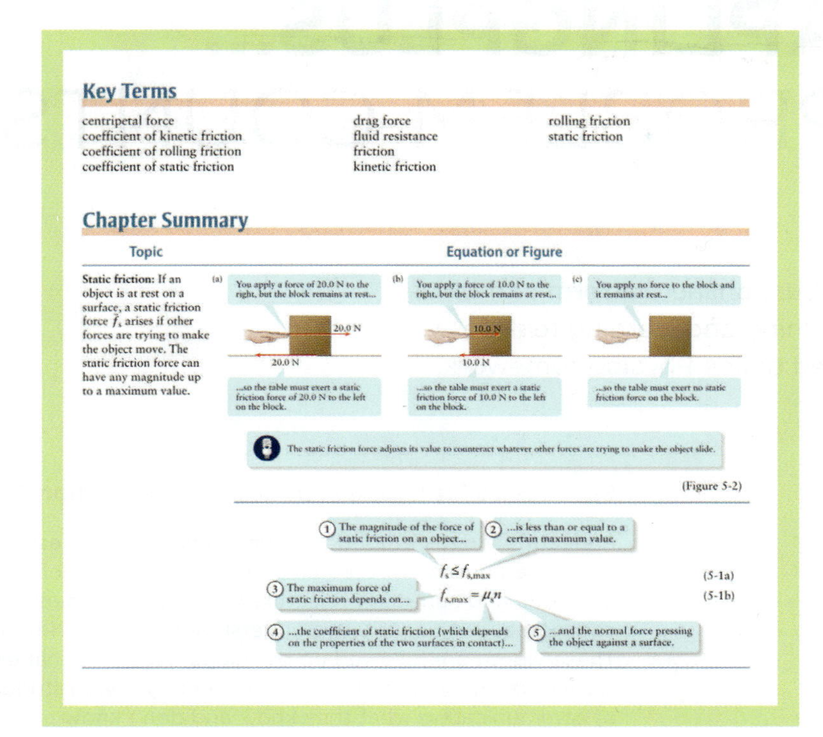

Key Terms

centripetal force
coefficient of kinetic friction
coefficient of rolling friction
coefficient of static friction
drag force
fluid resistance
friction
kinetic friction
rolling friction
static friction

Chapter Summary

Topic	Equation or Figure
Static friction: If an object is at rest on a surface, a static friction force $\vec{f}_s$ arises if other forces are trying to make the object move. The static friction force can have any magnitude up to a maximum value.	(a) You apply a force of 20.0 N to the right, but the block remains at rest... ...so the table must exert a static friction force of 20.0 N to the left on the block. (b) You apply a force of 10.0 N to the right, but the block remains at rest... ...so the table must exert a static friction force of 10.0 N to the left on the block. (c) You apply no force to the block and it remains at rest... ...so the table must exert no static friction force on the block. The static friction force adjusts its value to counteract whatever other forces are trying to make the object slide. (Figure 5-2)
	(1) The magnitude of the force of static friction on an object... (2) ...is less than or equal to a certain maximum value. $f_s \le f_{s,\max}$ (5-1a) (3) The maximum force of static friction depends on... $f_{s,\max} = \mu_s n$ (5-1b) (4) ...the coefficient of static friction (which depends on the properties of the two surfaces in contact)... (5) ...and the normal force pressing the object against a surface.

Student-centered framework

Key terms/visual summary
Help students synthesize ideas after reading or class time.

End-of-chapter questions tagged with related worked examples. Students can go back and review worked examples while working problems.

47. •• An object of mass $3M$, moving in the $+x$ direction at speed v_0, breaks into two pieces of mass M and $2M$ as shown in **Figure 7-26**. If $\theta_1 = 45°$ and $\theta_2 = 30°$, determine the final velocities of the resulting pieces in terms of v_0. SSM Example 7-4

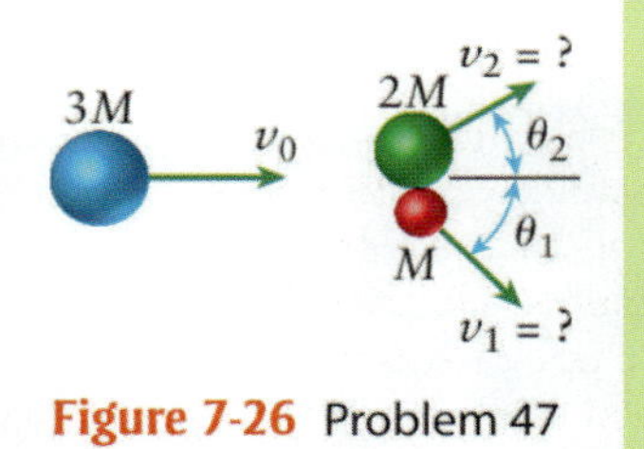

Figure 7-26 Problem 47

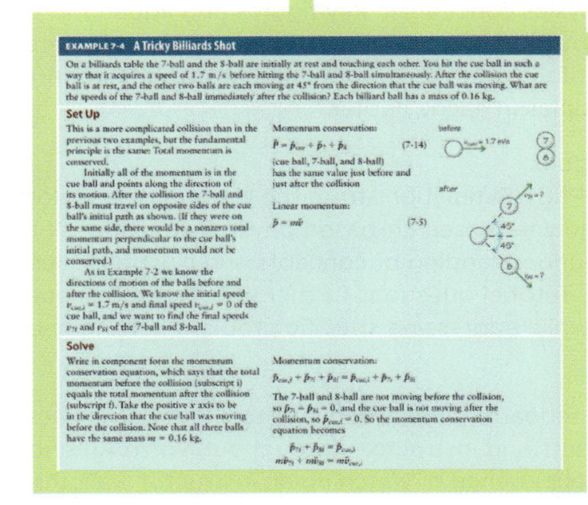

EXAMPLE 7-4 A Tricky Billiards Shot

On a billiards table the 7-ball and the 8-ball are initially at rest and touching each other. You hit the cue ball in such a way that it acquires a speed of 1.7 m/s before hitting the 7-ball and 8-ball simultaneously. After the collision the cue ball is at rest, and the other two balls are each moving at 45° from the direction that the cue ball was moving. What are the speeds of the 7-ball and 8-ball immediately after the collision? Each billiard ball has a mass of 0.16 kg.

Set Up

This is a more complicated collision than in the previous two examples, but the fundamental principle is the same: Total momentum is conserved.

Initially all of the momentum is in the cue ball and points along the direction of its motion. After the collision the 7-ball and 8-ball must travel on opposite sides of the cue ball's initial path as shown. (If they were on the same side, there would be a nonzero total momentum perpendicular to the cue ball's initial path, and momentum would not be conserved.)

As in Example 7-2 we know the directions of motion of the balls before and after the collision. We know the initial speed $v_{cue,i} = 1.7$ m/s and final speed $v_{cue,f} = 0$ of the cue ball, and we want to find the final speeds v_{7f} and v_{8f} of the 7-ball and 8-ball.

Momentum conservation:

$\vec{P} = \vec{p}_{cue} + \vec{p}_7 + \vec{p}_8$ (7-14)

(cue ball, 7-ball, and 8-ball) has the same value just before and just after the collision

Linear momentum:

$\vec{p} = m\vec{v}$ (7-5)

Solve

Write in component form the momentum conservation equation, which says that the total momentum before the collision (subscript i) equals the total momentum after the collision (subscript f). Take the positive x axis to be in the direction that the cue ball was moving before the collision. Note that all three balls have the same mass $m = 0.16$ kg.

Momentum conservation:

$\vec{p}_{cue,f} + \vec{p}_{7f} + \vec{p}_{8f} = \vec{p}_{cue,i} + \vec{p}_{7i} + \vec{p}_{8i}$

The 7-ball and 8-ball are not moving before the collision, so $\vec{p}_{7i} = \vec{p}_{8i} = 0$, and the cue ball is not moving after the collision, so $\vec{p}_{cue,f} = 0$. So the momentum conservation equation becomes

$\vec{p}_{7f} + \vec{p}_{8f} = \vec{p}_{cue,i}$
$m\vec{v}_{7f} + m\vec{v}_{8f} = m\vec{v}_{cue,i}$

Categorized end-of-chapter problems

- Conceptual questions
- Multiple-choice questions
- Estimation/numerical analysis
- Problems (organized by section)
- General problems

WITH SAPLINGPLUS... EVERY PROBLEM COUNTS

This comprehensive and robust online platform combines innovative, high-quality teaching and learning features with Sapling Learning's acclaimed online Physics homework.

SaplingPlus for *College Physics,* Second Edition, features:

Prelecture videos/embedded questions/bridge questions
Developed based on research and principles that defined smartPhysics, animated and narrated Prelecture videos give students both a conceptual and quantitative understanding of core Physics topics. These videos are followed up by *bridge questions* that extend student learning to in-class engagement by giving students a means of communicating what they know and don't know and providing instructors with access to valuable insights to tailor class time.

Interactive e-Book
New! For the first time, the e-Book is also available through an app that allows students to read offline and have the book read aloud to them, in addition to the highlighting, note taking, and keyword search that you have come to expect.

Sapling Learning problems
Where every problem counts—with hints, targeted feedback, and detailed solutions.

Balloon art concept checks
Designed to guide students through the process of identifying important physics concepts in key figures and equations. Mirroring the visual narrative in the form of word balloons, these interactive questions reinforce key ideas from the text by highlighting important physics principles in each chapter.

PhET simulations
New HTML5 PhET Simulations from the University of Colorado at Boulder's renowned research-based physics simulations help students gain a visual understanding of concepts and illustrate cause-and-effect relationships. Tutorial questions further encourage this quantitative exploration, while addressing specific problem-solving needs.

P'Casts
250 total whiteboard mobile-ready videos. Carefully selected by physics students and instructors throughout North America to help simulate the experience of watching an instructor walk through the steps and explanation of Physics concepts while solving a problem.

Pocket worked examples
All worked examples from *College Physics* are available as a downloadable item for mobile devices.

ANATOMY OF A SAPLING PROBLEM

Hints

Clues attached to every problem encourage critical thinking by providing suggestions for completing the problem, without giving away the answer.

Hint

The potential energy in this scenario is *gravitational potential energy*, which depends on the height of the car above some reference point. In this case, take the bottom of the hill to be the reference point, so the potential energy at the bottom of the hill is zero.

For this question, use $g = 10$ m/s^2 for the acceleration due to gravity.

Since the car is assumed to coast freely, without friction, you can make use of the principle of *energy conservation*.

A 2 kg toy car sits at the highest point of a 13 m high hill. The car is gently pushed forward until it begins to roll down the slope. Assuming the car coasts freely, without any friction or air resistance, how much kinetic energy (KE) and potential energy (PE) will it have at each of the indicated points? Complete the diagram by placing the correct label in each bin. Use $g = 10$ m/s^2 for the acceleration due to gravity. Note, the diagram is not drawn to scale.

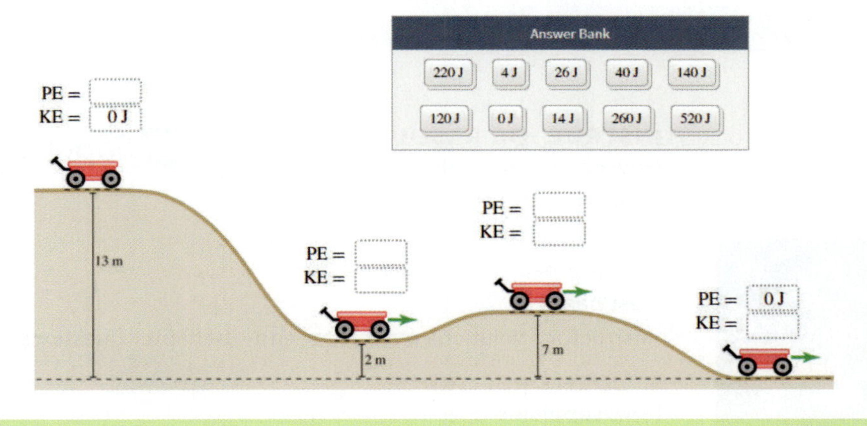

Targeted feedback

Each question includes wrong-answer specific feedback targeted to students' misconceptions.

Feedback

Remember that we are taking the bottom of the hill as our reference level. The potential energy at each point depends on the height h of that point relative to the bottom of the hill: PE $= mgh$.

A 3 kg toy car sits at the highest point of a 13 m high hill. The car is gently pushed forward until it begins to roll down the slope. Assuming the car coasts freely, without any friction or air resistance, how much kinetic energy (KE) and potential energy (PE) will it have at each of the indicated points? Complete the diagram by placing the correct label in each bin. Use $g = 10$ m/s^2 for the acceleration due to gravity. Note, the diagram is not drawn to scale.

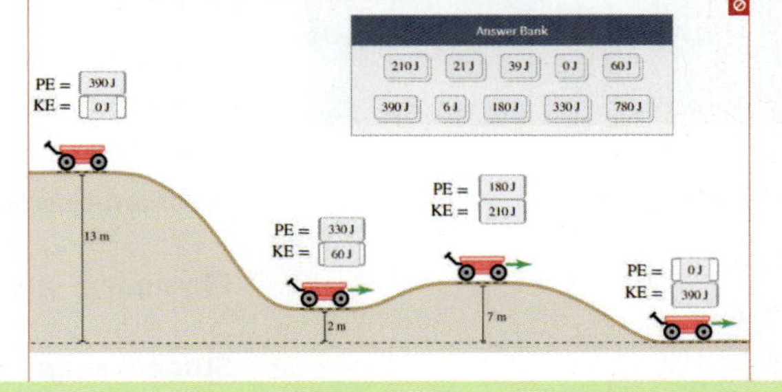

Detailed solutions

Fully worked solutions reinforce concepts and provide an in-product study guide for every problem in the Sapling Learning system.

Solution

The potential energy in this scenario is *gravitational potential energy*, which reference point. In this case, we take the bottom of the hill to be our reference At some height h above the bottom, the gravitational potential energy PE is g

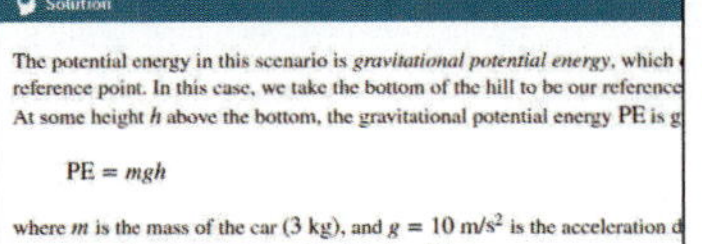

In the valley, at the point where $h = 2$ m, the potential energy is

$$\text{PE}_2 = (3 \text{ kg}) \times (10 \text{ m/s}^2) \times (2 \text{ m}) = 60 \text{ J}$$

Similarly, the potential energy at the top of the smaller hill can be found

$$\text{PE}_3 = 210 \text{ J}$$

A 3 kg toy car sits at the highest point of a 13 m high hill. The car is gently pushed forward until it begins to roll down the slope. Assuming the car coasts freely, without any friction or air resistance, how much kinetic energy (KE) and potential energy (PE) will it have at each of the indicated points? Complete the diagram by placing the correct label in each bin. Use $g = 10$ m/s^2 for the acceleration due to gravity. Note, the diagram is not drawn to scale.

The potential energy in this scenario is *gravitational potential energy*, which depends on the height of the car above some reference point. In this case, we take the bottom of the hill to be our reference point, so the potential energy at that point is zero. At some height h above the bottom, the gravitational potential energy PE is given by

$$\text{PE} = mgh$$

where m is the mass of the car (3 kg), and $g = 10$ m/s^2 is the acceleration due to gravity. We can now calculate the potential energy at each point. At the top of the hill, $h = 13$ m, so

RESOURCES TO SUPPORT *COLLEGE PHYSICS*, SECOND EDITION, AND SAPLINGPLUS

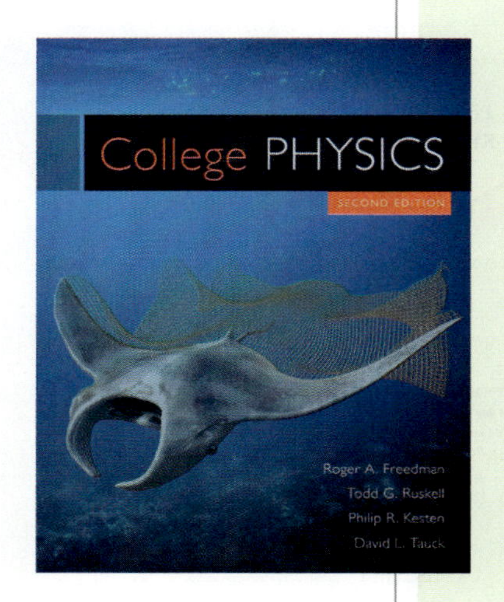

SaplingPlus

INSTRUCTOR RESOURCES

Housed in SaplingPlus

- iClicker questions
- Lecture slides
- Roger Freedman's in-class activities
- Test bank
- Instructor's solutions manual for end-of-chapter questions

STUDENT RESOURCES

Print supplement

- Student workbook problem-solving guide with solutions: ISBN: 1-319-16219-3

Housed in SaplingPlus

- Self-test concept checks
- Flash cards
- Interactive math tutorials
- Instructional videos
- PhET simulations
- Worked examples

PhysIndex—physical constants, conversion factors, and material data

Student support community

Course Solutions for Macmillan Physics

Custom Curriculum Solutions

SaplingPlus with e-Book

Textbooks

Available now for your Physics Lab...iOLab

The power of a lab in the palm of your hand

iOLab is a handheld data-gathering device that communicates wirelessly to its software and gives students a unique opportunity to see the concepts of physics in action. Students gain hands-on experience and watch their data graphed in real time. This can happen anywhere you have an iOLab device and a laptop: in the lab, in the classroom, in the dorm room, or in your basement. iOLab is flexible and makes it easy for instructors to design and implement virtually any experiment they want to assign their students or demonstrate in lecture.

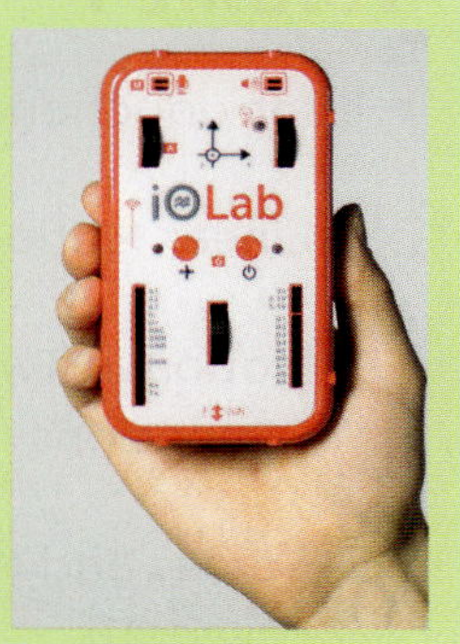

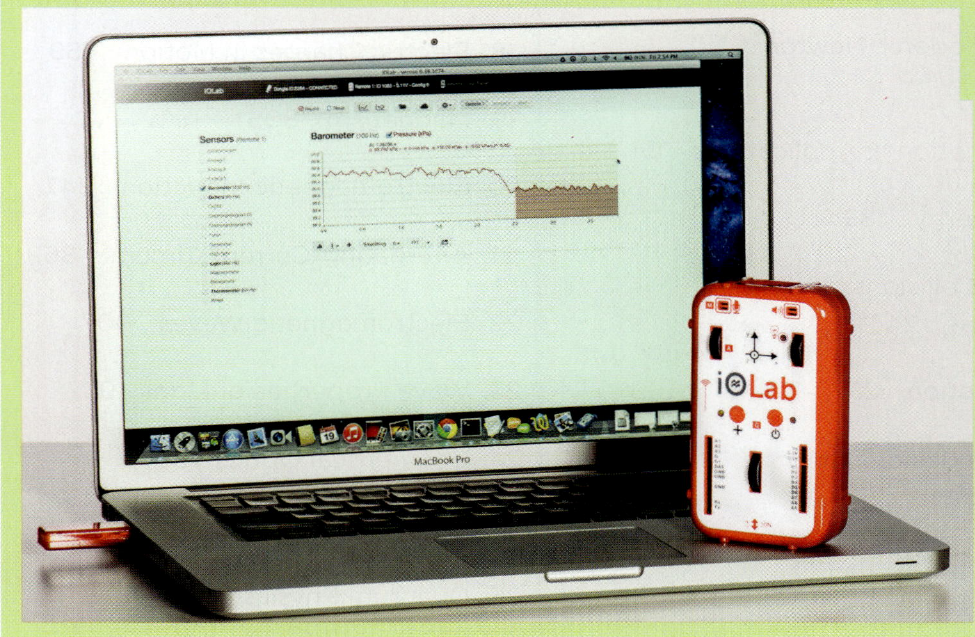

Reef In-Class Engagement

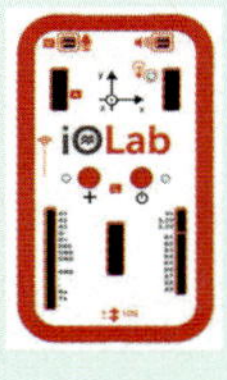

iOLab

Lab Solutions

BRIEF **CONTENTS**

Volume I

Volume II

Dear Students and Instructors,

Welcome to *College Physics!* We are excited to bring you this innovative text. No other college physics text presents material in quite this way. Our unique author team includes an experienced and highly successful textbook author, physicists who have spent years focusing on how students learn physics best, and even a biology professor who brings his perspective on what makes physics interesting to the students who take this course.

Our innovations, coupled with an engaging writing style, help students master concepts and succeed in developing and practicing the problem-solving skills they need to do well in this course. The visual impact of this text is something totally new: Word balloons throughout the text highlight key physics concepts in figures and equations so that students can learn the concepts in easy-to-manage pieces.

In addition, numerous features in the text are also designed to help students succeed in this course. Look for the **Chapter Goals** outlined at the beginning of each chapter; **Watch Out!** boxes that address misconceptions; **Got the Concept?** questions that test students' understanding of the material; **BIO-Medical** applications of physics in every chapter; **Take-Home Messages** that directly link to the chapter goals and help students focus on the important concepts presented in each chapter section; **Examples** that are broken down into key steps with an easy-to-follow *Set-Up*, *Solve*, *Reflect* structure; and a visual **Chapter Summary** at the end of each chapter, followed by **Questions and Problems**.

Our aim is to instill in students a deeper appreciation of physics—by showing how physics connects to their lives and their future careers, and by helping them succeed in the course. We hope you enjoy exploring and using this text!

Best Regards,

Roger, Todd, Phil, Dave

Contents

Volume II

Biological Applications

Unique and fully integrated physiological and biological applications are found throughout the text and are indicated by a **BIO-Medical** icon. Below is a list of select biological applications organized by chapter section for easy reference.

Chapter 1

Section 1-1: Hummingbirds and Physics
Section 1-3: Example 1-2 The World's Fastest Bird (speed)
Example 1-3 Hair Growth (unit conversion)

Chapter 2

Section 2-6: Example 2-10 Free Fall II: A Pronking Springbok (free fall)

Chapter 3

Section 3-3: Example 3-2 At What Angle Is Your Heart? (vector addition)
Section 3-8: Vestibular System (uniform circular motion)

Chapter 4

Section 4-2: Example 4-1 Small but Forceful (Newton's second law)
Section 4-5: Animal Propulsion (Newton's third law)

Chapter 5

Lubrication in Joints (static friction)
Section 5-2: Gecko Feet (static friction)
Example 5-1 Friction in Joints (static friction)
Section 5-5: Algal Spore (drag force)
Streamlined Animal Shapes (drag force)
Example 5-7 Terminal Speed (drag force)
Example 5-9 Making an Airplane Turn (apparent weight)

Chapter 6

Section 6-2: Example 6-2 Work Done by Actin (work)
Muscles and Doing Work (muscle function)
Section 6-5: Tendons (Hooke's Law)
Section 6-7: Arteries (energy transfer)
Muscles (potential energy)
Section 6-9: Exercise Rate (power)

Chapter 7

Section 7-1: Squid (momentum)
Section 7-7: Ballet Dancer (center of mass)

Chapter 8

Section 8-6: Lever Arms in Animal Anatomy (muscles)
Got the Concept? 8-8 Rotating the Human Jaw (moment of inertia)
Got the Concept? 8-9 Insects versus Birds (moment of inertia)
Section 8-9: Alligator Death Roll (angular momentum)
Got the Concept? 8-14 Danger! Falling Cat Zone (angular momentum)

Chapter 9

Section 9-1: Achilles Tendon (deformation)
Section 9-3: Example 9-1 Tensile Stress and Strain (sternoclavicular ligaments)
Section 9-4: Diving (compression)
Section 9-6: Arterial Blood Flow (elasticity)
Section 9-7: Example 9-4 Endothelial Cells and Shear (shear stress)
Section 9-8: Anterior Cruciate Ligament (stress)
Section 9-9: Example 9-6 Human ACL I: Maximum Force (tensile stress)
Example 9-7 Human ACL II: Breaking Strain (tensile strain)
Example 9-8 Human ACL III: The Point of No Return (tensile stress and strain)

Chapter 10

Section 10-5 Space Adaptation Sickness (weightlessness)
Muscle and Blood Loss (weightlessness)

Chapter 11

Section 11-5: Blood Pressure (gauge pressure)
Section 11-6: The Lungs (pressure differences)
Section 11-8: Fish Swim Bladder (density)
Example 11-10 Measuring Body Fat (Archimedes' principle)
Section 11-9: Measuring Blood Pressure (laminar versus turbulent flow)
Blood Flow (equation of continuity)
Example 11-12 How Many Capillaries? (equation of continuity)
Example 11-13 From Capillaries to the Vena Cavae (equation of continuity)
Section 11-10: Opercula (ram ventilation)
Section 11-11: Single-celled Alga *Chlamydomonas* (viscosity)
Blood Flow (Reynolds number)
Section 11-12: Gas Exchange (surface tension)

Chapter 12

Section 12-2: Got the Concept? 12-1 Heartbeats
Section 12-3: Intracellular Calcium Concentration (oscillations)
Section 12-4: Kangaroos and Dolphins (potential energy)
Section 12-5: Knee-jerk Reflex (physical pendulum)
Section 12-6: Example 12-7 Moment of Inertia of Swinging Leg (simple pendulum)
Section 12-7: Frog's Eardrum (underdamped oscillation)
Example 12-8 Oscillations of the Inner Ear (amplitude of oscillation)

Section 12-8: Flight of Flies, Wasps, and Bees (resonance in nature)
Got the Concept? 12-12 Mosquitos
Listening for Each Other (natural frequency)

Chapter 13
Section 13-3: Example 13-1 Wave Speed on a Sperm's Flagellum (wave speed)
Human Hearing (frequency)
Got the Concept? 13-2 Human Hearing (frequency)
Section 13-4: Example 13-3 Tension in a Sperm's Flagellum (transverse waves)
Section 13-7: Auditory Canals (sound waves)
Vocal Tract (closed pipes)
Example 13-7 An Amorous Frog (frequency)
Section 13-9: Eardrum (power)
Example 13-9 Loud and Soft (hearing)
Example 13-10 Delivering Energy to the Eardrum (energy)
Section 13-10: Ultrasonic Imaging (Doppler shift)
Example 13-12 Diagnostic Ultrasound (frequency)

Chapter 14
Section 14-3: Stomata (mean free path)
Section 14-4: Average Global Temperature (sea levels)
Section 14-6: Sweat (heat flow)
Section 14-7: Jackrabbits and Iguanas (heat transfer)
Temporal Artery Thermometer (heat transfer)
Radiation and Climate (global warming)
Polar Bear Fur (thermal conductivity)

Chapter 15
Section 15-2: Exercise (thermodynamic process)
Example 15-1 Cycling It Off (heat loss)
Section 15-5: Cellular Respiration (heat engine)
Section 15-6: Living Organisms (entropy)

Chapter 16
Section 16-3: Conductor of Electricity (biological systems)
Section 16-5: Electrophoresis (DNA profiling)

Chapter 17
Section 17-5: Example 17-8 Insulin Release
Section 17-6: Example 17-9 A Defibrillator Capacitor
Section 17-8: Example 17-12 Cell Membrane Capacitance

Chapter 18
Section 18-4: Potassium Channels (resistance)
Example 18-5 Resistance of a Potassium Channel
Nervous System (membrane potential)
Example 18-6 Giant Axons in Squid
Section 18-5: Example 18-7 Giant Axons in Squid Revisited
Section 18-7: Propagation of Action Potentials Along Axons (physics of capacitors)

Chapter 19
Section 19-3: Magnetic Field (MRI scanner)
Section 19-7: Electric Currents (magnetoencephalogram)
Solenoids (MRI scanner)
Earth's Magnetic Field (animal navigation)

Chapter 20
Section 20-1: Electromagnetic Induction [transcranial magnetic stimulation (TMS)]
Section 20-3: Eddy Currents (electromagnetic flow meter)
Eddy Currents (magnetic induction tomography)

Chapter 21
Section 21-3: Current (ventricular fibrillation)

Chapter 22
Section 22-2: Detecting Wavelengths (snakes)
Section 22-4: Electromagnetic Waves (skin)
Ionizing Radiation (DNA)

Chapter 23
Section 23-2: Index of Refraction (blood plasma)
Section 23-3: Light (endoscopy)
Section 23-5: Polarization of Light (bees)
Section 23-6: Interference of Light Waves (*Morpho menelaus*)
Wavelengths (*tapetum lucidum*)
Section 23-9: Value of uR (physician's eye chart)

Chapter 24
Section 24-2: Interpretation of Light (eye and brain)
Section 24-9: Focusing Images on the Retina (cornea and lens)

Chapter 25
Section 25-3: Got the Concept? 25-2 Speed of Sound in Water
Section 25-4: Watch Out! Time Dilation Occurs Whether or Not a Clock Is Present

Chapter 26
Section 26-2: Photoelectric Effect (photoemission electron microscopy)
Blackbodies (pupil)
Section 26-3: Cancer Radiation Therapy (x rays)

Chapter 27
Section 27-1: Magnetic Resonance Imaging (nuclear spin)
Section 27-2: MRI Scanner (spin of hydrogen nuclei)
Element Formation in Stars (fosters life)

Acknowledgments

Bringing a textbook into its second edition requires the coordinated effort of an enormous number of talented professionals. We are grateful for the dedicated support of our in-house team at W. H. Freeman; thank you for sustaining our concept and developing this book into a beautiful new edition.

We especially want to thank Senior Development Editor Blythe Robbins and our development editor, Meg Rosenburg, for their dedication to this book. We would also like to thank our program manager, Lori Stover, for encouraging us and leading our team; Content Director Tania Mlawer, Media Editor Victoria Garvey, Media Project Manager Daniel Comstock, and Physics Content Team Lead Josh Hebert for their thoughtful contributions; and Assistant Editor Kevin Davidson, Permissions Manager Jennifer MacMillan, and Senior Content Project Manager Kerry O'Shaughnessy for their patience and attention to detail. Of course, none of this would be possible without the support of Editorial Program Director Brooke Suchomel and Vice President for STEM Ben Roberts. Special thanks also go to our talented marketing team, Marketing Manager Maureen Rachford, Associate Digital Marketing Specialist Cate McCaffery, and Marketing Assistant Savannah DiMarco for their insight and assistance.

Friends and Family

One of us (RAF) thanks his wife, Caroline Robillard, for her patience with the seemingly endless hours that went into preparing this textbook. I also thank my students at the University of California, Santa Barbara, for giving me the opportunity to test and refine new ideas for making physics more accessible.

One of us (TGR) thanks his wife, Susan, and daughter, Allison, for their limitless patience and understanding with the countless hours spent working on this book. I also thank my parents who showed me how to live a balanced life.

One of us (PRK) would like to acknowledge valuable and insightful conversations on physics and physics teaching with Richard Barber, John Birmingham, and J. Patrick Dishaw of Santa Clara University, and to offer these colleagues my gratitude. Finally, I offer my gratitude to my wife Kathy and my children Sam and Chloe for their unflagging support during the arduous process that led to the book you hold in your hands.

One of us (DLT) thanks his family and friends for accommodating my tight schedule during the years that writing this book consumed. I especially want to thank my parents, Bill and Jean, for their boundless encouragement and support, and for teaching me everything I've ever really needed to know—they've shown me how to live a good life, be happy, and age gently. I greatly appreciate my sister for encouraging me not to abandon a healthy lifestyle just to write a book. I also want to thank my nonbiological family, Holly and Geoff, for leading me to Sonoma County and for making Sebastopol feel like home.

Solutions Manual authors:

Bryan Armentrout, *Nova Southeastern University*
Michael Dunham, *State University of New York at Fredonia*
Andrew Ekpenyong, *Creighton University*

Accuracy checker:

Jose Lozano, *Bradley University*

Student Workbook authors:

Perry Hilburn, *Gannon University*
Seong-Gon Kim, *Mississippi State University*
Garrett Yoder, *Eastern Kentucky University*
Linda McDonald, *North Park University*
Avishek Kumar, *Arizona State University*
Michael Dunham, *SUNY Fredonia*

Test Bank authors:

Anna Harlick, *University of Calgary*
Debashis Dasgupta, *University of Wisconsin at Milwaukee*
Elizabeth Holden, *University of Wisconsin at Platteville*

iClicker Slides author:

Adam Lark, *Hamilton College*

Lecture Slides author:

Adam Lark, *Hamilton College*

Questions and Problems editors:

Jonathan Bratt
Michael Scott
Fredrick "Mike" DeArmond
Syed "Asif" Hassan

Second-Edition Reviewers

We would also like to thank the many colleagues who carefully reviewed chapters for us. Their insightful comments significantly improved our book.

Miah Muhammad Adel, *University of Arkansas at Pine Bluff*
Mikhail M. Agrest, *The Citadel*
Vasudeva Rao Aravind, *Clarion University of Pennsylvania*
Yiyan Bai, *Houston Community College*
E. C. Behrman, *Wichita State University*
Antonia Bennie-George, *Green River College*
Ken Bolland, *The Ohio State University*
Matthew Joseph Bradley, *Santa Rosa Junior College*
Matteo Broccio, *University of Pittsburgh*
Daniel J. Costantino, *The Pennsylvania State University*
Adam Davis, *Wayne State College*
Sharvil Desai, *The Ohio State University*

Eric Deyo, *Fort Hays State University*
Diana I. Driscoll, *Case Western Reserve University*
Davene Eyres, *North Seattle College*
William Falls, *Erie Community College*
Sambandamurthy Ganapathy, *SUNY Buffalo*
Vladimir Gasparyan, *California State University, Bakersfield*
Frank Gerlitz, *Washtenaw Community College*
Svetlana Gladycheva, *Towson University*
Romulus Godang, *University of South Alabama*
Javier Gomez, *The Citadel*
Rick Goulding, *Memorial University of Newfoundland*
Ania Harlick, *Memorial University of Newfoundland*
Erik Helgren, *California State University, East Bay*
Perry G. Hillburn, *Gannon University*
Zdeslav Hrepic, *Columbus State University*
Patrick Huth, *Community College of Allegheny County*
Matthew Jewell, *University of Wisconsin, Eau Claire*
Wafaa Khattou, *Valencia College*
Seong-Gon Kim, *Mississippi State University*
Patrick Koehn, *Eastern Michigan University*
Ameya S. Kolarkar, *Auburn University*
Maja Krcmar, *Grand Valley State University*
Elena Kuchina, *Thomas Nelson Community College*
Avishek Kumar, *Arizona State University*
Chunfei Li, *Clarion University of Pennsylvania*
Jose Lozano, *Bradley University*
Dan MacIsaac, *SUNY Buffalo State College*
Linda McDonald, *North Park University*
Francis Mensah, *Virginia Union University*
Victor Migenes, *Texas Southern University*
Krishna Mukherjee, *Slippery Rock University of Pennsylvania*
Rumiana Nikolova-Genov, *College of DuPage*
Moses Ntam, *Tuskegee University*
Martin O. Okafor, *Georgia Perimeter College*
Gabriela Petculescu, *University of Louisiana at Lafayette*
Yuriy Pinelis, *University of Houston, Downtown*
Sulakshana Plumley, *Community College of Allegheny County*
Lawrence Rees, *Brigham Young University*
David Richardson, *Northwest Missouri State University*
Carlos Roldan, *Central Piedmont Community College*
Jeffrey Sabby, *Southern Illinois University Edwardsville*
Arun Saha, *Albany State University*
Haiduke Sarafian, *Pennsylvania State University*
Katrin Schenk, *Randolph College*
Surajit Sen, *SUNY Buffalo*
Jerry Shakov, *Tulane University*
Douglas Sharman, *San Jose State University*
Ananda Shastri, *Minnesota State University, Moorhead*
Marllin L. Simon, *Auburn University*
Stanley J. Sobolewski, *Indiana University of Pennsylvania*
Erin C. Sutherland, *Kennesaw State University*
Sarah F. Tebbens, *Wright State University*
Fiorella Terenzi, *Florida International University*
Dmitri Tsybychev, *Stony Brook University*
Laura Whitlock, *Georgia Perimeter College*
Pushpa Wijesinghe, *Arizona State University*
Fengyuan Yang, *The Ohio State University*
Garett Yoder, *Eastern Kentucky University*
Jiang Yu, *Fitchburg State University*
Ulrich Zurcher, *Cleveland State University*

First-Edition Reviewers

We would also like to acknowledge the many reviewers who reviewed the first edition of this text, as their contributions live on in this edition.

Don Abernathy, *North Central Texas College*
Elise Adamson, *Wayland Baptist University*
Miah Muhammad Adel, *University of Arkansas, Pine Bluff*
Ricardo Alarcon, *Arizona State University*
Z. Altounian, *McGill University*
Abu Amin, *Riverland Community College*
Sanjeev Arora, *Fort Valley State University*
Llani Attygalle, *Bowling Green State University*
Yiyan Bai, *Houston Community College*
Michael Bates, *Moraine Valley Community College*
Luc Beaulieu, *Memorial University*
Jeff J. Bechtold, *Austin Community College*
David Bennum, *University of Nevada*
Satinder Bhagat, *University of Maryland*
Dan Boye, *Davidson College*
Jeff Bronson, *Blinn College*
Douglas Brumm, *Florida State College at Jacksonville*
Mark S. Bruno, *Gateway Community College*
Brian K. Bucklein, *Missouri Western State University*
Michaela Burkardt, *New Mexico State University*
Kris Byboth, *Blinn College*
Joel W. Cannon, *Washington & Jefferson College*
Kapila Clara Castoldi, *Oakland University*
Paola M. Cereghetti, *Lehigh University*
Hong Chen, *University of North Florida*
Zengjun Chen, *Tuskegee University*
Uma Choppali, *Dallas County Community College*
Todd Coleman, *Century College*
José D'Arruda, *University of North Carolina, Pembroke*
Tinanjan Datta, *Georgia Regents University*
Chad L. Davies, *Gordon College*
Brett DePaola, *Kansas State University*
Sandra Doty, *Ohio University*
James Dove, *Metro Community College*
Carl T. Drake, *Jackson State University*
Rodney Dunning, *Longwood University*
Vernessa M. Edwards, *Alabama A & M University*
Davene Eyres, *North Seattle Community College*
Hasan Fakhruddin, *Ball State University*
Paul Fields, *Pima Community College*
Lewis Ford, *Texas A & M University*
J.A. Forrest, *University of Waterloo*
Scott Freedman, *Philadelphia Academy Charter High School*
Tim French, *Harvard University*
James Friedrichsen III, *Austin Community College*
Sambandamurthy Ganapathy, *SUNY Buffalo*
J. William Gary, *University of California, Riverside*
L. Gasparov, *University of North Florida*
Vladimir Gasparyan, *California State University, Bakersfield*
Brian Geislinger, *Gadsden State Community College*
Oommen George, *San Jacinto College*
Anindita Ghosh, *Suffolk County Community College*
Alan I. Goldman, *Iowa State University*
Richard Goulding, *Memorial University of Newfoundland*
Morris C. Greenwood, *San Jacinto College Central*

Thomas P. Guella, *Worcester State University*
Alec Habig, *University of Minnesota, Duluth*
Edward Hamilton, *Gonzaga University*
C. A. Haselwandter, *University of Southern California*
Zvonko Hlousek, *California State University, Long Beach*
Micky Holcomb, *West Virginia University*
Kevin M. Hope, *University of Montevallo*
J. Johanna Hopp, *University of Wisconsin, Stout*
Leon Hsu, *University of Minnesota*
Olenka Hubickyj Cabot, *San Jose State University*
Richard Ignace, *East Tennessee State University*
Elizabeth Jeffery, *James Madison University*
Yong Joe, *Ball State University*
Darrin Eric Johnson, *University of Minnesota, Duluth*
David Kardelis, *Utah State University, College of Eastern Utah*
Agnes Kim, *Georgia State College*
Ju H. Kim, *University of North Dakota*
Seth T. King, *University of Wisconsin, La Crosse*
Kathleen Koenig, *University of Cincinnati*
Olga Korotkova, *University of Miami*
Minjoon Kouh, *Drew University*
Tatiana Krivosheev, *Clayton State University*
Michael Kruger, *University of Missouri, Kansas City*
Jessica C. Lair, *Eastern Kentucky University*
Josephine M. Lamela, *Middlesex County College*
Patrick M. Len, *Cuesta College*
Shelly R. Lesher, *University of Wisconsin, La Crosse*
Zhujun Li, *Richland College*
Bruce W. Liby, *Manhattan College*
David M. Lind, *Florida State University*
Jeff Loats, *Metropolitan State College of Denver*
Susannah E. Lomant, *Georgia Perimeter College*
Jia Grace Lu, *University of Southern California*
Mark Lucas, *Ohio University*
Lianxi Ma, *Blinn College*
Aklilu Maasho, *Dyersburg State Community College*
Ron MacTaylor, *Salem State University*
Eric Mandell, *Bowling Green State University*
Maxim Marienko, *Hofstra University*
Mark Matlin, *Bryn Mawr College*
Dan Mattern, *Butler Community College*
Mark E. Mattson, *James Madison University*
Jo Ann Merrell, *Saddleback College*
Michael R. Meyer, *Michigan Technological University*
Karie A. Meyers, *Pima Community College*
Andrew Meyertholen, *University of Redlands*
John H. Miller Jr., *University of Houston*
Ronald C. Miller, *University of Central Oklahoma*
Ronald Miller, *Texas State Technical College System*
Hector Mireles, *California State Polytechnic University, Pomona*
Ted Monchesky, *Dalhousie University*
Steven W. Moore, *California State University, Monterey Bay*
Mark Morgan-Tracy, *University of New Mexico*
Dennis Nemeschansky, *University of Southern California*
Terry F. O'Dwyer, *Nassau Community College*
John S. Ochab, *J. Sargeant Reynolds Community College*
Umesh C. Pandey, *Central New Mexico Community College*
Archie Paulson, *Madison Area Technical College*
Christian Poppeliers, *Augusta State University*
James R. Powell, *University of Texas, San Antonio*
Michael Pravica, *University of Nevada, Las Vegas*
Kenneth M. Purcell, *University of Southern Indiana*
Kenneth Ragan, *McGill University*
Milun Rakovic, *Grand Valley State University*
Jyothi Raman, *Oakland University*
Natarajan Ravi, *Spelman College*
Lou Reinisch, *Jacksonville State University*
Sandra J. Rhoades, *Kennesaw State University*
John Rollino, *Rutgers University, Newark*
Rodney Rossow, *Tarrant County College*
Larry Rowan, *University of North Carolina*
Michael Sampogna, *Pima Community College*
Tumer Sayman, *Eastern Michigan University*
Jim Scheidhauer, *DePaul University*
Paul Schmidt, *Ball State University*
Morton Seitelman, *Farmingdale State College*
Saeed Shadfar, *Oklahoma City University*
Weidian Shen, *Eastern Michigan University*
Jason Shulman, *Richard Stockton College of New Jersey*
Michael J. Shumila, *Mercer County Community College*
R. Seth Smith, *Francis Marion University*
Frank Somer, *Columbia College*
Chad Sosolik, *Clemson University*
Brian Steinkamp, *University of Southern Indiana*
Narasimhan Sujatha, *Wake Technical Community College*
Maxim Sukharev, *Arizona State University*
James H. Taylor, *University of Central Missouri*
Richard Taylor, *University of Oregon*
E. Tetteh-Lartey, *Blinn Community College*
Fiorella Terenzi, *Brevard Community College*
Gregory B. Thompson, *Adrian College*
Marshall Thomsen, *Eastern Michigan University*
Som Tyagi, *Drexel University*
Vijayalakshmi Varadarajan, *Des Moines Area Community College*
John Vasut, *Baylor University*
Dimitrios Vavylonis, *Lehigh University*
Kendra L. Wallis, *Eastfield College*
Laura Weinkauf, *Jacksonville State University*
Heather M. Whitney, *Wheaton College*
Capp Yess, *Morehead State University*
Chadwick Young, *Nicholls State University*
Yifu Zhu, *Florida International University*
Raymond L. Zich, *Illinois State University*

1 Introduction to Physics

Anthony Mercieca/Science Source

What do you think?

The size of a hummingbird's head is closest to which of these lengths? (a) 1 micrometer (1 μm); (b) 1 millimeter (1 mm); (c) 1 centimeter (1 cm); (d) 1 meter (1 m).

The answer to the *What do you think?* question can be found at the end of the chapter.

In this chapter, your goals are to:

- (1-1) Explain the roles that concepts and equations play in physics.
- (1-2) Describe three key steps in solving any physics problem.
- (1-3) Identify the fundamental units used for measuring physical quantities and convert from one set of units to another.
- (1-4) Use significant figures in calculations.
- (1-5) Use dimensional analysis to check algebraic results.

To master this chapter, you should review:

- The principles of algebra and how to solve an equation for a particular variable.

1-1 Physicists use a special language—part words, part equations—to describe the natural world

A hummingbird is a beautiful sight to anyone. To a physicist, however, a hummingbird is also an illustration of several physical principles at work. As they beat, the hummingbird's wings push downward on the air, which causes the air to push upward in response and keep the bird aloft. The flow of blood through the hummingbird's body is governed by the properties of moving fluids, including the interplay of pressure and friction in the narrowest blood vessels. The bending of light rays as they enter the bird's eye forms an image on the retina, providing the hummingbird with vision. And the subtle interference of light waves reflecting off the feathers at the bird's throat gives them their characteristic iridescence. To truly understand and appreciate hummingbirds or any other living organism, we need to understand the principles of physics. The purpose of this book is to help you gain that understanding.

Learning physics is somewhat like learning a foreign language. If you're paying a brief visit to a foreign country, you may just memorize a few key phrases such as "Where is the hotel?" in the local language. But if you want to be fluent in the language, you don't memorize millions of phrases for the countless situations you might encounter. Instead, you learn a vocabulary of useful words and the grammar that allows you to combine those words into meaningful sentences. Using those basic tools, you can express an enormous variety of ideas, even those no one else has expressed before. In the same way, it may seem at the outset that physics is a long compilation of rules, equations, definitions, and concepts. But we will discover that this long list of physics "phrases" is created from a much smaller set of physics "grammar" and the key vocabulary of physics. Your goal is to become fluent in physics by understanding how to use the rules of physics grammar to connect the vocabulary of physics.

The vocabulary of physics is often expressed in terms of mathematical quantities, and these quantities, along with the grammar that relates them, are the equations of physics. Like most physics books, this one contains many equations. Not all of these equations are equally important, however. A much smaller set of *fundamental* equations expresses the key ideas of physics. By starting with these fundamental equations and combining them using the rules of algebra, you'll be able to solve all of the physics problems that you encounter in this book.

It's tempting to view the equations in this book, even the fundamental ones, as being all there is to learning physics. If you adopt this view, you'll get the idea that solving a physics problem is simply a hunt to find the correct equation. Not so! Each physics equation is shorthand for a *concept* that can be much more profound than the equation might suggest. One example is Albert Einstein's famous equation

$$E = mc^2$$

Figure 1-1 $E = mc^2$ When you burn natural gas in a kitchen range, you are converting some of the rest energy of methane and oxygen into light and heat.

In this equation E represents energy, m represents mass, and c represents the speed at which light travels in a vacuum. But what the equation *means* is something deeper. In this book we'll often write fundamental equations with explanations attached, such as this one:

Einstein's equation for rest energy (1-1)

$$E = mc^2$$

Rest energy of an object

Mass of the object

Speed of light in a vacuum

The greater the mass of an object, the greater its rest energy.

The *rest energy* of an object is the amount of energy it possesses simply as a consequence of having mass. (Einstein was the first physicist to deduce that energy of this kind existed.) Equation 1-1 shows that rest energy is *directly proportional* to mass: The greater the mass of an object, the greater its rest energy and vice versa. If the rest energy of an object decreases, its mass must decrease as well. You can find an example of this in many kitchens (**Figure 1-1**). In a gas range, natural gas (mostly methane, CH_4) combines with oxygen to produce carbon dioxide (CO_2) and water (H_2O). The chemical reaction that takes place is

$$CH_4 + 2O_2 \longrightarrow CO_2 + 2H_2O$$

√x See the Math Tutorial for more information on equations.

This process releases energy in the form of heat and light. This energy comes from rest energy, so the combined rest energy of the products (a CO_2 molecule and two H_2O molecules) is less than the rest energy of the reactants (a CH_4 molecule and two O_2 molecules). Equation 1-1 then tells us that the combined mass of the molecules present after the reaction is slightly *less* than the combined mass of the molecules present before the reaction. This is a remarkable statement, but sensitive measurements reveal it to be true. When you look at a flame, Equation 1-1 itself isn't the important part of this story: It's the *concept* behind this equation that's truly important. This is an important idea that can be easy to forget. When we come across an important idea like this one, or a situation that requires special care, we'll put it in a ***Watch Out!*** box like the one on the left side of this page.

WATCH OUT! You need to be careful.

When we come across a situation that requires special care, we'll put it in a *Watch Out!* box like this one.

TAKE-HOME MESSAGE FOR Section 1-1

✔ Like all other sciences, physics is a collection of interlinked concepts.

✔ The equations of physics help us summarize complex ideas in mathematical shorthand.

✔ It's essential to understand the concepts behind each fundamental equation.

1-2 Success in physics requires well-developed problem-solving skills

More than just broadening our understanding of the world, the process of learning physics involves learning to solve problems. Physics problems can appear challenging because they usually can't be done simply by selecting an equation and plugging in values. Solving physics problems is easier, however, if you build up a set of tools and techniques and then apply a consistent strategy. Our strategy for solving problems involves three steps that we refer to as ***Set Up, Solve,*** and ***Reflect.*** We'll use these steps in all of the worked examples in this book.

EXAMPLE 1-1 Strategy for Solving Problems

Set Up

At the beginning of any problem, ask yourself these questions: What is the problem asking for? Which quantities are known, and which are unknowns whose values you need to determine to solve the problem? What concepts will help solve the problem, and which equations should you use to apply these principles? Once you've answered these questions, you'll have a good idea of how to get started with solving the problem. Here are some helpful hints for setting up any problem:

Draw a picture. Physicists rely heavily on diagrams. You'll find one or more in many of the worked examples in this book. A good problem-solving picture should capture the motion, the process, the geometry, or whatever else defines the problem. You don't need an artistic masterpiece; it's perfectly okay to represent most complex objects as squares or circles or even dots.

Label all quantities. It makes sense to label all quantities such as length, speed, and temperature right on the picture you've drawn. But be careful! First, more than one of a given kind of variable is often in a problem. For example, if two objects are moving at different speeds, but you use the same symbol v for the speeds of both objects, confusion is bound to result. Subscripts are useful in such a case, for example, v_A and v_B for the speeds of the two objects. Also, don't hesitate to label lengths, velocities, and other quantities even if you haven't been given a value for them. You might be able to figure out the value of one of these quantities and use it later in the problem, or you might gain an insight that will guide you to a solution.

Look for connections. Often the best strategy for solving a physics problem is to look for connections between quantities. These connections could be relationships between two or more variables or a statement that a certain parameter has the same value for two objects. Furthermore, these connections will often enable you either to eliminate variables or to find a value for an unknown variable. As a result, you sometimes will have an equation that defines the variable of interest only in terms of variables that are known. So we'll always be on the lookout for connections such as these when starting a problem.

Solve

Once you've sized up the situation and selected the concepts and equations to use, finding the answer is often just an exercise in mathematics. During this step in the worked examples, we'll break down and show many intermediate steps as we describe our reasoning and thought processes. In this way, we'll model good problem solving, which will lead you to develop your own problem-solving skills and habits.

Reflect

An important but often overlooked part of problem solving is to review the answer to see what it tells you. If the problem calls for a numerical answer, it's important to ask, "Do the number and the units make sense?" This is often a good way for you to check the accuracy of your calculations. For example, if you've been asked to calculate the diameter of Earth and you come up with 40 centimeters, or if you come up with an answer in kilograms (the unit of mass), it's time to go back and check your calculations. If your answer is a formula rather than a single number, check that what the formula says is reasonable. As an example, if you've found a formula that relates the weight of a cat to its volume, your answer should show that a heavier cat has a larger volume than a lightweight one. If it doesn't, there's an error in your calculations that you need to find and fix.

Learning physics will be easier if you study the many worked examples in this book and get in the habit of following the same steps of ***Set Up, Solve,*** and ***Reflect*** in your own problem solving.

TAKE-HOME MESSAGE FOR Section 1-2

✔ Success in physics requires having a methodical approach to problem solving.

✔ When approaching any worked example, follow these three steps: ***Set Up*** a problem by selecting the appropriate concepts and equations, ***Solve*** by working through the mathematics, and ***Reflect*** on the results to see what they mean.

1-3 Measurements in physics are based on standard units of time, length, and mass

To the average person, physics calls to mind speculative theories about the nature of matter and the origins of space and time. But at its heart physics is an *experimental* science based on measurements of the natural world. Over a century ago the renowned Scottish physicist Lord Kelvin (1824–1907) expressed this viewpoint: "If you can measure that of which you speak, and can express it by a number, you know something of your subject; but if you cannot measure it, your knowledge is unsatisfactory."

TABLE 1-1 Fundamental Quantities and Their SI Units

Quantity	Unit	Abbreviation
time	second	s
length	meter	m
mass	kilogram	kg
temperature	kelvin	K
electric current	ampere	A
amount of substance	mole	mol
luminous intensity	candela	cd

Any measurement requires a standard of comparison. If you want to compare the heights of two children, for example, you might stand them side by side and use one child's height as a reference for the other. A better and more reliable method would be to use a vertical rod with equally spaced markings (in other words, a ruler) that acts as a standard of length.

In physics, three fundamental quantities for which it is essential to have standards are *time, length,* and *mass*. The standards, or **units**, for these three quantities are the *second*, the *meter*, and the *kilogram*, respectively, and the system of units based on these quantities is called the **Système International**, or **SI** for short. By understanding how these standards are defined, we'll get a sense of the precision toward which physicists strive in their measurements. **Table 1-1** lists all of the fundamental SI quantities, including others that we'll encounter in later chapters.

The definition of the **second** (abbreviated s) is based on the properties of the cesium atom. When such atoms are excited, they emit radio waves of a very definite frequency. The second is defined to be the amount of time required for cesium to emit 9,192,631,770 complete cycles of these waves. This may seem like a bizarre way to measure time, but modern radio technology is highly refined, and it's relatively easy to prepare a gas of cesium atoms with the right properties. Hence a physicist anywhere in the world could readily repeat these measurements.

In the past objects such as metal bars were used as standards of length. These were not very good standards, however, because their length changes slightly but measurably when the temperature changes. Today light waves are used to define the **meter** (abbreviated m), the SI standard of length. The speed of light in a vacuum is defined to be precisely 299,792,458 meters per second, so the meter is the distance that light travels in 1/299,792,458 of a second. This is a robust and repeatable standard because light in a vacuum travels at the same speed anywhere on Earth and, as best we can tell, everywhere in the universe. (We specify a vacuum because light travels more slowly in a material such as glass, water, or even air.)

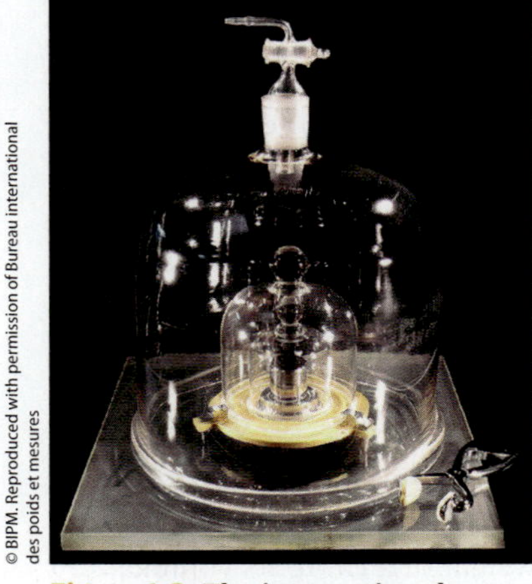

Figure 1-2 The international mass standard This cylinder is defined to have a mass of precisely 1 kilogram. To ensure that repeated measurements of its mass give the same value, the cylinder has to be treated very carefully to prevent even a small number of atoms from being scraped off.

Mass is a measure of the amount of material in an object. A feather has less mass than a baseball, which in turn has less mass than a cinder block. The ideal standard for mass would be one that is, like the standards for time and length, based on something that is the same throughout the universe—for example, the mass of a hydrogen atom. Unfortunately, physicists' present-day ability to measure masses on the atomic scale is not yet precise enough to give a really satisfactory standard of mass. As of this writing (2017) the **kilogram** (abbreviated kg) is defined to be the mass of a special cylinder of platinum-iridium alloy kept in a repository near Paris, France (**Figure 1-2**). Physicists hope that in the near

WATCH OUT! Mass and weight are different quantities.

It's important not to confuse *mass* and *weight*. While an object's mass tells you how much material that object contains, its weight tells you how strongly gravity pulls on that object's material. Consider a person who weighs 110 pounds on Earth, corresponding to a mass of 50 kilograms. Gravity is only about one-sixth as strong on the Moon as it is on Earth, so on the Moon this person would weigh only one-sixth of 110 pounds, or about 18 pounds. But that person's mass of 50 kilograms is the same on the Moon; wherever you go in the universe, you take all of your material along with you. We'll explore the relationship between mass and weight in Chapter 4.

future this archaic mass standard can be replaced by one based on the properties of atoms. (The *gram*, equal to 1/1000 of a kilogram, is not a fundamental SI unit.)

Units for many other physical quantities can be derived from the fundamental units of time, length, and mass. For example, speed is measured in meters per second (m/s), and weight is measured in kilograms times meters per second per second, or kilogram-meters per second squared ($\text{kg} \cdot \text{m/s}^2$). Some derived quantities are used so frequently that physicists have named a new unit to represent them. For example, the SI unit of weight is the newton (N), equal to $1\ \text{kg} \cdot \text{m/s}^2$, or about 0.225 pounds.

When we begin our study of heat and temperature in Chapter 14, we will need to introduce another unit called the *kelvin*, which is the fundamental unit of temperature. We'll also need another unit called the *ampere*, a unit of electric current, when we begin our study of electricity in Chapter 16. Until then, all of the units we'll work with will be based on seconds, meters, and kilograms.

Unit Conversions

In science and in ordinary life, a variety of other units are used for time, length, and mass. As we'll see, it's straightforward to convert between different units for the same quantity.

In addition to the second, some other common units of time include the following:

$$1 \text{ minute (min)} = 60 \text{ s}$$
$$1 \text{ hour (h)} = 3600 \text{ s}$$
$$1 \text{ day (d)} = 86{,}400 \text{ s}$$

Some other units of length that are handy when discussing sizes and distances include millimeters (mm), centimeters (cm), and kilometers (km). These units of length are related to the meter as follows:

$$1 \text{ mm} = 0.001 \text{ m}$$
$$1 \text{ cm} = 0.01 \text{ m}$$
$$1 \text{ km} = 1000 \text{ m}$$

Although the English system of inches (in), feet (ft), and miles (mi) is much older than the SI, today the English system is actually based on the SI system: The inch is *defined* to be exactly 2.54 cm. A useful set of conversions is as follows:

$$1 \text{ in} = 2.54 \text{ cm}$$
$$1 \text{ ft} = 0.3048 \text{ m}$$
$$1 \text{ mi} = 1.609 \text{ km}$$

You'll find more conversion factors in Appendix A.

We can use conversions such as these to convert any quantity from one set of units to another. For example, the glass skyscraper in London, England, called the Shard is 1016 feet tall (**Figure 1-3**). How can we convert this height to meters? **Figure 1-4** shows the technique. When doing unit conversions or other multiplication and division calculations involving units, you can treat the unit associated with the number as a variable. Units can get multiplied together, and they can also cancel if the same unit appears in the numerator and denominator.

Allan Baxter/Getty Images

Figure 1-3 From feet to meters The tallest building in the United Kingdom, the Shard, stands 1016 feet tall. How tall is this in meters?

Figure 1-4 Converting units To convert units, remember that a quantity does not change if you multiply by 1. The only thing to remember is to write 1 in an appropriate way.

(1) To convert units—for example, to convert a distance in feet (ft) to a distance in meters (m)—multiply by 1 in the appropriate form. Multiplying a number by 1 does not change the number.

$$1016 \text{ ft} = 1016 \text{ ft} \times 1$$

(2) To find the appropriate form of 1, begin with the conversion between the two units...

(3) ...and divide one side by the other. (If you divide a number by an equal number, the result is always equal to 1.) This gives two ways to express the conversion between feet and meters.

$$1 \text{ ft} = 0.3048 \text{ m, so } \frac{1 \text{ ft}}{0.3048 \text{ m}} = 1 \text{ and } \frac{0.3048 \text{ m}}{1 \text{ ft}} = 1$$

INCORRECT APPROACH:

(4) If you try multiplying 1016 ft by the first form of 1...

(5) ...the result is not in the desired units.

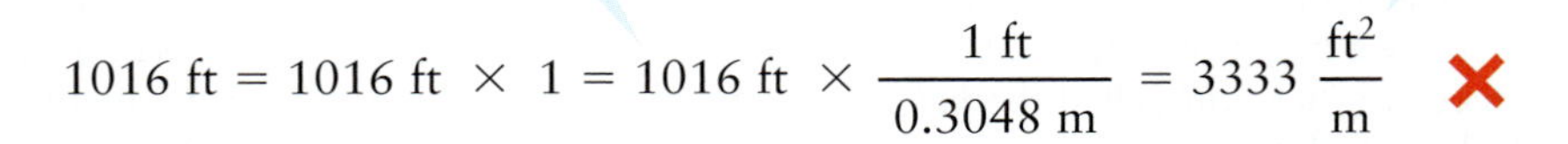

$$1016 \text{ ft} = 1016 \text{ ft} \times 1 = 1016 \text{ ft} \times \frac{1 \text{ ft}}{0.3048 \text{ m}} = 3333 \frac{\text{ft}^2}{\text{m}} \quad ✗$$

CORRECT APPROACH:

(6) If instead you multiply 1016 ft by the second form of 1...

(7) ...the units of ft cancel and the result is in m, as desired.

$$1016 \text{ ft} = 1016 \text{ ft} \times 1 = 1016 \cancel{\text{ft}} \times \frac{0.3048 \text{ m}}{1 \cancel{\text{ft}}} = 309.7 \text{ m} \quad \text{OK}$$

To convert a quantity from one set of units to another, multiply the quantity by 1 (equal to the ratio of the two units). Choose the ratio that enables the unwanted units to cancel out.

WATCH OUT! Take care with conversions that have multiple steps.

Some conversions cannot be easily carried out in a single step. As an illustration, let's verify the number of seconds in a day. Since 1 day (d) equals 24 hours (h), 1 h equals 60 minutes (min), and 1 min equals 60 seconds (s), we convert units three times:

$$1 \text{ d} = 1 \cancel{\text{d}} \left(\frac{24 \cancel{\text{h}}}{1 \cancel{\text{d}}}\right)\left(\frac{60 \cancel{\text{min}}}{1 \cancel{\text{h}}}\right)\left(\frac{60 \text{ s}}{1 \cancel{\text{min}}}\right) = 86{,}400 \text{ s}$$

Notice that we were careful to check the cancellation of each pair of like units (d and d, h and h, min and min). We strongly recommend that you carry out conversions such as this one in an equally methodical way.

BIO-Medical EXAMPLE 1-2 Converting Units: Speed

The world's fastest bird, the white-throated needletail (*Hirundapus caudacutus*), can move at speeds up to 47.0 m/s in level flight. (Falcons can go even faster, but they must dive to do so.) What is this speed in km/h? in mi/h?

Set Up

This is a problem in converting units. To find the speed in km/h, we will have to convert meters (m) to kilometers (km) and seconds (s) to hours (h). To then find the speed in mi/h, we will have to convert km to miles (mi). In each stage of the conversion, we'll follow the procedure shown in Figure 1-4.

Conversions:
1 h = 3600 s
1 km = 1000 m
1 mi = 1.609 km

Solve

We write the first two conversions above as a ratio equal to 1, such as (1 km)/(1000 m) = 1. We then multiply each of these by 47.0 m/s to find the speed in km/h. Note that we must set up the ratios so that the units of meters (m) and seconds (s) cancel as shown.

$$47.0 \text{ m/s} = \left(47.0 \frac{\cancel{\text{m}}}{\cancel{\text{s}}}\right)\left(\frac{3600 \cancel{\text{s}}}{1 \text{ h}}\right)\left(\frac{1 \text{ km}}{1000 \cancel{\text{m}}}\right)$$
$$= 169 \text{ km/h}$$

To convert this speed to mi/h, we use the ratio (1 mi)/(1.609 km) = 1.

$$169 \text{ km/h} = \left(169 \frac{\text{km}}{\text{h}}\right)\left(\frac{1 \text{ mi}}{1.609 \text{ km}}\right) = 105 \text{ mi/h}$$

Reflect

Since there are 3600 s in an hour, an object that moves at 1 m/s will travel 3600 m, or 3.6 km, in 1 h. So 1 m/s = 3.6 km/h. This says that *any* speed in km/h will be 3.6 times greater than the speed in m/s. You can check our first result by verifying that 169 is just 3.6 times 47.0.

To check our second result, note that a mile is a greater distance than a kilometer. Hence our fast-flying bird will travel fewer miles than kilometers in a one-hour time interval, and the speed in mi/h will be less than in km/h. This is just what we found.

WATCH OUT! Use the correct conversion factor.

You can get into trouble if you are careless in applying the method of taking the number whose units are to be converted and multiplying it by a conversion factor equal to 1. For example, suppose we tried to find the speed of the white-throated needletail (Example 1-2) in mi/h by multiplying 169 km/h by the conversion factor 1 = (1.609 km)/(1 mi).

$$169 \text{ km/h} = \left(169 \frac{\text{km}}{\text{h}}\right)\left(\frac{1.609 \text{ km}}{1 \text{ mi}}\right) = 272 \text{ km}^2/(\text{h}\cdot\text{mi})$$

We would then get a result with strange units (kilometers squared per hour per mile) because the unwanted units of kilometers didn't cancel. What's more, the speed in mi/h should be less than in km/h, not more. If you *always* keep track of units when doing conversions, you'll avoid mistakes of this kind. (See Figure 1-4 for another example of incorrect vs. correct unit conversion.)

Quantities versus Units

In this book we'll often use symbols to denote the values of physical quantities. We'll always write these symbols using *italics*. Some examples include the symbols v for speed, B for magnetic field, and E for energy. (As these examples show, the symbol is not always the first letter of the name of the quantity.) By contrast, the letters that we use to denote units, such as m (meters), s (seconds), and kg (kilograms), are not italicized. If the unit is named for a person, the letter is capitalized. An example is the unit of power, the watt (abbreviated W), named for the Scottish inventor James Watt.

In some cases the same symbol is used for more than one quantity. One example is the symbol T, which can denote a temperature, a period of time, or the tension in a stretched rope. The letter T (not italicized) is also used as an abbreviation for the unit of magnetic field, the tesla (named for the Serbian-American physicist Nikola Tesla). You'll often need to pay careful attention to the context in which a symbol is used to discern the quantity or unit it represents.

Scientific Notation

Physicists investigate objects that vary in size from the largest structures in the universe, including galaxies and clusters of galaxies, down to atomic nuclei and the particles found within nuclei. The time intervals that they analyze range from the age of the universe to a tiny fraction of a second. To describe such a wide range of phenomena, we need an equally wide range of both large and small numbers.

To avoid such confusing terms as "a million billion billion," physicists use a standard shorthand system called **scientific notation**. All the cumbersome zeros that accompany a large number are consolidated into one term consisting of 10 followed by an **exponent**, which is written as a superscript. The exponent indicates how many zeros you would need to write out the long form of the number. The exponent also tells you how many tens must be multiplied together to give the desired number, which is why the exponent is also called the **power of ten**. For example, ten thousand can be written as 10^4 ("ten to the fourth" or "ten to the fourth power") because $10^4 = 10 \times 10 \times 10 \times 10 = 10{,}000$. Some powers of ten are so common that we have created abbreviations for them that appear as a prefix to the actual unit. For example, rather than writing 10^9 watts (10^9 W), we can write 1 gigawatt, or 1 GW. Table 1-2 lists the prefix and its abbreviation for several common powers of ten.

TABLE 1-2 Prefixes

Factor	Prefix	Symbol
10^{-24}	yocto	y
10^{-21}	zepto	z
10^{-18}	atto	a
10^{-15}	femto	f
10^{-12}	pico	p
10^{-9}	nano	n
10^{-6}	micro	μ
10^{-3}	milli	m
10^{-2}	centi	c
10^{-1}	deci	d
10^{1}	deka	da
10^{2}	hecto	h
10^{3}	kilo	k
10^{6}	mega	M
10^{9}	giga	G
10^{12}	tera	T
10^{15}	peta	P
10^{18}	exa	E
10^{21}	zetta	Z
10^{24}	yotta	Y

GOT THE CONCEPT? 1-1
Smallest to Biggest

Each of these numbers represents the size of something. Arrange them from smallest to largest.
(a) 0.1 mm; (b) 7 μm;
(c) 6380 km; (d) 165 cm;
(e) 200 nm.

(Answers to *Got the Concept?* questions can be found at the end of the chapter.)

In scientific notation, numbers are usually written as a figure between 1 and 10 multiplied by the appropriate power of ten. The approximate distance between Earth and the Sun, for example, can be written as 1.5×10^8 km. This is far more convenient than writing 150,000,000 km or one hundred and fifty million kilometers. A power output of 2500 watts (2500 W) can be written in scientific notation as 2.5×10^3 W. (Alternatively, we can write this as 2.5 kW, where 1 kW = 1 kilowatt = 10^3 W.)

Note that most electronic calculators use a shorthand for scientific notation. To enter the number 1.5×10^8 you first enter 1.5, then press a key labeled EXP, EE, or 10^x, then enter the exponent 8. (The EXP, EE, or 10^x key takes care of the $\times$ 10 part of the expression.)

The number will then appear on your calculator's display as 1.5 E 8, 1.5 8, or some variation of these; typically the $\times$ 10 is not displayed as such. There are some variations from one calculator to another, so you should spend a few minutes to make sure you know the correct procedure for working with numbers in scientific notation. You will be using this notation continually in your study of physics, so this is time well spent.

WATCH OUT! Know how your calculator expresses scientific notation.

Confusion can result from the way calculators display scientific notation. Since 1.5×10^8 is displayed as 1.5 8 or 1.5 E 8, it is not uncommon to think that 1.5×10^8 is the same as 1.5^8. That is not correct, however; 1.5^8 is equal to 1.5 multiplied by itself eight times, or 25.63, which is not even close to 150,000,000 = 1.5×10^8. Another common mistake is to write 1.5×10^8 as 15^8. If you are inclined to do this, perhaps you are thinking that you can multiply 1.5 by 10, then tack on the exponent later. This also does not work; 15^8 is equal to 15 multiplied by itself eight times, or 2,562,890,625, which again is nowhere near 1.5×10^8. Reading over the manual for your calculator will help you to avoid these common errors.

GOT THE CONCEPT? 1-2
Comparing Sizes

For each pair, determine which quantity is bigger and find the ratio of the larger quantity to the smaller quantity.
(a) 1 mg, 1 kg
(b) 1 mm, 1 cm
(c) 1 MW, 1 kW
(d) 10^{-10} m, 10^{-14} m
(e) 10^{10} m, 10^{14} m

You can use scientific notation for numbers that are less than one by using a minus sign in front of the exponent. A negative exponent tells you to *divide* by the appropriate number of tens. For example, 10^{-2} ("ten to the minus two") means to divide by 10 twice, so $10^{-2} = (1/10) \times (1/10) = 1/100 = 0.01$. Table 1-2 also includes negative powers of ten.

You can also use scientific notation to express a number such as 0.00245: $0.00245 = 2.45 \times 0.001 = 2.45 \times 10^{-3}$. (Again, in scientific notation the first figure is a number between 1 and 10.) This notation is particularly useful when dealing with very small numbers. As an example, the diameter of a hydrogen atom is much more convenient to state in scientific notation (1.1×10^{-10} m) than as a decimal (0.00000000011 m) or a fraction (110 trillionths of a meter).

BIO-Medical EXAMPLE 1-3 Unit Conversion and Scientific Notation: Hair Growth

The hair on a typical person's head grows at an average rate of 1.5 cm per month (**Figure 1-5**). Approximately how far (in meters) will the ends of your hair move during a 50-min physics class? Express the answer using scientific notation.

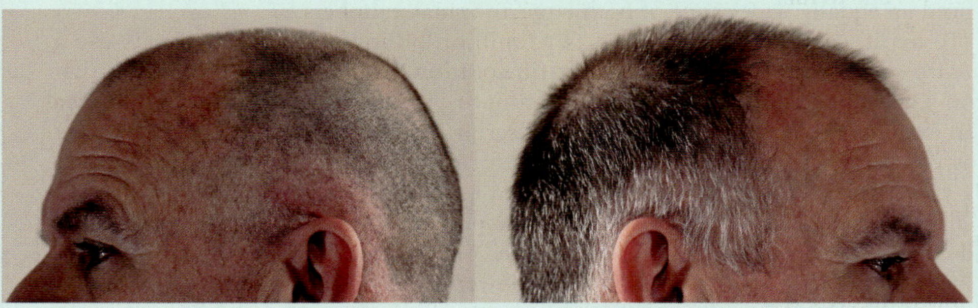

Figure 1-5 Hair growth How much does a person's hair grow during a 50-min class period?

Set Up

Like Example 1-2, this is a problem in converting units. We'll again follow the procedure shown in Figure 1-4. Not all months have the same number of days, so we'll assume an average 30-day month. We'll use the idea that since speed equals distance per time, the distance the hair moves equals the growth speed multiplied by the time in one class period.

Conversions:

$$1\ \text{min} = 60\ \text{s}$$
$$1\ \text{d} = 86{,}400\ \text{s}$$
$$1\ \text{month (average)} = 30\ \text{d}$$
$$1\ \text{cm} = 0.01\ \text{m}$$
$$\text{speed} = \frac{\text{distance}}{\text{time}}$$
$$\text{so distance} = \text{speed} \times \text{time}$$

Solve

Let's first express the speed at which hair grows in SI units (m/s). We use the conversion factors relating meters to centimeters, months to days, and days to seconds, and cancel units. We must move the decimal point to the right by nine steps to put the result in scientific notation, so the exponent is −9.

$$1.5\ \text{cm/month} = \left(1.5\ \frac{\text{cm}}{\text{month}}\right)\left(\frac{0.01\ \text{m}}{1\ \text{cm}}\right)\left(\frac{1\ \text{month}}{30\ \text{d}}\right)\left(\frac{1\ \text{d}}{86{,}400\ \text{s}}\right)$$
$$= 0.0000000058\ \text{m/s} = 5.8 \times 10^{-9}\ \text{m/s}$$

To find the distance that hair grows in 50 min, we multiply this speed by 50 min and include the conversion factor relating seconds and minutes. We then put the result into scientific notation. Note that the first number in the result is between 1 and 10. Note also that when multiplying together powers of ten, we simply add the exponents.

$$\text{distance} = \left(5.8 \times 10^{-9}\ \frac{\text{m}}{\text{s}}\right)(50\ \text{min})\left(\frac{60\ \text{s}}{1\ \text{min}}\right)$$
$$= (5.8 \times 10^{-9})(3 \times 10^{3})\ \text{m}$$
$$= 5.8 \times 3 \times 10^{-9} \times 10^{3}\ \text{m}$$
$$= 17 \times 10^{-9} \times 10^{3}\ \text{m}$$
$$= 1.7 \times 10^{1} \times 10^{-9} \times 10^{3}\ \text{m}$$
$$= 1.7 \times 10^{1-9+3}\ \text{m}$$
$$= 1.7 \times 10^{-5}\ \text{m}$$

Reflect

You can also express this result in micrometers (μm). Note that when dividing powers of ten, we subtract the exponents. A distance of 17 μm is about 1/40 mm. Do you think you could notice this change in length with the unaided eye?

$$1.7 \times 10^{-5}\ \text{m} = (1.7 \times 10^{-5}\ \text{m})\left(\frac{1\ \mu\text{m}}{10^{-6}\ \text{m}}\right)$$
$$= 1.7 \times \frac{10^{-5}}{10^{-6}}\ \mu\text{m}$$
$$= 1.7 \times 10^{-5-(-6)}\ \mu\text{m}$$
$$= 1.7 \times 10^{1}\ \mu\text{m} = 17\ \mu\text{m}$$

TAKE-HOME MESSAGE FOR Section 1-3

✔ Many physical quantities can be measured in units that are combinations of the SI units seconds, meters, and kilograms.

✔ To convert a physical quantity from one set of units to another, multiply it by an appropriate conversion factor equal to 1.

✔ In scientific notation, quantities are expressed as a number between 1 and 10 multiplied by 10 raised to a power.

✔ Prefixes can be used instead of scientific notation to indicate 10 raised to a power.

1-4 Correct use of significant figures helps keep track of uncertainties in numerical values

Although physicists strive to make their measurements as precise as possible, the unavoidable fact is that there is always some uncertainty in every measurement. This is not a result of sloppiness or of a lack of care on the part of the person doing the measurement; rather, it is due to fundamental limitations on how well measurements can be carried out.

If you try to measure your thumbnail using a ruler with markings in millimeters, you might find that it's between 15 and 16 mm in length. With keen eyes you might be able to tell that the length is very nearly halfway between those two markings on the ruler, so you might say that the length is approximately 15.5 mm. However, armed only with your ruler it would be impossible to say whether the length was 15.47, 15.51, or 15.54 mm. We would say that the fourth digit is *not significant*, that is, it does not contain meaningful information. Hence this measurement has only three **significant figures.** We specify this implicitly by giving the measured value as 15.5 mm, by which we mean that it's between 15.4 and 15.6 mm. More precise measurements have more significant figures; less precise measurements have fewer significant figures.

√x *See the Math Tutorial for more information on significant figures.*

An alternative way to present the number would be to give the length of your thumbnail as 15.5 ± 0.1 mm, with the number after the $\pm$ sign representing the uncertainty. This kind of presentation is very common in experimental work. In this book, however, we'll use significant figures to represent uncertainty.

Note that some numbers are *exact*, with no uncertainty whatsoever. Three examples are the number of meters in a kilometer (exactly 1000), the speed of light (defined to be exactly 299,792,458 m/s, as we learned in Section 1-3), and the value of π. The value of any measured quantity, however, will have some uncertainty.

Scientific notation is helpful in representing the number of significant figures in a measurement. When we say that the diameter of the Moon at its equator is 3476 km, we imply that there are four significant figures in the value and that the uncertainty in the diameter is only about 1 km. Since there are 1000 m in 1 km, we could also express the diameter as 3,476,000 m. This is misleading, however, since it gives the impression that there are *seven* significant figures in the value. That isn't correct; we do *not* know the value of the Moon's equatorial radius to within 1 meter. A better way to express the diameter in meters is as 3.476×10^6 m, which makes it explicit that there are only four significant figures in the value. The radius of the minor planet Huya is less well known and is about 460 km; only the first two digits are significant, so it's more accurate to express this dimension as 4.6×10^2 km.

GOT THE CONCEPT? 1-3
Significant Figures

? All of these numbers have the same value but are written with a different number of significant figures: (a) 123, (b) 1.230×10^2, (c) 123.000, (d) 1.23×10^2, (e) 0.12300×10^3. Rank them in order of least to most significant figures.

There are special rules to help determine whether zeroes in a number are significant. *Leading zeroes* are not significant, so the number 0.0035 has only two significant figures. *Trailing zeroes* to the right of the decimal point are significant, so the number 0.3500 has four significant figures. When we have trailing zeroes to the left of the decimal point, as in 350, things get ambiguous. To make it clear that the trailing zero is significant, it is best to write 3.50×10^2. However, there are times when it's more expedient to write exact numbers without scientific notation. For example, there are exactly 60 s in 1 min, so the 60 has as many significant figures as we need, and we won't bother writing 6.0000×10^1 s or 60.000 to indicate that the value should have five significant figures.

In most problems in this book, we'll use numbers with two, three, or four significant figures. For example, you might be asked about the motion of an ostrich that runs 43.9 m (three significant figures) in 2.3 s (two significant figures). If you use your calculator to find the speed of this ostrich (the distance traveled divided by the elapsed time), the result will be

$$\frac{43.9 \text{ m}}{2.3 \text{ s}} = 19.08695652 \text{ m/s} \quad \text{(incorrect)}$$

This answer is incorrect because there are *not* 10 significant figures in the answer. You do not know the result that precisely! The general rule for *multiplying or dividing* numbers is that the result has the same number of significant figures as the input number with the *fewest* significant figures. In our example the elapsed time of 2.3 s has fewer significant figures (two) than the distance of 43.9 m, so the result has only two significant figures. Rounding our result, we get

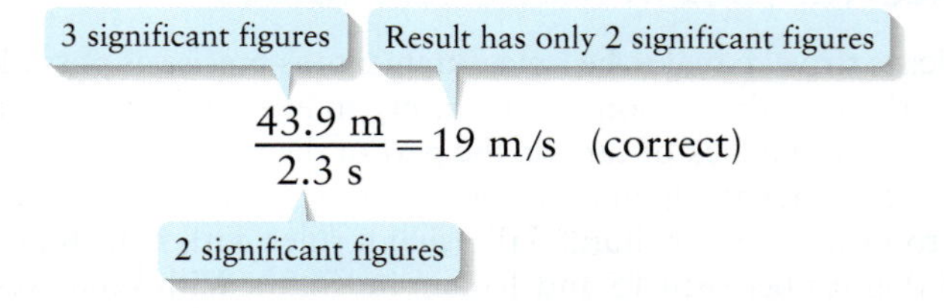

Following the same rule, if a strip of cloth is 35.65 cm long and 2.49 cm wide, the area (length times width) has only three significant figures: $(35.65 \text{ cm})(2.49 \text{ cm}) = 88.7685 \text{ cm}^2$, which rounds to 88.8 cm^2.

There's a slightly different rule for *adding or subtracting* numbers. Suppose that you drive 44.3 km (according to the odometer of your car) from Central City to the parking lot at the Metropolis shopping mall, then walk an additional 108 m or 0.108 km (according to your pedometer) to your favorite clothing store. The total distance you have traveled is (44.3 km) + (0.108 km), which according to your calculator is 44.408 km. This isn't quite right, however, because you know your driving distance only to the nearest tenth of a kilometer. As a result, you need to round off the answer to the nearest tenth of a kilometer, or 44.4 km. So

1 digit to the right of the decimal point | 3 digits to the right of the decimal point

$$44.3 \text{ km} + 0.108 \text{ km} = 44.4 \text{ km}$$

Result has only 1 digit to the right of the decimal point

Thus the general rule for adding or subtracting numbers is that the result can have no less uncertainty than the most uncertain input number. In our example that's the 44.3-km distance, which has an uncertainty of about 0.1 km. Stated another way, the answer may have no more digits to the right of the decimal point than the input number with the fewest digits to the right of the decimal point.

Let's summarize our general rules for working with significant figures:

Rule 1 When multiplying or dividing numbers, the result should have the same number of significant figures as the input number with the fewest significant figures. The same rule applies to squaring, taking the square root, calculating sines and cosines, and so on.

Rule 2 When adding or subtracting numbers, the result should have the same number of digits to the right of the decimal point as the input number with the fewest digits to the right of the decimal point.

We'll use these rules for working with significant figures in all of the worked examples in the book. (If you look back, you'll see that we actually used them in Examples 1-2 and 1-3 in Section 1-3.) Make sure that you follow these same rules in your own problem solving.

WATCH OUT! Be careful with significant figures and scientific notation.

For a number given in scientific notation, the exponent of 10 is known exactly; it doesn't affect the number of significant figures. So 1.45×10^{-2} has three significant figures, and 2.2×10^{-4} has two significant figures. If you *multiply* these together, by Rule 1 their product has only two significant figures: $(1.45 \times 10^{-2})(2.2 \times 10^{-4}) = 3.19 \times 10^{-2+(-4)} = 3.19 \times 10^{-6}$, which rounds to 3.2×10^{-6}. To *add* these two numbers together, first write them using the same exponent of 10: $1.45 \times 10^{-2} + 2.2 \times 10^{-4} = 1.45 \times 10^{-2} + 0.022 \times 10^{-2} = (1.45 + 0.022) \times 10^{-2}$. In this sum, there are two significant figures after the decimal point for the first number versus three significant figures for the second number. By Rule 2 the sum can have only two significant figures after the decimal point, so we must round 1.45 + 0.022 = 1.472 to 1.47, and the final answer is 1.47×10^{-2}.

WATCH OUT! Be mindful of significant figures when rounding numbers.

If you're doing a calculation with several steps, it's best to retain *all* of the digits in your calculation until the very end. Then you can round off the result as necessary to give the answer to the correct number of significant figures. For example, suppose you want to calculate $32.1 \times 4.998 \times 4.87$. Rule 1 above tells us that the final answer has only three significant figures. The product of the first two numbers is $32.1 \times 4.998 = 160.4358$, which if you round off at this point becomes 160. If you multiply this rounded number by 4.87, you get 779.2, which rounds to 779. If, however, you calculate $32.1 \times 4.998 \times 4.87$ directly using your calculator *without* rounding the intermediate answer, you get 781.322346. Rounding this to three significant figures gives the correct answer, which is 781. Moral: Don't round until the end of your calculation!

EXAMPLE 1-4 Significant Figures: Combining Volumes

What is the volume (in cubic kilometers) of Earth? What is the volume of the Moon? What is the combined volume of the two worlds together? The radius of Earth is 6378 km, and the radius of the Moon is 1738 km.

Set Up

From the Math Tutorial the volume of a sphere of radius R is $4\pi R^3/3$. Since we multiply the radius by itself, the answer can have no more significant figures than the value of the radius.

Radius of Earth $R_E = 6378$ km
Radius of Moon $R_M = 1738$ km

Solve

For Earth R_E has four significant figures, so the volume has only four significant figures.

Volume of Earth:

$$V_E = \frac{4\pi(6378 \text{ km})^3}{3} = 1.086781293 \times 10^{12} \text{ km}^3$$

Round to four significant figures:

$$V_E = 1.087 \times 10^{12} \text{ km}^3$$

The radius of the Moon also has four significant figures, and so its volume does as well.

Volume of Moon:

$$V_M = \frac{4\pi(1738 \text{ km})^3}{3} = 2.199064287 \times 10^{10} \text{ km}^3$$

Round to four significant figures:

$$V_M = 2.199 \times 10^{10} \text{ km}^3$$

To find the combined volume we add V_E and V_M. To make sure we retain the correct number of significant figures in our answer, we express both numbers in scientific notation with the same exponent. According to Rule 2 for addition, our answer must have only three significant figures to the right of the decimal point.

$$V_E = 1.087 \times 10^{12} \text{ km}^3$$
$$V_M = 2.199 \times 10^{10} \text{ km}^3 = 0.02199 \times 10^{12} \text{ km}^3$$

To avoid intermediate rounding errors we will use the unrounded values and then round the result to three digits to the right of the decimal point at the end of the calculation.

$$V_E + V_M = (1.086781293 \times 10^{12} \text{ km}^3) + (0.02199064287 \times 10^{12} \text{ km}^3)$$
$$= (1.086781293 + 0.02199064287) \times 10^{12} \text{ km}^3$$
$$= 1.10877193587 \times 10^{12} \text{ km}^3$$

This has too many significant figures, so we round to the final answer of

$$V_E + V_M = 1.109 \times 10^{12} \text{ km}^3$$

Reflect

In this problem we had to use both the significant figure rule for multiplication and the significant figure rule for addition. You should be prepared to use both of these rules in solving problems on your own. In this example if we had used the rounded values of the volumes, we would have gotten $V_E + V_M = 1.087 \times 10^{12} \text{ km}^3 + 0.02199 \times 10^{12} \text{ km}^3 = 1.10899 \times 10^{12} \text{ km}^3$, which still rounds to $1.109 \times 10^{12} \text{ km}^3$. But this is not always the case, so don't round until you have finished all your calculations.

TAKE-HOME MESSAGE FOR Section 1-4

✔ Significant figures characterize the precision of a measurement or the uncertainty of a numerical value.

✔ When multiplying or dividing numbers the result has the same number of significant figures as the input number with the fewest significant figures.

✔ When adding or subtracting numbers the result has the same number of digits to the right of the decimal point as the input number with the fewest number of digits to the right of the decimal point *when all numbers are expressed in scientific notation with the same exponent.*

✔ When writing a number all trailing zeroes to the right of the decimal point are significant figures.

1-5 Dimensional analysis is a powerful way to check the results of a physics calculation

Once you've solved a problem, it's useful to be able to quickly determine how likely it is that you've solved it correctly. One valuable approach is to do a **dimensional analysis.** This approach is particularly useful when you calculate an algebraic result and want to check it before substituting numerical values.

To understand what we mean by dimensional analysis, first note that any physical quantity has a *dimension.* Three physical quantities that we introduced in Section 1-3 are mass, length, and time. (Table 1-1 in Section 1-3 lists additional physical quantities.) We say that the mass of a hydrogen atom has dimensions of mass, the diameter of the atom has dimensions of length, and the amount of time it takes light to travel that diameter has dimensions of time. Mass, length, and time are *fundamental* dimensions because they cannot be expressed in terms of other, more fundamental quantities. (We'll encounter a few other fundamental dimensions in later chapters.)

Dimensions *have* units but are not units themselves. Units are used to help express the numerical value of a dimension. A dimension of length may be expressed in units of meters; for example, the length of a typical house cat is 0.65 m. The dimension of time might be expressed in units of nanoseconds, and the dimension of mass might be expressed in units of kilograms.

The dimensions of many quantities are made up of a combination of fundamental dimensions. One example is speed, which has dimensions of distance per time, or (distance)/(time) for short. Another is volume, which has dimensions of distance cubed, or $(\text{distance})^3$.

Here's the key to using dimensional analysis: *In any equation, the dimensions must be the same on both sides of the equation.* As an example, suppose you hold a ball at height h above the ground and then let it fall. The ball takes a time t to reach the ground. You calculate that the relationship between the height h and the time t is

$$h = vt^2 \tag{1-2}$$

where v is the speed of the ball just before it hits the ground. Is Equation 1-2 dimensionally correct? To find out, replace each symbol by the dimensions of the quantity that it represents. The height h has dimensions of length or distance, speed v has dimensions of (distance)/(time), and time t has dimensions of time. Dimensions cancel just like algebraic quantities, so if we simplify this, we get

$$\begin{aligned}\text{distance} &= \frac{\text{distance}}{\text{time}} \times (\text{time})^2 \\ &= \frac{\text{distance}}{\text{time}} \times \text{time} \times \text{time} \\ &= \text{distance} \times \text{time}\end{aligned}$$

This is *inconsistent*: The left-hand side of Equation 1-2 (h) has dimensions of distance, while the right-hand side (vt^2) has dimensions of distance multiplied by time. So Equation 1-2 *cannot* be correct.

It turns out that the correct relationship between the height h from which the ball is dropped and the time t it takes to reach the ground is

$$h = \frac{1}{2}gt^2 \tag{1-3}$$

where g is a constant. Let's use dimensional analysis to figure out the dimensions of g. Again h has units of distance and t has units of time. The number 1/2 has no dimensions at all, so it can be ignored (it is a pure number with no units, so it is *dimensionless*). We rewrite Equation 1-3 in terms of dimensions and solve for the dimensions of g:

$$\begin{aligned}\text{distance} &= (\text{dimensions of } g) \times (\text{time})^2 \\ (\text{dimensions of } g) &= \frac{\text{distance}}{(\text{time})^2}\end{aligned}$$

WATCH OUT! Dimensions are not the same as units.

! Be careful to make the distinction between *dimensions* and *units.* For example, volume may be expressed in any number of different units such as m^3, cm^3, mm^3, or km^3. But no matter what units are used, volume *always* has dimensions of $(\text{distance})^3$.

The SI units of g must then be m/s^2 (meters per second squared, or meters per second per second). We'll learn the significance of the quantity g (called the *acceleration due to gravity*) in Chapter 2.

WATCH OUT! Dimensional analysis can't tell you everything.

! Suppose that as a result of an algebraic error, we failed to include the factor of 1/2 in Equation 1-3 and instead found $h = gt^2$. Because the pure number 1/2 has no dimensions, it wouldn't change the dimensional analysis we did above. So a dimensional analysis would not catch this mistake. Performing dimensional analysis can tell you whether an answer is wrong, but even if the dimensions of your answer are correct, the numerical value might still be wrong.

GOT THE CONCEPT? 1-4 Dimensional Analysis

? A block of mass m oscillates back and forth on the end of a spring. (Flip forward in the book and look at Figure 12-2.) You'll find that the time T that the block takes for one full back-and-forth cycle depends on a constant k that has dimensions of mass divided by time squared according to

$$T = 2\pi\sqrt{\frac{m}{k}}$$

Use dimensional analysis to determine whether this result could be correct.

TAKE-HOME MESSAGE FOR Section 1-5

✔ Dimensional analysis is the technique of checking the dimensions of your algebraic answer before you substitute values to compute a numeric result.

✔ The dimensions of any quantity can be put together from a few fundamental dimensions, including mass, distance, and time.

✔ The key to using dimensional analysis: *In any equation the dimensions must be the same on both sides of the equation.*

Key Terms

dimensional analysis
exponent
kilogram
mass
meter
power of ten
scientific notation
second
significant figures
Système International
units

Chapter Summary

Topic	Equation or Figure
Problem-solving strategy: We'll solve all problems using a strategy that incorporates three steps.	*Set Up* the problem *Solve* for the desired quantities *Reflect* on the answer
Unit conversion: Units can be converted by multiplying any value by an appropriate representation of 1.	$47.0\ \text{m/s} = \left(47.0\frac{\cancel{\text{m}}}{\cancel{\text{s}}}\right)\left(\frac{3600\ \cancel{\text{s}}}{1\ \text{h}}\right)\left(\frac{1\ \text{km}}{1000\ \cancel{\text{m}}}\right) = 169\ \text{km/h}$
Scientific notation: Any number can be written in scientific notation, which usually consists of a number between 1 and 10 multiplied by 10 raised to some power. Prefixes can be used as abbreviations for many powers of ten.	$150{,}000{,}000\ \text{km} = 1.5 \times 10^8\ \text{km}$ $2500\ \text{watts} = 2.5 \times 10^3\ \text{W} = 2.5\ \text{kilowatts} = 2.5\ \text{kW}$

Significant figures: Any measured quantity has a finite number of significant digits. You need to keep track of the number of significant digits in all your calculations. There are different rules for addition/subtraction and multiplication/division.

$$\frac{43.9\text{ m}}{2.3\text{ s}} = 19\text{ m/s} \quad \text{(correct)}$$

3 significant figures | Result has only 2 significant figures | 2 significant figures

Rule 1 When multiplying or dividing numbers, the result should have the same number of significant figures as the input number with the fewest significant figures. The same rule applies to squaring, taking the square root, calculating sines and cosines, and so on.

Rule 2 When adding or subtracting numbers, the result should have the same number of digits to the right of the decimal point as the input number with the fewest digits to the right of the decimal point.

$$44.3\text{ km} + 0.108\text{ km} = 44.4\text{ km}$$

1 digit to the right of the decimal point | 3 digits to the right of the decimal point | Result has only 1 digit to the right of the decimal point

Dimensional analysis: Analyzing the dimensions of an equation can provide a clue as to the correctness of that equation. Dimensions and units are not the same thing.

$$h = \frac{1}{2}gt^2 \tag{1-3}$$

$$\text{distance} = (\text{dimensions of } g) \times (\text{time})^2$$

$$(\text{dimensions of } g) = \frac{\text{distance}}{(\text{time})^2}$$

Answer to What do you think? Question

(c) 1 cm. The length of a typical hummingbird is 7.5 to 13 cm, and the head is 1 or 2 cm in length.

Answers to Got the Concept? Questions

1-1 From smallest to largest, the order is (e), (b), (a), (d), and (c). Some examples of objects of about the size of the five distances include: 200 nanometers (e) is the diameter of a typical virus; 7 μm (b) is the diameter of a human red blood cell; 0.1 mm (a) is the diameter of a typical human hair; 165 cm (d) is the average height of a woman in the United States; and (c) 6380 km is the mean radius of Earth.

1-2 (a) One kilogram is bigger than one milligram by a factor of 1 million; 1 mg equals 0.001 g equals 0.000001 kg. (b) One centimeter is 10 times bigger than 1 mm; 1 mm equals 0.001 m, and 1 cm equals 0.01 m. (c) One megawatt is 1000 times bigger than 1 kW; 1 MW equals 1,000,000 W, and 1 kW equals 1000 W. (d) A distance of 10^{-10} m is 10^4 (10,000) times bigger than 10^{-14} m: $(10^{-10}\text{ m})/(10^{-14}\text{ m}) = 10^{-10-(-14)} = 10^{-10+14} = 10^4$. (e) A distance 10^{14} m is 10^4 (10,000) times bigger than 10^{10} m; $(10^{14}\text{ m})/(10^{10}\text{ m}) = 10^{14-10} = 10^4$.

1-3 123 has three significant digits, as does 1.23×10^2. 1.230×10^2 is written with four significant digits, 0.12300×10^3 has five significant digits, and 123.000 has six.

1-4 Since m has dimensions of mass, k has dimensions of mass per time squared, and 2π is dimensionless, the dimensions of the right-hand side of the equation $T = 2\pi\sqrt{m/k}$ are

$$\sqrt{\frac{\text{mass}}{\text{mass}/(\text{time})^2}} = \sqrt{\text{mass} \times \frac{(\text{time})^2}{\text{mass}}}$$
$$= \sqrt{(\text{time})^2} = \text{time}$$

The left-hand side of the equation, T, also has dimensions of time. Hence you can have some confidence that this relationship is correct. You cannot say, however, that this equation is *exactly* correct because the pure number 2π has no dimensions and therefore does not play a role in the dimensional analysis. Had the equation been, say, $T = 5\sqrt{m/k}$ the dimensional analysis would be exactly the same.

Questions and Problems

In a few problems you are given more data than you actually need; in a few other problems you are required to supply data from your general knowledge, outside sources, or informed estimate.

For all problems use $g = 9.80\ m/s^2$ for the free-fall acceleration due to gravity. Neglect friction and air resistance unless instructed to do otherwise.

• Basic, single-concept problem
•• Intermediate-level problem; may require synthesis of concepts and multiple steps
••• Challenging problem
SSM *Solution is in Student Solutions Manual*
Example *See worked example for a similar problem*

Conceptual Questions

1\. • Define the basic quantities in physics (length, mass, and time) and list the appropriate SI unit that is used for each quantity.

2\. • Is it possible to define a system of units in which length is not one of the fundamental properties?

3\. • What properties should an object, system, or process have for it to be a useful standard of measurement of a physical quantity such as length or time?

4\. • Why do physicists and other scientists prefer to use metric units and prefixes (for example, micro, milli, and centi) to American Standard Units, for example, inches and pounds? SSM

5\. • All valid equations in physics have consistent units. Are all equations that have consistent units valid? Support your answer with examples. SSM

6\. • If you use a calculator to divide 3411 by 62.0, you will get something like 55.016129. (Exactly what you get will depend on your calculator.) Of course, you know that all those decimal places aren't significant. How should you write the answer?

7\. • Consider the number 61,000. (a) What is the least number of significant figures this might have? (b) The greatest number? (c) If the same number is expressed as 6.10×10^4, how many significant figures does it have?

8\. • Acceleration has dimensions L/T^2, where L is length and T is time. What are the SI units of acceleration? SSM

Multiple-Choice Questions

9\. • Which of the following are fundamental quantities?
A. density (mass per volume)
B. length
C. area
D. resistance
E. all of the above

10\. • Which length is the largest?
A. 10 nm
B. 10 cm
C. 10^2 mm
D. 10^{-2} m
E. 1 m

11\. • One nanosecond is
A. 10^{-15} s
B. 10^{-6} s
C. 10^{-9} s
D. 10^{-3} s
E. 10^9 s

12\. • How many cubic meters are there in a cubic centimeter?
A. 10^2
B. 10^6
C. 10^{-2}
D. 10^{-3}
E. 10^{-6}

13\. • How many square centimeters are there in a square meter?
A. 10
B. 10^2
C. 10^4
D. 10^{-2}
E. 10^{-4} SSM

14\. • Calculate 1.4 + 15 + 7.15 + 8.003 using the proper number of significant figures.
A. 31.553
B. 31.550
C. 31.55
D. 31.6
E. 32

15\. • Calculate 0.688/0.28 using the proper number of significant figures.
A. 2.4571
B. 2.457
C. 2.46
D. 2.5
E. 2

16\. • Which of the following relationships is dimensionally consistent with a value for acceleration that has dimensions of distance per time per time? In these equations x is distance, t is time, and v is speed.
A. v^2/t
B. v/t
C. v/t^2
D. v/x^2
E. v^2/x^2

17\. • Calculate 25.8×70.0 using the proper number of significant figures.
A. 1806.0
B. 1806
C. 1810
D. 1800
E. 2000 SSM

Estimation/Numerical Analysis

18\. • Estimate the distance from your dorm or apartment to your physics lecture room. Give your answer in meters.

19\. • Estimate the time required for a baseball to travel from home plate to the center field fence in a major league baseball game.

20\. • Estimate the number of laptop computers on your campus.

21\. • Estimate the volume (in liters) of water used daily by each student at your school. SSM

22\. • Estimate the amount (in kg) of nonrecyclable and noncompostable garbage that each student at your school produces each day.

23. • Estimate the height (in m) of the tallest building on your campus.

24. • Enrico Fermi once estimated the length of all the sidewalks in Chicago in the 1940s. Estimate the length (in km) of sidewalks in present-day Chicago.

25. • **Biology** Although human cells vary in size, the volume of a typical cell is equivalent to the volume of a sphere that has an approximately 10^{-5} m radius. Estimate the number of cells in your body. (*Hint*: Consider your body as a collection of cylinders.) SSM

26. • **Biology** Estimate the volume flow rate (in units of m^3/s) of the air that fills your lungs as you take a deep breath.

Problems

1-3 Measurements in physics are based on standard units of time, length, and mass

27. • Write the following numbers in scientific notation:
A. 237
B. 0.00223
C. 45.1
D. 1115
E. 14,870
F. 214.78
G. 0.00000442
H. 12,345,678 Example 1-3

28. • Write the following numbers using decimals:
A. 4.42×10^{-3}
B. 7.09×10^{-6}
C. 8.28×10^{2}
D. 6.02×10^{6}
E. 456×10^{3}
F. 22.4×10^{-3}
G. 0.375×10^{-4}
H. 138×10^{-6} Example 1-3

29. • Write the metric prefixes for the following powers of ten:
A. 10^3 = ________
B. 10^9 = ________
C. 10^6 = ________
D. 10^{12} = ________
E. 10^{-3} = ________
F. 10^{-12} = ________
G. 10^{-6} = ________
H. 10^{-9} = ________ SSM Example 1-3

30. • What are the numerical values for the following prefixes:
A. p = ________
B. m = ________
C. M = ________
D. μ = ________
E. f = ________
F. G = ________
G. T = ________
H. c = ________ Example 1-3

31. • Complete the following conversions:
A. 125 cm = ________ m
B. 233 g = ________ kg
C. 786 ms = ________ s
D. 454 kg = ________ mg
E. 208 cm^2 = ________ m^2
F. 444 m^2 = ________ cm^2
G. 12.5 cm^3 = ________ m^3
H. 144 m^3 = ________ cm^3 Example 1-2

32. • Complete the following conversions:
A. 238 ft = ________ m
B. 772 in = ________ cm
C. 1220 in^2 = ________ cm^2
D. 559 oz = ________ L
E. 973 lb = ________ g
F. 122 ft^3 = ________ m^3
G. 1.28 mi^2 = ________ km^2
H. 442 in^3 = ________ cm^3 Example 1-2

33. • Complete the following conversions:
A. 328 cm^3 = ________ L
B. 112 L = ________ m^3
C. 220 hectares = ________ m^2
D. 44,300 m^2 = ________ hectares
E. 225 L = ________ m^3
F. 17.2 hectares · m = ________ L
G. 225,300 L = ________ hectares · m
H. 2000 m^3 = ________ mL SSM Example 1-2

34. • Complete the following conversions:
A. 33.5 gal = ________ L
B. 62.8 L = ________ gal
C. 216 acre · ft = ________ L
D. 1770 gal = ________ m^3
E. 22.8 fl oz = ________ cm^3
F. 54.2 cm^3 = ________ cups
G. 1.25 hectares = ________ acre
H. 664 mm^3 = ________ qt Example 1-2

35. • Write the following quantities in scientific notation without prefixes: (a) 300 km, (b) 33.7 μm, and (c) 77.5 GW. Example 1-3

36. • Write the following with prefixes and not using scientific notation: (a) 3.45×10^{-4} s, (b) 2.00×10^{-11} W, (c) 2.337×10^{8} m, and (d) 6.54×10^{4} g. Example 1-3

37. •• The United States is about the only country that still uses the units feet, miles, and gallons. However, you might see some car specifications that give fuel efficiency as 7.6 km per kilogram of fuel. Given that a mile is 1.609 km, a gallon is 3.785 liters, and a liter of gasoline has a mass of 0.729 kg, what is the car's fuel efficiency in miles per gallon? SSM Example 1-2

1-4 Correct use of significant figures helps keep track of uncertainties in numerical values

38. • Give the number of significant figures in each of the following numbers:
A. 112.4
B. 10
C. 3.14159
D. 700
E. 1204.0
F. 0.0030
G. 9.33×10^{3}
H. 0.02240 Example 1-4

39. • Complete the following operations using the correct number of significant figures:
A. 5.36×2.0 = ________
B. $\frac{14.2}{2}$ = ________
C. 2×3.14159 = ________
D. 4.040×5.55 = ________
E. 4.444×3.33 = ________
F. $\frac{1000}{333.3}$ = ________
G. 2.244×88.66 = ________
H. 133×2.000 = ________ Example 1-4

40. • Complete the following operations using the correct number of significant figures:

A.	B.	C.	D.
4.55	80.00	71.1	200
+ 21.6	− 112.3	+ 3.70	+ 33.7

SSM

Example 1-4

1-5 Dimensional analysis is a powerful way to check the results of a physics calculation

41. • One equation that describes motion of an object is $x = \frac{1}{2}at^2 + v_0 t + x_0$, where x is the position of the object, a is the acceleration, t is time, v_0 is the initial speed, and x_0 is the initial position. Show that the dimensions in the equation are consistent.

42. • One equation that describes motion of an object is $x = vt + x_0$, where x is the position of the object, v is its speed, t is time, and x_0 is the initial position. Show that the dimensions in the equation are consistent. SSM

43. • The motion of a vibrating system is described by $y(x, t) = A_0 e^{-\alpha t} \sin(kx - \omega t)$. Find the SI units for k, ω, and α.

44. • The kinetic energy of a particle is $K = \frac{1}{2}mv^2$, where m is the mass of the particle and v is its speed. Show that 1 joule (J), the SI unit of energy, is equivalent to 1 $\text{kg} \cdot \text{m}^2/\text{s}^2$.

45. • The period T of a simple pendulum, the time for one complete oscillation, is given by $T = 2\pi\sqrt{(L/g)}$, where L is the length of the pendulum and g is the acceleration due to gravity. Show that the dimensions in the equation are consistent.

46. • Which of the following could be correct based on a dimensional analysis?

A. The volume flow rate is 64 m^3/s.
B. The height of the Transamerica building is 332 m^2.
C. The time required for a fortnight is 66 m/s.
D. The speed of the train is 9.8 m/s^2.
E. The weight of a standard kilogram mass is 2.2 lb.
F. The density of gold is 19.3 kg/m^2. SSM

General Problems

47. ••• You hire a printer to print concert tickets. He delivers them in circular rolls labeled as 1000 tickets each. You want to check the number of tickets in each roll without counting thousands of tickets. You decide to do it by measuring the diameter of the rolls. If the tickets are 2 in long and 0.22 mm thick and are rolled on a core 3 cm in diameter, what should be the diameter of a roll of 1000 tickets? Example 1-4

48. • **Biology** A typical human cell is approximately 10 μm in diameter and enclosed by a membrane that is 5.0 nm thick. (a) What is the volume of the cell? (b) What is the volume of the cell membrane? (c) What percentage of the cell volume does its membrane occupy? Model the cell as a sphere. Example 1-3

49. • **Medical** Express each quantity in the standard SI units requested. (a) An adult should have no more than 2500 mg of sodium per day. What is the limit in kg? (b) A 240-mL cup of whole milk contains 35 mg of cholesterol. Express the cholesterol concentration in the milk in kg/m^3 and in mg/mL. (c) A typical human cell is about 10 μm in diameter, modeled as a sphere. Express its volume in cubic meters. (d) A low-strength aspirin tablet (sometimes called a baby aspirin) contains 81 mg of the active ingredient. How many kg of the active ingredient does a 100-tablet bottle of baby aspirin contain? (e) The average flow rate of urine out of the kidneys is typically 1.2 mL/min. Express the rate in m^3/s. (f) The density of blood proteins is about 1.4 g/cm^3. Express the density in kg/m^3. Example 1-2

50. • **Medical** The concentration of PSA (prostate-specific antigen) in the blood is sometimes used as a screening test for possible prostate cancer in men. The value is normally reported as the number of nanograms of PSA per milliliter of blood. A PSA of 1.7 (ng/mL) is considered low. Express that value in (a) g/L, (b) standard SI units of kg/m^3, and (c) μg/L. Example 1-2

51. • The acceleration g (units m/s^2) of a falling object near a planet is given by the following equation $g = GM/R^2$. If the planet's mass M is expressed in kg and the distance of the object from the planet's center R is expressed in m, determine the units of the gravitational constant G. Example 1-2

52. ••• **Medical** At a resting pulse rate of 75 beats per minute, the human heart typically pumps about 70 mL of blood per beat. Blood has a density of 1060 kg/m^3. Circulating all of the blood in the body through the heart takes about 1 min in a person at rest. (a) How much blood (in L and m^3) is in the body? (b) On average, what mass of blood (in g and kg) does the heart pump each beat? SSM Example 1-2

53. • During a certain experiment, light is found to take 37.1 μs to traverse a measured distance of 11.12 km. Determine the speed of light from the data. Express your answer in SI units and in scientific notation, using the appropriate number of significant figures. Example 1-3

54. •• **Medical** The body mass index (BMI) estimates the amount of fat in a person's body. It is defined as the person's mass m in kg divided by the square of the person's height h in m. (a) Write the formula for BMI in terms of m and h. (b) In the United States, most people measure weight in pounds and height in feet and inches. Show that with weight W in pounds and height h in inches, the BMI formula is BMI $= 703\ W/h^2$. (c) A person with a BMI between 25.0 and 30.0 is considered overweight. If a person is 5′11″ tall, for what range of mass will he be considered overweight? Example 1-2

55. •• **Medical** A typical prostate gland has a mass of about 20 g and is about the size of a walnut. The gland can be modeled as a sphere 4.50 cm in diameter. (a) What is the density (mass per volume) of the prostate? Express your answer in g/cm^3 and in standard SI units. (b) How does the density compare to that of water? (c) During a biopsy of the prostate, a thin needle is used to remove a series of cylindrical tissue samples. If the cylinders have a total length of 28.0 mm and a diameter of 0.100 mm, what is the total mass (in g) of tissue taken? (d) What percentage of the mass of the prostate is removed during the biopsy? Example 1-2

Linear Motion

Boris Austin/Getty Images

What do you think?

When a jumper is at the high point of her trajectory, is she accelerating? If so, is she accelerating upward or downward?

The answer to the *What do you think?* question can be found at the end of the chapter.

In this chapter, your goals are to:

- (2-1) State and explain the definition of linear motion.
- (2-2) Explain the meanings of and relationships among displacement, average velocity, and constant velocity.
- (2-3) Explain the distinction between velocity and acceleration, and interpret graphs of velocity versus time.
- (2-4) Use and interpret the equations and graphs for linear motion with constant acceleration.
- (2-5) Solve constant-acceleration linear motion problems.
- (2-6) Solve problems involving objects in free fall.

To master this chapter, you should review:

- (1-2) The three key steps in solving any physics problem.
- (1-4) How to use significant figures in calculations.

2-1 Studying motion in a straight line is the first step in understanding physics

We live in a universe of motion. We are surrounded by speeding cars, scampering animals, and moving currents of air. Our planet is in motion as it spins on its axis and orbits the Sun. There is motion, too, within our bodies, including the pulsations of the heart and the flow of blood through the circulatory system. One of the principal tasks of physics is to describe motion in all of its variety, and it is with the description of motion—a subject called *kinematics*—that we begin our study of physics.

In general, objects can move in all three dimensions: forward and back, left and right, and up and down. In this chapter, however, we'll concentrate on the simpler case of **linear motion** (also called *one-dimensional motion*), or motion in a straight line. Examples of linear motion include an airliner accelerating down the runway before it takes off, a blood cell moving along a capillary, and a rocket climbing upward from the launch pad. All of the concepts that we'll study in this chapter apply directly to motion in two or three dimensions, which we'll study in Chapter 3. You'll use the ideas of this chapter and the next throughout your study of physics, so the time you spend understanding them now will be an excellent investment for the future.

TAKE-HOME MESSAGE FOR Section 2-1

✔ Linear motion (also called *one-dimensional motion*) is motion in a straight line.

✔ We'll build on the ideas of linear motion in later chapters to understand more complex kinds of motion, such as motion in two or three dimensions.

2-2 Constant velocity means moving at a steady speed in the same direction

Figure 2-1 shows objects in linear motion: three swimmers traveling down the lanes of a swimming pool. Their motion is particularly simple because each swimmer has a **constant velocity**: The swimmer's speed stays the same, and the swimmer always moves in the same direction. Let's look at this kind of motion more closely.

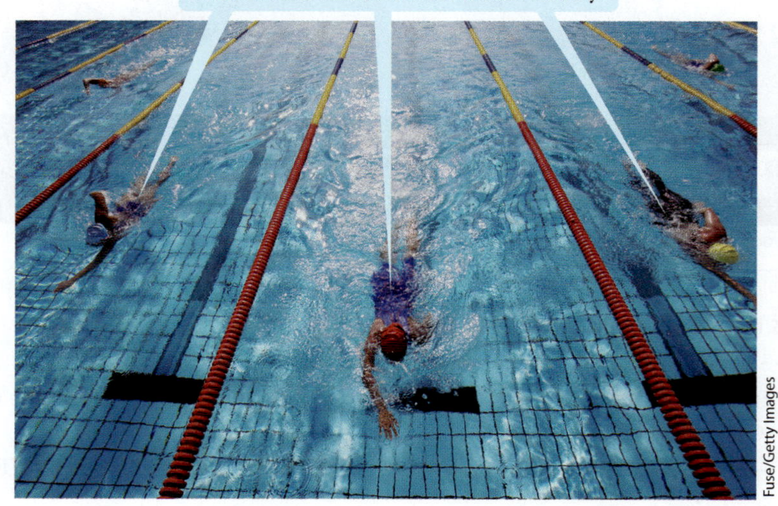

Each swimmer moves with a steady speed and always moves in the same direction. Hence each swimmer has a constant velocity.
Fuse/Getty Images

Figure 2-1 Constant velocity An object has a constant velocity when its speed remains the same (it neither speeds up nor slows down) and its direction of motion remains the same. These three swimmers all move in the same direction yet have different constant velocities because their speeds are different: The middle swimmer (who reaches the end of the pool first) is the fastest.

Coordinates, Displacement, and Average Velocity

A useful way to keep track of the changing position of a moving object (such as a swimmer) is a **coordinate system** as shown in **Figure 2-2a**. We use the symbol x to denote the position of the object at a given time t. The value of x can be positive or negative, depending on where the object is relative to an **origin** or a **reference position** that we take to be at $x = 0$. We can choose the reference position to be anywhere that's convenient. For example, we might choose $x = 0$ to be the point where we're sitting alongside a swimming pool. If our seat is 10.0 meters (10.0 m for short) from one end of a swimming pool with a length of 50.0 m, then one end of the pool is at $x = -10.0$ m and the other end is at $x = +40.0$ m. We are free to choose the positive x direction (the direction in which x increases) as we wish; in Figure 2-2a, we've chosen it to be to the right.

The **coordinates**, or **position**, of an object relative to an origin are what we call **vector** quantities. Vectors are special quantities that contain two kinds of information: size, or **magnitude**, and direction. For instance, from $x = 0$ the magnitude of the position of the right-hand end of the pool in Figure 2-2a is 40.0 m, and to get there from $x = 0$ you must travel in the positive x direction. We summarize these statements by saying that this end is at $x = +40.0$ m. The left-hand end of the pool in Figure 2-2a is 10.0 m away from $x = 0$, and to get there you must travel in the negative x direction (the direction in which x decreases). So the position of this end is $x = -10.0$ m.

Distance, by comparison, is *not* a vector quantity: It does not contain information about direction. (If you say something is 50.0 m away, you're stating its distance; if you say it's 50.0 m away to the east, you're stating its position.) That's why distance, unlike position, is always given as a positive number. If you were asked, "How far is it from New York to Boston?" you would never answer, "Negative 300 km!"

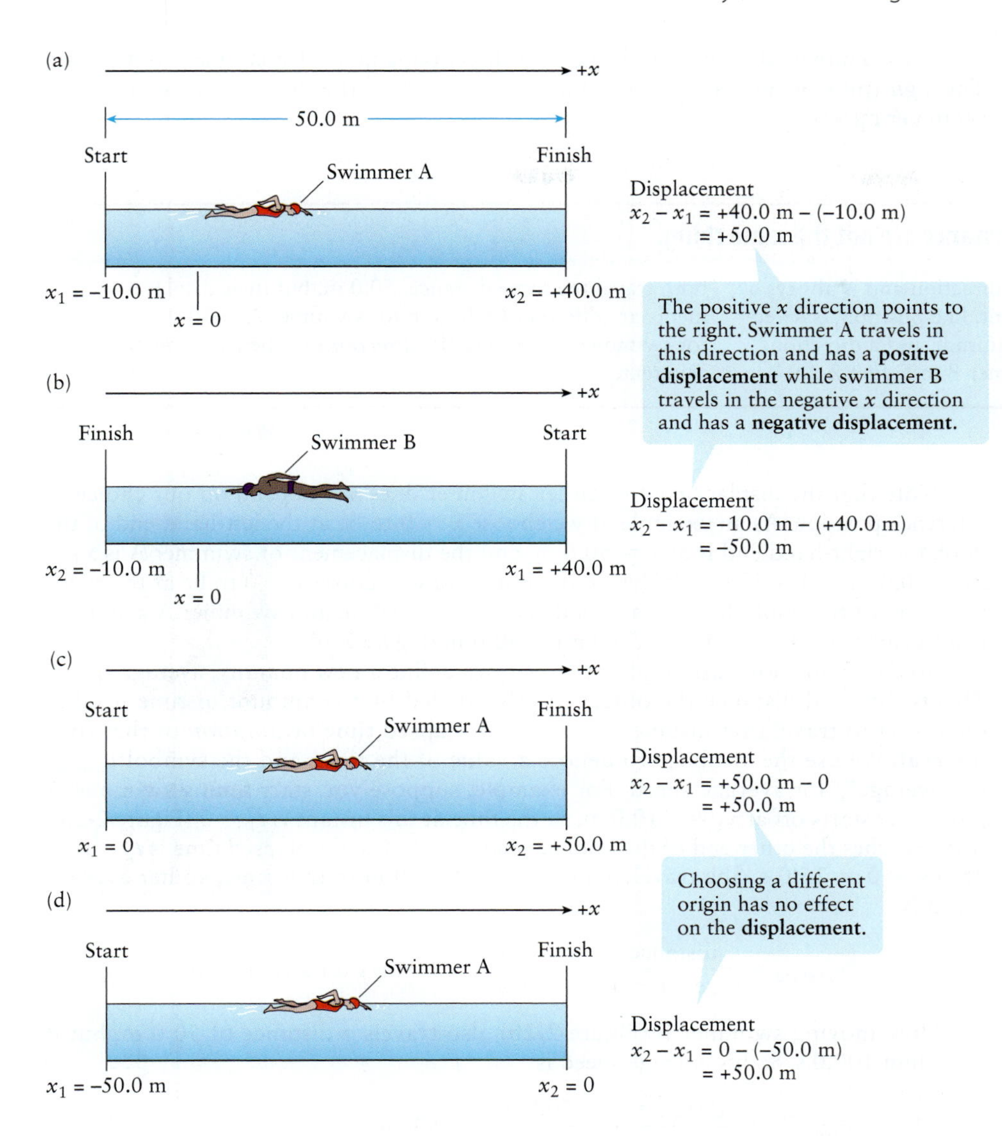

Figure 2-2 Coordinate systems and displacement (a) With our choice of coordinate system, swimmer A has a positive displacement and (b) swimmer B has a negative displacement. (c), (d) Displacement is independent of the coordinate system.

WATCH OUT! A quantity that can be positive or negative might not be a vector.

Some quantities that can be either positive or negative are *not* vectors. For instance, the temperature might be +35°C at noon in midsummer but −10°C on a cold winter night. But there is no direction such as "up" or "left" associated with temperature, so it is not a vector quantity. You need to rely on the context to decide whether the sign of a quantity refers to a direction.

The **displacement** of an object is the difference between its positions at two different times. It tells us how far and in what direction the object moves, so displacement, like position, is a vector. For example, if swimmer A travels the length of the swimming pool from left to right (Figure 2-2a), she starts at $x_1 = -10.0$ m and ends at $x_2 = +40.0$ m. (The subscripts "1" and "2" help us keep track of which position is which.) Her displacement is the *later* position minus the *earlier* position: $x_2 - x_1 = +40.0 \text{ m} - (-10.0 \text{ m}) = +50.0 \text{ m}$. The positive value of displacement means that her value of x increases and she moves 50.0 m in the positive x direction. By comparison, swimmer B (**Figure 2-2b**) travels the length of the pool in the opposite direction. For him $x_1 = +40.0$ m and $x_2 = -10.0$ m, and his displacement is $x_2 - x_1 = (-10.0 \text{ m}) - 40.0 \text{ m} = -50.0 \text{ m}$.

Swimmer B moves 50.0 m in the negative x direction, and his displacement is negative. The *sign* (plus or minus) of the displacement tells you the *direction* in which each swimmer moves.

WATCH OUT! Displacement and distance are not the same thing.

Remember that distance has no direction and is always a positive number. But displacement, like position, is a vector quantity that has both a magnitude and a direction. Swimmer A in Figure 2-2a and swimmer B in Figure 2-2b both travel the same distance, 50.0 m, but their displacements are different (+50.0 m for swimmer A, −50.0 m for swimmer B) because the *directions* of their motion are different.

Note that the displacement of either swimmer does *not* depend on our choice of reference position. As an example, if we choose $x = 0$ to be at the left-hand end of the pool, the right-hand end is at $x = 50.0$ m, and the displacement of swimmer A is $x_2 - x_1 = 50.0 \text{ m} - 0 = 50.0$ m (**Figure 2-2c**). If instead we choose $x = 0$ to be at the right-hand end of the pool, the left-hand end is at $x = -50.0$ m, and swimmer A again has displacement $x_2 - x_1 = 0 - (-50.0 \text{ m}) = 50.0$ m (**Figure 2-2d**).

To determine how *fast* an object moves, we define a new quantity, **average speed.** This is the total distance the object travels divided by the amount of time it takes the object to travel that distance—that is, the **elapsed time** or *duration* of that **time interval.** We use the symbol t to denote a value of the time, and the symbol v_{average} ("v-average") for average speed. For example, suppose you start timing swimmer A just as she starts off at $x_1 = -10.0$ m, so the time at this instant is $t_1 = 0$ (Figure 2-2a). If she reaches the other end of the pool at time $t_2 = 25.0$ s, the elapsed time is $t_2 - t_1 = 25.0 \text{ s} - 0 = 25.0$ s. She traveled a distance of 50.0 m in that time, so her average speed is

$$v_{\text{average}} = \frac{\text{distance}}{\text{time interval}} = \frac{50.0 \text{ m}}{25.0 \text{ s}} = 2.00 \text{ m/s for swimmer A}$$

Slow-moving swimmer B (Figure 2-2b) also travels a distance of 50.0 m, but it takes him 100.0 s. So his average speed is

$$v_{\text{average}} = \frac{\text{distance}}{\text{time interval}} = \frac{50.0 \text{ m}}{100.0 \text{ s}} = 0.500 \text{ m/s for swimmer B}$$

Note that average speed, like distance, tells you nothing about the direction of motion. Like distance, average speed is never negative.

WATCH OUT! Speed and velocity are not the same thing.

Average speed and average velocity are related but are *not* the same quantity. Average speed is equal to the *distance* traveled divided by the elapsed time. Distances are always positive, so average speed is always positive. Average velocity can be positive or negative, depending on the direction in which the object moves. Average velocity is a vector.

To get direction information, we need to calculate the **average velocity** of an object, which is a vector quantity. We use the symbol $v_{\text{average},x}$ ("v-average-x") for average velocity. This is very close to the symbol for average speed but includes the subscript x to remind us that the object is moving in either the positive or negative x direction. The average velocity equals the displacement divided by the elapsed time for that displacement. For swimmer A in Figure 2-2a, who starts at position $x = -10.0$ m and ends at position $x = +40.0$ m in 25.0 s, the average velocity is

$$v_{\text{average},x} = \frac{x_2 - x_1}{t_2 - t_1} = \frac{40.0 \text{ m} - (-10.0 \text{ m})}{25.0 \text{ s} - 0}$$

$$= \frac{50.0 \text{ m}}{25.0 \text{ s}} = +2.00 \text{ m/s for swimmer A}$$

So swimmer A moves in the positive x direction (as shown by the positive sign of $v_{\text{average},x}$), or from left to right in Figure 2-2a, at an average *speed* of 2.00 meters per second. The standard SI units for average velocity and average speed are meters per second (m/s). **Table 2-1** lists some other common units for these quantities.

TABLE 2-1 Units of Velocity and Speed

The SI unit of velocity and speed is meters per second (m/s). Here are some other common units:
1 kilometer per hour (km/h) = 0.2778 m/s
1 mile per hour (mi/h) = 0.4470 m/s = 1.609 km/h
1 foot per second (ft/s) = 0.3048 m/s = 1.097 km/h = 0.6818 mi/h
1 knot = 1 nautical mile per hour = 0.5148 m/s = 1.853 km/h = 1.152 mi/h = 1.689 ft/s

By comparison, swimmer B in Figure 2-2b starts at $x_1 = +40.0$ m at $t_1 = 0$, and at $t_2 = 100.0$ s he reaches $x_2 = -10.0$ m. Hence the average velocity of Swimmer B is

$$v_{\text{average},x} = \frac{x_2 - x_1}{t_2 - t_1} = \frac{-10.0\text{ m} - 40.0\text{ m}}{100.0\text{ s} - 0}$$
$$= \frac{-50.0\text{ m}}{100.0\text{ s}} = -0.500\text{ m/s for swimmer B}$$

Swimmer B moves at only 0.500 m/s. The minus sign of $v_{\text{average},x}$ says that he travels in the negative x direction, or from right to left in Figure 2-2b. Equation 2-1 summarizes these ideas about average velocity:

Average velocity of an object in linear motion

Displacement (change in position) of the object over a certain time interval: The object moves from x_1 to x_2, so $\Delta x = x_2 - x_1$.

$$v_{\text{average},x} = \frac{x_2 - x_1}{t_2 - t_1} = \frac{\Delta x}{\Delta t}$$

For both the displacement and the elapsed time, subtract the earlier value (x_1 or t_1) from the later value (x_2 or t_2).

Elapsed time for the time interval: The object is at x_1 at time t_1 and x_2 at time t_2, so the elapsed time is $\Delta t = t_2 - t_1$.

Average velocity for linear motion equals displacement divided by elapsed time (2-1)

In Equation 2-1 Δx ("delta-x") is an abbreviation for the change $x_2 - x_1$ in the position x, and Δt ("delta-t") is an abbreviation for the change $t_2 - t_1$ in the time t; that is, the elapsed time. Throughout our study of physics we'll use the symbol Δ ("delta") to represent the change in a quantity.

WATCH OUT! Be sure you understand the meaning of Δ.

The symbol Δx does *not* mean a quantity Δ multiplied by a quantity x! Instead, it means the *change* in the value of x.

WATCH OUT! Average speed is not always the magnitude of the average velocity.

The example of the swimmers may make you think that average speed is simply the magnitude of the average velocity. As long as the object travels in a straight line and does not turn around, this is true. But consider the average speed and average velocity of a swimmer who does a complete lap in the pool. This swimmer starts and ends at the same location. Her total displacement is then zero. This means her average velocity is also zero. But does it mean her average speed was also zero? No! She traveled two lengths of the pool—this distance is definitely *not* zero. So her average speed is not zero either.

Constant Velocity, Motion Diagrams, and Graphs of *x* versus *t*

We've calculated the average velocity for each swimmer's entire trip from one end of the pool to the other. But we can also calculate the average velocity for any *segment* of a swimmer's trip between the two ends. A swimmer has *constant velocity* if the average velocity calculated for *any* segment of the trip has the same value as for any other segment. As an example, in **Figure 2-3a** swimmer A is at $x_1 = 0.0$ m at $t_1 = 5.0$ s and at $x_2 = 30.0$ m at $t_2 = 20.0$ s. Her average velocity for this part of her motion is

$$\begin{aligned} v_{\text{average},x} &= \frac{x_2 - x_1}{t_2 - t_1} = \frac{30.0\text{ m} - 0.0\text{ m}}{20.0\text{ s} - 5.0\text{ s}} \\ &= \frac{30.0\text{ m}}{15.0\text{ s}} = +2.00\text{ m/s for swimmer A} \end{aligned}$$

That's the same as we calculated above for the entire trip from one end of the pool to the other. If the average velocity has the same value for all segments, we call it simply the *velocity* and give it the symbol v_x ("v-x").

Figure 2-3a shows how to depict swimmer A's entire motion in a way that makes it clear that her velocity is constant. We draw a dot to represent her position at equal time intervals (say, $t = 0$, $t = 5.0$ s, $t = 10.0$ s, $t = 15.0$ s, and so on). Taken together, these dots make up a **motion diagram**. The spacing between adjacent dots shows you the distance that the object traveled during the corresponding time interval. The spacing is the same between any two adjacent dots in Figure 2-3a, which tells you that swimmer A travels equal distances in the $+x$ direction—that is, has equal displacement—in equal time intervals. That's just what we mean by saying that swimmer A has constant velocity.

Figure 2-3b shows a motion diagram for swimmer B. Since he moves more slowly than swimmer A (speed 0.5 m/s compared to 2.00 m/s), he travels a smaller distance in the same time interval. Hence the dots in swimmer B's motion diagram are more closely spaced than those for swimmer A.

Yet another useful way to depict motion is in terms of a graph of position versus time, also known as an ***x–t* graph**. In **Figure 2-4a**, we've turned swimmer A's motion diagram on its side so that the x axis now runs upward. Then to emphasize that each dot in the diagram corresponds to a specific time t, we've moved each dot to the right by an amount that corresponds to the value of t. We add a horizontal axis for time to make it easy to read the value of t that corresponds to each dot. The result is an x–t graph, a graph of the swimmer's coordinate x versus the time t. To finish the graph, we draw a smooth curve connecting the dots. For this constant-velocity motion, the graph is a straight line.

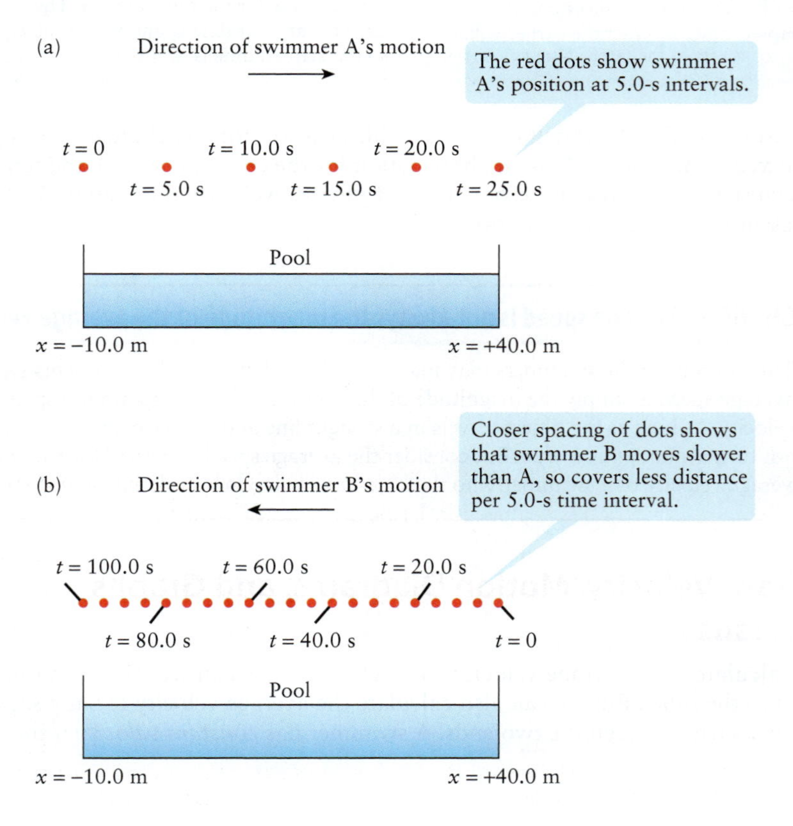

Figure 2-3 Motion diagrams for constant velocity A series of dots shows the positions of (a) swimmer A and (b) swimmer B at equal time intervals.

WATCH OUT! Horizontal motion can be graphed on the vertical axis.

Although the swimmer's motion in the x direction in Figure 2-3a is *horizontal*, we've drawn the x–t graph of Figure 2-4a with the x axis *vertical*. We do this because in graphs that show how a quantity varies with time, we always put time on the horizontal axis. For example, a graph of how a stock mutual fund performs always has time (that is, different dates) on the horizontal axis and the value of the fund on the vertical axis. That's why we put the coordinate x on the vertical axis, no matter how the object's motion is actually oriented in space.

Figure 2-4b shows the x–t graph for swimmer B. This swimmer also has a constant velocity, and the x–t graph is again a straight line. However, this graph differs from the x–t graph for swimmer A (Figure 2-4a) in two ways: where the line touches the vertical axis, and the *slope* of the line (how steep it is and whether it slopes up or down).

The point on the graph where the line touches the vertical (x) axis shows you the position x of the object at $t = 0$. We give this the symbol x_0 ("x-zero"). If the motion starts at $t = 0$, then x_0 represents the *initial position* of the object. The slope of the line tells you the value of the object's *velocity*. If you consider two points on the line, corresponding to times t_1 and t_2 and positions x_1 and x_2, the slope is the vertical difference $x_2 - x_1$ between those two points (the "rise" of the graph) divided by the horizontal

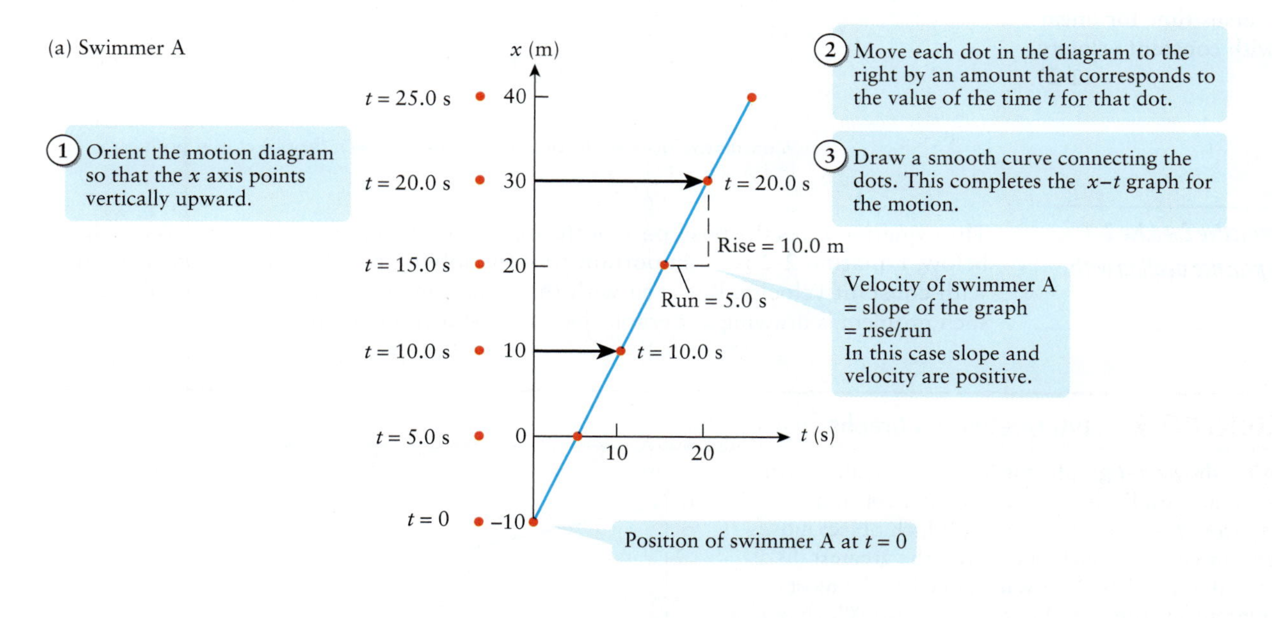

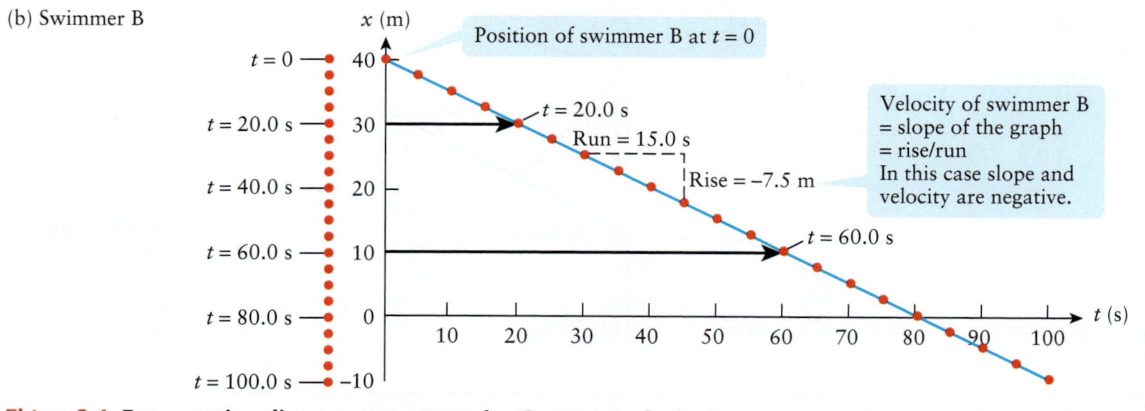

Figure 2-4 From motion diagrams to x–t graphs: Constant velocity How to convert the motion diagrams from Figure 2-3 for (a) swimmer A and (b) swimmer B into graphs of position x versus time t. If the velocity is constant, the x–t graph is a straight line. The slope (rise divided by run) of the straight line equals the velocity.

difference $t_2 - t_1$ (the "run" of the graph). But that's just our definition of the average velocity, Equation 2-1. The x–t graph in Figure 2-4a has a steep slope (so swimmer A is moving rapidly) and slopes upward (so she is moving in the positive x direction). In Figure 2-4b the x–t graph has a shallow slope (so swimmer B is moving slowly) and slopes downward (so he is moving in the negative x direction).

The Equation for Constant-Velocity Linear Motion

We can write a simple and useful equation for constant-velocity motion by starting with Equation 2-1, $v_{\text{average},x} = (x_2 - x_1)/(t_2 - t_1)$. Since the velocity is constant, we replace $v_{\text{average},x}$ by v_x. We choose the earlier time t_1 in this equation to be $t = 0$ and use x_0 instead of x_1 as the symbol for the object's position at this time. We also use the symbols t and x instead of t_2 and x_2 to represent the later time and the object's position at that time. Then Equation 2-1 becomes

$$v_x = \frac{x - x_0}{t - 0} = \frac{x - x_0}{t}$$

We multiply both sides of this equation by t then add x_0 to both sides. The result is

Position versus time for linear motion with constant velocity (2-2)

Position at time t of an object in linear motion with constant velocity

Position at time $t = 0$ of the object

$$x = x_0 + v_x t$$

Constant velocity of the object

Time at which the object is at position x

Go to Interactive Exercise 2-1 for more practice dealing with velocity.

This equation gives the position x of the object as a function of the time t. As we'll see below, Equation 2-2 is an important tool for solving problems about linear motion with constant velocity. But even with this equation, an essential part of solving any such problem is drawing x–t graphs like those shown in Figure 2-4.

GOT THE CONCEPT? 2-1 Interpreting x–t Graphs

? Figure 2-5 shows x–t graphs for four objects, all drawn on the same axes. Each object moves at a constant velocity from $t = 0$ to $t = 5.0$ s. (a) Which object moves at the greatest speed? (b) Which one travels the greatest distance from $t = 0$ to $t = 5.0$ s? (c) Which one has the most positive displacement from $t = 0$ to $t = 5.0$ s? (d) Which one has the most negative displacement from $t = 0$ to $t = 5.0$ s? (e) Which one ends up at the most positive value of x? (f) Which one ends up at the most negative value of x?

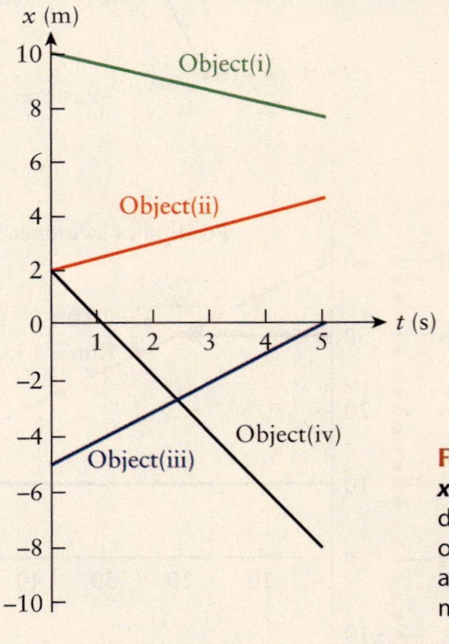

(Answers to *Got the Concept?* questions can be found at the end of the chapter.)

x (m)
10
8
6
4
2
0
−2
−4
−6
−8
−10
1 2 3 4 5
t (s)
Object(i)
Object(ii)
Object(iii)
Object(iv)

Figure 2-5 Four different x–t graphs Each graph depicts a different example of straight-line motion. What are the properties of each motion?

EXAMPLE 2-1 Average Velocity

You drive from Bismarck, North Dakota, to Fargo, North Dakota, on Interstate Highway 94, an approximately straight road 315 km in length. You leave Bismarck and travel 210 km at constant velocity in 2.50 h, then stop for 0.50 h at a rest area. You then drive the remaining 105 km to Fargo at constant velocity in 1.00 h. You then turn around immediately and drive back to Bismarck nonstop at constant velocity in 2.75 h. (a) Draw an x–t graph for the entire round trip. Then calculate the average velocity for (b) the trip from Bismarck to the rest area, (c) the trip from the rest area to Fargo, (d) the entire outbound trip from Bismarck to Fargo, (e) the return trip from Fargo to Bismarck, and (f) the round trip from Bismarck to Fargo and back to Bismarck.

Set Up

This is an example of linear motion along the straight highway between Bismarck and Fargo. We set up our coordinates as shown, with $x = 0$ at the starting position (Bismarck) and the positive x direction from Bismarck toward Fargo. We also choose $t = 0$ to be when the car leaves Bismarck. Using Equation 2-1 will tell us the average velocity for each portion of the trip.

Average velocity for linear motion:

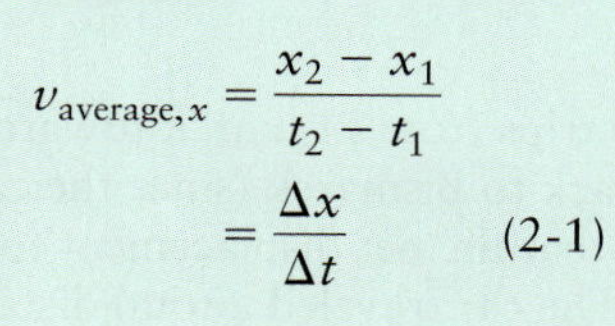

$$v_{\text{average},x} = \frac{x_2 - x_1}{t_2 - t_1} = \frac{\Delta x}{\Delta t} \quad (2\text{-}1)$$

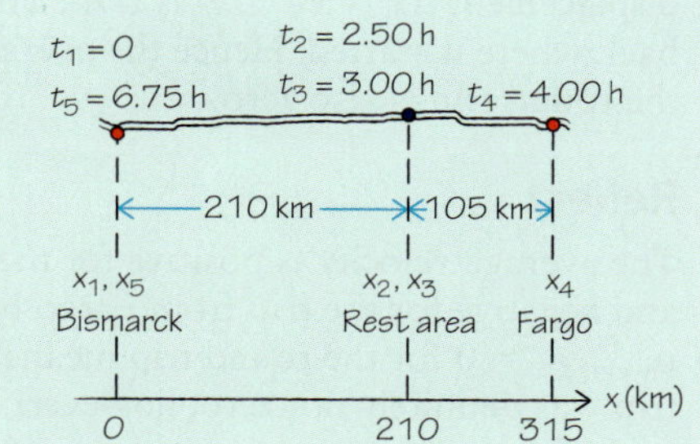

Solve

(a) To draw the x–t graph for the car's motion, we first put a dot on the graph for the car's positions and the corresponding times at the beginning and end of each leg of the trip. The velocity is constant on each leg, so we draw straight lines connecting successive points. The slope of the graph is positive (upward) when the car is moving in the positive x direction toward Fargo and negative (downward) when the car is moving in the negative x direction back toward Bismarck. The slope is *zero* (the line is flat) while the car is at the rest area and not moving.

(b) Calculate the average velocity for the trip from Bismarck (position $x_1 = 0$ at time $t_1 = 0$) to the rest area (position $x_2 = 210$ km at time $t_2 = 2.50$ h).

Bismarck to rest area:

$$v_{\text{average},x} = \frac{\Delta x}{\Delta t} = \frac{x_2 - x_1}{t_2 - t_1} = \frac{210\text{ km} - 0}{2.50\text{ h} - 0} = +84.0\text{ km/h}$$

(c) Calculate the average velocity for the trip from the rest area (position $x_3 = 210$ km at time $t_3 = 2.50\text{ h} + 0.50\text{ h} = 3.00$ h) to Fargo (position $x_4 = 315$ km at time $t_4 = 3.00\text{ h} + 1.00\text{ h} = 4.00$ h).

Rest area to Fargo:

$$v_{\text{average},x} = \frac{\Delta x}{\Delta t} = \frac{x_4 - x_3}{t_4 - t_3} = \frac{315\text{ km} - 210\text{ km}}{4.00\text{ h} - 3.00\text{ h}} = +105\text{ km/h}$$

(d) Calculate the average velocity for the entire outbound trip from Bismarck (position $x_1 = 0$ at time $t_1 = 0$) to Fargo (position $x_4 = 315$ km at time $t_4 = 4.00$ h).

Bismarck to Fargo

$$v_{\text{average},x} = \frac{\Delta x}{\Delta t} = \frac{x_4 - x_1}{t_4 - t_1} = \frac{315\text{ km} - 0}{4.00\text{ h} - 0} = +78.8\text{ km/h}$$

(e) Calculate the average velocity for the return trip from Fargo (position $x_4 = 315$ km at time $t_4 = 4.00$ h) to Bismarck (position $x_5 = 0$ at time $t_5 = 4.00\text{ h} + 2.75\text{ h} = 6.75$ h).

Fargo back to Bismarck:

$$v_{\text{average},x} = \frac{\Delta x}{\Delta t} = \frac{x_5 - x_4}{t_5 - t_4} = \frac{0 - 315\text{ km}}{6.75\text{ h} - 4.00\text{ h}} = -115\text{ km/h}$$

(f) For the round trip from Bismarck (position $x_1 = 0$ at time $t_1 = 0$) to Fargo and back to Bismarck (position $x_5 = 0$ at time $t_5 = 6.75$ h), the net displacement $\Delta x = x_5 - x_1$ is *zero*: The car ends up back where it started. Hence the average velocity for the round trip is also zero.

Bismarck to Fargo and back to Bismarck:

$$v_{\text{average},x} = \frac{\Delta x}{\Delta t} = \frac{x_5 - x_1}{t_5 - t_1} = \frac{0 - 0}{6.75\text{ h} - 0} = 0$$

Reflect

The average velocity is positive for the trips from Bismarck toward Fargo (since the car moves in the positive x direction) and negative for the trip from Fargo back to Bismarck (since the car moves in the negative x direction). The result $v_{\text{average},x} = 0$ for the round trip means that the net displacement for the round trip was zero. The average *speed* for the trip was definitely not zero, however: The car traveled a total distance of $2(315\text{ km}) = 630$ km in a total elapsed time of 6.75 h, so the average speed was $(630\text{ km})/(6.75\text{ h}) = 93.3$ km/h. Displacement and distance are not the same thing; likewise, average velocity and average speed are not the same thing.

EXAMPLE 2-2 When Swimmers Pass

Consider again swimmers A and B, whose motion is depicted in Figure 2-3. Both of them begin swimming at $t = 0$. (a) Write the position-versus-time equation, Equation 2-2, for each swimmer. Use the coordinate system shown in Figure 2-3. (b) Calculate the position of each swimmer at $t = 10.0$ s. (c) Calculate the time when each swimmer is at $x = 20.0$ m. (d) At what time and position do the two swimmers pass each other?

Set Up

We can use Equation 2-2 because each swimmer is in linear motion with constant velocity. Figure 2-3 shows where each swimmer starts and each swimmer's constant velocity, which are just the quantities we need to substitute into Equation 2-2. (We use subscripts "A" and "B" to denote the quantities that pertain to each swimmer.)

$$x = x_0 + v_x t \qquad (2\text{-}2)$$

Swimmer A starts at $x = -10.0$ m at $t = 0$ and has velocity $+2.00$ m/s, so

$$x_{A0} = -10.0\text{ m}$$
$$v_{Ax} = +2.00\text{ m/s}$$

Swimmer B starts at $x = +40.0$ m at $t = 0$ and has velocity -0.500 m/s, so

$$x_{B0} = +40.0\text{ m}$$
$$v_{Bx} = -0.500\text{ m/s}$$

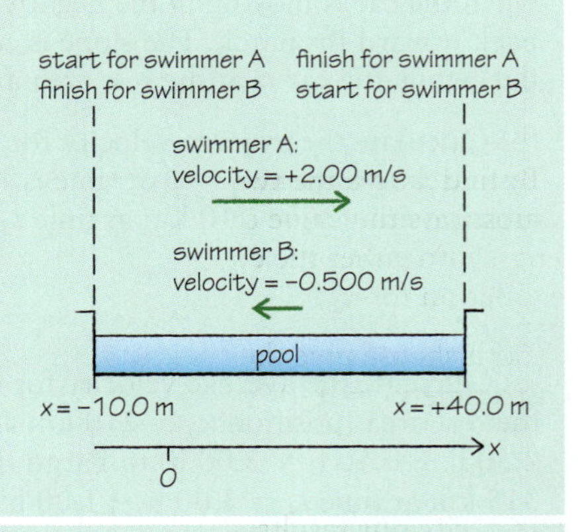

Solve

(a) Use the given information to write down Equation 2-2 for each swimmer. It's useful to draw the x–t graphs for both swimmers on the same graph (see Figure 2-4).

Swimmer A:

$$x_A = x_{A0} + v_{Ax}t \quad \text{so}$$
$$x_A = -10.0\text{ m} + (2.00\text{ m/s})t$$

Swimmer B:

$$x_B = x_{B0} + v_{Bx}t \quad \text{so}$$
$$x_B = +40.0\text{ m} + (-0.500\text{ m/s})t$$

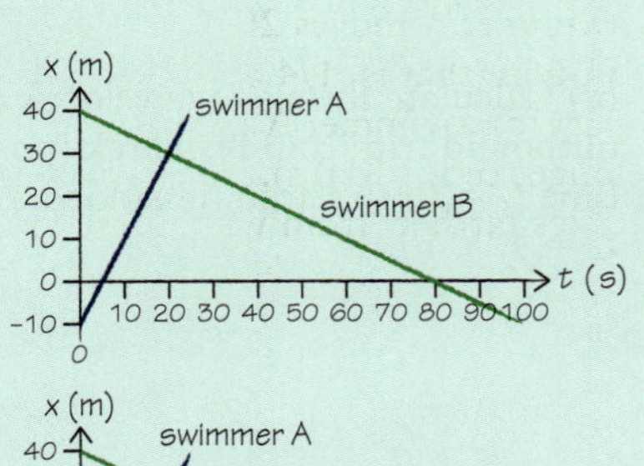

(b) To find the position of each swimmer at $t = 10.0$ s, just substitute this value of t into the equations for x_A and x_B from (a).

Swimmer A:

$$x_A = -10.0\text{ m} + (2.00\text{ m/s})(10.0\text{ s})$$
$$= -10.0\text{ m} + 20.0\text{ m}$$
$$= +10.0\text{ m}$$

Swimmer B:

$$x_B = +40.0\text{ m} + (-0.500\text{ m/s})(10.0\text{ s})$$
$$= +40.0\text{ m} + (-5.00\text{ m})$$
$$= +35.0\text{ m}$$

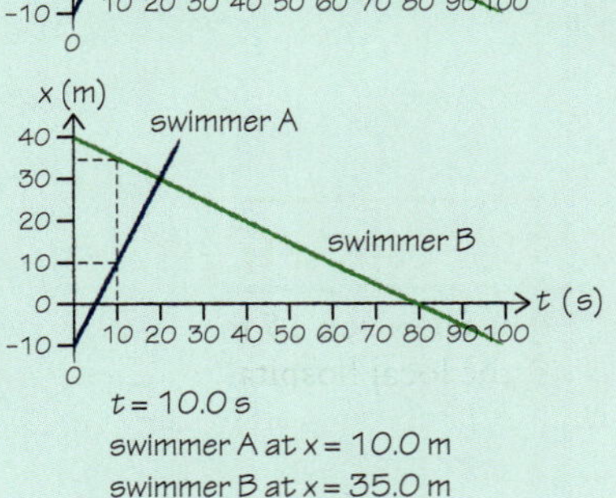

(c) In order to solve for the times when the two swimmers are at $x = 20.0$ m, we must first rearrange the equations from (a) so that t (the quantity we are trying to find) is by itself on one side of each equation. We then substitute 20.0 m for both x_A and x_B.

Swimmer A:

$$x_A = -10.0 \text{ m} + (2.00 \text{ m/s})t \quad \text{so}$$
$$x_A + 10.0 \text{ m} = (2.00 \text{ m/s})t$$
$$t = \frac{x_A + 10.0 \text{ m}}{2.00 \text{ m/s}}$$

Substitute $x_A = 20.0$ m:

$$t = \frac{20.0 \text{ m} + 10.0 \text{ m}}{2.00 \text{ m/s}} = 15.0 \text{ s for A}$$

Swimmer B:

$$x_B = +40.0 \text{ m} + (-0.500 \text{ m/s})t \quad \text{so}$$
$$x_B - 40.0 \text{ m} = (-0.500 \text{ m/s})t$$
$$t = \frac{x_B - 40.0 \text{ m}}{(-0.500 \text{ m/s})}$$

Substitute $x_B = 20.0$ m:

$$t = \frac{20.0 \text{ m} - 40.0 \text{ m}}{(-0.500 \text{ m/s})} = 40.0 \text{ s for B}$$

x (m)
swimmer A
swimmer B
t (s)
x = 20.0 m
swimmer A at t = 15.0 s
swimmer B at t = 40.0 s

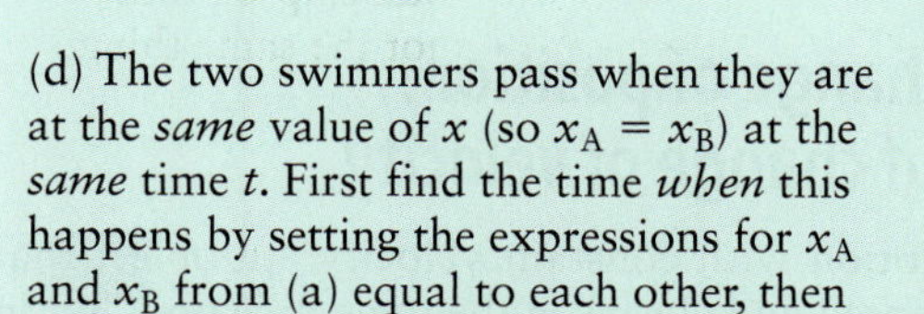

(d) The two swimmers pass when they are at the *same* value of x (so $x_A = x_B$) at the *same* time t. First find the time *when* this happens by setting the expressions for x_A and x_B from (a) equal to each other, then solving for the corresponding value of t.

Swimmer A:

$$x_A = -10.0 \text{ m} + (2.00 \text{ m/s})t$$

Swimmer B:

$$x_B = +40.0 \text{ m} + (-0.500 \text{ m/s})t$$

When the two swimmers pass, $x_A = x_B$

$$-10.0 \text{ m} + (2.00 \text{ m/s})t = 40.0 \text{ m} + (-0.500 \text{ m/s})t$$
$$(2.00 \text{ m/s})t - (-0.500 \text{ m/s})t = 40.0 \text{ m} - (-10.0 \text{ m})$$
$$(2.50 \text{ m/s})t = 50.0 \text{ m}$$
$$t = \frac{50.0 \text{ m}}{2.50 \text{ m/s}} = 20.0 \text{ s}$$

point where swimmers A and B pass each other:
x = 30.0 m at t = 20.0 s
x (m)
swimmer A
swimmer B
t (s)

To find *where* the two swimmers pass, substitute the value of t for when they pass into either the equation for x_A or the equation for x_B.

At $t = 20.0$ s swimmer A's position is

$$x_A = -10.0 \text{ m} + (2.00 \text{ m/s})t$$
$$= -10.0 \text{ m} + (2.00 \text{ m/s})(20.0 \text{ s})$$
$$= 30.0 \text{ m}$$

(Both should give the same answer, since at this time $x_A = x_B$.)

$t = 20.0$ s is the time when the two swimmers pass, so the position where they pass is $x = 30.0$ m.

Reflect

To verify our results, note that swimmer B moves 1/4 as fast as swimmer A. This checks out for part (b): In 10.0 s, swimmer A moves 20.0 m in the positive x direction from her starting point at $x_{A0} = -10.0$ m, and swimmer B moves 5.00 m (that is, 1/4 as far) in the negative x direction from his starting point at $x_{B0} = +40.0$ m. It also checks out for part (c): swimmer A travels 30.0 m in 15.0 s, while slow-moving swimmer B travels a shorter distance (20.0 m) in a longer time (40.0 s). And in part (d), the two swimmers meet 10.0 m from where swimmer B started but 40.0 m (four times farther) from where swimmer A started.

You can also check the answer to part (d) by substituting $t = 20.0$ s into the equation for x_B. You should find $x_B = 30.0$ m, the same as x_A (since $t = 20.0$ s is when the two swimmers pass). Do you?

GOT THE CONCEPT? 2-2 Average Velocity

It takes you 15 min to drive 6.0 mi in a straight line to the local hospital. It takes 10 min to go the last 3.0 mi, 2.0 min to go the last mile, and only 30 s (= 0.50 min) to go the last 0.50 mi. What is your average velocity for the trip? Take the positive x direction to be from your starting point toward the hospital.

TAKE-HOME MESSAGE FOR Section 2-2

✔ The displacement Δx of an object in linear motion is the difference between the object's position x at two different times (displacement Δx is the *later* position minus the *earlier* position).

✔ The average velocity of the object between those two different times, $v_{\text{average},x}$, is the displacement Δx (the difference between the two positions) divided by the elapsed time Δt (the difference between the two times).

✔ An object has constant velocity if $v_{\text{average},x}$ has the same value for any time interval. If the velocity is constant, the object always moves in the same direction and maintains a steady speed.

✔ Motion diagrams and x–t graphs are important tools for interpreting what happens during linear motion. The slope of an x–t graph equals the object's velocity.

✔ You can solve many problems that involve straight-line motion by using Equation 2-1, which relates average velocity, displacement, and elapsed time.

✔ If the velocity is constant, you can also use Equation 2-2 to relate velocity, position, and time.

2-3 Velocity is the rate of change of position, and acceleration is the rate of change of velocity

In Section 2-2 we considered linear motion with constant velocity: an object that moves in a straight line with unvarying speed and always in the same direction. But in many important situations the speed, the direction of motion, or both, can change as an object moves. Some examples include a car speeding up to get through an intersection before the light turns red; the same car slowing down as a police officer signals the driver to pull over; or a dog that runs away from you to pick up a thrown stick, then changes direction and returns the stick to you.

Whenever an object changes its speed or direction of motion—that is, whenever its velocity changes—we say that the object *accelerates* and undergoes an *acceleration*. Before we study acceleration, however, we first need to take a closer look at the idea of velocity.

Instantaneous Velocity

Figure 2-6 is a motion diagram of an object that moves with a changing velocity: a jogger who starts at rest, speeds up, then slows down toward the end of a 100-m run that takes him 38.0 s. The dots show his position at equal time intervals of 2.0 s. Because he speeds up and slows down, these dots are *not* equally spaced: His displacement, and so his average velocity, is *not* the same for all 2.0-s intervals.

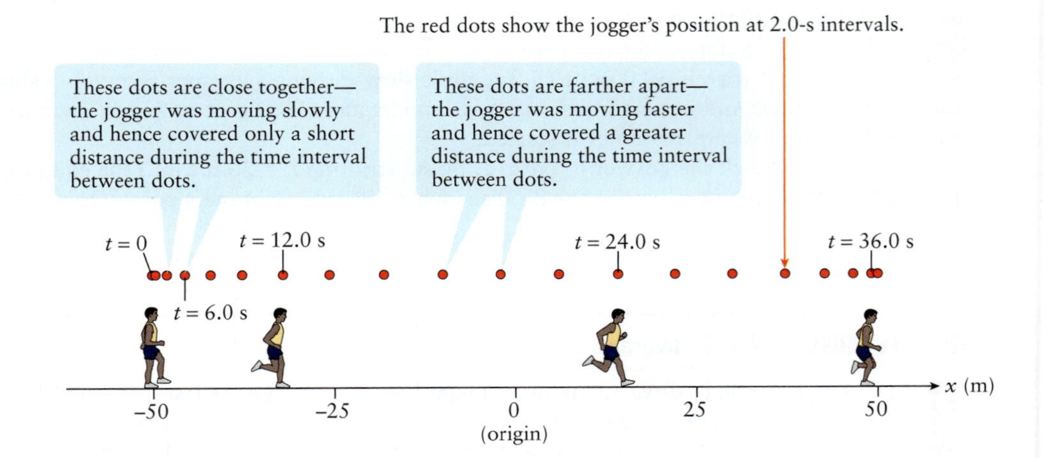
The red dots show the jogger's position at 2.0-s intervals.
These dots are close together—the jogger was moving slowly and hence covered only a short distance during the time interval between dots.
These dots are farther apart—the jogger was moving faster and hence covered a greater distance during the time interval between dots.
t = 0
t = 6.0 s
t = 12.0 s
t = 24.0 s
t = 36.0 s
–50
–25
0
(origin)
25
50
x (m)

Figure 2-6 Motion with varying velocity A jogger's velocity changes as he speeds up and slows down. This motion diagram shows these velocity changes.

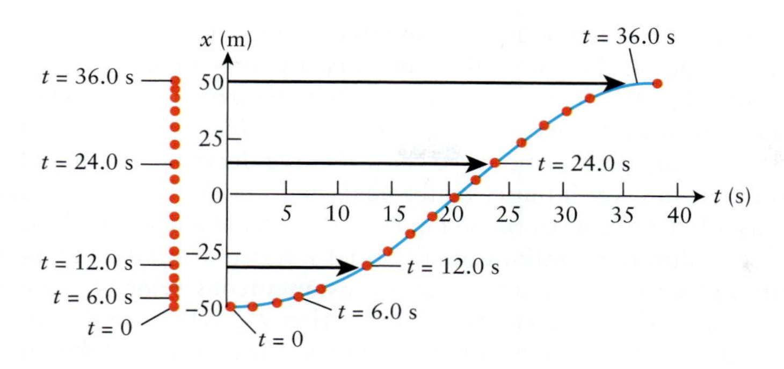

Figure 2-7 From motion diagram to x–t graph: Varying velocity Converting the jogger's motion diagram from Figure 2-6 to an x–t graph. Compare to Figure 2-4.

Figure 2-7 shows the x–t graph for the jogger. The graph is not a straight line because the velocity is not constant. In Section 2-2 we saw that if the velocity *is* constant, as for the two swimmers whose motion is graphed in Figure 2-4, the slope of the straight-line x–t graph tells us the value of the velocity. We'll now show that the same is true when the velocity is not constant.

In Figure 2-8a, we've picked out two points during the jogger's motion, representing a time interval from $t = 6.0$ s to $t = 24.0$ s. The rise of a straight line connecting these points on the x–t graph equals the displacement Δx during the interval, and the run of this line equals the duration Δt of the interval. From Equation 2-1, the average velocity for this interval is $v_{\text{average},x} = \Delta x/\Delta t$; this is the rise divided by the run, or the *slope* of this straight line. For the case shown in Figure 2-8a, $\Delta x = +59.5$ m, and $\Delta t = 18.0$ s, so $v_{\text{average},x} = (+59.5 \text{ m})/(18.0 \text{ s}) = +3.31$ m/s. The plus sign means that the straight line slopes upward, which tells us that x increased during the time interval, and the jogger moved in the positive x direction.

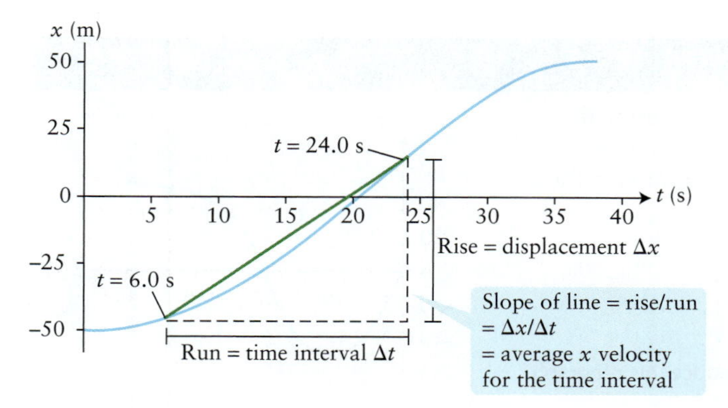

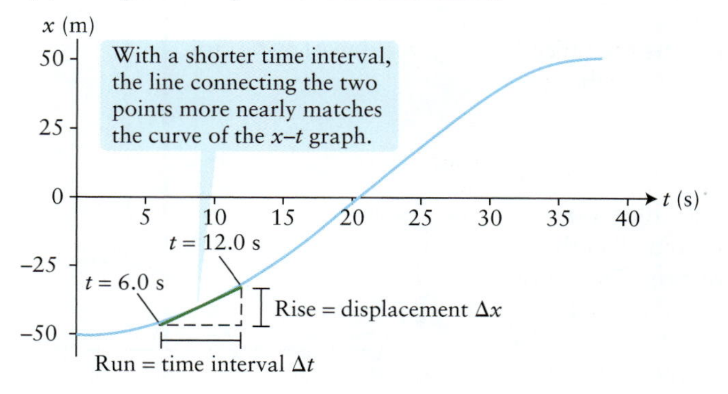

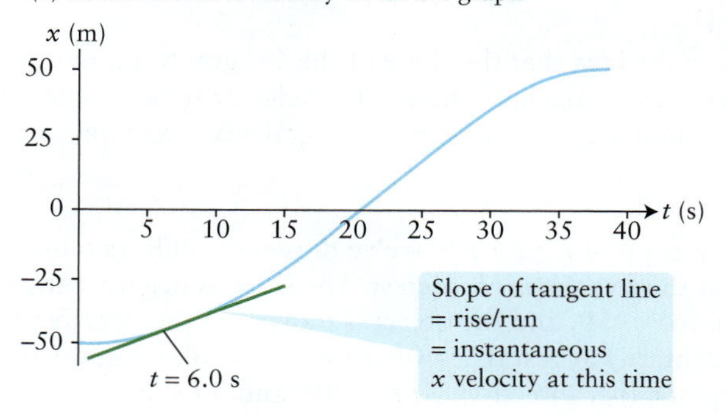

Figure 2-8 Average and instantaneous velocity Finding the average velocity of the jogger over shorter and shorter time intervals. The slope of a line that is tangent to the x–t graph at a given time t gives the instantaneous velocity at time t.

WATCH OUT! Instantaneous speed is the magnitude of instantaneous velocity.

Unlike average speed, which is generally not the absolute value of the average velocity, the *instantaneous* speed is always equal to the absolute value of the *instantaneous* velocity.

The straight line in Figure 2-8a doesn't exactly match the shape of the x–t curve between the two points. However, if we decrease the time interval from $t = 6.0$ s to $t = 12.0$ s, as in **Figure 2-8b**, the straight line is a better match to the x–t curve. And if we make the time interval *very* short—say, from $t = 6.000$ s to 6.001 s—the straight line becomes a nearly perfect match to the x–t curve during that interval. Indeed, if we make the time interval infinitesimally small in duration, the straight line has the same slope as a line tangent to the x–t curve, as in **Figure 2-8c**. This slope equals the average velocity during an infinitesimally brief interval around an instant. We call this quantity v_x (with no "average"), or the **instantaneous velocity** at the instant in question. (We used the same symbol v_x in Section 2-2 when we discussed objects that move with constant velocity. If the velocity is constant, v_x has the same value at all times.)

The sign of the instantaneous velocity tells us what direction the object is moving at the instant in question: v_x is positive if the object is moving in the positive x direction, and v_x is negative if the object is moving in the negative x direction. The absolute value or magnitude of v_x is the **instantaneous speed** of the object, which we denote by the symbol v without a subscript. Instantaneous speed is always positive or zero, never negative. A car's speedometer is a familiar device for measuring instantaneous speed.

As an example, consider the jogger's x–t graph in Figure 2-8c and the tangent line at the point for $t = 6.0$ s. If you carefully measure the rise and the run of this tangent line, then take their quotient, you'll find that the slope of this line is $+1.50$ m/s. So at $t = 6.0$ s the jogger's instantaneous velocity is $v_x = +1.50$ m/s; at this instant he is moving in the positive x direction at speed $v = 1.50$ m/s. Henceforth we'll use the terms **"velocity"** to refer to instantaneous velocity (which we'll use much more often than average velocity) and **"speed"** to refer to instantaneous speed.

Mathematically, an object's velocity v_x is the *rate of change* of the object's position as given by the coordinate x. The faster the position changes, the greater the magnitude of the velocity v_x and the greater the speed of the object. Positive v_x means that x is increasing, and the object is moving in the positive x direction; negative v_x means that x is decreasing, and the object is moving in the negative x direction.

EXAMPLE 2-3 Decoding an x–t Graph

Figure 2-9 shows an x–t graph for the motion of an object. (a) Determine the direction in which the object is moving at $t = 0$, 10 s, 20 s, 30 s, and 40 s. (b) Describe the object's motion in words and draw a motion diagram for the object for the period from $t = 0$ to $t = 40$ s.

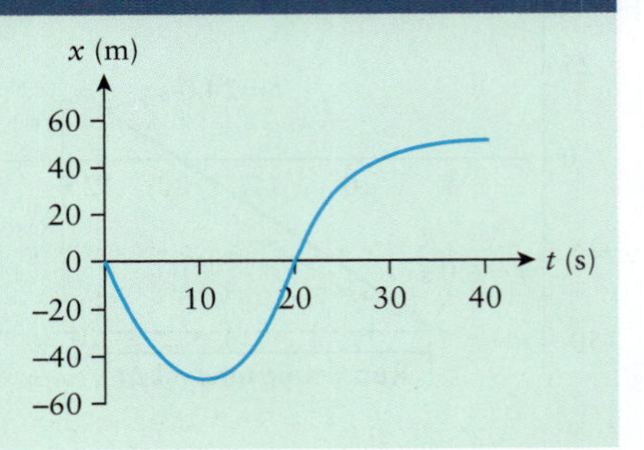

Figure 2-9 An x–t graph to interpret How must an object move to produce this x–t graph?

Set Up

We'll use the idea that the slope of the x–t graph for an object tells us the object's velocity v_x. The algebraic sign of the velocity (plus or minus) tells us whether the object is moving in the positive or negative x direction.

v_x = slope of x–t graph

Solve

(a) On a copy of Figure 2-9, we've drawn green lines tangent to the blue x–t curve at the five times of interest. The slope is negative at $t = 0$ (point 1 in the figure), so $v_x < 0$, and the object is moving in the negative x direction. The object is momentarily at rest ($v_x = 0$) at $t = 10$ s (point 2 in the figure), so it is not moving in either direction. At $t = 20$ s and 30 s (points 3 and 4 in the figure) the object is moving in the positive x direction ($v_x > 0$); at $t = 40.0$ s (point 5 in the figure) the object is again at rest ($v_x = 0$).

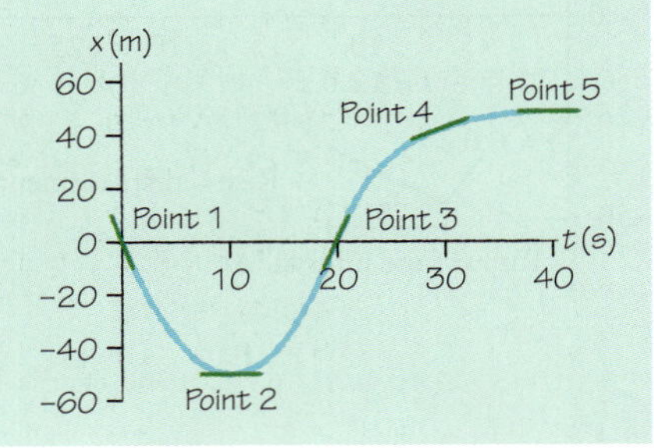

(b) At time $t = 0$ the object is at $x = 0$ and moving in the negative x direction. The x–t graph has a negative slope until $t = 10$ s, so the object continues to move in the negative x direction; because the slope of the x–t graph becomes shallower, the object is slowing down so that the dots are closer together. At $t = 10$ s, the object comes to rest momentarily at $x = -50$ m. After $t = 10$ s, the x–t graph has a positive slope, so the object moves in the positive x direction. From $t = 10$ s to 20 s, the object is speeding up because the slope of the x–t graph is increasing. The object has the fastest speed (because the graph has the greatest slope) at $t = 20$ s when the object again passes through $x = 0$. After $t = 20$ s the slope of the x–t graph is still positive but getting shallower; so the object is still moving in the positive x direction but its speed is decreasing so that the dots again get closer together. The object finally comes to rest at $x = +50$ m at $t = 40$ s.

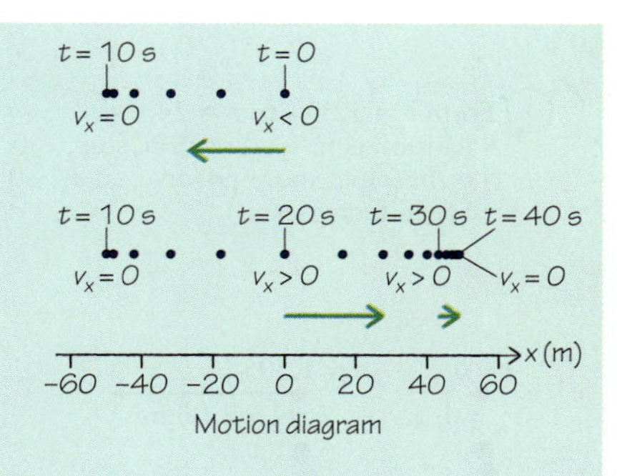

Reflect

It's a common mistake to look at Figure 2-9 and say that the velocity at $t = 10$ s is negative. For this value of time, the curve is below the horizontal axis and so the *coordinate* x is negative. However, the x–t graph has zero *slope* at $t = 10$ s, so at this instant the coordinate x is neither increasing nor decreasing, and the velocity is *zero*. Remember that the *value* of an x–t graph shows you the object's *position* (in other words, where the object is), while the *slope* of the x–t graph tells you its *velocity* (that is, how fast and in which direction the object is moving).

Average Acceleration and Instantaneous Acceleration

Just as velocity tells you how rapidly position changes, acceleration tells you how rapidly velocity changes. Suppose an object like the jogger in Figure 2-6 has velocity v_{1x} at time t_1 and velocity v_{2x} at time t_2. We define the object's **average acceleration** $a_{\text{average},x}$ over this time interval to be the change in velocity $\Delta v_x = v_{2x} - v_{1x}$ divided by the elapsed time $\Delta t = t_2 - t_1$:

Average acceleration of an object in linear motion

Change in **velocity** of the object over a certain time interval: The velocity changes from v_{1x} to v_{2x}.

$$a_{\text{average},x} = \frac{\Delta v_x}{\Delta t} = \frac{v_{2x} - v_{1x}}{t_2 - t_1}$$

Elapsed time for the time interval: The object has velocity v_{1x} at time t_1, and has velocity v_{2x} at time t_2.

Average acceleration for linear motion
(2-3)

If a race car initially at rest blasts away from the starting line and rapidly comes to its top speed, its velocity changes substantially (Δv_x is large) in a short time (Δt is small), and its average acceleration $a_{\text{average},x}$ has a large magnitude. If, instead, an elderly horse starts from rest and gradually speeds up to a slow amble, the horse's velocity changes only a little (Δv_x is small) and takes a long time to change (Δt is large). In this case the horse's average acceleration $a_{\text{average},x}$ has only a small magnitude.

Velocity has units of meters per second (m/s) and time has units of seconds (s), so the SI units of acceleration are meters per second per second, or meters per second squared (m/s^2). Saying that an object has an acceleration of $2.0\ \text{m/s}^2$ means that the object's velocity becomes more positive by 2.0 m/s every second; an acceleration of $-4.5\ \text{m/s}^2$ means that the velocity becomes more negative by 4.5 m/s every second.

Like velocity, average acceleration is a vector and contains direction information, as we explain below.

Interpreting Positive and Negative Acceleration

We saw in Section 2-2 that a positive velocity means that the object is moving in the positive x direction; a negative velocity means the object is moving in the negative x direction. In a similar way, the sign of the average acceleration $a_{\text{average},x}$ tells us the direction in which the object is *accelerating*: in the positive x direction if $a_{\text{average},x}$ is positive and in the negative direction if $a_{\text{average},x}$ is negative. To see what this

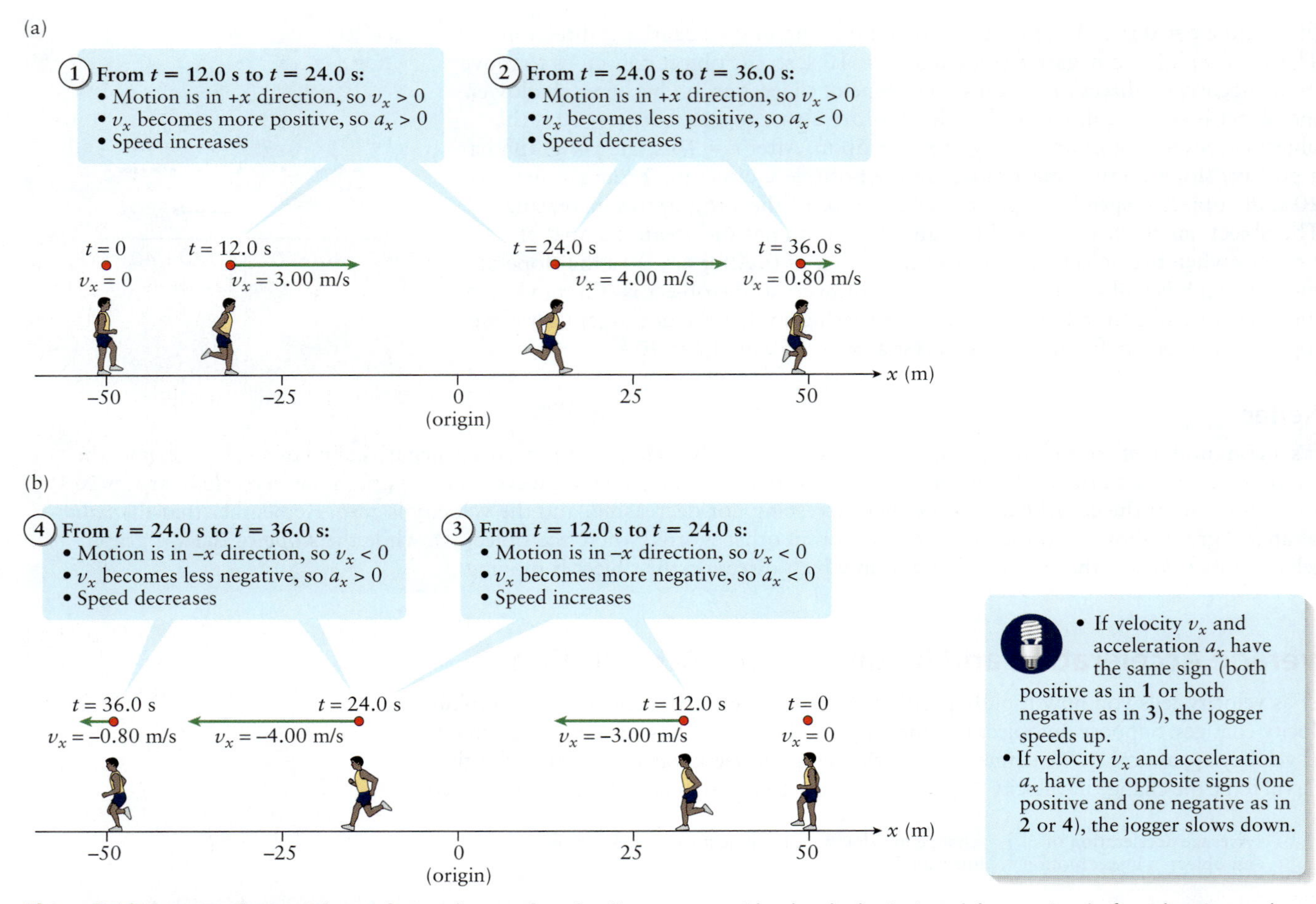

Figure 2-10 Interpreting positive and negative acceleration You must consider the algebraic signs (plus or minus) of acceleration *and* velocity to determine whether an object is slowing down or speeding up.

means, let's look again at our jogger (**Figure 2-10a**). He moves in the positive x direction at all times. During the time interval from $t_1 = 12.0$ s to $t_2 = 24.0$ s, his velocity increases from $+3.00$ m/s to 4.00 m/s, and the change in his velocity is positive: $\Delta v_x = v_{2x} - v_{1x} = (+4.00 \text{ m/s}) - (+3.00 \text{ m/s}) = +1.00 \text{ m/s}$. The jogger's average acceleration for this time interval is also positive: $a_{\text{average},x} = \Delta v_x/\Delta t = (+1.00 \text{ m/s})/(24.0 \text{ s} - 12.0 \text{ s}) = +0.0833 \text{ m/s}^2$. We say that he accelerates in the positive x direction. The jogger's acceleration is in the *same direction* as his velocity, and he *speeds up*.

By contrast, for the time interval from $t_2 = 24.0$ s to $t_3 = 36.0$ s the jogger's velocity decreases from $+4.00$ m/s to $+0.80$ m/s, and the change in his velocity is *negative:* $\Delta v_x = v_{3x} - v_{2x} = (+0.80 \text{ m/s}) - (+4.00 \text{ m/s}) = -3.20 \text{ m/s}$. This means that the average acceleration for this time interval is also negative: $a_{\text{average},x} = \Delta v_x/\Delta t = (-3.20 \text{ m/s})/(36.0 \text{ s} - 24.0 \text{ s}) = -0.267 \text{ m/s}^2$. We say that the jogger accelerates in the negative x direction. In this case the jogger's acceleration is *opposite* to his velocity, and he *slows down*.

It's common to think that positive acceleration always corresponds to speeding up and that negative acceleration always corresponds to slowing down. That's *not* always true, however! **Figure 2-10b** shows an example: The jogger from Figure 2-10a repeats the same motion but in the negative x direction, so his velocity is negative at all times. The figure shows that the jogger has *negative* average acceleration when he is speeding up and *positive* acceleration when he is slowing down. Here's a simple rule to help you understand when an object is speeding up and when it is slowing down:

When an object moving in a straight line speeds up, its velocity and acceleration have the same sign (both plus or both minus). When an object moving in a straight line slows down, its velocity and acceleration have opposite signs (one plus and the other minus).

WATCH OUT! Acceleration doesn't have to mean increasing speed.

In everyday language, "acceleration" is used to mean "speeding up" and "deceleration" is used to mean "slowing down." In physics, however, acceleration refers to *any* change in velocity and so includes *both* speeding up and slowing down (**Figure 2-11**). Speeding up is "accelerating forward" (that is, in the direction of motion) and slowing down is "accelerating backward" (that is, opposite to the direction of motion).

Figure 2-11 Two pedals that cause acceleration The right-hand pedal in a car is called the accelerator because it's used to make the car go faster. But the left-hand pedal, which controls the brakes, is also an accelerator pedal because it's used to make the car slow down.

EXAMPLE 2-4 Calculating Average Acceleration

The object described in Example 2-3 has velocity $v_x = -10$ m/s at $t = 0$, $v_x = 0$ at $t = 10$ s, $v_x = +10$ m/s at $t = 20$ s, $v_x = +1.4$ m/s at $t = 30$ s, and $v_x = 0$ at $t = 40$ s. Find the object's average acceleration for the time intervals (a) $t = 0$ to 10 s, (b) $t = 10$ s to 20 s, (c) $t = 20$ s to 30 s, and (d) $t = 30$ s to 40 s.

Set Up

The figures in Example 2-3 show the x–t graph and motion diagram for this object. For each time interval we'll use the definition of average acceleration, Equation 2-3.

Average acceleration for linear motion:

$$a_{\text{average},x} = \frac{\Delta v_x}{\Delta t} = \frac{v_{2x} - v_{1x}}{t_2 - t_1} \quad (2\text{-}3)$$

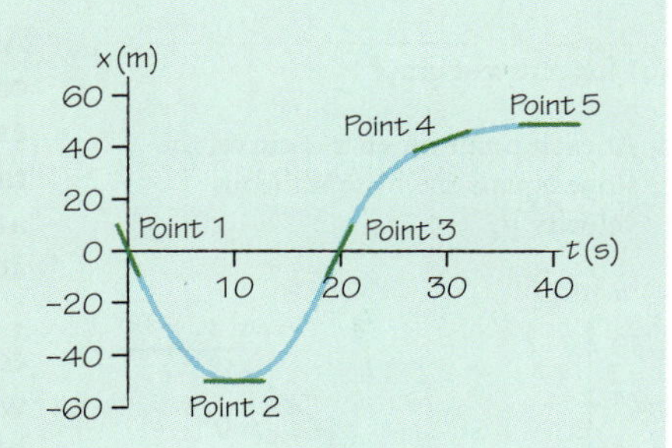

Solve

(a) For the time interval $t_1 = 0$ to $t_2 = 10$ s (from point 1 to point 2 in the x–t graph), we have $v_{x1} = -10$ m/s and $v_{x2} = 0$. Because the velocity becomes more positive (it goes from negative to zero), the average x acceleration is positive.

$$a_{\text{average},x} = \frac{v_{2x} - v_{1x}}{t_2 - t_1} = \frac{0 - (-10\text{ m/s})}{10\text{ s} - 0} = \frac{+10\text{ m/s}}{10\text{ s}} = +1.0\text{ m/s}^2$$

(b) For the time interval $t_2 = 10$ s to $t_3 = 20$ s (from point 2 to point 3 in the x–t graph), we have $v_{x2} = 0$ and $v_{x3} = +10$ m/s. The velocity again becomes more positive (it goes from zero to a positive value), so the average x acceleration is again positive.

$$a_{\text{average},x} = \frac{v_{3x} - v_{2x}}{t_3 - t_2} = \frac{(+10\text{ m/s}) - 0}{20\text{ s} - 10\text{ s}} = \frac{+10\text{ m/s}}{10\text{ s}} = +1.0\text{ m/s}^2$$

(c) For the time interval $t_3 = 20$ s to $t_4 = 30$ s (from point 3 to point 4 in the x–t graph), we have $v_{x3} = +10$ m/s and $v_{x4} = +1.4$ m/s. The velocity becomes more negative (less positive), so the average x acceleration is negative.

$$a_{\text{average},x} = \frac{v_{4x} - v_{3x}}{t_4 - t_3} = \frac{(+1.4\text{ m/s}) - (+10\text{ m/s})}{30\text{ s} - 20\text{ s}} = \frac{-8.6\text{ m/s}}{10\text{ s}} = -0.86\text{ m/s}^2$$

(d) For the time interval $t_4 = 30$ s to $t_5 = 40$ s (from point 4 to point 5 in the x–t graph), we have $v_{x4} = +1.4$ m/s and $v_{x5} = 0$. Again the velocity becomes more negative (less positive) and the average acceleration is negative.

$$a_{\text{average},x} = \frac{v_{5x} - v_{4x}}{t_5 - t_4} = \frac{0 - (+1.4\text{ m/s})}{40\text{ s} - 30\text{ s}} = \frac{-1.4\text{ m/s}}{10\text{ s}} = -0.14\text{ m/s}^2$$

Reflect

Let's check these results to see how they agree with the general rule about the algebraic signs of velocity v_x and average acceleration $a_{\text{average},x}$. From $t_1 = 0$ to $t_2 = 10$ s, v_x is negative (the x–t graph slopes downward), but $a_{\text{average},x} = +1.0\text{ m/s}^2$ is positive. Because the signs are different, the speed must decrease during this interval, and indeed it does (from 10 m/s to zero). From $t_2 = 10$ s to $t_3 = 20$ s, v_x is positive (the x–t graph slopes upward), and $a_{\text{average},x} = +1.0\text{ m/s}^2$ is also positive. During this interval the signs of v_x and $a_{\text{average},x}$ are the same, and the speed increases (from 0 to 10 m/s), in accordance with the rule. During the intervals from $t_3 = 20$ s to $t_4 = 30$ s (for which $a_{\text{average},x} = -0.86\text{ m/s}^2$) and from $t_4 = 30$ s to $t_5 = 40$ s (for which $a_{\text{average},x} = -0.14\text{ m/s}^2$), v_x and $a_{\text{average},x}$ have opposite signs (v_x is positive and $a_{\text{average},x}$ is negative), and the speed decreases (from 10 m/s to 1.4 m/s to zero) as it should. When looking at changes in speed, the sign of the acceleration alone isn't what's important—what matters is how the signs of velocity and acceleration compare.

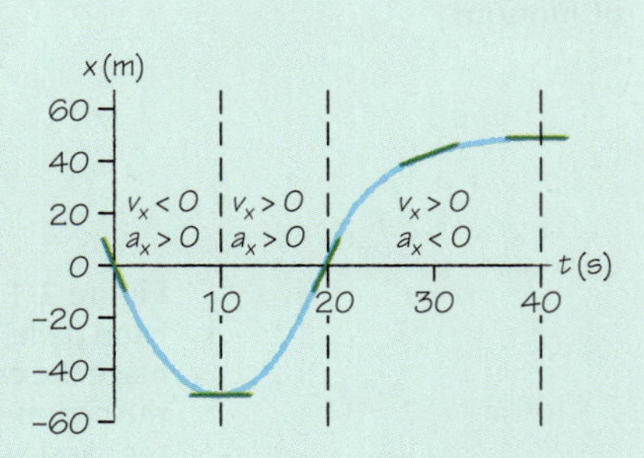

Instantaneous Acceleration and the v_x–t Graph

An object's acceleration can change from one moment to the next. For example, a car can accelerate forward and gain speed, cruise at a steady velocity with zero acceleration, then accelerate backward when the driver steps on the brakes. The **instantaneous acceleration**, or simply **acceleration**, describes how the velocity is changing at a given instant. We use the symbol a_x (without the word "average") to denote acceleration.

Just as an object's velocity v_x is the rate of change of its position as expressed by its coordinate x, the object's acceleration a_x is the rate of change of its velocity v_x. In other words, *acceleration is to velocity as velocity is to position.* The slope of a graph of an object's position versus time (an x–t graph) indicates its velocity v_x. In the same way, the slope of a graph of the object's velocity versus time—that is, a **v_x–t graph**—tells us its acceleration a_x. If the v_x–t graph has an upward (positive) slope, the velocity is becoming more positive, and the acceleration is positive; if the v_x–t graph has a downward (negative) slope, the velocity is becoming more negative, and the acceleration is negative. If the velocity is not changing, so that the acceleration is zero, the v_x–t graph is horizontal and has zero slope.

You can see these properties of the v_x–t graph by inspecting **Figure 2-12**, which shows both the x–t graph and the v_x–t graph for the jogger whose motion we depicted in Figures 2-6, 2-7, and 2-10a. From $t = 0$ to $t = 16$ s his velocity becomes more positive, which we can see in two ways: The x–t graph has an *increasing* slope (which means v_x is becoming more positive), and the v_x–t graph has a *positive* slope (which means a_x is positive). From $t = 16$ s to $t = 28$ s, the jogger has a constant velocity, which is why the x–t graph has a *constant* slope (which means v_x is not changing), and the v_x–t graph is a horizontal line with *zero* slope (which means a_x is zero) during these times. Finally, from $t = 28$ s to $t = 38$ s, the jogger's velocity is becoming more negative, so the x–t graph has a *decreasing* slope (which means v_x is becoming more negative), and the v_x–t graph has a *negative* slope (which means a_x is negative). **Table 2-2** summarizes these key features of x–t and v_x–t graphs.

(a) Jogger's x–t graph

At each point on an x–t curve, the slope equals the instantaneous velocity v_x.

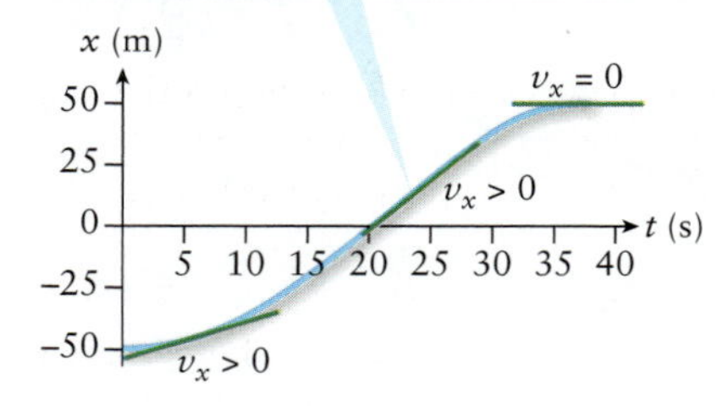

(b) Jogger's v_x–t graph

At each point on a v_x–t curve, the slope equals the instantaneous acceleration a_x.

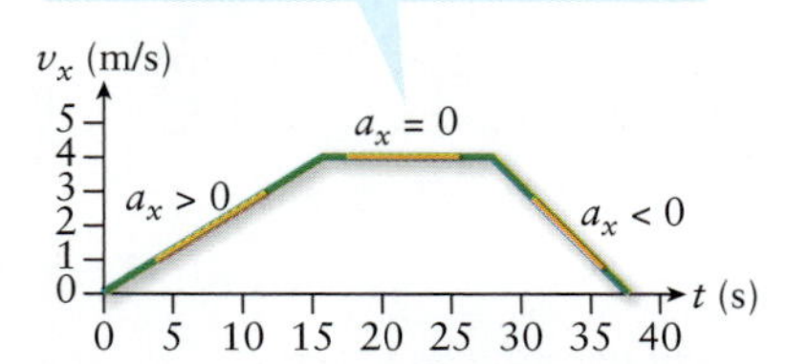

Figure 2-12 From x–t graph to v_x–t graph Graphs of a jogger's (a) position versus time and (b) velocity versus time.

TABLE 2-2 Interpreting x–t Graphs and v_x–t Graphs

A graph of coordinate x versus time t (x–t graph) and a graph of x velocity v_x versus time t (v_x–t graph) are different ways to depict the motion of an object along a straight line.

Type of graph	x–t graph	v_x–t graph
The *value* of the graph tells you...	the coordinate x of the object at a given time t	the velocity v_x of the object at a given time t
The *slope* of the graph tells you...	the velocity v_x of the object at a given time t	the acceleration a_x of the object at a given time t
Changes in the slope of the graph tell you...	the acceleration a_x of the object at a given time t	whether the acceleration is changing

EXAMPLE 2-5 Decoding a v_x–t Graph

Figure 2-13 shows both the x–t graph and the v_x–t graph for the motion of the same object that we examined in Examples 2-3 and 2-4. Using these graphs alone, determine whether the object's acceleration a_x is positive, negative, or zero at (a) $t = 0$, (b) $t = 10$ s, (c) $t = 20$ s, and (d) $t = 30$ s.

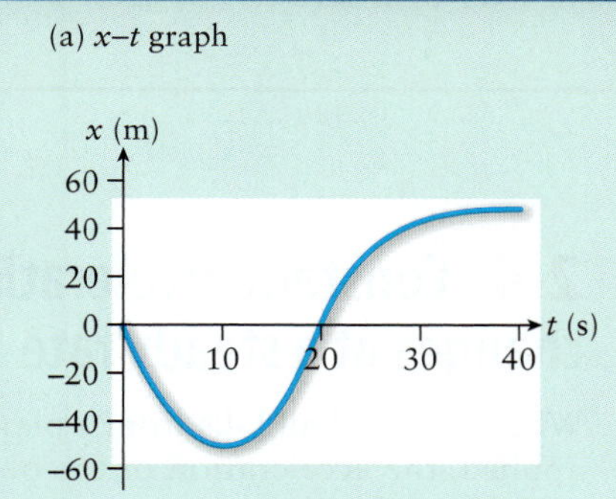

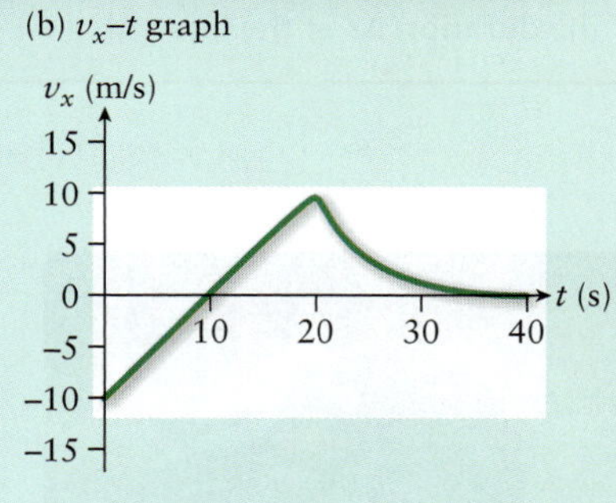

Figure 2-13 Two graphs to interpret The x–t graph and v_x–t graph for the object from Examples 2-3 and 2-4.

Set Up

We use two ideas: (i) acceleration is the rate of change of velocity, and (ii) the value of the acceleration equals the slope of the v_x–t graph (Table 2-2).

Solve

(a) At $t = 0$ the v_x–t graph has a positive (upward) slope, which means that v_x is increasing and hence that a_x is positive. The x–t graph shows the same thing. This graph has a steep negative slope at $t = 0$, which means v_x is negative, but as time increases the slope becomes shallower as the slope and v_x become closer to zero. Because v_x is changing from a negative value toward zero, the velocity is increasing, and a_x is positive.

(b) At $t = 10$ s the v_x–t graph has the same upward slope as at $t = 0$, which means that the acceleration a_x is again positive. The velocity is zero at this instant, so the object is momentarily at rest. It's still accelerating, however; the object has a negative velocity just before $t = 10$ s, and the velocity is positive just after $t = 10$ s. This is the time at which the object turns around. You can also see this from the x–t graph, the slope of which is changing from negative to zero to positive around $t = 10$ s.

(c) At $t = 20$ s the v_x–t graph has zero slope, so the acceleration is zero. At this instant the velocity is neither increasing nor decreasing. That's why the x–t graph at this time is nearly a straight line of constant slope, indicating that the velocity isn't changing at this instant.

(d) At $t = 30$ s the v_x–t graph has a negative (downward) slope, so a_x is negative. The x–t graph has a positive slope at this instant (v_x is positive) but is flattening out, so the slope and v_x are both becoming more negative, which is another way of saying that the acceleration is negative.

Reflect

We can check our answers by seeing what they tell us about how the *speed* of the object is changing. At $t = 0$ the object has negative velocity and positive acceleration. Because v_x and a_x have opposite signs, the object is slowing down. At $t = 20$ s the velocity is positive but the acceleration is zero; the velocity is not changing at this instant, so the object is neither speeding up nor slowing down. Finally, at $t = 30$ s, the velocity is positive and the acceleration is negative, so the object is once again slowing down. You can confirm these conclusions by comparing with the motion diagram in Example 2-3. Are they consistent? Can you also use Figure 2-13 to show that at $t = 40$ s the object is at rest (it has zero velocity) and is remaining at rest (its acceleration is zero, so the velocity remains equal to zero)?

GOT THE CONCEPT? 2-3 Interpreting the Sign of Acceleration

Which statement about the motion depicted in Figure 2-13 is correct? (a) At $t = 5$ s, a_x is positive and the object is speeding up; (b) at $t = 5$ s, a_x is positive and the object is slowing down; (c) at $t = 5$ s, a_x is negative and the object is speeding up; (d) at $t = 5$ s, a_x is negative and the object is slowing down.

TAKE-HOME MESSAGE FOR Section 2-3

✔ Instantaneous velocity v_x (or *velocity* for short) tells you an object's speed and direction of motion at a given instant. It is equal to the rate of change of the object's position and equals the slope of the object's x–t graph.

✔ The average acceleration of an object, $a_{\text{average},x}$, is the change in its velocity v_x over a given time interval divided by the duration Δt of the interval.

✔ Instantaneous acceleration a_x (or *acceleration* for short) equals the rate of change of an object's velocity at a given instant. It also equals the slope of the object's v_x–t graph.

✔ An object speeds up when its velocity and acceleration have the same sign and slows down when they have opposite signs.

2-4 Constant acceleration means velocity changes at a steady rate

Yauhen_D/Shutterstock

Figure 2-14 An object with constant acceleration As this car brakes to a halt, its velocity is changing at a steady rate. Hence its acceleration—the rate of change of velocity—remains constant.

When a basketball falls from a player's hand or a car comes to a halt with the brakes applied, the acceleration of the object turns out to be nearly *constant*—that is, the velocity of the object changes at a steady rate (**Figure 2-14**). This situation is so common that it's worth deriving a set of kinematic equations that describe motion in a straight line with **constant acceleration**. These equations turn out to be useful even in situations where the acceleration isn't strictly constant.

If the acceleration is constant, the *instantaneous* acceleration a_x at any instant of time is the same as the *average* acceleration $a_{\text{average},x}$ for any time interval. Thus we can use a_x in place of $a_{\text{average},x}$ in any equation. In particular let's consider a time interval from $t = 0$ to some other time t. We use the symbol v_{0x} for the velocity at $t = 0$, which we call the initial velocity, and the symbol v_x (with no zero) for the velocity at time t. From the definition of average acceleration given in Equation 2-3,

$$a_x = \frac{v_x - v_{0x}}{t - 0} \quad (2\text{-}4)$$

We can rewrite Equation 2-4 to get an expression for the velocity v_x at any time t. Multiply both sides of the equation by t, add v_{0x} to both sides, and rearrange:

Velocity, acceleration, and time for constant acceleration only (2-5)

$$v_x = v_{0x} + a_x t$$

Velocity at time t of an object in linear motion with constant acceleration (v_x); Velocity at time $t = 0$ of the object (v_{0x}); Constant acceleration of the object (a_x); Time at which the object has velocity v_x (t)

Equation 2-5 tells us how the object's velocity v_x varies with time. If $a_x = 0$ so that the object is not accelerating, v_x at any time t is the same as the velocity v_{0x} at $t = 0$: In other words, the velocity is constant. If a_x is not zero and the object is accelerating, the term $a_x t$ in this equation says that the velocity changes with time.

We'd also like to have an expression that shows how the object's x coordinate varies with time. To obtain such an expression first note that if the object is at coordinate x_0 at $t = 0$ and at coordinate x at time t, its x displacement during the time interval from 0 to t is $\Delta x = x - x_0$. From Equation 2-1, Δx divided by the duration $\Delta t = t - 0$ of the time interval equals the average velocity for this time interval:

$$v_{\text{average},x} = \frac{\Delta x}{\Delta t} = \frac{x - x_0}{t - 0}$$

We can rewrite this expression using the same steps we used in writing Equation 2-5:

$$x = x_0 + v_{\text{average},x}t \tag{2-6}$$

Equation 2-6 tells us that to determine where the object is at time t—that is, what its x coordinate is—we need to know the average velocity $v_{\text{average},x}$ for the time interval from 0 to t.

The average velocity is easy to find if the acceleration is constant. For example, suppose that the initial velocity is $v_{0x} = 3.0$ m/s and the acceleration is $a_x = 2.0$ m/s^2. From Equation 2-5, the velocity is

$$v_x = (3.0 \text{ m/s}) + (2.0 \text{ m/s}^2)(0 \text{ s}) = 3.0 \text{ m/s at } t = 0$$
$$v_x = (3.0 \text{ m/s}) + (2.0 \text{ m/s}^2)(1.0 \text{ s}) = 5.0 \text{ m/s at } t = 1.0 \text{ s}$$
$$v_x = (3.0 \text{ m/s}) + (2.0 \text{ m/s}^2)(2.0 \text{ s}) = 7.0 \text{ m/s at } t = 2.0 \text{ s}$$

To find the average of these three velocities—that is, the average velocity $v_{\text{average},x}$ for the time interval from $t = 0$ to $t = 2.0$ s—we add them together and divide by 3:

$$v_{\text{average},x} = \frac{3.0 \text{ m/s} + 5.0 \text{ m/s} + 7.0 \text{ m/s}}{3} = \frac{15.0 \text{ m/s}}{3} = 5.0 \text{ m/s}$$

Notice that $v_{\text{average},x} = 5.0$ m/s is also the value of the instantaneous velocity v_x at the midpoint of the interval ($t = 1.0$ s). Furthermore, $v_{\text{average},x}$ is equal to the average of the instantaneous velocities at the *beginning and end* of the time interval ($v_x = 3.0$ m/s at $t = 0$ and $v_x = 7.0$ m/s at $t = 2.0$ m/s; the average of these is $(3.0 \text{ m/s} + 7.0 \text{ m/s})/2 = (10.0 \text{ m/s})/2 = 5.0$ m/s). You will *always* get this result if the acceleration is constant. So for constant a_x, the average velocity for the time interval from time 0 to time t is the average of v_{0x} and v_x:

$$v_{\text{average},x} = \frac{v_{0x} + v_x}{2} \tag{2-7}$$

If we substitute the expression for v_x from Equation 2-5 into Equation 2-7, we get

$$v_{\text{average},x} = \frac{v_{0x} + v_{0x} + a_x t}{2} = \frac{2v_{0x} + a_x t}{2} = v_{0x} + \frac{1}{2}a_x t \tag{2-8}$$

Equation 2-8 gives the average velocity of an object moving with constant acceleration for the interval from 0 to t.

WATCH OUT! Instantaneous velocity and average velocity are different quantities.

Equations 2-5 and 2-8 look quite similar, but they describe two very different things. The first of these, $v_x = v_{0x} + a_x t$, tells us the *instantaneous velocity* at the *end* of the time interval from 0 to t. The second of these, $v_{\text{average},x} = v_{0x} + (1/2)a_x t$, is a formula for the *average velocity during* this time interval.

If we substitute the expression for $v_{\text{average},x}$ from Equation 2-8 into Equation 2-6, we get an equation for the coordinate of the object x at time t:

$$x = x_0 + v_{\text{average},x}\, t = x_0 + \left(v_{0x} + \frac{1}{2}a_x t\right)t$$

or, simplifying,

Position, acceleration, and time for constant acceleration only (2-9)

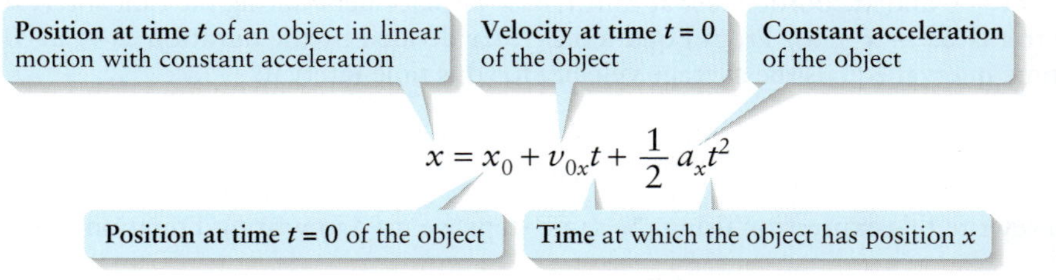

If we know the object's initial position x_0, its initial velocity v_{0x}, and its constant acceleration a_x, Equation 2-9 tells us its position x at any time t.

A good way to check Equation 2-9 is to consider a couple of special cases. If $v_{0x} = 0$ (the object is initially at rest) and $a_x = 0$ (the object's velocity doesn't change), Equation 2-9 says that $x = x_0$ at any time t. In other words, the object doesn't move! If v_{0x} is not zero (the object is moving initially) but $a_x = 0$ (the acceleration is zero), the object has a constant velocity $v_x = v_{0x}$ and its position at time t is $x = x_0 + v_{0x}t = x_0 + v_xt$. This relationship is the same as Equation 2-2 for linear motion with constant velocity, which we found in Section 2-2. Note that if there is a nonzero acceleration a_x, Equation 2-9 shows that there is an additional displacement of $(1/2)a_xt^2$ during the time interval from 0 to t compared to the case of constant velocity.

Graphing Motion with Constant Acceleration

A good way to interpret Equations 2-5 and 2-9 is to draw the associated motion diagram, x–t graph, and v_x–t graph. **Figure 2-15** shows three examples of these motion diagrams and graphs. In each case the spacing between the dots of the motion diagram changes as the object's velocity changes. In cases (a) and (b) the acceleration is positive, so the velocity is becoming more positive. And in cases (a) and (b) the x–t graph is a special curve called a **parabola** that curves upward (the slope, which denotes v_x, is increasing), and the v_x–t graph is a straight line that slants upward (the slope, which denotes a_x, is positive). In case (c) the acceleration is negative: The x–t graph is a parabola that curves downward (because v_x is getting more negative, the slope of this curve is also becoming more negative), and the v_x–t graph is a straight line that slants downward (because a_x, and thus the slope of this curve, is negative).

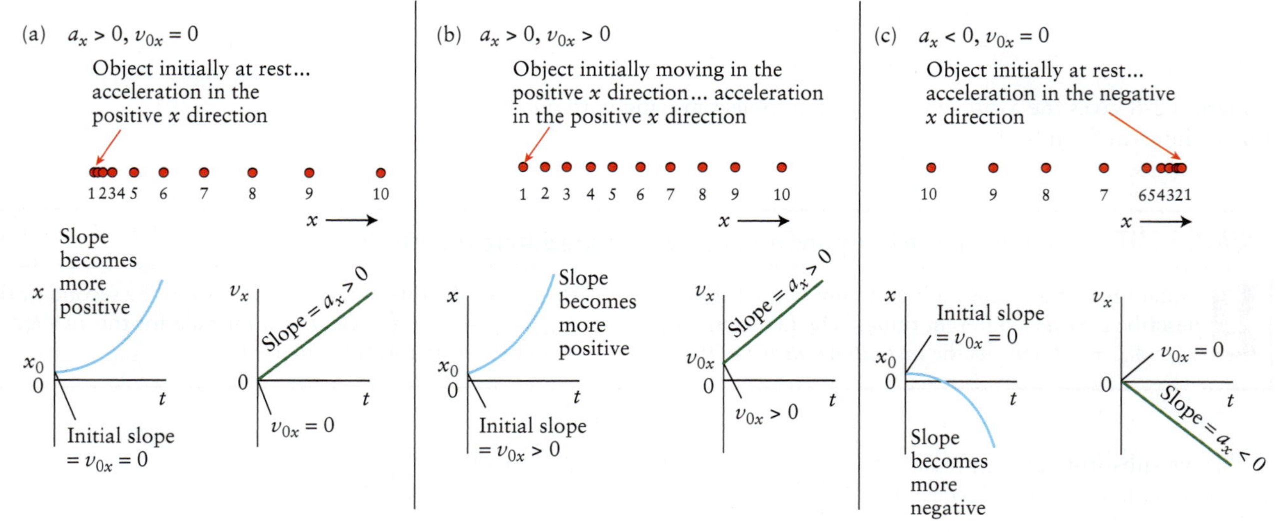

Figure 2-15 Motion with constant acceleration: Three examples Each of these three sets of an x–t graph and a v_x–t graph depicts a different example of linear motion with constant acceleration.

In each of the cases shown in Figure 2-15, we have assumed that x_0 is positive. You should redraw each of the x–t graphs assuming instead that x_0 is negative. (This change has no effect on the v_x–t graphs. Can you see why not?)

WATCH OUT! Know your graphs.

Be careful not to confuse the x–t graph and the v_x–t graph for constant acceleration. When the acceleration is not zero, the x–t graph is *curved* because its slope at each point represents the velocity at the corresponding time t. The velocity is continually changing, so the slope is different at each point and the graph is curved. (Remember that acceleration means a change in velocity, so constant acceleration means the velocity changes at all times at the same rate.) By contrast, the v_x–t graph is a *straight line* because its slope at each point represents the *acceleration* at the corresponding time. The acceleration is constant, so the v_x–t graph has a constant slope (which makes the graph a straight line).

GOT THE CONCEPT? 2-4 Drawing x–t and v_x–t Graphs

? Draw the x–t graph and v_x–t graph for motion with constant acceleration for each of the following cases: (a) $a_x > 0, v_{0x} < 0$; (b) $a_x < 0, v_{0x} > 0$; and (c) $a_x < 0, v_{0x} < 0$. (Assume that x_0 is positive in each case.)

The Kinematic Equations for Motion with Constant Acceleration

Equations 2-5 and 2-9 give the object's x coordinate and velocity at a given time t. There is also a third equation that is often useful in analyzing linear motion with constant acceleration. This equation expresses the object's velocity when the object is at a given x coordinate, without reference to time. To create this equation, first solve Equation 2-5 for the time at which the object has velocity v_x:

$$t = \frac{v_x - v_{0x}}{a_x}$$

$\sqrt{x}$ See the Math Tutorial for more information on solving simultaneous equations and quadratic equations.

Then substitute this expression into Equation 2-9, $x = x_0 + v_{0x}t + \frac{1}{2}a_x t^2$, every place you see the quantity t:

$$x = x_0 + v_{0x}\left(\frac{v_x - v_{0x}}{a_x}\right) + \frac{1}{2}a_x\left(\frac{v_x - v_{0x}}{a_x}\right)^2 \quad (2\text{-}10)$$

You should fill in the algebraic steps needed to rewrite this equation as

$$v_x^2 = v_{0x}^2 + 2a_x(x - x_0) \quad (2\text{-}11)$$

Velocity at position x of an object in linear motion with constant acceleration

Velocity at position x_0 of the object

Constant acceleration of the object

Two positions of the object

Velocity, acceleration, and position for constant acceleration only (2-11)

Equations 2-5, 2-9, and 2-11 are tremendously important because they allow you to solve *any* problem in kinematics that involves straight-line motion with constant acceleration. Situations with constant or nearly constant acceleration are actually quite common in nature and technology, so these equations will be useful to us throughout our study of physics. In the following section we'll see how to tackle problems of this kind.

GOT THE CONCEPT? 2-5 Interpreting Motion with Constant Acceleration

At $t = 0$ an object has velocity $+2.00$ m/s. Its acceleration is constant and equal to -1.00 m/s^2. State whether the object is (i) speeding up, (ii) slowing down, or (iii) neither at each of the following times: (a) $t = 1.00$ s, (b) $t = 2.00$ s, (c) $t = 3.00$ s. (*Hint*: Use Equation 2-5.)

TAKE-HOME MESSAGE FOR Section 2-4

✔ If an object in linear motion has a constant acceleration a_x, its velocity v_x changes at a steady rate.

✔ The v_x–t graph for an object in linear motion with constant acceleration a_x is a straight line. The slope of the line tells you the acceleration.

✔ The x–t graph for an object in linear motion with constant acceleration a_x is a parabola. The parabola curves upward if a_x is positive and curves downward if a_x is negative. The slope of the parabola where it meets the vertical axis tells you the velocity at $t = 0$.

2-5 Solving straight-line motion problems: Constant acceleration

Table 2-3 summarizes the properties of Equations 2-5, 2-9, and 2-11 for linear motion with constant acceleration. The three equations in Table 2-3 involve six physical quantities: the time t, the object's initial position coordinate x_0 at time 0, the object's position x at time t, its velocity v_{0x} at time 0, its velocity v_x at time t, and its constant acceleration a_x. However, not all of these quantities appear in any one of the three constant-acceleration equations. In the following worked examples, we'll show how to decide which equation or equations are the proper ones to use for a given problem.

As we'll see, there are situations where you'll need to use more than one of the equations in Table 2-3 to solve the problem. In these situations you'll need to use your skills at solving simultaneous equations. In addition Equation 2-9 is a *quadratic* equation that involves the square of the time t. This quadratic equation can have two, one, or no solutions—you'll have to decide which of these is the case for the particular problem at hand. See the Math Tutorial to review how to work with simultaneous equations and with quadratic equations.

TABLE 2-3 Equations for Linear Motion with Constant Acceleration

By using one or more of these equations, you can solve any problem involving motion in a straight line for which the acceleration is constant. The check marks indicate which of the quantities t (time), x_0 (the x coordinate at time 0), x (the x coordinate at time t), v_{0x} (the velocity at time 0), v_x (the velocity at time t), and a_x (the constant acceleration) appear in each equation.

Equation		Which quantities the equation includes					
		t	x	x_0	v_{0x}	v_x	a_x
$v_x = v_{0x} + a_x t$	(2-5)	✓			✓	✓	✓
$x = x_0 + v_{0x}t + \frac{1}{2}a_x t^2$	(2-9)	✓	✓	✓	✓		✓
$v_x^2 = v_{0x}^2 + 2a_x(x - x_0)$	(2-11)		✓	✓	✓	✓	✓

EXAMPLE 2-6 Motion with Constant Acceleration I: Cleared for Takeoff!

After winning the lottery you are shopping for a corporate jet. During a demonstration flight with a salesperson, you notice that a particular type of jet takes off after traveling 1.20 km down the runway and after reaching a speed of 226 km/h (140 mi/h). (a) Sketch a motion diagram, an x–t graph, and a v_x–t graph for the jet's motion as it rolls down the runway. (b) What is the magnitude of the airplane's acceleration (assumed constant) during the takeoff roll? (c) How long does it take the airplane to reach takeoff speed?

Set Up

We take the positive x axis to point along the runway in the direction that the jet travels, choose the origin at the point where the jet starts to move, and choose $t = 0$ to be when it starts to move. We are told the jet's initial velocity ($v_{0x} = 0$ because the jet starts at rest) as well as its velocity $v_x = 226$ km/h at position $x = 1.20$ km. Our goals in parts (b) and (c) are to find the jet's acceleration a_x and the time t when the jet attains $v_x = 226$ km/h. To summarize what we know and what we don't, we make a table of the six physical quantities involved in this problem, with a question mark for each of the quantities we want to find.

Since we want to determine the values of *two* unknown quantities, a_x and t, we need two equations. Table 2-3 tells us that to find a_x from x_0, x, v_x, and v_{0x}, we should use Equation 2-11. Once we know the value of a_x, we can use Equation 2-5 to find the time t at which $v_x = 226$ km/h.

To find a_x from x_0, x, v_{0x}, and v_x:

$$v_x^2 = v_{0x}^2 + 2a_x(x - x_0) \quad \text{(2-11)}$$

To find t from v_{0x}, v_x, and a_x:

$$v_x = v_{0x} + a_x t \quad \text{(2-5)}$$

start of takeoff roll $v_{0x} = 0$ — O

takeoff $v_x = 226$ km/h — 1.20 km

Quantity	Know/Don't Know
t	?
x_0	0
x	1.20 km
v_{0x}	0
v_x	226 km/h
a_x	?

Solve

(a) Because the jet speeds up as it rolls down the runway, successive dots in its motion diagram are spaced increasingly far apart. The x–t graph is a parabola (because a_x is constant) that starts at $x_0 = 0$ at $t = 0$ (the jet is initially at the origin), has zero slope at $t = 0$ (the initial velocity is zero), and curves upward (the slope increases as v_x increases). The v_x–t graph starts at $v_{0x} = 0$ at $t = 0$ (the initial velocity is zero) and is a straight line that slopes upward (the acceleration is constant and positive). (Compare Figure 2-15a.) We draw the graphs only to the point of takeoff because after that point the jet is flying and no longer simply moving along the x axis.

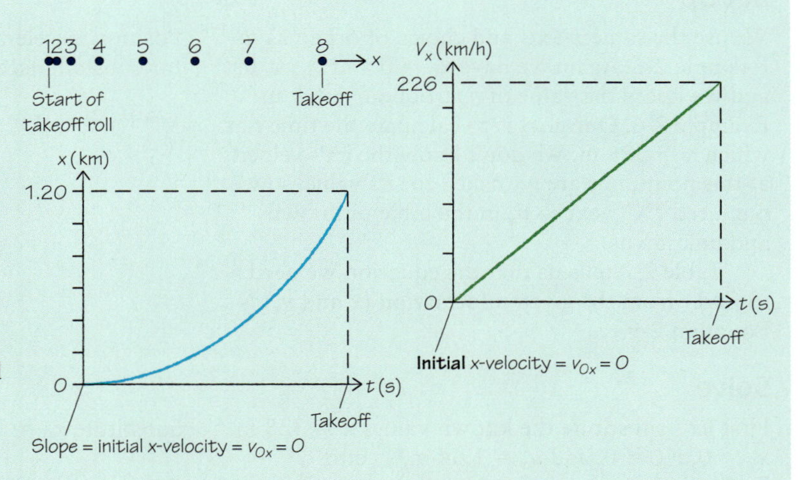

(b) We first convert $v_x = 226$ km/h to m/s and convert $x = 1.20$ km to m. We then solve Equation 2-11 for a_x and substitute in the known values.

$$v_x = (226 \text{ km/h})\left(\frac{1000 \text{ m}}{1 \text{ km}}\right)\left(\frac{1 \text{ h}}{60 \text{ min}}\right)\left(\frac{1 \text{ min}}{60 \text{ s}}\right) = 62.8 \text{ m/s}$$

$$x = (1.20 \text{ km})\left(\frac{1000 \text{ m}}{1 \text{ km}}\right) = 1.20 \times 10^3 \text{ m}$$

Substitute into Equation 2-11:

$$a_x = \frac{v_x^2 - v_{0x}^2}{2(x - x_0)} = \frac{(62.8 \text{ m/s})^2 - (0 \text{ m/s})^2}{2(1.20 \times 10^3 \text{ m} - 0 \text{ m})} = 1.64 \text{ m/s}^2$$

(c) We now know the value of a_x from part (b), so we can solve Equation 2-5 for the time t at which $v_x = 226$ km/h $= 62.8$ m/s. When you enter the values for v_x and a_x into your calculator, retain the extra digits from your calculations above and round the value for the time t only at the end.

Substitute a_x from part (b) into Equation 2-5:

$$t = \frac{v_x - v_{0x}}{a_x} = \frac{(62.8 \text{ m/s}) - (0 \text{ m/s})}{1.64 \text{ m/s}^2} = 38.2 \text{ s}$$

Reflect

Both of our numerical answers have the correct units: a_x is in m/s^2 and t is in s. We can check our answers to part (b) and part (c) by substituting $x = 0$, $v_{0x} = 0$, $a_x = 1.64\ m/s^2$, and $t = 38.2$ s into Equation 2-9, the one constant-acceleration equation that we haven't used yet. We get $x = 1.20$ km, which means that when the jet reaches takeoff speed, it is 1.20 km down the runway—just as we were told in the statement of the problem. That tells us our results are consistent.

Substitute a_x from part (b) and t from part (c) into Equation 2-9:

$$x = x_0 + v_{0x}t + \frac{1}{2}a_x t^2$$

$$= 0 + (0)(38.2\ s) + \frac{1}{2}(1.64\ m/s^2)(38.2\ s)^2$$

$$= 1.20 \times 10^3\ m = 1.20\ km$$

EXAMPLE 2-7 Motion with Constant Acceleration II: Which Solution Is Correct?

Consider again the jet takeoff described in Example 2-6. There is a marker alongside the runway 328 m from the point where the jet starts to move. How long after beginning its takeoff roll does the jet pass the marker?

Set Up

We use the same x axis and choice of origin as in Example 2-6. Again we have $x_0 = 0$ and $v_{0x} = 0$, and we know the value of a_x from our work in Example 2-6. Our goal is to calculate the time t at which $x = 328$ m. We don't know the jet's velocity at this point, nor are we asked for its value, so we put a red "X" next to v_x in the table of knowns and unknowns.

Table 2-3 tells us that the equation we need to find t from the given information (x and v_x) is Equation 2-9.

Position, acceleration, and time for constant acceleration only:

$$x = x_0 + v_{0x}t + \frac{1}{2}a_x t^2 \quad (2\text{-}9)$$

Quantity	Know/Don't Know
t	?
x_0	0
x	328 m
v_{0x}	0
v_x	X
a_x	$1.64\ m/s^2$

Solve

First let's substitute the known values $x = 328$ m, $x_0 = 0$, $v_{0x} = 0$, and $a_x = 1.64\ m/s^2$ into Equation 2-9.

Substitute $x_0 = 0$ and $v_{0x} = 0$, then simplify:

$$x = 0 + (0)t + \frac{1}{2}a_x t^2 = \frac{1}{2}a_x t^2$$

Then substitute $x = 328$ m, $a_x = 1.64\ m/s^2$:

$$328\ m = \frac{1}{2}(1.64\ m/s^2)t^2$$

Rearrange the equation to solve for t.

$$t^2 = \frac{2(328\ m)}{1.64\ m/s^2} = 4.00 \times 10^2\ s^2$$

$$t = \pm\sqrt{4.00 \times 10^2\ s^2} = \pm 20.0\ s$$

The $\pm$ symbol in front of 20.0 s means that the square of either $t = +20.0$ s or $t = -20.0$ s is equal to $t^2 = 4.00 \times 10^2\ s^2$. In this case we clearly want a positive value of time: The jet must reach the marker at $x = 328$ m some time *after* beginning its takeoff roll at $t = 0$. So the correct answer is $t = 20.0$ s.

Choose the value that corresponds to a time after the jet started moving:

$t = 20.0$ s

Reflect

In Example 2-6 we found that the jet takes 38.2 s to travel 1.20 km. Our answer says that it takes less time (20.0 s) to travel a shorter distance (328 m), which is reasonable.

What may not seem reasonable is that we got *two* answers to the problem, only one of which made sense. To see the meaning of the other answer, $t = -20.0$ s, remember that we assumed that the jet has *constant* acceleration $a_x = 1.64\ \text{m/s}^2$. As far as the equations are concerned, the jet has *always* had this acceleration, even before $t = 0$! According to these equations, before $t = 0$ the jet had the same positive value of a_x but had a *negative* velocity v_x and was moving *backward* along the runway in the negative x direction. The answer $t = -20.0$ s refers to the time during this fictitious motion when the backward-moving jet passed through the point $x = 328$ m.

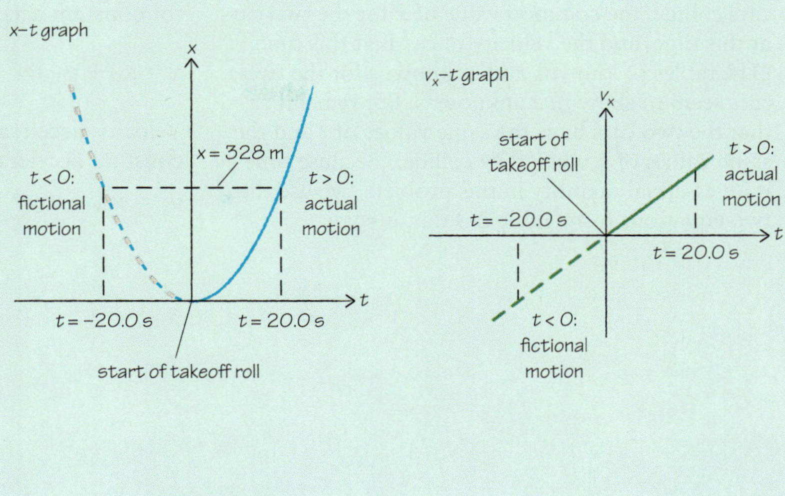

EXAMPLE 2-8 Motion with Constant Acceleration III: Demolition Derby

As a stunt for a movie, two cars are to collide with each other head-on (**Figure 2-16**). The two cars initially are 125 m apart, car A is heading straight for car B at 30.0 m/s, and car B is at rest. Car A maintains the same velocity, while car B accelerates toward car A at a constant 4.00 m/s^2. (a) When do the cars collide? (b) Where do the cars collide? (c) How fast is car B moving at the moment of collision?

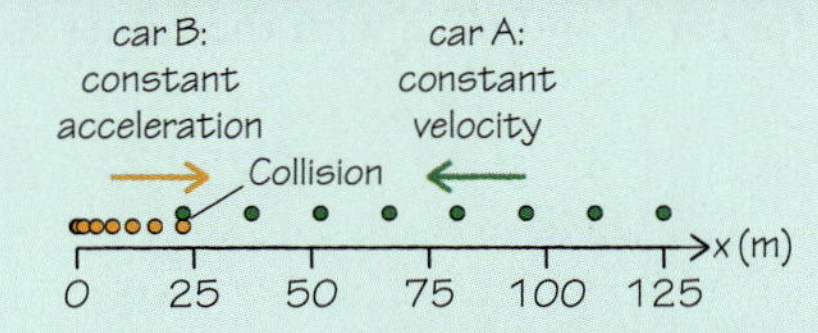

Figure 2-16 **A head-on collision** When and where will these two cars collide?

Set Up

This situation is similar to the two swimmers in Example 2-2 in Section 2-2. The two cars collide when they are at the same place—in other words, at the time t when they have the same x coordinate, as the motion diagram shows. We choose the x axis so that car B is initially at the origin and car A is initially at $x = x_{A0} = 125$ m. Then car A has initial velocity $v_{A0x} = -30.0$ m/s (it is moving in the negative x direction, toward the origin) and zero acceleration, while car B has zero initial velocity (it starts at rest) and constant acceleration $a_{Bx} = +4.00\ \text{m/s}^2$ (it accelerates in the positive x direction).

The two cars collide at the time t when their x–t graphs cross. Note from the v_x–t graph that the two cars never have the same *velocity* because they move in opposite directions.

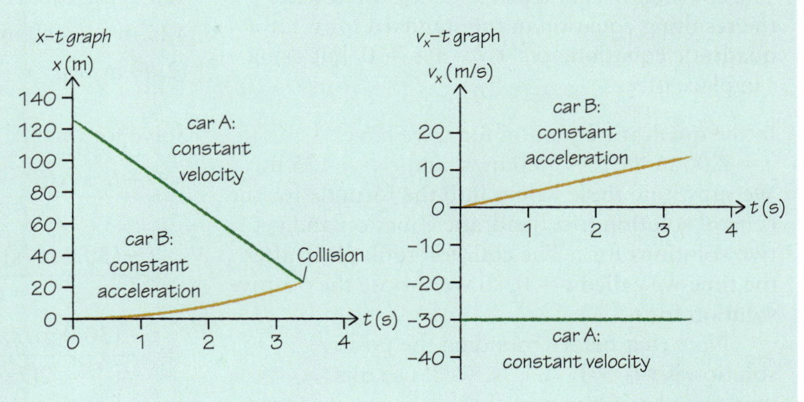

Our goals are to find the value of t when the two cars collide, the common value of x for the two cars at this time, and the velocity of car B at this time. (The tables of knowns and unknowns for the two cars seem to show *five* unknowns. But remember that the two cars have the same values of t and the same values of x when they collide.) Because time is an essential quantity in this problem, we use the two equations from Table 2-3 that involve t.

Position, acceleration, and time for constant acceleration only:

$$x = x_0 + v_{0x}t + \frac{1}{2}a_x t^2 \qquad (2\text{-}9)$$

Velocity, acceleration, and time for constant acceleration only:

$$v_x = v_{0x} + a_x t \qquad (2\text{-}5)$$

For Car A:

Quantity	Know/Don't Know
t	?
x_0	125 m
x	?
v_{0x}	−30.0 m/s
v_x	−30.0 m/s
a_x	0

For Car B:

Quantity	Know/Don't Know
t	?
x_0	0
x	?
v_{0x}	0
v_x	?
a_x	4.00 m/s^2

Solve

(a) Write Equation 2-9 twice, once with the values for car A and once with the values for car B.

Car A ($x_{A0} = 125$ m, $v_{A0x} = -30.0$ m/s, $a_{Ax} = 0$):

$$x_A = 125\text{ m} + (-30.0\text{ m/s})t + \frac{1}{2}(0)t^2$$
$$= 125\text{ m} - (30.0\text{ m/s})t$$

Car B ($x_{B0} = 0$, $v_{B0x} = 0$, $a_{Bx} = +4.00\text{ m/s}^2$):

$$x_B = 0 + (0)t + \frac{1}{2}(4.00\text{ m/s}^2)t^2 = (2.00\text{ m/s}^2)t^2$$

The collision occurs when $x_A = x_B$. We rewrite the resulting equation in the standard form for a quadratic equation, $ax^2 + bx + c = 0$, but using t in place of x.

When the cars collide, $x_A = x_B$, so
$125\text{ m} - (30.0\text{ m/s})t = (2.00\text{ m/s}^2)t^2$ or
$(2.00\text{ m/s}^2)t^2 + (30.0\text{ m/s})t - 125\text{ m} = 0$

In the quadratic equation for t, we have $a = 2.00\text{ m/s}^2$, $b = 30.0$ m/s, and $c = -125$ m. We substitute these values into the formula for the general solution to a quadratic equation and get two solutions for t. The collision took place after the time we called $t = 0$, so we choose the positive solution for t.

Note that before rounding the positive solution is $t = 3.397247$ s. We'll use this unrounded result in part (b).

Solve for time t:

$$t = \frac{-b \pm \sqrt{b^2 - 4ac}}{2a}$$
$$= \frac{-(30.0\text{ m/s}) \pm \sqrt{(30.0\text{ m/s})^2 - 4(2.00\text{ m/s}^2)(-125\text{ m})}}{2(2.00\text{ m/s}^2)}$$
$$= \frac{-(30.0\text{ m/s}) \pm \sqrt{1900\text{ m}^2/\text{s}^2}}{2(2.00\text{ m/s}^2)}$$
$$= \frac{-(30.0\text{ m/s}) \pm (43.6\text{ m/s})}{4.00\text{ m/s}^2}$$
$$= 3.40\text{ s or } -18.4\text{ s}$$

The answer must be positive, so desired answer is $t = 3.40$ s.

(b) To find where the collision took place, we can plug t from part (a) into either the equation for x_A or the equation for x_B because at this time x_A and x_B are equal. Let's choose the first of these. Make sure to use the unrounded value of $t = 3.397247$ s in your calculation and round only at the end.

Solve for the position of car A at $t = 3.40$ s:

$$x_A = 125\text{ m} - (30.0\text{ m/s})t$$
$$= 125\text{ m} - (30.0\text{ m/s})(3.397247\text{ s})$$
$$= 23.1\text{ m}$$

So the collision takes place 23.1 m from the origin (the place where car B started).

(c) To calculate how fast car B was moving when the collision occurred, we will use Equation 2-5, now that we know from part (a) the time at which the collision occurred.

Solve for the velocity of car B at $t = 3.40$ s. Its initial velocity is 0, and its acceleration is 4.00 m/s².

$$v_x = v_{0x} + a_x t = 0 + (4.00\ \text{m/s}^2)(3.40\ \text{s})$$
$$= 13.6\ \text{m/s}$$

Reflect

You can check the answer for part (b) by verifying that you get the same result if you substitute t from part (a) into the equation for x_B. Be sure to try this!

Here's another way to look at the answer to part (b). When the two cars collide, car B has moved 23.1 m (from the origin to $x = 23.1$ m) while car A has moved more than four times as far (from $x = 125$ m to $x = 23.1$ m). These results make sense: Car B started at rest and reached a speed of only 13.6 m/s, while car A moved at a faster steady speed of 30.0 m/s and so covered more distance in the same amount of time.

GOT THE CONCEPT? 2-6 Don't Trust Your Calculator Too Much

A calculator tells you that $(1/2)(1.64\ \text{m/s}^2)\ (38.2\ \text{s})^2 = 1196.5768$ m, or 1.1965768 km. Explain why we said in Example 2-6 that this product equals 1.20 km.

GOT THE CONCEPT? 2-7 Distance Traveled While Slowing

A racing cyclist on a straight track slows down at a steady rate from 15.0 m/s to 10.0 m/s in 4.00 s. How far do the cyclist and her bicycle travel while she is slowing down? (a) 40.0 m; (b) 50.0 m; (c) 60.0 m; (d) not enough information given to decide

TAKE-HOME MESSAGE FOR Section 2-5

✔ To solve problems that involve linear motion with constant acceleration, you need just three equations (see Table 2-3) that involve six quantities: time t, initial position x_0, position x at time t, velocity v_{0x} at time $t = 0$, velocity v_x at time t, and (constant) acceleration a_x.

✔ Using a table of known and unknown quantities as part of solving any problem helps you select which equations will help you find the desired quantity of interest.

2-6 Objects falling freely near Earth's surface have constant acceleration

Perhaps the most important case of constant acceleration is the motion of falling objects near the surface of Earth (**Figure 2-17**). Experiment shows that if air resistance isn't important, an object falling near the surface has a constant, downward acceleration. We will learn later that this acceleration is due to a gravitational force exerted by Earth on the object.

When is air resistance important? To help understand this, drop a sheet of notebook paper. It speeds up for a second or so then wafts downward at a slow, roughly constant speed. The speed doesn't keep increasing, so this is definitely *not* a case of constant acceleration. But now take a second, identical piece of notebook paper and wad it into a ball. If you drop the first sheet and the second, wadded sheet side by side, you'll find that the wadded one falls more rapidly and *does* keep on gaining speed as it falls. The difference is that by wadding up the second sheet, you've given it a smaller cross-sectional area and thus made it less susceptible to air resistance.

If we can ignore the effects of the air, the players and the ball are all in **free fall:** Their acceleration is constant and downward.
y
g
g
g
• An object in free fall has a constant downward acceleration.
• The magnitude g of the acceleration is the same no matter what the size of the object.
• The acceleration is the same whether the object is moving upward, moving downward, or momentarily at rest.

Figure 2-17 Free fall If an airborne object is moving relatively slowly, the effects of the air on the object can be ignored. Then the object moves with a constant acceleration due to gravity.

Competition bicyclists use this same trick when they bend low over the handlebars. This reduces their cross section and the retarding effects of air resistance so that they can go faster for the same effort.

But even the wadded-up paper experiences some air resistance. Imagine, though, that you could somehow crush the paper to an infinitesimally small sphere. There would be essentially no air resistance on this tiny sphere, and only the pull of gravity would affect its fall. In this idealized situation, called **free fall**, the sphere would have a constant downward acceleration.

Free fall is an idealization, but many real-life situations come very close to this ideal. Some examples of nearly free fall include the basketball and players shown in Figure 2-17, a high diver descending toward the water, and a leaping frog in midair. In each of these cases, the falling object experiences minimal air resistance because it has a relatively slow speed, a small cross section, or both, and so falls with constant downward acceleration.

WATCH OUT! Be careful when falling isn't free.

If an object is falling at high speed, or has a relatively large cross section, we *cannot* ignore air resistance and so the motion is *not* free fall. Some examples include a sky diver descending under a parachute, a falling leaf, and a hawk with folded wings plummeting at high speed downward onto its prey. In each of these cases, the falling object reaches a maximum speed, or *terminal speed*, at which the retarding effects of air resistance balance the downward pull of gravity. We'll return to situations of this kind in Chapter 5.

Measuring Free-Fall Acceleration

You might be wondering how we can make the claim that the acceleration of a freely falling object near the surface of Earth is constant. The answer is simple—by experiment! **Figure 2-18** shows a strobe photograph of a billiard ball dropped from rest. The image of the ball is recorded at intervals of 0.10 s. It's conventional to call the vertical axis the y axis (the horizontal axis is often called the x axis). It's also conventional to choose the positive y direction to point upward so that increasing values of y correspond to increasing height. After the ball is dropped from rest, it accelerates downward (in the negative y direction) and acquires a negative velocity v_y (the subscript "y" reminds us that we are now discussing motion along the y axis). We choose the initial position of the ball to be $y_0 = 0$ so that as the ball descends its y coordinate becomes a larger and larger negative number.

Careful measurements of the y coordinate and velocity of the ball yield the values shown in **Table 2-4**. We plot the values of v_y and y versus time in **Figure 2-19**. We learned

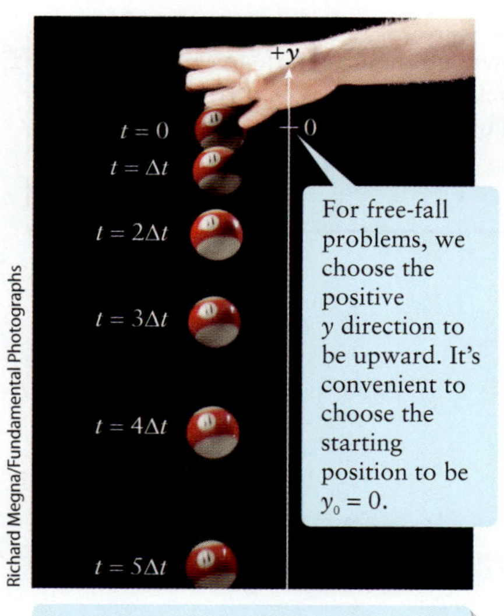

Figure 2-18 Analyzing a freely falling object Measurements of this falling ball confirm that it has a constant acceleration.

TABLE 2-4 Motion of a Falling Ball

This table lists the y coordinate and y velocity at 0.10-s intervals for the falling ball shown in Figure 2-17. Initially the ball is at rest at $y = 0$. The positive y direction is upward, so as the ball descends it has a negative y velocity and it moves to increasingly negative values of y.

Time t (s)	y coordinate (m)	y velocity (m/s)
0	0.00	0.00
0.10	−0.05	−0.98
0.20	−0.20	−1.96
0.30	−0.44	−2.94
0.40	−0.78	−3.92
0.50	−1.23	−4.91
0.60	−1.76	−5.89

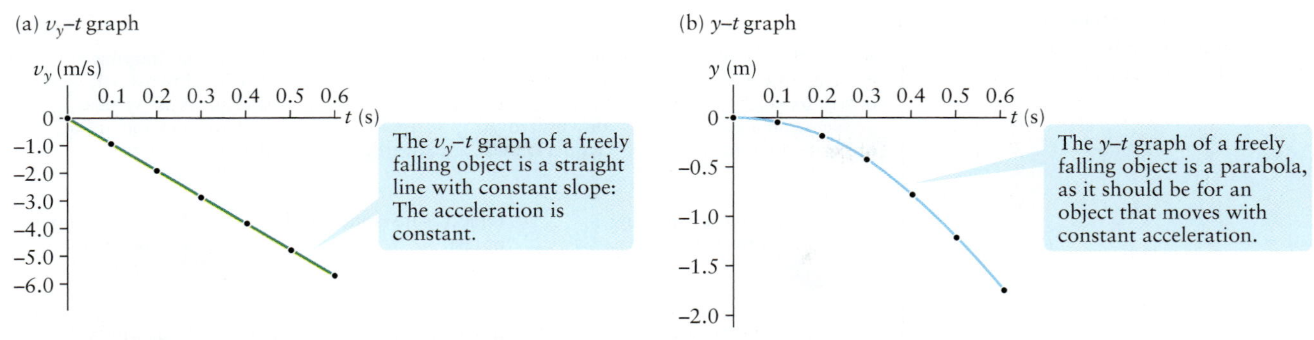

Figure 2-19 Graphing free fall Graphs of velocity versus time and position versus time for the freely falling object shown in Figure 2-18.

in Section 2-3 that the slope of an object's velocity–time graph tells us that object's acceleration. Because a straight line has a constant slope, we can conclude from **Figure 2-19a** that the acceleration of the falling ball is constant. We can determine the value of this acceleration by measuring the slope of the v_x–t graph. Table 2-4 says that the velocity is $v_{1y} = -1.96$ m/s at $t_1 = 0.20$ s, and the velocity is $v_{2y} = -3.92$ m/s at $t_2 =$ 0.40 s, so the acceleration a_y of the falling ball is

$$a_y = \text{slope of } v_y\text{–}t \text{ graph} = \frac{v_{2y} - v_{1y}}{t_2 - t_1} = \frac{-3.92 \text{ m/s} - (-1.96 \text{ m/s})}{0.40 \text{ s} - 0.20 \text{ s}} = -9.8 \text{ m/s}^2$$

We can confirm this value of acceleration in free fall by comparing the graph of position versus time in **Figure 2-19b** with the positions we can calculate by using the value $a_y = -9.8$ m/s^2. To make these calculations, we use Equation 2-9 for motion with constant acceleration, $x = x_0 + v_{0x}t + \frac{1}{2}a_x t^2$, but change every "$x$" in this equation to a "y." Using the initial velocity $v_{0y} = 0$ and initial y coordinate $y_0 = 0$, this equation becomes $y = \frac{1}{2}a_y t^2$. If we substitute $t = 0.30$ s and $a_y = -9.8$ m/s^2, our predicted position for the falling ball is

$$y = \frac{1}{2}(-9.8 \text{ m/s}^2)(0.30 \text{ s})^2 = -0.44 \text{ m}$$

which agrees with the value in Table 2-4.

Acceleration Due to Gravity

Experiments such as that depicted in Figure 2-18 show that an object in free fall near Earth's surface has a downward acceleration of approximate magnitude 9.80 m/s^2. (There are slight variations in this value from one place to another, related to differences in elevation and in the density of material beneath the surface. Keep in mind that 9.80 m/s^2 is an *average* value, which we state to three significant figures for convenience in calculation.) Physicists refer to this magnitude as the **acceleration due to gravity** and denote it by the symbol g:

$$g = 9.80 \text{ m/s}^2 \quad (2\text{-}12)$$

The value of g decreases very slightly with altitude: It is about 0.3% less at 10,000 m (32,800 ft) above sea level than at sea level. As we will learn in Chapter 10, a spacecraft journeying far beyond Earth encounters very different values of g. The value of g is also totally different on the surfaces of the Moon and other planets.

The downward acceleration due to gravity also acts on objects that are moving upward or are even momentarily stationary. If a ball is tossed straight up, as it ascends its velocity is positive (upward), and its acceleration is negative (downward). Since the velocity and acceleration have opposite signs, the ball slows down (see our discussion in Section 2-3). After the ball has reached the high point of its motion and is descending, its velocity and acceleration are both negative (downward). The signs of velocity and acceleration are the same, so the ball speeds up as it falls. At all times that the ball is in free fall, it has the *same* magnitude and direction of acceleration (**Figure 2-20**).

WATCH OUT! g is always positive, never negative.

Note that g is a *positive* number because it is the *magnitude* (absolute value) of the acceleration due to gravity. If we take the positive y direction to point upward, as we did in Figures 2-17 and 2-18, the acceleration due to gravity is negative: $a_y = -g = -9.80$ m/s^2. But no matter whether you take the positive direction to be upward or downward, the quantity g is always positive.

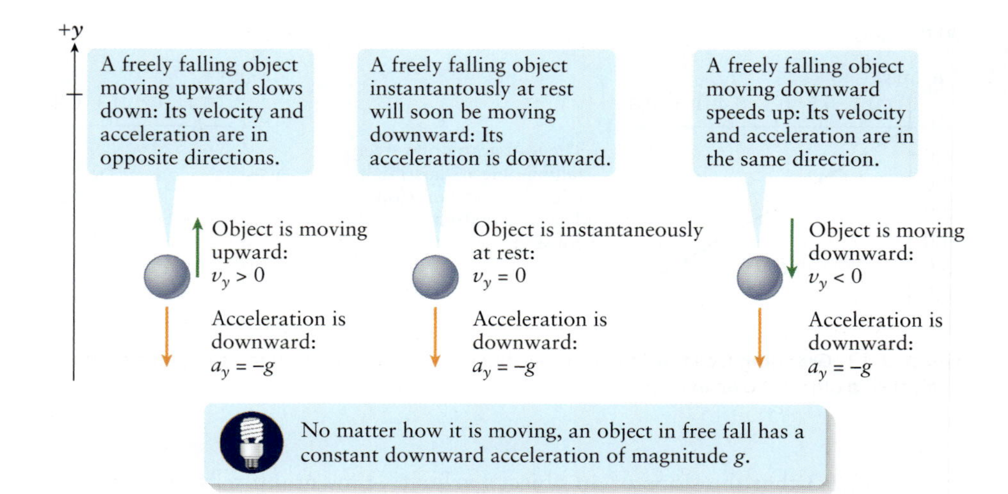
+y
A freely falling object moving upward slows down: Its velocity and acceleration are in opposite directions.
A freely falling object instantantously at rest will soon be moving downward: Its acceleration is downward.
A freely falling object moving downward speeds up: Its velocity and acceleration are in the same direction.
Object is moving upward: $v_y > 0$
Object is instantaneously at rest: $v_y = 0$
Object is moving downward: $v_y < 0$
Acceleration is downward: $a_y = -g$
Acceleration is downward: $a_y = -g$
Acceleration is downward: $a_y = -g$
No matter how it is moving, an object in free fall has a constant downward acceleration of magnitude g.

Figure 2-20 Free-fall acceleration The acceleration due to gravity does not depend on how the object is moving.

WATCH OUT! What happens when an object in free fall reaches its highest point?

It's a common error to think that if a ball is tossed upward, its acceleration at the high point of its motion is zero. It's true that the *velocity* is zero at this point because the ball is momentarily at rest. The velocity is changing, however; the ball's velocity was upward a moment before reaching the high point, and the ball's velocity is downward a moment after. Whenever an object's velocity is changing, that object has an acceleration. Thus the ball is still accelerating at the high point of the motion, with the same value of acceleration as on the way up and as on the way down (see Figure 2-20). If the acceleration at the high point of the motion actually *were* zero, the tossed ball would reach the high point of its motion and simply stop in midair: With no acceleration and therefore no change in velocity, the ball's zero velocity would remain zero forever! That doesn't happen in the real world, which tells us that the acceleration can't be zero at the high point.

The Equations of Free Fall

Let's see how to describe free fall that involves motion that is straight up and down—that is, *one-dimensional* free fall or *linear* free fall. As we did in Figures 2-17, 2-18, and 2-20, it's conventional to call the vertical axis the y axis and to choose the positive y direction to point upward. Then the acceleration in free fall points in the negative y direction, so the acceleration is $a_y = -g$. (Remember from Equation 2-12 that g is a *positive* number.) We can write the equations for free fall by replacing every "x" in Equations 2-5, 2-9, and 2-11 with a "y" and by using $-g$ for the acceleration a_y:

Velocity, acceleration, and time for free fall (2-13)

Velocity at time t of an object in free fall — **Velocity at time $t = 0$** of the object

$$v_y = v_{0y} - gt$$

Acceleration due to gravity (g is positive) — **Time** at which the object has position v_y

Position, acceleration, and time for free fall (2-14)

Position at time t of an object in free fall — **Velocity at time $t = 0$** of the object — **Acceleration due to gravity** (g is positive)

$$y = y_0 + v_{0y}t - \frac{1}{2}gt^2$$

Position at time $t = 0$ of the object — **Time** at which the object is at position y

Velocity at position y of an object in free fall

Velocity at position y_0 of the object

Position of object when it has velocity v_{0y}

$$v_y^2 = v_{0y}^2 - 2g\,(y - y_0)$$

Acceleration due to gravity (g is positive)

Position of object when it has velocity v_y

Velocity, acceleration, and position for free fall
(2-15)

Note the minus sign in front of each term containing g, which is a reminder that the acceleration is always *downward*, in the negative y direction. If a freely falling object is rising, it slows down; if the object is descending, it speeds up (Figure 2-20). When solving problems that involve free fall, make sure that you always check your answers to see that they agree with these commonsense principles! If they don't—if, say, your answers tell you that a ball thrown straight up goes faster and faster as it climbs—you need to go back and check your work.

GOT THE CONCEPT? 2-8
Interpreting Graphs for Free Fall

? Figure 2-15 on page 40 shows three possible combinations of x–t graphs and v_x–t graphs. If we interpret these as y–t graphs and v_y–t graphs, with the positive y direction chosen to be upward, which of these graph combinations are *not* possible for an object in free fall?

Solving Free-Fall Problems

Table 2-5 summarizes the three equations we can use to analyze the motion of an object experiencing free fall. We use the same approach to solve problems that we used in Section 2-5. After all, the equations in Table 2-5 are the same equations as in Table 2-3, but for the special case of motion along the y axis, and with $a_y = -g$. As you read through the following examples, keep in mind that the process we use to solve each problem is no less important than the final answer.

Go to Interactive Exercise 2-2 for more practice dealing with free fall.

TABLE 2-5 Equations for Linear Free Fall

By using one or more of these equations, you can solve any problem involving free-fall motion that is straight up and down. The check marks indicate which of the quantities t (time), y_0 (the y coordinate at time 0), y (the y coordinate at time t), v_{0y} (the y velocity at time 0), v_y (the y velocity at time t), and $a_y = -g$ (the constant, downward y acceleration) appear in each equation. Note that g, the magnitude of the acceleration due to gravity, is a *positive* quantity.

Equation		Which quantities the equation includes					
		t	y	y_0	v_{0y}	v_y	$a_y = -g$
$v_y = v_{0y} - gt$	(2-13)	✓			✓	✓	✓
$y = y_0 + v_{0y}t - \frac{1}{2}gt^2$	(2-14)	✓	✓	✓	✓		✓
$v_y^2 = v_{0y}^2 - 2g(y - y_0)$	(2-15)		✓	✓	✓	✓	✓

WATCH OUT! Up or down, it's still free fall.

! In many free-fall problems an object like a tossed ball first moves upward then falls back downward (see Example 2-10 below). Many students make such problems more complicated by writing separate sets of equations for the upward and downward parts of the motion. But remember that the acceleration is the same, $a_y = -g$, the whole time the object is in free fall. One set of equations is all you need for both the upward and downward motion, as Example 2-10 shows.

EXAMPLE 2-9 Free Fall I: A Falling Monkey

A monkey drops from a tree limb to grab a piece of fruit on the ground 1.80 m below (**Figure 2-21**). (a) How long does it take the monkey to reach the ground? (b) How fast is the monkey moving just before it reaches the ground? Neglect the effects of air resistance, so you can treat the monkey as being in free fall.

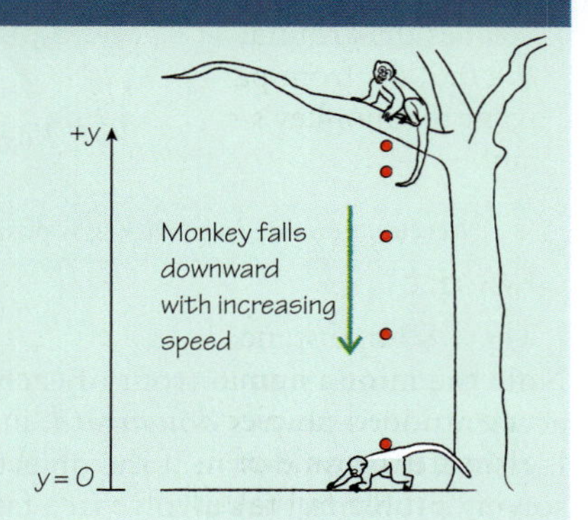

Figure 2-21 A monkey in free fall What are the monkey's position and velocity as functions of time?

Set Up

To describe our freely falling monkey, we choose the positive y direction to be upward (see Figure 2-21). Then the monkey's constant downward acceleration is $a_y = -g = -9.80\ \text{m/s}^2$ and we can use the equations in Table 2-5 as needed. Our goals are to find the time when he reaches the ground and the magnitude of his velocity just *before* reaching the ground. (Once the monkey is on the ground, his speed is zero.)

As Figure 2-21 shows, we choose the origin ($y = 0$) to be on the ground. Then the monkey's initial y coordinate is $y_0 = 1.80$ m, his y coordinate when he reaches the ground is zero, and his initial velocity is $v_{0y} = 0$ (because he starts at rest).

We've drawn the y–t graph and v_y–t graph for the monkey's motion using these values of y_0 and v_{0y}. Because the acceleration is negative, the y–t graph curves downward (the slope, which represents the velocity v_y becomes increasingly negative as time goes by), and the v_y–t graph has a downward (negative) slope.

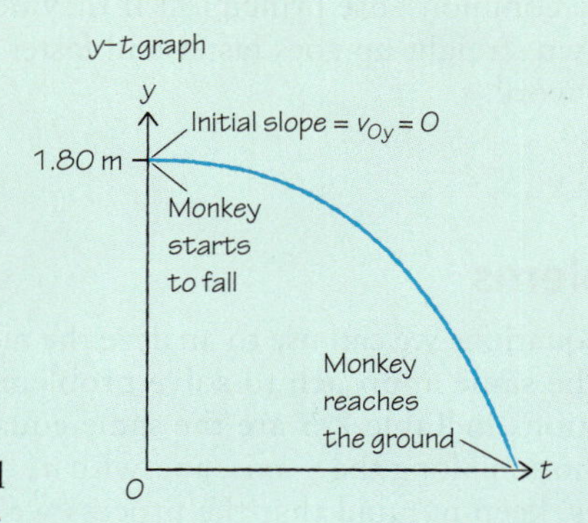

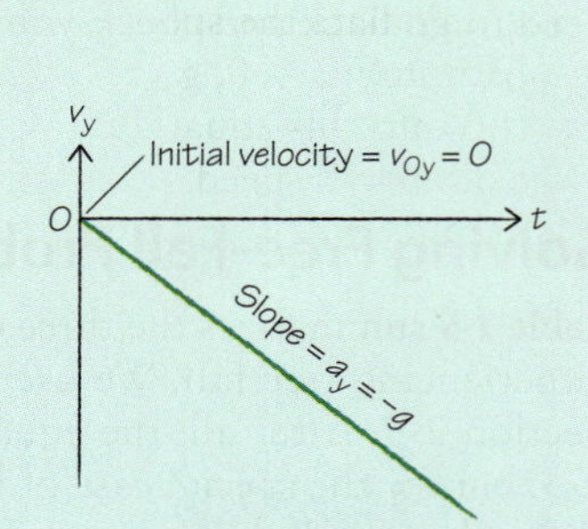

As we did for the examples in Section 2-5, we summarize the problem with a table of known and unknown quantities. We know the values of y_0, y, v_{0y}, and a_y, and want to find the values of (a) t and (b) v_y. To solve for t from the known information, we use Equation 2-14. Once we know the value of t, we can substitute it into Equation 2-13 to solve for the monkey's velocity v_y.

Position, acceleration, and time for free fall:

$$y = y_0 + v_{0y}t - \frac{1}{2}gt^2 \quad (2\text{-}14)$$

Velocity, acceleration, and time for free fall:

$$v_y = v_{0y} - gt \quad (2\text{-}13)$$

Quantity	Know/Don't Know
t	?
y_0	1.80 m
y	0
v_{0y}	0
v_y	?
$a_y = -g$	$-9.80\ \text{m/s}^2$

Solve

(a) Substitute the known values from our table into Equation 2-14. Then solve for t by rearranging the equation so that t^2 is by itself on the left-hand side of the equation and by taking the square root. The equation for t has both a positive solution and a negative solution. The monkey started to fall at $t = 0$, and we are looking for a time after this, so we must choose the positive value of t.

To find the time t when the monkey reaches the ground, substitute values into Equation 2-14:

$$0 = 1.80\ \text{m} + (0)t - \frac{1}{2}(9.80\ \text{m/s}^2)t^2$$

Solve for t:

$$\frac{1}{2}(9.80\ \text{m/s}^2)t^2 = 1.80\ \text{m}$$

$$t^2 = \frac{1.80\ \text{m}}{(1/2)(9.80\ \text{m/s}^2)} = 0.367\ \text{s}^2$$

$$t = \pm\sqrt{0.367\ \text{s}^2} = \pm\ 0.606\ \text{s}$$

Choose the positive value of t:
$t = 0.606$ s

(b) To find the monkey's velocity v_y just before it reaches the ground, substitute the known values and $t = 0.606$ s from part (a) into Equation 2-13. We want the monkey's *speed*, which is a positive number and is just the absolute value of v_y.

To find the velocity just before reaching the ground, substitute into Equation 2-13:

$$v_y = 0 - (9.80\ \text{m/s}^2)(0.606\ \text{s})$$
$$= -5.94\ \text{m/s so}$$

Speed = 5.94 m/s

Reflect

The 1.80-m distance that the monkey falls is about the height of a human adult. Try dropping a pencil or a wadded-up piece of paper from this height and estimate the time of its fall. Our result in part (a) says that the fall takes less than a second. Can you verify that this is about right?

As a check on our result in part (b), we can also calculate the speed using Equation 2-15. If you substitute $v_{0y} = 0$, $g = 9.80\ \text{m/s}^2$, $y_0 = 1.80$ m, and $y = 0$ into this equation, you should again find that the monkey's speed is 5.94 m/s. Do you? Check and find out!

Velocity, acceleration, and position for free fall:

$$v_y^2 = v_{0y}^2 - 2g(y - y_0) \qquad (2\text{-}15)$$

BIO-Medical EXAMPLE 2-10 Free Fall II: A Pronking Springbok

When startled, a springbok (*Antidorcas marsupialis*) like the one shown in **Figure 2-22** leaps upward several times in succession, reaching a height of several meters. (This behavior, called *pronking*, may inform predators that the springbok knows of their presence.) During one such jump, a springbok leaves the ground at a speed of 6.00 m/s. (a) What maximum height above the ground does the springbok attain? (b) How long does it take the springbok to attain this height? (c) What is the springbok's velocity when it is 1.25 m above the ground? (d) At what times is the springbok 1.25 m above the ground?

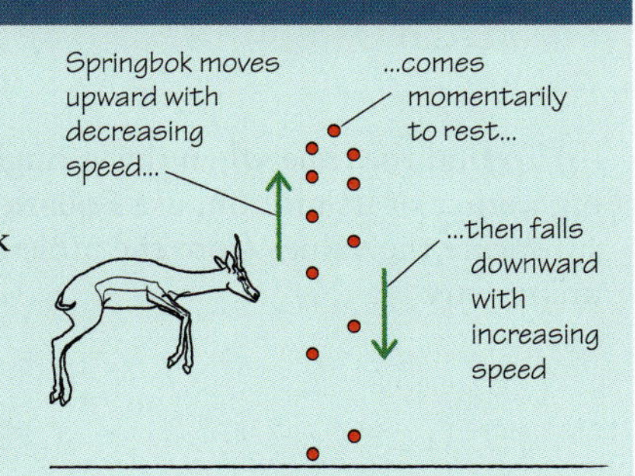

Figure 2-22 Pronking When a springbok jumps into the air, it is in free fall.

Set Up

Like the monkey in Example 2-9, the springbok is in free fall once it leaves the ground. Unlike the monkey, however, the springbok is moving initially and so v_{0y} is *not* zero.

At the moment when the springbok attains its maximum height, its velocity is zero. So parts (a) and (b) are really asking for the springbok's vertical coordinate and the time t when $v_y = 0$. In part (c) our goal is to find the springbok's velocity v_y at a certain vertical coordinate, and in part (d) we wish to find the times t at which the springbok is located at this coordinate.

We again choose the y axis to point upward and again place the origin at ground level. Then the springbok has initial y coordinate $y_0 = 0$ and initial velocity $v_{0y} = +6.00$ m/s (positive because it's moving upward). The drawings show the springbok's y–t graph and v_y–t graph. Because of the initial upward velocity, the y–t graph has an upward slope at $t = 0$, and the v_y–t graph starts with a positive value at $t = 0$.

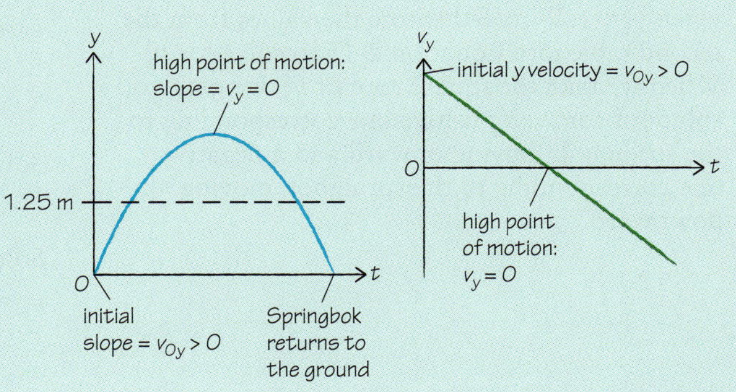

The tables of known and unknown quantities tell us that in parts (a) and (b) we want to calculate the value of y and the time t when the springbok reaches its highest point. We can use Equation 2-15 to find the position when the velocity is zero and Equation 2-13 to determine how much time is required for the velocity to equal zero. In part (c) we want the value of v_y at a height corresponding to $y = 1.25$ m. We can again use Equation 2-15. In part (d) we'll use Equation 2-13 to find the time t when the springbok has this value of y. We expect the springbok to pass through this height twice, once moving upward and once moving back downward.

For parts (a) and (c):

$$v_y^2 = v_{0y}^2 - 2g(y - y_0) \qquad (2\text{-}15)$$

For parts (b) and (d):

$$v_y = v_{0y} - gt \qquad (2\text{-}13)$$

For parts (a) and (b)

Quantity	Know/Don't Know
t	?
y_0	0
y	?
v_{0y}	+6.00 m/s
v_y	0
$a_y = -g$	-9.80 m/s^2

For parts (c) and (d)

Quantity	Know/Don't Know
t	?
y_0	0
y	1.25 m
v_{0y}	+6.00 m/s
v_y	?
$a_y = -g$	-9.80 m/s^2

Solve

(a) To determine the value of y corresponding to the high point of the motion, we substitute the values from our first table into Equation 2-15. Then we solve this equation for y, the height of the springbok above its launch point ($y_0 = 0$).

Substitute into Equation 2-15:

$$0^2 = (6.00 \text{ m/s})^2 - 2(9.80 \text{ m/s}^2)(y - 0)$$

Solve for y at the high point of the motion:

$$2(9.80 \text{ m/s}^2)y = (6.00 \text{ m/s})^2$$

$$y = \frac{(6.00 \text{ m/s})^2}{2(9.80 \text{ m/s}^2)} = 1.84 \text{ m}$$

(b) To find the time when the springbok is at the high point of its motion, use Equation 2-13 and substitute the values from the table of knowns and unknowns.

Substitute into Equation 2-13:

$$0 = 6.00 \text{ m/s} - (9.80 \text{ m/s}^2)t$$

Solve for t at the high point of the motion:

$$(9.80 \text{ m/s}^2)t = 6.00 \text{ m/s}$$

$$t = \frac{6.00 \text{ m/s}}{9.80 \text{ m/s}^2} = 0.612 \text{ s}$$

(c) Now we want to determine the value of v_y when $y = 1.25$ m. Substitute the values from the second table into Equation 2-15 to solve for v_y. When we take the square root of v_y^2, we get two solutions for v_y: a positive one corresponding to the springbok moving upward and a negative one corresponding to the springbok moving downward.

Substitute into Equation 2-15:

$$v_y^2 = (6.00 \text{ m/s})^2 - 2(9.80 \text{ m/s}^2)(1.25 \text{ m} - 0)$$
$$= 36.0 \text{ m}^2/\text{s}^2 - 24.5 \text{ m}^2/\text{s}^2$$
$$= 11.5 \text{ m}^2/\text{s}^2$$

Solve for v_y when $y = 1.25$ m:

$$v_y = \pm\sqrt{11.5 \text{ m}^2/\text{s}^2}$$

Solutions:

$v_y = +3.39$ m/s (springbok moving up)

$v_y = -3.39$ m/s (springbok moving down)

(d) From part (c) we know the two values of v_y for when the springbok passes through $y = 1.25$ m. To find the times t when it passes through this height, we substitute each value of v_y into Equation 2-13, along with $v_{0y} = 6.00$ m/s and $g = 9.80$ m/s^2.

Substitute into Equation 2-13:
Springbok moving up ($v_y = +3.39$ m/s):

$$+3.39 \text{ m/s} = +6.00 \text{ m/s} - (9.80 \text{ m/s}^2)t$$

Solve for t:

$$(9.80 \text{ m/s}^2)t = 6.00 \text{ m/s} - 3.39 \text{ m/s}$$

$$t = \frac{6.00 \text{ m/s} - 3.39 \text{ m/s}}{9.80 \text{ m/s}^2} = \frac{2.61 \text{ m/s}}{9.80 \text{ m/s}^2}$$

$$= 0.266 \text{ s}$$

Springbok moving down ($v_y = -3.39$ m/s):

$$-3.39 \text{ m/s} = +6.00 \text{ m/s} - (9.80 \text{ m/s}^2)t$$

Solve for t:

$$(9.80 \text{ m/s}^2)t = 6.00 \text{ m/s} - (-3.39 \text{ m/s})$$

$$t = \frac{6.00 \text{ m/s} - (-3.39 \text{ m/s})}{9.80 \text{ m/s}^2} = \frac{9.39 \text{ m/s}}{9.80 \text{ m/s}^2}$$

$$= 0.958 \text{ s}$$

Reflect

A professional basketball player can jump to a height of about 1 m; our result in part (a) says a springbok can jump even higher. This makes sense because a springbok has four powerful legs to propel it upward rather than just two!

We can check our result in part (c) by noticing that 3.39 m/s is less than 6.00 m/s. In other words, when the springbok is rising through $y = 1.25$ m, it is going more slowly than when it left the ground—as it should be because gravity slows the springbok as it ascends.

You can also check the times that we calculated in part (d) by solving for them directly without first calculating the values of v_y. You can do this with Equation 2-14: Substitute $y = 1.25$ m, $y_0 = 0$, $v_{0y} = 6.00$ m/s, and $g = 9.80$ m/s^2 and solve for t in the same way that we did in Example 2-8 in Section 2-5. Try this! You should get the same two answers for t, 0.266 s and 0.958 s.

GOT THE CONCEPT? 2-9 How Many Times Does It Pass?

You throw a ball straight upward at an initial speed of 5.00 m/s. How many times does the ball pass through a point 2.00 m above the point where it left your hand? Ignore the effects of air resistance. (a) Two; (b) one; (c) zero; (d) not enough information given to decide.

TAKE-HOME MESSAGE FOR Section 2-6

✔ An object is in free fall if its motion is influenced by gravity alone (that is, if air resistance is so small that it can be ignored). Free fall includes moving upward, moving downward, and the high point of the motion.

✔ Free-fall motion near the surface of Earth is constant-acceleration motion, with a downward acceleration of magnitude $g = 9.80$ m/s^2. The magnitude g is always a positive number.

✔ When discussing free fall, we always choose the positive y direction to be upward. Then the acceleration of an object in free fall is $a_y = -g$. For an object in free fall, the y–t graph is a parabola with a downward curvature, and the v_y–t graph is a straight line with downward (negative) slope.

✔ You can use the three equations in Table 2-5 to solve any problems involving linear motion in free fall.

✔ The same techniques that we used in Section 2-6 for general constant-acceleration problems can also be used for free-fall problems.

Key Terms

acceleration
acceleration due to gravity
average acceleration
average speed
average velocity
constant acceleration
constant velocity
coordinates
coordinate system
displacement
elapsed time
free fall
instantaneous acceleration
instantaneous speed
instantaneous velocity
linear motion
magnitude
motion diagram
negative displacement
origin
parabola
position
positive displacement
reference position
speed
time interval
v_x–t graph
x–t graph
vector
velocity

Chapter Summary

Topic	Equation or Figure

Linear motion: We describe the motion of an object along a straight line by stating how its position x changes with the passage of time t. The change in the object's position over a time interval is called the displacement.

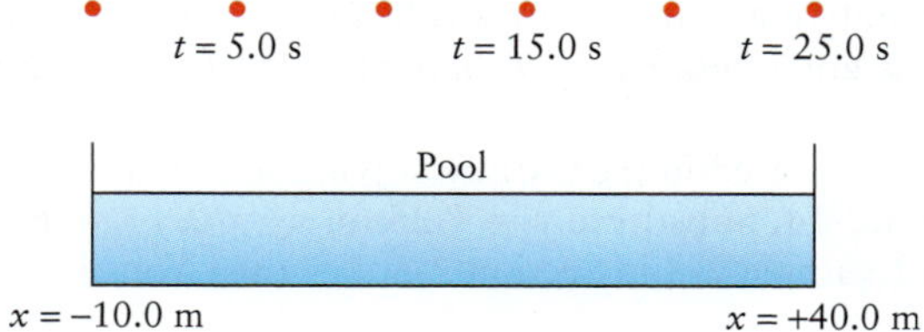

(Figure 2-3a)

Average velocity and instantaneous velocity: The average velocity of an object during a time interval describes how fast the object moved and in what direction. If the time interval is very short, the average velocity becomes the instantaneous velocity v_x (velocity for short).

Average velocity of an object in linear motion

Displacement (change in position) of the object over a certain time interval: The object moves from x_1 to x_2, so $\Delta x = x_2 - x_1$.

$$v_{\text{average},x} = \frac{x_2 - x_1}{t_2 - t_1} = \frac{\Delta x}{\Delta t} \qquad (2\text{-}1)$$

For both the displacement and the elapsed time, subtract the earlier value (x_1 or t_1) from the later value (x_2 or t_2).

Elapsed time for the time interval: The object is at x_1 at time t_1 and x_2 at time t_2, so the elapsed time is $\Delta t = t_2 - t_1$.

Constant-velocity linear motion: If an object has the same velocity v_x at all times, it maintains a steady speed and always moves in the same direction. A graph of position x versus time t for constant-velocity motion is a straight line with slope v_x.

Position at time t of an object in linear motion with constant velocity

Position at time $t = 0$ of the object

$$x = x_0 + v_x t \qquad (2\text{-}2)$$

Constant velocity of the object

Time at which the object is at position x

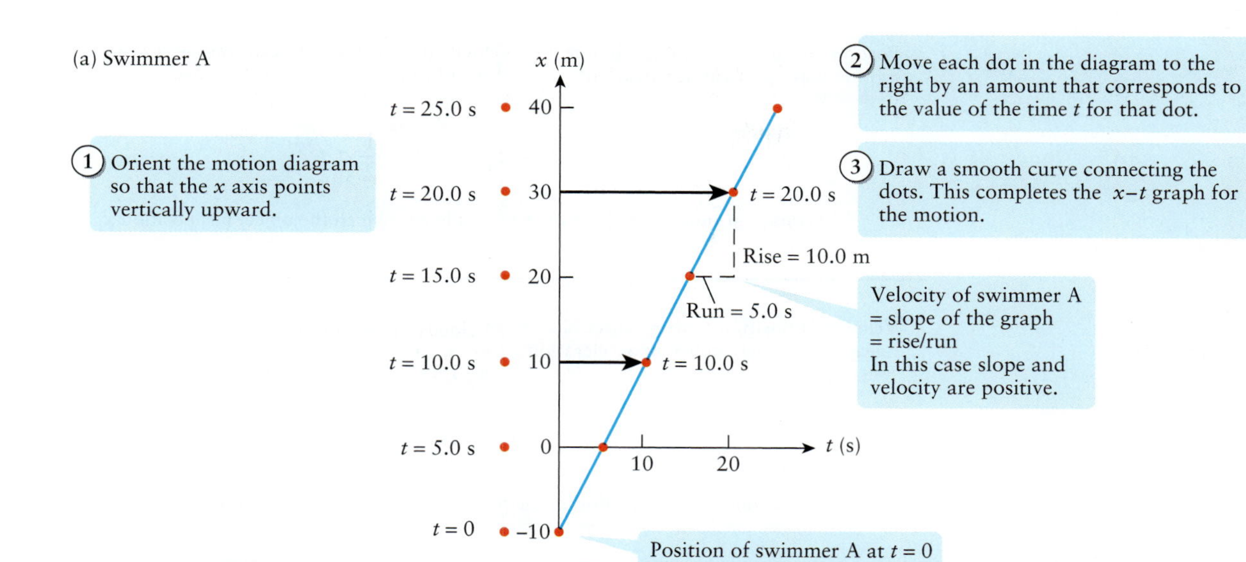

(Figure 2-4a)

Average acceleration and instantaneous acceleration: An object's average acceleration over a time interval equals the average rate of change of the object's velocity during that interval. If the time interval is very short, average acceleration becomes instantaneous acceleration a_x (acceleration for short).

Average acceleration of an object in linear motion

Change in **velocity** of the object over a certain time interval: The velocity changes from v_{1x} to v_{2x}.

$$a_{\text{average},x} = \frac{\Delta v_x}{\Delta t} = \frac{v_{2x} - v_{1x}}{t_2 - t_1} \tag{2-3}$$

Elapsed time for the time interval: The object has velocity v_{1x} at time t_1, and has velocity v_{2x} at time t_2.

Interpreting velocity and acceleration: The slope of an object's x–t graph (graph of position vs. time) is its velocity v_x. The slope of its v_x–t graph (graph of velocity vs. time) is its acceleration a_x. An object speeds up if v_x and a_x have the same sign, and slows down if v_x and a_x have opposite signs.

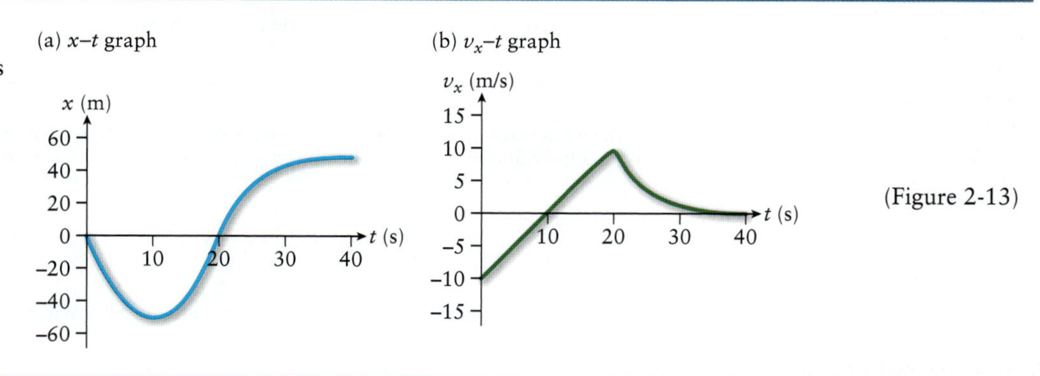

(Figure 2-13)

Motion with constant acceleration: The velocity of an object changes at the same rate at all times. The x–t graph for such an object is a parabola, and the v_x–t graph is a straight line.

Velocity at time t of an object in linear motion with constant acceleration

Velocity at time $t = 0$ of the object

$$v_x = v_{0x} + a_x t \tag{2-5}$$

Constant acceleration of the object

Time at which the object has velocity v_x

Position at time t of an object in linear motion with constant acceleration

Velocity at time $t = 0$ of the object

Constant acceleration of the object

$$x = x_0 + v_{0x}t + \frac{1}{2}a_x t^2 \quad (2\text{-}9)$$

Position at time $t = 0$ of the object

Time at which the object has position x

Velocity at position x of an object in linear motion with constant acceleration

Velocity at position x_0 of the object

$$v_x^2 = v_{0x}^2 + 2a_x(x - x_0) \quad (2\text{-}11)$$

Constant acceleration of the object

Two positions of the object

Free fall: An object that moves upward or downward under the influence of gravity alone is in free fall. It has a constant downward acceleration of magnitude g. Near the surface of Earth, $g = 9.80\ \text{m/s}^2$.

Velocity at time t of an object in free fall

Velocity at time $t = 0$ of the object

$$v_y = v_{0y} - gt \quad (2\text{-}13)$$

Acceleration due to gravity (g is positive)

Time at which the object has position v_y

Position at time t of an object in free fall

Velocity at time $t = 0$ of the object

Acceleration due to gravity (g is positive)

$$y = y_0 + v_{0y}t - \frac{1}{2}gt^2 \quad (2\text{-}14)$$

Position at time $t = 0$ of the object

Time at which the object is at position y

Velocity at position y of an object in free fall

Velocity at position y_0 of the object

Position of object when it has velocity v_{0y}

$$v_y^2 = v_{0y}^2 - 2g(y - y_0) \quad (2\text{-}15)$$

Acceleration due to gravity (g is positive)

Position of object when it has velocity v_y

Answer to What do you think? Question

The athlete is accelerating downward at this point and at *all* times from when her feet leave the ground to when they again touch the ground. This is a characteristic of free fall, discussed in Section 2-6.

Answers to Got the Concept? Questions

2-1 (a) (iv), (b) (iv), (c) (iii), (d) (iv), (e) (i), (f) (iv). Object (iv) has the steepest slope, so this object moves fastest *and* travels the greatest distance. The slope is negative, so the object moves in the negative x direction, but speed and distance are both positive quantities. Object (iii) has the most positive (upward) slope, so this object has the most positive velocity and hence the most positive displacement (from $x_1 = -5.0$ m to $x_2 = 0$ m, so $x_2 - x_1 = +5.0$ m). Object (iv) has the most negative (downward) slope, so this object has the most negative velocity and hence the most negative displacement (from $x_1 = +2.0$ m to $x_2 = -8.0$ m, so $x_2 - x_1 = -10.0$ m). The graphs show that at $t = 5.0$ s object (i) ends up at the most positive value of x (7.5 m), and object (iv) ends up at the most negative value of x (−8.0 m).

2-2 24 mi/h. Average velocity equals the net displacement (6.0 mi) divided by the elapsed time (15 min = 0.25 h), or $v_x = (6.0 \text{ mi})/(0.25 \text{ h}) = 24.0$ mi/h. The other information tells you how the velocity varied during the trip, but isn't relevant to the average velocity of the trip as a whole.

2-3 (b) We can determine the sign of the acceleration a_x from the slope of the v_x–t graph in Figure 2-13b. At $t = 5$ s, this graph has a positive slope, so a_x is positive and the velocity is becoming more positive. The v_x–t graph also shows that the value of the velocity v_x at $t = 5$ s is negative. Because v_x and a_x have opposite signs at $t = 5$ s, it follows that the object is slowing down at this time. You can confirm this result from Figure 2-13b: The object has $v_x = -10$ m/s at $t = 0$ and so is moving at 10 m/s in the negative x direction, while the object has $v_x = 0$ at $t = 10$ s and so is at rest. So the speed decreases from $t = 0$ to $t = 10$ s, and the object slows down.

2-4 (a) Because the acceleration is positive, the x–t graph is a parabola that curves upward, and the v_x–t graph slopes upward. The slope of the x–t graph is negative at $t = 0$ because $v_{0x} < 0$. (b) Because the acceleration is negative, the x–t graph is a parabola that curves downward, and the v_x–t graph slopes downward. The slope of the x–t graph is positive at $t = 0$ because $v_{0x} > 0$. (c) Because the acceleration is negative, the x–t graph is a parabola that curves downward, and the v_x–t graph slopes downward. The slope of the x–t graph is negative at $t = 0$ because $v_{0x} < 0$.

2-5 (a) (ii), (b) (iii), (c) (i) To decide how the speed of an accelerating object is changing, we need to compare the signs of its velocity and its acceleration. (a) At $t = 1.00$ s, $v_x = v_{0x} + a_x t = +2.00 \text{ m/s} + (-1.00 \text{ m/s}^2)(1.00 \text{ s}) = +1.00$ m/s. The velocity is positive, but the acceleration $a_x = -1.00 \text{ m/s}^2$ is negative, so the object is slowing down. (b) At $t = 2.00$ s, $v_x = +2.00 \text{ m/s} + (-1.00 \text{ m/s}^2)(2.00 \text{ s}) = 0$. The object is instantaneously at rest. It has just come to a stop (slowing down), but it is also speeding up because of the positive acceleration. (c) At $t = 3.00$ s, $v_x = +2.00 \text{ m/s} + (-1.00 \text{ m/s}^2)(3.00 \text{ s}) = -1.00$ m/s. The velocity and acceleration are both negative, so the object is speeding up.

2-6 We learned in Section 1-4 that when two numbers are multiplied together, the number with the smaller number of significant figures determines the number of significant figures in the product. Because 1.64 m/s^2 and 38.2 s each have only three significant figures, their product can only have three significant figures. (The factor of 1/2 is an exact number, so it doesn't affect how many significant figures are in the result.)

2-7 (b) "Slowing down at a steady rate" means that the cyclist has a constant acceleration. In this situation, the average velocity over a certain time interval is just the average of the instantaneous velocities at the beginning and end of the interval (see Equation 2-7). Because $v_{0x} = 15.0$ m/s and $v_x = 10.0$ m/s, the average velocity over the 4.00-s time interval is

$$v_{\text{average},x} = \frac{v_{0x} + v_x}{2} = \frac{15.0 \text{ m/s} + 10.0 \text{ m/s}}{2} = 12.5 \text{ m/s}$$

The cyclist's displacement during the time interval equals the average velocity multiplied by the elapsed time, or (12.5 m/s)(4.00 s) = 50.0 m. You can find this same result using the equations in Table 2-3, but it's much easier to use this basic fact about the average velocity when acceleration is constant.

2-8 (a), (b) The acceleration in free fall is $a_y = -g$, which is negative. Hence the impossible graphs are (a) and (b), which assume positive acceleration. The possible case is (c): The y–t graph is a downward-curved parabola, and the v_y–t graph is a straight line with a negative slope.

2-9 (c) The ball *never* reaches a height 2.00 m above the release point. You can see this in more than one way. One is to calculate the maximum vertical height of the ball (where the ball comes momentarily to rest) by using Equation 2-15 and setting $v_y = 0$. Let the point where you release the ball be at $y = 0$.

$$v_y^2 = v_{0y}^2 - 2g(y - y_0)$$

$$0 = (5.00 \text{ m/s})^2 - 2(9.80 \text{ m/s}^2)(y - 0)$$

$$y = \frac{(5.00 \text{ m/s})^2}{2(9.80 \text{ m/s}^2)} = 1.28 \text{ m}$$

The maximum height attained is 1.28 m, so the ball never reaches a height of 2.00 m above the release point. Another way to reach the same conclusion is to use Equation 2-14 to calculate the time t when the ball is at $y = 2.00$ m:

$$y = y_0 + v_{0y}t - \frac{1}{2}gt^2$$

$$2.00 \text{ m} = 0 + (5.00 \text{ m/s})t - \frac{1}{2}(9.80 \text{ m/s}^2)t^2$$

$$(4.90 \text{ m/s}^2)t^2 + (-5.00 \text{ m/s})t + 2.00 \text{ m} = 0$$

$$t = \frac{-(-5.00 \text{ m/s}) \pm \sqrt{(-5.00 \text{ m/s})^2 - 4(4.90 \text{ m/s}^2)(2.00 \text{ m})}}{2(4.90 \text{ m/s}^2)}$$

The quantity inside the square root is

$$(-5.00 \text{ m/s})^2 - 4(4.90 \text{ m/s}^2)(2.00 \text{ m})$$
$$= 25.0 \text{ m}^2/\text{s}^2 - 39.2 \text{ m}^2/\text{s}^2 = -14.2 \text{ m}^2/\text{s}^2$$

The square root of a negative number is *imaginary:* There is no real number that is equal to $\sqrt{-14.2}$. So this problem has no solution, and we conclude that the ball never reaches a height of 2.00 m.

Questions and Problems

In a few problems you are given more data than you actually need; in a few other problems you are required to supply data from your general knowledge, outside sources, or informed estimate.

Interpret as significant all digits in numerical values that have trailing zeros and no decimal points. For all problems use $g = 9.80\ \text{m/s}^2$ for the free-fall acceleration due to gravity. Neglect friction and air resistance unless instructed to do otherwise.

•	Basic, single-concept problem
••	Intermediate-level problem; may require synthesis of concepts and multiple steps
•••	Challenging problem
SSM	*Solution is in Student Solutions Manual*
Example	*See worked example for a similar problem*

Conceptual Questions

1. • Under what circumstances is it acceptable to omit the units during a physics calculation? What is the advantage of using SI units in *all* calculations, no matter how trivial?

2. • Compare the concepts of *speed* and *velocity*. Do these two quantities have the same units? When can you interchange these two with no confusion? When would it be problematic? SSM

3. • Explain the difference between average *speed* and average *velocity*.

4. • Under what circumstances will the magnitude of the displacement and the distance traveled be the same? When will they be different?

5. • Which speed gives the largest straight-line displacement in a fixed time: 1 m/s, 1 km/h, or 1 mi/h?

6. • Under what circumstance(s) will the average velocity of a moving object be the same as the instantaneous velocity? SSM

7. • What happens to an object's velocity when the object's acceleration is in the opposite direction to the velocity? SSM

8. • Discuss the direction and magnitude of the velocity and acceleration of a ball that is thrown straight up, from the time it leaves your hand until it returns and you catch it.

9. •The manufacturer of a high-end sports car plans to present its latest model's acceleration in units of m/s^2 rather than the customary units of miles/hour/second. Discuss the advantages and disadvantages of such an ad campaign in the global marketplace. Would you suggest making any modifications to this plan?

10. • What are the units of the slopes of the following graphs: (a) displacement versus time? (b) velocity versus time? (c) distance fallen by a dropped rock versus time?

11. • The upper limit of the braking acceleration for most cars is about the same magnitude as the acceleration due to gravity on Earth. Compare the braking motion of a car with a ball thrown straight upward. Both have the same initial speed. Ignore air resistance.

12. • A video is made of a ball being thrown up into the air and then falling. Is there any way to tell whether the video is being played backward? Explain your answer.

13. • The velocity of a ball thrown straight up decreases as it rises. Does its acceleration increase, decrease, or remain the same as the ball rises? Explain your answer.

14. • A device launches a ball straight up from the edge of a cliff so that the ball falls and hits the ground at the base of the cliff. The device is then turned so that a second, identical ball is launched straight down from the same height. Does the second ball hit the ground with a velocity that is higher than, lower than, or the same as the first ball? Explain your answer. SSM

15. • Is there any consistent reason why "up" can't be labeled as "negative" or "left" as "positive"? Explain why many physics professors and textbooks recommend choosing *up* and *right* as the positive directions in a description of motion.

Multiple-Choice Questions

16. • Figure 2-23 shows a position versus time graph for a moving object. At which lettered point is the object moving the slowest?
    A. A
    B. B
    C. C
    D. D
    E. E SSM

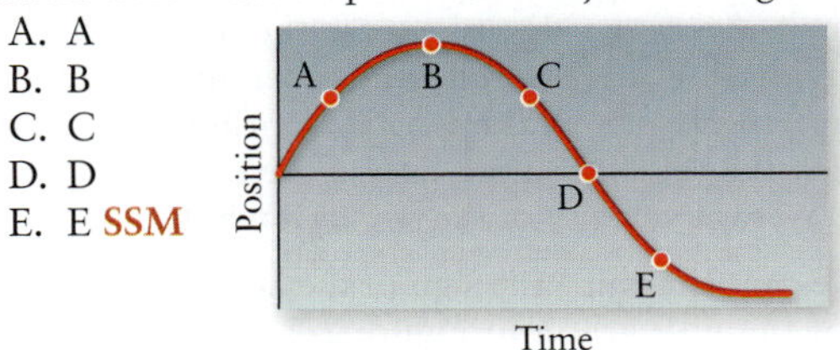

Figure 2-23 Problem 16

17. • Figure 2-24 shows a position versus time graph for a moving object. At which lettered point is the object moving the fastest?
    A. A
    B. B
    C. C
    D. D
    E. E

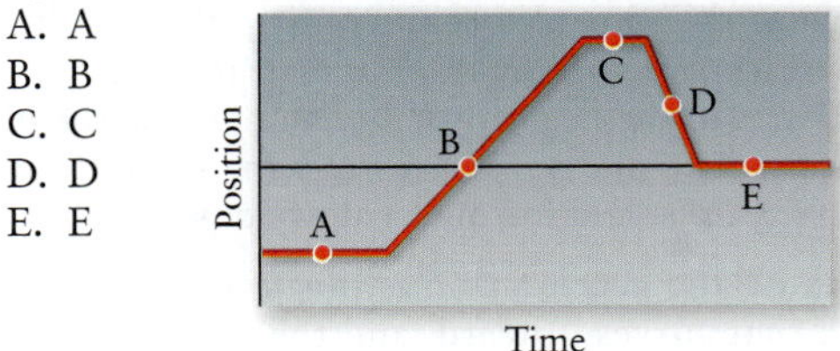

Figure 2-24 Problem 17

18. • A person is driving a car down a straight road. The instantaneous acceleration is constant and in the direction of the car's motion. The speed of the car is
    A. increasing.
    B. decreasing.
    C. constant.
    D. increasing but will eventually decrease.
    E. decreasing but will eventually increase.

19. • A person is driving a car down a straight road. The instantaneous acceleration is constant and directed opposite the direction of the car's motion. The speed of the car is
    A. increasing.
    B. decreasing.
    C. constant.
    D. increasing but will eventually decrease.
    E. decreasing but will eventually increase.

20. • A 1-kg ball and a 10-kg ball are dropped from a height of 10 m at the same time. In the absence of air resistance,
    A. the 1-kg ball hits the ground first.
    B. the 10-kg ball hits the ground first.
    C. the two balls hit the ground at the same time.
    D. the 10-kg ball will take 10 times the amount of time to reach the ground.
    E. there is not enough information to determine which ball hits the ground first.

21. • When throwing a ball straight up, which of the following are correct about the magnitudes of its velocity (v) and its acceleration (a) at the highest point in its path?
   A. both $v = 0$ and $a = 0$
   B. $v \neq 0$ but $a = 0$
   C. $v = 0$ but $a = 9.80\ \text{m/s}^2$
   D. $v \neq 0$ but $a = 9.80\ \text{m/s}^2$
   E. not enough information to determine the velocity (v) and acceleration (a)

Estimation/Numerical Analysis

22. • **Medical** Estimate the acceleration, upon hitting the ground, of a painter who loses his balance and falls from a step stool. Compare this to the acceleration he would experience if he bends his knees as he hits the ground.

23. • Estimate the time required for an average car to reach freeway speeds. Estimate the time required for an average car to slow down to zero from freeway speeds.

24. • Estimate the average speed of a runner competing in a marathon race.

25. •• **Medical** The severity of many sports injuries is related to the magnitude of the acceleration that an athlete's body undergoes as it comes to rest, especially when joints (such as ankles and knees) are not properly aligned during falls. (a) Estimate the acceleration of a woman who falls while running at full speed and comes to rest on a muddy field. Compare the acceleration to that of the same woman who falls while running at full speed and comes to rest on a running track made from a synthetic material. (b) Repeat these estimates for a male athlete. Assume they both are world-class athletes. SSM

26. • Estimate the average speed of a commercial airliner in flight.

27. • **Biology** Estimate the acceleration that a cat undergoes as it jumps from the floor to a countertop.

28. • **Astronomy** Estimate the speed of Earth as it orbits the Sun.

29. • Estimate the acceleration of a large cruise ship that is leaving port to head out to sea. SSM

30. • (a) Estimate the displacement of a swimmer during a typical workout. (b) Estimate the distance swam during the same workout.

31. • **Sports** (a) Estimate the acceleration of a thrown baseball. (b) Estimate the acceleration of a kicked soccer ball.

32. • A trainer times his racehorse as it completes a workout on a long, straight track. The position versus time data are given below. Plot an x–t graph and calculate the average speed of the horse between (a) 0 and 10 s, (b) 10 and 30 s, and (c) 0 and 50 s.

x(m)	t(s)	x(m)	t(s)
0	0	500	30
90	5	550	35
180	10	600	40
270	15	650	45
360	20	700	50
450	25		

33. • Write the equations for $x(t)$ for each interval of constant acceleration motion for the object whose position as a function of time is shown in the following table. (*Hint*: Graph the data on a graphing calculator or in a spreadsheet.) SSM

x(m)	t(s)	x(m)	t(s)	x(m)	t(s)
−12	0	18	9	85	18
−6	1	24	10	90	19
0	2	33	11	95	20
6	3	45	12	100	21
12	4	60	13	90	22
15	5	65	14	80	23
15	6	70	15	70	24
15	7	75	16	70	25
15	8	80	17		

34. • A coconut is dropped from a tall tree. Complete the following table. You may wish to program a spreadsheet to calculate these values more quickly.

t(s)	y(m)	v(m/s)	a(m/s^2)
0	0	0	9.8
1			
2			
3			
4			
5			
10			

Problems

2-2 Constant velocity means moving at a steady speed in the same direction

35. • Convert the following speeds:
   A. 30.0 m/s = ______ km/h
   B. 14.0 mi/h = ______ km/h
   C. 90.0 km/s = ______ mi/h
   D. 88.0 ft/s = ______ mi/h
   E. 100 mi/h = ______ m/s

36. • A bowling ball moves from $x_1 = 3.50$ cm to $x_2 = -4.70$ cm during the time interval from $t_1 = 3.00$ s to $t_2 = 5.50$ s. What is the ball's average velocity? Example 2-1

37. • What must a jogger's average speed be in order to travel 13.0 km in 3.25 h? Example 2-1

38. • **Sports** The Olympic record for the marathon set in 2008 is 2 h, 6 min, 32 s. The marathon distance is 26.2 mi. What was the average speed of the record-setting runner in km/h? Example 2-1

39. •• **Sports** Kevin completes his morning workout at the pool. He swims 4.00×10^3 m (80 laps in the 50.0-m-long pool) in 1.00 h. (a) What is the average velocity of Kevin during his workout? (b) What is his average speed? (c) With a burst of speed, Kevin swims one 25.0-m stretch in 9.27 s. What is Kevin's average speed over those 25 m? SSM Example 2-1

40. •• A student rides her bicycle home from physics class to get her physics book and then heads back to class. It takes her 21.0 min to make the 12.2 km trip home and 13.0 min to get back to class. (a) Calculate the average velocity of the student for the round-trip (from the lecture hall to home and back to the lecture hall). (b) Calculate her average speed for the trip from the lecture hall to her home. (c) Calculate the average speed of the girl for the trip from her home back to the lecture hall. (d) Calculate her average speed for the round trip. Example 2-1

41 • A school bus takes 0.700 h to reach the school from your house. If the average speed of the bus is 56.0 km/h, how far does the bus have to travel? Example 2-1

42. •• A car traveling at 80.0 km/h is 1500 m behind a truck traveling at 70.0 km/h. How long will it take the car to catch up with the truck? Example 2-2

43. • **Medical** Alcohol consumption slows people's reaction times. In a controlled government test, it takes a certain driver 0.320 s to hit the brakes in a crisis when unimpaired and 1.00 s when drunk. When the car is initially traveling at 90.0 km/h, how much farther does the car travel before coming to a stop when the person is drunk compared to sober? SSM Example 2-1

44. •• A jet takes off from SFO (San Francisco, CA) and flies to ORD (Chicago, IL). The distance between the airports is 3.00×10^3 km. After a 1.00 h layover, the jet returns to San Francisco. The total time for the round-trip (including the layover) is 9 h, 52 min. If the westbound trip (from ORD to SFO) takes 24 more minutes than the eastbound portion, calculate the time required for each leg of the trip. What is the average speed for the overall trip? What is the average speed *without* the layover? Example 2-1

2-3 Velocity is the rate of change of position, and acceleration is the rate of change of velocity

45. • **Biology** The position versus time graph for a red blood cell leaving the heart is shown in **Figure 2-25**. Determine the instantaneous speed of the red blood cell when t = 10 ms. Recall, 1 ms = 0.001 s, 1 mm = 0.001 m. Example 2-3

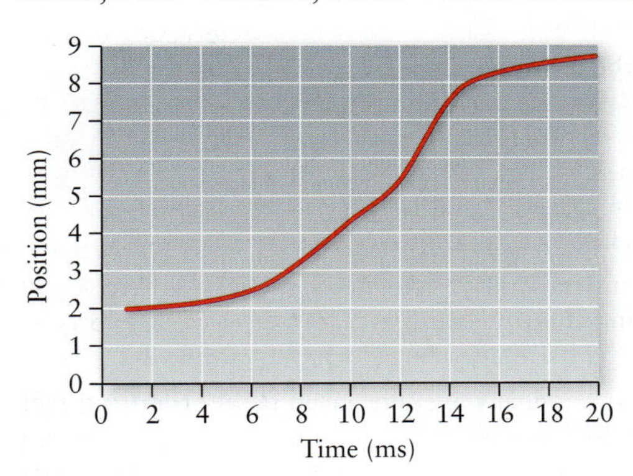

Figure 2-25
Problem 45

46. • **Figure 2-26** shows an x–t graph for some object. (a) At t = 4 s how far is the object from its position at t = 0? (b) At about what time is the object's *speed* greatest? (c) What is the velocity of the object between t = 2 s and t = 4 s? Example 2-3

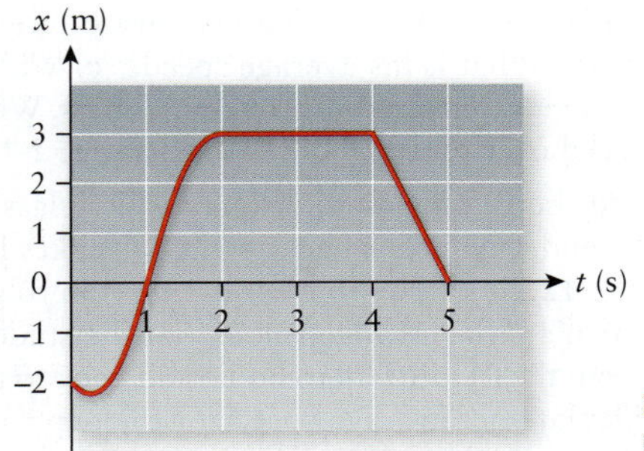

Figure 2-26
Problems 46–47

47. • Figure 2-26 shows an x–t graph for some object. (a) Over what time interval is the acceleration negative? (b) At about what time is the object's *speed* the greatest? (c) What is the acceleration between t = 4 s and t = 5 s? Example 2-4

48. • **Figure 2-27** shows a v_x–t graph for some object. (a) What is the object's maximum velocity? (b) How many times does the object change direction? (c) What is the acceleration of the object between t = 2 s and t = 3 s? (d) What is the object's average acceleration between t = 0 and t = 6 s? Example 2-5

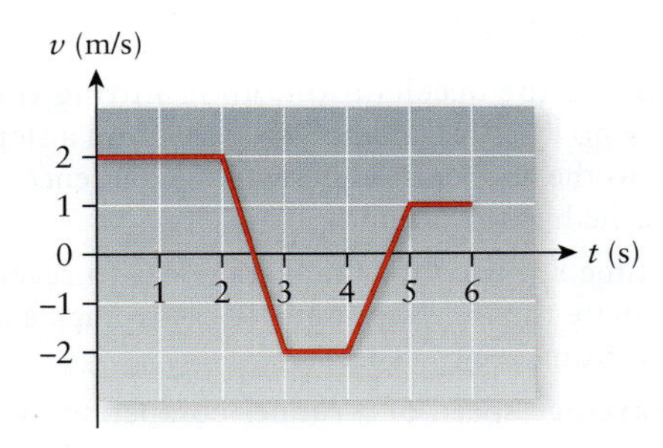

Figure 2-27
Problem 48

49. •• Consider the x–t plot in **Figure 2-28**. For each defined interval identify the *sign* (that is, + or −) of velocity v and acceleration a. Example 2-5

A. 0 s ≤ t ≤ 2 s
B. 2 s ≤ t ≤ 3 s
C. 3 s ≤ t ≤ 4 s
D. 4 s ≤ t ≤ 5 s
E. 5 s ≤ t ≤ 5.8 s
F. t ≥ 5.8 s

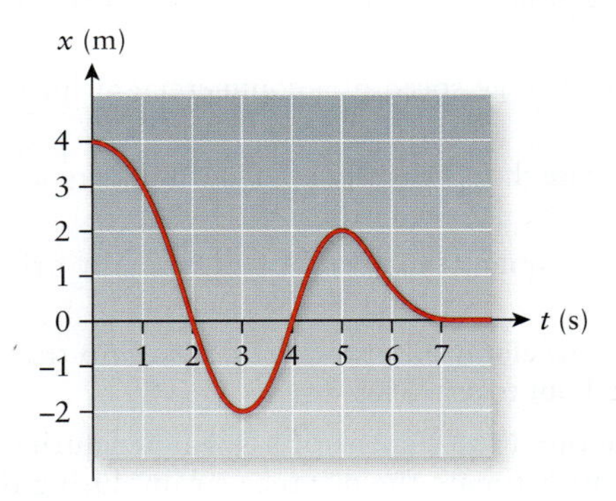

Figure 2-28
Problem 49

50. •• A driver is traveling along a straight road at the speed limit V. After 2 min, she slows at a constant rate to a stop at a traffic light. Two minutes later the light turns green, and she accelerates at a constant rate back up to the speed limit. Five minutes later she again slows at a constant rate to a stop. After 3 min she performs a u-turn then accelerates at a constant rate back up to speed V. Two minutes later she reaches her destination and slows at a constant rate to a stop. Sketch a graph of velocity versus time that represents her trip. (Since you are not given values for acceleration, your graph need only be qualitatively correct.) Example 2-5

51. ••• A car is stopped at a traffic light, defined as position x = 0. At t = 0 the light turns green, and the car accelerates constantly at 3 m/s^2 until it reaches 15 m/s at t = 5 s, at which time it continues on at that velocity. At t = 2 s a speeding truck passes through the traffic light in the same direction traveling

at a constant velocity of 20 m/s. On the *same* set of axes, draw the position versus time graphs for both the car and the truck out to $t = 6$ s. Example 2-5

2-5 Solving straight-line motion problems: Constant acceleration

52. •• Two trains, traveling toward one another on a straight track, are 300 m apart when the engineers on both trains become aware of the impending collision and hit their brakes. The eastbound train, initially moving at 98.0 km/h, slows down at 3.50 m/s^2. The westbound train, initially moving at 120 km/h, slows down at 4.20 m/s^2. Will the trains stop before colliding? If so, what is the distance between them once they stop? If not, what initial separation would have been needed to avert a disaster? **SSM** Example 2-8

53. • Paola can flex her legs from a bent position through a distance of 20.0 cm. Paola leaves the ground when her legs are straight, at a speed of 4.43 m/s. Calculate the magnitude of her acceleration, assuming that it is constant. Example 2-6

54. • A runner starts from rest and achieves a maximum speed of 8.97 m/s. If her acceleration is 9.77 m/s^2, how long does it take her to reach that speed? **SSM** Example 2-7

55. • A car traveling at 35.0 km/h speeds up to 45.0 km/h in a time of 5.00 s. The same car later speeds up from 65.0 km/h to 75.0 km/h in a time of 5.00 s. (a) Calculate the magnitude of the constant acceleration for each of these intervals. (b) Determine the distance traveled by the car during each of these time intervals. Example 2-6

56. • A car starts from rest and reaches a maximum speed of 34.0 m/s in a time of 12.0 s. Calculate the magnitude of its average acceleration. Example 2-4

57. • The world's fastest cars are rated by the time required for them to accelerate from 0 to 60.0 mi/h. Convert 60.0 mi/h to km/h and then calculate the acceleration in m/s^2 for each of the cars on the following list: Example 2-4

1. Bugatti Veyron 16.4	2.4 s
2. Caparo T1	2.5 s
3. Ultima GTR	2.6 s
4. SSC Ultimate Aero TT	2.7 s
5. Saleen S7 Twin Turbo	2.8 s

58. •• Using the information for the Bugatti Veyron in the previous problem, calculate the distance that the car would travel in the time it takes to reach 90.0 km/h, which is the speed limit on many Canadian roads. Compare your answer to the distance that the Saleen S7 Twin Turbo would travel while accelerating to the same final speed. **SSM** Example 2-6

59. •A driver is traveling at 30.0 m/s when he sees a moose crossing the road 80.0 m ahead. The moose becomes distracted and stops in the middle of the road. If the driver of the car slams on the brakes, what is the minimum constant acceleration he must undergo to stop short of the moose and avert an accident? (In British Columbia, Canada, alone, there are over 4000 moose–car accidents each year.) Example 2-6

60. • **Biology** A sperm whale can accelerate at about 0.100 m/s^2 when swimming on the surface. How far will a whale travel if it starts at a speed of 1.00 m/s and accelerates to a speed of 2.25 m/s? Assume the whale travels in a straight line. **SSM** Example 2-6

61. • At the start of a race, a horse accelerates out of the gate at a rate of 3.00 m/s^2. How long does it take the horse to cover the first 25.0 m of the race? Example 2-7

62. •• Derive the equation that relates position to speed and acceleration but in which the time variable does not appear. Start with the basic equation for the definition of acceleration, $a = (v - v_0)/t$; solve for t; and substitute the resulting expression into the position versus time equation, $x = x_0 + v_0 t + \frac{1}{2}at^2$.

2-6 Objects falling freely near Earth's surface have constant acceleration

63. • A ball is dropped from rest at a height of 25.0 m above the ground. (a) How fast is the ball moving when it is 10.0 m above the ground? (b) How much time is required for it to reach the ground level? Ignore the effects of air resistance. Example 2-9

64. •• Alex climbs to the top of a tall tree while his friend Gary waits on the ground below. Gary throws a ball up to Alex, who allows the ball to go past him before catching the ball on its way down. The ball has an initial speed of 10 m/s and is caught 3.5 m above where it was thrown. How long after the ball was thrown does Alex catch it? Ignore the effects of air resistance. Example 2-10

65. •• **Biology** A fox locates its prey, usually a mouse, under the snow by slight sounds the rodent makes. The fox then leaps straight into the air and burrows its nose into the snow to catch its next meal. If a fox jumps to a height of 85.0 cm, calculate (a) the speed at which the fox leaves the snow and (b) how long the fox is in the air. Ignore the effects of air resistance. Example 2-10

66. • **Medical** More people end up in U.S. emergency departments because of fall-related injuries than from any other cause. At what speed would someone hit the ground who accidentally stepped off the top rung of a 6-ft-tall stepladder? (That step is usually embossed with the phrase "Warning! Do not stand on this step.") Ignore the effects of air resistance. **SSM** Example 2-9

67. •• A ball is thrown straight up at 18.0 m/s. How fast is the ball moving after 1.00 s? After 2.00 s? After 5.00 s? When does the ball reach its maximum height? Ignore the effects of air resistance. Example 2-10

68. •• A tennis ball is hit straight up at 20.0 m/s from the edge of a sheer cliff. Sometime later, the ball passes the original height from which it was hit. (a) How fast is the ball moving at that time? (b) If the cliff is 30.0 m high, how long will it take the ball to reach the ground level? (c) What total distance did the ball travel? Ignore the effects of air resistance. Example 2-10

General Problems

69. •• Katie serves a ping pong ball by tossing the ball straight up and hitting the ball once it returns to the same point where it was tossed. If the ball is in the air for 0.45 s, what is the ball's initial speed after being tossed? Ignore the effects of air resistance. Example 2-10

70. •• **Biology** Cheetahs can accelerate to a speed of 20.0 m/s in 2.50 s and can continue to accelerate to reach a top speed of 29.0 m/s. Assume the acceleration is constant until the top speed is reached and is zero thereafter. (a) Express the cheetah's top speed in mi/h. (b) Starting from a crouched position, how long does it take a cheetah to reach its top speed, and how

far does it travel in that time? (c) If a cheetah sees a rabbit 120 m away, how long will it take to reach the rabbit, assuming the rabbit does not move? Example 2-6

71. ••• **Medical** Very large accelerations can injure the body, especially if they last for a considerable length of time. One model used to gauge the likelihood of injury is the severity index (SI), defined as $SI = a^{5/2}t$, where a is the acceleration in multiples of g, and t is the time the acceleration lasts (in seconds). In one set of studies of rear-end motor vehicle collisions, a person's velocity increases by 15.0 km/h with an acceleration of 35.0 m/s^2. (a) What is the severity index for the collision? (b) How far does the person travel during the collision if the car was initially moving forward at 5.00 km/h? Example 2-6

72. •• Randy and Deborah are standing at one end of a basketball court, each with a basketball. Randy rolls his basketball so that it moves at a constant speed of 5.0 m/s toward the other end of the court. Deborah waits 2.0 s before she rolls her basketball. With what speed must Deborah's ball move if it is to arrive at the other end of the court at the same time as Randy's basketball? The length of the court is 26 m. Example 2-8

73. ••• Blythe and Geoff compete in a 1.00-km race. Blythe's strategy is to run the first 600 m of the race at a constant speed of 4.00 m/s, and then accelerate to her maximum speed of 7.50 m/s, which takes her 1.00 min, and then finish the race at that speed. Geoff decides to accelerate to his maximum speed of 8.00 m/s at the start of the race and to maintain that speed throughout the rest of the race. It takes Geoff 3.00 min to reach his maximum speed. Assuming all accelerations are constant, who wins the race?

74. •• **Figure 2-29** is a graph of the velocity of a moving car. What is its instantaneous acceleration at times $t = 2$ s, $t = 4.5$ s, $t = 6$ s, and $t = 8$ s? Example 2-5

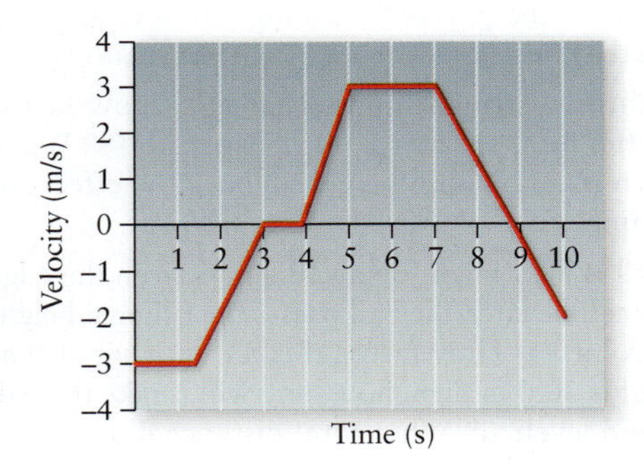

Figure 2-29
Problem 74

75. ••• A ball is dropped from an upper floor, some unknown distance above your apartment. As you look out of your window, which is 1.50 m tall, you observe that it takes the ball 0.180 s to traverse the length of the window. Determine how high above the top of your window the ball was dropped. Ignore the effects of air resistance. SSM Example 2-6

76. •• A ball is thrown straight up at 15.0 m/s. (a) How much time does it take for the ball to be 5.00 m above its release point? (b) How fast is the ball moving when it is 7.00 m above its release point? (c) How much time is required for the ball to reach a point that is 7.00 m above its release point? Why are there two answers to part (c)? Example 2-10

77. •• In the book and film *Coraline*, the title character drops a rock into a deep well and hears the sounds of it hitting the bottom 5.50 s later. If the speed of sound is 340 m/s, determine the depth of the well. Example 2-6

78. ••• A rocket is fired straight up from the ground. It contains a payload and two stages of solid rocket fuel that are designed to burn for 10.0 s and 5.00 s, respectively, with no time interval between them. In stage 1 the upward acceleration is 15.0 m/s^2. In stage 2 the acceleration is 12.0 m/s^2. Neglecting air resistance, (a) how fast is the payload moving at the end of stage 1? (b) How fast is it moving at the end of stage 2? (c) What is the altitude of the rocket at the end of stage 1? (d) What is the altitude of the rocket at the end of stage 2? (e) What is the maximum altitude obtained by the payload? (f) What is the total travel time of the payload from launch to landing? Assume the acceleration due to gravity is constant. Example 2-6

79. •• **Biology** A black mamba snake has a length of 4.30 m and a top speed of 8.90 m/s! Suppose a mongoose and a black mamba find themselves nose to nose. In an effort to escape the snake accelerates past the mongoose at 18 m/s^2 from rest. (a) How much time does it take for the snake to reach its top speed? (b) How far does the snake travel in that time? (c) How much reaction time does the mongoose have before the tail of the mamba snake passes by? Example 2-6

80. ••• **Sports** Kharissia wants to complete a 1000-m race with an average speed of 8.00 m/s. After 750 m, she has averaged 7.20 m/s. What average speed must she maintain for the remainder of the race in order to attain her goal? Example 2-1

81. •• A ball is thrown straight down at 1.50 m/s from a tall tower. Two seconds (2.00 s) later a second ball is thrown straight up at 4.00 m/s. How far apart are the two balls 4.00 s after the second ball is thrown? Ignore the effects of air resistance. Example 2-9

82. •• A two-stage rocket blasts off vertically from rest on a launch pad. During the first stage, which lasts for 15.0 s, the acceleration is a constant 2.00 m/s^2 upward. At the end of the first stage, the second stage engine fires, producing an upward acceleration of 3.00 m/s^2 that lasts for 12.0 s. At the end of the second stage, the engines no longer fire and therefore cause no acceleration, so the rocket coasts to its maximum altitude. (a) What is the maximum altitude of the rocket? (b) Over the time interval from blastoff at the launch pad to the instant that the rocket falls back to the launch pad, what are its (i) average speed and (ii) average velocity? Ignore the effects of air resistance. Example 2-6

3 Motion in Two or Three Dimensions

Tom Pfeiffer/VolcanoDiscovery/Getty Images

What do you think?

If air resistance isn't a factor, does a clump of lava emerging from a volcano take more time to (a) climb to its maximum height or (b) descend from its maximum height back to the elevation from which it was launched? Or (c) do the climb and descent take the same amount of time?

In this chapter, your goals are to:

- (3-1) Recognize the differences among one-, two-, and three-dimensional motion.
- (3-2) Describe the properties of a vector and how to find the sum or difference of two vectors.
- (3-3) Relate the components of a vector to its magnitude and direction, and use components in calculations involving vectors.
- (3-4) Explain how displacement, velocity, and acceleration are described in terms of vectors.
- (3-5) Identify the key features of projectile motion and how to interpret this kind of motion.
- (3-6) Solve problems involving projectile motion.
- (3-7) Describe why an object moving in a circle is always accelerating.
- (3-8) Explain how the structure of the ear helps us sense acceleration.

To master this chapter, you should review:

- (2-2, 2-3) The ideas of displacement, velocity, and acceleration for motion in a straight line.
- (2-4, 2-5) The concepts and equations of straight-line motion with constant acceleration.
- (2-6) The concepts and equations of free fall.

3-1 The ideas of linear motion help us understand motion in two or three dimensions

The clumps of lava shown in the above photo follow *curved* paths. That means that we can't describe them simply by using the ideas of Chapter 2 for motion in a straight line (one-dimensional motion). But we can extend the same concepts of displacement, velocity, and acceleration to motion along curved paths in two or three dimensions—that is, to motion in a plane or to generalized motion in space (**Figure 3-1a**).

We'll look in detail at an important kind of two-dimensional motion called *projectile motion*. This kind of motion in a plane occurs when the only source of an object's acceleration is the downward pull of gravity (**Figure 3-1b**). Projectile motion is not a perfect description of the flight of batted baseballs or kicked footballs, for which air resistance can be important, but it can still give useful approximate results.

We'll also examine the important case of motion in a circle (**Figure 3-1c**). We'll discover that even if an object travels around a circle at a constant speed, it is nonetheless accelerating. That's because the object is *turning*—the direction of its motion is continuously changing. In later chapters we'll use this observation to help explain why birds bank their wings when they turn, why cars sometimes skid when turning on a rain-slicked road, and other aspects of the natural and technological world.

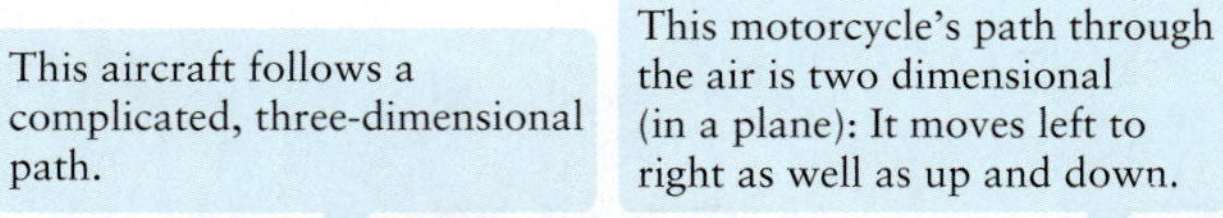

This aircraft follows a complicated, three-dimensional path.
This motorcycle's path through the air is two dimensional (in a plane): It moves left to right as well as up and down.

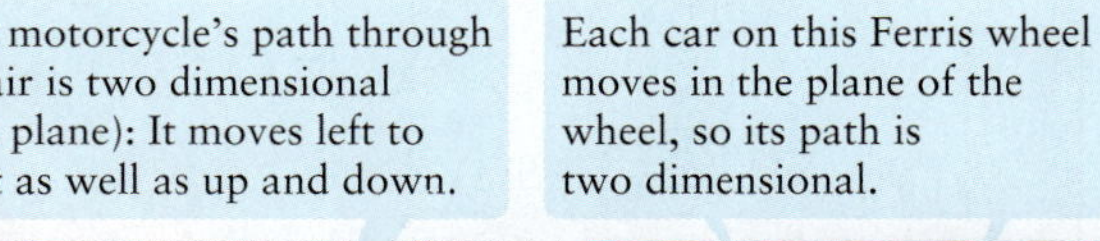

Each car on this Ferris wheel moves in the plane of the wheel, so its path is two dimensional.

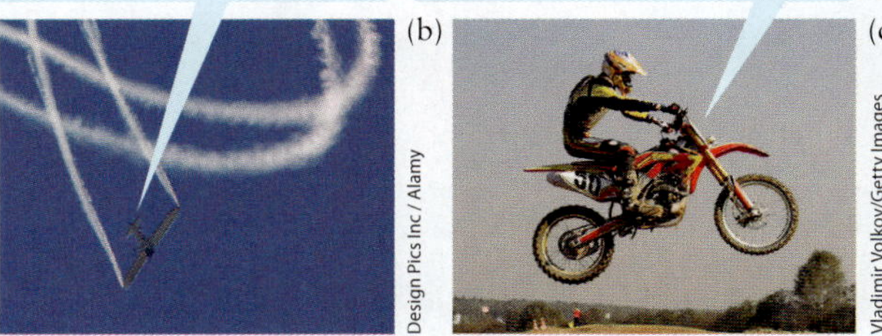

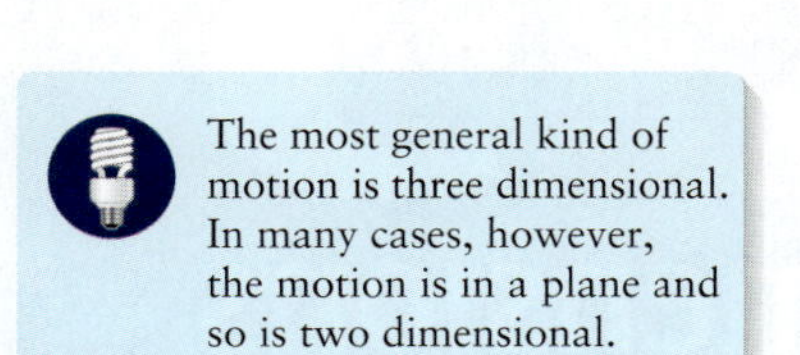

The most general kind of motion is three dimensional. In many cases, however, the motion is in a plane and so is two dimensional.

Figure 3-1 Motion in three and two dimensions Three examples of motion.

You might worry that motion in two or three dimensions is two or three times harder to analyze than one-dimensional motion along a straight line. Not so! By using the idea of a *vector* (a quantity with both a magnitude and a direction) that we introduced in Section 2-2, we'll be able to use all of the same techniques that we used in Chapter 2 for studying straight-line motion. The only difference is that in two- or three-dimensional motion, the vectors for position, displacement, velocity, and acceleration do *not* always point along the same straight line. In the following section we'll see how to work with vectors in two and three dimensions.

TAKE-HOME MESSAGE FOR Section 3-1

✔ Motion in two or three dimensions uses the same ideas of velocity and acceleration as linear motion.

✔ In two- and three-dimensional motion, velocity and acceleration have to be treated as vector quantities.

3-2 A vector quantity has both a magnitude and a direction

One afternoon you decide to take your textbook to study at your favorite coffee house. As **Figure 3-2a** shows, you could take a number of different paths to the coffee house. No matter which path you choose, however, the net result is that you end up 520 m from your apartment in a direction 60° north of east (**Figure 3-2b**). This change in your position is your *displacement*, and it has both a *magnitude* (520 m) and a *direction* (60° north of east). This is an extension of the idea of displacement that we introduced in Chapter 2 for motion along a straight line. In that case the sign of the displacement (positive or negative) told us whether the displacement was in the positive or negative x direction. Because your motion from apartment to coffee house is in a *plane*, however, we need to expand our definition of displacement.

Vectors and Scalars

Any quantity that has both a magnitude and a direction is called a **vector**. An example is the **displacement vector** in Figure 3-2b, which we depict as an arrow, that points from your starting point (your apartment) to your destination (the coffee house). The length of the arrow denotes the **magnitude** of the vector, which in this case is 520 m, the straight-line distance from starting point to destination. The **direction** of the displacement vector in this case is 60° north of east. Two vectors are equal *only* if they have the same magnitude and the same direction (**Figure 3-2c**). If two vectors differ in their magnitude, direction, or both, they are not equal to each other (**Figure 3-2d**).

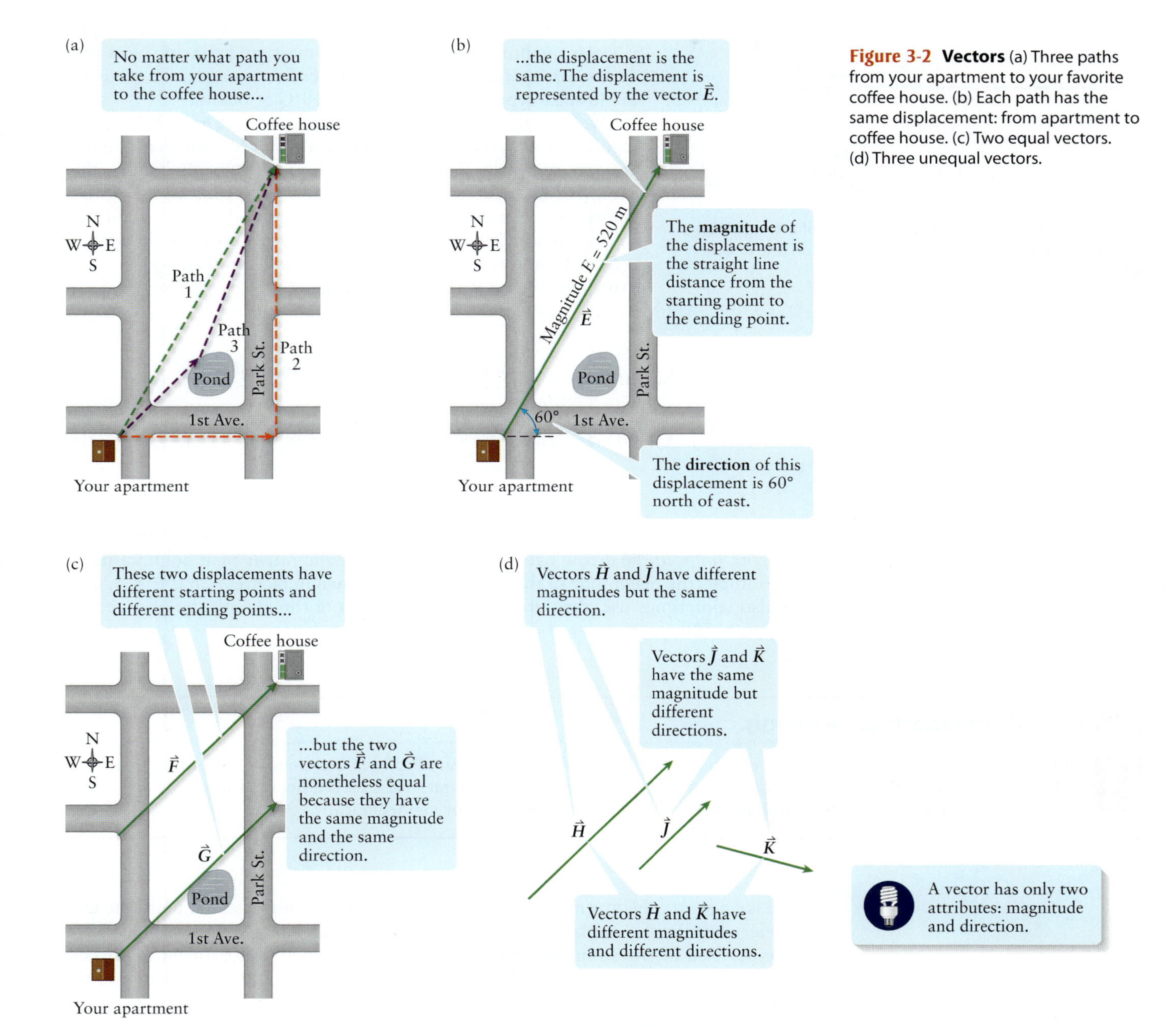
(a)
No matter what path you take from your apartment to the coffee house...
Coffee house
N
W
E
S
Path 1
Path 3
Path 2
Pond
Park St.
1st Ave.
Your apartment
(b)
...the displacement is the same. The displacement is represented by the vector $\vec{E}$.
Coffee house
Magnitude E = 520 m
$\vec{E}$
The **magnitude** of the displacement is the straight line distance from the starting point to the ending point.
Pond
Park St.
60°
1st Ave.
Your apartment
The **direction** of this displacement is 60° north of east.
(c)
These two displacements have different starting points and different ending points...
Coffee house
$\vec{F}$
$\vec{G}$
...but the two vectors $\vec{F}$ and $\vec{G}$ are nonetheless equal because they have the same magnitude and the same direction.
Pond
Park St.
1st Ave.
Your apartment
(d)
Vectors $\vec{H}$ and $\vec{J}$ have different magnitudes but the same direction.
Vectors $\vec{J}$ and $\vec{K}$ have the same magnitude but different directions.
$\vec{H}$
$\vec{J}$
$\vec{K}$
Vectors $\vec{H}$ and $\vec{K}$ have different magnitudes and different directions.

Figure 3-2 Vectors (a) Three paths from your apartment to your favorite coffee house. (b) Each path has the same displacement: from apartment to coffee house. (c) Two equal vectors. (d) Three unequal vectors.

A vector has only two attributes: magnitude and direction.

Many important physical quantities have both a magnitude and a direction, and hence are vectors. When we say that a pigeon is flying at 10 m/s due west, we are stating its *velocity* vector (**Figure 3-3**). The pigeon's *speed* is the magnitude of this vector. Just as for the displacement vector, we use an arrow to denote the vector, and we use the length of the vector to denote the magnitude. That's why we draw the velocity vector for a hummingbird flying at 20 m/s as having twice the length of the velocity vector for the 10-m/s pigeon in Figure 3-3: The hummingbird's speed (the magnitude of its velocity) is twice as great. Other vectors that we will encounter in our

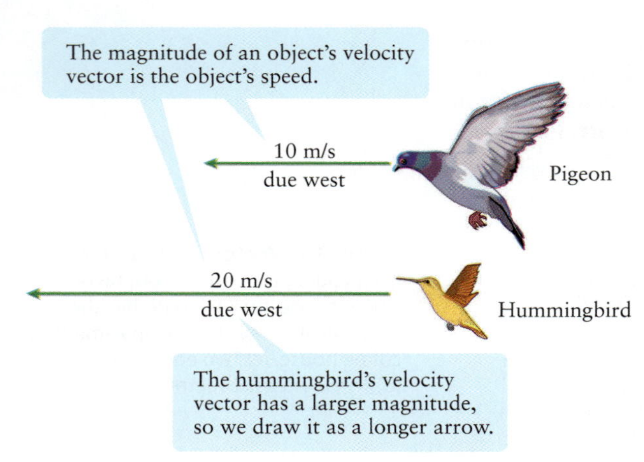
The magnitude of an object's velocity vector is the object's speed.
10 m/s
due west
Pigeon
20 m/s
due west
Hummingbird
The hummingbird's velocity vector has a larger magnitude, so we draw it as a longer arrow.

Figure 3-3 Velocity vectors The velocity of a flying bird is a vector: It has both a magnitude and a direction (the direction of the bird's travel).

study of physics include force (a push or pull), acceleration (which describes changes in velocity), and electric and magnetic fields.

Other physical quantities have *no* direction associated with them, so they are *not* vectors. These quantities are called **scalars** and can be described simply by stating a number and a unit. For example, the average temperature inside a home is 20°C, and the duration of a typical university lecture is 50 min. Other examples of scalar quantities are area, volume, mass, and density. Note that some scalar quantities can be negative: For example, the temperature inside a typical home freezer is −18°C or below.

We use different symbols for scalars and vectors to help us distinguish between them. We always use an *italic* letter to denote a scalar quantity, such as T for temperature or t for time. In handwriting you use ordinary letters for these quantities. By contrast, we always denote vectors using ***boldface italic*** letters with an arrow on top. Thus we write displacement vectors like those shown in Figure 3-2 as $\vec{E}$, $\vec{F}$, $\vec{G}$, and so on. In handwriting you should *always* draw an arrow over the symbol for a vector quantity.

Scalar: C — The symbol for a scalar is shown in *italic*.

Vector: $\vec{A}$ — The symbol for a vector is shown in ***boldface italic*** with an arrow.

The symbol for the magnitude of a vector is the same symbol as for the vector itself, but *without* the arrow over the symbol and in italic rather than boldface italic. For example, the magnitude of the displacement $\vec{E}$ shown in Figure 3-2b is $E = 520$ m. We also sometimes use absolute value signs to denote the magnitude of a vector: $|\vec{E}| = E = 520$ m.

WATCH OUT! Use vector notation correctly.

Note that it is *never* correct to write an equation such as $\vec{E} = 520$ m. A vector is not simply equal to its magnitude; its direction is just as important. You must instead say, "$\vec{E}$ has magnitude 520 m and points 60° north of east." (Later in this chapter we'll see a different, shorthand way to represent a vector mathematically.)

Like a scalar, the magnitude of a vector is given by a number and a unit. The unit is meters (m) for the displacement shown in Figure 3-2b. The magnitude of a vector can *never* be a negative number. If asked the distance from your apartment to the coffee house in Figure 3-2b, you would never say, "It's negative 520 m from here"! In the same way, speed (the magnitude of velocity) is never negative, which is why there are no negative numbers on a speedometer.

Adding Vectors

Scalars add together according to the rules of ordinary arithmetic: If your house has a floor area of 200 m^2 and you build an extra room with a floor area of 15 m^2, the net result is a house with a floor area of 200 m^2 + 15 m^2 = 215 m^2. Vectors behave differently, however. As an example, suppose you walk due east from your apartment to the corner of 1st Avenue and Park Street, a displacement $\vec{A}$, then walk due north to the coffee house, a displacement $\vec{B}$ (**Figure 3-4a**). Your net displacement is from your apartment to the coffee house; the combined effect of these two displacements is the same as the single displacement $\vec{E}$. Combining $\vec{A}$ and $\vec{B}$ to get $\vec{E}$ is an example of **vector addition**, and we say that $\vec{E}$ is the **vector sum** of $\vec{A}$ and $\vec{B}$. That is, $\vec{E} = \vec{A} + \vec{B}$. Figure 3-4a shows that another way to get to the coffee house is to travel first from your apartment to the duck pond (displacement $\vec{C}$) and then from the duck pond to the coffee house (displacement $\vec{D}$). So it's also true that $\vec{E} = \vec{C} + \vec{D}$.

Figure 3-4a illustrates that to perform vector addition we draw the two vectors in sequence, with the tail of the second vector up against the tip of the first vector. The sum of these two vectors then points from the tail of the first vector to the tip of the second vector. Even if the two vectors are not originally tip to tail, you can make them that way by moving or *translating* one vector's tail to the other vector's tip while keeping the directions of the vectors the same, as **Figure 3-4b** shows. This figure also shows that the order in which you add vectors doesn't matter, so $\vec{A} + \vec{B} = \vec{B} + \vec{A}$. Ordinary scalar addition behaves the same way: for example, $3 + 4 = 4 + 3$.

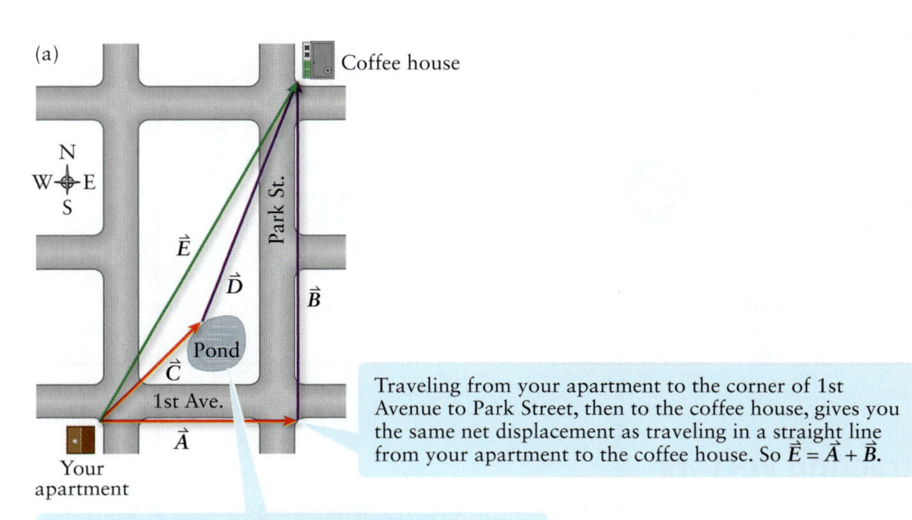
(a)
Coffee house
N
W
E
S
Park St.
$\vec{E}$
$\vec{D}$
$\vec{B}$
Pond
$\vec{C}$
1st Ave.
$\vec{A}$
Your apartment
Traveling from your apartment to the corner of 1st Avenue to Park Street, then to the coffee house, gives you the same net displacement as traveling in a straight line from your apartment to the coffee house. So $\vec{E} = \vec{A} + \vec{B}$.
A trip via the duck pond also gives you the same net displacement as a straight-line trip. So $\vec{E} = \vec{C} + \vec{D}$.

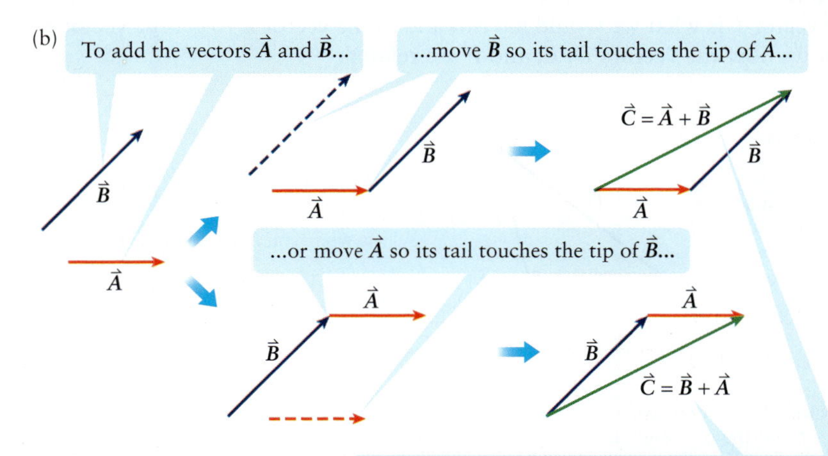
(b)
To add the vectors $\vec{A}$ and $\vec{B}$...
...move $\vec{B}$ so its tail touches the tip of $\vec{A}$...
$\vec{C} = \vec{A} + \vec{B}$
$\vec{B}$
$\vec{A}$
...or move $\vec{A}$ so its tail touches the tip of $\vec{B}$...
$\vec{C} = \vec{B} + \vec{A}$
...and the sum $\vec{C}$ of the two vectors runs from the tail of the first vector to the tip of the second vector. You get the same result no matter which order you add the vectors: $\vec{C} = \vec{A} + \vec{B} = \vec{B} + \vec{A}$.

Figure 3-4 Vector addition
(a) All three paths from your apartment to the coffee house give the same total displacement vector: $\vec{E} = \vec{A} + \vec{B} = \vec{C} + \vec{D}$. (b) How to place the vectors $\vec{A}$ and $\vec{B}$ so they can be added.

When adding two vectors, place the tip of the first vector so it touches the tail of the second vector. The vector sum then extends from the tail of the first vector to the tip of the second vector.

WATCH OUT! Vector addition is not the same as ordinary addition.

It's a common error to conclude that the magnitude of $\vec{A} + \vec{B}$, the sum of two vectors, is equal to the sum of the magnitude of $\vec{A}$ and the magnitude of $\vec{B}$. This usually gives the *wrong* answer when adding vectors. As an example, in Figure 3-5a the vectors $\vec{A}$ and $\vec{B}$ are perpendicular, so these two vectors and their sum $\vec{C} = \vec{A} + \vec{B}$ make up three sides of a right triangle. In this case the Pythagorean theorem tells us that the magnitude of $\vec{C}$ is given by $C^2 = A^2 + B^2$, so $C = \sqrt{A^2 + B^2}$. This is less than $A + B$. The one situation in which the magnitude $|\vec{A} + \vec{B}|$ *is* equal to the sum of magnitudes $A + B$ is when $\vec{A}$ and $\vec{B}$ point in the same direction (Figure 3-5b). Figure 3-5c shows that if $\vec{A}$ and $\vec{B}$ point in *opposite* directions, the magnitude $|\vec{A} + \vec{B}|$ is equal to the *difference* of A and B. In Section 3-3 we'll learn a technique for easily calculating the magnitude and direction of a vector sum.

(a)

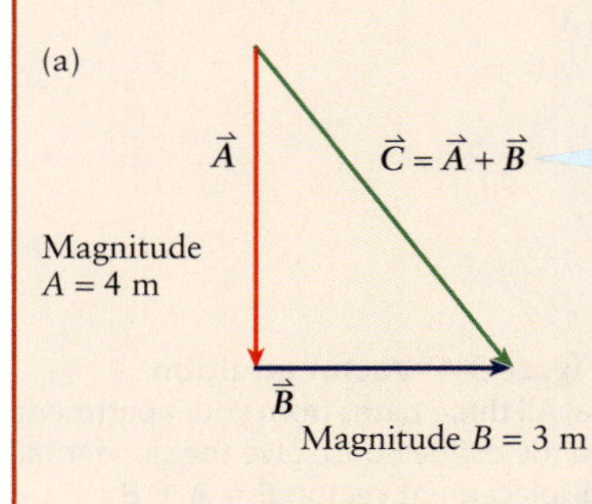

Magnitude of this vector sum:
$$C = \sqrt{A^2 + B^2} = \sqrt{(4\ \text{m})^2 + (3\ \text{m})^2} = \sqrt{16\ \text{m}^2 + 9\ \text{m}^2} = \sqrt{25\ \text{m}^2} = 5\ \text{m}$$

(b)

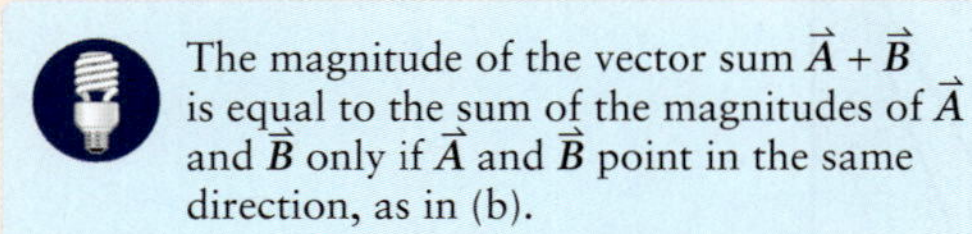

Magnitude of this vector sum:
$$C = A + B = 4\ \text{m} + 3\ \text{m} = 7\ \text{m}$$

(c)

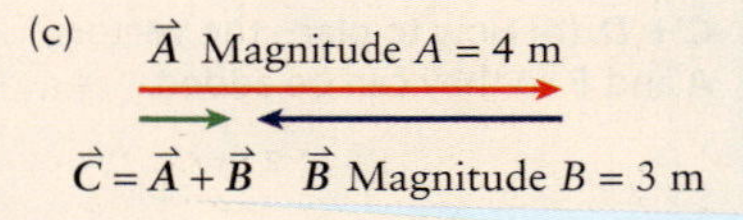

Magnitude of this vector sum:
$$C = A - B = 4\ \text{m} - 3\ \text{m} = 1\ \text{m}$$

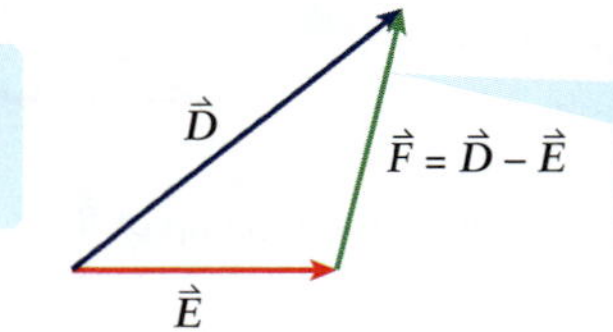

The magnitude of the vector sum $\vec{A} + \vec{B}$ is equal to the sum of the magnitudes of $\vec{A}$ and $\vec{B}$ only if $\vec{A}$ and $\vec{B}$ point in the same direction, as in (b).

Figure 3-5 Special cases of vector addition The sum $\vec{A} + \vec{B}$ of two vectors that point (a) perpendicular to each other, (b) in the same direction, and (c) in opposite directions.

Subtracting Vectors

Go to Picture It 3-1 for more practice dealing with adding and subtracting vectors.

A simple problem in arithmetic is "What is 7 minus 4?" What this problem is really asking is, "What do I have to add to the second number (in our example, 4) to get the first number (in our example, 7)?" The answer is $7 - 4 = 3$ because 3 added to 4 gives 7. **Vector subtraction** works in much the same way. When we say that $\vec{D} - \vec{E} = \vec{F}$, we mean that the vector $\vec{F}$ is what we would have to add to the vector $\vec{E}$ to get the vector $\vec{D}$. Figure 3-6a shows how the vectors $\vec{D}$, $\vec{E}$, and $\vec{F}$ are related.

(a) Subtracting vectors:

The difference $7 - 4 = 3$ is what you must add to 4 to get 7.

$\vec{D}$ $\vec{F} = \vec{D} - \vec{E}$ $\vec{E}$

Similarly, the vector difference $\vec{F} = \vec{D} - \vec{E}$ is what you must add to $\vec{E}$ to get $\vec{D}$.

(b) How to calculate $\vec{F} = \vec{D} - \vec{E}$:

1. Form the vector $-\vec{E}$ (with the same magnitude as $\vec{E}$ but the opposite direction).

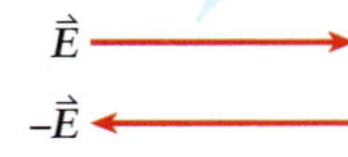

2. Add $\vec{D}$ and $-\vec{E}$ by placing them tip to tail. Their sum is $\vec{D} + (-\vec{E}) = \vec{D} - \vec{E}$.

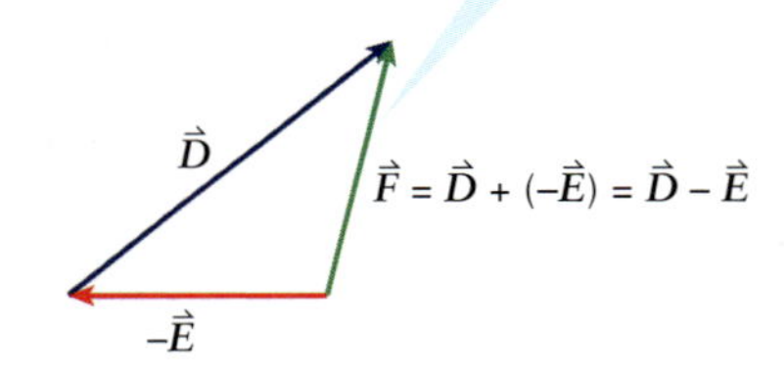

Figure 3-6 Vector subtraction Subtracting $\vec{E}$ from $\vec{D}$ is the same as adding $-\vec{E}$ to $\vec{D}$. (a) Subtracting one vector from another. (b) How to place the vectors $\vec{D}$ and $\vec{E}$ so they can be subtracted.

To see how to carry out vector subtraction, let's first notice something about ordinary subtraction: When we *subtract* 4 from 7, we're really *adding* -4 to 7, so that $7 - 4 = 7 + (-4) = 3$. In the same way, when we *subtract* the vector $\vec{E}$ from the vector $\vec{D}$ to calculate the **vector difference** $\vec{D} - \vec{E}$, we're really *adding* the vector $-\vec{E}$ to the vector $\vec{D}$: that is, $\vec{D} - \vec{E} = \vec{D} + (-\vec{E})$. To take the *negative of a vector*, we keep its magnitude the same and reverse its direction (**Figure 3-6b**). Thus if $\vec{E}$ is a displacement vector of magnitude 600 m that points due *east*, the vector $-\vec{E}$ also has magnitude 600 m but points due *west*. Figure 3-6b shows how to carry out the subtraction $\vec{D} - \vec{E}$ by adding $\vec{D}$ and $-\vec{E}$. If you know how to add two vectors, you also know how to subtract them!

A number of different animal species use vector addition and subtraction to navigate. As an example, honeybees (**Figure 3-7a**) fly along a jagged path from flower to flower in search of nectar, but they fly straight back—on a beeline—to their hive. A bee manages this feat by keeping track of the individual distances and directions it travels and then computing the vector needed to go home. That is, if a honeybee's individual displacements from the hive are $\vec{A}$, $\vec{B}$, and $\vec{C}$, it knows that its net displacement from the hive is $\vec{A} + \vec{B} + \vec{C}$ and that it must travel through an additional displacement $-(\vec{A} + \vec{B} + \vec{C})$ to return to the hive (**Figure 3-7b**). Fiddler crabs, desert ants, and hamsters also navigate using vectors. We'll use vector subtraction in Section 3-4 to help us extend the ideas of velocity and acceleration to motion in two or three dimensions.

(a)

Rob Flynn/USDA ARS

(b)

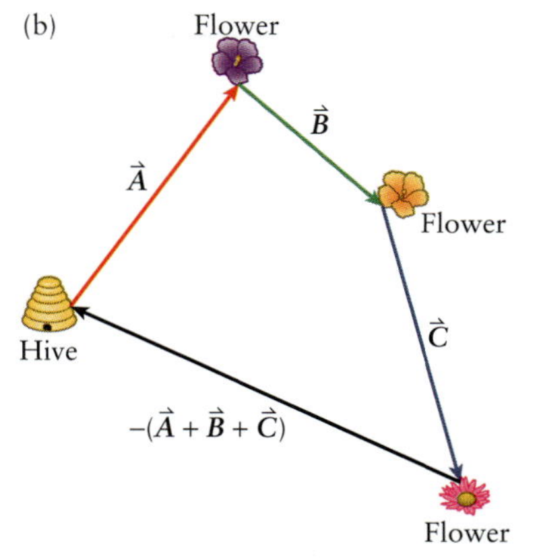
Flower
$\vec{B}$
$\vec{A}$
Flower
Hive
$\vec{C}$
$-(\vec{A} + \vec{B} + \vec{C})$
Flower

Figure 3-7 Honeybee vectors (a) A honeybee (genus *Apis*) visiting a flower. (b) Honeybees keep track of their vector displacements $\vec{A}$, $\vec{B}$, $\vec{C}$, . . . and use vector subtraction to calculate the direct route back to the hive.

Multiplying a Vector by a Scalar

A third useful bit of mathematics involving vectors is **vector multiplication by a scalar.** **Figure 3-8** shows the simple rules for this. If c is a *positive* scalar, then the product of c and a vector $\vec{A}$ is a new vector $c\vec{A}$ that points in the same direction as $\vec{A}$ but has a different magnitude, equal to cA (the product of c and the magnitude of $\vec{A}$). If c is a *negative* scalar, then $c\vec{A}$ points in the direction opposite to $\vec{A}$ and has magnitude $|c|A$ (the absolute value of c multiplied by the magnitude of $\vec{A}$). We actually used this idea already when we defined the negative of a vector (Figure 3-6b): The vector $-\vec{E}$ equals the product of the negative scalar -1 and the vector $\vec{E}$, so $-\vec{E}$ points in the direction opposite to $\vec{E}$ and has magnitude $|-1|E = (1)E = E$ (that is, the same magnitude as $\vec{E}$).

We can also *divide* a vector by a scalar. This is really the same thing as multiplying by a scalar. As an example, dividing the vector $\vec{A}$ by 2 is the same as multiplying $\vec{A}$ by 1/2: $\vec{A}/2 = (1/2)\vec{A}$. So $\vec{A}/2$ is a vector that points in the same direction as $\vec{A}$ and has one-half the magnitude (Figure 3-8).

We'll multiply and divide a vector by a scalar frequently in our study of physics. In Section 3-4 we'll use these ideas to help define the velocity and acceleration vectors.

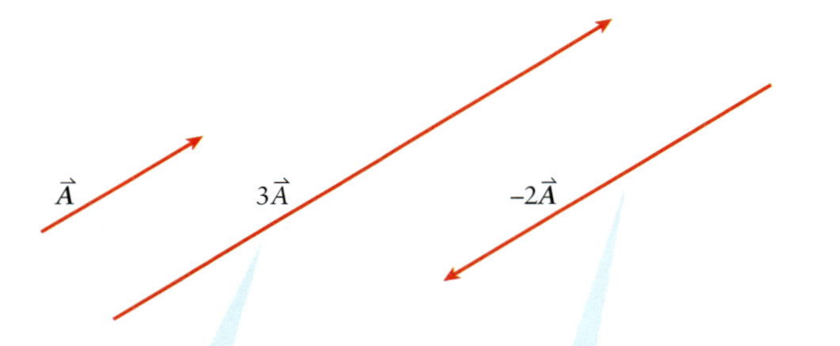
$\vec{A}$
$3\vec{A}$
$-2\vec{A}$
$\frac{\vec{A}}{2} = \frac{1}{2}\vec{A}$
The number 3 is positive, so the vector $3\vec{A}$ points in the same direction as $\vec{A}$ but has 3 times the magnitude.
The number –2 is negative, so the vector $-2\vec{A}$ points in the direction opposite to $\vec{A}$ and has 2 times the magnitude.
Dividing a vector by 2 is the same as multiplying it by 1/2, so the vector $\vec{A}/2$ points in the same direction as $\vec{A}$ (since 1/2 is positive) and has 1/2 the magnitude.

Figure 3-8 Multiplying a vector by a scalar How to multiply a vector $\vec{A}$ by a positive scalar or a negative scalar, and how to divide a vector by a scalar.

GOT THE CONCEPT? 3-1 Adding and Subtracting Vectors

Vector $\vec{A}$ points due north, and vector $\vec{B}$ points due west. Both vectors have the same magnitude. Which of the following vectors points southwest? (a) $\vec{A} + \vec{B}$; (b) $\vec{A} - \vec{B}$; (c) $\vec{B} - \vec{A}$; (d) $-\vec{A} - \vec{B}$; (e) none of these.

TAKE-HOME MESSAGE FOR Section 3-2

✔ Vectors are quantities that have both a magnitude and a direction. Scalars are ordinary numbers.

✔ To find the vector sum $\vec{A} + \vec{B}$ of two vectors, place the tail of the second vector ($\vec{B}$) against the tip of the first vector ($\vec{A}$). The vector sum points from the tail of the first vector to the tip of the second vector.

✔ The vector difference of two vectors, $\vec{A} - \vec{B}$, is the sum of $\vec{A}$ and $-\vec{B}$ (the negative of vector $\vec{B}$, which has the same magnitude as $\vec{B}$ but points in the opposite direction).

✔ The product of a scalar c and a vector $\vec{A}$ is a new vector that has magnitude $|c|A$ (the absolute value of c multiplied by the magnitude of $\vec{A}$). This vector points in the same direction as $\vec{A}$ if c is positive but in the opposite direction if c is negative.

3-3 Vectors can be described in terms of components

We saw in Figure 3-5 that adding two vectors $\vec{A}$ and $\vec{B}$ is straightforward if the vectors are perpendicular, point in the same direction, or point in opposite directions. But what can we do in the majority of situations, in which vectors are not so conveniently arranged?

A powerful technique that we can use for all kinds of calculations with vectors, no matter how they're oriented, is the method of **components**. As an application, consider the arc of a kicked football in the absence of air resistance. As we'll see in Chapter 4, gravity pulls only in the downward vertical direction. As a result, it does not affect the football's horizontal motion. Likewise, wind blowing horizontally does not affect the vertical motion of the ball. As a result we can analyze the two-dimensional motion of the football as two separate, one-dimensional motions—vertical motion and horizontal motion. We can describe the football's velocity as a vector, so it makes sense to break that vector into horizontal and vertical components.

Figure 3-9a shows a pair of mutually perpendicular coordinate axes labeled x and y. These two axes define a plane, and any point in the plane can be identified by its x and y coordinates. For instance, a point located at $x = 3$ and $y = 5$ has coordinates (3, 5). The figure also shows a vector $\vec{A}$ whose direction lies in the x–y plane. If we place the tail of this vector at the **origin**—that is, at the point $x = 0$, $y = 0$—the

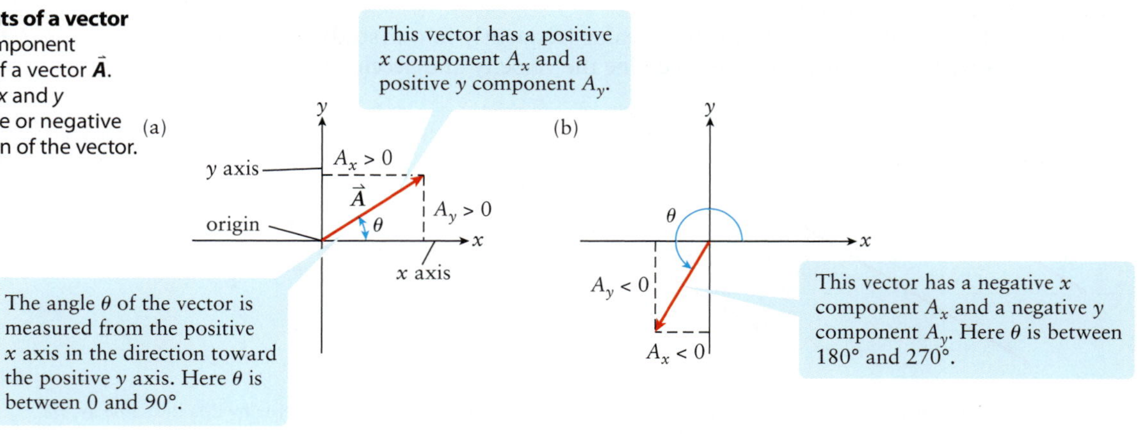

(a)
This vector has a positive x component A_x and a positive y component A_y.
y
y axis
$A_x > 0$
$\vec{A}$
$A_y > 0$
origin
θ
x
x axis
The angle θ of the vector is measured from the positive x axis in the direction toward the positive y axis. Here θ is between 0 and 90°.
(b)
y
θ
x
$A_y < 0$
$A_x < 0$
This vector has a negative x component A_x and a negative y component A_y. Here θ is between 180° and 270°.

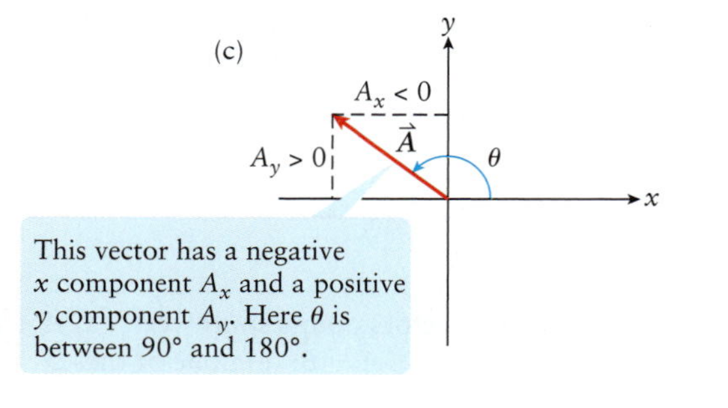

(c)
y
$A_x < 0$
$\vec{A}$
$A_y > 0$
θ
x
This vector has a negative x component A_x and a positive y component A_y. Here θ is between 90° and 180°.

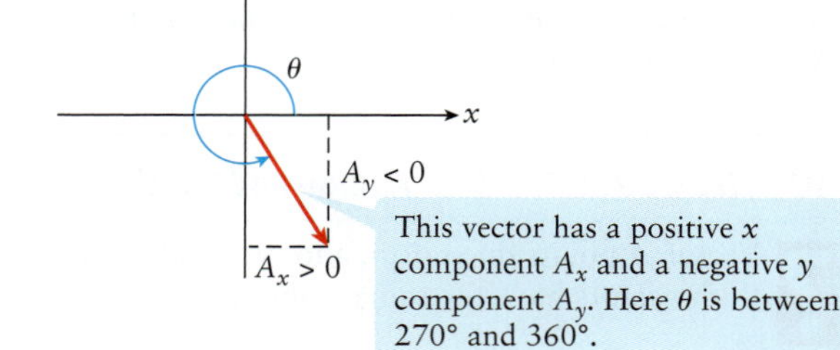

(d)
y
θ
x
$A_y < 0$
$A_x > 0$
This vector has a positive x component A_x and a negative y component A_y. Here θ is between 270° and 360°.

Figure 3-9 Components of a vector (a) Determining the x component A_x and y component A_y of a vector $\vec{A}$. (b), (c), (d) Whether the x and y components are positive or negative depends on the direction of the vector.

coordinates of the tip of the vector $\vec{A}$ are (A_x, A_y), as shown in Figure 3-9a. The quantities A_x (we say "A-sub-x") and Ay (we say "A-sub-y") are called the components of the vector $\vec{A}$: A_x is called the **x component**, and A_y is called the **y component.**

As Figure 3-9a shows, to find the x component A_x we first draw the vector $\vec{A}$ with its tail at the origin, then draw a line perpendicular to the x axis from the tip of $\vec{A}$ to the x axis. Where this line intersects the x axis tells you the value of A_x. In a similar way, to find the value of A_y, we draw a line perpendicular to the y axis from the tip of $\vec{A}$ to the y axis. Depending on the direction of the vector, the components can be both positive (Figure 3-9a), both negative (**Figure 3-9b**), or of different signs (**Figures 3-9c** and **3-9d**).

Once you have stated the x component and y component of a vector, you have defined the vector completely. So you can describe a vector such as $\vec{A}$ in terms of either (1) its magnitude A and the angle θ or (2) its components A_x and A_y. You can use whichever description is more convenient. **Figure 3-10** shows how to convert between these two descriptions. Note that the vector $\vec{A}$ of magnitude A, its x component A_x, and its y component A_y form three sides of a right triangle. From trigonometry the cosine of the angle θ equals the adjacent side A_x divided by the hypotenuse A, the sine of θ equals the opposite side A_y divided by the hypotenuse A, and the tangent of θ equals the opposite side A_y divided by the adjacent side A_x. Furthermore, the Pythagorean theorem tells us that $A^2 = A_x^2 + A_y^2$.

Rewriting these relationships, we get two equations that tell us how to find the components from the magnitude and direction, and two that tell us how to find the magnitude and direction from the components:

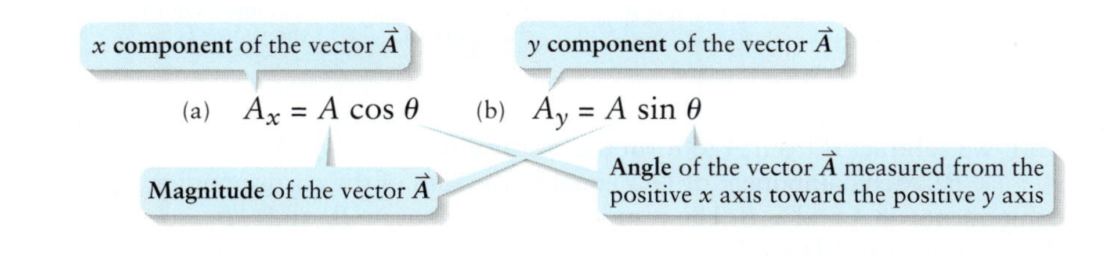

Finding vector components from vector magnitude and direction
(3-1)

Magnitude of the vector $\vec{A}$

Angle of the vector $\vec{A}$ measured from the positive x axis toward the positive y axis

(a) $A = \sqrt{A_x^{\,2} + A_y^{\,2}}$ (b) $\tan\theta = \dfrac{A_y}{A_x}$

x component of the vector $\vec{A}$

y component of the vector $\vec{A}$

Finding vector magnitude and direction from vector components
(3-2)

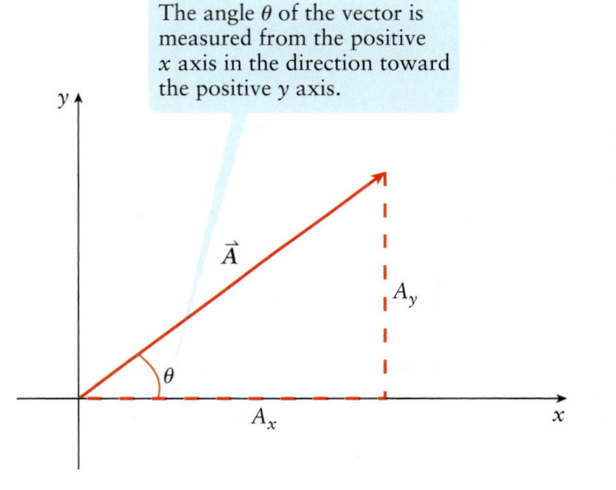

To find the components A_x and A_y from the magnitude A and angle θ:

$$A_x = A\cos\theta$$
$$A_y = A\sin\theta$$

To find the magnitude A and angle θ from the components A_x and A_y:

$$A = \sqrt{A_x^{\,2} + A_y^{\,2}}$$
$$\tan\theta = \frac{A_y}{A_x}$$

Figure 3-10 Two ways to describe a vector You can describe a vector $\vec{A}$ in terms of its magnitude A and direction θ or in terms of its components A_x and A_y.

WATCH OUT! Be careful with angles when doing calculations with vector components.

Many students get frustrated when working with Equations 3-1 and 3-2 because they didn't realize their calculator was set to "radian" mode for angles. If you're given an angle θ in degrees, or are trying to calculate an angle in degrees, always check that your calculator is in "degrees" mode. In addition, note that Equations 3-1 and 3-2 are valid *only* if the angle θ is measured from the positive x axis in the direction toward the positive y axis, as in Figures 3-9 and 3-10. We recommend that you *always* measure θ in that way. If you choose to measure the angle θ in some other way, you'll need to apply the trigonometric relations carefully to properly relate the magnitude, x component, and y component to your chosen angle.

Vector Arithmetic with Components

Describing vectors in terms of their components greatly simplifies the vector arithmetic that we described in Section 3-2. Here are the rules:

(1) **Figure 3-11** shows that if you add the vectors $\vec{A}$ and $\vec{B}$ to form the vector sum $\vec{C} = \vec{A} + \vec{B}$, each component of $\vec{C}$ is just the sum of the corresponding components of $\vec{A}$ and $\vec{B}$:

Rules for vector addition using components (3-3)

$$\text{(a)}\quad C_x = A_x + B_x$$

$$\text{(b)}\quad C_y = A_y + B_y$$

x component of the vector $\vec{C} = \vec{A} + \vec{B}$; x component of the vector $\vec{A}$; x component of the vector $\vec{B}$

y component of the vector $\vec{C} = \vec{A} + \vec{B}$; y component of the vector $\vec{A}$; y component of the vector $\vec{B}$

(2) If we subtract $\vec{B}$ from $\vec{A}$ to form the vector difference $\vec{D} = \vec{A} - \vec{B} = \vec{A} + (-\vec{B})$, each component of $\vec{D}$ is equal to the sum of the corresponding components of $\vec{A}$ and $-\vec{B}$. The components of $-\vec{B}$ are $-B_x$ and $-B_y$, so the components of $\vec{D}$ are $D_x = A_x + (-B_x)$ and $D_y = A_y + (-B_y)$, or

Rules for vector subtraction using components (3-4)

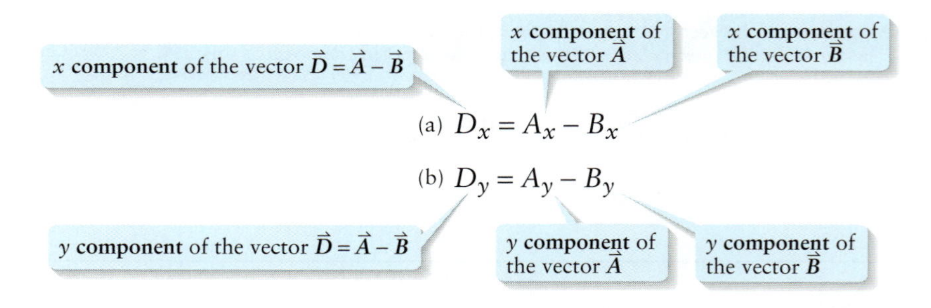

Figure 3-11 Vector addition using components The simplest way to add two vectors is in terms of their components.

The y component of the vector $\vec{C} = \vec{A} + \vec{B}$ equals the sum of the y components of $\vec{A}$ and $\vec{B}$: $C_y = A_y + B_y$

The x component of the vector $\vec{C} = \vec{A} + \vec{B}$ equals the sum of the x components of $\vec{A}$ and $\vec{B}$: $C_x = A_x + B_x$

y, x, B_y, C_y, A_y, $\vec{C}$, $\vec{B}$, $\vec{A}$, A_x, B_x, C_x

Each component of a vector sum $\vec{C} = \vec{A} + \vec{B}$ equals the sum of the corresponding components of the vectors $\vec{A}$ and $\vec{B}$.

That is, each component of $\vec{D} = \vec{A} - \vec{B}$ equals the difference between the corresponding components of $\vec{A}$ and $\vec{B}$.

(3) Finally, multiplying a vector $\vec{A}$ by a scalar c gives a new vector $\vec{E} = c\vec{A}$ whose components are just the components of $\vec{A}$ multiplied by c:

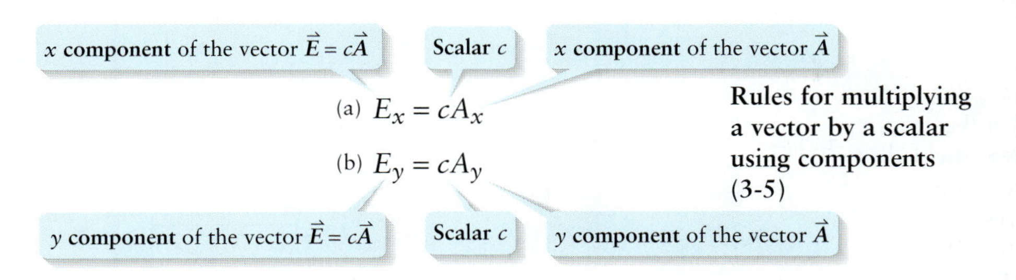

$$\text{(a)}\quad E_x = cA_x$$
$$\text{(b)}\quad E_y = cA_y$$

Rules for multiplying a vector by a scalar using components (3-5)

Note that if c is negative, the sign of each vector component is reversed—which means that the direction of the vector as a whole is reversed. That's just what we saw in Figure 3-8.

To use these rules you need to know the vectors $\vec{A}$ and $\vec{B}$ in terms of their components. If instead you're given $\vec{A}$ and $\vec{B}$ in terms of their magnitudes and directions, you can find their components using Equations 3-1. Once you've found the components of the desired result (such as a vector sum or a vector difference), you can determine its magnitude and direction using Equations 3-2.

GOT THE CONCEPT? 3-2 Vector Components

Which of the vectors in the accompanying figure has a positive x component and a negative y component? (a) $\vec{A}$; (b) $\vec{B}$; (c) $\vec{C}$; (d) $\vec{D}$; (e) none of these.

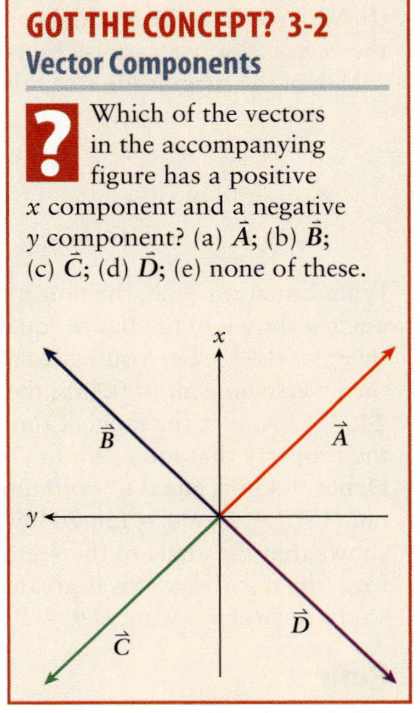

EXAMPLE 3-1 Different Descriptions of a Vector

(a) Find the x and y components of a velocity vector $\vec{v}$ that has magnitude 35.0 m/s and points in a direction 36.9° west of north. (b) Find the magnitude and direction of a displacement vector $\vec{r}$ that has x component −24.0 m and y component −11.0 m. Choose the positive x direction to point east and the positive y direction to point north.

Set Up

Our drawing shows the two vectors and the x–y axes. We'll use Equations 3-1 to find the x and y components of the vector $\vec{v}$ in part (a) and Equations 3-2 to find the magnitude and direction of the vector $\vec{r}$ in part (b).

Finding vector components from vector magnitude and direction:

$$A_x = A\cos\theta \qquad (3\text{-}1a)$$
$$A_y = A\sin\theta \qquad (3\text{-}1b)$$

Finding vector magnitude and direction from vector components:

$$A = \sqrt{A_x^2 + A_y^2} \qquad (3\text{-}2a)$$
$$\tan\theta = \frac{A_y}{A_x} \qquad (3\text{-}2b)$$

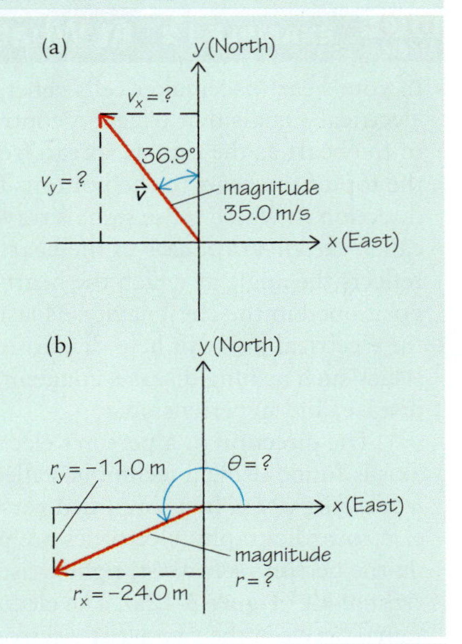

Solve

(a) The angle 36.9° is measured from the positive y axis. To use Equations 3-1, the angle of the vector $\vec{v}$ must be measured from the positive x axis. The figure shows that the angle we need is 90° + 36.9° = 126.9°. Use this angle in Equations 3-1 along with the magnitude v = 35.0 m/s of the vector $\vec{v}$.

$$v_x = (35.0\text{ m/s})\cos 126.9° = (35.0\text{ m/s})(-0.600) = -21.0\text{ m/s}$$

$$v_y = (35.0\text{ m/s})\sin 126.9° = (35.0\text{ m/s})(0.800) = +28.0\text{ m/s}$$

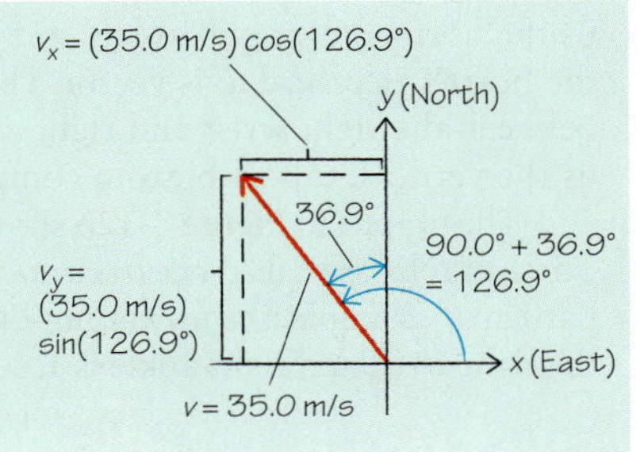

(b) We are given the x and y components of the vector $\vec{r}$, so we can use Equation 3-2a to calculate the magnitude r of this vector.

$$r = \sqrt{r_x^2 + r_y^2}$$
$$= \sqrt{(-24.0\text{ m})^2 + (-11.0\text{ m})^2}$$
$$\text{so } r = 26.4\text{ m}$$

From Equation 3-2b, the tangent of the angle θ shown in the figure equals r_y divided by r_x, or 0.458. Use your calculator to find the inverse tangent of 0.458; the result is 24.6°. However, the tangent function has the property that $\tan \phi = \tan(180° + \phi)$. Hence 0.458 is equal to *both* tan 24.6° and $\tan(180° + 24.6°) = \tan 204.6°$. The figure shows that the angle of the vector $\vec{r}$ measured from the positive x axis is greater than 180°, so the answer we want is $\theta = 204.6°$.

$$\tan\theta = \frac{r_y}{r_x} = \frac{-11.0\text{ m}}{-24.0\text{ m}} = 0.458$$
$$\theta = \tan^{-1} 0.458 = 24.6° \text{ or}$$
$$= 205$$

Reflect

The drawing of $\vec{v}$ in part (a) shows that this vector points in the negative x direction and the positive y direction. This agrees with our results, which show that v_x is negative and v_y is positive. The drawing of $\vec{r}$ in part (b) helped us decide which value of θ was the correct one. This shows why it's important to *always* draw a picture for any problem that involves vectors.

BIO-Medical EXAMPLE 3-2 At What Angle Is Your Heart?

In your heart pacemaker cells generate electrical signals that trigger a contraction of the heart as the signals spread from the top of the heart to the bottom. The direction in which these signals travel is called the *electrical axis* of the heart, which reflects the angle at which the heart is positioned in the chest cavity. Measuring the electrical axis can help diagnose health issues such as lung disease, congenital heart disease, and hypertension.

The direction of a person's electrical axis is found using a technique called *electrocardiography*. In a simplified version of electrocardiography, electrodes are placed on the person's left wrist, right wrist, and right ankle (Figure 3-12a). The electrical signal between the two wrist electrodes tells us the horizontal, right-to-left component of the heart's electrical axis vector. The signal between the right wrist and right ankle gives us the vertical, top-to-bottom component.

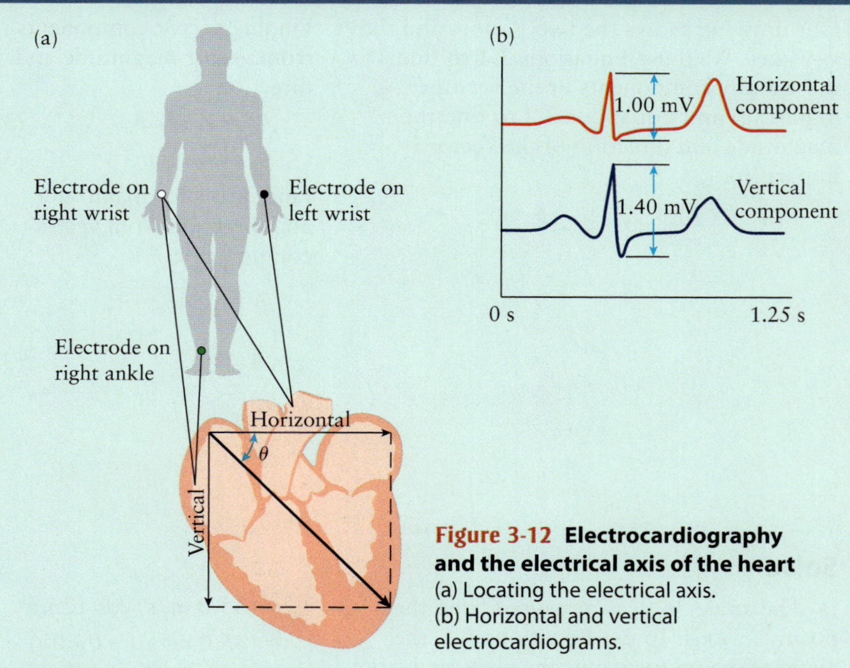

Figure 3-12 Electrocardiography and the electrical axis of the heart (a) Locating the electrical axis. (b) Horizontal and vertical electrocardiograms.

The graphs in Figure 3-12b show sample recordings of the electrical signal versus time for these two pairs of electrodes. These graphs are called *electrocardiograms*, abbreviated ECG or EKG (after their German name *Elektrokardiogramm*). In this particular example, the maximum ECG signal between the wrists is 1.00 millivolt (mV), and the maximum ECG signal between the right wrist and right ankle is 1.40 mV. At what angle θ does the electrical signal propagate across the heart?

Set Up

Our goal is to find the angle θ shown in the figure, which is the direction of the heart's electrical axis measured relative to the horizontal. We choose the x axis to be horizontal and choose the y axis to be vertically downward. (We can choose the axes to be whatever we want, provided they're mutually perpendicular.) The sketch shows the electrical axis as a green vector; it has an x component of 1.00 mV (shown in red) and a y component of 1.40 mV (shown in purple). As in Example 3-1, we'll use Equation 3-2b to find this angle.

Finding vector direction from vector components:

$$\tan\theta = \frac{A_y}{A_x} \qquad (3\text{-}2b)$$

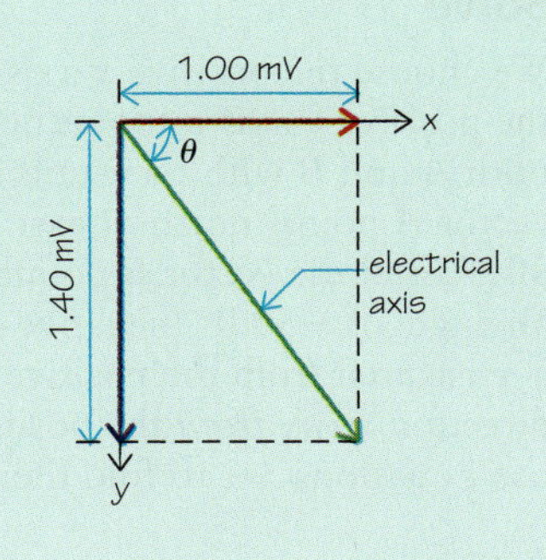

Solve

The angle θ is measured from the positive x axis toward the positive y axis, so we can safely use Equation 3-2b. Using your calculator you'll find that $\tan^{-1} 1.40 = 54.5°$. We saw in Example 3-1 that a second solution is 180° plus the result from your calculator, or 180° + 54.5°, or 234.5°. However, the figure above shows that θ is between 0 and 90°, so the correct answer is 54.5°.

$$\tan\theta = \frac{y\text{ component}}{x\text{ component}} = \frac{1.40\text{ mV}}{1.00\text{ mV}} = 1.40$$

$$\theta = \tan^{-1} 1.40 = 54.5° \text{ or } 234.5°$$

Correct answer: $\theta = 54.5°$

Reflect

If the horizontal and vertical signals were equally strong, the electrical axis would be tilted at an angle of 45° from the horizontal. In this example the vertical component is somewhat larger than the horizontal component, so the angle is a bit more than 45°. Normal clinical values are between −30° (that is, pointing 30° above the horizontal) and +90° (pointing straight down), so our result is reasonable.

EXAMPLE 3-3 Adding Vectors Using Components

You travel 62.0 km in a direction 30.0° north of east then turn to a direction 50.0° south of east and travel an additional 23.0 km (Figure 3-13). How far and in what direction are you from your starting point?

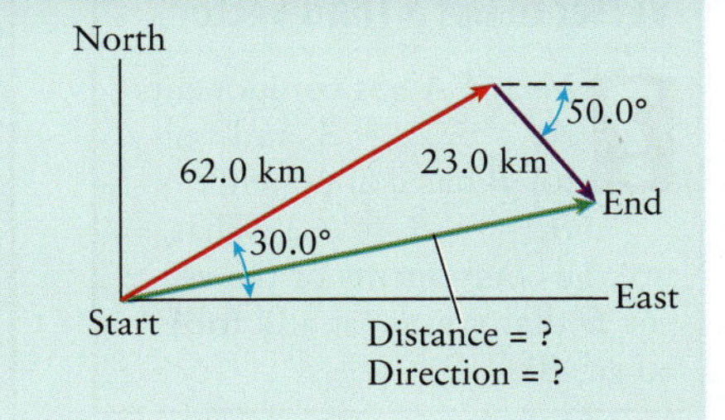

Figure 3-13 A problem in vector addition What is your net displacement?

Set Up

Each leg of the trip constitutes a displacement vector, which we show as $\vec{A}$ and $\vec{B}$ in the sketch. To find the magnitude and direction of the total displacement, which is just the vector sum $\vec{C} = \vec{A} + \vec{B}$, we will first find the x and y components of $\vec{A}$ and $\vec{B}$ using Equations 3-1. We'll then find the components of the vector sum $\vec{C}$ using Equations 3-3. Finally, we'll find the magnitude and direction of $\vec{C}$ using Equations 3-2.

$$A_x = A\cos\theta \qquad (3\text{-}1a)$$
$$A_y = A\sin\theta \qquad (3\text{-}1b)$$
$$C_x = A_x + B_x \qquad (3\text{-}3a)$$
$$C_y = A_y + B_y \qquad (3\text{-}3b)$$
$$A = \sqrt{A_x^2 + A_y^2} \qquad (3\text{-}2a)$$
$$\tan\theta = \frac{A_y}{A_x} \qquad (3\text{-}2b)$$

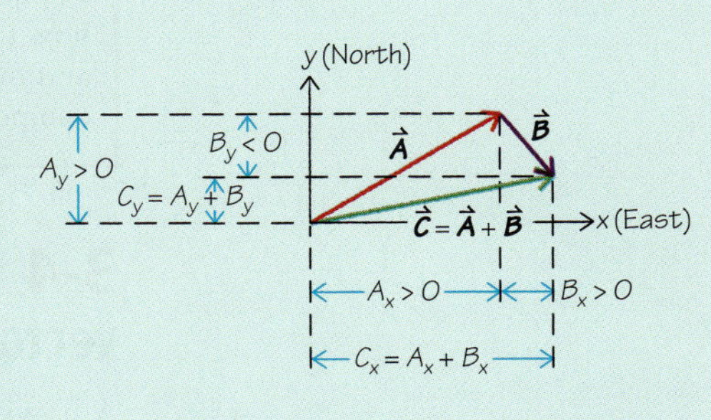

Solve

We choose the positive x axis to point east and the positive y axis to point north, and draw both $\vec{A}$ and $\vec{B}$ with their tails at the origin. The vector $\vec{A}$ points north of east at an angle of 30.0°, and the vector $\vec{B}$ points south of east at an angle of −50.0° (negative since this angle is measured from the positive x axis in the direction *away from* the positive y axis). We use Equations 3-1 to find their components.

$$A_x = (62.0\text{ km})\cos 30.0° = 53.7\text{ km}$$
$$A_y = (62.0\text{ km})\sin 30.0° = 31.0\text{ km}$$
$$B_x = (23.0\text{ km})\cos(-50.0°) = 14.8\text{ km}$$
$$B_y = (23.0\text{ km})\sin(-50.0°) = -17.6\text{ km}$$

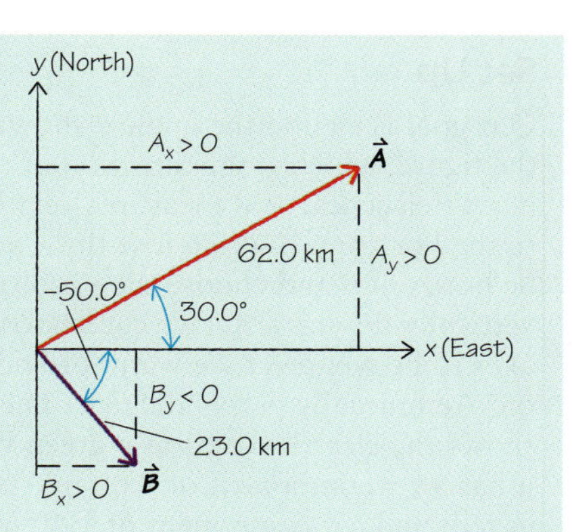

Given the components of $\vec{A}$ and $\vec{B}$, we can now find the components of $\vec{C} = \vec{A} + \vec{B}$ using Equations 3-3.

$$C_x = A_x + B_x = 53.7\text{ km} + 14.8\text{ km} = 68.5\text{ km}$$
$$C_y = A_y + B_y = 31.0\text{ km} + (-17.6\text{ km}) = 13.4\text{ km}$$

Now calculate the magnitude and direction of $\vec{C}$ using Equations 3-2. Both components of $\vec{C}$ are positive, so the vector points north of east, and the angle θ_C is between 0 and 90°. Hence the desired angle is 11.1°, not 180° + 11.1° = 191.1°.

$$C = \sqrt{C_x^2 + C_y^2} = \sqrt{(68.5\text{ km})^2 + (13.4\text{ km})^2}$$
$$C = 69.8\text{ km}$$
$$\tan\theta_C = \frac{C_y}{C_x} = \frac{13.4\text{ km}}{68.5\text{ km}} = 0.195$$
$$\theta_C = \tan^{-1} 0.195 = 11.1°$$

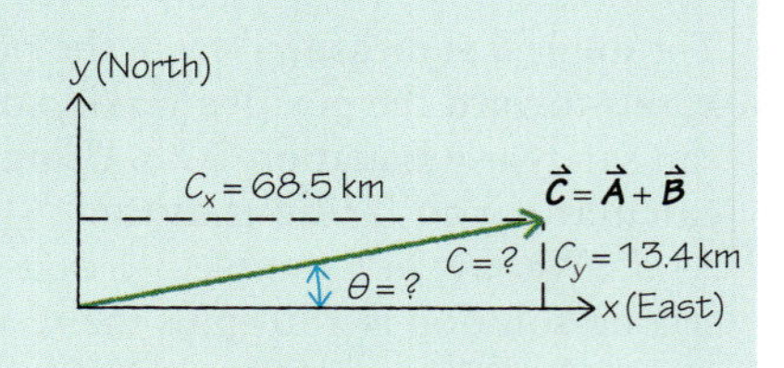

Reflect

The net displacement takes you 69.8 km from the starting point at an angle of 11.1° north of east. These numbers would be very difficult to get simply from the drawing of the two vectors shown above.

We now have all the mathematical tools that we need to study motion in two or three dimensions. In the next section we'll begin this study by seeing how to relate the position, velocity, and acceleration of an object using vectors.

GOT THE CONCEPT? 3-3
Adding One Vector to Another Vector to Get a Third Vector

? Vector $\vec{A}$ has components $A_x = 5$, $A_y = 3$, and vector $\vec{B}$ has components $B_x = 7$, $B_y = -2$. What are the components of the vector $\vec{D}$ that you must add to $\vec{A}$ to get $\vec{B}$?

TAKE-HOME MESSAGE FOR Section 3-3

✔ The easiest way to do arithmetic with vectors (adding, subtracting, and multiplying by a scalar) is to use components.

✔ Equations 3-1 let you find the components of a vector from its magnitude and direction, and Equations 3-2 let you find the magnitude and direction of a vector from its components.

✔ Equations 3-3, 3-4, and 3-5 show how to add vectors, subtract vectors, and multiply a vector by a scalar using components.

✔ For any problem involving vectors, it's essential to draw a sketch of the situation and to measure angles from the positive x direction toward the positive y direction.

✔ Be cautious using Equation 3-2b, $\tan\theta = A_y/A_x$. This equation actually gives two answers for the direction θ of a vector $\vec{A}$; your sketch will help you decide which answer is correct.

3-4 For motion in a plane, velocity and acceleration are vector quantities

A hawk glides over a meadow at a shallow angle then steepens its flight path to dive onto a mouse before sailing upward again (**Figure 3-14a**). How can an ornithologist describe the hawk's motion using the language of physics?

Just as for the straight-line motion that we discussed in Chapter 2, to have a complete description of the hawk's motion means that we must know its position at every instant of time. This knowledge requires that we have a reference point, or origin, from which to measure positions. Let's choose the origin to be at the point where the ornithologist stands and observes the motion. Then at each instant we can visualize a **position vector** $\vec{r}$ that extends from the origin to the hawk's position at that instant. **Figure 3-14b** shows one such position vector. If we know what $\vec{r}$ is at each instant of time t, we know the path or **trajectory** that the hawk follows through space as well as what time the hawk passes through each point along that trajectory. In Figure 3-14b we have drawn dots to show the hawk's position at 1-s intervals, so this figure also shows the hawk's motion diagram.

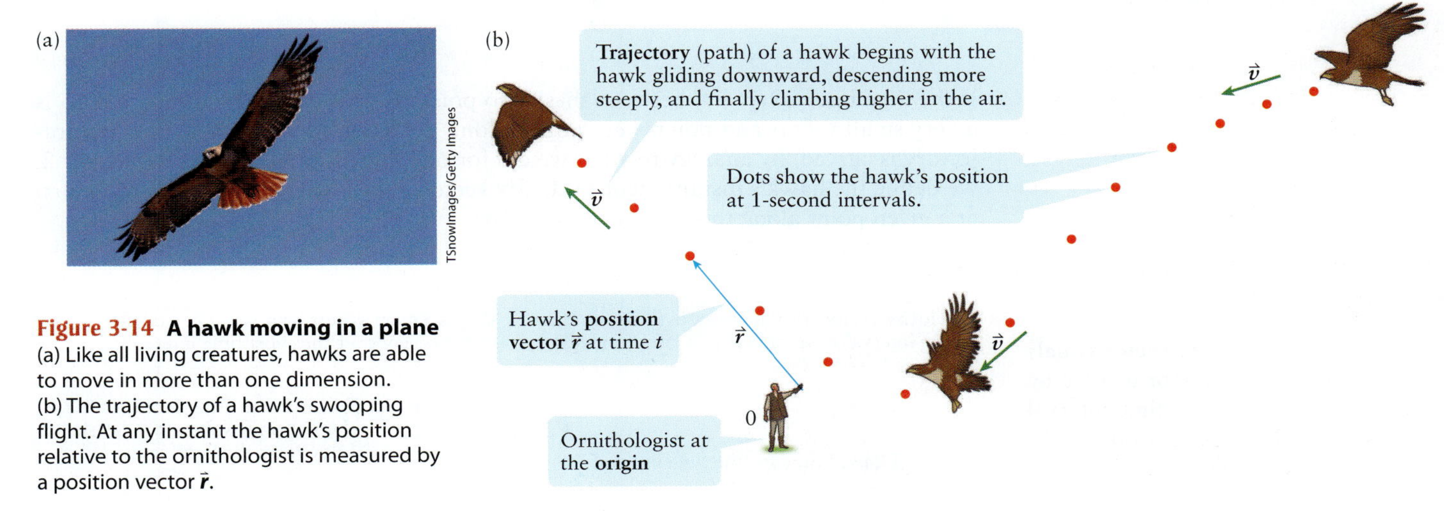
(a)
TSnowImages/Getty Images
(b)
Trajectory (path) of a hawk begins with the hawk gliding downward, descending more steeply, and finally climbing higher in the air.
Dots show the hawk's position at 1-second intervals.
$\vec{v}$
$\vec{v}$
$\vec{v}$
Hawk's position vector $\vec{r}$ at time t
$\vec{r}$
0
Ornithologist at the origin

Figure 3-14 A hawk moving in a plane (a) Like all living creatures, hawks are able to move in more than one dimension. (b) The trajectory of a hawk's swooping flight. At any instant the hawk's position relative to the ornithologist is measured by a position vector $\vec{r}$.

For simplicity let's assume that our hawk moves in only two of the three dimensions of space, so it is in **two-dimensional motion**. This means that the hawk may move up and down as well as forward and back, but doesn't turn left or right. One key reason for making this simplifying assumption is that two-dimensional motion is a *lot* easier to draw than three-dimensional motion! More importantly, many real-life motions are two-dimensional—for example, the flight of a thrown baseball and Earth's orbit around the Sun. In these situations we need only two coordinate axes, which we typically call x and y. In mathematics two such axes define a plane, so another name for two-dimensional motion is **motion in a plane**. If we use x and y to define the two perpendicular directions within the plane, we'll refer to this plane as the **x–y plane**.

The Velocity Vector

Just as for the one-dimensional motion that we studied in Chapter 2, understanding two- or three-dimensional motion means knowing how rapidly and in what direction the object moves along its trajectory. To see how to describe this type of trajectory, consider how the hawk shown in Figure 3-14b moves during a time interval between two instants of time that we call t_1 and t_2 (see **Figure 3-15**). The change of the hawk's position during that interval is its displacement vector $\Delta\vec{r}$ for that time interval. This vector points from the object's position at t_1 to its position at t_2 and is the difference between the object's position vector $\vec{r}_2$ at t_2 and its position vector $\vec{r}_1$ at t_1:

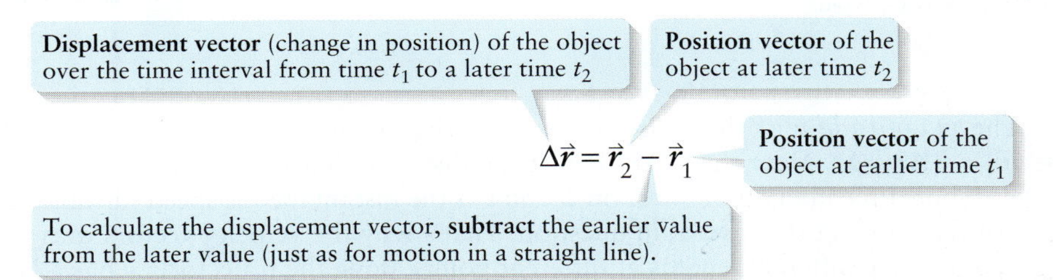
Displacement vector (change in position) of the object over the time interval from time t_1 to a later time t_2
Position vector of the object at later time t_2
$\Delta\vec{r} = \vec{r}_2 - \vec{r}_1$
Position vector of the object at earlier time t_1
To calculate the displacement vector, subtract the earlier value from the later value (just as for motion in a straight line).

Displacement vector equals the change in position vector over a time interval
(3-6)

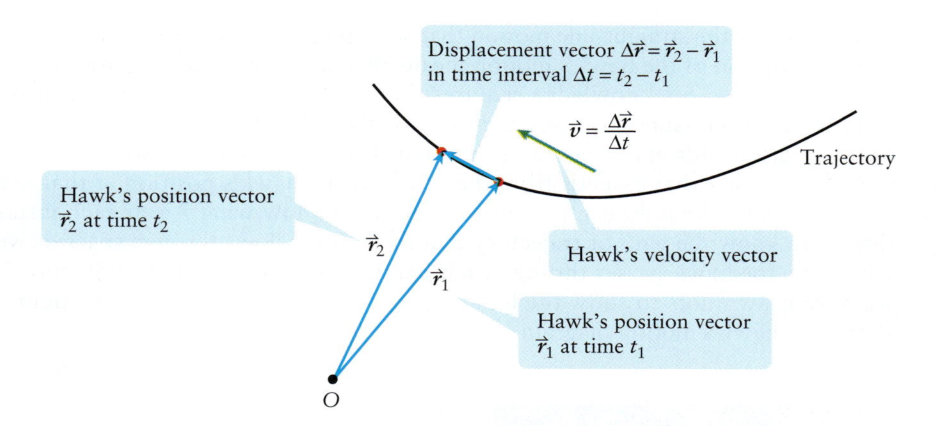

Figure 3-15 The displacement vector and velocity vector The displacement and velocity vectors both point along the trajectory.

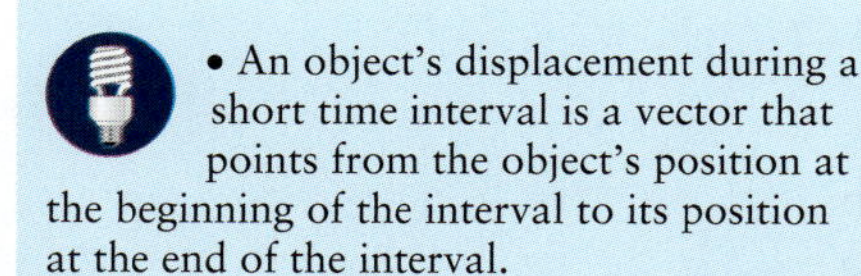

- An object's displacement during a short time interval is a vector that points from the object's position at the beginning of the interval to its position at the end of the interval.
- The object's velocity vector and displacement vector for that short time interval both point along the object's trajectory.

If the time interval $t_2 - t_1$ between these two points is very small, $\Delta\vec{r}$ in Figure 3-15 is a very small vector and points very nearly along the hawk's trajectory, even if the trajectory is curved. By analogy to what we did for one-dimensional motion in Chapter 2, we define the hawk's **instantaneous velocity vector** $\vec{v}$ (or just **velocity vector** for short) at a given point along the trajectory as follows:

Velocity vector equals displacement vector divided by time interval

$$\vec{v} = \frac{\Delta\vec{r}}{\Delta t} = \frac{\vec{r}_2 - \vec{r}_1}{t_2 - t_1} \quad (3\text{-}7)$$

Velocity vector for the object over a very short time interval from time t_1 to a later time t_2

Displacement vector (change in position) of the object over the short time interval

Elapsed time for the time interval

For both the displacement and the elapsed time, **subtract** the earlier value from the later value.

Equation 3-7 tells us to multiply the displacement vector $\Delta\vec{r}$ by $1/\Delta t$ (see Figure 3-8). Since $1/\Delta t$ is positive, the velocity vector $\vec{v}$ defined by Equation 3-7 points in the same direction as $\Delta\vec{r}$. So $\vec{v}$ at each point also points along the trajectory (**Figure 3-16**).

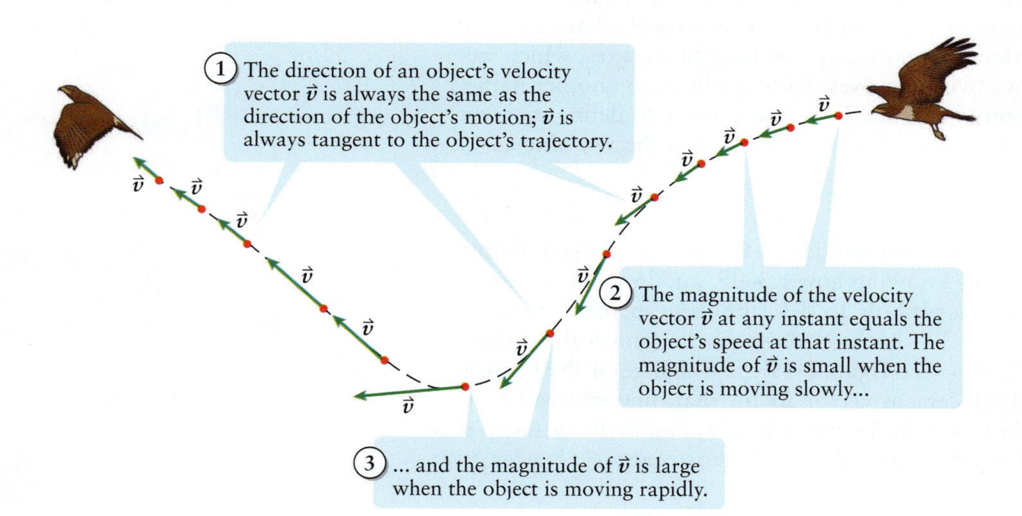

(1) The direction of an object's velocity vector $\vec{v}$ is always the same as the direction of the object's motion; $\vec{v}$ is always tangent to the object's trajectory.
(2) The magnitude of the velocity vector $\vec{v}$ at any instant equals the object's speed at that instant. The magnitude of $\vec{v}$ is small when the object is moving slowly...
(3) ... and the magnitude of $\vec{v}$ is large when the object is moving rapidly.
$\vec{v}$

Figure 3-16 Velocity vectors on a curved trajectory The varying velocity of the swooping hawk depicted in Figure 3-14.

The direction and magnitude at any instant of an object's velocity vector $\vec{v}$ tells you the direction of the object's motion and the object's speed at that instant.

The *magnitude* v of the velocity vector is the magnitude of $\Delta\vec{r}$ divided by Δt: $v = |\vec{v}| = |\Delta\vec{r}|/\Delta t$. Note that $|\Delta\vec{r}|$ is just the distance that the hawk travels during the very short time interval of duration Δt. So v is the instantaneous speed (distance per time) of the hawk at the instant in question. That's why the velocity vectors drawn in Figure 3-16 have different lengths at different points along the trajectory: $\vec{v}$ has a large magnitude where the hawk is moving rapidly and a small magnitude where the hawk is moving slowly (compare Figure 3-3). Like velocity and speed for motion in a straight line, the magnitude of the velocity vector $\vec{v}$ has units of meters per second (m/s).

Velocity Components

Like any other vector, we can describe an object's velocity vector $\vec{v}$ in terms of its components. The x component of $\vec{v}$, which we call v_x, equals the x component of displacement (the change in x coordinate during the time interval, $\Delta x = x_2 - x_1$) divided by the time interval $\Delta t = t_2 - t_1$; we define v_y, the y component of $\vec{v}$, in a similar way. Thus

$$\text{(a) } v_x = \frac{\Delta x}{\Delta t} = \frac{x_2 - x_1}{t_2 - t_1}$$

$$\text{(b) } v_y = \frac{\Delta y}{\Delta t} = \frac{y_2 - y_1}{t_2 - t_1}$$

x component of the **velocity vector** $\vec{v}$

y component of the **velocity vector** $\vec{v}$

x component of the **displacement** during a short time interval

y component of the **displacement** during a short time interval

Elapsed time for the time interval

Components of the velocity vector (3-8)

Equation 3-8a is the same expression that we wrote in Section 2-2 for the average velocity along the x axis in straight-line motion. When the time interval is short, this expression also gives the x component of the instantaneous velocity. Equation 3-8b is the corresponding equation for straight-line motion along the y axis. So our two-dimensional vector definition of velocity is a natural extension of our definition of straight-line motion.

The Acceleration Vector and Its Direction

You are driving a car along a country road. You speed up along a straight part of the road, then slow down when you see a curve up ahead, and finally go around the curve at a constant speed (**Figure 3-17**). In which of these three cases are you accelerating?

The answer is that you are accelerating in *all three* of these cases: speeding up, slowing down, and turning. Remember from Section 2-4 that an object undergoes an acceleration whenever its velocity changes. Because velocity is a vector, this means that an object accelerates if there is a change in *any* aspect of the velocity vector $\vec{v}$. When your car speeds up (**Figure 3-18a**) or slows down (**Figure 3-18b**) while moving in a straight line, the *magnitude* of $\vec{v}$ changes, and there is an acceleration. When your car is turning (**Figures 3-18c** and **3-18d**), the *direction* of $\vec{v}$ changes and so there is also an acceleration, even if the car's speed doesn't change.

Vladimir Ovchinnikov/Getty Images

Figure 3-17 An accelerating car Although a car's throttle control or gas pedal is commonly called the accelerator, a car really has *three* "accelerator" controls: the gas pedal, which makes the car speed up; the brake pedal, which makes the car slow down; and the steering wheel, which changes the car's direction of motion. All three of these change either the magnitude or the direction of the car's velocity vector $\vec{v}$.

The direction of a car's acceleration depends on whether it is speeding up, slowing down, turning, or a combination of these. To explore this, first recall from Section 2-4 our definition of the acceleration at a given time: the change in velocity during a very short time interval around that time divided by the duration of the time interval. For an object that moves in a plane, we have to consider the change in its velocity *vector* during a time interval. We again consider an infinitesimally short time interval from time t_1, when the object's velocity is $\vec{v}_1$, to a later time t_2, when the object has velocity $\vec{v}_2$. The object's **acceleration vector** $\vec{a}$ (short for **instantaneous acceleration vector**) at the time of this infinitesimal time interval is equal to the change in velocity divided by the duration of the interval:

$$\vec{a} = \frac{\Delta\vec{v}}{\Delta t} = \frac{\vec{v}_2 - \vec{v}_1}{t_2 - t_1}$$

Acceleration vector for the object over a very short time interval from time t_1 to a later time t_2

Change in velocity of the object over the short time interval

For both the velocity change and the elapsed time, **subtract** the earlier value from the later value.

Elapsed time for the time interval

Acceleration vector equals change in velocity vector divided by time interval (3-9)

(a) Speeding up in a straight line

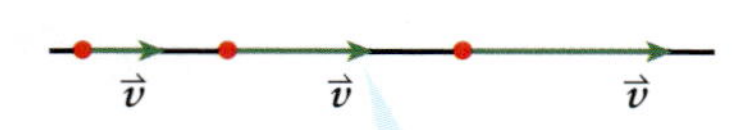

(b) Slowing down in a straight line

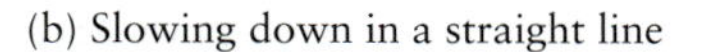

(c) Turning right at a steady speed

(d) Turning left at a steady speed

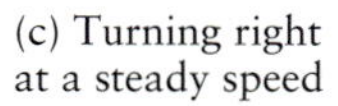
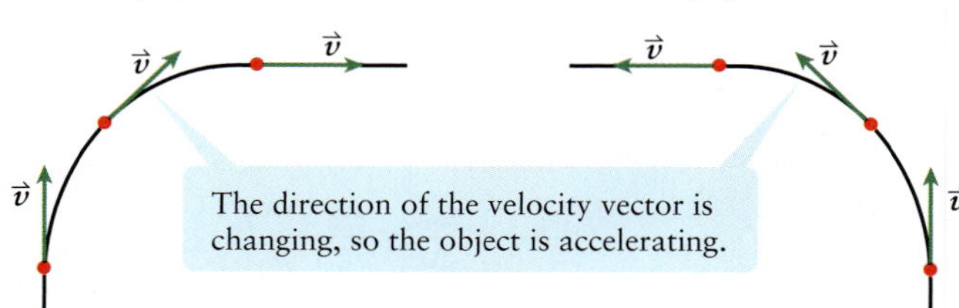

An object accelerates if it speeds up, slows down, or changes direction. That's because in each case the velocity vector changes in either its magnitude or its direction.

Figure 3-18 Four different examples of acceleration An object is accelerating if *any* aspect of its velocity vector $\vec{v}$ is changing.

Figure 3-19 Finding the direction of the acceleration vector An object's acceleration vector $\vec{a}$ for a given brief time interval points in the same direction as the difference $\Delta\vec{v} = \vec{v}_2 - \vec{v}_1$ between the object's velocities at the end and beginning of the interval. The four examples shown here correspond to the four situations in Figure 3-18.

The acceleration vector $\vec{a}$ points in the same direction as $\Delta\vec{v}$, and its magnitude equals the magnitude of $\Delta\vec{v}$ divided by Δt: $|\vec{a}| = |\Delta\vec{v}|/\Delta t$. Just like acceleration for motion in a straight line, the magnitude of the acceleration vector has units of meters per second squared (m/s^2).

While an object's velocity vector $\vec{v}$ always points along its trajectory, its acceleration vector may not. Figure 3-19 shows how to find the direction of the acceleration vector $\vec{a} = \Delta\vec{v}/\Delta t$ for each situation shown in Figure 3-18. In Figure 3-19a a car is

(a) Speeding up in a straight line.

1. Add $\vec{v}_2$ and $-\vec{v}_1$ to form $\Delta\vec{v} = \vec{v}_2 - \vec{v}_1$:

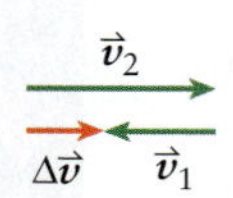

2. Average acceleration $\vec{a}_{\text{average}}$ is in the same direction as $\Delta\vec{v}$—in this case, in the direction of motion.

(b) Slowing down in a straight line.

1. Add $\vec{v}_2$ and $-\vec{v}_1$ to form $\Delta\vec{v} = \vec{v}_2 - \vec{v}_1$:

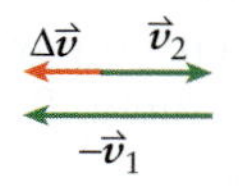

2. Average acceleration $\vec{a}_{\text{average}}$ is in the same direction as $\Delta\vec{v}$—in this case, opposite to the direction of motion.

(c) Turning right at a steady speed.

1. Add $\vec{v}_2$ and $-\vec{v}_1$ to form $\Delta\vec{v} = \vec{v}_2 - \vec{v}_1$:

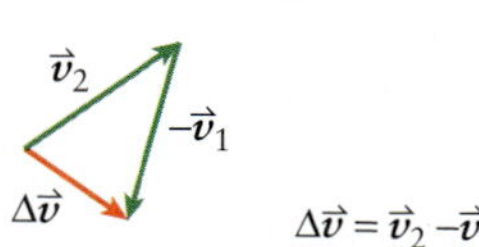

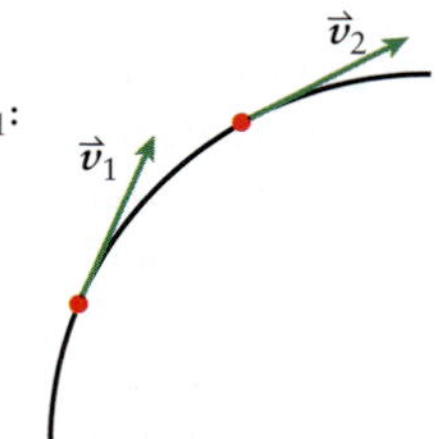

2. Average acceleration $\vec{a}_{\text{average}}$ is in the same direction as $\Delta\vec{v}$—in this case, toward the inside of the turn.

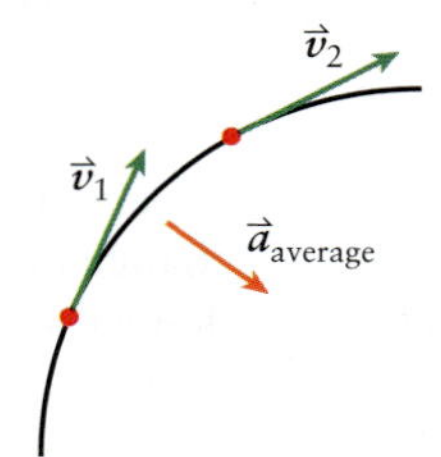

(d) Turning left at a steady speed.

1. Add $\vec{v}_2$ and $-\vec{v}_1$ to form $\Delta\vec{v} = \vec{v}_2 - \vec{v}_1$:

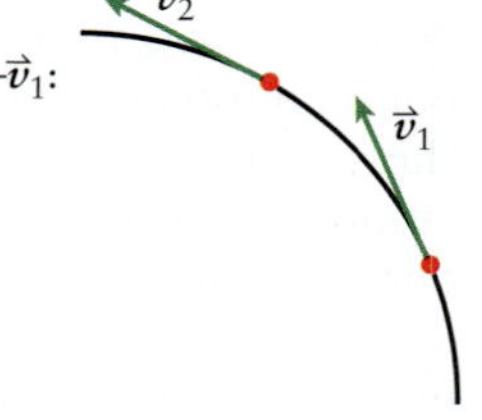

$\vec{v}_2$ $-\vec{v}_1$ $\Delta\vec{v}$

2. Average acceleration $\vec{a}_{\text{average}}$ is in the same direction as $\Delta\vec{v}$—in this case, toward the inside of the turn.

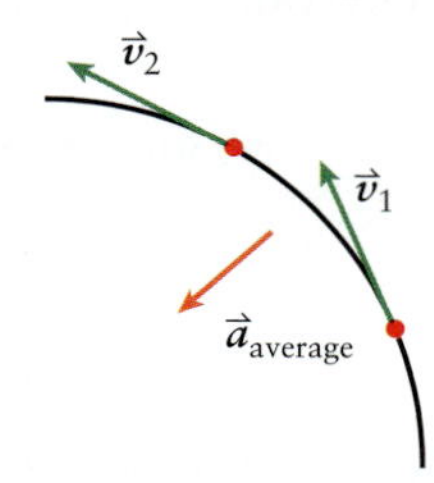

speeding up as it moves in a straight line. The velocity change $\Delta\vec{v}$ points in the same direction as the car's motion, and so its acceleration vector $\vec{a}$ is likewise in the direction of motion (the car "accelerates forward"). If instead the car slows down as it moves in a straight line, as shown in **Figure 3-19b**, $\Delta\vec{v}$ and the acceleration both point opposite to the car's motion (it "accelerates backward"). In **Figures 3-19c** and **3-19d** the car is going around a curve at a constant speed. In this case the acceleration vector $\vec{a}$ is *perpendicular* to the car's trajectory and points toward the *inside* of the curve. If the car is turning to the right, as in Figure 3-19c, the acceleration $\vec{a}$ is to the right; if it is turning to the left, as in Figure 3-19d, the acceleration $\vec{a}$ is to the left. We can say that a car going around a curve "accelerates sideways."

To convince yourself that Figures 3-19c and 3-19d correctly show the direction of acceleration while turning, note that *when you are riding in a vehicle that accelerates, you feel a push in the direction opposite the vehicle's acceleration.* (We'll see why this is so in Chapter 4.) If you're riding in a car that speeds up, you feel pushed *backward* into your seat. This push is opposite to the car's *forward* acceleration vector $\vec{a}$ (see Figure 3-19a). If the car brakes suddenly, you feel thrown *forward* in a direction opposite to the *backward* acceleration vector of the car (see Figure 3-19b). And if the car makes a turn at constant speed, you feel pushed to the *outside* of the turn. The acceleration vector $\vec{a}$ is opposite to this direction and hence must point to the *inside* of the turn, as Figures 3-19c and 3-19d show.

Figure 3-20 shows another useful way to think about the acceleration vector $\vec{a}$. At a given point along an object's trajectory, $\vec{a}$ has two components: one that is perpendicular to the trajectory and one that points along the trajectory in the direction that the object is moving. The acceleration component perpendicular to the trajectory tells us about the change in the object's *direction*, and the component along the trajectory tells us about the change in the object's *speed*.

1. Find the direction of $\vec{a}$ for (a) the interval from point 1 to point 2 and for (b) the interval from point 2 to point 3.

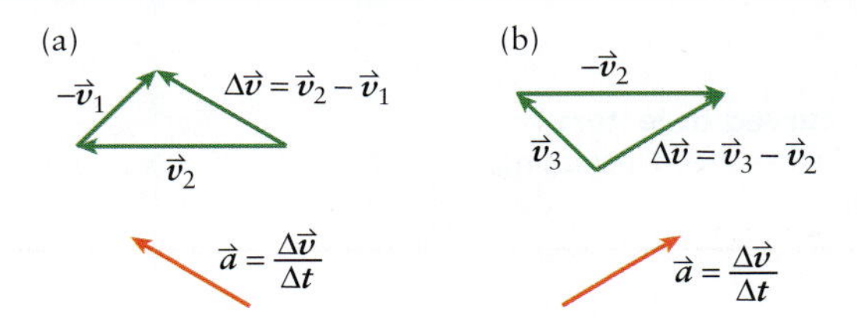

A hawk speeds up as it moves from point 1 to point 2 and slows down as it moves from point 2 to point 3.

For each interval we can break the acceleration $\vec{a}$ into a component along the trajectory and a component perpendicular to the trajectory.

2. For both intervals there is a component of acceleration that points perpendicular to the trajectory (toward the inside of the turn). This describes changes in the hawk's direction of motion.

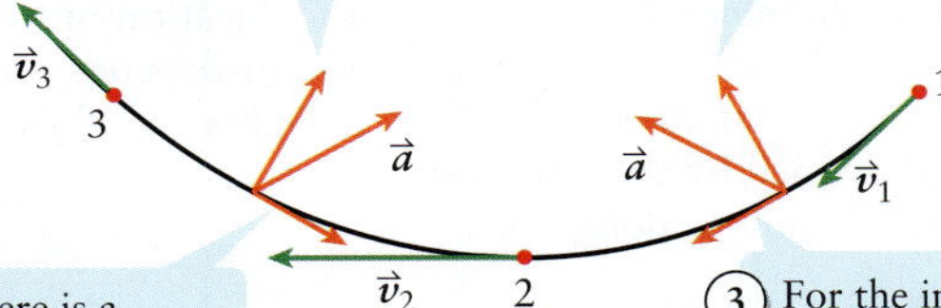

4. For the interval from point 2 to point 3, there is a component of acceleration opposite to the direction of the hawk's motion. This describes the decrease in the hawk's speed.

3. For the interval from point 1 to point 2, there is a component of acceleration in the same direction as the hawk's motion. This describes the increase in the hawk's speed.

If an object's speed is increasing, $\vec{a}$ has a component in the direction of motion.
If an object's speed is decreasing, $\vec{a}$ has a component opposite to the direction of motion.
If an object's direction of motion is changing (the trajectory is curved), $\vec{a}$ has a component perpendicular to its trajectory that points *toward* the inside of the turn.

Figure 3-20 Finding the acceleration of a swooping hawk How to find the acceleration vector of an object from its velocity vectors.

It's also often useful to express the acceleration vector $\vec{a}$ of an object in terms of its x and y components. Use the symbols v_{1x} and v_{1y} for the velocity components of the object at time t_1 and the symbols v_{2x} and v_{2y} for the velocity components at time t_2. Then a_x, the x component of $\vec{a}$, equals the x component of velocity change ($\Delta v_x = v_{x2} - v_{x1}$) divided by the time interval $\Delta t = t_2 - t_1$, and the y component of $\vec{a}$ equals the y component of velocity change ($\Delta v_y = v_{y2} - v_{y1}$) divided by Δt:

Components of the acceleration vector (3-10)

$$\text{(a)}\ a_x = \frac{\Delta v_x}{\Delta t} = \frac{v_{x2} - v_{x1}}{t_2 - t_1}$$

$$\text{(b)}\ a_y = \frac{\Delta v_y}{\Delta t} = \frac{v_{y2} - v_{y1}}{t_2 - t_1}$$

x component of the **acceleration vector $\vec{a}$**

y component of the **acceleration vector $\vec{a}$**

Change in the x component of the **velocity** during a short time interval

Change in the y component of the **velocity** during a short time interval

Elapsed time for the time interval

In the next section we'll use the component description of velocity and acceleration to help us analyze the motion of a *projectile*—an object moving through the air that is acted on by gravity alone.

GOT THE CONCEPT? 3-4 Acceleration and Changing Speed

? Figure 3-21 shows the acceleration vector $\vec{a}$ at several points along an object's curved trajectory. At each of points 1, 2, 3, and 4, is the object (a) speeding up, (b) slowing down, or (c) maintaining a constant speed?

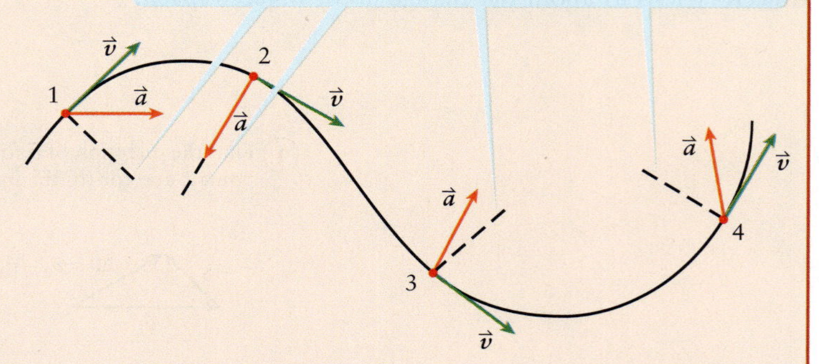

At each point, the dashed line is perpendicular to the trajectory.
1
2
3
4

Figure 3-21 Acceleration on a curved trajectory At each of the points shown, is the object's speed increasing, decreasing, or staying the same?

TAKE-HOME MESSAGE FOR Section 3-4

✔ If an object moves during a time interval, its displacement $\Delta\vec{r}$ is a vector that points from its position at the beginning of the interval to its position at the end of the interval.

✔ The velocity $\vec{v}$ of an object is a vector. It equals the displacement $\Delta\vec{r}$ for a very short interval divided by the duration of the interval. The velocity vector always points along the object's trajectory.

✔ The acceleration $\vec{a}$ of an object is also a vector. It equals the change in velocity $\Delta\vec{v}$ for a very short interval divided by the duration of the interval. The direction of the acceleration vector depends on how the object's speed is changing and on whether the object is changing direction.

3-5 A projectile moves in a plane and has a constant acceleration

We now have all the tools we need to analyze the kind of motion shown in the photograph that opens this chapter, in which clumps of molten lava are ejected from a volcano. The curved paths followed by the lava are the same as those followed by streams of water from a fountain and by a leaping ballet dancer (**Figure 3-22**). The direction of motion of an object that follows such a curved path is constantly changing. Hence the object is *accelerating* at all points during the motion. What is the magnitude and direction of this acceleration?

If the object is moving through the air at a relatively slow speed so that air resistance can be neglected, the object's acceleration is due to gravity alone. Then the object is in *free fall* and has a constant downward acceleration, just as we discussed in Section 2-6. In that section we looked only at motion that was straight up-and-down; now we want to consider **projectile motion**, which is free fall with both vertical motion *and* horizontal motion. (We often refer to an object undergoing this kind of motion as a **projectile**.) Just as for vertical free fall, the downward acceleration of a projectile has a magnitude g, called the *acceleration due to gravity*. We'll use the value $g = 9.80\ \text{m/s}^2$.

Projectile motion is always *two*-dimensional, so the motion lies in a plane. To see why, think about tossing a basketball upward and toward the east (**Figure 3-23**). When the ball leaves your hand, its velocity vector has a horizontal component toward the east, a vertical component that points upward, and zero component along the north–south direction. Because the acceleration vector points straight downward, there is *no* north–south component of acceleration. Therefore, the north–south velocity doesn't change—this component of velocity starts at zero and remains zero. So the basketball always remains in a vertical plane that's oriented east–west, as Figure 3-23 shows.

(a)

Jim Jurica/Getty Images

(b)

NVC ARTS, a Warner Music Group company

Figure 3-22 Two examples of projectile motion (a) Water in a fountain. (b) A leaping ballet dancer.

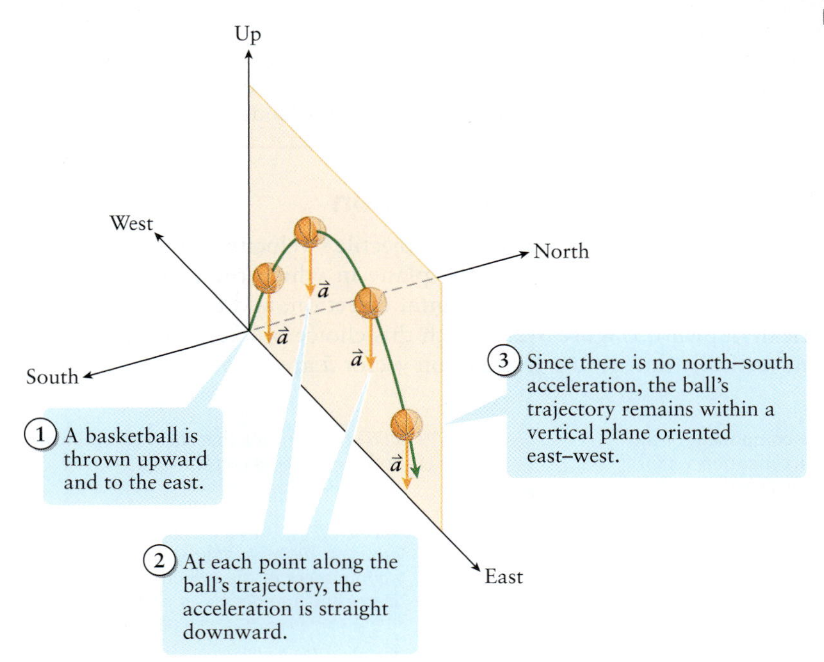
Up
West
North
South
East
$\vec{a}$
1 A basketball is thrown upward and to the east.
2 At each point along the ball's trajectory, the acceleration is straight downward.
3 Since there is no north–south acceleration, the ball's trajectory remains within a vertical plane oriented east–west.

Figure 3-23 Why projectile motion is two-dimensional motion The trajectory of a projectile is in a vertical plane because the acceleration due to gravity points downward.

Figure 3-24 shows the acceleration vector $\vec{a}$ at several points along a projectile's trajectory as well as the components of $\vec{a}$ along and perpendicular to the trajectory. There is a perpendicular component of $\vec{a}$ at *all* points, so the projectile's direction of motion is changing at all points, and the trajectory is curved all along its length. When the projectile is ascending there is also a component of acceleration opposite to its motion, so the projectile is slowing down. When the projectile is descending there is an acceleration component in the direction of motion, which tells us that the projectile is speeding up. The only point at which the speed is instantaneously *not* changing is at the highest point of the trajectory, where the acceleration vector is entirely perpendicular to the trajectory. At this point the projectile has its *minimum* speed.

Figure 3-24 Projectile motion Acceleration components along and perpendicular to the trajectory.

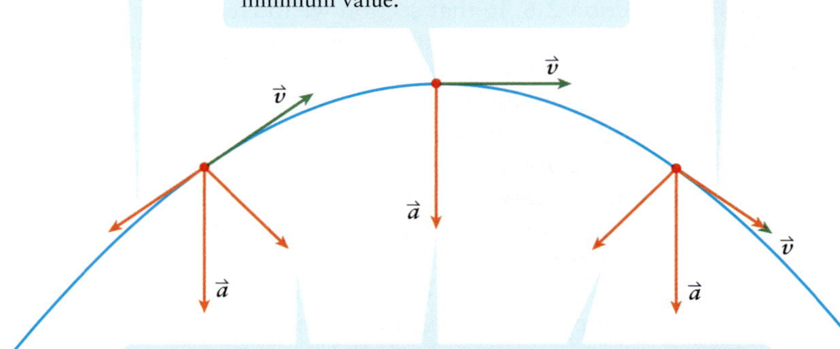
There is a component of acceleration opposite to the direction of motion: The projectile is slowing down.
There is a component of acceleration in the direction of motion: The projectile is speeding up.
There is no component of acceleration along the direction of motion: The speed of the projectile has its minimum value.
$\vec{v}$
$\vec{v}$
$\vec{v}$
$\vec{a}$
$\vec{a}$
$\vec{a}$
At all points there is a component of acceleration perpendicular to the trajectory: The direction of motion is changing.

• The acceleration $\vec{a}$ of a projectile is constant.
• Because the projectile's path is curved, $\vec{a}$ has varying components along and perpendicular to the direction of motion.

WATCH OUT! A projectile is a special case.

Not every object in flight can be regarded as a projectile. As an example, a gliding bird is definitely *not* a projectile: The air exerts lift and drag forces on the bird that can't be ignored because these forces are comparable to the force of gravity. A batted baseball also feels a substantial drag force exerted by the air because of its high speed (much greater than the speed of the basketball in Figure 3-23), so this object isn't really a projectile either. Nonetheless, studying projectile motion is an important first step toward understanding these more complex kinds of motion.

The Equations of Projectile Motion

Let's be more quantitative about how a projectile's velocity changes during its flight. We use the x and y axes to define the plane in which the projectile moves, with the positive x axis pointing in a horizontal direction, and the positive y axis pointing vertically upward (**Figure 3-25**). With this choice of axes, the components of the constant, vertically downward acceleration vector $\vec{a}$ are

Components of the acceleration vector in projectile motion (3-11)

$$\text{(a)}\quad a_x = 0$$

$$\text{(b)}\quad a_y = -g$$

x **component** of the **acceleration** vector $\vec{a}$ for **projectile motion**

Zero value means that the x component of the velocity is constant.

y **component** of the **acceleration** vector $\vec{a}$ for **projectile motion**

Constant negative value means that the y component of the velocity is always decreasing and decreases at a steady rate.

WATCH OUT! Projectile motion means constant vertical acceleration.

Just as for vertical free fall, it's important to remember that the vertical, y, component of acceleration is the same ($a_y = -g$) at all times during projectile motion. This is true whether the projectile is moving upward, at the high point of its motion, or moving downward.

Because both the x and y components of acceleration are constant, we can use our results from Chapter 2 to easily write the equations that show how the x and y components of the projectile's velocity $\vec{v}$ vary with time. We'll let $t = 0$ represent the instant when the projectile begins its flight. At this moment the projectile has x component of velocity v_{0x} and y component of velocity v_{0y}. (The "0" in the subscripts reminds us that these are the components at $t = 0$.) Equation 2-5 from Section 2-4 tells us that if the

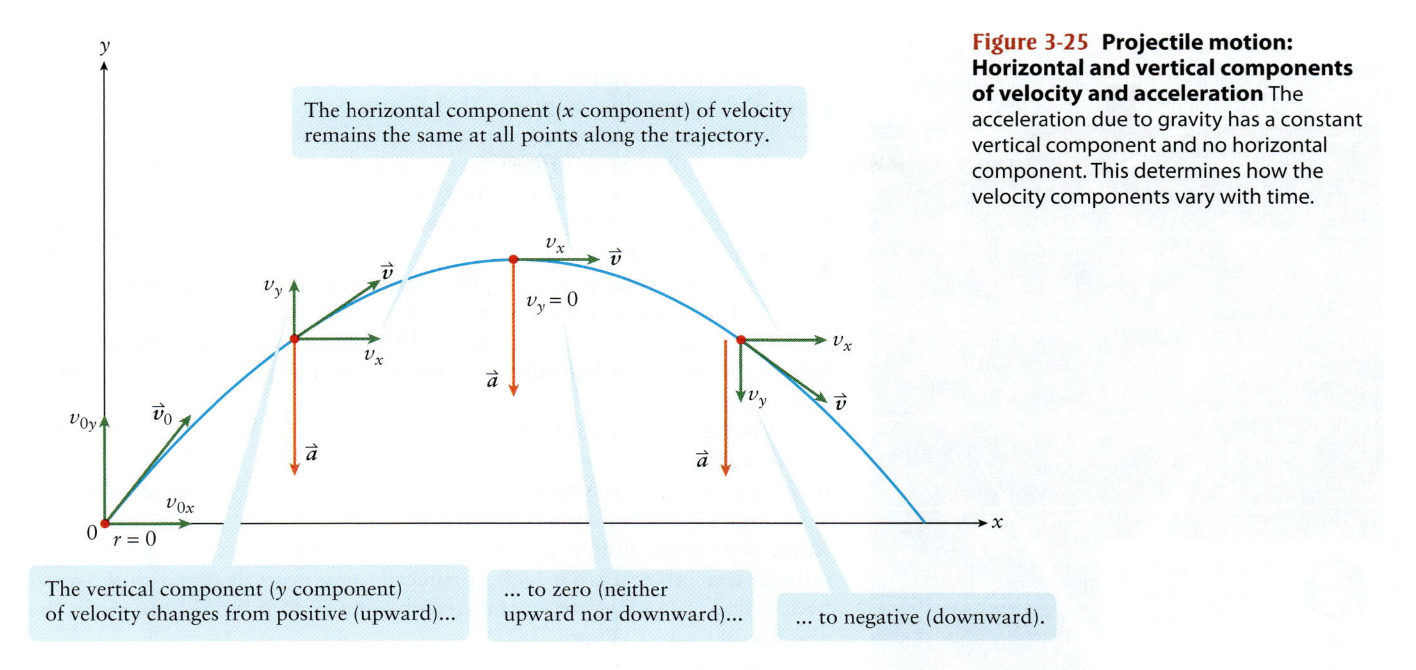

Figure 3-25 Projectile motion: Horizontal and vertical components of velocity and acceleration The acceleration due to gravity has a constant vertical component and no horizontal component. This determines how the velocity components vary with time.

x component of acceleration is constant, the x component of velocity v_x as a function of time is $v_x = v_{0x} + a_x t$. The same equation rewritten for the y direction is $v_y = v_{0y} + a_y t$. If we substitute the values $a_x = 0$ and $a_y = -g$ from Equations 3-11, we get

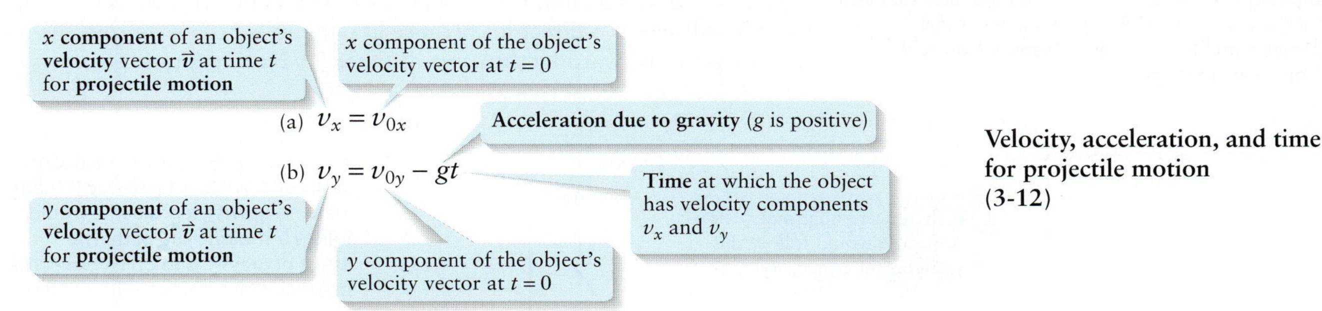

Velocity, acceleration, and time for projectile motion (3-12)

Equation 3-12a tells us that *in projectile motion the horizontal component of velocity doesn't change*. The vertical component of velocity *does* change with time, however; as Equation 3-12b shows, the y component of velocity decreases at a steady rate. If v_{0y} is positive as shown in Figure 3-25, v_y decreases to zero (at the high point of the trajectory) and then continues to decrease or becomes increasingly negative (that is, the projectile moves downward at an ever-faster rate).

We can use another equation from Chapter 2 to write the equations for the projectile's x and y coordinates at any time t. We let x_0 and y_0 be the projectile's coordinates at time $t = 0$. Because the x component of acceleration a_x is constant, we know that $x = x_0 + v_{0x}t + (1/2)a_x t^2$ from Equation 2-9 in Section 2-4. The same equation rewritten for the y direction is $y = y_0 + v_{0y}t + (1/2)a_y t^2$. For projectile motion $a_x = 0$ and $a_y = -g$, so these equations become

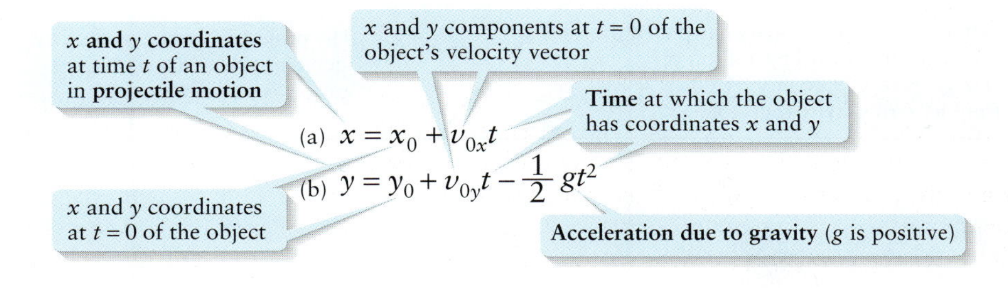

Position, acceleration, and time for projectile motion (3-13)

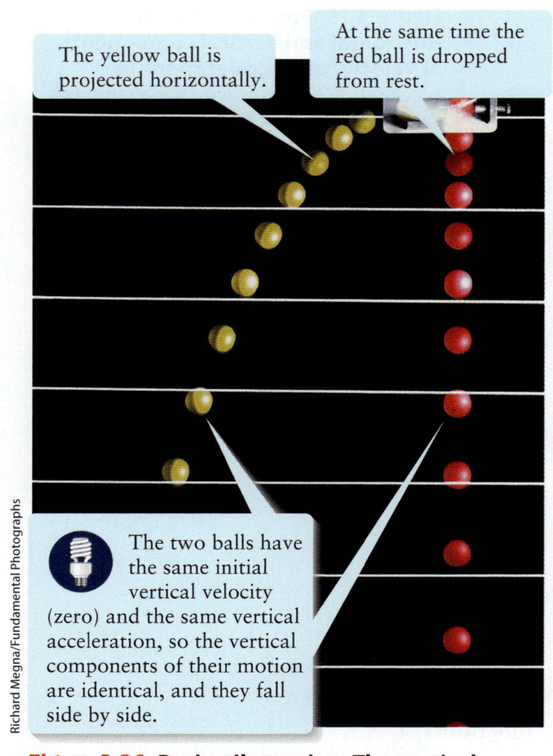

The yellow ball is projected horizontally.
At the same time the red ball is dropped from rest.
The two balls have the same initial vertical velocity (zero) and the same vertical acceleration, so the vertical components of their motion are identical, and they fall side by side.
Richard Megna/Fundamental Photographs

Figure 3-26 Projectile motion: The vertical and horizontal components of motion are independent Although these two balls have different initial horizontal velocities, their initial vertical velocities—and hence their subsequent vertical motions—are identical.

The curved path that the projectile follows in accordance with Equations 3-13 is called a **parabola**. You can see this shape in the photograph that opens this chapter and in Figure 3-22.

You probably recognize Equations 3-12b and 3-13b from our discussion of free fall in Section 2-6; they're exactly the same as Equations 2-13 and 2-14, respectively. So *we can think of projectile motion as a combination of horizontal motion with constant velocity and vertical free fall.* What's more, *the horizontal and vertical motions of a projectile are independent.* To see this, notice that the expressions for v_x and x in Equations 3-12 and 3-13 involve only x quantities, while the expressions for y and v_y involve only y quantities. The multiflash photograph in **Figure 3-26** helps to show the independence of the horizontal and vertical motions of a projectile.

Another way to think about projectile motion is to first imagine what would happen if we could turn gravity off so that $g = 0$. Then a thrown projectile would have zero acceleration, its velocity would remain constant, and its trajectory would be a straight line. However, a projectile's trajectory curves downward because gravity *does* exist. Equation 3-13b tells us that after a time t, the projectile has been dragged downward by a distance $(1/2)gt^2$ from the straight-line path it would follow in the absence of gravity (**Figure 3-27**).

The demonstration depicted in **Figure 3-28** illustrates this way of thinking about projectile motion. A projectile is aimed directly at a hanging target. Just as the projectile starts its flight toward the target, the target is released and begins to fall. If gravity did not exist, the target wouldn't fall, the projectile would follow a straight-line path, and the projectile would score a direct hit on the target. Because gravity does exist, however, the target does fall, and the projectile follows a curved path. But the projectile still scores a direct hit because gravity pulls *both* the projectile and the target downward by the *same* distance $(1/2)gt^2$ in a time t.

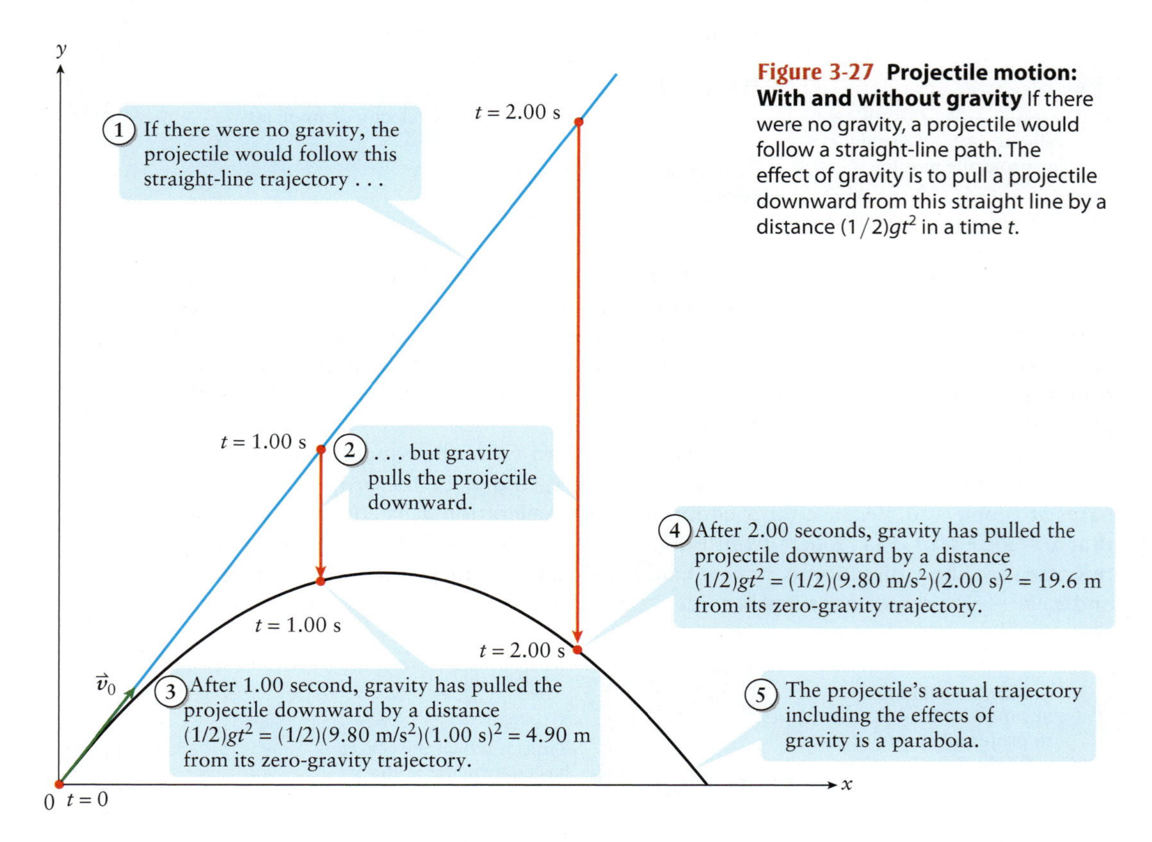

Figure 3-27 Projectile motion: With and without gravity If there were no gravity, a projectile would follow a straight-line path. The effect of gravity is to pull a projectile downward from this straight line by a distance $(1/2)gt^2$ in a time t.

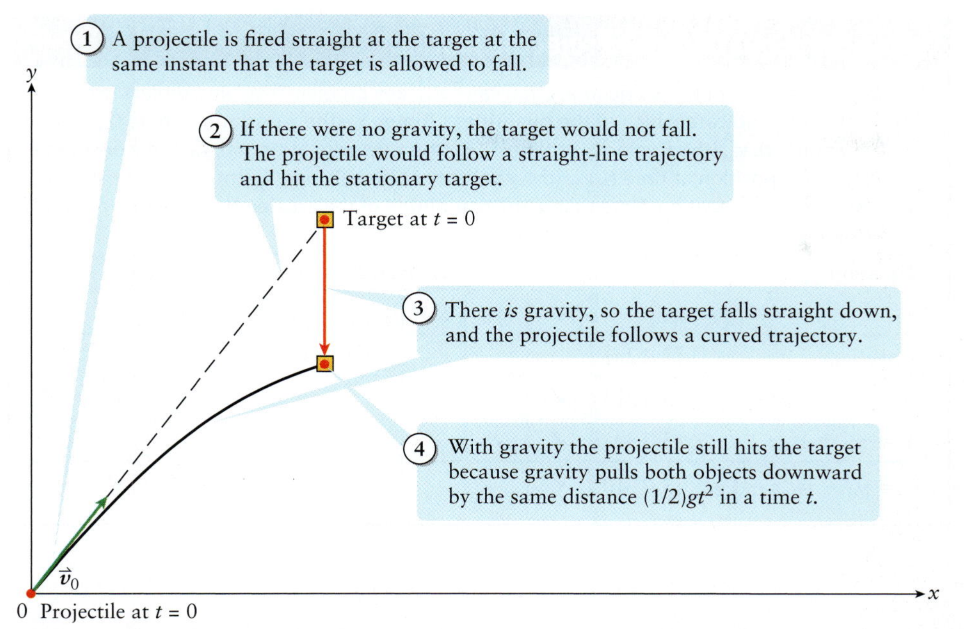

1 A projectile is fired straight at the target at the same instant that the target is allowed to fall.
2 If there were no gravity, the target would not fall. The projectile would follow a straight-line trajectory and hit the stationary target.
Target at t = 0
3 There *is* gravity, so the target falls straight down, and the projectile follows a curved trajectory.
4 With gravity the projectile still hits the target because gravity pulls both objects downward by the same distance $(1/2)gt^2$ in a time t.
y
x
$\vec{v}_0$
0 Projectile at t = 0

Figure 3-28 Hitting a dropped target This experiment demonstrates that all objects in free fall have the same downward acceleration.

Thus we have two alternative ways of thinking about projectile motion: (1) a combination of horizontal, constant-velocity motion and vertical free-fall or (2) straight-line motion in the direction of the initial velocity, but pulled downward by a distance $(1/2)gt^2$ in time t. As we'll see in the next section, both of these interpretations can be useful when solving problems with projectile motion.

GOT THE CONCEPT? 3-5
The Changing Velocity of a Projectile

? An object launched at some initial speed and angle follows a parabolic arc. At what point during the trajectory is the magnitude of its velocity the smallest? At what point, if any, will the velocity of the object be zero?

TAKE-HOME MESSAGE FOR Section 3-5

✔ A projectile is an object that is launched and then falls freely (that is, moves without air resistance).

✔ Only gravity acts on a projectile, so the acceleration is downward and has the same magnitude throughout the motion.

✔ Projectile motion can be thought of as a combination of horizontal motion with constant velocity and vertical free fall.

✔ Projectile motion can also be thought of as straight-line motion in the direction of the initial velocity with an additional downward motion due to gravity.

3-6 You can solve projectile motion problems using techniques learned for straight-line motion

Equations 3-12 and 3-13 are the principal equations we'll use for solving problems in projectile motion. Since these equations are identical to those for constant-velocity motion (along the x direction) and free fall (along the y direction) from Chapter 2, we'll be able to use many of the same problem-solving techniques learned in that chapter.

Go to Interactive Exercise 3-1 for more practice dealing with projectile motion.

In Table 3-1 we've collected Equations 3-12 and 3-13 along with information about which quantities are involved in each equation. In the same way that you used Table 2-3 in Section 2-5, you can use this table to decide which equations to use for a given projectile problem.

TABLE 3-1 Equations for Projectile Motion

By using one or more of these equations, you can solve any problem involving projectile motion. The check marks indicate which of the quantities t (time), x_0 (the x position at time 0), x (the x position at time t), v_{0x} (the x velocity at time 0), v_x (the x velocity at time t), y_0 (the y position at time 0), y (the y position at time t), v_{0y} (the y velocity at time 0), and v_y (the y velocity at time t) are present in each equation. Note that g, the magnitude of the acceleration due to gravity, is a *positive* quantity.

Equation		Which quantities the equation includes								
		t	x_0	x	v_{0x}	v_x	y_0	y	v_{0y}	v_y
$v_x = v_{0x}$	(3-12a)				✓	✓				
$v_y = v_{0y} - gt$	(3-12b)	✓							✓	✓
$x = x_0 + v_{0x}t$	(3-13a)	✓	✓	✓	✓					
$y = y_0 + v_{0y}t - \frac{1}{2}gt^2$	(3-13b)	✓					✓	✓	✓	

EXAMPLE 3-4 Cliff Diving I

A diver leaps from a cliff in La Quebrada (Acapulco), Mexico. The cliff is 30.0 m above the surface of the water. He leaves the cliff moving at 2.00 m/s at 40.0° above the horizontal. If air resistance can be neglected, (a) how long does it take him to hit the water, and (b) how far does he travel horizontally before reaching the ocean?

Set Up

We can treat the diver as a projectile because air resistance is negligible. We choose the origin ($x_0 = 0$, $y_0 = 0$) to be where the diver leaves the cliff and starts to move as a projectile. The diver's initial velocity (of magnitude $v_0 = 2.00$ m/s and angle $\theta_0 = 40.0°$) has positive x and y components that we can find using trigonometry. In part (a) we'll find the time t when he reaches the water at $y = -30.0$ m. We'll use this value of t in part (b) to find his x position at this time. This tells us how far he traveled horizontally during the dive. Table 3-1 tells us which equations to use for each part.

$$v_{0x} = v_0 \cos\theta_0$$
$$v_{0y} = v_0 \sin\theta_0$$

For part (a):

$$y = y_0 + v_{0y}t - \frac{1}{2}gt^2 \qquad \text{(3-13b)}$$

For part (b):

$$x = x_0 + v_{0x}t \qquad \text{(3-13a)}$$

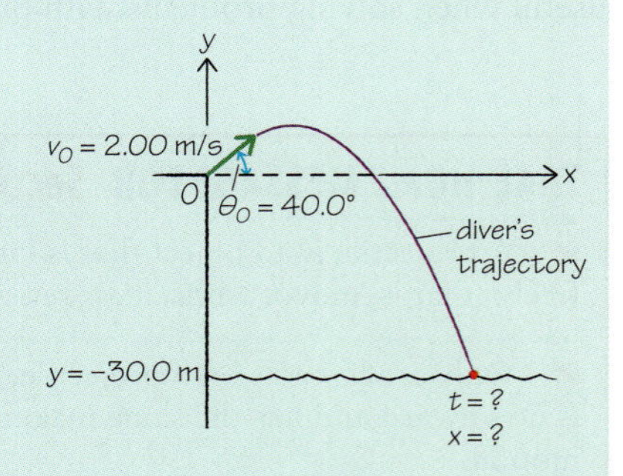

Solve

(a) First find the x and y components of the diver's initial velocity.

$$\begin{aligned} v_{0x} &= v_0 \cos\theta_0 \\ &= (2.00 \text{ m/s}) \cos 40.0° \\ &= 1.53 \text{ m/s} \end{aligned}$$

$$\begin{aligned} v_{0y} &= v_0 \sin\theta_0 \\ &= (2.00 \text{ m/s}) \sin 40.0° \\ &= 1.29 \text{ m/s} \end{aligned}$$

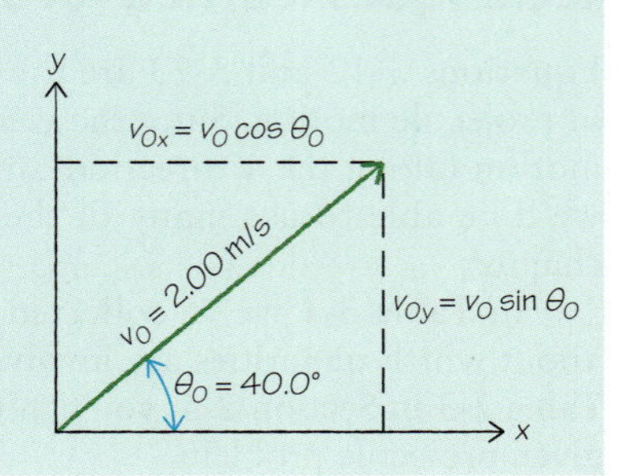

Then solve Equation 3-13b for time t and insert the known values $y_0 = 0$, $y = -30.0$ m, $v_{0y} = 1.29$ m/s, and $g = 9.80$ m/s^2.

$$y = y_0 + v_{0y}t - \frac{1}{2}gt^2 \text{ or } \frac{1}{2}gt^2 - v_{0y}t + (y - y_0) = 0$$

This is a quadratic equation of the form $at^2 + bt + c = 0$, with $a = g/2$, $b = -v_{0y}$, and $c = y - y_0$.

$$t = \frac{-b \pm \sqrt{b^2 - 4ac}}{2a} = \frac{-(-v_{0y}) \pm \sqrt{(-v_{0y})^2 - 4\left(\frac{g}{2}\right)(y - y_0)}}{2\left(\frac{g}{2}\right)}$$

$$= \frac{v_{0y} \pm \sqrt{v_{0y}^2 - 2g(y - y_0)}}{g}$$

$$= \frac{(1.29 \text{ m/s}) \pm \sqrt{(1.29 \text{ m/s})^2 - 2(9.80 \text{ m/s}^2)(-30.0 \text{ m})}}{9.80 \text{ m/s}^2}$$

$$= \frac{(1.29 \text{ m/s}) \pm \sqrt{590 \text{ m}^2/\text{s}^2}}{9.80 \text{ m/s}^2} = \frac{(1.29 \text{ m/s}) \pm (24.3 \text{ m/s})}{9.80 \text{ m/s}^2}$$

$$= 2.61 \text{ s or } -2.35 \text{ s}$$

The diver reaches the water after leaving the cliff at $t = 0$, so the solution for t that we want is the positive one.

$$t = 2.61 \text{ s}$$

(b) We substitute the diver's constant x component of velocity $v_{0x} = 1.53$ m/s along with $t = 2.61$ s from part (a) into Equation 3-13a to determine the diver's final horizontal position.

$$x - x_0 = v_{0x}t = (1.53 \text{ m/s})(2.61 \text{ s}) = 4.00 \text{ m}$$

Reflect

We can check our results by using one of our interpretations of projectile motion. If there were no gravity, the diver wouldn't fall but would continue along a straight line at an angle of 40.0° above the horizontal. Then in 2.61 s the diver would rise 3.37 m above the cliff. In fact the diver falls a distance $\frac{1}{2}gt^2 = 33.4$ m below his zero-gravity trajectory in 2.61 s. The net result is that the diver goes through a vertical displacement of -30.0 m—exactly the displacement from the height of the cliff to sea level.

If gravity did not exist, in time $t = 2.61$ s the diver would travel horizontally: $x - x_0 = v_{0x}t$

$$= (1.53 \text{ m/s})(2.61 \text{ s}) = 4.00 \text{ m}$$

vertically: $y - y_0 = v_{0y}t$

$$= (1.29 \text{ m/s})(2.61 \text{ s}) = 3.35 \text{ m}$$

When gravity is considered the diver falls a distance

$$\frac{1}{2}gt^2 = \frac{1}{2}(9.80 \text{ m/s}^2)(2.61 \text{ s})^2 = 33.4 \text{ m}$$

so the diver's net vertical displacement in 2.61 s is 3.35 m − 33.4 m = −30.0 m

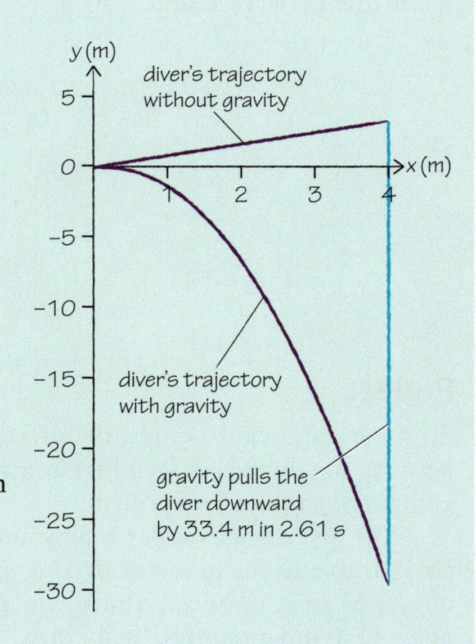

Part (a) of this example illustrates that projectile motion problems can have more than one solution, just like problems in linear motion with constant acceleration (see Example 2-8 in Section 2-5) or free-fall problems (see Example 2-10 in Section 2-6). Be careful!

Go to Interactive Exercise 3-2 for more practice dealing with projectile motion.

EXAMPLE 3-5 Cliff Diving II

What is the velocity (magnitude and direction) of the diver in Example 3-4 when he enters the water?

Set Up

From Example 3-4 we know the diver's initial ($t = 0$) velocity components, $v_{0x} = 1.53$ m/s and $v_{0y} = 1.29$ m/s, as well as the time $t = 2.61$ s that it takes for him to fall to the surface of the water. We'll use this information and Equations 3-12 to calculate his velocity components v_x and v_y when he reaches the water. We'll then use trigonometry to find the magnitude and direction of his velocity at that point.

Velocity, acceleration, and time for projectile motion:

$$v_x = v_{0x} \quad (3\text{-}12a)$$

$$v_y = v_{0y} - gt \quad (3\text{-}12b)$$

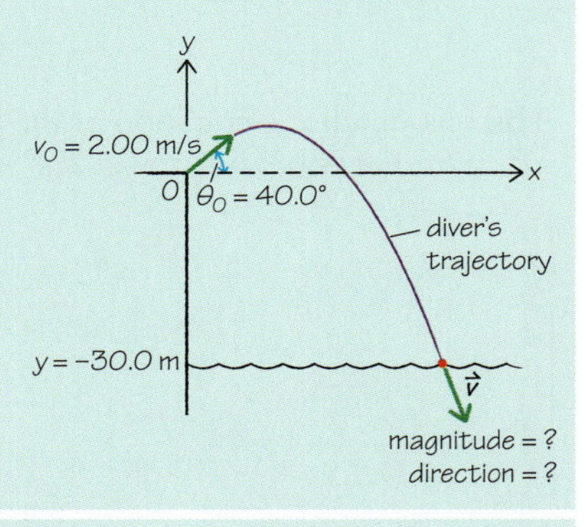

Solve

Equation 3-12a tells us that v_x doesn't change as the diver plummets toward the water, so it has the same value as when he left the cliff at $t = 0$. By contrast, Equation 3-12b tells us that v_y decreases by an amount gt.

$$v_x = v_{0x} = 1.53 \text{ m/s}$$

$$v_y = v_{0y} - gt$$
$$= 1.29 \text{ m/s} - (9.80 \text{ m/s}^2)(2.61 \text{ s})$$
$$= -24.3 \text{ m/s}$$

The diver's x and y components of velocity and his speed v make up the three sides of a right triangle, so we can find v from v_x and v_y using the Pythagorean theorem. Note from the sketch that because v_x is positive and v_y is negative, the angle θ must be between 0 and $-90°$.

Magnitude of $\vec{v}$:

$$v = \sqrt{v_x^2 + v_y^2}$$
$$= \sqrt{(1.53 \text{ m/s})^2 + (-24.3 \text{ m/s})^2}$$
$$= 24.3 \text{ m/s}$$

Direction of $\vec{v}$:

$$\tan\theta = \frac{v_y}{v_x}$$
$$= \frac{-24.3 \text{ m/s}}{1.53 \text{ m/s}} = -15.8$$

$$\theta = \tan^{-1}(-15.9)$$
$$= -86.4°$$

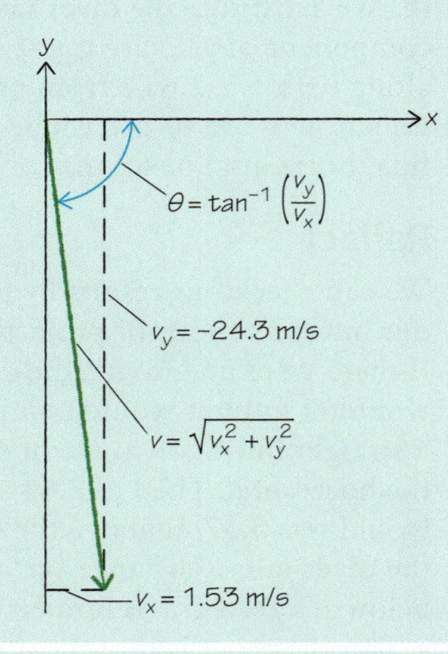

Reflect

To three significant figures, the diver's speed v is the same as the absolute value of his y component of velocity v_y. That's because the diver is descending at a very steep angle of almost $-90°$, so the x component of his velocity is very small compared to the y component.

The diver is moving at a substantial speed when he enters the water. You can show that $v = 87.6$ km/h or 54.4 mi/h, comparable to the highway driving speed of an automobile. (In fact the diver's speed will be somewhat less due to the effects of air resistance.) That's why it's essential that the diver enter the water with his body oriented at just the right angle to avoid a painful "belly flop."

EXAMPLE 3-6 How High Does It Go?

A frog hops so that it leaves the ground at speed v_0 and at an angle θ_0 above the horizontal. What is the maximum height that the frog reaches during its leap? Ignore air resistance.

Set Up

We considered a similar problem in Example 2-10 (Section 2-6), in which a springbok jumped straight up. For both the springbok and the frog, the y

Velocity, acceleration, and time for projectile motion:

$$v_y = v_{0y} - gt \quad (3\text{-}12b)$$

component of velocity is zero at the peak of the motion (when the animal is moving neither up nor down). Unlike the springbok, the frog also has an x component of velocity, but this has no effect on the up-and-down motion.

We use a coordinate system as shown in the sketch, with the origin at the frog's starting point so that its initial coordinates are $x_0 = 0$, $y_0 = 0$. We'll use Equation 3-12b to find the time t_1 when the frog's y component of velocity is zero so that it is at the highest point of its leap. We'll then use Equation 3-13b to find the frog's y coordinate y_1 at this time.

Position, acceleration, and time for projectile motion:

$$y = y_0 + v_{0y}t - \frac{1}{2}gt^2 \quad \text{(3-13b)}$$

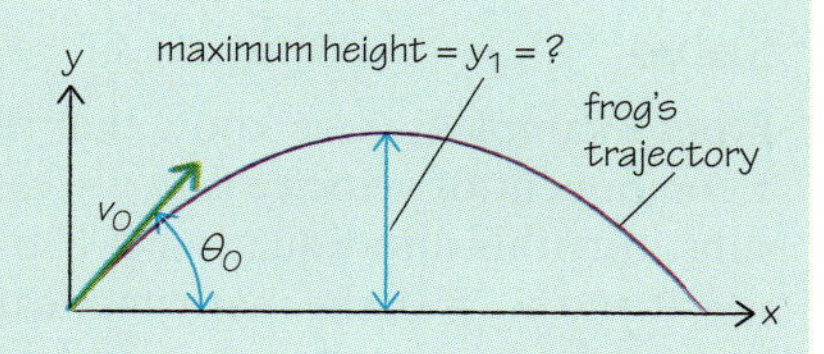

Solve

First express the frog's initial y component of velocity v_{0y} in terms of its initial speed v_0 and launch angle θ_0. Then, substitute this expression into Equation 3-12b. Set $v_y = 0$ (corresponding to the high point of the motion) at time $t = t_1$, then solve for t_1.

$$v_{0y} = v_0 \sin\theta_0 \text{ so}$$
$$v_y = v_0 \sin\theta_0 - gt$$
$$0 = v_0 \sin\theta_0 - gt_1$$
$$gt_1 = v_0 \sin\theta_0$$
$$t_1 = \frac{v_0 \sin\theta_0}{g}$$

Now use Equation 3-13b to solve for the frog's y coordinate at time t_1. Substitute $y_0 = 0$, $v_{0y} = v_0 \sin\theta_0$, and $t_1 = (v_0 \sin\theta_0)/g$ from above.

$$y_1 = 0 + (v_0 \sin\theta_0)\left(\frac{v_0 \sin\theta_0}{g}\right) - \frac{1}{2}g\left(\frac{v_0 \sin\theta_0}{g}\right)^2$$
$$= \frac{v_0^2 \sin^2\theta_0}{g} - \frac{v_0^2 \sin^2\theta_0}{2g} \text{ so}$$
$$y_1 = \frac{v_0^2 \sin^2\theta_0}{2g}$$

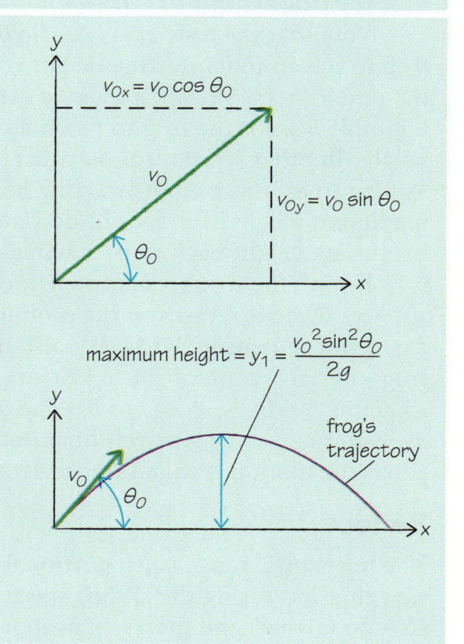

Reflect

In this example our result is a formula rather than a number. Can you verify that our answer for y_1 has the correct units?

To interpret what this formula tells us, let's look at how the maximum height y_1 depends on the values of the frog's initial speed v_0 and launch angle θ_0 as well as on the value of g. Because y_1 is proportional to the square of v_0, doubling the frog's launch speed makes it go *four* times as high. Note also that y_1 is proportional to $\sin^2\theta_0$. The sine function has its maximum value for $\theta_0 = 90°$, which corresponds to the frog jumping straight up. Both the dependence on v_0 and the dependence on θ_0 make sense; the frog leaps higher if its launch speed is greater or if it jumps at a more vertical angle.

Finally, note that y_1 is *inversely* proportional to g (the factor of g is in the denominator). Hence, if the value of g were smaller, you would divide by a smaller number to calculate y_1 and you would get a larger answer. This result also makes sense: If gravity were weaker, the frog would reach a greater height.

EXAMPLE 3-7 How Far Does It Go?

When the frog in Example 3-6 returns to the ground, how far from its launch point does it land?

Set Up

The frog started its flight at $x_0 = 0$, $y_0 = 0$. When the frog lands, it returns to the same height $y = 0$ at horizontal coordinate x_2. The frog's distance from its launch point when it lands is the x component of its displacement, $x_2 - x_0$. To determine x_2, we'll first use Equation 3-13b to calculate the time t_2 when the frog lands and returns to $y = 0$. We'll then substitute this time into Equation 3-13a to calculate the frog's x coordinate when it lands, x_2.

Position, acceleration, and time for projectile motion:

$$x = x_0 + v_{0x}t \quad \text{(3-13a)}$$

$$y = y_0 + v_{0y}t + \frac{1}{2}gt^2 \quad \text{(3-13b)}$$

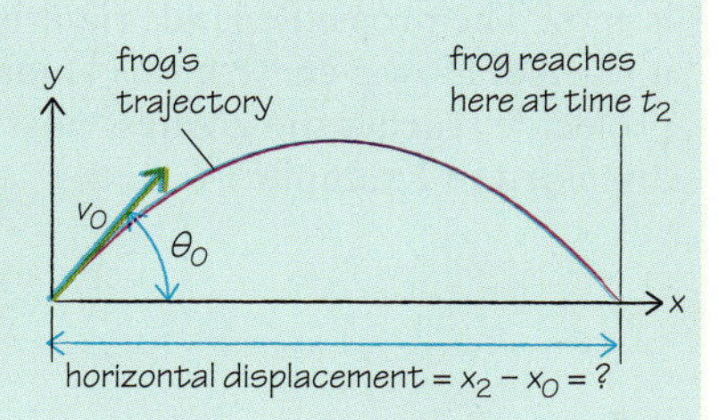

Solve

From Example 3-6 we know that the frog's initial y component of velocity is $v_{0y} = v_0 \sin\theta_0$. We substitute this into Equation 3-13b along with $y = 0$ and $y_0 = 0$, the mathematical statement that the frog has returned to ground level, and set $t = t_2$ (the time when the frog returns to ground level).

Clearly one solution to the above equation is $t_2 = 0$. But we know the frog lands after it takes off. So we divide this expression by t_2, eliminating the quadratic, and making it much simpler to find the second solution for t_2, which will be the time the frog *returns* to the ground.

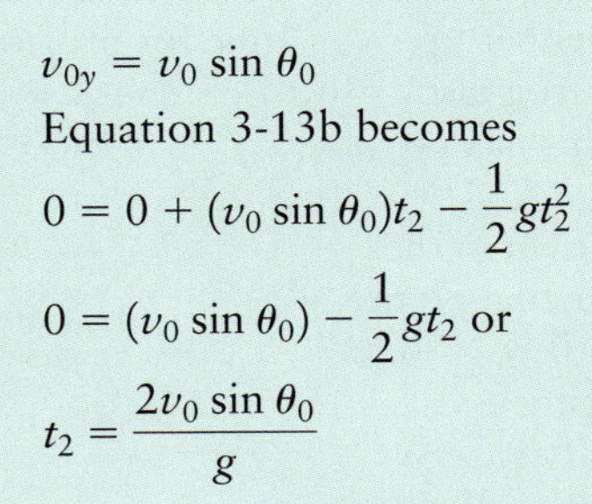

$$v_{0y} = v_0 \sin\theta_0$$

Equation 3-13b becomes

$$0 = 0 + (v_0 \sin\theta_0)t_2 - \frac{1}{2}gt_2^2$$

$$0 = (v_0 \sin\theta_0) - \frac{1}{2}gt_2 \text{ or}$$

$$t_2 = \frac{2v_0 \sin\theta_0}{g}$$

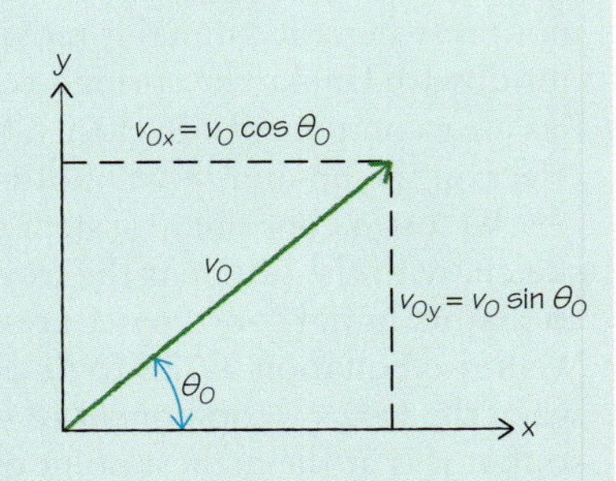

Note that the time $t_2 = (2v_0 \sin\theta_0)/g$ for the frog to return to its starting height is exactly *twice* the time $t_1 = (v_0 \sin\theta_0)/g$ that we calculated in Example 3-6 for the frog to reach its maximum height. In other words, a projectile takes just as long to ascend from its starting height to its maximum height as it does to descend from its maximum height back to its starting height.

To find the frog's x component of displacement $x_2 - x_0$ when it returns to the ground, substitute $t = t_2$ into Equation 3-13a. We also need the frog's initial x component of velocity, which from Example 3-6 is $v_{0x} = v_0 \cos\theta_0$. The distance $x_2 - x_0$ that the frog travels horizontally during its projectile motion is called the *horizontal range*.

$$x_2 - x_0 = v_{0x}t_2$$

At time $t_2 = (2v_0 \sin\theta_0)/g$, find the displacement $x_2 - x_0$:

Horizontal range:

$$x_2 - x_0 = (v_0 \cos\theta_0)\left(\frac{2v_0 \sin\theta_0}{g}\right) = \frac{2v_0^2 \sin\theta_0 \cos\theta_0}{g}$$

Reflect

The horizontal range is proportional to v_0^2 (which says that increasing the launch speed makes the frog go farther) and inversely proportional to g (which says that the frog would also go farther if gravity were weaker).

To understand how the horizontal range depends on the launch angle θ_0, it's helpful to rewrite our result (using a formula from trigonometry) as $v_0^2 \sin 2\theta_0/g$. This expression shows that horizontal range is greatest for $\theta_0 = 45°$, so that $2\theta_0 = 90°$ and the function $\sin 2\theta_0$ has its greatest value ($\sin 90° = 1$). For launch angles less than 45°, the horizontal range is shorter because the projectile is in the air for a relatively short time and so cannot travel as far horizontally. For launch angles greater than 45°, the projectile spends more time in the air yet has a shorter horizontal range because its initial velocity is mostly upward (see the sketch).

Notice that for $\theta_0 = 90°$ the horizontal range is *zero*: The projectile lands right back where it started. However, with this launch angle the projectile reaches the greatest possible height, as we discussed in Example 3-6.

Horizontal range

$$= \frac{2v_0^2 \sin\theta_0 \cos\theta_0}{g}$$

From trigonometry,

$$2 \sin\theta_0 \cos\theta_0 = \sin 2\theta_0$$

so

$$\text{horizontal range} = \frac{v_0^2 \sin 2\theta_0}{g}$$

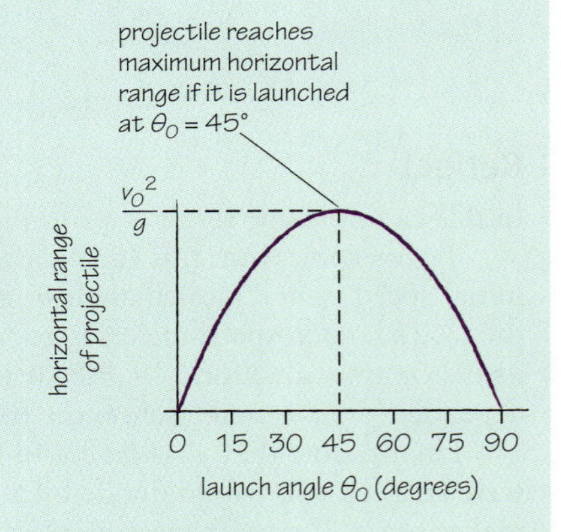

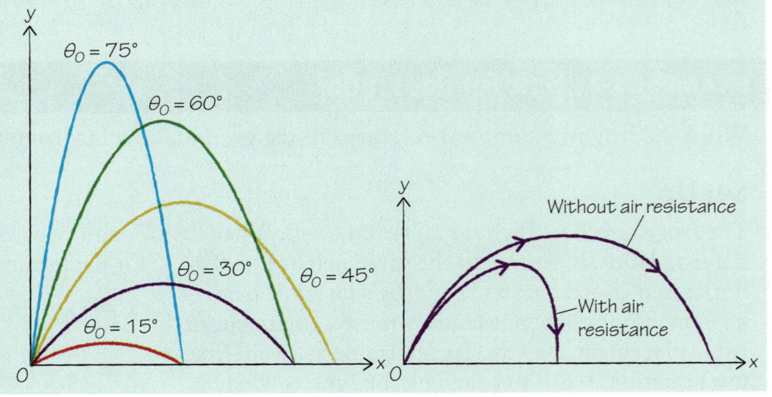

WATCH OUT! Air resistance affects range.

The rule that a launch angle of 45° gives the greatest horizontal range doesn't work for objects such as kicked footballs or batted baseballs. The reason is that air resistance plays a significant role in the flight of these objects, so their acceleration is not constant, their trajectories are not parabolic, and the formulas we've developed in this section don't apply. Detailed calculations that account for air resistance show that a batted baseball has the greatest horizontal range for a launch angle of about 35° above the horizontal.

GOT THE CONCEPT? 3-6 How Fast Is a Flying Frog?

Suppose the frog in Examples 3-6 and 3-7 leaves the ground at a launch angle $\theta_0 = 45°$. If the frog's speed is v_0 when it leaves the ground, what is its speed when it is at its maximum height? (a) v_0; (b) between v_0 and $v_0/2$; (c) $v_0/2$; (d) between $v_0/2$ and zero; (e) zero; (f) not enough information given to decide.

TAKE-HOME MESSAGE FOR Section 3-6

✔ In solving problems that involve projectile motion, you can use many of the same techniques that we used in Section 2-7 to solve one-dimensional free-fall problems.

✔ Equations 3-12 and 3-13 are essential tools for solving any projectile motion problem.

3-7 An object moving in a circle is accelerating even if its speed is constant

It takes a harder tug on a car's steering wheel to round a tight corner at high speed than it does to take a gentle curve at low speed. We saw in Section 3-4 that a car following a curved path experiences an acceleration directed toward the inside of the curve and that the driver causes this acceleration by turning the steering wheel. So the car's acceleration must be greater for a high-speed, tight turn than for a low-speed, gentle turn. But *how much* greater? What is the relationship between the car's acceleration, its speed, and the radius of its circular path?

To see the answer let's look at the case of an object going around a circular path at a constant speed, also called **uniform circular motion**. Because the object's speed is constant, its acceleration vector $\vec{a}$ has no component tangent to its trajectory. Hence $\vec{a}$ is perpendicular to the trajectory and points toward the *center* of the circle (Figure 3-29). An acceleration that points in this direction is called a **centripetal acceleration**, which comes from a Latin term meaning "seeking the center." Let's see how to calculate the magnitude of the centripetal acceleration vector.

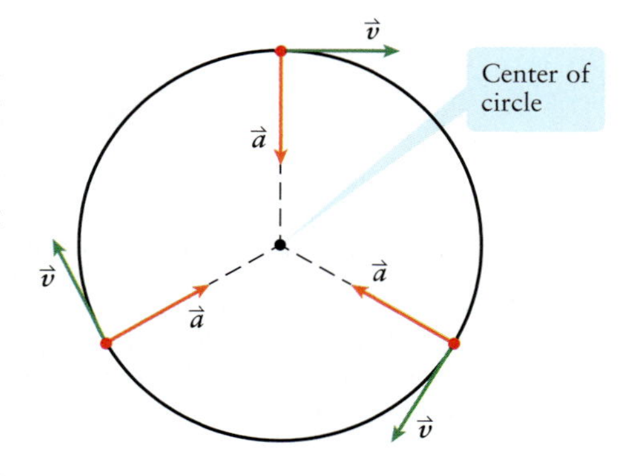

• In uniform circular motion the speed is *constant*, but the object accelerates because the velocity $\vec{v}$ is always changing direction.
• Because the speed is constant, the acceleration $\vec{a}$ is always perpendicular to the velocity $\vec{v}$.
• $\vec{a}$ always points toward the center of the circle.

Figure 3-29 Uniform circular motion An object moving around a circle at a constant speed.

Analyzing Motion in a Circle

Figure 3-30a shows two points along the trajectory of an object in uniform circular motion around a circular path of radius r. From point 1 to point 2, the object travels a distance Δs around the circle, and an imaginary line drawn from the center of the circle to the object rotates through an angle $\Delta\theta$. We can write a simple relationship between Δs and $\Delta\theta$ provided that we express $\Delta\theta$ in *radians*:

$$\Delta\theta = \frac{\Delta s}{r} \; (\Delta\theta \text{ in radians}) \tag{3-14}$$

You can check this formula by thinking about the case in which the object goes all the way around the circle so that the distance Δs equals $2\pi r$, the circumference of the circle. Then Equation 3-14 says that $\Delta\theta = (2\pi r)/r = 2\pi$, which is just the number of radians that make up a complete circle.

If the time interval between points 1 and 2 is very short, the angle $\Delta\theta$ is small, and the curved arc of length Δs is nearly a straight line. Then the lines from the center of the circle to points 1 and 2 and the curved arc form a triangle with two sides of length r separated by an angle $\Delta\theta$ and a third side of length Δs given by Equation 3-14 (see **Figure 3-30b**). Thus Equation 3-14 is a relationship between the length Δs of the short side of the triangle, the length r of the two long sides, and the angle $\Delta\theta$.

Figure 3-30c shows why Equation 3-14 is useful: The velocity vector rotates through the same angle $\Delta\theta$ as the object moves from point 1 (velocity $\vec{v}_1$) to point 2 (velocity $\vec{v}_2$). Because the speed is constant, $\vec{v}_1$ and $\vec{v}_2$ have the same magnitude v. In **Figure 3-30d**, we've converted $\vec{v}_1$ into $-\vec{v}_1$ by reversing its direction and placed the head of $-\vec{v}_1$ against the tail of $\vec{v}_2$ to form the vector difference $\Delta\vec{v} = \vec{v}_2 - \vec{v}_1$. This vector diagram is equivalent to a triangle with two sides of length v separated by an angle $\Delta\theta$ and a third side of length Δv, which is the magnitude of the velocity change between points 1 and 2 (**Figure 3-30e**).

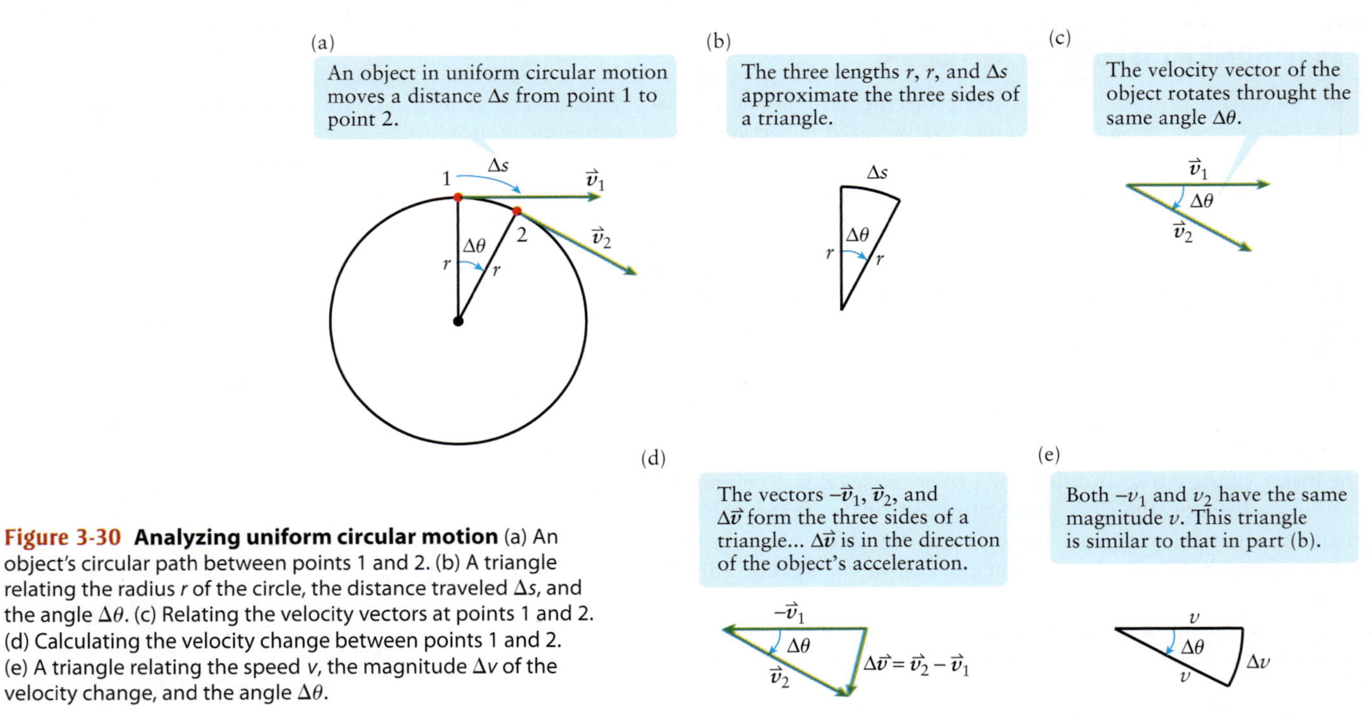

(a)
An object in uniform circular motion moves a distance Δs from point 1 to point 2.
Δs
1
2
v₁
v₂
Δθ
r
r
(b)
The three lengths r, r, and Δs approximate the three sides of a triangle.
Δs
Δθ
r
r
(c)
The velocity vector of the object rotates throught the same angle Δθ.
v₁
Δθ
v₂
(d)
The vectors −v₁, v₂, and Δv form the three sides of a triangle... Δv is in the direction of the object's acceleration.
−v₁
Δθ
v₂
Δv = v₂ − v₁
(e)
Both −v₁ and v₂ have the same magnitude v. This triangle is similar to that in part (b).
v
Δθ
v
Δv

Figure 3-30 Analyzing uniform circular motion (a) An object's circular path between points 1 and 2. (b) A triangle relating the radius r of the circle, the distance traveled Δs, and the angle $\Delta\theta$. (c) Relating the velocity vectors at points 1 and 2. (d) Calculating the velocity change between points 1 and 2. (e) A triangle relating the speed v, the magnitude Δv of the velocity change, and the angle $\Delta\theta$.

You can see that the triangle in Figure 3-30e is *similar* to the triangle in Figure 3-30b. Both triangles have two equal sides separated by the same angle, so the relationship between the lengths of their sides is the same. Comparing the two figures shows that Δv and v in Figure 3-30e correspond to Δs and r, respectively, in Figure 3-30b. So we can write an equation involving Δv and v that's the equivalent of Equation 3-14:

$$\Delta\theta = \frac{\Delta v}{v} \tag{3-15}$$

We'd like to calculate the magnitude of the object's acceleration, which is equal to $a_{cent} = \Delta v/\Delta t$. (We use the subscript "cent" to remind us that the acceleration is *centripetal*.) To do this calculation we first set our two expressions for $\Delta\theta$, Equations 3-14 and 3-15, equal to each other. This gives a new equation that relates Δv, v, Δs, and r:

$$\frac{\Delta v}{v} = \frac{\Delta s}{r}$$

To get an expression for $a_{cent} = \Delta v/\Delta t$, multiply both sides of this equation by v and divide both sides by Δt. Then we get

$$a_{\text{cent}} = \frac{\Delta v}{\Delta t} = \left(\frac{\Delta s}{r}\right)\left(\frac{v}{\Delta t}\right) = \left(\frac{\Delta s}{\Delta t}\right)\left(\frac{v}{r}\right) \quad (3\text{-}16)$$

Now the object's speed v is equal to the distance Δs that it travels divided by the time interval Δt required to travel this distance: $v = \Delta s/\Delta t$. Therefore, we can replace $\Delta s/\Delta t$ in Equation 3-16 by v, and our expression for the object's centripetal acceleration becomes $a_{\text{cent}} = v(v/r)$, or

Centripetal acceleration: magnitude of the acceleration of an object in uniform circular motion

Speed of an object as it moves around the circle

$$a_{\text{cent}} = \frac{v^2}{r}$$

Radius of the circle

Centripetal acceleration for motion in a circle (3-17)

The units of v and r are m/s and m, respectively. Hence v^2/r in Equation 3-17 has units $(\text{m/s})^2/\text{m}$ or m/s^2, which are the correct units for acceleration.

Equation 3-17 tells us that the magnitude of the centripetal acceleration is proportional to the square of the speed and inversely proportional to the radius of the circle. So there is greater centripetal acceleration the greater the object's speed and the smaller the radius (and hence the tighter the turn). That's entirely consistent with our discussion at the beginning of this section.

Although we derived Equation 3-17 for the case of an object moving around a circle at constant speed, the same equation applies to an object that travels at constant speed around any *portion* of a circle. As an example, think of a bird initially flying north that makes a 90° right turn along a semicircular arc at constant speed until it is heading east. The part of the bird's flight during which it is turning constitutes uniform circular motion, with an acceleration directed toward the center of the circle of which the arc is part (**Figure 3-31**).

Equation 3-17 is also valid when the object's speed changes (a case called *nonuniform* circular motion). In this case, as we saw in Section 3-4, the acceleration vector has both a component along the direction of motion (which describes changes in speed) and a component perpendicular to the motion (which describes changes in direction). It turns out that the magnitude of this perpendicular component at each point around the circular path is given by Equation 3-17, $a_{\text{cent}} = v^2/r$, provided that we set v equal to the instantaneous value of the speed at that point (**Figure 3-32**).

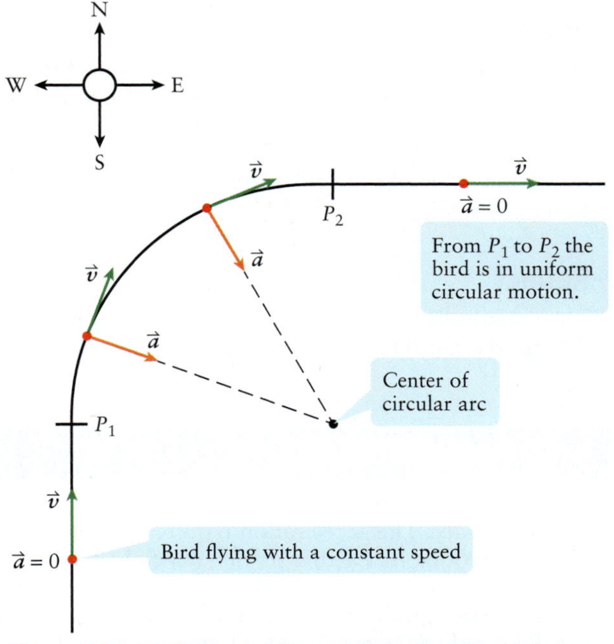

Figure 3-31 Uniform circular motion along an arc An object doesn't have to travel all the way around a complete circle to be in uniform circular motion.

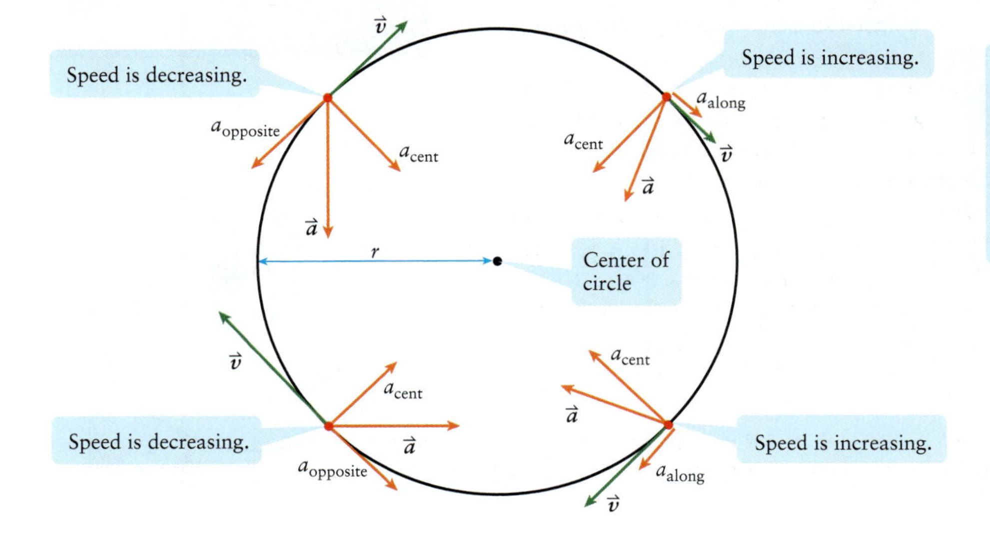

• If an object moves around a circle with varying speed, the acceleration $\vec{a}$ has a component along or opposite the direction of motion.
• Whether the speed is increasing, decreasing, or remaining constant, the acceleration always has a component $a_{\text{cent}} = v^2/r$ directed toward the center of the circle.

Figure 3-32 Nonuniform circular motion If the speed varies in circular motion, the acceleration vector is not directed toward the center of the circle.

EXAMPLE 3-8 Around the Bend

A certain curve in a level road is signposted for a maximum recommended speed of 65 km/h. The curve is a circular arc with a radius of 95 m. What is the centripetal acceleration of a car that takes this curve at the maximum recommended speed?

Set Up

We are given the car's speed $v = 65$ km/h and the radius $r = 95$ m of its trajectory, so we can calculate its centripetal acceleration a_{cent} using Equation 3-17.

Centripetal acceleration:

$$a_{cent} = \frac{v^2}{r} \qquad (3\text{-}17)$$

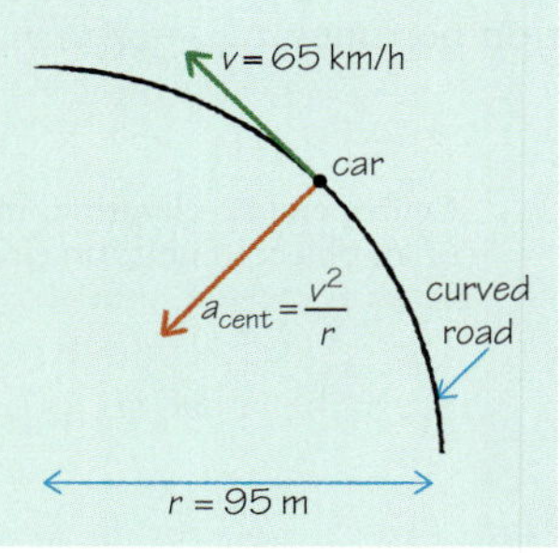

Solve

We are given r in m and v in km/h. To calculate the car's acceleration in m/s^2, we must first convert v to m/s.

$$v = 65 \text{ km/h} \times \left(\frac{1000 \text{ m}}{1 \text{ km}}\right) \times \left(\frac{1 \text{ h}}{3600 \text{ s}}\right)$$
$$= 18 \text{ m/s}$$

Given the value of v we calculate a_{cent} using Equation 3-17.

$$a_{cent} = \frac{v^2}{r} = \frac{(18 \text{ m/s})^2}{95 \text{ m}} = 3.4 \text{ m/s}^2$$

Reflect

The magnitude of the centripetal acceleration is just over one-third of the magnitude of the acceleration due to gravity during free fall. This certainly seems like a reasonable acceleration.

EXAMPLE 3-9 Orbital Speed

The International Space Station (ISS) orbits Earth at an altitude of 360 km above Earth's surface. The acceleration required to keep the ISS in orbit along a circular path is provided by Earth's gravity, which is directed toward the center of Earth and hence toward the center of the orbit as shown in Figure 3-33. (Earth's gravitational pull decreases gradually with altitude, so the acceleration due to gravity that acts on the ISS is only 8.78 m/s^2.) (a) What is the orbital speed of the ISS in meters per second, kilometers per hour, and miles per hour? (b) How long (in minutes) does it take the ISS to complete an orbit? Note that the radius of Earth is 6378 km.

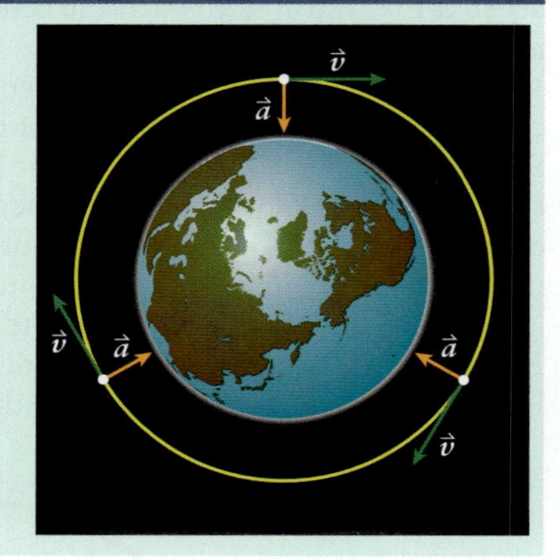

Figure 3-33 An orbiting space station The centripetal acceleration of an orbiting space station or other satellite is provided by Earth's gravity.

Set Up

We are given the ISS's centripetal acceleration a_{cent} and orbital radius r (the sum of Earth's radius and the altitude of the ISS, as the sketch shows). Using this information we can calculate the speed v using Equation 3-17. Once we know v we can calculate the time T for the ISS to complete one orbit (that is, to travel a distance equal to the circumference of the orbit) using the relationship (speed) = (distance)/(time).

Centripetal acceleration:

$$a_{cent} = \frac{v^2}{r} \qquad (3\text{-}17)$$

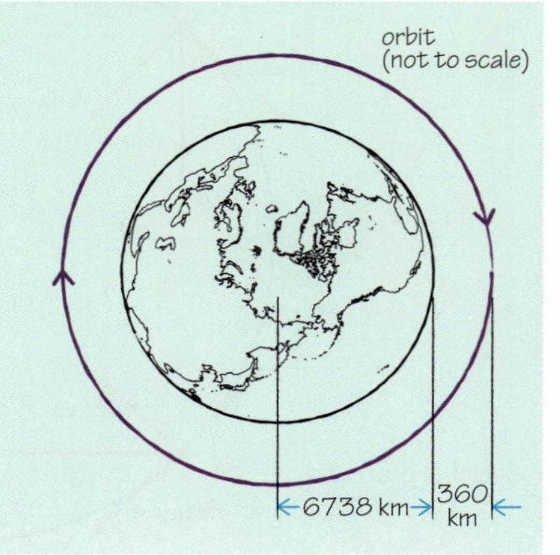

Solve

(a) We want an expression for the speed v of the space station, so we rearrange Equation 3-17 to put v by itself on the left-hand side.

$$a_{\text{cent}} = \frac{v^2}{r}$$
$$v^2 = ra_{\text{cent}}$$
$$v = \sqrt{ra_{\text{cent}}}$$

We calculate the orbital radius r by adding Earth's radius and the altitude of the ISS then convert the result to meters.

$$r = 6378 \text{ km} + 360 \text{ km}$$
$$= 6738 \text{ km} \times \left(\frac{10^3 \text{ m}}{1 \text{ km}}\right) = 6.738 \times 10^6 \text{ m}$$

Substitute our calculated value of r into the expression for v.

$$v = \sqrt{(6.738 \times 10^6 \text{ m})(8.78 \text{ m/s}^2)}$$
$$= \sqrt{5.92 \times 10^7 \text{ m}^2/\text{s}^2} = 7.69 \times 10^3 \text{ m/s}$$

Convert the speed to km/h and mi/h.

$$v = 7.69 \times 10^3 \text{ m/s} \times \left(\frac{1 \text{ km}}{10^3 \text{ m}}\right) \times \left(\frac{3600 \text{ s}}{1 \text{ h}}\right)$$
$$= 2.77 \times 10^4 \text{ km/h}$$
$$v = 2.77 \times 10^4 \text{ km/h} \times \left(\frac{1 \text{ mi}}{1.609 \text{ km}}\right)$$
$$= 1.72 \times 10^4 \text{ mi/h}$$

(b) The distance that the ISS travels in one orbit is equal to the orbit circumference $2\pi r$. Because (speed) = (distance)/(time), the time T for one orbit equals the orbit circumference divided by the speed. We convert this answer to minutes.

$$\text{distance} = 2\pi(6.738 \times 10^6 \text{ m})$$
$$= 4.234 \times 10^7 \text{ m}$$
$$\text{time } T = \frac{\text{distance}}{\text{speed}} = \frac{4.234 \times 10^7 \text{ m}}{7.69 \times 10^3 \text{ m/s}}$$
$$= 5.50 \times 10^3 \text{ s} = (5.50 \times 10^3 \text{ s}) \times \left(\frac{1 \text{ min}}{60 \text{ s}}\right)$$
$$= 91.7 \text{ min}$$

Reflect

The sketch shows why there is one particular speed for a circular orbit of a given altitude. If you drop a ball, it falls straight downward. If you throw it horizontally, it follows a curved path before hitting the ground. The faster you throw the ball, the farther it travels prior to impact, and the more gradually curved its trajectory. If you throw the ball with a sufficiently great speed, the curvature of its trajectory precisely matches the curvature of Earth. Then the ball remains at the same height above Earth's surface and never falls to the ground. Putting a satellite such as the International Space Station into orbit works the same way, except that a rocket rather than a strong pitching arm is used to provide the necessary speed. Gravity is making the ISS "fall," but because of its high speed, it falls *around* Earth rather than *toward* Earth. As we calculated in part (b), the ISS makes a complete orbit about every hour and a half.

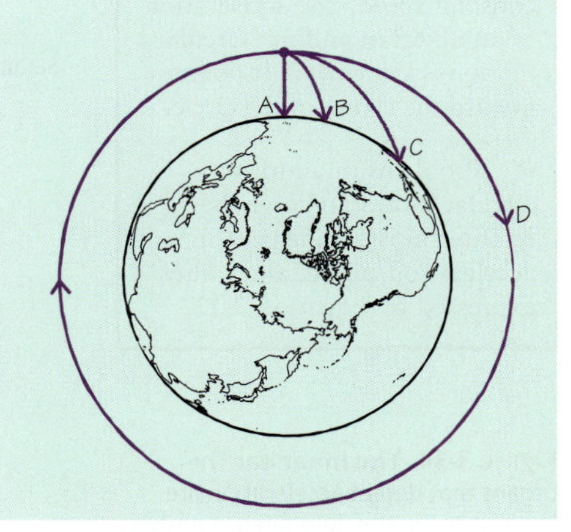

WATCH OUT! What causes weightlessness—and what doesn't.

It's a common misconception that an object in orbit is "beyond the pull of Earth's gravity." It's true that an astronaut on board the International Space Station floats around the cabin *as though* she were weightless and unaffected by gravity. But if gravity didn't act on an orbiting satellite or its occupants, the satellite would have *zero* acceleration and hence couldn't stay in orbit: It would maintain a constant velocity and simply fly off into space in a straight line. In fact a satellite such as the ISS remains in orbit precisely because it *does* feel the pull of gravity, which provides the acceleration necessary to make the satellite follow a circular trajectory around Earth. An astronaut on board the ISS *seems* to be weightless because she is really an independent satellite of Earth following the same orbit as the ISS does. Because the astronaut and the ISS are both following the same trajectory, there is nothing pushing the astronaut toward any of the station's walls. As a result, even though gravity is acting on the astronaut, she *feels* weightless.

GOT THE CONCEPT? 3-7 Acceleration in Uniform Circular Motion

If an object is in uniform circular motion, its speed is constant. Is its *acceleration* constant?

GOT THE CONCEPT? 3-8 Finding the Road Radius

If the car in Example 3-8 went around a curve at twice the speed assumed in that example, how much larger would the radius of the curve have to be in order for the centripetal acceleration to be the same? (a) $\sqrt{2}$ times as large; (b) twice as large; (c) $2\sqrt{2}$ times as large; (d) 4 times as large; (e) 16 times as large.

3-8 The vestibular system of the ear allows us to sense acceleration

TAKE-HOME MESSAGE FOR Section 3-7

✔ If an object follows a curved trajectory, it has an acceleration even if its speed remains constant. Motion in a circle is an important example of motion of this kind.

✔ Uniform circular motion is motion around a circle at a constant speed. The acceleration of an object in uniform circular motion is centripetal: It points toward the center of the circle.

✔ Problems in uniform circular motion involve the relationships among centripetal acceleration, speed, and radius expressed by Equation 3-17.

You can sense acceleration when you are thrown forward when the car you are riding in stops suddenly or pushed into your seat when riding in a roller coaster car that makes a sharp turn. But you can also detect much more subtle accelerations, such as the gentle swaying of a hammock, even with your eyes closed. How do our bodies perceive such small accelerations?

The answer lies in the *vestibular system* of the ear (Figure 3-34). This is a series of tiny, interconnected canals and chambers in the skull.

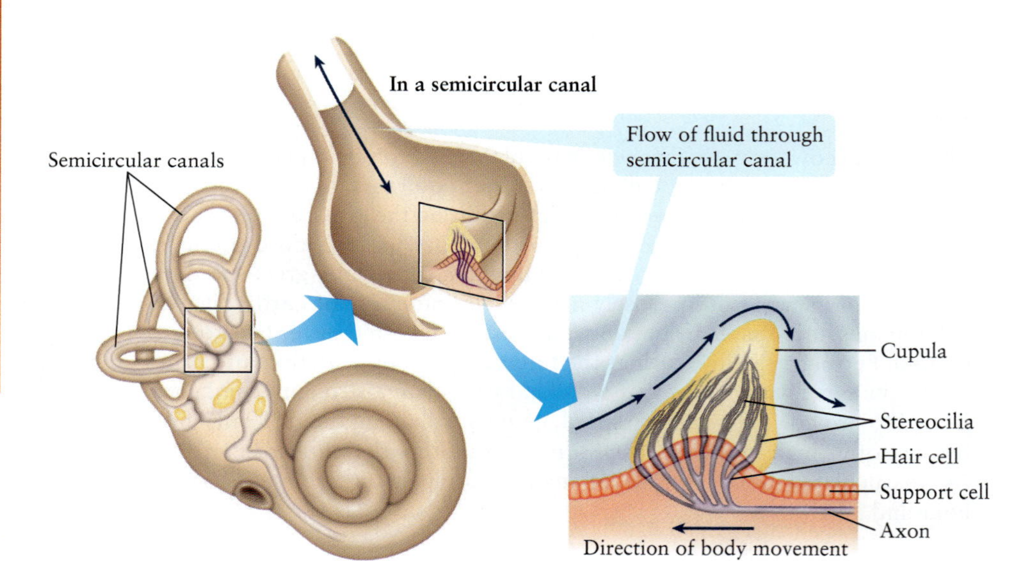

Figure 3-34 The inner ear The organs that detect acceleration are located inside the fluid-filled chambers of the inner ear.

Any change in the motion of the head means that the head has experienced an acceleration, and two chambers at the center of the vestibular system detect the horizontal and vertical components of the acceleration vector. A layer of sensory cells is oriented vertically in one chamber and horizontally in the other. A bundle of microscopic, hair-like extensions called *cilia* stick up from each sensory cell into a gelatinous substance coated with a layer of calcium carbonate crystals. The rock-like appearance of these crystals gives them their name, *otoliths* ("ear stones" in Greek).

When the head accelerates, the relatively massive otoliths lag behind due to their inertia, which causes the cilia to bend. This bending causes the sensory cells to transmit an electrical signal to the brain. The sensory cells produce a maximum signal when the tip of a hair-cell bundle moves by less than 0.5 μm, so these cells are exquisitely sensitive to even gentle accelerations. In addition, each individual hair cell responds best to being bent in one particular direction. Any acceleration of the head therefore results in a unique pattern of nerve signals being sent to the brain, which enables you to tell both the magnitude and direction of the head's acceleration.

The other organs of the vestibular system detect *angular* acceleration (that is, the rate of change of your head's rotational motion). This is accomplished by three *semicircular canals* that are oriented perpendicular to one another (see Figure 3-34). The interior of each canal is filled with fluid. Whenever the head rotates around the axis of one of the canals, the fluid inside it moves, bending a structure inside the canal called a *cupula* whose hair cells function like the cilia. The response of the brain to the resulting electrical signals it receives gives you the feeling of angular acceleration. Rotations of the head as small as 0.005° can be sensed in this way.

When the vestibular system functions normally, you are oblivious to its existence. If it malfunctions, however, you feel dizzy and may not be able to walk or even stand upright. In addition, if information from the vestibular system doesn't match the information from the eyes, the result is a sense of disorientation, like the feeling of seasickness some people experience in a small boat on choppy water.

TAKE-HOME MESSAGE FOR Section 3-8

✔ The vestibular system of the ear detects the acceleration vector of your head. It separately measures the horizontal and vertical components of acceleration.

Key Terms

acceleration vector (or instantaneous acceleration vector)
centripetal acceleration
component
direction
displacement vector
magnitude
motion in a plane
origin
parabola
position vector
projectile
projectile motion
scalar
trajectory
two-dimensional motion
uniform circular motion
vector
vector addition
vector difference
vector multiplication by a scalar
vector subtraction
vector sum
velocity vector (or instantaneous velocity vector)
x component
y component
x-*y* plane

Chapter Summary

Topic	Equation or Figure
Vectors and scalars: A vector quantity such as displacement or velocity has both a magnitude and a direction. By contrast, a scalar quantity such as temperature or time has no direction. We write a vector in boldface italic with an arrow above it. The magnitude of a vector $\vec{A}$ is written as A.	(d) Vectors $\vec{H}$ and $\vec{J}$ have different magnitudes but the same direction. Vectors $\vec{J}$ and $\vec{K}$ have the same magnitude but different directions. 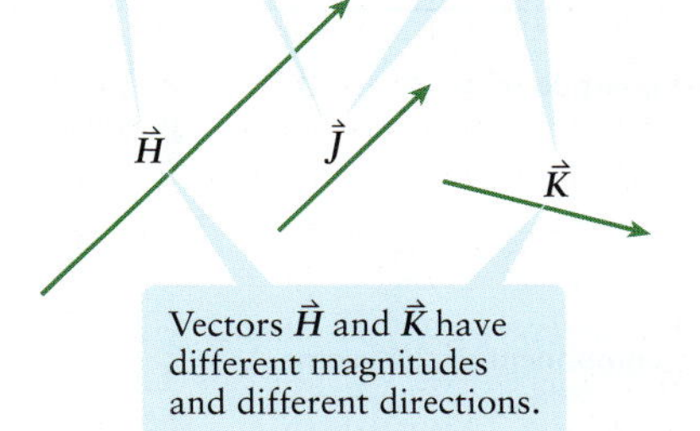  A vector has only two attributes: magnitude and direction. (Figure 3-2d)

Vector arithmetic: To find the sum of two vectors $\vec{A}$ and $\vec{B}$, place them tip to tail. Then $\vec{A} + \vec{B}$ points from the tail of the first vector to the tip of the second vector. To find the difference $\vec{A} - \vec{B}$ of two vectors, add $\vec{A}$ and $-\vec{B}$ (a vector with the same magnitude as $\vec{B}$ that points opposite to $\vec{B}$). Multiplying a vector by a scalar can change the vector's magnitude, direction, or both.

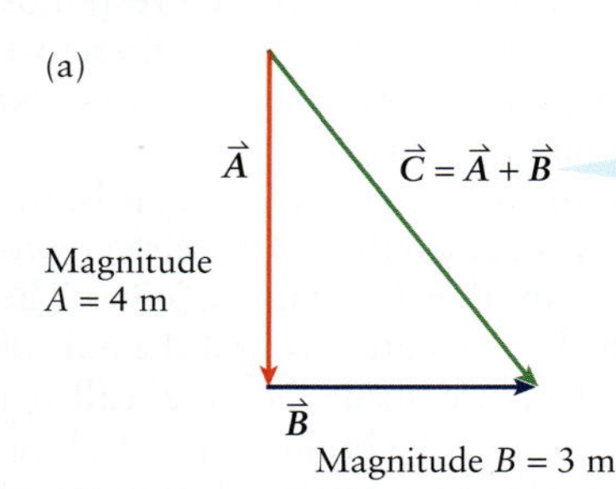

Magnitude of this vector sum:

$$C = \sqrt{A^2 + B^2} = \sqrt{(4\text{ m})^2 + (3\text{ m})^2} = \sqrt{16\text{ m}^2 + 9\text{ m}^2} = \sqrt{25\text{ m}^2} = 5\text{ m}$$

(Figure 3-5a)

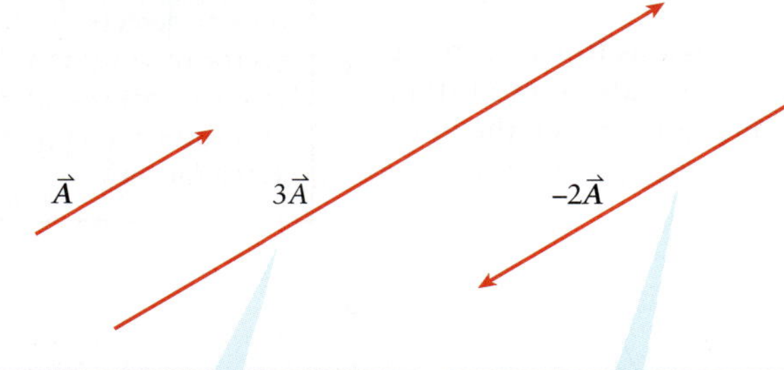

The number 3 is positive, so the vector $3\vec{A}$ points in the same direction as $\vec{A}$ but has 3 times the magnitude.

The number −2 is negative, so the vector $-2\vec{A}$ points in the direction opposite to $\vec{A}$ and has 2 times the magnitude.

Dividing a vector by 2 is the same as multiplying it by 1/2, so the vector $\vec{A}/2$ points in the same direction as $\vec{A}$ (since 1/2 is positive) and has 1/2 the magnitude.

(Figure 3-8)

Vector components: A vector can be expressed in terms of either its magnitude and direction or its x and y components. The components can be calculated from the magnitude and direction, and the magnitude and direction can be calculated from the components.

This vector has a positive x component A_x and a positive y component A_y.

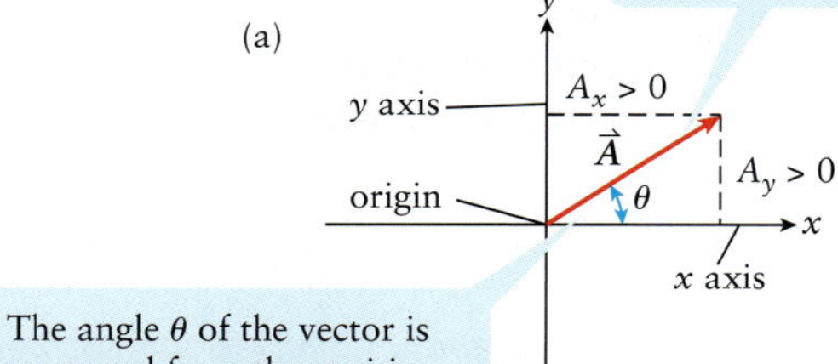

The angle θ of the vector is measured from the positive x axis in the direction toward the positive y axis. Here θ is between 0 and 90°.

(Figure 3-9a)

x component of the vector $\vec{A}$ — **y component** of the vector $\vec{A}$

$$\text{(a)}\quad A_x = A\cos\theta \qquad \text{(b)}\quad A_y = A\sin\theta \tag{3-1}$$

Magnitude of the vector $\vec{A}$ — **Angle** of the vector $\vec{A}$ measured from the positive x axis toward the positive y axis

Magnitude of the vector $\vec{A}$ — **Angle** of the vector $\vec{A}$ measured from the positive x axis toward the positive y axis

$$\text{(a)}\quad A = \sqrt{A_x^2 + A_y^2} \qquad \text{(b)}\quad \tan\theta = \frac{A_y}{A_x} \tag{3-2}$$

x component of the vector $\vec{A}$ — **y component** of the vector $\vec{A}$

Vector arithmetic with components: To find the components of a vector sum $\vec{A}+\vec{B}$, add the respective components. To find the components of a vector difference $\vec{A}-\vec{B}$, subtract the respective components. To find the components of a vector $\vec{A}$ multiplied by a scalar c, multiply each component of $\vec{A}$ by c.

$$\text{(a) } C_x = A_x + B_x \qquad \text{(b) } C_y = A_y + B_y \qquad (3\text{-}3)$$

x component of the vector $\vec{C}=\vec{A}+\vec{B}$; x component of the vector $\vec{A}$; x component of the vector $\vec{B}$

y component of the vector $\vec{C}=\vec{A}+\vec{B}$; y component of the vector $\vec{A}$; y component of the vector $\vec{B}$

$$\text{(a) } D_x = A_x - B_x \qquad \text{(b) } D_y = A_y - B_y \qquad (3\text{-}4)$$

x component of the vector $\vec{D}=\vec{A}-\vec{B}$; x component of the vector $\vec{A}$; x component of the vector $\vec{B}$

y component of the vector $\vec{D}=\vec{A}-\vec{B}$; y component of the vector $\vec{A}$; y component of the vector $\vec{B}$

$$\text{(a) } E_x = cA_x \qquad \text{(b) } E_y = cA_y \qquad (3\text{-}5)$$

x component of the vector $\vec{E}=c\vec{A}$; Scalar c; x component of the vector $\vec{A}$

y component of the vector $\vec{E}=c\vec{A}$; Scalar c; y component of the vector $\vec{A}$

The displacement and velocity vectors: The displacement $\Delta\vec{r}$ of an object during a time interval is a vector that points from the object's position at the earlier time t_1 to its position at the later time t_2. The velocity vector $\vec{v}$ at a given time points along the object's trajectory; its magnitude is the object's speed. The components of $\vec{v}$ are given by the same expressions we used in Chapter 2 for velocity in linear motion.

$$\Delta\vec{r} = \vec{r}_2 - \vec{r}_1 \qquad (3\text{-}6)$$

Displacement vector (change in position) of the object over the time interval from time t_1 to a later time t_2

Position vector of the object at later time t_2

Position vector of the object at earlier time t_1

To calculate the displacement vector, **subtract** the earlier value from the later value (just as for motion in a straight line).

$$\vec{v} = \frac{\Delta\vec{r}}{\Delta t} = \frac{\vec{r}_2 - \vec{r}_1}{t_2 - t_1} \qquad (3\text{-}7)$$

Velocity vector for the object over a very short time interval from time t_1 to a later time t_2

Displacement vector (change in position) of the object over the short time interval

Elapsed time for the time interval

For both the displacement and the elapsed time, **subtract** the earlier value from the later value.

$$\text{(a) } v_x = \frac{\Delta x}{\Delta t} = \frac{x_2 - x_1}{t_2 - t_1} \qquad \text{(b) } v_y = \frac{\Delta y}{\Delta t} = \frac{y_2 - y_1}{t_2 - t_1} \qquad (3\text{-}8)$$

x component of the velocity vector $\vec{v}$

y component of the velocity vector $\vec{v}$

x component of the **displacement** during a short time interval

y component of the **displacement** during a short time interval

Elapsed time for the time interval

The acceleration vector: An object has a nonzero acceleration vector $\vec{a}$ whenever it is speeding up, slowing down, *or* changing direction. The x and y components of $\vec{a}$ are given by the same expressions we used in Chapter 2 for acceleration in linear motion. Alternatively, we can express $\vec{a}$ in terms of its component along the direction of motion (which describes changes in speed) and its component perpendicular to the direction of motion (which describes changes in direction).

Acceleration vector for the object over a very short time interval from time t_1 to a later time t_2

Change in velocity of the object over the short time interval

$$\vec{a} = \frac{\Delta\vec{v}}{\Delta t} = \frac{\vec{v}_2 - \vec{v}_1}{t_2 - t_1} \quad (3\text{-}9)$$

For both the velocity change and the elapsed time, **subtract** the earlier value from the later value.

Elapsed time for the time interval

x component of the acceleration vector $\vec{a}$

Change in the x component of the velocity during a short time interval

$$\text{(a)}\ a_x = \frac{\Delta v_x}{\Delta t} = \frac{v_{x2} - v_{x1}}{t_2 - t_1}$$

$$\text{(b)}\ a_y = \frac{\Delta v_y}{\Delta t} = \frac{v_{y2} - v_{y1}}{t_2 - t_1} \quad (3\text{-}10)$$

Change in the y component of the velocity during a short time interval

Elapsed time for the time interval

y component of the acceleration vector $\vec{a}$

Projectile motion: A projectile is a moving object acted on by gravity alone. It has a constant downward acceleration of magnitude g. In the equations for the motion of a projectile, we always choose the x axis to be horizontal and the y axis to point vertically upward.

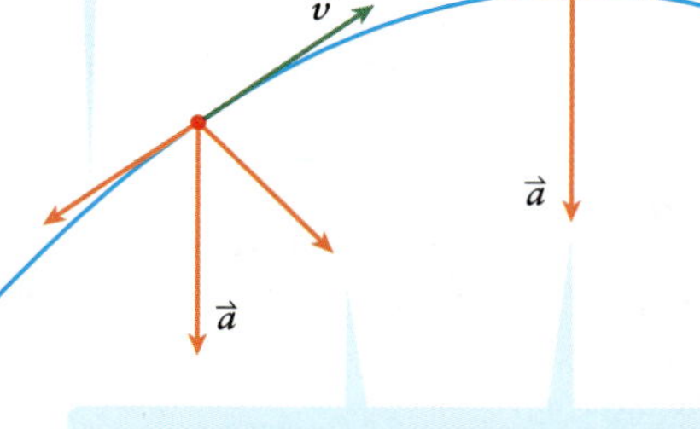

There is a component of acceleration opposite to the direction of motion: The projectile is slowing down.
There is a component of acceleration in the direction of motion: The projectile is speeding up.
There is no component of acceleration along the direction of motion: The speed of the projectile has its minimum value.
$\vec{v}$
$\vec{a}$
At all points there is a component of acceleration perpendicular to the trajectory: The direction of motion is changing.

(Figure 3-24)

• The acceleration $\vec{a}$ of a projectile is constant.
• Because the projectile's path is curved, $\vec{a}$ has varying components along and perpendicular to the direction of motion.

x component of an object's **velocity vector $\vec{v}$** at time t for **projectile motion**

x component of the object's velocity vector at $t = 0$

$$\text{(a)}\ v_x = v_{0x}$$

$$\text{(b)}\ v_y = v_{0y} - gt \quad (3\text{-}12)$$

Acceleration due to gravity (g is positive)

Time at which the object has velocity components v_x and v_y

y component of an object's **velocity vector $\vec{v}$** at time t for **projectile motion**

y component of the object's velocity vector at $t = 0$

x and y coordinates at time t of an object in **projectile motion**

x and y components at $t = 0$ of the object's velocity vector

Time at which the object has coordinates x and y

$$\text{(a)}\ x = x_0 + v_{0x}t$$

$$\text{(b)}\ y = y_0 + v_{0y}t - \frac{1}{2}gt^2 \quad (3\text{-}13)$$

x and y coordinates at $t = 0$ of the object

Acceleration due to gravity (g is positive)

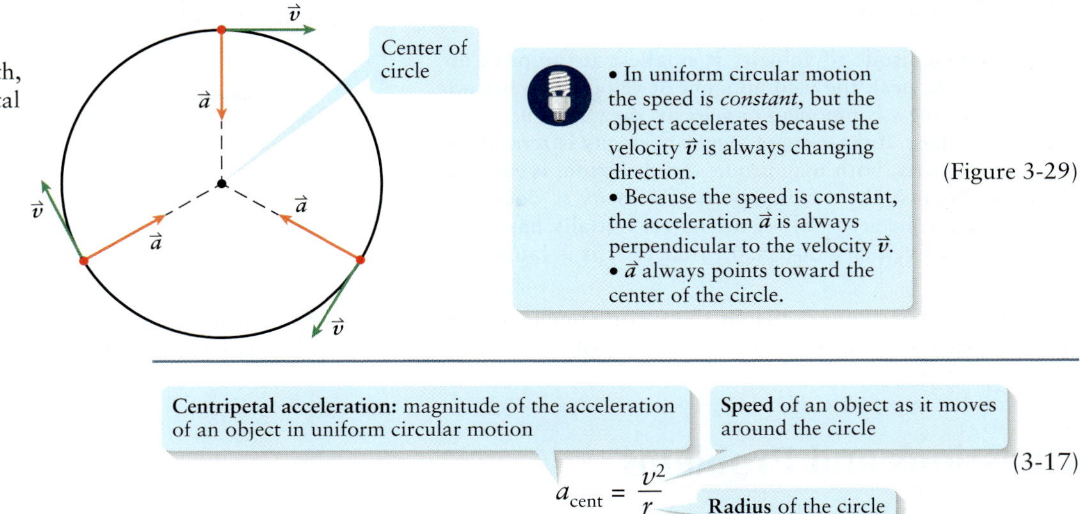

Motion in a circle: For an object moving at constant speed around a circular path, the acceleration is centripetal (directed toward the center of the circle).
Center of circle
• In uniform circular motion the speed is constant, but the object accelerates because the velocity $\vec{v}$ is always changing direction.
• Because the speed is constant, the acceleration $\vec{a}$ is always perpendicular to the velocity $\vec{v}$.
• $\vec{a}$ always points toward the center of the circle.

(Figure 3-29)

Centripetal acceleration: magnitude of the acceleration of an object in uniform circular motion

Speed of an object as it moves around the circle

$$a_{\text{cent}} = \frac{v^2}{r} \tag{3-17}$$

Radius of the circle

Answer to What do you think? Question

(c) The lava clump takes the *same* amount of time to climb to its maximum height as it does to descend from its maximum height back to its starting height. See Example 3-7 in Section 3-6.

Answers to Got the Concept? Questions

3-1 (c) The illustration shows the vectors $\vec{A}$ and $\vec{B}$ and the four combinations. One way to end up with a vector pointing southwest is to add a vector pointing west to a vector pointing south. Vector $\vec{B}$ points west. If we multiply vector $\vec{A}$ (which points north) by -1, the vector $-\vec{A}$ will point south. So $\vec{B} + (-\vec{A}) = \vec{B} - \vec{A}$ points southwest.

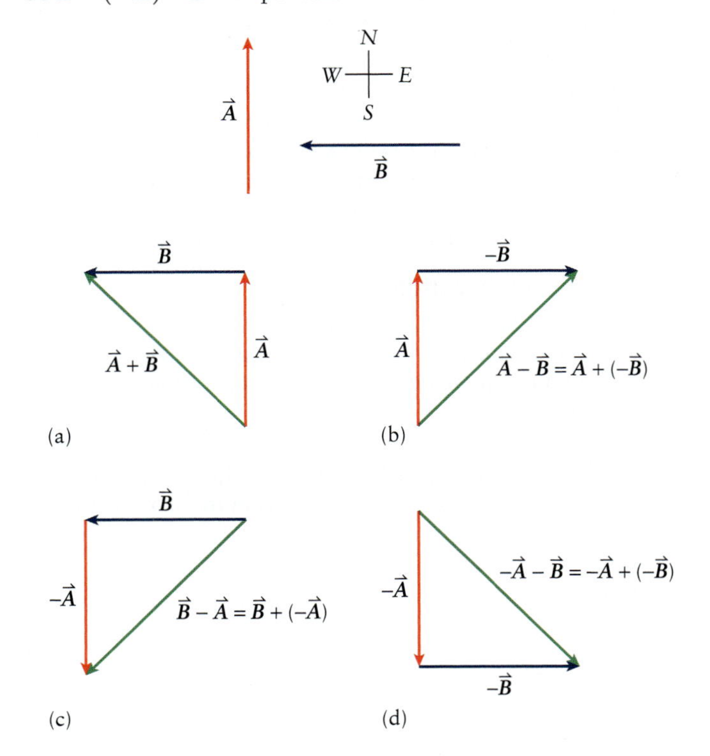

3-2 (a) You may have answered (d) if you assumed that the positive x direction points to the right and the positive y direction points upward. But the figure shows that in this case we've chosen the positive x direction to point upward and the positive y direction to point to the left. Hence the vector we want is $\vec{A}$, which points upward (in the positive x direction) and to the right (in the negative y direction). When thinking about vector components, always be mindful of the directions of the axes!

3-3 $D_x = 2$, $D_y = -5$. Our goal is to find a vector $\vec{D}$ with the property that $\vec{A} + \vec{D} = \vec{B}$, so $\vec{D} = \vec{B} - \vec{A}$. From Equations 3-4, the components of $\vec{D}$ are

$$D_x = B_x - A_x = 7 - 5 = 2$$
$$D_y = B_y - A_y = (-2) - 3 = -5$$

In other words, the x component of $\vec{D} = \vec{B} - \vec{A}$ is the x component of $\vec{B}$ minus the x component of $\vec{A}$, and similarly for the y component of $\vec{D} = \vec{B} - \vec{A}$. Our result tells us that the vector $\vec{D}$ has components $D_x = 2$, $D_y = -5$. If you wanted to, you could express $\vec{D}$ in terms of its magnitude and direction. But it's perfectly fine to write $\vec{D}$ in terms of its components as we have done here.

3-4 Point 1, (a); point 2, (c); point 3, (b); point 4, (a). At points 1 and 4 the object is speeding up because its acceleration vector $\vec{a}$ points partially in the direction of motion (that is, partially in the direction of the velocity vector $\vec{v}$). At point 3 the acceleration vector points partially opposite to the direction of $\vec{v}$, so at this point the object is slowing down. At point 2 the object is neither speeding up nor slowing down but is maintaining a constant speed. The acceleration vector is perpendicular to the velocity vector. Note that at all four points on this trajectory, the acceleration vector $\vec{a}$ points in whole or in part toward the inside of the trajectory's curve.

3-5 The magnitude of velocity is smallest at the peak of motion. At the peak the y component of velocity is zero, so regardless of the x component of velocity, the magnitude of $\vec{v}$ must be smallest. If the x component of velocity is zero, then the total velocity, both magnitude and direction, is zero at the peak. Because the x component of velocity v_x does not change, v_x can be zero only if the object initially has no x component of velocity ($v_{0x} = 0$); that is, if it is launched straight up.

3-6 (b) At its maximum height the y component of the frog's velocity is zero, and the x component of its velocity equals the x component of its initial velocity: $v_x = v_{0x} = v_0 \cos \theta_0$. Since $\cos 45° = 1/\sqrt{2} = 0.707$, at the frog's maximum height $v_x = 0.707v_0$ and its speed is $0.707v_0$—that is, between v_0 and $v_0/2$.

3-7 No. The *magnitude* of the acceleration $a_{cent} = v^2/r$ is constant, since the speed v is constant, but the *direction* of the acceleration vector is continually changing so that $\vec{a}$ points toward the center of the circle at all times (see Figure 3-29). Therefore, the acceleration is *not* constant.

3-8 (d) The centripetal acceleration is given by Equation 3-17, $a_{cent} = v^2/r$. If the speed v is doubled, the numerator v^2 in the expression for a_{cent} increases by a factor of $2^2 = 4$. The radius r must also increase by a factor of 4 in order to maintain the same value of a_{cent}.

Questions and Problems

In a few problems you are given more data than you actually need; in a few other problems, you are required to supply data from your general knowledge, outside sources, or informed estimate.

Interpret as significant all digits in numerical values that have trailing zeros and no decimal points. For all problems use $g = 9.80$ m/s^2 for the free-fall acceleration due to gravity.

•	Basic, single-concept problem
••	Intermediate-level problem; may require synthesis of concepts and multiple steps
•••	Challenging problem
SSM	*Solution is in Student Solutions Manual*
Example	*See worked example for a similar problem*

Conceptual Questions

1. • What is the difference between a scalar and a vector? Give an example of a scalar and an example of a vector.

2. • Explain what is meant by the magnitude of a vector.

3. • (a) Can the sum of two vectors that have different magnitudes ever be equal to zero? If so, give an example. If not, explain why the sum of two vectors cannot be equal to zero. (b) Can the sum of three vectors that have different magnitudes ever be equal to zero? SSM

4. • Describe a situation in which the average velocity and the instantaneous velocity vectors are identical. Describe a situation in which these two velocity vectors are different.

5. • **Astronomy** If you were playing tennis on the Moon, what adjustments would you need to make in order for your shots to stay within the boundaries of the court? Would the trajectories of the balls look different on the Moon compared to on Earth?

6. • Consider the effects of air resistance on a projectile. Describe qualitatively how the projectile's velocities and accelerations in the vertical and horizontal directions differ when the effects of air resistance are ignored and when the effects are considered. SSM

7. • During the motion of a projectile, which of the following quantities are constant during the flight: x, y, v_x, v_y, a_x, a_y? (Neglect any effects due to air resistance.)

8. • For a given, fixed launch speed, at what angle should you launch a projectile to achieve (a) the longest range, (b) the longest time of flight, and (c) the greatest height? (Neglect any effects due to air resistance.) SSM

9. • A rock is thrown from a bridge at an angle 20° below horizontal. At the instant of impact, is the rock's speed greater than, less than, or equal to the speed with which it was thrown? Explain your answer. (Neglect any effects due to air resistance.)

10. • **Sports** A soccer player kicks a ball at an angle 60° from the ground. The soccer ball hits the ground some distance away. Is there any point at which the velocity and acceleration vectors are perpendicular to each other? Explain your answer. (Neglect any effects due to air resistance.)

11. • **Sports** Suppose you are the coach of a champion long jumper. Would you suggest that she take off at an angle less than 45°? Why or why not?

12. • (a) Explain the difference between an object undergoing uniform circular motion and an object experiencing projectile motion. (b) In what ways are these kinds of motion similar?

13. • An ape swings through the jungle by hanging from a vine. At the lowest point of its motion, is the ape accelerating? If so, what is the direction of its acceleration? SSM

14. • A cyclist rides around a flat, circular track at constant speed. Is his acceleration vector zero? Explain your answer.

15. • You are driving your car in a circular path on flat ground with a constant speed. At the instant you are driving north and turning right, are you accelerating? If so, what is the direction of your acceleration at that moment? If not, why not?

Multiple-Choice Questions

16. • Which of the following is not a vector?
A. average velocity
B. instantaneous velocity
C. distance
D. displacement
E. acceleration

17. • Vector $\vec{A}$ has an x component and a y component that are equal in magnitude. Which of the following is the angle that vector $\vec{A}$ makes with respect to the x axis in the same x–y coordinate system?
A. 0°
B. 45°
C. 60°
D. 90°
E. 120° SSM

18. • The vector in Figure 3-35 has a length of 4.00 units and makes a 30.0° angle with respect to the y axis as shown. What are the x and y components of the vector?

A. 3.46, 2.00
B. −2.00, 3.46
C. −3.46, 2.00
D. 2.00, −3.46
E. −3.46, −2.00

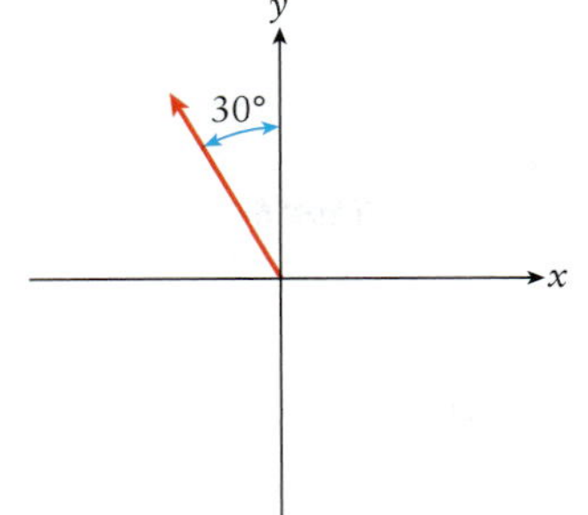

Figure 3-35 Problem 18

19. • The acceleration of a particle in projectile motion

A. points along the parabolic path of the particle.
B. is directed horizontally.
C. vanishes at the particle's highest point.
D. is vertically downward.
E. is zero.

20. • Adam drops a ball from rest from the top floor of a building at the same time that Bob throws a ball horizontally from the same location. Which ball hits the level ground below first? (Neglect any effects due to air resistance.)

A. Adam's ball
B. Bob's ball
C. They both hit the ground at the same time.
D. It depends on how fast Bob throws the ball.
E. It depends on how fast the ball falls when Adam drops it.

21. • **Sports** Two golf balls are hit from the same point on a flat field. Both are hit at an angle of 30° above the horizontal. Ball 2 has twice the initial speed of ball 1. If ball 1 lands a distance d_1 from the initial point, at what distance d_2 does ball 2 land from the initial point? (Neglect any effects due to air resistance.)

A. $d_2 = 0.5d_1$
B. $d_2 = d_1$
C. $d_2 = 2d_1$
D. $d_2 = 4d_1$
E. $d_2 = 8d_1$ SSM

22. • A zookeeper is trying to shoot a monkey sitting at the top of a tree with a tranquilizer gun. If the monkey drops from the tree at the same instant that the zookeeper fires, where should the zookeeper aim if he wants to hit the monkey? (Neglect any effects due to air resistance.)

A. Aim straight at the monkey.
B. Aim lower than the monkey.
C. Aim higher than the monkey.
D. Aim to the right of the monkey.
E. It's impossible to determine.

23. • The acceleration vector of a particle in uniform circular motion

A. points along the circular path of the particle and in the direction of motion.
B. points along the circular path of the particle and opposite the direction of motion.
C. is zero.
D. points toward the center of the circle.
E. points outward from the center of the circle.

24. • If the speed of an object in uniform circular motion remains constant while the radial distance is doubled, the magnitude of the radial acceleration decreases by what factor?

A. 2
B. 3
C. 4
D. 6
E. 1

25. • You toss a ball into the air at an initial angle 40° from the horizontal. At what point in the ball's trajectory does the ball have the smallest speed? (Neglect any effects due to air resistance.)

A. just after it is tossed
B. at the highest point in its flight
C. just before it hits the ground
D. halfway between the ground and the highest point on the rise portion of the trajectory
E. halfway between the ground and the highest point on the fall portion of the trajectory

Estimation/Numerical Analysis

26. • If $\vec{r}$ has a magnitude of 24 and points in a direction 36° south of west, find the vector components of $\vec{r}$. Use a protractor and some graph paper to verify your answer by drawing $\vec{r}$ and measuring the length of the lines representing its components.

27. • A vector $\vec{r}$ has a magnitude of 18 units and makes a 30° angle with respect to the x axis. Find the vector components of $\vec{r}$ using a protractor and some graph paper to verify your answer by drawing $\vec{r}$ and measuring the length of the lines representing its components.

28. • **Sports** In the 1970 National Basketball Association championship, Jerry West made a 60-ft shot from beyond half court to lead the Los Angeles Lakers to an improbable tie at the buzzer with the New York Knicks. Neglecting air resistance, estimate the initial speed of the ball. (The Knicks won the game in overtime.)

29. • **Sports** In Detroit in 1971, Reggie Jackson hit one of the most memorable home runs in the history of the Major League Baseball All-Star Game. The approximate trajectory is plotted in Figure 3-36. (The asymmetry is due to air resistance.) Using the information in the graph, estimate the initial speed of the ball as it left Reggie's bat. SSM

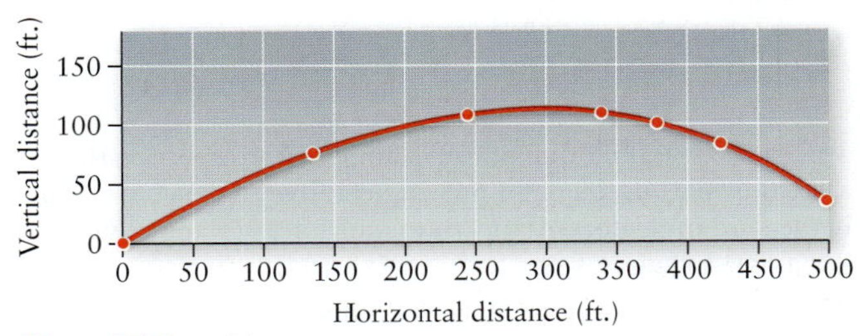

Figure 3-36 Problem 29

30. •• Mary and Kelly are running side by side on a 400-m track. As they go around one of the turns, Mary is running on the inside lane, and Kelly is running in the outer lane. They both enter and leave the turn at the same time. (a) Estimate the difference in their centripetal accelerations. (b) Calculate the ratio of their centripetal accelerations.

31. • Use a spreadsheet program or a graphing calculator to make (a) a graph of v_x versus time, (b) a graph of v_y versus

time, (c) a graph of a_x versus time, and (d) a graph of a_y versus time for an object that undergoes projectile motion. Identify the points where the object reaches its highest point and where it hits the ground at the end of its flight.

Problems

3-2 A vector quantity has both a magnitude and a direction

3-3 Vectors can be described in terms of components

32. • Vector $\vec{A}$ has components $A_x = 6$ and $A_y = 9$. Vector $\vec{B}$ has components $B_x = 7$, $B_y = -3$, and vector $\vec{C}$ has components $C_x = 0$, $C_y = -6$. Determine the components of the following vectors: (a) $\vec{A} + \vec{B}$, (b) $\vec{A} - 2\vec{C}$, (c) $\vec{A} + \vec{B} - \vec{C}$, and (d) $\vec{A} + \frac{1}{2}\vec{B} - 3\vec{C}$. Example 3-3

33. • Calculate the magnitude and direction of the vector $\vec{r}$ using Figure 3-37. Example 3-1

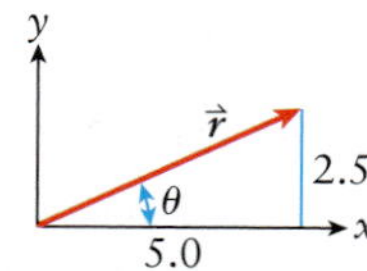

Figure 3-37 Problem 33

34. • What are the components A_x and A_y of vector $\vec{A}$ in the three coordinate systems shown in Figure 3-38? SSM Example 3-1

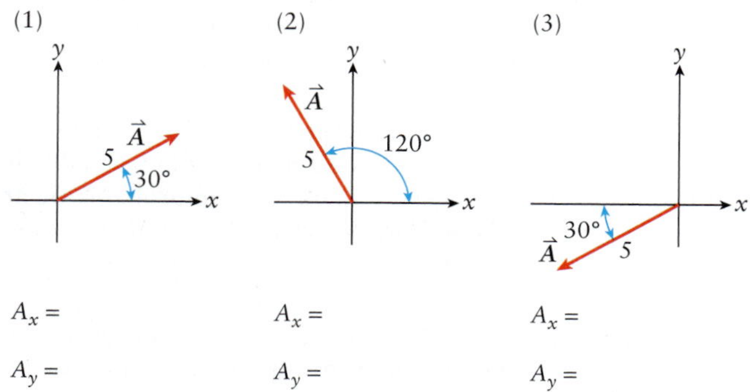

Figure 3-38 Problem 34

35. • Each of the following vectors is given in terms of its x and y components. Find the magnitude of each vector and the angle it makes with respect to the $+x$ axis. Example 3-1

A. $A_x = 3, A_y = -2$
B. $A_x = -2, A_y = 2$
C. $A_x = 0, A_y = -2$

36. •• $\vec{A}$ is 66.0 m long at a 28° angle with respect to the $+x$ axis. $\vec{B}$ is 40.0 m long at a 56° angle above the $-x$ axis. What is the sum of vectors $\vec{A}$ and $\vec{B}$ (magnitude and angle with the $+x$ axis)? Example 3-3

37. •• Given the vector $\vec{A}$ with components $A_x = 2.00$ and $A_y = 6.00$, and the vector $\vec{B}$ with components $B_x = 3.00$ and $B_y = -2.00$, calculate the magnitude and angle with respect to the $+x$ axis of the vector sum $\vec{C} = \vec{A} + \vec{B}$. Example 3-3

38. •• Given the vector $\vec{A}$ with components $A_x = 2.00$, $A_y = 6.00$, the vector $\vec{B}$ with components $B_x = 2.00$, $B_y = -2.00$, and the vector $\vec{D} = \vec{A} - \vec{B}$, calculate the magnitude and angle with the $+x$ axis of the vector $\vec{D}$. SSM Example 3-3

39. •• Two velocity vectors are given as follows: $\vec{A} = 30$ m/s, 45° north of east and $\vec{B} = 40$ m/s, due north. Calculate each of the resultant velocity vectors: (a) $\vec{A} + \vec{B}$, (b) $\vec{A} - \vec{B}$, (c) $2\vec{A} + \vec{B}$. Example 3-3

40. •• Consider the set of vectors in Figure 3-39. Nathan says the magnitude of the resultant vector is 7, and the resultant vector points in a direction 37° in the northeasterly direction. What, if anything, is wrong with his statement? If something is wrong, explain the error(s) and how to correct it (them). Example 3-3

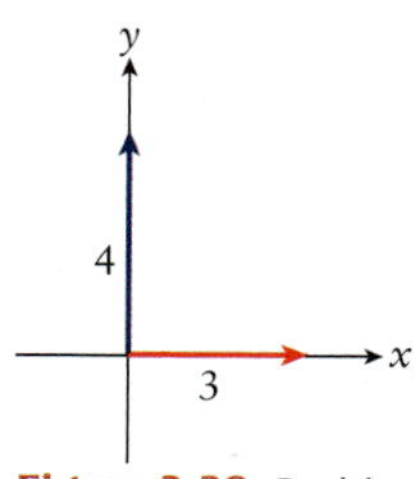

Figure 3-39 Problem 40

3-4 For motion in a plane, velocity and acceleration are vector quantities

41. •• What are the magnitude and direction of the change in velocity if the initial velocity is 30 m/s south and the final velocity is 40 m/s west? Example 3-3

42. •• The two vectors shown in Figure 3-40 represent the initial and final velocities of an object during a trip that took 5 s. Calculate the average acceleration during this trip. Is it possible to determine whether the acceleration was uniform from the information given in the problem? SSM Example 3-3

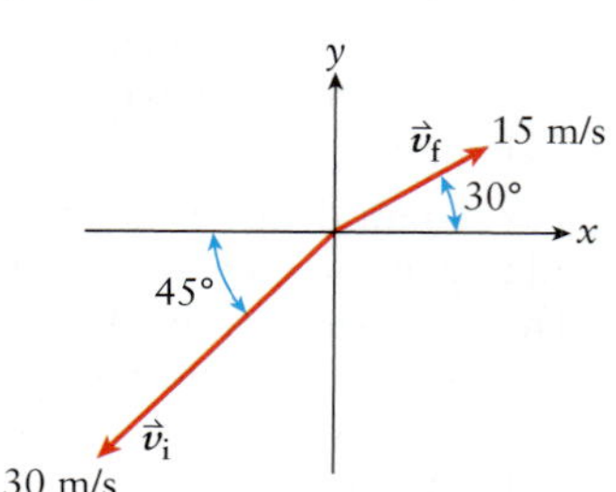

Figure 3-40 Problem 42

43. •• An object travels with a constant acceleration for 10 s. The vectors in Figure 3-41 represent the final and initial velocities. Carefully graph the x component of the velocity versus time, the y component of the velocity versus time, and the y component of the acceleration versus time. Example 3-3

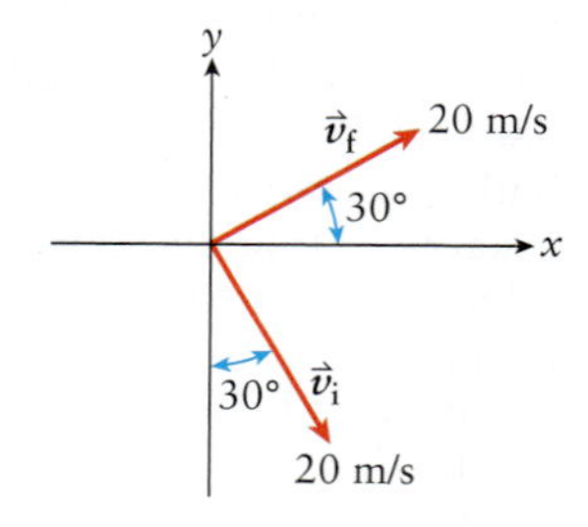

Figure 3-41 Problem 43

44. •• An object experiences a constant acceleration of 2.00 m/s^2 along the $-x$ axis for 2.70 s, attaining a velocity of 16.0 m/s in a direction 45° from the $+x$ axis. Calculate the initial velocity vector of the object. Example 3-3

45. •• Cody starts at a point 6.00 km to the east and 4.00 km to the south of a location that represents the origin of a coordinate system for a map. He ends up at a point 10.0 km to the west and 6.00 km to the north of the map origin. (a) What was his average velocity if the trip took him 4.00 h to complete? (b) Cody walks to his destination at a constant rate. His friend Marcus covers the distance with a combination of jogging, walking, running, and resting so that the total trip time is also 4.00 h. How do their average velocities compare? Example 3-3

3-5 A projectile moves in a plane and has a constant acceleration

3-6 You can solve projectile motion problems using techniques learned for straight-line motion

46. • An object is undergoing projectile motion as shown from the side in Figure 3-42. Assume the object starts its motion at ground level. For the five positions shown, draw to scale vectors representing the magnitudes of (a) the x components of the velocity, (b) the y components of the velocity, and

(c) the accelerations. (Neglect any effects due to air resistance.) Example 3-5

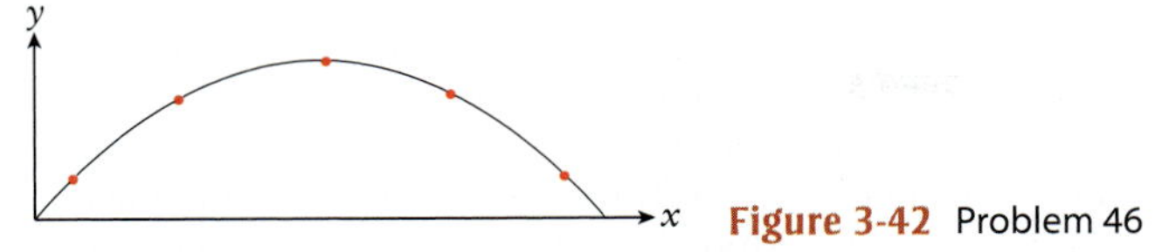

Figure 3-42 Problem 46

47. •• An object undergoing projectile motion travels 1.00×10^2 m in the horizontal direction before returning to its initial height. If the object is thrown initially at a 30.0° angle from the horizontal, determine the x component and the y component of the initial velocity. (Neglect any effects due to air resistance.) Example 3-7

48. •• Five balls are thrown off a cliff at the angles shown in Figure 3-43. Each has the same initial speed. Rank (a) the time required for each to hit the ground, (b) the vertical displacement of each, and (c) the vertical component of final velocity. (Neglect any effects due to air resistance.) SSM Example 3-5

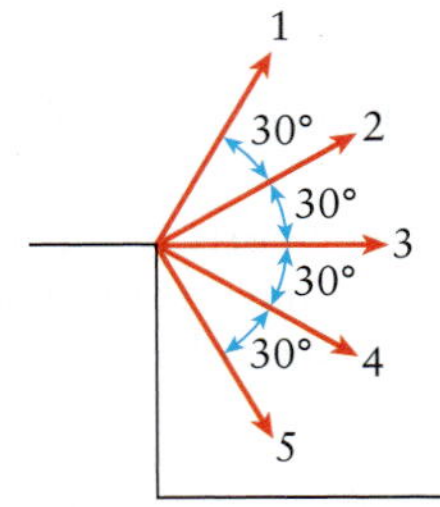

Figure 3-43 Problem 48

49. • **Biology** A Chinook salmon can jump out of water with a speed of 6.30 m/s. How far horizontally can a Chinook salmon travel through the air if it leaves the water with an initial angle of 40°? (Neglect any effects due to air resistance.) Example 3-7

50. • **Biology** A tiger leaps horizontally out of a tree that is 4.00 m high. If he lands 5.00 m from the base of the tree, calculate his initial speed. (Neglect any effects due to air resistance.) SSM Example 3-4

51. • A football is punted at 25.0 m/s at an angle of 30.0° above the horizon. What is the velocity vector of the ball when it is 5.00 m above ground level? Assume it starts 1.00 m above ground level. (Neglect any effects due to air resistance.) Example 3-5

52. •• A dart is thrown at a dartboard 2.37 m away. When the dart is released at the same height as the center of the dartboard, it hits the center in 0.447 s. At what angle relative to the floor was the dart thrown? (Neglect any effects due to air resistance.) Example 3-5

3-7 An object moving in a circle is accelerating even if its speed is constant

53. • A ball attached to a string is twirled in a circle of radius 1.25 m. If the constant speed of the ball is 2.25 m/s, what is the time required to travel once around the circle? Example 3-9

54. • A ball is twirled on a 0.870-m-long string with a constant speed of 3.36 m/s. Calculate the acceleration of the ball. Be sure to specify the direction of the acceleration. SSM Example 3-8

55. • Riders on a Ferris wheel of diameter 16.0 m move in a circle with a centripetal acceleration of 2.00 m/s^2. What is the speed of the Ferris wheel? Example 3-9

56. •• In 1892 George W. G. Ferris designed a carnival ride in the shape of a large wheel. This Ferris wheel had a diameter of 76 m and rotated one revolution every 20 min. What was the magnitude of the acceleration that riders experienced? SSM Example 3-8

57. • A car races at a constant speed of 330 km/h around a flat, circular track 1.00 km in diameter. What is the car's centripetal acceleration in m/s^2? Example 3-8

58. • **Astronomy** We know that the Moon revolves around Earth during a period of 27.3 days. The average distance from the center of Earth to the center of the Moon is 3.84×10^8 m. What is the acceleration of the Moon due to its motion around Earth? Example 3-9

59. •• Calculate the accelerations of (a) Earth as it orbits the Sun with an average orbital radius of 1.5×10^{11} m and (b) a car traveling along a circular path that has a radius of 50.0 m at a speed of 20.0 m/s. Example 3-9

60. • **Biology** In a vertical dive, a peregrine falcon can accelerate at 0.6 times the free-fall acceleration (that is, at $0.6g$) in reaching a speed of about 100 m/s. If a falcon pulls out of a dive into a circular arc at this speed and can sustain a centripetal acceleration of $0.6g$, what is the minimum radius of the turn? SSM Example 3-9

61. •• Commercial ultracentrifuges can rotate at rates of 100,000 rpm (revolutions per minute). As a consequence, they can create accelerations on the order of $800{,}000g$. (A "g" represents an acceleration of 9.80 m/s^2.) Calculate the distance from the rotation axis of the sample chamber in such a device. What is the speed of an object traveling under the given conditions? Example 3-9

General Problems

62. ••• **Sports** Aaron Rodgers stands on the 20-yard line, poised to throw long. He throws the ball at initial velocity $v_0 = 15.0$ m/s and releases it at an angle $\theta = 45.0°$. Having faked an end around, Jordy Nelson comes racing past Aaron at a constant velocity $V_J = 8.00$ m/s, heading straight down the field. Assuming that Jordy catches the ball at the same height above the ground that Aaron throws it, how long must Aaron wait to throw, after Jordy goes past, so that the ball falls directly into Jordy's hands? SSM Example 3-5

63. ••• You throw a ball from the balcony onto the court in the basketball arena. You release the ball at a height of 7.00 m above the court, with an initial velocity equal to 9.00 m/s at 33.0° above the horizontal. A friend of yours, standing on the court 11.0 m from the point directly beneath you, waits for a period of time after you release the ball and then begins to move directly away from you at an acceleration of 1.80 m/s^2. (She can only do this for a short period of time!) If you throw the ball in a line with her, how long after you release the ball should she wait to start running directly away from you so that she'll catch the ball exactly 1.00 m above the floor of the court? Example 3-5

64. •• Marcus and Cody want to hike to a destination 12.0 km north of their starting point. Before heading directly to the destination, Marcus walks 10.0 km in a direction that is 30.0° north of east and Cody walks 15.0 km in a direction that is 45.0° north of west. How much farther must each hike on the second part of the trip? Example 3-3

65. • Nathan walks due east a certain distance and then walks due south twice that distance. He finds himself 15.0 km from his starting position. How far east and how far south does Nathan walk? Example 3-3

66. •• A water balloon is thrown horizontally at a speed of 2.00 m/s from the roof of a building that is 6.00 m above the

ground. At the same instant the balloon is released, a second balloon is thrown straight down at 2.00 m/s from the same height. Determine which balloon hits the ground first and how much sooner it hits the ground than the other balloon. (Neglect any effects due to air resistance.) SSM Example 3-5

67. •• You throw a rock from the upper edge of a 75.0-m vertical dam with a speed of 25.0 m/s at 65.0° above the horizon. How long after throwing the rock will it (a) return to its initial height, and (b) hit the water flowing out at the base of the dam? The speed of sound in the air is 344 m/s. (Neglect any effects due to air resistance.) Example 3-4

68. • An airplane releases a ball as it flies parallel to the ground at a height of 255 m (Figure 3-44). If the ball lands on the ground at a horizontal displacement of exactly 255 m from the release point, calculate the airspeed of the plane. (Neglect any effects due to air resistance.) Example 3-4

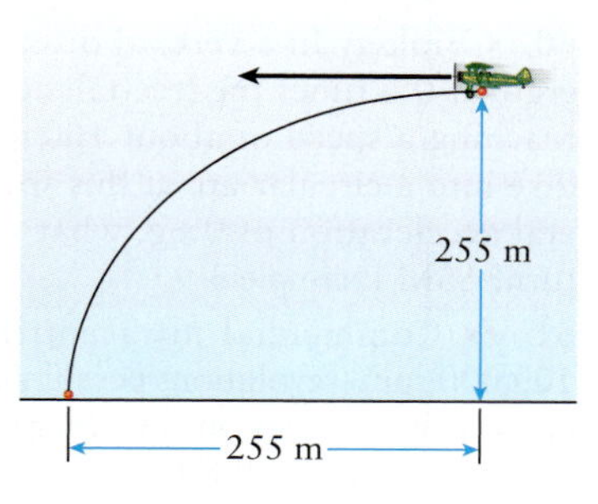

Figure 3-44 Problem 68

69. •• An airplane flying upward at 35.3 m/s and an angle of 30.0° relative to the horizontal releases a ball when it is 255 m above the ground. Calculate (a) the time it takes the ball to hit the ground, (b) the maximum height of the ball, and (c) the horizontal distance the ball travels from the release point to the ground. (Neglect any effects due to air resistance.) Example 3-4

70. •• **Sports** In 1993, Javier Sotomayor set a world record of 2.45 m in the men's outdoor high jump. He is 193 cm (6 ft 4 in) tall. By treating his body as a point located at half his height, and given that he left the ground a horizontal distance from the bar of 1.5 m at a takeoff angle of 65.0°, determine Javier Sotomayor's takeoff speed. (Neglect any effects due to air resistance.) Example 3-6

71. •• **Sports** A boy runs straight off the end of a diving platform at a speed of 5.00 m/s. The platform is 10.0 m above the surface of the water. (a) Calculate the boy's speed when he hits the water. (b) How much time is required for the boy to reach the water? (c) How far horizontally will the boy travel before he hits the water? (Neglect any effects due to air resistance.) Example 3-5

72. • **Sports** Gabriele Reinsch threw a discus 76.80 m on July 9, 1988, to set the women's world record. Assume that she launched the discus with an elevation angle of 45.0° and that her hand was 2.00 m above the ground at the instant of launch. What was the initial speed of the discus required to achieve that range? (Neglect any effects due to air resistance.) SSM Example 3-5

73. •• **Astronomy** The froghopper, a tiny insect, is a remarkable jumper. Suppose you raised a colony of the little critters on the Moon, where the acceleration due to gravity is only 1.62 m/s^2. If on Earth a froghopper's maximum jump height is h and maximum horizontal range is R, what would its maximum height and range be on the Moon in terms of h and R? Assume a froghopper's takeoff speed is the same on the Moon and on Earth. Example 3-7

74. •• **Sports** In 1998, Jason Elam kicked a record field goal. The football started on the ground a horizontal distance of 63.0 yards from the goal posts and just barely cleared the 10-ft-high bar. If the initial trajectory of the football was 40.0° above the horizontal, (a) what was its initial speed and (b) how long after the ball was struck did it pass through the goal posts? (Neglect any effects due to air resistance.) Example 3-4

75. ••• **Sports** In the hope that the Moon and Mars will one day become tourist attractions, a golf course is built on each. An average golfer on Earth can drive a ball from the tee about 63% of the distance to the hole. If this is to be true on the Moon and on Mars, by what factor should the dimensions of the golf courses on the Moon and Mars be changed relative to a course on Earth? (Neglect any effects due to air resistance.) Example 3-7

76. •• **Biology** Anne is working on a research project that involves the use of a centrifuge. Her samples must first experience an acceleration of 100g, but then the acceleration must increase by a factor of 8. By how much will the rotation speed have to increase? Express your answer as a multiple of the initial rotation rate. SSM Example 3-9

77. • **Medical** In a laboratory test of tolerance for high centripetal acceleration, pilots were swung in a circle 13.4 m in diameter. It was found that they blacked out when they were spun at 30.6 rpm (rev/min). (a) At what acceleration (in SI units and in multiples of g) did the pilots black out? (b) If you want to decrease the acceleration by 25.0% without changing the diameter of the circle, by what percent must you change the time for the pilot to make one spin? Example 3-8

78. • **Medical** Modern pilots can survive centripetal accelerations up to 9g (88 m/s^2). Can a fighter pilot flying at a constant speed of 500 m/s and in a circle that has a diameter of 8800 m survive to tell about his experience? Example 3-8

79. • **Sports** A girl's fast-pitch softball player does a windmill pitch, moving her hand through a circular arc with her arm straight. She releases the ball at a speed of 24.6 m/s. Just before the ball leaves her hand, the ball's centripetal acceleration is 1960 m/s^2. What is the length of her arm from the pivot point at her shoulder? Example 3-9

4 Forces and Motion I: Newton's Laws

Dmitri Ma/Shutterstock

What do you think?

The tug is pulling the airliner with just enough force to maintain a constant velocity. If the tug instead pulls with double this force, what will happen to the airliner? (a) It will move with a faster constant speed; (b) it will move for a while with a constant speed and then its speed will increase; (c) its speed will increase continuously.

In this chapter, your goals are to:

- (4-1) Describe the importance of forces in determining how an object moves.
- (4-2) Use Newton's second law to relate the net force on an object to the object's acceleration.
- (4-3) Recognize the distinctions among mass, weight, and inertia.
- (4-4) Draw and use free-body diagrams in problems that involve forces.
- (4-5) Describe how Newton's third law relates forces that act on different objects.
- (4-6) Apply the sequence of steps used in solving all problems involving forces, including those that also involve kinematics.

To master this chapter, you should review:

- (2-2, 2-3) The ideas of displacement, velocity, and acceleration for motion in a straight line.
- (2-4, 2-5, and 2-6) The concepts and equations of straight-line motion with constant acceleration, including free fall.
- (3-2, 3-3) How to add vectors and how to do vector calculations using components.
- (3-4) The ideas of displacement, velocity, and acceleration in two dimensions.

4-1 How objects move is determined by the forces that act on them

In the language of physics, the tug in the above photo exerts a *force* on the airliner. Force is a vector: It has both magnitude and direction. For example, the force the tug exerts on the airliner points in the direction of the airliner's motion, while the ground exerts a friction force on the airliner that points backward (opposite to the airliner's motion). The magnitude of each force is a measure of how strong it pushes or pulls on the airliner.

What determines how the airliner moves is not any one single force but rather the combined effect of *all* the forces that act on it. If the forward force from the tug has the same magnitude as the backward friction force, these two oppositely directed force vectors cancel and add to zero. In this case the airliner moves with a constant velocity. But if the tug pulls harder than the force of friction, the forces do not cancel, and there is a *net* force in the forward direction. Then the airliner accelerates.

Experiment shows that no matter how hard the tug pulls on the airliner, the airliner pulls back on the tug with an equally strong force. (This force on the tug is why

TAKE-HOME MESSAGE FOR Section 4-1

✔ What determines how an object moves is the combined effect of all of the forces (pushes or pulls) that act on the object.

✔ If one object exerts a force on a second object, the second object must also exert a force on the first object.

the tug's engine has to work harder with the airliner attached.) This turns out to be true for forces of all kinds: If one object exerts a force on a second object, the second object necessarily exerts a force on the first one.

These observations are at the heart of *Newton's laws of motion*, a time-tested set of physical principles that have a tremendous range of applicability. We'll use Newton's laws throughout the remainder of this book. We'll devote this chapter and the next to understanding these laws and some of their most important applications.

4-2 If a net external force acts on an object, the object accelerates

If you want to start a soccer ball rolling along the ground, you have to give it a push or kick. If you're a hockey goalie and want to deflect a puck away from the goal that you're defending, you have to hit the puck with your hockey stick. And if your dog runs off in pursuit of the neighbor's cat, you have to pull on the dog's leash to slow it down and bring it to a halt. In each of these cases you're changing the velocity of an object, either making it speed up (for the soccer ball), changing the direction of its motion (for the hockey puck), or making it slow down (for your dog). And in each case to cause the object's velocity to change, you have to exert a **force**—that is, a push or a pull—on that object.

Forces are *vectors* because they have both magnitude and direction. (You might pull a glass of your favorite beverage gently toward you or push a plate of cafeteria food forcefully away from you.) When more than one force acts on an object, what determines how the object moves is the vector *sum* of all of the forces acting on that object, also called the **net external force** on the object (or, for short, the **net force**). If the individual forces acting on an object are $\vec{F}_1$, $\vec{F}_2$, $\vec{F}_3$, and so on, we can write the net force on that object as

Net external force on an object (4-1)

The **net external force** acting on an object... ...equals the **vector sum** of all of the individual forces that act on the object.

$$\sum \vec{F}_{\text{ext}} = \vec{F}_1 + \vec{F}_2 + \vec{F}_3 + \ldots$$

The sum includes only external forces (forces exerted on the object by other objects).

For example, three forces act on a baseball player as he slides to get safely to base (**Figure 4-1**). One of these forces is the downward pull of Earth's gravity, called the **gravitational force** (denoted by a lowercase $\vec{w}$). The other two forces are exerted by the surface on which the player slides. The **friction force** on the player (which we denote by a lowercase $\vec{f}$) acts parallel to the surface and opposite to the player's motion. The

(a)

(b)

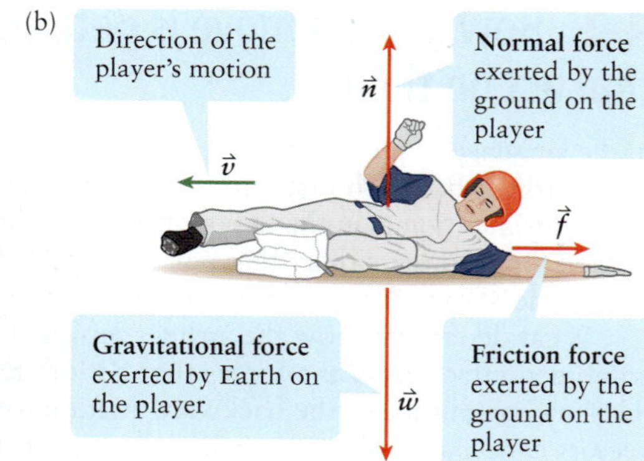

Figure 4-1 Forces on a baseball player (a) A baseball player sliding to get safely on base. (b) The red vectors indicate the directions of each external force that acts on the player.

The only forces that affect the player's motion are *external* forces (forces exerted on the player by other objects). These include Earth's gravitational force (the player's weight) and forces from objects that the player is touching (in this case the normal force and friction force exerted by the ground).

friction force arises from chemical bonds that form between atoms on the ground and atoms on the player's uniform. These bonds resist being broken, so the player feels a force slowing him down as he slides. The **normal force** on the player, by contrast, is directed perpendicular to the surface. (In physics and mathematics "normal" is not the opposite of "abnormal"; rather, it's another word for "perpendicular.") Like the friction force, the normal force arises from interactions between atoms: The atoms in the player's uniform resist being squeezed into the atoms on the ground. This force, which we denote by a lowercase $\vec{n}$, prevents the player from falling through the ground below him. If you're sitting in a chair right now, you're feeling an upward normal force exerted by the chair on your rear end; if you're standing, you're feeling an upward normal force exerted on your feet by the ground.

The three forces that act on the baseball player in Figure 4-1 are called **external forces** because they are exerted by other objects outside of the player's body. The gravitational force is exerted by Earth as a whole, while the friction force and normal force are exerted by the ground that's in contact with the player.

Internal forces are also exerted by one part of the player's body on another. One example is the force that the player's shoulder exerts on his left arm to keep it from falling off. While internal forces are important for understanding the interactions of one part of the player's body with another, these forces are *not* included in the sum in Equation 4-1. That's why the left-hand side of that equation has the subscript "ext" for "external." In a moment we'll see why it's appropriate to include only external forces in Equation 4-1.

Newton's Second Law

In the late seventeenth century the English physicist and mathematician Sir Isaac Newton (1642–1727) published a treatise in which he described three fundamental relationships between force and motion. These relationships, called **Newton's laws of motion**, explain nearly all physical phenomena in our everyday experience. One of these relationships, commonly known as **Newton's second law,** is the most important for our discussion of forces. It states:

If a net external force acts on an object, the object accelerates. The net external force is equal to the product of the object's mass and the object's acceleration:

If a net external force acts on an object...

...the object accelerates. The acceleration is in the same direction as the net force.

$$\sum \vec{F}_{ext} = m\vec{a} \qquad (4\text{-}2)$$

The magnitude of acceleration that the net external force causes depends on the mass m of the object (the quantity of material in the object). The greater the mass, the smaller the acceleration.

Newton's second law of motion (4-2)

Newton's second law is simple to state but can be challenging to fully understand. Here are its three essential features:

(1) *A net force in a certain direction causes acceleration in that direction.* The baseball player shown in Figure 4-1 has a backward acceleration (he slows down) because the net force acting on him points opposite to the direction of his motion (**Figure 4-2a**). If we ignore air resistance, a falling basketball is acted on by only a single force, the downward gravitational force. So the net force on the ball is downward, and it accelerates downward (**Figure 4-2b**). This relationship between net force and acceleration is the reason we devoted so much effort in Chapters 2 and 3 to understanding the nature of acceleration.

(2) *The magnitude of acceleration caused by the net force depends on the object's mass.* **Mass** is a measure of how much matter an object has. The SI unit of mass is the **kilogram**, abbreviated kg. A liter of water has a mass of 1 kg. Newton's second law says that the product of an object's mass and the object's acceleration equals the net external force on the object. So if you deliver identically strong kicks to a tennis ball (mass $m = 0.058$ kg) and to a soccer ball (mass $m = 0.43$ kg), the more

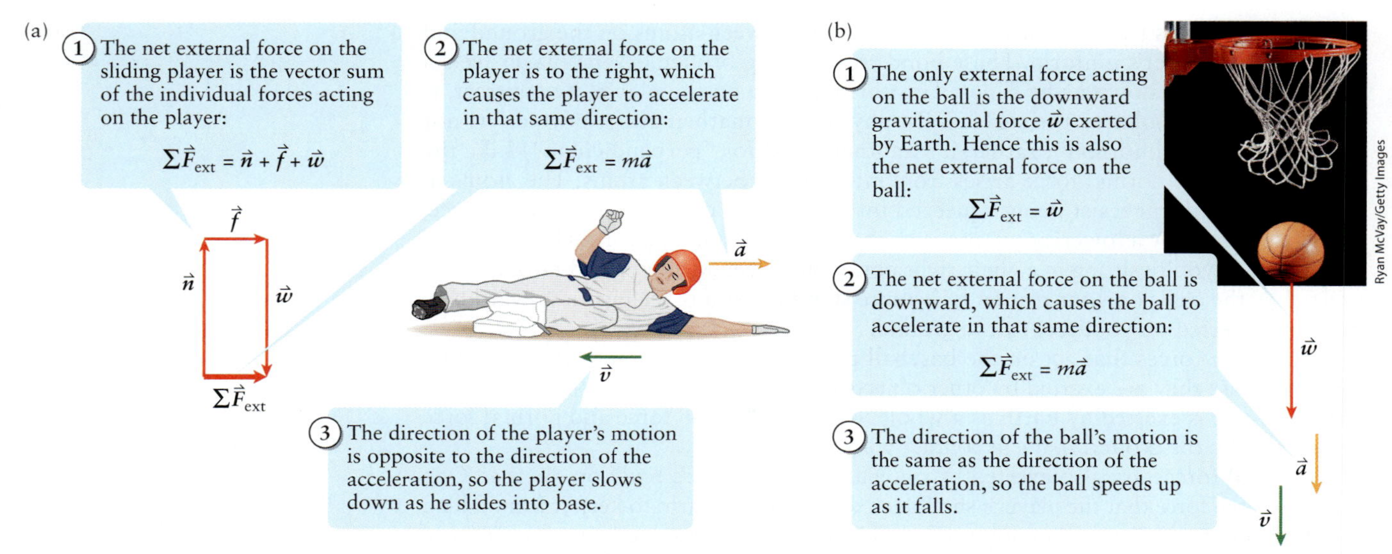

(a)
1 The net external force on the sliding player is the vector sum of the individual forces acting on the player:
$\sum\vec{F}_{ext} = \vec{n} + \vec{f} + \vec{w}$
2 The net external force on the player is to the right, which causes the player to accelerate in that same direction:
$\sum\vec{F}_{ext} = m\vec{a}$
$\vec{f}$
$\vec{n}$
$\vec{w}$
$\sum\vec{F}_{ext}$
$\vec{a}$
$\vec{v}$
3 The direction of the player's motion is opposite to the direction of the acceleration, so the player slows down as he slides into base.
(b)
1 The only external force acting on the ball is the downward gravitational force $\vec{w}$ exerted by Earth. Hence this is also the net external force on the ball:
$\sum\vec{F}_{ext} = \vec{w}$
2 The net external force on the ball is downward, which causes the ball to accelerate in that same direction:
$\sum\vec{F}_{ext} = m\vec{a}$
$\vec{w}$
3 The direction of the ball's motion is the same as the direction of the acceleration, so the ball speeds up as it falls.
$\vec{a}$
$\vec{v}$
Ryan McVay/Getty Images

Figure 4-2 The net external force determines acceleration The net external force on an object is the vector sum $\sum\vec{F}_{ext}$ of the individual forces that act on that object. The acceleration $\vec{a}$ produced by the net external force is in the same direction as $\sum\vec{F}_{ext}$.

massive soccer ball will experience less acceleration while it's in contact with your foot and will fly off with a slower speed.

(3) *Only external forces acting on an object affect that object's acceleration.* As an example, sit in your chair with your feet off the ground and pull upward on your belt with both hands. No matter how hard you try, you can't lift yourself out of the chair! Your body has to accelerate to rise out of the chair (it has to go from being stationary to being in motion), but the force of your hands on your belt is an *internal* force (one part of your body pulls on another part). Thus this force can't produce an acceleration. A helpful friend could lift you out of your chair, but that would happen because your friend exerts an *external* force (one that originates outside your body).

To see the significance of mass more clearly, divide both sides of Equation 4-2 by the object's mass m:

Newton's second law, alternative form (4-3)

The acceleration of an object...

...is proportional to and in the same direction as the net external force on the object...

$$\vec{a} = \frac{1}{m}\sum\vec{F}_{ext}$$

...and inversely proportional to the mass of the object.

Figure 4-3 illustrates the ideas of Equation 4-3.

WATCH OUT! Mass and weight are not the same thing.

An object's mass is the same no matter where the object is; for example, an astronaut walking on the Moon has the same mass as she has on Earth (the amount of matter in her body is exactly the same on either world). However, the astronaut has less *weight* on the Moon because gravity is weaker on the Moon than on Earth. We'll discuss the distinction between mass and weight in more detail in Section 4-3.

The SI unit of force is the **newton**, abbreviated N. (An uppercase abbreviation is used for units that bear a person's name.) A net force of one newton applied to an object with a mass of one kilogram gives the object an acceleration of 1 m/s^2. Because $\sum\vec{F}_{ext} = m\vec{a}$ from Equation 4-2, it follows that

$$1\ \text{N} = 1\ \text{kg} \times 1\ \text{m/s}^2 = 1\ \text{kg}\cdot\text{m/s}^2$$

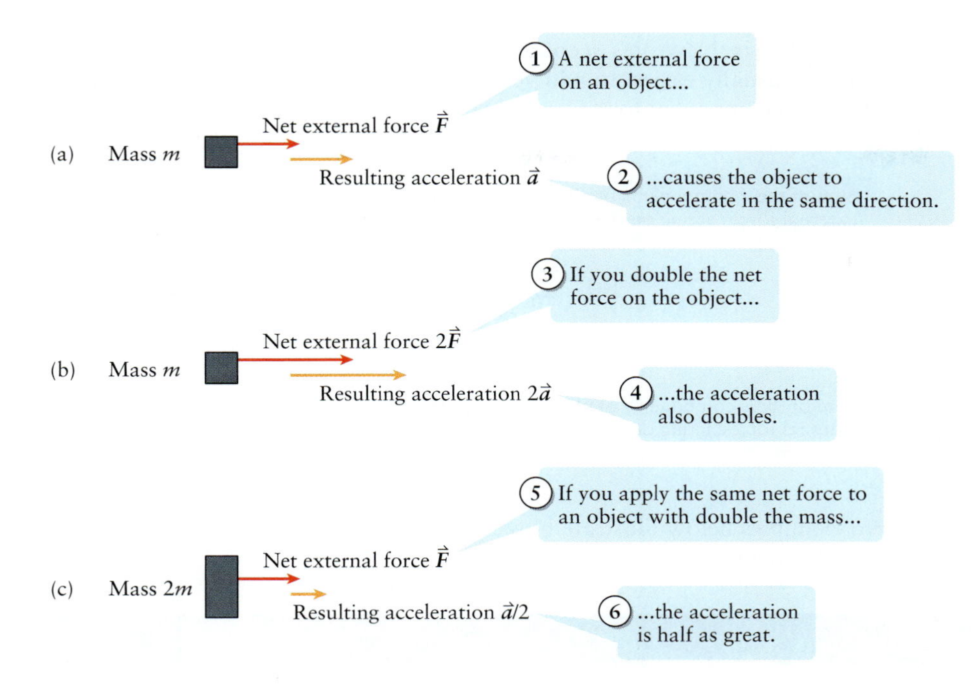
1 A net external force on an object...
Net external force $\vec{F}$
(a) Mass m
Resulting acceleration $\vec{a}$
2 ...causes the object to accelerate in the same direction.
3 If you double the net force on the object...
Net external force $2\vec{F}$
(b) Mass m
Resulting acceleration $2\vec{a}$
4 ...the acceleration also doubles.
5 If you apply the same net force to an object with double the mass...
Net external force $\vec{F}$
(c) Mass $2m$
Resulting acceleration $\vec{a}/2$
6 ...the acceleration is half as great.

Figure 4-3 Net force, mass, and acceleration Newton's second law, $\sum \vec{F}_{ext} = m\vec{a}$ or $\vec{a} = (1/m)\sum \vec{F}_{ext}$, tells us how much acceleration results when a net external force $\sum \vec{F}_{ext}$ is applied to an object of mass m.

The acceleration of an object is directly proportional to the net external force that acts on an object. The acceleration of an object is inversely proportional to the mass m of the object.

$$\vec{a} = \frac{1}{m}\Sigma\vec{F}_{ext}$$

The English unit of force is the **pound** (abbreviated lb). To three significant figures, 1 lb equals 4.45 N, and 1 N = 0.225 lb. One newton is a bit less than a quarter of a pound.

WATCH OUT! It's the net force that matters.

Newton's second law tells us that a net force is required to make an object accelerate, that is, to *change* its velocity. However, your experience may suggest that a force has to act on an object to make it move even at a *constant* velocity. After all, you might reason, you have to exert a force to make a book slide across a table (**Figure 4-4a**). In fact there's no contradiction between these two statements! The explanation is that the *net* force on the sliding book is zero. The forward force that you exert on the book just balances the backward force of friction that the table exerts on the book, just as the upward normal force exerted on the book by the table just balances the downward gravitational force. Thus the vector sum of *all* forces on the book is zero or $\sum \vec{F}_{ext} = 0$. From Newton's second law this means that $\vec{a} = 0$, so the book moves over the table with a constant velocity (Figure 4-4a). If you push harder on the book, the force that you exert is greater in magnitude than the friction force and so the net force $\sum \vec{F}_{ext}$ points forward; then the book has a forward acceleration and speeds up (**Figure 4-4b**). If you let go of the sliding book, the net force $\sum \vec{F}_{ext}$ points backward, so the book has a backward acceleration and slows to a stop (**Figure 4-4c**). Just remember that it's the *net* force on an object that determines its acceleration, not any one particular force.

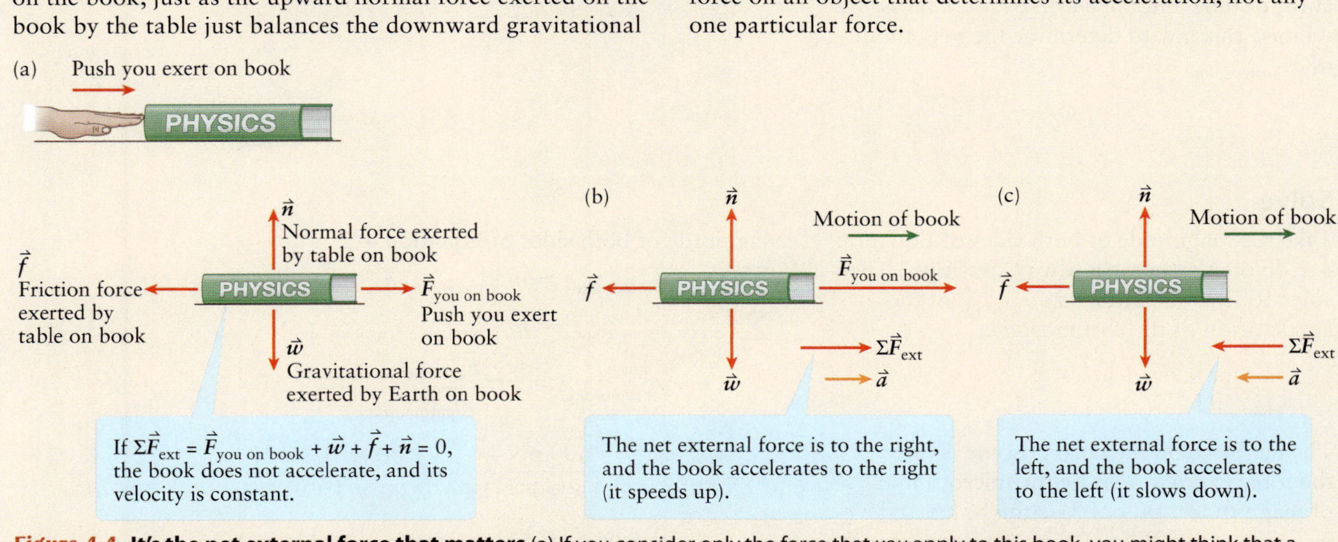
(a) Push you exert on book
PHYSICS
$\vec{n}$ Normal force exerted by table on book
$\vec{f}$ Friction force exerted by table on book
$\vec{F}_{you\ on\ book}$ Push you exert on book
$\vec{w}$ Gravitational force exerted by Earth on book
If $\Sigma\vec{F}_{ext} = \vec{F}_{you\ on\ book} + \vec{w} + \vec{f} + \vec{n} = 0$, the book does not accelerate, and its velocity is constant.
(b)
Motion of book
$\vec{F}_{you\ on\ book}$
$\Sigma\vec{F}_{ext}$
$\vec{a}$
The net external force is to the right, and the book accelerates to the right (it speeds up).
(c)
Motion of book
$\Sigma\vec{F}_{ext}$
$\vec{a}$
The net external force is to the left, and the book accelerates to the left (it slows down).

Figure 4-4 It's the net external force that matters (a) If you consider only the force that *you* apply to this book, you might think that a force is required to maintain constant velocity. If instead you consider *all* of the forces that act on the book, you'll see that a zero *net* external force keeps the book's velocity constant. If the net external force on the book is not zero, the book accelerates in the direction of the net force.

The sum $\sum \vec{F}_{\text{ext}}$ in Newton's second law involves vector addition. We saw in Chapter 3 that it's usually easiest to add vectors if we use components, so we'll often use Equation 4-2 in component form:

Newton's second law in component form (4-4)

$$\text{(a)}\quad \sum F_{\text{ext},x} = ma_x$$
$$\text{(b)}\quad \sum F_{\text{ext},y} = ma_y$$

Sum of the x components of all external forces acting on the object

x component of the object's acceleration

Mass of the object

Sum of the y components of all external forces acting on the object

y component of the object's acceleration

The two examples below illustrate how to use Newton's second law to relate net force, mass, and acceleration.

BIO-Medical EXAMPLE 4-1 Small but Forceful

Microtubules are assembled from protein molecules (**Figure 4-5**). Microtubules help cells maintain their shape and are responsible for various kinds of movements within cells, such as pulling apart chromosomes during cell division. Measurements show that microtubules can exert forces from a few pN (1 pN = 1 piconewton = 10^{-12} N) up to hundreds of nN (1 nN = 1 nanonewton = 10^{-9} N). A particular bacterial chromosome has a mass of 2.00×10^{-17} kg. If a microtubule applies a force of 1.00 pN to the chromosome, what is the magnitude of the chromosome's acceleration? (Ignore any other forces that might act on the chromosome.)

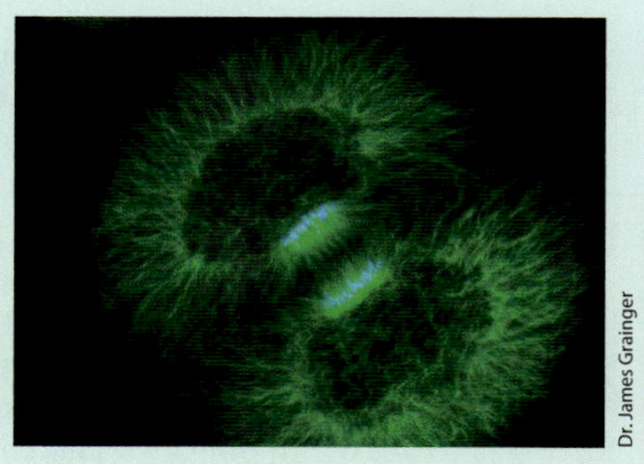

Figure 4-5 Microtubule Microtubules (shown here in green within a fertilized sea urchin egg undergoing division) are protein molecules found within cells. They help cells hold their shape and are responsible for various kinds of movements within cells.

Set Up

Newton's second law tells us that the acceleration $\vec{a}_{\text{chromosome}}$ of the chromosome is determined by the net force acting on the chromosome and the chromosome's mass. We will use this law to determine the magnitude of $\vec{a}_{\text{chromosome}}$.

Newton's second law of motion:

$$\sum \vec{F}_{\text{ext on chromosome}} = m_{\text{chromosome}} \vec{a}_{\text{chromosome}} \quad (4\text{-}2)$$

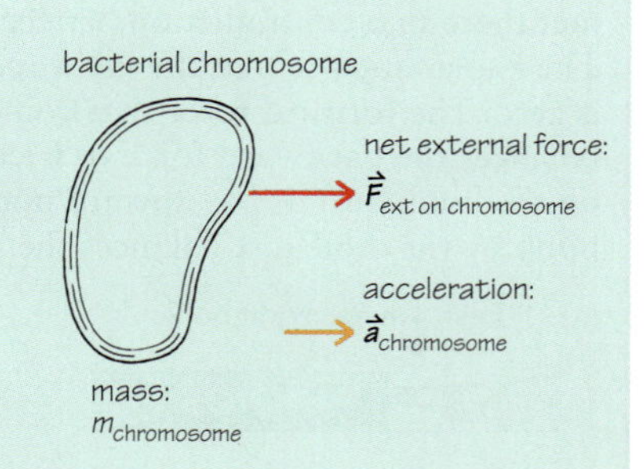

Solve

Take the magnitude of both sides of Equation 4-2. (Note that mass is always positive.) Then solve for the magnitude $a_{\text{chromosome}}$ of the acceleration of the chromosome.

Magnitude of both sides of Equation 4-2:

$$\left|\sum \vec{F}_{\text{ext on chromosome}}\right| = m_{\text{chromosome}} \left|\vec{a}_{\text{chromosome}}\right| \text{ so}$$
$$\left|\sum \vec{F}_{\text{ext on chromosome}}\right| = m_{\text{chromosome}} a_{\text{chromosome}}$$
$$a_{\text{chromosome}} = \frac{\left|\sum \vec{F}_{\text{ext on chromosome}}\right|}{m_{\text{chromosome}}}$$

The net force on the chromosome is just the force exerted on it by the microtubule, of magnitude 1.00 pN. Use this value and $m_{\text{chromosome}} = 2.00 \times 10^{-17}$ kg to solve for the magnitude $a_{\text{chromosome}}$.

$$a_{\text{chromosome}} = \frac{1.00 \text{ pN}}{2.00 \times 10^{-17} \text{ kg}} = \frac{1.00 \times 10^{-12} \text{ N}}{2.00 \times 10^{-17} \text{ kg}}$$
$$= 5.00 \times 10^{4} \text{ N/kg}$$

Use the definition $1\ \text{N} = 1\ \text{kg}\cdot\text{m/s}^2$:

$$a_{\text{chromosome}} = (5.00 \times 10^4\ \text{N/kg})\left(\frac{1\ \text{kg}\cdot\text{m/s}^2}{1\ \text{N}}\right)$$
$$= 5.00 \times 10^4\ \text{m/s}^2$$

Reflect

This acceleration is about 5000 times g, the acceleration due to gravity! It is important to note, however, that the force exerted by the microtubule is opposed by resistive forces that we chose to neglect. The acceleration would not be quite this big in reality. Note also that the forces exerted by microtubules act for only very short periods of time.

EXAMPLE 4-2 A Three-Person Tug-of-War

Jesse, Karim, and Luis, three hungry fraternity brothers, are each pulling on a cafeteria tray laden with desserts. The mass of the tray and its contents is 2.50 kg. The tray sits on a horizontal dining table, and each of the fraternity brothers pulls horizontally on the tray. Jesse pulls due south with a force of magnitude 85.0 N, and Karim pulls with a force of magnitude 90.0 N in a direction 35.0° north of east. If the tray accelerates due north at 20.0 m/s^2, how hard and in what direction does Luis pull on the tray? Ignore the effects of friction.

Set Up

The tray accelerates due to the net force that acts on it, which is the vector sum of the individual forces exerted by Jesse, Karim, and Luis. We've drawn these individual force vectors with their tails together. (We're told to ignore friction, so no other horizontal forces act on the tray. The tray doesn't accelerate vertically, so there's no net vertical force—the upward normal force exerted by the table top balances the downward gravitational force.) Because we know the mass and acceleration of the tray, we can use Newton's second law to calculate the net force on the tray. We'll then use vector addition and subtraction to determine the unknown force exerted by Luis.

$$\sum \vec{F}_{\text{ext on tray}} = m_{\text{tray}}\vec{a}_{\text{tray}} \qquad (4\text{-}2)$$

In component form:

$$\sum F_{\text{ext on tray},x} = m_{\text{tray}}a_{\text{tray},x} \qquad (4\text{-}4a)$$
$$\sum F_{\text{ext on tray},y} = ma_{\text{tray},y} \qquad (4\text{-}4b)$$

Net force on tray = sum of forces exerted by Jesse, Karim, and Luis:

$$\sum \vec{F}_{\text{ext on tray}}$$
$$= \vec{F}_{\text{Jesse on tray}} + \vec{F}_{\text{Karim on tray}} + \vec{F}_{\text{Luis on tray}}$$

Solve

Vector arithmetic is easiest if we use vector components. The forces the three brothers exert are all in the horizontal plane, so we choose x and y axes that lie in this plane with $+x$ pointing east and $+y$ pointing north. We've drawn all three force vectors with their tails at the origin. The tray's acceleration vector $\vec{a}_{\text{tray}}$ points in the positive y direction.

$$a_{\text{tray},x} = 0$$
$$a_{\text{tray},y} = 20.0\ \text{m/s}^2$$

Use Newton's second law in component form, Equations 4-4, to calculate the x and y components of the net force on the tray. Recall that $1\ \text{kg}\cdot\text{m/s}^2 = 1\ \text{N}$.

$$\sum F_{\text{ext on tray},x} = m_{\text{tray}}a_{\text{tray},x} = m_{\text{tray}}(0) = 0$$
$$\sum F_{\text{ext on tray},y} = m_{\text{tray}}a_{\text{tray},y} = (2.50\ \text{kg})\,(20.0\ \text{m/s}^2)$$
$$= 50.0\ \text{kg}\cdot\text{m/s}^2 = 50.0\ \text{N}$$

The x component of the net force on the tray is the sum of the x components of the individual forces on the tray, and similarly for the y component.

$$\sum F_{\text{ext on tray},x} = 0 = F_{\text{Jesse on tray},x} + F_{\text{Karim on tray},x} + F_{\text{Luis on tray},x}$$

$$\sum F_{\text{ext on tray},y} = 50.0\text{ N} = F_{\text{Jesse on tray},y} + F_{\text{Karim on tray},y} + F_{\text{Luis on tray},y}$$

Solve for the components of the force that Luis exerts on the tray:

$$F_{\text{Luis on tray},x} = -F_{\text{Jesse on tray},x} - F_{\text{Karim on tray},x}$$
$$F_{\text{Luis on tray},y} = 50.0\text{ N} - F_{\text{Jesse on tray},y} - F_{\text{Karim on tray},y}$$

To proceed we need to know the x and y components of the forces that Jesse and Karim exert on the tray.

$\vec{F}_{\text{Jesse on tray}}$:
magnitude 85.0 N, points due south (in the negative y direction), so

$$F_{\text{Jesse on tray},x} = 0$$
$$F_{\text{Jesse on tray},y} = -85.0\text{ N}$$

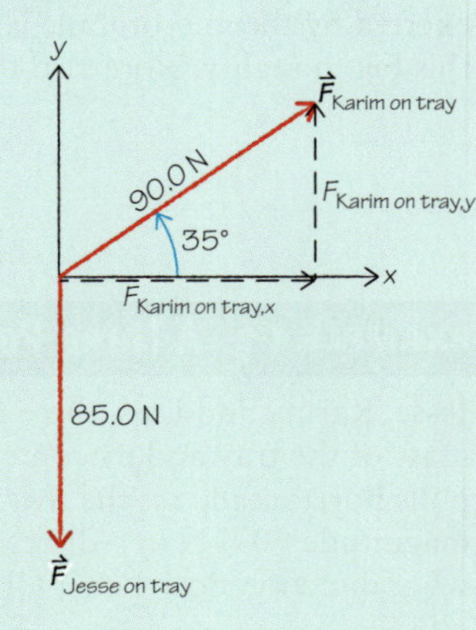

$\vec{F}_{\text{Karim on tray}}$:
magnitude 90.0 N, points 35.0° north of east, so $F_{\text{Karim on tray},x}$ and $F_{\text{Karim on tray},y}$ are both positive

$$F_{\text{Karim on tray},x} = (90.0\text{ N})\cos 35.0° = 73.7\text{ N}$$
$$F_{\text{Karim on tray},y} = (90.0\text{ N})\sin 35.0° = 51.6\text{ N}$$

Substitute these values into the expressions for the components $F_{\text{Luis on tray},x}$ and $F_{\text{Luis on tray},y}$ and solve.

$$F_{\text{Luis on tray},x} = -(0) - (73.7\text{ N}) = -73.7\text{ N}$$
$$F_{\text{Luis on tray},y} = 50.0\text{ N} - (-85.0\text{ N}) - (51.6\text{ N}) = 83.4\text{ N}$$

The force that Luis exerts, $\vec{F}_{\text{Luis on tray}}$, has a negative x component and positive y component, so it points west of north. Calculate the magnitude of $\vec{F}_{\text{Luis on tray}}$ using the Pythagorean theorem and the direction of $\vec{F}_{\text{Luis on tray}}$ using trigonometry.

Magnitude of $\vec{F}_{\text{Luis on tray}}$:

$$F_{\text{Luis on tray}} = \sqrt{(F_{\text{Luis on tray},x})^2 + (F_{\text{Luis on tray},y})^2} = \sqrt{(-73.7\text{ N})^2 + (83.4\text{ N})^2} = 111\text{ N}$$

Angle θ of $\vec{F}_{\text{Luis on tray}}$ measured west of north:

$$\tan\theta = \frac{73.7\text{ N}}{83.4\text{ N}} = 0.884$$
$$\theta = \tan^{-1} 0.884 = 41.5°$$

Reflect

Luis pulls on the tray with a force of 111 N at an angle of 41.5° west of north. In terms of components, Luis pulls west (in the negative x direction) with a force of 73.7 N, thus canceling Karim's 73.7-N pull to the east. Luis also pulls north (in the positive y direction) with a force of 83.4 N; combined with Karim's 51.6-N northward pull, Luis's pull overwhelms Jesse's 85.0-N pull toward the south. Hence the net force on the tray and its delicious contents is northward, and its acceleration is northward as well.

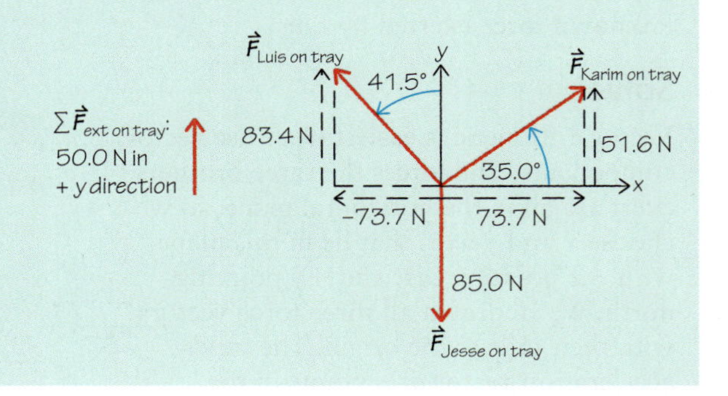

Example 4-2 illustrates an important point: *It's always wise to choose coordinate axes that align with one or more of the vectors.* In this example we chose the axes to be east–west and north–south so that the tray's acceleration $\vec{a}_{\text{tray}}$ and the force $\vec{F}_{\text{Jesse on tray}}$ exerted by Jesse were both along one of the coordinate axes. This choice made it easy to express these vectors in terms of components and so simplified the calculation. Using a different choice of coordinate axes (say, one in which the positive x axis pointed 30°

north of west and the positive y axis pointed 30° east of north), you would have ended up with the same result, but with a good deal more effort.

GOT THE CONCEPT? 4-1 Newton's Second Law

? Rank the following objects in order of the magnitude of the net force that acts on the object, from greatest magnitude to smallest magnitude. (a) A 1250-kg automobile gaining speed at 2.00 m/s^2. (b) A 200,000-kg airliner flying in a straight line at a constant 280 m/s. (c) A 4500-kg truck slowing down at 0.600 m/s^2. (d) An 80.0-kg hiker climbing up a 4.00° slope at a constant 1.20 m/s.

GOT THE CONCEPT? 4-2 Measuring Mass in Zero Gravity

? Imagine you are aboard a spacecraft far from any planet or star, so there is no gravitational force on you or anything else in the spacecraft. Floating in front of you are two spheres that are identical in size and appearance. One is made of lead, and the other is made of plastic; the lead sphere is considerably more massive than the plastic one. Can you devise a simple experiment to determine which sphere is lead and which is plastic?

TAKE-HOME MESSAGE FOR Section 4-2

✔ The net external force on an object is the vector sum of all of the individual forces that act on it from other objects.

✔ Newton's second law states that an object accelerates if the net external force on the object is not zero.

✔ The acceleration is in the same direction as the net force, is directly proportional to the magnitude of the net force, and is inversely proportional to the mass of the object (the quantity of material in the object).

4-3 Mass, weight, and inertia are distinct but related concepts

How much do you weigh? If you grew up in the United States, chances are your answer will be in pounds. If not, you'll probably use kilograms. We've already declared that the SI units of mass are kilograms. Does that mean that pounds and kilograms are both units of the same quantity? Are weight and mass the same?

The answer to both questions is "no!" Although people often use the terms "mass" and "weight" interchangeably in everyday conversation, they have very different meanings in science.

As we discussed in Section 4-2, the mass of an object describes how much matter is contained in the object. By contrast the **weight** of that object is the magnitude of the gravitational force that acts on the object. Because it's a force, weight is measured in newtons (*not* kilograms!) in the SI system and pounds in the English system. We'll use the symbols w for weight and $\vec{w}$ for gravitational force.

Consider an object that has mass m and is acted on by only the gravitational force. The object could be a ball falling in a vacuum, so there is no air resistance (**Figure 4-6**). Newton's second law, Equation 4-2, tells us that the gravitational force $\vec{w}$ equals the object's mass multiplied by its acceleration $\vec{a}$:

$$\vec{w} = m\vec{a}$$

The magnitude of the gravitational force is the weight w, so

$$w = ma$$

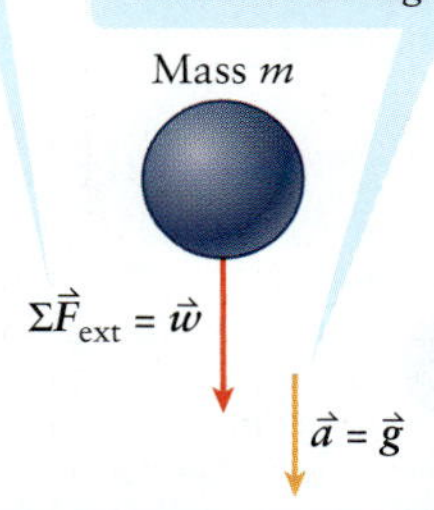
(1) If an object falls without air resistance, the net external force on the object equals the downward gravitational force on that object, which has magnitude w.
(2) An object falling without air resistance accelerates downward. The magnitude of the acceleration is g.
Mass m
$\Sigma\vec{F}_{ext} = \vec{w}$
$\vec{a} = \vec{g}$
(3) Newton's second law, $\Sigma\vec{F}_{ext} = m\vec{a}$, tells us that $\vec{w} = m\vec{g}$ and so $w = mg$.

Figure 4-6 Gravitational force Applying Newton's second law to a freely falling object shows that the magnitude of the gravitational force is $w = mg$.

where a is the magnitude of the object's acceleration. We know from Section 2-6 that if only gravity acts on an object, the object's acceleration has magnitude g. So we get the following expression for the weight of an object of mass m:

Weight of an object of mass m (4-5)

Weight of an object (equal to the magnitude of the gravitational force on that object)

$$w = mg$$

Mass of the object

Magnitude of the **acceleration due to gravity**

That is, *the weight of an object is equal to the object's mass multiplied by g, the acceleration due to gravity*. This statement makes it clear that the units of weight cannot be the same as the units of mass. In Chapters 2 and 3 we used an average value of $g = 9.80\ \text{m/s}^2$ but recognized that the exact value of g varies slightly from place to place on Earth. Therefore, the weight of an object must change as it is moved from place to place. *Hence weight is not an intrinsic property of an object.* Not only does the weight of an object change depending on where you are on Earth, but as we will see in Chapter 10, it can be significantly different on other worlds such as the Moon and Mars.

WATCH OUT! Objects have weight even if they are not accelerating.

We arrived at Equation 4-5 for the weight of an object by assuming that the object was falling freely. However, Equation 4-5 is true even if the object is not accelerating. A 10.0-kg object near Earth's surface has a weight of 98.0 N whether it's sitting on a table, falling off the table, or being pulled across the floor.

GOT THE CONCEPT? 4-3 Most Massive to Least Massive

Rank the following objects according to their mass, from largest to smallest. The acceleration due to gravity at the equator of Mars is $3.69\ \text{m/s}^2$. (a) A rock on the Martian equator that weighs 10.0 N. (b) A rock at Earth's equator that weighs 10.0 N. (c) A rock on the Martian equator that has a mass of 10.0 kg. (d) A rock at Earth's equator that has a mass of 10.0 kg.

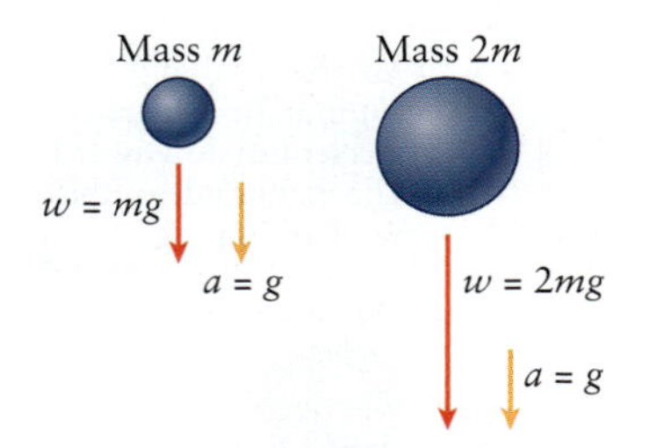

The gravitational force on an object is proportional to its mass. The more massive object experiences a greater gravitational force than the less massive object but has the same acceleration.

Figure 4-7 Greater mass means greater gravitational force Both of these objects are falling in a vacuum, so only the gravitational force acts on each object.

Consider dropping two different objects side by side in a vacuum so that the only force acting on each object is the gravitational force (**Figure 4-7**). The heavier object weighs twice as much as the other, so it is subjected to twice as much gravitational force and hence twice as much net force as the lighter object.

So you might expect that the heavier object would have a greater downward acceleration $\vec{a}$. But the heavier object also has a greater mass and Equation 4-5 tells us that weight and mass are directly proportional. If the heavier object has twice the weight of the lighter one, it also has twice the mass. An object's acceleration is equal to the net force that acts on it divided by its mass (recall the alternative form of Newton's second law, $\vec{a} = \frac{1}{m}\sum \vec{F}_{\text{ext}}$, from Equation 4-3). Hence the quotient $\frac{1}{m}\sum \vec{F}_{\text{ext}}$ has the *same* value for both objects, and they both fall with the same acceleration in a vacuum.

Can you lift an object that weighs 10 N? A 1000-N object? In Section 4-2 we mentioned the following conversions, which are valid to three significant figures:

$$1\ \text{N} = 0.225\ \text{lb}$$
$$1\ \text{lb} = 4.45\ \text{N}$$

(You can find more precise conversion factors in Appendix A.) According to Equation 4-5 the mass of an object that weighs 10.0 N is $m = w/g = (10.0\ \text{N})/(9.80\ \text{m/s}^2) = 1.02\ \text{kg}$. A 1-L bottle of water has a mass of 1 kg; that is, it weighs about 10 N. In pounds, that's 10 multiplied by 0.225 lb or about 2.2 lb. You'd have no problem lifting that amount. Many people can lift an object that weighs 1000 N (225 lb); that's a mass of about 100 kg. Remembering that 1 liter of water has a mass of 1 kg and a weight of 10 N can help you get an intuitive feel for the numerical values of masses and forces.

Mass, Inertia, and Newton's First Law

The concept of mass is intertwined with the observation that all objects *resist changes* in their state of motion. We use the term "**inertia**" for the tendency of an object to resist a change in motion. For example, a *stationary* object (such as a rock lying on the ground or a roommate sleeping on the couch) tends to remain stationary. An object *in motion*, like a fast-moving hockey puck sliding on the ice, tends to keep moving in the same direction with the same speed.

WATCH OUT! Don't confuse mass, weight, and inertia.

Note the differences among mass, weight, and inertia. Mass is the *quantity of material* in an object; no matter where in the universe you take the object, its mass remains the same. Weight is the *magnitude of the gravitational force* on an object; it's proportional to the mass but depends on the value of g at the object's location. Inertia is the *tendency* of *all* objects to maintain the same motion. Unlike mass or weight, we do not give numerical values to the inertia of an object.

Newton described inertia as an intrinsic, unchanging property of mass and of objects. This property is defined in **Newton's first law:**

An object at rest tends to stay at rest, and an object in uniform motion tends to stay in motion with the same speed and in the same direction unless acted upon by a net force.

Colette Sutton

Figure 4-8 An object at rest Kjeragbolten is a boulder suspended above an abyss in Norway that is 984 m (3228 ft) deep. It remains at rest because the vector sum of the external forces acting on it—the downward force exerted on it by Earth's gravity, the additional downward force exerted on it by a brave tourist, and the upward forces exerted on it by the surrounding rock—is zero.

Newton's first law contradicts the older theories of motion developed by the ancient Greeks, principally by the philosopher Aristotle (384 B.C.–322 B.C.). Aristotle believed that every object had a natural place in the world. For example, heavy objects, such as rocks, are naturally at rest on Earth, and light objects, such as clouds, are naturally at rest in the sky. In Aristotle's view, a moving object tended to stop when it found its natural rest position. To keep an object moving, reasoned Aristotle, a force had to be applied to it; to make it move faster, a greater force was required. Newton saw more deeply. He identified the concept of the *net* force on an object and realized that the net force on an object determines its acceleration rather than its speed. (This realization is Newton's second law.) If the net force on an object is zero, the object has zero acceleration and moves with a constant velocity—that is, with the same speed and in the same direction. (You should review the discussion of Figure 4-4 in Section 4-2.)

At this point you can think of Newton's first law as a special case of the *second* law that applies when the net force on an object is zero. We'll nonetheless call these laws "first" and "second" in the same manner in which Newton numbered them. In equation form we can write the first law as follows:

If the net external force on an object is zero... ...the object does not accelerate...

$$\text{If } \sum \vec{F}_{\text{ext}} = 0, \text{ then } \vec{a} = 0 \text{ and } \vec{v} = \text{constant} \tag{4-6}$$

...and the velocity of the object remains constant. If the object is at rest, it remains at rest; if it is in motion, it continues in motion in a straight line at a constant speed.

Newton's first law of motion (4-6)

We say that an object is in **equilibrium** if the net external force on it is zero (**Figure 4-8**). A chandelier hanging from the ceiling is in equilibrium: The chain from which the chandelier is suspended exerts an upward force that exactly

See the Math Tutorial for more information on trigonometry.

balances the downward gravitational force on the chandelier. The chandelier's velocity is zero and remains zero. But a moving airliner is also in equilibrium if it flies in a straight line at a constant speed. The forward thrust provided by the airliner's engines balances the backward drag force that the air exerts on the airliner, and the upward lift provided by air flowing around the wings balances the downward gravitational force. The airliner has a nonzero velocity, but its *acceleration* is zero and so the airliner is in equilibrium.

The following example illustrates how to use Newton's first law to analyze an object at rest.

EXAMPLE 4-3 Let Sleeping Cats Lie

A 40.0-N cat is asleep on a ramp that is tilted by an angle of 15.0° from the horizontal. Three forces act on the cat: the downward gravitational force, a normal force perpendicular to the ramp, and a friction force directed uphill parallel to the ramp. The cat remains at rest while sleeping. Determine the magnitude of each force that acts on the cat.

Set Up

As in Example 4-2, we've drawn all of the external force vectors on the cat with their tails touching. The cat's weight is $w = 40.0$ N; this is just the magnitude of the gravitational force on the cat ($\vec{w}_{cat}$). We know the directions of the normal force $\vec{n}$ (perpendicular to the ramp) and the friction force $\vec{f}$ (uphill parallel to the ramp). Our task is to find the magnitudes of $\vec{n}$ and $\vec{f}$. We'll do this using Newton's first law: Because the cat remains at rest, the sum of $\vec{w}_{cat}$, $\vec{n}$, and $\vec{f}$ must be zero. So we can solve this problem by using vector addition.

Newton's first law of motion:

$$\sum \vec{F}_{\text{ext on cat}} = \vec{w}_{cat} + \vec{n} + \vec{f} = 0 \qquad (4\text{-}6)$$

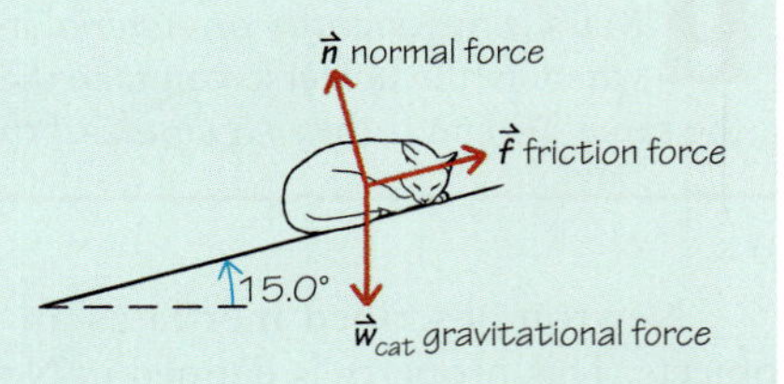

Solve

Like most vector addition problems, this one is most easily solved using components. If $\sum \vec{F}_{\text{ext on cat}} = 0$, then each of the components of $\sum \vec{F}_{\text{ext on cat}}$ must also be equal to zero. It's convenient to choose the x axis to be along the tilted ramp and the y axis to be perpendicular to the ramp. (The x and y axes must be perpendicular to each other but don't have to be horizontal or vertical.) Using this choice, two of the vectors, $\vec{n}$ and $\vec{f}$, lie either directly along or directly opposite to one of the axes.

Newton's first law in component form:

$$\sum F_{\text{ext on cat},x} = w_{\text{cat},x} + n_x + f_x = 0$$

$$\sum F_{\text{ext on cat},y} = w_{\text{cat},y} + n_y + f_y = 0$$

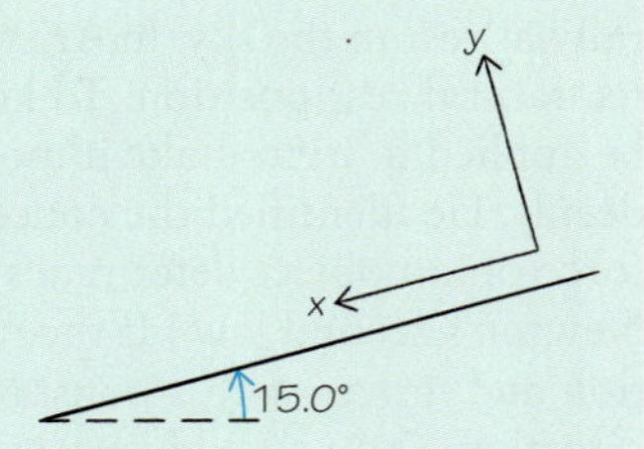

Write the x and y components of each of the external force vectors in terms of their magnitudes w, n, and f and the angle $\theta = 15.0°$ of the ramp. Note that $\vec{w}_{cat}$ has a positive x component (down the ramp) and a negative y component (into the ramp), $\vec{n}$ has only a positive y component (perpendicular to the ramp), and $\vec{f}$ has only a negative x component (up the ramp).

Components of gravitational force $\vec{w}_{cat}$:

$w_{\text{cat},x} = +w_{\text{cat}} \sin\theta$

$w_{\text{cat},y} = -w_{\text{cat}} \cos\theta$

Components of normal force $\vec{n}$:

$n_x = 0$

$n_y = +n$

Components of $\vec{f}$:

$f_x = -f$

$f_y = 0$

$n_y = n$
$f_x = -f$
w_{cat}
$w_{cat,y} = -w_{cat} \cos\theta$
$\theta = 15.0°$
$w_{cat,x} = +w_{cat} \sin\theta$

Substitute the expressions for $w_{cat,x}$, $w_{cat,y}$, n_x, n_y, f_x, and f_y into Newton's first law and solve for the magnitudes n and f.

Substitute into Newton's first law in component form:

$$\sum F_{\text{ext on cat},x} = w_{cat} \sin\theta + 0 + (-f) = 0$$

$$\sum F_{\text{ext on cat},y} = -w_{cat} \cos\theta + n + 0 = 0$$

From the y equation:

$$n = w_{cat} \cos\theta = (40.0 \text{ N}) \cos 15.0° = 38.6 \text{ N}$$

From the x equation:

$$f = w_{cat} \sin\theta = (40.0 \text{ N}) \sin 15.0° = 10.4 \text{ N}$$

Reflect

The downward gravitational force has magnitude 40.0 N, the normal force acting perpendicular to the ramp has magnitude 38.6 N, and the friction force acting uphill has magnitude 10.4 N. Note that no single force "balances" any of the other forces: All three forces are needed to mutually balance each other and keep the cat at rest.

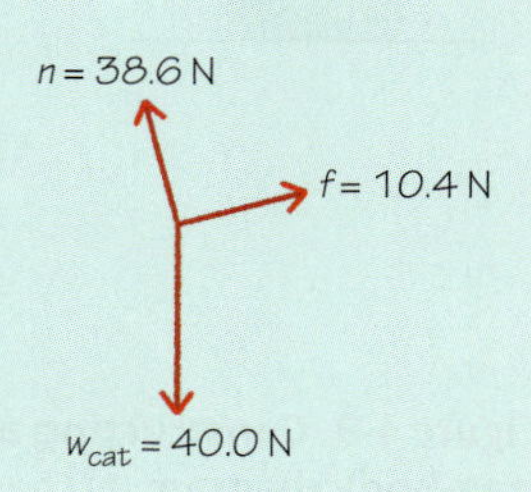

WATCH OUT! The normal force isn't always equal to the weight.

It's a common mistake to assume that the normal force on an object (in this case, the sleeping cat) has the same magnitude as the object's weight. That's clearly *not* true in this case: The normal-force magnitude $n = 38.6$ N is *less* than the cat's 40.0-N weight. As a general rule it's *always* safest to treat the magnitude of the normal force as an unknown, as we have done in this example.

GOT THE CONCEPT? 4-4 Tilt the Cat, but Don't Wake It Up

You increase the angle θ of the ramp in Example 4-3 to a new value greater than 15.0°. The cat remains blissfully asleep and at rest at the new angle. What effect does this have on the magnitude n of the normal force and the magnitude f of the friction force? (a) n and f both increase; (b) n and f both decrease; (c) n increases and f decreases; (d) n decreases and f increases; (e) none of these; (f) The answer depends on the new value of θ.

TAKE-HOME MESSAGE FOR Section 4-3

✔ The weight of an object is the magnitude of the gravitational force that acts on it.

✔ Newton's first law states that if the external forces on an object sum to zero, the object does not accelerate. If the object is stationary, it remains at rest, and if it is in motion, it continues to move in a straight line with constant velocity.

4-4 Making a free-body diagram is essential in solving any problem involving forces

You need the right key to open a locked door. And you need the proper map to navigate in a foreign city. The key that unlocks physics problems involving forces and the map that allows you to navigate these problems is the *free-body diagram*.

A **free-body diagram** is a graphical representation of all of the external forces acting on an object. It's useful because Newton's second law and Newton's first law both involve the sum of all external forces on an object, $\sum \vec{F}_{ext}$. The term "free body" means that we draw only the object on which the forces act, *not* the other objects that exert those forces. We drew free-body diagrams for the tray in Example 4-2 and the cat in Example 4-3 (see the Set Up step in each example). As another example, **Figure 4-9a** shows a block on a horizontal table. A wind is blowing from left to right, and the wind pushes on the block and makes it slide. **Figure 4-9b** lists the steps involved in constructing this block's free-body diagram.

(a) A block on a horizontal table.

(b) Drawing the free-body diagram for the block.

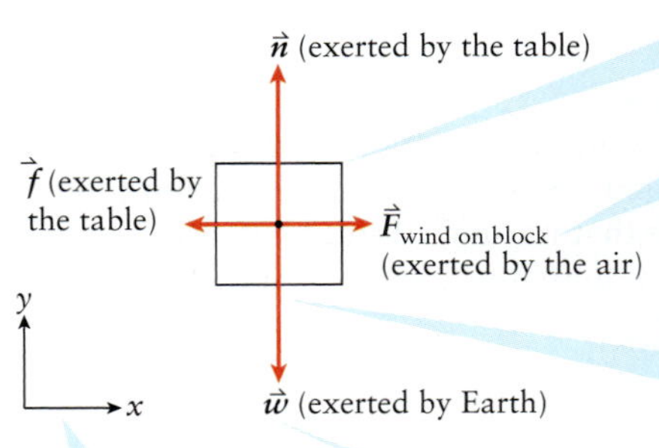

1. Sketch the object on which the forces act.
2. Identify what other objects exert forces on this object. This includes Earth (which exerts a gravitational force) and anything that touches the object.
3. For each force that acts on the object, draw the force vector with its tail at the center of the object. Make sure each vector points in the correct direction. Do not include forces that the object exerts on the other objects.
4. Label each force with its symbol. Be certain that you can identify what other object exerts each force. If you're not sure what exerts a force, it's probably not a real force!
5. Choose, draw, and label the directions of the positive x and y axes. Choose the axes so that as many force vectors as possible lie along one of the axes.

Figure 4-9 Constructing a free-body diagram (a) The wind pushes to the right on this block, which slides on a horizontal table. (b) How to create a free-body diagram that depicts all of the external forces that act on the block.

Often we'll label a force to indicate what object is exerting the force and what object the force acts on. For example, if a girl pushes on a book, we will label that force $\vec{F}_{\text{girl on book}}$. (We used a label of this kind in Figure 4-9b for the force of the wind blowing on the block.) However, we'll label some commonly encountered forces in a different way. We'll use $\vec{w}$ for the gravitational force that Earth exerts on an object. If the object is in contact with a surface, we'll use $\vec{f}$ for the friction force on the object that acts parallel to the surface and $\vec{n}$ for the normal force on the object that acts perpendicular to the surface (see Figure 4-9b). We'll refer to the friction force and normal force as **contact forces**: They arise only when two objects are in contact with each other. By contrast the gravitational force is *not* a contact force: A basketball in midair is pulled downward by the gravitational force even though it's not in contact with the ground.

(a)

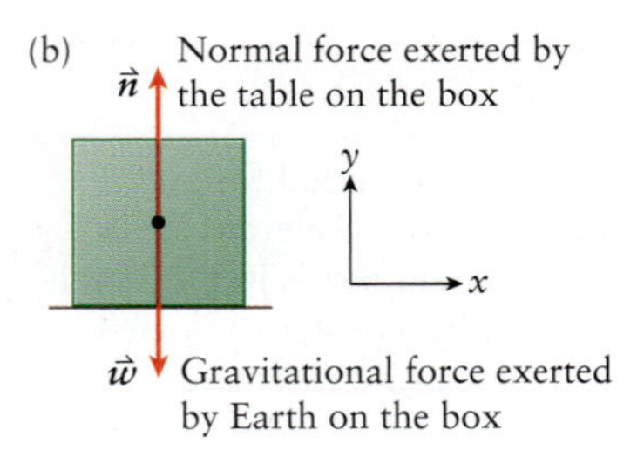

(b)

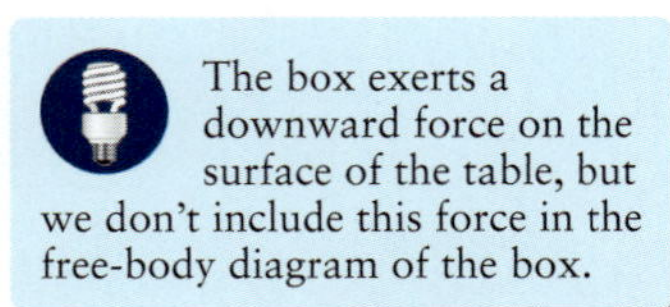

The box exerts a downward force on the surface of the table, but we don't include this force in the free-body diagram of the box.

Figure 4-10 A free-body diagram: A stationary box atop a table (a) A box rests on a horizontal table top. (b) The free-body diagram for the box.

Examining Free-Body Diagrams

Let's look at some examples of free-body diagrams and what we can learn from them. In **Figure 4-10a**, a stationary box rests on a table with a horizontal top. The free-body diagram in **Figure 4-10b** shows the two forces that act on the box: the downward gravitational force $\vec{w}$ and the upward normal force $\vec{n}$ exerted by the table.

We choose the y axis to be vertical; then both forces lie along this axis. The sum of $\vec{w}$ and $\vec{n}$ must be zero in order for the box to remain at rest, so the free-body diagram tells us that $\vec{w}$ and $\vec{n}$ must be opposite in direction and have the same magnitude. Note that the *box* exerts a downward force on the surface of the *table*, but we don't include that force in the free-body diagram for the box. A diagram for the box includes only those forces that act *on* the box.

In **Figure 4-11a** a box is placed on an inclined ramp with a rough surface. There is friction between the box and the ramp, so the free-body diagram in **Figure 4-11b** includes a friction force $\vec{f}$ as well as the gravitational force $\vec{w}$ and the normal force $\vec{n}$. Because the ramp is tilted by an angle θ from the horizontal, the normal force (which is perpendicular to the ramp) is tilted from the vertical by the same angle θ. We've chosen the x axis to be parallel to the ramp and the y axis to be perpendicular to the ramp;

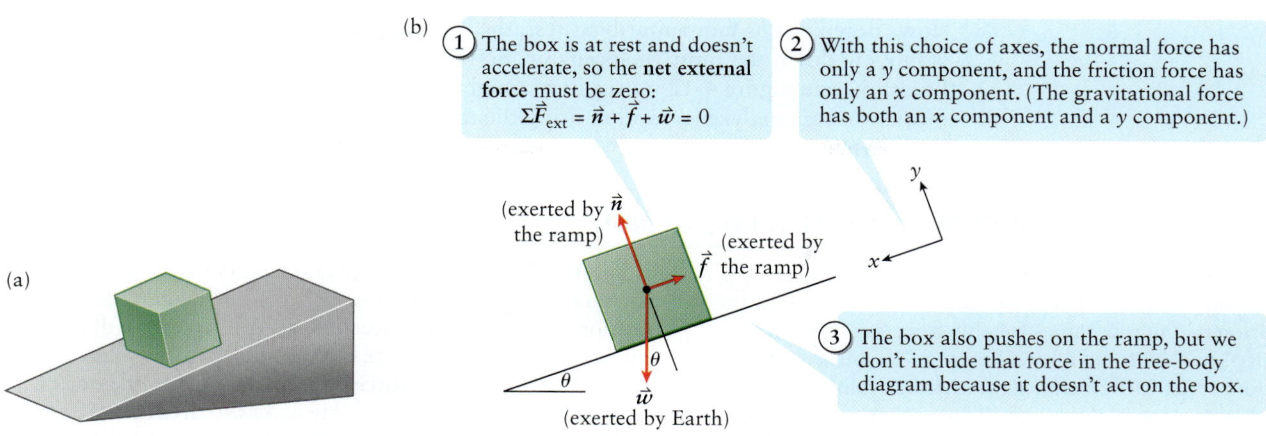

(a)
(b)
1 The box is at rest and doesn't accelerate, so the **net external force** must be zero:
$\Sigma\vec{F}_{\text{ext}} = \vec{n} + \vec{f} + \vec{w} = 0$
2 With this choice of axes, the normal force has only a y component, and the friction force has only an x component. (The gravitational force has both an x component and a y component.)
(exerted by the ramp) $\vec{n}$
(exerted by the ramp) $\vec{f}$
y
x
3 The box also pushes on the ramp, but we don't include that force in the free-body diagram because it doesn't act on the box.
θ
θ
$\vec{w}$
(exerted by Earth)

Figure 4-11 A free-body diagram: A stationary box on a ramp (a) Friction keeps the box from sliding down the ramp. (b) The free-body diagram for the box.

then $\vec{f}$ has only an x component, and $\vec{n}$ has only a y component. By assumption the block remains at rest, so the net external force $\sum \vec{F}_{\text{ext}}$ on it—that is, the vector sum of $\vec{w}$, $\vec{n}$, and $\vec{f}$—must be zero. So the friction force $\vec{f}$ must point in the negative x direction in order to cancel the (positive) x component of the gravitational force $\vec{w}$; the normal force $\vec{n}$ points in the positive y direction and must cancel the (negative) y component of $\vec{w}$. (This is the same situation as in Example 4-3 in Section 4-3.) Note that the x and y components of the gravitational force $\vec{w}$ are less than the magnitude of the force, which is the box's weight w. The free-body diagram therefore tells us something useful: The magnitudes of the friction force and normal force are both *less* than the weight of the box. Compare this situation to the box on a horizontal surface shown in Figure 4-9, for which the magnitude of the normal force is just equal to the weight.

What happens if we grease the surface of the ramp so that there is no more friction? In this case the friction force $\vec{f}$ goes to zero, and the free-body diagram is as shown in **Figure 4-12**. Now the vector sum $\sum \vec{F}_{\text{ext}}$ of the external forces on the box cannot be zero: For the normal force and gravitational forces to cancel each other, they would have to point in opposite directions, which the free-body diagram in Figure 4-12 shows they do not. Instead the net external force points down the ramp in the positive x direction. According to Newton's second law, $\sum \vec{F}_{\text{ext}} = m\vec{a}$ (Equation 4-2), the box must therefore have a nonzero acceleration $\vec{a}$ down the ramp. We learn this simply by examining the free-body diagram for the box.

Note that neither the normal force nor the gravitational force in Figure 4-12 depends on how the box is moving. Whether the box is sliding up the ramp, momentarily at rest, or sliding down the ramp, these forces are the same. This means the acceleration is also the same in each of these cases. If the box is moving up the ramp, it

WATCH OUT! Draw the force vectors in a free-body diagram with their tails touching.

In Figures 4-9, 4-10, 4-11, and 4-12, we've drawn the individual force vectors that act on an object so that their tails all touch within the object. (We also did this in Examples 4-2 and 4-3.) This makes it easier to calculate the x and y components of these vectors.

Figure 4-12 also includes vectors for the net external force $\sum \vec{F}_{\text{ext}}$ and the acceleration $\vec{a}$. If you add vectors like this to your free-body diagram, draw them *alongside* (not touching) the object, as in Figure 4-12. In your diagram only the individual forces that act on the object should be touching it.

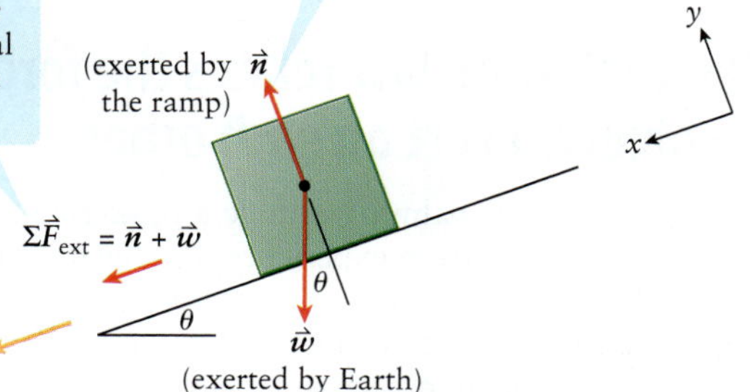

Figure 4-12 A free-body diagram: A box sliding on a ramp The free-body diagram for a box sliding on a frictionless ramp.

slows down; if it is momentarily at rest, the box starts moving down the ramp; and if it is moving down the ramp, it speeds up. So the free-body diagram also tells us that the box shown in Figure 4-12 moves in a straight line with a constant acceleration—which means we can analyze its motion using the techniques that we learned in Chapter 2.

WATCH OUT! Don't add nonexistent forces to a free-body diagram.

Students are sometimes tempted to add a force in a free-body diagram that is somehow associated with the motion. Common names for this force are "the force of velocity" or "the force of acceleration." If you feel this temptation, resist it! Remember that neither velocity nor acceleration is a force. Keep in mind also that acceleration is the *result* of force: For example, the acceleration of the box in Figure 4-12 is a result of the nonzero force in the x direction. (It's true that Newton's second law says that $\sum \vec{F}_{\text{ext}} = m\vec{a}$, which may mislead you into thinking that the quantity $m\vec{a}$ is itself a force. But remember the real meaning of Newton's second law: A net external force $\sum \vec{F}_{\text{ext}}$ on an object makes the object accelerate. So $m\vec{a}$ is *not* a force, and equation $\sum \vec{F}_{\text{ext}} = m\vec{a}$ just tells you the magnitude and direction of that resulting acceleration $\vec{a}$.) If you're still not convinced, ask yourself what external object could exert a "force of velocity" or a "force of acceleration" on the object you're analyzing. (Remember only *external* forces can affect an object's motion. You may want to review the discussion in Section 4-2.) In Figure 4-12 we've already accounted for Earth's gravity, and we've already accounted for the one and only thing that touches the box (the surface of the ramp). So there can be no external forces on the box other than the two we've already drawn in Figure 4-12.

We'll use free-body diagrams throughout our study of forces and motion, and you should as well. Before we delve more deeply into how to solve problems using Newton's first and second laws, we'll introduce Newton's *third* law. This will give us deeper insight into the way in which objects exert forces on each other.

GOT THE CONCEPT? 4-5 Choose the Correct Free-Body Diagram

When a box is placed on an inclined ramp and released, it moves down the ramp with increasing speed. There is friction between the ramp and the box. Which of the free-body diagrams shown in **Figure 4-13** correctly depicts the forces on the box as it slides down the ramp?

(a)
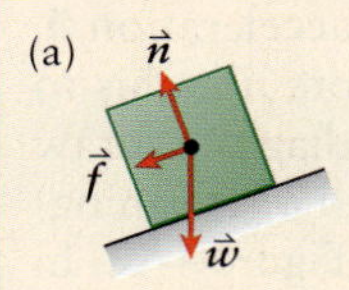

(b)
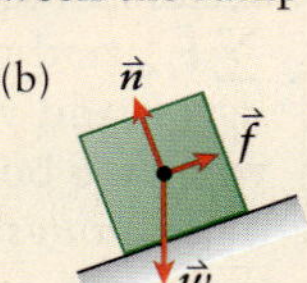

(c)
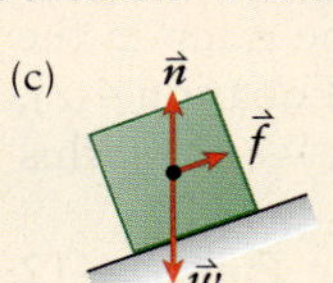

(d)
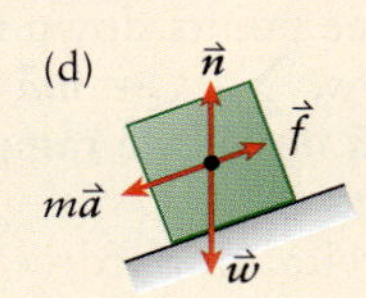

(e)
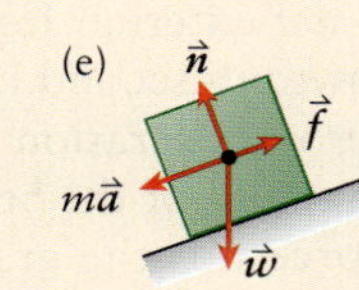

Figure 4-13 A box sliding down a ramp with friction Which of these free-body diagrams is correct?

TAKE-HOME MESSAGE FOR Section 4-5

✔ In order to solve a problem about how an object moves and the forces that act on it, you must draw a free-body diagram for the object. This diagram depicts all of the external forces that act on the object.

✔ Choose the x and y axes in a free-body diagram so that most of the forces point along one or the other of the axes.

4-5 Newton's third law relates the forces that two objects exert on each other

If you hold a book in your palm or dribble a basketball, you're exerting a force on an object. However, you know from experience that the *object* exerts a force on *you* as well. You can feel the book pushing back against your palm, and you can feel the slap of the basketball against your hand as you push down on it. What is the connection between the force that you exert on an object and the force that the object exerts back on you?

Here's an experiment that offers an answer to this question. Take two spring scales (such as you might use for weighing a fish) and connect their upper ends together (**Figure 4-14a**). You hold the free end of one scale, and you attach the free end of the other scale to a hook mounted to the wall (**Figure 4-14b**). When you pull on the free end of the first scale, the wall pulls on the free end of the other scale. Remarkably, no matter how hard you pull, the reading on the scale that you hold (which measures the amount of force you exert) is *identical* to the reading on the scale attached to the wall (which measures the amount of force that the wall exerts). In other words, the force that you exert on the wall has the *same* magnitude as the force that the wall exerts back on you (**Figure 4-14c**).

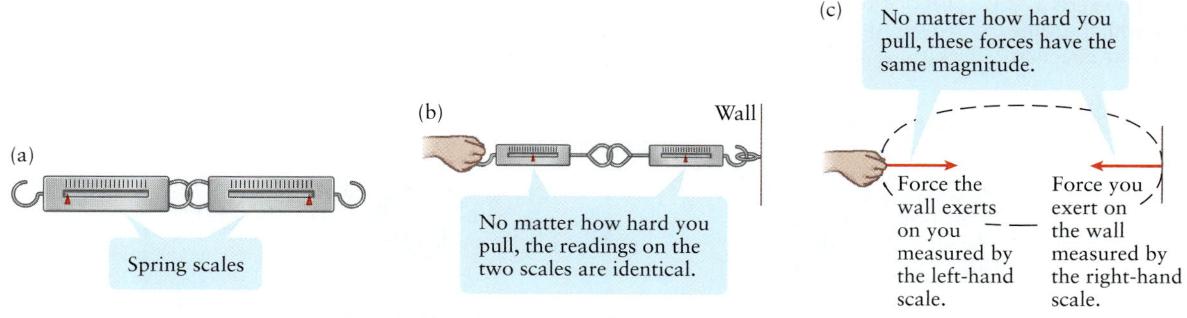
(a)
Spring scales
(b)
Wall
No matter how hard you pull, the readings on the two scales are identical.
(c)
No matter how hard you pull, these forces have the same magnitude.
Force the wall exerts on you measured by the left-hand scale.
Force you exert on the wall measured by the right-hand scale.

Figure 4-14 Comparing the forces that objects exert on each other (a) Two spring scales connected together. Each scale measures the force exerted on it. (b) One scale is connected to a wall, and you pull on the other end. (c) The force you exert on the wall with the scales is equal in magnitude to the force that the wall exerts back on you.

In Figure 4-14 both you and the wall are stationary. We can try the same experiment with two objects in motion—say, a car and a trailer connected together by the same arrangement of two spring scales (**Figure 4-15**). The readings on each scale depend on whether the car and trailer are moving at a constant speed, speeding up, or slowing down. In all of these cases, however, experiment shows that the spring scale attached to the car reads *exactly* the same as the spring scale attached to the trailer. So the force that the car exerts on the trailer always has the same magnitude as the force that the trailer exerts on the car. The only difference between the two forces is that they are in *opposite* directions: The car pulls forward on the trailer, while the trailer pulls backward on the car.

We can summarize these observations in **Newton's third law:**

If object A *exerts a force on object* B, *object* B *exerts a force on object* A *that has the same magnitude but is in the opposite direction. These two forces act on different objects.*

In declaring this law Newton added: "If you press a stone with your finger, the finger is also pressed by the stone." This quote reminds us that Newton's third law refers to two forces that act on *different* objects. In Newton's example these two forces are your finger pressing on the stone and the stone pressing on your finger. Physicists speak of the *interaction* between two objects and refer to the two forces involved in an interaction as a **force pair.** (These two forces are sometimes called the *action* and the *reaction.*)

We can express Newton's third law as an equation relating the force that an object A exerts on another object B (written as $\vec{F}_{A \text{ on } B}$) to the **reaction force** that object B exerts on object A (written as $\vec{F}_{B \text{ on } A}$):

The car and trailer are linked together by two spring scales (like those shown in Figure 4-14) connected end to end.

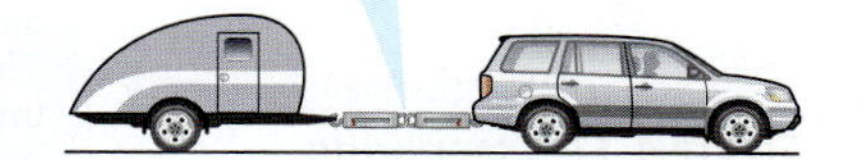

Each spring scale has the same reading: The force that the car exerts on the trailer always has the same magnitude as the force that the trailer exerts on the car. This is true no matter how the car and trailer move (and is also true if they aren't moving at all).

Figure 4-15 Comparing the forces that moving objects exert on each other A pair of spring scales like those shown in Figure 4-14a are used to connect a car and trailer.

Force that object A exerts on object B

Force that object B exerts on object A

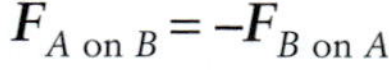

$$\vec{F}_{A \text{ on } B} = -\vec{F}_{B \text{ on } A} \quad (4\text{-}7)$$

Newton's third law of motion

If object A exerts a force on object B, object B must exert a force on object A with the same magnitude but opposite direction.

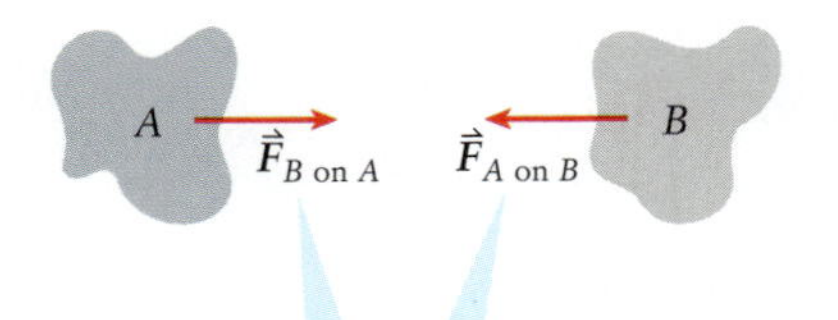
A
$\vec{F}_{B \text{ on } A}$
$\vec{F}_{A \text{ on } B}$
B
These two forces are equal in magnitude but opposite in direction.

Figure 4-16 Newton's third law If object A exerts a force $\vec{F}_{A \text{ on } B}$ on object B, then object B necessarily exerts a force $\vec{F}_{B \text{ on } A}$ on object A with the same magnitude but the opposite direction.

The minus sign means that $\vec{F}_{B \text{ on } A}$ has the same magnitude as $\vec{F}_{A \text{ on } B}$ but is in the opposite direction (**Figure 4-16**). The two forces in a force pair are sometimes called "equal and opposite," which means that they are equal *in magnitude* and opposite *in direction*.

Comparing Newton's Three Laws

We've emphasized that Newton's third law is fundamentally different from the first and second laws because it tells us about forces that act on *two different* objects. By contrast, Newton's first and second laws tell us about the effect of the net external force acting on a *single* object. Here's an example that illustrates this important distinction. **Figure 4-17a** shows a ball held at rest in a person's hand. The drawing shows three forces: the gravitational force of Earth pulling down on the ball (the ball's weight), the force of the person's hand on the ball, and the force of the ball on the person's hand. All three forces have equal magnitudes but for different reasons!

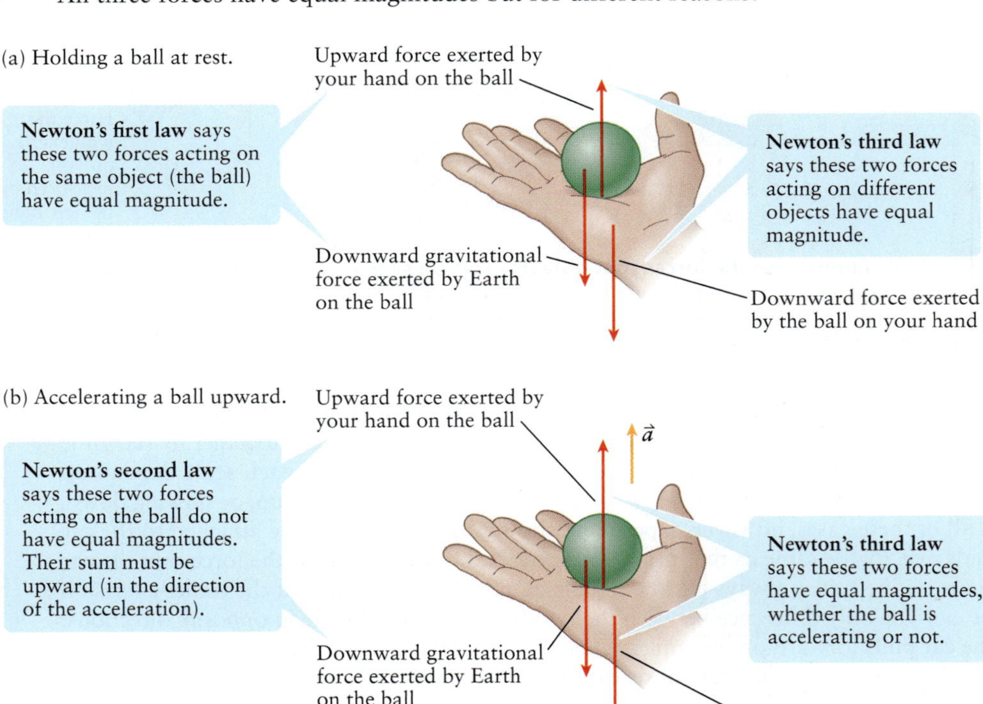
(a) Holding a ball at rest.
Upward force exerted by your hand on the ball
Newton's first law says these two forces acting on the same object (the ball) have equal magnitude.
Newton's third law says these two forces acting on different objects have equal magnitude.
Downward gravitational force exerted by Earth on the ball
Downward force exerted by the ball on your hand
(b) Accelerating a ball upward.
Upward force exerted by your hand on the ball
$\vec{a}$
Newton's second law says these two forces acting on the ball do not have equal magnitudes. Their sum must be upward (in the direction of the acceleration).
Newton's third law says these two forces have equal magnitudes, whether the ball is accelerating or not.
Downward gravitational force exerted by Earth on the ball
Downward force exerted by the ball on your hand

Figure 4-17 Comparing Newton's three laws (a) Newton's first law and third law applied to a ball at rest in your palm. (b) Newton's second law and third law applied to a ball that you accelerate upward.

- The force of Earth on the ball and the force of the person's hand on the ball have equal magnitudes and opposite directions according to Newton's *first* law. These are the only two external forces acting *on* the ball, so their vector sum $\sum \vec{F}_{\text{ext}}$ must be zero (that is, the two forces must balance each other) in order that the ball remain at rest and not accelerate.
- The force of the person's hand on the ball and the force of the ball on the person's hand have equal magnitudes and opposite directions according to Newton's *third* law. These two forces are a force pair: If A represents the hand and B represents the ball, these are the forces of A on B and of B on A.

Figure 4-17b shows the situation when the person's hand is accelerating the ball upward. Now the vector sum $\sum \vec{F}_{\text{ext}}$ of external forces on the ball is not zero, so the force of Earth on the ball does *not* have the same magnitude as the force of the person's hand on the ball. However, the force of the person's hand on the ball still *does* have the same magnitude of the force of the ball on the hand because Newton's third law works at all times, even when the objects are accelerating.

Here's a general rule to keep in mind:

Use Newton's first law or second law when dealing with the forces that act on a given object. Use Newton's third law when relating the forces that act between two objects.

WATCH OUT! Newton's third law involves only two objects.

Two forces can be a force pair only if they are exerted between the *same* two objects. Consider the situation in which you are holding a rock in your hand. The weight force exerted by Earth on the rock cannot ever be a force pair to the normal force of your hand on the same rock. These two forces involve three objects: Earth, rock, and hand. So these two forces cannot be a force pair.

GOT THE CONCEPT? 4-6 Is Motion Impossible?

You ask a friend to push a large crate across a floor (**Figure 4-18**). Your friend declares that this feat isn't possible because whatever force he applies to the box will be met with an equal but opposite force. The vector sum of these forces, he argues, will be zero, so the crate will have zero acceleration and won't move. What's wrong with this argument?

Figure 4-18 Pushing a crate Do Newton's laws make it impossible to push this crate across the floor?

GOT THE CONCEPT? 4-7 Which Forces Act?

Two crates are at rest, one touching the other, on a horizontal surface. You push horizontally on crate 1 as shown by the red arrow in **Figure 4-19**. Which of the following is a force that acts on crate 2? (a) A force exerted by crate 1; (b) a force exerted by you; (c) a reaction force exerted by crate 2; (d) a force due to the acceleration of crate 2; (e) more than one of these.

1
2

Figure 4-19 Pushing two crates If you push on crate 1 as shown by the red arrow, what forces act on crate 2?

Newton's Third Law and Tension

We can use Newton's third law to help us understand another important type of contact force. **Tension** is the force exerted by a rope (or thread, string, cable, or wire) on an object to which the rope is connected. As an example, the rope shown in **Figure 4-20a** exerts a tension force on the sailor who holds on to the left-hand end of the rope and exerts a tension force on the boat tied to its right-hand end. We'll use the symbol $\vec{T}$ for a tension force.

Here's an important fact about tension: If a rope has sufficiently small mass, the tension forces that the rope exerts at its two ends have the *same* magnitude. To see why

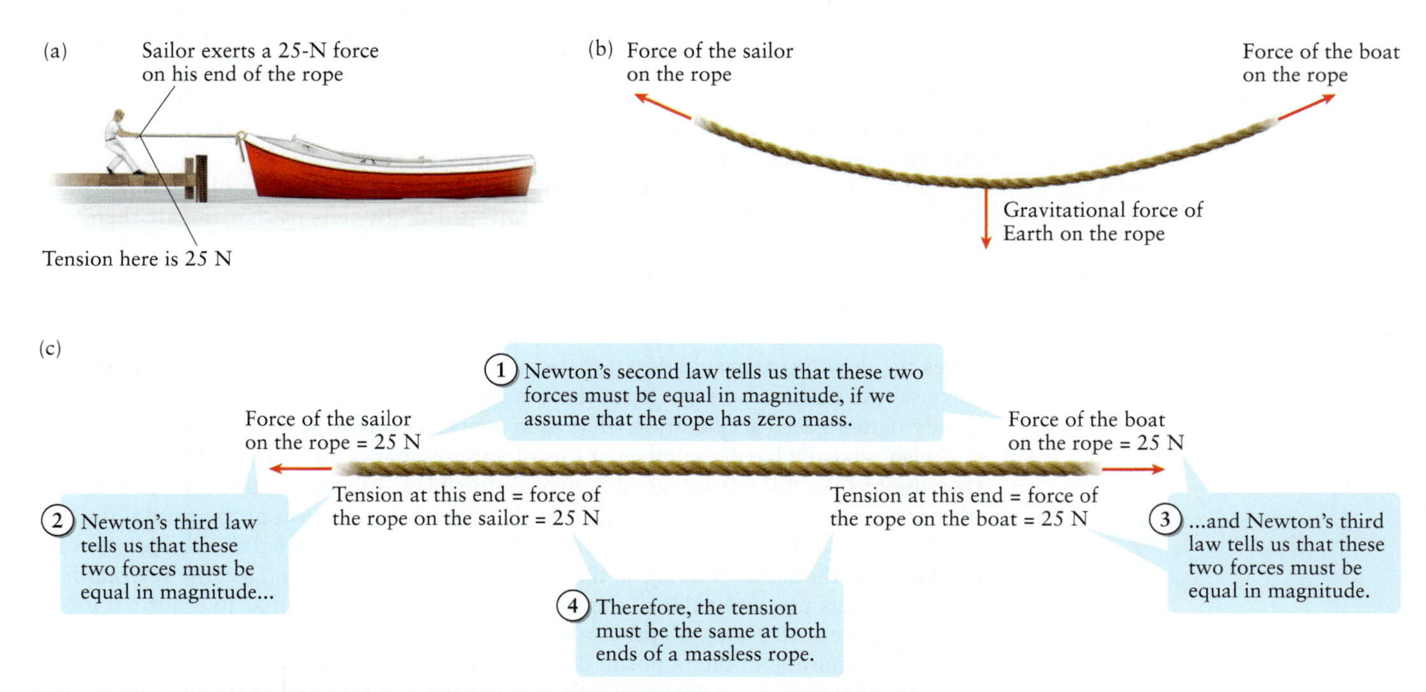
(a)
Sailor exerts a 25-N force on his end of the rope
Tension here is 25 N
(b)
Force of the sailor on the rope
Force of the boat on the rope
Gravitational force of Earth on the rope
(c)
1 Newton's second law tells us that these two forces must be equal in magnitude, if we assume that the rope has zero mass.
Force of the sailor on the rope = 25 N
Force of the boat on the rope = 25 N
Tension at this end = force of the rope on the sailor = 25 N
Tension at this end = force of the rope on the boat = 25 N
2 Newton's third law tells us that these two forces must be equal in magnitude...
3 ...and Newton's third law tells us that these two forces must be equal in magnitude.
4 Therefore, the tension must be the same at both ends of a massless rope.

Figure 4-20 Comparing the tensions at the two ends of a rope (a) A sailor docking a boat pulls on one end of a rope. (b) The forces that act on the rope. (c) The forces that act on a massless rope. In this case the tension force must have the same magnitude at either end of the rope.

these forces have the same magnitude, consider a sailor who docks a boat by pulling on one end of a rope whose other end is tied to the boat (Figure 4-20a). If he pulls with a given force—say, 25 N—on his end of the rope, Newton's third law tells us that the rope pulls back on him with a force of 25 N. How much force does the *other* end of the rope exert on the boat?

To see the answer, consider **Figure 4-20b**, which shows the forces that act *on* the rope. The weight of the rope causes the rope to sag so that the angle of the rope is different at the two ends. However, in many situations the weight of the rope is very small compared to the pulling forces exerted on the rope by the objects attached to its ends. That's the case in **Figure 4-20c**. So we can safely approximate the rope as having *zero* mass and hence *zero* weight. Using this approximation the rope won't sag at all under its own weight (because it has none) and will be perfectly straight. Then the only forces acting on the rope are those at either end. Because the rope is straight, these two forces point in opposite directions (Figure 4-20c).

Newton's *second* law says that the sum of external forces on the rope, $\sum\vec{F}_{ext}$, is equal to $m\vec{a}$, the product of the rope's mass m and its acceleration $\vec{a}$. However, if the rope has zero mass ($m = 0$), the product $m\vec{a}$ is always equal to zero, even if the rope is accelerating. Therefore, the sum of external forces on the rope must be zero ($\sum\vec{F}_{ext} = 0$) under all circumstances. For this to be the case, the oppositely directed forces acting on either end of the rope (Figure 4-20c) must have the same magnitude whether the rope is accelerating or not.

Newton's *third* law tells us that the force exerted *on* each end of the rope has the same magnitude as the tension forces exerted *by* each end of the rope. Because equal-magnitude forces act on each end of a massless rope, it follows that the tension is also the same at *both* ends: If the tension is 25 N at the left-hand end, it is 25 N at the right-hand end (Figure 4-20c). So we are left with the following general rule:

The tension is the same at both ends of a rope or string, provided the weight of the rope or string is much less than the forces exerted on the ends of the rope or string.

TAKE-HOME MESSAGE FOR Section 4-5

✔ Newton's third law tells us that if object *A* exerts a force on object *B*, object *B* has to exert a force on object *A* of the same magnitude but in the opposite direction.

✔ A rope exerts the same tension at both ends, provided the mass of the rope can be ignored.

Newton's Third Law and Propulsion

Newton's third law plays an essential role in biology: All living systems (including you) use it for *propulsion*. As an example, suppose you start running from a standing start. It's common to say that you propel *yourself* forward, but this isn't strictly correct: It takes an external force to accelerate you forward from rest, and an external force has to come from outside your body. What happens is that to start running, you use your feet to exert a *backward* force on the surface of the running track. By Newton's third law, the track exerts a *forward* force of the same magnitude on your feet. It's this force exerted by the track that's responsible for propelling you forward.

Bird flight also depends on Newton's third law. As they flap, a bird's wings push air both backward and downward. Newton's third law tells us that the air therefore exerts forward and upward components of force on the bird's wings. The forward component of force exerted on the bird's wings by the air balances the backward force of air resistance that acts on the bird's body, so the net horizontal external force on the bird is zero. The upward component of force that the air exerts on the bird's wings balances the downward gravitational force on the bird's body, so the net vertical external force on the bird is also zero. Since there is no net external force on the bird, it continues flying with zero acceleration (that is, in a straight line with constant speed).

Fish typically have a number of fins of various shapes and sizes (**Figure 4-21a**) that they use to take full advantage of Newton's third law. In almost all species of fish, the primary means of forward propulsion is the side-to-side motion of the caudal fin and the rear part of the fish's body. **Figure 4-21b** shows the fish from above as it sweeps its caudal fin to the right. In doing so the fish exerts a force $\vec{F}_{\text{fish on water}}$ on the surrounding water that acts perpendicular to the fish's body, in a manner similar to a normal force. In accordance with Newton's third law, the water exerts a force on the fish, $\vec{F}_{\text{water on fish}}$, of equal magnitude in the opposite direction. The forward component of this force on the fish, labeled F_{thrust} in Figure 4-21b, propels the fish forward against the backward

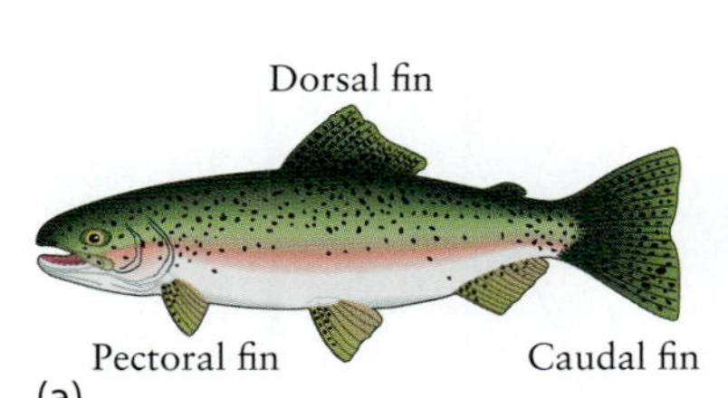

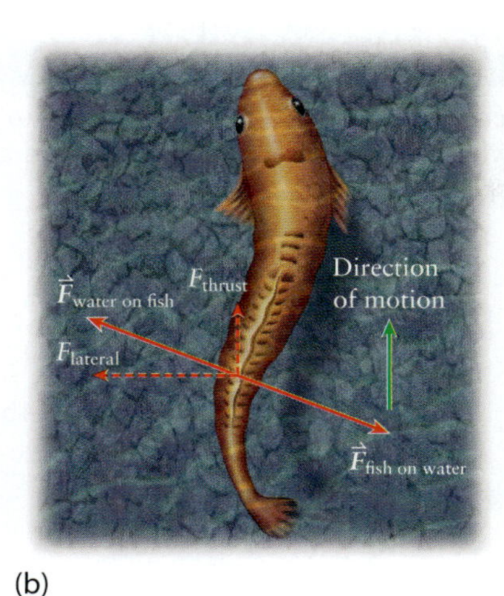

Figure 4-21 Fish propulsion (a) Fish have a variety of fins for propulsion and directional control. The caudal fin is primarily responsible for pushing the fish forward through the water. (b) When a fish swishes its caudal fin back and forth, it exerts a force $\vec{F}_{\text{fish on water}}$ on the water. The reaction to this is the force $\vec{F}_{\text{water on fish}}$ that the water exerts on the fish; the forward component of this force, F_{thrust}, propels the fish forward.

resistance of water on the front of the fish. In some species the motion of the dorsal fin (Figure 4-21a) also contributes to forward propulsion.

Figure 4-21b shows that there is also a force component F_{lateral} on the fish that acts perpendicular to the net motion of the fish. These lateral forces contribute to the stability of the fish and help it turn and maneuver. Most fish can also make subtle adjustments to the orientation of the pectoral fins on the sides of the fish, enabling it to stop, start, and change direction quickly.

4-6 All problems involving forces can be solved using the same series of steps

In this section we'll use Newton's laws to solve a variety of problems that involve forces and acceleration. We can set up *any* force problem using just two steps.

(1) Make a free-body diagram for the object or objects of interest using the rules that we outlined in Section 4-4. Choose the coordinate axes and label the positive directions. (As we've mentioned before, it's best to choose these axes so that one or more of the vectors in the problem—the acceleration and the individual forces—lie along one of the axes.)

(2) Write Newton's second law for each object and in each direction separately. Use the free-body diagram to make sure that for each object, all of the forces that act on that object are included in the sum of external forces. Then solve for the unknowns.

Problems Involving a Single Object

Our first few examples involve just a single object. In some problems the goal is to find the object's acceleration; in other cases you need to determine one or more of the forces acting on the object. Sometimes determining the acceleration or unknown force is not the goal itself but just a necessary step to arrive at the final answer.

$\sqrt{x}$ ***See the Math Tutorial for more information on trigonometry.***

One thing to think about in these problems is the choice of x and y axes for your force diagram. If the object is on an incline, choose one of the axes to point along the incline. If the object is moving in a known direction, it's best to choose one of the axes to point in that direction. If you're not sure which way the object will accelerate (say, whether to the left or to the right), don't worry about it—choose an axis to point in one direction or the other. If it turns out that the acceleration is in the direction opposite to the one you chose, you'll know because the acceleration will turn out to be negative.

EXAMPLE 4-4 What's the Angle?

A pair of fuzzy dice hangs on a lightweight thread from the rearview mirror of your high-performance sports car. At what angle from the vertical do the dice hang when the car is accelerating forward at a constant 6.50 m/s^2?

Set Up

The dice move along with the car and so have the same forward acceleration of 6.50 m/s^2. So there must be a net forward force on the dice. The individual forces on the dice are the gravitational force $\vec{w}_{dice}$ exerted by Earth and the tension force $\vec{T}$ exerted by the thread. Because the thread is lightweight, the gravitational force on it is negligible, and the thread will be straight when taut. So the angle of the thread from the vertical—which is the quantity we're trying to find—is the same as the angle θ of the tension force $\vec{T}$. We'll find this angle by applying Newton's second law to the dice.

Newton's second law in component form:

$$\sum F_{\text{ext on dice},x} = m_{\text{dice}} a_{\text{dice},x} \quad (4\text{-}4a)$$

$$\sum F_{\text{ext on dice},y} = m_{\text{dice}} a_{\text{dice},y} \quad (4\text{-}4b)$$

Net force on dice = sum of tension force and gravitational force:

$$\sum \vec{F}_{\text{ext on dice}} = \vec{T} + \vec{w}_{\text{dice}}$$

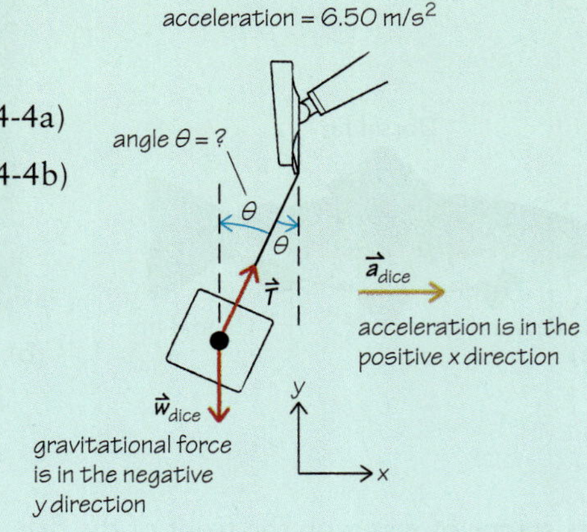

Solve

Write the x and y components of the external forces that act on the dice and the acceleration of the dice.

Components of tension force $\vec{T}$:

$$T_x = T \sin\theta$$
$$T_y = T \cos\theta$$

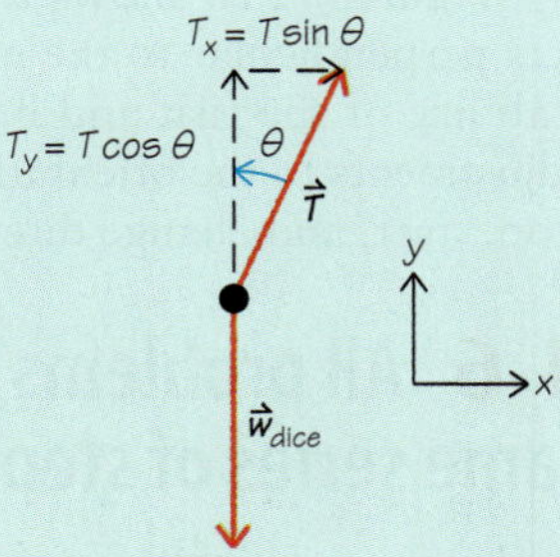

Components of gravitational force $\vec{w}_{\text{dice}}$ with magnitude $w_{\text{dice}} = m_{\text{dice}} g$:

$$w_{\text{dice},x} = 0$$
$$w_{\text{dice},y} = -w_{\text{dice}} = -m_{\text{dice}} g$$

Components of acceleration $\vec{a}_{\text{dice}}$:

$$a_{\text{dice},x} = a_{\text{dice}} = 6.50 \text{ m/s}^2$$
$$a_{\text{dice},y} = 0$$

Substitute the expressions for the components of $\vec{w}$, $\vec{T}$, and $\vec{a}_{\text{dice}}$ into Newton's second law. This gives two equations, one from the x components and one from the y components.

Newton's second law in component form:

$$x\text{: } \sum F_{\text{ext on dice},x} = T_x + w_{\text{dice},x} = m_{\text{dice}} a_{\text{dice},x}, \text{ so}$$
$$T \sin\theta + 0 = m_{\text{dice}} a_{\text{dice}}$$
$$y\text{: } \sum F_{\text{ext on dice},y} = T_y + w_{\text{dice},y} = m_{\text{dice}} a_{\text{dice},y}, \text{ so}$$
$$T \cos\theta + (-m_{\text{dice}} g) = 0$$

We don't know either the magnitude T or the angle θ of the tension force. All we are asked to solve for is the value of θ, so we eliminate T from the two equations above. We also use the definition of the tangent function, $\tan\theta = (\sin\theta)/(\cos\theta)$.

From the y equation:

$$T \cos\theta = m_{\text{dice}} g, \text{ so}$$
$$T = \frac{m_{\text{dice}} g}{\cos\theta}$$

Substitute this into the x equation:

$$T \sin\theta = m_{\text{dice}} a_{\text{dice}}, \text{ so}$$
$$\frac{m_{\text{dice}} g}{\cos\theta} \sin\theta = m_{\text{dice}} a_{\text{dice}}$$

Divide through by m_{dice} and g and apply the definition of tangent.

$$\frac{\sin\theta}{\cos\theta} = \tan\theta = \frac{a_{\text{dice}}}{g}$$

Solve for θ:

$$\theta = \tan^{-1}\left(\frac{a_{dice}}{g}\right) = \tan^{-1}\left(\frac{6.50 \text{ m/s}^2}{9.80 \text{ m/s}^2}\right) = \tan^{-1}(0.663)$$

$$= 33.6°$$

Reflect

Note that our answer doesn't depend on the mass m_{dice}. The angle θ would be the same whether the dice were made of lightweight plastic or solid gold.

We can check our result by using a range of values of acceleration. When the car is not accelerating (as would be the case if its velocity is constant), $a_{dice} = 0$ and $\theta = \tan^{-1} 0 = 0$. The dice hang straight down, which is expected. The greater the acceleration a_{dice}, the greater the value of θ—which is also what we would expect.

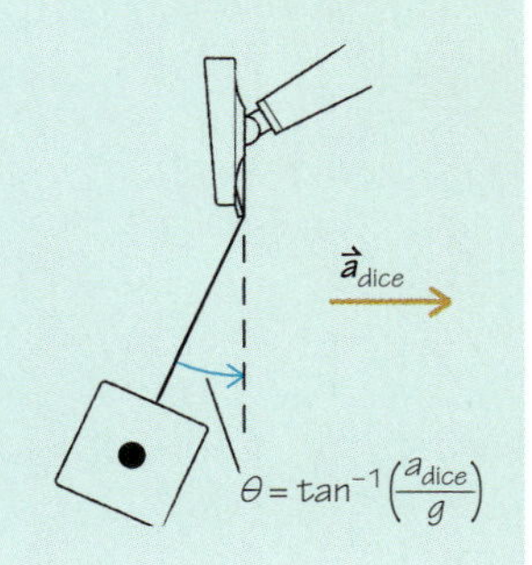

WATCH OUT! Newton's laws apply only in inertial frames of reference.

You might argue that the dice in the preceding example accelerate *backward* rather than forward. After all, from your perspective as the driver, the dice swing backward when you step on the accelerator pedal. The driver's seat isn't a good place from which to apply Newton's laws, however, because you are yourself accelerating! It turns out that Newton's laws work *only* if you observe from a vantage point, or *frame of reference,* that is *not* accelerating. Such a vantage point is called an **inertial frame of reference.** (A more detailed study of Newton's first law than we can present in this book usually considers Newton's first law as a *definition* of inertial frames of reference. That's really why Newton's first law gets its own number.) One such frame of reference would be a person standing by the road as you accelerate away. The dice initially will lag behind the car due to their inertia, which is why the dice swing backward relative to the driver. Once the dice have stabilized, however, they move with the same acceleration as the car. The same inertial effect explains why you feel pushed back in your seat when the car accelerates forward, and you feel thrown forward against your safety belt if the car brakes rapidly. No real force acts on you in either case: It's just your inertia. (If you think inertia *is* a real force, ask yourself what could be exerting it. If you think it's your own body, remember that only external forces affect motion; if you think it's "the force of acceleration" or "the force of velocity," remember from Section 4-4 that these forces don't exist.)

EXAMPLE 4-5 Slipping and Sliding I

One wintry day you accidentally drop your physics book (mass 2.50 kg) on an icy, essentially frictionless sidewalk that tilts at an angle of 10.0° below the horizontal. What is the book's acceleration as it slides downhill?

Set Up

We'll find the book's acceleration using Newton's second law. Because there is no friction in this situation, the only forces acting on the book are the gravitational force $\vec{w}_{book}$ (which points straight down) and the normal force $\vec{n}$ (which points perpendicular to the surface of the inclined sidewalk). This situation is similar to the block shown in Figure 4-12 (Section 4-4). Because the motion is down the ramp, we choose the x axis to be in that direction.

Newton's second law in component form applied to the book:

$$\sum F_{ext\ on\ book,x} = m_{book}a_{book,x} \quad (4\text{-}4a)$$
$$\sum F_{ext\ on\ book,y} = m_{book}a_{book,y} \quad (4\text{-}4b)$$

Net force on book = sum of normal force and gravitational force:

$$\sum \vec{F}_{ext\ on\ book} = \vec{n} + \vec{w}_{book}$$

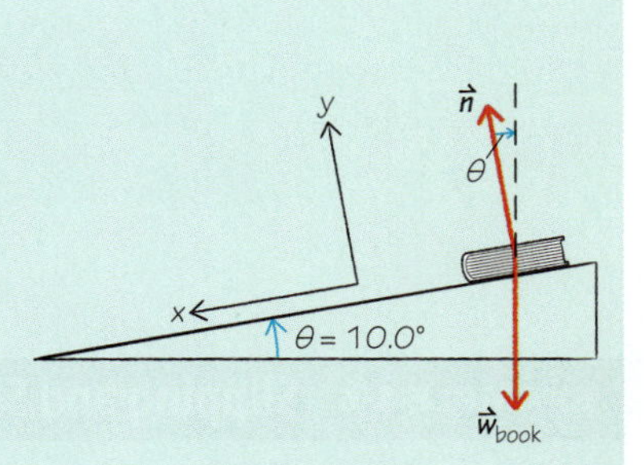

Solve

Write the forces and the acceleration in component form. The book doesn't move perpendicular to the ramp, so the y component of its acceleration is zero.

Components of normal force $\vec{n}$:

$$n_x = 0$$
$$n_y = n$$

Components of gravitational force $\vec{w}_{\text{book}}$:

magnitude $w_{\text{book}} = m_{\text{book}}g$

$$w_{\text{book},x} = w_{\text{book}} \sin\theta = m_{\text{book}}g \sin\theta$$

$$w_{\text{book},y} = -w_{\text{book}} \cos\theta = -m_{\text{book}}g \cos\theta$$

Components of acceleration $\vec{a}_{\text{book}}$:

$a_{\text{book},x}$: to be determined

$a_{\text{book},y} = 0$

Use the components to write Newton's second law for the book in component form.

Newton's second law in component form:

$$x\text{: } \sum F_{\text{ext on book},x} = n_x + w_x = 0 + m_{\text{book}}g \sin\theta = m_{\text{book}}a_{\text{book},x}\text{, so}$$

$$m_{\text{book}}g \sin\theta = m_{\text{book}}a_{\text{book},x}$$

$$y\text{: } \sum F_{\text{ext on book},y} = n_y + w_y = n - m_{\text{book}}g \cos\theta = m_{\text{book}}a_{\text{book},y}\text{, so}$$

$$n - m_{\text{book}}g \cos\theta = 0$$

All we are asked to find is the acceleration of the book, so we need only the x equation.

Dividing the x equation by m_{book}:

$$a_{\text{book},x} = g \sin\theta = (9.80 \text{ m/s}^2) \sin 10.0° = 1.70 \text{ m/s}^2$$

Reflect

Our result tells us that the book accelerates in the positive x direction (downhill). This result makes sense: When no friction opposes its motion, the book should gain speed as it slides downhill. If the slope were steeper, θ and $\sin\theta$ would be greater, and the acceleration $a_{\text{book},x} = g \sin\theta$ would be greater as well. Just as for an object in free fall, the acceleration doesn't depend on the book's mass.

Note also that our analysis would be exactly the same if you put the book near the bottom of the incline and gave it an upward push to start it moving. The gravitational force and normal force would be unchanged, so the magnitude and direction of the acceleration would likewise be unchanged. Because the acceleration is downhill and the velocity uphill, the book would slow down as it moved up the incline—again, just as we would expect.

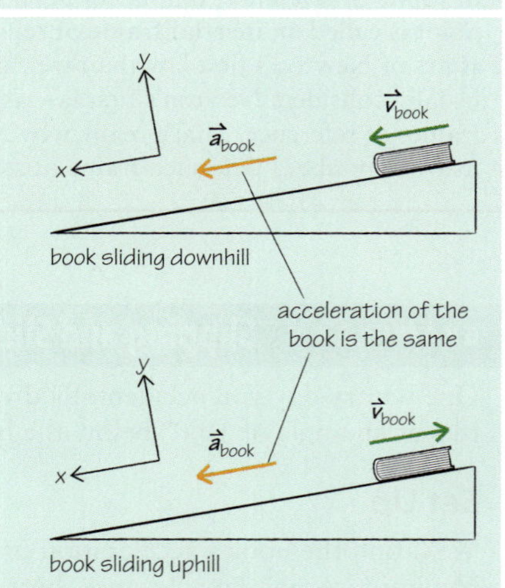

EXAMPLE 4-6 Slipping and Sliding II

If there is friction between the book and the incline in the previous example, the book's downhill acceleration will be less than we calculated above. If the book gains speed at only 0.200 m/s^2 as it slides downhill, what is the magnitude of the friction force on the 2.50 kg book?

Set Up

The situation is much the same as in Example 4-5 but with an additional friction force $\vec{f}$ which points up the incline, opposing the book's motion. We are given the book's acceleration, so we'll use Newton's second law to determine the magnitude of the friction force.

Newton's second law in component form applied to the book:

$$\sum F_{\text{ext on book},x} = m_{\text{book}} a_{\text{book},x} \quad \text{(4-4a)}$$

$$\sum F_{\text{ext on book},y} = m_{\text{book}} a_{\text{book},y} \quad \text{(4-4b)}$$

Net force on book = sum of normal force, gravitational force, and friction force:

$$\sum \vec{F}_{\text{ext on book}} = \vec{n} + \vec{w}_{\text{book}} + \vec{f}$$

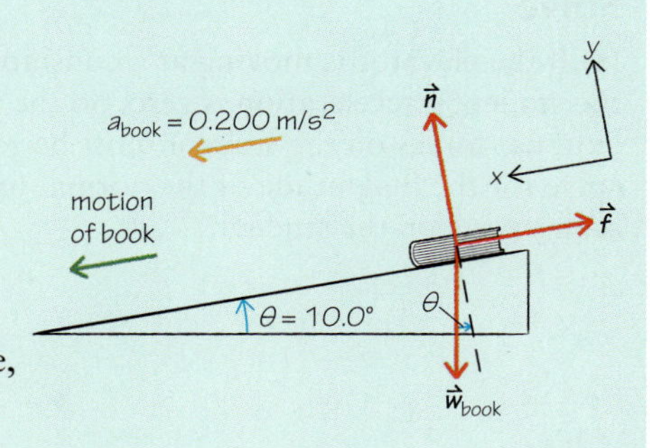

Solve

The vectors $\vec{n}$, $\vec{w}_{\text{book}}$, and $\vec{a}_{\text{book}}$ break into components the same way that they did in Example 4-5. We complete the list with the vector components of $\vec{f}$.

Components of friction force $\vec{f}$:

$f_x = -f$

$f_y = 0$

Use the components to write Newton's second law for the book in component form.

Newton's second law in component form:

x: $\sum F_{\text{ext on book},x} = n_x + w_x + f_x = 0 + m_{\text{book}} g \sin\theta - f = m_{\text{book}} a_{\text{book},x}$ so $m_{\text{book}} g \sin\theta - f = m_{\text{book}} a_{\text{book},x}$

y: $\sum F_{\text{ext on book},y} = n_y + w_y + f_y = n - m_{\text{book}} g \cos\theta + 0 = m_{\text{book}} a_{\text{book},y}$ so $n - m_{\text{book}} g \cos\theta = 0$

Solve the x equation for the magnitude f of the friction force.

$m_{\text{book}} g \sin\theta - f = m_{\text{book}} a_{\text{book},x}$ so

$$\begin{aligned} f &= m_{\text{book}} g \sin\theta - m_{\text{book}} a_{\text{book},x} \\ &= m_{\text{book}}(g \sin\theta - a_{\text{book},x}) \\ &= (2.50\ \text{kg})[(9.80\ \text{m/s}^2)\sin 10.0° - 0.200\ \text{m/s}^2] \\ &= 3.75\ \text{kg}\cdot\text{m/s}^2 = 3.75\ \text{N} \end{aligned}$$

Reflect

This example is very much like Example 4-3 in Section 4-3. The only difference is that the net force on the cat in Example 4-3 is zero (the cat is at rest), while the net force on the sliding book in this example is *not* zero.

EXAMPLE 4-7 How to "Lose Weight" the Easy Way

A 70.0 kg student stands on a bathroom scale in an elevator. What is the reading on the scale (a) when the elevator is moving upward at a constant 0.500 m/s? (b) When the elevator is accelerating downward at 1.00 m/s²?

Set Up

The reading on a bathroom scale tells you how hard you are pushing on it, which may not be the same as your weight. (Try pushing on a bathroom scale with your hand. The reading measures how strong you are, not how much you weigh.) Newton's third law tells us that the downward normal force of the student on the scale is equal in magnitude to the upward normal force exerted by the scale on the student, $\vec{n}_{\text{scale on student}}$. So what we want to know is the magnitude of $\vec{n}_{\text{scale on student}}$ in each situation, which we'll find by applying Newton's second law to the student. The only other force acting on the student is the gravitational force $\vec{w}_{\text{student}}$. Both forces and the acceleration in part (b) are in the vertical direction, so we only need the y component.

y component of Newton's second law applied to the student:

$$\sum F_{\text{ext on student},y} = m_{\text{student}} a_{\text{student},y} \quad \text{(4-4b)}$$

Net force on student = sum of normal force and gravitational force:

$$\sum \vec{F}_{\text{ext on student}} = \vec{n}_{\text{scale on student}} + \vec{w}_{\text{student}}$$

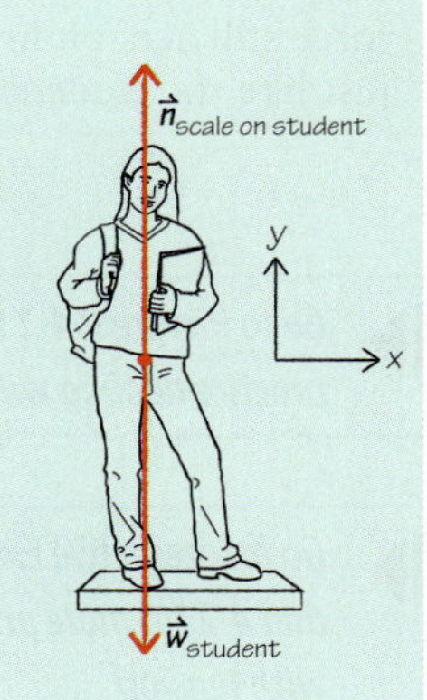

Solve

(a) If the elevator is moving at a constant velocity, the student's acceleration is zero. So the sum of the external forces on the student must be zero. We then solve for the magnitude of the normal force that the scale exerts on the student.

Student's y velocity is constant, so

$a_{\text{student},y} = 0$

Newton's second law in component form:

$$y: \sum F_{\text{ext on student},y} = n_{\text{scale on student}} + (-w_{\text{student}}) = m_{\text{student}} a_{\text{student},y} = 0 \text{ so}$$

$$n_{\text{scale on student}} = w_{\text{student}} = m_{\text{student}} g = (70.0 \text{ kg})(9.80 \text{ m/s}^2) = 686 \text{ N}$$

constant velocity: $a_{\text{student},y} = 0$

$\vec{n}_{\text{scale on student}}$

net external force = 0

$\vec{w}_{\text{student}}$

(b) If the elevator, and so also the student, are accelerating downward (in the negative y direction) at 1.00 m/s^2, the net force on the student must be downward. The net force can be downward only if the upward normal force exerted by the scale is smaller than the downward gravitational force. Again we solve for the magnitude $n_{\text{scale on student}}$ using Newton's second law.

The student's acceleration has a downward (negative) y component:

$a_{\text{student},y} = -1.00 \text{ m/s}^2$

Newton's second law in component form:

$$y: \sum F_{\text{ext on student},y} = n_{\text{scale on student}} + (-w_{\text{student}}) = m_{\text{student}} a_{\text{student},y} \text{ so}$$

$$n_{\text{scale on student}} = w_{\text{student}} + m_{\text{student}} a_{\text{student},y} = 686 \text{ N} + (70.0 \text{ kg})(-1.00 \text{ m/s}^2) = 616 \text{ N}$$

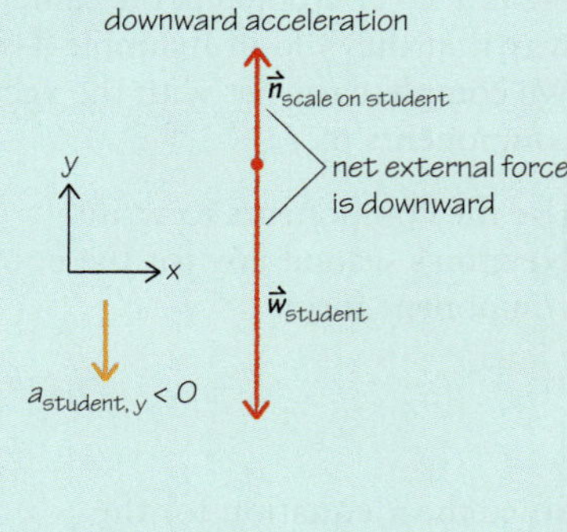

Reflect

According to the scale the student "weighs" 70 N less when the elevator accelerates downward. You've probably experienced this phenomenon in an elevator when it starts moving downward from rest or when it comes to a halt when moving upward. A similar phenomenon occurs in an elevator that accelerates *upward*. Then $a_{\text{student},y}$ is positive, $n_{\text{scale on student}}$ is greater than the student's weight w_{student}, and the student feels heavier than normal.

If the elevator is in free fall,

$a_{\text{student},y} = -g$

$$n_{\text{scale on student}} = w_{\text{student}} + m_{\text{student}} a_{\text{student},y} = m_{\text{student}} g + m_{\text{student}}(-g) = 0$$

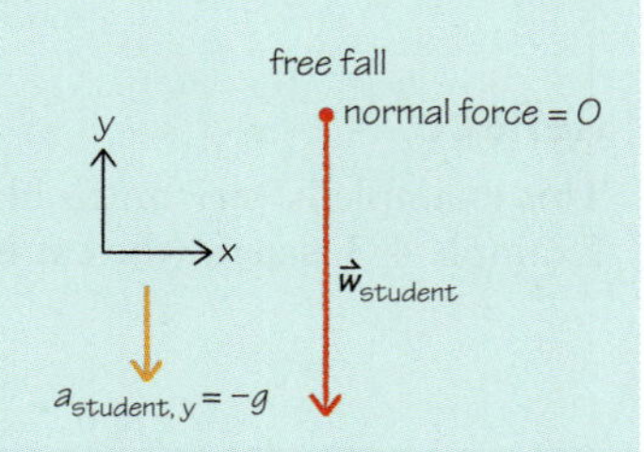

What if the elevator is in free fall? Then $a_{\text{student},y} = -g = -9.80 \text{ m/s}^2$ (negative because the acceleration is downward). Substituting this value into our result gives $n_{\text{scale on student}} = 0$. The scale reads zero, and the student will feel weightless! This situation is called *apparent weightlessness*. (The student isn't truly weightless because the gravitational force still acts on her.) We'll see in Chapter 10 that astronauts in Earth orbit are also in free fall, so they *feel* weightless just like the student in the free-falling elevator.

Go to Picture It 4-1 for more practice dealing with weight

Go to Interactive Exercises 4-1 and 4-2 for more practice dealing with tension

Problems Involving Ropes and Tension

What object is pulling the water skier shown in **Figure 4-22**? You might be tempted to say that it's the boat—but that can't be correct because the boat doesn't touch him. Instead the boat exerts a force on the tow rope, and it's the tow rope that pulls on the skier. The force that pulls the skier is therefore a *tension* force.

As we learned in Section 4-5, if the weight of the tow rope is small compared to the forces exerted by the objects to which it's attached (in this case, the boat and the water skier), we can ignore its mass altogether. Then the magnitude of the tension force

Karl Weatherly/Photodisc/Getty Images

Figure 4-22 Towing a water skier The boat exerts a force on the tow rope, and the tow rope pulls on the skier.

is the *same* at both ends of the rope no matter how the boat, skier, and rope are moving. We'll use this principle in the next two examples, both of which involve moving objects that are connected by a lightweight string.

EXAMPLE 4-8 Tension in a String

You pull two blocks connected by a lightweight string across a horizontal table. Block 1 has a mass of 1.00 kg, and block 2 has a mass of 2.00 kg. There is negligible friction between the blocks and the table. The string will break if the tension is greater than 6.00 N. What is the maximum pull that you can apply to block 1 without breaking the string?

Set Up

As in the previous examples we start by drawing a free-body diagram. Note that here there are *two* bodies (block 1 and block 2), so we have to be careful and draw *two* free-body diagrams to include the forces that act on each block. We choose the positive x axis to be in the direction of the blocks' horizontal motion and choose the positive y axis to be vertically upward.

$$\sum \vec{F}_{\text{ext on 1}} = m_1\vec{a}_1$$

$$\sum \vec{F}_{\text{ext on 2}} = m_2\vec{a}_2 \qquad (4\text{-}2)$$

Net force on block 1 = sum of force you exert, normal force, gravitational force, and tension:

$$\sum \vec{F}_{\text{ext on 1}} = \vec{F}_{\text{you on 1}} + \vec{n}_1 + \vec{w}_1 + \vec{T}_{\text{string on 1}}$$

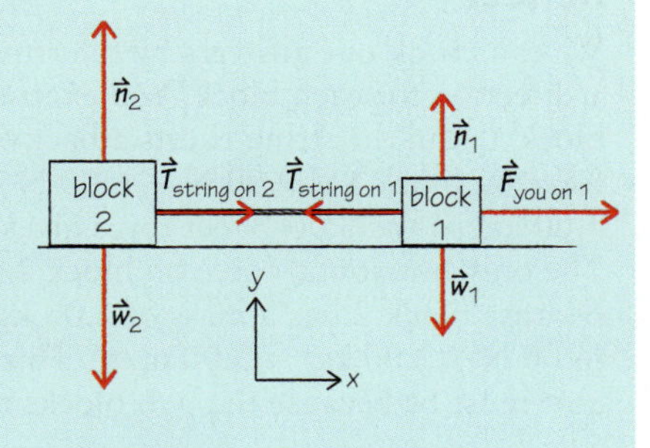

We are told that the string connecting the blocks is lightweight, so the tension at either end of the string has the same magnitude: $T_{\text{string on 1}} = T_{\text{string on 2}}$. We'll call this magnitude T for short. Our goal is to find the maximum value of $F_{\text{you on 1}}$ (the magnitude of the force that you exert on block 1) so that T is no greater than 6.00 N.

Net force on block 2 = sum of normal force, gravitational force, and tension:

$$\sum \vec{F}_{\text{ext on 2}} = \vec{n}_2 + \vec{w}_2 + \vec{T}_{\text{string on 2}}$$

Tension is the same at both ends of the string:

$$T_{\text{string on 1}} = T_{\text{string on 2}} = T$$

Solve

We then write Newton's second law for each block in component form. If we assume that the string doesn't stretch, the two blocks move together and have the same x component of acceleration, which we call a_x.

Both blocks have the same x acceleration:
$a_{1x} = a_{2x} = a_x$
Both blocks have zero y acceleration.

Newton's second law in component form for block 1:

$$x: \sum F_{\text{ext on }1,x} = F_{\text{you on }1} + (-T) = m_1 a_x$$

$$y: \sum F_{\text{ext on }1,y} = n_1 + (-w_1) = 0$$

Newton's second law in component form for block 2:

$$x: \sum F_{\text{ext on }2,x} = T = m_2 a_x$$

$$y: \sum F_{\text{ext on }2,y} = n_2 + (-w_2) = 0$$

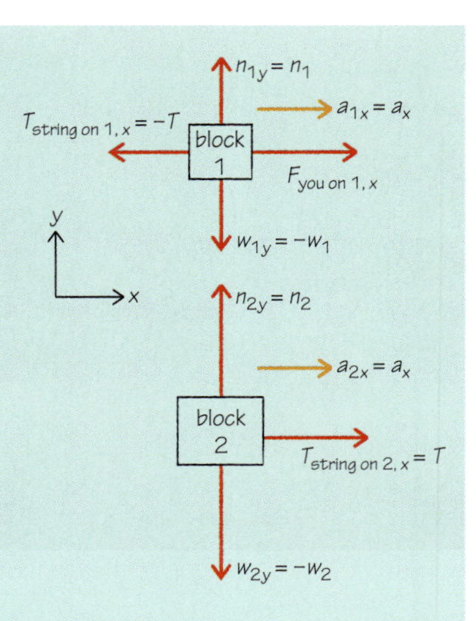

We want to find the value of $F_{\text{you on }1}$ when $T = 6.00$ N (the maximum tension that will prevent the string from breaking). We use the x equation for block 2 to find the x acceleration a_x in this situation and then we use this value of a_x in the x equation for block 1 to solve for $F_{\text{you on }1}$.

First solve for the value of a_x when $T = 6.00$ N. Use the x equation for block 2:

$$T = m_2 a_x$$

$$a_x = \frac{T}{m_2} = \frac{6.00\text{ N}}{2.00\text{ kg}} = 3.00\text{ N/kg} = 3.00\text{ m/s}^2$$

Now substitute $a_x = 3.00\text{ m/s}^2$ into the x equation for block 1 and solve for $F_{\text{you on }1}$:

$$F_{\text{you on }1} - T = m_1 a_x$$

$$F_{\text{you on }1} = T + m_1 a_x = 6.00\text{ N} + (1.00\text{ kg})(3.00\text{ m/s}^2)$$

$$= 6.00\text{ N} + 3.00\text{ kg}\cdot\text{m/s}^2 = 6.00\text{ N} + 3.00\text{ N} = 9.00\text{ N}$$

Reflect

We can check our answers by referring back to Newton's second law in the x direction for each block. You exert a forward force $F_{\text{you on }1} = 9.00$ N on block 1, and the string exerts a backward force of 6.00 N. So the net forward force on block 1 is 9.00 N − 6.00 N = 3.00 N. This block has a mass of 1.00 kg, so the 3.00-N net force gives block 1 an acceleration of 3.00 m/s^2. The only horizontal force on block 2 is the forward tension force of 6.00 N; because block 2 has a mass of 2.00 kg, the forward acceleration produced is (6.00 N)/(2.00 kg) = 3.00 m/s^2. This acceleration is the same as for block 1, as it must be because the two blocks move together.

T = 6.00 N
Fyou on 1 = 9.00 N
2.00 kg
1.00 kg
acceleration $= \frac{6.00\text{ N}}{2.00\text{ kg}}$ = 3.00 m/s^2
acceleration $= \frac{(9.00\text{ N} - 6.00\text{ N})}{1.00\text{ kg}}$ = 3.00 m/s^2

Because the two blocks move together, we could have treated the two blocks together as a single unit with a mass equal to $m_1 + m_2 = 3.00$ kg. That wouldn't have helped us solve this problem, however. The reason is that the tension forces are *internal* forces (they involve one part of the combined unit pulling on another part) in this approach. As a result there would be no way to apply the condition that the tension cannot exceed 6.00 N.

Here's the lesson: When solving problems that involve forces and accelerations for more than one object, it's safest to *always* apply Newton's laws to each object separately.

However, the combined system of two blocks can be used as another way to find the acceleration of the two blocks moving together. Can you solve the equations for the two blocks as a unit to show again that this acceleration is 3.00 m/s^2?

Newton's second law in component form for the two blocks as a unit:

$$x: \sum F_{\text{ext on unit},x} = F_{\text{you on }1} = (m_1 + m_2) a_{\text{on unit},x}$$

$$y: \sum F_{\text{ext on }1,y} = n + (-w) = 0$$

We can't solve for $F_{\text{you on }1}$ because the tension T does not appear in these equations.

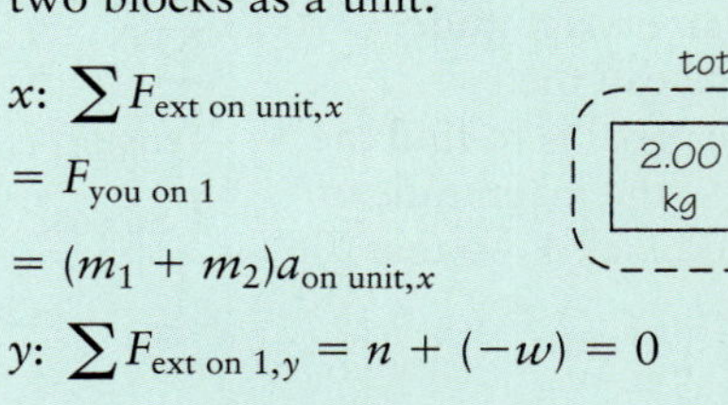

EXAMPLE 4-9 Atwood's Machine

Two blocks are connected by a lightweight string that passes over a pulley as shown in **Figure 4-23**. (This arrangement is known as *Atwood's machine* after Rev. George Atwood, who devised it in 1784 as a way to demonstrate motion with constant acceleration.) Block 1 is more massive than block 2, so when released block 1 accelerates downward, and block 2 accelerates upward. Assume that the mass of the pulley is so small that we can neglect it. Derive expressions for (a) the acceleration of each block and (b) the tension in the string.

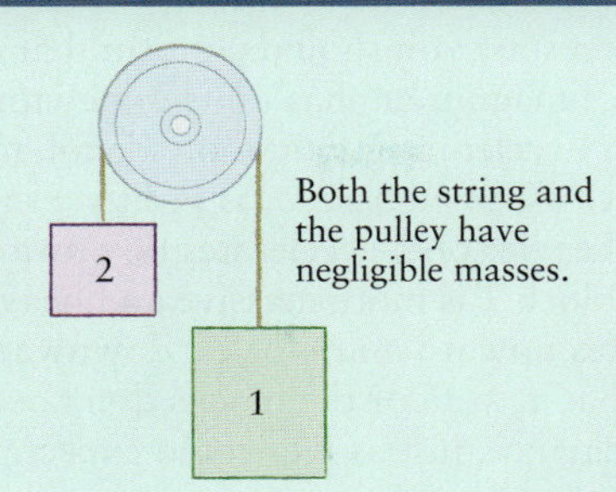

2
1
Both the string and the pulley have negligible masses.

Figure 4-23 Atwood's machine When the blocks are released from rest, block 1 falls and, block 2 rises. What are their accelerations, and what is the tension in the string?

Set Up

Each block has only two forces acting on it: a downward gravitational force and an upward tension force. We choose the positive y direction for block 1 to be downward and the positive y direction for block 2 to be upward. Using this choice both blocks accelerate in the positive y direction and both blocks have the same y component of acceleration (because we assume that the string doesn't stretch). One of our goals is to calculate this acceleration, which we call a_y.

Because we can neglect the mass of the pulley, the magnitude of the tension force has the same value T at both ends of the lightweight string. Our other goal is to calculate the value of T.

Newton's second law in the y direction applied to each block:

$$\sum F_{\text{ext on 1}, y} = m_1 a_{1y}$$
$$\sum F_{\text{ext on 2}, y} = m_2 a_{2y} \qquad (4\text{-}2)$$

Net force on block 1 = sum of gravitational force and tension:

$$\sum \vec{F}_{\text{ext on 1}} = \vec{w}_1 + \vec{T}_{\text{string on 1}}$$

Net force on block 2 = sum of gravitational force and tension:

$$\sum \vec{F}_{\text{ext on 2}} = \vec{w}_2 + \vec{T}_{\text{string on 2}}$$

Both blocks have the same y acceleration:

$a_{1y} = a_{2y} = a_y$

Magnitude of tension is the same at both ends of the string:

$T_{\text{string on 1}} = T_{\text{string on 2}} = T$

Solve

Write the y component of Newton's second law for each block.

Newton's second law in component form for block 1:

$$y\text{: } \sum F_{\text{ext on 1},y} = w_1 + (-T) = m_1 a_y \text{ or } m_1 g - T = m_1 a_y$$

Newton's second law in component form for block 2:

$$y\text{: } \sum F_{\text{ext on 2},y} = T + (-w_2) = m_2 a_y \text{ or } T - m_2 g = m_2 a_y$$

We have two equations (one for block 1 and one for block 2) and two quantities that we want to find (a_y and T). One way to proceed is to first solve for one unknown in terms of the other. Let's solve the block 1 equation for T in terms of a_y. Then substitute this expression for T into the block 2 equation and solve for a_y.

Solve the block 1 equation for T:

$m_1 g - T = m_1 a_y$ so $T = m_1 g - m_1 a_y$

Substitute this into the block 2 equation:

$(m_1 g - m_1 a_y) - m_2 g = m_2 a_y$

Solve for a_y:

$m_1 g - m_2 g = m_1 a_y + m_2 a_y$

$(m_1 - m_2)\, g = (m_1 + m_2)\, a_y$ or

$$a_y = \left(\frac{m_1 - m_2}{m_1 + m_2}\right) g$$

Now that we have an expression for a_y, substitute it back into the equation for T that we found from the block 1 equation. This resulting expression gives us our final answer for T.

From the block 1 equation for T:

$T = m_1 g - m_1 a_y$

Substitute our expression for a_y and simplify:

$$T = m_1 g - m_1\left(\frac{m_1 - m_2}{m_1 + m_2}\right) g = m_1 g\left(1 - \frac{m_1 - m_2}{m_1 + m_2}\right)$$

$$= m_1 g\left(\frac{m_1 + m_2}{m_1 + m_2} - \frac{m_1 - m_2}{m_1 + m_2}\right) = m_1 g\left(\frac{m_1 + m_2 - m_1 + m_2}{m_1 + m_2}\right) \text{ or}$$

$$T = \frac{2 m_1 m_2 g}{m_1 + m_2}$$

Reflect

Do these results make sense? As a start you should confirm that our expression for a_y (an acceleration) has units of m/s^2, and our expression for T (a force magnitude) has units of newtons. Do they?

To check our expression for acceleration a_y in more detail, notice that it depends on the difference $m_1 - m_2$ between the two masses. If $m_1 > m_2$ as we assumed here, a_y is positive; then both blocks accelerate in the positive y direction, so block 1 (the more massive and heavier one) accelerates downward, and block 2 accelerates upward.

If instead $m_1 < m_2$ so that block 2 is more massive and heavier, the acceleration a_y is negative. Then block 1 (which is now the lighter one) accelerates upward and block 2 downward. Finally, if the two blocks have equal mass so that $m_1 = m_2$, our expression says that $a_y = 0$, or the blocks don't accelerate at all. If the two equal-mass blocks begin at rest, they will remain at rest and in balance, just as we would expect.

What about our expression for the tension T? To check this out let's choose some specific values of m_1 and m_2. With $m_1 = 4.00$ kg and $m_2 = 2.00$ kg, the tension is $T = 26.1$ N. That's less than the 39.2-N weight of block 1 but greater than the 19.6-N weight of block 2. This result makes sense because for block 1 the downward gravitational force overwhelms the upward tension force and the block accelerates downward, while for block 2 the upward tension force overwhelms gravity, and the block accelerates upward. Note that because the tension force "holds back" the downward acceleration of block 1, the value of a_y is less than g.

Let $m_1 = 4.00$ kg and $m_2 = 2.00$ kg

Weight of block 1 $= m_1 g$
$= (4.00 \text{ kg})(9.80 \text{ m/s}^2)$
$= 39.2$ N

Weight of block 2 $= m_2 g$
$= (2.00 \text{ kg})(9.80 \text{ m/s}^2)$
$= 19.6$ N

Tension in string:

$$T = \frac{2m_1 m_2 g}{m_1 + m_2} = \frac{2(4.00 \text{ kg})(2.00 \text{ kg})(9.80 \text{ m/s}^2)}{4.00 \text{ kg} + 2.00 \text{ kg}} = 26.1 \text{ N}$$

You should repeat these calculations for different values of the two masses. Try $m_1 = 1.00$ kg and $m_2 = 3.00$ kg: How does the tension compare to the weights of the two blocks in this case? Is the acceleration positive or negative in this case?

Acceleration of either block:

$$a_y = \left(\frac{m_1 - m_2}{m_1 + m_2}\right)g = \left(\frac{4.00 \text{ kg} - 2.00 \text{ kg}}{4.00 \text{ kg} + 2.00 \text{ kg}}\right)(9.80 \text{ m/s}^2) = 3.27 \text{ m/s}^2$$

Kinematics and Newton's Laws

All of the objects that we considered in Examples 4-4 through 4-9 move in straight lines. What's more, the forces acting on each object are constant. So the *accelerations* that these forces produce are constant as well. That means we can apply all of the kinematic formulas for straight-line motion with constant acceleration that we learned in Chapter 2. If you can solve problems that use these formulas, and can also solve problems that involve Newton's laws, it's straightforward to solve problems that involve both of these. Here's an example.

EXAMPLE 4-10 Down the Slopes

A 60.0-kg skier starts from rest at the top of a hill with a 30.0° slope. She reaches the bottom of the slope 4.00 s later. If there is a constant 72.0-N friction force that resists her motion, how long is the hill?

Set Up

All three of the forces that act on the skier—the normal force $\vec{n}$, the gravitational force $\vec{w}_{skier}$, and the friction force $\vec{f}$—are constant. So the net force on the skier is constant, as is her acceleration. We can therefore use a constant-acceleration formula to find the length of the hill. We'll find the skier's downhill acceleration using Newton's second law.

$$x = x_0 + v_{0x}t + \frac{1}{2}a_x t^2 \quad (2\text{-}9)$$

Newton's second law in component form for the skier:

$$x: \sum F_{\text{ext on skier},x} = m_{\text{skier}} a_x$$

$$y: \sum F_{\text{ext on skier},y} = m_{\text{skier}} a_y$$

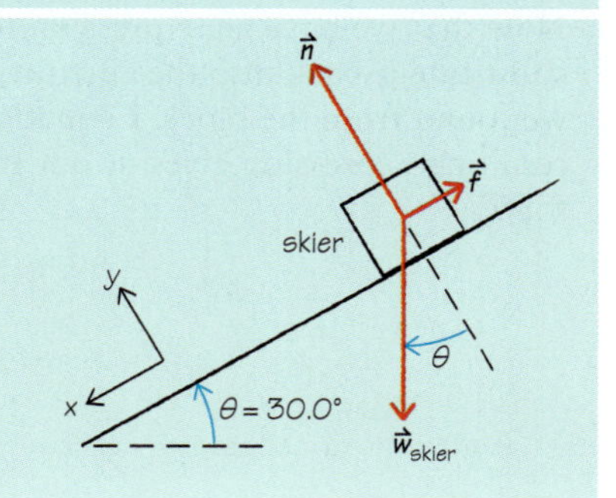

We choose the positive x axis to point down the slope and the positive y axis to point perpendicular to the slope as shown. (Compare Example 4-6.) Then the skier's acceleration points along the x axis only.

Net force on skier = sum of normal force, gravitational force, and friction force:

$$\sum \vec{F}_{\text{ext on skier}} = \vec{n} + \vec{w}_{\text{skier}} + \vec{f}$$

Solve

The normal force $\vec{n}$ has only a positive y component. The gravitational force $\vec{w}_{\text{skier}}$ has a positive (downhill) x component and a negative y component, while the friction force has only a negative (uphill) x component. Use these components and Newton's second law to solve for the skier's x component of acceleration a_x.

Find the skier's x component of acceleration from the x component of Newton's second law:

$$\sum F_{\text{ext on skier},x} = w_{\text{skier}} \sin\theta + (-f) = m_{\text{skier}} a_x$$

Weight of skier:

$w_{\text{skier}} = m_{\text{skier}} g$, so

$$m_{\text{skier}} a_x = m_{\text{skier}} g \sin\theta - f$$

$$a_x = g \sin\theta - \frac{f}{m_{\text{skier}}}$$

$$= (9.80\ \text{m/s}^2) \sin 30.0° - \frac{72.0\ \text{N}}{60.0\ \text{kg}}$$

$$= 4.90\ \text{m/s}^2 - 1.20\ \text{m/s}^2$$

$$= 3.70\ \text{m/s}^2$$

Now that we know the skier's acceleration in the x direction, we can use the kinematic equation to solve for her final position, which will be equal to the length of the hill as long as she starts at $x_0 = 0$. Because the skier starts at rest, her initial velocity is $v_{0x} = 0$.

Find the skier's final position after 4.00 s, assuming she starts at $x_0 = 0$.

$$x = x_0 + v_{0x} t + \frac{1}{2} a_x t^2$$

$$= 0 + 0 + \frac{1}{2}(3.70\ \text{m/s}^2)(4.00\ \text{s})^2$$

$$= 29.6\ \text{m}$$

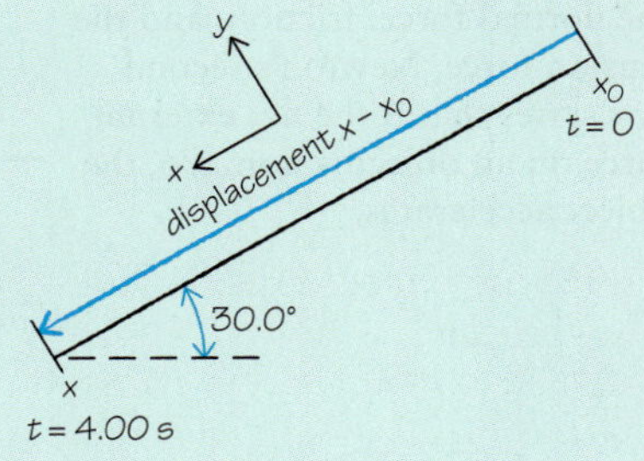

Reflect

Is our answer consistent? With an acceleration of $3.70\ \text{m/s}^2$, our skier reaches a velocity of 14.8 m/s after 4.00 s. Because she started at rest, her average x component of velocity for the trip is half the final velocity, or 7.40 m/s. If you travel at 7.40 m/s for 4.00 s, you cover a distance of 29.6 m—which is just the answer that we obtained.

Skier's velocity at $t = 4.00$ s:

$$v_x = v_{0x} + a_x t \qquad (2\text{-}5)$$

The skier's initial velocity is $v_{0x} = 0$, so

$$v_x = 0 + (3.70\ \text{m/s}^2)(4.00\ \text{s}) = 14.8\ \text{m/s}$$

Average velocity for the skier's motion with constant acceleration:

$$v_{\text{average},x} = \frac{v_{0x} + v_x}{2} = \frac{0 + 14.8\ \text{m/s}}{2} = 7.40\ \text{m/s} \qquad (2\text{-}7)$$

Distance traveled in 4.00 s:

$$x = x_0 + v_{\text{average},x} t = 0 + (7.40\ \text{m/s})(4.00\ \text{s}) = 29.6\ \text{m}$$

By now you should have a pretty good sense of how to attack problems that involve force and acceleration. In the next chapter we'll use the same techniques to tackle problems involving a more detailed description of friction as well as motion in a circle.

TAKE-HOME MESSAGE FOR Section 4-6

✔ To solve a problem that involves forces, begin by making a free-body diagram for the object (or objects) of interest and choosing coordinate axes.

✔ Once you have drawn the free-body diagram, write Newton's second law for each coordinate axis and solve for the unknowns.

Key Terms

contact force
equilibrium
external force
force
force pair
free-body diagram
friction force
gravitational force
inertia
inertial frame of reference
internal force
kilogram
mass
net external force (or net force)
newton
Newton's first law
Newton's laws of motion
Newton's second law
Newton's third law
normal force
pound
reaction force
tension
weight

Chapter Summary

Topic	Equation or Figure
Forces and Newton's second law of motion: The net external force on an object is the vector sum of all forces exerted on that object by other objects. Common forces include the gravitational force, the normal force, friction, and the tension force. Newton's second law states that if the net external force on an object is nonzero, the object accelerates.	The **net external force** acting on an object... ...equals the **vector sum** of all of the individual forces that act on the object. $$\sum \vec{F}_{ext} = \vec{F}_1 + \vec{F}_2 + \vec{F}_3 + \ldots \quad (4\text{-}1)$$ The sum includes only external forces (forces exerted on the object by other objects).
	If a net external force acts on an object... ...the object accelerates. The acceleration is in the same direction as the net force. $$\sum \vec{F}_{ext} = m\vec{a} \quad (4\text{-}2)$$ The magnitude of acceleration that the net external force causes depends on the mass m of the object (the quantity of material in the object). The greater the mass, the smaller the acceleration.
	The acceleration of an object... ...is proportional to and in the same direction as the net external force on the object... $$\vec{a} = \frac{1}{m}\sum \vec{F}_{ext} \quad (4\text{-}3)$$ ...and inversely proportional to the mass of the object.
Mass, weight, inertia, and Newton's first law of motion: The mass of an object is the quantity of material that it contains. Its weight is the magnitude of the gravitational force on the object. Inertia is the tendency of an object to maintain the same motion. Newton's first law states that if the net external force on an object is zero, it maintains a constant velocity.	**Weight** of an object (equal to the magnitude of the gravitational force on that object) $$w = mg \quad (4\text{-}5)$$ **Mass** of the object; **Magnitude** of the **acceleration due to gravity**
	If the net external force on an object is zero... ...the object does not accelerate... $$\text{If } \sum \vec{F}_{ext} = 0, \text{ then } \vec{a} = 0 \text{ and } \vec{v} = \text{constant} \quad (4\text{-}6)$$ ...and the velocity of the object remains constant. If the object is at rest, it remains at rest; if it is in motion, it continues in motion in a straight line at a constant speed.

Free-body diagrams: A free-body diagram depicts all of the external forces that act on an object. The coordinate axes are chosen so that as many of the vectors as possible (force and acceleration) point along one of the axes.

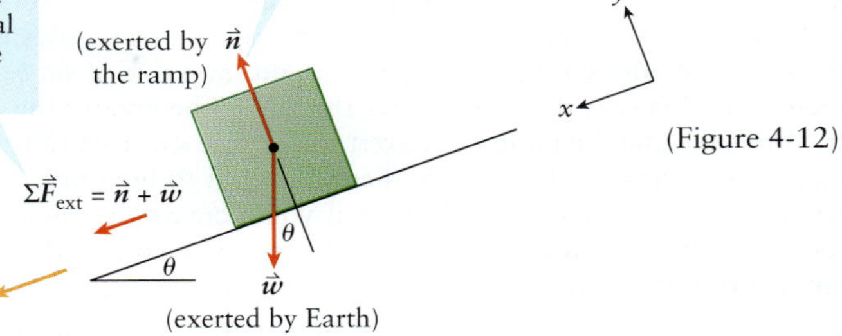

(1) With the friction force absent, the remaining normal force and gravitational force do not cancel. So the net external force $\Sigma\vec{F}_{ext} = \vec{n} + \vec{w}$ is not zero.
(2) The box neither leaps off the ramp nor sinks down into it, so it doesn't accelerate in the y direction. So the normal force must cancel the (negative) y component of the gravitational force.
(3) Nothing cancels the positive x component of the gravitational force, so the net force is in the positive x direction...
(4) ...and the acceleration of the block is in the positive x direction, the same as the direction of the net force.
(exerted by the ramp)
$\vec{n}$
$\Sigma\vec{F}_{ext} = \vec{n} + \vec{w}$
$\vec{a}$
θ
$\vec{w}$
(exerted by Earth)
y
x

(Figure 4-12)

Newton's third law of motion: The third law relates the forces that two objects exert on each other. This law is valid no matter how the objects move.

Force that object A exerts on object B — Force that object B exerts on object A

$$\vec{F}_{A \text{ on } B} = -\vec{F}_{B \text{ on } A} \quad (4\text{-}7)$$

If object A exerts a force on object B, object B must exert a force on object A with the same magnitude but opposite direction.

Solving force problems: All problems that involve forces can be solved by (1) first drawing a free-body diagram for the object in question and (2) then writing out Newton's second law in component form and solving for the unknowns.

$$\text{(a)} \quad \sum F_{ext,x} = ma_x$$
$$\text{(b)} \quad \sum F_{ext,y} = ma_y \quad (4\text{-}4)$$

Sum of the x components of all external forces acting on the object; x component of the object's acceleration; Mass of the object; Sum of the y components of all external forces acting on the object; y component of the object's acceleration

Answer to What do you think? Question

(c) The forward force exerted on the airliner from the tug now exceeds the backward force of friction on the airliner. Hence there is a net forward force on the airliner, and it accelerates forward—that is, it speeds up.

Answers to Got the Concept? Questions

4-1 (c), (a), (b) and (d) [tie]. From Newton's second law, $\sum\vec{F}_{ext} = m\vec{a}$, the magnitude of the net force $\sum\vec{F}_{ext}$ equals the object's mass m multiplied by the magnitude of the object's acceleration $\vec{a}$. For case (a) this product is $(1250 \text{ kg})(2.00 \text{ m/s}^2) = 2500 \text{ kg}\cdot\text{m/s}^2 = 2500 \text{ N}$; for case (c) it is $(4500 \text{ kg})(0.600 \text{ m/s}^2) = 2700 \text{ N}$. (The truck is slowing down rather than speeding up, but all that counts is the *magnitude* of the acceleration.) So there is a greater net force on the truck in (c) than on the automobile in (a). In cases (b) and (d) the objects are not accelerating (the velocities are constant), so $\vec{a} = 0$ and the net force on the object is zero.

4-2 Newton's second law holds everywhere in the universe, even in the absence of gravity. The second law says that if you apply the same force to each sphere, the massive lead sphere will have a smaller acceleration than the less massive plastic sphere. You can use this same principle here on Earth: Hold this textbook in one hand and a pencil in the other; then try to shake both hands side to side. The hand holding this textbook will move far less than the other hand because this book is quite a bit more massive than a pencil. So the acceleration of the book (and the hand holding it) will be much smaller than the acceleration of the pencil, assuming that each hand exerts roughly the same force.

4-3 (c) and (d) [tie], (a), (b). From Equation 4-5 an object's mass m is related to its weight w and the acceleration due to gravity g by $m = w/g$. So the mass of rock (a) is $m_a = (10.0\text{ N})/(3.69\text{ m/s}^2) = 2.71\text{ kg}$, and the mass of rock (b) is $m_b = (10.0\text{ N})/(9.80\text{ m/s}^2) = 1.02\text{ kg}$. Both rock (c) and rock (d) have a mass of 10.0 kg.

4-4 (d) Example 4-3 shows that $n = w_{cat} \cos\theta$ and $f = w_{cat} \sin\theta$, where w_{cat} is the weight of the cat. As the value of θ increases, the values of both $\cos\theta$ and n decrease, and the value of $\sin\theta$ increases. This says that the steeper the angle, the greater the friction force that the ramp must exert to keep the cat from sliding. As we'll learn in Chapter 5, there is a limit to how much friction force a surface can exert. So if you increase the angle too much, the ramp will no longer be able to provide the needed amount of friction, and the unhappy cat will start to slide.

4-5 (b) Three external forces act on the box: the gravitational force $\vec{w}$ (exerted by Earth), the normal force $\vec{n}$ (exerted by the surface of the ramp), and the friction force $\vec{f}$ (also exerted by the surface of the ramp). Acceleration is not a force, so the quantity $m\vec{a}$ doesn't belong in the free-body diagram. This rules out choices (d) and (e). The normal force $\vec{n}$ is not vertical in this situation but points perpendicular to the inclined surface of the ramp; this rules out choices (c) and (d). Finally, the friction force $\vec{f}$ opposes sliding, so if the box is sliding down the ramp it must be that $\vec{f}$ points up the ramp. This rules out choice (a). Hence only choice (b) is correct. Note that this is the same free-body diagram as shown in Figure 4-11b for a block at rest on the ramp. The difference here is that friction is weaker than in Figure 4-11b, so the block slides down the ramp rather than remaining stationary.

4-6 It is true that when your friend pushes on the crate, the crate pushes back on him according to Newton's third law. However, these two forces act on two *different* objects. The force exerted by the crate acts on your friend, *not* on the crate. So we don't include it in the sum of external forces acting on the crate. (Remember that one of the keys to applying Newton's second law is considering *only* those external forces that act *on* the object you're considering—in this case, the crate.) The only horizontal forces that act on the crate are your friend's forward push and the backward push of friction exerted by the floor. If your friend can overcome that friction force, the crate will accelerate and start moving!

4-7 (a) Crate 1 is in contact with crate 2 and so exerts a force on it. Answer (b) is incorrect: You do *not* exert a force on crate 2 because you are not in direct contact with it. (It's true that crate 1 exerts a force on crate 2 only because you push on crate 1, but crate 2 doesn't "know" that: All it "feels" is the push from crate 1.) Crate 2 does exert a reaction force to the push it receives from crate 1. But Newton's third law tells us that this reaction acts on crate 1, *not* on crate 2, so answer (c) is incorrect. As we discussed in Section 4-4, acceleration is caused by force, not the other way around, and there is no such thing as "the force of acceleration." So answer (d), too, is incorrect.

Questions and Problems

In a few problems you are given more data than you actually need; in a few other problems you are required to supply data from your general knowledge, outside sources, or informed estimate.

Interpret as significant all digits in numerical values that have trailing zeros and no decimal points.

For all problems use $g = 9.80\text{ m/s}^2$ for the free-fall acceleration due to gravity.

•	Basic, single-concept problem
••	Intermediate-level problem; may require synthesis of concepts and multiple steps
•••	Challenging problem
SSM	*Solution is in Student Solutions Manual*
Example	*See worked example for a similar problem*

Conceptual Questions

1. • According to Newton's second law, does the direction of the net force always equal the direction of the acceleration? SSM

2. • If the sum of the forces acting on an object equals zero, does this imply that the object is at rest?

3. • What are the basic SI units (kg, m, s) for force according to Newton's second law $\sum \vec{F}_{ext} = m\vec{a}$?

4. • Explain why the force that a horizontal surface exerts on an object at rest on the surface is called the normal force.

5. • What is the net force on a bathroom scale when a 75-kg person stands on it? SSM

6. • Two forces of 30 N and 70 N act on an object. What are the minimum and maximum values for the sum of these two forces?

7. • You apply a 60-N force to push a box across the floor at constant speed. If you increase the applied force to 80 N, will the box speed up to some new constant speed, or will it continue to speed up until it hits the wall? Assume that the floor is horizontal and the surface is uniform. Explain your answer.

8. • A definition of the inertia of an object is that it is a measure of the quantity of matter. How does this definition compare with the definition discussed in the chapter?

9. • **Astronomy** Why would it be easier to lift a truck on the Moon's surface than it is on Earth? SSM

10. • When constructing a free-body diagram, why is it a good idea to choose your coordinate system so that the motion of an object is along one of the axes?

11. • **Sports** How can a fisherman land a 5-lb fish using fishing line that is rated at 4 lb?

12. • List all the forces acting on a bottle of water if it were sitting on your desk.

13. • **Sports** A boxer claims that Newton's third law helps him while boxing. He says that during a boxing match the force that his jaw feels is the same as the force that his opponent's fist feels (when the opponent is doing the punching). Therefore, his opponent will feel the same force as he feels and he will be able to fight on without any problems, no matter how many punches he receives or gives. It will always be an "even fight." Discuss any flaws in his reasoning.

14. • **Astronomy** We know that the Sun pulls on Earth. Does Earth also pull on the Sun? Why or why not? SSM

15. • **Medical** Use Newton's third law to explain the forces involved in walking.

16. • **Biology** Use Newton's third law to explain how birds are able to fly forward. SSM

17. • A certain rope will break under any tension greater than 800 N. How can it be used to lower an object weighing 850 N over the edge of a cliff without the rope breaking? SSM

18. • Tension is a very common force in day-to-day life. Identify five ordinary situations that involve the force of tension.

19. • A chair is mounted on a scale inside an elevator in the physics building. Describe the variation in the scale reading as the elevator begins to ascend, goes up at constant speed, stops, begins to descend, goes down at a constant speed, and stops again.

20. • **Medical** Why does the American Academy of Pediatrics recommend that all infants sit in rear-facing car seats starting with their first ride home from the hospital? Explain your answer.

21. • **Medical** Why should the driver and passengers in a car wear seat belts? Explain your answer.

22. • **Medical, Sports** Gymnastics routines are done over a padded floor to protect athletes who fall. Why is falling on padding safer than falling on concrete?

23. • Using physics principles, explain why your hand hurts after punching a solid wall. Assuming your punches are all identical, use physics principles to explain why it hurts less if the wall is padded.

Multiple-Choice Questions

24. • The net force on a moving object suddenly becomes zero and remains zero. The object will
    A. stop abruptly.
    B. reduce speed gradually.
    C. continue at constant velocity.
    D. increase speed gradually.
    E. reduce speed abruptly.

25. • Which has greater monetary value, a newton of gold on Earth or a newton of gold on the Moon?
    A. the newton of gold on Earth
    B. the newton of gold on the Moon
    C. The value is the same, regardless of location.
    D. One cannot say without checking the weight on the Moon.
    E. the newton of gold on the Moon but only when inside a spaceship

26. • According to Newton's second law of motion, when a net force acts on an object, the acceleration is
    A. zero.
    B. inversely proportional to the object's mass.
    C. independent of mass.
    D. inversely proportional to the net force.
    E. directly proportional to the object's mass.

27. • In the absence of a net force, an object can be
    A. at rest.
    B. in motion with a constant velocity.
    C. accelerating.
    D. at rest or in motion with a constant velocity.
    E. It's not possible to know without more information. SSM

28. • **Medical** A car stops suddenly during a head-on collision, causing the driver's brain to slam into the skull. The resulting injury would most likely be to which part of the brain?
    A. frontal portion of the brain
    B. rear portion of the brain
    C. middle portion of the brain
    D. left side of the brain
    E. right side of the brain

29. • **Medical** During the sudden impact of a car accident, a person's neck can experience abnormal forces, resulting in an injury commonly known as *whiplash*. If a victim's head and neck move in the manner shown in **Figure 4-24**, his car was hit from the
    A. front.
    B. rear.
    C. right side.
    D. left side.
    E. top during a rollover. SSM

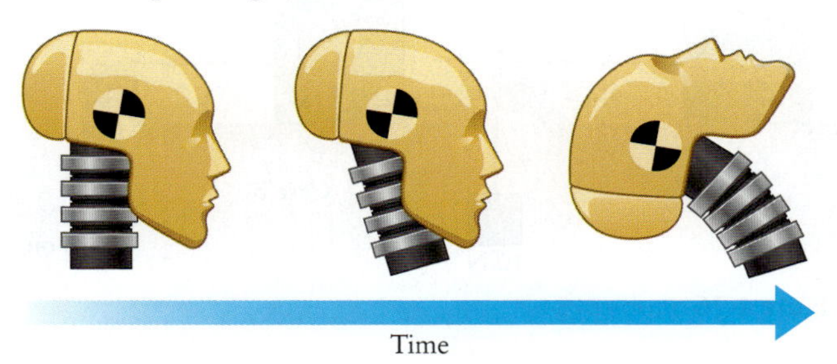

Figure 4-24 Problem 29

30. • When a net force acts on an object, the object
    A. is at rest.
    B. is in motion with a constant velocity.
    C. has zero speed.
    D. is accelerating.
    E. is at rest or in motion with a constant velocity.

31. • In the absence of a net force, an object cannot be
    A. at rest.
    B. in motion with a constant velocity.
    C. accelerating.
    D. moving with an acceleration of zero.
    E. experiencing opposite but equal forces.

32. • How does the magnitude of the normal force exerted by the ramp in **Figure 4-25** compare to the weight of the block? The normal force is
    A. equal to the weight of the block.
    B. greater than the weight of the block.
    C. less than the weight of the block.
    D. possibly equal to or less than the weight of the block, depending on whether or not the ramp surface is smooth.
    E. possibly greater than or equal to the weight of the block, depending on whether or not the ramp surface is smooth.

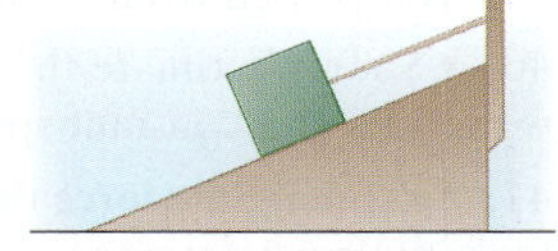

Figure 4-25 Problem 32

33. • While an elevator traveling upward slows down to stop, the normal force on the feet of a passenger is _____ her weight. While an elevator traveling downward slows down to stop, the normal force on the feet of a passenger is _____ his weight.
    A. larger than; smaller than
    B. larger than; larger than
    C. smaller than; smaller than
    D. smaller than; larger than
    E. equal to; equal to

34. • Case (a) in **Figure 4-26** shows block A accelerated across a frictionless table by a hanging 10-N block (1.02 kg). In case

(b) the same block A is accelerated by a steady 10-N tension in the string. Treat the masses of the strings, as well as the masses and friction of the pulleys, as negligible. The acceleration of block A in case (b) is

A. greater than its acceleration in case (a).
B. less than its acceleration in case (a).
C. equal to its acceleration in case (a).
D. twice its acceleration in case (a).
E. half its acceleration in case (a).

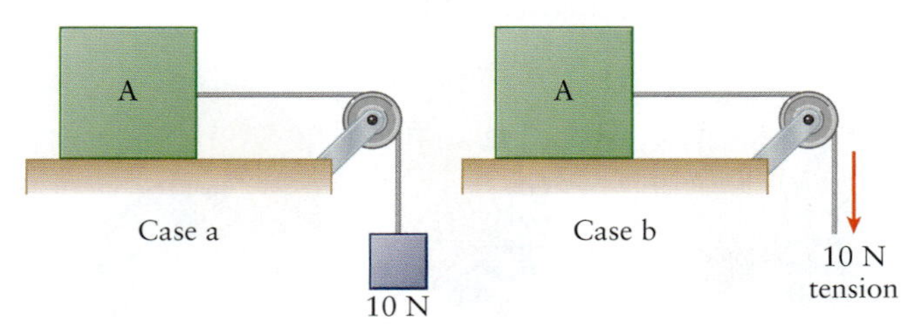

Figure 4-26 Problem 34

Estimation/Numerical Analysis

35. • Estimate the weight of five common objects using newtons (*not* pounds or ounces).

36. • Estimate the normal force acting on an apple that rests on a flat surface. What would this estimate be if the surface were tilted at an incline of 30° with the horizontal?

37. • **Sports** Estimate the average force that a major league baseball pitcher exerts on a baseball when he throws it. SSM

38. • Estimate the maximum tension in the cable that supports a typical elevator.

39. • **Sports** Estimate the average force imparted on a tennis ball as it is fired from a tennis ball machine.

40. • **Sports** Estimate the tension in a rope that pulls you on water skis at a constant speed behind a speedboat.

41. • Estimate the force that is produced with the "jaws of life" (the pneumatic tool used to rip open the jammed doors of a vehicle that is involved in an accident).

42. • Estimate the force in newtons required to flip on a light switch.

43. • Estimate the force in newtons associated with snapping your fingers.

44. • **Medical** During an arthroscopic surgery repair of a patient's knee, a portion of healthy tendon is grafted in place of a damaged anterior cruciate ligament (ACL). To test the results the surgeon applies a known force to the ligament and increases it at 1-s intervals for 10 s. Plot the data listed in the table as a function of time in a spreadsheet or on a graphing calculator and extrapolate the best-fit line to estimate the force that would correspond to a time of 15 s.

Time (s)	Force (mN)
0	0
1	2.28
2	5.73
3	11.28
4	12.27
5	12.54
6	12.38
7	14.11
8	16.79
9	21.08
10	28.51

45. •• Use a spreadsheet or graphing calculator to plot the velocity versus height of an express elevator. The elevator starts from rest on the ground floor, has constant acceleration until reaching its maximum speed of 10 m/s at the 4th floor (the distance between each floor is 3.5 m), remains at this speed until the 20th floor, and then has constant acceleration until it stops at the 30th floor. Identify the point(s) on the graph when the weight of a rider as measured by a spring scale will reach its maximum value. SSM

Problems

4-2 If a net external force acts on an object, the object accelerates

46. • What is the acceleration of a 2.00×10^3-kg car if the net force on the car is 4.00×10^3 N? Example 4-1

47. • What net force is needed to accelerate a 2.00×10^3-kg car at 2.00 m/s^2? Example 4-1

48. •• Applying a constant net force to an object causes it to accelerate at 10 m/s^2. What will the acceleration of the object be if (a) the force is doubled, (b) the mass is halved, (c) the force is doubled and the mass is doubled, (d) the force is doubled and the mass is halved, (e) the force is halved, (f) the mass is doubled, (g) the force is halved and the mass is halved, and (h) the force is halved and the mass is doubled? Example 4-1

49. • Suppose that the engine of a certain car can result in a maximum force of 15,000 N being applied to the car from the road. In the absence of any other forces, what is the maximum acceleration this engine can produce in a 1250-kg car? Example 4-1

50. •• Three rugby players are pulling horizontally on ropes attached to a box, which remains stationary. Player 1 exerts a force F_1 equal to 1.00×10^2 N at an angle θ_1 equal to 60.0° with respect to the $+x$ direction (**Figure 4-27**). Player 2 exerts a force F_2 equal to 2.00×10^2 N at an angle θ_2 equal to 37.0° with respect to the $+x$ direction. The view in the figure is from above. Ignore friction and note that gravity can be ignored in this problem. (a) Determine the force $\vec{F}_3$ exerted by player 3. State your answer by giving the components of $\vec{F}_3$ in the directions perpendicular to and parallel to the positive x direction. (b) Redraw the diagram and add the force $\vec{F}_3$ as carefully as you can. (c) Player 3's rope breaks, and player 2 adjusts by pulling with a force of magnitude F_2' equal to 1.50×10^2 N at the same angle as before. In which direction is the acceleration of the box relative to the $+x$ direction shown? (d) In part (c) the magnitude of the acceleration is measured to be 10.0 m/s^2. What is the mass of the box? SSM Example 4-2

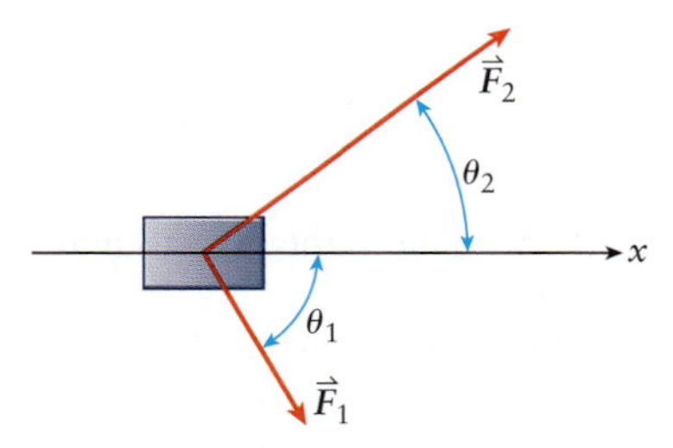

Figure 4-27 Problem 50

51. •• Three forces act on a 2.00-kg object at angles $\theta_1 = 40.0°$, $\theta_2 = 60.0°$, and $\theta_3 = 20.0°$, as shown in **Figure 4-28**. Find the magnitude of $\vec{F}_2$ and $\vec{F}_3$ if the magnitude of F_1 is 1.00 N and the acceleration of the object is 1.50 m/s^2 toward the $+x$ direction. Example 4-2

Figure 4-28 Problem 51

4-3 Mass, weight, and inertia are distinct but related concepts

52. • What is the weight on Earth of a wrestler who has a mass of 120 kg? Example 4-3

53. • A bluefin tuna has a mass of 250 kg. What is its weight? SSM Example 4-3

54. •• **Astronomy** An astronaut has a mass of 80.0 kg. How much would the astronaut weigh on Mars where surface gravity is 38.0% of that on Earth? Example 4-3

55. • What is the net force on an apple that weighs 3.5 N when you hold it at rest in your hand? Example 4-3

56. •• A 1300-kg car is capable of a maximum acceleration of 5.0 m/s^2. If this car is required to push a stalled car of mass 1700 kg, what is the maximum magnitude of acceleration of the two-car system? Example 4-3

4-4 Making a free-body diagram is essential in solving any problem involving forces

57. • Draw a free-body diagram for a heavy crate being lowered by a steel cable straight down at a constant speed. Example 4-3

58. • Draw a free-body diagram for a box being pushed horizontally by a person across a smooth, frictionless floor at a steadily increasing speed. SSM Example 4-3

59. • Draw a free-body diagram for a bicycle rolling down a hill. Ignore the friction between the bicycle wheels and the hill, but consider any air resistance. Example 4-3

60. •• A tugboat uses its winch to pull up a sinking sailboat with an upward force of 4500 N. The mass of the boat is 200 kg and the water acts on the sailboat with a drag force of 2000 N. Draw a free-body diagram for the sailboat and describe the motion of the sailboat. Example 4-3

4-6 All problems involving forces can be solved using the same series of steps

61. •• Box A weighs 80 N and rests on a table (**Figure 4-29**). A rope that connects boxes A and B drapes over a pulley so that box B hangs above the table, as shown in the figure. The pulley and rope are massless, and the pulley is frictionless. What force does the table exert on box A if box B weighs (a) 35 N, (b) 70 N, (c) 90 N? Example 4-9

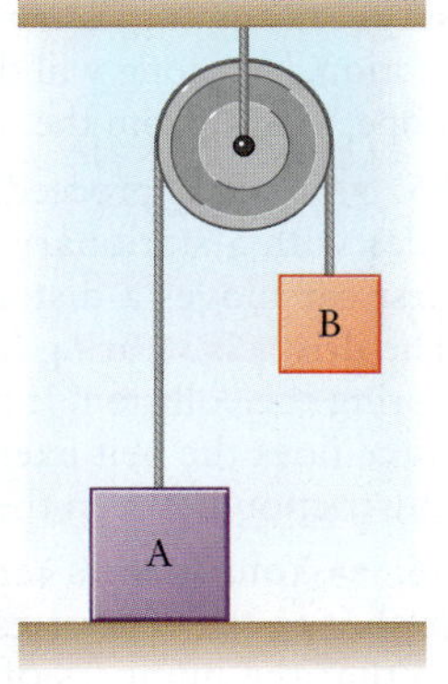

Figure 4-29 Problem 61

62. •• Two forces act on an object of mass $M = 3.00$ kg as shown in **Figure 4-30**. Because of these forces, the object experiences an acceleration of $a = 7.00$ m/s^2 in the $+x$ direction. If $\theta_1 = 30.0°$, (a) calculate the magnitude of $\vec{F}_2$ and (b) make a careful drawing to show its direction, given that $F_1 = 20.0$ N. Example 4-2

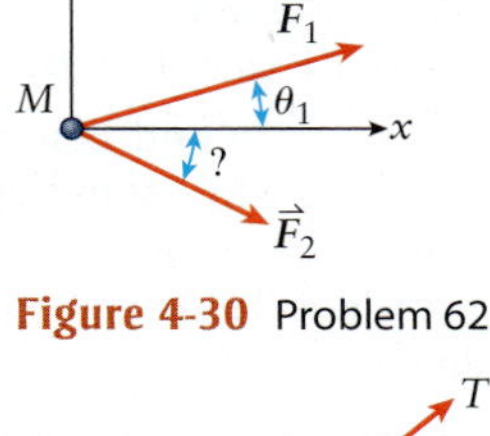

Figure 4-30 Problem 62

63. ••A 1.00×10^2-kg streetlight is supported equally by two ropes as shown in **Figure 4-31**. One rope pulls up and to the right, 40.0° above the horizontal; the other rope pulls up and to the left, 40.0° above the horizontal. What is the tension in each rope? SSM Example 4-3

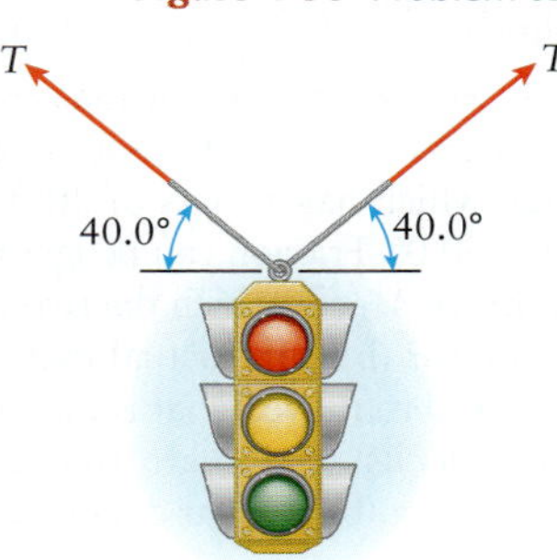

Figure 4-31 Problem 63

64. •••A 2.00×10^2-N sign is supported by two ropes as shown in **Figure 4-32**. If $\theta_L = 45.0°$ and $\theta_R = 30.0°$, what is the tension in each rope? Example 4-3

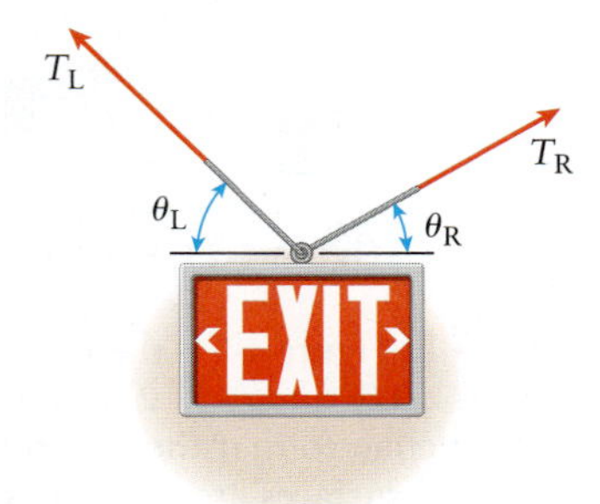

Figure 4-32 Problem 64

65. •• The distance between two telephone poles is 50.0 m. When a 0.500-kg bird lands on the telephone wire midway between the poles, the wire sags 0.15 m. How much tension does the bird produce in the wire? Ignore the weight of the wire. Example 4-3

66. • A locomotive pulls 10 identical freight cars with an acceleration of 2.0 m/s^2. How does the force between the third and fourth cars compare to the force between the seventh and eighth cars? Example 4-8

67. • A locomotive pulls 10 identical freight cars. The force between the locomotive and the first car is 1.00×10^5 N, and the acceleration of the train is 2.00 m/s^2. There is no friction to consider. Find the force between the ninth and tenth cars. SSM Example 4-8

68. • A 0.0100-kg block and a 2.00-kg block are attached to the ends of a rope. A student holds the 2.00-kg block and lets the 0.0100-kg block hang below it; then he lets go. What is the tension in the rope while the blocks are falling, before either hits the ground? Air resistance can be neglected. Example 4-8

69. • While parachuting, a 66.0-kg person experiences a downward acceleration of 2.50 m/s^2. What is the downward force on the parachute from the person? Example 4-4

70. • A bicycle and 50.0-kg rider accelerate at 1.00 m/s^2 up an incline of 10.0° above the horizontal. What is the magnitude of the force that the bicycle exerts on the rider? Example 4-6

71. •• A car uniformly accelerates from 0 to 28.0 m/s. A 60.0-kg passenger experiences a horizontal force of 4.00×10^2 N. How much time does it take for the car to reach 28.0 m/s? SSM Example 4-10

72. •• A car accelerates from 0 to 1.00×10^2 km/h in 4.50 s. What force does a 65.0-kg passenger experience during this acceleration? Example 4-10

73. • Adam and Ben pull hand over hand on opposite ends of a rope while standing on a frictionless frozen pond. Adam's mass is 75.0 kg, and Ben's mass is 50.0 kg. If Adam's acceleration is 1.00 m/s^2 to the east, what are the magnitude and direction of Ben's acceleration? Example 4-8

74. •• Two blocks of masses M_1 and M_2 are connected by a massless string that passes over a massless pulley (Figure 4-33). M_2, which has a mass of 20.0 kg, rests on a long ramp of angle $\theta = 30.0°$. Friction can be ignored in this problem. (a) What is the value of M_1 for which the two blocks are in equilibrium (no acceleration)? (b) If the actual mass of M_1 is 5.00 kg and the system is allowed to move, what is the magnitude of the acceleration of the two blocks? (c) In part (b) does M_2 move up or down the ramp? (d) In part (b) how far does block M_2 move in 2.00 s? Example 4-9

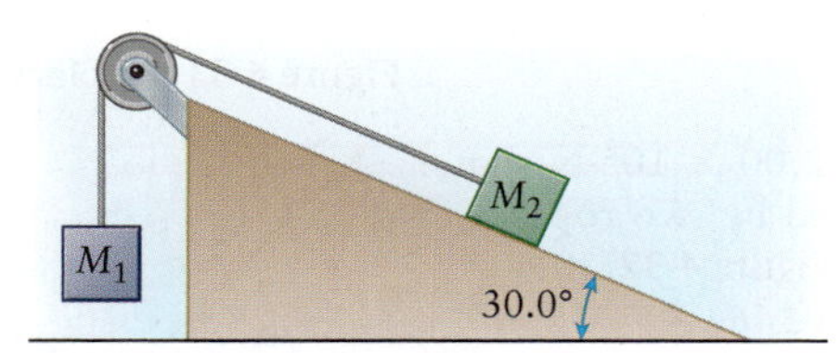

Figure 4-33 Problem 74

75. •• In Figure 4-34, the block on the left incline is 6.00 kg. If $\theta_1 = 60.0°$ and $\theta_2 = 25.0°$, find the mass of the block on the right incline so that the system is in equilibrium (no acceleration). All surfaces are frictionless. SSM Example 4-9

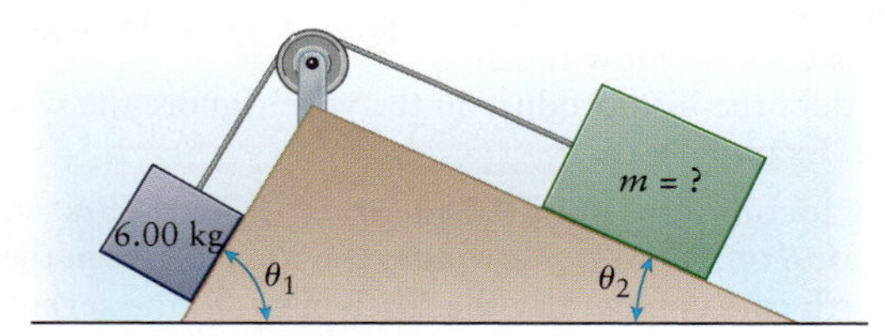

Figure 4-34 Problem 75

General Problems

76. • A reckless coyote engineer straps on a pair of ice skates and attaches a rocket capable of producing 5.0×10^3 N of thrust to his back. Together, the coyote and the rocket have a mass of 120 kg. If the coyote starts at rest on level, frictionless ice and bends over such that the rocket thrust is directed parallel to the ice, what is his final speed if the rocket burns for 5.0 s? Example 4-5

77. • You stand at the base of a 4.00-m long frictionless ramp that is inclined at an angle of 9.00°. You want to slide a 2.00-kg object up the ramp so that it stops *just* as it reaches the top. What initial velocity must you give the object? Example 4-5

78. • You and a friend are ice skating. Standing in the middle of the rink, you give your 85-kg friend a push with a force of 3.0×10^2 N. Assuming that the ice surface is frictionless, what magnitude of acceleration does your friend experience? Example 4-5

79. • A skier starts at rest atop a ski slope that has a slope of exactly 40.0°. Her total mass is 72.0 kg. Approximating the slope to be frictionless and assuming that the skier skis straight down the slope, what will her speed be after 25.0 s? Example 4-5

80. • A 5.0-kg block slides in a straight line. The velocity of the block as a function of time is displayed in the v_x–t graph in Figure 4-35. Calculate the net force on the block for the time intervals t = 0 to 1 s, 1 to 3 s, 3 to 5 s, 5 to 6 s, and 6 to 7 s.

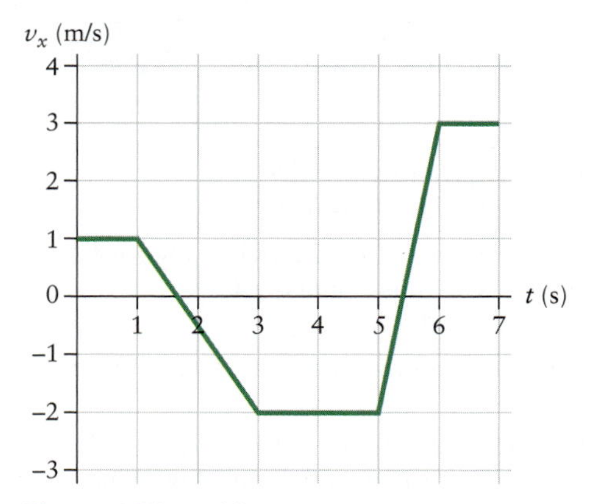

Figure 4-35 Problem 80

81. • In an Atwood's machine, box A of unknown mass is attached to box B that has a mass of 2.00 kg (Figure 4-36). The two boxes are attached by a massless rope that hangs over a massless, frictionless pulley. When in motion, it is found that box B moves upward with an acceleration of 3.00 m/s^2. (a) What is the tension in the rope? (b) What is the mass of box A? Example 4-9

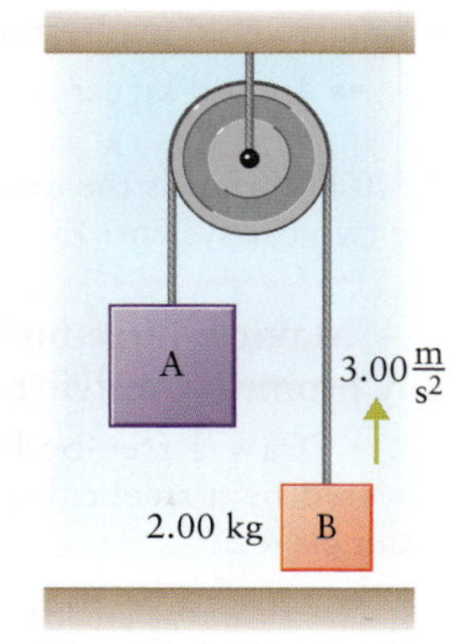

Figure 4-36 Problem 81

82. • **Medical, Astronomy** A spaceship takes off vertically from rest with an acceleration of 29.0 m/s^2. What force is exerted on a 75.0-kg astronaut during takeoff? Express your answer in newtons and also as a multiple of the astronaut's weight on Earth. Example 4-1

83. • A girl of mass 50.0 kg is standing on a weight scale in an elevator that is initially at rest. As the elevator begins to move, the scale displays a value of 350 N. (a) What is the magnitude of the acceleration that the girl experiences? (b) Is the elevator moving up or down? Example 4-7

84. •• A child on a sled starts from rest at the top of a 20.0° slope. Assuming that there are no forces resisting the sled's motion, how long will the child take to reach the bottom of the slope, 210 m from the top? Example 4-10

85. •• A car is proceeding at a speed of 14.0 m/s when it collides with a stationary car in front. During the collision, the first car moves a distance of 0.300 m as it comes to a stop. The driver is wearing her seat belt, so she remains in her seat during the collision. If the driver's mass is 52.0 kg, how much force does the belt exert on her during the collision? Neglect any friction between the driver and the seat. Example 4-10

86. •• Your friend's car runs out of fuel, and you volunteer to push it to the nearest gas station. You carefully drive your car so that the bumpers of the two cars are in contact and then slowly accelerate to a speed of 2.00 m/s over the course of

1.00 min. If the mass of your friend's car is 1200 kg, what is the contact force between the two bumpers? Example 4-10

87. •• A 30.0-kg golden retriever stands on a scale in an elevator. Calculate the reading on the scale when the elevator (a) accelerates at 3.50 m/s^2 downward, (b) when the elevator cruises down at a steady speed, and (c) when the elevator accelerates at 4.00 m/s^2 upward. SSM Example 4-7

88. •• A person weighs 588 N. If she stands on a scale while riding on the Inclinator (the lift at the Luxor Hotel in Las Vegas), what will be the reading on the scale? Assume the Inclinator moves at a constant acceleration of 1.25 m/s^2, in a direction 39.0° above the horizontal, and that the rider stands vertically in the elevator car. Example 4-7

89. •• A rider in an elevator weighs 700 N. If this person stands on a scale in an elevator, describe the variation in the scale readings as the elevator initially starts from rest, accelerates upward at 3.00 m/s^2, cruises upward at 4.00 m/s, slows to a stop at 2.00 m/s^2, then free-falls all the way to the bottom of the elevator shaft before striking springs that bring the car to a safe stop. Example 4-7

90. •• A fuzzy die that has a weight of 1.80 N hangs from the ceiling of a car by a massless string. The car travels with a forward acceleration of 2.70 m/s^2 on a horizontal road. The string makes an angle θ with respect to the vertical, shown in **Figure 4-37**. What is the angle θ? Example 4-4

Figure 4-37 Problem 90

91. •• **Biology** On average, froghopper insects have a mass of 12.3 mg and jump to a height of 428 mm. The takeoff velocity is achieved as the little critter flexes its legs over a distance of approximately 2.00 mm. Assume a vertical jump with constant acceleration. (a) How long does the jump last (the jump itself, not the time in the air), and what is the froghopper's acceleration during that time? (b) Make a free-body diagram of the froghopper during its leap (but before it leaves the ground). (c) What force did the ground exert on the froghopper during the jump? Express your answer in millinewtons and as a multiple of the insect's weight. SSM Example 4-10

92. •• **Medical** A car traveling at 28.0 m/s hits a bridge abutment. A passenger in the car, who has a mass of 45.0 kg, moves forward a distance of 55.0 cm while being brought to rest by an inflated air bag. Assuming that the force that stops the passenger is constant, what is the magnitude of this force? Example 4-10

93. •• Two mountain climbers are working their way up a glacier when one falls into a crevasse (**Figure 4-38**). The icy slope, which makes an angle of θ = 45.0° with the horizontal, can be considered frictionless. Sue's weight is pulling Paul up the 45.0° slope. If Sue's mass is 66.0 kg and she falls 2.00 m in 10.0 s starting from rest, calculate (a) the tension in the rope joining them and (b) Paul's mass. SSM Example 4-9

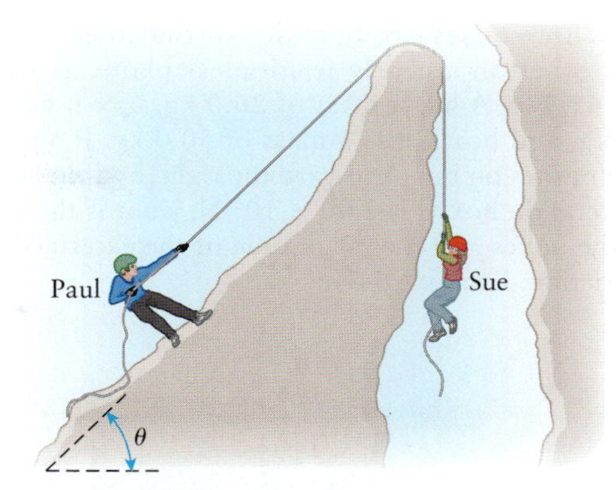

Figure 4-38 Problem 93

94. •• **Medical** During a front-end car collision, the acceleration limit for the chest is 60*g*. If a car was initially traveling at 48.0 km/h, (a) how much time does it take for the car to come to rest, assuming a constant acceleration equal to the threshold acceleration for damage to the chest? (b) Draw a free-body diagram of a person during the crash. (c) What force in newtons does the air bag exert on the chest of a 72.0-kg person? The trunk of the body comprises about 43.0% of body weight. (d) Why doesn't this force injure the person? Example 4-10

95. •• A window washer sits in a bosun's chair that dangles from a massless rope that runs over a massless, frictionless pulley and back down to the man's hand. The combined mass of man and chair is 95.0 kg. With how much force must he pull downward to raise himself (a) at constant speed and (b) with an upward acceleration 1.50 m/s^2? SSM Example 4-9

96. •• A 2.00×10^2-kg block is hoisted by pulleys that are massless and frictionless, as shown in **Figure 4-39**. If a force of 1.50×10^3 N is applied to the massless rope, what is the acceleration of the suspended mass? Example 4-9

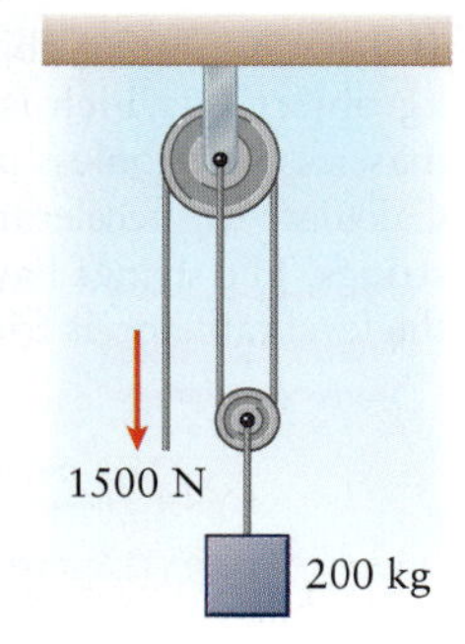

Figure 4-39 Problem 96

97. ••• Two blocks connected by a light string are being pulled across a frictionless horizontal tabletop by a hanging 10.0-N weight (block C) (**Figure 4-40**). Block A has a mass of 2.00 kg. The mass of block B is only 1.00 kg. The strings remain taut at all times. Assuming the pulley is massless and frictionless, what are the values of the tensions T_1 and T_2? SSM Example 4-8

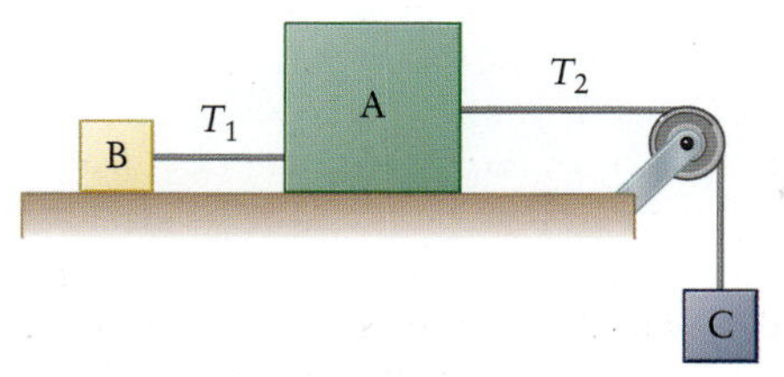

Figure 4-40 Problem 97

98. ••• Three boxes are lined up so that they are touching each other on a nearly frictionless plane, as shown in **Figure 4-41**. Box A has a mass of 20.0 kg, box B has a mass of 30.0 kg, and box C has a mass of 50.0 kg. If an external force (F) pushes on box A toward the right, and the force that box B exerts on box C is 2.00×10^2 N, what is the acceleration of the boxes and the magnitude of the external force F? Ignore friction. Example 4-8

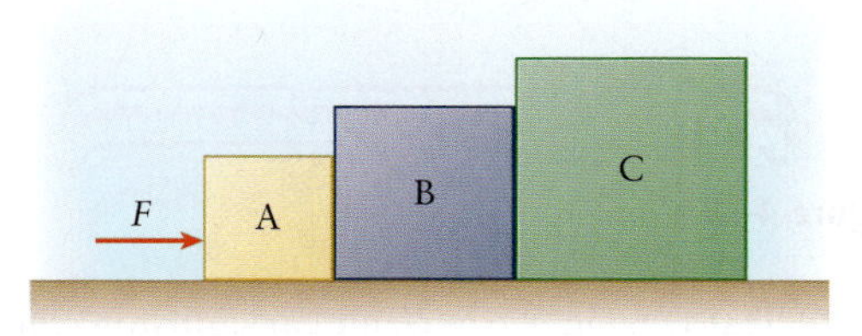

Figure 4-41 Problem 98

99. •• A 2.00-kg object A is connected with a massless string across a massless, frictionless pulley to a 3.00-kg object B (**Figure 4-42**). The smaller object rests on a nearly frictionless plane, which is tilted at an angle of $\theta = 40.0°$ as shown. What are the acceleration of the system and the tension in the string? Example 4-9

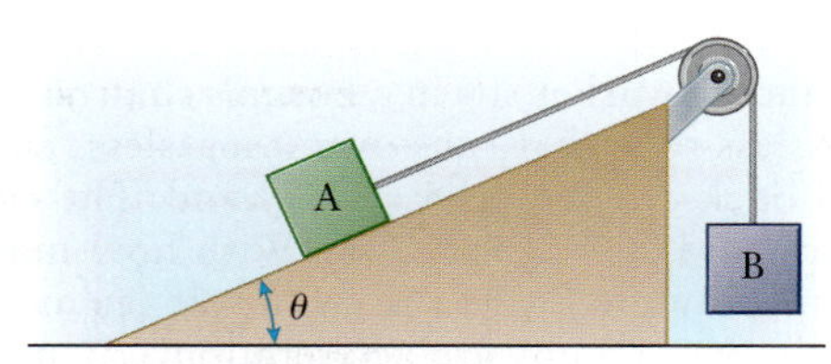

Figure 4-42 Problem 99

100. ••• A 1.00-kg object A is connected with a string to a 2.00-kg object B, which is connected with a second string over a massless, frictionless pulley to a 4.00-kg object C (**Figure 4-43**). Calculate the acceleration of the system and the tension in both strings. The strings have negligible mass and do not stretch, and the level tabletop is frictionless. Example 4-8

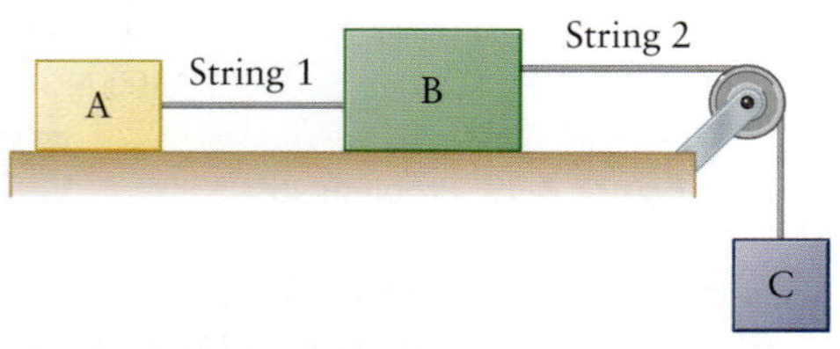

Figure 4-43 Problem 100

101. ••• A 1.00-kg object A is connected with a string to a 2.00-kg object B, which is connected with a second string over a massless, frictionless pulley to a 4.00-kg object C (**Figure 4-44**). The first two objects are placed onto a frictionless inclined plane that makes an angle $\theta = 30.0°$ as shown. Calculate the acceleration of the masses and the tensions in both strings. Example 4-8

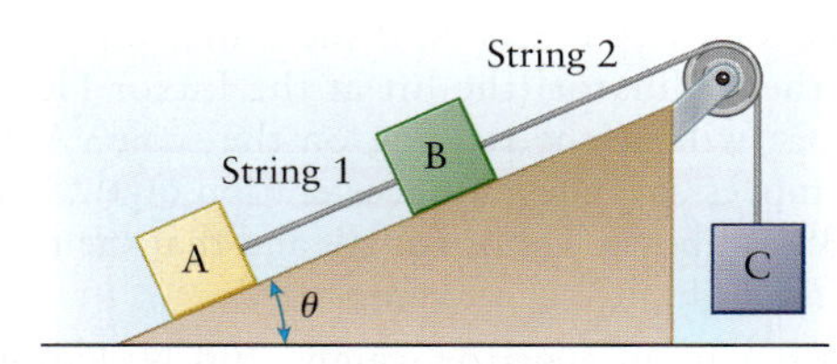

Figure 4-44 Problem 101

102. ••• **Sports** An athlete drops from rest from a platform 10.0 m above the surface of a 5.00-m-deep pool. Assuming that the athlete enters the water vertically and moves through the water with constant acceleration, what is the minimum average force the water must exert on a 62.0-kg diver to prevent her from hitting the bottom of the pool? Express your answer in newtons and also as a multiple of the diver's weight. Air resistance during the athlete's dive can be ignored in this problem. Example 4-10

103. ••• A person pulls three crates over a smooth horizontal floor as shown in **Figure 4-45**. The crates are connected to each other by identical horizontal strings A and B, each of which can support a maximum tension of 45.0 N before breaking. (a) What is the largest pulling force that can be exerted without breaking either of the strings? (b) What are the tensions in A and B just before one of the strings breaks? Example 4-8

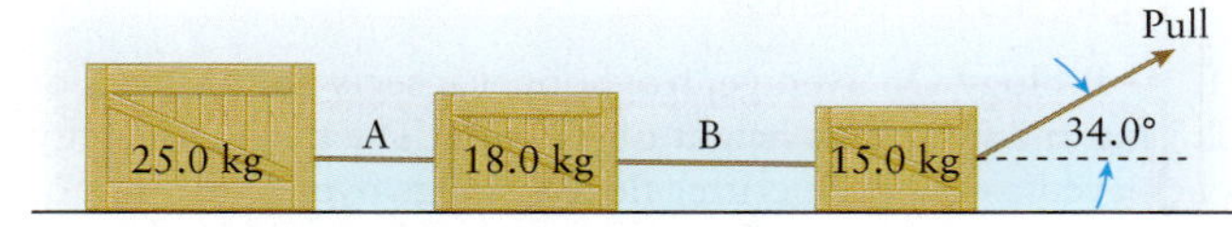

Figure 4-45 Problem 103

5 Forces and Motion II: Applications

Elena Lelikova/Shutterstock

What do you think?

When an airplane or a bird flies around a circular path at a constant speed, is there a nonzero *net* force on the airplane or bird? If so, does the net force point (a) inward toward the center of the circle, (b) outward away from the center of the circle, or (c) in some other direction?

In this chapter, your goals are to:

- (5-1) Appreciate that Newton's laws apply to all forces and in all situations.
- (5-2) Recognize what determines the magnitude of the static friction force.
- (5-3) Be able to find the magnitude and direction of the force of kinetic friction.
- (5-4) Solve problems in which the forces on an object include static or kinetic friction.
- (5-5) Analyze situations in which fluid resistance is important.
- (5-6) Apply Newton's laws to objects in uniform circular motion.

To master this chapter, you should review:

- (3-4) The concepts and equations of motion in two or three dimensions.
- (3-7) The concepts and equations of uniform circular motion.
- (4-2 to 4-6) How to solve problems using Newton's laws of motion.

5-1 We can use Newton's laws in situations beyond those we have already studied

In Chapter 4 we learned the basic laws that describe the motions of objects, and we saw how external forces acting on an object determine the magnitude and direction of the object's acceleration. In this chapter we'll see how to apply these laws to a variety of very common situations. In particular, we'll take a much closer look at situations involving friction forces and drag forces—two forces that resist the motion of one object relative to another. We'll also analyze the direction of the forces acting on an object moving in a circular path and describe the resulting acceleration of that object.

The force of *friction* plays an important role in many physical situations. Friction acts in almost every case when two surfaces are in contact, and it can be present whether or not the two surfaces are sliding past each other. In Chapter 4 we used the friction force in some examples; in this chapter we'll learn more about the nature of friction and see how to solve more complex problems that involve friction. Another important situation arises when objects move through a fluid (a gas or liquid), such as a football flying through the air or a submarine cruising beneath the surface of the ocean. The fluid exerts a *drag force* on each of these objects. The drag force makes it more difficult for an object to move through a fluid and is greater the faster an object travels.

The forces of friction and drag are special *kinds* of forces. Another important situation is one that can involve forces of *any* kind: an object moving in a circular path

TAKE-HOME MESSAGE FOR Section 5-1

✔ Newton's laws also apply in situations that involve frictional forces, drag forces, and motion in a circle.

(a)

omgimages/Getty Images

The ground exerts a static friction force on the stationary children's feet. This force prevents their feet from sliding.

(b)

JGI/Jamie Grill/Getty Images

The water slide exerts a kinetic friction force on the moving child that slows her down. The water reduces the friction but doesn't eliminate it.

Figure 5-1 Static friction versus kinetic friction Both (a) static friction and (b) kinetic friction arise from the interaction between two surfaces in contact.

at constant speed, like the airplane shown in the photograph that opens this chapter. In this situation, no matter what forces act on the object, the net force must give the object an acceleration toward the center of the circle.

5-2 The static friction force changes magnitude to offset other applied forces

Friction is a force that resists the sliding of one object past another. If you push or pull on an object to make it slide but it doesn't move, we call the force that is preventing sliding a **static friction** force (**Figure 5-1a**). If the object is moving, we call the force that opposes this motion a **kinetic friction** force (**Figure 5-1b**). Both static and kinetic friction forces result from interactions between the microscopic irregularities (bumps) of the two surfaces in contact. Even surfaces that appear or feel smooth can be rough at the microscopic level. Irregularities on one surface can catch on or be impeded by irregularities on the other surface.

We'll concentrate on static friction in this section and analyze kinetic friction in Section 5-3.

Properties of Static Friction

To demonstrate some properties of static friction, gently rest your open palm and fingertips against a wall and then try to drag your hand downward while maintaining contact with the wall. You'll find that it doesn't take much effort to make your hand slide. If you repeat the same experiment, but push a bit harder against the wall, you'll notice that it takes a bit more effort to get your hand to move. If you push as hard as you can on the wall, you'll find it very challenging to slide your hand down the wall. If the wall has a smooth surface, however, you'll find that your hands slip more easily than on a rough-surfaced wall.

Here's what we learn from this simple experiment:

(1) In each case the force that tries to keep your stationary hand from sliding is the force of static friction. We'll use the symbol $\vec{f}_s$ ("f-sub-s") for this force. (The subscript "s" reminds us that we're talking about *static* friction.) The static friction force $\vec{f}_s$ exerted on your hand by the wall acts parallel to your hand and parallel to the wall, that is, parallel to the surfaces in contact.

(2) You can overcome static friction by pushing against it with sufficient force. This says that in a given situation, the static friction force has a maximum magnitude $f_{s,max}$. You overcome the static friction force, and so start your hand sliding, if you apply a force in a direction parallel to the surfaces, whose magnitude is greater than $f_{s,max}$.

(3) The value of $f_{s,max}$ depends on how hard the two surfaces are pushed together—that is, on the *normal force* that acts perpendicular to the two surfaces. The greater the magnitude n of the normal force, the greater the maximum static friction force and the more difficult it is to make the surfaces slide past each other.

(4) The force of static friction depends on the properties of the surfaces in contact.

We can combine these observations into a simple mathematical relationship:

Magnitude of the static friction force

(5-1a)

(5-1b)

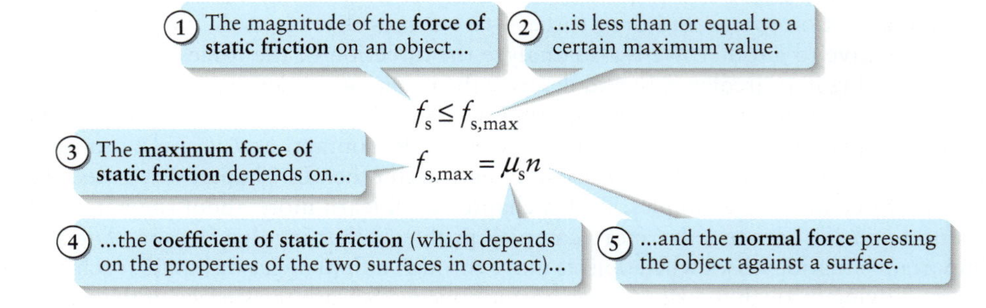

The quantity μ_s in Equation 5-1b is called the **coefficient of static friction**. Its value depends on the properties of both surfaces in contact; for example, μ_s has different values for steel on steel, for steel on glass, and for glass on glass. Because both f_s and n have units of force, Equation 5-1b tells us that μ_s must be dimensionless.

Typical values of the coefficient of static friction are between about 0.1 and 0.8. For example, μ_s is approximately 0.1 for a ski on snow, approximately 0.5 for two pieces of wood in contact, and approximately 0.7 for two pieces of metal in contact.

BIO-
Medical

Lubrication reduces the frictional force between surfaces and hence the value of μ_s. In human joints, such as the elbow and knee, the presence of lubricating fluid results in a coefficient of static friction of about 0.01. This relatively low value of μ_s minimizes sticking and reduces wear and tear of the cartilage that supports these joints.

WATCH OUT! Static friction is described by an inequality, not an equality.

All of the mathematical relationships we've seen so far in our study of physics have been equalities, with an equals sign relating two quantities. But Equation 5-1a is an *inequality*. **Figure 5-2** helps show why this is so. Imagine that you apply a horizontal force to a stationary block on a horizontal table. You find that if you apply a 20.0-N force, the block remains at rest. Therefore, the static friction force must have the same 20.0-N magnitude to balance the force that you apply (**Figure 5-2a**). What happens if you apply a lesser force, say only 10.0 N? The block would still remain at rest, so the force of static friction is now only 10.0 N in magnitude to exactly cancel the applied force (**Figure 5-2b**). The block also remains at rest if you don't push on it at all, so in this case the static friction force is zero (**Figure 5-2c**). If you push enough to overcome the maximum force of static friction given by Equation 5-1b, the object will slide. In Section 5-3 we'll discuss what happens then.

(a) You apply a force of 20.0 N to the right, but the block remains at rest...

20.0 N

20.0 N

...so the table must exert a static friction force of 20.0 N to the left on the block.

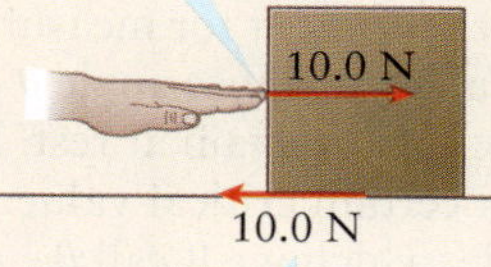
(b)
You apply a force of 10.0 N to the right, but the block remains at rest...
10.0 N
10.0 N
...so the table must exert a static friction force of 10.0 N to the left on the block.

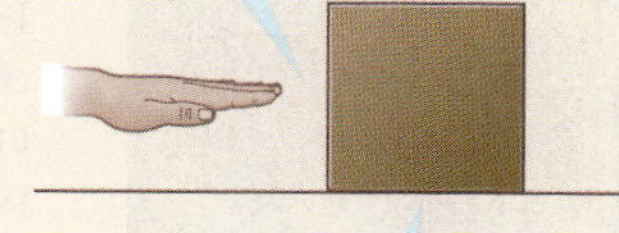
(c)
You apply no force to the block and it remains at rest...
...so the table must exert no static friction force on the block.

The static friction force adjusts its value to counteract whatever other forces are trying to make the object slide.

Figure 5-2 The magnitude of the static friction force For a given object at rest on a given surface, the magnitude of the static friction force is *not* always the same: It depends on the other forces acting on the object.

WATCH OUT! The direction of the static friction force is parallel to the contact surface.

Equations 5-1 relate only the *magnitudes* of the static friction force and the normal force. These forces are *not* in the same direction! The friction force always acts parallel to the surface of contact, while the normal force always acts perpendicular to the surface. (Remember that "normal" is another word for "perpendicular.")

Static friction explains why a nail does not slide out of the wood into which it is hammered, even if the nail is upside down (**Figure 5-3a**). The wood fits tightly around the nail and so exerts strong normal forces on the sides of the nail (**Figure 5-3b**). Equation 5-1b tells us that the maximum static friction force is therefore quite large, more than enough to keep the nail in place. If you live in a wood-frame house, static friction deserves much of the credit for holding it together! The same effect explains why it is difficult to take off a tight-fitting pair of boots or to pull an overstuffed pillow out of its pillowcase.

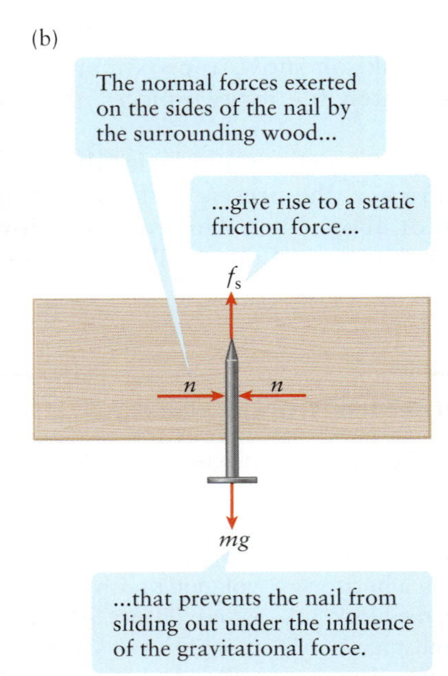

Figure 5-3 Static friction force on a nail (a) Hammering a nail upward into a piece of wood. (b) Forces on the nail.

BIO-Medical Geckos (tropical lizards of the family *Gekkonidae*) use static friction to walk on vertical surfaces, even on a smooth plate of glass (**Figure 5-4**). At first glance this seems impossible: On a vertical surface there is nothing pushing the gecko into the surface, so the normal force would be zero, and there would be zero static friction force (see Equation 5-1b). However, at the end of each of a gecko's toes are hundreds of thousands of tiny hairs or *setae*, each of which splits into hundreds of even smaller, flat pads. Each pad or *spatula* is only about 0.2 μm (2×10^{-7} m) across—so small that it can nestle in one of the tiny crevices on the surface of a pane of glass. Like what happens to the nail in Figure 5-3, the normal force between a spatula and the walls of its crevice give rise to a static friction force. Multiply that tiny force by the millions of spatulae on a gecko's four feet, and the result is a maximum force of friction as great as 20 N—more than adequate to support the weight of an average gecko, which is less than 0.5 N.

Figure 5-4 Walking on a glass wall A gecko uses static friction to support its weight while walking on a vertical surface. It can do this thanks to the specialized architecture of its feet.

Measuring the Coefficient of Static Friction

Here's a simple experiment for measuring the coefficient of static friction. Place a block of weight w on a ramp whose angle θ can be varied (**Figure 5-5a**). If the angle θ is small enough, the block will remain at rest. However, the block will start to slide if the angle is greater than a certain critical value. Let's see how to determine the value of μ_s from this critical angle, which we'll call θ_{slip}.

We'll follow the Method of Solving Force Problems that we used in Section 4-6. (You should review that strategy because we'll be using it a lot in this chapter.) **Figure 5-5b** shows the free-body diagram for the block when $\theta = \theta_{slip}$. Three forces act on the block: the normal force of magnitude n, which is perpendicular to the surface of the ramp; the gravitational force of magnitude w, which points downward; and the static friction force of magnitude f_s, which points opposite to the direction the block would move if it could. At this angle the block is just about to slip, which means the friction force must be at its maximum value. So from Equations 5-1, $f_s = f_{s,max} = \mu_s n$.

We set the sum of the external forces on the block in each direction equal to its mass multiplied by its acceleration in that direction (Newton's second law). We take the positive x direction to point down the ramp and the positive y direction to point perpendicular to the ramp, as Figure 5-5b shows. The block remains at rest, so $a_x = a_y = 0$. Then the equations of Newton's second law are

(5-2)

$$\text{(a)} \quad x: \sum F_{ext,x} = n_x + w_x + f_{s,max,x} = ma_x = 0$$

$$\text{(b)} \quad y: \sum F_{ext,y} = n_y + w_y + f_{s,max,y} = ma_y = 0$$

Figure 5-5b shows that the gravitational force has a positive (downhill) x component $w_x = w \sin \theta_{slip}$ and a negative y component $w_y = -w \cos \theta_{slip}$. The normal force points in the positive y direction (so $n_x = 0$ and $n_y = n$), and the static friction force points in the negative x direction (so $f_{s,max},x = -f_{s,max}$ and $f_{s,max,y} = 0$). So Equations 5-2 become

(5-3)

$$\text{(a)} \quad x: \sum F_{ext,x} = 0 + w \sin \theta_{slip} + (-f_{s,max}) = 0$$

$$\text{(b)} \quad y: \sum F_{ext,y} = n + (-w \cos \theta_{slip}) + 0 = 0$$

Our goal is to find the coefficient of static friction μ_s and so we need to incorporate this quantity into our equations. We do this by substituting Equation 5-1b, $f_{s,max} = \mu_s n$, into Equation 5-3a. Then, by rearranging Equations 5-3, we get the following expressions:

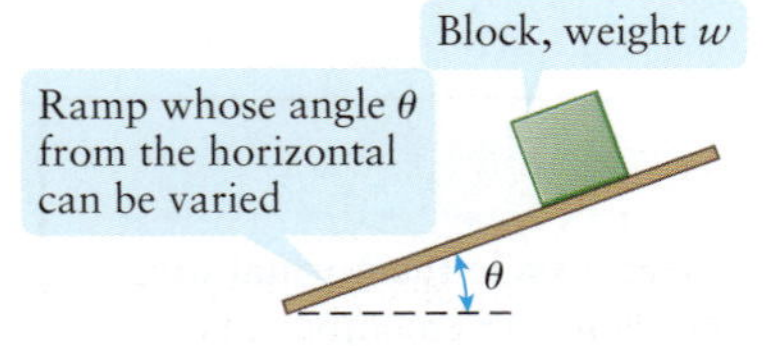

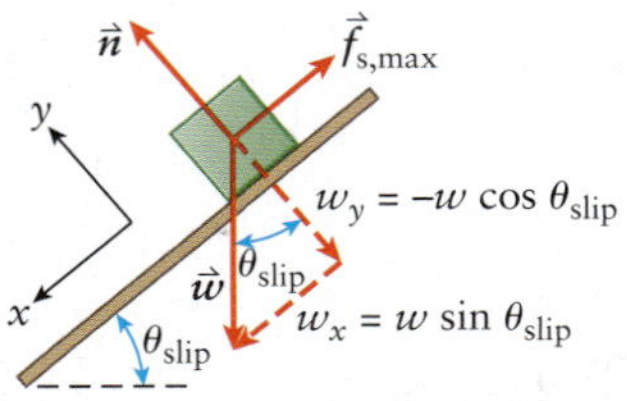

Figure 5-5 Measuring the coefficient of static friction The angle at which a block on an incline just begins to slip tells you the coefficient of static friction μ_s.

$$\text{(a)} \quad \mu_s n = w \sin \theta_{\text{slip}}$$
$$\text{(b)} \quad n = w \cos \theta_{\text{slip}} \qquad (5\text{-}4)$$

Equation 5-4b tells us the value of n, the magnitude of the normal force, so we can substitute this quantity into Equation 5-4a and solve for μ_s:

$$\mu_s(w \cos \theta_{\text{slip}}) = w \sin \theta_{\text{slip}}$$
$$\mu_s \cos \theta_{\text{slip}} = \sin \theta_{\text{slip}}$$

Or, using $\tan \theta = (\sin \theta)/(\cos \theta)$,

Coefficient of static friction for an object at rest on a surface

Angle of the surface at which the object **begins to slip**

$$\mu_s = \tan \theta_{\text{slip}}$$

Angle at which an object slips on an incline (5-5)

The greater the angle θ_{slip} beyond which the box begins to slide, the greater the value of $\tan \theta_{\text{slip}}$ and hence the greater the coefficient of static friction must be. Note that Equation 5-5 does *not* involve the weight of the block. A plastic block on a wooden ramp will begin to slide at the same angle θ regardless of its weight.

Go to the Math Tutorial for more information on trigonometry.

WATCH OUT! The coefficient of static friction can be greater than one.

Because typical values of the coefficient of static friction are less than 1, it is tempting to assume that μ_s is *always* less than 1. Not so! **Figure 5-6** shows a wooden block that rests without slipping on a ramp covered with sandpaper. The ramp is set at a 50° angle; if this is the angle beyond which the block will slip, then from Equation 5-5 we find that $\mu_s = \tan 50° = 1.2$. Whenever θ_{slip} is greater than 45°, the coefficient of static friction is greater than $\tan 45° = 1$.

David Tauck

Figure 5-6 A block on a sandpaper surface This block can remain at rest at an angle of up to 50°, so the coefficient of static friction for this block on sandpaper is tan 50° = 1.2 (see Equation 5-5).

WATCH OUT! The normal force exerted on an object is not always equal to the object's weight.

Since the maximum force of static friction is related to the normal force, it's important to remember that the normal force for an object on a surface is *not* always equal to the object's weight. (We cautioned you about this in Section 4-3). The situation shown in Figure 5-5 is an example of this caveat. Equation 5-4b shows that the normal force on this object is the object's weight multiplied by the cosine of the angle of the incline. Resist the temptation to always set the normal force equal to the weight!

BIO-Medical EXAMPLE 5-1 **Friction in Joints**

The wrist is made up of eight small bones called *carpals* that glide back and forth as you wave your hand from side to side. A thin layer of cartilage covers the surfaces of the carpals, making them smooth and slippery. In addition, the spaces between the bones contain synovial fluid, which provides lubrication. During a laboratory experiment, a physiologist applies a compression force to squeeze the bones together along their nearly planar bone surfaces. She then measures the force that must be applied parallel to the surface of contact to make them move. (**Figure 5-7** shows the contact region between these two carpal bone surfaces.) When the compression force is 11.2 N, the minimum force required to move the bones is 0.135 N. What is the coefficient of static friction in the joint?

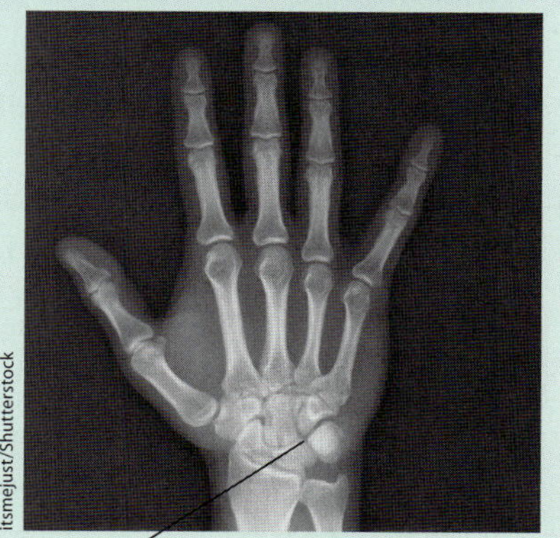

itsmejust/Shutterstock

Figure 5-7 Static friction in the wrist This X-ray image shows the carpal bones of the wrist.

Set Up

We begin by drawing the free-body diagram for one of the two carpals in question, which we call carpal A, when it is just about to slip relative to carpal B. The physiologist exerts two of the four forces acting on carpal A: the compression force $\vec{F}_{\text{compression}}$ (of magnitude 11.2 N) and the force $\vec{F}_{\text{slide}}$ applied parallel to the contact surface (of magnitude 0.135 N) to make bone A slide relative to bone B. The other two forces acting on carpal A are the normal force $\vec{n}$ and the static friction force $\vec{f}_{\text{s,max}}$, both exerted by carpal B on carpal A. (This friction force is the *maximum* force of static friction because the carpals are just about to slip.) Our goal is to find the value of μ_s, so we'll also use the relationship between $f_{\text{s,max}}$ and the normal force given by Equation 5-1b.

Newton's second law for carpal A:

$$\sum \vec{F}_{\text{ext on } A} = \vec{F}_{\text{compression}} + \vec{F}_{\text{slide}} + \vec{n} + \vec{f}_{\text{s,max}} = m_A \vec{a}_A$$

Magnitude of the static friction force:

$$f_{\text{s,max}} = \mu_s n \qquad (5\text{-}1\text{b})$$

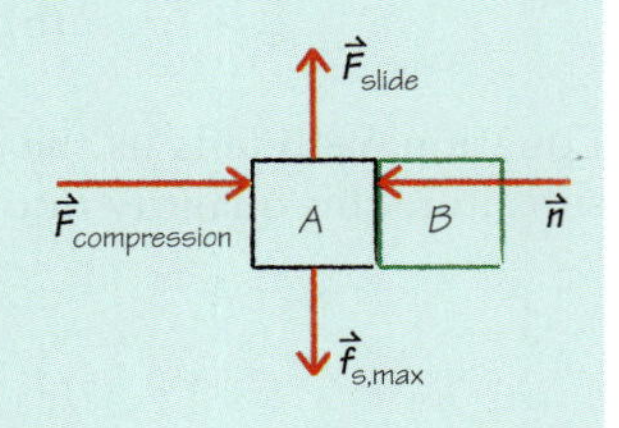

Solve

Let the positive x axis point up and the positive y axis point to the left. Since we are considering the situation in which the carpals have not yet begun to slide, carpal A is at rest and $a_{A,x} = a_{A,y} = 0$. We use this information to write Newton's second law for carpal A in component form.

Newton's second law for carpal A in component form:

$$x: \sum F_{\text{ext on } A,x} = 0 + F_{\text{slide}} + 0 + (-f_{\text{s,max}}) = m_A a_{A,x} = 0$$

$$y: \sum F_{\text{ext on } A,y} = (-F_{\text{compression}}) + 0 + n + 0 = m_A a_{A,y} = 0$$

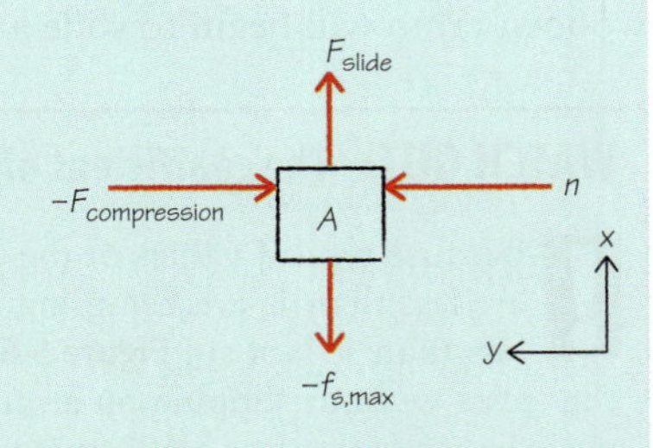

To find the value of μ_s using Equation 5-1b, we need to know the values of n and $f_{\text{s,max}}$. We find these quantities by using Newton's second law equations and the known values $F_{\text{compression}} = 11.2$ N and $F_{\text{slide}} = 0.135$ N.

Solve Equation 5-1b for μ_s:

$$\mu_s = \frac{f_{\text{s,max}}}{n}$$

Solve Newton's second law equations for $f_{\text{s,max}}$ and n. From the x component equation:

$$f_{\text{s,max}} = F_{\text{slide}} = 0.135 \text{ N}$$

From the y component equation:

$$n = F_{\text{compression}} = 11.2 \text{ N so}$$

$$\mu_s = \frac{0.135 \text{ N}}{11.2 \text{ N}} = 1.21 \times 10^{-2} = 0.0121$$

Reflect

This coefficient of static friction is very small, which means that very little effort is required to make your wrist move. The lubricating fluid between the bones is the key to keeping the friction so small. The fluid also helps reduce wear on the joints. If the bones or even the cartilage surfaces were in direct contact, our joints would wear down relatively quickly.

GOT THE CONCEPT? 5-1 Face to Face with Friction

? A rectangular block made of steel has dimensions 1.00 cm × 2.00 cm × 3.00 cm. You place it on a wooden board, and then you tilt the board until the block begins to slide down the board. Which face of the block should be in contact with the board so that it begins to slide at the *shallowest* angle? (a) The 1.00-cm × 2.00-cm face, (b) the 1.00-cm × 3.00-cm face, (c) the 2.00-cm × 3.00-cm face. (d) The block will begin to slide at the same angle in each case. (e) The answer depends on the type of steel and the type of wood.

TAKE-HOME MESSAGE FOR Section 5-2

✔ If two objects are at rest with respect to each other and an external force is applied to try to make them slide past each other, static friction acts to oppose that external force.

✔ In any situation, the magnitude of the static friction force on an object adjusts itself to balance the other forces acting on the object. However, there is a maximum possible magnitude of the static friction force, given by Equations 5-1.

✔ If an object is placed on a tilted ramp, it will slide if the angle of the ramp is greater than the slip angle given by Equation 5-5.

5-3 The kinetic friction force on a sliding object has a constant magnitude

In the previous section, we used a block on a tilted ramp (Figure 5-5) to help us explore the force of static friction. If the ramp is inclined by more than a certain critical angle, not even the maximum force of static friction will prevent the block from sliding. The block still experiences a resistive force as it slides. However, this force is no longer static friction but *kinetic* **friction**, the frictional force between two surfaces that move relative to each other.

The Magnitude of the Kinetic Friction Force

Like static friction, the force of kinetic friction $\vec{f}_k$ ("f-sub-k") between two objects is proportional to the normal force that one object exerts on the other perpendicular to their surface of contact—in other words, it is proportional to how hard the two surfaces are pushed together. However, the kinetic friction force is typically less than the maximum force of static friction.

We'll model the kinetic friction force as being independent of how fast the two objects slide past each other, as well as independent of the area of contact between the objects. (This model works well in many situations.) Then we can write an expression for the magnitude f_k of the kinetic friction force:

The **force of kinetic friction** depends on...

$$f_k = \mu_k n \qquad (5\text{-}6)$$

...the **coefficient of kinetic friction** (which depends on the properties of the two surfaces in contact)...

...and the **normal force** exerted on the object by the surface on which it slides.

Magnitude of the kinetic friction force (5-6)

Like the coefficient of static friction μ_s, the value of the **coefficient of kinetic friction** μ_k depends on the kinds of surfaces that are in contact. For typical surfaces μ_k is close to but less than μ_s. That's why it takes less force to keep a block sliding at constant speed over a horizontal surface than it does to start it moving in the first place (**Figure 5-8**). Values of the coefficient of kinetic friction for typical materials range between 0.1 and 0.8.

Go to Interactive Exercises 5-1 and 5-2 for more practice dealing with kinetic friction.

In **Figure 5-9a** the block is moving at a constant velocity, so its acceleration is zero and the net force exerted on the block is zero as well. In this case the horizontal force that you apply just balances the force of kinetic friction of magnitude f_k opposing the motion. If the horizontal force that you apply is greater in magnitude than f_k, the net force points forward and the block speeds up (**Figure 5-9b**). If you apply a horizontal force with a magnitude less than f_k to the sliding block, the block slows down because the net force points backward (**Figure 5-9c**).

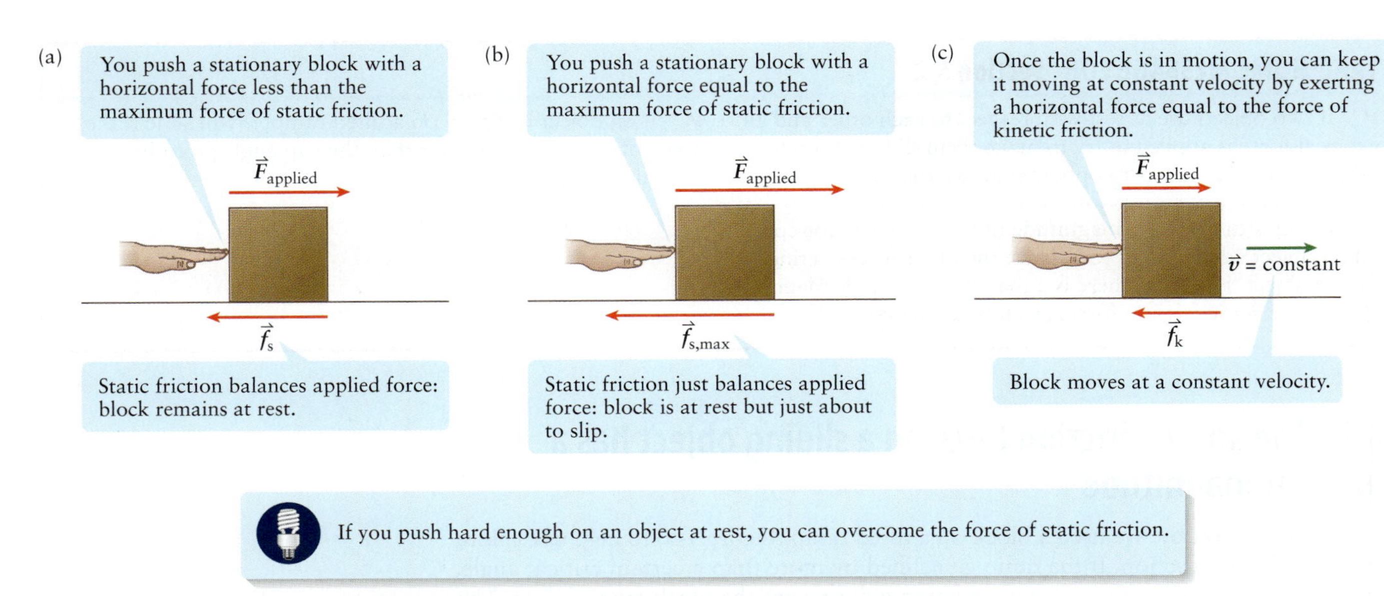

(a)
You push a stationary block with a horizontal force less than the maximum force of static friction.
$\vec{F}_{applied}$
$\vec{f}_s$
Static friction balances applied force: block remains at rest.
(b)
You push a stationary block with a horizontal force equal to the maximum force of static friction.
$\vec{F}_{applied}$
$\vec{f}_{s,max}$
Static friction just balances applied force: block is at rest but just about to slip.
(c)
Once the block is in motion, you can keep it moving at constant velocity by exerting a horizontal force equal to the force of kinetic friction.
$\vec{F}_{applied}$
$\vec{v}$ = constant
$\vec{f}_k$
Block moves at a constant velocity.
If you push hard enough on an object at rest, you can overcome the force of static friction.

Figure 5-8 From static friction to kinetic friction If you push hard enough on an object to overcome static friction and start the object sliding, the force of friction becomes the kinetic friction force.

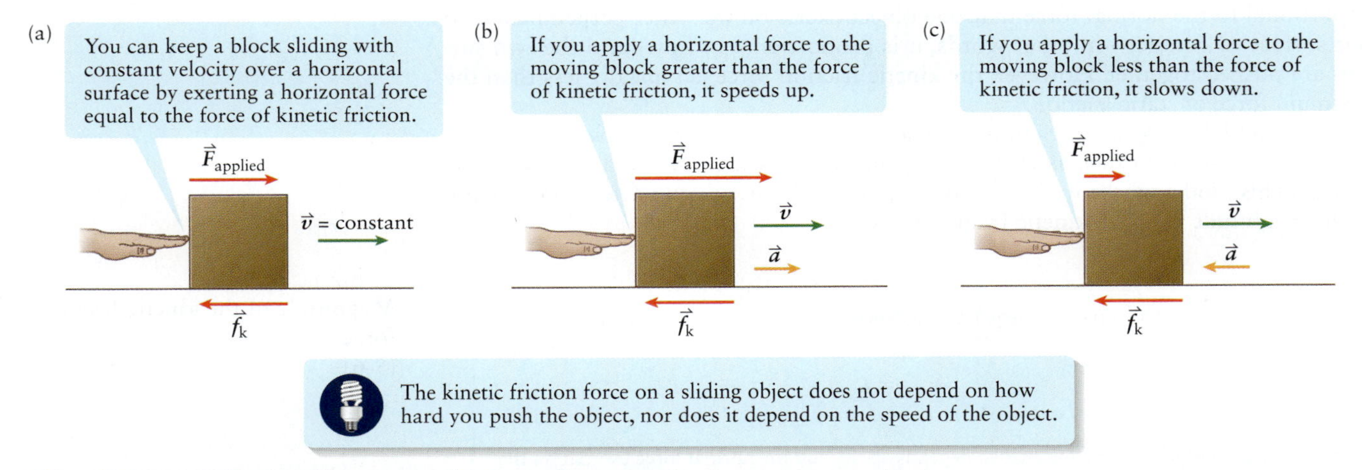

(a)
You can keep a block sliding with constant velocity over a horizontal surface by exerting a horizontal force equal to the force of kinetic friction.
$\vec{F}_{applied}$
$\vec{v}$ = constant
$\vec{f}_k$
(b)
If you apply a horizontal force to the moving block greater than the force of kinetic friction, it speeds up.
$\vec{F}_{applied}$
$\vec{v}$
$\vec{a}$
$\vec{f}_k$
(c)
If you apply a horizontal force to the moving block less than the force of kinetic friction, it slows down.
$\vec{F}_{applied}$
$\vec{v}$
$\vec{a}$
$\vec{f}_k$
The kinetic friction force on a sliding object does not depend on how hard you push the object, nor does it depend on the speed of the object.

Figure 5-9 Kinetic friction: Three examples Depending on how hard you push on an object sliding with kinetic friction, it will (a) maintain a constant velocity, (b) speed up, or (c) slow down.

EXAMPLE 5-2 Sliding with Kinetic Friction

In Example 4-6 (see Section 4-6) we considered a book sliding down an incline with friction. In a variation of this problem, suppose the coefficient of kinetic friction between the incline and the sliding 2.50-kg book is 0.350. (a) What is the downhill acceleration of the book if the incline is at an angle from the horizontal of 30.0°? (b) What is the acceleration if the angle is 10.0°?

Set Up

The book is moving down the incline of angle θ, so the kinetic friction force (which always opposes sliding) must point up the incline. To find the net external force on the book and hence the book's acceleration, we need to determine the magnitude of the friction force by using Equation 5-6. Once we've found a general expression for the book's acceleration in terms of θ, we'll substitute the values $\theta = 30.0°$ and $\theta = 10.0°$.

Newton's second law applied to the book:

$$\sum \vec{F}_{\text{ext on book}} = \vec{n} + \vec{w}_{\text{book}} + \vec{f}_{\text{k}} = m_{\text{book}}\vec{a}_{\text{book}}$$

Magnitude of the kinetic friction force:

$$f_{\text{k}} = \mu_{\text{k}} n \qquad (5\text{-}6)$$

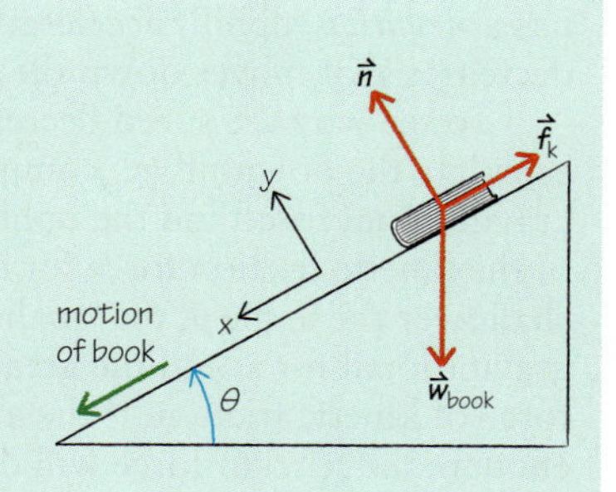

Solve

Write Newton's second law for the book in component form. Because the book's acceleration is along the incline (that is, along the x direction), it follows that $a_{\text{book},y} = 0$.

Newton's second law for the book in component form:

$$x: \sum F_{\text{ext on book},x} = 0 + w_{\text{book}} \sin\theta + (-f_{\text{k}}) = m_{\text{book}} a_{\text{book},x}$$

$$y: \sum F_{\text{ext on book},y} = n - w_{\text{book}} \cos\theta + 0 = m_{\text{book}} a_{\text{book},y} = 0$$

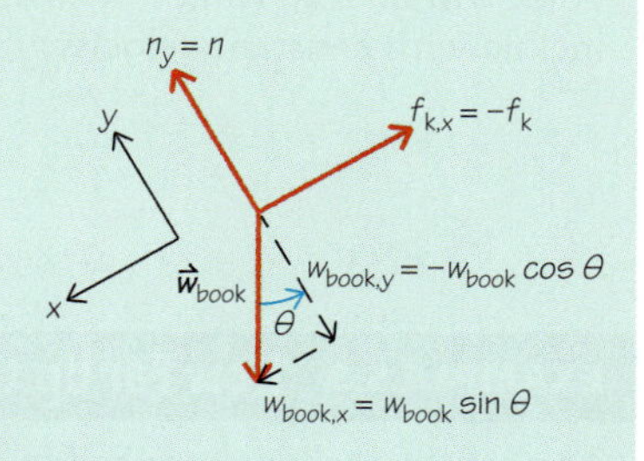

To find the magnitude of the kinetic friction force, we solve the y component equation for the magnitude of the normal force n and then substitute this expression into the equation for f_{k}. Next we substitute our result for f_{k} into the x component equation.

From the y component equation,

$n = w_{\text{book}} \cos\theta$

Substitute this into Equation 5-6:

$f_{\text{k}} = \mu_{\text{k}} n = \mu_{\text{k}}(w_{\text{book}} \cos\theta)$

Substitute this expression for f_{k} into the x equation:

$w_{\text{book}} \sin\theta - \mu_{\text{k}} w_{\text{book}} \cos\theta = m_{\text{book}} a_{\text{book},x}$

Now we can complete our solution for the book's acceleration $a_{\text{book},x}$. Use the relationship between gravitational force and mass.

Relationship between the mass m_{book} of the book and the gravitational force w_{book} on the book:

$w_{\text{book}} = m_{\text{book}} g$

Substitute into the x equation and solve for $a_{\text{book},x}$:

$m_{\text{book}} g \sin\theta - \mu_{\text{k}} m_{\text{book}} g \cos\theta = m_{\text{book}} a_{\text{book},x}$

The factor of m_{book} cancels, so

$a_{\text{book},x} = g \sin\theta - \mu_{\text{k}} g \cos\theta$ or

$a_{\text{book},x} = g(\sin\theta - \mu_{\text{k}} \cos\theta)$

Use this general formula for $a_{\text{book},x}$ for the given value of μ_{k} and the two given values of θ.

(a) Substitute $\theta = 30.0°$:

$$\begin{aligned} a_{\text{book},x} &= g(\sin\theta - \mu_{\text{k}} \cos\theta) \\ &= (9.80\ \text{m/s}^2)[\sin 30.0° - (0.350)(\cos 30.0°)] \\ &= (9.80\ \text{m/s}^2)[0.500 - (0.350)(0.866)] \\ &= 1.93\ \text{m/s}^2 \end{aligned}$$

(b) Substitute $\theta = 10.0°$:

$$\begin{aligned} a_{\text{book},x} &= g(\sin\theta - \mu_{\text{k}} \cos\theta) \\ &= (9.80\ \text{m/s}^2)[\sin 10.0° - (0.350)(\cos 10.0°)] \\ &= (9.80\ \text{m/s}^2)[0.174 - (0.350)(0.985)] \\ &= -1.68\ \text{m/s}^2 \end{aligned}$$

Reflect

The book has a positive (downhill) acceleration for $\theta = 30.0°$, which means the book speeds up as it slides downhill. However, when the incline is at a shallower angle of $\theta = 10.0°$, the book

Downhill component of gravitational force:

$w_{\text{book},x} = w_{\text{book}} \sin\theta = m_{\text{book}} g \sin\theta$

Uphill component of kinetic friction force:

$f_{\text{k}} = \mu_{\text{k}} n = \mu_{\text{k}} m_{\text{book}} g \cos\theta$

has a *negative* (uphill) acceleration: Its speed decreases as it moves down the incline.

To see why the speed decreases, let's calculate the downhill (x) component of the gravitational force and the uphill component of the kinetic friction force for each angle. The shallower the angle θ, the smaller the downhill gravitational force and the greater the uphill force of kinetic friction. If the angle is small enough, the friction force will dominate over the gravitational force—in which case the net force and acceleration point up the incline.

Can you show that the acceleration of the book will be zero (that is, it will slide down the incline with constant velocity) if $\theta = 19.3°$?

If $\theta = 30.0°$,

$$w_{\text{book},x} = (2.50\text{ kg})(9.80\text{ m/s}^2)\sin 30.0° = 12.3\text{ N}$$

$$f_k = (0.350)(2.50\text{ kg})(9.80\text{ m/s}^2)\cos 30.0° = 7.43\text{ N}$$

so gravity dominates over kinetic friction, and the net force is downhill.

If $\theta = 10.0°$,

$$w_{\text{book},x} = (2.50\text{ kg})(9.80\text{ m/s}^2)\sin 10.0° = 4.25\text{ N}$$

$$f_k = (0.350)(2.50\text{ kg})(9.80\text{ m/s}^2)\cos 10.0° = 8.44\text{ N}$$

Kinetic friction dominates over gravity and so the net force is uphill.

EXAMPLE 5-3 How Far Up the Incline?

You set the incline described in Example 5-2 at an angle of 30.0° from the horizontal and then push the 2.50-kg book up the incline with an initial speed of 4.20 m/s. The coefficient of kinetic friction between book and ramp is again 0.350. How far up the incline does the book travel before coming to a stop?

Set Up

As in Example 5-2, the three forces acting on the sliding book are the normal force, gravitational force, and kinetic friction force. The difference here is that the friction force points downhill rather than uphill because the book is now sliding uphill. (Remember that the kinetic friction force always opposes the slipping.)

This problem asks for a distance, so it's really a kinematics problem. Because all three forces that act on the book are constant, the net external force and acceleration are constant as well. Therefore we can use the formulas from Section 2-4 for linear motion with constant acceleration. We choose an equation that relates the book's initial velocity v_{0x}, its final velocity v_x, and its acceleration $a_{\text{book},x}$ to its displacement $x - x_0$, which is what we are asked to find.

Newton's second law applied to the book:

$$\sum \vec{F}_{\text{ext on book}} = \vec{n} + \vec{w}_{\text{book}} + \vec{f}_k = m_{\text{book}}\vec{a}_{\text{book}}$$

Magnitude of the kinetic friction force:

$$f_k = \mu_k n \quad (5\text{-}6)$$

Velocity, acceleration, and position for constant acceleration only:

$$v_x^2 = v_{0x}^2 + 2a_{\text{book},x}(x - x_0) \quad (2\text{-}11)$$

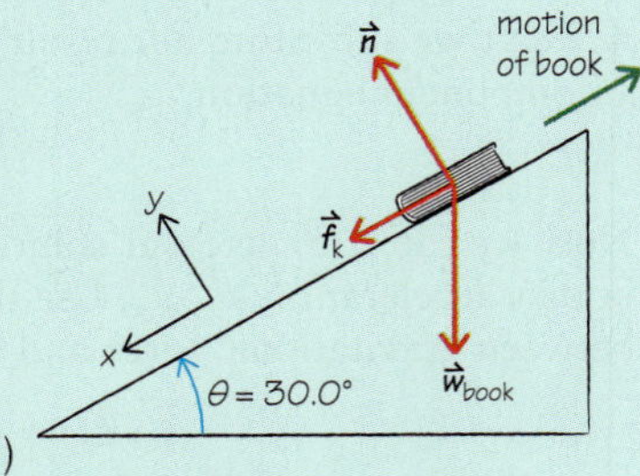

Solve

Write Newton's second law for the book in component form. The equations are the same as in Example 5-2 except that the friction force now has a positive (downhill) x component. The book's acceleration is along the x axis only, so $a_{\text{book},y} = 0$.

Newton's second law for the book in component form:

$$x: \sum F_{\text{ext on book},x} = 0 + w_{\text{book}}\sin\theta + f_k = m_{\text{book}}a_{\text{book},x}$$

$$y: \sum F_{\text{ext on book},y} = n - w_{\text{book}}\cos\theta + 0 = m_{\text{book}}a_{\text{book},y} = 0$$

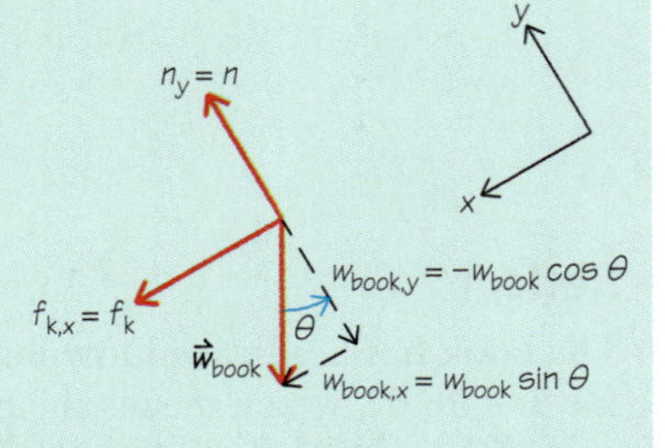

Use these equations and Equation 5-6 to solve for the acceleration of the book.

From the y component equation and $w_{\text{book}} = m_{\text{book}}g$,

$n = w_{\text{book}} \cos\theta = m_{\text{book}}\, g \cos\theta$

Substitute into Equation 5-6 and find the magnitude of the kinetic friction force:

$f_k = \mu_k n = \mu_k m_{\text{book}}\, g \cos\theta$

Substitute into x component equation and solve for $a_{\text{book},x}$:

$$m_{\text{book}}\, g \sin\theta + \mu_k m_{\text{book}}\, g \cos\theta = m_{\text{book}}\, a_{\text{book},x}$$

$$g \sin\theta + \mu_k\, g \cos\theta = a_{\text{book},x}$$

$$\begin{aligned} a_{\text{book},x} &= g(\sin\theta + \mu_k \cos\theta) \\ &= (9.80\ \text{m/s}^2)[\sin 30.0^\circ + (0.350)(\cos 30.0^\circ)] \\ &= (9.80\ \text{m/s}^2)[0.500 + (0.350)(0.866)] \\ &= 7.87\ \text{m/s}^2 \end{aligned}$$

(Acceleration is positive because it points downhill.)

Now we can solve for the displacement of the book from where it starts to where it stops. Its initial velocity v_{0x} is negative because the book starts off moving uphill. Its final velocity v_x is zero because it is at rest after traveling the maximum distance up the incline. Use Equation 2-11 to find $x - x_0$.

Equation 2-11:

$$v_x^2 = v_{0x}^2 + 2a_{\text{book},x}(x - x_0)$$

with $v_{0x} = -4.20$ m/s, $v_x = 0$, and $a_{\text{book},x} = 7.87\ \text{m/s}^2$

Solve for $x - x_0$:

$$\begin{aligned} x - x_0 &= \frac{v_x^2 - v_{0x}^2}{2a_{\text{book},x}} \\ &= \frac{0 - (-4.20\ \text{m/s})^2}{2(7.87\ \text{m/s}^2)} \\ &= -1.12\ \text{m} \end{aligned}$$

(Displacement is negative since the book moves uphill.)
So the book travels 1.12 m.

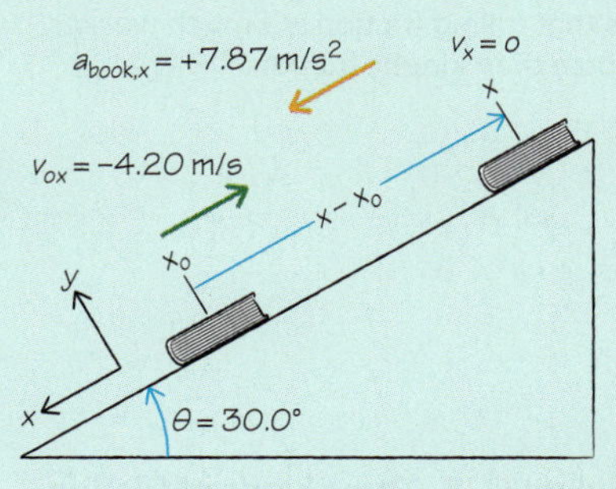

Reflect

The downhill acceleration of the book ($7.87\ \text{m/s}^2$) is much greater than in Example 5-2 ($1.93\ \text{m/s}^2$) because the force of kinetic friction now points downhill and so augments the downhill component of the gravitational force.

We can check our result by seeing how far the book would travel if the surface were horizontal, so $\theta = 0$. In this case the book travels 2.57 m, which is more than twice the distance we found for the book on the 30.0° incline. It travels a greater distance if the surface is horizontal because the force of gravity is no longer helping to slow the book down. This result is physically reasonable and gives us confidence that our answer is correct.

If the surface is horizontal ($\theta = 0$):

$$\begin{aligned} a_{\text{book},x} &= g(\sin 0 + \mu_k \cos 0) \\ &= g[0 + \mu_k(1)] \\ &= \mu_k g \\ &= (0.350)(9.80\ \text{m/s}^2) \\ &= 3.43\ \text{m/s}^2 \end{aligned}$$

horizontal surface
$a_{\text{book},x} = +3.43\ \text{m/s}^2$
$v_{0x} = -4.20$ m/s
$v_x = 0$
y x_0 $x - x_0$ x

Then the displacement of the book from where it starts to where it stops is

$$\begin{aligned} x - x_0 &= \frac{v_x^2 - v_{0x}^2}{2a_{\text{book},x}} = \frac{0 - (-4.20\ \text{m/s})^2}{2(3.43\ \text{m/s}^2)} \\ &= -2.57\ \text{m} \end{aligned}$$

So the book travels 2.57 m.

Rolling Friction

In addition to static friction and kinetic friction, a third important type of friction is **rolling friction**. If you start a billiard ball rolling across a horizontal surface, it will roll for quite a distance but will eventually come to a stop. A force must therefore act opposite to its motion, and it's this force that we call rolling friction (**Figure 5-10**). This force always acts opposite to the motion of a rolling object, whether it's rolling uphill, downhill, or on a horizontal surface.

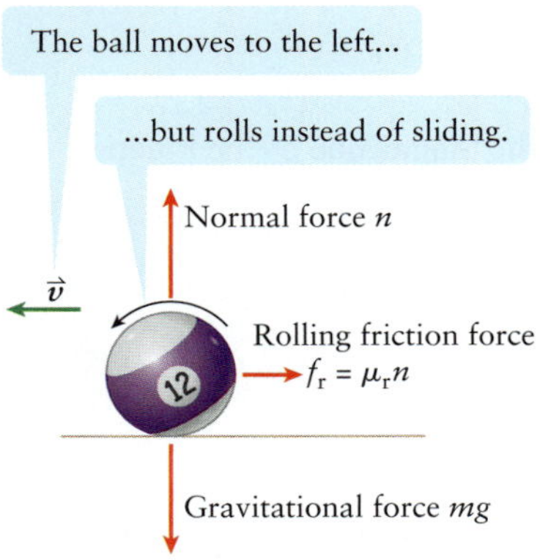

The ball moves to the left...

...but rolls instead of sliding.

Normal force n

$\vec{v}$

Rolling friction force $f_r = \mu_r n$

Gravitational force mg

Figure 5-10 A rolling billiard ball A ball can roll much farther on a horizontal surface than a block can slide on the same surface. The reason is that rolling friction is a much weaker force than kinetic friction.

Just as for static and kinetic friction, the force of rolling friction is proportional to the normal force that acts between the rolling object and the surface over which it rolls. So we can express the magnitude f_r of the rolling friction force as

$$f_r = \mu_r n \quad \text{(force of rolling friction)} \tag{5-7}$$

The **coefficient of rolling friction** μ_r can be a very small number—about 0.02 for rubber automobile tires on concrete and about 0.002 for the steel wheels of a railroad car on steel rails. As a result, rolling friction is generally much smaller than kinetic friction, so a round object will roll much farther than a brick-shaped object of the same mass will slide. In general, the more rigid the rolling object and the more rigid the surface over which it rolls, the smaller the value of μ_r. The tires of your car are an example: If you don't keep them properly inflated, the coefficient of rolling friction for your tires will be greater than it should be, and your fuel consumption will be greater as well.

Figure 5-11 summarizes the three varieties of friction force.

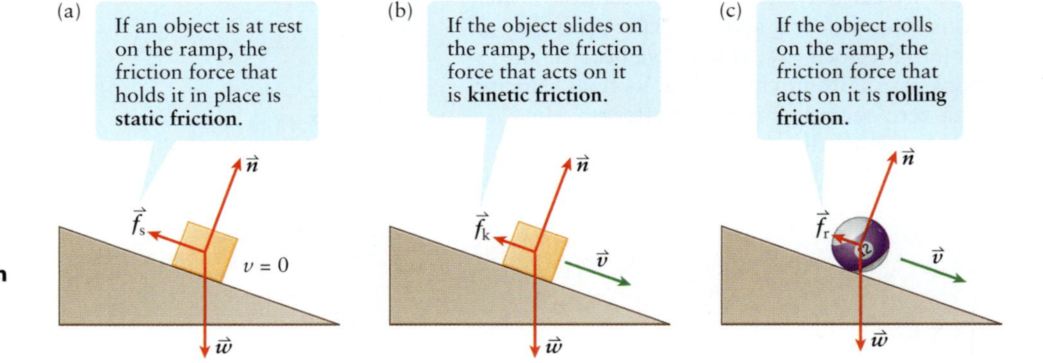
(a)
If an object is at rest on the ramp, the friction force that holds it in place is static friction.
$\vec{f}_s$
$\vec{n}$
$v = 0$
$\vec{w}$
(b)
If the object slides on the ramp, the friction force that acts on it is kinetic friction.
$\vec{f}_k$
$\vec{n}$
$\vec{v}$
$\vec{w}$
(c)
If the object rolls on the ramp, the friction force that acts on it is rolling friction.
$\vec{f}_r$
$\vec{n}$
$\vec{v}$
$\vec{w}$

Figure 5-11 Three kinds of friction Comparing (a) static friction, (b) kinetic friction, and (c) rolling friction.

GOT THE CONCEPT? 5-2 A Question of Mass

? Suppose the book in Examples 5-2 and 5-3 had a mass of 5.00 kg rather than 2.50 kg. How would this change the magnitude of the acceleration of the book as it moves along the ramp? (a) The magnitude would be greater going both downhill and uphill. (b) The magnitude would be less going both downhill and uphill. (c) The magnitude would be greater going downhill but less going uphill. (d) The magnitude would be less going downhill but greater going uphill. (e) None of these.

GOT THE CONCEPT? 5-3 Sliding Uphill to a Stop

Consider again the book sliding up the 30.0° incline in Example 5-3. For which of the following values of the coefficient of *static* friction will the book remain at rest after it stops? (a) 0.350 (the same as μ_k); (b) 0.600; (c) 0.750; (d) both (b) and (c); (e) all of (a), (b), and (c).

TAKE-HOME MESSAGE FOR Section 5-3

✔ If an object is sliding over a surface, the kinetic friction force on the object acts in the direction that opposes the sliding.

✔ The magnitude of the kinetic friction force is proportional to the normal force exerted on the object by the surface on which it is sliding (Equation 5-6).

✔ Like kinetic friction, rolling friction opposes motion and is proportional to the normal force, but the coefficient of friction for rolling is much smaller than that for sliding.

5-4 Problems involving static and kinetic friction are like any other problem with forces

In this section we'll look at a number of problems that involve static friction, kinetic friction, or both. We'll solve these examples using the same two-step approach that we introduced in Section 4-6. First, draw a free-body diagram for each object of interest; second, use the diagram to write down a Newton's second law equation for each object and solve for the unknowns. The only new wrinkle for problems that involve friction are the expressions for the magnitudes of the static friction force ($f_s \le f_{s,max} = \mu_s n$) and the kinetic friction force ($f_k = \mu_k n$).

EXAMPLE 5-4 Two Boxes, a String, and Friction

Two boxes are connected by a lightweight string that passes over a massless, frictionless pulley (**Figure 5-12**). Box 1 (mass $m_1 = 2.00$ kg) sits on a ramp inclined at 30.0° from the horizontal, while box 2 (mass $m_2 = 4.00$ kg) hangs from the other end of the string. When the boxes are released, the string remains taut, box 1 accelerates up the ramp, and box 2 accelerates downward. The coefficient of kinetic friction for box 1 sliding on the ramp is $\mu_k = 0.250$. What is the acceleration of each box?

Box 1

Box 2

Figure 5-12 A pulley problem with friction If there is friction between box 1 and the ramp, what is the acceleration of the boxes?

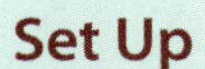

Set Up

Because box 2 has more mass than box 1, we expect that box 2 will fall, accelerating downward and pulling box 1 up the ramp. So we choose to have the positive x axis point up the ramp for box 1 and straight down for box 2. Box 1 has four forces acting on it: a normal force $\vec{n}_1$ from the surface of the ramp, its weight $\vec{w}_1$, the tension force from the string $\vec{T}_{\text{string on 1}}$, and the kinetic friction force from the ramp $\vec{f}_{k1}$. Two forces act on box 2: its weight $\vec{w}_2$ and the tension force from the string $\vec{T}_{\text{string on 2}}$. Because the string is lightweight and the pulley is massless and frictionless, the tensions $\vec{T}_{\text{string on 1}}$ and $\vec{T}_{\text{string on 2}}$ at either end have the same magnitude T. If the string doesn't stretch, the two boxes move together and their accelerations $\vec{a}_1$ and $\vec{a}_2$ have the same x component a_x. (Note that the positive x direction for each box is in the direction that the box moves.) Our goal is to calculate a_x.

Newton's second law applied to box 1:

$$\sum \vec{F}_{\text{ext on 1}} = \vec{n}_1 + \vec{w}_1 + \vec{T}_{\text{string on 1}} + \vec{f}_{k1} = m_1\vec{a}_1$$

Magnitude of the kinetic friction force on box 1:

$$f_{k1} = \mu_k n_1$$

Newton's second law applied to box 2:

$$\sum \vec{F}_{\text{ext on 2}} = \vec{w}_2 + \vec{T}_{\text{string on 2}} = m_2\vec{a}_2$$

Tensions have the same magnitude:

$$T_{\text{string on 1}} = T_{\text{string on 2}} = T$$

Accelerations have the same x component:

$$a_{1x} = a_{2x} = a_x$$

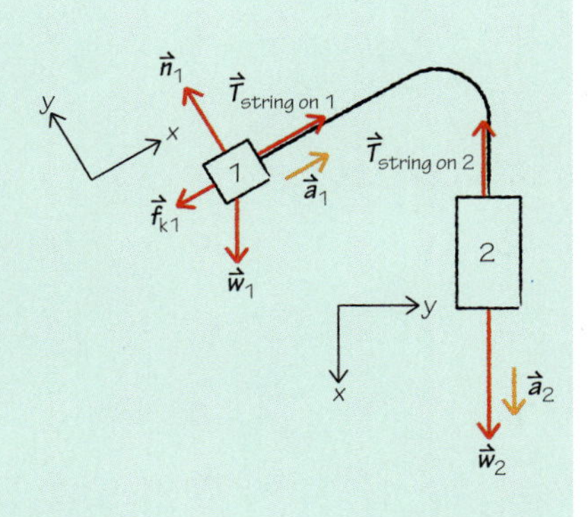

Solve

Write Newton's second law for each box in component form. Use T for the magnitude of each tension and a_x for the x component of each acceleration. Box 1 doesn't move in the y direction, so its y component of acceleration is zero and the net external force on box 1 has zero y component.

Newton's second law in component form for box 1:

$$x: 0 + (-w_1 \sin\theta) + T + (-f_k) = m_1 a_x$$

$$y: n_1 + (-w_1 \cos\theta) + 0 + 0 = 0$$

Newton's second law in component form for box 2:

$$x: w_2 + (-T) = m_2 g - T = m_2 a_x$$

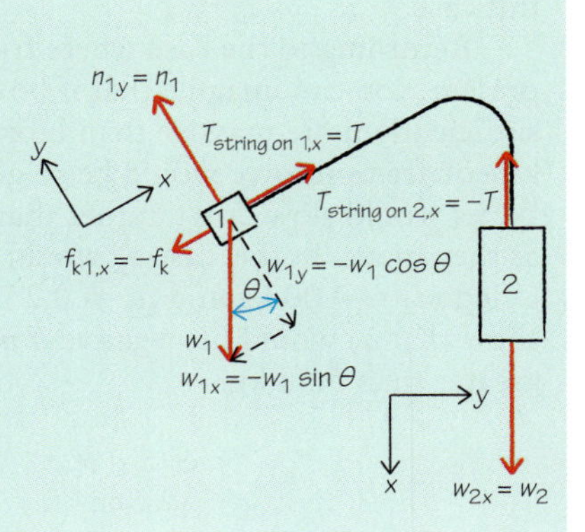

Calculate the normal force n_1 that acts on box 1 from the y component equation for that box. Use this expression to find the magnitude of the kinetic friction force on box 1 and substitute the result into the x equation for that box.

From the y component equation for box 1,

$$n_1 = w_1 \cos\theta = m_1 g \cos\theta$$

So the kinetic friction force on box 1 has magnitude

$$f_{k1} = \mu_k n_1 = \mu_k m_1 g \cos\theta$$

Substitute this expression and $w_1 = m_1 g$ into the x component equation for box 1:

$$-m_1 g \sin\theta + T - \mu_k m_1 g \cos\theta = m_1 a_x$$

We now have two equations that involve the unknown quantities T and a_x. We're only asked to find the value of a_x, so we eliminate T between the two equations.

x component equation for box 1:

$$-m_1 g \sin\theta + T - \mu_k m_1 g \cos\theta = m_1 a_x$$

x component equation for box 2:

$$m_2 g - T = m_2 a_x$$

Solve for T using the x component equation for box 1:

$$T = m_1 g \sin\theta + \mu_k m_1 g \cos\theta + m_1 a_x$$

Substitute this into the x component equation for box 2:

$$m_2 g - m_1 g \sin\theta - \mu_k m_1 g \cos\theta - m_1 a_x = m_2 a_x$$

Rearrange and solve for a_x:

$$m_2 g - m_1 g \sin\theta - \mu_k m_1 g \cos\theta = m_1 a_x + m_2 a_x$$

$$(m_2 - m_1 \sin\theta - \mu_k m_1 \cos\theta) g = (m_1 + m_2) a_x$$

$$\begin{aligned} a_x &= \frac{(m_2 - m_1 \sin\theta - \mu_k m_1 \cos\theta)}{(m_1 + m_2)} g \\ &= \frac{[4.00\text{ kg} - (2.00\text{ kg})\sin 30.0° - (0.250)(2.00\text{ kg})\cos 30.0°]}{(2.00\text{ kg} + 4.00\text{ kg})} \\ &\quad \times (9.80\text{ m/s}^2) \\ &= 4.19\text{ m/s}^2 \end{aligned}$$

Reflect

If there were no friction to slow the boxes down, we would expect the acceleration to be greater. We can check this conclusion by looking at the case $\mu_k = 0$ (which implies zero kinetic friction). Sure enough, a_x is greater in this case.

Returning to the case where friction is present, you can imagine that if box 1 were sufficiently more massive than box 2, the kinetic friction force would be so great that the boxes would slow down rather than speed up as they move. In this case a_x would be negative. Using $m_2 = 4.00$ kg and $\mu_k = 0.250$, can you show that a_x would be negative if m_1 were greater than 5.58 kg?

If there were no friction between box 1 and the ramp:

$\mu_k = 0$ so

$$\begin{aligned} a_x &= \frac{(m_2 - m_1 \sin\theta)}{(m_1 + m_2)} g \\ &= \frac{[4.00\text{ kg} - (2.00\text{ kg})\sin 30.0°]}{(2.00\text{ kg} + 4.00\text{ kg})}(9.80\text{ m/s}^2) \\ &= 4.90\text{ m/s}^2 \end{aligned}$$

EXAMPLE 5-5 Pinned Against a Wall

You place a block of plastic that weighs 33.0 N against a vertical wall and push it toward the wall with a force of 55.0 N (**Figure 5-13**). The coefficients of static and kinetic friction between the plastic and the wall are $\mu_s = 0.420$ and $\mu_k = 0.400$, respectively. Does the block remain at rest? If not, with what acceleration does it slip down the wall? Neglect any friction between you and the block.

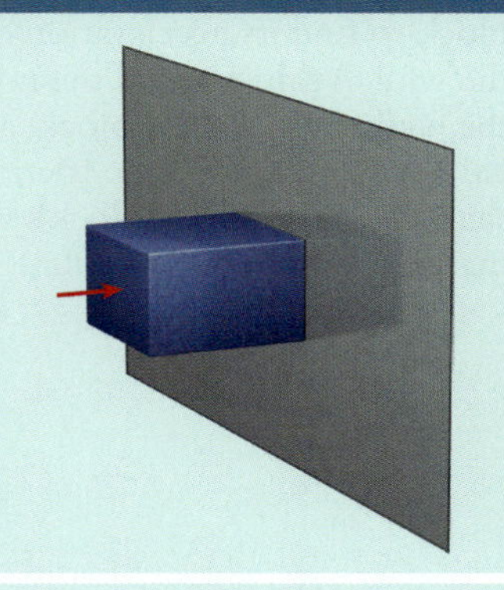

Figure 5-13 Pushing a block against a wall If you push a block against a wall with a force of a given magnitude, will the block remain at rest or slide down the wall?

Set Up

Four forces act on the block: the gravitational force $\vec{w}_{block}$, the force $\vec{F}_{you\ on\ block}$ you exert to push it against the wall, the normal force $\vec{n}_{block}$ that the wall exerts, and the upward friction force $\vec{f}$ (which opposes the downward gravitational force). The block will remain at rest only if the force of static friction is large enough to balance the 33.0-N gravitational force.

We'll use Equation 5-1b to determine the maximum force of static friction, which will help us decide whether the block will slip. If the block slips, we'll use Equation 5-6 for the force of kinetic friction to determine its acceleration.

Newton's second law applied to the block:

$$\sum \vec{F}_{ext\ on\ block} = \vec{w}_{block} + \vec{F}_{you\ on\ block} + \vec{n}_{block} + \vec{f} = m_{block}\vec{a}_{block}$$

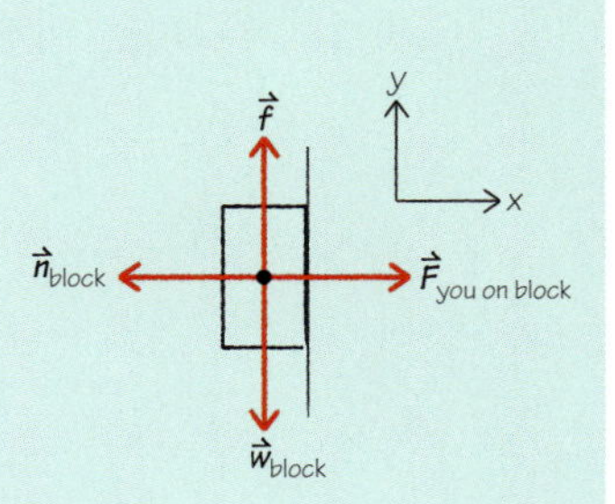

Magnitude of the static friction force:

$$f_{s,max} = \mu_s n \qquad (5\text{-}1b)$$

Magnitude of the kinetic friction force:

$$f_k = \mu_k n \qquad (5\text{-}6)$$

Solve

We use Newton's second law to determine the normal force that the wall exerts on the block and Equation 5-1b to find the maximum static friction force available. Only 23.1 N of static friction is available, which is less than the gravitational force on the block, so we conclude that the block will slip.

Newton's second law in component form for the block, assuming the block remains at rest (so the friction force is static friction):

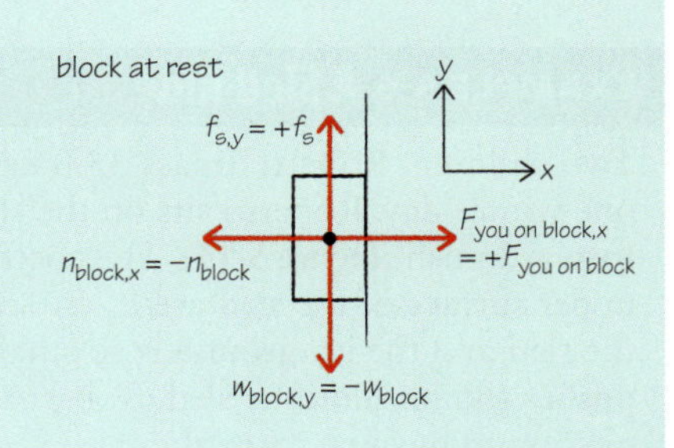

$$x: F_{you\ on\ block} + (-n_{block}) = 0$$

$$y: f_s + (-w_{block}) = 0$$

From the x component equation,

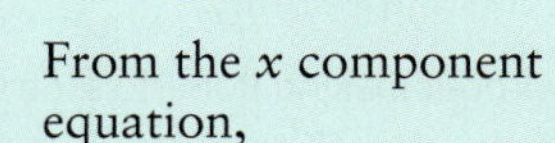

$$n_{block} = F_{you\ on\ block} = 55.0\ \text{N}$$

From Equation 5-1b, the maximum static friction force available is

$$f_{s,max} = \mu_s n_{block} = (0.420)(55.0\ \text{N}) = 23.1\ \text{N}$$

From the y component equation, the required static friction force is

$$f_s = w_{block} = 33.0\ \text{N}$$

This force is more than the maximum static friction force available, so the block can't remain at rest.

We again apply Newton's second law to the block to find its acceleration as it slides down the wall. Again we find the normal force that the wall exerts on the block, and we use this value to find the kinetic friction force. Note that we are given the block's weight, not its mass, so we have to calculate its mass m_{block}.

Newton's second law in component form for the block, assuming the block is sliding downward (so the friction force is kinetic friction):

$$x\text{: } F_{\text{you on block}} + (-n_{\text{block}}) = 0$$
$$y\text{: } f_{\text{k}} + (-w_{\text{block}}) = m_{\text{block}}a_{\text{block},y}$$

From the x component equation,

$$n_{\text{block}} = F_{\text{you on block}} = 55.0 \text{ N}$$

From Equation 5-6,

$$f_{\text{k}} = \mu_{\text{k}} n_{\text{block}} = (0.400)(55.0 \text{ N}) = 22.0 \text{ N}$$

Find the mass of the block:

$$w_{\text{block}} = m_{\text{block}}g$$
$$m_{\text{block}} = \frac{w_{\text{block}}}{g} = \frac{33.0 \text{ N}}{9.80 \text{ m/s}^2} = 3.37 \text{ kg}$$

Substitute into the y equation and solve for $a_{\text{block},y}$:

$$a_{\text{block},y} = \frac{f_{\text{k}} - w_{\text{block}}}{m_{\text{block}}} = \frac{22.0 \text{ N} - 33.0 \text{ N}}{3.37 \text{ kg}} = -3.27 \text{ m/s}^2$$

Reflect

We chose the positive y direction to be upward, so the negative value of $a_{\text{block},y}$ means that the block accelerates downward. The magnitude of the acceleration is somewhat less than $g = 9.80 \text{ m/s}^2$ because kinetic friction prevents the block from attaining free fall.

Can you show that the block would remain at rest if you pushed with a force of at least 78.6 N?

EXAMPLE 5-6 A Sled Ride

Two children, Roberto (mass 35.0 kg) and Mary (mass 30.0 kg), go out sledding one winter day. Roberto sits on the sled of mass 5.00 kg, and Mary gives the sled a forward push (**Figure 5-14**). The coefficients of friction between Roberto and the upper surface of the sled are $\mu_{\text{s}} = 0.300$ and $\mu_{\text{k}} = 0.200$; the friction force between the sled and the icy ground is so small that we can ignore it. Mary finds that if she pushes too hard on the sled, Roberto slides toward the back of the sled. What is the maximum pushing force she can exert without this happening?

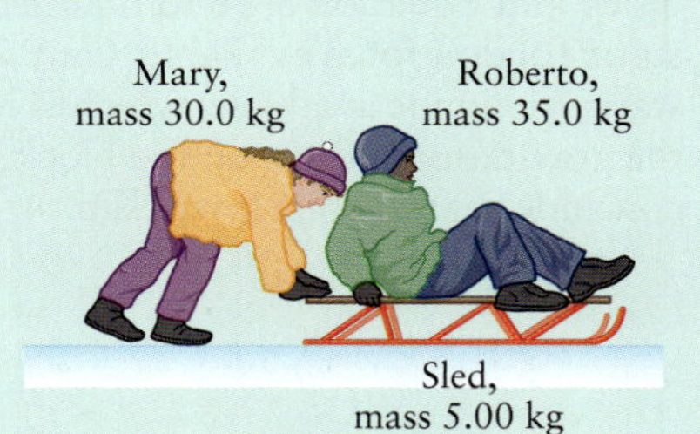

Figure 5-14 Pushing a sled How hard can Mary push without making Roberto slide toward the back of the sled?

Set Up

If Roberto doesn't slip, the friction force that acts between him and the sled is *static* friction. (The sled and Roberto are sliding over the ice, but they're not sliding relative to each other.) The maximum force of static friction sets a limit on how great the forward acceleration of Roberto and the sled together can be.

This suggests that we should consider two objects: (a) Roberto by himself and (b) Roberto and the sled considered as a unit. Roberto is pushed forward by the friction force exerted on him by the sled, while Roberto and the sled together are pushed forward by Mary's push $\vec{F}_{\text{Mary on unit}}$. (Note that Mary doesn't touch Roberto, so her push doesn't act on him directly.)

Newton's second law equation for Roberto:

$$\sum \vec{F}_{\text{ext on Roberto}} = \vec{n}_{\text{sled on Roberto}} + \vec{w}_{\text{Roberto}} + \vec{f}_{\text{s,sled on Roberto}} = m_{\text{Roberto}}\vec{a}_{\text{Roberto}}$$

Newton's second law equation for Roberto and the sled considered as a unit:

$$\sum \vec{F}_{\text{ext on unit}} = \vec{n}_{\text{ground on unit}} + \vec{w}_{\text{unit}} + \vec{F}_{\text{Mary on unit}} = m_{\text{unit}}\vec{a}_{\text{unit}}$$

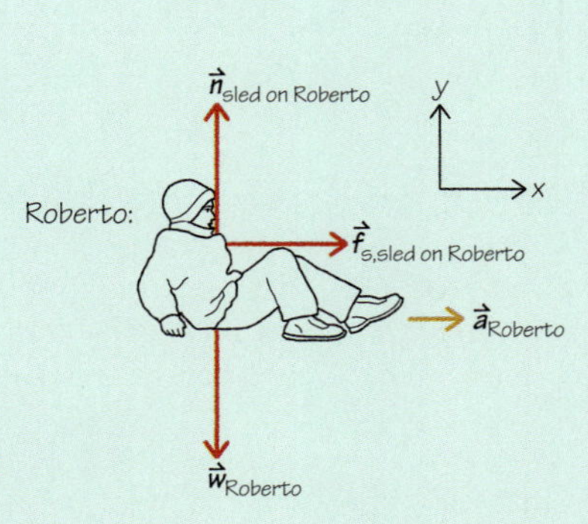

We'll determine the maximum acceleration that will prevent Roberto from slipping and then use the result to find the maximum forward force that Mary can exert.

Magnitude of the static friction force:

$$f_{s,max} = \mu_s n \quad (5\text{-}1b)$$

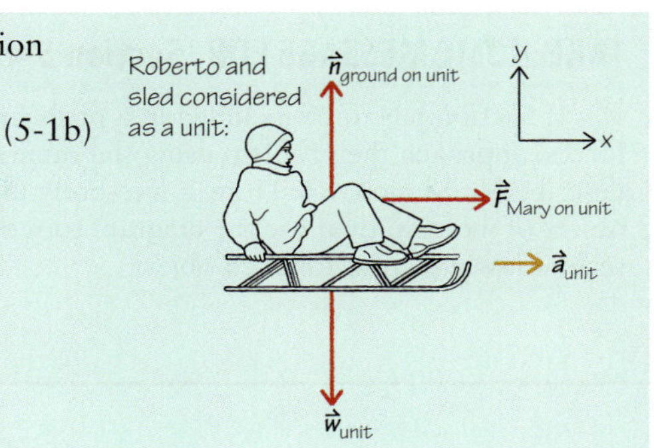

Solve

First let's consider just Roberto and the forces that act on him. He is pushed forward by the static friction force that the sled exerts on him. If Roberto is just about to slip, the static friction force has its maximum value. To determine this value we first find the normal force that the sled exerts on Roberto.

Newton's second law in component form for Roberto, assuming that the static friction force has its maximum value:

$$x\text{: } f_{s,max\ of\ sled\ on\ Roberto} = m_{Roberto}a_{Roberto,x}$$

$$y\text{: } n_{sled\ on\ Roberto} + (-w_{Roberto}) = 0$$

From the y component equation,

$$n_{sled\ on\ Roberto} = w_{Roberto} = m_{Roberto}g$$

So the maximum force of static friction that the sled exerts on Roberto is

$$f_{s,max\ of\ sled\ on\ Roberto} = \mu_s n_{sled\ on\ Roberto} = \mu_s m_{Roberto}g$$

So from the x component equation, Roberto's maximum acceleration is

$$a_{Roberto,x} = \frac{f_{s,max\ of\ sled\ on\ Roberto}}{m_{Roberto}} = \frac{\mu_s m_{Roberto}g}{m_{Roberto}} = \mu_s g$$

Because Roberto does not slip, he and the sled have the same acceleration. So we can treat Roberto and the sled as a single unit that accelerates forward due to Mary's push. (For this combined unit we don't have to consider the forces that Roberto and the sled exert on each other. That's because these are *internal* forces.) Because we know Roberto's (and the sled's) maximum acceleration that will prevent slipping, we can calculate the maximum force that Mary can exert.

Newton's second law in component form applied to Roberto and the sled considered as a unit:

$$x\text{: } F_{Mary\ on\ unit} = m_{unit}a_{unit,x}$$

$$y\text{: } n_{ground\ on\ unit} + (-w_{unit}) = 0$$

If Roberto has his maximum acceleration and does not slip on the sled,

$$a_{unit,x} = a_{Roberto,x} = \mu_s g$$

So the maximum force that Mary can exert without causing Roberto to slip is

$$\begin{aligned} F_{Mary\ on\ unit} &= m_{unit}\mu_s g \\ &= (m_{Roberto} + m_{sled})\mu_s g \\ &= [(35.0\ \text{kg}) + (5.00\ \text{kg})](0.300)(9.80\ \text{m/s}^2) \\ &= 118\ \text{kg} \cdot \text{m/s}^2 = 118\ \text{N} \end{aligned}$$

Reflect

What happens if Mary exerts a force greater than 118 N on the sled? There will not be enough static friction to prevent Roberto from sliding across the top of the sled. The net force on him will still be forward, but now it will be due to *kinetic* friction. This force has a smaller magnitude than the maximum force of static friction (we are told that μ_k has a smaller value than μ_s). So Roberto's forward acceleration will be less than that of the sled. From Mary's perspective, Roberto will slide backward relative to the sled.

GOT THE CONCEPT? 5-4 Speeding Up with Friction?

Can the force of friction cause an object to *speed up*? (a) Yes, but only the force of kinetic friction; (b) yes, but only the force of static friction; (c) yes, both kinetic and static friction can cause an object to speed up; (d) no.

TAKE-HOME MESSAGE FOR Section 5-4

✔ If friction has to be included in a problem that involves forces, approach the problem using the same steps as described in Section 4-6: Draw a free-body diagram for each object of interest; then use the diagram to write a Newton's second law equation for each object.

✔ If an object is not sliding, use Equations 5-1 to describe the static friction force. If the problem asks for a limiting case such as "when the object just begins to slip," assume the static friction force is maximum.

✔ If an object is sliding, use Equation 5-6 to describe the kinetic friction force.

5-5 An object moving through air or water experiences a drag force

When a solid object slides over another solid object, the force of kinetic friction provides resistance to its motion. Another important kind of resistance is the **fluid resistance** experienced by an object when it moves through a *fluid*—that is, a liquid or gas (**Figure 5-15**). The force that resists the motion of an object through a liquid such as water or a gas such as air is called a **drag force**.

(a)

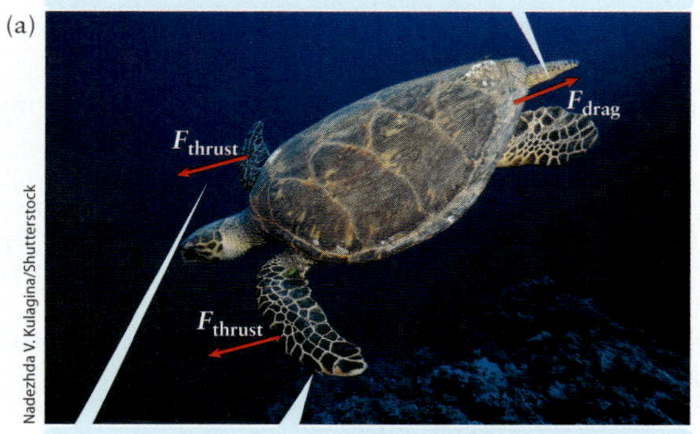

To maintain a constant forward velocity, the turtle pushes backward on the water. The water responds by exerting a forward force (thrust) on the flippers (Newton's third law).

(b)

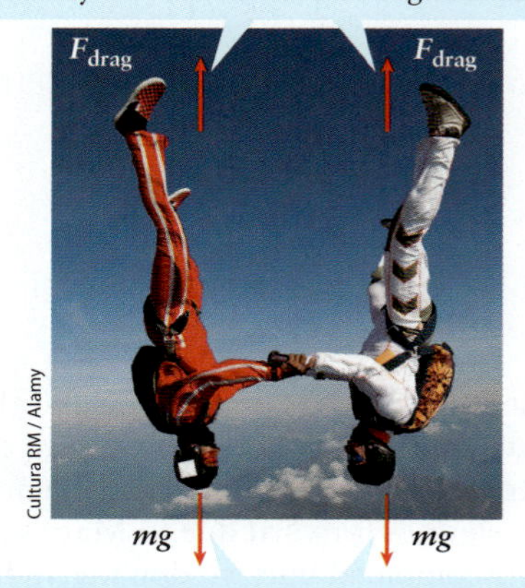

These skydivers feel an upward force of fluid resistance (drag) as they move downward through the air.
F_{drag}
F_{drag}
Cultura RM / Alamy
mg
mg
Fluid resistance opposes gravity. This slows their fall, giving them more time to enjoy the descent.

Figure 5-15 Fluid resistance When an object moves through either (a) a liquid such as water or (b) a gas such as air, it experiences fluid resistance (a drag force).

Drag Force on Microscopic Objects

Unlike the force of kinetic friction, the drag force depends on the object's *speed* v as it moves through the fluid. The faster the object's speed, the greater the magnitude of the drag force. For a very slow-moving object that is very small, such as a living cell or virus, the magnitude of the drag force is directly proportional to the speed v:

Drag force for small objects at low speeds (5-8)

Magnitude of the **drag force** on a **small object** moving at a **low speed**

$$F_{drag} = bv$$

Constant that depends on the properties of the object and of the fluid

Speed of the object relative to the fluid

The value of b in Equation 5-8 depends on the size, shape, and surface characteristics of the object and the properties of the liquid or gas through which the object is moving. As an example, for an algal spore (which has a spherical shape) of radius 0.020 mm moving in water, the coefficient b is 3.8×10^{-7} N·s/m. At a speed of 5.0×10^{-5} m/s, the drag force on this spore has magnitude

$$F_{\text{drag}} = bv = (3.8 \times 10^{-7}\ \text{N}\cdot\text{s/m})(5.0 \times 10^{-5}\ \text{m/s}) = 1.9 \times 10^{-11}\ \text{N}$$

BIO-Medical

This force is so small that you might think it would have no effect on the spore at all. However, the mass of a spore with this size is only 3.6×10^{-11} kg. If the only force acting on the spore is the drag force (**Figure 5-16**), the resulting acceleration is

$$a_{\text{spore},x} = \frac{F_{\text{drag},x}}{m_{\text{spore}}} = \frac{-1.9 \times 10^{-11}\ \text{N}}{3.6 \times 10^{-11}\ \text{kg}} = -0.53\ \text{m/s}^2$$

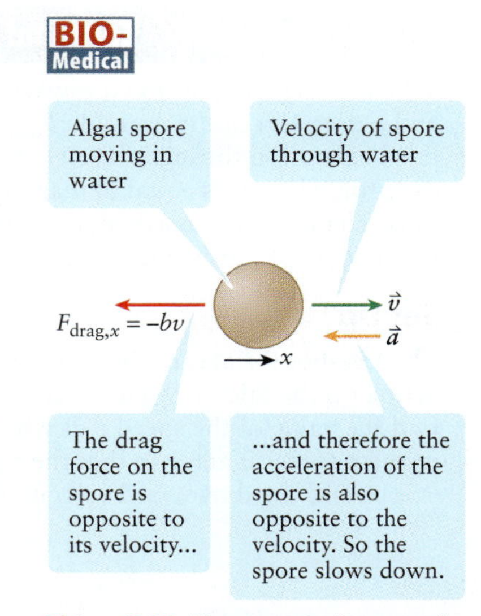

Figure 5-16 Drag on a microscopic spore A spherical algal spore moves through water in the positive x direction.

The negative values of $F_{\text{drag},x}$ and $a_{\text{spore},x}$ mean that the drag force acts opposite to the spore's motion, as in Figure 5-16. This substantial acceleration brings the slow-moving spore nearly to a halt relative to the water in a fraction of a second. Since such microscopic spores move very little *relative* to the water around them, they move readily from place to place if the water is itself in motion; the drag force exerted by the moving water carries the spores along with it. Thus, thanks to fluid resistance, the location where an algal spore finally germinates and produces a new organism depends crucially on water currents.

Drag Force on Larger Objects

For a larger object moving at a faster speed, such as the sea turtle or skydivers in Figure 5-15, the drag force has a different dependence on the speed v: Its magnitude is approximately proportional to the *square* of v. In this case we can write

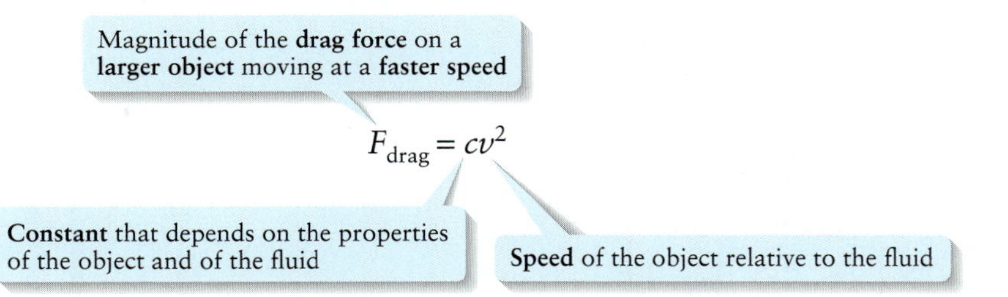

Drag force for larger objects at faster speeds (5-9)

Like the coefficient b in Equation 5-8, the quantity c in Equation 5-9 depends on the fluid through which the object moves and the size, shape, and surface properties of the object. For a baseball flying through the air, $c = 1.3 \times 10^{-3}$ N·s²/m². If the baseball is traveling at 25 m/s (a relatively slow pitching speed for a professional player), the magnitude of the drag force on the ball is

$$F_{\text{drag}} = (1.3 \times 10^{-3}\ \text{N}\cdot\text{s}^2/\text{m}^2)(25\ \text{m/s})^2 = 0.81\ \text{N}$$

This drag force is substantial: F_{drag} is a bit more than half the magnitude of the gravitational force on a regulation baseball, which weighs 1.4 N or 5 ounces. You can see that the drag force plays an important role in baseball and other ball sports.

Giving an object a streamlined shape can lower its value of c. For example, a dolphin (which has a streamlined body) moving in water has a value of c that is only about 1% as large as that of a sphere of the same cross-sectional area. Furthermore, because air is less dense than water, the value of c for a dolphin moving through air is only about 10^{-3} as great as for a dolphin moving through water. That explains why dolphins jump out of the water, or *porpoise*, when moving at high speeds: the dramatic reduction in drag force that they experience while airborne more than compensates for the effort required to leap clear of the water.

Another species that minimizes its value of the quantity c in Equation 5-9 in order to achieve high speed is the peregrine falcon (*Falco peregrinus*). This species of predatory bird preys on other, smaller birds, and attacks its prey by diving on it from high altitude. To achieve maximum speed in a dive, which maximizes the impact force it can impart when it strikes its prey, a peregrine falcon streamlines its shape by folding back its wings and tail and tucking in its feet. As a result a diving falcon can attain remarkable speeds, as the following example shows.

BIO-Medical EXAMPLE 5-7 Terminal Speed

When diving straight down toward its prey, a peregrine falcon is acted on by two forces: a downward gravitational force, and a drag force of magnitude F_{drag} given by Equation 5-9 directed vertically upward (opposite to the direction of the falcon's motion through the air). As the falcon falls and its speed v increases, the value of F_{drag} also increases. When the drag force becomes equal in magnitude to the gravitational force, the net force on the falcon is zero and the falcon ceases to accelerate. It has reached its *terminal speed*, so it no longer speeds up nor does it slow down. Find the terminal speed of a female peregrine falcon of mass 1.2 kg, for which the value of c is $1.6 \times 10^{-3}\ \text{N}\cdot\text{s}^2/\text{m}^2$.

Set Up

The free-body diagram shows the two forces acting on the falcon. We use Equation 5-9 to find the value of the speed v at which the sum of these forces is zero, so that the acceleration is zero and the downward velocity is constant.

$$\sum \vec{F}_{\text{ext on falcon}} = \vec{F}_{\text{drag on falcon}} + \vec{w}_{\text{falcon}} = m\vec{a}_{\text{falcon}} = 0$$

Drag force for larger objects at faster speeds:

$$F_{drag} = cv^2 \qquad (5\text{-}9)$$

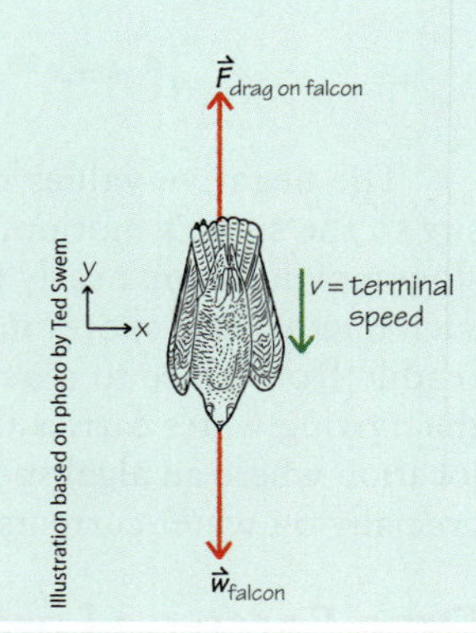

Solve

Write Newton's second law in component form and solve for the terminal speed v_{term}.

Newton's second law in component form applied to the falcon:

$$y\text{: } F_{\text{drag on falcon}} + (-w_{\text{falcon}}) = 0$$

At the terminal speed v_{term},

$$F_{\text{drag on falcon}} = cv_{term}^2 \text{ so}$$

$$cv_{term}^2 - w_{falcon} = 0$$

$$cv_{term}^2 = w_{falcon} = mg$$

$$v_{term}^2 = \frac{mg}{c}$$

$$v_{term} = \sqrt{\frac{mg}{c}}$$

Substitute the numerical values of m and c for the falcon.

Using $m = 1.2$ kg and $c = 1.6 \times 10^{-3}\ \text{N}\cdot\text{s}^2/\text{m}^2$,

$$v_{term} = \sqrt{\frac{(1.2\ \text{kg})(9.80\ \text{m/s}^2)}{1.6 \times 10^{-3}\ \text{N}\cdot\text{s}^2/\text{m}^2}}$$

$$= 86\ \text{m/s} = 310\ \text{km/h} = 190\ \text{mi/h}$$

Reflect

The high diving speeds attained by a peregrine falcon make it the fastest member of the animal kingdom.

The relationship $v_{term} = \sqrt{\frac{mg}{c}}$ explains the common notion that "heavier objects fall faster." A baseball and an iron ball of the same radius falling side by side in air have the same value of c (because they have the same shape and size), but the iron ball will have a greater terminal speed because its mass m is greater. So a heavier object *does* fall faster if we take the drag force into account. If the baseball and iron ball were dropped side by side in a vacuum, however, they would have the same acceleration of magnitude g and so would always have the same speed.

WATCH OUT! The drag force is not a constant force.

Unlike other forces we've considered, the magnitude of the drag force on an object changes as the object's speed changes. So if a drag force acts on an object, the net force on the object and hence the object's acceleration are *not* constant. As an example, a skydiver who falls without opening her parachute reaches half of her terminal speed ($v = v_{term}/2$) in 3 to 5 s. To achieve nearly terminal speed takes more than twice as long, about 15 to 20 s after beginning the fall. So her acceleration from rest to $v = v_{term}/2$ is much greater than that from $v = v_{term}/2$ to $v = v_{term}$. It would therefore be quite incorrect to try to apply the constant-acceleration formulas from Chapters 2 and 3.

GOT THE CONCEPT? 5-5 Comparing Drag Forces

? A baseball and a second object of the same mass are thrown into the air. When the second object is moving at one-half the speed of the baseball, the drag force on the second object is four times greater than the drag force on the baseball. Compared to the value of the quantity c for the baseball, the value of c for the second object is (a) 2 times greater, (b) 4 times greater, (c) 8 times greater, (d) 16 times greater, (e) none of these.

TAKE-HOME MESSAGE FOR Section 5-5

✔ Microscopic objects moving through a fluid at low speeds experience a drag force that is proportional to the speed (Equation 5-8).

✔ Larger objects moving through a fluid at faster speeds experience a drag force that is proportional to the square of the speed (Equation 5-9).

✔ A falling object reaches its terminal speed when the upward drag force just balances the downward gravitational force.

5-6 In uniform circular motion the net force points toward the center of the circle

Think of thrill seekers on a fast-moving amusement park ride or a motorcycle making a sharp turn (**Figure 5-17**). In both cases an object is following a curved path, so the direction of its velocity vector is changing. Because the direction of the velocity vector is changing, the object is *accelerating* (the object's speed can be constant or changing). In Section 3-5 we learned that when an object follows a curved path, the acceleration vector $\vec{a}$ must have a component directed toward the inside of the curve. We went on to discover in Section 3-8 that the magnitude of this component, called the *centripetal acceleration*, depends on the radius of the curve and on the object's speed. From Section 4-2 we know that for an object to accelerate there must be a net external force that points in the direction of the acceleration vector $\vec{a}$. In this section we'll put all of these ideas together and complete our analysis of objects that move along curved paths.

Each person on this amusement park ride moves in a circle at constant speed...

(a)

...so the acceleration of each person and the net force on each person both point toward the center of the person's circular path.

This motorcycle rider also moves along a circular path at constant speed...

(b)

...so his acceleration and the net force on him both point toward the center of the circular path.

Figure 5-17 Forces in uniform circular motion An object moving on a circular path at a constant speed is in uniform circular motion. Since the object's direction of motion is continuously changing, the object is accelerating and a net external force must act on it.

What kinds of force can provide the acceleration needed to make an object follow a curved path? The answer is simple: *any kind of force*, provided that it has a component perpendicular to the object's trajectory. In **Figure 5-17a**, for example, each person's acceleration arises from the normal force exerted on each person by the seat on which that person sits. For the motorcycle in **Figure 5-17b**, the acceleration is provided by the friction force that the ground exerts on the motorcycle's tires.

One of the simplest situations in which an object follows a curved trajectory is *uniform circular motion,* or motion around a circular path at a constant speed. We saw in Section 3-8 that in uniform circular motion, the acceleration at any point around the circle is directed toward the center of the circle (that is, it is *centripetal*) and has a magnitude a_{cent} that depends on the object's speed v and the radius r of the circle:

Centripetal acceleration (5-10)

$$a_{cent} = \frac{v^2}{r}$$

Centripetal acceleration: Magnitude of the acceleration of an object in uniform circular motion

Speed of the object as it moves around the circle

Radius of the circle

Because the acceleration is always toward the center of the circle, it follows from Newton's second law that at any instant the net external force acting on the object of mass m must likewise be directed toward the center of the circle. Furthermore, the magnitude of the net external force must be equal to ma_{cent}. We can write this statement in equation form as

Newton's second law equation for uniform circular motion (5-11)

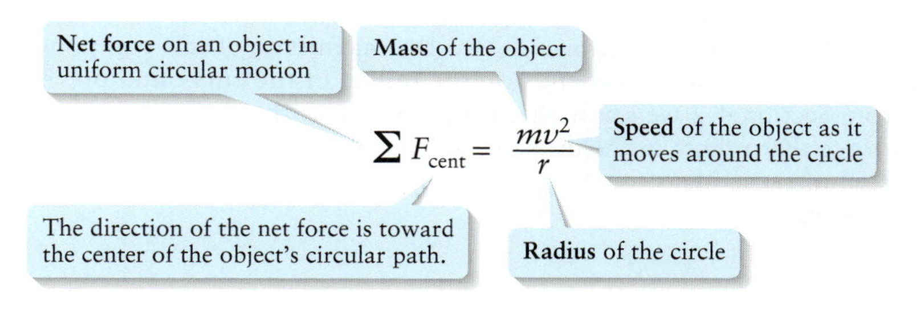

Go to Interactive Exercise 5-3 for more practice dealing with centripetal acceleration.

This equation says that in uniform circular motion the magnitude of the net external force on the object is equal to the object's mass m multiplied by the magnitude of the centripetal acceleration $a_{cent} = v^2/r$.

Let's look at several examples of how to apply these ideas to problems in which an object moves around a circle at constant speed.

WATCH OUT! "Centripetal force" is not a special *kind* of force—it describes the *direction* of the net force.

The force that points toward the inside of an object's curving trajectory and produces the centripetal acceleration is sometimes called the **centripetal force.** Keep in mind, however, that this is not a separate kind of force that you need to include in a free-body diagram. The word "centripetal" simply describes the direction of the force. Just as we might write F_x to denote the x component of a force, we write F_{cent} to denote the component of force directed toward the center of an object's circular trajectory. As we'll see in the following examples, in different circumstances different forces play the role of the centripetal force.

EXAMPLE 5-8 A Rock on a String

You tie one end of a lightweight string of length L around a rock of mass m. You hold the other end of the string in your hand and make the rock swing in a horizontal circle at constant speed. As you swing the rock the string makes an angle θ with the vertical (**Figure 5-18**). Derive an expression for the speed at which the rock moves around the circle. Ignore the effect of drag forces.

Figure 5-18 Moving in a circle How fast must the rock move to make the string hang at an angle θ from the vertical?

Set Up

Because the rock moves at constant speed in a horizontal circle, its acceleration vector points horizontally toward the center of the circle. We choose the positive x axis to point in this direction and the positive y axis to point upward.

Only two forces act on the rock: the tension force $\vec{T}$ exerted by the string and the downward gravitational force $\vec{w}_{\text{rock}}$. The tension force is directed along the string and so points at an angle θ to the vertical. The vertical component of $\vec{T}$ balances the downward force of gravity, while the horizontal component of $\vec{T}$ provides the centripetal acceleration. Our goal is to solve for the rock's speed v.

Newton's second law equation for the rock:

$$\sum \vec{F}_{\text{ext on rock}} = \vec{T} + \vec{w}_{\text{rock}} = m_{\text{rock}}\vec{a}_{\text{rock}}$$

Centripetal acceleration:

$$a_{\text{rock,cent}} = \frac{v^2}{r} \qquad (5\text{-}10)$$

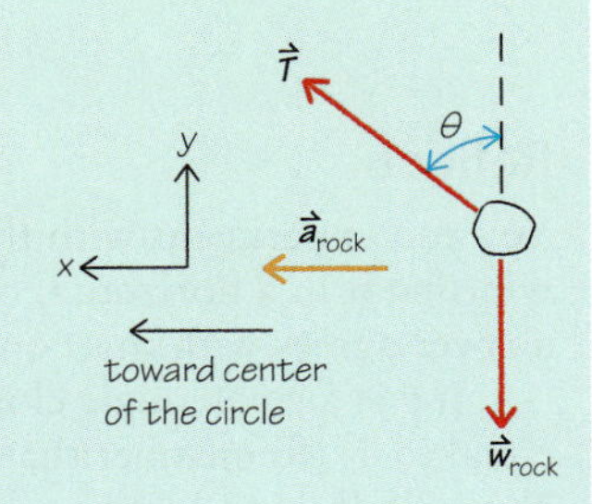

Solve

Write Newton's second law for the rock in component form. Note that the quantity v, which is what we're trying to find, doesn't appear in these equations. We'll introduce it in the next step through the expression for the rock's centripetal acceleration.

Newton's second law in component form applied to the rock:

$$x\text{: } T\sin\theta = m_{\text{rock}}a_{\text{rock,cent}}$$

$$y\text{: } T\cos\theta + (-w) = T\cos\theta - m_{\text{rock}}g = 0$$

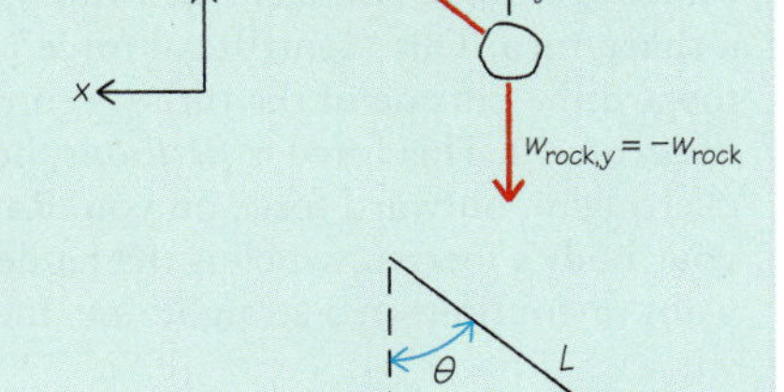

Use Equation 5-10 to express the acceleration of the rock. Note that because the string is at an angle, the radius of the rock's circular path is *not* equal to the length of the string.

Radius of the rock's circular path:

$$r = L\sin\theta$$

so the rock's centripetal acceleration is

$$a_{\text{rock,cent}} = \frac{v^2}{r} = \frac{v^2}{L\sin\theta}$$

Substitute the expression for centripetal acceleration into the Newton's second law equation for the x direction and solve for v.

Newton's second law equations for the rock become

$$x\text{: } T\sin\theta = m_{\text{rock}}a_{\text{rock,cent}} = \frac{m_{\text{rock}}v^2}{L\sin\theta}$$

$$y\text{: } T\cos\theta = m_{\text{rock}}g$$

Solve the y component equation for T:

$$T = \frac{m_{\text{rock}}g}{\cos\theta}$$

Substitute this expression into the x component equation and solve for v:

$$\frac{m_{\text{rock}}g}{\cos\theta}\sin\theta = \frac{m_{\text{rock}}v^2}{L\sin\theta}$$

$$\frac{g\sin\theta}{\cos\theta} = \frac{v^2}{L\sin\theta}$$

$$v^2 = \frac{gL\sin^2\theta}{\cos\theta}$$

$$v = \sqrt{\frac{gL\sin^2\theta}{\cos\theta}}$$

Reflect

You can experiment with this motion yourself by tying a piece of string around an eraser or other small object and whirling it in a horizontal circle. You'll find that the faster the object moves, the greater the angle θ. Let's check that our answer agrees with this conclusion.

If θ is a small angle close to zero so that the string hangs down almost vertically, $\sin\theta$ is small and $\cos\theta$ is nearly equal to 1. (Remember that $\sin 0 = 0$ and $\cos 0 = 1$.) So the ratio $(\sin^2\theta)/(\cos\theta)$ is a small number, and the speed v will be small. If θ is a large angle close to 90° so that the string is nearly horizontal, $\sin\theta$ is a little less than 1 and $\cos\theta$ is small (recall $\sin 90° = 1$ and $\cos 90° = 0$). In this case the ratio $(\sin^2\theta)/(\cos\theta)$ is a large number, and the speed v will be large. So our formula agrees with the experiment!

WATCH OUT! No "centrifugal force," please.

In Example 5-8 you may have been tempted to add a force that points toward the *outside* of the rock's curved trajectory (**Figure 5-19a**). Don't do it! If you felt this temptation, it's because you're thinking of the so-called centrifugal force you feel when riding in a car that makes a sharp turn. This "centrifugal force" seems to push you toward the outside of the turn ("centrifugal" means "fleeing the center"). This force is *fictitious,* however; nothing is exerting an outward force on you. Rather, you're just feeling your body's inertia, which is the tendency of all objects to want to continue in a straight line. Inertia explains what would happen if you let go of the string attached to the rock: The rock would fly off in a straight line as seen from above (**Figure 5-19b**) and so it would move away from the center of its circular path. That's not because of a mysterious centrifugal force but rather because of the *absence* of the inward tension force. The lesson here is that you should never include a centrifugal force in any of your free-body diagrams. (If you still feel that you have to add an outward force to "keep the object in balance," remember that an object going around a curve is *not* in balance: The object is accelerating, so the net external force on it is *not* zero!)

(a) *Incorrect* free-body diagram of the rock

There is no object that exerts $F_{\text{centrifugal}}$. So this force does not exist and should not be included in the free-body diagram.

$\vec{T}$, θ, $\vec{F}_{\text{centrifugal}}$, $\vec{a}_{\text{rock}}$, $\vec{w}_{\text{rock}}$, x, y

Toward center of the circle

(b) A top view of the rock's trajectory

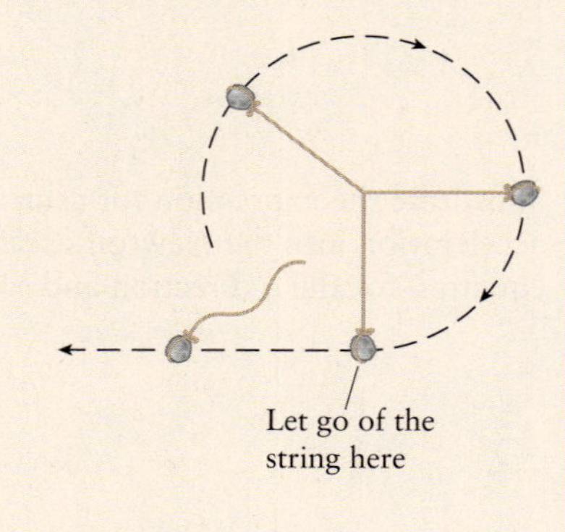

Figure 5-19 How *not* to draw a free-body diagram for uniform circular motion (a) A free-body diagram should never include a "centrifugal force." (b) If the string is let go, the rock flies off in a straight line due to its inertia—*not* a "centrifugal force."

EXAMPLE 5-9 Making an Airplane Turn

An airplane banks to one side in order to turn in that direction. By banking the plane, the *lift force*—a force that acts perpendicular to the direction of flight, caused by the motion of air over the airplane's wings—ends up with both a vertical component and a horizontal component (Figure 5-20). The horizontal component of this force provides the centripetal acceleration needed to make the airplane move around a circle. (a) If the airplane of mass m_{airplane} is traveling at speed v and is banked by an angle θ, derive an expression for the radius of the turn. (b) The pilot of the airplane has mass m_{pilot} and weight $w_{\text{pilot}} = m_{\text{pilot}}g$. If the pilot is sitting on a scale as the airplane makes its banked turn, what does the scale read?

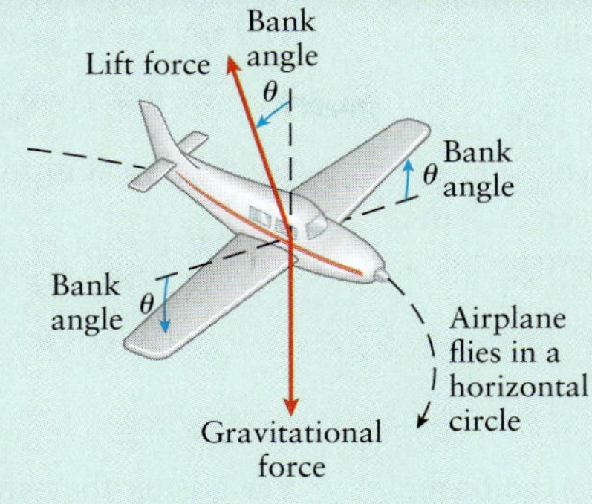

Figure 5-20 An airplane in a banked turn What mathematical expression describes the radius of the airplane's turn? The forward thrust force and the backward drag force cancel and are not shown.

Set Up

We draw two free-body diagrams, one for the airplane of mass m_{airplane} and one for the pilot of mass m_{pilot}. (In both cases the airplane and pilot are coming at us head on.) Both of these diagrams are almost identical to the free-body diagram for the rock in Example 5-8. Besides the gravitational force, a second force has a vertical component that balances the gravitational force (so that the airplane doesn't accelerate up or down) and a horizontal component that gives the object a centripetal acceleration.

In part (a) we want to find the radius r of the turn; in part (b) our goal is to determine the value of the normal force n (the force that the scale exerts on the pilot, which is the same as the reading on the scale). The value of n is the pilot's *apparent weight,* or how much she *feels* that she weighs during the turn.

Newton's second law equation for the airplane:

$$\sum \vec{F}_{\text{ext on airplane}} = \vec{L} + \vec{w}_{\text{airplane}} = m_{\text{airplane}}\vec{a}_{\text{airplane}}$$

Newton's second law equation for the pilot:

$$\sum \vec{F}_{\text{ext on pilot}} = \vec{n} + \vec{w}_{\text{pilot}} = m_{\text{pilot}}\vec{a}_{\text{pilot}}$$

Centripetal acceleration:

$$a_{\text{airplane}} = a_{\text{pilot}} = a_{\text{cent}} = \frac{v^2}{r} \quad (5\text{-}10)$$

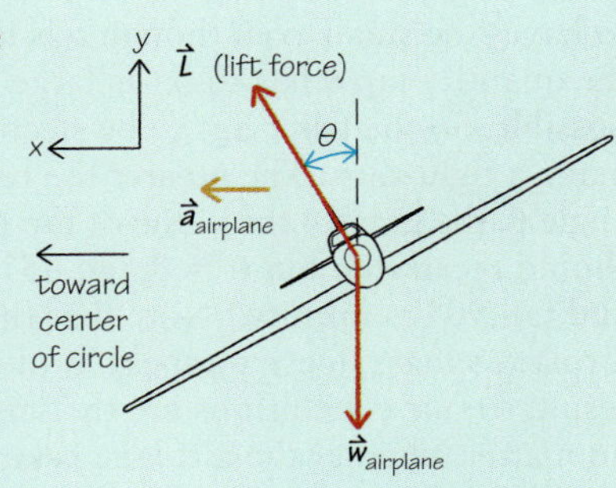

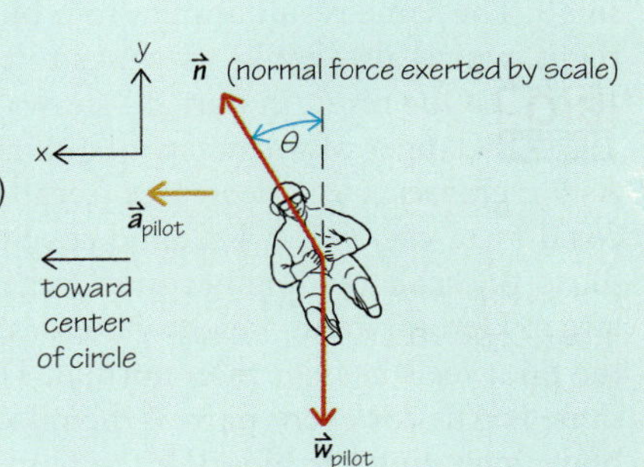

Solve

(a) Once again we begin by writing Newton's second law in component form. The x component equation involves the radius r of the turn, but we don't know the value of the lift force L. We find this force from the y component equation, and we then substitute the value of L into the x equation and solve for r. The airplane's acceleration is directed toward the center of the turn, so it is a centripetal acceleration.

Newton's second law in component form applied to the airplane:

$$x\text{: } L \sin\theta = m_{\text{airplane}}a_{\text{cent}} = m_{\text{airplane}}\frac{v^2}{r}$$

$$y\text{: } L \cos\theta + (-w_{\text{airplane}}) = L\cos\theta - m_{\text{airplane}}g = 0$$

From the y equation,

$$L\cos\theta = m_{\text{airplane}}g$$

$$L = \frac{m_{\text{airplane}}g}{\cos\theta}$$

Substitute this expression for L into the x equation and solve for r:

$$m_{\text{airplane}}\frac{v^2}{r} = L\sin\theta = \frac{m_{\text{airplane}}g}{\cos\theta}\sin\theta = m_{\text{airplane}}g\tan\theta$$

$$\frac{v^2}{r} = g\tan\theta$$

$$r = \frac{v^2}{g\tan\theta}$$

(b) The component form of Newton's second law for the pilot is very similar to that for the airplane—the main difference is that the lift force L is replaced by the normal force n. Here we need only the y component equation to solve for n. The pilot's acceleration is directed to the center of the turn, so it is a centripetal acceleration.

Newton's second law in component form applied to the pilot:

$$x\text{: } n \sin\theta = m_{\text{pilot}}a_{\text{cent}} = m_{\text{pilot}}\frac{v^2}{r}$$

$$y\text{: } n\cos\theta + (-w_{\text{pilot}}) = n\cos\theta - m_{\text{pilot}}g = 0$$

Solve the y component equation for n:

$$n\cos\theta = m_{\text{pilot}}g$$

$$n = \frac{m_{\text{pilot}}g}{\cos\theta}$$

Reflect

Our result in part (a) says that the faster the airplane's speed for a given bank angle θ, the larger the radius r of the turn. To make a tight turn at high speed (that is, to have r be small even though v is large), the quantity $\tan\theta$ has to be as large as possible. So the bank has to be steep to make a tight turn: The greater the bank angle θ, the greater the value of $\tan\theta$. (You should recall that $\tan 0 = 0$, $\tan 45° = 1$, and $\tan 90°$ = infinity.) Note that the airplane's mass doesn't appear in our result, so our conclusions are the same no matter whether the airplane is large or small. The same result applies to a bird in flight, which also banks its wings to turn.

BIO-Medical Our result in part (b) shows the danger of too steep a bank angle. As θ increases, $\cos\theta$ decreases (recall that $\cos 0 = 1$, $\cos 60° = 1/2$, and $\cos 90° = 0$), so the scale reading becomes larger and larger. Her apparent weight increases, and the pilot *feels* heavier than normal. The same is true for every part of the pilot's body, including her blood. If the bank angle is too steep, the pilot's heart can't pump this "heavy" blood up to her brain. As a result, a typical pilot will lose consciousness after five to ten seconds at an apparent weight of 4 to 5 times $m_{\text{pilot}}g$.

To counter these effects, fighter pilots (who often make very sharp turns at steep bank angles) wear *g-suits*. These are garments that apply pressure to the legs and lower abdomen, thus preventing blood from pooling in the lower body even during extreme aerial maneuvers.

The greater the bank angle θ, the smaller the turning radius r of the airplane:

$$r = \frac{v^2}{g\tan\theta}$$

For $\theta = 45°$: $\tan 45° = 1$

$$r = \frac{v^2}{g}$$

For $\theta = 60°$: $\tan 60° = 1.73$

$$r = \frac{v^2}{1.73g} = 0.577\frac{v^2}{g}$$

For $\theta = 80°$: $\tan 80° = 5.67$

$$r = \frac{v^2}{5.67g} = 0.176\frac{v^2}{g}$$

The greater the bank angle θ, the greater the pilot's scale reading and the heavier she feels:

Actual weight $= m_{\text{pilot}}g$

Apparent weight $= n = \dfrac{m_{\text{pilot}}g}{\cos\theta}$

For $\theta = 45°$: $\cos 45° = 0.707$

$$n = \frac{m_{\text{pilot}}g}{0.707} = 1.41m_{\text{pilot}}g$$

(The pilot feels "1.41 g.")

For $\theta = 60°$: $\cos 60° = 0.500$

$$n = \frac{m_{\text{pilot}}g}{0.500} = 2.00m_{\text{pilot}}g$$

(The pilot feels "2.00 g.")

For $\theta = 80°$: $\cos 80° = 0.174$

$$n = \frac{m_{\text{pilot}}g}{0.174} = 5.76m_{\text{pilot}}g$$

(The pilot feels "5.76 g" and might pass out.)

EXAMPLE 5-10 A Turn on a Level Road

A car of mass m_{car} exits the freeway and travels around a circular off-ramp at a constant speed of 15.7 m/s (56.5 km/h, or 35.1 mi/h). The ramp is level (it is not banked), so the centripetal force has to be provided by the force of friction. If the coefficient of static friction between the car's tires and the road is 0.550 (a typical value when the road surface is wet), what radius of curvature is required for the off-ramp so that the car does not skid?

Set Up

This example is very similar to Example 5-6 in Section 5-4. The only difference is that in Example 5-6 the static friction force gave Roberto a *forward* acceleration, while in this example the static friction force causes a *centripetal* acceleration.

The free-body diagram shows the car head-on. Note that even though the car is rolling on its tires, the force that provides the centripetal acceleration is *static* friction rather than rolling friction. That's because the force of interest is the one that prevents the car from sliding sideways away from the center of its circular path. (By contrast, rolling friction acts opposite to the car's forward motion and so doesn't contribute to the centripetal acceleration.) The minimum value of the radius r corresponds to the maximum centripetal acceleration $a_{cent} = v^2/r$, which in turn corresponds to the maximum value $f_{s,max}$ of the static friction force.

Newton's second law equation for the car:

$$\sum \vec{F}_{\text{ext on car}} = \vec{n} + \vec{w}_{car} + \vec{f}_s = m_{car}\vec{a}_{car}$$

Centripetal acceleration:

$$a_{car} = a_{cent} = \frac{v^2}{r} \quad (5\text{-}10)$$

Magnitude of the static friction force:

$$f_{s,max} = \mu_s n \quad (5\text{-}1b)$$

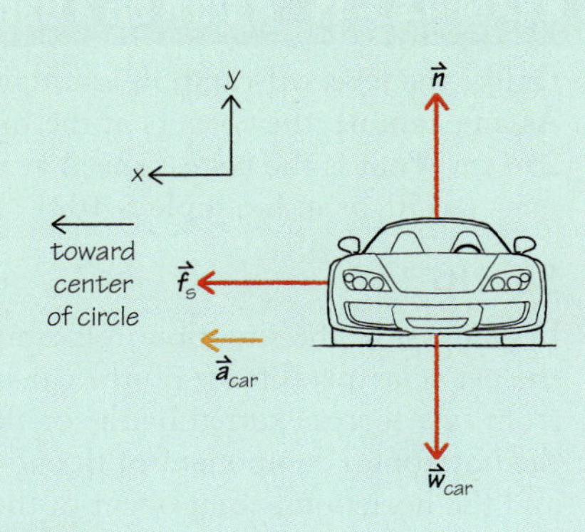

Solve

We write Newton's second law for the car in component form, with the static friction force set equal to its maximum value $f_{s,max}$. The horizontal (x) acceleration equals the centripetal acceleration; there is zero vertical (y) acceleration. We then solve for the radius r of the car's circular path.

Newton's second law in component form for the car, assuming that the static friction force has its maximum value:

$$x: f_{s,max} = m_{car}a_{cent}$$

$$y: n + (-w_{car}) = 0$$

y
x
$n_y = +n$
$f_{s,max,x} = +f_{s,max}$
$w_{car,y} = -w_{car}$

From the y component equation,

$$n = w_{car} = m_{car}g$$

So the maximum force of static friction of the tires is

$$f_{s,max} = \mu_s n = \mu_s m_{car} g$$

Substitute this expression and $a_{cent} = v^2/r$ into the x component equation, then solve for r:

$$\mu_s m_{car} g = m_{car}\frac{v^2}{r}$$

$$r = \frac{v^2}{\mu_s g}$$

Using $\mu_s = 0.550$ and $v = 15.7$ m/s,

$$r = \frac{(15.7 \text{ m/s})^2}{(0.550)(9.80 \text{ m/s}^2)} = 45.7 \text{ m}$$

Reflect

Our answer is the *minimum* radius of the off-ramp. If the radius is larger, the car will have a smaller centripetal acceleration $a_{cent} = v^2/r$ and the static friction force required will be less than the maximum force available.

Note that the minimum radius increases if the car is going faster (v is greater). Note also that for a given speed, the car can make a tighter turn (that is, r will be smaller) if the tires have a better "grip" on the road (so the coefficient of static friction μ_s is greater). The car would be able to make a tighter turn if the road were dry, in which case μ_s is about 1.0.

EXAMPLE 5-11 A Turn on a Banked Road

Unlike the level off-ramp in Example 5-10, many freeway off-ramps, as well as corners at racetracks, are banked. As an example, the corners at the Indianapolis Motor Speedway are banked by 9.20°. The radius of each corner is 256 m. What is the fastest speed at which a race car could take one of these corners if the coefficient of static friction is $\mu_s = 0.550$, as in Example 5-10?

Set Up

In contrast to the situation in Example 5-10, the net centripetal force on the car comes from *two* forces exerted by the banked road: the horizontal component of the normal force and the horizontal component of the static friction force. Because we want the car to move as fast as possible and hence have the maximum possible centripetal acceleration, it has to have the maximum available centripetal force exerted on it. In this extreme case the static friction force will be at its maximum value. Our goal is to find the value of v in this situation.

Newton's second law equation for the car:

$$\sum \vec{F}_{\text{ext on car}} = \vec{n} + \vec{w}_{\text{car}} + \vec{f}_s = m_{\text{car}}\vec{a}_{\text{car}}$$

Centripetal acceleration:

$$a_{\text{car}} = a_{\text{cent}} = \frac{v^2}{r} \qquad (5\text{-}10)$$

Magnitude of the static friction force:

$$f_{s,\max} = \mu_s n \qquad (5\text{-}1b)$$

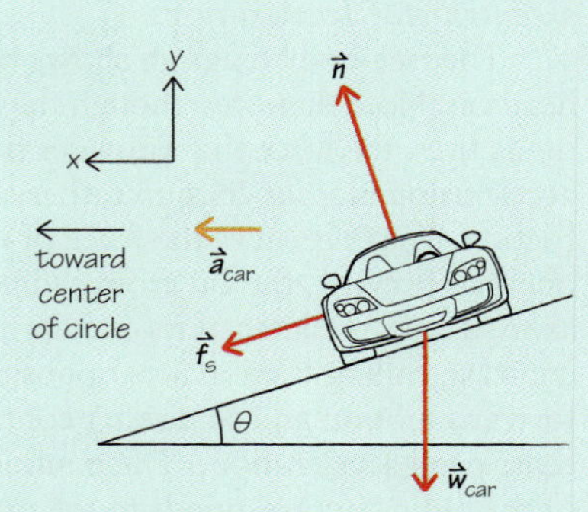

Solve

As in Example 5-10, we write Newton's second law for the car in component form, with the static friction force set equal to its maximum value $f_{s,\max}$. Again the horizontal (x) acceleration equals the centripetal acceleration; there is zero vertical (y) acceleration. The difference is that now the normal force and friction force have both x and y components.

Newton's second law in component form for the car, assuming that the static friction force has its maximum value:

$$x\text{: } n \sin\theta + f_{s,\max}\cos\theta = m_{\text{car}}a_{\text{cent}}$$

$$y\text{: } n\cos\theta + (-w_{\text{car}}) + (-f_{s,\max}\sin\theta) = 0$$

Substitute the expressions for the maximum force of static friction, the centripetal acceleration, and the weight of the car:

$$f_{s,\max} = \mu_s n \quad ; \quad a_{\text{cent}} = \frac{v^2}{r} \quad ; \quad w_{\text{car}} = m_{\text{car}}g$$

Then the Newton's law equations become

$$x\text{: } n\sin\theta + \mu_s n\cos\theta = m_{\text{car}}\frac{v^2}{r}$$

$$y\text{: } n\cos\theta - m_{\text{car}}g - \mu_s n\sin\theta = 0$$

Solve the y component equation for the normal force n.

Rearrange the y component equation to find n:

$$n\cos\theta - \mu_s n\sin\theta = m_{\text{car}}g$$

$$n(\cos\theta - \mu_s\sin\theta) = m_{\text{car}}g$$

$$n = \frac{m_{\text{car}}g}{\cos\theta - \mu_s\sin\theta}$$

Substitute this expression for n into the x component equation and solve for the speed v.

Rearrange the x component equation:

$$n\sin\theta + \mu_s n\cos\theta = m_{\text{car}}\frac{v^2}{r}$$

$$n(\sin\theta + \mu_s\cos\theta) = m_{\text{car}}\frac{v^2}{r}$$

Substitute the expression for n and solve for v:

$$\frac{m_{car}g}{\cos\theta - \mu_s \sin\theta}(\sin\theta + \mu_s\cos\theta) = m_{car}\frac{v^2}{r}$$

$$g\left(\frac{\sin\theta + \mu_s\cos\theta}{\cos\theta - \mu_s\sin\theta}\right) = \frac{v^2}{r}$$

$$v^2 = gr\left(\frac{\sin\theta + \mu_s\cos\theta}{\cos\theta - \mu_s\sin\theta}\right)$$

$$v = \sqrt{gr\left(\frac{\sin\theta + \mu_s\cos\theta}{\cos\theta - \mu_s\sin\theta}\right)}$$

Use the values of r, θ, and μ_s given for the Indianapolis Motor Speedway.

Using $r = 256$ m, $\theta = 9.20°$, and $\mu_s = 0.550$,

$\sin\theta = \sin 9.20° = 0.160$

$\cos\theta = \cos 9.20° = 0.987$

$$v = \sqrt{(9.80\text{ m/s}^2)(256\text{ m})\left[\frac{0.160 + (0.550)(0.987)}{0.987 - (0.550)(0.160)}\right]}$$

$$= 44.3\text{ m/s}$$

$$= (44.3\text{ m/s})\left(\frac{1\text{ km}}{1000\text{ m}}\right)\left(\frac{3600\text{ s}}{1\text{ h}}\right) = 159\text{ km/h}$$

$$= (159\text{ km/h})\left(\frac{1\text{ mi}}{1.61\text{ km}}\right) = 99.1\text{ mi/h}$$

Reflect

This speed is quite a bit less than that reached by race cars on the straightaway (200 to 300 km/h), so it's necessary to slow down before turning this corner.

We can check our result by looking at two special cases: (a) If the curve is *not* banked, then $\theta = 0$, $\sin\theta = 0$, and $\cos\theta = 1$. In this case our result reduces to what we found for the flat road in Example 5-10. (b) If the curve is banked but there is no friction, then $\mu_s = 0$ (because then the maximum force of friction is $f_{s,max} = \mu_s n = 0$). In this situation our result simplifies to what we found for the airplane making a level turn in Example 5-9. (The airplane isn't on a road, but the lift force $\vec{L}$ on a turning airplane plays exactly the same role as the normal force $\vec{n}$ on a car on a frictionless banked curve.)

You can show that in case (a) the car's maximum speed would be only 37.1 m/s, while in case (b) it would be only 20.2 m/s. A banked curve with friction allows the highest speeds!

(a) If the curve is not banked (so $\theta = 0$):

$$v = \sqrt{gr\left(\frac{\sin 0 + \mu_s\cos 0}{\cos 0 - \mu_s\sin 0}\right)}$$

$$= \sqrt{gr\left(\frac{0 + \mu_s(1)}{1 - \mu_s(0)}\right)}$$

$$= \sqrt{\mu_s g r}$$

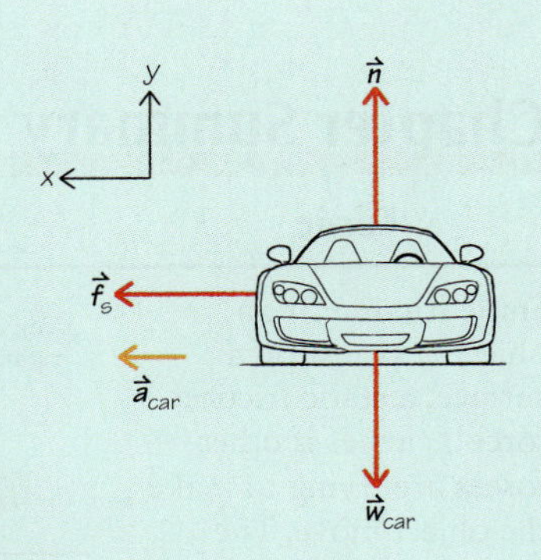

This equation is equivalent to what we found in Example 5-10 for a flat road:

$$r = \frac{v^2}{\mu_s g}, \text{ so } v^2 = \mu_s g r \text{ and } v = \sqrt{\mu_s g r}$$

(b) If the curve is banked but there is no friction (so $\mu_s = 0$):

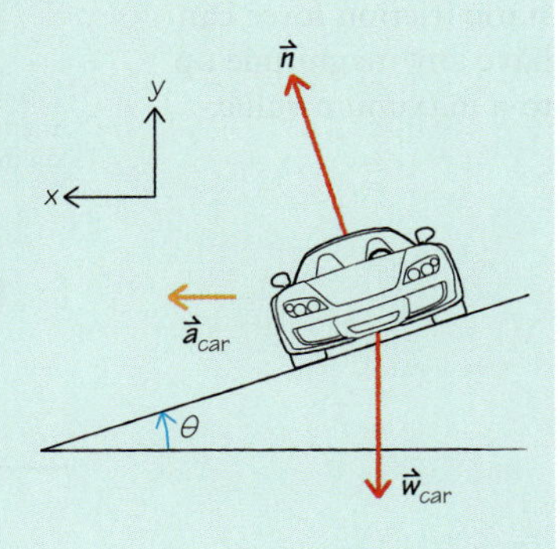

$$v = \sqrt{gr\left(\frac{\sin\theta + (0)\cos\theta}{\cos\theta - (0)\sin\theta}\right)}$$

$$= \sqrt{gr\left(\frac{\sin\theta}{\cos\theta}\right)}$$

$$= \sqrt{gr\tan\theta}$$

This equation is equivalent to what we found in Example 5-9 for an airplane making a banked turn:

$$r = \frac{v^2}{g\tan\theta}, \text{ so } v^2 = gr\tan\theta \text{ and}$$

$$v = \sqrt{gr\tan\theta}$$

GOT THE CONCEPT? 5-5 Comparing Circular Motions

? Five objects all move in uniform circular motion. The objects have different masses, move at different speeds, and travel in circular paths of different radii. Rank these objects in order of the magnitude of the centripetal force that acts on them, from greatest to least: (a) a 2.00-kg object moving at 10.0 m/s in a circle of radius 8.00 m, (b) a 2.00-kg object moving at 5.00 m/s in a circle of radius 4.00 m, (c) a 4.00-kg object moving at 5.00 m/s in a circle of radius 2.00 m, (d) a 4.00-kg object moving at 5.00 m/s in a circle of radius 8.00 m, and (e) an 8.00-kg object moving at 5.00 m/s in a circle of radius 1.00 m.

TAKE-HOME MESSAGE FOR Section 5-6

✔ If an object is in uniform circular motion, its acceleration is directed toward the center of the circle and given by Equation 5-10.

✔ The net force on an object in uniform circular motion is also directed toward the center of the circle (Equation 5-11). The net force in this case is called the centripetal force.

✔ Any kind of force (or combination of forces) can provide the centripetal force on an object in uniform circular motion.

Key Terms

centripetal force
coefficient of kinetic friction
coefficient of rolling friction
coefficient of static friction
drag force
fluid resistance
friction
kinetic friction
rolling friction
static friction

Chapter Summary

Topic	Equation or Figure
Static friction: If an object is at rest on a surface, a static friction force $\vec{f}_s$ arises if other forces are trying to make the object move. The static friction force can have any magnitude up to a maximum value.	(see figure below)

(a)
You apply a force of 20.0 N to the right, but the block remains at rest...
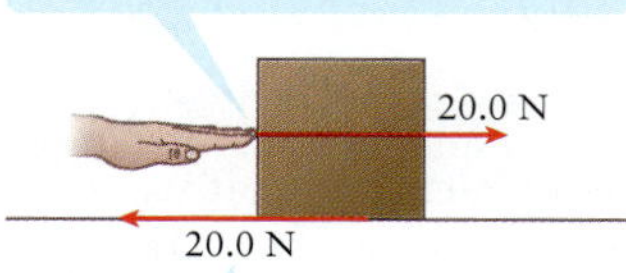
20.0 N
20.0 N
...so the table must exert a static friction force of 20.0 N to the left on the block.

(b)
You apply a force of 10.0 N to the right, but the block remains at rest...
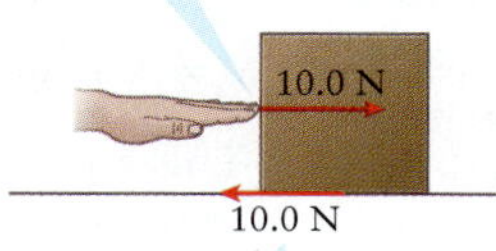
10.0 N
10.0 N
...so the table must exert a static friction force of 10.0 N to the left on the block.

(c)
You apply no force to the block and it remains at rest...
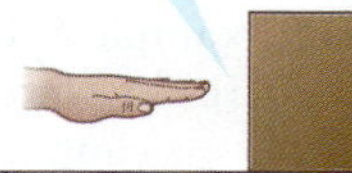
...so the table must exert no static friction force on the block.

The static friction force adjusts its value to counteract whatever other forces are trying to make the object slide.

(Figure 5-2)

① The magnitude of the **force of static friction** on an object...

② ...is less than or equal to a certain maximum value.

$$f_s \le f_{s,\max} \tag{5-1a}$$

③ The **maximum force of static friction** depends on...

$$f_{s,\max} = \mu_s n \tag{5-1b}$$

④ ...the **coefficient of static friction** (which depends on the properties of the two surfaces in contact)...

⑤ ...and the **normal force** pressing the object against a surface.

An object at rest on an incline: Static friction will keep an object at rest on an inclined surface provided the angle θ of the incline is less than a critical value θ_{slip}.

(a) A block at rest on an inclined ramp

Block, weight w

Ramp whose angle θ from the horizontal can be varied

θ

(Figure 5-5a)

Coefficient of static friction for an object at rest on a surface

Angle of the surface at which the object **begins to slip**

$$\mu_s = \tan \theta_{\text{slip}} \tag{5-5}$$

Kinetic friction: If an object is sliding over a surface, that surface exerts a kinetic friction force $\vec{f}_k$ that opposes sliding. We model this force as being independent of the object's speed. There is a similar expression for the force of rolling friction; the difference is that the coefficient for rolling has a smaller value than that for sliding.

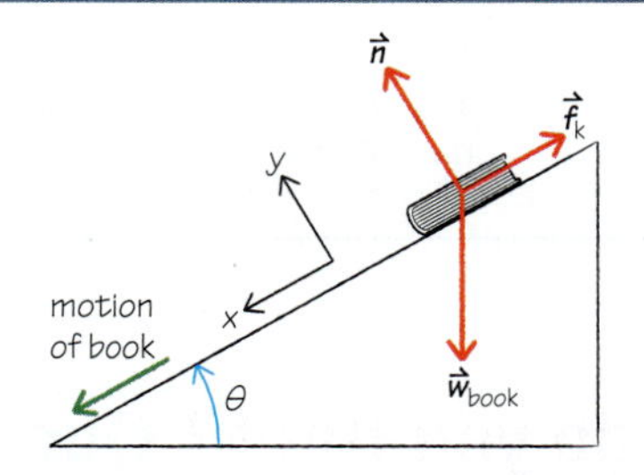

(Example 5-2)

The **force of kinetic friction** depends on...

$$f_k = \mu_k n \tag{5-6}$$

...the **coefficient of kinetic friction** (which depends on the properties of the two surfaces in contact)...

...and the **normal force** exerted on the object by the surface on which it slides.

Drag force: An object moving through a fluid (gas or liquid) experiences a resistive force that opposes its motion. If the object is small and moving at a low speed v relative to the fluid, the magnitude of the drag force is proportional to v. For larger objects moving at faster speeds, the drag force is proportional to the square of v.

Magnitude of the **drag force** on a **small object** moving at a **low speed**

$$F_{\text{drag}} = bv \tag{5-8}$$

Constant that depends on the properties of the object and of the fluid

Speed of the object relative to the fluid

Magnitude of the **drag force** on a **larger object** moving at a **faster speed**

$$F_{\text{drag}} = cv^2 \tag{5-9}$$

Constant that depends on the properties of the object and of the fluid

Speed of the object relative to the fluid

Uniform circular motion: If an object moves at constant speed around a circular path, the net force on the object points toward the center of the path. Any kind of force or combination of forces can contribute to this net force.

Centripetal acceleration: Magnitude of the acceleration of an object in uniform circular motion

Speed of the object as it moves around the circle

$$a_{\text{cent}} = \frac{v^2}{r} \quad (5\text{-}10)$$

Radius of the circle

Net force on an object in uniform circular motion

Mass of the object

$$\sum F_{\text{cent}} = \frac{mv^2}{r} \quad (5\text{-}11)$$

Speed of the object as it moves around the circle

The direction of the net force is toward the center of the object's circular path.

Radius of the circle

Answer to What do you think? Question

(a) There is a net force, and it points inward toward the center of the circle. Newton's second law tells us that the net force must be in the same direction as the airplane's acceleration, and in uniform circular motion the acceleration is inward. If you're riding in the airplane it may *feel* as though a force is pushing you outward. This outward "force" isn't real, however; it's just a consequence of your body's inertia (see Section 5-6).

Answers to Got the Concept? Questions

5-1 (d) As shown in Equation 5-1b, the maximum static friction force between the block and board depends only on the normal force n and the coefficient of static friction μ_s. It does *not* depend on the area of contact between the block and board. Therefore, the angle at which the block begins to slide will be the same no matter which face is in contact with the board.

5-2 (e) The acceleration of the book doesn't depend on its mass at all. The acceleration is caused by the kinetic friction force and the downhill component of the gravitational force. Both of these are proportional to the mass of the book. (The kinetic friction force has magnitude $f_k = \mu_k n$, and the normal force n is proportional to the block's weight and hence its mass.) The mass cancels out when we calculate the acceleration of the block, which is equal to the net external force on the block divided by its mass.

5-3 (d) We saw in Section 5-2 that an object at rest on a ramp with angle θ will just barely remain at rest (that is, the maximum force of static friction is just enough to balance the downhill component of the gravitational force), if $\mu_s = \tan\theta$. For an incline with $\theta = 30.0°$, $\mu_s = \tan 30.0° = 0.577$. If μ_s has a greater value than 0.577, the maximum static friction force available is greater than the downhill component of the gravitational force, and the object will still remain at rest. So both $\mu_s = 0.600$ and $\mu_s = 0.750$ will keep the object at rest. If $\mu_s = 0.350$, there isn't enough static friction and so the book will start sliding back downhill.

5-4 (c) Example 5-6 illustrates that both kinetic friction and static friction can cause an object to speed up. If Roberto doesn't slip on the upper surface of the sled, the force that accelerates him forward is the force of static friction. If he does slip, he nonetheless accelerates forward due to the force of kinetic friction. (In the latter case it's true that he slips backward relative to the sled. However, both he and the sled are accelerating forward; it's just that Roberto's acceleration is less than the acceleration of the sled.)

5-5 (d) The drag force for both objects is given by $F_{\text{drag}} = cv^2$, so $c = F_{\text{drag}}/v^2$. Because the drag force on the second object is four times greater and 1/2 the speed, the quantity c is $(4)/(1/2)^2 = 16$ times greater for the second object than for the baseball.

5-6 (e), (c), (a), (b) and (d) [tie] The centripetal force on an object of mass m moving at speed v in a circle of radius r is equal to mv^2/r—that is, to m multiplied by the object's centripetal acceleration $a_{\text{cent}} = v^2/r$. For the five cases this quantity is (a) $(2.00\text{ kg})(10.0\text{ m/s})^2/(8.00\text{ m}) = 25.0\text{ N}$; (b) $(2.00\text{ kg})(5.00\text{ m/s})^2/(4.00\text{ m}) = 12.5\text{ N}$; (c) $(4.00\text{ kg})(5.00\text{ m/s})^2/(2.00\text{ m}) = 50.0\text{ N}$; (d) $(4.00\text{ kg})(5.00\text{ m/s})^2/(8.00\text{ m}) = 12.5\text{ N}$; (e) $(8.00\text{ kg})(5.00\text{ m/s})^2/(1.00\text{ m}) = 2.00 \times 10^2\text{ N}$.

Questions and Problems

In a few problems you are given more data than you actually need; in a few other problems you are required to supply data from your general knowledge, outside sources, or informed estimate.

Interpret as significant all digits in numerical values that have trailing zeros and no decimal points. For all problems use $g = 9.80\ \text{m/s}^2$ for the free-fall acceleration due to gravity.

• Basic, single-concept problem
•• Intermediate-level problem; may require synthesis of concepts and multiple steps
••• Challenging problem
SSM *Solution is in Student Solutions Manual*
Example *See worked example for a similar problem*

Conceptual Questions

1. • Complete the sentence: The static frictional force between two surfaces is (never/sometimes/always) less than the normal force. Explain your answer.

2. • You want to push a heavy box across a rough floor. You know that the coefficient of static friction (μ_s) is larger than the coefficient of kinetic friction (μ_k). Should you push the box for a short distance, rest, push the box another short distance, and then repeat the process until the box is where you want it, or will it be easier to keep pushing the box across the floor once you get it moving?

3. • If the force of friction always opposes the sliding of an object, how then can a frictional force cause an object to increase in speed? SSM

4. • A solid rectangular block has sides of three different areas. You can choose to rest any of the sides on the floor as you apply a horizontal force to the block. Does the choice of side on the floor affect how hard it is to push the block? Explain your answer.

5. • You press a book against the wall with your hand. As you get tired you exert less force, but the book remains in the same spot on the wall. Do each of the following forces increase, decrease, or not change in magnitude when you reduce the force you are applying to the book: (a) weight of the book, (b) normal force of the wall on the book, (c) frictional force of the wall on the book, and (d) maximum static frictional force of the wall on the book?

6. • An antilock braking system (ABS) prevents wheels from skidding while drivers stomp on the brakes in emergencies. How far will a car with an ABS move before finally stopping, as compared to a car without an ABS?

7. • Describe two situations in which the normal force acting on an object is not equal to the object's weight.

8. •• You are sliding a piece of furniture across the floor at constant velocity when a friend jumps on top of it. Assume $\mu_k < \mu_s$ and use the ideas discussed in this chapter to describe what happens and why.

9. • *Synovial fluid* lubricates the surfaces where bones meet in joints, making the coefficient of friction between bones very small. Why is the minimum force required for moving bones in a typical knee joint different for different people, in spite of the fact that their joints have the same coefficient of friction? For simplicity assume the surfaces in the knee are flat and horizontal.

10. • At low speeds the drag force on an object moving through a fluid is proportional to its velocity. According to Newton's second law, force is proportional to acceleration. As acceleration and velocity aren't the same quantity, is there a contradiction here?

11. • As a skydiver falls faster and faster through the air, does the magnitude of his acceleration increase, decrease, or remain the same? Explain your answer. SSM

12. • Why do raindrops fall from the sky at different speeds? Explain your answer.

13. • James Bond leaps without a parachute from a burning airplane flying at 15,000 ft. Ten seconds later his assistant, who was following behind in another plane, dives after him, wearing her parachute and clinging to one for her hero. Is it possible for her to catch up with Bond and save him?

14. • What distinguishes the forces that act on a car driving over the top of a hill from those acting on a car driving through a dip in the road? Explain how the forces relate to the sensations passengers in the car experience during each situation.

15. • Explain how you might measure the centripetal acceleration of a car rounding a curve. SSM

16. • For an object moving in a circle, which of the following quantities are zero over one revolution: (a) displacement, (b) average velocity, (c) average acceleration, (d) instantaneous velocity, and (e) instantaneous centripetal acceleration?

17. • Why does water stay in a bucket that is whirled around in a vertical circle? Contrast the forces acting on the water when the bucket is at the lowest point on the circle to when the bucket is at the highest point on the circle. SSM

18. • Explain why curves in roads and cycling velodromes are banked. SSM

19. • Why might your car start to skid if you drive too fast around a curve?

Multiple-Choice Questions

20. • If a sport utility vehicle (SUV) drives up a slope of 45°, what must be the minimum coefficient of static friction between the SUV's tires and the road?
    A. 1.0
    B. 0.5
    C. 0.7
    D. 0.9
    E. 0.05

21. • A block of mass m slides down a rough incline with constant speed. If a similar block that has a mass of $4m$ were placed on the same incline, it would
    A. slide down at the same constant speed.
    B. accelerate down the incline.
    C. slowly slide down the incline and then stop.
    D. slide down with a faster constant speed.
    E. not move. SSM

22. • A 10-kg crate is placed on a horizontal conveyor belt moving with a constant speed. The crate does not slip. If the coefficients of friction between the crate and the belt are $\mu_s = 0.50$ and $\mu_k = 0.30$, what is the frictional force exerted on the crate?
   A. 98 N
   B. 49 N
   C. 29 N
   D. 9.8 N
   E. 0

23. • **Biology** The *Escherichia coli* (*E. coli*) bacterium propels itself through water by means of long, thin structures called flagella. If the force exerted by the flagella doubles, the velocity of the bacterium
   A. doubles.
   B. decreases by half.
   C. does not change.
   D. increases by a factor of four.
   E. cannot be determined without more information.

24. • A 1-kg wood ball and a 5-kg lead ball have identical sizes, shapes, and surface characteristics. They are dropped simultaneously from a tall tower. Air resistance is present. How do their accelerations compare?
   A. The 1-kg wood ball has the larger acceleration, but it cannot be calculated without more information.
   B. The 5-kg lead ball has the larger acceleration, but it cannot be calculated without more information.
   C. The accelerations are the same.
   D. The 5-kg ball accelerates at five times the acceleration of the 1-kg ball.
   E. The 1-kg ball accelerates at five times the acceleration of the 5-kg ball.

25. • A skydiver is falling at his terminal speed. Immediately after he opens his parachute
   A. his speed will be larger than his terminal speed.
   B. the magnitude of the drag force on the skydiver will decrease.
   C. the net force on the skydiver is in the downward direction.
   D. the magnitude of the drag force is larger than the skydiver's weight.
   E. the net force on the skydiver is zero. SSM

26. • Two identical rocks are tied to massless strings and whirled in nearly horizontal circles at the same speed. One string is twice as long as the other. What is the tension T_1 in the shorter string compared to the tension T_2 in the longer one?
   A. $T_1 = T_2/4$
   B. $T_1 = T_2/2$
   C. $T_1 = T_2$
   D. $T_1 = 2T_2$
   E. $T_1 = 4T_2$

27. • You are on a Ferris wheel moving in a vertical circle. When you are at the bottom of the circle, how does the magnitude of the normal force n exerted by your seat compare to your weight mg?
   A. $n = mg$
   B. $n > mg$, but cannot be exactly calculated without more information.
   C. $n < mg$, but cannot be exactly calculated without more information.
   D. $n = mg/2$
   E. $n = 2mg$ SSM

28. • Two rocks are tied to massless strings and whirled in nearly horizontal circles so that the time to travel around the circle once is the same for both. One string is twice as long as the other. The tension in the longer string is twice the tension in the shorter one. What is the mass m_1 of the rock at the end of the shorter string compared to the mass m_2 of the rock at the end of the longer one?
   A. $m_1 = m_2/4$
   B. $m_1 = m_2/2$
   C. $m_1 = m_2$
   D. $m_1 = 2m_2$
   E. $m_1 = 4m_2$

Estimation/Numerical Analysis

29. • Estimate the numerical value of the coefficient of static friction between a car's tires and dry pavement.

30 • Estimate the value of the coefficient of kinetic friction for a box of books on a carpeted floor.

31 • **Sports** Make a rough estimate of the value of the coefficient of kinetic friction between a baseball player's uniform and the infield surface as he slides into second base. SSM

32. • Give a numerical estimate of the minimum radius of curvature of an unbanked on-ramp for a typical freeway.

33. • **Sports** Estimate the size of the coefficient of kinetic friction between a hockey puck and the ice of a rink.

34. • (a) Estimate the magnitude of the coefficient of kinetic friction of a book as it slides across a tabletop. (b) Estimate the magnitude of the coefficient of static friction for the same book on the same tabletop.

35. • Estimate the magnitude of the coefficient of kinetic friction between you and the surface of a waterslide. SSM

36. • Estimate the magnitude of the coefficient of kinetic friction for a mug of root beer as it slides across a wooden bar.

37. • The following table lists the forces applied to a 1.00-kg crate at particular times. The crate begins at rest on a rough, horizontal surface until it begins to move at a constant speed in the horizontal direction. Using the data, estimate the coefficient of static friction and the coefficient of kinetic friction between the crate and the surface.

t (s)	F (N)	t (s)	F (N)
0	0	0.25	8.26
0.01	1.33	0.30	7.84
0.05	3.28	0.35	5.17
0.10	8.11	0.40	5.21
0.15	8.20	0.45	5.22
0.20	8.24	0.50	5.37

Problems

5-2 The static friction force changes magnitude to offset other applied forces

38. • What is the minimum horizontal force that will cause a 5.00-kg box to begin to slide on a horizontal surface when the coefficient of static friction is 0.670? Example 5-1

39. • A 7.60-kg object rests on a level floor with a coefficient of static friction of 0.550. What minimum horizontal force will cause the object to start sliding? SSM Example 5-1

40. • Draw a free-body diagram for the situation shown in **Figure 5-21**. An object of mass M rests on a ramp; there is friction between the object and the ramp. Example 5-1

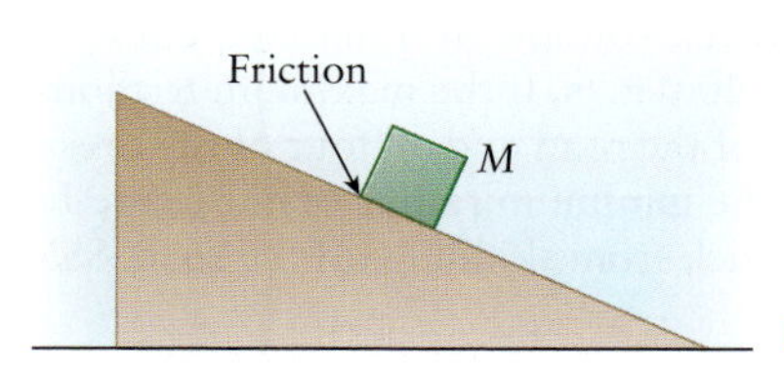

Figure 5-21 Problem 40

5-3 The kinetic friction force on a sliding object has a constant magnitude

41. • A book is pushed across a horizontal table at a constant speed. If the horizontal force applied to the book is equal to one-half of the book's weight, calculate the coefficient of kinetic friction between the book and the tabletop. SSM Example 5-2

42. • An object on a level surface experiences a horizontal force of 12.7 N due to kinetic friction. If the coefficient of kinetic friction is 0.37, what is the mass of the object? Example 5-2

43. • A 25.0-kg crate rests on a level floor. A horizontal force of 50.0 N accelerates the crate at 1.00 m/s^2. Calculate (a) the normal force on the crate, (b) the frictional force on the crate, and (c) the coefficient of kinetic friction between the crate and the floor. Example 5-2

44. • A mop is pushed across the floor with a force F of 50.0 N at an angle of $\theta = 50.0°$ (Figure 5-22). The mass of the mop head is 3.75 kg. Calculate the acceleration of the mop head if the coefficient of kinetic friction between the head and the floor is 0.400. Example 5-2

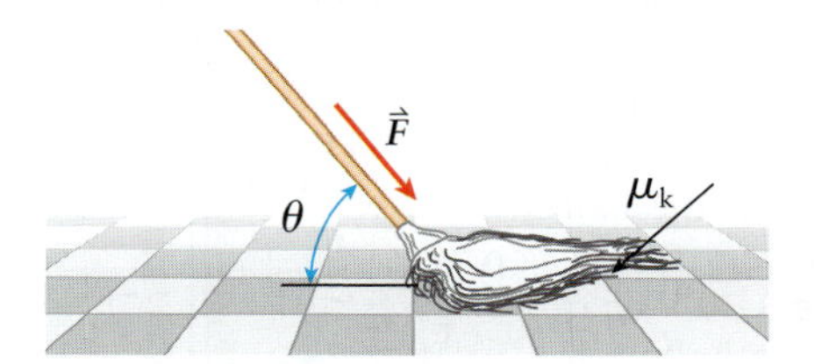

Figure 5-22 Problem 44

45. • If the coefficient of kinetic friction between an object with mass $M = 3.00$ kg and a flat surface is 0.400, what magnitude of force will cause the object to accelerate at 2.50 m/s^2? The force is applied at an angle of $\theta = 30.0°$ (Figure 5-23). Example 5-2

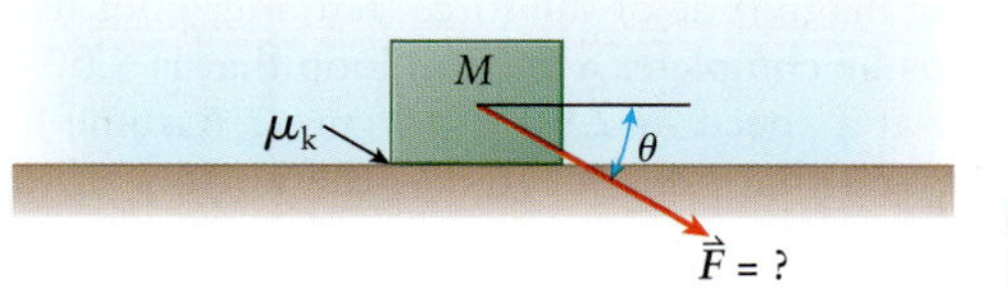

Figure 5-23 Problem 45

5-4 Problems involving static and kinetic friction are like any other problem with forces

46. •• A horizontal force $\vec{F}$ with magnitude 10.0 N is applied to a stationary block with a mass M of 2.00 kg as shown in Figure 5-24. The coefficient of static friction between the block and the floor is 0.750; the coefficient of kinetic friction is 0.450. Find the acceleration of the box. SSM Example 5-4

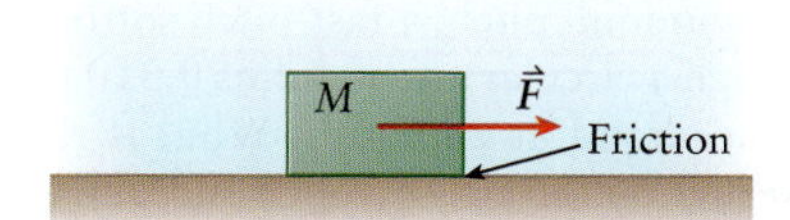

Figure 5-24 Problem 46

47. •• A taut string connects a crate with mass $M_1 = 5.00$ kg to a crate with mass $M_2 = 12.0$ kg (Figure 5-25). The coefficient of static friction between the smaller crate and the floor is 0.573; the coefficient of static friction between the larger crate and the floor is 0.443. What is the minimum magnitude of force $\vec{F}$ required to start the crates in motion? SSM Example 5-4

Figure 5-25 Problem 47

48. ••• A box of mass $M_{box} = 2.00$ kg rests on top of a crate with mass $M_{crate} = 5.00$ kg (Figure 5-26). The coefficient of static friction between the box and the crate is 0.667. The coefficient of static friction between the crate and the floor is 0.400. Calculate the minimum magnitude of force $\vec{F}$ that is required to move the crate to the right and the corresponding magnitude of the tension T in the rope that connects the box to the wall when the crate is moved. Example 5-6

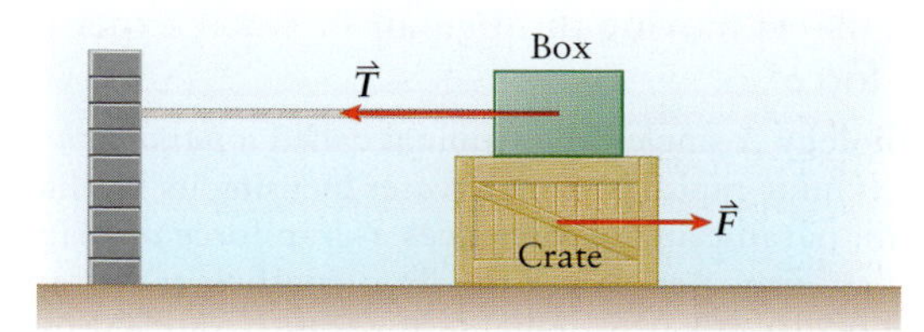

Figure 5-26 Problem 48

49. • Two blocks are connected over a massless, frictionless pulley (Figure 5-27). The mass of block 2 is 8.00 kg, and the coefficient of kinetic friction between block 2 and the incline is 0.220. The angle θ of the incline is 28.0°. Block 2 slides down the incline at constant speed. What is the mass of block 1? SSM Example 5-4

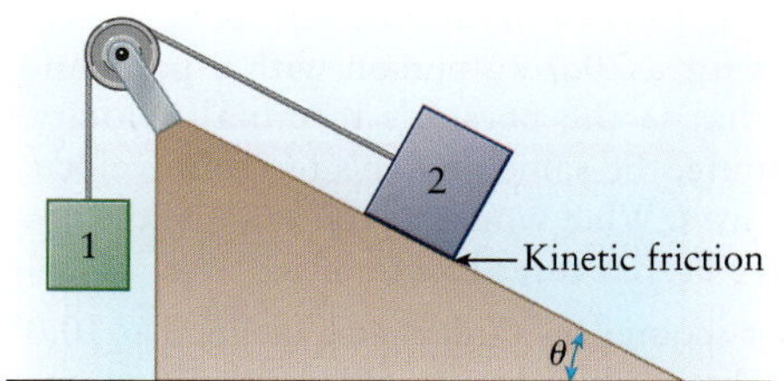

Figure 5-27 Problems 49 and 50

50. • Two blocks are connected over a massless, frictionless pulley (Figure 5-27). The mass of block 2 is 10.0 kg, and the coefficient of kinetic friction between block 2 and the incline is 0.200. The angle θ of the incline is 30.0°. If block 2 moves up the incline at constant speed, what is the mass of block 1? Example 5-4

51. •• An object with mass M_1 of 2.85 kg is held in place on an inclined plane that makes an angle θ of 40.0° with the horizontal (Figure 5-28). The coefficient of kinetic friction between the plane and the object is 0.552. A second object that has a mass M_2 of 4.75 kg is connected to the first object with a massless string over a massless, frictionless pulley. Calculate the magnitude and direction of the initial acceleration of the system

and the tension in the string once the objects are released. Example 5-4

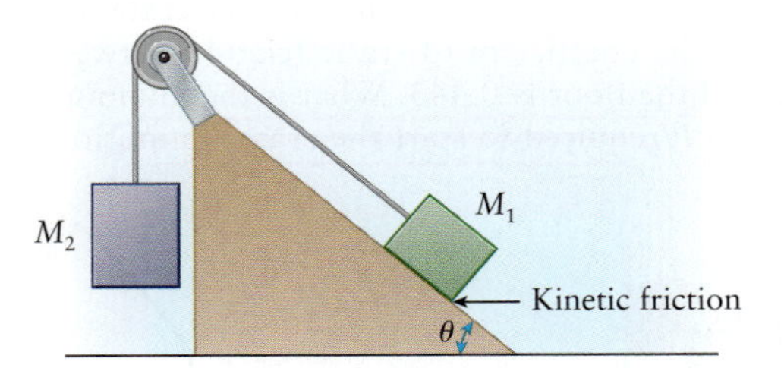

Figure 5-28 Problem 51

52. • Draw free-body diagrams for the situation shown in **Figure 5-29**. An object of mass M_2 rests on a frictionless table, and an object of mass M_1 sits on it; there is friction between the objects. A horizontal force $\vec{F}$ is applied to the lower object as shown. Example 5-6

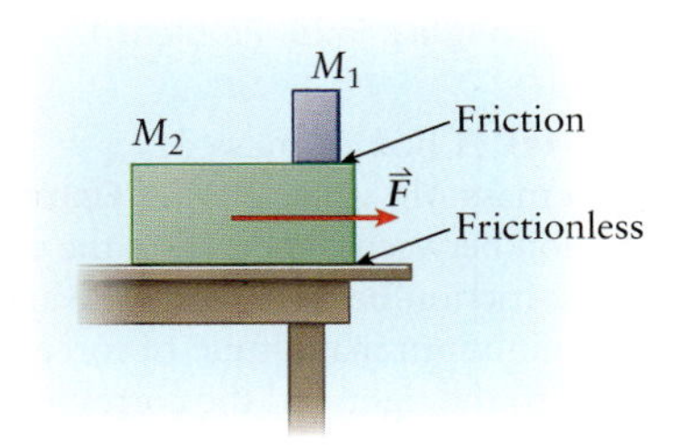

Figure 5-29 Problem 52

5-5 An object moving through air or water experiences a drag force

53. • **Biology** A single-celled animal called a *paramecium* propels itself quite rapidly through water by using its hairlike *cilia*. A certain paramecium experiences a drag force of magnitude $F_{drag} = cv^2$ in water, where the drag coefficient c is approximately 0.310. What propulsion force does this paramecium generate when moving at a constant (terminal) speed v of 0.150×10^{-3} m/s? Example 5-7

54. • **Biology** The bacterium *Escherichia coli* (*E. coli*) propels itself with long, thin structures called flagella. When its flagella exert a force of 1.50×10^{-13} N, the bacterium swims through water at a speed of 20.0 μm/s. Find the speed of the bacterium in water when the force exerted by its flagella is 3.00×10^{-13} N. Example 5-7

55. • We model the drag force of the atmosphere as $F_{drag} = cv^2$, where the value of c for a 70.0-kg person with a parachute is 18.0 kg/m. (a) What is the person's terminal velocity? (b) Without a parachute, the same person's terminal velocity would be about 50.0 m/s. What would be the value of the proportionality constant c in that case? Example 5-7

56. •• A girl rides her scooter on a hill that is inclined at 10.0° with the horizontal. The combined mass of the girl and scooter is 50.0 kg. On the way down, she coasts at a constant speed of 12.0 m/s, while experiencing a drag force that is proportional to the square of her velocity. What force, parallel to the surface of the hill, is required to increase her speed to 20.0 m/s? Neglect any other resistive forces, including the friction between the scooter and the hill. Example 5-7

5-6 In uniform circular motion, the net force points toward the center of the circle

57. • A hockey puck that has a mass of 0.170 kg is tied to a light string and spun in a circle of radius 1.25 m on frictionless ice. If the string breaks under a tension that exceeds 5.00 N, what is the maximum speed of the puck without breaking the string? Example 5-10

58. • A 1.50×10^3-kg truck rounds an unbanked curve on the highway at a speed of 20.0 m/s. If the maximum frictional force between the surface of the road and all four of the tires is 8.00×10^3 N, calculate the minimum radius of curvature for the curve to prevent the truck from skidding off the road. SSM Example 5-10

59. • A 25.0-g metal washer is tied to a 60.0-cm-long string and whirled around in a vertical circle at a constant speed of 6.00 m/s. Calculate the tension in the string (a) when the washer is at the bottom of the circular path and (b) when it is at the top of the path. Example 5-8

60. • A centrifuge spins small tubes in a circle of radius 10.0 cm at a rate of 1.20×10^3 rev/min. What is the centripetal force on a sample that has a mass of 1.00 g? SSM Example 5-10

61. • **Biology** Very high-speed ultracentrifuges are useful devices to sediment materials quickly or to separate materials. An ultracentrifuge spins a small tube in a circle of radius 10.0 cm at 6.00×10^4 rev/min. What is the centripetal force experienced by a sample that has a mass of 3.00 g? Example 5-10

62. • **Astro** What centripetal force is exerted on the Moon as it orbits about Earth at a center-to-center distance of 3.84×10^8 m with a period of 27.4 days? What is the source of the force? The mass of the Moon is equal to 7.35×10^{22} kg. Example 5-10

63. • At the Fermi National Accelerator Laboratory (Fermilab), a large particle accelerator makes protons travel in a circular orbit 6.3 km in circumference at a speed of nearly 3.0×10^8 m/s. What is the centripetal acceleration of one of the protons? Example 5-10

64. • In the game of tetherball a 1.25-m rope connects a 0.750-kg ball to the top of a vertical pole so that the ball can spin around the pole as shown in **Figure 5-30**. What is the speed of the ball as it rotates around the pole when the angle θ of the rope is 35.0° with the vertical? SSM Example 5-8

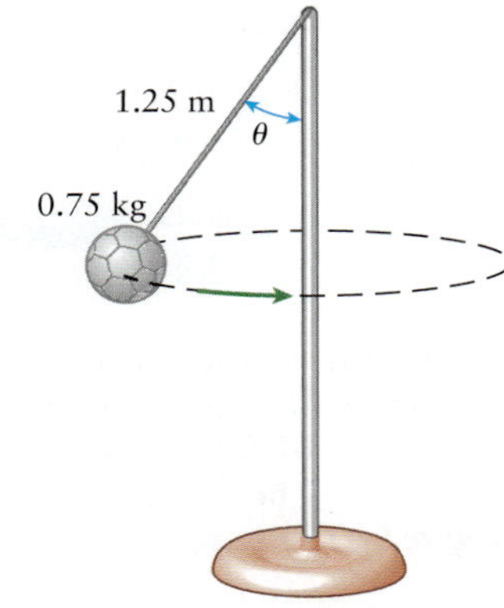

Figure 5-30 Problem 64

65. • What is the magnitude of the force that a jet pilot feels against his seat as he completes a vertical loop that is 5.00×10^2 m in radius at a speed of 2.00×10^2 m/s? Assume his mass is 70.0 kg and that he is located at the bottom of the loop. Example 5-8

66. • The radius of Earth is 6.38×10^6 m, and it completes one revolution in 1 day. What is the centripetal acceleration of an object located (a) on the equator and (b) at latitude 40.0° north? Example 5-10

67. •• A coin that has a mass of 25.0 g rests on a phonograph turntable that rotates at 78.0 rev/min. The center of the coin is 13.0 cm from the turntable axis. If the coin does not slip, what is the minimum value of the coefficient of static friction between the coin and the turntable surface? Example 5-10

68. • **Sports** In executing a windmill pitch, a fast-pitch softball player moves her hand through a circular arc of radius 0.310 m. The 0.190-kg ball leaves her hand at 24.0 m/s. What is the magnitude of the force exerted on the ball by her hand immediately before she releases it? Example 5-8

General Problems

69. ••• A block of mass M rests on a block of mass $M_1 = 5.00$ kg which is on a tabletop (**Figure 5-31**). A light string passes over a massless, frictionless pulley and connects the blocks. The coefficient of kinetic friction μ_k at both surfaces equals 0.330. A force of magnitude 60.0 N pulls the upper block to the left and the lower block to the right. The blocks are moving at a constant speed. Determine the mass of the upper block. Example 5-4

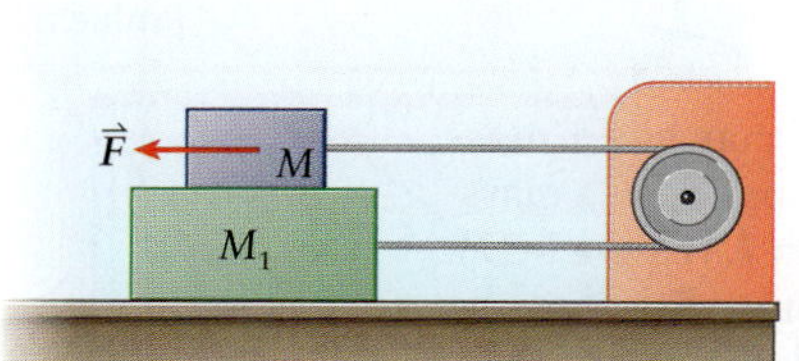

Figure 5-31
Problem 69

70. •• A wedding cake sits atop a table. A mischievous ring-bearer wants to show off his strength by lifting one end of the table. If the cake weighs 53.0 N and the coefficient of static friction between the cake base and the table is 0.400, to what angle can the table tilt before the cake starts to slide? Example 5-1

71. • You lean against a table such that your weight exerts a force of 1.50×10^2 N on the edge of the table, directed at an angle of 11.0° below a line drawn parallel to the table's surface (**Figure 5-32**). The table has a mass of 25.0 kg and the coefficient of static friction between its feet and the ground is 0.550. Does the table slide out from under you? Justify your answer with calculations. Example 5-10

Figure 5-32
Problem 71

72. • You find yourself pushing a 42.0-kg box up a 6.00° ramp into a moving truck. The coefficient of kinetic friction between the box and the metal ramp is 0.320. To save energy, you only push hard enough to move the box up the ramp at constant velocity. Assuming that your pushing force is directed parallel to the ramp, with what force do you push? Example 5-3

73. • While playing tug-of-war with your dog, you decide to pull him across the carpeted floor. Your dog has a mass of 18.5 kg, the rope toy you are pulling him with makes an angle of 23.0° with the floor, and the coefficient of kinetic friction between your dog's feet and the floor is 0.367. If you pull your dog at a constant velocity, with what force are you pulling on the rope? Example 5-3

74. • Luis absentmindedly leaves his physics textbook on the horizontal top of his 1825-kg car while getting in. The coefficients of friction between the 1.84-kg textbook and the top of the car are $\mu_s = 0.180$ and $\mu_k = 0.113$. If the car is on level ground, what is the magnitude of the maximum acceleration the car can have before Luis's textbook slides off the car? Example 5-6

75. • A librarian moonlighting as a magician wants to pull one book out of a stack of books. The stack is seven books high and all the books are of roughly equal mass M, of comparable size, and the coefficient of kinetic friction between the books is μ_k. The librarian needs the third book in the stack, counting from the bottom. The librarian gives the third book a hard, straight yank with a force of magnitude F perpendicular to the stack. In terms of F, M, g, and μ_k, what is the net horizontal force on the yanked book? Example 5-5

76. • A large block is being pushed against a smaller block such that the smaller block remains elevated while being pushed (**Figure 5-33**). The mass of the smaller block is $m = 0.75$ kg. The blocks need to have a minimum acceleration of $a = 15 \text{ m/s}^2$ in order for the smaller block to remain elevated and not slide down. What is the coefficient of static friction between the two blocks? Example 5-5

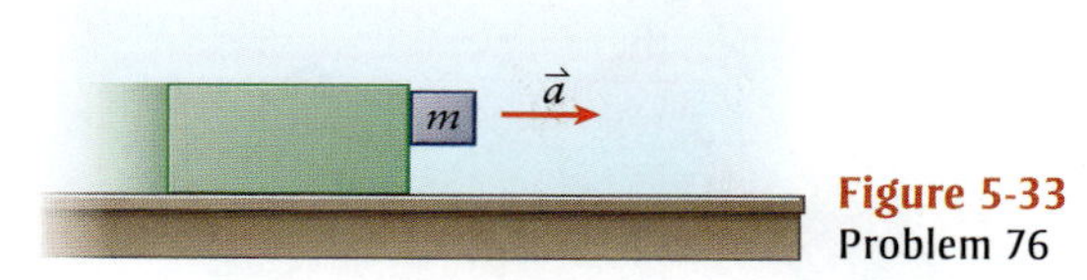

Figure 5-33
Problem 76

77. •• A 150-kg crate rests in the bed of a truck that slows from 50.0 km/h to a stop in 12.0 s. The coefficient of static friction between the crate and the truck bed is 0.655. (a) Will the crate slide during the braking period? Explain your answer. (b) What is the minimum stopping time for the truck that prevents the crate from sliding? Example 5-6

78. •• The coefficient of static friction between a rubber tire and dry pavement is about 0.800. Assume that a car's engine only turns the two rear wheels and that the weight of the car is uniformly distributed over all four wheels. (a) What limit does the coefficient of static friction place on the time required for a car to accelerate from rest to 60 mph (26.8 m/s)? (b) How can friction accelerate a car *forward* when friction *opposes* motion? SSM Example 5-2

79. •• Two blocks are connected over a massless, frictionless pulley (**Figure 5-34**). Block m_1 has a mass of 1.00 kg and block m_2 has a mass of 0.400 kg. The angle θ of the incline is 30.0°. The coefficients of static friction and kinetic friction between block m_1 and the incline are $\mu_s = 0.500$ and $\mu_k = 0.400$, respectively. What is the magnitude of the tension in the string? Example 5-5

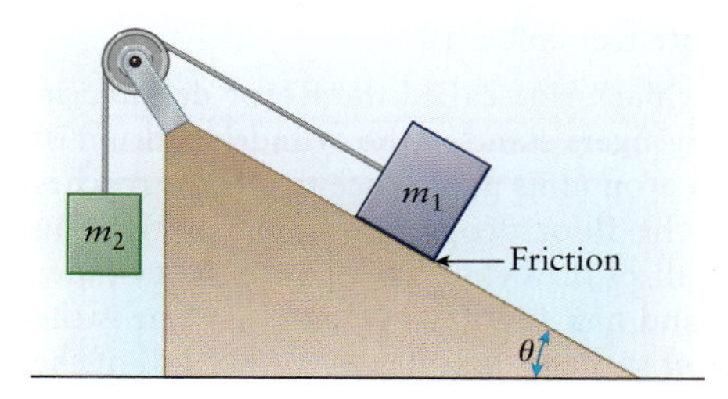

Figure 5-34
Problems 79 and 80

80. •• Two blocks are connected over a massless, frictionless pulley (Figure 5-34). Block 1 has a mass m_1 of 1.00 kg and Block 2 has a mass m_2 of 2.00 kg. The angle θ of the incline is 30.0°. The coefficients of static friction and kinetic friction between Block 1 and the incline are $\mu_s = 0.500$ and $\mu_k = 0.400$. What is the acceleration of Block 1? Example 5-4

81. • A runaway ski slides down a 250-m-long slope inclined at 37.0° with the horizontal. If the initial speed is 10.0 m/s,

how long does it take the ski to reach the bottom of the incline if the coefficient of kinetic friction between the ski and snow is (a) 0.100 and (b) 0.150? Example 5-2

82. • A 2.50-kg package slides down a 12.0-m long inclined plane that makes an angle of 20.0° with the horizontal. The package has an initial speed of 2.00 m/s at the top of the incline. What must the coefficient of kinetic friction between the package and the inclined plane be so that the package reaches the bottom with no speed? SSM Example 5-2

83. ••• In **Figure 5-35**, two blocks are connected to each other by a massless string over a massless and frictionless pulley. The mass of block 1 is $m_1 = 6.00$ kg. Assuming the coefficient of static friction $\mu_s = 0.542$ for all surfaces, find the range of values of the mass m_2 so that the system is in equilibrium. Example 5-5

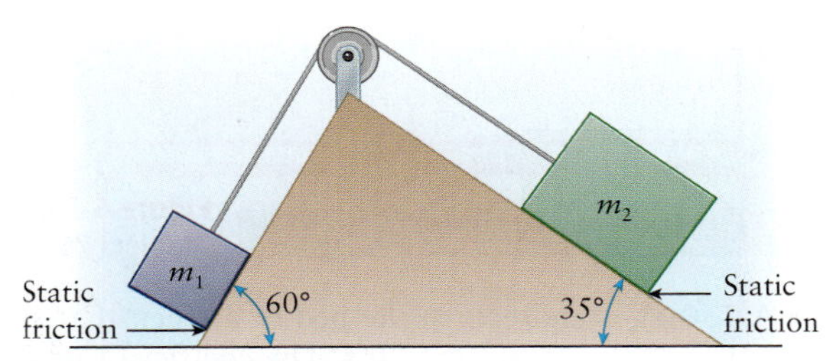

Figure 5-35 Problem 83

84. •• The terminal velocity of a raindrop that is 4.00 mm in diameter is approximately 8.50 m/s under controlled, windless conditions. The density of water is 1.00×10^3 kg/m^3. Recall that the density of an object is its mass divided by its volume. (a) If we model the air drag as being proportional to the square of the speed, $F_{\text{drag}} = cv^2$, what is the value of c? (b) Under the same conditions as above, what would be the terminal velocity of a raindrop that is 8.0 mm in diameter? Try to use your answer from part (a) to solve the problem by proportional reasoning instead of just doing the same calculation over again. Example 5-7

85.• Biomedical laboratories routinely use ultracentrifuges, some of which are able to spin at 1.00×10^5 rev/min about the central axis. The turning rotor in certain models is about 20.0 cm in diameter. At its top spin speed, what force does the rotor exert on a 2.00-g sample that is positioned at the greatest distance from the spin axis? Would the force be appreciably different if the sample were spun in a vertical or a horizontal circle? Why or why not? Example 5-10

86. •• An amusement park ride called the Rotor debuted in 1955 in Germany. Passengers stand in the cylindrical drum of the Rotor as it rotates around its axis. Once the Rotor reaches its operating speed, the floor drops but the riders remain pinned against the wall of the cylinder. Suppose the cylinder makes 25.0 rev/min and has a radius of 3.50 m. What is the minimum coefficient of static friction between the wall of the cylinder and the backs of the riders? Example 5-3

87. • An object of mass $m_1 = 0.125$ kg undergoes uniform circular motion and is connected by a massless string through a hole in a frictionless table to a larger object of mass $m_2 = 0.225$ kg (**Figure 5-36**). If the larger object is stationary, calculate the tension in the string and the speed of the circular motion of the smaller object. The radius R of the circular path is equal to 1.00 m. Example 5-10

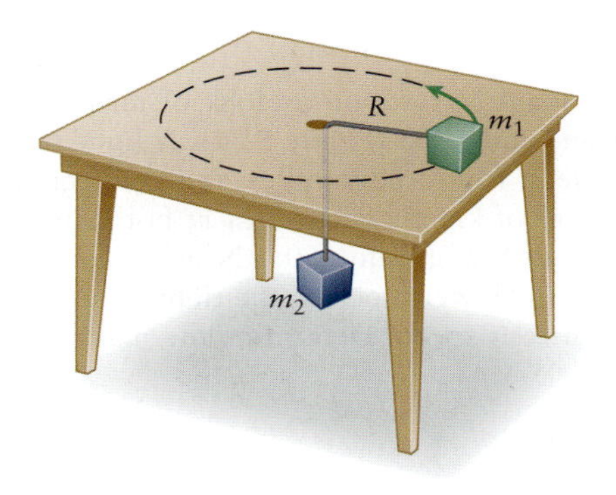

Figure 5-36 Problem 87

88. •• An object that has a mass M hangs from a support by a massless string of length L (**Figure 5-37**). The support is rotated so that the object follows a circular path at an angle θ from the vertical as shown. The object makes N revolutions per second. Derive an expression for the angle θ in terms of M, L, N, and any necessary physical constants. SSM Example 5-8

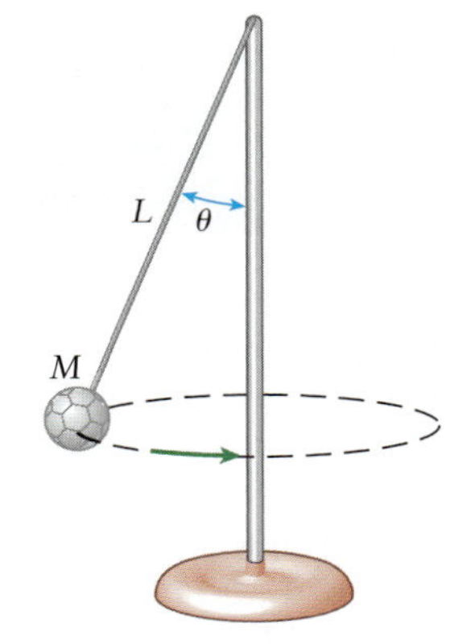

Figure 5-37 Problem 88

89. • **Medical** Occupants of cars hit from behind often suffer serious neck injury from whiplash. During a low-speed rear-end collision, a person's head suddenly pivots about the base of the neck through a 60.0° angle, a motion that lasts 250 ms. The distance from the base of the neck to the center of the head is typically about 0.20 m, and the head normally comprises about 6.0% of body weight. We can model the motion of the head as having uniform speed over the course of its pivot. (a) What is the acceleration of the head during the collision? (b) What force (in newtons and in pounds) does the neck exert on the head of a 75-kg person in the collision? (As a first approximation, neglect the force of gravity on the head.) (c) Would headrests mounted to the backs of the car seats help protect against whiplash? Why or why not? Example 5-10

90. • A curve that has a radius of 1.00×10^2 m is banked at an angle of 10.0° (**Figure 5-38**). If a 1.00×10^3-kg car navigates the curve at 65.0 km/h without skidding, what is the minimum coefficient of static friction between the pavement and the tires? Example 5-11

Figure 5-38 Problem 90

6 Work and Energy

John Warburton-Lee/Danita Delimont Stock Photography

What do you think?

The ostrich (*Struthio camelus*) is the world's largest bird, with twice the mass of an adult human. Despite its size, an ostrich can run at a steady 50 km/h (14 m/s, or 31 mi/h) for extended periods. Which anatomical structure is most important for sustaining a running ostrich's motion? (a) The ostrich's long neck; (b) its feathers; (c) its tendons; (d) its clawed feet.

In this chapter, your goals are to:

- (6-1) Explain the relationship between work and energy.
- (6-2) Calculate the work done by a constant force on an object moving in a straight line.
- (6-3) Describe what kinetic energy is and understand the work-energy theorem.
- (6-4) Apply the work-energy theorem to solve problems.
- (6-5) Recognize why the work-energy theorem applies even for curved paths and varying forces like the spring force.
- (6-6) Explain the meaning of potential energy and how conservative forces such as the gravitational force and the spring force give rise to gravitational potential energy and spring potential energy.
- (6-7) Explain the differences between, and when you can apply, the generalized law of conservation of energy and the conservation of total mechanical energy.
- (6-8) Identify which kinds of problems are best solved with energy conservation and the steps to follow in solving these problems.
- (6-9) Describe what power is and its relationship to work and energy.

To master this chapter, you should review:

- (2-5) The equations for constant-acceleration motion in a straight line.
- (4-2, 4-5) Newton's second and third laws of motion.

6-1 The ideas of work and energy are intimately related

Energy is used in nearly every living process—moving, breathing, circulating blood, digesting food, and absorbing nutrients. There are different kinds of energy, including kinetic, potential, and internal. Energy can be transformed from one of these kinds into another. But what *is* energy?

In this chapter we'll see that if an object or system has **energy**, then that object or system may be able to do *work*. Work has many different meanings in everyday life. In the language of physics, however, work has a very specific definition that describes what happens when a force is exerted on an object as it moves. One example of doing work is lifting a book from your desk to a high bookshelf; another is a football player pushing a blocking sled across the field (**Figure 6-1a**). By combining the definition of work with our knowledge of Newton's second law, we'll be led to the idea of *kinetic*

The football player applies a force F to the sled while it moves a distance d. Hence he does an amount of work on the sled equal to Fd.
(a)
F
d
KEVIN WOLF/AP Images

The girl does work on the basketball: She exerts a force on the ball as she pushes it away from her. As a result the ball acquires kinetic energy (energy of motion).
(b)
Jens Karlsson/Getty Images

(c)
Sampics/Corbis
As the ball hits its target and slows down, work is done on the target. Some kinetic energy goes into deforming the ball and giving it elastic potential energy, which converts again to kinetic energy when the ball springs back.

Figure 6-1 Work and energy In this chapter we'll explore the ideas of (a) work, (b) kinetic energy, and (c) potential energy.

energy, which is the energy that an object has due to its motion (**Figure 6-1b**). An object with kinetic energy has the ability to do work: For example, a moving ball has the ability to displace objects in its path (**Figure 6-1c**).

Sometimes kinetic energy is lost—it gets converted to a form of energy that can't be used to do work—such as when an egg thrown at high speed splatters against a wall. But if that egg is replaced with a rubber ball, the ball bounces off with nearly the same speed at which it was thrown and so we recover most of the ball's kinetic energy. We say that as the ball compresses against the wall, its kinetic energy is converted into *potential energy*—energy associated not with the ball's motion but with its shape and position. This potential energy is again converted into kinetic energy as the ball bounces back (Figure 6-1c).

In this chapter we'll look at the ways in which kinetic energy, potential energy, and *internal energy*—energy stored within an object, sometimes in a way that can't easily be extracted—can transform into each other. (As an example, the energy you need to make it through today is extracted from the internal energy of food that you consumed earlier.) In this way we'll discover the *law of conservation of energy*, which proves to be one of the great unifying principles of the natural sciences in general.

TAKE-HOME MESSAGE FOR Section 6-1

✔ A force acting on an object in motion can do work on that object.

✔ An object or system's energy is related to its ability to do work.

✔ There are different kinds of energy, including kinetic, potential, and internal. Energy can be transformed from one of these kinds into another.

6-2 The work that a constant force does on a moving object depends on the magnitude and direction of the force

The man depicted in **Figure 6-2** is doing *work* as he pushes a crate up a ramp. The amount of work that he does depends not only on how hard he pushes on the crate (that is, on the magnitude of the force that he exerts) but also on the distance over which he moves the crate (that is, the displacement of the crate). In a similar way, the football players shown in Figure 6-1a will be more exhausted if the coach asks them to push the blocking sled all the way down the field rather than a short distance.

These examples suggest how we should define the **work** done by a force on an object. Suppose a constant force $\vec{F}$ acts on an object as it moves through a straight-line displacement $\vec{d}$, and the force $\vec{F}$ is in the same direction as $\vec{d}$. Then the work done by

If the force $\vec{F}$ the man exerts on the crate is in the same direction as the displacement $\vec{d}$ of the crate, the work W that he does on the crate is the product of force and displacement: $W = Fd$.

Figure 6-2 Work is force times distance Calculating how much work the man does to push the crate up the ramp.

the force equals the product of the force magnitude F and the distance d over which the object moves:

Work done on an object by a constant force $\vec{F}$ that points in the **same direction** as the object's displacement $\vec{d}$

Magnitude of the constant force $\vec{F}$

$$W = Fd$$

Magnitude of the displacement $\vec{d}$

Work done by a constant force that points in the same direction as the straight-line displacement (6-1)

Note that Equation 6-1 refers only to situations in which the force acts in the *same* direction as the displacement. You've already seen two situations of this sort: The football players in Figure 6-1a push the sled backward as it moves backward, and the man in Figure 6-2 pushes the crate uphill as it moves uphill. Later we'll consider the case in which force and displacement are *not* in the same direction.

We saw in Chapters 4 and 5 that it's important to keep track of what object *exerts* a given force and on what object that force is exerted. It's equally important to keep track of both the object that exerts a force and the object on which the force does work. For example in Figure 6-2, the object exerting a force is the man, and the object on which the force does work is the crate.

The unit of work, the **joule** (J), is named after the nineteenth-century English physicist James Joule, who did fundamental research on the relationship between motion and work. From Equation 6-1,

$$1 \text{ J} = 1 \text{ N} \times 1 \text{ m or } 1 \text{ J} = 1 \text{ N} \cdot \text{m}$$

You do 1 J of work when you exert a 1-N push on an object as it moves through a distance of 1 m.

WATCH OUT! Work and weight have similar symbols.

! Since *work* and *weight* begin with the same letter, it's important to use different symbols to represent them in equations. We'll use an uppercase W for work and a lowercase w for weight (the magnitude of the gravitational force), and we recommend that you do the same to prevent confusion.

EXAMPLE 6-1 Lifting a Book

How much work must you do to lift a textbook with a mass of 2.00 kg—roughly the mass of the printed version of this book—by 5.00 cm? (You lift the book at a constant speed.)

Set Up

Newton's first law tells us that the net force on the book must be zero if it is to move upward at a constant speed. Hence the upward force you apply must be constant and equal in magnitude to the gravitational force on the book. Since the applied force is constant and in the direction in which the book moves, we can use Equation 6-1 to calculate the work done.

Work done on an object, force in the same direction as displacement:

$$W = Fd \quad (6\text{-}1)$$

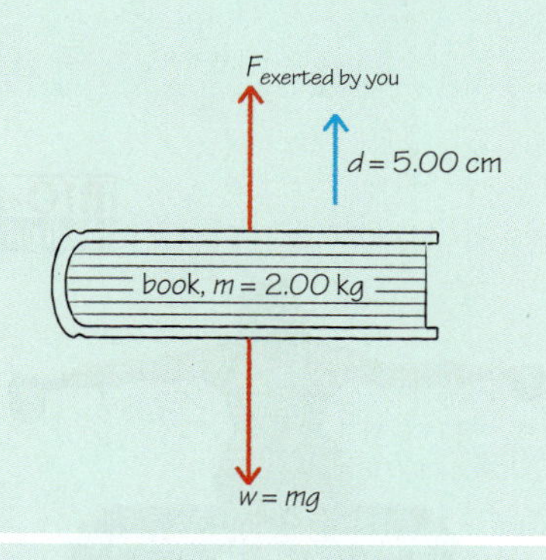

Solve

Calculate the magnitude F of the force that you exert, then substitute this and the displacement $d = 5.00$ cm into Equation 6-1 to determine the work that you do.

$$F = mg = (2.00 \text{ kg})(9.80 \text{ m/s}^2) = 19.6 \text{ N}$$

$$W = Fd = (19.6 \text{ N})(5.00 \text{ cm}) = (19.6 \text{ N})(0.0500 \text{ m}) = 0.980 \text{ N} \cdot \text{m} = 0.980 \text{ J}$$

Reflect

The work that you do is almost exactly 1 joule. The actual amount of energy that you would need to *expend* is several times more than 0.980 J, however. That's because your body isn't 100% efficient at converting energy into work. Some of the energy goes into heating your muscles. In fact, your muscles can consume energy even when they do *no* work, as we describe below. The value of 0.980 J that we calculated is just the amount of work that you do *on the book*.

BIO-Medical EXAMPLE 6-2 Work Done by Actin

In order to fertilize eggs, the sperm of the horseshoe crab (**Figure 6-3**) must penetrate two protective layers of the egg with a combined thickness of about 40 μm = 40×10^{-6} m. To achieve this, a bundle of the protein actin on the outer surface of the sperm pushes through the egg's protective layers with a constant force of 1.9×10^{-9} N. How much work does the actin bundle do in this process?

J Hindman/Shutterstock

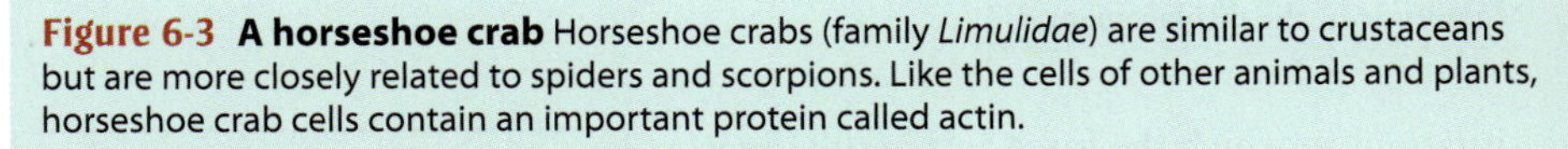
Figure 6-3 A horseshoe crab Horseshoe crabs (family *Limulidae*) are similar to crustaceans but are more closely related to spiders and scorpions. Like the cells of other animals and plants, horseshoe crab cells contain an important protein called actin.

Set Up

The force exerted is both constant and in the same direction as the motion of the end of the bundle. Hence we can once again use Equation 6-1.

$$W = Fd \qquad (6\text{-}1)$$

Solve

The work done by the actin bundle equals the force exerted by the actin bundle multiplied by the distance that it pushes through the outer layers of the egg.

$$W = Fd$$
$$= (1.9 \times 10^{-9}\ \text{N})(40 \times 10^{-6}\ \text{m})$$
$$= 7.6 \times 10^{-14}\ \text{J}$$

Reflect

The amount of work done by the actin bundle seems ridiculously small. But such a bundle is microscopic, with a length of only 60 μm (slightly greater than the thickness of the layers of the egg that it must penetrate) and a mass of only about 10^{-16} kg. Hence the actin bundle does about 10^3 J of work per kilogram of mass. If a 70-kg person could do that much work on a per-kilogram basis on the book in Example 6-1, he or she could lift the book 3.5 km (nearly 12,000 ft)! An actin bundle is small, but it packs a big punch.

$$\frac{\text{Work done by actin bundle}}{\text{Mass of actin bundle}} = \frac{7.6 \times 10^{-14}\ \text{J}}{10^{-16}\ \text{kg}} = 10^3\ \text{J/kg}$$

If a 70-kg person could do that much work on a per-kilogram basis,

$$\text{Work done} = W = (10^3\ \text{J/kg})(70\ \text{kg}) = 7.0 \times 10^4\ \text{J}$$

Height d to which this amount of work could raise a book of mass 2.00 kg and weight 19.8 N:

$$W = mgd, \text{ so } d = \frac{W}{mg} = \frac{7.0 \times 10^4\ \text{J}}{19.8\ \text{N}} = 3.5 \times 10^3\ \text{m} = 3.5\ \text{km}$$

Muscles and Doing Work

BIO-Medical

Figure 6-4 Getting tired while doing zero work Weights that you hold stationary in your outstretched arms undergo no displacement, so you do zero work on them. Why, then, do your arms get tired?

Pick up a heavy object and hold it in your hand at arm's length (**Figure 6-4**). After a while you'll notice your arm getting tired: It feels like you're doing work to hold the object in midair. But Equation 6-1 says that you're doing *no* work on the object because it isn't moving (its displacement d is zero). So why does your arm feel tired?

To see the explanation for this seeming paradox, notice that the muscles in your arm (and elsewhere in your body) exert forces on their ends by contracting. Skeletal muscles (those that control the motions of your arms, legs, fingers, toes, and other structures) are made up of bundles of muscle cells (**Figure 6-5a**). Muscle cells consist of bundles of myofibrils that are segmented into thousands of tiny structures called sarcomeres. Connected to each other, end to end, sarcomeres contain interdigitated filaments of the contractile proteins actin and myosin (**Figure 6-5b**). As the filaments slide past each other the sarcomeres can shorten, resulting in the muscle contracting; when the muscle relaxes, sarcomeres lengthen.

The reason your arm tires while holding a heavy object is that shortening the sarcomeres and maintaining the contraction against the load of the object requires energy. Metabolic pathways convert energy stored in the chemical bonds of fat and sugar to a form that your muscle cells can use, but these reactions generate waste products that change the conditions in the muscle. This leads to the feeling that you've been doing work—even though you're doing no work on the *object* you're holding.

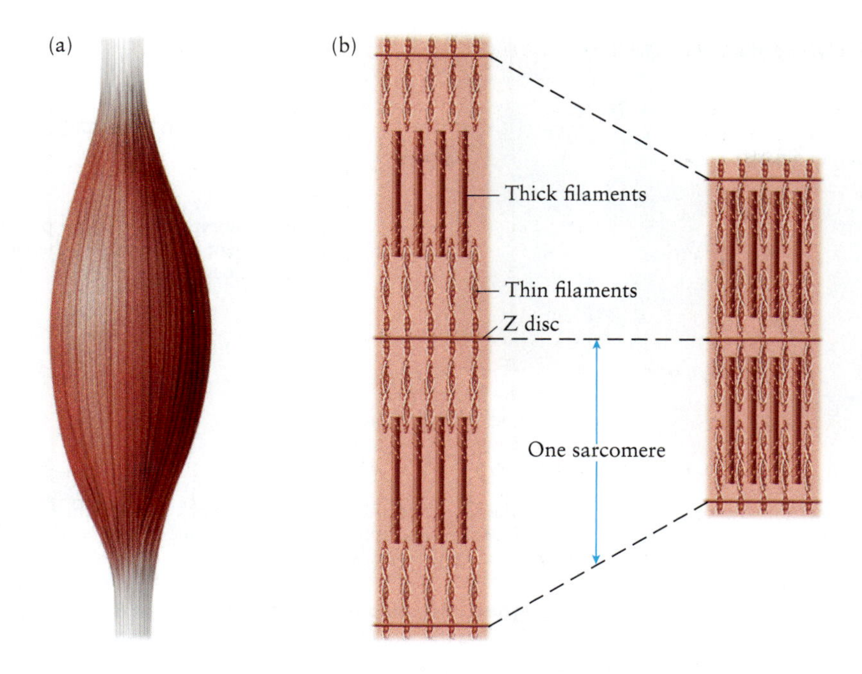

Figure 6-5 Muscles and muscle function (a) The muscles of your body are composed of many individual muscle fibers. (b) The internal structure of a muscle fiber when relaxed (left) and contracted (right).

If you again hold the object in one hand, but now rest your elbow on a table, you'll be able to hold the object for a longer time without fatigue. That's because in this case you're using only the muscles of your hand and forearm, not those of your upper arm. Fewer muscle fibers have to stay contracted, fewer crossbridges between filaments have to be reattached to maintain the contraction, and you expend less energy.

Work by Forces Not Parallel to Displacement and Negative Work

How can we calculate the work done by a constant force that is *not* in the direction of the object's motion? As an example, in **Figure 6-6a** a groundskeeper is using a rope to pull a screen across a baseball diamond to smooth out the dirt. The net tension force $\vec{F}$ that the rope exerts on the screen is at an angle θ with respect to the displacement $\vec{d}$ of the screen, so only the *component* of the force along the displacement does work on the screen (**Figure 6-6b**). This component is $F\cos\theta$, so the amount of work done by the force is

See the Math Tutorial for more information on trigonometry.

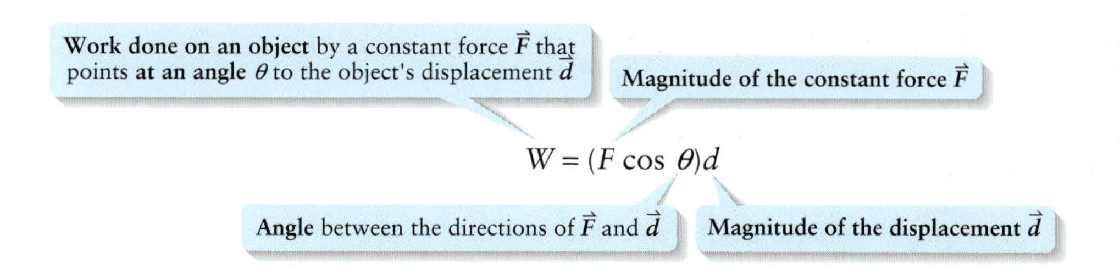

$$W = (F\cos\theta)d$$

Calculating the work done by a constant force at an angle θ to the straight-line displacement (6-2)

To understand Equation 6-2, let's look at some special cases.

- If $\vec{F}$ is in the same direction as the motion, $\theta = 0$ and $\cos\theta = \cos 0 = 1$. This is the same situation as shown in Figure 6-2, and in this case Equation 6-2 gives the same result as Equation 6-1: $W = Fd$.
- If the angle θ between the force and displacement is more than 0 but less than 90°, as in Figure 6-6b, then $\cos\theta$ is less than 1 but still positive. The work W done by the force $\vec{F}$ is positive but less than Fd.

(a)

The man exerts a force $\vec{F}$ on the screen. The direction of the force is not parallel to the ground.

$\vec{F}$

$\vec{d}$

As he exerts the force, the screen undergoes a displacement $\vec{d}$ along the ground.

(b)

Screen

$\vec{F}$

θ

$F \cos\theta$

$\vec{d}$

Only the component of force parallel to the displacement contributes to the work that the man does on the screen.

If a constant force $\vec{F}$ acts on an object at an angle θ to the object's displacement $\vec{d}$, the work done by the force equals the force component $F\cos\theta$ parallel to the displacement multiplied by the displacement d: $W = (F\cos\theta)\,d$.

Figure 6-6 When force is not aligned with displacement (a) A man exerts a force on a screen to pull it across a baseball field. (b) Finding the work that the man does on the screen (seen from the side).

A lazy dog exerts a downward force on the screen.

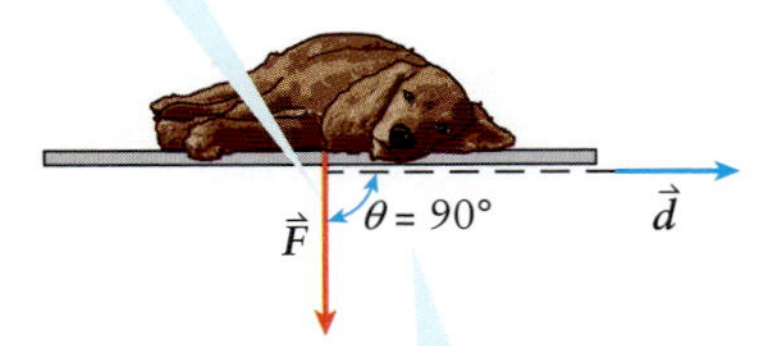

The angle between the force and the displacement $\vec{d}$ of the screen is 90° (the force and displacement are perpendicular). Hence the force has zero component in the direction of $\vec{d}$ and does zero work.

Figure 6-7 Doing zero work A dog resting atop the screen exerts a force on the screen as it slides but does zero work on the screen.

- If $\vec{F}$ is perpendicular to the direction of motion, $\theta = 90°$ and $\cos\theta = \cos 90° = 0$. In this case Equation 6-2 tells us that force $\vec{F}$ does *zero* work. An example is the force exerted by a lazy dog lying on top of the screen from Figure 6-6a (**Figure 6-7**). The dog exerts a downward force on the screen as the screen moves horizontally, so $\theta = 90°$ and the lazy dog does no work at all.
- If the angle θ in Equation 6-2 is greater than 90°, the value of $\cos\theta$ is negative. In this case force $\vec{F}$ does **negative work:** $W < 0$. As an example, **Figure 6-8** shows a cart rolling along the floor as a person tries to slow it down. The force $\vec{F}$ that the person exerts on the cart is directed opposite to the cart's displacement $\vec{d}$, so the angle θ in Equation 6-2 is 180°. Since $\cos 180° = -1$, this means that the work that the person does on the cart is $W_{\text{person on cart}} = -Fd$, which is negative.

What does it mean to do negative work on an object? Remember that when the force on an object (and hence its acceleration) points in the opposite direction of the object's velocity and displacement, the object slows down (see Figure 2-10). So if you want to slow an object down, you must do negative work on it. Conversely, if you do positive work on an object, you can speed it up.

(1) As the person tries to make the cart slow down, the cart and the person's hands move together to the right (they have the same displacement).

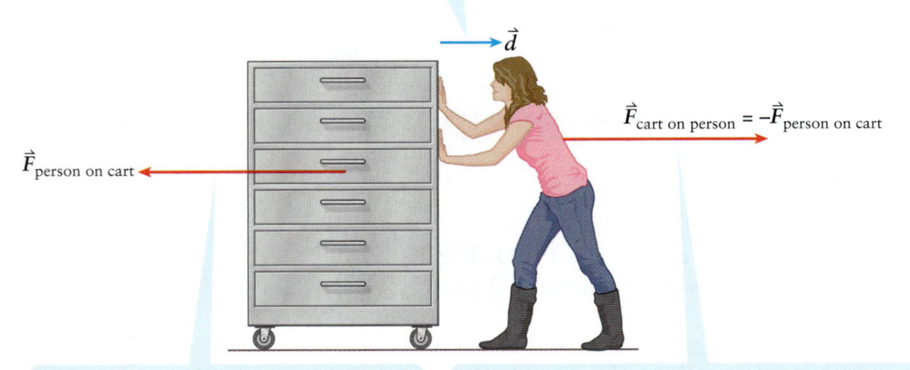

(2) The force of the person on the cart is opposite to the cart's displacement. Hence $\theta = 180°$, $\cos\theta = -1$, and the person does negative work on the cart.

(3) By Newton's third law, the cart exerts an equally strong force on the person, but in the opposite direction — that is, in the same direction of the displacement of the person's hands. So the cart does positive work on the person.

If one object (like the person) does negative work on a second object (like the cart), the second object does an equal amount of positive work on the first object.

Figure 6-8 Doing negative work As the cart rolls to the right, the person pushes on the cart to the left in order to make it slow down. As a result, she does negative work on the cart.

What else happens when you do negative work on an object? Newton's third law provides the answer: If the person in Figure 6-8 exerts a force $\vec{F}$ on the cart, the cart exerts a force $-\vec{F}$ on her (the same magnitude of force but in the opposite direction). The hands of the person have the same displacement $\vec{d}$ as the cart, and the force the cart exerts on her is in the *same* direction as her displacement ($\theta = 0$ and $\cos\theta = +1$ in Equation 6-2), so the work that the *cart* does on the *person* is $W_{\text{cart on person}} = +Fd$. *If object A does negative work on object B, then object B does an equal amount of positive work on object A.* For example, when a moving cue ball hits a stationary 8-ball on a pool table, the cue ball does *positive* work on the 8-ball: It pushes the 8-ball forward (in the direction that the 8-ball moves) and makes the 8-ball speed up. The 8-ball does *negative* work on the cue ball: It pushes back on the cue ball (in the direction opposite to the cue ball's motion) and makes the cue ball slow down.

Table 6-1 summarizes the relationship between the angle at which a force acts on an object and the work the force does on the object.

TABLE 6-1 Work Done By a Constant Force

If the angle between a force $\vec{F}$ on an object and the displacement $\vec{d}$ of the object is...	... then the work done by the force is...
Less than 90°	positive
90°	zero
More than 90°	negative

Calculating Work Done by Multiple Forces

What if more than one force acts on an object as it moves? Then the *total* work done on the object is the *sum* of the work done on the object by each force individually. For example, four forces act on the screen in Figure 6-6a: the tension in the rope, the gravitational force, the normal force, and the force of friction. The free-body diagram in **Figure 6-9** shows all of these forces, as well as the displacement vector $\vec{d}$. Note that we draw $\vec{d}$ to one side in the diagram to avoid confusing it with the forces and that the direction of motion is in the positive x direction as indicated by the dashed arrow.

We use Equation 6-2 to determine the work that each force does on the screen:

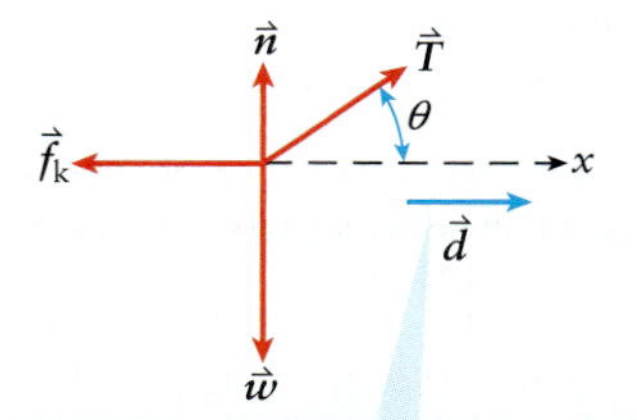

$\vec{n}$
$\vec{T}$
θ
$\vec{f}_k$
x
$\vec{d}$
$\vec{w}$
Always draw the displacement vector $\vec{d}$ to one side so you don't confuse it with the force vectors.

Figure 6-9 Calculating the work done by multiple forces The free-body diagram for the screen shown in Figure 6-6a. To find the work done by each force, draw the displacement vector $\vec{d}$ in the free-body diagram.

Tension force: This force $\vec{T}$ (which we labeled $\vec{F}$ in Figure 6-6a) acts at an angle θ with respect to the direction of motion, so $W_{\text{tension}} = Td\cos\theta$.

Gravitational force: The gravitational force $\vec{w}$ is perpendicular to the direction of motion, so the angle θ in Equation 6-2 is 90° for this force. Since $\cos 90° = 0$, the work done by the gravitational force is $W_{\text{gravity}} = 0$.

Normal force: The normal force $\vec{n}$ is also perpendicular to the direction of motion. So like the gravitational force, the normal force does no work at all: $W_{\text{normal}} = 0$.

Kinetic friction force: Because friction opposes sliding, the kinetic friction force vector $\vec{f}_k$ points in the direction opposite to the motion. The angle to use in Equation 6-2 is therefore 180°, and $W_{\text{friction}} = f_k d\cos 180° = -f_k d$. *When a force acts to oppose the motion of the object, the value of the work done by that force is negative.*

The total work done on the screen by all four forces is the sum of these:

$$W_{\text{total}} = W_{\text{tension}} + W_{\text{gravity}} + W_{\text{normal}} + W_{\text{friction}} = Td\cos\theta + 0 + 0 + (-f_k d)$$
$$= Td\cos\theta - f_k d = (T\cos\theta - f_k)d$$

If the horizontal component $T\cos\theta$ of the tension force (which does positive work) is greater in magnitude than the kinetic friction force f_k (which does negative work), the total work done on the screen is *positive* and the screen *speeds up*. But if $T\cos\theta$ is less than f_k, the total work done on the screen is *negative* and the screen *slows down*.

EXAMPLE 6-3 Up the Hill

You need to push a box of supplies (weight 225 N) from your car to your campsite, a distance of 6.00 m up a 5.00° incline. You exert a force of 85.0 N parallel to the incline, and a 56.0-N kinetic friction force acts on the box. Calculate (a) how much work you do on the box, (b) how much work the force of gravity does on the box, (c) how much work the friction force does on the box, and (d) the net work done on the box by all forces.

Set Up

Since this problem involves forces, we draw a free-body diagram for the box. The forces do not all point along the direction of the box's displacement $\vec{d}$, so we'll have to use Equation 6-2 to calculate the work done by each force.

Work done by a constant force, straight-line displacement:

$$W = Fd\cos\theta \quad (6\text{-}2)$$

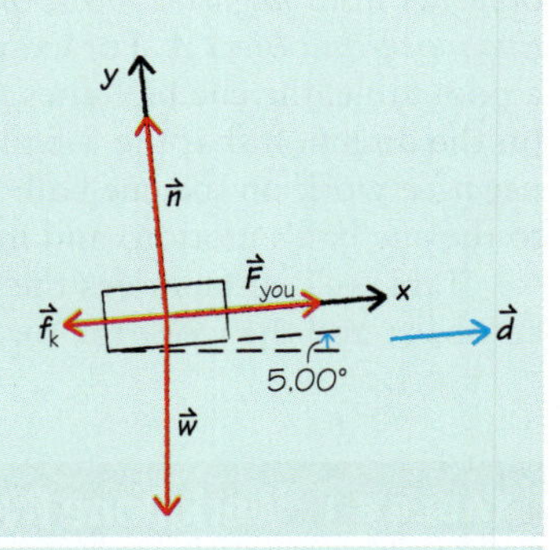

Solve

(a) The force you exert on the box is in the same direction as the box's displacement, so $\theta = 0°$ in Equation 6-2.

$$\begin{aligned} W_{you} &= F_{you}d\cos 0° \\ &= (85.0\ \text{N})(6.00\ \text{m})(1) \\ &= 5.10 \times 10^2\ \text{N} \cdot \text{m} \\ &= 5.10 \times 10^2\ \text{J} \end{aligned}$$

(b) The angle between the gravitational force and the displacement is $\theta = 90.0° + 5.00° = 95.0°$. Since this is more than 90°, $\cos\theta$ is negative and the gravitational force does negative work on the box.

$$\begin{aligned} W_{grav} &= wd\cos 95.0° \\ &= (225\ \text{N})(6.00\ \text{m})(-0.0872) \\ &= -1.18 \times 10^2\ \text{J} \end{aligned}$$

(c) The friction force points opposite to the displacement, so for this force $\theta = 180°$ and $\cos\theta = -1$.

$$\begin{aligned} W_{friction} &= f_k d\cos 180° \\ &= (56.0\ \text{N})(6.00\ \text{m})(-1) \\ &= -3.36 \times 10^2\ \text{J} \end{aligned}$$

(d) The net work is the sum of the work done by all four forces that act on the box. The normal force points perpendicular to the displacement, so it does zero work (for this force, $\theta = 90°$ and $\cos\theta = 0°$).

$$\begin{aligned} W_{total} &= W_{you} + W_{grav} + W_{friction} + W_n \\ &= (5.10 \times 10^2\ \text{J}) + (-1.18 \times 10^2\ \text{J}) \\ &\quad + (-3.36 \times 10^2\ \text{J}) + 0 \\ &= 0.56 \times 10^2\ \text{J} = 56\ \text{J} \end{aligned}$$

Reflect

To check our result, let's calculate how much work is done by the *net* force that acts on the box. Our discussion of Equation 6-2 tells us that we only need the component of the net force in the direction of the displacement, which in our figure is the positive x component. So the work done by the net force is just $\sum F_x$ multiplied by d.

We learn two things from this. One, the work done by the net force has the *same* value as the sum of the work done by the individual forces. Two, in this case the net force is in the direction of the displacement, which means that the box picks up speed as it moves *and* the net work done on the box is positive. We'll use both of these observations in the next section to relate the net work done on an object to the change in its speed.

Net force up the incline:

$$\begin{aligned} \sum F_x &= F_{you} - f_k - w\sin 5.00° \\ &= 85.0\ \text{N} - 56.0\ \text{N} - (225\ \text{N})(0.0872) \\ &= 9.4\ \text{N} \end{aligned}$$

(positive, so the net force is uphill)

Work done by net force:

$$\begin{aligned} W_{net} &= \left(\sum F_x\right)d \\ &= (9.4\ \text{N})(6.00\ \text{m}) \\ &= 56\ \text{J} \end{aligned}$$

This equals the sum of the work done by the four forces individually.

GOT THE CONCEPT? 6-1 Work: Positive, Negative, or Zero?

? For each of the following cases, state whether the work you do on the baseball is positive, negative, or zero. (a) You catch a baseball and your hand moves backward as you bring the ball to a halt; (b) you throw a baseball; (c) you carry a baseball in your hand as you ride your bicycle in a straight line at constant speed; (d) you carry a baseball in your hand as you ride your bicycle in a straight line at increasing speed.

TAKE-HOME MESSAGE FOR Section 6-2

✔ If a force acts on an object that undergoes a displacement, the force can do work on that object.

✔ For a constant force and straight-line displacement, the amount of work done equals the displacement multiplied by the component of the force parallel to that displacement.

✔ Whether the work done is positive, negative, or zero depends on the angle between the direction of the force and the direction of the displacement.

✔ If one object does negative work on a second object, the second object must do an equal amount of positive work on the first object.

6-3 Kinetic energy and the work-energy theorem give us an alternative way to express Newton's second law

In Example 6-3 we considered a box being pushed uphill that gained speed as it moved. We found that the total amount of work being done on this box was positive. Let's now show that there's a *general* relationship between the total amount of work done on an object and the change in that object's speed. To find this important relationship, we'll combine our definition of work from Section 6-2 with what we learned about one-dimensional motion in Chapter 2 and our knowledge of Newton's laws from Chapter 4.

Let's begin by considering an object that moves along a straight line that we'll call the x axis, traveling from initial coordinate x_i to final coordinate x_f with a constant acceleration a_x (**Figure 6-10**). Equation 2-11 gives us a relation between the object's velocity v_{ix} at x_i and its velocity v_{fx} at x_f:

$$v_{fx}^2 = v_{ix}^2 + 2a_x(x_f - x_i) \qquad (6\text{-}3)$$

In Equation 6-3 v_{fx}^2 is the square of the velocity at x_f, but it also equals the square of the object's *speed* v_f at x_f. That's because v_{fx} is equal to $+v_f$ if the object is moving in the positive x direction and equal to $-v_f$ if moving in the negative x direction. In either case, $v_{fx}^2 = v_f^2$. For the same reason $v_{ix}^2 = v_i^2$, where v_i is the object's speed at x_i. So we can rewrite Equation 6-3 as

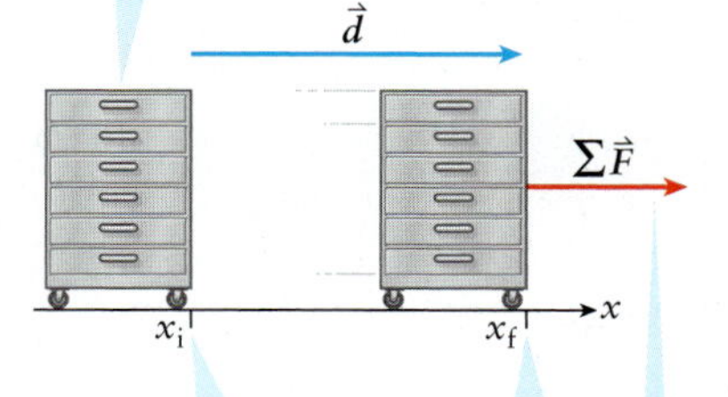

1 A cart of mass m rolls along a straight line that we'll call the x axis.
d
ΣF
x
x_i
x_f
2 The cart slides from initial position x_i to final position x_f.
3 As the cart moves, a constant net force in the x direction acts on the cart.

Figure 6-10 Deriving the work-energy theorem A cart moves along a straight line, traveling from initial coordinate x_i to final coordinate x_f. The net force is constant, so the cart has a constant acceleration a_x.

Relating speed, acceleration, and position for straight-line motion with constant acceleration

Speed at position x_f of an object in linear motion with constant acceleration

Speed at position x_i of the object

$$v_f^2 = v_i^2 + 2a_x\,(x_f - x_i) \qquad (6\text{-}4)$$

Constant acceleration of the object

Two positions of the object

In Equation 6-4 $x_f - x_i$ is the displacement of the object along the x axis. From Equation 6-2, if we multiply this by the x component of the net force on the object, $\sum F_x$—that is, by the net force component in the direction of the displacement—we get W_{net}, the work done by the net force on the object. (This assumes that the net force on the object is constant, which is consistent with our assumption that the object's acceleration is constant.) That is,

$$W_{net} = \left(\sum F_x\right)(x_f - x_i) \qquad (6\text{-}5)$$

Newton's second law tells us that $\sum F_x = ma_x$, where m is the mass of the object. So we can rewrite Equation 6-5 as

Calculating the work done by a constant net force, straight-line motion (6-6)

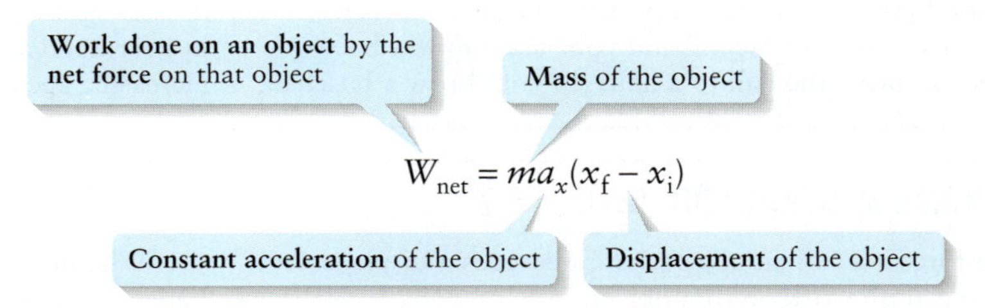

$$W_{\text{net}} = ma_x(x_f - x_i)$$

We can get another expression for the quantity $ma_x(x_f - x_i)$ by multiplying Equation 6-4 by $m/2$ and rearranging:

$$\frac{1}{2}mv_f^2 = \frac{1}{2}mv_i^2 + ma_x(x_f - x_i)$$

$$ma_x(x_f - x_i) = \frac{1}{2}mv_f^2 - \frac{1}{2}mv_i^2$$

The left-hand side of this equation is the work done on the object by the net force (see Equation 6-6). So

$$W_{\text{net}} = \frac{1}{2}mv_f^2 - \frac{1}{2}mv_i^2 \tag{6-7}$$

The right-hand side of Equation 6-7 is the *change* in the quantity $\frac{1}{2}mv^2$ over the course of the displacement (the value at the end of the displacement, where the speed is v_f, minus the value at the beginning where the speed is v_i). We call this quantity the **kinetic energy** K of the object:

Kinetic energy of a moving object (6-8)

Kinetic energy of an object
Mass of the object

$$K = \frac{1}{2}mv^2$$

Speed of the object

The units of kinetic energy are $\text{kg} \cdot \text{m}^2/\text{s}^2$. Since $1 \text{ J} = 1 \text{ N} \cdot \text{m}$ and $1 \text{ N} = 1 \text{ kg} \cdot \text{m/s}^2$, you can see that kinetic energy is measured in joules, the same as work. Using the definition of kinetic energy given in Equation 6-8, we can rewrite Equation 6-7 as

The work-energy theorem (6-9)

Work done on an object by the net force on that object

$$W_{\text{net}} = K_f - K_i$$

Kinetic energy of the object after the work is done on it
Kinetic energy of the object before the work is done on it

When an object undergoes a displacement, the work done on it by the net force equals the object's kinetic energy at the end of the displacement minus its kinetic energy at the beginning of the displacement. This statement is called the **work-energy theorem.** It is valid as long as the object acted on is rigid—that is, it doesn't deform like a rubber ball might. Although we have derived this theorem for the special case of straight-line motion with constant forces, it turns out to be valid even if the object follows a curved path and is acted on by varying forces. (We'll justify this claim later in Section 6-5.)

The Meaning of the Work-Energy Theorem

What is kinetic energy, and how does the work-energy theorem help us solve physics problems? To answer these questions let's first return to the cart from Figure 6-8 and imagine that it starts at rest on a horizontal frictionless surface. If you give the cart a push as in **Figure 6-11a**, the net force on the cart equals the force that you exert (the upward normal force on the cart cancels the downward gravitational force), so the work that you do is the work W_{net} done by the net force. The cart starts at rest, so $v_{\text{i}} = 0$ and the cart's initial kinetic energy $K_{\text{i}} = \frac{1}{2}mv_{\text{i}}^2$ is zero. After you've finished the push, the cart has final speed $v_{\text{f}} = v$ and kinetic energy $K_{\text{f}} = \frac{1}{2}mv^2$. So the work-energy theorem states that

$$W_{\text{you on cart}} = K_{\text{f}} - K_{\text{i}} = \frac{1}{2}mv^2 - 0 = \frac{1}{2}mv^2$$

In words, this gives us our first interpretation of kinetic energy: *An object's kinetic energy equals the work that was done to accelerate it from rest to its present speed.*

Now suppose your friend stands in front of the moving cart and brings it to a halt (**Figure 6-11b**). If we apply the work-energy theorem to the part of the motion where she exerts a force on the cart, the cart's initial kinetic energy is $K_{\text{i}} = \frac{1}{2}mv^2$, and its final kinetic energy is $K_{\text{f}} = 0$ (the cart ends up at rest). The net force on the cart is the force exerted by your friend, so the work she does equals the net work:

$$W_{\text{friend on cart}} = K_{\text{f}} - K_{\text{i}} = 0 - \frac{1}{2}mv^2 = -\frac{1}{2}mv^2$$

Our discussion in Section 6-2 tells us that the *cart* does an amount of work on your *friend* that's just the negative of the work that your friend does on the cart. So we can write

$$W_{\text{cart on friend}} = -W_{\text{friend on cart}} = \frac{1}{2}mv^2$$

This gives us a second interpretation of kinetic energy: *An object's kinetic energy equals the amount of work it can do in the process of coming to a halt from its present speed.*

(a) Making the cart speed up

(1) The cart rolls without friction and the normal force balances the gravitational force.

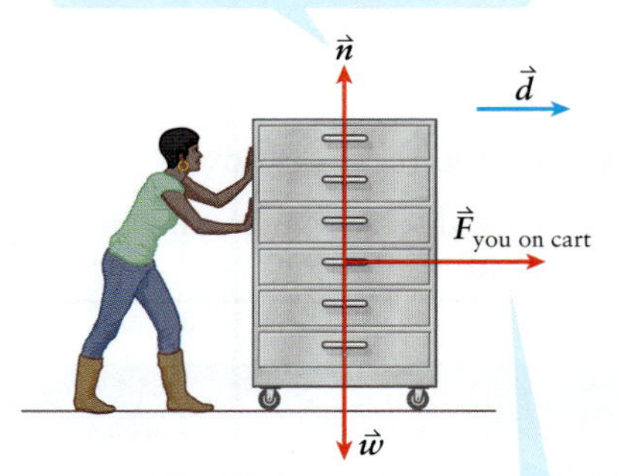

(2) The force you exert therefore equals the net force on the cart. You do positive work on the cart, and the cart gains speed and kinetic energy.

An object's kinetic energy equals the work that was done to accelerate it from rest to its present speed.

(b) Making the cart slow down

(1) The cart rolls without friction and the normal force balances the gravitational force.

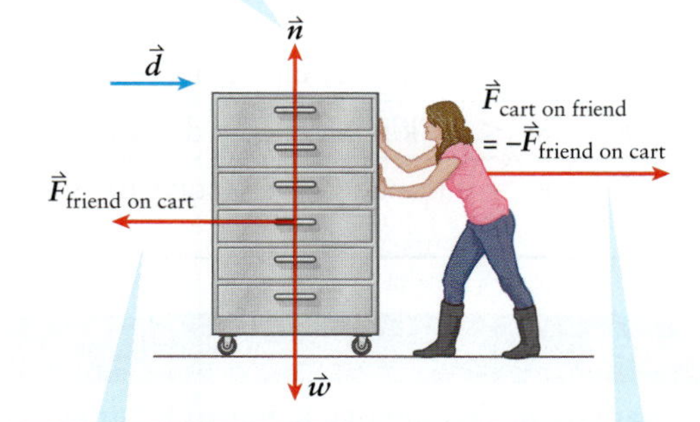

(2) The force your friend exerts therefore equals the net force on the cart. She does negative work on the cart, and the cart loses speed and kinetic energy.

(3) As the cart loses kinetic energy, it pushes on your friend and does positive work on her.

An object's kinetic energy equals the work it can do in the process of coming to a halt from its present speed.

Figure 6-11 The meaning of kinetic energy We can interpret kinetic energy in terms of (a) what it takes to accelerate an object from rest or (b) what the object can do in slowing to a halt.

This should remind you of the general definition of energy as the ability to do work, which we introduced in Section 6-1.

You already have a good understanding of this interpretation of kinetic energy. If you see a thrown basketball coming toward your head, you know intuitively that due to its mass and speed it has a pretty good amount of kinetic energy $K = \frac{1}{2}mv^2$, which means that it can do a pretty good amount of work on you (it can exert a force on your nose that pushes it inward a painful distance). You don't want this to happen, which is why you duck!

WATCH OUT! Kinetic energy is a scalar.

! We've become comfortable breaking an object's motion into components, so you might be tempted to resolve kinetic energy into components as well. However, kinetic energy depends on *speed*, not velocity. Because speed is not a vector, kinetic energy is a *scalar* quantity and *not* a vector. It wouldn't be meaningful to set up components of kinetic energy in different directions. Note also that we've only discussed the kinetic energy associated with the motion of an object as a whole, which we call **translational kinetic energy.** Kinetic energy is also associated with the rotation of an object around its own axis. We'll return to this *rotational kinetic energy* in a later chapter.

Net Work and Net Force

In Equation 6-9 we interpret W_{net} as the work done by the net force. But we saw in Example 6-3 in Section 6-2 that the work done by the net force equals the *sum* of the amount of work done by each individual force acting on the object. So we can also think of W_{net} in Equation 6-9 as the *net work* done by all of the individual forces. It turns out that this statement is true not just for the situation in Example 6-3; it applies to all situations.

This gives us a simplified statement of the work-energy theorem: *The net work done by all forces on an object, W_{net}, equals the difference between its final kinetic energy K_f and its initial kinetic energy K_i.* If the net work done is positive, then $K_f - K_i$ is positive, the final kinetic energy is greater than the initial kinetic energy, and the object gains speed. If the net work done is negative, then $K_f - K_i$ is negative, the final kinetic energy is less than the initial kinetic energy, and the object loses speed. If zero net work is done, the kinetic energy does not change, and the object maintains the same speed (**Table 6-2**). This agrees with the observations we made in Section 6-1.

Example 6-4 shows how to use the work-energy theorem.

TABLE 6-2 The Work-Energy Theorem

If the net work done on an object is . . .	. . . then the change in the object's kinetic energy $K_f - K_i$ is. . .	. . . and the speed of the object. . .
$W_{net} > 0$ (positive net work)	$K_f - K_i > 0$ (kinetic energy increases)	increases (object speeds up)
$W_{net} < 0$ (negative net work)	$K_f - K_i < 0$ (kinetic energy decreases)	decreases (object slows down)
$W_{net} = 0$ (zero net work)	$K_f - K_i = 0$ (kinetic energy stays the same)	is unchanged (object maintains the same speed)

EXAMPLE 6-4 Up the Hill Revisited

Consider again the box of supplies that you pushed up the incline in Example 6-3 (Section 6-2). Use the work-energy theorem to find how fast the box is moving when it reaches your campsite, assuming it is moving uphill at 0.75 m/s at the bottom of the incline.

Set Up

Our goal is to find the final speed v_f of the box. In Example 6-3 we were given the weight (and hence the mass m) of the box, and we found that the net work done on the box is $W_{net} = 56$ J. Since we're given $v_i = 0.75$ m/s, we can use Equations 6-8 and 6-9 to solve for v_f.

Work-energy theorem:

$$W_{net} = K_f - K_i \quad (6\text{-}9)$$

Kinetic energy:

$$K = \frac{1}{2}mv^2 \quad (6\text{-}8)$$

Solve

Use the work-energy theorem to solve for v_f in terms of the initial speed v_i, the net work done W_{net}, and the mass m.

Isolate the $\frac{1}{2}mv_f^2$ term on one side of the equation, then solve for v_f by multiplying through by $2/m$ and taking the square root.

Combine Equations 6-8 and 6-9:

$$W_{net} = \frac{1}{2}mv_f^2 - \frac{1}{2}mv_i^2$$

$$\frac{1}{2}mv_f^2 = \frac{1}{2}mv_i^2 + W_{net}$$

$$v_f^2 = v_i^2 + \frac{2W_{net}}{m}$$

$$v_f = \sqrt{v_i^2 + \frac{2W_{net}}{m}}$$

From Example 6-3 we know that $W_{net} = 56$ J. We find the mass m of the box from its known weight $w = 225$ N and the relationship $w = mg$. Use these to find the value of v_f.

$$m = \frac{w}{g} = \frac{225 \text{ N}}{9.80 \text{ m/s}^2} = 23.0 \text{ kg}$$

$$v_f = \sqrt{(0.75 \text{ m/s})^2 + \frac{2(56 \text{ J})}{23.0 \text{ kg}}}$$

$$= 2.3 \text{ m/s}$$

Reflect

We *could* have solved this problem by first finding the net force on the box, then calculating its acceleration from Newton's second law, and finally solving for v_f by using one of the kinematic equations from Chapter 2. But using the work-energy theorem is easier.

As a check on our result, note that we found in Example 6-3 that the net force on the box is 9.4 N uphill. Can you follow the procedure we just described to find the final velocity after moving 6.00 m up the incline?

Example 6-4 illustrates how using the work-energy theorem can simplify problem-solving. In the following section we'll see more examples of how to apply this powerful theorem to various cases of straight-line motion. In Section 6-5 we'll see how the work-energy theorem can be applied to problems in which the motion is along a curved path and in which the forces are not constant.

GOT THE CONCEPT? 6-2 Slap Shot

? A hockey player does work on a hockey puck in order to propel it from rest across the ice. When a constant force is applied over a certain distance, the puck leaves his stick at speed v. If instead he wants the puck to leave at speed $2v$, by what factor must he increase the distance over which he applies the same force? (a) $\sqrt{2}$; (b) 2; (c) $2\sqrt{2}$; (d) 4; (e) 8.

TAKE-HOME MESSAGE FOR Section 6-3

✔ The net work done on an object (the sum of the work done on it by all forces) as it undergoes a displacement is equal to the change in the object's kinetic energy during that displacement.

✔ The formula for the kinetic energy of an object of mass m and speed v is $K = \frac{1}{2}mv^2$.

✔ The kinetic energy of an object is equal to the amount of work that was done to accelerate the object from rest to its present speed.

✔ The kinetic energy of an object is also equal to the amount of work the object can do in the process of coming to a halt from its present speed.

6-4 The work-energy theorem can simplify many physics problems

In this section we'll explore the relationships among work, force, and speed by applying the definitions of work and kinetic energy (Equations 6-2 and 6-8) and the work-energy theorem (Equation 6-9) to a variety of physical situations. Even if the problem could be solved using Newton's second law, the work-energy theorem often makes the solution easier as well as giving additional insight.

Strategy: Problems with the Work-Energy Theorem

The work-energy theorem is a useful tool for problems that involve an object that moves a distance along a straight line while being acted on by constant forces. You can use this theorem to relate the forces, the distance traveled (the displacement), and the speed of the object at the beginning and end of the displacement.

Note that the work-energy theorem makes no reference to the *time* it takes the object to move through this displacement. If the problem requires you to use or find this time, you should use a different approach.

Set Up: Always draw a picture of the situation that shows the object's displacement. Include a free-body diagram that shows all of the forces that act on the object. Draw the direction of each force carefully, since the direction is crucial for determining how much work each force does on the object. Decide what unknown quantity the problem is asking you to determine (for example, the object's final speed or the magnitude of one of the forces).

Solve: Use Equation 6-2 to find expressions for the work done by each force. The sum of the work done by each force is the total or net work done on the object, W_{net}. Then use Equations 6-8 and 6-9 to relate this to the object's initial and final kinetic energies. Solve the resulting equations for the desired unknown.

Reflect: Always check whether the numbers have reasonable values and that each quantity has the correct units.

EXAMPLE 6-5 How Far to Stop?

The U.S. National Highway Traffic Safety Administration lists the minimum braking distance for a car traveling at 40.0 mi/h to be 101 ft. If the braking force is the same at all speeds, what is the minimum braking distance for a car traveling at 65.0 mi/h?

Set Up

Three forces act on the car—the gravitational force, the normal force exerted by the road, and the braking force—but if the road is level, only the braking force does work on the car. (The other two forces are perpendicular to the displacement, so $Fd \cos 90° = 0$.) We'll use Equations 6-2, 6-8, and 6-9 to find the relationship among the initial speed v_i, the braking force, and the distance the car must travel to have the final speed be $v_f = 0$.

Work done by a constant force, straight-line displacement:

$$W = Fd \cos \theta \qquad (6\text{-}2)$$

Kinetic energy:

$$K = \frac{1}{2}mv^2 \qquad (6\text{-}8)$$

Work-energy theorem:

$$W_{net} = K_f - K_i \qquad (6\text{-}9)$$

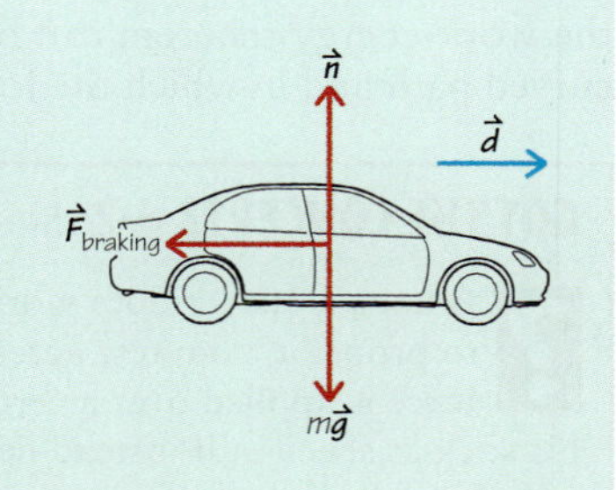

Solve

The car ends up at rest, so its final kinetic energy is zero. So the work-energy theorem tells us that the net work done on the car (equal to the work done on it by the braking force) equals the negative of the initial kinetic energy.

Work-energy theorem with final kinetic energy equal to zero:

$$W_{net} = 0 - K_i = -\frac{1}{2}mv_i^2$$

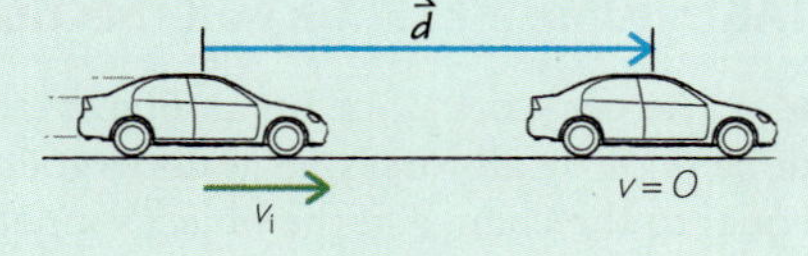

The braking force of magnitude $F_{braking}$ is directed opposite to the displacement, so $\theta = 180°$ in the expression for the work $W_{braking}$ done by this force.

Net work = work done by the braking force:

$$W_{net} = W_{braking}$$
$$= F_{braking}d \cos 180° = -F_{braking}d$$

Substitute $W_{net} = -F_{braking}d$ into the work-energy theorem and solve for the braking distance d. The result tells us that d is proportional to the square of the initial speed v_i.

Work-energy theorem becomes

$$-F_{braking}d = -\frac{1}{2}mv_i^2$$

Solve for d:

$$d = \frac{mv_i^2}{2F_{braking}}$$

We aren't given the mass m of the car or the magnitude $F_{braking}$ of the braking force, but we can set up a ratio between the values of distance for $v_{i1} = 40.0$ mi/h and $v_{i2} = 65.0$ mi/h.

Use this expression to set up a ratio for the value of d at two different initial speeds. The mass of the car and the braking force are the same for both initial speeds:

$$\frac{d_2}{d_1} = \frac{v_{i2}^2}{v_{i1}^2}$$

Solve for the stopping distance d_2 corresponding to $v_{i2} = 65.0$ mi/h:

$$d_2 = d_1\frac{v_{i2}^2}{v_{i1}^2} = (101 \text{ ft})\left(\frac{65.0 \text{ mi/h}}{40.0 \text{ mi/h}}\right)^2$$
$$= 267 \text{ ft}$$

Reflect

This example demonstrates what every driver training course stresses: The faster you travel, the more distance you should leave between you and the car ahead in case an emergency stop is needed. Increasing your speed by 62.5% (from 40.0 to 65.0 mi/h) increases your stopping distance by 164% (from 101 ft to 267 ft). The reason is that stopping distance is proportional to kinetic energy, which is proportional to the *square* of the speed.

EXAMPLE 6-6 Work and Kinetic Energy: Force at an Angle

At the start of a race, a four-man bobsleigh crew pushes their sleigh as fast as they can down the 50.0-m straight, relatively flat starting stretch (**Figure 6-12**). The net force that the four men together apply to the 325-kg sleigh has magnitude 285 N and is directed at an angle of 20.0° below the horizontal. As they push, a 60.0-N kinetic friction force also acts on the sleigh. What is the speed of the sleigh right before the crew jumps in at the end of the starting stretch?

sampics/Getty Images

Figure 6-12 Bobsleigh start The success of a bobsleigh team depends on the team members giving the sleigh a competitive starting speed.

Set Up

Again we'll use the work-energy theorem, this time to determine the final speed v_f of the bobsleigh (which starts at rest, so its initial speed is $v_i = 0$). The normal force and gravitational forces do no work on the sleigh since they act perpendicular to the displacement. The four men of the crew do positive work, while the friction force does negative work. The new wrinkle is that the force exerted by the crew points at an angle to the displacement. We'll deal with this using Equation 6-2.

The sleigh gains speed and kinetic energy during the motion, so the net work done by these two forces must be positive.

Work done by a constant force, straight-line displacement:

$$W = Fd\cos\theta \qquad (6\text{-}2)$$

Kinetic energy:

$$K = \frac{1}{2}mv^2 \qquad (6\text{-}8)$$

Work-energy theorem:

$$W_{net} = K_f - K_i \qquad (6\text{-}9)$$

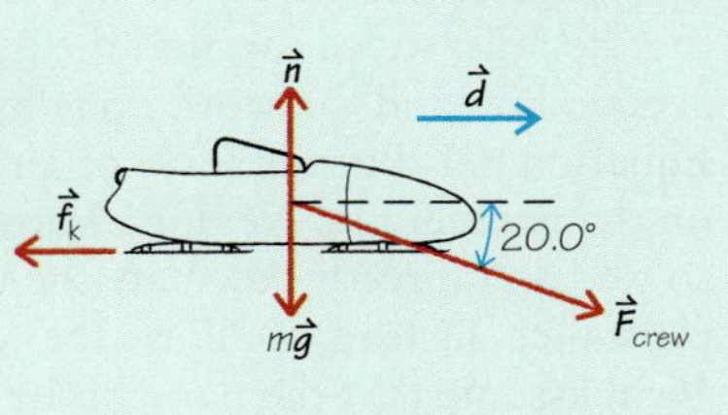

Solve

The sleigh starts at rest, so its initial kinetic energy is zero. From Equation 6-9 the net work done on the sleigh is therefore equal to its final kinetic energy, which is related to the final speed v_f.

Work-energy theorem with initial kinetic energy equal to zero:

$$W_{net} = K_f - 0 = \frac{1}{2}mv_f^2$$

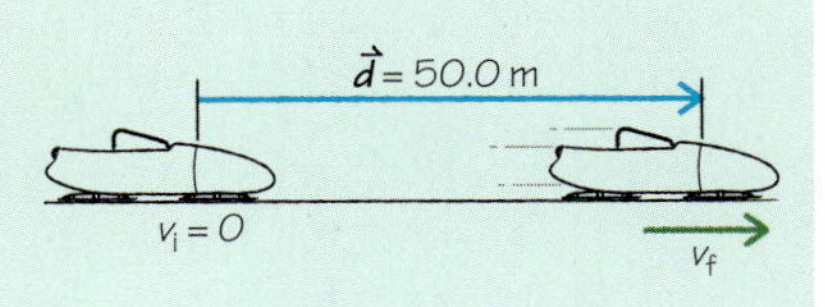

The force of the crew is at $\theta = 20.0°$ to the displacement, and the friction force is at $\theta = 180°$. Use these in Equation 6-2 to find the net work done on the sleigh.

Net work = work done by the crew plus work done by friction

$$\begin{aligned} W_{net} &= W_{crew} + W_{friction} \\ &= F_{crew}d \cos 20.0° + f_k d \cos 180° \\ &= (285 \text{ N})(50.0 \text{ m})(0.940) \\ &\quad + (60.0 \text{ N})(50.0 \text{ m})(-1) \\ &= 1.04 \times 10^4 \text{ J} \end{aligned}$$

Substitute W_{net} into the work-energy theorem and solve for the final speed v_f.

Work-energy theorem says $W_{net} = \frac{1}{2}mv_f^2$

Solve for v_f: $v_f = \sqrt{\frac{2W_{net}}{m}} = \sqrt{\frac{2(1.04 \times 10^4 \text{ J})}{325 \text{ kg}}}$

$= 8.00 \text{ m/s}$

Reflect

This answer for v_f is relatively close to the speed of world-class sprinters, so it seems reasonable. In reality the start of an Olympic four-man bobsleigh race is slightly downhill, so the sleigh's speed at the end of the starting stretch is typically even faster (11 or 12 m/s).

EXAMPLE 6-7 Find the Work Done by an Unknown Force

An adventurous parachutist of mass 70.0 kg drops from the top of Angel Falls in Venezuela, the world's highest waterfall (Figure 6-13). The waterfall is 979 m (3212 ft) tall and the parachutist deploys his chute after falling 295 m, at which point his speed is 54.0 m/s. During the 295-m drop, (a) what was the net work done on him and (b) what was the work done on him by the force of air resistance?

Figure 6-13 Falling with air resistance As the parachutist falls, air resistance does negative work on him.

Set Up

We are given the parachutist's mass and the distance that he falls, so we can find the gravitational force that acts on him and the work done by that force using Equation 6-2. But the force of air resistance also acts on the parachutist. We don't know its magnitude, so we can't directly calculate the work done by air resistance. Instead, we'll use the work-energy theorem. We'll assume that the parachutist starts at rest, so his initial kinetic energy is zero.

Work done by a constant force, straight-line displacement:

$$W = Fd \cos \theta \qquad (6\text{-}2)$$

Work-energy theorem:

$$W_{net} = K_f - K_i \qquad (6\text{-}9)$$

Kinetic energy:

$$K = \frac{1}{2}mv^2 \qquad (6\text{-}8)$$

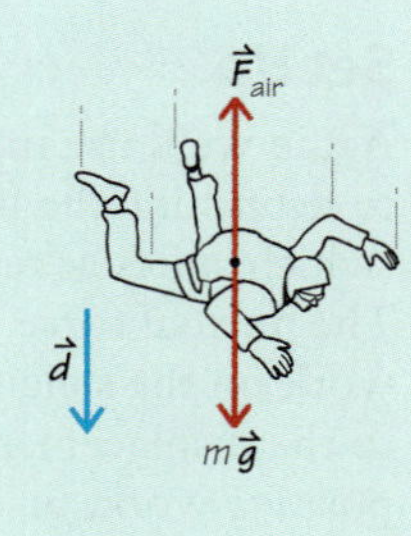

Solve

(a) Since $K_i = 0$, Equation 6-9 says that the net work done on the parachutist is just equal to his final kinetic energy.

$$\begin{aligned} W_{net} &= K_f = \frac{1}{2}mv_f^2 \\ &= \frac{1}{2}(70.0 \text{ kg})(54.0 \text{ m/s})^2 \\ &= 1.02 \times 10^5 \text{ J} \end{aligned}$$

$v_i = 0$

$d = 295$ m

$v_f = 54.0$ m/s

(b) The net work is the sum of the work done by the gravitational force, W_{grav}, and the work done by air resistance, W_{air}. Hence W_{air} is the difference between W_{net} and W_{grav}.

Work done by the gravitational force: Displacement is in the same direction (downward) as the force, so

$$\begin{aligned} W_{grav} &= mgd \\ &= (70.0 \text{ kg})(9.80 \text{ m/s}^2)(295 \text{ m}) \\ &= 2.02 \times 10^5 \text{ J} \end{aligned}$$

Hence the work done by air resistance is

$$\begin{aligned} W_{air} &= W_{net} - W_{grav} \\ &= 1.02 \times 10^5 \text{ J} - 2.02 \times 10^5 \text{ J} \\ &= -1.00 \times 10^5 \text{ J} \end{aligned}$$

Reflect

The work done by the force of air resistance is negative because the force is directed upward, opposite to the downward displacement. We can use the calculated value of W_{air} to find an average value of the air resistance force, considered to be a constant. This is only an *average* value because this force is *not* constant: The faster the parachutist falls, the greater the force of air resistance.

Average upward force of air resistance:

$$W_{air} = F_{air} d \cos 180° = -F_{air} d$$

Hence

$$\begin{aligned} F_{air} &= -\frac{W_{air}}{d} = -\frac{(-1.00 \times 10^5 \text{ J})}{295 \text{ m}} \\ &= 339 \text{ J/m} = 339 \text{ N} \end{aligned}$$

GOT THE CONCEPT? 6-3 Pushing Boxes

? You push two boxes from one side of a room to the other. Each box begins and ends at rest. The contact surfaces and areas between box and floor are the same for both, but one box is heavy while the other is light. There is friction between the box and the floor. (a) How does the work *you* do on each box compare? (i) You do more work on the heavy box; (ii) you do more work on the light box; (iii) you do the same amount of work on both boxes. (b) How does the *net* work done on each box compare? (i) There is more net work on the heavy box; (ii) there is more net work on the light box; (iii) there is the same net work on both boxes.

TAKE-HOME MESSAGE FOR Section 6-4

✔ The work-energy theorem allows us to explore the relationship between force, distance, and speed in a variety of physical situations.

✔ To solve problems using the work-energy theorem, begin by drawing a free-body diagram that shows all the forces that act on the object in question. Then write expressions for the kinetic energies at the beginning and end of the displacement and for the total work done on the object (the sum of the work done by each force). Relate these using the work-energy theorem and solve for the unknown quantity.

6-5 The work-energy theorem is also valid for curved paths and varying forces

We've derived the work-energy theorem only for the special case of straight-line motion with constant forces. Since we already know how to solve problems for that kind of motion, you may wonder what good this theorem is. Here's the answer: The work-energy theorem also works for motion along a *curved* path and in cases where the forces are *not* constant (**Figure 6-14a**).

To prove this, consider **Figure 6-14b**, which shows an object's curved path from an initial point i to a final point f. We mark a large number N of equally spaced intermediate points 1, 2, 3, . . . , N along the path and imagine breaking the path into short segments (from i to 1, from 1 to 2, from 2 to 3, and so on). If each segment is sufficiently short, we can treat it as a straight line. Furthermore, the forces on the object don't change very much during the brief time the object traverses one of these segments, so we can treat the forces as constant over that segment. (The forces may be different

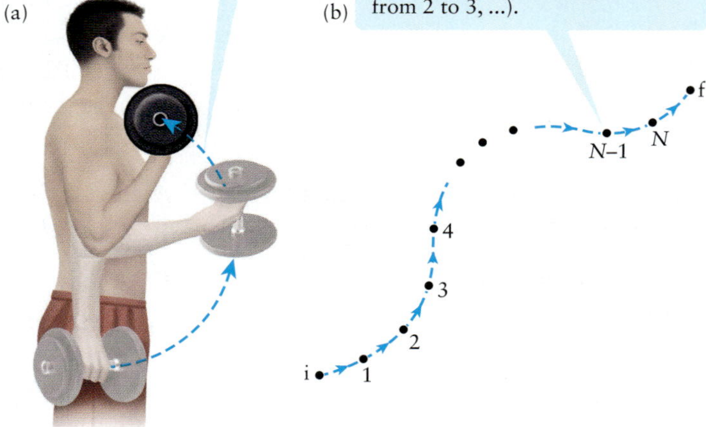
A weight lifter doing bicep curls makes the weight move through a curved path. The force that he exerts on the weight varies during its travel.
(a)
(b)
An object's curved path from an initial point i to a final point f. We mark a large number N of equally spaced intermediate points 1, 2, 3, ..., N along the path and imagine breaking the path into short segments (from i to 1, from 1 to 2, from 2 to 3, ...).
i
1
2
3
4
N–1
N
f

Figure 6-14 Motion along a curved path (a) The weight travels in a curved path during the bicep curl. (b) Analyzing motion along a general curved path.

from one segment to the next.) Since each segment is a straight line with constant forces, we can safely apply the work-energy theorem in Equation 6-9 to every segment:

$$
\begin{aligned}
W_{\text{net, i to 1}} &= K_1 - K_{\text{i}} \\
W_{\text{net, 1 to 2}} &= K_2 - K_1 \\
W_{\text{net, 2 to 3}} &= K_3 - K_2 \\
&\cdots \\
W_{\text{net, }N\text{ to f}} &= K_{\text{f}} - K_N
\end{aligned}
$$

Now add all of these equations together:

$$
\begin{aligned}
&W_{\text{net, i to 1}} + W_{\text{net, 1 to 2}} + W_{\text{net, 2 to 3}} + \cdots + W_{\text{net, }N\text{ to f}} \\
&\quad = (K_1 - K_{\text{i}}) + (K_2 - K_1) + (K_3 - K_2) + \cdots + (K_{\text{f}} - K_N)
\end{aligned}
\tag{6-10}
$$

The left-hand side of Equation 6-10 is the sum of the amounts of work done by the net force on each segment. This sum equals the *net* work done by the net force along the entire path from i to f, which we call simply W_{net}. On the right-hand side of the equation, $-K_1$ in the second term cancels K_1 in the first term, $-K_2$ in the third term cancels K_2 in the second term, and so on. The only quantities that survive on the right-hand side are $-K_{\text{i}}$ and K_{f}, so we are left with

$$W_{\text{net}} = K_{\text{f}} - K_{\text{i}}$$

This is *exactly* the same statement of the work-energy theorem as Equation 6-9. So the work-energy theorem is valid for *any* path and for *any* forces, whether constant or not. This is why the work-energy theorem is so important: You can apply it to situations where using forces and Newton's laws would be difficult or impossible.

EXAMPLE 6-8 A Swinging Spider

Figure 6-15 shows a South African kite spider (*Gasteracantha*) swinging on a strand of spider silk. Suppose a momentary gust of wind blows on a spider of mass 1.00×10^{-4} kg (0.100 gram) that's initially hanging straight down on such a strand. As a result, the spider acquires a horizontal velocity of 1.3 m/s. How high will the spider swing? Ignore air resistance.

Emil von Maltitz/Getty Images

Figure 6-15 A spider swinging on silk If we know the spider's speed at the low point of its arc, how do we determine how high it swings?

Set Up

Because the spider follows a curved path, this would be a very difficult problem to solve using Newton's laws directly. Instead, we can use the work-energy theorem. We are given the spider's mass and initial speed $v_i = 1.3$ m/s (which means we know its initial kinetic energy). We want to find its maximum height h at the point where the spider comes momentarily to rest (so its speed and kinetic energy are zero) before swinging back downward.

The free-body diagram shows that only two forces act on the swinging spider: the gravitational force and the tension force exerted by the silk. All we have to do is calculate the work done on the spider by these forces along this curved path. We'll do this by breaking the path into a large number of segments as in Figure 6-14.

Work-energy theorem:

$$W_{net} = K_f - K_i \quad (6\text{-}9)$$

Kinetic energy:

$$K = \frac{1}{2}mv^2 \quad (6\text{-}8)$$

Work done by a constant force, straight-line displacement:

$$W = Fd\cos\theta \quad (6\text{-}2)$$

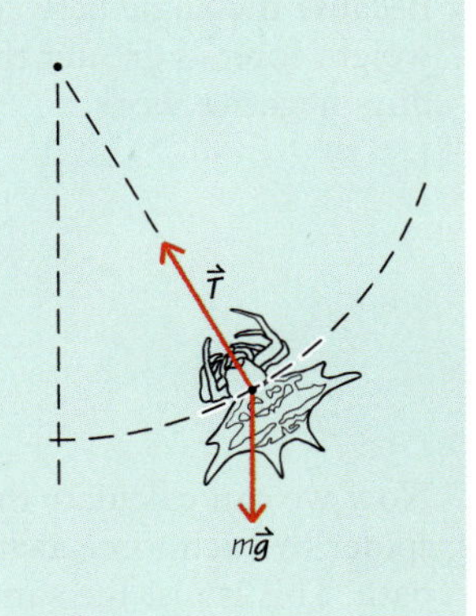

Solve

Rewrite the work-energy theorem in terms of the initial and final speeds of the spider and the work done by each force.

Combine Equations 6-9 and 6-8:

$$W_{net} = \frac{1}{2}mv_f^2 - \frac{1}{2}mv_i^2$$

Net work is the sum of the work done by the tension force T and the work done by the gravitational force:

$$W_{net} = W_{grav} + W_T$$

The final speed of the spider at the high point of its motion is $v_f = 0$, so the work-energy theorem becomes:

$$W_{grav} + W_T = -\frac{1}{2}mv_i^2$$

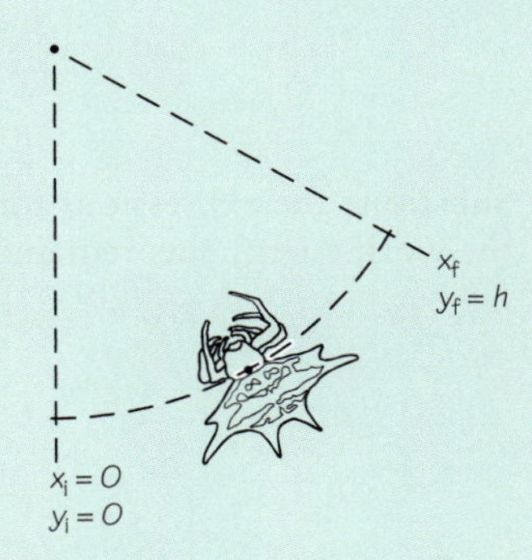

To calculate the work done by each force along the spider's curved path, break the path up into segments so short that each one can be considered as a straight line. On each segment the spider has a displacement $\Delta\vec{d}$ with horizontal component Δx and vertical component Δy. These are different for each individual segment.

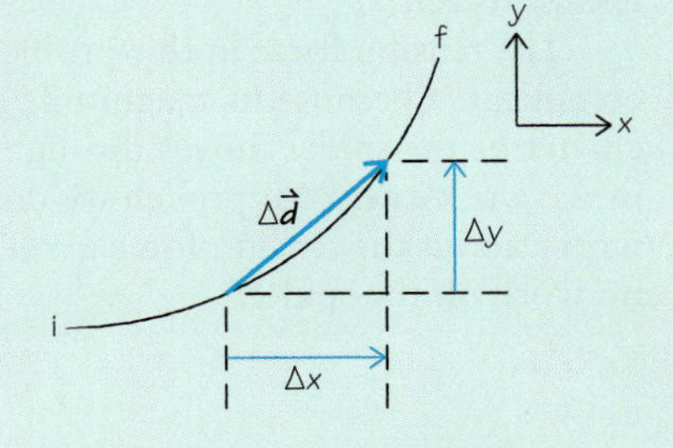

On each segment of the path the tension force $\vec{T}$ points radially inward, perpendicular to the path and hence perpendicular to the displacement $\Delta\vec{d}$. Hence $\theta = 90°$ in Equation 6-2, which means that the tension force does *no* work during this displacement.

Work done by the tension force as the spider moves through a short segment of its path:

$$\Delta W_T = T\,\Delta d\cos 90° = 0$$

The gravitational force $\vec{w}$ always points straight downward (in the negative y direction) and has constant magnitude $w = mg$. We apply Equation 6-2 to the displacement and weight force drawn as shown, and use the identity $\cos(90° + \phi) = -\sin\phi$.

Work done by the gravitational force as the spider moves through a short segment of its path:

$$\Delta W_{grav} = w\,\Delta d\cos\theta$$

Because the angle between the displacement and the weight force is greater than 90°, the weight force does negative work.

The angle between the force and displacement is $\theta = 90° + \phi$ so

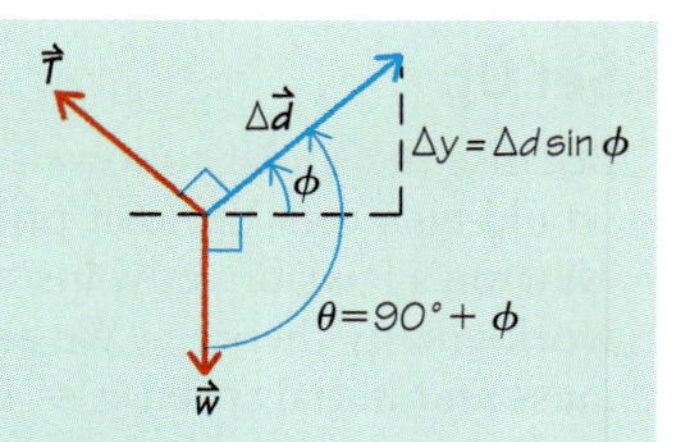

$$\Delta W_{grav} = w\,\Delta d \cos(90° + \phi)$$
$$= -w\,\Delta d \sin\phi$$

The figure shows

$\Delta d \sin\phi = \Delta y$, so

$$\Delta W_{grav} = -w\,\Delta y = -mg\,\Delta y$$

Now we can calculate the *net* work done on the spider by each force as it moves along its curved path. This is just the sum of the individual bits of work done on it along the short segments.

The tension force does zero work on each short segment, so

$$W_T = 0$$

Work done by the gravitational force over the entire path:

$$W_{grav} = (-mg\,\Delta y_1) + (-mg\,\Delta y_2) + (-mg\,\Delta y_3) + \cdots$$
$$= -mg(\Delta y_1 + \Delta y_2 + \Delta y_3 + \cdots)$$

The sum of all the Δy terms is the net vertical displacement $y_f - y_i = h - 0 = h$. So

$$W_{grav} = -mgh$$

Substitute the expressions for $W_T =$ and W_{grav} into the work-energy theorem and solve for h.

Work-energy theorem:

$$W_{grav} + W_T = -mgh + 0 = -\frac{1}{2}mv_i^2$$

$$h = \frac{v_i^2}{2g} = \frac{(1.3\text{ m/s})^2}{2(9.80\text{ m/s}^2)} = 0.086\text{ m} = 8.6\text{ cm}$$

Reflect

The maximum height reached is the *same* as if the spider had initially been moving *straight up* at 1.3 m/s without being attached to the silk. In both cases the gravitational force does the same amount of (negative) work to reduce the spider's kinetic energy to zero.

The tension force in this problem is complicated because its magnitude and direction change as the spider moves through its swing. But in the work-energy approach we don't have to worry about the tension force at all because it does no work on the spider.

If the spider is initially moving straight up:

$$W_{net} = \frac{1}{2}mv_f^2 - \frac{1}{2}mv_i^2$$

$$W_{grav} = -mgh = 0 - \frac{1}{2}mv_i^2$$

$$h = \frac{v_i^2}{2g} = 8.6\text{ cm}$$

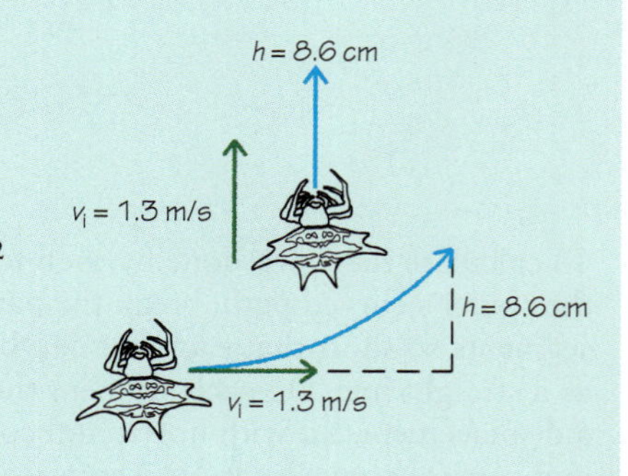

Example 6-8 illustrates an important point: *If a force always acts perpendicular to an object's curved path, it does zero work on the object.* This enabled us to ignore the effects of the tension force. We'll use this same idea again in Section 6-6.

Work Done by a Varying Force

In Example 6-8 the tension force acting on the spider varied in magnitude but did no work (because it acts perpendicular to the spider's displacement). In many situations, however, a force of variable magnitude *does* do work on an object. As an example, you must do work to stretch a spring. The force you exert to do this is not constant: The farther you stretch the spring, the greater the magnitude of the force you must exert. How can we calculate the amount of work that you do while stretching the spring?

To see the answer let's first consider a *constant* force F that acts on an object in the direction of its straight-line motion. **Figure 6-16** shows a graph of this force versus position as the object undergoes a displacement d. The *area* under the graph of force versus position equals the area of the colored rectangle in Figure 6-16: that is, the height of the rectangle (the force F) multiplied by its width (the displacement d). But the area Fd is just equal to the work done by the constant force. So *on a graph of force versus position, the work done by the force equals the area under the graph.*

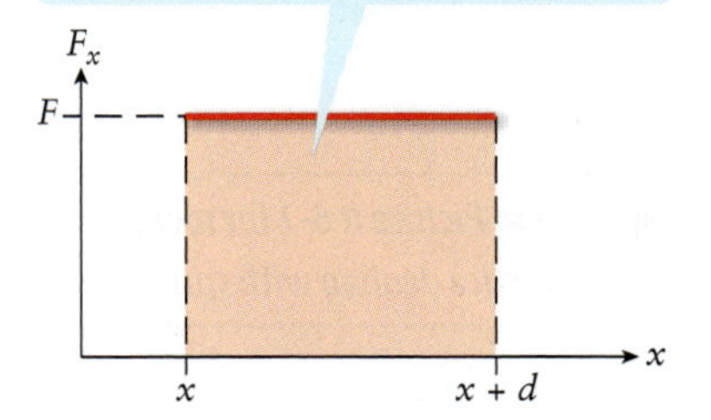
Work done by a constant force along direction of motion = Fd = area under graph of force versus position.
F_x
F
x
x
$x + d$

Figure 6-16 The "area rule" for work Finding the work done by a constant force using a graph of force versus position.

Let's apply this "area rule" to the work that you do in stretching a spring. Experiment shows that if you stretch a spring by a relatively small amount, the force that the *spring* exerts on you is directly proportional to the amount of stretch (**Figure 6-17**). We can write this relationship, known as **Hooke's law**, as

Force exerted by an **ideal spring**

Spring constant of the spring (a measure of its stiffness)

$$F_x = -kx$$

Extension of the spring ($x > 0$ if spring is stretched, $x < 0$ if spring is compressed)

Hooke's law for the force exerted by an ideal spring (6-11)

The minus sign in Equation 6-11 means that the force that the spring exerts on you is in the direction *opposite* to the stretch. If you pull one end of the spring in the positive x direction so $x > 0$, the spring pulls back on you in the negative x direction. (As we will see below, Hooke's law also describes situations in which the spring is compressed rather than stretched.) The quantity k, called the **spring constant**, depends on the stiffness of the spring: the greater the value of k, the stiffer the spring. If the force is measured in newtons and the extension in meters, the spring constant has units of N/m.

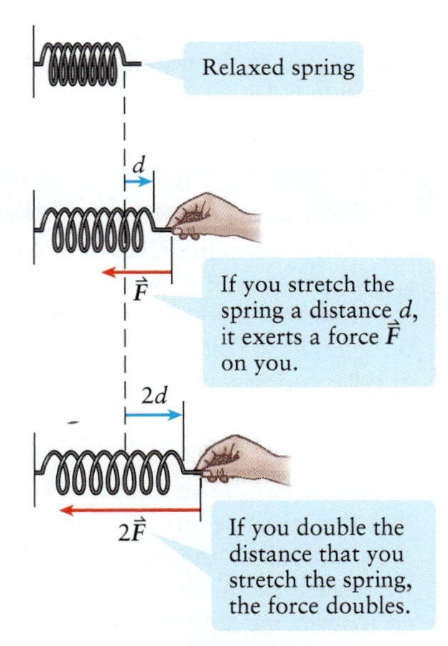
Relaxed spring
d
$\vec{F}$
If you stretch the spring a distance d, it exerts a force $\vec{F}$ on you.
$2d$
$2\vec{F}$
If you double the distance that you stretch the spring, the force doubles.

Figure 6-17 Hooke's law If you stretch an ideal spring, the force that it exerts on you is directly proportional to its extension.

Equation 6-11 gives the force that the *spring* exerts on *you* as you stretch it. By Newton's third law, the force that *you* exert on the *spring* has the same magnitude but the opposite direction. So the x component of the force you exert has the opposite sign to the force in Equation 6-11:

$$F_{\text{you on spring},x} = +kx \text{ (force that you exert on the spring)} \qquad \textbf{(6-12)}$$

Stretching the spring means $x > 0$, and to do this you must exert a force in the same direction. So $F_{\text{you on spring},x} > 0$ if $x > 0$, which is just what Equation 6-12 tells us.

Figure 6-18 graphs the force that you exert as a function of the distance x that the spring has been stretched. If the spring is initially stretched a distance x_1 and you

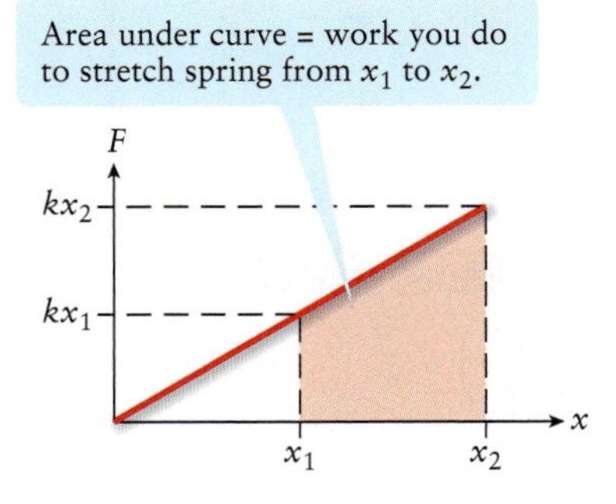
Area under curve = work you do to stretch spring from x_1 to x_2.
F
kx_2
kx_1
x
x_1
x_2

Figure 6-18 Applying the "area rule" to an ideal spring We can use the same technique as in Figure 6-16 to find the work required to stretch a spring.

stretch it further to x_2, the work W that you do is equal to the area of the colored trapezoid in Figure 6-18. From geometry, this area is equal to the *average* height of the graph multiplied by the width $x_2 - x_1$. Using Equation 6-12, we find

$$W = \left[\frac{(\text{force you exert at } x_1) + (\text{force you exert at } x_2)}{2}\right](x_2 - x_1)$$
$$= \frac{(kx_1 + kx_2)}{2}(x_2 - x_1) = \frac{1}{2}k(x_1 + x_2)(x_2 - x_1)$$

Go to Picture It 6-1 for more practice dealing with springs.

We can simplify this to

Work done to stretch a spring (6-13)

$$W = \frac{1}{2}kx_2^2 - \frac{1}{2}kx_1^2$$

Work that must be done on a spring to stretch it from $x = x_1$ to $x = x_2$

Spring constant of the spring (a measure of its stiffness)

x_1 = **initial stretch** of the spring
x_2 = **final stretch** of the spring

WATCH OUT! Who's doing the work?

Note that Equation 6-13 tells us the work that *you* do on the *spring*. The work that the *spring* does on *you* is equal to the negative of Equation 6-13.

Example 6-9 shows how to use Equation 6-13 to attack a problem that would have been impossible to solve with the force techniques from Chapters 4 and 5.

EXAMPLE 6-9 Work Those Muscles!

An athlete stretches a set of exercise cords 47 cm from their unstretched length. The cords behave like a spring with spring constant 860 N/m. (a) How much force does the athlete exert to hold the cords in this stretched position? (b) How much work did he do to stretch them? (c) The athlete loses his grip on the cords. If the mass of the handle is 0.25 kg, how fast is it moving when it hits the wall to which the other end of the cords is attached? (You can ignore gravity and assume that the cords themselves have only a small mass.)

Set Up

In part (a) we'll use Equation 6-12 to find the force that the athlete exerts, and in part (b) Equation 6-13 will tell us the work that he does. In part (c) we'll see how much work the *cords* do as they go from being stretched to relaxed. This work goes into the kinetic energy of the handle. We're ignoring the mass of the cords and therefore assuming that they have no kinetic energy of their own.

$$F_{\text{athlete on cords},x} = +kx \quad (6\text{-}12)$$

$$W = \frac{1}{2}kx_2^2 - \frac{1}{2}kx_1^2 \quad (6\text{-}13)$$

$$W_{\text{net}} = K_f - K_i \quad (6\text{-}9)$$

$$K = \frac{1}{2}mv^2 \quad (6\text{-}8)$$

relaxed
47 cm
stretched
released
$v_f = ?$

Solve

(a) The force that the athlete exerts is proportional to the distance $x = 47$ cm that the cords are stretched.

Force exerted by the athlete to hold the cords stretched:

$$F_{\text{athlete on cords},x} = +kx$$
$$= (860 \text{ N/m})(47 \text{ cm})\left(\frac{1 \text{ m}}{100 \text{ cm}}\right)$$
$$= 4.0 \times 10^2 \text{ N}$$

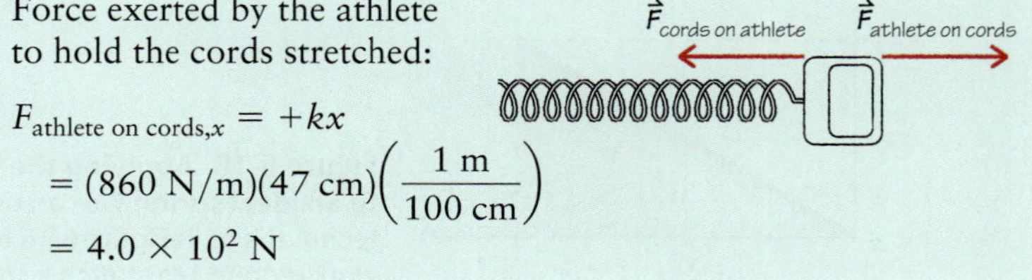

(b) The cords are not stretched at all to start (so $x_1 = 0$) and end up stretched by 47 cm (so $x_2 = 47$ cm $= 0.47$ m).

Work done by the athlete to stretch the cords:

$$W_{\text{athlete on cords}} = \frac{1}{2}kx_2^2 - \frac{1}{2}kx_1^2$$
$$= \frac{1}{2}(860\ \text{N/m})(0.47\ \text{m})^2 - \frac{1}{2}(860\ \text{N/m})(0\ \text{m})^2$$
$$= 95\ \text{N}\cdot\text{m} = 95\ \text{J}$$

(c) When the athlete releases the handle, the cords relax from their new starting position ($x_1 = 0.47$ m) to their final, unstretched position ($x_2 = 0$). The work that the cords do on the handle is given by the negative of Equation 6-13.

Work done by the cords on the handle as they relax:

$$W_{\text{cords on handle}}$$
$$= -\left(\frac{1}{2}kx_2^2 - \frac{1}{2}kx_1^2\right)$$
$$= -\left[\frac{1}{2}(860\ \text{N/m})(0\ \text{m})^2 - \frac{1}{2}(860\ \text{N/m})(0.47\ \text{m})^2\right]$$
$$= -(-95\ \text{N}\cdot\text{m}) = 95\ \text{J}$$

We'll ignore the force of gravity (that is, we assume the handle flies back horizontally and doesn't fall). Then the net work done on the handle equals the work done by the cords. This is equal to the change in the handle's kinetic energy. Use this to find the handle's speed when the cords are fully relaxed.

Net work done on handle:

$$W_{\text{net}} = W_{\text{cords on handle}} = 95\ \text{J}$$

Work-energy theorem applied to handle:

$$W_{\text{net}} = K_f - K_i = \frac{1}{2}mv_f^2 - \frac{1}{2}mv_i^2$$

Handle is initially at rest, so $v_i = 0$. Solve for final speed of the handle:

$$\frac{1}{2}mv_f^2 = W_{\text{net}} = 95\ \text{J}$$
$$v_f^2 = \frac{2W_{\text{net}}}{m}$$
$$v_f = \sqrt{\frac{2W_{\text{net}}}{m}} = \sqrt{\frac{2(95\ \text{J})}{0.25\ \text{kg}}} = 28\ \text{m/s}$$

Reflect

A common *incorrect* way to approach part (b) is to use the formula for the work done by a *constant* force: Multiply the force needed to hold the cords fully stretched by the cords' displacement. This gives the wrong answer because the force needed to pull the cords is *not* constant.

Incorrect way to calculate work that the athlete does on the cords:

$$W_{\text{athlete on cords}} = F_x d = (4.0 \times 10^2\ \text{N})(0.47\ \text{m}) = 190\ \text{J}$$

(actual value: $W_{\text{athlete on cords}} = 95$ J)

Notice that the amount of work that the athlete does to stretch the cords (95 J) is the *same* as the amount of work that the cords do on the handle when they relax. This suggests that the athlete stores energy in the cords by stretching them. We'll explore this idea in Section 6-6 as we introduce a new kind of energy, called *potential energy*.

More on Hooke's Law and Its Limits

The spring in Figure 6-17 exerts a force when it is stretched. A spring also exerts a force when it is *compressed* (**Figure 6-19a**). One example is a car's suspension, whose springs compress as you load passengers and luggage into the car. The force that an ideal spring exerts when compressed is given by Equation 6-11, $F_x = -kx$, the *same* equation that describes the force exerted by a *stretched* spring (**Figure 6-19b**). The only

(a) A compressed spring ($x < 0$) pushes in the $+x$ direction.

(b) A stretched spring ($x > 0$) pulls in the $-x$ direction.

$x < 0$ $\vec{F}_{\text{by spring}}$ x

$x > 0$ $\vec{F}_{\text{by spring}}$ x

F_x Force exerted by a spring: $F_x = -kx$ 0 x

Figure 6-19 Compressing and stretching an ideal spring Hooke's law, $F_x = -kx$, applies equally well to an ideal spring whether it is (a) compressed or (b) stretched.

difference is that x is negative if the spring is compressed. Hence if you compress the spring (push its end in the negative x direction), the spring pushes back on you in the positive x direction.

Like Equation 6-11, Equation 6-13 is also valid when a spring is compressed. If a spring with spring constant $k = 1000$ N/m is initially relaxed (so $x_1 = 0$) and you stretch it by 20 cm (so $x_2 = +20$ cm $= +0.20$ m), the work that you do is

$$W = \frac{1}{2}kx_2^2 - \frac{1}{2}k(0)^2 = \frac{1}{2}(1000 \text{ N/m})(0.20 \text{ m})^2 = 20 \text{ J}$$

The work that you do to *compress* the same spring by 20 cm (so $x_2 = -20$ cm $= -0.20$ m) is

$$W = \frac{1}{2}kx_2^2 - \frac{1}{2}k(0)^2 = \frac{1}{2}(1000 \text{ N/m})(-0.20 \text{ m})^2 = 20 \text{ J}$$

So you have to do positive work to compress a spring as well as to stretch it.

Hooke's law and Equations 6-11, 6-12, and 6-13 are only *approximate* descriptions of how real springs, elastic cords, and tendons behave. As an example, **Figure 6-20** is a graph of the force needed to stretch a human patellar tendon (which connects the kneecap to the shin). The curve isn't a straight line, which means that the force isn't directly proportional to the amount of stretch. The force you have to apply to the

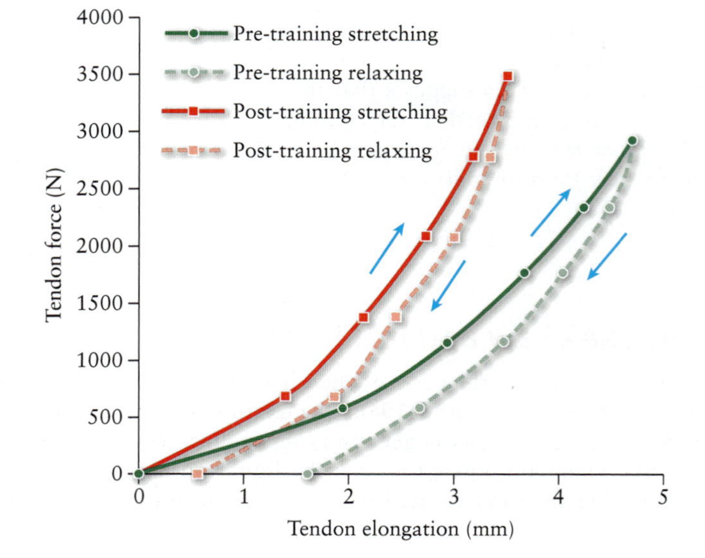

Figure 6-20 Tendons are not ideal springs This graph of force versus extension for a human patellar tendon is very different from that for an ideal spring (compare to Figure 6-19). The graph is not a straight line; there is a different graph for relaxing than for stretching the tendon; and the graph for tendons that have undergone exercise (square, red data points) is different from that for tendons that have not (circular, green data points).

tendon is also greater when you stretch than when you let it relax. What's more, the tendon can change its properties: The two sets of curves in Figure 6-20 are for males in their 70s before and after a 14-week course of physical training, which caused the patellar tendon to become much stronger and stiffer.

It's nonetheless true that Hooke's law is a useful approximation for the behavior of many materials when stretched or compressed, provided the amount of stretch or compression is small. We'll use this law in the next section.

GOT THE CONCEPT? 6-4 Double the Work

In Example 6-9 the athlete did 95 J of work to stretch the exercise cords by 47 cm. How far would he have to stretch the cords to do double the amount of work (2×95 J = 190 J)? (a) $\sqrt{2} \times 47$ cm = 66 cm; (b) 2×47 cm = 94 cm; (c) 4×47 cm = 188 cm.

TAKE-HOME MESSAGE FOR Section 6-5

✔ The work-energy theorem applies even when the object follows a curved path or the forces that act on the object are not constant.

✔ A force that always acts perpendicular to an object's curved path does zero work on the object.

✔ An ideal spring exerts a force proportional to the distance that it is stretched or compressed (Hooke's law). The work required to stretch or compress a spring by a given distance is proportional to the square of that distance.

6-6 Potential energy is energy related to an object's position

We've seen that an object in motion has the ability to do work, as measured by its kinetic energy. But a *stationary* object can also have the ability to do work. An example is the barbell in **Figure 6-21**. When held at rest above the weight lifter's head, the barbell has no kinetic energy. If the weight lifter should drop the barbell, however, it will fall to the ground with a resounding crash and leave a dent in the floor. In other words, the barbell will exert a downward force on the floor over a distance and so will do work. Thus the barbell has the *potential* to move and to do work simply because it's held so high. We use the term "**potential energy**" to refer to an ability to do work that's related to an object's *position*.

There are many examples of potential energy in your environment. The spring in a mousetrap has the potential to do very destructive work on any mouse foolish enough to release the trap. And the positively charged protons in a uranium nucleus, which want to push away from each other but are prevented from doing so by the strong force between the constituents of the nucleus, have the potential to do work if the nucleus is broken apart—a process that provides the energy released in a nuclear reactor.

Paul Bradbury/Getty Images

Figure 6-21 Potential energy The barbell in this photo is at rest and so has zero kinetic energy. But it has the *potential* to acquire kinetic energy if the weight lifter should drop it, so there is potential energy in this situation.

Gravitational Potential Energy

Let's quantify the amount of potential energy associated with the barbell in Figure 6-21. If the barbell has mass m and is initially a height h above the floor, as it falls the gravitational force does an amount of work $W = Fh = mgh$ on it. From the work-energy theorem this is equal to the kinetic energy that the dropped barbell has just before it hits the floor. (We're ignoring any effects of air resistance.) So we say that the potential energy (symbol U) of the Earth-barbell system before the barbell was dropped was $U = mgh$, and that this potential energy was converted to kinetic energy as the barbell fell. Note that potential energy has units of joules, the same as work and kinetic energy.

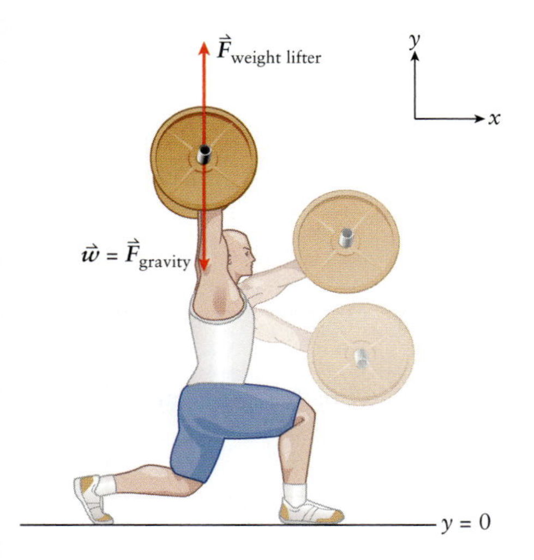

Figure 6-22 Adding potential energy The change in the kinetic energy of the barbell as it is raised by the weight lifter is zero. The work that he does to lift the barbell goes into increasing the potential energy.

Where did the potential energy come from? To see the answer, consider what happens as the weight lifter *raises* the barbell from the floor to a height h (**Figure 6-22**). During the lifting, the barbell begins with zero kinetic energy (sitting on the floor) and ends up with zero kinetic energy (at rest above the weight lifter's head). The net *change* in its kinetic energy is zero, so the net work done on the barbell is also zero. Hence the positive work that the weight lifter did to raise the barbell must just balance the negative work done by the gravitational force of magnitude mg. The work done by gravity is $W_{\text{grav}} = -(mg)(h) = -mgh$ (negative since the displacement is upward but the gravitational force is downward). Hence the work done by the weight lifter to raise the barbell to height h is $W_{\text{weight lifter}} = +mgh$, which is exactly equal to the potential energy mgh associated with the barbell when it is at height h. So the weight lifter is the source of the potential energy. We call $U = mgh$ the **gravitational potential energy** since it arises from the weight lifter doing work against the gravitational force.

WATCH OUT! To what object does gravitational potential energy "belong"?

We've been careful *not* to say "the barbell *has* gravitational potential energy." The reason is the gravitational potential energy is really a property of the *system* made up of the barbell and Earth (which exerts the gravitational force on the barbell). The weight lifter adds potential energy to this system by moving the barbell away from Earth. The gravitational potential energy mgh would be the same if the barbell remained at rest but Earth as a whole were pushed down a distance h. You should always be careful to define your system when using work and energy to solve problems.

In general, if the barbell of mass m is at a vertical coordinate y, the gravitational potential energy is

Gravitational potential energy associated with an object (6-14)

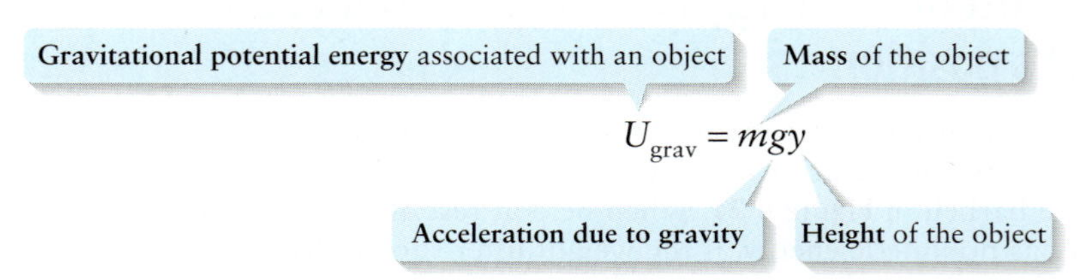

When the weight lifter in Figure 6-22 raises the barbell (so y increases), the gravitational potential energy increases; when he lowers or drops the barbell, y decreases, and the gravitational potential energy decreases.

Interpreting Potential Energy

Here's how the work-energy theorem describes what happens to a dropped barbell in the absence of air resistance: As the barbell falls, the gravitational force does work on it, and this work goes into changing the barbell's kinetic energy. In equation form,

(6-15) $$W_{\text{grav}} = \Delta K = K_{\text{f}} - K_{\text{i}} \text{ (only the gravitational force does work)}$$

Now let's *reinterpret* this statement in terms of gravitational potential energy and generalize to the case in which the barbell follows a curved path (as would be the case if the barbell was thrown rather than dropped) rather than falling straight down. The barbell in **Figure 6-23** moves along a curve from an initial y coordinate y_{i} to a final coordinate y_{f}. As we did for the swinging spider in Example 6-8 (Section 6-5), we divide the path into a large number of short segments, each of which is small enough that we can treat it as a straight line. The displacement vector along this segment is $\Delta\vec{d}$, with horizontal component Δx and vertical component Δy. The work done along this segment by the force of gravity $\vec{w}$ is exactly the same as it was in Example 6-8:

$$\Delta W_{\text{grav}} = -mg\,\Delta y$$

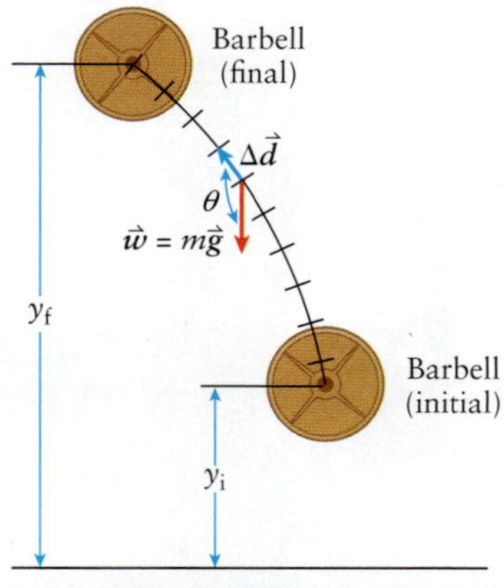

Figure 6-23 Raising a barbell along a curved path The work done on the barbell by the gravitational force $\vec{w}$ along a short segment $\Delta\vec{d}$ of this curved path is $\Delta W_{\text{grav}} = w\Delta d\cos\theta$. Adding up the work done along all such segments gives the net work done by the gravitational force between height y_{i} and height y_{f}.

The total work W_{grav} done by the force of gravity along the entire curved path is the sum of the ΔW_{grav} terms for each such element. The sum of all the Δy terms is the total change in the y coordinate, $y_f - y_i$, so

$$\begin{aligned} W_{grav} &= (-mg\,\Delta y_1) + (-mg\,\Delta y_2) + (-mg\,\Delta y_3) + \cdots \\ &= -mg(\Delta y_1 + \Delta y_2 + \Delta y_3 + \cdots) \\ &= -mg(y_f - y_i) = -mgy_f + mgy_i \end{aligned}$$

We can rewrite this in terms of the gravitational potential energy as given by Equation 6-14, $U_{grav} = mgy$:

$$\begin{aligned} W_{grav} &= -mgy_f + mgy_i = -(mgy_f - mgy_i) \\ &= -(U_{grav,f} - U_{grav,i}) = -\Delta U_{grav} \end{aligned} \tag{6-16}$$

In other words, *the work done by gravity on an object equals the negative of the change in gravitational potential energy of the Earth-object system.* If an object descends, the downward gravitational force does positive work on it and the gravitational potential energy of the system decreases (its change is negative). If an object rises, the downward gravitational force does negative work on it and the gravitational potential energy of the system increases (its change is positive). If an object begins and ends its motion at the same height, the gravitational force does zero net work on it and there is zero net change in the system's gravitational potential energy.

Go to Interactive Exercise 6-1 for more practice dealing with work, kinetic energy, and potential energy.

We can now restate the work-energy theorem for the falling barbell. If we substitute W_{grav} from Equation 6-16 into Equation 6-15, we get

If the **only force that does work** on a moving object is the **gravitational force...**

$$\Delta K = -\Delta U_{grav}$$

...then the **change in the object's kinetic energy** K...

...is equal to the **negative of the change in gravitational potential energy** U_{grav}.

If K increases, U_{grav} decreases and vice versa.

Change in kinetic energy if only the gravitational force does work (6-17)

If the object rises, gravitational potential energy increases and kinetic energy decreases (the object slows down). If the object descends, gravitational potential energy decreases and kinetic energy increases (the object speeds up). From this perspective we no longer need to talk about the *work* done by the gravitational force: That's accounted for completely by the change in the system's gravitational potential energy.

WATCH OUT! The choice of $y = 0$ for gravitational potential energy doesn't matter.

The value of $U_{grav} = mgy$ depends on what height you choose to be $y = 0$. But Equation 6-17 shows that what matters is the *change* in gravitational potential energy, and that does *not* depend on your choice of $y = 0$. That's because the change in gravitational potential energy depends only on the *difference* between the initial and final heights, not the heights themselves: $\Delta U_{grav} = U_f - U_i = mgy_f - mgy_i = mg(y_f - y_i)$. Example 6-10 illustrates this important point.

EXAMPLE 6-10 A Ski Jump

A skier of mass m starts at rest at the top of a ski jump ramp (**Figure 6-24**). The vertical distance from the top of the ramp to the lowest point is a distance H, and the vertical distance from the lowest point to where the skier leaves the ramp is D. The first part of the ramp is at an angle θ from the horizontal, and the second part is at an angle ϕ. Derive an expression for the speed of the skier when she leaves the ramp. Assume that her skis are well waxed, so that there is negligible friction between the skis and the ramp, and ignore air resistance.

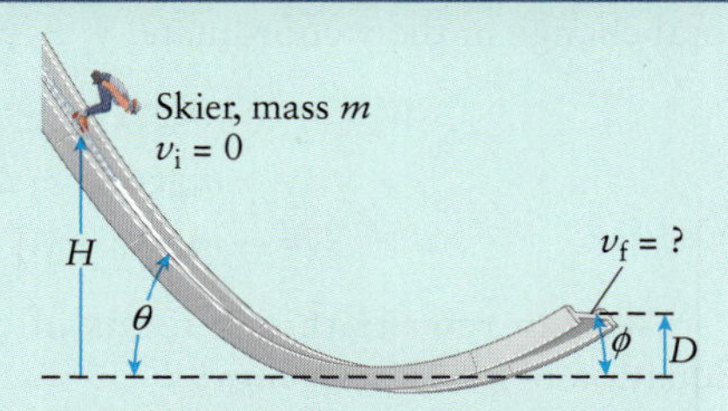

Figure 6-24 Flying off the ramp What is the skier's speed when she leaves the frictionless ramp?

Set Up

The only forces that act on the skier are the gravitational force and the normal force exerted by the ramp. The normal force does no work on her because it always acts perpendicular to the ramp and hence perpendicular to her direction of motion. This means that only the gravitational force does work, so we can use Equation 6-17 for the Earth-skier system to calculate the change in the skier's kinetic energy and hence her final speed v_f.

Change in kinetic energy if only the gravitational force does work:

$$\Delta K = -\Delta U_{grav} \quad (6\text{-}17)$$

Gravitational potential energy:

$$U_{grav} = mgy \quad (6\text{-}14)$$

Kinetic energy:

$$K = \frac{1}{2}mv^2 \quad (6\text{-}8)$$

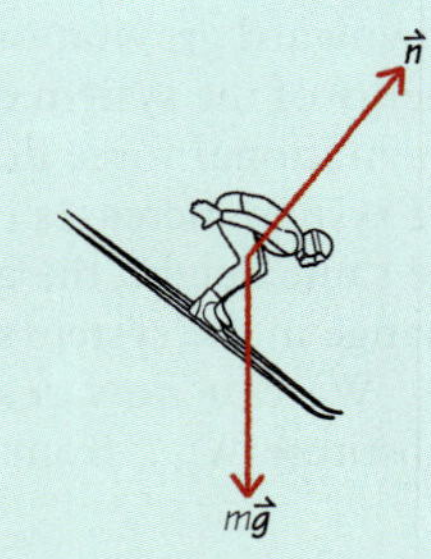

Solve

Let's take $y = 0$ to be at the low point of the ramp. The skier then begins at rest ($v_i = 0$) at $y_i = H$ and is at $y_f = D$ when she leaves the ramp. Use Equations 6–8, 6-14, and 6-17 to solve for v_f.

At starting point:

$$K_i = \frac{1}{2}mv_i^2 = 0$$

$$U_{grav,\,i} = mgy_i = mgH$$

At the point where skier leaves the ramp:

$$K_f = \frac{1}{2}mv_f^2$$

$$U_{grav,\,f} = mgy_f = mgD$$

Use Equation 6-17:

$-\Delta U_{grav} = \Delta K$, here

$$\Delta U_{grav} = U_{grav,\,f} - U_{grav,\,i} = mgD - mgH$$
$$= mg(D - H) = -mg(H - D)$$

(negative since $D < H$) and

$$\Delta K = K_f - K_i$$
$$= \frac{1}{2}mv_f^2 - 0 = \frac{1}{2}mv_f^2$$

(positive since skier is moving faster at the end of the ski jump than at the beginning). So

$$+\,mg(H - D) = \frac{1}{2}mv_f^2$$

$$v_f^2 = 2g(H - D)$$

$$v_f = \sqrt{2g(H - D)}$$

Reflect

The answer does *not* involve the angles θ or ϕ, or any other aspect of the ramp's shape. All that matters is the difference $H - D$ between the skier's initial and final heights. If the ramp had a different shape, the final speed v_f would be exactly the same.

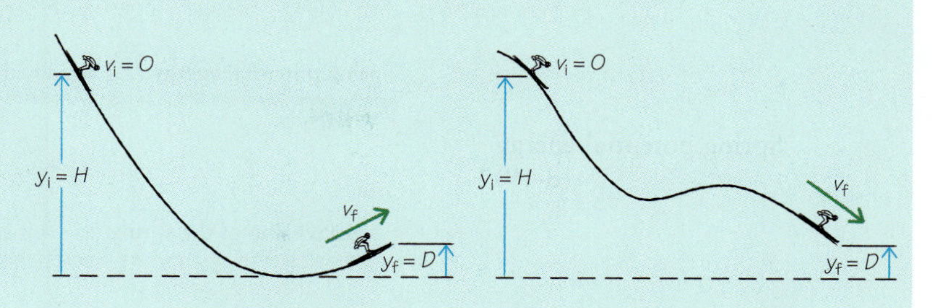

As we mentioned in the *Watch Out!* feature just before this example, our answer also shouldn't depend on our having chosen $y = 0$ to be at the low point of the ramp. For example, if we instead chose $y = 0$ to be where the skier leaves the ramp, the result for v_f would be the same. Try this yourself: Take $y = 0$ to be at the skier's starting point, so the end of the ramp is at $y = -(H - D)$. Do you get the same ΔU_{grav} with this choice?

With $y = 0$ at the point where the skier leaves the ramp:

At starting point:

$$U_{grav,\,i} = mgy_i = mg(H - D)$$

At the point where skier leaves the ramp:

$$U_{grav,\,f} = mgy_f = 0,\ \text{So}$$

$$\Delta U_{grav} = U_{grav,\,f} - U_{grav,\,i} = 0 - mg(H - D) = -mg(H - D)$$

This is the same as with our previous choice of $y = 0$, so we'll find the same value of v_f.

GOT THE CONCEPT? 6-5 Choosing a Point of Reference

? A ball is dropped from the top of an 11-story building, 45 m above the ground, to a balcony on the back of the ninth floor, 36 m above the ground. In which case is the change in the potential energy associated with the ball greatest? (a) If we choose the ground to be at $y = 0$; (b) if we choose the balcony to be at $y = 0$; (c) if we choose the top of the building to be at $y = 0$; (d) the change is the same in all three cases.

Spring Potential Energy

In Example 6-10 the work done by the gravitational force on the skier is $W_{grav} = -\Delta U_{grav}$ (see Equation 6-16). We saw that this work depends only on the skier's final and initial positions, *not* on the path that she took between those two points. That's why we can express the work done by the gravitational force in terms of the difference ΔU_{grav} between the gravitational potential energy values at the two points.

The same is true for the work done by an ideal spring. We saw in Section 6-5 that the work you must do to stretch a spring from an extension x_i to an extension x_f is $\frac{1}{2}kx_f^2 - \frac{1}{2}kx_i^2$ (this is Equation 6-13, with 1 replaced by i and 2 replaced by f). The work that the *spring* does is the negative of this: $W_{spring} = -(\frac{1}{2}kx_f^2 - \frac{1}{2}kx_i^2)$. Note that this also depends on only the initial and final extensions of the spring, not on the details of how you got from one to the other. Hence just as we did for the work done by the gravitational force, we can write the work done by an ideal spring in terms of a change in potential energy:

$$\begin{aligned} W_{spring} &= -\left(\frac{1}{2}kx_f^2 - \frac{1}{2}kx_i^2\right) \\ &= -(U_{spring,\,f} - U_{spring,\,i}) = -\Delta U_{spring} \end{aligned} \tag{6-18}$$

The quantity U_{spring} in Equation 6-18 is called the **spring potential energy:**

Spring potential energy (6-19)

Spring potential energy of a stretched or compressed spring

Spring constant of the spring

$$U_{spring} = \frac{1}{2}kx^2$$

Extension of the spring ($x > 0$ if spring is stretched, $x < 0$ if spring is compressed)

The spring potential energy is zero if the spring is relaxed ($x = 0$) and positive if the spring is stretched ($x > 0$) or compressed ($x < 0$) (**Figure 6-25**). This says that we have to do work to either stretch the spring or compress it, and the work that we do goes into the spring potential energy.

While a human tendon is not an ideal spring, we can think of it as storing spring potential energy when it is stretched. When you are running and one of your feet is in contact with the ground, the Achilles tendon at the back of that leg is stretched. The spring potential energy stored in that tendon literally "springs" you back in the air, helping you to sustain your running pace.

WATCH OUT! For spring potential energy, $x = 0$ means a relaxed spring.

For gravitational potential energy we're free to choose $y = 0$ to be anywhere we like (see Example 6-10). We don't have that kind of freedom for spring potential energy: In Equation 6-19, we *must* choose $x = 0$ to be where the spring is neither compressed nor stretched.

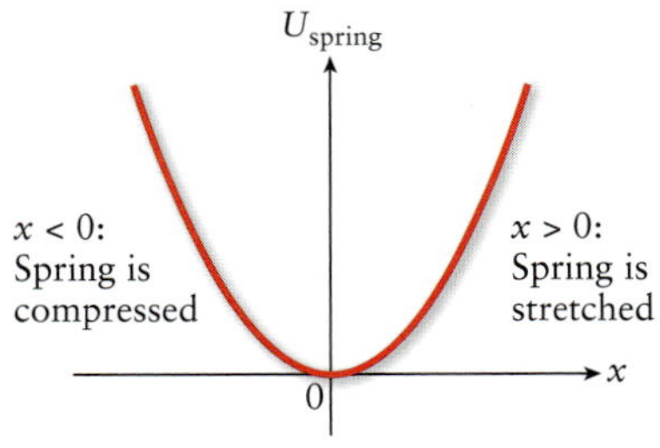

Figure 6-25 Spring potential energy The potential energy in a spring is proportional to the square of its extension x. (See Equation 6-19.)

Conservative and Nonconservative Forces

A force that can be associated with a potential energy, like the gravitational force or the force exerted by an ideal spring, is called a **conservative force.** (In Section 6-7 we'll see the reason for this term.) By contrast, the friction force is an example of a **nonconservative force** for which we *cannot* use the concept of potential energy. The reason is that unlike the gravitational force or the force exerted by an ideal spring, the work done by the friction force *does* depend on the path taken from the initial point to the final point. In **Figure 6-26a** we slide a book across a table top from an initial point to a final point along two different paths. Along either path the kinetic friction force has the same magnitude and points opposite to the direction of motion. Hence the friction force does more (negative) work along the curved path than along the straight path. Since the work done by friction depends on more than just the initial and final positions, we can't write it in terms of a change in potential energy. That's why there's no such thing as "friction potential energy."

Here's an equivalent way to decide whether a certain kind of force is conservative: If the work done by the force on a *round trip* (that is, one where the initial and final positions are the same) is *zero*, the force is conservative and we can use the idea of potential energy. This is the case for the gravitational force: if $y_f = y_i$, then $W_{grav} = -mgy_f + mgy_i = 0$. If you toss a ball straight up, the gravitational force does negative work on it as it rises and an equal amount of positive work as it falls back to your hand. The same is true for the spring force: If $x_f = x_i$, then $W_{spring} = \frac{1}{2}kx_f^2 - \frac{1}{2}kx_i^2 = 0$. But if you slide a book on a round trip on a table top, the total amount of work that the friction force does on the book is negative and *not* zero (**Figure 6-26b**). Hence the friction force is nonconservative. (To keep the book moving, you have to do an equal amount of positive work on the book as you push it.)

(a)

Friction does more negative work along path 2 than along path 1. Since the work done by friction depends on the path, we conclude that friction is **not** a conservative force.

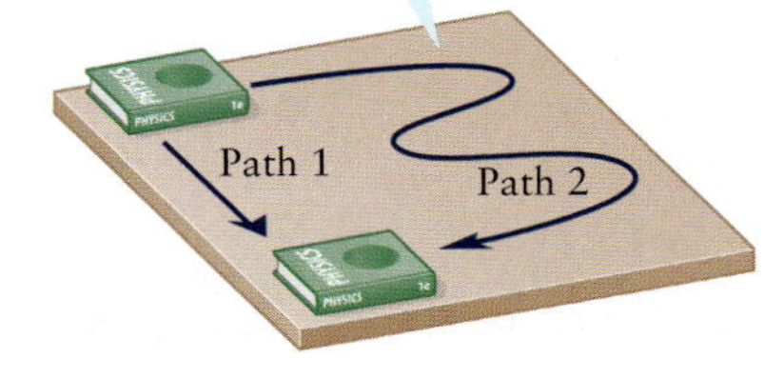

(b)

The book makes a round trip that begins and ends at this point. Friction does a nonzero, negative amount of work on the book for this round trip. So we again conclude that friction is **not** a conservative force.

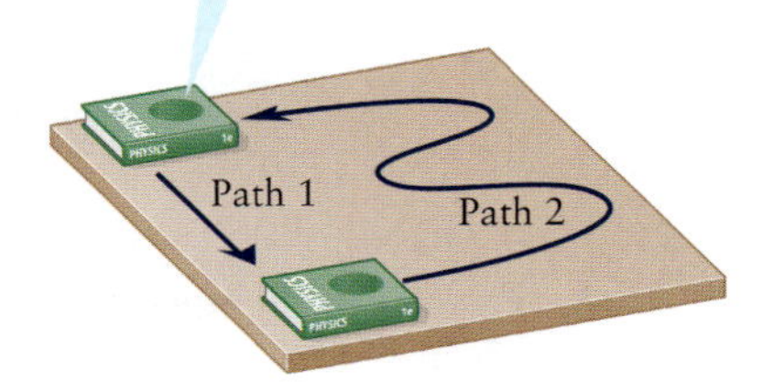

Figure 6-26 Kinetic friction is a nonconservative force Because the work done by kinetic friction depends on the path, and it does nonzero net work on a round trip, it is a nonconservative force.

Figure 6-20 shows that the force exerted by a human tendon is also nonconservative. The tendon exerts more force on its end while it is being stretched than when it is relaxing to its original length, so it does more negative work on the muscle attached to its end as it stretches than it does positive work as it relaxes. So the tendon does a nonzero (and negative) amount of work on this "round trip." To make the tendon go through a complete cycle, the muscle has to do a nonzero amount of *positive* work on the tendon, just like the positive work you must do to push the book around the path in Figure 6-26b.

What happens to the work that you do in these cases? Before we can answer this question, we'll combine the ideas of kinetic energy and potential energy into yet another kind of energy, called *total mechanical energy.* This will lead us to one of the central ideas not just of physics but of science as a whole, the idea of *conservation of energy*.

TAKE-HOME MESSAGE FOR Section 6-6

✔ The work done by a conservative force depends only on the initial and final positions of the object, not on the path the object followed from one position to the other. The gravitational force and the force exerted by an ideal spring are examples of conservative forces.

✔ A kind of energy, called potential energy, is associated with each kind of conservative force.

✔ The work done by a conservative force is equal to the negative of the change in the associated potential energy. If the force does positive work, the potential energy decreases; if the force does negative work, the potential energy increases.

6-7 If only conservative forces do work, total mechanical energy is conserved

Take a pencil in your hand and toss it upward. As the pencil ascends, it loses speed and its kinetic energy $K = \frac{1}{2}mv^2$ decreases. At the same time the pencil gains height so that the gravitational potential energy $U_{\text{grav}} = mgy$ increases. After the pencil reaches its maximum height and falls downward, its kinetic energy increases as it gains speed and the gravitational potential energy decreases as it loses height. This way of thinking about the pencil's up-and-down motion suggests that energy is *transformed* from one form (kinetic) into a different form (gravitational potential) as the pencil rises and is transformed back as the pencil descends.

Another example of energy transformation is happening inside your body right now. When your heart contracts it pushes blood into the arteries, stretching their walls to accommodate the increased volume. The stretched arterial walls behave like a stretched spring and so possess spring potential energy. In between heart contractions the arterial walls relax back to their equilibrium size and lose their potential energy, just as a stretched spring does when it relaxes to its unstretched length. As a result blood gains kinetic energy as the arterial walls push on it between heartbeats. Thus spring potential energy of the arteries is transformed into kinetic energy of the blood (**Figure 6-27**).

In both of these situations the *sum* of kinetic energy and potential energy—a sum that we call the *total mechanical energy*—keeps the same value and is *conserved.* When the system's kinetic energy K increases, its potential energy U decreases, and when K decreases, U increases. It's like having two bank accounts: You can transfer money from one account to the other, but the total amount of money in the two accounts remains the same. In this section we'll learn the special circumstances in which total mechanical energy is conserved, and we'll see how to express these ideas in equation form.

Suppose that as an object moves from an initial position to a final position, the only force that does work on the object is the gravitational force. [An example is the skier on a ski jump in Example 6-10 (Section 6-6). In addition to the gravitational

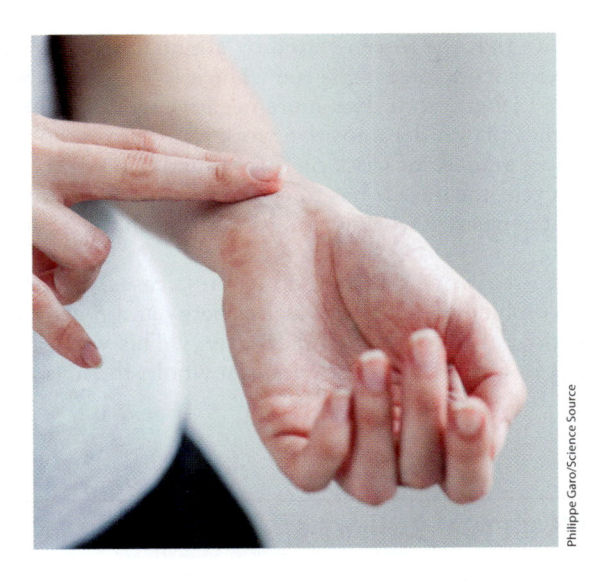

Figure 6-27 Kinetic and potential energy in the arteries When you feel the pulse in your radial artery, you are actually feeling spring potential energy of the arterial walls being converted to kinetic energy of the blood. This energy transformation, which occurs about once per second, is essential for life.

Philippe Garo/Science Source

force on the skier, there is a normal force exerted by the ramp, but it does no work on the skier because this force is always perpendicular to the skier's path.] According to Equation 6-17 in this situation the work-energy theorem tells us that the change in the object's kinetic energy is equal to the negative of the change in its potential energy:

$$\Delta K = -\Delta U_{\text{grav}} \text{ (if only the gravitational force does work)} \tag{6-20}$$

The change in kinetic energy equals the final value minus the initial value, and likewise for the gravitational potential energy. So we can rewrite Equation 6-20 as

$$K_{\text{f}} - K_{\text{i}} = -(U_{\text{grav, f}} - U_{\text{grav, i}}) = -U_{\text{grav, f}} + U_{\text{grav, i}}$$
$$\text{(if only the gravitational force does work)}$$

Let's rearrange this equation so that all the terms involving the initial situation are on the left-hand side of the equals sign and all the terms involving the final situation are on the right-hand side. We get

$$K_{\text{i}} + U_{\text{grav, i}} = K_{\text{f}} + U_{\text{grav, f}} \text{ (if only the gravitational force does work)} \tag{6-21}$$

If only the gravitational force does work on the object as it moves, its speed v and kinetic energy $K = \frac{1}{2}mv^2$ can change, and its height y and gravitational potential energy $U_{\text{grav}} = mgy$ can change. But the *sum* of K and U_{grav} has the same value at the end of the motion as at the beginning.

We can generalize these ideas and Equation 6-21 even more. Our results from Section 6-6 show that the work done by *any* conservative force can be written as the negative of the change in the associated potential energy. (Equations 6-18 and 6-19 show this for the case of the work done by an ideal spring: $W_{\text{spring}} = -\Delta U_{\text{spring}}$, where the spring potential energy is $U_{\text{spring}} = \frac{1}{2}kx^2$ and x is the extension of the spring.) If a number of conservative forces $\vec{F}_A, \vec{F}_B, \vec{F}_C, \ldots$ do work on an object as it moves, and the potential energies associated with each of these forces are $U_{\text{A}}, U_{\text{B}}, U_{\text{C}}, \ldots,$ then the total work done by these forces is

$$\begin{aligned} W_{\text{conservative}} &= W_A + W_B + W_C + \cdots \\ &= (-\Delta U_A) + (-\Delta U_B) + (-\Delta U_C) + \cdots \\ &= (-U_{A,\text{f}} + U_{A,\text{i}}) + (-U_{B,\text{f}} + U_{B,\text{i}}) + (-U_{C,\text{f}} + U_{C,\text{i}}) + \cdots \\ &= -(U_{A,\text{f}} + U_{B,\text{f}} + U_{C,\text{f}} + \cdots) + (U_{A,\text{i}} + U_{B,\text{i}} + U_{C,\text{i}} + \cdots) \\ &= -U_{\text{f}} + U_{\text{i}} = -\Delta U \end{aligned} \tag{6-22}$$

In Equation 6-22 the quantity U is the *total* potential energy, which is just the sum of the individual potential energies $U_{\text{A}}, U_{\text{B}}, U_{\text{C}}, \ldots$ If these conservative forces are the only forces that do work on the object, then $W_{\text{conservative}}$ in Equation 6-22 equals the net work done on the object, which from the work-energy theorem equals $\Delta K = K_f - K_i$. So if only conservative forces do work, we have

$$K_f - K_i = -U_f + U_i$$

or, rearranging,

Values of the **kinetic energy** K of an object at two points (i and f) during its motion

$$K_i + U_i = K_f + U_f$$

Values of the **potential energy** U at the same two points

Mechanical energy is conserved if only conservative forces do work (6-23)

If only conservative forces do work on an object, the sum $K + U$ maintains the same value throughout the motion. This sum is called the **total mechanical energy** E.

You can see that this is a more general form of Equation 6-21, which refers to the case in which only one conservative force—the gravitational force—does work. Note that we can take the initial point i and the final point f to be *any* two points during the motion, as long as only conservative forces do work.

The sum of kinetic energy K and potential energy U is called the **total mechanical energy** E. So Equation 6-23 tells us that if only conservative forces do work on an object, the total mechanical energy is *conserved;* that is, it maintains the same value during the motion (see **Figure 6-28**). In order for this energy conservation to take place, your system must include the sources of the conservative forces—for example, Earth and springs. You can now see the origin of the term "conservative" for forces like the gravitational force and the spring force: If only forces of this kind do work, the total mechanical energy of the system is conserved.

Another way to write Equation 6-23 is to move all of the terms in that equation to the same side of the equals sign:

$$K_f - K_i + U_f - U_i = 0$$

Since $K_f - K_i = \Delta K$ (the change in kinetic energy) and $U_f - U_i = \Delta U$ (the change in potential energy), we can rewrite this equation as

The kinetic energy K and the gravitational potential energy U_{grav} both change, but their sum E remains the same.

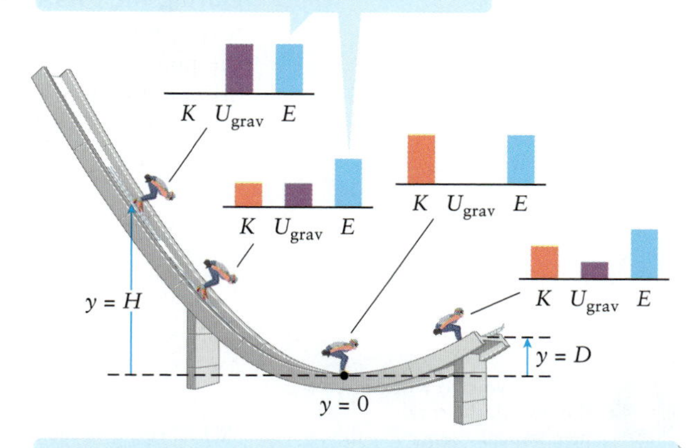

When only conservative forces like gravity act, the total mechanical energy E of the system is conserved.

Figure 6-28 Total mechanical energy is conserved As the skier moves on this frictionless ramp, the kinetic energy K and the gravitational potential energy U_{grav} both change, but their sum $E = K + U_{grav}$—the total mechanical energy—always has the same value.

Change in the kinetic energy K of an object during its motion

Change in the potential energy U associated with an object during its motion

$$\Delta K + \Delta U = 0$$

Conservation of mechanical energy—alternative version (6-24)

If only conservative forces do work on the object, the sum is zero: Total mechanical energy can transform between kinetic and potential forms, but there is no net change in the amount of total mechanical energy.

Equation 6-24 refers to the change in kinetic energy ΔK and the change in potential energy ΔU during the motion. If the only forces that do work on the object are conservative forces, the sum of these is zero: Any increase in kinetic energy comes with an equal decrease in potential energy, and vice versa.

Go to Picture It 6-2 for more practice dealing with mechanical energy.

WATCH OUT! Total mechanical energy is a property of a system, not a single object.

We saw in Section 6-6 that gravitational potential energy for an object like a baseball is really a shared property of a *system* of two objects, the baseball and Earth. Thus the total mechanical energy for a baseball in flight, which incorporates gravitational potential energy, is likewise a property of that system. Likewise, if a weight is attached to a horizontal spring and allowed to oscillate back and forth, the total mechanical energy is a shared property of the weight (which has kinetic energy) and the spring (which has spring potential energy). Whenever you think about total mechanical energy, you should always be able to state to what system that mechanical energy belongs.

In Section 6-8 we'll use the idea of conservation of mechanical energy to solve a number of physics problems. First, however, let's see how this idea has to be modified if there are also *nonconservative* forces that do work.

Generalizing the Idea of Conservation of Energy: Nonconservative Forces

There are many situations in which total mechanical energy is *not* conserved. As an example, at the beginning of this section we asked you to toss a pencil into the air. During the toss you had to do *positive* work on the pencil: You exerted an upward force on the pencil as you pushed it upward. As a result the total mechanical energy of the pencil increased (**Figure 6-29**). If you subsequently catch the pencil as it falls, you do *negative* work on the pencil because you exert an upward force on it as the pencil moves downward. As Figure 6-29 shows, the total mechanical energy decreases during the catch.

Because you do work on the pencil, the total mechanical energy E—the sum of kinetic energy K and gravitational energy U_{grav}—is not conserved throughout this process.

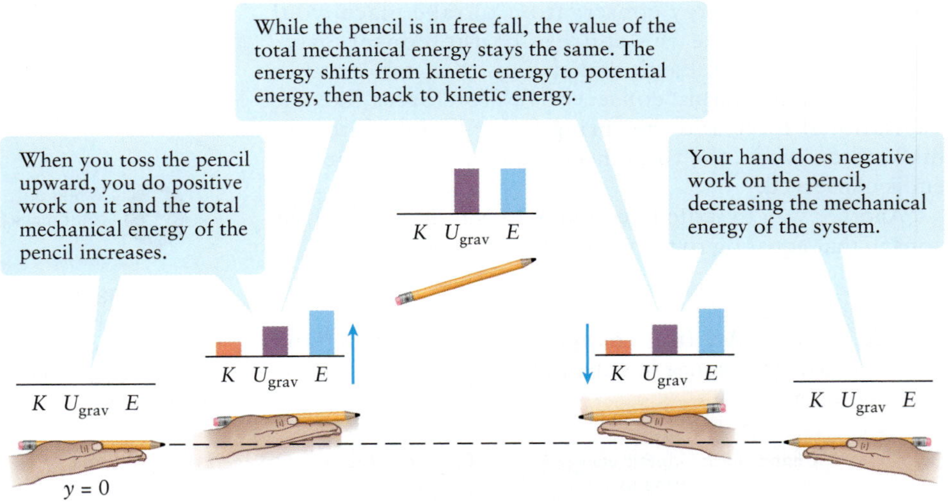

While the pencil is in free fall, the value of the total mechanical energy stays the same. The energy shifts from kinetic energy to potential energy, then back to kinetic energy.

When you toss the pencil upward, you do positive work on it and the total mechanical energy of the pencil increases.

Your hand does negative work on the pencil, decreasing the mechanical energy of the system.

K U_{grav} E

$y = 0$

Figure 6-29 Tossing and catching a pencil The total mechanical energy associated with the pencil increases as you toss it upward, remains constant while the pencil is in flight, then decreases as you catch it.

The total mechanical energy changes in these situations because the force exerted by your hand is not conservative. We can write a modified version of Equation 6-23 that can accommodate situations like these in which the total mechanical energy is *not* conserved. To do this let's go back to the work-energy theorem, $W_{net} = K_f - K_i$. If both conservative and nonconservative forces do work on an object, the net work is the sum of the work done by the two kinds of forces:

$$W_{net} = W_{conservative} + W_{nonconservative} \quad (6\text{-}25)$$

Equation 6-22 says that the work done by conservative forces is $W_{conservative} = -U_f + U_i = -\Delta U$. Substituting this and Equation 6-25 into the work-energy theorem, we get

$$W_{net} = (-U_f + U_i) + W_{nonconservative} = K_f - K_i$$

We can rearrange this to

Values of the **kinetic energy** K of an object at two points (i and f) during its motion

$$K_i + U_i + W_{\text{nonconservative}} = K_f + U_f$$

Values of the **potential energy** U at the same two points

Change in mechanical energy when nonconservative forces act (6-26)

If nonconservative forces do work on an object, the total mechanical energy $E = K + U$ changes its value.
If $W_{\text{nonconservative}} > 0$, $E_f > E_i$ and the total mechanical energy increases.
If $W_{\text{nonconservative}} < 0$, $E_f < E_i$ and the total mechanical energy decreases.

Equation 6-26 gives us more insight into the difference between conservative and nonconservative forces. In Figure 6-29 only the conservative force of gravity acts on the pencil from the time when it leaves your hand to when it returns to your hand. So the mechanical energy $E = K + U$ remains constant during this part of the motion. Kinetic energy decreases and changes into potential energy as the pencil ascends, but the potential energy is turned back into kinetic energy as the pencil falls. If the force of your hand were also a conservative force, all of the energy you expended to toss the pencil would be returned to your body when you caught the pencil; you could toss the pencil up and down all day and you would never get tired. This is *not* the case, however, and you *do* get tired. (If you don't believe this, try it with a book rather than a pencil and try tossing the book up and catching it a few dozen times.) Because the energy you expend in the toss is *not* returned to you in the catch, the force of your hand is nonconservative.

BIO-Medical

Here's an alternative way to think of the nonconservative forces depicted in Figure 6-29. To toss the pencil upward you actually use energy stored in the chemical bonds of adenosine triphosphate (ATP), carbohydrates, and fat in the muscles of your arm. We don't include this kind of energy in the quantity U in Equation 6-26, which refers only to potential energy associated with the pencil. Instead, we'll use the symbol E_{other} for **internal energy**, a catch-all kind of energy that is not included in K or U in Equation 6-26. Not all of the released chemical energy goes into moving your muscles; some goes into warming your muscles (which is why you get warm when you exercise) and hence into **thermal energy**, which is energy associated with the random motion of atoms and molecules (in this case inside your arm). Thermal energy is also considered part of E_{other}. So during the toss, some chemical energy is converted into thermal energy (and so remains within E_{other}), and the rest is used to do work on the pencil (**Figure 6-30**). Hence during the toss for which $W_{\text{nonconservative}} > 0$, E_{other} of your body decreases so $\Delta E_{\text{other}} < 0$.

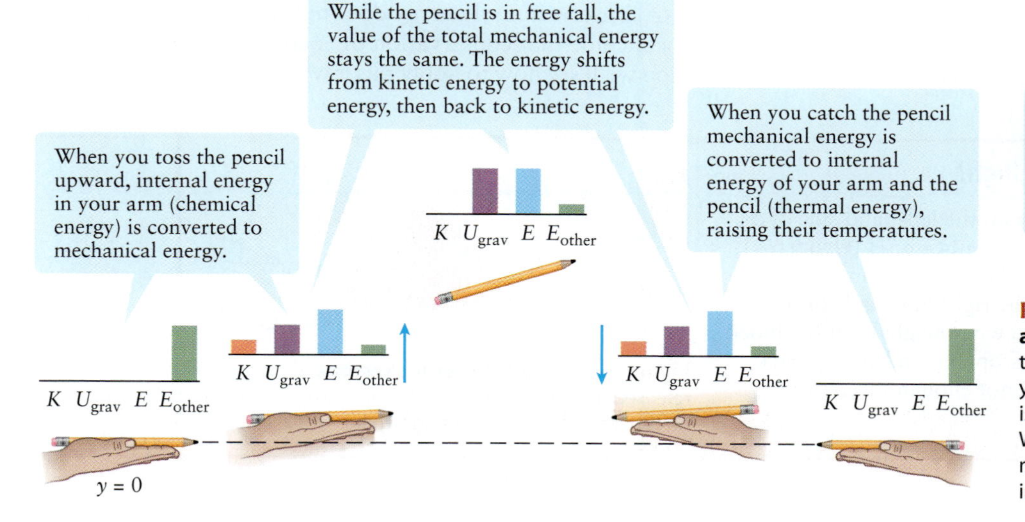
While the pencil is in free fall, the value of the total mechanical energy stays the same. The energy shifts from kinetic energy to potential energy, then back to kinetic energy.
When you toss the pencil upward, internal energy in your arm (chemical energy) is converted to mechanical energy.
When you catch the pencil mechanical energy is converted to internal energy of your arm and the pencil (thermal energy), raising their temperatures.
K Ugrav E Eother
y = 0

Total energy—the sum of total mechanical energy (K plus U_{grav}) and internal energy E_{other}—is conserved in this process.

Figure 6-30 Tossing and catching a pencil, revisited When you toss the pencil some internal energy of your body in chemical form (E_{other}) is converted to mechanical energy. When you catch the pencil this added mechanical energy is converted back into internal energy.

When you catch the pencil you again use your muscles and so you use more of the energy stored in chemical bonds. In the catch, however, both your arm and the pencil warm by a slight but measurable amount, corresponding to an increase in thermal energy, and this increase is greater than the loss of chemical energy (Figure 6-30). So during the catch, for which $W_{\text{nonconservative}} < 0$, E_{other} of you and the pencil increases so $\Delta E_{\text{other}} > 0$.

These observations suggest that the amount of nonconservative work done is related to the change in internal energy. Indeed, many careful measurements show that in *all* situations, $W_{\text{nonconservative}}$ is *exactly* equal to the negative of ΔE_{other}:

$$W_{\text{nonconservative}} = -\Delta E_{\text{other}} \tag{6-27}$$

With Equation 6-27 in mind we can rewrite Equation 6-26 as

$$K_{\text{i}} + U_{\text{i}} - \Delta E_{\text{other}} = K_{\text{f}} + U_{\text{f}}$$

If we move all of the terms in this equation to the same side of the equals sign, we get

$$K_{\text{f}} - K_{\text{i}} + U_{\text{f}} - U_{\text{i}} + \Delta E_{\text{other}} = 0$$

Now $K_{\text{f}} - K_{\text{i}}$ is the change in kinetic energy ΔK and $U_{\text{f}} - U_{\text{i}}$ is the change in potential energy ΔU. So we can rewrite this equation as

Change in the kinetic energy K of an object during its motion

Change in the potential energy U associated with an object during its motion

The law of conservation of energy (6-28)

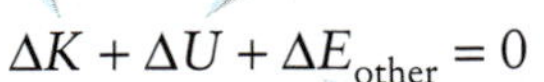

$$\Delta K + \Delta U + \Delta E_{\text{other}} = 0$$

Change in the internal energy E_{other} during the object's motion

Energy can transform between kinetic, potential, and internal forms, but there is no net change in the amount of energy.

Equation 6-28 suggests that we broaden our definition of energy to include both total mechanical energy ($E = K + U$) and internal energy E_{other}. In this case we have to expand the system to include not only the object and the sources of the conservative forces that produce ΔU but also the sources of the nonconservative forces that result in ΔE_{other} (like your hand in the case of the thrown pencil). As the object moves, K, U, and E_{other} can all change values, but the sum of these changes is zero: One kind of energy can transform into another, but the total amount of energy of all forms remains the same. This is the most general statement of the **law of conservation of energy**.

Unlike the similar-appearing Equation 6-24, which applies only if nonconservative forces do no work, Equation 6-28 is *always* true as long as the system contains all objects that are exerting or feeling forces. Scientists have made an exhaustive search for situations in which the law of conservation of energy does not hold. No such situation has ever been found. So we conclude that conservation of energy is an absolute law of nature. In the following section we'll see how to apply this law to solve physics problems.

GOT THE CONCEPT? 6-6 How High?

? A block is released from rest on the left-hand side of an asymmetric skateboard ramp at a height H (**Figure 6-31**). The angle from the horizontal of the left-hand side of the ramp is twice the angle of the right-hand side of the ramp. If the ramp is frictionless, to what height does the block rise on the right-hand side before stopping and sliding back down? (a) $2H$; (b) H; (c) $H/2$; (d) not enough information given to decide.

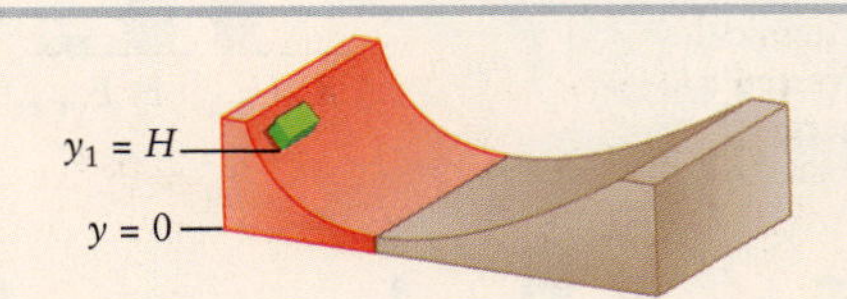

Figure 6-31 An asymmetric skateboard ramp If the block is released at height H on the left-hand side, to what height will it rise on the more gently sloped right-hand side?

TAKE-HOME MESSAGE FOR Section 6-7

✔ If the only forces that do work on an object are conservative forces, then the total mechanical energy E (the sum of the kinetic energy K and the potential energy U) is conserved. The values of K and U may change as the object moves, but the value of E remains the same.

✔ If nonconservative forces also do work on the object, the total mechanical energy changes. The value of E increases if positive nonconservative work is done and decreases if negative nonconservative work is done.

✔ Work done by a nonconservative force is associated with a change in internal energy. The total energy—the sum of kinetic, potential, and internal energies—is always conserved.

6-8 Energy conservation is an important tool for solving a wide variety of problems

Let's now see how to use the energy conservation equations valid when only conservative forces do work (Equations 6-23 and 6-24) and those valid when nonconservative forces also do work (Equations 6-26 and 6-28). As a rule, if a problem involves a conserved quantity (that is, a quantity whose value remains unchanged), it simplifies the problem tremendously. Often the conservation equation, which relates the value of the conserved quantity at one point in an object's motion to the value at a different point, is all you need to solve for the desired unknown. We'll see several problems of that kind in this section.

Go to Interactive Exercise 6-2 for more practice dealing with energy conservation.

EXAMPLE 6-11 A Ski Jump, Revisited

As in Example 6-10 (Section 6-6), a skier of mass m starts at rest at the top of a ski jump ramp (**Figure 6-32**). (a) Use an energy conservation equation to find an expression for the skier's speed as she flies off the ramp. There is negligible friction between the skis and the ramp, and you can ignore air resistance. (b) After the skier reaches the ground at point P in Figure 6-32, she begins braking and comes to a halt at point Q. Find an expression for the magnitude of the constant friction force that acts on her between points P and Q.

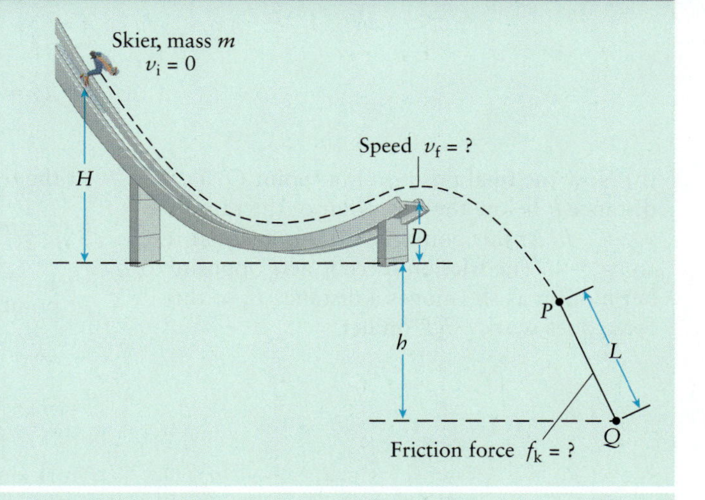

Figure 6-32 Flying off the ramp and coming to a halt What is the skier's speed as she leaves the ramp? How much friction is required to bring her to a halt?

Set Up

In part (a) of this problem the skier travels from an initial point i at the top of the ramp to a final point f at the end of the ramp. During this motion only the conservative force of gravity does work on her. (The ramp also exerts a normal force on her, but this force does no work since it is always perpendicular to her motion.) So we can use Equation 6-23. (There are no springs, so there is no spring potential energy.) Our goal for this part is to find the skier's speed v_f as she leaves the ramp.

Conservation of mechanical energy:

$$K_i + U_i = K_f + U_f \quad (6\text{-}23)$$

Gravitational potential energy:

$$U_{grav} = mgy \quad (6\text{-}14)$$

Change in mechanical energy when nonconservative forces act:

$$K_i + U_i + W_{nonconservative} = K_f + U_f \quad (6\text{-}26)$$

Kinetic energy:

$$K = \frac{1}{2}mv^2 \quad (6\text{-}8)$$

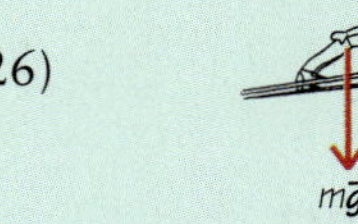

For part (b) we consider the skier's entire motion from the top of the ramp (which we again call point i) to where she finally comes to rest at point Q. A normal force also acts on her when she reaches the ground, but again this force does no work on her. Between points P and Q a nonconservative *friction* force also acts and does negative work on her (the force is opposite to her motion). So for the motion as a whole, we must use Equation 6-26. Our goal for part (b) is to find the magnitude f_k of the friction force.

$\vec{f}_k$, $\vec{n}$, $m\vec{g}$ between P and Q

Solve

(a) Write expressions for the skier's kinetic energy K and the gravitational potential energy U_{grav} at the top of the ramp (at $y_i = H$, where the speed is $v_i = 0$) and the end of the ramp (at $y_f = D$, where the speed is v_f).

At the top of the ramp:

$$K_i = \frac{1}{2}mv_i^2 = 0$$

$$U_{grav,\,i} = mgy_i = mgH$$

At the end of the ramp:

$$K_f = \frac{1}{2}mv_f^2$$

$$U_{grav,\,f} = mgy_f = mgD$$

Now substitute the energies into Equation 6-23 and solve for v_f.

Equation 6-23:

$K_i + U_i = K_f + U_f$, so

$$0 + mgH = \frac{1}{2}mv_f^2 + mgD$$

$$\frac{1}{2}mv_f^2 = mgH - mgD = mg\,(H - D)$$

$$v_f^2 = 2g(H - D)$$

$$v_f = \sqrt{2g(H - D)}$$

(b) Now the final position is at point Q, a distance h below the low point of the ramp so $y_Q = -h$. At this point the skier is again at rest, so $v_Q = 0$. The friction force f_k acts opposite to her motion as she moves a distance L, so this force does work $-f_kL$ on her.

At the top of the ramp:

$$K_i = \frac{1}{2}mv_i^2 = 0$$

$$U_{grav,\,i} = mgy_i = mgH$$

At point Q:

$$K_Q = \frac{1}{2}mv_Q^2 = 0$$

$$U_{grav,Q} = mgy_Q = mg(-h) = -mgh$$

Work done by the friction force, which points opposite to the skier's displacement:

$$W_{nonconservative} = f_kL\cos 180° = f_kL(-1) = -f_kL$$

Now use Equation 6-26 to solve for f_k.

Equation 6-26:

$K_i + U_i + W_{nonconservative} = K_Q + U_Q$, so

$$0 + mgH + (-f_kL) = 0 + (-mgh)$$

$$f_kL = mgH + mgh = mg\,(H + h)$$

$$f_k = mg\left(\frac{H + h}{L}\right)$$

Reflect

Our answer for part (a) is the same one that we found in Example 6-10, where we used the expression $W_{grav} = -\Delta U_{grav}$ for the work done by gravity. That's as it should be, since we used that expression in Section 6-7 to help derive Equation 6-23.

Our result for f_k in part (b) has the correct dimensions of force, the same as mg, since the dimensions of $H + h$ (the total vertical distance that the skier descends) cancel those of L (the skier's stopping distance). Note that the smaller the stopping distance L compared to $H + h$, the greater the friction force required to bring the skier to a halt.

An alternative way to solve part (b) would be to treat the second part of the skier's motion separately, with the initial point at the end of the ramp [at height D, where the skier's speed is $\sqrt{2g(H - D)}$ as found in part (a)] and the final point is at point Q (at height $-h$, where the skier's speed is zero). You should try solving the problem this way using Equation 6-26 and $W_{\text{nonconservative}} = -f_k L$; you should get the same answer for f_k. Do you?

EXAMPLE 6-12 Warming Skis and Snow

When the skier in Example 6-11 brakes to a halt, her skis and the snow over which she slides both warm up. What is the increase in the internal energy of the skis and snow in this process?

Set Up

This problem involves internal energy, so we'll use the most general statement of energy conservation. This says that during the skier's motion (from the top of the ramp to when she finally comes to rest at point Q) the sum of the changes in kinetic energy K, potential energy U, and internal energy E_{other} must be zero. We'll use this to find how much internal energy is added to the ski and snow during the braking.

The law of conservation of energy:

$$\Delta K + \Delta U + \Delta E_{\text{other}} = 0 \quad (6\text{-}28)$$

Solve

Substitute the change in kinetic energy ΔK and the change in potential energy ΔU into Equation 6-28 and solve for the increase in internal energy. We'll refer to Example 6-11 for the initial and final values of y and v.

At the top of the ramp: $K_i = \frac{1}{2}mv_i^2 = 0$

$$U_{\text{grav, i}} = mgy_i = mgH$$

At point Q: $K_Q = \frac{1}{2}mv_Q^2 = 0$

$$U_{\text{grav, }Q} = mgy_Q = mg(-h) = -mgh$$

Change in kinetic energy: $\Delta K = K_Q - K_i = 0 - 0 = 0$

Change in potential energy: $\Delta U = U_{\text{grav, }Q} - U_{\text{grav, i}}$

$$= -mgh - mgH = -mg\,(H + h)$$

From Equation 6-28, $\Delta K + \Delta U + \Delta E_{\text{other}} = 0$

$$0 + [-mg\,(H + h)] + \Delta E_{\text{other}} = 0$$

$$\Delta E_{\text{other}} = mg(H + h)$$

Reflect

The overall motion of the skier begins and ends with zero kinetic energy. The net result is that the (gravitational) potential energy decreases by an amount $mg(H + h)$ and the internal energy increases by the same amount. This represents the amount of thermal energy that is added to the skier's skis and the snow over which she slides as she brakes to a halt.

We can check our result by using Equation 6-27, which relates the change in internal energy to the amount of nonconservative work done. We get the same answer with this approach, as we must.

$$W_{\text{nonconservative}} = -\Delta E_{\text{other}} \quad (6\text{-}27)$$

From Example 6-11, $W_{\text{nonconservative}} = -f_k L$, and

$$f_k = mg\left(\frac{H + h}{L}\right), \text{ so}$$

$$W_{\text{nonconservative}} = -mg\left(\frac{H + h}{L}\right)L = -mg(H + h)$$

$$\Delta E_{\text{other}} = -W_{\text{nonconservative}} = mg(H + h)$$

EXAMPLE 6-13 A Ramp and a Spring

A child's toy uses a spring with spring constant $k = 36\ \text{N/m}$ to shoot a block up a ramp inclined at $\theta = 30°$ from the horizontal (**Figure 6-33**). The mass of the block is $m = 8.0\ \text{g}$. When the spring is compressed 4.2 cm and released, the block slides up the ramp, loses contact with the spring, and comes to rest a distance d along the ramp from where it started. Find the value of d. Neglect friction between the block and the surface of the ramp.

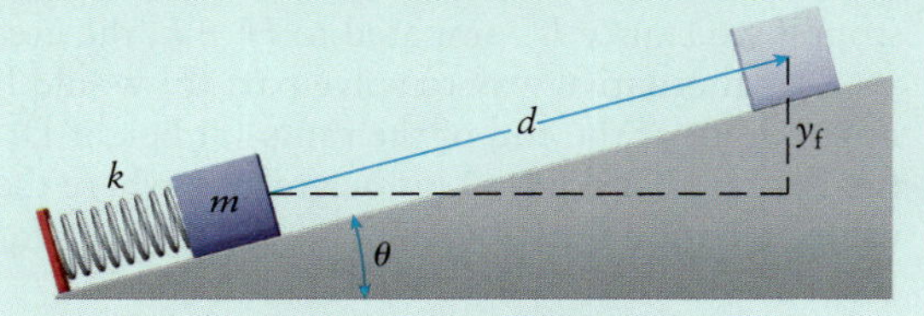

Figure 6-33 Spring potential energy and gravitational potential energy If the spring is compressed a certain distance, how far up the frictionless ramp will the block go?

Set Up

The forces on the block are a normal force, the gravitational force, and a spring force that acts while the block is in contact with the spring. The normal force is always perpendicular to the block's motion and so does no work. So only conservative forces (gravity and the spring force) do work, and total mechanical energy is conserved. We'll use Equation 6-23 to solve for the distance d. The initial point i is where the block is released, and the final point f is a distance d up the ramp from the initial point.

$$K_i + U_i = K_f + U_f \quad (6\text{-}23)$$

$$U_{grav} = mgy \quad (6\text{-}14)$$

$$U_{spring} = \frac{1}{2}kx^2 \quad (6\text{-}19)$$

$$K = \frac{1}{2}mv^2 \quad (6\text{-}8)$$

Solve

The block is at rest with zero kinetic energy at both the initial point and the final point. If we take $y = 0$ to be the height of the block where it is released, then $y_i = 0$; the final height y_f is related to the distance d by trigonometry. We use Equation 6-14 to write the initial and final gravitational potential energies. The initial spring potential energy is given by Equation 6-19 with $k = 36\ \text{N/m}$ and $x_i = -4.2\ \text{cm}$; the final spring potential energy is zero because the spring is relaxed.

Before block is released:

$$K_i = \frac{1}{2}mv_i^2 = 0$$

$$U_{grav,\,i} = mgy_i = 0$$

$$U_{spring,\,i} = \frac{1}{2}kx_i^2$$

When block has traveled a distance d up the ramp:

$$K_f = \frac{1}{2}mv_f^2 = 0$$

From Figure 6-33,

$y_f = d \sin\theta$, so

$$U_{grav,\,f} = mgy_f = mgd\sin\theta$$

$$U_{spring,\,f} = \frac{1}{2}kx_f^2 = 0$$

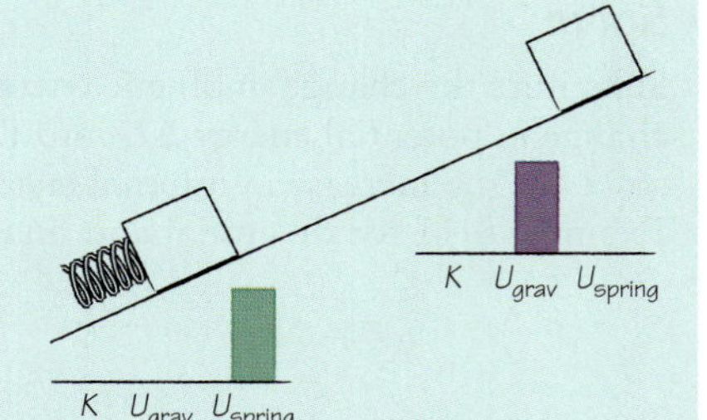

Substitute the energies into Equation 6-23. (Remember that the *total* potential energy U is the sum of the gravitational and spring potential energies.) Then solve for the distance d.

Equation 6-23:

$K_i + U_i = K_f + U_f$, or

$K_i + U_{grav,\,i} + U_{spring,\,i} = K_f + U_{grav,\,f} + U_{spring,\,f}$, so

$$0 + 0 + \frac{1}{2}kx_i^2 = 0 + mgd\sin\theta + 0$$

$$d = \frac{kx_i^2}{2mg\sin\theta}$$

Insert numerical values (being careful to express distances in meters and masses in kilograms) and find the value of d. Recall that $1\ \text{N} = 1\ \text{kg}\cdot\text{m/s}^2$.

$$x_i = (-4.2\ \text{cm})\left(\frac{1\ \text{m}}{100\ \text{m}}\right) = -0.042\ \text{m}$$

$$m = (8.0\ \text{g})\left(\frac{1\ \text{kg}}{1000\ \text{g}}\right) = 0.0080\ \text{kg}$$

$$d = \frac{(36 \text{ N/m})(-0.042 \text{ m})^2}{2(0.0080 \text{ kg})(9.80 \text{ m/s}^2) \sin 30^\circ}$$
$$= 0.81 \frac{\text{N} \cdot \text{s}^2}{\text{kg}} = 0.81 \frac{\text{kg} \cdot \text{m}}{\text{s}^2} \frac{\text{s}^2}{\text{kg}}$$
$$= 0.81 \text{ m} = 81 \text{ cm}$$

Reflect

In our solution we assumed that the final spring potential energy is zero, which means that the spring ends up neither compressed nor stretched. In other words, we assumed that the block moves so far up the ramp that it loses contact with the spring. That's consistent with our result: The block moves 81 cm up the ramp, much greater than the 4.2-cm distance by which the spring was originally compressed.

Our solution shows that the net effect of the block's motion is to convert spring potential energy at the initial position to gravitational potential energy at the final position. The block begins and ends with zero kinetic energy, just as the skier does in Examples 6-11 and 6-12. Note that the block *does* have kinetic energy at intermediate points during the motion, when it's moving up the ramp. However, we didn't need to worry about these intermediate stages of the motion.

This problem would have been impossible to solve using the techniques of Chapters 4 and 5 because the net force on the block isn't constant: The magnitude of the spring force starts off large, then decreases to zero as the block moves uphill and the spring relaxes. The energy approach is the *only* method that will lead us to the answer to this problem.

GOT THE CONCEPT? 6-7 Where Is It Zero?

In Example 6-13, which of the following are you free to choose as you see fit? (a) the point on the ramp where gravitational potential energy is zero; (b) the point on the ramp where spring potential energy is zero; (c) both of these; (d) neither of these.

TAKE-HOME MESSAGE FOR Section 6-8

✔ The kinds of problems that are best solved using the law of conservation of energy are those in which an object moves from one place to another under the action of well-defined forces.

✔ Drawing a free-body diagram is essential for deciding which forces act and do work on the object.

✔ If only conservative forces do work, total mechanical energy is conserved. If nonconservative forces also do work, you must use the more general statement of energy conservation.

6-9 Power is the rate at which energy is transferred

If you walk for a kilometer, your heart rate will increase above its resting value. But if you run for a kilometer, your heart rate will increase to an even higher value (**Figure 6-34**). Now that we've learned about energy, this may seem paradoxical: You covered the same distance and did roughly the same amount of work in both cases, so why is a higher heart rate needed for running? The answer, as we will see, is related to the *rate* of transferring energy and the rate of doing work.

The rate at which energy is transferred from one place to another, or from one form to another, is called **power**. The unit of power is the joule per second, or **watt** (abbreviated W): 1 W = 1 J/s. As an example, a 50-W light bulb is designed so that when it is in operation in an electric circuit, 50 J of energy is delivered to it every second by that circuit. Other common units of power are the *kilowatt* (1 kW = 1000 W) and the *horsepower* (1 hp = 746 W is a typical rate at which a horse does work by pulling on a plow). We'll usually denote power by the uppercase symbol P.

Let's apply the concept of power to the rate of *doing work*. If you are using a stationary exercise bike at the gym, a reading of 200 W on the bike's display means that

The AGE/Getty Images

Figure 6-34 Running versus walking Both running and walking involve doing work; the difference is the rate at which work is done.

you are doing 200 J of work on the bike every second. So the power you are applying to the pedals equals the amount of work that you do divided by the time it takes for you to do that work:

Power, work done, and time
(6-29)

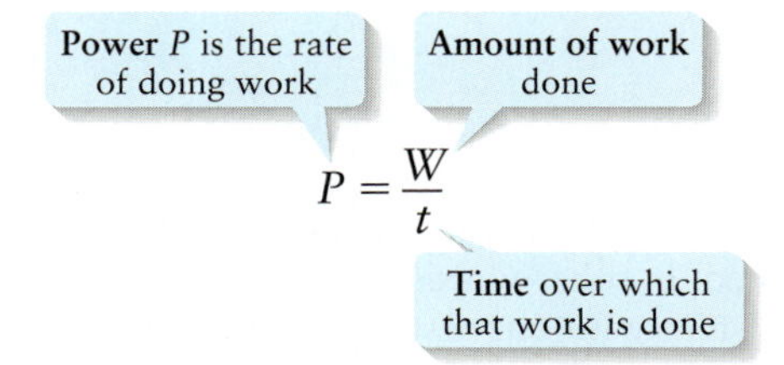
Power P is the rate of doing work
Amount of work done
$P = \frac{W}{t}$
Time over which that work is done

The quantity P in this equation is sometimes called the power *delivered* to the object on which work is being done.

To help interpret this equation, let's think again about the groundskeeper pulling a screen across a baseball diamond that we described in Section 6-2 (Figure 6-6). The groundskeeper exerted a force of magnitude F on the screen at an angle θ to the screen's displacement. So to move the screen a distance d over the ground at a steady speed, the groundskeeper did an amount of work $W = (F \cos \theta)d$, as described by Equation 6-2. If the groundskeeper took a time t to pull the screen this distance, the speed of the screen was $v = d/t$. So Equation 6-29 tells us that the power output that the groundskeeper delivered to the screen was

$$P = \frac{W}{t} = \frac{(F \cos \theta)d}{t} = (F \cos \theta)v \tag{6-30}$$

This equation says that the greater the force F that the groundskeeper exerts and the faster the speed v at which he makes the screen move, the more power he delivers to the screen.

The same equation also applies to the power that you deliver to the pedals of a bicycle or of a stationary exercise bike. When your foot pushes straight down on one of the pedals, as in **Figure 6-35a**, the angle θ in Equation 6-30 is zero (the force is in the direction that the pedal is moving). Since $\cos 0 = 1$, the power you deliver is then $P = Fv$. But if the force is at an angle of $\theta = 45°$, as in **Figure 6-35b**, the power you deliver is less: $P = (F \cos 45°)v = 0.707Fv$. So the power you deliver to the pedal is maximum when you push straight down on a downward-moving pedal. (It's also maximum when you pull straight up on an upward-moving pedal, so again $\theta = 0$). You can notice this most easily when you stand up on the pedals; you can feel the greater power being delivered when your feet move vertically.

We can now use the concept of power to explain the difference between walking and running that we mentioned at the beginning of this section. Whether you walk or run, you expend some of the internal energy E_{other} stored in your body and use it to do nonconservative work $W_{\text{nonconservative}}$ to propel yourself. (This work is nonconservative because you don't get the energy back after you stop.) From Equation (6-27) in Section 6-7,

$$W_{\text{nonconservative}} = -\Delta E_{\text{other}}$$

Here $W_{\text{nonconservative}}$ is positive (you do positive work) and ΔE_{other} is negative (your internal energy decreases). If we divide both sides of this equation by the time t that you walk or run, we get

$$\frac{W_{\text{nonconservative}}}{t} = \frac{-\Delta E_{\text{other}}}{t} \tag{6-31}$$

(a)

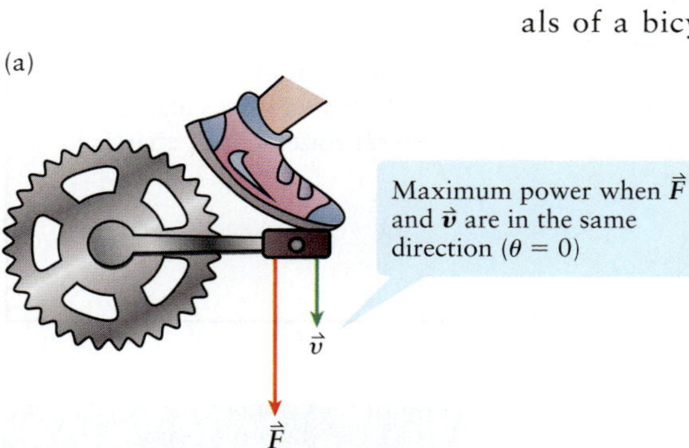
Maximum power when $\vec{F}$ and $\vec{v}$ are in the same direction ($\theta = 0$)
$\vec{v}$
$\vec{F}$

(b)

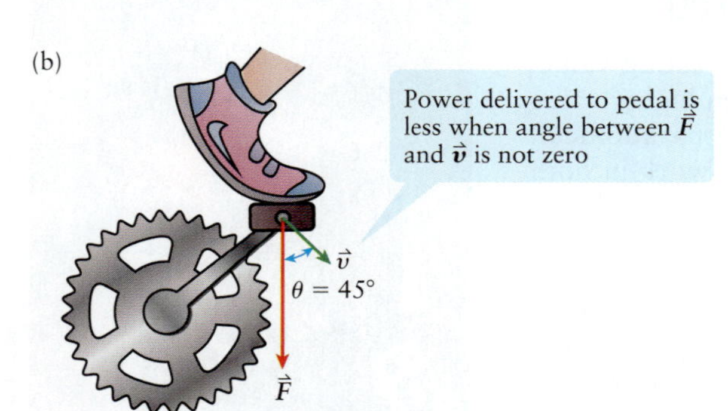
Power delivered to pedal is less when angle between $\vec{F}$ and $\vec{v}$ is not zero
$\vec{v}$
$\theta = 45°$
$\vec{F}$

Figure 6-35 Power in cycling The power you deliver to a bicycle pedal depends on the angle between the force you exert and the direction of the pedal's motion.

If you compare Equation 6-31 to Equation 6-30, you'll see that $W_{\text{nonconservative}}/t$ is just the power P that you deliver to make yourself move. The term $\Delta E_{\text{other}}/t$ is the rate at which you lose internal energy as you move. When you run rather than walk, you cover the same distance in a shorter time t, $P = W_{\text{nonconservative}}/t$ has a greater value, and you do work at a faster rate. To do work more rapidly your muscles consume internal energy more rapidly and also require more oxygen per second. Hence your respiration rate goes up in order to inhale more oxygen per second, and your heart rate goes up in order to more rapidly supply your muscles with oxygenated blood. This elevated heart rate is why running is superior to walking as cardiovascular exercise.

In walking, running, and all other forms of exercise, only part of the chemical energy that you expend goes into doing work. The rest goes into increasing the thermal energy of your body and your surroundings. The faster the rate at which you do work—that is, the greater your power output—the more rapidly you use up your chemical energy and the more rapidly the thermal energy increases (and so the more rapidly you warm up). Example 6-14 explores this for one particular type of vigorous exercise.

EXAMPLE 6-14 Power in a Rowing Race

Rowing demands one of the highest average power outputs of all competitive sports (**Figure 6-36**). In a typical rowing race a racing shell travels 2000 m in 6.00 min, during which each rower has an average power output of 4.00×10^2 W and expends 7.20×10^5 J = 720 kJ of chemical energy. (a) How much work does each rower do to propel the shell? (b) By how much does the thermal energy of each rower increase?

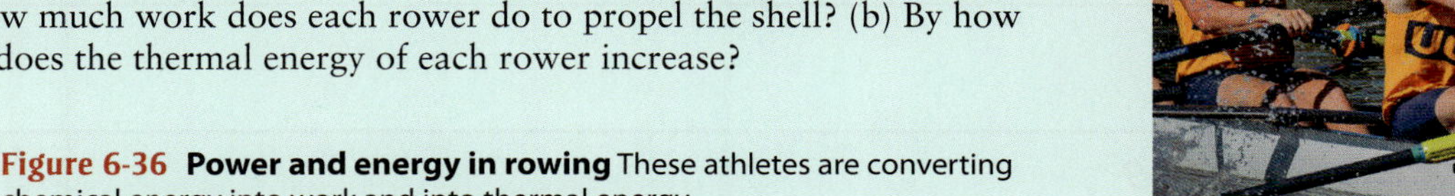

Figure 6-36 Power and energy in rowing These athletes are converting chemical energy into work and into thermal energy.

Sherene DuBois

Set Up

We are given the rower's power output P and the time t for which that power is delivered, so we can use Equation 6-29 to find the total (nonconservative) work that the rower does. We can then use Equation 6-31 to find the change ΔE_{other} in internal energy. This is the sum of the change in chemical energy (which we are given) and the change in thermal energy (which we want to find).

$$P = \frac{W_{\text{nonconservative}}}{t} \quad (6\text{-}29)$$

$$\frac{W_{\text{nonconservative}}}{t} = -\frac{\Delta E_{\text{other}}}{t} \quad (6\text{-}31)$$

Solve

(a) Solve Equation 6-29 for the work done by the rower. Remember that 1 W = 1 J/s, and be careful to express the time t in seconds.

Equation 6-29:

$$P = \frac{W_{\text{nonconservative}}}{t}, \text{ so}$$

$$W_{\text{nonconservative}} = Pt$$

$$P = 4.00 \times 10^2 \text{ W} = 4.00 \times 10^2 \text{ J/s}$$

$$t = (6.00 \text{ min})\left(\frac{60 \text{ s}}{1 \text{ min}}\right) = 3.60 \times 10^2 \text{ s}$$

$$W_{\text{nonconservative}} = (4.00 \times 10^2 \text{ W})(3.60 \times 10^2 \text{ s})$$
$$= 1.44 \times 10^5 \text{ J} = 144 \text{ kJ}$$

(b) We can now find the total internal change ΔE_{other}, which is the negative of the work that the rower does. We know the change in the rower's chemical energy is $\Delta E_{\text{chemical}} = -7.20 \times 10^5$ J (this is negative because the chemical energy decreases). So we can find the change in thermal energy $\Delta E_{\text{thermal}}$.

Multiply Equation 6-31 by time t:

$$W_{\text{nonconservative}} = -\Delta E_{\text{other}}, \text{ so}$$
$$\Delta E_{\text{other}} = -W_{\text{nonconservative}} = -1.44 \times 10^5 \text{ J}$$

This is the sum of $\Delta E_{\text{chemical}}$ and $\Delta E_{\text{thermal}}$:

$$\Delta E_{\text{other}} = \Delta E_{\text{chemical}} + \Delta E_{\text{thermal}}, \text{ so}$$
$$\Delta E_{\text{thermal}} = \Delta E_{\text{other}} - \Delta E_{\text{chemical}}$$
$$= (-1.44 \times 10^5 \text{ J}) - (-7.20 \times 10^5 \text{ J})$$
$$= +5.76 \times 10^5 \text{ J} = +576 \text{ kJ}$$

Reflect

In most of the world chemical energy or food energy is expressed in kilojoules, or kJ. In the United States it is expressed in food calories (symbol C), where 1 C = 4.186 kJ. So our rower "burned" 720 kJ or 172 food calories during the race

In accordance with the law of conservation of energy, we can account for 100% of the chemical energy used: 0.200 or 20.0% went into doing work, and 0.800 or 80.0% went into increasing the thermal energy.

Chemical energy used:

$$(720\text{ kJ})\left(\frac{1\text{ C}}{4.186\text{ kJ}}\right) = 172\text{ C}$$

Fraction of chemical energy that went into doing work:

$$\frac{144\text{ kJ}}{720\text{ kJ}} = 0.200$$

Fraction of chemical energy that went into thermal energy:

$$\frac{576\text{ kJ}}{720\text{ kJ}} = 0.800$$

GOT THE CONCEPT? 6-8

The power that a gasoline engine delivers to a car's wheels is roughly the same at all speeds. Is the forward force on the car greater at (a) low speeds or (b) high speeds? (*Hint*: Consider Equation 6-30.)

TAKE-HOME MESSAGE FOR Section 6-9

✔ Power is the rate of transferring energy from one place or form to another.

✔ The power delivered to an object is related to the speed of the object and the magnitude and direction of the applied force.

Key Terms

conservative force
energy
gravitational potential energy
Hooke's law
internal energy
joule
kinetic energy
law of conservation of energy
negative work
nonconservative force
potential energy
power
spring constant
spring potential energy
thermal energy
total mechanical energy
translational kinetic energy
watt
work
work-energy theorem

Chapter Summary

Topic	Equation or Figure
Work done by a force in the direction of displacement: If an object moves in a straight line while a constant force is applied in the same direction as the displacement, the work is just equal to the force times the displacement.	**Work done on an object** by a constant force $\vec{F}$ that points in the **same direction** as the object's displacement $\vec{d}$ — **Magnitude of the constant force** $\vec{F}$ — $$W = Fd \quad (6\text{-}1)$$ **Magnitude of the displacement** $\vec{d}$

Work done by a constant force at an angle θ to the straight-line displacement: If an object moves in a straight line while a constant force is applied at some angle θ to the displacement, the work is equal to the force times the displacement multiplied by the cosine of the angle between the force and displacement.

$$W = (F\cos\theta)d \quad (6\text{-}2)$$

- **Work done on an object** by a constant force $\vec{F}$ that points **at an angle** θ to the object's displacement $\vec{d}$
- **Magnitude of the constant force** $\vec{F}$
- **Angle** between the directions of $\vec{F}$ and $\vec{d}$
- **Magnitude of the displacement** $\vec{d}$

Kinetic energy and the work-energy theorem: The work-energy theorem states that the net work done in a displacement—the sum of the work done on the object by individual forces—is equal to the change in the object's kinetic energy, or energy of motion, during that displacement. This theorem is valid whether the path is curved or straight and whether the forces are constant or varying.

$$W_{net} = K_f - K_i \quad (6\text{-}9)$$

- **Work done on an object** by the **net force** on that object
- **Kinetic energy** of the object **after** the work is done on it
- **Kinetic energy** of the object **before** the work is done on it

$$K = \frac{1}{2}mv^2 \quad (6\text{-}8)$$

- **Kinetic energy** of an object
- **Mass** of the object
- **Speed** of the object

The spring force and Hooke's law: An ideal spring exerts a force that is proportional to how far it is stretched or compressed, and it is always opposite to the stretch or compression. The spring force is not constant, so the work needed to stretch or compress a spring is not simply the force magnitude multiplied by the displacement.

$$F_x = -kx \quad (6\text{-}11)$$

- **Force** exerted by an **ideal spring**
- **Spring constant** of the spring (a measure of its stiffness)
- **Extension** of the spring ($x > 0$ if spring is stretched, $x < 0$ if spring is compressed)

$$W = \frac{1}{2}kx_2^2 - \frac{1}{2}kx_1^2 \quad (6\text{-}13)$$

- **Work** that must be done on a spring to stretch it from $x = x_1$ to $x = x_2$
- **Spring constant** of the spring (a measure of its stiffness)
- x_1 = **initial stretch** of the spring
- x_2 = **final stretch** of the spring

Potential energy: Unlike kinetic energy, potential energy is associated with the position of an object. Gravitational potential energy increases with height. The potential energy of a spring increases with the stretch or compression of the spring. Only conservative forces, for which the work done does not depend on the path taken, are associated with a potential energy.

$$U_{grav} = mgy \quad (6\text{-}14)$$

- **Gravitational potential energy** associated with an object
- **Mass** of the object
- **Acceleration due to gravity**
- **Height** of the object

$$U_{spring} = \frac{1}{2}kx^2 \quad (6\text{-}19)$$

- **Spring potential energy** of a stretched or compressed spring
- **Spring constant** of the spring
- **Extension** of the spring ($x > 0$ if spring is stretched, $x < 0$ if spring is compressed)

Total mechanical energy and its conservation: If only conservative forces do work on an object, the total mechanical energy is conserved. (Other forces can act on the object, but they cannot do any work.) The kinetic and potential energies can change, but their sum remains constant.

Values of the **kinetic energy** K of an object at two points (i and f) during its motion

$$K_i + U_i = K_f + U_f \tag{6-23}$$

Values of the **potential energy** U at the same two points

If only conservative forces do work on an object, the sum $K + U$ maintains the same value throughout the motion. This sum is called the **total mechanical energy** E.

Change in the kinetic energy K of an object during its motion

Change in the potential energy U associated with an object during its motion

$$\Delta K + \Delta U = 0 \tag{6-24}$$

If only conservative forces do work on the object, the sum is zero: Total mechanical energy can transform between kinetic and potential forms, but there is no net change in the amount of total mechanical energy.

Generalized law of conservation of energy: If nonconservative forces do work, total mechanical energy is not conserved. If we broaden our definition of energy to include the internal energy of the interacting objects, then total energy is conserved in *all* cases.

Values of the **kinetic energy** K of an object at two points (i and f) during its motion

$$K_i + U_i + W_{\text{nonconservative}} = K_f + U_f \tag{6-26}$$

Values of the **potential energy** U at the same two points

If nonconservative forces do work on an object, the total mechanical energy $E = K + U$ changes its value.
If $W_{\text{nonconservative}} > 0$, $E_f > E_i$ and the total mechanical energy increases.
If $W_{\text{nonconservative}} < 0$, $E_f < E_i$ and the total mechanical energy decreases.

Change in the kinetic energy K of an object during its motion

Change in the potential energy U associated with an object during its motion

$$\Delta K + \Delta U + \Delta E_{\text{other}} = 0 \tag{6-28}$$

Change in the internal energy E_{other} during the object's motion

Energy can transform between kinetic, potential, and internal forms, but there is no net change in the amount of energy.

Power: Power is the rate at which energy is transferred from one place or form to another place or form. If work is being done on an object, the power delivered equals the rate at which work is done.

Power P is the rate of doing work

Amount of work done

$$P = \frac{W}{t} \tag{6-29}$$

Time over which that work is done

Answer to What do you think? Question

(c) The long, thick Achilles tendon on the back of an ostrich's leg acts like a spring. When a running ostrich lands on one foot, this tendon stretches and converts kinetic energy into spring potential energy. When this foot subsequently leaves the ground, the tendon relaxes, and this stored spring energy is turned back into kinetic energy. Were it not for this form of energy storage, the ostrich's leg muscles would have to push much harder with each step to maintain the animal's motion.

Answers to Got the Concept? Questions

6-1 (a) negative, (b) positive, (c) zero, (d) positive. In (a) you exert a force opposite to the ball's displacement to slow it down, so the work you do on the ball is negative. In (b) the force you exert on the ball as you throw it is in the same direction as the ball's displacement, so you do positive work on the ball. In (c) you exert an upward force on the ball to keep it from falling, but there is no forward or backward force on the ball. Its displacement is forward, so the force you exert is perpendicular to the ball's displacement and no work is done. In (d) you must push the ball forward (in the direction of its displacement) to make it speed up, so you do positive work on the ball.

6-2 (d) The work-energy theorem equates the work done on an object to the change in its kinetic energy. The work done is proportional to the distance over which the force is applied, so this distance is directly proportional to the change in kinetic energy. When an object starts from rest, the change in kinetic energy (from Equation 6-9) equals the final kinetic energy, which is proportional to the square of the final speed. So the distance is proportional to the square of the final speed, and doubling the speed therefore requires that the force be applied over $2^2 = 4$ times the distance.

6-3 (a) (i), (b) (iii) The floor exerts a greater normal force n on the heavy box, so the kinetic friction force $f_k = \mu_k n$ is greater for the heavy box. Hence you must push harder on the heavy box to balance the greater kinetic friction force. You therefore do more work on the heavy box than on the light box because you exert a greater force over the same distance. The work-energy theorem tells us that that *net* work done on *both* boxes is the same and equal to zero. That's because both boxes have initial kinetic energy $K_i = 0$ (they start at rest) and final kinetic energy $K_f = 0$ (they end at rest), so $W_{net} = K_f - K_i = 0 - 0 = 0$. You do positive work on each box, but the floor does an equal amount of negative work through the kinetic friction force.

6-4 (a) From Example 6-10 the work that the athlete must do to stretch the cords by a distance x_2 from their relaxed state ($x_1 = 0$) is $W_{\text{athlete on cords}} = \frac{1}{2}kx_2^2 - \frac{1}{2}kx_1^2 = \frac{1}{2}kx_2^2 - 0 = \frac{1}{2}kx_2^2$. This says that the work done is proportional to the square of the stretch. Equivalently, this says that the stretch is proportional to the square root of the work done. So to increase the work done by a factor of 2, the stretch must increase by a factor of $\sqrt{2}$.

6-5 (d) Equation 6-16 tells us that the change in gravitational potential energy as the ball moves from height y_i (the top of the building) to height y_f (the height of the balcony) is $\Delta U_{grav} = U_f - U_i = mgy_f - mgy_i = mg(y_f - y_i)$. The difference $y_f - y_i$ is 36 m -45 m $= -9$ m; that is, the balcony is 9 m below the top of the building. This distance doesn't depend at all on where you choose $y = 0$! (If $y = 0$ is at the balcony, $y_i = 9$ m and $y_f = 0$, so $y_f - y_i = 0 - 9$ m $= -9$ m; if $y = 0$ is at the top of the building, $y_i = 0$ and $y_f = -9$ m, and $y_f - y_i = -9$ m $- 0 = -9$ m.) As far as *changes* in gravitational potential energy are concerned, you can choose the level $y = 0$ to be wherever you like.

6-6 (b) There is no friction, and the normal force does no work on the block (it always acts perpendicular to the path of the block). Hence only the conservative gravitational force does work and so total mechanical energy is conserved. Let point i be where the block is released and point f be where it stops momentarily on the right-hand side of the ramp. The block is not moving at either point, so $K_i = K_f = 0$. Equation 6-23 then says that $U_i = U_f$, which means that the gravitational potential energy is the same at both points. Since $U_{grav} = mgy$, this means that the block is at the same height y at both points. Hence the block reaches the same height $y = H$ on the right-hand side of the ramp as on the left-hand side.

6-7 (a) We learned in Section 6-6 that we can choose $y = 0$ to be wherever we like when calculating gravitational potential energy $U_{grav} = mgy$. However, we do not have the same freedom when calculating spring potential energy $U_{spring} = \frac{1}{2}kx^2$; we *must* choose $x = 0$ to be where the spring is relaxed (neither stretched nor compressed).

6-8 (a) The forward force is in the same direction as the car's motion, so the power delivered is given by Equation 6-30 with $\theta = 0$: $P = (F \cos 0)v = Fv$. The power P, and hence the product Fv, is the same at low speed and high speed, so the force F has a large value when the speed v has a small value. As a result a car powered by a gasoline engine has a greater forward force (and greater acceleration) when it is first starting from rest than it does when it's already moving fast.

Questions and Problems

In a few problems you are given more data than you actually need; in a few other problems you are required to supply data from your general knowledge, outside sources, or informed estimate.

Interpret as significant all digits in numerical values that have trailing zeros and no decimal points. For all problems use $g = 9.80$ m/s^2 for the free-fall acceleration due to gravity.

•	Basic, single-concept problem
••	Intermediate-level problem; may require synthesis of concepts and multiple steps
•••	Challenging problem
SSM	*Solution is in Student Solutions Manual*
Example	*See worked example for a similar problem*

Conceptual Questions

1. • Using Equation 6-2, explain how the work done on an object by a force can be equal to zero.

2. • Why do seasoned hikers step *over* logs that have fallen in their path rather than stepping *onto* them? SSM

3. • Your roommate lifts a cement block, carries it across the room, and sets it back down on the floor. Is the net work she did on the block positive, negative, or zero? Explain your answer.

4. • Can the normal force ever do work on an object? Explain your answer. SSM

5. • One of your classmates in physics reasons, "If there is no displacement, then a force will perform no work." Suppose the student pushes with all of her might against a massive boulder, but the boulder doesn't move. (a) Does she expend energy pushing against the boulder? (b) Does she do work on the boulder? Explain your answer.

6. • Can kinetic energy ever have a negative value? Explain your answer.

7. • Can a change in kinetic energy ever have a negative value? Explain your answer.

8. • Define the concept of "nonconservative force" in your own words. Give three examples.

9. • A satellite orbits around Earth in a circular path at a high altitude. Explain why the gravitational force does zero work on the satellite.

10. • Model rockets are propelled by an engine containing a combustible propellant. Two rockets, one twice as heavy as the other, are launched using engines that contain the same amount of propellant. (a) Is the maximum kinetic energy of the heavier rocket less than, equal to, or more than the maximum kinetic energy of the lighter one? (b) Is the speed of the heavier rocket just after the propellant is used up less than, equal to, or greater than the launch speed of the lighter one?

11. • When does the kinetic energy of a rock that is dropped from the edge of a high cliff reach its maximum value? Answer the question (a) when the air resistance is negligible and (b) when there is significant air resistance. SSM

12. • Can gravitational potential energy have a negative value? Explain your answer.

13. • Bicycling to the top of a hill is much harder than coasting back down to the bottom. Do you have more gravitational potential energy at the top of the hill or at the bottom? Explain your answer.

14. • Analyze the types of energy that are associated with the "circuit" that a snowboarder follows from the bottom of a mountain, to its peak, and back down again. Be sure to explain all changes in energy. SSM

15. • A common classroom demonstration involves holding a bowling ball attached by a rope to the ceiling close to your face and releasing it from rest. In theory, you should not have to worry about being hit by the ball as it returns to its starting point. However, some of the most "exciting" demonstrations *have* involved the professor being hit by the ball! (a) Why should you expect not to be hit by the ball? (b) Why *might* you be hit if you actually perform the demonstration?

16. • Why are the ramps for people with disabilities quite long instead of short and steep?

17. • A rubber dart can be launched over and over again by the spring in a toy gun. Where is the energy generated to launch the dart?

18. • The energy provided by your electric company is *not* sold by the joule. Instead, you are charged by the kilowatt-hour (typically 1 kWh ≅ \$0.25 including taxes). Explain why this makes sense for the average consumer. It might be helpful to read the electrical specifications of an appliance (such as a hair dryer or a blender).

Multiple-Choice Questions

19. • Box 2 is pulled up a rough incline by box 1 in an arrangement as shown in **Figure 6-37**. How many forces are doing work on box 2?

A. 1
B. 2
C. 3
D. 4
E. 5

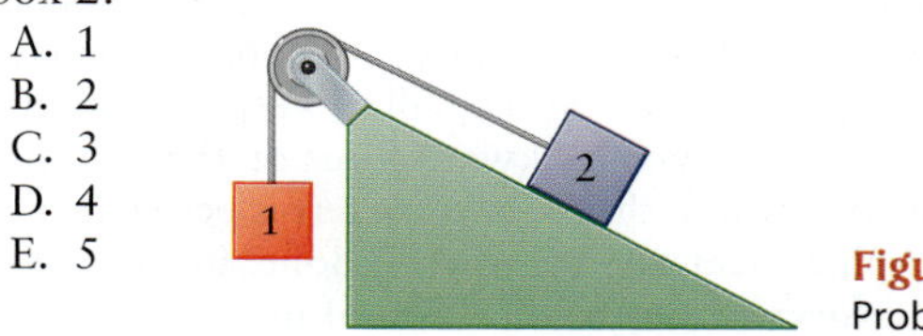

Figure 6-37 Problem 19

20. • **Figure 6-38** shows four situations in which a box slides to the right a distance d across a frictionless floor as a result of applied forces, *one* of which is shown. The magnitudes of the forces shown are identical. Rank the four cases in order of increasing work done on the box by the force shown during the displacement.

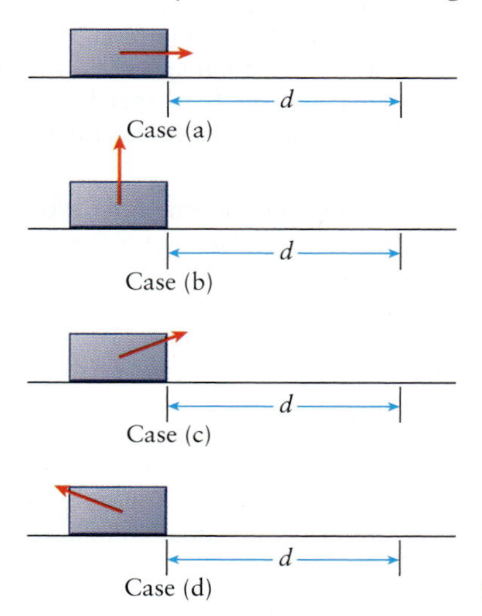

Figure 6-38 Problem 20

21. • A box is dragged a distance d across a floor by a force $\vec{F}$ which makes an angle θ with the horizontal (**Figure 6-39**). If the magnitude of the force is held constant but the angle θ is increased up to 90°, the work done by the force in dragging the box

A. remains the same.
B. increases.
C. decreases.
D. first increases, then decreases.
E. first decreases, then increases. SSM

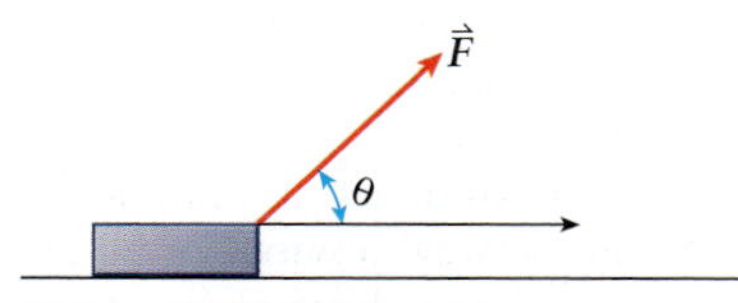

Figure 6-39 Problem 21

22. • A car moves along a straight, level road. If the car's velocity changes from 60 mi/h due east to 60 mph due west, the kinetic energy of the car
   A. remains the same.
   B. increases.
   C. decreases.
   D. first increases, then decreases.
   E. first decreases, then increases.

23. • A boy swings a ball on a string at constant speed in a horizontal circle that has a circumference equal to 6 m. What is the work done on the ball by the 10-N tension force in the string during one revolution of the ball?
   A. 190 J D. 15 J
   B. 60 J E. 0
   C. 30 J

24. • As part of a lab experiment, Allison uses an air-track cart of mass m to compress a spring of constant k by an amount x from its equilibrium length. The air-track has negligible friction. When Allison lets go, the spring launches the cart. What cart velocity should she expect after it is launched by the spring?

   A. $\sqrt{2kx/m}$ B. $\sqrt{kx/m}$ C. $(\sqrt{k/m})x$ D. $(\sqrt{2k/m})x$

25. •Adam and Bobby are twins who have the same weight. Adam drops to the ground from a tree at the same time that Bobby begins his descent down a frictionless slide. If they both start at the same height above the ground, how do their kinetic energies compare when they hit the ground?
   A. Adam has greater kinetic energy than Bobby.
   B. Bobby has greater kinetic energy than Adam.
   C. They have the same kinetic energy.
   D. Bobby has twice the kinetic energy as Adam.
   E. More information is required to compare their kinetic energies. SSM

26. • Three balls are thrown off a tall building with the same speed but in different directions. Ball A is thrown in the horizontal direction; ball B starts out at 45° above the horizontal; ball C begins its flight at 45° below the horizontal. Which ball has the greatest speed just before it hits the ground? Ignore any effects due to air resistance.
   A. Ball A
   B. Ball B
   C. Ball C
   D. All balls have the same speed.
   E. Balls B and C have the same speed, which is greater than the speed of ball A.

Estimation/Numerical Analysis

27. • About how many joules are required to lift a suitcase into the trunk of a car?

28. • **Biology** Describe a method for estimating the amount of energy required for an average adult human to take a single step at a typical walking speed, then use it to make the estimate. Express your answer in joules.

29. • **Biology** Estimate the amount of energy required for an average human to hike to the top of a 1.5-km mountain along a path 5.8 km in total distance. Express your answer in Joules. SSM

30. • (a) What is the kinetic energy of a commercial jet airplane when it "touches down" on a runway to land? (b) Where does the energy "go" when the plane stops?

31. • Estimate the kinetic energy of a cue ball during a typical billiards shot.

32. • Estimate the work done in carrying your backpack up one flight of stairs.

33. • Estimate the maximum kinetic energy of a puma, one of the best leapers of the animal kingdom. SSM

34. • Thor works as an express elevator attendant in the famed Empire State Building in New York City. If the potential energy that Thor gains during each ascent throughout his career were added and converted into kinetic energy, what would his speed be?

35. • Estimate the rate, in watts, at which your body metabolizes energy if your weight does not change. Note that 1 food calorie equals approximately 4200 J.

36. • Estimate how much power is required to throw a fastball in the game of baseball. Given that the human body is only about 25% efficient, how much power would a pitcher actually have to expend to throw a fastball?

37. • The data below represent a changing force that acts on an object in the x direction (the force is parallel to the displacement). Graph F versus x using a graphing calculator or a spreadsheet. Calculate the work done by the force (a) in the first 0.100 m, (b) in the first 0.200 m, (c) from 0.100 to 0.200 m, and (d) for the entire motion (for $0 < x < 0.250$ m).

x (m)	F (N)	x (m)	F (N)	x (m)	F (N)
0	0.00	0.0900	12.00	0.180	12.50
0.0100	2.00	0.100	12.48	0.190	12.50
0.0200	4.00	0.110	12.48	0.200	12.50
0.0300	6.00	0.120	12.48	0.210	12.48
0.0400	8.00	0.130	12.60	0.220	9.36
0.0500	10.00	0.140	12.60	0.230	6.24
0.0600	10.50	0.150	12.70	0.240	3.12
0.0700	11.00	0.160	12.70	0.250	0.00
0.0800	11.50	0.170	12.60		

Problems

6-2 The work that a constant force does on a moving object depends on the magnitude and direction of the force

38. • A crane very slowly lifts a 2.00×10^2-kg crate a vertical distance of 15.0 m. How much work does the crane do on the crate? How much work does gravity do on the crate? Example 6-1

39. • **Sports** In the men's weight-lifting competition of the 2008 Beijing Olympics, Matthias Steiner made his record lift of 446 kg from the floor to over his head (2.0 m). How much work is done on the weight by Steiner? SSM Example 6-1

40. • A 350-kg box is pulled 7.00 m up a 30.0° inclined plane by an external force of 5.00×10^3 N that acts parallel to the frictionless plane. Calculate the work done by (a) the external force, (b) gravity, and (c) the normal force. Example 6-3

41. • Three clowns try to move a 3.00×10^2-kg crate 12.0 m to the right across a smooth, low-friction floor. Moe pushes to the right with a force of 5.00×10^2 N, Larry pushes to the left with 3.00×10^2 N, and Curly pushes straight down with 6.00×10^2 N. Calculate the work done by each of the clowns. Example 6-3

42. • An assistant for the football team carries a 30.0-kg cooler of water from the top row of the stadium, which is 20.0 m above the field level, down to the bench area on the field. (a) If the speed of the cooler is constant throughout the trip, calculate the work done by the assistant on the cooler of water. (b) How much work is done by the force of gravity on the cooler of water? Example 6-1

43. •• A statue is crated and moved for cleaning. The mass of the statue and the crate is 150 kg. As the statue slides down a ramp inclined at 40.0°, the curator pushes up, parallel to the ramp's surface, so that the crate does not accelerate (**Figure 6-40**). If the statue slides 3.0 m down the ramp, and the coefficient of kinetic friction between the crate and the ramp is 0.54, calculate the work done on the crate by each of the following: (a) the gravitational force, (b) the curator, (c) the friction force, and (d) the normal force between the ramp and the crate. SSM Example 6-3

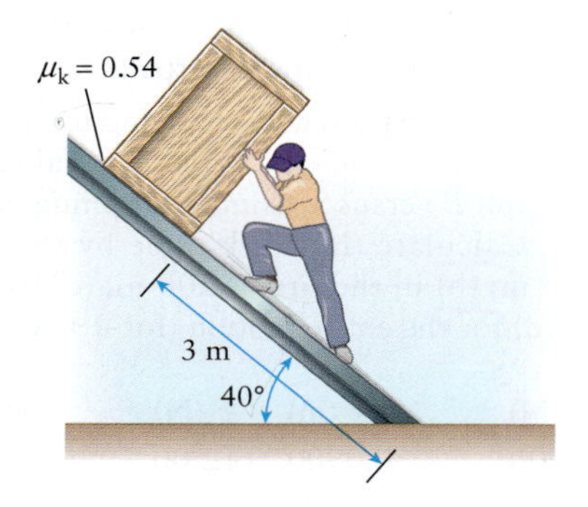

Figure 6-40 Problem 43

6-3 Kinetic energy and the work-energy theorem give us an alternative way to express Newton's second law

44. • A 1250-kg car moves at 20.0 m/s. How much work must be done on the car to increase its speed to 30.0 m/s? Example 6-4

45. • A bumblebee has a mass of about 0.25 g. If its speed is 10.0 m/s, calculate its kinetic energy. SSM Example 6-4

46. • A small truck has a mass of 2100 kg. How much work is required to decrease the speed of the vehicle from 22.0 m/s to 12.0 m/s on a level road? Example 6-4

47. • An 8500-metric ton freight train is out of control and moving at 90 km/h on level track. How much work must a superhero do on the train to bring it to a halt? Example 6-4

48. • A 10.0-kg box starts at rest on a level floor. An external, horizontal force of 2.00×10^2 N is applied to the box for a distance of 4.00 m. If the coefficient of kinetic friction between the box and the floor is 0.440, how fast is the block moving at the end of the 4.00 m? Assume it starts from rest. Example 6-4

6-4 The work-energy theorem can simplify many physics problems

49. • A force of 1200 N pushes a man on a bicycle forward. Air resistance pushes against him with a force of 800 N. If he starts from rest and is on a level road, how fast will he be moving after 20.0 m? The mass of the bicyclist and his bicycle is 90.0 kg. Example 6-5

50. • Starting from rest a 75.0-kg skier skis down a slope 8.00×10^2 m long that has an average incline of 40.0°. The speed of the skier at the bottom of the slope is 20.2 m/s. How much work was done by nonconservative forces? Example 6-5

51. •• A book slides across a level, carpeted floor at an initial speed of 4.00 m/s and comes to rest after 3.25 m. Calculate the coefficient of kinetic friction between the book and the carpet. Assume the only forces acting on the book are friction, weight, and the normal force. Example 6-7

52. •• Calculate the final speed of the 2.00-kg object that is pushed for 22.0 m by a 40.0-N force directed 20.0° below the horizontal on a level, frictionless floor (**Figure 6-41**). Assume the object starts from rest. SSM Example 6-6

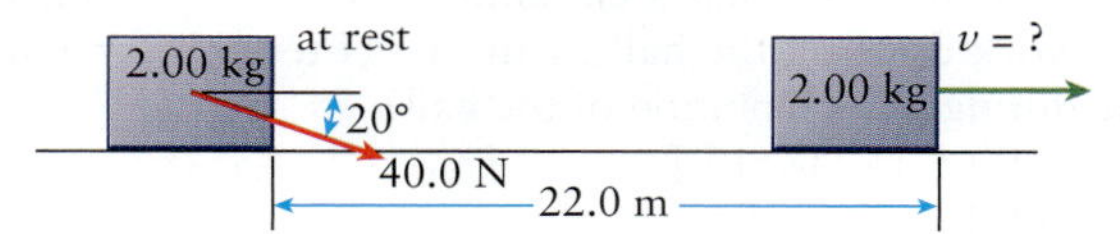

Figure 6-41 Problem 52

53. • A 325-g model boat facing east floats on a pond. The wind in its sail provides a force of 1.85 N that points 25° north of east. The force on its keel is 0.782 N pointing south. The drag force of the water on the boat is 0.750 N toward the west. If the boat starts from rest and heads east, how fast is it moving after it travels for a distance of 3.55 m? Example 6-6

54. • **Sports** A catcher in a baseball game stops a pitched ball originally moving at 44.0 m/s at the moment it first came in contact with the catcher's glove. After contacting the glove, the ball traveled an additional 12.5 cm before coming to a complete stop. The mass of the ball is 0.145 kg. What is the average force that the glove imparts to the ball during the catch? Comment on the force that the catcher's hand experiences during the catch. SSM Example 6-5

6-5 The work-energy theorem is also valid for curved paths and varying forces

55. • A sled of mass m = 22.0 kg is accelerated from rest on a frictionless surface to a final velocity of v = 12.5 m/s. **Figure 6-42** displays a graph of force vs. distance for the sled in terms of F_{max}. Use the graph to help you solve for F_{max} in terms of m, v, and numerical factors.

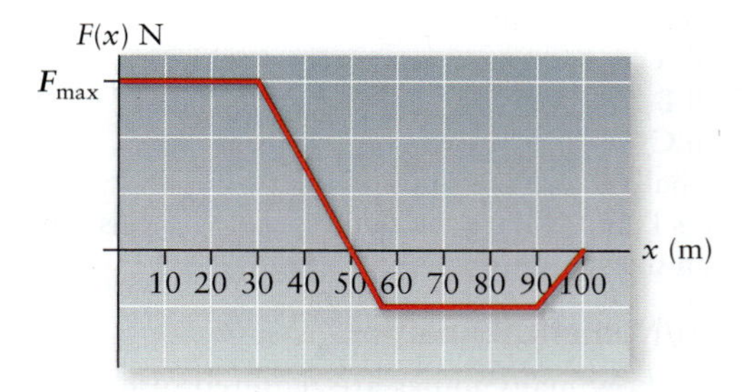

Figure 6-42 Problem 55

56. • An object attached to the free end of a horizontal spring of constant 450 N/m is pulled from a position 12 cm beyond equilibrium to a position 18 cm beyond equilibrium. Calculate the work the spring does on the object. SSM Example 6-9

57. • A 5.00-kg object is attached to one end of a horizontal spring that has a negligible mass and a spring constant of 250 N/m. The other end of the spring is fixed to a wall. The spring is compressed by 10.0 cm from its equilibrium position and released from rest. (a) What is the speed of the object when it is 8.00 cm from equilibrium? (b) What is the

speed when the object is 5.00 cm from equilibrium? (c) What is the speed when the object is at the equilibrium position? Example 6-9

58. • A pendulum is constructed by attaching a small metal ball to one end of a 1.25-m-long string that hangs from the ceiling (**Figure 6-43**). The ball is released when it is raised high enough for the string to make an angle of 30.0° with the vertical. How fast is it moving at the bottom of its swing? Does the mass of the ball affect the answer? SSM Example 6-8

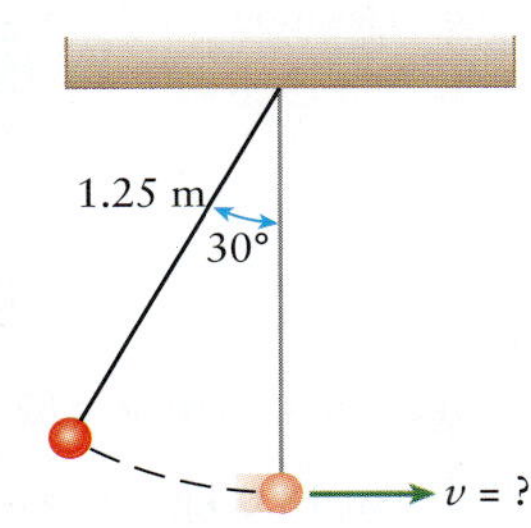

Figure 6-43 Problem 58

59. • Logan is standing on a dock holding onto a rope swing that is 4 m long and suspended from a tree branch above (**Figure 6-44**). The rope is taut and makes a 30° angle with the vertical direction. Logan swings in a circular arc until he releases the rope when it makes an angle of 12° from vertical, but on the other side. If Logan's mass is 75 kg, how much work does gravity do on him up to the point where he releases the rope? Example 6-8

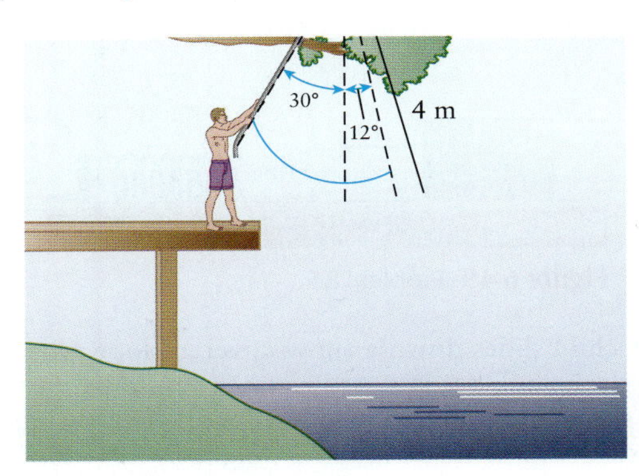

Figure 6-44 Problem 59

60. • Jack and Jill, whose masses are identical, go up a 300-m hill to fetch a pail of water. Jack climbs a sheer rock face to reach the top, while Jill follows a meandering path 835 m long. What is the magnitude of the difference between the work Jack and Jill do against gravity to get to the top? Example 6-8

61. • Wei drags a heavy piece of driftwood for 910 m along an irregular path. If Wei ends 750 m from where he started and exerted a constant force of 625 N, parallel to his path, the entire time, how much work did he do? Example 6-8

62. • Earth orbits the Sun at a radius of about 1.5×10^8 km. At this distance the force of gravity on Earth due to the Sun is 3.6×10^{22} N. Assuming Earth's orbit to be perfectly circular, how much work does the Sun's gravity do on Earth in one year? Example 6-8

63. • Mikaela is out for a bike ride on a breezy day. The wind blows out of the west such that it exerts a constant drag force of 115 N pointing east. Initially riding north on flat roads, Mikaela traverses a 1.2-km-long *circular arc* at a constant speed that ends with her heading directly into the wind; the arc is a quarter circle that starts pointing north and ends pointing west (**Figure 6-45**). How much work does the wind do on her as she rounds this curve from point A to point B? Example 6-8

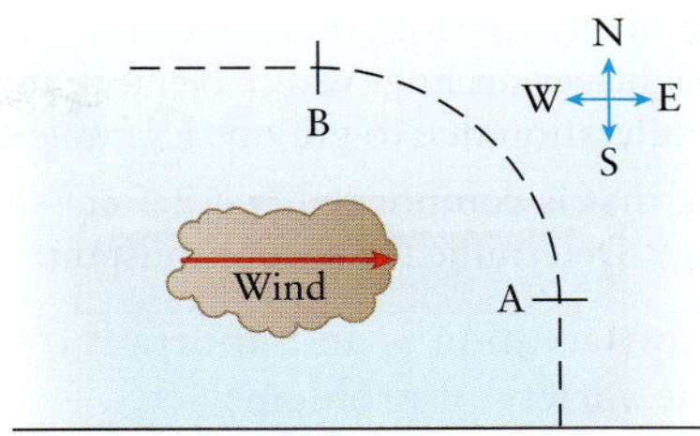

Figure 6-45 Problem 63

6-6 Potential energy is energy related to an object's position

64. • What is the gravitational potential energy relative to the ground associated with a 1.00-N Gravenstein apple hanging from a limb 2.50 m above the ground? Example 6-10

65. • Pilings are driven into the ground at a building site by dropping a 2000-kg object onto them. What change in gravitational potential energy does the object undergo if it is released from rest 18.0 m above the ground and ends up 2.00 m above the ground? Example 6-10

66. • A 40.0-kg boy steps on a skateboard and pushes off from the top of a hill. What change in gravitational potential energy takes place as the boy glides down to the bottom of the hill, 4.35 m below the starting level? SSM Example 6-10

67. • How much additional potential energy is stored in a spring that has a spring constant of 15.5 N/m if the spring starts 10.0 cm from the equilibrium position and ends up 15.0 cm from the equilibrium position?

68. • A spring that has a spring constant of 2.00×10^2 N/m is oriented vertically with one end on the ground. (a) What distance must the spring compress for a 2.00-kg object placed on its upper end to reach equilibrium? (b) By how much does the potential energy stored in the spring increase during the compression?

69. • An external force moves a 3.50-kg box at a constant speed up a frictionless ramp (**Figure 6-46**). The force acts in a direction parallel to the ramp. (a) Calculate the work done on the box by this force as it is pushed up the 5.00-m ramp to a height of 3.00 m. (b) Compare the value with the change in gravitational potential energy that the box undergoes as it rises to its final height. Example 6-10

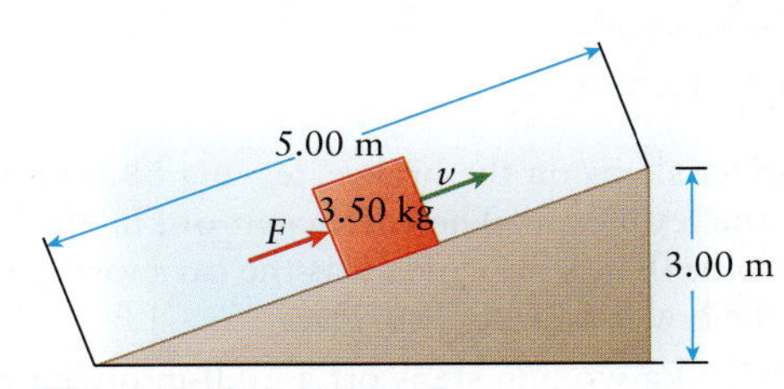

Figure 6-46 Problem 69

70. • Over 630 m in height, the Burj Khalifa is the world's tallest skyscraper. What is the change in gravitational potential energy of a $20 gold coin (33.5 g) when it is carried from ground level up to the top of the Burj Khalifa? Neglect any slight variations in the acceleration due to gravity. SSM Example 6-10

71. • For a great view and a thrill, check out EdgeWalk at the CN Tower in Toronto, where you can walk on the roof of the tower's main pod, 356 m above the ground. What is the

gravitational potential energy relative to the surface of Earth of a 65.0-kg sightseer on EdgeWalk? Neglect any slight variations in the acceleration due to gravity. Example 6-10

72. • A spring that is compressed by 12.5 cm stores 3.33 J of potential energy. Determine the spring constant.

6-8 Energy conservation is an important tool for solving a wide variety of problems

73. • **Sports** A 0.145-kg baseball rebounds off of a wall. The rebound speed is one-third of the original speed. By what percent does the kinetic energy of the baseball change in the collision with the wall? Where does the energy go? Example 6-12

74. • A ball is thrown straight up with an initial speed of 15.0 m/s. At what height will the ball have one-half of its initial speed? Example 6-11

75. • A water balloon is thrown straight down with an initial speed of 12.0 m/s from a second floor window, 5.00 m above ground level. How fast is the balloon moving when it hits the ground? SSM Example 6-11

76. • A gold coin (33.5 g) is dropped from the top of the Burj Khalifa building, 630 m above ground level. In the absence of air resistance, how fast would it be moving when it hit the ground? Example 6-11

77. •• Starting from rest, a 30.0-kg child rides a 9.00-kg sled down a frictionless ski slope. At the bottom of the hill, her speed is 7.00 m/s. If the slope makes an angle of 15.0° with the horizontal, how far did she slide on the sled? Example 6-11

78. •• **Sports** During a long jump Olympic champion Carl Lewis's center of mass rose 1.2 m from the launch point to the top of the arc. What minimum speed did he need at launch if he was traveling at 6.6 m/s at the top of the arc? Example 6-11

79. • An ice cube starts at rest at point A, and slides down a frictionless track as shown in **Figure 6-47**. Calculate the speed of the cube at points B, C, D, and E. Example 6-11

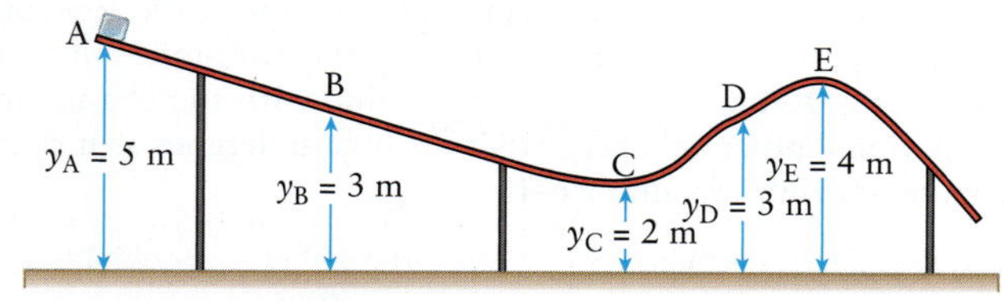

Figure 6-47 Problem 79

80. • A driver slams on the brakes, leaving 88.0-m-long skid marks on the level road. The coefficient of kinetic friction is estimated to be 0.480. How fast was the car moving when the driver hit the brakes? Example 6-11

81. •• A 65.0-kg woman steps off a 10.0-m diving platform and drops straight down into the water. If she reaches a depth of 4.50 m, what is the average resistance force exerted on her by the water? Ignore air resistance. Example 6-11

82. •• A skier leaves the starting gate at the top of a ski jump with an initial speed of 4.00 m/s (**Figure 6-48**). The starting position is 120 m higher than the end of the ramp, which is 3.00 m above the snow. Find the final speed of the skier if he lands 145 m down the 20.0° slope. Assume there is no friction on the ramp, but air resistance causes a 50% loss in the final kinetic energy. The GPS reading of the elevation of the skier is 4212 m at the top of the jump and 4039 m at the landing point. SSM Example 6-11

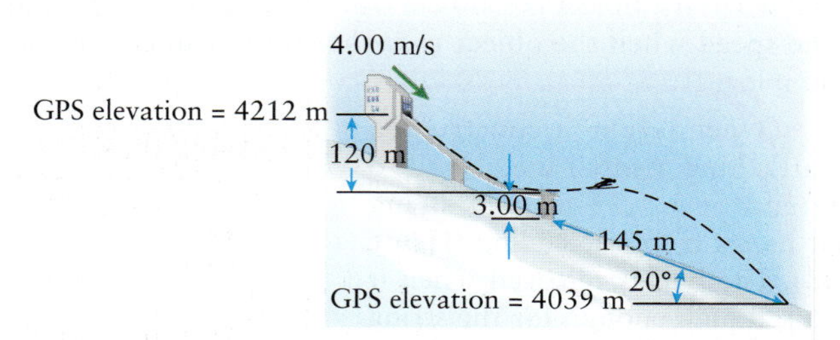

Figure 6-48 Problem 82

83. • An 18.0-kg suitcase falls from a hot-air balloon that is at rest at a height of 385 m above the surface of Earth. The suitcase reaches a speed of 30.0 m/s just before it hits the ground. Calculate the percentage of the initial energy that is "lost" to air resistance. Example 6-12

84. •• An ideal spring is used to stop blocks as they slide along a table without friction (**Figure 6-49**). A 0.85-kg block traveling at a speed of 2.1 m/s can be stopped over a distance of 0.15 m, once it makes contact with the spring. What distance would a 1.3-kg block travel after making contact with the spring, if the block were traveling at a speed of 3.3 m/s? Example 6-13

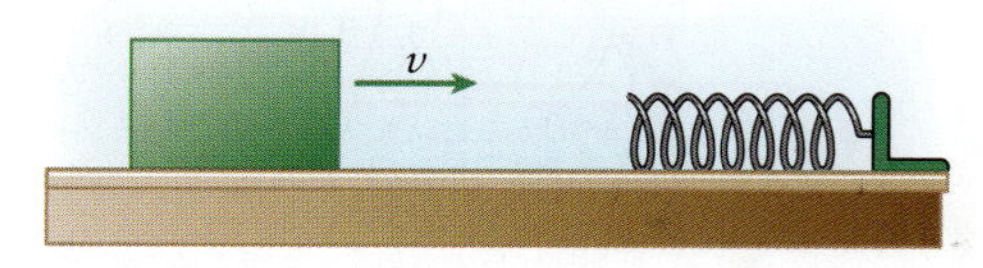

Figure 6-49 Problem 84

85. • A child slides down a snow-covered slope on a sled. At the top of the hill, her mother gives her a push to start her off with a speed of 1.00 m/s. The frictional force acting on the sled is one-fifth of the combined weight of the child and the sled. If she travels for a distance of 25.0 m and her speed at the bottom is 4.00 m/s, calculate the angle that the hill makes with the horizontal. Example 6-12

86. • Neil and Gus are having a competition to see who can launch a marble highest in the air using their own spring. Neil has a firm spring (k_{Neil} = 50.8 N/m), but it can be compressed only a maximum of $y_{Neil, max}$ = 14.0 cm. Gus's spring is less firm than Neil's (k_{Gus} = 12.7 N/m), but its maximum compression is greater ($y_{Gus, max}$ = 27.0 cm). Given that the mass of the marble is 4.5 g and ignoring the effects of air resistance, who can launch the marble the highest and what is the winning height? Assume the marble starts from the same height, in both cases when the springs are at maximum compression. Example 6-13

87. • Continuing with the previous problem, what is the speed of the marble as it leaves each spring? Example 6-13

6-9 Power is the rate at which energy is transferred

88. • The power output of professional cyclists averages about 350 W when climbing mountains. How much energy does a typical 70-kg pro cyclist expend climbing a 12.3-km-long mountain road with a 3.9° average slope at an average speed of 22.5 km/h? Example 6-14

89. • One of the fastest elevators in the world is found in the Taipei 101 building. It ascends at 16.83 m/s and rises 382.2 m.

Each elevator has a maximum load capacity of 1600 kg. What power is required to raise a maximum load from the lowest level to the top floor? Example 6-14

90. • Neglecting extraneous factors like wind resistance, and given that the human body is only about 25% efficient, how much power must an 80.0-kg runner produce in order to run at a constant speed of 3.75 m/s up an 8.0° incline? Example 6-14

91. • A tower crane has a hoist motor rated at 167 hp. Assuming the crane is limited to using 70% of its maximum hoisting power for safety reasons, what is the shortest time in which the crane can lift a 5700-kg load over a distance of 85 m? Example 6-14

General Problems

92. • You push a 20.0-kg crate at constant velocity up a ramp inclined at an angle of 33.0° to the horizontal. The coefficient of kinetic friction between the ramp and the crate, μ_k is equal to 0.200. How much work must you do to push the crate a distance of 2.00 m? Example 6-7

93. •• A 12.0-kg block (M) is released from rest on a frictionless incline that makes an angle of 28.0°, as shown in **Figure 6-50**. Below the block is a spring that has a spring constant of 13,500 N/m. The block momentarily stops when it compresses the spring by 5.50 cm. How far does the block move down the incline from its release point to the stopping point? SSM Example 6-13

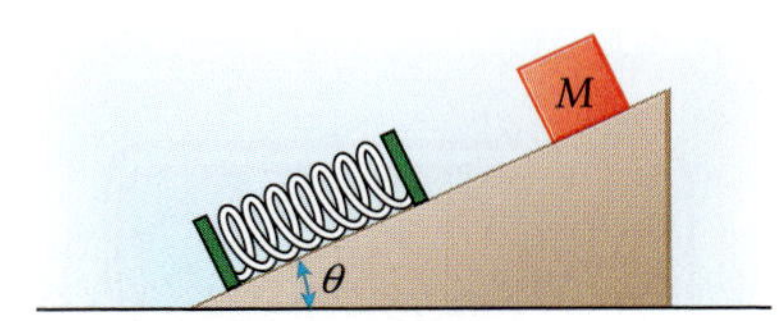

Figure 6-50 Problem 93

94. •• **Sports** A man on his luge (total mass of 88.0 kg) emerges onto the horizontal straight track at the bottom of the hill with a speed of 28.0 m/s. If the luge and rider slow at a constant rate of 2.80 m/s^2, what is the total work done on them by the force that slows them to a stop? Example 6-5

95. • **Biology** An adult dolphin is about 5.00 m long and weighs about 1600 N. How fast must he be moving as he leaves the water in order to jump to a height of 2.50 m? Ignore any effects due to air resistance. Example 6-10

96. • A 3.00-kg block is placed at the top of a track consisting of two frictionless quarter circles of radius R = 2.00 m connected by a 7.00-m-long, straight, horizontal surface (**Figure 6-51**). The coefficient of kinetic friction between the block and the horizontal surface is μ_k = 0.100. The block is released from rest. What maximum vertical height does the block reach on the right-hand section of the track? Example 6-11

Figure 6-51 Problem 96

97. • An object is released from rest on a frictionless ramp of angle θ_1 = 60.0°, at a (vertical) height H_1 = 12.0 m above the base of the ramp (**Figure 6-52**). The bottom end of the ramp *merges smoothly* with a second frictionless ramp that rises at angle θ_2 = 37.0°. (a) How far along the second ramp does the object slide before coming to a momentary stop? (b) When the object is on its way back down the second ramp, what is its speed at the moment that it is a (vertical) height H_2 = 7.00 m above the base of the ramp? SSM Example 6-10

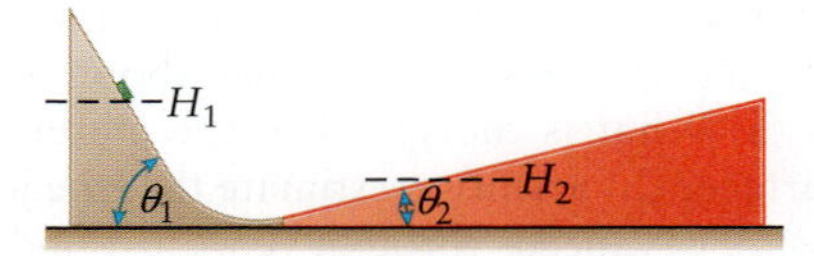

Figure 6-52 Problem 97

98. •• **Biology** An average froghopper insect has a mass of 12.3 mg and reaches a maximum height of 290 mm when its takeoff angle is 58.0° above the horizontal. What is the takeoff speed of the froghopper? Example 6-10

99. ••• A 20.0-g object is placed against the free end of a spring (k equal to 25.0 N/m) that is compressed 10.0 cm (**Figure 6-53**). Once released, the object slides 1.25 m across the tabletop and eventually lands 1.60 m from the edge of the table on the floor, as shown. Is there friction between the object and the tabletop? If there is, what is the coefficient of kinetic friction? The sliding distance on the tabletop includes the 10.0-cm compression of the spring, and the tabletop is 1.00 m above the floor level. Example 6-13

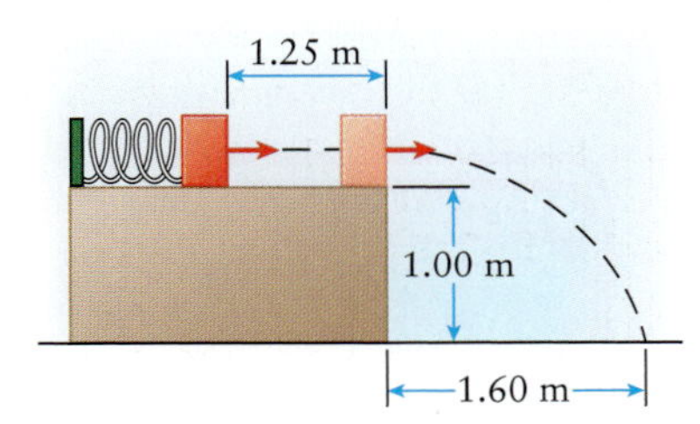

Figure 6-53 Problem 99

100. •• A 1.00-kg object is attached by a thread of negligible mass, which passes over a pulley of negligible mass, to a 2.00-kg object. The objects are positioned so that they are the same height from the floor and then released from rest. What are the speeds of the objects when they are separated vertically by 1.00 m? Example 6-10

101. •• Gravel-filled runaway truck lanes are designed to stop trucks that have lost their brakes on mountain grades. Typically such a lane is horizontal (if possible) and about 35.0 m long. We can think of the ground as exerting a frictional drag force on the truck. If a truck enters the gravel lane with a speed of 55.0 mph (24.6 m/s), use the work-energy theorem to find the minimum coefficient of kinetic friction between the truck and the lane to be able to stop the truck. Example 6-5

102. ••• In 2006 the United States produced 282 × 10^9 kilowatt-hours (kWh) of electrical energy from 4138 hydroelectric plants (1.00 kWh = 3.60 × 10^6 J). On average each plant is 90% efficient at converting mechanical energy to electrical energy, and the average dam height is 50.0 m. (a) At 282 × 10^9 kWh of electrical energy produced in one

year, what is the average power output per hydroelectric plant? (b) What total mass of water flowed over the dams during 2006? (c) What was the average mass of water per dam and the average volume of water per dam that provided the mechanical energy to generate the electricity? (The density of water is 1000 kg/m^3.) (d) A gallon of gasoline contains 45.0×10^6 J of energy. How many gallons of gasoline did the 4138 dams save? Example 6-14

103. ••• **Astronomy** Violent gas eruptions have been observed on Mars, where the acceleration due to gravity is 3.7 m/s^2. The jets throw sand and dust about 75.0 m above the surface. (a) What is the speed of the material just as it leaves the surface? (b) Scientists estimate that the jets originate as high-pressure gas speeds through vents just underground at about 160 km/h. How much energy per kilogram of material is lost due to nonconservative forces as the high-speed matter forces its way to the surface and into the air? Example 6-12

104. • A 3.00-kg block is sent up a ramp of angle θ equal to 37.0° with an initial speed $v_0 = 20.0$ m/s. Between the block and the ramp, the coefficient of kinetic friction is $\mu_k = 0.50$, and the coefficient of static friction is $\mu_s = 0.80$. How far up the ramp (measured along the ramp) does the block go before it comes to a stop? Example 6-5

105. ••• A small block of mass M is placed halfway up on the inside of a frictionless, circular loop of radius R (**Figure 6-54**). The size of the block is very small compared to the radius of the loop. Determine an expression for the minimum downward speed with which the block must be released in order to guarantee that it will make a full circle. Example 6-10

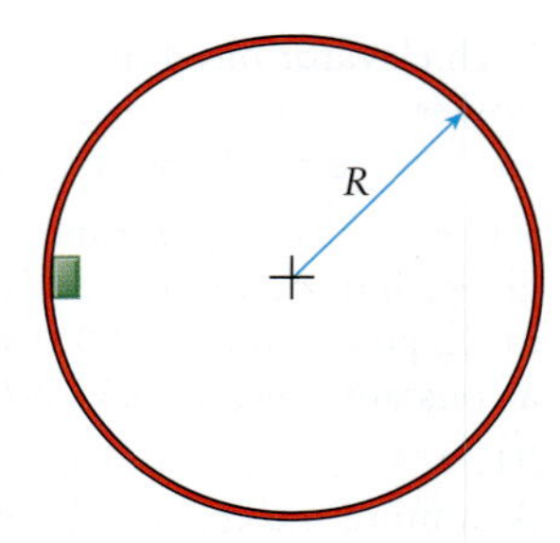

Figure 6-54 Problem 105

106. •• Niagara Falls in Canada has the highest flow rate of any waterfall in the world with a height of more than 50 m. The total average flow rate of the falls is 2400 m^3/s and its average height is 51 m. Given that the density of water is 1000 kg/m^3, calculate the average power output of Niagara Falls.

107. •• The Carmatech SAR12 pneumatic air rifle can accelerate a 3.0-g paintball from zero to 280 ft/s over a barrel length of 20 in. Calculate the average amount of power delivered to the paintball when fired.

7 Momentum, Collisions, and the Center of Mass

Jim Zipp/Science Source

What do you think?

Cooper's hawk (*Accipiter cooperi*) is a species of raptor that hunts and captures other birds in flight. Suppose a fast-moving Cooper's hawk attacks a slow-moving pigeon from behind and captures the pigeon in its talons. Compared to the total kinetic energy of the two separate birds immediately before the capture, the kinetic energy of the system of two birds moving together immediately after the capture is (a) greater; (b) the same; (c) less.

In this chapter, your goals are to:

- (7-1) Comprehend the significance of momentum and the center of mass.
- (7-2) Define the linear momentum of an object and explain how it differs from kinetic energy.
- (7-3) Explain the conditions under which the total momentum of a system is conserved and why total momentum is conserved in a collision.
- (7-4) Identify the differences and similarities between elastic, inelastic, and completely inelastic collisions.
- (7-5) Apply momentum conservation and energy conservation to problems about elastic collisions.
- (7-6) Relate the momentum change of an object, the force that causes the change, and the time over which the force acts.
- (7-7) Explain the physical significance of the center of mass and describe how the net force on a system affects the motion of the system's center of mass.

To master this chapter, you should review:

- (3-2) Multiplying a vector by a scalar.
- (3-5) The definitions of the velocity and acceleration vectors.
- (4-2, 4-5) Newton's second and third laws of motion.
- (6-2, 6-3) The ideas of work, kinetic energy, and the work-energy theorem.

7-1 Newton's third law helps lead us to the idea of momentum

The ideas of kinetic energy and the work-energy theorem that we introduced in Chapter 6 gave us new ways to think about motion. These ideas derive fundamentally from Newton's second law, which states that the net external force on an object determines the object's acceleration. In this chapter we'll learn even more new physics by reconsidering Newton's *third* law: When two objects interact with each other, they exert forces of equal magnitude but opposite direction on each other.

(a) Stephen Mcsweeny/Shutterstock (b) David Fleetham/Alamy (c) Gino's Premium Images/Alamy

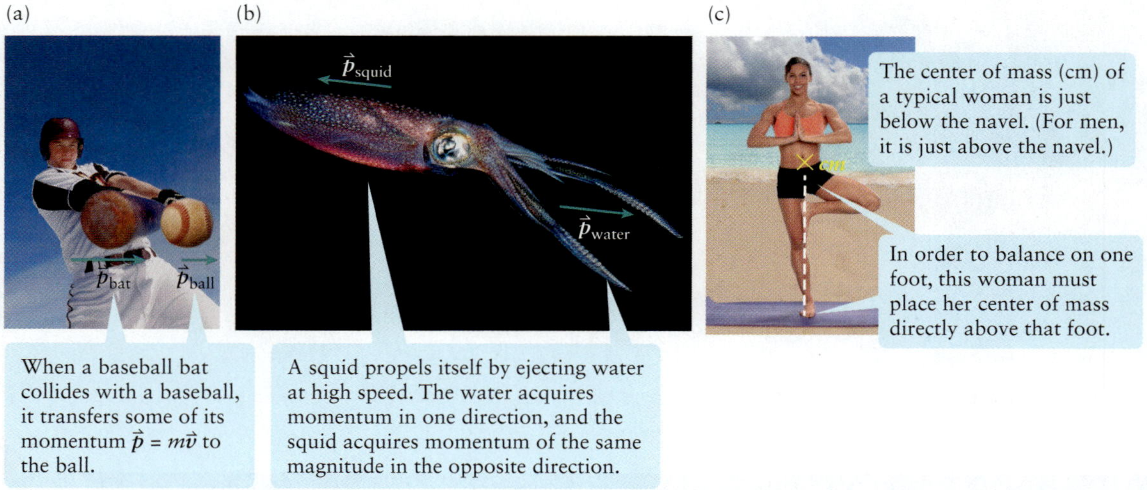

(a)

$\vec{p}_{bat}$ $\vec{p}_{ball}$

When a baseball bat collides with a baseball, it transfers some of its momentum $\vec{p} = m\vec{v}$ to the ball.

(b)

$\vec{p}_{squid}$

$\vec{p}_{water}$

A squid propels itself by ejecting water at high speed. The water acquires momentum in one direction, and the squid acquires momentum of the same magnitude in the opposite direction.

(c)

The center of mass (cm) of a typical woman is just below the navel. (For men, it is just above the navel.)

cm

In order to balance on one foot, this woman must place her center of mass directly above that foot.

Figure 7-1 Momentum and center of mass In this chapter we'll explore (a) momentum, (b) conservation of momentum, and (c) the center of mass of a system.

An important way to express this concept is in terms of the *momentum* of an object (written as $\vec{p}$), which is the product of an object's mass and its velocity. Since velocity is a vector, momentum is as well. We'll see that we can describe the behavior of two interacting objects, such as a baseball bat and a ball (**Figure 7-1a**) or a squid and the water it ejects to propel itself (**Figure 7-1b**), in terms of momentum: Each object in the pair undergoes a change in velocity and hence in momentum, and the momentum changes are equal in magnitude but opposite in direction. This important observation will lead us to a physical principle called the *law of conservation of momentum*. This law will turn out to be essential for analyzing what happens when two objects interact with each other.

Intimately related to the idea of momentum is the notion of *center of mass*. This is a special point associated with a system of objects that moves as though all of the mass of the system were concentrated there and as though all of the forces on the system act at that point. You've used the idea of the center of mass if you've ever balanced on one foot (**Figure 7-1c**). If all of the mass of your body were concentrated into a small blob placed on the ground, that blob would remain at rest because it would be supported from below. In the same way, to have your body remain at rest while standing on the ground on one foot, you lean so that your center of mass (located near your navel) is above your foot and so is supported from below. We'll develop the idea of the center of mass by expanding on what we'll have learned about momentum.

We'll begin our exploration of the momentum concept by considering a simple system: a person standing on a skateboard who decides to jump off.

TAKE-HOME MESSAGE FOR Section 7-1

✔ The idea of momentum is useful in situations where two or more objects interact with each other.

7-2 Momentum is a vector that depends on an object's mass, speed, and direction of motion

You're standing atop a stationary skateboard (**Figure 7-2a**). You then jump straight to the left off the skateboard, and the skateboard rolls away to the right (**Figure 7-2b**). If you try this, you'll find that the skateboard rolls to the right much faster than you fly through the air to the left. Why is this? What determines how much faster the skateboard moves than you do?

Like any question about motion, the best way to answer this question is by using Newton's laws. **Figure 7-2c** shows the free-body diagrams for you and your skateboard while your foot is in contact with the skateboard and you're pushing off. The vertical forces on the skateboard cancel, as do the vertical forces on you, so the net force on the skateboard and the net force on you are both horizontal. There is negligible friction between the ground and the wheels of the skateboard, so the only horizontal forces

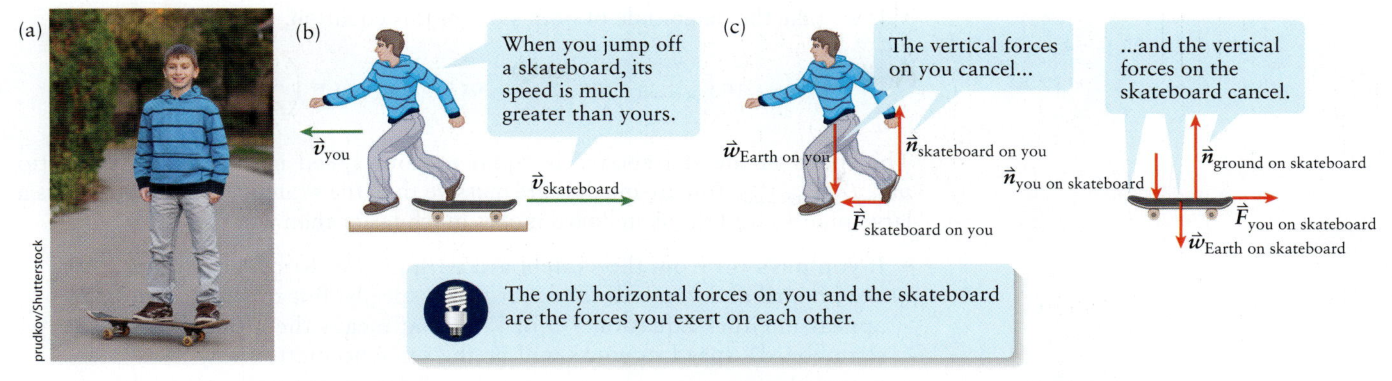

Figure 7-2 Jumping off a skateboard You are (a) initially at rest atop a skateboard then (b) jump off horizontally. (c) Free-body diagrams for you and the skateboard during the push-off.

that act are the forces that you and the skateboard exert on each other. If we apply Newton's second law to the skateboard and to you, we get the equations

$$\begin{aligned}\sum \vec{F}_{\text{ext on skateboard}} &= \vec{F}_{\text{you on skateboard}} = m_{\text{skateboard}}\vec{a}_{\text{skateboard}} \\ \sum \vec{F}_{\text{ext on you}} &= \vec{F}_{\text{skateboard on you}} = m_{\text{you}}\vec{a}_{\text{you}}\end{aligned} \tag{7-1}$$

The vectors $\vec{a}_{\text{skateboard}}$ and $\vec{a}_{\text{you}}$ are the accelerations of the skateboard and you, respectively, during the push-off.

Equations 7-1 involve the *accelerations* of the skateboard and you, but we're interested in the *velocities* of the skateboard and you after the push-off. To see how to get these, let's go back to the definition of the acceleration vector in Section 3-5:

Acceleration vector for the object over a very short time interval from time t_1 to a later time t_2

Change in velocity of the object over the short time interval

$$\vec{a} = \frac{\Delta\vec{v}}{\Delta t} = \frac{\vec{v}_2 - \vec{v}_1}{t_2 - t_1}$$

For both the velocity change and the elapsed time, subtract the earlier value from the later value.

Elapsed time for the time interval

Acceleration vector equals change in velocity vector divided by time interval
(3-9)

In Equation 3-9 the elapsed time Δt is how long the push-off lasts. The initial velocity before the push-off is $\vec{v}_1 = 0$ for both the skateboard and you, since both begin at rest. We'll use the symbols $\vec{v}_{\text{skateboard}}$ and $\vec{v}_{\text{you}}$ for the velocities of the skateboard and you just after the push-off (corresponding to $\vec{v}_2$ in Equation 3-9). When we substitute these and Equation 3-9 into Equations 7-1, then multiply both sides of these equations by Δt (the duration of the push-off), we get

$$\begin{aligned}\vec{F}_{\text{you on skateboard}}\Delta t &= m_{\text{skateboard}}\vec{v}_{\text{skateboard}} \\ \vec{F}_{\text{skateboard on you}}\Delta t &= m_{\text{you}}\vec{v}_{\text{you}}\end{aligned} \tag{7-2}$$

Now remember that from Newton's third law, the force that the skateboard exerts on you has the same magnitude as the force that you exert on the skateboard but points in the opposite direction (Figure 7-2c): $\vec{F}_{\text{skateboard on you}} = -\vec{F}_{\text{you on skateboard}}$. If we substitute this into Equations 7-2, you can see that the quantities $m_{\text{skateboard}}\vec{v}_{\text{skateboard}}$ and $m_{\text{you}}\vec{v}_{\text{you}}$ likewise have the same magnitude but point in opposite directions:

$$m_{\text{skateboard}}\vec{v}_{\text{skateboard}} = -m_{\text{you}}\vec{v}_{\text{you}} \tag{7-3}$$

This relationship is a direct consequence of Newton's second and third laws of motion. Here's how to interpret Equation 7-3:

- The minus sign in Equation 7-3 tells us that $\vec{v}_{\text{skateboard}}$ and $\vec{v}_{\text{you}}$ are in opposite directions. As Figure 7-2b shows, if you push the skateboard to the right, you must move to the left.

- If we take the magnitude of both sides of this equation, we get

$$m_{\text{skateboard}} v_{\text{skateboard}} = m_{\text{you}} v_{\text{you}} \quad \text{or} \quad v_{\text{skateboard}} = \left(\frac{m_{\text{you}}}{m_{\text{skateboard}}}\right) v_{\text{you}} \tag{7-4}$$

The speed of the skateboard is equal to your speed multiplied by the ratio $m_{\text{you}}/m_{\text{skateboard}}$. You are much more massive than the skateboard, so this ratio is a large number and the skateboard moves much faster than you do.

- If you push off from the skateboard with greater force, both the skateboard and you will fly off at faster speeds. But this force doesn't appear in either Equation 7-3 or 7-4. That means the *ratio* of the skateboard's speed to yours will be the same no matter how hard you push.

Our analysis of the skateboard problem, and especially Equation 7-3, suggests that we think about a quantity that's equal to the product of an object's mass m and its velocity vector $\vec{v}$. We'll call this quantity the object's **linear momentum:**

Linear momentum (7-5)

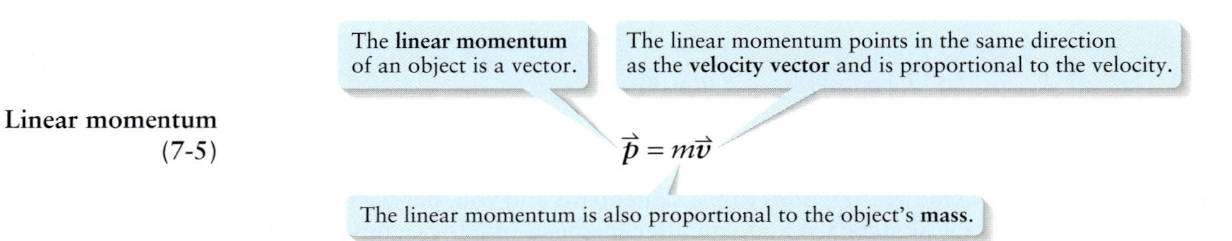

The **linear momentum** of an object is a vector.
The linear momentum points in the same direction as the **velocity vector** and is proportional to the velocity.
$\vec{p} = m\vec{v}$
The linear momentum is also proportional to the object's **mass.**

We'll usually call $\vec{p}$ simply the **momentum** of the object, which is how most physicists refer to it. Note that since the mass m is a positive scalar quantity, the momentum $\vec{p}$ points in the same direction as the velocity $\vec{v}$ (see Section 3-2). **Figure 7-3** compares the momentum vectors for three different objects.

Equation 7-5 shows that the units of momentum are the units of mass multiplied by the units of velocity, or kg·m/s. This combination of units doesn't have a special name in the SI system, so we simply say that the momentum of a 1000-kg car driving at 20 m/s has magnitude $p = mv = (1000 \text{ kg})(20 \text{ m/s}) = 20{,}000 \text{ kg}\cdot\text{m/s}$, or "twenty thousand kilogram-meters per second."

We can use the definition of momentum, Equation 7-5, to rewrite Equation 7-3 for the skateboarder and you: $m_{\text{skateboard}}\vec{v}_{\text{skateboard}} = -m_{\text{you}}\vec{v}_{\text{you}}$ becomes

$$\vec{p}_{\text{skateboard}} = -\vec{p}_{\text{you}} \tag{7-6}$$

Just after the push-off, you and the skateboard each have the same *magnitude* of momentum, but your momentum is directly opposite to the skateboard's momentum. We'll use this observation in Example 7-1.

(a)
These two objects have the same momentum $\vec{p} = m\vec{v}$: The magnitude of momentum is the same, and they are both moving in the same direction.
$v = 6.0$ m/s
$p = mv = 6.0$ kg·m/s
$m = 1.0$ kg

(b)
These two objects have different momentum: The magnitude $p = mv$ is the same for both, but they move in different directions so the momentum vector $\vec{p} = m\vec{v}$ is different.
$v = 3.0$ m/s
$p = mv = 6.0$ kg·m/s
$m = 2.0$ kg

(c)
$v = 3.0$ m/s
$p = mv = 6.0$ kg·m/s
$m = 2.0$ kg

Figure 7-3 The momentum vector Comparing the momentum vectors for three different objects.

WATCH OUT! Don't confuse momentum and kinetic energy.

Like momentum, the kinetic energy K of an object (introduced in Chapter 6) depends on its mass m and its speed v: $K = \frac{1}{2}mv^2$. But kinetic energy is a *scalar* quantity (it has no direction), while momentum $\vec{p} = m\vec{v}$ is a *vector* quantity. Furthermore, two objects with the same magnitude of momentum can have very different kinetic energies. Example 7-1 explores this further.

EXAMPLE 7-1 Momentum and Kinetic Energy

Suppose that you have a mass of 50.0 kg. Your skateboard has a mass of 2.50 kg. Just after you push off as in Figure 7-2, you are moving to the left at 0.600 m/s. (a) What is the magnitude of your momentum and the magnitude of the skateboard's momentum just after you push off? (b) What is the speed of the skateboard just after you push off? (c) What is your kinetic energy and the skateboard's kinetic energy just after you push off?

Set Up

We use the definitions of linear momentum and kinetic energy. Our discussion above and Equation 7-6 tell us how to find the speed of the skateboard from your speed.

$$\vec{p} = m\vec{v} \qquad (7\text{-}5)$$

$$K = \frac{1}{2}mv^2 \qquad (6\text{-}8)$$

$$\vec{p}_{\text{skateboard}} = -\vec{p}_{\text{you}} \qquad (7\text{-}6)$$

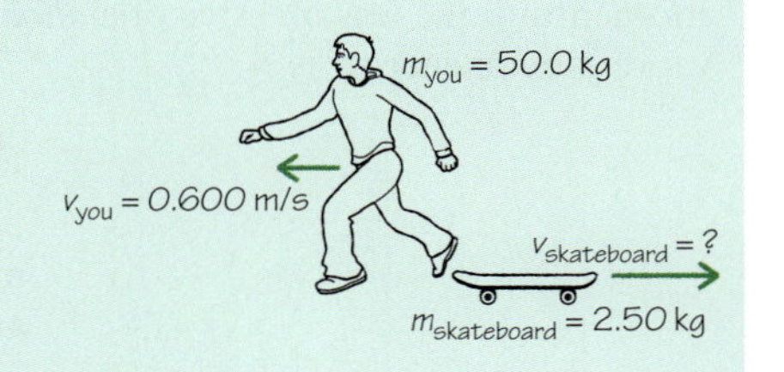

Solve

(a) Equation 7-5 gives us the magnitude of your momentum. Equation 7-6 tells us that the skateboard has as much momentum to the right as you have to the left.

Your momentum vector:

$$\vec{p}_{\text{you}} = m_{\text{you}}\vec{v}_{\text{you}}$$

Magnitude of this vector:

$$\begin{aligned} p_{\text{you}} &= m_{\text{you}}v_{\text{you}} \\ &= (50.0\ \text{kg})(0.600\ \text{m/s}) \\ &= 30.0\ \text{kg}\cdot\text{m/s} \end{aligned}$$

Since $\vec{p}_{\text{skateboard}} = -\vec{p}_{\text{you}}$, you and the skateboard have the same magnitude of momentum:

$$p_{\text{skateboard}} = m_{\text{skateboard}}v_{\text{skateboard}} = p_{\text{you}} = m_{\text{you}}v_{\text{you}} = 30.0\ \text{kg}\cdot\text{m/s}$$

(b) We know the skateboard's mass and momentum, so we can solve for its speed.

$$p_{\text{skateboard}} = m_{\text{skateboard}}v_{\text{skateboard}} \quad \text{so}$$

$$\begin{aligned} v_{\text{skateboard}} &= \frac{p_{\text{skateboard}}}{m_{\text{skateboard}}} \\ &= \frac{30.0\ \text{kg}\cdot\text{m/s}}{2.50\ \text{kg}} \\ &= 12.0\ \text{m/s} \end{aligned}$$

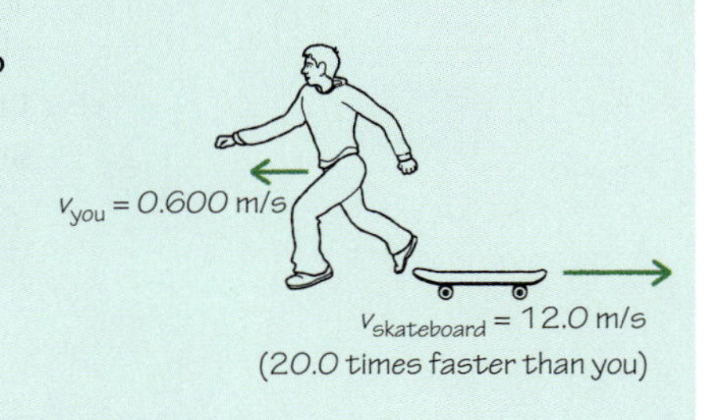

(c) Using Equation 6-8, we find the kinetic energies from the masses and speeds of the two objects.

Your kinetic energy:

$$\begin{aligned} K_{\text{you}} &= \frac{1}{2}m_{\text{you}}v_{\text{you}}^2 \\ &= \frac{1}{2}(50.0\ \text{kg})(0.600\ \text{m/s})^2 \\ &= 9.00\ \text{kg}\cdot\text{m}^2/\text{s}^2 = 9.00\ \text{J} \end{aligned}$$

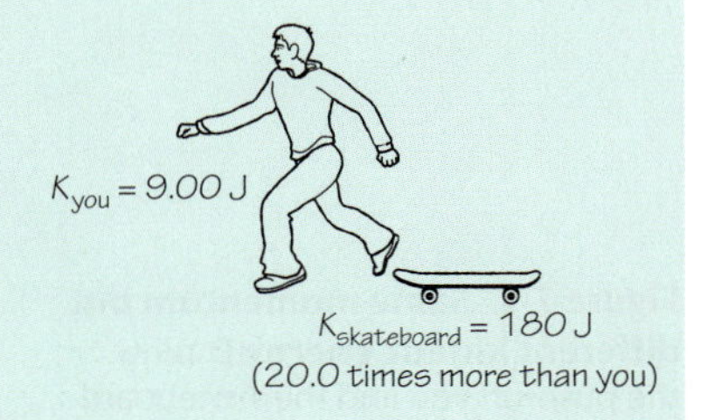

$$K_{\text{skateboard}} = \frac{1}{2}m_{\text{skateboard}}v_{\text{skateboard}}^2 = \frac{1}{2}(2.50\ \text{kg})(12.0\ \text{m/s})^2 = 180\ \text{J}$$

Reflect

The skateboard's speed is 12.0 m/s, about the same as the speed limit for cars in a residential neighborhood. It's not difficult to shove a skateboard hard enough to give it that speed.

As we discussed above, the skateboard moves much faster than you do after the push-off because it has much less mass. The ratio of your mass to that of the skateboard is 20.0 to 1: This same factor of 20.0 appears in the ratio of speeds (the skateboard is faster). It also appears in the ratio of kinetic energies: The skateboard has 20.0 times more kinetic energy than you do. If two objects have the same magnitude of momentum, the less massive one *always* has more kinetic energy.

$$\text{Skateboard speed: } v_{\text{skateboard}} = \left(12.0\frac{\text{m}}{\text{s}}\right)\left(\frac{1\text{ km}}{1000\text{ m}}\right)\left(\frac{3600\text{ s}}{1\text{ h}}\right)$$
$$= 43.2\text{ km/h} = 26.8\text{ mi/h}$$

$$\text{Ratio of masses: } \frac{m_{\text{you}}}{m_{\text{skateboard}}} = \frac{50.0\text{ kg}}{2.50\text{ kg}} = 20.0$$

$$\text{Ratio of speeds: } \frac{v_{\text{you}}}{v_{\text{skateboard}}} = \frac{0.600\text{ m/s}}{12.0\text{ m/s}} = 0.0500 = \frac{1}{20.0}$$

$$\text{Ratio of kinetic energies: } \frac{K_{\text{you}}}{K_{\text{skateboard}}} = \frac{9.00\text{ J}}{180\text{ J}} = 0.0500 = \frac{1}{20.0}$$

We can get more insight into why the two objects in Example 7-1 have the same magnitude of momentum but different kinetic energies. Let's combine Equations 7-2 and 7-3 with the definition $\vec{p} = m\vec{v}$ for momentum:

$$\vec{F}_{\text{you on skateboard}}\Delta t = m_{\text{skateboard}}\vec{v}_{\text{skateboard}} = \vec{p}_{\text{skateboard}}$$
$$\vec{F}_{\text{skateboard on you}}\Delta t = m_{\text{you}}\vec{v}_{\text{you}} = \vec{p}_{\text{you}} \qquad (7\text{-}7)$$

Equations 7-7 say that the momentum that each object acquires during the push-off is equal to the net force that acts on it multiplied by the *time* over which the force acts. Newton's third law tells us that you and the skateboard exert forces of equal magnitude on each other and that when one force is present the other one must be as well. Hence both forces act for the same time Δt, and the skateboard and you end up with equal magnitudes of momentum (although in opposite directions). The kinetic energy that each object acquires, however, is equal to the work done on that object during the push-off, and work equals force multiplied by the *distance* over which the force acts. The skateboard moves faster than you do during the push-off, so it travels a greater distance than you do and so has more work done on it by the same magnitude of force. Therefore, the skateboard ends up with more kinetic energy (**Figure 7-4**).

We've seen how the idea of momentum is useful for the special case of a person pushing off a skateboard. In the next section we'll see how to apply this idea to the general case in which two objects exert forces on each other.

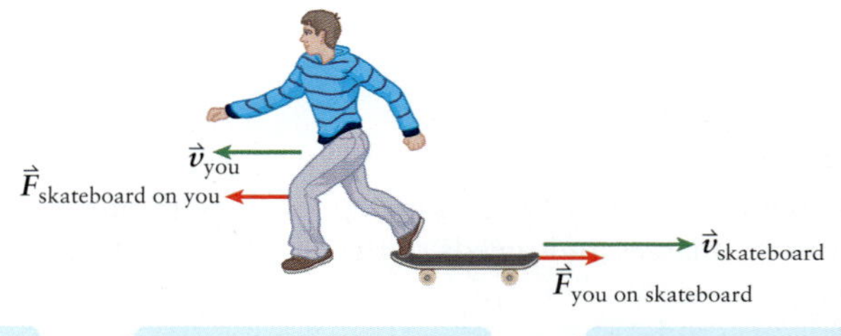

① $\vec{F}_{\text{skateboard on you}}$ and $\vec{F}_{\text{you on skateboard}}$ have the same magnitude.

② These forces act on you and the skateboard for the same amount of time, so you and the skateboard acquire the same magnitude of momentum.

③ The skateboard moves faster and covers a greater distance while the forces act, so more work is done on the skateboard and it acquires more kinetic energy.

Figure 7-4 Same momentum but different kinetic energy During the push-off you and the skateboard acquire the same magnitude of momentum, but the skateboard acquires more kinetic energy.

GOT THE CONCEPT? 7-1 Armed and Dangerous

You are challenged to a skeet shooting competition while wearing rollerblades. You fire the rifle, and you recoil backward. Compared to the kinetic energy of the bullet just after the shot, the kinetic energy of you and the rifle together is (a) greater; (b) the same; (c) less; (d) the answer depends on how powerful the rifle is.

TAKE-HOME MESSAGE FOR Section 7-2

✔ The linear momentum of an object is a vector. It is equal to the product of the object's mass and its velocity.

✔ If two objects at rest push away from each other, each object acquires an equal amount of momentum in opposite directions (if no other net forces act on the objects).

✔ If two objects of different masses have the same magnitude of momentum, the object with the smaller mass has more kinetic energy than the other object.

7-3 The total momentum of a system of objects is conserved under certain conditions

In Section 7-2 we found that just after you pushed off a skateboard, your momentum and the momentum of the skateboard had the same magnitude but were in opposite directions. In terms of vectors (Equation 7-6), we wrote this as

$$\vec{p}_{\text{skateboard}} = -\vec{p}_{\text{you}}$$

Let's rearrange this equation so that both momentum vectors are on the same side of the equals sign:

$$\vec{p}_{\text{you}} + \vec{p}_{\text{skateboard}} = 0 \tag{7-8}$$

The left-hand side of Equation 7-8 is the **total momentum** just after the push-off of you and the skateboard together. Since momentum is a vector, we have to use the rules of vector arithmetic to add them together. Since these two vectors have equal magnitudes but opposite directions, they add to zero. So the system of you and the skateboard has zero total momentum just after the push-off.

Note also that the total momentum just *before* the push-off is also zero: Initially neither you nor the skateboard is moving, so both objects have zero momentum and the sum of these is likewise zero. So for this special case the total momentum of the system of you and the skateboard is *conserved*: It has the same value (in this case zero) before and after the push-off (**Figure 7-5**).

We saw in Chapter 6 that the *total mechanical energy E* of a system is conserved only under special circumstances (if no work is done by nonconservative forces). Is something similar true for the total momentum of a system? And if so, what are the special circumstances under which total momentum is conserved? Let's find the answers to these questions, using our discussion from Section 7-2 as a guide.

① Before the push-off: No motion, so total momentum of the system of you and skateboard is zero: $\vec{p}_{\text{you}} + \vec{p}_{\text{skateboard}} = 0$

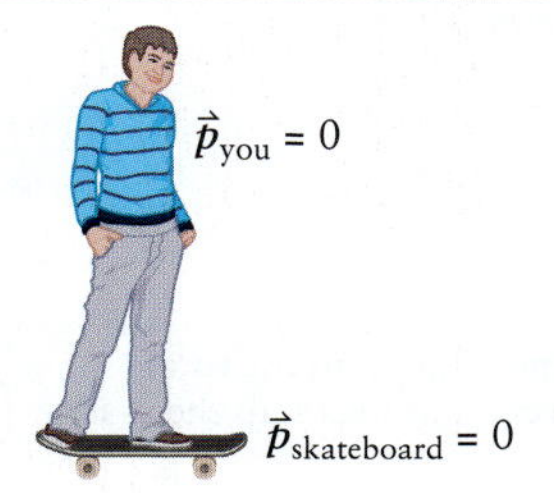

② After the push-off: Both objects have momentum, but the total momentum is still zero: $\vec{p}_{\text{you}} + \vec{p}_{\text{skateboard}} = 0$

In this situation the total momentum of the system is conserved: The push-off does not affect its value.

Figure 7-5 Total momentum When you push off from the skateboard, your momentum and the momentum of the skateboard both change. But the *total* momentum of the system of you and the skateboard is unchanged.

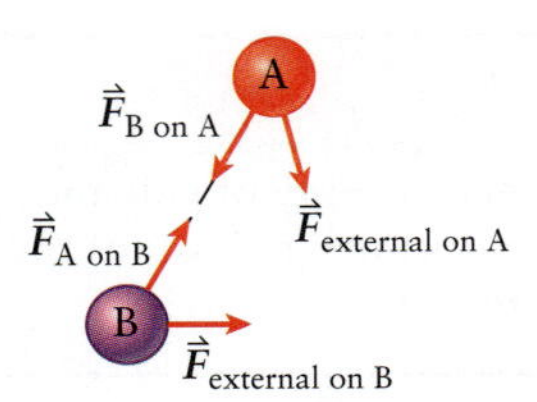

Figure 7-6 Internal forces and external forces If we treat the two objects A and B as a system, the forces that these objects exert on each other are *internal* forces. Forces exerted on A and B by other objects outside the system are *external* forces. The internal forces can either pull the objects together or push them apart.

A System of Several Objects: Internal and External Forces

Figure 7-6 shows a system of two objects. These objects could be billiard balls on a billiard table, oxygen molecules in the air around you, or a planet and its moon. Two kinds of forces can act on each object: forces exerted by other members of the system, which we call **internal forces**, and forces exerted by objects outside the system, which we call **external forces**. For two billiard balls the internal forces are the forces that one ball exerts on another when they collide; the external forces are the normal and friction forces exerted by the billiard table, the gravitational force exerted by Earth, and the force you exert on the cue ball with the cue. For the two objects A and B shown in Figure 7-6, we can write Newton's second law as

$$\begin{aligned}\sum \vec{F}_{\text{on A}} &= \sum \vec{F}_{\text{external on A}} + \vec{F}_{\text{B on A}} = m_{\text{A}}\vec{a}_{\text{A}} \\ \sum \vec{F}_{\text{on B}} &= \sum \vec{F}_{\text{external on B}} + \vec{F}_{\text{A on B}} = m_{\text{B}}\vec{a}_{\text{B}}\end{aligned} \tag{7-9}$$

We can reduce Equations 7-9 to a single equation by adding them together. This has several advantages. First, it leaves us with a single equation to analyze. Second, from Newton's third law, we know that $\vec{F}_{\text{B on A}} = -\vec{F}_{\text{A on B}}$. So when we add Equations 7-9, these two terms cancel each other out. Mathematically, we end up with

$$\sum \vec{F}_{\text{external on A}} + \sum \vec{F}_{\text{external on B}} = m_{\text{A}}\vec{a}_{\text{A}} + m_{\text{B}}\vec{a}_{\text{B}} \tag{7-10}$$

The internal forces that one object in our system exerts on another have disappeared from Equation 7-10. This will prove to be incredibly important for analyzing collisions: We'll be able to calculate the effects of a collision without knowing the details of those internal forces.

Over a time interval Δt the velocity of A changes from $\vec{v}_{\text{Ai}}$ to $\vec{v}_{\text{Af}}$ ("i" for initial, "f" for final), and the velocity of B changes from $\vec{v}_{\text{Bi}}$ to $\vec{v}_{\text{Bf}}$. Using Equation 3-9, $\vec{a} = \Delta\vec{v}/\Delta t$, we can rewrite each of the accelerations in Equation 7-10 in terms of the changes in velocity:

$$\sum \vec{F}_{\text{external on A}} + \sum \vec{F}_{\text{external on B}} = \frac{m_{\text{A}}(\vec{v}_{\text{Af}} - \vec{v}_{\text{Ai}})}{\Delta t} + \frac{m_{\text{B}}(\vec{v}_{\text{Bf}} - \vec{v}_{\text{Bi}})}{\Delta t} \tag{7-11}$$

We can further rewrite Equation 7-11 by multiplying both sides of each equation by Δt, using the definition of momentum $\vec{p} = m\vec{v}$, and rearranging terms on the right-hand side:

$$\left(\sum \vec{F}_{\text{external on A}} + \sum \vec{F}_{\text{external on B}}\right)\Delta t = m_{\text{A}}\vec{v}_{\text{Af}} - m_{\text{A}}\vec{v}_{\text{Ai}} + m_{\text{B}}\vec{v}_{\text{Bf}} - m_{\text{B}}\vec{v}_{\text{Bi}}$$

$$\left(\sum \vec{F}_{\text{external on A}} + \sum \vec{F}_{\text{external on B}}\right)\Delta t = (\vec{p}_{\text{Af}} + \vec{p}_{\text{Bf}}) - (\vec{p}_{\text{Ai}} + \vec{p}_{\text{Bi}}) \tag{7-12}$$

In Equation 7-12 $\vec{p}_{\text{Ai}}$ and $\vec{p}_{\text{Bi}}$ are the momenta of A and B, respectively, at the beginning of the time interval Δt, and $\vec{p}_{\text{Af}}$ and $\vec{p}_{\text{Bf}}$ are the momenta of A and B, respectively, at the end of the time interval.

The quantity in parentheses on the left-hand side of Equation 7-12 is the sum of all of the *external* forces that act on the *system* of objects A and B. We'll call this $\sum \vec{F}_{\text{external on system}}$ for short. The right-hand side of the equation is the difference between the *total* momentum of the system after the time interval Δt, $\vec{P}_{\text{f}} = \vec{p}_{\text{Af}} + \vec{p}_{\text{Bf}}$, and the *total* momentum of the system before the time interval, $\vec{P}_{\text{i}} = \vec{p}_{\text{Ai}} + \vec{p}_{\text{Bi}}$. So we can rewrite Equation 7-12 as

External force and total momentum change for a system of objects

$$\left(\sum \vec{F}_{\text{external on system}}\right)\Delta t = \vec{P}_{\text{f}} - \vec{P}_{\text{i}} = \Delta\vec{P} \tag{7-13}$$

The **sum of all external forces** acting on a system of objects

Duration of a time interval over which the external forces act

Change during that time interval **in the total momentum** of the objects that make up the system

Equation 7-13 says something quite remarkable: *Only the external forces acting on a system can affect the system's total momentum.* The internal forces of one object on another allow momentum to be transferred between the objects (for example, when a moving cue ball hits an eight-ball at rest and sends the eight-ball flying), but they don't affect the value of the *total* momentum.

We now have the answer to the question "When is momentum conserved?" Equation 7-13 says that the total momentum does not change over the time interval Δt if the net external force on the system is zero. Then the left-hand side of Equation 7-13 is zero and so there is zero difference between the final total momentum $\vec{P}_f = \vec{p}_{Af} + \vec{p}_{Bf}$ and the initial total momentum $\vec{P}_i = \vec{p}_{Ai} + \vec{p}_{Bi}$:

If the net external force on a system of objects is zero, **the total momentum of the system** is conserved.

Law of conservation of momentum

$$\vec{P}_f = \vec{P}_i \tag{7-14}$$

Then the total momentum of the system at the end of a time interval...

...is equal to the total momentum of the system at the beginning of that time interval.

Equation 7-14 is the **law of conservation of momentum.** It explains why the total momentum was conserved for the system of you and the skateboard in Section 7-2: There were external forces acting on you and on the skateboard, but the vector sum of these external forces was zero. Hence the total momentum of you and the skateboard had the same value (zero) both before and after the push-off.

Momentum Conservation and Collisions

The law of conservation of momentum turns out to be useful even when the net external force on a system is *not* zero. One example is a collision between two automobiles (**Figure 7-7**). During the collision the net vertical force on each car is zero: The upward normal force exerted by the ground balances the downward gravitational force. But the ground also exerts a horizontal friction force on each car, and there is nothing to balance the friction forces. So there is a net external force on the system of two cars. To see why we can nonetheless ignore this net external force, let's rewrite Equation 7-12 the way it would appear if we didn't immediately cancel the internal forces when we added Equations 7-9 (the internal forces still appear on the left-hand side of the equation):

$$\left(\sum \vec{F}_{\text{external on A}} + \sum \vec{F}_{\text{external on B}}\right)\Delta t + (\vec{F}_{\text{B on A}} + \vec{F}_{\text{A on B}})\Delta t = \vec{p}_{Af} + \vec{p}_{Bf} - \vec{p}_{Ai} - \vec{p}_{Bi} \tag{7-15}$$

1 When the two cars collide they exert strong forces on each other.
2 The vertical forces on each car cancel.
Car A
Car B
$\vec{n}_A$
$\vec{n}_B$
$\vec{F}_{\text{B on A}}$
$\vec{F}_{\text{A on B}}$
$\vec{f}_{kA}$
$\vec{f}_{kB}$
$\vec{w}_A$
$\vec{w}_B$
3 There is a kinetic friction force on each car, but these are very small compared to the forces that car A exerts on car B and car B exerts on car A.
AP Photo/Dale Davi

Figure 7-7 Forces in a collision During a collision between two cars, the internal forces (the forces of one car on another) are so great that we can ignore the external forces on the system.

So we can say that:
- The net force on car A is $\vec{F}_{\text{B on A}}$ and the net force on car B is $\vec{F}_{\text{A on B}}$.
- These forces are internal to the system of two cars.

The forces $\vec{F}_{B \text{ on } A}$ and $\vec{F}_{A \text{ on } B}$ that the cars exert on each other are *very* large: They can deform metal and shatter both plastic and glass. By comparison, the external friction forces are much smaller in magnitude. So it's a very good approximation to ignore the external friction forces during the brief duration Δt of the collision and set the terms $\sum \vec{F}_{\text{external on A}}$ and $\sum \vec{F}_{\text{external on B}}$ in Equation 7-15 equal to zero:

$$(\vec{F}_{\text{B on A}} + \vec{F}_{\text{A on B}})\Delta t = \vec{p}_{\text{Af}} + \vec{p}_{\text{Bf}} - \vec{p}_{\text{Ai}} - \vec{p}_{\text{Bi}} \tag{7-16}$$

But by Newton's third law we still have $\vec{F}_{\text{B on A}} = -\vec{F}_{\text{A on B}}$, so the left-hand side of Equation 7-16 for the two colliding cars is zero. Thus the change in total momentum during the collision is still zero, or

$$\vec{p}_{\text{Af}} + \vec{p}_{\text{Bf}} = \vec{p}_{\text{Ai}} + \vec{p}_{\text{Bi}} \tag{7-17}$$

This is the same as Equation 7-14, the law of conservation of momentum. So we conclude that

If the internal forces during a collision are much greater in magnitude than the external forces, the total momentum of the colliding objects has the same value just before and just after the collision.

In *most* collisions the internal forces are much larger in magnitude than the external forces. For example, when two hockey players collide on the ice, the internal forces of one player on the other are strong enough to knock the wind out of the players or even cause injury. These internal forces are much greater than the friction forces that the ice exerts on the players. Similarly, the forces that act between a tennis ball and tennis racquet in a serve are so large that the ball distorts noticeably. The external forces that act on the ball and racquet during the time they are in contact—the gravitational forces and the force of the player's hand on the racquet—are very feeble by comparison. So it's almost always safe to say that the total momentum of a system of colliding objects just *after* the collision is the same as just *before* the collision.

Go to Interactive Exercise 7-1 for more practice dealing with conservation of momentum.

WATCH OUT! In a collision momentum is conserved only *during* the collision.

The general rule that we have discovered says that the total momentum of colliding objects is conserved *only* during the very brief period during which the collision takes place. Once the collision is over the strong internal forces no longer act, the external forces become dominant, and the total momentum is no longer conserved. For example, the two colliding cars shown in Figure 7-7 continued to move after the collision (with a total momentum just after the collision that equals their total momentum just before the collision). Once the collision has ended, however, friction with the track and grass soon caused the cars to come to a halt and lose all of their momentum.

WATCH OUT! Remember, momentum is a vector.

It's important to note that Equation 7-14, the law of conservation of momentum, is a *vector* equation. If the vectors $\vec{P}_f$ and $\vec{P}_i$ are equal, it must be that the x components of the two vectors are equal *and* the y components of the two vectors are equal. Examples 7-2, 7-3, and 7-4 illustrate how to solve problems that involve the vector nature of momentum.

EXAMPLE 7-2 Conservation of Momentum: A Collision on the Ice

Gordie, a 100-kg hockey player, is initially moving to the right at 5.00 m/s directly toward Mario, a stationary 80.0-kg player. After the two players collide head-on, Mario is moving to the right at 3.75 m/s. (a) In what direction and at what speed is Gordie moving after the collision? (b) What was the change in Gordie's momentum in the collision? What was the change in Mario's momentum?

Set Up

The system that we're considering is made up of the two players. The vertical forces on each player (the normal force exerted by the ice and the gravitational force) cancel each other, so there is no net external force in the vertical direction. The friction forces between the players and the ice are small compared to the forces that Gordie and Mario exert on each other. So we can treat the total momentum of the system as conserved during the collision. We'll use this to find Gordie's final velocity and the changes in momentum of each player. We choose the positive x direction to be to the right, as shown.

Momentum conservation:

$$\vec{P} = \vec{p}_{\text{G}} + \vec{p}_{\text{M}} \quad (7\text{-}14)$$

(G for Gordie, M for Mario) has the same value just before and just after the collision

$$\vec{p} = m\vec{v} \quad (7\text{-}5)$$

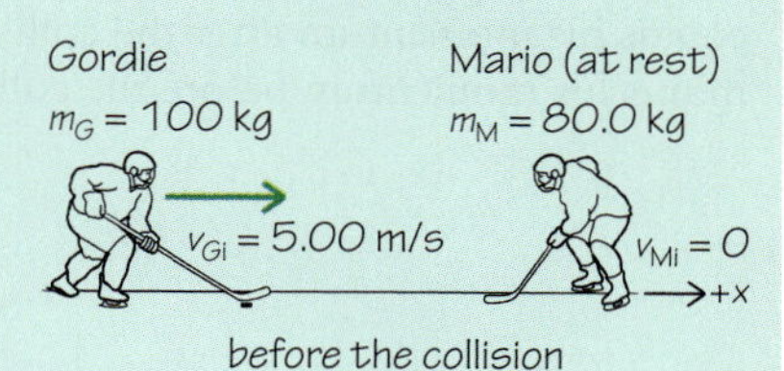

Solve

(a) Write the equation of momentum conservation, using the subscript i (for intial) for values just before the collision and subscript f (for final) for values just after.

Just before the collision the total momentum is

$$\vec{P}_{\text{i}} = \vec{p}_{\text{Gi}} + \vec{p}_{\text{Mi}} = m_{\text{G}}\vec{v}_{\text{Gi}} + m_{\text{M}}\vec{v}_{\text{Mi}}$$

Just after the collision the total momentum is

$$\vec{P}_{\text{f}} = \vec{p}_{\text{Gf}} + \vec{p}_{\text{Mf}} = m_{\text{G}}\vec{v}_{\text{Gf}} + m_{\text{M}}\vec{v}_{\text{Mf}}$$

Momentum is conserved:

$$\vec{P}_{\text{f}} = \vec{P}_{\text{i}}$$

$$m_{\text{G}}\vec{v}_{\text{Gf}} + m_{\text{M}}\vec{v}_{\text{Mf}} = m_{\text{G}}\vec{v}_{\text{Gi}} + m_{\text{M}}\vec{v}_{\text{Mi}}$$

Note that $\vec{v}_{\text{Mi}} = 0$ since Mario is originally at rest. We need to find $\vec{v}_{\text{Gf}}$ (Gordie's velocity just after the collision).

Since the motion is entirely along the x axis, we only need the x component of the momentum conservation equation. Solve for Gordie's final x velocity, $v_{\text{Gf}x}$; then substitute the values of the players' masses, Gordie's velocity before the collision, and Mario's velocity after the collision.

$$m_{\text{G}}v_{\text{Gf}x} + m_{\text{M}}v_{\text{Mf}x} = m_{\text{G}}v_{\text{Gi}x} + m_{\text{M}}v_{\text{Mi}x}$$

Mario is originally at rest, so $v_{\text{Mi}x} = 0$ and

$$m_{\text{G}}v_{\text{Gf}x} + m_{\text{M}}v_{\text{Mf}x} = m_{\text{G}}v_{\text{Gi}x}$$

$$m_{\text{G}}v_{\text{Gf}x} = m_{\text{G}}v_{\text{Gi}x} - m_{\text{M}}v_{\text{Mf}x}$$

$$v_{\text{Gf}x} = v_{\text{Gi}x} - \frac{m_{\text{M}}v_{\text{Mf}x}}{m_{\text{G}}}$$

Players' masses: $m_{\text{G}} = 100$ kg, $m_{\text{M}} = 80.0$ kg

Gordie's initial x velocity: $v_{\text{Gi}x} = +5.00$ m/s

Mario's final x velocity: $v_{\text{Mf}x} = +3.75$ m/s

$$\begin{aligned} v_{\text{Gf}x} &= (+5.00 \text{ m/s}) - \frac{(80.0 \text{ kg})(+3.75 \text{ m/s})}{100 \text{ kg}} \\ &= +5.00 \text{ m/s} - 3.00 \text{ m/s} \\ &= +2.00 \text{ m/s} \end{aligned}$$

Gordie ends up moving at 2.00 m/s to the right (in the positive x direction).

(b) The change in each player's momentum equals his momentum after the collision minus his momentum before the collision.

$$\begin{aligned}\text{Gordie: } \Delta p_{\mathrm{G}x} &= m_{\mathrm{G}} v_{\mathrm{Gf}x} - m_{\mathrm{G}} v_{\mathrm{Gi}x} \\ &= (100\ \mathrm{kg})(+2.00\ \mathrm{m/s}) - (100\ \mathrm{kg})(+5.00\ \mathrm{m/s}) \\ &= 200\ \mathrm{kg \cdot m/s} - 500\ \mathrm{kg \cdot m/s} \\ &= -300\ \mathrm{kg \cdot m/s}\end{aligned}$$

$$\begin{aligned}\text{Mario: } \Delta p_{\mathrm{M}x} &= m_{\mathrm{M}} v_{\mathrm{Mf}x} - m_{\mathrm{M}} v_{\mathrm{Mi}x} \\ &= (80.0\ \mathrm{kg})(+3.75\ \mathrm{m/s}) - (80.0\ \mathrm{kg})(0\ \mathrm{m/s}) \\ &= +300\ \mathrm{kg \cdot m/s}\end{aligned}$$

Reflect

Our answer to part (a) tells us that after the collision Gordie is still moving in the positive x direction but with reduced speed: He has lost x momentum, while Mario (who was originally at rest) has gained x momentum. In fact, as part (b) shows, the amount of x momentum that Gordie loses ($300\ \mathrm{kg \cdot m/s}$) is exactly the same as the amount of x momentum that Mario gains. Thus we can think of the collision between the two players as a *transfer* of momentum between Gordie and Mario. No momentum is lost in the collision; it simply changes hands from one player to the other.

Notice that in this problem we needed to know Mario's final x velocity $v_{\mathrm{Mf}x}$ in order to find Gordie's final x velocity $v_{\mathrm{Gf}x}$. That's because the statement that momentum is conserved in the x direction gave us only one equation which relates $v_{\mathrm{Mf}x}$ and $v_{\mathrm{Gf}x}$. Hence we were able to solve for only one unknown quantity. If we didn't know either of the players' final velocities, we would have needed additional information—in particular, the duration of the collision and the magnitude of the force that one player exerts on the other during the collision—which we unfortunately do not have. The law of conservation of momentum is a great tool, but by itself it can't tell you everything!

Although we used conservation of momentum in this example, note that neither player's *individual* momentum is conserved: Gordie's momentum after the collision is different than before the collision and likewise for Mario's momentum. It's only the *total* momentum of Gordie and Mario's system that is conserved. *In general the momentum of any particular object within a system will not be conserved.*

EXAMPLE 7-3 A Collision at the Bowl-a-Rama

The sequence of images in **Figure 7-8** shows a bowling ball striking a stationary pin. The second image shows that when the collision occurs the ball strikes a bit to the left of the horizontal center of the pin. The third image shows what happens after the collision: The ball moves on a path to the left of its original direction, and the pin shoots off to the right. Just after one such collision, the ball is moving off horizontally at a 10.0° angle and the pin is moving off horizontally at a 60.0° angle (measured relative to the original direction of motion of the ball). The mass of the ball is 3.50 times greater than the mass of the pin. Just after the collision is the pin moving faster or slower than the ball, and by what factor?

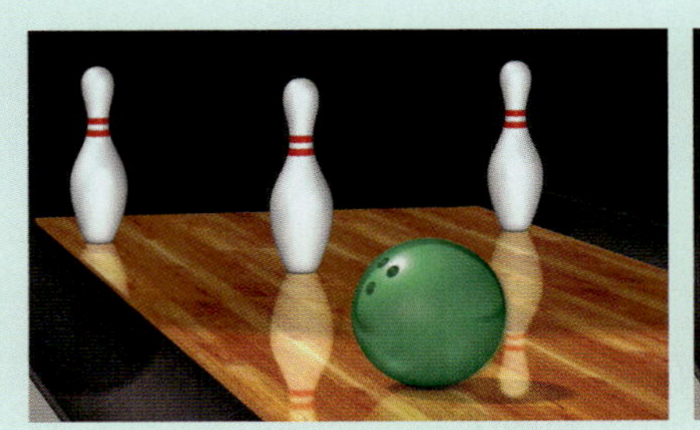

Figure 7-8 Hitting a bowling pin off-center After the bowling ball strikes the left-hand side of the pin, the ball deflects to the left and the pin moves off to the right. How do their speeds compare?

Set Up

The forces that the ball and pin exert on each other are very strong (imagine what it would feel like if your finger were between the ball and pin when they hit!). Compared to these we can ignore the external forces exerted on the system of ball and pin, so the total momentum of this system is conserved. The collision is two-dimensional, so we must account for both the x and y components of momentum.

Momentum conservation:

$$\vec{P} = \vec{p}_{\mathrm{B}} + \vec{p}_{\mathrm{P}} \quad (7\text{-}14)$$

(B for ball, P for pin) has the same value just before and just after the collision

Linear momentum:

$$\vec{p} = m\vec{v} \quad (7\text{-}5)$$

ball (B)
10.0°
60.0°
pin (P)
y
x

We are given the directions of motion of the ball and pin before and after the collision, and we want to find the *ratio* of the pin's final speed v_{Pf} to the ball's final speed v_{Bf}. We don't know the mass of either the ball or the pin, but we do know the ratio of their masses.

Ratio of the mass of the ball to the mass of the pin:

$$\frac{m_B}{m_P} = 3.50$$

Solve

The total momentum before the collision (subscript i) equals the total momentum after the collision (subscript f). Write this for both the x component and the y component of momentum.

Momentum conservation:

$$\vec{p}_{Bf} + \vec{p}_{Pf} = \vec{p}_{Bi} + \vec{p}_{Pi}, \text{ or}$$

$$m_B\vec{v}_{Bf} + m_P\vec{v}_{Pf} = m_B\vec{v}_{Bi} + m_P\vec{v}_{Pi}, \text{ or}$$

$$m_B v_{Bfx} + m_P v_{Pfx} = m_B v_{Bix} + m_P v_{Pix}$$

(total x momentum is conserved)

$$m_B v_{Bfy} + m_P v_{Pfy} = m_B v_{Biy} + m_P v_{Piy}$$

(total y momentum is conserved)

Just prior to the collision, the ball has no component of velocity in the y direction (so $v_{Biy} = 0$) and the pin is at rest (so $v_{Pix} = v_{Piy} = 0$). Just after the collision the ball is moving at a 10.0° angle with positive x and y components of velocity, while the pin is moving at a 60.0° angle with a positive x velocity and a negative y velocity. Use these facts to rewrite the equations for the conservation of x and y momentum.

x equation:

$$m_B v_{Bf} \cos 10.0° + m_P v_{Pf} \cos 60.0° = m_B v_{Bi}$$

y equation:

$$m_B v_{Bf} \sin 10.0° + (-m_P v_{Pf} \sin 60.0°) = 0$$

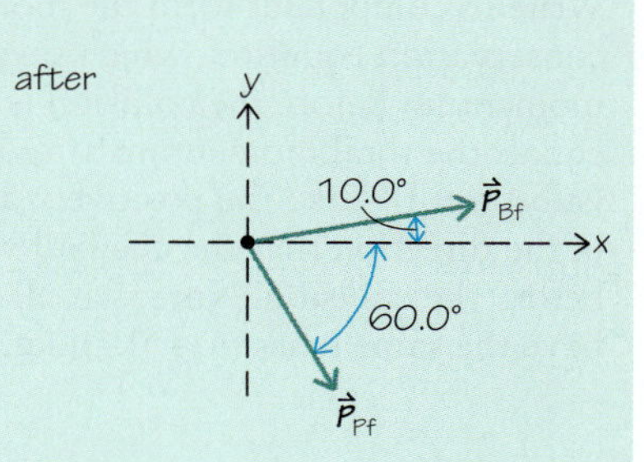

Our goal is to find the ratio of the speed of the pin just after the collision (v_{Pf}) to the speed of the ball just after the collision (v_{Bf}). The equation for x momentum isn't useful for this because we aren't given the value of the ball's speed v_{Bi} before the collision. Instead we use the equation for y momentum to solve for v_{Pf}/v_{Bf}, and then substitute the ratio of the masses of the two objects.

$$m_B v_{Bf} \sin 10.0° = m_P v_{Pf} \sin 60.0°$$

$$\frac{v_{Pf}}{v_{Bf}} = \frac{m_B}{m_P}\frac{\sin 10.0°}{\sin 60.0°} = (3.50)\frac{0.174}{0.866} = 0.702$$

After the collision the speed of the pin is 0.702 times that of the ball.

Reflect

The pin moves relatively slowly because it suffered only a glancing blow from the ball. With a more head-on impact the pin can end up moving away from the collision at a much faster speed than the ball. We'll explore this in Section 7-5.

EXAMPLE 7-4 A Tricky Billiards Shot

On a billiards table the 7-ball and the 8-ball are initially at rest and touching each other. You hit the cue ball in such a way that it acquires a speed of 1.7 m/s before hitting the 7-ball and 8-ball simultaneously. After the collision the cue ball is at rest, and the other two balls are each moving at 45° from the direction that the cue ball was moving. What are the speeds of the 7-ball and 8-ball immediately after the collision? Each billiard ball has a mass of 0.16 kg.

Set Up

This is a more complicated collision than in the previous two examples, but the fundamental principle is the same: Total momentum is conserved.

Initially all of the momentum is in the cue ball and points along the direction of its motion. After the collision the 7-ball and 8-ball must travel on opposite sides of the cue ball's initial path as shown. (If they were on the same side, there would be a nonzero total momentum perpendicular to the cue ball's initial path, and momentum would not be conserved.)

As in Example 7-2 we know the directions of motion of the balls before and after the collision. We know the initial speed $v_{\text{cue,i}} = 1.7$ m/s and final speed $v_{\text{cue,f}} = 0$ of the cue ball, and we want to find the final speeds $v_{7\text{f}}$ and $v_{8\text{f}}$ of the 7-ball and 8-ball.

Momentum conservation:

$$\vec{P} = \vec{p}_{\text{cue}} + \vec{p}_7 + \vec{p}_8 \qquad (7\text{-}14)$$

(cue ball, 7-ball, and 8-ball) has the same value just before and just after the collision

Linear momentum:

$$\vec{p} = m\vec{v} \qquad (7\text{-}5)$$

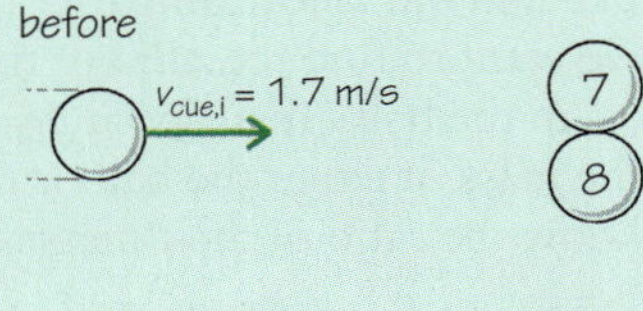

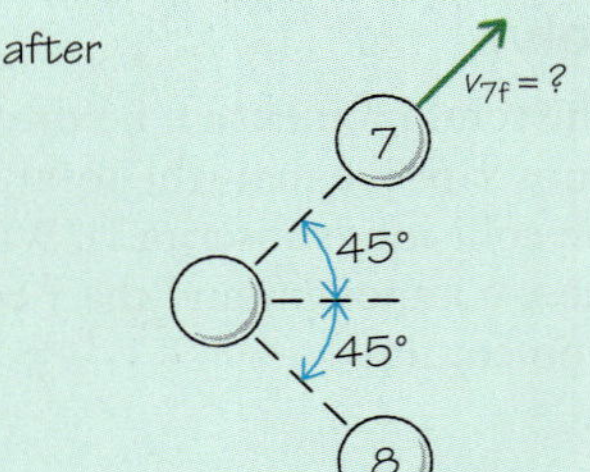

Solve

Write in component form the momentum conservation equation, which says that the total momentum before the collision (subscript i) equals the total momentum after the collision (subscript f). Take the positive x axis to be in the direction that the cue ball was moving before the collision. Note that all three balls have the same mass $m = 0.16$ kg.

Momentum conservation:

$$\vec{p}_{\text{cue,f}} + \vec{p}_{7\text{f}} + \vec{p}_{8\text{f}} = \vec{p}_{\text{cue,i}} + \vec{p}_{7\text{i}} + \vec{p}_{8\text{i}}$$

The 7-ball and 8-ball are not moving before the collision, so $\vec{p}_{7\text{i}} = \vec{p}_{8\text{i}} = 0$, and the cue ball is not moving after the collision, so $\vec{p}_{\text{cue,f}} = 0$. So the momentum conservation equation becomes

$$\vec{p}_{7\text{f}} + \vec{p}_{8\text{f}} = \vec{p}_{\text{cue,i}}$$
$$m\vec{v}_{7\text{f}} + m\vec{v}_{8\text{f}} = m\vec{v}_{\text{cue,i}}$$

or in component form

$$mv_{7\text{f}x} + mv_{8\text{f}x} = mv_{\text{cue,i}x} \text{ (total } x \text{ momentum is conserved)}$$
$$mv_{7\text{f}y} + mv_{8\text{f}y} = mv_{\text{cue,i}y} \text{ (total } y \text{ momentum is conserved)}$$

The cue ball's initial velocity had zero y component (so $v_{\text{cue,i}y} = 0$). Of the 7-ball and 8-ball, one has positive final y velocity and the other negative final y velocity in order to satisfy the condition that total y momentum is conserved. Use these observations to simplify the momentum equations.

x equation:

$$mv_{7\text{f}} \cos 45° + mv_{8\text{f}} \cos 45° = mv_{\text{cue,i}}$$

y equation:

$$mv_{7\text{f}} \sin 45° + (-mv_{8\text{f}} \sin 45°) = 0$$

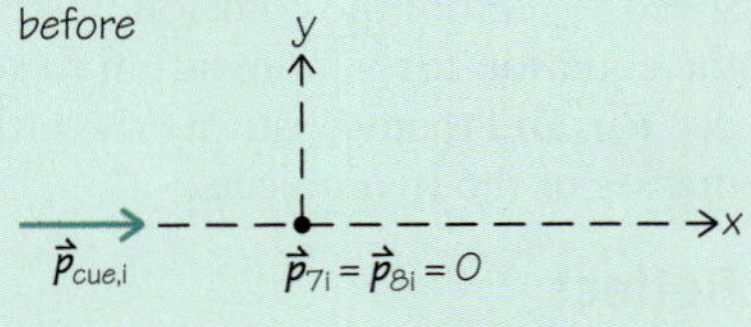

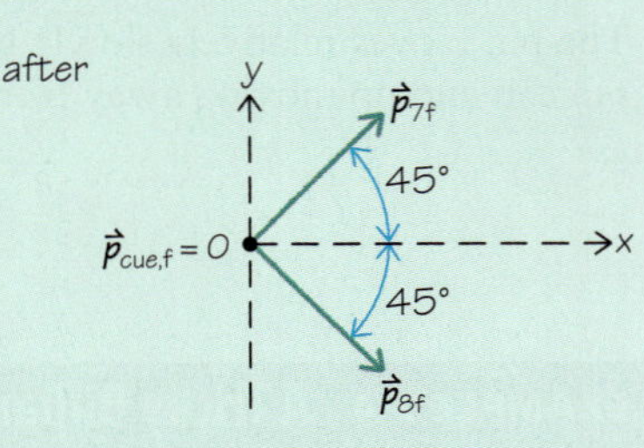

Solve the momentum conservation equations for the final speeds of the 7-ball and 8-ball.

From the y equation

$$mv_{7\text{f}} \sin 45° = mv_{8\text{f}} \sin 45° \text{ so}$$

$$v_{7\text{f}} = v_{8\text{f}}$$

After the collision the 7-ball and the 8-ball move with the same speed. To find this common speed, replace v_{8f} in the x equation with v_{7f} and then solve for v_{7f}:

$$mv_{7f} \cos 45° + mv_{7f} \cos 45° = mv_{\text{cue,i}}$$

$$2mv_{7f} \cos 45° = mv_{\text{cue,i}}$$

$$v_{7f} = \frac{v_{\text{cue,i}}}{2 \cos 45°} = \frac{1.7 \text{ m/s}}{2 \cos 45°} = 1.2 \text{ m/s}$$

After the collision both the 7-ball and the 8-ball are moving at 1.2 m/s.

Reflect

Note that we didn't need the value of the billiard ball mass. That's because all three balls have the same mass m, so m canceled out in the equations. That didn't happen in Example 7-3 because the bowling ball and pin had *different* masses.

There were three objects in this collision rather than two as in Examples 7-2 and 7-3. But the same ideas of momentum conservation apply no matter how many objects are involved in the collision.

GOT THE CONCEPT? 7-2 Which Way Did the Other One Go?

An object moving due west collides with a second object that is initially at rest. Just after the collision the first object is moving toward the southeast. Just after the collision the second object must be moving toward the (a) northeast; (b) southeast; (c) southwest; (d) northwest; (e) west.

TAKE-HOME MESSAGE FOR Section 7-3

✔ In a system of objects internal forces are those that one object in the system exerts on another object in the system. External forces are those exerted on objects in the system by other, outside objects.

✔ If the net external force on a system of objects is zero, the total momentum of the system is conserved. Objects within the system can exchange momentum with each other, but the vector sum of the momentum of all members of the system remains constant.

✔ During a collision the internal forces are typically much larger in magnitude than any external forces. Hence the external forces can be ignored, and the total momentum just after the collision equals the total momentum just before the collision.

7-4 In an inelastic collision some of the mechanical energy is lost

We've learned that momentum is conserved in collisions of all kinds. We also learned in Chapter 6 that *mechanical energy* is conserved in certain circumstances. Are *both* mechanical energy and momentum conserved in collisions?

The answer to this question is "sometimes." In Section 6-7 we learned that mechanical energy is conserved if *only* conservative forces do work. An example of this is two objects that behave like ideal springs when they collide: They compress, converting some of the kinetic energy of the colliding objects into spring potential energy; then they relax and convert the potential energy back into kinetic energy (**Figure 7-9a**). A collision of this kind, in which the forces between the colliding objects are conservative, is called an **elastic collision.** In an elastic collision both total momentum *and* total mechanical energy are conserved.

Figure 7-9 Collision variations (a) The ideal springs between these colliding objects exert conservative forces. Mechanical energy is conserved in this elastic collision. (b) The forces between colliding billiard balls are very nearly conservative, and the collision is very nearly elastic. (c) The forces between colliding cars are not conservative. Mechanical energy is lost, and the collision is inelastic.

(a)

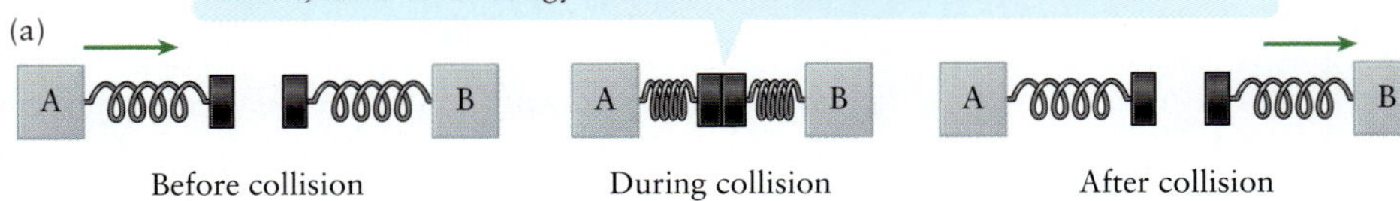
Because the ideal springs between these colliding objects exert conservative forces, mechanical energy is conserved in this **elastic collision.**
A
B
Before collision
During collision
After collision

(b)

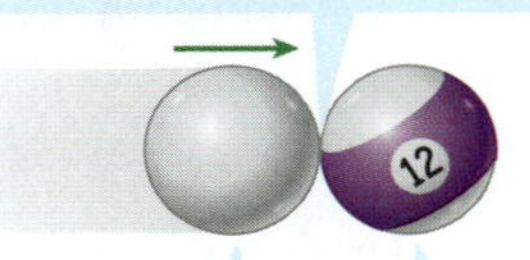
The collision of the billiard balls is elastic: It does not cause any permanent deformation.
12
The forces between colliding ideal billiard balls are conservative. No mechanical energy is lost in this **elastic collision.**

(c)

The collision of the two cars is inelastic: It causes a permanent deformation.
The forces between colliding cars are not conservative. Mechanical energy is lost in this **inelastic collision.**

In elastic collisions, mechanical energy is conserved. In inelastic collisions, mechanical energy is lost.

Elastic collisions are happening all around you: When oxygen and nitrogen molecules in the air collide with each other, the collisions are almost always elastic. That's because the forces between molecules (which are fundamentally electric in nature) are conservative forces. On a larger scale a collision between billiard balls is very nearly elastic (**Figure 7-9b**). A billiard ball deforms slightly when hit but immediately springs back to its original shape with very little of the energy being lost.

Something very different usually happens when two automobiles collide: The bodies of the automobiles deform and do *not* spring back (**Figure 7-9c**). The forces involved in this deformation are *nonconservative* so mechanical energy is lost and is converted to internal energy. (This is by design. By absorbing mechanical energy as it deforms, the structure of an automobile prevents that energy from being used to do potentially harmful work on the automobile's occupants.) A collision in which mechanical energy is *not* conserved is called an **inelastic collision.**

If we know the masses of the colliding objects and their velocities before and after the collision, it's straightforward to determine whether the collision is elastic or inelastic. Here's the idea: By analogy to a collision between ideal springs (Figure 7-9a), there is zero potential energy just before and just after the collision (corresponding to the springs being relaxed). So just before and just after the collision, the total mechanical energy of the system is equal to the sum of the *kinetic energies* of the colliding objects. If the total kinetic energy has the same value before and after the collision, the collision is elastic; if there is less total kinetic energy after the collision, the collision is inelastic. Example 7-5 illustrates this technique for analyzing collisions.

EXAMPLE 7-5 Elastic or Inelastic?

Determine whether the following collisions are elastic or inelastic: (a) The collision of two hockey players in Example 7-2 (Section 7-3); (b) the collision of three billiard balls in Example 7-4 (Section 7-3).

Set Up

For each collision we calculate the total kinetic energy (the sum of the individual kinetic energies of the colliding objects) just before and just after the collision. If these are equal, total mechanical energy is conserved and the collision is elastic; if they are not equal, the collision is inelastic.

Kinetic energy:

$$K = \frac{1}{2}mv^2 \quad (6\text{-}8)$$

Solve

(a) The collision between the two hockey players in Example 7-2 is one-dimensional (that is, along a straight line). So the square of each player's speed v is the same as the square of each player's x velocity v_x.

Players' masses:

$m_G = 100$ kg, $m_M = 80.0$ kg

Before the collision Gordie has x velocity $v_{Gix} = +5.00$ m/s and Mario is at rest. So the initial total kinetic energy is

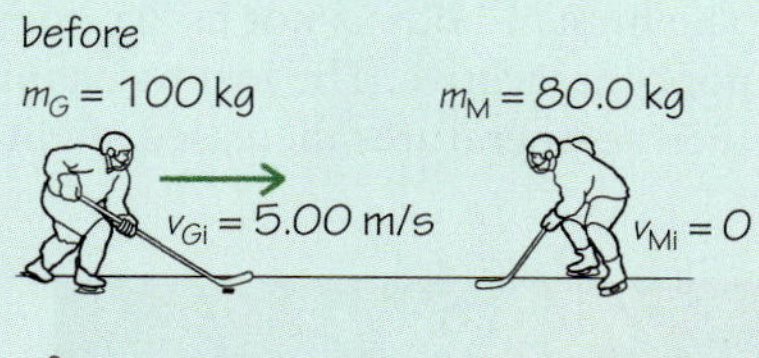

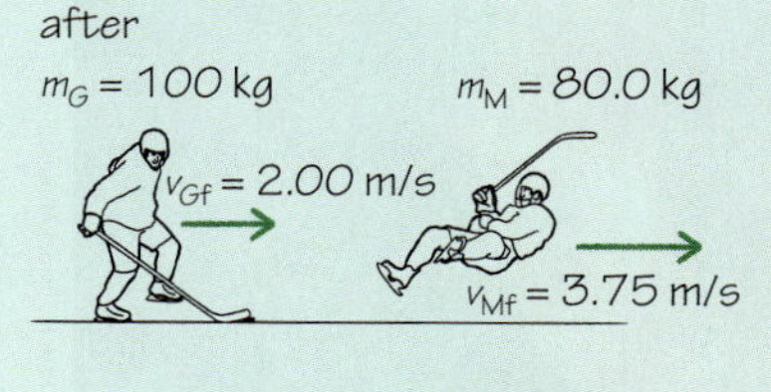

$$\begin{aligned}K_{Gi} + K_{Mi} &= \frac{1}{2}m_G v_{Gix}^2 + \frac{1}{2}m_M v_{Mix}^2 \\ &= \frac{1}{2}(100 \text{ kg})(5.00 \text{ m/s})^2 \\ &\quad + \frac{1}{2}(80.0 \text{ kg})(0 \text{ m/s})^2 \\ &= 1.25 \times 10^3 \text{ J}\end{aligned}$$

After the collision, Gordie has x velocity $v_{Gfx} = +2.00$ m/s and Mario has x velocity $v_{Mfx} = +3.75$ m/s. So the final total kinetic energy is

$$\begin{aligned}K_{Gf} + K_{Mf} &= \frac{1}{2}m_G v_{Gfx}^2 + \frac{1}{2}m_M v_{Mfx}^2 \\ &= \frac{1}{2}(100 \text{ kg})(2.00 \text{ m/s})^2 + \frac{1}{2}(80.0 \text{ kg})(3.75 \text{ m/s})^2 \\ &= 763 \text{ J}\end{aligned}$$

The total kinetic energy is less after the collision than before, so this collision is inelastic.

(b) Repeat the calculation for the three billiard balls in Example 7-4, for which we know the masses, the speeds before the collision, and the speeds after the collision.

Each ball has mass $m = 0.16$ kg. Before the collision the cue ball is moving at speed $v_{cue,i} = 1.7$ m/s, and both the 7-ball and 8-ball are at rest. So the initial total kinetic energy is

$$\begin{aligned}&K_{cue,i} + K_{7i} + K_{8i} \\ &= \frac{1}{2}mv_{cue,i}^2 + \frac{1}{2}mv_{7i}^2 + \frac{1}{2}mv_{8i}^2 \\ &= \frac{1}{2}(0.16 \text{ kg})(1.7 \text{ m/s})^2 \\ &\quad + \frac{1}{2}(0.16 \text{ kg})(0 \text{ m/s})^2 \\ &\quad + \frac{1}{2}(0.16 \text{ kg})(0 \text{ m/s})^2 = 0.23 \text{ J}\end{aligned}$$

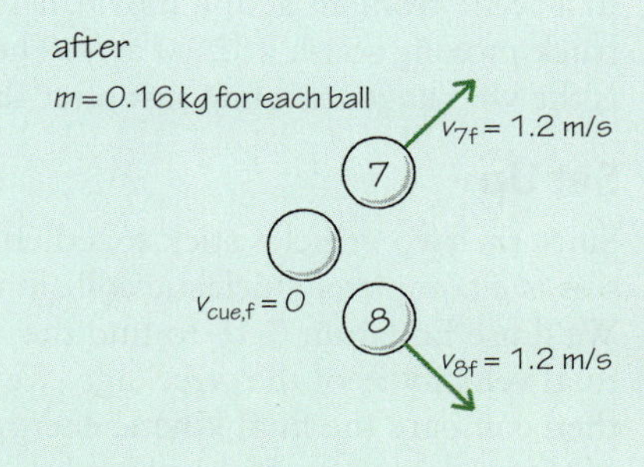

After the collision the cue ball is at rest, the 7-ball is moving at 1.2 m/s, and the 8-ball is also moving at 1.2 m/s. So the final total kinetic energy is

$$\begin{aligned}K_{cue,f} + K_{7f} + K_{8f} &= \frac{1}{2}mv_{cue,f}^2 + \frac{1}{2}mv_{7f}^2 + \frac{1}{2}mv_{8f}^2 \\ &= \frac{1}{2}(0.16 \text{ kg})(0)^2 + \frac{1}{2}(0.16 \text{ kg})(1.2 \text{ m/s})^2 \\ &\quad + \frac{1}{2}(0.16 \text{ kg})(1.2 \text{ m/s})^2 \\ &= 0.23 \text{ J}\end{aligned}$$

The total kinetic energy is unchanged by the collision, so this collision is elastic.

Reflect

The human body is not as "springy" as a billiard ball, so the forces between Gordie and Mario when they collide are nonconservative. The lost mechanical energy went into increasing the two players' internal energies, which means that their temperatures increased slightly as a result of the collision.

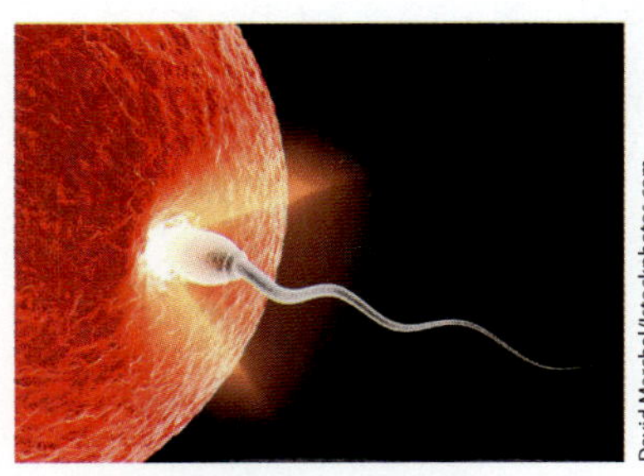

Figure 7-10 A completely inelastic collision A sperm cell and an ovum move together after they collide.

Completely Inelastic Collisions

The type of collision in which the *most* mechanical energy is lost is a **completely inelastic collision**, in which two objects stick together after they collide. A collision between two cars is completely inelastic if the cars lock together and don't separate. (If the cars deform and then bounce apart, the collision is inelastic but not *completely* inelastic.) You are the result of a completely inelastic collision in which a sperm cell fused with an ovum, which nine months later led to your birth (**Figure 7-10**).

When an object of mass m_A and velocity $\vec{v}_{Ai}$ undergoes a completely inelastic collision with a second object of mass m_B and velocity $\vec{v}_{Bi}$, we can regard what remains after the collision as a single object of mass $m_A + m_B$. Momentum conservation then gives us an equation for the velocity $\vec{v}_f$ of this combined object just after the collision:

Momentum conservation in a completely inelastic collision (7-18)

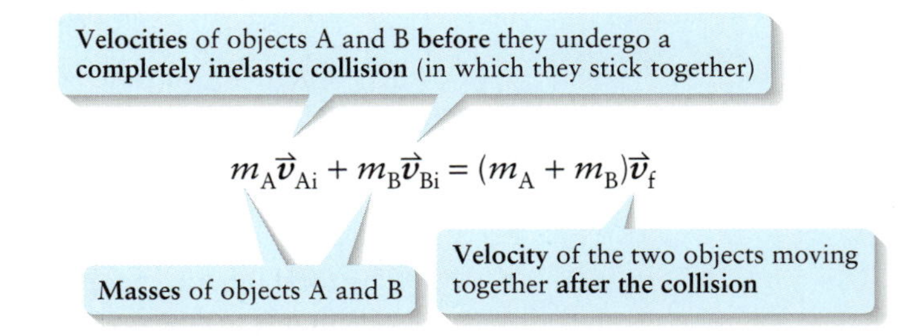

$$m_A\vec{v}_{Ai} + m_B\vec{v}_{Bi} = (m_A + m_B)\vec{v}_f$$

Go to Interactive Exercise 7-2 for more practice dealing with inelastic collisions.

Examples 7-6 and 7-7 illustrate how to use this equation.

EXAMPLE 7-6 A Head-On Collision

In a scene from an action movie, a 1.50×10^3-kg car moving north at 35.0 m/s collides head-on with a 7.50×10^3-kg truck moving south at 25.0 m/s. The car and truck stick together after the collision. (a) How fast and in what direction is the wreckage traveling just after the collision? (b) How much mechanical energy is lost in the collision?

Set Up

Since the two vehicles stick together, this is a completely inelastic collision. We'll use Equation 7-18 to find the final velocity $\vec{v}_f$ of the wreckage. We'll then compare the final kinetic energy of the wreckage to the combined kinetic energies of the car and truck before the collision; the difference is the amount of mechanical energy that's lost in the collision.

Momentum conservation in a completely inelastic collision:

$$m_{car}\vec{v}_{car,i} + m_{truck}\vec{v}_{truck,i} = (m_{car} + m_{truck})\vec{v}_f \quad (7\text{-}18)$$

Kinetic energy:

$$K = \frac{1}{2}mv^2 \quad (6\text{-}8)$$

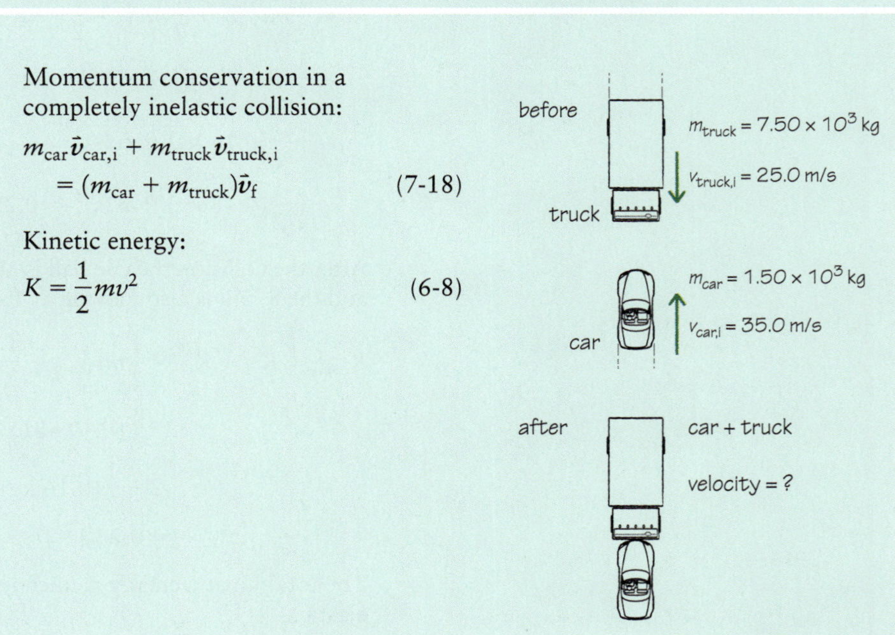

Solve

(a) The collision is along a straight line, which we call the x axis. We take the positive x axis to be to the north, in the direction of the car's motion before the collision. We use the x component of Equation 7-18 to solve for the final velocity.

Equation for conservation of x momentum in a completely inelastic collision:

$$m_{car}v_{car,ix} + m_{truck}v_{truck,ix} = (m_{car} + m_{truck})v_{fx}$$

Solve for final velocity of wreckage:

$$v_{fx} = \frac{m_{car}v_{car,ix} + m_{truck}v_{truck,ix}}{m_{car} + m_{truck}}$$

With our choice of x axis, we have

$$v_{car,ix} = +35.0 \text{ m/s}$$
$$v_{truck,ix} = -25.0 \text{ m/s}$$

Substitute values:

$$v_{fx} = \frac{\begin{bmatrix}(1.50 \times 10^3 \text{ kg})(+35.0 \text{ m/s}) \\ +(7.50 \times 10^3 \text{ kg})(-25.0 \text{ m/s})\end{bmatrix}}{1.50 \times 10^3 \text{ kg} + 7.50 \times 10^3 \text{ kg}} = -15.0 \text{ m/s}$$

The wreckage moves at 15.0 m/s to the south (in the negative x direction).

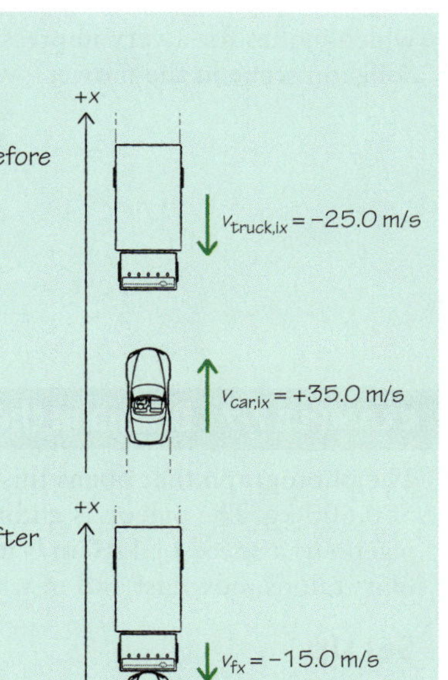

(b) Calculate the kinetic energies before and after the collision and compare.

Total kinetic energy before the collision:

$$K_{car,i} + K_{truck,i} = \frac{1}{2}m_{car}v_{car,i}^2 + \frac{1}{2}m_{truck}v_{truck,i}^2$$
$$= \frac{1}{2}(1.50 \times 10^3 \text{ kg})(35.0 \text{ m/s})^2$$
$$+ \frac{1}{2}(7.50 \times 10^3 \text{ kg})(25.0 \text{ m/s})^2$$
$$= 9.19 \times 10^5 \text{ J} + 2.34 \times 10^6 \text{ J}$$
$$= 3.26 \times 10^6 \text{ J}$$

Total kinetic energy after the collision:

$$K_{car,f} + K_{truck,f} = \frac{1}{2}(m_{car} + m_{truck})v_f^2$$
$$= \frac{1}{2}(1.50 \times 10^3 \text{ kg} + 7.50 \times 10^3 \text{ kg})(15.0 \text{ m/s})^2$$
$$= 1.01 \times 10^6 \text{ J}$$

The amount of mechanical energy lost in the collision is the difference between the initial and final energies:

$$(K_{car,i} + K_{truck,i}) - (K_{car,f} + K_{truck,f})$$
$$= 3.26 \times 10^6 \text{ J} - 1.01 \times 10^6 \text{ J}$$
$$= 2.25 \times 10^6 \text{ J}$$

Reflect

It makes sense that the wreckage moves in the direction of the truck's initial motion. Before the collision the magnitude of the southbound truck's momentum was much greater than that of the northbound car, so the total momentum was to the south.

Our result in part (b) shows that more than two-thirds of the total mechanical energy is lost in the collision. This lost energy goes into doing work to deform the two vehicles,

Momentum of car before the collision:

$$p_{car,ix} = m_{car}v_{car,ix}$$
$$= (1.50 \times 10^3 \text{ kg})(+35.0 \text{ m/s})$$
$$= +5.25 \times 10^4 \text{ kg} \cdot \text{m/s}$$

Momentum of truck before the collision:

$$p_{truck,ix} = m_{truck}v_{truck,ix}$$
$$= (7.50 \times 10^3 \text{ kg})(-25.0 \text{ m/s})$$
$$= -1.88 \times 10^5 \text{ kg} \cdot \text{m/s}$$

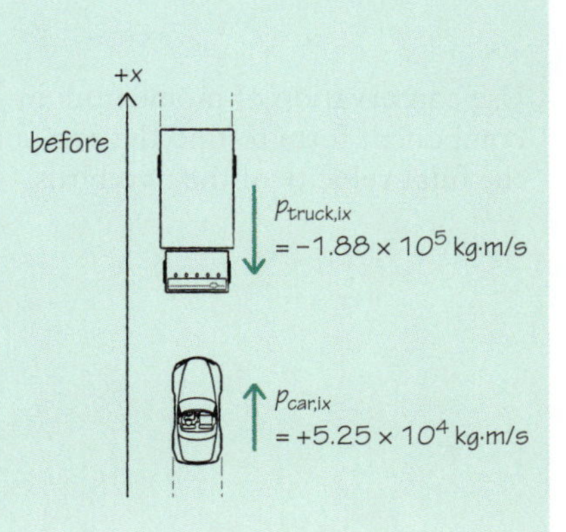

which makes for a very impressive collision scene in the movie.

Total momentum before the collision:

$$P_{ix} = p_{car,ix} + p_{truck,ix}$$
$$= +5.25 \times 10^4 \text{ kg} \cdot \text{m/s} + (-1.88 \times 10^5 \text{ kg} \cdot \text{m/s})$$
$$= -1.35 \times 10^5 \text{ kg} \cdot \text{m/s}$$

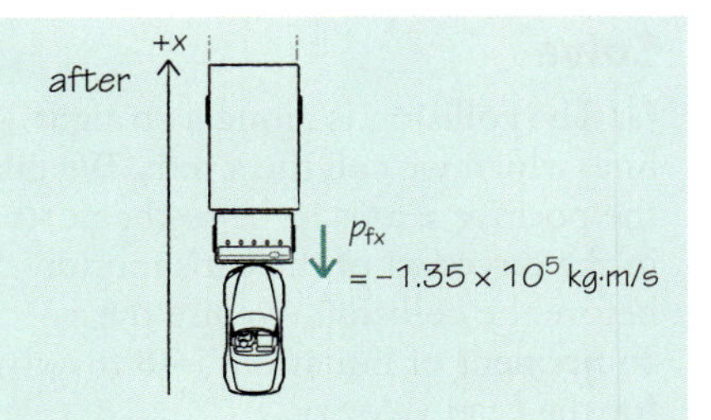

EXAMPLE 7-7 Drama in the Skies

The photograph that opens this chapter shows a Cooper's hawk attacking a pigeon from behind. Each bird has a mass of 0.600 kg. The pigeon is gliding due west at 8.00 m/s at a constant altitude, and the Cooper's hawk is diving on the pigeon at a speed of 12.0 m/s at an angle of 30.0° below the horizontal. Just after the Cooper's hawk grabs the pigeon in its talons, how fast and in what direction are the two birds moving?

Set Up

This is a completely inelastic collision because the two birds move together afterward. Unlike Example 7-6 this collision takes place in two dimensions, so we will need to consider more than one component of Equation 7-18 to find the final velocity $\vec{v}_f$ of the two birds after the collision.

Momentum conservation in a completely inelastic collision:

$$m_{pigeon}\vec{v}_{pigeon,i} + m_{hawk}\vec{v}_{hawk,i} = (m_{pigeon} + m_{hawk})\vec{v}_f \quad (7\text{-}18)$$

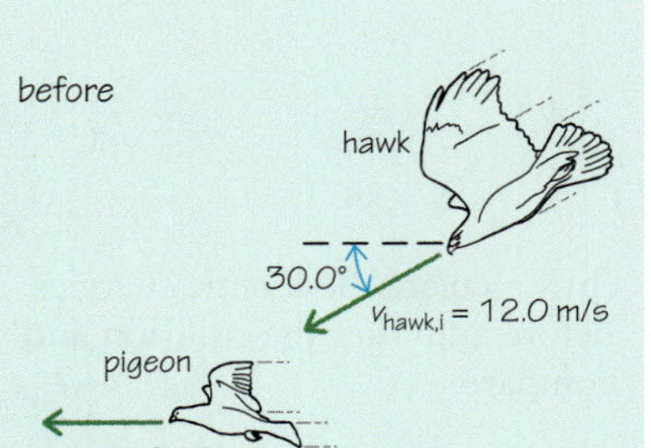

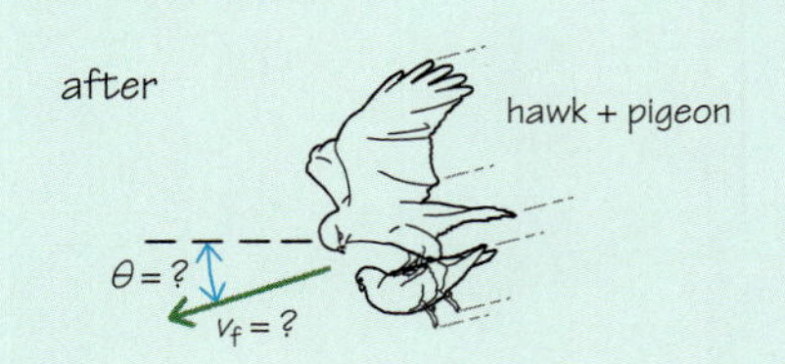

Solve

We choose the positive x axis to be horizontal and to the west, and we choose the positive y axis to be upward. We write the components of the initial velocities of the hawk and pigeon.

Pigeon is initially moving in the positive x direction, so

$$v_{pigeon,ix} = v_{pigeon,i} = +8.00 \text{ m/s}$$
$$v_{pigeon,iy} = 0$$

Hawk is initially moving in the positive x direction and the negative y direction, so

$$v_{hawk,ix} = v_{hawk,i} \cos 30.0° = (12.0 \text{ m/s}) \cos 30.0° = +10.4 \text{ m/s}$$

$$v_{hawk,iy} = -v_{hawk,i} \sin 30.0° = -(12.0 \text{ m/s}) \sin 30.0° = -6.00 \text{ m/s}$$

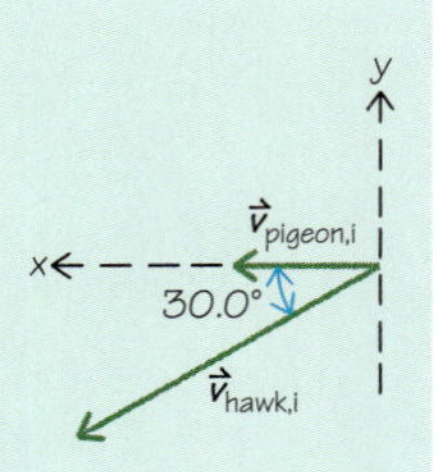

Use conservation of momentum in component form to find the components of the final velocity of the two birds.

Equation for conservation of x momentum in the collision:

$$m_{pigeon}v_{pigeon,ix} + m_{hawk}v_{hawk,ix} = (m_{pigeon} + m_{hawk})v_{fx}$$

Solve for the final x velocity of the two birds:

$$v_{fx} = \frac{m_{pigeon}v_{pigeon,ix} + m_{hawk}v_{hawk,ix}}{m_{pigeon} + m_{hawk}}$$
$$= \frac{(0.600 \text{ kg})(+8.00 \text{ m/s}) + (0.600 \text{ kg})(+10.4 \text{ m/s})}{0.600 \text{ kg} + 0.600 \text{ kg}}$$
$$= +9.20 \text{ m/s}$$

Equation for conservation of y momentum in the collision:

$$m_{\text{pigeon}}v_{\text{pigeon,iy}} + m_{\text{hawk}}v_{\text{hawk,iy}} = (m_{\text{pigeon}} + m_{\text{hawk}})v_{\text{fy}}$$

Solve for the final y velocity of the two birds:

$$\begin{aligned} v_{\text{fy}} &= \frac{m_{\text{pigeon}}v_{\text{pigeon,iy}} + m_{\text{hawk}}v_{\text{hawk,iy}}}{m_{\text{pigeon}} + m_{\text{hawk}}} \\ &= \frac{(0.600\ \text{kg})(0\ \text{m/s}) + (0.600\ \text{kg})(-6.00\ \text{m/s})}{0.600\ \text{kg} + 0.600\ \text{kg}} \\ &= -3.00\ \text{m/s} \end{aligned}$$

Given the components of the final velocity vector $\vec{v}_{\text{f}}$, use trigonometry to find the magnitude and direction of $\vec{v}_{\text{f}}$.

Speed after collision = magnitude of final velocity vector:

$$\begin{aligned} v_{\text{f}} &= \sqrt{v_{\text{fx}}^2 + v_{\text{fy}}^2} \\ &= \sqrt{(+9.20\ \text{m/s})^2 + (-3.00\ \text{m/s})^2} \\ &= 9.67\ \text{m/s} \end{aligned}$$

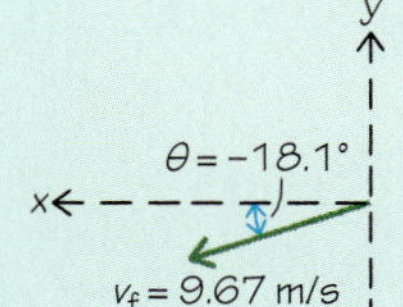

Direction of final velocity vector:

$$\tan\theta = \frac{v_{\text{fy}}}{v_{\text{fx}}} = \frac{-3.00\ \text{m/s}}{+9.20\ \text{m/s}} = -0.326$$

$$\theta = \arctan(-0.326) = -18.1°$$

After the collision the two birds move at 9.67 m/s in a direction 18.1° below the horizontal.

Reflect

The impact causes the hawk to slow down from 12.0 m/s to 9.67 m/s and causes the pigeon to speed up from 8.00 m/s to 9.67 m/s. This is just what we would expect when a slow-moving pigeon is hit from behind by a fast-moving hawk.

WATCH OUT! An inelastic collision doesn't have to be *completely* inelastic.

A common misconception is that a collision is inelastic if two objects come together and stick, and it is elastic if they bounce off each other. Only the first part of this statement is true! Part (a) of Example 7-5 describes a collision between two hockey players that is inelastic even though the players bounce off each other. For a collision to be inelastic, some mechanical energy has to be lost in the collision, and this can happen even if the colliding objects don't stick together.

Three Special Cases of Completely Inelastic Collisions

Figure 7-11a shows an object of mass m_{A} moving with velocity $\vec{v}_{\text{Ai}}$ and about to collide with a second object of mass m_{B} that is initially at rest, so $\vec{v}_{\text{Bi}} = 0$. If the collision is

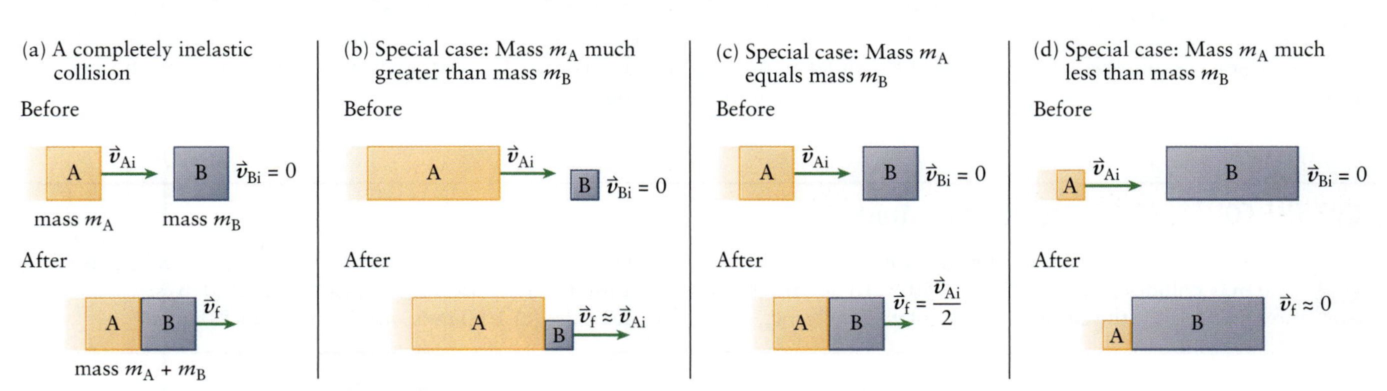

Figure 7-11 Analyzing a completely inelastic collision (a) Object A has a completely inelastic collision with object B (initially at rest). (b) If A is much more massive than B, the final velocity is almost the same as the initial velocity of A. (c) If A and B have the same mass, the final velocity is one-half the initial velocity of A. (d) If A is much less massive than B, the final velocity is nearly zero.

completely inelastic, we can use Equation 7-18 to find the velocity $\vec{v}_f$ after the collision. Since $\vec{v}_{Bi} = 0$,

$$m_A\vec{v}_{Ai} = (m_A + m_B)\vec{v}_f$$

and so

$$\vec{v}_f = \frac{m_A\vec{v}_{Ai}}{m_A + m_B} \tag{7-19}$$

(completely inelastic collision, object B initially at rest)

Let's consider three important special cases of this equation.

(1) *The moving object has much greater mass than the object at rest* (**Figure 7-11b**). Whenever one quantity in an expression is significantly larger than another, it's a good approximation to ignore the smaller quantity if the two quantities are added or subtracted. So when the mass of object A is much larger than the mass of object B, we can safely replace $m_A + m_B$ in Equation 7-19 with m_A:

$$\vec{v}_f \approx \frac{m_A\vec{v}_{Ai}}{m_A} = \vec{v}_{Ai}$$

In this case the stationary object is so small that the moving object is essentially unaffected by the collision, and its velocity remains the same. If a fast-moving car runs into a hovering fly, the car slows down only imperceptibly.

(2) *The two objects have the same mass* (**Figure 7-11c**). If $m_A = m_B$, Equation 7-19 becomes

$$\vec{v}_f = \frac{\vec{v}_{Ai}}{2}$$

This says that when a moving object collides with and sticks to a stationary one of the same mass, the combined system moves at half the initial speed.

(3) *The moving object has much less mass than the object at rest* (**Figure 7-11d**). In this case the mass of object A is much less than the mass of object B, so the quantity m_A in the numerator of Equation 7-19 is very much less than the sum $m_A + m_B$ in the denominator. Therefore $m_A/(m_A + m_B)$ is close to zero, and in this approximation Equation 7-19 becomes

$$\vec{v}_f \approx 0$$

This says that because object A has so little mass, object B hardly moves at all when it is struck. An example is using an axe to split logs for the fireplace (**Figure 7-12**). Each impact is a completely inelastic collision between the axe (object A) and a large, massive, stationary object B made up of the log, the stump on which the log rests, and the ground below the stump. The final velocity of the axe and log is zero, which means *all* of the kinetic energy of the axe is lost. This lost energy goes into heating the axe and heating and ripping apart the log.

Oleksandr Kulichenko/Getty Images

Figure 7-12 Firewood physics In splitting this log, why does the axe blade get hot as it repeatedly strikes the log?

GOT THE CONCEPT? 7-3 How to Stop a Truck

In Example 7-6 a 1.50×10^3-kg car moving north at 35.0 m/s collides head-on with a 7.50×10^3-kg truck moving south. If the collision is completely inelastic, how fast must the truck be moving so that the wreckage is at rest immediately after the collision? (a) 20.0 m/s; (b) 14.0 m/s; (c) 10.0 m/s; (d) 7.0 m/s; (e) 5.0 m/s.

TAKE-HOME MESSAGE FOR Section 7-4

✔ In an elastic collision the forces between the colliding objects are conservative. Both momentum and mechanical energy are conserved in an elastic collision.

✔ An inelastic collision is one in which there are nonconservative forces between the colliding objects. In this case momentum is conserved in the collision, but mechanical energy is not conserved.

✔ In a completely inelastic collision the colliding objects stick together. Momentum is conserved, but mechanical energy is not.

7-5 In an elastic collision both momentum and mechanical energy are conserved

In Section 7-4 we introduced the idea of an *elastic* collision in which the forces between colliding objects are conservative and no mechanical energy is lost in the collision. Hence *both* total momentum (a vector) and total mechanical energy (a scalar) are conserved in an elastic collision. We call the two colliding objects A and B, with masses m_A and m_B. We then have two conservation equations that relate the velocities $\vec{v}_{Af}$ and $\vec{v}_{Bf}$ of the objects after the collision to the velocities $\vec{v}_{Ai}$ and $\vec{v}_{Bi}$ before the collision:

$$\text{(a)} \quad m_A\vec{v}_{Af} + m_B\vec{v}_{Bf} = m_A\vec{v}_{Ai} + m_B\vec{v}_{Bi}$$
(momentum is conserved in an elastic collision)

$$\text{(b)} \quad \frac{1}{2}m_A v_{Af}^2 + \frac{1}{2}m_B v_{Bf}^2 = \frac{1}{2}m_A v_{Ai}^2 + \frac{1}{2}m_B v_{Bi}^2 \tag{7-20}$$
(mechanical energy is conserved in an elastic collision)

Equations 7-20 require some effort to solve, and even after that they often don't tell the whole story of an elastic collision. As an example, consider a collision on a pool table between a moving cue ball (A) with known initial velocity $\vec{v}_{Ai}$ and an 8-ball (B) that is initially at rest so $\vec{v}_{Bi} = 0$. Such a collision is almost perfectly elastic. If you're a pool player, you would like to know how fast and in what direction each ball will go after the collision, which means that you want to know the x and y components of their final velocities. So there are four unknowns in this problem: v_{Afx}, v_{Afy}, v_{Bfx}, and v_{Bfy}. But Equations 7-20 give us only *three* equations for these four unknowns: two from Equation 7-20a for the x and y components of momentum and one from Equation 7-20b. So these equations by themselves aren't enough to solve the problem.

Although we don't have enough information to solve Equations 7-20 for a two-dimensional collision, we can solve these equations completely for a *one-dimensional* elastic collision, so all of the motions are along a straight line. To make the problem even simpler, we'll consider the case in which object B is initially at rest (**Figure 7-13**). An example is a moving cue ball that hits a stationary 8-ball head on. We call that axis along which the motion takes place the x axis. Then $v_{Bix} = 0$ (the initial velocity of object B is zero) and Equations 7-20 become

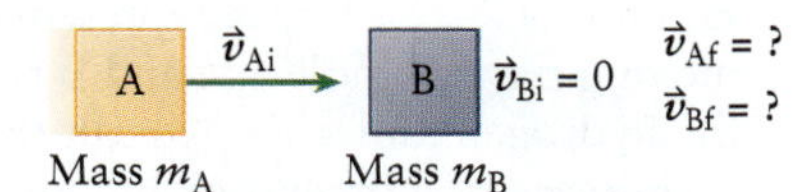

Figure 7-13 A head-on elastic collision If momentum and mechanical energy are both conserved in this head-on collision, what are the final velocities of the two objects?

$$\text{(a)} \quad m_A v_{Afx} + m_B v_{Bfx} = m_A v_{Aix}$$
(momentum is conserved, head-on elastic collision, object B initially at rest)

$$\text{(b)} \quad \frac{1}{2}m_A v_{Afx}^2 + \frac{1}{2}m_B v_{Bfx}^2 = \frac{1}{2}m_A v_{Aix}^2 \tag{7-21}$$
(mechanical energy is conserved, head-on elastic collision, object B initially at rest)

Since each object moves along the x axis only, in Equation 7-21b we've replaced the square of each object's speed with the square of its x velocity.

The two unknowns in Equations 7-21 are the final velocities v_{Afx} and v_{Bfx} of the two objects. Solving these equations takes quite a few steps, so we'll just present the results:

$$v_{Afx} = v_{Aix}\left(\frac{m_A - m_B}{m_A + m_B}\right)$$
$$v_{Bfx} = v_{Aix}\left(\frac{2m_A}{m_A + m_B}\right) \tag{7-22}$$

(final velocities in a head-on elastic collision, object B initially at rest)

You can verify Equations 7-22 by substituting them into Equations 7-21 and simplifying.

Equations 7-22 show that the final velocity of each object is proportional to the initial velocity v_{Aix} of object A, and also depends on a factor in parentheses that involves the masses of the two objects. The *direction* of each object's motion is given by the sign of its final velocity. If the sign of an object's final velocity is the same as the sign of v_{Aix}, the object leaves the collision in the same direction as object A was moving initially. If the final velocity has the opposite sign from v_{Aix}, the object is moving in the direction opposite to the initial motion. For example, if v_{Afx} is negative while v_{Aix} is positive, object A has bounced backward after colliding with initially stationary object B.

WATCH OUT! Some equations are less important than others.

Equations 7-22 are useful for solving a certain, very special kind of collision problem, but they are not *essential* equations that you should worry about trying to memorize or even remember in detail. What *is* worth remembering is the key physics of an elastic collision: Both momentum and kinetic energy are conserved.

EXAMPLE 7-8 When Molecules Collide Elastically

In air, at room temperature, a typical molecule of oxygen (O_2) moves at about 500 m/s. (In Chapter 14 we'll explore the relationship between the temperature of a gas and the speeds of molecules in that gas.) The mass of an O_2 molecule is 32.0 u, where 1 u = 1 atomic mass unit = 1.66×10^{-27} kg. Suppose an O_2 molecule is initially moving at 5.00×10^2 m/s in the positive x direction and has a head-on elastic collision with a molecule of hydrogen (H_2, of mass 2.02 u) that is at rest. Determine the final velocity of each molecule, the initial and final momentum of each molecule, and the initial and final kinetic energy of each molecule.

Set Up

Since this is a one-dimensional elastic collision in which one object is initially at rest, we can find the final velocities using Equations 7-22. Here A represents the moving oxygen molecule and B represents the hydrogen molecule. We can then use the definitions of momentum and kinetic energy to find the initial and final values of these quantities.

Final velocities in a head-on elastic collision, object B initially at rest:

$$v_{Afx} = v_{Aix}\left(\frac{m_A - m_B}{m_A + m_B}\right)$$

$$v_{Bfx} = v_{Aix}\left(\frac{2m_A}{m_A + m_B}\right) \quad (7\text{-}22)$$

$$\vec{p} = m\vec{v} \quad (7\text{-}5)$$

$$K = \frac{1}{2}mv^2 \quad (6\text{-}8)$$

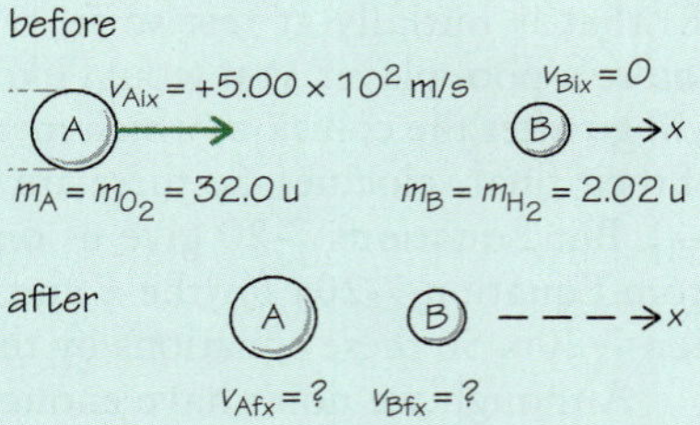

Solve

Calculate the final velocities using Equations 7-22. Since these equations involve the ratios of the masses, we can use the mass values in atomic mass units without needing to convert to kilograms.

Oxygen molecule A is initially moving in the positive x direction, so

$$v_{Aix} = +5.00 \times 10^2 \text{ m/s}$$

Mass of molecule A (O_2) is $m_A = 32.0$ u
Mass of molecule B (H_2) is $m_B = 2.02$ u

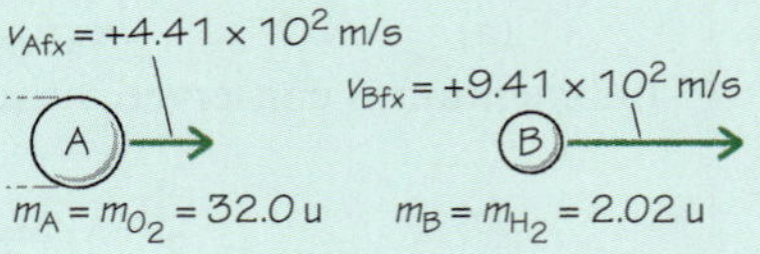

Final velocity of O_2 molecule:

$$v_{Afx} = v_{Aix}\left(\frac{m_A - m_B}{m_A + m_B}\right)$$

$$= (+5.00 \times 10^2 \text{ m/s})\left(\frac{32.0 \text{ u} - 2.02 \text{ u}}{32.0 \text{ u} + 2.02 \text{ u}}\right)$$

$$= +4.41 \times 10^2 \text{ m/s}$$

Final velocity of H_2 molecule:

$$v_{Bfx} = v_{Aix}\left(\frac{2m_A}{m_A + m_B}\right)$$
$$= (+5.00 \times 10^2 \text{ m/s})\left[\frac{2(32.0 \text{ u})}{32.0 \text{ u} + 2.02 \text{ u}}\right]$$
$$= +9.41 \times 10^2 \text{ m/s}$$

Both final velocities are positive, so after the collision both molecules are moving in the same direction as the initial motion of the O_2 molecule. The H_2 molecule is moving almost twice as fast as the initial velocity of the O_2 molecule.

We now know both the initial and final velocities, so we can calculate the initial and final values of momentum and kinetic energy for both molecules. For these calculations we need the masses in kilograms.

Masses of the molecules in kilograms:

$$m_A = \text{mass of } O_2 \text{ molecule}$$
$$= (32.0 \text{ u})(1.66 \times 10^{-27} \text{ kg/u}) = 5.31 \times 10^{-26} \text{ kg}$$

$$m_B = \text{mass of } H_2 \text{ molecule}$$
$$= (2.02 \text{ u})(1.66 \times 10^{-27} \text{ kg/u}) = 3.35 \times 10^{-27} \text{ kg}$$

Calculate the initial and final values of x momentum:

For molecule A (O_2):

$$p_{Aix} = m_A v_{Aix} = (5.31 \times 10^{-26} \text{ kg})(+5.00 \times 10^2 \text{ m/s})$$
$$= +2.66 \times 10^{-23} \text{ kg} \cdot \text{m/s}$$

$$p_{Afx} = m_A v_{Afx} = (5.31 \times 10^{-26} \text{ kg})(+4.41 \times 10^2 \text{ m/s})$$
$$= +2.34 \times 10^{-23} \text{ kg} \cdot \text{m/s} = 0.881 p_{Aix}$$

For molecule B (H_2):

$$p_{Bix} = m_B v_{Bix} = 0 \text{ (molecule B initially at rest)}$$
$$p_{Bfx} = m_B v_{Bfx} = (3.35 \times 10^{-27} \text{ kg})(+9.41 \times 10^2 \text{ m/s})$$
$$= +3.15 \times 10^{-24} \text{ kg} \cdot \text{m/s} = 0.119 p_{Aix}$$

In the collision the O_2 molecule transfers 0.119 of its initial momentum to the H_2 molecule, leaving 0.881 for itself.

Calculate the initial and final values of kinetic energy:

For molecule A (O_2):

$$K_{Ai} = \frac{1}{2} m_A v_{Aix}^2 = \frac{1}{2}(5.31 \times 10^{-26} \text{ kg})(+5.00 \times 10^2 \text{ m/s})^2$$
$$= 6.64 \times 10^{-21} \text{ J}$$

$$K_{Af} = \frac{1}{2} m_A v_{Afx}^2 = \frac{1}{2}(5.31 \times 10^{-26} \text{ kg})(+4.41 \times 10^2 \text{ m/s})^2$$
$$= 5.16 \times 10^{-21} \text{ J} = 0.777 K_{Ai}$$

For molecule B (H_2):

$$K_{Bi} = \frac{1}{2} m_B v_{Bix}^2 = 0 \text{ (molecule B initially at rest)}$$

$$K_{Bf} = \frac{1}{2} m_B v_{Bfx}^2 = \frac{1}{2}(3.35 \times 10^{-27} \text{ kg})(9.41 \times 10^2 \text{ m/s})^2$$
$$= 1.48 \times 10^{-21} \text{ J} = 0.223 K_{Ai}$$

In the collision the O_2 molecule transfers 0.223 of its initial kinetic energy to the H_2 molecule, leaving 0.777 for itself.

Reflect

Both the *total* momentum and the *total* kinetic energy of the system of two molecules are conserved in this collision: Each of these quantities has the same numerical value after the collision as before the collision.

In the case of an O_2 molecule (object A) hitting a stationary H_2 molecule (object B), moving object A is much more massive than stationary target object B. We found that object A slows down only slightly and loses only a little momentum and kinetic energy. Object B flies off moving faster than the original speed of object A.

We invite you to repeat these calculations for the case in which the moving O_2 molecule collides with a second O_2 molecule that is at rest. You'll find that the moving O_2 molecule ends up at rest and the stationary O_2 molecule ends up moving with the original velocity of the first molecule, 5.00×10^2 m/s. In this case the first O_2 molecule transfers all of its momentum and all of its kinetic energy to the second O_2 molecule.

You should also repeat these calculations for the case in which the moving O_2 molecule collides with a protein molecule of mass 3.20×10^4 u that is at rest. In this case you'll find that the O_2 molecule has a final velocity of -4.99×10^2 m/s. The negative sign means that the O_2 molecule *bounces back*, and the magnitude means that the O_2 molecule ends up with nearly the same speed as it had originally (5.00×10^2 m/s). You'll also find that the massive protein molecule ends up with a final velocity of just $+0.999$ m/s, so it moves very slowly in the initial direction of motion of the O_2 molecule. Be sure to also calculate the final momentum and kinetic energy of the protein molecule; you'll find that it ends up with about twice the initial momentum of the O_2 molecule but has little kinetic energy.

Below we'll see that these results are just what we would expect for elastic collisions.

Example 7-8 suggests that just as we did in our exploration of completely inelastic collisions (Section 7-4), we can gain insight into the physics of elastic collisions by considering three special cases involving the relative masses of the two objects. Again we consider a moving object A that has a head-on elastic collision with a stationary object B as is shown in **Figure 7-14**, so we can use Equations 7-22.

(1) *The moving object has much greater mass than the object at rest* (**Figure 7-14a**). If the mass of object A is much larger than the mass of object B, we can safely replace both $m_A - m_B$ and $m_A + m_B$ in Equations 7-22 with m_A:

$$v_{Afx} = v_{Aix}\left(\frac{m_A - m_B}{m_A + m_B}\right) \approx v_{Aix}\left(\frac{m_A}{m_A}\right) = v_{Aix}$$

$$v_{Bfx} = v_{Aix}\left(\frac{2m_A}{m_A + m_B}\right) \approx v_{Aix}\left(\frac{2m_A}{m_A}\right) = 2v_{Aix}$$

Go to Picture It 7-1 for more practice dealing with elastic collisions.

In this extreme case the motion of the massive object A is unaffected by the elastic collision, but the second, much smaller object B flies off at twice the initial speed of object A. That's what happens in the nearly elastic collision between a bowling ball and a stationary bowling pin, which has only about one-quarter the mass of the ball: The ball continues with nearly the same speed, while the pin flies off at a faster speed.

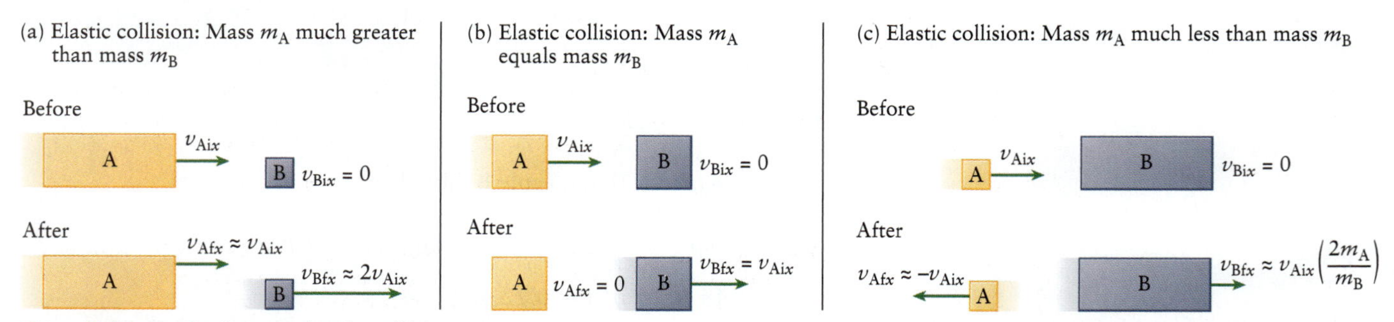

Figure 7-14 Analyzing an elastic collision Object A has a head-on elastic collision with object B (initially at rest). (a) If A is much more massive than B, A is almost unaffected and B flies off with twice the initial velocity of A. (b) If A and B have the same mass, A comes to rest and B flies off with the initial velocity of A. (c) If A is much less massive than B, A bounces back with nearly its initial speed and B moves forward very slowly.

(2) *The two objects have the same mass* (**Figure 7-14b**). If $m_A = m_B$, Equations 7-22 become

$$v_{Afx} = v_{Aix}\left(\frac{m_A - m_B}{m_A + m_B}\right) = 0$$

$$v_{Bfx} = v_{Aix}\left(\frac{2m_A}{m_A + m_B}\right) = v_{Aix}\left(\frac{2m_A}{2m_A}\right) = v_{Aix}$$

In this case object A comes to a stop during the elastic collision and object B leaves the collision with the initial velocity of A. If a cue ball hits an 8-ball head-on, the result of this nearly elastic collision is that the cue ball stops and the 8-ball rolls away with the speed that the cue ball had before the collision.

(3) *The moving object has much less mass than the object at rest* (**Figure 7-14c**). In this case the mass of object A is small compared to the mass of object B, so $m_A - m_B$ is approximately equal to $-m_B$ and $m_A + m_B$ is approximately equal to m_B. Then Equations 7-22 become

$$v_{Afx} = v_{Aix}\left(\frac{m_A - m_B}{m_A + m_B}\right) \approx v_{Aix}\left(\frac{-m_B}{m_B}\right) = -v_{Aix}$$

$$v_{Bfx} = v_{Aix}\left(\frac{2m_A}{m_A + m_B}\right) \approx v_{Aix}\left(\frac{2m_A}{m_B}\right)$$

In this case object A is reflected back from the elastic collision, moving with the same speed it had initially but in the opposite direction. Hence object A has lost almost no kinetic energy, but its momentum has changed from $p_{Aix} = +m_A v_{Aix}$ to $p_{Afx} = -m_A v_{Aix}$. Object B is moving very slowly (because m_A is much less than m_B, the ratio $2m_A/m_B$ is much less than 1). But because its mass is so great, object B has *twice* as much momentum as object A had before the collision:

$$p_{Bfx} = m_B v_{Bfx} = m_B v_{Aix}\left(\frac{2m_A}{m_B}\right) = 2m_A v_{Aix}$$

Then the total momentum of the two objects after the collision is

$$p_{Afx} + p_{Bfx} = -m_A v_{Aix} + 2m_A v_{Aix} = m_A v_{Aix}$$

which is the same as the momentum of object A before the collision. So total momentum is conserved in this case, as it must be. An example is a pebble hitting a boulder: The pebble bounces back, while the recoil of the boulder is so little as to be imperceptible. You should review Example 7-8, which describes what happens when a moving O_2 molecule collides elastically with a stationary molecule of smaller mass (an H_2 molecule), the same mass (another O_2 molecule), or greater mass (a protein molecule).

Figure 7-15 summarizes the differences between elastic collisions, inelastic collisions, and completely inelastic collisions.

GOT THE CONCEPT? 7-4 Classifying Collisions

? For each of the following collisions, state whether it is elastic or inelastic. (a) A ball is dropped from rest and hits the ground. It rebounds to half of its original height. (b) After a ball with 200 J of kinetic energy hits a stationary ball on a horizontal surface, the first ball has 125 J of kinetic energy, and the second has 75 J of kinetic energy. (c) A dog leaps in the air and catches a ball in its teeth.

(a) Elastic collision: Momentum conserved, mechanical energy conserved

Before

A $\vec{v}_{Ai}$ $\vec{v}_{Bi}$ B

After

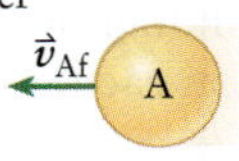

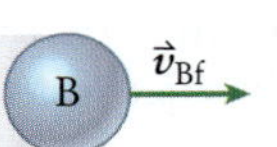

(b) Inelastic collision: Momentum conserved, mechanical energy not conserved

Before

A $\vec{v}_{Ai}$ $\vec{v}_{Bi}$ B

After

$\vec{v}_{Af}$ A B $\vec{v}_{Bf}$

(c) Completely inelastic collision: Momentum conserved, mechanical energy not conserved, colliding objects stick together

Before

A $\vec{v}_{Ai}$ $\vec{v}_{Bi}$ B

After

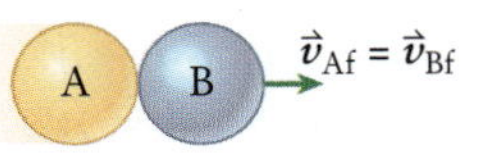

Figure 7-15 Three types of collision Comparing (a) elastic, (b) inelastic, and (c) completely inelastic collisions.

TAKE-HOME MESSAGE FOR Section 7-5

✔ A collision is elastic if mechanical energy as well as linear momentum is conserved.

✔ If a moving object has a head-on elastic collision with a second, stationary object, how the objects move after the collision depends on the ratio of masses of the two objects.

7-6 What happens in a collision is related to the time the colliding objects are in contact

fStop Images GmbH/Alamy

Figure 7-16 A sudden stop with airbags How does a deployed airbag help prevent serious injury?

During a collision that brings a car to a sudden stop, the air bags installed in most cars inflate (**Figure 7-16**). Their purpose is to prevent serious injury by cushioning the impact between the car's occupants and hard surfaces in the vehicle.

The principle of the air bag is to minimize the *force* on the occupants by maximizing the *time* that it takes to bring the occupants to rest. To see how this works, let's look at the relationship between the net force on an object, the time that the net force acts, and the change in the object's momentum caused by the net force.

Collision Force, Contact Time, and Momentum Change

We learned in Section 7-3 that if external forces act on a *system* of objects for a time Δt, the result is a change in the momentum of the system from an initial value $\vec{P}_i$ to a final value $\vec{P}_f$. This change is given by Equation 7-13:

$$\left(\sum \vec{F}_{\text{external on system}}\right)\Delta t = \vec{P}_f - \vec{P}_i$$

Now suppose the system is made up of only a single object. Then we can replace the symbol $\vec{P}$ (for the momentum of a system) by $\vec{p}$ (for the momentum of a single object), and Equation 7-13 becomes

External force and momentum change for an object (7-23)

The **sum of all external forces** acting on an object

Duration of the time interval over which the external forces act

$$\left(\sum \vec{F}_{\text{external on object}}\right)\Delta t = \vec{p}_f - \vec{p}_i = \Delta\vec{p} \tag{7-23}$$

Change in the momentum $\vec{p}$ of the object during that time interval

The left-hand side of Equation 7-23 is called the **impulse** that acts on the object. This equation tells us that to change the momentum of an *object* from $\vec{p}_i$ to $\vec{p}_f$, a certain amount of impulse is required. This impulse can be the result of a large external force $\sum \vec{F}_{\text{external on object}}$ that acts for a short time Δt or a smaller external force that acts for a longer time.

If the momentum change occurs as a result of a collision, the net external force on an object is predominantly due to the other object with which it is colliding; any other forces are very weak by comparison (see Section 7-3). So for the special case of a collision, Equation 7-23 becomes

Collision force, contact time, and momentum change (7-24)

Force exerted on an object during a collision

Duration of the collision = **contact time**

$$\vec{F}_{\text{collision}}\Delta t = \vec{p}_f - \vec{p}_i = \Delta\vec{p} \tag{7-24}$$

Change in the momentum $\vec{p}$ of the object during the collision

The **contact time** in Equation 7-24 is the amount of time that the colliding objects are in contact and hence the amount of time that the objects exert forces on each other.

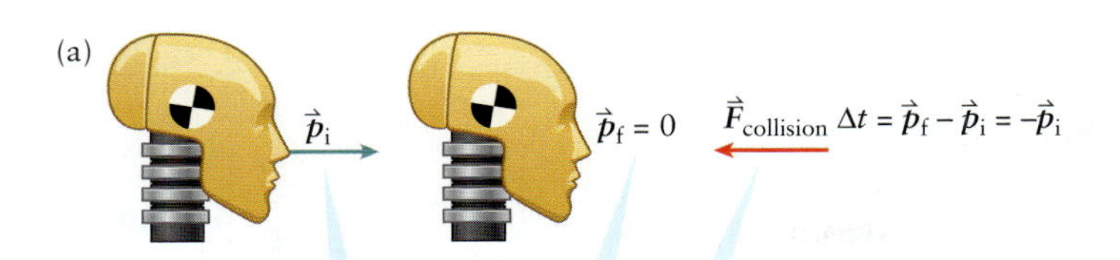
(a)
$\vec{p}_i$
$\vec{p}_f = 0$
$\vec{F}_{collision}\,\Delta t = \vec{p}_f - \vec{p}_i = -\vec{p}_i$
As the crash test dummy comes to rest, its momentum changes from an initial value $\vec{p}_i$ to a final value $\vec{p}_f = 0$ (zero momentum). The product $\vec{F}_{collision}\,\Delta t$ equals the negative of $\vec{p}_i$.
(b)
$\vec{F}_{collision}$
If the crash test dummy strikes a hard surface, Δt is very short and $\vec{F}_{collision}$ is very large.

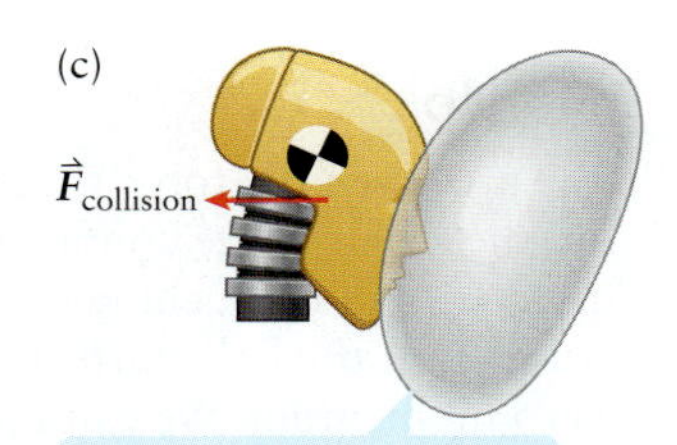
(c)
$\vec{F}_{collision}$
If the crash test dummy strikes an airbag, Δt is longer and $\vec{F}_{collision}$ is reduced.

Figure 7-17 Contact time (a) To bring a crash test dummy to rest requires that $\vec{F}_{collision}\,\Delta t$ have a certain large value. (b) A short contact time Δt means that the force $\vec{F}_{collision}$ will be large. (c) A longer contact time Δt reduces the magnitude of $\vec{F}_{collision}$.

Equation 7-24 explains the principle of the air bag shown in Figure 7-16. If a car comes to a sudden stop, the momentum of one of its occupants changes from a large initial value $\vec{p}_i$ to a final value of zero ($\vec{p}_f = 0$). So the momentum change $\vec{p}_f - \vec{p}_i$ is large, which means the product $\vec{F}_{collision}\Delta t$ also has a large value (**Figure 7-17a**). If there are no air bags, the occupant has a "hard" collision with the structure of the car and also comes to a sudden stop. In this case the contact time Δt is very short and so $\vec{F}_{collision}$ must have a *very* large value: The car's structure exerts a tremendous force on the occupant that is likely to cause injury, as **Figure 7-17b** shows. But if the car is equipped with air bags, the occupant covers a greater distance as it comes to rest and so is in contact with the air bag for a longer time Δt (**Figure 7-17c**). Hence $\vec{F}_{collision}$ (which represents the force exerted on the occupant by the air bag) is greatly reduced, and injury is avoided. For the same reason, if you jump down from a table it's less painful if you bend your knees when landing. Flexing your legs during the collision between your feet and the ground maximizes the contact time Δt that it takes for you to come to a stop.

WATCH OUT! Constant force or average force.

In deriving Equations 7-23 and 7-24, we have assumed that the forces acting on the object are *constant* forces that do not change during the time Δt. We can still use these equations even if the forces are not constant, however: Just treat $\sum \vec{F}_{external\ on\ object}$ in Equation 7-23 and $\vec{F}_{collision}$ in Equation 7-24 as the *average* values of these forces over the time interval Δt.

Example 7-9 shows how to apply the idea of contact time to a collision in which an object speeds up rather than slowing down.

EXAMPLE 7-9 Follow-through

Tennis players are taught to *follow through* as they serve the ball—that is, to continue to swing the racket after first striking the ball. The rationale is that by increasing the time that the ball and racket are in contact, the ball might have a larger speed as it leaves the racket. Using her racket, a player can exert an average force of 560 N on a 57-g tennis ball during an overhand serve (**Figure 7-18**). At what speed does the ball leave her racket if the contact time between them is 0.0050 s? What would the speed of the ball be if she improved her follow-through and increased the contact time by a factor of 1.3 to 0.0065 s? During an overhand serve, the tennis ball is struck when it is essentially motionless.

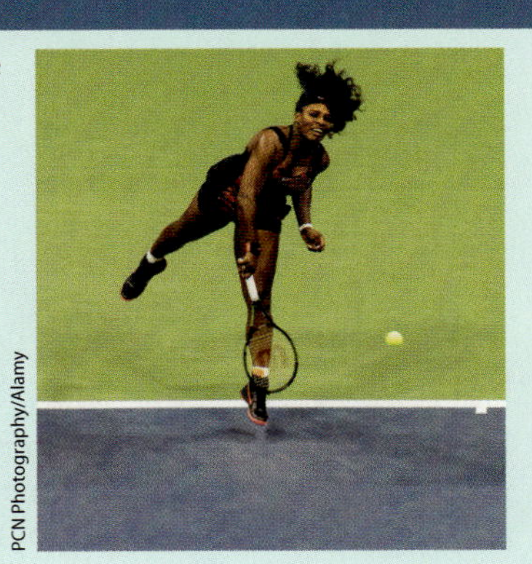

Figure 7-18 Tennis physics How does an improved follow-through affect the speed of a tennis ball?

Set Up

We are given the force that the racket exerts on the ball and the contact time (the time that the ball is touching the racket). We also know that the ball starts at rest and so with zero momentum. We can use Equation 7-24 to calculate the final momentum of the ball, and then use Equation 7-5 to determine the final speed.

Collision force, contact time, and momentum change:

$$\vec{F}_{\text{collision}}\Delta t = \vec{p}_{\text{f}} - \vec{p}_{\text{i}} \quad (7\text{-}24)$$

$$\vec{p} = m\vec{v} \quad (7\text{-}5)$$

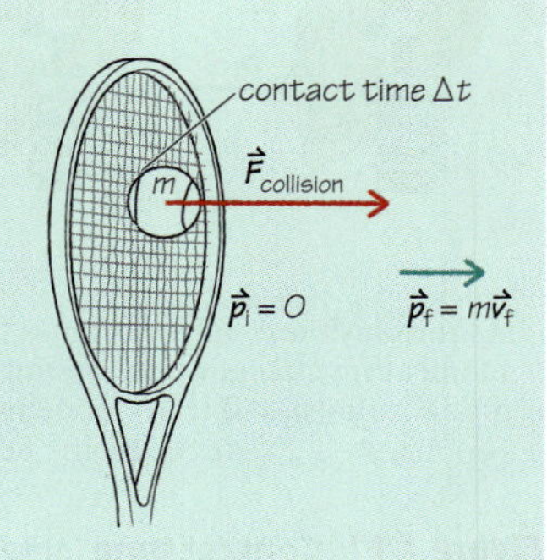

Solve

The racket follows a curved path during the serve. But since the collision between racket and ball lasts such a short time, the ball is in contact with the racket for just a short segment of that path, which we can treat as a straight line. We choose the positive x axis to be in the direction of the motion of the ball.

The collision is one-dimensional, so we just need the x component of Equation 7-24:

$$F_{\text{collision},x}\,\Delta t = p_{\text{f}x} - p_{\text{i}x}$$

The ball is initially at rest, so $v_{\text{i}x} = 0$ and

$$p_{\text{i}x} = 0$$

After the collision with the racket, the ball has (unknown) velocity $v_{\text{f}x}$, so

$$p_{\text{f}x} = mv_{\text{f}x}$$

Then the x component of Equation 7-24 becomes

$$F_{\text{collision},x}\,\Delta t = mv_{\text{f}x} - 0 = mv_{\text{f}x}$$

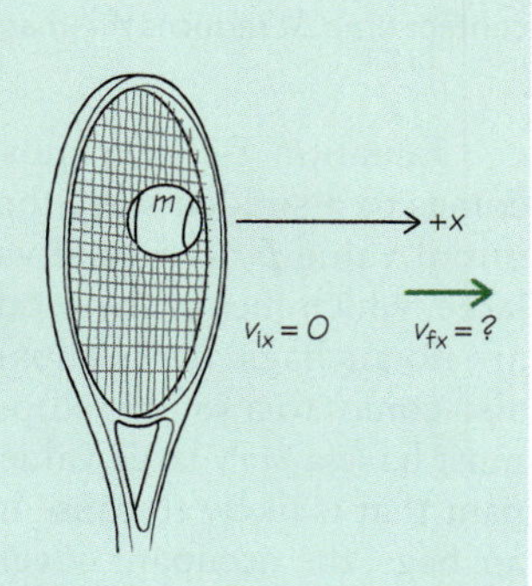

Solve for the final velocity of the ball in each case.

Rewrite the x component of Equation 7-24 to solve for $v_{\text{f}x}$:

$$v_{\text{f}x} = \frac{F_{\text{collision},x}\Delta t}{m}$$

In both cases $F_{\text{collision},x} = 560$ N

$$m = 57\text{ g}\left(\frac{1\text{ kg}}{1000\text{ g}}\right) = 0.057\text{ kg}$$

Calculate $v_{\text{f}x}$ if the contact time is 0.0050 s. Recall that $1\text{ N} = 1\text{ kg}\cdot\text{m/s}^2$:

$$v_{\text{f}x} = \frac{F_{\text{collision},x}\Delta t}{m} = \frac{(560\text{ N})(0.0050\text{ s})}{0.057\text{ kg}}$$

$$= 49\frac{\text{N}\cdot\text{s}}{\text{kg}}\left(\frac{1\text{ kg}\cdot\text{m/s}^2}{1\text{ N}}\right) = 49\text{ m/s}$$

With improved follow-through and a contact time of 0.0065 s,

$$v_{\text{f}x} = \frac{F_{\text{collision},x}\Delta t}{m} = \frac{(560\text{ N})(0.0065\text{ s})}{0.057\text{ kg}} = 64\text{ m/s}$$

Reflect

For a given force the change in momentum and therefore the change in velocity during a collision are directly proportional to the contact time. Increasing the contact time Δt from 0.0050 s to (1.3)(0.0050 s) = 0.0065 s results in an increase in the final speed of the tennis ball from 49 m/s to (1.3)(49 m/s) = 64 m/s.

Impulse and Momentum Versus Work and Kinetic Energy

Equation 7-23 says that the product of the net force on an object and the time that the force acts equals the change in momentum of the object. This should remind you of the work-energy theorem from Section 6-3, which we can write as

$$W_{\text{net}} = Fd\cos\theta = K_{\text{f}} - K_{\text{i}} \qquad (7\text{-}25)$$

This says that the net work on an object—which is the product of the net force on an object parallel to its displacement and the *displacement* of the object during the time that the force acts—equals the change in *kinetic energy* of the object. Equation 7-25 gives us an alternative way to think about the principle of the air bag shown in Figure 7-16. To bring a moving crash test dummy to rest, its kinetic energy $K = \frac{1}{2}mv^2$ must change from a large initial value K_{i} to a final value $K_{\text{f}} = 0$, so the required change in kinetic energy is $K_{\text{f}} - K_{\text{i}} = 0 - K_{\text{i}} = -K_{\text{i}}$. Hence a certain amount of (negative) work has to be done on the dummy to bring it to rest. Without an air bag, the dummy stops in a very short distance when it strikes the hard surfaces of the car, so the magnitude of the displacement d is small and the force F on the dummy must be large. With an air bag, however, the dummy travels a larger distance as it slows to a stop. In this case the magnitude of the displacement d is larger, and the force F is smaller and less injurious.

GOT THE CONCEPT? 7-5
Force of the Kick

? A soccer player's foot is in contact with the 0.43-kg ball for 8.6×10^{-3} s. The ball is initially at rest. What is the speed of the ball immediately after being kicked if the player exerts a force of 1700 N on it? (a) 1.5 m/s; (b) 3.4 m/s; (c) 15 m/s; (d) 34 m/s.

TAKE-HOME MESSAGE FOR Section 7-6

✔ The change in momentum of an object during a collision equals the force exerted on the object during the collision multiplied by the contact time—the length of time that the object touches another during the collision.

✔ The same change in momentum can be produced by a large force acting for a short time or by a small force acting for a longer time.

7-7 The center of mass of a system moves as though all of the system's mass were concentrated there

The concept of momentum can help us answer a question that may have been lurking in the back of your mind since Chapter 4. We learned in that chapter that the acceleration of an object is determined by the external forces on the object as described by Newton's second law: $\sum \vec{F}_{\text{ext}} = m\vec{a}$. But if the object is rotating, like a gymnast tumbling in midair or a barrel rolling downhill, different parts of the object have different accelerations. So the quantity $\vec{a}$ in $\sum \vec{F}_{\text{ext}} = m\vec{a}$ must refer to the acceleration of a specific point on the object. But what point is that?

The point we are looking for is called the **center of mass** of the object. As we will see, the center of mass moves as though all of the object's mass were squeezed into a tiny blob at that point and all of the external forces acted on that blob. We'll begin by showing how to calculate the position of the center of mass; then we'll use ideas about momentum to justify why the center of mass behaves as it does.

Averages and Weighted Averages

The position of an object's center of mass is a kind of *average* that takes into account the masses and positions of all of the pieces that make up that object. If the object is a car, you can think of these pieces as the various components of a car, or the individual atoms that comprise the car. To see how this average is defined, let's first remind ourselves how to calculate an ordinary average.

To find the average of a set of N numbers, add them together and divide by N. For example, the average of the three numbers ($N = 3$) 4, 8, and 18 is $(4 + 8 + 18)/3 = 30/3 = 10$. The average of the set of six numbers $n_1 = 4, n_2 = 4, n_3 = 8, n_4 = 8, n_5 = 8, n_6 = 10$ is

$$\overline{n} = \frac{1}{6}\sum_{i=1}^{6} n_i = \frac{4 + 4 + 8 + 8 + 8 + 10}{6} = \frac{42}{6} = 7$$

Lightweight mint

Heavier soup bowl

The system behaves as though all of its mass were concentrated at the center of mass. That's where the waiter supports the tray.

Figure 7-19 Center of mass at the restaurant The tray balances if it is supported at the center of mass of the system of tray, soup bowl, and mint.

Here n_i represents the ith value in the set, and the symbol $\overline{n}$ (n with a bar on top) denotes the average value of the numbers n_i. Notice that because some of the values in the set occur more than once, we could write this average as

$$\overline{n} = \frac{2(4) + 3(8) + 10}{6} = \frac{2}{6}(4) + \frac{3}{6}(8) + \frac{1}{6}(10) = 7 \tag{7-26}$$

Each term in this sum includes the fraction of the entire set comprised of the value that follows it: 2/6 of the set has value 4, 3/6 of the set has value 8, and 1/6 of the set has value 10. In writing the sum this way, we have defined a **weighted average**, in which each value affects the result to a greater or lesser extent depending on how often it appears in the set. Here *weight* does not refer to a force due to gravity but rather to the amount that each value contributes to the result.

Let's see how to use a weighted average to define the center of mass of a complex object. Imagine a waiter who supports a tray that has a bowl of soup on one side and a small mint on the other (**Figure 7-19**). To balance the combination of tray, soup bowl, and mint, the waiter supports the tray at a position much closer to the soup than the mint. This position is the center of mass of the three objects: If the tray, soup bowl, and mint were all squeezed into a single blob, the waiter would put his hand directly under the blob to support it. The position of the center of mass isn't the ordinary average of the positions of tray, soup bowl, and mint, which would be a point at the geometrical center of the tray. (If the waiter tried to balance the tray there, the result would be soup everywhere.) Instead, the center of mass is a *weighted* average that puts more emphasis on the position of the high-mass soup bowl than on the position of the low-mass mint.

Equation 7-26 provides some guidance as to how to write such a weighted average. Suppose we have a system of N objects with different masses $m_1, m_2, m_3, \ldots, m_N$. The total mass M_{tot} of the system is the sum of these N individual masses:

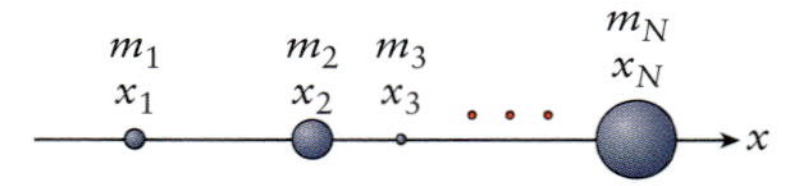

Figure 7-20 A system of objects The position of the center of mass of this collection of objects depends on the masses and positions of the individual objects (Equation 7-28).

$$M_{\text{tot}} = m_1 + m_2 + m_3 + \ldots + m_N = \sum_{i=1}^{N} m_i \tag{7-27}$$

The mass of each individual object represents a fraction m_i/M of the total mass. Now suppose the N objects are at different positions along the x axis: $x_1, x_2, x_3, \ldots, x_N$. (**Figure 7-20**). By analogy to Equation 7-26 we write the position of the center of mass as a weighted average of these positions:

Position of the center of mass of a system (7-28)

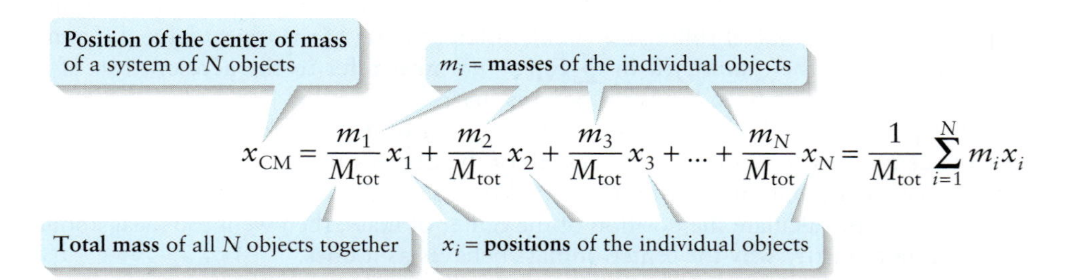

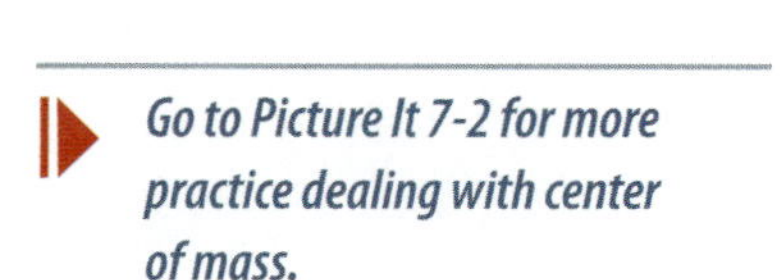

Go to Picture It 7-2 for more practice dealing with center of mass.

The more massive a given object, the greater the ratio m_i/M_{tot} and the greater the importance of that object's position x_i in the sum given by Equation 7-28. Example 7-10 illustrates this property of the position of the center of mass.

EXAMPLE 7-10 Locating the Center of Mass

Two objects are located on the x axis at the positions $x_1 = 2.0$ m and $x_2 = 8.0$ m. For the system made up of these two objects, find the position of the center of mass (a) if m_1 and m_2 are both equal to 3.0 kg and (b) if $m_1 = 3.0$ kg and $m_2 = 33$ kg.

Set Up

Since there are just two objects, the total mass of the system is just the sum of their two masses. We use Equation 7-28 in each case to locate the center of mass.

Position of the center of mass of a system:

$$x_{CM} = \frac{m_1}{M_{tot}}x_1 + \frac{m_2}{M_{tot}}x_2 + \frac{m_3}{M_{tot}}x_3 + \ldots + \frac{m_N}{M_{tot}}x_N = \frac{1}{M_{tot}}\sum_{i=1}^{N} m_i x_i \quad (7\text{-}28)$$

Total mass of the system of two objects:

$$M_{tot} = m_1 + m_2 \quad (7\text{-}27)$$

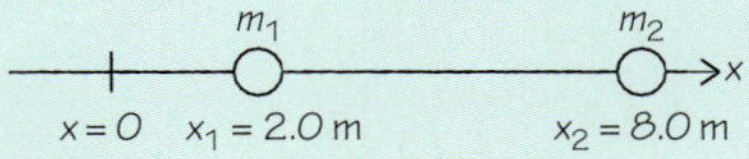

m_1 m_2 $x = 0$ $x_1 = 2.0$ m $x_2 = 8.0$ m

Solve

(a) Calculate the position of the center of mass for the case in which both objects have the same mass.

The two objects have equal masses:

$m_1 = m_2 = 3.0$ kg

Total mass of the system:

$$M_{tot} = m_1 + m_2 = 3.0 \text{ kg} + 3.0 \text{ kg} = 6.0 \text{ kg}$$

Position of the center of mass:

$$x_{CM} = \frac{m_1}{M_{tot}}x_1 + \frac{m_2}{M_{tot}}x_2 = \frac{3.0 \text{ kg}}{6.0 \text{ kg}}(2.0 \text{ m}) + \frac{3.0 \text{ kg}}{6.0 \text{ kg}}(8.0 \text{ m}) = 5.0 \text{ m}$$

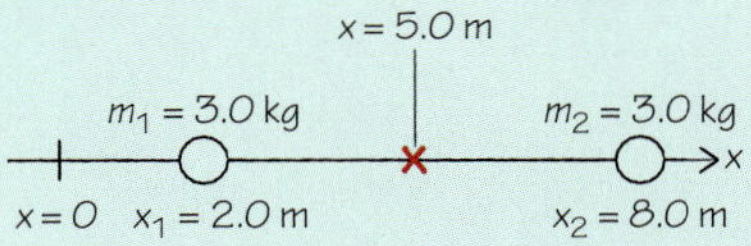

$x = 5.0$ m; $m_1 = 3.0$ kg; $m_2 = 3.0$ kg; $x = 0$ $x_1 = 2.0$ m; $x_2 = 8.0$ m

(b) Repeat part (a) for the case in which m_2 is much larger than m_1.

The two objects have different masses:

$m_1 = 3.0$ kg, $m_2 = 33$ kg

Total mass of the system:

$$M_{tot} = m_1 + m_2$$
$$3.0 \text{ kg} + 33 \text{ kg} = 36 \text{ kg}$$

Position of the center of mass:

$$x_{CM} = \frac{m_1}{M_{tot}}x_1 + \frac{m_2}{M_{tot}}x_2 = \frac{3.0 \text{ kg}}{36 \text{ kg}}(2.0 \text{ m}) + \frac{33 \text{ kg}}{36 \text{ kg}}(8.0 \text{ m}) = 7.5 \text{ m}$$

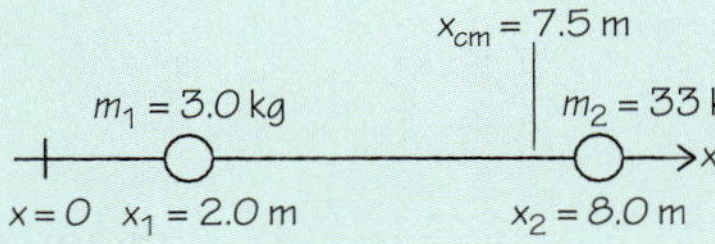

$x_{cm} = 7.5$ m; $m_1 = 3.0$ kg; $m_2 = 33$ kg; $x = 0$ $x_1 = 2.0$ m; $x_2 = 8.0$ m

Reflect

In part (a) the position $x_{CM} = 5.0$ m is 3.0 m from object 1 and 3.0 m from object 2—in other words exactly halfway between the two identical objects. In part (b), by contrast, the center of mass is 5.5 m from the less massive object 1 ($m_1 = 3.0$ kg) and only 0.5 m from the more massive object 2 ($m_2 = 33$ kg). If one object in the system is much more massive than the others, the center of mass is closest to that most massive object. In both cases if the two objects were connected by a rigid lightweight rod, you could balance the system in your hand if you supported it at the center of mass just like the tray shown in Figure 7-19.

It's important to note that the relative position of the center of mass does *not* depend on the choice of origin. As an example, suppose we choose $x = 0$ to be at the position of object 1 so that object 2 is at position $x_2 = 6.0$ m. In this case we find the center of mass in case (b) to be at $x_{CM} = 5.5$ m. That looks different from the value of 7.5 m that we found above but is still 5.5 m from object 1 and 0.5 m from object 2—exactly as we found before.

Find the center of mass in case (b), with two objects of different mass, but now with the origin chosen to be at the position of object 1:

$m_1 = 3.0$ kg, $m_2 = 33$ kg so $M_{tot} = 36$ kg

$x_1 = 0$, $x_2 = 6.0$ m

Position of the center of mass:

$$x_{CM} = \frac{m_1}{M_{tot}}x_1 + \frac{m_2}{M_{tot}}x_2 = \frac{3.0 \text{ kg}}{36 \text{ kg}}(0) + \frac{33 \text{ kg}}{36 \text{ kg}}(6.0 \text{ m}) = 5.5 \text{ m}$$

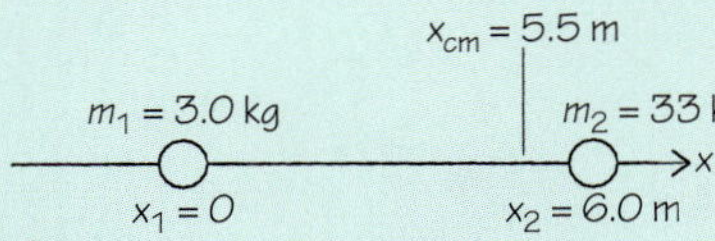

$x_{cm} = 5.5$ m; $m_1 = 3.0$ kg; $m_2 = 33$ kg; $x_1 = 0$; $x_2 = 6.0$ m

More on the Position of the Center of Mass

In writing Equation 7-28 we assumed that all of the objects are arranged along a line that we call the x axis. If the objects are arranged in a plane defined by the x and y axes, we need to specify the x and y coordinates of the position of the center of mass. Equation 7-28 gives the x coordinate, and a very similar equation gives us the y coordinate:

$$y_{\mathrm{CM}} = \frac{m_1}{M_{\mathrm{tot}}}y_1 + \frac{m_2}{M_{\mathrm{tot}}}y_2 + \frac{m_3}{M_{\mathrm{tot}}}y_3 + \ldots + \frac{m_N}{M_{\mathrm{tot}}}y_N = \frac{1}{M_{\mathrm{tot}}}\sum_{i=1}^{N} m_i y_i \quad (7\text{-}29)$$

(y coordinate of the position of the center of mass)

If the objects are arranged in three dimensions, you would also need an equation similar to Equations 7-28 and 7-29 for the z coordinate z_{CM} of the center of mass.

The center of mass of the system in part (a) of Example 7-10 is midway between the two objects, at the *geometrical* center of the system. This is true whenever the system is *symmetrical* so that the mass is distributed in the same way on both sides of the center of mass. For example, the center of mass of a billiard ball is at its geometrical center, as is the center of mass of a uniform metal rod. In part (b) of Example 7-10 the center of mass is *not* at the geometrical center because the system is not symmetrical: Masses m_1 and m_2 have very different values.

WATCH OUT! There doesn't have to be any mass at the center of mass.

Note that in both parts (a) and (b) of Example 7-10, there's nothing physically located *at* the center of mass. The same is true for any symmetrical object with a hole at its center, like a Blu-ray disc, a doughnut, or a ping-pong ball: In each case the center of mass is located in the center of the empty hole.

Momentum, Force, and the Motion of the Center of Mass

It's now time to justify the statement that the center of mass moves as though all of the mass of the system were concentrated there. To begin let's combine Equations 7-28 and 7-29 into a single *vector* equation for the position of the center of mass:

$$\vec{r}_{\mathrm{CM}} = \frac{m_1}{M_{\mathrm{tot}}}\vec{r}_1 + \frac{m_2}{M_{\mathrm{tot}}}\vec{r}_2 + \frac{m_3}{M_{\mathrm{tot}}}\vec{r}_3 + \ldots + \frac{m_N}{M_{\mathrm{tot}}}\vec{r}_N = \frac{1}{M_{\mathrm{tot}}}\sum_{i=1}^{N} m_i \vec{r}_i \quad (7\text{-}30)$$

(vector position of the center of mass)

In Equation 7-30 $\vec{r}_{\mathrm{CM}}$ is the position vector of the center of mass, and the vectors $\vec{r}_1, \vec{r}_2, \vec{r}_3, \ldots, \vec{r}_N$ represent the positions of the N individual objects that make up the system. Each vector $\vec{r}_i$ has components x_i and y_i (and, if a third dimension is required, z_i), and $\vec{r}_{\mathrm{CM}}$ has components x_{CM} and y_{CM} (and z_{CM} if required).

Let's now allow the objects that make up the system to move. (The objects need not be connected, so they may or may not move together.) During a time interval Δt each of the vectors $\vec{r}_i$ in Equation 7-30 changes by an amount $\Delta\vec{r}_i$, which can have a different value for each object. From Equation 7-30 the change $\Delta\vec{r}_{\mathrm{CM}}$ in the position vector of the center of mass during Δt is

$$\Delta\vec{r}_{\mathrm{CM}} = \frac{m_1}{M_{\mathrm{tot}}}\Delta\vec{r}_1 + \frac{m_2}{M_{\mathrm{tot}}}\Delta\vec{r}_2 + \frac{m_3}{M_{\mathrm{tot}}}\Delta\vec{r}_3 + \ldots + \frac{m_N}{M_{\mathrm{tot}}}\Delta\vec{r}_N = \frac{1}{M_{\mathrm{tot}}}\sum_{i=1}^{N} m_i \Delta\vec{r}_i \quad (7\text{-}31)$$

(change in vector position of the center of mass)

Equation 3-7 in Section 3-4 tells us that if an object changes its position by $\Delta\vec{r}$ during a time Δt, the velocity of the object is $\vec{v} = \Delta\vec{r}/\Delta t$. If we divide Equation 7-31 through by Δt, we get an expression for the velocity of the center of mass:

$$\begin{aligned}\vec{v}_{\mathrm{CM}} &= \frac{\Delta\vec{r}_{\mathrm{CM}}}{\Delta t} \\ &= \frac{1}{\Delta t}\left(\frac{m_1}{M_{\mathrm{tot}}}\Delta\vec{r}_1 + \frac{m_2}{M_{\mathrm{tot}}}\Delta\vec{r}_2 + \frac{m_3}{M_{\mathrm{tot}}}\Delta\vec{r}_3 + \ldots + \frac{m_N}{M_{\mathrm{tot}}}\Delta\vec{r}_N\right) \quad (7\text{-}32) \\ &= \frac{m_1}{M_{\mathrm{tot}}}\vec{v}_1 + \frac{m_2}{M_{\mathrm{tot}}}\vec{v}_2 + \frac{m_3}{M_{\mathrm{tot}}}\vec{v}_3 + \ldots + \frac{m_N}{M_{\mathrm{tot}}}\vec{v}_N = \frac{1}{M_{\mathrm{tot}}}\sum_{i=1}^{N} m_i \vec{v}_i\end{aligned}$$

(vector velocity of the center of mass)

Just as the position of the center of mass is a weighted average of the positions of the objects that make up the system, the *velocity* of the center of mass is a weighted average of the *velocities* of the objects.

The quantity $m_i\vec{v}_i$ on the right-hand side of Equation 7-32 is just the *momentum* of the ith object in the system. So $\sum_{i=1}^{N} m_i\vec{v}_i$ is the vector sum of the momentum of all objects that make up the system. This is just the *total* momentum of the system, which we denote as $\vec{P}$. If we multiply Equation 7-32 by the total mass of the system M_{tot}, we get

The **total momentum** of a system...

...equals the **vector sum** of the **momentum of all objects** in the system...

$$\vec{P} = \sum_{i=1}^{N} m_i\vec{v}_i = M_{tot}\vec{v}_{CM}$$

...and also equals the **total mass of the system** multiplied by the **velocity of the center of mass.**

Total momentum and the velocity of the center of mass (7-33)

In other words the total momentum of a system of objects of total mass M_{tot} is the *same* as if all of the objects were squeezed into a single blob moving at the velocity of the center of mass (**Figure 7-21**). This means that the total linear momentum of a football doesn't depend on how fast the football is spinning but only on the velocity of the center of mass of the football (which is at the football's geometrical center because the football is symmetrical).

We can now use Equation 7-33 to see what affects the motion of the center of mass. This equation shows that in order for the velocity of the center of mass to change by an amount $\Delta\vec{v}_{CM}$, there must be a change in the total momentum of the system:

$$\Delta\vec{P} = M_{tot}\Delta\vec{v}_{CM}$$

(The mass M_{tot} doesn't change since the system always includes the same set of objects.) But Equation 7-13 in Section 7-3 tells us that the total momentum can change during a time Δt *only* if a net external force acts on the system during that time:

$$\left(\sum\vec{F}_{\text{external on system}}\right)\Delta t = \Delta\vec{P}$$

Combining these two equations, we get

$$\left(\sum\vec{F}_{\text{external on system}}\right)\Delta t = M_{tot}\Delta\vec{v}_{CM}$$

Divide both sides of this equation by Δt, and recall that the acceleration $\vec{a}$ of an object equals the change in its velocity divided by the time over which the velocity changes: $\vec{a} = \Delta\vec{v}/\Delta t$ (Equation 3-9 in Section 3-5). The result is

The **net external force** on a system...

...causes the **center of mass** of the system to **accelerate.**

$$\sum\vec{F}_{\text{external on system}} = M_{tot}\frac{\Delta\vec{v}_{CM}}{\Delta t} = M_{tot}\vec{a}_{CM}$$

How the net external force on a system affects the center of mass (7-34)

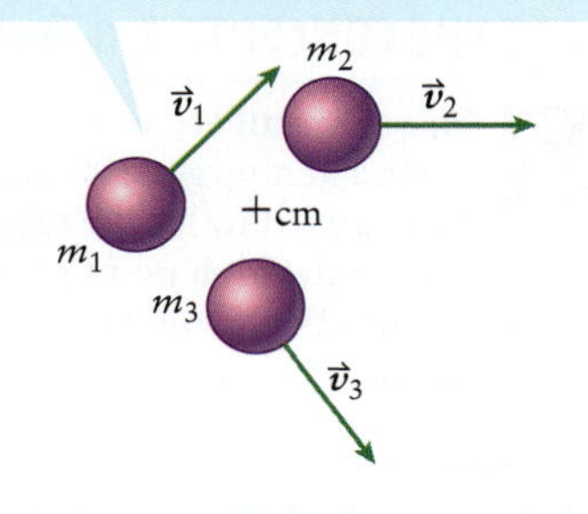
The total momentum of the system is $\vec{P} = m_1\vec{v}_1 + m_2\vec{v}_2 + m_3\vec{v}_3...$
m_2
$\vec{v}_1$
$\vec{v}_2$
+cm
m_1
m_3
$\vec{v}_3$
$M_{tot} = m_1 + m_2 + m_3 + ...$
$\vec{v}_{cm}$
...which is the same as the momentum of all the mass moving together at the velocity of the center of mass: $\vec{P} = M_{tot}\vec{v}_{cm}$.

Figure 7-21 Total momentum of a system Two ways to represent the total momentum of a system of objects.

Equation 7-34 looks almost exactly like Newton's second law for a single object of mass m:

$$\sum\vec{F}_{ext} = m\vec{a}$$

So what Equation 7-34 tells us is that *the center of mass of a system of objects moves exactly as if the entire mass M_{tot} of the system were concentrated at the center of mass, and all of the external forces on the system acted on that concentrated mass.* This justifies the statements we made at the beginning of this section about the significance of the center of mass.

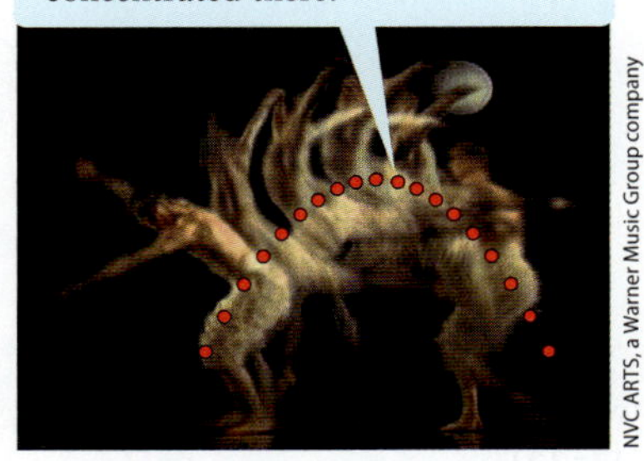

The dancer's center of mass follows a parabolic trajectory, just as if all of the dancer's mass were concentrated there.
NVC ARTS, a Warner Music Group company

Figure 7-22 Center of mass of a ballet dancer The motion of the ballet dancer's body is very complex, but the motion of the dancer's center of mass is very simple.

BIO-Medical Equation 7-34 also says that only *external* forces affect the motion of the center of mass. If various parts of the system exert forces on each other, these forces can affect how those parts move relative to each other, but they have *no* effect on the motion of the center of mass. As an example, consider the ballet dancer shown in Figure 7-22. As he leaps from right to left, his arms, legs, head, and torso move in complicated ways relative to each other. But his center of mass, shown by the orange dots, moves in a simple parabolic trajectory. This is the same trajectory that would be followed by a blob with the same mass as the dancer that was acted on by only the external force of gravity.

If the net external force on a system is zero, Equation 7-34 says that the center of mass cannot accelerate even if the various parts of the system move relative to each other. As an example, if the dancer in Figure 7-22 were in gravity-free outer space, his center of mass would continue moving in a straight line at a constant speed no matter how he moved the various parts of his body.

The center of mass will play an important role when we discuss rotational motion in Chapter 8. We'll see that if an object is rotating as it moves through space, like a spinning football in flight or a spinning tire on a moving car, its motion can naturally be thought of as being in two pieces: the motion of the object's center of mass, plus rotation of the object around its center of mass.

GOT THE CONCEPT? 7-6 An Exploding Cannon Shell

? An army artillery unit fires a cannon shell over level ground at a target 200 m away. The cannon is perfectly aimed to hit the target, and air resistance can be neglected. At the high point of the shell's trajectory, the shell explodes into two identical halves, both of which hit the ground at the same time. One half falls vertically downward from the point of the explosion. Where does the other half land? (a) On the target; (b) 50 m short of the target; (c) 50 m beyond the target; (d) 100 m beyond the target; (e) 150 m beyond the target.

TAKE-HOME MESSAGE FOR Section 7-7

✔ The center of mass of a system of objects is a weighted average of the positions of the individual objects.

✔ The total momentum of the system equals the system's total mass multiplied by the velocity of the center of mass.

✔ If there is a net external force on the system, the center of mass accelerates as though all of the mass of the system was concentrated into a blob at the center of mass. If there is zero net external force on the system, the center of mass does not accelerate.

Key Terms

center of mass
completely inelastic collision
contact time
elastic collision
external forces
impulse
inelastic collision
internal forces
law of conservation of momentum
linear momentum (momentum)
momentum (linear momentum)
total momentum
weighted average

Chapter Summary

Topic	Equation or Figure
Linear momentum: The momentum of an object is a vector that points in the same direction as its velocity. It depends on both the mass and velocity of the object. Do not confuse momentum with kinetic energy, which is a scalar quantity.	The **linear momentum** of an object is a vector. The linear momentum points in the same direction as the **velocity vector** and is proportional to the velocity. $\vec{p} = m\vec{v}$ (7-5) The linear momentum is also proportional to the object's **mass**.

External force and total momentum change for a system of objects: The total momentum of a system of objects is the vector sum of the individual momentum of each object in the system. Only external forces (which originate from objects outside the system) affect the total momentum; internal forces (exerted by one member of the system on another) do not.

The **sum of all external forces** acting on a system of objects

Duration of a time interval over which the external forces act

$$\left(\sum \vec{F}_{\text{external on system}}\right)\Delta t = \vec{P}_{\text{f}} - \vec{P}_{\text{i}} = \Delta\vec{P} \tag{7-13}$$

Change during that time interval **in the total momentum** of the objects that make up the system

Collisions and the law of conservation of momentum: The total momentum of a system is conserved (maintains the same value) if there is zero net external force on the system. If the internal forces during a collision are much greater in magnitude than the external forces, the total momentum of the colliding objects is conserved.

If the net external force on a system of objects is zero, **the total momentum of the system** is conserved.

$$\vec{P}_{\text{f}} = \vec{P}_{\text{i}} \tag{7-14}$$

Then the total momentum of the system at the end of a time interval...

...is equal to the total momentum of the system at the beginning of that time interval.

Kinds of collisions: Momentum is conserved in collisions of all types. In an elastic collision mechanical energy is also conserved. Mechanical energy is not conserved in an inelastic collision. In a completely inelastic collision the colliding objects stick together after the collision.

(a) Elastic collision: Momentum conserved, mechanical energy conserved

Before

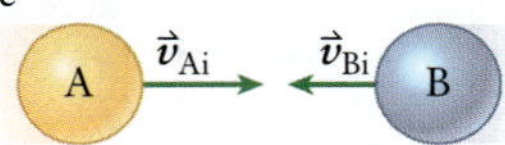

After

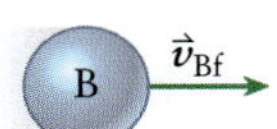

(b) Inelastic collision: Momentum conserved, mechanical energy not conserved

Before

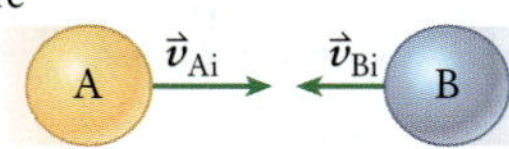

After

B $\vec{v}_{\text{Bf}}$

(c) Completely inelastic collision: Momentum conserved, mechanical energy not conserved, colliding objects stick together

Before

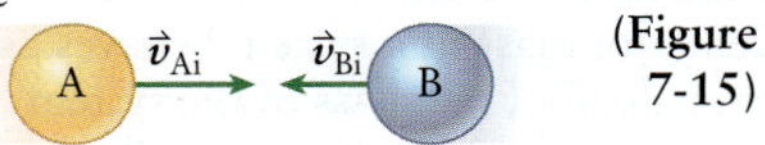

After

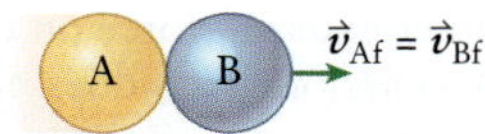

(Figure 7-15)

Completely inelastic collisions: The common final velocity in a completely inelastic collision depends on the masses and initial velocities of the colliding objects.

Velocities of objects A and B **before** they undergo a **completely inelastic collision** (in which they stick together)

$$m_{\text{A}}\vec{v}_{\text{Ai}} + m_{\text{B}}\vec{v}_{\text{Bi}} = (m_{\text{A}} + m_{\text{B}})\vec{v}_{\text{f}} \tag{7-18}$$

Masses of objects A and B

Velocity of the two objects moving together **after the collision**

Impulse and external force: To cause a certain change in an object's momentum requires a certain impulse (the product of the net external force on an object and the time that those forces act).

The **sum of all external forces** acting on an object

Duration of the time interval over which the external forces act

$$\left(\sum \vec{F}_{\text{external on object}}\right)\Delta t = \vec{p}_{\text{f}} - \vec{p}_{\text{i}} = \Delta\vec{p} \tag{7-23}$$

Change in the momentum $\vec{p}$ of the object during that time interval

Contact time: The impulse in a collision is determined by the force in the collision and the contact time (the time that the colliding objects interact with each other). The same momentum change can be caused by a strong force with a short contact time or a weak force with a long contact time.

Force exerted on an object during a collision

Duration of the collision = **contact time**

$$\vec{F}_{\text{collision}}\Delta t = \vec{p}_{\text{f}} - \vec{p}_{\text{i}} = \Delta\vec{p} \qquad (7\text{-}24)$$

Change in the momentum $\vec{p}$ of the object during the collision

Center of mass: The center of mass of a system of objects is a weighted average of the positions of the individual objects.

Position of the center of mass of a system of N objects

m_i = **masses** of the individual objects

$$x_{\text{CM}} = \frac{m_1}{M_{\text{tot}}}x_1 + \frac{m_2}{M_{\text{tot}}}x_2 + \frac{m_3}{M_{\text{tot}}}x_3 + \ldots + \frac{m_N}{M_{\text{tot}}}x_N = \frac{1}{M_{\text{tot}}}\sum_{i=1}^{N} m_i x_i \qquad (7\text{-}28)$$

Total mass of all N objects together

x_i = **positions** of the individual objects

Total momentum and the motion of the center of mass: The total momentum of the system is the same as if all of the mass were concentrated at the center of mass and moving with the center-of-mass velocity.

The **total momentum** of a system...

...equals the **vector sum** of the **momentum of all objects** in the system...

$$\vec{P} = \sum_{i=1}^{N} m_i\vec{v}_i = M_{\text{tot}}\vec{v}_{\text{CM}} \qquad (7\text{-}33)$$

...and also equals the **total mass of the system** multiplied by the **velocity of the center of mass.**

Motion of the center of mass: The center of mass of a system behaves as though all of the mass of the system were concentrated there and all of the external forces acted at that point. If there is no net external force on the system, the center of mass does not accelerate.

The **net external force** on a system...

...causes the **center of mass** of the system to **accelerate.**

$$\sum \vec{F}_{\text{external on system}} = M_{\text{tot}}\frac{\Delta\vec{v}_{\text{CM}}}{\Delta t} = M_{\text{tot}}\vec{a}_{\text{CM}} \qquad (7\text{-}34)$$

Answer to What do you think? Question

(c) If two objects moving at different velocities collide and stick together, kinetic energy is always lost. That's because some portion of one or both of the objects is compressed in the collision. Hence negative work is done on the object or objects, and kinetic energy is lost. We discuss collisions of this kind (called *completely inelastic* collisions) in Section 7-4.

Answers to Got the Concept? Questions

7-1 (c) This situation is exactly like the case of you pushing off the skateboard. The rifle and the bullet exert forces of equal magnitude on each other, and these forces act for the same amount of time. So just after the shot, the rifle (with you and your rollerblades attached) has the same magnitude of *momentum* as does the bullet. Because the bullet has such a small mass, however, it ends up with much more kinetic energy than the combination of the rifle, you, and your rollerblades. Since kinetic energy is a measure of how much work an object can do in coming to rest (see Section 6-3), it follows that the bullet will do much more damage to whatever it hits than will you recoiling on your rollerblades. (That's an important reason a sharpshooter holds the rifle and lets the bullet fly rather than the other way around!)

7-2 (d) Momentum is conserved in this collision. The collision is two dimensional, so we need both x and y axes; we'll take the positive x axis to point west (the direction in which the first object is moving in initially) and the positive y axis to point north. Then before the collision the total momentum (which is equal to the momentum of the first object) has a positive x component and a zero y component. Because momentum is conserved, the total momentum after the collision must also point in the positive x direction. After the collision the first object has negative x momentum and negative y momentum, so the second object must have a large positive x momentum (so that the total x momentum can be positive) and a positive y momentum (so that the total y momentum can be zero). Therefore, the second object is moving west and north.

7-3 (d) In order for the system of car and truck to be at rest after this completely inelastic collision, the total momentum of the system must be zero. The car has a northward momentum of magnitude $p_{car,i} = m_{car}v_{car,i} = (1.50 \times 10^3 \text{ kg})(35.0 \text{ m/s}) = 5.25 \times 10^4 \text{ kg} \cdot \text{m/s}$, so the truck must have a southward momentum of the same magnitude for the total momentum to be zero. Therefore, $p_{truck,i} = m_{truck}v_{truck,i} = (7.50 \times 10^3 \text{ kg})v_{truck,i} = 5.25 \times 10^4 \text{ kg} \cdot \text{m/s}$, and $v_{truck,i} = (5.25 \times 10^4 \text{ kg} \cdot \text{m/s})/(7.50 \times 10^3 \text{ kg}) = 7.00 \text{ m/s}$.

7-4 (a) inelastic, (b) elastic, (c) inelastic. In (a) the ball collides with Earth, which is very much more massive than the ball. If the collision were elastic, the ball would rebound with nearly the same kinetic energy as it had just before hitting the ground and so it would bounce back to its original height. Since this does not happen, mechanical energy must have been lost and the collision must have been inelastic. In (b) the total kinetic energy just after the collision is 125 J + 75 J = 200 J, the same as just before the collision. So mechanical energy is conserved, and the collision is elastic. In (c) the colliding objects (the dog and the ball) move together after the collision, so this must be a *completely* inelastic collision as described in Section 7-4.

7-5 (d) This is the same situation as Example 7-9, so we can use the same equation for the final velocity of the ball: $v_{fx} = F_{collision,x}\Delta t/m = (1700 \text{ N})(8.6 \times 10^{-3} \text{ s})/(0.43 \text{ kg}) = 34 \text{ m/s}$ (about 120 km/h or 76 mi/h).

7-6 (d) If the shell did not explode, its center of mass would hit the target. The force that blows the cannon shell apart is *internal* to the shell, so it does not affect the motion of the center of mass. So the center of mass still hits the target. The explosion happens when the shell is halfway along its trajectory, so the half that falls vertically lands 100 m short of the target (half the horizontal distance from cannon to target). We saw in Example 7-10 that the center of mass of a system of two equal masses is halfway between the two masses. So the other mass must land 100 m on the other side of the target.

Questions and Problems

In a few problems you are given more data than you actually need; in a few other problems you are required to supply data from your general knowledge, outside sources, or informed estimate.

Interpret as significant all digits in numerical values that have trailing zeros and no decimal points.

For all problems use $g = 9.80 \text{ m/s}^2$ for the free-fall acceleration due to gravity.

- • Basic, single-concept problem
- •• Intermediate-level problem; may require synthesis of concepts and multiple steps
- ••• Challenging problem
- SSM *Solution is in Student Solutions Manual*
- Example *See worked example for a similar problem*

Conceptual Questions

1. • If the mass of a basketball is 18 times that of a tennis ball, can they ever have the same momentum? Explain your answer. SSM

2. • Starting from Newton's second law explain how a collision that is free from external forces conserves momentum. In other words, explain how the momentum of the system remains constant with time.

3. • The classic collision problem involves two hard spheres colliding. When you define the system as *both* spheres, momentum is conserved, but when you define the system as only *one* of the spheres, momentum is *not* conserved. Explain why momentum is conserved in the first case but not the second.

4. • Two objects have equal kinetic energies. Are the magnitudes of their momenta equal? Explain your answer.

5. • How would you determine if a collision is elastic or inelastic?

6. • Cite two examples of totally inelastic collisions that occur in your daily life.

7. •• Why is conservation of energy alone not sufficient to explain the motion of a Newton's cradle toy, shown in Figure 7-23? Consider the case when two balls are raised and released together.

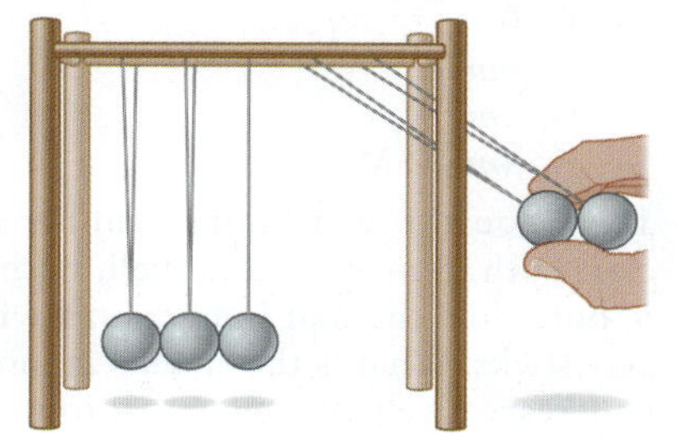

Figure 7-23 Problem 7

8. • An arrow shot into a straw target penetrates a distance that depends on the speed with which it strikes the target. How does the penetration distance change if the arrow's speed is doubled? Be sure to list all the assumptions you make while arriving at your answer.

9. • Using the common definition of the word *impulse*, comment on physicists' choice to define impulse as the change in momentum.

10. • A recent U.S. patent application describes a "damage avoidance system" for cell phones, which, upon detecting an impending, uncontrolled impact with a surface, deploys an air-filled bag around the phone. Explain how this could protect the cell phone from damage.

11. • A glass will break if it falls onto a hardwood floor but not if it falls from the same height onto a padded, carpeted floor. Describe the different outcomes in the collision between a glass and the floor in terms of fundamental physical quantities.

12. • A child stands on one end of a long wooden plank that rests on a frictionless icy surface. (a) Describe the motion of the plank when she runs to the other end of the plank. (b) Describe the motion of the center of mass of the system. (c) How would your answers change if she had walked the plank rather than run down it?

13. • Based on what you know about center of mass, why is it potentially dangerous to step off of a small boat before it is secured to the dock? SSM

14. • A man and his large dog sit at opposite ends of a rowboat, floating on a still pond. You notice from shore that the boat moves as the dog walks toward his owner. Describe the motion of the boat from your perspective.

15. • After being thrown into the air, a lit firecracker explodes at the apex of its parabolic flight. (a) Is momentum conserved before or after the explosion? (b) Is the mechanical energy conserved? (c) What is the path of the center of mass? Neglect the effects of air resistance and explain your answers.

Multiple-Choice Questions

16. • A large semitrailer truck and a small car have equal momentum. How do their speeds compare?
   A. The truck has a much higher speed than the car.
   B. The truck has only a slightly higher speed than the car.
   C. Both have the same speed.
   D. The truck has only a slightly lower speed than the car.
   E. The truck has a much lower speed than the car.

17. • A tennis player smashes a ball of mass m horizontally at a vertical wall. The ball rebounds at the same speed v with which it struck the wall. Has the momentum of the ball changed, and if so, what is the magnitude of the change?
   A. mv
   B. 0
   C. $mv/2$
   D. $2mv$
   E. $4mv$ SSM

18. • You throw a bouncy rubber ball and a wet lump of clay, both of mass m, at a wall. Both strike the wall at speed v, but while the ball bounces off with no loss of speed, the clay sticks. What is the change in momentum of the clay and ball, respectively, assuming that toward the wall is the positive direction?
   A. 0; mv
   B. mv; 0
   C. 0; $-2mv$
   D. $-mv$; $-mv$
   E. $-mv$; $-2mv$

19. • Consider a completely inelastic, head-on collision between two particles that have equal masses and equal speeds. Describe the velocities of the particles after the collision.
   A. The velocities of both particles are zero.
   B. Both of their velocities are reversed.
   C. One of the particles continues with the same velocity and the other comes to rest.
   D. One of the particles continues with the same velocity, and the other reverses direction at twice the speed.
   E. More information is required to determine the final velocities.

20. • An object is traveling in the positive x direction with speed v. A second object that has half the mass of the first is traveling in the opposite direction with the same speed. The two experience a completely inelastic collision. The final x component of the velocity is
   A. 0
   B. $v/2$
   C. $v/3$
   D. $2v/3$
   E. v

21. • Consider a completely elastic head-on collision between two particles that have the same mass and the same speed. What are the velocities after the collision?
   A. Both are zero.
   B. The magnitudes of the velocities are the same, but the directions are reversed.
   C. One of the particles continues with the same velocity, and the other comes to rest.
   D. One of the particles continues with the same velocity, and the other reverses direction at twice the speed.
   E. More information is required to determine the final velocities. SSM

22. • Two small, identical steel balls collide completely elastically. Initially ball 1 is moving with velocity v_1, and ball 2 is stationary. After the collision, the final velocities of ball 1 and ball 2 are
   A. $\frac{1}{2}v_1$; $\frac{1}{2}v_1$
   B. v_1; $2v_1$
   C. 0; v_1
   D. $-v_1$; 0
   E. $-v_1$; $2v_1$

23. • Two ice skaters, Lilly and John, face each other while stationary and push against each other's hands. John's mass is twice that of Lilly. How do their speeds compare after the push-off?
   A. Lilly's speed is one-fourth of John's speed.
   B. Lilly's speed is one-half of John's speed.
   C. Lilly's speed is the same as John's speed.
   D. Lilly's speed is twice John's speed.
   E. Lilly's speed is four times John's speed.

24. • A friend throws a heavy ball toward you while you are standing on smooth ice. You can either catch the ball or deflect it back toward your friend. Which of the following options will maximize your speed right after your interaction with the ball?
   A. You should catch the ball.
   B. You should deflect the ball back toward your friend at the same speed with which it hit your hand.
   C. You should let the ball go past you without touching it.
   D. It doesn't matter—your speed is the same regardless of what you do.
   E. You should deflect the ball back toward your friend at half the speed with which it hit your hand.

25. ••• Two blocks are released from rest on either side of a frictionless half-pipe (**Figure 7-24**). Block B is more massive than block A. The height H_B from which block B is released is less than H_A, the height from which block A is released. The blocks collide elastically on the flat section. After the collision, which is correct?
   A. Block A rises to a height greater than H_A, and block B rises to a height less than H_B.
   B. Block A rises to a height less than H_A, and block B rises to a height greater than H_B.
   C. Block A rises to height H_A, and block B rises to height H_B.
   D. Block A rises to height H_B, and block B rises to height H_A.
   E. The heights to which the blocks rise depends on where along the flat section they collide. SSM

Figure 7-24 Problem 25

Estimation/Numerical Analysis

26. • **Astronomy** A 50-m wide asteroid collides with Mars. Estimate the ratio of the magnitude of the momentum transfer from the asteroid impact to the momentum of Mars's orbital motion.

27. • Estimate the momentum of a car driving the speed limit on a freeway.

28. • **Sports** Estimate the momentum of a fastball thrown by a major league pitcher.

29. • **Sports** Estimate the momentum of a tennis ball served by a professional tennis player.

30. • **Medical, Biology** Estimate the location of the center of mass of your body. SSM

31. • Estimate the momentum of a bumblebee when it strikes a motorcycle rider.

32. • **Astronomy** Estimate the momentum of Earth as it orbits the Sun.

33. • **Sports** Compare the momentum of a fast-pitch softball to a major league fastball.

34. • Estimate the impulse delivered to a tennis ball that rebounds from a practice wall. SSM

35. •• A car moving at 20 m/s slams into the back end of a car stopped at a red light. After the collision, the two cars stick together. If the cars have the same mass, estimate the distance the two cars travel before coming to rest. Assume that neither driver applies his brakes during the collision.

36. ••• The following gives the force (in newtons) acting on a 2-kg object as a function of time. (a) Make a graph of force versus time. (b) If the object starts from rest, what is its speed after 25 s?

t (s)	F (N)	t (s)	F (N)	t (s)	F (N)
0	−20	9	25	18	25
1	−20	10	25	19	25
2	−10	11	25	20	25
3	0	12	25	21	20
4	10	13	25	22	15
5	15	14	25	23	10
6	18	15	25	24	5
7	20	16	25	25	0
8	25	17	25		

Problems

7-2 Momentum is a vector that depends on an object's mass, speed, and direction of motion

37. • A 1.00×10^4-kg train car moves east at 15 m/s. Determine the momentum of the train car. Example 7-1

38. • **Sports** The magnitude of the instantaneous momentum of a 57-g tennis ball is 2.6 kg·m/s. What is its speed? SSM Example 7-1

39. • Determine the initial momentum, final momentum, and change in momentum of a 1250-kg car initially backing up at 5.00 m/s, then moving forward at 14.0 m/s. Example 7-1

40. • **Sports** What is the magnitude of momentum of a 135-kg defensive lineman running at 7.00 m/s? Example 7-1

41. • One ball has four times the mass and twice the speed of another. (a) How does the magnitude of momentum of the more massive ball compare to the magnitude of momentum of the less massive one? (b) How does the kinetic energy of the more massive ball compare to the kinetic energy of the less massive one? Example 7-1

42. • A girl who has a mass of 55.0 kg rides the skateboard to class at a speed of 6.00 m/s. (a) What is the magnitude of her momentum? (b) If the magnitude of momentum of the skateboard itself is 30.0 kg·m/s, what is its mass? SSM Example 7-1

43. • Blythe and Geoff are ice skating together. Blythe has a mass of 50.0 kg, and Geoff has a mass of 80.0 kg. Blythe pushes Geoff in the chest when both are at rest, causing him to move away at a speed of 4.00 m/s. (a) Determine Blythe's speed after she pushes Geoff. (b) In what direction does she move? Example 7-1

7-3 The total momentum of a system of objects is conserved under certain conditions

44. • A 2.00-kg object is moving east at 4.00 m/s when it collides with a 6.00-kg object that is initially at rest. After the collision the larger object moves east at 1.00 m/s. What is the final velocity of the smaller object after the collision? Assume no external forces act on the objects. Example 7-2

45. • A 3.00-kg object is moving toward the right at 6.00 m/s. A 5.00-kg object moves to the left at 4.00 m/s. After the two objects collide the 3.00-kg object moves toward the left at 2.00 m/s. What is the final velocity of the 5.00-kg object? Assume no external forces act on the objects. Example 7-2

46. • Two hockey players collide on the ice and go down together in a tangled heap. Player 1 has a mass of 105 kg, and player 2 has a mass of 92 kg. Before the collision player 1 had a velocity of $v_1 = 6.3$ m/s in the $+x$ direction, and player 2 had a velocity of $v_2 = 5.6$ m/s at an angle of 72° with respect to the $+x$ axis as shown in **Figure 7-25**. At what speed and in what direction do they slide together on the ice after the collision? Example 7-7

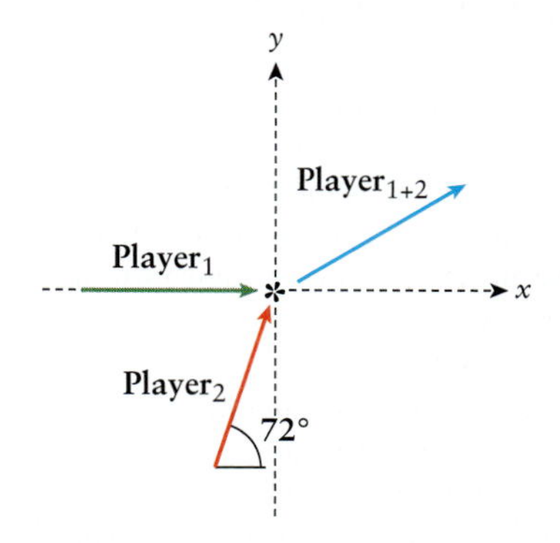

Figure 7-25 Problem 46

47. •• An object of mass $3M$, moving in the $+x$ direction at speed v_0, breaks into two pieces of mass M and $2M$ as shown in **Figure 7-26**. If $\theta_1 = 45°$ and $\theta_2 = 30°$, determine the final velocities of the resulting pieces in terms of v_0. SSM Example 7-4

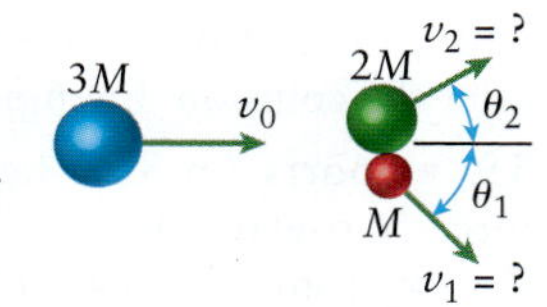

Figure 7-26 Problem 47

48. •• In a game of pool the cue ball is rolling at 2.00 m/s in a direction 30.0° north of east when it collides with the eight ball (initially at rest). The mass of the cue ball is 170 g, but the mass of the eight ball is only 156 g. After the collision the cue ball heads off at 10.0° north of east, and the eight ball moves off due north. What are the final speeds of each ball after the collision? Example 7-3

49. • **Biology** During mating season, male bighorn sheep establish dominance with head-butting contests which can be heard up to a mile away. When two males butt heads, the "winner" is the one that knocks the other backward. In one contest a sheep of mass 95.0 kg moving at 10.0 m/s runs directly into a sheep of mass 80.0 kg moving at 12.0 m/s. Which ram wins the head-butting contest? SSM Example 7-2

50. • A superhero and a supervillain are having a battle. The hero is flying parallel to the ground at a speed of 51.8 m/s when the villain on the ground picks up and hurls a car at him. The 1860-kg car is moving parallel to the ground at 19.1 m/s when it strikes the 102-kg hero at an angle of −46° with respect to the hero's initial direction of motion, as shown in **Figure 7-27**. The hero and the car are stuck together after the collision. What is their speed, and in what direction relative to the hero's initial velocity do they travel after the collision? Example 7-7

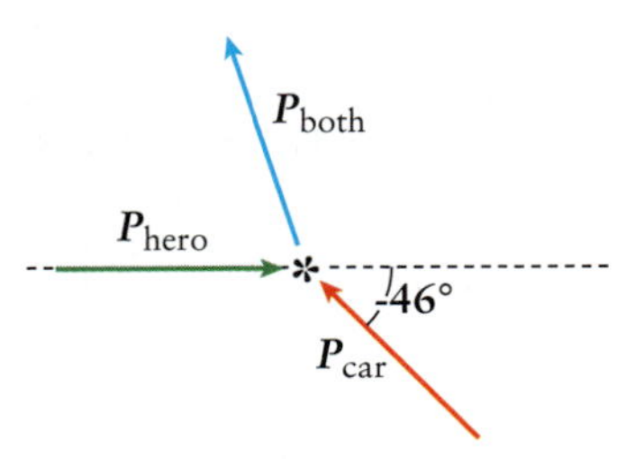

Figure 7-27 Problem 50

7-4 In an inelastic collision, some of the mechanical energy is lost

51. • A 1.00×10^4-kg train car moving due east at 20.0 m/s collides with and couples to a 2.00×10^4-kg train car that is initially at rest. (a) What is the common velocity of the two-car train after the collision? (b) What is the total kinetic energy of the two train cars before and after the collision? Example 7-6

52. • A large fish has a mass of 25.0 kg and swims at 1.00 m/s toward and then swallows a smaller fish that is not moving. If the smaller fish has a mass of 1.00 kg, what is the speed of the larger fish immediately after it finishes lunch? Example 7-6

53. • A 5.00-kg howler monkey is swinging due east on a vine. It overtakes and grabs onto a 6.00-kg monkey also moving east on a second vine. The first monkey is moving at 12.0 m/s at the instant it grabs the second, which is moving at 8.00 m/s. (a) After they join on the same vine, what is their common speed? (b) What is the total kinetic energy of the two monkeys before and after they join the same vine? SSM Example 7-6

54. • A 1200-kg car is moving at 20.0 m/s due north. A 1500-kg car is moving at 18.0 m/s due east. The two cars simultaneously approach an icy intersection where, with no brakes or steering, they collide and stick together. Determine the speed and direction of the combined two-car wreck immediately after the collision. Example 7-7

55. • **Sports** An 85.0-kg linebacker is running at 8.00 m/s directly toward the sideline of a football field. He tackles a 75.0-kg running back moving at 9.00 m/s straight toward the goal line (perpendicular to the original direction of the linebacker). Determine their common speed and direction immediately after they collide. Example 7-7

7-5 In an elastic collision both momentum and mechanical energy are conserved

56. •• One way that scientists measure the mass of an unknown particle is to bounce a known particle, such as a proton or an electron, off the unknown particle in a bubble chamber. The initial and rebound velocities of the known particle are measured from photographs of the bubbles it creates as it moves; the information is used to determine the mass of the unknown particle. (a) If a known particle of mass m and initial speed v_0 collides elastically, head-on with a stationary unknown particle and then rebounds with speed v, find an expression for the mass of the unknown particle in terms of m, v, and v_0. (b) If the known particle is a proton and the unknown particle is a neutron, what will be the recoil speed of the proton and the final speed of the neutron? Example 7-8

57. • A 2.00-kg ball moves at 3.00 m/s toward the right. It collides elastically with a 4.00-kg ball that is initially at rest. Determine the velocities of the balls after the collision. Example 7-8

58. • A 10.0-kg block of ice is sliding due east at 8.00 m/s when it collides elastically with a 6.00-kg block of ice that is sliding in the same direction at 4.00 m/s. Determine the velocities of the blocks of ice after the collision. SSM Example 7-8

59. • A 0.170-kg ball moves at 4.00 m/s toward the right. It collides elastically with a 0.155-kg ball moving at 2.00 m/s toward the left. Determine the final velocities of the balls after the collision. Example 7-8

60. • A neutron traveling at 2.00×10^5 m/s collides elastically with a deuteron that is initially at rest. Determine the final speeds of the two particles after the collision. The mass of a neutron is 1.67×10^{-27} kg, and the mass of a deuteron is 3.34×10^{-27} kg. Example 7-8

7-6 What happens in a collision is related to the time the colliding objects are in contact

61. •• A sudden gust of wind exerts a force of 20.0 N for 1.20 s on a bird that had been flying at 5.00 m/s. As a result the bird ends up moving in the opposite direction at 7.00 m/s. What is the mass of the bird? Example 7-9

62. • Determine the average force exerted on your hand as you catch a 0.200-kg ball moving at 20.0 m/s. Assume the time of contact is 0.0250 s. SSM Example 7-9

63. • An expert boxer delivers a 3.20×10^3 N punch to the head of his opponent. If the punch contacts the 4.82-kg head for 0.35 s, and the unfortunate boxer's head has a mass of 4.82 kg, what is the speed of the head when the punch loses contact? Assume no other forces act on the opponent's head. (In reality, the force of the neck on the head plays an important role.) Example 7-9

64. • **Sports** A baseball of mass 0.145 kg is thrown at a speed of 40.0 m/s. The batter strikes the ball with a force of 25,000 N; the bat and ball are in contact for 0.500 ms. Assuming that the force is exactly opposite to the original direction of the ball, determine the final speed of the ball. Example 7-9

65. • Bean bag rounds used by police are nonlethal projectiles fired from shotguns. If a 40.0-g round strikes at 70.0 m/s and delivers all its momentum over a time of 0.15 s, what is the average force of impact? Example 7-9

66. •• **Sports** A baseball bat strikes a ball when both are moving at 31.3 m/s (relative to the ground) toward each other. The bat and ball are in contact for 1.20 ms, after which the ball is traveling opposite its initial direction at a speed of 42.5 m/s. The mass of the bat and the ball are 850 and 145 g, respectively.

Calculate the magnitude and direction of the impulse given to (a) the ball by the bat and (b) the bat by the ball. (c) What average force does the bat exert on the ball? (d) Why doesn't the force shatter the bat? Example 7-9

7-7 The center of mass of a system moves as though all of the system's mass were concentrated there

67. • Find the coordinates of the center of mass of the three objects shown in **Figure 7-28** if $m_1 = 4.00$ kg, $m_2 = 2.00$ kg, and $m_3 = 3.00$ kg. Distances are in meters. Example 7-10

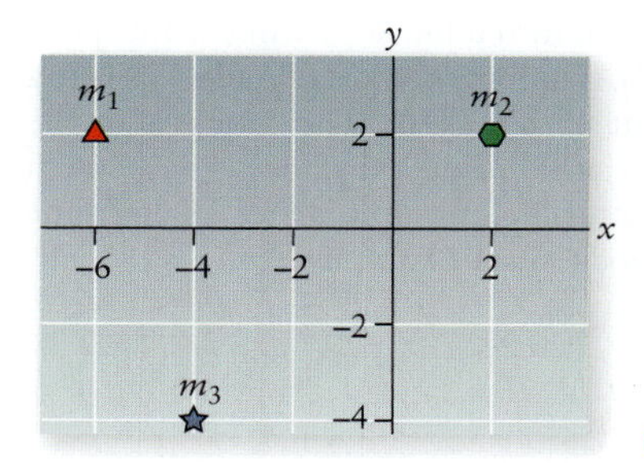

Figure 7-28 Problem 67

68. • What are the coordinates of the center of mass for the combination of the three objects shown in **Figure 7-29**? The uniform rod has a mass of 10.0 kg, has a length of 30.0 cm, and is located at $x = 50.0$ cm. The oval football has a mass of 2.00 kg, a semimajor axis of 15.0 cm, has a semiminor axis of 8.00 cm, and is centered at $x = -50.0$ cm. The spherical volleyball has a mass of 1.00 kg, has a radius of 10.0 cm, and is centered at $y = -30$ cm. Assume both balls are of uniform mass density. SSM Example 7-10

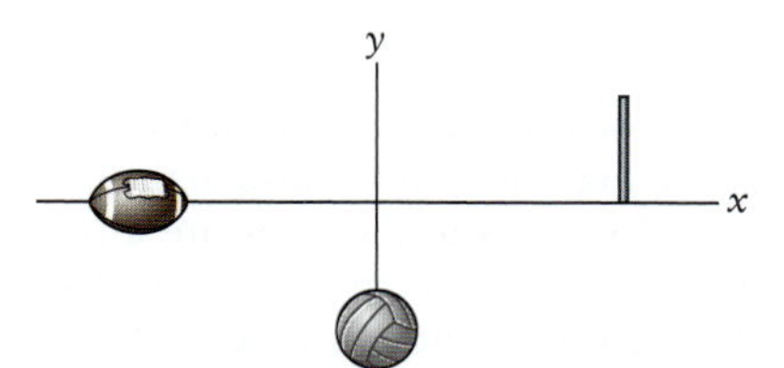

Figure 7-29 Problem 68

69. •• Four beads each of mass M are attached at various locations to a hoop of mass M and radius R (**Figure 7-30**). Find the center of mass of the hoop and beads. Example 7-10

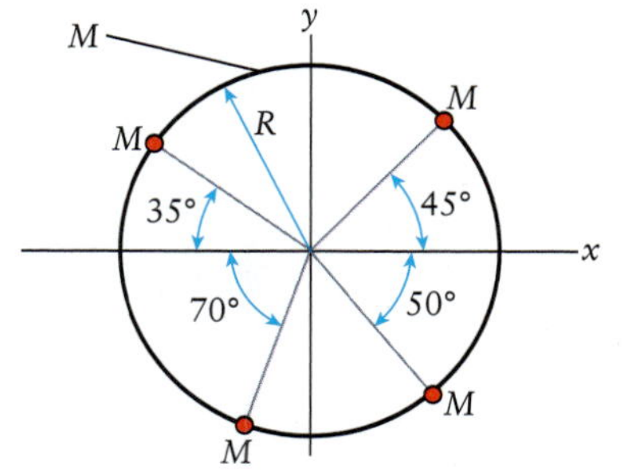

Figure 7-30 Problem 69

70. •• A 3.0-kg block (A) is attached to a 1.0-kg block (B) by a massless spring that is compressed and locked in place, as shown in **Figure 7-31**. The blocks slide without friction along the x direction at an initial constant speed of 2.0 m/s. At time $t = 0$, the positions of blocks A and B are $x = 1.0$ m and $x = 1.2$ m, respectively, at which point a mechanism releases the spring, and the blocks begin to oscillate as they slide. If 2.0 s later block B is located at $x = 6.0$ m, where will block A be located? Example 7-10

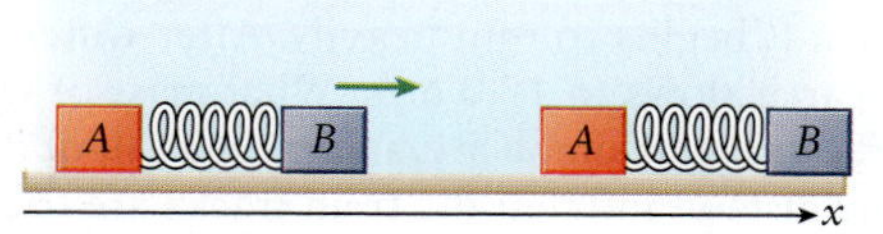

Figure 7-31 Problem 70

General Problems

71. •• **Sports** A major league baseball has a mass of 0.145 kg. Neglecting the effects of air resistance, determine the momentum of the ball when it hits the ground if it falls from rest on the roof of the Willis Tower in Chicago, Illinois, a height of 442.0 m. Example 7-1

72. ••• Forensic scientists can determine the speed at which a rifle fires a bullet by shooting into a heavy block hanging by a wire. As the bullet embeds itself in the block, the block and embedded bullet swing up; the impact speed is determined from the maximum angle of the swing. (a) Which would make the block swing higher, a 0.204 Ruger bullet of mass 2.14 g and muzzle speed 1290 m/s or a 7-mm Remington Magnum bullet of mass 9.71 g and muzzle speed 948 m/s? Assume the bullets enter the block right after leaving the muzzle of the rifle. (b) Using your answer in part (a), determine the mass of the block so that when hit by the bullet it will swing through a 60.0° angle. The block hangs from wire of length 1.25 m and negligible mass. (c) What is the speed of an 8.41-g bullet that causes the block to swing upward through a 30.0° angle? Example 7-6

73. •• A friend suggests that if all the people in the United States dropped down from a 1-m-high table at the same time, Earth would move in a noticeable way. To test the credibility of this proposal, (a) determine the momentum imparted to Earth by 300 million people, of an average mass of 65.0 kg, dropping from 1.00 m above the surface. Assume no one bounces. (b) What change in Earth's speed would result? SSM Example 7-6

74. •• Sally finds herself stranded on a frozen pond so slippery that she can't stand up or walk on it. To save herself, she throws one of her heavy boots horizontally, directly away from the closest shore. Sally's mass is 60 kg, the boot's mass is 5 kg, and Sally throws the boot with speed equal to 30.0 m/s. (a) What is Sally's speed immediately after throwing the boot? (b) Where is the center of mass of the Sally–boot system, relative to where she threw the boot, after 10.0 s? (c) How long does it take Sally to reach the shore, a distance of 30.0 m away from where she threw the boot? For all parts, assume the ice is frictionless. Example 7-1

75. ••• You have been called to testify as an expert witness in a trial involving a head-on collision. Car A weighs 680 kg and was traveling eastward. Car B weighs 500 kg and was traveling westward at 72.0 km/h. The cars locked bumpers and slid eastward with their wheels locked for 6.00 m before stopping. You have measured the coefficient of kinetic friction between the tires and the pavement to be 0.750. How fast (in miles per hour) was car A traveling just before the collision? Example 7-6

76. • A 5000-kg open train car is rolling at a speed of 20.0 m/s when it begins to rain heavily, and 200 kg of water collects

quickly in the car. If only the total mass has changed, what is the speed of the flooded train car? For simplicity, assume that all of the water collects at one instant and that the train tracks are frictionless. Example 7-2

77. • An 8000-kg open train car is rolling at a speed of 20.0 m/s when it begins to rain heavily. After water has collected in the car, it slows to 19.0 m/s. What mass of water has collected in the car? For simplicity, assume that all of the water collects at one instant and that the train tracks are frictionless. SSM Example 7-2

78. •• An open rail car of initial mass 10,000 kg is moving at 5.00 m/s when rocks begin to fall into it from a conveyor belt. The rate at which the mass of rocks increases is 500 kg/s. Find the speed of the train car after rocks have fallen into the car for a total of 3.00 s. Example 7-2

79. • A 65-kg novice skier stops to rest partway down a slope. An inattentive snowboarder with the same 65-kg mass is barreling down the same hill at 9.6 m/s and crashes right into the back of the skier. Miraculously, the collision is perfectly elastic and nobody falls over. What are the velocities of the skier and snowboarder after the collision? Example 7-8

80. •• **Sports** The sport of curling is quite popular in Canada. A curler slides a 19.1-kg stone so that it strikes a competitor's stationary stone at 6.40 m/s before moving at an angle of 120° from its initial direction. The competitor's stone moves off at 5.60 m/s. Determine the final speed of the first stone and the final direction of the second one. Example 7-3

81. •• **Biology** Lions can run at speeds up to approximately 80.0 km/h. A hungry 135-kg lion running northward at top speed attacks and holds onto a 29.0-kg Thomson's gazelle running eastward at 60.0 km/h. Find the speed and direction of travel of the lion–gazelle system just after the lion attacks. SSM Example 7-7

82. •• **Biology** The mass of a pigeon hawk is twice that of the pigeons it hunts. Suppose a pigeon is gliding north at a speed of v_P = 23.0 m/s when a hawk swoops down, grabs the pigeon, and flies off (**Figure 7-32**). The hawk was flying north at speed of v_H = 35.0 m/s, at an angle θ = 45° below the horizontal, at the instant of the attack. Find the final velocity vector of the birds just after the attack. Example 7-7

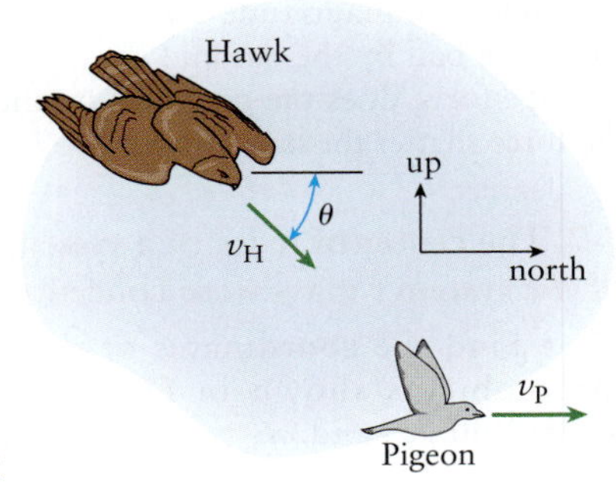

Figure 7-32 Problem 82

83. •• A 12.0-g bullet is fired into a block of wood with speed v = 250 m/s (**Figure 7-33**). The block is attached to a spring that has a spring constant of 200 N/m. The block with the embedded bullet compresses the spring a distance d = 30.0 cm to the right, before momentarily coming to a stop. Determine the mass of the wooden block. SSM Example 7-6

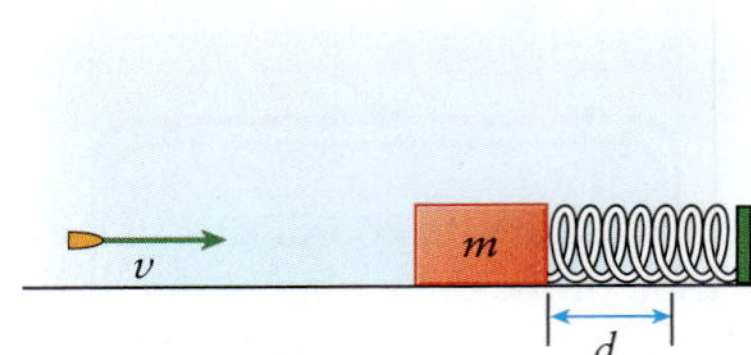

Figure 7-33 Problem 83

84. •• In a ballistic pendulum experiment, a small marble is fired into a cup attached to the end of a pendulum. If the mass of the marble is 0.00750 kg and the mass of the pendulum is 0.250 kg, how high will the pendulum swing if the marble has an initial speed of 6.00 m/s? Assume that the mass of the pendulum is concentrated at its end. Example 7-6

85. ••• A 0.0750-kg ball is thrown at 25.0 m/s toward a brick wall. (a) Determine the impulse that the wall imparts to the ball when it hits and rebounds at 25.0 m/s in the opposite direction. (b) Determine the impulse that the wall imparts to the ball when it hits and rebounds at 25.0 m/s at an angle of 45.0°. (c) If the ball thrown in part (b) contacts the wall for 0.0100 s, determine the magnitude and direction of the average force that the wall exerts on the ball. Example 7-9

8 Rotational Motion

(a) Kevin Lee - ISU/Getty Images (b) ASSOCIATED PRESS

In this chapter, your goals are to:

- (8-1) Define translation and rotation.
- (8-2) Explain what is meant by the moment of inertia of an object and how to use it to calculate the rotational kinetic energy of a rotating object.
- (8-3) Describe the techniques for finding the moment of inertia of an object, including use of the parallel-axis theorem.
- (8-4) Apply the conservation of mechanical energy to rotating objects, including objects that roll without slipping.
- (8-5) Identify the equations of rotational kinematics and know how to use them to solve problems.
- (8-6) Define the concept of lever arm and know how to use it to calculate the torque generated by a force.
- (8-7) Explain the meaning of Newton's second law for rotational motion.
- (8-8) Define what is meant by the angular momentum of a rotating object and of a moving particle, and explain the circumstances under which angular momentum is conserved.
- (8-9) Explain how to find the direction of the angular momentum, angular velocity, and torque vectors.

To master this chapter, you should review:

- (2-2, 2-4) The definitions of average and instantaneous velocity and acceleration.
- (3-7) The equation for the length of an arc in terms of its radius and angle.
- (4-2) Newton's second law of motion.
- (6-3, 6-7) The definition of kinetic energy and the principle of conservation of energy.
- (7-2, 7-3) The concept of linear momentum and when linear momentum is conserved.
- (7-7) The concept of the center of mass of an object.

What do you think?

It's easier for a figure skater to rotate her body if she holds her arms by her side than if she holds her arms straight out. (You can try this yourself—spin around on your feet with your arms held in and with your arms held out.) Why does the orientation of her arms make a difference?

8-1 Rotation is an important and ubiquitous kind of motion

Up to this point we've considered the motions of objects such as a thrown baseball, a speeding car, and a soaring bird. We use the word "**translation**" to refer to these kinds of motion, in which an object as a whole moves through space. However, we haven't worried

A Ferris wheel rotates around a fixed (stationary) axis.

(a)

Holly Kuchera/Getty Images

This mother-of-pearl moth caterpillar (*Pleurotya ruralis*) wraps itself into a circle for self-defense: It can roll away from predators 40 times faster than it can walk. As it rolls, the caterpillar moves as a whole (it translates) as well as rotates.

(b)

Dr. John Brackenbury/Science Source

Our planet rotates to the east once per day around an axis that extends from the north pole to the south pole. Earth also translates: It takes one year to move in an orbit around the Sun.

(c)

NASA/NOAA/GSFC/Suomi NPP/VIIRS/Norman Kuring

Figure 8-1 **Rotational motion** Three examples of rotating objects.

TAKE-HOME MESSAGE FOR Section 8-1

✔ Translation refers to the motion of an object as a whole through space.

✔ Rotation refers to the spinning motion of an object.

✔ An object can rotate around a fixed axis (rotation without translation), or its axis can move through space as the object rotates (rotation with translation).

much about the kind of motion called **rotation**, in which an object spins around an axis. (One example is the figure skater in the photographs on the previous page, who rotates around an axis that extends upward from the skate that touches the ground. **Figure 8-1** shows three more examples.) In this chapter we'll investigate rotational motion.

In some situations, an object undergoes rotation but not translation. An example is the Ferris wheel shown in **Figure 8-1a**: The Ferris wheel as a whole remains at the same location in the amusement park, but the wheel rotates continuously to entertain its riders. Other objects that rotate without translation are a spinning ceiling fan and the rotating platter in a microwave oven. In other situations, like a rolling caterpillar (**Figure 8-1b**) or the wheels of a fast-moving bicycle, an object undergoes rotation and translation at the same time. You've spent your entire life on a planet that experiences both rotation and translation (**Figure 8-1c**). In this chapter we'll look at rotation both without and with translation.

We saw in Chapter 6 that it can be easier to describe motion in terms of work and energy than in terms of forces. With this in mind, let's first ask the following question: How can we find the kinetic energy due to an object's rotation—that is, its *rotational kinetic energy*? Once we've answered this question, we'll be able to use conservation of energy to solve many problems involving rotational motion. Later in the chapter we'll look at an alternative approach involving *torque* and *angular acceleration*—quantities that bear the same relationship to rotational motion that force and acceleration do to translational motion.

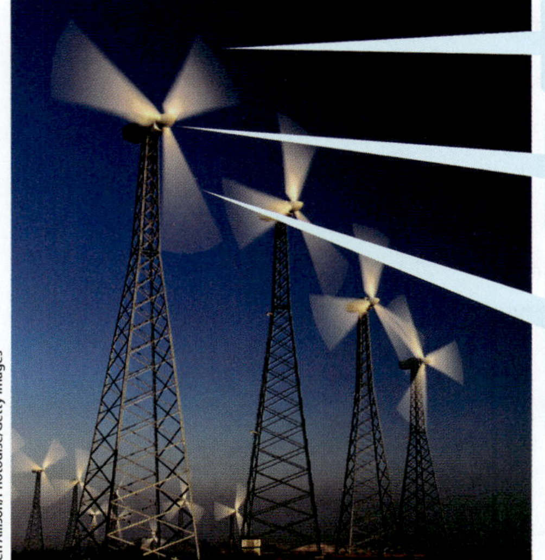

Each blade of this rotating wind turbine is rigid: It maintains the same shape at all times.
The wind turbine rotates around a fixed axis: The axis remains in the same place and keeps the same orientation.
While all parts of the blade rotate together, different parts move at different speeds: The blurring shows that the outer parts (at a greater distance r from the axis) move faster than the inner parts of the blade.
Glen Allison/Photodisc/Getty Images

Figure 8-2 **A rotating rigid object** The blades of a wind turbine undergo rotation but do not undergo translation (the wind turbine as a whole doesn't go anywhere).

8-2 An object's rotational kinetic energy is related to its angular velocity and how its mass is distributed

To keep things simple, we'll begin our study of rotational motion by considering rotation without translation. To be specific, we'll examine a **rigid object**—that is, one whose shape doesn't change as it rotates. The spinning wind turbines shown in **Figure 8-2** are examples of rotating rigid objects. (By contrast, the figure skater in the photographs that open this chapter can change her shape as she spins.) We'll also assume that the rotation axis is *fixed*—that is, it keeps the same orientation in space. The wind turbines in Figure 8-2 rotate around fixed axes.

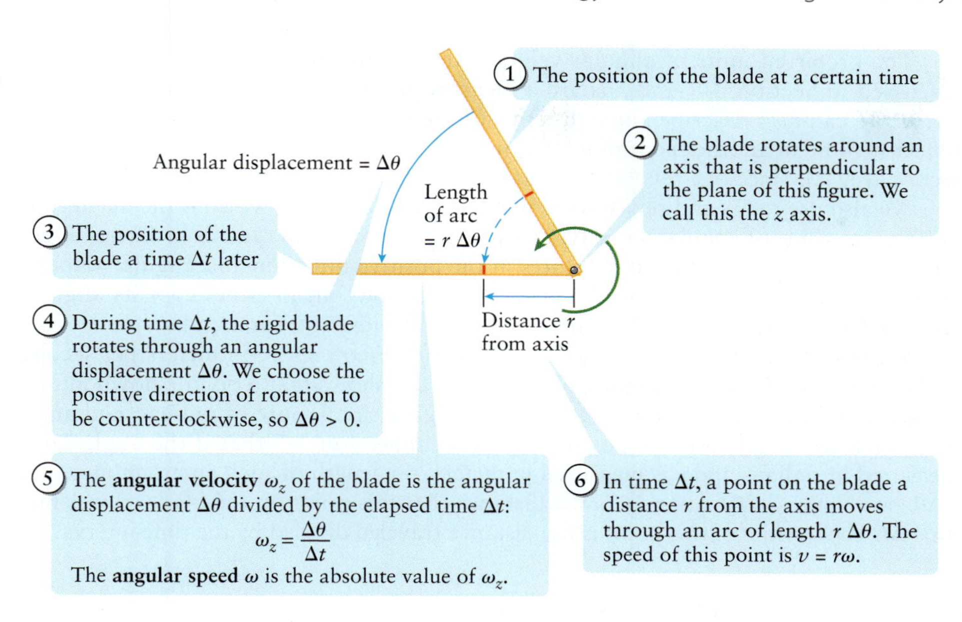

1 The position of the blade at a certain time
2 The blade rotates around an axis that is perpendicular to the plane of this figure. We call this the z axis.
Angular displacement = Δθ
Length of arc = r Δθ
3 The position of the blade a time Δt later
Distance r from axis
4 During time Δt, the rigid blade rotates through an angular displacement Δθ. We choose the positive direction of rotation to be counterclockwise, so Δθ > 0.
5 The angular velocity ω_z of the blade is the angular displacement Δθ divided by the elapsed time Δt:
ω_z = Δθ/Δt
The angular speed ω is the absolute value of ω_z.
6 In time Δt, a point on the blade a distance r from the axis moves through an arc of length r Δθ. The speed of this point is v = rω.

Figure 8-3 Angular velocity We define angular velocity by analogy to the way that we defined velocity for straight-line motion in Chapter 2.

Angular Velocity and Angular Speed

Let's see how to determine the rotational kinetic energy of the turbine shown in Figure 8-2. We'll begin by considering the motion of one of its blades, shown in **Figure 8-3**.

Recall from Section 6-3 that an object's kinetic energy depends on its mass m and its speed v: $K = \frac{1}{2}mv^2$. However, we *cannot* use this formula directly to find the rotational kinetic energy of a turbine blade. Here's why: During a short time interval Δt, the entire blade rotates through a small angle $\Delta\theta$ (pronounced "delta-theta"), which we call the **angular displacement** of the blade. However, during this time interval a point near the rotation axis travels only a short distance, while a point farther from the axis travels a greater distance. Hence a point farther out along the blade moves faster and has a greater speed than a point closer to the axis. You can see this in Figure 8-2: The blade is more blurred the farther it is from the axis. So the speed v is *different* for different parts of the blade. Hence we can't immediately use the formula $K = \frac{1}{2}mv^2$ from Section 6-3 to find the blade's kinetic energy because it's not clear which speed v we should use.

See the Math Tutorial for more information on trigonometry.

To see how to account for different speeds at different locations, let's look in more detail at how the blade moves during a time interval Δt (Figure 8-3). Let's choose angles to increase in the counterclockwise direction; the blade in Figure 8-3 rotates counterclockwise, so with this choice the angular displacement $\Delta\theta$ in Figure 8-3 is positive. By analogy to how we defined average velocity for straight-line motion in Section 2-2, we define the **average angular velocity** of the blade to be

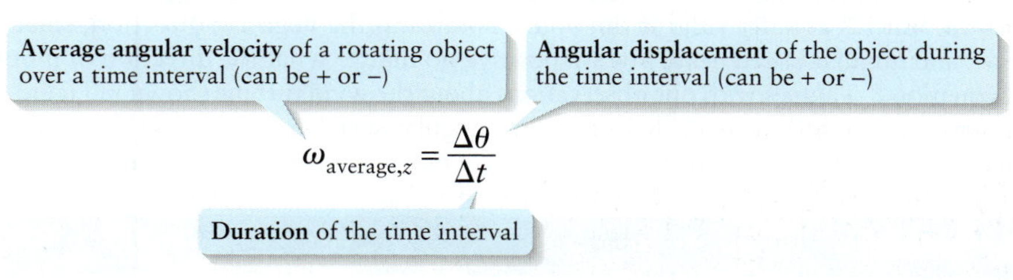

$$\omega_{\text{average},z} = \frac{\Delta\theta}{\Delta t}$$

Average angular velocity (8-1)

Here ω is the Greek letter omega, and the z in the subscript tells us that the blade is rotating around an axis that we call the z axis (Figure 8-3). For any rotating rigid object, the value of $\omega_{\text{average},z}$ is the same for all pieces of the object. The average angular velocity can be positive or negative, depending on the direction in which the object rotates. A common choice is to take counterclockwise rotation to be positive and clockwise rotation to be negative, but the choice is up to you.

Just as we did for ordinary velocity in Section 2-4, we'll define the *instantaneous* angular velocity ω_z, or just **angular velocity** for short, as the rate at which the object is rotating at a given instant. That is, the instantaneous angular velocity is the average angular velocity for a very short time interval.

The preferred units of angular velocity are radians per second, or rad/s. As we discussed in Section 3-8, radians are a measure of angle; there are 2π radians in a circle. We can also describe angle in terms of the number of revolutions (rev) that an object makes, so another common set of units for angular velocity is revolutions per second (rev/s). Since one revolution equals 2π radians, $1\text{ rev/s} = 2\pi\text{ rad/s}$.

We also use rad/s and rev/s as the units for **angular speed**, which is the magnitude or absolute value of angular velocity. An object that rotates 5.0 radians in 1 s has angular velocity $\omega_z = +5.0$ rad/s if it rotates in the positive direction and angular velocity $\omega_z = -5.0$ rad/s if it rotates in the negative direction, but in either case its angular speed is 5.0 rad/s. We'll use the symbol ω (with no subscript) for angular speed.

We now have the tools we need to find the *ordinary* speed in meters per second (m/s) of a point along the rotating blade. Figure 8-3 shows that a point (shown in red) a distance r from the rotation axis moves through a circular arc of radius r and angle $\Delta\theta$, which we've taken to be positive. We saw in Section 3-7 that if the angle $\Delta\theta$ is measured in radians and r is measured in meters, the length of such an arc in meters is $r\,\Delta\theta$ (see Figure 3-30). So, $r\,\Delta\theta$ is the distance that this point on the blade travels in a time Δt. The speed v of this point is the distance traveled divided by the time interval, or

$$v = \frac{r\,\Delta\theta}{\Delta t} \tag{8-2}$$

But $\Delta\theta/\Delta t$ is just the angular speed ω of the rotating blade. (Since the blade is rotating in the positive direction, its angular velocity $\omega_z = \Delta\theta/\Delta t$ is positive and so the angular speed ω—which is always positive—is the same as ω_z.) So we can rewrite Equation 8-2 for the speed of a point on the blade as

Speed of a point on a rotating rigid object (8-3)

Speed (in m/s) of a point on a rotating rigid object

Distance (in m) from the axis of rotation to the point

$$v = r\omega$$

Angular speed (in rad/s) of the rotating rigid object

WATCH OUT! Use correct units when relating speed and angular speed.

We derived Equation 8-3 from the statement that the length of the arc in Figure 8-3 is $r\,\Delta\theta$. This is true *only* if the angle $\Delta\theta$ is measured in radians (see Section 3-7). Hence you can safely use Equation 8-3 only if the angular speed ω is measured in radians per second (rad/s). If you are given the angular speed in other units, such as revolutions per second (rev/s), revolutions per minute (rev/min), or degrees per second, you must convert it to rad/s before you can use Equation 8-3.

We derived Equation 8-3 assuming that the object is rotating in the positive direction. But it's equally valid if the object rotates in the negative direction, since the speed v and angular speed ω are always positive no matter what the direction of motion.

Equation 8-3 agrees with our observations about the wind turbine shown in Figure 8-2. All points on each turbine blade have the same angular speed ω, but points that are farther from the axis are at a greater distance r and so are moving at a greater speed $v = r\omega$.

EXAMPLE 8-1 Speed versus Angular Speed

A turbine blade is rotating at 25.0 rev/min. How fast (in m/s) is a point on the blade moving that is (a) 0.500 m from the rotation axis and (b) 1.00 m from the rotation axis?

Set Up

For each point we are given the angular speed ω of the blade and the distance r from the rotation axis. We'll use Equation 8-3 to find the speed v of that piece.

Speed of a point on a rotating object:

$$v = r\omega \tag{8-3}$$

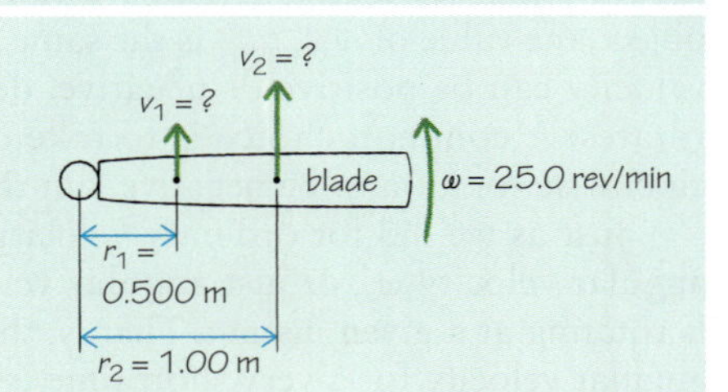

Solve

We are given the angular speed in revolutions per minute. We first need to convert this to radians per second so we can safely use Equation 8-3.

$\omega = 25.0$ rev/min

Convert revolutions to radians and minutes to seconds:

$$\omega = \left(25.0\,\frac{\text{rev}}{\text{min}}\right)\left(\frac{2\pi\text{ rad}}{1\text{ rev}}\right)\left(\frac{1\text{ min}}{60\text{ s}}\right)$$

$$= \frac{(25.0)(2\pi)}{60}\frac{\text{rad}}{\text{s}} = 2.62\text{ rad/s}$$

For each case substitute $\omega = 2.62$ rad/s and the value of the distance r of the point from the axis.

(a) $v_1 = r_1\omega = (0.500\text{ m})(2.62\text{ rad/s}) = 1.31\text{ m/s}$

(b) $v_2 = r_2\omega = (1.00\text{ m})(2.62\text{ rad/s}) = 2.62\text{ m/s}$

Reflect

Note that piece (b) is twice as far from the rotation axis as is piece (a) and has twice the speed. That's because in the same amount of time, piece (b) travels in a circle that has twice the radius and twice the circumference of the circle traveled by piece (a); so piece (b) must cover twice as much distance as piece (a).

Note that the units of radians disappeared when we calculated the speed of each piece. We did this because a radian isn't truly a unit but simply a way of counting the number of revolutions or fractions of a revolution through which an object has rotated.

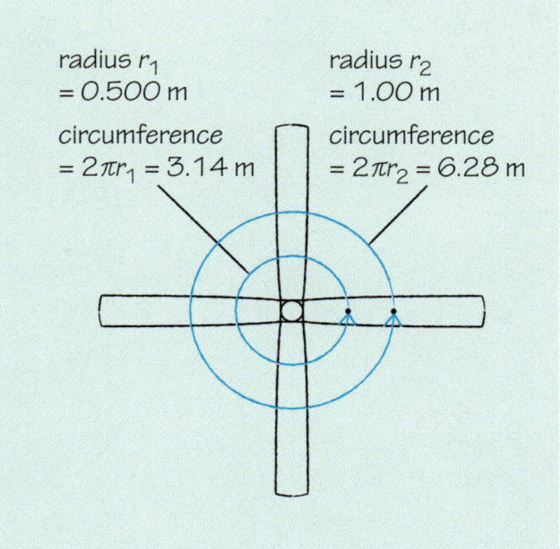
radius r_1
= 0.500 m
circumference
= $2\pi r_1$ = 3.14 m
radius r_2
= 1.00 m
circumference
= $2\pi r_2$ = 6.28 m

Moment of Inertia and Rotational Kinetic Energy

Let's see how to use Equation 8-3 to find the kinetic energy of a rotating object such as a wind turbine blade. We'll assume the object is rigid, so all parts of the object rotate at the same angular speed ω. The speed v is different at different points along the blade, however. To account for this let's imagine that the rigid blade is divided up into many small pieces (**Figure 8-4**). There are N such small pieces, where N is a large number. We label each piece using the subscript "i," where i can equal any integer from 1 to N: $i = 1, 2, 3, \ldots, N$. The ith piece has mass m_i and is a distance r_i from the rotation axis. The reason that we have to use so many pieces is that we want each piece to be very small so that there is only one value of the distance r_i for each piece.

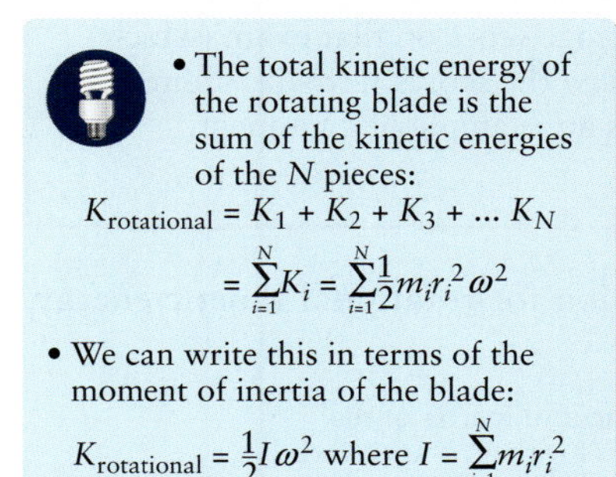
• The total kinetic energy of the rotating blade is the sum of the kinetic energies of the N pieces:
$K_{\text{rotational}} = K_1 + K_2 + K_3 + \ldots K_N$
$= \sum_{i=1}^{N} K_i = \sum_{i=1}^{N} \frac{1}{2} m_i r_i^2 \omega^2$
• We can write this in terms of the moment of inertia of the blade:
$K_{\text{rotational}} = \frac{1}{2} I\omega^2$ where $I = \sum_{i=1}^{N} m_i r_i^2$

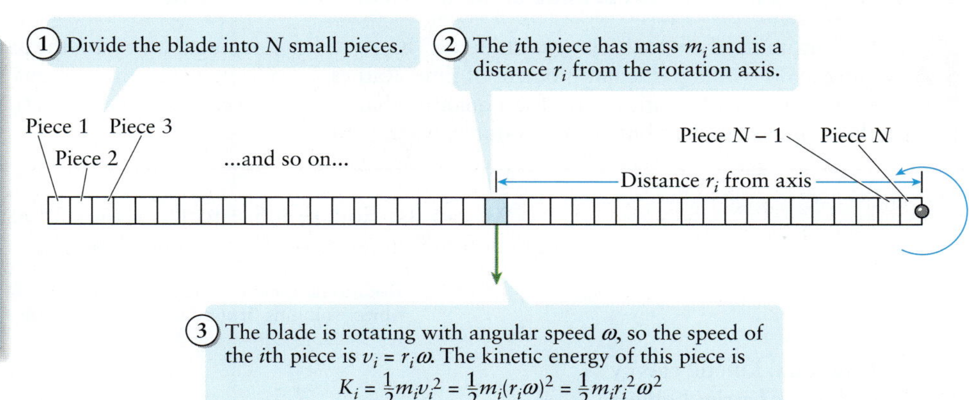
1 Divide the blade into N small pieces.
2 The ith piece has mass m_i and is a distance r_i from the rotation axis.
Piece 1
Piece 2
Piece 3
...and so on...
Piece $N - 1$
Piece N
Distance r_i from axis
3 The blade is rotating with angular speed ω, so the speed of the ith piece is $v_i = r_i\omega$. The kinetic energy of this piece is
$K_i = \frac{1}{2} m_i v_i^2 = \frac{1}{2} m_i (r_i\omega)^2 = \frac{1}{2} m_i r_i^2 \omega^2$

Figure 8-4 Calculating rotational kinetic energy and moment of inertia The kinetic energy of a rotating object is the sum of the kinetic energies of all of its component pieces.

From Equation 8-3 the speed of the ith small piece of the blade is $v_i = r_i\omega$. If we measure ω in radians per second and r_i in meters, v_i will be in meters per second (m/s). The kinetic energy of the ith piece of the blade is therefore

$$K_i = \frac{1}{2}m_i v_i^2 = \frac{1}{2}m_i(r_i\omega)^2 = \frac{1}{2}m_i r_i^2\omega^2$$

We can find the total kinetic energy of the entire rotating turbine blade by simply adding together the kinetic energies of each piece of the blade:

$$K_{\text{rotational}} = K_1 + K_2 + K_3 + \ldots + K_N = \sum_{i=1}^{N} K_i = \sum_{i=1}^{N}\frac{1}{2}m_i r_i^2\omega^2 \tag{8-4}$$

The subscript "rotational" in Equation 8-4 reminds us that this is the kinetic energy of the turbine blade due to its rotation. The quantities $\frac{1}{2}$ and ω^2 have the same values for each term in the sum, so we can factor them out of the sum. (If you're not sure whether it's correct to factor constant terms out of a sum, try it with a few numbers; for example, 2(3) + 2(4) + 2(5) = 2 (3 + 4 + 5).) The result is

$$K_{\text{rotational}} = \frac{1}{2}\left(\sum_{i=1}^{N} m_i r_i^2\right)\omega^2 \tag{8-5}$$

What is the quantity in parentheses in Equation 8-5, $\sum_{i=1}^{N} m_i r_i^2$? Although it involves a sum over all the little pieces into which we've divided the turbine blade, it is *not* simply the total mass M of the blade. That sum would be $M = \sum_{i=1}^{N} m_i$, without the factor of r_i^2. Instead the sum $\sum_{i=1}^{N} m_i r_i^2$ is a new quantity that tells us how the mass of the blade is *distributed*: It depends on both the mass of each small piece (m_i) and how far away from the rotation axis that piece is (r_i). This quantity is called the **moment of inertia** (sometimes called *rotational inertia*) of the turbine blade. We represent it by the symbol I:

To find the **moment of inertia** I of an object...

...we imagine dividing the object into **N small pieces,** and calculate the sum...

Moment of inertia of an object (8-6)

$$I = \sum_{i=1}^{N} m_i r_i^2 \tag{8-6}$$

...of the product of the **mass m_i of the ith piece...**

...and the square of the **distance r_i** from the ith piece to the rotation axis.

The SI units of moment of inertia are kilograms multiplied by meters squared (kg · m²).

WATCH OUT! Take a moment to understand the meaning of "moment."

! Despite its name, the term "*moment* of inertia" has nothing to do with a particular moment of time. Rather, "moment" is a mathematical term for a quantity that tells you how things are distributed. For example, when your class takes an exam, the average score on that exam (which tells you something about how the scores were distributed between zero and perfect) is an example of a moment.

We can use Equation 8-6 to write the expression for **rotational kinetic energy,** Equation 8-5, in terms of the moment of inertia:

Rotational kinetic energy of a rigid object spinning around an axis

Moment of inertia of the object for that rotation axis

Rotational kinetic energy of a rigid object (8-7)

$$K_{\text{rotational}} = \frac{1}{2}I\omega^2 \tag{8-7}$$

Angular speed of the object

We've derived this formula for a rotating turbine blade, but it applies to *any* rotating rigid object. The rotational kinetic energy of a rigid object is equal to one-half the object's moment of inertia multiplied by the square of the object's angular speed. Note that in Equation 8-7 the value of the moment of inertia I is for a specific rotation axis. As we'll see in the next section, the value of I for a given object can be different depending on the axis around which that object rotates.

WATCH OUT! Use correct units in calculating rotational kinetic energy.

In deriving Equation 8-7 we used Equation 8-3, $v = r\omega$, which is valid only if angles are measured in radians. To have the units of rotational kinetic energy work out properly with Equation 8-7, you *must* express angular speed ω in radians per second (rad/s). The following example illustrates the need to use correct units.

EXAMPLE 8-2 Turbine Kinetic Energy

The blades of a wind turbine have a combined moment of inertia of $2.00 \times 10^3\ \text{kg} \cdot \text{m}^2$ for rotation around the turbine axis. If the blades make 4.00 complete revolutions every minute, what is the rotational kinetic energy of the blades?

Set Up

We use the definition of rotational kinetic energy from Equation 8-7. We are given the value of the moment of inertia I, but we are not told the value of the angular speed ω. Instead we are told that the turbine makes 4.00 complete revolutions (4.00 rev for short) in 1 min. We can use this value to calculate the angular velocity from its definition (Equation 8-1). If we take the direction of rotation to be positive, the angular speed (which is the magnitude of angular velocity) is the same as $\omega_{\text{average},z}$.

Rotational kinetic energy:

$$K_{\text{rotational}} = \frac{1}{2} I\omega^2 \quad (8\text{-}7)$$

Angular velocity:

$$\omega_{\text{average},z} = \frac{\Delta\theta}{\Delta t} \quad (8\text{-}1)$$

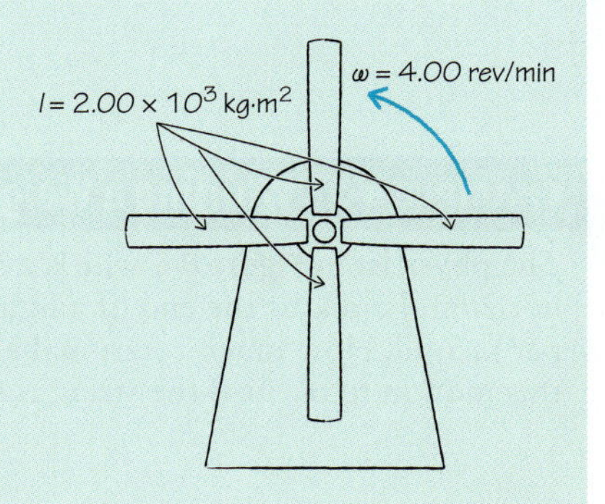

Solve

First calculate the angular velocity and angular speed. Be careful to convert revolutions per minute to radians per second.

$$\omega_{\text{average},z} = \frac{\Delta\theta}{\Delta t} = \left(\frac{4.00\ \text{rev}}{1\ \text{min}}\right)\left(\frac{2\pi\ \text{rad}}{1\ \text{rev}}\right)\left(\frac{1\ \text{min}}{60\ \text{s}}\right)$$

$$= \frac{(4.00)(2\pi)}{60}\frac{\text{rad}}{\text{s}} = 0.419\ \text{rad/s}$$

The angular speed is the magnitude of this, which is equal to $\omega_{\text{average},z}$ since its value is positive: $\omega = 0.419$ rad/s

Substitute this value of ω and the given value of I into the expression for rotational kinetic energy. We can drop the "rad" from the units of the final answer because radians aren't a true unit.

$$K_{\text{rotational}} = \frac{1}{2} I\omega^2 = \frac{1}{2}(2.00 \times 10^3\ \text{kg} \cdot \text{m}^2)(0.419\ \text{rad/s})^2$$

$$= 175\ \text{kg} \cdot \text{m}^2/\text{s}^2 = 175\ \text{J}$$

Reflect

This answer means that 175 J of work must be done on the turbine blades to bring them up to this angular speed. The blades are shaped so that the wind pushes on them and does the work for you. To make the blades spin this fast on a day with no wind, you would have to push on the blades yourself to do 175 J of work. Remember that work equals force multiplied by distance, so if you pushed the tip of one blade a distance of 1.00 m to start it moving, you would have to exert a force of 175 N or 39.4 lb. (You'd actually have to exert even more force than that to overcome friction in the bearings of the turbine.)

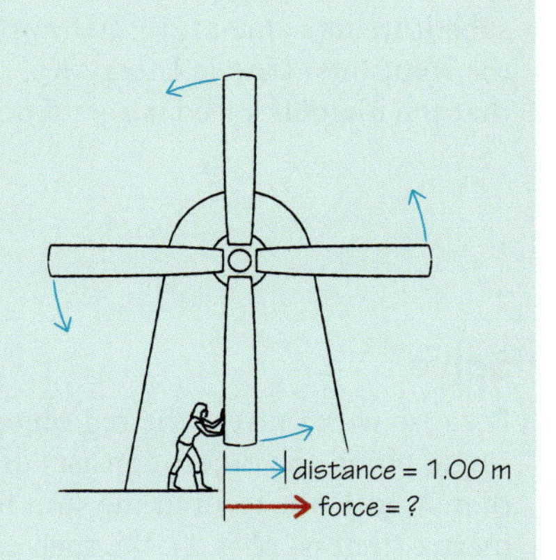

Rotational Kinetic Energy versus Translational Kinetic Energy

It's useful to compare the formula for *rotational* kinetic energy from Equation 8-7 with the formula for *translational* kinetic energy that we learned in Section 6-3:

Translational kinetic energy

$$K_{\text{translational}} = \frac{1}{2}mv^2$$

Rotational kinetic energy

$$K_{\text{rotational}} = \frac{1}{2}I\omega^2$$

Notice that mass m in the translational kinetic energy equation corresponds to moment of inertia I in the rotational kinetic energy equation. Mass is a property of matter that represents the resistance of an object to a change in its velocity $\vec{v}$; by analogy, an object's moment of inertia represents the resistance of the object to a change in its *angular* velocity ω_z.

Both translational and rotational kinetic energy depend on the mass of the moving object. However, the moment of inertia (Equation 8-6) and therefore the rotational kinetic energy both depend on how the mass is *distributed* with respect to the axis of rotation. A bit of mass far from the rotation axis has a larger effect on the value of the moment of inertia than the same amount of mass close to the axis. The following example illustrates this effect.

EXAMPLE 8-3 Whirling a Weight

The physicist in **Figure 8-5** whirls a small red object of mass 0.200 kg in a nearly horizontal circle at the end of a lightweight string. The object makes 5.00 revolutions per second. How much rotational kinetic energy must the physicist supply to cause this motion to occur if the string is (a) 0.300 m long or (b) 0.600 m long?

Figure 8-5 A weight moving in a circle What is the rotational kinetic energy of the small red weight moving in a horizontal circle?

David Tauck

Set Up

The statement that the string is "lightweight" means that we can ignore its mass and kinetic energy, so we need only calculate the kinetic energy of the red object. Figure 8-5 shows that the object is small compared to the length of the string, so it's reasonable to treat the object as though all its mass were concentrated at a single point. We'll use Equation 8-6 to find the moment of inertia I of the object in each case, then substitute its value into Equation 8-7 to calculate the rotational kinetic energy $K_{\text{rotational}}$. We're told that the angular speed is $\omega = 5.00$ rev/s.

Moment of inertia:

$$I = \sum_{i=1}^{N} m_i r_i^2 \qquad (8\text{-}6)$$

Rotational kinetic energy:

$$K_{\text{rotational}} = \frac{1}{2}I\omega^2 \qquad (8\text{-}7)$$

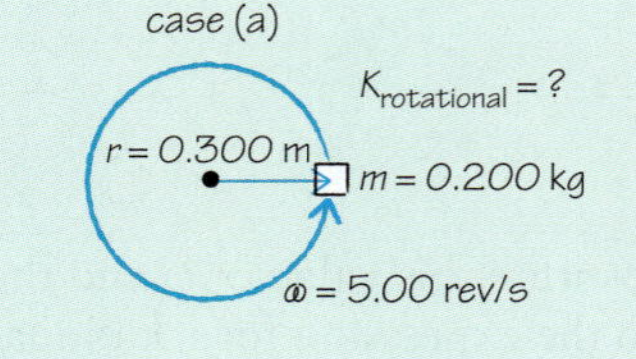

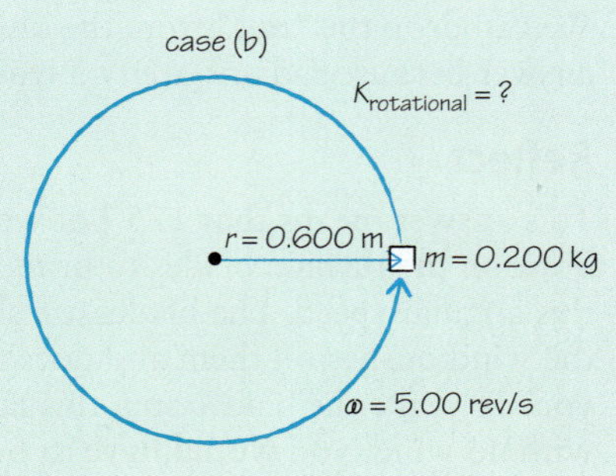

Solve

Because we can treat the red object as just a single point, the object consists of one piece. So there's only one term in the sum for moment of inertia (that is, $N = 1$). We then calculate the moment of inertia for each case.

$$I = \sum_{i=1}^{N} m_i r_i^2 = m_{\text{object}}\, r_{\text{object}}^2$$

(a) String 0.300 m long: $I = (0.200\text{ kg})(0.300\text{ m})^2 = 0.0180\text{ kg}\cdot\text{m}^2$

(b) String 0.600 m long: $I = (0.200\text{ kg})(0.600\text{ m})^2 = 0.0720\text{ kg}\cdot\text{m}^2$

We are given the angular speed in revolutions per second, so we must convert it to radians per second.

$$\omega = \left(5.00\ \frac{\text{rev}}{\text{s}}\right)\left(\frac{2\pi\ \text{rad}}{1\ \text{rev}}\right) = (5.00)(2\pi)\frac{\text{rad}}{\text{s}} = 31.4\ \text{rad/s}$$

Finally, insert the values of I and ω into the formula for rotational kinetic energy.

(a) String 0.300 m long:

$$K_{\text{rotational}} = \frac{1}{2}I\omega^2 = \frac{1}{2}(0.0180\ \text{kg}\cdot\text{m}^2)(31.4\ \text{rad/s})^2$$
$$= 8.88\ \text{kg}\cdot\text{m}^2/\text{s}^2 = 8.88\ \text{J}$$

(b) String 0.600 m long:

$$K_{\text{rotational}} = \frac{1}{2}I\omega^2 = \frac{1}{2}(0.0720\ \text{kg}\cdot\text{m}^2)(31.4\ \text{rad/s})^2$$
$$= 35.5\ \text{kg}\cdot\text{m}^2/\text{s}^2 = 35.5\ \text{J}$$

Reflect

The angular velocity ω is the same in both situations. However, the physicist must supply four times more kinetic energy to the rotating object—that is, he has to do four times as much work—when he uses the longer string: $4 \times 8.88\ \text{J} = 35.5\ \text{J}$. That's because doubling the length of the string makes the moment of inertia four times greater.

GOT THE CONCEPT? 8-1 Angular Speed of a DVD

? On a digital video disc (DVD), video and audio data are stored in a series of tiny pits that are evenly spaced along a long spiral that spans most of the surface of the disc and extends from the inner edge of the DVD to the outer edge. The scanning laser in a DVD player reads this information at a constant rate. As the player reads information recorded closer and closer to the outer edge of the DVD, should the disc (a) rotate faster, (b) rotate slower, or (c) rotate at the same speed?

TAKE-HOME MESSAGE FOR Section 8-2

✔ Angular velocity is a measure of how rapidly and in what direction an object rotates. Angular speed is the magnitude of angular velocity.

✔ The speed of a point on a rotating object depends on the point's distance from the rotation axis and the object's angular speed.

✔ The moment of inertia of an object depends on the object's mass and how that mass is distributed relative to the object's rotation axis.

✔ The rotational kinetic energy of an object spinning around an axis depends on its angular speed and its moment of inertia for that axis.

8-3 An object's moment of inertia depends on its mass distribution and the choice of rotation axis

We saw in Section 8-2 that the moment of inertia I of an object plays the same role for rotational motion that the object's mass m does for translational motion. The moment of inertia of an object depends on how its mass is *distributed* throughout the object. An example is the door of a bank vault (**Figure 8-6**). The door takes effort to move in part because it's so massive. But an additional factor is that a substantial fraction of the door's mass is in its locking mechanism, and this mechanism is located a good distance away from the hinges around which the door rotates. The greater the fraction of the door's mass that's located far from the hinges, the more difficult it is to start the door rotating if it's at rest or to stop it moving if it's already in motion.

Because the moment of inertia of an object is such an important property, we'll devote this section to some examples of calculating its value. We'll see how the moment of inertia depends on the way in which the object's mass is distributed and how the moment of inertia of a given object can change depending on the particular axis around which the object rotates.

Imagestate Media Partners Limited - Impact Photos/Alamy Stock Photo

Figure 8-6 Moving a bank vault door How easy or difficult it is to start this door moving depends on its moment of inertia for rotation around the hinges. The moment of inertia is determined by the mass of the door and how that mass is distributed.

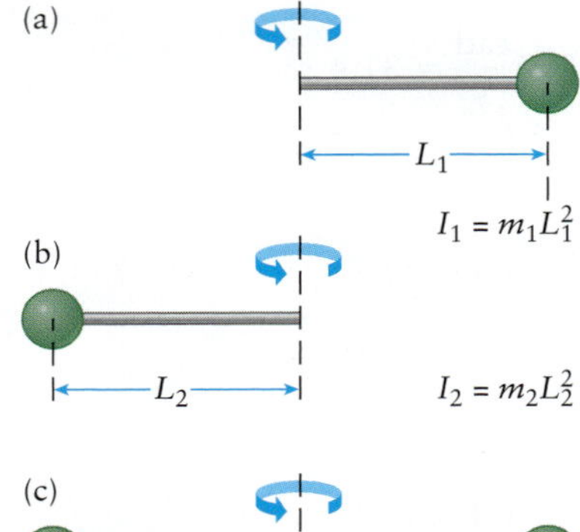

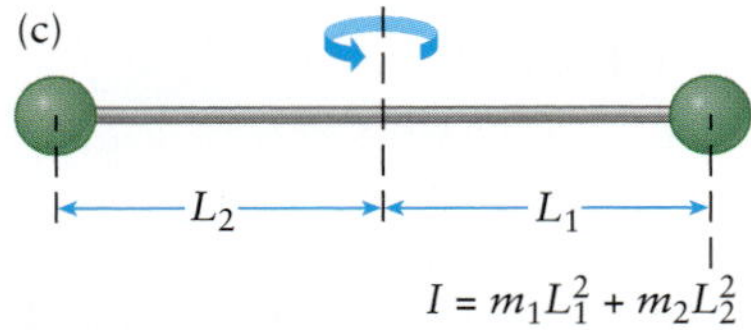

Figure 8-7 Calculating moment of inertia The sum of the moments of inertia of the objects in (a) and (b) equals the moment of inertia of the composite object in (c).

Go to Picture It 8-1 for more practice dealing with moment of inertia.

Moment of Inertia of a Collection of Small Pieces

From Section 8-2, if we imagine dividing an object into a large number of small objects of masses $m_1, m_2, m_3, \ldots m_N$, the moment of inertia I of the object is

$$I = \sum_{i=1}^{N} m_i r_i^2 \tag{8-6}$$

In this expression r_i is the distance from the axis around which the object rotates to the position of the ith small piece.

Equation 8-6 reveals three important properties of the moment of inertia:

(1) *The moment of inertia is additive.* Each term $m_i r_i^2$ in Equation 8-6 represents the moment of inertia of the ith piece of the object, and the total moment of inertia is the sum of the terms for all pieces. It follows that if you know the moments of inertia of two objects around some rotation axis and you attach them to form a single object, the moment of inertia of the new object—around the same axis—is the sum of the moments of inertia of the two separate objects (**Figure 8-7**).

(2) *The farther away from the rotation axis an object's mass lies, the greater the object's moment of inertia.* In Equation 8-6, the farther the small pieces of the object are from the rotation axis, the greater the values of r_i and the greater the moment of inertia. Example 8-4 demonstrates this idea.

(3) *The moment of inertia depends on the rotation axis.* If a different rotation axis is used, the quantities r_i in Equation 8-6 change and so the moment of inertia changes as well. In general, the moment of inertia for a given object is different for each different rotation axis. We'll see how this works in Example 8-5.

Let's illustrate these statements by looking at objects that we can describe as being composed of just a few pieces.

EXAMPLE 8-4 Moment of Inertia for Dumbbells I

A dumbbell is made up of two small, massive spheres connected by a lightweight rod. For a dumbbell made with two identical 50.0-kg spheres, find the moment of inertia for rotation around the midpoint of the dumbbell if the spheres are separated by (a) 1.20 m or (b) 2.40 m.

Set Up

The problem describes the rod connecting the spheres as "lightweight," so we can ignore the mass of the rod. The two spheres are also described as "small," so we can treat each sphere as an infinitesimally small point object. Because the dumbbell is made up of only two small pieces, the sum in Equation 8-6 has only two terms, one for each sphere. That is, the moment of inertia of the dumbbell is the sum of the moments of inertia of the two spheres.

Moment of inertia:

$$I = \sum_{i=1}^{N} m_i r_i^2 = m_1 r_1^2 + m_2 r_2^2 \tag{8-6}$$

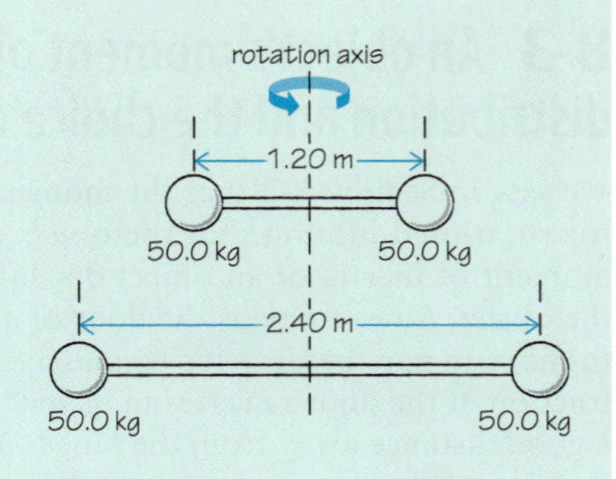

Solve

Each sphere has mass 50.0 kg, so $m_1 = m_2 =$ 50.0 kg in each case. For the first dumbbell each sphere is one-half of 1.20 m from the rotation axis, so $r_1 = r_2 = 0.600$ m; similarly, in the second case each sphere is one-half of 2.40 m from the rotation axis, so $r_1 = r_2 = 1.20$ m. Substitute these values into the formula for moment of inertia I.

For both dumbbells,

$$I = m_1 r_1^2 + m_2 r_2^2 = (50.0\ \text{kg})\, r_1^2 + (50.0\ \text{kg})\, r_2^2$$

(a) For the first dumbbell:

$$I = (50.0\ \text{kg})(0.600\ \text{m})^2 + (50.0\ \text{kg})(0.600\ \text{m})^2 = 36.0\ \text{kg} \cdot \text{m}^2$$

(b) For the second dumbbell:

$$I = (50.0\ \text{kg})(1.20\ \text{m})^2 + (50.0\ \text{kg})(1.20\ \text{m})^2 = 144\ \text{kg} \cdot \text{m}^2$$

Reflect

Although both dumbbells have the same total mass $m_1 + m_2 = 50.0\ \text{kg} + 50.0\ \text{kg} = 100.0\ \text{kg}$, the mass is distributed in different ways. Hence, they have very different values of the moment of inertia. The farther the massive spheres are from the rotation axis, the greater the value of the moment of inertia.

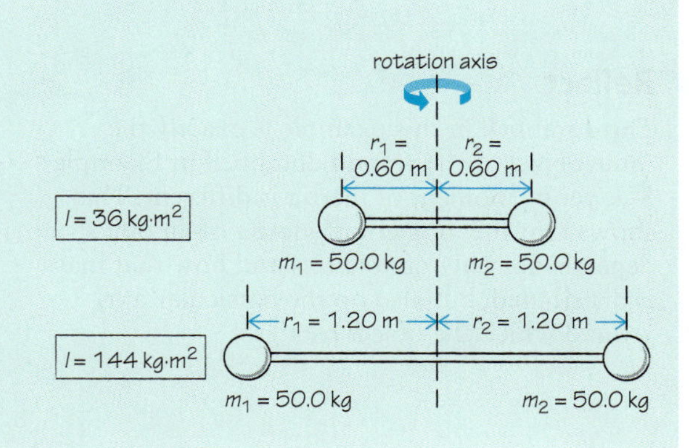

EXAMPLE 8-5 Moment of Inertia for Dumbbells II

Consider again the second dumbbell from Example 8-4, which has two small 50.0-kg spheres separated by 2.40 m. (a) Calculate the moment of inertia of this same dumbbell if it rotates around the center of one of the spheres. (b) Determine the kinetic energy of this dumbbell if it rotates at 1.00 rad/s around its midpoint and if it rotates at 1.00 rad/s around the center of one of its spheres.

Set Up

As in Example 8-4 the moment of inertia I of the dumbbell is the sum of two terms, one for each sphere. We expect a different answer for I in part (a) because the values of r_1 and r_2 are different from those in Example 8-4. Once we know the value of I for each choice of axis, we can calculate the rotational kinetic energy using Equation 8-7.

Moment of inertia:

$$I = \sum_{i=1}^{N} m_i r_i^2 = m_1 r_1^2 + m_2 r_2^2 \qquad (8\text{-}6)$$

Rotational kinetic energy:

$$K_{\text{rotational}} = \frac{1}{2} I \omega^2 \qquad (8\text{-}7)$$

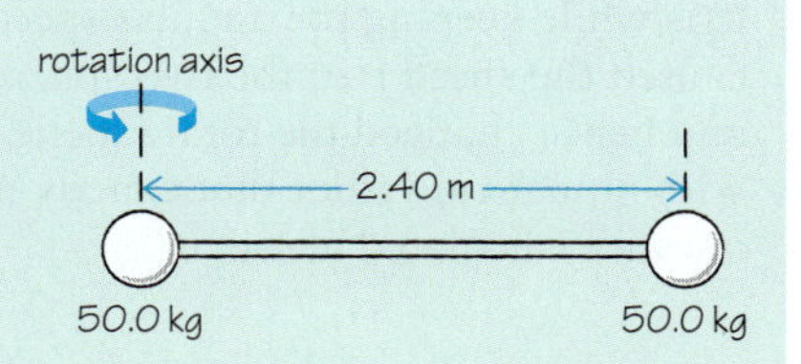

Solve

(a) Again each sphere has mass 50.0 kg, so $m_1 = m_2 = 50.0$ kg in each case. Now one sphere is *on* the rotation axis, so $r_1 = 0$; the other sphere is 2.40 m from the rotation axis, so $r_2 = 2.40$ m. Substitute these values into the formula for I.

$$I = m_1 r_1^2 + m_2 r_2^2 = (50.0\ \text{kg})(0)^2 + (50.0\ \text{kg})(2.40\ \text{m})^2 = 288\ \text{kg} \cdot \text{m}^2$$

(b) In Example 8-4 we found that the moment of inertia of this dumbbell for rotation around its midpoint was $I = 144\ \text{kg} \cdot \text{m}^2$. Use this in

For rotation around the midpoint of the dumbbell,

$$I = 144\ \text{kg} \cdot \text{m}^2$$

Equation 8-7 to find the rotational kinetic energy when the dumbbell rotates around this axis at 1.00 rad/s.

Calculate the rotational kinetic energy for $\omega = 1.00$ rad/s:

$$K_{\text{rotational}} = \frac{1}{2}I\omega^2 = \frac{1}{2}(144 \text{ kg} \cdot \text{m}^2)(1.00 \text{ rad/s})^2 = 72.0 \text{ J}$$

Repeat the calculation for rotation around one of the spheres. We found above that for this axis, $I = 288 \text{ kg} \cdot \text{m}^2$.

For rotation around one of the spheres of the dumbbell,

$I = 288 \text{ kg} \cdot \text{m}^2$

Calculate the rotational kinetic energy for $\omega = 1.00$ rad/s:

$$K_{\text{rotational}} = \frac{1}{2}I\omega^2 = \frac{1}{2}(288 \text{ kg} \cdot \text{m}^2)(1.00 \text{ rad/s})^2 = 144 \text{ J}$$

Reflect

The dumbbell in this example is exactly the same object as the second dumbbell in Example 8-4, yet its moment of inertia is different. This shows that the moment of inertia of an object depends not only on its mass and how that mass is distributed, but also on the particular axis around which the object rotates.

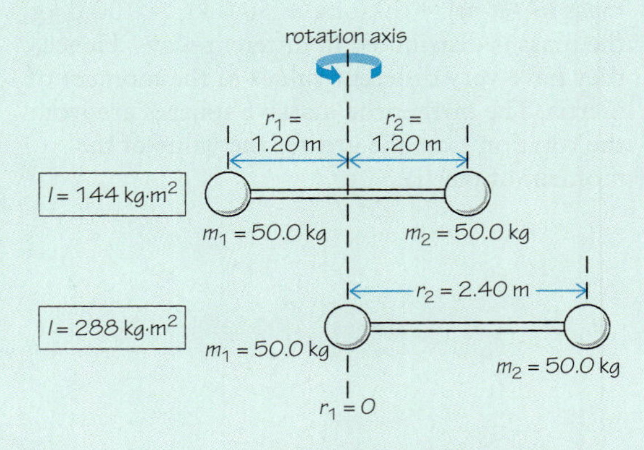

We can check our results for rotational energy in the two cases by finding the *translational* kinetic energies of the individual spheres and adding them together. (Remember from Example 8-4 that we can treat each sphere as a point object, so we don't have to worry about the kinetic energy of a sphere spinning on its axis.) Our calculations show that changing the rotation axis while keeping the angular speed ω the same caused the speed v of the two spheres to change and hence changed the total kinetic energy—with numerical values that exactly match the ones we calculated above.

For rotation around the midpoint of the dumbbell at angular speed $\omega = 1.00$ m/s: Each sphere is a distance $r = 1.20$ m from the rotation axis, so the speed of each sphere is

$$v = r\omega = (1.20 \text{ m})(1.00 \text{ rad/s}) = 1.20 \text{ m/s}$$

Each 50.0-kg sphere has translational kinetic energy

$$K_{\text{sphere}} = \frac{1}{2}mv^2 = \frac{1}{2}(50.0 \text{ kg})(1.20 \text{ m/s})^2 = 36.0 \text{ J}$$

So the total kinetic energy of the system of two spheres rotating around its midpoint is 36.0 J + 36.0 J = 72.0 J.

For rotation around one of the spheres at angular speed $\omega = 1.00$ rad/s: One sphere is at the rotation axis and is essentially at rest, so it has zero kinetic energy. The other sphere is a distance $r = 2.40$ m from the rotation axis, so the speed of this sphere is

$$v = r\omega = (2.40 \text{ m})(1.00 \text{ rad/s}) = 2.40 \text{ m/s}$$

The translational kinetic energy of this 50.0-kg sphere is

$$K_{\text{sphere}} = \frac{1}{2}mv^2 = \frac{1}{2}(50.0 \text{ kg})(2.40 \text{ m/s})^2 = 144 \text{ J}$$

So the total kinetic energy of the system of two spheres rotating around one of the spheres is 0 J + 144 J = 144 J.

WATCH OUT! An object has more than one moment of inertia.

If your friend points to an object and asks, "What is the moment of inertia of that object?" it could be a trick question! Example 8-5 shows that objects do *not* have a single moment of inertia; rather, the value of the moment of inertia depends on the specific rotation axis. Any object, even one like a DVD that commonly rotates around one particular axis (**Figure 8-8a**), can be made to rotate around any number of axes (**Figure 8-8b**). The axis does not even have to pass through the object—for example, imagine tying a string to the edge of a DVD and swinging it around in a circle (**Figure 8-8c**). The DVD would be rotating around an axis that lies completely outside the DVD itself. The moment of inertia of the DVD will be different for each of the axes shown in Figure 8-8. *Always* make sure that you identify the axis of rotation before determining the moment of inertia of an object!

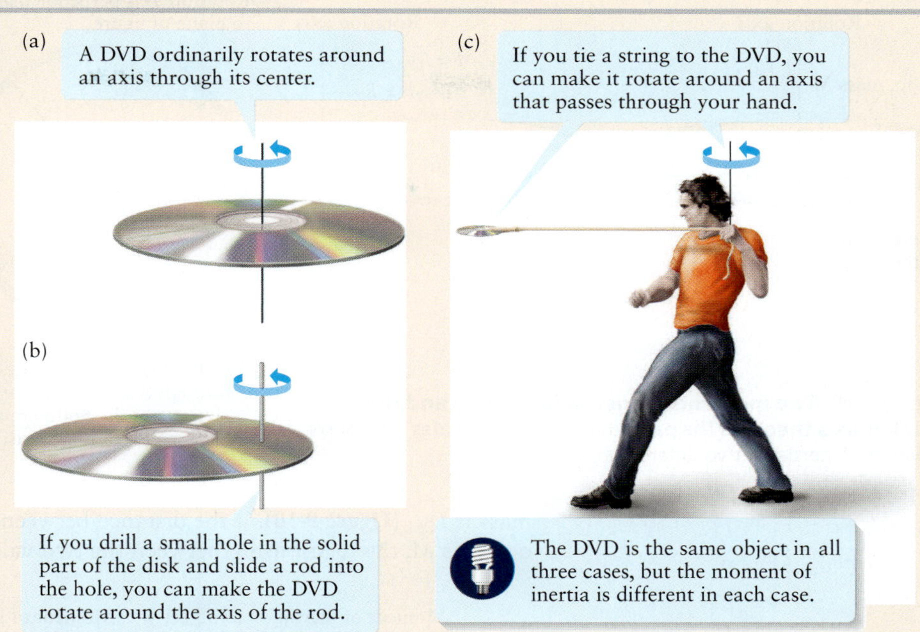

(a)
A DVD ordinarily rotates around an axis through its center.
(b)
If you drill a small hole in the solid part of the disk and slide a rod into the hole, you can make the DVD rotate around the axis of the rod.
(c)
If you tie a string to the DVD, you can make it rotate around an axis that passes through your hand.
The DVD is the same object in all three cases, but the moment of inertia is different in each case.

Figure 8-8 Three rotation axes for the same object Changing the rotation axis of an object changes the value of its moment of inertia.

The Parallel-Axis Theorem

Finding the moment of inertia of an object can be challenging for certain choices of rotation axis. An example is a gymnast's hoop of mass M and radius R. The gymnast rotates the hoop around an axis that is perpendicular to the plane of the hoop and is near the hoop's rim (**Figure 8-9a**). If we imagine dividing the hoop into a large number of small pieces labeled $i = 1, 2, 3, \ldots, N$, every piece is a different distance r_i from the rotation axis, so it's very difficult to calculate the moment of inertia using Equation 8-6.

However, a remarkable relationship exists between the moment of inertia of an object for an axis through its *center of mass*, and the moment of inertia when the object rotates around any other parallel axis. (We introduced the idea of center of mass in Section 7-7. This would be a good time to review that section.) This relationship, called the **parallel-axis theorem**, is useful because it's often relatively easy to determine an object's moment of inertia around its center of mass.

Here's the statement of the parallel-axis theorem: Suppose the moment of inertia of an object for a certain rotation axis is I. The moment of inertia of the same object for a second axis that's parallel to the first one but passes through the object's center of

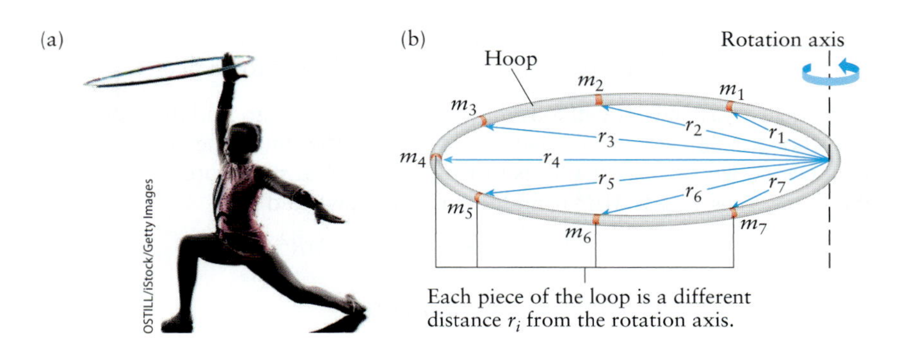

Figure 8-9 A gymnast's hoop (a) A hoop rotating around an axis that passes through its rim. (b) Calculating the hoop's moment of inertia for this axis.

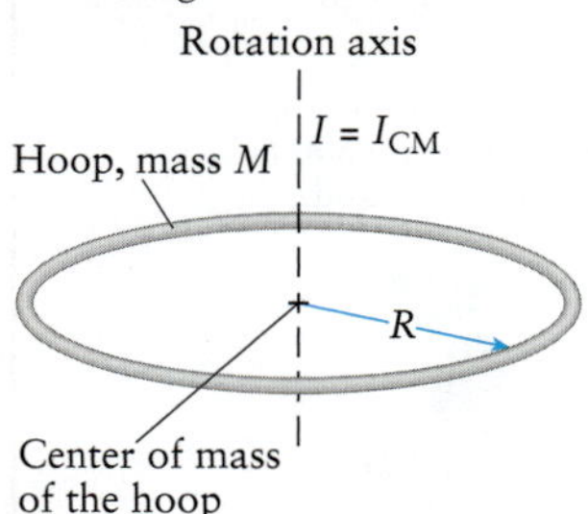
(a) Hoop rotating around an axis through its center
Rotation axis
$I = I_{CM}$
Hoop, mass M
R
Center of mass of the hoop

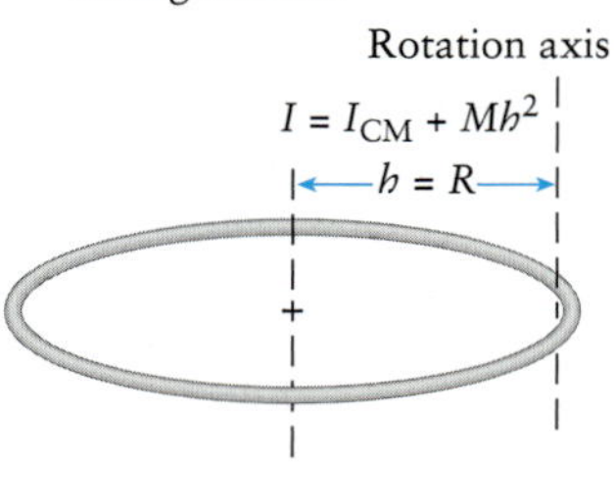
(b) Hoop rotating around an axis through its rim
Rotation axis
$I = I_{CM} + Mh^2$
$h = R$

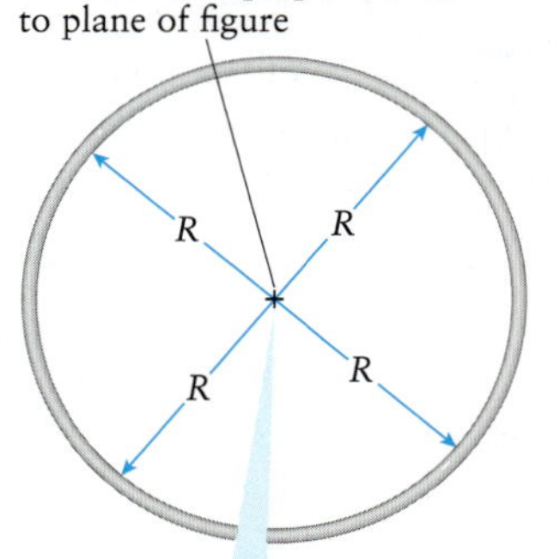
(c) Top view of the hoop
Rotation axis perpendicular to plane of figure
R
R
R
R
If the rotation axis passes through the center of the hoop, all parts of the hoop are a distance R from the axis.

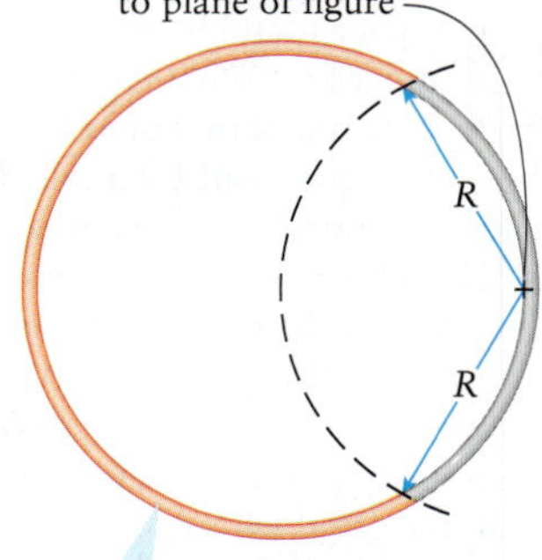
Rotation axis perpendicular to plane of figure
R
R
With this choice of axis, most of the hoop is farther than R from the axis.

Figure 8-10 Two moments of inertia for a hoop and the parallel-axis theorem The parallel-axis theorem relates the hoop's moment of inertia for two different axes.

mass is I_{CM} (**Figure 8-10**). If the distance between the two axes is h and the mass of the object is M, the relationship between the two values of moment of inertia is

The parallel-axis theorem (8-8)

$$I = I_{CM} + Mh^2$$

Moment of inertia of an object for a **certain rotation axis** (I)

Moment of inertia of the same object for a second, **parallel axis through its center of mass** (I_{CM})

Mass of the object (M)

Distance between the two parallel axes (h)

To see the parallel-axis theorem in action, let's return to the gymnast's hoop of mass M and radius R shown in **Figure 8-9b**. The theorem tells us that to determine the hoop's moment of inertia for an axis at its rim, we should first calculate its moment of inertia for a parallel axis that passes through its center of mass (**Figure 8-10a**). The center of mass of a symmetrical object such as a hoop is at the geometrical center of the object. (Recall from Section 7-7 that there doesn't actually have to be any mass *at* an object's center of mass.) If we divide the hoop into many small pieces, each of those pieces is the *same* distance R from this rotation axis. So in Equation 8-6 for the moment of inertia we can replace r_i for each value of i by R. This equation then becomes

$$I_{CM} = \sum_{i=1}^{N} m_i r_i^2 = \sum_{i=1}^{N} m_i R^2 = \left(\sum_{i=1}^{N} m_i\right) R^2 \quad (8\text{-}9)$$

The quantity $\sum_{i=1}^{N} m_i$ in Equation 8-9 is just the sum of the masses of all of the individual pieces that make up the hoop. This sum is just the mass M of the hoop as a whole, so we can rewrite Equation 8-9 as

$$I_{CM} = MR^2 \quad (8\text{-}10)$$

(moment of inertia of a hoop of mass M and radius R, axis through its center of mass and perpendicular to its plane)

Now we can use the parallel-axis theorem, Equation 8-8, to find the moment of inertia of this hoop for a parallel axis that passes through the rim of the hoop (**Figure 8-10b**). The distance between the two axes is just the radius R of the hoop, so we let $h = R$ in Equation 8-8. Using $I_{CM} = MR^2$ from Equation 8-10, we find

$$I = I_{CM} + MR^2 = MR^2 + MR^2$$

or

$$I = 2MR^2 \quad (8\text{-}11)$$

(moment of inertia of a hoop of mass M and radius R, axis through its rim and perpendicular to its plane)

This result would have been extremely difficult to obtain without using the parallel-axis theorem.

Why is the moment of inertia of the hoop of radius R for an axis through its rim (Equation 8-11) greater than that for an axis through its center (Equation 8-10)? The explanation is that with the rotation axis at the rim, the average distance from the axis to the pieces that make up the hoop is greater than R (**Figure 8-10c**). (A portion of the hoop is closer than R, but a larger portion is farther away than R.) As we have seen before the farther an object's mass lies from the rotation axis, the greater its moment of inertia.

EXAMPLE 8-6 Using the Parallel-Axis Theorem I

Use the parallel-axis theorem and the results of Example 8-4 to find the moment of inertia of the dumbbell in Example 8-5 rotated about one end of the dumbbell.

Set Up

The two spheres at the ends of the dumbbell have the same mass, so the center of mass of the dumbbell is at its midpoint. In Example 8-4 we found the moment of inertia of this dumbbell for an axis that passes through the midpoint, so this expression is I_{CM}. The dumbbell in Example 8-5 is identical but has a different, parallel rotation axis, so we can find its moment of inertia using the parallel-axis theorem, Equation 8-8.

Parallel-axis theorem:

$$I = I_{CM} + Mh^2 \qquad (8\text{-}8)$$

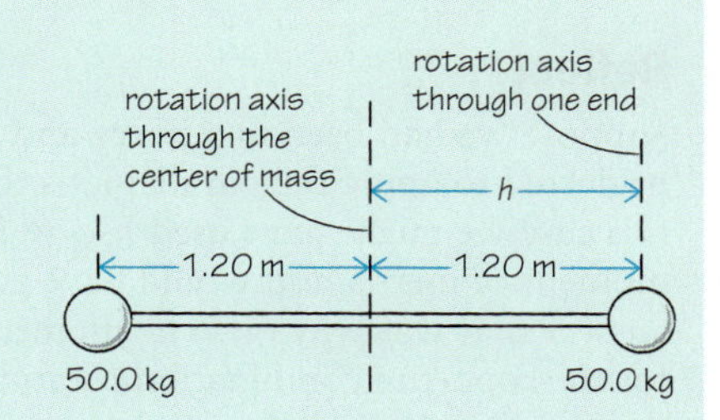

Solve

From Example 8-4, the moment of inertia of the dumbbell through its center of mass is $I_{CM} = 144\ \text{kg}\cdot\text{m}^2$. The axis of rotation for the dumbbell in Example 8-5 is a distance $h = 1.20$ m from the center of mass, and the total mass M of the dumbbell is the sum of the masses of the two 50.0-kg spheres. Substitute these values into the parallel-axis theorem.

$$I_{CM} = 144\ \text{kg}\cdot\text{m}^2$$

$$h = 1.20\ \text{m}$$

$$M = 50.0\ \text{kg} + 50.0\ \text{kg} = 100.0\ \text{kg}$$

$$I = I_{CM} + Mh^2 = 144\ \text{kg}\cdot\text{m}^2 + (100.0\ \text{kg})(1.20\ \text{m})^2$$

$$= 288\ \text{kg}\cdot\text{m}^2$$

Reflect

We get the same answer as in Example 8-5, which is a nice check that the parallel-axis theorem works.

EXAMPLE 8-7 Using the Parallel-Axis Theorem II

A child's swing at a rustic amusement park is made up of an old tire hanging from a lightweight rope tied to a tree limb. The tire has a radius of 0.310 m and a mass of 11.0 kg, and hangs from a rope 2.50 m long. If the rope–tire combination is swinging around the tree limb at 1.20 rad/s at the low point of its motion, what is its kinetic energy?

Set Up

To find the kinetic energy due to the rotation of the rope–tire combination, we need to know its moment of inertia. The rope is lightweight, so we can ignore its mass. The tire is like a hoop that rotates around an axis at the position of the tree limb. We know the moment of inertia of a hoop for an axis through its center of mass (Equation 8-10), so we can find the tire's moment of inertia around the tree limb using the parallel-axis theorem, Equation 8-8. Then we can find the rotational kinetic energy using Equation 8-7.

Moment of inertia of a hoop of mass M and radius R, axis through its center of mass and perpendicular to its plane:

$$I_{CM} = MR^2 \qquad (8\text{-}10)$$

Parallel-axis theorem:

$$I = I_{CM} + Mh^2 \qquad (8\text{-}8)$$

Rotational kinetic energy:

$$K_{rotational} = \frac{1}{2}I\omega^2 \qquad (8\text{-}7)$$

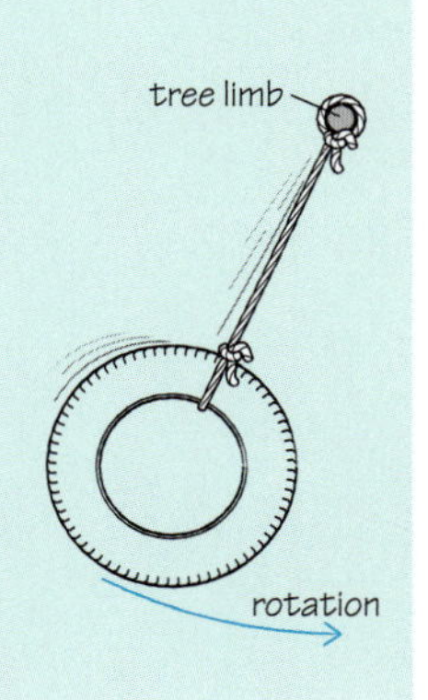

Solve

The tire has mass $M = 11.0$ kg and radius $R = 0.310$ m. The distance from the tree-limb axis to a parallel axis that passes through the center of the hoop (its center of mass) is $h = 0.310\text{ m} + 2.50\text{ m} = 2.81$ m. Substitute these values into the formula for the tire's moment of inertia I for rotation around the tree limb.

$$\begin{aligned} I &= I_{CM} + Mh^2 \\ &= MR^2 + Mh^2 \\ &= (11.0\text{ kg})(0.310\text{ m})^2 + (11.0\text{ kg})(2.81\text{ m})^2 \\ &= 87.9\text{ kg}\cdot\text{m}^2 \end{aligned}$$

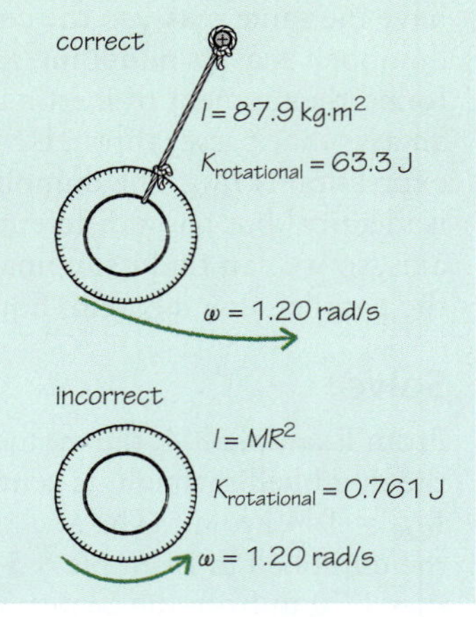

Then substitute the moment of inertia and the angular speed $\omega = 1.20$ rad/s into the expression for rotational kinetic energy.

$$\begin{aligned} K_{\text{rotational}} &= \frac{1}{2}I\omega^2 \\ &= \frac{1}{2}(87.9\text{ kg}\cdot\text{m}^2)(1.20\text{ rad/s})^2 \\ &= 63.3\text{ J} \end{aligned}$$

Reflect

Suppose we had been in a hurry and had neglected to apply the parallel-axis theorem. In this case we might have used $I_{CM} = MR^2$ for the moment of inertia and would have gotten an answer that was way off. The moment of inertia for the rope–tire combination rotating around the tree limb is *much* larger than that for the tire rotating around its center of mass. Hence, for a given angular speed, the kinetic energy for the rope–tire combination is much greater.

Incorrect calculation:

$$\begin{aligned} K_{\text{rotational}} &= \frac{1}{2}I_{CM}\,\omega^2 = \frac{1}{2}MR^2\omega^2 \\ &= \frac{1}{2}(11.0\text{ kg})(0.310\text{ m})^2(1.20\text{ rad/s})^2 \\ &= 0.761\text{ J (wrong!)} \end{aligned}$$

correct
$I = 87.9$ kg·m^2
$K_{\text{rotational}} = 63.3$ J
$\omega = 1.20$ rad/s
incorrect
$I = MR^2$
$K_{\text{rotational}} = 0.761$ J
$\omega = 1.20$ rad/s

The Moment of Inertia for Common Shapes

It was straightforward to calculate the moment of inertia for the hoop shown in Figure 8-10 because all parts of the hoop are the same distance from its rotation axis. For solid objects with other shapes, calculating the moment of inertia is more difficult and requires the use of calculus, which is beyond our scope. Instead, we'll just present the results for some common shapes that show up in various situations involving rotation (**Table 8-1**). We assume that these objects are *uniform*; that is, each object has the same composition throughout its volume.

You can see from Table 8-1 that the moment of inertia of a thin cylindrical shell of mass M and radius R around its central axis ($I = MR^2$) is twice that of a solid cylinder of the same mass and radius ($I = \frac{1}{2}MR^2$). You should expect the cylindrical shell to have a larger moment of inertia because the moment of inertia of an object is strongly influenced by how far the mass is from the rotation axis. In the case of the thin cylindrical shell, all of the mass is located a distance R from the axis, while for the solid cylinder only a fraction of the mass is that far from the axis. Hence the moment of inertia of the shell must be larger than the moment of inertia of the solid cylinder.

We can use the parallel-axis theorem to verify some of the results shown in Table 8-1. For example, consider a thin rod that has mass M and length L. The center of mass of this rod is at its geometric center. From Table 8-1, the moment of inertia of such a rod rotating around an axis perpendicular to the rod and through its center is

$$I_{CM} = \frac{ML^2}{12}$$

(moment of inertia of a thin uniform rod, rotation axis through its center)

What is the moment of inertia of this same rod for an axis through one end and perpendicular to the rod? This axis is parallel to an axis through the center of mass

TABLE 8-1 Moments of Inertia of Uniform Objects of Various Shapes

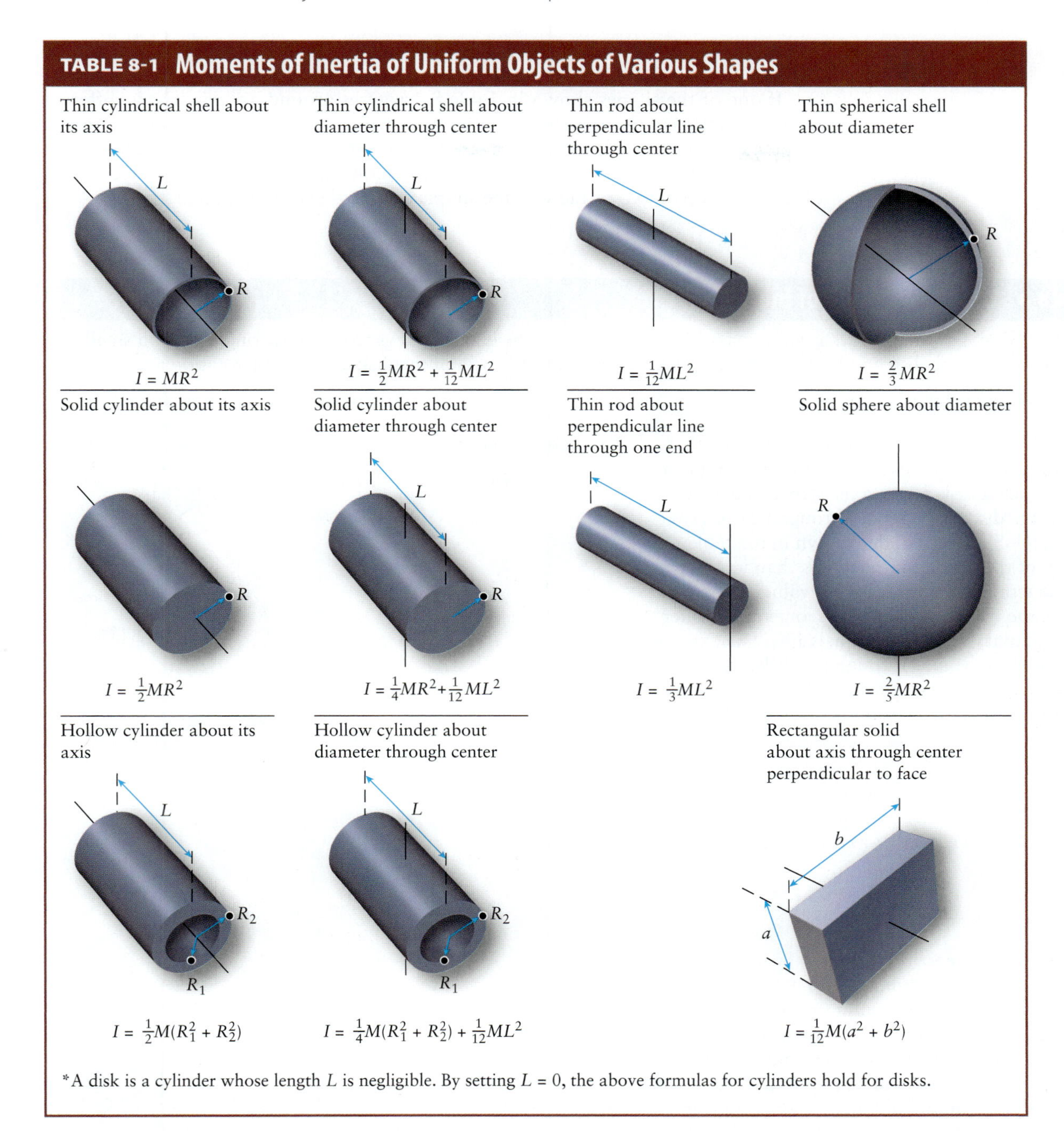

Thin cylindrical shell about its axis
L
R
$I = MR^2$
Thin cylindrical shell about diameter through center
L
R
$I = \frac{1}{2}MR^2 + \frac{1}{12}ML^2$
Thin rod about perpendicular line through center
L
$I = \frac{1}{12}ML^2$
Thin spherical shell about diameter
R
$I = \frac{2}{3}MR^2$
Solid cylinder about its axis
R
$I = \frac{1}{2}MR^2$
Solid cylinder about diameter through center
L
R
$I = \frac{1}{4}MR^2 + \frac{1}{12}ML^2$
Thin rod about perpendicular line through one end
L
$I = \frac{1}{3}ML^2$
Solid sphere about diameter
R
$I = \frac{2}{5}MR^2$
Hollow cylinder about its axis
L
R_2
R_1
$I = \frac{1}{2}M(R_1^2 + R_2^2)$
Hollow cylinder about diameter through center
L
R_2
R_1
$I = \frac{1}{4}M(R_1^2 + R_2^2) + \frac{1}{12}ML^2$
Rectangular solid about axis through center perpendicular to face
b
a
$I = \frac{1}{12}M(a^2 + b^2)$

*A disk is a cylinder whose length L is negligible. By setting $L = 0$, the above formulas for cylinders hold for disks.

of the rod and a distance $h = L/2$ from that axis. From Equation 8-8, the moment of inertia for an axis through the end is

$$I = I_{CM} + Mh^2 = \frac{ML^2}{12} + M\left(\frac{L}{2}\right)^2$$
$$= \left(\frac{1}{12} + \frac{1}{4}\right)ML^2 = \left(\frac{1}{12} + \frac{3}{12}\right)ML^2 = \frac{4ML^2}{12}$$

or, simplifying,

$$I = \frac{ML^2}{3}$$

(moment of inertia of a thin uniform rod, axis through one end, perpendicular to rod)

This expression is just the result shown in Table 8-1 for the rod rotating around its end.

If one of the shapes shown in Table 8-1 rotates around an axis that is different than but parallel to the axis shown in the table, we can use the parallel-axis theorem to find the new moment of inertia. Because the moment of inertia is additive, you can also use the results shown in Table 8-1 to find the moment of inertia of a more complex object made up of two or more of the simple objects shown in the table.

EXAMPLE 8-8 Earring Moment of Inertia

An earring is a thin, uniform disk that has a mass M and a radius R. The earring hangs from the earring post by a small hole near the edge of the disk and is free to rotate. Find the moment of inertia of the disk around this rotation axis.

Set Up

A solid disk is an example of a solid cylinder like that shown in Table 8-1. This table shows that the moment of inertia of such a disk for an axis perpendicular to the plane of the disk and passing through its center is $I = MR^2/2$. (Unlike the cylinder shown in the table, the length L of the earring is much less than its radius R. But the length has no effect on the value of I.)

Because the disk is uniform, its geometrical center is its center of mass, so $MR^2/2$ equals I_{CM}. The axis we want is parallel to the axis through the disk's center, so we can use the parallel-axis theorem, Equation 8-8, to find the moment of inertia around the axis at the rim.

Parallel-axis theorem:

$$I = I_{CM} + Mh^2 \quad (8\text{-}8)$$

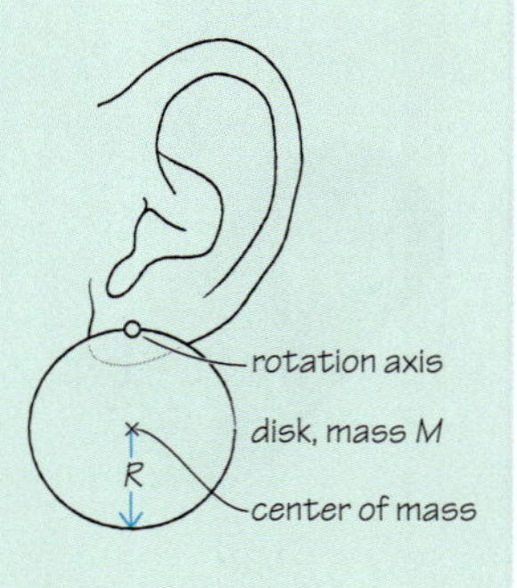

Solve

The edge of the disk is a distance $h = R$ from the center of mass of the disk. Substitute this value and $I_{CM} = MR^2/2$ into the parallel-axis theorem.

$$I = \frac{MR^2}{2} + MR^2$$

or

$$I = \frac{3MR^2}{2}$$

Reflect

If the wearer of these earrings nods her head up and down while dancing at a club, she can make the disks rotate back and forth around the posts in an eye-catching way. The more massive the disks and the larger their radius, the greater their moment of inertia and the more effort will be required to get them moving (recall that rotational kinetic energy is given by $K_{rotational} = \frac{1}{2}I\omega^2$). Making the earrings small and lightweight will require less effort from the wearer!

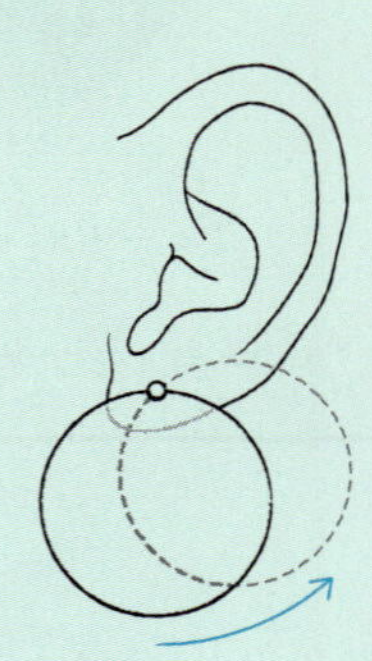

GOT THE CONCEPT? 8-2 "X" Marks the Moment of Inertia, Part I

Two thin, uniform rods, each of which has mass M and length L, are attached at their centers to form an "X" or "+" shape. What is the moment of inertia when this configuration is rotated around an axis that passes through their centers, perpendicular to the plane of the two rods? (a) $ML^2/24$; (b) $ML^2/12$; (c) $ML^2/6$; (d) $ML^2/4$; (e) $ML^2/3$.

GOT THE CONCEPT? 8-3 "X" Marks the Moment of Inertia, Part II

Four thin, uniform rods, each of which has mass $M/2$ and length $L/2$, are attached at their ends to form an "X" or "+" shape. What is the moment of inertia when this configuration is rotated around the axis that passes through the center of the "X" perpendicular to the plane of the rods? (a) $ML^2/24$; (b) $ML^2/12$; (c) $ML^2/6$; (d) $ML^2/4$; (e) $ML^2/3$.

TAKE-HOME MESSAGE FOR Section 8-3

✔ The moment of inertia of an object is additive and increases as the object's mass is moved farther from the rotation axis.

✔ An object's moment of inertia depends on the axis around which the object rotates.

✔ The parallel-axis theorem relates the moment of inertia for an axis through an object's center of mass to the moment of inertia for a second, parallel axis.

8-4 Conservation of mechanical energy also applies to rotating objects

Every year thousands of girls and boys race in the All-American Soap Box Derby in homemade cars powered only by the pull of gravity (**Figure 8-11**). In building their cars competitors have the flexibility to be creative, but they can be disqualified for using unapproved wheels—specifically, wheels for which too much of the mass is concentrated toward the center of the wheel. How would such wheels affect the outcome of the race? We can answer this and many related questions by using the conservation of energy.

Joe Raedle/Getty Images News/ Getty Images

Figure 8-11 Soap box derby racers These unpowered race cars roll downhill propelled only by gravity. How do the properties of a car's spinning wheels affect the car's speed?

We learned in Section 6-7 that the mechanical energy of a system—the sum of kinetic energy and potential energy—is *conserved* if no nonconservative forces do work on that system. In Sections 8-2 and 8-3 we introduced the *rotational* kinetic energy of a rigid body with moment of inertia I rotating with angular velocity ω: $K_{\text{rotational}} = \frac{1}{2} I\omega^2$. Does conservation of energy still hold true if we include rotational kinetic energy? And in particular does it hold true for objects that are *both* translating (moving through space) *and* rotating, like a ball rolling downhill or the spinning wheels of a soap box derby car?

Happily, the answer to both questions is "yes" because rotational kinetic energy isn't really a new kind of energy. As we saw in Section 8-2, the rotational kinetic energy of a rigid body is just the combined kinetic energy of all of its component pieces due to the motion of those pieces around the rotation axis. So the principle of conservation of energy also holds if we include rotational kinetic energy.

We can also use the principle of conservation of energy for a rigid object that's both moving through space as a whole and rotating. In such a situation it turns out that we can write the object's *total* kinetic energy as the sum of two terms: the *translational* kinetic energy associated with the motion of the object's center of mass (see Section 7-7), and the *rotational* kinetic energy associated with the object's rotation around its center of mass (**Figure 8-12**). If the object has mass M and its center of mass is moving with speed v_{CM}, its translational kinetic energy is $K_{\text{translational}} = \frac{1}{2} Mv_{\text{CM}}^2$; if the object's moment of inertia for an axis through its center of mass is I_{CM} and it

1. The center of mass of this wrench (shown as a red cross) moves in a straight line...

2. ...and the wrench rotates around the center of mass.

Berenice Abbott/Science Source

Figure 8-12 Combined translation and rotation In this time-lapse image, you are looking down on a spinning wrench as it slides across a table. The wrench naturally rotates around its center of mass (shown by the red "cross"), so we can think of its motion as the translation of the center of mass plus the rotation of the wrench around the center of mass.

rotates with angular speed ω, its rotational kinetic energy is $K_{\text{rotational}} = \frac{1}{2} I_{\text{CM}}\omega^2$. The total kinetic energy (translational plus rotational) of the object is then

Total kinetic energy for a rigid object undergoing both translation and rotation (8-12)

$$K = K_{\text{translational}} + K_{\text{rotational}} = \frac{1}{2} M v_{\text{CM}}^2 + \frac{1}{2} I_{\text{CM}}\omega^2$$

① **Total kinetic energy** of a system that is both translating and rotating

② The **translational kinetic energy** is the same as if all of the mass of the system were moving at the speed of the center of mass.

③ The **rotational kinetic energy** is the same as if the system were not translating, and all of the mass were rotating around the center of mass.

Many objects that undergo both translation and rotation will be symmetrical, like a rolling bowling ball or bicycle tire. In such a case the center of mass is at the object's geometrical center, and Table 8-1 in Section 8-3 will help you determine the value of I_{CM} for that object. (The wrench shown in Figure 8-12 is not symmetric: It has more mass at one end than the other, so its center of mass is *not* at its geometrical center.)

Whenever we have a situation involving a rigid object that undergoes both translation and rotation, we can make use of the energy relationships from Section 6-7 provided that we use Equation 8-12 as the expression for kinetic energy. If only conservative forces do work, mechanical energy (the sum of kinetic energy K and potential energy U) is conserved, so the value of $K + U$ is the same at any two times during the motion:

(6-23)

$$K_{\text{i}} + U_{\text{i}} = K_{\text{f}} + U_{\text{f}}$$

(if only conservative forces do work)

If nonconservative forces also do work, the final mechanical energy $K_{\text{f}} + U_{\text{f}}$ is equal to the initial mechanical energy $K_{\text{i}} + U_{\text{i}}$ plus $W_{\text{nonconservative}}$, the amount of nonconservative work that is done during the motion:

(6-26)

$$K_{\text{i}} + U_{\text{i}} + W_{\text{nonconservative}} = K_{\text{f}} + U_{\text{f}}$$

(if nonconservative forces do work)

In both Equations 6-23 and 6-26, the kinetic energy K is now given by Equation 8-12 and the potential energy U includes a term for each conservative force that acts (for example, the gravitational force or a spring force).

Example 8-9 illustrates how to use these ideas to solve a simple problem that involves both translational and rotational motion.

EXAMPLE 8-9 Flying Disc Energy

A flying disc with a mass of 0.175 kg and a diameter 0.266 m is used in the team sport called Ultimate. A player takes a disc at rest and does 1.00 J of work on it, causing the disc to fly off in a horizontal direction. When the disc leaves the player's hand, nine-tenths of its kinetic energy is translational and one-tenth is rotational. Find (a) the speed of the flying disc's center of mass and (b) the angular speed of the rotating disc. Treat the disc as uniform.

Set Up

The work done by the player counts as nonconservative work, so we use Equation 6-26 to describe how the disc's energy changes. We use Equation 8-12 to describe the kinetic energy of the disc, which involves its moment of inertia I_{CM} for an axis through its center of mass (the same as its geometrical center). As in Example 8-8 in Section 8-3, this moment of inertia is $I_{\text{CM}} = MR^2/2$, where $M = 0.175$ kg and R is one-half of the disc's diameter (see the first entry in the second row of Table 8-1). Our goal is to find the values of v_{CM} and ω just after the disc leaves the player's hand.

If nonconservative forces do work:

$$K_{\text{i}} + U_{\text{i}} + W_{\text{nonconservative}} = K_{\text{f}} + U_{\text{f}} \qquad (6\text{-}26)$$

Total kinetic energy for a rigid object undergoing both translation and rotation:

$$K = K_{\text{translational}} + K_{\text{rotational}}$$
$$= \frac{1}{2}Mv_{\text{CM}}^2 + \frac{1}{2}I_{\text{CM}}\omega^2 \qquad (8\text{-}12)$$

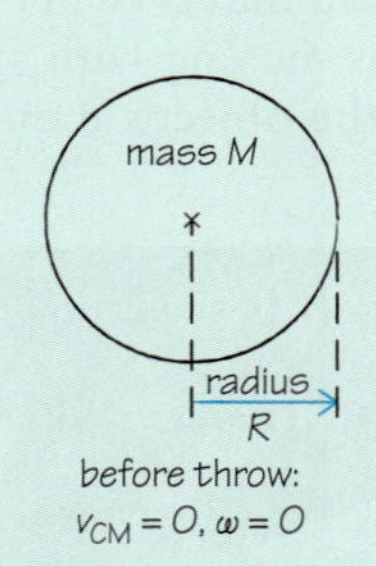

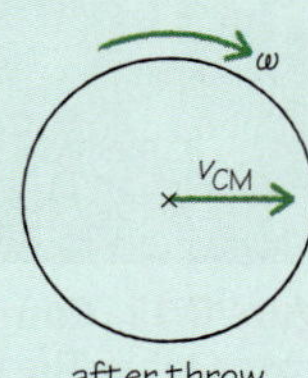

Solve

During the throw the disc moves horizontally, so its height stays the same and its gravitational potential energy is constant; that is, $U_f = U_i$. The disc starts at rest, so $K_i = 0$ (it has zero initial kinetic energy). So the work done on the disc by the player is transformed into the disc's final kinetic energy K_f.

$$K_i + U_i + W_{\text{nonconservative}} = K_f + U_f$$

Since $U_f = U_i$, $K_i = 0$, and $W_{\text{nonconservative}} = 1.00$ J, this becomes

$$1.00 \text{ J} = K_f$$

The disc's 1.00 J of kinetic energy is divided so that 9/10 is translational and 1/10 is rotational.

$$K_{\text{translational}} = \frac{1}{2}Mv_{\text{CM}}^2 = (9/10)(1.00 \text{ J}) = 0.900 \text{ J}$$

$$K_{\text{rotational}} = \frac{1}{2}I_{\text{CM}}\omega^2 = (1/10)(1.00 \text{ J}) = 0.100 \text{ J}$$

(a) Find the speed of the disc's center of mass.

From the expression for translational kinetic energy,

$$v_{\text{CM}}^2 = \frac{2K_{\text{translational}}}{M}$$

$$v_{\text{CM}} = \sqrt{\frac{2K_{\text{translational}}}{M}} = \sqrt{\frac{2(0.900 \text{ J})}{0.175 \text{ kg}}} = \sqrt{\frac{2(0.900 \text{ kg} \cdot \text{m}^2/\text{s}^2)}{0.175 \text{ kg}}}$$

$$= 3.21 \text{ m/s}$$

(b) Calculate the disc's moment of inertia for the axis of rotation through its center, then solve for its angular speed ω. Note that if $K_{\text{rotational}}$ is in joules and I_{CM} is in kg $\cdot$ m^2, ω is in radians per second.

Calculate the disc's moment of inertia through its center of mass:

$$I_{\text{CM}} = \frac{1}{2}MR^2 = \frac{1}{2}(0.175 \text{ kg})\left(\frac{0.266 \text{ m}}{2}\right)^2$$

$$= 1.55 \times 10^{-3} \text{ kg} \cdot \text{m}^2$$

From the expression for rotational kinetic energy,

$$\omega^2 = \frac{2K_{\text{rotational}}}{I_{\text{CM}}}$$

$$\omega = \sqrt{\frac{2K_{\text{rotational}}}{I_{\text{CM}}}} = \sqrt{\frac{2(0.100 \text{ J})}{1.55 \times 10^{-3} \text{ kg} \cdot \text{m}^2}}$$

$$= \sqrt{\frac{2(0.100 \text{ kg} \cdot \text{m}^2/\text{s}^2)}{1.55 \times 10^{-3} \text{ kg} \cdot \text{m}^2}} = 11.4 \text{ rad/s}$$

Reflect

This is a problem for which the conservation of energy equation is essential but not sufficient. In order to solve for the two unknowns, v_{CM} and ω, we needed a second relationship telling us how the kinetic energy was distributed—the statement that the disc's kinetic energy was 9/10 translational and 1/10 rotational.

Note that v_{CM} is greater than a typical walking speed (1 to 2 m/s) but less than the speed of a world-class sprinter (about 10 m/s). Since 2π (just over 6) radians represents a full revolution, the value of ω tells us that the disc spins just under two revolutions per second. Are these values what you might expect for a hand-thrown flying disc?

Rolling Without Slipping

Let's return to one of the soap box derby cars that we thought about at the beginning of this section (Figure 8-11) and ask how we can use energy concepts to analyze its motion. For simplicity let's think about a single wheel rolling downhill (**Figure 8-13a**) rather than the entire car. After the wheel has traveled a certain distance down the hill, how fast is it moving as a whole, and how rapidly is it spinning?

Before the start of the race, the wheel has gravitational potential energy relative to the bottom of the hill but no kinetic energy. The wheel picks up speed as it rolls down the race course, a transformation of potential energy into kinetic. Unlike the flying disc

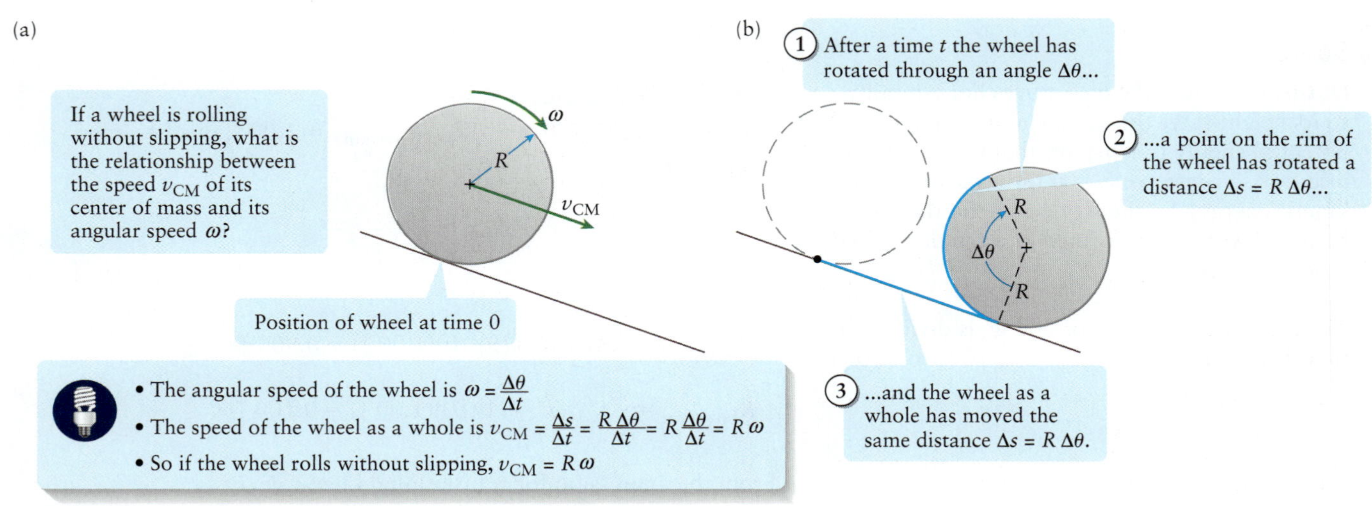

Figure 8-13 A wheel rolling downhill If a wheel rolls without slipping, there is a definite relationship between its angular speed and the speed of its center of mass.

in Example 8-9, however, we need to figure out how much of that kinetic energy is translational energy and how much is rotational. Furthermore, there must be friction between the wheel and the road in order for the wheel to roll. (If the road were frictionless, the wheel would simply slide downhill.) So we also need to know how much mechanical energy is lost due to friction as the wheel rolls.

To make further progress, we use the idea that the wheel is **rolling without slipping**—that is, the wheel does not skid or slide over the road. In this situation there is a specific relationship between the speed of the center of the wheel, v_{CM}, and the angular speed ω of the wheel's rotation. To see what this relationship is, consider what happens when the wheel rolls through a small angle $\Delta\theta$ over a time interval Δt (**Figure 8-13b**). Imagine that a thread has been wrapped around the circumference of the wheel and unwinds as the wheel rolls, marking the distance traveled. During the interval Δt a point on the rim of the wheel of radius R rotates a distance $\Delta s = R\Delta\theta$. Since the wheel doesn't slip on the road, $\Delta s = R\Delta\theta$ is also the length of thread that the wheel unwinds onto the road and hence the distance that the wheel as a whole moves down the road. Hence v_{CM}, the speed of the wheel as a whole, is equal to the distance Δs divided by the duration Δt of the time interval:

$$v_{CM} = \frac{\Delta s}{\Delta t} = \frac{R\Delta\theta}{\Delta t} = R\frac{\Delta\theta}{\Delta t}$$

But $\Delta\theta/\Delta t$ is just equal to ω, the angular speed of the wheel, so

Condition for rolling without slipping (8-13)

Speed of the center of mass of a rolling object

$$v_{CM} = R\omega$$

Radius of the object

Angular speed of the object's rotation

Equation 8-13 says that there is a direct proportionality between the linear speed v_{CM} and angular speed ω of a circular object that rolls without slipping. As the wheel in Figure 8-13 rolls downhill it gains speed so v_{CM} increases, and it rotates faster so ω increases, but these increases are always proportional to each other so that v_{CM} and ω are always related by Equation 8-13.

How much mechanical energy is lost due to friction if an object rolls without slipping, like the wheel in Figure 8-13? If the object and the surface are perfectly rigid, the answer is simple: The force of friction does *no* work at all, so *no* energy is lost. If an object *slides* over a surface, we saw in Section 6-7 that the kinetic friction force does work on the object and changes its mechanical energy. But there is no sliding if the object rolls without slipping, so friction does no work. The friction force still plays an important role: It's what makes the object roll. So as the wheel in Figure 8-13 rolls downhill and gravitational potential energy is converted to kinetic energy, the friction force ensures that a portion of the kinetic energy goes into the rotational form.

Nonrigid Objects and Rolling Friction

While friction does no work on an object that rolls without slipping, if the object or surface is not perfectly rigid—that is, if it can *deform*—there *is* a related force that does work on the object as it rolls. An example is a bicycle tire or automobile tire, which flexes and flattens on its bottom where it contacts the road. As the tire rolls, different parts of the tire successively get compressed as they touch the ground and relax as they lose contact with the ground. This flexing causes the tire to get warmer, which expends mechanical energy. The energy comes from the mechanical energy of the rolling tire, which means that left to itself on a horizontal surface the tire will lose kinetic energy and slow down. The net effect is the same as if a force were acting to oppose the rolling tire's motion. This is just the force that we called *rolling friction* in Section 5-3. Rolling friction is also present if the surface on which the object rolls is deformable, like a wrestling mat or the green baize that covers a billiards table.

You can minimize rolling friction on your bicycle or automobile by keeping tires properly inflated. Underinflated tires flex more, which means that they absorb more energy and reduce the car's mileage. That's also why transporting goods via railroad is more efficient than using trucks: Railroad cars have very rigid steel wheels, which flex much less than even properly inflated truck tires.

WATCH OUT! Rolling doesn't have to imply rolling friction.

An object rolling on a surface experiences rolling friction *only* if the object, the surface, or both are deformable. In this chapter we'll assume that all rolling objects and the surface on which they roll are perfectly rigid. Then there's no rolling friction. A friction force may act on the rolling object, but it's the force of *static* friction (since the object rolls rather than slides over the surface). Static friction does no work on the object, so in this case mechanical energy is conserved.

We now see how to use the energy approach to solve problems involving objects that roll without slipping. Since no nonconservative work is done by the force of friction, we use Equation 6-23, $K_i + U_i = K_f + U_f$, which states that mechanical energy is conserved as the object rolls. The kinetic energy of the rolling object is given by Equation 8-12, $K = \frac{1}{2}Mv_{CM}^2 + \frac{1}{2}I_{CM}\omega^2$, and the speed v_{CM} and angular velocity ω are related by Equation 8-13, $v_{CM} = R\omega$. The following example makes use of these equations.

EXAMPLE 8-10 Downhill Race: Disk versus Hoop

A uniform disk and a hoop are both allowed to roll without slipping down a ramp of height H (**Figure 8-14**). Both objects have the same radius R and the same mass M. If both objects start from rest at the same time, which one reaches the bottom of the ramp first?

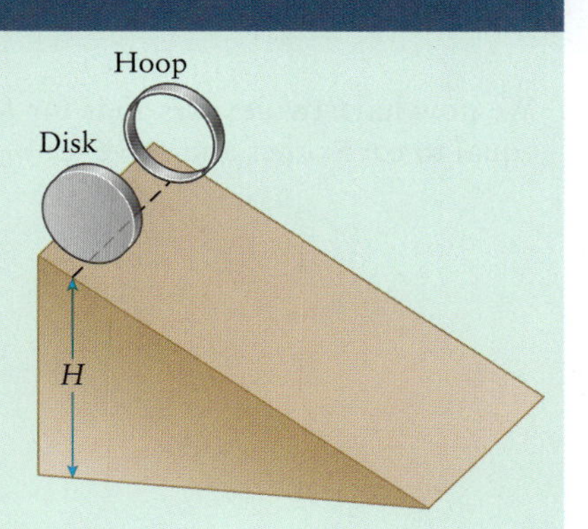

Figure 8-14 Who wins the race? A competition between two objects rolling down a ramp.

Set Up

Both objects roll without slipping, so we can use Equations 6-23, 8-12, and 8-13 to describe their motion. The only real difference between the two objects is their moment of inertia: From Table 8-1 $I_{CM} = MR^2/2$ for the uniform disk, while $I_{CM} = MR^2$ for the hoop. To decide which one wins the race, we'll calculate v_{CM} at the bottom of the ramp for each object. The one with the faster speed will get to the bottom first and be declared the winner. We'll solve for $v_{CM,f}$ (the speed at the bottom of the ramp) for a general object that rolls down the ramp starting from rest: Only at the end will we plug in the value of I_{CM} for each object.

Conservation of mechanical energy:

$$K_i + U_i = K_f + U_f \qquad (6\text{-}23)$$

Kinetic energy for an object that undergoes both translation and rotation:

$$K = \frac{1}{2}Mv_{CM}^2 + \frac{1}{2}I_{CM}\omega^2 \qquad (8\text{-}12)$$

Condition for rolling without slipping:

$$v_{CM} = R\omega \qquad (8\text{-}13)$$

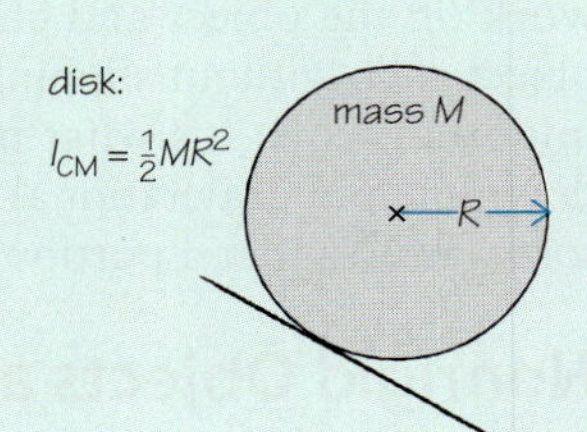

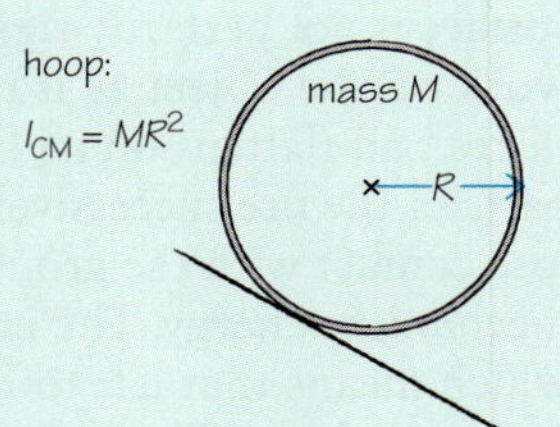

Solve

Initially the object is at rest with zero kinetic energy, so $K_i = 0$. The initial gravitational potential energy is $U_i = Mgy_i$, and the final gravitational potential energy is $U_f = Mgy_f$. The figure shows that $y_i - y_f = H$, so we can solve for the object's final kinetic energy K_f at the bottom of the ramp.

Conservation of energy equation:

$$0 + Mgy_i = K_f + Mgy_f, \text{ or}$$

$$K_f = Mgy_i - Mgy_f = Mg(y_i - y_f)$$

Since $y_i - y_f = H$,

$$K_f = MgH$$

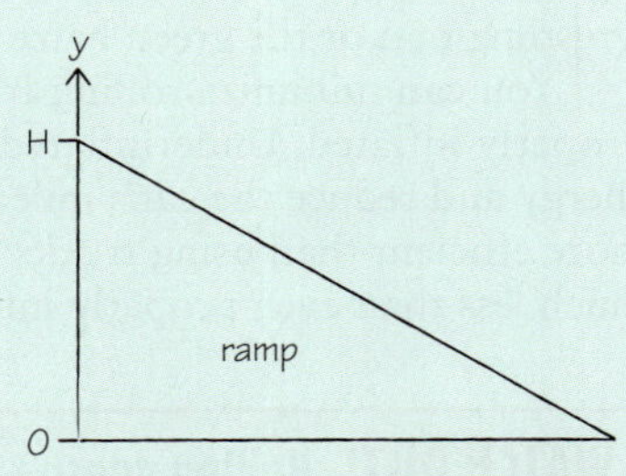

The kinetic energy is part translational and part rotational. We can use Equation 8-13 to write ω_f (the angular velocity at the bottom of the ramp) in terms of $v_{CM,f}$ (the linear speed at the bottom of the ramp) and so can express the final kinetic energy in terms of the linear speed $v_{CM,f}$ only.

Kinetic energy at the bottom of the ramp:

$$K_f = \frac{1}{2}Mv_{CM,f}^2 + \frac{1}{2}I_{CM}\omega_f^2$$

Rolling without slipping:

$$v_{CM,f} = R\omega_f, \text{ so}$$

$$\omega_f = \frac{v_{CM,f}}{R}$$

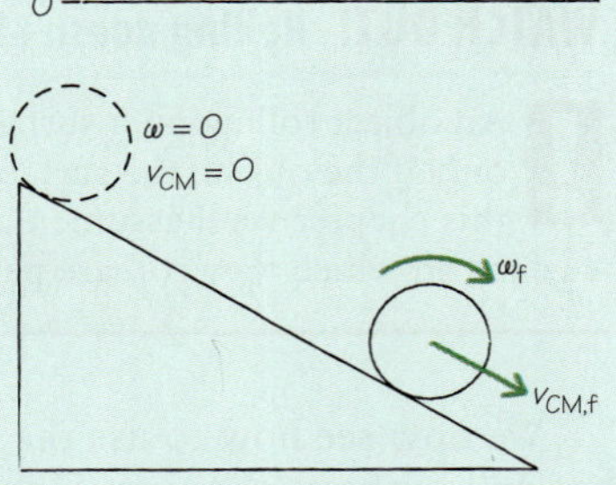

Substitute into kinetic energy equation:

$$K_f = \frac{1}{2}Mv_{CM,f}^2 + \frac{1}{2}I_{CM}\left(\frac{v_{CM,f}}{R}\right)^2 = \frac{1}{2}Mv_{CM,f}^2 + \frac{1}{2}\left(\frac{I_{CM}}{R^2}\right)v_{CM,f}^2$$

$$= \frac{1}{2}\left(M + \frac{I_{CM}}{R^2}\right)v_{CM,f}^2$$

We now have two expressions for K_f. Set them equal to each other and solve for $v_{CM,f}$.

$$\frac{1}{2}\left(M + \frac{I_{CM}}{R^2}\right)v_{CM,f}^2 = MgH$$

$$v_{CM,f}^2 = \frac{2MgH}{M + \frac{I_{CM}}{R^2}}$$

$$v_{CM,f} = \sqrt{\frac{2MgH}{M + \frac{I_{CM}}{R^2}}}$$

Finally, substitute $I_{CM} = MR^2/2$ for the disk and $I_{CM} = MR^2$ for the hoop and find the final speed for each object.

The speed of the disk at the bottom of the ramp is faster than that of the hoop: $\sqrt{(4/3)gH}$ compared to $\sqrt{gH}$. Both objects will accelerate down the ramp, but at any given position down the ramp, the disk will be moving faster than the hoop. In a race the disk would win.

For the uniform disk:

$$I_{CM} = \frac{MR^2}{2}$$

$$M + \frac{I_{CM}}{R^2} = M + \frac{M}{2} = \frac{3M}{2}$$

$$v_{CM,f} = \sqrt{\frac{2MgH}{3M/2}} = \sqrt{\left(\frac{2}{3M}\right)(2MgH)} = \sqrt{\frac{4gH}{3}}$$

For the hoop:

$$I_{CM} = MR^2$$

$$M + \frac{I_{CM}}{R^2} = M + M = 2M$$

$$v_{CM,f} = \sqrt{\frac{2MgH}{2M}} = \sqrt{gH}$$

Reflect

Why does the disk win? The only difference between the two objects is that the uniform disk has a smaller moment of inertia than the hoop ($MR^2/2$ compared to MR^2), so a smaller fraction of the gravitational potential energy is converted to rotational kinetic energy as it rolls down the ramp. That leaves more energy available for translational motion, which means that the disk travels faster down the ramp and wins the race.

We can verify this by calculating the ratio of rotational kinetic energy $K_{rotational}$ to translational kinetic energy $K_{translational}$ for a rolling object. This ratio is I_{CM}/MR^2, so the smaller the moment of inertia I_{CM} for a given mass M and radius R—like the uniform disk compared to the hoop—the smaller the rotational kinetic energy and the more energy is available for translation.

Ratio of rotational kinetic energy to translational kinetic energy:

$$\frac{K_{rotational}}{K_{translational}} = \frac{\frac{1}{2}I_{CM}\omega^2}{\frac{1}{2}Mv_{CM}^2} = \frac{I_{CM}\omega^2}{Mv_{CM}^2}$$

If the object is rolling without slipping,

$$v_{CM} = R\omega$$

$$\omega = \frac{v_{CM}}{R}, \text{ so}$$

$$\frac{K_{rotational}}{K_{translational}} = \frac{I_{CM}\omega^2}{Mv_{CM}^2} = \frac{I_{CM}}{Mv_{CM}^2}\left(\frac{v_{CM}}{R}\right)^2 = \frac{I_{CM}}{MR^2}$$

At the beginning of this section we mentioned that soap box derby cars are not allowed to have wheels with too much of their mass concentrated toward their centers. We can now explain this using Example 8-10, which shows that a wheel in the form of a solid disk rolls down an incline more rapidly than a wheel shaped like a hoop (like a bicycle tire). The reason for this difference is that the solid disk has more of its mass located near the rotation axis and so has a smaller moment of inertia around its rotation axis than does the hoop ($I_{CM} = MR^2/2$ for the solid disk versus $I_{CM} = MR^2$ for the hoop). A smaller moment of inertia I_{CM} means less energy is required to rotate the wheel, so more energy is available for translational kinetic energy and the speed of the car as a whole is greater. Having wheels with too much mass toward the central rotation axis, and hence too small a moment of inertia, would give a soap box derby car an unfair advantage!

GOT THE CONCEPT? 8-4 Does Size Matter?

In Example 8-10 suppose the solid disk had half the radius and half the mass of the hoop. In a downhill race between these two, would the disk still win? Why or why not?

GOT THE CONCEPT? 8-5 A Three-Way Downhill Race

A uniform disk, a uniform sphere, and a block are released from rest down adjacent ramps of identical angle. The disk and sphere roll without slipping, while the block slides without friction. All three objects start from the same height. Rank the three objects in the order in which they reach the bottom of the ramp, starting with the first to arrive.

TAKE-HOME MESSAGE FOR Section 8-4

✔ The total kinetic energy K of an object of mass M is the sum of the translational kinetic energy of its center of mass, $K_{\text{translational}} = \frac{1}{2} Mv_{\text{CM}}^2$, and the rotational kinetic energy of the object's rotation around its center of mass, $K_{\text{rotational}} = \frac{1}{2} I_{\text{CM}}\omega^2$.

✔ The same energy conservation equations we learned in Chapter 6 also apply when there is rotational motion. The only difference is that the kinetic energy K now includes both translational and rotational kinetic energies.

✔ If an object rolls without slipping on a surface, the speed of its center of mass is directly proportional to its angular speed. If the rolling object is perfectly rigid, a friction force is present but does no work on the object as it rolls.

8-5 The equations for rotational kinematics are almost identical to those for linear motion

Up to this point we've discussed rotation by using the concepts of energy and energy conservation almost exclusively. However, in many situations we need to use **rotational kinematics**: a description not just of how rapidly an object rotates but also how rapidly its angular velocity is changing, that is, its *angular acceleration*.

Angular acceleration is an important factor in the design of rotating machinery. When you turn on an electric fan on a sweltering hot day, you don't just want the fan to start spinning but to come up to speed right away. But if you're riding on a carnival merry-go-round, you want it to gain speed gradually and smoothly at the beginning of the ride (and lose speed in the same way at the end of the ride) rather than in a sudden jerk. In this section we'll see how angular acceleration is defined and how to do calculations using it. Just as motion in a straight line is easiest to analyze when the acceleration is constant, we'll see that the simplest kind of rotational motion is that with constant *angular* acceleration.

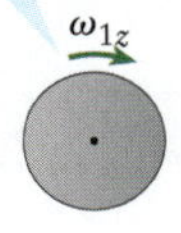
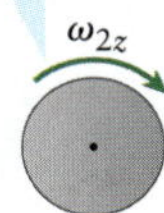

At time t_1 an object rotates with angular velocity ω_{1z}.
At a later time t_2 the object rotates with a different angular velocity ω_{2z}.
ω_{1z}
ω_{2z}
The object's average angular acceleration equals the change in angular velocity divided by the elapsed time:
$\alpha_{\text{average},z} = \frac{\omega_{2z} - \omega_{1z}}{t_2 - t_1} = \frac{\Delta\omega_z}{\Delta t}$

Figure 8-15 Defining angular acceleration The greater the magnitude of the average angular acceleration $\alpha_{\text{average},z}$ of a rotating object, the more rapidly the angular velocity changes.

Defining Angular Acceleration

We'll define angular acceleration in much the same way that we defined the acceleration of an object moving in a straight line in Section 2-4. In that section we described the average acceleration $a_{\text{average},x}$ of an object moving along the x axis as the change Δv_x in its x velocity during a time interval divided by the duration Δt of the time interval:

$$a_{\text{average},x} = \frac{\Delta v_x}{\Delta t} \text{ (average } x \text{ acceleration)}$$

If the duration Δt of the time interval becomes very short, the ratio $\Delta v_x/\Delta t$ becomes the *instantaneous* acceleration a_x or, for short, simply the acceleration.

Let's follow the same steps for rotational motion. Suppose a rigid object is rotating around the z axis with angular velocity ω_{1z} at time t_1 and with angular velocity ω_{2z} at a later time t_2 (**Figure 8-15**). The **average angular acceleration** $\alpha_{\text{average},z}$ (the Greek letter α, or alpha) for the time interval between t_1 and t_2 is the change in angular velocity, $\Delta\omega_z = \omega_{2z} - \omega_{1z}$, divided by the duration of the time interval, $\Delta t = t_2 - t_1$:

Average angular acceleration (8-14)

$$\alpha_{\text{average},z} = \frac{\omega_{2z} - \omega_{1z}}{t_2 - t_1} = \frac{\Delta\omega_z}{\Delta t}$$

Average angular acceleration of a rotating object

Change in angular velocity of the object over a certain time interval: The angular velocity changes from ω_{1z} to ω_{2z}.

Elapsed time for the time interval: The object has angular velocity ω_{1z} at time t_1, and has angular velocity ω_{2z} at time t_2.

If the two times t_1 and t_2 are very close to each other, so that Δt is very short, this becomes the *instantaneous* angular acceleration α_z, or just **angular acceleration** for short. At any instant angular acceleration is equal to the rate of change of angular velocity at that instant. Since angular velocity is measured in rad/s, angular acceleration is measured in rad/s^2.

Just as all pieces of a rigid object rotate with the same angular velocity ω_z at any time t, all pieces have the same angular acceleration. The value of α_z is positive if the angular velocity ω_z is becoming more positive and negative if ω_z is becoming more negative.

WATCH OUT! The sign of angular acceleration can be misleading.

Just as for the acceleration of an object moving along a straight line, a positive value of angular acceleration α_z does *not* necessarily correspond to speeding up and a negative value of α_z does not necessarily correspond to slowing down. An object's rotation speeds up if ω_z and α_z have the same algebraic sign—if both are positive, the object is rotating in the positive direction and speeding up, while if both are negative, the object is rotating in the negative direction and speeding up. If ω_z and α_z have opposite signs (one positive and one negative), the object's rotation is slowing down (**Figure 8-16**).

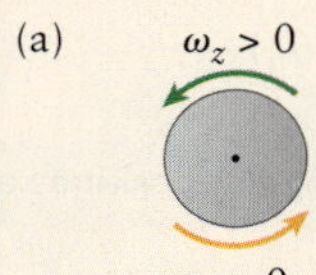

Figure 8-16 The sign of angular acceleration A rotating object is speeding up if its angular velocity ω_z and angular acceleration α_z are either (a) both positive or (b) both negative. It is slowing down if (c) ω_z is positive and α_z is negative or (d) ω_z is negative and α_z is positive.

Motion with Constant Angular Acceleration

We saw in Chapter 2 that an important case of motion in a straight line is when the acceleration a_x is constant so that the velocity v_x changes at a steady rate. Let's look at the analogous situation of rotational motion with constant *angular* acceleration. A ball or wheel rolling downhill from rest moves with nearly constant angular acceleration, as do many types of rotating machinery as they start up or slow down.

Go to Interactive Exercise 8-1 for more practice dealing with rotational kinematics.

It's straightforward to write down the equations for rotational motion with constant angular acceleration. To see how, take a look at the equations for linear velocity v_x and angular velocity ω_z, and the equations for linear acceleration a_x and angular acceleration α_z:

Linear motion	Rotational motion
$v_x = \dfrac{\Delta x}{\Delta t}$ (linear velocity)	$\omega_z = \dfrac{\Delta \theta}{\Delta t}$ (angular velocity)
$a_x = \dfrac{\Delta v_x}{\Delta t}$ (linear acceleration)	$\alpha_z = \dfrac{\Delta \omega_z}{\Delta t}$ (angular acceleration)

Comparing these equations shows that the rotational quantities θ, ω_z, and α_z are related to each other in exactly the same way that x, v_x, and a_x are related to each other. So we can take the equations for constant linear acceleration and convert them to the equations for constant angular acceleration by replacing x with θ, v_x with ω_z, and a_x with α_z. The equations for linear motion are

$$v_x = v_{0x} + a_x t \text{ (constant } x \text{ acceleration only)} \quad (2\text{-}5)$$

$$x = x_0 + v_{0x}t + \frac{1}{2}a_x t^2 \text{ (constant } x \text{ acceleration only)} \quad (2\text{-}9)$$

$$v_x^2 = v_{0x}^2 + 2a_x(x - x_0) \text{ (constant } x \text{ acceleration only)} \quad (2\text{-}11)$$

Hence the equations for constant angular acceleration are

$$\omega_z = \omega_{0z} + \alpha_z t$$

Angular velocity at time t of a rotating object with constant angular acceleration

Angular velocity at time $t = 0$ of the object

Constant angular acceleration of the object

Time at which the object has angular velocity ω_z

Angular velocity, angular acceleration, and time for constant angular acceleration only (8-15)

Angular position, angular acceleration, and time for constant angular acceleration only (8-16)

$$\theta = \theta_0 + \omega_{0z}t + \frac{1}{2}\alpha_z t^2$$

- θ: **Angular position** at time t of a rotating object with constant angular acceleration
- θ_0: **Angular position** at time $t = 0$ of the object
- ω_{0z}: **Angular velocity** at time $t = 0$ of the object
- t: **Time** at which the object is at angular position θ
- α_z: **Constant angular acceleration** of the object

Angular velocity, angular acceleration, and angular position for constant angular acceleration only (8-17)

$$\omega_z^2 = \omega_{0z}^2 + 2\alpha_z(\theta - \theta_0)$$

- ω_z: **Angular velocity** at angular position θ of a rotating object with constant angular acceleration
- ω_{0z}: **Angular velocity** at angular position θ_0 of the object
- α_z: **Constant angular acceleration** of the object
- θ, θ_0: Two **angular positions** of the object

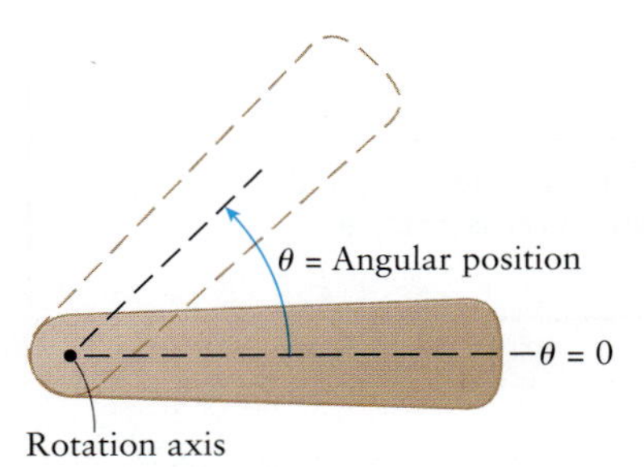

Figure 8-17 Angular position Imagine a line drawn on this rotating object from its rotation axis outward. The angle of this line at a given time is the object's angular position θ at that time.

To see what we mean by *angular position*, imagine a line drawn on the object outward from the rotation axis (**Figure 8-17**). The angle of this line from a reference direction changes as the object rotates, and it's this angle that we call the **angular position** θ. The angular displacement from time 0 to time t is $\theta - \theta_0$.

We can use Equations 8-15, 8-16, and 8-17 to solve a wide range of problems in rotational motion, just as we used Equations 2-5, 2-9, and 2-11 for linear motion problems in Chapter 2. Here's a representative example.

WATCH OUT! Equations 8-15, 8-16, and 8-17 are for rigid objects only.

The three equations we've just presented can be used only if *all parts* of a rotating object have the same angular velocity and angular acceleration at any given time—in other words, only if the object is *rigid*. They *cannot* be used for a rotating object that isn't rigid, such as water in a bathtub as it swirls down the drain or a cake mix being stirred in a bowl. In Example 8-11, and in the problems at the end of the chapter, we'll always assume that the rotating objects are rigid.

EXAMPLE 8-11 A Stopping Top

A top spinning at 4.00 rev/s comes to a complete stop in 64.0 s. Assuming the top slows down at a constant rate, how many revolutions does it make before coming to a stop?

Set Up

The statement that the top slows down at a constant rate means that its angular acceleration is constant, so we can use Equations 8-15 and 8-16. (We won't use Equation 8-17 because that equation doesn't involve time, which is one of the quantities we're given.) These two equations involve six quantities (t, θ_0, θ, ω_{0z}, ω_z, and α_z).

To find constant angular acceleration α_z from ω_{0z}, ω_z, and t:

$$\omega_z = \omega_{0z} + \alpha_z t \qquad (8\text{-}15)$$

In this case we are given the initial angular velocity $\omega_{0z} = 4.00$ rev/s, the final angular velocity $\omega_z = 0$, and the elapsed time $t = 64.0$ s. If we let the top start at an angular position of $\theta_0 = 0$, we want to find the top's final angular position θ during this time but don't know the value of the angular acceleration α_z. We'll use Equation 8-15 to determine α_z from the known information then substitute this value into Equation 8-16 to find the value of θ.

To find the final angular position θ from θ_0, ω_{0z}, α_z, and t:

$$\theta = \theta_0 + \omega_{0z}t + \frac{1}{2}\alpha_z t^2 \qquad (8\text{-}16)$$

Solve

Rewrite Equation 8-15 to solve for α_z. Then substitute the values of ω_z, ω_{0z}, and t. (Note that we don't need to convert revolutions to radians: We just have to be consistent with units throughout our solution.)

Rewrite Equation 8-15: $\omega_z - \omega_{0z} = \alpha_z t$

$$\alpha_z = \frac{\omega_z - \omega_{0z}}{t}$$

The resulting value of α_z is negative while ω_{0z} is positive, which correctly says that the rotation is slowing down.

Substitute values: $\alpha_z = \dfrac{0 - 4.00 \text{ rev/s}}{64.0 \text{ s}} = -0.0625 \text{ rev/s}^2$

Solve for θ by plugging this value of α_z into Equation 8-16 along with the given values of ω_{0z} and t.

Equation 8-16: $\theta = \theta_0 + \omega_{0z}t + \frac{1}{2}\alpha_z t^2$

$$= 0 \text{ rev} + (4.00 \text{ rev/s})(64.0 \text{ s}) + \frac{1}{2}(-0.0625 \text{ rev/s}^2)(64.0 \text{ s})^2$$

$$= 256 \text{ rev} - 128 \text{ rev}$$

$$= 128 \text{ rev}$$

Reflect

We can check our answer by substituting the values for ω_z, ω_{0z}, and α_z into Equation 8-17 (which we did not use above) and solving for θ. This gives us the same answer for θ as above.

Note that by writing $\omega_{0z} = 4.00$ rev/s, we made the assumption that the top is initially rotating in the positive direction. Can you show that if you instead make $\omega_{0z} = -4.00$ rev/s, so that the top is initially rotating in the negative direction, you get $\theta - \theta_0 = -128$ rev? (This means that the top rotates 128 rev but in the negative direction.)

Equation 8-17: $\omega_z^2 = \omega_{0z}^2 + 2\alpha_z(\theta - \theta_0)$

Rewrite to solve for θ: $\omega_z^2 - \omega_{0z}^2 = 2\alpha_z(\theta - \theta_0)$

$$\theta = \frac{\omega_z^2 - \omega_{0z}^2}{2\alpha_z} + \theta_0 = \frac{(0 \text{ rev/s})^2 - (4.00 \text{ rev/s})^2}{2(-0.0625 \text{ rev/s}^2)} + 0 \text{ rev}$$

$$= 128 \text{ rev}$$

GOT THE CONCEPT? 8-6 Rate of Change of Angular Speed

Rank the following rotating objects in order of the rate at which the speed of their rotation is changing, from speeding up at the fastest rate to slowing down at the fastest rate. (a) A disk with $\omega_z = 2.00$ rad/s and $\alpha_z = -1.00$ rad/s^2; (b) a wheel with $\omega_z = 3.00$ rad/s and $\alpha_z = -2.00$ rad/s^2; (c) a ceiling fan with $\omega_z = -2.00$ rad/s and $\alpha_z = -1.00$ rad/s^2; (d) a flywheel with $\omega_z = 0$ and $\alpha_z = -0.750$ rad/s^2; (e) a circular saw with $\omega_z = 3.00$ rad/s and $\alpha_z = 0.750$ rad/s^2.

TAKE-HOME MESSAGE FOR Section 8-5

✔ Just as acceleration in a straight line is the rate of change of velocity, angular acceleration is the rate of change of angular velocity.

✔ If the angular acceleration is constant for a rigid object, there are three simple equations that relate time, angular position, angular velocity, and angular acceleration.

8-6 Torque is to rotation as force is to translation

Where do you push or pull on a door to open it most easily? In what direction should you exert that push or pull? These may seem like odd questions to ask in a chapter about rotational motion. But keep in mind that when you open a door you are giving it an *angular acceleration*: The door starts at rest and begins rotating around its hinges, so you're changing the door's angular velocity. So we can rephrase our questions like this: Where, and in what direction, should we push or pull on a door to give it an angular acceleration around its hinges?

Experience tells you that it's nearly impossible to open a door by pushing or pulling near its hinges, as with force $\vec{F}_1$ in **Figure 8-18**. A much better place to apply a force is on the opposite end of a door from its hinges, which is why the doorknob is located there. Even then, it's ineffective to push or pull on the doorknob in a direction parallel to the plane of the door (as with force $\vec{F}_2$ in Figure 8-18). The easiest way to open the door is to exert a force far from the hinges (that is, at the doorknob) in a direction perpendicular to the plane of the door, as with force $\vec{F}_3$ in Figure 8-18.

This example shows that in order to give an object an angular acceleration, what matters is not just how hard you push or pull on the object, but also *where* and *in what direction* that push or pull is applied. A physical quantity that relates all of these aspects of an applied force is the *torque* associated with that force. In this section we'll find that just as the net force acting on an object determines its translational or linear acceleration (Newton's second law), the net *torque* acting on an object determines its *angular* acceleration.

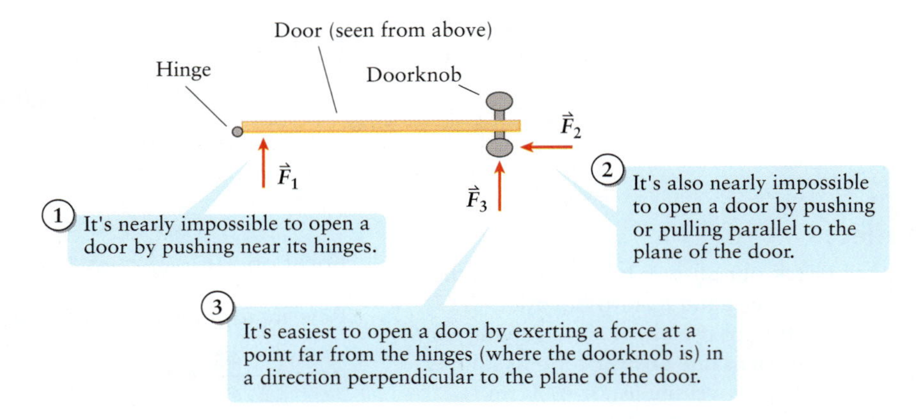

Door (seen from above)
Hinge
Doorknob
$\vec{F}_1$
$\vec{F}_2$
$\vec{F}_3$
1 It's nearly impossible to open a door by pushing near its hinges.
2 It's also nearly impossible to open a door by pushing or pulling parallel to the plane of the door.
3 It's easiest to open a door by exerting a force at a point far from the hinges (where the doorknob is) in a direction perpendicular to the plane of the door.

Figure 8-18 Opening a door When you push or pull on a door to open it, it matters where and in what direction the force is applied.

Defining the Magnitude and Direction of Torque

Suppose that you exert a force $\vec{F}$ on an object as shown in **Figure 8-19a**. We use the symbol $\vec{r}$ to denote the vector from the rotation axis to the point where the force is applied, and we use the symbol ϕ (the Greek letter phi) for the angle between the directions of $\vec{r}$ and $\vec{F}$. The component of $\vec{F}$ that points straight out from the rotation axis, $F \cos \phi$, doesn't have any tendency to make the object rotate. But the perpendicular component of $\vec{F}$, $F \sin \phi$, *does* tend to make the object rotate—in this case in a clockwise direction. We describe the rotational effect of the force $\vec{F}$ by a quantity called the **torque** τ (the Greek letter tau) associated with the force. This is given by

Magnitude of torque (8-18)

$$\tau = rF \sin \phi$$

Magnitude of the torque produced by a force acting on an object

Magnitude of the force

Distance from the rotation axis of the object to where the force is applied

Angle between the vector $\vec{r}$ (from the rotation axis to where the force is applied) and the force vector $\vec{F}$

Equation 8-18 tells us the *magnitude* of the torque, but torque also has *direction*. For the force shown in Figure 8-19a, we say that the associated torque is clockwise because it tends to make the object rotate clockwise. By contrast, the force depicted in **Figure 8-19b** tends to make the object rotate counterclockwise, so we say that the torque associated with this force is counterclockwise. (In Section 8-9 we'll see that

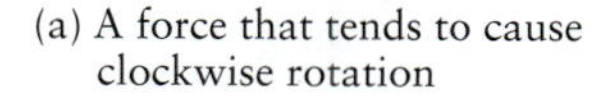

Views of object from along its rotation axis
(a) A force that tends to cause clockwise rotation

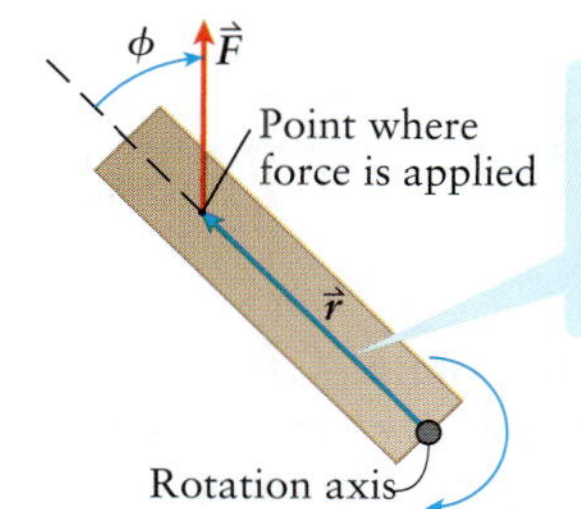

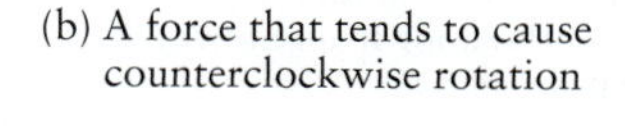

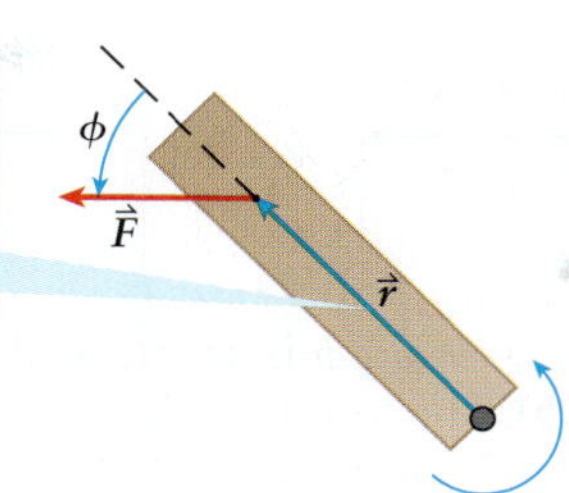

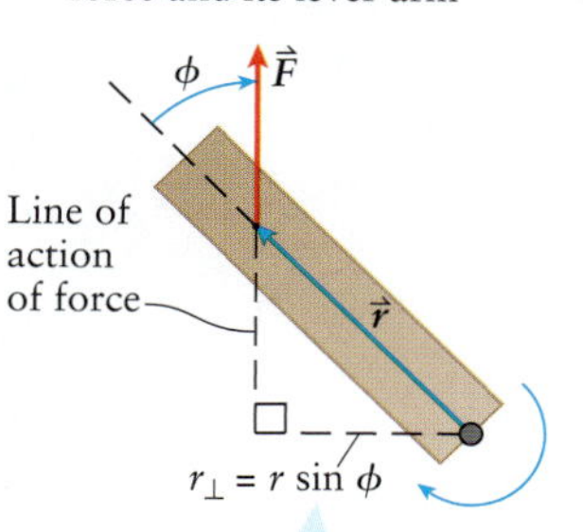

Figure 8-19 Torque (a) The vector $\vec{r}$ points from an object's rotation axis to where a force $\vec{F}$ is applied to the object. The torque produced by the force has magnitude $\tau = rF \sin\phi$. In this case the torque is clockwise because it tends to make the object rotate clockwise. (b) This torque is counterclockwise because it tends to make the object rotate counterclockwise. (c) Finding the torque in terms of the lever arm $r_\perp$.

torque itself can be regarded as a vector. But for rotation around an axis with fixed direction, like a door rotating around its hinges, all we need is the idea that torque can be clockwise or counterclockwise.)

In Section 8-2 we denoted the rotation axis as the z axis and used the symbol ω_z to denote angular velocity, a quantity that can be positive or negative depending on the direction of rotation. We did this to distinguish ω_z from the angular speed ω, which is the magnitude of angular velocity. In the same way we'll use τ_z as the symbol for torque and τ (with no subscript) to denote the magnitude of torque, as in Equation 8-18. If we choose the positive rotation direction to be counterclockwise, then in Figure 8-19a the torque is negative (it tends to cause clockwise rotation) and $\tau_z < 0$. In Figure 8-19b the torque has the same magnitude τ (the distance r, force magnitude F, and angle ϕ are the same as in Figure 8-19a), but the torque tends to cause counterclockwise rotation and so $\tau_z > 0$.

Go to Interactive Exercise 8-2 for more practice dealing with torque.

Figure 8-19c shows another way to calculate the magnitude of the torque for the situation in Figure 8-19a. As this figure shows, the **line of action** of the force is just an extension of the force vector $\vec{F}$ through the point where the force is applied. The **lever arm** of the force (also called the *moment arm*) is the perpendicular distance from the rotation axis to the line of action of the force, which is why we denote it by the symbol $r_\perp$ ($\perp$ is mathematical shorthand for "perpendicular"). Trigonometry shows that the lever arm $r_\perp$ equals $r \sin\phi$, the same quantity that appears in Equation 8-18. Hence we can rewrite that equation as

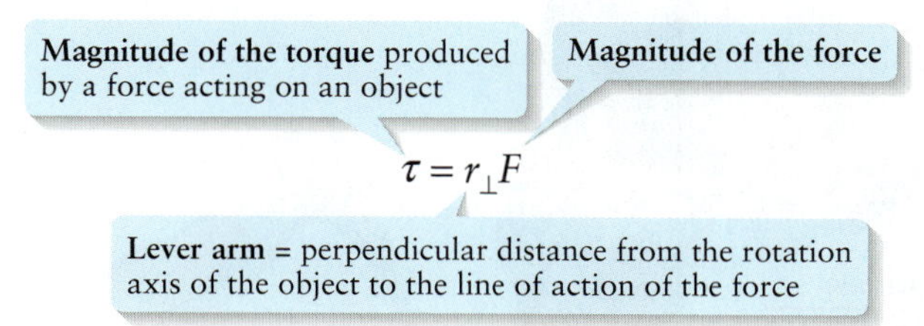

Magnitude of torque in terms of lever arm (8-19)

Equation 8-19 is mathematically equivalent to Equation 8-18 but defines torque in a slightly different way: It says that the magnitude of the torque produced by a force equals the lever arm of the force multiplied by the force magnitude. The longer the lever arm for a given force magnitude, the greater the torque.

Figure 8-20 illustrates these ideas about torque using the door that we discussed previously (see Figure 8-18). For a given force magnitude F, the lever arm $r_\perp$ and hence the torque magnitude τ will be as large as possible if r is as large as possible and if $\sin\phi$ has its maximum value of 1, which occurs if $\phi = 90°$. In other words a force produces the maximum torque—and hence the maximum rotating effect—if it is applied as far from the rotation axis as possible and in a direction perpendicular to a line from the rotation axis to the point of application (see **Figure 8-20a**). That's just what we said in our discussion of the door. By contrast if the force is applied close to the rotation axis so the distance r is small—like pushing on a door at a point close to the hinges—the

Figure 8-20 Torques on a door (a) For a force $\vec{F}$ of a given magnitude, the torque is greatest when r is large and $\phi = 90°$. The torque is small if (b) r is small or (c) $\sin\phi$ is small.

(a) Pulling on a door far from the hinge (view as seen looking down on the door from above): Large lever arm $r_\perp$.

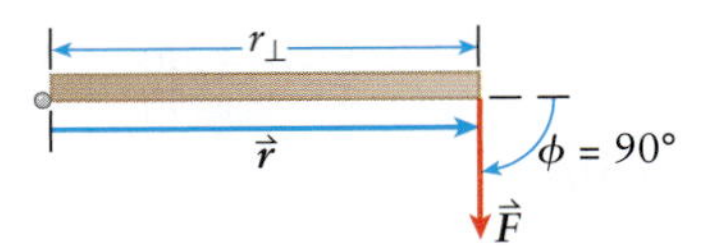

(b) Pulling on a door close to the hinge: Small lever arm $r_\perp$.

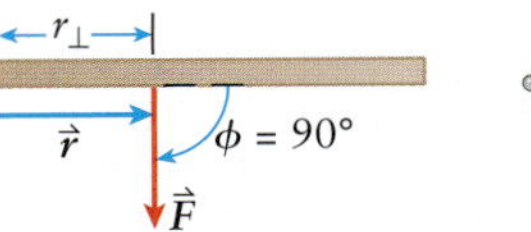

(c) Pushing on a door directly toward the hinge: Zero lever arm $r_\perp$.

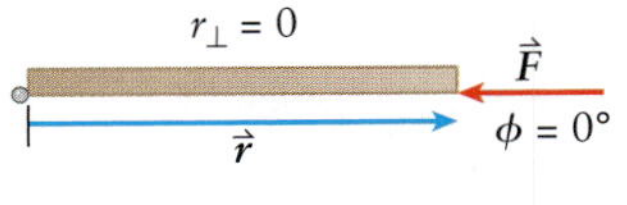

lever arm $r_\perp = r \sin\phi$ is short, so the torque magnitude and rotating effect are small (**Figure 8-20b**). The torque is also small if the force is applied in nearly the same direction as the vector $\vec{r}$ as in **Figure 8-20c**. In that case the angle ϕ is nearly zero, $\sin\phi$ has a small value, and the lever arm $r_\perp = r \sin\phi$ is again short. This is like pulling on a door in a direction nearly parallel to the door's plane. (The same is true if the force is directed nearly opposite to $\vec{r}$, so ϕ is close to 180° and again $\sin\phi$ is close to zero.)

From Equation 8-19, the SI units of torque are newtons multiplied by meters, or newton-meters (abbreviated N · m).

WATCH OUT! Torque and work have the same units but are very different quantities.

Both torque and work have units of newton-meters. But you can't set one equal to the other. Work involves the product of the component of a force parallel to a displacement. Torque is the product of the component of a force perpendicular to a lever arm. Because physically they represent two very different things, it makes no sense to set one equal to the other.

GOT THE CONCEPT? 8-7 Using Torque to Loosen a Bolt

A socket wrench, like the one in **Figure 8-21**, can be used to loosen a bolt. If a bolt has become frozen in place and is especially difficult to loosen, a common trick is to slide a section of pipe over the handle of the wrench and turn the bolt while gripping the end of the pipe. Why does this work? Why do many experienced craftsmen tend to avoid using this trick?

Figure 8-21 Wrench torque How does lengthening the handle of a wrench make it easier to loosen a bolt?

Lever Arms in Anatomy

BIO-Medical As Equation 8-19 shows even a small force F can generate a large torque τ if the lever arm $r_\perp$ is long enough. Humans and other animals take advantage of this principle through the arrangement of their muscles and bones. As an example, **Figure 8-22a** shows that the point at which the masseter muscle is attached to the lower jawbone is far from the joint around which the jaw rotates. Hence the force exerted by the masseter muscle has a long lever arm, and so produces a large torque on the lower jawbone. This large torque enables you to crack a nut between your back teeth or use your front teeth to tear into a raw carrot.

A very different arrangement is used in your forearm. One end of the

(a)

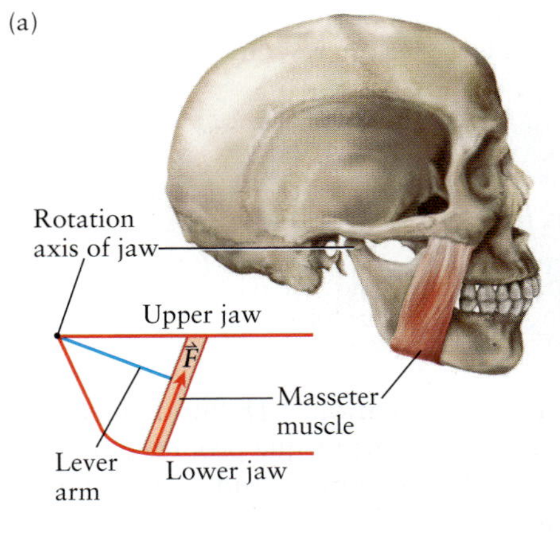

(b)

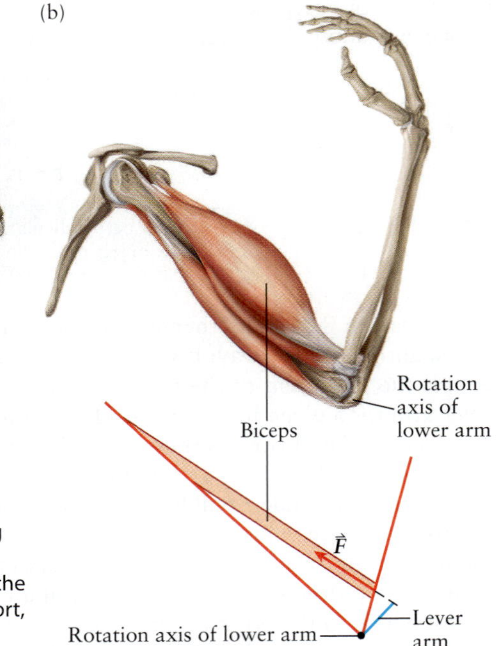

Figure 8-22 Lever arms in the human jaw and arm (a) The masseter muscle exerts a force on the lower jaw with a relatively long lever arm, so the torque on the lower jaw is quite large. (b) The lever arm for the force of the biceps muscle on the forearm is relatively short, so the torque is relatively small.

biceps muscle attaches to the bone of the upper arm and the other to the lower arm just below the elbow (**Figure 8-22b**). Even though the biceps muscle can exert a large force, this muscle–joint arrangement doesn't generate a large torque because the lever arm is relatively short. In chimpanzees the bicep is attached farther from the elbow joint, giving it the ability to generate more forearm torque than a human despite the chimpanzee's smaller size and smaller muscles.

Torque, Angular Acceleration, and Newton's Second Law for Rotation

We have suggested that a torque acting on an object (like the torque produced by pushing or pulling on a door) affects its rotation. What is the precise relationship between torque and the rotational effect that it causes? To find out let's recall Newton's second law for *linear* motion along the x axis:

$$\sum F_{\text{ext},x} = ma_x \quad (4\text{-}4\text{a})$$

This equation states that an object accelerates in response to the sum of all the external forces that act on the object: This sum $\sum F_{\text{ext},x}$ equals the object's mass m multiplied by its acceleration a_x. To find the rotational equivalent of this equation, remember from Section 8-2 that moment of inertia I plays the same role for rotational motion that mass m does for linear motion and recall from Section 8-5 that angular acceleration α is the rotational equivalent of linear acceleration a_x. If torque plays the same role for rotational motion as force plays for translational motion, then we should be able to replace F_x with τ_z, m with I, and a_x with α_z in Equation 4-4a to get the rotational version of Newton's second law:

The **net torque,** or sum of all external torques acting on an object for the axis about which it can rotate

The **angular acceleration** produced by the net torque

$$\sum \tau_{\text{ext},z} = I\alpha_z$$

The object's **moment of inertia** for the axis about which it can rotate

Newton's second law for rotational motion
(8-20)

Experiment shows that Equation 8-20 is, in fact, correct. An object acquires an *angular* acceleration α_z in response to the sum $\sum \tau_{\text{ext},z}$ of all of the external *torques* that act on the object, and this angular acceleration is directly proportional to the net torque. The magnitude of the angular acceleration also depends on the object's moment of inertia I around the rotation axis. The basic physics is the same as the original form of Newton's second law, Equation 4-4a.

BIO-Medical GOT THE CONCEPT? 8-8 Rotating the Human Jaw

? As **Figure 8-23** shows, the human jaw can rotate around three axes as shown. Chewing is accomplished primarily by rotations around the axis labeled y, while moving your jaw from side to side involves rotating it around the axis labeled z. Measurements show that for a lower jaw with a mass of approximately 0.4 kg, typical values for the moments of inertia are $I_y = 3 \times 10^{-4}\ \text{kg}\cdot\text{m}^2$ and $I_z = 9 \times 10^{-4}\ \text{kg}\cdot\text{m}^2$. (a) Does the anatomy of the jaw favor (i) up-down motions or (ii) side-to-side motion? (b) Is more torque required (i) to open your jaw or (ii) to rotate it from side to side?

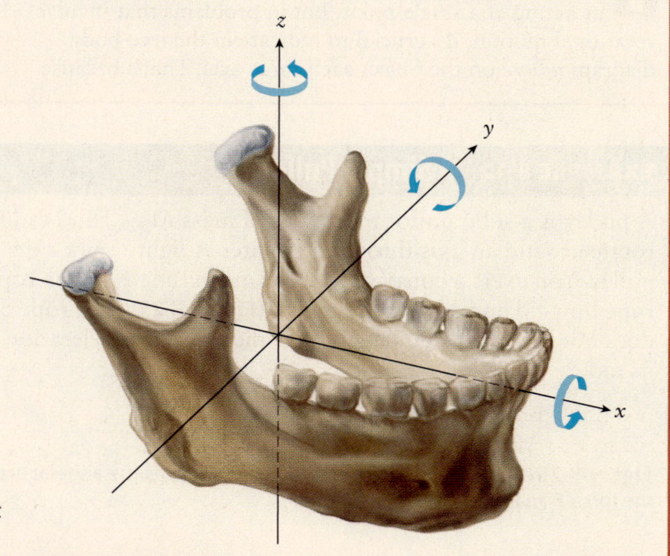

Figure 8-23 The human jaw The human jaw has a different moment of inertia for each of the three axes x, y, and z.

BIO-Medical **GOT THE CONCEPT? 8-9 Insects versus Birds**

The maneuverability of a flying insect or a bird depends in large part on the contributions their wings make to the moments of inertia around their roll, pitch, and yaw axes (**Figure 8-24**). As much as 15% of the total body mass of a bird can be in its wings, while the wings of a typical insect are a considerably smaller fraction of its total mass. Which would you expect to be able to maneuver more quickly in flight: (a) a flying insect or (b) a bird?

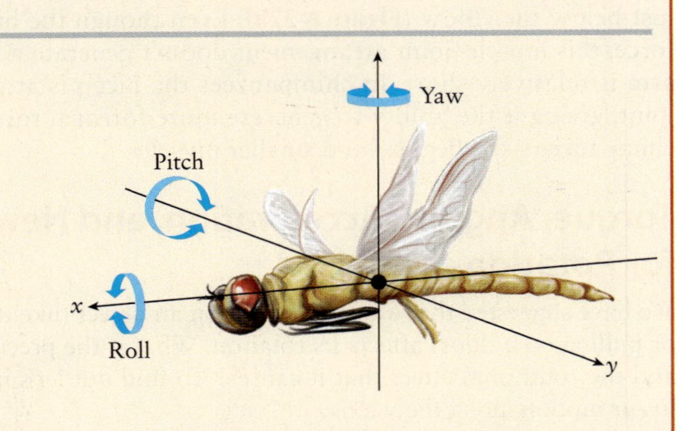

Figure 8-24 Rotation axes of an insect An insect can rotate in three ways: It can roll around its longitudinal *x* axis, pitch around its lateral *y* axis, or yaw around its *z* axis.

TAKE-HOME MESSAGE FOR Section 8-6

✔ If a force acting on an object tends to change that object's rotation, the force generates a torque. The magnitude of the torque depends on where the force is applied relative to the rotation axis and on the orientation of the force.

✔ The magnitude of a torque is equal to the magnitude of the force that causes the torque multiplied by the lever arm of the force (the perpendicular distance from the rotation axis to the line of action of the force).

✔ Newton's second law for rotation says that if a net external torque $\sum \tau_z$ acts on an object, the object acquires an angular acceleration α_z. These are related via the moment of inertia I: $\sum \tau_{\text{ext},z} = I\alpha_z$.

8-7 The techniques used for solving problems with Newton's second law also apply to rotation problems

The examples in this section show how to use the rotational form of Newton's second law. Just as for the translational form of this law, in each example we'll begin our solution by drawing a free-body diagram.

In many problems it's necessary to use *both* the translational and rotational forms of Newton's second law. Examples 8-13 and 8-14 show how to tackle problems like these.

WATCH OUT! In rotational problems, it matters *where* the force acts.

In problems that involve Newton's second law for translational motion, we draw all of the forces on an object as acting at a single point. But in problems that involve rotational motion, it's crucial to indicate in the free-body diagram *where* on the object each force acts. That's because the torque generated by a force depends on where the force is applied relative to the rotation axis. In the examples below, study closely how forces are depicted in free-body diagrams and how the corresponding torques are calculated.

EXAMPLE 8-12 A Simple Pulley I

A pulley is a solid uniform cylinder of mass M_{pulley} and radius R that is free to rotate around an axis through its center. A lightweight rope is wound around the pulley. You exert a constant force of magnitude F on the rope, which makes the rope unwind and rotates the pulley (**Figure 8-25**). The rope does not stretch and does not slip on the pulley. What is the angular acceleration of the pulley about its axle?

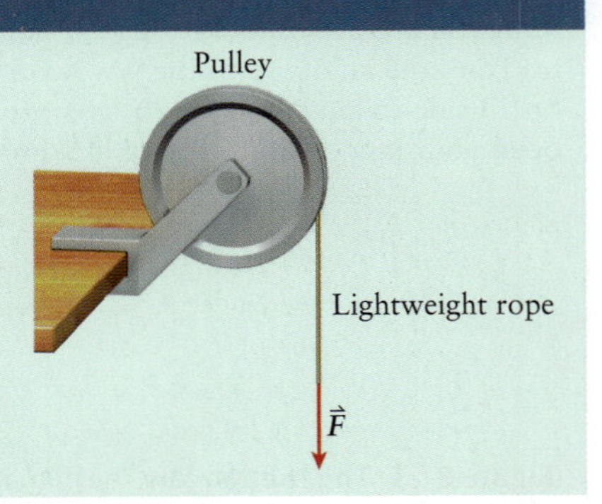

Figure 8-25 Exerting a torque on a pulley What is the angular acceleration of the pulley as the force $\vec{F}$ makes the rope unwind?

Set Up

We begin by drawing a free-body diagram for the pulley, taking care to draw each force at the point where it acts. We are told that the rope is lightweight (that is, it has much less mass than the pulley), so the force that you exert on the free end of the rope has the same magnitude F as the force that the rope exerts on the pulley. This force exerts a torque on the pulley and causes an angular acceleration. We'll use the rotational form of Newton's second law to determine this angular acceleration.

Newton's second law for rotational motion:

$$\sum \tau_{\text{ext},z} = I_{\text{pulley}}\alpha_{\text{pulley},z} \qquad (8\text{-}20)$$

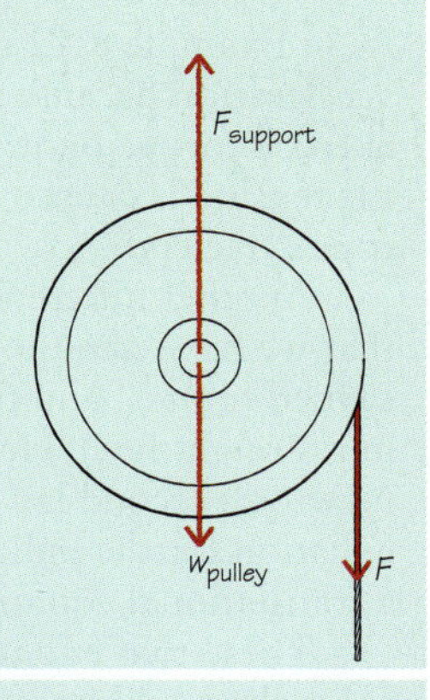

Solve

The free-body diagram shows that only the tension force F exerts a torque on the pulley causing it to rotate around its axle. (The support force F_{support} and the weight of the pulley W_{pulley} both act at its rotation axis, so the lever arm is zero for both of these forces.) The lever arm for the tension force is R, so the tension torque is R multiplied by F.

Torque due to tension force:

$$\tau_z = r_\perp F = RF$$

(This torque makes the pulley rotate in the clockwise direction, so we take clockwise to be the positive rotation direction.)

This is the only torque acting on the pulley, so

$$\sum \tau_{\text{ext},z} = RF$$

From Table 8-1 in Section 8-3, the moment of inertia of a solid cylinder around its central axis is $I = MR^2/2$. Insert this into Newton's second law for rotational motion and solve for $\alpha_{\text{pulley},z}$.

$$\sum \tau_{\text{ext},z} = I_{\text{pulley}}\alpha_{\text{pulley},z}$$

Substitute values of $\sum \tau_{\text{ext},z}$ and $I_{\text{pulley}} = M_{\text{pulley}}R^2/2$:

$$RF = \frac{1}{2}M_{\text{pulley}}R^2\alpha_{\text{pulley},z}$$

Solve for angular acceleration α_z: $\alpha_{\text{pulley},z} = \dfrac{2RF}{M_{\text{pulley}}R^2} = \dfrac{2F}{M_{\text{pulley}}R}$

Reflect

Our result for the angular acceleration $\alpha_{\text{pulley},z}$ depends on the pulling force F, the pulley mass M_{pulley}, and the pulley radius R. It makes sense that our result is proportional to the ratio F/M_{pulley}: A greater force F means a stronger pull and a greater angular acceleration, while a greater mass M_{pulley} means the pulley is more difficult to rotate and gives a smaller angular acceleration.

Our result also shows that the larger the pulley radius R, the smaller the angular acceleration $\alpha_{\text{pulley},z}$. This may seem backward, since a larger radius means that the force F causes a larger torque $\tau = RF$. However, increasing the radius increases the moment of inertia $I = M_{\text{pulley}}R^2/2$ by a larger factor than it increases the torque. (Doubling the radius doubles the torque but quadruples the moment of inertia.) So the moment of inertia plays a more important role, which is why $\alpha_{\text{pulley},z}$ decreases with increasing pulley radius.

EXAMPLE 8-13 A Simple Pulley II

You attach a block of mass M_{block} to the free end of the rope in Example 8-12 (**Figure 8-26**). When the block is released from rest and falls downward, the rope unwinds, and the pulley rotates around its central axis. As in Example 8-12 the rope neither stretches nor slips on the pulley. Derive expressions for the acceleration of the block, the angular acceleration of the pulley, and the tension in the string.

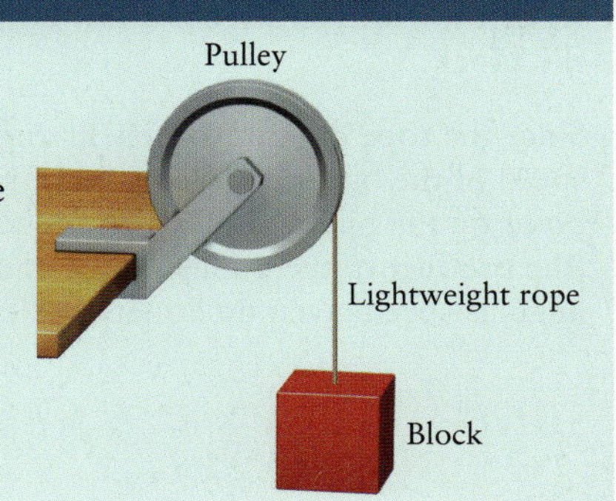

Figure 8-26 Block and pulley What is the downward acceleration of the block as it falls, and what is the angular acceleration of the pulley as the rope unwinds?

Set Up

As in Example 8-12 the pulley has an angular acceleration because the tension of the rope exerts a torque on it. The difference here is that the tension is caused by the block attached to the rope's free end.

Another difference from Example 8-12 is that we have *two* moving objects: the pulley (which rotates) and the block (which moves in a straight line). Hence we have to write two Newton's second law equations: a rotational equation for the pulley, as in Example 8-12, and a translational equation for the block.

These two equations won't be enough to solve the problem, since we're trying to find *three* quantities: the block's acceleration $a_{block,x}$, the pulley's angular acceleration $\alpha_{pulley,z}$, and the string tension T. Happily, we can get a third equation because the string doesn't stretch. This tells us that the speed of the block equals the speed of a point on the pulley rim. We'll use this to find a relationship between $a_{block,x}$ and $\alpha_{pulley,z}$.

Newton's second law for rotational motion applies to the pulley:

$$\sum \tau_{ext,z} = I_{pulley}\alpha_{pulley,z} \quad (8\text{-}20)$$

Newton's second law for translational motion applies to the block:

$$\sum F_{ext,x} = M_{block}a_{block,x} \quad (4\text{-}4a)$$

Statement that rope doesn't stretch:

$$v_{block} = v_{rim\ of\ pulley}$$

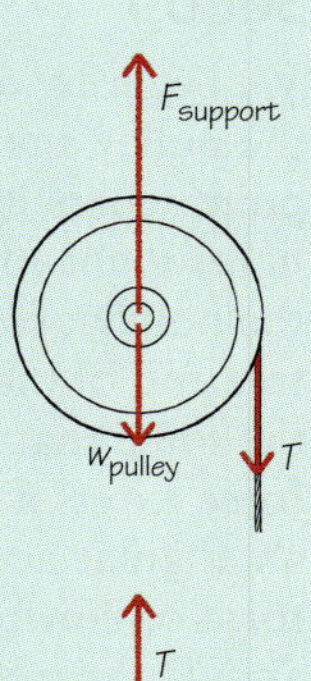

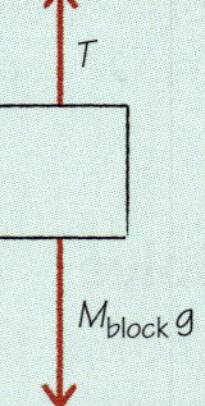

Solve

As in Example 8-12, the only force that exerts a torque on the pulley is the rope tension, which we call T. The lever arm of this force is R, so the net torque is $\sum\tau_{ext,z} = RT$. (As in Example 8-12 this torque makes the pulley rotate in the clockwise direction. So we take clockwise to be the positive rotation direction.) Insert this and $I_{pulley} = M_{pulley}R^2/2$ into the rotational Newton's second law equation and solve for the tension.

For the pulley,

$$\sum \tau_{ext,z} = I_{pulley}\alpha_{pulley,z}$$

Substitute values of $\sum\tau_z$ and I_{pulley}:

$$RT = \frac{1}{2}M_{pulley}R^2\alpha_{pulley,z}$$

Divide through by R:

$$T = \frac{1}{2}M_{pulley}R\alpha_{pulley,z}$$

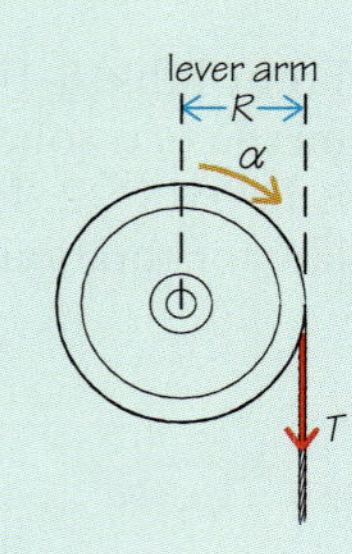

We take the positive x direction to point downward, which is the direction in which the block will move when the pulley rotates clockwise. With this choice the acceleration $a_{block,x}$ of the block and the angular acceleration $\alpha_{pulley,z}$ of the pulley are both positive, the gravitational force $M_{block}g$ is in the positive x direction, and the tension force T on the block is in the negative x direction. (The rope has negligible mass, so the tension is the same throughout its length.) Use this to write the Newton's second law equation for the block.

For the block,

$$\sum F_{ext,x} = M_{block}a_{block,x}$$

Substitute the individual forces:

$$M_{block}g - T = M_{block}a_{block,x}$$

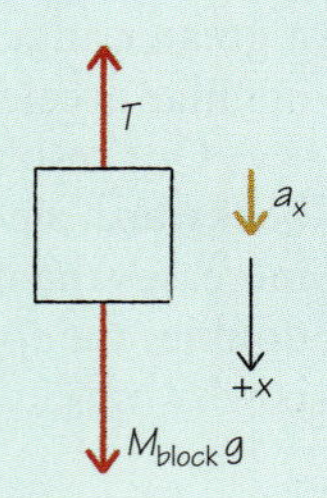

Since the rope doesn't stretch, at any instant the speed of the block v_{block} equals the speed of a point on the pulley rim, which is equal to $R\omega$ (the product of the pulley radius and the pulley angular speed given by Equation 8-3). As the

At a certain time, $v_{block} = v_{rim\ of\ pulley}$ or

$$v_{block,x} = R\omega_{pulley} = R\omega_{pulley,z}$$

block falls it gains speed and the pulley gains angular speed, but the speed of the block remains equal to the speed of the pulley's rim. Use this to relate the linear acceleration of the block to the angular acceleration of the pulley.

(The block is moving downward in the positive x direction, so its x velocity $v_{\text{block},x}$ is positive and equal to its speed v_{block}. The pulley is rotating in the positive direction, so its angular velocity $\omega_{\text{pulley},z}$ is positive and equal to the pulley's angular speed ω_{pulley}.)

$v_{\text{rim of pulley}}$

same speed

v_{block}

A time Δt later,

$$v_{\text{block},x} + \Delta v_{\text{block},x} = R(\omega_{\text{pulley},z} + \Delta\omega_{\text{pulley},z})$$

so

$$v_{\text{block},x} + \Delta v_{\text{block},x} = R\omega_{\text{pulley},z} + R\,\Delta\omega_{\text{pulley},z}$$

Since $v_{\text{block},x} = R\omega_{\text{pulley},z}$ from above, it follows that

$$\Delta v_{\text{block},x} = R\,\Delta\omega_{\text{pulley},z}$$

Divide both sides by Δt:

$$\frac{\Delta v_{\text{block},x}}{\Delta t} = R\frac{\Delta\omega_{\text{pulley},z}}{\Delta t}$$

The ratios on the two sides are just the acceleration of the block and the angular acceleration of the pulley. So

$$a_{\text{block},x} = R\alpha_{\text{pulley},z} \text{ or } \alpha_{\text{pulley},z} = \frac{a_{\text{block},x}}{R}$$

Combine the three equations to solve for the three unknowns.

Substitute $\alpha_{\text{pulley},z} = a_{\text{block},x}/R$ into the pulley equation:

$$T = \frac{1}{2}M_{\text{pulley}}R\alpha_{\text{pulley},z} = \frac{1}{2}M_{\text{pulley}}R\left(\frac{a_{\text{block},x}}{R}\right) = \frac{1}{2}M_{\text{pulley}}a_{\text{block},x}$$

Substitute this expression for T into the block equation:

$$M_{\text{block}}g - T = M_{\text{block}}a_{\text{block},x} \text{ so}$$

$$M_{\text{block}}g - \frac{1}{2}M_{\text{pulley}}a_{\text{block},x} = M_{\text{block}}a_{\text{block},x}$$

Solve for $a_{\text{block},x}$:

$$M_{\text{block}}g = M_{\text{block}}a_{\text{block},x} + \frac{1}{2}M_{\text{pulley}}a_{\text{block},x}$$

$$= \left(M_{\text{block}} + \frac{1}{2}M_{\text{pulley}}\right)a_{\text{block},x}$$

$$a_{\text{block},x} = \left(\frac{M_{\text{block}}}{M_{\text{block}} + M_{\text{pulley}}/2}\right)g$$

Now find $\alpha_{\text{pulley},z}$ by substituting this into $\alpha_{\text{pulley},z} = a_{\text{block},x}/R$:

$$\alpha_{\text{pulley},z} = \frac{a_{\text{block},x}}{R} = \left(\frac{M_{\text{block}}}{M_{\text{block}} + M_{\text{pulley}}/2}\right)\frac{g}{R}$$

Finally, solve for T from the pulley equation:

$$T = \frac{1}{2}M_{\text{pulley}}R\alpha_{\text{pulley},z} = \frac{1}{2}M_{\text{pulley}}R\left(\frac{M_{\text{block}}}{M_{\text{block}} + M_{\text{pulley}}/2}\right)\frac{g}{R}$$

$$= M_{\text{block}}g\left(\frac{M_{\text{pulley}}}{2M_{\text{block}} + M_{\text{pulley}}}\right)$$

Reflect

The downward acceleration of the block is less than g because the tension of the rope exerts an upward pull on the block that partially cancels the downward pull of gravity. The expression for tension T confirms this: T is less than $M_{block}g$, which says that the tension force doesn't completely balance out the gravitational force. If it did, the block would just hang there when released and wouldn't move at all!

Here's a way to check our results: If the pulley is very much lighter than the block, so M_{pulley} is nearly zero, it's as though the block is connected to nothing at all. In this case the block's downward acceleration would be g (it would be in free fall), and there would be no tension in the rope. Can you use our results for $a_{block,x}$ and T to show that this is the case?

$$a_{block,x} = \left(\frac{M_{block}}{M_{block} + M_{pulley}/2}\right)g$$

Inside the parentheses the numerator M_{block} is less than the denominator $M_{block} + M_{pulley}/2$, so the ratio inside the parentheses is less than 1 and $a_{block,x} < g$.

$$T = M_{block}\,g\left(\frac{M_{pulley}}{2M_{block} + M_{pulley}}\right)$$

Inside the parentheses the numerator M_{pulley} is less than the denominator $2M_{block} + M_{pulley}$, so the ratio inside the parentheses is less than 1 and $T < M_{block}g$.

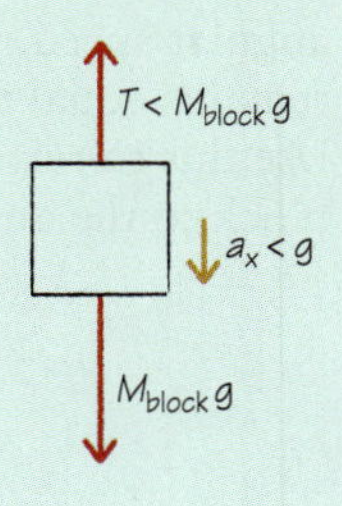

EXAMPLE 8-14 Rolling Without Slipping Revisited

In Example 8-10 in Section 8-4, we considered a uniform disk of mass M and radius R that rolls down a ramp without slipping. Use Newton's second law for rotational motion and for linear motion to determine the downhill acceleration of the disk's center of mass. The ramp is inclined at an angle θ to the horizontal.

Set Up

Unlike Example 8-13, in which we had one object (the pulley) undergoing purely rotational motion and another object (the block) undergoing purely translational motion, here we have *one* object that undergoes *both* rotational and translational motion. The principles are the same, however. We'll write down equations for both the rotational and translational forms of Newton's second law for the disk.

We also know from Section 8-4 that because the disk rolls without slipping, there's a relationship between the speed of the disk's center of mass v_{CM} and its angular speed ω. In a manner similar to what we did in Example 8-13, we'll relate the disk's center of mass acceleration and its angular acceleration.

Newton's second law for rotational motion of the disk around its central axis (the z axis):

$$\sum \tau_{ext,z} = I\alpha_z \qquad (8\text{-}20)$$

Newton's second law for linear motion of the disk—note that we include both x and y equations:

$$\sum F_{ext,x} = Ma_{CM,x}$$
$$\sum F_{ext,y} = Ma_{CM,y} \qquad (4\text{-}4)$$

Disk rolls without slipping:

$$v_{CM} = R\omega \qquad (8\text{-}13)$$

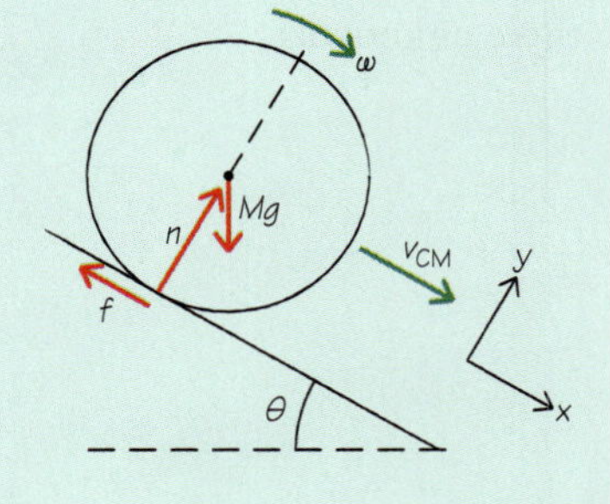

Solve

Three forces act on the disk as it rolls: the downward force of gravity Mg, the normal force n that acts perpendicular to the surface of the ramp, and the uphill force of friction f that keeps the disk from sliding downhill.

The force of gravity exerts no torque around the center of mass because it acts at the disk's center of mass, so its lever arm is $r_\perp = 0$. The normal force acts along a line that passes through the center of mass, so its lever arm is also zero and it exerts no torque. For the friction force, $r_\perp = R$ and $\tau_{friction,z} = Rf$. Substitute this and $I = MR^2/2$ into Newton's second law for rotation.

Newton's second law for rotation:

$$\sum \tau_{ext,z} = I\alpha_z$$

Substitute $\sum \tau_z = \tau_{friction,z} = Rf$ and $I = MR^2/2$:

$$Rf = \frac{1}{2}MR^2\alpha_z$$

Divide both sides of this equation by R:

$$f = \frac{1}{2}MR\alpha_z$$

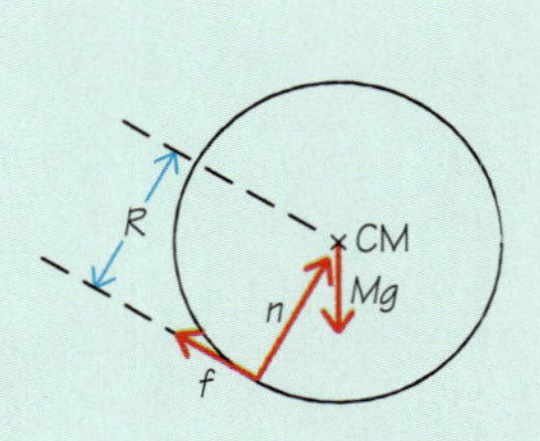

For translational motion, write the equations for the net force in the x and y directions. There is no acceleration in the y direction (perpendicular to the ramp) so $a_{CM,y} = 0$. The acceleration downhill in the x direction, $a_{CM,x}$, results from the forces of gravity and friction.

Net force in each direction:

$$\sum F_x = Mg \sin \theta - f$$

$$\sum F_y = n - Mg \cos \theta$$

Substitute into Newton's second law for translational motion:

$$Mg \sin \theta - f = Ma_{CM,x}$$

$$n - Mg \cos \theta = 0$$

(since $a_{CM,y} = 0$)

The y equation for linear motion isn't useful (it would only be of interest if we wanted to calculate the normal force n). The other two equations—one for rotational motion and one for linear motion along the x axis—involve three unknowns: f, α_z, and $a_{CM,x}$. So we need a third equation. This comes from the condition for rolling without slipping, $v_{CM} = R\omega$.

Rotation: $f = \frac{1}{2}MR\alpha_z$

Translational motion along x axis: $Mg \sin \theta - f = Ma_{CM,x}$

Rolling without slipping: $v_{CM} = R\omega$

In Example 8-13 the similar equation $v_{block,x} = R\omega_{pulley,z}$ led to $a_{block,x} = R\alpha_{pulley,z}$. The same mathematical steps (which we won't repeat here) lead us to the same relationship between $a_{CM,x}$ and α_z for the disk, which is our third equation.

Hence $a_{CM,x} = R\alpha_z$ or $\alpha_z = \dfrac{a_{CM,x}}{R}$

Combine the three equations to solve for $a_{CM,x}$.

Substitute $\alpha_z = a_{CM,x}/R$ into the rotational equation:

$$f = \frac{1}{2}MR\alpha_z = \frac{1}{2}MR\left(\frac{a_{CM,x}}{R}\right) = \frac{1}{2}Ma_{CM,x}$$

Substitute this expression for f into the linear motion equation and solve for $a_{CM,x}$:

$$Mg \sin \theta - f = Ma_{CM,x} \text{ so } Mg \sin \theta - \frac{1}{2}Ma_{CM,x} = Ma_{CM,x}$$

$$Mg \sin \theta = Ma_{CM,x} + \frac{1}{2}Ma_{CM,x} = \frac{3}{2}Ma_{CM,x}$$

$$g \sin \theta = \frac{3}{2}a_{CM,x}$$

$$a_{CM,x} = \frac{2}{3}g \sin \theta$$

Reflect

If there were no friction, the disk would slide downhill rather than roll and its acceleration would be $g \sin \theta$ (the component along the ramp of the acceleration due to gravity), as in Example 4-5. Friction slows the disk's downhill motion, which is why its acceleration is only two-thirds of $g \sin \theta$. Can you show that the magnitude of the friction force is $(Mg/3) \sin \theta$?

We can check our result for acceleration by asking how fast the disk will be moving at the bottom of the ramp if it starts from rest at

Motion with constant x acceleration:

$$v_{CM,x}^2 = v_{CM,0x}^2 + 2a_{CM,x}(x - x_0)$$

Disk starts at rest: $v_{CM,0x} = 0$

Displacement down the ramp:

$$x - x_0 = \frac{H}{\sin \theta}$$

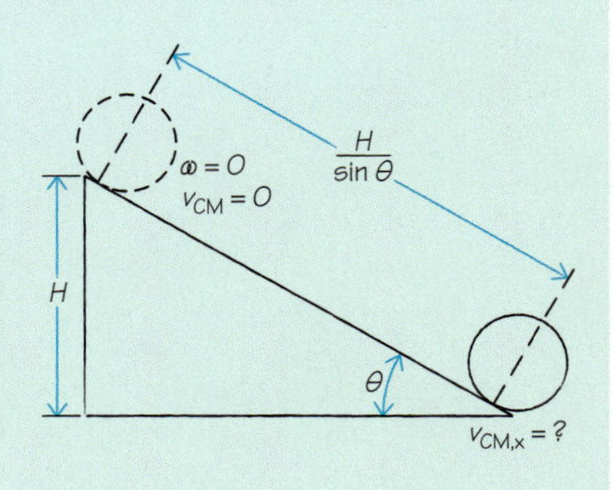

a height H. The answer we get is $\sqrt{(4/3)gH}$. This is the same as we found in Example 8-10 in Section 8-4 using energy conservation—which is encouraging!

We calculated the acceleration above: $a_{\text{CM},x} = \dfrac{2}{3} g \sin\theta$, so

$$v_{\text{CM},x}^2 = 0 + 2\left(\frac{2}{3} g \sin\theta\right)\left(\frac{H}{\sin\theta}\right) = \frac{4gH}{3}$$

$$v_{\text{CM},x} = \sqrt{\frac{4gH}{3}}$$

GOT THE CONCEPT? 8-10 Which Friction?

What kind of friction force acts on the disk in Example 8-14 as it rolls downhill: (a) static friction, (b) kinetic friction, or (c) a mixture of static and kinetic friction?

GOT THE CONCEPT? 8-11 A Uniform Sphere Rolling Downhill

Suppose we replace the disk in Example 8-14 with a uniform solid sphere like a billiard ball. Compared to the linear acceleration of the disk, what will be the linear acceleration of the sphere? (a) greater; (b) the same; (c) less.

TAKE-HOME MESSAGE FOR Section 8-7

✔ Always draw a free-body diagram as part of the solution of any problem that involves forces and/or torques. In order to be able to calculate torques correctly, draw each force at the point where it is applied.

✔ In problems of this kind apply Newton's second law for translation to any object that changes position and Newton's second law for rotation to any object that rotates. Both laws must be applied to any object that undergoes both translation and rotation.

✔ In problems that involve both translation and rotation you will need to relate the translational acceleration to the rotational acceleration.

8-8 Angular momentum is conserved when there is zero net torque on a system

When a spinning figure skater pulls her arms and legs in close to her body, the rate at which she spins automatically increases (see the photographs that open this chapter). You can demonstrate this same effect by sitting in an office chair. Sit with your arms outstretched and hold a weight, like a brick or a full water bottle, in either hand. Now use your feet to start your body and the chair rotating, lift your feet off the ground, and then pull your arms inward. Your rotation will speed up quite noticeably!

Why does this happen? For both cases, the figure skater and you in the office chair, there's no torque acting to make the object rotate faster. Instead, both of these situations are examples of the *conservation of angular momentum*. This principle is the rotational analog of the conservation of *linear* momentum that we introduced in Chapter 7.

Angular Momentum and Angular Momentum Conservation

To see what we mean by angular momentum, consider an object rotating around a fixed z axis that doesn't change its orientation. The object's moment of inertia around this axis is I and its angular velocity is ω_z. We've seen that these quantities are analogous to the mass m and velocity v_x of an object in straight-line motion along the x axis. From Section 7-2, the linear momentum for such straight-line motion is

$$p_x = mv_x \tag{8-21}$$

By analogy to Equation 8-21 we'll define a new quantity called the **angular momentum** of an object rotating around the z axis. We'll give this the symbol L_z:

The **angular momentum** of a rigid object rotating around a fixed axis...

...is proportional to the **angular velocity** at which the object rotates around that axis.

$$L_z = I\omega_z$$

The angular momentum is also proportional to the object's **moment of inertia** for that rotation axis.

Angular momentum for rotation around a fixed axis
(8-22)

The greater the moment of inertia and the faster the angular velocity, the greater the value of angular momentum. Note that like angular velocity, angular momentum can be positive or negative depending on which way the object is rotating and which direction of rotation you choose to be positive.

WATCH OUT! Angular momentum and linear momentum are different quantities.

Although linear momentum p_x and angular momentum L_z are analogous to each other, they are *not* the same quantity and do not have the same units. Linear momentum has units of mass times velocity, or kg · m/s. The units of angular momentum are the units of moment of inertia (kg · m^2) multiplied by the units of angular velocity (rad/s), which we can write as kg · m^2/s. (We can eliminate the "rad" since radians are not a true unit.)

Why is angular momentum important? We saw in Section 7-3 that the total *linear* momentum of a system of two or more objects is *conserved* if there is no net external force on the system. The analogous statement about angular momentum is

If there is no net external torque on a system, the angular momentum of the system is conserved.

This is the principle of **conservation of angular momentum.** It explains what happens to a spinning figure skater when she pulls her limbs in. As she does so her moment of inertia I decreases because more of her body's mass is closer to her rotation axis. There are no external torques on her—the force of gravity and the normal force don't have any effect on her rotation, and the friction force on her skates is very small. Hence her angular momentum L_z keeps the same value, and as I decreases her angular velocity ω_z has to increase so that the product $L_z = I\omega_z$ doesn't change. The same thing happens in the office chair experiment we described at the beginning of this section.

GOT THE CONCEPT? 8-12 Tetherball Physics

In the child's game of tetherball, a rope attached to the top of a tall pole is tied to a ball. Players hit the ball in opposite directions in an attempt to wrap the rope around the pole. As the ball wraps around the pole, does the angular speed of the ball (a) decrease, (b) stay the same, or (c) increase? Explain in terms of angular momentum. Treat the rope as having negligible mass and neglect any resistive forces (such as air resistance and friction).

EXAMPLE 8-15 Spinning Figure Skater

A figure skater executing a "scratch spin" gradually pulls her arms in toward her body while spinning on one skate. During the spin her angular speed increases from 1.50 rad/s (approximately one revolution every four seconds) to 15.0 rad/s (approximately 2.5 rev/s). (a) By what factor does her moment of inertia around her central axis change as she pulls in her arms? (b) By what factor does her rotational kinetic energy change?

Set Up

There is very little friction between the skater and the ice, so the net torque on her is essentially zero and her angular momentum can be considered constant. We'll use this to determine her moment of inertia after she pulls her arms in, I_{after}, in terms of her moment of inertia before pulling them in, I_{before}. (Note that we are given the values $\omega_{before} = 1.50$ rad/s and $\omega_{after} = 15.0$ rad/s.) Then we'll compare her rotational kinetic energy before and after pulling her arms in by using Equation 8-7.

Angular momentum of the skater:

$$L_z = I\omega_z \quad (8\text{-}22)$$

Rotational kinetic energy of the skater:

$$K_{rotational} = \frac{1}{2}I\omega^2 \quad (8\text{-}7)$$

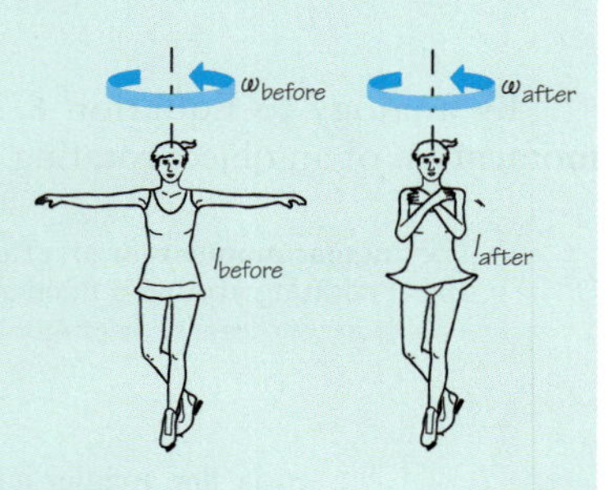

Solve

(a) Since angular momentum is conserved, we write an equation which says that the angular momentum after she pulls her arms in equals the angular momentum before she pulls them in. Then we solve for I_{after} in terms of I_{before}.

Conservation of angular momentum:

$$L_{after,z} = L_{before,z}$$

From Equation 8-22, this says

$$I_{after}\omega_{after,z} = I_{before}\omega_{before,z}$$

Let the direction of the skater's rotation be positive. Then her angular velocity ω_z at any instant is positive and equal to her angular speed ω at that instant. So the angular momentum conservation equation says

$$I_{after}\omega_{after} = I_{before}\omega_{before}$$

Rearrange to find the ratio of I_{after} to I_{before}:

$$\frac{I_{after}}{I_{before}} = \frac{\omega_{before}}{\omega_{after}} = \frac{1.50 \text{ rad/s}}{15.0 \text{ rad/s}} = 0.100$$

The moment of inertia after she pulls her arms in is only 0.100 times (one-tenth) as large as the value before she pulls her arms in.

(b) Now we can use Equation 8-7 to compare the skater's rotational kinetic energy before and after she pulls her arms in.

$$K_{rotational,before} = \frac{1}{2}I_{before}\omega_{before}^2$$

$$K_{rotational,after} = \frac{1}{2}I_{after}\omega_{after}^2$$

Calculate the ratio of her rotational kinetic energy before and after:

$$\frac{K_{rotational,after}}{K_{rotational,before}} = \frac{(1/2)I_{after}\omega_{after}^2}{(1/2)I_{before}\omega_{before}^2} \text{ so}$$

$$\frac{K_{rotational,after}}{K_{rotational,before}} = \left(\frac{I_{after}}{I_{before}}\right)\left(\frac{\omega_{after}}{\omega_{before}}\right)^2 = (0.100)\left(\frac{15.0 \text{ rad/s}}{1.50 \text{ rad/s}}\right)^2$$

$$= (0.100)\,(10.0)^2 = 10.0$$

Her rotational kinetic energy after she pulls her arms in is 10.0 times as great as the value before she pulls her arms in.

Reflect

The skater's moment of inertia changes by a factor of 0.100 = 1/10.0. This might seem large, because her arms are a relatively small fraction of her total mass. However, the moment of inertia

Moment of inertia of the skater:

$$I = \sum_{i=1}^{N} m_i r_i^2 \quad (8\text{-}6)$$

depends on the square of distance (see Equation 8-6), so holding her arms close to her body rather than extended has a significant effect on the skater's moment of inertia.

Although the skater's angular momentum maintains the same value as she pulls her arms in, her rotational kinetic energy does not. You can see why by rewriting the expression for rotational kinetic energy in terms of the angular momentum magnitude $L = I\omega$ and the angular speed ω. Since L is unchanged and ω increases when the skater pulls her arms in, her rotational kinetic energy must increase. The extra kinetic energy comes from the skater herself: Chemical energy from food she ate is used to flex muscles and bring her arms in.

Rotational kinetic energy:

$$K_{\text{rotational}} = \frac{1}{2}I\omega^2 = \frac{1}{2}(I\omega)\omega = \frac{1}{2}L\omega$$

The skater's angular speed ω increases while the magnitude L of her angular momentum stays the same, so $K_{\text{rotational}}$ increases.

Angular Momentum of a Particle

Example 8-15 demonstrates how the angular momentum concept helps us solve problems about rotating objects. It can also be useful to think about the angular momentum of a single *particle*. We've already seen the utility of this concept for the example of a tetherball moving around a pole (see Got the Concept? 8-12). Let's look at the general case of a particle of mass m moving with velocity $\vec{v}$ so that its linear momentum is $\vec{p} = m\vec{v}$. As an example, let's consider a girl (who we regard as a particle) running at constant speed toward a playground merry-go-round (**Figure 8-27**). We take the rotation axis to be the axis of the merry-go-round. At a given instant the vector $\vec{r}$ points directly away from the rotation axis to the position of the girl, and there is an angle ϕ between the vectors $\vec{r}$ and $\vec{p}$. What is the girl's angular momentum around this rotation axis?

In Equation 8-19 we defined the magnitude of the torque due to a force $\vec{F}$ as $\tau = r_\perp F$, where $r_\perp$ is the perpendicular distance from the rotation axis to the point

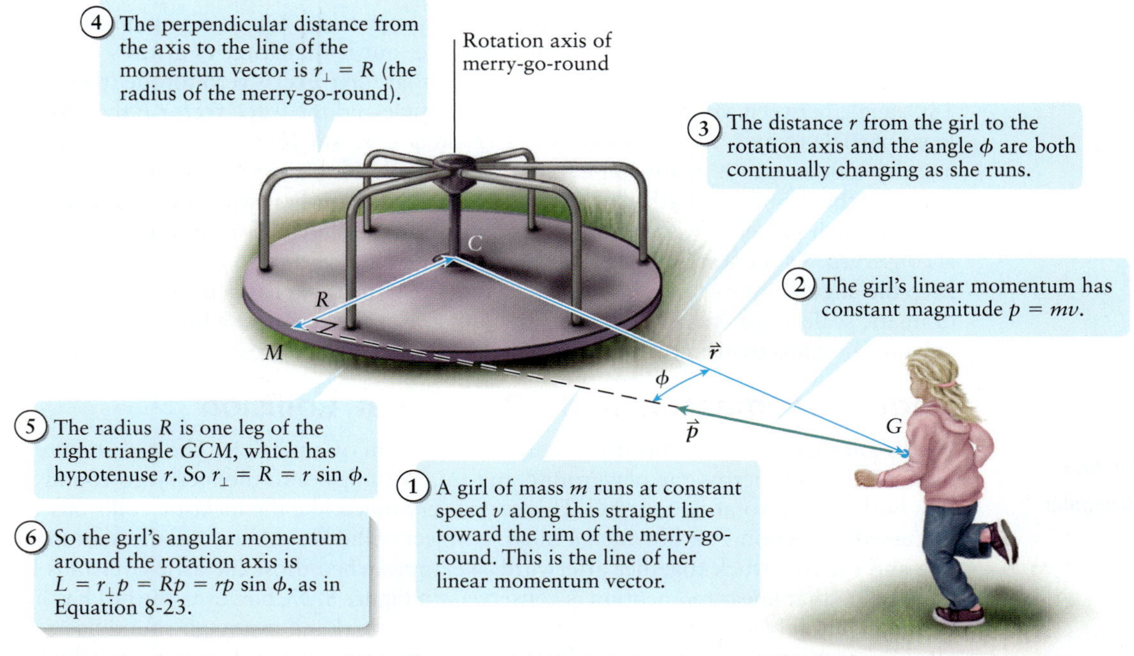

4 The perpendicular distance from the axis to the line of the momentum vector is $r_\perp = R$ (the radius of the merry-go-round).
Rotation axis of merry-go-round
3 The distance r from the girl to the rotation axis and the angle ϕ are both continually changing as she runs.
C
R
M
2 The girl's linear momentum has constant magnitude $p = mv$.
$\vec{r}$
ϕ
$\vec{p}$
G
5 The radius R is one leg of the right triangle GCM, which has hypotenuse r. So $r_\perp = R = r \sin\phi$.
1 A girl of mass m runs at constant speed v along this straight line toward the rim of the merry-go-round. This is the line of her linear momentum vector.
6 So the girl's angular momentum around the rotation axis is $L = r_\perp p = Rp = rp \sin\phi$, as in Equation 8-23.

Figure 8-27 A girl running with linear momentum and angular momentum The angular momentum of the girl around the rotation axis is $L = r_\perp p = rp \sin\phi$ (see Equation 8-23).

where the force is applied. We define the angular momentum of a particle with linear momentum $\vec{p}$ in an analogous way:

Magnitude of the angular momentum of a particle (8-23)

$$L = r_{\perp} p = rp \sin \phi$$

For example, in Figure 8-27 the perpendicular distance $r_{\perp}$ equals the radius R of the merry-go-round.

Does Equation 8-23 agree with our earlier definition $L_z = I\omega_z$ (Equation 8-22)? To find out, note that $p\sin\phi = mv\sin\phi = mv_{\perp}$, where $v_{\perp}$ is the component of the particle's velocity that's perpendicular to $\vec{r}$ in Figure 8-27. We can also write $v_{\perp} = r\omega$, where ω is the angular speed of the particle around the rotation axis. If we substitute these into Equation 8-23 for the magnitude of the particle's angular momentum, we get

(8-24) $$L = rp\sin\phi = rmv\sin\phi = rmv_{\perp} = rm(r\omega) = mr^2\omega$$

Equation 8-6 tells us that if there's only a single particle of mass m a distance r from the rotation axis, its moment of inertia is $I = mr^2$. So Equation 8-24 for the magnitude of the particle's angular momentum becomes $L = I\omega$. But this is just the magnitude of the angular momentum $L_z = I\omega_z$ as given by Equation 8-22. So Equation 8-23 for the angular momentum of a particle is completely consistent with our earlier work.

WATCH OUT! Angular momentum doesn't require the motion to be in a circle.

Note that a particle doesn't have to move in a circle, or even along a curved path, to have angular momentum. Consider the girl in Figure 8-27, who moves along a straight line at constant velocity. You can think of her as rotating around a fixed point (the axis of the merry-go-round) with an ever-changing radius from the axis.

The girl in Figure 8-27 has a constant linear momentum because the net external force acting on her is zero. She also has a constant *angular* momentum: Figure 8-27 shows that her angular momentum is $L = r_{\perp}p = Rp$, where R is the radius of the merry-go-round. Since R and her linear momentum p are both constant, her angular momentum L is constant. That happens because there is also zero net *torque* acting on her.

If there *is* a net force that acts on a particle, but if that force is always directed toward a fixed point that acts as a rotation axis, that force exerts zero torque (there is a 180° angle between the vector $\vec{r}$ from axis to particle and the force vector $\vec{F}$ that points from particle to axis, so the torque is $\tau = rF \sin \phi = rF \sin 180° = 0$). In that case the angular momentum is again conserved. One example of this is the motion of water as it circles the drain of a bathtub. The water is pulled directly toward the drain, so the force on each parcel of water exerts no torque and the angular momentum of the water is conserved. As the water approaches the drain, r decreases and so $p = mv$ increases to keep the angular momentum $L = rp_{\perp}$ constant. The same thing happens on a much more dramatic scale as air is drawn into the low pressure at the center of a hurricane. The circulating air gains tremendous speed, which gives the hurricane great destructive power.

Angular Momentum and a Rotational "Collision"

Go to Picture It 8-2 for more practice dealing with angular momentum.

Here's another application of the angular momentum of a particle. The girl shown in Figure 8-27 runs across the playground and finally jumps onto the merry-go-round so that both end up rotating together. This seems very much like the inelastic collisions we discussed in Section 7-4, in which a moving object collided with an initially stationary one and the two stuck together afterward. We approached those inelastic collisions by demanding that linear momentum is conserved. In Figure 8-27, however, the girl has linear momentum before the collision, and the system of girl and merry-go-round has *zero* total linear momentum after the collision (the system rotates but does not move bodily from one place to another). So *linear* momentum is not conserved in this collision.

However, the *angular* momentum of the system of girl and merry-go-round *is* conserved because there are no external torques acting on the system around the merry-go-round's rotation axis. (The vertical force of gravity and normal force don't affect the rotation, and there is very little friction in the bearings of the merry-go-round.) During the collision the girl and the merry-go-round exert torques on each other, but these torques are *internal* to the system of girl plus merry-go-round and so don't affect the *total* angular momentum. All that happens in the collision is that angular momentum is transferred between the girl and the merry-go-round. Example 8-16 shows how to analyze what happens.

EXAMPLE 8-16 A Girl on a Merry-Go-Round

The girl in Figure 8-27 has a mass of 30.0 kg. She runs toward the merry-go-round at 3.0 m/s, then jumps on. The merry-go-round is initially at rest, has a mass of 100 kg and a radius of 2.0 m, and can be treated as a uniform disk. Find the rotation speed of the merry-go-round after the girl jumps on.

Set Up

As we described above, angular momentum is conserved in this process. Before she jumps on, the girl has all of the angular momentum, with a magnitude given by Equation 8-23. After she jumps on, she and the merry-go-round rotate together with the same angular speed ω. Together they behave like a single object made up of a rotating disk with an extra mass (the girl) at its rim.

Angular momentum of the girl before jumping on:

$$L = rp_{\perp} = rp \sin \phi \qquad (8\text{-}23)$$

Angular momentum of the merry-go-round with the girl riding on it:

$$L_z = I\omega_z \qquad (8\text{-}22)$$

Moment of inertia of the merry-go-round plus girl:

$$I = I_{\text{merry-go-round}} + I_{\text{girl}} \qquad (8\text{-}6)$$

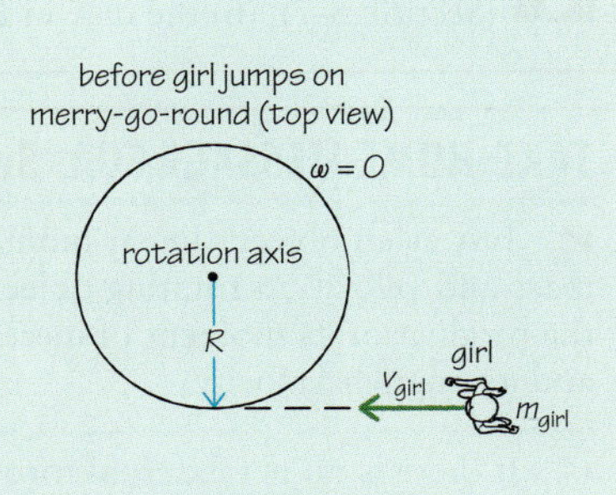

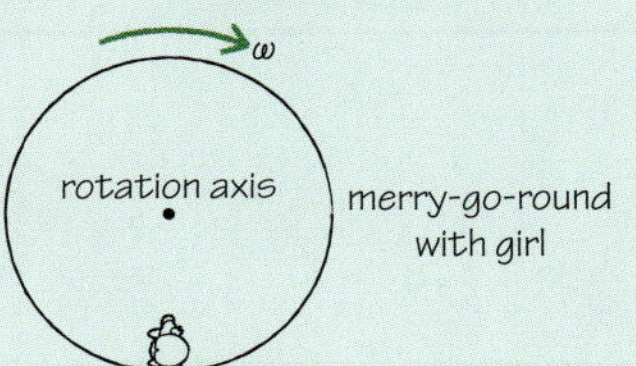

Solve

Figure 8-27 shows that the distance $r \sin \phi$ is just equal to the radius of the merry-go-round, $R = 2.0$ m. Use this to find the magnitude of the angular momentum before she jumps on.

Before the girl jumps on her linear momentum has magnitude

$$p = m_{\text{girl}}v_{\text{girl}} = (30 \text{ kg})(3.0 \text{ m/s}) = 90 \text{ kg}\cdot\text{m/s}$$

Her angular momentum is

$$L = rp \sin \phi = (r \sin \phi)p$$
$$= Rp = (2.0 \text{ m})(90 \text{ kg}\cdot\text{m/s}) = 180 \text{ kg}\cdot\text{m}^2/\text{s}$$

After she jumps on, the magnitude of the angular momentum is $L = I\omega$, where ω is what we are trying to find. Calculate the value of I for the system of merry-go-round plus girl.

Moment of inertia of the merry-go-round alone, considered as a uniform cylinder: From Table 8-1,

$$I_{\text{merry-go-round}} = \frac{1}{2}M_{\text{merry-go-round}}R^2 = \frac{1}{2}(100 \text{ kg})(2.0 \text{ m})^2 = 200 \text{ kg}\cdot\text{m}^2$$

Moment of inertia of the girl alone, considered as an object of mass m_{girl} rotating a distance R from the axis:

$$I_{\text{girl}} = m_{\text{girl}}R^2 = (30 \text{ kg})(2.0 \text{ m})^2 = 120 \text{ kg}\cdot\text{m}^2$$

Moment of inertia of the merry-go-round plus girl:

$$I = I_{\text{merry-go-round}} + I_{\text{girl}} = 200 \text{ kg}\cdot\text{m}^2 + 120 \text{ kg}\cdot\text{m}^2 = 320 \text{ kg}\cdot\text{m}^2$$

Angular momentum is conserved, so set the magnitude of angular momentum before the girl jumps on equal to the magnitude after she jumps on. Then solve for the final angular speed ω. Note that if L is in kg · m²/s and I is in kg · m², then ω is in rad/s.

Before the girl jumps on $L = 180 \text{ kg}\cdot\text{m}^2/\text{s}$

After she jumps on angular momentum is $L_z = I\omega_z$ with magnitude $L = I\omega = (320 \text{ kg}\cdot\text{m}^2)\omega$

Angular momentum is conserved, so the magnitude L is the same before and after:

$$180 \text{ kg}\cdot\text{m}^2/\text{s} = (320 \text{ kg}\cdot\text{m}^2)\omega$$

$$\omega = \frac{180 \text{ kg}\cdot\text{m}^2/\text{s}}{320 \text{ kg}\cdot\text{m}^2} = 0.56 \text{ rad/s}$$

Reflect

To get a better sense of our result, we convert it to rev/min. Our answer says that the merry-go-round spins 5.4 times per minute. In other words, it makes one revolution every (1/5.4) minute, or about once every 11 s—a reasonable pace for a merry-go-round.

Convert ω to revolutions per minute:

$$\omega = (0.56 \text{ rad/s})\left(\frac{1 \text{ rev}}{2\pi \text{ rad}}\right)\left(\frac{60 \text{ s}}{1 \text{ min}}\right) = \frac{(0.56)(60)}{2\pi} \frac{\text{rev}}{\text{min}}$$

$$= 5.4 \text{ rev/min}$$

GOT THE CONCEPT? 8-13 Conserving angular momentum

For which of the following objects is angular momentum conserved as it rotates: (a) the pulley in Example 8-12 (Section 8-7), (b) the disk in Example 8-14 (Section 8-7), (c) both the pulley and the disk, or (d) neither the pulley nor the disk?

TAKE-HOME MESSAGE FOR Section 8-8

✔ Just as an object's linear momentum is the product of its mass and velocity, a rotating object's angular momentum is the product of its moment of inertia around its rotation axis and its angular velocity.

✔ If there is no net external torque on a system, the angular momentum of the system is conserved. This holds true even if the system includes an object whose shape and moment of inertia change.

✔ A particle moving past a point has angular momentum relative to that point, even if the particle is not following a curved path.

8-9 Rotational quantities such as angular momentum and torque are actually vectors

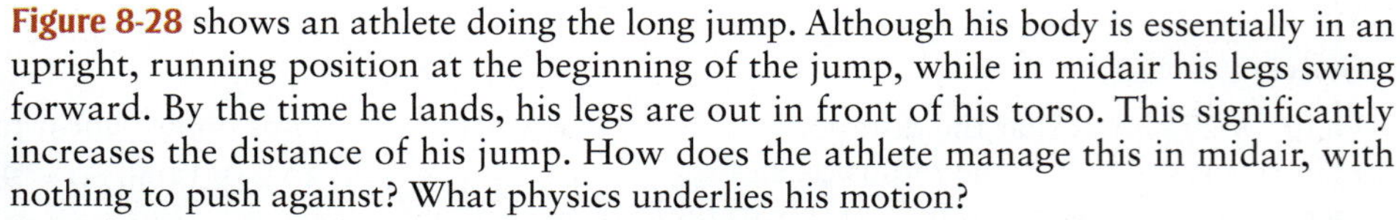

Phil O'Brien/AP Images

Figure 8-28 Angular momentum during a long jump This long jumper rotates his legs counterclockwise (upward) by swinging his arms clockwise. In this way he keeps the total angular momentum of his body equal to zero.

Figure 8-28 shows an athlete doing the long jump. Although his body is essentially in an upright, running position at the beginning of the jump, while in midair his legs swing forward. By the time he lands, his legs are out in front of his torso. This significantly increases the distance of his jump. How does the athlete manage this in midair, with nothing to push against? What physics underlies his motion?

The answer is that there was no net torque acting on the athlete during his leap, so his angular momentum was conserved. (The force of gravity acted on him, but this acted at his center of mass and so exerted no torque around the center of mass.) As the athlete left the ground, the net angular momentum of his body around its center of mass was zero, or nearly so. Once in the air he rotated his legs up and forward (counterclockwise in Figure 8-28), giving them a nonzero angular momentum. In order for angular momentum to be conserved, some other part of his body had to rotate in such a way that the two contributions to his *net* angular momentum canceled. He rotated his arms rapidly clockwise while in the air, which helped rotate his upper torso clockwise in Figure 8-28. The angular momentum of the arms in one direction thus canceled the angular momentum of his legs in the opposite direction.

Angular Velocity and Angular Momentum as Vectors

We can understand this property of angular momentum most easily by considering angular momentum as a *vector* $\vec{L}$. In what direction does $\vec{L}$ point? We need to account for the direction of rotation, but there is no way to align a single vector in the plane of the rotation to indicate this direction (**Figure 8-29a**). Hence we take the angular momentum vector to point along the axis of rotation, which we call the z axis. By convention the specific direction of $\vec{L}$ is given by a **right-hand rule**. Curl the fingers of your right hand in the direction of motion and stick your thumb straight out. By the right-hand rule, your thumb points in the direction of the angular momentum vector (**Figure 8-29b**). We use this same rule to determine the direction of the angular

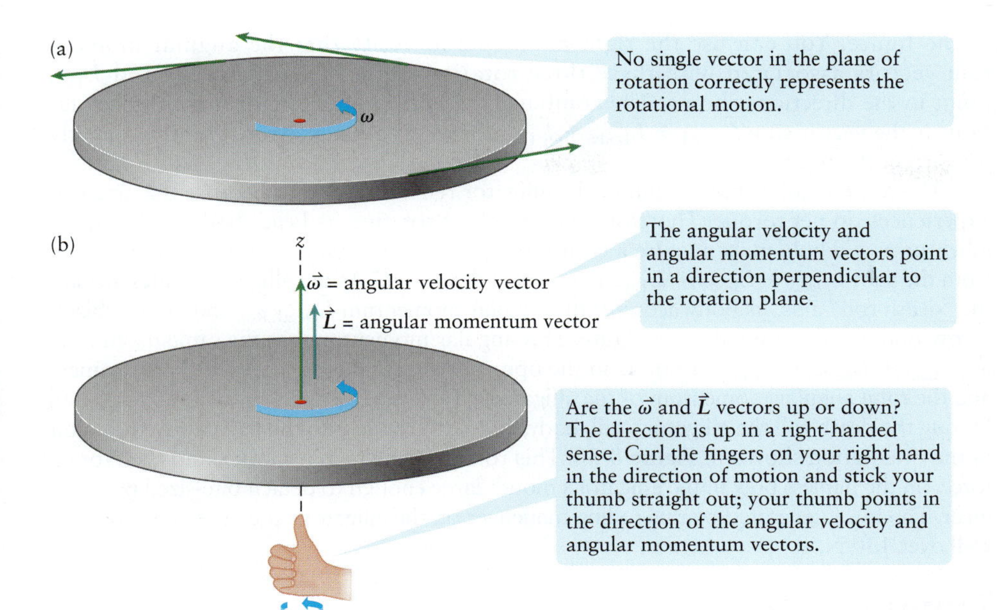
(a)
ω
No single vector in the plane of rotation correctly represents the rotational motion.
(b)
z
$\vec{\omega}$ = angular velocity vector
$\vec{L}$ = angular momentum vector
The angular velocity and angular momentum vectors point in a direction perpendicular to the rotation plane.
Are the $\vec{\omega}$ and $\vec{L}$ vectors up or down? The direction is up in a right-handed sense. Curl the fingers on your right hand in the direction of motion and stick your thumb straight out; your thumb points in the direction of the angular velocity and angular momentum vectors.

Figure 8-29 The angular velocity and angular momentum vectors Rather than lying in (a) the plane of rotation, the angular velocity and angular momentum vectors of a rotating object are (b) oriented along the rotation axis. The directions of $\vec{L}$ and $\vec{\omega}$ are given by a right-hand rule.

velocity vector $\vec{\omega}$: This also points along the rotation axis. If the object rotates in the direction shown in Figure 8-29b, $\vec{\omega}$ and $\vec{L}$ point in the positive z direction and their components ω_z and L_z are positive; if the object rotates in the opposite direction, the vectors $\vec{\omega}$ and $\vec{L}$ point in the negative z direction and their components ω_z and L_z are negative. That's the origin of the positive and negative signs for ω_z and L_z that we introduced in Sections 8-2 and 8-8, respectively. (Strictly speaking, $\vec{L}$ points along the rotation axis only if the object's mass is arranged symmetrically around the rotation axis. We'll consider only situations of this kind, however, so you needn't worry about this point.)

How does this apply to the long jumper? If more than one element of a system is rotating, the net angular momentum $\vec{L}$ of the system is the vector sum of the angular momenta of each individual element. Therefore, the angular momentum of the jumper's arms and upper torso cancels out that of his legs. This stabilizes his motion so that both arms and legs are forward at the end of the jump, resulting in a longer flight.

Alligator Angular Momentum

BIO-Medical Let's explore this physics more deeply by considering the feeding behavior of alligators, which involves rotations of their head, body, and tail. The structure of an alligator's jaws and teeth makes it impossible for these reptiles to cut large prey into chunks small enough to swallow. To make matters worse, alligators have no leverage when they clamp onto their prey while swimming—their legs are too short to hold the food, and they can't push against the river bottom. However, by tightly biting the prey and then executing a spinning maneuver, they can exert forces large enough to remove bite-sized pieces of food. This "death roll" arises from the conservation of angular momentum.

① This upward vector is the combined angular momentum due to the alligator rotating its head, body, and tail.

Death roll axis

$\vec{L}_{\text{Head}} + \vec{L}_{\text{Body}} + \vec{L}_{\text{Tail}}$

$\vec{L}_{\text{Head}}$

$\vec{L}_{\text{Death Roll}}$

$\vec{L}_{\text{Body}}$

$\vec{L}_{\text{Tail}}$

② This downward vector is the "death roll" angular momentum of the alligator as a whole. This makes the total angular momentum of the alligator equal to zero.

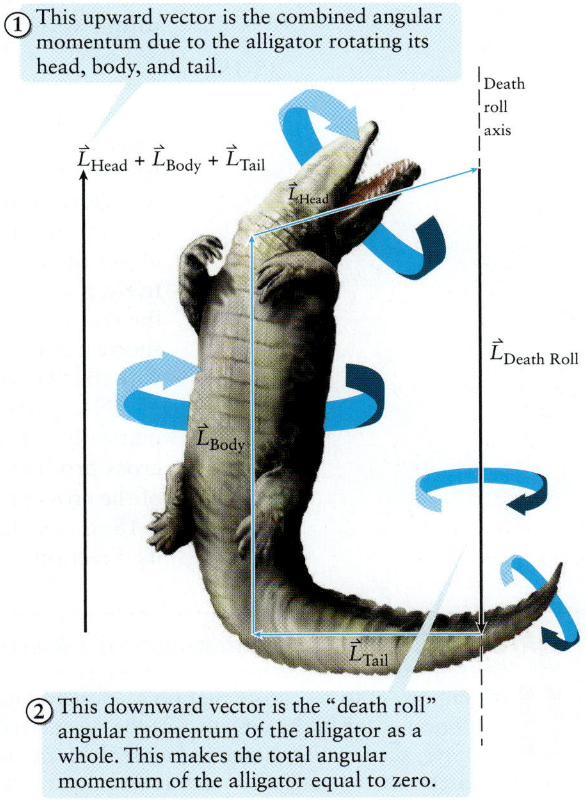

Figure 8-30 The alligator "death roll" Alligators use the conservation of angular momentum to rotate in a "death roll" that helps them consume their prey. *(Information from Frank E. Fish and colleagues.)*

After the alligator has securely grabbed hold of its prey, it bends its tail and head to one side, forming a C-shape (**Figure 8-30**). This step is crucial to the execution of the "death roll" because it enables the animal to spin its head, body, and tail around different axes. The directions of these three rotations are shown as large blue arrows in the figure. You can use the right-hand rule to verify that the angular momentum vectors associated with these three rotations, labeled $\vec{L}_{\text{Head}}$, $\vec{L}_{\text{Body}}$, and $\vec{L}_{\text{Tail}}$, point in the directions shown. The combined angular momentum of these three rotations is the vector sum $\vec{L}_{\text{Head}} + \vec{L}_{\text{Body}} + \vec{L}_{\text{Tail}}$, shown as a black arrow pointing directly upward in the figure.

The *total* angular momentum of the alligator remains constant because the alligator experiences no net torque. Therefore, as a result of rotating its head, body, and tail, the alligator as a whole must acquire a compensatory, net rotation *in the opposite direction* from the sum $\vec{L}_{\text{Head}} + \vec{L}_{\text{Body}} + \vec{L}_{\text{Tail}}$. This is the "death roll." As the alligator rotates around the "death roll" axis, its body acquires the angular momentum $\vec{L}_{\text{Death Roll}}$ shown as a black arrow pointing downward in the figure. This angular momentum has the same magnitude as $\vec{L}_{\text{Head}} + \vec{L}_{\text{Body}} + \vec{L}_{\text{Tail}}$ but points in the opposite direction, so these two vectors cancel and the *total* angular momentum of the alligator is $\vec{L}_{\text{Head}} + \vec{L}_{\text{Body}} + \vec{L}_{\text{Tail}} + \vec{L}_{\text{Death Roll}} = 0$. During the death roll the alligator's tail, body, and head rotate into the page away from you in the orientation shown in Figure 8-30. This rotation results in a torque, and therefore a force, on the hapless prey in the alligator's mouth large enough to detach bite-sized pieces of meat. The force generated by this rolling maneuver on the alligator's prey is estimated to be well over 130 N.

Torque as a Vector

Like angular momentum $\vec{L}$, torque can be expressed as a vector $\vec{\tau}$ that has both a magnitude and a direction. Equation 8-18 from Section 8-6 tells us the magnitude of the torque vector:

$$\tau = rF \sin \phi \tag{8-18}$$

In Equation 8-18 r is the position from the rotation axis to the point where a force of magnitude F is applied. Both the position $\vec{r}$ and the force $\vec{F}$ are vectors, and ϕ is the angle between $\vec{r}$ and $\vec{F}$. The direction of $\vec{\tau}$ is given by a right-hand rule that's similar to the one we described to find the direction of the angular momentum or angular velocity vectors: Curl the fingers of your right hand in the direction that the torque would tend to make the object rotate, and your right thumb will point in the direction of $\vec{\tau}$ (**Figure 8-31**). Alternatively, point the fingers of your right hand along the direction of $\vec{r}$ so that your palm faces the vector $\vec{F}$, then curl your fingers from the direction of $\vec{r}$ to the direction of $\vec{F}$ along the shortest path. Your extended right thumb then points in the direction of $\vec{\tau}$. (Practice this with the two situations shown in Figure 8-31.)

Note that $\vec{\tau}$ is perpendicular to the plane defined by the vectors $\vec{r}$ and $\vec{F}$. Mathematically, a combination of two vectors $\vec{r}$ and $\vec{F}$ that has these properties is called the **cross product** or **vector product** of $\vec{r}$ and $\vec{F}$. You can learn more about the properties of the cross product in the Math Tutorial. In this text, though, we will rely on Equation 8-18 to calculate the magnitude of the torque and on the right-hand rule to determine its direction.

WATCH OUT! There is no motion along the direction of torque.

It's important to recognize that no motion and no change in motion occur in the direction of the torque vector. For example, the torques shown in Figure 8-31 are directed either straight up or straight down, but they don't cause the bar on which the torque acts to *move* either up or down: Instead the bar rotates around the rotation axis. The direction of the torque vector is simply a mathematical tool.

If the net torque on an object is zero—that is, if the *vector* sum $\sum \vec{\tau}$ of all external torques acting on the object is zero—the angular momentum of the object is conserved and its angular acceleration is zero. A special case of this is when the object is not rotating initially. In this case the object will remain at rest with zero angular momentum if the vector sum of all the torques on the object is zero. The following example illustrates how to use this principle.

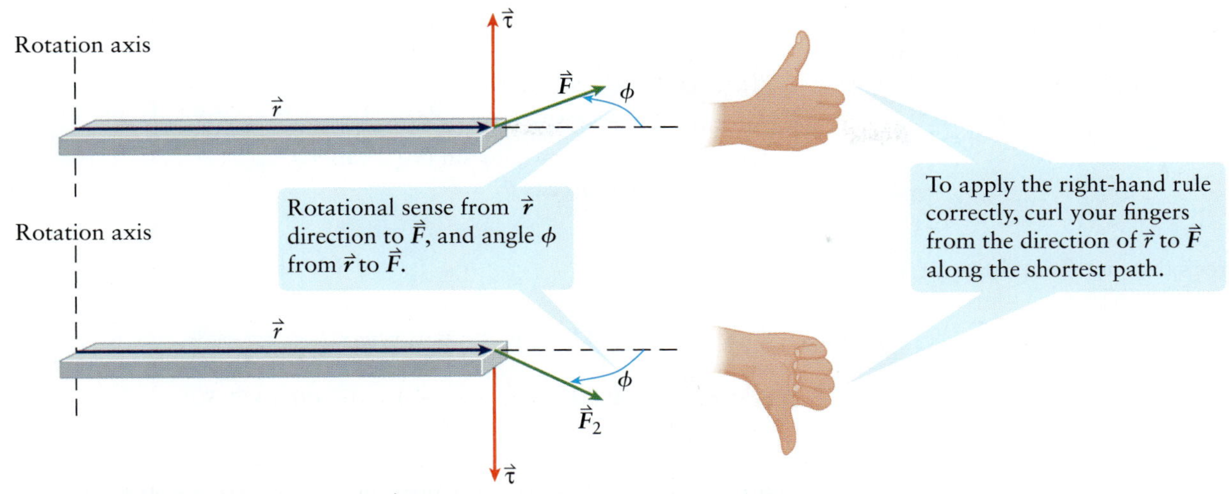

Figure 8-31 The torque vector A force $\vec{F}$ acts on an object at a point that is separated from the rotation axis by a vector $\vec{r}$. The torque $\vec{\tau}$ produced by this force is a vector that is perpendicular to both $\vec{r}$ and $\vec{F}$. The magnitude of $\vec{\tau}$ is $\tau = rF \sin \phi$, and the direction of $\vec{\tau}$ is given by a right-hand rule.

EXAMPLE 8-17 Balancing on a Seesaw

A seesaw is a uniform plank supported at its midpoint. Akeelah and her little sister Bree sit on opposite sides of the seesaw. Akeelah weighs 1.50 times as much as Bree, and Akeelah sits 80.0 cm from the midpoint of the seesaw. If the seesaw is tilted by an angle θ from the horizontal, where should Bree sit so that they just balance?

Set Up

When the system made up of Akeelah, Bree, and the seesaw is balanced, the angular acceleration of the system is zero: It has no tendency to start rotating one way or the other. Hence the net torque on the system must be zero.

We are told that Akeelah's distance from the pivot is $r_A = 80.0$ cm $= 0.800$ m and that Akeelah's weight is 1.50 times that of Bree: that is, $w_A = 1.50 w_B$. Our goal is to find the distance r_B from the pivot to Bree that will satisfy the condition of zero torque. We'll write this condition in terms of the torque vector, using Equation 8-18 to calculate the magnitudes of the torques, and the right-hand rule to determine their directions.

Condition that the net torque is zero:

$$\sum \vec{\tau} = 0$$

Magnitude of the torque

$$\tau = rF \sin \phi \qquad (8\text{-}18)$$

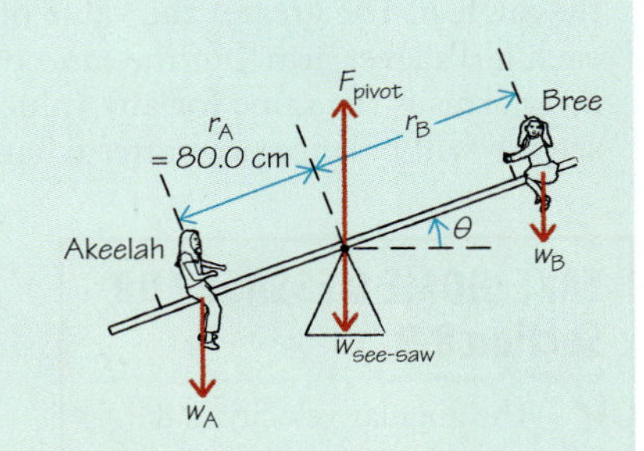

Solve

Neither the weight of the seesaw nor the support force exerts any torque on the system around the pivot point, since both of these forces act at the pivot and have $\vec{r} = 0$. The only torques are due to the weight of Akeelah, $\vec{w}_A$, and the weight of Bree, $\vec{w}_B$. The right-hand rule tells us that the corresponding torques $\vec{\tau}_A$ and $\vec{\tau}_B$ are in opposite directions, into and out of the page, respectively, so their vector sum can be zero (as it must be to keep the seesaw in balance).

Net torque on the system of Akeelah, Bree, and the seesaw:

$$\sum \vec{\tau} = \vec{\tau}_A + \vec{\tau}_B = 0$$

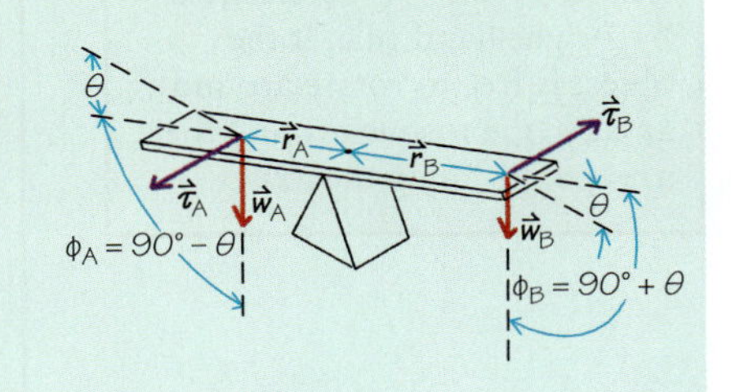

For the net torque to be zero, the opposite torques $\vec{\tau}_A$- and $\vec{\tau}_B$ must have the same magnitude. The angle between $\vec{r}_A$ and $\vec{w}_A$ is $\phi_A = 90° - \theta$, and the angle between $\vec{r}_B$ and $\vec{w}_B$ is $\phi_B = 90° + \theta$. From trigonometry, $\sin(90° - \theta) = \sin(90° + \theta) = \cos\theta$.

Opposite torques $\vec{\tau}_A$ and $\vec{\tau}_B$ have the same magnitude:

$$\tau_A = \tau_B$$

$$r_A w_A \sin\phi_A = r_B w_B \sin\phi_B$$

Substitute $\phi_A = 90° - \theta$, $\phi_B = 90° + \theta$:

$$r_A w_A \sin(90° - \theta) = r_B w_B \sin(90° + \theta)$$

$$r_A w_A \cos\theta = r_B w_B \cos\theta$$

$$r_A w_A = r_B w_B$$

Solve for Bree's distance from the pivot, r_B.

From $r_A w_A = r_B w_B$, $r_B = r_A \dfrac{w_A}{w_B}$

Substitute $r_A = 0.800$ m, $w_A = 1.50 w_B$:

$$r_B = r_A \frac{w_A}{w_B} = (0.800 \text{ m})\left(\frac{1.50 w_B}{w_B}\right) = (0.800 \text{ m})(1.50) = 1.20 \text{ m}$$

Bree must sit 1.20 m from the pivot in order to balance Akeelah.

Reflect

Notice the significance of the lever arm in this problem. Bree, who is lighter than Akeelah, can create a balance by sitting farther from the center. A small force can give rise to a large torque when the lever arm is large. Bree has 1/1.50 the weight of Akeelah, so Bree's lever arm must be 1.50 times greater than Akeelah's and she must sit 1.50 times as far from the pivot as Akeelah.

Note also that the answer doesn't depend on the angle θ. The greater the value of θ, the shorter each girl's lever arm, but the *ratio* of the two lever arms is the same for any value of θ. So the seesaw will balance no matter what the angle θ.

Lever arm for Akeelah:

$$r_{A\perp} = r_A \cos\theta$$

Lever arm for Bree:

$$r_{B\perp} = r_B \cos\theta$$

Ratio of the two lever arms:

$$\frac{r_{B\perp}}{r_{A\perp}} = \frac{r_B \cos\theta}{r_A \cos\theta} = \frac{r_B}{r_A} = 1.50$$

This is the same for any value of θ.

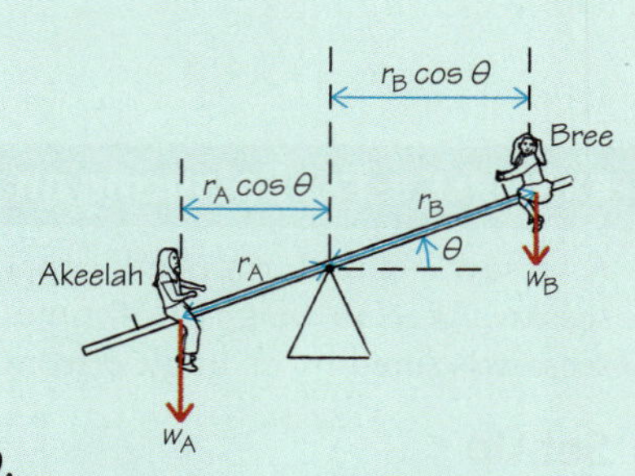

TAKE-HOME MESSAGE FOR Section 8-9

✔ The angular velocity and angular momentum of a rotating object are vectors that lie along the object's rotation axis. The direction of these vectors is determined by a right-hand rule.

✔ Torque is also a vector whose direction is determined by a right-hand rule. If the object is free to rotate around an axis, the torque around that axis points along the axis.

BIO-Medical GOT THE CONCEPT? 8-14 Danger! Falling Cat Zone

Everyone knows that a falling cat usually lands on its feet. This requires the cat to rotate itself into an upright position while falling (see Figure 8-32). The key to a cat's ability to start and stop rotating in midair is bending its body in the middle at the start of the fall. Explain how this helps the cat right itself while falling.

Figure 8-32 A falling cat How does this cat manage to right itself in midair so that it lands on its feet?

Key Terms

angular acceleration
angular displacement
angular momentum
angular position
angular speed
angular velocity
average angular acceleration
average angular velocity
conservation of angular momentum
cross product (vector product)
lever arm
line of action
moment of inertia
parallel-axis theorem
right-hand rule
rigid object
rolling without slipping
rotation
rotational kinematics
rotational kinetic energy
torque
translation
vector product (cross product)

Chapter Summary

Topic | **Equation or Figure**

Angular velocity and angular speed: The angular velocity ω_z of a rigid object is the rate at which it rotates. (The z axis is the rotation axis of the object.) Angular velocity can be positive or negative, depending on which direction of rotation you choose to be positive. Angular speed ω is the magnitude of angular velocity. A point a distance r from the rotation axis has speed $v = r\omega$.

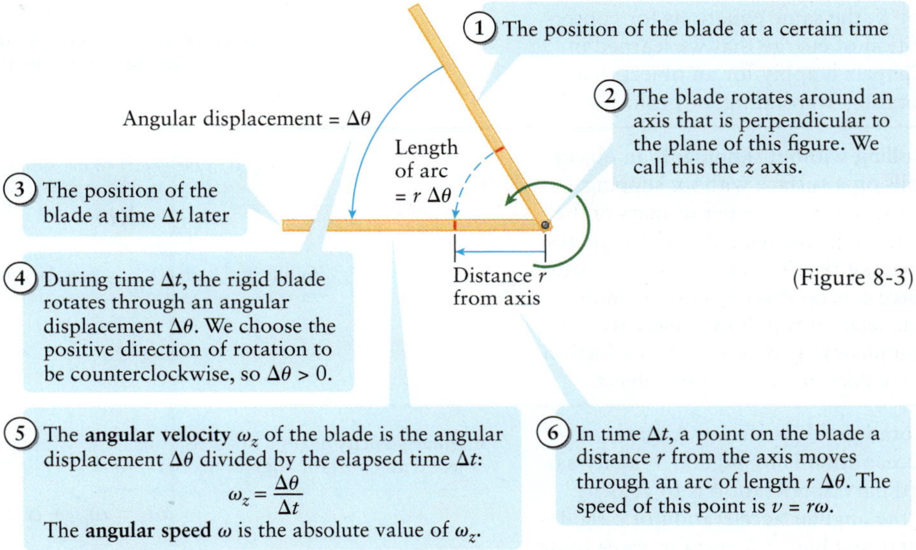

(1) The position of the blade at a certain time

(2) The blade rotates around an axis that is perpendicular to the plane of this figure. We call this the z axis.

Angular displacement = $\Delta\theta$

Length of arc = $r\,\Delta\theta$

(3) The position of the blade a time Δt later

Distance r from axis

(4) During time Δt, the rigid blade rotates through an angular displacement $\Delta\theta$. We choose the positive direction of rotation to be counterclockwise, so $\Delta\theta > 0$.

(5) The **angular velocity** ω_z of the blade is the angular displacement $\Delta\theta$ divided by the elapsed time Δt:

$$\omega_z = \frac{\Delta\theta}{\Delta t}$$

The **angular speed** ω is the absolute value of ω_z.

(6) In time Δt, a point on the blade a distance r from the axis moves through an arc of length $r\,\Delta\theta$. The speed of this point is $v = r\omega$.

(Figure 8-3)

Moment of inertia and rotational kinetic energy: The quantity that plays the role of mass for rotational motion is an object's moment of inertia. This depends on the mass of the object but also on how the components of the object are positioned relative to the rotation axis. (Table 8-1 lists the moments of inertia for various common shapes.) The rotational kinetic energy of an object is related to its moment of inertia and angular speed.

To find the **moment of inertia** I of an object... ...we imagine dividing the object into **N small pieces,** and calculate the sum... ...of the product of the **mass m_i of the ith piece**... ...and the square of the **distance r_i** from the ith piece to the rotation axis.

$$I = \sum_{i=1}^{N} m_i r_i^2 \qquad (8\text{-}6)$$

Rotational kinetic energy of a rigid object spinning around an axis

Moment of inertia of the object for that rotation axis

Angular speed of the object

$$K_{\text{rotational}} = \frac{1}{2} I\omega^2 \qquad (8\text{-}7)$$

Parallel-axis theorem: If an object's moment of inertia for an axis through its center of mass is I_{CM}, its moment of inertia for a second axis parallel to the center-of-mass axis and separated by a distance h is $I = I_{CM} + Mh^2$. Here M is the mass of the object.

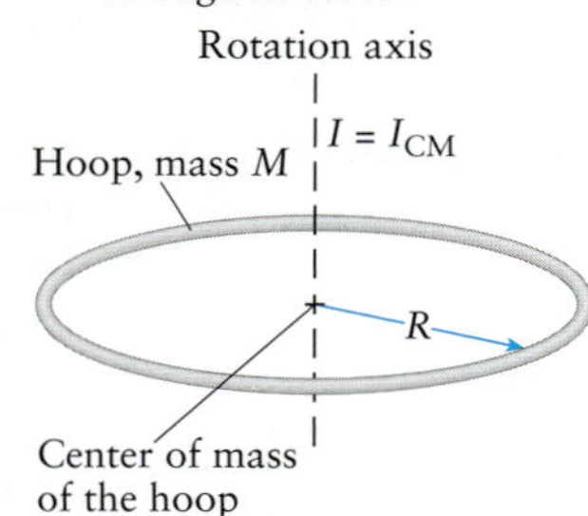

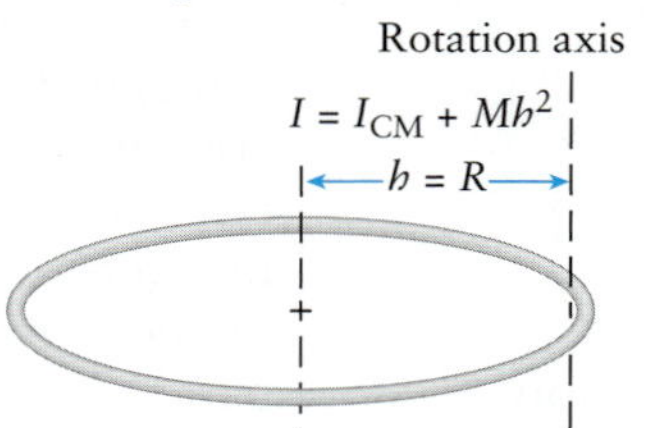

(Figure 8-10 a and b)

Total kinetic energy and conservation of energy: The total kinetic energy K of an object is the sum of its translational kinetic energy and rotational kinetic energy. With this expression for K, the same equations for conservation of energy that we learned in Chapter 6 apply for an object that is rotating, translating, or both.

(1) **Total kinetic energy** of a system that is both translating and rotating

(2) The **translational kinetic energy** is the same as if all of the mass of the system were moving at the speed of the center of mass.

$$K = K_{\text{translational}} + K_{\text{rotational}} = \frac{1}{2} Mv_{CM}^2 + \frac{1}{2} I_{CM}\omega^2 \tag{8-12}$$

(3) The **rotational kinetic energy** is the same as if the system were not translating, and all of the mass were rotating around the center of mass.

Rolling without slipping: If an object rolls on a surface without slipping, the speed of the center of mass of the object must be related to the angular speed of the object's rotation. A friction force acts on this object to maintain this relationship. If the object is completely rigid, however, this friction force does no work on the object.

Speed of the center of mass of a rolling object

$$v_{CM} = R\omega \tag{8-13}$$

Radius of the object

Angular speed of the object's rotation

Rotational kinematics: Angular acceleration is to angular velocity as ordinary acceleration is to velocity. If the angular acceleration of a rigid rotating object is constant, three basic equations relate the time, angular position, angular velocity, and angular acceleration of the object.

Angular velocity at time t of a rotating object with constant angular acceleration

Angular velocity at time $t = 0$ of the object

$$\omega_z = \omega_{0z} + \alpha_z t \tag{8-15}$$

Constant angular acceleration of the object

Time at which the object has angular velocity ω_z

Angular position at time t of a rotating object with constant angular acceleration

Angular velocity at time $t = 0$ of the object

Constant angular acceleration of the object

$$\theta = \theta_0 + \omega_{0z}t + \frac{1}{2}\alpha_z t^2 \tag{8-16}$$

Angular position at time $t = 0$ of the object

Time at which the object is at angular position θ

Angular velocity at angular position θ of a rotating object with constant angular acceleration

Angular velocity at angular position θ_0 of the object

$$\omega_z^2 = \omega_{0z}^2 + 2\alpha_z(\theta - \theta_0) \tag{8-17}$$

Constant angular acceleration of the object

Two **angular positions** of the object

Views of object from along its rotation axis

(a) A force that tends to cause clockwise rotation

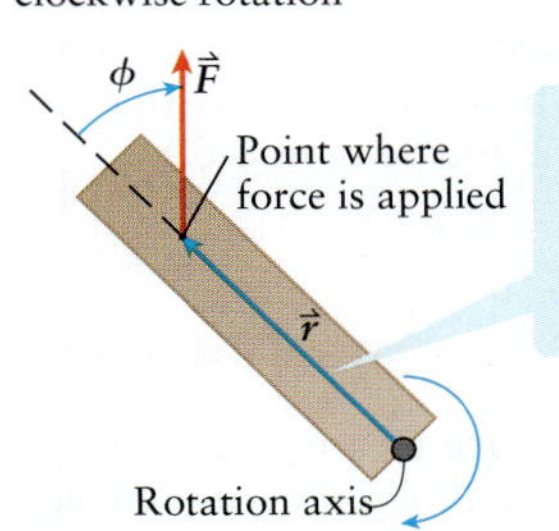

The vector $\vec{r}$ points from an object's rotation axis to where a force $\vec{F}$ is applied to the object.

(b) A force that tends to cause counterclockwise rotation

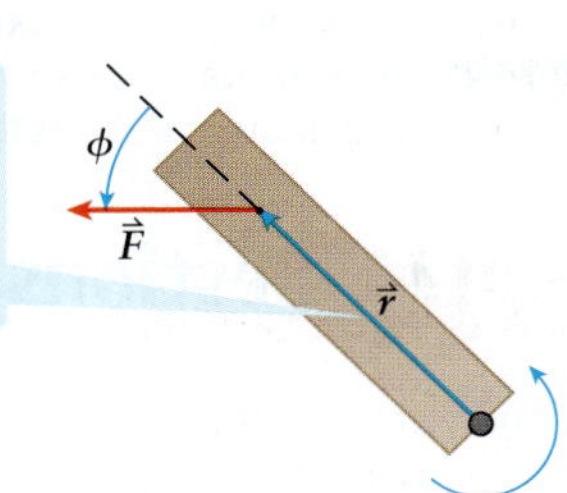

(c) Determining the line of action of a force and its lever arm

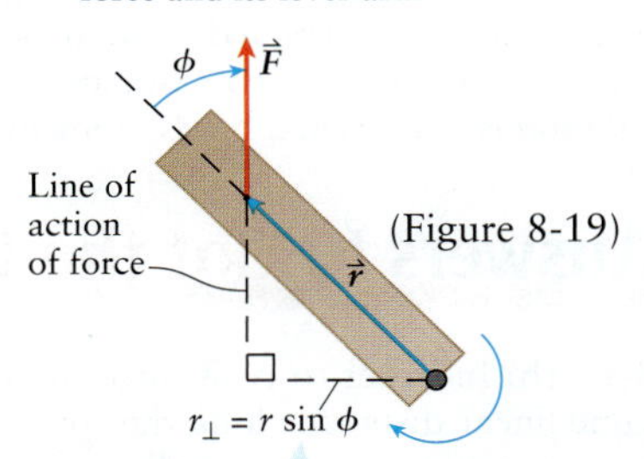

(Figure 8-19)

The lever arm $r_\perp$ is perpendicular to the line of action of the force, and extends from the rotation axis to the line of action of the force.

Torque: If a force $\vec{F}$ acts at a position $\vec{r}$ from an object's rotation axis, it can change the rotation of the object. The torque associated with the force has magnitude $\tau = rF \sin \phi = r_\perp F$, where $r_\perp$ is the lever arm of the force. The torque τ_z can be positive or negative.

Rotational form of Newton's second law: A net external torque on an object gives the object an angular acceleration. To analyze an object that is able to translate as well as rotate (such as one that rolls without slipping), both Newton's second law in its original form $\left(\sum \vec{F} = m\vec{a}\right)$ and this rotational form must be used.

The **net torque**, or sum of all external torques acting on an object for the axis about which it can rotate

The **angular acceleration** produced by the net torque

$$\sum \tau_{\text{ext},z} = I\alpha_z \tag{8-20}$$

The object's **moment of inertia** for the axis about which it can rotate

Angular momentum and the conservation of angular momentum: A rotating rigid object has an angular momentum L_z whose sign depends on the direction of rotation. A particle moving relative to an axis also has angular momentum around that axis. If there is no net external torque on an object or system of objects, its angular momentum remains constant.

The **angular momentum** of a rigid object rotating around a fixed axis...

...is proportional to the **angular velocity** at which the object rotates around that axis.

$$L_z = I\omega_z \tag{8-22}$$

The angular momentum is also proportional to the object's **moment of inertia** for that rotation axis.

Magnitude of the **angular momentum of a particle**

Magnitude of the **linear momentum of a particle**

Angle between the momentum vector and the vector from axis to particle

$$L = r_\perp p = rp \sin \phi \tag{8-23}$$

Perpendicular distance from the axis to the line of the momentum vector

Distance from the axis to the particle

Rotational quantities as vectors: The angular velocity $\vec{\omega}$, angular momentum $\vec{L}$, and torque $\vec{\tau}$ are actually all vector quantities that lie along the rotation axis. Their directions are given by a right-hand rule.

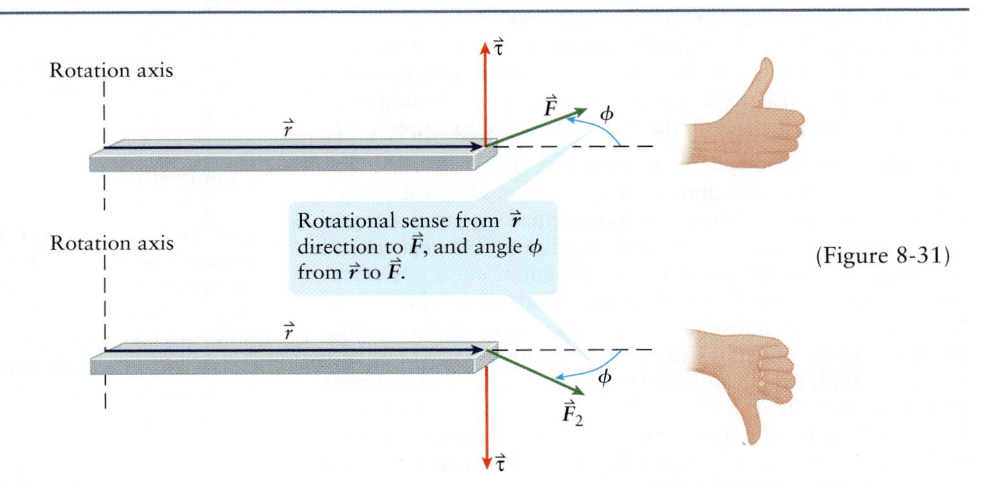

(Figure 8-31)

Answer to What do you think? Question

The resistance of an object to changes in its velocity—that is, its inertia—is determined by its mass: More mass means more inertia. By contrast, the resistance of an object to changes in its *rotation* is determined by its moment of inertia, which depends on both the object's mass and how that mass is distributed. The skater has a smaller moment of inertia with her arms pulled in, so it's easier to make her body spin rapidly than if her arms are extended. (See Section 8-2 and Section 8-6.)

Answers to Got the Concept? Questions

8-1 (b) In order to read information at a constant rate, the same linear distance along the spiral must pass the reader in a given amount of time regardless of radius. Distance in this case is the length of a circular arc, which is given by the product of angular displacement $\Delta\theta$ and radius r. In order for the product $r\Delta\theta$ to remain constant as r increases, the angular displacement $\Delta\theta$ in any fixed amount of time Δt must decrease. So as the scanning laser gets closer to the outer edge of the disc, the angular speed $\omega = \Delta\theta/\Delta t$ must decrease. Hence the motor that rotates the DVD inside the player must slow down as the scanning laser moves from the inner part of the disc to the outer part.

8-2 (c) Table 8-1 shows that the moment of inertia of a thin, uniform rod rotating around its center is $I_{\text{rod}} = ML^2/12$. Because the moment of inertia is additive, the moment of inertia of the two rods together is $I = 2I_{\text{rod}} = 2\,(ML^2/12) = ML^2/6$.

8-3 (c) The moment of inertia of a thin, uniform rod that has a mass m and a length l rotating around its end is $ml^2/3$. Each of the four rods in this problem (with $m = M/2, l = L/2$) rotates around its end, and so each rod has a moment of inertia

$$I_{\text{rod}} = \frac{(M/2)(L/2)^2}{3} = \frac{ML^2}{24}$$

when rotating around its end. Because the moment of inertia is additive, the moment of inertia of the four rods together is

$$I = 4I_{\text{rod}} = 4\left(\frac{ML^2}{24}\right) = \frac{ML^2}{6}$$

Notice that the final configuration of the four rods in this *Got the Concept?* problem is identical to that of the two rods in the previous one. So it's not surprising that the result is the same.

8-4 The solid disk would still win the race. As Example 8-10 shows the speeds after moving a vertical distance H down the ramp are $v_{\text{CM}} = \sqrt{4gH/3}$ for the solid disk and $v_{\text{CM}} = \sqrt{gH}$ for the hoop. These don't depend at all on either the mass or the radius of either object. We saw a similar result in Section 6-6 for the purely translational motion of a skier on a ski jump: The final speed didn't depend on the skier's mass (see Example 6-10).

8-5 Block, uniform sphere, uniform disk. Because it does not rotate and has no work done on it by friction, *all* of the block's gravitational potential energy is transformed into translational kinetic energy. By contrast, as both the disk and the sphere roll downhill, part of the gravitational potential energy is transformed into rotational kinetic energy. That means that at any location, the hoop and sphere will be moving more slowly than the block. Hence the block reaches the bottom first. The sphere arrives second, since from Table 8-1 it has a smaller moment of inertia than the disk ($I_{\text{CM}} = 2MR^2/5$ for the sphere versus $I_{\text{CM}} = MR^2/2$ for the disk) and so less of the sphere's kinetic energy goes into rotation (see Example 8-10). The disk, with its large moment of inertia, comes in last.

8-6 (c), then (d) and (e) are tied for second place, then (a), then (b). The angular velocity ω_z and angular acceleration α_z have the same sign for objects (c) and (e), so the rotation is speeding up for both objects; (c) is speeding up at a greater rate because the absolute value of α_z is greater than for (e). Object (d) is at rest and has a nonzero α_z, so it must also be speeding up; its α_z has the same absolute value as object (e), so these two objects are speeding up at the same rate. Objects (a) and (b) are slowing down because ω_z and α_z have opposite signs; object (b) is slowing down at a faster rate because it has a larger absolute value of α_z.

8-7 Extending the handle of the wrench with the pipe lengthens the lever arm, because it increases the distance between the rotation axis and the point where the force is applied. Equation 8-21 shows that torque increases as the lever arm increases, even when the force remains the same, making it more likely that the bolt will rotate. The downside is that with this arrangement, applying even a small force may result in a torque large enough to break the bolt or the wrench!

8-8 (a) (i), (b) (ii) The torque required to create a certain angular acceleration is proportional to the moment of inertia (Equation 8-23). Because I_y is only about one-third as great as I_z, up-and-down motions (rotation around the y axis) are easier than side-to-side motions (rotation around the z axis). About three times as much torque is required to rotate the jaw from side to side (rotation around the z axis) as compared to the torque required to cause the jaw to open and close (rotation around the y axis).

8-9 (a) A typical bird has more of its mass farther from the roll axis compared to the insect, so its moment of inertia around that axis will be larger compared to its mass. Hence proportionally more effort is required to give the bird an angular acceleration around that axis. This makes the bird less responsive to the torques exerted by the aerodynamic forces on its wings. Thus a typical bird is less maneuverable than a flying insect, which can make rapid and abrupt turns.

8-10 (a) The disk moves down the ramp but doesn't *slide* down the ramp. Instead, as the disk rolls, each part of its rim sets down on the ramp and then lifts off again. Hence the friction force that acts on the disk is purely static friction, not kinetic (sliding) friction.

8-11 (a) The situation is the same as Example 8-14 but with a moment of inertia $I = (2/5)MR^2$ rather than $I = (1/2)MR^2$ (see Table 8-1). Since the sphere has a moment of inertia smaller than that of the disk, it requires a smaller torque to give it an angular acceleration. Hence the uphill friction force f acting on the sphere will be less than that on the disk. As a result, the net downhill force on the sphere, and thus its downhill acceleration, will be greater for the sphere than for the disk.

You can also get this result by redoing the calculation in Example 8-14 using $I = (2/5)MR^2$. The rotational form of Newton's second law becomes

$$Rf = \frac{2}{5}MR^2\alpha_z \text{ or } f = \frac{2}{5}MR\alpha_z$$

With $\alpha_z = a_{CM,x}/R$, this becomes

$$f = \frac{2}{5}MR\left(\frac{a_{CM,x}}{R}\right) = \frac{2}{5}Ma_{CM,x}$$

Substitute this into the Newton's second law equation for the sphere's motion along the x axis:

$$Mg\sin\theta - f = Ma_{CM,x}$$

$$Mg\sin\theta - \frac{2}{5}Ma_{CM,x} = Ma_{CM,x}$$

$$Mg\sin\theta = Ma_{CM,x} + \frac{2}{5}Ma_{CM,x} = \frac{7}{5}Ma_{CM,x}$$

$$a_{CM,x} = \frac{5}{7}g\sin\theta$$

Since $5/7$ (about 0.714) is greater than $2/3$ (about 0.667), this is in fact greater than the disk's acceleration $a_{CM,x} = (2/3)g\sin\theta$.

8-12 (c) The angular speed of the ball increases. Once the ball has been hit and set in motion it has angular momentum around the pole, and because we neglect resistive forces there is no external torque on it. (Yes, there is a force on the ball due to tension in the rope. This force is straight inward toward the pole, however, so the angle ϕ between $\vec{r}$ and $\vec{F}$ in Equation 8-20, the definition of torque, is 180°. Hence the magnitude of the torque is zero: $\tau = rF\sin\phi = rF\sin 180° = 0$.) Angular momentum L_z is therefore conserved. As the rope loops around the pole, the ball gets closer to the rotational axis and its moment of inertia I decreases. Hence the angular speed (which is the magnitude of the angular velocity ω_z) must increase in order for the magnitude of $L_z = I\omega_z$ to remain constant.

8-13 (d) The angular momentum L of a system is conserved only if no net torque acts on the system. This isn't the case for *either* the pulley or the disk: A net torque due to the tension force acts on the pulley as it rotates, and a net torque due to the force of friction acts on the disk as it rolls downhill. For both objects the moment of inertia I remains unchanged (the objects don't change shape) and the angular velocity ω_z increases, so $L_z = I\omega_z$ increases. Conservation laws are great when they apply, but remember that they do *not* apply in all situations!

8-14 By bending in the middle the cat divides its body into two segments, each of which can rotate around a different axis. The cat uses its muscles to rotate its front section into an upright position while simultaneously rotating its hind quarters slightly in the same direction. This small twist cancels the angular momentum generated by the rotation of the front part of his body. Next, the opposite occurs: As the rear of the animal rotates into alignment with the front, it slightly twists its forequarters. Because the net angular momentum of the cat must remain zero, the cat's body must twist in order to cancel the angular momentum associated with the front and rear of its body separately.

Questions and Problems

In a few problems you are given more data than you actually need; in a few other problems you are required to supply data from your general knowledge, outside sources, or informed estimate. Interpret as significant all digits in numerical values that have trailing zeros and no decimal points.

For all problems use $g = 9.80\ \text{m/s}^2$ for the free-fall acceleration due to gravity.

•	Basic, single-concept problem
••	Intermediate-level problem; may require synthesis of concepts and multiple steps
•••	Challenging problem
SSM	*Solution is in Student Solutions Manual*
Example	*See worked example for a similar problem*

Conceptual Questions

1. • Define the SI unit radian. The unit appears in some physical quantities (for example, the angular velocity of a turntable is 3.5 rad/s) and it is omitted in others (for example, the translational velocity at the rim of a turntable is 0.35 m/s). Because the formula relating rotational and translational quantities involves multiplying by a radian ($v = r\omega$), discuss when it is appropriate to include radians and when the unit should be dropped.

2. • Why is it critical to define the axis of rotation when you set out to find the moment of inertia of an object? SSM

3. • The four solids shown in Figure 8-33 have equal heights, widths, and masses. The axes of rotation are located at the center of each object and are perpendicular to the plane of the paper. Rank the moments of inertia from greatest to least.

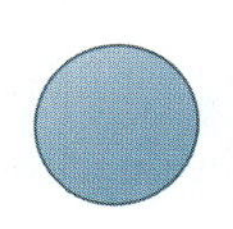

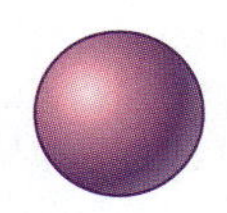

Figure 8-33 Problem 3

4. • What is the ratio of rotational kinetic energy for two balls, each tied to a light string and spinning in a circle with a radius equal to the length of the string? The first ball has a mass m and a string of length L, and rotates at a rate of ω. The second ball has a mass $2m$ and a string of length $2L$, and rotates at a rate of 2ω.

5. • In Chapter 4 you learned that the mass of an object determines how that object responds to an applied force. Write a rotational analog to that idea based on the concepts of this chapter.

6. • Describe what a "torque wrench" is (look up the definition, if not known) and discuss any difficulties that a Canadian auto or bicycle mechanic might have working with an American mechanic's tools (and vice versa).

7. • In describing rotational motion it is often useful to develop an analogy with translational motion. First, write a set of equations describing translational motion. Then write the rotational analogs (for example, $\theta = \theta_0 \ldots$) of the translational equations (for example, $x = x_0 + v_{0x}t + \frac{1}{2}a_xt^2$) using the following legend:

$$x \Leftrightarrow \theta \quad v_x \Leftrightarrow \omega_z \quad a_x \Leftrightarrow \alpha_z \quad F_x \Leftrightarrow \tau_z \quad m \Leftrightarrow I$$

$$p_x \Leftrightarrow L_z \quad K_{\text{translational}} \Leftrightarrow K_{\text{rotational}}$$

8. • A student cannot open a door at her school. She pushes with ever-greater force, and still the door will not budge! Knowing that the door does push open, is not locked, and a minimum torque is required to open the door, give a few reasons why this might be occurring.

9. • Rank the torques exerted on the bolts in A–D (Figure 8-34) from least to greatest. Note that the forces in B and D make an angle of 45° with the wrench. Assume the wrenches and the magnitude of the force F are identical.

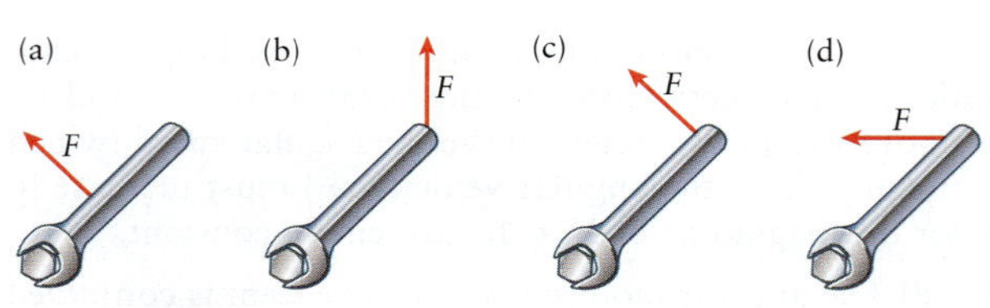

Figure 8-34 Problem 9

10. • A hollow cylinder rolls without slipping up an incline, stops, and then rolls back down. Which of the following graphs in **Figure 8-35** shows the (a) angular acceleration and (b) angular velocity for the motion? Assume that up the ramp is the positive direction.

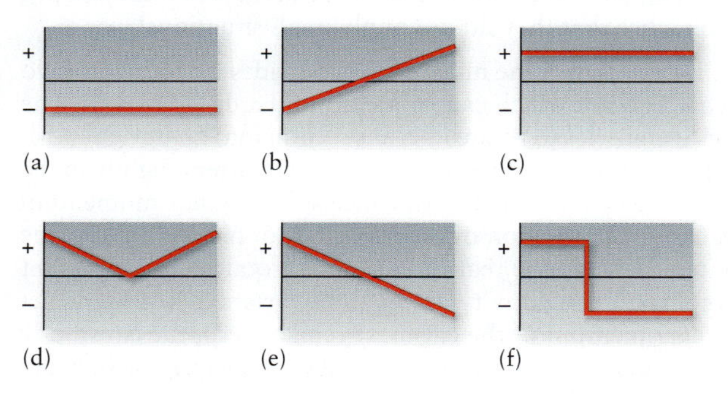

Figure 8-35 Problem 10

11. • Consider a situation in which a merry-go-round, starting from rest, speeds up in the counterclockwise direction. It eventually reaches and maintains a maximum angular velocity. After a short time the merry-go-round then starts to slow down and eventually stops. Assume the accelerations experienced by the merry-go-round have constant magnitudes.(a) Which graph in **Figure 8-36** describes the angular velocity as the merry-go-round speeds up? (b) Which graph describes the angular position as the merry-go-round speeds up? (c) Which graph describes the angular velocity as the merry-go-round travels at its maximum velocity? (d) Which graph describes the angular position as the merry-go-round travels at its maximum velocity? (e) Which graph describes the angular velocity as the merry-go-round slows down? (f) Which graph describes the angular position as the merry-go-round slows down? (g) Draw a graph of the torque experienced by the merry-go-round as a function of time during the scenario described in the problem. SSM

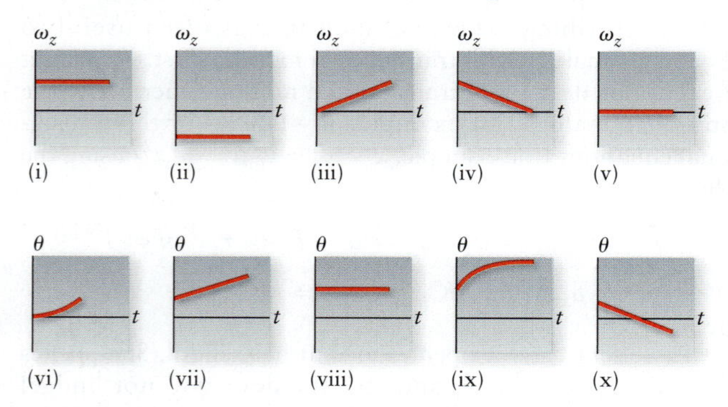

Figure 8-36 Problem 11

12. • Using the rotational concepts of this chapter, explain why a uniform solid sphere beats a uniform solid cylinder which beats a ring when the three objects "race" down an inclined plane while rolling without slipping. SSM

13. • Describe any inconsistencies in the following statement: "The units of torque are N · m, but that's not the same as the units of energy."

14. • What are the units of the following quantities: (a) rotational kinetic energy, (b) moment of inertia, and (c) angular momentum?

15. • Explain which physical quantities change when an ice skater moves her arms in and out as she rotates in a pirouette. What causes her angular velocity to change, if it changes at all?

16. • Which quantity is larger: the angular momentum of Earth rotating on its axis each day or the angular momentum of Earth revolving about the Sun each year? Try to determine the answer without using a calculator.

17. • Explain how an object moving in a straight line can have a nonzero angular momentum.

18. • Analyze the following statement and determine if there are any physical inconsistencies: While rotating a ball on the end of a string of length L, the rotational kinetic energy remains constant as long as the length and angular speed are fixed. When the ball is pulled inward and the length of the string is shortened, the rotational kinetic energy will remain constant due to conservation of energy, but the angular momentum will not because there is an external force acting on the ball to pull it inward. The moment of inertia and angular speed will, of course, remain the same throughout the process because the ball is rotating in the same plane throughout the motion.

19. • A freely rotating turntable moves at a steady angular velocity. A glob of cookie dough falls straight down and attaches to the very edge of the turntable. Describe which quantities (angular velocity, angular acceleration, torque, rotational kinetic energy, moment of inertia, or angular momentum) are conserved during the process and describe qualitatively what happens to the motion of the turntable. SSM

20. • What are the units of angular velocity ($\vec{\omega}$)? Why are factors of 2π present in many equations describing rotational motion?

21. • While watching two people on a seesaw, you notice that the person at the top always leans backward, while the person at the bottom always leans forward. (a) Why do the riders do this? (b) Assuming they are sitting equidistant from the pivot point of the seesaw, what, if anything, can you say about the relative masses of the two riders? SSM

22. • Referring to the time-lapse photograph of a falling cat in Figure 8-32, do you think that a cat will fall on her feet if she does not have a tail? Explain your answer using the concepts of this chapter.

Multiple-Choice Questions

23. • A solid sphere of radius R, a solid cylinder of radius R, and a rod of length R all have the same mass, and all three are rotating with the same angular velocity. The sphere is rotating around an axis through its center. The cylinder is rotating around its long axis, and the rod is rotating around an axis through its center but perpendicular to the rod. Which one has the greatest rotational kinetic energy?

A. the sphere
B. the cylinder
C. the rod
D. the rod and cylinder have the same rotational kinetic energy
E. they all have the same kinetic energy

24. • How would a flywheel's (spinning disk's) kinetic energy change if its moment of inertia were five times larger but its angular speed were five times smaller?
A. 0.1 times as large as before
B. 0.2 times as large as before
C. same as before
D. 5 times as large as before
E. 10 times as large as before

25. •• You have two steel spheres; sphere 2 has twice the radius of sphere 1. What is the ratio of the moment of inertia $I_2:I_1$ measured about an axis through the center of the spheres?
A. 2
B. 4
C. 8
D. 16
E. 32

26. •• A solid ball, a solid disk, and a hoop, all with the same mass and the same radius, are set rolling without slipping up an incline, all with the same initial energy. Which goes farthest up the incline?
A. the ball
B. the disk
C. the hoop
D. the hoop and the disk roll to the same height, farther than the ball
E. they all roll to the same height

27. •• A solid ball, a solid disk, and a hoop, all with the same mass and the same radius, are set rolling without slipping up an incline, all with the same initial linear speed. Which goes farthest up the incline?
A. the ball
B. the disk
C. the hoop
D. the hoop and the disk roll to the same height, farther than the ball
E. they all roll to the same height SSM

28. • Todd and Susan are riding on a merry-go-round. Todd rides on a horse toward the outside of the circular platform, and Susan rides on a horse toward the center of the circular platform. When the merry-go-round is rotating at a constant angular speed, Todd's angular speed is
A. exactly half as much as Susan's.
B. larger than Susan's.
C. smaller than Susan's.
D. the same as Susan's.
E. exactly twice as much as Susan's.

29. • Todd and Susan are riding on a merry-go-round. Todd rides on a horse toward the outer edge of a circular platform. and Susan rides on a horse toward the center of the circular platform. When the merry-go-round is rotating at a constant angular speed ω, Todd's speed v is
A. exactly half as much as Susan's.
B. larger than Susan's.
C. smaller than Susan's
D. the same as Susan's.
E. exactly twice as much as Susan's.

30. • While a gymnast is in the air during a leap, which of the following quantities must remain constant for her?
A. position
B. velocity
C. momentum
D. angular velocity
E. angular momentum about her center of mass

31. • The moment of inertia of a thin ring about its symmetry axis is $I_{CM} = MR^2$. What is the moment of inertia if you twirl a large ring around your finger, so that in essence it rotates about a point on the ring, about an axis parallel to the symmetry axis?
A. $5MR^2$
B. $2MR^2$
C. MR^2
D. $1.5MR^2$
E. $0.5MR^2$ SSM

32. • You give a quick push to a ball at the end of a massless, rigid rod, causing the ball to rotate clockwise in a horizontal circle (**Figure 8-37**). The rod's pivot is frictionless. After the push has ended, the ball's angular velocity
A. steadily increases.
B. increases for a while, then remains constant.
C. decreases for a while, then remains constant.
D. remains constant.
E. steadily decreases.

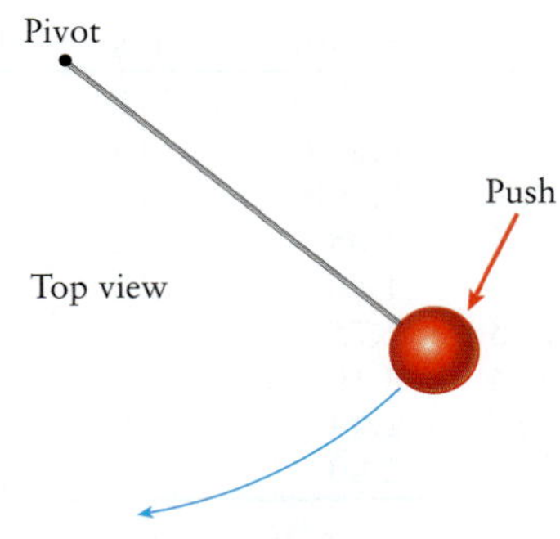

Figure 8-37 Problem 32

Estimation/Numerical Analysis

33. • Estimate the angular speed of a car moving around a cloverleaf on-ramp of a typical freeway. Cloverleaf ramps extend through approximately three-quarters of a circle to connect two perpendicular freeways. SSM

34. • A fan is designed to last for a certain time before it will have to be replaced (planned obsolescence). The fan has one speed (at a maximum of 750 rpm), which it reaches in 2 s starting from rest. It takes the fan 10 s to stop rotating once it is turned off. If the manufacturer specifies that the fan will operate up to 1 billion rotations, estimate how many days you will be able to use the fan.

35. • Estimate the torque you apply when you open a door in your house, and specify the axis to which your estimate refers.

36. • Make a rough estimate of the moment of inertia of a pencil that is spun about its center by a nervous student during an exam.

37. • Estimate the moment of inertia of a figure skater as she rotates about the vertical axis that passes straight down through the center of her body into the ice. Make this estimation for the extreme parts of a pirouette (arms fully extended and arms drawn in tightly).

38. • Estimate the angular displacement (in radians *and* degrees) of Earth in one day of its orbit around the Sun.

39. • Estimate the angular speed of the apparent passage of the Sun across the sky of Earth (from dawn until dusk).

40. • Estimate the angular acceleration of a lone sock that is inside a washing machine that starts from rest and reaches the maximum speed of its spin cycle in typical fashion.

41. • Estimate the angular momentum about the center of rotation for a "skip-it ball" that is spun around on the ankle of a small child (the child hops over the ball as it swings around and around her feet). SSM

42. • Using a spreadsheet and the data below, calculate the average angular speed of the rotating object over the first 10 s. Calculate the average angular acceleration from 15 to 25 s. If the object has a moment of inertia of 0.25 kg · m^2 about the axis of rotation, calculate the average torque during the following time intervals: $0 < t < 10$ s, 10 s $< t <$ 15 s, and 15 s $< t <$ 25 s.

t (s)	θ (rad)	t (s)	θ (rad)	t (s)	θ (rad)
0	0	9	3.14	18	6.48
1	0.349	10	3.50	19	8.53
2	0.700	11	3.50	20	11.0
3	1.05	12	3.49	21	14.1
4	1.40	13	3.50	22	17.6
5	1.75	14	3.51	23	21.6
6	2.10	15	3.51	24	26.2
7	2.44	16	3.98	25	31.0
8	2.80	17	5.01		

Problems

8-2 An object's rotational kinetic energy is related to its angular velocity and how its mass is distributed

43. • What is the angular speed, in rad/s, of an object that completes 2.00 rev every 12.0 s? Example 8-2

44. • A car rounds a curve with a translational speed of 12.0 m/s. If the radius of the curve is 7.00 m, calculate the angular speed in rad/s. Example 8-1

45. • Convert the following: Example 8-2

45.0 rev/min = ______rad/s

$33\frac{1}{3}$ rpm = ______rad/s

2π rev/s = ______rad/s

46. • Calculate the angular speed of the Moon as it orbits Earth. The Moon completes one orbit about Earth in 27.4 days and the Earth–Moon distance is 3.84×10^8 m. SSM Example 8-1

47. • If a 0.250-kg point object rotates at 3.00 rev/s about an axis that is 0.500 m away, what is the kinetic energy of the object? Example 8-3

48. • What is the rotational kinetic energy of an object that has a moment of inertia of 0.280 kg · m^2 about the axis of rotation when its angular speed is 4.00 rad/s? Example 8-2

49. • What is the moment of inertia of an object that rotates at 13.0 rev/min about an axis and has a rotational kinetic energy of 18.0 J? Example 8-2

50. • What is the angular speed of a rotating wheel that has a moment of inertia of 0.330 kg · m^2 and a rotational kinetic energy of 2.75 J? Give your answer in both rad/s and rev/min. SSM Example 8-2

8-3 An object's moment of inertia depends on its mass distribution and the choice of rotation axis

51. • What is the combined moment of inertia for the three point objects about the axis O in Figure 8-38? Example 8-5

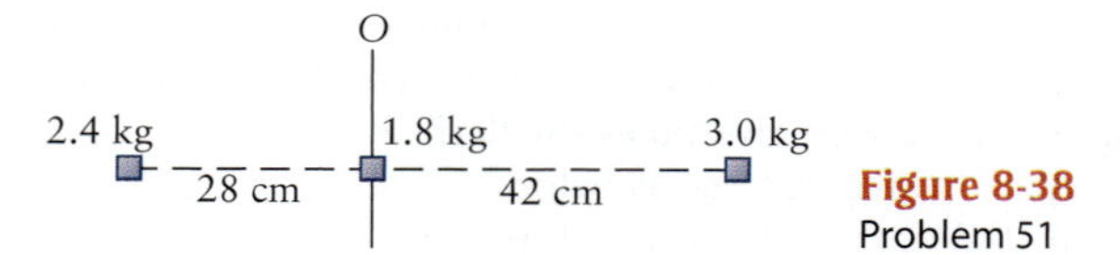

Figure 8-38 Problem 51

52. • What is the combined moment of inertia of three point objects (m_1 = 1.00 kg, m_2 = 1.50 kg, m_3 = 2.00 kg) tied together with massless strings and rotating about the axis O as shown in Figure 8-39? Example 8-4

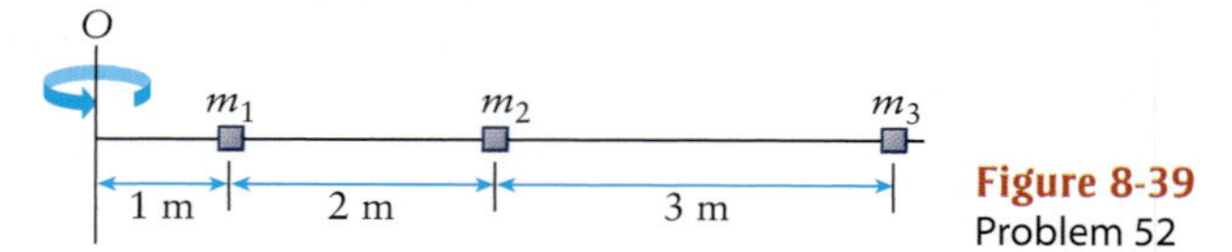

Figure 8-39 Problem 52

53. •• A baton twirler in a marching band complains that her baton is defective (Figure 8-40). The manufacturer specifies that the baton should have an overall length of L = 60.0 cm and a total mass between 940 and 950 g (there is one 350-g object on each end). Also according to the manufacturer, the moment of inertia about the central axis passing through the baton should fall between 0.0750 and 0.0800 kg · m^2. The twirler (who has completed a class in physics) claims this is impossible. Who's right? Explain your answer. Example 8-8

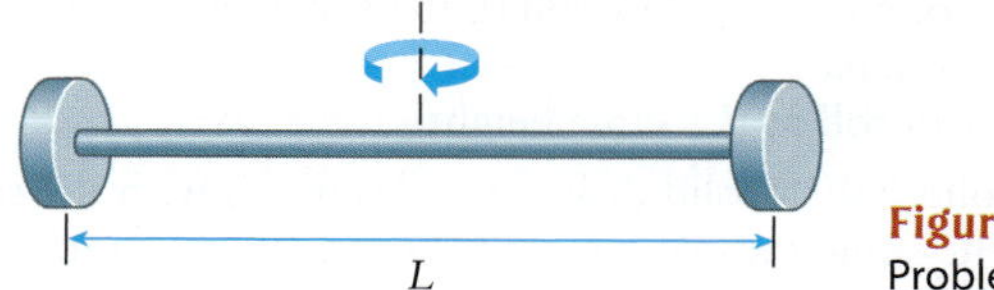

Figure 8-40 Problem 53

54. • What is the moment of inertia of a steering wheel about the axis that passes through its center? Assume the rim of the wheel has a radius R and a mass M. Assume that there are five radial spokes that connect in the center as shown in Figure 8-41. The spokes are thin rods of uniform mass density with length R and mass $\frac{1}{2}M$, evenly spaced around the wheel. SSM Example 8-8

Figure 8-41 Problem 54

55. • Using the parallel-axis theorem, calculate the moment of inertia for a solid, uniform sphere about an axis that is tangent to its surface (Figure 8-42). Example 8-8

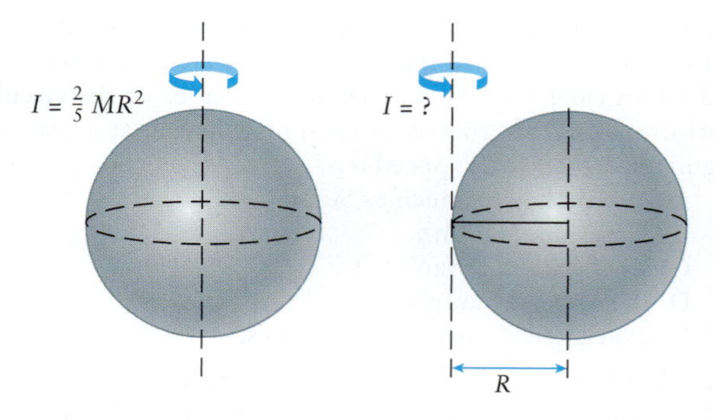

Figure 8-42 Problem 55

56. • Calculate the moment of inertia for a uniform, solid cylinder (mass M, radius R) if the axis of rotation is tangent to the side of the cylinder as shown in **Figure 8-43**. Example 8-8

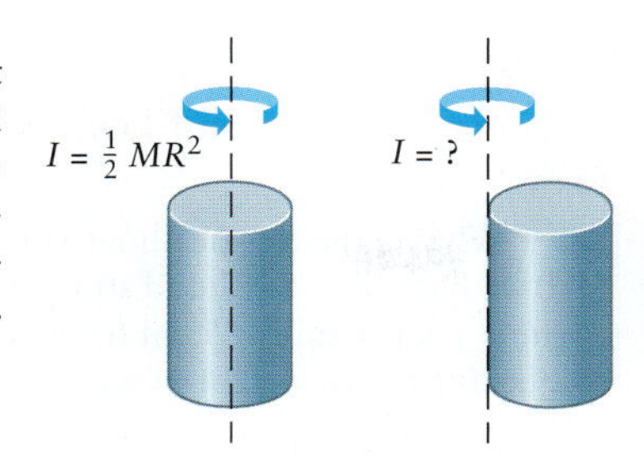

Figure 8-43 Problem 56

57. • Calculate the moment of inertia for a thin uniform rod that is 1.25 m long and has mass of 2.25 kg. The axis of rotation passes through the rod at a point one-third of the way from the left end (**Figure 8-44**). Example 8-7

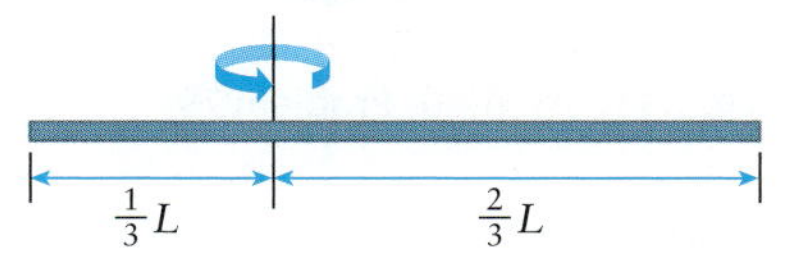

Figure 8-44 Problem 57

58. •• Calculate the moment of inertia of a thin plate that is 5.00 cm × 7.00 cm in area and has a uniform mass density of 1.50 g/cm^2. The axis of rotation is located at the left side, as shown in **Figure 8-45**. Example 8-8

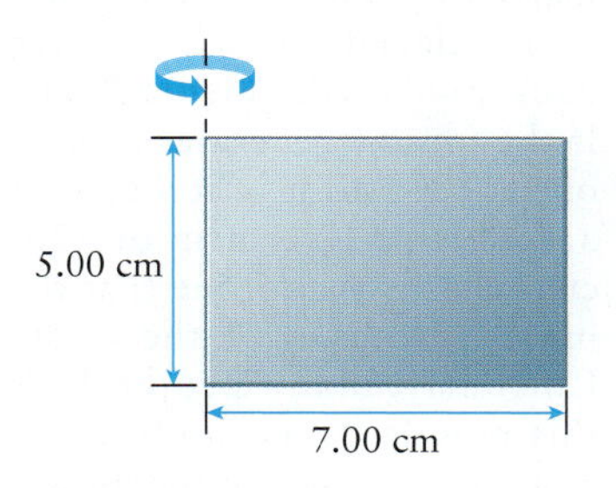

Figure 8-45 Problem 58

59. •• Calculate the radius of a solid uniform sphere of mass M that has the same moment of inertia about an axis through its center of mass as a second solid uniform sphere of radius R and mass M which has the axis of rotation passing tangent to the surface and parallel to the center of mass axis (**Figure 8-46**). Example 8-8

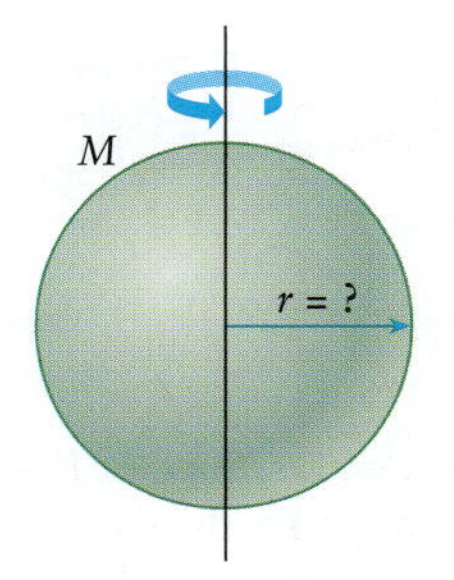

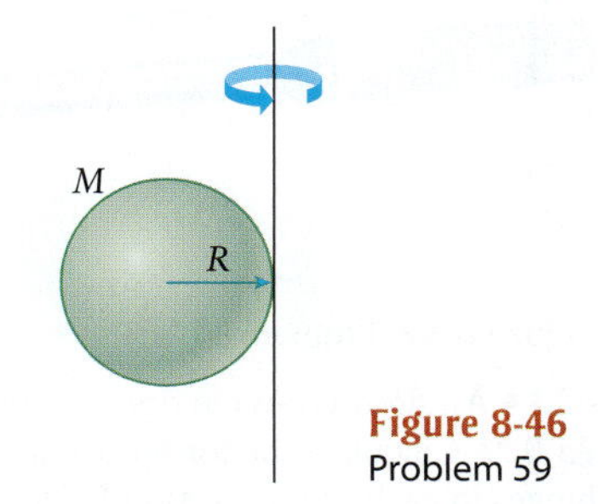

Figure 8-46 Problem 59

60. •• Two uniform, solid spheres (one has a mass M and a radius R and the other has a mass M and a radius $2R$) are connected by a thin, uniform rod of length $3R$ and mass M (**Figure 8-47**). Find the moment of inertia about the axis through the center of the rod. SSM Example 8-7

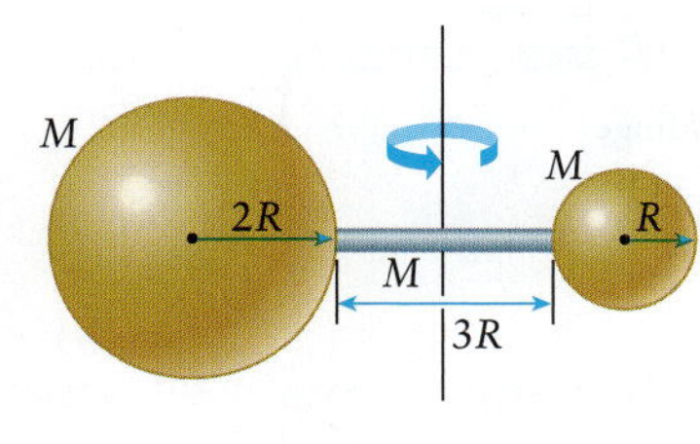

Figure 8-47 Problem 60

61. •• What is the moment of inertia of the sphere–rod system shown in **Figure 8-48** where the sphere is of uniform mass density and has a radius R and a mass M and the rod is thin and massless, and has a length L? The sphere–rod system is spun about an axis A. Example 8-7

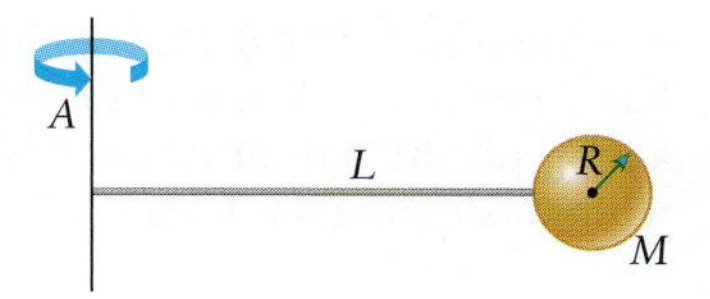

Figure 8-48 Problem 61

8-4 Conservation of mechanical energy also applies to rotating objects

62. • **Sports** A bowling ball that has a radius of 11.0 cm and a mass of 5.00 kg rolls without slipping on a level lane at 2.00 rad/s. Calculate the ratio of the translational kinetic energy to the rotational kinetic energy of the bowling ball. SSM Example 8-10

63. • **Astronomy** Earth is approximately a solid sphere, has a mass of 5.98×10^{24} kg, a radius of 6.38×10^6 m, and completes one rotation about its central axis each day. Calculate the rotational kinetic energy of Earth as it spins on its axis. Example 8-9

64. • **Astronomy** Calculate the translational kinetic energy of Earth as it orbits the Sun once each year (the Earth–Sun distance is 1.50×10^{11} m). Calculate the ratio of the translational kinetic energy to the rotational kinetic energy calculated in the previous problem. Example 8-9

65. • A potter's flywheel is made of a 5.00-cm-thick, round slab of concrete that has a mass of 60.0 kg and a diameter of 35.0 cm. This disk rotates about an axis that passes through its center, perpendicular to its round area. Calculate the angular speed of the slab about its center if the rotational kinetic energy is 15.0 J. Express your answer in both rad/s and rev/min. Example 8-9

66. •• **Sports** A flying disk (160 g, 25.0 cm in diameter) spins at a rate of 3.00×10^2 rpm with its center balanced on a fingertip. What is the rotational kinetic energy of the Frisbee if the disc has 70.0% of its mass on the outer edge (basically a thin ring 25.0-cm in diameter) and the remaining 30.0% is a nearly flat disk 25.0-cm in diameter? SSM Example 8-9

67. • A uniform, solid cylinder of radius 5.00 cm and mass 3.00 kg starts from rest at the top of an inclined plane that is 2.00 m long and tilted at an angle of 25.0° with the horizontal and rolls without slipping down the ramp. What is the cylinder's speed at the bottom of the ramp? Example 8-10

68. • A uniform, solid sphere of radius 5.00 cm and mass 3.00 kg starts with a translational speed of 2.00 m/s at the top of an inclined plane that is 2.00 m long and tilted at an angle of 25.0° with the horizontal and rolls without slipping down the ramp. What is the sphere's speed at the bottom of the ramp? Example 8-10

69. ••• A spherical marble that has a mass of 50.0 g and a radius of 0.500 cm rolls without slipping down a loop-the-loop track that has a radius of 20.0 cm. The marble starts from rest and *just barely* clears the loop to emerge on the other side of the track. What is the minimum height that the marble must start from to make it around the loop? Example 8-10

70. ••• A billiard ball of mass 160 g and radius 2.50 cm starts with a translational speed of 2.00 m/s at point A on the track

as shown in **Figure 8-49**. If point B is at the top of a hill that has a radius of curvature of 60 cm, what is the normal force acting on the ball at point B? Assume the billiard ball rolls without slipping on the track. Example 8-10

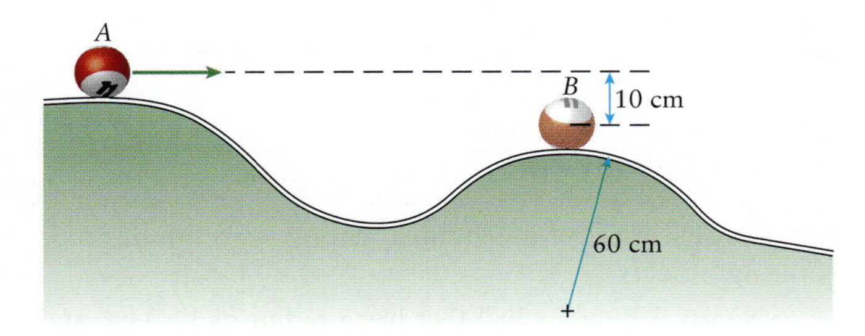

Figure 8-49 Problem 70

8-5 The equations for rotational kinematics are almost identical to those for linear motion

71. • Suppose a roulette wheel is spinning at 1 rev/s. (a) How long will it take for the wheel to come to rest if it experiences an angular acceleration of -0.02 rad/s^2? (b) How many rotations will it complete in that time? Example 8-11

72. • A spinning top completes 6.00×10^3 rotations before it starts to topple over. The average speed of the rotations is 8.00×10^2 rpm. Calculate how long the top spins before it begins to topple. Example 8-11

73. • A child pushes a merry-go-round that has a diameter of 4.00 m and goes from rest to an angular speed of 18.0 rpm in a time of 43.0 s. (a) Calculate the average angular acceleration (in rad/s^2) of the merry-go-round. (b) Calculate the angular displacement (in rad) of the merry-go-round during this time interval. (c) What is the maximum tangential speed of the child if she rides on the edge of the platform? Example 8-11

74. • Allison twirls an umbrella around its central axis so that it completes 24.0 rotations in 30.0 s. (a) If the umbrella starts from rest, calculate the angular acceleration (in rad/s^2) of a point on the outer edge. (b) What is the maximum tangential speed of a point on the edge if the umbrella has a radius of 55.0 cm? Example 8-11

75. • Prior to the music CD, stereo systems had a phonographic turntable on which vinyl disk recordings were played. A particular phonographic turntable starts from rest and achieves a final constant angular speed of $33\frac{1}{3}$ rpm in a time of 4.5 s. (a) How many rotations did the turntable undergo during that time? (b) The classic Beatles album *Abbey Road* is 47 min and 7 s in duration. If the turntable requires 8 s to come to rest once the album is over, calculate the total number of rotations for the complete start-up, playing, and slow-down of the album. Example 8-11

76. • A CD player varies its speed as it changes circular tracks on the CD. A CD player is rotating at 300 rpm. To read another track the angular speed is increased to 450 rpm in a time of 0.75 s. Calculate the average angular acceleration in rad/s^2 during the change. SSM Example 8-11

77. • **Astronomy** A communication satellite circles Earth in a geosynchronous orbit such that the satellite remains directly above the same point on the surface of Earth. (a) What angular displacement (in radians) does the satellite undergo in 1 h of its orbit? (b) Calculate the angular speed of the satellite in rev/min and rad/s. Example 8-11

8-7 The techniques used for solving problems with Newton's second law also apply to rotation problems

78. • What is the torque about your shoulder axis if you hold a 10.0-kg barbell in one hand straight out and at shoulder height? Assume your hand is 75 cm from your shoulder and neglect the torque due to the weight of your arm. SSM Example 8-12

79. • A driver applies a horizontal force of 20.0 N (to the right) to the top of a steering wheel, as shown in **Figure 8-50**. The steering wheel has a radius of 18.0 cm and a moment of inertia of 0.0970 kg · m^2. Calculate the angular acceleration of the steering wheel about the central axis due to this force. Example 8-12

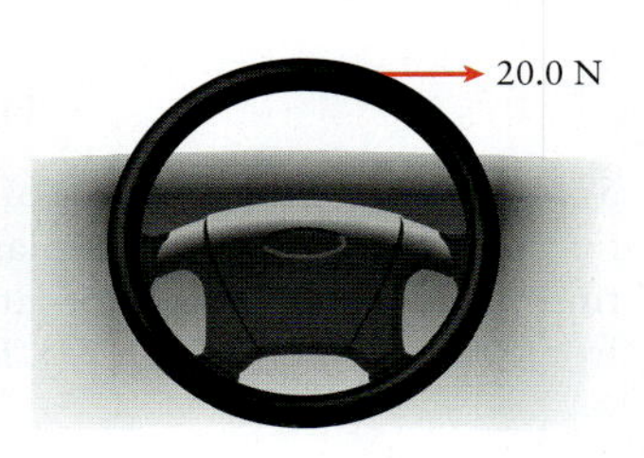

Figure 8-50 Problem 79

80. • **Medical** When the palmaris longus muscle in the forearm is flexed, the wrist moves back and forth (**Figure 8-51**). If the muscle generates a force of 45.0 N and it is acting with an effective lever arm of 22.0 cm, what is the torque that the muscle produces on the wrist? Curiously, many people lack this muscle. Some studies correlate the absence of the muscle with carpal tunnel syndrome. Example 8-12

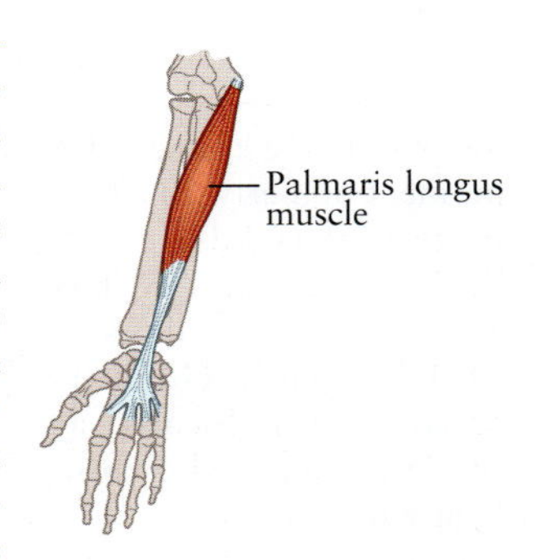

Figure 8-51 Problem 80

81. • A torque wrench is used to tighten a nut on a bolt. The wrench is 25 cm long, and a force of 120 N is applied at the end of the wrench as shown in **Figure 8-52**. Calculate the torque about the axis that passes through the bolt. Example 8-12

Figure 8-52 Problem 81

82. • An 85.0-cm-wide door is pushed open with a force of F = 75.0 N. Calculate the torque about an axis that passes through the hinges in each of the cases in **Figure 8-53**. SSM Example 8-12

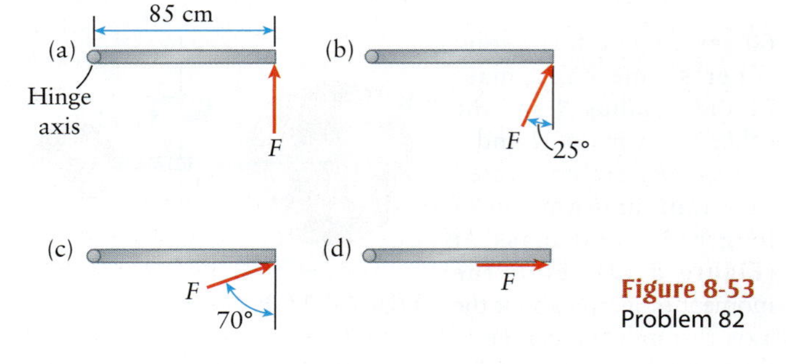

Figure 8-53 Problem 82

83. • A robotic arm lifts a barrel of radioactive waste (**Figure 8-54**). If the maximum torque delivered by the arm about the axis O is 3.00×10^3 N · m and the distance r in the diagram is 3.00 m, what is the maximum mass of the barrel? Example 8-12

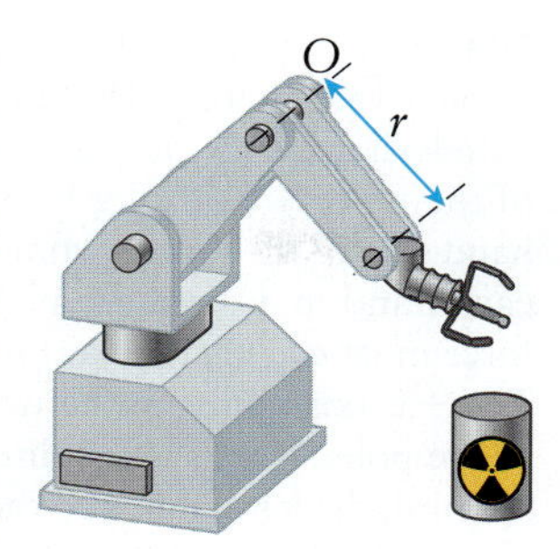

Figure 8-54 Problem 83

84. • A typical adult can deliver about 10 N · m of torque when attempting to open a twist-off cap on a bottle. What is the maximum force that the average person can exert with his fingers if most bottle caps are about 2 cm in diameter? Example 8-12

85. •• A potter's wheel is initially at rest. A constant external torque of 75.0 N · m is applied to the wheel for 15.0 s, giving the wheel an angular speed of 5.00×10^2 rev/min. (a) What is the moment of inertia of the wheel? (b) The external torque is then removed, and a brake is applied. If it takes the wheel 2.00×10^2 s to come to rest after the brake is applied, what is the magnitude of the torque exerted by the brake? Example 8-12

86. •• A solid cylindrical pulley with a mass of 1.00 kg and a radius of 0.25 m is free to rotate about its axis. An object of mass 0.250 kg is attached to the pulley with a light string (**Figure 8-55**). Assuming the string does not stretch or slip, calculate the tension in the string and the angular acceleration of the pulley. Example 8-13

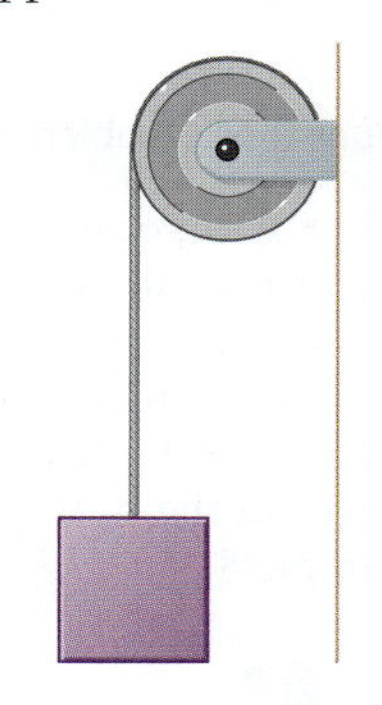
Figure 8-55 Problem 86

87. •• **Figure 8-56** shows a solid, uniform cylinder of mass 7.00 kg and radius 0.450 m with a light string wrapped around it. A 3.00-N tension force is applied to the string, causing the cylinder to roll without slipping across a level surface as shown. (a) What is the angular acceleration of the cylinder? (b) Calculate the magnitude and direction of the frictional force that acts on the cylinder. Example 8-14

Figure 8-56 Problem 87

88. • A crane winch is lifting a 2300-kg mass. The winch can be modeled as a cable of negligible mass wound around a solid uniform cylinder with a 12-cm radius and a mass of 320 kg that rotates around its central axis. How much torque is required to accelerate the 2300-kg mass straight upward at 0.35 m/s^2? Assume there is no friction in the system and that the winch radius remains constant. Example 8-12

89. ••• A block with mass $m_1 = 2.00$ kg rests on a frictionless table. It is connected with a light string over a pulley to a hanging block of mass $m_2 = 4.00$ kg. The pulley is a uniform disk with a radius of 4.00 cm and a mass of 0.500 kg (**Figure 8-57**). (a) Calculate the acceleration of each block and the tension in each segment of the string. (b) How long does it take the blocks to move a distance of 2.25 m? (c) What is the angular speed of the pulley at this time? Example 8-13

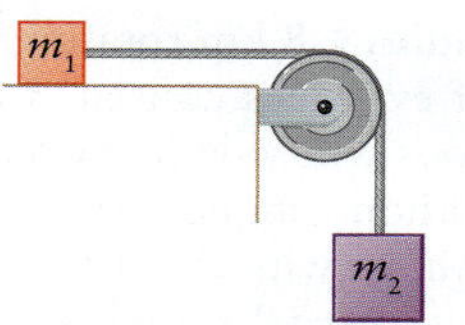

Figure 8-57 Problem 89

90. •• A yo-yo with a mass of 0.0750 kg and a rolling radius of $r = 2.50$ cm rolls down a string with a linear acceleration of 6.50 m/s^2 (**Figure 8-58**). (a) Calculate the tension in the string and the angular acceleration of the yo-yo. (b) What is the moment of inertia of this yo-yo? Example 8-14

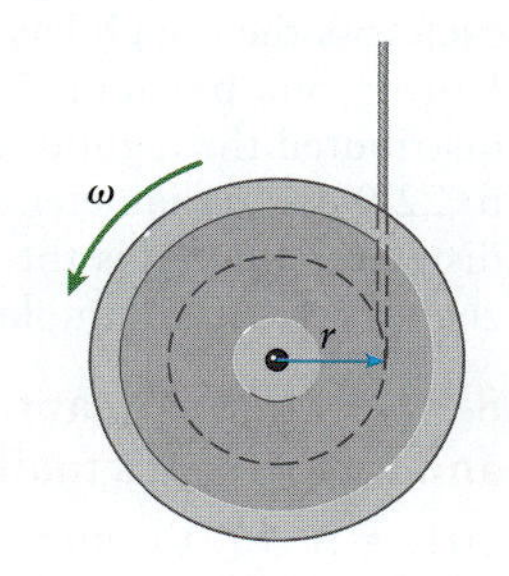

Figure 8-58 Problem 90

8-8 Angular momentum is conserved when there is zero net torque on a system

91. • What is the angular momentum about the central axis of a thin disk that is 18.0 cm in diameter, has a mass of 2.50 kg, and rotates at a constant 1.25 rad/s? Example 8-15

92. • What is the angular momentum of a 0.300-kg tetherball when it whirls around the central pole at 60.0 rpm and at a radius of 125 cm? Example 8-16

93. • **Astronomy** Calculate the angular momentum of Earth as it orbits the Sun. Recall that the mass of Earth is 5.98×10^{24} kg, the distance between Earth and the Sun is 1.50×10^{11} m, and the time for one orbit is 365.3 days. Example 8-16

94. • **Astronomy** Calculate the angular momentum of Earth as it spins on its central axis once each day. Assume Earth is approximately a uniform, solid sphere that has a mass of 5.98×10^{24} kg and a radius of 6.38×10^6 m. Example 8-15

95. • What is the speed of an electron in the lowest energy orbital of hydrogen, of radius equal to 5.29×10^{-11} m? The mass of an electron is 9.11×10^{-31} kg, and its angular momentum in this orbital is 1.055×10^{-34} J · s. Example 8-16

96. • What is the angular momentum of a 70.0-kg person riding on a Ferris wheel that has a diameter of 35.0 m and rotates once every 25.0 s? SSM Example 8-16

97. • A professor sits on a rotating stool that spins at 10.0 rpm while she holds a 1.00-kg weight in each of her hands. Her outstretched arms are 0.750 m from the axis of rotation, which passes through her head into the center of the stool. When she draws the weights in toward her body, her angular speed increases to 20.0 rpm. Neglecting the mass of her arms, how far are the weights from the rotational axis at the increased speed? Example 8-15

98. • A 2.15-kg, 16.0-cm radius, high-end turntable is rotating freely at 33.3 rpm when a naughty child drops 11.0 g of chewing gum onto it 10.0 cm from the rotation axis. Assuming that the gum sticks where it lands, and that the turntable can be modeled as a solid, uniform disk, what is the new angular speed of the turntable? Example 8-16

99. • A giant toroidal space station is being evacuated. Model the 1.0×10^{10}-kg station as a cylindrical hoop of

radius 1.8 km rotating at a rate of 2π rad/min. If each wave of evacuees consists of one thousand 4.0×10^5-kg escape pods whose launchers are pointed radially outward from the station rim, how much does the angular speed of the station *change* with the first wave of evacuees? Express your answer in rad/sec. Example 8-16

100. A chef is tossing 0.500 kg of pizza crust dough. With each toss the dough has an initial angular speed of 5.20 rad/s. During one particular toss, the dough starts out uniformly distributed throughout a 20.0-cm diameter disk and expands to 22.0 cm in diameter. Assuming the mass remains uniformly distributed, what is the angular speed of the dough when the chef catches it? Example 8-15

8-9 Rotational quantities such as angular momentum and torque are actually vectors

101. • A 50.0-g meter stick is balanced at its midpoint (50.0 cm). Then a 0.100-kg and a 0.200-kg mass are hung with light string from the 10.0-cm and 70.0-cm points, respectively. (**Figure 8-59**). Calculate the clockwise and counterclockwise torques acting on the board due to the four forces shown about an axis pointing out of the page at the following points: (a) the 0-cm point, (b) the 50-cm point, and (c) the 100-cm point. (d) Is the meter stick still balanced after the two masses have been added? Example 8-17

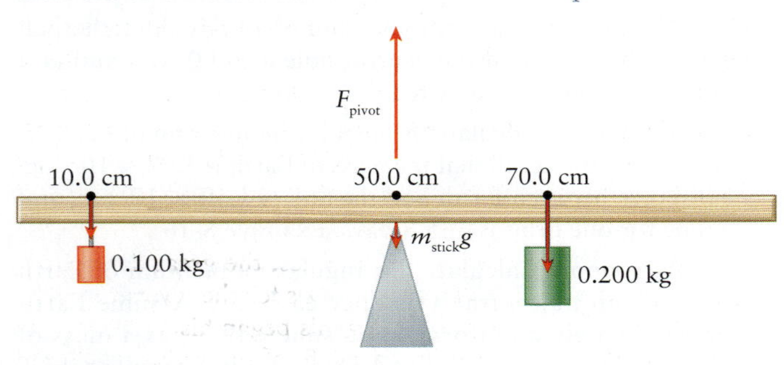

Figure 8-59 Problem 101

102. • A 325-kg merry-go-round with a radius of 1.40 m is spinning counterclockwise as viewed from above at 4.70 rad/s. A 36.0-kg child is hanging on tightly 1.25 m from the rotation axis. Her father applies friction to the outer rim to slow the merry-go-round to a stop in 5 s. How much torque must he apply? Model the merry-go-round as a solid disk. Give both the magnitude and the direction of the torque. Example 8-17

103. • A circular revolving door has a radius of R = 1.00 m, total moment of inertia of $I = 119\ \text{kg}\cdot\text{m}^2$ and is free to rotate on frictionless bearings as shown in **Figure 8-60**. Each of the door "arms" makes a tight seal with the frame, so that each arm provides f_k = 45.0 N of frictional force when the door rotates. If an 85.0-kg adult pushes perpendicularly to one of the doors at a distance r = 0.600 m from the rotation axis, what force must the person exert to make the door spin at a constant rate, assuming all four arms are in contact with the frame? Example 8-12

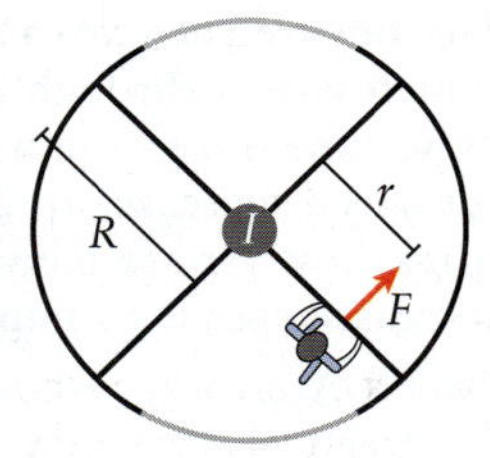

Figure 8-60 Problem 103

104. •• A tightrope walker is walking between two buildings using a 15.0-m long, 18.0-kg pole for balance (**Figure 8-61**). The daredevil grips the pole with each hand 0.600 m from the center of the pole. A 0.540-kg bird lands on the very end of the left-hand side of the pole. Assuming the daredevil applies forces with each hand in a direction perpendicular to the pole, how much force must *each* hand exert to counteract the torque of the bird? The $+x$ axis points in the walking direction, through the center of the pole. What are the directions of the torque vectors due to the bird, the left hand, and the right hand? Example 8-17

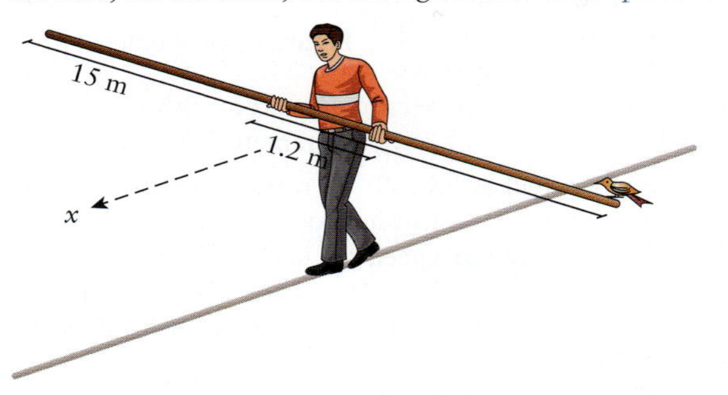

Figure 8-61 Problem 104

105. • A rigid, uniform, 4.00-m long, 40.0-kg beam rests on a pivot placed 3.00 m from one end (**Figure 8-62**). An 80.0-kg man walks up the beam from the end resting on the ground. How far along the beam must the man walk before the low end lifts off the ground? What is the direction of the net torque vector on the beam until the man reaches that critical point? Example 8-17

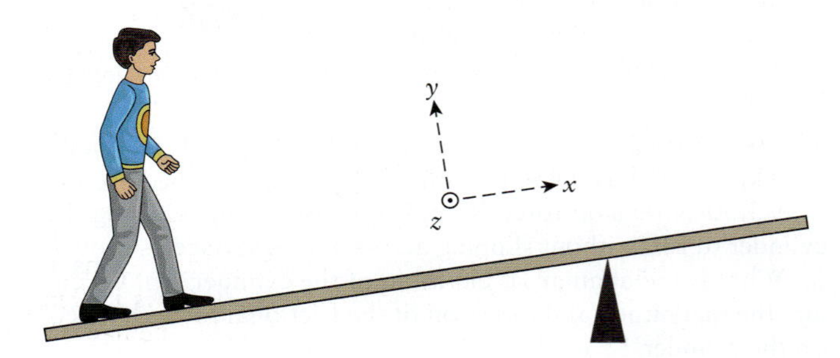

Figure 8-62 Problem 105

106. • A yo-yo, with an inner cylinder radius of r = 2.5 cm and outer disk radius of R = 7.5 cm, is at rest on a 30.0° incline. It is held in place by friction and the tension in the string that is wrapped around its inner cylinder, as shown in **Figure 8-63**. If the mass of the yo-yo is 420 g, what is the direction and magnitude of the friction force? Example 8-14

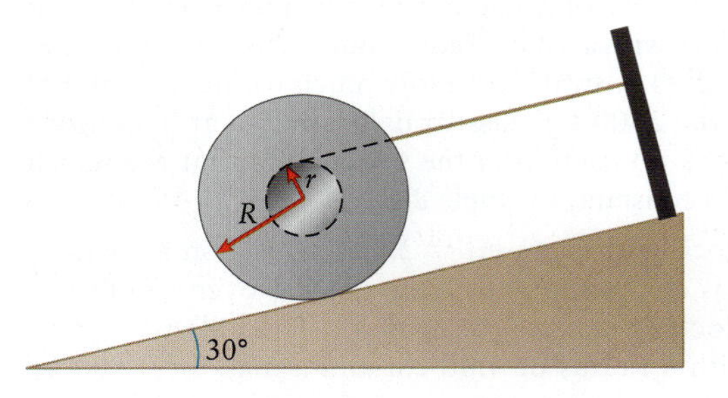

Figure 8-63 Problem 106

General Problems

107. •• A baton is constructed by attaching two small objects that each have a mass M to the ends of a uniform rod that has a length L and a mass M. Find an expression for the moment of inertia of the baton when it is rotated around a point $(3/8)$ L from one end. Example 8-8

108. • The outside diameter of the playing area of an optical Blu-ray disc is 11.75 cm, and the inside diameter is 4.50 cm. When viewing movies, the disc rotates so that a laser maintains a constant linear speed relative to the disc of 7.50 m/s as it tracks over the playing area. (a) What are the maximum and minimum angular speeds (in rad/s and rpm) of the disc? (b) At which location of the laser on the playing area do these speeds occur? (c) What is the average angular acceleration of a Blu-ray disc as it plays an 8.0-h set of movies? Example 8-11

109. • A table saw has a 25.0-cm-diameter blade that rotates at a rate of 7000 rpm. It is equipped with a safety mechanism that can stop the blade within 5.00 ms if something like a finger is accidentally placed in contact with the blade. (a) What average angular acceleration occurs if the saw starts at 7000 rpm and comes to rest in this time? (b) How many rotations does the blade complete during the stopping period? Example 8-11

110. •• In 1932 Albert Dremel of Racine, Wisconsin, created his rotary tool that has come to be known as a dremel. (a) Suppose a dremel starts from rest and achieves an operating speed of 35,000 rev/min. If it requires 1.20 s for the tool to reach operating speed and it is held at that speed for 45.0 s, how many rotations has the bit made? (b) Suppose it requires another 8.50 s for the tool to return to rest. What are the average angular accelerations for the start-up and the slow-down periods? (c) How many rotations does the tool complete from start to finish? Example 8-11

111. •• **Medical, Sports** On average, both arms and hands together account for 13% of a person's mass, while the head is 7.0% and the trunk and legs account for 80%. We can model a spinning skater with his arms outstretched as a vertical cylinder (head + trunk + legs) with two solid uniform rods (arms + hands) extended horizontally. Suppose a 62.0-kg skater is 1.80 m tall, has arms that are each 65.0 cm long (including the hands), and a trunk that can be modeled as being 35.0 cm in diameter. If the skater is initially spinning at 70.0 rpm with his arms outstretched, what will his angular velocity be (in rpm) when he pulls in his arms until they are at his sides parallel to his trunk? Example 8-15

112. •• Because of your success in physics class you are selected for an internship at a prestigious bicycle company in its research and development division. Your first task involves designing a wheel made of a hoop that has a mass of 1.00 kg and a radius of 50.0 cm, and spokes with a mass of 10.0 g each. The wheel should have a total moment of inertia 0.280 kg · m^2. (a) How many spokes are necessary to construct the wheel? (b) What is the mass of the wheel? SSM Example 8-8

113. •• Two beads that each have a mass M are attached to a thin rod that has a length $2L$ and a mass $M/8$ (**Figure 8-64**). Each bead is initially a distance $L/4$ from the center of the rod. The whole system is set into uniform rotation about the center of the rod, with initial angular frequency $\omega_i = 20\pi$ rad/s. If the beads are then allowed to slide to the ends of the rod, what will the angular frequency become? Example 8-15

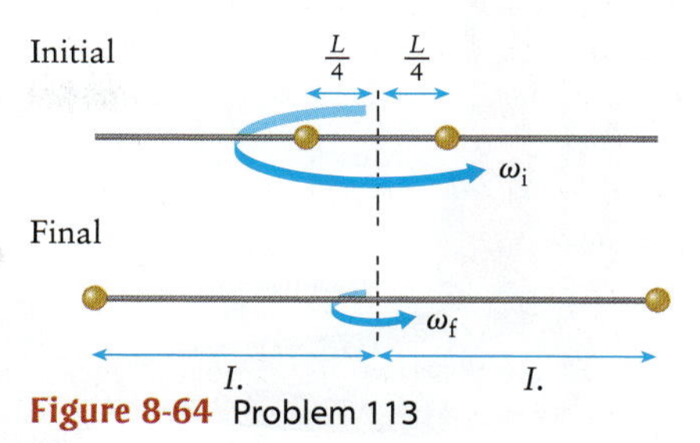

Figure 8-64 Problem 113

114. •• A uniform disk that has a mass $M = 0.300$ kg and a radius $R = 0.270$ m rolls up a ramp of angle $\theta = 55.0°$ with initial speed $v = 4.8$ m/s. If the disk rolls without slipping, how far up the ramp does it go? Example 8-14

115. •• In a new model of a machine, a spinning solid spherical part of radius R must be replaced by a ring of the same mass which is to have the same kinetic energy. Both parts need to spin at the same rate, the sphere about an axis through its center and the ring about an axis perpendicular to its plane at its center. (a) What should the radius of the ring be in terms of R? (b) Will both parts have the same angular momentum? If not, which one will have more?

116. • Many 2.5-in-diameter (6.35-cm) computer hard disks spin at a constant 7200 rpm operating speed. The disks have a mass of about 7.50 g and are essentially uniform throughout with a very small hole at the center. If they reach their operating speed 2.50 s after being turned on, what average torque does the disk drive supply to the disk during the acceleration? Example 8-12

117. ••• **Sports** At the 1984 Olympics, the great diver Greg Louganis won one of his 10 gold medals for the reverse $3\frac{1}{2}$ somersault tuck dive. In the dive, Louganis began his $3\frac{1}{2}$ turns with his body tucked in at a maximum height of approximately 2.0 m above the platform, which itself was 10.0 m above the water. He spun uniformly $3\frac{1}{2}$ times and straightened out his body just as he reached the water. A reasonable approximation is to model the diver as a thin uniform rod 2.0 m long when he is stretched out and as a uniform solid cylinder of diameter 0.75 m when he is tucked in. (a) What was Louganis's average angular speed as he fell toward the water with his body tucked in? *Hint:* How long did it take him to reach the water from his highest point? (b) What was his angular speed just after he stretched out? (c) How much did Louganis's rotational kinetic energy change while extending his body if his mass was 75 kg? Example 8-15

118. ••• **Medical** The bones of the forearm (radius and ulna) are hinged to the humerus at the elbow (**Figure 8-65**). The biceps muscle connects to the bones of the forearm about 2 cm beyond the joint. Assume the forearm has a mass of 2 kg and a length of 0.4 m. When the humerus and the biceps are nearly vertical and the forearm is horizontal, if a person wishes to hold an object of mass M so that her forearm remains motionless, what is the relationship between the force exerted by the biceps muscle and the mass of the object? Example 8-17

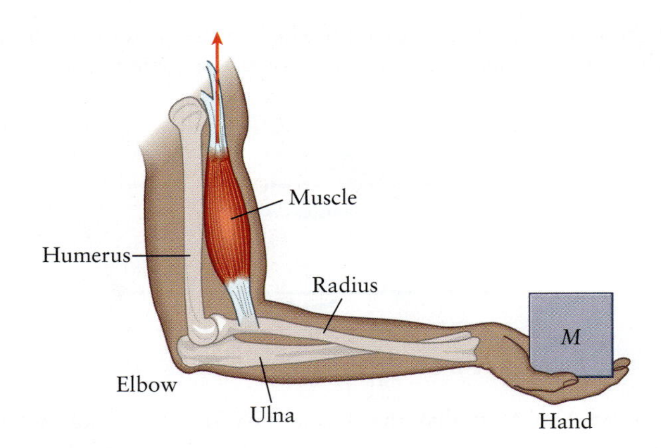

Figure 8-65 Problem 118

119. ••• **Medical** The femur of a human leg (mass 10 kg, length 0.9 m) is in traction (**Figure 8-66**). The center of gravity of the leg is one-third of the distance from the pelvis to the bottom of the foot. Two objects, with masses m_1 and m_2, are hung at the ends of the leg using pulleys to provide upward support. A third object of 8 kg is hung to provide tension along the leg. The body provides tension as well. (a) What is the mathematical relationship between m_1 and m_2? Is this relationship unique in the sense that there is only one combination of m_1 and m_2 that maintains the leg in static equilibrium? (b) How does the relationship change if the tension force due to m_1 is applied at the leg's center of mass? Example 8-17

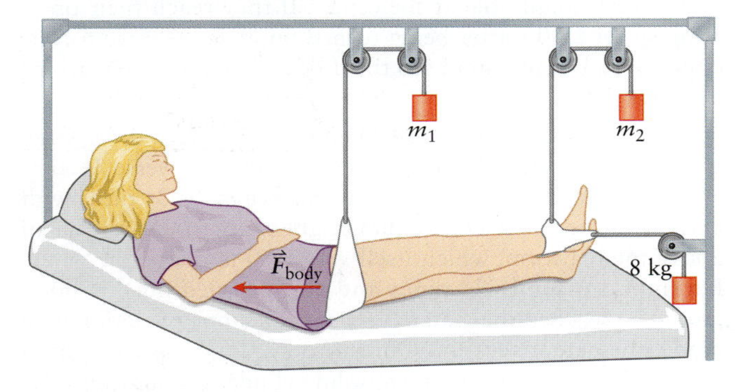

Figure 8-66 Problem 119

120. • **Astronomy** It is estimated that 60,000 tons of meteors and other space debris accumulate on Earth each year. Assume the debris is accumulated uniformly across the surface of Earth. (a) How much does Earth's rotation rate change per year as a result of this accumulation? (That is, find the change in angular velocity.) (b) How long would it take the accumulation of debris to change the rotation period by 1 s? SSM Example 8-15

121. •• **Astronomy** Suppose we decided to use the rotation of Earth as a source of energy. (a) What is the maximum amount of energy we could obtain from this source? (b) By the year 2025 the projected rate at which the world uses energy is expected to be 6.6×10^{20} J/y. If energy use continues at that rate, for how many years would the spin of Earth supply our energy needs? Does this seem long enough to justify the effort and expense involved? (c) How long would it take before our day was extended to 48 h instead of 24 h? Assume that Earth is uniform throughout. Example 8-2

122. •• **Astronomy** In a little over 5 billion years, our Sun will collapse to a white dwarf approximately 16,000 km in diameter. (Ignore the fact that the Sun will lose mass as it ages.) (a) What will our Sun's angular momentum and rotation rate be as a white dwarf? (Express your answers as multiples of its present-day values.) (b) Compared to its present value, will the Sun's rotational kinetic energy increase, decrease, or stay the same when it becomes a white dwarf? If it does change, by what factor will it change? The radius of the Sun is presently 6.96×10^8 m. Example 8-15

123. • **Astronomy** (a) If all the people in the world (~7 billion) lined up along the equator, would Earth's rotation rate increase or decrease? Justify your answer. (b) How would the rotation rate change if all people were no longer on Earth? Assume the average mass of a human is 70.0 kg. Example 8-15

124. • A 1.00×10^3-kg merry-go-round (a flat, solid cylinder) supports 10 children, each with a mass of 50.0 kg, located at the axis of rotation (thus you may assume the children have no angular momentum at that location). Describe a plan to move the children such that the angular velocity of the merry-go-round decreases to one-half its initial value. Example 8-15

125. •• One way for pilots to train for the physical demands of flying at high speeds is with a device called the "human centrifuge." It involves having the pilots travel in circles at high speeds so that they can experience forces greater than their own weight. The diameter of the NASA device is 17.8 m. (a) Suppose a pilot starts at rest and accelerates at a constant rate so that he undergoes 30 rev in 2 min. What is his angular acceleration (in rad/s^2)? (b) What is his angular velocity (in rad/s) at the end of that time? (c) After the 2-min period, the centrifuge moves at a constant speed. The *g*-force experienced is the centripetal force keeping the pilot moving along a circular path. What is the *g*-force experienced by the pilot? ($1\,g$ = mass $\times$ 9.80 m/s^2) (d) The pilot can tolerate 12 *g*'s in the horizontal direction. How long would it take the centrifuge to reach that state if it starts at the angular speed found in part (c) and accelerates at the rate found in part (a)? Example 8-5

126. •• A flywheel of mass 35.0 kg and diameter 60.0 cm spins at 400 rpm when it experiences a sudden power loss. The flywheel slows due to friction in its bearings during the 20.0 s the power is off. If the flywheel makes 200 complete revolutions during the power failure, (a) at what rate is the flywheel spinning when the power comes back on? (b) How long would it have taken for the flywheel to come to a complete stop?

127. • A 620-g basketball rolls without slipping across a level floor at a speed of 1.0 m/s. At what speed would a 40.0-g glass marble travel, as it rolls without slipping, if it has the same kinetic energy as the basketball? Model the basketball as a thin spherical shell and the marble as a uniform solid sphere. Example 8-10

9

Elastic Properties of Matter: Stress and Strain

Vladimir Pcholkin/Exactostock-1672/Superstock

What do you think?

This athlete's arms actually shorten slightly as they support her weight. If the diameter of her arms was only half as much, would the amount that her arms shorten be (a) the same, (b) greater, or (c) less?

In this chapter, your goals are to:

- (9-1) Explain the meaning of stress and strain in physics.
- (9-2) Define Hooke's law for an object under tension or compression.
- (9-3) Solve problems that involve the relationships between tensile stress and strain and between compressive stress and strain.
- (9-4) Explain the importance of the minus sign in the relationship between volume stress and volume strain.
- (9-5) Solve problems in which an object is under volume stress.
- (9-6) Define what is meant by shear, and explain what shear stress and shear strain are.
- (9-7) Solve problems that involve the relationship between shear stress and shear strain.
- (9-8) Explain what must happen to a material for it to go from elastic to plastic to failure, and why biological materials do not become plastic.
- (9-9) Solve problems concerning the stresses and strains on materials that may not obey Hooke's law.

To master this chapter, you should review:

- (1-4) The rules for working with significant figures.
- (6-5) Hooke's law for the force required to stretch or compress an ideal spring.

9-1 When an object is under stress, it deforms

BIO-Medical So far in our study of physics we've treated objects as perfectly rigid. That is, an object's shape and size don't change no matter what forces we apply to it. But what about real objects that *do* change shape when forces are applied? As an example, look at the Achilles tendon that connects the heel bone to the muscles of the calf (**Figure 9-1**). When you bend your ankle to point

Figure 9-1 Stress and strain When an object such as an Achilles tendon is put under stress, it undergoes strain.

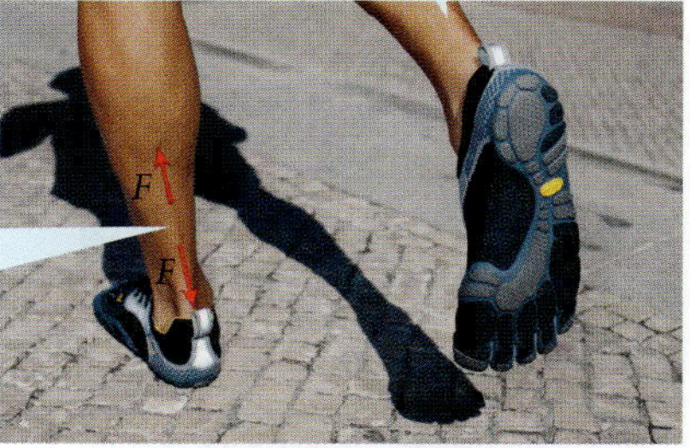

This Achilles tendon is relaxed.
This Achilles tendon is under **stress:** The muscles at its upper end exert an upward force F, and the heel bone at its lower end exerts an equally strong downward force. The tendon responds by undergoing **strain** (it stretches).
F
F
Rachel Torres/Alamy

Stress refers to forces that act to deform an object (change its size or shape). **Strain** refers to the deformation caused by these forces.

(a)

(b)

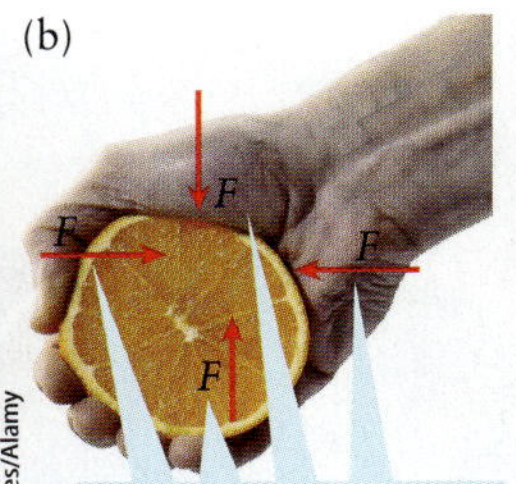

(c)

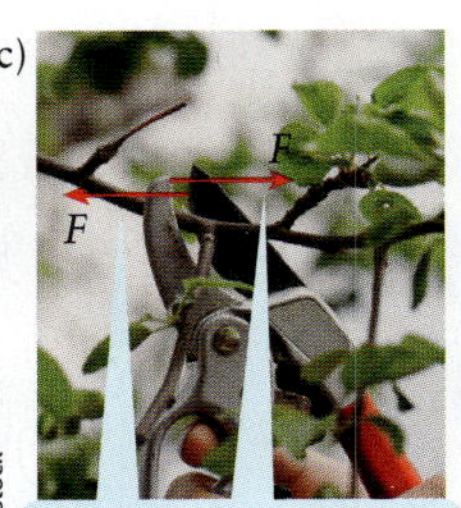

Figure 9-2 Three kinds of stress You apply (a) tensile stress when you stretch an object along its length, (b) volume stress when you squeeze it on all sides, and (c) shear stress when you apply offset forces (as with a pair of shears).

your toes upward, the calf muscles pull upward on the top of the Achilles tendon just as hard as the heel bone pulls downward on the bottom of the tendon. Hence the net force on the tendon is zero, and the tendon as a whole doesn't accelerate. However, individual *parts* of the tendon move as the tendon stretches and deforms. In this chapter, we'll examine the physics of stretching, squeezing, compressing, and twisting materials.

We say that the applied forces in Figure 9-1 exert a *stress* on the tendon that makes it deform, and we call the resulting deformation the *strain*. We'll see that there are three distinct kinds of stress, which differ in the manner in which they are applied (**Figure 9-2**). One of our principal goals will be to see how much an object deforms when we apply a particular amount of stress—for example, how much a tendon stretches when you flex a muscle, or how much a skin diver's body compresses due to water pressure as she dives to the bottom of a lake.

Some materials, such as your Achilles tendon, your earlobes, or the tip of your nose, are *elastic*: If you apply a stress to stretch or squeeze them, they snap back to their original shape after the stress is removed. Other materials, like a well-chewed piece of bubble gum, are *plastic*: They exhibit a permanent change of shape when stretched. Even an elastic material like rubber can remain irreversibly deformed if large enough forces are applied. The material might even break apart and *fracture*, like a plastic fork that's bent too far. We'll look at the differences among these kinds of behavior later in this chapter.

WATCH OUT! Stress and strain are different quantities.

In everyday language the words "stress" and "strain" can mean pretty much the same thing. You might say, "I'm under a lot of stress" or "I'm under a lot of strain." In physics, however, **stress** refers to forces that make a body deform, and **strain** refers to how much deformation those forces cause.

TAKE-HOME MESSAGE FOR Section 9-1

✔ An object is said to be under stress when forces act on the object to deform it (change its shape or size). The amount of deformation that results is called the strain.

✔ Elastic materials return to their original shape and size when the stress is removed, while plastic materials exhibit a permanent change in shape.

9-2 An object changes length when under tensile or compressive stress

The Achilles tendon shown in Figure 9-1 and the rope shown in **Figure 9-2a** are under a kind of stress called **tension** or *tensile stress*: A force is applied at each end that makes the tendon or rope stretch. **Figure 9-3a** shows a gymnast whose upper legs are under tension. (Don't confuse this with the emotion of feeling "under tension," such as may happen before taking a physics exam!) A closely related kind of stress is **compression** or *compressive stress*, which occurs when a force is applied to each end of an object that tends to squeeze the object. The gymnast's arms in **Figure 9-3b** are under compression. If you grab each end of a pencil and try to pull the ends apart, you're putting the pencil under tension; if you push the ends toward each other, you're putting the pencil under compression.

(a)

(b)

F
F
This gymnast's arm is under compression: The shoulder pushes down on her arm and the ring in her hand pushes up on her arm.
Christoph Wilhelm/Getty Images

Figure 9-3 Tension and compression (a) An object under tension tends to stretch along its length; (b) an object under compression tends to shorten along its length.

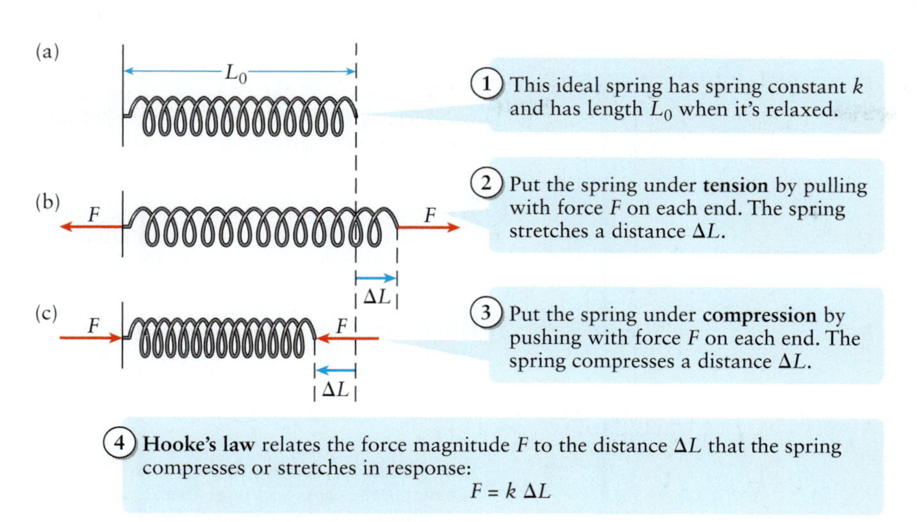

(a)
L_0
1 This ideal spring has spring constant k and has length L_0 when it's relaxed.
(b)
F
F
2 Put the spring under **tension** by pulling with force F on each end. The spring stretches a distance ΔL.
ΔL
(c)
F
F
3 Put the spring under **compression** by pushing with force F on each end. The spring compresses a distance ΔL.
ΔL
4 **Hooke's law** relates the force magnitude F to the distance ΔL that the spring compresses or stretches in response:
$F = k\,\Delta L$

Figure 9-4 Hooke's law for an ideal spring When an ideal spring is under tension or compression, the change in the spring's length is directly proportional to the forces applied to its ends.

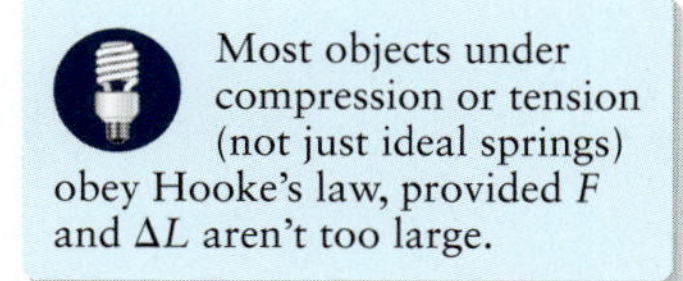

Most objects under compression or tension (not just ideal springs) obey Hooke's law, provided F and ΔL aren't too large.

Hooke's Law

How do objects behave when they are under tension or compression? We learned the answer to this question in Chapter 6 for one kind of object: an ideal spring (**Figure 9-4a**). When we stretch a spring along its length, we apply forces of the same magnitude F to both ends (**Figure 9-4b**). (This ensures that the net external force on the spring is zero, so the spring as a whole doesn't accelerate from rest.) According to **Hooke's law**, which we first encountered in Section 6-5, the force magnitude F is proportional to the distance ΔL that the spring stretches, provided that ΔL is small compared to the relaxed length of the spring, L_0:

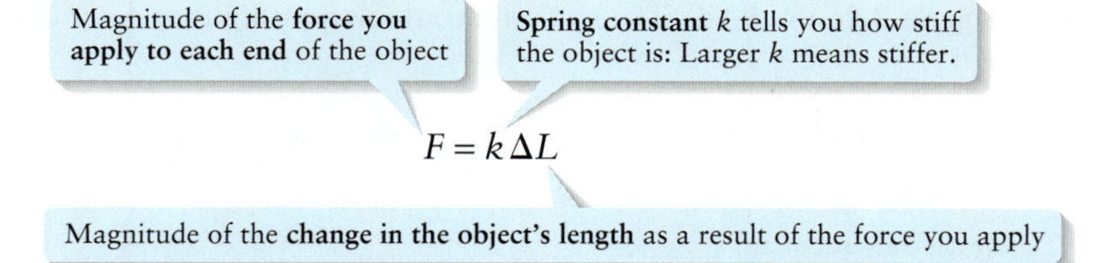

Hooke's law (9-1)

(In Chapter 6 we used the symbol x for the amount that the spring stretches. We've changed this to ΔL to emphasize that this distance is the *change* in the overall length of the spring.)

The quantity k in Equation 9-1 is called the **spring constant**: It's a measure of the stiffness of the spring. To see this, note that we can rewrite Equation 9-1 as $k = F/\Delta L$. The stiffer the spring, the shorter the amount of stretch ΔL for a given force magnitude F and so the greater the value of $k = F/\Delta L$. If the spring is loose and floppy, there is more stretch for a given force magnitude and $k = F/\Delta L$ has a smaller value. The SI units of k are newtons per meter (N/m).

Hooke's law also applies to an ideal spring under compression (**Figure 9-4c**): When compressed by forces of magnitude F, an ideal spring decreases in length by a distance ΔL given by Equation 9-1. The value of k for an ideal spring is the same for compression as for tension. Hence if stretching an ideal spring with forces of magnitude F makes its length increase by 1.00 mm, squeezing that spring with forces of the same magnitude F will make its length decrease by 1.00 mm.

Hooke's law is important because it doesn't apply to just ideal springs. Experiment shows that it holds true for most objects when they are stretched, provided that the amount of stretch is small. An example is a rubber strip made by cutting apart a rubber band (**Figure 9-5a**). When put under tension, the strip stretches by a distance ΔL that's directly proportional to the force magnitude F (**Figure 9-5b**). Likewise, when the strip is compressed, it shrinks by a distance ΔL that's directly proportional to F.

Experiment shows that the spring constant k for a given object depends on three things: its relaxed length L_0, its cross-sectional area A, and the material of which it is made (for example, the particular kind of rubber). The last of these is given by a

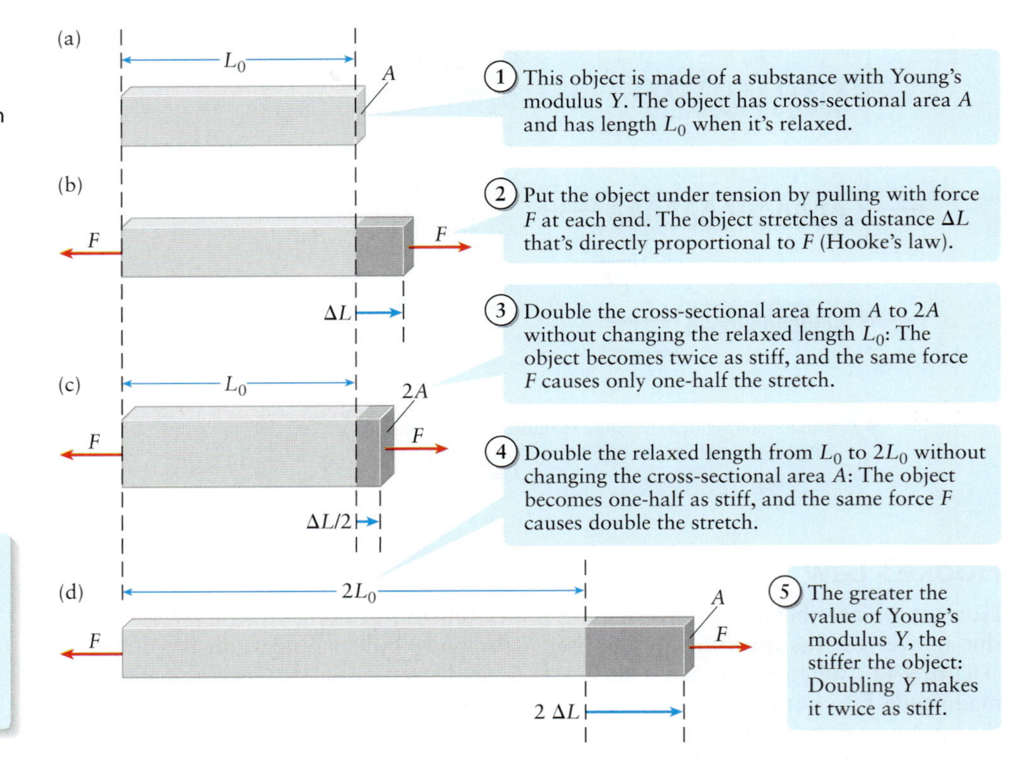

Figure 9-5 Hooke's law for tension and compression An object under tension obeys Hooke's law if the stress is not too great. The same is true for an object under compression.

The object's spring constant is proportional to its Young's modulus and its cross-sectional area but inversely proportional to its length: $k = Y\dfrac{A}{L_0}$.

quantity called **Young's modulus** (or elastic modulus) for the material, denoted by the symbol Y. The greater the value of Y, the greater the intrinsic stiffness of the material. Experiment shows that k is directly proportional to Y and A but inversely proportional to L_0 (**Figures 9-5c** and **9-5d**). The mathematical relationship that describes these observations is

Spring constant for an object under tension or compression (9-2)

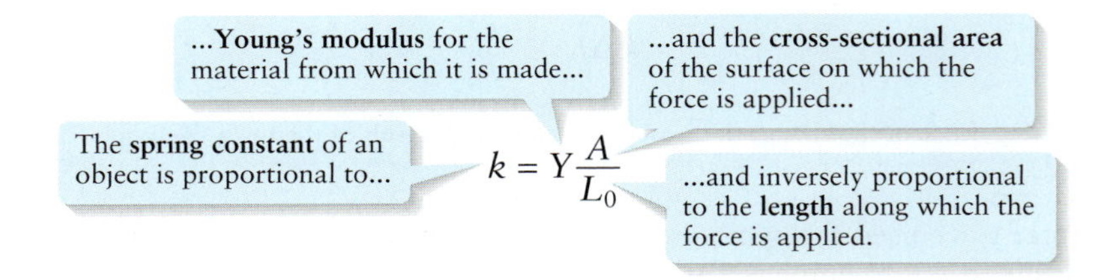

...**Young's modulus** for the material from which it is made...
...and the **cross-sectional area** of the surface on which the force is applied...
The **spring constant** of an object is proportional to...
...and inversely proportional to the **length** along which the force is applied.

$$k = Y\frac{A}{L_0} \tag{9-2}$$

This says that a short, thick object (small L_0, large A) is stiffer than a long, thin object (large L_0, small A) made of the same material and hence with the same value of Y. That's why a rubber eraser is harder to stretch than a rubber band. Equation 9-2 also says that the stiffer the material used to make an object—that is, the greater the value of Y—the stiffer the object will be.

WATCH OUT! Young's modulus is not the same as the spring constant.

Don't confuse Young's modulus Y with the spring constant k. Young's modulus tells you how stiff a given *material* is; the spring constant refers to a particular *object* made of that material. If two objects have the same shape and size (and so the same values of A and L_0) but are made of different materials, Equation 9-2 says the object made of the material with greater Y will be harder to stretch or compress and will have a greater value of k. Equation 9-2 also says that if two objects are made of the same material (so Y is the same) but have different dimensions (so A, L_0, or both are different), their spring constants k can be different.

The first column of **Table 9-1** gives specific values of Young's modulus Y for various materials. The SI units of Y are newtons per square meter, or N/m^2. This unit is also

TABLE 9-1 Elastic Properties of Selected Materials

Material	Young's Modulus (10^9 N/m^2)	Bulk Modulus (10^9 N/m^2)	Shear Modulus (10^9 N/m^2)
aluminum	70	76	26
brass	100	80	40
concrete	30	13	15
iron	211	170	82
nylon	3		4.1
rubber band	0.005		0.003
steel	200	140	78
air		1.41×10^{-4}	
ethyl alcohol		1.0	
water		2.2	
human ACL	0.1		
human lung		$1.5\text{–}9.8 \times 10^{-6}$	
pig endothelial cell			2×10^{-5}

called the **pascal**, abbreviated Pa: 1 Pa = 1 N/m^2. As an example, $Y = 5 \times 10^6$ Pa for the rubber in a typical rubber band, while $Y = 2 \times 10^{11}$ Pa for steel. So compared to the force required to stretch a rubber band, the force required to stretch by the same amount a steel rod of the same cross-sectional area and length would be $(2 \times 10^{11})/(5 \times 10^6) = 4 \times 10^4$ times greater. Materials found in the human body have a wide range of values of Y (**Figure 9-6**).

GOT THE CONCEPT? 9-1 Compare the Spring Constants

Rank the following four rods in order of how easy they are to stretch, from easiest to most difficult. All four rods are made of the same material. If any two rods are equally easy to stretch, say so. (a) A rod 5.00 cm long with cross-sectional area 2.00 mm^2; (b) a rod 2.50 cm long with cross-sectional area 2.00 mm^2; (c) a rod 5.00 cm long with cross-sectional area 1.00 mm^2; (d) a rod 2.50 cm long with cross-sectional area 1.00 mm^2.

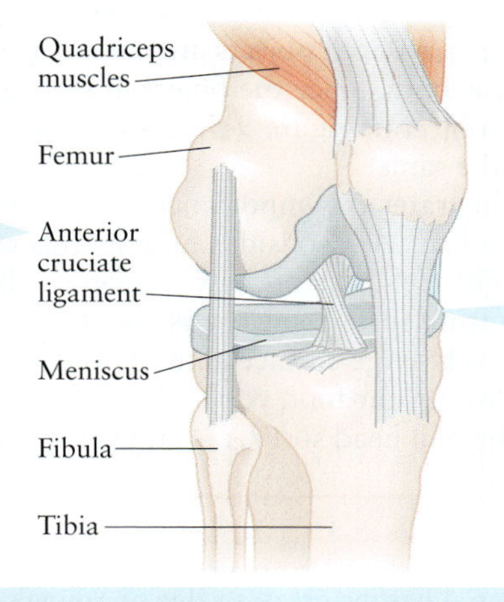

The anterior cruciate ligament (ACL) has a moderate value of Young's modulus: Y = about 10^8 Pa for tension.

The meniscus, which is compressed between the femur and the tibia, has a small value of Young's modulus: Y = about 10^6 Pa for compression, comparable to the silicone rubber nipple of a baby's pacifier.

An ACL is about 1/2000 as stiff as a steel rod ($Y = 2 \times 10^{11}$ Pa) of the same dimensions, but 20 times stiffer than a rubber band ($Y = 5 \times 10^6$ Pa). Hence your knee can bend freely (which an ACL of steel would not allow) without letting your tibia flop around loosely (as would happen with an ACL of rubber).

Figure 9-6 Young's modulus in the knee Comparing the values of Young's modulus for some common materials to the values for two components of the human knee.

Relating Stress and Strain

Here's a useful way to rewrite Hooke's law for an object under tension or compression. This will also help us define *stress* and *strain* more precisely. First we substitute k from Equation 9-2 into Equation 9-1.

$$F = \left(Y\frac{A}{L_0}\right)\Delta L$$

Then we divide both sides by the cross-sectional area A:

Hooke's law for an object under tension or compression (9-3)

Tensile or compressive stress on the object: applied force F divided by the object's cross-sectional area A

Young's modulus Y tells you how stiff the material is of which the object is made: Larger Y means stiffer.

$$\frac{F}{A} = Y\frac{\Delta L}{L_0} \tag{9-3}$$

Tensile or compressive strain of the object: resulting change in length ΔL divided by the object's relaxed length L_0

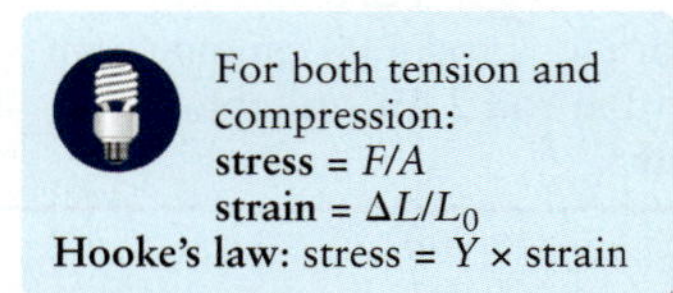

Figure 9-7 Stress and strain for tension or compression Hooke's law relates the stress on an object to the resulting strain of the object.

Figure 9-7 depicts the quantities included in Equation 9-3. The ratio F/A, equal to the force *per area* exerted on each end of the object, is called the **tensile stress** (if the object is being stretched) or **compressive stress** (if the object is being squeezed). Like Young's modulus, stress has units of newtons per square meter (N/m^2), or pascals (Pa).

The ratio $\Delta L/L_0$ that appears on the right-hand side of Equation 9-3 is called the **tensile strain** or **compressive strain**: It equals the distance that the object stretches or shrinks expressed as a fraction of its relaxed length. Tensile or compressive strain is a length divided by another length, so it's a dimensionless quantity.

Equation 9-3 tells us that we can express Hooke's law for tension or compression as follows:

(9-4) (tensile or compressive stress) = (Young's modulus) × (tensile or compressive strain)

That is, *stress and strain are directly proportional,* and the constant of proportionality is Young's modulus. This gives us an alternative way to interpret Figure 9-5. Compared to the object under tension in Figure 9-5b, the object in Figure 9-5c is under one-half the tensile stress (the force is the same but the cross-sectional area is twice as great). Hence the tensile strain is also reduced by one-half: Both objects have the same relaxed length L_0, so the object in Figure 9-5c stretches only half as far as the object in Figure 9-5b. The objects in Figures 9-5b and 9-5d have the same force at each end and the same cross-sectional area, so both objects are under the same tensile stress and the tensile strain is the same for both. Since the object in Figure 9-5d is twice as long, it stretches twice as far as the object in Figure 9-5b so that the ratio of the stretch distance to the relaxed length is the same.

Here's an analogy that illustrates the importance of the ratio of applied force (F) to the area over which it's applied (A). Consider how it feels to stand barefoot on a floor compared to how it might feel to stand barefoot on the head of a nail. In both cases the force F of the supporting surface on you is equal to your weight. When that force is applied over the relatively large surface area of your feet, so F/A is small and there is little compressive stress on your foot, you will be comfortable. But if you stand on the small surface area of the nail head so F/A is large, it will definitely hurt!

GOT THE CONCEPT? 9-2 Stress, Strain, and Young's Modulus

You pull with equal force on each end of two rods of identical cross-sectional area. Both rods obey Hooke's law. Rod 1, made of material P, stretches 1.00 mm. Rod 2, made of material Q, stretches 2.00 mm. Which material has the greater value of Young's modulus? (a) Material P; (b) material Q; (c) they both have the same Young's modulus; (d) not enough information given to decide. *Hint:* Read the details of this question carefully.

WATCH OUT! Hooke's law has important limitations.

Hooke's law in Equation 9-3 applies *only* when the tensile strain is small, that is, when the change in length is small relative to the overall initial length. In Section 9-8 we'll investigate what can happen if the strain is *not* small. Hooke's law also applies only to materials that are of a uniform composition, such as a solid bar of iron. Furthermore, for metals the value of Young's modulus is the same for compression as for tension; for other materials the values for compression and tension may be very different. In Section 9-8 we'll also look at biological materials like ligaments and tendons, which are not of uniform composition and so don't obey Hooke's law very faithfully. For many purposes, however, it's a reasonable approximation to apply Hooke's law even to biological materials like these.

TAKE-HOME MESSAGE FOR Section 9-2

✔ Young's modulus Y is a measure of how easy it is to stretch or compress an object.

✔ Tensile stress equals the stretching force exerted on each end of an object divided by the object's cross-sectional area. The resulting change in length relative to the relaxed length, $\Delta L/L_0$, is called the tensile strain.

✔ We use the terms "compressive stress" and "compressive strain" for the case in which an object is squeezed along a single direction rather than stretched.

✔ If an object obeys Hooke's law, stress and strain are directly proportional.

9-3 Solving stress–strain problems: Tension and compression

Let's use the ideas of Section 9-2 to solve a problem that involves an object under tension.

BIO-Medical EXAMPLE 9-1 Tensile Stress and Strain

You are doing a laboratory experiment with a specimen cut from the sternoclavicular ligament in the shoulder of a cadaver (Figure 9-8). The segment is 2.00 mm by 2.00 mm in cross section and 30.0 mm long. When you apply a force of 1.00 N to each end of the specimen, the specimen stretches by 1.20 mm. Determine (a) the tensile stress on the specimen, (b) the tensile strain of the specimen, and (c) Young's modulus for the specimen. (d) If you repeated the experiment applying the same forces to a second specimen of ligament that has the same length but is 4.00 mm by 4.00 mm in cross section, how would the answers to (a), (b), and (c) change? How far would this specimen stretch?

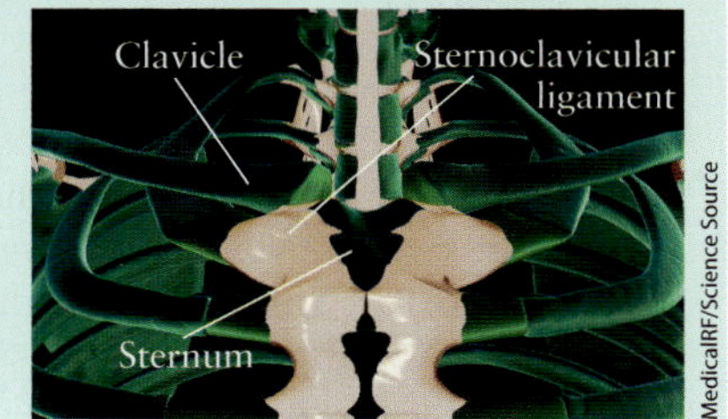

Figure 9-8 Sternoclavicular ligament Ligaments connect one bone to another in the body; the sternoclavicular ligaments connect the sternum (breastbone) to the clavicles.

Set Up

We are given the dimensions of the specimen and the force $F = 1.00$ N applied to each end. We'll find the tensile stress and tensile strain using the definitions of these quantities, and we'll find Young's modulus Y using Hooke's law for tension.

$$\text{tensile stress} = \frac{F}{A}$$

$$\text{tensile strain} = \frac{\Delta L}{L_0}$$

$$(\text{tensile stress}) = (\text{Young's modulus}) \times (\text{tensile strain}) \quad (9\text{-}3)$$

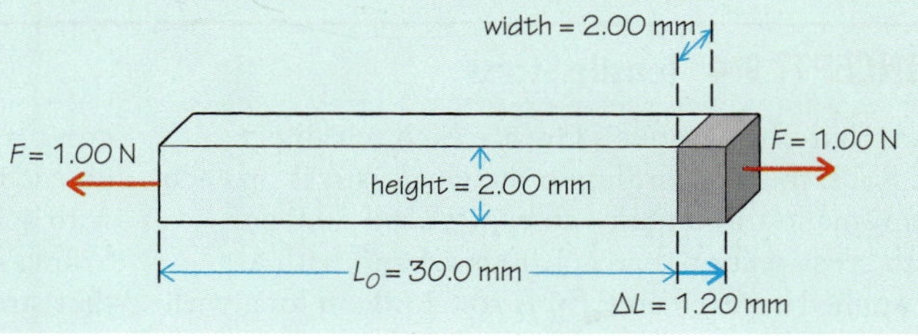

Solve

(a) The tensile stress is the applied force $F = 1.00$ N divided by the cross-sectional area. We convert millimeters to meters.

$$\text{area } A = (2.00 \text{ mm}) \times (2.00 \text{ mm})$$
$$= (2.00 \times 10^{-3} \text{ m}) \times (2.00 \times 10^{-3} \text{ m}) = 4.00 \times 10^{-6} \text{ m}^2$$
$$\text{tensile stress} = \frac{F}{A} = \frac{1.00 \text{ N}}{4.00 \times 10^{-6} \text{ m}^2}$$
$$= 2.50 \times 10^5 \text{ N/m}^2 = 2.50 \times 10^5 \text{ Pa}$$

(b) The tensile strain is the change in length $\Delta L = 1.20$ mm divided by the relaxed length $L_0 = 30.0$ mm.

$$\text{tensile strain} = \frac{\Delta L}{L_0} = \frac{1.20 \text{ mm}}{30.0 \text{ mm}}$$
$$= 0.0400 = 4.00 \times 10^{-2}$$

(c) Rearranging Hooke's law tells us that Young's modulus for this specimen equals the tensile stress divided by the tensile strain.

$$\text{Young's modulus } Y = \frac{\text{tensile stress}}{\text{tensile strain}} = \frac{2.50 \times 10^5 \text{ Pa}}{4.00 \times 10^{-2}}$$
$$= 6.25 \times 10^6 \text{ Pa}$$

(d) If each of the cross-sectional dimensions doubles from 2.00 to 4.00 mm, the cross-sectional area A increases by a factor of $2 \times 2 = 4$, so the tensile stress is $\frac{1}{4}$ as much as the previous value of 2.50×10^5 Pa. The value of Young's modulus is *unaffected*, since this depends on the type of material and not on its dimensions. From Hooke's law, the new tensile strain equals the new tensile stress divided by Y. Since Y is the same but the tensile stress is $\frac{1}{4}$ the previous value, the tensile strain is likewise $\frac{1}{4}$ of the previous value of 4.00×10^{-2}. The new specimen therefore stretches by $\frac{1}{4}$ as much as the first one.

New cross-sectional area A

$$= (4.00 \text{ mm}) \times (4.00 \text{ mm})$$
$$= (4.00 \times 10^{-3} \text{ m}) \times (4.00 \times 10^{-3} \text{ m})$$
$$= 1.60 \times 10^{-5} \text{ m}$$

New tensile stress

$$= \frac{F}{A} = \frac{1.00 \text{ N}}{1.60 \times 10^{-5} \text{ m}^2} = 6.25 \times 10^4 \text{ N/m}^2 = 6.25 \times 10^4 \text{ Pa}$$

Young's modulus $Y = 6.25 \times 10^6$ Pa

New tensile strain

$$= \frac{\text{new tensile stress}}{\text{Young's modulus}} = \frac{6.25 \times 10^4 \text{ Pa}}{6.25 \times 10^6 \text{ Pa}} = 1.00 \times 10^{-2}$$

$$\Delta L = (\text{new tensile strain}) \times L_0 = (1.00 \times 10^{-2})(30.0 \text{ mm})$$
$$= 0.300 \text{ mm}$$

Reflect

This example emphasizes the key message of Hooke's law: For relatively small stresses the stress is proportional to the strain. Note that the value of Young's modulus for the sternoclavicular ligament is quite a bit less than that for the ACL ($Y = 1 \times 10^8$ Pa). This is because different ligaments in different parts of the body are made of different materials.

For this 2.00 mm × 2.00 mm × 30.0 mm segment of sternoclavicular ligament, can you show that a stretching force of 0.025 N would be required to increase the 30.0-mm length of the segment by 0.10%?

GOT THE CONCEPT? 9-3 Compressive Stress and Strain

A brass rod ($Y = 1 \times 10^{11}$ Pa) and a steel rod ($Y = 2 \times 10^{11}$ Pa) of the same dimensions are subjected to identical squeezing forces on their ends. (a) Which experiences the greater compressive stress, the brass rod or the steel rod? (b) Which undergoes the greater compressive strain, the brass rod or the steel rod? (c) Which undergoes the greatest change in length, the brass rod or the steel rod?

GOT THE CONCEPT? 9-4 Tensile Stress

?

Four rods that obey Hooke's law are each put under tension. Rank them according to the tensile stress on each rod, from smallest to largest value. (a) A rod 50.0 cm long with cross-sectional area 1.00 mm^2 and with a 200-N force applied on each end. (b) A rod 25.0 cm long with cross-sectional area 1.00 mm^2 and with a 200-N force applied on each end. (c) A rod with cross-sectional area 2.00 mm^2 with a 100-N force applied on each end. (d) A rod with Young's modulus 2.00×10^{10} Pa and relaxed length 50.0 cm that stretches 0.025 mm due to forces on its ends.

TAKE-HOME MESSAGE FOR Section 9-3

✔ In problems involving tension or compression, use Hooke's law, Equation 9-3, to relate the stress (force divided by cross-sectional area) to the strain (change in length divided by relaxed length) and Young's modulus for the material. Then solve for the unknown quantity.

9-4 An object expands or shrinks when under volume stress

We saw in Sections 9-2 and 9-3 that applied forces can cause an object to stretch or compress along its length. Applied forces can also cause changes in an object's *volume*. You can see this effect when a fire extinguisher is discharged (**Figure 9-9a**). Inside the fire extinguisher, the gas (which is the object in question) is under high pressure, and this pressure keeps the gas confined within the small volume of the extinguisher cylinder. When you trigger the fire extinguisher, the released gas moves into the low-pressure outside air and expands to a much larger volume. The *volume stress* on the extinguisher gas is the *change* in the pressure that pushes on the gas, and the *volume strain* is a measure of how much the volume of this gas changes in response to that pressure change.

BIO-Medical For the gas that escapes from the fire extinguisher in Figure 9-9a, the volume stress is negative (the pressure on the gas decreases), and the volume strain is positive (the volume of the gas increases). The reverse is true for a descending scuba diver (**Figure 9-9b**). The deeper in the ocean she dives, the greater the pressure on her body, so the volume stress is positive (the pressure on her body increases). This causes gases within her body to compress, so their volume strain is negative (the volume of the gas decreases). These compressed gases dissolve in her body fluids. If a diver ascends too quickly, the gases may pop out of solution and form bubbles that can cause excruciating pain (a condition called *decompression sickness*). To avoid this, scuba divers are trained to return to the surface at a gradual pace so that the pressure on their bodies decreases gradually. The dissolved gases then expand slowly and can safely be exhaled.

(a) Triggering a fire extinguisher releases the high-pressure gases inside the tank. The pressure is much lower outside the tank, so the gases expand.

Michael Blann/Getty Images

(b)

Exactostock/SuperStock

As this scuba diver descends, the pressure on her increases. This compresses gases inside her body and makes them dissolve in her body fluids.

Volume stress refers to the change in pressure on an object.
Volume strain refers to the change in the object's volume caused by this pressure change.

Figure 9-9 Volume stress and strain Volume stress involves a change in pressure. (a) A pressure decrease makes the volume increase. (b) A pressure increase makes the volume decrease.

Relating Volume Stress and Volume Strain: Hooke's Law Revisited

Let's be more quantitative about volume stress and volume strain. First notice that every object around you has forces exerted on it by the surrounding material (**Figure 9-10a**). Your body has forces exerted on it by the air around you; a submerged fish has forces exerted on it by the water in which it swims; and a rock buried below Earth's surface has forces exerted on it by the surrounding dirt. The magnitude of the force per unit area on the surface of the object is called the **pressure** p. (Unfortunately, we used this same symbol in Chapter 7 for momentum. We promise never to use these two quantities in the same equation!) The units of pressure are newtons per square meter (N/m^2) or pascals (Pa). This is the same unit we used for the tensile or compressive stress: F/A, which can also be thought of as a pressure.

As **Figure 9-10** illustrates, an object's volume changes when the pressure p changes. We call the pressure change Δp the **volume stress**, and we call the resulting change in volume ΔV divided by the original volume V_0—that is, the quantity $\Delta V/V_0$—the **volume strain**. If the pressure increases so that the volume stress Δp is positive, then the object shrinks, and the volume strain is negative (**Figure 9-10b**). If the pressure decreases so that the volume stress is negative, then the object expands, and the volume strain is positive (**Figure 9-10c**). Note that volume stress, like pressure, has units of N/m^2 or Pa, while volume strain is dimensionless (it's a volume divided by another volume). These are the same units as the quantities tensile stress and tensile strain that we studied in Sections 9-2 and 9-3.

Experiment shows that if the volume stress on an object is not too large, the resulting volume strain is *directly proportional* to the volume stress. That is, Hooke's law relates volume stress and strain in the same manner as it does for tensile stress and strain:

$$\text{(volume stress)} = -\text{(bulk modulus)} \times \text{(volume strain)} \qquad (9\text{-}5)$$

That is, *volume stress and volume strain are directly proportional*. The proportionality constant in Equation 9-5 is called the **bulk modulus** of the material of which the object is made. Like Young's modulus, the bulk modulus B has units of N/m^2 or Pa. The second

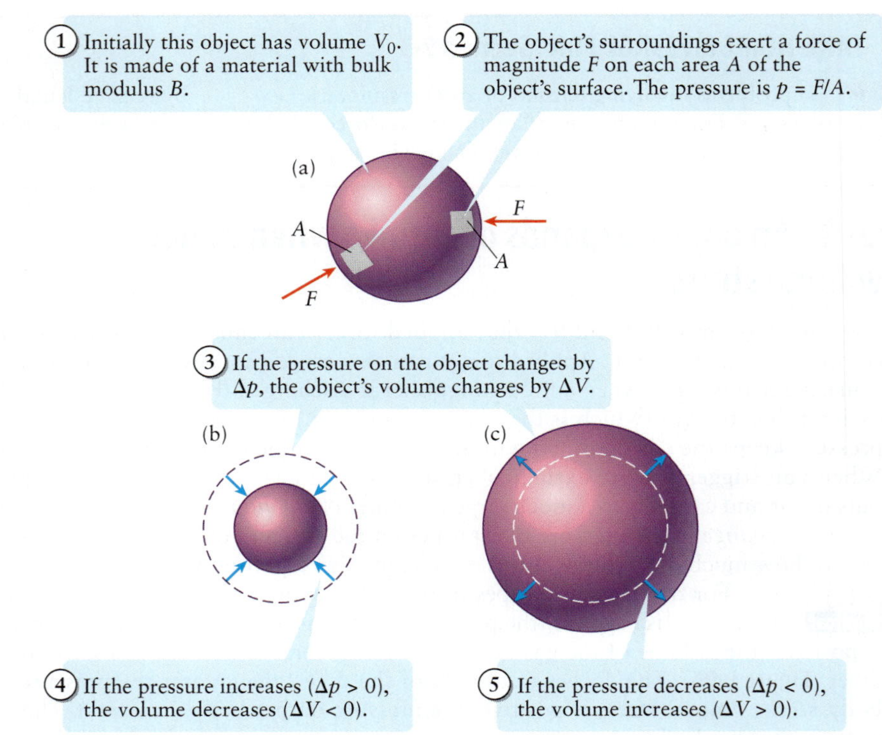

1 Initially this object has volume V_0. It is made of a material with bulk modulus B.
2 The object's surroundings exert a force of magnitude F on each area A of the object's surface. The pressure is $p = F/A$.
(a)
A
F
F
A
3 If the pressure on the object changes by Δp, the object's volume changes by ΔV.
(b)
(c)
4 If the pressure increases ($\Delta p > 0$), the volume decreases ($\Delta V < 0$).
5 If the pressure decreases ($\Delta p < 0$), the volume increases ($\Delta V > 0$).

Figure 9-10 Hooke's law for volume stress An object under volume stress obeys Hooke's law if the stress is not too great.

volume stress = Δp
volume strain = $\Delta V/V_0$
Hooke's law:
volume stress = $-B$ × volume strain

column of Table 9-1 lists the values of the bulk modulus of various substances. The minus sign in Equation 9-5 tells us that volume stress and volume strain have opposite signs; if one is positive, the other is negative (see parts b and c of Figure 9-10).

In terms of the quantities Δp, V_0, and ΔV depicted in Figure 9-10, we can rewrite Equation 9-5 as

Volume stress on the object: change in the pressure p on all sides of the object

Bulk modulus B: How difficult it is to compress the material of which the object is made; larger B means harder to compress.

Hooke's law for an object under volume stress (9-6)

$$\Delta p = -B\frac{\Delta V}{V_0}$$

Important minus sign! If Δp is positive, then ΔV is negative; if Δp is negative, then ΔV is positive.

Volume strain of the object: resulting change in volume ΔV divided by the object's initial volume V_0

To get more insight into the meaning of Equation 9-6, multiply both sides by $-V_0/B$ to rewrite it so that ΔV is by itself on one side of the equation:

$$\Delta V = -\frac{\Delta p}{B}V_0 \tag{9-7}$$

Equation 9-7 says that for a given pressure change Δp and initial volume V_0, making the bulk modulus B larger makes the change in volume ΔV smaller. A material that has a relatively large bulk modulus is relatively *incompressible*, which means that its volume doesn't change much even when it experiences a large volume stress. From the entries in Table 9-1 you can see that materials such as iron and concrete have very large bulk moduli; they are hard to compress, which makes them good materials for constructing buildings. By contrast, air has a bulk modulus more than a million times smaller than that of steel and is relatively easy to compress. The human lung is made of even more compressible material: Under physiologic conditions, the bulk modulus of the human lung varies between about 1500 and 10,000 Pa in young adults and increases with age. These low values of B minimize the effort needed to make the lungs expand and contract during respiration.

Another interpretation of the bulk modulus can be gained by examining Equation 9-7 in the case when Δp is equal to B. When this happens we see that ΔV is equal to $-V_0$. That is, the object has shrunk to zero volume! So we can think of B as the *increase* in pressure required to completely shrink an object made of that material. It's pretty challenging to shrink most objects to a point that has zero volume. That's one reason B is so large. If Δp is equal to $-B$, then ΔV is equal to V_0. That is, the volume of the object has doubled. So bulk modulus can also be interpreted as the *decrease* in pressure required for an object to double its volume.

GOT THE CONCEPT? 9-5 Volume Stress and Volume Strain

Two identical balloons are filled to the same size, one with water and one with air. When the two balloons are subjected to the same volume stress (the same force per unit surface area pushing in on the balloon), which of the two will exhibit the larger change in volume: (a) the water-filled balloon or (b) the air-filled balloon?

GOT THE CONCEPT? 9-6 Pressure and Force

You apply a pressure of 5.00×10^6 N/m^2 on all sides of a sphere of radius 2.00 m. What is the net force that is exerted on the sphere as a result? (a) 6.28×10^7 N; (b) 1.68×10^8 N; (c) 2.51×10^8 N; (d) zero.

TAKE-HOME MESSAGE FOR Section 9-4

✔ Volume stress equals the *change* in pressure on an object.

✔ An object under volume stress either expands (if the stress is negative) or contracts (if the stress is positive); the volume strain is the change in the object's volume divided by its initial volume.

✔ If an object obeys Hooke's law, volume stress and volume strain are directly proportional; the constant of proportionality is the bulk modulus B for the material of which the object is made.

9-5 Solving stress–strain problems: Volume stress

In the following examples we'll look at two problems that involve volume stress and strain. Note that the ideas of volume stress and strain apply equally well to objects that are gases, liquids, and solids.

See the Math Tutorial for more information on geometry.

EXAMPLE 9-2 A Submerged Cannonball

A solid iron cannonball falls into the Pacific near the Galapagos Islands, where the ocean is 2400 m deep and the pressure is 2.40×10^7 Pa at the bottom. (By contrast, air pressure at the surface is 1.01×10^5 Pa.) By what percentage does the cannonball shrink in volume as it sinks to the bottom?

Set Up

We use Hooke's law, Equation 9-6, to find the volume strain $\Delta V/V_0$ of the cannonball due to the additional pressure exerted on it (the volume stress Δp). Since $\Delta V/V_0$ expresses the change in volume as a fraction, we can convert it to a percentage by multiplying it by 100.

Hooke's law for an object under volume stress:

$$\Delta p = -B\frac{\Delta V}{V_0} \qquad (9\text{-}6)$$

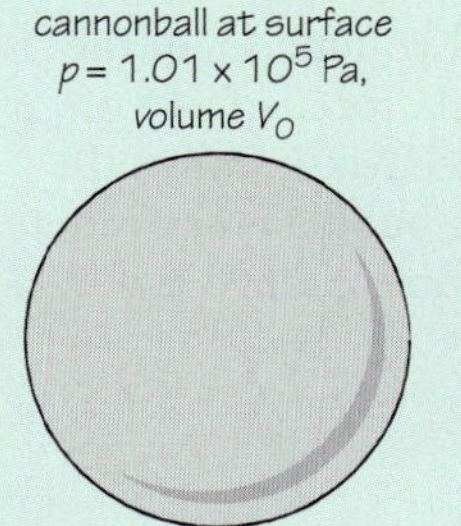

Solve

The volume stress is the difference between the pressure at the bottom of the ocean and the pressure at the surface.

Volume stress Δp:

$$= (\text{pressure at bottom of ocean}) - (\text{pressure at surface})$$
$$= 2.40 \times 10^7 \text{ Pa} - 1.01 \times 10^5 \text{ Pa} = 2.39 \times 10^7 \text{ Pa}$$

Use the value of B for iron from Table 9-1 and solve for $\Delta V/V_0$. Then convert this to a percentage.

$$\frac{\Delta V}{V_0} = -\frac{\Delta p}{B} = -\frac{2.39 \times 10^7 \text{ Pa}}{1.7 \times 10^{11} \text{ Pa}}$$
$$= -1.4 \times 10^{-4}$$

Percentage change in volume:

$$\frac{\Delta V}{V_0} \times 100\% = -1.4 \times 10^{-4} \times 100\%$$
$$= -0.014\%$$

Reflect

The minus sign tells us that $\Delta V < 0$ and so the volume of the cannonball decreases. Because iron has such a large bulk modulus B, even this immense pressure increase causes only a miniscule change in the cannonball's volume, just 14/1000 of 1%.

Note that while the volume stress Δp is given to three significant figures, the answer has only two significant figures. Can you see why? (See the discussion of significant figures in Section 1-4 if you're not sure.)

EXAMPLE 9-3 Bubbles Rising

A scuba diver 10.0 m below the surface of a lake releases a spherical air bubble 2.00 cm in radius. As it rises to the surface, the pressure exerted by the surrounding water on the bubble decreases by 9.80×10^4 Pa due to the decreasing amount of water above the bubble. This pressure decrease causes the bubble to expand. Use Hooke's law to find the radius of the bubble as it breaks the surface.

Set Up

As in Example 9-2, we use the Hooke's law relationship between volume stress and volume strain, Equation 9-6. This gives us the change in volume ΔV versus the initial volume V_0. We assume that the amount of air in the bubble doesn't change as it rises. Thus the change in volume of the bubble is due only to the volume stress, which equals the change in water pressure pushing on the outside of the bubble.

$$\Delta p = -B\frac{\Delta V}{V_0} \quad (9\text{-}6)$$

volume stress $= \Delta p$
$= -9.80 \times 10^4$ Pa
(negative because the pressure on the outside of the bubble decreases)

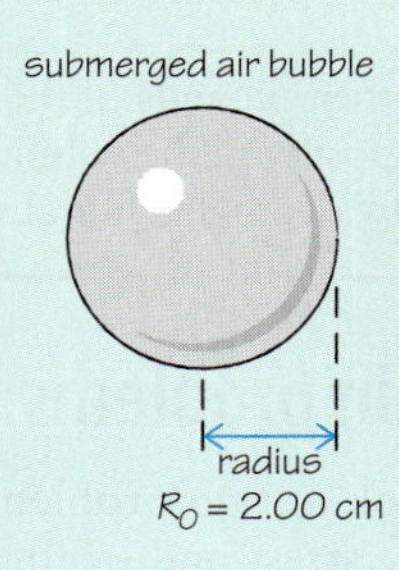

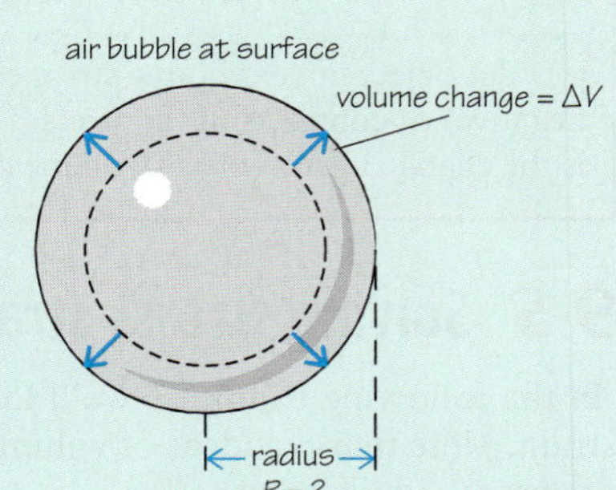

Solve

Use the bulk modulus of air from Table 9-1 and the value of Δp to solve for the volume strain $\Delta V/V_0$. The volume strain is positive because the bubble expands.

Bulk modulus of air from Table 9-1:
$B = 1.41 \times 10^5$ Pa, so

$$\frac{\Delta V}{V_0} = -\frac{\Delta p}{B} = -\frac{(-9.80 \times 10^4 \text{ Pa})}{1.41 \times 10^5 \text{ Pa}} = 0.695$$

Next we find the initial volume V_0 of the bubble using the formula for a sphere of radius $R_0 = 2.00$ cm (see Math Tutorial). The final volume of the bubble is the initial volume V_0 plus the volume change ΔV.

Initial volume of a spherical bubble:

$$V_0 = \frac{4\pi}{3}R_0^3 = \frac{4\pi}{3}(2.00 \text{ cm})^3$$
$$= 33.5 \text{ cm}^3$$

Final volume:

$$V = V_0 + \Delta V = V_0\left(1 + \frac{\Delta V}{V_0}\right)$$
$$= (33.5 \text{ cm}^3)(1 + 0.695) = (33.5 \text{ cm}^3)(1.695)$$
$$= 56.8 \text{ cm}^3$$

Solve for the final radius R of the bubble using the formula $V = (4\pi/3)R^3$ for the volume of a sphere of radius R.

$$V = \frac{4\pi}{3}R^3$$

$$R = \sqrt[3]{\frac{3V}{4\pi}} = \sqrt[3]{\frac{3(56.8 \text{ cm}^3)}{4\pi}} = 2.38 \text{ cm}$$

Reflect

Our calculation shows that as the bubble rises to the surface, its volume increases by a factor of 1.695, and its radius increases from 2.00 to 2.38 cm. Consider the implications of this result on decompression sickness (described in Section 9-4). When a diver ascends too quickly, not only do bubbles form, but also they get bigger as the diver rises to the surface.

It turns out that the answers we calculated here aren't exactly correct. The reason is that the air pressure *inside* the bubble also decreases as it expands, and as a result the bubble expands a bit less than we have calculated. A more detailed calculation shows that the volume actually increases by a factor of 1.623, and the final radius is 2.35 cm. Our calculations with Hooke's law are pretty close, however, which shows that this law can give quite useful results.

TAKE-HOME MESSAGE FOR Section 9-5

✔ In problems involving volume stress and strain, use Hooke's law, Equation 9-6, to relate the stress (the change in pressure on an object) to the strain (change in the object's volume divided by its original volume) and the bulk modulus of the material. Then solve for the unknown quantity.

9-6 A solid object changes shape when under shear stress

When you run, play sports, or do any sort of physical activity, blood flow increases to your active tissues. The arteries need to expand or *dilate* to accommodate this increased blood flow. In addition to the dilation caused by increased blood pressure pushing outward against arterial walls, the deformation of endothelial cells that line the inside of the artery also leads to dilation of the vessel (**Figure 9-11a, b**). The force of blood pushing on the exposed surfaces of the cells deforms them in the direction of blood flow (**Figure 9-11c**). Because the cells are firmly attached to the arterial wall, the change in shape is neither a stretch nor a compression. Instead, the cells deform in a way similar to a cube of gelatin dessert being pushed parallel to the top face. In response to this deformation, the endothelial cells release nitric oxide gas. This gas diffuses to the muscle cells in the arterial wall and causes them to relax. As a result the artery dilates, thus relieving the stress on the endothelial cells (**Figure 9-11d**).

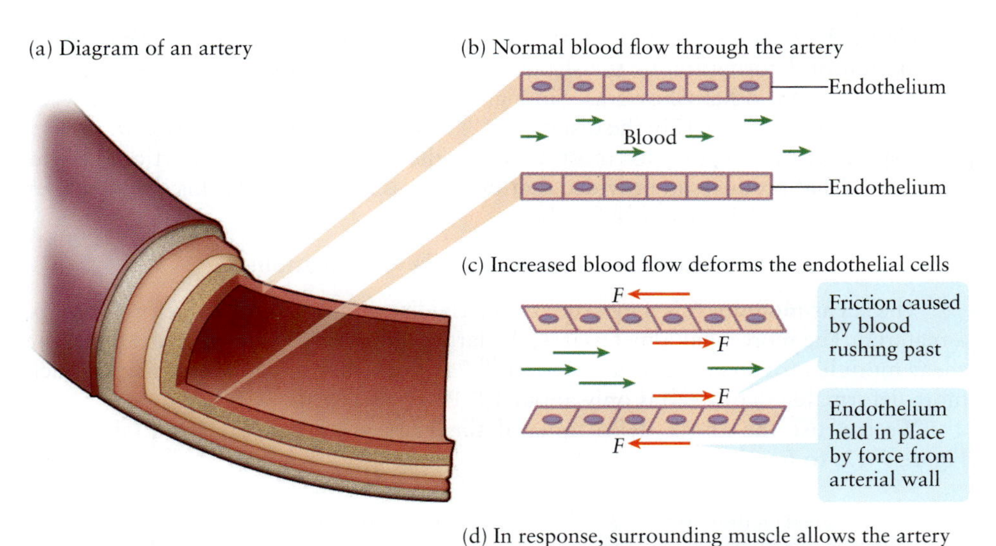
(a) Diagram of an artery
(b) Normal blood flow through the artery
Endothelium
Blood
Endothelium
(c) Increased blood flow deforms the endothelial cells
F
F
Friction caused by blood rushing past
F
F
Endothelium held in place by force from arterial wall
(d) In response, surrounding muscle allows the artery to expand, relieving stress on the endothelium

Shear stress refers to forces that deform an object [like the endothelium in (c)] without making it expand or contract. **Shear strain** refers to the change in the object's shape caused by this stress.

Figure 9-11 Shear stress and strain When blood flow through an artery increases due to physical exertion, the endothelium deforms without becoming larger or smaller. This is an example of shear stress.

The kind of stress shown in Figure 9-11c is called **shear**. Unlike tensile stress and compressive stress (which are applied along a line within the object) or volume stress (which is applied over the entire surface of the object), shear stress is caused by forces applied *parallel* (tangential) to the faces of an object. As Figure 9-11c shows, these forces are *offset* from each other: They do not act along the same line. An object is said to experience shear when one face is made to move or slide relative to the opposite face.

Like the other types of stress we've encountered, **shear stress** is defined as a force per unit area. As an example, consider the object shown in **Figure 9-12a**. In **Figure 9-12b** opposite forces of the same magnitude $F_{\parallel}$ are applied to the upper and lower surfaces of this object. The subscript "$\parallel$" means "parallel," and reminds us that these forces act parallel to the surfaces as in Figure 9-12b. (For the endothelial cells depicted in Figure 9-11c, one force is exerted by the blood that flows past the cells. The other force is exerted by the arterial wall to which the cells are attached.) The shear stress associated with these forces is the force magnitude $F_{\parallel}$ divided by the area A over which this force is applied, or $F_{\parallel}/A$. Just like tensile and volume stress, the units of shear stress are N/m^2 or Pa.

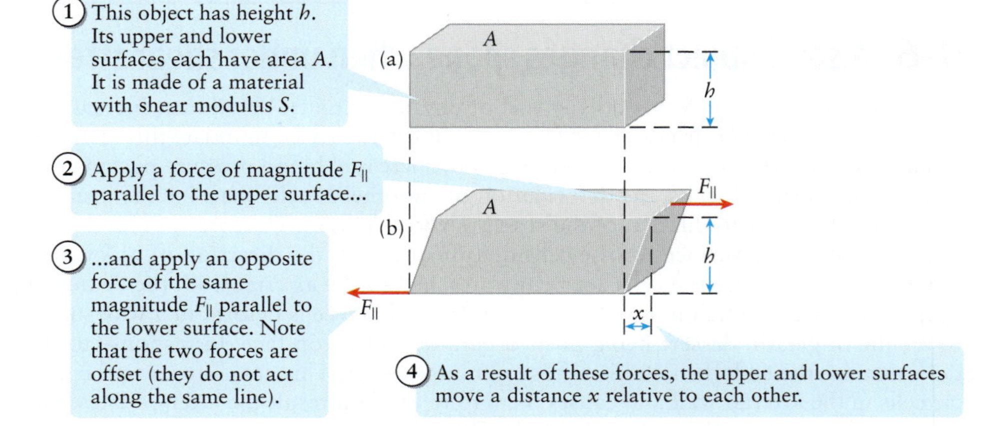

1 This object has height h. Its upper and lower surfaces each have area A. It is made of a material with shear modulus S.
(a)
A
h
2 Apply a force of magnitude F‖ parallel to the upper surface...
(b)
A
F‖
h
3 ...and apply an opposite force of the same magnitude F‖ parallel to the lower surface. Note that the two forces are offset (they do not act along the same line).
F‖
x
4 As a result of these forces, the upper and lower surfaces move a distance x relative to each other.

Figure 9-12 Shear stress and strain Shear stress involves oppositely directed forces that act on different surfaces of an object. These offset forces cause a twisting deformation (shear strain).

shear stress = $F_{\parallel}/A$
strain = x/h
Hooke's law:
shear stress = $S \times$ shear strain

Figure 9-12 shows that this shear stress causes the object to deform. We measure the amount of deformation by the **shear strain**, which is the displacement of one surface relative to the other (shown as x in Figure 9-12b) divided by the distance between the surfaces (in Figure 9-12b, the distance h). Thus the shear strain equals x/h.

Experiment shows that for most materials the shear strain is proportional to the shear stress that produces it, provided that the stress is not too great. This is yet another version of Hooke's law, this time for shear:

$$\text{shear stress} = \text{shear modulus} \times \text{shear strain} \tag{9-8}$$

The **shear modulus** S is a measure of the rigidity and resistance to deformation of a material. The more rigid a material is, the larger its shear modulus. For example, the shear modulus for steel is about 7.8×10^{10} Pa, while that of rubber (which is much more deformable than steel) is only about 10^6 Pa.

We can use the above definitions of shear stress and shear strain (shown in Figure 9-12) to rewrite Equation 9-8 as

Hooke's law for an object under shear stress (9-9)

$$\frac{F_{\parallel}}{A} = S\frac{x}{h}$$

Shear stress on the object: force $F_{\parallel}$ applied to each of two opposite surfaces divided by the area A of each surface

Shear modulus S: how difficult it is to deform the material of which the object is made: larger S means more difficult

Shear strain of the object: distance x that the two surfaces move relative to each other divided by the distance h between the surfaces

WATCH OUT! There is no shear modulus for a liquid or gas.

While liquids and gases can sustain a *volume* stress that causes a change in volume, they cannot sustain a *shear* stress due to forces applied tangentially to their surfaces. Instead of deforming like the object shown in Figure 9-12, a force tangential to the surface of a liquid or gas causes it to flow rather than deform. This means that the shear modulus is defined *only* for solids.

GOT THE CONCEPT? 9-7 Shear Stress and Shear Strain

Figure 9-13 shows three objects that are all made of the same material. The forces applied to the three objects all have the same magnitude. (a) Rank the objects in order of the shear stress acting on them, from greatest to least. (b) Rank the objects in order of the shear strain that they will undergo, from greatest to least.

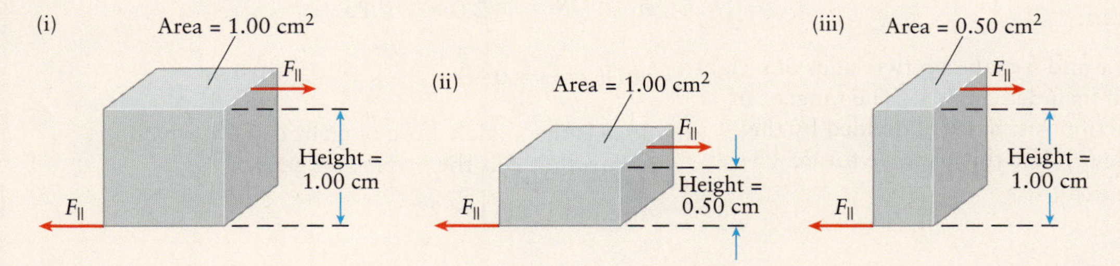

Figure 9-13 Three objects under shear stress The same forces act on all three objects, all of which are made of material with the same shear modulus. Which one experiences the greatest strain?

TAKE-HOME MESSAGE FOR Section 9-6

✔ Shear stress occurs when a force is applied parallel to the face of an object and causes that face to move relative to the opposite face. Shear strain describes the extent to which an object is deformed by shear stress.

✔ The shear modulus measures the rigidity of a material and is only defined for solids.

9-7 Solving stress–strain problems: Shear stress

Here are two examples of how to solve problems involving shear stress and shear strain.

BIO-Medical EXAMPLE 9-4 Endothelial Cells and Shear

Endothelial cells in an artery experience shear stress as a result of blood flow. In an *in vitro* study of arterial endothelial cells from pigs, the shear strain was observed to be proportional to the shear stress when this stress was 8.6×10^3 N/m². The shear modulus of a typical endothelial cell is 2.0×10^4 N/m². Figure 9-14 shows a normal endothelial cell and a cell experiencing shear. At what angle ϕ does one surface of a cell move relative to the opposite surface? (The angle ϕ is labeled in Figure 9-14.)

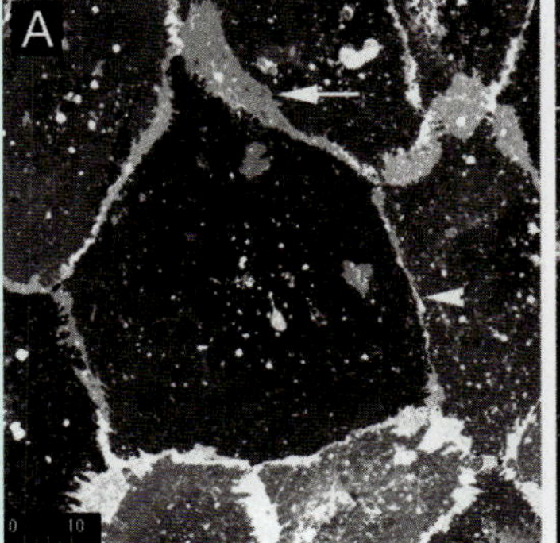

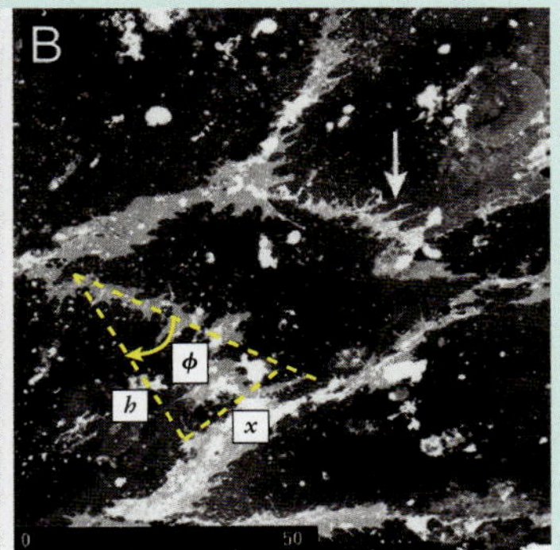

Figure 9-14 Shear stress on endothelial cells A normal endothelial cell (left) and a cell experiencing shear stress (right).

Set Up

The observation that shear stress and shear strain are proportional means that Hooke's law applies in this situation, so we can use Equation 9-9. We are given the values of the shear stress and the shear modulus, so we can find the ratio x/h (the shear strain). Once we know this ratio we'll use trigonometry to determine the angle ϕ shown in Figure 9-14.

Hooke's law for shear:

$$\frac{F_{\parallel}}{A} = S\frac{x}{h} \qquad (9\text{-}9)$$

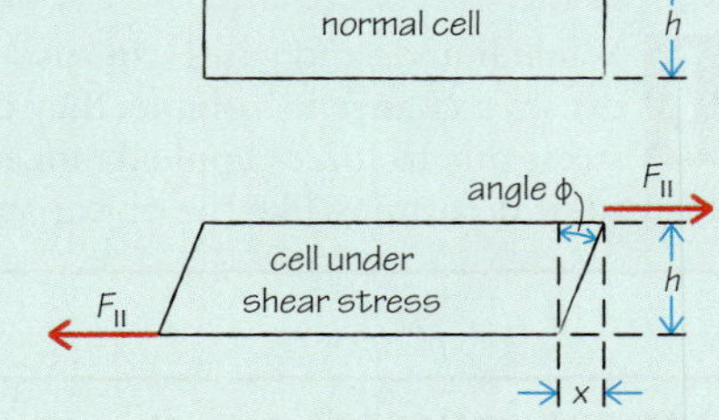

Solve

Rearrange Equation 9-9 to calculate the shear strain. Then substitute $F_{\parallel}/A = 8.6 \times 10^3$ Pa for the shear stress and $S = 2.0 \times 10^4$ Pa for the shear modulus.

Divide both sides of Equation 9-9 by S:

$$\frac{x}{h} = \frac{(F_{\parallel}/A)}{S} = \frac{8.6 \times 10^3 \text{ Pa}}{2.0 \times 10^4 \text{ Pa}} = 0.43$$

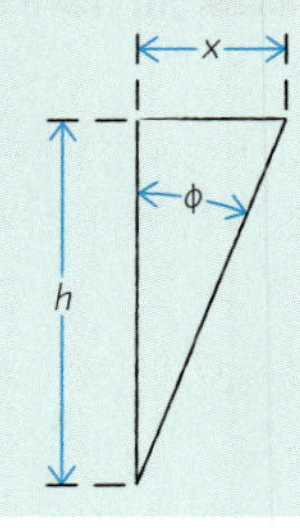

The lengths x and h make up two sides of a right triangle with included angle ϕ. The tangent of ϕ equals the opposite side (x) divided by the adjacent side (h). Use this to solve for ϕ.

$$\tan\phi = \frac{x}{h} = 0.43$$

So ϕ is the angle whose tangent is 0.43, which means that ϕ is the inverse tangent of 0.43:

$$\phi = \tan^{-1} 0.43 = 23°$$

Reflect

The value of $\phi = 23°$ is quite large, which means that these endothelial cells undergo a dramatic change in shape (see Figure 9-14). This illustrates that endothelial cells in arteries, and the vascular system in general, are *very* sensitive to shear stress due to fluid flow. Note that, as we mention in Section 9-6, endothelial cells release nitric oxide gas when under shear stress; this gas release causes the underlying muscle to relax, thus expanding the artery. This increases blood flow while reducing the flow speed and relieving the shear stress on the endothelial cells—a classic example of negative feedback in physiology.

EXAMPLE 9-5 Earthquake Damage

The dedication plate mounted to the base of a building was deformed during an earthquake. The plate, made from a metal alloy of shear modulus 4.00×10^{10} Pa, was originally 80.0 cm high, 50.0 cm long, and 5.00 mm thick. The earthquake shifted the top surface of the plate 0.100 mm relative to the bottom surface. What shear force did the plate experience during the earthquake?

Set Up

The displacement of the two surfaces of the plate ($x = 0.100$ mm) is small compared to the distance between them ($h = 80.0$ cm), so the shear strain x/h is very small. This suggests that we can use Hooke's law for shear, Equation 9-9. We are given the value of the shear modulus ($S = 4.00 \times 10^{10}$ Pa) and the dimensions of the plate, so we can solve for the shear force $F_{\parallel}$:

Hooke's law for shear:

$$\frac{F_{\parallel}}{A} = S\frac{x}{h} \qquad (9\text{-}9)$$

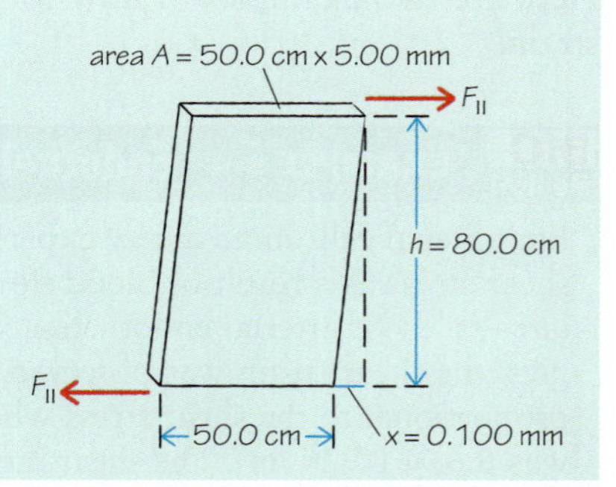

Solve

First calculate the shear stress $F_{\parallel}/A$ using Equation 9-9, making sure to convert to SI units.

$$\frac{F_{\parallel}}{A} = S\frac{x}{h} = (4.00 \times 10^{10} \text{ Pa})\left(\frac{0.100 \text{ mm}}{80.0 \text{ cm}}\right)$$
$$= (4.00 \times 10^{10} \text{ N/m}^2)\left(\frac{1.00 \times 10^{-4} \text{ m}}{0.800 \text{ m}}\right)$$
$$= (4.00 \times 10^{10} \text{ N/m}^2)(1.25 \times 10^{-4})$$
$$= 5.00 \times 10^6 \text{ N/m}^2$$

The area A over which the force is applied is the area of the top or bottom surface of the plate. Multiply the shear stress by A to calculate the force magnitude $F_{\|}$:

$$A = (50.0 \text{ cm})(5.0 \text{ mm}) = (0.500 \text{ m})(5.00 \times 10^{-3} \text{ m})$$
$$= 2.50 \times 10^{-3} \text{ m}$$
$$F_{\|} = \frac{F_{\|}}{A}A = (5.00 \times 10^{6} \text{ N/m}^2)(2.50 \times 10^{-3} \text{ m})$$
$$= 1.25 \times 10^{4} \text{ N}$$

Reflect

The shear force was 12,500 N, or about 1.25 tons! This force is about the same as the weight of a small car. It takes a *lot* of force to deform metal.

TAKE-HOME MESSAGE FOR Section 9-7

✔ In problems involving shear stress and strain, use Hooke's law, Equation 9-9, to relate the shear stress, the shear strain, and the shear modulus of the material. (It's very helpful to draw a picture of the situation.) Then solve for the unknown quantity.

9-8 Objects deform permanently or fail when placed under too much stress

Your arteries, ears, nose, and ligaments, as well as many other structures in your body, can be stretched, compressed, or bent without deforming them permanently. Somewhat like a rubber band, after you stretch your earlobe or squeeze the skin on your forearm by applying a modest force, it quickly returns to its original shape. Structures with this property are called **elastic**. (We used the term *elastic* in Chapter 7 to refer to collisions in which mechanical energy is conserved. Our new definition is really the same. If two elastic objects collide with each other, they may deform during the collision but will spring back to their original shapes afterward. Hence no energy is used up during the collision, and the total mechanical energy is conserved.) But if a structure is deformed too much, it will not spring back and may even tear or fracture. This is what can happen when too much force is applied to a human knee, making the anterior cruciate ligament (ACL) stretch too far and creating a tear in the ACL. How can we describe what happens when a structure or material is deformed so much that it is no longer elastic?

From Elastic to Plastic to Failure

If the tensile stress on an object is increased so that the change in the object's length ΔL becomes large relative to its initial length L_0, the deformation of the object becomes permanent. This is what happens if you stretch a metal spring too far: The spring is permanently deformed and can no longer return to its original shape (**Figure 9-15**). The tensile stress at which this occurs is called the **yield strength**. If the tensile stress exceeds the yield strength and the object deforms permanently, we say that the object has become **plastic**.

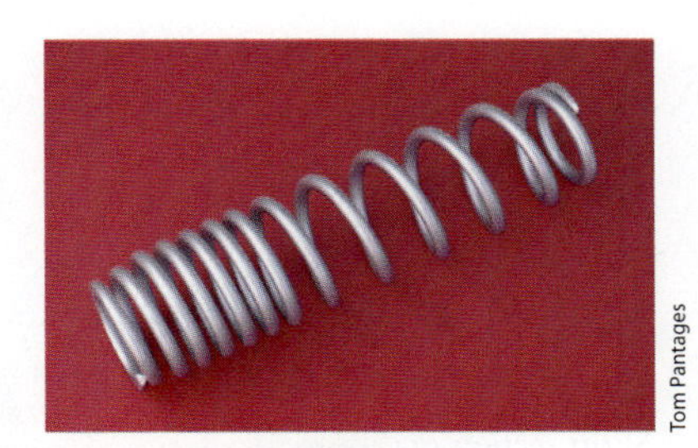

Tom Pantages

Figure 9-15 A permanently deformed spring This spring has been stretched so far that it no longer returns to its original coiled shape. The metal of which the spring is made became plastic during the stretch.

WATCH OUT! When is a plastic object plastic?

! The term *plastic* may seem like a strange one to use to describe objects that don't spring back to their original shape. After all, most objects that we think of as "plastic"—such as a plastic fork or plastic bottle—are actually quite springy. But to make these objects the plastic material is heated so that it can flow into a mold, then cooled so that it solidifies into the desired shape. Strictly speaking, the material is only "plastic" when it is flowing into its new shape.

Figure 9-16 shows elastic and plastic behavior in an idealized graph of tensile stress versus tensile strain. In the elastic regime the tensile strain is *reversible:* If you increase the amount of stress by a small amount, then release that extra stress (say, by pulling a little harder on the ends of a spring, then easing off on the pull), the object will relax to

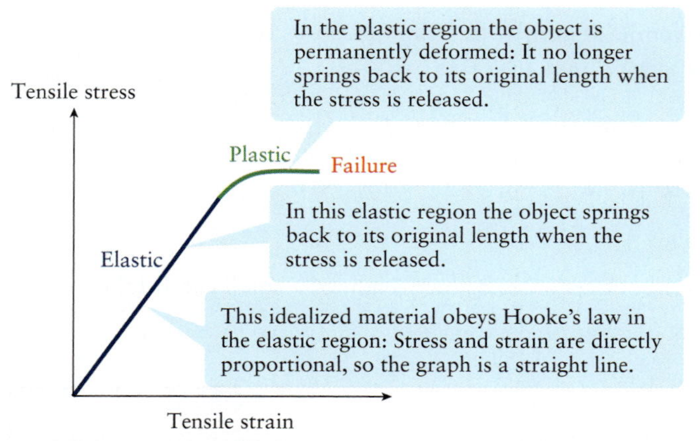

In the plastic region the object is permanently deformed: It no longer springs back to its original length when the stress is released.
Tensile stress
Plastic
Failure
In this elastic region the object springs back to its original length when the stress is released.
Elastic
This idealized material obeys Hooke's law in the elastic region: Stress and strain are directly proportional, so the graph is a straight line.
Tensile strain

Figure 9-16 Elastic behavior, plastic behavior, and failure This graph shows tensile stress versus tensile strain for an idealized material.

the length it had before you increased the stress. By contrast, in the plastic region the tensile strain is *irreversible*. If you pull a little harder on the ends of the object, it stretches a bit more and stays stretched even after you stop pulling.

When the strain on an object is large enough, it undergoes **failure**: The structure of the material starts to lose its integrity, which eventually leads to the object breaking apart. The maximum tensile stress that the material can withstand before failure is called the **ultimate strength**.

Biological Tissue: From Elastic to Failure

BIO-Medical Most biological tissue, including your skin and ligaments—such as the anterior cruciate ligament, or ACL, in your knee (Figure 9-6) and the sternoclavicular ligament in your shoulder (Figure 9-8)—responds to increased tensile stress in a different way than the idealized material graphed in Figure 9-16. This tissue does *not* have a plastic regime: Once a stress is applied that exceeds the ability of biological tissue to spring back to its initial shape, it undergoes partial or complete failure. You can demonstrate this by holding each end of a string bean—a simple biological structure—with your fingertips and trying to pull the two ends apart. If you pull too hard, the string bean will break in two rather than acquire a permanent stretch.

The reason why biological tissue is different from metals has to do with its composition. As we discussed in Section 9-2, stretchable biological materials are *not* uniform but are composed of a combination of tiny fibrils of collagen and elastin (see **Figure 9-17a**). As a result these materials do *not* obey Hooke's law if the tensile stress is small. If the tensile stress is too great, the fibrils do not acquire a permanent stretch but start breaking apart (**Figure 9-17b**).

Figure 9-18 shows an idealized curve of stress versus strain for biological tissue. The curve is not a straight line for small stresses (region I in Figure 9-18), which means that stress and strain are not directly proportional, and Hooke's law is not strictly obeyed. Hooke's law *is* more valid for moderate stresses (region II in Figure 9-18). If the stress is too great and the elastin fibers begin to tear, the tissue fails (region III in Figure 9-18). This may be referred to as *fracture*, *rupture*, or *tearing*. That's what happens when an athlete suffers a knee injury and ends up with a partially or completely torn ACL.

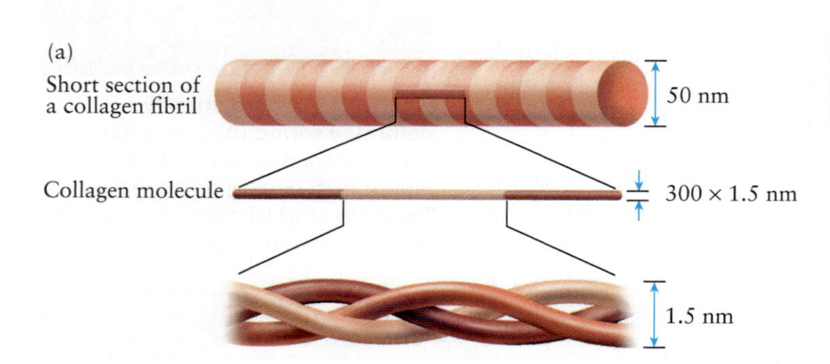

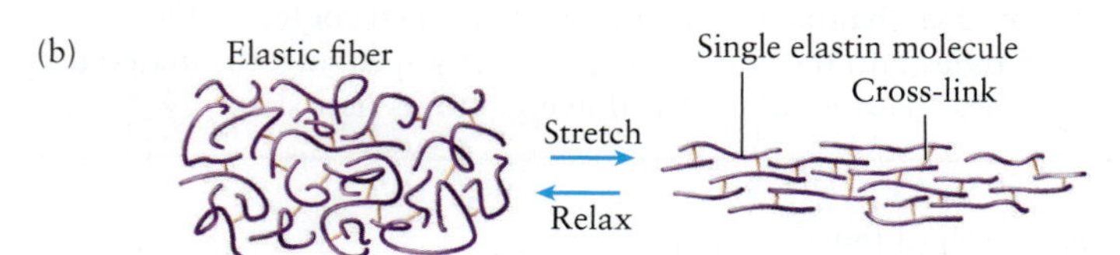

Figure 9-17 Microscopic structure of biological tissue Connective tissues get their rigidity from the protein collagen and their elastic properties from the protein elastin.

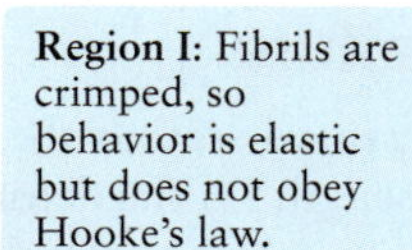

Region I: Fibrils are crimped, so behavior is elastic but does not obey Hooke's law.

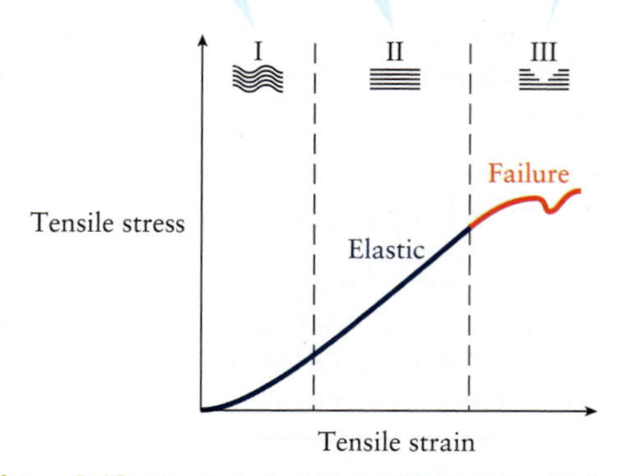

Figure 9-18 Elastic behavior and failure for biological materials This graph shows tensile stress versus tensile strain for typical biological materials. Because such materials are made of elastin and collagen fibrils, this behavior is very different from that of the idealized material shown in Figure 9-16.

WATCH OUT! Materials can be elastic yet not obey Hooke's law.

While biological tissue does not obey Hooke's law in the region of low tensile stress (region I in Figure 9-18), it is nonetheless elastic in this region. That's because it springs back to its original length when the tensile stress is removed. The same is true for the larger stresses within region II in Figure 9-18, in which Hooke's law is more nearly valid. An object that obeys Hooke's law is necessarily elastic, but an elastic object does *not* necessarily obey Hooke's law!

In this section we've considered only the effects of *tensile* stress. Failure can also occur when an object is under compression (think of what would happen if you were to push the two ends of a peeled banana together). It can also happen under shear (which is what happens when you hold a piece of paper in your hands and tear it in two).

GOT THE CONCEPT? 9-8 Hooke's Law and Elastic Behavior

To stretch a certain object by 1.00 cm takes a force of 25.0 N applied at each end. To stretch the same object by 2.00 cm takes a force of 60.0 N applied at each end. In both cases the object returns to its initial length when the force is released. (a) Does the object obey Hooke's law? (b) Does the object display elastic behavior when the 60.0-N forces are applied?

TAKE-HOME MESSAGE FOR Section 9-8

✔ When an object is subjected to a stress that results in a relatively small strain, it deforms but returns to its initial shape after the stress is removed (the elastic regime).

✔ If the stress on an object is increased so that the change in length becomes large relative to the object's initial length, the deformation of the object becomes irreversible (the plastic regime) and represents the onset of failure, which eventually leads to the object breaking apart.

✔ Biological materials do not have a plastic regime: As the stress is increased, they go from the elastic regime directly to failure.

9-9 Solving stress–strain problems: From elastic behavior to failure

In the following three examples we'll examine the stress–strain behavior of a typical biological material, the anterior cruciate ligament (ACL) found in the human knee.

BIO-Medical EXAMPLE 9-6 Human ACL I: Maximum Force

A study of the properties of human ACLs revealed that the ultimate strength for an ACL in a younger person is, on average, 3.8×10^7 Pa. A typical cross-sectional area of the ACL is 4.4×10^{-5} m^2. What is the force exerted on the ACL when the tensile stress is at the maximum value that the ACL can withstand?

Set Up

The ultimate strength of an object is the maximum tensile stress, or maximum force per area, that the object can withstand. Use F_{max} to denote this maximum force, which is what we want to find. We use the definition of tensile stress from Equation 9-3 to relate the ultimate strength to the cross-sectional area A and the force F_{max}.

$$\text{tensile stress} = \frac{F}{A}$$

so

$$\text{ultimate strength} = \text{maximum tensile stress} = \frac{F_{max}}{A}$$

Solve

Rearrange the equation for ultimate strength to solve for F_{max}. Then substitute the known values and use 1 Pa = 1 N/m^2.

$$\begin{aligned} F_{max} &= (\text{ultimate strength}) \times A \\ &= (3.8 \times 10^7 \text{ Pa}) \times (4.4 \times 10^{-5} \text{ m}^2) \\ &= (3.8 \times 10^7 \text{ N/m}^2) \times (4.4 \times 10^{-5} \text{ m}^2) \\ &= 1.7 \times 10^3 \text{ N} \end{aligned}$$

Reflect

Our result shows that a single knee ligament can withstand a sizeable force of 1.7×10^3 N. That's greater than the weight of the heaviest professional players in American football, who have a mass of about 150 kg. In normal use (walking, running, and so on) the forces that act on the ACL are only a fraction of F_{max}, the maximum force that the ACL can withstand.

Offensive tackle in American football:

$$\text{mass } m = 150 \text{ kg}$$
$$\text{weight } mg = (150 \text{ kg})(9.80 \text{ m/s}^2) = 1.5 \times 10^3 \text{ N}$$

BIO-Medical EXAMPLE 9-7 Human ACL II: Breaking Strain

A study of the properties of human anterior cruciate ligaments found that at the point of ACL failure in a younger person, the tensile strain of the ACL is approximately 0.60. The typical length of the ACLs studied was 2.7 cm. How far can the ACL stretch before it breaks?

Set Up

When an object is under tensile stress, its tensile strain equals how far it stretches ΔL divided by its total unstretched length L_0 (Equation 9-3). We know the tensile strain at the point at which the ACL fails (when the ACL is stretched as far as possible) as well as its unstretched length, so we can solve for its maximum stretch ΔL_{max}.

$$\text{tensile strain} = \frac{\Delta L}{L_0}$$

so

tensile strain at failure

$$= \text{maximum tensile strain} = \frac{\Delta L_{max}}{L_0}$$

area $A = 4.4 \times 10^{-5}$ m^2

F_{max} F_{max}

$L_0 = 2.7$ cm ΔL

Solve

Rearrange the equation for the tensile strain at failure and solve for ΔL_{max}.

$$\Delta L_{max} = (\text{tensile strain at failure}) \times L_0 = (0.60)(2.7 \text{ cm}) = 1.6 \text{ cm}$$

Reflect

If 1.6 cm (almost two-thirds of an inch) seems like a large stretch, it is! A maximum tensile strain of 0.60 means that the ACL can stretch by 60% before failing.

BIO-Medical EXAMPLE 9-8 Human ACL III: The Point of No Return

A study of the properties of human anterior cruciate ligaments found that the **yield strength** of the ACL of a younger person is approximately 3.3×10^7 Pa. By what percentage of its initial length can the ACL stretch before it will no longer return to its original length intact? Use $Y_{ACL} = 1.1 \times 10^8$ Pa as the value for Young's modulus for an ACL.

Set Up

The yield strength of an object is the tensile stress beyond which the object is no longer elastic. In Figure 9-17, which shows a representative stress–strain curve for biological tissue, this corresponds to the boundary between region II and region III. While Hooke's law (Equation 9-4) is not strictly valid throughout the elastic regions I and II, we can use it to estimate the tensile strain at the point where the tensile stress equals the yield strength. This and the definition of tensile strain will tell us the percentage increase in the length of the ACL.

Hooke's law for tension expressed in terms of stress and strain:

tensile stress

$$= \text{Young's modulus} \times \text{tensile strain} \quad (9\text{-}4)$$

$$\text{tensile strain} = \frac{\Delta L}{L_0}$$

Solve

We first solve Hooke's law, Equation 9-4, for the value of the tensile strain of the ACL when the tensile stress equals the yield strength.

$$\text{tensile strain} = \frac{\text{tensile stress}}{\text{Young's modulus}}$$

$$\text{tensile stress} = \text{yield strength} = 3.3 \times 10^7 \text{ Pa}$$

$$\text{Young's modulus} = Y_{ACL} = 1.1 \times 10^8 \text{ Pa}$$

Convert the tensile strain to a percentage increase in the length of the ACL.

Hence when the tensile stress equals the yield strength,

$$\text{tensile strain} = \frac{3.3 \times 10^7 \text{ Pa}}{1.1 \times 10^8 \text{ Pa}} = 0.30$$

$$\text{tensile strain} = \frac{\Delta L}{L_0} = \text{fractional change in the length of the ACL}$$

To convert a fraction to a percentage, multiply by 100%:
tensile strain = (0.30)(100%) = 30%

Reflect

We have found that the human ACL is elastic enough to withstand being stretched by a factor of 30% while still being able to return undamaged to its normal length. In Example 9-7 we discovered that a human ACL tears completely when stretched by about 60%. For strains between 30% and 60% some damage—for example, a partial tear—can occur.

TAKE-HOME MESSAGE FOR Section 9-9

✔ The ideas of stress and strain are valid even for materials that do not obey Hooke's law. In this case we can still use Hooke's law to get rough estimates of the strain that corresponds to a given stress.

Key Terms

bulk modulus
compression
compressive strain
compressive stress
elastic
failure
Hooke's law
pascal
plastic
pressure
shear
shear modulus
shear strain
shear stress
spring constant
strain
stress
tensile strain
tensile stress
tension
ultimate strength
volume strain
volume stress
yield strength
Young's modulus

Chapter Summary

Topic	Equation or Figure
Stress and strain: An object is under stress when forces act to deform it (change its size or shape). Stress has units of force per area and is measured in pascals (1 Pa = 1 N/m^2). The amount of deformation that results is the strain, which is a dimensionless quantity.	(Figure 9-1)

This Achilles tendon is relaxed.

This Achilles tendon is under **stress**: The muscles at its upper end exert an upward force *F*, and the heel bone at its lower end exerts an equally strong downward force. The tendon responds by undergoing **strain** (it stretches).

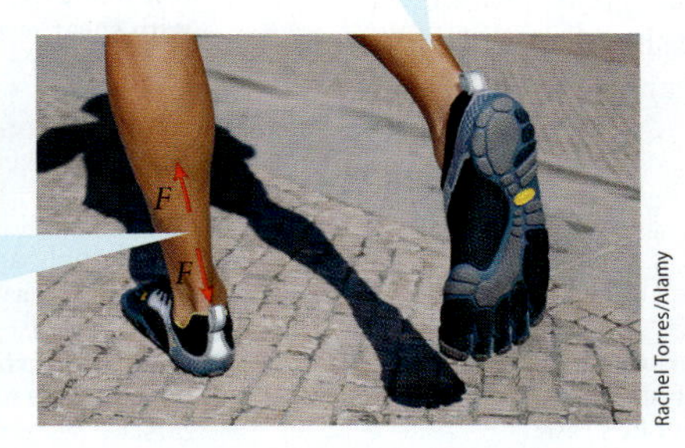

Rachel Torres/Alamy

Stress refers to forces that act to deform an object (change its size or shape). **Strain** refers to the deformation caused by these forces.

Tension, compression, and Hooke's law: An object is under tension if forces act to lengthen it along one dimension; it is under compression if forces act to shorten it along one dimension. If the object obeys Hooke's law, the strain is directly proportional to the stress; the constant of proportionality, called Young's modulus, depends on what the object is made of but not on its dimensions. The spring constant k for an object like that shown here equals YA/L_0.

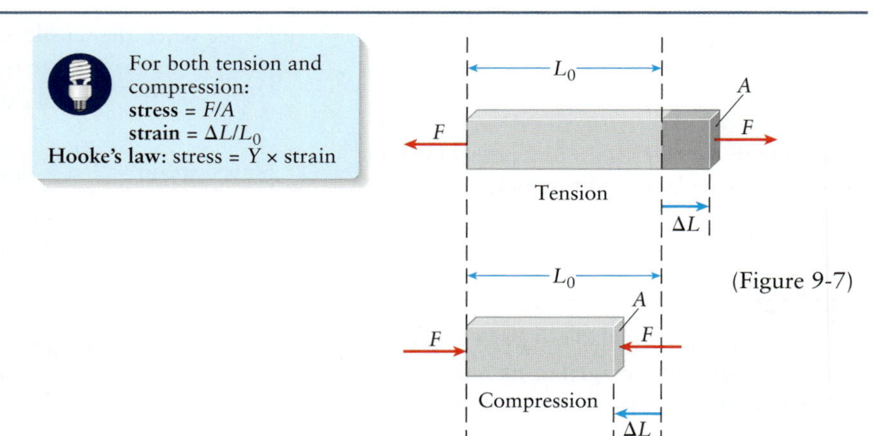

(Figure 9-7)

Volume stress and strain: An object is under volume stress if there is a *change* in the pressure p on the object. The stress is equal to the pressure change Δp. If Δp is small, Hooke's law applies and the strain (the fractional change in volume, $\Delta V/V_0$) is proportional to the stress. The proportionality constant is the bulk modulus of the material. The signs of stress and strain are opposite: Positive volume stress ($\Delta p > 0$) causes a negative volume strain ($\Delta V/V_0 < 0$) and vice versa.

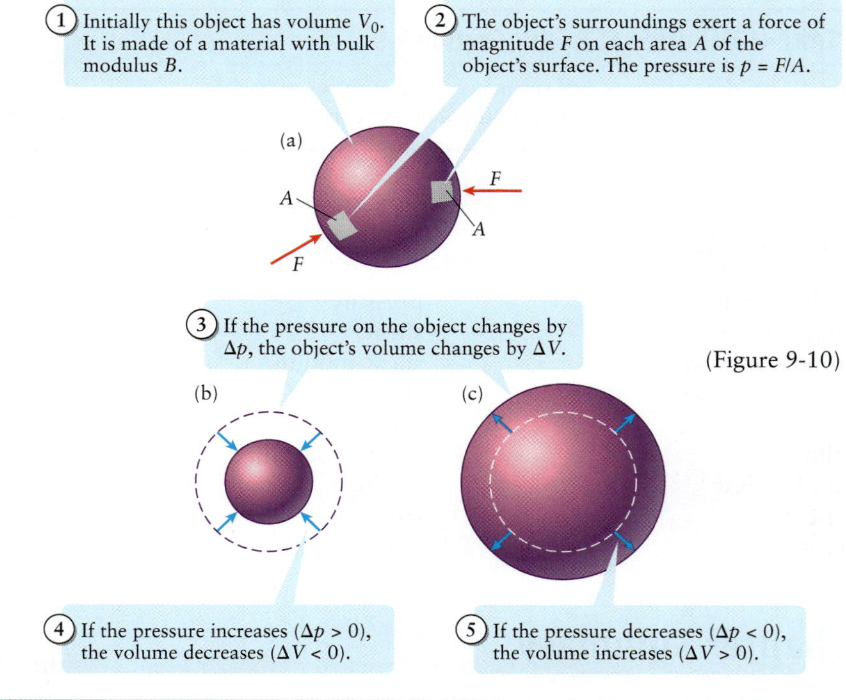

1 Initially this object has volume V_0. It is made of a material with bulk modulus B.
2 The object's surroundings exert a force of magnitude F on each area A of the object's surface. The pressure is $p = F/A$.
(a)
A
F
F
A
3 If the pressure on the object changes by Δp, the object's volume changes by ΔV.
(b)
(c)
4 If the pressure increases ($\Delta p > 0$), the volume decreases ($\Delta V < 0$).
5 If the pressure decreases ($\Delta p < 0$), the volume increases ($\Delta V > 0$).

(Figure 9-10)

volume stress = Δp
volume strain = $\Delta V/V_0$
Hooke's law:
volume stress = $-B$ × volume strain

Shear stress and strain: An object is under shear stress if offset forces $F_{\parallel}$ act to change the object's shape. If the shear stress is not too great, Hooke's law applies and the shear strain is proportional to the shear stress.

1 This object has height h. Its upper and lower surfaces each have area A. It is made of a material with shear modulus S.

2 Apply a force of magnitude $F_{\parallel}$ parallel to the upper surface...

3 ...and apply an opposite force of the same magnitude $F_{\parallel}$ parallel to the lower surface. Note that the two forces are offset (they do not act along the same line).

4 As a result of these forces, the upper and lower surfaces move a distance x relative to each other.

(a) A h (b) A $F_{\parallel}$ h $F_{\parallel}$ x

(Figure 9-12)

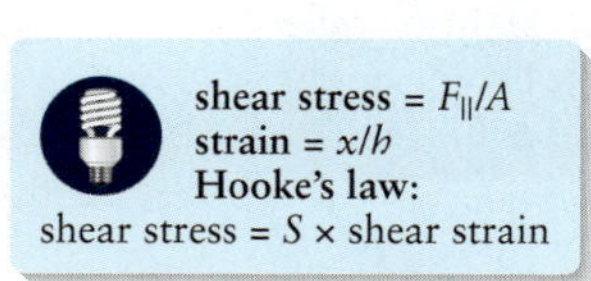

Elasticity, plasticity, and failure: Most materials return to their original size and shape when the stress is removed, provided the stress is not too great. This is called elastic behavior. If the stress is greater than a certain value, the material becomes plastic: It retains the deformation even after the stress is removed. Beyond a certain even greater value, the material undergoes failure (it ruptures or fractures). Biological materials do not become plastic, but go from elastic behavior to failure if the stress is made too great.

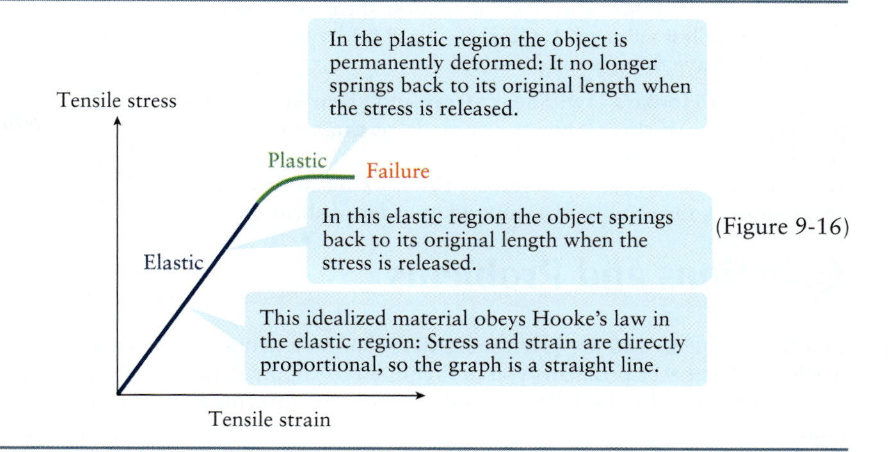

Tensile stress
In the plastic region the object is permanently deformed: It no longer springs back to its original length when the stress is released.
Plastic
Failure
In this elastic region the object springs back to its original length when the stress is released.
Elastic
This idealized material obeys Hooke's law in the elastic region: Stress and strain are directly proportional, so the graph is a straight line.
Tensile strain

(Figure 9-16)

Answer to What do you think? Question

(b) In Section 9-2 we learn that the amount that an object's length decreases due to compression is proportional to the compressive stress F/A, which is the force F applied (in this case to the athlete's arms) divided by the object's cross-sectional area A. If the arms had one-half the diameter, they would have one-quarter the area A and the compressive stress F/A would be four times as much (assuming that F is unchanged). Hence her arms would shorten by four times the distance.

Answers to Got the Concept? Questions

9-1 (c), (a) and (d) [tie], (b) The easier a rod is to stretch, the smaller its spring constant k. Equation 9-2 tells us that $k = YA/L_0$. All four rods have the same Young's modulus Y (since they are all made of the same material). So the rod with the smallest value of k is the one with the smallest ratio A/L_0—that is, the one with the smallest cross-sectional area and the longest length. The values of this ratio for each of the four rods are (a) $(2.00\text{ mm}^2)/(5.00\text{ cm}) = 0.400\text{ mm}^2/\text{cm}$, (b) $(2.00\text{ mm}^2)/(2.50\text{ cm}) = 0.800\text{ mm}^2/\text{cm}$, (c) $(1.00\text{ mm}^2)/(5.00\text{ cm}) = 0.200\text{ mm}^2/\text{cm}$, (d) $(1.00\text{ mm}^2)/(2.50\text{ cm}) = 0.400\text{ mm}^2/\text{cm}$.

9-2 (d) Not enough information is given to decide. To see why not, rewrite Equation 9-3 as

$$Y = \frac{F/A}{\Delta L/L_0} = \frac{\text{tensile stress}}{\text{tensile strain}}$$

To compare the value of Y for the two materials, we need to compare the ratio of tensile stress to tensile strain for the two rods. The force F and cross-sectional area A are the same for both rods, so both are under the same tensile stress. We also know that ΔL is twice as great for rod 2 as for rod 1. But we don't know how the tensile *strain* compares for the two rods because we aren't told anything about the relaxed lengths L_0 of the two rods. (That was the detail we asked you to look for.) So we don't have enough information to decide which material has the greater value of Y.

9-3 (a) Same for both, (b) brass rod, (c) brass rod. Each rod has the same cross-sectional area A and is subjected to forces of the same magnitude F, so the compressive *stress* F/A is the same for both. From Hooke's law the compressive strain equals the compressive stress divided by Young's modulus Y: $\Delta L/L = (F/A)(1/Y)$. The brass rod has a smaller value of Y, so it will experience the greater compressive strain. The distance ΔL that each rod contracts equals the original length (which is the same for both rods) multiplied by the compressive strain, so the brass rod (which experiences greater strain) contracts more than the steel rod.

9-4 (c), then (d), then (a) and (b) (tie) For the first three cases use the definition of tensile stress as force divided by cross-sectional area:

$$\text{(a) tensile stress} = \frac{F}{A} = \frac{200\text{ N}}{(1.00\text{ mm}^2)}\left(\frac{100\text{ mm}}{1\text{ m}}\right)^2$$
$$= 2.0 \times 10^6\text{ N/m}^2$$

(b) Same as (a); the length is different, but this has no effect on calculating the tensile stress

$$\text{(c) tensile stress} = \frac{F}{A} = \frac{100\text{ N}}{(2.00\text{ mm}^2)}\left(\frac{100\text{ mm}}{1\text{ m}}\right)^2$$
$$= 5.0 \times 10^5\text{ N/m}^2$$

(d) For the fourth case we use Hooke's law, which states that

$$\text{tensile stress} = (\text{Young's modulus}) \times (\text{tensile strain}) = Y\frac{\Delta L}{L_0} =$$
$$(2.00 \times 10^{10}\text{ Pa})\left(\frac{0.025\text{ mm}}{50.0\text{ cm}}\right)\left(\frac{1\text{ cm}}{10\text{ mm}}\right) = 1.0 \times 10^6\text{ N/m}^2$$

9-5 (b) The volume of the balloon filled with air will change the most. From Table 9-1 you can see that the bulk modulus of water is about 2×10^4 times larger than that of air. Water is therefore about 2×10^4 times *less* compressible than air.

9-6 (d) Because the pressure is the same on all sides, the forces pushing on opposite sides of the sphere are of equal magnitude but opposite direction. Hence the *net* force is zero!

9-7 (a) (iii), (i) and (ii) (tie); (b) (iii), (i) and (ii) (tie). In part (a) the shear stress equals $F_{\parallel}/A$. The force $F_{\parallel}$ is the same in each case. Hence object (iii) has the greatest shear stress since it has the smallest value of A. Objects (i) and (ii) have the same area and so have the same shear stress.

Part (b) uses Hooke's law: Shear stress and shear strain are proportional. All three objects are made of the same material with shear modulus S, so the constant or proportionality is the same in each case. Therefore, the ranking in order of shear strain is the same as the ranking in order of shear stress.

9-8 (a) No; (b) yes. If an object obeys Hooke's law, doubling the stretch (and hence doubling the tensile strain) would require double the force (and hence double the tensile stress). This object does *not* obey Hooke's law since doubling the stretch from 1.00 to 2.00 cm requires more than double the force (from 25.0 N to 60.0 N). The object *is* nonetheless elastic because it returns to its unstretched length when the tensile forces go away.

Questions and Problems

In a few problems you are given more data than you actually need; in a few other problems you are required to supply data from your general knowledge, outside sources, or informed estimate.

Interpret as significant all digits in numerical values that have trailing zeros and no decimal points. For all problems use $g = 9.80 \text{ m/s}^2$ for the free-fall acceleration due to gravity. Neglect friction and air resistance unless instructed to do otherwise.

•	Basic, single-concept problem
••	Intermediate-level problem; may require synthesis of concepts and multiple steps
•••	Challenging problem
SSM	*Solution is in Student Solutions Manual*
Example	*See worked example for a similar problem*

Conceptual Questions

1. • Describe the small-stretch limit of Hooke's law for a spring.

2. • Give a few reasons why Hooke's law is intuitively obvious and a few reasons why it is counterintuitive.

3. • Is it possible for a long cable hung vertically to break under its own weight? Explain your answer. SSM

4. • A 2" × 4" pine stud oriented horizontally is securely clamped at one end to an immovable object. A heavy weight hangs from the free end of the wood, causing it to bend. (a) Which part of the plank is under compression? (b) Which part of the plank is under tension? (c) Is there any part that is neither stretched nor compressed?

5. • A steel wire and a brass wire, each of length L and diameter D, are joined together to form a wire of length $2L$. If this wire is then used to hang an object of mass m, describe the amount of stretch in the two segments of wire.

6. • What can cause nylon tennis racket strings to break when they are hit by the ball?

7. • Human skin is under tension like a rubber glove that has had air blown into it. Why does skin acquire wrinkles as people get older?

8. • Balloons expand as they rise through the atmosphere. Why does this happen?

9. • One way of determining the volume of an irregular solid is to submerge it in a liquid to measure the volume change of the liquid in its container. What assumptions are being made about the object and the object's bulk modulus when performing such experiments?

10. • Marshmallows and iron have vastly different bulk moduli. One has a much larger value than the other. Suppose a marshmallow cube and an iron cube of approximately the same size were placed in a vacuum chamber. Explain what would happen to each of the cubes as the air was slowly removed from the chamber.

11. • (a) What is the difference between Young's modulus and bulk modulus? (b) What are the units of these two physical quantities?

12. • Is it possible, when tightening the lug nuts on the wheel of your car, to use too much torque and break off one of the bolts? Explain your answer.

13. • Shear modulus (S) is sometimes known as *rigidity*. Can you explain why rigidity is an appropriate synonym? SSM

14. • Devise a simple way of determining which modulus (Young's, bulk, or shear) is appropriate for any given stress–strain problem.

15. • Why are tall mountains typically shaped like cones rather than a straight vertical columnlike structure?

16. • (a) Describe some common features of strain that were defined in this chapter. (b) We encountered three types of strain (tensile, volume, and shear). What are some distinguishing features of these quantities?

17. • Define the terms yield strength and ultimate strength.

18. • **Biology** In some recent studies it has been shown that women are more susceptible to torn ACLs than men when competing in similar sports (most notably in soccer and basketball). What are some reasons why this disparity might exist?

19. • **Biology** The leg bone of a cow has an ultimate strength of about $150 \times 10^6 \text{ N/m}^2$ and a maximum strain of about 1.5%. The antler of a deer has an ultimate strength of about $160 \times 10^6 \text{ N/m}^2$ and a maximum strain of about 12%. Explain the relationship between structure and function in these data.

Multiple-Choice Questions

20. • The units for strain are
A. N/m.
B. N/m^2.
C. N.
D. $\text{N} \cdot \text{m}^2$.
E. none of the above.

21. • The units for stress are
A. N/m.
B. N/m^2.
C. N.
D. $\text{N} \cdot \text{m}$.
E. $\text{N} \cdot \text{m}^2$.

22. • When tension is applied to a metal wire of length L, it stretches by ΔL. If the same tension is applied to a wire of the

same material with the same cross-sectional area, but of length $2L$, by how much will it stretch?

A. ΔL
B. $2\ \Delta L$
C. $0.5\ \Delta L$
D. $3\ \Delta L$
E. $4\ \Delta L$

23. • A steel cable lifting a heavy box stretches by ΔL. If you want the cable to stretch by only half of ΔL, by what factor must you increase its diameter?

A. 2
B. 4
C. $\sqrt{2}$
D. 1/2
E. 1/4 SSM

24. • A wire is stretched just to its breaking point by a force F. A longer wire made of the same material has the same diameter. The force that will stretch it to its breaking point is

A. larger than F.
B. smaller than F.
C. equal to F.
D. much smaller than F.
E. much larger than F.

25. • Two solid rods have the same length and are made of the same material with circular cross sections. Rod 1 has a radius r, and rod 2 has a radius $r/2$. If a compressive force F is applied to both rods, their lengths are reduced by ΔL_1 and ΔL_2, respectively. The ratio $\Delta L_1/\Delta L_2$ is

A. 1/4.
B. 1/2.
C. 1.
D. 2.
E. 4.

26. • A wall mount for a television consists in part of a mounting plate screwed or bolted flush to the wall. Which kinds of stress play a role in keeping the mount securely attached to the wall?

A. compression stress
B. tension stress
C. shear stress
D. bulk stress
E. A, B, and C

27. • When choosing building construction materials, what kinds of materials would you choose, all other things being equal?

A. materials with a relatively large bulk modulus
B. materials with a relatively small bulk modulus
C. either materials with a large or a small bulk modulus
D. it doesn't matter as long as the building is not too tall
E. materials with a relatively small shear modulus SSM

28. • A book is pushed sideways, deforming it as shown in Figure 9-19. To describe the relationship between stress and strain for the book in this situation, you would use

A. Young's modulus.
B. bulk modulus.
C. shear modulus.
D. both Young's modulus and bulk modulus.
E. both shear modulus and bulk modulus.

$\vec{F}$
$\vec{f}_s$

Figure 9-19 Problem 28

29. • A steel cable supports an actor as he swings onto the stage. The weight of the actor stretches the steel cable. To describe the relationship between stress and strain for the steel cable, you would use

A. Young's modulus.
B. bulk modulus.
C. shear modulus.
D. both Young's modulus and bulk modulus.
E. both shear modulus and bulk modulus.

Estimation/Numerical Analysis

30. • Estimate Young's modulus for (a) an elastic bungee cord and (b) a wooden pencil.

31. • Estimate Young's modulus for a strip of paper. SSM

32. • Estimate the shear modulus for a chilled stick of butter taken from a refrigerator. Describe how this value would change as the butter warms to room temperature.

33. • Estimate the force needed to break a bone in your arm.

34. • Estimate the force needed to bend a bar of 0.5-in. rebar. (Rebar is made from iron and is used to reinforce concrete.)

35. • Estimate the force needed to puncture a 0.5-cm-thick sheet of aluminum with a 1-cm diameter rivet.

36. • Estimate the shear strain experienced by a typical athletic shoe in a basketball game.

37. • The following data are associated with an alloy of steel. Plot a graph of stress versus strain for the alloy. What are (a) the yield strength, (b) the ultimate strength, and (c) Young's modulus, for the material? SSM

Strain (%)	Stress (10^9 N/m^2)	Strain (%)	Stress (10^9 N/m^2)
0	0	1.0	300
0.1	125	1.5	325
0.2	250	2.0	350
0.3	230	2.5	375
0.4	230	3.0	400
0.5	235	3.5	375
0.6	240	4.0	350
0.7	250	4.5	325
0.8	260	5.0	300
0.9	270		

38. • **Biology** A galloping horse experiences the following stresses and corresponding strains on its front leg bone. Plot a graph of the stress versus strain for the bone and identify (a) the elastic region, (b) the yield strength, and (c) Young's modulus for this type of bone.

Stress (10^6 N/m^2)	Strain (%)
35	0.2
70	0.4
105	0.6
140	1.0
175	1.5

Problems

9-1 When an object is under stress, it deforms
9-2 An object changes length when under tensile or compressive stress
9-3 Solving stress–strain problems: Tension and compression

39. • A cylindrical steel rod is originally 250 cm long and has a diameter of 0.254 cm. A force is applied longitudinally and the rod stretches 0.85 cm. What is the magnitude of the force? Example 9-1

40. • A solid band of rubber, which has a circular cross section of radius 0.25 cm, is stretched a distance of 3.0 cm by a force of 87 N. Calculate the original length of the cylinder of rubber. Example 9-1

41. • A bar of aluminum has a cross section of 1.0 cm × 1 .0 cm and a length of 88 cm. (a) What force would be needed to stretch the bar to 1.00 m? (b) What is the tensile strain of the bar of aluminum at that point? SSM Example 9-1

42. • The tensile stress on a concrete block is 0.52×10^9 N/m^2. What is its tensile strain under this force? Example 9-1

43. • A physicist examines a metal sample and measures the ratio of the tensile stress to the tensile strain to be 95×10^9 N/m^2. From what material might the sample be made? Example 9-1

44. • A 10.0-m-long copper wire is pulled with a force of 1200 N and it stretches 10.0 cm. Calculate the radius of the wire if the value of Young's modulus is 110×10^9 N/m^2. Example 9-1

45. • (a) Calculate the ratio of the tensile strain on an aluminum bar to that on a steel bar if both bars have the same cross-sectional area and the same force is applied to each bar. (b) Does the original length of each bar affect your answer to part (a)? SSM Example 9-1

46. • **Biology** What is the compressive stress on your feet if your weight is spread out evenly over both soles of your shoes? Assume your mass is 55.0 kg and each shoe has an area of 200.0 cm^2. Example 9-1

47. • **Biology** Compare the answer to Problem 9-46 to the case where each of your shoes only has an area of 10.0 cm^2 in contact with the floor (as might be the case for high-heeled shoes). Example 9-1

48. • **Medical** The anterior cruciate ligament in a woman's knee is 2.5 cm long and has a cross-sectional area of 0.54 cm^2. If a force of 3.0×10^3 N is applied longitudinally, how much will the ligament stretch? Example 9-1

49. •• **Medical** One model for the length of a person's ACL (L_{ACL}, in millimeters) relates it to the person's height (h, in centimeters) with the linear function $L_{ACL} = 0.4606h - 41.29$. Age, gender, and weight do not significantly influence the relationship. If a basketball player has a height of 2.29 m, (a) approximately how long is his ACL? (b) If a pressure of 10.0×10^6 N/m^2 is applied longitudinally to his ligament, how much will it stretch? Example 9-1

9-4 An object expands or shrinks when under volume stress

9-5 Solving stress–strain problems: Volume stress

50. • A cube of lead (each side is 5.0 cm) is pressed equally on all six sides with forces of 1.0×10^5 N. What will the new dimensions of the cube be after the forces are applied ($B = 46 \times 10^9$ N/m^2 for lead)? Example 9-2

51. • A rigid cube (each side is 0.10 m) is filled with water and frozen solid. When water freezes its volume expands about 9%. How much pressure is exerted on the sides of the cube? *Hint:* Imagine trying to squeeze the block of ice back into the original cube. SSM Example 9-2

52. • A sphere of copper that has a radius of 5.00 cm is compressed uniformly by a force of 2.00×10^8 N. Calculate (a) the change in volume of the sphere and (b) the sphere's final radius. The bulk modulus for copper is 140×10^9 N/m^2. Example 9-2

9-6 A solid object changes shape when under shear stress

9-7 Solving stress–strain problems: Shear stress

53. • Shear forces act on a steel door during an earthquake (**Figure 9-20**). If the shear strain is 0.005, calculate the force acting on the door with dimensions 0.044 m × 0.81 m × 2.03 m. Example 9-5

Figure 9-20 Problem 53

54. • A brass nameplate is 2.00 cm × 10.0 cm × 20.0 cm in size. If a force of 2.00×10^5 N acts on the upper left side and the bottom right side (**Figure 9-21**), find the shear strain (x/h) and the angle ϕ. SSM Example 9-4

Figure 9-21 Problem 54

55. • **Medical** In patients with asthma, an increased thickness of the airways causes a local reduction in stress through the airway walls. The effect can be as much as a 50% reduction in the local shear modulus of the airways of an asthmatic patient compared to those of a healthy person. Calculate the ratio of the shear strain in an asthmatic airway to that of a healthy airway. Example 9-4

56. •• A force of 5.0×10^6 N is applied tangentially at the center of one side of a brass cube. The angle of shear ϕ is measured to be 0.65°. Calculate the volume of the original cube. Example 9-4

57. • An enormous piece of granite 2.00×10^2 m thick with a shear modulus of 5.00×10^{10} N/m^2 is sheared from its geologic formation by an earthquake force of 275×10^9 N. The area on which the force acts is a square of side x as shown in **Figure 9-22**. Find the value of x if the shear force produces a shear strain of 0.125. Example 9-4

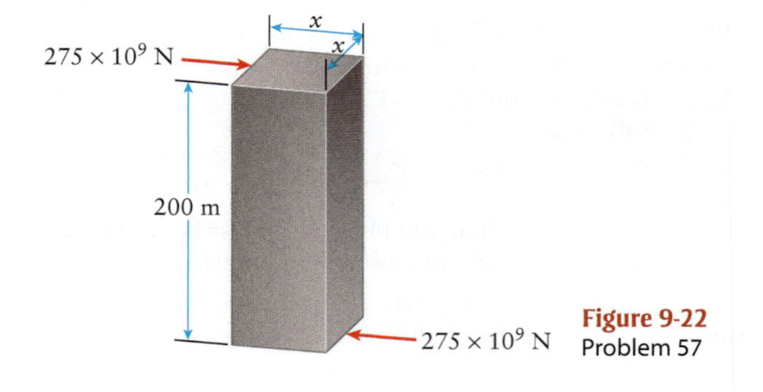

Figure 9-22 Problem 57

58. • **Medical** In regions of the cardiovascular system where there is steady laminar blood flow, the shear stress on cells lining the walls of the blood vessels is about 20 dyne/cm^2. If the shear strain is about 0.008, estimate the shear modulus for the affected cells. Note 1 dyne = 1 g·cm/s^2 and 1 N = 10^5 dyne. SSM Example 9-5

9-8 Objects deform permanently or fail when placed under too much stress

9-9 Solving stress–strain problems: From elastic behavior to failure

59. • A piece of steel piano wire is 1.60 m long and has a diameter of 0.20 cm. What is the magnitude of the tension required to break it? The ultimate strength of steel is 5.0×10^8 N/m^2. SSM Example 9-6

60. • Steel will ultimately fail if the shear stress exceeds 4.00×10^8 N/m^2. Determine the force required to shear a steel bolt that is 0.50 cm in diameter. Example 9-6

61. • A theater rigging company uses a safety factor of 10 for all its ropes, which means that all ultimate breaking strengths will be underestimated by a factor of 10 just to be safe. Suppose a rope with an ultimate breaking strength of 1.0×10^4 N is tied with a knot that decreases rope strength by 50%. (a) If the rope is used to support a load of 1.0×10^3 N, what is the safety factor? (b) Will the rigging company be able to use the rope with the knot? SSM Example 9-6

62. • Standard 12-gauge copper wiring (commonly used in home electrical wiring) is 2.053 mm in diameter. Copper has a yield strength of 70 MPa (1 MPa = 10^6 Pa). (a) What is the maximum amount of weight that a 12-gauge copper wire can hold before it can no longer return to its original length intact? (b) The yield strength of silk produced by a silkworm is 500 MPa. What minimum diameter of thread of a silkworm's silk would be needed to hold the same weight as the 12-gauge copper wire, provided that the elasticity of the thread is preserved? Example 9-8

General Problems

63. •• The elastic limit of an alloy is 0.6×10^9 N/m^2. What is the minimum radius of a 4-m-long wire made from the alloy if a single strand is designed to support a commercial sign that has a weight of 8000 N and hangs from a fixed point? To stay within safety codes, the wire cannot stretch more than 5 cm. Example 9-8

64. • A 50.0-kg air-conditioning unit slips from its window mount, but the end of the electrical cord gets caught in the mounting bracket. In the process the cord (which is 0.50 cm in diameter) stretches from 3.0 to 4.5 m. What is Young's modulus for the cord? Example 9-1

65. • **Medical** The largest tendon in the body, the Achilles tendon connects the calf muscle to the heel bone of the foot. This tendon is typically 25.0 cm long, is 5.0 mm in diameter, and has a Young's modulus of 1.47×10^9 N/m^2. If an athlete has stretched the tendon to a length of 26.1 cm, what is the tension (in newtons and pounds) in the tendon? Example 9-1

66. • The strongest man-made fiber is a type of carbon fiber that has an ultimate strength of 6.37 GPa (1 GPa = 10^9 Pa). Those fibers, however, have an approximate diameter of only 5.0×10^{-6} m. If the fibers could be made thicker while maintaining the same strength, what minimum diameter of fiber would be needed to lift an adult human (m = 75 kg)? Example 9-6

67. •• **Biology** Many caterpillars construct cocoons from silk, one of the strongest naturally occurring materials known. Each thread is typically 2.0 μm in diameter and the silk has a Young's modulus of 4.0×10^9 N/m^2. (a) How many strands would be needed to make a rope 9.0 m long that would stretch only 1.00 cm when supporting a pair of 85-kg mountain climbers? (b) Assuming that there is no appreciable space between the parallel strands, what would be the diameter of the rope? Does the diameter seem reasonable for a rope that mountain climbers might carry? SSM Example 9-1

68. •• A 2.8-carat diamond is grown under a high pressure of 58×10^9 N/m^2. (a) By how much does the volume of a spherical 2.8-carat diamond expand once it is removed from the chamber and exposed to atmospheric pressure? (b) What is the increase in the diamond's radius? One carat is 0.200 g, and you can use 3.52 g/cm^3 for the density of diamond, and 4.43×10^{11} N/m^2 for the bulk modulus of diamond. Example 9-3

69. • A glass marble that has a diameter of 1.00 cm is dropped into a graduated cylinder that contains 20.0 cm of mercury. (a) By how much does the volume of the marble shrink while at the bottom of the mercury, where the pressure is 2.7×10^4 N/m^2 greater than at the surface? (b) What is the corresponding change in radius associated with the compression? The bulk modulus of glass is 50×10^6 N/m^2. Example 9-2

70. • **Sports** During the 2004 Olympic clean-and-jerk weight-lifting competition, Hossein Rezazadeh lifted 263.5 kg. Mr. Rezazadeh himself had a mass of 163 kg. Ultimately the weight is all supported by the tibia (shin bone) of the lifter's legs. The average length of a tibia is 385 mm, and its diameter (modeling it as having a round cross section) is about 3.0 cm. Young's modulus for bone is typically about 2.0×10^{10} N/m^2. (a) By how much did the lift compress the athlete's tibia, assuming that the bone is solid? (b) Does this seem to be a significant compression? (c) Is it necessary to include the lifter's weight in your calculations? Why or why not? SSM Example 9-1

71. • When a house is moved it is gradually raised and supported on wooden blocks. A typical house averages about 54,000 kg. The house is supported uniformly on six stacks of blocks of Douglas fir wood (which has a Young's modulus of 13×10^9 N/m^2). Each block is 25 by 75 cm. If the wood is stacked 1.5 m high, by how much will the house compress the supporting stack of blocks? Example 9-1

72. •• (a) What diameter steel cable is needed to support a large diesel engine with a mass of 4.0×10^3 kg? (b) By how much will the 10.0-m-long cable stretch once the engine is raised up off the ground? Assume the ultimate breaking strength of steel is 4.00×10^8 N/m^2. Example 9-6

73. • **Sports** A runner's foot pushes backward on the ground as shown in **Figure 9-23**. This results in a 25-N shearing force exerted in the forward direction by the ground, and in the backward direction by the foot, distributed over an area of 15 cm^2 and a 1.0-cm-thick sole. If the shear angle θ is 5.0°, what is the shear modulus of the sole? Example 9-4

Figure 9-23 Problem 73

74. • The spherical bubbles near the surface of a glass of water are 2.5 mm in diameter at sea level where the atmosphere exerts a pressure of 1.01×10^5 N/m^2 over the surface of each bubble. If the glass of water is taken to the elevation of Golden, CO, where the atmosphere exerts a pressure of 8.22×10^4 N/m^2 over the bubble surface, what will be the diameter of the bubbles? SSM Example 9-3

75. •• **Biology** A particular human hair has a Young's modulus of 4.0×10^9 N/m^2 and a diameter of 150 μm. (a) If a 250-g object is suspended by the single strand of hair that is originally 20.0 cm long, by how much will the hair stretch? (b) If the same object were hung by an aluminum wire of the same dimensions as the hair, by how much would the aluminum stretch? (Try to do this part without repeating the previous calculation but use proportional reasoning instead.) (c) If we think of the strand of hair as a spring, what is its spring constant? (d) How does the hair's spring constant compare with that of ordinary springs in your physics laboratory? Example 9-1

76. • **Medical** The femur bone in the human leg has a minimum effective cross section of 3.0 cm^2. How much compressive force can it withstand before breaking? Assume the ultimate strength of the bone to be 1.7×10^8 N/m^2. Example 9-6

77. ••• A beam is attached to a vertical wall with a hinge. The mass of the beam is 1000 kg, and it is 4 m long. A steel support wire is tied from the end of the beam to the wall, making an angle of 30° with the beam (**Figure 9-24**). (a) By summing the torque about the axis passing through the hinge, calculate the tension in the support wire. Assume the beam is uniform so that the weight acts at its exact center. (b) What is the minimum cross-sectional area of the steel wire so that it is not permanently stretched? The yield strength (elastic limit) for steel is 290×10^6 N/m^2, and the ultimate breaking strength is 400×10^6 N/m^2. Example 9-8

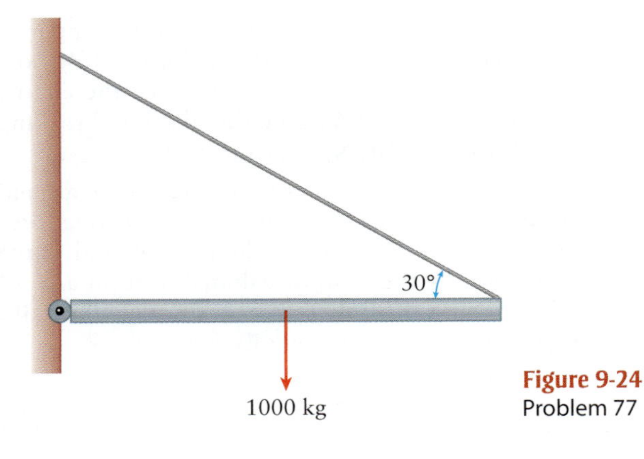

Figure 9-24 Problem 77

78. •• **Biology** The typical compressive ultimate stress in the transverse direction is 133×10^6 N/m^2 for human bones and 178×10^6 N/m^2 for cow bones. (a) Is a human or a cow more likely to suffer a transverse break in a bone? Why? (b) The compressive longitudinal yield stress of human bones is approximately 182×10^6 N/m^2, and the compressive longitudinal ultimate stress is about 195×10^6 N/m^2. For cows, the compressive longitudinal yield stress of bone is about 196×10^6 N/m^2, and the compressive longitudinal ultimate stress is 237×10^6 N/m^2. Explain how a cow's bones are much more capable of supporting their extreme weight in comparison to a human's bones. (c) A bone in a woman's leg has an effective cross-sectional area of 3.00 cm^2. If the bone is 35 cm long, how much compressive force can it withstand before breaking? How much will her bone compress if it is subjected to a force one-tenth the magnitude of the force that breaks it? The longitudinal elastic modulus of human bone is about 9.6×10^9 N/m^2. Example 9-6

79. •• An 85-kg mass falls 15 m from rest before it is jerked to a stop by a safety cable in 0.07 s. The cable is attached to a 9-mm diameter eyebolt sunk into a rigid wooden beam (**Figure 9-25**). Assuming the beam does not deform at all, what is the average shear stress on the bolt? Example 9-5

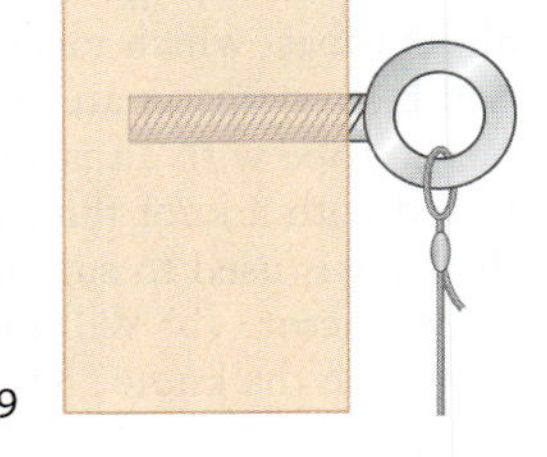

Figure 9-25 Problem 79

80. •• **Sports** Bicycle caliper brakes consist of rubber pads mounted in rigid metal "shoes." The normal force applied by a bicycle caliper brake pad to the wheel rim is about 4.00×10^2 N. The contact area of the brake pad is approximately 4.0 cm × 0.70 cm, with the long dimension parallel to the rim's rotation. The pad is 1.0 cm thick. The coefficient of kinetic friction between the rubber pad and the aluminum rim is 0.50. Given that the shear modulus of the rubber is 0.30 MPa, calculate the deformation angle φ of the brake pad. Example 9-4

81. • A warehouse worker accidentally slides into a shelving unit, stopping himself with his hands by exerting a force of 3700 N on the top shelf. The shelving unit is 1.5 m tall, 2.5 m long, and 0.67 m deep. During the collision the shelving unit is knocked into a parallelogram shape with the top overhanging the bottom by 13 cm along the long axis before it springs back into shape. Assuming all the force of the collision was translated into shear forces acting parallel to the top surface, calculate the shear modulus of the shelving unit. Example 9-5

NASA

Gravitation

> **What do you think?**
>
> The International Space Station (ISS) orbits Earth about 350 km (about 220 mi) above the surface of our planet, which has a radius of about 6370 km (3960 mi). Astronauts aboard the ISS feel as though they are weightless. Compared to the gravitational force that Earth exerts on an astronaut standing on our planet's surface, how great is the gravitational force that Earth exerts on the same astronaut when she is aboard the ISS?
> (a) The same; (b) about 10% less; (c) about 50% less; (d) about 75% less; (e) essentially zero.

In this chapter, your goals are to:

- (10-1) Identify what it means to say that gravitation is universal.
- (10-2) Explain how Newton's law of universal gravitation describes the attractive gravitational force between any two objects.
- (10-3) Describe the general expression for gravitational potential energy and how to relate it to the expression used near Earth's surface.
- (10-4) Apply the law of universal gravitation and the expression for gravitational potential energy to analyze the orbits of satellites and planets.
- (10-5) Explain the origin of apparent weightlessness.

To master this chapter, you should review:

- (3-2, 3-3) How to add vectors using components.
- (3-7) The properties of uniform circular motion.
- (3-8) How the inner ear senses acceleration.
- (4-3) The nature of the gravitational force near Earth's surface.
- (6-3, 6-6, and 6-7) The ideas of kinetic energy, potential energy, and the conservation of total mechanical energy.
- (8-2, 8-6, and 8-8) The concepts of angular speed, torque, and angular momentum.

10-1 Gravitation is a force of universal importance

Eighty years ago the idea of humans orbiting Earth or sending spacecraft to other worlds was regarded as science fiction. Today science fiction has become commonplace reality. Humans live and work in Earth orbit aboard the International Space Station (see photograph above), have ventured as far as the Moon, and have sent robotic spacecraft to explore all the planets of the solar system.

While we think of spaceflight as an innovation of the twentieth century, we can trace its origins to Isaac Newton's revolutionary seventeenth-century notion of *universal gravitation*—the idea that all objects in the universe attract each other through the force of gravity. In Chapters 3 and 4 we learned that Earth's gravitational force acts on all objects near our planet's surface. In this chapter we'll extend our discussion to consider the gravitational forces exerted by any massive object on any other massive object. Gravitational forces between the components of our planet are partly responsible for holding Earth together (**Figure 10-1a**), gravitational forces

Figure 10-1 Gravitation is universal The force of gravitation is responsible for (a) keeping our Earth from flying apart, (b) keeping moons and rings in orbit around their planets, and (c) holding galaxies—among the largest structures in the universe—together.

(a) Earth

Earth is held together by the mutual gravitational attraction of all its parts.

NASA/NOAA

exerted by Saturn keep its moons and other small bodies in orbit (**Figure 10-1b**), and gravitational forces between stars and the material between the stars hold entire galaxies of stars together (**Figure 10-1c**).

(b) Saturn, its rings, and three of its moons
Saturn's rings are made of countless small objects that, like Saturn's moons, are held in orbit by the planet's gravitation.
NASA/JPL/USGS

(c) Galaxy NGC 6744
This galaxy (a near-twin of our Milky Way) contains more than 10^{11} stars. The stars' gravitational attraction for each other holds the galaxy together.
European Southern Observatory

In this chapter we will learn about the properties of the gravitational force, including how Newton deduced that this force must grow weaker as objects move farther apart. We will see how to treat the gravitational potential energy of a system of two objects interacting with each other, such as a satellite and Earth. We'll then use these ideas about gravitational force and gravitational potential energy to understand the nature of orbits. We'll find that the same ideas that apply to a satellite orbiting Earth also apply to planets orbiting the Sun. Indeed, we will see how Newton's idea of universal gravitation allowed him to explain three laws of planetary motion that had been discovered, but not understood, decades before. We'll also use the idea of universal gravitation to understand why astronauts in orbit feel weightless, and learn about some of the physiological challenges that astronauts face as a result.

TAKE-HOME MESSAGE FOR Section 10-1

✔ Isaac Newton deduced that gravitation is universal—it is a force that acts between any two massive objects in the universe.

✔ The ideas of universal gravitation will help us understand the motions of satellites around Earth and of planets around the Sun.

10-2 Newton's law of universal gravitation explains the orbit of the Moon

Newton deduced much of what we now know about gravitation by considering the motion of Earth's Moon (**Figure 10-2**). It had been understood for centuries that the Moon orbits around Earth. Newton knew from his laws of motion that there must therefore be a force acting on the Moon to keep it from flying off into space, and that Earth must exert this force on the Moon. This is just the force of Earth's gravitation. But how strong is that force? Is this force the same, stronger, or weaker than it would be if the Moon were closer? Let's see how Newton answered these questions by using the ideas of circular motion.

The Image Bank/Getty Images

Figure 10-2 The Moon Isaac Newton used the known distance to the Moon and the Moon's speed to deduce its acceleration, which he concluded was caused by Earth's gravitational attraction. This gave him an important clue about how the gravitational force depends on distance.

Newton knew that the Moon orbits around Earth at a nearly constant speed in a nearly circular path. The evidence for this is that the Moon changes its position against the background stars. It moves against this background at a nearly constant rate (which tells us that its speed is nearly constant), and as it moves its apparent size changes very little (which tells us that the Moon maintains a roughly constant distance from Earth, and so its orbit is nearly circular).

The radius of the Moon's orbit—that is, the distance from Earth to the Moon—was also well known in Newton's time, even though travel to the Moon was centuries in the future. This was possible because as seen from different locations on Earth, the Moon appears to be in slightly different positions relative to the background of stars

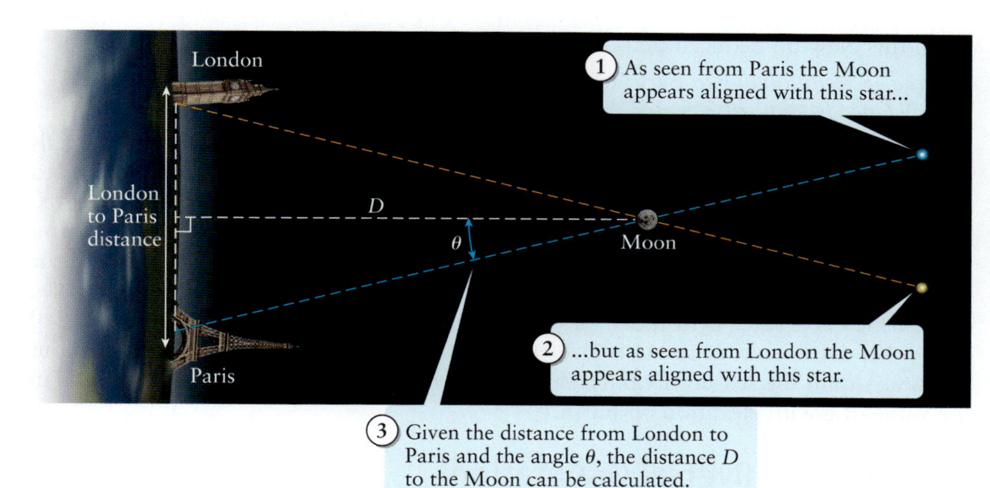

London
1 As seen from Paris the Moon appears aligned with this star...
London to Paris distance
D
θ
Moon
2 ...but as seen from London the Moon appears aligned with this star.
Paris
3 Given the distance from London to Paris and the angle θ, the distance D to the Moon can be calculated.

Figure 10-3 Measuring the distance to the Moon The Moon is close enough that it appears to be in slightly different positions against the backdrop of distant stars when viewed from different locations on Earth. This made it possible to calculate the distance to the Moon many centuries before humans were able to send a spacecraft there.

(**Figure 10-3**). By measuring the differences in apparent position and using trigonometry, it's possible to measure the distance D in Figure 10-3. From these measurements, the distance from the center of Earth to the Moon (equal to Earth's radius $R = 6370$ km, which was also known in Newton's time, plus the distance D in Figure 10-3 from the surface of Earth to the Moon) turns out to be 384,000 km $= 3.84 \times 10^8$ m. This is the radius r of the Moon's orbit.

As we learned in Section 3-7, an object that travels at constant speed v around a circular path of radius r—that is, an object in *uniform circular motion*—has an acceleration toward the center of the circle of magnitude

Centripetal acceleration: magnitude of the acceleration of an object in uniform circular motion

Speed of an object as it moves around the circle

$$a_{\text{cent}} = \frac{v^2}{r}$$

Radius of the circle

Centripetal acceleration (3-17)

The Moon's acceleration is due to the gravitational pull of Earth on the Moon, so a_{cent} is equal to the value of g at the position of the Moon. Equation 3-17 shows that to determine the Moon's acceleration Newton had to know the values of the radius r of the Moon's orbit—that is, the distance from the center of Earth to the Moon—and the speed v of the Moon in its orbit.

From observations of the Moon's apparent motion relative to the stars, it was known in Newton's time that the Moon makes one complete orbit around Earth in a time $T = 27.3$ days, or 2.36×10^6 s. The speed v of the Moon in its orbit is therefore the circumference of its orbit, equal to 2π times the radius $r = 3.84 \times 10^8$ m of its orbit, divided by the time $T = 2.36 \times 10^6$ s to complete an orbit. The result is $v = 1.02 \times 10^3$ m/s (about 3670 km/h or 2280 mi/h). If you substitute these values for v and r into Equation 3-17, you'll find that the Moon's acceleration is, to two significant digits,

$$a_{\text{cent}} = \frac{v^2}{r} = \frac{(1.02 \times 10^3 \text{ m/s})^2}{3.84 \times 10^8 \text{ m}} = 2.7 \times 10^{-3} \text{ m/s}^2$$

Note that a_{cent} is *very* much less than the value $g = 9.80$ m/s^2 at the surface of Earth. This implies that Earth's gravitational attraction—which is what exerts a force on the Moon to give it its centripetal acceleration—gets weaker with increasing distance. In fact the Moon's acceleration is less than the value of g at Earth's surface by a factor of

$$\frac{a_{\text{cent, Moon}}}{g_{\text{surface of Earth}}} = \frac{2.7 \times 10^{-3} \text{ m/s}^2}{9.80 \text{ m/s}^2} = 2.8 \times 10^{-4} = \frac{1}{3600} = \frac{1}{60^2}$$

As it happens, the distance from the center of Earth to Earth's surface is smaller than the distance from the center of Earth to the Moon by a factor of

$$\frac{r_{\text{Earth center to surface}}}{r_{\text{Earth center to Moon}}} = \frac{6370 \text{ km}}{384{,}000 \text{ km}} = \frac{1}{60}$$

From this Newton deduced that

> *the acceleration due to Earth's gravity decreases in proportion to the square of the distance from Earth's center.*

If you could stand atop a tower 6370 km tall—that is, a tower whose height equals the radius of Earth—you would be twice as far from Earth's center as a person at sea level, and the acceleration due to gravity would be $1/(2)^2 = 1/4$ as great as at sea level. By contrast if you were to stand atop Earth's tallest mountain, Mount Everest, you would be a little less than 9 km above sea level. The distance from Earth's center to the peak of Mount Everest is 6379 km as compared to 6370 km from Earth's center to sea level, which is greater by a factor of $(6379 \text{ km})/(6370 \text{ km}) = 1.0014$. So the value of g atop Mount Everest is about $1/(1.0014)^2 = 0.997$ of the value at sea level, or about 0.3% less than the sea-level value. This shows why there is only a small difference in the value of g with elevation on Earth.

The Law of Universal Gravitation

The following equation expresses Newton's observation that the value of the acceleration g due to Earth's gravitation decreases in proportion to the square of the distance r from the center of Earth:

$$g = \frac{\text{(a constant)}}{r^2}$$

The value of the constant in this equation is chosen so that at a point on Earth's surface, where r equals Earth's radius of 6370 km, the value of g equals 9.80 m/s^2. We learned in Section 4-3 that the magnitude of Earth's gravitational force on an object of mass m is mg. So if an object of mass m is located a distance r from Earth's center, the gravitational force of Earth on this object is

$$F_{\text{Earth on object}} = \frac{\text{(a constant)}m}{r^2}$$

That is, Earth's gravitational force on an object of mass m is directly proportional to the mass m of the object that experiences the force.

Newton generalized this to say that *any* object of mass m_1 exerts a gravitational force on a second object of mass m_2, that this force is directly proportional to the mass m_2 of the object that experiences the force, and that this force is inversely proportional to the square of the distance r between the centers of the objects (**Figure 10-4a**):

$$F_{1 \text{ on } 2} = \frac{\text{(a constant)}m_2}{r^2} \quad (10\text{-}1)$$

For the force of Earth on the Moon, m_2 is the Moon's mass, and r is the Earth–Moon distance. If Equation 10-1 is true for *any* two objects, then it must also be true that the object of mass m_2 exerts a gravitational force on the object of mass m_1, and this force is directly proportional to the mass m_1. But Newton's third law (Section 4-5) tells us that the forces that the two objects exert on each other have opposite directions and the same magnitude: $F_{2 \text{ on } 1} = F_{1 \text{ on } 2}$. Hence the gravitational force that each object exerts on the other must be directly proportional to *both* m_1 and m_2. This chain of reasoning leads us to **Newton's law of universal gravitation:**

Newton's law of universal gravitation (10-2)

$$F_{1 \text{ on } 2} = F_{2 \text{ on } 1} = \frac{Gm_1m_2}{r^2}$$

Gravitational constant (same for any two objects)

Masses of the two objects

Any two objects (1 and 2) exert **equally strong gravitational forces** on each other.

Center-to-center distance between the two objects

The gravitational forces are attractive: $\vec{F}_{1 \text{ on } 2}$ pulls object 2 toward object 1 and $\vec{F}_{2 \text{ on } 1}$ pulls object 1 toward object 2.

As Figure 10-4a shows the forces that the two objects exert on each other are attractive and are directed along the line that connects the centers of the two objects.

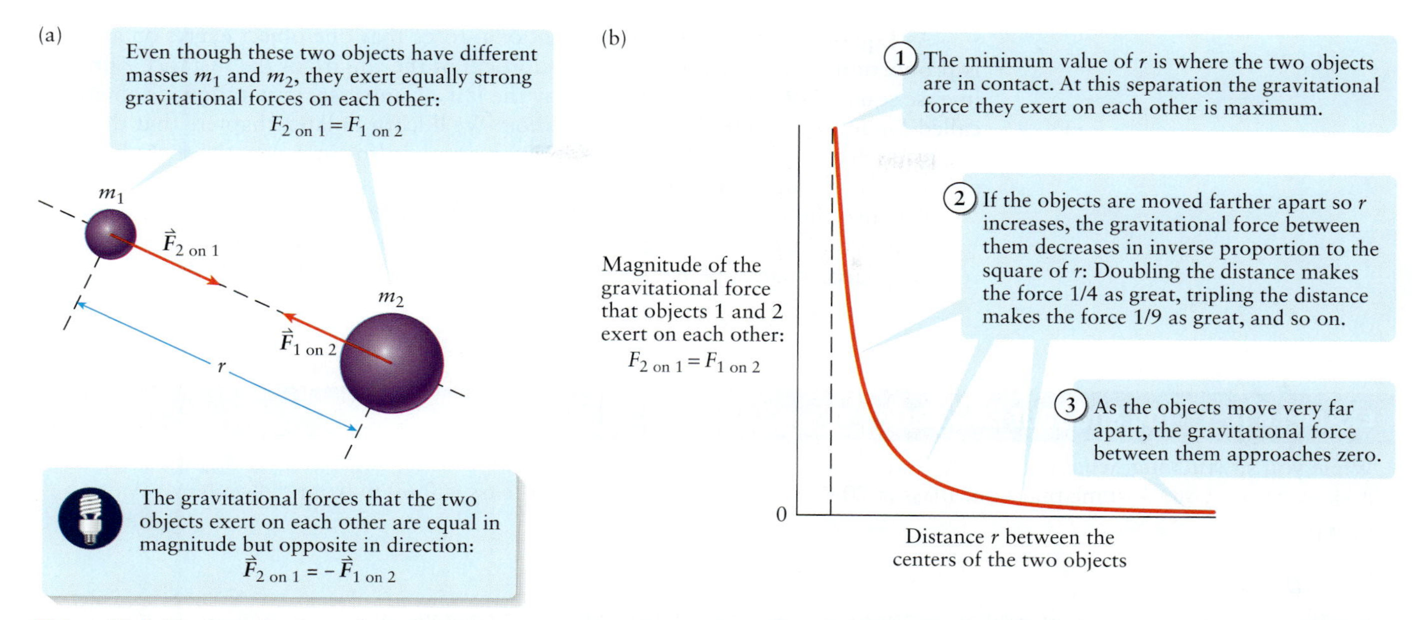

(a)
Even though these two objects have different masses m_1 and m_2, they exert equally strong gravitational forces on each other: $F_{2 \text{ on } 1} = F_{1 \text{ on } 2}$
m_1
$\vec{F}_{2 \text{ on } 1}$
m_2
$\vec{F}_{1 \text{ on } 2}$
r
The gravitational forces that the two objects exert on each other are equal in magnitude but opposite in direction: $\vec{F}_{2 \text{ on } 1} = -\vec{F}_{1 \text{ on } 2}$
(b)
1 The minimum value of r is where the two objects are in contact. At this separation the gravitational force they exert on each other is maximum.
2 If the objects are moved farther apart so r increases, the gravitational force between them decreases in inverse proportion to the square of r: Doubling the distance makes the force 1/4 as great, tripling the distance makes the force 1/9 as great, and so on.
3 As the objects move very far apart, the gravitational force between them approaches zero.
Magnitude of the gravitational force that objects 1 and 2 exert on each other: $F_{2 \text{ on } 1} = F_{1 \text{ on } 2}$
0
Distance r between the centers of the two objects

Figure 10-4 The law of universal gravitation (a) Any two objects exert gravitational forces on each other. These forces are equal in magnitude but opposite in direction: $\vec{F}_{2 \text{ on } 1} = -\vec{F}_{1 \text{ on } 2}$. (b) The magnitude of the gravitational force is inversely proportional to the square of the center-to-center distance r.

The proportionality constant G is called the **gravitational constant.** We'll see later in this section how the value of G (which Newton did not know) is determined. Its currently accepted value, to three significant figures, is

$$G = 6.67 \times 10^{-11} \text{ N} \cdot \text{m}^2/\text{kg}^2$$

Figure 10-4b graphs the magnitude of the gravitational force between the two objects as a function of the distance r between their centers.

WATCH OUT! Be sure to understand the limitations of Equation 10-2.

Strictly speaking, Equation 10-2 applies to two infinitesimally small objects with masses m_1 and m_2. It's much more challenging to calculate the gravitational force between two extended objects with complicated shapes, such as between Earth and the Moon (neither of which is a perfect sphere) or between Earth and the International Space Station (the shape of which you can see in the photograph that opens this chapter). However, in practice we can safely use Equation 10-2 if the *distance* between the two objects is large compared to the *size* of either object. That's the case for Earth and the Moon, which are separated by a distance of 384,000 km—much larger than the radius of either Earth (6370 km) or the Moon (1740 km). It's also safe to use Equation 10-2 if one object is very much smaller than the other, as is the case for a satellite (like the International Space Station) orbiting Earth, as long as we can approximate the large object (in this case Earth) as being roughly uniform. In other situations, such as calculating the force between the two galaxies shown in Figure 10-1c (which are comparable in size and whose separation is comparable to the size of either galaxy), Equation 10-2 still gives useful estimates of the magnitude of the gravitational force. Note also that Equation 10-2 is not valid if one object is *inside* the other. For example, we can't use this equation to find the gravitational force on you when you're in a tunnel deep inside Earth's interior.

WATCH OUT! The distance r in Equation 10-2 is the center-to-center distance.

Always remember that r in Equation 10-2 is the distance between the *centers* of the two objects. If you're calculating the gravitational force that Earth exerts on a person standing on Earth's surface, the distance r equals the radius of Earth—that is, the distance from the center of Earth to that person. For nearly symmetrical objects like Earth or the Moon, the geometrical center is the same as the center of mass, so we'll often use the terms "center" and "center of mass" interchangeably when discussing gravitational forces.

As Equation 10-2 shows the gravitational force that one object exerts on another is proportional to $1/r^2$, which is the reciprocal of the square of the distance r or "the inverse square" of r for short. That's why the law of universal gravitation is sometimes called *the inverse-square law for gravitation*. We'll learn in later chapters that the *electric* force that one charged particle exerts on another (for instance, the force between an electron and a proton) also obeys an inverse-square law.

Because the gravitational constant is such a small value, gravitational forces between two objects also tend to be pretty small unless at least one of the objects is relatively large (that is, planet-sized). The following examples illustrate this.

EXAMPLE 10-1 You and Your Backpack

While you sit studying, what is the magnitude of the gravitational force that your 10.0-kg backpack sitting on your desk exerts on you? Assume that your mass is 70.0 kg and that your center of mass is 0.60 m from the center of your backpack.

Set Up

We'll use Newton's law of universal gravitation, Equation 10-2, to find the magnitude of the gravitational force.

Newton's law of universal gravitation:

$$F_{\text{bag on you}} = \frac{Gm_{\text{bag}}m_{\text{you}}}{r^2} \quad (10\text{-}2)$$

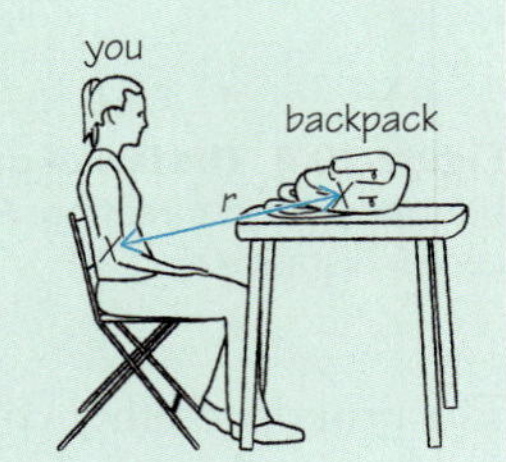

Solve

Substitute the given values of G, $m_{\text{bag}} = 10.0$ kg, $m_{\text{you}} = 70.0$ kg, and $r = 0.60$ m into the expression for gravitational force. The distance is given to only two significant figures, so our result has only two significant figures.

$$F_{\text{bag on you}} = \frac{Gm_{\text{bag}}m_{\text{you}}}{r^2}$$

$$= \frac{(6.67 \times 10^{-11}\ \text{N}\cdot\text{m}^2/\text{kg}^2)(10.0\ \text{kg})(70.0\ \text{kg})}{(0.60\ \text{m})^2}$$

$$= 1.3 \times 10^{-7}\ \text{N}$$

Reflect

The gravitational force your bag exerts on you is *very* small, about one ten-millionth of a newton. This is equivalent to the weight of a few specks of dust. This force is far smaller than the gravitational force Earth exerts on you (see the next example). It's also far smaller than the force of friction between you and your chair should you start to slide. That's why we generally neglect the gravitational force between everyday objects.

EXAMPLE 10-2 Earth's Gravitational Force on You

Calculate the magnitude of the gravitational force that Earth exerts on you, again assuming you have a mass of 70.0 kg. Earth has mass 5.97×10^{24} kg and radius 6370 km.

Set Up

Again we'll find the magnitude of the gravitational force using Newton's law of universal gravitation, Equation 10-2. The distance r between the centers of the two objects (Earth and you) equals Earth's radius, 6370 km. The extra meter or so distance from Earth's surface to the center of your body is insignificant compared to the size of Earth's radius, so we ignore it.

Newton's law of universal gravitation:

$$F_{\text{Earth on you}} = \frac{Gm_{\text{Earth}}m_{\text{you}}}{r^2} \quad (10\text{-}2)$$

r = distance from center of Earth to you
$= R_{\text{Earth}}$ = radius of Earth

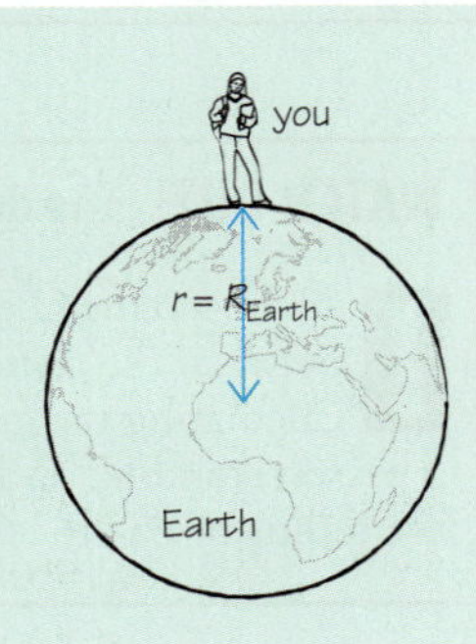

Solve

To use the distance $r = R_{\text{Earth}} = 6370$ km in Equation 10-2, we first convert it to meters.

$$r = R_{\text{Earth}} = (6370 \text{ km}) \times \frac{1000 \text{ m}}{1 \text{ km}} = 6.37 \times 10^6 \text{ m}$$

Substitute the values of G, m_{Earth}, m_{you}, and r into Equation 10-2.

$$F_{\text{Earth on you}} = \frac{Gm_{\text{Earth}}m_{\text{you}}}{r^2} = \frac{Gm_{\text{Earth}}m_{\text{you}}}{R_{\text{Earth}}^2}$$
$$= \frac{(6.67 \times 10^{-11} \text{ N}\cdot\text{m}^2/\text{kg}^2)(5.97 \times 10^{24} \text{ kg})(70.0 \text{ kg})}{(6.37 \times 10^6 \text{ m})^2}$$
$$= 687 \text{ N} = 690 \text{ N to two significant figures}$$

Reflect

We can check our result by using the expression $w = mg$ for gravitational force that we introduced in Section 4-3. We get the same result as above to two significant figures, which shows that the values of G and m_{Earth} are compatible with the value of g at Earth's surface.

Using the expression $w = mg$ for gravitational force,

$$F_{\text{Earth on you}} = m_{\text{you}}g = (70.0 \text{ kg})(9.80 \text{ m/s}^2) = 686 \text{ N} = 690 \text{ N to two significant figures}$$

Finding the Value of *G* and the Mass of Earth

Example 10-2 shows that there is a connection between the value of g at Earth's surface and the values of G and m_{Earth}, the gravitational constant and Earth's mass. To see this more clearly take the two expressions for the gravitational force on you and set them equal to each other:

$$F_{\text{Earth on you}} = m_{\text{you}}g = \frac{Gm_{\text{Earth}}m_{\text{you}}}{R_{\text{Earth}}^2}$$

Then divide both sides of the equation by your mass m_{you} to get an expression for g:

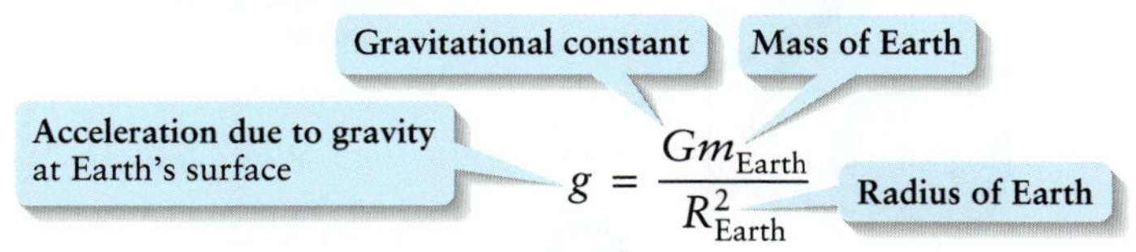

$$g = \frac{Gm_{\text{Earth}}}{R_{\text{Earth}}^2}$$

Value of g at Earth's surface (10-3)

This equation tells us that whenever we calculate a weight or an acceleration using g, we are simply using a shorthand form of the law of universal gravitation applied near Earth's surface. So the familiar expression $w = mg$ is just a special form of the general law of universal gravitation, Equation 10-2.

Equation 10-3 relates the value of g (which we can measure by observing a freely falling object) to the values of Earth's radius R_{Earth}, Earth's mass m_{Earth}, and the gravitational constant G. The value of R_{Earth} has been known since the third century B.C., when Eratosthenes, a scholar from Cyrene in modern-day Libya, first determined its value. (**Figure 10-5** shows how Eratosthenes was able to do this.) But that leaves two unknown values in Equation 10-3, G and m_{Earth}. With only one equation to relate them, it seems impossible to determine the values of both of these unknowns.

The solution to this problem was provided by the British scientist Henry Cavendish in 1798, more than a century after Newton published the law of universal gravitation. Cavendish conducted the elegant experiment shown schematically in **Figure 10-6**, now known simply as the **Cavendish experiment**. The gravitational force of each large sphere on the nearby small sphere makes the wooden rod rotate, which twists the wire from which the rod is suspended. The wire resists being twisted and exerts a torque on the rod that tries to return the wire to its relaxed state, much as a stretched spring exerts a force that tries to return the spring to its unstretched length. In equilibrium the torque exerted by the wire just balances the gravitational torque. By measuring the wire beforehand Cavendish knew how much torque the wire exerted for a given rotation angle, and thus he could determine the equal-magnitude gravitational torque. He could then determine the gravitational force on each small sphere required to cause

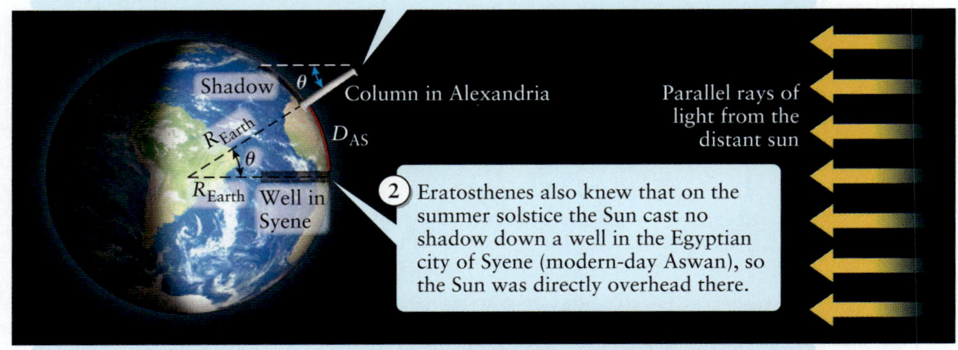

1 As the story is told, on the day of the summer solstice (when the Sun is highest in the sky as seen from the northern hemisphere), Eratosthenes observed the shadow cast by a column in the city of Alexandria.

2 Eratosthenes also knew that on the summer solstice the Sun cast no shadow down a well in the Egyptian city of Syene (modern-day Aswan), so the Sun was directly overhead there.
3 Given the known distance D_{AS} from Alexandria to Syene and his measurement of the angle θ, Eratosthenes was able to calculate the radius R_{Earth}.

Figure 10-5 How Eratosthenes measured the radius of Earth Around 240 B.C. the scholar Eratosthenes measured Earth's radius by analyzing the shadows cast by the Sun at two different locations.

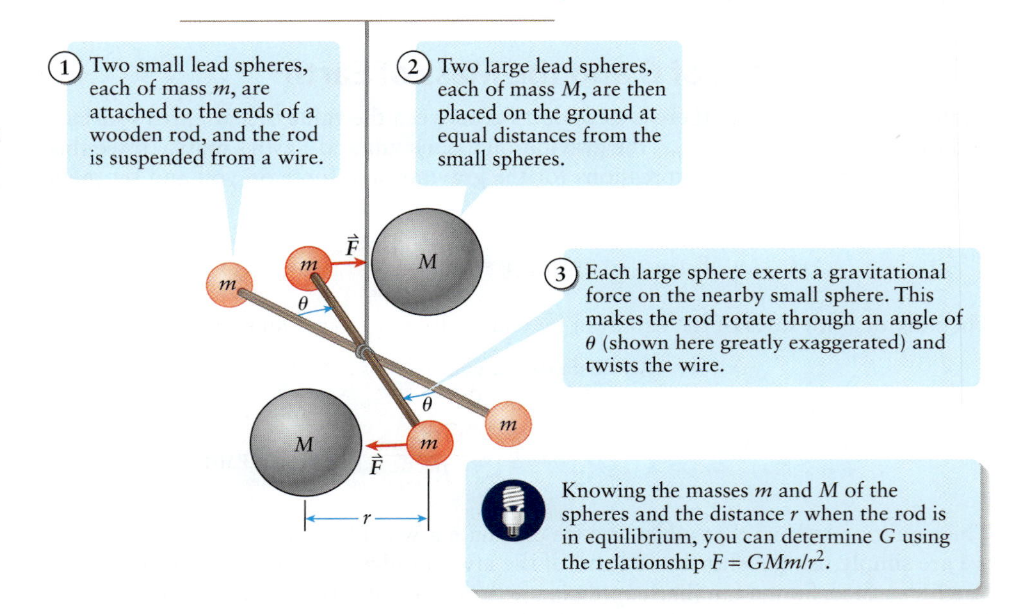

1 Two small lead spheres, each of mass m, are attached to the ends of a wooden rod, and the rod is suspended from a wire.
2 Two large lead spheres, each of mass M, are then placed on the ground at equal distances from the small spheres.
3 Each large sphere exerts a gravitational force on the nearby small sphere. This makes the rod rotate through an angle of θ (shown here greatly exaggerated) and twists the wire.
Knowing the masses m and M of the spheres and the distance r when the rod is in equilibrium, you can determine G using the relationship $F = GMm/r^2$.

Figure 10-6 The Cavendish experiment This experiment is used to determine the value of G. In the original (1798) version of the experiment, the wooden rod was 1.8 m in length, each small lead sphere had a mass of 0.73 kg, each large lead sphere had a mass of 158 kg, and the distance r was about 23 cm.

this torque on the rod. With this technique Cavendish could measure extremely small forces on the order of 10^{-7} N, which made his experiment possible.

Given the magnitude of the force F, the masses m and M of the small and large spheres, respectively, and the distance r between neighboring large and small spheres, we can use Newton's law of universal gravitation to solve for the value of G:

$$F = \frac{GMm}{r^2} \quad \text{so} \quad G = \frac{Fr^2}{Mm}$$

Go to Picture It 10-1 for more practice dealing with gravitational force.

Modern experiments to measure G to great precision still use variations on Cavendish's original apparatus. Once the value of G is known, we can use Equation 10-3 to determine the value of Earth's mass m_{Earth} (now known to be 5.97×10^{24} kg).

Armed with knowledge of the value of G, we have all the tools we need to apply the law of universal gravitation to the orbits of satellites and planets. To analyze these orbits it's important to first consider the potential energy associated with the gravitational force, which is the topic of the next section. First, however, here are two examples that make use of the law of universal gravitation.

EXAMPLE 10-3 At What Altitude Is g Cut in Half?

At what height above Earth's surface is the value of g equal to one-half the value at the surface?

Set Up

Equation 10-3 tells us the value of g at Earth's surface, which we'll call $g_{surface}$. Our goal is to find the height h above the surface at which g has half this value. We'll use the law of universal gravitation and the relationship between the gravitational force on an object and the value of g. Remember that the distance r in the law of universal gravitation is *not* the height above the surface but rather the distance to Earth's center. This equals the radius of Earth plus the height of the object.

Value of g at Earth's surface:

$$g_{surface} = \frac{Gm_{Earth}}{R_{Earth}^2} \quad (10\text{-}3)$$

Newton's law of universal gravitation:

$$F_{Earth\ on\ object} = \frac{Gm_{Earth}m_{object}}{r^2} \quad (10\text{-}2)$$

Gravitational force on an object in terms of g:

$$F_{Earth\ on\ object} = m_{object}g$$

Distance from Earth's center to a height h above the surface:

$$r = R_{Earth} + h$$

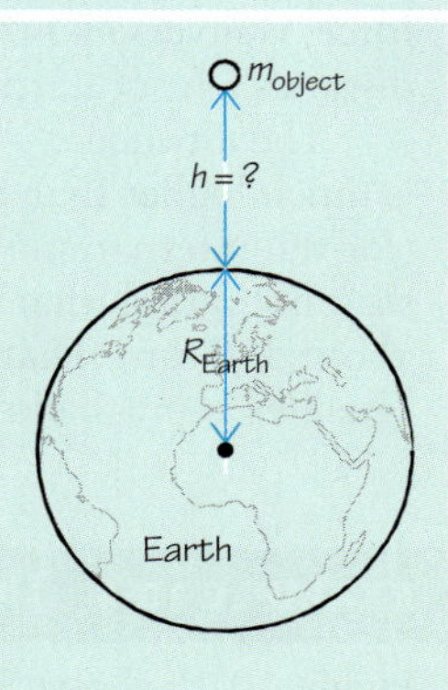

Solve

Use the two expressions for the gravitational force on an object to solve for g at a height h above the surface.

Set the two expressions for $F_{Earth\ on\ object}$ equal to each other:

$$F_{Earth\ on\ object} = m_{object}g = \frac{Gm_{Earth}m_{object}}{r^2}$$

To solve for g divide through by m_{object}:

$$g = \frac{Gm_{Earth}}{r^2}$$

Substitute $r = R_{Earth} + h$ in the expression for g:

$$g = \frac{Gm_{Earth}}{(R_{Earth} + h)^2}$$

We want to find the value of h at which g equals one-half of $g_{surface}$.

The value of g at height h equals $1/2$ the value of $g_{surface}$:

$$g = \frac{1}{2}g_{surface}$$

Substitute the expression for g we found above and the expression for $g_{surface}$ from Equation 10-3:

$$\frac{Gm_{Earth}}{(R_{Earth} + h)^2} = \frac{1}{2}\frac{Gm_{Earth}}{R_{Earth}^2}$$

Note that Gm_{Earth} cancels, leaving

$$\frac{1}{(R_{Earth} + h)^2} = \frac{1}{2R_{Earth}^2}$$

Take the reciprocal of both sides of this equation:

$$(R_{Earth} + h)^2 = 2R_{Earth}^2$$

To eliminate the squares take the square root of both sides:

$$R_{Earth} + h = \sqrt{2}\,R_{Earth}$$

To get an expression for h, subtract R_{Earth} from both sides of this equation:

$$h = \sqrt{2}R_{Earth} - R_{Earth}$$
$$= (\sqrt{2} - 1)R_{Earth}$$

Substitute $R_{Earth} = 6370$ km:

$$h = (\sqrt{2} - 1)(6370 \text{ km}) = (1.414 - 1)(6370 \text{ km})$$
$$= 2640 \text{ km}$$

Reflect

Our calculation shows that g, like the gravitational force, is inversely proportional to the square of the distance r from Earth's center.

The altitude $h = (\sqrt{2} - 1)R_{\text{Earth}} = 2640$ km is much higher than that of most Earth satellites (for the International Space Station shown in the photograph that opens this chapter, h is only about 350 km). That's why the value of g for these satellites is much closer to g_{surface} than to $g_{\text{surface}}/2$.

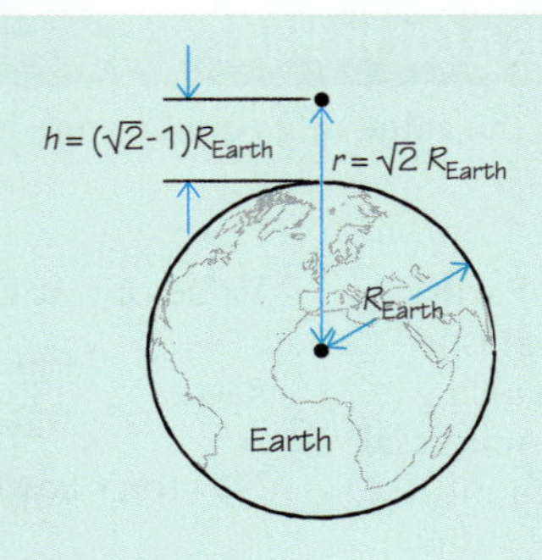

EXAMPLE 10-4 The Moons of Saturn

Figure 10-1b shows Saturn's moons Tethys, Dione, and Rhea. (Saturn's moons, of which 62 were known as of this writing, are named for mythological figures associated with the Roman god Saturn.) Their masses are 6.2×10^{20} kg for Tethys, 1.1×10^{21} kg for Dione, and 2.3×10^{21} kg for Rhea. All three of these moons are quite small compared to Earth's moon (mass 7.36×10^{22} kg). Tethys, Dione, and Rhea move in different orbits around Saturn and at different speeds, and on occasion they form a right triangle (Figure 10-7). Calculate the *net* gravitational force on Tethys due to Dione and Rhea when the three moons are in this configuration. Compare this to the force that Saturn (mass 5.7×10^{26} kg) exerts on Tethys, which orbits 2.95×10^5 km from Saturn's center.

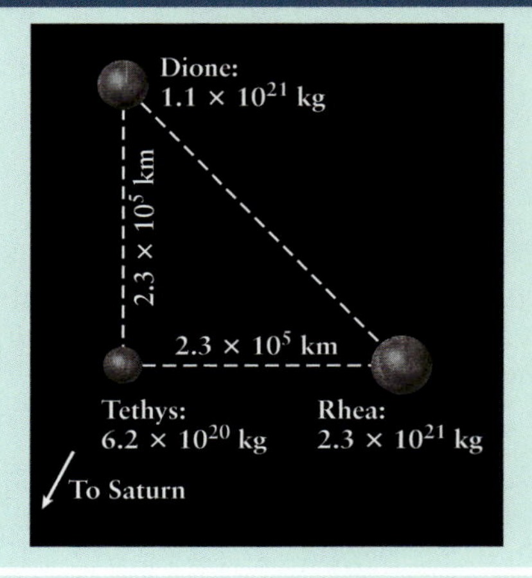

Figure 10-7 Three moons of Saturn What is the net gravitational force on Saturn's moon Tethys due to the other moons Dione and Rhea?

Set Up

Both Dione and Rhea exert gravitational forces on Tethys as given by Equation 10-2. As the figure shows these forces act at right angles to each other. The *net* gravitational force on Tethys due to the two other moons is the vector sum of the two individual forces. We'll use the ideas of vector addition and vector components from Sections 3-2 and 3-3 to find this vector sum and determine the magnitude and direction of the net force.

Newton's law of universal gravitation:

$$F_{\text{moon on Tethys}} = \frac{Gm_{\text{moon}}m_{\text{Tethys}}}{r^2_{\text{moon to Tethys}}} \quad (10\text{-}2)$$

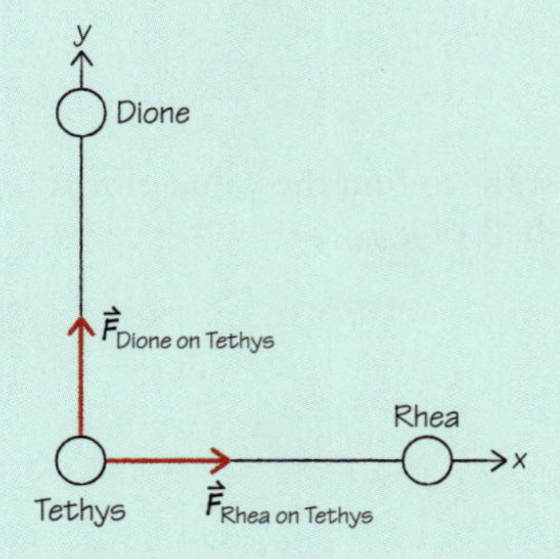

Solve

The easiest way to add vectors is by using components, so we choose x and y axes as shown. With this choice of axes, the force of Dione on Tethys has only a y component, and the force of Rhea on Tethys has only an x component. Calculate these components using the values of r shown in Figure 10-7.

Convert the Dione-Tethys distance and Rhea–Tethys distance from kilometers to meters:

$$r_{\text{Dione to Tethys}} = r_{\text{Rhea to Tethys}} = 2.3 \times 10^5 \text{ km}$$

$$= (2.3 \times 10^5 \text{ km})\left(\frac{1000 \text{ m}}{1 \text{ km}}\right) = 2.3 \times 10^8 \text{ m}$$

The force of Dione on Tethys has zero x component and a positive y component:

$$F_{\text{Dione on Tethys},x} = 0$$

$$F_{\text{Dione on Tethys},y} = +\frac{Gm_{\text{Dione}}m_{\text{Tethys}}}{r^2_{\text{Dione to Tethys}}}$$

$$= +\frac{(6.67 \times 10^{-11} \text{ N}\cdot\text{m}^2/\text{kg}^2)(1.1 \times 10^{21} \text{ kg})(6.2 \times 10^{20} \text{ kg})}{(2.3 \times 10^8 \text{ m})^2}$$

$$= +8.6 \times 10^{14} \text{ N}$$

The force of Rhea on Tethys has a positive x component and zero y component:

$$F_{\text{Rhea on Tethys},x} = +\frac{Gm_{\text{Rhea}}m_{\text{Tethys}}}{r^2_{\text{Rhea to Tethys}}}$$

$$= +\frac{(6.67 \times 10^{-11}\ \text{N}\cdot\text{m}^2/\text{kg}^2)(2.3 \times 10^{21}\ \text{kg})(6.2 \times 10^{20}\ \text{kg})}{(2.3 \times 10^{8}\ \text{m})^2}$$

$$= +1.8 \times 10^{15}\ \text{N}$$

$$F_{\text{Rhea on Tethys},y} = 0$$

The net force on Tethys is the vector sum of the two individual forces. Find the components of the net force.

The x component of the net force on Tethys is the sum of the x components of the individual forces:

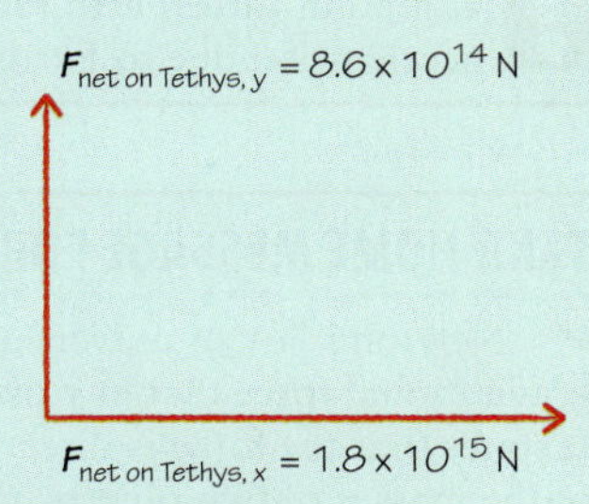

$$F_{\text{net on Tethys},x}$$
$$= F_{\text{Dione on Tethys},x} + F_{\text{Rhea on Tethys},x}$$
$$= 0\ \text{N} + 1.8 \times 10^{15}\ \text{N}$$
$$= +1.8 \times 10^{15}\ \text{N}$$

Similarly, the y component of the net force on Tethys is the sum of the y components of the individual forces:

$$F_{\text{net on Tethys},y} = F_{\text{Dione on Tethys},y} + F_{\text{Rhea on Tethys},y}$$
$$= +8.6 \times 10^{14}\ \text{N} + 0\ \text{N}$$
$$= +8.6 \times 10^{14}\ \text{N}$$

From the components of the net force, find its magnitude and direction using the techniques learned in Section 3-3.

Find the magnitude of the net force:

$$F_{\text{net on Tethys}} = \sqrt{(F_{\text{net on Tethys},x})^2 + (F_{\text{net on Tethys},y})^2}$$
$$= \sqrt{(+1.8 \times 10^{15}\ \text{N})^2 + (+8.6 \times 10^{14}\ \text{N})^2}$$
$$= 2.0 \times 10^{15}\ \text{N}$$

Find the angle of the net force relative to the x axis:

$$\tan\theta = \frac{F_{\text{net on Tethys},y}}{F_{\text{net on Tethys},x}}$$
$$= \frac{+8.6 \times 10^{14}\ \text{N}}{+1.8 \times 10^{15}\ \text{N}} = 0.48$$

so

$$\theta = \tan^{-1} 0.48 = 26°$$

Reflect

The net gravitational force on Tethys from Dione and Rhea is 2.0×10^{15} N, which seems tremendous. But the gravitational force that Saturn exerts on Tethys is about 1.4×10^5 (140,000) times greater than the force from the other moons. This means the orbit of Tethys around Saturn is affected hardly at all by the presence of the other moons.

Note that the net force on Tethys from the other two moons is not directed at either of those moons but rather somewhere between the two. This isn't surprising: In general the sum of two vectors doesn't point in the direction of either individual vector.

Convert the distance from the center of Saturn to Tethys from kilometers to meters:

$$r_{\text{Saturn to Tethys}} = (2.95 \times 10^5\ \text{km})\left(\frac{1000\ \text{m}}{1\ \text{km}}\right) = 2.95 \times 10^8\ \text{m}$$

Then the gravitational force that Saturn exerts on Tethys is

$$F_{\text{Saturn on Tethys}} = \frac{Gm_{\text{Saturn}}m_{\text{Tethys}}}{r^2_{\text{Saturn to Tethys}}}$$

$$= \frac{(6.67 \times 10^{-11}\ \text{N}\cdot\text{m}^2/\text{kg}^2)(5.7 \times 10^{26}\ \text{kg})(6.2 \times 10^{20}\ \text{kg})}{(2.95 \times 10^8\ \text{m})^2}$$

$$= 2.7 \times 10^{20}\ \text{N}$$

This force is greater than the net force that Dione and Rhea together exert on Tethys by a factor of

$$\frac{F_{\text{Saturn on Tethys}}}{F_{\text{net on Tethys}}} = \frac{2.7 \times 10^{20}\ \text{N}}{2.0 \times 10^{15}\ \text{N}} = 1.4 \times 10^5$$

GOT THE CONCEPT? 10-1 Gravitational Force

Compared to the gravitational force that Earth exerts on a 50-kg object located three Earth radii above the north pole, the gravitational force that Earth exerts on a 50-kg object located at the north pole is (a) the same; (b) 3 times greater; (c) 4 times greater; (d) 9 times greater; (e) 16 times greater.

GOT THE CONCEPT? 10-2 Gravitational Acceleration

Compared to the acceleration due to Earth's gravity at a point three Earth radii above the north pole, the acceleration due to Earth's gravity at the north pole is (a) the same; (b) 3 times greater; (c) 4 times greater; (d) 9 times greater; (e) 16 times greater.

TAKE-HOME MESSAGE FOR Section 10-2

✔ Newton's law of universal gravitation describes the gravitational force that any object exerts on another object. It states that the force is directly proportional to the product of the masses of two objects and inversely proportional to the square of the distance between their centers.

✔ The relationship we use for weight near Earth's surface, $w = mg$, is simply an application of the law of universal gravitation.

10-3 The gravitational potential energy of two objects is negative and increases toward zero as the objects are moved farther apart

In Chapter 6 we introduced the ideas of kinetic energy (Section 6-3) and potential energy (Section 6-6). We found that if a force is *conservative*—so that the work done by that force on an object as it moves depends only on the object's starting point and ending point, not on the path the object takes between them—then we can associate that force with potential energy. If only conservative forces do work on an object as it moves, the total mechanical energy—the sum of the kinetic energy and potential energy of the system—always keeps the same value and is *conserved* (Section 6-7).

One particularly important conservative force is the gravitational force on an object of mass m near Earth's surface, $w = mg$. We found that the potential energy associated with this force is $U_{grav} = mgy$, where y is the height of the object above some level that we choose to call $y = 0$. But because this simple expression for U_{grav} is based on the expression $w = mg$ for gravitational force, it *cannot* be correct for the more general form of the gravitational force between two objects $F = Gm_1m_2/r^2$. What, then, is the correct expression for the gravitational potential energy of two objects of mass m_1 and m_2 separated by a distance r?

Deriving this expression from the law of universal gravitation is a task that requires calculus, which is beyond our scope. Instead we'll just present the answer then examine it to see why it makes sense:

Gravitational potential energy (10-4)

$$U_{grav} = -\frac{Gm_1m_2}{r}$$

Gravitational constant (same for any two objects)

Masses of the two objects

Gravitational potential energy of a system of two objects (1 and 2)

Center-to-center distance between the two objects

The gravitational potential energy is zero when the two objects are infinitely far apart. If the objects are brought closer together (so r is made smaller), U_{grav} decreases (it becomes more negative).

The expression in Equation 10-4 looks nothing like the familiar formula $U_{grav} = mgy$. To understand the difference let's consider two different cases in which gravitational potential energy changes (Figure 10-8).

In **Figure 10-8a** a woman on Earth's surface raises a book of mass m_{book} a vertical distance d. There is a gravitational force on the book of magnitude $m_{\text{book}}g$, so she must exert an upward force of the same magnitude. So she does an amount of work $m_{\text{book}}gd$ in the process. The book begins and ends at rest, so the work $m_{\text{book}}gd$ that she does goes into increasing the gravitational potential energy associated with the book: $\Delta U_{\text{grav}} = m_{\text{book}}gd$. If the woman stands on a ladder and once again lifts the book by a distance d, the gravitational force she works against is still $m_{\text{book}}g$ since the gravitational force exerted by Earth changes hardly at all near Earth's surface. So she again does work $m_{\text{book}}gd$, and the gravitational potential energy increases by the same amount $m_{\text{book}}gd$ as before. As **Figure 10-8b** shows, the gravitational potential energy increases in direct proportion to the object's height y, so the graph of U_{grav} versus height is a straight line: $U_{\text{grav}} = m_{\text{book}}gy$.

Now the same person dons a spacesuit and brings the book with her to a point in space a distance r_A from the center of Earth (**Figure 10-8c**). At this point the gravitational force that Earth exerts on the book has magnitude $F_A = Gm_{\text{Earth}}m_{\text{book}}/r_A^2$. If she moves the book an additional distance d away from Earth, she has to exert a force of the same magnitude on the book and so does an amount of work $F_A d = (Gm_{\text{Earth}}m_{\text{book}}/r_A^2)d$. Just as on Earth, this work goes into increasing the gravitational potential energy. If she now moves to a greater distance r_B from the center of Earth and repeats the process, Earth's gravitational force has a smaller magnitude $F_B = Gm_{\text{Earth}}m_{\text{book}}/r_B^2$. So at this greater distance she does a smaller amount of work $F_B d = (Gm_{\text{Earth}}m_{\text{book}}/r_B^2)d$ and so adds a smaller amount of gravitational potential energy. This means that at great distances the gravitational potential energy does *not* increase in direct proportion to the distance r. The graph of U_{grav} versus distance is not a straight line, but a curve that becomes shallower with increasing values of r (**Figure 10-8d**). This is just what Equation 10-4 tells us, which helps to justify this expression for gravitational potential energy.

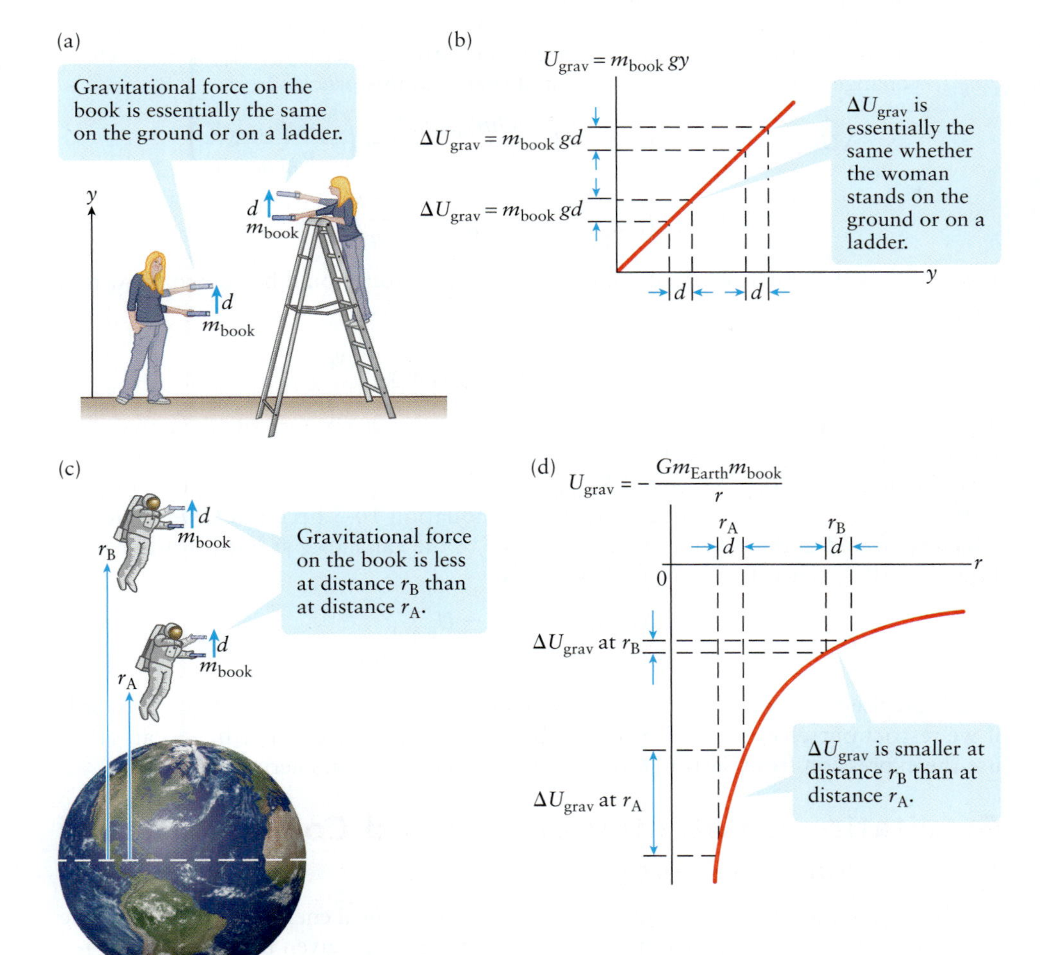
(a)
Gravitational force on the book is essentially the same on the ground or on a ladder.
y
d
m_{book}
d
m_{book}
(b)
$U_{\text{grav}} = m_{\text{book}}gy$
$\Delta U_{\text{grav}} = m_{\text{book}}gd$
$\Delta U_{\text{grav}} = m_{\text{book}}gd$
ΔU_{grav} is essentially the same whether the woman stands on the ground or on a ladder.
d
d
y
(c)
d
m_{book}
r_B
Gravitational force on the book is less at distance r_B than at distance r_A.
d
m_{book}
r_A
(d)
$U_{\text{grav}} = -\frac{Gm_{\text{Earth}}m_{\text{book}}}{r}$
r_A
r_B
d
d
r
0
ΔU_{grav} at r_B
ΔU_{grav} at r_A
ΔU_{grav} is smaller at distance r_B than at distance r_A.

Figure 10-8 Gravitational potential energy near Earth's surface and far away (a) A woman lifts a book of mass m_{book} a distance d while standing on the ground. She then repeats the process while standing on a ladder. (b) The gravitational force and the change in gravitational potential energy is basically the same in both cases, so the graph of U_{grav} versus height is a straight line. (c) The woman again raises the book a distance d, first at distance r_A from Earth's center, then at distance r_B. (d) The gravitational force and the change in gravitational potential energy are less at distance r_B than at distance r_A, so the graph of U_{grav} versus distance is a curve that gets shallower as r increases.

More on Gravitational Potential Energy

Here are three important attributes of Equation 10-4 for gravitational potential energy.

(1) *The gravitational potential energy of two objects interacting is a shared property of the two objects.* We learned in Section 6-6 that the gravitational potential energy $U_{grav} = mgy$ for an object of mass m near Earth's surface actually "belongs" to both Earth and the object, since it's associated with their mutual gravitational attraction. In the same way $U_{grav} = -Gm_1m_2/r$ as given by Equation 10-4 "belongs" to the *system* made up of an object of mass m_1 and an object of mass m_2.

(2) *The gravitational potential energy as given by Equation 10-4 is never positive.* The minus sign in Equation 10-4 means that the value of U_{grav} given by this expression is always negative, except when the two objects are infinitely far apart so $r \to \infty$ and $U_{grav} = 0$ (see Figure 10-8d). Don't be concerned by this! We learned in Section 6-6 that what's really physically meaningful are not the values of potential energy, but rather the *differences* between the values of potential energy at two different points. That means we can choose the point where potential energy equals zero to be wherever it's convenient. Our choice is that $U_{grav} = 0$ when the two objects are infinitely far apart. With this choice, U_{grav} as given by Equation 10-4 decreases and becomes more and more negative as an object of mass m_1 and an object of mass m_2 move toward each other, just as gravitational potential energy decreases as a ball drops toward the ground (that is, toward Earth).

(3) *The gravitational potential energy as given by Equation 10-4 is consistent with the formula $U_{grav} = mgy$ from Section 6-6.* To see how this can be, imagine an object of mass m a height y_i above Earth's surface—that is, at a distance $r = R_{Earth} + y_i$ from Earth's center. If you lift this object to a new height y_f above the surface, its new distance from Earth's center is $r = R_{Earth} + y_f$. The change in the gravitational potential energy in this process is

$$\Delta U_{grav} = U_{grav,f} - U_{grav,i} = \left(-\frac{Gm_{Earth}m}{R_{Earth} + y_f}\right) - \left(-\frac{Gm_{Earth}m}{R_{Earth} + y_i}\right)$$
$$= Gm_{Earth}m\left(\frac{1}{R_{Earth} + y_i} - \frac{1}{R_{Earth} + y_f}\right)$$

If we express this in terms of a common denominator, you should be able to show that we get

$$\Delta U_{grav} = m\left[\frac{Gm_{Earth}}{(R_{Earth} + y_i)(R_{Earth} + y_f)}\right](y_f - y_i) \qquad (10\text{-}5)$$

This looks like a horrible mess! But if the heights y_i and y_f are both small compared to Earth's radius R_{Earth}—that is, if the object of mass m remains close to Earth's surface—then the quantities $R_{Earth} + y_i$ and $R_{Earth} + y_f$ are both approximately equal to R_{Earth}. In that case the quantity in square brackets in Equation 10-5 becomes Gm_{Earth}/R_{Earth}^2, which we found in the previous section is equal to the value of g at Earth's surface (see Equation 10-3). Then Equation 10-5 becomes

$$\Delta U_{grav} = U_{grav,f} - U_{grav,i}$$
$$= mg(y_f - y_i) = mgy_f - mgy_i$$

That's precisely the result we would have obtained using $U_{grav} = mgy$. In other words if we restrict ourselves to motion of the object near the surface of Earth, we can safely use the expression from Section 6-6 for gravitational potential energy.

Gravitational Potential Energy and Conservation of Total Mechanical Energy

If only the gravitational force does work, the total mechanical energy E—the sum of the kinetic energy K and the gravitational potential energy U_{grav} given by Equation 10-4—is *conserved*. An example is a rocket that's launched from Earth's surface with a certain initial speed. Earth is so massive compared to the rocket that we can assume that there is no effect on our planet. Then, if we ignore the effects of air resistance during the rocket's initial climb through the atmosphere, only gravity does work on the rocket, and the total mechanical energy of the Earth-rocket system is conserved. The following example demonstrates how to use this idea.

EXAMPLE 10-5 How High Does It Go?

You launch a rocket of mass 2.40×10^4 kg straight upward from Earth's surface. The rocket's engines burn for a short time, giving the rocket an initial speed of 9.00 km/s (32,400 km/h or 20,100 mi/h), then shut off (Figure 10-9). To what maximum height above Earth's surface will this rocket rise? Earth's mass is 5.97×10^{24} kg, and its radius is 6370 km = 6.37×10^6 m.

U.S. Air Force/Science Source

Figure 10-9 A rocket launch What is the relationship between the speed imparted to the rocket at launch and the maximum height that it reaches?

Set Up

We'll ignore the force of air resistance on the rocket as it ascends through the atmosphere (this force, like kinetic friction, is nonconservative). Then Earth's gravity is the only force that does work on the rocket, and total mechanical energy is conserved.

With such a tremendous launch speed, we expect the rocket to reach a very high altitude. So the rocket will not remain close to Earth's surface, and we must use the more general form for gravitational potential energy given by Equation 10-4.

Initially the rocket has speed v_i = 9.00 km/s = 9.00×10^3 m/s and is a distance r_i from Earth's center equal to the radius of Earth. At its maximum height the rocket is at rest ($v_f = 0$) and is a distance h above the surface. Our goal is to determine h.

Total mechanical energy is conserved:

$$K_i + U_{grav,i} = K_f + U_{grav,f} \quad (6\text{-}23)$$

Gravitational potential energy:

$$U_{grav} = -\frac{Gm_{Earth}m_{rocket}}{r} \quad (10\text{-}4)$$

Kinetic energy of rocket:

$$K = \frac{1}{2}m_{rocket}v^2 \quad (6\text{-}8)$$

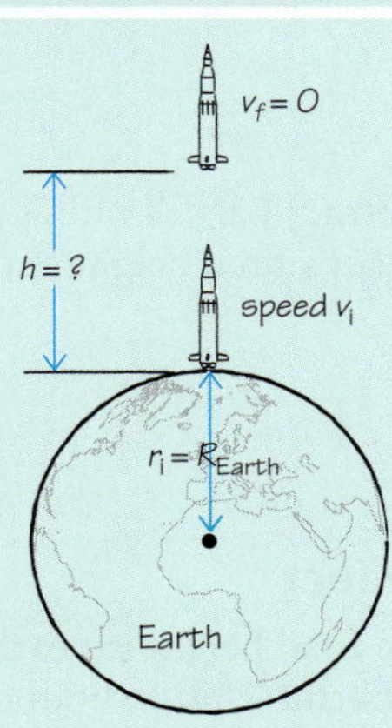

Solve

We first determine the total mechanical energy of the system using the given values for the rocket's initial speed and position.

When the rocket is initially launched at the surface, its kinetic energy is

$$K_i = \frac{1}{2}m_{rocket}v_i^2 = \frac{1}{2}(2.40 \times 10^4 \text{ kg})(9.00 \times 10^3 \text{ m/s})^2$$
$$= 9.72 \times 10^{11} \text{ kg} \cdot \text{m}^2/\text{s}^2 = 9.72 \times 10^{11} \text{ J}$$

(Recall that $1 \text{ J} = 1 \text{ kg} \cdot \text{m}^2/\text{s}^2$.)

The gravitational potential energy is

$$U_{grav,i} = -\frac{Gm_{Earth}m_{rocket}}{r_i} = -\frac{Gm_{Earth}m_{rocket}}{R_{Earth}}$$
$$= -\frac{(6.67 \times 10^{-11} \text{ N} \cdot \text{m}^2/\text{kg}^2)(5.97 \times 10^{24} \text{ kg})(2.40 \times 10^4 \text{ kg})}{6.37 \times 10^6 \text{ m}}$$
$$= -1.50 \times 10^{12} \text{ N} \cdot \text{m} = -1.50 \times 10^{12} \text{ J}$$

(Recall that $1 \text{ J} = 1 \text{ N} \cdot \text{m}$.)

The total mechanical energy is

$$E_i = K_i + U_{grav,i} = (9.72 \times 10^{11}\ J) + (-1.50 \times 10^{12}\ J) = -5.28 \times 10^{11}\ J$$

When the rocket is at the high point of its trajectory, it is momentarily at rest, and its kinetic energy is zero. At this point the (conserved) total mechanical energy is equal to the gravitational potential energy. Use this to solve for the rocket's distance r_f from Earth's center at its high point.

At the high point of the trajectory, rocket speed is $v_f = 0$, and its kinetic energy is

$$K_f = \frac{1}{2} m_{rocket} v_f^2 = 0$$

The total mechanical energy has the same value as when the rocket was launched:

$$E_f = K_f + U_{grav,f} = E_i = -5.28 \times 10^{11}\ J$$

Since $K_f = 0$, this means that

$$U_{grav,f} = -\frac{Gm_{Earth} m_{rocket}}{r_f} = E_i = -5.28 \times 10^{11}\ J$$

Solve for r_f: $r_f = -\dfrac{Gm_{Earth} m_{rocket}}{E_i}$

$$= -\frac{(6.67 \times 10^{-11}\ N \cdot m^2/kg^2)(5.97 \times 10^{24}\ kg)(2.40 \times 10^4\ kg)}{(-5.28 \times 10^{11}\ J)}$$

$$= 1.81 \times 10^7\ m = 1.81 \times 10^4\ km$$

Subtract Earth's radius from r_f to find the rocket's final height above the surface.

The rocket's final distance from Earth's center, r_f, equals the radius of Earth (R_{Earth}) plus the rocket's final height h above Earth's surface:

$$r_f = R_{Earth} + h$$

Solve for h:

$$h = r_f - R_{Earth} = 1.81 \times 10^4\ km - 6.37 \times 10^3\ km = 1.17 \times 10^4\ km$$

Reflect

The final height is nearly twice the radius of Earth! This justifies our decision to use Equation 10-4 for gravitational potential energy. Had we used the expression from Chapter 6, $U_{grav} = m_{rocket}gy$, we would have gotten an incorrect answer because that expression assumes that g has the same value at all heights. In fact g decreases substantially in value at greater distances from Earth, which means that the rocket climbs much higher than the expression $U_{grav} = m_{rocket}gy$ would predict.

From Example 10-3 (Section 10-2) the value of g at a distance r from Earth's center is

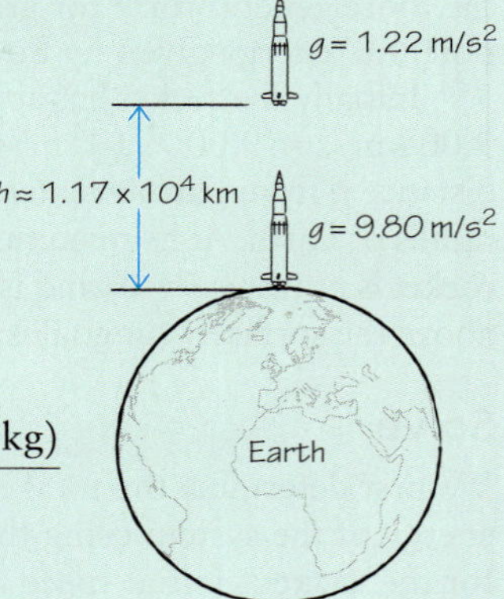

$$g = \frac{Gm_{Earth}}{r^2}$$

At the rocket's maximum distance

$$g = \frac{Gm_{Earth}}{r_f^2}$$

$$= \frac{(6.67 \times 10^{-11}\ N \cdot m^2/kg^2)(5.97 \times 10^{24}\ kg)}{(1.81 \times 10^7\ m)^2}$$

$$= 1.22\ m/s^2$$

which is *much* less than $g = 9.80\ m/s^2$ at Earth's surface.

A common *incorrect* way to calculate h would be to use $U_{grav} = m_{rocket}gy$, which is based on the (false!) assumption that $g = 9.80\ m/s^2$ at all altitudes. Then conservation of total mechanical energy would have told us that

$$\frac{1}{2} m_{rocket} v_i^2 + m_{rocket} g y_i = \frac{1}{2} m_{rocket} v_f^2 + m_{rocket} g y_f$$

With $y_i = 0$ (rocket at surface), $y_f = h$ (rocket at maximum height), and $v_f = 0$ (rocket momentarily at rest at maximum height), this becomes

$$\frac{1}{2}m_{\text{rocket}}v_i^2 + 0 = 0 + m_{\text{rocket}}gh$$

$$h = \frac{v_i^2}{2g} = \frac{(9.00 \times 10^3 \text{ m/s})^2}{2(9.80 \text{ m/s}^2)} = 4.13 \times 10^6 \text{ m} = 4.13 \times 10^3 \text{ km}$$

(Incorrect answer!)

The actual height that the rocket reaches is almost three times higher, 1.17×10^4 km, because in fact the value of g decreases with altitude.

Escape Speed

Like a ball thrown vertically upward, the rocket in Example 10-5 reaches a peak height and eventually returns to the surface. If the rocket is launched with a great enough speed, however, it will *never* return. That's because the gravitational force that Earth exerts on the rocket decreases with increasing distance from Earth and approaches zero as the rocket moves to infinitely great distances (see Equation 10-2 and Figure 10-4b). As in Example 10-5, the rocket engines give it a certain initial speed then shut off. The rocket then coasts upward and loses speed due to Earth's gravity. As the rocket climbs, the gravitational force on the rocket decreases, so the rocket loses an ever-smaller amount of speed per second. If the initial speed of the rocket is great enough, there will still be some speed remaining when the rocket is very far from Earth, and it will keep on going forever. We say that the rocket has *escaped* from Earth.

WATCH OUT! You can never be "beyond the pull of Earth's gravity."

There's a common misconception that if a rocket travels far enough from Earth, it reaches a point where it's "beyond" the pull of our planet's gravity. But Equation 10-2 shows that in fact the gravitational force that Earth exerts on an object of mass m, $F = Gm_{\text{Earth}}m/r^2$, becomes zero only when the distance r from Earth to the object is *infinity*. The force gets progressively weaker as the object moves farther away from Earth, but it never entirely disappears. A rocket can escape from Earth if it has enough speed to overcome the attraction of Earth's gravity and travel to infinity. But the rocket still feels that attraction at all points along its infinite voyage.

We can find the minimum speed for a projectile to escape from Earth by using Equation 10-4 for gravitational potential energy. As in Example 10-5, we'll ignore the effects of air resistance during the time that the projectile is passing through Earth's atmosphere. The projectile (mass m) is launched from Earth's surface (that is, from a distance $r_i = R_{\text{Earth}}$ from Earth's center) with a speed v_i. The total mechanical energy just as the projectile is launched is then

$$E_i = K_i + U_{\text{grav,i}} = \frac{1}{2}mv_i^2 + \left(-\frac{Gm_{\text{Earth}}m}{R_{\text{Earth}}}\right) \tag{10-6}$$

When the projectile is a distance r from Earth's center and moving at a speed v, the total mechanical energy is

$$E = K + U_{\text{grav}} = \frac{1}{2}mv^2 + \left(-\frac{Gm_{\text{Earth}}m}{r}\right) \tag{10-7}$$

Since total mechanical energy is conserved, E_i as given by Equation 10-6 is equal to E as given by Equation 10-7:

$$\frac{1}{2}mv_i^2 + \left(-\frac{Gm_{\text{Earth}}m}{R_{\text{Earth}}}\right) = \frac{1}{2}mv^2 + \left(-\frac{Gm_{\text{Earth}}m}{r}\right) \tag{10-8}$$

Now let's suppose that the projectile just barely escapes Earth, so its speed is zero when it is infinitely far away. That corresponds to replacing v with zero and r with infinity in Equation 10-8. The reciprocal of infinity is zero, so in this case both terms on the right-hand side of Equation 10-8 are zero: The projectile ends up with zero kinetic energy (it is at rest), and the gravitational potential energy ends up with a zero value (the projectile is infinitely far from Earth; see Figure 10-8d). Then Equation 10-8 becomes

$$\frac{1}{2}mv_i^2 + \left(-\frac{Gm_{\text{Earth}}m}{R_{\text{Earth}}}\right) = 0 \quad (10\text{-}9)$$

The initial speed v_i in Equation 10-9 is called the **escape speed**, which we denote by the symbol v_{escape}. This is the minimum speed at which an object must be launched from Earth's surface to escape to infinity. To find its value replace v_i in Equation 10-9 with v_{escape}, add $Gm_{\text{Earth}}m/R_{\text{Earth}}$ to both sides, and multiply both sides by $2/m$.

$$\frac{1}{2}mv_{\text{escape}}^2 = \frac{Gm_{\text{Earth}}m}{R_{\text{Earth}}}$$

$$v_{\text{escape}}^2 = \frac{2Gm_{\text{Earth}}}{R_{\text{Earth}}}$$

Finally, take the square root of both sides:

Escape speed (10-10)

$$v_{\text{escape}} = \sqrt{\frac{2Gm_{\text{Earth}}}{R_{\text{Earth}}}}$$

Speed that a projectile must have at Earth's surface in order to **escape Earth**

Gravitational constant

Mass of Earth

Radius of Earth

If you substitute $G = 6.67 \times 10^{-11}\ \text{N}\cdot\text{m}^2/\text{kg}^2$, $m_{\text{Earth}} = 5.97 \times 10^{24}$ kg, and $R_{\text{Earth}} = 6.37 \times 10^6$ m into Equation 10-10, you'll find that $v_{\text{escape}} = 1.12 \times 10^4$ m/s = 11.2 km/s (about 40,300 km/h or 25,000 mi/h). Rockets designed to send spacecraft to other planets must have powerful rocket engines (see Figure 10-9) in order to accelerate their payload to such high speeds.

Notice that the launch speed of the rocket in Example 10-5 was only 9.00 km/s, which is less than escape speed. That's why the rocket in that example did *not* escape but reached a maximum distance from Earth before falling back. To have the rocket escape it would have to be launched at 11.2 km/s or faster.

We can also use Equation 10-10 to find the escape speed from *any* planet or satellite: Just replace m_{Earth} and R_{Earth} with the mass and radius of the object from which the projectile is launched.

EXAMPLE 10-6 Escaping from a Martian Moon

You have landed your spacecraft on Phobos, the larger of the two airless moons of Mars (Figure 10-10). Phobos is roughly spherical with an average radius of 11.1 km = 1.11×10^4 m. By dropping a baseball while standing on Phobos, you find that g has a very small value at the moon's surface, only 0.00580 m/s^2. At what speed would you have to throw the baseball to have it escape Phobos?

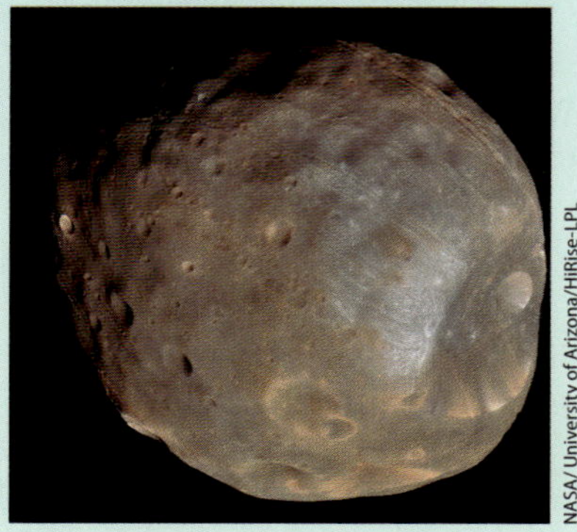

Figure 10-10 Phobos What is the escape speed from this miniature moon of Mars, just 11.1 km in radius?

Set Up

To find the escape speed v_{escape} using Equation 10-10, we need the radius of Phobos, which we are given, and the mass of Phobos, which we are *not* given. However, we recall from Equation 10-3 in Section 10-2 that the value of g on Earth's surface is related to Earth's mass and radius. We can write this same equation for Phobos, which will let us use the measured value of g on Phobos to determine its mass.

Escape speed from Phobos:

$$v_{\text{escape}} = \sqrt{\frac{2Gm_{\text{Phobos}}}{R_{\text{Phobos}}}} \qquad (10\text{-}10)$$

Value of g at the surface of Phobos:

$$g = \frac{Gm_{\text{Phobos}}}{R_{\text{Phobos}}^2} \qquad (10\text{-}3)$$

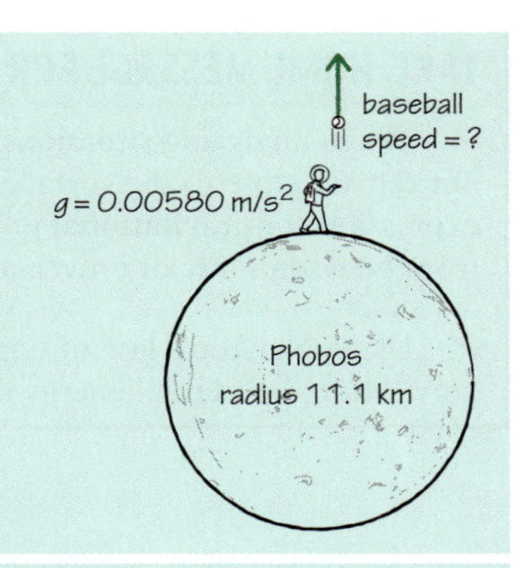

Solve

First use Equation 10-3 to determine the mass of Phobos.

From Equation 10-3 $g = \dfrac{Gm_{\text{Phobos}}}{R_{\text{Phobos}}^2}$

Solve this for the mass of Phobos:

$$m_{\text{Phobos}} = \frac{gR_{\text{Phobos}}^2}{G} = \frac{(0.00580 \text{ m/s}^2)(1.11 \times 10^4 \text{ m})^2}{6.67 \times 10^{-11} \text{ N} \cdot \text{m}^2/\text{kg}^2}$$

$$= 1.07 \times 10^{16} \frac{\text{kg}^2 \cdot \text{m/s}^2}{\text{N}} = 1.07 \times 10^{16} \text{ kg}$$

(Recall that $1 \text{ N} = 1 \text{ kg} \cdot \text{m/s}^2$.)

Then use Equation 10-10 to find the escape speed from Phobos.

Escape speed:

$$v_{\text{escape}} = \sqrt{\frac{2Gm_{\text{Phobos}}}{R_{\text{Phobos}}}}$$

$$= \sqrt{\frac{2(6.67 \times 10^{-11} \text{ N} \cdot \text{m}^2/\text{kg}^2)(1.07 \times 10^{16} \text{ kg})}{1.10 \times 10^4 \text{ m}}}$$

$$= \sqrt{129 \frac{\text{N} \cdot \text{m}}{\text{kg}}} = \sqrt{129 \frac{\text{kg} \cdot \text{m}}{\text{s}^2} \frac{\text{m}}{\text{kg}}} = \sqrt{129 \frac{\text{m}^2}{\text{s}^2}}$$

$$= 11.3 \text{ m/s} = 40.9 \text{ km/h} = 25.4 \text{ mi/h}$$

(We once again used $1 \text{ N} = 1 \text{ kg} \cdot \text{m/s}^2$.)

Reflect

The escape speed on Phobos is almost exactly 1/1000 the escape speed on Earth! In general the smaller the planet or moon, the lower the escape speed.

A typical high school baseball pitcher can easily throw a baseball at more than 25 m/s (90 km/h or 56 mi/h), so a well-trained astronaut should have no problem throwing a baseball at the 11.3-m/s escape speed on Phobos.

GOT THE CONCEPT? 10-3 Which Direction for Escape?

? In Example 10-6 suppose you throw the baseball with a speed faster than 11.3 m/s. Figure 10-11 shows four possible directions in which you could throw the ball at that speed. In which of these directions would the ball escape from Phobos? (a) Direction A only; (b) direction A or B; (c) direction A, B, or C; (d) direction A, B, C, or D; (e) none of these.

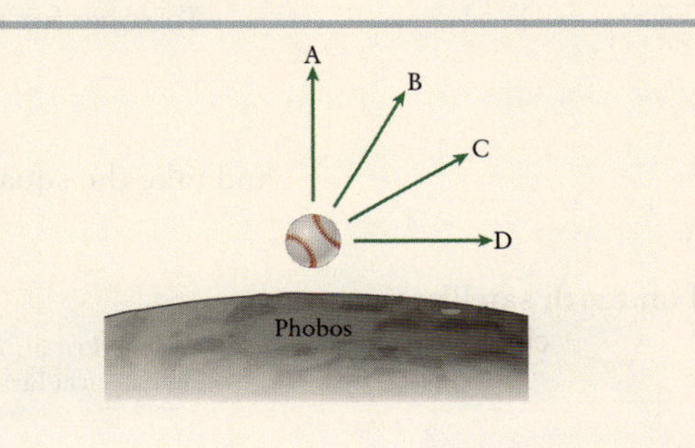

Figure 10-11 Four directions for escaping Phobos If a ball is thrown at escape speed, in what direction(s) can it be thrown so that it *will* escape?

TAKE-HOME MESSAGE FOR Section 10-3

✔ When analyzing situations in which an object does not remain close to the surface of Earth, you must use the expression for gravitational potential energy that is derived from Newton's law of universal gravitation.

✔ Using Newton's law of universal gravitation, the gravitational potential energy of two objects is always negative or zero. It is zero when the two objects are infinitely far apart, and it becomes more and more negative as the objects move closer together.

✔ Escape speed is the speed at which a projectile must be launched from a planet or moon so that it never falls back to the world from which it was launched.

10-4 Newton's law of universal gravitation explains Kepler's laws for the orbits of planets and satellites

We saw in Section 10-2 that Newton deduced the law of universal gravitation from the properties of the Moon's orbit around Earth. Because gravitation is universal, the same principles apply to the orbit of *any* celestial object around another.

Circular Orbits: Orbital Speed and Orbital Period

The simplest type of orbit to analyze is a circular orbit. Many Earth satellites, including the International Space Station (see the photograph that opens this chapter) and the satellites that provide signals for Global Positioning System (GPS) navigation, are in circular or nearly circular orbits.

Although the first Earth satellite was put in orbit in 1957, Newton understood what was required to put a satellite in a circular orbit almost three centuries earlier (Figure 10-12). If a cannonball is dropped from a great height such as the top of a mountain, it will fall straight down. If it is thrown horizontally at a moderate speed, the cannonball will follow a curved arc before hitting the ground. But if it is thrown horizontally with just the right speed, the surface of Earth will fall away below the cannonball so that the cannonball always remains at the same height above the surface. Put another way, Earth's gravitational attraction will cause the cannonball to accelerate toward Earth's center. If the cannonball's speed is just right, the result will be uniform circular motion of the sort that we studied in Section 3-7, in which the acceleration is always directed toward the center of the cannonball's circular path (Figure 10-13).

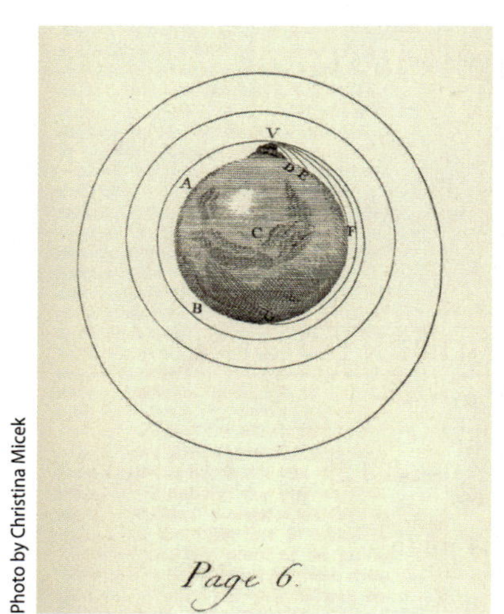

Photo by Christina Micek

Figure 10-12 Newton's recipe for an Earth satellite Isaac Newton imagined that if a cannonball were fired horizontally from the top of a mountain with sufficient speed, the rate at which it fell could be made to match the rate at which Earth's surface fell away. The cannonball would therefore end up orbiting Earth. Newton created this illustration for the same 1687 book in which he presented the law of universal gravitation.

We can find the speed v required for a circular orbit of radius r from Equation 3-17 for the acceleration in uniform circular motion: $a_{cent} = v^2/r$. For a satellite of mass m orbiting Earth, the acceleration is provided by Earth's gravitational force. From Newton's second law, this says

$$F_{\text{Earth on satellite}} = ma_{\text{cent}}$$

or, from Equation 3-17 for uniform circular motion and Equation 10-2 for the law of universal gravitation,

$$\frac{Gm_{\text{Earth}}m}{r^2} = m\frac{v^2}{r}$$

To solve for v, multiply both sides of this equation by r/m

$$\frac{Gm_{\text{Earth}}}{r} = v^2$$

and take the square root:

Speed of an Earth satellite in a circular orbit (10-11)

$$v = \sqrt{\frac{Gm_{\text{Earth}}}{r}}$$

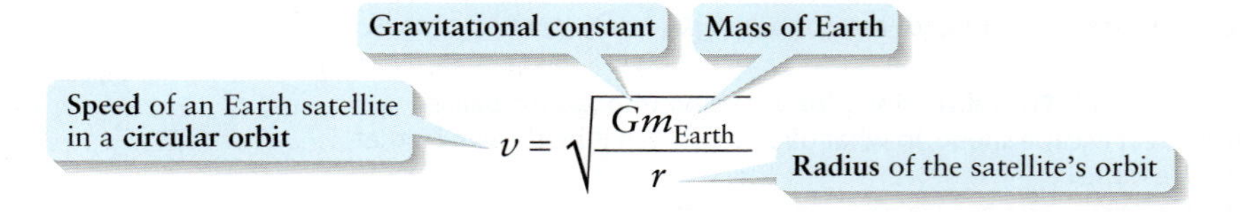

Many satellites, including the International Space Station, are in *low Earth orbit:* Their height above the surface is a few hundred kilometers, which is a short distance compared to Earth's radius of 6370 km. So we can find the approximate speed of a satellite in low Earth orbit by replacing r in Equation 10-11 with the radius of Earth and substituting the values $G = 6.67 \times 10^{-11}\ \text{N} \cdot \text{m}^2/\text{kg}^2$, $m_{\text{Earth}} = 5.97 \times 10^{24}$ kg, and $R_{\text{Earth}} = 6.37 \times 10^6$ m:

$$v = \sqrt{\frac{Gm_{\text{Earth}}}{R_{\text{Earth}}}} = 7.91 \times 10^3 \text{ m/s} = 7.91 \text{ km/s}$$

(orbital speed for low Earth orbit)

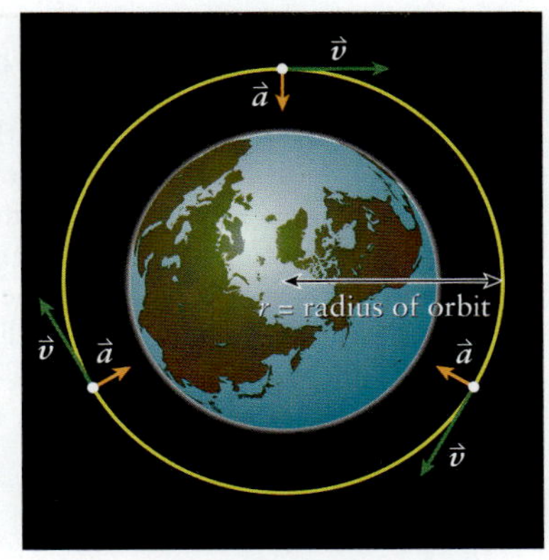

Figure 10-13 A circular orbit A satellite in a circular orbit is in uniform circular motion: The acceleration (due to Earth's gravitational force) has a constant magnitude and is always directed toward Earth's center, and the velocity has a constant magnitude.

This speed (about 28,500 km/h or 17,700 mi/h) is slower than the escape speed of 11.2 km/s (which we discussed in Section 10-3) by a factor of $1/\sqrt{2} = 0.707$.

Equation 10-11 shows that increasing the orbital radius r decreases the speed for a circular orbit. We saw an example of this in Section 10-2: The Moon orbits Earth at an average distance of 3.84×10^8 m, about 60 times Earth's radius, and its orbital speed is only 1.02×10^3 m/s $= 1.02$ km/s. You can understand why the speed decreases with increasing orbital radius by considering a ball on the end of a string. Imagine whirling the ball on a string around your hand so that the ball makes a circular orbit around your hand. To make the ball move at high speed around a small circle, you have to exert a substantial pull on the string. But if you lengthen the string and make the same ball move at low speed around a large circle, much less pull is required. Similarly, a satellite that orbits close to Earth experiences a substantial gravitational pull and so moves at high speed, while a planet in a larger orbit experiences less gravitational force and moves at a lower speed.

Another useful way to describe how rapidly a satellite moves around its circular orbit is in terms of the **orbital period**, which is the time required to complete one orbit. In uniform circular motion the speed v is constant, so the orbital period T is just the circumference $2\pi r$ of the orbit (the distance around the circular orbit of radius r) divided by the speed: $T = 2\pi r/v$. Using Equation 10-11 for v this becomes

$$T = \frac{2\pi r}{v} = 2\pi r\sqrt{\frac{r}{Gm_{\text{Earth}}}}$$

It's convenient to square both sides of this equation so as to eliminate the square root. Note that the square of r is r^2 and the square of $\sqrt{r}$ is r, so we end up with a factor of $r^2 \times r = r^3$ on the right-hand side of the equation:

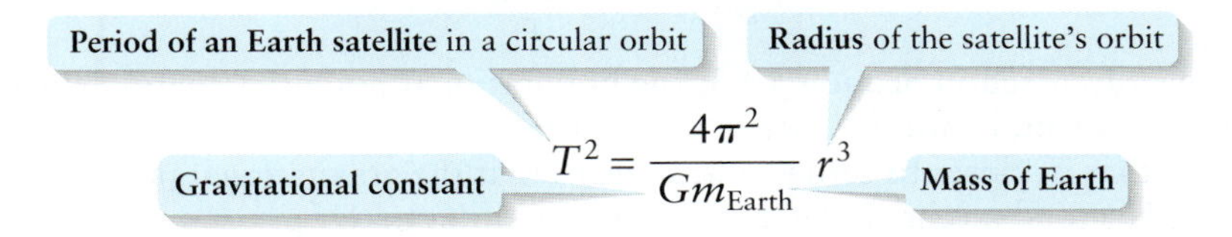

$$T^2 = \frac{4\pi^2}{Gm_{\text{Earth}}} r^3$$

Relationship between orbital period and radius for a circular orbit (10-12)

For an object in low Earth orbit, for which r equals R_{Earth}, this tells us that $T = 5.06 \times 10^3$ s $= 84.4$ min, or a little less than an hour and a half.

In words Equation 10-12 says that the *square* of the orbital period T is directly proportional to the *cube* of the radius r of the orbit. By comparison Equation 10-11 tells us that the orbital speed v is inversely proportional to the *square root* of the orbital radius r. As an example, consider a satellite that orbits at a distance $r = 4R_{\text{Earth}}$ from Earth's center (that is, at an altitude of $3R_{\text{Earth}}$ above Earth's surface). Since r is four times greater than for a satellite in low Earth orbit, the quantity r^3 is $4^3 = 4 \times 4 \times 4 = 64$ times greater. So the quantity T^2 is also 64 times greater than for an object in low Earth orbit, which means the period T is $\sqrt{64} = 8$ times longer. The orbital speed is proportional to $1/\sqrt{r}$, which is $1/\sqrt{4} = 1/2$ as great as for a satellite in low Earth orbit. These proportionality rules therefore tell us that for a satellite with $r = 4R_{\text{Earth}}$, the orbital period is 8×84.4 min $= 675$ min (11.2 hours), and the orbital speed is $(1/2) \times 7.91$ km/s $= 3.95$ km/s.

WATCH OUT! When a satellite orbits Earth, Earth moves too.

One aspect of orbital motion that we've ignored in this discussion is Newton's third law: If Earth exerts a gravitational force on a satellite, the satellite exerts an equally strong force on Earth. So Earth also moves in a small orbit. Strictly speaking Earth and the satellite both orbit around the center of mass of the Earth–satellite system. However, even the largest satellite humans have ever placed in orbit has a tiny mass compared to the mass of Earth. So for all practical purposes the center of mass of the Earth–satellite system is at Earth's center. By contrast the Moon's mass is a reasonable fraction of Earth's mass (about 1.2%), so for detailed calculations this effect must be taken into account (Figure 10-14).

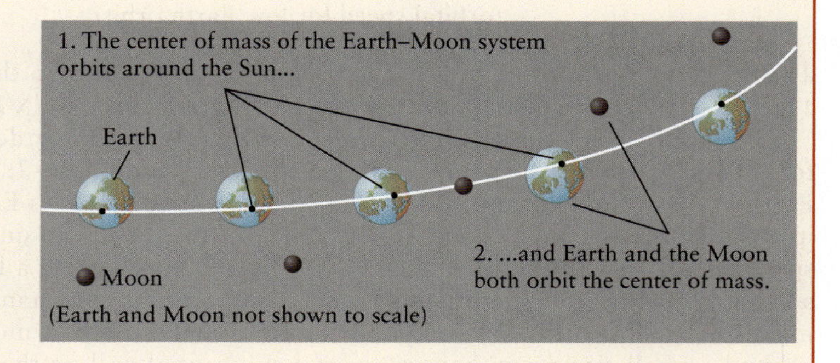
1. The center of mass of the Earth–Moon system orbits around the Sun...
Earth
Moon
2. ...and Earth and the Moon both orbit the center of mass.
(Earth and Moon not shown to scale)

Figure 10-14 The motions of Earth and Moon Because Earth and the Moon both exert gravitational forces on each other, it's not correct to say that the Moon orbits Earth. Rather, both objects move around their common center of mass.

Circular Orbits: Energy

Before placing a satellite into a circular orbit, an important question to ask is "How much energy will it take to put the satellite in the desired orbit?" The answer determines how powerful a rocket will be needed for the task. We have all the tools we need to answer this question: Equation 10-11 tells us the speed the satellite must have in its circular orbit, which allows us to determine the required kinetic energy, and Equation 10-4 tells us the gravitational potential energy. So the kinetic energy for a satellite of mass m in a circular orbit of radius r is

$$K = \frac{1}{2}mv^2 = \frac{1}{2}m\left(\sqrt{\frac{Gm_{\text{Earth}}}{r}}\right)^2 = \frac{Gm_{\text{Earth}}m}{2r}$$

Making the orbital radius r larger means the satellite moves more slowly, and the kinetic energy decreases. The gravitational potential energy is

$$U_{\text{grav}} = -\frac{Gm_{\text{Earth}}m}{r}$$

As we discussed in Section 10-3, the gravitational potential energy is negative. Making the orbital radius r larger makes the gravitational potential energy less negative—that is, closer to zero—which means that the gravitational potential energy increases. The total mechanical energy is the sum of K and U_{grav}:

$$E = K + U_{\text{grav}} = \frac{1}{2}mv^2 + \left(-\frac{Gm_{\text{Earth}}m}{r}\right) = \frac{Gm_{\text{Earth}}m}{2r} + \left(-\frac{Gm_{\text{Earth}}m}{r}\right)$$

or

Total mechanical energy for a circular orbit (10-13)

$$E = -\frac{Gm_{\text{Earth}}m}{2r}$$

Gravitational constant

Mass of Earth

Mass of satellite

Total mechanical energy for a satellite in a circular orbit around Earth

Radius of the satellite's orbit

The greater the radius r of the circular orbit, the greater (less negative) the total mechanical energy.

The total mechanical energy E is always negative because of the way we defined gravitational potential energy. A zero value of E corresponds to an orbit with an infinitely large radius r. In this limiting case the gravitational potential energy is zero, and the satellite would have zero speed and zero kinetic energy.

As an example, for a satellite in low Earth orbit (with $r = R_{\text{Earth}}$) the total mechanical energy is $E = -Gm_{\text{Earth}}m/2R_{\text{Earth}}$. Before the spacecraft was launched and was sitting on Earth's surface, its kinetic energy was zero, and E was equal to the gravitational potential energy: $E = U_{\text{grav}} = -Gm_{\text{Earth}}m/R_{\text{Earth}}$. So the amount of energy that had to be imparted to the satellite to put it into low Earth orbit was

$$\Delta E = E_{\text{in orbit}} - E_{\text{on surface}} = \left(-\frac{Gm_{\text{Earth}}m}{2R_{\text{Earth}}}\right) - \left(-\frac{Gm_{\text{Earth}}m}{R_{\text{Earth}}}\right) = +\frac{Gm_{\text{Earth}}m}{2R_{\text{Earth}}}$$

For a 1-kg satellite ($m = 1.00$ kg) $\Delta E = 3.13 \times 10^7$ J. So 3.13×10^7 J of energy has to be imparted to each kilogram of a satellite placed into low Earth orbit. In practice the energy requirements are much greater because the rocket and its fuel (which are much more massive than the satellite itself) also have to be given kinetic energy. One way to *reduce* the energy requirements is to take advantage of Earth's rotation. Every point on Earth's surface is rotating to the east (which is why we see the Sun rise in the east), and the speed in meters per second is greatest near the equator. That's why NASA and the European Space Agency launch satellites into orbit from locations relatively close to the equator (Florida and French Guiana, respectively) and launch them toward the east (Figure 10-15).

Figure 10-15 Launching toward the east This aerial photograph shows a space shuttle launch. Having climbed through the clouds, the spacecraft climbs upward and to the east (to the left in the photograph) from Florida. In this way the spacecraft takes advantage of Earth's eastward rotation, which gives the spacecraft 406 m/s of speed even before it lifts off.

WATCH OUT! When a satellite falls to Earth, it's not just gravity's fault.

Orbiting satellites do sometimes fall out of orbit and crash back to Earth. When this happens, however, the real culprit is not gravity but air resistance. A satellite in a relatively low orbit is actually flying through the tenuous outer wisps of Earth's atmosphere. Air resistance does negative work on the spacecraft, just like a kinetic friction force. This causes the total mechanical energy E to decrease and become more negative. From Equation 10-13 this means that the orbital radius r becomes smaller, and the satellite sinks to a lower altitude, where it encounters more air resistance and sinks even lower. Eventually the satellite either strikes Earth or burns up in flight due to air friction. By contrast the Moon and planets orbit in the near-vacuum of interplanetary space. Hence, they are unaffected by this kind of air resistance, and they have remained in orbit around the Sun since the solar system formed 4.56 billion years ago.

EXAMPLE 10-7 A Satellite for Satellite Television

The broadcasting satellites used in a satellite television system orbit Earth's equator with a period of exactly 24 hours, the same as Earth's rotation period. As a result, these satellites are *geostationary*: They always remain over the same spot on the equator, so they always appear to be in the same position in the sky as seen from anywhere on Earth's surface. (A satellite TV receiver "dish" is aimed to receive the signal broadcast from one of these satellites.) (a) At what distance from Earth's center must a television satellite orbit? (b) What must be its orbital speed? (c) If a television satellite has a mass of 3.50×10^3 kg, how much energy must it be given to place it in orbit?

Set Up

A geostationary satellite has $T = 24$ h; we'll use this information and Equation 10-12 to find the radius r of the orbit. Once we find the value of r, we'll use Equation 10-11 to find the orbital speed of the satellite.

Equation 10-13 tells us the total mechanical energy of the spacecraft in orbit, and Equation 10-4 will tell us the total mechanical energy when the spacecraft is at rest on Earth's surface before being launched. The difference between these two values is the energy that must be given to the satellite.

Relationship between period and radius for a circular orbit:

$$T^2 = \frac{4\pi^2}{Gm_{\text{Earth}}}r^3 \qquad (10\text{-}12)$$

Speed of an Earth satellite in a circular orbit:

$$v = \sqrt{\frac{Gm_{\text{Earth}}}{r}} \qquad (10\text{-}11)$$

Total mechanical energy for a circular orbit:

$$E = -\frac{Gm_{\text{Earth}}m}{2r} \qquad (10\text{-}13)$$

Gravitational potential energy:

$$U_{\text{grav}} = -\frac{Gm_1m_2}{r} \qquad (10\text{-}4)$$

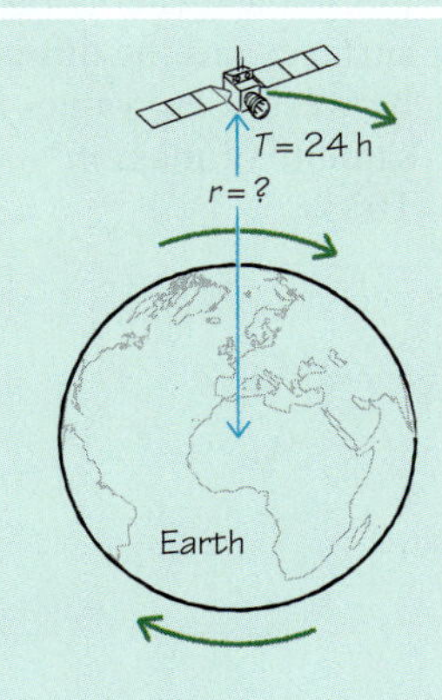

Solve

(a) To use Equation 10-12 to find the orbital radius r, first convert the orbital period T to seconds.

Period of the satellite orbit:

$$T = 24\text{ h}\left(\frac{60\text{ min}}{1\text{ h}}\right)\left(\frac{60\text{ s}}{1\text{ min}}\right) = 8.64 \times 10^4\text{ s}$$

Solve Equation 10-12 for r^3:

$$r^3 = \frac{Gm_{\text{Earth}}T^2}{4\pi^2}$$

$$= \frac{(6.67 \times 10^{-11}\text{ N}\cdot\text{m}^2/\text{kg}^2)(5.97 \times 10^{24}\text{ kg})(8.64 \times 10^4\text{ s})^2}{4\pi^2}$$

$$= 7.53 \times 10^{22}\,\frac{\text{N}\cdot\text{m}^2\cdot\text{s}^2}{\text{kg}} = 7.53 \times 10^{22}\text{ m}^3$$

(We used $1\text{ N} = 1\text{ kg}\cdot\text{m/s}^2$.)

Take the cube root to find r:

$$r = \sqrt[3]{r^3} = \sqrt[3]{7.53 \times 10^{22}\text{ m}^3}$$

$$= 4.22 \times 10^7\text{ m} = 4.22 \times 10^4\text{ km}$$

This is 6.63 times Earth's radius ($R_{\text{Earth}} = 6.37 \times 10^3$ km).

(b) Then use Equation 10-11 to find the orbital speed of the satellite.

Orbital speed:

$$v = \sqrt{\frac{Gm_{\text{Earth}}}{r}}$$

$$= \sqrt{\frac{(6.67 \times 10^{-11}\text{ N}\cdot\text{m}^2/\text{kg}^2)(5.97 \times 10^{24}\text{ kg})}{4.22 \times 10^7\text{ m}}}$$

$$= \sqrt{9.43 \times 10^6\,\frac{\text{N}\cdot\text{m}}{\text{kg}}} = \sqrt{9.43 \times 10^6\,\frac{\text{kg}\cdot\text{m}}{\text{s}^2}\,\frac{\text{m}}{\text{kg}}}$$

$$= \sqrt{9.43 \times 10^6\,\frac{\text{m}^2}{\text{s}^2}}$$

$$= 3.07 \times 10^3\text{ m/s} = 1.11 \times 10^4\text{ km/h} = 6.87 \times 10^3\text{ mi/h}$$

(Again we used $1\text{ N} = 1\text{ kg}\cdot\text{m/s}^2$.)

(c) Find the total mechanical energy in orbit, the total mechanical energy on Earth's surface, and the difference between these values for the satellite of mass $m = 3.50 \times 10^3$ kg.

When the satellite is in orbit, the total mechanical energy is

$$E_{\text{in orbit}} = -\frac{Gm_{\text{Earth}}m}{2r}$$

$$= -\frac{(6.67 \times 10^{-11}\text{ N}\cdot\text{m}^2/\text{kg}^2)(5.97 \times 10^{24}\text{ kg})(3.50 \times 10^3\text{ kg})}{2(4.22 \times 10^7\text{ m})}$$

$$= -1.65 \times 10^{10}\text{ N}\cdot\text{m}$$

$$= -1.65 \times 10^{10}\text{ J}$$

With the spacecraft at rest on Earth's surface, the total mechanical energy is just the gravitational potential energy:

$$E_{\text{on surface}} = U_{\text{grav, on surface}} = -\frac{Gm_{\text{Earth}}m}{r}$$

$$= -\frac{(6.67 \times 10^{-11}\text{ N}\cdot\text{m}^2/\text{kg}^2)(5.97 \times 10^{24}\text{ kg})(3.50 \times 10^3\text{ kg})}{6.37 \times 10^6\text{ m}}$$

$$= -2.19 \times 10^{11}\text{ N}\cdot\text{m} = -2.19 \times 10^{11}\text{ J}$$

The amount of energy that must be given to the satellite to put it into orbit is

$$E_{\text{in orbit}} - E_{\text{on surface}} = (-1.65 \times 10^{10}\text{ J}) - (-2.19 \times 10^{11}\text{ J})$$

$$= 2.02 \times 10^{11}\text{ J}$$

Reflect

Our answer to part (a) shows that geostationary broadcast satellites orbit at a tremendous distance from Earth. Part (c) shows that it takes a tremendous amount of energy to put them there.

You can check the result for part (b) by confirming that the satellite's orbital speed v equals $2\pi r$ (the circumference of the circular orbit of radius r) divided by T (the time to complete one orbit). Does it?

Kepler's Laws of Planetary Motion

Decades before Newton published his law of universal gravitation, astronomers had carefully observed and recorded the positions of the planets as they traced out their orbits. By analyzing these observations, the German mathematician and astronomer Johannes Kepler discovered three laws that summarize the motions of all of the planets.

Kepler's laws describe the shape of planetary orbits, the speed at which a planet moves along its orbit, and the time it takes a planet to complete an orbit. One of Newton's great accomplishments was to show that his law of universal gravitation, in conjunction with his laws of motion, explained *why* the planets move according to Kepler's laws. Let's look at Kepler's three laws in turn.

The first of Kepler's laws is the law of orbits:

The orbit of each planet is an ellipse with the Sun located at one focus of the ellipse.

You can draw an ellipse by using a loop of string, two thumbtacks, and a pencil as Figure 10-16a shows. Each thumbtack in the figure is at a *focus* (plural foci) of the ellipse; an ellipse has two foci. The longest diameter of an ellipse, called the *major axis*, passes through both foci. Half of that distance is called the **semimajor axis** and is usually designated by the symbol a. (Unfortunately this is also the symbol for acceleration. We promise never to use semimajor axis and acceleration in the same equation!) A circle is a special case of an ellipse in which the two foci are at the same point; this corresponds to using only a single thumbtack in Figure 10-16a. The semimajor axis of a circle is equal to its radius.

The **eccentricity** (symbol e) of an ellipse describes how elongated the ellipse is. The value of e can range from 0 (a circle) to just under 1 (nearly a straight line). The greater the eccentricity, the more elongated the ellipse (Figure 10-16b). All of the objects that

(a) The geometry of an ellipse

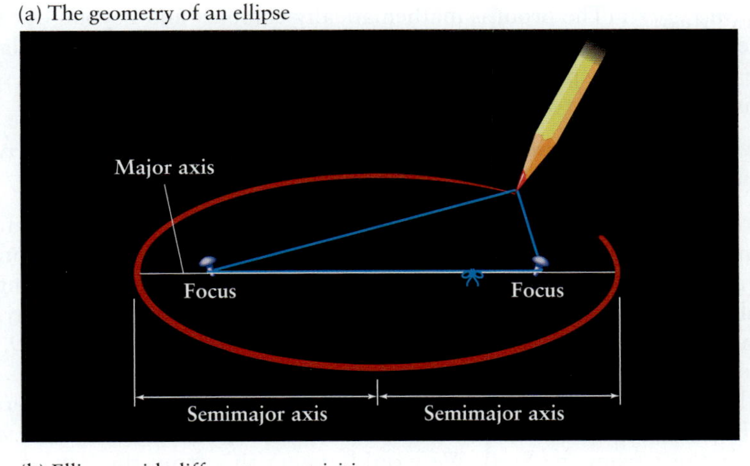

(b) Ellipses with different eccentricities

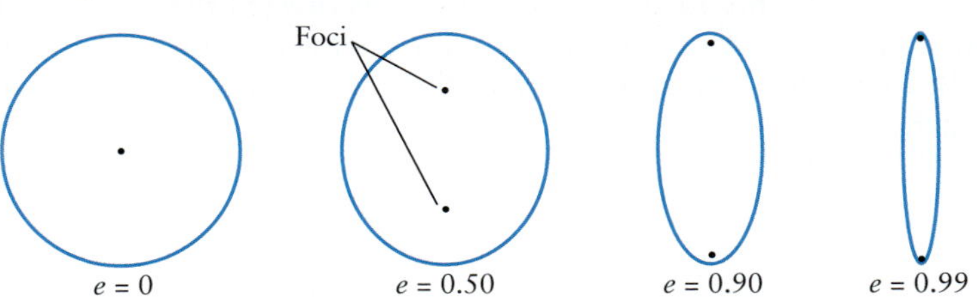

Figure 10-16 Ellipses (a) To draw an ellipse, use two thumbtacks to secure the ends of a piece of string then use a pencil to pull the string taut. If you move the pencil while keeping the string taut, the pencil traces out an ellipse. The thumbtacks are located at the two foci of the ellipse. The major axis is the greatest distance across the ellipse; the semimajor axis is half of this distance. (b) These ellipses have the same major axis but different eccentricities. An ellipse can have any eccentricity from $e = 0$ (a circle) to just under $e = 1$ (virtually a straight line).

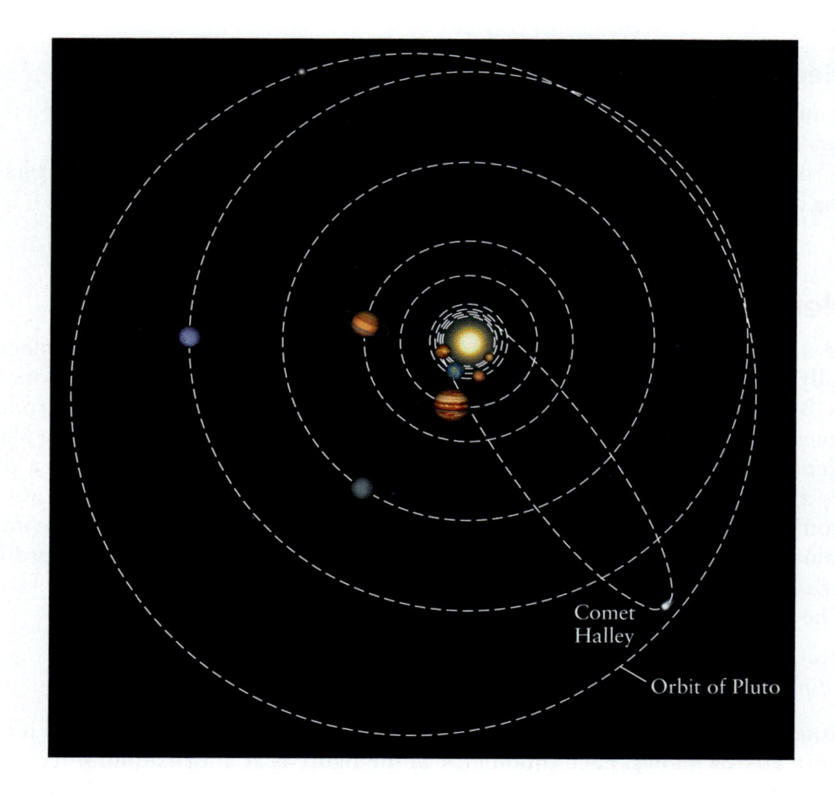

Figure 10-17 Orbits in the solar system The orbits of the planets and the minor planet Pluto are ellipses with relatively small eccentricities. Comet Halley, which moves around its elliptical orbit once every 76 years, has an eccentricity of 0.967.

orbit the Sun have orbits that are at least slightly elliptical, with e greater than zero. The most circular of any planetary orbit is that of Venus, with an eccentricity e of just 0.007; Mercury's orbit has $e = 0.206$. The minor planet Pluto has an orbit with $e = 0.249$, and a number of small bodies called comets move in very elongated orbits with eccentricities just less than 1 (Figure 10-17). For any elliptical orbit, the Sun is at one focus; there is nothing at the other focus.

One of Newton's triumphs was to show that the planets can have elliptical orbits only if the force F attracting them toward the Sun is in inverse proportion to the square of the distance r from Sun to planet. In other words the elliptical shape of the planets is a verification of the law of universal gravitation, which states that $F = Gm_{\text{sun}}m_{\text{planet}}/r^2$. (The proof is mathematically complex and beyond our scope.)

Kepler's second law states that unlike the case of a circular orbit, a planet does *not* move at a constant speed on an elliptical orbit. His law of areas describes how the speed changes around the orbit:

> *A line joining the Sun and a planet sweeps out equal areas in equal intervals of time, regardless of the position of the planet in the orbit.*

Figure 10-18 illustrates this law. Suppose that it takes 30 days for a planet to go from point A to point B. During that time an imaginary line joining the Sun and the planet sweeps out a nearly triangular area. Kepler discovered that a line joining the Sun and the planet also sweeps out exactly the same area during any other 30-day interval. In other words, if the planet also takes 30 days to go from point C to point D, then the two shaded segments in Figure 10-18 are equal in area.

The law of areas tells us that the planet moves fastest at *perihelion* (the point in its orbit closest to the Sun) and slowest at *aphelion* (the point when the planet is farthest from the Sun). We would expect this from the law of universal gravitation: A planet should speed up as it moves toward the Sun and approaches perihelion, and it should slow down as it moves away from the Sun and toward aphelion. But why does the speed vary in the particular manner described by the law of areas?

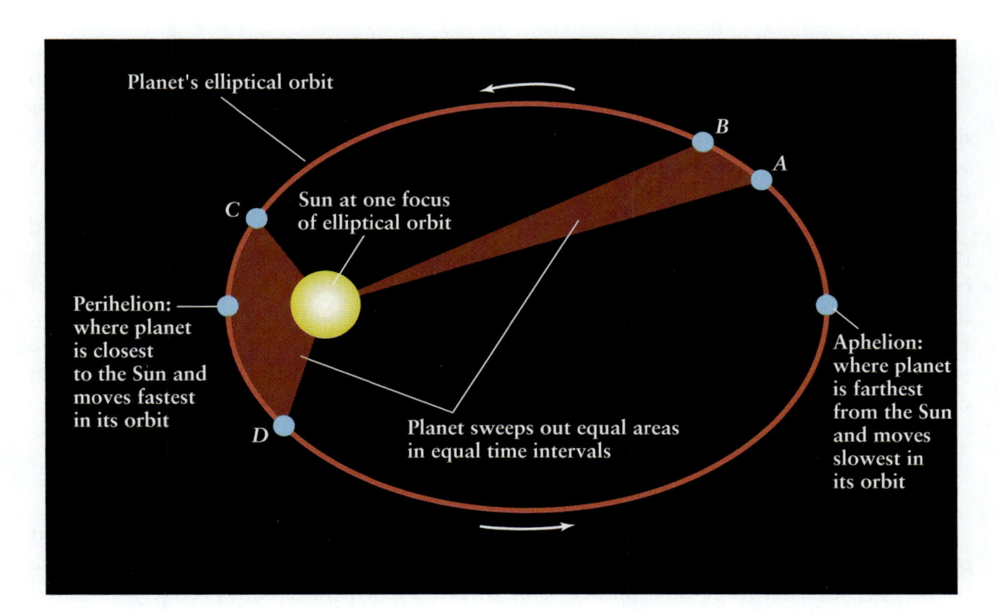
Planet's elliptical orbit
B
A
C
Sun at one focus of elliptical orbit
Perihelion: where planet is closest to the Sun and moves fastest in its orbit
D
Planet sweeps out equal areas in equal time intervals
Aphelion: where planet is farthest from the Sun and moves slowest in its orbit

Figure 10-18 Kepler's law of orbits and law of areas According to Kepler's law of orbits, a planet travels around the Sun along an elliptical orbit with the Sun at one focus. (There is nothing at the other focus.) According to his law of areas, as the planet moves an imaginary line joining the planet and the Sun sweeps out equal areas in equal intervals of time (from A to B or from C to D). By using these laws in his calculations, Kepler found a perfect fit to the apparent motions of the planets.

To answer this question let's consider how a planet moves during a very short time interval Δt. Figure 10-19 shows that if Δt is short, an imaginary line from the Sun to the planet moves through an angle $\Delta\theta$ and sweeps out an area $\Delta A = \frac{1}{2}r^2\Delta\theta$, where r is the distance from Sun to planet. Hence the *rate* at which this line sweeps out area is ΔA divided by Δt, or

$$\text{rate at which area is swept out} = \frac{\Delta A}{\Delta t} = \frac{1}{2}r^2\frac{\Delta\theta}{\Delta t} = \frac{1}{2}r^2\omega$$

See the Math Tutorial for more information on geometry.

In this expression $\omega = \Delta\theta/\Delta t$ is the *angular speed* of the planet around the Sun (see Section 8-2). Kepler's law of areas says that $\Delta A/\Delta t$ has the same value at all points along the orbit, so

$$\frac{1}{2}r^2\omega = \text{constant (Kepler's law of areas)} \tag{10-14}$$

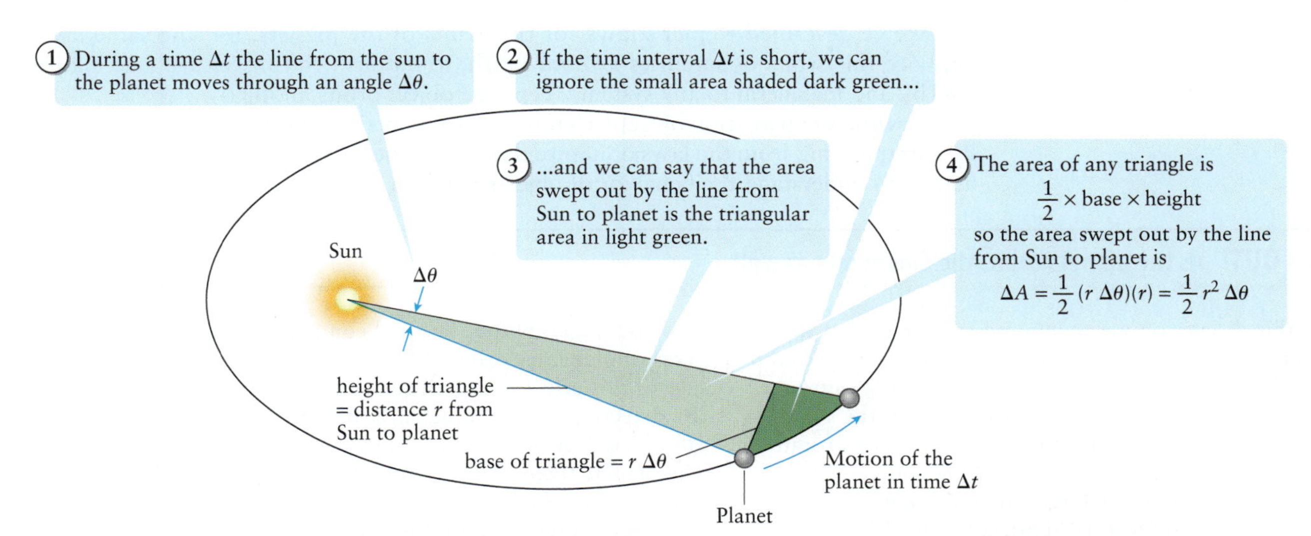
1 During a time Δt the line from the sun to the planet moves through an angle $\Delta\theta$.
2 If the time interval Δt is short, we can ignore the small area shaded dark green...
3 ...and we can say that the area swept out by the line from Sun to planet is the triangular area in light green.
4 The area of any triangle is $\frac{1}{2}$ × base × height so the area swept out by the line from Sun to planet is $\Delta A = \frac{1}{2}(r\,\Delta\theta)(r) = \frac{1}{2}r^2\,\Delta\theta$
Sun
$\Delta\theta$
height of triangle = distance r from Sun to planet
base of triangle = $r\,\Delta\theta$
Motion of the planet in time Δt
Planet

Figure 10-19 Explaining Kepler's law of areas As a planet moves around its elliptical orbit, a line from the Sun to the planet sweeps out the same area $\Delta A = \frac{1}{2}r^2\Delta\theta$ in any time interval Δt. This turns out to be equivalent to the statement that the planet's angular momentum is conserved.

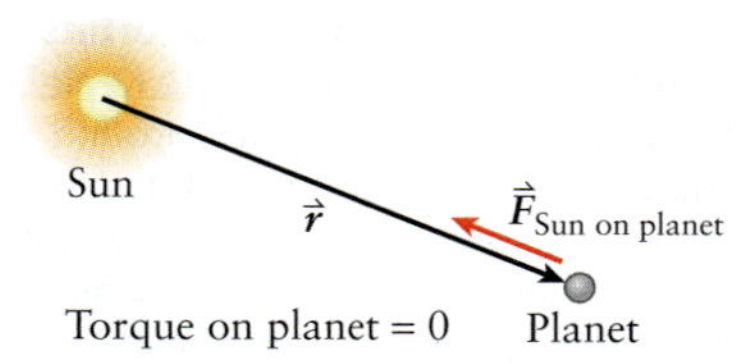

Figure 10-20 The Sun's gravity produces zero torque The gravitational force of the Sun on a planet is directed toward the Sun. Hence this force produces no torque around the Sun, so the planet's angular momentum remains constant.

Now that we have the law of areas in equation form, we can understand how it arises. Figure 10-20 shows that the gravitational force $\vec{F}$ exerted on a planet by the Sun is always directed opposite to the vector that points from the Sun to the planet. This means that the gravitational force exerts zero *torque* on the planet (see Section 8-6, where we learned that a force that acts either in the direction of $\vec{r}$ or opposite to $\vec{r}$—as is the case in Figure 10-20—produces zero torque).

We learned in Section 8-8 that the *angular momentum* of a system is conserved (that is, remains constant) if there is no net torque on it. Hence the angular momentum of a planet is conserved as it orbits the Sun. For a rotating object with moment of inertia I and angular speed ω, the magnitude of the angular momentum is $L = I\omega$. Furthermore, the moment of inertia of an object of mass m at a distance r from the rotation axis is $I = mr^2$ (see Section 8-2). So the statement that the angular momentum of an orbiting planet is conserved is

$$L = mr^2\omega = \text{constant} \tag{10-15}$$

(angular momentum of a planet is conserved)

If we divide Equation 10-15 by the planet's mass m (which does not change as it orbits) and also divide by 2, we get Equation 10-14. So the law of areas is a consequence of the gravitational force on a planet being directed toward the Sun, precisely as stated in the law of universal gravitation (see Figure 10-4a).

The first two of Kepler's laws describe how a given planet moves in its orbit. His third law, the law of periods, compares the orbital periods of orbits of different sizes:

> *The square of the period of a planet's orbit is proportional to the cube of the semimajor axis of the orbit.*

This is precisely the relationship that we found for *circular* orbits in Equation 10-12. Newton was able to show that the law of periods follows from the law of universal gravitation even for elliptical orbits, if we replace the orbital radius r in Equation 10-12 with the semimajor axis a. For objects orbiting the Sun we must also replace Earth's mass m_{Earth} with the mass of the Sun:

Newton's form of Kepler's law of periods (10-16)

Period of an object in an elliptical orbit **around the Sun**
Semimajor axis of the object's orbit

$$T^2 = \frac{4\pi^2}{Gm_{\text{Sun}}}a^3 \tag{10-16}$$

Gravitational constant
Mass of the Sun

We've described Kepler's laws for the orbits of the planets. But the same laws apply to the orbits of satellites around Earth, to moons orbiting Saturn (see Figure 10-1b), and in general to any system where one object orbits another.

Newton's explanation of Kepler's three laws was perhaps his ultimate achievement. By extrapolating from the laws of nature that he saw here on Earth, he became the first human to understand the behavior of objects in the heavens.

WATCH OUT! As the planets orbit, the Sun moves as well.

! In this discussion we've ignored the forces that the planets exert on the Sun, and we have assumed that the Sun remains at rest. In fact these forces cause the Sun to "wobble" around the center of mass of the solar system in the same way the Moon causes Earth to rotate around the center of mass of the Earth-Moon system. Although this "wobble" is larger in amplitude than the radius of the Sun (6.96×10^5 km), we've neglected it because the amplitude is small compared to the semimajor axes of planetary orbits. However, astronomers have made use of this effect to search for planets orbiting other stars. By detecting the "wobble" of a star, astronomers have been able to detect the presence of that star's planets even though the planets themselves are too faint to be seen with even the most powerful telescopes. Hundreds of *extrasolar planets* have been discovered in this way.

The following example demonstrates one application of Equation 10-16.

EXAMPLE 10-8 A Comet's Orbit

Distances in the solar system are typically measured not in meters or kilometers but in *astronomical units* (au), where 1 au is the semimajor axis of Earth's orbit around the Sun (1 au = 1.50×10^8 km). Earth's orbital period around the Sun is one year. Find the orbital period of a comet that is 0.50 au from the Sun at perihelion and 17.50 au from the Sun at aphelion.

Set Up

We'll use the data about the comet to find its semimajor axis a. From this we'll use Newton's form of Kepler's law of periods to determine the period T.

Newton's form of Kepler's law of periods:

$$T^2 = \frac{4\pi^2}{Gm_{\text{Sun}}}a^3 \qquad (10\text{-}16)$$

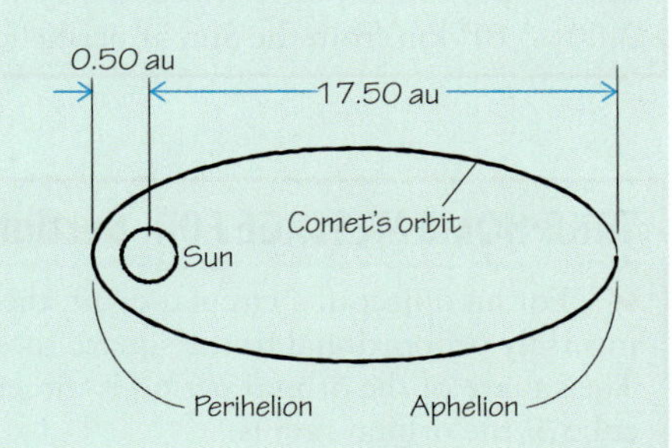

Solve

The drawing shows that the length of the major axis of the comet's orbit is the sum of the perihelion and aphelion distances. The semimajor axis a is one-half the length of the major axis.

Length of the major axis: (distance from Sun to comet at perihelion) + (distance from Sun to comet at aphelion)

$= 0.50 \text{ au} + 17.50 \text{ au} = 18.0 \text{ au}$

Semimajor axis:

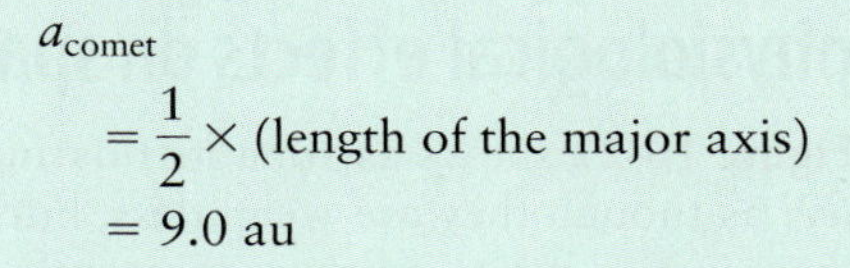

$$a_{\text{comet}} = \frac{1}{2} \times (\text{length of the major axis}) = 9.0 \text{ au}$$

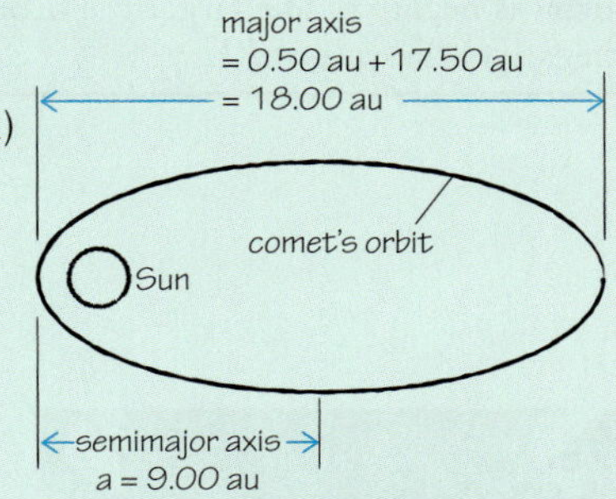

We could calculate T directly from Equation 10-16 by substituting the values of G, m_{Sun}, and the semimajor axis of the comet's orbit (which we would have to convert to meters). But a much simpler approach is to compare the comet's orbit to that of Earth, for which the semimajor axis is one au and the period is one year.

For Earth: $T_{\text{Earth}}^2 = \dfrac{4\pi^2}{Gm_{\text{Sun}}}a_{\text{Earth}}^3$

For the comet: $T_{\text{comet}}^2 = \dfrac{4\pi^2}{Gm_{\text{Sun}}}a_{\text{comet}}^3$

If we divide the second equation by the first one, the factor of $4\pi^2/Gm_{\text{Sun}}$ cancels out, leaving

$$\frac{T_{\text{comet}}^2}{T_{\text{Earth}}^2} = \frac{a_{\text{comet}}^3}{a_{\text{Earth}}^3}$$

Substitute values:

$$\frac{T_{\text{comet}}^2}{(1 \text{ y})^2} = \frac{(9.0 \text{ au})^3}{(1 \text{ au})^3} = 729$$

$$T_{\text{comet}} = \sqrt{729} \times 1 \text{ y} = 27 \text{ y}$$

Reflect

Our technique for solving this problem really used Kepler's original form of the law of periods: The square of the period is directly proportional to the cube of the semimajor axis. The semimajor axis for the comet's orbit is 9.0 times larger than that of Earth's orbit, so the square of the orbital period for the comet must be $(9.0)^3 = 729$ times larger than the square of the 1-year orbital period of Earth. Therefore the comet's orbital period is $\sqrt{729}$ years, or 27 years.

GOT THE CONCEPT? 10-4 Comparing Orbits

? Rank the following five objects in order of their orbital period around the Sun, from shortest period to longest period. (a) An asteroid in a circular orbit with radius 3.00×10^8 km; (b) an asteroid in a circular orbit with radius 4.00×10^8 km; (c) an asteroid in an elliptical orbit that is 3.00×10^8 km from the Sun at perihelion and 4.00×10^8 km from the Sun at aphelion; (d) an asteroid in an elliptical orbit that is 2.00×10^8 km from the Sun at perihelion and 5.00×10^8 km from the Sun at aphelion; and (e) an asteroid in an elliptical orbit that is 2.00×10^8 km from the Sun at perihelion and 4.00×10^8 km from the Sun at aphelion.

TAKE-HOME MESSAGE FOR Section 10-4

✔ For an object in a circular orbit, the orbital speed is inversely proportional to the square root of the orbital radius. The square of the orbital period is directly proportional to the cube of the orbital radius.

✔ The total mechanical energy E for an object in a circular orbit is negative. The larger the orbital radius, the greater (less negative) the value of E.

✔ Kepler's three laws of planetary motion—the law of orbits, the law of areas, and the law of periods—were deduced from observations of how the planets move, but can all be explained by Newton's law of universal gravitation.

10-5 Apparent weightlessness can have major physiological effects on space travelers

Education Images/UIG/Getty Images

Figure 10-21 Astronauts in orbit In the apparent weightlessness aboard an orbiting spacecraft, there is no up or down.

Figure 10-21 shows astronauts floating inside an orbiting spacecraft. The astronauts *feel* as though they are weightless, but this is a misnomer: Earth's gravitation acts on them just as it acts on their spacecraft. As Figure 10-22 shows because the astronauts are the same distance from Earth as their spacecraft, they have the same acceleration as the spacecraft and so "fall together" with their orbiting spacecraft. Since the astronauts have no tendency to move toward the floor, ceiling, or any other part of the spacecraft, they are in a state of apparent weightlessness.

One of the ways in which we perceive our weight is through the force the floor exerts on our feet. If you ride in an elevator that accelerates rapidly upward, you'll feel the floor pushing upward harder on your feet as though your weight had increased. If instead the elevator accelerates rapidly downward, the force that the floor exerts on your feet is reduced, and you feel as though your weight had decreased. In the extreme case where the elevator was falling freely with a downward acceleration of magnitude $g = 9.80\ \text{m/s}^2$, the force of the floor on your feet goes to zero and you have a sense of apparent weightlessness, even though gravity still acts on you. As long as both you and your surroundings experience the same acceleration, you will feel "weightless."

So it's not necessary for a spacecraft to be in orbit around a planet for the occupants to experience apparent weightlessness. All that's needed is for the spacecraft to be falling freely, so the only external force on the spacecraft is the gravitational force. Astronauts will feel weightless whether they are orbiting Earth aboard the International Space Station, on a mission from Earth to the Moon, or on an interplanetary voyage to Mars. (For a mission to Mars or any other planet, the spacecraft follows an elliptical orbit with the Sun at one focus, just as in Figure 10-18. This orbit is chosen so that it intersects the orbit of the destination planet.) The only times during the flight when the astronauts do not feel weightless are when the rockets are firing, either to put the spacecraft into the desired orbit or to take it out of orbit for landing.

At first glance it might seem that being in a state of apparent weightlessness would be relaxing. In fact there are many very negative effects on an astronaut's body. One common problem is *space adaptation syndrome* (also called *space sickness*). About half of all astronauts suffer from this condition during their first few days of spaceflight. The symptoms include motion sickness and disorientation. Research suggests that space adaptation syndrome is related to the vestibular system of the inner ear, which senses

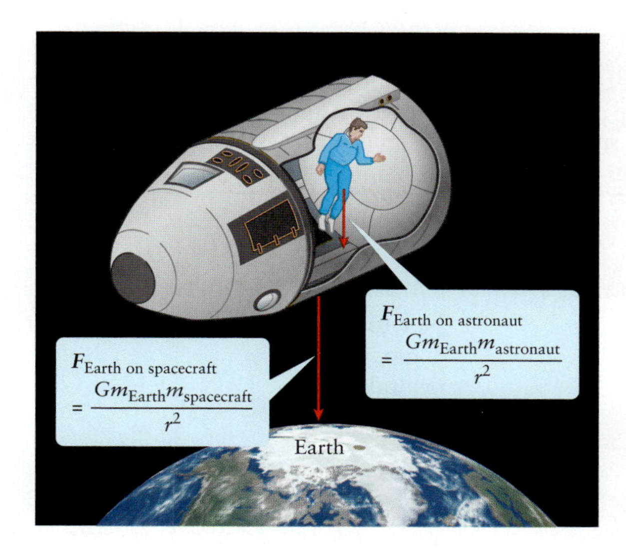

Figure 10-22 The origin of apparent weightlessness If the only external force acting on a spacecraft is the gravitational force, then the spacecraft and its occupant have the same acceleration $a = F/m$. As a result the spacecraft and the astronaut fall freely along the same trajectory, and the astronaut feels "weightless."

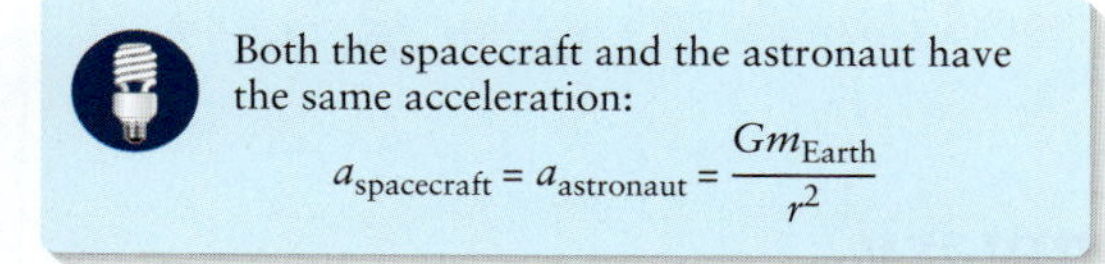

gravity and acceleration (see Section 3-8). If an astronaut has a more sensitive vestibular system than normal and there is an asymmetry between the vestibular systems in the left and right ears, the astronaut is likely to be plagued by space adaptation syndrome.

Both the spacecraft and the astronaut have the same acceleration:
$a_{\text{spacecraft}} = a_{\text{astronaut}} = \dfrac{Gm_{\text{Earth}}}{r^2}$

Two other problems that bedevil all astronauts are *loss of muscle mass* and *blood loss*. You exercise the muscles in your calves and spine simply by walking and standing. But in apparent weightlessness these muscles are not exercised and can lose as much as 5% of their mass per week. During a very long-duration spaceflight, such as a mission to Mars (which could take 10 months each way), astronauts might lose up to 40% of their capacity to do physical work, equivalent to a 40-year-old astronaut's muscles deteriorating to those of an 80-year-old. For this reason astronauts exercise as much as possible in space using special equipment designed to mimic gravity. (Building muscle mass before going into space does not appear to help: Astronauts who begin their missions with the greatest muscle mass also show the greatest decline while in space.)

The cause of blood loss is more subtle. When you stand upright on Earth, blood pools in your feet due to gravity, so your blood pressure is higher in your feet than in your brain. (That's why blood pressure is always measured at the same vertical position in the body, typically at your upper arm at the same level as your heart.) In apparent weightlessness, however, blood distributes itself more evenly through the body: Blood moves out of the legs and into the large veins in the upper body where stretch receptors respond to the apparent increase in blood volume. (This gives astronauts in space a characteristic puffy-faced appearance.) Activation of these stretch receptors initiates a reflex to reduce blood volume. The reflex involves the release of a hormone from the heart, inhibition of the release of a hormone from the brain, partial inhibition of the neural signals that cause arteries to contract, and the production of copious amounts of dilute urine. As a result astronauts can lose up to 22% of their blood volume within three days of apparent weightlessness. The weightless environment reduces demand on the heart because it does not have to overcome the effects of gravity to provide the brain with blood, so the heart muscle begins to atrophy as well. This is another reason for astronauts to do cardiovascular exercise while in space.

Astronauts recover muscle strength relatively quickly after returning from space (at a rate of about one day of recovery per day in space), and blood volume can be restored within a few days by drinking fluids. More problematic is recovery from the *bone loss* that occurs in space. Astronauts' bones atrophy at a rate of about 1% per month in apparent weightlessness, and recovery from a six-month mission may require two or three years back on Earth coupled with a program of strenuous exercise.

Another threat to astronauts' well-being is that while human muscle and bone tend to atrophy in apparent weightlessness, microbes such as *E. coli* and *Staphylococcus* actually reproduce more rapidly under these conditions. This leads to increased risk for contamination and serious infection during a long-duration spaceflight. Such are the biological challenges that confront the future of human exploration of space.

TAKE-HOME MESSAGE FOR Section 10-5

✔ Astronauts in space feel apparent weightlessness because their acceleration is the same as that of the spacecraft.

✔ The physiological challenges of apparent weightlessness include space sickness; loss of muscle mass, blood, and bone; and increased risk of microbial infection.

Key Terms

apparent weightlessness
Cavendish experiment
eccentricity
escape speed
gravitational constant
law of areas
law of orbits
law of periods
Newton's law of universal gravitation
orbital period
semimajor axis

Chapter Summary

Topic	Equation or Figure	
Newton's law of universal gravitation: Any two objects attract each other with a gravitational force. This force is proportional to the product of the masses of the objects and inversely proportional to the square of the distance between the centers of the two objects.	$$F_{1\text{ on }2} = F_{2\text{ on }1} = \frac{Gm_1m_2}{r^2}$$ **Gravitational constant** (same for any two objects) **Masses** of the two objects **Any two objects** (1 and 2) exert **equally strong gravitational forces** on each other. **Center-to-center distance** between the two objects **The gravitational forces are attractive:** $\vec{F}_{1\text{ on }2}$ pulls object 2 toward object 1 and $\vec{F}_{2\text{ on }1}$ pulls object 1 toward object 2.	(10-2)
Value of g at Earth's surface: The acceleration due to gravity at our planet's surface is related to Earth's mass and radius. At greater distances r from Earth's center, the value of g decreases in inverse proportion to the square of r.	$$g = \frac{Gm_{\text{Earth}}}{R^2_{\text{Earth}}}$$ **Gravitational constant** **Mass of Earth** **Acceleration due to gravity** at Earth's surface **Radius of Earth**	(10-3)
The Cavendish experiment: The value of the gravitational constant G in the law of universal gravitation can be determined by measuring the gravitational attraction between objects of known mass.	(1) Two small lead spheres, each of mass m, are attached to the ends of a wooden rod, and the rod is suspended from a wire. (2) Two large lead spheres, each of mass M, are then placed on the ground at equal distances from the small spheres. (3) Each large sphere exerts a gravitational force on the nearby small sphere. This makes the rod rotate through an angle of θ (shown here greatly exaggerated) and twists the wire.	(Figure 10-6)

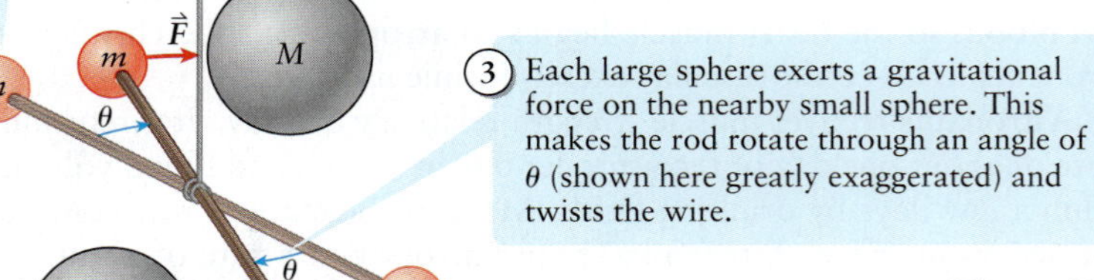

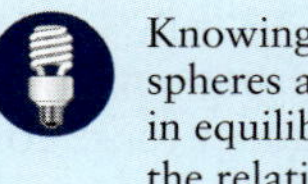

Knowing the masses m and M of the spheres and the distance r when the rod is in equilibrium, you can determine G using the relationship $F = GMm/r^2$.

Gravitational potential energy: The general expression for the gravitational potential energy of two objects follows from the law of universal gravitation. In this expression the potential energy is zero when the two objects are infinitely far apart; for any finite separation, U_{grav} is negative.

$$U_{grav} = -\frac{Gm_1m_2}{r} \quad (10\text{-}4)$$

Gravitational constant (same for any two objects)

Masses of the two objects

Gravitational potential energy of a system of two objects (1 and 2)

Center-to-center distance between the two objects

The gravitational potential energy is zero when the two objects are infinitely far apart. If the objects are brought closer together (so r is made smaller), U_{grav} decreases (it becomes more negative).

Escape speed: If an object is launched from Earth's surface at the escape speed or faster, it will never fall back to Earth but will escape to infinity.

$$v_{escape} = \sqrt{\frac{2Gm_{Earth}}{R_{Earth}}} \quad (10\text{-}10)$$

Gravitational constant

Mass of Earth

Speed that a projectile must have at Earth's surface in order to **escape Earth**

Radius of Earth

Circular orbits: A satellite in a circular orbit around Earth moves with a constant speed v. The orbital period T and the total mechanical energy E in a circular orbit both increase with increasing orbital radius r.

$$v = \sqrt{\frac{Gm_{Earth}}{r}} \quad (10\text{-}11)$$

Gravitational constant

Mass of Earth

Speed of an Earth satellite in a **circular orbit**

Radius of the satellite's orbit

$$T^2 = \frac{4\pi^2}{Gm_{Earth}} r^3 \quad (10\text{-}12)$$

Period of an Earth satellite in a circular orbit

Radius of the satellite's orbit

Gravitational constant

Mass of Earth

$$E = -\frac{Gm_{Earth}m}{2r} \quad (10\text{-}13)$$

Gravitational constant

Mass of Earth

Mass of satellite

Total mechanical energy for a satellite in a **circular orbit** around Earth

Radius of the satellite's orbit

The greater the radius r of the circular orbit, the greater (less negative) the total mechanical energy.

Elliptical orbits: A circular orbit is actually a special case of an elliptical orbit. The planets move in elliptical orbits with the Sun at one focus. The speed varies in accordance with Kepler's law of areas, which is equivalent to the statement that the angular momentum of a planet is conserved. The orbital period is given by the same expression as for circular orbits around Earth, with Earth's mass replaced by the mass of the Sun and the orbital radius replaced by

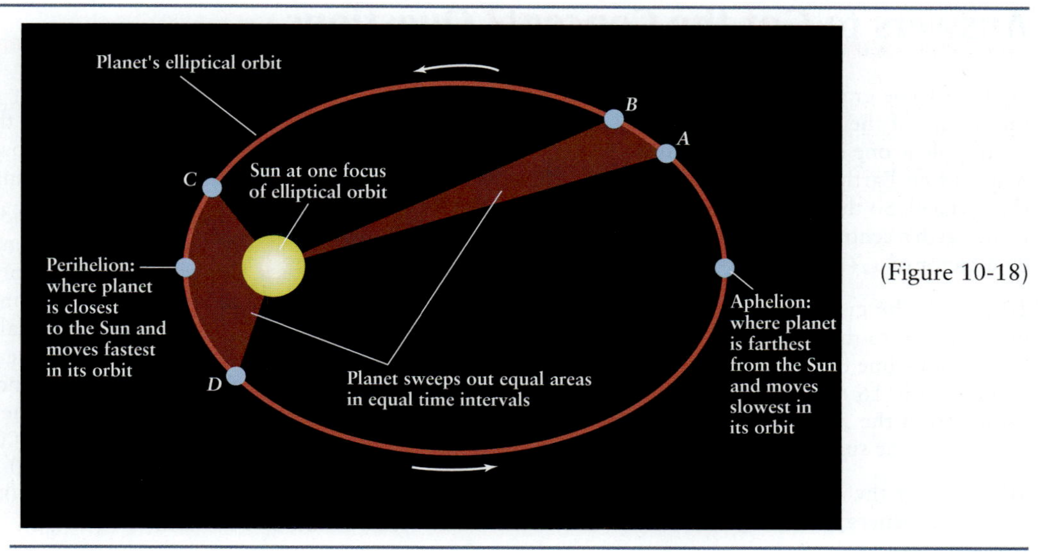

(Figure 10-18)

the semimajor axis (one-half the length of the long axis of the ellipse).

$$T^2 = \frac{4\pi^2}{Gm_{\text{Sun}}} a^3 \quad (10\text{-}16)$$

Period of an object in an elliptical orbit around the Sun

Semimajor axis of the object's orbit

Gravitational constant

Mass of the Sun

Apparent weightlessness: Because gravitation is universal, astronauts riding in a spacecraft (with the rockets off) have the same acceleration as the spacecraft. As a result they "fall" along with the spacecraft and feel as though they are weightless.

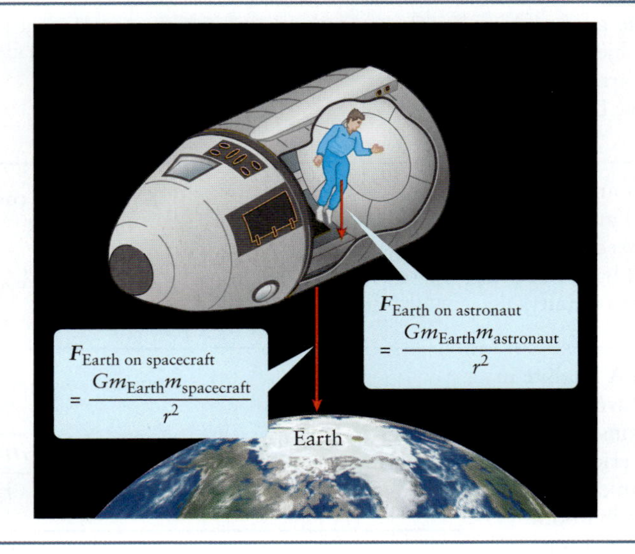

$F_{\text{Earth on astronaut}} = \frac{Gm_{\text{Earth}}m_{\text{astronaut}}}{r^2}$
$F_{\text{Earth on spacecraft}} = \frac{Gm_{\text{Earth}}m_{\text{spacecraft}}}{r^2}$
Earth

(Figure 10-22)

Both the spacecraft and the astronaut have the same acceleration:

$$a_{\text{spacecraft}} = a_{\text{astronaut}} = \frac{Gm_{\text{Earth}}}{r^2}$$

Answer to What do you think? Question

(b) Newton's law of universal gravitation states that Earth's gravitational force is inversely proportional to the square of the distance from Earth's center. At the altitude of the ISS, the distance from Earth's center is 6370 km + 350 km = 6720 km, which is (6720 km)/(6370 km) = 1.055 times Earth's radius. Hence the gravitational force at the altitude of the ISS is $1/(1.055)^2 = 0.90$ as great as at Earth's surface, or 10% less than at Earth's surface. It's true that an astronaut aboard the ISS *feels* weightless, but that's not because there is no gravitational force on her. Rather, there are gravitational forces on both the astronaut and the ISS, which give both objects the same acceleration. Hence both objects fall together around Earth in the same orbit. (See Sections 10-2 and 10-5.)

Answers to Got the Concept? Questions

10-1 (e) The gravitational force is inversely proportional to the square of the distance from Earth's center. A point at the north pole is one Earth radius from the center, while the other point is four Earth radii from the center (three Earth radii from the surface). So the point at the north pole is 1/4 the distance from Earth's center, and the gravitational force there is greater by a factor of $1/(1/4)^2 = 4^2 = 16$.

10-2 (e) The gravitational acceleration g is also inversely proportional to the square of the distance from Earth's center. By the same line of reasoning as in the previous question, the value of g is 16 times greater at the north pole (one Earth radius from the planet's center) than at a point three Earth radii above the surface (four Earth radii from the center).

10-3 (d) If the ball has escape speed or greater, the total mechanical energy is enough for the ball to travel to infinity. The total mechanical energy is a scalar, not a vector, so the direction in which the ball is thrown doesn't matter. Changing the direction in which the ball is thrown simply affects the path that the ball follows to infinity.

10-4 (a) and (e) [tie], (c) and (d) [tie], (b) The square of the orbital period T is proportional to the cube of the semimajor axis a. So a ranking in order of the orbital period is the same as a ranking in order of the semimajor axis. For a circular orbit the semimajor axis is the orbital radius; for an elliptical orbit the semimajor axis is half the major (long) axis, and the long axis is the sum of the distances at perihelion and aphelion. So the semimajor axes of the orbits are (a) 3.00×10^8 km, (b) 4.00×10^8 km, (c) $\frac{1}{2} \times (3.00 \times 10^8 \text{ km} + 4.00 \times 10^8 \text{ km}) = 3.50 \times 10^8$ km, (d) $\frac{1}{2} \times (2.00 \times 10^8 \text{ km} + 5.00 \times 10^8 \text{ km}) = 3.50 \times 10^8$ km, and (e) $\frac{1}{2} \times (2.00 \times 10^8 \text{ km} + 4.00 \times 10^8 \text{ km}) = 3.00 \times 10^8$ km.

Questions and Problems

In a few problems you are given more data than you actually need; in a few other problems you are required to supply data from your general knowledge, outside sources, or informed estimate.

Interpret as significant all digits in numerical values that have trailing zeros and no decimal points. For all problems use $g = 9.80\ \text{m/s}^2$ for the free-fall acceleration due to gravity. Neglect friction and air resistance unless instructed to do otherwise. Refer to Appendix B for any required astronomical data like the masses or orbital radii of the planets.

• Basic, single-concept problem
•• Intermediate-level problem; may require synthesis of concepts and multiple steps
••• Challenging problem
SSM *Solution is in Student Solutions Manual*
Example *See worked example for a similar problem*

Conceptual Questions

1. • Isaac Newton probably gained no insight by being hit on the head with a falling apple, but he did believe that "the Moon is falling." Explain this statement. SSM

2. • Would the magnitude of the acceleration due to gravity near Earth's surface increase more if Earth's mass were doubled or if Earth's radius were cut in half? Justify your answer.

3. • According to Equation 10-2 what happens to the force between two objects if (a) the mass of one object is doubled and (b) if both masses are halved? SSM

4. • Every object in the universe with mass experiences an attractive gravitational force due to every other object with mass. Why don't you feel a pull from objects close to you?

5. • The gravitational force acts on all objects in proportion to their mass. Why don't heavy objects fall faster than light ones, if you neglect air resistance?

6. • When powering a trip from Earth to the Moon, the thrusters of a ship don't have to get the ship all the way to the Moon. Why do the thrusters have to get the ship only partway there?

7. • The Sun gravitationally attracts the Moon. Does the Moon orbit the Sun?

8. • The fabrication of precision ball bearings may be better performed on the International Space Station than on the surface of Earth. Explain why.

9. • Why is the gravitational potential energy of two objects negative?

10. • When a frog is at the bottom of a well, the frog–Earth system has a negative amount of gravitational potential energy with respect to the ground level where we choose to set potential energy equal to zero. Can the frog escape from the well if it jumps upward with a positive amount of kinetic energy that is larger than the negative potential energy when the frog is at the bottom of the well? Explain.

11. • Describe some ways that Newton's law of gravity may have affected human evolution. SSM

12. • Imagine a world where the force of gravity were proportional to the inverse *cube* of the distance between two objects ($F_{\text{gravity}} \propto 1/r^3$). How would Kepler's law of periods be changed?

13. • Much attention has been devoted to the exact numerical power in Newton's law of universal gravitation ($F_{\text{gravity}} \propto 1/r^2$). Some theorists have investigated whether the dependence might be slightly larger or smaller than 2. What would be the significance (what impact would there be) if the power was not exactly 2?

14. • Earth moves faster in its orbit around the Sun during the winter in the northern hemisphere than it does during the summer in the northern hemisphere. Is Earth closer to the Sun during the northern hemisphere's winter or during the northern hemisphere's summer? Explain your answer. SSM

15. • A satellite is to be raised from one circular orbit to one farther from Earth's surface. What will happen to its period?

16. • Geostationary satellites remain stationary over one point on Earth. How is this accomplished?

Multiple-Choice Questions

17. • Planet A orbits a star with a period T. Planet B circles the same star at four times the distance of planet A. Planet B is four times as massive as planet A. Planet B orbits the star with a period of
   A. $4T$.
   B. $8T$.
   C. $16T$.
   D. T.
   E. $T/4$.

18. • According to Newton's universal law of gravitation, $F = \dfrac{Gm_1m_2}{r^2}$, if both masses are doubled, the force is
   A. four times as much as the original value.
   B. twice as much as the original value.
   C. the same as the original value.
   D. one-half of the original value.
   E. one-fourth of the original value.

19. • According to Newton's universal law of gravitation, $F = \dfrac{Gm_1m_2}{r^2}$, if the distance r is doubled, the force is
   A. four times as much as the original value.
   B. twice as much as the original value.
   C. the same as the original value.
   D. one-half of the original value.
   E. one-fourth of the original value.

20. • Which is larger, the Sun's pull on Earth or Earth's pull on Sun?
   A. The Sun's pull on Earth is larger.
   B. Earth's pull on the Sun is larger.
   C. They pull on each other equally.
   D. The Sun's pull on Earth is twice as large as Earth's pull on the much larger Sun.
   E. There is no pull or force between Earth and the Sun.

21. • Compare the weight of a mountain climber when she is at the bottom of a mountain with her weight when she is at the top of the mountain. In which case is her weight larger?
   A. She weighs more at the bottom.
   B. She weighs more at the top.
   C. Both are the same.
   D. She weighs twice as much at the top.
   E. She weighs four times as much at the top. SSM

22. • A satellite and the International Space Station have the same mass and are going around Earth in concentric orbits. The distance of the satellite from Earth's center is twice that of the International Space Station's distance. What is the ratio of the centripetal force acting on the satellite compared to that acting on the International Space Station?
   A. 1/4
   B. 1/2
   C. 1
   D. 2
   E. 4

23. • A satellite in a low-altitude circular orbit just above a planet's surface has speed v. To escape the planet, its speed must be
   A. v.
   B. $\sqrt{2}v$.
   C. $2v$.
   D. $v/2$.
   E. $v/\sqrt{2}$. SSM

24. • The escape speed from planet X is v. Planet Y has the same radius as planet X but is twice as dense. The escape speed from planet Y is
   A. v.
   B. $\sqrt{2}v$.
   C. $2v$.
   D. $v/2$.
   E. $v/\sqrt{2}$.

25. • Two satellites having equal masses are in circular orbits around Earth. Satellite A has a smaller orbital radius than satellite B. Which statement is true?
   A. Satellite A has more kinetic energy, less potential energy, and less mechanical energy (kinetic energy plus potential energy) than satellite B.
   B. Satellite A has less kinetic energy, less potential energy, and less mechanical energy than satellite B.
   C. Satellite A has more kinetic energy, more potential energy, and less mechanical energy than satellite B.
   D. Satellite A and satellite B have the same amount of mechanical energy.
   E. Satellite A and satellite B have the same amount of kinetic energy and no potential energy because they are in motion. SSM

Estimation/Numerical Analysis

26. • Estimate the gravitational force of attraction between two people having a conversation.

27. • Estimate how close two 1-kg weights must be in an ordinary science lab in order to measure the gravitational force one exerts on the other. SSM

28. • Estimate the gravitational force from Earth that passengers would experience on a commercial airliner at cruising altitude. How much "lighter" would a person be in flight as compared to at sea level?

29. • Estimate the gravitational force of attraction between two skyscrapers built on adjacent city blocks. Would this force need to be taken into account in the engineering of the buildings?

30. • **Biology** Estimate the difference in height due to compression of the vertebrae when a 160-cm-tall sherpa at the peak of Mount Everest travels to sea level.

31. • Suppose there are two identical spheres made out of bowling ball material. Estimate how large each sphere would have to be in order for there to be an easily measured force of gravitational attraction between the spheres when in contact. What would be the escape speed for such a sphere?

32. •• The eight planets, as well as Pluto and Comet Halley, orbit the Sun according to Newton's law of universal gravitation. The table has data for the periods (T) and semimajor axes (a) for each orbital ellipse. (a) Derive a best-fit constant, similar to that referred to in Kepler's law of periods. (b) Predict the value of a for the elliptical orbit of Comet Halley.

Object	T (days)	a (au)
Mercury	87.97	0.3871
Venus	224.7	0.7233
Earth	365.2	1.000
Mars	687.0	1.5234
Jupiter	4332	5.204
Saturn	10,832	9.582
Uranus	30,799	19.23
Neptune	60,190	30.10
Pluto	90,613.3	9.48
Halley	27,507	?

Problems

10-1 Gravitation is a force of universal importance
10-2 Newton's law of universal gravitation explains the orbit of the Moon

33. • A 5.00×10^2-kg tree stump is located 1.00×10^3 m from a 12,000-kg boulder. Determine the magnitude and direction of the gravitational force exerted by the tree stump on the boulder. Example 10-1

34. • Kramer goes bowling and decides to employ the force of gravity to "pick up a spare." He rolls the 7.0-kg bowling ball very slowly so that it comes to rest a center-to-center distance of 0.20 m from the one remaining 1.5-kg bowling pin. Determine the force of gravity between the ball and the pin at this point and comment on the efficacy of the technique. Treat the ball and pin as point objects for this problem. Example 10-1

35. • A 150-g baseball is 1.00×10^2 m from a 935-g bat. What is the force of gravitational attraction between the two objects? Ignore the size of the bat and ball (treat them as point objects) for this problem. SSM Example 10-1

36. • In 1994 the performer Rod Stewart drew over 3 million people to a concert in Rio de Janeiro, Brazil. (a) If the people in the group had an average mass of 80.0 kg, what collective gravitational force would the group have on a 4.50-kg eagle soaring 3.00×10^2 m above the throng? If you treat the group as a point object, you will get an upper limit for the gravitational force. (b) What is the ratio of that force of attraction to the force between Earth and the eagle? Example 10-2

37. • Compare the gravitational force on a 1-kg apple that is on the surface of Earth versus the gravitational force due to the Moon on the same apple in the same location on the surface of Earth. Assume that Earth and the Moon are spherical and that both have their masses concentrated at their respective centers. Example 10-2

38. • Determine the average magnitude of the force of gravity between the Sun and Earth. Example 10-4

39. • Determine the net force of gravity acting on the Moon during an eclipse when it is directly between Earth and the Sun. SSM Example 10-4

40. • A star that has a mass equal to the mass of our Sun is located 7.50×10^9 km from another star that has a mass that is one-half of the Sun's mass. At what point(s) will the gravitational force from the two stars on a 1.00×10^5-kg space probe be equal to zero? Example 10-4

41. • At what point between Earth and the Moon will a 5.00×10^4-kg space probe experience no net gravitational force? SSM Example 10-4

42. • Compare the weight of a 5.00-kg object on Earth's surface to the gravitational force between a 5.00-kg object that is one Earth radius from another object of mass equal to 5.98×10^{24} kg. Use Newton's universal law of gravitation for the second part of the question. Example 10-2

10-3 The gravitational potential energy of two objects is negative and increases toward zero as the objects are moved farther apart

43. • How much energy would be required to move the Moon from its present orbit around Earth to a location that is twice as far away? Assume the Moon's orbit around Earth is nearly circular and has a radius of 3.84×10^8 m. Example 10-5

44. • How much work is done by the force of gravity as a 10.0-kg object moves from a point that is 6.00×10^3 m above sea level to a point that is 1.00×10^3 m above sea level? SSM Example 10-5

45. • (a) What is the escape speed of a space probe that is launched from the surface of Earth? (b) Would the answer change if the launch occurs on top of a very tall mountain? Explain your answer. Example 10-6

46. • A small asteroid that has a mass of 1.00×10^2 kg is moving at 2.00×10^2 m/s when it is 1.00×10^3 km above the Moon. (a) How fast will the asteroid be traveling when it impacts the lunar surface if it is heading straight toward the center of the Moon? (b) How much work does the Moon do in stopping the asteroid if neither the Moon nor the asteroid heats up in the process? The radius of the Moon is 1.737×10^6 m. SSM Example 10-5

47. •• The volume of water in the Pacific Ocean is about 7.0×10^8 km^3. The density of seawater is about 1030 kg/m^3. (a) Determine the gravitational potential energy of the Moon–Pacific Ocean system when the Pacific is facing away from the Moon. (b) Repeat the calculation when Earth has rotated so that the Pacific Ocean faces toward the Moon. (c) Estimate the maximum speed of the water in the Pacific Ocean due to the tidal influence of the Moon. For the sake of the calculations, treat the Pacific Ocean as a pointlike object (obviously a very rough approximation). Example 10-5

10-4 Newton's law of universal gravitation explains Kepler's laws for the orbits of planets and satellites

48. • The orbit of Mars around the Sun has a radius that is 1.524 times greater than the radius of Earth's orbit. Determine the time required for Mars to complete one revolution. Example 10-8

49. • The space shuttle usually orbited Earth at altitudes of around 3.00×10^5 m. (a) Determine the time for one orbit of the shuttle about Earth. (b) How many sunrises per day did the astronauts witness? Example 10-7

50. • What was the speed of a space shuttle that orbited Earth at an altitude of 3.00×10^5 m? Example 10-7

51. • A satellite orbits Earth at an altitude of 8.00×10^4 km. Determine the time for the satellite to orbit Earth once. Example 10-7

52. • A satellite requires 86.5 min to orbit Earth once. Assuming a circular orbit, what is the circumference of the satellite's orbit? SSM Example 10-7

53. • The orbital period of Saturn is 29.46 years. Determine the semimajor axis of its orbit. Example 10-7

54. • Determine the altitude of a geosynchronous satellite. Such a satellite will have precisely the same orbital period as Earth's rotational period. Example 10-7

55. • Given that Earth orbits the Sun with a semimajor axis of 1.000 au and an approximate orbital period of 365.24 days, determine the mass of the Sun. Example 10-7

56. • The Moon orbits Earth in a nearly circular orbit that lasts 27.32 days. Determine the distance from the surface of the Moon to the surface of Earth. SSM Example 10-7

57. • A planet orbits a star with an orbital radius of 1.00 au. If the star has a mass that is 1.75 times our own Sun's mass, determine the time for one revolution of the planet around the star. Example 10-7

General Problems

58. •• A 425-kg satellite is launched into a circular orbit that has a period of 702 min and a radius of 20,100 km around Earth. (a) Determine the gravitational potential energy of the satellite's orbit. (b) Estimate the energy required to place the satellite in orbit around Earth. Example 10-7

59. •• Astronomy The four largest of Jupiter's moons are listed in the table below. Using these data, Kepler's three laws, and the law of universal gravitation, (a) complete the table and (b) determine the mass of Jupiter. Example 10-8

Moon	Semimajor Axis (km)	Orbital Period (days)
Io	421,700	1.769
Europa	671,034	?
Ganymede	?	7.155
Callisto	?	16.689

60. •• (a) What speed is needed to launch a rocket due east near the equator into a low Earth orbit? Assume the rocket skims along the surface of Earth, so $r \approx R_E$. (b) Repeat the calculation for a rocket fired due west. SSM Example 10-7

61. •• The former Soviet Union launched the first artificial Earth satellite, *Sputnik*, in 1957. Its mass was 84 kg, and it made one orbit every 96 min. (a) Determine the altitude of

Sputnik's orbit above Earth's surface, assuming circular orbits. (b) What was *Sputnik*'s weight in orbit and at Earth's surface? Example 10-7

62. • **Astronomy** The 2004 landings of the Mars rovers *Spirit* and *Opportunity* involved many stages, resulting in each probe having zero vertical velocity about 12 m above the surface of Mars. Determine (a) the time required for the final free-fall descent of the probes and (b) the vertical velocity at impact. The mass of Mars is 6.419×10^{23} kg, and its radius is 3.397×10^6 m. Example 10-2

63. ••• **Biology, Astronomy** On Earth froghoppers can jump upward with a takeoff speed of 2.8 m/s. Suppose you took some of the insects to an asteroid. If it is small enough, they can jump free of it and escape into space. (a) What is the diameter (in kilometers) of the largest spherical asteroid from which they could jump free? Assume a typical asteroid density of 2.0 g/cm^3. (b) Suppose that one of the froghoppers jumped horizontally from a small hill on an asteroid. What would the diameter (in kilometers) of the asteroid need to be so that the insect could go into a circular orbit just above the surface? Example 10-6

64. • **Astronomy** The International Space Station (ISS) orbits Earth in a nearly circular orbit that is 345 km above Earth's surface. (a) How many hours does it take for the ISS to make each orbit? (b) Some of the experiments performed by astronauts in the ISS involve the effects of "weightlessness" on objects. What gravitational force does Earth exert on a 10.0-kg object in the ISS? Express your answer in newtons and as a fraction of the force that Earth would exert on the object at Earth's surface. (c) Considering your answer in part (b), how can an object be considered *weightless* in the ISS? Example 10-7

65. •• **Astronomy** Measurements on the asteroid Apophis have shown that its aphelion (farthest distance from the Sun) is 1.099 au, its perihelion (closest distance from the Sun) is 0.746 au, and its mass is 2.7×10^{10} kg. (a) Determine the semimajor axis of Apophis in astronomical units and in meters. (b) How many days does it take Apophis to orbit the Sun? (c) At what point in its orbit is Apophis traveling fastest, and at what point is it traveling slowest? (d) Determine the ratio of its maximum speed to its minimum speed. Example 10-8

66. •• A rack of seven spherical bowling balls (each 8.00 kg, radius of 11.0 cm) is positioned along a line 1.00 m from a point *P*, as shown in Figure 10-23. Determine the gravitational force the bowling balls exert on a ping-pong ball of mass 2.70 g centered at point *P*. Example 10-4

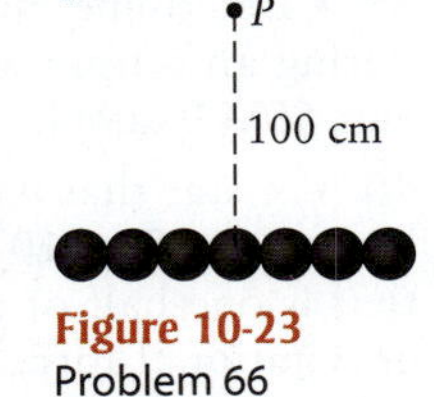

Figure 10-23 Problem 66

67. • **Astronomy** The Sun and solar system actually are not at rest in our Milky Way galaxy. We orbit around the center of the Milky Way galaxy once every 225,000,000 years, at a distance of 27,000 light-years. (One light-year is the distance that light travels in one year: 1 ly = 9.46×10^{12} km = 9.46×10^{15} m.) If the mass of the Milky Way were concentrated at the center of the galaxy, what would be the mass of the galaxy? Example 10-7

68. •• Locate the point(s) along the line $\overline{AB}$ where a small, 1.00-kg object could rest such that the net gravitational force on it due to the two objects shown is exactly zero (Figure 10-24). SSM Example 10-4

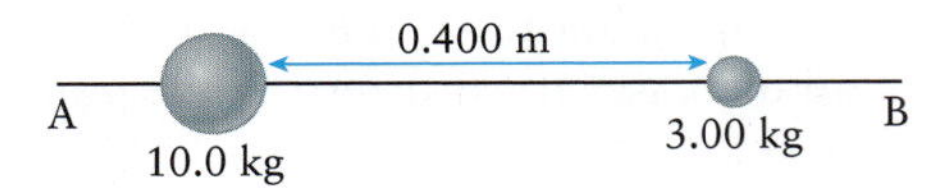

Figure 10-24 Problem 68

11 Fluids

Andrey Nekrasov / Alamy

What do you think?

The dolphin in the photo experiences an upward buoyant force exerted by the water surrounding it. If you replaced the dolphin with a life-size sculpture of a dolphin made of iron (which is denser than a dolphin's body), would the magnitude of the buoyant force (a) increase, (b) decrease, or (c) remain the same?

In this chapter, your goals are to:

- (11-1) Describe the similarities and differences between liquids and gases.
- (11-2) Recognize how to apply the definition of density.
- (11-3) Explain the origin of fluid pressure in terms of molecular motion.
- (11-4) Calculate the pressure at a given depth in a fluid in hydrostatic equilibrium.
- (11-5) Explain the difference between absolute pressure and gauge pressure.
- (11-6) Calculate the force on an object due to a difference in pressure on its sides.
- (11-7) Explain how to apply Pascal's principle to a fluid at rest.
- (11-8) Apply Archimedes' principle to find the buoyant force on an object in a fluid.
- (11-9) Use the equation of continuity to analyze the flow of an incompressible fluid.
- (11-10) Apply Bernoulli's principle to relate fluid pressure and flow speed in an incompressible fluid.
- (11-11) Explain what happens in flows where viscosity is important.
- (11-12) Describe the role of surface tension in the behavior of liquids.

To master this chapter, you should review:

- (6-7) What happens to the total mechanical energy of a system when nonconservative forces do work.
- (9-4) How pressure is defined.

11-1 Liquids and gases are both examples of fluids

Liquids and gases—collectively known as *fluids* for their ability to flow—are the most common states of matter in the universe. Indeed, part of our planet's interior, the entirety of our Sun, and most of your body are composed of fluids (**Figure 11-1**). Solid objects in our daily environment such as rocks and trees are exceptional cases in a universe dominated by fluids.

Solids are substances whose individual molecules cannot move freely but remain in essentially fixed positions relative to one another. That's why a solid object maintains its shape, though it may bend or deform if you apply forces to it. In contrast, **fluids** are substances that can flow because their molecules can move freely with respect to each other and are not tied to fixed locations.

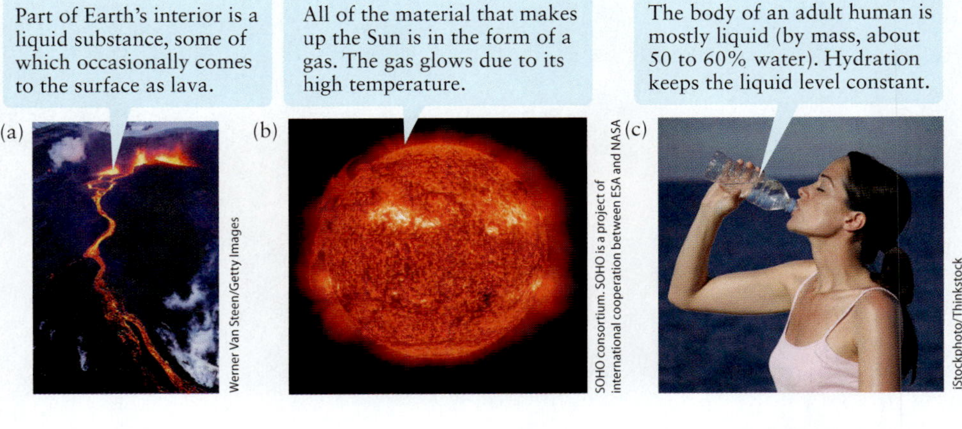

Part of Earth's interior is a liquid substance, some of which occasionally comes to the surface as lava.
All of the material that makes up the Sun is in the form of a gas. The gas glows due to its high temperature.
The body of an adult human is mostly liquid (by mass, about 50 to 60% water). Hydration keeps the liquid level constant.
(a)
(b)
(c)
Werner Van Steen/Getty Images
SOHO consortium. SOHO is a project of international cooperation between ESA and NASA
iStockphoto/Thinkstock

Figure 11-1 Fluids Materials that can flow make up (a) part of our Earth, (b) all of our Sun (as well as all of the stars visible in the night sky, which are objects like the Sun), and (c) most of our bodies. Note that the photo in (a) shows at least three different fluids: molten lava flowing down the rocks, liquid water in the ocean, and the air in our atmosphere.

Fluids fall into two broad categories: liquids and gases. **Liquids** tend to maintain the same volume regardless of the shape and size of their container. For example, 1 cubic meter (1 m^3) of water taken from a storage tank will exactly fill a 1-m^3 container into which it is poured (**Figure 11-2a**). **Gases**, however, expand to fill whatever volume is available to them. If you place a cylinder containing 1 m^3 of compressed oxygen gas in a large room and open the valve, the oxygen gas spreads out over the entire room so that its volume is much greater than 1 m^3 (**Figure 11-2b**). It can also be compressed further to fill an even smaller container.

What explains the difference between these two types of fluids? In a liquid, molecules are close to one another (almost touching). At close range the attractive forces between molecules are strong, keeping the molecules together so that the volume of the liquid stays the same. In a gas, however, the molecules are much farther apart; in the air you're breathing now, the average distance between molecules is about 10 times the size of a single molecule. At these greater separations the molecules exert little or no attractive forces on each other, so nothing prevents the molecules from spreading out to occupy all of the available volume.

In this chapter we will touch on many of the key aspects of fluids. The central concepts of *density* and *pressure*, coupled with the notion of what it means for a fluid to be in *equilibrium*, will help us understand why water pressure increases with increasing

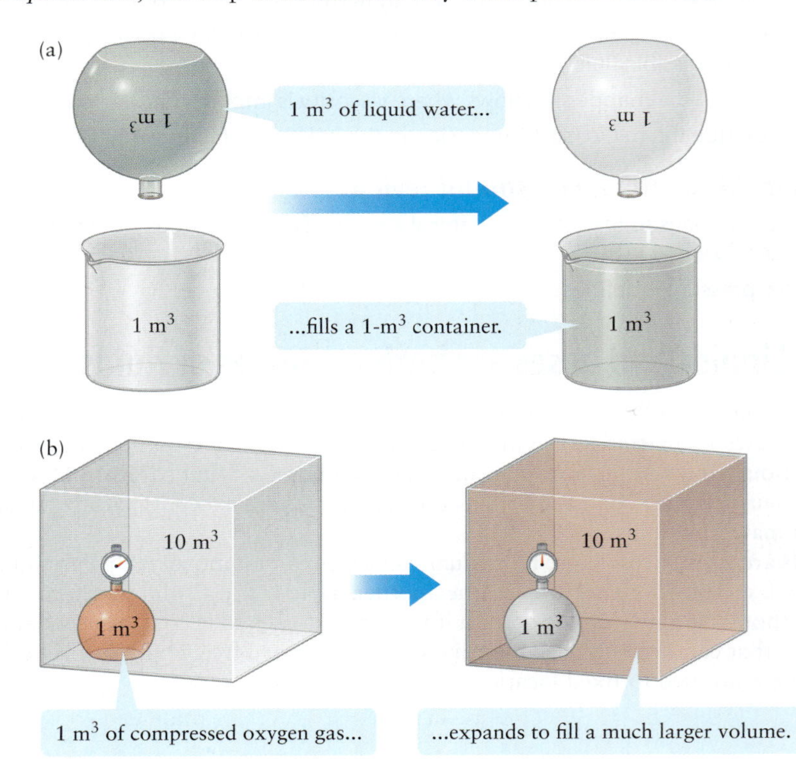

Figure 11-2 Liquids versus gases (a) A liquid maintains the same volume when you pour it from one container to another. (b) A gas expands to fill whatever volume is available.

depth. They also explain the phenomenon of *buoyancy* that enables marine animals to float under water without sinking or rising. Fluids in *motion* are influenced not only by pressure and density but also by *viscous* forces, which are the frictional forces of one part of the fluid rubbing past another. By examining some of the myriad ways in which fluids flow, we will gain insight into the flight of airplanes and birds, the behavior of blood in the circulatory system, and the peculiar challenges that face very small swimming organisms. Finally, we will see how the mutual attraction of molecules in a liquid gives rise to surface tension, a phenomenon that impacts the function of your lungs.

TAKE-HOME MESSAGE FOR Section 11-1

✔ Fluids (which include liquids and gases) do not have a definite shape. They can flow to take on the shape of their container because their molecules are in constant motion.

✔ Liquids maintain essentially the same volume as they flow, but gases will expand or contract as necessary to fill the available space.

✔ The attractive forces between molecules are strong in liquids but are weak in gases because the molecules are farther apart.

11-2 Density measures the amount of mass per unit volume

One way to describe how tightly packed the molecules are in a gas, liquid, or solid is in terms of *density*. The **density** of a substance, denoted by the Greek letter ρ (rho), is the mass of the substance divided by the volume that it occupies. The greater the density or *mass per volume* of a substance, the more kilograms of that substance are packed into a given number of cubic meters of volume:

The **density** ρ ("rho") of a certain substance...

...equals the **mass** m of a given quantity of that substance...

$$\rho = \frac{m}{V} \tag{11-1}$$

...divided by the **volume** V of that quantity of the substance.

Definition of density (11-1)

A given number of water molecules of combined mass m occupy a smaller volume V in liquid water than they do in water vapor (a gas). Hence the density $\rho = m/V$ of liquid water is greater than that of water vapor.

WATCH OUT! Density doesn't tell you how closely packed the molecules are.

A certain number of oxygen gas molecules have 14% greater density than the same number of nitrogen gas molecules with the same spacing between molecules. That's because density is *mass* per unit volume, and the mass of an oxygen molecule is 14% greater than that of a nitrogen molecule. To find the spacing between molecules in a given substance, you need to know both the density of the substance *and* the mass per molecule.

Density is usually stated in kilograms per cubic meter (kg/m^3). A common alternative choice of units is grams per cubic centimeter, or g/cm^3 (1 g/cm^3 = 1000 kg/m^3). As an example, one cubic meter of liquid water at a temperature of 4°C has a mass of 1000 kg and therefore a density $\rho = (1000 \text{ kg})/(1 \text{ m}^3) = 1000 \text{ kg/m}^3$. Twice the mass of liquid water (2000 kg) occupies twice the volume (2 m^3) and so its density is the same: $\rho = (2000 \text{ kg})/(2 \text{ m}^3) = 1000 \text{ kg/m}^3$. This example shows that density depends only on the nature of the substance, *not* on how much of the substance is present. When we say that gold has a density of 19,300 kg/m^3 = 1.93×10^4 kg/m^3, we mean that *any* quantity of gold has this same density. **Table 11-1** lists the densities of some common substances.

TABLE 11-1 Densities of Various Substances

Substance	Density (kg/m^3)
air at sea level at 15°C	1.23
dry timber (white pine)	370
fresh wood (red oak)	600
gasoline	680
ethanol (ethyl alcohol)	806
lipids (fats and oils)	915–945
ice at 0°C	917
fresh water at 4°C	1.000×10^3
fresh water at 20°C	998
seawater at 4°C	1.025×10^3
muscle	1.06×10^3
deer antler (low-density bone)	1.86×10^3
cow femur (typical bone)	2.06×10^3
tympanic bulla of a fin whale (densest bone)	2.47×10^3
aluminum (Al)	2.70×10^3
mollusk shell	2.7×10^3
calcite (mineral of shell)	2.8×10^3
tooth enamel, human	2.9×10^3
apatite (mineral of bone, tooth)	3.2×10^3
planet earth (average density)	5.52×10^3
iron (Fe)	7.8×10^3
mercury (Hg)	13.595×10^3
gold (Au)	19.3×10^3

An alternative way to describe the density of a substance is to compare its density to that of another substance that serves as a standard. One common standard is liquid water at 4°C ($\rho = 1000\ kg/m^3$), and the **specific gravity** of a substance is equal to its density divided by that of 4°C liquid water. Human blood, for instance, has a specific gravity of 1.06, which means it is 1.06 times as dense as water: $\rho_{blood} = (1.06)(1000\ kg/m^3) = 1.06 \times 10^3\ kg/m^3$. The term "specific gravity" in this case means that a certain volume of blood weighs 1.06 times as much as the same volume of water—that is, it experiences 1.06 times as much gravitational force. On average, red blood cells make up about 38% of the volume of blood in women and 42% in men; these cells are only slightly denser than water, which is why the specific gravity of blood is just a little greater than 1.

In general, the density of a liquid depends on temperature. For example, if you take water from the refrigerator (temperature about 4°C) and let it warm to room temperature (about 20°C), the water expands slightly. Because the mass m of the water is the same but its volume V has increased, the density $\rho = m/V$ of 20°C water is slightly less (by about 0.2%) than that of 4°C water. Most liquids expand and become less dense when the temperature increases. Between 0°C and 4°C, water is a notable exception to this general rule: Within this temperature range, water *contracts* and becomes *more* dense as the temperature increases. That's why we give the densities of water and salt water in Table 11-1 at 4°C, the temperature at which these liquids are densest.

The density of gases is sensitive to both temperature and pressure. Heated air expands, which lowers its density; applying pressure to air makes the air occupy a smaller volume, which raises its density. Because gases can easily be compressed by squeezing, they are **compressible fluids**. By contrast, for most practical purposes liquids are **incompressible fluids**: The volume and the density of a liquid change very little when it is squeezed because the molecules of a liquid are almost touching and resist being squeezed much closer together. You can easily compress air to a higher density using a hand pump of the sort used for inflating bicycle tires, but to compress water appreciably takes tremendous pressures like those found at the bottom of the ocean.

While the main focus of this chapter is on fluids, the idea of density applies to solids as well. The following examples show some applications of this idea.

EXAMPLE 11-1 Density of Chicken

A typical whole chicken on sale at a supermarket has a mass of 2.3 kg and is approximately spherical in shape, with a radius of about 8.0 cm. What is the approximate density of such a chicken in kg/m^3? Compare the result to the densities of the biological materials listed in Table 11-1.

Set Up

We are given the chicken's mass and its radius, and our goal is to calculate its density ρ. We use the definition of density in Equation 11-1, $\rho = m/V$.

The chicken is very nearly spherical, so to determine its volume we use the expression for the volume of a sphere of radius R (see the Math Tutorial).

Definition of density:

$$\rho = \frac{m}{V} \qquad (11\text{-}1)$$

Volume of a sphere of radius R:

$$V = \frac{4\pi R^3}{3}$$

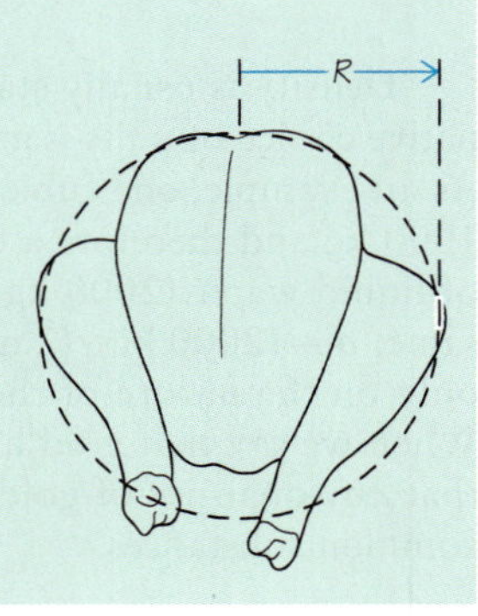

Solve

To calculate the density in the desired units, we need to find the chicken's volume in cubic meters (m^3). We first express its radius in meters.

$$R = (8.0 \text{ cm})\left(\frac{1 \text{ m}}{100 \text{ cm}}\right) = 8.0 \times 10^{-2} \text{ m}$$

We then calculate the volume of the chicken in cubic meters.

$$V = \frac{4\pi R^3}{3} = \frac{4\pi(8.0 \times 10^{-2} \text{ m})^3}{3} = 2.1 \times 10^{-3} \text{ m}^3$$

The density is the mass of the chicken ($m = 2.3$ kg) divided by this volume.

$$\rho = \frac{m}{V} = \frac{2.3 \text{ kg}}{2.1 \times 10^{-3} \text{ m}^3} = 1.1 \times 10^3 \text{ kg/m}^3$$

Reflect

Our answer doesn't match up precisely with any of the values given in Table 11-1. That's because a chicken's body is not a pure substance but a mixture of substances of different densities: fats, water, muscle, and bone. Our calculated density of $1.1 \times 10^3 \text{ kg/m}^3$ is actually the *average* density of the chicken. We explore the idea of average density further in Example 11-2.

You may be wondering if it's reasonable to approximate a chicken as being spherical. If you look at the sketch, some parts of the chicken extend beyond the sphere, and some of the sphere is empty. But the volume of the chicken outside the sphere is nearly equal to the volume of the "empty" part of the sphere, so the volume of the chicken is actually pretty close to the volume of the sphere.

EXAMPLE 11-2 Average Density

A submersible vessel used for deep-sea research has a metal hull with a large cavity in its interior for the crew and their equipment. Suppose that the hull is made of $1.130 \times 10^3 \text{ m}^3$ of a particular steel alloy of density $7.910 \times 10^3 \text{ kg/m}^3$ and that the cavity in the hull has volume $7.600 \times 10^3 \text{ m}^3$ and is filled with air of density 1.200 kg/m^3. Find the *average* density of the submersible, that is, the density of the submersible as a whole.

Set Up

The submersible contains two materials—the steel of the hull and the air in the interior spaces—which have different densities and occupy different volumes. We want to find a third density, the *average* density of the entire submersible, equal to its total mass m_{sub} divided by its total volume V_{sub}. We again use the definition of density, Equation 11-1. We'll apply this definition three times, once for each density.

Definition of density:

$$\rho = \frac{m}{V} \quad (11\text{-}1)$$

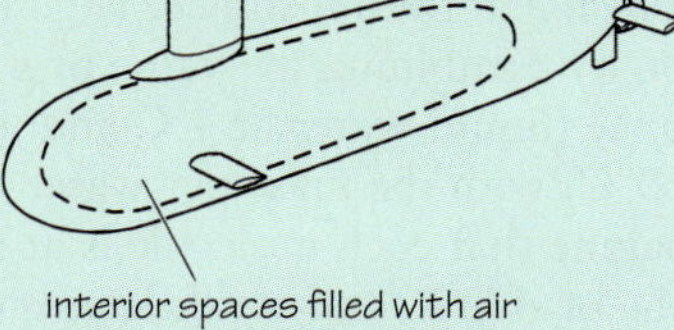

Solve

The submersible's average density $\rho_{average}$ is its total mass m_{sub} divided by its total volume V_{sub}, so we need to calculate both m_{sub} and V_{sub}. The volume of the submersible (V_{sub}) is the sum of the volume of the hull (V_{hull}) and the volume of the cavity (V_{cavity}).

Average density of the sub as a whole:

$$\rho_{average} = \frac{m_{sub}}{V_{sub}}$$

Total volume of the sub:

$$V_{sub} = V_{hull} + V_{cavity} = 1.130 \times 10^3 \text{ m}^3 + 7.600 \times 10^3 \text{ m}^3 = 8.730 \times 10^3 \text{ m}^3$$

From Equation 11-1 the mass of steel in the hull (m_{hull}) and the mass of air in the cavity (m_{cavity})

Rearrange Equation 11-1:

$$m = \rho V$$

are given by their respective densities multiplied by their respective volumes.

Solve for m_{hull} and m_{cavity}:

$$m_{hull} = \rho_{steel} V_{hull} = (7.910 \times 10^3 \text{ kg/m}^3)(1.130 \times 10^3 \text{ m}^3) = 8.938 \times 10^6 \text{ kg}$$

$$m_{cavity} = \rho_{air} V_{cavity} = (1.200 \text{ kg/m}^3)(7.600 \times 10^3 \text{ m}^3) = 9.120 \times 10^3 \text{ kg}$$

The total mass of the submersible (m_{sub}) is the sum of the mass of steel in the hull and the mass of air in the cavity.

$$m_{sub} = m_{hull} + m_{cavity} = 8.938 \times 10^6 \text{ kg} + 9.120 \times 10^3 \text{ kg} = 8.947 \times 10^6 \text{ kg}$$

To get the average density, substitute the total volume V_{sub} and the total mass m_{sub} into the expression for $\rho_{average}$.

$$\rho_{average} = \frac{m_{sub}}{V_{sub}} = \frac{8.947 \times 10^6 \text{ kg}}{8.730 \times 10^3 \text{ m}^3} = 1.025 \times 10^3 \text{ kg/m}^3$$

Reflect

No one component of the submersible has a density of 1.025×10^3 kg/m^3; instead, this represents an average of the densities of the two components. Note that the average density is intermediate between the density of air (1.200 kg/m^3) and the density of steel (7.910×10^3 kg/m^3).

This particular value of average density is greater than the density of fresh water (1.000×10^3 kg/m^3) but is the same as that of seawater (1.025×10^3 kg/m^3). As we will see in Section 11-8, this means that the submersible will sink in fresh water but will float when submerged in seawater—which is a useful thing for a deep-sea research vessel to do.

EXAMPLE 11-3 How Much Space for a Molecule?

A single molecule of water (H_2O) has a mass of 2.99×10^{-26} kg. Find the average volume per water molecule (a) in liquid water at 4°C and (b) in water vapor (that is, water in gas form) at 120°C, which has a density of 0.559 kg/m^3. In each case, what is the length of each side of a cube with this volume? Compare to the size of a water molecule, which is about 2.0×10^{-10} m across.

Set Up

The density ρ of a substance tells us the mass per volume. We want to find the volume per *molecule* in the substance. We'll do this calculation by combining the value of ρ for each substance (liquid water at 4°C and water vapor at 120°C) with the mass per water molecule. We'll imagine that each molecule is at the center of a cube of side L, and we'll determine L from the formula for the volume of a cube.

Definition of density:

$$\rho = \frac{m}{V} \quad (11\text{-}1)$$

Volume of a cube of side L:

$$V = L \times L \times L = L^3$$

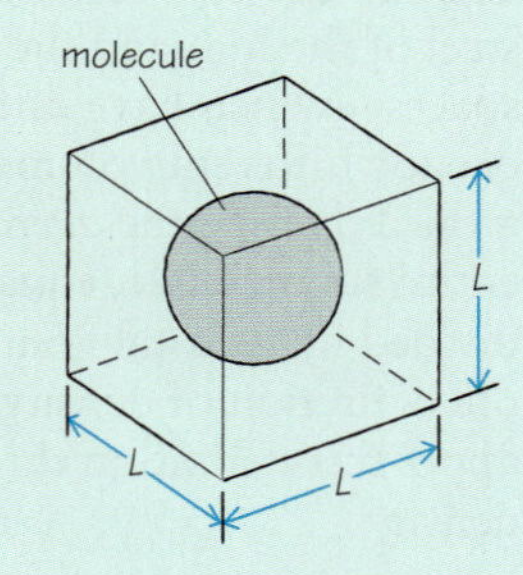

Solve

Rewrite Equation 11-1 to solve for the volume occupied by a quantity m of the substance.

From Equation 11-1,

$$V = \frac{m}{\rho}, \text{ or}$$

$$\text{volume occupied by a molecule} = \frac{\text{(mass of a molecule)}}{\text{(density)}}$$

(a) Table 11-1 tells us that water at 4°C has a density of 1.000×10^3 kg/m^3. Use this to find the volume per molecule and the length L of a cube with this volume.

For water at 4°C,

volume occupied by a molecule = V

$$= \frac{2.99 \times 10^{-26} \text{ kg}}{1.00 \times 10^3 \text{ kg/m}^3} = 2.99 \times 10^{-29} \text{ m}^3$$

The volume V is equal to L^3; that is, V is the cube of L. So L is the cube *root* of V:

$$L = \sqrt[3]{V} = \sqrt[3]{2.99 \times 10^{-29}\ \text{m}^3}$$
$$= 3.10 \times 10^{-10}\ \text{m}$$

This is about 50% larger than the size of a water molecule (roughly 2.0×10^{-10} m).

(b) Repeat the calculation for water vapor at 120°C. The molecule is the same and so has the same mass, but the density is far less than that for the liquid at 4°C:

For water vapor at 120°C,

volume occupied by a molecule = V

$$= \frac{2.99 \times 10^{-26}\ \text{kg}}{0.559\ \text{kg/m}^3}$$
$$= 5.35 \times 10^{-26}\ \text{m}^3$$

The length L of a cube with this volume is

$$L = \sqrt[3]{V} = \sqrt[3]{5.35 \times 10^{-26}\ \text{m}^3}$$
$$= 3.77 \times 10^{-9}\ \text{m}$$

This is larger than the size of a water molecule by a factor of $(3.77 \times 10^{-9}\ \text{m})/(2.0 \times 10^{-10}\ \text{m}) = 19$.

Reflect

In both liquids and gases, the molecules are in constant motion. That's why the volumes we calculated in this example are the *average* volume per molecule.

Our results show that in the liquid state a water molecule almost fills the volume that it occupies on average. So there is very little extra room to move the molecules closer together, which means that it is very difficult to compress a liquid. In the gaseous state, however, on average each molecule is surrounded by lots of empty space. In this state it's much easier to push the molecules closer together, and so a gas is relatively easy to compress.

GOT THE CONCEPT? 11-1 Weight and Density

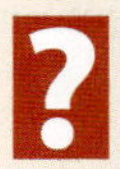

Which is heavier, a block of iron or a block of ice? (a) The block of iron; (b) the block of ice; (c) they have the same weight; (d) not enough information given to decide.

TAKE-HOME MESSAGE FOR Section 11-2

✔ The density of a substance describes how much mass of a particular substance occupies a given volume.

✔ The specific gravity of a substance equals the density of that substance relative to the density of fresh water at 4°C.

✔ An object's average density equals its total mass divided by its overall volume. It is intermediate in value between the densities of the least dense and most dense constituents of the object.

11-3 Pressure in a fluid is caused by the impact of molecules

Pressure is a commonplace idea. Everyone has had their blood pressure measured during a visit to the physician. An air-filled balloon bursts if the pressure within it is too great. And weather reports refer to low-pressure and high-pressure weather systems in the atmosphere. But what *is* pressure?

Figure 11-3 shows how pressure is defined. If you place a small, thin, flat plate in a fluid—a gas or a liquid—you'll find that the fluid exerts a force on either face of the plate. You can feel these forces if you stick your hand into a swimming pool or a sink full of water. The same kind of forces act on your hand when it's surrounded by air, but are very feeble compared to the forces exerted by water. The net force that the fluid exerts on either face of the plate acts perpendicular to the face. The **pressure** of the fluid acting on the face is the magnitude of this perpendicular force divided by the area of the face:

Definition of pressure (11-2)

$$p = \frac{F_\perp}{A}$$

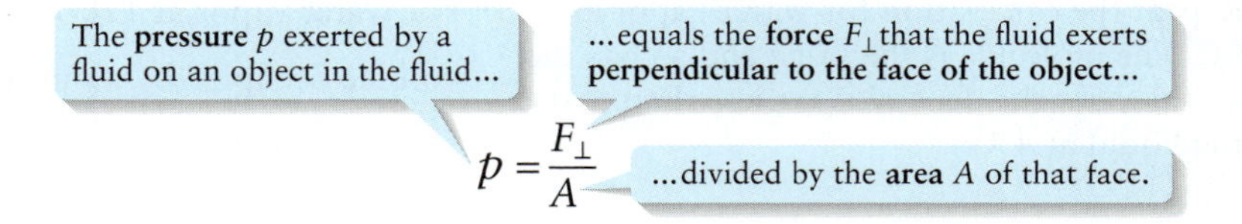

The pressure p exerted by a fluid on an object in the fluid...
...equals the force $F_\perp$ that the fluid exerts perpendicular to the face of the object...
...divided by the area A of that face.

WATCH OUT! The symbols for pressure and density are not the same.

We use a lowercase p to represent pressure and the lower case Greek letter ρ (rho) to represent density. These variables look a lot alike, so you need to be careful, especially when you write them, that you properly distinguish between the two.

To understand why a fluid exerts forces like those shown in Figure 11-3, keep in mind that the molecules of a liquid or gas are in constant motion. Any object placed in a fluid is struck repeatedly by these moving molecules. The force of each individual impact is miniscule, but there are so many molecules and so many impacts that their combined effect is appreciable. It's that combined force that gives rise to the forces labeled $F_\perp$ in Figure 11-3. As an example, more than 10^{23} air molecules collide with a square centimeter of your skin every second, exerting a combined force of about 10 N on that square centimeter. In a fluid at rest (such as water in the kitchen sink that is not flowing or draining out, or the air in a room with no wind currents), the molecules of the fluid move in entirely random directions. At any point in the fluid there are as many molecules moving to the left as to the right, as many moving upward as downward, and so on. Hence a small, thin, flat plate placed in the fluid (Figure 11-3) experiences the same number of impacts, and hence the same magnitude of force, on both of its sides no matter which way the plate is oriented.

If we double the surface area A exposed to the fluid, the force magnitude $F_\perp$ also doubles because there is twice as much area to be hit by the molecules of the fluid. However, the pressure $p = F_\perp/A$ remains the same. Pressure is a property of the fluid, not of the object that's immersed in the fluid. Furthermore, because the pressure is the same no matter how the disk in Figure 11-3 is oriented, pressure has no direction: Unlike force, pressure is *not* a vector.

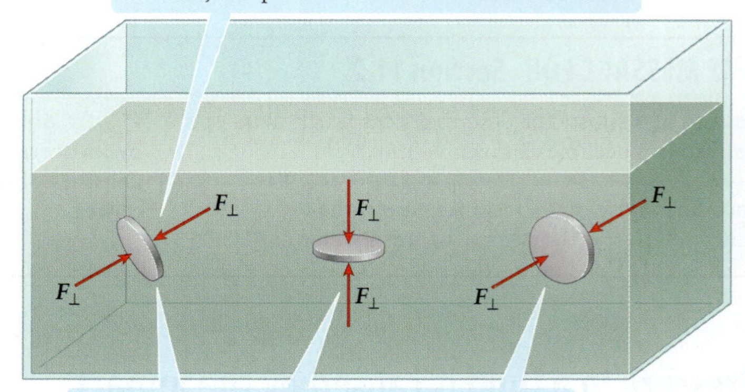

Figure 11-3 Pressure on a thin plate in a fluid The fluid pressure on this plate does not depend on its orientation.

The SI units of pressure are newtons per square meter (N/m^2). As we discussed in Section 9-2, one newton per square meter is also called a **pascal** (abbreviated Pa): 1 Pa = 1 N/m^2. We'll discuss other common units of pressure in Section 11-5.

WATCH OUT! Don't confuse force and pressure.

Although force and pressure are related, they are *not* the same quantity. Here's an example. If you take off your shoes and stand up, your mass is distributed evenly over the surface area of your two feet. The pressure on the soles of your feet is your mass divided by the surface area of the bottom of your two feet and is probably not uncomfortable. But if you stood on a bed of nails, it would hurt! Your weight hasn't changed, so the force you exert on the surface supporting you hasn't changed. However, because the supporting surface—the tips of a few nails—has a much smaller surface area, the pressure on your feet is much greater (and much more painful) in this case. However, people can comfortably *lie* on a suitably prepared bed of nails. By lying flat, a person's weight is distributed over the area of the tips of many nails, resulting in a tolerable pressure (**Figure 11-4**).

Figure 11-4 A bed of nails It doesn't hurt to lie on this bed of nails. That's because the force that the nails exert is distributed over the area of many nails, resulting in a tolerable pressure on the person's back.

At sea level on Earth, the pressure of the atmosphere is about 1.01×10^5 N/m^2, or 1.01×10^5 Pa. By contrast, the pressure at an altitude of 15,000 m (41,000 ft, about the highest altitude at which airliners fly) is only 1.78×10^4 Pa, or 17.6% of the pressure at sea level. One reason the pressure is lower at high altitude is that the air is less dense there (0.289 kg/m^3 at 15,000 m versus 1.23 kg/m^3 at sea level). Hence there are fewer air molecules to produce pressure through collisions. A second reason is that the molecules at high altitude move more slowly than those near sea level, so when collisions do take place they are relatively gentle. We'll learn in Chapter 14 that gas molecules move more slowly in a cold gas than a warm gas. The air at 15,000 m is indeed quite cold: The average temperature there is −56°C (−69°F) compared to 15°C (59°F) at sea level. The air pressure at 15,000 m is too low to sustain human life, which is why airliners have *pressurized* cabins: Air from the jet engine intakes is pumped into the cabin, increasing the pressure inside the cabin to what it would be at an altitude of 1800−2400 m (6000−8000 ft) above sea level.

The variation of pressure with elevation isn't unique to our atmosphere. In oceans, in lakes, and even in a glass of water there is greater pressure as you go deeper and lower pressure as you ascend. In the next section we'll learn more about how and why pressure increases with depth in a fluid.

GOT THE CONCEPT? 11-2 Something's Fishy

A tropical fish is at rest in the middle of an aquarium. Compared to the pressure on the left side of the fish, the pressure on the right side of the fish has (a) the same magnitude and the same direction; (b) the same magnitude and opposite direction; (c) a different magnitude and the same direction; (d) a different magnitude and opposite direction; (e) none of these.

TAKE-HOME MESSAGE FOR Section 11-3

✔ A fluid exerts pressure on any object with which it is in contact.

✔ The pressure equals the force that the fluid exerts perpendicular to the object's face divided by the area of that face.

(a)

Roger Freedman

(b)

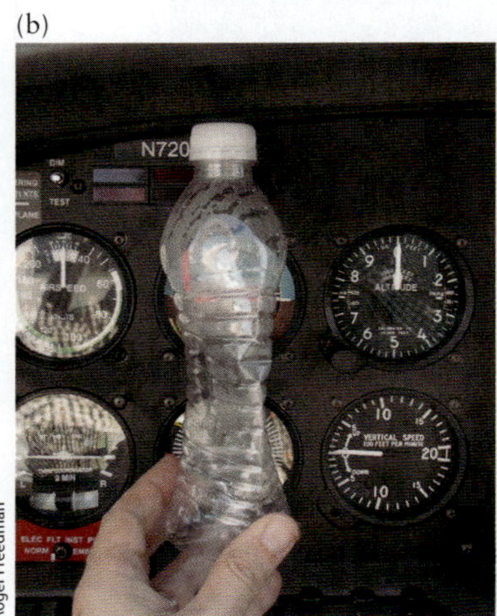

Roger Freedman

Figure 11-5 Air pressure varies with elevation (a) This empty plastic bottle was opened to the air in a non-pressurized airplane at an altitude of 9000 ft (2740 m) then tightly capped. (b) Outside air pressure made the bottle collapse when it was returned to sea level.

11-4 In a fluid at rest pressure increases with increasing depth

Figure 11-5 shows the results of an experiment with air in our atmosphere. An empty plastic bottle is opened at an altitude of 9000 ft (2740 m) above sea level so that the bottle fills with air at that altitude. The bottle is then tightly capped (**Figure 11-5a**). As the bottle is brought back to sea level, it compresses as though it had been squeezed by an invisible hand (see **Figure 11-5b**).

Why does this happen? The answer is that the pressure in a fluid *increases* as you go *deeper* into the fluid. For example, air pressure in the atmosphere is more than 25% greater at sea level than it is at an altitude of 2000 m. The air inside the bottle in Figure 11-5 was at the pressure found at 2740 m, so it compressed when it was brought to sea level and was surrounded by air at higher pressure. In this section we'll explore the relationship between pressure and depth in a fluid.

Hydrostatic Equilibrium

Let's consider a large tank filled with a fluid at rest (**Figure 11-6**). When a fluid is at rest (that is, not flowing), we say it is in **hydrostatic equilibrium**. The term *hydrostatic* specifically refers to water at rest, but it's common to use this term to refer to *any* kind of fluid in equilibrium.

Now imagine a box-shaped volume of fluid within the tank, as in Figure 11-6. The area of the top and bottom of the box is A, and the height of the box is d. We won't put an actual box in the fluid; instead we simply imagine a boundary that separates the box-shaped volume from the rest of the fluid.

The weight of the fluid above the box exerts a downward force of magnitude F_{down} on the box. Because fluids exert a force in all directions, the fluid below the box pushes upward on the box with a force of magnitude F_{up}. (We'll see in a moment that F_{up} is *not* equal to F_{down}.) The fluid also exerts forces on the sides of the box, but the *net* force on the sides is zero: There is as much force on the left side as on the right side, and these forces cancel. Finally, we must consider the force of gravity on the fluid inside the box. The fluid has weight. This weight appears as a downward force $\vec{w}$ on the fluid volume. Figure 11-6 shows the three remaining forces on the box—an upward force on the bottom, a downward force on the top, and the downward force of gravity.

Consider a box-shaped volume of fluid that is part of a larger quantity of fluid at rest.

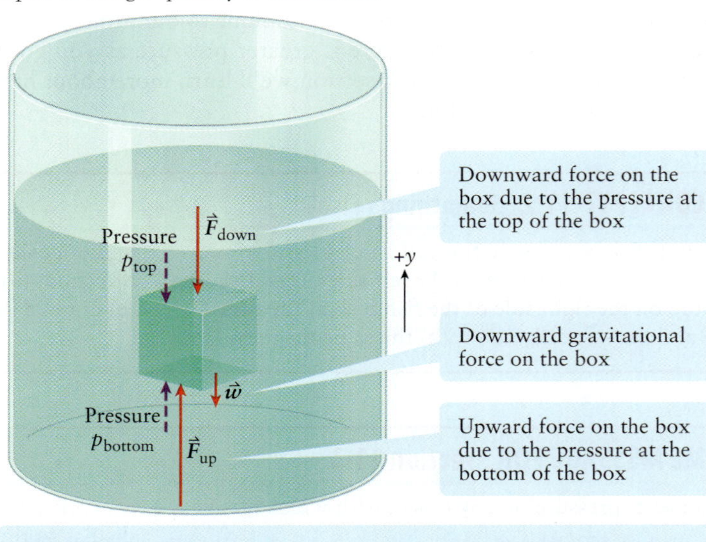

In order for the fluid to be in **hydrostatic equilibrium**, the net force on the box must be zero: $\Sigma\vec{F} = \vec{F}_{up} + \vec{F}_{down} + \vec{w} = 0$

Figure 11-6 Hydrostatic equilibrium We can derive the equation of hydrostatic equilibrium by considering a box-shaped portion of a fluid at rest. For the box to be in equilibrium, the pressure at its bottom must be greater than the pressure at its top.

We started with the assumption that the fluid in the tank is at rest. This requires that there be no *net* force on the fluid volume; that is, the vector sum of the three forces acting on the box must be zero:

$$\sum \vec{F} = \vec{F}_{\text{up}} + \vec{F}_{\text{down}} + \vec{w} = 0 \tag{11-3}$$

All of these forces are vertically upward or downward, so we need only the y components of these forces. Taking the positive y direction to be upward, Equation 11-3 becomes

$$\sum F_y = F_{\text{up}} - F_{\text{down}} - w = 0 \tag{11-4}$$

We can see that because the fluid in the box has a nonzero weight w, the forces F_{up} and F_{down} are *not* equal. From Equation 11-2, $p = F_{\perp}/A$, we can express the magnitude F_{up} of the upward force (which acts perpendicular to the underside of the box) as the pressure p_{bottom} at the bottom of the box multiplied by the area A of the bottom of the box. Similarly the magnitude F_{down} of the downward force equals the pressure p_{top} at the top of the box multiplied by the area A of the top of the box. Furthermore the weight w of the fluid in the box equals the product of the mass m of fluid in the box and the acceleration due to gravity g. So Equation 11-4 becomes

$$p_{\text{bottom}}A - p_{\text{top}}A - mg = 0 \quad \text{or} \quad p_{\text{bottom}}A = p_{\text{top}}A + mg \tag{11-5}$$

Equation 11-5 tells us that because of the weight mg of the fluid in the box, the pressure p_{bottom} has to be greater than the pressure p_{top}. In other words the pressure of the fluid is greater at a point that's deep in the fluid (the bottom of the box) than at a point that's less deep (the top of the box) and so pressure increases with increasing depth in the fluid.

A Uniform-Density Fluid: The Equation of Hydrostatic Equilibrium

We can simplify Equation 11-5 even further if we assume that the fluid has **uniform density**; that is, the density of the fluid has the same value throughout the fluid. The mass m of the volume of fluid in the box is the product of the density of the fluid, ρ, and the volume of the box. The height of the rectangular box is d, so its volume is $V = Ad$, and the mass of the fluid in the box is $m = \rho V = \rho Ad$. So Equation 11-5, the statement that the fluid in the box remains at rest, becomes

$$p_{\text{bottom}}A = p_{\text{top}}A + \rho Adg \tag{11-6}$$

Now divide each term in Equation 11-6 by A:

$$p_{\text{bottom}} = p_{\text{top}} + \rho gd \tag{11-7}$$

In other words the pressure at the bottom of the box is greater than the pressure at the top of the box. Furthermore, the box is just an imaginary construct, so the real meaning of Equation 11-7 is as follows: At *any* two points in the fluid with vertical separation d, the pressure at the lower point is greater than the pressure at the upper point by an amount ρgd. In particular let p_0 be the pressure at the upper point (Figure 11-6). Then the pressure p at a point a distance d below the upper point is

$$p = p_0 + \rho gd \tag{11-8}$$

Pressure at a certain point in a **fluid at rest**

Density of the fluid (same at all points in the fluid)

Pressure at a second, lower point in the fluid

Acceleration due to gravity

Depth of the second point where the pressure is p below the point where the pressure is p_0

Variation of pressure with depth in a fluid with uniform density (equation of hydrostatic equilibrium)
(11-8)

Equation 11-8 must be satisfied for a fluid to remain at rest, so this is also called the **equation of hydrostatic equilibrium.**

WATCH OUT! Understand what Equation 11-8 tells you.

Be careful with the sign of the $\rho g d$ term in Equation 11-8: Remember that the symbol d means *depth*. If the second point is *below* the point where the pressure is p_0, d is positive and p is greater than p_0. If the second point is *above* the point where the pressure is p_0, d is *negative* and p is less than p_0. In other words pressure increases as you go deeper in a fluid such as the ocean but decreases as you ascend. Note also that Equation 11-8 is based on the assumption that the fluid has the same density ρ at all points. If the density is noticeably different at different depths (as is the case for the air in our atmosphere), Equation 11-8 gives only approximate results.

EXAMPLE 11-4 Air Pressure versus Water Pressure

You dig a swimming pool in your backyard that is 2.00 m deep. What is the pressure at the bottom of the pool (a) before you fill it with water and (b) after it is filled with water? At the time that you make the measurements, the air pressure at ground level is 1.0100×10^5 Pa.

Set Up

We are given the pressure $p_0 = 1.0100 \times 10^5$ Pa at the top of the pool and want to find the pressure p at a depth $d = 2.00$ m below the top of the pool. We'll use Equation 11-8 for this purpose. The only difference between parts (a) and (b) is the density ρ of the fluid: In part (a) we use the density of air, while in part (b) we use the density of water.

Variation of pressure with depth:

$$p = p_0 + \rho g d \qquad (11\text{-}8)$$

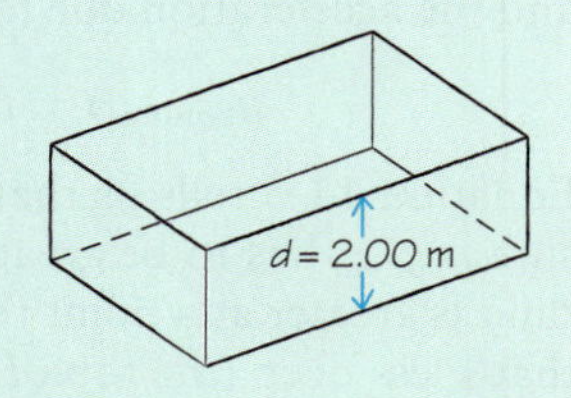

Solve

(a) From Table 11-1 the density of air at sea level is $\rho = 1.23$ kg/m^3. Substitute this into Equation 11-8.

Pressure at the bottom of the pool filled with air:

$$p = (1.0100 \times 10^5 \text{ Pa}) + (1.23 \text{ kg/m}^3)(9.80 \text{ m/s}^2)(2.00 \text{ m})$$
$$= 1.0100 \times 10^5 \text{ Pa} + 24.1 \text{ kg/(m}\cdot\text{s}^2)$$

Convert units: Recall that

$$1 \text{ N} = 1 \text{ kg}\cdot\text{m/s}^2 \text{ so}$$
$$1 \text{ Pa} = 1 \text{ N/m}^2 = 1 \text{ kg/(m}\cdot\text{s}^2)$$

So the pressure at the bottom of the air-filled pool is

$$p = 1.0100 \times 10^5 \text{ Pa} + 24.1 \text{ Pa}$$
$$= 1.0100 \times 10^5 \text{ Pa} + 0.000241 \times 10^5 \text{ Pa}$$
$$= 1.0102 \times 10^5 \text{ Pa}$$

(b) When the pool is filled with water, the calculation is the same as in part (a) except that we use the density of water from Table 11-1, $\rho = 1.000 \times 10^3$ kg/m^3.

Pressure at the bottom of the pool filled with water:

$$p = (1.0100 \times 10^5 \text{ Pa}) + (1.000 \times 10^3 \text{ kg/m}^3)(9.80 \text{ m/s}^2)(2.00 \text{ m})$$
$$= 1.0100 \times 10^5 \text{ Pa} + 1.96 \times 10^4 \text{ kg/(m}\cdot\text{s}^2)$$
$$= 1.0100 \times 10^5 \text{ Pa} + 0.196 \times 10^5 \text{ Pa}$$
$$= 1.206 \times 10^5 \text{ Pa}$$

Reflect

The pressure at the bottom of the air-filled pool is only 0.02% greater than at the top of the pool. That's such a tiny difference that we can ignore it. It takes a much greater variation in altitude than 2.00 m for the pressure difference in the air to be noticeable. (Your ears—which are sensitive pressure sensors—may "pop" when you drive up into the mountains, but they won't "pop" when you climb a ladder or a single flight of stairs.)

There *is* a noticeable pressure difference between the top and bottom of the water-filled pool: The pressure is about 20% greater at the bottom than at the surface. The difference is that water is almost a thousand times denser than air. So a swimming pool full of water weighs almost a thousand times more than a swimming pool full of air and so it produces an additional pressure almost a thousand times greater.

GOT THE CONCEPT? 11-3 Watching a U-tube

? You pour water into a U-shaped tube, as shown in Figure 11-7. The left-hand leg of the tube is 2.00 cm in radius, while the right-hand leg is 1.00 cm in radius. When the water is in equilibrium, how will the height of the water in the left-hand leg and the pressure at the bottom of the left-hand leg compare to the height of water in the right-hand leg and the pressure at the bottom of the right-hand leg? (a) Greater height, greater pressure; (b) same height, same pressure; (c) lower height, lower pressure; (d) same height, greater pressure; (e) none of these.

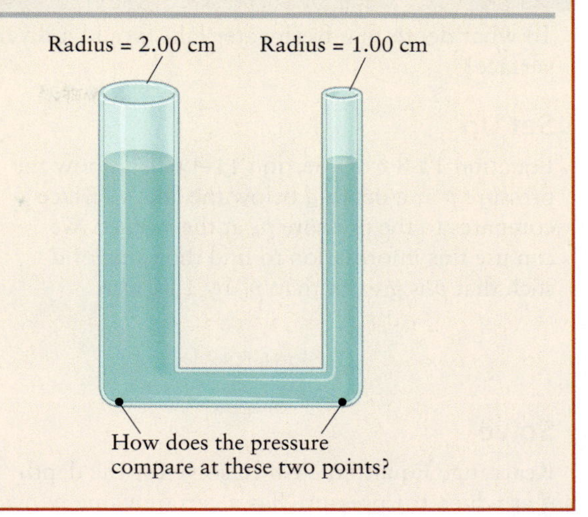
Radius = 2.00 cm
Radius = 1.00 cm
How does the pressure compare at these two points?

Figure 11-7 A hollow U-tube If water is poured into this hollow tube, how will the water level in each leg and the pressure at the bottom of each leg compare?

TAKE-HOME MESSAGE FOR Section 11-4

✔ In a fluid in hydrostatic equilibrium (at rest), pressure increases as you go deeper into the fluid.

✔ The pressure difference between any two points within a uniform fluid in hydrostatic equilibrium depends only on the difference in the vertical height of the two points and the density of the fluid.

11-5 Scientists and medical professionals use various units for measuring fluid pressure

You might not be aware of it, but all of the air in the atmosphere directly above your head has weight, and that weight is pushing down on you right now. To be more precise, it's pushing in on you and exerting a considerable pressure from every direction.

The air pressure at sea level varies somewhat due to the passage of weather systems, but on average is equal to 1.01325×10^5 Pa in SI units or about 14.7 pounds per square inch (lb/in^2) in English units. This average value of atmospheric pressure at sea level is defined to be one **atmosphere,** abbreviated atm:

$$1 \text{ atm} = 1.01325 \times 10^5 \text{ Pa} = 14.7 \text{ lb/in}^2$$

To three significant figures, 1 atm = 1.01×10^5 Pa. Since 1 Pa = 1 N/m^2, this means that if you paint a square 1 m on a side onto the ground at sea level, the weight of the column of air that sits atop that square and extends to the upper limit of Earth's atmosphere is 1.01×10^5 N. A metric ton has a mass $m = 1000$ kg and a weight $mg = (1000 \text{ kg})(9.80 \text{ m/s}^2) = 9.80 \times 10^3$ N; the weight of air above that 1-m square is more than 10 times greater (to be precise, 10.3 metric tons). If you prefer to think in English units, paint a 1-in. square onto the ground at sea level: The weight of air above that square is 14.7 lb.

The atmosphere (atm) is a convenient unit for dealing with relatively large pressures. For example, the pressures found at the bottom of the ocean can exceed 10^3 atm, and high-pressure gas systems can operate at pressures of hundreds or even thousands of atmospheres. Very *low* pressures are often measured in pascals rather than atmospheres; for example, a high-quality vacuum pump can reduce the pressure inside a container to 10^{-5} Pa or lower.

EXAMPLE 11-5 Diver's Rule of Thumb

To what depth in a freshwater lake would a diver have to descend for the pressure to be 1.00 atm greater than at the surface?

Set Up

Equation 11-8 from Section 11-4 tells us how the pressure p at a depth d below the lake's surface compares to the pressure p_0 at the surface. We can use this information to find the value of d such that p is greater than p_0 by 1.00 atm.

Variation of pressure with depth:

$$p = p_0 + \rho g d \qquad (11\text{-}8)$$

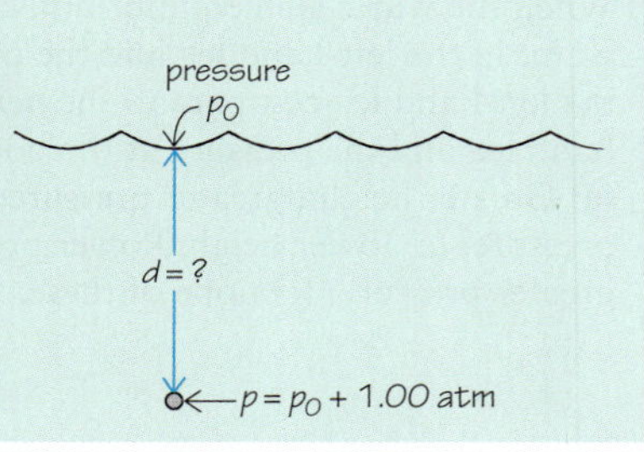

Solve

Rearrange Equation 11-8 to solve for the depth d at which the pressure has a certain value p.

From Equation 11-8,

$$p - p_0 = \rho g d$$

$$d = \frac{p - p_0}{\rho g}$$

We want the pressure p at depth d to be greater than the pressure p_0 at the surface by 1.00 atm, so $p - p_0 = 1.00$ atm. Substitute this value as well as the value of density ρ for fresh water from Table 11-1. Then use the conversion factors 1 atm = 1.01×10^5 Pa = 1.01×10^5 N/m^2 and 1 N = 1 kg·m/s^2.

Pressure difference:

$$p - p_0 = 1.00 \text{ atm}$$

Density of fresh water from Table 11-1:

$$\rho = 1.000 \times 10^3 \text{ kg/m}^3 \text{ so}$$

$$d = \frac{p - p_0}{\rho g}$$

$$= \frac{1.00 \text{ atm}}{(1.000 \times 10^3 \text{ kg/m}^3)(9.80 \text{ m/s}^2)} \times \left(\frac{1.01 \times 10^5 \text{ N/m}^2}{1 \text{ atm}}\right)$$

$$= 10.3 \frac{\text{m}^3}{\text{kg}} \frac{\text{s}^2}{\text{m}} \frac{\text{N}}{\text{m}^2} \times \left(\frac{1 \text{ kg} \cdot \text{m/s}^2}{1 \text{ N}}\right)$$

$$= 10.3 \text{ m}$$

Reflect

Our answer is a little bit more than 10 m, so it's a good approximation (a "rule of thumb" for divers) to say that our diver experiences a pressure change of 1 atm going from the surface to a depth of 10 m. You can see that descending an additional 10 m would increase the pressure by another atmosphere, and so on.

Measuring Pressure

You've almost certainly encountered pressure measurements when you've gone to the doctor or listened to a weather report. However, the value of pressure was probably not presented to you with units of atmospheres, Pa, or lb/in^2, but rather in either inches or millimeters of mercury. To understand these units of pressure, we'll examine how to construct a simple device called a *barometer* that's used to measure air pressure.

Fill a tube closed at one end with mercury (Hg), which has density ρ_{Hg}. While keeping the open end sealed, turn the tube upside down into a pan partially filled with mercury. (***Warning:*** Mercury is poisonous, so do *not* try this experiment at home!) Now release the seal on the end of the tube, which is beneath the surface of the mercury in the pan. No air can get into the tube, so an equal volume of vacuum must remain in the place of whatever volume of fluid drains from the tube. **Figure 11-8** shows this process.

Let's consider what happens once the system has come to hydrostatic equilibrium and a column of mercury of height H (measured from the surface of the mercury of the

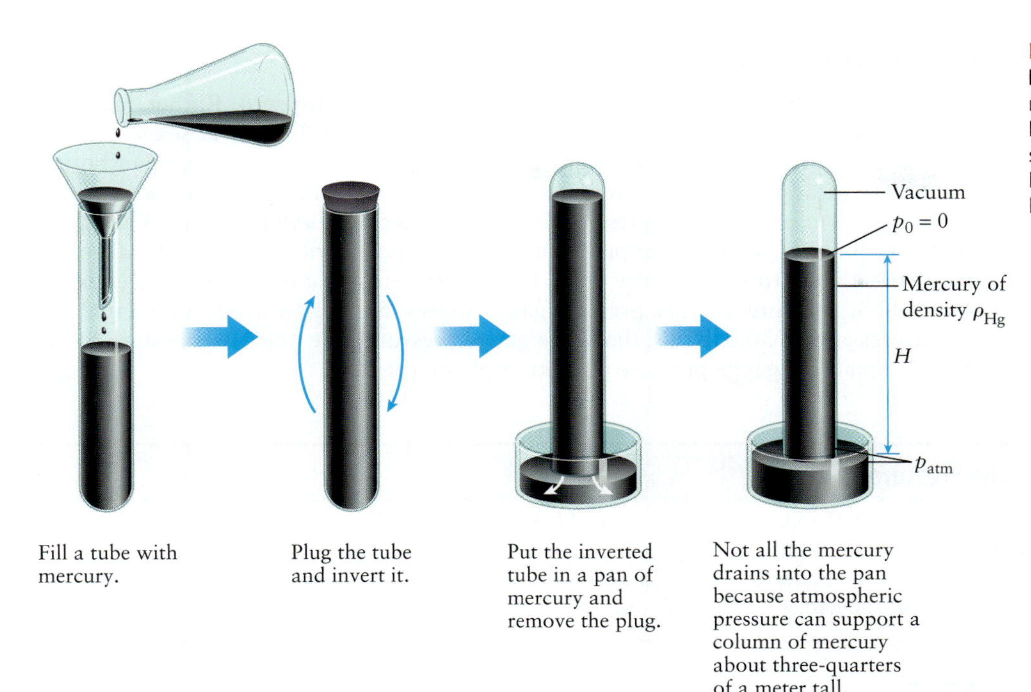

Vacuum
$p_0 = 0$
Mercury of density ρ_{Hg}
H
p_{atm}
Fill a tube with mercury.
Plug the tube and invert it.
Put the inverted tube in a pan of mercury and remove the plug.
Not all the mercury drains into the pan because atmospheric pressure can support a column of mercury about three-quarters of a meter tall.

Figure 11-8 Constructing a simple barometer A simple barometer to measure atmospheric pressure can be made by inverting a tube of fluid such as mercury in a pan. A height H of liquid remains in the tube, supported by the outside air pressure.

pan) remains in the tube. The pressure at the top of this column is $p_0 = 0$ because the pressure in a vacuum is zero. The pressure at the exposed surface of the mercury in the pan equals atmospheric pressure p_{atm}. Since the mercury is in hydrostatic equilibrium, the pressure at a point inside the mercury column at the same elevation as the exposed mercury surface is *also* equal to p_{atm}. From the equation of hydrostatic equilibrium, Equation 11-8, the pressure p_{atm} is greater than p_0 by $\rho_{Hg}gH$:

$$p_{atm} = p_0 + \rho_{Hg}gH$$

Since $p_0 = 0$ in the vacuum at the top of the mercury column, this equation becomes

$$p_{atm} = \rho_{Hg}gH \tag{11-9}$$

(atmospheric pressure measured with a mercury barometer)

As the atmospheric pressure p_{atm} rises and falls, the height H of the mercury column rises and falls along with it. Hence the height of the mercury column is a direct measure of atmospheric pressure.

A mercury barometer like that shown in Figure 11-8 is not a perfect instrument for measuring pressure: The density of mercury ρ_{Hg} changes with temperature, and the value of the acceleration due to gravity g depends on location. But at a temperature of 0°C so that the density of mercury is $\rho_{Hg} = 1.35951 \times 10^3\ \text{kg/m}^3$ and at a location where $g = 9.80665\ \text{m/s}^2$ (a "standard" value of g), the height of a column of mercury supported by exactly 1 atm of pressure is found to be 760 mm. For that reason pressures are sometimes given in **millimeters of mercury,** or mmHg for short. Thus 1 atm = 760 mmHg. An equivalent (and preferred) unit is the **torr,** defined as 1/760 atm. This unit is named for the Italian scientist and mathematician Evangelista Torricelli, who invented the mercury barometer in the 1640s. The torr is a more appropriate unit than mmHg since most modern barometers work on different principles that don't involve columns of mercury.

If mercury is poisonous, why was it used in barometers? It's a case of convenience triumphing over safety. Equation 11-9 shows that the greater the density of the fluid, the lower the height H of the fluid column in the barometer. Because mercury is so dense—13.6 times denser than water—a mercury barometer is relatively compact and convenient to use. If you wanted to make a barometer like that shown in Figure 11-8 but using water rather than mercury, it would have to be 13.6 times taller than a mercury barometer. Can you show that if $p_{atm} = 1$ atm, the height of the liquid column in a water barometer would be 10.3 m—far too tall to be convenient?

Gauge Pressure

What is the air pressure inside a flat automobile tire? It's not zero: Rather, it's equal to atmospheric pressure (about 1 atm, 1.01×10^5 Pa, or 14.7 lb/in^2 at sea level). The same is true for a balloon that hasn't been inflated and for an empty, open bottle, because air at atmospheric pressure is inside these vessels. But if you use a tire pressure gauge to measure the air pressure in the deflated tire, it *will* read zero. This **gauge pressure** shows how much the pressure *exceeds* atmospheric pressure. The **absolute pressure**—that is to say, the true value of pressure—inside a deflated tire is equal to atmospheric pressure. If a tire pressure gauge shows that the air inside an inflated tire is at a pressure of 30.0 lb/in^2, that's the gauge pressure: The absolute pressure inside the tire equals the gauge pressure plus atmospheric pressure.

GOT THE CONCEPT? 11-4 Rank the Pressures

? You have two balloons at sea level. One is deflated, and the other is inflated to 0.50 atm according to a pressure gauge connected to the mouth of the balloon. Rank the following from highest to lowest: (a) the gauge pressure inside the deflated balloon; (b) the absolute pressure inside the deflated balloon; (c) the gauge pressure inside the inflated balloon; (d) the absolute pressure inside the inflated balloon.

BIO-Medical Blood Pressure

Most of us encounter a pressure gauge when we visit the doctor and have our blood pressure measured (**Figure 11-9**). Because the heart is a pump, the pressure in the system changes over the period of each heartbeat. Blood pressure is usually given as two numbers, for example, "120 over 80." These values are measured in units of mmHg and represent the high and low values of the *gauge* pressure in the arteries at the level of the heart.

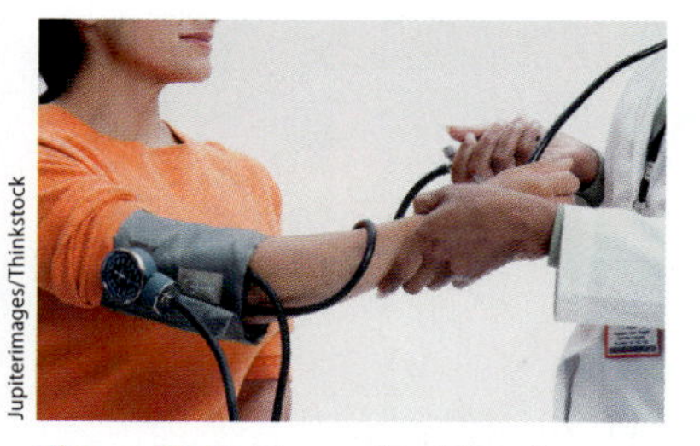
Jupiterimages/Thinkstock

Figure 11-9 Measuring blood pressure A blood pressure cuff is applied around the upper arm, at the same elevation as the patient's heart. If the pressure were measured at a different elevation, the values would be either lower or higher than the actual pressures produced by the heart.

Blood pressure is usually measured using a cuff around your upper arm, at the same elevation as your heart. We can understand why this is so by using the relationship between pressure and depth given by Equation 11-8. The blood within the body behaves rather like the fluid in the container shown in Figure 11-6: The farther down you go within the fluid, the greater the pressure. Since the heart is the principal organ of the circulatory system, let's have p_0 be the pressure at the position of the heart. Then Equation 11-8 tells us that the pressure of blood in the feet will be higher than the pressure in the heart, and the pressure in the head will be lower. To measure the actual pressure produced by the heart's pumping action, it's important to measure that pressure at the same elevation as the heart itself—which explains the placement of a blood pressure cuff.

The average blood pressure (gauge pressure) in an upright person's feet is about 100 mmHg higher than when it leaves the heart, or about twice as great as the gauge pressure measured in a person's arm. In contrast, the pressure in the head is about 50 mmHg less than that at the heart. If you could somehow stretch your neck upward by a meter or so, the blood pressure in your head would drop to zero and no blood would reach your brain. So it's all for the best that you can't stretch your neck that far! (Some animals do have necks that are even longer: The head of an adult giraffe is about 2 m higher than his heart. To get blood to the brain, the giraffe heart has to generate much higher pressure than a human heart does—about 260 mmHg!)

These differences in pressure between different regions of the body affect the flow of blood in the veins returning to the heart from the head and neck. Blood pressure is always lower in veins than in arteries; if you can feel a pulse in your neck, you're feeling the carotid *artery* with its relatively high pressure. The gauge pressure of blood entering the heart is very close to zero. Since the pressure above the heart is even lower, the veins of the head and neck tend to collapse! As blood continues to enter the veins, pressure builds up, and the vessels reopen. But as the blood flows through them, the pressure will again drop, and the veins will collapse again. The result is intermittent venous blood flow in the head and neck.

You may be wondering why the gauge pressure of the blood entering the heart is so much lower than the gauge pressure of the blood leaving the heart, when the blood is at essentially the same height. Equation 11-8 seems to indicate that these two pressures should be the same. However, you need to remember that Equation 11-8 is valid only under the conditions of *hydrostatic* equilibrium. Because blood is pumped by the heart and flows through the circulatory system, the blood is certainly *not* in hydrostatic equilibrium and we cannot use Equation 11-8. In Section 11-11 we'll discuss viscous flow, which helps to explain the difference in arterial and venous pressure.

TAKE-HOME MESSAGE FOR Section 11-5

✔ The pressure exerted by air at sea level is approximately equal to 1 atm. This pressure is due to the weight of a column of air that extends to the top of our atmosphere.

✔ Gauge pressure is measured relative to some reference pressure. Usually this reference pressure is atmospheric pressure.

✔ Blood pressure is measured in mmHg, which is the height of a column of mercury the pressure could support. An equivalent unit to mmHg which physicists prefer is the torr.

11-6 A difference in pressure on opposite sides of an object produces a net force on the object

What makes a cork fly out of a champagne bottle? What makes the pistons move in the cylinders of an automobile engine? And what makes your beverage move up the straw into your mouth as you sip? In each case the cause is not strictly pressure but rather pressure *difference*. The cork pops out because carbon dioxide gas inside the champagne bottle is at higher pressure than the air outside the bottle. Burning gasoline in the cylinder produces hot gas on one side of the piston that's at much higher pressure than the air on the other side, forcing the piston to move. And when you suck on a straw, you reduce the pressure in your mouth to a lower value than in the outside air, and the pressure difference drives liquid up the straw.

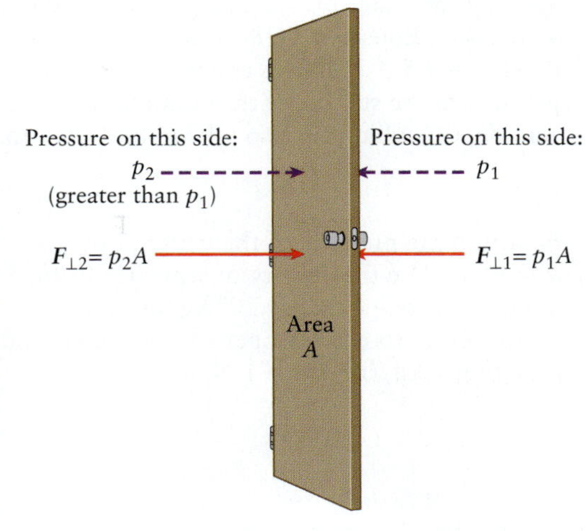

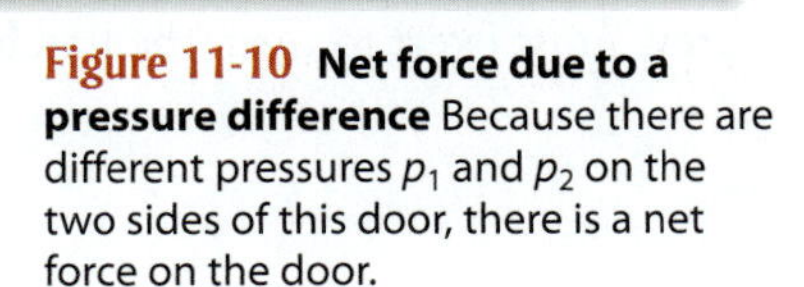

Figure 11-10 Net force due to a pressure difference Because there are different pressures p_1 and p_2 on the two sides of this door, there is a net force on the door.

As a simple example of the force produced by a pressure difference, consider an object like a door with fluid on each side. The object has area A, there is pressure p_1 on one side of the object, and there is a greater pressure p_2 on the other side (**Figure 11-10**). By rearranging Equation 11-2, $p = F_\perp/A$, we can calculate the magnitude of the force that each fluid exerts perpendicular to the object's surface:

$$\text{Force on side 1: magnitude } F_{\perp 1} = p_1 A$$
$$\text{Force on side 2: magnitude } F_{\perp 2} = p_2 A$$

Since $p_2 > p_1$, the force $F_{\perp 2}$ on side 2 is greater than the force $F_{\perp 1}$ on side 1. These forces are in opposite directions, so the *net* force on the object has magnitude

$$F_{net} = F_{\perp 2} - F_{\perp 1} = p_2 A - p_1 A$$

A **pressure difference** on two opposite sides of an object **produces a net force.**

Area of each side of the object

$$F_{net} = (p_2 - p_1)\,A \qquad (11\text{-}10)$$

High pressure on one side of the object

Lower pressure on the other side of the object

Net force on an object due to a pressure difference (11-10)

That is, the net force on the object due to the pressures acting on it is proportional to the difference in pressure between the object's two sides. Note that Equation 11-10 refers *only* to the net force on an object due to a difference in pressure on its two sides. It doesn't include other forces such as gravity that may act on the object.

EXAMPLE 11-6 Submarine Hatch

The air in the crew compartment of a research submarine is maintained at a pressure of 1.00 atm so that the crew can breathe normally. The sub operates at a depth of 85.0 m below the surface of the ocean. At this depth what force would be required to push open a hatch of area 2.00 m^2?

Set Up

There is 1.00 atm of air pressure on the inside of the hatch (call this p_{air}) pushing the hatch outward, but this is much less than the outside water pressure (call this p_{water}) pushing the hatch inward. The net inward force F_{net} due to this pressure difference is given by Equation 11-10; $A = 2.00$ m^2 is the surface area of the hatch. To open the hatch the crew would have to exert an outward force of the same magnitude F_{net}.

To calculate F_{net} we need to know the value of the water pressure p_{water}. We'll find this value using Equation 11-8 for the pressure at depth $d = 85.0$ m. In this equation p_0 equals the pressure at the surface of the ocean (that is, at zero depth), which we also take to be 1.00 atm.

Net force on the hatch due to the pressure difference:

$$F_{net} = (p_{water} - p_{air})A \qquad (11\text{-}10)$$

Variation of pressure with depth:

$$p_{water} = p_0 + \rho g d \qquad (11\text{-}8)$$

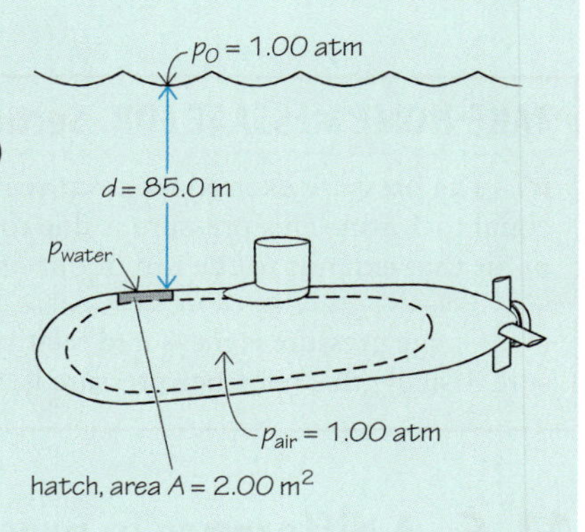

Solve

First find the pressure of the water at depth $d = 85$ m. Use the density of seawater from Table 11-1, $\rho = 1.025 \times 10^3$ kg/m^3; convert atmospheres to newtons per square meter; and note that 1 kg/(m · s^2) = 1 N/m^2.

Water pressure on the outside of the hatch:

$$p_{water} = p_0 + \rho g d$$

In this, $p_0 = 1.00 \text{ atm} = 1.01 \times 10^5 \text{ N/m}^2$, so

$$\begin{aligned} p_{water} &= 1.01 \times 10^5 \text{ N/m}^2 \\ &\quad + (1.025 \times 10^3 \text{ kg/m}^3)(9.80 \text{ m/s}^2)(85.0 \text{ m}) \\ &= 1.01 \times 10^5 \text{ N/m}^2 + 8.54 \times 10^5 \text{ kg/(m} \cdot \text{s}^2) \\ &= 1.01 \times 10^5 \text{ N/m}^2 + 8.54 \times 10^5 \text{ N/m}^2 \\ &= 9.55 \times 10^5 \text{ N/m}^2 \end{aligned}$$

Now we can calculate the magnitude of the net force on the hatch due to the pressure difference. This is the same as the force that the crew must exert to open the hatch.

Air pressure on the inside of the hatch:

$p_{air} = 1.00 \text{ atm} = 1.01 \times 10^5 \text{ N/m}^2$, so

$$\begin{aligned} F_{net} &= (p_{water} - p_{air})A \\ &= (9.55 \times 10^5 \text{ N/m}^2 - 1.01 \times 10^5 \text{ N/m}^2)(2.00 \text{ m}^2) \\ &= (8.54 \times 10^5 \text{ N/m}^2)(2.00 \text{ m}^2) \\ &= 1.71 \times 10^6 \text{ N} \end{aligned}$$

Reflect

This is an immense amount of force, equivalent to nearly 200 U.S. tons! Clearly the crew would not be able to open the hatch. In order for a crew member in diving gear to exit the submerged vessel, a portion of the crew compartment would have to be flooded with water from the outside. This action will equalize the inside and outside pressures and make it possible to open the hatch.

BIO-Medical The Lungs

Pressure differences drive air into and out of the lungs. Frogs, for example, force air into their lungs by swallowing mouthfuls of air. In contrast we humans suck air into our lungs by contracting the diaphragm and the muscles of the chest wall to increase the volume of our chest cavity. This pulls the lungs open, increasing their volume and therefore decreasing the pressure inside them. As this pressure becomes less than atmospheric pressure, the pressure difference pushes air through the airways and into the expanding lungs.

It takes energy for us to contract the muscles required for *inhalation*. However, normal *exhalation* does not require our muscles to do additional work on our chest wall. Relaxing the diaphragm and chest muscles allows the volume of the chest cavity to decrease. This process causes the pressure in the chest to rise above atmospheric pressure and forces air out of the lungs. Because atmospheric pressure is constant, changes in lung pressure determine the direction of airflow.

Breathing becomes more difficult if the chest is under pressure, such as if you're at the bottom of a football pile-on. The outside pressure on the chest makes it more difficult to expand the chest in order to inhale. Measurements show that if the pressure on a person's chest exceeds atmospheric pressure by 0.05 atm, the chest can't expand and inhalation is no longer possible. Example 11-7 shows that this places a limit on how deep underwater a person can be and still breathe through a hollow tube that opens above the surface.

EXAMPLE 11-7 Breathing Underwater

Secret agent Steele Branson dives into a shallow freshwater pond to avoid capture by his nemesis. He intends to lie flat on his back on the bottom of the pond and breathe through a hollow reed. What is the length of the longest reed for which this could work?

Set Up

Branson needs to ensure that the difference between the water pressure p on his chest and the pressure p_0 of the atmosphere at the surface (where the open end of the reed is) does not exceed 0.05 atm. Equation 11-8 tells us how the water pressure varies with depth d, so our goal is to find the depth such that $p - p_0$ equals 0.05 atm.

Variation of pressure with depth:

$$p = p_0 + \rho g d \qquad (11\text{-}8)$$

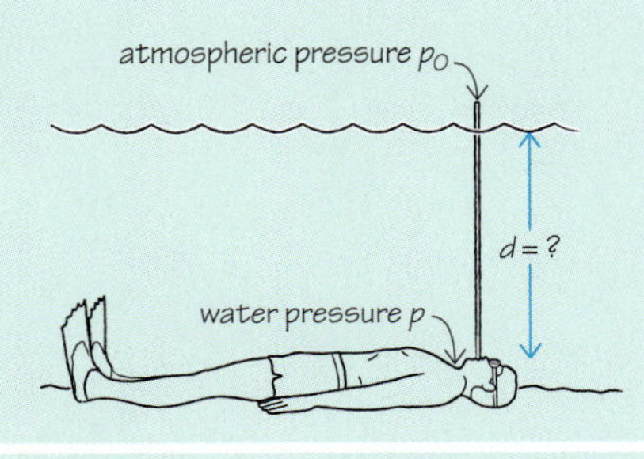

Solve

Solve Equation 11-8 to find the depth d in terms of the water pressure on Branson's chest (p) and atmospheric pressure (p_0). Then find the depth d for which $p - p_0 = 0.05$ atm. Use the density of fresh water given in Table 11-1 as well as the definitions 1 atm $= 1.01 \times 10^5$ N/m^2 and 1 N $= 1$ kg$\cdot$m/s^2.

Rearrange Equation 11-8:

$$p - p_0 = \rho g d$$

Solve for d:

$$d = \frac{p - p_0}{\rho g}$$

$$= \frac{0.05 \text{ atm}}{(1.000 \times 10^3 \text{ kg/m}^3)(9.80 \text{ m/s}^2)} \times \frac{1.01 \times 10^5 \text{ N/m}^2}{1 \text{ atm}}$$

$$= 0.5 \frac{\text{N} \cdot \text{s}^2}{\text{kg}} \times \frac{1 \text{ kg} \cdot \text{m/s}^2}{1 \text{ N}} = 0.5 \text{ m}$$

Reflect

The longest hollow reed through which Branson can breathe underwater is about half a meter—not that long, really. To check this result let's calculate the weight of water half a meter deep on the area of a person's chest, about 30 cm × 60 cm. The answer is about 900 N or 200 lb. Certainly if a 200-lb person were sitting on your chest, you'd find it hard to breathe!

Scuba divers can descend to depths much greater than 0.5 m, but to do so they carry tanks of pressurized air. The regulator on a scuba tank delivers this air to the diver's mouth at the same pressure as the surrounding water, so $p - p_0 = 0$. Hence it's just as easy for a scuba diver to breathe underwater as on dry land.

Area of chest:

$$A = 30 \text{ cm} \times 60 \text{ cm} = 0.30 \text{ m} \times 0.60 \text{ m} = 0.18 \text{ m}^2$$

Mass of water in a rectangular volume with base 0.18 m^2 and height 0.50 m:

$$m = \rho V = (1.000 \times 10^3 \text{ kg/m}^3) \times (0.18 \text{ m}^2)(0.50 \text{ m}) = 90 \text{ kg}$$

Weight of that water:

$$mg = (90 \text{ kg})(9.8 \text{ m/s}^2) = 900 \text{ N} = 200 \text{ lb}$$

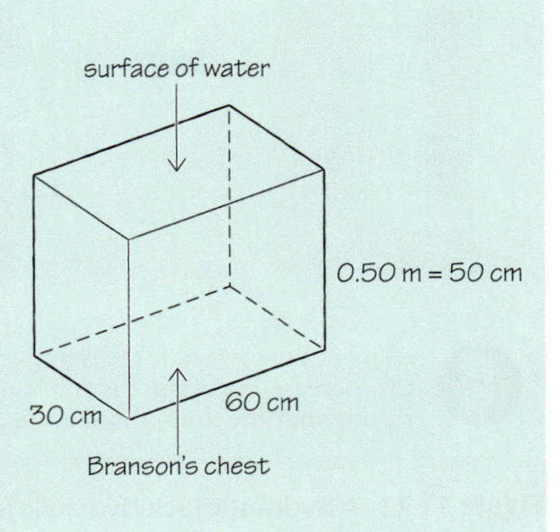

GOT THE CONCEPT? 11-5 Paper and Pressure

Hold a piece of paper in your two hands so that the plane of the paper is horizontal. Considering the air pressure on the top of the paper and the air pressure on the bottom of the paper, is the force F_{net} on the paper described by Equation 11-10 (a) upward, (b) downward, or (c) zero?

TAKE-HOME MESSAGE FOR Section 11-6

✔ If the fluid pressure is different on two sides of an object, the result is a net force on the object. This net force is proportional to the pressure difference and to the area of the object.

11-7 A pressure increase at one point in a fluid causes a pressure increase throughout the fluid

When you squeeze one end of a tube of toothpaste, the pressure you apply is transmitted throughout the tube, and toothpaste comes out the other end. This is a simple example of a more general principle first proposed by the seventeenth-century French philosopher, mathematician, and scientist Blaise Pascal, who did pioneering investigations of the nature of pressure in fluids:

> ***Pascal's principle:*** *Pressure applied to a confined, static fluid is transmitted undiminished to every part of the fluid as well as to the walls of the container.*

We've already seen **Pascal's principle** in action in Section 11-4. There we learned that if a fluid is at rest—that is, in hydrostatic equilibrium—and the pressure at a certain point in the fluid is p_0, the pressure at a second point a distance d deeper in the fluid is $p = p_0 + \rho gd$ (Equation 11-8). If we were to increase the value of the pressure p_0 at the first point by, say, 50 Pa, the value of the pressure p at the second point would also have to increase by 50 Pa to maintain hydrostatic equilibrium.

A common practical application of Pascal's principle is a *hydraulic jack*. This device makes it possible to lift heavy objects using a force much less than the object's weight. You'll find these devices in operation wherever cars are being worked on: A hydraulic lift raises the car so that the mechanic can work on the car's underside. Smaller versions are used to raise and lower the chairs you sit in at the dentist and the barber.

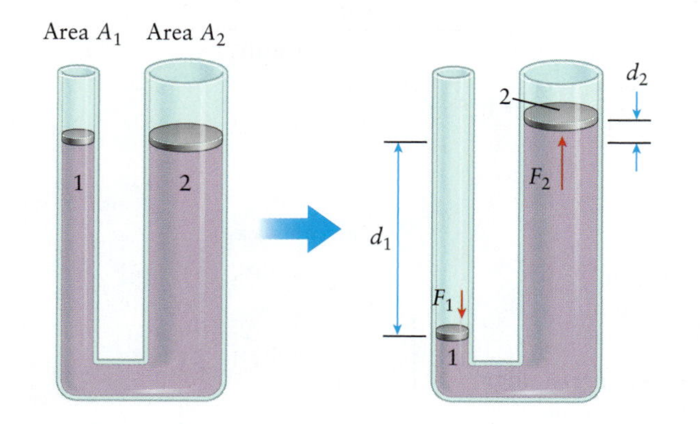

When a downward force is applied to the cap on side 1, the increase in pressure is transmitted undiminished throughout the fluid, forcing the cap on side 2 to rise.

Figure 11-11 A hydraulic jack Hydraulic jacks are used in workshops to elevate cars under repair and in dental offices to elevate the patient's chair. They operate using Pascal's principle.

Figure 11-11 shows the construction of a simplified hydraulic lift. Start with a tube bent into a "U" shape. On the side labeled "1" the tube is much narrower than on the side labeled "2"; we'll call the cross-sectional areas of the two sides of the tube A_1 and A_2, respectively. We partially fill the tube with an incompressible liquid such as oil so that its density does not change when pressure is applied. The liquid rises to the same height on both sides of the tube. (If you're not sure why, see the "Got the Concept" question at the end of Section 11-4.) We'll put a moveable cap at each end of the U-tube to keep the fluid from leaking out.

Now apply a downward force of magnitude F_1 to the cap on side 1. This causes the pressure in the fluid under the cap to increase by an amount $\Delta p = F_1/A_1$. According to Pascal's principle, this change in pressure is transmitted throughout the tube. We will therefore see this same pressure increase Δp below the cap on side 2, and that will cause an upward force $F_2 = \Delta p \times A_2$ on that cap. Substituting $\Delta p = F_1/A_1$ into this expression for F_2 gives

$$F_2 = \Delta p \times A_2 = \left(\frac{F_1}{A_1}\right)A_2$$

or, rearranging,

$$F_2 = F_1\left(\frac{A_2}{A_1}\right) \quad (11\text{-}11)$$

Since A_2 is greater than A_1, F_2 is greater than F_1, which means that a small force applied to side 1 results in a larger force applied to the cap on side 2. This is why a hydraulic jack is useful to auto mechanics: A relatively small force applied to side 1 can produce enough force on side 2 to raise a heavy car off the ground.

Equation 11-11 may make it seem like we get extra force "for free." However, there is a trade-off: The cap on side 1 must be pushed down a large distance to make the cap on side 2 move up a short distance. Here's why: When force F_1 is applied to the cap on side 1, that cap moves down a distance d_1 and displaces a volume of liquid $V_1 = d_1A_1$. The increase in pressure throughout the tube causes liquid to rise on side 2, pushing up the cap on that side by a distance d_2. The increase in liquid volume under cap 2 is $V_2 = d_2A_2$. Since the liquid used in the U-tube is assumed to be incompressible, these two volumes of liquid must be equal. Hence

$$d_2A_2 = d_1A_1$$

or

$$d_2 = d_1\left(\frac{A_1}{A_2}\right) \quad (11\text{-}12)$$

The factor A_1/A_2 in Equation 11-12 is the reciprocal of the factor that relates F_2 to F_1 in Equation 11-11. Although the *force* that is applied on side 2 is larger than the force on side 1, the *distance* the cap on side 2 moves is smaller than the distance the cap on side 1 was moved.

$\sqrt{x}$ *See the Math Tutorial for more information on direct and inverse proportions.*

GOT THE CONCEPT? 11-6 Work and a Hydraulic Lift

Consider a hydraulic lift like that shown in Figure 11-11. The cap on side 1 has a radius of 1.00 cm, and the cap on side 2 has a radius of 5.00 cm. As you push down on the cap on side 1, you do a certain amount of work W_1. How much work is done on the cap on side 2 as it rises up? (a) $25W_1$; (b) $5W_1$; (c) W_1; (d) $W_1/5$; (e) $W_1/25$.

TAKE-HOME MESSAGE FOR Section 11-7

✔ Pascal's principle states that pressure applied to a confined, static fluid is transmitted undiminished to every part of the fluid as well as to the walls of the container.

✔ Pascal's principle explains the hydraulic lift, which allows a small force exerted on a fluid at one location to translate into a larger force exerted by the fluid at another location.

11-8 Archimedes' principle helps us understand buoyancy

Drop a plastic toy into a pond, and you're not surprised when it pops back up to the surface. A boat anchor won't come back up—which is also not surprising. But why does a boat, which weighs much more than its anchor, float on the surface rather than sink?

To answer these questions, think of a fluid (gas or liquid) at rest. To maintain hydrostatic equilibrium of any portion of the fluid, like the shaded blob shown in **Figure 11-12a**, the net force on that portion of fluid must be zero. Hence the combined force due to the pressure of the surrounding fluid—greater pressure on the bottom, less pressure on top—must be an upward force that exactly balances the weight of the shaded blob of fluid. We call this combined upward force the **buoyant force**; it "buoys up" the blob of fluid just enough to keep it from sinking.

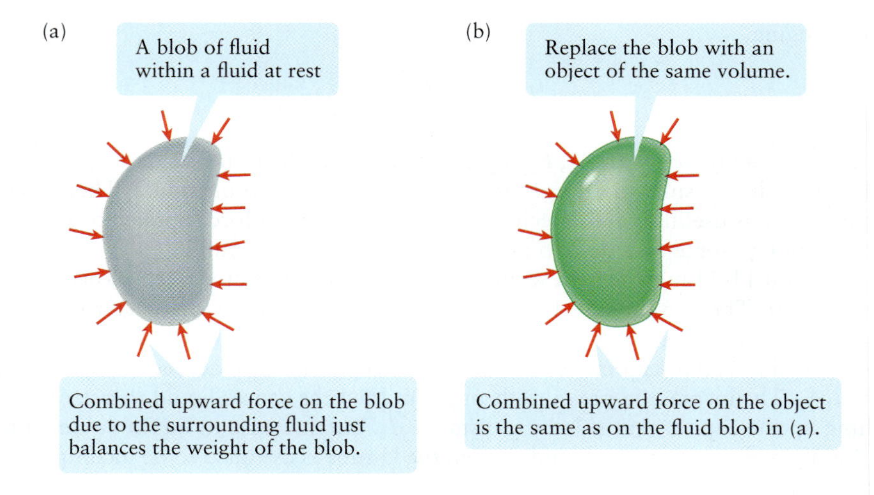
(a)
A blob of fluid within a fluid at rest
Combined upward force on the blob due to the surrounding fluid just balances the weight of the blob.
(b)
Replace the blob with an object of the same volume.
Combined upward force on the object is the same as on the fluid blob in (a).

Figure 11-12 Archimedes' principle (a) The blob is a portion of a fluid at rest. The pressure from the surrounding fluid is greater at the bottom than at the top. (b) If the blob is replaced by some other object of the same volume, that object feels the same upward (buoyant) force as did the blob.

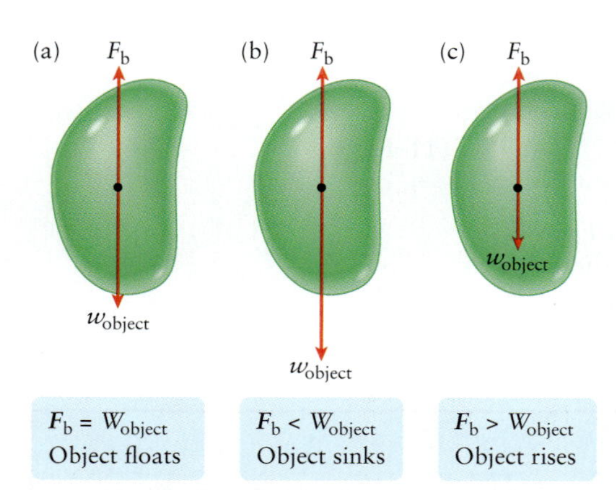
(a) F_b
w_{object}
(b) F_b
w_{object}
(c) F_b
w_{object}
$F_b = W_{object}$ Object floats
$F_b < W_{object}$ Object sinks
$F_b > W_{object}$ Object rises

Figure 11-13 Buoyancy: Floating, sinking, and rising All three of these objects are immersed in the same fluid. All three have the same shape and size and so experience the same buoyant force F_b. Whether the object floats, sinks, or rises depends on its weight.

Now we replace the blob of fluid in Figure 11-12a with an object of exactly the same dimensions (**Figure 11-12b**). We say that this object *displaces* a volume of fluid just equal to the volume of the object that's immersed in the fluid. For a boat floating in water, the displaced volume equals that portion of the boat's volume that's underwater. For a submarine under water, a balloon in air, or any other object that's totally immersed in a fluid, the displaced volume equals the total volume of the object.

Here's the critical observation: At each point on the object's surface, the *pressure* of the surrounding fluid (caused by fluid molecules colliding with the object) is *exactly* the same as it was before we swapped the object for the blob of fluid. So the submerged object feels the *same* buoyant force as the blob of fluid did, with a magnitude just equal to the weight of that fluid blob. This statement about buoyant force is called **Archimedes' principle**:

The buoyant force on an object immersed in a fluid is equal to the weight of the fluid that the object displaces.

To be specific, suppose the object displaces a volume $V_{displaced}$ of fluid. If the fluid has density ρ_{fluid}, the fluid that the object displaces has mass $m_{fluid} = \rho_{fluid} V_{displaced}$ and weight $m_{fluid} g = \rho_{fluid} V_{displaced} g$. Hence the magnitude F_b of the buoyant force that acts on the object is

Buoyant force (11-13)

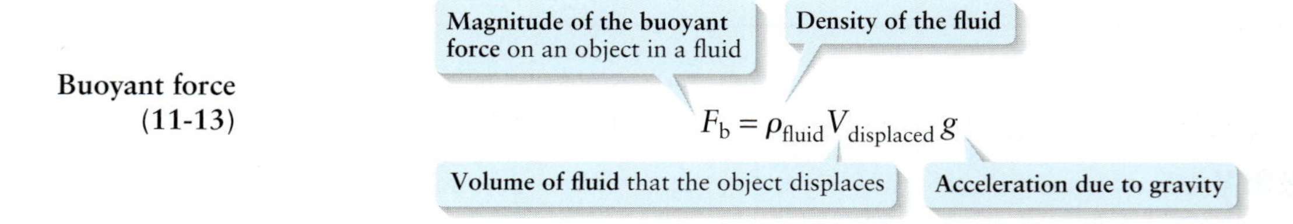
Magnitude of the buoyant force on an object in a fluid
Density of the fluid
Volume of fluid that the object displaces
Acceleration due to gravity

$$F_b = \rho_{fluid} V_{displaced}\, g$$

If the object's weight is exactly the same as the magnitude of the buoyant force F_b—that is, exactly the same as the weight of the displaced fluid—the net force on the object is zero and the object neither sinks nor rises; instead, it floats, remaining at the same height within the fluid (**Figure 11-13a**). If the object weighs more than the displaced fluid, the object's weight overwhelms the buoyant force; hence the net force is downward and the object sinks (**Figure 11-13b**). If the object weighs less than the displaced fluid, the buoyant force overwhelms the object's weight so that the net force is upward and the object rises (**Figure 11-13c**).

Floating: Submarines, Fish, Ships, and Balloons

Many types of fish, as well as submarines, are able to float while submerged. In this case the volume $V_{\text{displaced}}$ of displaced fluid in Equation 11-13 is the same as the volume of the object. If the object has density ρ, its mass is $m = \rho V_{\text{displaced}}$ and its weight is $\rho V_{\text{displaced}} g$. Then the condition that the buoyant force on the object has the same magnitude as the object's weight is

$$\rho_{\text{fluid}} V_{\text{displaced}} g = \rho V_{\text{displaced}} g$$

or, dividing through by $V_{\text{displaced}} g$,

$$\rho_{\text{fluid}} = \rho$$

(condition that an object of density ρ floats while submerged in a fluid of density ρ_{fluid})

In words, *a submerged object can float only if its density is the same as the fluid in which it is immersed.*

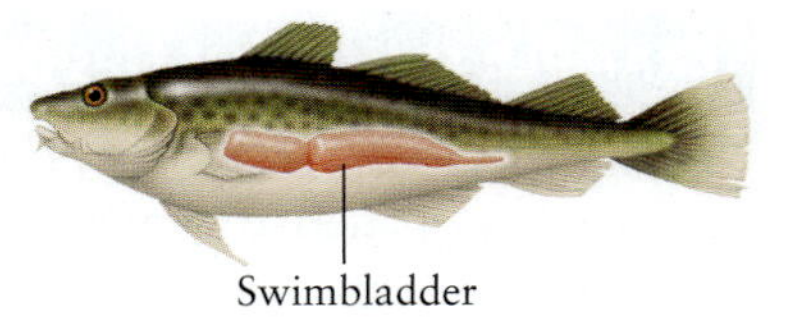

Figure 11-14 A fish with a swimbladder The gas-filled swimbladder gives this fish the same overall density as the surrounding water so that it floats while submerged.

Submarines and deep-sea research vessels are made of steel, which is far denser than water. They are nonetheless able to float underwater because they are hollow, with much of the internal volume filled with low-density air. The amount of air can be adjusted by filling ballast tanks with seawater or emptying the ballast tanks so that the submarine's *overall* density (the mass of the submarine divided by its volume) is equal to that of water (see Example 11-2 in Section 11-2). For the same reason, many species of fish (including salmon, herring, mackerel, and cod) have a flexible gas-filled sac called a *swimbladder* (**Figure 11-14**). These fish are able to float while submerged because they can regulate the amount of gas in the swimbladder to make their overall density equal to that of water. (In cod and other species this regulation is done by exchanging gas between the swimbladder and the blood.)

Common aquarium fish also have swimbladders and are able to float while submerged. But they move their fins constantly even if they are standing still. The reason is that the equilibrium provided by a swimbladder is *unstable.* If a water current displaces the fish slightly upward, the pressure at its new position is slightly less than before, and Pascal's principle (Section 11-7) tells us that this reduced pressure will be transmitted throughout the fish to the swimbladder. The reduced pressure makes the gas in the swimbladder expand, which decreases the average density of the fish as a whole. As a result the fish is now less dense than the surrounding water and will continue to move upward. Similarly, if the fish should be displaced below its equilibrium position, the swimbladder will shrink in response to the increased pressure, thus increasing the density of the fish and making it sink even further. The continuous motions of the fish's fins are an effort to maintain a constant depth.

Fish that lack swimbladders are denser than water and hence cannot float; they must swim continuously to avoid sinking to the bottom. (The same is true for sharks.) This is not a hardship for fish such as flounder and dogfish, which lack a swimbladder but whose natural habitat is at the ocean bottom.

An object that is *less* dense than water floats on the surface with only part of its volume submerged; the lower the object's density relative to that of water, the higher it floats (see Example 11-8). Steel is denser than water, but ships made of steel are hollow inside; this makes their overall density less than that of water and so allows them to remain afloat while only partially submerged (**Figure 11-15**).

We've mostly been discussing objects in liquids such as water, but there is also a buoyant force on an object immersed in a gas such as the atmosphere. The buoyant force on a balloon filled with helium (which is less dense than air) is greater than the balloon's weight, which is why it rises if released. By contrast, a balloon filled with room-temperature air falls if released: The air inside the balloon is at higher pressure and has a higher density than the outside air, so the balloon's weight is more than the buoyant force acting on it. Hot air is less dense than cold air, so an air-filled balloon can be made to float if it's equipped with a burner to warm the balloon's contents. That's the principle of a hot-air balloon.

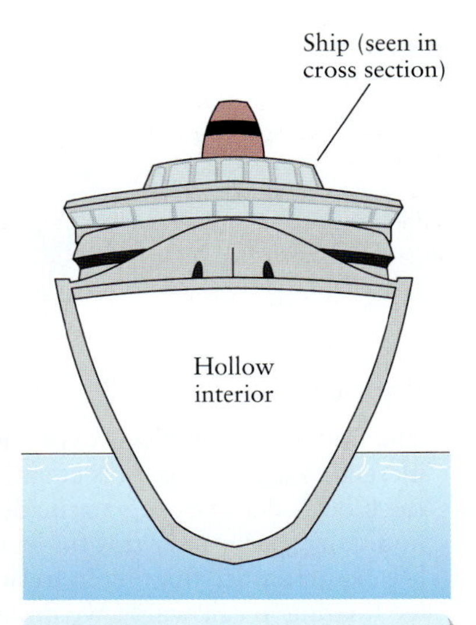

Steel is denser than water, but ships made of steel are hollow inside; this makes their overall density less than that of water and so allows them to remain afloat while only partially submerged.

Figure 11-15 How steel ships float What determines whether an object floats is how its overall density compares to the density of water.

EXAMPLE 11-8 What Lies Beneath the Surface?

A solid block of density ρ_{block} is placed in a liquid of density ρ_{liquid}. If the block is less dense than the liquid (that is, $\rho_{\text{block}} < \rho_{\text{liquid}}$), derive an expression for the fraction of the block's volume that is submerged.

Set Up

Archimedes' principle tells us that the buoyant force exerted on the block by the liquid equals the weight of the displaced liquid. In order for the block to float, the upward buoyant force must balance the weight of the block. We'll use this principle to compare the volume of displaced liquid $V_{\text{displaced}}$—equal to the volume of the block that's submerged in the liquid—to the total volume of the block.

Buoyant force:

$$F_{\text{b}} = \rho_{\text{liquid}} V_{\text{displaced}} g \qquad (11\text{-}13)$$

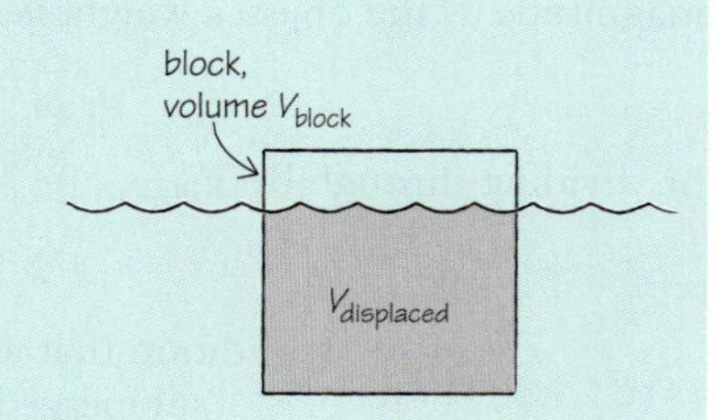

Solve

First write expressions for the block's mass m_{block} and weight w_{block}. When the block is in equilibrium, the upward buoyant force on the block equals the block's weight. Use this and the expression for w_{block} to solve for the ratio of $V_{\text{displaced}}$ to V_{block}.

volume of block = V_{block}

mass of block = (density of block) × (volume of block):

$$m_{\text{block}} = \rho_{\text{block}} V_{\text{block}}$$

weight of block = (mass of block) × g:

$$w_{\text{block}} = m_{\text{block}} g = \rho_{\text{block}} V_{\text{block}} g$$

Net force on the block in equilibrium is zero:

$$\sum F_y = F_{\text{b}} - w_{\text{block}} = 0 \text{ so}$$

$$F_{\text{b}} = w_{\text{block}}$$

From Equation 11-13,

$F_{\text{b}} = \rho_{\text{liquid}} V_{\text{displaced}} g$, so

$$\rho_{\text{liquid}} V_{\text{displaced}} g = \rho_{\text{block}} V_{\text{block}} g$$

Divide through by g:

$$\rho_{\text{liquid}} V_{\text{displaced}} = \rho_{\text{block}} V_{\text{block}}$$

Solve for ratio of $V_{\text{displaced}}$ to V_{block}:

$$\text{fraction of block's volume that is submerged} = \frac{V_{\text{displaced}}}{V_{\text{block}}} = \frac{\rho_{\text{block}}}{\rho_{\text{liquid}}}$$

Reflect

The fraction of the block's volume that is submerged, $V_{\text{displaced}}/V_{\text{block}}$, is equal to the ratio of the block's density ρ_{block} to the density of the liquid, ρ_{liquid}. Let's try this for a couple of real-life examples: ice floating in fresh water and ice floating in salt water. We find that a cube of pure ice in a glass of water floats with 91.7% of its volume submerged, while an iceberg in salt water floats a bit higher with 89.5% of its volume below the surface.

Block of ice floating in fresh water:

ρ_{block} = density of ice = 917 kg/m^3

ρ_{liquid} = density of fresh water = 1.000×10^3 kg/m^3

$$\text{fraction of ice that's submerged} = \frac{V_{\text{displaced}}}{V_{\text{block}}} = \frac{\rho_{\text{block}}}{\rho_{\text{liquid}}}$$

$$= \frac{917 \text{ kg/m}^3}{1.000 \times 10^3 \text{ kg/m}^3} = 0.917$$

So 91.7% of the ice is submerged.

Block of ice floating in salt water:

ρ_{block} = density of ice = 917 kg/m^3

ρ_{liquid} = density of salt water = 1.025×10^3 kg/m^3

$$\text{fraction of ice that's submerged} = \frac{V_{\text{displaced}}}{V_{\text{block}}} = \frac{\rho_{\text{block}}}{\rho_{\text{liquid}}}$$

$$= \frac{917 \text{ kg/m}^3}{1.025 \times 10^3 \text{ kg/m}^3} = 0.895$$

So 89.5% of the ice is submerged.

EXAMPLE 11-9 Underwater Float

A solid plastic ball of density 6.00×10^2 kg/m^3 and radius 2.00 cm is attached by a lightweight string to the bottom of an aquarium filled with fresh water. What is the tension in the string?

Set Up

Three forces act on the ball: the downward force of gravity, the upward buoyant force exerted by the water, and the downward tension force exerted by the string (which is what we want to find). We'll first use Equation 11-13 to determine the buoyant force on the ball and then use Newton's first law to solve for the tension force.

Buoyant force:

$$F_b = \rho_{water} V_{displaced} g \qquad (11\text{-}13)$$

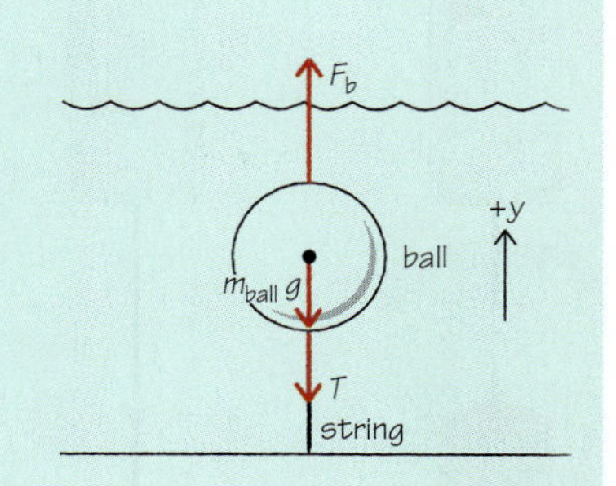

Solve

Newton's first law tells us that since the ball is at rest, the net force on the ball must be zero. Solve for the tension T in the string in terms of the buoyant force F_b and the weight of the ball $m_{ball}g$.

Newton's first law:

$$\sum F_y = F_b - T - m_{ball}g = 0 \text{ so}$$

$$T = F_b - m_{ball}g$$

The entire ball is submerged, so the volume of water that it displaces is equal to the volume of the ball of radius $R_{ball} = 2.00$ cm $= 2.00 \times 10^{-2}$ m. Use this to calculate the buoyant force.

Buoyant force:

$$F_b = \rho_{water} V_{displaced} g$$

$$V_{displaced} = V_{ball}$$

$$= \frac{4}{3}\pi R_{ball}^3 = \frac{4}{3}\pi(2.00 \times 10^{-2}\text{ m})^3$$

$$= 3.35 \times 10^{-5}\text{ m}^3 \text{ so}$$

$$F_b = (1.000 \times 10^3\text{ kg/m}^3)(3.35 \times 10^{-5}\text{ m}^3)(9.80\text{ m/s}^2)$$

$$= 0.328\text{ kg}\cdot\text{m/s}^2 = 0.328\text{ N}$$

Use the density of the ball to calculate its weight.

$$\text{Weight of ball} = m_{ball}g = \rho_{ball}V_{ball}g$$

$$= (6.00 \times 10^2\text{ kg/m}^3)(3.35 \times 10^{-5}\text{ m}^3)(9.80\text{ m/s}^2)$$

$$= 0.197\text{ kg}\cdot\text{m/s}^2 = 0.197\text{ N}$$

Finally, use the relationship that we derived from Newton's first law to solve for the tension T.

$$T = F_b - m_{ball}g = 0.328\text{ N} - 0.197\text{ N}$$

$$= 0.131\text{ N}$$

Reflect

The ball is less dense than water, so the buoyant force ($F_b = 0.328$ N) is greater than the weight of the ball ($m_{ball}g = 0.197$ N). Hence the string must exert a downward tension force to keep the ball from rising. If the string were to break, the ball would rise to the surface of the water. Can you see from Example 11-8 that the ball would end up floating with 60.0% of its volume submerged?

Apparent Weight

In Example 11-9 the submerged plastic ball is less dense than water and so has to be tethered to keep it from floating upward. Let's now think about a submerged object that's denser than water. The problem now is how to keep the object from sinking, which we solve by suspending it from above. **Figure 11-16a** shows such a submerged object hanging from a spring balance, along with an identical object that isn't submerged. Although both objects have the same mass M and the same volume V, the weights registered on the spring balances are *not* the same. Why? The difference is that for the submerged object on the left, the buoyant force due to the displaced water opposes the force of gravity and makes the object seem to weigh less. Thus the scale

An object submerged in a fluid that is denser than air has an apparent weight less than its weight when submerged in air.

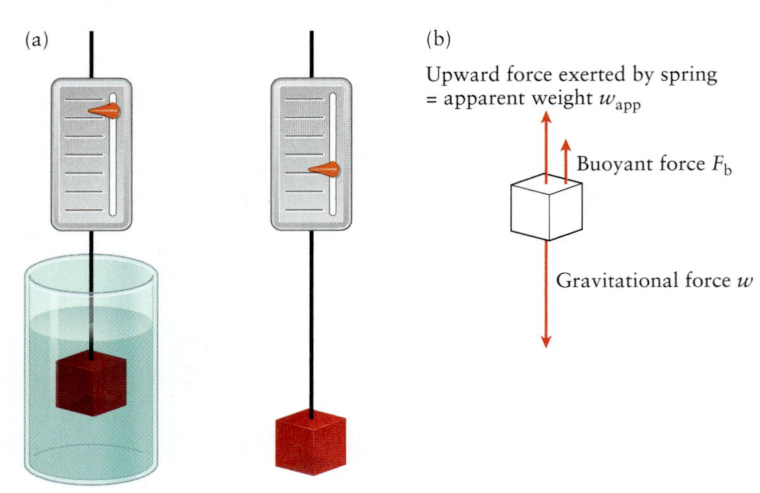

Figure 11-16 **Apparent weight** (a) An object hung from a spring balance has an apparent weight that depends on the fluid in which it is submerged. The reading on the scale indicates the force exerted by the spring, which is equal in magnitude to the object's apparent weight. (b) A free-body diagram showing the three forces acting on the submerged object in (a).

reads the object's **apparent weight** rather than its true weight. (In Section 10-5 we saw that an object in orbit has an apparent weight of zero due to its acceleration. The kind of apparent weight we're discussing here is for an object that's *not* accelerating—instead it's submerged in a fluid and acted on by a buoyant force.)

You can experience the difference between apparent weight and true weight in a swimming pool. If you swim below the surface of the water, you feel as though you weigh less than normal. Just as for the plastic ball in Figure 11-16, the upward buoyant force on you partially cancels the force of gravity. Astronauts use this effect to train for working in low gravity: They practice by working while submerged in an immense pool of water.

How is the apparent weight related to the true weight? **Figure 11-16b** shows the free-body diagram for the submerged object on the left in Figure 11-16a. The three forces acting on the object are the downward force of gravity (whose magnitude is the object's true weight w), the upward buoyant force F_b exerted by the water, and the upward force exerted by the spring balance. The scale on the spring balance measures how much force the balance exerts, so this force is just equal to the apparent weight w_{app}. The net force on the hanging object is zero (because the object is at rest), so

Go to Picture It 11-1 for more practice dealing with the buoyant force.

$$\sum F_y = w_{app} + F_b - w = 0$$

and

$$w_{app} = w - F_b \quad (11\text{-}14)$$

(apparent weight of a submerged object)

The apparent weight of a submerged object equals its true weight minus the buoyant force exerted on the object by the surrounding fluid.

If the object is completely submerged, the volume of fluid that it displaces equals the volume V of the object. Then we can replace $V_{displaced}$ in Equation 11-13 (the statement of Archimedes' principle) by V, and Equation 11-14 becomes

$$w_{app} = w - \rho_{fluid} V g \quad (11\text{-}15)$$

(apparent weight of an object of weight w and volume V completely submerged in a fluid of density ρ_{fluid})

So by measuring the apparent weight of an object while submerged (w_{app}) and its true weight when not submerged (w), you can determine the object's volume. This method can be quite handy when the object is irregularly shaped!

We can also write the apparent weight in terms of the density ρ of the object. Since the object has volume V, its mass is $m = \rho V$ and its true weight is $w = mg = \rho V g$. Then Equation 11-15 becomes

$$w_{app} = \rho V g - \rho_{fluid} V g = (\rho - \rho_{fluid}) V g \quad (11\text{-}16)$$

(apparent weight of an object of density ρ and volume V completely submerged in a fluid of density ρ_{fluid})

Go to Interactive Exercise 11-1 and 11-2 for more practice dealing with the buoyant force.

Equation 11-16 says that if we know an object's volume V, we can determine its density ρ by measuring its apparent weight w_{app} when submerged in a fluid of density ρ_{fluid}. According to legend, Archimedes came to this realization while he was himself submerged in his bathtub—a discovery that led him to his understanding of buoyancy.

A useful application of Equation 11-16 is to find the *average* density ρ of an object composed of two materials of known density. This makes it possible to determine what fraction of the object is composed of each material. Example 11-10 shows how to use this technique to find the ratio of lean tissue to fat tissue in a person's body.

BIO-Medical **EXAMPLE 11-10 Measuring Body Fat**

In the human body the density of lean muscle is about 1.06×10^3 kg/m^3, and the density of fat tissue is about 9.30×10^2 kg/m^3. If a person who weighs 833 N in air has an apparent weight of 27.4 N when submerged in water, what percent of his body mass is fat? (Assume that the person is made of muscle and fat only.)

Set Up

To determine what fraction of this person's body mass is fat, we'll first determine the average density ρ of his body. To do this we'll first find his volume V using Equation 11-15 and the measured values $w = 833$ N and $w_{app} = 27.4$ N. Then we can use Equation 11-16 to calculate his density ρ. We'll next compare ρ to the given densities of muscle and fat to determine the fat fraction.

Two equations for the apparent weight w_{app} of an object of weight w, volume V, and density ρ submerged in fluid of density ρ_{fluid}:

$$w_{app} = w - \rho_{fluid}Vg \qquad (11\text{-}15)$$

$$w_{app} = (\rho - \rho_{fluid})Vg \qquad (11\text{-}16)$$

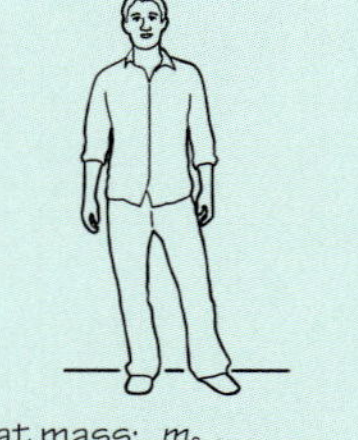

Solve

Solve Equation 11-15 for the person's volume V then substitute the given values of w_{app}, w, g, and $\rho_{fluid} = 1.000 \times 10^3$ kg/m^3 (the density of water).

Rearrange Equation 11-15 to solve for V:

$$\rho_{fluid}Vg = w - w_{app}$$

$$V = \frac{w - w_{app}}{\rho_{fluid}g} = \frac{833\text{ N} - 27.4\text{ N}}{(1.000 \times 10^3\text{ kg/m}^3)(9.80\text{ m/s}^2)}$$

$$= 0.0822\text{ m}^3$$

Now use this value of V in Equation 11-16 and solve for the person's density ρ. We'll keep an extra significant figure for this intermediate calculation and adjust them at the end.

Rearrange Equation 11-16 to solve for ρ:

$$(\rho - \rho_{fluid})Vg = w_{app}$$

$$\rho - \rho_{fluid} = \frac{w_{app}}{Vg}$$

$$\rho = \rho_{fluid} + \frac{w_{app}}{Vg}$$

$$= 1.000 \times 10^3\text{ kg/m}^3 + \frac{27.4\text{ N}}{(0.0822\text{ m}^3)(9.80\text{ m/s}^2)}$$

$$= 1.034 \times 10^3\text{ kg/m}^3$$

The overall density ρ is less than the density of lean muscle but greater than the density of fat, so the body is a mixture of these two quantities. Use the definition of density, $\rho = m/V$, to calculate the fraction x of the person's mass that is fat.

Let x = fraction of body mass m that is fat.

Then the mass of fat in the body is

$$m_{fat} = xm$$

The remaining mass in the body is muscle:

$$m_{muscle} = m - m_{fat} = m - xm = (1 - x)m$$

volume of body fat = (mass of fat)/(density of fat), so

$$V_{fat} = \frac{m_{fat}}{\rho_{fat}} = \frac{xm}{\rho_{fat}}$$

volume of body muscle = (mass of muscle)/(density of muscle), so

$$V_{muscle} = \frac{m_{muscle}}{\rho_{muscle}} = \frac{(1 - x)m}{\rho_{muscle}}$$

overall volume of body = (mass of body)/(overall density of body), so

$$V = \frac{m}{\rho}$$

The total volume of the body equals the sum of the volume of fat and the volume of muscle:

$$V = V_{\text{fat}} + V_{\text{muscle}}, \text{ so}$$

$$\frac{m}{\rho} = \frac{xm}{\rho_{\text{fat}}} + \frac{(1-x)m}{\rho_{\text{muscle}}}$$

Rearrange this equation and solve for x:

$$\frac{1}{\rho} = \frac{x}{\rho_{\text{fat}}} + \frac{(1-x)}{\rho_{\text{muscle}}}$$

$$\frac{1}{\rho} - \frac{1}{\rho_{\text{muscle}}} = \frac{x}{\rho_{\text{fat}}} - \frac{x}{\rho_{\text{muscle}}} = x\left(\frac{1}{\rho_{\text{fat}}} - \frac{1}{\rho_{\text{muscle}}}\right)$$

$$x = \frac{\left(\frac{1}{\rho} - \frac{1}{\rho_{\text{muscle}}}\right)}{\left(\frac{1}{\rho_{\text{fat}}} - \frac{1}{\rho_{\text{muscle}}}\right)}$$

$$= \frac{\left(\frac{1}{1.034 \times 10^3 \text{ kg/m}^3} - \frac{1}{1.06 \times 10^3 \text{ kg/m}^3}\right)}{\left(\frac{1}{9.30 \times 10^2 \text{ kg/m}^3} - \frac{1}{1.06 \times 10^3 \text{ kg/m}^3}\right)}$$

$$= 0.180$$

Reflect

Our result tells us that 0.180 (or 18.0%) of this person's mass is fat. Two-thirds of Americans have a percent body fat of 25% or more and are therefore overweight or obese. Adult men with between 18 and 24% body fat are considered healthy.

We can check our result for the person's density by calculating it in a different way: Take the person's weight of 833 N and divide by g to get the mass, then divide that by the volume that we calculated above from Equation 11-15. Happily we get the same result as we did using Equation 11-16.

Note that the method outlined here is only an approximation of an actual body fat calculation; we ignored the lungs and bones.

Second calculation of density:

$$\rho = \frac{m}{V}$$

Determine mass from weight:

$$w = mg, \text{ so } m = \frac{w}{g} = \frac{833 \text{ N}}{9.80 \text{ m/s}^2} = 85.0 \text{ kg}$$

From above, $V = 0.0822 \text{ m}^3$. So the person's density is

$$\rho = \frac{85.0 \text{ kg}}{0.0822 \text{ m}^3} = 1.03 \times 10^3 \text{ kg/m}^3$$

GOT THE CONCEPT? 11-7 Buoyancy I

? A plastic cube with a coin taped to its top surface is floating partially submerged in water. Mark the level of the water on the cube then remove the coin and tape it to the *bottom* of the cube. Will the cube sit (a) higher in the water, (b) lower in the water, or (c) the same as it was when the coin was on the top of the cube? *Hint:* The buoyant force on an object is the weight of the fluid that it displaces, and in this problem water is displaced by whatever is below the surface.

GOT THE CONCEPT? 11-8 Buoyancy II

? An open, glass soda bottle will float in a tub of water as long as it is empty. Place an empty, open glass bottle into a tub of water so that it floats and mark the level of the water on the wall of the tub. Now submerge the bottle so that it fills with water and sinks to the bottom. Is the level of the water in the tub (a) higher, (b) lower, or (c) the same as it was when the bottle was floating?

TAKE-HOME MESSAGE FOR Section 11-8

✔ An object surrounded by a fluid experiences a buoyant force equal to the weight of the fluid displaced by the object.

✔ An object immersed in a fluid has an apparent weight equal to its true weight minus the magnitude of the buoyant force on the object.

11-9 Fluids in motion behave differently depending on the flow speed and the fluid viscosity

In the preceding two sections we have considered fluids at rest. All around us, however, we find fluids in *motion*. Masses of air shift through the atmosphere to bring today's weather, river water courses downhill, and the oceans move in and out with the tides. Within our bodies moving fluids—blood, lymph, and the air used in respiration—play essential roles in sustaining life.

In a wind tunnel smoke trails display the smooth, steady pattern of laminar flow around this automobile.
(a)
Andy Sacks/Getty Images

The smoke rising from these incense sticks changes from laminar flow to chaotic, turbulent flow.
(b)
iStockphoto/Thinkstock

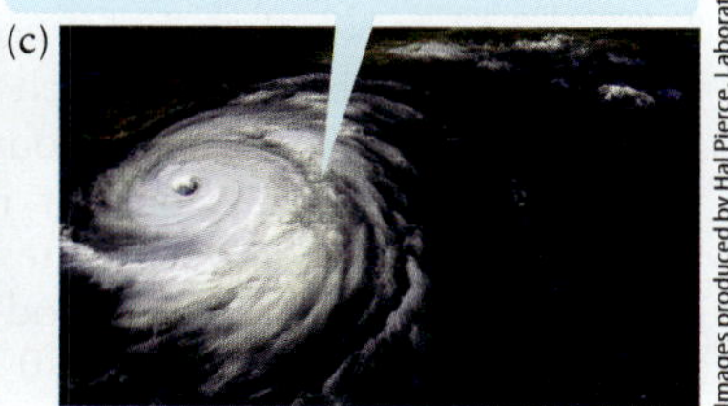

Although air rotates around the center of a hurricane, the flow is largely irrotational (a physicist's way of saying that the flow velocity changes in a special and gradual way from one point to another).
(c)
Images produced by Hal Pierce, Laboratory for Atmospheres, NASA Goddard Space Flight Center

Figure 11-17 Some examples of fluid flow Three very different situations in which air (a fluid) is in motion.

Fluid flows in both nature and technology are breathtaking in their diversity (**Figure 11-17**). We'll begin our study of fluids in motion by examining how physicists classify different types of fluid flow.

Steady Flow and Unsteady Flow

The simplest type of fluid flow is one in which the flow pattern doesn't change with time, like a stream in which water moves very smoothly with no variations. Such fluid motion is called **steady flow**. The direction and speed of the flow can be different from one point in the fluid to another (**Figure 11-18**). At any one point, however, the flow velocity remains constant from one moment to the next.

In **unsteady flow** the velocity at a given point can change with time. You experience the unsteady flow of air when you stand outside on a gusty day: The direction and speed of the wind at your position changes erratically. A less erratic example of unsteady flow is the pulsing motion of blood as it exits your heart through the aorta. Ocean waves, too, involve an unsteady pulsing motion that carries water (and anyone bobbing in the water) alternately up and down.

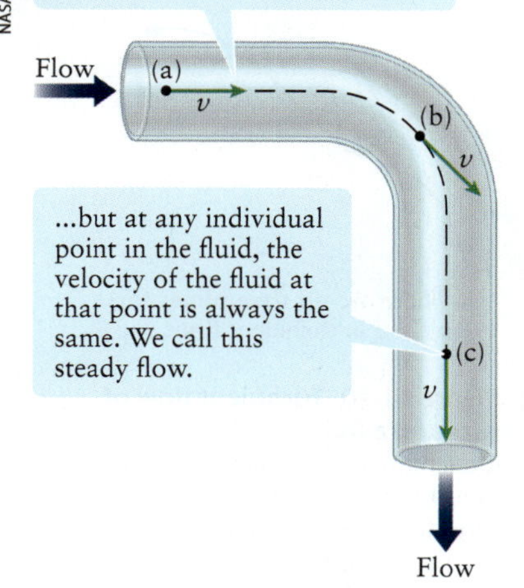

Figure 11-18 Steady flow of fluid in a pipe In steady fluid flow the velocity can be different at different points in the flow, but the velocity at any one point maintains the same value at all times.

WATCH OUT! Steady flow doesn't mean that the fluid velocity is the same everywhere.

Note that even though the fluid velocity at any one position in steady flow remains constant, the velocity of any given quantity of fluid *can* change as it moves. For example, water flowing through a curved pipe such as the one shown in Figure 11-18 changes its direction of motion as it follows the bends of the pipe. When we say that the flow is steady, we mean that when the next quantity of fluid passes through the same point in the pipe, its velocity at that given point will be the same as the previous quantity of fluid that passed through that same point.

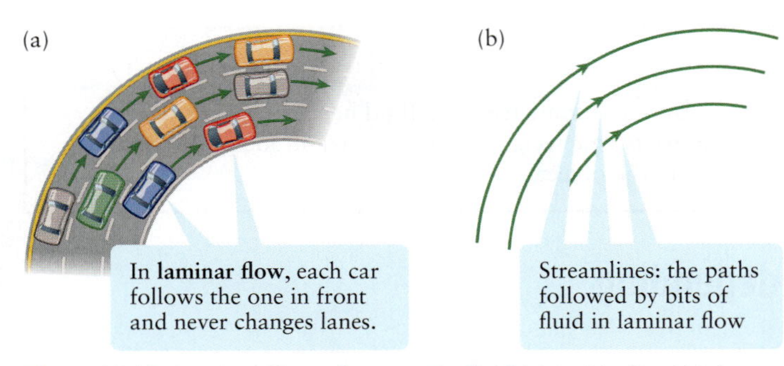

Figure 11-19 Laminar flow of cars and a fluid (a) An idealized highway in which cars move in laminar flow. (b) Streamlines in a fluid with laminar flow.

Laminar Flow and Turbulent Flow

Imagine a multi-lane highway filled with moving cars. If none of the cars ever changed lanes, you'd have an efficient, smooth-running transportation system in which each car would follow exactly the same trajectory as the one in front of it and the one behind it (**Figure 11-19a**). If you now replace the cars with bits of fluid, you would have a type of fluid motion called **laminar flow**. Each bit of fluid follows a path called a **streamline** that is the direct equivalent of one of the lanes of our idealized highway (**Figure 11-19b**). The smoke trails in Figure 11-17a show the streamlines of laminar flow in a wind tunnel.

Real highways are not as well organized as those in our imaginary example. Some cars change lanes, some cars move faster than others in the same lane, and occasionally there are collisions (**Figure 11-20a**). A fluid that behaves in this way is undergoing **turbulent flow** (**Figure 11-20b**). There are no streamlines in this case, since adjacent bits of fluid can follow very different paths. Figure 11-17b shows the turbulent flow of smoke.

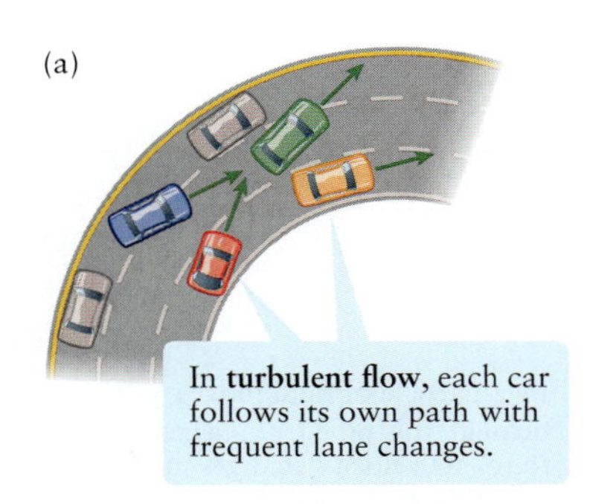

(a)
In turbulent flow, each car follows its own path with frequent lane changes.

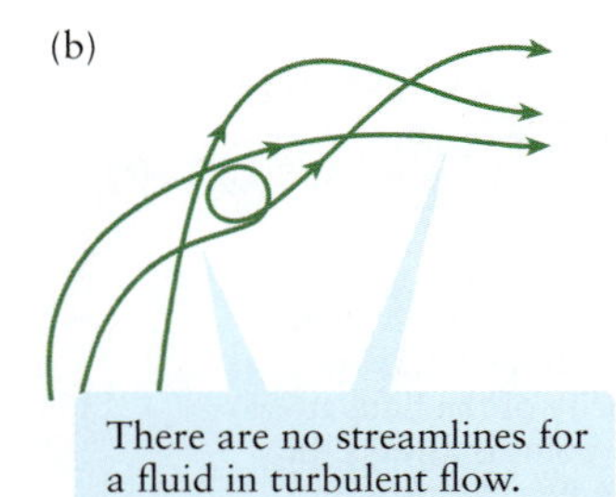

Figure 11-20 Turbulent flow of cars and a fluid (a) A less-than-ideal highway in which cars move in turbulent flow. (b) In a fluid with turbulent flow, bits of fluid move in a haphazard and seemingly unpredictable way.

Generally speaking a given type of flow changes from laminar to turbulent if the flow speed exceeds some critical value (which depends on the particular type of fluid). That's why airliners approaching Denver, which is downwind of the Rocky Mountains, can have a smooth ride when the wind is light and the airflow is laminar but a much bumpier ride when the wind is howling and the airflow turbulent.

BIO-Medical Turbulence is much noisier than laminar flow. As an example, when you make a hissing or "S" sound, you blow air past your teeth in a way that produces turbulence. By contrast, if you form your lips into an "o" and blow with equal force, your teeth are out of the way, the airflow is more laminar, and the sound is much softer. Your diastolic blood pressure—the second of the two numbers in a blood pressure report such as 120/80 (see Section 11-5)—is measured by putting a high-pressure cuff around your upper arm, gradually lowering the pressure of the cuff, and listening for a change in sound. At pressures just above the diastolic value, the artery is partially compressed, making the flow turbulent and noisy; at the diastolic pressure and below, the noise disappears because the artery is fully expanded and the flow is once again in its normal laminar state.

Viscous Flow and Inviscid Flow

When adjacent parts of a fluid move at different velocities, the parts rub and exert frictional forces on each other. Just as kinetic friction opposes the sliding motion of a block on a ramp, this rubbing opposes the sliding of one bit of fluid past another. This intrinsic resistance to flow is called **viscosity**. Motor oil is more *viscous*—that is, it has a greater viscosity—than water, which in turn is more viscous than air. Many liquids are less viscous at high temperatures; for example, warm maple syrup flows more easily than cold. Most automobiles with gasoline engines use a *multi-viscosity* oil designed to flow (and hence lubricate the engine's moving parts) equally well over a broad range of temperatures.

Fluids also experience friction when they flow past a solid surface. This friction is so great that it leads to what is called the **no-slip condition**: Right next to the solid surface, the velocity of the fluid is *zero* so the fluid does not "slip" over the surface. Instead, if fluid is flowing past the surface with a speed v, there is a **boundary layer** next to the surface within which the fluid speed increases from zero at the solid surface to the full speed v at the edge of the layer (**Figure 11-21**). A boundary layer of this kind develops around an automobile in motion, which is why driving even at freeway speeds doesn't blow dirt off the car. Any dirt particles lie well within the boundary layer, where the air is hardly moving at all. Even running water over the car won't dislodge all of the dirt particles, since there is also a boundary layer for water flow; only scrubbing with a sponge will complete the job.

Figure 11-21 shows that at different depths within the boundary layer, the fluid flows at different speeds. Hence frictional (viscous) forces act in the boundary layer, and these forces oppose the flow past the solid surface. This effect is called *viscous*

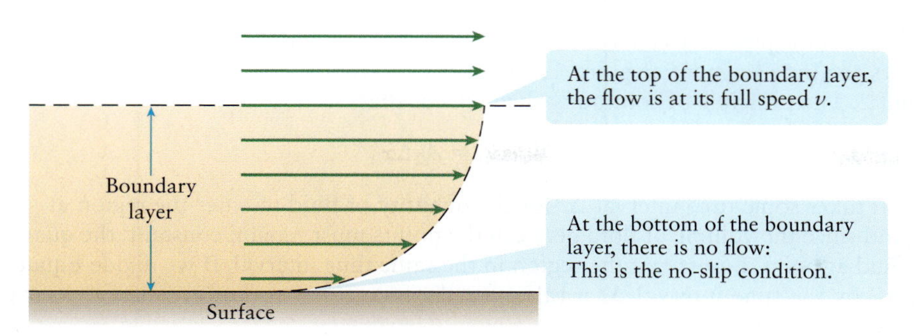
At the top of the boundary layer, the flow is at its full speed v.
Boundary layer
At the bottom of the boundary layer, there is no flow: This is the no-slip condition.
Surface

Figure 11-21 A boundary layer in a viscous fluid A boundary layer in a fluid moving past a flat surface.

drag, and it makes an important contribution to the force of air resistance felt by a moving car. If the flow in the boundary layer is turbulent, there is a greater difference in flow speeds between bits of fluid close to the car's skin, so the viscous forces are greater and the vehicle experiences more drag. Giving a car a "streamlined" shape is a way to make the flow of air around the car less turbulent and more laminar so that the air follows streamlines, as in Figure 11-17a. This reduces the drag so that less power has to be provided by the engine, and the car gets better mileage.

While every fluid has some viscosity, in many physical situations the viscosity is relatively unimportant. (An example is the airflow around a bird in flight. The viscous forces are much less important than the forces due to the pressure of air on the bird.) In these situations it's reasonable to simplify the problem by imagining that the moving fluid has *zero* viscosity. Such **inviscid flow** is an idealization, just like the frictionless ramps and massless strings that we considered in Chapter 4. We'll make use of this idealization in Section 11-10.

Finally, as we did in our earlier consideration of static fluids, we'll examine only the special case of incompressible fluids—that is, fluids in which changes in pressure do not affect the density of the fluid. Flowing water, blood, and air usually behave as incompressible fluids. All of these fluids *can* be compressed by exerting pressure on them, but in many practical situations the pressures are low enough that the compression can be ignored.

The Equation of Continuity

No matter what other properties a moving fluid may have, it must obey the following restriction: Mass can neither be created nor destroyed as the fluid flows. Therefore, in a *steady* flow, if a certain mass of fluid flows *into* a given region of space (say, the interior of a certain segment of a pipe) in a given time interval, the same amount of mass must flow *out of* that region during that same time interval. This is called the *principle of continuity*. For example, if 1 kg of fluid flows into one end of a pipe each second, then 1 kg/s must flow out the other end. (If the principle of continuity were violated, the amount of fluid within the pipe would either increase or decrease. But in this case the flow wouldn't be steady as we originally assumed it to be.)

We can make a simple mathematical statement of the principle of continuity for the case of an *incompressible* fluid. The principle of continuity tells us that in a given time interval the same *volume* of incompressible fluid must flow into a certain region of space as flows out of that region. **Figure 11-22** illustrates this principle for a pipe whose cross-sectional area varies along its length. Let 1 and 2 be two different points along the pipe, and consider the

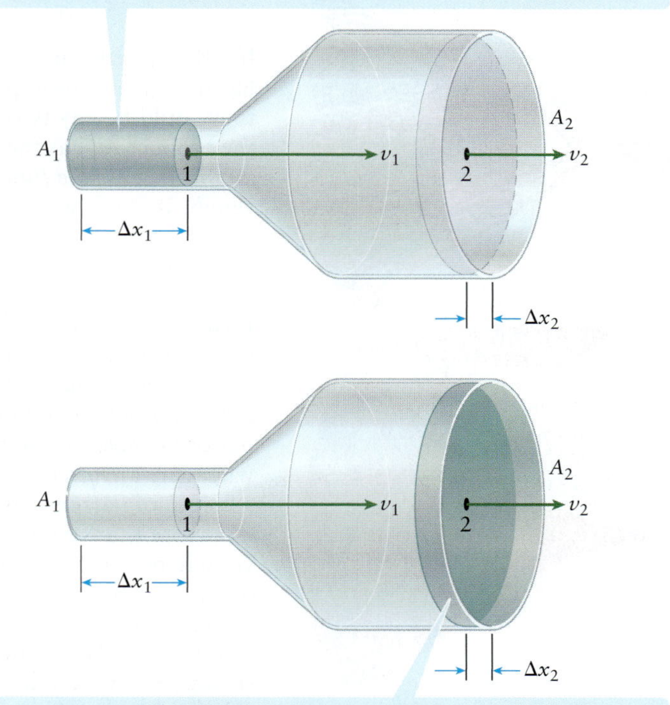
During the time it takes this quantity of fluid to move a distance Δx_1...
A_1
A_2
v_1
v_2
1
2
Δx_1
Δx_2
A_1
A_2
v_1
v_2
1
2
Δx_1
Δx_2
...this identical quantity of fluid moves a distance Δx_2. The amount of fluid between points 1 and 2 remains constant.

Figure 11-22 The equation of continuity This illustration shows the flow of an incompressible fluid through a pipe of varying diameter. A slug of fluid of volume $A_1\Delta x_1$ moving at speed v_1 enters the region between points 1 and 2, which causes a slug of fluid of volume $A_2\Delta x_2$ moving at speed v_2 to exit the region. The equation of continuity says that fluid enters the region at the same volume flow rate that it leaves the region.

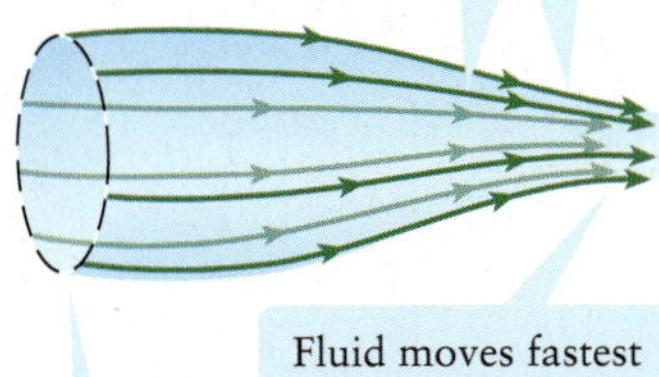

In laminar flow, fluid enclosed by these streamlines remains in the enclosed volume.
Fluid moves fastest where streamlines are closest together.
Fluid moves slowest where streamlines are far apart.

Figure 11-23 Streamline spacing and flow speed In laminar flow of an incompressible fluid, the spacing between streamlines tells you about the flow speed.

segment of the pipe between these two points. While a quantity of fluid of volume $A_1\Delta x_1$ enters the pipe segment at point 1, a quantity of volume $A_2\Delta x_2$ exits the segment at point 2. The principle of continuity then tells us that

$$A_1\Delta x_1 = A_2\Delta x_2 \tag{11-17}$$

It takes some time interval Δt for the quantity of fluid to enter the region at point 1. And since the volume of fluid between the points must remain constant, the quantity of fluid at point 2 must exit the region in the same time interval. If we divide Equation 11-17 by the time interval Δt it takes for the quantities of fluid to enter or exit the region, we arrive at

$$A_1\frac{\Delta x_1}{\Delta t} = A_2\frac{\Delta x_2}{\Delta t} \tag{11-18}$$

To see why we divided through by Δt, note that the fluid at point 1 moves a distance Δx_1 during the time interval Δt and so has speed $v_1 = \Delta x_1/\Delta t$. Similarly, the fluid at point 2 (which moves a distance Δx_2 during the same time interval) has speed $v_2 = \Delta x_2/\Delta t$. So we can rewrite Equation 11-18 as the following relationship, called the **equation of continuity:**

Equation of continuity for steady flow of an incompressible fluid (11-19)

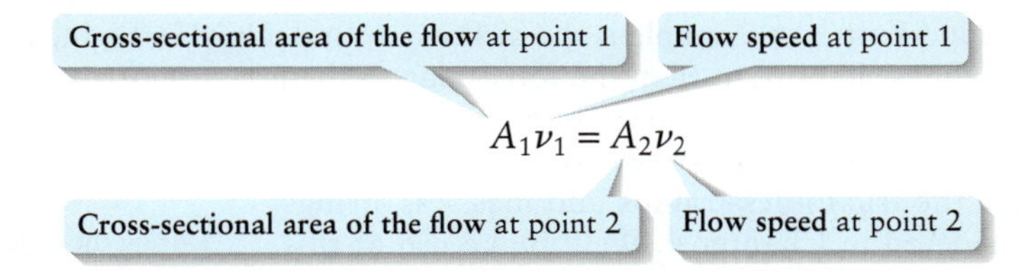

$$A_1v_1 = A_2v_2$$

In other words, the product of the pipe's cross-sectional area A and the flow speed v has the same value at point 1 as at point 2. The choice of these points is quite arbitrary; they could be any two points along the pipe's length. So Equation 11-19 tells us that the product Av has the same value *everywhere* along the pipe. *Where a pipe is narrow, an incompressible fluid moves rapidly; where the pipe is broad, the fluid moves more slowly.* If the flow is laminar, the "pipe" doesn't have to be a solid object with walls; it can be a volume enclosed by a set of streamlines (**Figure 11-23**). It follows that *where streamlines are close together, an incompressible fluid moves rapidly; where streamlines are far apart, the fluid moves more slowly.* The airflow around the car shown in Figure 11-17a is mostly incompressible, so the flow is rapid over the roof of the car where streamlines are close together.

The equation of continuity explains what happens when you put your thumb over the end of a garden hose. This action reduces the area through which water can flow out of the hose and so the water emerges at a faster speed. The same principle explains why there are often strong winds through mountain passes, where the air is forced into a "pipe" of narrow cross section.

Note that the product Av in Equation 11-19 has units of $(m^2)(m/s) = m^3/s$, or volume per unit time. This quantity, called the **volume flow rate**, tells you the number of cubic meters of fluid that pass a given point each second. For example, the average volume flow rate of blood through the aorta of a resting human is about $10^{-4}\ m^3/s$, or 6 L/min. (This is an *average* volume flow rate since the flow is in pulses rather than steady.) So the equation of continuity says that *the volume flow rate of an incompressible fluid moving through a pipe is the same at all points.*

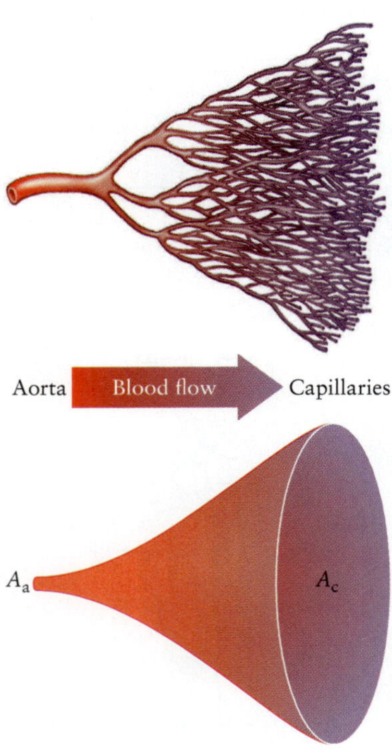

Figure 11-24 The human circulatory system Blood leaving the heart flows first through a single vessel, the aorta. Branches off the aorta eventually lead to billions of capillaries. Although each capillary is very small, the total cross-sectional area of all capillaries taken together is much larger than that of the aorta.

BIO-Medical This principle needs a little restatement if the pipe branches into a number of small pipes. An example is the human circulatory system, where the flow of oxygenated blood from the aorta is spread out into an enormous number of narrow capillaries (**Figure 11-24**). In this case the volume flow rate in the aorta is the same as the *combined* volume flow rate through all of the tiny capillaries. If the combined cross-sectional area of the capillaries, A_c, were the same as the cross-sectional area A_a of the aorta, blood would flow at the same speed throughout the system. In fact, in the human circulatory system A_c for the capillaries is greater than A_a for the aorta, so the flow speed in the capillaries is *slower* than in the aorta. This slower speed gives the blood in the capillaries more time for exchange of nutrients, gases, and waste products between the blood and the surrounding tissues.

WATCH OUT! The continuity equation is for incompressible flow only.

Note that Equation 11-19 applies only if the flow is *incompressible*, so its density is the same in all circumstances. In certain cases of fluid motion, however, the fluid does change density, and we say that the flow is *compressible*. Air flowing faster than sound behaves like a compressible fluid: When such fast-moving air enters a constriction such as a narrow pipe, it slows down and the molecules of the fluid get squeezed together so that the density of the fluid increases. In such a case Equation 11-19 doesn't apply. An analogy for a compressible fluid is how automobiles behave on a highway. You can think of the cars as "molecules" that make up a "fluid" that "flows" along the highway. In light traffic the cars in this "fluid" are far apart, but when this "fluid" passes through a narrower "pipe"—for example, a section of a two-lane highway where one lane is closed due to construction—the flow slows down, the cars get squeezed together, and a traffic jam results.

EXAMPLE 11-11 Flow in a Constriction

An incompressible fluid in a pipe of cross-sectional area 4.0×10^{-4} m^2 flows at a speed of 3.0 m/s. What is the flow speed in a part of the pipe that is constricted and has a cross-sectional area of 2.0×10^{-4} m^2?

Set Up

We are given the cross-sectional area and flow speed at one point in the pipe, and we wish to determine the flow speed at another point in the same pipe that has a different cross-sectional area. Since the fluid is incompressible, we can use the equation of continuity, Equation 11-19.

Equation of continuity for steady flow of an incompressible fluid:

$$A_1 v_1 = A_2 v_2 \qquad (11\text{-}19)$$

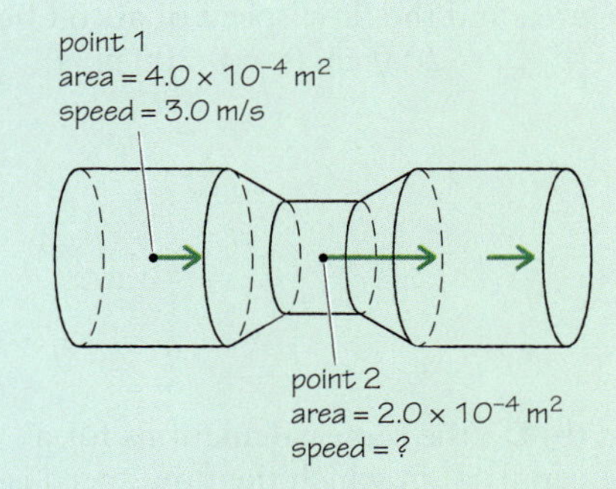

Solve

Let point 1 be a location where the pipe has its full cross-sectional area. Then $A_1 = 4.0 \times 10^{-4}$ m^2 and $v_1 = 3.0$ m/s. Choose point 2 to be in the constriction, where $A_2 = 2.0 \times 10^{-4}$ m^2. Use the equation of continuity to solve for v_2.

Rearrange Equation 11-19 to solve for v_2:

$$v_2 = \frac{A_1}{A_2} v_1$$

Substitute values:

$$v_2 = \frac{(4.0 \times 10^{-4}\ \text{m}^2)}{(2.0 \times 10^{-4}\ \text{m}^2)}(3.0\ \text{m/s})$$
$$= 6.0\ \text{m/s}$$

Reflect

Our result shows that the speed in the narrow constriction is faster than in the broad part of the pipe, just as the equation continuity tells us it must be.

BIO-Medical EXAMPLE 11-12 How Many Capillaries?

The inner diameter of the human aorta is about 2.50 cm, while that of a typical capillary is about 6.00 μm = 6.00×10^{-6} m (see Figure 11-24). In a person at rest, the average flow speed of blood is about 20.0 cm/s in the aorta and about 1.00 mm/s in a capillary. Calculate (a) the volume flow rate of blood in the aorta, (b) the volume flow rate in a single capillary, and (c) the total number of open capillaries into which blood from the aorta is distributed at any one time.

Set Up

Figure 11-24 shows the situation. We are given the dimensions of the aorta and each capillary as well as the flow speed in each of these pipes. Our goal is to determine the volume flow rate (in volume per unit time, or m^3/s) in the aorta and in a capillary as well as the number of capillaries into which the aorta empties. The volume flow rate in a pipe is related to its cross-sectional area and the speed of the fluid moving in the pipe. Like water, blood acts like an incompressible fluid. (It will compress appreciably only under pressures much higher than those found in the body.) So we can use the ideas of the equation of continuity, including the idea that the volume flow rate through the aorta must be equal to the flow rate through all of the open capillaries combined.

Equation of continuity for steady flow of an incompressible fluid:

$$A_1 v_1 = A_2 v_2 \qquad (11\text{-}19)$$

1 = aorta
2 = all open capillaries combined

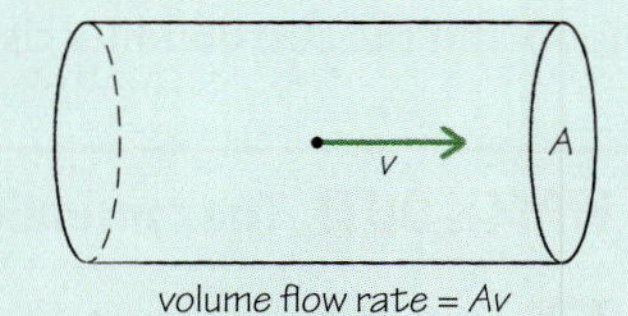

Solve

(a) The volume flow rate in the aorta is equal to the product of its cross-sectional area and the flow speed of aortal blood ($v_{\text{aorta}} = 20.0 \text{ cm/s} = 0.200 \text{ m/s}$).

Radius of aorta:

$$r_{\text{aorta}} = 1/2 \times (\text{diameter of aorta}) = 1/2 \times 2.50 \text{ cm} = 1.25 \text{ cm} = 1.25 \times 10^{-2} \text{ m}$$

Cross-sectional area of aorta:

$$A_{\text{aorta}} = \pi r_{\text{aorta}}^2 = \pi(1.25 \times 10^{-2} \text{ m})^2 = 4.91 \times 10^{-4} \text{ m}^2$$

volume flow rate in aorta:

$$A_{\text{aorta}} v_{\text{aorta}} = (4.91 \times 10^{-4} \text{ m}^2)(0.200 \text{ m/s}) = 9.82 \times 10^{-5} \text{ m}^3/\text{s}$$

(b) Do the same calculations for a capillary, in which the flow speed is $v_{\text{capillary}} = 1.00 \text{ mm/s} = 1.00 \times 10^{-3} \text{ m/s}$.

Radius of a capillary:

$$r_{\text{capillary}} = 1/2 \times (\text{diameter of capillary}) = 1/2 \times 6.00 \times 10^{-6} \text{ m} = 3.00 \times 10^{-6} \text{ m}$$

Cross-sectional area of a capillary:

$$A_{\text{capillary}} = \pi r_{\text{capillary}}^2 = \pi(3.00 \times 10^{-6} \text{ m})^2 = 2.83 \times 10^{-11} \text{ m}^2$$

volume flow rate in a capillary:

$$A_{\text{capillary}} v_{\text{capillary}} = (2.83 \times 10^{-11} \text{ m}^2)(1.00 \times 10^{-3} \text{ m/s}) = 2.83 \times 10^{-14} \text{ m}^3/\text{s}$$

(c) Our results from (a) and (b) show that compared to the volume flow rate through a *single* capillary, the volume flow rate through the aorta is 3.47×10^9 times greater. The idea of continuity tells us that the combined volume flow rate through *all* the open capillaries must be equal to the volume flow rate through the aorta. We therefore learn the total number of open capillaries.

$$\frac{\text{volume flow rate in aorta}}{\text{volume flow rate in a capillary}} = \frac{9.82 \times 10^{-5} \text{ m}^3/\text{s}}{2.83 \times 10^{-14} \text{ m}^3/\text{s}} = 3.47 \times 10^9$$

(volume flow rate in aorta) = (total volume flow rate in all open capillaries combined)

...so there must be 3.47×10^9 open capillaries.

Reflect

Our results show that the human circulatory system is truly extensive!

As a check on our results, note that the *combined* cross-sectional areas of all capillaries is $9.82 \times 10^{-2} \text{ m}^2$, which is 200 times

Total cross-sectional area of all open capillaries combined:

$$A_{\text{all open capillaries}} = (3.47 \times 10^9) A_{\text{capillary}} = (3.47 \times 10^9)(2.83 \times 10^{-11} \text{ m}^2) = 9.82 \times 10^{-2} \text{ m}^2$$

greater than the cross-sectional area of the aorta. By the equation of continuity, the flow speed in the capillaries should therefore be *slower* than in the aorta by a factor of $1/(2.00 \times 10^2)$; that is, $v_{\text{capillary}} = v_{\text{aorta}}/(2.00 \times 10^2) = (0.200 \text{ m/s})/(2.00 \times 10^2) = 1.00 \times 10^{-3} \text{ m/s}$. This gives us back one of the numbers we started with, so our calculation is consistent.

$$\frac{\text{area of all open capillaries combined}}{\text{area of aorta}} = \frac{9.82 \times 10^{-2} \text{ m}^2}{4.91 \times 10^{-4} \text{ m}^2} = 2.00 \times 10^2$$

BIO-Medical EXAMPLE 11-13 From Capillaries to the Vena Cavae

Blood returns to the heart from the capillaries through two veins known as the *vena cavae*. The combined cross-sectional area of the vena cavae is 10.0 cm^2. At what average speed does blood move through these veins?

Set Up

We know the net volume flow rate of blood in the aorta from Example 11-12, and we're given the cross-sectional area $A_{\text{vc}} = 10.0 \text{ cm}^2$ for the vena cavae. We'll use the equation of continuity to find the flow speed of blood in the vena cavae, v_{vc}.

Equation of continuity for steady flow of an incompressible fluid:

$$A_1 v_1 = A_2 v_2 \qquad (11\text{-}19)$$

1 = aorta
2 = vena cavae

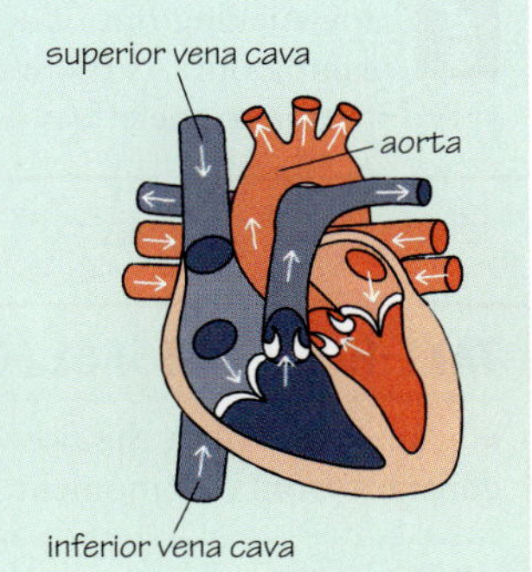

Solve

If we assume no blood volume is lost as it circulates through the body, the volume flow rate in the aorta is the same as in the vena cavae.

From Example 11-12,

$$\text{volume flow rate in aorta} = A_{\text{aorta}} v_{\text{aorta}} = (4.91 \times 10^{-4} \text{ m}^2)(0.200 \text{ m/s}) = 9.82 \times 10^{-5} \text{ m}^3/\text{s}$$

volume flow rate in venae cavae = $A_{\text{vc}} v_{\text{vc}}$, where

$$A_{\text{vc}} = (10.0 \text{ cm}^2)\left(\frac{1 \text{ m}}{100 \text{ cm}}\right)^2 = 1.00 \times 10^{-3} \text{ m}^2$$

The two volume flow rates are the same, so

$$A_{\text{aorta}} v_{\text{aorta}} = A_{\text{vc}} v_{\text{vc}}$$

$$v_{\text{vc}} = \frac{A_{\text{aorta}} v_{\text{aorta}}}{A_{\text{vc}}} = \frac{9.82 \times 10^{-5} \text{ m}^3/\text{s}}{1.00 \times 10^{-3} \text{ m}^2} = 9.82 \times 10^{-2} \text{ m/s} = 9.82 \text{ cm/s}$$

Reflect

Our results tell us that blood flows more slowly in the vena cavae (9.82×10^{-2} m/s) than in the aorta (0.200 m/s). That makes sense, for the venae cavae have a larger cross-sectional area than the aorta.

It's a common misconception that the slow-moving blood in the capillaries continues to move slowly as it returns to the heart. This example shows that this isn't the case! Compared to an average speed of 1.00 mm/s in the capillaries, blood moves 98.2 times faster in the vena cavae. The reason is that the cross-sectional area of the vena cavae is 1/98.2 as great as the combined cross-sectional area of the open capillaries, and the blood has to speed up as it returns to the heart to compensate for the decreased cross-sectional area.

Speed of blood in vena cavae compared to speed of blood in capillaries is

$$\frac{v_{vc}}{v_{\text{all open capillaries}}} = \frac{9.82 \times 10^{-2} \text{ m/s}}{1.00 \times 10^{-3} \text{ m/s}} = 98.2$$

Explanation: Cross-sectional area of vena cavae compared to cross-sectional area of all open capillaries combined is

$$\frac{A_{vc}}{A_{\text{all open capillaries}}} = \frac{1.00 \times 10^{-3} \text{ m}^2}{9.82 \times 10^{-2} \text{ m}^2} = \frac{1}{98.2}$$

GOT THE CONCEPT? 11-9 How Diameter Goes with the Flow

? While walking past a construction site, you notice a pipe sticking out of a second floor window, with water rushing out. As the water flows to the ground, it must speed up due to the effect of gravity. How does the diameter of the flowing stream of water change as it descends? If we assume that the flow remains laminar, (a) the diameter increases; (b) the diameter stays the same; (c) the diameter decreases.

TAKE-HOME MESSAGE FOR Section 11-9

✔ In steady flow, the flow velocity at any one point remains constant from one moment to the next.

✔ In laminar flow, each bit of fluid follows a streamline. In turbulent flow, adjacent bits of fluid can follow very different paths.

✔ The viscosity of a fluid is a measure of its resistance to flow. It results in the formation of a boundary layer next to a solid surface. Fluid speed increases from zero at the solid surface to the full speed v of the flow at the edge of the boundary layer.

✔ In steady, incompressible flow, the product of the cross-sectional area of the flow and the speed of the flow is constant: If the cross-sectional area increases, the speed of the flow decreases, and vice versa.

11-10 Bernoulli's equation helps us relate pressure and speed in fluid motion

Hold a piece of notebook paper by two corners so that the paper droops downward. Then blow on the top of the paper as shown in **Figure 11-25**. You might expect that the force of the air expelled from your mouth would push the paper downward. But remarkably, the paper actually lifts *up* where it is struck by the moving air.

What's happened is that by making air move over the top of the paper, you've lowered the pressure of that air. Hence there is greater air pressure on the underside of the paper than on the top, and the paper lifts up. This simple experiment illustrates that certain kinds of fluid flow have a property described by **Bernoulli's principle: In a moving fluid, the pressure is low where the fluid is moving rapidly.** This principle was first identified by the eighteenth-century mathematician Daniel Bernoulli.

Bernoulli's principle explains why an open door may swing closed on a windy day. The pressure in the moving air outside the house is lower than the pressure of the still air inside the house. The difference in air pressure on the two sides of the door pulls the door toward the outside, slamming it shut. An umbrella bulges upward in the wind for the same reason; there is low-pressure, fast-moving air on top of the umbrella, but high-pressure still air in the space underneath.

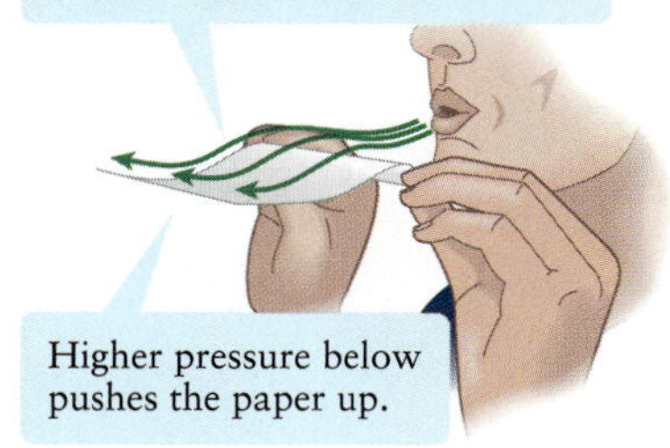
Blowing on the top of a sheet of paper lowers the pressure there.
Higher pressure below pushes the paper up.

Figure 11-25 Bernoulli's principle A piece of paper lifts up if you blow over the top of the paper.

Figure 11-26 illustrates the origins of Bernoulli's principle. A pipe of varying diameter carries an incompressible fluid from left to right. According to the equation of continuity that we discussed in Section 11-9, the fluid moves fastest in the narrow part of the pipe (point 2). Hence a parcel of fluid must speed up as it enters the narrow part. A net force must act on the fluid to change its velocity (that is, cause it to accelerate). This net force must be to the right as a parcel of fluid enters the narrow part. If we assume that there is negligible viscosity (see Section 11-9), the only forces that could be acting are those due to differences in pressure on the left and right sides of the parcel. As Figure 11-26 shows, to produce the required acceleration, the pressure must be lower in the narrow part of the pipe (where the fluid moves rapidly) than in the wide part (where the fluid moves slowly). This is just what Bernoulli's principle says: The pressure is lowest where the fluid moves the fastest.

WATCH OUT! Pressure differences cause velocity changes, not the other way around.

It's *not* correct to say that the pressure differences in Figure 11-26 are *caused* by the changing velocity of the fluid. In fact, just the reverse is true: The changes in fluid velocity are caused by the pressure differences! Recall the meaning of Newton's second law: An object accelerates (that is, changes its velocity) in response to a net force acting on it. In other words, a change in velocity is a result of a net force, not the other way around. In the same way the meaning of Bernoulli's principle is that a fluid undergoes a change in velocity as a result of pressure differences, not the reverse.

Bernoulli's Equation

Let's expand on these ideas and see how to express Bernoulli's principle as a rather simple and useful equation. By seeing where this equation comes from, we'll be able to understand why Bernoulli's principle holds only under certain special conditions. Many kinds of flow satisfy these conditions, but many others—including, for example, the flow of blood in the circulatory system—do not.

The shaded volume in **Figure 11-27a** shows a quantity of a fluid in motion. If we assume that the flow is *laminar* (see Section 11-9), we can think of this quantity of fluid as being enclosed within streamlines just as though it was flowing inside a pipe. As time goes by this quantity of fluid moves along its "pipe," displacing other fluid at its front end and being displaced by fluid at its back end. **Figure 11-27b** shows our quantity of fluid a brief time interval Δt after the instant shown in Figure 11-27a. The fluid has vacated a volume $A_1\Delta x_1$ and has moved into a volume $A_2\Delta x_2$.

Let's further assume that the flow is *steady* so that the fluid velocity at any fixed position in the fluid remains the same. In this case the motion of the shaded volume

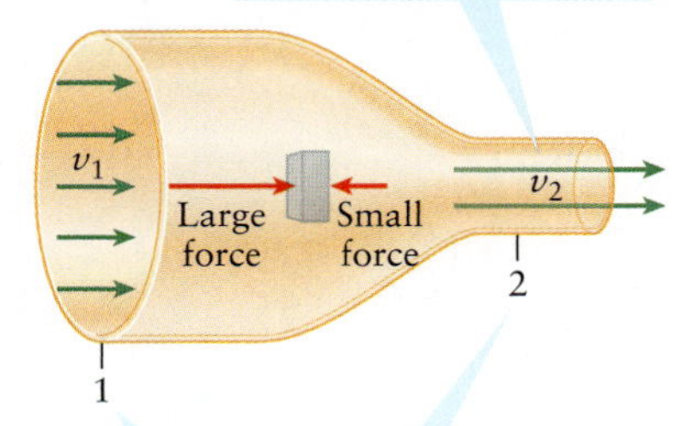
Fluid must move faster at point 2 than at point 1 to satisfy the equation of continuity.
v_1
v_2
Large force
Small force
1
2
To make the fluid speed up between point 1 and point 2, there must be a higher pressure at 1 than at 2.

Figure 11-26 Interpreting Bernoulli's principle Pressure differences drive a fluid through a constriction. The larger pressure on the left means that a parcel of fluid entering the constriction has a stronger force pushing it to the right than to the left, so the parcel speeds up as it approaches the constriction.

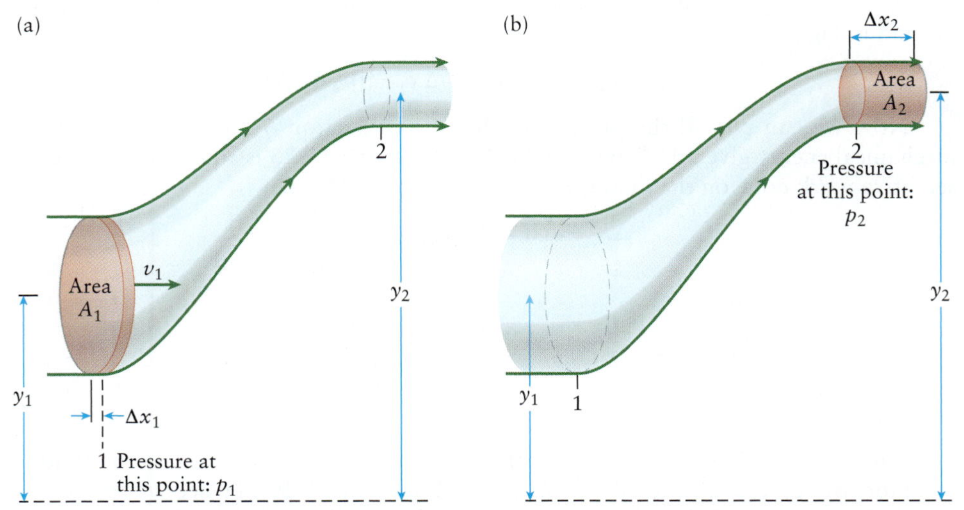
(a)
(b)
Area A_1
v_1
y_1
Δx_1
1 Pressure at this point: p_1
2
y_2
Δx_2
Area A_2
2 Pressure at this point: p_2
1

Figure 11-27 Deriving Bernoulli's equation (a) A slug of incompressible fluid of volume $A_1\Delta x_1$ moving at speed v_1 enters the region between points 1 and 2. (b) A short time later, a slug of fluid of volume $A_2\Delta x_2$ moving at speed v_2 exits the region. The equation of continuity says that fluid enters the region at the same rate that it exits. (Compare Figure 11-26.)

of fluid from the situation in Figure 11-27a to that in Figure 11-27b is the same as if the fluid had "lost" the volume $A_1\Delta x_1$ at its back end and "gained" the volume $A_2\Delta x_2$ at its front end, with no other changes. If we make an additional assumption that the fluid is *incompressible*—so that its density ρ is the same at all points in the fluid—then the volume "lost" at the back end must be the same as the volume "gained" at the front end:

$$\text{volume "lost"} = A_1\Delta x_1 = A_2\Delta x_2 = \text{volume "gained"} \tag{11-20}$$

While the volume of the moving incompressible fluid remains the same, its *energy* can change. Over the time interval Δt the shaded volume of fluid has lost the kinetic energy and gravitational potential energy associated with the "lost" volume at point 1 but has gained the kinetic and potential energies associated with the "gained" volume at point 2. The total energy change associated with the shaded volume of fluid is

$$\Delta E = \Delta K + \Delta U \tag{11-21}$$

We can write the mass of this volume of fluid as $\rho A_1\Delta x_1$ (when it is at point 1) or $\rho A_2\Delta x_2$ (when it is at point 2). We'll use the symbols v_1 and v_2 to denote the fluid speed at points 1 and 2, respectively, and use the symbols y_1 and y_2 for the heights at points 1 and 2. Then, using the familiar formulas $K = \frac{1}{2}mv^2$ for kinetic energy and $U = mgy$ for gravitational potential energy of the fluid-Earth system, we can write ΔK and ΔU in Equation 11-21 as

$$\Delta K = \text{change in kinetic energy} = +\frac{1}{2}(\rho A_2\Delta x_2)v_2^2 - \frac{1}{2}(\rho A_1\Delta x_1)v_1^2 \tag{11-22}$$

$$\Delta U = \text{change in gravitational potential energy} = +(\rho A_2\Delta x_2)gy_2 - (\rho A_1\Delta x_1)gy_1 \tag{11-23}$$

We learned in Section 6-7 that the total mechanical energy of a system (in this case Earth and the shaded volume of fluid) changes only when work is done on it by nonconservative forces. For simplicity we'll assume that there's no friction of any kind so that the fluid is *inviscid* (that is, it has zero viscosity). Then the only nonconservative forces that could act on the shaded volume of fluid are forces from the pressure of the surrounding fluid. There is pressure on all sides of the shaded volume of fluid, but work is done only by those forces that act on the *moving* parts of the fluid. From Figure 11-27 we can determine that the forces (pressure times area) on the back and front ends of the shaded volume of fluid are p_1A_1 and p_2A_2, respectively. The force on the back end pushes in the same direction as the displacement Δx_1 and so does positive work; the force on the front end does negative work because it pushes opposite to the displacement Δx_2. Hence the total nonconservative work done on the shaded volume of fluid in Figure 11-27 is

$$W_{\text{nonconservative}} = +(p_1A_1)\Delta x_1 - (p_2A_2)\Delta x_2 \tag{11-24}$$

Now we can put all the pieces together. From Section 6-7 the total change in mechanical energy given by Equations 11-21, 11-22, and 11-23 is equal to the nonconservative work done on the fluid given by Equation 11-24. That is,

$$\Delta K + \Delta U = W_{\text{nonconservative}}$$

or

$$+\frac{1}{2}(\rho A_2\Delta x_2)v_2^2 - \frac{1}{2}(\rho A_1\Delta x_1)v_1^2 + (\rho A_2\Delta x_2)gy_2 - (\rho A_1\Delta x_1)gy_1$$
$$= +(p_1A_1)\Delta x_1 - (p_2A_2)\Delta x_2 \tag{11-25}$$

Equation 11-25 looks like a horrible mess. But because we assumed the fluid is incompressible, $A_1\Delta x_1$ is equal to $A_2\Delta x_2$ (see Equation 11-20). Hence we can divide out all the factors of $A_1\Delta x_1$ and $A_2\Delta x_2$, leaving the simpler expression

$$+\frac{1}{2}\rho v_2^2 - \frac{1}{2}\rho v_1^2 + \rho gy_2 - \rho gy_1 = p_1 - p_2$$

We can rearrange this expression to read

$$p_1 + \frac{1}{2}\rho v_1^2 + \rho g y_1 = p_2 + \frac{1}{2}\rho v_2^2 + \rho g y_2 \tag{11-26}$$

Equation 11-26 says that the quantity $p + \frac{1}{2}\rho v^2 + \rho g y$ has the same value at point 1 (the back end of the shaded volume of fluid) as at point 2 (the front end). But we could have chosen the back and front ends to be *anywhere* along the length of the shaded volume of fluid in Figure 11-27. So the quantity $p + \frac{1}{2}\rho v^2 + \rho g y$ must have the same value at *any* point within the streamlines that enclose this shaded volume.

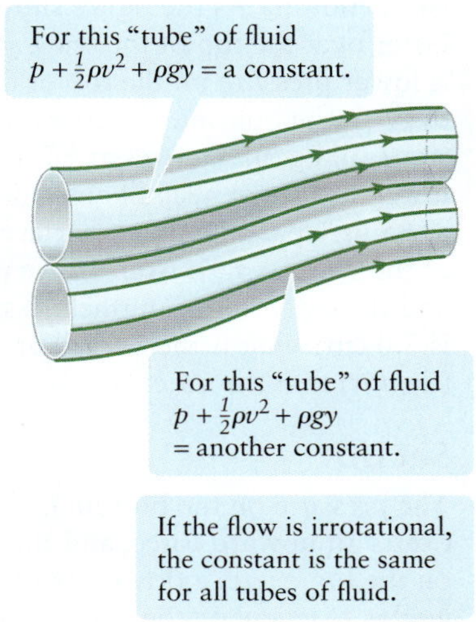

What about a volume of fluid immediately adjacent to the shaded one we've been discussing (**Figure 11-28**)? In principle this fluid could be moving at a very different speed than the fluid in the shaded volume. But in real fluids in which the flow is laminar, there is enough viscosity—that is, enough frictional force within the fluid—that adjoining bits of fluid move at nearly the same speed. Then the speed varies gradually from one part of the fluid to another, with no abrupt jumps. Physicists say that such a flow is **irrotational**. This does *not* mean that the fluid can't rotate—it certainly can, like the hurricane in Figure 11-17c or like water swirling as it goes down the drain. Rather, the curious term "irrotational" refers to what would happen to a little paddlewheel that was put in the fluid and allowed to move with it. If the fluid speed on one side of the paddlewheel were sharply different from that on the other side, the paddlewheel would start turning. In the idealized case we're discussing, there are no such differences, and the paddlewheel wouldn't rotate—which is why we call the flow irrotational.

For this "tube" of fluid
$p + \frac{1}{2}\rho v^2 + \rho g y$ = a constant.

For this "tube" of fluid
$p + \frac{1}{2}\rho v^2 + \rho g y$
= another constant.

If the flow is irrotational, the constant is the same for all tubes of fluid.

Figure 11-28 Irrotational flow and Bernoulli's equation Two adjacent tubes of fluid. If the flow is irrotational, the quantity $p + \frac{1}{2}\rho v^2 + \rho g y$ has the same value for both tubes and for *all* parts of the fluid.

Let's now make a final assumption that the fluid flow shown in Figure 11-28 is irrotational. This assumption is actually in contradiction to our earlier approximation that the flow is inviscid; *some* viscosity must be present to make the flow irrotational. So our approximation is really that the fluid has a little viscosity but not too much! With this additional assumption the quantity $p + \frac{1}{2}\rho v^2 + \rho g y$ has the same value in the shaded volume in Figure 11-27 as it does in any adjacent volume. It also has the same value in the next volume over, and so on. So our idealized fluid has the same value of $p + \frac{1}{2}\rho v^2 + \rho g y$ at *all* points in the fluid. Put another way, points 1 and 2 in Equation 11-26 could be any two points in the fluid.

Our final result is **Bernoulli's equation**:

Fluid pressure at a given point in the fluid

Speed of the fluid at that point

Vertical coordinate of that point

$$p + \frac{1}{2}\rho v^2 + \rho g y = \text{a constant with the same value throughout the fluid} \tag{11-27}$$

Density of the fluid (uniform throughout the fluid)

Acceleration due to gravity

Bernoulli's equation (11-27)

Note that if the fluid is at rest, so that $v = 0$ at all points, Equation 11-27 becomes $p + \rho g y$ = constant; that is, as y decreases and you go deeper in a static fluid, the quantity $\rho g y$ decreases and the pressure p increases. This is just the relationship for pressure at various depths in a static fluid that we found in Section 11-4. Note also that if we compare two points at the same height y, Equation 11-27 tells us that the pressure p is high where the fluid speed v is low—which is just the statement of Bernoulli's principle that we made at the beginning of this section.

In order to derive Bernoulli's equation, we had to make several approximations: The fluid flow had to be *laminar, steady, incompressible, inviscid,* and *irrotational.* In many real-life situations these assumptions aren't valid. For example, the flow of water down a waterfall is turbulent, not laminar; viscosity is important for blood flow in capillaries; and the air flowing around a supersonic airplane undergoes substantial compression. But there are a number of situations where Bernoulli's equation gives reasonably good results. In the examples that follow we'll examine how to use Bernoulli's equation in some of these situations.

EXAMPLE 11-14 Lift on a Wing

Figure 11-29 shows a computer simulation of air flowing around an airplane wing (seen end-on). As the figure shows, air flows faster over the top of the wing so that there is lower pressure on the top of the wing than on the bottom. Suppose the air (density 1.20 kg/m^3) moves at 75 m/s past the lower surface of a small airplane's wing and at 85 m/s past the upper surface. If the area of the wing as seen from above is 10.0 m^2 and the top-to-bottom thickness of the wing is 7.0 cm, what is the overall upward force (the *lift*) that the air exerts on the wing?

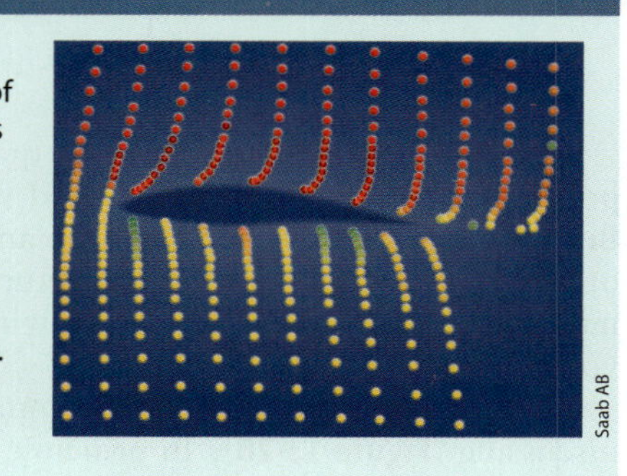

Figure 11-29 A computer simulation of airflow around a wing A vertical column of parcels of air (shown by colored dots) starts at the left and moves to the right, flowing around a wing. The parcels passing over the top of the wing go faster than those passing below the bottom of the wing (they're spaced farther apart horizontally), so there must be lower pressure on the top. (The dots are colored red in regions of low pressure.)

Set Up

The pressure on the bottom of the wing exerts an upward force, and the pressure on the top exerts a downward force. Our goal is to calculate the *combined* vertical force that these pressures exert on the wing. We use Bernoulli's equation to relate the pressure on the two surfaces of the wing. Pressure is force per area, so the force on either surface of the wing is the pressure multiplied by the wing area.

Bernoulli's equation:

$$p + \frac{1}{2}\rho v^2 + \rho g y = \text{a constant} \tag{11-27}$$

Definition of pressure:

$$p = \frac{F_\perp}{A} \tag{11-2}$$

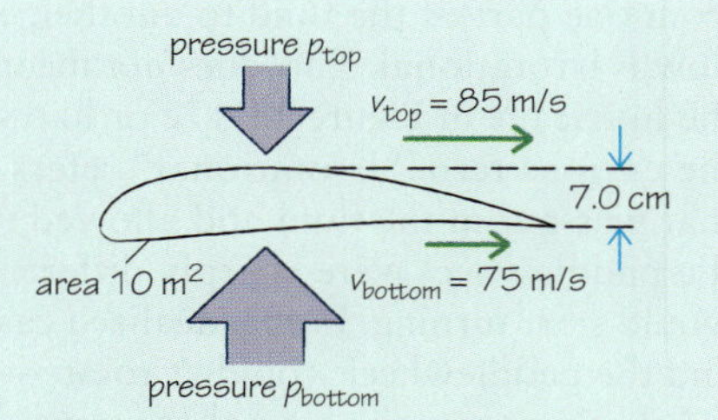

Solve

We want to calculate the net upward force on the wing, or the lift L. Let p_{bottom} and p_{top} be the pressures on the bottom and top of the wing respectively. Express L in terms of these pressures and the wing area A.

From Equation 11-2 the upward force on the bottom of the wing is

$$F_{\text{bottom}} = p_{\text{bottom}} A$$

and the downward force on the top of the wing is

$$F_{\text{top}} = p_{\text{top}} A$$

Combined upward force on the wing:

$$\begin{aligned} L &= F_{\text{bottom}} - F_{\text{top}} \\ &= p_{\text{bottom}} A - p_{\text{top}} A \\ &= (p_{\text{bottom}} - p_{\text{top}}) A \end{aligned}$$

F_{top}
area A = 10.0 m²
lift $L = F_{bottom} - F_{top}$
F_{bottom}

Use Bernoulli's equation to write an expression for the pressure difference $p_{\text{bottom}} - p_{\text{top}}$.

$p + \frac{1}{2}\rho v^2 + \rho g y$ has the same value on the top and bottom of the wing, so

$$p_{\text{bottom}} + \frac{1}{2}\rho v_{\text{bottom}}^2 + \rho g y_{\text{bottom}} = p_{\text{top}} + \frac{1}{2}\rho v_{\text{top}}^2 + \rho g y_{\text{top}}$$

Solve for $p_{\text{bottom}} - p_{\text{top}}$:

$$p_{\text{bottom}} - p_{\text{top}} = \frac{1}{2}\rho v_{\text{top}}^2 - \frac{1}{2}\rho v_{\text{bottom}}^2 + \rho g y_{\text{top}} - \rho g y_{\text{bottom}}$$

Calculate the pressure difference using $\rho = 1.20 \text{ kg/m}^3$, $v_{\text{top}} = 85 \text{ m/s}$, $v_{\text{bottom}} = 75 \text{ m/s}$, and $y_{\text{top}} - y_{\text{bottom}} = 7.0 \text{ cm} = 7.0 \times 10^{-2} \text{ m}$ (the top-to-bottom thickness of the wing).

$$\begin{aligned} \frac{1}{2}\rho v_{\text{top}}^2 - \frac{1}{2}\rho v_{\text{bottom}}^2 &= \frac{1}{2}\rho (v_{\text{top}}^2 - v_{\text{bottom}}^2) \\ &= \frac{1}{2}(1.20 \text{ kg/m}^3)[(85 \text{ m/s})^2 - (75 \text{ m/s})^2] \\ &= 9.6 \times 10^2 \text{ kg/(m} \cdot \text{s}^2) = 9.6 \times 10^2 \text{ N/m}^2 \end{aligned}$$

$$\rho g y_{top} - \rho g y_{bottom} = \rho g(y_{top} - y_{bottom})$$
$$= (1.20 \text{ kg/m}^3)(9.80 \text{ m/s}^2)(7.0 \times 10^{-2} \text{ m})$$
$$= 0.82 \text{ kg/(m} \cdot \text{s}^2) = 0.82 \text{ N/m}^2 \text{ so}$$

$$p_{bottom} - p_{top} = 9.6 \times 10^2 \text{ N/m}^2 + 0.82 \text{ N/m}^2$$
$$= 9.6 \times 10^2 \text{ N/m}^2 \text{ to two significant figures}$$

Calculate the lift L by multiplying the pressure difference by the wing area.

$$L = (p_{bottom} - p_{top})A = (9.6 \times 10^2 \text{ N/m}^2)(10.0 \text{ m}^2)$$
$$= 9.6 \times 10^3 \text{ N}$$

Reflect

In straight-and-level flight the airplane is not accelerating vertically, so the upward force of lift must exactly balance the weight of the airplane. So our wing can keep an airplane of weight 9.6×10^3 N (about 2200 lb) in the air.

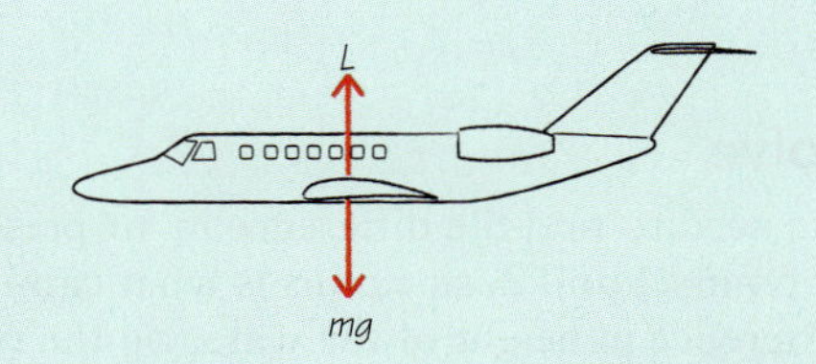

Our result shows that for given speeds of airflow along the top and bottom of the wing, the lift L is proportional to the wing area A. The heavier the airplane, the larger the wing required to maintain flight. The same principle applies to gliding birds: An eagle or vulture weighs more than a hawk and so has a larger wing.

Even if the airplane were not moving, there would still be a small pressure difference of $\rho g y_{top} - \rho g y_{bottom} = 0.82 \text{ N/m}^2$ between the upper and lower surfaces of the wing. This pressure difference means that the air exerts a small buoyant force on the wing. Our calculations show that this small pressure difference is totally negligible compared to the pressure difference $\frac{1}{2}\rho v_{top}^2 - \frac{1}{2}\rho v_{bottom}^2 = 9.6 \times 10^2 \text{ N/m}^2$ due to air traveling faster past the top of the wing than past the bottom of the wing.

WATCH OUT! Air molecules don't meet at the trailing edge of a wing.

Why does air travel at different speeds over the two surfaces of the wing? A common misconception is that air traveling along the top of the wing takes just as much time to go from the wing's leading edge to its trailing edge as does the air that travels along the bottom. According to this misconception, air has to move faster along the top of the wing because the upper surface is curved more than the lower, and molecules that parted company at the leading edge must somehow meet up at the trailing edge. But the computer simulation in Figure 11-29 shows that this isn't the case at all. Air travels over the upper surface of the wing *much* faster than the common misconception would have us believe, and the molecules *don't* meet up at the trailing edge. A better explanation is that air follows the curvature of the wing due to its (slight) viscosity. This curvature is such that the wing pushes air downward. By Newton's third law the air must push the wing *upward* equally hard. To produce this upward force, or lift, there has to be a pressure difference between the two surfaces of the wing, and this pressure difference is what causes the air to flow at different speeds over the top and bottom of the wing.

EXAMPLE 11-15 A Venturi Meter

Figure 11-30 shows a simple device called a *Venturi meter* for measuring fluid velocity in a gas such as air. When gas passes from left to right through the horizontal pipe, it speeds up as it passes through the constriction at point 2. Bernoulli's principle tells us that the gas pressure must be lower at point 2 than at point 1, and the pressure difference causes the liquid in the U-tube to drop on the left-hand side and rise on the right-hand side. Suppose the gas is air (density 1.20 kg/m^3) that enters the left-hand side of the Venturi meter at 25.0 m/s. The horizontal tube has cross-sectional area 2.00 cm^2 at point 1 and cross-sectional area 1.00 cm^2 at point 2. If the liquid in the U-tube is water, what is the difference in height between the water columns on the left-hand and right-hand sides of the tube?

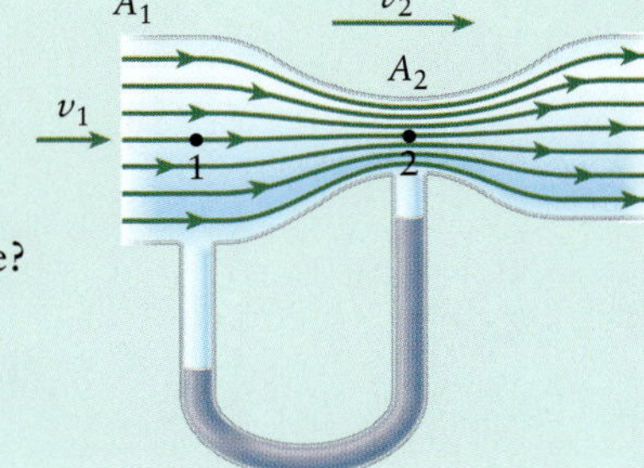

Figure 11-30 A Venturi meter A Venturi meter, or flow meter, can be used to measure the flow speed of a gas.

Set Up

In this problem there are *two* fluids, the air that flows through the horizontal pipe and the water in the U-tube. Hence we'll use Bernoulli's equation twice: once to relate the moving air at point 1 to the moving air at point 2, and once to relate the heights of the water on the two sides of the U-tube (which is what we're trying to find). We'll also use the equation of continuity to relate the speeds of the air at points 1 and 2.

Bernoulli's equation:

$$p + \frac{1}{2}\rho v^2 + \rho g y = \text{a constant} \quad (11\text{-}27)$$

Equation of continuity for steady flow of an incompressible fluid:

$$A_1 v_1 = A_2 v_2 \quad (11\text{-}19)$$

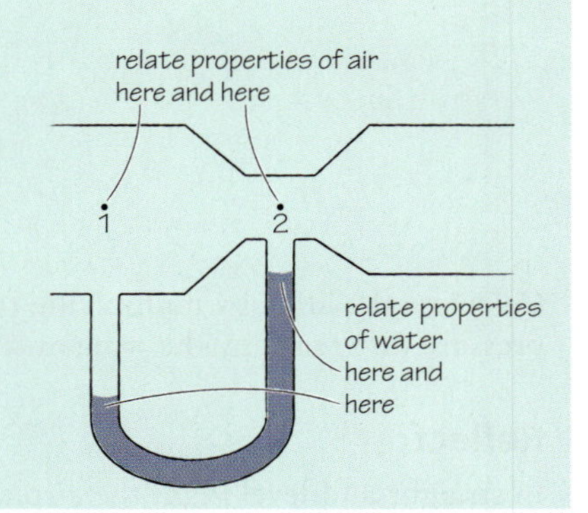

Solve

We need to find the difference in air pressure at points 1 and 2, since this is what causes the difference in height of the water on the two sides of the U-tube. Find this difference using Bernoulli's equation and the equation of continuity applied to the air in the horizontal pipe, keeping in mind that we know the values of v_1, A_1, and A_2.

Bernoulli's equation for the air at points 1 and 2:

$$p_1 + \frac{1}{2}\rho_{\text{air}} v_1^2 + \rho_{\text{air}} g y_1 = p_2 + \frac{1}{2}\rho_{\text{air}} v_2^2 + \rho_{\text{air}} g y_2$$

At the center of the horizontal pipe, $y_1 = y_2$, so

$$p_1 + \frac{1}{2}\rho_{\text{air}} v_1^2 = p_2 + \frac{1}{2}\rho_{\text{air}} v_2^2$$

$$p_1 - p_2 = \frac{1}{2}\rho_{\text{air}} v_2^2 - \frac{1}{2}\rho_{\text{air}} v_1^2$$

From the equation of continuity,

$$v_2 = \frac{A_1}{A_2} v_1 \text{ so}$$

$$p_1 - p_2 = \frac{1}{2}\rho_{\text{air}}(v_2^2 - v_1^2) = \frac{1}{2}\rho_{\text{air}}\left[\left(\frac{A_1}{A_2}\right)^2 v_1^2 - v_1^2\right]$$

$$= \frac{1}{2}\rho_{\text{air}} v_1^2\left[\left(\frac{A_1}{A_2}\right)^2 - 1\right]$$

$$= \frac{1}{2}(1.20 \text{ kg/m}^3)(25.0 \text{ m/s})^2\left[\left(\frac{2.00 \text{ cm}^2}{1.00 \text{ cm}^2}\right)^2 - 1\right]$$

$$= 1.13 \times 10^3 \text{ kg/(m} \cdot \text{s}^2) = 1.13 \times 10^3 \text{ N/m}^2$$

The pressure difference between points 1 and 2 is also the pressure difference between the water at two points: the top of the water column on the left-hand side of the U-tube and the water column on the right-hand side. The water is at rest ($v = 0$) at both points. Use this technique to find the height difference between the two water columns.

Bernoulli's principle for the water at the tops of the two columns:

$$p_1 + \frac{1}{2}\rho_{\text{water}} v_{\text{water},1}^2 + \rho_{\text{water}} g y_{\text{water},1}$$

$$= p_2 + \frac{1}{2}\rho_{\text{water}} v_{\text{water},2}^2 + \rho_{\text{water}} g y_{\text{water},2}$$

Water is at rest: $v_{\text{water},1} = v_{\text{water},2} = 0$

Solve for the height difference of the two water columns:

$$p_1 + \rho_{\text{water}} g y_{\text{water},1} = p_2 + \rho_{\text{water}} g y_{\text{water},2}$$

$$\rho_{\text{water}} g y_{\text{water},2} - \rho_{\text{water}} g y_{\text{water},1} = p_1 - p_2$$

$$y_{\text{water},2} - y_{\text{water},1} = \frac{p_1 - p_2}{\rho_{\text{water}} g} = \frac{1.13 \times 10^3 \text{ kg/(m} \cdot \text{s}^2)}{(1.000 \times 10^3 \text{ kg/m}^3)(9.80 \text{ m/s}^2)}$$

$$= 0.115 \text{ m} = 11.5 \text{ cm}$$

Reflect

This height difference is large enough to easily measure, so the Venturi meter is a practical device.

We made an approximation in our solution that the pressure difference between points 1 and 2 within the horizontal pipe is the same as the pressure difference between the two water columns. This isn't exactly correct because there is a greater weight of air above the column on the left-hand side. However, the resulting additional pressure difference is so small that we can ignore it.

Additional pressure difference between the tops of the two water columns due to the extra weight of air on the left-hand side:

$$\begin{aligned}\rho_{\text{air}} g(y_{\text{water},2} - y_{\text{water},1}) &= (1.20\ \text{kg/m}^3)(9.80\ \text{m/s}^2)(0.115\ \text{m}) \\ &= 1.35\ \text{kg/(m} \cdot \text{s}^2) = 1.35\ \text{N/m}^2\end{aligned}$$

This is 0.12% of the pressure difference calculated above ($p_1 - p_2 = 1.13 \times 10^3\ \text{N/m}^2$), so we can neglect it.

Applications of Bernoulli's Principle

Can a thrown baseball be a "curveball"—that is, can it be made to follow a path that curves left or right? In the early days of baseball, most people thought a curveball was just an optical illusion. It wasn't until 1941, when *Life* magazine published photographs of a curveball taken with a strobe light, that baseball fans (and everyone else) had proof that a properly thrown baseball can be made to curve. Bernoulli's principle helps to explain why.

Imagine a baseball thrown without any spin, what ball players call a "knuckleball." As the ball flies, the air rushes past the left and right sides of the ball at the same speed. Since the speed of the air is the same on both sides, according to Bernoulli's equation the air pressure in the air is also the same on both sides. Hence there is no tendency for the ball's trajectory to curve either right or left.

Now imagine the ball spinning as it moves (**Figure 11-31**). In this figure we're looking at the ball from above, and it is rotating counterclockwise as seen from this vantage point. The ball is moving in the direction shown by the arrow, and air rushes past it in the opposite direction. Because the surface of the ball is rough, and because the baseball has raised seams that hold the leather cover of the ball together, viscosity drags a layer of air around in the same direction as the spin of the ball (compare Figure 11-21). As a result the net speed of air relative to the ball is slower on the right-hand side of the ball than on the left-hand side. Bernoulli's principle tells us that a speed difference corresponds to a pressure difference, and the slower speed to the right of the ball means that there is higher pressure there. This pressure difference means that as viewed from above, this baseball feels a force to the left and will curve to the left!

BIO-Medical Bernoulli's principle can help baseball pitchers win games; it also helps fish such as mackerel stay alive. A mackerel swims with its mouth open, allowing water to enter through the mouth, pass through the gills where oxygen is extracted, and exit through an aperture called the *operculum*. The streamlines of water flow around the mackerel to follow the contours of its body. **Figure 11-32** shows that these streamlines are close together near the operculum. As we learned in Section 11-9, flow in an incompressible fluid such as water is rapid where streamlines are close together but slow where the streamlines are far apart. Hence the water flow is *faster* at the operculum than at the mouth, and so according to Bernoulli's principle the pressure at the operculum is *lower* than at the mouth. This pressure difference helps to force water from mouth to operculum by way of the mackerel's gills. This effect, called *ram ventilation*, is essential to the mackerel's ability to extract sufficient oxygen from the water. Other species of fish use ram ventilation only at high swimming speeds; at lower speeds, where ram ventilation is less effective, they use muscular action to pump water through their gills.

As the ball rotates counterclockwise (in this picture) a layer of air close to the ball is also dragged around counterclockwise. This boundary layer moves in the same direction as the air rushing past the ball on the left but opposes the air flow on the right.

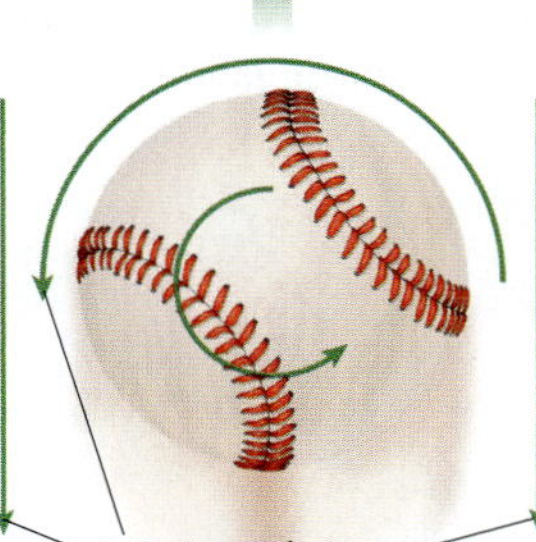

According to Bernoulli's equation, higher net air speed on the left results in lower pressure. The pressure difference between the right and left sides of the ball results in a net force, so this ball will curve to the left!

Figure 11-31 **What makes a curveball curve** If a baseball rotates, viscosity drags a layer of air around the ball in the same direction as the ball's rotation.

GOT THE CONCEPT? 11-10 Bernoulli in Action

Imagine holding two pieces of paper vertically, with a small gap between them, and then blowing gently into the gap. Would you expect the pieces of paper to (a) be drawn together, (b) be pushed apart, or (c) be unaffected? (Try it!)

TAKE-HOME MESSAGE FOR Section 11-10

✔ Bernoulli's equation relates the pressure, speed, and height of a fluid whose flow is *laminar, steady, incompressible, inviscid,* and *irrotational.*

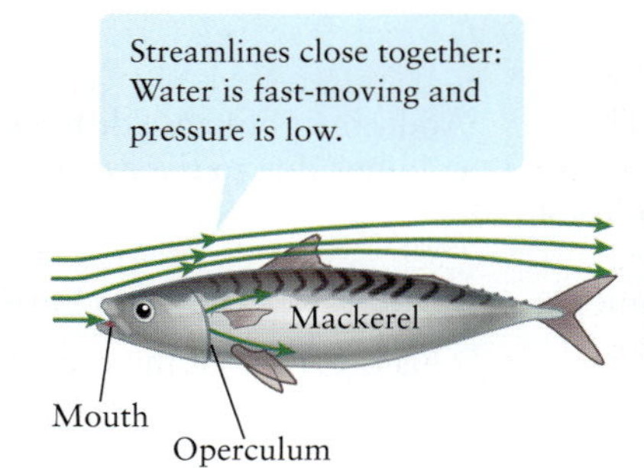

Streamlines close together: Water is fast-moving and pressure is low.
Mackerel
Mouth
Operculum

Figure 11-32 Bernoulli's principle and fish respiration Mackerel breathe by letting water flow from the mouth through the gills (where oxygen is extracted from the water) and out the operculum. Bernoulli's principle tells us that the water pressure at the operculum is lower when the fish is in motion, thus facilitating flow through the gills.

11-11 Viscosity is important in many types of fluid flow

We saw in Section 11-10 that Bernoulli's equation holds true only under certain very special assumptions. In particular we saw that the fluid had to have a little bit of viscosity to make the laminar flow be irrotational. But we couldn't allow the fluid to have too much viscosity, since that would mean that *viscous forces*—that is, forces on fluid due to friction—would play an important role in determining the acceleration of a bit of moving fluid.

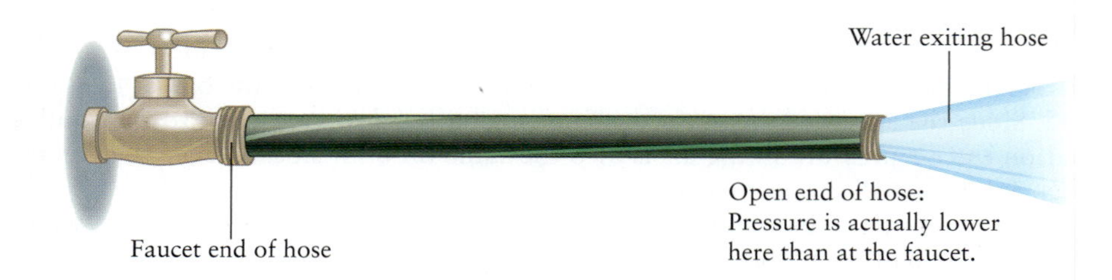

Figure 11-33 When Bernoulli's equation doesn't work A horizontal garden hose attached to a faucet. If Bernoulli's equation held true, the water pressure would be the same everywhere along the length of the hose. Bernoulli's equation does not apply here because viscous forces are important in this situation.

For many situations involving fluid flow, there is "too much" viscosity, and Bernoulli's equation is *not* valid. One example is water flowing in a garden hose (**Figure 11-33**). The hose has the same cross-sectional area A throughout its length, so in order to satisfy the equation of continuity (which says that the product Av is a constant), the flow speed v must be the same at all points along the hose's length. Furthermore, if the hose is horizontal, then all points along the hose are at the same height y. If Bernoulli's equation, Equation 11-27 (which says that $p + \frac{1}{2}\rho v^2 + \rho g y$ is a constant), applied to this situation, then the water pressure p would also have to have the same value everywhere inside the hose. But in fact the pressure at the open end of the hose is substantially *lower* than at the end attached to the faucet. The reason is that frictional forces oppose the flow of water through the hose, so there has to be additional pressure at the faucet to sustain the flow.

The human circulatory system is another situation where Bernoulli's equation doesn't hold. The speed of blood flow in capillaries (about 10^{-3} m/s) is hundreds of times slower than in the aorta (about 0.2 m/s), so the term $\frac{1}{2}\rho v^2$ in Equation 11-27 has a much smaller value in the capillaries. Bernoulli's equation would therefore predict that blood pressure is much higher in the capillaries than in the aorta. In fact, the pressure is higher in the *aorta*; this is necessary to push against the frictional forces that the blood encounters on its way to the capillaries.

Figure 11-34 Measuring viscosity An experiment to measure the viscosity of a fluid.

Let's be more quantitative about viscosity and its consequences. **Figure 11-34** shows one way that physicists measure the viscosity of a fluid. Two glass plates, each of area A, are separated by a distance d and have a quantity of fluid filling the space between them. The lower plate is held stationary while the upper plate is pulled sideways at a constant speed v. Experiment shows that if the speed v is not too great, the fluid at the top remains attached to the upper plate and the fluid at the bottom remains attached to the lower plate, while the fluid in between slides sideways in sheets like a deck of cards. The force (magnitude F) that must be applied to the upper plate to make the fluid move in this way is relatively large for a viscous fluid like oil but relatively low for a fluid like water. The force magnitude F is also proportional to the speed v and the area A but inversely proportional to the plate spacing d. We can summarize these observations in the following expression for F, which defines what we mean by the viscosity η (the Greek letter eta):

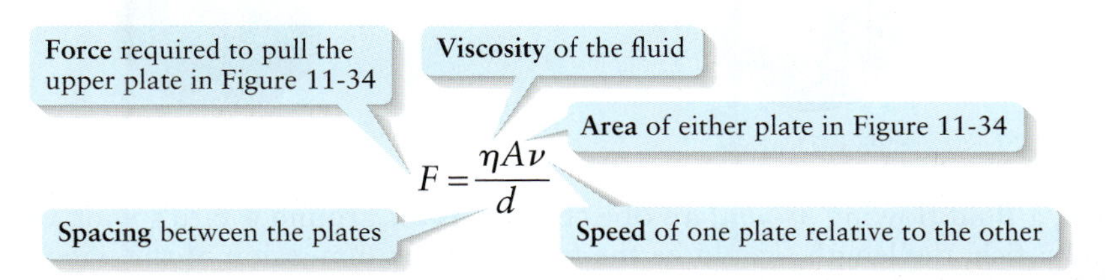

Definition of viscosity (11-28)

Equation 11-28 says that the greater the viscosity η, the greater the force that must be applied to move one plate relative to the other. The units of viscosity are pascals times seconds, or Pa · s for short. Viscosities are also sometimes given in *poise*, named for the nineteenth-century French scientist Jean Poiseuille; 1 poise equals 0.1 Pa · s.

Table 11-2 lists values of viscosity for several common fluids. Note that the viscosity of a fluid generally depends on temperature; liquids such as maple syrup and motor oil become less viscous and so flow more easily at higher temperatures.

TABLE 11-2 Viscosities of Various Substances

Substance	Viscosity (Pa · s)
air at 0°C	1.709×10^{-5}
air at 20°C	1.808×10^{-5}
air at 40°C	1.904×10^{-5}
fresh water at 0°C	1.787×10^{-3}
fresh water at 20°C	1.002×10^{-3}
fresh water at 40°C	6.53×10^{-4}
seawater at 20°C	1.072×10^{-3}
mercury (Hg) at 20°C	1.554×10^{-3}
blood at 37°C	0.003–0.004
typical motor oil at 20°C	0.1
glycerin at 20°C	1.490
chocolate syrup at 20°C	1.5
ketchup at 20°C	5.1
honey at 20°C	10
peanut butter at 20°C	25.5

Reynolds Number: Comparing Viscous Forces and Forces Due to Pressure Differences

The experiment shown in Figure 11-34 is useful for measuring viscosity because the only forces making the fluid move are the frictional (viscous) forces that act between parts of the fluid and between the fluid and plates. In most real-life situations, however, there are also forces due to pressure differences acting on the fluid (for example, the pressure difference between the two ends of a garden hose as you water the lawn). To describe the relative importance of forces due to pressure difference and viscous forces in fluid flow, physicists use a dimensionless quantity—that is, a pure number with no units—called the **Reynolds number:**

Reynolds number of a certain fluid flow
Density of the fluid
A length characteristic of the situation

$$Re = \frac{\rho l v}{\eta}$$

Viscosity of the fluid
Speed of the flow

Reynolds number (11-29)

Roughly speaking, the Reynolds number Re (*not* R_e) is the ratio of the forces due to pressure differences acting on a small piece of the fluid to the viscous forces acting on the same piece of fluid. Because the viscosity η is in the denominator in this expression, the Reynolds number increases as the viscosity *decreases*, and vice versa. Thus Re has a large value if viscosity is relatively unimportant, but it has a small value if viscosity plays a dominant role.

Definition of the length l in Equation 11-29 depends on the situation. For a fluid flowing in a tube, such as water in a pipe or air through your nostrils, l could be the diameter of the tube; the narrower the pipe, the smaller the Reynolds number, the more important the role of viscosity, and the less likely it is that Bernoulli's equation will be valid.

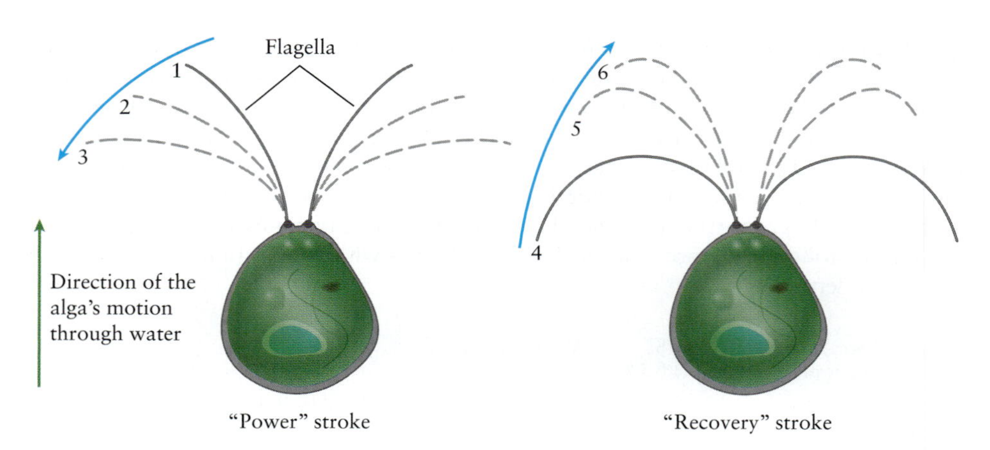

Figure 11-35 Viscous forces on swimming algae (a) The tiny flagellate *Chlamydomonas*, which is just a few micrometers in diameter, moves forward through water during a power stroke. (b) Due to the viscosity of water, *Chlamydomonas* actually moves backward during the recovery stroke.

For a fluid flowing *around* an object, such as air around a bird's wing or water around a fish, the length l could be the front-to-back dimension of the wing or fish. Hence the Reynolds number Re is smaller, and viscosity more important, for a small organism moving through water than for a larger organism moving through water at the same speed. If the small organism moves at a slow speed v, the Reynolds number is even smaller and viscosity can become extremely important. As an example, the viscosity of water is relatively unimportant for a human doing the breaststroke underwater. The swimmer accelerates forward when she pulls her arms back in a "power stroke," pushing against the water, and continues forward even when she pulls her arms forward in the "recovery stroke." By contrast, viscosity is a huge effect for the single-celled alga *Chlamydomonas*, which propels itself slowly through water in a breaststroke-like way using two flagella (**Figure 11-35**). When the cell pulls its flagella back in the "recovery stroke," the cell actually *backs up* by about a third of the distance that it moves forward during a "power stroke." You can better empathize with the plight of *Chlamydomonas* by imagining yourself swimming in a tub of molasses!

Low Reynolds Number: Laminar Flow

If the viscosity of a flowing fluid is sufficiently high, and hence Re is sufficiently low, the flow will be laminar but will not obey Bernoulli's equation. An important example of this is the flow of fluids within pipes, such as the flow of vaccine in a hypodermic needle or of blood in your circulatory system. (Blood vessels can curve and can vary in diameter along their length, but for simplicity we'll consider straight pipes with circular cross sections.) To sustain a steady flow there must be a pressure difference between the ends of the pipe; fluid flows from the high-pressure end to the low-pressure end. That's why a hypodermic syringe is equipped with a plunger; the force of a thumb on the plunger provides the necessary high pressure on one end of the needle.

Figure 11-36 shows the laminar flow of fluid within a circular pipe seen in cross section. Due to friction the fluid that's right up against the inside walls of the pipe doesn't move at all. This is the *no-slip condition* that we described in Section 11-9. The fluid a small distance from the walls moves a little, the fluid a bit farther away moves a bit faster, and so on. The fastest flow is at the very center of the pipe. This flow pattern is called *parabolic* (since the dashed curve in Figure 11-36 is a parabola lying on its side). A derivation using calculus gives the following expression, called the **Hagen–Poiseuille equation**, relating the pressure difference between the two ends of the pipe to the resulting flow rate. (Non-French speakers can pronounce Poiseuille as "pwa-zoo-yah.")

Cross section of a pipe

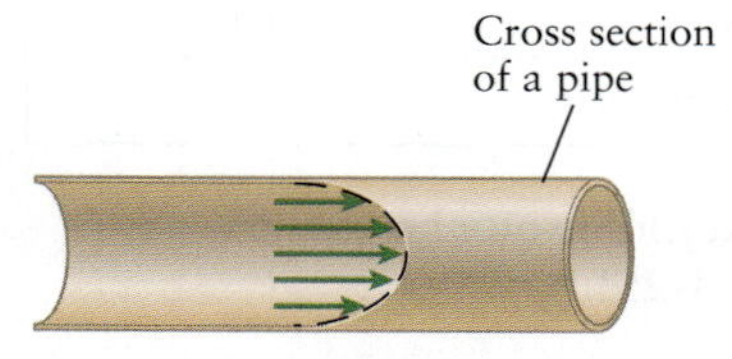

Figure 11-36 Laminar flow of a viscous liquid in a circular pipe The viscous fluid moves fastest at the center of the pipe and does not move at all at the pipe walls.

Hagen–Poiseuille equation for laminar flow of a viscous fluid (11-30)

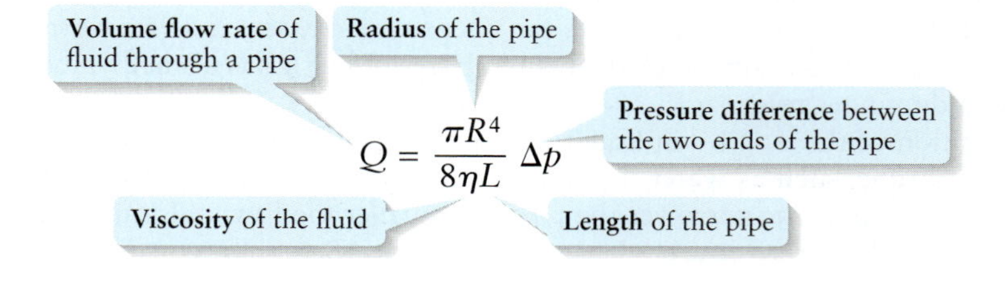

$$Q = \frac{\pi R^4}{8\eta L}\Delta p \qquad (11\text{-}30)$$

The longer the pipe (greater L) and the smaller its radius R, the less the volume flow rate Q caused by a given pressure difference Δp. You can understand this by comparing a volume of fluid that fills the inside of a short, wide pipe to the same volume of fluid that fills a long, narrow pipe. More of the fluid in the long, narrow pipe is in contact with the pipe walls and hence is not in motion, so the flow rate is less.

BIO-Medical Equation 11-30 states that for the same pressure difference Δp, using a slightly wider pipe gives a much larger flow rate. Doubling the radius increases the flow rate by a factor of $2^4 = 16$! This equation also tells us that if we want to get the same volume flow rate Q through a narrower pipe (smaller R), we have to apply a greater pressure difference Δp between the two ends of the pipe. This helps explain how a diet high in cholesterol can lead to *hypertension*, or high blood pressure; cholesteric plaque building up within arteries decreases their inner radius R. Also, because the plaque build-up makes the arteries less elastic, hardened arteries expand less than healthy ones to accommodate blood pumped into them by the heart. Therefore, the heart must supply higher pressure to sustain the same blood flow.

Turbulent flow over the top of the bird's wing makes these feathers deflect upward.
iStockphoto/Thinkstock

Figure 11-37 Turbulent air flow over a buzzard's wings The airflow over the top of this buzzard's wings is largely turbulent, causing some of the small feathers near mid-span to be pushed up.

High Reynolds Number: Turbulent Flow

Suppose fluid is moving through a circular pipe with the flow pattern shown in Figure 11-36. If there's a small imperfection in the pipe walls (biological pipes are never perfectly circular) or if something should disturb the flow, what happens? If the Reynolds number is low enough (that is, if the flow is sufficiently slow or the viscosity sufficiently high), the answer is "not much": The flow pattern is deformed a little, but thanks to viscous forces the pattern recovers its parabolic shape a short distance downstream of the disturbance. If the Reynolds number is greater than a certain value, however, the flow loses its regularity and becomes *turbulent* like the smoke trails in Figure 11-17b. For fluid flow in circular pipes, the critical value of the Reynolds number is about $Re = 2000$, where l in Equation 11-29 is the diameter (not the radius) of the pipe.

Unlike laminar flow, turbulent flow is messy as well as difficult to describe mathematically. It is also very common in nature. The flow of air around the wing of a gliding bird is laminar near the wing's leading edge but becomes turbulent as the air moves along the wing's upper surface. The effect is more pronounced if the bird glides at a slower speed, in which case its wings are inclined at a greater angle to the direction of flight to provide enough lift. The turbulent air moving over the wing's upper surface can cause individual wing feathers to deflect upward (**Figure 11-37**). If the bird flies too slowly and tilts the wing up too sharply, the air on the upper surface becomes so turbulent that it no longer follows the contours of the wing. Since the wing produces lift by forcing air downward, this onset of turbulence means that the lifting force of the wing decreases dramatically (a phenomenon called an *aerodynamic stall*) and the bird drops suddenly. Birds recover from this drop by dipping their wings slightly to restore a more laminar flow. Airplane wings are subject to the same effect, which pilots deal with in the same way.

Blood flow in the human circulatory system is often approximately laminar. But the Reynolds number in the larger blood vessels is close to the critical value that separates laminar and turbulent flows. Hence a relatively small obstruction in the aorta—the large artery that carries blood from the heart to the rest of the body—can trigger turbulence. Turbulent flow in the aorta generates noise, which is one thing that a physician listens for (and hopes not to hear) when using a stethoscope to check your heart. A totally benign noise caused by turbulence is the sound you hear on a windy day as turbulent air blows past your ears.

EXAMPLE 11-16 Laminar or Turbulent?

Is the flow of water at 20°C laminar or turbulent if (a) the pipe has radius $0.500 \text{ mm} = 5.00 \times 10^{-4}$ m and the average flow speed is 0.800 m/s; (b) the radius is $0.125 \text{ mm} = 1.25 \times 10^{-4}$ m and the average flow speed is 12.8 m/s? At this temperature the density of water is 998 kg/m^3.

Set Up

The key to solving this problem is the Reynolds number, which for flow in a pipe of circular cross section is given by Equation 11-29. In this equation l is the diameter of the pipe. The flow is laminar if Re is less than 2000; otherwise it is turbulent.

Reynolds number:

$$Re = \frac{\rho l v}{\eta} \qquad (11\text{-}29)$$

Solve

(a) Use the diameter of the pipe (twice the radius) and the flow speed to calculate the Reynolds number. We use Table 11-2 to find the viscosity of water.

$$l = 2(5.00 \times 10^{-4}\ \text{m}) = 1.00 \times 10^{-3}\ \text{m}$$

Reynolds number of the flow:

$$Re = \frac{(9.98 \times 10^{2}\ \text{kg/m}^3)(1.00 \times 10^{-3}\ \text{m})(0.800\ \text{m/s})}{1.002 \times 10^{-3}\ \text{Pa} \cdot \text{s}}$$

$$= 797$$

Since the Reynolds number is less than 2000, the flow is laminar.

(b) Repeat the calculation for the new pipe diameter and flow speed.

$$l = 2(1.25 \times 10^{-4}\ \text{m}) = 2.50 \times 10^{-4}\ \text{m}$$

Reynolds number of the flow:

$$Re = \frac{(9.98 \times 10^{2}\ \text{kg/m}^3)(2.50 \times 10^{-4}\ \text{m})(12.8\ \text{m/s})}{1.002 \times 10^{-3}\ \text{Pa} \cdot \text{s}}$$

$$= 3.19 \times 10^{3}$$

In this case the Reynolds number is greater than 2000, so the flow is turbulent.

Reflect

Notice that compared to the pipe in (a), the pipe in (b) has 1/4 the radius and hence 1/16 the cross-sectional area (which is proportional to the square of the radius). Notice also that the average flow speed in (b) is 16 times that in (a). So the volume flow rate (the product of cross-sectional area A and flow speed v) is the *same* in both (a) and (b), but the flow is turbulent in the narrower pipe of (b). In a similar way an obstruction in a blood vessel (which reduces the vessel's radius) doesn't change the flow rate but does increase the Reynolds number and so can induce turbulence.

Volume flow rate in first pipe:

$$A_1 v_1 = \pi R_1^2 v_1$$
$$= \pi(5.00 \times 10^{-4}\ \text{m})^2 \times (0.800\ \text{m/s})$$
$$= 6.28 \times 10^{-7}\ \text{m}^3/\text{s}$$

Volume flow rate in second pipe:

$$A_2 v_2 = \pi R_2^2 v_2$$
$$= \pi(1.25 \times 10^{-4}\ \text{m})^2(12.8\ \text{m/s})$$
$$= 6.28 \times 10^{-7}\ \text{m}^3/\text{s}$$

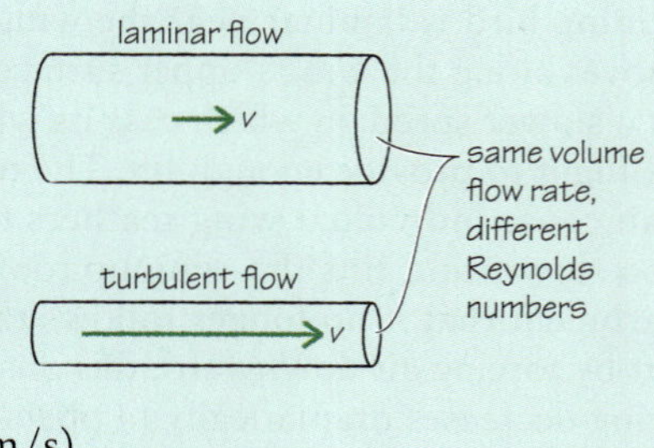

EXAMPLE 11-17 Viscous Laminar Flow in a Long, Narrow Pipe

Water at 20°C moves in laminar flow at an average flow speed of 0.800 m/s through a pipe of radius 0.500 mm = 5.00×10^{-4} m. The pipe is 2.00 m long. Determine what the pressure difference must be between the two ends of the pipe.

Set Up

We saw in part (a) of Example 11-16 that the flow is laminar for water at this temperature moving through such a pipe at this speed. Because the pipe is long and narrow, viscosity is important and laminar flow is an example of Hagen–Poiseuille flow. (The density and viscosity are given in Example 11-16.) Our goal is to find the pressure difference Δp between the ends of the pipe, and we use the Hagen–Poiseuille equation, Equation 11-30.

Hagen–Poiseuille equation for laminar flow of a viscous fluid:

$$Q = \frac{\pi R^4}{8 \eta L} \Delta p \qquad (11\text{-}30)$$

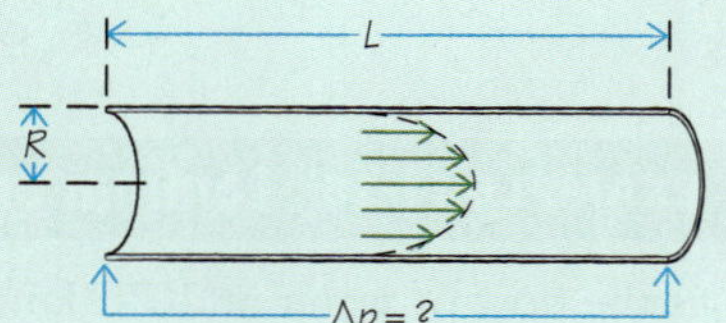

Solve

In Example 11-16 we calculated the volume flow rate.

Volume flow rate:

$$Q = 6.28 \times 10^{-7}\ \text{m}^3/\text{s}$$

Solve the Hagen–Poiseuille equation for the pressure difference Δp and substitute values.

$$\Delta p = \frac{8\eta L Q}{\pi R^4}$$

$$= \frac{8(1.002 \times 10^{-3}\ \text{Pa}\cdot\text{s})(2.00\ \text{m})(6.28 \times 10^{-7}\ \text{m}^3/\text{s})}{\pi(5.00 \times 10^{-4}\ \text{m})^4}$$

$$= 5.13 \times 10^4\ \text{Pa}$$

Reflect

Even though the viscosity of water is low, the pressure required to sustain the flow is substantial (about half an atmosphere) because the pipe is so narrow and long.

$$\Delta p = (5.13 \times 10^4\ \text{Pa})\left(\frac{1\ \text{atm}}{1.01325 \times 10^5\ \text{Pa}}\right)$$

$$= 0.506\ \text{atm}$$

GOT THE CONCEPT? 11-11 When Is Viscosity More Important?

Two identical tubes have fluid flowing through them at the same speed. In one tube the fluid is air, and in the other tube the fluid is fresh water. In which tube is viscosity more important? (a) the tube with water; (b) the tube with air; (c) viscosity is equally important in each tube; (d) not enough information given to decide.

TAKE-HOME MESSAGE FOR Section 11-11

✔ The Reynolds number for a fluid flow is large if viscous forces are relatively unimportant in the flow and small if viscous forces play a dominant role.

✔ If a fluid flow has a sufficiently low Reynolds number, the flow is laminar. A sufficiently high Reynolds number means that the flow is turbulent.

✔ The Hagen–Poiseuille equation describes the laminar flow of a viscous fluid. This equation relates the pressure difference between the two ends of a pipe to the resulting flow rate.

11-12 Surface tension explains the shape of raindrops and how respiration is possible

Why does rain fall in droplets? Why does water roll off a duck's back? And why is it useful to use detergent when washing clothes? We can answer all of these questions by examining an important property of liquids called *surface tension.*

A liquid maintains an essentially constant volume because its molecules exert strong attractive forces on each other. As a result the molecules try to huddle close to each other. The potential energy associated with these forces is lowest if as many of the molecules as possible are completely surrounded by other molecules, and if as few as possible are on the surface (where they are surrounded by other molecules on one side only). Therefore, to minimize this potential energy, the molecules arrange themselves so as to minimize the surface area of the liquid (Figure 11-38). A sphere has the smallest surface area for a given volume, which is why raindrops are spherical. The drop behaves as though its surface were a membrane that resists being stretched, which is why this behavior is called **surface tension**.

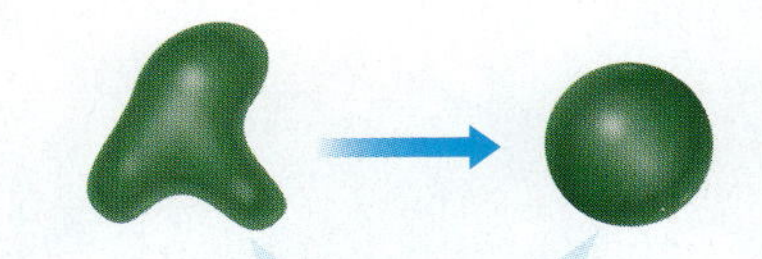

An irregular blob of liquid tends to become spherical due to surface tension.

Figure 11-38 Surface tension The effect of surface tension on a liquid blob or drop.

WATCH OUT! Raindrops aren't shaped like "teardrops."

It's a common belief that raindrops are teardrop-shaped, with a rounded lower half and a pointed upper half. But the aerodynamic forces that would force a raindrop into this shape are quite weak compared to the attractive forces between molecules. In fact, due to surface tension, raindrops are nearly spherical in shape.

A spherical drop of radius R has volume $(4\pi/3)R^3$ and surface area $4\pi R^2$. The importance of surface tension is proportional to the ratio of the drop's surface area (which is the size of the "membrane" enclosing the drop) to its volume (which tells you the total amount of fluid in the drop). This ratio is $4\pi R^2/[(4\pi/3)R^3] = 3/R$, which *increases* as the radius R *decreases*. Hence surface tension is more important for small drops than for large ones. If you pour enough liquid into a container, surface tension is overwhelmed by gravitational forces and the liquid assumes a shape that fills the container and has a horizontal upper surface. In zero gravity even large quantities of liquid form spherical drops (**Figure 11-39**).

The surface tension of water enables ducks to remain warm and dry even after submerging themselves in search of a meal. A duck's feather has a crisscross grid of tiny barbules that form a fine mesh (**Figure 11-40a**). To push through this mesh the water surface must stretch as shown in **Figure 11-40b**. Surface tension resists such stretching of the surface "membrane," so water does not penetrate the mesh, and the duck's skin remains dry. The same effect makes it difficult for water to penetrate between the closely spaced fibers of clothing. That's why clothing is best washed using a detergent, which is a substance that reduces the surface tension of water.

BIO-Medical

Surface tension also plays an important role in human respiration. In the lungs oxygen exchange takes place across the walls of tiny, balloon-like structures called *alveoli* (**Figure 11-41**). The inside surface of the alveoli is wet, and the surface tension of this fluid generates a pressure that tends to pull the walls of the alveoli inward. The magnitude of this pressure is proportional to the surface tension and inversely proportional to the radius of the alveolus:

$$\text{pressure} = (2 \times \text{surface tension})/(\text{alveolar radius})$$

The problem is that the alveoli do not all have the same radius, so smaller alveoli experience a greater pressure than larger ones. As a result, small alveoli would tend to collapse into larger ones if not for a substance called a *surfactant* that coats the inside surface of each alveolus. This remarkable substance decreases surface tension as a function of its concentration in the fluid lining the inside of the alveoli. As the radius of the alveolus decreases during exhalation, the concentration of surfactant increases on the surface of the fluid lining the alveolus and therefore decreases surface tension and the resulting pressure in the alveolus. This prevents smaller alveoli from collapsing into larger ones. Also, during inhalation the reduced surface tension makes it easier to inflate the lungs with the relatively small pressure difference between inhaled air and the low-pressure air in the thoracic cavity—about 130 Pa, or 0.13% of atmospheric pressure. A condition called respiratory distress syndrome occurs in premature infants whose lungs are deficient in surfactant; the increased surface tension pressure can make their alveoli collapse. Administering a replacement surfactant helps these newborn babies survive.

Figure 11-39 Surface tension on a weightless water drop In the apparent weightlessness of an orbiting spacecraft, surface tension allows rather large quantities of water to form spherical drops.

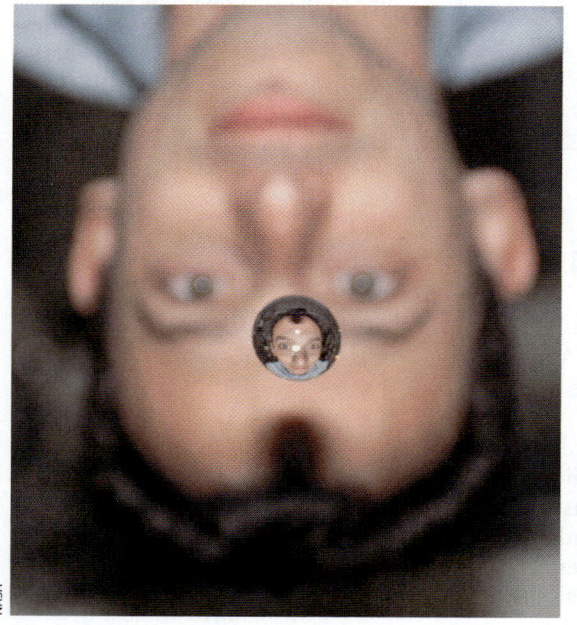

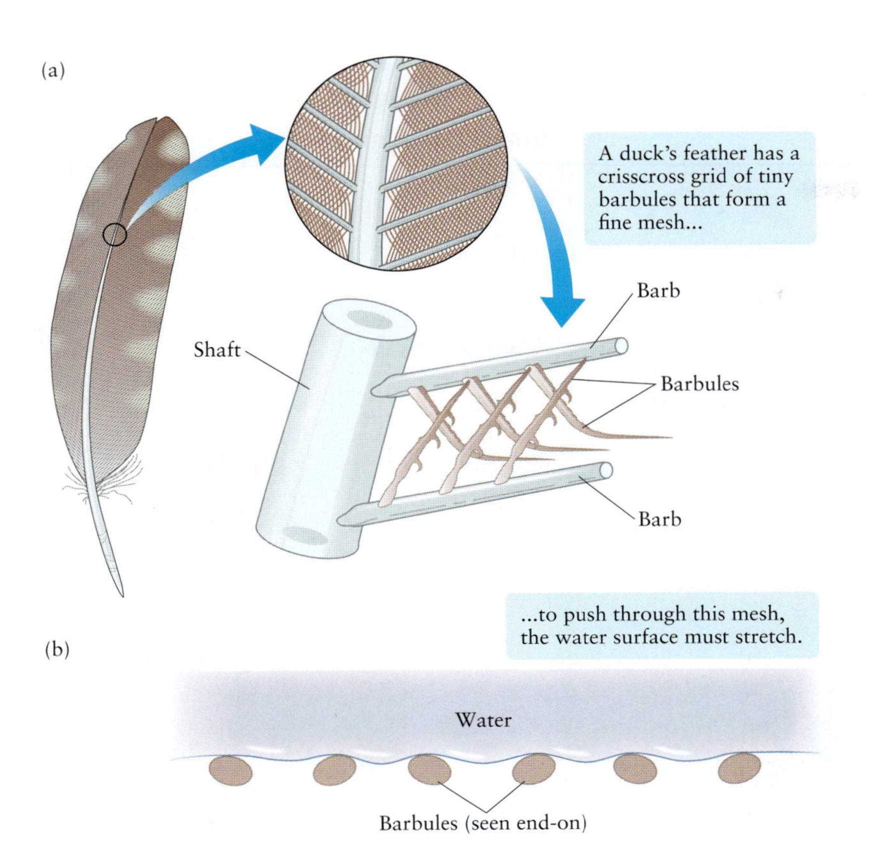
(a)
A duck's feather has a crisscross grid of tiny barbules that form a fine mesh...
Barb
Shaft
Barbules
Barb
...to push through this mesh, the water surface must stretch.
(b)
Water
Barbules (seen end-on)

Figure 11-40 Surface tension and bird feathers (a) This drawing of a bird feather shows the tiny interlocking barbules. (b) Due to surface tension, water cannot easily pass through the tiny spaces between barbules.

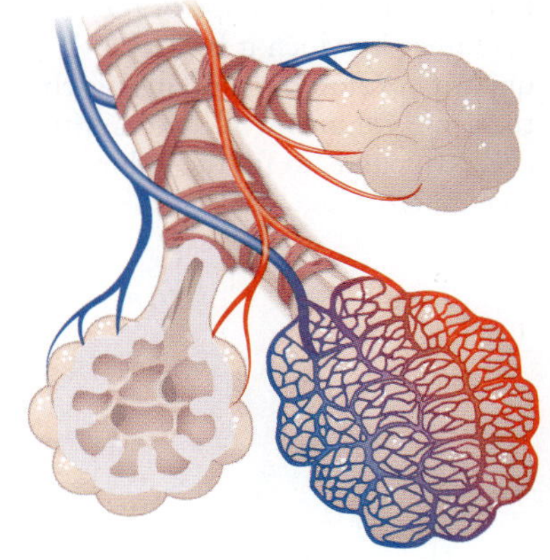

Figure 11-41 Alveoli This illustration shows a cluster of the tiny balloon-like sacs called alveoli. A human lung contains about 3×10^8 alveoli.

TAKE-HOME MESSAGE FOR Section 11-12

✔ The molecules of a liquid minimize their energy by surrounding themselves with their fellow molecules. As a result, liquids tend to form spherical drops.

✔ The surface of a liquid drop resists being stretched. This behavior is called surface tension.

Key Terms

absolute pressure
apparent weight
Archimedes' principle
atmosphere (unit of pressure)
Bernoulli's equation
Bernoulli's principle
boundary layer
buoyant force
compressible fluid
density
equation of continuity
equation of hydrostatic equilibrium
fluid
gas
gauge pressure
Hagen–Poiseuille equation
hydrostatic equilibrium
incompressible fluid
inviscid flow
irrotational flow
laminar flow
liquid
millimeters of mercury
no-slip condition
pascal
Pascal's principle
pressure
Reynolds number
solid
specific gravity
steady flow
streamline
surface tension
torr
uniform density
unsteady flow
turbulent flow
viscosity
volume flow rate

Chapter Summary

Topic | **Equation or Figure**

Fluids: The molecules in a fluid are not arranged in any organized structure. A fluid does not have any shape of its own, and flows to conform to the shape of the container in which it is placed. A quantity of liquid maintains nearly the same volume, while a quantity of gas expands to fill the volume available.

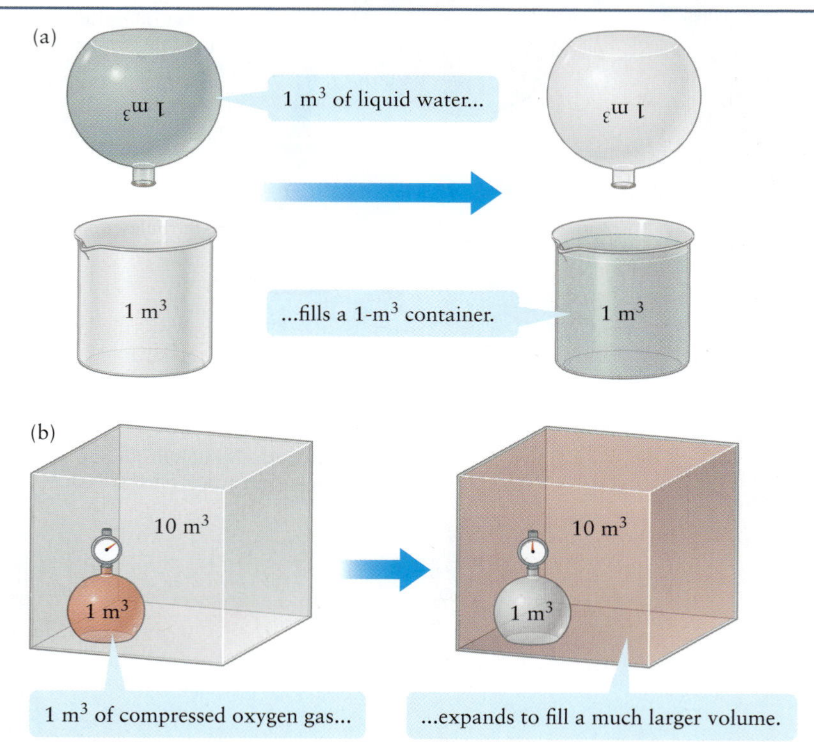

(Figure 11-2)

Density: The density ρ of a substance is its mass per volume. Density depends on how closely packed the molecules of the substance are, as well as on the mass of the individual molecules.

The **density** ρ ("rho") of a certain substance...

...equals the **mass** m of a given quantity of that substance...

...divided by the **volume** V of that quantity of the substance.

$$\rho = \frac{m}{V} \quad (11\text{-}1)$$

Pressure: The pressure p exerted by a fluid is the perpendicular force it exerts on a given area in the fluid divided by the area. Unlike force, pressure is not a vector. Pressure measured relative to atmospheric pressure is called gauge pressure.

The **pressure** p exerted by a fluid on an object in the fluid...

...equals the **force** $F_\perp$ that the fluid exerts **perpendicular to the face of the object...**

...divided by the **area** A of that face.

$$p = \frac{F_\perp}{A} \quad (11\text{-}2)$$

Fluids at rest: A stationary fluid must be in hydrostatic equilibrium so that the sum of the forces on any parcel of fluid is zero. In a fluid of uniform density, the pressure at a given point increases with depth to support the weight of the fluid above that point.

Pressure at a certain point in a **fluid at rest**

Density of the fluid (same at all points in the fluid)

Pressure at a second, lower point in the fluid

Acceleration due to gravity

$$p = p_0 + \rho g d \quad (11\text{-}8)$$

Depth of the second point where the pressure is p below the point where the pressure is p_0

Pressure difference: If there are different pressures on the two sides of an object, the different pressures produce a net force on the object.

A **pressure difference** on two opposite sides of an object **produces a net force.**

Area of each side of the object

$$F_{net} = (p_2 - p_1)A \quad (11\text{-}10)$$

High pressure on one side of the object

Lower pressure on the other side of the object

Pascal's principle: Pressure applied to a confined, static fluid is transmitted undiminished to every part of the fluid as well as to the walls of the container. A hydraulic jack, which converts a small force acting over a large distance into a large force acting over a small distance, relies on Pascal's principle.

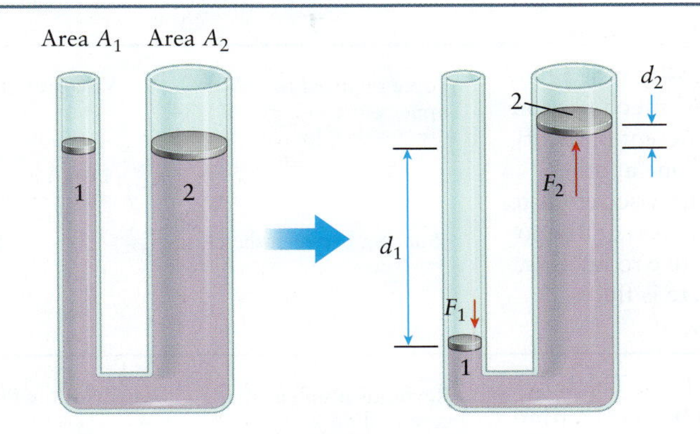

When a downward force is applied to the cap on side 1, the increase in pressure is transmitted undiminished throughout the fluid, forcing the cap on side 2 to rise.

(Figure 11-11)

Buoyancy: An object in a fluid feels an upward buoyant force F_b equal to the weight of fluid that the object displaces. A submerged object will neither sink nor rise if its average density equals that of the fluid.

Magnitude of the buoyant force on an object in a fluid

Density of the fluid

$$F_b = \rho_{fluid} V_{displaced}\, g \quad (11\text{-}13)$$

Volume of fluid that the object displaces

Acceleration due to gravity

Fluid flow: The motion of a fluid is classified as being steady or unsteady; and as laminar or turbulent; viscous (with internal friction) or inviscid (without internal friction).

(b)

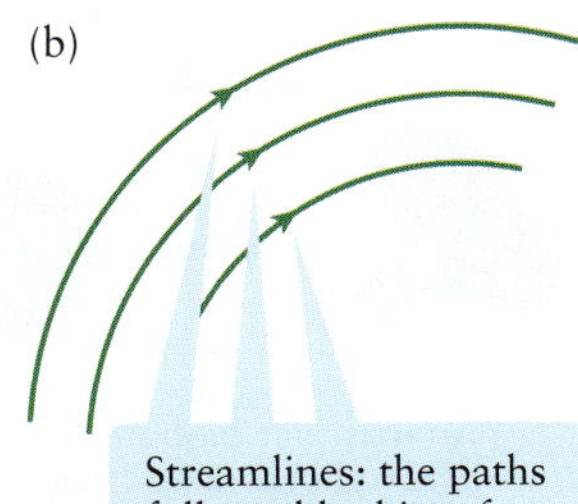

Streamlines: the paths followed by bits of fluid in laminar flow

(Figure 11-19b)

(b)

There are no streamlines for a fluid in turbulent flow.

(Figure 11-20b)

Equation of continuity: If an incompressible (constant-density) fluid flows through a pipe, the size of the pipe and the flow speed may change, but the volume flow rate (the cross-sectional area of the flow multiplied by the flow speed) remains the same.

Cross-sectional area of the flow at point 1

Flow speed at point 1

$$A_1 v_1 = A_2 v_2 \quad (11\text{-}19)$$

Cross-sectional area of the flow at point 2

Flow speed at point 2

Bernoulli's principle and Bernoulli's equation: The pressure in a moving fluid is reduced at locations where the fluid moves rapidly. The mathematical statement of this principle, called Bernoulli's equation, is only accurate for idealized laminar, inviscid flows.

$$p + \frac{1}{2}\rho v^2 + \rho g y = \text{a constant with the same value throughout the fluid} \quad (11\text{-}27)$$

Fluid pressure at a given point in the fluid; **Speed** of the fluid at that point; **Vertical coordinate** of that point; **Density** of the fluid (uniform throughout the fluid); **Acceleration due to gravity**

Viscosity: The viscosity η of a fluid is a measure of how much friction is present when different parts of the fluid slide past one another. One way to measure viscosity is to see how much force is required to slide one plate relative to another, parallel plate if there is fluid between the plates.

$$F = \frac{\eta A v}{d} \quad (11\text{-}28)$$

Force required to pull the upper plate in Figure 11-34; **Viscosity** of the fluid; **Area** of either plate in Figure 11-34; **Spacing** between the plates; **Speed** of one plate relative to the other

Turbulent viscous flow: The flow of a viscous liquid becomes turbulent if the Reynolds number Re of the flow is too large. This happens if the flow speed is too great, the characteristic dimension of the flow (for example, the diameter of the pipe through which the fluid moves) is too large, or both.

$$Re = \frac{\rho l v}{\eta} \quad (11\text{-}29)$$

Reynolds number of a certain fluid flow; **Density** of the fluid; A **length** characteristic of the situation; **Viscosity** of the fluid; **Speed** of the flow

Laminar viscous flow: To make a viscous fluid flow through a pipe, there must be a pressure difference between the two ends of the pipe. A greater pressure difference is required for long, thin pipes than for short, broad ones.

$$Q = \frac{\pi R^4}{8\eta L}\Delta p \quad (11\text{-}30)$$

Volume flow rate of fluid through a pipe; **Radius** of the pipe; **Pressure difference** between the two ends of the pipe; **Viscosity** of the fluid; **Length** of the pipe

Surface tension: Attractive forces between the molecules of a liquid make it favorable for the liquid to minimize its surface area. This effect causes water to form droplets and makes it difficult to force water through small apertures.

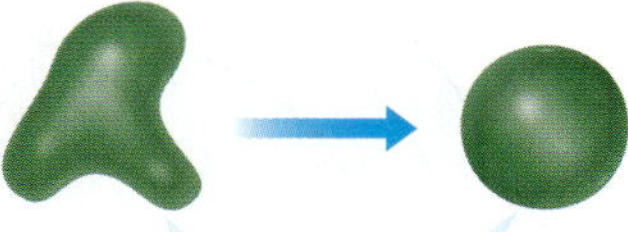

An irregular blob of liquid tends to become spherical due to surface tension.

(Figure 11-38)

Answer to What do you think? Question

(c) The upward buoyant force on an object immersed in a fluid, such as a dolphin immersed in water, is equal in magnitude to the weight of the fluid that the object displaces. A life-size sculpture of a dolphin displaces the same volume of water, and hence the same weight of water, as an actual dolphin. So the buoyant force would not change if you replaced the dolphin with a sculpture of the same size and shape, even if the sculpture is made of iron. The difference is that a dolphin's weight is about the same as the upward buoyant force exerted by the surrounding water, so a dolphin can float with little effort. By contrast, the weight of an iron dolphin sculpture is much greater than the buoyant force, so the sculpture would sink. (See Section 11-8.)

Answers to Got the Concept? Questions

11-1 (d) If you chose the block of iron, be careful! Table 11-1 tells us that iron has a much greater density ρ (7.8×10^3 kg/m^3) than does ice (0.917×10^3 kg/m^3), but the mass m (and hence the weight) of each block depends on its volume V. From Equation 11-1, $m = \rho V$. If the two blocks have the same volume, the block of iron will weigh more than the block of ice. But if the block of ice is the size of a house while the block of iron is the size of your fingernail, the ice will definitely outweigh the iron.

11-2 (e) Remember that, unlike force, pressure is *not* a vector and so has no direction. Therefore any statement that refers to the direction of pressure is an incorrect and misleading one! The pressure on each side of the fish is the same.

11-3 (b) In hydrostatic equilibrium the pressure in the water depends only on the vertical position. To convince yourself of this assertion, let p_0 be the pressure at the bottom of the left-hand leg, and p be the pressure at the bottom of the right-hand leg. These two pressures are related by Equation 11-8, $p = p_0 + \rho g d$. Since both points are at the same height, $d = 0$, and $p = p_0$. That is, the pressure p at the bottom of the left-hand tube must be the *same* as the pressure at the bottom of the right-hand tube, since these points are at the same level. (If the pressure was higher on one side or the other, water would flow from the region of high pressure to the region of low pressure until the pressures equalized.) To see that the height of each column is the same, now let the pressure at the top of each column of water, which is equal to air pressure, be written as p_0. Since the pressure p_0 at the top of the water column is the same on both sides and the pressure p at the bottom of the water column is the same on both sides, it follows from $p = p_0 + \rho g d$ that the depth of water d must be the same on both sides as well.

11-4 (d), (b), (c), (a). The deflated balloon has air at atmospheric pressure inside, so its gauge pressure is zero and its absolute pressure is about 1.0 atm. The inflated balloon has a gauge pressure of 0.50 atm, and its absolute pressure is about (0.50 atm) + (1.0 atm) = 1.5 atm.

11-5 (a) We know from Section 11-4 that pressure increases with increasing depth in a fluid, including the atmosphere. So the air pressure on the bottom of the paper is slightly greater than the pressure on the top, and the net force due to these pressures is upward (although very small). It's true that the sum of *all* forces acting on the paper—including the force of gravity and the force exerted by your hands—is zero, because the paper is at rest. But that's not what the question is asking.

11-6 (c) The cap on side 2 has 5 times the radius and so 25 times the area of the cap on side 1 (that is, $A_2/A_1 = 25$). So from Equation 11-11 the force on side 2 (F_2) is 25 times greater than the force on side 1 (F_1). But from Equation 11-12, the displacement of the cap on side 2 (d_2) is only 1/25 as much as the displacement of the cap on side 1 (d_1). Since work equals the product of force and displacement, the work done on side 2 is the *same* as the work done on side 1, because $25 \times (1/25) = 1$. The hydraulic jack doesn't increase the amount of work that you can do: It just makes it possible for you to do that work by applying a relatively small force F_1.

11-7 (a) Regardless of whether the coin is on the top or bottom of the cube, since the cube and coin are floating, the buoyant force of the water displaced must equal the weight of the cube plus the weight of the coin. (If that weren't true, the cube and coin would sink.) So the amount of water displaced is the same in both cases. However, when the coin is on the top, only the cube is displacing water, but when the coin is on the bottom, both the cube and the coin displace water. In this second case, then, the volume of the coin is contributing to the buoyant force, which means that the cube itself won't displace as much water as when the buoyant force was due totally to the amount of the cube that was submerged. For that reason, more of the cube will be above the surface of the water when the coin is taped to the bottom, and the cube will sit higher in the water.

11-8 (b) When the bottle was floating it displaced a volume of water equal to its weight. That's because in order to float, the buoyant force must equal the bottle's weight, and according to Archimedes' principle, the buoyant force is also equal to the weight of the displaced water. However, when the bottle has sunk to the bottom of the tub, the volume of water it displaces is equal only to the volume of the glass that forms the bottle. Since the bottle sank, we know that the glass must be denser than water, which means the glass itself takes up less volume than the equivalent weight of water. The net result is that when the bottle sinks, the water level goes down.

11-9 (c) According to the equation of continuity, the product of the cross-sectional area of a flow of fluid and the speed of flow is constant. For that reason, as the speed of the flow increases, the cross-sectional area must decrease. This "neck-down" effect is visible even for height differences of about a meter and is quite noticeable in waterfalls before the flow turns turbulent.

11-10 (a) According to Bernoulli's principle when two regions in a flowing fluid have different flow speed, the pressure must be lower in the region in which the flow is faster. In the experiment in which you blow between two papers, because air flows faster between the papers than in the region surrounding the papers, the pressure between them is lower than in the surrounding region, causing the papers to be drawn together—not pushed apart.

11-11 (b) The Reynolds number $Re = \rho l v/\eta$ is a measure of the relative importance of pressure forces and viscous forces in a flowing fluid. The smaller the value of Re, the more important viscosity is. The two tubes have the same dimensions (so the length l is the same), and the fluid flows at the same speed (so v is the same). Hence what's different between the two flows is the value of ρ/η, the ratio of density to viscosity. From Tables 11-1 and 11-2, the values of this ratio for the two fluids are as follows.

Water at 20°C:

$$\begin{aligned}\rho/\eta &= (998\ \text{kg/m}^3)/(1.002 \times 10^{-3}\ \text{Pa}\cdot\text{s}) \\ &= 9.96 \times 10^5\ \text{kg/(m}^3\cdot\text{Pa}\cdot\text{s)}\end{aligned}$$

Air at around 15 to 20°C:

$$\begin{aligned}\rho/\eta &= (1.23\ \text{kg/m}^3)/(1.808 \times 10^{-5}\ \text{kg/m}^3) \\ &= 6.80 \times 10^4\ \text{kg/(m}^3\cdot\text{Pa}\cdot\text{s)}\end{aligned}$$

The value of ρ/η for air is less than one-tenth of the value for water, so viscous effects are actually more important for air than for water.

Questions and Problems

In a few problems you are given more data than you actually need; in a few other problems you are required to supply data from your general knowledge, outside sources, or informed estimate.

Interpret as significant all digits in numerical values that have trailing zeros and no decimal points.

For all problems use $g = 9.80 \text{ m/s}^2$ for the free-fall acceleration due to gravity. Neglect friction and air resistance unless instructed to do otherwise.

• Basic, single-concept problem

•• Intermediate-level problem; may require synthesis of concepts and multiple steps

••• Challenging problem

SSM *Solution is in Student Solutions Manual*

Example *See worked example for a similar problem*

Conceptual Questions

1. • A dam holds back a very long, deep lake. If the lake were twice as long, but still just as deep, how much thicker would the dam need to be? Why?

2. • **Medical** When you cut your finger badly, why might it be wise to hold it high above your head?

3. • **Medical** When you donate blood, is the collection bag held below or above your body? Why?

4. • **Medical** Usually blood pressure is measured on the arm at the same level as the heart. How would the results differ if the measurement were made on the leg instead?

5. • **Medical** After sitting for many hours during a trans-Pacific flight, a passenger jumps up quickly as soon as the "fasten seat belt" light is turned off and immediately falls to the floor, unconscious! Within seconds of being in the horizontal position, however, consciousness is restored. What happened?

6. • A dam will be built across a river to create a reservoir. Does the pressure in the reservoir at the base of the dam depend on the shape of the dam, given that the depth of the reservoir will be the same regardless of the choice of dam shape? Explain your answer. SSM

7. • Two identically shaped containers in the shape of a truncated cone are placed on a table, but one is inverted such that the small end is resting on the table. The containers are filled with the same height of water. The pressure at the bottom of each container is the same. However, the weight of the water in each container is different. Explain why this statement is correct.

8. • An ice cube floats in a glass of water so that the water level is exactly at the rim. After the ice cube melts will all the water still be in the glass? Explain your answer.

9. • Aluminum is more dense than plastic. You are given two identical, closed, and opaque boxes. One contains a piece of aluminum, and the other contains a piece of plastic. Without opening the boxes is it possible to tell which box contains the plastic and which box contains the aluminum? Explain your answer.

10. • The salinity of the Great Salt Lake varies from place to place ranging from about two to as much as eight times the salinity of ocean water. Is it easier or harder for a person to float in the Great Salt Lake compared to floating in ocean water? Why? SSM

11. • A wooden boat floats in a small pond, and the level of the water at the edge of the pond is marked. Will the level of the water rise, fall, or stay the same when the boat is removed from the pond?

12. • You are given two objects of identical size, one made of aluminum and the other of lead. You hang each object separately from a spring balance. Because lead is denser than aluminum, the lead object weighs more. Now you weigh each object while it is submerged in water. Will the difference between the measured weights of the aluminum object and the lead object be greater than, less than, or the same as it was when the objects were weighed in air? Explain your answer.

13. • A river runs through a wide valley and then through a narrow channel. How do the velocities of the flows of water compare between the wide valley and the narrow channel?

14. • Why does the stream of water from a faucet become narrower as it falls?

15. • The wind is blowing from west to east. Should landing airplanes approach the runway from the west or the east? Explain your answer.

16. • Use Bernoulli's equation to explain why a house roof is easily blown off during a tornado or hurricane. SSM

17. • A cylindrical container is filled with water. If a hole is cut on the side of the container so that the water shoots out, what is the direction of the water flow the instant it leaves the container?

Multiple-Choice Questions

18. • Object A has density ρ_1. Object B has the same shape and dimensions as object A, but it is three times as massive. Object B has density ρ_2 such that

A. $\rho_2 = 3\rho_1$.
B. $\rho_2 = \dfrac{\rho_1}{3}$.
C. $\rho_2 = \rho_1$.
D. $\rho_2 = 2\rho_1$.
E. $\rho_2 = \dfrac{\rho_1}{2}$.

19. • If the gauge pressure is doubled, the absolute pressure will

A. be halved.
B. be doubled.
C. be unchanged.
D. be increased, but not necessarily doubled.
E. be decreased, but not necessarily halved. SSM

20. • A toy floats in a swimming pool. The buoyant force exerted on the toy depends on the volume of

A. water in the pool.
B. the pool.
C. the toy under water.
D. the toy above water.
E. none of the above choices.

21. • An object floats in water with 5/8 of its volume submerged. The ratio of the density of the object to that of water is
A. 8/5.
B. 5/8.
C. 1/2.
D. 2/1.
E. 3/8.

22. • An ice cube floats in a glass of water. As the ice melts what happens to the water level?
A. It rises.
B. It remains the same.
C. It falls by an amount that cannot be determined from the information given.
D. It falls by an amount proportional to the volume of the ice cube.
E. It falls by an amount proportional to the volume of the ice cube that was initially above the water line.

23. • Water flows through a 0.5-cm-diameter pipe connected to a 1-cm-diameter pipe. Compared to the speed of the water in the 0.5-cm pipe, the speed in the 1-cm pipe is
A. one-quarter the speed in the 0.5-cm pipe.
B. one-half the speed in the 0.5-cm pipe.
C. the same as the speed in the 0.5-cm pipe.
D. double the speed in the 0.5-cm pipe.
E. quadruple the speed in the 0.5-cm pipe. SSM

24. • **Medical** Blood flows through an artery that is partially blocked. As the blood moves from the wider region into the narrow region, the blood speed
A. increases.
B. decreases.
C. stays the same.
D. drops to zero.
E. alternately increases and then decreases.

25. • Two wooden boxes of equal mass but different density are held beneath the surface of a large container of water. Box A has smaller average density than box B. When the boxes are released, they accelerate upward to the surface. Which box has the greater acceleration?
A. Box A
B. Box B
C. They are the same.
D. We need to know the actual densities of the boxes in order to answer the question.
E. It depends on the contents of the boxes. SSM

Estimation/Numerical Analysis

26. • Estimate the volume of water consumed in a major Canadian city each day.

27. • Estimate the lift that a human body might experience in hurricane-force winds. SSM

28. • Estimate the mass of Earth if all the water in the oceans were replaced with mercury.

29. • Estimate the total hydrostatic *force* on the Hoover Dam.

30. • Estimate the difference in buoyancy for a boat in fresh water versus the same boat in seawater.

31. • Estimate the maximum load that you might be able to lift with a simple hydraulic jack.

32. • **Medical** Estimate the pressure (in pascals) generated by the left ventricle of a human heart during contraction.

Problems

11-1 Liquids and gases are both examples of fluids
11-2 Density measures the amount of mass per unit volume

33. • Determine the mass of a cube of iron that is 2 cm $\times$ 2 cm $\times$ 2 cm in size. Example 11-1

34. • What is the radius of a sphere made of aluminum, if its mass is 24 kg? Example 11-1

35. • (a) Determine the average density of a cylinder that is 20 cm long, 1.0 cm in radius, and has a mass of 37 g. (b) Referring to Table 11-1, of what material might the cylinder be made? (c) What is the specific gravity of the material? SSM Example 11-1

36. • How long are the sides of an ice cube if its mass is 0.35 kg? Example 11-3

37. • **Biology** Approximately 65% of a person's body weight is water. What is the volume of water in a 65-kg man? Example 11-3

38. • A regulation men's basketball has an inflated circumference of 75 cm and an uninflated mass of 623.69 g. What is the average density of the basketball when it is inflated with air at an absolute pressure of 1.544 atm? Air at this pressure has a density of 1.89 kg/m^3. You may assume the vinyl shell has negligible thickness. Example 11-2

39. • A 20 ft $\times$ 8.0 ft $\times$ 8.5 ft shipping container has a mass of 2350 kg when empty and has about 33 m^3 of cargo space. The container is 2/3 full of bags of cat litter that have a density of 0.54 g/cm^3. What is the average density of the container? Will it float if it falls off a ship? Example 11-2

40. • **Astronomy** When a massive star reaches the end of its life, it is possible for a supernova to occur. This may result in the formation of a very small, but very dense, neutron star, the density of which is about the same as a neutron. Determine the radius of a neutron star that has the mass of our Sun and the same density as a neutron. A neutron has a mass of 1.7×10^{-27} kg and an approximate radius of 1.2×10^{-15} m. SSM Example 11-3

41. • Earth has radius 6380 km and mass 5.98×10^{24} kg. Determine the average density of our planet. Example 11-1

11-3 Pressure in a fluid is caused by the impact of molecules

42. • An elephant that has a mass of 6000 kg evenly distributes her weight on all four feet. (a) If her feet are approximately circular and each has a diameter of 50 cm, estimate the pressure on each foot. (b) Compare the answer in part (a) with the pressure on each of your feet when you are standing up. Make some rough but reasonable assumptions about the area of your feet.

43. • The head of a nail is 0.32 cm in diameter. You hit it with a hammer with a force of 25 N. (a) What is the pressure on the head of the nail? (b) If the pointed end of the nail, opposite to the head, is 0.032 cm in diameter, what is the pressure on that end? SSM

11-4 In a fluid at rest, pressure increases with increasing depth

44. • Calculate the pressure at the bottom of a 0.25-m tall graduated cylinder that is half full of mercury and half full of water. Don't forget about atmospheric pressure. Example 11-4

45. • A 5.0 m × 10 m swimming pool is filled to a depth of 10 m. What is the pressure on the bottom of the pool? Example 11-4

46. • What is the difference in blood pressure (mmHg) between the top of the head and bottom of the feet of a 1.75-m-tall person standing vertically? The density of blood is 1.06×10^3 kg/m^3. SSM Example 11-4

47. • At 25°C the density of ether is 72.7 kg/m^3 and the density of iodine is 4930 kg/m^3. A cylinder is filled with iodine to a depth of 1.5 m. How tall would a cylinder filled with ether need to be so that the pressure at the bottom is the same as the pressure at the bottom of the cylinder filled with iodine? Example 11-4

11-5 Scientists and medical professionals use various units for measuring fluid pressure

48. • A diver is 10.0 m below the surface of the ocean. The surface pressure is 1 atm. What are the absolute pressure and gauge pressure he experiences? Example 11-5

49. • **Medical** Suppose that your pressure gauge for determining the blood pressure of a patient measured absolute pressure instead of gauge pressure. How would you write the normal value of systolic blood pressure, 120 mmHg, in such a case? SSM

50. • Convert the following pressures to the SI unit of pascals (Pa), where 1 Pa = 1 N/m^2: (a) 1500 kPa; (b) 35 psi; (c) 2.85 atm; (d) 883 torr.

51. • Assuming the atmospheric pressure is exactly 1 atm at sea level, determine the atmospheric pressure in Death Valley, California, 85 m below sea level. Example 11-5

52. • Elaine wears her wide-brimmed hat at the beach. If the atmospheric pressure at the beach is exactly 1 atm, determine the weight of the imaginary column of air that "rests" on her hat if its diameter is 45 cm. Example 11-5

53. • What is the *absolute* pressure in pascals (Pa) of the air inside a bicycle tire that is inflated to 65 psi?

11-6 A difference in pressure on opposite sides of an object produces a net force on the object

54. •• A rectangular swimming pool is 8.0 m × 35 m in area. The depth varies uniformly from 1.0 m in the shallow end to 2.0 m in the deep end. (a) Determine the pressure at the bottom of the deep end of the pool and at the shallow end. (b) What is the net force on the bottom of the pool due to the water in the pool? (Ignore the effects of the air above the pool for this part.) Example 11-6

55. • What is the net force on an airplane window of area 1000 cm^2 if the pressure inside the cabin is 0.95 atm and the pressure outside is 0.85 atm? SSM Example 11-6

56. • Suppose that the hatch on the side of a Mars lander is built and tested on Earth so that the internal pressure just balances the external pressure. The hatch is a disk 50.0 cm in diameter. When the lander goes to Mars, where the external pressure is 650 N/m^2, what will be the net force (in newtons and pounds) on the hatch, assuming that the internal pressure is the same in both cases? Will it be an inward or an outward force? SSM Example 11-6

57. • What is the net force on the walls of a 55-gal drum that is on the bottom of the ocean at a depth of 250 m? Assume the drum is a cylinder with a height of $34\frac{1}{2}$ in and a diameter of $21\frac{5}{8}$ in. (Remember to convert inches to meters!) The interior pressure of the drum is exactly 1 atm. Example 11-6

11-7 A pressure increase at one point in a fluid causes a pressure increase throughout the fluid

58. • What is the maximum weight that can be raised by the hydraulic lift shown in **Figure 11-42**?

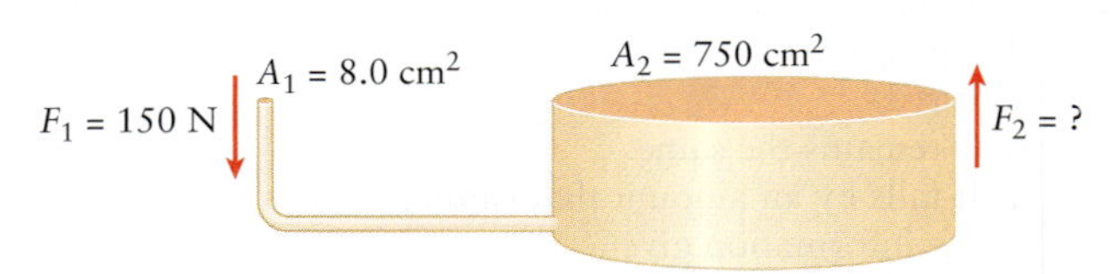

Figure 11-42 Problem 58

59. • A hydraulic lift is designed to raise a 9.00×10^2-kg car. If the "large" piston has a radius of 35.0 cm and the "small" piston has a radius of 2.00 cm, determine the minimum force exerted on the small piston to accomplish the task.

60. • (a) What force will the large piston provide if the small piston in a hydraulic lift is moved down as shown in **Figure 11-43**? (b) If the small piston is depressed a distance of Δy_1, by how much will the large piston rise? (c) How much work is done in pushing down the small piston compared to the work done in raising the large piston if Δy_1 = 0.20 m? SSM

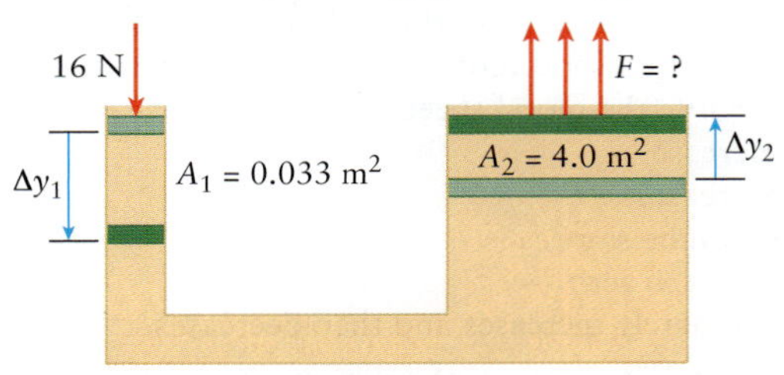

Figure 11-43 Problem 60

61. • A hydraulic lift has a leak so that it is only 75.0% efficient in raising its load. If the large piston exerts a force of 150 N when the small piston is depressed with a force of 15.0 N and the radius of the small piston is 5.00×10^{-2} m, what is the radius of the large piston?

11-8 Archimedes' principle helps us understand buoyancy

62. • A rectangular block of wood, 10.0 cm × 15.0 cm × 40.0 cm, has a specific gravity of 0.600. (a) Determine the buoyant force that acts on the block when it is placed in a pool of fresh water. *Hint:* Draw a free-body diagram labeling all of the forces on the block. (b) What fraction of the block is submerged? (c) Determine the weight of the water that is displaced by the block. Example 11-8

63. • A rectangular block of wood floats in fresh water with its lower 10.0 cm submerged. What distance will be submerged when it floats in seawater (specific gravity 1.025)? Example 11-8

64. •• A cube of side *s* is completely submerged in a pool of fresh water. (a) Derive an expression for the pressure difference between the bottom and top of the cube. (b) After drawing a free-body diagram, derive an algebraic expression for the net force on the cube. (c) What is the weight of the displaced water when the cube is submerged? Your expressions may include some or all of the following

quantities: atmospheric pressure, the density of fresh water, the length of the side of the cube, the mass of the cube, and the acceleration due to gravity. Example 11-9

65. • A log raft is 3.00 m × 4.00 m × 0.150 m and is made from trees that have an average density of 7.00×10^2 kg/m^3. How many people can stand on the raft and keep their feet dry, assuming an average person has a mass of 70.0 kg? Example 11-8

66. • A crown that is supposed to be made of solid gold is under suspicion. When the crown is weighed in air, it has a weight of 5.15 N. When it is suspended from a digital balance and lowered into water, its apparent weight is measured to be 4.88 N. Given that the specific gravity of gold is 19.3, comment on the authenticity of the crown. SSM Example 11-10

67. • A woman floats in a region of the Great Salt Lake where the water is about four times saltier than the ocean and has a density of about 1130 kg/m^3. The woman has a mass of 55 kg, and her density is 985 kg/m^3 after exhaling as much air as possible from her lungs. Determine the percentage of her volume that will be above the waterline. Example 11-8

11-9 Fluids in motion behave differently depending on the flow speed and the fluid viscosity

68. • (a) A hose is connected to a faucet and used to fill a 5.0-L container in a time of 45 s. Determine the volume flow rate in m^3/s. (b) Determine the velocity of the water in the hose in part (a) if it has a radius of 1.0 cm. SSM Example 11-11

69. • At what speed is the water leaving the 7.5-mm-diameter nozzle of a hose with a volume flow rate of 0.45 m^3/s? Example 11-11

70. • Determine the time required for a 50.0-L container to be filled with water when the speed of the incoming water is 25.0 cm/s and the cross-sectional area of the hose carrying the water is 3.00 cm^2. Example 11-11

71. • **Medical** A cylindrical blood vessel is partially blocked by the buildup of plaque. At one point, the plaque decreases the diameter of the vessel by 60.0%. The blood approaching the blocked portion has speed v_0. Just as the blood enters the blocked portion of the vessel, what will be its speed in terms of v_0? Example 11-11

72. • **Medical** You inject your patient with 2.50 mL of medicine. If the inside diameter of the 31-gauge needle is 0.114 mm and the injection lasts 0.650 s, determine the average speed of the fluid as it leaves the needle. SSM Example 11-11

73. • In July 1995 a spillway gate broke at the Folsom Dam in California. During the uncontrolled release the flow rate through the gate peaked at 40,000 ft^3/s, and about 1.35 billion gallons of water were lost (nearly 40% of the reservoir). Estimate the time that the gate was open. Example 11-11

74. • The return-air ventilation duct in a home has a cross-sectional area of 9.0×10^2 cm^2. The air in a room with dimensions 7.0 m × 10.0 m × 2.4 m is to be completely circulated in a 30-min cycle. What is the speed of the air in the duct? Example 11-11

11-10 Bernoulli's equation helps us relate pressure and speed in fluid motion

75. •• Water flows from a fire truck through a hose that is 11.7 cm in diameter and has a nozzle that is 2.00 cm in diameter. The firefighters stand on a hill 5.00 m above the level of the truck. When the water leaves the nozzle, it has a speed of 20.0 m/s. Determine the minimum gauge pressure in the truck's water tank. Example 11-14

76. • At one point Hurricane Katrina had maximum sustained winds of 175 mi/h (240 km/h) and a low pressure in the eye of 666.52 mmHg (0.877 atm). Using the given air speed, compare the atmospheric pressure that would be predicted by Bernoulli's equation to the measured value. Comment on any discrepancies. Assume the pressure of the air is normally 1.00 atm. SSM Example 11-15

77. • When the atmospheric pressure is 1.00 atm, a water fountain ejects a stream of water that rises to a height of 5.00 m. There is a 1.00-cm-radius pipe that leads from a pressurized tank to the opening that ejects the water. Calculate the height of the fountain if it were operational while the eye of a hurricane passed by. Assume that the atmospheric pressure in the eye is 0.877 atm and the tank's pressure remains the same. Example 11-14

78. •• A cylinder that is 20 cm tall is filled with water (**Figure 11-44**). If a hole is made in the side of the cylinder, 5 cm below the top level, how far from the base of the cylinder will the stream land? Assume that the cylinder is large enough so that the level of the water in the cylinder does not drop significantly. Example 11-15

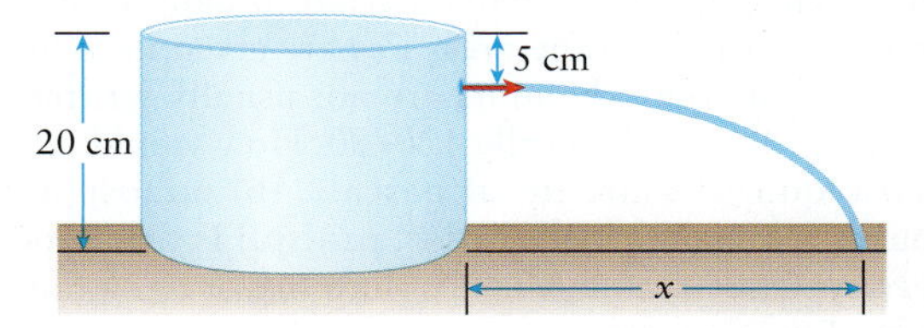

Figure 11-44 Problem 78

11-11 Viscosity is important in many types of fluid flow

79. • Water flows at 0.500 mL/s through a horizontal tube that is 30.0 cm long and has an inside diameter of 1.50 mm. Determine the pressure difference required to drive this flow if the viscosity of water is 1.00 mPa · s. Assume laminar flow. Example 11-17

80. • **Biology** Blood takes about 1.50 s to pass through a 2.00-mm-long capillary. If the diameter of the capillary is 5.00 μm and the pressure drop is 2.60 kPa, calculate the viscosity of blood. Assume laminar flow. SSM Example 11-17

81. •• A very large tank is filled to a depth of 250 cm with oil that has a density of 860 kg/m^3 and a viscosity of 180 mPa · s. If the container walls are 5.00 cm thick and a cylindrical hole of radius 0.750 cm has been bored through the base of the container, what is the initial volume flow rate (in L/s) of the oil through the hole? Example 11-17

82. ••• Water flows to an outlet from a pumping station 5.00 km away. From the pumping station to the outlet there is a net vertical rise of 19 m. Take the coefficient of viscosity of water to be 0.0100 poise. The pipe from the pumping station to the outlet is 1.00 cm in diameter, and the gauge pressure in the pipe where it exits the pumping station is 520 kPa. At what volume flow rate does water flow from the outlet? Example 11-17

General Problems

83. •• **Medical** The human body contains about 5.0 L of blood that has a density of 1060 kg/m^3. Approximately 45% (by

mass) of the blood is cells and the rest is plasma. The density of blood cells is approximately 1125 kg/m^3, and about 1% of the cells are white blood cells, the rest being red blood cells. The red blood cells are about 7.5 μm across (modeled as spheres). (a) What is the mass of the blood in a typical human? (b) Approximately how many blood cells (of both types) are in the blood? Example 11-3

84. • In 2009 a person in Italy won a lottery prize of 146.9 million euros, which at the time was worth 211.8 million U.S. dollars. Gold at the time was worth $953 per troy ounce, and silver was worth $14.16 per troy ounce. A troy ounce is 31.1035 g, the density of gold is 19.3 g/cm^3, and the density of silver is 10.5 g/cm^3. (a) If the lucky lottery winner opted to be paid in a single cube of pure gold, how high would the cube be? Could he carry it home? (b) What would be the height of a silver cube of the same value?. Example 11-3

85. • A hybrid car travels about 50.0 mi/gal of gasoline. The density of gasoline is 737 kg/m^3 and 1 gal equals 3.788 L. Express the car's mileage in miles per kilogram (mi/kg) of gas. Example 11-3

86. • **Medical** Blood pressure is normally expressed as the ratio of the *systolic* pressure (when the heart just ejects blood) to the *diastolic* pressure (when the heart is relaxed). The measurement is made at the level of the heart (usually at the middle of the upper arm), and the pressures are given in millimeters of mercury, although the units are not usually written. Normal blood pressure is typically 120/80. How would you write normal blood pressure in (a) pascals, (b) atmospheres, or (c) pounds per square inch (lb/in^2, psi)? (d) Is the blood pressure, as typically stated, the absolute pressure or the gauge pressure? Explain your answer. SSM

87. • **Astronomy** Oceans as deep as 0.50 km once may have existed on Mars. The acceleration due to gravity on Mars is 0.379g. (a) If there were any organisms in the Martian ocean in the distant past, what pressure (absolute and gauge) would they have experienced at the bottom, assuming the surface pressure was the same as it is on present-day Earth? Assume that the salinity of Martian oceans was the same as oceans on Earth. (b) If the bottom-dwelling organisms in part (a) were brought from Mars to Earth, how deep could they go in our ocean without exceeding the maximum pressure they experienced on Mars? Example 11-4

88. • **Medical** Blood pressure is normally taken on the upper arm at the level of the heart. Suppose, however, that a patient has both arms in casts so that you cannot take his blood pressure in the usual way. If you have him stand up and take the blood pressure at his calf, which is 95.0 cm below his heart, what would normal blood pressure be? The density of blood is 1060 kg/m^3. Example 11-4

89. • **Medical** A syringe which has an inner diameter of 0.60 mm is attached to a needle with an inner diameter of 0.25 mm. A nurse uses the syringe to inject an ideal fluid into a patient's artery where blood pressure is 140/100. What minimum force must the nurse apply to the syringe? SSM Example 11-15

90. •• A water tank is being filled by a 1.0-cm-diameter hose. The water in the hose has a uniform speed of 15 cm/s. Meanwhile, the tank springs a leak at the bottom. The hole has a diameter of 0.50 cm. Determine the equilibrium level of the water in the tank if water continues flowing into the tank at the same rate. Example 11-15

91. • A large water tank is 18 m above the ground (Figure 11-45). Suppose a pipe with a diameter of 8.0 cm is connected to the tank and leads down to the ground. How fast does the water rush out of the pipe at ground level? Assume that the tank is open to the atmosphere. Example 11-5

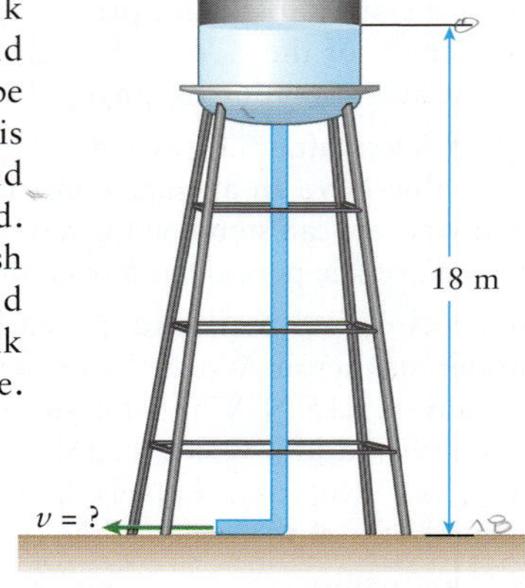

Figure 11-45 Problem 91

92. • **Medical** The aorta is approximately 25 mm in diameter. The mean pressure there is about 100 mmHg, and the blood flows through the aorta at approximately 60 cm/s. Suppose that at a certain point a portion of the aorta is blocked so that the cross-sectional area is reduced to 3/4 of its original area. The density of blood is 1060 kg/m^3. (a) How fast is the blood moving just as it enters the blocked portion of the aorta? (b) What is the gauge pressure (in mmHg) of the blood just as it enters the blocked portion? Example 11-15

93. ••• An equilateral prism of wood with specific gravity 0.6 is placed into a freshwater pool (Figure 11-46). Determine the ratio y_d/y_u of the depth of submersion when the prism is pointed down to the depth of submersion when the prism is pointed up. The prism has a side of s. Example 11-18

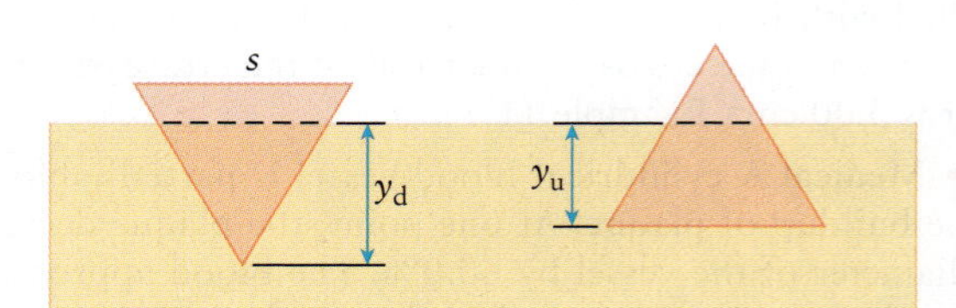

Figure 11-46 Problem 93

94. •• **Sports** The air around the spinning baseball in Figure 11-47 (top view) experiences a faster speed on the left than the right side and hence a smaller pressure on the left than the right. Suppose the ball is traveling in the x direction at 38.0 m/s and it breaks 15.0 cm to the left in the $-y$ direction. Determine the pressure difference between the right and left sides of the ball. Assume the ball has a mass of 142 g and a radius of 3.55 cm. In addition, assume it travels at a constant speed of 38 m/s in the x direction from the pitcher's mound to home plate (60.5 ft = 18.44 m). *Hint:* The pressure difference equals the leftward force multiplied by the cross-sectional area of the ball, πr^2. Example 11-14

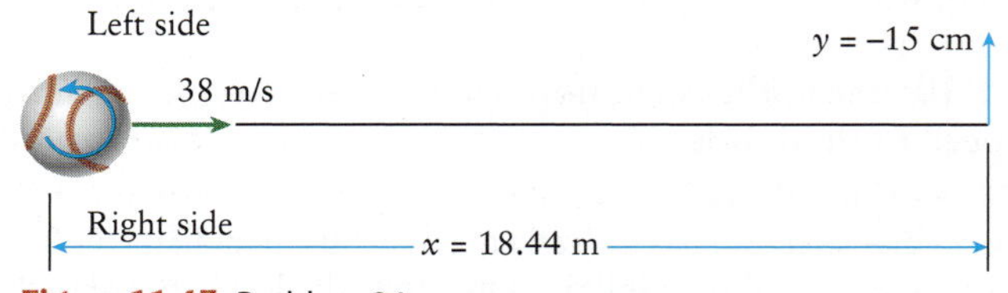

Figure 11-47 Problem 94

Oscillations

What do you think?

The destructive force of an earthquake results from the side-to-side and up-and-down oscillation of the ground. Consider an earthquake that moves the ground 2.0 mm up and 2.0 mm down, making one up-and-down cycle every 0.80 s. Compared to the maximum acceleration of the ground in this earthquake, what is the maximum acceleration of the ground in a second earthquake that makes the ground move up and down the same distance but takes only 0.40 s per cycle? (a) $\frac{1}{4}$ as great; (b) $\frac{1}{2}$ as great; (c) the same; (d) twice as great; (e) four times as great.

In this chapter, your goals are to:

- (12-1) Define oscillation.
- (12-2) Describe the key properties of oscillations, including what makes oscillation happen.
- (12-3) Explain the connection between Hooke's law and the special kind of oscillation called simple harmonic motion.
- (12-4) Discuss how kinetic energy and potential energy vary during an oscillation.
- (12-5) Explain what determines the period, frequency, and angular frequency of a simple pendulum.
- (12-6) Explain what determines the period, frequency, and angular frequency of a physical pendulum.
- (12-7) Describe what happens in underdamped, critically damped, and overdamped oscillations.
- (12-8) Identify the circumstances under which resonance occurs in a forced oscillation.

To master this chapter, you should review:

- (2-4) How to interpret $x-t$, v_x-t, and a_x-t graphs.
- (6-5, 6-6) The force exerted by an ideal spring that obeys Hooke's law and the potential energy stored in such a spring.
- (3-7, 8-2) The relationships among speed, angular speed, and centripetal acceleration in circular motion.
- (8-6) Newton's second law for rotational motion and how to calculate the torque exerted by a force.

12-1 We live in a world of oscillations

Try sitting by yourself in your room and keeping absolutely still. If no one else is in the room, it may seem as though there is no motion anywhere around you. Not so! Your heart is in continuous motion, pulsing at a rate of about one cycle per second. Your eardrums are also vibrating softly in response to the sound of your breathing. And in the electrical wires within the walls of your room, electrons move back and forth about 60 times per second as part of the processes that supply energy to the room lights, your computer, and other appliances. These are just three examples of a kind of motion called **oscillation**, in which an object moves back and forth around a position of *equilibrium*—that is, a point at which it experiences zero net force. (We introduced

The wings of common hummingbirds oscillate up and down about 50 times per second.
(a)
Paul Piebinga/Vetta/Getty Images

The timekeeper at the heart of most wristwatches is a tiny piece of quartz that oscillates 32,768 times per second.
(b)
Photodisc/Thinkstock

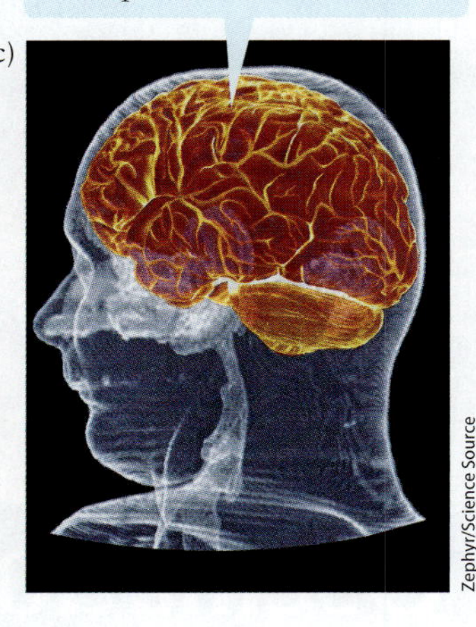

This MRI image is made by mapping how protons oscillate within molecules in the brain in response to radio waves.
(c)
Zephyr/Science Source

Figure 12-1 Oscillations The kind of motion that we call oscillation is commonplace (a) in nature, (b) in technology, and (c) in medicine.

the idea of equilibrium in Section 4-3.) As an example, your eardrum is in equilibrium when it is relaxed and displaced neither inward nor outward. **Figure 12-1** shows three examples of oscillations in nature and technology.

We'll begin this chapter by studying the causes of oscillation. We'll find that oscillation is possible only when an object experiences a *restoring force*—one that always pushes or pulls the object toward the equilibrium position. Oscillations are particularly simple in nature if the restoring force is directly proportional to how far the object is displaced from equilibrium. This proves to be an important special case that's a very good approximation to many kinds of oscillation.

Just as for other kinds of motion, we'll find that it's useful to describe oscillations in terms of kinetic and potential energy. The energy approach will also help us to understand the role of *damping* in oscillations—for example, what happens to make the oscillations of a swinging pendulum die out so that the pendulum eventually ends up at rest. Finally, we'll examine *forced* oscillations, in which a rhythmic force is applied to an object to sustain its oscillations even when damping is present. The ideas of forced oscillation will help us understand how the right kind of sound can break a wine glass, how some insects fly, and why architects must design buildings to avoid catastrophic vibrations.

TAKE-HOME MESSAGE FOR Section 12-1

✔ Oscillation is a kind of motion in which an object moves back and forth around an equilibrium position.

12-2 Oscillations are caused by the interplay between a restoring force and inertia

Figure 12-2 shows one of the simplest ways to make an object oscillate. A block is attached to a horizontal spring and is free to slide back and forth. (For simplicity we'll assume that there's no friction between the block and the surface over which it slides.) We'll call the horizontal direction along which the block can move the x axis and let $x = 0$ be the point at which the spring is relaxed (**Figure 12-2a**). Then the quantity x represents the *position* of the block as well as the *displacement* of the block from equilibrium.

When the spring is stretched, corresponding to $x > 0$, it exerts a force on the block that points in the negative x direction so $F_x < 0$ (**Figure 12-2b**). If instead the spring is compressed, corresponding to $x < 0$, it exerts a force on the block that points in the positive x direction so $F_x > 0$ (**Figure 12-2c**). In other words, no matter which way the block is displaced from equilibrium, the spring exerts a force that tends to pull or push the block back toward the equilibrium position. A force of this kind is called a **restoring force**.

To start the block in Figure 12-2 oscillating, move the block so as to stretch the spring by a distance A and then release it (**Figure 12-3a**). The restoring force of the spring pulls the block back toward the equilibrium position at $x = 0$, so the block accelerates and continues to gain speed as it moves back toward equilibrium (**Figure 12-3b**). The block reaches its maximum speed as it passes through equilibrium (**Figure 12-3c**). Even though there is zero net force on the block at the equilibrium position, the block's

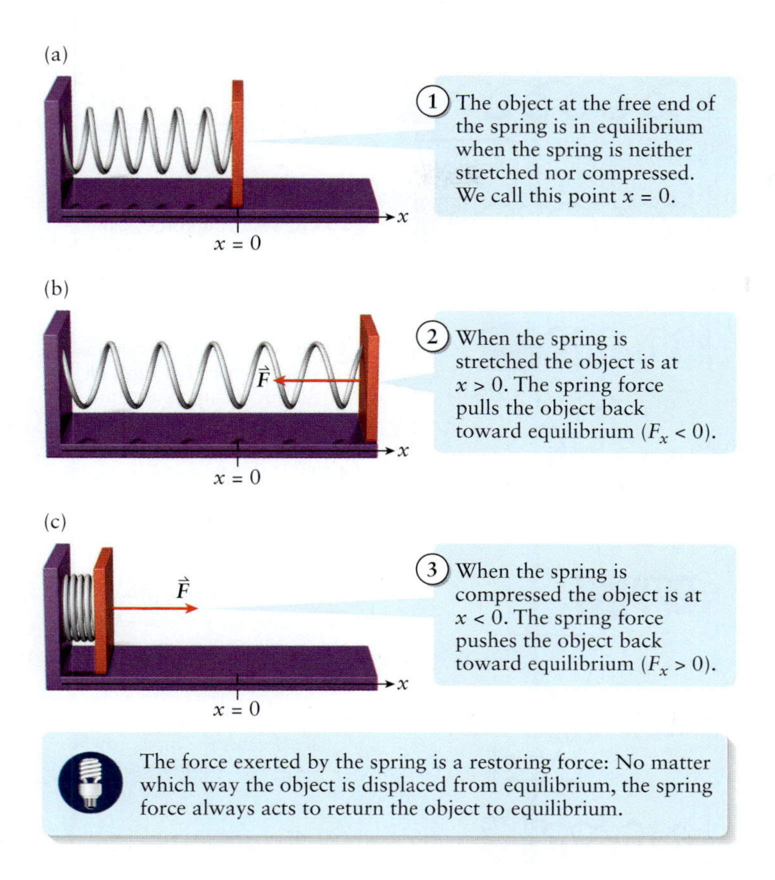

(a)
① The object at the free end of the spring is in equilibrium when the spring is neither stretched nor compressed. We call this point $x = 0$.
$x = 0$
(b)
$\vec{F}$
② When the spring is stretched the object is at $x > 0$. The spring force pulls the object back toward equilibrium ($F_x < 0$).
$x = 0$
(c)
$\vec{F}$
③ When the spring is compressed the object is at $x < 0$. The spring force pushes the object back toward equilibrium ($F_x > 0$).
$x = 0$
The force exerted by the spring is a restoring force: No matter which way the object is displaced from equilibrium, the spring force always acts to return the object to equilibrium.

Figure 12-2 A restoring force In order for oscillations to take place, there must be a force that always acts to return the oscillating object to equilibrium.

inertia causes it to overshoot this point. As a result the block compresses the spring, so the restoring force is now directed opposite to the block's motion as it slows down (**Figure 12-3d**). Thus the block eventually comes momentarily to rest with the spring compressed (**Figure 12-3e**). The restoring force of the spring still acts, so the block again starts moving and gains speed as it heads back toward the equilibrium position (**Figure 12-3f**). The object reaches equilibrium a second time and again overshoots the equilibrium position (**Figure 12-3g**). The block stretches the spring as it overshoots, and the restoring force causes the block to slow down (**Figure 12-3h**). The block ends up once again at its initial position with zero speed (**Figure 12-3i**). The cycle then repeats. If there is no friction or air drag, the oscillation will go on forever!

Not every oscillation involves a spring, but *every* oscillation follows the same general sequence of steps shown in Figure 12-3. This figure shows that the two key ingredients for oscillation are a restoring force that always draws the oscillating object back toward the equilibrium position and inertia that causes the object to overshoot equilibrium.

Oscillation Period and Frequency

The time for one complete cycle of an oscillation is called the **period** of the oscillation. We'll use the symbol T for period. Intuition should tell you that a stiffer spring (which will exert a stronger restoring force) will push and pull the block more rapidly through the steps shown in Figure 12-3 and so will cause the period to be shorter. That same intuition will tell you that a more massive block, which has more inertia, will go through the steps more slowly and hence have a longer period. We'll verify that these ideas are correct later in this section.

A quantity related to the oscillation period is the **frequency** f of the oscillation. The frequency is equal to the number of cycles per unit time. The unit of frequency is the **hertz** (abbreviated Hz): 1 Hz = 1 cycle per second. A pendulum that oscillates once per second has a frequency $f = 1$ Hz. If you play the A4 key on a piano (A above middle C, also called "concert A"), the piano string vibrates 440 times per second, so the frequency is $f = 440$ Hz. The vibrating string compresses the air at this same frequency, which is the frequency of sound that you hear.

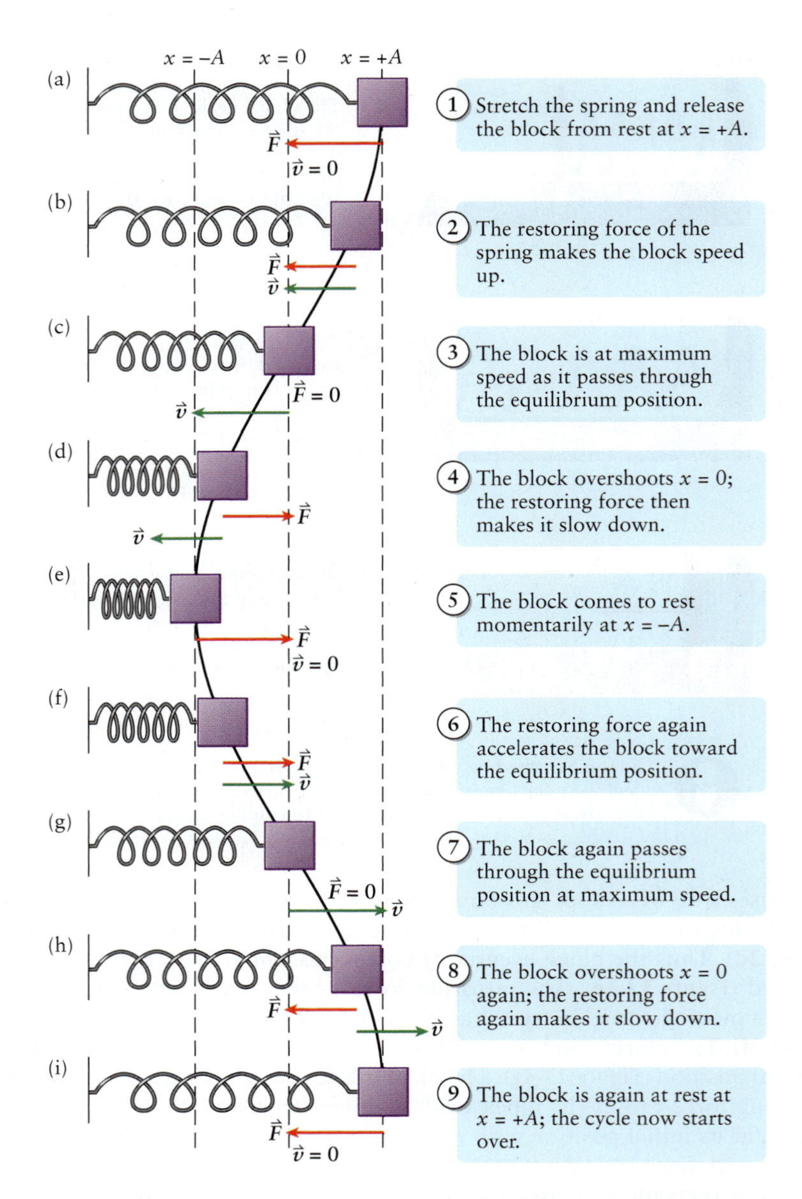

$x = -A$
$x = 0$
$x = +A$
(a)
(b)
(c)
(d)
(e)
(f)
(g)
(h)
(i)
$\vec{F}$
$\vec{v} = 0$
$\vec{v}$
$\vec{F} = 0$
1 Stretch the spring and release the block from rest at $x = +A$.
2 The restoring force of the spring makes the block speed up.
3 The block is at maximum speed as it passes through the equilibrium position.
4 The block overshoots $x = 0$; the restoring force then makes it slow down.
5 The block comes to rest momentarily at $x = -A$.
6 The restoring force again accelerates the block toward the equilibrium position.
7 The block again passes through the equilibrium position at maximum speed.
8 The block overshoots $x = 0$ again; the restoring force again makes it slow down.
9 The block is again at rest at $x = +A$; the cycle now starts over.

Figure 12-3 A cycle of oscillation Nine stages in the oscillation of a block attached to a spring.

Since period T is the number of seconds that elapse per cycle and frequency f is the number of cycles that happen per second, it follows that these quantities are the reciprocals of each other:

Frequency and period (12-1)

$$f = \frac{1}{T} \quad \text{and} \quad T = \frac{1}{f} \tag{12-1}$$

Period T of an oscillation

Frequency f of the oscillation

EXAMPLE 12-1 Frequency and Period

(a) What is the oscillation period of your eardrum when you are listening to the A4 note on a piano (frequency 440 Hz)? (b) A bottle floating in the ocean bobs up and down once every 2.00 min. What is the frequency of this oscillation?

Set Up

We'll use Equation 12-1 to find the period T from the frequency f and vice versa.

Frequency and period:

$$f = \frac{1}{T} \quad \text{and} \quad T = \frac{1}{f} \tag{12-1}$$

Solve

(a) The period of this oscillation is the reciprocal of the frequency. Note that a cycle is not a true unit but simply a way of counting, so we can remove it from the final answer as needed.

$$T = \frac{1}{f} = \frac{1}{440 \text{ Hz}} = \frac{1}{440 \text{ cycles/s}} = \frac{1}{440}\frac{\text{s}}{\text{cycle}}$$
$$= 2.3 \times 10^{-3} \text{ s}$$

(b) The frequency of this oscillation is the reciprocal of the period. To get a result in Hz, we must first convert the period to seconds. Note that the period is the number of seconds per cycle, so we can add "cycle" to the calculation as needed.

$$T = 2.00 \text{ min} \times \frac{60 \text{ s}}{1 \text{ min}} = 120 \text{ s}$$
$$f = \frac{1}{T} = \frac{1}{120 \text{ s}} = \frac{1}{120 \text{ s/cycle}} = \frac{1}{120}\frac{\text{cycle}}{\text{s}}$$
$$= 8.33 \times 10^{-3} \text{ Hz}$$

Reflect

Since frequency and period are the reciprocals of each other, a large value of f (high frequency) corresponds to a small value of T (short period) as in (a) and a small value of f (low frequency) corresponds to a large value of T (long period) as in (b).

Oscillation Amplitude

If the spring in Figures 12-2 and 12-3 exerts the same magnitude of force when it is stretched a certain distance as when it is compressed the same distance, the block in Figure 12-3 will move as far to the left of equilibrium ($x < 0$) as it does to the right of equilibrium ($x > 0$). In this chapter we'll consider only restoring forces that have this symmetric property. In this case the block will oscillate between $x = +A$ and $x = -A$, as shown in Figure 12-3. The distance A is called the **amplitude** of the oscillation. It's equal to the maximum displacement of the object from equilibrium. The amplitude is *always* a positive number, never a negative one.

WATCH OUT! The amplitude of oscillation is the maximum displacement from equilibrium.

It's tempting to think of amplitude as the distance from where the spring is stretched the most (Figure 12-3a) to where the spring is compressed the most (Figure 12-3e). That's not correct: The amplitude A equals the distance from the *equilibrium* position to the point of maximum stretch, and it also equals the distance from the equilibrium position to the point of maximum compression. Keep this principle in mind to avoid giving answers that are wrong by a factor of 2.

Period, frequency, and amplitude are just part of a complete description of an oscillation. In the following section we'll see how to get such a complete description for an important special kind of oscillation.

BIO-Medical GOT THE CONCEPT? 12-1 Heartbeats

Jess has a resting pulse of 50 heartbeats per minute. When Jess sprints down the track, her pulse increases to 150 beats per minute. Compared to when she is at rest, the oscillation period of her heart when she is sprinting is (a) 9 times as great; (b) 3 times as great; (c) the same; (d) 1/3 as great; (e) 1/9 as great.

TAKE-HOME MESSAGE FOR Section 12-2

✔ Oscillation is caused by the interplay between (i) a restoring force that returns an object to its equilibrium position (where it experiences zero net force) and (ii) inertia that makes the object overshoot equilibrium.

✔ The period of an oscillation is the time for one complete cycle. The frequency, which is the reciprocal of the period, is the number of cycles per second.

✔ The amplitude of an oscillation equals the maximum displacement from equilibrium.

12-3 The simplest form of oscillation occurs when the restoring force obeys Hooke's law

To have a complete description of what happens during an oscillation, we need to know the position, velocity, and acceleration of the oscillating object at all times during its motion. We can actually find these for the case in which the restoring force is *directly proportional* to the distance that the object is displaced from equilibrium. This relationship is called **Hooke's law** for the force exerted by an ideal spring (Section 6-5). Hooke's law is important because experiment shows that if you stretch a spring by a relatively small amount, the force that the spring exerts on you is directly proportional to the amount of stretch:

Hooke's law (6-11)

Force exerted by an **ideal spring** | **Spring constant** of the spring (a measure of its stiffness)

$$F_x = -kx$$

Extension of the spring ($x > 0$ if spring is stretched, $x < 0$ if spring is compressed)

The minus sign in Equation 6-11 tells us that if the spring is stretched, $x > 0$, it exerts a force in the negative x direction so $F_x < 0$. It also tells us that if the spring is compressed, $x < 0$, it exerts a force in the positive x direction so $F_x > 0$. That's exactly the relationship between displacement x and spring force F_x shown in Figure 12-2.

As we did in Section 6-5, we'll call a spring that obeys Equation 6-11 an *ideal* spring. Real springs deviate from Hooke's law a little. But experiment shows that if the value of x is kept small enough, so that the oscillations have small amplitude, almost every spring obeys Hooke's law quite faithfully.

If the spring in Figures 12-2 and 12-3 is an ideal one, then the net force on the block in Figures 12-2 and 12-3 is equal to the spring force given by Equation 6-11. (There's also a downward gravitational force on the block plus an upward normal force exerted by the surface over which it slides. But these forces cancel each other so that the net force on the block has zero vertical component.) Newton's second law for the x motion of the block tells us that this net force is equal to the mass m of the block multiplied by the block's acceleration a_x:

$$-kx = ma_x$$

If we divide both sides of this equation by the mass m of the block, we get

$$a_x = -\left(\frac{k}{m}\right)x \tag{12-2}$$

Equation 12-2 tells us that *if a block oscillates because it is attached to an ideal spring that obeys Hooke's law, the acceleration of the block is proportional to the negative of the displacement.* The acceleration is negative when the displacement x is positive, positive when the displacement is negative, and zero when the displacement is zero (that is, when the block is at equilibrium).

Because the acceleration of the block is not constant, we *cannot* describe the motion of the block in Figure 12-3 using the equations from Chapter 2 for straight-line motion with constant acceleration. (Those are nice, simple equations, but they simply don't apply here.) Instead we'll have to solve Equation 12-2 to find x, v_x, and a_x as functions of time t. This sounds like a complicated mathematical problem, but it turns out that we can solve it by using what we know about another, seemingly unrelated, kind of motion: an object moving around a circle at constant speed.

Uniform Circular Motion and Hooke's Law

Figure 12-4a shows an object of mass m moving at constant speed v around a circular path of radius A. (We'll see shortly why we choose this symbol for radius.) The object travels once around the circle, an angle of 2π radians, in a time T, so the object's constant *angular* speed is $\omega = 2\pi/T$.

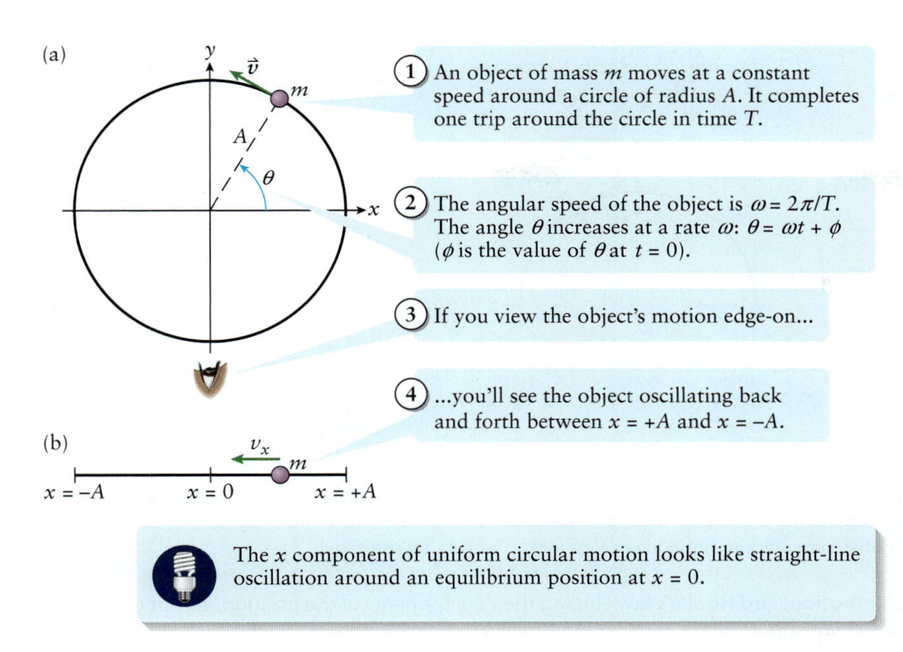

(a)
y
v
m
A
θ
x
1 An object of mass m moves at a constant speed around a circle of radius A. It completes one trip around the circle in time T.
2 The angular speed of the object is ω = 2π/T. The angle θ increases at a rate ω: θ = ωt + ϕ (ϕ is the value of θ at t = 0).
3 If you view the object's motion edge-on...
4 ...you'll see the object oscillating back and forth between x = +A and x = –A.
(b)
v_x
m
x = –A
x = 0
x = +A
The x component of uniform circular motion looks like straight-line oscillation around an equilibrium position at x = 0.

Figure 12-4 Oscillation and uniform circular motion If you look at one component of uniform circular motion, what you see is an oscillation.

How does this uniform circular motion relate to oscillation? Imagine that you view the circular path edge-on so that all you can see is the x component of the object's motion (**Figure 12-4b**). From this vantage point you'll see the object oscillating back and forth along a straight-line path, moving a distance A to the left of center and the same distance A to the right of center (Figure 12-4b). The object will take a time T for one back-and-forth cycle. So what you'll see is the object moving in the same way as the block on a spring in Figure 12-3, with period T and amplitude A.

Let's show that the x component of the uniform circular motion obeys Equation 12-2 so that the x component of the object's acceleration is proportional to the object's x coordinate. To do this we'll use three important facts about uniform circular motion. From Section 8-2 the *linear* speed v of an object in uniform circular motion is equal to the *angular* speed ω multiplied by the radius of the circle. The radius of the circle in Figure 12-4a is A, so the speed of the object's circular motion is

$$v = \omega A \tag{12-3}$$

We also know from Section 3-7 that an object in uniform circular motion has a centripetal acceleration (that is, an acceleration directed toward the center of the circle). This acceleration has a constant magnitude a_{cent} equal to the square of the speed v divided by the radius of the circle, which in this case is equal to A. Using Equation 12-3 we can write the magnitude of the centripetal acceleration as

$$a_{\text{cent}} = \frac{v^2}{A} = \frac{(\omega A)^2}{A} = \frac{\omega^2 A^2}{A} = \omega^2 A \tag{12-4}$$

Finally, from Equation 8-16 we have the following equation for the angle θ from the positive x axis for the angular position of an object at time t, assuming that the object has constant angular acceleration:

$$\theta = \theta_0 + \omega_{0z} t + \frac{1}{2}\alpha_z t^2 \tag{8-16}$$

In Equation 8-16 α_z is the angular acceleration of the object, which is zero in this case because the motion is uniform and the angular velocity is constant. The quantity ω_{0z} is the initial z component of the angular velocity of the object, which is equal to the object's constant angular speed ω. (Figure 12-4a shows that the object moves around the circle in the positive, counterclockwise direction, so the angular velocity is positive and equal to the angular speed.) The angle θ_0 is the value of θ at time $t = 0$; we'll use the symbol ϕ (the Greek letter "phi") for this quantity. Then we can write Equation 8-16 as $\theta = \phi + \omega t$, or

$$\theta = \omega t + \phi \tag{12-5}$$

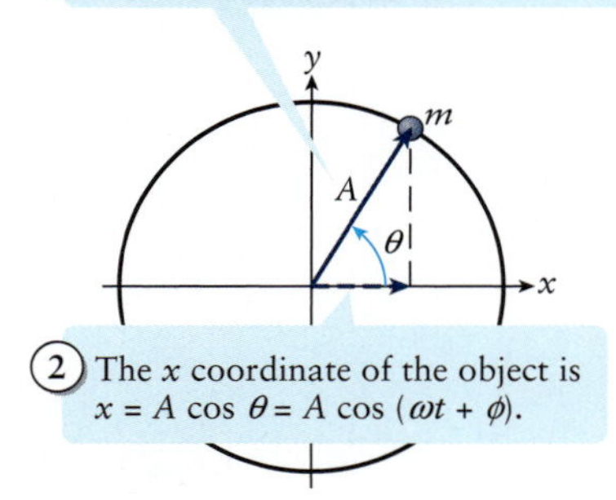

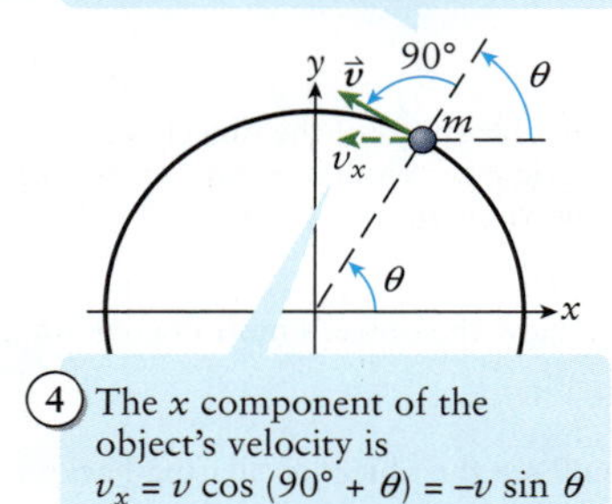

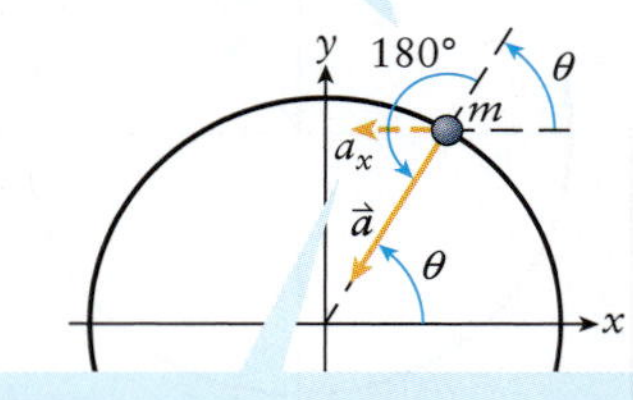

Compare the equations for x and a_x: You'll see that a_x = −ω²x. That's the same relationship as for Hooke's law, a_x = −(k/m)x, with ω² = k/m. So the x component of uniform circular motion is the same as the simple harmonic motion of an object oscillating on the end of an ideal, Hooke's law spring.

Figure 12-5 Oscillation, uniform circular motion, and Hooke's law Finding the x components of the position, velocity, and acceleration in uniform circular motion reveals an important connection to simple harmonic motion.

Now let's put all of these pieces together. **Figure 12-5a** shows that the x coordinate of the object as it moves around the circle is $x = A\cos\theta$. From Equation 12-5 we can write this as

$$x = A\cos(\omega t + \phi) \tag{12-6}$$

Figure 12-5b shows that the x velocity of the object is $v_x = -v\sin\theta$. From Equation 12-3 and Equation 12-5, we can write this as

$$v_x = -\omega A\sin(\omega t + \phi) \tag{12-7}$$

From **Figure 12-5c** the x acceleration of the object is $a_x = -a_{\text{cent}}\cos\theta$. Equation 12-4 and Equation 12-5 together tell us that we can write the x acceleration as

$$a_x = -\omega^2 A\cos(\omega t + \phi) \tag{12-8}$$

$\sqrt{x}$ See the Math Tutorial for more information on trigonometry.

Here's the payoff for all of this mathematics: If we compare Equation 12-6 and Equation 12-8, we see that the x acceleration of an object in uniform circular motion is directly proportional to the object's x coordinate:

$$a_x = -\omega^2 x \tag{12-9}$$

That's *precisely* the relationship between acceleration a_x and displacement x that we found for an object of mass m attached to a spring with spring constant k that obeys Hooke's law: $a_x = -(k/m)x$ (Equation 12-2). The two equations are identical if we set the square of the angular speed ω in Figure 12-4 and Figure 12-5 equal to the ratio of the quantities k/m:

$$\omega^2 = \frac{k}{m} \quad \text{or} \quad \omega = \sqrt{\frac{k}{m}} \tag{12-10}$$

In other words, Equation 12-9 and Equation 12-10 tell us something quite profound and unexpected: *A block of mass m attached to an ideal, Hooke's law spring of spring constant k oscillates in precisely the same way as the x coordinate of an object in uniform circular motion with angular speed ω, provided that $\omega = \sqrt{k/m}$.* We made the detour into analyzing uniform circular motion so that we could arrive at this insight.

With this observation in mind, we can now go back and write a complete description of how an object oscillates when acted on by a restoring force that obeys Hooke's law.

Simple Harmonic Motion: Angular Frequency, Period, and Frequency

We've seen that the period T of the oscillation in Figure 12-4b is the same as the time T that it takes the object in uniform circular motion in Figure 12-4a to travel once around the circle. The angular speed ω equals $2\pi/T$, so T equals $2\pi/\omega$. Using Equation 12-10 for the value of ω along with Equation 12-1 for the relationship between period T and frequency f, we get the following results for the oscillations of a block of mass m attached to an ideal, Hooke's law spring of spring constant k:

Angular frequency: $$\omega = \sqrt{\frac{k}{m}}$$

Period: $$T = \frac{2\pi}{\omega} = 2\pi\sqrt{\frac{m}{k}}$$

Frequency: $$f = \frac{1}{T} = \frac{1}{2\pi}\sqrt{\frac{k}{m}} = \frac{\omega}{2\pi}$$

k = **spring constant** of the spring

m = **mass** of the object connected to the spring

Angular frequency, period, and frequency for a block attached to an ideal spring
(12-11)

In the first of Equations 12-11, we've introduced a new quantity called the **angular frequency**. This has the same symbol and the same value as the angular *speed* of the circular motion in Figure 12-4. That's because our analysis of Figure 12-4 showed that the back-and-forth motion of a block attached to an ideal spring is equivalent to the x component of uniform circular motion. If you find this confusing to remember, just keep in mind that the ordinary frequency f equals $\omega/2\pi$, or equivalently $\omega = 2\pi f$ (see the third of Equations 12-11). As we'll see, the quantity ω appears naturally in the equations for oscillation, which is why we use it. Like angular speed, angular frequency has units of radians per second (rad/s).

The second of Equations 12-11 tells us that the period of oscillation T is proportional to the square root of the mass m and inversely proportional to the square root of the spring constant k. This means that increasing the mass m makes the period longer, while increasing the spring constant k makes the period shorter. That's exactly what we predicted in Section 12-2 for the oscillations of a block attached to a spring. The last of Equations 12-11 is just the statement that the frequency f is the reciprocal of the period T. This equation says that increasing the mass causes the frequency to decrease, while increasing the spring constant makes the frequency increase.

Note that the amplitude A does *not* appear in Equations 12-11, which means that the angular frequency, period, and frequency of an oscillation are *independent* of the amplitude if the restoring force obeys Hooke's law. This is called the **harmonic property**. This may come as a surprise, since doubling the amplitude means the oscillating object has to cover twice as much distance during an oscillation and so should take a longer time to complete an oscillation. But doubling the amplitude also means that the object reaches larger displacements x and so experiences a stronger Hooke's law restoring force $F_x = -kx$. This effect makes the object move faster during the oscillation. The net result is that the period T, the time for one oscillation, is unaffected by changing the amplitude. That means the frequency $f = 1/T$ and the angular frequency $\omega = 2\pi f = 2\pi/T$ are unaffected as well. This is the case *only* if the restoring force obeys Hooke's law. If the restoring force depends on displacement in a different way, the oscillations do not have the harmonic property.

The word "harmonic" sounds like a musical term, and indeed the harmonic property is important in musical instruments. We'll learn in Chapter 13 that the pitch of a musical sound is determined primarily by the frequency of oscillation of the musical instrument that makes the sound, while the loudness of the sound is determined primarily by the amplitude. If changing the amplitude caused a change in frequency, playing the same key on a piano would make a different pitch if you pressed the key softly or pushed hard on the key. That would render a piano almost completely useless. But because the strings of a piano have the harmonic property, they vibrate at the same frequency and produce sounds of the same pitch whether they are played soft or loud.

As we've described, the harmonic property is a direct result of the restoring force being directly proportional to the displacement from equilibrium (Hooke's law, Equation 6-11). For this reason the kind of oscillation that results from a Hooke's law restoring force is called **simple harmonic motion (SHM)**. Later in this chapter we'll encounter kinds of oscillations that are not simple harmonic motion; the oscillations of a pendulum are one example. We'll see that in many such cases, however, these kinds of oscillations are *approximately* SHM if the amplitude of oscillation is sufficiently small.

WATCH OUT! Simple harmonic motion is not the same as circular motion.

Even though we describe simple harmonic motion in terms of an *angular* frequency, the oscillating object is *not* actually moving in a circle! As we've seen we can *visualize* simple harmonic motion as one component of the motion of an object moving in a circle with an angular speed ω, but there doesn't need to be an object that's actually moving in a circle.

EXAMPLE 12-2 SHM I: Angular Frequency, Period, and Frequency

An object of mass 0.80 kg is attached to an ideal horizontal spring of spring constant 1.8×10^2 N/m and set into oscillation as in Figure 12-3. (a) Calculate the angular frequency, period, and frequency of the object. (b) Calculate the angular frequency, period, and frequency if the mass of the object is quadrupled to 4×0.80 kg $= 3.2$ kg.

Set Up

An ideal spring obeys Hooke's law, so the oscillations are simple harmonic motion. This means that in part (a) we can use Equations 12-11 to calculate the angular frequency ω, period T, and frequency f from the spring constant $k = 1.8 \times 10^2$ N/m and the mass $m = 0.80$ kg. In part (b) we'll use these equations to determine how the values of ω, T, and f change when we change the value of the mass m.

Angular frequency, period, and frequency:

$$\omega = \sqrt{\frac{k}{m}}$$

$$T = \frac{2\pi}{\omega} = 2\pi\sqrt{\frac{m}{k}}$$

$$f = \frac{1}{T} = \frac{1}{2\pi}\sqrt{\frac{k}{m}} = \frac{\omega}{2\pi} \qquad (12\text{-}11)$$

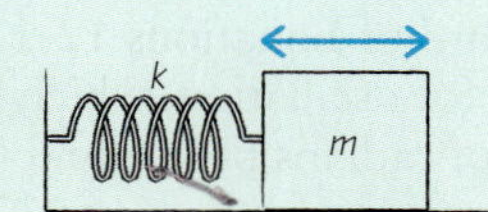

Solve

(a) Substitute the given values for the mass m and spring constant k into the first of Equations 12-11 to calculate the angular frequency.

$$\omega = \sqrt{\frac{k}{m}} = \sqrt{\frac{1.8 \times 10^2 \text{ N/m}}{0.80 \text{ kg}}}$$

Recall that $1 \text{ N} = 1 \text{ kg} \cdot \text{m/s}^2$, so

$$\frac{1 \text{ N/m}}{\text{kg}} = \frac{1 \text{ kg/s}^2}{\text{kg}} = \frac{1}{\text{s}^2} = \frac{1 \text{ rad}^2}{\text{s}^2}$$

(A radian is a way of counting, not a true unit, so we can insert it or remove it as needed.) Then

$$\omega = \sqrt{\frac{k}{m}} = \sqrt{\frac{1.8 \times 10^2}{0.80}\frac{\text{rad}^2}{\text{s}^2}} = 15 \text{ rad/s}$$

Then use the second and third of Equations 12-11 to find the period and frequency.

If the angular frequency ω is in rad/s, the period T will be in s:

$$T = \frac{2\pi}{\omega} = \frac{2\pi}{15 \text{ rad/s}} = 0.42 \text{ s}$$

The reciprocal of the period in s is the frequency in Hz:

$$f = \frac{1}{T} = \frac{1}{0.42 \text{ s}} = 2.4 \text{ Hz}$$

(b) We could substitute the new value of the mass m into Equations 12-11 to find the new values of ω, T, and f. Instead we use proportional reasoning to find the new values from the old ones.

The relationships $\omega = \sqrt{\frac{k}{m}}$ and $f = \frac{1}{2\pi}\sqrt{\frac{k}{m}}$ tell us that the angular frequency and the frequency are inversely proportional to the square root of the mass. The new mass is four times greater than the old mass, so the new values of ω and f are $1/\sqrt{4} = 1/2$ the old values:

$$\omega_{new} = \frac{\omega_{old}}{\sqrt{4}} = \frac{\omega_{old}}{2} = \frac{15 \text{ rad/s}}{2} = 7.5 \text{ rad/s}$$

$$f_{new} = \frac{f_{old}}{2} = \frac{2.4 \text{ Hz}}{2} = 1.2 \text{ Hz}$$

In the same way, the relationship $T = 2\pi\sqrt{\frac{m}{k}}$ tells us that the period T is directly proportional to the square root of the mass. Again the mass has increased by a factor of 4, so the new period is

$$T_{new} = \sqrt{4}\,T_{old} = 2T_{old} = 2(0.42 \text{ s}) = 0.84 \text{ s}$$

Reflect

Our results show that when the mass of the oscillating object increases, the frequency decreases and the period increases. To check our results, can you confirm that f_{new} is equal to $\omega_{new}/2\pi$ and that T_{new} is equal to $1/f_{new}$?

Simple Harmonic Motion: Position, Velocity, and Acceleration

Let's now look in more detail at the position x, velocity v_x, and acceleration a_x of an object moving in simple harmonic motion. From Equation 12-6, the position is given by

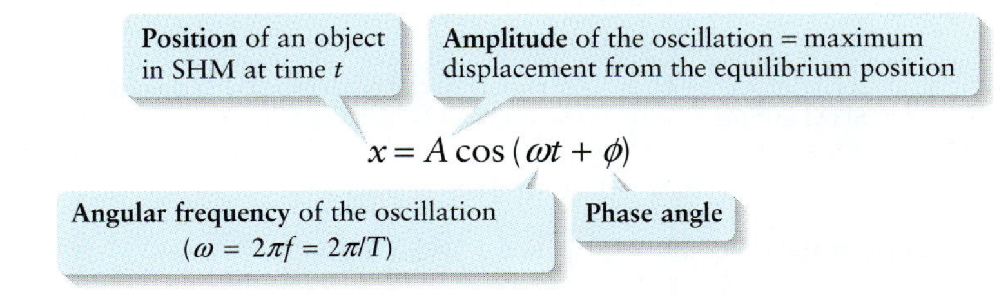

Position as a function of time for simple harmonic motion (12-12)

The cosine function has values from $+1$ to -1. So Equation 12-12 says that the value of x varies from $x = A$ (when the oscillating object is at its most positive displacement from equilibrium) through $x = 0$ (the equilibrium position) to $x = -A$ (when the object is at its most negative displacement from equilibrium). The cosine is one of the simplest of all oscillating functions, which helps justify the name *simple* harmonic motion. The cosine and sine functions are called **sinusoidal functions** (from the Latin word *sinus*, meaning "bent" or "curved") because of their shape.

The angular frequency ω is measured in rad/s and time t is measured in s, so the product ωt in Equation 12-12 is in radians—just as the argument of the cosine function should be. One cycle of oscillation lasts a time T, over which time the product ωt varies from 0 at $t = 0$ to ωT at $t = T$. But since $T = 2\pi/\omega$ from the second of Equations 12-11, it follows that $\omega T = \omega(2\pi/\omega) = 2\pi$. So over the course of one cycle the value of ωt varies from 0 to $\omega T = 2\pi$. The cosine function goes through one complete cycle when its value increases by 2π, so this tells us that the position x goes through one complete cycle when the time increases by T. That's just what we mean by saying that T is the period of the oscillation.

The one quantity in Equation 12-12 that seems a bit mysterious is the **phase angle** ϕ, which is measured in radians. To explain its significance recall from our comparison between uniform circular motion and oscillation that ϕ represents the angular position at $t = 0$ of an object in uniform circular motion. For oscillation the value of ϕ tells us where in the oscillation cycle the object is at $t = 0$. As an example, if $\phi = 0$ Equation 12-12 becomes

$$x = A\cos\omega t \text{ (simple harmonic motion, phase angle } \phi = 0) \quad \textbf{(12-13)}$$

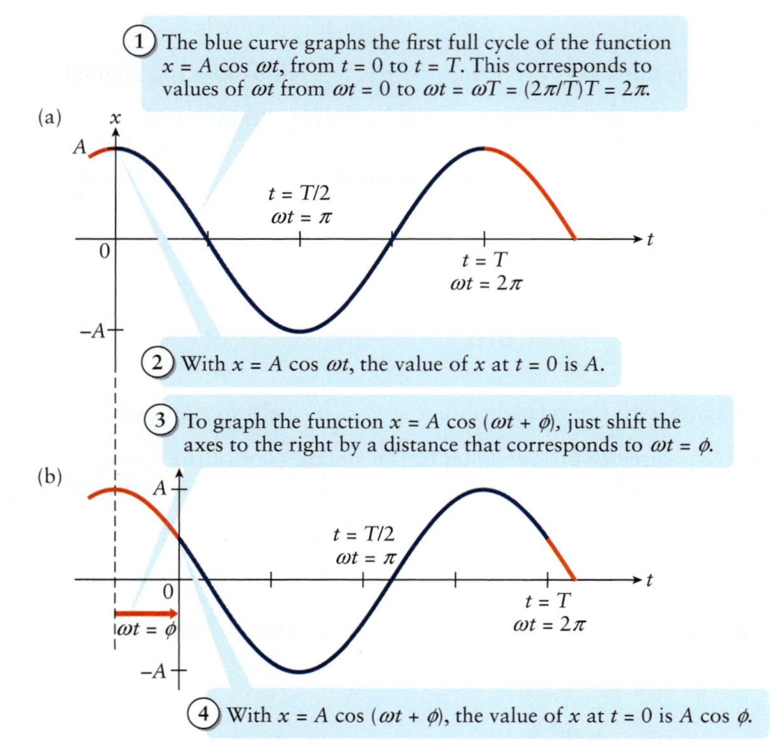

(a)
x
A
t = T/2
ωt = π
0
t
t = T
ωt = 2π
−A
(1) The blue curve graphs the first full cycle of the function x = A cos ωt, from t = 0 to t = T. This corresponds to values of ωt from ωt = 0 to ωt = ωT = (2π/T)T = 2π.
(2) With x = A cos ωt, the value of x at t = 0 is A.
(3) To graph the function x = A cos (ωt + ϕ), just shift the axes to the right by a distance that corresponds to ωt = ϕ.
(b)
A
t = T/2
ωt = π
0
t
t = T
ωt = 2π
ωt = ϕ
−A
(4) With x = A cos (ωt + ϕ), the value of x at t = 0 is A cos ϕ.

Figure 12-6 **Phase angle** The two cycles shown in blue differ by a phase angle ϕ.

Figure 12-6a is the x–t graph that corresponds to Equation 12-13. This graph of x versus t is an ordinary cosine function, which has its most positive value ($x = +A$) when the argument of the function is zero (at $t = 0$) and its most negative value ($x = -A$) one-half cycle later (at $t = T/2$). You can see that the slope of this x–t graph is zero at $t = 0$, which means that the object has zero velocity at $t = 0$. In other words, the value $\phi = 0$ corresponds to starting the object at rest at $t = 0$ at the position $x = A$. That's just the motion that we depicted in Figure 12-3.

By comparison **Figure 12-6b** is the graph that corresponds to Equation 12-12, $x = A\cos(\omega t + \phi)$, for a case in which ϕ is *not* zero but has a positive value. As the figure shows, you can think of the phase angle as shifting the axes to the right if ϕ is positive, as in Figure 12-6b, or to the left when ϕ is negative. Note that in this x–t graph the value of x at $t = 0$ is *not* $+A$; rather, from Equation 12-12 it is equal to

$$x(0) = A\cos(0 + \phi) = A\cos\phi$$

So for an object in simple harmonic motion the value of the phase angle ϕ tells you the point in the oscillation cycle that corresponds to $t = 0$. Note that Figures 12-6a and 12-6b really depict the *same* oscillation, with the same amplitude A and period T; the only difference is the point in the cycle that we call $t = 0$.

The phase angle also appears in the expressions for velocity and acceleration in simple harmonic motion (Equations 12-7 and 12-8):

Velocity as a function of time for simple harmonic motion (12-14)

$$v_x = -\omega A\sin(\omega t + \phi)$$

Velocity of an object in SHM at time t

Amplitude of the oscillation = maximum displacement from the equilibrium position

Angular frequency of the oscillation ($\omega = 2\pi f = 2\pi/T$)

Phase angle

Acceleration as a function of time for simple harmonic motion (12-15)

$$a_x = -\omega^2 A\cos(\omega t + \phi)$$

Acceleration of an object in SHM at time t

Amplitude of the oscillation = maximum displacement from the equilibrium position

Angular frequency of the oscillation ($\omega = 2\pi f = 2\pi/T$)

Phase angle

In the case $\phi = 0$, Equations 12-14 and 12-15 become

(12-16)

$$v_x = -\omega A\sin\omega t$$

$$a_x = -\omega^2 A\cos\omega t$$

(simple harmonic motion, phase angle $\phi = 0$)

Figure 12-7 graphs x versus t, v_x versus t, and a_x versus t for the case $\phi = 0$. In this case the x–t graph is a cosine function, the v_x–t graph is a negative sine function, and the a_x–t graph is a negative cosine function. You can confirm that these graphs are consistent with each other by recalling from Section 2-4 that the slope of the x–t graph equals the velocity v_x and the slope of the v_x–t graph equals the acceleration a_x. Note also that the a_x–t graph looks like the negative of the x–t graph. That's what we expect

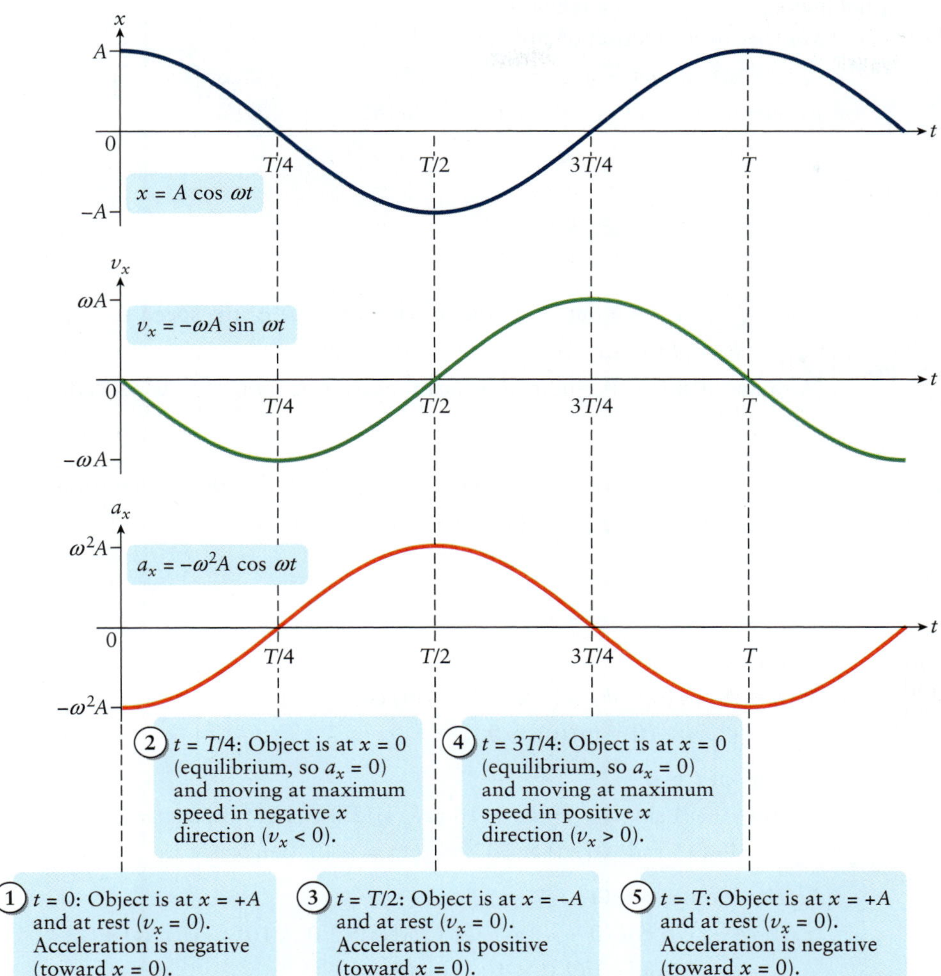

x
A
0
$-A$
$T/4$
$T/2$
$3T/4$
T
t
$x = A \cos \omega t$
v_x
ωA
$-\omega A$
$v_x = -\omega A \sin \omega t$
a_x
$\omega^2 A$
$-\omega^2 A$
$a_x = -\omega^2 A \cos \omega t$
(2) $t = T/4$: Object is at $x = 0$ (equilibrium, so $a_x = 0$) and moving at maximum speed in negative x direction ($v_x < 0$).
(4) $t = 3T/4$: Object is at $x = 0$ (equilibrium, so $a_x = 0$) and moving at maximum speed in positive x direction ($v_x > 0$).
(1) $t = 0$: Object is at $x = +A$ and at rest ($v_x = 0$). Acceleration is negative (toward $x = 0$).
(3) $t = T/2$: Object is at $x = -A$ and at rest ($v_x = 0$). Acceleration is positive (toward $x = 0$).
(5) $t = T$: Object is at $x = +A$ and at rest ($v_x = 0$). Acceleration is negative (toward $x = 0$).

Figure 12-7 Displacement, velocity, and acceleration in SHM These graphs show displacement x (Equation 12-13) as well as velocity v_x and acceleration a_x (Equations 12-16) for simple harmonic motion for the case $\phi = 0$. Compare to Figure 12-3, which shows the same motion.

from Hooke's law, $a_x = -(k/m)x$ (Equation 12-2), which tells us that the acceleration in simple harmonic motion is proportional to the negative of the displacement x.

If the phase angle ϕ is not equal to zero, the vertical axes of all of the graphs in Figure 12-7 are shifted to the left or right as in Figure 12-6. In this case the velocity at $t = 0$ is equal to

$$v_x(0) = -\omega A \sin(0 + \phi) = -\omega A \sin \phi$$

If $\phi = 0$, the velocity at $t = 0$ is zero and the oscillating object is at rest at that instant (see Figure 12-7). If ϕ is not zero, at $t = 0$ the velocity may not be zero. Figure 12-6b illustrates this: In this case the x–t graph has a negative slope at $t = 0$, which shows that the velocity at that instant is not zero.

Go to Interactive Exercise 12-1 for more practice dealing with simple harmonic motion.

EXAMPLE 12-3 SHM II: Position, Velocity, and Acceleration

As in Example 12-2, an object of mass 0.80 kg is attached to an ideal horizontal spring of spring constant 1.8×10^2 N/m and set into oscillation (see Figure 12-3). The amplitude of the motion is 2.0×10^{-2} m, and the phase angle is $\phi = \pi/2$. (a) Find the maximum speed of the object and the maximum magnitude of its acceleration during the oscillation. (b) Find the position, velocity, and acceleration of the object at $t = 0$. (c) Sketch graphs of the position, velocity, and acceleration of the object as functions of time.

Set Up

We can use the simple harmonic motion equations because the spring is ideal

Position as a function of time in SHM:

$$x = A \cos(\omega t + \phi) \qquad (12\text{-}12)$$

(it obeys Hooke's law). From Example 12-2 we know that the angular frequency is $\omega = \sqrt{k/m} = 15$ rad/s, and we are given the values of the amplitude $A = 2.0 \times 10^{-2}$ m and the phase angle $\phi = \pi/2$. Notice that a_x from Equation 12-15 is equal to $-\omega^2$ times x from Equation 12-12.

Velocity as a function of time in SHM:

$$v_x = -\omega A \sin(\omega t + \phi) \qquad (12\text{-}14)$$

Acceleration as a function of time in SHM:

$$a_x = -\omega^2 A \cos(\omega t + \phi) \qquad (12\text{-}15)$$

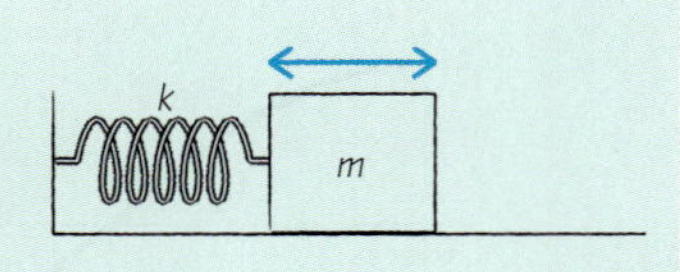

Solve

(a) Equation 12-14 tells us the velocity v_x as a function of time. The sine function has values from $+1$ to -1, so v_x has values from $+\omega A$ to $-\omega A$. The maximum speed is therefore ωA.

Since $v_x = -\omega A \sin(\omega t + \phi)$, the maximum value of the speed is

$$v_{max} = \omega A = (15 \text{ rad/s})(2.0 \times 10^{-2} \text{ m}) = 0.30 \text{ m/s}$$

(A radian is not a true unit, so we can insert it or remove it as needed.)

In the same way Equation 12-15 tells us the acceleration a_x as a function of time. The cosine varies from $+1$ to -1, so a_x has values from $+\omega^2 A$ to $-\omega^2 A$ and the maximum magnitude of a_x is $\omega^2 A$.

Since $a_x = -\omega^2 A \cos(\omega t + \phi)$, the maximum value of the acceleration is

$$a_{max} = \omega^2 A = (15 \text{ rad/s})^2 (2.0 \times 10^{-2} \text{ m}) = 4.5 \text{ m/s}^2$$

(b) To find the values of x, v_x, and a_x at $t = 0$, substitute the values of A, ω, and ϕ as well as $t = 0$ into Equations 12-12, 12-14, and 12-15. Note that $\cos(\pi/2) = 0$ and $\sin(\pi/2) = 1$.

Position at $t = 0$:

$$x(0) = A\cos(0 + \phi) = (2.0 \times 10^{-2} \text{ m})\cos(\pi/2)$$
$$= (2.0 \times 10^{-2} \text{ m})(0) = 0$$

Velocity at $t = 0$:

$$v_x(0) = -\omega A \sin(0 + \phi) = -(15 \text{ rad/s})(2.0 \times 10^{-2} \text{ m})\sin(\pi/2)$$
$$= -(15 \text{ rad/s})(2.0 \times 10^{-2} \text{ m})(1) = -0.30 \text{ m/s}$$

Acceleration at $t = 0$:

$$a_x(0) = -\omega^2 A \cos(0 + \phi) = -(15 \text{ rad/s})^2(2.0 \times 10^{-2} \text{ m}) \cos(\pi/2)$$
$$= -(15 \text{ rad/s})^2(2.0 \times 10^{-2} \text{ m})(0) = 0$$

(c) To draw the x–t graph, recognize that this graph must be sinusoidal and that the slope of the x–t graph is the velocity v_x. The values of x and v_x at $t = 0$ that we found in part (b) tell us that at $t = 0$ the x–t graph has value 0 and a negative slope. The value of x oscillates between $+A = +2.0 \times 10^{-2}$ m and $-A = -2.0 \times 10^{-2}$ m. The graph repeats itself after a time t equal to the period $T = 0.42$ s.

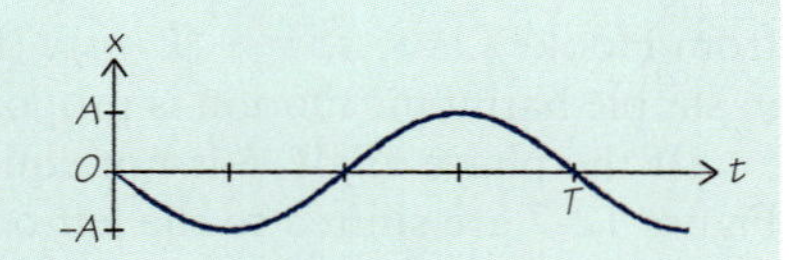

The v_x–t graph is also sinusoidal, and its slope is the acceleration a_x. The values of v_x and a_x at $t = 0$ tell us that at $t = 0$ the v_x–t graph has a negative value and zero slope. From part (a) the value of v_x oscillates between $+v_{max} = +0.30$ m/s and $-v_{max} = -0.30$ m/s. The graph repeats itself after a time t equal to the period $T = 0.42$ s.

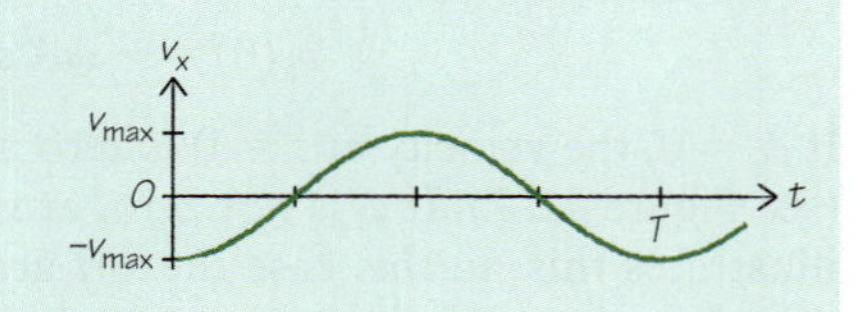

Finally, comparing Equations 12-12 and 12-15 shows that $a_x = -\omega^2 x$, so the a_x–t graph looks like the negative of the x–t graph (compare Figure 12-7). From part (a) the value of a_x oscillates between $+a_{max} = +4.5$ m/s^2 and $-a_{max} = -4.5$ m/s^2. The graph repeats itself after a time t equal to the period $T = 0.42$ s.

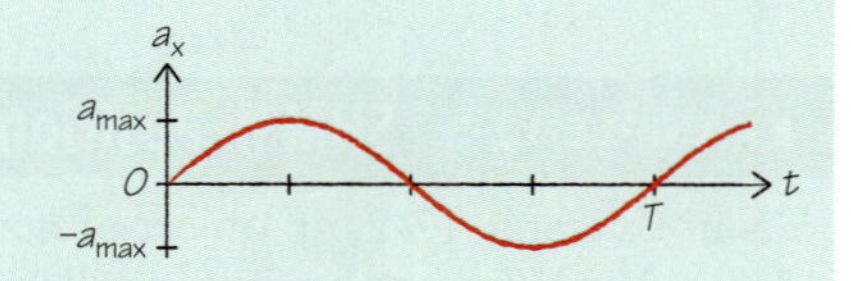

Reflect

To start an oscillation going so that the graphs look like what we have drawn, begin with the block at rest at equilibrium ($x = 0$) and give it a sharp hit with your hand in the negative x direction so that it starts with a negative x velocity v_x. Note that the x–t, v_x–t, and a_x–t graphs look like those in Figure 12-7 but with the vertical axis shifted to the right by a quarter-cycle. That's what we would expect: Figure 12-6 shows that the effect of a phase angle ϕ is to shift the vertical axis horizontally by ϕ, and $\pi/2$ is one-quarter of 2π (the number of radians in a complete cycle).

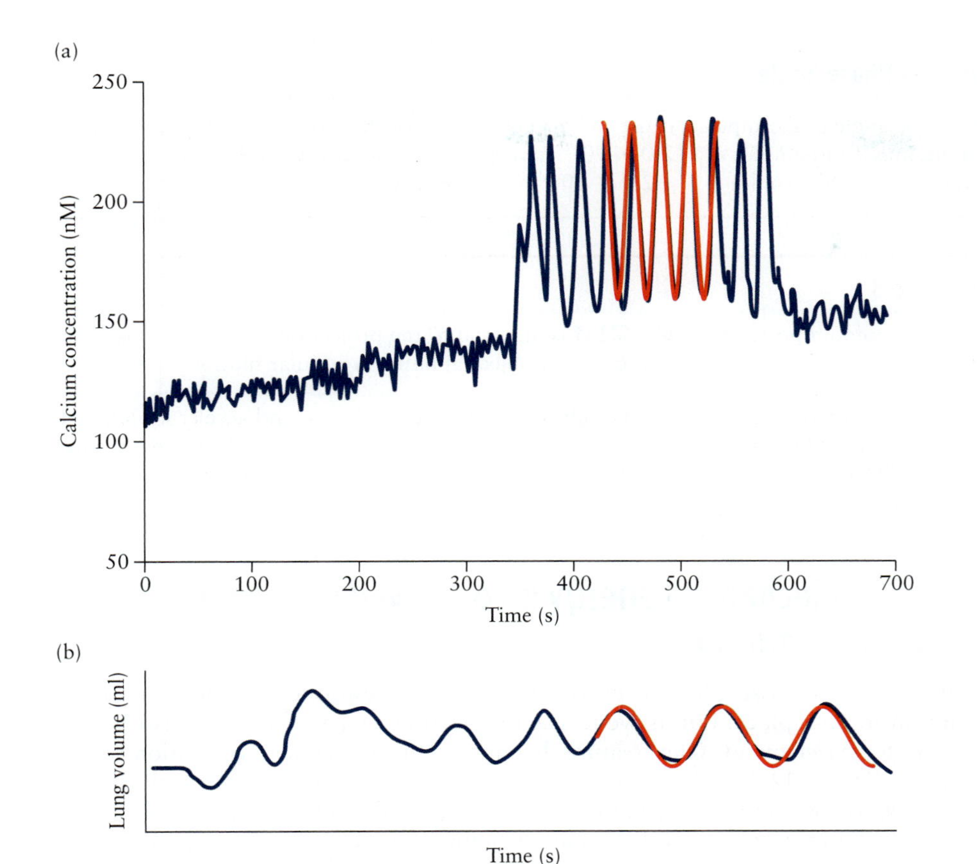

Figure 12-8 Simple harmonic motion in physiology (a) The concentration of calcium ions (Ca^{2+}) within a cell can oscillate under certain conditions. A sinusoidal curve (red) is a good match to these oscillations. (b) This graph shows the volume of air in a person's lungs as a function of time. The oscillations in lung volume are also well matched to a sinusoidal curve (red).

Our discussion of simple harmonic motion has concentrated on blocks attached to springs. But the same basic equations, and the sinusoidal graph of the oscillation as a function of time, apply to many different oscillating systems. **Figure 12-8a** shows an example from cell biology. In some cells the intracellular concentration of calcium ions (Ca^{2+}) oscillates sinusoidally. (These oscillations can regulate enzyme activity, mitochondrial metabolism, or gene expression.) The red sinusoidal curve drawn on top of the data matches these oscillations well. **Figure 12-8b** shows another biological example, the volume of air in a person's lungs as a function of time. Again we've drawn a red sinusoidal curve on top of the data; the close match between the curves shows that the oscillations are essentially simple harmonic motion.

GOT THE CONCEPT? 12-2 Changing the Amplitude

? Suppose you were to increase the amplitude of the oscillation in Example 12-3 by a factor of 4, from 2.0×10^{-2} m to 8.0×10^{-2} m. The spring constant and mass remain the same as in Example 12-3. This would cause the frequency of the oscillation to (a) increase by a factor of 4; (b) increase by a factor of 2; (c) decrease by a factor of 1/4; (d) decrease by a factor of 1/2; (e) remain unchanged.

GOT THE CONCEPT? 12-3 Changing the Spring Constant

? Suppose you were to increase the spring constant of the spring in Example 12-3 by a factor of 4, from 1.8×10^2 N/m to 7.2×10^2 N/m. The mass and amplitude remain the same as in Example 12-3. This would cause the maximum speed of the block during the oscillation to (a) increase by a factor of 4; (b) increase by a factor of 2; (c) decrease by a factor of 1/4; (d) decrease by a factor of 1/2; (e) remain unchanged.

GOT THE CONCEPT? 12-4 Changing the Phase Angle

Suppose you were to change the phase angle in Example 12-3 to $\phi = \pi$. The spring constant, mass, and amplitude remain the same as in Example 12-3. This would cause the displacement x of the block at $t = 0$ to be (a) $A = 2.0 \times 10^{-2}$ m; (b) positive, but less than A; (c) zero; (d) $-A = -2.0 \times 10^{-2}$ m; (e) negative, but between zero and $-A$.

TAKE-HOME MESSAGE FOR Section 12-3

✔ The amplitude of an oscillation is the maximum positive displacement from equilibrium.

✔ An object oscillating under the influence of a Hooke's law restoring force is called simple harmonic motion (SHM). The angular frequency, period, and frequency in SHM are unaffected by changes in the oscillation amplitude.

✔ SHM is identical to the projection onto the x axis of the motion of an object in uniform circular motion.

✔ Graphs of the position, velocity, and acceleration in SHM are sinusoidal curves. In each of these graphs the phase angle shifts the location of the vertical axis (where time equals zero) to the right or left.

12-4 Mechanical energy is conserved in simple harmonic motion

We've been discussing how a spring can provide the restoring force necessary for oscillation to take place. But as we learned in Section 6-6, a spring can also be used to store *potential energy*. One creature that applies this principle to oscillation is the kangaroo (**Figure 12-9**). A kangaroo's tendons act like springs to store spring potential energy, which is transformed into kinetic energy as the animal hops; this increases the efficiency of movement. Dolphins use a similar mechanism in swimming. Even when a dolphin increases its speed by beating its tail faster, there is hardly any change in the rate at which it consumes oxygen (an indication of the dolphin's power output). The explanation is that the tissue in a dolphin's tail acts much like a spring, storing potential energy as the tail flips either up or down and then transforming the potential energy into kinetic energy. As a result, the dolphin can swim efficiently at high speeds.

These observations suggest that it's worthwhile to look at oscillations again from an energy perspective. To do this we'll study the interplay between kinetic energy and potential energy for a block connected to the free end of an ideal, Hooke's law spring, oriented horizontally as in Figures 12-2 and 12-3.

Our starting point is the expression from Section 6-6 for the potential energy stored in an ideal spring:

Spring potential energy (6-19)

Spring potential energy of a stretched or compressed spring

Spring constant of the spring

$$U_{\text{spring}} = \frac{1}{2}kx^2$$

Extension of the spring ($x > 0$ if spring is stretched, $x < 0$ if spring is compressed)

The tendons on the backs of this kangaroo's legs are fully stretched. As such they store spring potential energy...

...which is released during a hop, as the kangaroo straightens its legs and lets its tendons relax...

...and the released spring potential energy is converted into kinetic energy of the kangaroo.

Figure 12-9 Spring potential energy When stretched, tendons in a kangaroo's legs store potential energy in much the same way as a spring.

This equation tells us that the potential energy stored in a spring varies during a cycle of oscillation. When the object is at equilibrium ($x = 0$), the spring is neither stretched nor compressed, and the spring potential energy is zero: $U_{spring} = 0$. Whenever the spring is stretched ($x > 0$) or compressed ($x < 0$), the spring potential energy is greater than zero. The potential energy is greatest when the spring is either at its maximum extension (when $x = A$) or maximum compression ($x = -A$). At either of those points $U_{spring} = \frac{1}{2}kA^2$.

The *kinetic* energy of the oscillating block, $K = \frac{1}{2}mv^2$, also varies during an oscillation cycle. The block has its maximum speed v_{max} when passing through the equilibrium position, so the kinetic energy has its maximum value $K = \frac{1}{2}mv_{max}^2$ there as well. At any other point in the oscillation the block is moving more slowly than when it passes through equilibrium, so the kinetic energy is less than its maximum value. The kinetic energy has its minimum value $K = 0$ when the oscillating block is momentarily at rest; this happens when the spring is at maximum extension ($x = A$) or maximum compression ($x = -A$).

This analysis shows that for a block oscillating on an ideal spring, the kinetic energy is maximum where the spring potential energy is minimum (at $x = 0$), and the kinetic energy is minimum where the spring potential energy is maximum (at $x = A$ and $x = -A$). This is just what we would expect! If there are no nonconservative forces such as friction acting, the total mechanical energy E (the sum of the kinetic energy of the block and the potential energy of the spring) should remain constant, and energy will be transformed back and forth between its kinetic and potential forms as the block oscillates.

To analyze this in more detail, let's look at the expression for the total mechanical energy E of the system of block and spring:

$$E = K + U_{spring} = \frac{1}{2}mv^2 + \frac{1}{2}kx^2 \tag{12-17}$$

Let's use our results from Section 12-3 to see how E, K, and U_{spring} change during the course of an oscillation. From Equation 12-12 the position as a function of time is $x = A\cos(\omega t + \phi)$, so the spring potential energy as a function of time is

$$U_{spring} = \frac{1}{2}kx^2 = \frac{1}{2}kA^2\cos^2(\omega t + \phi) \tag{12-18}$$

Since the value of $\cos(\omega t + \phi)$ ranges from $+1$ through zero to -1, you can see that the value of $\cos^2(\omega t + \phi)$ ranges from 0 to 1. Hence the value of U_{spring} ranges from 0 to a maximum value $\frac{1}{2}kA^2$, just as we mentioned above.

Equation 12-14 tells us that the velocity as a function of time is $v_x = -\omega A\sin(\omega t + \phi)$. The velocity can be negative, but the square of the velocity is positive and equal to the square of the speed. Hence the kinetic energy as a function of time is

$$K = \frac{1}{2}mv^2 = \frac{1}{2}m\omega^2A^2\sin^2(\omega t + \phi)$$

From the first of Equations 12-11, we know that the angular frequency for a block oscillating on the end of an ideal spring is $\omega = \sqrt{k/m}$. So $m\omega^2 = m(k/m) = k$, and we can rewrite our expression for the kinetic energy as a function of time as

$$K = \frac{1}{2}mv^2 = \frac{1}{2}kA^2\sin^2(\omega t + \phi) \tag{12-19}$$

Like the cosine function the value of $\sin(\omega t + \phi)$ ranges from $+1$ through zero to -1, and the value of $\sin^2(\omega t + \phi)$ ranges from 0 to 1. So just like the spring potential energy, the kinetic energy of the block ranges in value from 0 to a maximum value $\frac{1}{2}kA^2$.

See the Math Tutorial for more information on trigonometry.

We can now insert our expressions for K from Equation 12-19 and U_{spring} from Equation 12-18 into Equation 12-17 for the total mechanical energy E. We get

$$\begin{aligned} E = K + U_{spring} &= \frac{1}{2}kA^2\sin^2(\omega t + \phi) + \frac{1}{2}kA^2\cos^2(\omega t + \phi) \\ &= \frac{1}{2}kA^2[\sin^2(\omega t + \phi) + \cos^2(\omega t + \phi)] \end{aligned} \tag{12-20}$$

Equation 12-20 looks like a complicated function of time. But we can simplify it thanks to an important result from trigonometry:

$$\sin^2\theta + \cos^2\theta = 1 \qquad \text{for any value of } \theta$$

Then Equation 12-20 becomes

Total mechanical energy of a mass oscillating on an ideal spring (12-21)

$$E = K + U_{\text{spring}} = \frac{1}{2}kA^2$$

Total mechanical energy of the oscillating mass–spring system

Spring constant

Amplitude of the oscillation

During the oscillation, the kinetic energy K and spring potential energy U_{spring} both change...

...but their sum, the total mechanical energy E, always has the same value.

Because both k and A are constant for a specific motion of a given oscillator, the total energy of the oscillating system is constant. This is a statement of the conservation of energy: The energy of the system transforms between the kinetic and potential forms, but the total mechanical energy remains constant (**Figure 12-10**).

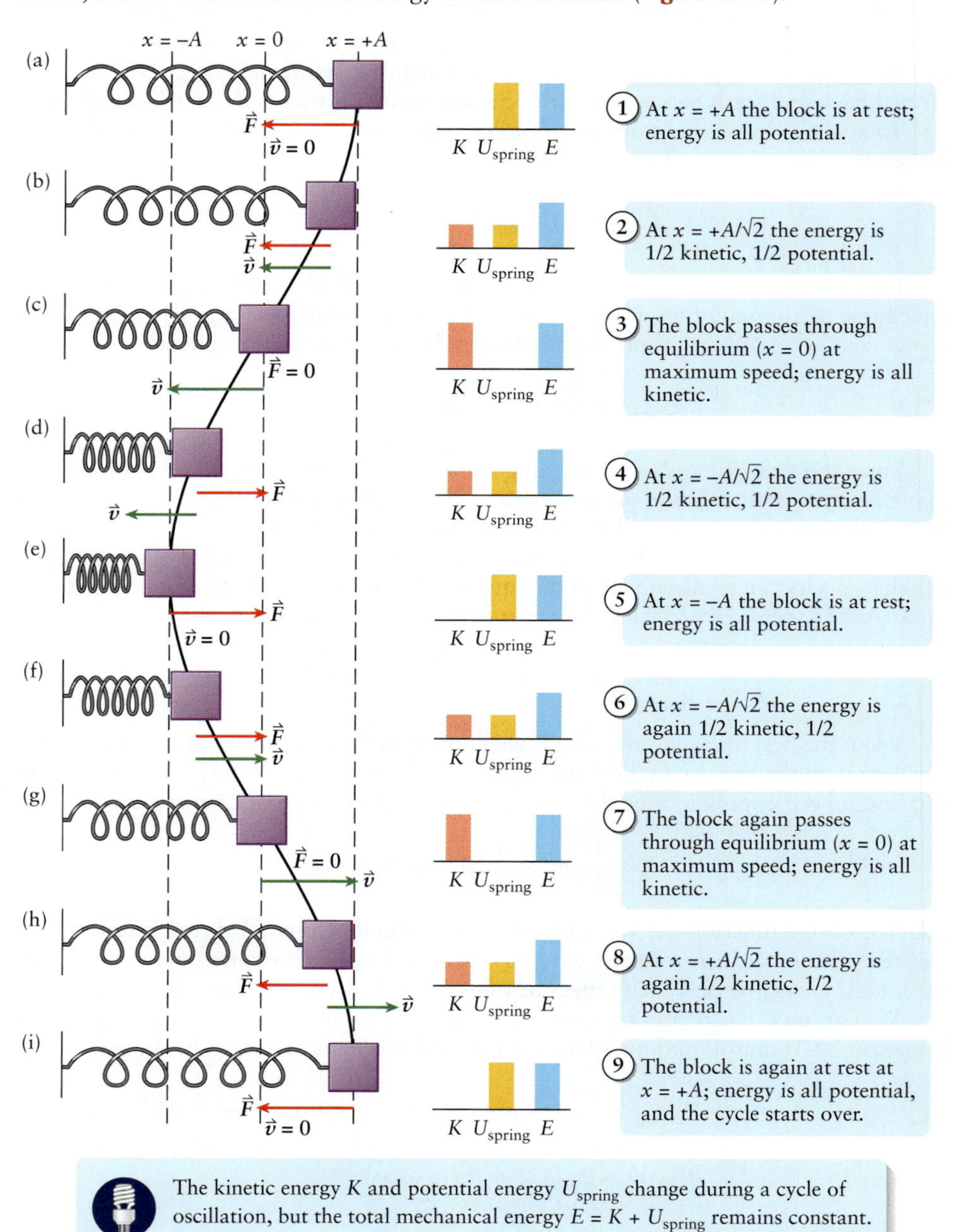

x = −A
x = 0
x = +A
(a)
(b)
(c)
(d)
(e)
(f)
(g)
(h)
(i)
F
v = 0
F = 0
v
K U_spring E
1 At x = +A the block is at rest; energy is all potential.
2 At x = +A/√2 the energy is 1/2 kinetic, 1/2 potential.
3 The block passes through equilibrium (x = 0) at maximum speed; energy is all kinetic.
4 At x = −A/√2 the energy is 1/2 kinetic, 1/2 potential.
5 At x = −A the block is at rest; energy is all potential.
6 At x = −A/√2 the energy is again 1/2 kinetic, 1/2 potential.
7 The block again passes through equilibrium (x = 0) at maximum speed; energy is all kinetic.
8 At x = +A/√2 the energy is again 1/2 kinetic, 1/2 potential.
9 The block is again at rest at x = +A; energy is all potential, and the cycle starts over.
The kinetic energy K and potential energy U_spring change during a cycle of oscillation, but the total mechanical energy E = K + U_spring remains constant.

Figure 12-10 Energy during a cycle of oscillation The bar graphs show the kinetic energy K, spring potential energy U_{spring}, and total mechanical energy $E = K + U_{\text{spring}}$ at nine stages in the oscillation of a block attached to a spring (compare Figure 12-3).

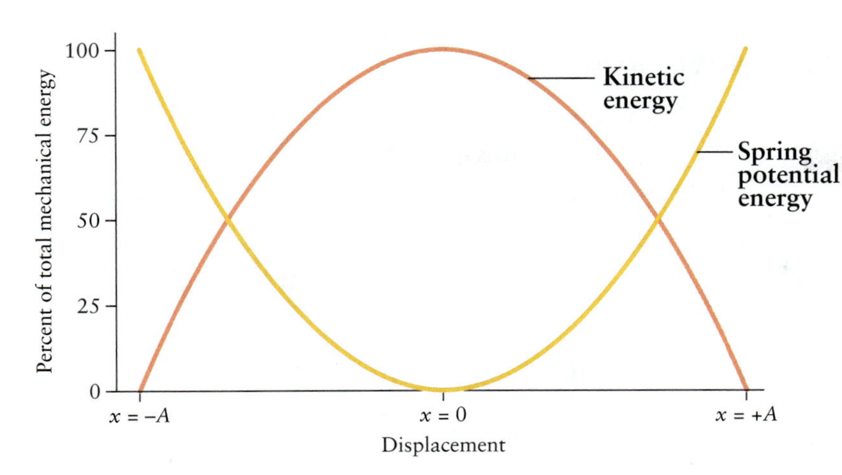

Figure 12-11 Kinetic energy and potential energy versus displacement The total mechanical energy—the sum of the kinetic energy and spring potential energy—associated with an object in simple harmonic motion is constant. The percentage of the total mechanical energy that is in the kinetic and potential forms depends on the object's displacement from equilibrium.

Equations 12-18 and 12-19 give the spring potential energy and kinetic energy as functions of time. It's also helpful to write these relationships as functions of the displacement x. We know from Equation 6-19 that the spring potential energy is $U_{\text{spring}} = (1/2)kx^2$; the graph of this as a function of x is a parabola with its minimum value at $x = 0$ (**Figure 12-11**). From Equation 12-21 the kinetic energy K is equal to the total mechanical energy $E = (1/2)kA^2$ (a constant) minus the potential energy U_{spring}:

$$K = E - U_{\text{spring}} = \frac{1}{2}kA^2 - \frac{1}{2}kx^2 \quad (12\text{-}22)$$

This is an "upside-down" parabola that has its *maximum* value at $x = 0$, as Figure 12-11 shows.

Note that U_{spring} is greatest at the two extremes of displacement ($x = +A$ and $x = -A$) and zero at equilibrium ($x = 0$). The kinetic energy K is zero at the two extremes, where the block momentarily comes to a stop, and greatest as the object passes through $x = 0$. So at $x = 0$ the energy is 100% kinetic and 0% potential, while at $x = +A$ and $x = -A$ the energy is 0% kinetic and 100% potential.

EXAMPLE 12-4 SHM III: Kinetic and Potential Energy

As in Example 12-3 an object of mass 0.80 kg is attached to an ideal horizontal spring of spring constant 1.8×10^2 N/m. The object is initially at rest at equilibrium. You then start the object oscillating with amplitude 2.0×10^{-2} m. (a) How much work do you have to do on the object to set it into oscillation? (b) What is the speed of the object when the spring is compressed by 1.0×10^{-2} m? (c) How far is the object from equilibrium when the kinetic energy of the object equals the potential energy in the spring?

Set Up

We'll use energy ideas to answer these questions. Equation 12-21 will let us find the (constant) total mechanical energy E of the system from the given spring constant and amplitude. Equation 6-19 will let us find the spring potential energy for any displacement x; we can then find the kinetic energy K using Equation 12-21, and from that find the speed for that value of x.

Total mechanical energy of a mass oscillating on an ideal spring:

$$E = K + U_{\text{spring}} = \frac{1}{2}kA^2 \quad (12\text{-}21)$$

Spring potential energy:

$$U_{\text{spring}} = \frac{1}{2}kx^2 \quad (6\text{-}19)$$

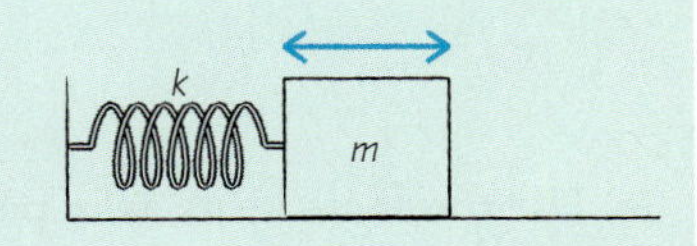

Solve

(a) Initially the system has zero kinetic energy and zero potential energy. You can start the oscillation by pulling the object so that you stretch the spring by a distance $A = 2.0 \times 10^{-2}$ m from equilibrium. The work you do goes into the potential energy of the spring.

(Work you do) = (Change of potential energy of the spring as you stretch it from $x = 0$ to $x = A$)

Initial potential energy

$$= \frac{1}{2}k(0)^2 = 0$$

Final potential energy $= \frac{1}{2}kA^2$

So the work you do is

$$W = \frac{1}{2}kA^2 - 0$$

$$= \frac{1}{2}(1.8 \times 10^2 \text{ N/m})(2.0 \times 10^{-2} \text{ m})^2$$

$$= 3.6 \times 10^{-2} \text{ N} \cdot \text{m} = 3.6 \times 10^{-2} \text{ J}$$

x = O

x = A = 2.0 × 10⁻² m

(b) When the spring is compressed by 1.0×10^{-2} m, the displacement is $x = -1.0 \times 10^{-2}$ m. The first step in determining the block's speed at this value of x is to determine its kinetic energy.

From (a) the total mechanical energy is

$$E = \frac{1}{2}kA^2 = 3.6 \times 10^{-2} \text{ J}$$

From Equation 6-19 the spring potential energy for $x = -1.0 \times 10^{-2}$ m is

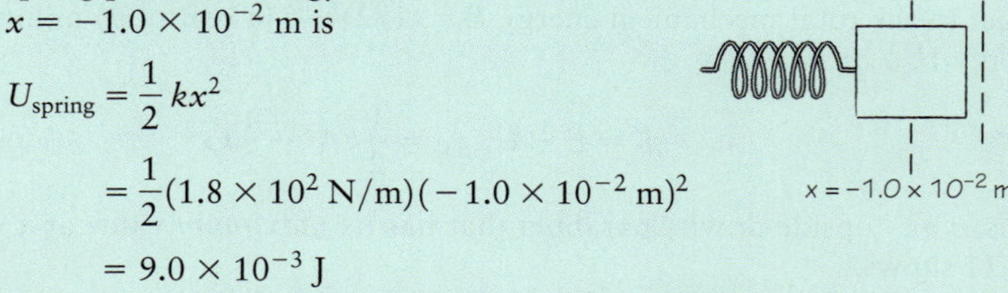

$$U_{\text{spring}} = \frac{1}{2}kx^2$$

$$= \frac{1}{2}(1.8 \times 10^2 \text{ N/m})(-1.0 \times 10^{-2} \text{ m})^2$$

$$= 9.0 \times 10^{-3} \text{ J}$$

So from Equation 12-21 the kinetic energy at this value of x is

$$K = E - U_{\text{spring}} = 3.6 \times 10^{-2} \text{ J} - 9.0 \times 10^{-3} \text{ J}$$
$$= 2.7 \times 10^{-2} \text{ J}$$

Given the kinetic energy K of the block and its mass $m = 0.80$ kg, calculate the speed of the block.

Solve for the speed v:

Kinetic energy is $K = \frac{1}{2}mv^2$, so

$$v^2 = \frac{2K}{m}$$

$$v = \sqrt{\frac{2K}{m}} = \sqrt{\frac{2(2.7 \times 10^{-2} \text{ J})}{0.80 \text{ kg}}} = 0.26 \text{ m/s}$$

(Recall from Chapter 6 that if the kinetic energy is in joules and the mass is in kilograms, the speed is in meters per second.)
The *velocity* of the block could be +0.26 m/s or −0.26 m/s, depending on what direction it is moving as it passes through this point.

(c) Use the same ideas as in part (b) to solve for the value of x at which the kinetic energy K equals the spring potential energy U_{spring}.

From above the total mechanical energy is

$$E = \frac{1}{2}kA^2 = 3.6 \times 10^{-2} \text{ J}$$

The spring potential energy is

$$U_{\text{spring}} = \frac{1}{2}kx^2$$

and the kinetic energy is

$$K = E - U_{\text{spring}}$$

At the value of x for which $K = U_{spring}$,

$$\frac{1}{2}kA^2 - \frac{1}{2}kx^2 = \frac{1}{2}kx^2$$

Multiply both sides by $2/k$ and solve for x:

$$A^2 - x^2 = x^2 \quad \text{so} \quad 2x^2 = A^2 \quad \text{and} \quad x^2 = \frac{A^2}{2}$$

$$x = \pm\sqrt{\frac{A^2}{2}} = \pm\frac{A}{\sqrt{2}}$$

$$= \pm\frac{2.0 \times 10^{-2}\ \text{m}}{\sqrt{2}} = \pm 1.4 \times 10^{-2}\ \text{m}$$

Note that x can be either positive (the spring is stretched by 1.4×10^{-2} m) or negative (the spring is compressed by 1.4×10^{-2} m).

Reflect

The result for part (b) would have been very difficult to find without using the energy approach. You would have needed to use the equations from Section 12-3 to solve for the time t at which the block passes through $x = -1.0 \times 10^{-2}$ m, then use this value of t to find the velocity of the block at that time. The energy approach makes this much easier.

Note that the points $x = \pm A/\sqrt{2} = \pm 0.71A$ where the kinetic and potential energies are equal are shown in Figure 12-10 (see parts (b), (d), (f), and (h) of that figure). These points are *not* halfway between the equilibrium position ($x = 0$) and the extremes of the motion ($x = \pm A$); they are actually closer to the extremes. Can you show that at $x = \pm A/2$, the energy is 75% kinetic and 25% potential? (*Hint:* See the solution to part (b).)

GOT THE CONCEPT? 12-5 Rank the Energies

? Consider four different systems, each made of a block attached to an ideal horizontal spring. Rank them in order of their total mechanical energy, from largest to smallest value. (a) Block mass 0.50 kg and spring constant 5.0×10^2 N/m, with amplitude 0.020 m; (b) block mass 0.60 kg and spring constant 3.0×10^2 N/m, with speed 1.0 m/s when passing through equilibrium; (c) block mass 1.2 kg and spring constant 4.0×10^2 N/m, with speed 0.50 m/s when passing through $x = -0.010$ m; (d) block mass 2.0 kg and spring constant 2.0×10^2 N/m, with speed 0.20 m/s when passing through $x = 0.050$ m.

TAKE-HOME MESSAGE FOR Section 12-4

- ✔ Energy is transformed back and forth from kinetic energy to potential energy in systems which contain mechanical or biological springs.
- ✔ When an object attached to a spring is at its maximum displacement, the energy of the system of spring and object is entirely potential energy.
- ✔ When the object passes through equilibrium, the energy of the system of spring and object is entirely kinetic energy.
- ✔ The total energy of the system, the maximum potential energy, and the maximum kinetic energy are all equal to $(1/2)kA^2$.

12-5 The motion of a pendulum is approximately simple harmonic

You've probably had a physician strike the patellar tendon just below your kneecap with a hammer then watched your lower leg swing upward and back (**Figure 12-12a**). For a physician this is a test of your patellar reflex: Hitting that tendon sends a signal to your spinal cord, which in turn sends a signal to the quadriceps muscle

(a)

BSIP SA/Alamy

(b)

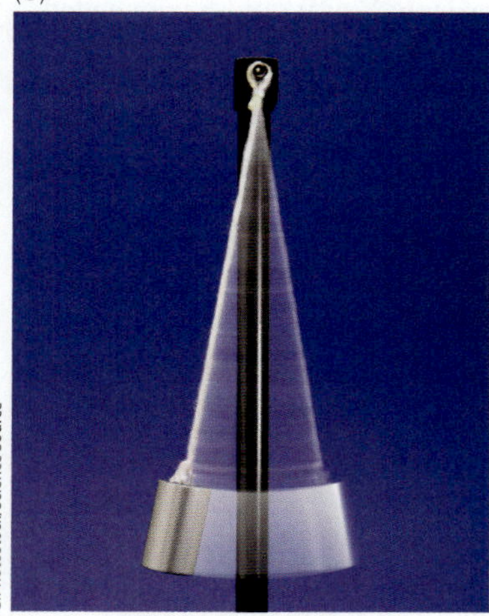

GIPhotoStock/Science Source

Figure 12-12 Swinging leg, swinging pendulum The motions of both (a) your lower leg in a physician's office and (b) a mass hanging from a string have the ingredients for oscillation: a restoring force (gravity) that tends to pull the object toward equilibrium (hanging straight down), and inertia that causes the object to overshoot equilibrium.

on your upper thigh. This makes that muscle flex and makes the lower leg move. But to a physicist this is the same kind of motion as a mass swinging at the end of a string (**Figure 12-12b**). When allowed to move freely, both your leg and the mass hang straight down in their equilibrium positions thanks to the influence of gravity. When either your leg or the mass is displaced from equilibrium, released, and allowed to move freely, the force of gravity pulls it back toward equilibrium—that is, gravity serves as a restoring force. And in either case inertia causes the object (leg or mass) to overshoot the equilibrium position, resulting in an oscillation. Both a swinging leg and a mass on the end of a string are examples of **pendulums**—systems that oscillate back and forth due to the restoring force of gravity.

Since pendulums actually *rotate* around a point, we'll describe them using the language of rotational motion that we developed in Chapter 8. So instead of relating a restoring force to the acceleration that it produces, as we did in Section 12-3 for a block attached to an ideal spring, we'll describe pendulum motion in terms of a restoring *torque* that produces an *angular* acceleration.

In this section we'll explore oscillations of a **simple pendulum**. This is one in which all of the mass is concentrated at a single point. It's an idealized version of the mass-and-string arrangement shown in Figure 12-12b in which the string of length L has zero mass and all of the mass m is compressed to a point. (In Section 12-6 we'll look at other, less idealized pendulums.)

Figure 12-13a shows a simple pendulum. In equilibrium the string hangs straight down; the figure shows the pendulum displaced by an angle θ from the vertical. The blue dashed curve shows the path that the object of mass m at the end of the pendulum will take to return to equilibrium. **Figure 12-13b** shows the free-body diagram for this object. The two forces that act on this object are a tension force of magnitude T exerted by the string and a gravitational force of magnitude mg. The tension force points along the string, so the line of action of the tension force passes through the pivot point. Hence the tension force exerts no torque around the pivot (it does nothing to make the pendulum rotate around the pivot). The gravitational force does exert a torque, however. From Section 8-6 its magnitude is the magnitude mg of the gravitational force multiplied by the lever arm of the force (the perpendicular distance from the line of action of the gravitational force to the

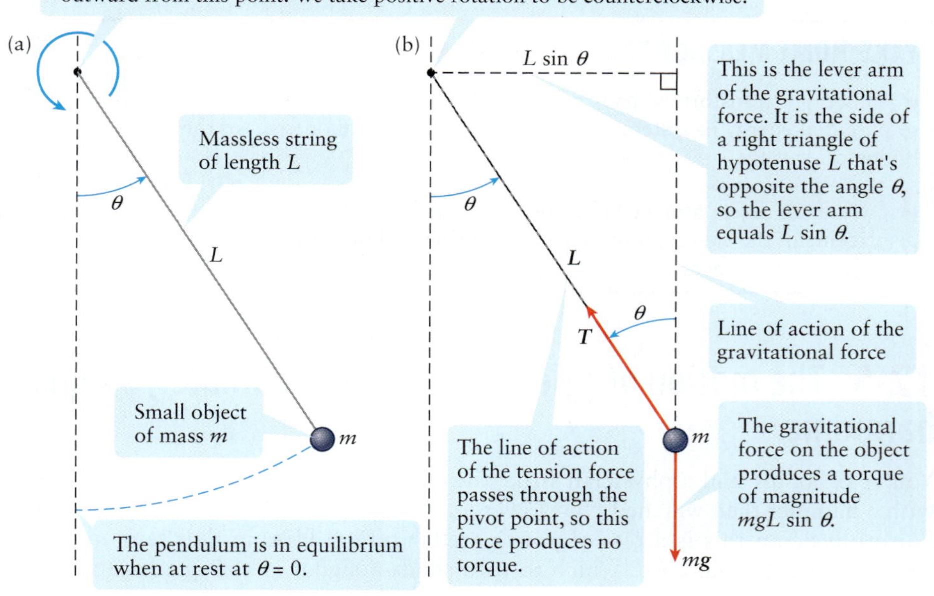

Pivot point: The positive z axis (the rotation axis of the pendulum) points outward from this point. We take positive rotation to be counterclockwise.
(a)
Massless string of length L
θ
L
Small object of mass m
m
The pendulum is in equilibrium when at rest at θ = 0.
(b)
L sin θ
This is the lever arm of the gravitational force. It is the side of a right triangle of hypotenuse L that's opposite the angle θ, so the lever arm equals L sin θ.
θ
L
θ
T
Line of action of the gravitational force
m
The line of action of the tension force passes through the pivot point, so this force produces no torque.
The gravitational force on the object produces a torque of magnitude mgL sin θ.
mg

Figure 12-13 A simple pendulum (a) Layout of a simple pendulum. (b) The forces on a simple pendulum.

pivot). Figure 12-13b shows that if the pendulum is at an angle θ from the vertical, this lever arm is $L \sin \theta$. Hence we can write the torque as:

$$\tau_z = -mgL \sin \theta \tag{12-23}$$

The subscript z indicates that the pendulum tends to rotate around the z axis shown in Figure 12-13a. The minus sign in Equation 12-23 indicates that this is a *restoring* torque. If the pendulum swings to the right of vertical (counterclockwise) so that θ and $\sin \theta$ are positive, as in Figure 12-13a, the torque will be negative (clockwise): That means it acts to decrease θ and so return the pendulum to the equilibrium position $\theta = 0$. If the pendulum swings instead to the left of vertical (clockwise) so that θ and $\sin \theta$ are negative, the torque will be positive (counterclockwise) and so will again act to restore the pendulum to equilibrium.

The torque in Equation 12-23 is the only torque that acts on the pendulum, so this is also the *net* torque. Newton's second law for rotational motion (Section 8-6) says that the net torque on an object is equal to the moment of inertia multiplied by the angular acceleration. For an object made up of a single point of mass m a distance L from the rotation axis, the moment of inertia is $I = mL^2$. Using the torque from Equation 12-23, we have

$$-mgL \sin \theta = mL^2 \alpha_z$$

If we divide both sides of this equation by mL^2 and rearrange, we get

$$\alpha_z = -\frac{g}{L} \sin \theta \tag{12-24}$$

Let's see whether Equation 12-24 is equivalent to Hooke's law. If it is, that means the oscillations of simple pendulums are simple harmonic motion and we can use all of our results from Section 12-3. We saw in Section 12-3 that we can write Hooke's law for straight-line oscillation as

$$a_x = -\omega^2 x$$

That is, the acceleration is directly proportional to the displacement, and the proportionality constant is the negative of the square of the angular frequency ω. From Section 8-5 the quantities x and a_x for straight-line motion correspond to the quantities θ and α_z, respectively, for rotational motion. So the rotational version of Hooke's law is

$$\alpha_z = -\omega^2 \theta \tag{12-25}$$

The difference between Equation 12-24 for the pendulum and Equation 12-25, the rotational version of Hooke's law, is that Equation 12-24 involves $\sin \theta$ rather than θ. So in general the oscillations of a simple pendulum do *not* obey Hooke's law, and so the motion of the pendulum is *not* simple harmonic motion.

If, however, the angle θ is relatively small and we measure θ in radians, it turns out that $\sin \theta$ is *approximately* equal to θ (**Figure 12-14**):

$$\sin \theta \approx \theta \quad \text{when } \theta \text{ is small}$$

(You can confirm this with your calculator. Try calculating $\sin \theta$ for $\theta =$ 1 rad, 0.5 rad, 0.2 rad, 0.1 rad, and 0.01 rad. You'll find that as you try smaller values of θ, the value of $\sin \theta$ gets closer and closer to θ.) With this approximation, Equation 12-24 for the simple pendulum becomes

$$\alpha_z \approx -\frac{g}{L}\theta \quad \text{if } \theta \text{ is small} \tag{12-26}$$

Compare this to Hooke's law for rotational motion, Equation 12-25, and you'll see that the oscillations of a pendulum *with small amplitude* (so that θ is always small) obey Hooke's law and that the angular frequency of the simple pendulum's oscillations is given by

$$\omega^2 = \frac{g}{L} \quad \text{or} \quad \omega = \sqrt{\frac{g}{L}}$$

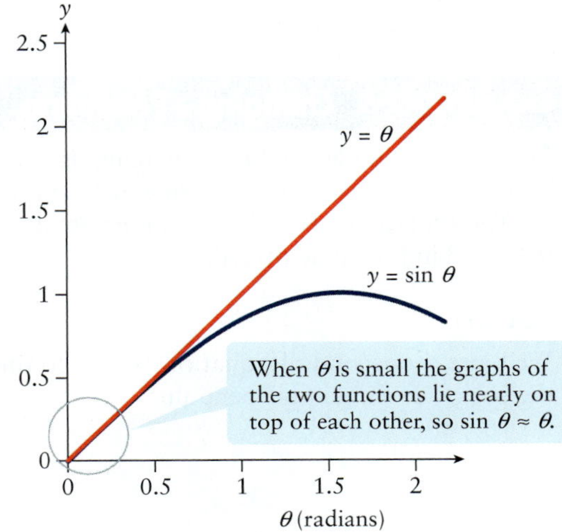

Figure 12-14 Approximating sin θ If the value of θ is small, $\sin \theta$ is approximately equal to θ (measured in radians).

As in Section 12-3 the period T is equal to $2\pi/\omega$, and the frequency f is equal to $1/T$ or $\omega/2\pi$. So we can write the following results for the small-amplitude oscillations of a simple pendulum:

Angular frequency, period, and frequency for a simple pendulum (small amplitude) (12-27)

Angular frequency $$\omega = \sqrt{\frac{g}{L}}$$

Period $$T = \frac{2\pi}{\omega} = 2\pi\sqrt{\frac{L}{g}}$$

Frequency $$f = \frac{1}{T} = \frac{1}{2\pi}\sqrt{\frac{g}{L}} = \frac{\omega}{2\pi}$$

g = acceleration due to gravity

L = length of the simple pendulum

Go to Picture It 12-1 for more practice dealing with simple pendulums.

Equations 12-27 show that the angular frequency, period, and frequency for a simple pendulum do *not* depend on the mass of the object at the end of the pendulum. That's because the restoring torque is provided by the gravitational force. Doubling the mass doubles the moment of inertia, which by itself would make the oscillations happen more slowly, but it also doubles the restoring torque, which by itself would make the oscillations happen more rapidly. The two effects cancel each other out so that ω, T, and f are unaffected by changes in the pendulum mass.

Because the small-amplitude oscillations of a simple pendulum obey Hooke's law, it also follows that the angular frequency, period, and frequency do not depend on the *amplitude* of the oscillations. Here the amplitude is the maximum angle from the vertical that the pendulum attains. If you were to pull the simple pendulum in Figure 12-13a to an angle $\theta = 0.1$ rad—about 6°—and let it go, the pendulum would oscillate between $\theta = +0.1$ rad (to the right of vertical) and $\theta = -0.1$ rad (to the left of vertical), and the amplitude of the oscillation would be 0.1 rad.

WATCH OUT! Large-amplitude oscillations of a simple pendulum are *not* simple harmonic motion.

Equations 12-27 apply only if the angle θ reached by the pendulum is always small enough that $\sin\theta$ is approximately equal to θ. Figure 12-14 shows that this is not a good approximation if θ is greater than about 0.5 rad (about 30°). If the amplitude is larger than this, the motion is *not* simple harmonic and the values of ω, T, and f *do* depend on the amplitude. As an example, if the amplitude is $\pi/2$ (90°), the period of oscillation is about 18% greater than the value given by the second of Equations 12-27.

EXAMPLE 12-5 Changing a Pendulum

You make a simple pendulum by hanging a small 60-g marble from an elastic thread of negligible mass. With the marble attached the thread is 0.40 m long. (a) Find the period when the marble is pulled a small angle to one side and released. (b) You replace the marble by another small one of mass 260 g. When you do this the thread stretches an additional 0.10 m. Find the new period.

Set Up

We'll use the second of Equations 12-27 to find the period of this simple pendulum.

Period of a simple pendulum:

$$T = 2\pi\sqrt{\frac{L}{g}} \qquad (12\text{-}27)$$

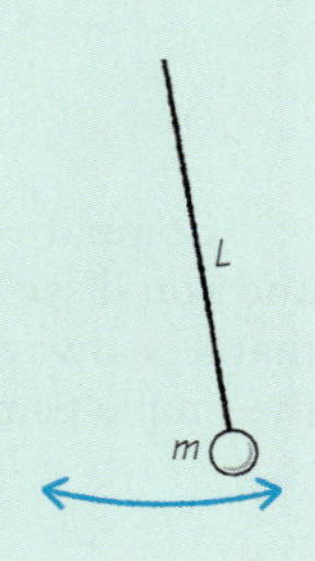

Solve

(a) Find the period of the pendulum with the initial length $L = 0.40$ m.

Then the period is

$$T = 2\pi\sqrt{\frac{L}{g}} = 2\pi\sqrt{\frac{0.40 \text{ m}}{9.80 \text{ m/s}^2}} = 2\pi\sqrt{\frac{0.40 \text{ s}^2}{9.80}} = 1.3 \text{ s}$$

(b) Repeat the calculation with the new value of the length.

The new length of the pendulum is

$$L_{\text{new}} = 0.40 \text{ m} + 0.10 \text{ m} = 0.50 \text{ m}$$

So the new period is

$$T_{\text{new}} = 2\pi\sqrt{\frac{L_{\text{new}}}{g}} = 2\pi\sqrt{\frac{0.50 \text{ m}}{9.80 \text{ m/s}^2}} = 1.4 \text{ s}$$

Reflect

Although the mass of the marble changed, we did *not* have to use its mass in the calculation. As we've seen, the period of a simple pendulum depends only on its length, not on its mass.

Note that we know the value of g to three significant figures, but our answers for the period are given to only two significant figures. Can you see why? (*Hint:* Review Section 1-4 if you're not sure.)

GOT THE CONCEPT? 12-6 Simple Pendulum I

? A small marble is attached to a thread that has a negligible mass and is hung from a support. When the marble is pulled back a small distance and released, it swings in simple harmonic motion with a frequency of 0.60 Hz. What is the frequency of the pendulum after the length of the thread is increased by a factor of 4 but the marble is released in the same way? (a) 2.4 Hz; (b) 1.2 Hz; (c) 0.60 Hz; (d) 0.30 Hz; (e) 0.15 Hz.

GOT THE CONCEPT? 12-7 Simple Pendulum II

? A small marble is attached to a thread that has a negligible mass and is hung from a support. When the marble is pulled back a small distance and released, it swings in simple harmonic motion with a frequency of 0.60 Hz. If the marble is replaced by one that has four times the mass of the original one, and the new marble is pulled back by half the distance of the original marble before being released, what is the new frequency? Assume that the length of the thread remains the same. (a) 2.4 Hz; (b) 1.2 Hz; (c) 0.60 Hz; (d) 0.30 Hz; (e) 0.15 Hz.

GOT THE CONCEPT? 12-8 Pendulum on an Asteroid

? A small marble is attached to a thread that has a negligible mass and is hung from a support. When the marble is pulled back a small distance and released, it swings in simple harmonic motion with a frequency of 0.60 Hz. What is the frequency of the pendulum if it is transported to the surface of a small asteroid, where the acceleration due to gravity is only 1/100 that on Earth's surface? (a) 60 Hz; (b) 6.0 Hz; (c) 0.60 Hz; (d) 0.060 Hz; (e) 0.0060 Hz.

TAKE-HOME MESSAGE FOR Section 12-5

✔ In a simple pendulum all of the mass of the pendulum is concentrated at a fixed distance from a rotation point.

✔ The angular frequency ω, the period T, and the frequency f of a simple pendulum depend on the length of the pendulum and the acceleration due to gravity. They do not depend on the pendulum mass.

12-6 A physical pendulum has its mass distributed over its volume

We began the previous section by comparing two examples of pendulum motion: the swing of your lower leg when a physician tests your patellar reflex (Figure 12-12a) and the motion of a mass on the end of a string (Figure 12-12b). The mass on a string is nearly an ideal simple pendulum, since all of the mass is concentrated into a very small volume. But your lower leg is *not* a simple pendulum because its mass is distributed along the entire distance from the knee to the foot. A pendulum like your lower leg whose mass is distributed throughout its volume is called a **physical pendulum.** Other examples of physical pendulums include a swinging church bell, a chandelier swaying back and forth after an earth tremor, and the pendulum of an old-fashioned grandfather clock. Let's see how to find the angular frequency, period, and frequency for the oscillations of a physical pendulum.

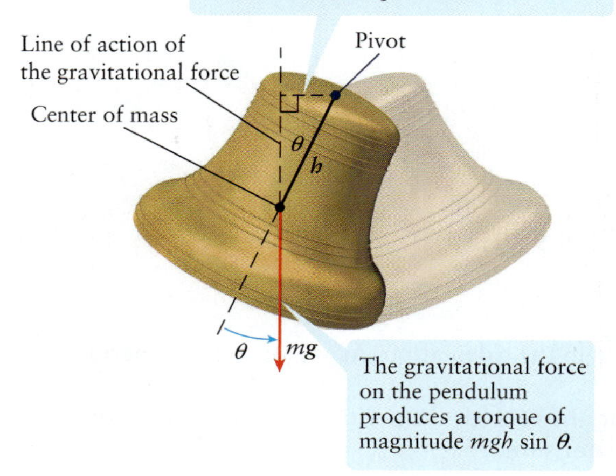
The z axis points perpendicular to the plane of the pendulum's rotation.
This is the lever arm of the gravitational force. It is the side of a right triangle of hypotenuse h that's opposite the angle θ, so the lever arm equals $h \sin \theta$.
Line of action of the gravitational force
Pivot
Center of mass
θ
h
θ
mg
The gravitational force on the pendulum produces a torque of magnitude $mgh \sin \theta$.

Figure 12-15 A physical pendulum The forces on a physical pendulum (one whose mass is not all concentrated in a single small blob, unlike the simple pendulum shown in Figure 12-13).

Figure 12-15 shows an example of a physical pendulum of mass m. Two forces act on the pendulum, a gravitational force of magnitude mg that acts at the center of mass of the pendulum, a distance h from the pivot point, and a support force that acts at the pivot. We haven't drawn the support force because it acts at the pivot point and so exerts no torque around that point. The torque exerted by the gravitational force is equal to the magnitude mg of the gravitational force multiplied by the lever arm of the gravitational force, which is the perpendicular distance from the line of action of the gravitational force to the pivot. Figure 12-15 shows that if the pendulum is displaced from the vertical by an angle θ, the lever arm equals $h \sin \theta$. Hence the torque on the pendulum is

$$\tau_z = -mgh \sin \theta \tag{12-28}$$

Just as for the torque on a simple pendulum, the minus sign in Equation 12-28 means that this is a restoring torque that always acts to pull the pendulum back toward its equilibrium position $\theta = 0$, with the center of mass directly below the pivot.

The torque in Equation 12-28 is the only torque on the pendulum and so is equal to the net torque $\Sigma\tau_z$. From Newton's second law for rotation (Section 8-6), $\Sigma\tau_z = I\alpha_z$, we have

$$-mgh \sin \theta = I\alpha_z \tag{12-29}$$

In Equation 12-29 the quantity I is the moment of inertia of the pendulum around the pivot point. Just as for the simple pendulum, if the angle θ is small then $\sin \theta$ is approximately equal to θ (measured in radians). With this approximation we can rewrite Equation 12-29 as

$$\alpha_z \approx -\left(\frac{mgh}{I}\right)\theta \quad \text{if } \theta \text{ is small} \tag{12-30}$$

Compare Equation 12-30 to the corresponding equation for the simple pendulum that we derived in Section 12-5, $\alpha_z \approx -(g/L)\theta$ (Equation 12-26): The equation is identical except that g/L has been replaced by mgh/I. So we conclude that just as for the simple pendulum, the oscillations of a physical pendulum are simple harmonic motion provided that the amplitude is relatively small. To find the angular frequency, period, and frequency of a physical pendulum, we simply take Equations 12-27 for a simple pendulum and replace g/L with mgh/I:

Angular frequency, period, and frequency for a physical pendulum (small amplitude) (12-31)

Angular frequency $\omega = \sqrt{\dfrac{mgh}{I}}$

Period $T = \dfrac{2\pi}{\omega} = 2\pi\sqrt{\dfrac{I}{mgh}}$

Frequency $f = \dfrac{1}{T} = \dfrac{1}{2\pi}\sqrt{\dfrac{mgh}{I}} = \dfrac{\omega}{2\pi}$

m = **mass** of physical pendulum

g = **acceleration due to gravity**

h = **distance** from pivot point to center of mass of physical pendulum

I = **moment of inertia** of physical pendulum about pivot

Here's a check on our results for a physical pendulum. If all of the mass of the physical pendulum is concentrated at the center of mass, then the moment of inertia of the physical pendulum around the pivot is $I = mh^2$. Then the quantity mgh/I in Equations 12-31 becomes

$$\frac{mgh}{I} = \frac{mgh}{mh^2} = \frac{g}{h} \quad (12\text{-}32)$$

If we change the symbol for the distance from the pivot to the center of mass from h to L, the quantity in Equation 12-32 becomes g/L. Then, from the first of Equations 12-31, the angular frequency for a physical pendulum with all of its mass concentrated at the center of mass becomes $\omega = \sqrt{g/L}$. That's exactly the result we found in Section 12-5 for a simple pendulum, which is just a physical pendulum with all of its mass concentrated a distance L from the pivot (see Figure 12-13 and the first of Equations 12-27). So our results for a *physical* pendulum give us the correct answers for the special case of a *simple* pendulum, just as they should.

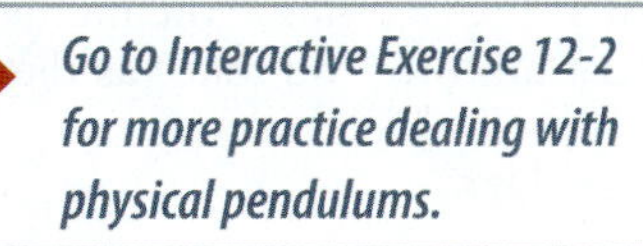

Go to Interactive Exercise 12-2 for more practice dealing with physical pendulums.

EXAMPLE 12-6 An Oscillating Rod

A uniform rod has a length L and a mass m, and is supported so that it can swing freely from one end. Derive an expression for the period of oscillation when the rod is pulled slightly from the vertical and released.

Set Up

Equations 12-31 tell us the period of a physical pendulum. The center of mass of a uniform rod is at its center, and its moment of inertia around one end is given in Table 8-1 (Section 8-3).

Period of a physical pendulum:

$$T = 2\pi\sqrt{\frac{I}{mgh}} \quad \text{(from Equations 12-31)}$$

Moment of inertia for a uniform rod of mass m and length L rotating about one end:

$$I = \frac{1}{3}mL^2$$

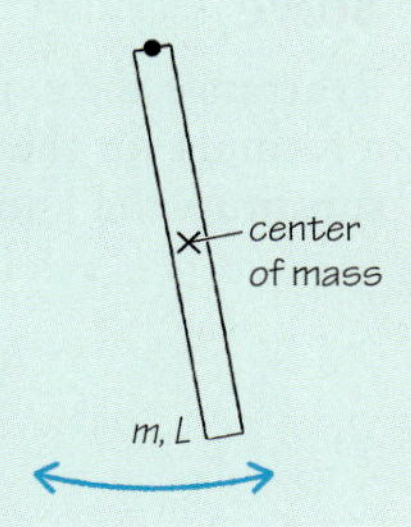

Solve

Given the moment of inertia I and the distance h, solve for the period.

The center of mass is a distance $L/2$ from the pivot, so

$$h = \frac{L}{2}$$

Substitute I and h into the expression for the period T from Equations 12-31:

$$T = 2\pi\sqrt{\frac{(1/3)mL^2}{mg(L/2)}} = 2\pi\sqrt{\frac{2mL^2}{3mgL}}$$

$$= 2\pi\sqrt{\frac{2L}{3g}}$$

Reflect

Compared to the period of a simple pendulum of length L, $T_{\text{simple pendulum}} = 2\pi\sqrt{L/g}$, the period of a uniform rod free to rotate about its end is smaller by a factor of $\sqrt{2/3} = 0.816$. In other words, the physical pendulum oscillates more quickly. That may seem surprising if you just compare the torques on the two objects. The center of mass of the rod is a distance $h = L/2$ from the pivot, so from Equation 12-28 the torque on the rod is $\tau_z = -mgh \sin\theta = -(mgL/2)\sin\theta$. By contrast, for a simple pendulum all of the mass m is a distance L from the pivot, so $h = L$ and the torque on the pendulum is $\tau_z = -mgh \sin\theta = -mgL \sin\theta$. So the rod experiences half as much torque pulling it back toward equilibrium as does the simple pendulum. Why, then, does the rod oscillate more rapidly?

To see the explanation we also need to compare the moment of inertia of the rod ($I = mL^2/3$) to the moment of inertia of the simple pendulum ($I = mL^2$). So while the rod experiences half as much torque as the simple pendulum, the rod has only one-third as much inertia as does the simple pendulum. Hence the rod ends up oscillating faster.

BIO-Medical EXAMPLE 12-7 Moment of Inertia of a Human Leg

How you walk is affected in part by how your legs swing around the rotation axis created by your hip joints. Because your leg has a complex shape and contains bone, muscle, skin, and other materials, it would be very difficult to calculate the moment of inertia I of a leg around the hip. Instead I can be found experimentally by measuring the period of the leg when allowed to swing freely. In a clinical study a person's leg of length 0.88 m is estimated to have a mass of 6.5 kg and a center of mass 0.37 m from the rotation axis through the hip. When allowed to swing freely it oscillates with a period of 1.2 s. Determine the moment of inertia of the leg in rotation around the axis through the hip. Compare your answer to a uniform rod that has the same mass and length (see Example 12-6).

Set Up

The expression for the period T of a physical pendulum depends on the moment of inertia I for rotation around the pivot, the mass m, and the distance h from pivot to center of mass. We're given the values of $T = 1.2$ s, $m = 6.5$ kg, and $h = 0.37$ m, so we can solve for I. We'll then compare our result to a uniform rod with the same mass ($m = 6.5$ kg) and length ($L = 0.88$ m) as a leg.

Period of a physical pendulum:

$$T = 2\pi\sqrt{\frac{I}{mgh}} \quad \text{(from Equations 12-31)}$$

Moment of inertia for a uniform rod of mass m and length L rotating about one end:

$$I = \frac{1}{3}mL^2$$

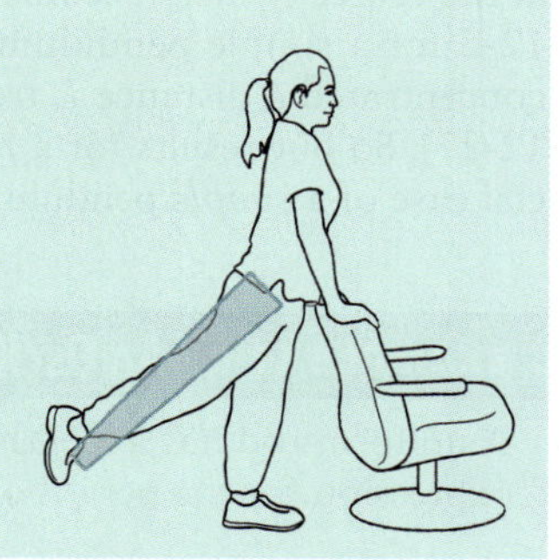

Solve

Rearrange the expression for period T to find a formula for the moment of inertia I, then substitute the known values.

Begin with the expression

$$T = 2\pi\sqrt{\frac{I}{mgh}}$$

Square both sides to get rid of the square root, then solve for I:

$$T^2 = \frac{4\pi^2 I}{mgh} \text{ so } I = \frac{mghT^2}{4\pi^2}$$

Substitute numerical values:

$$I = \frac{(6.5 \text{ kg})(9.80 \text{ m/s}^2)(0.37 \text{ m})(1.2 \text{ s})^2}{4\pi^2} = 0.86 \text{ kg} \cdot \text{m}^2$$

As a comparison, find the moment of inertia of a uniform rod with the same mass and length of the leg rotating about one end:

For a rod rotating around one end,

$$I = \frac{1}{3}mL^2 = \frac{1}{3}(6.5 \text{ kg})(0.88 \text{ m})^2 = 1.7 \text{ kg} \cdot \text{m}^2$$

Reflect

The moment of inertia of the uniform rod is *twice* as large as the moment of inertia of that leg. Is that reasonable? Consider that your upper leg (above the knee) is more massive than your lower leg, as suggested by the sketch above, so more of the leg's mass is above the knee than below the knee. By contrast, the mass of a uniform rod is distributed uniformly along its length. We learned in Section 8-3 that the farther an object's mass lies from the rotation axis, the greater the moment of inertia for that axis. So it's not surprising that the uniform rod, which has more of its mass farther from the rotation axis, has a greater moment of inertia than the leg.

GOT THE CONCEPT? 12-9 Period of a Rod

A uniform rod has the same length and mass as the leg in the previous example. When supported at one end and allowed to rotate, would the period of this rod be (a) greater than the period of the leg; (b) the same as the period of the leg; (c) less than the period of the leg; or (d) not enough information given to decide?

GOT THE CONCEPT? 12-10 Frequency of a Physical Pendulum

You measure the oscillation frequency of a physical pendulum when it is pivoted at one end. You then rearrange the experiment so that the physical pendulum is pivoted around its center of mass. Compared to the frequency when it is pivoted at one end, the frequency when it is pivoted at the center of mass is (a) greater; (b) the same; (c) less, but not zero; (d) zero; (e) not enough information given to decide. (*Hint:* Think about how the pendulum will move when supported at its center of mass.)

TAKE-HOME MESSAGE FOR Section 12-6

✔ A physical pendulum has its mass distributed throughout its volume. Unlike a simple pendulum, it cannot be treated as a point object suspended at the end of a string of negligible mass.

✔ The gravitational torque that arises when the pendulum is displaced from equilibrium acts as a restoring torque that causes the pendulum to return to its equilibrium position.

✔ The angular frequency, period, and frequency of a physical pendulum depend on its mass, the distance from the pivot point to the pendulum's center of mass, and the pendulum's moment of inertia around that point.

12-7 When damping is present, the amplitude of an oscillating system decreases over time

Stand up and let one arm hang limp by your side. With the other hand pull the limp arm forward then let it go. The limp arm will swing back and forth like a pendulum, but within a few swings the oscillations die out, and your arm is once again hanging by your side. In the same way the vibrations of your eardrum, in response to a source of sound, quickly die away when the source is cut off, as do the vibrations of a cymbal after it has been struck with a drumstick. In all of these cases the oscillations are diminished or *damped* by some kind of friction force. Let's examine the properties of these **damped oscillations**.

In Section 5-3 we learned about the kinetic friction force that arises when one object slides over another. In that section we assumed that the friction force did not depend on the speed of the sliding. For many oscillating systems, however, it turns out to be a better description to say that the force that causes damping *does* depend on speed. The simplest such damping force is one that is proportional to the speed of the oscillating object:

$$\vec{F}_{\text{damping}} = -b\vec{v} \tag{12-33}$$

The minus sign says that the damping force always opposes the motion of the object: If the object moves to the left, the damping force is to the right, and so on. The quantity b is called the **damping coefficient**. Its value depends on various physical characteristics of a particular system. For example, think of an oscillating object that's immersed in a fluid such as air, water, or oil. The value of b in this case depends on the viscosity of the fluid as well as on the size of the oscillating object. The units of force are newtons, where $1\ \text{N} = 1\ \text{kg} \cdot \text{m/s}^2$, and the units of velocity are m/s, so it follows that the units of the damping coefficient b are $\text{N/(m/s)} = (\text{kg} \cdot \text{m/s}^2)(\text{s/m}) = \text{kg/s}$.

Let's consider an object of mass m attached to an ideal, Hooke's law spring of spring constant k that oscillates along the x axis as in Figure 12-2. The force exerted by the spring has only an x component, equal to $-kx$. If the object is also subject to a damping force given by Equation 12-33, that force also has only an x component (because the velocity has only an x component): $F_{\text{damping},\,x} = -bv_x$. The net force on the object is the sum of the spring and damping forces, so Newton's second law for the object is

$$ma_x = -kx - bv_x \tag{12-34}$$

Equation 12-34 is a difficult equation to solve, so we'll just present the solution for x as a function of time. It turns out that the character of the solution depends on the size of the damping coefficient b and that there are three distinct solutions that depend on whether the damping coefficient is relatively small, equal to a certain critical value, or greater than that critical value.

Underdamped Oscillations

If b is relatively small (less than $2\sqrt{km}$), the system still oscillates but with ever-decreasing amplitude. (This is what you observed in the experiment we described at the beginning of this section about letting your arm swing.) This is called **underdamped oscillation.** If the amplitude when the oscillation begins at $t = 0$ is A, the amplitude at a later time t is

Amplitude of an underdamped oscillation (12-35)

$$A(t) = Ae^{-(b/2m)t}$$

The **amplitude** of an underdamped oscillator at time t after it is set in motion... ...equals the **initial amplitude** A at $t = 0$... ...multiplied by an **exponential function** that decreases with time. **Damping coefficient** · **Mass** of the oscillating object

Figure 12-16 is a graph of this function. The decreasing amplitude means that the oscillating object makes smaller and smaller excursions around the equilibrium position as time goes by.

The presence of the damping force also changes the oscillation angular frequency from the value $\omega = \sqrt{k/m}$ without damping. With damping present the angular frequency is

Angular frequency of an underdamped oscillation (12-36)

$$\omega_{\text{damped}} = \sqrt{\frac{k}{m} - \left(\frac{b}{2m}\right)^2}$$

Angular frequency of an underdamped mass–spring combination · **Spring constant** of the spring · **Damping coefficient** · **Mass** of the oscillating object

The presence of the term $(b/2m)^2$ inside the square root in Equation 12-36 means that the angular frequency is less with damping present than if there is no damping. The presence of the damping force acts to slow down the motion, so it makes sense that damping should decrease the angular frequency and so increase the oscillation period.

If we combine the information in Equations 12-35 and 12-36, we get the following expression for the displacement x as a function of time for a mass–spring combination that is underdamped, that is, that has a relatively small damping force:

$$x = Ae^{-(b/2m)t}\cos(\omega_{\text{damped}}t + \phi) \quad (12\text{-}37)$$

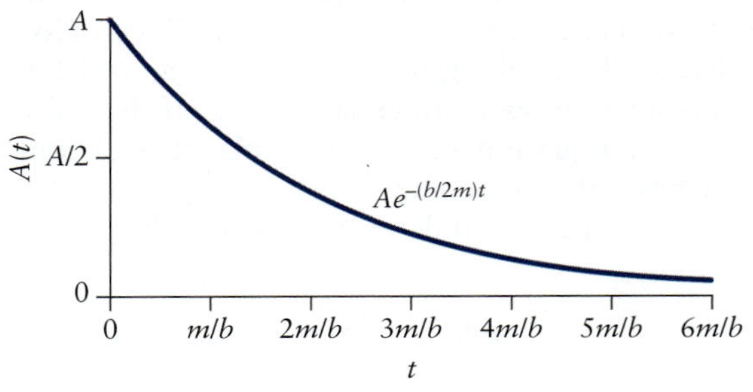

Figure 12-16 Amplitude of an underdamped oscillation If an oscillating system has relatively little damping, its oscillations are like simple harmonic motion but with an amplitude $A(t)$ that decreases with time.

The cosine function in Equation 12-37 tells us that an underdamped mass–spring combination oscillates with angular frequency ω_{damped}; the exponential function tells us that the amplitude of those oscillations decreases with time. Equation 12-35 tells us that the greater the value of the damping coefficient b, the more rapidly the amplitude decreases; Equation 12-36 tells us that a greater value of b makes the value of the angular frequency ω_{damped} smaller and so makes the oscillations happen more slowly (smaller frequency and greater period).

BIO-Medical **Figure 12-17** shows a real-life underdamped oscillation. The blue curve shows measurements of the displacement of a frog's eardrum after it is set into oscillation by a burst of sound. The red curve is an exponential function like that in Equations 12-35 and 12-37. There is a close match between the theoretical red curve and the peaks of the experimental blue curve, which shows that the amplitude decays just as we would expect in an underdamped oscillation. Indeed, Equation 12-37 turns out to be an accurate description of the shape of the blue curve.

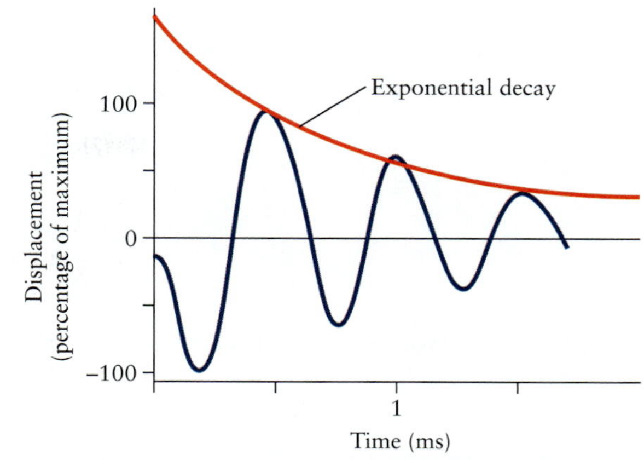

Figure 12-17 Damped oscillations of a frog's eardrum When stimulated by a burst of sound, the eardrum of a frog oscillates with decreasing amplitude (blue curve). The red curve is a decreasing exponential function like that shown in Figure 12-16.

BIO-Medical EXAMPLE 12-8 Oscillations in the Inner Ear

As we discussed in Section 3-8, special sensory cells in the human inner ear play a role in detecting acceleration and in enabling you to balance. In response to a prolonged stimulus, these cells send out an oscillating signal that can resemble the blue curve in **Figure 12-18**. The red dashed line follows Equation 12-35; for the data shown $b/2m = 42\ \text{s}^{-1}$ and the period of the oscillations is about 3.2×10^{-3} s. Calculate the time it takes for the amplitude of the oscillations to fall to 20% of the initial amplitude. Express the answer in seconds and as a multiple of the oscillation period.

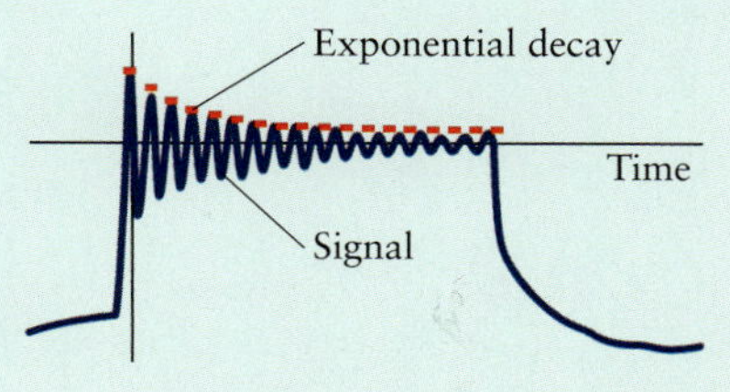

Figure 12-18 Damped oscillations in the inner ear The blue curve shows a signal sent out by sensory cells in the human inner ear. The red dashed line shows the decay in the amplitude of these signals.

Set Up

The signal oscillates with a decaying amplitude, so this is an underdamped oscillation. Our goal is to find the time t when $A(t)$ equals 0.20 times the initial amplitude A. We'll use Equation 12-35 for amplitude as a function of time.

Amplitude of an underdamped oscillation:

$$A(t) = Ae^{-(b/2m)t} \qquad (12\text{-}35)$$

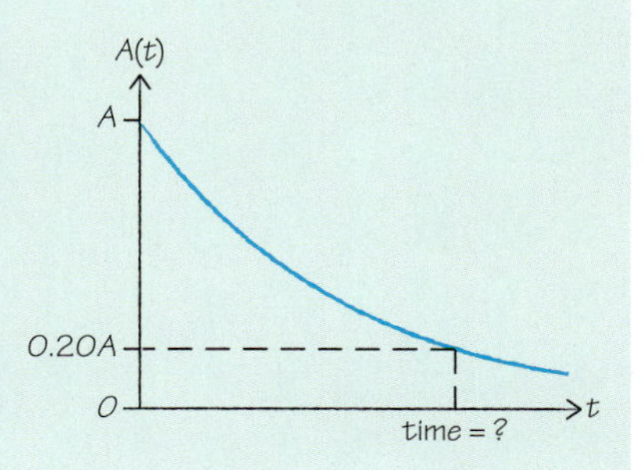

Solve

Set the time-dependent amplitude equal to 0.20 of the original amplitude and solve for t. To eliminate the exponential, use the properties of the natural logarithm function.

We want the value of t that satisfies the equation

$$A(t) = 0.20A = Ae^{-(b/2m)t}$$

Divide through by the initial amplitude A:

$$0.20 = e^{-(b/2m)t}$$

To get an expression for the time t, we need to "undo" the exponential function. We can do this using the natural logarithm function (ln), which has the property that

$$\ln e^x = x \text{ for any } x$$

With this in mind take the natural logarithm of both sides of the above equation:

$$\ln 0.20 = \ln e^{-(b/2m)t}, \text{ so}$$
$$\ln 0.20 = -(b/2m)t$$

Then solve for t:

$$t = -\frac{\ln 0.20}{(b/2m)} = -\frac{(-1.6)}{(42\ \text{s}^{-1})} = \frac{1.6}{42}\ \text{s} = 0.038\ \text{s}$$

Express the answer in terms of the period of the oscillation.

The period of the oscillation is 3.2×10^{-3} s, so we can write the time t as

$$t = (0.038\ \text{s}) \times \left(\frac{1\ \text{period}}{3.2 \times 10^{-3}\ \text{s}}\right) = 12\ \text{periods}$$

Reflect

Our results show that the time for the signal amplitude to decay to 0.20 of its initial value corresponds to about 12 complete oscillations. You can use a ruler to make measurements on Figure 12-18 to verify the height of the peak (that is, the amplitude) of the twelfth cycle in the data is close to 20% of the height of the first peak.

Note that the signal shown in Figure 12-18 is an *electrical* signal, not a mechanical one, yet it obeys the same rules as mechanical oscillations. This suggests that the key physics behind oscillation also applies to nonmechanical oscillations. In later chapters we will discover the tremendous importance of electrical oscillations for understanding how power is transmitted over wires and how radio communication works.

Critically Damped and Overdamped Oscillations

Equation 12-36 shows that the angular frequency of oscillation gets smaller and smaller as the damping coefficient b is increased. You can see that the angular frequency becomes *zero* if the quantity under the square root in Equation 12-36 becomes zero. Zero angular frequency means that there are *no* cycles of oscillation per second, which means that the system does not oscillate at all! In this situation we say that the oscillations are **critically damped**: When displaced from equilibrium, the system returns smoothly to equilibrium with *no* overshoot, so there is no oscillation. The suspension of an automobile is designed to give nearly critical damping. (The damping is provided by the shock absorbers.) If the suspension were underdamped, as can happen if the shock absorbers are worn out, an automobile that drove over a bump would oscillate up and down on its suspension several times before settling down, which would make for a very uncomfortable ride.

The value of b that corresponds to critical damping is the value that makes $\omega_{\text{damping}} = 0$ in Equation 12-36. This will happen if

$$\frac{k}{m} - \left(\frac{b}{2m}\right)^2 = 0 \quad \text{or} \quad \left(\frac{b}{2m}\right)^2 = \frac{k}{m}$$

To solve this equation for the value of b for which $\omega_{\text{damped}} = 0$, take the square root of both sides and multiply both sides by $2m$:

$$\frac{b}{2m} = \sqrt{\frac{k}{m}} \quad \text{and so} \quad b = 2m\sqrt{\frac{k}{m}}$$

Simplifying, we get

$$b = 2\sqrt{km} \text{ (condition for critical damping)} \tag{12-38}$$

As an example, in Examples 12-2, 12-3, and 12-4 (Section 12-3) we considered an object of mass 0.80 kg oscillating on a spring of spring constant 1.8×10^2 N/m. From Equation 12-38, this system will be critically damped if we add a damping force for which the damping coefficient b is

$$\begin{aligned} b &= 2\sqrt{km} = 2\sqrt{(1.8 \times 10^2\ \text{N/m})(0.80\ \text{kg})} \\ &= 2\sqrt{(1.8 \times 10^2\ \text{kg/s}^2)(0.80\ \text{kg})} \\ &= 24\ \text{kg/s} \end{aligned}$$

(We used the definition of the newton: $1\ \text{N} = 1\ \text{kg} \cdot \text{m/s}^2$, so $1\ \text{N/m} = 1\ \text{kg/s}^2$.)

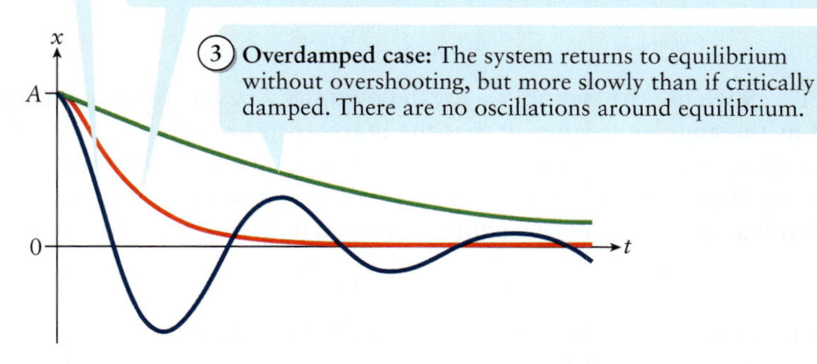

1 Underdamped case: The system oscillates around the equilibrium position, and the amplitude continually decreases.
2 Critically damped case: The system returns to equilibrium without overshooting. There are no oscillations around equilibrium.
3 Overdamped case: The system returns to equilibrium without overshooting, but more slowly than if critically damped. There are no oscillations around equilibrium.
x
A
0
t

Figure 12-19 Underdamped, critically damped, and overdamped oscillations The motion of a damped oscillating system depends on whether it is underdamped ($b < 2\sqrt{km}$), critically damped ($b = 2\sqrt{km}$), or overdamped ($b > 2\sqrt{km}$). In each case the system is displaced from equilibrium by a distance A and then released.

If b has a value greater than $2\sqrt{km}$, the oscillations are **overdamped**. If an overdamped system is displaced from equilibrium, it again returns to equilibrium without overshooting, so again there are no oscillations. But now the damping is so great that the system takes even longer to return to equilibrium than in the critically damped case. If a system is critically damped, it returns to equilibrium in the shortest possible time. **Figure 12-19** illustrates underdamped, critically damped, and overdamped oscillations.

Go to Picture It 12-2 for more practice dealing with damped oscillations.

GOT THE CONCEPT? 12-11 Underdamped, Critically Damped, or Overdamped?

A block is attached to an ideal spring. The system of block and spring is critically damped: When the block is displaced from equilibrium and released, the block returns smoothly to equilibrium in the minimum time possible and does not overshoot equilibrium. If you replace the block with a new one of twice the mass, but the damping coefficient and spring constant remain the same, what kind of oscillations will result? (a) Underdamped; (b) critically damped; (c) overdamped; (d) not enough information given to decide.

TAKE-HOME MESSAGE FOR Section 12-7

✔ A damping force is proportional to the velocity of an oscillating object and opposes its motion.

✔ If the damping is light so that the system is underdamped, the object oscillates in periodic motion with an amplitude that decreases with time. The frequency is less than if damping were absent.

✔ Heavier damping can cause critical damping or overdamping, in which case the system returns to equilibrium without overshooting.

12-8 Forcing a system to oscillate at the right frequency can cause resonance

As you did at the beginning of Section 12-7, stand up and let one arm hang limply by your side. With the other hand, again pull the limp arm forward and let it go, and note the frequency at which the arm tends to oscillate on its own. Now put the other hand in your pocket and swing the limp arm back and forth as you might do while walking. Try making it swing at a low frequency, a moderate frequency, and a high frequency. You'll find that it's easiest to make the arm swing at a frequency close to that at which it oscillates on its own, and that it's harder to make it swing at a slower frequency or higher frequency.

What you've just demonstrated are some of the properties of **forced oscillations**. In a forced oscillation a periodic driving force causes a system to oscillate at the frequency of that driving force. The experiment with your arm shows that it's easiest to force the system to oscillate if the *driving* frequency is near the system's *natural* frequency, which is the frequency at which it would oscillate on its own.

A periodic driving force can take many forms. In some cases it's a single push per cycle, such as what happens when a parent pushes a child on a swing. It turns out that the simplest case to analyze is a *sinusoidal* driving force. Let's see what happens when such a driving force acts on a damped mass–spring combination.

We saw in Section 12-7 that if an underdamped mass–spring combination is displaced from equilibrium, it oscillates with an ever-decreasing amplitude and an angular frequency

(12-39)
$$\omega_{\text{damped}} = \sqrt{\frac{k}{m} - \left(\frac{b}{2m}\right)^2} = \sqrt{\omega_0^2 - \left(\frac{b}{2m}\right)^2}$$

In Equation 12-39 we use the symbol ω_0 for the quantity $\sqrt{k/m}$. We recognize this from Section 12-3 as the angular frequency that the mass–spring combination would have if there were *no* damping (see the first of Equations 12-11). We'll call ω_0 the **natural angular frequency** for the combination of mass and spring. (The natural *frequency* measured in Hz is ω_0 divided by 2π.) Now let's apply a sinusoidal driving force to the mass:

(12-40)
$$F(t) = F_0 \cos \omega t$$

The angular frequency ω of the driving force, called the **driving angular frequency,** can have any value. As an example, consider the human eardrum, which behaves like a damped mass–spring system. Sound entering the ear acts as a driving force that makes the eardrum oscillate; the sound can have any of a wide range of frequencies.

With the driving force $F(t)$ from Equation 12-40 acting along with the force of the spring and the damping force, Newton's second law for the object of mass m attached to the spring is

(12-41)
$$ma_x = -kx - bv_x + F_0 \cos \omega t$$

As was the case for Equation 12-34, the equation for damped oscillations without a driving force that we presented in Section 12-7, solving Equation 12-41 is beyond our scope. But here's what the solutions are like: When the driving force is first applied, the motion is a complicated combination of an oscillation at the angular frequency ω_{damped} given by Equation 12-39 and an oscillation at the driving angular frequency ω. The oscillation at ω_{damped} quickly dies away (it's a damped oscillation), and the system is left oscillating at the driving angular frequency ω. The amplitude of these oscillations is

Amplitude of a damped, driven oscillation
(12-42)

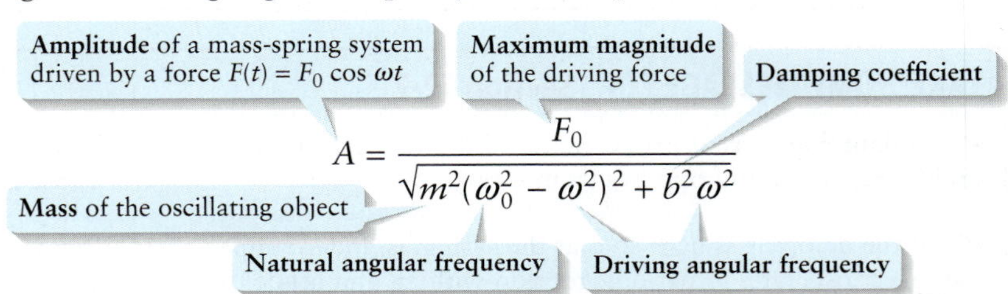

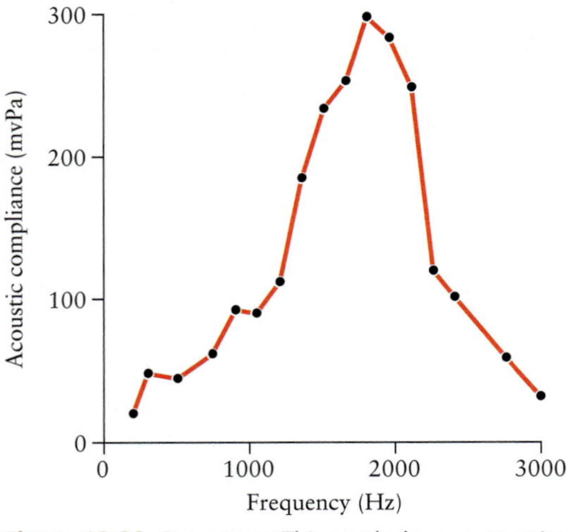

Figure 12-20 Resonance This graph shows a quantity related to the oscillation amplitude of a frog's eardrum (on the vertical axis) as a function of the frequency of sound waves causing the oscillation. The peak around 1800 Hz corresponds to the eardrum being forced at the eardrum's natural frequency of 1800 Hz.

Although we've discussed the case of an underdamped system only, Equation 12-42 turns out to be valid no matter how large the value of b.

Equation 12-42 shows that for a given driving force magnitude F_0, the oscillation amplitude A depends on the driving angular frequency ω. The amplitude is largest when the denominator in Equation 12-42 is smallest. That happens when b is small (so there is very little damping) and when ω is close to the natural angular frequency ω_0 so $\omega_0^2 - \omega^2$ is close to zero. In other words, a system oscillates with large amplitude if you drive at close to its natural angular frequency, and the amplitude is particularly large if the system is only lightly damped. This phenomenon is called **resonance.** **Figure 12-20** shows resonance in the eardrum of a frog.

Why does resonance happen? It turns out that if the driving angular frequency ω is equal to the natural angular frequency ω_0, the velocity v_x of the oscillating object is *in phase* with the driving force given by Equation 12-40. That is, the object is passing through equilibrium and has its maximum positive velocity v_x at the time when the force $F(t) = F_0 \cos \omega t$ has its maximum positive value; the object is at its maximum displacement and momentarily at rest ($v_x = 0$) when $F(t)$ equals zero; and so on. The net result is that when ω and ω_0 are equal, the driving force $F_0 \cos \omega t$ (which is in the same direction of the velocity) is the *negative* of the damping force

$-bv_x$ (which is always opposite to the velocity). So at this driving angular frequency the driving force *cancels* the damping force! In this case Equation 12-41 becomes

$$ma_x = -kx$$

That's the same equation obeyed by a mass attached to an ideal spring with *no* damping and *no* driving force. Such a system oscillates freely at angular frequency $\omega_0 = \sqrt{k/m}$.

Figure 12-21 shows a dramatic example of resonance. A crystal wine glass is placed directly in front of a loudspeaker, and the tone emitted by the speaker is tuned—that is, its angular frequency ω is adjusted—to match the natural angular frequency ω_0 of the glass. The oscillations of the glass have little damping, as you can verify by holding such a glass by the stem and tapping the glass with a fingernail: The glass rings for quite a long time, showing that the oscillations are only lightly damped by the internal friction of the glass and so the damping coefficient b is small. As a result the amplitude of the oscillations when driven at the natural angular frequency is very large—so large that the glass literally tears itself apart. (This doesn't work if you use an ordinary drinking glass. Even if the loudspeaker is tuned to the natural angular frequency of the glass, the larger internal friction makes b greater than for a crystal wine glass, and the amplitude of oscillation of the driven drinking glass isn't large enough to break the glass.)

Joe Kirby/Alamy Stock Photo

Figure 12-21 Destructive resonance When forced to oscillate at its natural angular frequency, a crystal wine glass can vibrate with such a large amplitude that it shatters.

BIO-Medical One example of resonance in nature is the flight of flies, wasps, and bees. These insects use resonance to help them beat their wings. Their bodies are equipped with special flight muscle cells that can apply cyclic forces at frequencies as high as 1000 Hz or more. These high frequencies correspond to the natural frequencies of the system of an insect's thorax and wings. The special flight muscle cells force the wings of the insect to oscillate at these natural frequencies. The result is that the wings oscillate with large amplitude A even though the force exerted by the flight muscle cells has a small magnitude F_0 (see Equation 12-42).

BIO-Medical GOT THE CONCEPT? 12-12 Mosquitoes Listening for Each Other

? The sounds made by male mosquitoes in flight are close to 650 Hz, while those made by females are close to 400 Hz. The antennae of mosquitoes differ from male to female (**Figure 12-22**): The natural frequency of a male's antennae is about 400 Hz, while that of the female's is about 200 Hz. Biologists suspect that some mosquitoes can detect the presence of others based on nerve signals generated when their antennae vibrate. Would you expect that (a) male mosquitoes can detect other males, (b) females can detect other females, (c) males can detect females, (d) females can detect males; (e) more than one of these, or (f) none of these?

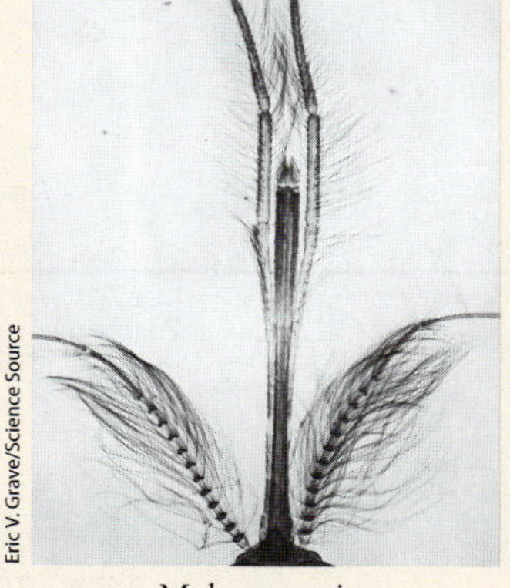

Eric V. Grave/Science Source

Male mosquito

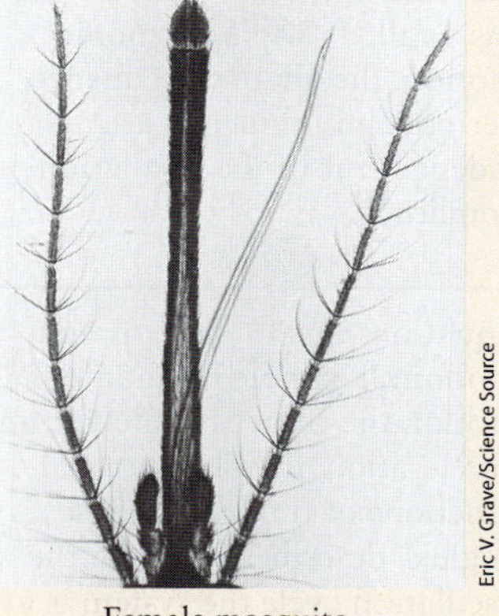

Eric V. Grave/Science Source

Female mosquito

Figure 12-22 Mosquito antennae Mosquitoes may be able to use their antennae to detect oscillations produced by other mosquitoes. What does the physics of resonance tell us about this?

TAKE-HOME MESSAGE FOR Section 12-8

✔ An oscillating system can be forced to oscillate at any frequency.

✔ When the driving angular frequency is the same as the natural angular frequency of the system, the amplitude of oscillations becomes large, and especially large if the damping force is small. This is called resonance.

Key Terms

amplitude
angular frequency
critically damped oscillations
damped oscillations
damping coefficient
driving angular frequency
forced oscillations
frequency
harmonic property
hertz
Hooke's law
natural angular frequency
oscillation
overdamped oscillations
pendulum

period
phase angle
physical pendulum
resonance
restoring force
simple harmonic motion (SHM)
simple pendulum
sinusoidal function
underdamped oscillation

Chapter Summary

Topic	Equation or Figure
Oscillations: A system will oscillate if there is (a) a restoring force that always pulls the system back toward equilibrium and (b) inertia that causes the system to overshoot equilibrium. The period T is the time for one oscillation cycle; the frequency $f = 1/T$ is the number of cycles per second (measured in Hz); the angular frequency ω equals $2\pi f$; and the amplitude A is the maximum displacement from equilibrium.	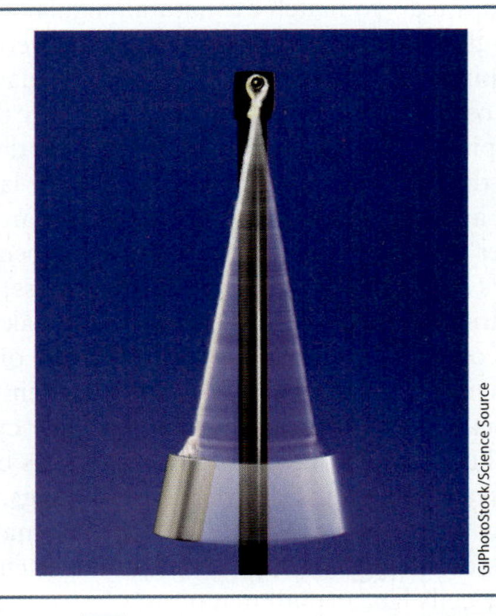 GIPhotoStock/Science Source (Figure 12-12b)
Simple harmonic motion: Oscillations are particularly simple if the restoring force obeys Hooke's law, $F_x = -kx$. In this case, called simple harmonic motion, the angular frequency, period, and frequency are independent of the oscillation amplitude.	**Angular frequency** $\omega = \sqrt{\frac{k}{m}}$ (k = **spring constant** of the spring; m = **mass** of the object connected to the spring) **Period** $T = \frac{2\pi}{\omega} = 2\pi\sqrt{\frac{m}{k}}$ **Frequency** $f = \frac{1}{T} = \frac{1}{2\pi}\sqrt{\frac{k}{m}} = \frac{\omega}{2\pi}$ (12-11)
Equations of simple harmonic motion: In simple harmonic motion the position, velocity, and acceleration are all sinusoidal functions of time. The phase angle ϕ describes where in the oscillation cycle the system is at $t = 0$. We can understand this by comparing uniform circular motion to the motion of a block attached to an ideal spring.	$x = A\cos(\omega t + \phi)$ (12-12) **Position** of an object in SHM at time t; **Amplitude** of the oscillation = maximum displacement from the equilibrium position; **Angular frequency** of the oscillation ($\omega = 2\pi f = 2\pi/T$); **Phase angle**
	$v_x = -\omega A\sin(\omega t + \phi)$ (12-14) **Velocity** of an object in SHM at time t; **Amplitude** of the oscillation = maximum displacement from the equilibrium position; **Angular frequency** of the oscillation ($\omega = 2\pi f = 2\pi/T$); **Phase angle**
	$a_x = -\omega^2 A\cos(\omega t + \phi)$ (12-15) **Acceleration** of an object in SHM at time t; **Amplitude** of the oscillation = maximum displacement from the equilibrium position; **Angular frequency** of the oscillation ($\omega = 2\pi f = 2\pi/T$); **Phase angle**

Energy in simple harmonic motion: If there is no friction, the total mechanical energy of a system in simple harmonic motion is conserved. At equilibrium the kinetic energy is maximum, and the spring potential energy is zero; at the extremes of the motion, the kinetic energy is zero, and the spring potential energy is maximum.

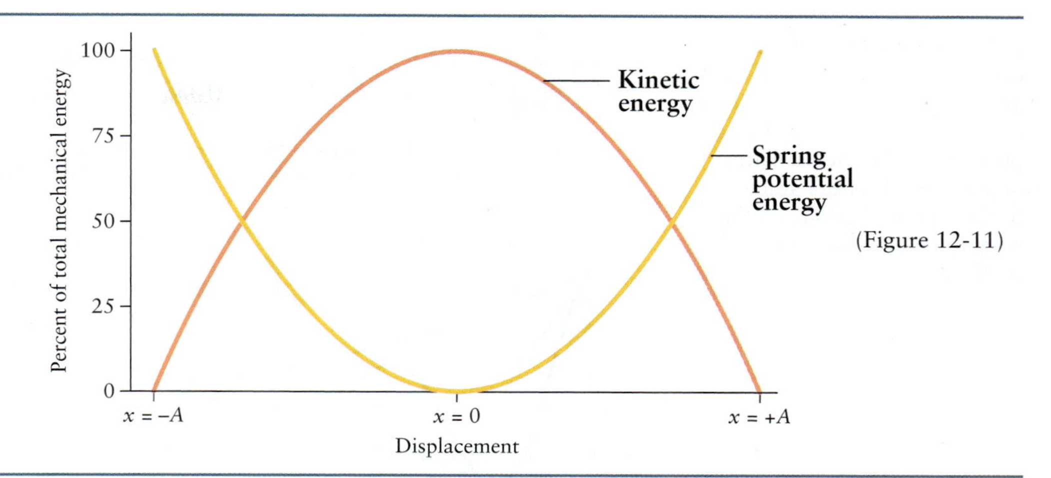

Percent of total mechanical energy
100
75
50
25
0
Kinetic energy
Spring potential energy
$x = -A$
$x = 0$
$x = +A$
Displacement

(Figure 12-11)

The simple pendulum: The oscillations of a simple pendulum are simple harmonic motion if the amplitude of oscillation is small. In this case the angular frequency of oscillation of a simple pendulum of length L is $\omega = \sqrt{g/L}$; this does not depend on the mass m.

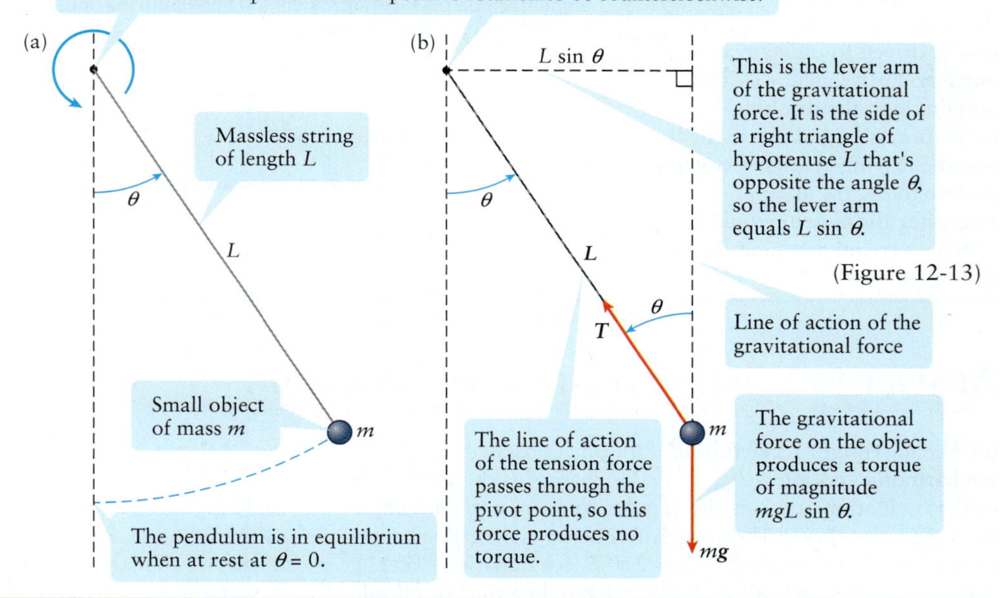

Pivot point: The positive z axis (the rotation axis of the pendulum) points outward from this point. We take positive rotation to be counterclockwise.
(a)
(b)
Massless string of length L
θ
L
Small object of mass m
m
The pendulum is in equilibrium when at rest at $\theta = 0$.
$L \sin \theta$
This is the lever arm of the gravitational force. It is the side of a right triangle of hypotenuse L that's opposite the angle θ, so the lever arm equals $L \sin \theta$.
T
Line of action of the gravitational force
The line of action of the tension force passes through the pivot point, so this force produces no torque.
The gravitational force on the object produces a torque of magnitude $mgL \sin \theta$.
mg

(Figure 12-13)

The physical pendulum: A pendulum of arbitrary shape also oscillates in simple harmonic motion if the amplitude is small. The angular frequency of oscillation in this case is $\omega = \sqrt{mgh/I}$ for a physical pendulum with mass m whose center of mass is a distance h from the pivot. The pendulum's moment of inertia around the pivot is I.

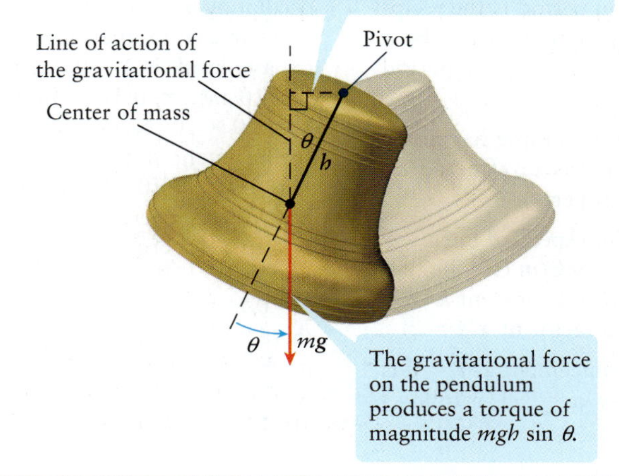

The z axis points perpendicular to the plane of the pendulum's rotation.
This is the lever arm of the gravitational force. It is the side of a right triangle of hypotenuse h that's opposite the angle θ, so the lever arm equals $h \sin \theta$.
Line of action of the gravitational force
Pivot
Center of mass
θ
h
mg
The gravitational force on the pendulum produces a torque of magnitude $mgh \sin \theta$.

(Figure 12-15)

Damped oscillations: In most cases a damping force acts on an oscillating system. This force can be approximated as being proportional to the oscillating object's speed. The character of the motion depends on the magnitude of the damping coefficient b.

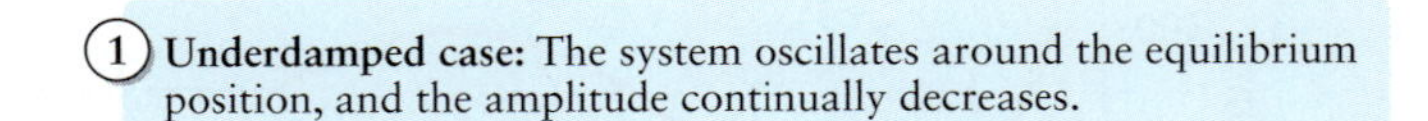

1 Underdamped case: The system oscillates around the equilibrium position, and the amplitude continually decreases.

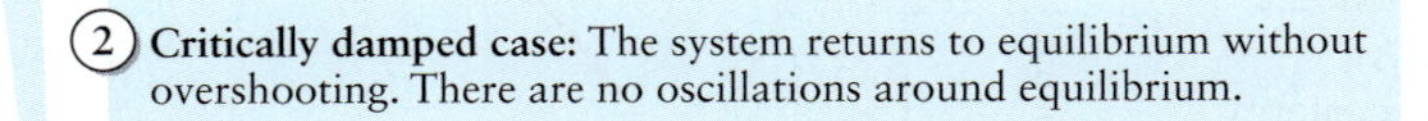

2 Critically damped case: The system returns to equilibrium without overshooting. There are no oscillations around equilibrium.

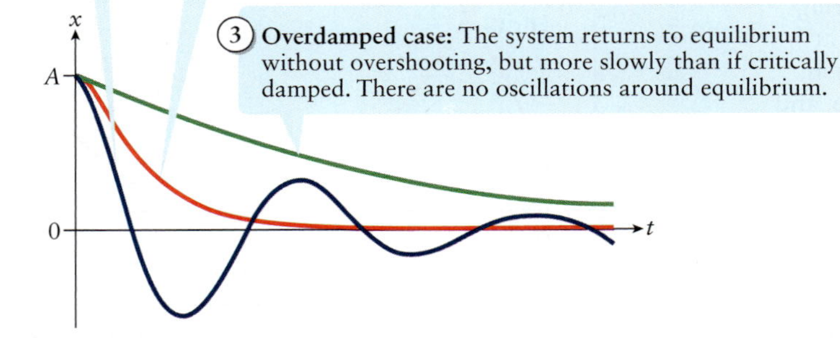

3 Overdamped case: The system returns to equilibrium without overshooting, but more slowly than if critically damped. There are no oscillations around equilibrium.
x
A
0
t

(Figure 12-19)

Driven oscillations: Even if damping is present, a system can be driven to oscillate at any angular frequency ω. The amplitude of the resulting oscillation is largest if b is small and the driving angular frequency is close to ω_0, the angular frequency the system would have if there were no damping. This is called resonance.

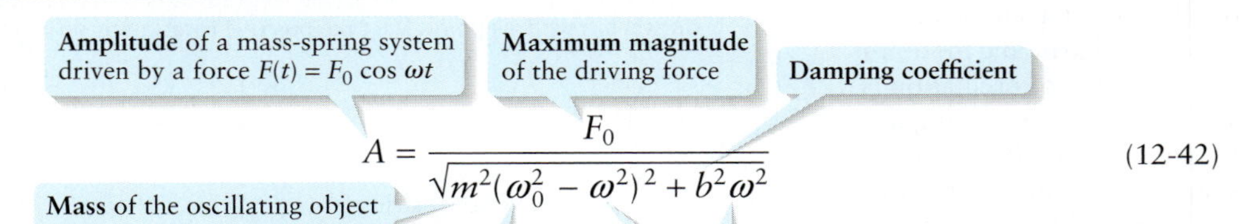

$$A = \frac{F_0}{\sqrt{m^2(\omega_0^2 - \omega^2)^2 + b^2\omega^2}} \quad (12\text{-}42)$$

Answer to What do you think? Question

(e) The maximum acceleration of an object oscillating in simple harmonic motion is $a_{max} = \omega^2 A$, where A is the amplitude of the oscillation and ω is the angular frequency (see Example 12-3 in Section 12-3). The angular frequency is in turn inversely proportional to the period T: $\omega = 2\pi/T$ (see Equations 12-11). The two oscillations in the problem have the same amplitude $A = 2.0$ mm, but the second one has one-half the period ($T = 0.40$ s versus $T = 0.80$ s). Because ω is inversely proportional to T, the second oscillation has double the angular frequency of the first one. Since a_{max} is proportional to the square of ω, it follows that the second oscillation has $2^2 =$ four times the maximum acceleration of the first one.

Answers to Got the Concept? Questions

12-1 (d) The number of beats per unit time is just the oscillation frequency of the heart. The oscillation period is the reciprocal of the frequency, so when the frequency increases by a factor of 3 (from 50 beats/min to 150 beats/min) the period decreases to 1/3 of its original value.

12-2 (e) This oscillation is an example of simple harmonic motion, which means that the frequency is unaffected by changes in the amplitude (the harmonic property).

12-3 (b) From Example 12-3 the maximum speed of the block is $v_{max} = \omega A$. The angular frequency is $\omega = \sqrt{k/m}$ (see the first of Equations 12-11). So increasing the spring constant k by a factor of 4 increases ω and $v_{max} = \omega A$ by a factor of $\sqrt{4} = 2$.

12-4 (d) From Equation 12-12, the displacement at $t = 0$ is $x(0) = A \cos(0 + \phi) = A \cos \pi$. Since $\cos \pi = -1$, it follows that $x(0) = -A$.

12-5 (b), (d), (c), (a) The total mechanical energy of a block oscillating on an ideal spring is $E = K + U_{spring} = \frac{1}{2}mv^2 + \frac{1}{2}kx^2 = \frac{1}{2}kA^2$. In case (a) the spring constant is $k = 5.0 \times 10^2$ N/m, and the amplitude is $A = 0.020$ m, so $E = \frac{1}{2}kA^2 = \frac{1}{2}(5.0 \times 10^2 \text{ N/m})(0.020 \text{ m})^2 = 0.10$ J. In case (b) the mass is $m = 0.60$ kg, $k = 3.0 \times 10^2$ N/m, and $v = 1.0$ m/s at $x = 0$. At $x = 0$ the spring potential energy is zero, so the energy is purely kinetic: $E = K = \frac{1}{2}mv^2 = \frac{1}{2}(0.60 \text{ kg})(1.0 \text{ m/s})^2 = 0.30$ J. In case (c) there is both kinetic energy ($m = 1.2$ kg and $v = 0.50$ m/s) and spring potential energy ($k = 4.0 \times 10^2$ N/m and $x = -0.010$ m). So $E = K + U_{spring} = \frac{1}{2}mv^2 + \frac{1}{2}kx^2 = \frac{1}{2}(1.2 \text{ kg})(0.50 \text{ m/s})^2 + \frac{1}{2}(4.0 \times 10^2 \text{ N/m})(-0.010 \text{ m})^2 = 0.17$ J. Similarly, in case (d) there is both kinetic energy ($m = 2.0$ kg and $v = 0.20$ m/s) and potential energy ($k = 2.0 \times 10^2$ N/m and $x = 0.050$ m), so $E = K + U_{spring} = \frac{1}{2}mv^2 + \frac{1}{2}kx^2 = \frac{1}{2}(2.0 \text{ kg})(0.20 \text{ m/s})^2 + \frac{1}{2}(2.0 \times 10^2 \text{ N/m})(0.050 \text{ m})^2 = 0.29$ J.

12-6 (d) The last of Equations 12-27 tells us that $f = (1/2\pi)\sqrt{g/L}$, so the frequency is inversely proportional to the square root of the length L of the pendulum. If L is increased by a factor of 4, the frequency changes by a factor of $1/\sqrt{4} = 1/2$. So the frequency is halved from 0.60 Hz to 0.30 Hz when the length of the pendulum increases by a factor of 4.

12-7 (c) The frequency $f = (1/2\pi)\sqrt{g/L}$ is independent of the mass of the pendulum and independent of the amplitude. So the frequency of the new pendulum is the same as before: $f =$ 0.60 Hz.

12-8 (d) The frequency $f = (1/2\pi)\sqrt{g/L}$ is proportional to the square root of g, the acceleration due to gravity. The value of g on the asteroid is 1/100 as great as on Earth, so the frequency of oscillation there is different by a factor of $\sqrt{1/100} = 1/10$. So the new frequency on the surface of the asteroid is $(1/10)(0.60 \text{ Hz}) = 0.060$ Hz.

12-9 (a) We know from Equations 12-31 that the period of a physical pendulum is $T = 2\pi\sqrt{I/mgh}$, so it is proportional to the square root of the ratio I/h. For a uniform rod of length 0.88 m (the same as the length of the leg), the center of mass is at the center of the rod. So the distance from the pivot point at one end of the rod to the center of mass is one-half of 0.88 m, or 0.44 m. For the leg the corresponding distance is 0.37 m, so h is greater for the rod by a factor of $(0.44 \text{ m})/(0.37 \text{ m}) = 1.2$. In Example 12-7 we found that the moment of inertia I of the rod is twice as large as that for the leg when both rotate around one end. So the period of the rod is greater than that of the leg by a factor of $\sqrt{(2/1.2)} = 1.3$. The effect of the greater moment of inertia overwhelms the effect of the greater distance h, so the rod has a greater period of oscillation and takes longer to complete one back-and-forth swing.

12-10 (d) A physical pendulum supported at its center of mass is in equilibrium no matter what its orientation. That's because the gravitational force acts at the support point, so there is no net torque around that point. (In terms of Equations 12-31, that means $h = 0$: There is zero distance from the pivot point to the center of mass.) Hence the pendulum does not oscillate at all. It goes through *zero* cycles per second, and the frequency is zero.

12-11 (a) From Equation 12-38 the condition for critical damping is $b = 2\sqrt{km}$. This condition was satisfied for the original block. With the new block the mass m is twice as great, so the quantity $2\sqrt{km}$ has increased by a factor of $\sqrt{2}$. However, the damping coefficient still has the same value, so now b is less than $2\sqrt{km}$ and the damping is less than the critical value. So the oscillations are now underdamped.

12-12 (c) The frequency of sound made by the female in flight is well tuned to the natural frequency of a male's antennae, so his antennae will exhibit large-amplitude vibrations (resonance) in the presence of a female mosquito. This matching of natural frequency of antennae and frequency of flight sounds does not occur between pairs of male or pairs of female mosquitoes.

Questions and Problems

In a few problems you are given more data than you actually need; in a few other problems you are required to supply data from your general knowledge, outside sources, or informed estimate. Interpret as significant all digits in numerical values that have trailing zeros and no decimal points.

For all problems use $g = 9.80 \text{ m/s}^2$ for the free-fall acceleration due to gravity. Neglect friction and air resistance unless instructed to do otherwise.

- • Basic, single-concept problem
- •• Intermediate-level problem, may require synthesis of concepts and multiple steps
- ••• Challenging problem

SSM *Solution is in Student Solutions Manual*

Example *See worked example for a similar problem*

Conceptual Questions

1. • One fundamental premise of simple harmonic motion is that a force must be proportional to an object's displacement. Is anything else required?

2. • List several examples of simple harmonic motion that you have observed in everyday life.

3. • Not all oscillatory motion is simple harmonic, but simple harmonic motion is always oscillatory. Explain this statement and give an example to support your explanation. SSM

4. • If the rise and fall of your lungs is considered to be simple harmonic motion, how would you relate the period of the motion to your breathing rate (breaths per minute)? SSM

5. • (a) What are the units of ω? (b) What are the units of ωt?

6. • Explain how *either* a cosine or a sine function can describe simple harmonic motion.

7. • Compare $x(t) = A \cos(\omega t)$ to $x(t) = A \cos(\omega t + \phi)$. What is the phase angle ϕ, and how does it change the simple harmonic motion?

8. • Why is mechanical energy always conserved in an ideal simple harmonic oscillator?

9. • In simple harmonic motion what phase angle separates the kinetic and potential energy?

10. • Galileo was one of the first scientists to observe that the period of a simple harmonic oscillator is independent of its amplitude. Explain what it means that the period is independent of the amplitude. Be sure to mention how the requirement that simple harmonic motion undergo small oscillations is affected by this supposition.

11. • Explain the difference between a simple pendulum and a physical pendulum.

12. • What are three factors that can help you distinguish between a simple pendulum and a physical pendulum?

13. • Explain how you could do an experiment to measure the elevation of your location through the use of a simple pendulum. SSM

14. • In the case of the damped harmonic oscillator, what are the units of the damping constant, b?

15. • Explain the difference between the frequency of the driving force and the natural frequency of an oscillator. SSM

16. • The application of an external force other than gravity on a simple pendulum can create many different outcomes, depending on how frequently the force is applied. Explain what will happen to the amplitude of the motion if an external

force is applied to a simple pendulum at the same frequency as the natural frequency of the pendulum.

Multiple-Choice Questions

17. • A block is attached to a horizontal spring, the spring is stretched so the block is located at $x = A$, and the spring is released. At what point in the resulting simple harmonic motion is the speed of the block at its maximum?
   A. $x = A$ and $x = -A$
   B. $x = 0$
   C. $x = 0$ and $x = A$
   D. $x = 0$ and $x = -A$
   E. $x = 0, x = -A$, and $x = A$ SSM

18. • A block is attached to a horizontal spring, the spring is stretched so the block is located at $x = A$, and the spring is released. At what point in the resulting simple harmonic motion is the magnitude of the acceleration of the block at its maximum?
   A. $x = A$ and $x = -A$
   B. $x = 0$
   C. $x = 0$ and $x = A$
   D. $x = 0$ and $x = -A$
   E. $x = 0, x = -A$, and $x = A$

19. • A small object is attached to a horizontal spring, pushed to position $x = -A$, and released. In one full cycle of its motion, the total distance traveled by the object is
   A. $A/2$.
   B. $A/4$.
   C. A.
   D. $2A$.
   E. $4A$.

20. • A small object is attached to a horizontal spring and set in simple harmonic motion with amplitude A and period T. How long does it take for the object to travel a total distance of $6A$?
   A. $T/2$
   B. $3T/4$
   C. T
   D. $3T/2$
   E. $2T$

21. • An object–spring system undergoes simple harmonic motion. If the mass of the object is doubled, what will happen to the period of the motion?
   A. The period will increase.
   B. The period will decrease by an unknown amount.
   C. The period will not change.
   D. The period will decrease by a factor of 2.
   E. The period will decrease by a factor of 4. SSM

22. • An object–spring system undergoes simple harmonic motion. If the amplitude increases but the mass of the object is not changed, the total energy of the system
   A. increases.
   B. decreases.
   C. doesn't change.
   D. undergoes a sinusoidal change.
   E. decreases exponentially.

23. • You can double the maximum speed of a simple harmonic oscillator by
   A. doubling the amplitude.
   B. reducing the mass to one-fourth its original value.
   C. increasing the spring constant to four times its original value.
   D. all of the above.
   E. none of the above.

24. • Replacing an object on a spring with an object having one-quarter the original mass will have the result of changing the frequency of the vibrating spring by a factor of
   A. $\frac{1}{4}$.
   B. $\frac{1}{2}$.
   C. 1 (no change).
   D. 2.
   E. 4.

25. • A uniform rod of length L hangs from one end and oscillates with a small amplitude. The moment of inertia for a rod rotating about one end is $I = \frac{1}{3}ML^2$. What is the period of the rod's oscillation?
   A. $2\pi\sqrt{\frac{L}{g}}$
   B. $2\pi\sqrt{\frac{2L}{3g}}$
   C. $2\pi\sqrt{\frac{L}{2g}}$
   D. $2\pi\sqrt{\frac{L}{3g}}$
   E. $2\pi\sqrt{\frac{L}{6g}}$

26. • If the period of a simple pendulum is T and we increase its length so that it's four times longer, what will the new period be?
   A. $T/2$
   B. T
   C. $2T$
   D. $4T$
   E. It is unchanged.

Estimation/Numerical Analysis

27. • (a) Plot a graph of the cosine function, cos (x), using a graphing calculator or software program. (b) How does the plot change when a phase angle of 30° is introduced, that is, cos $(x + 30°)$?

28. • Estimate the period of motion for your car's windshield wipers on (a) the low setting and (b) the high setting.

29. • Estimate the period of motion for a child swinging on a playground swing. SSM

30. • Estimate the period for normal, human eye blinking.

31. • Estimate the frequency of a hummingbird's wings.

32. • Estimate the frequency of a dog's panting on a hot summer day.

33. • Estimate the period of motion for a yo-yo that has been let down all the way and swings in simple harmonic motion.

34. • Estimate the frequency of a copy machine making (a) one-sided copies and (b) two-sided copies.

35. • Estimate the effect on the frequency of a grandfather clock pendulum if a 0.01-g fly lands on the very end of the pendulum's bob. SSM

36. • Estimate the energy (both potential and kinetic) associated with the pendulum of a ticking grandfather clock.

37. •• Estimate the period of a kitchen broom allowed to swing in small oscillations while suspended from the end of the handle.

38. • A force is measured with a force sensor at the times listed in the accompanying table. (a) Make a plot of force versus time and determine if the force can be modeled using a single sine or cosine function. (b) Can a period be identified for this force? If so, what is it?

t (s)	F (N)	t (s)	F (N)
0	−20	1.3	+10
0.1	−10	1.4	0
0.2	0	1.5	−10
0.3	+10	1.6	−20
0.4	+20	1.7	−10
0.5	+10	1.8	0
0.6	0	1.9	+10
0.7	−10	2.0	+20
0.8	−20	2.1	+10
0.9	−10	2.2	0
1.0	0	2.3	−10
1.1	+10	2.4	−20
1.2	+20	2.5	−10

39. •• The "ENSO" cycle (El Niño/Southern Oscillation) describes how the subsurface temperature of the Pacific Ocean changes. **Figure 12-23** indicates cold episodes with blue shading and warm episodes with red shading for one region of the Pacific. Estimate the periodic pattern for the effect. SSM

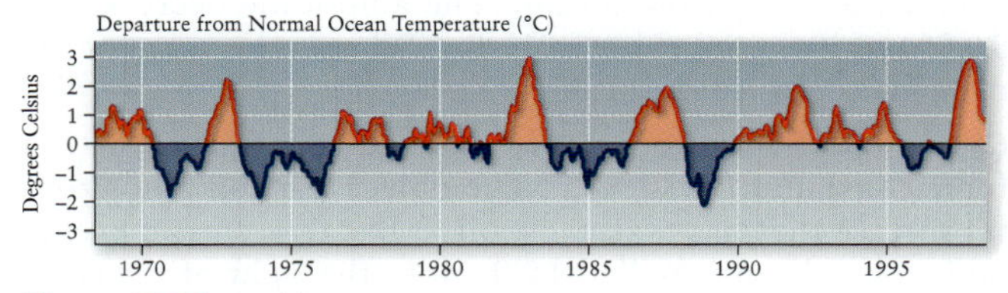

Figure 12-23 Problem 39

40. • A simple pendulum that has a length of 1.25 m and a bob with a mass of 4.0 kg is pulled an angle θ from its natural equilibrium position (hanging straight down). Use a spreadsheet or programmable calculator to calculate the maximum potential energy of the Earth–pendulum system and complete the table.

Maximum Angle from Vertical	Maximum Potential Energy of Bob (J)	Maximum Speed of Bob (m/s)
5°		
10°		
15°		
20°		
25°		

Problems

12-2 Oscillations are caused by the interplay between a restoring force and inertia

41. • The period of a simple harmonic oscillator is 0.0125 s. What is its frequency? Example 12-1

42. • A simple harmonic oscillator completes 1250 cycles in 20.0 min. Calculate (a) the period and (b) the frequency of the motion. Example 12-1

43. • An object on the end of a spring oscillates with a frequency of 15 Hz. Calculate (a) the period of the motion and (b) the number of oscillations that the object undergoes in 120 s. SSM Example 12-1

12-3 The simplest form of oscillation occurs when the restoring force obeys Hooke's law

44. •• High tide occurs at 8:00 A.M. and is 1 m above sea level. Six hours later, low tide is 1 m below sea level. After another 6 h high tide occurs (again 1 m above sea level) then finally one last low tide (6 h later, 1 m below sea level). (a) Write a mathematical expression that would predict the level of the ocean at this beach at any time of day. (b) Find the times in the day when the ocean level is exactly at sea level. Example 12-3

45. • A spring of unstretched length L and spring constant k is attached to a wall and an object of mass M resting on a frictionless surface (**Figure 12-24**). The object is pulled such that the spring is stretched a distance A then released. Let the $+x$ direction be to the right. (a) What is the position function $x(t)$ for the object? Take $x = 0$ to be the position of the object when the spring is relaxed. (b) What is the *velocity* of the object at $t = (7/6)T$, where T is the period? (c) What is the acceleration of the object at $t = T/4$? Example 12-3

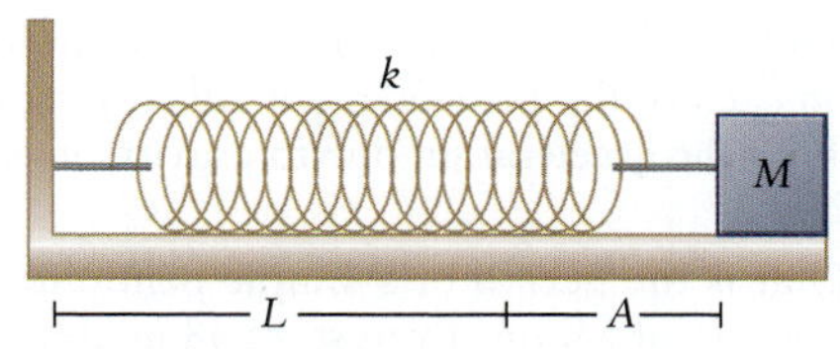

Figure 12-24 Problem 45

46. • What is the mass of an object that is attached to a spring with a force constant of 2.00×10^2 N/m if 14 oscillations occur each 16 s? Example 12-2

47. • A 0.200-kg object is attached to the end of a 55.0 N/m spring. It is displaced 10.0 cm to the right of equilibrium and released on a horizontal, frictionless surface. Calculate the period of the motion. Example 12-2

12-4 Mechanical energy is conserved in simple harmonic motion

48. • An object attached to the end of a spring slides on a horizontal frictionless surface with simple harmonic motion. Determine the location at which the object's kinetic energy and potential energy are the same. Assume the maximum displacement of the object is A so that it oscillates between $+A$ and $-A$. Example 12-4

49. •• A 0.250-kg object attached to a spring oscillates on a frictionless horizontal table with a frequency of 4.00 Hz and an amplitude of 20.0 cm. Calculate (a) the maximum potential energy of the system, (b) the displacement of the object when the potential energy is one-half of the maximum, and (c) the potential energy when the displacement is 10.0 cm. SSM Example 12-4

50. •• The potential energy of an object on a spring is 2.4 J at a location where the kinetic energy is 1.6 J. If the amplitude of the simple harmonic motion is 20.0 cm, (a) calculate the spring constant and (b) find the largest force that the object experiences. Example 12-4

51. •• The potential energy of a simple harmonic oscillator is given by $U = \frac{1}{2}kx^2$. (a) If $x(t) = A\cos(\omega t)$, plot the potential energy versus time for three full periods of motion. (b) Write the expression for the velocity, $v(t)$, and (c) add the plot of the kinetic energy, $K = \frac{1}{2}mv^2$, to your graph. SSM Example 12-4

12-5 The motion of a pendulum is approximately simple harmonic

52. • A simple pendulum on the surface of Earth is 1.24 m long. What is the period of its oscillation? Example 12-5

53. • The period of a simple pendulum on the surface of Earth is 2.25 s. Determine its length. Example 12-5

54. • In 1851 Jean Bernard Léon Foucault suspended a pendulum (later named the Foucault pendulum) from the dome of the Panthéon in Paris. The mass of the pendulum was 28.00 kg and the length of the rope was 67.00 m. The acceleration due to gravity in Paris is 9.809 m/s^2. Calculate the period of the pendulum. Example 12-5

55. • Geoff counts the number of oscillations of a simple pendulum at a location where the acceleration due to gravity is 9.80 m/s^2, and finds that it takes 25.0 s for 14 complete cycles. Calculate the length of the pendulum. SSM Example 12-5

56. • **Astronomy** What is the period of a 1.00-m-long simple pendulum on each of the planets in our solar system? You will need to look up the acceleration due to gravity on each planet. Example 12-5

57. • (a) What is the period of a simple pendulum of length 1.000 m at the top of Mount Everest, 8848 m above sea level? (b) Express your answer as a number times T_0, the period at sea level where h equals 0. The acceleration due to gravity in terms of elevation is $g = g_0\left(\frac{R_E}{R_E + h}\right)^2$ where g_0 is the average acceleration due to gravity at sea level, R_E is Earth's radius, and h is the elevation above sea level. Take g_0 to be 9.800 m/s^2 and Earth's radius R_E to be 6.380×10^6 m. Example 12-5

58. • A simple pendulum of length 0.350 m starts from rest at a maximum displacement of 10.0° from the equilibrium position on the surface of Earth. (a) At what time will the pendulum be located at an angle of displacement of 8.00°? (b) What about 5.00°? (c) When will the pendulum return to its starting position? Example 12-5

59. •• A simple pendulum oscillates between ±8.0° (as measured from the vertical) on the surface of Earth. The length of the pendulum is 0.50 m. Compare the time intervals between ±8.0° and ±4.0°. SSM Example 12-5

12-6 A physical pendulum has its mass distributed over its volume

60. • A pendulum made of a uniform rod with a length of 30.0 cm is set into harmonic motion about one end, on the surface of Earth. Calculate the period of its motion. Example 12-6

61. • A physical pendulum on the surface of Earth consists of a uniform spherical bob that has a mass M of 1.0 kg and a radius R of 0.50 m suspended from a massless string that has a length L of 1.0 m. What is the period T of small oscillations of the pendulum? Example 12-6

62. •• A thin, round disk made of acrylic plastic with a density of 1.1 g/cm^3 is 20.0 cm in diameter and 1.0 cm thick (**Figure 12-25**). A very small hole is drilled through the disk at a point d = 8.0 cm from the center. The disk is hung from the hole on a nail and set into simple harmonic motion on the surface of Earth with a maximum angular displacement (measured from vertical) of θ = 7.0°. Calculate the period of the motion. Example 12-6

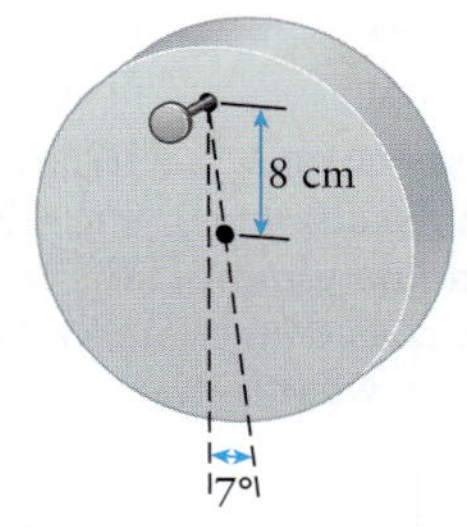

Figure 12-25 Problem 62

63. • A solid sphere, made of acrylic plastic with a density of 1.1 g/cm^3, has a radius of 5.0 cm (**Figure 12-26**). A very small "eyelet" is screwed into the surface of the sphere and a horizontal support rod is passed through the eyelet, allowing the sphere to pivot around this fixed axis. If the sphere is displaced slightly from equilibrium on the surface of Earth, find the period of its harmonic motion when it is released. SSM Example 12-6

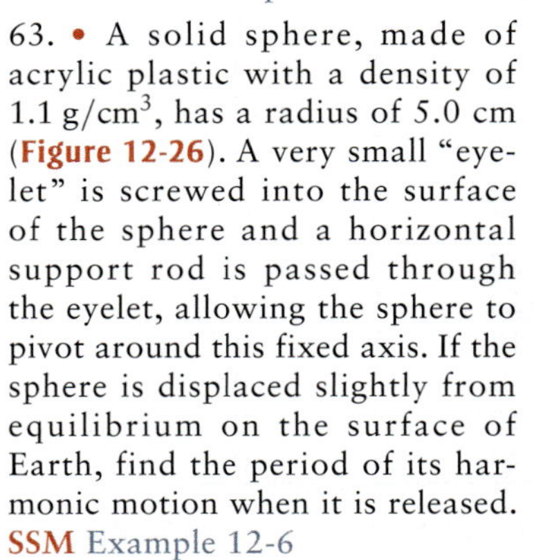

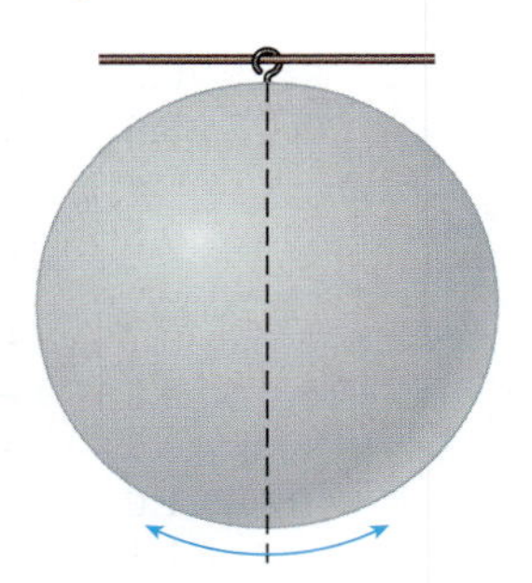
Figure 12-26 Problem 63

12-7 When damping is present, the amplitude of an oscillating system decreases over time

64. • An oscillator that has a mass of 0.100 kg experiences damped harmonic oscillation. The amplitude decreases to 36.8% of its initial value in 10.0 s. What is the value of the damping coefficient b? Example 12-8

65. •• A 0.500-kg object is attached to a spring with a force constant of 2.50 N/m. The object rests on a horizontal surface that has a viscous, oily substance spread evenly on it. The object is pulled 15.0 cm to the right of the equilibrium position and set into harmonic motion. After 3.00 s the amplitude has fallen to 7.00 cm due to frictional losses in the oil. (a) Calculate the natural frequency of the system, (b) the damping constant for the oil, and (c) the frequency of oscillation that will be observed for the motion. (d) How much time will it take before the oscillations have died down to one-tenth of the original amplitude (1.50 cm)? Example 12-8

12-8 Forcing a system to oscillate at the right frequency can cause resonance

66. • A forced oscillator is driven at a frequency of 30.0 Hz with a peak force of 16.5 N. The natural frequency of the physical system is 28.0 Hz. If the damping constant is 1.25 kg/s and the mass of the oscillating object is 0.750 kg, calculate the amplitude of the motion.

67. •• An oscillating system has a natural frequency of 50.0 rad/s. The damping coefficient is 2.0 kg/s. The system is driven by a force $F(t)$ = (100 N) cos ((50 rad/s)t). What is the amplitude of the oscillations? SSM

68. • A 5.0-kg object oscillates on a spring with a force constant of 180 N/m. The damping coefficient is 0.20 kg/s. The system is driven by a sinusoidal force of maximum value 50.0 N, and an angular frequency of 20.0 rad/s. (a) What is the amplitude of the oscillations? (b) If the driving frequency is varied, at what frequency will resonance occur? Example 12-8

General Problems

69. • The acceleration of an object that has a mass of 0.025 kg and exhibits simple harmonic motion is given by $a(t) = (10\text{ m/s}^2)\cos(\pi t + \pi/2)$. Calculate its velocity at $t = 2.0$ s, assuming the velocity of the object is $v = 3.18$ m/s at $t = 0$. SSM Example 12-3

70. • Show that the formulas for the period of an object on a spring ($T = 2\pi\sqrt{m/k}$) and a simple pendulum ($T = 2\pi\sqrt{L/g}$) are dimensionally correct. Example 12-1

71. • The position for a particular simple harmonic oscillator is given by $x(t) = (0.15\text{ m})\cos(\pi t + \pi/3)$. What are (a) the velocity of the oscillator at $t = 1.0$ s and (b) its acceleration at $t = 2.0$ s? SSM Example 12-3

72. • A simple harmonic oscillator is observed to start its oscillations at the maximum amplitude when $t = 0$. Devise a function for the position that is consistent with this initial condition. Repeat when the oscillations start at the equilibrium position when $t = 0$. Example 12-3

73. •• A 0.200-kg object is attached to a spring that has a force constant of 75.0 N/m. The object is pulled 8.00 cm to the right of equilibrium and released from rest to slide on a horizontal, frictionless table. (a) Calculate the maximum speed of the object. (b) Find the location of the object when it has one-third of the maximum speed, is moving to the right, and is speeding up. Example 12-3

74. •• A 0.100-kg object is fixed to the end of a spring that has a spring constant of 15.0 N/m. The object is displaced 15.0 cm to the right and released from rest at $t = 0$ to slide on a horizontal, frictionless table. (a) Calculate the first three times when the object is at the equilibrium position. (b) What are the first three times when the object is 10.0 cm to the left of equilibrium? (c) What is the first time that the object is 5.00 cm to the right of equilibrium, moving toward the left? Example 12-3

75. •• A damped oscillator with a period of 30.0 s shows a reduction of 30.0% in amplitude after 1.0 min. Calculate the percent loss in mechanical energy per cycle. SSM Example 12-8

76. •• A system consisting of a small 1.20-kg object attached to a light spring oscillates on a smooth, horizontal surface. A graph of the position x of the object as a function of time is shown in Figure 12-27. Use the graph to answer the following questions. (a) What are the amplitude, period, frequency, and angular frequency of the motion? (b) What is the spring constant of the spring? (c) What is the maximum speed of the object? (d) What is the maximum acceleration of the object? Example 12-3

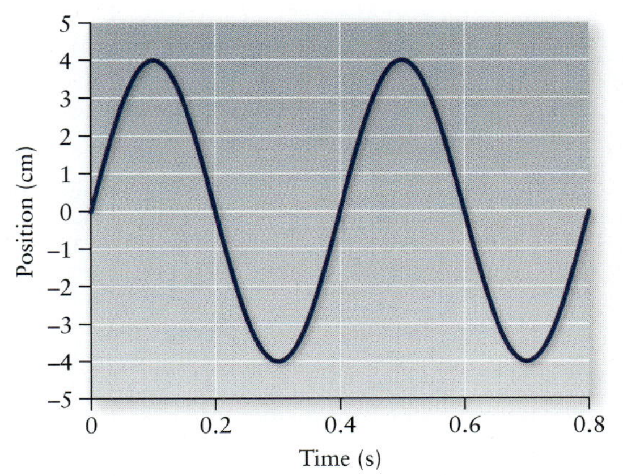

Figure 12-27 Problem 76

77. •• A pendulum made of a uniform rod with a length L_0 of 85 cm hangs from one end and is allowed to oscillate (Figure 12-28). (a) What length of a simple pendulum L_1 will have the same period of simple harmonic motion? (b) A second pendulum made of a uniform rod hangs from a point that is 5.0 cm from its end. What total length L_2 should it have to give it the same period as the first pendulum? Example 12-6

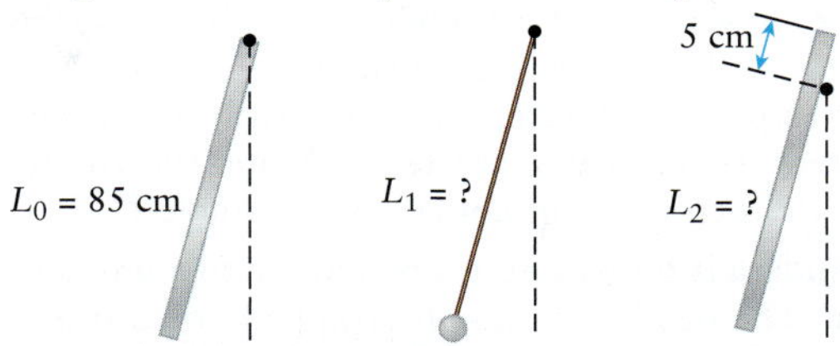

Figure 12-28 Problem 77

78. ••• (a) Using conservation of energy, derive a formula for the speed of an object that has a mass M, is on a spring that has a force constant k, and is oscillating with an amplitude of A, as a function of position $v(x)$. (b) If M has a value of 250 g, the spring constant is 85 N/m, and the amplitude is 10.0 cm, use the formula to calculate the speed of the object at $x = 0$ cm, 2.0 cm, 5.0 cm, 8.0 cm, and 10.0 cm. Example 12-4

79. •• When a small ball swings at the end of a very light, uniform bar, the period of the pendulum is 2.00 s. What will be the period of the pendulum if the bar has the same mass as the ball and the ball is removed and only the bar swings? Example 12-6

80. ••• Devise a "correction" to the formula for the period of a pendulum due to a change in altitude (and the corresponding change in the gravitational field). Your answer should provide the percent change in the period for each 1000-m increase (or decrease) in elevation. Example 12-5

81. •• A block with a mass of 0.750 kg resting on a frictionless surface is attached to an unstretched spring with a length of 15.0 cm and a spring constant of $k = 9.80 \times 10^3$ N/m. The spring is attached to a wall at its other end. A 7.50-g, 9-mm-diameter bullet is fired into the block at a speed of 355 m/s and comes to a halt inside the block. Letting the initial position of the block be $x = 0$ and the positive direction be toward the wall, what is the position function $x(t)$ for the bullet–block system after the collision? Example 12-3

82. ••• **Astronomy** Your spaceship lands on a moon of a planet around a distant star. As you initially circled the moon, you measured its diameter to be 5480 km. After landing you observe that a simple pendulum that had a frequency of 3.50 Hz on Earth now has a frequency of 1.82 Hz. (a) What is the mass of the moon? Express your answer in kilograms and as a multiple of our Moon's mass. (b) Could you have used the vibrations of a spring–object system to determine the moon's mass? Explain your reasoning. Example 12-5

83. •• A bungee jumper who has a mass of 80.0 kg leaps off a very high platform. A crowd excitedly watches as the jumper free-falls, reaches the end of the bungee cord, then gets "yanked" up by the elastic cord, again and again. One observer measures the time between the low points for the jumper to be 9.5 s. Another observer realizes that simple harmonic motion can be used to describe the process because several of the subsequent bounces for the jumper require 9.5 s also. Finally, the jumper comes to rest a distance of 40.0 m below the jump point. Calculate (a) the effective spring constant for the elastic bungee cord and (b) its unstretched length. Example 12-2

84. • **Biology** Spiderwebs are quite elastic, so when an insect gets caught in the web, its struggles cause the web to vibrate. This alerts the spider to a potential meal. The frequency of vibration of the web gives the spider an indication of the mass of the insect. (a) Would a rapidly vibrating web indicate a large (massive) or a small insect? Explain your reasoning. (b) Suppose that a 15-mg insect lands on a horizontal web and depresses it 4.5 mm. If we model the web as a spring, what would be its effective spring constant? (c) At what rate would the web in part (b) vibrate, assuming that its mass is negligible compared to that of the insect? (d) Would the vibration rate differ if the web were not horizontal? Example 12-2

85. • Andrea is transporting a pendulum to a physics demonstration. The pendulum base is strapped down in the bed of her truck. The pendulum itself is a small 450-g object suspended from a 1.6-m-long, lightweight inflexible wire. While driving at 13.5 m/s Andrea is forced to come to an abrupt stop in 1.2 s. Assuming her acceleration is constant and that the pendulum bob was stationary before she slammed on the brakes, what is the amplitude of the resulting oscillation of the pendulum in radians? Will the oscillations be *simple* harmonic? Example 12-5

86. •• A 2.0-kg object is attached to a spring and undergoes simple harmonic motion. At $t = 0$ the object starts from rest, 10.0 cm from the equilibrium position. If the force constant of the spring k is equal to 75 N/m, calculate (a) the maximum speed and (b) the maximum acceleration of the object. (c) What is the velocity of the object at $t = 5.0$ s? Example 12-3

87. ••• A 475-kg piece of delicate electronic equipment is to be hung by a 2.80-m-long steel cable. If the equipment is disturbed (such as by being bumped), the vertical vibrational frequency must not exceed 25.0 Hz. (a) What is the maximum diameter the cable can have? (b) Would the piece of equipment cause the cable to stretch appreciably? Example 12-3

13 Waves

Republished with permission of Elsevier, from J. Cosson. "A Moving Image of Flagella: News and Views on the Mechanisms Involved in Axonemal Beating." Cell Biology International, Vol. 20, No. 2, 83–94; permission conveyed through Copyright Clearance Center, Inc.

In this chapter, your goals are to:

- (13-1) Describe what a mechanical wave is.
- (13-2) Explain the key properties of transverse, longitudinal, and surface waves.
- (13-3) Define the relationship between simple harmonic motion and what happens in a sinusoidal wave.
- (13-4) Explain what determines the propagation speed of a mechanical wave.
- (13-5) Describe what happens when two sinusoidal waves from different sources interfere with each other.
- (13-6) Define the properties of standing waves on a string that is fixed at both ends.
- (13-7) Explain the nature of standing sound waves in open and closed pipes.
- (13-8) Describe how beats arise from combining two sound waves of slightly different frequencies.
- (13-9) Explain what is meant by the intensity and sound intensity level of a sound wave.
- (13-10) Recognize why the frequency of a sound changes if the source and listener are moving relative to each other.

To master this chapter, you should review:

- (6-2, 6-9) The relationship between work, force, and displacement, and the definition of power.
- (9-2, 9-4) The nature of the Young's modulus and bulk modulus of a substance.
- (11-2) The definition of density.
- (12-2, 12-3, 12-4, and 12-8) The properties of simple harmonic motion and forced oscillations.

What do you think?

A spermatozoon moves by beating its tail-like flagellum (typically about 40 μm in length) against the surrounding fluid. Because the flagellum is flexible, waves travel along its length. If a spermatozoon were to beat its flagellum at a faster rate, would the distance along the flagellum from one wave crest to the next be (a) longer, (b) shorter, or (c) unaffected?

13-1 Waves are disturbances that travel from place to place

If you point your index finger downward and flex your wrist repeatedly, your finger oscillates up and down. That's the kind of motion we discussed in Chapter 12: Your finger has an equilibrium position, and it oscillates on either side of that position. But if you dip that finger in a pool of water and repeat this motion, something new happens: A series of ripples spreads away from your finger. The disturbance that you create in this way is called a **wave**. It's actually made up of countless miniature oscillations, since each part of the water's surface oscillates as the wave passes by (**Figure 13-1a**).

What's remarkable about water waves is that the *disturbance* travels a substantial distance across the water's surface, but the individual water molecules stay in pretty much

(a)
A disturbance at one place in the water, such as a falling drop...
...causes a wave to spread outward over the water's surface.
Don Farrall/Getty Images
(b)
Sound waves from the lecturer travel through the classroom, but the air doesn't move along with them.
moodboard/Fotolia.com

Figure 13-1 Mechanical waves Two examples of mechanical waves: (a) one in a liquid and (b) one in a gas. Mechanical waves can also travel through solids.

the same place. The same thing is true for *sound* waves. When you sit in an hour-long lecture, sound waves travel from the lecturer to the back of the lecture hall, but there's no overall flow of air from the front to the back of the hall (**Figure 13-1b**). If there were, by the end of the hour the air would be denser at the back of the lecture hall than at the front!

In this chapter our subject is disturbances of this kind, called **mechanical waves.** In a mechanical wave a disturbance propagates through a material substance that we call the **medium** for the wave. The medium could be a liquid like water (Figure 13-1a), a gas like air (Figure 13-1b), or a solid. (If you live in a residence hall or an apartment, you certainly know that sound waves can travel through solid walls.) Not all waves are mechanical: As we'll learn in Chapter 22, *electromagnetic* waves such as radio signals, light rays, and x-ray beams don't require a medium to propagate and actually propagate fastest in a vacuum.

We'll begin this chapter by looking at the basic properties of mechanical waves. We'll see how to describe waves using mathematics, and we'll discover what determines the speed at which mechanical waves propagate. We'll go on to investigate what happens when two waves overlap and interfere with each other. In some cases the result is a *standing* wave in which the disturbance no longer propagates; in other cases we get a wave that propagates but whose frequency varies with time in a curious way. We'll study the manner in which we perceive sound waves, and we'll see how the properties of sound change when the source, the listener, or both are in motion.

TAKE-HOME MESSAGE FOR Section 13-1

✔ In a mechanical wave a disturbance travels through and displaces a material medium (gas, liquid, or solid). There is no net flow of matter with the disturbance.

13-2 Mechanical waves can be transverse, longitudinal, or a combination of these

Most mechanical waves are of one of the two types shown in **Figure 13-2**. The wave shown in **Figure 13-2a** is called a **transverse wave** because the individual parts of the wave medium (the rope) move in a direction *perpendicular* to the direction in which

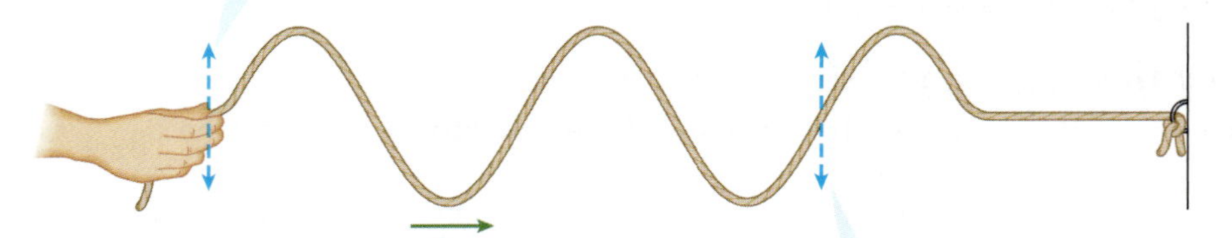

(a)
1 If the hand holding the rope moves up and down, a **transverse wave** travels along the rope.
2 The wave disturbance propagates (moves along the rope) horizontally...
3 ...but individual parts of the rope (the wave medium) move up and down, perpendicular (transverse) to the propagation direction.

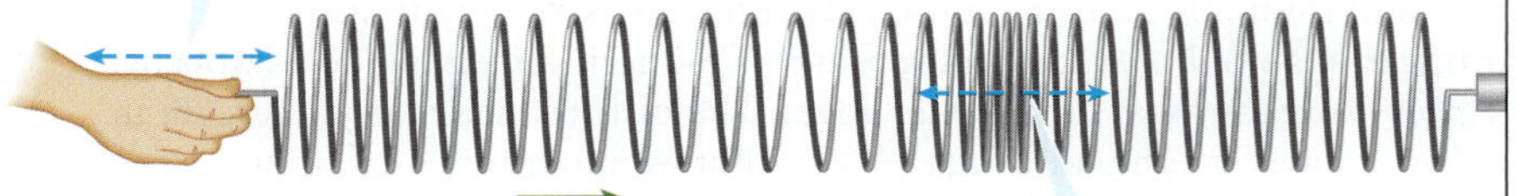

(b)
1 If the hand holding the spring moves back and forth, a **longitudinal wave** travels along the spring.
2 The wave disturbance propagates (moves along the spring) horizontally...
3 ...but individual parts of the spring (the wave medium) move back and forth, parallel (longitudinal) to the propagation direction.

Figure 13-2 Transverse and longitudinal waves These two types of waves differ in how the parts of the wave medium (the material carrying the wave) move relative to the direction that the wave propagates.

the wave propagates. By contrast, in the **longitudinal wave** shown in **Figure 13-2b**, the individual parts of the wave medium (the spring) move in the direction *parallel* to the direction of wave propagation.

You've participated in a kind of transverse wave if you've been to a crowded sports stadium and "done the wave," in which fans jump up section by section and then sit back down (**Figure 13-3**). The disturbance moves horizontally through the crowd, but individual parts of the wave medium—that is, the individual fans—move up and down, perpendicular to the direction that the "wave" propagates. (Note that this "wave" has the key property of waves that we described in Section 13-1: The disturbance moves through the crowd, but the members of the crowd don't move along with it.)

The "wave" propagates horizontally through the stadium...
...but individual members of the crowd (the wave medium) move up and down, transverse to the propagation direction.
Stock Photos/Glow Images

Figure 13-3 A transverse "wave" Sports fans call this "doing the wave," but a more descriptive name would be "doing the transverse wave."

You can easily make a longitudinal wave just by clapping your hands (**Figure 13-4**). If you clap your hands together sharply at the opening of a tube, your hands squeeze on the air between them. As a result, the air pressure between your hands increases. When you release your hands this higher-pressure air pushes on the air around it, causing the surrounding air to compress and undergo a pressure increase. The air behind the newly compressed air then relaxes back to its normal pressure. The result is that a pulse of increased pressure travels down the tube. (There's no net flow of air down the tube, just a disturbance in pressure.) If you clap your hands rhythmically, a series of pressure pulses propagates down the tube. We call this propagating disturbance a **sound wave**. Sound waves are also called **pressure waves** because they consist of regions of higher and lower pressure. (You've probably noticed that your ears, which you use to detect sound waves, are sensitive to changes in air pressure.)

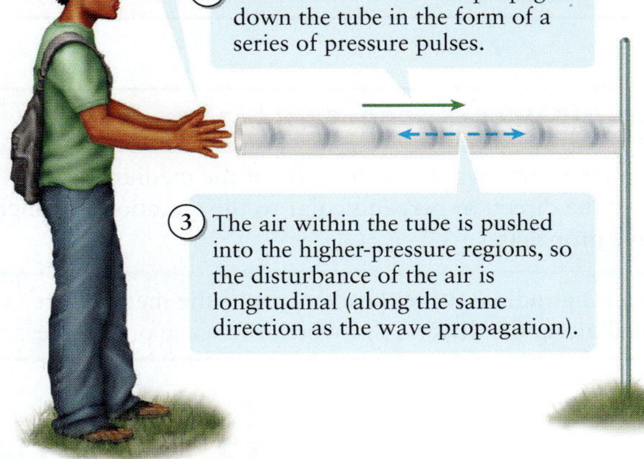
1 Clapping your hands rhythmically at the opening of a tube filled with air...
2 ...makes a sound wave propagate down the tube in the form of a series of pressure pulses.
3 The air within the tube is pushed into the higher-pressure regions, so the disturbance of the air is longitudinal (along the same direction as the wave propagation).

Figure 13-4 A longitudinal wave A sound wave is a longitudinal wave.

Here's how you can see that a sound wave is a longitudinal wave. Within each pressure pulse of the sound wave, air molecules are squeezed together; as a pulse passes, the molecules move apart again. So within the tube air molecules slosh back and forth parallel to the axis of the tube and so parallel to the direction of wave propagation. That's just our definition of a longitudinal wave (see Figure 13-2b).

An earthquake produces *both* transverse *and* longitudinal waves that propagate through the body of Earth. The longitudinal waves travel at about twice the speed of the transverse waves, so a seismic monitoring station some distance away from the earthquake site (the *epicenter*) will receive the longitudinal waves before the transverse waves. The more distant the location from the epicenter, the greater the time delay, so by measuring the time delay at a given location scientists can determine how far from that location the earthquake took place. By correlating such measurements made at three or more locations, they can triangulate the position of the epicenter.

A third variety of wave is a **surface wave**, which is a wave that propagates on the surface of a medium. Two common examples are waves on the surface of the ocean and the waves that spread away from a disturbance in a pond (Figure 13-1a). A surface wave is actually a *combination* of a transverse wave and a longitudinal wave. As a surface wave propagates parallel to the (horizontal) surface of the water, particles near the surface move both vertically and horizontally—that is, both transverse and longitudinal to the direction of wave propagation. You can see this behavior by watching a buoy floating in a harbor: As waves pass by the buoy, the buoy bobs up and down (transverse motion) and moves back and forth (longitudinal motion). Surface waves can occur on the surface of solids as well as liquids. As an example, earthquakes produce not only transverse and longitudinal waves that travel through the body of Earth but also waves that travel along Earth's surface. The surface waves are often the most destructive (see the photo that opens Chapter 12, which shows the devastation caused by the surface waves of a powerful earthquake.)

While there are essential differences among transverse, longitudinal, and surface waves, they all have this characteristic in common: The wave propagates through its medium thanks to *restoring forces* that tend to return the medium to its undisturbed state. As an example, for the transverse wave on a rope shown in Figure 13-2a, the restoring force is due to the tension in the rope. To see the importance of tension, imagine detaching the rope from the post on the right-hand side of Figure 13-2a and placing the rope on top of a table. Since the rope is now slack, there is zero tension. If you now wiggle one end of the rope, that end will move in response but *no* wave will travel along the length of the rope. With no restoring force due to tension, the medium (the rope) doesn't resist being disturbed and there is no wave propagation.

The restoring force is different for different types of waves. For the longitudinal waves on a spring shown in Figure 13-2b, it's the Hooke's law force that opposes the spring being either stretched or compressed. In the case of sound waves in air (Figure 13-4), the pressure of the air itself provides the restoring force. A region of higher pressure pushes against the neighboring regions, thus expanding and lowering its pressure back to the equilibrium value; a region of lower pressure contracts due to pressure from its neighbors, thus raising its pressure toward equilibrium. And for surface waves in water the restoring force is the gravitational force, which tends to make the water surface smooth and level and so opposes the kind of disturbance shown in Figure 13-1a.

We saw in Section 12-2 that the presence of a restoring force was also an essential ingredient for *oscillation*. This suggests that there are deep connections between the physics of oscillation that we studied in Chapter 12 and the physics of waves. We'll explore these connections in the next several sections.

GOT THE CONCEPT? 13-1 Restoring Forces in Waves

? Choose the selection that correctly fills in the blanks in the following sentence: In the wave shown in Figure 13-2a the restoring force on a piece of the rope acts in the __________ direction, and in the wave shown in Figure 13-2b the restoring force on a piece of the spring acts in the __________ direction. (a) Vertical, vertical; (b) horizontal, vertical; (c) vertical, horizontal; (d) horizontal, horizontal; (e) any of these, depending on circumstances.

TAKE-HOME MESSAGE FOR Section 13-2

✔ In a transverse wave the elements of the medium are disturbed in the direction perpendicular to the direction in which the wave propagates.

✔ In a longitudinal wave the elements of the medium are disturbed along the direction in which the wave propagates.

✔ Surface waves propagate on the surface of a medium and are a combination of a transverse and longitudinal wave.

✔ In mechanical waves of all types, a restoring force must be present to make the wave propagate.

13-3 Sinusoidal waves are related to simple harmonic motion

In general waves can be very complicated. For example, when you're having a conversation with a friend, the sound waves coming from your mouth spread out in all directions, and the character of the wave changes from one moment to another as you vary the pitch and volume of your voice and pronounce different vowels and consonants. Instead of beginning our mathematical description of waves with complicated cases such as these, we'll start by considering a simple but important kind of wave.

Our first simplification is to restrict ourselves to **sinusoidal waves**, in which the wave pattern at any instant is a sinusoidal function (a sine or cosine). This restriction may seem very limiting, but in fact, *any* wave pattern can be formed by combining sinusoidal waves. That's the principle behind some types of musical *synthesizer*. By combining sinusoidal sound waves with different characteristics, a synthesizer can simulate the sound of other musical instruments or create wholly new sounds.

We'll also restrict ourselves to waves that travel along a straight line, like the transverse and longitudinal waves shown in Figure 13-2 and the sound wave in a tube in Figure 13-4. We call these **one-dimensional waves** since they propagate along a single dimension of space.

To be specific let's consider a sinusoidal, one-dimensional wave on a rope. This is actually what's drawn in Figure 13-2a. If the hand holding the left-hand end of

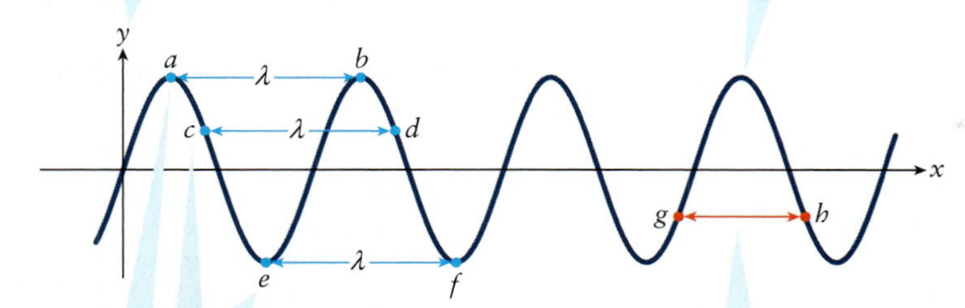

1 The "snapshot" of the wave at a given instant is a graph at time t of the displacement y of the medium versus the coordinate x along the length of the medium.
2 This is a sinusoidal wave: The graph of displacement y versus coordinate x is a sinusoidal function (sine or cosine).
y
x
a
b
c
d
e
f
g
h
λ
λ
λ
3 Points a and b are one wavelength λ apart: The distance between them corresponds to one complete cycle of the wave. The same is true for points c and d and for points e and f.
4 The displacement is the same at points g and h, but these points are not one wavelength apart: The distance between them does not correspond to a complete cycle.

Figure 13-5 A wave "snapshot": Wavelength The wavelength of a wave is the distance over which the wave repeats itself.

the rope in Figure 13-2a moves up and down in simple harmonic motion with frequency f and period T, the position of the hand as a function of time is a sinusoidal function (see Section 12-3). The wave that propagates down the rope will then also be a sinusoidal function. Let's look at the properties of such a sinusoidal, one-dimensional wave.

Sinusoidal Waves: Wavelength, Amplitude, Period, Frequency, and Propagation Speed

There are two ways to visualize a wave. One is to imagine taking a "snapshot" or "freeze-frame" of the wave to see what the wave as a whole looks like at a given time. The other is to concentrate on a given piece of the medium and see how that piece moves as a function of time. (For the wave on a rope shown in Figure 13-2a, that would mean focusing on a single small piece of the rope and watching its motion.)

In **Figure 13-5** we show a "snapshot" of a sinusoidal wave on a rope at a given instant of time. This "snapshot" is a graph of the displacement y of the medium (the rope) from equilibrium versus position x along the length of the medium. For a transverse wave on the rope, the displacement y is indeed perpendicular to x as shown in Figure 13-5. Figure 13-5 could also represent a "snapshot" of a longitudinal wave like that in Figure 13-2b; in that case y represents the displacement *in the x direction* of pieces of the medium from their undisturbed, equilibrium positions.

Figure 13-5 shows that the **wavelength** λ (the Greek letter "lambda") is the distance the disturbance travels over one full cycle of the wave. You can measure λ from one *crest* (high point) of the wave to the next crest, one *trough* (low point) to the next trough, or any specific point in the cycle to the analogous point in the next cycle. A wavelength is a distance, so its units in the SI system are meters.

Figure 13-6, another "snapshot" of a sinusoidal wave, shows the **amplitude** A of the wave. Just like the amplitude of an oscillation, A is the maximum displacement from equilibrium that occurs as the wave moves through its medium. It is *not* the difference between the maximum positive and maximum negative displacement.

To help us understand how a wave propagates over time, imagine a series of "snapshots" of a sinusoidal wave equally spaced in time like the frames of a movie. **Figure 13-7** shows such a series. As the wave

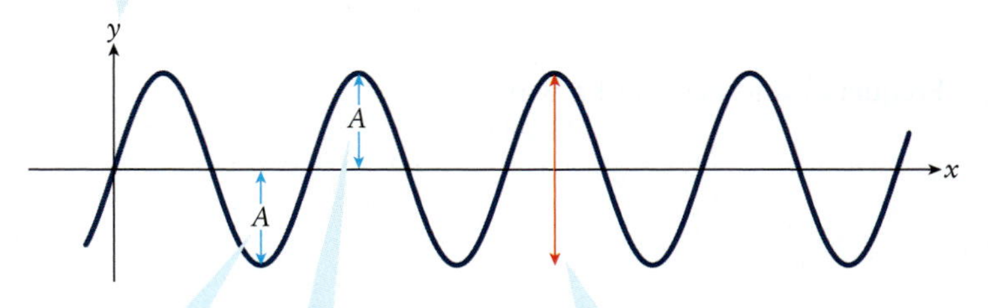

Figure 13-6 A wave "snapshot": Amplitude The amplitude of a wave is the maximum displacement from equilibrium.

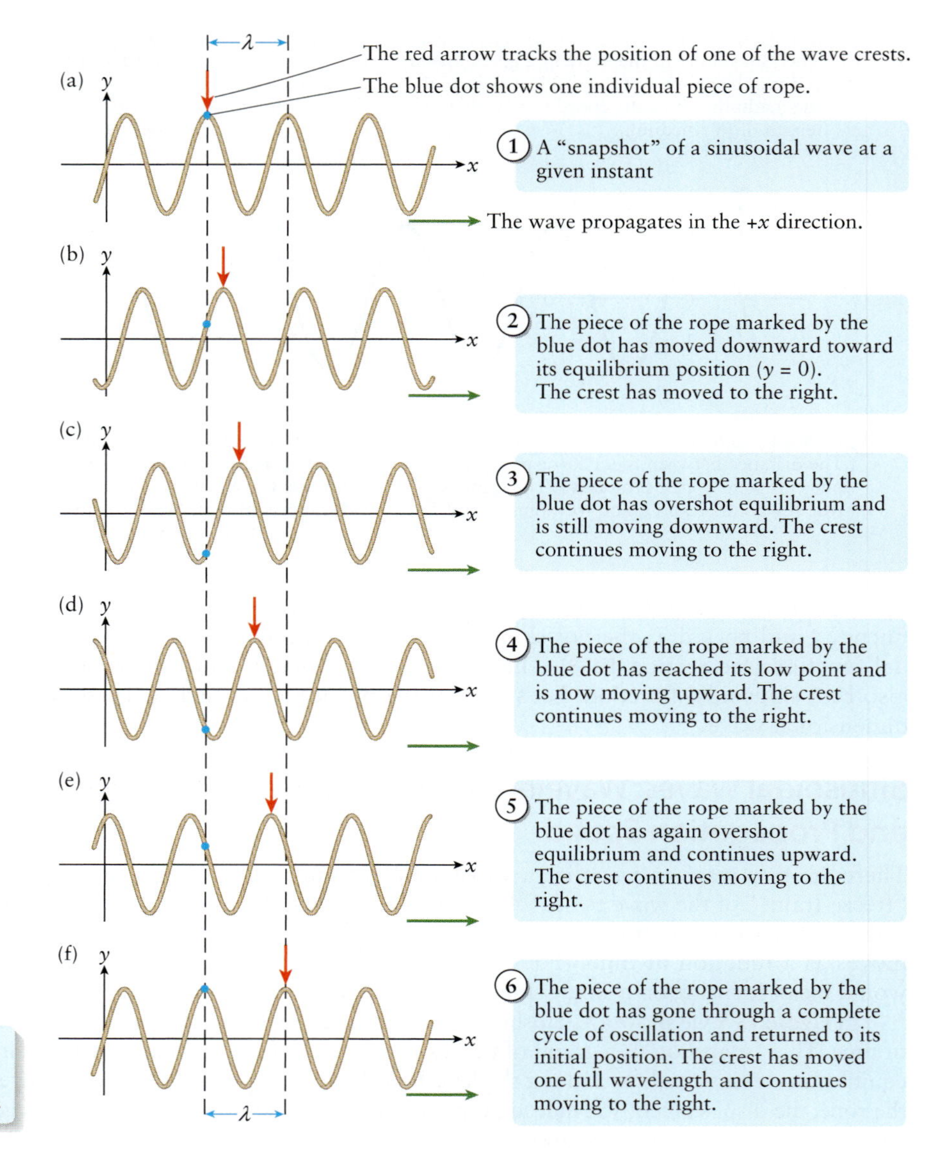

λ
The red arrow tracks the position of one of the wave crests.
The blue dot shows one individual piece of rope.
(a) y
x
1 A "snapshot" of a sinusoidal wave at a given instant
The wave propagates in the +x direction.
(b) y
x
2 The piece of the rope marked by the blue dot has moved downward toward its equilibrium position (y = 0). The crest has moved to the right.
(c) y
x
3 The piece of the rope marked by the blue dot has overshot equilibrium and is still moving downward. The crest continues moving to the right.
(d) y
x
4 The piece of the rope marked by the blue dot has reached its low point and is now moving upward. The crest continues moving to the right.
(e) y
x
5 The piece of the rope marked by the blue dot has again overshot equilibrium and continues upward. The crest continues moving to the right.
(f) y
x
6 The piece of the rope marked by the blue dot has gone through a complete cycle of oscillation and returned to its initial position. The crest has moved one full wavelength and continues moving to the right.
λ

Figure 13-7 Scenes from a wave "movie" These successive "snapshots" of a wave in motion reveal the connection between the period of a wave and the wavelength. The elapsed time from (a) to (f) is one period T.

During one period of oscillation T of the wave medium, the wave travels a distance of one wavelength λ.

moves through the medium, an individual piece of the medium (like the blue dot in Figure 13-7) oscillates up and down. Just as in Section 12-2 we use the term **"period"** and the symbol T for the time required for each piece of the medium to go through a complete oscillation cycle. Furthermore, just as in Sections 12-2 and 12-3 the **frequency** f (in hertz, or Hz) is the number of cycles that a piece of the medium goes through per second, and the **angular frequency** ω (in rads/s) equals the frequency multiplied by 2π:

Frequency and period of a wave (12-1)

Period of an oscillation

$$f = \frac{1}{T} \quad \text{and} \quad T = \frac{1}{f}$$

Frequency of the oscillation

Angular frequency of a wave (13-1)

Angular frequency of a wave

$$\omega = 2\pi f = \frac{2\pi}{T}$$

The values of T, f, and ω are the same *everywhere* in the medium where the wave is present, so these quantities are properties of the wave as a whole. The frequency of a sound wave helps determine its pitch: The higher the frequency, the higher the pitch.

Figure 13-7 also shows something remarkable about waves: *During the time that it takes an individual piece of the medium to complete one cycle of oscillation, the wave travels a distance of one wavelength through the medium.* We can use this observation to relate the wavelength and frequency to the *speed* of the wave. Since the wave travels a distance λ in a time T, the **propagation speed** v_p of the wave is

$$v_p = \frac{\lambda}{T}$$

From Equations 12-1, the reciprocal $1/T$ of the period is just the frequency f. So we can rewrite the propagation speed as

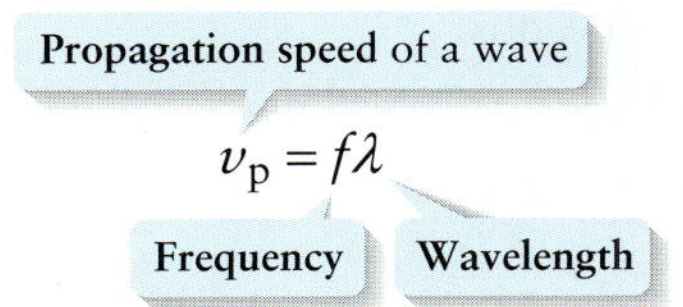

Propagation speed, frequency, and wavelength of a wave (13-2)

WATCH OUT! Be careful relating propagation speed, frequency, and wavelength.

Equation 13-2 may give you the impression that by changing the frequency or wavelength of a wave, you can change the propagation speed. In general this is *not* true! For many common types of waves, including sound waves in air and waves on a rope, the propagation speed is the *same* no matter what the frequency. If sound waves of different frequencies traveled at different speeds, a concertgoer sitting in the back row of a theater would hear a time delay between notes of low frequency (like those from a tuba) and notes of high frequency (like those from a flute); experience tells us that this isn't the case. For waves that propagate at a given speed v_p, increasing the frequency f decreases the wavelength λ and vice versa so that the product $f\lambda$ keeps the same value.

WATCH OUT! Propagation speed is not the same as the speed of pieces of the medium.

Be careful not to confuse v_p, the speed of the *wave*, with the speed of individual pieces of the medium. As Figure 13-7 shows, such pieces oscillate up and down in a transverse wave (or back and forth in a longitudinal wave), so their speed is continuously changing in magnitude. By contrast, the speed of a wave through a uniform medium (one that has the same properties throughout) stays the same at all times. The magnitude of the propagation speed can also be very different from the speed of pieces of the wave medium. As an example, sound travels through dry air at a temperature of 20°C at $v_p = 343$ m/s (about 1230 km/h, or 767 mi/h). But for a typical sound wave produced by a person speaking, the maximum speed of the air due to the wave passing through is less than 10^{-4} m/s, or one-tenth of a millimeter per second.

BIO-Medical EXAMPLE 13-1 Wave Speed on a Sperm's Flagellum

The photograph that opens this chapter shows a sea urchin spermatozoon (sperm cell) in motion. These cells move by beating a long, tail-like flagellum against the surrounding fluid. Although some aspects of the motor that drives the flagellum are distributed along the length of the flagellum, we can approximate the flagellum as behaving like a uniform rope with a transverse wave moving along it. The flagellum is 40 μm long, the period of oscillation of the flagellum is 0.030 s, and there are approximately two complete cycles of the wave in the length of the flagellum. Estimate the speed of the transverse wave as it propagates along the flagellum.

Set Up

We are given the oscillation period T and information about the wavelength λ. We use Equations 12-1 to determine the frequency of oscillation of the flagellum. Given this, we use Equation 13-2 to find the wave speed.

Period and frequency:

$$f = \frac{1}{T} \qquad (12\text{-}1)$$

Propagation speed, frequency, and wavelength of a wave:

$$v_p = f\lambda \qquad (13\text{-}2)$$

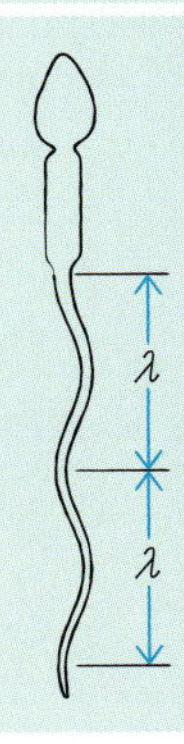

Solve

Find the frequency from the period.

From $T = 0.030$ s, the frequency is

$$f = \frac{1}{T} = \frac{1}{0.030\ \text{s}} = 33\ \text{Hz}$$

Two wavelengths fit into the length of the flagellum, so the wavelength is one-half of the 40-μm length. Use this to determine v_p.

Wavelength of the wave:

$$\lambda = \frac{1}{2} \times 40\ \mu\text{m} \times \frac{10^{-6}\ \text{m}}{1\ \mu\text{m}} = 2.0 \times 10^{-5}\ \text{m}$$

Then the wave speed is

$$v_p = f\lambda = (33\ \text{Hz})(2.0 \times 10^{-5}\ \text{m}) = 7 \times 10^{-4}\ \text{m/s}$$

to one significant figure.

Reflect

A wave moving at this propagation speed would travel about 2 m per hour. To put this in perspective, the world record holder in the World Snail Racing Championships (held annually in Congham, England) covered the 13-in. course in 2 min, which is just under 10 m/h. The propagation speed on the flagellum is only a factor of 5 slower—not bad for something 1000 times smaller than a snail.

Convert v_p to meters per hour:

$$v_p = 7 \times 10^{-4}\ \text{m/s} \times \frac{3600\ \text{s}}{1\ \text{h}} = 2\ \text{m/h}$$

Sinusoidal Waves: Displacement as a Function of Position and Time

For a complete description of a wave, we need to know the displacement y of the medium at every position x and every time t. A function that provides this description is called the **wave function** $y(x,t)$. As its symbol suggests, y is a function of *both* x and t.

To obtain this function for a sinusoidal wave, let's start with a "snapshot" of such a wave with amplitude A and wavelength λ. **Figure 13-8a** shows such a "snapshot" at a given instant of time, which we'll call $t = 0$. Since the wave is sinusoidal and repeats itself over the distance from $x = 0$ to $x = \lambda$, we can write the wave function at $t = 0$ as

$$y(x,0) = A \cos\left(\frac{2\pi x}{\lambda} + \phi\right) \tag{13-3}$$

See the Math Tutorial for more information on trigonometry.

The factor $2\pi x/\lambda$ varies from 0 at $x = 0$ to 2π at $x = \lambda$, so the cosine function goes through a complete cycle over a distance of one wavelength. The **phase angle** ϕ tells us what point in the cycle corresponds to $x = 0$. (We used a phase angle ϕ in a similar way in our description of simple harmonic motion in Section 12-3.) In Figure 13-8a $\phi = 0$, so the function is a cosine that has its maximum value at $x = 0$. In Figure 13-7a, which shows a different sinusoidal wave at $t = 0$, $\phi = -\pi/2$ (the function is shifted by $\pi/2$ radians, or one-quarter cycle, compared to the wave function in Figure 13-8a).

How can we get the wave function for *any* time t from Equation 13-3? Let's suppose that the wave is propagating in the positive x direction at speed v_p. As **Figure 13-8b** shows we can get the wave function at time t by replacing x in the wave function at $t = 0$ with $x - v_p t$:

$$y(x,t) = A \cos\left[\frac{2\pi}{\lambda}(x - v_p t) + \phi\right]$$

$$= A \cos\left(\frac{2\pi x}{\lambda} - \frac{2\pi v_p t}{\lambda} + \phi\right)$$

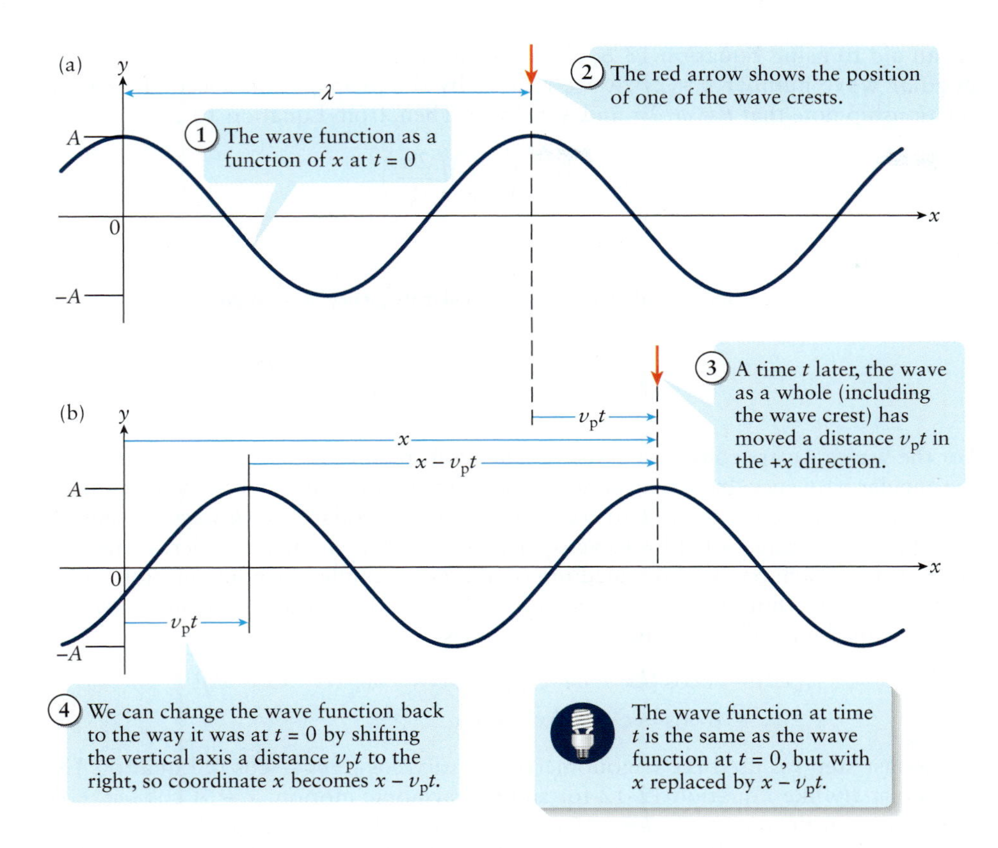

(a)
y
λ
A
0
−A
x
1 The wave function as a function of x at t = 0
2 The red arrow shows the position of one of the wave crests.
(b)
y
$v_p t$
x
$x - v_p t$
A
0
−A
x
3 A time t later, the wave as a whole (including the wave crest) has moved a distance $v_p t$ in the +x direction.
4 We can change the wave function back to the way it was at t = 0 by shifting the vertical axis a distance $v_p t$ to the right, so coordinate x becomes $x - v_p t$.
The wave function at time t is the same as the wave function at t = 0, but with x replaced by $x - v_p t$.

Figure 13-8 A sinusoidal wave function The wave function describes the displacement y as a function of position x and time t. If all parts of the wave move at the same propagation speed v_p, we can relate (a) the wave function at time $t = 0$ and (b) the wave function at a later time t.

We can simplify this expression by using Equation 13-2 for the propagation speed, $v_p = f\lambda$. Dividing both sides of Equation 13-2 by λ tells us that $v_p/\lambda = f$. Then the wave function becomes

$$y(x,t) = A\cos\left(\frac{2\pi x}{\lambda} - 2\pi f t + \phi\right) \qquad (13\text{-}4)$$

The quantity $2\pi f$ is the angular frequency ω of the wave (see Equations 13-1), and the combination $2\pi/\lambda$ in Equation 13-4 is called the **angular wave number**. We use the symbol k for this quantity:

Angular wave number (13-5)

$$k = \frac{2\pi}{\lambda}$$

Angular wave number; Wavelength

Since 2π represents the number of radians in one cycle and wavelength λ is in meters, the angular wave number k is measured in radians per meter (rad/m). We use the adjective "angular" since the term "wave number" is typically used for $1/\lambda$, the reciprocal of the wavelength. This quantity multiplied by 2π is the angular wave number $k = 2\pi/\lambda$, in the same fashion that frequency f multiplied by 2π is the angular frequency $\omega = 2\pi f$.

In terms of angular wave number and angular frequency, we can rewrite Equation 13-4 as

Wave function for a sinusoidal wave propagating in the $+x$ direction (13-6)

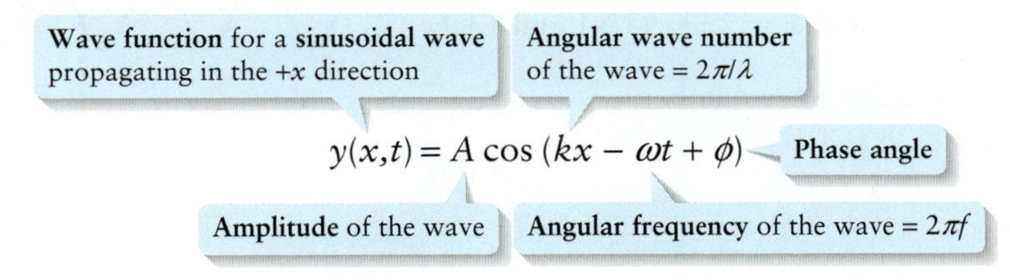

As an aid to using Equation 13-6, note that the angular frequency $\omega = 2\pi f$ and the angular wave number $k = 2\pi/\lambda$ are related by the propagation speed. To see this relationship note that $f = \omega/2\pi$ and $\lambda = 2\pi/k$. Then, from Equation 13-2,

$$v_p = f\lambda = \left(\frac{\omega}{2\pi}\right)\left(\frac{2\pi}{k}\right)$$

or

Propagation speed, angular frequency, and angular wave number (13-7)

$$v_p = \frac{\omega}{k}$$

Propagation speed of a wave; **Angular frequency** of the wave (ω); **Angular wave number** of the wave (k)

For the waves we are considering, the propagation speed v_p is the same no matter what the angular frequency: Increasing ω causes the angular wave number k to increase (that is, causes the wavelength $\lambda = 2\pi/k$ to decrease) so that the ratio $v_p = \omega/k$ stays the same.

We can use Equation 13-6 to verify a statement that we made earlier: If the wave is sinusoidal, each party of the medium oscillates in simple harmonic motion. To see that this is true, think of x as a constant (so that we are considering a single piece of the wave medium). Then Equation 13-6 becomes

$$\begin{aligned} y(x,t) &= A\cos(kx - \omega t + \phi) = A\cos[-\omega t - (-kx - \phi)] \\ &= A\cos[\omega t + (-kx - \phi)] \end{aligned} \quad (13\text{-}8)$$

In the last step we used the trigonometric identity $\cos(-\theta) = \cos\theta$. Equation 13-8 looks *exactly* like Equation 12-12 for simple harmonic motion, $x = A\cos(\omega t + \phi)$, if we interpret the quantity $-kx - \phi$ as the phase angle of the simple harmonic oscillation of the wave medium at position x. Equation 13-8 shows that all parts of the medium oscillate with the same amplitude A and the same angular frequency ω, but with a phase angle that depends on the position x within the medium. Two pieces of the medium oscillate together—that is, they are *in phase*—if they are a whole number (1, 2, 3,...) of wavelengths apart and oscillate opposite to each other—that is, they are *out of phase*—if they are $\frac{1}{2}$, $1\frac{1}{2}$, $2\frac{1}{2}$,... wavelengths apart.

We developed Equation 13-6 for a wave propagating in the positive x direction. If the wave propagates instead in the *negative* x direction, in time t it moves a distance $v_p t$ to the *left* in Figure 13-8, and the wave function at time t is the same as the wave function at time $t = 0$, with x replaced by $x + v_p t$. The result is that the wave function for a sinusoidal wave propagating in the negative x direction is the same as Equation 13-6, but with $kx - \omega t$ replaced by $kx + \omega t$.

EXAMPLE 13-2 A Wave on a Rope

A wave travels down a stretched rope. The wave has amplitude 2.00 cm and wavelength 15.0 cm, and each part of the rope goes through a complete cycle once every 0.400 s. (a) Find the frequency, angular frequency, angular wave number, and propagation speed for this wave. (b) If the wave has phase angle $\phi = 0$, find the displacement of the rope at $x =$ 0.450 m (a point 0.450 m from the end of the rope) at time $t = 3.40$ s.

Set Up

We are given the amplitude $A = 2.00$ cm, the oscillation period $T = 0.400$ s, and the wavelength $\lambda = 15.0$ cm. We use Equations 12-1 and 13-1 to determine the frequency and angular frequency, Equation 13-5 to find the angular wave number, and either Equation 13-2 or Equation 13-7 to find the propagation speed. We can find the displacement at any position x and any time t using Equation 13-6.

Period, frequency, and angular frequency:

$$f = \frac{1}{T} \text{ and } \omega = 2\pi f \quad (12\text{-}1 \text{ and } 13\text{-}1)$$

Angular wave number:

$$k = \frac{2\pi}{\lambda} \quad (13\text{-}5)$$

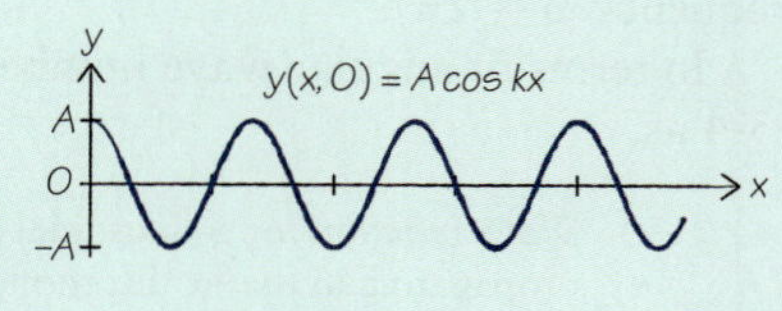

Propagation speed of a wave:

$$v_{\mathrm{p}} = f\lambda \qquad (13\text{-}2)$$

$$v_{\mathrm{p}} = \frac{\omega}{k} \qquad (13\text{-}7)$$

Wave function of a sinusoidal wave:

$$y(x,t) = A\cos(kx - \omega t + \phi) \qquad (13\text{-}6)$$

Solve

(a) Calculate the frequency f, angular frequency ω, angular wave number k, and propagation speed v_{p}.

We know $T = 0.400$ s, so the frequency and angular frequency are

$$f = \frac{1}{T} = \frac{1}{0.400 \text{ s}} = 2.50 \text{ Hz}$$

$$\omega = 2\pi f = (2\pi)(2.50 \text{ Hz}) = 15.7 \text{ rad/s}$$

The angular wave number is

$$k = \frac{2\pi}{\lambda} = \frac{2\pi}{15.0 \text{ cm}}\left(\frac{100 \text{ cm}}{1 \text{ m}}\right) = 41.9 \text{ rad/m}$$

From Equation 13-2, the propagation speed is

$$v_{\mathrm{p}} = f\lambda = (2.50 \text{ Hz})(15.0 \text{ cm})\left(\frac{1 \text{ m}}{100 \text{ cm}}\right) = 0.375 \text{ m/s}$$

Alternatively, from Equation 13-7

$$v_{\mathrm{p}} = \frac{\omega}{k} = \frac{15.7 \text{ rad/s}}{41.9 \text{ rad/m}} = 0.375 \text{ m/s}$$

(b) Find the displacement y at $x = 0.450$ m at time $t = 3.40$ s.

We take the direction of wave propagation to be the $+x$ direction. Then, since $\phi = 0$,

$$\begin{aligned} y(x,t) &= A\cos(kx - \omega t) \\ &= (2.00 \text{ cm})\cos[(41.9 \text{ rad/m})(0.450 \text{ m}) - (15.7 \text{ rad/s})(3.40 \text{ s})] \\ &= (2.00 \text{ cm})\cos(18.8 \text{ rad} - 53.4 \text{ rad}) \\ &= (2.00 \text{ cm})\cos(-34.6 \text{ rad}) = (2.00 \text{ cm})(-1.00) \\ &= -2.00 \text{ cm} = -0.0200 \text{ m} \end{aligned}$$

Reflect

The displacement at $x = 0.450$ m and $t = 3.40$ s is equal to the negative of the amplitude A, which means that at this point and this time the rope is at its maximum negative displacement. Here's why this makes sense: $x = 0.450$ m is 3 wavelengths away from $x = 0$, so there are 3 complete wave cycles between this point and the end of the rope. So at any time the displacement at $x = 0.450$ m is the same as at $x = 0$ (these points oscillate in phase). Furthermore, the time $t = 3.40$ s is equal to eight and a half periods of oscillation. At $x = 0$ the wave function is most positive ($y = A = +2.00$ cm) at $t = 0$, and most negative ($y = -A = -2.00$ cm) at times equal to a half period, one and a half periods, and so on. So at $t = 3.40$ s the wave function is at its most negative at $x = 0$ and at $x = 0.450$ m, which is just what we found.

Compared to the wavelength $\lambda = 15.0$ cm $= 0.150$ m, the distance from the end of the rope ($x = 0$) to the point $x = 0.450$ m is

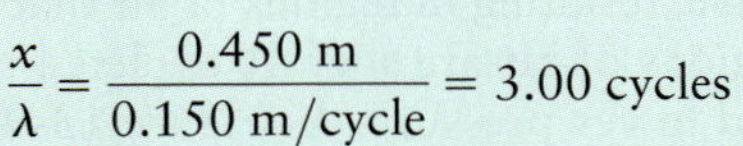

$$\frac{x}{\lambda} = \frac{0.450 \text{ m}}{0.150 \text{ m/cycle}} = 3.00 \text{ cycles}$$

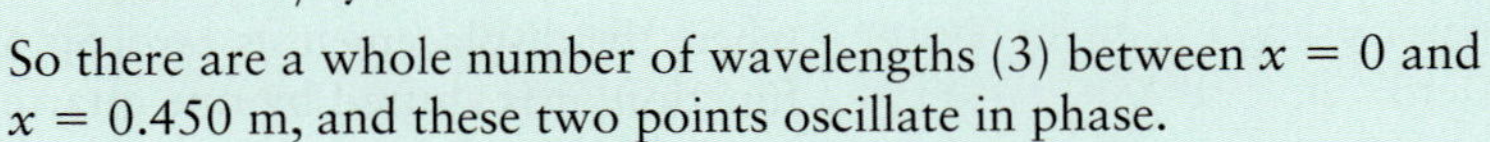

So there are a whole number of wavelengths (3) between $x = 0$ and $x = 0.450$ m, and these two points oscillate in phase.

Compared to the period $T = 0.400$ s, the time $t = 3.40$ s is

$$\frac{t}{T} = \frac{3.40 \text{ s}}{0.400 \text{ s/cycle}} = 8.50 \text{ cycles}$$

So the wave has gone through $8\frac{1}{2}$ cycles of oscillation since $t = 0$, and points that were at their most positive displacement at $t = 0$ are at their most negative displacement at $t = 3.40$ s.

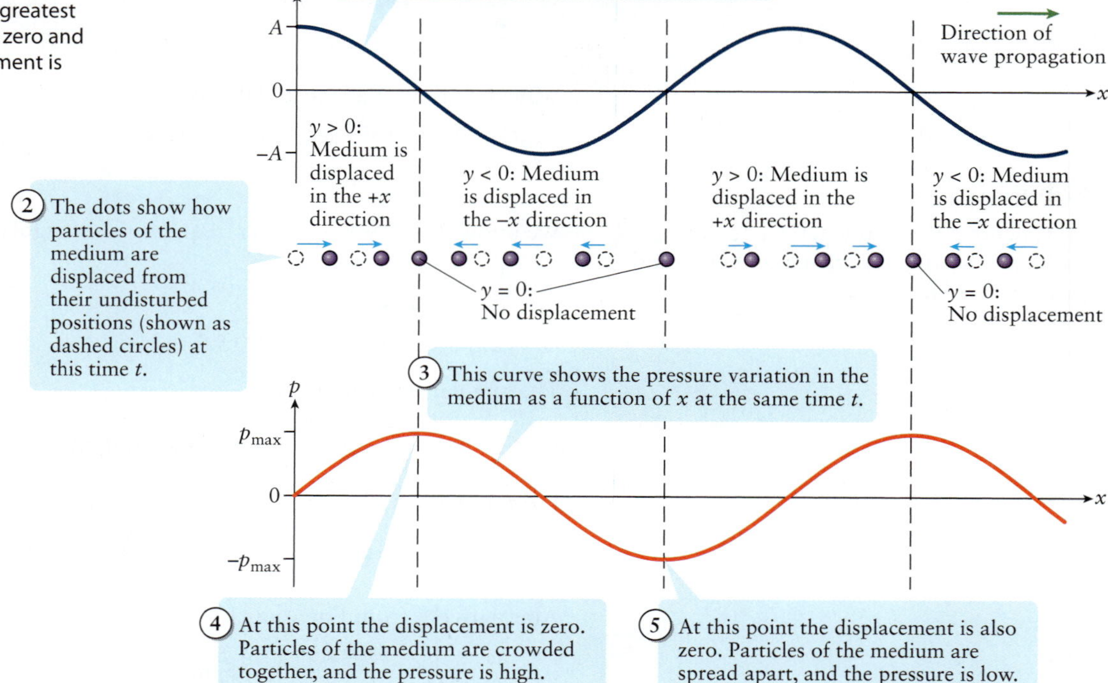
1 This curve shows the displacement of the medium as a function of x at a certain time t.
y
A
0
$-A$
x
Direction of wave propagation
$y > 0$: Medium is displaced in the $+x$ direction
$y < 0$: Medium is displaced in the $-x$ direction
$y > 0$: Medium is displaced in the $+x$ direction
$y < 0$: Medium is displaced in the $-x$ direction
2 The dots show how particles of the medium are displaced from their undisturbed positions (shown as dashed circles) at this time t.
$y = 0$: No displacement
$y = 0$: No displacement
3 This curve shows the pressure variation in the medium as a function of x at the same time t.
p
p_{max}
0
$-p_{max}$
x
4 At this point the displacement is zero. Particles of the medium are crowded together, and the pressure is high.
5 At this point the displacement is also zero. Particles of the medium are spread apart, and the pressure is low.

Figure 13-9 Pressure and displacement in a sinusoidal sound wave The pressure variation from normal in a sound wave is greatest where the displacement is zero and is zero where the displacement is greatest.

Sinusoidal Sound Waves

BIO-Medical

We've described sinusoidal waves in terms of their displacement y as a function of position x and time t. For sound waves in air it's usually more convenient to describe them in terms of *pressure variations* as a function of x and t. Humans hear by detecting very small variations of pressure on the eardrum. The ear can detect pressure changes as small as 2×10^{-5} Pa, equivalent to a change in atmospheric pressure of 1 part in *5 billion* (5×10^{9}). If you've ever had difficulty getting your ears to "pop" when driving from the mountains down to sea level or on an airliner descending to land, you know how sensitive to pressure differences your ears can be.

Since a sound wave is a longitudinal wave, a positive value of $y(x,t)$ means that the medium is displaced in the positive x direction, while a negative value of $y(x,t)$ means that the medium is displaced in the negative x direction (**Figure 13-9**). As a result as a sound wave propagates it squeezes together air molecules at some places along the wave, resulting in regions of air that have higher density and higher pressure. Adjacent regions of air are slightly depleted of air molecules and therefore have lower density and lower pressure. As Figure 13-9 shows the pressure is greatest or least at points where the displacement is *zero*. So while the pressure is also described by a sinusoidal function, it is shifted by one quarter-cycle from the wave function for displacement $y(x,t)$. That's why we've drawn the displacement graph in Figure 13-9 as a *cosine* curve but the pressure variation graph as a *sine* curve. In particular, if the displacement is described by Equation 13-6,

$$y(x,t) = A\cos(kx - \omega t + \phi)$$

the pressure variation is given by

Pressure variation for a **sinusoidal sound wave** propagating in the $+x$ direction

Angular wave number of the wave $= 2\pi/\lambda$

$$p(x,t) = p_{max}\sin(kx - \omega t + \phi)$$

Phase angle

Pressure amplitude of the wave

Angular frequency of the wave $= 2\pi f$

Pressure variation for a sinusoidal sound wave propagating in the $+x$ direction (13-9)

The quantity p_{max} is the **pressure amplitude.** It represents the maximum pressure variation above or below the pressure of the undisturbed air. A positive value of $p(x,t)$ corresponds to regions of compression (higher density and pressure), while negative values indicate regions of expansion (lower density and pressure).

WATCH OUT! Pressure variation is not the same as pressure.

Note that $p(x, t)$ in Equation 13-9 is not the total air pressure but rather represents the pressure *variation* in the sound wave relative to the pressure of the undisturbed air. At any point in the air, the pressure is equal to the pressure of the undisturbed air (normally around 10^5 Pa) plus $p(x,t)$.

BIO-Medical GOT THE CONCEPT? 13-2 Human Hearing

A person with normal hearing can hear sound waves that range in frequency from approximately 20 Hz to 20 kHz (1 kHz = 1 kilohertz = 10^3 Hz). Compared to a 20-Hz sound wave, a 20-kHz sound wave has (a) a shorter wavelength and a faster propagation speed; (b) a longer wavelength and a faster propagation speed; (c) a shorter wavelength and a slower propagation speed; (d) a longer wavelength and a faster propagation speed; (e) none of these.

TAKE-HOME MESSAGE FOR Section 13-3

✔ In a sinusoidal wave each part of the wave medium goes through simple harmonic motion.

✔ The wavelength λ is the distance the disturbance travels over one full cycle of the wave. In one period T the wave travels (propagates) a distance of one wavelength.

✔ For many common types of wave, the propagation speed is the same no matter what the frequency. Increasing the wave frequency decreases the wavelength and vice versa.

✔ The wave function $y(x,t)$ describes the displacement of the wave medium at any position x and at any time t. For a sinusoidal wave $y(x,t)$ is a sinusoidal function.

13-4 The propagation speed of a wave depends on the properties of the wave medium

As we discussed in the previous section, for many common kinds of mechanical waves the propagation speed does not depend on the frequency or wavelength. So what *does* it depend on? We can see the answer by remembering that for a mechanical wave to propagate through a medium, there must be a *restoring force* that tends to return the medium to equilibrium when it's disturbed. The greater the magnitude of this restoring force, the more rapidly the medium goes back to equilibrium and the faster the wave will propagate. At the same time the *inertia* of the medium acts to slow down the wave propagation. If the same restoring force is applied to a more sluggish medium with more inertia (think of water compared to maple syrup), the medium will return to equilibrium more slowly and the wave will propagate at a lower speed.

Let's look at how the competition between restoring force and inertia determines the propagation speed of transverse waves on a rope and of longitudinal waves in a fluid or solid.

Speed of a Transverse Wave on a String

For a transverse wave on a string, the restoring force is provided by the *tension* in the rope. We'll denote this by the symbol F. (We used T as the symbol for tension in earlier chapters, but in this chapter we're using T to denote the period of a wave.) If you pluck a rope, the wave that you create will move more rapidly if you pull on the end of the rope to increase the tension.

The inertia of the rope depends on its *mass per unit length* or **linear mass density** (SI units kg/m), which we denote by the Greek letter μ ("mu"). If the rope is uniform, μ is just equal to the mass of the rope divided by its length. A thick rope has more mass

Go to Interactive Exercise 13-1 for more practice dealing with tension.

per unit length than does a piece of ordinary string and so has more inertia. As the mass per unit length of a rope increases, the wave propagation speed decreases.

If we take both restoring force and inertia into account, we find that the propagation speed v_p of a transverse wave on a rope is

Speed of a transverse wave on a rope (13-10)

$$v_p = \sqrt{\frac{F}{\mu}}$$

Propagation speed of a **transverse wave on a rope**; **Tension** in the rope; **Mass per unit length** of the rope

Equation 13-10 can be derived by applying Newton's second law to a segment of a rope; however, the derivation is beyond our scope.

BIO-Medical EXAMPLE 13-3 Tension in a Sperm's Flagellum

We found in Example 13-1 (Section 13-3) that transverse waves propagate along the flagellum of a spermatozoon at a speed of about 7×10^{-4} m/s. A flagellum is about 40 μm in length and 0.25 μm in radius, and the material of which the flagellum is made has a density of 1100 kg/m^3 (slightly greater than the density of water). Estimate the tension in the flagellum.

Set Up

We can determine the tension F in the flagellum from the propagation speed v_p and the linear mass density μ. To determine μ (in kg/m), we'll use the definition of *ordinary* density ρ (mass per unit volume introduced in Section 11-2) to find the total mass of the flagellum, then divide it by the flagellum length $L = 40$ μm.

Speed of a transverse wave on a rope (or flagellum):

$$v_p = \sqrt{\frac{F}{\mu}} \qquad (13\text{-}10)$$

Definition of density:

$$\rho = \frac{m}{V} \qquad (11\text{-}1)$$

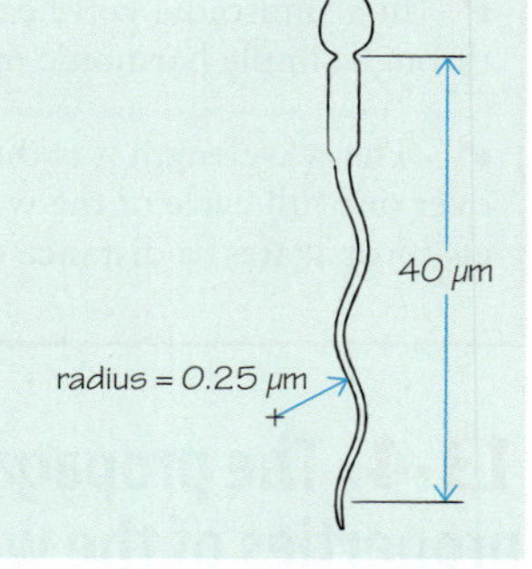

Solve

The flagellum is like a circular cylinder of radius $r = 0.25$ μm and length $L = 40$ μm. Its volume is the cross-sectional area *multiplied* by the length. From the definition of density the mass of the flagellum is its density multiplied by its volume. The linear mass density equals the mass *divided* by the flagellum's length. So the length drops out of the calculation.

Cross-sectional area of a cylinder of radius $r = 0.25$ μm:

$$A = \pi r^2 = \pi\left[0.25\ \mu\text{m} \times \left(\frac{10^{-6}\ \text{m}}{1\ \mu\text{m}}\right)\right]^2 = 2.0 \times 10^{-13}\ \text{m}^2$$

The volume of the flagellum (a cylinder with this cross-sectional area and length $L = 40$ μm) is $V = AL$.

Mass of the flagellum:

$$m = \rho V = \rho A L$$

The linear mass density of the flagellum is its mass divided by its length L:

$$\mu = \frac{m}{L} = \frac{\rho A L}{L} = \rho A$$

$$= (1.1 \times 10^3\ \text{kg/m}^3)(2.0 \times 10^{-13}\ \text{m}^2)$$

$$= 2.2 \times 10^{-10}\ \text{kg/m}$$

Given the propagation speed v_p and the linear mass density μ, we can calculate the tension F.

The speed of transverse waves on the flagellum is

$$v_p = \sqrt{\frac{F}{\mu}}$$

To solve for F, square both sides:

$$v_p^2 = \frac{F}{\mu}$$

$$F = \mu v_p^2 = (2.2 \times 10^{-10} \text{ kg/m})(7 \times 10^{-4} \text{ m/s})^2$$
$$= 1 \times 10^{-16} \text{ kg}\cdot\text{m/s}^2 = 1 \times 10^{-16} \text{ N}$$

to one significant figure.

Reflect

The tension in the flagellum is truly miniscule, about 1/10 of a millionth of a billionth of a newton. Trying to measure such an infinitesimal force directly is almost impossible. But as this example shows, we can determine such forces indirectly by using the properties of waves.

The Speed of Longitudinal Waves

Just as for a transverse wave on a rope, the speed of a *longitudinal* wave in a medium depends on the restoring force and the inertia of the medium. For a longitudinal wave in a fluid (a gas or a liquid), we call a longitudinal wave a *sound* wave. The disturbance associated with the wave corresponds to changes in the pressure of the fluid. We saw in Section 9-4 that the *bulk modulus* B of a material is a measure of how it responds to pressure changes. A larger bulk modulus means the material is more difficult to compress, which means that it has a greater tendency to return to its original volume when the pressure is released. In other words, the value of B tells us about the restoring force that acts within a fluid when disturbed by the pressure changes associated with a longitudinal wave. A measure of the inertia of the material is its density ρ, equal to its mass per volume (see Example 13-3 above). A detailed analysis using Newton's laws shows that the speed of sound waves in a fluid is

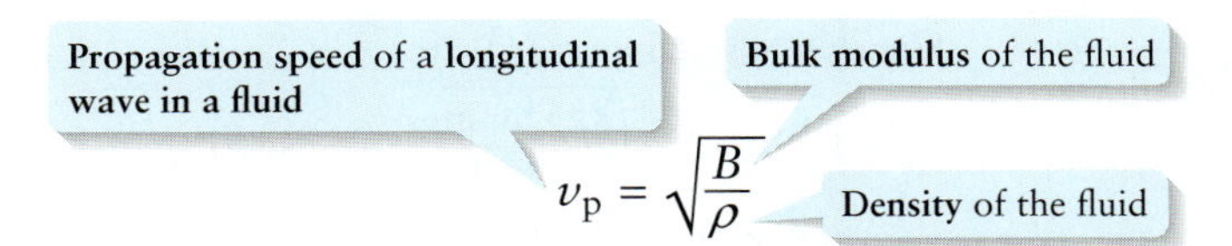

Speed of a longitudinal wave in a fluid (13-11)

The greater the bulk modulus B, the faster the propagation speed; the greater the density ρ, the slower the propagation speed.

Sound waves in *air* are particularly important. The bulk modulus B and density ρ of air both depend on temperature and humidity, so Equation 13-11 tells us that the speed of sound in air depends on both of these quantities. For most calculations we will use the speed of sound in dry air at 20°C:

$$v_{\text{sound}} = 343 \text{ m/s} \tag{13-12}$$

The speed of sound is slower in colder air. For example, at the altitude where most jetliners fly (about 11,000 m, or 36,000 ft, above sea level), the average temperature is about −55°C and the speed of sound is only about 295 m/s.

The speed of longitudinal waves in a *solid* is given by a different expression than Equation 13-11. If we just consider one-dimensional waves such as those that might travel the length of a solid rod, what matters is not the *bulk* modulus B but rather *Young's modulus* Y. We learned in Section 9-2 that Young's modulus tells us how difficult it is to stretch or compress a piece of material along its length; those are just the kind of stresses experienced in a rod with a longitudinal wave propagating down its long axis. The propagation speed of such a wave is

Propagation speed of a **longitudinal wave along a solid rod**

Young's modulus of the solid

$$v_p = \sqrt{\frac{Y}{\rho}}$$

Density of the solid

Speed of a longitudinal wave in a solid rod (13-13)

We've stated that the propagation speed of waves is independent of frequency or wavelength in many important cases. One notable exception to this general rule is the speed of *surface* waves in water: For such waves the propagation speed turns out to be greater for waves with longer wavelength λ. As an example, waves with $\lambda = 1.0$ m travel at about 1.2 m/s, while longer-wavelength waves with $\lambda = 10$ m move at about 3.9 m/s.

EXAMPLE 13-4 Propagation Speeds

Calculate the speed of propagation of two longitudinal waves: (a) A sound wave in room-temperature water, which has bulk modulus 2.2×10^9 N/m^2 and density 1.0×10^3 kg/m^3; (b) a wave traveling along a steel rail of a railroad track, with Young's modulus 2.1×10^{11} N/m^2 and density 8.0×10^3 kg/m^3.

Set Up

We use Equation 13-11 to calculate the wave speed in water (a fluid) and Equation 13-13 for the wave speed along the steel rod (an example of a solid rod).

Speed of a longitudinal wave in water (a fluid):

$$v_p = \sqrt{\frac{B_{water}}{\rho_{water}}} \quad (13\text{-}11)$$

Speed of a longitudinal wave along a solid steel rod:

$$v_p = \sqrt{\frac{Y_{steel}}{\rho_{steel}}} \quad (13\text{-}13)$$

Solve

(a) For water use $B_{water} = 2.2 \times 10^9$ N/m^2 and $\rho_{water} = 1.0 \times 10^3$ kg/m^3.

Speed of sound in water:

$$v_p = \sqrt{\frac{B_{water}}{\rho_{water}}} = \sqrt{\frac{2.2 \times 10^9 \text{ N/m}^2}{1.0 \times 10^3 \text{ kg/m}^3}}$$

$$= \sqrt{2.2 \times 10^6 \frac{\text{N} \cdot \text{m}^3}{\text{kg} \cdot \text{m}^2}} = 1.5 \times 10^3 \sqrt{\frac{\text{N} \cdot \text{m}}{\text{kg}}}$$

Since $1 \text{ N} = 1 \text{ kg} \cdot \text{m/s}^2$,

$$v_p = 1.5 \times 10^3 \sqrt{\frac{\text{kg} \cdot \text{m}}{\text{s}^2} \frac{\text{m}}{\text{kg}}} = 1.5 \times 10^3 \sqrt{\frac{\text{m}^2}{\text{s}^2}}$$

$$= 1.5 \times 10^3 \text{ m/s}$$

For the steel rod use $Y_{steel} = 2.1 \times 10^{11}$ N/m^2 and $\rho_{steel} = 8.0 \times 10^3$ kg/m^3.

Speed of longitudinal waves on the rail:

$$v_p = \sqrt{\frac{Y_{steel}}{\rho_{steel}}} = \sqrt{\frac{2.1 \times 10^{11} \text{ N/m}^2}{8.0 \times 10^3 \text{ kg/m}^3}}$$

$$= \sqrt{2.6 \times 10^7 \frac{\text{m}^2}{\text{s}^2}} = 5.1 \times 10^3 \text{ m/s}$$

Reflect

Even though water is almost a thousand times denser than air ($\rho_{water} = 1.0 \times 10^3$ kg/m^3 versus $\rho_{air} = 1.2$ kg/m^3), sound travels over four times faster in water (1.5×10^3 m/s versus 343 m/s in air). The explanation is that the bulk modulus B of water is over *ten thousand* times greater than that for air, 2.2×10^9 N/m^2 versus 1.4×10^5 N/m^2 (which says that water is much more difficult to compress than air). The bulk modulus factor thus overwhelms the density factor so that the wave speed $v_p = \sqrt{B/\rho}$ is greater in water than in air.

In the same way, Young's modulus for steel (2.1×10^{11} N/m^2) is about 100 times greater than the bulk modulus for water (2.2×10^9 N/m^2), which says that steel is *very* difficult to compress, and there is a tremendous restoring force when it is disturbed by a wave. This more than makes up for steel being eight times denser (and so having eight times more inertia) than water. So longitudinal waves travel even faster along a steel rail than they do in water.

GOT THE CONCEPT? 13-3 Crack of a Bat

The speed of sound in air is about 343 m/s, while the speed of *light* in air is 3.0×10^8 m/s. While sitting in the stands at a baseball game, you hear the crack of the bat hitting the ball 0.20 s after you see the batter swing. About how far is your seat from the batter? (a) About 70 m; (b) about 140 m; (c) about 340 m; (d) about 680 m; (e) about 1700 m.

TAKE-HOME MESSAGE FOR Section 13-4

✔ The propagation speed of a mechanical wave depends on the strength of the restoring force in the medium (which makes the wave propagate faster) and on the inertia of the medium (which makes the wave propagate more slowly).

✔ For transverse waves on a rope, the propagation speed depends on the ratio of the tension in the rope to its linear mass density (mass per unit length).

✔ The propagation speed of longitudinal waves (sound waves) in a fluid depends on the ratio of the bulk modulus of the fluid to the density of the fluid. For waves traveling along a solid rod, the speed is determined by the ratio of Young's modulus to the density.

13-5 When two waves are present simultaneously, the total disturbance is the sum of the individual waves

So far in our study of waves, we've considered only single waves in isolation. But in the real world we are bombarded by multiple waves simultaneously. If you are reading this book outside, you may hear sound waves from the conversations of others, from the wind blowing through the leaves, and from vehicles such as buses or airplanes. How can we describe what happens when more than one wave is present in a given medium?

In most cases what happens when more than one wave is present in a medium is this: Nothing unusual happens! If you hear the sound of a singer's voice at the same time as the sound of a piano, each sound wave is unaffected by the other. What you hear is simply the *sum* of the two sound waves. That observation is at the heart of an important physical principle called the *principle of superposition*:

When two waves are present simultaneously, the total wave is the sum of the two individual waves.

Where the principle of superposition can lead to surprising results is if the two waves have the *same frequency*. As we discussed in Section 13-3, we can write the wave function of a one-dimensional sinusoidal wave as $y(x,t) = A\cos(kx - \omega t + \phi)$, where A is the amplitude, $k = 2\pi/\lambda$ is the angular wave number, $\omega = 2\pi f$ is the angular frequency, and ϕ is the phase angle (Equation 13-6). Since the propagation speed is $v_p = \omega/k$ (Equation 13-7), two waves with the same frequency f and hence the same angular frequency ω will have the same angular wave number k and wavelength λ. Suppose we add two waves of the same amplitude and angular frequency but different phase angles. If wave 1 has phase angle zero and wave 2 has phase angle ϕ, we can think of ϕ as the **phase difference** for the two waves (that is, the difference between their phase angles). The wave function of the total wave is

$$\begin{aligned} y_{\text{total}}(x,t) &= y_1(x,t) + y_2(x,t) \\ &= A\cos(kx - \omega t) + A\cos(kx - \omega t + \phi) \\ &= A[\cos(kx - \omega t) + \cos(kx - \omega t + \phi)] \end{aligned}$$

√x *See the Math Tutorial for more information on trigonometry.*

To add the two cosine functions in this equation, we use the trigonometric identity for the sum of two cosines found in the Math Tutorial in the back of the book. After a little simplification we find

$$y_{\text{total}}(x,t) = 2A\cos\left(\frac{\phi}{2}\right)\cos\left(kx - \omega t + \frac{\phi}{2}\right) \tag{13-14}$$

In other words, the total wave has the *same* angular frequency ω and angular wave number k as each individual wave but has amplitude $2A\cos(\phi/2)$. Let's look at three examples.

(1) *Two waves in phase* ($\phi = 0$). If the phase difference is zero, then the two waves are in phase with each other: Their crests, or high points, occur at the same time t as do their troughs, or low points (**Figure 13-10a**). Since $\cos 0 = 1$, Equation 13-14 tells us that the total wave has amplitude $2A\cos 0 = 2A$. In other words, the waves reinforce each other so that the amplitude of the total wave is twice

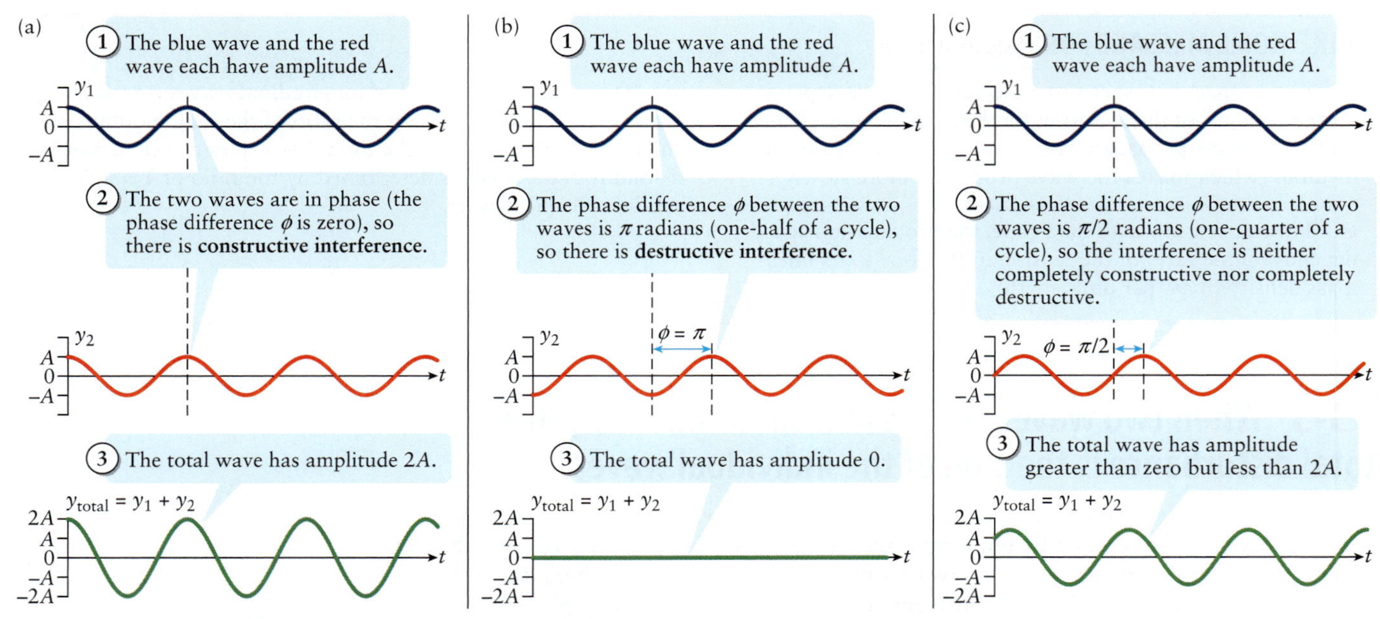
(a)
1 The blue wave and the red wave each have amplitude A.
2 The two waves are in phase (the phase difference φ is zero), so there is constructive interference.
3 The total wave has amplitude 2A.
(b)
1 The blue wave and the red wave each have amplitude A.
2 The phase difference φ between the two waves is π radians (one-half of a cycle), so there is destructive interference.
φ = π
3 The total wave has amplitude 0.
(c)
1 The blue wave and the red wave each have amplitude A.
2 The phase difference φ between the two waves is π/2 radians (one-quarter of a cycle), so the interference is neither completely constructive nor completely destructive.
φ = π/2
3 The total wave has amplitude greater than zero but less than 2A.
y1
y2
ytotal = y1 + y2
t

Figure 13-10 Superposition: Combining two waves with the same frequency When two waves arrive at the same location, the total wave is the sum of the two individual waves. The properties of the total wave depend on whether the individual waves are (a) in phase, (b) out of phase, or (c) something in between.

that of either individual wave. We call this **constructive interference.** Since each wave repeats itself once per cycle, constructive interference also happens if the phase difference corresponds to a whole number of cycles: $\phi = 2\pi, 4\pi, 6\pi,...$

(2) *Two waves π out of phase* ($\phi = \pi$). Since 2π radians corresponds to one complete cycle, a phase difference of π radians means that the two individual waves are one-half cycle out of step with each other (**Figure 13-10b**). The waves are *out of phase*: A crest of the first wave occurs at the same time as a trough of the second wave, and vice versa. In this case the two waves cancel each other, so from Equation 13-14 the amplitude of the total wave is $2A \cos \pi/2 = 0$ (recall that $\cos \pi/2 = 0$). This situation is called **destructive interference.** Destructive interference also occurs if the phase difference corresponds to a half cycle plus any whole number of cycles, or π plus an integer multiple of 2π: $\phi = 3\pi, 5\pi, 7\pi,...$

(3) *Two waves with an intermediate phase difference.* If the two waves are neither in phase nor out of phase (**Figure 13-10c**), we simply say that there is **interference** between the two waves: It is neither completely constructive nor completely destructive.

Figure 13-11 shows a real-life example of such interference. Two oscillating fingers dip in unison into a tank of water, producing two sets of surface waves like those in Figure 13-1a with the same amplitude, frequency, and same wavelength. In certain locations some of which are shown by the yellow dots in Figure 13-11, the two

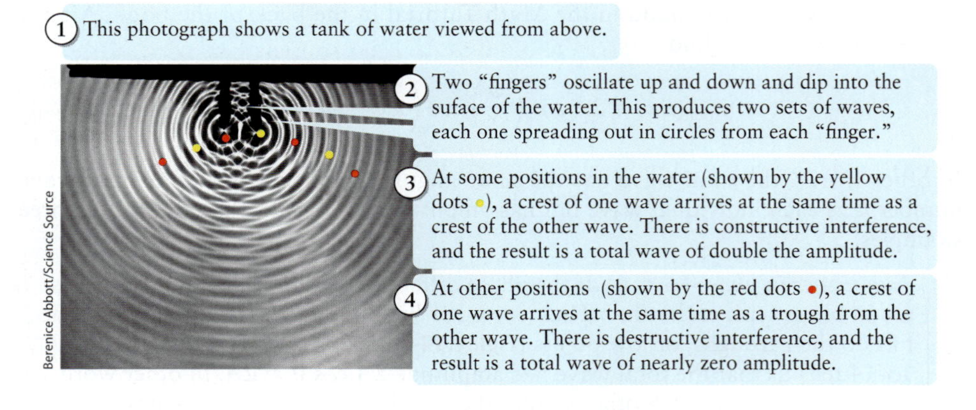
1 This photograph shows a tank of water viewed from above.
2 Two "fingers" oscillate up and down and dip into the suface of the water. This produces two sets of waves, each one spreading out in circles from each "finger."
3 At some positions in the water (shown by the yellow dots ●), a crest of one wave arrives at the same time as a crest of the other wave. There is constructive interference, and the result is a total wave of double the amplitude.
4 At other positions (shown by the red dots ●), a crest of one wave arrives at the same time as a trough from the other wave. There is destructive interference, and the result is a total wave of nearly zero amplitude.
Berenice Abbott/Science Source

Figure 13-11 Combining water waves Circular surface waves in water spread outward from two sources that oscillate in unison.

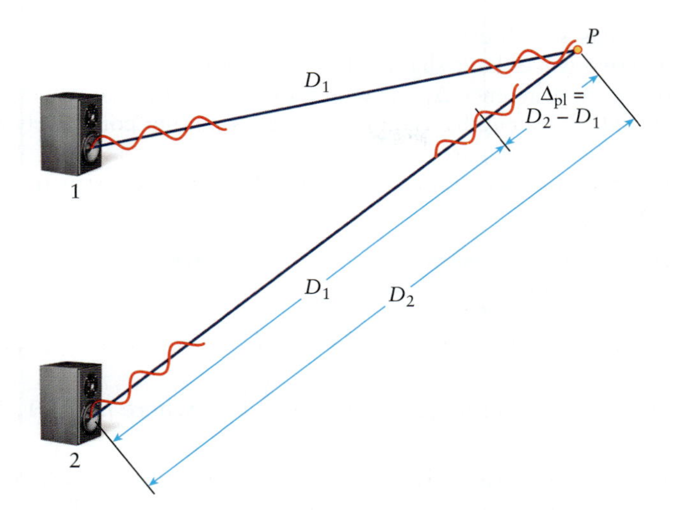

Figure 13-12 Combining sound waves Two sinusoidal sound waves of the same amplitude and wavelength travel different distances D_1 and D_2 to point *P*. How they interfere at *P* depends on how many wavelengths fit into the path length difference Δ_{pl}.

waves arrive in phase: A crest of the first wave arrives at the same time as a crest of the second wave. There is constructive interference at these locations, and the total wave has a large amplitude. At other locations, some of which are shown by the red dots, a crest of one wave arrives at the same time as a trough of the other wave. The two waves are out of phase, and there is destructive interference. At these locations the two waves essentially cancel each other, so that the total wave is nearly zero. Because interference of the surface waves occurs at all points in Figure 13-11, we call the overall pattern in that figure an **interference pattern.**

The interference effect shown in Figure 13-11 may seem surprising, since the two wave *sources* are oscillating in phase. It turns out that in this case what matters are the *distances* from each source to the point where the waves meet and interfere. The same principle applies to waves of all kinds, so in **Figure 13-12** we've drawn two loudspeakers that emit sinusoidal sound waves of the same amplitude A, the same wavelength λ, and the same frequency f. We've drawn the sound waves as sinusoidal functions that indicate the displacement of air along the direction of propagation. The two loudspeakers oscillate *in phase*, so they both emit the exact same waves at any time t. A listener is at point P, a distance D_1 from speaker 1 and a distance D_2 from speaker 2. We call these distances the *path lengths* from either speaker to point P. The **path length difference** Δ_{pl} ("delta-sub-p-l") for the two waves is the difference between these two distances: $\Delta_{pl} = D_2 - D_1$, so $D_2 = D_1 + \Delta_{pl}$.

At the instant shown in Figure 13-12, wave 1 happens to reach a peak of a cycle at point P. Notice that wave 2 travels farther to reach P. The additional distance is the path length difference Δ_{pl}. Because the two waves are emitted in phase, each wave hits a peak a distance D_1 from its source. So to determine what happens when the two waves overlap at P, we need only to ask how many cycles of wave 2 fit into the extra distance Δ_{pl}.

If exactly one wavelength λ fits into Δ_{pl}, then wave 2 will be at a peak at point P, just as if it were at distance D_1 from the source. The same is true if 2λ, 3λ, 4λ, or *any* integer number of full wavelengths fit into Δ_{pl}. (It's also true if $\Delta_{pl} = 0$, which is the case if point P is equal distances from the two speakers.) And because we've already established that wave 1 is at the same point in its cycle at point P as wave 2 is at D_1 from source 2, we can make this comparison regardless of where in the cycles the waves are at P. If an integer number of wavelengths fit into Δ_{pl}, the two waves will rise and fall together as they arrive at point P. The two waves are therefore *in phase* at point P, and constructive interference takes place there (see Figure 13-10a).

Suppose instead that exactly one-half of a wavelength, or an odd number of half wavelengths ($\lambda/2$, $3\lambda/2$, $5\lambda/2$,...), fits into the path length difference Δ_{pl} shown in Figure 13-12. In this case when wave 2 is at a peak at distance D_1 from the source, it will be at a trough in its cycle at point P. So waves 1 and 2 will be *out of phase* at point P, resulting in destructive interference. If the amplitudes of the two waves at P are equal, the total wave will be zero as in Figure 13-10b. (The cancellation isn't complete for the situation shown in Figure 13-11. That's because the amplitude of each surface wave decreases with distance from the source. So at points where D_1 and D_2 are different, the two waves have slightly different amplitudes.)

If the path length difference Δ_{pl} is neither a whole number of wavelengths nor an odd number of half-wavelengths, the interference is neither completely constructive nor completely destructive (Figure 13-10c).

Here's a summary of the conditions that must be met for constructive and destructive interference:

Path length difference from the two wave sources to the point where interference occurs

Wavelength

Conditions for constructive and destructive interference of waves from two sources (13-15)

Constructive interference: $\Delta_{pl} = n\lambda$ where $n = 0, 1, 2, 3, \ldots$

Destructive interference: $\Delta_{pl} = \left[n + \frac{1}{2}\right]\lambda$ where $n = 0, 1, 2, 3, \ldots$

Note that for constructive interference, $n = 0$ refers to the case where there is *zero* path length difference, so the two waves naturally arrive in phase. This is the situation for the center yellow dot in Figure 13-11.

WATCH OUT! Interference effects require special conditions.

A sound system with two speakers resembles the setup in Figure 13-12. Based on our discussion of interference, you might expect that such a system would have "dead spots" where there is destructive interference between the sound waves coming from the two speakers (like the red dots in Figure 13-11) and "loud spots" where there is constructive interference (like the yellow dots in Figure 13-11). But even if you have such a system, you've probably never noticed any such "dead spots" or "loud spots." One reason is that the positions where destructive interference occurs change if the wavelength changes (see Equations 13-15). Music contains sounds of many different frequencies and hence many different wavelengths, and places that are "dead" or "loud" spots for one of the wavelengths will not be "dead" or "loud" for other wavelengths. Another reason is that in a stereo system the signals coming from the left-hand speaker are not identical to those from the right-hand speaker. A final reason is that you also receive sound waves that come from the speakers and bounce off the walls or ceiling before reaching you. These have different path lengths than the waves that reach you directly from the speakers, and so they can "smooth out" any interference between the waves that reach you directly.

EXAMPLE 13-5 Stereo Interference

Your sound system consists of two speakers 2.50 m apart. You sit 2.50 m from one of the speakers so that the two speakers are at the corners of a right triangle. As a test you have both speakers emit the same *pure tone* (that is, a sinusoidal sine wave). The speakers emit in phase. At your location what is the lowest frequency for which you will get (a) *destructive* interference? (b) *constructive* interference?

Set Up

There is a path length difference for the two waves that reach your ear: One travels a distance $D_1 = 2.50$ m; the other travels a longer distance D_2 equal to the hypotenuse of the right triangle. Since the product of frequency and wavelength equals the constant speed of sound, the *lowest* frequency for each kind of interference corresponds to the *longest* wavelength.

Constructive interference:

$\Delta_{pl} = n\lambda$ where $n = 0, 1, 2, 3, \ldots$

Destructive interference:

$$\Delta_{pl} = \left(n + \frac{1}{2}\right)\lambda$$

where $n = 0, 1, 2, 3, \ldots$ (13-15)

Propagation speed of a sound wave:

$v_{sound} = f\lambda$ (13-2)

Speed of sound in dry air at 20°C:

$v_{sound} = 343$ m/s (13-12)

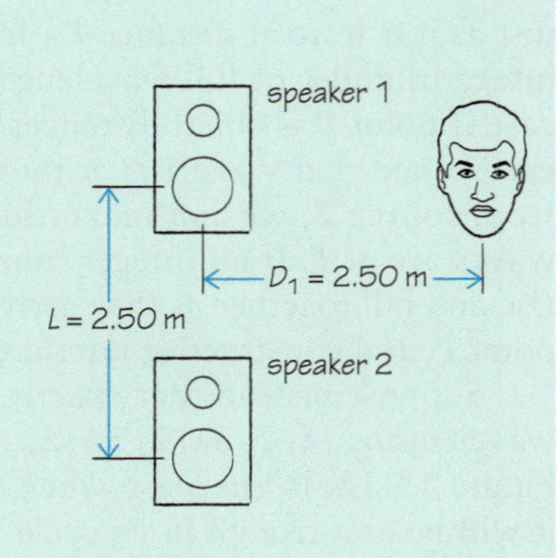

Solve

(a) The path length difference Δ_{pl} is equal to the difference between the two path lengths D_2 and D_1.

Path length from speaker 2 to the listener:

$$D_2 = \sqrt{D_1^2 + L^2} = \sqrt{(2.50\text{ m})^2 + (2.50\text{ m})^2} = 3.54\text{ m}$$

Path length difference:

$$\Delta_{pl} = 3.54\text{ m} - 2.50\text{ m} = 1.04\text{ m}$$

The longest wavelength for which there is destructive interference is the one for which one half-wavelength fits into the distance Δ_{pl}. This corresponds to $n = 0$ in the destructive-interference relation in Equations 13-15.

If one half-wavelength fits into the distance Δ_{pl},

$$\Delta_{pl} = \frac{1}{2}\lambda$$

$$\lambda = 2\Delta_{pl} = 2(1.04\text{ m}) = 2.08\text{ m}$$

The frequency of a sound wave with this wavelength is

$$f = \frac{v_{\text{sound}}}{\lambda} = \frac{343\text{ m/s}}{2.08\text{ m}} = 165\text{ s}^{-1} = 165\text{ Hz}$$

If the sound has this frequency, there will be destructive interference at the listener's position and the sound level will be diminished.

(b) The longest wavelength for which there is constructive interference is the one for which one full wavelength fits into the distance Δ_{pl}. This corresponds to $n = 1$ in the constructive-interference relation from Equations 13-15. Note that $n = 0$ is possible only if the path length difference is zero, which in this case it is not.

If one wavelength fits into the distance Δ_{pl},

$$\Delta_{pl} = \lambda = 1.04\text{ m}$$

The frequency of a sound wave with this wavelength is

$$f = \frac{v_{\text{sound}}}{\lambda} = \frac{343\text{ m/s}}{1.04\text{ m}} = 330\text{ s}^{-1} = 330\text{ Hz}$$

If the sound has this frequency, there will be constructive interference at the listener's position and the sound level will be elevated.

Reflect

The frequencies that we found are of musical importance: 165 Hz is E below middle C (E3 in musical nomenclature), and 330 Hz is E above middle C (or E4). Note that the higher frequency is exactly double that of the lower frequency. In music two such tones are said to be an *octave* apart.

Notice that we didn't consider the number of wavelengths that fit into the distance between you and either speaker. All that matters is the *difference* between the two paths. Can you show that in the destructive case the listener is 1.20 wavelengths away from speaker 1 and 1.70 wavelengths away from speaker 2? Can you also show that in the constructive case the numbers are 2.40 wavelengths and 3.40 wavelengths?

GOT THE CONCEPT? 13-4 Interference Pattern

If you increase the frequency at which the fingers in Figure 13-11 dip into the water, would the interference pattern (a) move outward away from the fingers, (b) move inward toward the fingers, (c) remain unchanged, or (d) any of these, depending on the value of the frequency?

TAKE-HOME MESSAGE FOR Section 13-5

✔ When two waves from different sources interfere with each other, they obey the principle of superposition. At every point where more than one wave passes simultaneously, the net disturbance of the medium equals the sum of the displacements that each wave would have caused individually.

✔ If two waves are in phase at some location, the two waves reinforce each other and there is constructive interference. If the two waves are out of phase at some location, they cancel each other and there is destructive interference.

13-6 A standing wave is caused by interference between waves traveling in opposite directions

Each of the six strings of a guitar is the same length, and yet each string produces a sound of different pitch when plucked. Why is this? When you turn one of the tuning keys to tighten a string, its pitch goes up. Why? And when one of the strings is pinched against the fingerboard, the pitch also goes up. Again, why?

The answers to these questions will help explain the operation of stringed musical instruments of all kinds, including guitars, violins, harps, and pianos. The basic physics is interference, which we introduced in the last section. But instead of looking at interference between waves coming from two different sources, we'll consider the interference between waves traveling in *opposite directions* along a stretched string.

(a) As the pulse arrives at the pole, the string exerts an upward force on the pole...

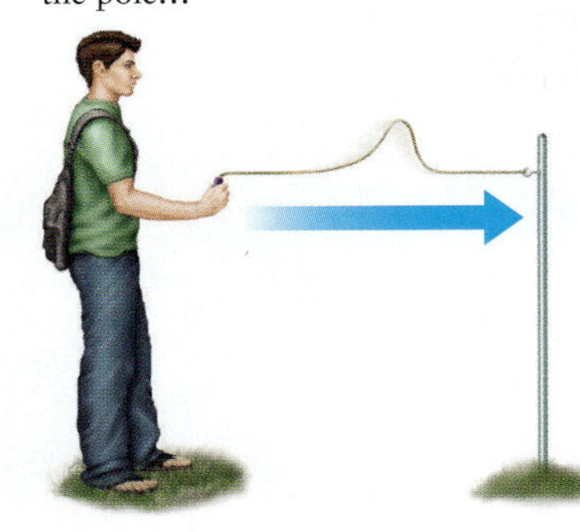

(b) ...so the pole exerts a downward force on the string. This causes the pulse to become inverted as it is reflected.

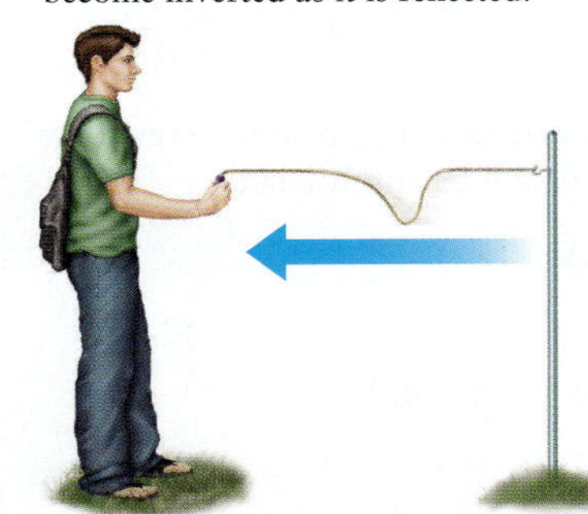

Figure 13-13 Reflecting a wave (a) A single pulse moves down a stretched string tied to a pole. (b) The pulse reflects back from the pole.

Standing Waves on a String: Modes

To see how such a situation could arise, consider what happens when you use your hand to send a single pulse down a string tied to a pole (**Figure 13-13a**). When the pulse reaches the fixed end, the string exerts an upward force on the pole. The pole must then exert a downward force on the string, as required by Newton's third law. This action reflects an inverted pulse back along the string (**Figure 13-13b**).

If you were to wiggle the end of the string up and down periodically, instead of a pulse you would create a sinusoidal wave on the string. The reflection would be an inverted sinusoidal wave traveling back toward you with the same amplitude, wavelength, and frequency as the incoming wave.

Let's see what happens when we add together two sinusoidal waves traveling in opposite directions. From Equation 13-6 the wave function for such a wave traveling in the positive x direction is $y(x,t) = A\cos(kx - \omega t + \phi)$. For simplicity we'll choose the phase angle ϕ to be zero. As we discussed in Section 13-3, a wave traveling in the *negative* x direction has the same wave function but with $kx - \omega t$ replaced by $kx + \omega t$. We'll also use $\phi = 0$ for this wave but add a minus sign in front of the amplitude to indicate that the wave is inverted. From the principle of superposition the total wave is the sum of these two sinusoidal waves with the same amplitude A, angular wave number $k = 2\pi/\lambda$, and angular frequency $\omega = 2\pi f$:

$$\begin{aligned} y_{\text{total}}(x,t) &= A\cos(kx - \omega t) + [-A\cos(kx + \omega t)] \\ &= A[\cos(kx - \omega t) - \cos(kx + \omega t)] \end{aligned}$$

We can again use a trigonometric identity to simplify the difference of the two cosine functions (see the Math Tutorial), arriving at

$$y_{\text{total}}(x,t) = 2A(\sin kx)(\sin \omega t) \tag{13-16}$$

The wave function in Equation 13-16 describes a wave pattern unlike any we've seen yet. Note that the x and t terms do not appear in the same sinusoidal function. As a result the crests of the wave do *not* move from place to place: The wave is maximum at points where $\sin kx$ equals either $+1$ or -1, and at those points the value of y_{total} oscillates between $+2A$ and $-2A$ as the function $\sin \omega t$ varies between $+1$ and -1. At the points where $\sin kx = 0$, the wave is *always* zero! Because the wave does not move but simply stays in the same place, we call it a **standing wave**. By contrast, a wave of the form $y(x,t) = A\cos(kx - \omega t)$ is referred to as a **traveling wave**. Note that we formed the standing wave by combining two traveling waves that propagate in opposite directions. The two traveling waves each carry a disturbance; when we add them together the result is that the disturbance stays in the same place. Pieces of the string still move, but the *pattern* created by the interference between two traveling sinusoidal waves—the standing wave—remains stationary.

Now think again about the string in Figure 13-13. Let L be the length of the string, take $x = 0$ to be at the end of the string you hold in your hand, and let $x = L$ be at the end of the string attached to the pole. Note from Equation 13-16 that at $x = 0$, $y_{\text{total}} = 0$ because $\sin k(0) = \sin 0 = 0$. The end that you hold in your hand moves very little compared to the amplitude of the wave on the string, so it's quite accurate to say that the amplitude of the wave at your hand is zero. So waves will reflect from *both* ends of the string, and a steady standing wave can be set up. At the other end of the string at $x = L$, the transverse displacement y must also be zero at all times because that end is rigidly attached to the pole. So $y_{\text{total}}(L,t) = 0$ for all values of t. Equation 13-16 tells us that this will be true only if $\sin kL = 0$. Now $\sin\theta$ is equal to zero if $\theta = 0, \pi, 2\pi, 3\pi, \ldots$ So it must be true that

$$kL = n\pi \text{ where } n = 1, 2, 3, \ldots \tag{13-17}$$

(We can't have $kL = 0$, since that would mean either that the string has zero length or the angular wave number $k = 2\pi/\lambda$ is zero, corresponding to the wave having an infinite wavelength.) Equation 13-17 is the condition that the displacement of the string is zero at both ends. Since $k = 2\pi/\lambda$, we can rewrite Equation 13-17 as

$$\frac{2\pi L}{\lambda} = n\pi$$

or, simplifying,

Length of a string held at both ends

Wavelength of a standing wave on the string

$$L = \frac{n\lambda}{2} \quad \text{where } n = 1, 2, 3, \ldots$$

Wavelengths for a standing wave on a string (13-18)

Here λ represents the wavelength of both the standing wave and also the original wave we created on the string. So according to Equation 13-18 a whole number n of half-wavelengths must fit onto the string to generate the standing wave pattern described by Equation 13-16. **Figure 13-14** shows the patterns for $n = 1$ (for which half of a wavelength fits), $n = 2$ (for which two half-wavelengths, or one full wavelength, fits), and $n = 3$ (for which three half-wavelengths, or one and a half full wavelengths, fits). Each of these patterns is called a **standing wave mode.**

Figure 13-15 shows photographs of the $n = 1$, 2, and 3 standing wave modes on a real string. In these photographs the movement of the end of the string that creates the standing waves is extremely slight. This justifies treating the end of this string as fixed, as we described above.

Figure 13-14 and Figure 13-15 show that for all modes except $n = 1$, there are other positions besides the ends at which the displacement of the string is zero for all times. These are the points where $\sin kx = 0$ in Equation 13-16, so the wave function $y_{\text{total}}(x,t)$ equals zero. Any point where the displacement is always zero is called a **node** of the standing wave. For the $n = 1$ standing wave mode, there are nodes at each end of the string; for the $n = 2$ mode there is an additional node at the center of the string, $x = L/2$; and for the $n = 3$ mode there are two nodes at $x = L/3$ and $x = 2L/3$.

Positions along the standing wave at which the oscillation of the string is maximal are called **antinodes.** These lie halfway between adjacent nodes. For the $n = 1$ mode in Figure 13-14 and Figure 13-15 there is one antinode at the center of the string ($x = L/2$); for the $n = 2$ mode there are two antinodes, at $x = L/4$ and $x = 3L/4$; and for the $n = 3$ mode there are three antinodes, at $x = L/6$, $x = L/2$, and $x = 5L/6$.

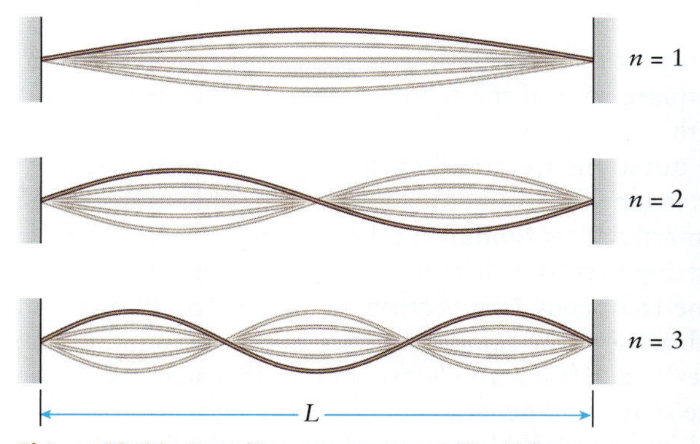

Figure 13-14 Standing waves on a string I These illustrations show the first three standing wave modes of a string connected to two supports. The mode number n counts the number of half-wavelengths that fit into the length of the string.

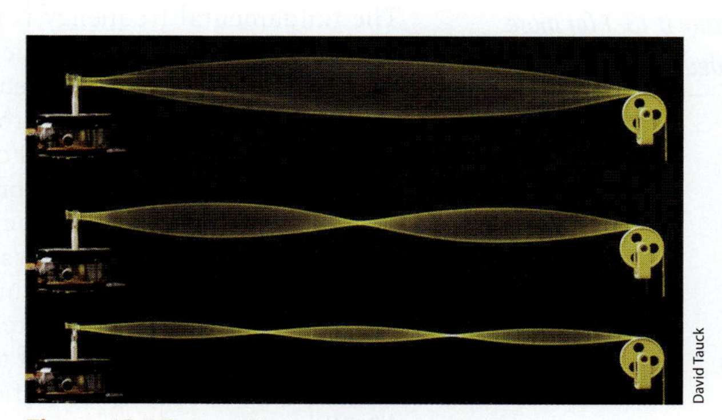

Figure 13-15 Standing waves on a string II These photographs show the first three standing wave modes on a string. The left-hand end of the string moves up and down (slightly) to create each standing wave.

Standing Waves on a String: Frequencies and Musical Sound

We learned in Section 13-4 that the propagation speed of a *traveling* wave on a string under tension F and with linear mass density μ is $v_p = \sqrt{F/\mu}$ (Equation 13-10). Furthermore, the frequency f and wavelength λ of a traveling wave are related by $v_p = f\lambda$ (Equation 13-2). Because a standing wave on a string is a superposition of two traveling waves, we can use these two relationships to find the *frequencies* associated with each of the standing wave modes on a string fixed at both ends.

By combining Equations 13-2 and 13-10, we get the following relationship between the frequency f and wavelength λ of a wave on a string:

$$f\lambda = \sqrt{\frac{F}{\mu}} \quad \text{or} \quad f = \frac{1}{\lambda}\sqrt{\frac{F}{\mu}}$$

To make this specific to *standing* waves on a string, we use the relation $L = n\lambda/2$ from Equation 13-18, which we can rewrite as $(1/\lambda) = n/(2L)$. Substituting this into the above equation, we get

Frequencies for a standing wave on a string (13-19)

$$f_n = \frac{n}{2L}\sqrt{\frac{F}{\mu}} \quad \text{where } n = 1, 2, 3, \ldots$$

*n*th possible frequency of a standing wave on a string held at both ends

Tension in the string

Length of the string

Linear mass density of the string

The subscript n on the symbol f_n in Equation 13-19 denotes the frequency of the nth standing wave mode (see Figure 13-14 or Figure 13-15). Each of these frequencies represents a *natural* frequency of the string, at which it will oscillate if displaced from equilibrium and released. To make the string oscillate at the $n = 1$ frequency, displace it into the $n = 1$ shape shown in Figure 13-14 and let it go; to make it oscillate at the $n = 2$ frequency, displace it into the $n = 2$ shape; and so on. They are also the *resonant* frequencies of the string. If one end of the string is set into oscillation at one of the frequencies given by Equation 13-19, the string will oscillate with large amplitude (see Figure 13-15). If the end of the string is forced to oscillate at a frequency other than those given by Equation 13-19, the result will be a jumble of small-amplitude wiggles rather than a well-behaved standing wave.

The frequency f_1, which is the lowest natural frequency of the string, is called the **fundamental frequency**. The corresponding ($n = 1$) standing wave mode is called the **fundamental mode**. Equation 13-19 shows that the fundamental frequency is equal to

$$f_1 = \frac{1}{2L}\sqrt{\frac{F}{\mu}} \quad (13\text{-}20)$$

Go to Picture It 13-1 for more practice dealing with overtones.

The fundamental frequency is proportional to the square root of the string tension F, inversely proportional to the square root of the string's linear mass density μ, and inversely proportional to the length L of the string.

Equation 13-20 answers the questions we posed at the beginning of this section about the strings of a guitar. For stringed instruments like a guitar, plucking a string normally causes it to vibrate primarily at its fundamental frequency. If it's an acoustic guitar, the oscillation of the string acts as a driving force that causes the body of the guitar to oscillate at the same frequency (see Section 12-8). The oscillations of the body in turn produce a sound wave of the same frequency in the surrounding air, which is what reaches the listener's ear. (If it's an electric guitar, the pickups on the guitar detect the oscillation of the string and generate an electrical signal of the same frequency. This is then amplified in a loudspeaker to produce sound.) So Equation 13-20 is really an equation for the sound frequencies produced by a stringed musical instrument. It shows that for strings of a given length L, the frequency and hence the pitch of the sound is greater for strings with less mass per unit length μ. That's why the first string on a guitar, which has a fundamental frequency of 329.6 Hz, is much thinner than the sixth string, which has a fundamental frequency of 82.4 Hz. It also shows that if you increase the tension F on a string, the frequency and pitch increase. And it explains what happens when you pinch one of the strings against the fingerboard: This shortens the length of string that's free to vibrate, thereby causing the frequency and pitch to go up.

Equation 13-19 shows that all of the other natural frequencies of a string held at both ends are integer multiples of f_1:

$$f_n = \frac{n}{2L}\sqrt{\frac{F}{\mu}} = n\left(\frac{1}{2L}\sqrt{\frac{F}{\mu}}\right) = nf_1$$

The $n = 2$ frequency f_2, called the *second harmonic* (or *first overtone*), is equal to $2f_1$; the $n = 3$ frequency f_3, called the *third harmonic* (or *second overtone*), is equal to $3f_1$; and so on.

The word "harmonic" should make you think of music, and indeed stringed musical instruments are an important application of the standing wave modes of a string. When you pluck a guitar string, the string actually vibrates in a *superposition* of standing waves with $n = 1$, $n = 2$, $n = 3$, and so on. (Just as two sinusoidal traveling waves can be present simultaneously on a string and combine to make a sinusoidal standing wave, more than one standing wave can be present on a string at the same time.) If it's an acoustic guitar, the guitar body is forced to oscillate simultaneously at the frequencies $f_1, f_2, f_3, \ldots$ and so produces a sound wave that is a superposition of sinusoidal waves at these frequencies. You might think that this would produce a hopeless jumble of sound. In fact, the result is a *nonsinusoidal* sound wave with the same frequency f_1 as the fundamental frequency of the string (**Figure 13-16**). The shape of the wave function of this nonsinusoidal wave—which depends on the relative amplitudes and phases of the sinusoidal waves at frequencies $f_1, f_2, f_3, \ldots$—determines the *tone quality* or *timbre* of the sound. Different musical instruments playing the same note all generate sound waves of the same frequency, but with different amounts of the various harmonics. As a result, different instruments produce sound waves with differently shaped wave functions and hence different tone qualities, which is why they sound different.

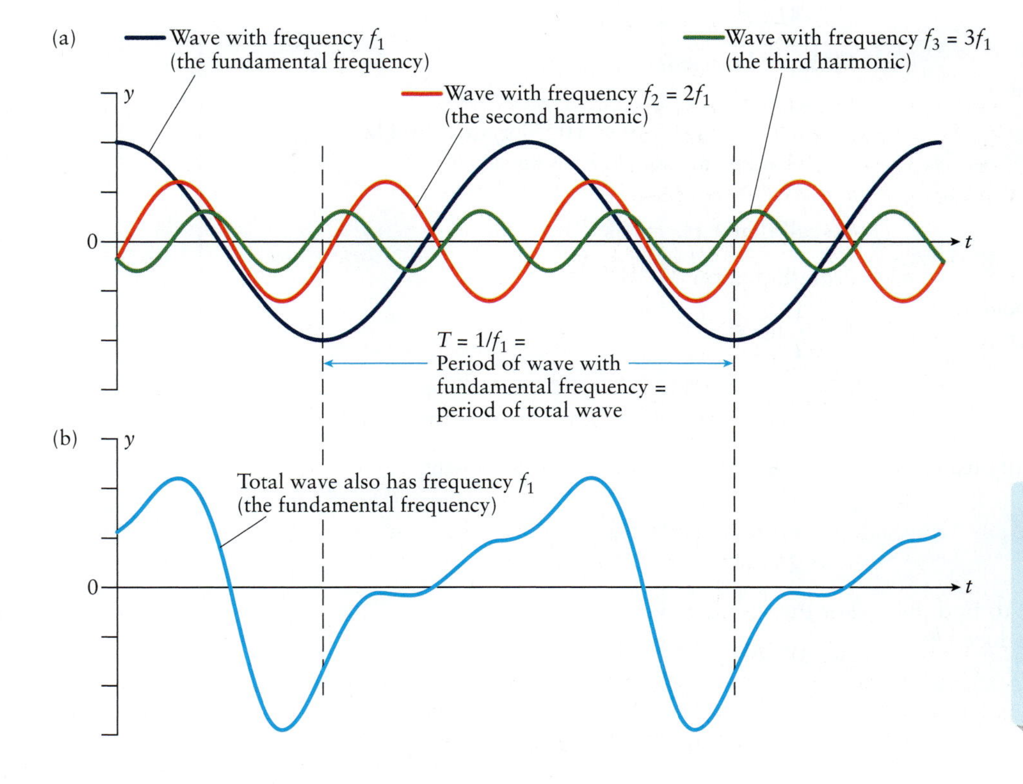

Figure 13-16 Adding harmonics (a) A sound wave is produced by a string oscillating simultaneously at its fundamental frequency f_1, its second harmonic f_2, and its third harmonic f_3. (b) The total sound wave has frequency f_1 but is not a sinusoidal wave.

• Adding waves of frequencies $f_1, f_2, f_3, \ldots$ (the harmonics of f_1) produces a wave of frequency f_1 with a non-sinusoidal wave function.
• The shape of the sound wave produced by a musical instrument determines the tone quality of the sound.

EXAMPLE 13-6 Guitar String Physics

The third string on a guitar is 0.640 m long and has a linear mass density of 1.14×10^{-3} kg/m. This string is to play a note of frequency 196 Hz (to a musician, this note is G3 or G below middle C). (a) What must be the tension in the string? (b) What is the wavelength of the fundamental standing wave on the string? (c) What is the speed of transverse waves on this string? (d) What is the wavelength of the sound wave produced by the wave?

Set Up

We are given the string length $L = 0.640$ m and the mass per length of the string $\mu = 1.14 \times 10^{-3}$ kg/m. We'll use Equation 13-19 to find the tension F that gives a fundamental frequency $f_1 = 196$ Hz. We'll use Equation 13-18 to find the wavelength λ of the standing wave and use this along with Equation 13-2 to find the speed of waves on the string. The sound wave produced by the guitar has the same *frequency* as the standing wave on the string. We'll find the *wavelength* of the sound wave using Equations 13-2 and 13-12.

Frequencies for a standing wave on a string:

$$f_n = \frac{n}{2L}\sqrt{\frac{F}{\mu}} \text{ where } n = 1, 2, 3, \ldots \quad (13\text{-}19)$$

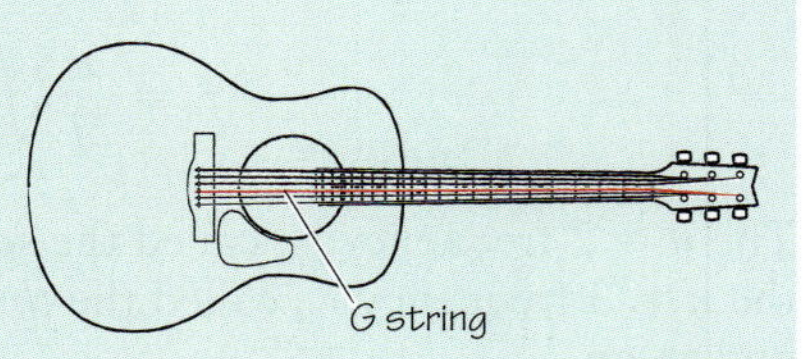

Wavelengths for a standing wave on a string:

$$L = \frac{n\lambda}{2} \text{ where } n = 1, 2, 3, \ldots \quad (13\text{-}18)$$

Propagation speed, frequency, and wavelength of a wave:

$$v_p = f\lambda \quad (13\text{-}2)$$

Speed of sound in dry air at 20°C:

$$v_{sound} = 343 \text{ m/s} \quad (13\text{-}12)$$

Solve

(a) The fundamental frequency is given by Equation 13-19 with $n = 1$. Solve this equation for the string tension F.

The fundamental frequency ($n = 1$) is

$$f_1 = \frac{1}{2L}\sqrt{\frac{F}{\mu}}$$

To solve for the string tension F, first square both sides to get rid of the square root:

$$f_1^2 = \frac{1}{4L^2}\frac{F}{\mu}$$

Multiply both sides by $4L^2\mu$:

$$\begin{aligned} F &= 4L^2\mu f_1^2 \\ &= 4(0.640 \text{ m})^2(1.14 \times 10^{-3} \text{ kg/m})(196 \text{ Hz})^2 \\ &= 71.8 \text{ kg} \cdot \text{m} \cdot \text{Hz}^2 = 71.8 \text{ kg} \cdot \text{m/s}^2 \\ &= 71.8 \text{ N} \end{aligned}$$

(Recall that 1 Hz = 1 s^{-1}.)

(b) For the fundamental mode (the $n = 1$ standing wave) of a string held at both ends, the length of the string equals one half-wavelength.

From Equation 13-18,

$$L = \frac{\lambda}{2} \text{ for } n = 1$$

$$\lambda = 2L = 2(0.640 \text{ m}) = 1.28 \text{ m}$$

(wavelength of the standing wave on the string)

(c) Find the wave speed on the string using Equation 13-2.

Speed of transverse waves on this string:

$$\begin{aligned} v_p &= f\lambda \\ &= (196 \text{ Hz})(1.28 \text{ m}) \\ &= 251 \text{ m/s} \end{aligned}$$

(d) Use Equations 13-2 and 13-12 to find the wavelength of the sound wave produced by the guitar.

For the sound wave,

$$v_{sound} = f\lambda$$

The frequency of the sound wave is the same as the frequency of the standing wave on the spring, but the wavelength is different because the wave speed is different:

$$\lambda = \frac{v_{\text{sound}}}{f} = \frac{343 \text{ m/s}}{196 \text{ Hz}} = \frac{343 \text{ m/s}}{196 \text{ s}^{-1}}$$
$$= 1.75 \text{ m}$$

Reflect

There are six strings on the guitar, each of which is under about the same tension. So the total force that acts on the guitar at the points where the strings are attached is about $6 \times 71.8 \text{ N} = 431 \text{ N}$ (about 97 lb). The guitar must be of sturdy construction to withstand these forces.

You can check our result for the speed of transverse waves on the string by using Equation 13-10, $v_p = \sqrt{F/\mu}$, and the value of F that we found in part (a). Do you get the same answer this way?

Note that our answer for part (d) is reasonable. The wavelength λ of the sound wave is greater than the wavelength of the standing wave on the string because, although the frequency f is the same for both waves, the propagation speed v_p is greater for the sound wave and $v_p = f\lambda$.

GOT THE CONCEPT? 13-5 Changing the String Tension I

If the tension in a guitar string is increased by a factor of 2, the wavelength of the $n = 1$ (fundamental) standing wave mode will (a) increase to twice its previous value; (b) increase, but by a factor different than 2; (c) decrease to 1/2 of its previous value; (d) decrease but by a factor different than 1/2; (e) none of these.

GOT THE CONCEPT? 13-6 Changing the String Tension II

If the tension in a guitar string is increased by a factor of 2, the frequency of the $n = 1$ (fundamental) standing wave mode will (a) increase to twice its previous value; (b) increase, but by a factor different than 2; (c) decrease to 1/2 of its previous value; (d) decrease, but by a factor different than 1/2; (e) none of these.

TAKE-HOME MESSAGE FOR Section 13-6

✔ Two identical waves traveling in opposite directions can interfere and form a standing wave. In a standing wave the points where there is zero displacement (the nodes) always remain at the same places in the medium.

✔ For a stretched string held down at both ends, the only allowed standing waves are those that have an integer number of half-wavelengths in the length of the string.

✔ The length, linear mass density, and tension of a string determine the standing wave frequencies for that string.

13-7 Wind instruments, the human voice, and the human ear use standing sound waves

You've probably noticed that your singing voice sounds much better in the shower than in the open. You may have also noticed that a flute (a musical instrument based on a long tube) produces lower notes than a piccolo (which uses a shorter tube). And you know that the sound of fingernails on a chalkboard is a particularly unpleasant one—so much so that just thinking about that sound may make you cringe.

We can help to explain all of these effects by again invoking the idea of *standing waves*, which we introduced in the last section. The standing waves we need to consider are not waves on a string, but rather standing *sound* waves that result from traveling sound waves that reflect from the ends of an enclosure or tube.

Figure 13-17 shows an example of such a standing sound wave. Sound waves travel horizontally through the transparent tube and are reflected when they strike the ends,

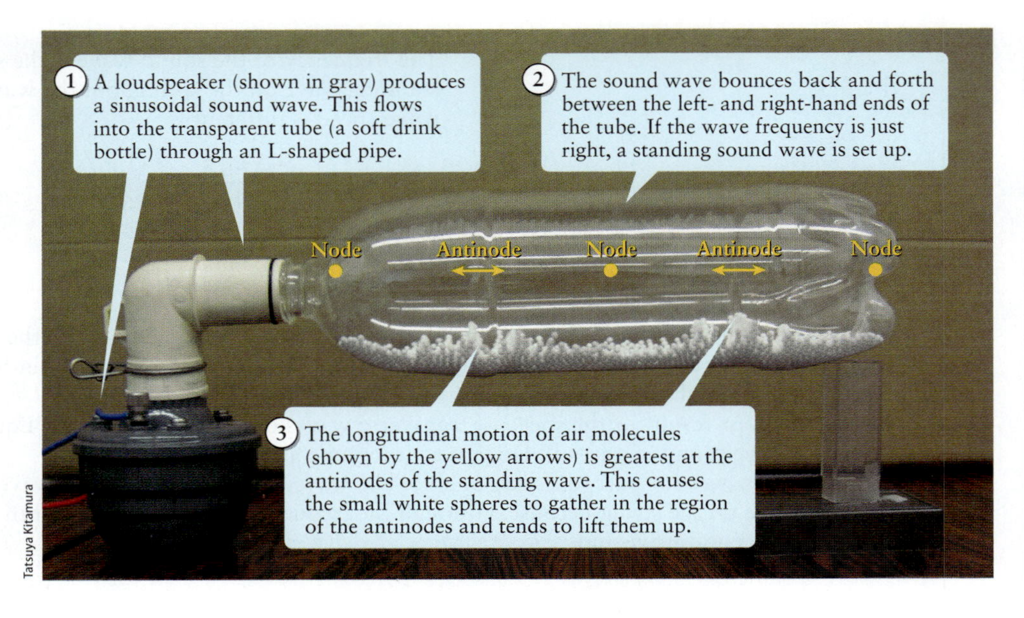

1 A loudspeaker (shown in gray) produces a sinusoidal sound wave. This flows into the transparent tube (a soft drink bottle) through an L-shaped pipe.
2 The sound wave bounces back and forth between the left- and right-hand ends of the tube. If the wave frequency is just right, a standing sound wave is set up.
Node
Antinode
Node
Antinode
Node
3 The longitudinal motion of air molecules (shown by the yellow arrows) is greatest at the antinodes of the standing wave. This causes the small white spheres to gather in the region of the antinodes and tends to lift them up.
Tatsuya Kitamura

Figure 13-17 Visualizing a standing sound wave The small white spheres inside this apparatus (known as *Kundt's tube* after the German physicist who devised it in the nineteenth century) help to identify the locations of the antinodes of a standing sound wave in air, which would otherwise be invisible.

just as transverse waves on a string are reflected at the fixed ends of the string. Sound is a longitudinal wave, so the motion of the air caused by the wave moving horizontally through the tube is also in the horizontal direction. The solid ends of the tube prevent this motion from taking place there, so there is zero displacement at the ends. This is just like the situation for a string with both ends fixed (see Figure 13-14). So just as for a standing wave on a string, a standing *sound* wave will be set up if a whole number of half-wavelengths of the wave fit into the length of the tube (Equation 13-18). In Figure 13-17 the frequency of sound waves provided by the loudspeaker has been adjusted so that two half-wavelengths (one full wavelength) fit into the tube's length, corresponding to $n = 2$ in Figure 13-14 and Equation 13-18.

Note that we can use Figure 13-14 for the displacement in a standing wave on a string to depict the displacement in a standing *sound* wave in a tube like the one shown in Figure 13-17. The difference is that because sound is a longitudinal wave, the displacement is now measured *along* the direction of wave propagation (see Figure 13-9).

Note also that the *pressure variation* in a traveling sound wave is a quarter-cycle out of phase with the displacement. Figure 13-9 shows this for a traveling sound wave, but the same is true for a standing sound wave. So where there is zero displacement in a standing sound wave (a displacement node) is where the pressure variation is maximum (a pressure antinode); where the displacement oscillates with maximum amplitude (a displacement antinode) is where the pressure variation is zero, so the pressure is always equal to the undisturbed air pressure (a pressure node).

Each standing wave sound frequency corresponds to a natural frequency of the tube. That means that if you produce a sound in the tube at one of those natural frequencies, *resonance* will take place and a strong standing wave will be set up at that frequency. The net result will be that the sound wave you produce will be enhanced, just like the oscillation of a mass–spring combination is enhanced if you force it to oscillate at its natural frequency (see Section 12-8). That explains what happens when you sing in the shower. In the $n = 1$ standing wave mode, half a wavelength will fit between the walls of a shower stall. A typical shower stall is about 1 m wide, so the wavelength of the $n = 1$ mode is about 2 m. The frequency f_1 of this mode is related to the wavelength λ and the propagation speed v_{sound} for sound waves by $v_{\text{sound}} = f_1\lambda$ (Equation 13-2), so

$$f_1 = \frac{v_{\text{sound}}}{\lambda} = \frac{343 \text{ m/s}}{2 \text{ m}} \approx 170 \text{ Hz}$$

So if you sing at a frequency near 170 Hz (E or F below middle C), the shower stall will resonate and your voice will sound much fuller than it would outside the shower. The same will happen at frequencies that are multiples of 170 Hz (340 Hz, 510 Hz, 680 Hz, etc.).

Standing Waves and Wind Instruments: Closed Pipes

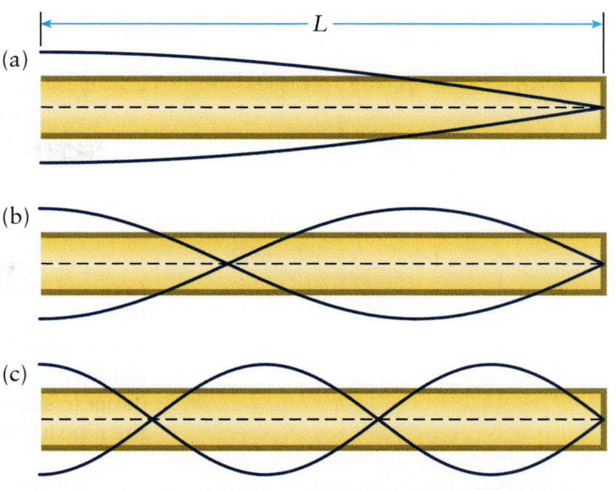

Figure 13-18 Standing sound waves in a closed pipe I (a), (b), and (c) show the displacement patterns for the first three standing sound wave modes of a pipe closed at its right-hand end and open at its left-hand end. There is a displacement node at the closed end and a displacement antinode at the open end.

Just as a guitar or other stringed instrument produces tones by using standing waves on a string, a *wind* instrument makes use of standing waves in an air-filled tube. These instruments are played by setting the air within the tube into oscillation. In a brass instrument like a trumpet or trombone, the musician does this by making a buzzing sound with her lips; in a woodwind like a clarinet or saxophone, the musician blows into a reed that vibrates the air in the instrument. In order for the sound to get out of the instrument, it has to be *open* at one or both ends. (When a sound wave traveling down the tube reaches the open end, part of it is reflected back. So the ingredients for making a standing wave are still present.) This means that a wind instrument has a different set of standing wave modes than those shown in Figure 13-14.

Most wind instruments behave like a **closed pipe**, in which one end is closed and the other is open to the air. The end that the musician blows into behaves like a closed end: The displacement is essentially zero, and the pressure variation has its maximum amplitude. (This is analogous to what happens at the left-hand end of the string in Figure 13-15. It's at that end that a force is applied to make the string oscillate, but the displacement there is very small. In the same way there is very little displacement of the air at the musician's end of a brass instrument or woodwind, but there is a large pressure variation due to the musician either buzzing her lips or making a reed vibrate.) The other end of the tube is open to the air. As a result the pressure there is essentially equal to the pressure of the surrounding air, so the pressure variation is close to zero and the displacement amplitude is large. In other words, a closed tube has a displacement node at the closed end but a displacement antinode at the open end.

Figure 13-18 shows the first three standing wave modes in a closed pipe. We represent the magnitude of the displacement of the air from equilibrium by the separation of the blue curves from the dashed centerline. **Figure 13-18a** shows that in the fundamental mode of this closed pipe, only *one-quarter* of a wavelength fits in the length of the pipe. You can see this more clearly in **Figure 13-19a**, in which we've extended the sinusoidal representation of the fundamental mode out beyond the open end to show one full wavelength. For a closed pipe, then, the relationship between the length of the tube L and the wavelength λ of the sound wave that sets up the fundamental mode of the standing wave interference is

$$L = \frac{1}{4}\lambda$$

WATCH OUT! The standing sound wave in a closed pipe does not extend far beyond the pipe.

This extension of the wave pattern that we've drawn in Figure 13-19a does *not* represent the standing wave; the standing wave exists inside the pipe only. We've drawn this only to allow you to compare the wavelength of the sound with the length of the pipe.

Figures 13-18b and **13-18c** show the next two standing wave modes of a closed pipe, and **Figure 13-19** shows the displacement patterns extended outside the pipe. You can see that three quarters (3/4) and five quarters (5/4) of a wavelength, respectively, fit into the pipe for these modes. You can also see that the general relationship between L and λ for a closed pipe is

Length of a closed pipe (open at one end, closed at the other)

Wavelength of a standing wave in the pipe

$$L = \frac{n\lambda}{4} \quad \text{where } n = 1, 3, 5, \ldots$$

Wavelengths for a standing sound wave in a closed pipe (13-21)

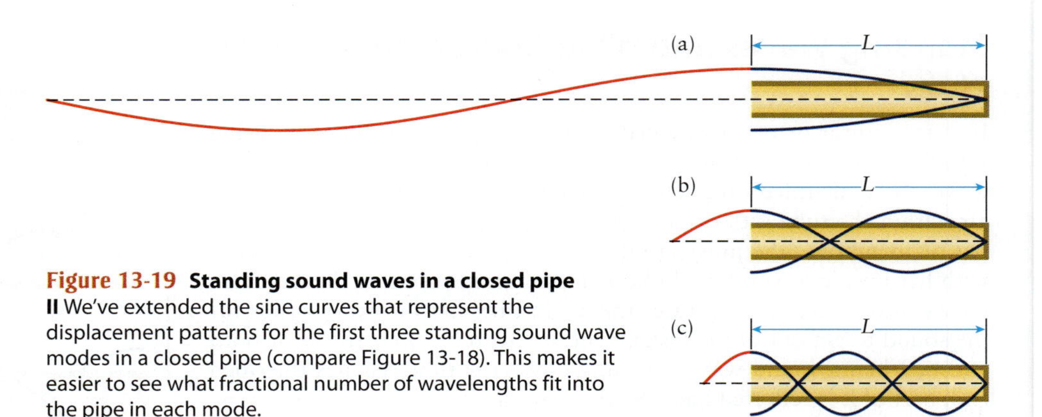

Figure 13-19 Standing sound waves in a closed pipe II We've extended the sine curves that represent the displacement patterns for the first three standing sound wave modes in a closed pipe (compare Figure 13-18). This makes it easier to see what fractional number of wavelengths fit into the pipe in each mode.

Notice that only the *odd* harmonics ($n = 1, 3, 5, \ldots$) can exist in a pipe closed at one end. From $v_{\text{sound}} = f\lambda$ (Equation 13-2 for sound) the corresponding natural frequencies for a closed pipe are

$$f_n = \frac{v_{\text{sound}}}{\lambda} = \frac{v_{\text{sound}}}{(4L/n)} = n\left(\frac{v_{\text{sound}}}{4L}\right) \text{ where } n = 1, 3, 5, \ldots \tag{13-22}$$

You can see from Equation 13-22 that the fundamental frequency of a closed pipe of length L is $f_1 = v_{\text{sound}}/4L$. The next harmonic is the *third* harmonic, $f_3 = 3(v_{\text{sound}}/4L) = 3f_1$; the next harmonic after that is the *fifth* harmonic, $f_5 = 5f_1$; and so on. For a closed pipe the even harmonics are absent.

Just like what happens when you pluck a guitar string, blowing into a closed-pipe wind instrument produces standing sound waves at more than one of the natural frequencies given by Equation 13-22. As a result the sound that emanates from the instrument will be a combination of sinusoidal sound waves at frequencies $f_1, f_3, f_5, \ldots$, and the net result will be a nonsinusoidal wave with frequency f_1 (see Figure 13-16). The particular combination of harmonics is different for different wind instruments playing the same note, which gives each instrument its own distinctive tone quality. Equation 13-22 tells us that to change the fundamental frequency of a closed pipe, you must change the pipe length L. For a brass instrument this is done either with a slide (as in a trombone) or with valves that open and close air passages (as in a trumpet or tuba). For a woodwind like a clarinet or oboe, this is done by opening holes in the pipe that effectively shorten the length of the vibrating air column.

You actually have closed pipes on either side of your head. These are the *auditory canals* of your left and right ears (**Figure 13-20**), which extend a distance of about

Figure 13-20 The human ear Sound waves entering the human ear via the auditory canal cause the eardrum to vibrate, which in turn cause vibrations of the three ossicles (the smallest bones in the human body). These induce vibrations in the fluid that fills the inner ear. These vibrations are detected and converted to an electrical signal that is carried by the auditory nerve to the brain. The auditory canal acts like a closed pipe, which makes hearing especially sensitive for frequencies around 3430 Hz.

2.5 cm = 0.025 m from the opening (an open end) to the eardrum (a closed end). From Equation 13-22 the fundamental frequency ($n = 1$) of this short closed pipe is

$$f_1 = \frac{v_{\text{sound}}}{4L} = \frac{343 \text{ m/s}}{4(0.025 \text{ m})} = 3430 \text{ Hz}$$

This suggests that if a sound wave with a frequency of about 3430 Hz enters your ear, it will cause resonance to take place and the sound wave will be enhanced. In fact, a person with normal hearing is most sensitive to sounds around this frequency. And an important reason why the sound of fingernails scraping across a chalkboard is so unpleasant is that this is a sound in the range from 2000 to 4000 Hz—precisely where your hearing is the most sensitive.

Your vocal tract, which extends from the vocal folds in your throat to your lips, is also a closed pipe (closed at the vocal folds, which behave rather like the reed in a clarinet or oboe, and open at the lips). Unlike other musical instruments, you can reshape your vocal tract by the way that you hold your tongue and your lips. This changes the natural frequencies of the vocal tract (so that they no longer obey Equation 13-22) and makes each spoken sound distinct. To demonstrate this try saying "a, e, i, o, u" and notice how your tongue and lips change position. Then use your fingers to hold your lips together and repeat the experiment; without the ability to reshape your vocal tract, all five vowels will now sound almost the same.

BIO-Medical **EXAMPLE 13-7 An Amorous Frog**

The male tree-hole frog of Borneo (*Metaphrynella sundana*) attracts females by croaking out a simple call dominated by a tone of a single frequency. To amplify his call and enhance his attractiveness to potential mates, he finds a tree with a cavity in its trunk, sits inside the cavity, and adjusts the frequency of his call to match the fundamental frequency of the cavity. If the cavity is cylindrical and 11 cm deep, at what frequency should the frog make his mating call?

Set Up

The cavity in the tree trunk acts like a closed pipe (open on the outside of the tree, closed on the inside). Equation 13-22 gives the frequency of the fundamental ($n = 1$) mode of this cavity.

Frequencies for a standing sound wave in a closed pipe:

$$f_n = n\left(\frac{v_{\text{sound}}}{4L}\right) \text{ where } n = 1, 3, 5, \ldots \quad (13\text{-}22)$$

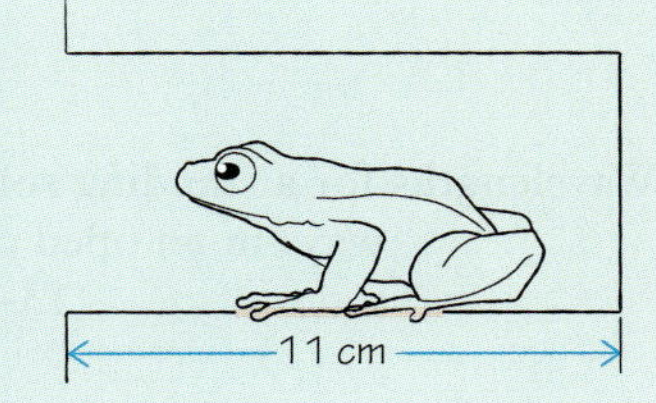

Solve

Calculate the fundamental frequency using $n = 1$ and $v_{\text{sound}} = 343$ m/s.

The fundamental frequency ($n = 1$) is

$$f_1 = \frac{v_{\text{sound}}}{4L} = \frac{343 \text{ m/s}}{4(11 \text{ cm})}\left(\frac{100 \text{ cm}}{1 \text{ m}}\right)$$
$$= 780 \text{ s}^{-1} = 780 \text{ Hz}$$

Reflect

Our calculated result is close to the measured value of the frequency of a male tree-hole frog's mating call in a cavity 11 cm deep. These frogs have been observed to match the natural frequency of tree cavities from 10 to 15 cm in depth.

Standing Waves and Wind Instruments: Open Pipes

Unlike most other wind instruments, a *flute* is a tube that is open at *both* ends. This is called an **open pipe**. A flautist plays this instrument by blowing across the top of a hole in the pipe; this causes a pressure variation that causes the air inside the pipe to oscillate.

Because this pipe is open at each end, there is a displacement antinode (pressure node) at each end. **Figure 13-21** shows the displacement patterns for the first three standing wave modes of such a pipe. (Just as at the open end of a closed pipe, a sound

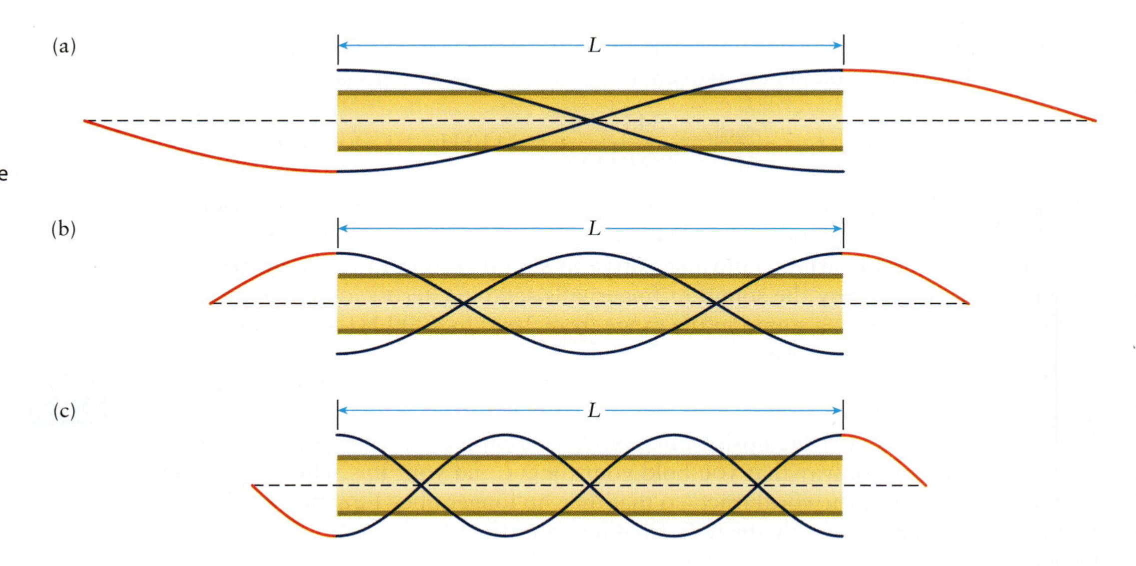

Figure 13-21 Standing sound waves in an open pipe (a), (b), and (c) show the displacement patterns for the first three standing sound wave modes of a pipe open at both ends. There is a displacement antinode at each open end.

wave traveling through an open pipe is partially reflected at each open end. The traveling waves moving in opposite directions through the pipe give rise to a standing wave.) You can see that in **Figure 13-21a** one half-wavelength fits in the length L of the pipe. This represents the fundamental mode of the pipe, so the wavelength λ of the fundamental mode is given by

$$L = \frac{1}{2}\lambda$$

Similarly, the standing wave modes in **Figures 13-21b** and **13-21c** have two half-wavelengths (one full wavelength) and three half-wavelengths, respectively, in the length of the pipe. In general, the relationship between L and λ for an open pipe is

Wavelengths for a standing sound wave in an open pipe (13-23)

Length of an open pipe (open at both ends)

Wavelength of a standing wave in the pipe

$$L = \frac{n\lambda}{2} \quad \text{where } n = 1, 2, 3, \ldots$$

This is the same relationship as for standing waves on a string held at both ends (Equation 13-18) or for a pipe *closed* at both ends (see Figure 13-17). Although the wavelengths are the same as in those two cases, an open pipe has antinodes, not nodes, at its two ends.

The corresponding natural frequencies are

$$f_n = \frac{v_{\text{sound}}}{\lambda} = \frac{v_{\text{sound}}}{(2L/n)} = n\left(\frac{v_{\text{sound}}}{2L}\right) \text{ where } n = 1, 2, 3, \ldots \qquad (13\text{-}24)$$

Unlike a closed pipe (Equation 13-22), an open pipe has *all* harmonics of its fundamental ($n = 1$) frequency. Note that for a given length L, the fundamental frequency of a closed pipe (from Equation 13-22, $f_1 = v_{\text{sound}}/4L$) is one-half of the fundamental frequency of an open pipe (from Equation 13-24, $f_1 = v_{\text{sound}}/2L$). If you cover one end of an open pipe while blowing into it, you change it into a closed pipe and the sound it produces will be lower in frequency and pitch.

GOT THE CONCEPT? 13-7 Filling a Water Bottle

Imagine that you gently tap the side of a water bottle as you fill it. As the water level rises, the pitch of the sound made by the bottle as you tap it (a) increases; (b) decreases; (c) first increases, then decreases; (d) first decreases, then increases; (e) stays the same.

TAKE-HOME MESSAGE FOR Section 13-7

✔ Traveling sound waves in a pipe can interfere to form a standing sound wave.

✔ A displacement node always appears at a closed end of the pipe, and a displacement antinode always appears at an open end.

✔ For a closed pipe (open at one end, closed at the other) the standing wave frequencies are odd multiples of the fundamental frequency. For an open pipe (open at both ends) the standing wave frequencies include both even and odd multiples of the fundamental frequency.

13-8 Two sound waves of slightly different frequencies produce beats

You might hear a sound something like "wah wah wah" when a guitar player tunes her instrument by plucking two strings at the same time. As she adjusts the tension on one string, the time between the "wahs" gets so long that they can no longer be heard. This phenomenon, known as **beats**, arises from the interference between sound waves. Beats are most pronounced when two waves of nearly identical frequencies interfere.

To see how beats arise let's see what happens when we combine two sinusoidal waves with the same amplitude A but with slightly different frequencies f_1 and f_2. **Figure 13-22** shows the result: The total wave is also a sinusoidal wave whose frequency f is the average of f_1 and f_2, but with an amplitude that varies between 0 and $2A$. We use the term beats for this up-and-down variation in amplitude. The frequency of the beats, also called the **beat frequency**, is equal to the *absolute value* of the difference between the two frequencies:

Beat frequency heard when sound waves of two similar frequencies interfere

Frequencies of the individual sound waves

$$f_{\text{beats}} = |f_2 - f_1|$$

Beat frequency (13-25)

(The absolute value ensures that we get a positive value for f_{beats}. We do this because a negative frequency has no meaning.) For example, if the two individual sound waves have frequencies $f_1 = 200$ Hz and $f_2 = 202$ Hz, what you will hear is a tone at a frequency of 201 Hz (the average of 200 Hz and 202 Hz) with an amplitude that rises and falls at frequency $f_{\text{beats}} = |f_2 - f_1| = |202 \text{ Hz} - 200 \text{ Hz}| = 2$ Hz. That is, the amplitude will rise and fall twice per second, or once every 1/2 s.

The greater the difference between the frequencies f_1 and f_2, the greater the beat frequency and the more rapid the beats. If the frequency difference is large enough, the beats are no longer perceptible and you will hear two distinct frequencies.

WATCH OUT! Don't confuse the beat frequency with the frequency of the sound.

When you hear two waves with slightly different frequencies f_1 and f_2, the beat frequency given by Equation 13-25 is the frequency at which the amplitude of the sound rises and falls. The frequency of the sound itself is different: It's just the average of f_1 and f_2, or $(f_1 + f_2)/2$.

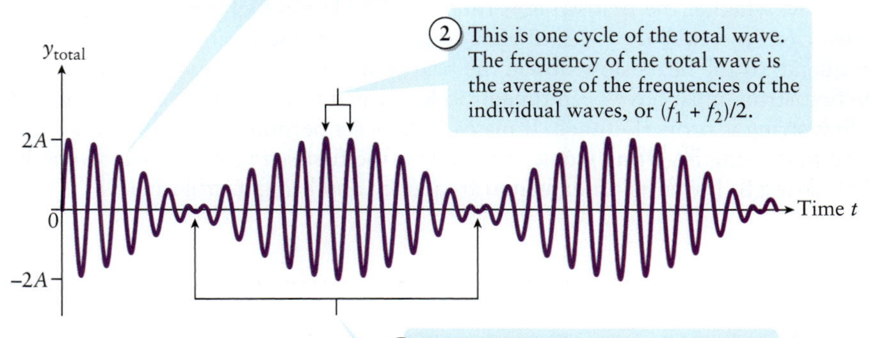

1 This is the total wave that results when two individual sine waves of similar frequencies f_1 and f_2 are combined.
2 This is one cycle of the total wave. The frequency of the total wave is the average of the frequencies of the individual waves, or $(f_1 + f_2)/2$.
y_{total}
2A
0
−2A
Time t
3 The amplitude of the total wave rises and falls periodically. The frequency of this rising and falling is the beat frequency, $f_{\text{beats}} = |f_2 - f_1|$.

Figure 13-22 Beats When two sinusoidal waves of nearly the same frequency are added together, the total wave has a rising and falling amplitude, or beats.

EXAMPLE 13-8 Guitar Tuning

You have just replaced the first string on your guitar and wish to tune it to the correct frequency of 329.6 Hz (the note E4). To do this you pluck both this string and the sixth (E2) string, which you know is properly tuned to 82.4 Hz. You hear one beat every 2.50 s. How far out of tune (in Hz) is the first string?

Set Up

The frequencies 82.4 and 329.6 Hz are far apart, so it may seem surprising that you would get beats when both are sounded. But remember that a plucked string oscillates not only at its fundamental frequency but also at the higher *harmonics* of the string. So what you hear is beats between the sound made by the first string and one of the harmonics of the sixth string.

Beat frequency:

$$f_{\text{beats}} = |f_2 - f_1| \qquad (13\text{-}25)$$

Frequencies for a standing wave on a string:

$$f_n = \frac{n}{2L}\sqrt{\frac{F}{\mu}} \text{ where } n = 1, 2, 3, \ldots \qquad (13\text{-}19)$$

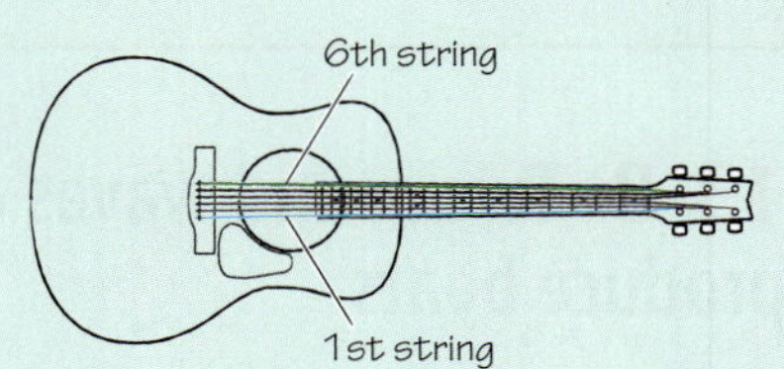

Solve

The harmonics of the sixth string are at integer multiples of its fundamental frequency.

For the sixth string, $f_1 = 82.4$ Hz. The harmonics are

$$f_2 = 2f_1 = 164.8 \text{ Hz}$$

$$f_3 = 3f_1 = 247.2 \text{ Hz}$$

$$f_4 = 4f_1 = 329.6 \text{ Hz}$$

So if the guitar is properly tuned, the fundamental frequency of the first string is the same as the fourth harmonic frequency of the sixth string. If the first string is mistuned, the frequencies will be different and you will hear beats.

Find the beat frequency and how far out of tune the first string is.

There is one beat per 2.50 s, so the beat frequency (the number of beats per second) is

$$f_{\text{beats}} = \frac{1}{2.50 \text{ s}} = 0.400 \text{ s}^{-1} = 0.400 \text{ Hz}$$

This is equal to the difference between the fundamental frequency of the first string and the fourth harmonic frequency of the sixth string:

$$f_{\text{beats}} = 0.400 \text{ Hz} = |f_{1 \text{ for first string}} - f_{4 \text{ for sixth string}}|$$

$$= |f_{1 \text{ for first string}} - 329.6 \text{ Hz}|$$

So the first string is out of tune by 0.400 Hz.

Reflect

With just the information given, we don't know whether the first string is tuned 0.400 Hz too high (in which case its frequency is 330.0 Hz) or too low (in which case its frequency is 329.2 Hz). To find out try increasing the tension on the first string to raise its fundamental frequency. If the beats slow down, the beat frequency is getting closer to zero as you bring the first string into tune, so the first string must have been tuned too low. Keep increasing the tension until the beats stop altogether, at which point the first string is properly tuned. If instead the beats become more rapid, the beat frequency is increasing, so the tuning of the first string is getting worse. This means the first string was tuned too high, so you should decrease the tension on the first string to lower its fundamental frequency and bring it into tune.

GOT THE CONCEPT? 13-8 Tuning Forks

When you strike two tuning forks simultaneously, you hear 3 beats per second. The frequency of the first tuning fork is 440 Hz. What is the frequency of the second tuning fork? (a) 446 Hz; (b) 443 Hz; (c) 437 Hz; (d) 434 Hz; (e) not enough information is given to decide.

TAKE-HOME MESSAGE FOR Section 13-8

✔ When two waves of nearly identical frequencies and wavelengths interfere, the result is a total wave whose amplitude rises and falls periodically. This phenomenon is called beats.

✔ The frequency of the beats equals the absolute value of the difference in frequency between the two waves.

13-9 The intensity of a wave equals the power that it delivers per square meter

Humans hear sounds over a wide range of volumes, from very faint sounds (such as leaves rustling in a gentle breeze) to sounds so loud that they can cause pain (such as the sound of the siren on an emergency vehicle driving nearby). Let's analyze the physical differences between loud sounds and faint sounds.

Wave Energy, Power, and Intensity

An important factor in determining the loudness of a sound is the energy carried by the sound wave. To see why this is so, note that in order for the ear to detect a sound, the sound wave must exert a force on the eardrum that makes the eardrum move (be displaced). This means that the sound wave must do *work* on the eardrum (recall from Section 6-2 that work is done when a force acts over a distance). Doing work involves the transfer of energy, and in this case the energy comes from the wave itself. So we can think of the ear as a device for detecting the energy in a sound wave.

The ear is particularly sensitive to the *rate* at which sound wave energy is delivered to it. As an analogy, imagine having someone pour a pitcher of water over your outstretched hand. You'll feel very little effect if the water falls on you very slowly, a drop at a time. But if the pitcher is turned over so that all of the water hits your hand in a short time, your hand will feel pushed down. In the same way your ear has a greater response if sound wave energy is delivered to your ear at a rapid rate.

As we learned in Section 6-9, the rate at which energy is transferred is called **power.** The unit of power is the joule per second, or **watt** (abbreviated W): 1 W = 1 J/s. As an example, a 50-W light bulb is designed so that when it is in operation 50 J of energy is delivered to it every second by the electric circuit to which the bulb is connected. If you are using a stationary exercise bicycle at the gym, a reading of 200 W on the bike's display means that you are doing 200 joules of work on the pedals every second. We'll usually denote power by the symbol P.

Let's apply these ideas about energy and power to a sinusoidal sound wave. We've seen that in a sinusoidal wave, pieces of the wave medium undergo simple harmonic motion. In Section 12-4 we learned that if an object of mass m is attached to an ideal spring of spring constant k and oscillates in simple harmonic motion with amplitude A, the total mechanical energy of the mass–spring system is

$$E = \frac{1}{2}kA^2 \qquad \text{(12-21)}$$

We also learned in Section 12-3 that the angular frequency for an object oscillating on an ideal spring is $\omega = \sqrt{k/m}$, so $\omega^2 = k/m$ and $k = m\omega^2$. If we insert this into Equation 12-21, we get

$$E = \frac{1}{2}m\omega^2 A^2 \qquad \text{(13-26)}$$

Let's see what this says about a sinusoidal sound wave in air with angular frequency ω and displacement amplitude A. (Remember that for a sound wave A represents the maximum longitudinal displacement of the air as the wave passes through it.) According to Equation 13-26 if this wave is present in a chunk of air of mass m, the *wave energy* present in that chunk is $E = (1/2)m\omega^2 A^2$. The wave carries that energy along with it as it propagates. To get a measure of the *power* associated with the wave, let's imagine a cubical volume in the air with sides of length L (**Figure 13-23**). If the

This volume of air has mass $m = \rho L^3$. A sound wave of angular frequency ω and amplitude A is present throughout the volume.

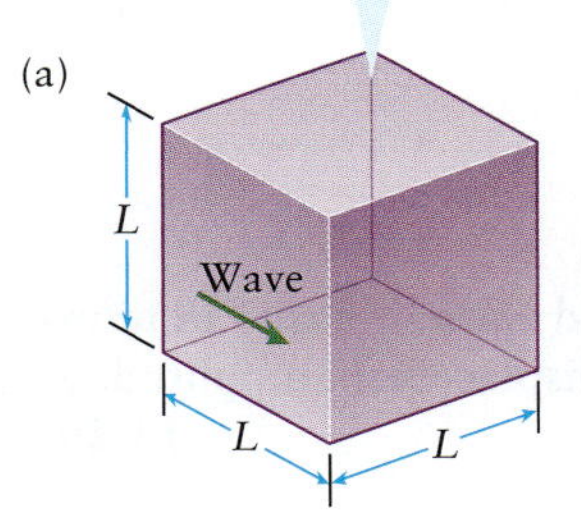

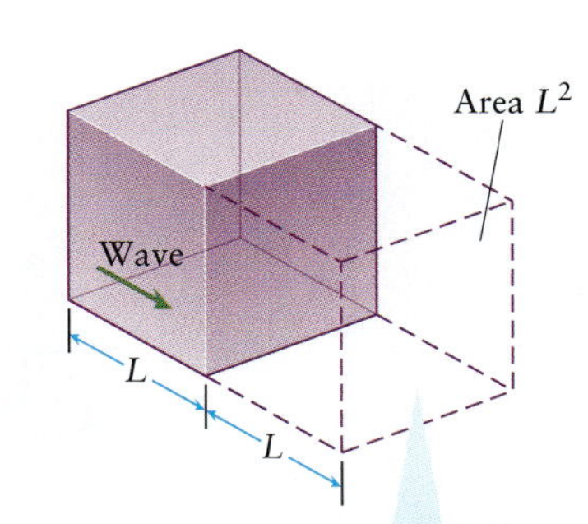

The volume L^3 of wave energy travels a distance L in time $t = L/v_{sound}$. The power equals the wave energy E in the volume divided by t. The intensity is the power divided by L^2.

Figure 13-23 Power and intensity in a sound wave (a) This cubical volume encloses a quantity of air with a sound wave propagating through it. (b) If we imagine the volume moving along with the wave, we can find the power and intensity associated with the wave.

air has density (mass per volume) ρ, the mass of air in this volume equals the density ρ multiplied by the volume L^3 of the cube: $m = \rho L^3$. From Equation 13-26 the wave energy contained in this volume is

$$E = \frac{1}{2}\rho L^3 \omega^2 A^2 \tag{13-27}$$

The energy in this volume moves along with the wave at propagation speed v_{sound}, so the time required for this volume of energy to move through its own length L is $t = L/v_{\text{sound}}$. So the rate at which this energy crosses through an imaginary square of side L (Figure 13-23)—that is, the *power* P delivered across this surface—equals the energy E given by Equation 13-27 divided by the time $t = L/v_{\text{sound}}$:

$$P = \frac{E}{t} = \frac{(1/2)\rho L^3 \omega^2 A^2}{L/v_{\text{sound}}} = \frac{1}{2}\rho L^2 v_{\text{sound}} \omega^2 A^2 \tag{13-28}$$

WATCH OUT! In a wave there is a flow of energy, not of material.

Note that in Figure 13-23 what moves a distance L in time t is the *wave energy* contained within the cubical volume of side L. The wave and its associated energy move together, but the wave *medium* doesn't go anywhere: Each piece of the medium just oscillates around its equilibrium position.

Equation 13-28 says that the power associated with the wave is proportional to the square of the wave amplitude A. This turns out to be a general attribute of waves of all kinds, not just sound waves. Note also that the power is proportional to L^2, which is the cross-sectional area through which the wave energy passes. The power *divided* by the cross-sectional area L^2 is called the **intensity** I of the wave, measured in watts per square meter (W/m^2). From Equation 13-28 this is

Sound wave intensity in terms of displacement amplitude A

$$I = \frac{1}{2}\rho v_{\text{sound}} \omega^2 A^2 \tag{13-29}$$

Intensity of a sound wave (I)
Density of the wave medium (usually air) (ρ)
Displacement amplitude of the wave (A)
Speed of sound waves in the medium (v_{sound})
Angular frequency of the wave (ω)

If you increase the displacement amplitude A of a sound wave of a given frequency, the intensity of the sound will increase. For an ear or other sound detector of a given size, the amount of sound wave power that is received equals the intensity I multiplied by the area of the detector. So increasing the intensity means more sound power enters your ear, and you will perceive the sound to be louder.

BIO-Medical

Note that the total power (intensity times area) reaching your eardrum would increase if it had a larger cross-sectional area. That's why many nocturnal animals such as bats and coyotes have large external ears: This maximizes the cross-sectional area over which they can collect low-amplitude sound waves coming from potential prey or predators. The larger collected power is then directed to their eardrums, improving their sensitivity. Intensity is a useful measure of the strength of a sound wave because the area factor has been divided out: A coyote may have more sensitive ears than yours (because of the increased collection area of its outer ears), but if you both are exposed to the same sound wave, both your external ears experience the same intensity.

Equation 13-29 is expressed in terms of the displacement amplitude A. However, our ears are more directly sensitive to the *pressure* variations in a sound wave. The displacement amplitude A in a sound wave is related to the pressure amplitude p_{max}, or maximum pressure variation, by

$$A = \frac{p_{\text{max}}}{\rho v_{\text{sound}} \omega} \tag{13-30}$$

If we substitute Equation 13-30 for A into Equation 13-29, we get an alternative expression for the intensity of a sound wave:

$$I = \frac{1}{2}\rho v_{sound}\omega^2\left(\frac{p_{max}}{\rho v_{sound}\omega}\right)^2 = \frac{p_{max}^2}{2}\frac{\rho v_{sound}\omega^2}{\rho^2 v_{sound}^2\omega^2}$$

or, simplifying,

Intensity of a sound wave

Pressure amplitude of the wave

$$I = \frac{p^2{}_{max}}{2\rho v_{sound}}$$

Density of the wave medium (usually air)

Speed of sound waves in the medium

Sound wave intensity in terms of pressure amplitude p_{max} (13-31)

Equation 13-31 says that for sound waves in a medium with density ρ for which the speed of sound is v_{sound}, the intensity is proportional to the square of the pressure amplitude.

A typical human ear can detect sounds with frequencies from 20 to 20,000 Hz. If they were equally sensitive to all frequencies in this range, greater intensity would imply greater loudness. However, our ears are *not* equally sensitive across the audible range: As we learned in Section 13-7, a person with normal hearing is most sensitive to frequencies around 3400 Hz and less sensitive at higher and lower frequencies. At 1000 Hz the lowest intensity that a person with normal hearing can sense (the *threshold of hearing*) is about 10^{-12} W/m^2. The threshold of hearing is substantially higher at frequencies to which the ear is less sensitive, about 10^{-11} W/m^2 at 10,000 Hz and about 10^{-8} W/m^2 at 100 Hz. At most frequencies the greatest intensity that a person can tolerate without pain (the *threshold of pain*) is about 1 W/m^2.

BIO-Medical EXAMPLE 13-9 Loud and Soft

The amplitude of the motion of your eardrum in response to a sound wave more or less matches the amplitude of the motion of nearby air molecules. (a) What is the oscillation amplitude of your eardrum when you hear a tone that has a frequency of 1000 Hz and an intensity of 1.00 W/m^2, near the threshold of pain? What is the pressure amplitude of this sound? (b) What about a 1000-Hz tone with an intensity of only 1.00×10^{-12} W/m^2, at the threshold of hearing? Use $\rho = 1.20$ kg/m^3 for the density of air and $v_{sound} = 343$ m/s for the speed of sound in air.

Set Up

We'll use Equations 13-29 and 13-31 to relate the intensity of a sound wave to its displacement amplitude A and pressure amplitude p_{max}. The problem statement tells us that the displacement amplitude of the eardrum is about the same as that of the air.

Sound wave intensity in terms of displacement amplitude A:

$$I = \frac{1}{2}\rho v_{sound}\omega^2 A^2 \quad (13\text{-}29)$$

Sound wave intensity in terms of pressure amplitude p_{max}:

$$I = \frac{p_{max}^2}{2\rho v_{sound}} \quad (13\text{-}31)$$

Solve

(a) Rearrange Equation 13-29 to find an expression for the displacement amplitude.

From Equation 13-29,

$$A^2 = \frac{2I}{\rho v_{sound}\omega^2}$$

We are given $f = 1000$ Hz, so substitute $\omega = 2\pi f$:

$$A^2 = \frac{2I}{\rho v_{sound}(2\pi f)^2} = \frac{2I}{4\pi^2\rho v_{sound}f^2} = \frac{I}{2\pi^2\rho v_{sound}f^2}$$

Take the square root of both sides and substitute $I = 1.00\ \text{W/m}^2$, $\rho = 1.20\ \text{kg/m}^3$, $v_{\text{sound}} = 343\ \text{m/s}$, and $f = 1000\ \text{Hz}$:

$$A = \sqrt{\frac{I}{2\pi^2 \rho v_{\text{sound}} f^2}} = \sqrt{\frac{1.00\ \text{W/m}^2}{2\pi^2(1.20\ \text{kg/m}^3)(343\ \text{m/s})(1000\ \text{Hz})^2}}$$

$$= 1.11 \times 10^{-5} \sqrt{\frac{\text{W} \cdot \text{s}}{\text{kg} \cdot \text{Hz}^2}}$$

Note that $1\ \text{W} = 1\ \text{J/s}$ and $1\ \text{J} = 1\ \text{N} \cdot \text{m} = 1\ \text{kg} \cdot \text{m}^2/\text{s}^2$, and $1\ \text{Hz} = 1/\text{s} = 1\ \text{s}^{-1}$. So

$$1\sqrt{\frac{\text{W} \cdot \text{s}}{\text{kg} \cdot \text{Hz}^2}} = 1\sqrt{\frac{\text{J} \cdot \text{s}^2}{\text{kg}}} = 1\sqrt{\frac{\text{kg} \cdot \text{m}^2}{\text{kg}}} = 1\ \text{m}$$

The units of the displacement amplitude A are therefore meters, as they should be:

$A = 1.11 \times 10^{-5}\ \text{m}$

Use the same approach with Equation 13-31 to find the pressure amplitude $p_{\max}$.

From Equation 13-31,

$$I = \frac{p_{\max}^2}{2\rho v_{\text{sound}}}$$

Solve for $p_{\max}$:

$$p_{\max}^2 = 2\rho v_{\text{sound}} I$$

$$p_{\max} = \sqrt{2\rho v_{\text{sound}} I} = \sqrt{2(1.20\ \text{kg/m}^3)(343\ \text{m/s})(1.00\ \text{W/m}^2)}$$

$$= 28.7\sqrt{\frac{\text{kg} \cdot \text{W}}{\text{m}^4 \cdot \text{s}}}$$

Again note that $1\ \text{W} = 1\ \text{J/s}$, $1\ \text{J} = 1\ \text{N} \cdot \text{m}$, and $1\ \text{N} = 1\ \text{kg} \cdot \text{m/s}^2$. So $1\ \text{W} = 1\ \text{N} \cdot \text{m/s}$ and $1\ \text{kg} = 1\ \text{N} \cdot \text{s}^2/\text{m}$, and the units of our answer are

$$1\sqrt{\frac{\text{kg} \cdot \text{W}}{\text{m}^4 \cdot \text{s}}} = 1\sqrt{\frac{1}{\text{m}^4 \cdot \text{s}} \frac{\text{N} \cdot \text{s}^2}{\text{m}} \frac{\text{N} \cdot \text{m}}{\text{s}}} = 1\sqrt{\frac{\text{N}^2}{\text{m}^4}} = 1\ \text{N/m}^2$$

The units of $p_{\max}$ are newtons per square meter or pascals ($1\ \text{N/m}^2 = 1\ \text{Pa}$), which is correct for pressure:

$$p_{\max} = 28.7\ \text{N/m}^2 = 28.7\ \text{Pa}$$

(b) To find the displacement amplitude A and pressure amplitude $p_{\max}$ for the new intensity, we don't have to redo the above calculations. Instead, we notice from our calculations that A and $p_{\max}$ are both proportional to the square root of the intensity I.

The new intensity is $1.00 \times 10^{-12}\ \text{W/m}^2$, which is 10^{-12} times the previous intensity. Since A and $p_{\max}$ are both proportional to $\sqrt{I}$, the new amplitudes are equal to the previous values multiplied by $\sqrt{10^{-12}} = 10^{-6}$:

$$A = (1.11 \times 10^{-5}\ \text{m}) \times 10^{-6} = 1.11 \times 10^{-11}\ \text{m}$$

$$p_{\max} = (28.7\ \text{N/m}^2) \times 10^{-6} = 2.87 \times 10^{-5}\ \text{N/m}^2 = 2.87 \times 10^{-5}\ \text{Pa}$$

Reflect

For the painfully loud sound at 1000-Hz, the eardrum oscillates with an amplitude slightly larger than 10 μm, or about 50% larger than the diameter of a human red blood cell. At the threshold of hearing the oscillation amplitude of the eardrum to a 1000 Hz tone is about 10^{-11} m, about one-tenth the diameter of a hydrogen atom. The pressure amplitudes are also very small: 28.9 Pa is about the difference in atmospheric pressure between the floor and ceiling of a bedroom, and 2.89×10^{-5} Pa is about the same pressure that would be exerted on your eardrum by the weight of a single wing of a fly. These remarkable numbers illustrate the exquisite sensitivity of human hearing.

Sound and the Inverse-Square Law

If you've ever convinced a teacher to hold class outside on a warm spring day, you may have noticed that the teacher's voice is much more difficult to hear outdoors than indoors. That's because when a source of sound (like the teacher's voice) radiates sound waves in all directions without anything getting in the way of the waves, the *intensity* of the waves decreases with increasing distance. In contrast, inside the lecture hall, sound bounces off the walls, back toward the audience, so the intensity does not decrease as quickly with increasing distance from the source.

Figure 13-24 shows how this comes about. Since waves carry energy, a source of sound waves is a source of energy. The rate at which the source emits energy in the form of waves is the *power* of the source. None of the power is lost as the waves spread away from the source, but at greater distances the power is distributed over a greater area. Hence the wave intensity (power divided by area) and the perceived loudness of the sound both decrease with increasing distance. If the source has power P_0 and emits equally in all directions as shown in Figure 13-24, at a distance r from the source the power is spread uniformly over a spherical surface of radius r. The area of such a surface is $4\pi r^2$, so the intensity equals

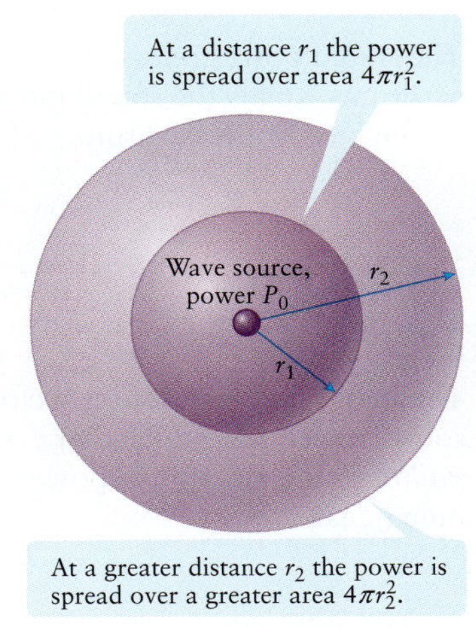

At a distance r_1 the power is spread over area $4\pi r_1^2$.
Wave source, power P_0
r_2
r_1
At a greater distance r_2 the power is spread over a greater area $4\pi r_2^2$.

Figure 13-24 The inverse-square law for sound waves A source of sound waves has power P_0. At greater distances from the source, this power is spread over a greater area, so the intensity is less.

$$I = \frac{P_0}{4\pi r^2} \qquad (13\text{-}32)$$

Intensity of a sound wave at a distance r from a source that emits in all directions

Power output of the sound source

Distance from the source to the point where the intensity is measured

Inverse-square law for sound waves (13-32)

Equation 13-32 says that the intensity is inversely proportional to the square of the distance from the source, so this is called the **inverse-square law** for waves.

The inverse-square law explains what happens when class is held outdoors: Compared to the intensity of the teacher's voice at a distance of 1 m, the intensity at 2 m is $(1/2)^2 = 1/4$ as great, and the intensity at 3 m is $(1/3)^2 = 1/9$ as great. So with increasing distance the teacher becomes increasingly difficult to hear. The inverse-square law doesn't apply indoors because sound reflects off the walls and ceiling, so sound energy that would be lost outdoors is reflected back toward you and the other members of the audience.

BIO-Medical EXAMPLE 13-10 Delivering Energy to an Eardrum

You are at an outdoor concert, standing 20 m from a speaker tower that generates 2500 W of sound power. How much sound energy from the speaker tower strikes one of your eardrums per second? A typical human eardrum has a surface area of 55 mm^2. Assume that the speaker tower emits sound uniformly in all directions.

Set Up

The inverse-square law for sound waves tells us the intensity of the sound that reaches you. To find the power (energy per time) delivered to your eardrum, we use the idea that intensity equals power per area, so power equals intensity times area.

Inverse-square law for sound waves:

$$I = \frac{P_0}{4\pi r^2} \qquad (13\text{-}32)$$

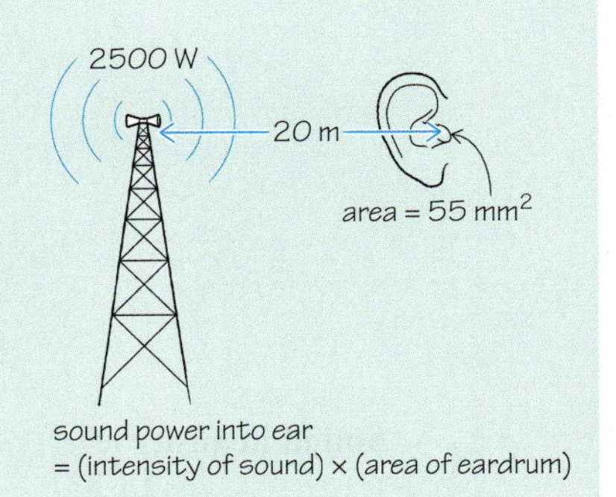

2500 W
20 m
area = 55 mm^2
sound power into ear
= (intensity of sound) × (area of eardrum)

Solve

Use Equation 13-32 to find the sound intensity at your position.

At a distance $r = 20$ m from a source with power $P_0 = 2500$ W,

$$I = \frac{P_0}{4\pi r^2} = \frac{2500\ \text{W}}{4\pi (20\ \text{m})^2} = 0.50\ \text{W/m}^2$$

The sound power that strikes one of your eardrums equals the sound intensity multiplied by the area of the eardrum.

Power (energy per time) on your eardrum:

$$P = I \times (\text{area of eardrum})$$
$$= (0.50\ \text{W/m}^2)(55\ \text{mm}^2)\left(\frac{1\ \text{m}}{10^3\ \text{mm}}\right)^2$$
$$= 2.7 \times 10^{-5}\ \text{W} = 2.7 \times 10^{-5}\ \text{J/s}$$

Reflect

The intensity of the sound is close to the threshold of pain, so you would be advised to stand farther away from the sound tower. But notice that even at this very high volume, the power that reaches the ear is very small (about one *millionth* the amount of power needed to light a 25-watt light bulb). This is another testament to the sensitivity of the human ear.

Sound Intensity Level

The range of sound intensities that we can hear spans the range from about 10^{-12} to about 1 W/m^2. It's convenient to compress this broad range of numbers by using *logarithms*. This leads to an alternative way to describe the intensity of a sound wave in terms of *sound intensity level.*

The idea behind logarithms is that any positive number x can be written as $x = b^y$, where b is called the *base*. In this language we say that y is "the logarithm of x to the base b." As an example, 3 is the logarithm of 8 to the base 2: $8 = 2^3$. We write this relationship as follows:

$$\text{If } x = b^y, \text{ then } y = \log_b x$$

So if x equals b raised to the power y, then y is the logarithm to the base b of x.

A common choice for the base b is 10. With this choice the above relationship becomes

$$\text{If } x = 10^y, \text{ then } y = \log_{10} x$$

For example, the logarithm to the base 10 of $1000 = 10^3$ is $\log_{10} 10^3 = 3$, and the logarithm to the base 10 of $0.01 = 10^{-2}$ is $\log_{10} 10^{-2} = -2$. (Any number raised to power 0 equals 1, so the logarithm of 1 is 0 in *any* base.) On your calculator the base-10 logarithm is probably denoted by a function key named log.

Note that the logarithm isn't just for numbers that are multiples of 10. As an example, a number between 1 and 10 has a base-10 logarithm that is between 0 (the logarithm of 1) and 1 (the logarithm of $10 = 10^1$). So $\log_{10} 2 = 0.30$, $\log_{10} 3 = 0.48$, and so on.

Let's now see how the logarithm is used for describing the intensity of a sound. The argument of the logarithm must be a pure number without dimensions or units. We satisfy this requirement by taking the ratio of the intensity I of a sound to $10^{-12}\ \text{W/m}^2$, the intensity of a 1000-Hz tone at the threshold of hearing. The **sound intensity level** of a sound with intensity I equals the base-10 logarithm of this ratio multiplied by 10:

Sound intensity level (13-33)

$$\beta = (10\ \text{dB}) \log_{10}\left[\frac{I}{10^{-12}\ \text{W/m}^2}\right]$$

Sound intensity level of a sound wave

Intensity of the sound

Reference intensity

The factor of 10 that multiplies the logarithm in Equation 13-33 is there for convenience. The units of sound intensity level are the **decibel** (dB), named after the

Scottish-born scientist and inventor Alexander Graham Bell. The sound intensity level for a 1000-Hz tone at the threshold of hearing is then

$$\beta = (10 \text{ dB}) \log_{10}\left(\frac{10^{-12} \text{ W/m}^2}{10^{-12} \text{ W/m}^2}\right) = (10 \text{ dB}) \log_{10}(1) = 0 \text{ dB}$$

The sound intensity level for a tone at the threshold of pain is

$$\beta = (10 \text{ dB}) \log_{10}\left(\frac{1 \text{ W/m}^2}{10^{-12} \text{ W/m}^2}\right) = (10 \text{ dB}) \log_{10}(10^{12}) = 120 \text{ dB}$$

So by using sound intensity level we express the intensities of sound in human hearing by the range from 0 to 120 dB rather than from 10^{-12} to 1 W/m^2. **Table 13-1** lists the sound intensity level of a range of sounds.

TABLE 13-1
Sound Intensity Levels

Sound	Sound level (dB)
jet engine at 25 m	150
live amplified music	120
car horn at 1 m	110
jackhammer	100
city street	90
vacuum cleaner	70
quiet conversation	50
rustling leaves	20
breathing	10

WATCH OUT! A sound intensity level of 0 dB is not the absence of sound.

We are used to thinking of a quantity with a value of zero as being nothing at all. But a sound with sound intensity level 0 dB, while faint, is still a sound. The zero value just means that this is the faintest sound that someone with normal hearing can hear at a frequency of 1000 Hz.

EXAMPLE 13-11 Double the Intensity

When the intensity of a sound doubles, how does the sound intensity level change?

Set Up

Equation 13-33 relates the sound intensity I to the sound intensity level β. We'll use this to compare two sounds, one of intensity I_1 and the other of intensity $I_2 = 2I_1$.

Sound intensity level:

$$\beta = (10 \text{ dB}) \log_{10}\left(\frac{I}{10^{-12} \text{ W/m}^2}\right) \quad (13\text{-}33)$$

Solve

Write expressions for the two sound intensity levels.

For sound 1:

$$\beta_1 = (10 \text{ dB}) \log_{10}\left(\frac{I_1}{10^{-12} \text{ W/m}^2}\right)$$

For sound 2:

$$\beta_2 = (10 \text{ dB}) \log_{10}\left(\frac{I_2}{10^{-12} \text{ W/m}^2}\right) = (10 \text{ dB}) \log_{10}\left(\frac{2I_1}{10^{-12} \text{ W/m}^2}\right)$$

Use an important property of logarithms to compare β_1 and β_2.

The logarithm of the *product* of two numbers A and B is equal to the *sum* of their logarithms:

$$\log_{10}(AB) = \log_{10} A + \log_{10} B$$

Using this property we can express β_2 as

$$\beta_2 = (10 \text{ dB}) \log_{10}\left(\frac{2I_1}{10^{-12} \text{ W/m}^2}\right) = (10 \text{ dB}) \log_{10}\left[2\left(\frac{I_1}{10^{-12} \text{ W/m}^2}\right)\right]$$

$$= (10 \text{ dB})\left[\log_{10} 2 + \log_{10}\left(\frac{I_1}{10^{-12} \text{ W/m}^2}\right)\right]$$

$$= (10 \text{ dB}) \log_{10} 2 + (10 \text{ dB}) \log_{10}\left(\frac{I_1}{10^{-12} \text{ W/m}^2}\right)$$

$$= (10 \text{ dB}) \log_{10} 2 + \beta_1$$

So *multiplying* the intensity by 2 is the same as *adding* (10 dB) $\log_{10} 2$ to the sound intensity level.

Find the *difference* of the two sound intensity levels.

The difference of the two values of β is

$$\beta_2 - \beta_1 = (10 \text{ dB})\log_{10} 2 = (10 \text{ dB})(0.30) = 3.0 \text{ dB}$$

Reflect

A factor of 2 increase in intensity corresponds to adding 3.0 dB to the sound intensity level. Increasing the intensity by a factor of 4 = 2 × 2 means adding 3.0 dB + 3.0 dB = 6.0 dB to the sound intensity level; increasing intensity by 8 = 2 × 2 × 2 means adding 3.0 dB + 3.0 dB + 3.0 dB = 9.0 dB to the sound intensity level; and so on.

GOT THE CONCEPT? 13-9 From One Fan to Two

You measure the sound intensity level of an electric fan to be 60 dB. What is the sound intensity level when a second identical fan is turned on, at the same distance from you as the first one? (a) 120 dB; (b) 90 dB; (c) 66 dB; (d) 63 dB; (e) 62 dB.

TAKE-HOME MESSAGE FOR Section 13-9

✔ The intensity of a sound wave is the power (energy per time) that the wave delivers per unit area. The intensity does not depend on the ear or device used to detect the sound.

✔ The power delivered by a sound wave is equal to the intensity of the wave multiplied by the collecting area.

✔ In the absence of obstructions or reflections, the intensity of sound from a source decreases in inverse proportion to the distance from the source.

✔ The sound intensity level involves the logarithm of a sound wave's intensity divided by a reference value of intensity.

13-10 The frequency of a sound depends on the motion of the source and the listener

Until now we have considered only waves generated by stationary emitters and observed by stationary listeners. Everyday experience, however, suggests that something curious happens when a moving object creates a sound. As a police car with sirens blaring zooms by you, for example, the frequency of the sound is higher as the car approaches and lower after it passes. Even without a siren, a fast-moving car generates a characteristic high-to-low frequency sound (something like "neee-urrrr") as it approaches and passes you. And although we don't often get the chance to experience this phenomenon in reverse, if you were to move at high speed toward a stationary police car with its sirens blaring, the sound of the siren would follow a similar high-to-low frequency shift as you approached and then passed it. This effect is known as the **Doppler effect**, named for the Austrian physicist Christian Doppler who first proposed this phenomenon associated with waves in 1842.

The Doppler Effect and Frequency Shift

In **Figure 13-25a**, a police siren emits a periodic sound wave that has a fixed frequency and a fixed wavelength. Because the car is stationary, each wave crest—that is, each region of highest pressure along the wave, shown as arcs of circles—is centered on the siren. On the left side of **Figure 13-25b** the car moves from left to right. The wave crest

(a)

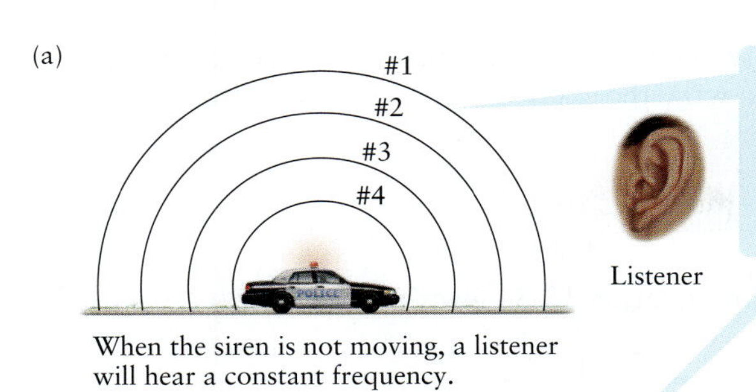
#1
#2
#3
#4
POLICE
Listener
Wave crests spread outward from the siren. The numbers on the crests show the sequence in which they were emitted: First #1, then #2, then #3, then #4, so the most recently emitted crest is closest to the siren.
When the siren is not moving, a listener will hear a constant frequency.

(b)

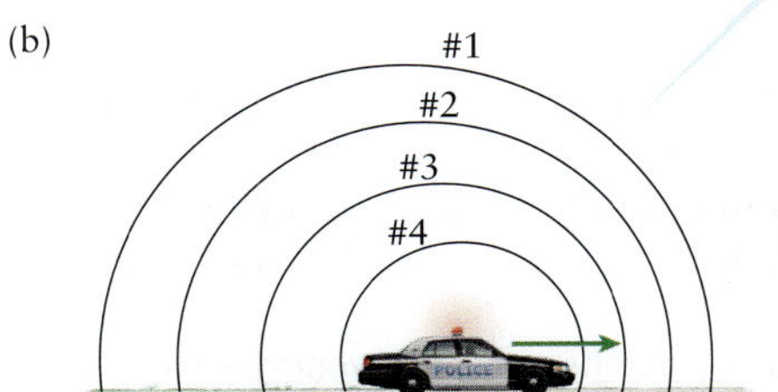
#1
#2
#3
#4
POLICE
Listener
As the siren approaches the listener, the sound frequency is higher.

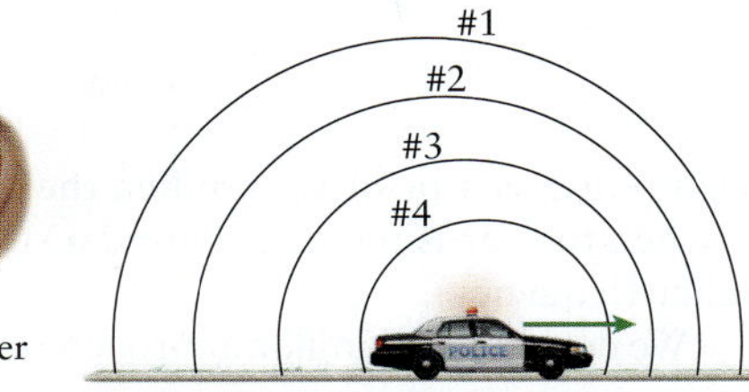
#1
#2
#3
#4
POLICE
The frequency is lower as the siren moves away from the listener.

Figure 13-25 The Doppler effect: A moving source A police car siren emits a periodic sound wave that has a fixed frequency and a fixed wavelength. (a) The police car is stationary, so each region of highest pressure along the wave, shown as arcs of circles, is centered on the siren. (b) The car moves from left to right. Because the source of the sound waves moves as the waves propagate, the wave crests are closer together as the car approaches a listener's ear and farther apart as the car recedes from the listener.

associated with the largest circle originated first, when the car was somewhere to the left, and the circle of the smallest radius represents the most recently generated wave crest. Because the car moves to the right as the sound wave propagates, the siren emits each new wave crest closer to the previous one than if the car were stationary. Because the distance between wave crests determines the wavelength heard by a stationary listener, someone standing in front of the car detects a shorter wavelength than the siren generates. The opposite is true on the right-hand side of Figure 13-25b, in which the car moves away from the listener.

The relationship $f = v_{\text{sound}}/\lambda$ (Equation 13-2 for sound) tells us that a shorter wavelength results in a higher frequency. So the listener in Figure 13-25b hears a higher frequency than the siren generates as the car approaches. Analogously, because the wave crests are spread farther apart behind the car, the observed wavelength lengthens, resulting in the listener hearing a lower frequency.

In **Figure 13-26** we expand the view of the moving police car to show the car at two instants in time separated by the period T of the tone created by the siren. A crest of the sound wave was created when the car was at the location marked by the red dashed line; in the time T the wave crest has propagated as indicated by the red semicircle. That distance is given by the product of the speed of sound v_{sound} and T, as shown. The distance the car moves in that time is $v_{\text{car}}T$, where v_{car} is the speed of the car. This new location is where the car will emit the next wave crest. So a stationary listener in front of the car detects $\lambda_{\text{listener}}$ (the distance between two successive wave crests) as the wavelength of the siren's tone. As seen in the figure these three distances—the distance the sound moves, the distance the car moves, and the distance between two successive wave crests at the place where the listener sits—are related by

$$v_{\text{car}}T + \lambda_{\text{listener}} = v_{\text{sound}}T$$

or

$$\lambda_{\text{listener}} = v_{\text{sound}}T - v_{\text{car}}T \qquad (13\text{-}34)$$

Although this equation mathematically describes the Doppler effect, by convention we write the relationship as

GOT THE CONCEPT? 13-10
A Car Alarm

? As you run toward your parked car to turn off the blaring alarm system, will the frequency of the tone you hear be (a) higher than, (b) the same as, or (c) lower than the frequency you hear when standing still?

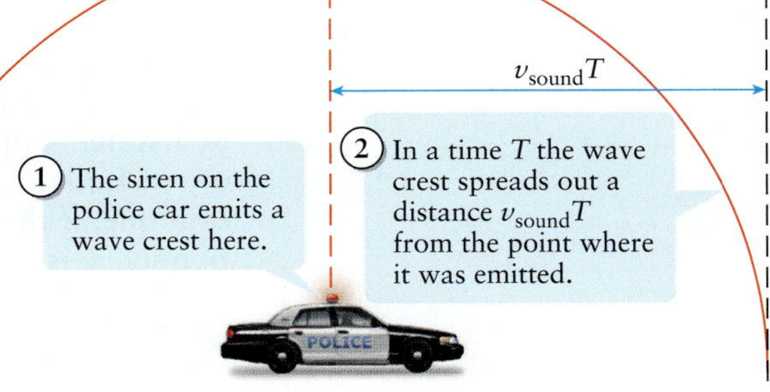
$v_{\text{sound}}T$
1 The siren on the police car emits a wave crest here.
2 In a time T the wave crest spreads out a distance $v_{\text{sound}}T$ from the point where it was emitted.
POLICE
3 In the same time T the police car travels a distance $v_{\text{car}}T$.

$v_{\text{car}}T$
$\lambda_{\text{listener}}$

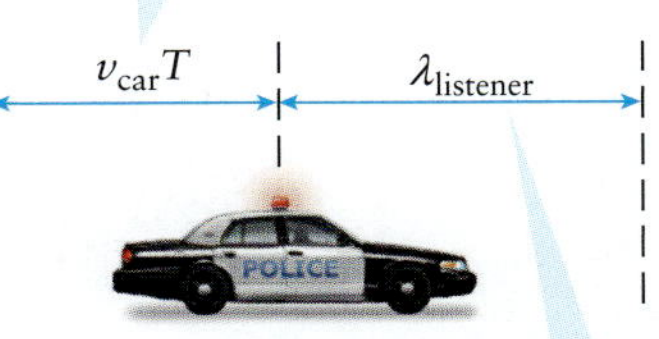
POLICE

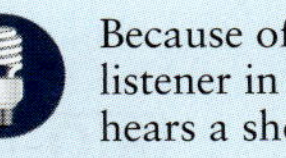
Because of the car's motion, a listener in front of the police car hears a shortened wavelength.

Figure 13-26 The Doppler effect: A moving source II The wavelength of the sound heard by a stationary listener as a source of sound approaches depends on the distance the wave travels in a certain time and the distance the source travels in that same time.

one between the actual and the observed frequency relative to the speed of the emitter rather than the speed of sound v. To write the expression we use $T = 1/f$ and Equation 13-2 for sound waves:

$$\lambda_{\text{listener}} = \frac{v_{\text{sound}}}{f_{\text{listener}}}$$

Combining these equations with Equation 13-34 gives

$$\frac{v_{\text{sound}}}{f_{\text{listener}}} = \frac{v_{\text{sound}}}{f} - \frac{v_{\text{car}}}{f}$$

or, rearranging,

$$f_{\text{listener}} = \left(\frac{v_{\text{sound}}}{v_{\text{sound}} - v_{\text{car}}}\right)f$$

Because v_{car} is a positive number, the fraction in parentheses must be greater than 1. So when the car is moving toward a stationary listener, $f_{\text{listener}} > f$ (the listener hears a higher frequency).

We can apply a similar approach to determine the observed frequency as the source moves away from a stationary listener, as well as for the cases in which the listener moves relative to a stationary source. Using "source" instead of "car" as a more general way to indicate the source of the sound, here's a single equation for the frequency heard by the listener no matter how the source and listener are moving:

The Doppler effect (13-35)

Frequency of sound detected by the listener

Frequency of sound emitted by the source

$$f_{\text{listener}} = \left[\frac{v_{\text{sound}} \pm v_{\text{listener}}}{v_{\text{sound}} \mp v_{\text{source}}}\right]f$$

v_{sound} = speed of sound
v_{listener} = speed of the listener
v_{source} = speed of the source

In both numerator and denominator:
- Use upper sign if source and listener are approaching.
- Use lower sign if source and listener are moving apart.

Notice that the plus sign is above the minus sign in the numerator but below it in the denominator. In Equation 13-35 we use the upper sign when the source or listener approaches the other and the lower sign when it retreats from the other. Equation 13-35 can be used for more than one case simultaneously; for example, if the source and listener are both approaching each other, the observed frequency is

$$f_{\text{listener}} = \left(\frac{v_{\text{sound}} + v_{\text{listener}}}{v_{\text{sound}} - v_{\text{source}}}\right)f$$

In this case the numerator is larger than the speed of sound v_{sound}, and the denominator is smaller than v_{sound}; both factors result in the listener hearing a higher frequency than the one emitted by the source.

The speed of a moving object can be determined by measuring the shift in frequency associated with the Doppler effect. A common way to do this is to create a wave of known frequency and then bounce it off the moving object to be studied. A *radar speed gun* uses this principle: It sends out a *radio* wave of a known frequency and compares it to the frequency of the wave after it is reflected by a moving car. (Because this device uses radio waves rather than sound waves, v_{sound} in Equation 13-35 is replaced by the speed of light c.) Bats and dolphins use the same technique with sound waves to track the motion of prey.

Measuring velocity using a Doppler shift has become important in medical applications, too—for example, to measure blood flow and to record real-time images of a moving fetus. Because the frequencies of sound used for this kind of imaging, between 2 and 10 MHz, are well above the range of human hearing, this technique is called *ultrasonic imaging.*

Consider the process by which ultrasound is used to determine the speed of blood flow through a heart valve. (An obstructed valve is often characterized by an increase in the speed of the blood flow.) A probe that emits a low-intensity sound wave of (high) frequency f_{emitted} is placed over the chest and focused on a region localized around one heart valve. Because the blood is moving, it experiences a wave of frequency f' that has been shifted according to the Doppler effect for a moving listener. Using Equation 13-35 we write

$$f' = \left(\frac{v_{\text{sound}} \pm v}{v_{\text{sound}}}\right) f_{\text{emitted}}$$

In this expression v is the speed of the blood, and the upper sign applies to motion toward the probe and the lower sign to motion away from the probe. The speed of sound in human tissue is about $v_{\text{sound}} = 1540$ m/s.

The sound wave reflects off the blood and returns to the probe. This reflected wave leaves the blood with frequency f', so it's equivalent to the blood emitting sound at frequency f'. So we again apply Equation 13-35, treating the blood as a moving source. The frequency f'' of the reflected ultrasound detected by the probe is then

$$\begin{aligned} f'' &= \left(\frac{v_{\text{sound}}}{v_{\text{sound}} \mp v}\right) f' \\ &= \left(\frac{v_{\text{sound}}}{v_{\text{sound}} \mp v}\right)\left(\frac{v_{\text{sound}} \pm v}{v_{\text{sound}}}\right) f_{\text{emitted}} \\ &= \left(\frac{v_{\text{sound}} \pm v}{v_{\text{sound}} \mp v}\right) f_{\text{emitted}} \end{aligned} \tag{13-36}$$

Equation 13-36 says that when a wave of frequency f_{emitted} from a probe hits blood or another object moving with speed v, the wave that is reflected back to the probe is received with frequency f''. As for Equation 13-35 the upper signs apply to motion toward the probe, and the lower signs to motion away from the probe.

BIO-Medical EXAMPLE 13-12 Diagnostic Ultrasound

A certain ultrasound device can measure a fetal heart rate as low as 50 beats per minute. This corresponds to the surface of the heart moving at about 4.0×10^{-4} m/s. If the probe generates ultrasound that has a frequency of 2.0 MHz (1 MHz = 1 megahertz = 10^6 Hz), what frequency shift must the machine be able to detect? Use $v_{\text{sound}} = 1540$ m/s for the speed of sound in human tissue.

Set Up

We are given the frequency $f_{\text{emitted}} = 2.0$ MHz of the sound waves that the probe emits. The frequency *shift* is the difference between f_{emitted} and the frequency f'' of the waves that the probe receives after they reflect from the heart.

Frequency received after reflecting from a moving object:

$$f'' = \left(\frac{v_{\text{sound}} \pm v}{v_{\text{sound}} \mp v}\right) f_{\text{emitted}} \tag{13-36}$$

Solve

Let's begin by considering a region of the heart that is moving toward the probe. Then we use the upper sign in the numerator and denominator of Equation 13-36.

If the heart wall is approaching the probe,

$$f'' = \left(\frac{v_{\text{sound}} + v}{v_{\text{sound}} - v}\right) f_{\text{emitted}}$$

The numerator $v_{\text{sound}} + v$ is greater than the denominator $v_{\text{sound}} - v$, so the quantity in parentheses is greater than 1 and f'' will be greater than f_{emitted}.

Write an expression for the difference between f'' and f_{emitted}.

The frequency shift is

$$f'' - f_{\text{emitted}} = \left(\frac{v_{\text{sound}} + v}{v_{\text{sound}} - v}\right) f_{\text{emitted}} - f_{\text{emitted}}$$

$$= \left(\frac{v_{\text{sound}} + v}{v_{\text{sound}} - v} - 1\right) f_{\text{emitted}}$$

Substitute $v = 4.0 \times 10^{-4}$ m/s, $v_{\text{sound}} = 1540$ m/s, and $f_{\text{emitted}} = 2.0 \times 10^6$ Hz:

$$f'' - f_{\text{emitted}} = \left(\frac{1540 \text{ m/s} + 4.0 \times 10^{-4} \text{ m/s}}{1540 \text{ m/s} - 4.0 \times 10^{-4} \text{ m/s}} - 1\right)(2.0 \times 10^6 \text{ Hz})$$

$$= 1.0 \text{ s}^{-1} = 1.0 \text{ Hz}$$

The wave received by the probe is 1.0 Hz higher in frequency than the wave emitted by the probe.

Reflect

You should repeat the above calculation for a region of the heart that is moving *away* from the probe. Then you'll use the *lower* sign in the numerator and denominator of Equation 13-36. You should find that the frequency shift in this case is $f'' - f_{\text{emitted}} = -1.0$ Hz (that is, the wave received by the probe is 1.0 Hz *lower* in frequency than the wave emitted by the probe). So an ultrasound device must be able to measure a frequency shift that is rather small compared to the emitted frequency $f_{\text{emitted}} = 2.0 \times 10^6$ Hz.

Our calculation shows that the frequency shift is proportional to the frequency emitted by the probe. Larger frequency shifts make it easier to detect smaller velocities, so an ultrasound device that operates at a higher frequency has a greater sensitivity. One constraint on this is that absorption of ultrasound waves by human tissue increases with frequency, so higher frequencies cannot penetrate as deeply into the body.

Sound from a Supersonic Source

Something curious occurs when a source of sound moves faster than the speed of sound (that is, the source is *supersonic*). In **Figure 13-27a** a jet airplane sits on the ground before taking off; the concentric circles centered on the airplane represent the spherical wave crests of the sound it generates. The three drawings in parts (b), (c), and (d) of Figure 13-27 show the airplane flying at increasingly faster speeds. As the airplane increases its speed, it catches up to the sound it generated earlier in the motion, squeezing the wave crests closer together ahead of the airplane. But notice that if the airplane moves at the same speed as the sound, as in **Figure 13-27c**, the airplane and the sound it generates travel the same distance in any time interval. In this case the crests of all of the sound waves bunch together. If the airplane is supersonic, so that its speed exceeds the speed of sound as in **Figure 13-27d**, the airplane moves farther in any time interval than the sound that it emits.

Figure 13-28a shows an expanded view of what happens when the airplane is supersonic. Each circle represents the spherical wave crest of a sound generated when the airplane was at the center of the circle. Notice that all of the sound wave crests generated over the time Δt that the airplane has moved from its initial position (the lighter-colored image to the left) to its final position (the image on the right) interfere constructively. This constructive interference takes place on the surface of a cone, known as the **Mach cone** after the Austrian physicist Ernst Mach, who first explained this phenomenon in 1877. In Figure 13-28a (which is only two dimensional), we represent the cone by the two black lines. Note that there is *no* sound to the right of the Mach cone in Figure 13-28a. An observer to the right of the cone will be able to see the airplane but will not be able to hear it: No sound from the airplane will have reached him yet.

Because all of the wave crests pile up at the Mach cone, this is a region of substantially higher pressure than the surrounding air. If you are standing on the ground when a supersonic airplane flies over, you experience this as a **shock wave** (so called because there is no advance warning that it is coming) or as a **sonic boom**. The pressure increase in a shock wave from an airplane is in the range of 5 to 50 Pa. The shock

Crests of sound waves are expanding spheres, drawn as circles.

(a)

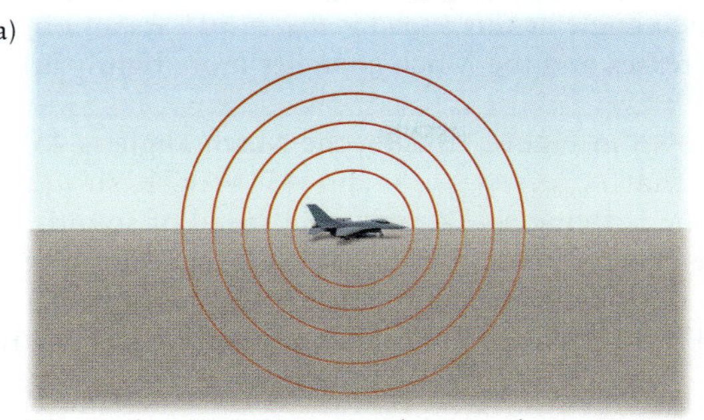

The airplane is stationary on the ground.

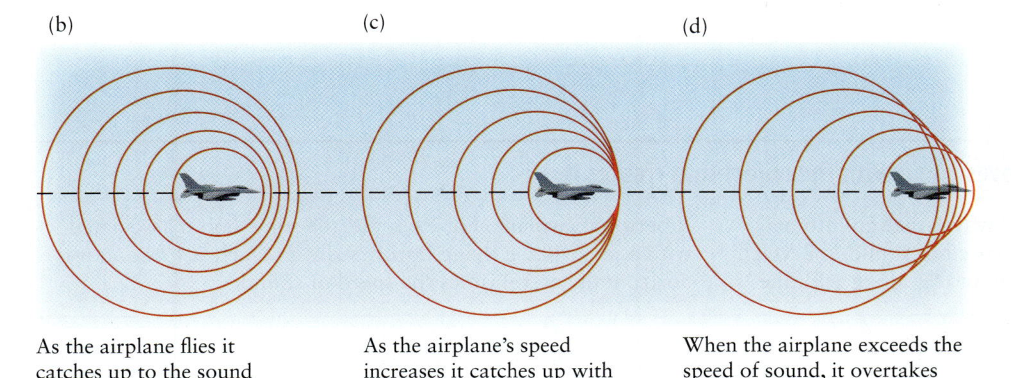

(b)
(c)
(d)
As the airplane flies it catches up to the sound crests it created earlier.
As the airplane's speed increases it catches up with more sound crests.
When the airplane exceeds the speed of sound, it overtakes previously created sound crests.

Figure 13-27 To the speed of sound—and beyond (a) Concentric circles centered on a stationary jet plane represent the spherical wave fronts of the sound it generates. As the speed of the plane increases after takeoff, the plane catches up to the sound wave fronts. (b) The plane is not moving as fast as the sound waves propagate. (c) When the plane moves at the speed of the sound waves, it just exactly catches up to the wave fronts. (d) When the speed of the plane exceeds the speed of sound, the plane moves farther in any time interval than the corresponding sound wave front.

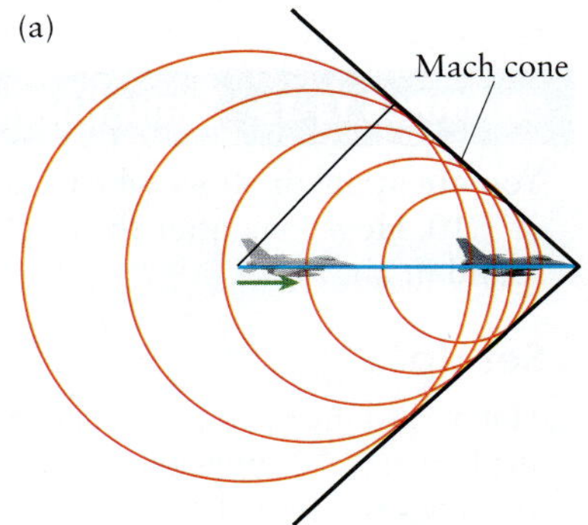

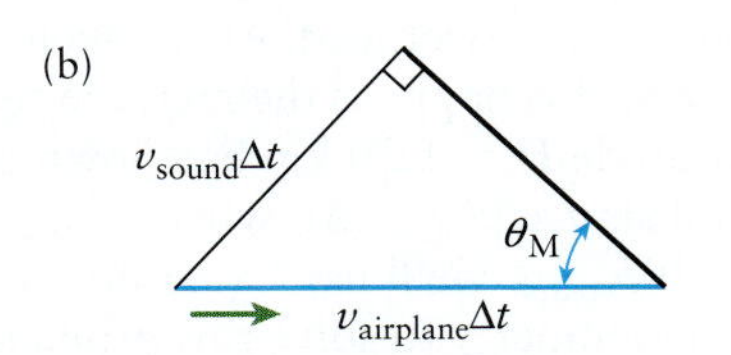

Figure 13-28 Faster than sound (a) In an expanded view of a plane flying faster than the speed of sound, each circle represents the spherical wave crest of a sound generated when the plane was at the center of the circle. The crests of all of the sound waves generated during the time that the plane has moved from its initial position to its final position interfere constructively on the surface of the Mach cone. (b) Relating the Mach angle to the speed of sound and the speed of the airplane.

wave is relatively narrow in extent, as you can see in **Figure 13-29**. This is a photograph of an X-15 aircraft in a wind tunnel with air moving at more than twice the speed of sound. In the figure a number of separate shock waves, each originating from various discontinuities on the airplane's surface, are made evident by a photographic technique that reveals regions of differing air density. Notice that each of these regions is only a fraction of the length of the airplane. For this reason you would hear a sharp, explosive sound from each shock wave as this aircraft passed overhead.

Note that the sonic boom is *not* the amplified sound of the aircraft engine. As Figure 13-29 shows, shock waves spread out from several parts of the airplane and are not associated with the engine at all. If an airplane is still flying supersonically after it runs out of fuel, it still produces a sonic boom.

The angle of the Mach cone depends on the airplane's speed, which we call $v_{airplane}$. To see how to calculate this angle, inspect parts (a) and (b) of Figure 13-28. In a time Δt the airplane travels a distance $v_{airplane}\Delta t$; this is shown by a blue line in Figure 13-28a and **Figure 13-28b**. The distance $v_{airplane}\Delta t$ forms the hypotenuse of a right triangle, one leg of which is the distance $v_{sound}\Delta t$ that the sound travels in time Δt. The sine of the angle θ_M in Figure 13-28b, known as the **Mach angle**, equals the side opposite this angle ($v_{sound}\Delta t$) divided by the hypotenuse ($v_{airplane}\Delta t$). The factors of Δt cancel, leaving

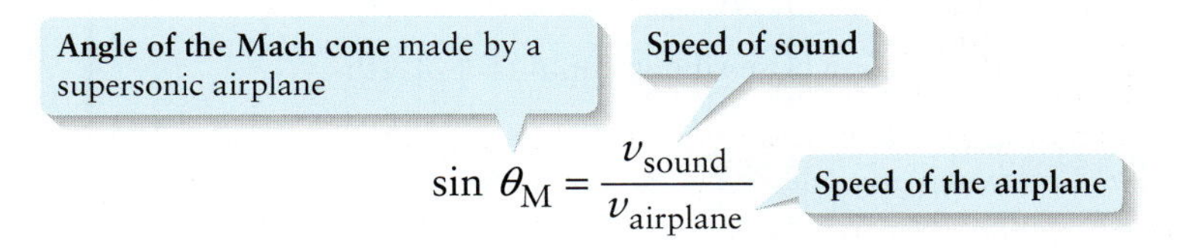

$$\sin \theta_M = \frac{v_{sound}}{v_{airplane}} \quad (13\text{-}37)$$

Mach angle (13-37)

NASA/Science Photo Library

Figure 13-29 **Shock waves** This photograph shows air moving over an X-15 aircraft at more than twice the speed of sound in a supersonic wind tunnel. Shock waves originate from various discontinuities on its surface.

If an airplane is flying at the speed of sound, the ratio $v_{sound}/v_{airplane}$ equals 1. Since $\sin 90° = 1$, the Mach angle at this speed is $\theta_M = 90°$. As an object's speed increases above v_{sound}, θ_M decreases and the Mach cone lies more tightly around the path of the airplane.

For the case shown in Figure 13-28b, the Mach angle is 45°; in this case Equation 13-37 tells us that $v_{sound}/v_{airplane} = \sin 45° = 0.71$, so $v_{airplane} = v_{sound}/0.71 = 1.4v_{sound}$ (the airplane is flying at 1.4 times the speed of sound). The ratio of an airplane's speed to the speed of sound is called the **Mach number**: $M = v_{airplane}/v_{sound}$. So you can see that $M = 1.4$ for the airplane shown in Figure 13-28b (it is said to be flying "at Mach 1.4"). In terms of the Mach number, we can rewrite Equation 13-37 as $\sin \theta_M = 1/M$.

WATCH OUT! The Mach cone moves along with the object that creates it.

Although the individual sound waves that go into making a sonic boom travel at the speed of sound, the Mach cone on which the crests pile up moves along with the supersonic airplane. For example, the sonic boom associated with a jet airplane flying horizontally at Mach 1.4 also moves horizontally at 1.4 times the speed of sound.

EXAMPLE 13-13 Sonic Boom

You are at an air show when a supersonic airplane flies directly over you at an altitude of 1.20 km and a Mach number of 1.10. How long after the airplane was directly overhead do you hear the sonic boom? Use 343 m/s for the speed of sound in air.

Set Up

Mach 1.10 means that the speed of the airplane is 1.10 times the speed of sound. You hear the sonic boom when the Mach cone passes over you, a time Δt after the airplane is overhead. As the sketch shows, when this happens the airplane flying at altitude $H = 1.20$ km has flown past you a distance $v_{airplane}\Delta t$, where $v_{airplane} = 1.10v_{sound}$. We'll use Equation 13-37 and trigonometry to solve this problem.

Mach angle:

$$\sin \theta_M = \frac{v_{sound}}{v_{airplane}} \qquad (13\text{-}37)$$

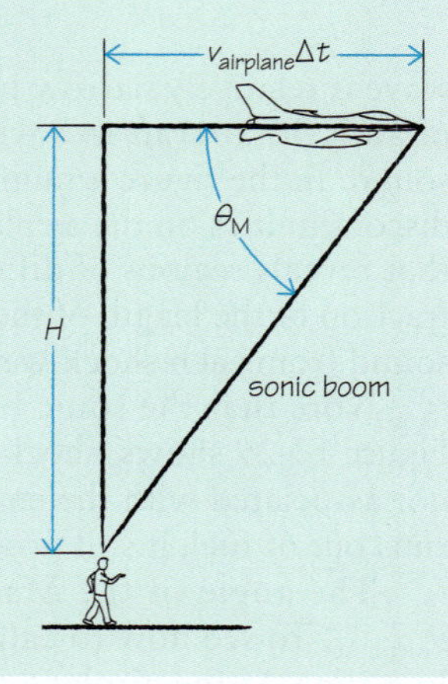

Solve

Since we know the speed of the airplane, we can use Equation 13-37 to find the Mach angle θ_M.

The speed of the airplane is $v_{airplane} = 1.10v_{sound}$, so Equation 13-37 becomes

$$\sin \theta_M = \frac{v_{sound}}{v_{airplane}} = \frac{v_{sound}}{1.10v_{sound}} = \frac{1}{1.10}$$

So θ_M is the angle whose sine is equal to $1/1.10$:

$$\theta_M = \sin^{-1}\left(\frac{1}{1.10}\right) = \sin^{-1} 0.909 = 65.4°$$

The figure above shows that the distances H and $v_{\text{airplane}}\,\Delta t$ are two sides of a right triangle. Since H is opposite to the angle θ_M and $v_{\text{airplane}}\Delta t$ is adjacent to θ_M, their ratio is the tangent of θ_M.

From the figure

$$\tan\theta_M = \frac{H}{v_{\text{airplane}}\Delta t}$$

We know the values of $\theta_M = 65.4°$, $H = 1.20$ km, $v_{\text{airplane}} = 1.10v_{\text{sound}}$, and $v_{\text{sound}} = 343$ m/s, so we can solve for Δt:

$$\Delta t = \frac{H}{v_{\text{airplane}}\tan\theta_M} = \frac{H}{1.10v_{\text{sound}}\tan\theta_M}$$
$$= \frac{1.20\text{ km}}{1.10(343\text{ m/s})\tan 65.4°}\left(\frac{10^3\text{ m}}{1\text{ km}}\right)$$
$$= 1.46\text{ s}$$

The sonic boom reaches you 1.46 s after the airplane flies overhead.

Reflect

Note that if the airplane emits a sound wave when it is directly above you at altitude $H = 1.20$ km $= 1200$ m, that sound would need a time $H/v_{\text{sound}} = (1200\text{ m})/(343\text{ m/s}) = 3.49$ s to reach you. The sonic boom reaches you *before* that, just 1.46 s after the airplane is directly above you. That's because the shock wave is made up of wave crests that were emitted by the airplane *before* the airplane was directly above you (see Figure 13-28a), so the wave crests have had more time to reach you.

GOT THE CONCEPT? 13-11 Sound from a Supersonic Airplane

When an airplane is traveling at Mach 2.0 (twice the speed of sound), at what speed do sound waves spread away from the point where the airplane emits them?

(a) 2.0 times the speed of sound; (b) faster than the speed of sound but slower than 2.0 times the speed of sound; (c) the speed of sound; (d) slower than the speed of sound.

TAKE-HOME MESSAGE FOR Section 13-10

✔ The frequency of a sound wave heard by a listener is higher if the listener and source are moving toward each other and lower if they are moving apart. This is called the Doppler effect.

✔ If a source of sound waves moves through the air faster than the speed of sound, a shock wave (sonic boom) is produced.

Key Terms

amplitude
angular frequency
angular wave number
antinode
beat frequency
beats
closed pipe
constructive interference
decibel
destructive interference
Doppler effect
frequency
fundamental frequency
fundamental mode
intensity
interference
interference pattern
inverse-square law
linear mass density
longitudinal wave (pressure wave)
Mach angle
Mach cone
Mach number
mechanical wave
medium
node
one-dimensional wave
open pipe
path length difference
period
phase angle
phase difference
power
pressure amplitude
pressure wave (longitudinal wave)
propagation speed
shock wave
sinusoidal wave
sonic boom
sound intensity level
sound wave
standing wave
standing wave mode
surface wave
transverse wave
traveling wave
watt
wave
wave function
wavelength

Chapter Summary

Topic	Equation or Figure

Mechanical waves: A mechanical wave is a disturbance that propagates through a material medium. There is no net flow of material through the medium. Mechanical waves can be transverse, longitudinal, or a combination of the two.

(a)

① If the hand holding the rope moves up and down, a **transverse wave** travels along the rope.

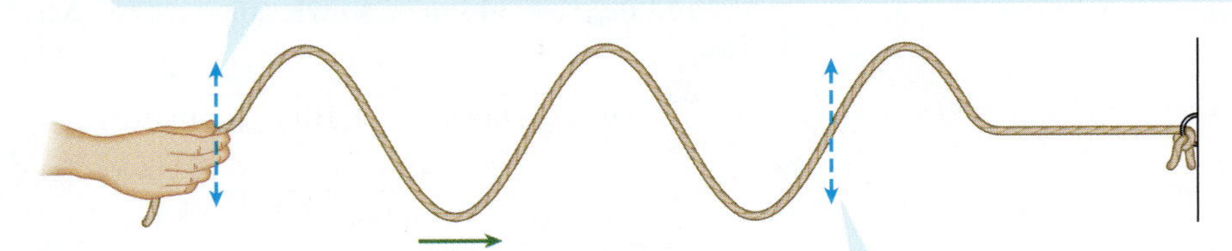

② The wave disturbance propagates (moves along the rope) horizontally...

③ ...but individual parts of the rope (the wave medium) move up and down, perpendicular (transverse) to the propagation direction.

(Figure 13-2)

(b)

① If the hand holding the spring moves back and forth, a **longitudinal wave** travels along the spring.

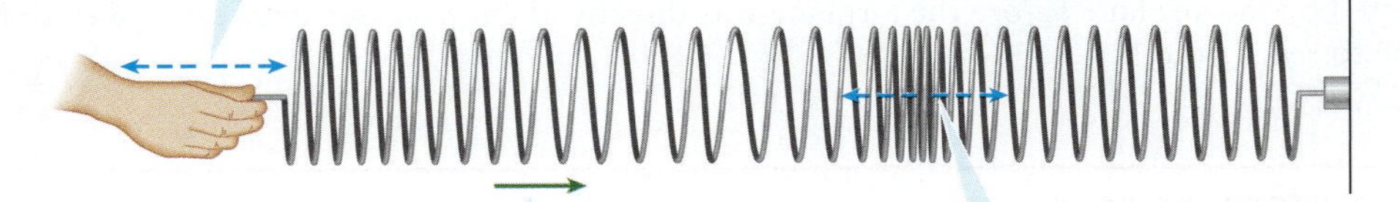

② The wave disturbance propagates (moves along the spring) horizontally...

③ ...but individual parts of the spring (the wave medium) move back and forth, parallel (longitudinal) to the propagation direction.

Sinusoidal waves: In a sinusoidal wave, each piece of the medium undergoes simple harmonic motion with the same amplitude and frequency. At any instant the wave pattern repeats itself over a distance λ known as the wavelength. The product of the frequency and wavelength equals the propagation speed of the wave.

Propagation speed of a wave

$$v_p = f\lambda \quad (13\text{-}2)$$

Frequency — **Wavelength**

Wave function for a **sinusoidal wave** propagating in the $+x$ direction

Angular wave number of the wave $= 2\pi/\lambda$

$$y(x,t) = A\cos(kx - \omega t + \phi) \quad (13\text{-}6)$$

Phase angle

Amplitude of the wave

Angular frequency of the wave $= 2\pi f$

Propagation speed of a wave: The speed at which a wave travels through a medium depends on the type of wave and the properties of the medium.

Propagation speed of a **transverse wave on a rope**

Tension in the rope

$$v_p = \sqrt{\frac{F}{\mu}} \quad (13\text{-}10)$$

Mass per unit length of the rope

Propagation speed of a **longitudinal wave in a fluid**

Bulk modulus of the fluid

$$v_p = \sqrt{\frac{B}{\rho}} \quad (13\text{-}11)$$

Density of the fluid

Propagation speed of a **longitudinal wave along a solid rod**

Young's modulus of the solid

$$v_p = \sqrt{\frac{Y}{\rho}} \quad (13\text{-}13)$$

Density of the solid

Superposition and interference: When two waves are present simultaneously in a medium, the total wave is the sum of the two individual waves. If the two waves have the same frequency but emanate from different places, there will be positions where constructive and destructive interference occur.

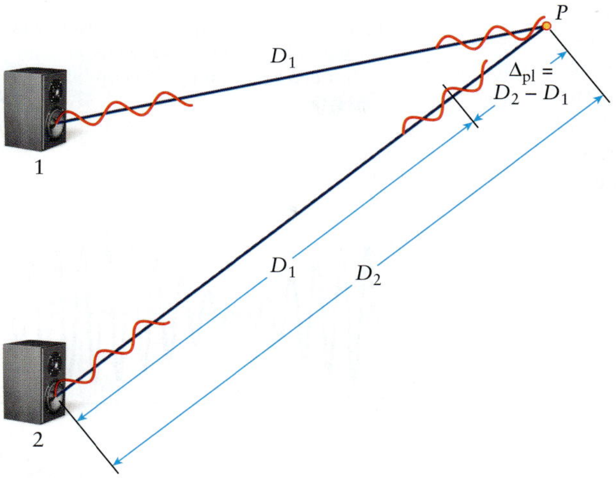

(Figure 13-12)

Standing waves on a string: When sinusoidal waves reflect back and forth from the ends of a string, the result can be a standing wave. Standing waves are possible only if a whole number of half-wavelengths fit into the length of the string. Each standing wave mode has its own natural frequency.

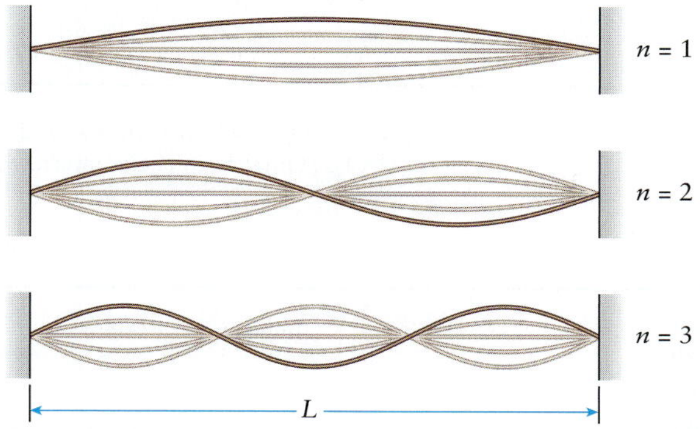

(Figure 13-14)

***n*th possible frequency** of a standing wave on a string held at both ends

Tension in the string

$$f_n = \frac{n}{2L}\sqrt{\frac{F}{\mu}} \quad \text{where } n = 1, 2, 3, \ldots \qquad (13\text{-}19)$$

Length of the string

Linear mass density of the string

Standing sound waves in a pipe: Standing waves can also occur when sinusoidal sound waves reflect back and forth from the ends of a pipe. The allowed wavelengths and frequencies depend on whether the pipe is closed (closed at one end and open at the other) or open (open at both ends).

Closed pipe:

(a)

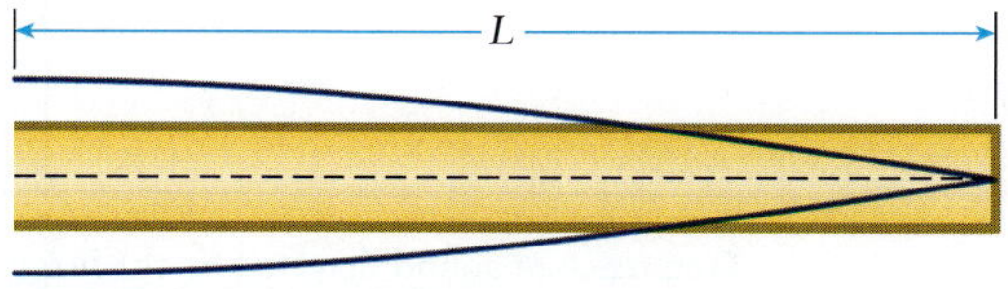

(Figure 13-18a)

Open pipe:

(a)

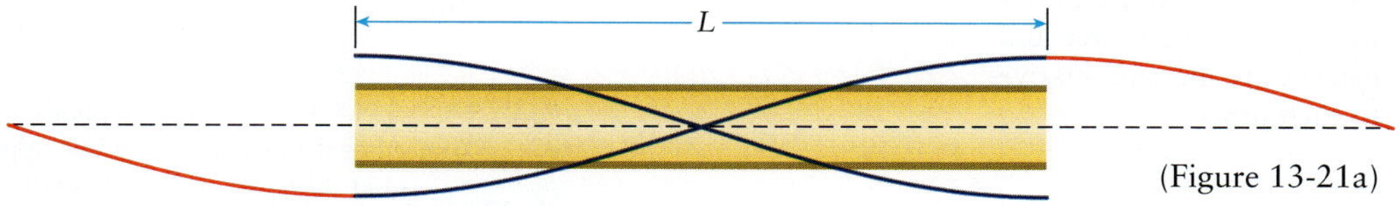

(Figure 13-21a)

Beats: When two sound waves of similar frequencies f_1 and f_2 interfere, the result is a total wave with a frequency $(f_1 + f_2)/2$ and an amplitude that rises and falls at the beat frequency $|f_2 - f_1|$.

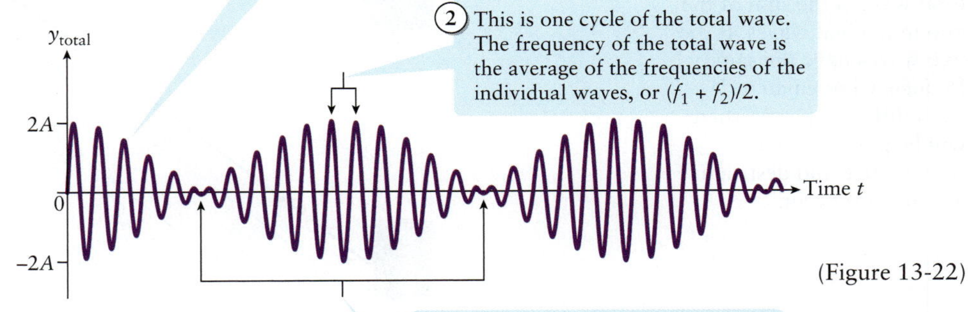

1 This is the total wave that results when two individual sine waves of similar frequencies f_1 and f_2 are combined.
2 This is one cycle of the total wave. The frequency of the total wave is the average of the frequencies of the individual waves, or $(f_1 + f_2)/2$.
y_{total}
2A
0
–2A
Time t
3 The amplitude of the total wave rises and falls periodically. The frequency of this rising and falling is the beat frequency, $f_{beats} = |f_2 - f_1|$.

(Figure 13-22)

Beat frequency heard when sound waves of two similar frequencies interfere

Frequencies of the individual sound waves

$$f_{beats} = |f_2 - f_1| \tag{13-25}$$

Wave intensity: The intensity of a sound wave is the amount of power (energy per unit time) that the wave delivers per unit area. It can be expressed in terms of the displacement amplitude or pressure amplitude of the wave. If the wave source emits equally in all directions, the intensity decreases with distance according to an inverse-square law. Sound intensity level is an alternative way to express intensity using a logarithmic scale.

Intensity of a sound wave

Pressure amplitude of the wave

$$I = \frac{p^2_{max}}{2\rho v_{sound}} \tag{13-31}$$

Density of the wave medium (usually air)

Speed of sound waves in the medium

Intensity of a sound wave at a distance r from a source that emits in all directions

Power output of the sound source

$$I = \frac{P_0}{4\pi r^2} \tag{13-32}$$

Distance from the source to the point where the intensity is measured

The Doppler effect: If a source of sound emits waves with frequency f, a listener will hear a different frequency if the source and listener are moving relative to each other.

Frequency of sound detected by the listener

Frequency of sound emitted by the source

$$f_{listener} = \left[\frac{v_{sound} \pm v_{listener}}{v_{sound} \mp v_{source}}\right] f \tag{13-35}$$

v_{sound} = speed of sound
$v_{listener}$ = speed of the listener
v_{source} = speed of the source

In both numerator and denominator:
- Use upper sign if source and listener are approaching.
- Use lower sign if source and listener are moving apart.

Sonic booms: If an object moves through the air faster than the speed of sound, the sound waves from the object combine to form a conical shock wave. The angle of this cone depends on the speed of the object.

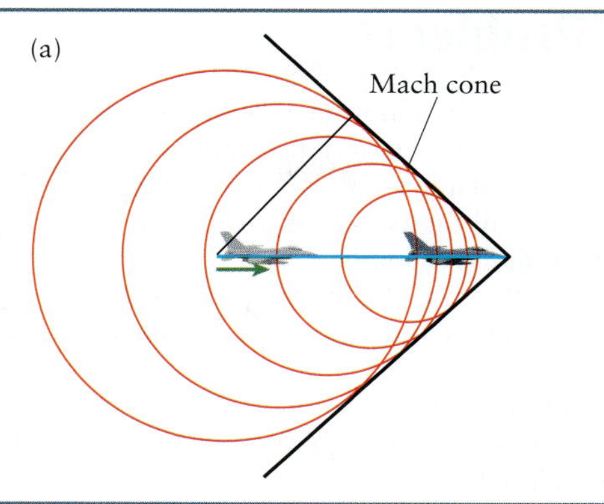

(Figure 13-28a)

Answer to What do you think? Question

(b) We learn in Section 13-3 that the speed of waves on a string (or a flagellum) does not depend on the frequency of the wave. If the frequency increases, the speed remains the same but the wavelength (the distance from one crest to the next) decreases.

Answers to Got the Concept? Questions

13-1 (c) In the transverse wave shown in Figure 13-2a, the individual pieces of rope (the wave medium) move vertically. So the restoring force that influences their motion must also act vertically. In the same way, since the pieces of the spring in Figure 13-2b move horizontally, the restoring force that acts on them must be in the horizontal direction.

13-2 (e) The propagation speed v_p of sound waves does not depend on the frequency f. The frequency and the wavelength λ are related by Equation 13-2, $v_p = f\lambda$, so increasing the frequency makes the wavelength shorter but has *no* effect on the propagation speed.

13-3 (a) The speed of light is so great that you *see* the batter swing a tiny fraction of a second after the swing took place. The speed of sound is relatively slow by comparison. From the relationship $d = v_{\text{sound}}t$ with distance d, propagation speed v_{sound}, and time t, the distance that sound travels in $t = 0.20$ s is $d = (343 \text{ m/s})(0.20 \text{ s}) = 69$ m. Note that the time for light to travel this same distance is $t_{\text{light}} = d/v_{\text{light}} = (69 \text{ m})/(3.0 \times 10^8 \text{ m/s}) = 2.3 \times 10^{-7}$ s, or about 1/4 of a microsecond (1 microsecond $= 1\ \mu\text{s} = 10^{-6}$ s).

13-4 (b) Increasing the frequency decreases the wavelength. The wavelength sets the scale for the entire interference pattern (the positions of destructive and constructive interference depend on the wavelength), so decreasing the wavelength will make all of the distances smaller, and the interference pattern will shrink inward toward the two fingers.

13-5 (e) From Equation 13-18 the wavelength of the $n = 1$ standing wave mode on a string of length L is given by $L = \lambda/2$, so $\lambda = 2L$. This does not depend on the tension in the string, so the wavelength remains the same.

13-6 (b) From Equation 13-19 the frequency of each standing wave mode on the string is proportional to the square root of the tension F. So doubling the tension will cause the frequency to increase not by a factor of 2 but by a factor of $\sqrt{2} = 1.41$.

13-7 (a) The bottle is like an air-filled pipe that is open at the top and closed on the bottom (at the surface of the water). The fundamental frequency of this pipe is inversely proportional to the length of the pipe (see Equation 13-22). As you fill the bottle the length of the air-filled pipe decreases so that the fundamental frequency goes up. The pitch of the sound from the bottle depends on the fundamental frequency, so the pitch goes up as well.

13-8 (e) The beat frequency equals the difference between the frequencies of the two waves that interfere. The beat frequency here is 3 Hz, so the second tuning fork has a frequency of either 440 Hz + 3 Hz = 443 Hz or 440 Hz − 3 Hz = 437 Hz. Without more information we can't tell which of these two is correct.

13-9 (d) The sound waves from each fan carry the same power, so at any distance from the fans the intensity from each fan is the same. Thus the total intensity doubles when both fans are running. We saw in Example 13-11 that doubling the intensity corresponds to adding 3.0 dB to the sound intensity level. So the sound intensity level due to both fans is 60 dB + 3.0 dB = 63 dB.

13-10 (a) The frequency you hear is higher while you're running compared to when you're standing still. Because you are moving toward the source of the sound, you will intercept more wave crests in a given amount of time than you would were you standing still. The qualitative approach we took using Figure 13-25 for a moving source of sound and a stationary listener applies equally well for a stationary source and a moving listener.

13-11 (c) No matter how a source of sound waves is moving, the waves spread outward at the speed of sound from the point where they were emitted. We used this idea in drawing Figure 13-27.

Questions and Problems

In a few problems you are given more data than you actually need; in a few other problems you are required to supply data from your general knowledge, outside sources, or informed estimate. Interpret as significant all digits in numerical values that have trailing zeros and no decimal points.

For all problems use $g = 9.80 \text{ m/s}^2$ for the free-fall acceleration due to gravity. Neglect friction and air resistance unless instructed to do otherwise.

•	Basic, single-concept problem
••	Intermediate-level problem; may require synthesis of concepts and multiple steps
•••	Challenging problem
SSM	*Solution is in Student Solutions Manual*
Example	*See worked example for a similar problem*

Conceptual Questions

1. • Explain the difference between longitudinal waves and transverse waves and give two examples of each.

2. • When you talk to your friend, are the air molecules that reach his ear the same ones that were in your lungs? Explain your answer. SSM

3. • Draw a sketch of a transverse wave and label the amplitude, wavelength, a crest, and a trough.

4. • Are water waves longitudinal or transverse? Explain your answer.

5. • Earthquakes produce several types of wave. The most significant are the primary wave (or P wave) and the secondary wave (or S wave). The primary wave is a longitudinal wave that can travel through liquids and solids. The secondary wave is a transverse wave that can travel only through solids. The speed of a P wave usually falls between 1000 and 8000 m/s; the speed of an S wave is around 60 to 70% of the P wave speeds. For any given seismic event discuss the damage that might be due to P waves and how that would differ when compared to the damage due to S waves.

6. • Explain the differences and similarities between the concepts of frequency f and angular frequency ω. SSM

7. • If a tree falls in the forest and no humans are there to hear it, was any sound produced?

8. • A sound wave passes from air into water. Give a qualitative explanation of how these properties change: (a) wave speed, (b) frequency, and (c) wavelength of the wave. SSM

9. • Two solid rods have the same Young's modulus, but one has larger density than the other. In which rod will the speed of longitudinal waves be greater? Explain your answer, making reference to the variables that affect the wave speed.

10. • Search the Internet for the words "rarefaction," "phonon," and "compression" in the context of longitudinal waves. Describe how a longitudinal wave is made up of phonons and explain the connection between rarefaction and compression in such a wave.

11. • Explain the concept of phase difference, ϕ, and predict the outcome of two identical waves interfering when $\phi = 0°$, $\phi = 90°$, $\phi = 180°$, $\phi = 270°$, and $\phi = 300°$.

12. • A car radio is tuned to receive a signal from a particular radio station. While the car slows to a stop at a traffic signal, the reception of the radio seems to fade in and out. Use the concept of interference to explain the phenomenon. *Hint:* In broadcast technology, the phenomenon is known as multipathing.

13. • (a) What is a transverse standing wave? (b) For a string stretched between two fixed points, describe how a disturbance on the string might lead to a standing wave.

14. • How does the length of an organ pipe determine the fundamental frequency?

15. • Why do the sounds emitted by organ pipes that are closed on one end and open on the other *not* have even harmonics? Include in your explanation a sketch of the resonant waves that are formed in the pipe.

16. • Two pianists sit down to play two identical pianos. However, a string is out of tune on Elizabeth's piano. The $G_3^\#$ key (208 Hz) appears to be the problem. When Greg plays the note on his piano and Elizabeth plays hers, a beat frequency of 6 Hz is heard. Luckily a piano tuner is present, and she is ready to correct the problem. However, in all the confusion she inadvertently increases the tension in Greg's $G_3^\#$ string by a factor of 1.058. Now both pianos are out of tune! Yet oddly, when Elizabeth plays her $G_3^\#$ note and Greg plays his $G_3^\#$, there is no beat frequency. Explain what happened.

17. • (a) Explain the differences and similarities among the concepts of intensity, sound level, and power. (b) What happens to intensity as the source of sound moves closer to the observer? (c) What happens to sound level? (d) What happens to power? SSM

18. • (a) Describe in words the nature of a sonic boom. (b) Now, referring to the formula for the Doppler shift, explain the phenomenon.

19. • If you stand beside a railroad track as a train sounding its whistle moves past, you will experience the Doppler effect. Describe any changes in the perceived sound that a person riding on the train will hear.

20. • Discuss several ways that the human body creates or responds to waves.

Multiple-Choice Questions

21. • A visible disturbance propagates around a crowded soccer stadium when fans, section by section, jump up and then sit back down. What type of wave is this?
    A. longitudinal wave
    B. transverse wave
    C. surface wave
    D. spherical wave
    E. pressure wave

22. • Two point sources produce waves of the same wavelength that are in phase. At a point midway between the sources, you would expect to observe
    A. constructive interference.
    B. destructive interference.
    C. alternating constructive and destructive interference.
    D. constructive or destructive interference depending on the wavelength.
    E. no interference.

23. • Standing waves are set up on a string that is fixed at both ends so that the ends are nodes. How many nodes are there in the fourth mode?
    A. 2
    B. 3
    C. 4
    D. 5
    E. 6 SSM

24. • Which of the following frequencies are higher harmonics of a string with fundamental frequency of 80 Hz?
    A. 80 Hz
    B. 120 Hz
    C. 160 Hz
    D. 200 Hz
    E. 220 Hz

25. • A trombone has a variable length. When a musician blows air into the mouthpiece and causes air in the tube of the horn to vibrate, the waves set up by the vibrations reflect back and forth in the horn to create standing waves. As the length of the horn is made shorter, what happens to the frequency?
    A. The frequency remains the same.
    B. The frequency will increase.
    C. The frequency will decrease.
    D. The frequency will increase or decrease depending on how hard the horn player blows.
    E. The frequency will increase or decrease depending on the diameter of the horn.

26. • Two tuning forks of frequency 480 Hz and 484 Hz are struck simultaneously. What is the beat frequency resulting from the two sound waves?
    A. 964 Hz
    B. 482 Hz
    C. 4 Hz
    D. 2 Hz
    E. 0 Hz

27. • If a source radiates sound uniformly in all directions, and you triple your distance from the sound source, what happens to the sound intensity at your new position?
    A. The sound intensity increases to three times its original value.
    B. The sound intensity does not change.
    C. The sound intensity drops to 1/3 its original value.
    D. The sound intensity drops to 1/9 its original value.
    E. The sound intensity drops to 1/27 its original value. SSM

28. • If the amplitude of a sound wave is tripled, the intensity will
    A. decrease by a factor of 3.
    B. increase by a factor of 3.
    C. remain the same.
    D. decrease by a factor of 9.
    E. increase by a factor of 9.

29. • A certain sound level is increased by an additional 20 dB. By how much does the intensity increase?
    A. The intensity increases by a factor of 2.
    B. The intensity increases by a factor of 20.
    C. The intensity increases by a factor of 100.
    D. The intensity increases by a factor of 200.
    E. The intensity does not increase.

30. • A person sitting in a parked car hears an approaching ambulance siren at a frequency f_1, and as it passes him and moves away he hears a frequency f_2. The actual frequency f of the source is
    A. $f > f_1$
    B. $f < f_2$
    C. $f_2 < f < f_1$
    D. $f = f_2 + f_1$
    E. $f = f_2 - f_1$

Estimation/Numerical Analysis

31. • (a) Estimate the distance that sound waves travel in 10 s. (b) Explain why you might get ±5% of the calculated distance on any given day that you perform an experiment to measure the speed of sound.

32. • Estimate the number of beats per second of a hummingbird's wings.

33. • (a) Estimate the speed of a human wave like those seen at a large sports venue. (b) How would you define the corresponding concepts of wavelength, frequency, and amplitude for the human wave?

34. • Describe how you might estimate the speed of the transverse waves on a violin string. SSM

35. • **Biology** Peristalsis is the rhythmic, wavelike contraction of smooth muscles to propel material through the digestive tract. A typical peristaltic wave will only last for a few seconds in the small intestine, traveling at only a few centimeters per second. Estimate the wavelength of the digestive wave.

36. • Estimate the amount of time it takes for a sound wave to travel directly *through* Earth, that is, from one point on the surface to another along a diameter. See Table 9-1 for typical bulk moduli of solids, Table 11-1 for typical densities of solids, and Appendix B for the radius of Earth.

37. • (a) Estimate the amplitude, frequency, and wavelength of a typical ocean wave, far at sea, away from coastlines. (b) Based on your estimations what is the typical speed of ocean waves?

38. • (a) Using a graphing calculator, spreadsheet, or graphing program, graph the wave function, $y(x)$ described by the data shown. Assume the data were taken at time $t = 0$ s. (b) Write a mathematical function that describes the traveling wave if the period of the motion is 2 s.

x (m)	y (m)	x (m)	y (m)
0	0	7	−4.5
1	4.5	8	−7.8
2	7.8	9	−9
3	9	10	−7.8
4	7.8	11	−4.5
5	4.5	12	0
6	0		

Problems

13-1 Waves are disturbances that travel from place to place
13-2 Mechanical waves can be transverse, longitudinal, or a combination of these
13-3 Sinusoidal waves are related to simple harmonic motion

39. • The period of a sound wave is 0.0100 s. Calculate the frequency f and the angular frequency ω. Example 13-1

40. • A wave on a string propagates at 22 m/s. If the frequency is 24 Hz, calculate the wavelength and angular wave number. SSM Example 13-1

41. •• A transverse wave on a string has an amplitude of 20 cm, a wavelength of 35 cm, and a frequency of 2.0 Hz. Write the mathematical description of the displacement from equilibrium for the wave if (a) at $t = 0$, $x = 0$ and $y = 0$; (b) at $t = 0$, $x = 0$ and $y = +20$ cm; (c) at $t = 0$, $x = 0$ and $y = -20$ cm; and (d) at $t = 0$, $x = 0$ and $y = 12$ cm. Example 13-2

42. • Show that the dimensions of speed (distance/time) are consistent with both versions of the expression for the propagation speed of a wave: $v_p = \frac{\omega}{k}$ and $v_p = \lambda f$. Example 13-1

43. •• Write the wave equation for a periodic transverse wave traveling in the positive x direction at a speed of 20 m/s if it has a frequency of 10 Hz and an amplitude in the y direction of 0.10 m. Example 13-2

44. • A wave on a string is described by the equation $y(x,t) = (0.5 \text{ m}) \cos [(1.0 \text{ rad/m})x - (10 \text{ rad/s})t]$. What are (a) the frequency, (b) wavelength, and (c) speed of the wave? SSM Example 13-2

45. •• A pressure wave traveling through air is described by the function $p(x,t) = (1.0 \text{ atm}) \sin [(6.0 \text{ rad/m})x - (4.0 \text{ rad/s})t]$. (a) What is the pressure amplitude of the wave? (b) What is the wave number of the wave? (c) What is the frequency of the wave? (d) What is the speed of the wave? Example 13-2

46. •• The equation for a particular wave is $y(x,t) = (0.10 \text{ m}) \cos (kx - \omega t)$. If the frequency of the wave is 2.0 Hz, what is the value of y at $x = 0$ when $t = 4.0$ s? Example 13-2

47. • Using the graph in **Figure 13-30**, write the mathematical description of the wave if the period of the motion is 4 s and the wave moves to the right (toward the positive x direction). Example 13-2

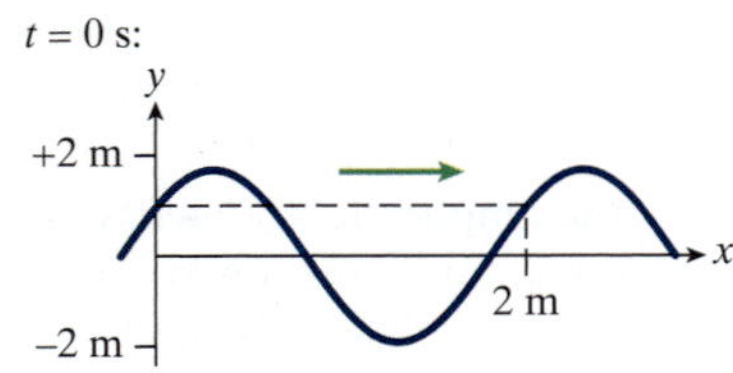

Figure 13-30 Problem 47

48. • Write a mathematical description of the wave represented by the graphs in **Figure 13-31**. SSM Example 13-2

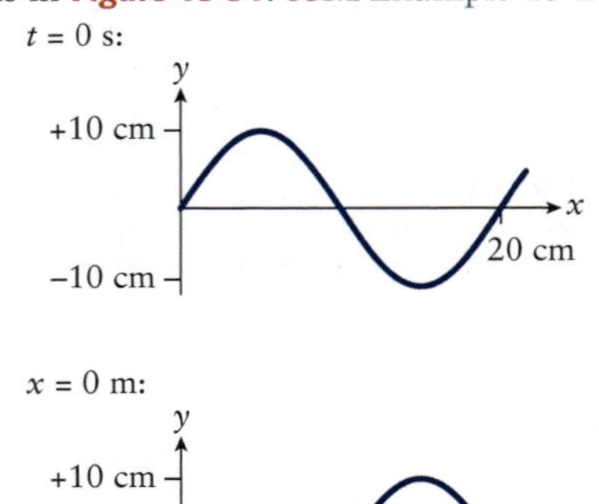

Figure 13-31 Problem 48

13-4 The propagation speed of a wave depends on the properties of the wave medium

49. • If the bulk modulus for liquid A is twice that of liquid B, and the density of liquid A is one-half of the density of liquid B, what does the ratio of the speeds of sound in the two liquids (v_A/v_B) equal? Example 13-4

50. • A string that has a mass of 5.0 g and a length of 2.2 m is pulled taut with a tension of 78 N. Calculate the speed of transverse waves on the string. SSM Example 13-3

51. • A long rope is shaken up and down by a rodeo contestant. The transverse waves travel 12.8 m in a time of 2.1 s. If the tension in the rope is 80.0 N, calculate the mass per unit length for the rope. Example 13-3

52. •• The violin is a four-stringed instrument tuned so that the ratio of the frequencies of adjacent strings is 3 to 2. (This is the ratio when taken as high frequency to lower frequency.) If the diameter of the E string (the highest frequency) on a violin is 0.25 mm, find the diameters of the remaining strings (A, D, and G), assuming they are tuned as indicated (what musicians call intervals of a perfect fifth), they are made of the same material, and they all have the same tension. Example 13-3

53. • At room temperature the bulk modulus of glycerine is about $4.35 \times 10^9 \text{ N/m}^2$, and the density of glycerine is about 1260 kg/m^3. Calculate the speed of sound in glycerin. Example 13-4

54. • The bulk modulus of water is $2.2 \times 10^9 \text{ N/m}^2$. The density of water is 1000 kg/m^3. Calculate the speed of sound in water. SSM Example 13-4

55. • When sound travels through the ocean, where the bulk modulus is $2.34 \times 10^9 \text{ N/m}^2$, the wavelength associated with 1000-Hz waves is 1.51 m. Calculate the density of seawater. Example 13-4

56. • What is the speed of sound in gasoline? The bulk modulus for gasoline is $1.3 \times 10^9 \text{ N/m}^2$. The density of gasoline is 0.74 kg/L. Example 13-4

13-5 When two waves are present simultaneously, the total disturbance is the sum of the individual waves

57. • Two waves interfere at the point x in **Figure 13-32**. The resultant wave is shown. Draw the three possible shapes for the two waves that interfere to produce this outcome. In each case one wave should head toward the right, and one wave should head toward the left.

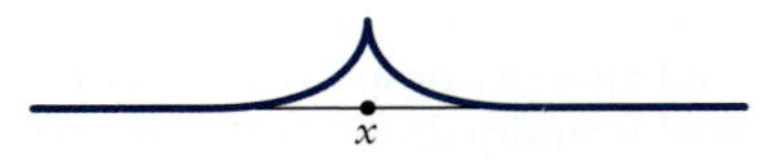

Figure 13-32 Problem 57

58. • In **Figure 13-33**, rectangular waveforms approach each other. For each case use the ideas of interference to predict the superposed wave that results when the two waves are coincident. Describe the superposed wave graphically and with text. SSM

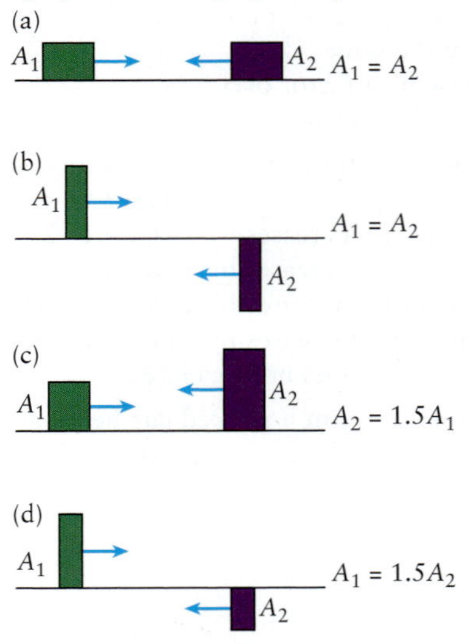

Figure 13-33 Problem 58

59. • Construct the resultant wave that is formed when the two waves shown in each case occupy the same space and interfere (**Figure 13-34**).

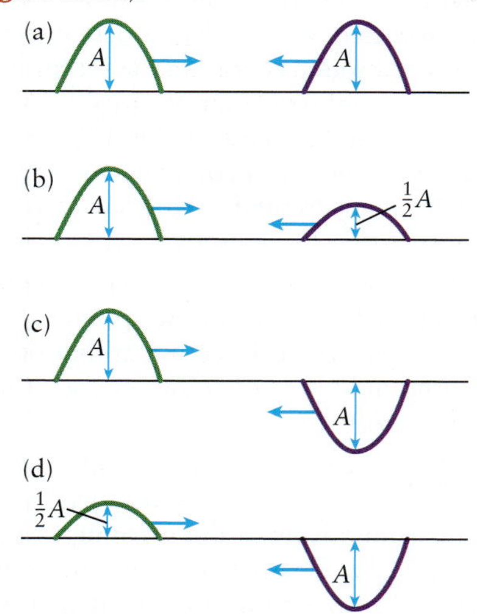

Figure 13-34 Problem 59

60. •• Two identical speakers (1 and 2) are playing a tone with a frequency of 171.5 Hz, in phase (**Figure 13-35**). The speakers are located 6.00 m apart. Determine what points (A, B, C, D, or E, all separated by 1.00 m) will experience constructive interference along the line that is 6.00 m in front of the speakers. Point A is directly in front of speaker 1. The speed of sound is 343 m/s. SSM Example 13-5

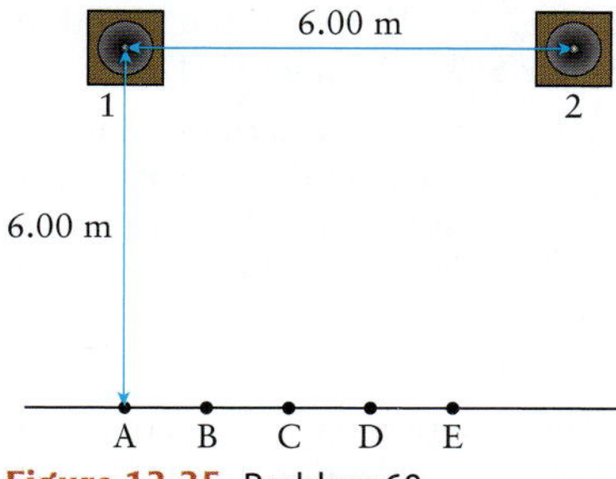

Figure 13-35 Problem 60

61. • Two 60-Hz tone generators are set up such that they interfere constructively at a point midway between them along a line drawn through both speakers. How far must one speaker be moved along the line such that *destructive* interference occurs at the exact same point? Example 13-5

13-6 A standing wave is caused by interference between waves traveling in opposite directions

62. • A string is tied at both ends, and a standing wave is established. The length of the string is 2.00 m, and it vibrates in the fundamental mode ($n = 1$). If the speed of waves on the string is 60.0 m/s, calculate the frequency and wavelength of the waves. Example 13-6

63. • A string is fixed on both ends with a standing wave vibrating in the fourth harmonic. Draw the shape of the wave and label the location of antinodes (A) and nodes (N). Example 13-6

64. • A 2.35-m-long string is tied at both ends, and it vibrates with a fundamental frequency of 24.0 Hz. Find the frequencies and make a sketch of the next four harmonics. What is the speed of waves on the string? Example 13-6

65. •• A string that is 1.25 m long has a mass of 0.0548 kg and a tension of 2.00×10^2 N. The string is tied at both ends and vibrated at various frequencies. (a) What frequencies would you need to apply to excite the first four harmonics? (b) Make a sketch of the first four harmonics for these standing waves. SSM Example 13-6

66. •• A string of length L is tied at both ends, and a harmonic mode is created with a frequency of 40.0 Hz. If the next successive harmonic is at 48.0 Hz and the speed of transverse waves on the string is 56.0 m/s, find the length of the string. Note that the fundamental frequency is not necessarily 40.0 Hz. Example 13-6

67. •• An object of mass M is used to provide tension in a 4.50-m-long string that has a mass of 0.252 kg, as shown in **Figure 13-36.** A standing wave that has a wavelength equal to 1.50 m is produced by a source that vibrates at 30.0 Hz. Calculate the value of M. Example 13-6

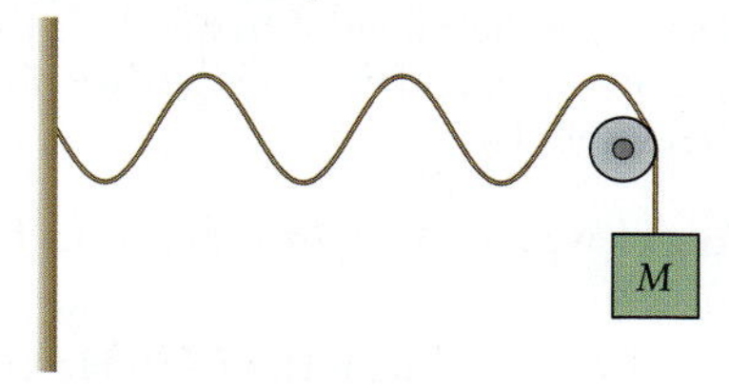

Figure 13-36 Problem 67

68. •• A guitar string has a mass per unit length of 2.35 g/m. If the string is vibrating between points that are 60.0 cm apart, find the tension when the string is designed to play a note of 440 Hz (A4). Example 13-6

69. •• An E string on a violin has a diameter of 0.25 mm and is made from steel (density of 7800 kg/m^3). The string is designed to sound a fundamental note of 660 Hz, and its unstretched length is 32.5 cm. Calculate the tension in the string. SSM Example 13-6

13-7 Wind instruments, the human voice, and the human ear use standing sound waves

70. • An organ pipe of length L sounds its fundamental tone at 40.0 Hz. The pipe is open on both ends, and the speed of sound in air is 343 m/s. Calculate (a) the length of the pipe and (b) the frequency of the first four harmonics. Example 13-7

71. • An organ pipe sounds two successive tones at 228.6 Hz and 274.3 Hz. Determine whether the pipe is open at both ends or open at one end and closed at the other. SSM Example 13-7

72. • A narrow glass tube is 0.40 m long and sealed on the bottom end. It is held under a loudspeaker that sounds a tone at 220 Hz, causing the tube to radically resonate in its first harmonic. Find the speed of sound in the room. Example 13-7

73. •• The third harmonic of an organ pipe that is open at both ends and is 2.25 m long excites the fourth harmonic in another organ pipe. Determine the length of the other pipe and whether it is open at both ends or open at one end and closed at the other. Assume the speed of sound is 343 m/s in air. Example 13-7

74. • **Biology** If the human ear canal with a typical length of about 2.8 cm is regarded as a tube open at one end and closed at the eardrum, what is the fundamental frequency that we should hear best? Assume that the speed of sound is 343 m/s in air. Example 13-7

75. • **Biology** A male alligator emits a subsonic mating call that has a frequency of 18 Hz by taking a large breath into his chest cavity and then releasing it. If the hollow chest cavity of an alligator behaves approximately as a pipe open at only one end, estimate its length. Is your answer consistent with the typical size of an alligator? Example 13-7

76. • **Biology** The trunk of a very large elephant may extend up to 3 m! It acts much like an organ pipe open only at one end when the elephant blows air through it. (a) Calculate the fundamental frequency of the sound the elephant can create with its trunk. (b) If the elephant blows even harder, the next harmonic may be sounded. What is the frequency of this first overtone? Example 13-7

77. • The longest pipe in the Mormon Tabernacle Organ in Salt Lake City has a "speaking length" of 9.75 m; the smallest pipe is 1.91 cm long. Assuming that the speed of sound is 343 m/s, determine the range of frequencies that the organ can produce if both open and closed pipes are used with these lengths. SSM Example 13-7

13-8 Two sound waves of slightly different frequencies produce beats

78. • The sound from a tuning fork of 440 Hz produces beats against the unknown emissions from a vibrating string. If beats are heard at a frequency of 4 Hz, what is the vibrational frequency of the string? Example 13-8

79. • A guitar string is "in tune" at 440 Hz. When a standardized tuning fork rated at 440 Hz is simultaneously sounded with the guitar string, a beat frequency of 5 Hz is heard. (a) How far out of tune is the string? (b) What should you do to correct this? Example 13-8

80. • Two tuning forks are both rated at 256 Hz, but when they are struck at the same time, a beat frequency of 4 Hz is created. If you know that one of the tuning forks is in tune (but you are not sure which one), what are the possible values of the "out-of-tune" fork? Example 13-8

81. •• A guitar string has a tension of 1.00×10^2 N and is supposed to have a frequency of 1.10×10^2 Hz. When a standard tone of that value is sounded while the string is plucked, a beat frequency of 2.00 Hz is heard. The peg holding the string is loosened (decreasing the tension), and the beat frequency increases. What should the tension in the string be in order to achieve perfect pitch? SSM Example 13-8

82. •• Two identical guitar strings under 2.00×10^2 N of tension produce sound with frequencies of 290 Hz. If the tension in one string drops to 196 N, what will be the frequency of beats when the two strings are plucked at the same time? Example 13-8

13-9 The intensity of a wave equals the power that it delivers per square meter

83. • By what factor should you move away from a point source of sound waves in order to lower the intensity by (a) a factor of 10? (b) a factor of 3? (c) a factor of 2? (d) Discuss how this problem would change if it were a point source that was radiating into a hemisphere instead of a sphere. Example 13-10

84. • Calculate the ratio of the acoustic power generated by a blue whale (190 dB) compared to the sound generated by a jackhammer (105 dB). Assume that the receiver of the sound is the same distance from the two sources of sound. Example 13-11

85. •• The steady drone of rush hour traffic persists on a stretch of freeway for 4 h each day. A nearby resident undertakes a plan to harness the wasted sound energy and use it for her home. She mounts a 1-m^2 "microphone" that absorbs 30% of the sound that hits it. If the ambient sound level is 100 dB at the microphone, calculate the amount of sound energy that is collected. Do you think her plan is "sound"? How could you improve it? Explain, with supporting calculations. SSM Example 13-11

86. • A longitudinal wave has a measured sound level of 85 dB at a distance of 3.0 m from the speaker that created it. (a) Calculate the intensity of the sound at that point. (b) If the speaker were a point source of sound energy, find its power output. Example 13-11

87. • **Biology** A single goose sounds a loud warning when an intruder enters the farmyard. Some distance from the goose, you measure the sound level of the warning to be 88 dB. If a gaggle of 30 identical geese is present, and they are all approximately the same distance from you, what will the collective sound level be if they all sound off simultaneously? Neglect any interference effects. Example 13-11

88. •• Two sound levels, β_1 and β_2, can be compared relative to each other (rather than to some standardized intensity threshold). Find a formula for this by calculating $\Delta\beta = \beta_2 - \beta_1$. You will need to employ your knowledge of basic logarithm operations to derive the formula. Example 13-11

89. • By what factor must you increase the intensity of a sound in order to hear (a) a 1-dB rise in the sound level? (b) What about a 20-dB rise? SSM Example 13-11

90. • **Biology** The intensity of a certain sound at your eardrum is 0.0030 W/m^2. (a) Calculate the rate at which sound energy hits your eardrum. Assume that the area of your eardrum is about 55 mm^2. (b) What power output is required from a point source that is 2.0 m away in order to create that intensity? Example 13-10

91. • **Biology** The area of a typical eardrum is 5.0×10^{-5} m^2. Find the sound power incident on an eardrum at the threshold of pain. Example 13-9

13-10 The frequency of a sound depends on the motion of the source and the listener

92. • A fire engine's siren is 1600 Hz when at rest. What frequency do you detect if you move with a speed of 28 m/s (a) toward the fire engine, and (b) away from it? Assume that the speed of sound is 343 m/s in air. Example 13-12

93. • **Medical** An ultrasound machine can measure blood flow speeds. Assume the machine emits acoustic energy at a frequency of 2.0 MHz, and the speed of the wave in human tissue is taken to be 1500 m/s. What is the beat frequency between the transmitted and reflected waves if blood is flowing in large arteries at 0.020 m/s directly away from the sound source? Example 13-12

94. • A car sounding its horn (rated by the manufacturer at 600 Hz) is moving at 20 m/s toward the east. A stationary observer is standing due east of the oncoming car. (a) What frequency will he hear assuming that the speed of sound is 343 m/s? (b) What if the observer is standing due west of the car as it drives away? Example 13-12

95. •• A bicyclist is moving toward a wall while holding a tuning fork rated at 484 Hz. If the bicyclist detects a beat frequency of 6 Hz (between the waves coming directly from the tuning fork and the echo waves coming from the wall), calculate the speed of the bicycle. Assume the speed of sound is 343 m/s. SSM Example 13-12

96. •• **Medical** An ultrasonic scan uses the echo waves coming from something moving (such as the beating heart of a fetus) inside the body and the waves that are directly received from the transmitter to form a measurable beat frequency. This allows the speed of the internal structure to be isolated and analyzed. What is the beat frequency detected when waves with a frequency of 5.00 MHz are used to scan a fetal heartbeat (moving at a speed of ± 10.0 cm/s)? The speed of the ultrasound waves in tissue is about 1540 m/s. Example 13-12

97. •• **Biology** A bat emits a high-pitched squeal at 50.0 kHz as it approaches an insect at 10.0 m/s. The insect flies away from the bat, and the reflected wave that echoes off the insect returns to the bat at a frequency of 50.05 kHz. Calculate the speed of the insect as it tries to avoid being the bat's next meal. (A bat can eat over 3000 mosquitoes in one night!) SSM Example 13-12

General Problems

98. •• A large volcanic eruption triggers a tsunami. At a seismic station 250 km away, the instruments record that the time *difference* between the arrival of the tidal wave and the arrival of the sound of the explosion is 9.25 min. Tsunamis typically travel at approximately 800 km/h. (a) Which sound arrives first, the sound in the air or in the water? Prove your answer numerically. (b) How long after the explosion does it take for the first sound wave to reach the seismic station? (c) How long after the explosion does it take for the tsunami to reach the seismic station? Example 13-4

99. • A transverse wave is propagating according to the following wave function:

$y(x,t) = (1.25 \text{ m}) \cos [(5.00 \text{ rad/m})x - (4.00 \text{ rad/s})t]$ (a) Plot a graph of y versus x when $t = 0$ s. (b) Repeat when $t = 1.00$ s. Example 13-2

100. • **Medical** A diagnostic *sonogram* produces a picture of internal organs by passing ultrasound through the tissue. In one application it is used to find the size, location, and shape of the prostate in preparation for surgery or other treatment. The speed of sound in the prostate is 1540 m/s, and a diagnostic sonogram uses ultrasound of frequency 2.00 MHz. The density of the prostate is 1060 kg/m^3. (a) What is the wavelength of the sonogram ultrasound? (b) What is the bulk modulus for the prostate gland? Example 13-4

101. •• **Biology** You may have seen a demonstration in which a person inhales some helium and suddenly speaks in a high-pitched voice. Let's investigate the reason for the change in pitch. At 0°C the density of air is 1.40 kg/m^3, the density of helium is 0.1786 kg/m^3, the speed of sound in helium is 972 m/s, and the speed of sound in air is 331 m/s. (a) What is the bulk modulus of helium at 0°C? (b) If a person produces a sound of frequency 0.500 kHz while speaking with his lungs full of air, what frequency sound will that person produce if his respiratory tract is filled with helium instead of air? (c) Use your result to explain why the person sounds strange when he breathes in helium. SSM Example 13-7

102. • **Biology** When an insect ventures onto a spiderweb, a slight vibration is set up, alerting the spider. The density of spider silk is approximately 1.3 g/cm^3, and its diameter varies considerably depending on the type of spider, but 3.0 mm is typical. If the web is under a tension of 0.50 N when a small beetle crawls onto it 25 cm from the spider, how long will it take for the spider to receive the first waves from the beetle? Example 13-3

103. •• If two musical notes are an *octave* apart, the frequency of the higher note is twice that of the lower note. The note concert A usually has a frequency of 440 Hz (although there is some variation). (a) What is the frequency of a note that is two octaves above (higher than) concert A in pitch? (b) If a certain string on a viola is tuned to middle C by adjusting its tension to T, what should be the tension (in terms of T) of the string so that it plays a note one octave below middle C? SSM Example 13-6

104. ••• In Western music the octave is divided into 12 notes as follows: C, C$^\#$/D^b, D, D$^\#$/E^b, E, F, F$^\#$/G^b, G, G$^\#$/A^b, A, A$^\#$/B^b, C′. Note that some of the notes are the same, such as C$^\#$ and D^b. Each of the 12 notes is called a *semitone*. In the ideal *tempered* scale, the ratio of the frequency of any semitone to the frequency of the note below it is the same for all pairs of adjacent notes. So, for example, $f_D/f_{C\#}$ is the same as $f_{A\#}/f_A$. (a) Show that the ratio of the frequency of any semitone to the frequency of the note just below it is $2^{1/12}$. (b) If A is 440 Hz, what is the frequency of F$^\#$ in the tempered scale? (c) If you want to tune a string from B^b to B by changing only its tension, by what ratio should you change the tension? Should you increase or decrease the tension? Example 13-6

105. •• Find the temperature in an organ loft in Vancouver, British Columbia, if the 5th *overtone* associated with the pipe that is resonating corresponds to 1500 Hz. The pipe is 0.70 m long, and its type (open–open or open–closed) is not specified. The speed of sound in air depends on the centigrade temperature (T) according to the following: Example 13-7

$$v(T) = \sqrt{109{,}700 + 402T}$$

106. • **Biology** An adult female ring-necked duck is typically 16 in. long, and the length of her bill plus neck is about 5.0 cm. (a) Calculate the expected fundamental frequency of the quack of the duck. For a rough but reasonable approximation, assume that the sound is produced only in the neck and bill. (b) An adult male ring-necked duck is typically 18 in. long. If its other linear dimensions are scaled up in the same ratio from those of the female, what would be the fundamental frequency of its quack? (c) Which would produce a higher-pitch quack, the male or female? Example 13-7

107. • When a worker puts on earplugs, the sound level of a jackhammer decreases from 105 dB to 75 dB. Then the worker moves twice as far from the sound. Determine the sound level with the earplugs at the new location. Example 13-11

108. ••• **Medical** High-intensity focused ultrasound (HIFU) is one treatment for certain types of cancer. During the procedure a narrow beam of high-intensity ultrasound is focused on the tumor, raising its temperature to nearly 90°C to kill it. A range of frequencies and intensities can be used, but in one treatment a beam of frequency 4.0 MHz produced an intensity of 1500 W/cm^2. The energy was delivered in short pulses for a total time of 2.5 s over an area measuring 1.4 mm by

5.6 mm. The speed of sound in the soft tissue was 1540 m/s, and the density of that tissue was 1058 kg/m^3. (a) What was the wavelength of the ultrasound beam? (b) How much energy was delivered to the tissue during the 2.5-s treatment? (c) What was the maximum displacement of the molecules in the tissue as the beam passed through? Example 13-10

109. •• Many natural phenomena produce very high-energy, but inaudible, sound waves at frequencies below 20 Hz (*infrasound*). During the 2003 eruption of the Fuego volcano in Guatemala, sound waves of frequency 10 Hz (and even less) with a sound level of 120 dB were recorded. (a) What was the maximum displacement of the air molecules produced by the waves? (b) How much energy would such a wave deliver to a 2.0 m by 3.0 m wall in 1.0 min? Assume the density of air is 1.2 kg/m^3. SSM Example 13-11

110. ••• Two identical 375-g speakers are mounted on parallel springs, each having a spring constant of 50.0 N/cm. Both speakers face in the same direction and produce a steady tone of 1.00 kHz. Both sounds have an amplitude of 35.0 cm, but they oscillate 180° out of phase with each other. What is the highest frequency of the beat that a person will hear if she stands in front of the speakers? Example 13-12

111. • A jogger hears a car alarm and decides to investigate. While running toward the car, she hears an alarm frequency of 869.5 Hz. After passing the car, she hears the alarm at a frequency of 854.5 Hz. If the speed of sound is 343 m/s, calculate the speed of the jogger. Example 13-12

112. ••• A rescuer in an all-terrain vehicle (ATV) is tracking two injured hikers in the desert, each of whom has an emergency locator transmitter (ELT) stored in his backpack (**Figure 13-37**). The beacons give off radio signals at 121.5 MHz, in phase, and there is a receiver in the ATV that is tuned to that frequency. The speed of the radio waves is 3.00×10^8 m/s. The ATV is traveling due east, 2.00×10^2 m north of the hikers, and the hikers are 1.00×10^2 m apart. What is the spacing between the points at which the driver detects constructive interference between the two signals? Example 13-5

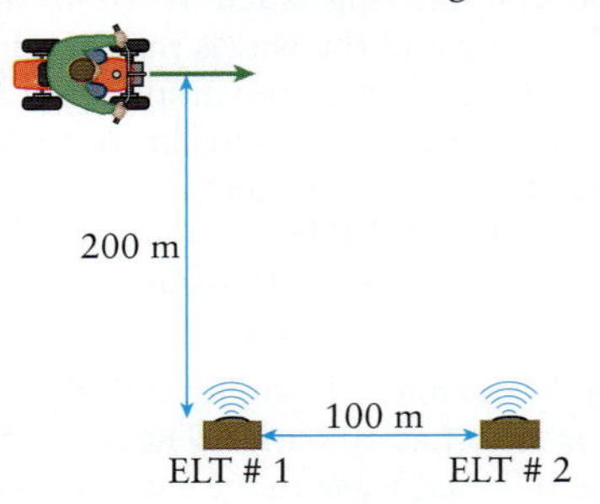

Figure 13-37 Problem 112

113. •• Two identical speakers that face each other and that are separated by a distance of 2.00 m emit a constant tone in phase. You stand in line with the speakers such that your right ear is exactly halfway between the two speakers, resulting in your left ear being closer to one speaker than the other. You notice that what you hear in your right ear is loud, but in your left ear you hear almost nothing. Given that your hearing is fine in both ears and the distance between your ears is 22.0 cm, what are the two lowest frequencies that the speakers could possibly be emitting? Example 13-5

14 Thermodynamics I

Helmut Schmitz, University of Bonn

What do you think?

Fire beetles (*Melanophila acuminata*) converge on burning forests from distances of up to 10 km to lay their eggs in trees weakened by fires. These insects can detect a fire from so far away because they are equipped with special chambers that serve as sensitive thermometers. When heated even by a small amount, the fluid inside the chambers expands and causes the pressure within to increase; sensory neurons respond to the pressure change. Compared to how much the volume of the fluid increases in response to the temperature increase, by how much does the volume of the chambers increase? (a) A greater amount; (b) the same amount; (c) by a lesser but nonzero amount; (d) the volume actually remains the same; (e) the volume actually decreases.

In this chapter, your goals are to:

- (14-1) Define what thermodynamics is.
- (14-2) Explain the meaning of temperature and thermal equilibrium.
- (14-3) Describe the origin of pressure in an ideal gas and explain the relationship between molecular kinetic energy and gas temperature.
- (14-4) Explain how objects change in size when their temperature changes.
- (14-5) Examine the relationship between the quantity of heat that flows into or out of an object and the temperature change of that object.
- (14-6) Explain how heat must flow in order to cause a substance to change between the solid, liquid, and gas phases.
- (14-7) Describe the key properties of heat transfer by radiation, convection, and conduction.

To master this chapter, you should review:

- (7-3) The relationship between force, collision time, and momentum change in a collision.
- (9-2) The meaning of tensile and compressive stress.
- (11-8) Archimedes' principle for buoyancy.

14-1 A knowledge of thermodynamics is essential for understanding almost everything around you—including your own body

Thermodynamics, broadly defined, is the branch of physics that deals with relationships among properties of substances such as temperature, pressure, and volume, as well as the energy and flow of energy associated with these properties. If that sounds pretty abstract, that's because thermodynamics is such a general subject that it applies to almost *everything* in your surroundings. Thermodynamics explains how the ice cubes slowly melting in your beverage keep the beverage cold and how the clothing you wear keeps you warm. Your body is an example of a complex thermodynamic system: Some of the energy released by digesting food goes into maintaining the temperature of your body at a healthy value, and some of the energy is transferred to your surroundings and warms the air around you. Indeed, to understand any aspect of nature or technology that involves the idea of *temperature*, we have to invoke thermodynamic concepts.

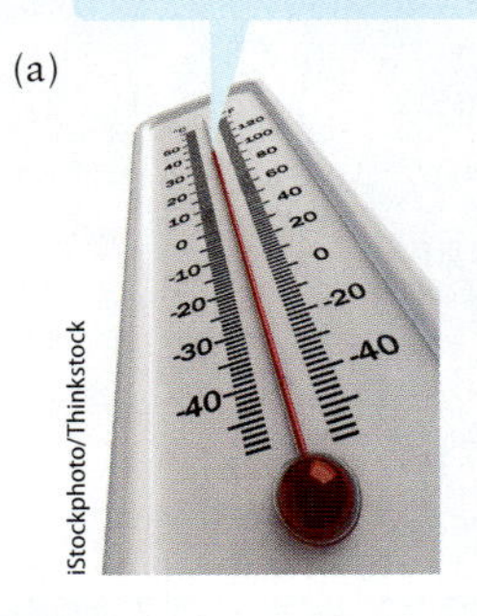

A liquid-in-glass thermometer works on the principle that substances like the liquid expand as the temperature increases.
(a)
iStockphoto/Thinkstock

This ice cube melts because heat flows into it, but its temperature remains at a constant 0°C until it has completely melted.
(b)
Sebastian Duda/Shutterstock.com

These cold-blooded iguanas lose heat by convection but take heat in by radiation from the Sun and by conduction from the hot rock beneath them.
(c)
BlueGreen Pictures/BlueGreen Pictures/Superstock

Figure 14-1 Thermodynamics The subject matter of thermodynamics includes (a) the meaning of temperature and how it is measured, (b) what heat is and what happens when heat flows into or out of an object, and (c) the ways in which heat is transferred from one object to another.

TAKE-HOME MESSAGE FOR Section 14-1

✔ Thermodynamics is the study of the relationships among properties of substances, including temperature, pressure, and volume, as well as the energy and flow of energy associated with these properties.

Thermodynamics is such a broad subject that it will take us this chapter and the next to introduce and apply its key concepts. We'll begin this chapter by defining what we mean by the temperature of an object, and we'll see that the temperature of a gas has a particularly simple interpretation. We'll go on to examine how objects change their dimensions when the temperature changes (**Figure 14-1a**). We'll define *heat* as energy that flows from one place to another due to a temperature difference, and we'll see how to analyze the energy flow that takes place when a substance changes phase, such as from solid to liquid (**Figure 14-1b**). We'll conclude this chapter by looking at the ways in which heat can flow from one object to another (**Figure 14-1c**).

14-2 Temperature is a measure of the energy within a substance

One of the most fundamental quantities in thermodynamics is the *temperature* of an object. We are used to the idea that we can raise the temperature of an object like a piece of chicken or a slice of eggplant by heating it, say in a frying pan. We also know that we can lower the temperature of an object by cooling it, as in a refrigerator or freezer. But these everyday experiences fail to answer an important question: What *is* temperature?

One way to define what we mean by the temperature of a substance is in terms of what happens on the *microscopic* level—that is, how the individual molecules that make up the substance behave. (In *monatomic* substances, such as helium, the molecules are actually individual atoms. In *polyatomic* substances the molecules are made up of combinations of atoms; for example, in water each molecule is made up of two hydrogen atoms and an oxygen atom. We'll use the term *molecule* to refer to the constituents of both monatomic and polyatomic substances.)

Molecules in any substance are always in motion. In a gas or liquid, molecules move in every direction with a range of speeds. A molecule in a solid has an equilibrium position, but it constantly wiggles back and forth around that equilibrium position. We can define **temperature** as a measure of the kinetic energy associated with molecular motion. As the temperature of a given substance increases, so does the average kinetic energy of a molecule in that substance.

This microscopic definition of temperature isn't very helpful, however. For one thing, it isn't very precise: The relationship between the average kinetic energy of a molecule and the temperature is different for gases, liquids, and solids, and it often depends on what type of molecule makes up the substance. What's more, this definition doesn't tell us how to *measure* temperature, since it's not very practical to analyze individual molecules in a substance to determine the temperature of that substance.

A practical approach to measuring temperature is to say that *temperature is the property that you measure with a* **thermometer**. There are many kinds of thermometers that work in different ways, but they're all based on the principle that certain substances change their properties when the temperature changes. For example, nearly all liquids increase in volume as temperature increases. A liquid thermometer uses a column of alcohol or mercury inside a glass tube, and the height of that column goes up as the temperature increases and the liquid expands (see Figure 14-1a).

Strictly speaking, a thermometer measures its *own* temperature. The reason you can use a thermometer to measure the temperature of an object is the phenomenon of *thermal equilibrium*. Experiment shows that when two objects that have different temperatures are in **thermal contact**, so that energy can flow from one to the other, energy flows until both objects reach the same temperature. This final temperature lies between the two original temperatures of the objects. This kind of energy flow that happens due to a temperature difference is what we call *heat transfer*. (We'll present a careful definition of what we mean by *heat* in Section 14-5.) For example, when you take a scoop of ice cream from the freezer and put it in a room-temperature dish, energy flows from the dish into the ice cream. As a result, the temperature of the dish decreases and the temperature of the ice cream rises. The energy flow stops when the two objects are at the *same* temperature. At this point we say the two objects are in **thermal equilibrium**, and we say that *two objects in thermal equilibrium are at the same temperature*. If one of those objects is a thermometer, the reading on the thermometer indicates the temperature of both the thermometer *and* the object with which it is in contact (**Figure 14-2**).

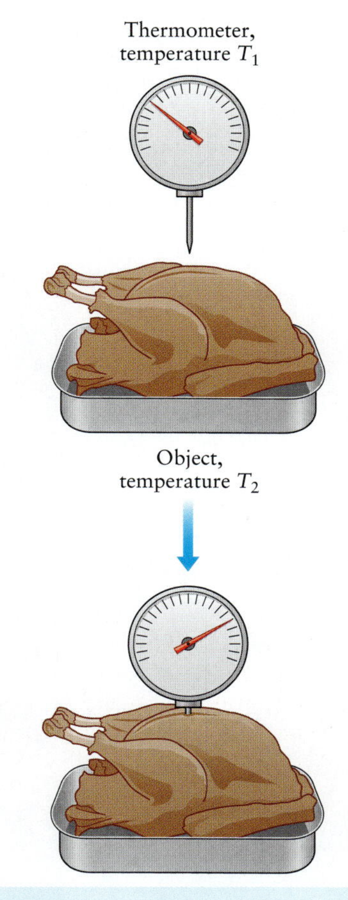
Thermometer, temperature T_1
Object, temperature T_2

Thermometer and object are both at the same intermediate temperature T_{final}; they are in thermal equilibrium.

Figure 14-2 Using a thermometer When a thermometer is placed in contact with an object and allowed to come to thermal equilibrium, the thermometer and the object end up at the same temperature.

A good measurement of any property is one that doesn't change the property being measured. (If you want to know what your pet cat does when you're not around, you won't find out by following it around the house.) Unfortunately, any thermometer causes a change in the temperature of the object being measured. That's because some heat transfer has to take place between the object and the thermometer, and this will cause the object's temperature to either increase or decrease. A good thermometer is one that minimizes the amount of heat transfer so that the temperature of the object hardly changes before the thermometer and object are in thermal equilibrium. Then the thermometer reading is very close to the temperature that the object had before you made it interact with the thermometer.

Now imagine we put a thermometer C in thermal contact with object A (**Figure 14-3a**). Heat flows until they reach thermal equilibrium and the reading on the thermometer tells us the temperature of object A. We now put the thermometer C in thermal contact with object B, allowing heat transfer between the thermometer and object B until they come to thermal equilibrium (**Figure 14-3b**). The reading on the thermometer then tells us the temperature of *both* A and B, so at this point the two objects have the same temperature. Finally, we put A and B into thermal contact with each other (**Figure 14-3c**). When we do this last step, we find that nothing changes: There is no heat flow between A and B, which means that these two objects at the same temperature must have been in thermal equilibrium *before* we put them in direct contact with each other. In other words, *two objects at the same temperature must be in thermal equilibrium, even if they do not interact*. If we combine this observation with our previous observations about thermal equilibrium, we end up with a single statement called the **zeroth law of thermodynamics**:

If two objects are each in thermal equilibrium with a third object, they are also in thermal equilibrium with each other.

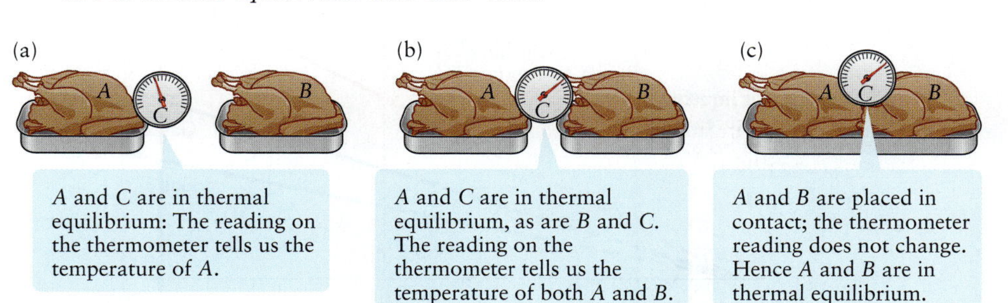
(a)
A
C
B
A and C are in thermal equilibrium: The reading on the thermometer tells us the temperature of A.
(b)
A
C
B
A and C are in thermal equilibrium, as are B and C. The reading on the thermometer tells us the temperature of both A and B.
(c)
A
C
B
A and B are placed in contact; the thermometer reading does not change. Hence A and B are in thermal equilibrium.

The **zeroth law of thermodynamics:** If two objects A and B are in thermal equilibrium with a third object C, then A and B are also in thermal equilibrium with each other.

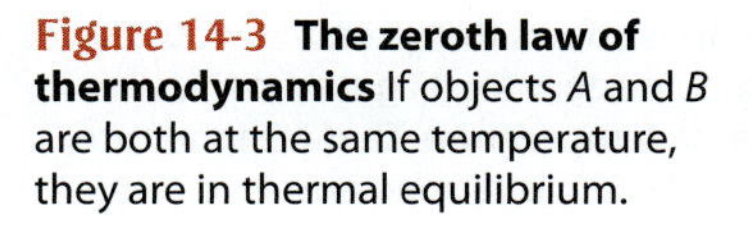

Figure 14-3 The zeroth law of thermodynamics If objects A and B are both at the same temperature, they are in thermal equilibrium.

From the zeroth law of thermodynamics, we can draw the important conclusion that *temperature is a property of objects in general*, not just of special devices called thermometers. (There are also a *first* and a *second* law of thermodynamics; we'll study these in Chapter 15.) To measure temperature, the tube of a liquid thermometer has numbers marked on it to help measure the volume of the liquid. The values associated with the markings are arbitrary, and throughout history a variety of schemes have been used to determine what the numbers on a thermometer should be.

The most common temperature scale in everyday life is the **Celsius scale**, based on the work of the eighteenth-century Swedish astronomer Anders Celsius. In this scale the freezing point of water is approximately 0°C and the boiling point is approximately 100°C (both values are at a standard pressure of 1 atm). By contrast, the official temperature scale used in the United States, the Cayman Islands, and Belize is the **Fahrenheit scale**, originated by the German scientist Daniel Fahrenheit (who also invented the alcohol thermometer and mercury thermometer). On this scale water freezes at 32°F and boils at 212°F at a pressure of 1 atm.

The range of Fahrenheit temperatures between freezing and boiling (212°F − 32°F = 180°F) corresponds almost exactly to a difference of 100°C on the Celsius scale. So a change of 1°C is nearly equivalent to (100/180)°F, or (5/9)°F. If we combine this with the observation that 0°C corresponds to 32°F, we get an approximate conversion between Fahrenheit and Celsius scales:

$$T_C = \frac{5}{9}(T_F - 32) \quad \text{and} \quad T_F = \frac{9}{5}T_C + 32 \tag{14-1}$$

You should use Equations 14-1 to verify that 0°C corresponds to 32°F and 100°C corresponds to 212°F.

The Celsius and Fahrenheit scales were based on the properties of a particular substance (water) under particular circumstances (normal atmospheric pressure on Earth). A scale that has its basis in much more fundamental physics is the **Kelvin scale**, first proposed by the nineteenth-century Scottish physicist William Thomson (1st Baron Kelvin). The Kelvin scale is based on the relationship between pressure and temperature of low-density gases. (In Section 14-3 we'll see why it's important that the density be low.) Experiment shows that the pressure in a sealed volume of gas decreases as temperature decreases and that a graph of pressure versus temperature is a straight line (**Figure 14-4**). For a given volume both the slope of this line and the pressure at a given temperature depend on the quantity of gas in the volume. But no matter what kind of gas or what quantity of gas is used, if we extrapolate the lines in Figure 14-4 to low temperatures and pressures, we find that the pressure goes to zero at the *same* temperature. (We have to *extrapolate* each line because at sufficiently low temperatures any gas becomes a liquid.) This temperature is called **absolute zero** because a lower temperature is not physically possible. Zero temperature in the Kelvin scale is absolute zero, so temperatures in the Kelvin scale are never negative.

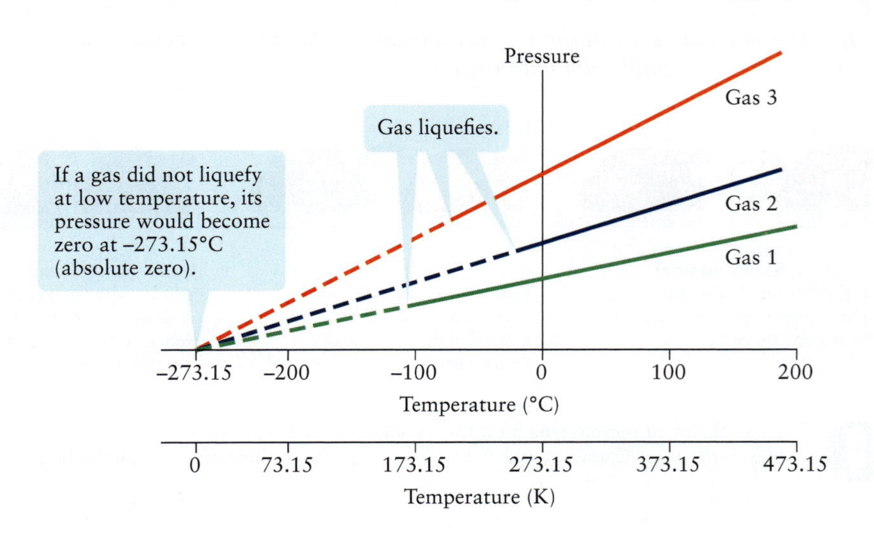

Figure 14-4 Ideal gases and absolute zero The rate at which pressure in a sealed volume of gas decreases as temperature decreases depends on the amount of gas present. However, extrapolating the graphs shows that pressure would become zero for *all* gases at the same temperature. That temperature, −273.15°C, is termed absolute zero.

The unit of temperature used in the Kelvin scale is the **kelvin**, abbreviated K. The value of the kelvin is based on an easily reproduced temperature that can be precisely measured, which is the temperature of the **triple point** of water. The triple point of a substance is the pressure and temperature of that substance at which the solid, liquid, and vapor phases of the substance all coexist. (We'll learn more about the phases of matter in Section 14-6.) For water the triple point pressure is 0.00603 atm, and the triple point temperature of water is defined to be 273.16 K. With this choice of the value of the kelvin, a change of 1 degree Celsius is *exactly* equal to a change of 1 kelvin. The triple point of water turns out to be at 0.01°C, so 0°C corresponds to 273.15 K, and 0 K (absolute zero) corresponds to −273.15°C. To convert from a temperature in kelvins to degrees Celsius, you simply *subtract* 273.15, and to convert a Celsius temperature to a Kelvin temperature, you *add* 273.15:

$$T_C = T_K - 273.15 \text{ and } T_K = T_C + 273.15 \tag{14-2}$$

Water freezes at approximately 0°C = 273.15 K and boils at approximately 100°C = 373.15 K. (The precise values of these temperatures depend on the atmospheric pressure.)

WATCH OUT! Temperature is in kelvins, not "degrees kelvin."

It's correct to say that an object has a temperature of a certain number of degrees Celsius or of degrees Fahrenheit, as in "20 degrees Celsius." But in the Kelvin scale we say an object has a temperature of a certain number of kelvins, as in "293 kelvins." We do *not* say, "293 degrees kelvin." Note also that the name of the scale (Kelvin) is capitalized, but the unit (kelvin) is not.

As we will see in the following section, the Kelvin scale is a natural one to use for many purposes in physics. Here's an example of how to convert among temperature values in Fahrenheit, Celsius, and Kelvin.

Go to Picture It 14-1 for more practice dealing with temperature scales.

EXAMPLE 14-1 Cold, Hot, and In Between

On the Fahrenheit scale, a cold day might be 5°F. A comfortable temperature is 68°F, which is widely accepted as "room temperature," and a hot day might be 95°F. Express all three temperatures in degrees Celsius and kelvins.

Set Up

We can use the first of Equations 14-1 to convert from degrees Fahrenheit to degrees Celsius and the second of Equations 14-2 to convert from degrees Celsius to temperatures on the Kelvin scale.

Conversion between Celsius and Fahrenheit:

$$T_C = \frac{5}{9}(T_F - 32) \text{ and } T_F = \frac{9}{5}T_C + 32 \tag{14-1}$$

Conversion between Celsius and Kelvin:

$$T_C = T_K - 273.15 \text{ and } T_K = T_C + 273.15 \tag{14-2}$$

Solve

Convert the cold temperature of 5°F to degrees Celsius and to kelvins.

$$T_{C,\text{cold}} = \frac{5}{9}(T_{F,\text{cold}} - 32) = \frac{5}{9}(5 - 32)$$
$$= -15°C$$
$$T_{K,\text{cold}} = -15 + 273.15$$
$$= 258 \text{ K}$$

Do the conversions for the comfortable room temperature of 68°F.

$$T_{C,\text{room}} = \frac{5}{9}(T_{F,\text{room}} - 32) = \frac{5}{9}(68 - 32)$$
$$= 20°C$$
$$T_{K,\text{room}} = 20 + 273.15$$
$$= 293 \text{ K}$$

Finally, convert the hot temperature of 95°F to the Celsius and Kelvin scales.

$$T_{C,hot} = \frac{5}{9}(T_{F,hot} - 32) = \frac{5}{9}(95 - 32)$$
$$= 35°C$$
$$T_{K,hot} = 35 + 273.15$$
$$= 308 \text{ K}$$

Reflect

Our calculations show that typical temperatures on Earth's surface are close to 300 K. That's a good reference number to keep in mind. Another more precise equivalence to remember is that a typical room temperature is 68°F = 20°C = 293 K.

GOT THE CONCEPT? 14-1 Ranking Temperatures

Rank these temperatures from coldest to warmest: (a) 280 K, (b) −13°C, (c) −13°F.

TAKE-HOME MESSAGE FOR Section 14-2

✔ The temperature of an object is a measure of the kinetic energy of the molecules of which that object is made.

✔ If two objects with different temperatures are in thermal contact so that energy can flow from one to the other, energy will flow until the two objects are at the same temperature. This final state is called thermal equilibrium.

✔ Two objects are in thermal equilibrium if, and only if, they are at the same temperature.

14-3 In a gas, the relationship between temperature and molecular kinetic energy is a simple one

We said in Section 14-2 that the temperature of an object is related to the kinetic energy of its individual molecules. Let's take a look at this relationship in more detail for the case of *gases*, which are in many ways the simplest form of matter. In a liquid or solid, adjacent molecules nearly touch each other, so the forces between molecules are strong. But in a low-density gas the average distance between molecules is large compared to the size of a molecule (see Example 11-3 in Section 11-2). Because gas molecules are very far apart on average, the forces they exert on each other are so weak that we can ignore them. As we'll see, that leads to tremendous simplifications in the relationship between the temperature of a gas and the kinetic energy of its molecules.

The Ideal Gas Law

Figure 14-4 shows that for a low-density gas at a constant volume, the graph of pressure versus temperature is a straight line, and the pressure p is zero when the Kelvin temperature T is zero. This means that the pressure is directly proportional to the Kelvin temperature. We can write this in equation form as

$$p = (\text{constant}) \times T \quad (14\text{-}3)$$

(low-density gas at constant volume)

√x *See the Math Tutorial for more information on direct and inverse proportionality.*

The pressure of a given quantity of gas also depends on the volume V that the gas occupies. For example, think of a quantity of gas in a flexible container whose volume we can change. Experiment shows that if the temperature of the gas is held constant and the volume is decreased, the pressure increases. In fact, the pressure turns out to be *inversely proportional* to the volume, or in equation form

$$p = \frac{(\text{constant})}{V} \quad (14\text{-}4)$$

(low-density gas at constant temperature)

We also find that for a given kind of gas, if the volume and temperature are held fixed but the *mass* of gas in a container is increased, the pressure increases in direct proportion to the mass. Increasing the mass means increasing the number N of molecules present in the gas, so we can write this relationship as

$$p = (\text{constant}) \times N \qquad (14\text{-}5)$$

(low-density gas at constant volume and temperature)

We can combine all three equations, 14-3, 14-4, and 14-5, together into a single relationship:

$$p = (\text{constant}) \times \frac{NT}{V}$$

This relationship is most commonly written in the form

Ideal gas law in terms of number of molecules (14-6)

$$pV = NkT$$

- **Pressure** of an ideal gas
- **Volume** occupied by the gas
- **Number of molecules** present in the gas
- **Temperature** of the gas on the Kelvin scale
- **Boltzmann constant**

The constant k in Equation 14-6, called the **Boltzmann constant**, turns out to have the same value for *all* gases. To four significant figures,

$$k = 1.381 \times 10^{-23}\ \text{J/K}$$

Equation 14-6 is called the **ideal gas law**, and a gas that would obey this equation exactly is called an **ideal gas**. Real gases are *not* ideal: They do not obey this equation exactly, especially at high pressures or at temperatures close to the point at which the gases become liquids. At relatively low gas densities, however, the ideal gas law does a good job of representing the relationship between pressure, volume, and temperature for real gases. That's why an ideal gas is a useful idealization.

Given the very large number of molecules in most samples of gas, it's often more convenient to write Equation 14-6 in terms of the number of *moles* of a gas. One **mole** of a substance contains as many molecules as there are atoms in exactly 12 grams of carbon-12, the variety of carbon that has six protons and six neutrons in its nucleus. To four significant figures the mole, abbreviated mol, contains 6.022×10^{23} molecules. This value is called Avogadro's number N_A:

$$N_A = 6.022 \times 10^{23}\ \text{molecules/mol}$$

The number N of molecules in a substance is therefore equal to the number of moles n multiplied by Avogadro's number: $N = nN_A$. Then Equation 14-6 becomes

$$pV = nN_AkT$$

The product N_Ak has a special name: We call it the **ideal gas constant** and give it the symbol R. To four significant figures,

$$R = N_Ak = 8.314\ \text{J/(mol}\cdot\text{K)}$$

With this definition the ideal gas law (Equation 14-6) becomes

Ideal gas law in terms of number of moles (14-7)

$$pV = nRT$$

- **Pressure** of an ideal gas
- **Volume** occupied by the gas
- **Number of moles** present in the gas
- **Temperature** of the gas on the Kelvin scale
- **Ideal gas constant**

The quantities pressure p, volume V, and temperature T characterize the physical state of a system. A relationship among these quantities, like Equation 14-6 or Equation 14-7 for an ideal gas, is called an **equation of state**. The equation of state for a real gas, a liquid, or a solid is substantially more complicated than that for an ideal gas.

EXAMPLE 14-2 An Ideal Gas at Room Temperature

At ordinary room temperature (20°C) and a pressure of 1 atm (1.013×10^5 Pa = 1.013×10^5 N/m^2), how many molecules are there in 1.00 m^3 of an ideal gas? How many moles are there?

Set Up

We can use the two forms of the ideal gas law to determine the number of molecules N and number of moles n. Note that the temperature in the ideal gas law is the *Kelvin* temperature, but we're given the value of the *Celsius* temperature. We'll have to convert the temperature using the second of Equations 14-2.

Ideal gas law in terms of number of molecules:

$$pV = NkT \quad (14\text{-}6)$$

Ideal gas law in terms of number of moles:

$$pV = nRT \quad (14\text{-}7)$$

Conversion between Celsius and Kelvin:

$$T_C = T_K - 273.15 \text{ and}$$
$$T_K = T_C + 273.15 \quad (14\text{-}2)$$

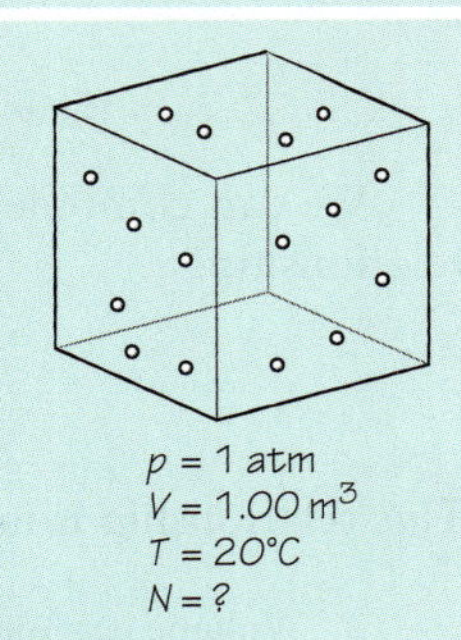

Solve

First convert the temperature to the Kelvin scale.

From the second of Equations 14-2, a temperature of 20°C corresponds to a Kelvin temperature of

$$T = 20 + 273.15 = 293 \text{ K}$$

Solve Equation 14-6 for the number of molecules N and substitute numerical values.

From Equation 14-6

$pV = NkT$, so

$$N = \frac{pV}{kT} = \frac{(1.013 \times 10^5 \text{ N/m}^2)(1.00 \text{ m}^3)}{(1.381 \times 10^{-23} \text{ J/K})(293 \text{ K})}$$
$$= 2.50 \times 10^{25} \frac{\text{N} \cdot \text{m}}{\text{J}}$$

Since 1 N · m = 1 J, N has *no* units: It is a pure number (in this case the number of molecules in 1.00 m^3). So

$$N = 2.50 \times 10^{25} \text{ molecules}$$

In a similar way, find the number of moles using Equation 14-7.

From Equation 14-7

$pV = nRT$, so

$$n = \frac{pV}{RT} = \frac{(1.013 \times 10^5 \text{ N/m}^2)(1.00 \text{ m}^3)}{(8.314 \text{ J/(mol} \cdot \text{K))}(293 \text{ K})}$$
$$= 41.6 \frac{\text{N} \cdot \text{m} \cdot \text{mol}}{\text{J}} = 41.6 \text{ mol}$$

Reflect

We can check our results by confirming that the number of molecules equals the number of moles multiplied by Avogadro's number (the number of molecules per mole). To put our result into perspective, the number of *stars* in the observable universe is estimated to be between 3×10^{22} and 10^{24}. The number of molecules in a cubic meter of gas is truly astronomical!

We can also calculate N from the number of moles n and Avogadro's number N_A:

$$N = nN_A = (41.6 \text{ mol})(6.022 \times 10^{23} \text{ molecules/mol})$$
$$= 2.50 \times 10^{25} \text{ molecules}$$

Temperature and Translational Kinetic Energy

With the ideal gas law in hand, let's now see the relationship between pressure, volume, and the *average translational kinetic energy* of molecules in a gas in a container. We'll see that the pressure arises from the forces that individual molecules exert when they collide with the walls of the container.

Figure 14-5 shows a cubical box that has sides of length L and contains N molecules of an ideal gas. Each molecule has mass m. When the gas is in equilibrium, the gas as a whole exhibits no net motion. Although every molecule is moving and has momentum, the *total* momentum of all molecules in the gas averages to zero. In Figure 14-5 one representative molecule has just collided with and bounced off the left wall. We'll follow this molecule for the time Δt that it takes for the single molecule to return to this same position after colliding with the wall on the right side of the container. We'll assume that the molecule doesn't collide with any other molecules during this round trip. (Later in this section we'll see why this is a reasonable assumption to make.) For simplicity we'll initially consider the molecule to be moving in the x direction only. We'll also make the simplifying assumptions that the walls of the container are rigid and the gas molecules behave like hard spheres so that the collisions are *elastic* and no energy is lost when a molecule collides with a wall.

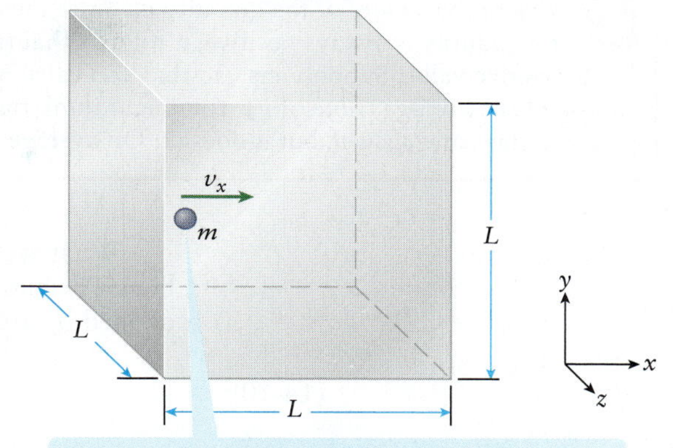
The gas molecule shown has just collided with and been reflected from the wall on the left.
v_x
m
L
L
L
y
x
z
The molecule, moving in the positive x direction, travels a distance L, collides with the right wall, and travels a distance L again before returning to this same position. This round trip takes a time Δt.

Figure 14-5 The origin of gas pressure A cubical container contains a gas. This figure shows one representative gas molecule that has just bounced off the left-hand wall. The combined effect of all molecules bouncing off the walls gives rise to the pressure of the gas.

When the molecule in Figure 14-5 hits the right-hand wall of the box, it reverses direction but loses no kinetic energy, so its velocity changes from $+v_x$ (to the right) to $-v_x$ (to the left) and its momentum changes from $+mv_x$ to $-mv_x$. So the *change* in the molecule's momentum is $(-mv_x) - (+mv_x) = -2mv_x$. This momentum change occurs thanks to the force that the wall exerts on the molecule during the hit. The hit lasts just a short time, and there is one such hit by the right-hand wall on the molecule per time Δt (the time needed for the molecule to make a round trip and return to the right-hand wall). If we call $F_{\text{wall on molecule},x}$ the *average* force that the right-hand wall exerts on the molecule over the time Δt, Equation 7-24 (Section 7-6) says that this force multiplied by Δt equals the change in momentum that the molecule undergoes due to its interaction with the right-hand wall:

$$F_{\text{wall on molecule},x}\Delta t = -2mv_x \quad \text{so} \quad F_{\text{wall on molecule},x} = -\frac{2mv_x}{\Delta t}$$

This is negative because the right-hand wall pushes the molecule to the left, in the negative x direction. By Newton's third law, the average force that the *molecule* exerts on the *wall* is the opposite of this and is *positive* (the molecule exerts a force on the wall in the positive x direction):

$$F_{\text{molecule on wall},x} = -F_{\text{wall on molecule},x} = \frac{2mv_x}{\Delta t}$$

We emphasize that this is the *average* force that this molecule exerts on the wall: During an actual hit the force is large, and the rest of the time (when the molecule is not in contact with the wall) the force is zero. Now Δt is the time it takes the molecule moving at speed v_x to move a distance $2L$ (back and forth across the length L of the box). Since time equals distance divided by speed, it follows that $\Delta t = 2L/v_x$. Substituting this into our expression for $F_{\text{molecule on wall},x}$ we get

$$F_{\text{molecule on wall},x} = \frac{2mv_x}{(2L/v_x)} = \frac{mv_x^2}{L} \qquad \text{(14-8)}$$

Equation 14-8 tells us the average force on the right-hand wall from *one* molecule. However, there are N molecules of mass m in the box shown in Figure 14-5, and each of them collides periodically with the right-hand wall. If every molecule had the same value of v_x, the total force from all gas molecules combined would just be N times the force in Equation 14-8. This is extremely unlikely to be the case, however. So we get the total force that the gas as a whole exerts on the wall by replacing v_x^2 in Equation 14-8 with the *average value* of v_x^2 (that is, averaged over its value for each individual molecule) and then multiplying by N:

$$F_{\text{gas on wall},x} = \frac{Nm(v_x^2)_{\text{average}}}{L} \qquad \text{(14-9)}$$

WATCH OUT! The average of the square is not the square of the average.

The quantity $(v_x^2)_{\text{average}}$ in Equation 14-9 is the *average value* of the *square* of the quantity v_x. Since the square of a quantity is always positive, it follows that $(v_x^2)_{\text{average}}$ has a positive value. Suppose instead that you calculated the *square* of the *average value* of v_x. You might think this would give you the same answer, but it doesn't: On average there are as many molecules moving to the left in Figure 14-5 as there are moving to the right, so there are as many with a positive value of v_x as with a negative value. So the average value of v_x is zero, and the square of that is also zero. When we say to take the average value of the square, we mean square first, *then* average!

What we actually measure is the *pressure* that the gas exerts on the walls of the box. The right-hand wall is a square of side L and area L^2, so the pressure (force divided by area) on the right-hand wall is

$$p = \frac{F_{\text{gas on wall},x}}{L^2} = \frac{Nm(v_x^2)_{\text{average}}}{L^3} = \frac{Nm(v_x^2)_{\text{average}}}{V} \tag{14-10}$$

In Equation 14-10 we've used the symbol V for the volume L^3 of the cubical container of side L.

In general the gas molecules are moving in the y and z directions as well as the x direction. Since there's no preferred direction inside the box, we expect that the average values of v_y^2 and v_z^2 are the same as the average value of v_x^2: $(v_x^2)_{\text{average}} = (v_y^2)_{\text{average}} = (v_z^2)_{\text{average}}$. Furthermore, the square of the *speed* v of a given molecule is the sum of the squares of its velocity components: $v^2 = v_x^2 + v_y^2 + v_z^2$. If we put these observations together, we can write an expression for the average value of the square of the speed of gas molecules:

$$(v^2)_{\text{average}} = (v_x^2)_{\text{average}} + (v_y^2)_{\text{average}} + (v_z^2)_{\text{average}} = 3(v_x^2)_{\text{average}} \tag{14-11}$$

Equation 14-11 tells us that the quantity $(v_x^2)_{\text{average}}$ in Equation 14-10 is equal to one-third of the average value of the square of the speed: $(v_x^2)_{\text{average}} = (v^2)_{\text{average}}/3$. Then we can rewrite Equation 14-10 for the gas pressure as

$$p = \frac{Nm(v^2)_{\text{average}}}{3V} \tag{14-12}$$

Now $K_{\text{translational}} = (1/2)mv^2$ is the translational kinetic energy of a single molecule. (We emphasize *translational* since the molecule could also be rotating around its axis and have rotational kinetic energy. Its constituent atoms could also be vibrating within the molecule, so there could be vibrational kinetic energy as well.) So $K_{\text{translational,average}} = (1/2)m(v^2)_{\text{average}}$ is the *average* translational kinetic energy of a single molecule in the gas. In terms of this we can rewrite Equation 14-12 as

$$p = \frac{Nm(v^2)_{\text{average}}}{3V} = \frac{2N}{3V}\left[\frac{1}{2}m(v^2)_{\text{average}}\right] = \frac{2N}{3V}K_{\text{translational,average}} \tag{14-12}$$

The pressure of a given quantity of ideal gas is directly proportional to the average translational kinetic energy of a molecule of the gas. The more kinetic energy that the molecules have on average, the greater the pressure.

We can learn something equally important by comparing Equation 14-12 with the ideal gas law in terms of number of molecules, Equation 14-6: $pV = NkT$, or $p = (N/V)kT$. In order for both Equations 14-6 and 14-12 to be correct, it must be true that $(2/3)K_{\text{translational,average}} = kT$, or

Temperature and average translational kinetic energy of an ideal gas molecule (14-13)

Average translational kinetic energy of a molecule in an ideal gas

Average value of the square of a gas molecule's speed

Temperature of the gas on the Kelvin scale

$$K_{\text{translational,average}} = \frac{1}{2}m(v^2)_{\text{average}} = \frac{3}{2}kT \tag{14-13}$$

Mass of a single gas molecule

Boltzmann constant

In words, Equation 14-13 says that *the average translational kinetic energy of a molecule in an ideal gas is directly proportional to the Kelvin temperature of the gas*. This justifies the statement we made in Section 14-2 about the physical meaning of temperature. Equation 14-13 relates a *microscopic* property of an ideal gas (the average kinetic energy of an individual gas molecule) to a *macroscopic* property (the temperature of the gas, something you might measure with an ordinary thermometer).

Equation 14-13 also gives us a microscopic interpretation of absolute zero. If the temperature is 0 K, then the average translational kinetic energy of a single gas molecule must also be zero. This is possible only if the speed of the molecule is zero. So absolute zero is the temperature at which all molecular motion would cease.

Note from Equation 14-13 that the average translational kinetic energy of a gas molecule does *not* depend on the *mass* of the molecule, just on the temperature. For example, the air around you is composed almost entirely of molecules of nitrogen (N_2) and oxygen (O_2). A mole of N_2 molecules has a mass of 28.0 g, and a mole of O_2 molecules has a mass of 32.0 g, so the mass m of an individual N_2 molecule is $(28.0/32.0) = 0.875$ as much as the mass of an individual O_2 molecule. Nonetheless, on average each kind of molecule in the air has the *same* translational kinetic energy: An average nitrogen molecule must therefore be moving faster than an average oxygen molecule.

A measure of how fast gas molecules move on average is the **root-mean-square speed or rms speed.** To see how this is defined, first imagine finding the average value of v^2, the square of the speed v, for all the molecules in a gas that have a given mass m. This is the quantity that we've called $(v^2)_{\text{average}}$. Another word for average is *mean*, which is why $(v^2)_{\text{average}}$ is called the *mean-square* of the speed. The *root*-mean-square speed is the square root of the mean-square:

Go to Interactive Exercise 14-1 for more practice dealing with rms speed.

$$v_{\text{rms}} = \sqrt{(v^2)_{\text{average}}} \qquad (14\text{-}14)$$

Note that v_{rms} has units of m/s. From Equation 14-13, $(1/2)m(v^2)_{\text{average}} = (3/2)kT$, so $(v^2)_{\text{average}} = 3kT/m$. If we substitute this into Equation 14-14, we get

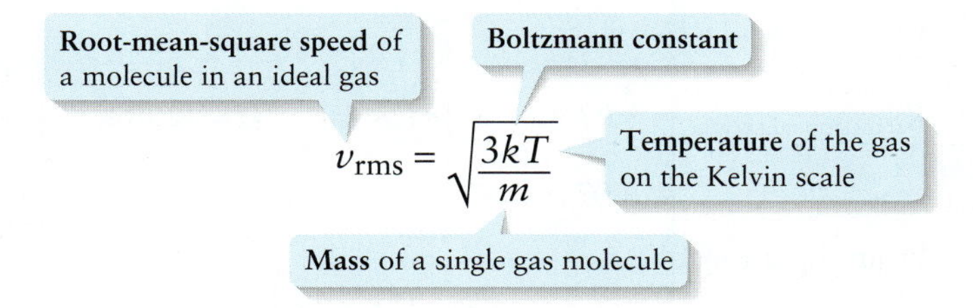

Root-mean-square speed of molecules in an ideal gas (14-15)

WATCH OUT! The root-mean-square value is not the same as the average value.

We use the root-mean-square speed v_{rms} as a measure of the typical molecular speed in a gas because it follows naturally from Equation 14-13, which relates temperature to translational kinetic energy. It is *not*, however, the same as the *average* molecular speed. To see the difference consider a gas with just five molecules that have speeds 1.00, 2.00, 3.00, 4.00, and 5.00 m/s. The average of these speeds is

$$v_{\text{average}} = \frac{\left[\begin{array}{l}1.00\text{ m/s} + 2.00\text{ m/s} + 3.00\text{ m/s}\\ + 4.00\text{ m/s} + 5.00\text{ m/s}\end{array}\right]}{5} = 3.00\text{ m/s}$$

but the rms speed is

$$v_{\text{rms}} = \sqrt{(v^2)_{\text{average}}} = \sqrt{\frac{\left[\begin{array}{l}(1.00\text{ m/s})^2 + (2.00\text{ m/s})^2 + (3.00\text{ m/s})^2\\ + (4.00\text{ m/s})^2 + (5.00\text{ m/s})^2\end{array}\right]}{5}} = 3.32\text{ m/s}$$

It turns out that the average value of any collection of numbers is *always* less than the rms value. For the special case of an ideal gas, a detailed analysis shows that the average speed is $\sqrt{8/(3\pi)} = 0.921$ of the rms speed.

EXAMPLE 14-3 Oxygen at Room Temperature

Calculate (a) the average translational kinetic energy and (b) the root-mean-square speed of an oxygen molecule in air at room temperature (20°C). One mole of oxygen molecules has a mass of 32.0 g = 32.0×10^{-3} kg.

Set Up

Equation 14-13 tells us the average translational kinetic energy, and Equation 14-15 tells us the rms speed of molecules. Note that we're given the mass per mole of O_2, but we'll need to convert this to m, the mass per molecule. We'll also need to convert the temperature from Celsius to Kelvin.

Temperature and average translational kinetic energy of an ideal gas molecule:

$$K_{\text{translational,average}} = \frac{1}{2}m(v^2)_{\text{average}} = \frac{3}{2}kT \qquad (14\text{-}13)$$

Root-mean-square speed of molecules in an ideal gas:

$$v_{\text{rms}} = \sqrt{\frac{3kT}{m}} \qquad (14\text{-}15)$$

Conversion between Celsius and Kelvin:

$$T_C = T_K - 273.15 \text{ and } T_K = T_C + 273.15 \qquad (14\text{-}2)$$

oxygen molecules

Solve

(a) First find the mass per molecule m and the Kelvin temperature T.

The number of molecules per mole is Avogadro's number, $N_A = 6.022 \times 10^{23}$ molecules/mol. So the mass per O_2 molecule is

$$m = \frac{(32.0 \times 10^{-3}\text{ kg/mol})}{(6.022 \times 10^{23}\text{ molecules/mol})} = 5.31 \times 10^{-26}\text{ kg/molecule}$$

From the second of Equations 14-2, the Kelvin temperature that corresponds to 20°C is

$$T = 20 + 273.15 = 293\text{ K}$$

Use Equation 14-13 to find the average kinetic energy per molecule.

From Equation 14-13

$$K_{\text{translational,average}} = \frac{3}{2}kT = \frac{3}{2}(1.381 \times 10^{-23}\text{ J/K})(293\text{ K}) = 6.07 \times 10^{-21}\text{ J}$$

(b) Use Equation 14-15 to find the rms speed of an O_2 molecule.

From Equation 14-15

$$v_{\text{rms}} = \sqrt{\frac{3kT}{m}} = \sqrt{\frac{3(1.381 \times 10^{-23}\text{ J/K})(293\text{ K})}{5.31 \times 10^{-26}\text{ kg}}} = 478\text{ m/s}$$

Reflect

The rms speed of O_2 molecules is tremendous, 478 m/s = 1720 km/h = 1070 mi/h. Note that very few O_2 molecules in the air travel at precisely that speed: Their speeds range from nearly zero to many times faster than 478 m/s.

Note that the nitrogen (N_2) molecules in air at 20°C have the *same* translational kinetic energy (which does not depend on the mass m of the molecules) but a *different* rms speed (which is proportional to the reciprocal of $\sqrt{m}$). As we discussed above, an N_2 molecule has 0.875 the mass of an O_2 molecule, so the rms speed for N_2 is $1/\sqrt{0.875} = 1.07$ times faster than the rms speed for O_2.

Degrees of Freedom

The factor of three in Equation 14-13 for the average translational kinetic energy of a molecule, $K_{\text{translational,average}} = (1/2)m(v^2)_{\text{average}} = (3/2)kT$, is actually rather significant. It arises because $(v^2)_{\text{average}}$ is the sum of three terms, one for each component of molecular motion (Equation 14-11): $(v^2)_{\text{average}} = (v_x^2)_{\text{average}} + (v_y^2)_{\text{average}} +$

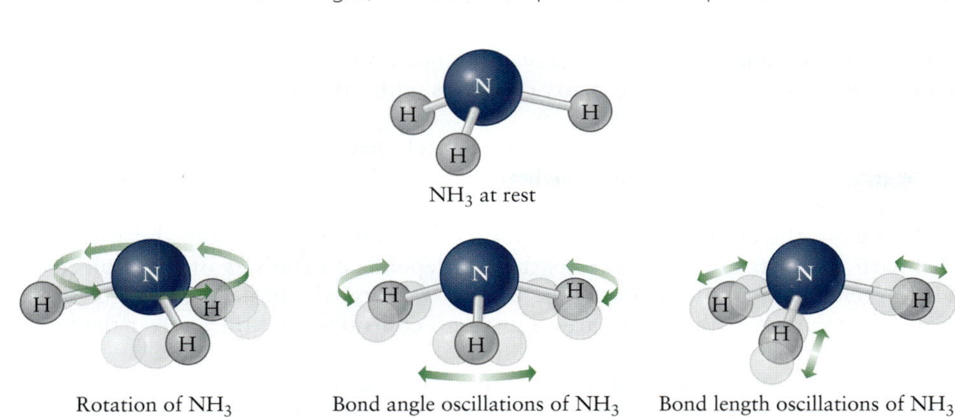

Figure 14-6 Motions of an ammonia molecule If a gas is made of molecules with two or more atoms, a molecule can have energy of motion even if the translational velocity of the molecule is zero. For an ammonia (NH_3) molecule, the molecule can rotate, and both the bond angle and the bond length can oscillate. This means that NH_3 has more degrees of freedom than just the three translational degrees of freedom.

$(v_z^2)_{\text{average}}$. As we discussed above, each of these terms has the same value. That means we can write the average kinetic energy of a molecule as the sum of three equal terms:

$$\frac{1}{2}m(v^2)_{\text{average}} = \frac{1}{2}m(v_x^2)_{\text{average}} + \frac{1}{2}m(v_y^2)_{\text{average}} + \frac{1}{2}m(v_z^2)_{\text{average}} = \frac{3}{2}kT$$

In other words, on average the translational kinetic energy is shared equally between energies associated with the motions in each of the three possible directions. Each direction of motion available to the molecule contributes an average translational kinetic energy of one-third of $(3/2)kT$, or $(1/2)kT$. We refer to each of these possible directions of motion as a **degree of freedom** of the system. (An object that could move only along a straight line would have one degree of freedom, and one that could move only in a plane would have two degrees of freedom. An object that can move in all three dimensions of space has three degrees of freedom.)

The number of degrees of freedom of a gas can be more than simply the number of directions in which gas molecules can translate. If the molecules contain more than one atom, the molecule can have energy of motion even if its translational velocity is zero. As an example **Figure 14-6** shows possible motions of the ammonia molecule (NH_3). The entire ammonia molecule can rotate, and it can vibrate in two ways: The nitrogen (N) and hydrogen (H) atoms can oscillate along the length of the bond, and the hydrogen atoms can be disturbed so that the angle of the bonds oscillates. Each of these also represents a possible degree of freedom of the gas molecule, and each can contribute an additional $(1/2)kT$ to the average energy per molecule. Thus the average energy per molecule is likely to be more than $(3/2)kT$ for a gas that is composed of polyatomic molecules. For example, for O_2 and N_2 at room temperature, the total energy per molecule is actually close to $(5/2)kT$ because the molecules can both translate (for which there are three degrees of freedom) and rotate (for which there are two degrees of freedom: If the long axis of the O_2 molecule is the z axis, one degree of freedom corresponds to rotation around the x axis and the other to rotation around the y axis). Quantum mechanics tells us that there's not enough energy at room temperature for bond length oscillations to contribute significantly to the energy of these diatomic molecules. The fact that the energy of a molecule is shared equally among each degree of freedom is called the **equipartition theorem.**

Mean Free Path

In our analysis of what happens in an ideal gas at the microscopic level, we considered what happens when a gas molecule hits one of the container walls, but we ignored the effects of collisions between gas molecules. How frequently do such collisions between molecules occur? One way to answer this question is in terms of the **mean free path** of a gas molecule. This is the average distance that a molecule travels from the time at which it collides with one molecule to when it collides with another molecule. If this distance is large compared to the size of the container that holds the gas, we can conclude that a molecule is unlikely to have any collisions with other molecules as it bounces back and forth between the container walls. But if the mean free path is short, a molecule may undergo many collisions with other molecules during a single trip across the container.

The mean free path depends on how densely packed the gas molecules are. If there are very few gas molecules in the volume of the container, collisions between molecules are unlikely and the mean free path is long. If, however, there are many molecules in the container, the likelihood of a collision is greater and the mean free path is short. The mean free path also depends on the *size* of the molecules: The larger the gas molecules are, the bigger "targets" the molecules are for other molecules, the more likely they are to undergo collisions, and the shorter the mean free path.

If there are N gas molecules in a volume V, and if we model the molecules as spheres of radius r, the mean free path (to which we give the symbol λ, Greek letter lambda) turns out to be

$$\lambda = \frac{1}{4\sqrt{2}\pi r^2 (N/V)} \tag{14-16}$$

Just as we expected, the mean free path is short if r is large (the molecules are large in size) or if N/V is large (there are many molecules present per volume, so the molecules are close together). Equation 14-6 tells us that $pV = NkT$, which we can rearrange to read $N/V = p/kT$, so Equation 14-16 becomes

$$\lambda = \frac{kT}{4\sqrt{2}\pi r^2 p} \tag{14-17}$$

We invite you to use Equation 14-17 to calculate the mean free path for oxygen (O_2) molecules, which are about 2.0×10^{-10} m in radius, at ordinary room temperature (20°C or 293 K) and a pressure of 1 atm $= 1.013 \times 10^5$ N/m^2. You'll find that $\lambda = 5.6 \times 10^{-8}$ m, a distance that is large compared to the size of a molecule but hundreds of millions of times smaller than the size of a typical container for air. An O_2 molecule trying to travel from one side to another of a room 2 m wide would undergo tens of millions of collisions along the way! How, then, can we justify the assumption we made earlier in this section that we can *ignore* collisions between molecules? The explanation is that a given molecule collides with other molecules that are moving in *all* directions. Some collisions make the molecule speed up, and others make it slow down; some deflect it to the left, and others deflect it to the right. The result is that the effects of all these collisions largely cancel out, so on average we can treat molecules as if they were unaffected by collisions. Note that collisions also tend to *randomize* the velocities of the molecules that make up the gas, which justifies our assumption that the average values of v_x^2, v_y^2, and v_z^2 are all the same.

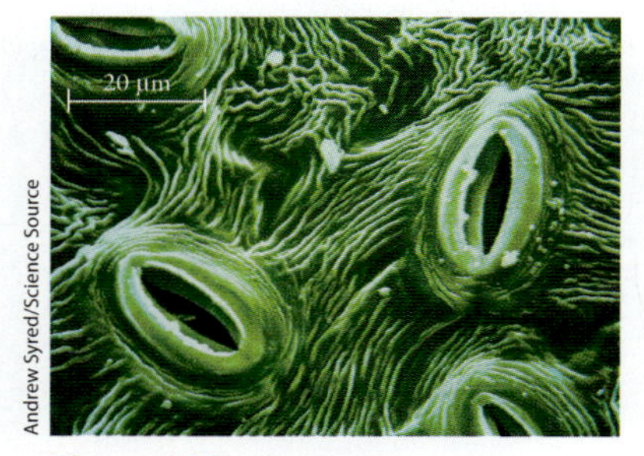

Figure 14-7 Stomata on the epidermis of a leaf This photomicrograph shows fine details on the leaf of an elder tree (*Sambucus nigra*). The size of the stomata, through which the leaf exchanges gases with the atmosphere, is related to the mean free path of gas molecules.

BIO-Medical The concept of mean free path plays an important role in botany. The leaves of plants "breathe" through their pores or *stomata*, tiny apertures through which they exchange gas molecules with the surrounding air (**Figure 14-7**). In photosynthesis carbon dioxide (CO_2) is taken in through the pores and oxygen (O_2) is released, in respiration O_2 is taken in and CO_2 is released, and in transpiration water vapor (H_2O) is released. Each pore is like a narrow tube, and gas molecules passing through the tube experience collisions with the walls of the tube as well as with each other. When the diameter of the tube is large compared to the mean free path of the gas molecules, gas molecules collide mostly with each other and only infrequently with the tube walls. As a result, gas can flow relatively easily through the tube. But if the tube diameter is about the same as the mean free path, collisions with the walls become more frequent and the rate of gas flow slows dramatically. For this reason the pores are able to expand when conditions require increased gas flow (for example, when intense sunlight shines on the leaf and the rate of photosynthesis goes up).

GOT THE CONCEPT? 14-2 Comparing Gases

A small fraction of the molecules in air are atoms of helium. A mole of helium has a mass of 4.00 g, and a helium atom has a radius of about 3×10^{-11} m. State whether each of the following quantities will be greater, less, or the same for helium than for oxygen (O_2, with a mass of 32.0 g/mol and a radius of about 2.0×10^{-10} m) in a given quantity of air: (a) the average translational kinetic energy per molecule; (b) the total energy of all kinds per molecule; (c) the rms speed; (d) the mean free path.

TAKE-HOME MESSAGE FOR Section 14-3

✔ The ideal gas law relates the pressure, volume, temperature, and number of molecules (or number of moles) for a low-density gas.

✔ The pressure that a gas exerts on the walls of its container is due to collisions that the gas molecules make with the walls.

✔ The temperature of a gas is a measure of the average translational kinetic energy per molecule. In a gas at Kelvin temperature T, this energy is $(3/2)kT$. This does not depend on the mass of the molecule. Molecules with more than one atom can have additional energy associated with rotation or vibration.

✔ The average distance that a molecule travels between collisions with other molecules is called the mean free path. The larger the molecule and the more densely molecules are packed, the shorter the mean free path.

14-4 Most substances expand when the temperature increases

Nearly all objects expand when heated and contract when cooled. For example, the lid on a jar of pickles may be too tight to unscrew when you first take the jar from the refrigerator, but when the jar warms up the lid expands and is easier to remove. This is known as **thermal expansion**. Thermal expansion is the basis of many thermometers, including those that use alcohol or mercury (see Section 14-2 and Figure 14-1a).

In a solid, thermal expansion happens because with increasing temperature the molecules that comprise the solid not only oscillate with greater amplitude around their equilibrium positions but also shift their equilibrium positions so that they are farther away from their neighbors. In a liquid or gas increasing temperature means that adjacent molecules collide with each other with greater momentum; this pushes adjacent molecules farther apart and increases the volume of the liquid or gas.

Linear Expansion

Experiment shows that if the temperature change ΔT of a solid object is not too great, the change in each dimension of the object is proportional to ΔT. In particular, if a solid object initially has length L_0, the change ΔL in its length when the temperature changes by ΔT is

Change in length of an object

Length of the object before the temperature change

$$\Delta L = \alpha L_0 \Delta T$$

Coefficient of linear expansion of the substance of which the object is made

Temperature change of the object that causes the length change

Change in length due to a temperature change (14-18)

TABLE 14-1 Coefficients of Linear Expansion

Substance	Coefficient of linear expansion (α) $K^{-1} \times 10^{-6}$	Substance	Coefficient of linear expansion (α) $K^{-1} \times 10^{-6}$
aluminum	22.2	iron, cast	10.4
antimony	10.4	iron, forged	11.3
beryllium	11.5	iron, pure	12.0
brass	18.7	lead	28.0
brick	5.5	marble	12
bronze	18.0	plaster	25
carbon–diamond	1.2	platinum	9.0
cement	10.0	porcelain	3.0
concrete	14.5	quartz, fused	0.59
copper	16.5	rubber	77
glass, hard	5.9	silver	19.5
glass, plate	9.0	solder	24.0
glass, pyrex	4.0	steel	13.0
gold	14.2	wood, oak parallel to grain	4.9
graphite	7.9		

The quantity α (Greek letter "alpha") in Equation 14-18 is called the **coefficient of linear expansion**. It depends on what the object is made of but not on the shape or size of the object. **Table 14-1** lists the coefficient of linear expansion for a number of substances. Notice in Equation 14-18 that α has dimensions of inverse temperature (often K^{-1}), so the dimensions of $\alpha L_0 \, \Delta T$ on the right side of the equation match those of ΔL on the left side.

WATCH OUT! The coefficient of linear expansion can have different units.

The units of α can be $(°C)^{-1}$ as well as K^{-1}. That's because the sizes of 1°C and 1 K are defined to be the same, so the temperature *change* ΔT has the same value whether the temperature is measured on the Celsius scale or the Kelvin scale.

Equation 14-18 and Table 14-1 explain why you can loosen a tight-fitting steel lid on a glass jar of pickles by running hot water over the lid. Initially the steel lid and the glass mouth of the jar have the same diameter L_0. The hot water makes both the steel and the glass undergo the same temperature change ΔT, but the steel lid expands more (ΔL is greater) because the coefficient of thermal expansion is greater for steel ($\alpha = 13.0 \times 10^{-6}\ K^{-1}$) than for glass ($\alpha = 4 \times 10^{-6}$ to $9 \times 10^{-6}\ K^{-1}$). That makes the fit less snug and makes it easier to unscrew the lid.

Equation 14-18 should remind you of Equation 9-3 (Section 9-2), which says that the change ΔL in length of an object due to a tensile or compressive *stress* (force per unit area) is proportional to the initial length L_0 and the stress F/A:

Go to Picture It 14-2 for more practice dealing with linear expansion.

$$\frac{F}{A} = Y\frac{\Delta L}{L_0} \quad \text{or} \quad \Delta L = \left(\frac{1}{Y}\right)L_0\frac{F}{A}$$

We saw in Section 9-2 that this direct proportionality between stress and length change is valid if the stress is not too great. In the same fashion the direct proportionality between ΔL and ΔT given by Equation 14-18 is valid if ΔT is not too great. (If the temperature change is too great, the direct proportionality between ΔL and ΔT breaks down, and we have to treat α as a function of temperature. The character of this function is different for different substances.)

EXAMPLE 14-4 An Expanding Bridge

Bridges are constructed with expansion joints (**Figure 14-8**) to allow the bridge to expand on hot days without buckling. Suppose the bridge is made in two halves, each of which is made of steel and is 10.0 m long when the temperature is 18.0°C. The left-hand end of the left half is held in place and cannot move, as is the right-hand end of the right half. How large a gap should there be between the two halves at 18.0°C so that the structure does not buckle on an exceptionally hot day when the temperature is 45.0°C?

peacepics/Fotolia.com

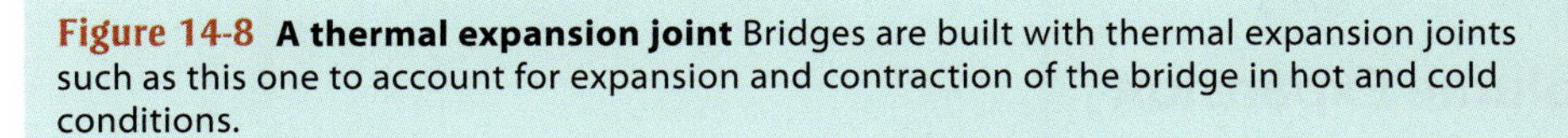

Figure 14-8 A thermal expansion joint Bridges are built with thermal expansion joints such as this one to account for expansion and contraction of the bridge in hot and cold conditions.

Set Up

At 18.0°C each half of the bridge has length $L_0 = 10.0$ m. We want to choose the gap d between them so that at 45.0°C the gap just closes. For this to happen each half must increase in length by a distance $d/2$ as the temperature increases. We use Equation 14-18 for the length increase of each half of the bridge. Table 14-1 gives the coefficient of thermal expansion α for steel.

Change in length due to a temperature change:

$$\Delta L = \alpha L_0 \Delta T \quad (14\text{-}18)$$

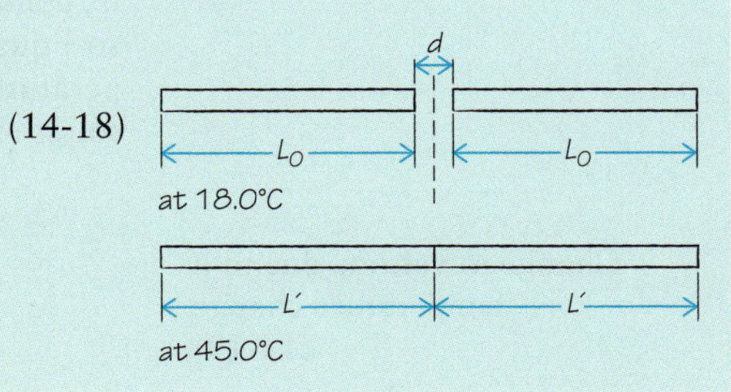

Solve

Write an expression for the length change of each bridge half when the temperature increases from 18.0°C to 45.0°C.

From Table 14-1

$\alpha = 13.0 \times 10^{-6}\ \text{K}^{-1}$ for steel

The temperature change is

$$\Delta T = 45.0°\text{C} - 18.0°\text{C} = 27.0\ \text{K}$$

(Recall that a temperature change of 1°C is the same as a temperature change of 1 K.)

Then from Equation 14-18 the length change of each bridge half is

$$\begin{aligned}\Delta L &= \alpha L_0\, \Delta T \\ &= (13.0 \times 10^{-6}\ \text{K}^{-1})(10.0\ \text{m})(27.0\ \text{K}) \\ &= 3.51 \times 10^{-3}\ \text{m} = 3.51\ \text{mm}\end{aligned}$$

The length change ΔL of each bridge half is equal to $d/2$ (half the gap at 18.0°C), so d is equal to $2\Delta L$.

The necessary gap between the two bridge halves at 18.0°C is

$$\begin{aligned}d &= 2\ \Delta L = 2(3.51\ \text{mm}) \\ &= 7.02\ \text{mm} = 0.276\ \text{in.}\end{aligned}$$

Reflect

This gap is small enough that it offers no resistance to a car's tire or a pedestrian's foot. Without a sufficiently large expansion joint, the bridge halves in this problem would push against each other on a hot day and eventually warp or buckle as shown.

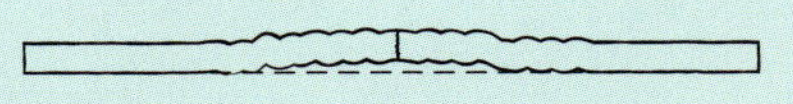

WATCH OUT! Holes in an object expand when heated.

Example 14-4 may lead you to conclude that when the temperature of an object increases, the substance of which it is made expands to fill any holes in the object. Not so! *Every* dimension of the object increases, which means the size of the hole also increases (**Figure 14-9**). That's what happens when you warm the metal lid on a jar to loosen it: The inner diameter of the lid (the size of the "hole" on the inside of the lid) increases so that it no longer fits so tightly on the mouth of the glass jar.

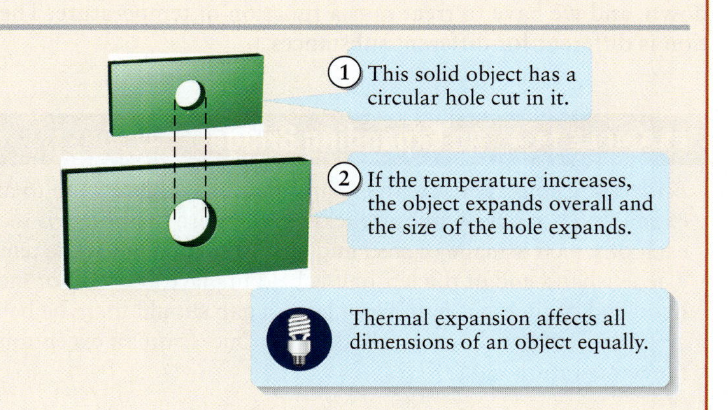

1 This solid object has a circular hole cut in it.
2 If the temperature increases, the object expands overall and the size of the hole expands.
Thermal expansion affects all dimensions of an object equally.

Figure 14-9 Heating a hole When a block with a hole is heated, the hole and the block both expand.

Volume Expansion

When the temperature of a solid object changes, its length, width, and height change according to the linear thermal expansion relationship of Equation 14-18. As a result, its *volume* expands as well. A fluid (a liquid or gas) has no fixed dimensions, however, so Equation 14-18 doesn't apply, but fluids also change their volume when the temperature changes. An equation that describes the volume change with temperature for solids and fluids alike is

Change in volume due to a temperature change (14-19)

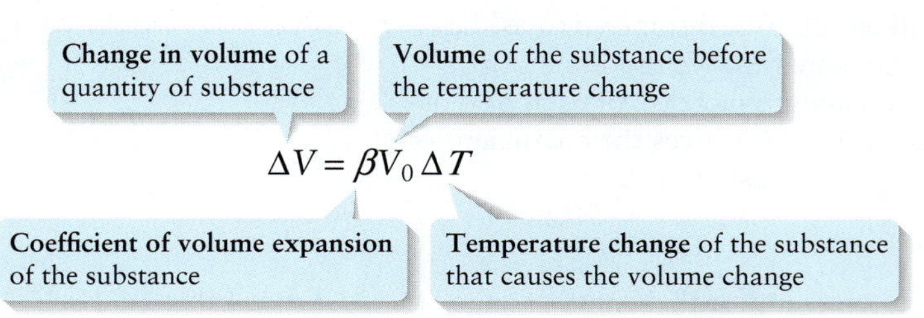

TABLE 14-2 Coefficients of Volume Expansion for Liquids

Substance	Coefficient of volume expansion (β) $K^{-1} \times 10^{-6}$
gasoline	905
ethanol	750
water	207
mercury	182

The **coefficient of volume expansion** β (Greek letter "beta") also has units of K^{-1} or $(°C)^{-1}$.

For solids it turns out that the coefficient of volume expansion is just three times the coefficient of linear expansion: $\beta = 3\alpha$. (The factor of 3 arises because volume expansion involves all three dimensions of space, while linear expansion involves just one dimension.) So you can find the value of β for solids by multiplying the values of α in Table 14-1 by 3.

Table 14-2 lists values of β for selected liquids. Note that for ethanol, the liquid used in most alcohol thermometers, $\beta = 750 \times 10^{-6}\ K^{-1}$. Contrast this to the volume expansion coefficient for glass from Table 14-1: α is between $4.0 \times 10^{-6}\ K^{-1}$ and $9.0 \times 10^{-6}\ K^{-1}$ for glass, so $\beta = 3\alpha$ is between $12 \times 10^{-6}\ K^{-1}$ and $27 \times 10^{-6}\ K^{-1}$. That explains how an alcohol thermometer works: When the temperature goes up the glass tube of the thermometer and the ethanol inside the tube both expand, but the ethanol expands much more (because it has such a large value of β) and so the ethanol rises within the tube. The same is true for a mercury thermometer, since mercury also has a large value of β (see Table 14-2).

We don't list the value of β for gases in Table 14-2. That's because it's most convenient to use Equation 14-7 (the ideal gas law, $pV = nRT$) rather than Equation 14-19 to keep track of the changes in the volume of a gas with temperature.

For nearly all substances β is positive, so the volume increases (the substance expands, and $\Delta V > 0$) when the temperature increases ($\Delta T > 0$), and the volume decreases (the substance contracts, and $\Delta V < 0$) when the temperature decreases ($\Delta T < 0$). But liquid water is a conspicuous exception to this rule at temperatures below 4°C. Below that temperature liquid water actually *expands* as the temperature decreases further (**Figure 14-10**). So water is less dense at 0°C (the freezing point) than at slightly warmer temperatures. As a result, as the water in rivers and lakes gets cold during the

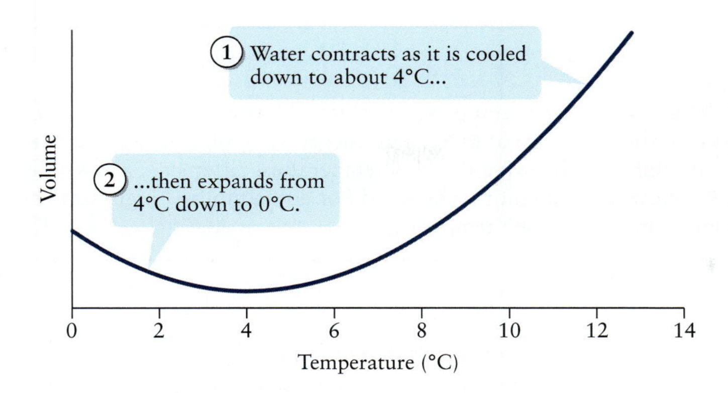

Figure 14-10 The strange case of cold water Although most substances expand when heated and contract when cooled, water is different. As water is cooled below 4°C, it stops contracting and actually expands as the temperature decreases from 4°C to the freezing point.

winter, the coldest water (closest to the freezing point) floats to the surface, while denser, warmer water sinks to the bottom. This means that the surface water freezes first and that rivers and lakes freeze from the surface down. Were water like most substances, the coldest water would be denser than warmer water and would sink to the bottom, and bodies of water would freeze from the bottom up. In this case rivers and lakes could freeze throughout their volume during extended periods of cold temperatures, possibly destroying aquatic life. Life on Earth might not have been able to survive if water did not exhibit the unusual property seen in Figure 14-10.

Figure 14-10 shows another aspect of water that affects life on Earth: Above 4°C an increase in temperature causes liquid water to expand. The average global temperature on Earth has been increasing for decades due to the burning of fossil fuels such as coal and gasoline, which releases carbon dioxide (CO_2) in the atmosphere and causes a temperature increase. (We'll discuss the physics of this process in Section 14-7.) As a result, the water in the oceans is expanding, which causes the sea level to rise. At present the rate of rise is about 3 mm/y (millimeters per year), about half of which is due to thermal expansion of seawater; the rest is caused by the melting of land-based ice such as glaciers, causing water to run off into the ocean.

GOT THE CONCEPT? 14-3 Warming a Cube

A piece of solid aluminum is in the shape of a cube. If the temperature of the cube is increased so that the length of each side increases by 0.0010 (that is, by 0.10%), by how much does the volume of the cube increase? (a) 0.0010; (b) $2 \times 0.0010 = 0.0020$; (c) $3 \times 0.0010 = 0.0030$; (d) $(0.0010)^2 = 1.0 \times 10^{-6}$; (e) $(0.0010)^3 = 1.0 \times 10^{-9}$.

TAKE-HOME MESSAGE FOR Section 14-4

✔ Nearly all substances and objects undergo thermal expansion: They expand when their temperature increases, and they contract when their temperature decreases. Liquid water below 4°C is a notable exception.

✔ If the change in temperature is relatively small, the change in the length of an object or the volume of a quantity of substance is proportional to the change in temperature.

14-5 Heat is energy that flows due to a temperature difference

We've seen that the temperature of an object is a measure of the average kinetic energy of its molecules. We've also seen that if two objects at different temperatures are placed in contact, their temperatures eventually come to the same value. So there must be a *flow* of energy between objects at different temperatures. We use the term **heat** and the symbol Q for the energy that flows from one object to another as a result of a temperature difference. Since heat is a form of energy, it has units of joules (J). If heat flows into an object, the value of Q for that object is positive; if heat flows out of an object, the value of Q for that object is negative (**Figure 14-11**).

WATCH OUT! Objects do not contain heat.

We emphasize that the term *heat* applies only to energy that is in the process of flowing between objects at different temperatures. We use a different term, **internal energy**, for the energy within an object due to the kinetic *and* potential energies associated with the individual molecules that comprise the object. Note also that *temperature* (a measure of the kinetic energy per molecule) and *heat* (energy flowing due to a temperature difference) are *not* the same quantity. The word *hot* doesn't mean "high heat" but rather "high temperature."

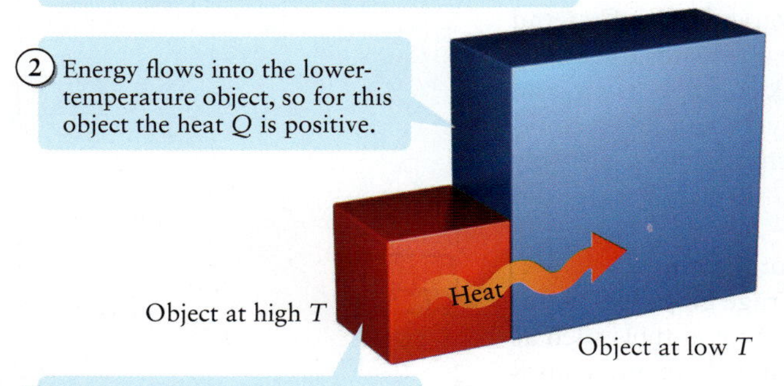

1 If two objects at different temperatures T are in contact, energy flows between them due to the temperature difference. This kind of energy flow is called heat (symbol Q).
2 Energy flows into the lower-temperature object, so for this object the heat Q is positive.
Heat
Object at high T
Object at low T
3 Energy flows out of the higher-temperature object, so for this object the heat Q is negative.

Figure 14-11 Heat flow in thermal contact Energy in the form of heat (Q) flows from the higher-temperature object to the lower-temperature one until the two objects are in thermal equilibrium at the same temperature.

You can understand how heat flows between two objects in contact by considering what happens on the molecular level. Where the two objects touch, molecules on the surfaces of the two objects can collide with each other. The molecules in the higher-temperature object have more kinetic energy than those in the lower-temperature object. As a result, collisions between surface molecules in the two objects end up transferring kinetic energy from the molecules in the higher-temperature object to those in the lower-temperature object. Collisions between neighboring molecules in each object share these energy transfers among all of the object's molecules. The net result is that energy is transferred from the higher-temperature object to the lower-temperature one.

Experiment shows that in most circumstances the flow of heat into or out of an object changes the object's temperature. (The exception is when the object undergoes a *phase change* such as from solid to liquid, liquid to solid, liquid to gas, or gas to liquid. In the next section we'll discuss what happens in a phase change.) If the quantity of heat Q that flows into an object is relatively small, the resulting temperature change ΔT turns out to be *directly* proportional to Q and *inversely* proportional to the mass m of the object. In other words, the greater the quantity of heat Q that flows into an object, the more its temperature changes; the more massive the object and so the more material that makes up the object, the smaller the temperature change for a given quantity of heat Q. (The same quantity of heat that will cook a single meatball will cause hardly any temperature change in a pot roast.) We can write this relationship as

$$\Delta T = (\text{constant}) \times \frac{Q}{m}$$

The constant in this equation depends on the substance of which the object is made. It's conventional to express this equation as

Quantity of heat and the resulting temperature change (14-20)

$$Q = mc\Delta T$$

Quantity of heat that flows into (if $Q > 0$) or out of ($Q < 0$) an object

Specific heat of the substance of which the object is made

Temperature change of the object that results from the heat flow

Mass of the object

Go to Picture It 14-3 for more practice dealing with specific heat.

The quantity c is called the **specific heat** of the material that makes up the object. Its units are joules per kilogram per Kelvin (J/(kg·K) or $\text{J}\cdot\text{kg}^{-1}\cdot\text{K}^{-1}$). For example, the value of c for aluminum is 910 J/(kg·K); this means that 910 J of heat must flow into a 1-kg block of aluminum to raise its temperature by 1 K (or, equivalently, 1°C). **Table 14-3** lists the values of specific heats for a range of substances. The value of the specific heat for any substance varies somewhat with the temperature; however, over the range of temperatures we typically experience, these variations are small enough that we'll ignore them.

Equation 14-20 says that if heat flows *into* an object, so $Q > 0$, it will cause the temperature to increase so $\Delta T > 0$. That's what happens when you warm a saucepan of water on the stove: Heat flows from the stove's electric heating element or gas flame into the saucepan (so $Q > 0$ for the saucepan), and the saucepan's temperature increases (so $\Delta T > 0$ for the saucepan). If heat flows *out* of an object, so $Q < 0$, it will cause the temperature to decrease, so $\Delta T < 0$. This happens to the saucepan when you take it off the stove and allow it to cool: Heat flows from the saucepan to its surroundings (so $Q < 0$ for the saucepan), and the saucepan's temperature decreases (so $\Delta T < 0$ for the saucepan). Example 14-5 shows how to use Equation 14-20 to find the temperature change of an object due to a certain amount of heat flow.

A slightly more complicated application of Equation 14-20 is to find the final temperature of two objects that begin at different temperatures and are placed in contact (for example, hot coffee poured into a cold container). Example 14-6 shows how to solve this sort of problem. In problems of this kind we'll make the simplifying assumption that the two objects are *thermally isolated*; that is, they can exchange energy with each other but don't exchange energy with anything in their environment. So the amount of energy that flows out of the higher-temperature object is equal to the amount of energy that flows into the lower-temperature object.

TABLE 14-3 Specific Heats

Substance	Specific heat ($J \cdot kg^{-1} \cdot K^{-1}$)
air (50°C)	1046
aluminum	910
benzene	1750
copper	387
glass	840
gold	130
ice	(−10°C to 0°C) 2093
iron/steel	452
lead	128
marble	858
mercury	138
methyl alcohol	2549
silver	236
steam (100°C)	2009
water (0°C to 100°C)	4186
wood	1700

EXAMPLE 14-5 Camping Thermodynamics

A certain camping stove releases 5.00×10^4 J of energy per minute from burning propane. (This requires that it uses up the propane at a rate of about 1 g/min.) If half of the released energy is transferred to 2.00 kg of water in a pot above the flame, how much does the temperature of the water change in 1.00 min? (The other half of the energy released by the stove goes into warming the surrounding air and the material of the pot.)

Set Up

Energy flows from the stove into the water because of the temperature difference between them, so this is the kind of energy that we call heat. In 1 min heat Q equal to one-half of 5.00×10^4 J flows into the water. Our goal is to find the resulting temperature change ΔT of the water.

Quantity of heat and the resulting temperature change:

$$Q = mc\,\Delta T \qquad (14\text{-}20)$$

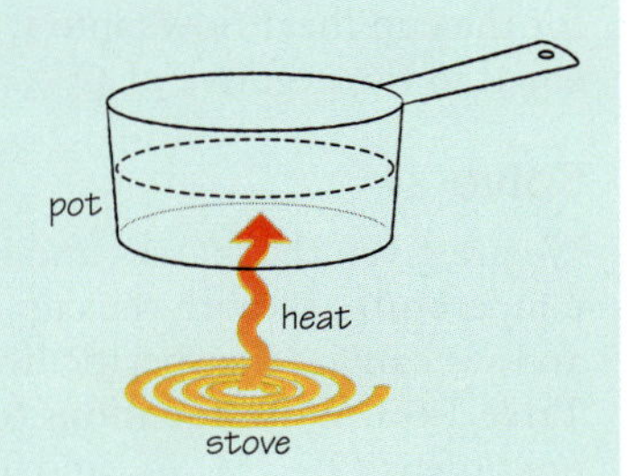

Solve

We are given the values of Q and m, and Table 14-3 lists the value of c for water. Use this and Equation 14-20 to find the temperature change of water in 1.00 min.

Solve Equation 14-20 for the temperature change ΔT:

$$\Delta T = \frac{1}{mc}Q$$

The mass is $m = 2.00$ kg, $c = 4186\ J \cdot kg^{-1} \cdot K^{-1}$ for water from Table 14-3, and $Q = (1/2) \times (5.00 \times 10^4\ J)$. So

$$\Delta T = \frac{(1/2)(5.00 \times 10^4\ J)}{(2.00\ kg)(4186\ J \cdot kg^{-1} \cdot K^{-1})} = 2.99\ K = 2.99°C$$

If the water starts at a temperature of 20.0°C, after 1.00 min its temperature will be 20.0°C + 2.99°C = 23.0°C. After another 1.00 min its temperature will be 23.0°C + 2.99°C = 26.0°C, and so on.

Reflect

If you've ever used a camping stove, you know that it takes quite a while to bring water to a boil. Using the numbers in this problem, we see it would take roughly 27 min to increase the temperature of 2.00 kg of water (corresponding to a volume of 2.00 L) from 20°C to 100°C. That's because water has a very large specific heat, so it takes a lot of heat to raise its temperature by a given amount.

EXAMPLE 14-6 Cooling Coffee

Your expensive coffee maker produces coffee at a temperature of 95.0°C (203°F), which you find is a bit too warm to drink comfortably. To cool the coffee you pour 0.350 kg of brewed coffee (mostly water) at 95.0°C into a 0.250-kg aluminum cup that is initially at room temperature (20.0°C). What is the final temperature of the coffee and cup? Assume that the coffee and cup are thermally isolated.

Set Up

There are three unknown quantities in this problem: the quantity of heat Q_{cup} that flows into the cup, the quantity of heat Q_{coffee} that flows into the coffee (which will be negative since heat flows *out* of the coffee), and the final temperature T_f of the two objects (which is what we want to find). So we need three equations that relate these quantities.

We can get two equations by writing Equation 14-20 twice, once for the cup and once for the coffee. To get a third equation we'll use the idea that the two objects are thermally isolated, so the energy that flows *out* of the hot coffee must equal the energy that flows *into* the cool aluminum cup. So Q_{coffee} and Q_{cup} both involve the same number of joules. However, Q is negative for the coffee (heat flows out of it) and positive for the cup (heat flows into it), so Q_{coffee} is equal to the negative of Q_{cup}.

Quantity of heat and the resulting temperature change:

$$Q = mc\,\Delta T \tag{14-20}$$

Energy is conserved:

$$Q_{coffee} = -Q_{cup}$$

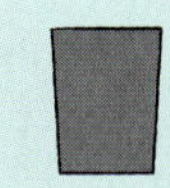

Solve

We are given the masses and initial temperatures of both the cup and the coffee, and we can get the specific heats from Table 14-3. Write equations for the three unknowns Q_{cup}, Q_{coffee}, and T_f. Note that for each object ΔT is the difference between the final and initial temperatures for that object.

The cup has mass $m_{cup} = 0.250$ kg and initial temperature $T_{cup,i} =$ 20.0°C, is made of aluminum with $c_{Al} = 910\ \text{J}\cdot\text{kg}^{-1}\cdot\text{K}^{-1}$, and ends up at temperature T_f. Equation 14-20 for the cup says

$$Q_{cup} = m_{cup}c_{Al}\,\Delta T_{cup} = m_{cup}c_{Al}(T_f - T_{cup,i})$$

The coffee has mass $m_{coffee} = 0.350$ kg and initial temperature $T_{coffee,i} =$ 95.0°C, is made almost completely of water with $c_{water} = 4186\ \text{J}\cdot\text{kg}^{-1}\cdot\text{K}^{-1}$, and ends up at the same final temperature T_f as the cup. Equation 14-20 for the coffee says

$$Q_{coffee} = m_{coffee}c_{water}\,\Delta T_{coffee} = m_{coffee}c_{water}(T_f - T_{coffee,i})$$

The equation of energy conservation is

$$Q_{coffee} = -Q_{cup}$$

Combine these three equations to get a single equation for T_f, the quantity we are trying to find.

Substitute Q_{cup} and Q_{coffee} from the first two equations into the third equation for energy conservation:

$$Q_{coffee} = -Q_{cup}$$

$$m_{coffee}c_{water}(T_f - T_{coffee,i}) = -m_{cup}c_{Al}(T_f - T_{cup,i})$$

The only unknown quantity in this equation is the final temperature T_f.

Solve for the final temperature T_f. Recall that a temperature change of 1°C is equivalent to a temperature change of 1 K, so we can use kelvins and degrees Celsius interchangeably in our calculation.

Multiply out both sides of the equation:

$$m_{coffee}c_{water}T_f - m_{coffee}c_{water}T_{coffee,i} = -m_{cup}c_{Al}T_f + m_{cup}c_{Al}T_{cup,i}$$

Rearrange so that all the terms with T_f are on the same side of the equation:

$$m_{coffee}c_{water}T_f + m_{cup}c_{Al}T_f = m_{coffee}c_{water}T_{coffee,i} + m_{cup}c_{Al}T_{cup,i}$$

or

$$(m_{coffee}c_{water} + m_{cup}c_{Al})\,T_f = m_{coffee}c_{water}T_{coffee,i} + m_{cup}c_{Al}T_{cup,i}$$

Solve for T_f:

$$T_f = \frac{m_{coffee}c_{water}T_{coffee,i} + m_{cup}c_{Al}T_{cup,i}}{m_{coffee}c_{water} + m_{cup}c_{Al}}$$

$$= \frac{\left[\begin{array}{r}(0.350\text{ kg})(4186\text{ J}\cdot\text{kg}^{-1}\cdot\text{K}^{-1})(95.0°\text{C}) \\ +\,(0.250\text{ kg})(910\text{ J kg}^{-1}\text{ K}^{-1})(20.0°\text{C})\end{array}\right]}{(0.350\text{ kg})(4186\text{ J}\cdot\text{kg}^{-1}\cdot\text{K}^{-1}) + (0.250\text{ kg})(910\text{ J}\cdot\text{kg}^{-1}\cdot\text{K}^{-1})}$$

$$= 84.9°\text{C}$$

The coffee cools from 95.0°C to 84.9°C, and the aluminum cup warms from 20.0°C to 84.9°C.

Reflect

The temperature of the coffee decreases and the temperature of the cup increases, just as we expected. We can check our results by calculating Q_{cup} and Q_{coffee}. This calculation shows that $Q_{coffee} = -Q_{cup}$, which must be true for energy to be conserved.

Note that the final temperature of 84.9°C is closer to the initial temperature of the coffee (95.0°C) than to the initial temperature of the cup (20.0°C). That's because the mass and the specific heat are both greater for the coffee than for the cup, so a given quantity of heat produces a smaller temperature change in the coffee than in the cup.

The temperature change for the cup is

$$\Delta T_{cup} = T_f - T_{cup,i} = 84.9°\text{C} - 20.0°\text{C} = +64.9°\text{C} = +64.9\text{ K}$$

(Recall that 1°C and 1 K represent the same temperature change.) The heat that flows into the cup is

$$\begin{aligned} Q_{cup} &= m_{cup}c_{Al}(T_f - T_{cup,i}) \\ &= (0.250\text{ kg})(910\text{ J}\cdot\text{kg}^{-1}\cdot\text{K}^{-1})(+64.9\text{ K}) \\ &= 1.48 \times 10^4\text{ J} \end{aligned}$$

The temperature change for the coffee is

$$\Delta T_{coffee} = T_f - T_{coffee,i} = 84.9°\text{C} - 95.0°\text{C} = -10.1°\text{C} = -10.1\text{ K}$$

The heat that flows into the coffee is

$$\begin{aligned} Q_{coffee} &= m_{coffee}c_{water}\Delta T_{coffee} \\ &= (0.350\text{ kg})(4186\text{ J kg}^{-1}\cdot\text{K}^{-1})(-10.1\text{ K}) \\ &= -1.48 \times 10^4\text{ J} \end{aligned}$$

This is negative because heat flows *out* of the coffee.

In Examples 14-5 and 14-6 we used the SI unit of heat, the joule. However, other units are also commonly used. The **calorie** (cal), equal to 4.186 J, is defined as the quantity of heat required to increase the temperature of one gram (1 g) of pure water from 14.5°C to 15.5°C. So in terms of calories, the specific heat of water at 14.5°C is $c = 1$ $\text{cal}\cdot\text{g}^{-1}\cdot\text{K}^{-1}$. The energy content of foods are given in *food* calories, often denoted by a capital "C," equal to 1000 calories or 1 kilocalorie: 1 C = 1 kcal = 4186 J = 4.186 kJ. In countries other than the United States, the energy content on food labels is given in units of kilojoules (**Figure 14-12**).

The unit of heat in the English system is the **British thermal unit** (BTU), defined as the quantity of heat required to increase the temperature of 1 lb of pure water from 63°F to 64°F. The energy flow of air conditioners and heaters is often given in BTUs. One BTU is equal to 1055 J, 252 cal, or 0.252 kcal.

GOT THE CONCEPT? 14-4 Final Temperature I

Two objects that are made of the same substance are placed in thermal contact. Object A is far more massive than object B. Will the final temperature of the two objects be (a) close to the initial temperature of object A, (b) close to the initial temperature of object B, or (c) about midway between the two initial temperatures?

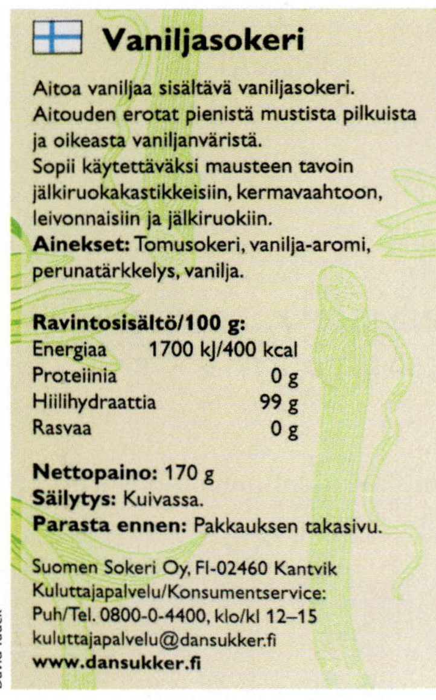

Vaniljasokeri
Aitoa vaniljaa sisältävä vaniljasokeri.
Aitouden erotat pienistä mustista pilkuista ja oikeasta vaniljanväristä.
Sopii käytettäväksi mausteen tavoin jälkiruokakastikkeisiin, kermavaahtoon, leivonnaisiin ja jälkiruokiin.
Ainekset: Tomusokeri, vanilja-aromi, perunatärkkelys, vanilja.
Ravintosisältö/100 g:
Energiaa 1700 kJ/400 kcal
Proteiinia 0 g
Hiilihydraattia 99 g
Rasvaa 0 g
Nettopaino: 170 g
Säilytys: Kuivassa.
Parasta ennen: Pakkauksen takasivu.
Suomen Sokeri Oy, FI-02460 Kantvik
Kuluttajapalvelu/Konsumentservice:
Puh/Tel. 0800-0-4400, klo/kl 12–15
kuluttajapalvelu@dansukker.fi
www.dansukker.fi

David Tauck

Figure 14-12 Counting calories and kilojoules The label on a package of flavored sugar from Finland lists the energy content in both kilojoules (kJ) and food calories (1 food calorie = 1 kilocalorie = 1 kcal).

GOT THE CONCEPT? 14-5 Final Temperature II

Two objects that have the same mass are placed in thermal contact. Object A is made of a substance that has a much larger specific heat than the substance from which object B is made. Will the final temperature of the two objects be (a) close to the initial temperature of object A, (b) close to the initial temperature of object B, or (c) about midway between the two initial temperatures?

TAKE-HOME MESSAGE FOR Section 14-5

✔ Heat is energy that flows from one object to another as a result of a temperature difference. Heat Q is positive if energy flows into an object and negative if it flows out.

✔ The specific heat c of a substance is the amount of heat (in J) required to raise the temperature of 1 kg of the substance by 1 K.

14-6 Energy must enter or leave an object in order for it to change phase

If you add energy to an object, will its temperature increase? The best answer is "Maybe." If you warm (add heat to) an ice cube to its melting point of 0°C, then continue to warm it, the temperature of the ice cube will *remain* at 0°C until the ice has all melted (Figure 14-1b). The solid, liquid, and gaseous states of water are called its **phases**, and when ice melts the water undergoes a **phase change**. Energy must be either absorbed or released in order for a substance to change from one phase to another, but the temperature of the substance stays the same during a phase change. For example, in boiling water at 100°C the energy that the water absorbs goes entirely into rearranging the organization of the water molecules to effect the phase change, with none left over to raise the temperature. You can verify this by putting a thermometer in a pot of water and heating the pot on a stove. The water temperature increases until it reaches 100°C and the water begins to boil. As the water boils away the temperature will remain at 100°C; at this temperature all of the heat flowing into the water goes into changing the phase of the water.

You're familiar with several kinds of phase change. When a liquid becomes a solid, the process is called **freezing**; when a solid becomes a liquid, the process is called melting or **fusion**. Other common phase changes include from liquid to gas, or **vaporization**, and from gas to liquid, or **condensation**. Some substances can change from solid to gas without an intermediate liquid phase. A common example is carbon dioxide (CO_2); at room temperature atmospheric pressure CO_2 goes from a solid form, commonly known as *dry ice*, directly to CO_2 gas. A phase change from solid directly to gas is called **sublimation**; the reverse process, from gas directly to solid, is called **deposition**.

For a given substance at a given pressure, there is a specific temperature at which a phase change occurs. For example, at 1 atm of pressure water can change between its solid and liquid forms at 0°C. This is the temperature at which these two phases can *coexist*. If you allow heat to flow into ice at 0°C, it will begin to melt and the just-melted water will be at the same temperature of 0°C. As heat continues to flow the temperature of the ice–water mixture will remain at 0°C until all of the ice has melted. Only then will an additional heat flow cause a temperature increase in the liquid water that remains. If you reverse the process and allow heat to flow out of water at 0°C, it will begin to freeze and you'll have a mixture of liquid water and solid ice at 0°C. If heat continues to flow out of the system, the mixture stays at 0°C until the water is converted completely to ice, after which point the temperature can drop below 0°C.

Latent Heat

Suppose a quantity of substance is at the temperature at which a phase change can occur. The amount of heat per unit mass that must flow into or out of the substance to cause the phase change is called the **latent heat.** For liquid water at 100°C the latent heat is the amount of heat per kilogram that must flow into the water to vaporize it; for liquid water at 0°C the latent heat is the amount of heat per kilogram that must flow *out* of the water to freeze it. The units of latent heat are joules per kilogram (J/kg).

The value of the latent heat depends on the substance as well as on the kind of phase change. For example, 2.47×10^5 J of heat must flow into a 1-kg block of iron to melt it at its melting temperature of 1811 K, but even more heat (3.34×10^5 J) must flow into a 1-kg block of ice to melt it at its melting temperature of 273 K. We say that the **latent heat of fusion** (that is, melting) is $L_F = 2.47 \times 10^5$ J/kg for iron and $L_F = 3.34 \times 10^5$ J/kg for water. The **latent heat of vaporization** for water is $L_V = 2.26 \times 10^6$ J/kg, so 2.26×10^6 J of heat must flow into 1 kg of liquid water at its vaporization temperature of 100°C in order to vaporize it. The values of L_F and L_V for water show that it takes about seven times more energy to vaporize a kilogram of water at 100°C than it does to melt a kilogram of ice at 0°C.

Note that the latent heat of fusion is also the energy *released* per unit mass when a substance changes from a liquid to a solid, and the latent heat of vaporization is the energy *released* per unit mass when a substance changes from gas to liquid. **Table 14-4** lists the latent heat of fusion and the latent heat of vaporization for various substances.

TABLE 14-4 Melting and Boiling Temperatures (T_F and T_V) and Latent Heats of Fusion and Vaporization (L_F and L_V) for Various Substances at 1 atm of Pressure

Substance	T_F (K)	T_V (K)	L_F (J/kg) × 10^3	L_V (J/kg) × 10^3
alcohol (ethyl)	159	351	104	830
aluminum	933	2792	397	10,900
copper	1357	2835	209	4730
gold	1337	3129	63.7	1645
hydrogen (H_2)	14	20	59.5	445
iron	1811	3134	247	6090
lead	600	2022	24.5	866
nitrogen (N_2)	63	77	25.3	199
oxygen (O_2)	54	90	13.7	213
water	273	373	334	2260

The amount of energy needed to cause a phase change to occur is proportional to the mass m of substance. The more massive a block of ice or iron, for example, the more energy is required to melt it. We can summarize these relationships in a single equation:

Quantity of heat that flows into or out of a quantity of substance to cause a phase change

Use + sign if heat flows into the substance; use − sign if heat flows out of the substance.

$$Q = \pm mL \quad (14\text{-}21)$$

Latent heat for the phase change

Mass of the substance

Heat required for a phase change (14-21)

Equation 14-21 applies to all phase changes. For example, if the phase change is fusion (a solid becomes a liquid), then L is the latent heat of fusion L_F and we use the plus sign so Q is positive (heat is added to the substance to make it melt). If the phase change is condensation (a gas becomes a liquid), L is the latent heat of vaporization L_V and we use the minus sign so Q is negative (heat is released by the substance as it condenses).

WATCH OUT! Equation 14-21 describes what happens in the phase change, not what it takes to get there.

Remember that a phase change can happen only if the substance is at the appropriate temperature (for example, it must be at the vaporization temperature for vaporization or condensation to take place). Equation 14-21 does *not* include the heat that must flow into or out of a substance to get the substance to that temperature. To calculate that you must use Equation 14-20, $Q = mc\Delta T$. So if you want to know how much heat would be needed to vaporize a quantity of liquid water initially at 20°C, you would first use Equation 14-20 to find the amount of heat needed to raise the water temperature from 20°C to 100°C, then use Equation 14-21 to find the amount of heat needed to vaporize that water at 100°C.

Equation 14-21 says that energy in the form of heat must flow into a substance to cause a phase change from solid to liquid or from liquid to gas. Often the source of that energy is a second object in thermal contact with the first. For example, as the sweat on our skin evaporates on a hot day, energy flows from our bodies to the beads of perspiration. There is heat flow *into* the sweat, so $Q_{\text{sweat}} > 0$, and heat flow *out of* the body, so $Q_{\text{body}} < 0$. Energy conservation says that the number of joules of heat that flows out of the body equals the number of joules that flows into the sweat, so $Q_{\text{body}} = -Q_{\text{sweat}}$. The energy lost from the body to the sweat causes a decrease in the temperature of the body (which, unlike the sweat, does not undergo a phase change). In other words, perspiration helps us to cool off. A typical person perspires about 0.60 kg of sweat per day, and sweat has a latent heat of vaporization of about 2.43×10^6 J/kg. From Equation 14-21 the heat that flows into one day's worth of sweat to make it evaporate is

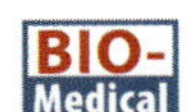

$$Q_{\text{sweat}} = mL_V = (0.60 \text{ kg})(2.43 \times 10^6 \text{ J/kg}) = 1.5 \times 10^6 \text{ J}$$

Go to Interactive Exercise 14-2 for more practice dealing with phase changes.

The heat flow into the body is $Q_{\text{body}} = -Q_{\text{sweat}} = -1.5 \times 10^6$ J to two significant figures, so the body *loses* 1.5×10^6 J of energy by sweating. Recall from Section 14-5 that one food calorie (1 kcal) equals 4186 J, so the amount of energy that the body loses by sweating is $(1.5 \times 10^6 \text{ J})(1 \text{ kcal}/4186 \text{ J}) = 350$ kcal. So you must consume at least 350 kcal daily to compensate for the energy lost by sweating. (You also need to drink at least 0.60 kg of water, with a volume of 0.60 L, to compensate for the water lost.)

We can apply the same approach to what happens when an ice cube is dropped into a glass of water. Heat is transferred from the water (so $Q_{\text{water}} < 0$) into the ice (so $Q_{\text{ice}} > 0$) in order for the ice to melt: The number of joules of energy that leaves the water equals the number of joules that enters the ice, so $Q_{\text{water}} = -Q_{\text{ice}}$. The temperature of the water decreases, and if the water loses enough energy, some or all of it could even freeze.

The following examples illustrate how to solve some typical problems that involve phase changes.

EXAMPLE 14-7 Melting Ice I

A 1.00-kg block of ice initially at a temperature of 0°C is placed inside an experimental apparatus that allows 250 kJ of heat to flow into the ice. What mass of ice melts as a result? What is the final temperature of the melted ice?

Set Up

The ice starts at the melting temperature, so no heat has to be added to the ice to get it to that temperature. Heat must be added to melt the ice, so $Q_{\text{ice}} > 0$. From Table 14-4 the latent heat of fusion for water is $L_F = 334$ kJ/kg $= 334 \times 10^3$ J/kg, so it would take 334 kJ of heat to completely melt 1.00 kg of ice. The heat supplied is less than that, so only part of the ice will melt. We'll use Equation 14-21 to determine the mass that melts.

Heat required to melt a mass m of ice:

$$Q_{\text{ice}} = +mL_F \qquad (14\text{-}21)$$

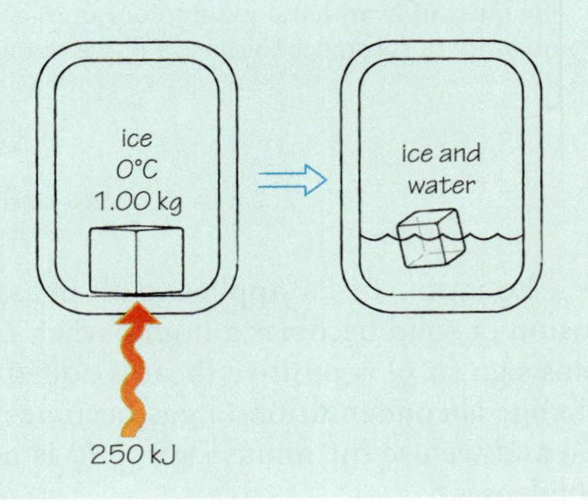

Solve

Find the mass of ice that melts.

From Equation 14-21 the mass that melts is

$$m = \frac{Q_{\text{ice}}}{L_F} = \frac{250 \text{ kJ}}{334 \text{ kJ/kg}} = 0.75 \text{ kg}$$

Almost three-quarters of the 1.00 kg of ice melts. Because all of the heat goes into melting the ice, we end up with solid ice and liquid water that are both at the initial temperature of 0°C.

Reflect

If 334 kJ of heat were transferred to the block of ice, it would completely melt. Any more heat added would go into raising the temperature of the resulting liquid water to a value greater than 0°C.

EXAMPLE 14-8 Melting Ice II

A chunk of ice at −10.0°C is placed in an insulated container that holds 1.00 kg of water at 20.0°C. When the system comes to thermal equilibrium, all of the ice has melted and the entire system is at a temperature of 0.00°C. What was the initial mass of the ice?

Set Up

Three processes occur as the system reaches thermal equilibrium: (a) The temperature of the water decreases from 20.0°C to 0.0°C, so heat flows out of the water; (b) the temperature of the ice increases from −10.0°C to 0.0°C, so heat flows into the ice; and (c) the ice melts at 0°C, so additional heat flows into the ice. Equation 14-20 describes the heat flows associated with the temperature changes, and Equation 14-21 describes the heat flow associated with melting the ice. Energy conservation says that the heat that flows into the ice equals the heat that flows out of the water. Our goal is to find the initial mass of ice, m_{ice}.

Quantity of heat and the resulting temperature change:

$$Q = mc\,\Delta T \qquad (14\text{-}20)$$

Heat required to melt a mass m of ice:

$$Q_{\text{ice}} = +mL_F \qquad (14\text{-}21)$$

Energy is conserved:

$$Q_{\text{ice}} = -Q_{\text{water}}$$

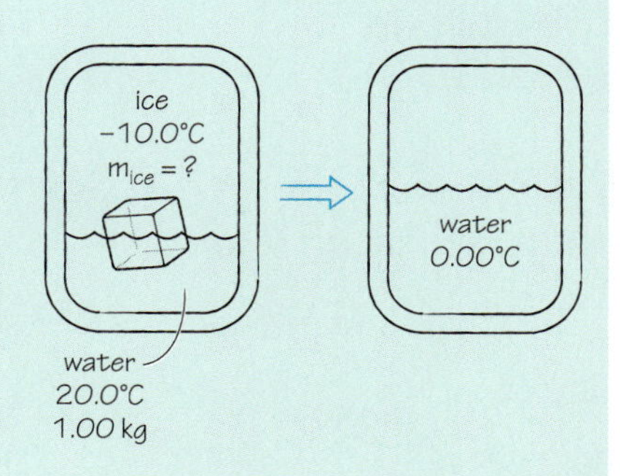

Solve

Write Equation 14-20 for the water and for the ice, and write Equation 14-21 for the ice. The specific heats for water and ice are given in Table 14-3, and the latent heat of fusion is given in Table 14-4.

Q_{water} is negative: Heat flows *out* of the water to make its temperature decrease.

Q_{ice} is positive: Both processes require that heat flows *into* the ice. Note that the term that involves melting the ice is about 16 times larger than the term that involves raising the temperature of the ice from −10.0°C to 0.0°C.

(We again use the idea that a temperature change in degrees Celsius is equivalent to a temperature change in kelvins.)

The heat flow into the water that is required to lower its temperature from 20.0°C to 0.0°C:

$$\begin{aligned} Q_{\text{water}} &= m_{\text{water}} c_{\text{water}}\,\Delta T_{\text{water}} \\ &= (1.00 \text{ kg})(4186 \text{ J}\cdot\text{kg}^{-1}\cdot\text{K}^{-1})(0.0°\text{C} - 20.0°\text{C}) \\ &= (1.00 \text{ kg})(4186 \text{ J}\cdot\text{kg}^{-1}\cdot\text{K}^{-1})(-20.0 \text{ K}) \\ &= -8.37 \times 10^4 \text{ J} \end{aligned}$$

The heat flow into the ice is the sum of the heat required to raise its temperature by −10.0°C to 0.0°C and the heat required to melt it at 0.0°C:

$$\begin{aligned} Q_{\text{ice}} &= m_{\text{ice}} c_{\text{ice}}\,\Delta T_{\text{ice}} + m_{\text{ice}} L_F = m_{\text{ice}}(c_{\text{ice}}\,\Delta T_{\text{ice}} + L_F) \\ &= m_{\text{ice}}[(2093 \text{ J}\cdot\text{kg}^{-1}\cdot\text{K}^{-1})[0.0°\text{C} - (-10.0°\text{C})] \\ &\quad + 3.34 \times 10^5 \text{ J/kg}] \\ &= m_{\text{ice}}(2.093 \times 10^4 \text{ J/kg} + 3.34 \times 10^5 \text{ J/kg}) \\ &= m_{\text{ice}}(3.55 \times 10^5 \text{ J/kg}) \end{aligned}$$

Use energy conservation to relate Q_{water} and Q_{ice}, then solve for m_{ice}.

The energy flow into the ice equals the negative of the (negative) energy flow into the water:

$$Q_{ice} = -Q_{water}$$
$$m_{ice}(3.55 \times 10^5 \text{ J/kg}) = -(-8.37 \times 10^4 \text{ J})$$

Solve for m_{ice}:

$$m_{ice} = \frac{8.37 \times 10^4 \text{ J}}{3.55 \times 10^5 \text{ J/kg}} = 0.236 \text{ kg}$$

Reflect

The amount of ice required has a mass of 0.236 kg. In more familiar terms, 1 kg is the mass of 1 L of water, and an ice cube from a typical ice cube tray has a mass of about 0.04 kg (40 g). So it would take about six ice cubes to reproduce the process described in this problem, which doesn't seem unreasonable.

Note that if the container that holds the ice and water were *not* insulated, we would also have to worry about heat flow between the water–ice mixture and the container. Using an insulated container allows us to avoid this complication.

Phase Diagrams

A useful tool for visualizing the possible phase changes for a particular substance is a **phase diagram.** This is a graph of pressure versus temperature for the substance. For example, **Figure 14-13** is the phase diagram for water. The three red curves in the figure show the boundaries in terms of pressure and temperature between phases. Each point on one of the red curves represents a combination of pressure and temperature at which two phases can coexist and a phase change can take place. For example, at a pressure of 1 atm solid ice and liquid water can coexist at 0°C; this is the melting temperature of ice at that pressure. Similarly, at 1 atm pressure liquid water and water vapor can coexist at 100°C, the temperature at which water boils and vaporizes. The phase diagram shows that at different pressures the phase changes occur at different temperatures. For example, if the pressure is less than 1 atm, the boundary between liquid water and water vapor is at a temperature less than 100°C. That's why water

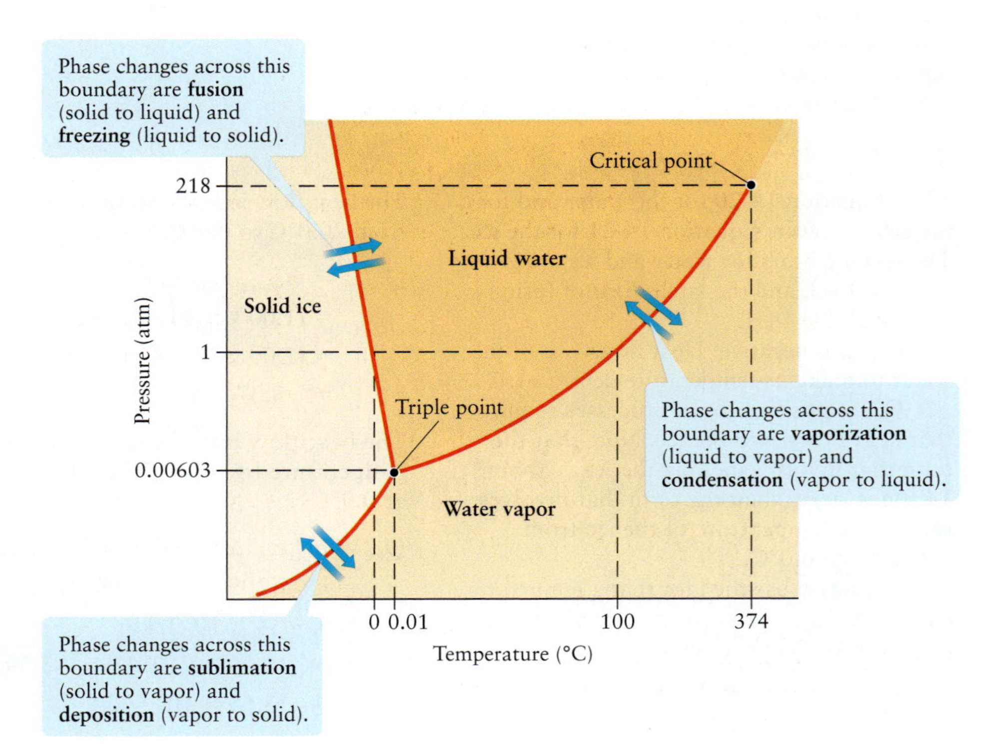
Phase changes across this boundary are **fusion** (solid to liquid) and **freezing** (liquid to solid).
Critical point
218
Liquid water
Solid ice
Pressure (atm)
1
Triple point
0.00603
Water vapor
Phase changes across this boundary are **vaporization** (liquid to vapor) and **condensation** (vapor to liquid).
0 0.01
100
374
Temperature (°C)
Phase changes across this boundary are **sublimation** (solid to vapor) and **deposition** (vapor to solid).

Figure 14-13 The phase diagram for water A phase diagram shows the relationship between pressure, temperature, and the phase of the substance. Each of the three red curves in the figure marks a boundary between two phases and shows the temperature at which a phase change is possible as a function of pressure.

boils at only 95°C at the elevation of Denver or Albuquerque (1610 m or 5280 ft), where air pressure is only about 0.82 atm. Due to the reduced boiling temperature, foods prepared in boiling water at such elevations have to be cooked longer than they would be at sea level.

In Section 14-2 we introduced the idea of the *triple point* of a substance, which is the particular combination of pressure and temperature at which the solid, liquid, and gas phases can all coexist. In Figure 14-13 the triple point is where the three red curves meet. If the pressure is greater than the triple-point pressure, the substance can exist in either the solid, liquid, or gas (vapor) phase depending on the temperature. That's the case for water at sea level on Earth: The triple point of water is 0.00603 atm and 0.01°C, so at normal atmospheric pressure (1 atm) water can exist in any of the three phases. Pressure decreases with increasing altitude, so very high in the atmosphere the pressure is near or below the triple-point pressure. That's why stratus and cumulus clouds, which are made of droplets of liquid water, are found mostly at low altitudes where the pressure of the air is well above the triple-point pressure. By contrast, the most common clouds in the upper atmosphere are cirrus clouds made of ice crystals; the low pressures at such altitudes make it more difficult for water to be in the liquid phase.

Figure 14-13 shows that the red curve that forms the boundary between the liquid and gas (vapor) phases comes to an end at a certain point, which for water occurs at 218 atm and 374°C. This special combination of pressure and temperature is called the **critical point**. If the pressure is greater than the critical-point pressure, the temperature is greater than the critical-point temperature, or both, there is no sharp dividing line between gas and liquid: The substance exists in a single phase called a *supercritical fluid*. Above the critical temperature, unlike below it, no increase in pressure will cause a gas to condense into liquid form.

GOT THE CONCEPT? 14-6 Ranking Energies for Water

? Use the information in Tables 14-3 and 14-4 to rank the following from largest to smallest: (a) the energy needed to raise the temperature of 1 kg of ice from −5°C to −4°C; (b) the energy needed to melt 1 kg of ice at 0°C; (c) the energy needed to raise the temperature of 1 kg of liquid water from 20°C to 21°C; (d) the energy needed to boil 1 kg of water at 100°C; and (e) the energy needed to raise the temperature of 1 kg of steam from 110°C to 111°C.

TAKE-HOME MESSAGE FOR Section 14-6

✔ Energy in the form of heat must either be absorbed or released in order for a substance to change from one phase to another.

✔ The latent heat of fusion is the amount of heat per unit mass required to cause a substance to undergo a phase change between the solid and liquid phases. Similarly, the latent heat of vaporization is the amount of heat per unit mass required to cause a substance to undergo a phase change between the liquid and gaseous phases.

✔ The triple point is the particular combination of pressure and temperature at which the solid, liquid, and gas phases of a substance can coexist.

14-7 Heat can be transferred by radiation, convection, or conduction

Animals have evolved to live in a broad range of climates. The jackrabbit in **Figure 14-14a** has a scrawny body, long skinny legs, enormous ears, and thin fur, all of which promote heat loss and enables the animal to survive in the desert. In contrast, the arctic hare's compact body, stubby legs, relatively small ears, and thick insulating fur help prevent heat loss (**Figure 14-14b**).

The iguanas shown in Figure 14-1c use different strategies for controlling their body temperature: After a cold swim in the Pacific Ocean, these reptiles warm

(a)

(b)

Figure 14-14 Controlling heat transfer Both (a) jackrabbits and (b) arctic hares belong to the same genus (*Lepus*). However, jackrabbits evolved to radiate heat in the hot desert environment, while arctic hares evolved to conserve heat in the cold conditions of the Arctic.

themselves by lying in the tropical sun atop rocks that have already been heated by the Sun. In this section we'll look at how heat is transferred from one place to another, such as from a jackrabbit to its environment or from a hot rock to a cold iguana.

Heat can be transferred from one object to another through three processes: *radiation*, *convection*, and *conduction*.

Radiation is energy transfer by the emission (or absorption) of electromagnetic waves. Some examples of electromagnetic waves are visible light, infrared radiation, and microwaves. (We'll see in Chapter 23 that all of these are different manifestations of the same physical phenomenon.) You use radiation when you lie out on a sunny day or warm yourself under a heat lamp. The iguanas in Figure 14-1c are using radiation to gather energy directly from the sun.

Convection is energy transfer by the motion of a liquid or gas (such as air). Convection is what happens in a pot of water boiling on the stove: Warm water from the bottom of the pot rises, carrying energy to the cooler water at the top of the pot. Convection carries energy away from the arctic hare in Figure 14-14b, as the air warmed by the hare's body rises away from the animal.

Conduction is energy transfer by the collision of particles in one object with the particles in another. It requires physical contact between the two objects. If you use an electric blanket to keep warm on a cold winter's night, the energy flows from the blanket into your body by conduction. The iguanas in Figure 14-1c also use conduction to take in energy from the hot rock on which they are lying.

Let's look at each of these processes of heat transfer individually.

Radiation

Both convection and conduction rely on molecules bumping against one another, but radiation does not. The energy emitted by the Sun reaches Earth after passing through 150 million kilometers of almost completely empty space.

Experiment shows that *any* object emits energy in the form of radiation. The rate at which radiation is emitted by an object—that is, the radiated *power* P in joules per second or watts—is proportional to the object's surface area A and to the fourth power of the Kelvin temperature T of the object:

Rate of energy flow in radiation (14-22)

Rate at which an object emits energy in the form of radiation

Emissivity of the object (a number between 0 and 1)

$$P = e\sigma A T^4$$

Temperature of the object on the Kelvin scale

Stefan-Boltzmann constant $= 5.6704 \times 10^{-8}\ \text{W} \cdot \text{m}^{-2} \cdot \text{K}^{-4}$

Surface area of the object

The **Stefan–Boltzmann constant** σ (the Greek letter sigma) has the same value for all objects. The quantity e is the **emissivity** of the surface; its value indicates how well or how poorly a surface radiates. A surface with a value of e close to 1 is a good radiator of thermal energy.

The factor of T^4 in Equation 14-22 means that the radiated power can change substantially with a change in temperature. For example, a kiln for making ceramic pots can easily reach temperatures of 600 K; a pot at this temperature will radiate energy at $2^4 = 16$ times the rate as when the pot is at room temperature of about 300 K. (The red glow from the heating element in an electric toaster is an example of radiation of this kind.) The factor A in Equation 14-22 shows that for a given surface temperature T, an object will radiate energy at a greater rate if it has a larger surface area. That's why it's useful for the jackrabbit in Figure 14-14a to have a lanky frame and big ears; this maximizes its surface area and makes it easier for the jackrabbit to get rid of excess energy in the hot desert. The more compact arctic hare in Figure 14-14b has much less surface area for its volume, so it radiates away less of its internal energy to its frigid surroundings.

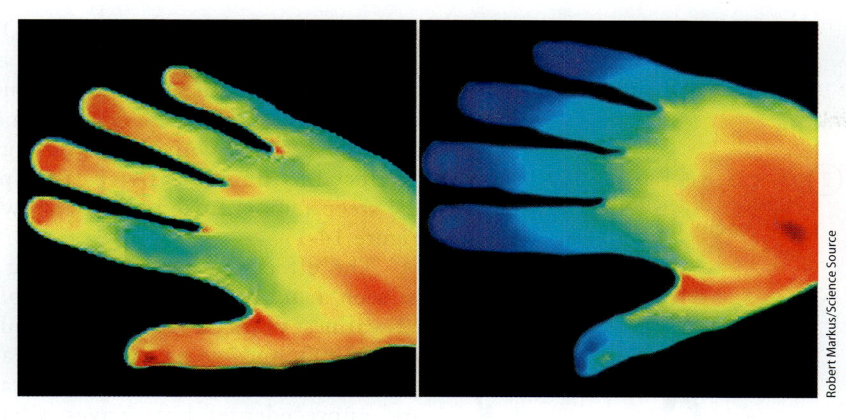

Robert Markus/Science Source

Figure 14-15 Radiation from a human hand A special infrared camera was used to record these images of a person's right hand before (left) and after (right) smoking a cigarette. The intensity of the radiation from different parts of the hand, and hence the temperature of those parts, is indicated by colors from red (warmest) through yellow, green, and blue (coldest). The temperature of the fingers drops after smoking because the nicotine in the tobacco causes blood vessels to contract and reduces blood circulation to the extremities.

Just as for jackrabbits and arctic hares, radiation plays a significant role in heat loss from the human body. Under typical conditions approximately half of the energy transferred from the body to the environment is in the form of radiation. This radiation is emitted almost entirely in the form of infrared light to which your eye is not sensitive, so you can't see yourself glowing in the dark. But specialized devices can detect this radiation (**Figure 14-15**). One such device now in common use in medicine is the *temporal artery thermometer*. The nurse runs the thermometer over your head in the region of the temporal artery, and the circuitry in the thermometer detects the power radiated by the blood in the artery. This gives a very accurate measurement of body temperature while being noninvasive.

Radiation and Climate

Radiation is also the mechanism that keeps Earth from freezing. Our planet's interior releases very little heat, so what keeps the surface warm is energy that reaches us from the Sun in the form of electromagnetic radiation. In equilibrium the rate at which Earth *absorbs* solar energy (about 1.21×10^{17} W) must be equal to the rate at which it *emits* energy into space as given by Equation 14-22. Since the emission rate depends on the surface temperature, this is what determines our planet's average surface temperature T. If Earth had no atmosphere, T would be about 254 K (= −19°C = −2°F), so cold that oceans and lakes around the world should be frozen over. In fact, Earth's actual average surface temperature is a much more livable 287 K (= 14°C = 57°F). The explanation for this discrepancy is called the **greenhouse effect**: Our atmosphere prevents some of the radiation emitted by Earth's surface from escaping into space. Certain gases in our atmosphere called **greenhouse gases**, among them water vapor and carbon dioxide, are transparent to visible light but not to infrared radiation. Consequently, visible sunlight has no trouble entering our atmosphere and warming the surface. But the infrared radiation coming from the heated surface is partially trapped by the atmosphere and cannot escape into space. This lowers Earth's net emissivity e in Equation 14-22. In order to have the power radiated into space equal to the power received from the Sun, the factor of T^4 in Equation 14-22 must be greater to compensate for the reduced value of e. The net effect is that Earth's surface is some 33°C (59°F) warmer than it would be without the greenhouse effect, and water remains unfrozen over most of the planet.

The warming caused by the greenhouse effect gives our planet the moderate temperatures needed for the existence of life. For more than a century, however, our technological civilization has been adding greenhouse gases to the atmosphere at an unprecedented rate. **Figure 14-16** shows how the concentration of carbon dioxide (CO_2) has varied in our atmosphere over the past 400,000 years. (Data from past centuries come from analyzing bubbles of air trapped in ancient ice in the Antarctic.) While there is natural variation in the CO_2 concentration, the value has skyrocketed since the beginning of the Industrial Revolution thanks to our burning fossil fuels such as coal and petroleum. The result has been an amplification of the greenhouse effect and an increase in the average surface temperature, an effect known

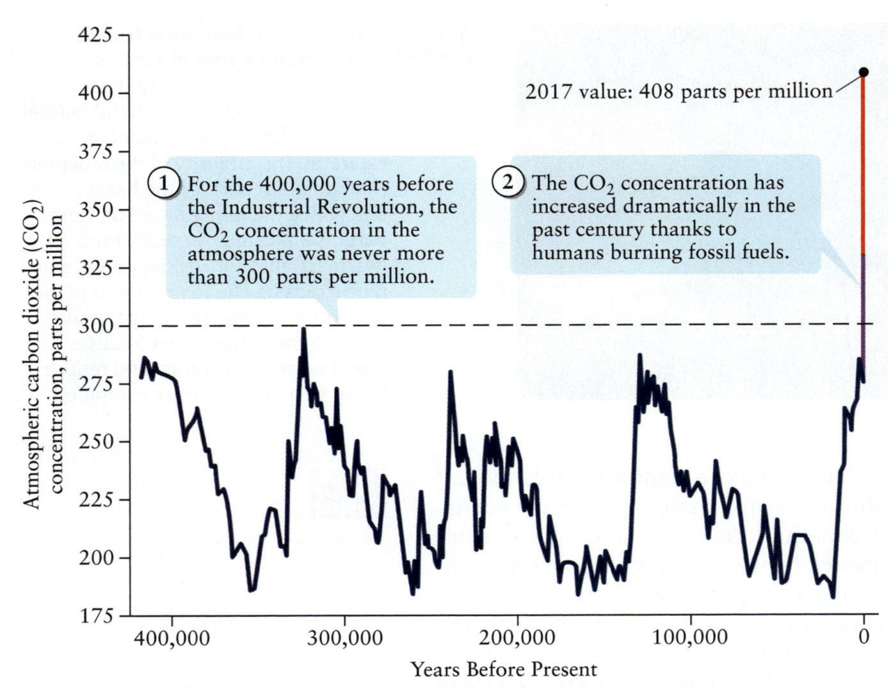

Figure 14-16 Carbon dioxide in the atmosphere Because carbon dioxide (CO_2) absorbs infrared radiation emitted by Earth's surface, its presence in our atmosphere decreases Earth's emissivity and causes the average surface temperature to increase. Thanks to the burning of fossil fuels, the level of CO_2 in our atmosphere is now greater than it has been any time in the past 400,000 years.

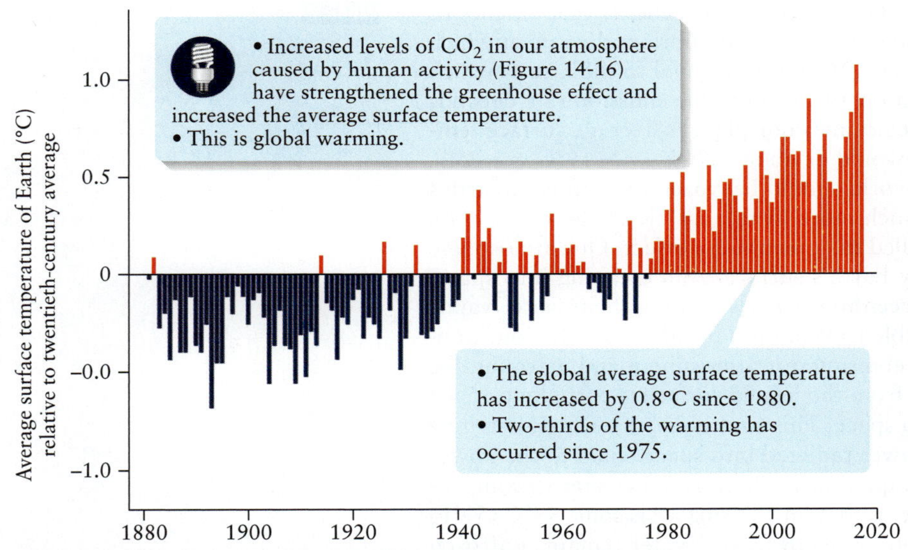

Figure 14-17 Earth's average surface temperature is increasing The increase in greenhouse gases shown in Figure 14-16 has led to elevated temperatures averaged over our planet's surface. The rate of increase has accelerated over recent decades.

as **global warming** (**Figure 14-17**). Other explanations for global warming have been proposed, such as changes in the sun's brightness, but these do not stand up to close scrutiny. Only greenhouse gases produced by human activity can explain the steep temperature increase shown in Figure 14-17.

The effects of global warming can be seen around the world. Eleven of the 12 warmest years on record have occurred since 2000, and we have seen increasing numbers of droughts, water shortages, and unprecedented heat waves. Glaciers worldwide are receding, Arctic sea ice is decreasing by 13% per decade, and a portion of the Antarctic ice shelf has broken off. Unfortunately, global warming is predicted to intensify in the decades to come. The UN Intergovernmental Panel on Climate Change predicts that if nothing is done to decrease the rate at which we add greenhouse gases to our atmosphere, the average surface temperature will continue to rise by an additional 1.4 to 5.8°C during the twenty-first century. What is worse, temperatures will rise in some regions and decline in others and the patterns of rainfall will be substantially altered. Agriculture depends on rainfall, so these changes in rainfall patterns can cause major disruptions in the world food supply. Studies suggest that the climate changes caused by a 3°C increase in the average surface temperature would cause a worldwide drop in cereal crops of 20 to 400 million tons, putting 400 million more people at risk of hunger.

The solution to global warming will require concerted and thoughtful action. Global warming cannot be stopped completely: Even if we were to immediately halt all production of greenhouse gases, the average surface temperature would increase an additional 2°C by 2100 thanks to the natural inertia of Earth's climate system. Instead, our goal is to minimize the effects of global warming by changing the ways in which we produce energy, making choices about how to decrease our requirements for energy, and searching for ways to remove CO_2 from the atmosphere and trap it in the oceans or beneath our planet's surface. Confronting global warming is perhaps the greatest challenge to face our civilization in the twenty-first century.

Convection

You probably learned long ago that "hot air rises." It does, and the rising air carries energy with it from one region to another in a process called convection. **Figure 14-18**

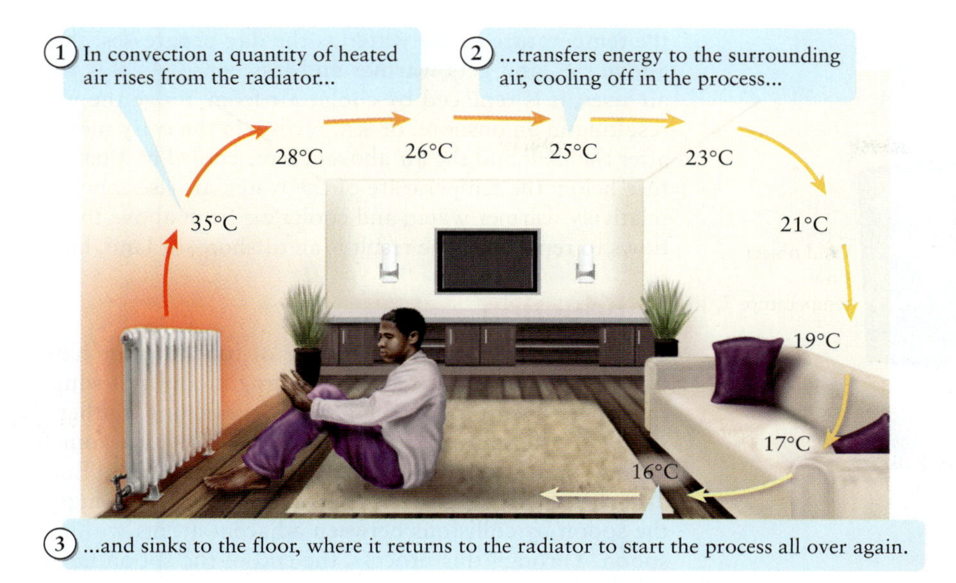

1 In convection a quantity of heated air rises from the radiator...
2 ...transfers energy to the surrounding air, cooling off in the process...
28°C
26°C
25°C
23°C
35°C
21°C
19°C
17°C
16°C
3 ...and sinks to the floor, where it returns to the radiator to start the process all over again.

Figure 14-18 A convection current A continuous circulating flow forms when warm air rises, forcing the air above it to move out of the way and then downward somewhere else.

shows a *convection current* in a room: a continuously circulating flow that forms when rising air forces the air above it to move out of the way and then downward somewhere else.

Convection is possible because fluids (gases and liquids) expand when heated. As a quantity of fluid expands, it becomes less dense than its surroundings and so is buoyed up according to Archimedes' principle (Section 11-8). The hot air "floats" in cooler air, and the cooler air sinks.

Convection plays an important role in driving motions in the atmosphere. People who live near the coasts of large bodies of water often experience a breeze that blows toward the shore in late afternoon and toward the water in early morning (**Figure 14-19**). Because the specific heat of land is much lower than the specific heat of water, the land heats up and cools down more quickly than the water. Especially when it is hot and the sky is clear, the temperature of the land becomes higher than

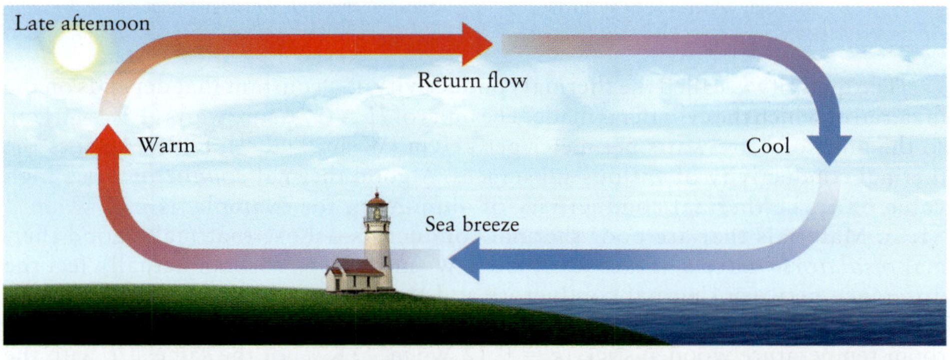

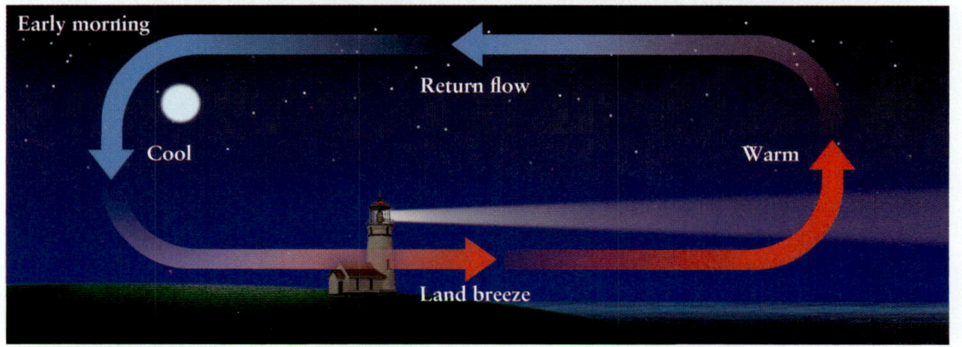

Figure 14-19 Seaside convection Because the specific heat of land is much lower than the specific heat of water, the land heats up and cools down more quickly than the water. This creates onshore sea breezes in late afternoon and offshore land breezes during early morning.

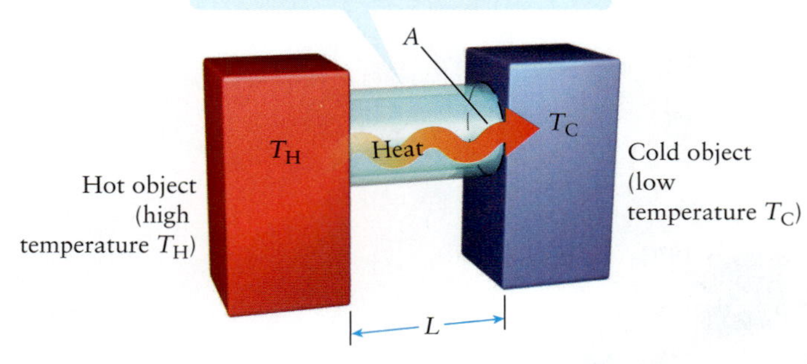
1 The rate H of heat transfer through the cylinder is proportional to the temperature difference $T_H - T_C$.
A
T_H
Heat
T_C
Hot object (high temperature T_H)
Cold object (low temperature T_C)
L
2 The rate of heat transfer through the cylinder is greater when its length L is small and its cross-sectional area A is large.

Figure 14-20 Conduction The rate of heat exchange by conduction between two objects depends on the temperature difference as well as on the cross-sectional area and length of the contact region between them.

the temperature of the water as the day progresses; the air above the land gets warmer and less dense. As this warm air rises, it is replaced by cooler air from above the water, resulting in an onshore, or sea, breeze. In the early morning, after the land and the air above it have cooled to a temperature below the temperature of the water, air rises above the relatively warmer water, and cooler air from above the land flows to replace it. The result is an offshore, or land, breeze.

Conduction

Perhaps you've made the mistake of touching the handle of a metal spoon that has been resting in a pot of soup on the stove. Even though the handle is touching neither the stove nor the soup, it can get hot enough to hurt your fingers if you touch it. The explanation is *conduction*: Energy absorbed from the soup is transmitted along the length of the spoon by collisions between adjacent atoms within the spoon. Through this process, the end of the spoon farthest from the soup eventually will also be hot.

Figure 14-20 shows an idealized situation in which conduction takes place. The cylinder of length L and cross-sectional area A is in thermal contact at one end with a hot object at temperature T_H and in thermal contact at the other end with a cold object at temperature T_C. Experiment shows that the rate of heat transfer is proportional to the temperature difference $T_H - T_C$ (the greater the temperature difference, the more rapidly heat flows). The rate of heat transfer is also greater if the cylinder is short (L is small) and wide (A is large). We can put these observations together into a single equation:

Rate of energy flow in conduction (14-23)

Rate of heat transfer H in conduction = quantity of heat Q that flows divided by the time Δt it takes to flow

Temperature difference of the two places between which heat flows

$$H = \frac{Q}{\Delta t} = k\frac{A}{L}(T_H - T_C)$$

Thermal conductivity of the material through which the heat flows

Cross-sectional area and length of the material through which the heat flows

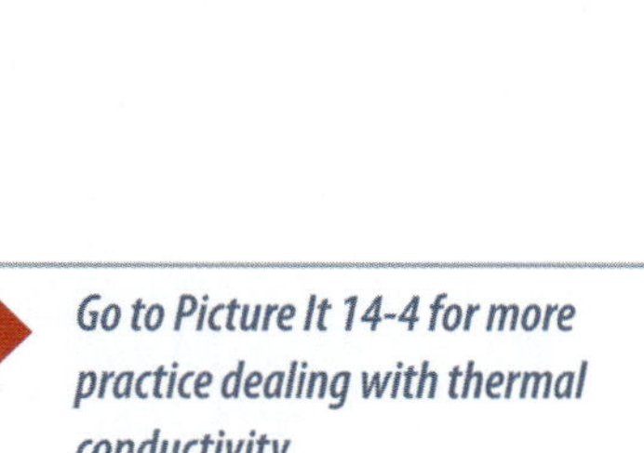
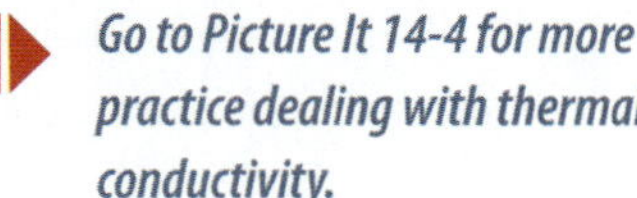

Go to Picture It 14-4 for more practice dealing with thermal conductivity.

The quantity k, called the **thermal conductivity**, is a constant that depends on the material of which the cylinder is made. The units of $H = Q/\Delta t$ are watts ($1 \text{ W} = 1 \text{ J/s}$), so the units of k are watts per meter per kelvin ($\text{W}\cdot\text{m}^{-1}\cdot\text{K}^{-1}$). **Table 14-5** lists the thermal conductivity of various substances. A good thermal *conductor* has a high value of k; the thermal conductivity of aluminum, for example, is $235 \text{ W}\cdot\text{m}^{-1}\cdot\text{K}^{-1}$. Materials that are poor thermal conductors—those that make good thermal *insulators*—have k values less than $1 \text{ W}\cdot\text{m}^{-1}\cdot\text{K}^{-1}$. You can actually feel the difference between thermal conductors and thermal insulators. If you pick up a room-temperature aluminum rod ($k = 235 \text{ W}\cdot\text{m}^{-1}\cdot\text{K}^{-1}$) with one hand and a room-temperature wooden stick ($k = 0.12 \text{ W}\cdot\text{m}^{-1}\cdot\text{K}^{-1}$) of the same size with the other hand, the aluminum rod will feel colder. Both objects are at the same temperature (which is lower than the temperature of your hand), but the much higher thermal conductivity of the aluminum means that heat can flow into it from your hand at a faster rate.

BIO-
Medical

Table 14-5 shows that the thermal conductivity of air is relatively low. This small value makes air, in particular air trapped between other materials, a good insulator. Most animals rely on air trapped between fur or feathers to slow the rate of heat loss in cold conditions. This phenomenon is even more developed in polar bears. The hair that makes up their fur is hollow; the air trapped within the hair shafts makes it a good thermal insulator. Clothing, particularly that made from cotton, wool, and other woven cloth, keeps us warm primarily because of the air trapped between the fibers. Animal fat is also a good insulator. A typical value of thermal conductivity for human

TABLE 14-5 Thermal Conductivities

Material	Thermal conductivity k ($W \cdot m^{-1} \cdot K^{-1}$)	Material	Thermal conductivity k ($W \cdot m^{-1} \cdot K^{-1}$)
air	0.024	hydrogen	0.168
aluminum	235	nitrogen	0.024
brick	0.9	plywood	0.13
copper	401	sand (dry)	0.35
cotton	0.03	silver	429
earth (dry)	1.5	steel	46
human fat	0.2	styrofoam	0.033
fiberglass	0.04	water	0.58
glass (window)	0.96	wood (white pine)	0.12
granite	1.7 – 4.0	wool	0.04
gypsum (plaster) board	0.17		

fat is $0.2\ W \cdot m^{-1} \cdot K^{-1}$, which is about the same as whale blubber and nearly a third smaller than the value of water ($k = 0.58\ W \cdot m^{-1} \cdot K^{-1}$). That's one reason why blubber is so important to marine mammals in frigid polar waters.

EXAMPLE 14-9 Heat Loss Through a Window

Windows are a major source of heat loss from a house. That's because of the materials typically used in construction, glass has one of the highest thermal conductivities. Determine the rate of heat loss for a house on an evening when the temperatures of the outer and inner surfaces of the windows are 14.0°C and 15.0°C, respectively. Take the total window area of the house to be 28.0 m^2 and the thickness of the windows to be 3.80 mm.

Set Up

The rate of heat flow H through all of the windows combined is the same as if there were a single window with area $A = 28.0\ m^2$ and thickness $L = 3.80\ mm = 3.80 \times 10^{-3}\ m$. We use Equation 14-23 to calculate the value of H.

Rate of energy flow in conduction:

$$H = k\frac{A}{L}(T_H - T_C) \qquad (14\text{-}23)$$

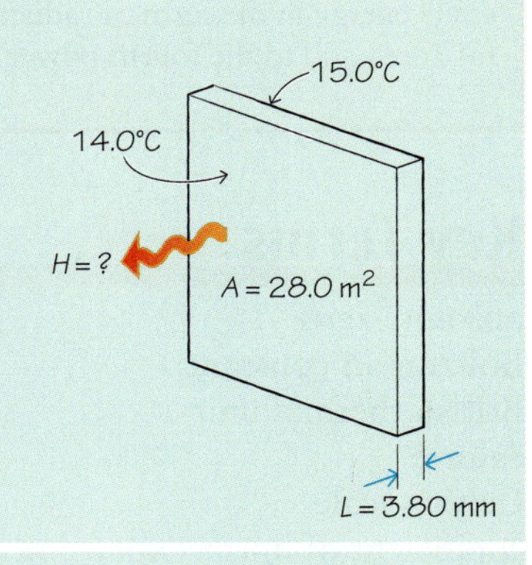

Solve

Substitute values into Equation 14-23.

From Table 14-5 the thermal conductivity of window glass is $k = 0.96\ W \cdot m^{-1} \cdot K^{-1}$. The rate of energy flow through the windows is

$$H = k\frac{A}{L}(T_H - T_C)$$

$$= (0.96\ W \cdot m^{-1} \cdot K^{-1})\left(\frac{28.0\ m^2}{3.80 \times 10^{-3}\ m}\right)(15.0°C - 14.0°C)$$

$$= 7.1 \times 10^3\ W = 7.1\ kW$$

(Recall that a temperature difference of 1.0°C is the same as a temperature difference of 1.0 K.)

Reflect

To maintain the interior of the house at the same temperature, energy has to be provided to the interior by the heating system at a rate of 7.1×10^3 W or 7.1 kW. To put that into perspective, a typical microwave oven or portable hair dryer requires about 1 kW of power to operate. So keeping this house warm with electric heat will require the same amount of electric power as seven microwave ovens or seven hair dryers running continuously.

One way to keep heating costs down is to replace the windows in this house with *dual-pane* or *triple-pane* windows. These windows have two or three sheets of glass separated by a gap of 1 to 2 cm, with gas filling the gap. The greater thickness of glass increases the value of L in Equation 14-23, and the gas between the panes reduces the overall thermal conductivity k.

GOT THE CONCEPT? 14-7 Which Plate Is Which?

The photograph in **Figure 14-21** was taken 30 s after identical ice cubes were placed on each of two black plates. The plates are the same size and were initially at the same temperature. Which of the following is the most likely composition of the two plates? (a) Left-hand plate is wood, right-hand plate is aluminum; (b) left-hand plate is aluminum, right-hand plate is wood; (c) left-hand plate is brick, right-hand plate is glass; (d) left-hand plate is glass, right-hand plate is brick; (e) not enough information to decide.

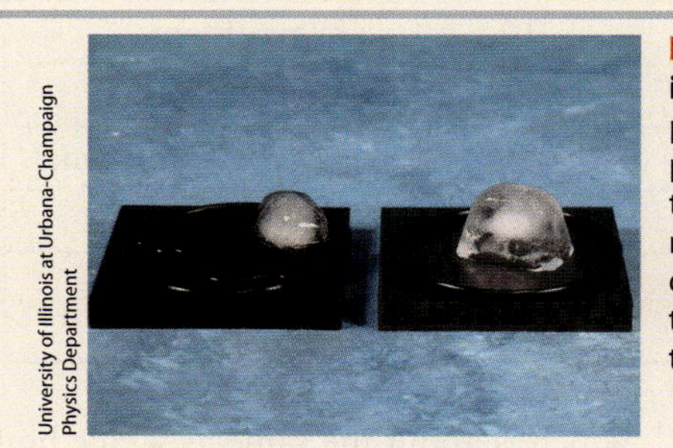

University of Illinois at Urbana-Champaign Physics Department

Figure 14-21 Melting ice cubes An ice cube placed on the left-hand plate melts more rapidly than one placed on the right-hand plate. What does this tell you about the compositions of the two plates?

TAKE-HOME MESSAGE FOR Section 14-7

✔ Heat transfer from one place or object to another can occur through three processes: radiation, convection, and conduction.

✔ In radiation, energy flows from one object to a cooler one in the form of electromagnetic waves. The rate at which an object emits energy in the form of radiation is proportional to its surface area and to the fourth power of its Kelvin temperature.

✔ In convection, energy is carried through a liquid or gas by the motion of that liquid or gas.

✔ In conduction, energy flows within an object through collisions between the particles that make up that object. The rate of heat flow through an object by conduction depends on the dimensions of the object and on the temperature difference between its ends.

Key Terms

absolute zero
Boltzmann constant
British thermal unit
calorie
Celsius scale
coefficient of linear expansion
coefficient of volume expansion
condensation
conduction
convection
critical point
degree of freedom
deposition
emissivity
equation of state
equipartition theorem
Fahrenheit scale
freezing
fusion
global warming
greenhouse effect
greenhouse gas
heat
ideal gas
ideal gas constant
ideal gas law
internal energy
kelvin
Kelvin scale
latent heat (of fusion or vaporization)
mean free path
mole
phase
phase change
phase diagram
radiation
root-mean-square speed (rms speed)
specific heat
Stefan–Boltzmann constant
sublimation
temperature
thermal conductivity
thermal contact
thermal equilibrium
thermal expansion
thermodynamics
thermometer
triple point
vaporization
zeroth law of thermodynamics

Chapter Summary

Topic	Equation or Figure
Temperature and thermal quilibrium: The temperature of an object is a measure of the kinetic energy of its molecules. Two objects are in thermal equilibrium (so no energy flows from one to the other) if and only if they are at the same temperature. We use both the Celsius and Kelvin temperature scales; on the Kelvin scale zero temperature is absolute zero.	(Figure 14-2)
Ideal gases: In an ideal gas there are no interactions between molecules of the gas. The ideal gas law relates the pressure, volume, and Kelvin temperature of the gas; it can be expressed in terms of either the number of molecules or the number of moles.	$pV = NkT$ (14-6) $pV = nRT$ (14-7)
Molecular motion in an ideal gas: The average translational kinetic energy of a molecule of the gas is directly proportional to the Kelvin temperature T. The root-mean-square molecular speed is proportional to the square root of T. If the gas molecules contain more than one atom, there can also be energy in the vibration or rotation of the molecules.	$K_{\text{translational,average}} = \frac{1}{2}m(v^2)_{\text{average}} = \frac{3}{2}kT$ (14-13) $v_{\text{rms}} = \sqrt{\frac{3kT}{m}}$ (14-15)

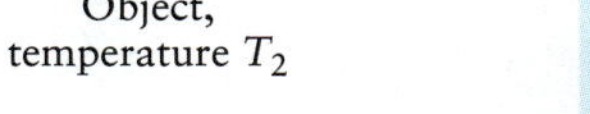

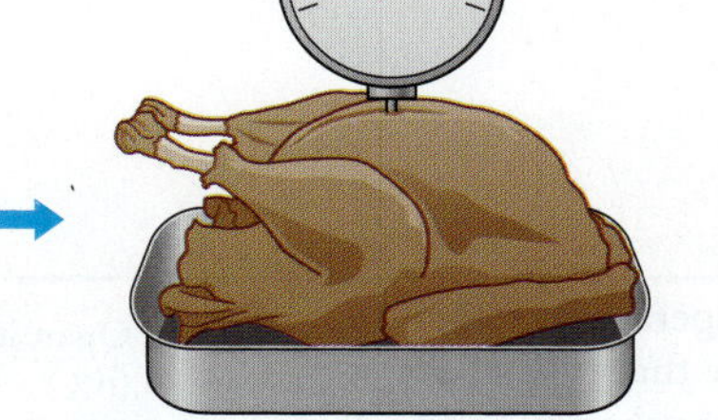

Thermometer and object are both at the same intermediate temperature T_{final}; they are in thermal equilibrium.

Equation (14-6) labels: **Volume** occupied by the gas; **Number of molecules** present in the gas; **Pressure** of an ideal gas; **Temperature** of the gas on the Kelvin scale; **Boltzmann constant**

$$pV = NkT \quad (14\text{-}6)$$

Equation (14-7) labels: **Volume** occupied by the gas; **Number of moles** present in the gas; **Pressure** of an ideal gas; **Temperature** of the gas on the Kelvin scale; **Ideal gas constant**

$$pV = nRT \quad (14\text{-}7)$$

Equation (14-13) labels: **Average translational kinetic energy** of a molecule in an ideal gas; Average value of the square of a gas molecule's speed; **Temperature** of the gas on the Kelvin scale; **Mass** of a single gas molecule; **Boltzmann constant**

$$K_{\text{translational,average}} = \frac{1}{2}m(v^2)_{\text{average}} = \frac{3}{2}kT \quad (14\text{-}13)$$

Equation (14-15) labels: **Root-mean-square speed** of a molecule in an ideal gas; **Boltzmann constant**; **Temperature** of the gas on the Kelvin scale; **Mass** of a single gas molecule

$$v_{\text{rms}} = \sqrt{\frac{3kT}{m}} \quad (14\text{-}15)$$

Thermal expansion: The length of a solid object changes with temperature, as does the volume of a solid or liquid. In most cases the dimensions increase with increasing temperature; liquid water between 0°C and 4°C is an exception.

Change in length of an object

Length of the object before the temperature change

$$\Delta L = \alpha L_0 \Delta T \tag{14-18}$$

Coefficient of linear expansion of the substance of which the object is made

Temperature change of the object that causes the length change

Change in volume of a quantity of substance

Volume of the substance before the temperature change

$$\Delta V = \beta V_0 \Delta T \tag{14-19}$$

Coefficient of volume expansion of the substance

Temperature change of the substance that causes the volume change

Heat and temperature change: Heat is energy that flows from one place to another due to a temperature difference. If there is no change in the phase of an object, heat flow into or out of the object causes a temperature change ΔT.

Quantity of heat that flows into (if $Q > 0$) or out of ($Q < 0$) an object

Specific heat of the substance of which the object is made

$$Q = mc\Delta T \tag{14-20}$$

Temperature change of the object that results from the heat flow

Mass of the object

Heat and phase change: Solid, liquid, and gas are the different possible phases of a substance. At a given pressure there is a specific temperature at which a substance can change from one phase to another. Each phase change requires heat to flow into or out of the substance. For one particular combination of pressure and temperature, called the triple point, the three phases can coexist.

Quantity of heat that flows into or out of a quantity of substance to cause a phase change

Use + sign if heat flows into the substance; use – sign if heat flows out of the substance.

$$Q = \pm mL \tag{14-21}$$

Latent heat for the phase change

Mass of the substance

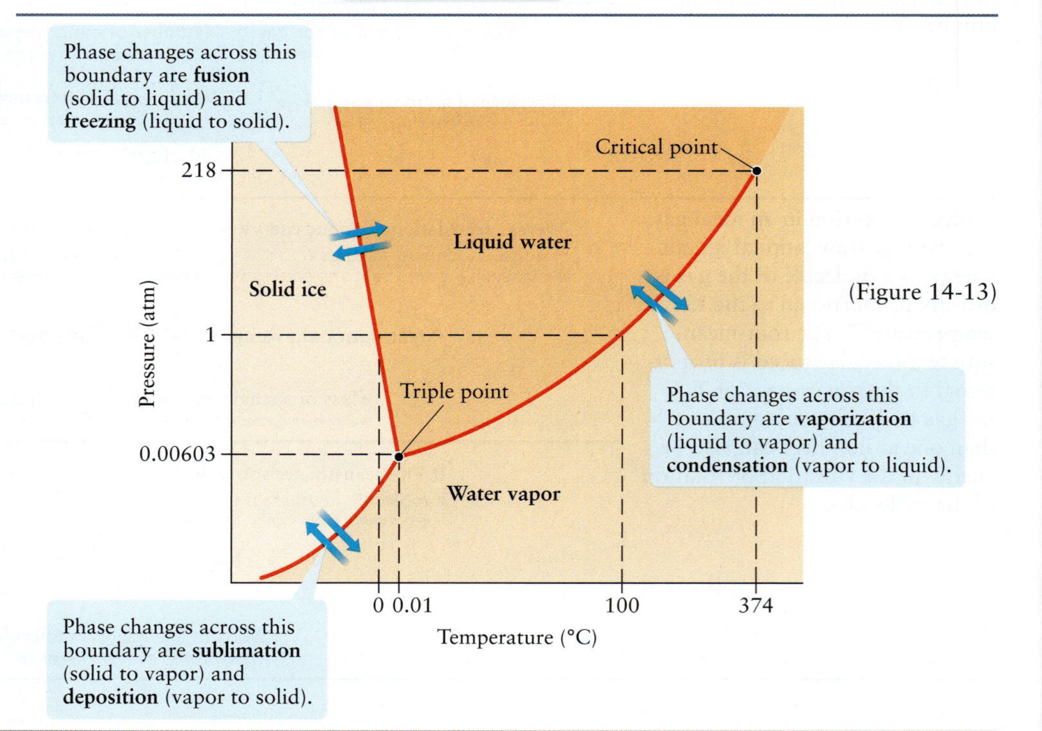
Phase changes across this boundary are **fusion** (solid to liquid) and **freezing** (liquid to solid).
Critical point
218
Liquid water
Solid ice
Pressure (atm)
1
Triple point
0.00603
Water vapor
Phase changes across this boundary are **vaporization** (liquid to vapor) and **condensation** (vapor to liquid).
0 0.01
100
374
Temperature (°C)
Phase changes across this boundary are **sublimation** (solid to vapor) and **deposition** (vapor to solid).

(Figure 14-13)

Radiation, convection, and conduction: Energy can flow from place to place by radiation (electromagnetic waves), convection (circulation of a liquid or gas), or conduction (direct contact between solid objects). The rate at which an object emits energy by radiation is proportional to the fourth power of the object's Kelvin temperature. The rate at which energy flows through an object by conduction depends on the temperature difference between the ends of the object, the dimensions of the object, and the object's thermal conductivity.

Rate at which an object emits energy in the form of radiation

Emissivity of the object (a number between 0 and 1)

$$P = e\sigma AT^4 \qquad (14\text{-}22)$$

Temperature of the object on the Kelvin scale

Stefan-Boltzmann constant $= 5.6704 \times 10^{-8}\ \text{W}\cdot\text{m}^{-2}\cdot\text{K}^{-4}$

Surface area of the object

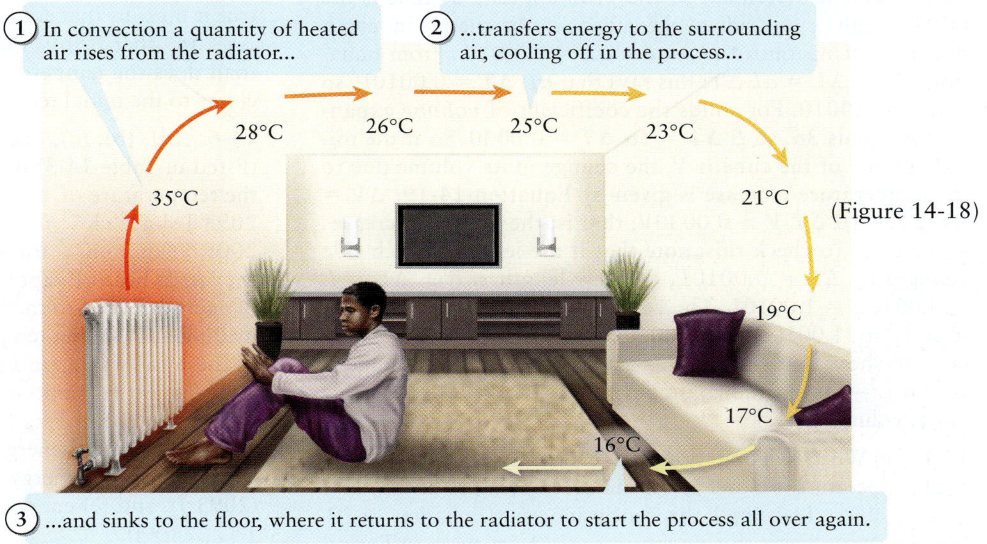

1 In convection a quantity of heated air rises from the radiator...

2 ...transfers energy to the surrounding air, cooling off in the process...

35°C 28°C 26°C 25°C 23°C 21°C 19°C 17°C 16°C

(Figure 14-18)

3 ...and sinks to the floor, where it returns to the radiator to start the process all over again.

Rate of heat transfer H in conduction = quantity of heat Q that flows divided by the time Δt it takes to flow

Temperature difference of the two places between which heat flows

$$H = \frac{Q}{\Delta t} = k\frac{A}{L}(T_H - T_C) \qquad (14\text{-}23)$$

Thermal conductivity of the material through which the heat flows

Cross-sectional area and length of the material through which the heat flows

Answer to What do you think? Question

(c) Almost every substance expands in response to a temperature increase, so we expect that the chamber walls will expand. If they expanded as much or more than the fluid inside the chambers, we would expect that the pressure of the fluid against the chamber walls would stay the same or decrease. Because the pressure increases, it must be that the chamber walls expand by a smaller amount than the fluid. A liquid-in-glass thermometer uses the same basic principle: A temperature increase makes the liquid expand more than the glass tube that contains the liquid, so the liquid level rises in the tube.

Answers to Got the Concept? Questions

14-1 (c), (b), (a) To compare these three temperatures, convert them all to the Kelvin scale. Temperature (b) is $-13°\text{C}$ or, with the second of Equations 14-2, $T_K = T_C + 273.15 = (-13) + 273.15 = 260$ K. Temperature (c) is $-13°\text{F}$, which from the first of Equations 14-1 is $T_C = \frac{5}{9}(T_F - 32) = \frac{5}{9}(-13 - 32) = \frac{5}{9}(-45) = -25°\text{C}$ and from the second of Equations 14-2 is $T_K = T_C + 273.15 = (-25) + 273.15 = 248$ K. So temperature (c) (248 K) is coldest, temperature (b) (260 K) is second coldest, and temperature (a) (280 K) is warmest.

14-2 (a) the same, (b) less, (c) greater, (d) greater. The average translational kinetic energy per molecule $K_{\text{translational,average}} = (3/2)kT$ (Equation 14-13) depends on the temperature but not on the kind of molecule, so this is the same for helium as for oxygen in air at a given temperature. Because oxygen is a diatomic molecule, it can rotate as well as translate, so the average energy per O_2 molecule is about $(5/2)kT$; helium is monatomic, so it has only translational kinetic energy with an average value of $(3/2)kT$ per molecule. The rms speed $v_{rms} = \sqrt{3kT/m}$ (Equation 14-15) is greater for helium because helium has a smaller mass per molecule m. Finally, the mean free path λ (Equation 14-16) is inversely proportional to the square of the molecular radius r, so λ is greater for helium than for oxygen.

14-3 (c) The statement of the problem tells us that if the initial length of one side of the cube is L, the change in length due to the temperature increase is $\Delta L = 0.0010L$. From Equation 14-18, $\Delta L = \alpha L \Delta T$; this says that $\alpha L \Delta T = 0.0010L$, so $\alpha \Delta T = 0.0010$. For solids the coefficient of *volume* expansion β equals 3α, so $\beta \Delta T = 3\alpha \Delta T = 0.0030$. So if the initial volume of the cube is V, the change in its volume due to the temperature increase is given by Equation 14-19: $\Delta V = \beta V \Delta T = (\beta \Delta T)V = 0.0030V$; that is, the volume increases by 0.0030. To check this, note that if the length of each side changes by $\Delta L = 0.0010L$, the new length is $L + \Delta L = L + 0.0010L = 1.0010L$. The volume of the cube then changes from L^3 to $(1.0010L)^3 = (1.0010)^3L^3$. You can use your calculator to show that $(1.0010)^3 = 1.0030$, so the new volume is $1.0030L^3$—that is, the volume has increased by 0.0030 of its initial volume L^3.

14-4 (a) When two objects are made of the same substance, the final temperature is closer to the initial temperature of the more massive object. Let's check this conclusion using Equation 14-20. For the two objects this equation says $Q_A = m_A c_A \Delta T_A$ and $Q_B = m_B c_B \Delta T_B$. If the two objects exchange energy with each other only, $Q_A = -Q_B$ (see Example 14-5), so $m_A c_A \Delta T_A = -m_B c_B \Delta T_B$. The two objects are made of the same substance and so have the same specific heat ($c_A = c_B$); thus our equation simplifies to $m_A \Delta T_A = -m_B \Delta T_B$. Since m_A is much greater than m_B, for this equation to be true it must be that ΔT_A is much smaller in absolute value than ΔT_B. In other words, the temperature of A changes much less than does the temperature of B, and the final temperature is closer to the initial temperature of A.

14-5 (a) When two objects have the same mass, the final temperature is closer to the initial temperature of the object with the greater specific heat. Let's check this conclusion using Equation 14-20. For the two objects this equation says $Q_A = m_A c_A \Delta T_A$ and $Q_B = m_B c_B \Delta T_B$. If the two objects exchange energy with each other only, $Q_A = -Q_B$ (see Example 14-5), so $m_A c_A \Delta T_A = -m_B c_B \Delta T_B$. The two objects have the same mass ($m_A = m_B$), so our equation simplifies to $c_A \Delta T_A = -c_B \Delta T_B$. Since c_A is much greater than c_B, for this equation to be true it must be that ΔT_A is much smaller in absolute value than ΔT_B. In other words, the temperature of A changes much less than does the temperature of B, and the final temperature is closer to the initial temperature of A.

14-6 (d), (b), (c), (a), (e) The specific heat of a substance (listed in Table 14-3) is the amount of energy needed to raise the temperature of 1 kg of substance by 1°C (or 1 K). This is $2093 \text{ J} \cdot \text{kg}^{-1} \cdot \text{K}^{-1}$ for ice, $4186 \text{ J} \cdot \text{kg}^{-1} \cdot \text{K}^{-1}$ for water, and $2009 \text{ J} \cdot \text{kg}^{-1} \cdot \text{K}^{-1}$ for steam. The latent heat of fusion and the latent heat of vaporization are, respectively, the amounts of energy needed to melt and vaporize 1 kg of substance at its melting and boiling temperatures. From Table 14-4 the values of these for water are $L_F = 3.34 \times 10^5$ J/kg and $L_V = 2.26 \times 10^6$ J/kg. So the rank of the energies from largest to smallest is (d) energy to boil 1 kg (2.26×10^6 J), (b) energy to melt 1 kg (3.34×10^5 J), (c) energy to raise the temperature of water by 1°C (4186 J), (a) energy to raise the temperature of ice by 1°C (2093 J), and (e) energy to raise the temperature of steam by 1°C (2009 J).

14-7 (b) The ice cube on the left melts much more rapidly because it is placed on a plate with a high thermal conductivity. Table 4-5 shows that the thermal conductivity of aluminum ($k = 235 \text{ W} \cdot \text{m}^{-1} \cdot \text{K}^{-1}$) is much greater than that of wood ($k = 0.12 \text{ W} \cdot \text{m}^{-1} \cdot \text{K}^{-1}$), so this is a possibility. By contrast, the thermal conductivities of brick ($k = 0.9 \text{ W} \cdot \text{m}^{-1} \cdot \text{K}^{-1}$) and glass ($k = 0.96 \text{ W} \cdot \text{m}^{-1} \cdot \text{K}^{-1}$) are almost identical, so we wouldn't expect much difference between the melting rate of ice cubes placed on a brick plate and a glass plate.

Questions and Problems

In a few problems you are given more data than you actually need; in a few other problems you are required to supply data from your general knowledge, outside sources, or informed estimate. Interpret as significant all digits in numerical values that have trailing zeros and no decimal points. For all problems use $g = 9.80 \text{ m/s}^2$ for the free-fall acceleration due to gravity. Neglect friction and air resistance unless instructed to do otherwise.

•	Basic, single-concept problem
••	Intermediate-level problem; may require synthesis of concepts and multiple steps
•••	Challenging problem
SSM	*Solution is in Student Solutions Manual*
Example	*See worked example for a similar problem*

Conceptual Questions

1. • State the zeroth law of thermodynamics and explain how it is used in physics.

2. • Search the Internet for *temperature scales*. How many different temperature scales can you find?

3. • Explain the physical significance of the value −273.15°C.

4. • A physics student has decided that reading the textbook is very time-consuming and has further decided that she will simply attempt the problems in the book without any prior background reading. During the completion of one problem that involves a temperature conversion, the student concludes that the answer is −508°F. Discuss the validity of the answer.

5. • A careful physics student is reading about the concept of temperature and has what she thinks is a bright idea. While reading the text, she discovers that the Celsius temperature scale and the Kelvin temperature scale both have the *same* increments for equal temperature differences. However, the Fahrenheit scale was described as having *larger* increments for the same temperature differences. From this information the student determines that Celsius (or Kelvin) thermometers will always be *shorter* than Fahrenheit thermometers. Explain the parts of her idea that are valid and the parts that are not.

6. • The average normal human body temperature falls within the range of 37 ± 0.5°C or 98.6 ± 0.9°F. Discuss (a) the difficulties in determining an exact value for normal body temperature and (b) why there is a range of values.

7. • Describe some possible uses for the following thermometers whose temperature ranges are provided in kelvins.

Thermometer A: 200–270 K
Thermometer B: 230–370 K
Thermometer C: 300–550 K
Thermometer D: 300–315 K

8. • "Temperature is the physical quantity that is measured with a thermometer." Discuss the limitations of this working definition.

9. • A very old mercury thermometer is discovered in a physics lab. All the markings on the glass have worn away. How could you recalibrate the thermometer?

10. • The phrase "temperature measures kinetic energy" is somewhat difficult to understand and can be seen as contradictory. Give a few reasons why the explanation actually makes sense and a few reasons why it seems contradictory.

11. • If ideal gas A is thermally in contact with ideal gas B for a significant time, what can you say (if anything) about the following variables: P_A versus P_B, V_A versus V_B, and T_A versus T_B? (P, V, and T stand for pressure, volume, and temperature, respectively.)

12. • A skunk is threatened by a great horned owl and emits its foul-smelling fluid in order to get away. Will the odor be more easily detected on a day when it is cooler or when it is warmer? Explain your answer. SSM

13. • The ideal gas law can be written as $PV = NkT$ or $PV = nRT$. (a) Explain the different contexts in which you might use one or the other and (b) define all the variables and constants.

14. • In warm regions highway repair work is often completed in the summer months. Describe why this is thermodynamically sound.

15. • Why are some sidewalks often formed in small segments, separated by a small gap between the segments?

16. • Older mechanical thermostats use a bimetallic strip to open and close a mercury switch that turns the heat on or off. A bimetallic strip is made of two different metals fastened together. One side is made of one metal, and the other side is made of a different metal. Describe how a bimetallic strip functions in such a thermostat. SSM

17. • A brick wall is composed of 19.0-cm-long bricks ($\alpha = 5.5 \times 10^{-6}\ K^{-1}$) and 1.00-cm-long sections of mortar ($\alpha = 8.0 \times 10^{-6}\ K^{-1}$) in between the bricks. Describe the expansion effects on a 20.0-m-long section of wall that undergoes a temperature change of 25.0°C. SSM

18. •• In Section 14-3 we show that $K_{translational,average} = (3/2)kT$. The result is valid for a three-dimensional collection of atoms. (a) Discuss any changes in the formula that would correspond to a one-dimensional system. (b) What change would be necessary for a 10-dimensional space?

19. • Using the concepts of heat, temperature changes, and heat capacity, give a simplistic explanation of the global demographic factoid that 90% of the world's population lives within 100 km of a coastline.

20. • (a) Which is hotter: a kilogram of boiling water or a kilogram of steam? (b) Which one can cause a more severe burn? Why? Assume the steam is at the lowest temperature possible for the corresponding pressure. SSM

21. • Explain the term "latent" as it applies to phase changes.

22. • Describe the sequence of thermodynamic steps that a very cold block of ice (below 0°C) will undergo as it transforms into steam at a temperature above 100°C.

23. • You have probably heard the old saying "It ain't the heat, it's the humidity that's so unbearable." Discuss what this *really* means and try to include some reference to the definitions of heat and temperature that we have established in this chapter.

24. • There are several different units besides the joule used to measure thermal energy. (a) List the various units associated with thermal energy and (b) indicate in which area of science or technology they are used.

25. • Write a brief definition of the three different modes of heat transfer: radiation, convection, and conduction.

26. • Describe how convection causes hot air to rise in a room.

27. • Explain how a light material, such as fiberglass insulation, makes an effective barrier to keep a home warm in the winter and cool in the summer.

Multiple-Choice Questions

28. • Two objects that have different sizes, masses, and temperatures come into close contact with each other. Thermal energy is transferred

A. from the larger object to the smaller object.
B. from the object that has more mass to the one that has less mass.
C. from the object that has the higher temperature to the object that has the lower temperature.
D. from the object that has the lower temperature to the object that has the higher temperature.
E. back and forth between the two objects until they come to equilibrium.

29. • If you halve the value of the root-mean-square speed or v_{rms} of an ideal gas, the absolute temperature must be

A. reduced to one-half its original value.
B. reduced to one-quarter its original value.
C. unchanged.
D. increased to twice its original value.
E. increased to four times its original value. SSM

30. • Two gases each have the same number of molecules, same volume, and same atomic radius, but the atomic mass of gas B is twice that of gas A. Compare the mean free paths.

A. The mean free path of gas A is four times larger than gas B.
B. The mean free path of gas A is four times smaller than gas B.
C. The mean free path of gas A is the same as gas B.
D. The mean free path of gas A is two times larger than gas B.
E. The mean free path of gas A is two times smaller than gas B.

31. • If you heat a thin, circular ring so that its temperature is twice what it was originally, the ring's hole

A. becomes larger.
B. becomes smaller by an unknown amount.
C. remains the same size.
D. becomes four times smaller.
E. becomes two times smaller.

32. • If you add heat to water at 0°C, the water will decrease in volume until it reaches

A. 1°C.
B. 2°C.
C. 3°C.
D. 4°C.
E. 100°C.

33. • Two objects that are not initially in thermal equilibrium are placed in close contact. After a while,

A. the specific heats of both objects will be equal.
B. the thermal conductivity of each object will be the same.
C. the temperature of the cooler object will rise the same amount that the hotter one drops.
D. the temperature of each object will be the same.
E. the temperature of the cooler object will rise twice as much as the temperature of the hotter one drops. SSM

34. • When a substance goes directly from a solid state to a gaseous form, the process is known as

A. vaporization.
B. fusion.
C. melting.
D. condensation.
E. sublimation.

35. • Which heat transfer process(es) is (are) important in the transfer of energy from the Sun to Earth?

A. radiation
B. convection
C. conduction
D. conduction and radiation
E. conduction, radiation, and convection

36. • A clay pot at room temperature is placed in a kiln, and the pot's temperature doubles. How much more heat per second is the pot radiating when hot compared to when cool?

A. 2 times
B. 4 times
C. 8 times
D. 16 times
E. 32 times

37. • If the thickness of a uniform wall is doubled, the rate of heat transfer through the wall is

A. quadrupled.
B. doubled.
C. halved.
D. unchanged.
E. one-fourth as much. SSM

Estimation/Numerical Analysis

38. • Estimate how much energy is required to warm Earth's oceans by 1°C.

39. • Estimate how much energy is radiated by Earth in one 24-hour period.

40. • Estimate the amount of energy required to vaporize Lake Superior.

41. • A modern 20-story, climate-controlled office building is held at a uniform temperature of 23°C. Estimate how much thermal energy is conducted through its windows in 1 hour if the exterior temperature is a toasty 36°C.

42. • Estimate how much heat is required to melt a typical ice cube taken from your freezer.

43. • Estimate the time you have before you succumb to hypothermia if you fall through the ice into a lake. SSM

44. • Estimate the time required for a kettle of water to boil on a natural gas stove.

45. • Estimate how much more heat a person loses on a cold winter day when she doesn't wear a hat compared to when she does wear one.

46. • If 100 physics students each give off 100 W as they sit in the classroom taking an exam, estimate the temperature increase of the room during the hour-long test. Assume that 1% of the heat goes into raising the air temperature.

47. • In the final stages in the life of a star (such as our Sun), the radiated power decreases by one-half. Estimate the temperature change that would accompany such a shift.

48. •• **Biology** Although many mechanisms help large mammals (such as elephants) keep cool in hot climates, the size of ears is a big factor. The Asian elephant's ears are about 20% smaller than the ears of the African elephant. If the radiant loss of heat through the billions of capillaries in elephant ears is the primary means of staying cool, estimate the relative temperature difference between the territory of Asian elephants versus the territory of African elephants.

49. • Estimate the fraction of heat that is given off by radiation versus conduction when a person lies down in the snow. SSM

50. • A 1.0-g sample of metal is slowly heated. The temperature increase for each increment of heat added is tabulated in the table. Plot a graph for the data and predict the specific heat for this common metal. If the measurements are ±10%, make a guess as to the type of metal it might be.

Heat (J)	Temperature change (°C)
0	0
3.9	10
7.9	20
20	50
40	100

51. • Ordinary ice (frozen H_2O) is cold, dry ice (solid CO_2) is very cold, liquid nitrogen is extremely cold, and liquid helium is almost unimaginably cold! The cost of each is approximately exponentially proportional to the temperature at which they change states. The temperature of dry ice is −78.5°C, and it costs about \$1/kg. The temperature of liquid nitrogen is −196°C, and it costs about \$2/kg. The temperature of liquid helium is −268°C, and it costs about \$100/kg. (a) Predict the cost of 1 kg of ordinary ice. Before you make the estimate, consider the most appropriate temperature scale to use! (b) Check the price of a bag or block of ice at your local grocery or convenience store. How close is your estimate?

52. • (a) Plot water density as a function of temperature for the data listed in the table. (b) Using the graph, discuss how changes in water's density affect the freezing of lakes in the winter.

Density (g/cm^3)	Temperature (°C)
0.99990	8.0
0.99996	6.0
1.00000	4.0
0.99996	2.0
0.99988	0.0

Problems

14-1 A knowledge of thermodynamics is essential for understanding almost everything around you—including your own body

14-2 Temperature is a measure of the energy within a substance

53. • Convert the following temperatures: Example 14-1
 A. 28°C = ________ °F
 B. 58°F = ________ °C
 C. 128 K = ________ °F
 D. 78°F = ________ K
 E. 37°C = ________ °F

54. • Convert the following temperatures and comment on any physical significance of each: Example 14-1
 A. 0°C = ______ °F
 B. 212°F = _____ °C
 C. 273 K = _____ °F
 D. 68°F = ______ K

55. • Starting from $T_C = 5/9(T_F - 32)$ (the first of Equations 14-1), derive a formula for converting from °C to °F (the second of Equations 14-1).

56. • The highest temperature ever recorded on Earth is 56.7°C, in Death Valley, California, in 1913. The lowest temperature on record is −89.2°C, measured at Vostok Research Station in Antarctica in 1983. Convert these extreme temperatures to °F and kelvin. Example 14-1

57. • In adults the normal range for oral (under the tongue) temperature is approximately 36.7°C to 37.0°C. Calculate the range and differences in °F and in kelvins. Example 14-1

58. • Explain why there must exist a numerical value that is the same on the Celsius scale as on the Fahrenheit scale. Show the calculation that yields this special value. SSM

59. • A thermally isolated system has a temperature of T_A. The temperature of a second isolated system is T_B. When the two systems are placed in thermal contact with each other, they come to an equilibrium temperature of T_C. Describe all the possibilities regarding the relative magnitudes of T_A and T_B when (a) $T_C < T_A$, and (b) $T_C > T_A$.

14-3 In a gas, the relationship between temperature and molecular kinetic energy is a simple one

60. • One mole of an ideal gas is at a pressure of 1.00 atm and occupies a volume of 1.00 L. (a) What is the temperature of the gas? (b) Convert the temperature to °C and °F. Example 14-2

61. • Calculate the energy of a sample of 1.00 mol of ideal oxygen (O_2) gas molecules at a temperature of 300 K. Assume that the molecules are free to rotate and move in three dimensions. Example 14-3

62. • A 55.0-g sample of a certain gas occupies 4.13 L at 20.0°C and 10.0 atm pressure. What is the gas? SSM Example 14-2

63. • The boiling point of water at 1.00 atm is 373 K. What is the volume occupied by water gas due to evaporation of 10.0 g of liquid water at 1.00 atm and 373 K? Example 14-2

64. • An ideal gas is confined to a container at a temperature of 300 K. What is the average translational kinetic energy of an atom of the gas? Example 14-3

65. • Calculate v_{rms} for a helium atom if 1.00 mol of the gas is confined to a 1.00-L container at a pressure of 10.0 atm. Example 14-3

66. • Two gases present in the atmosphere are water vapor (H_2O) and oxygen (O_2). What is the ratio of their rms speeds? SSM Example 14-3

67. • State-of-the-art vacuum equipment can attain pressures as low as 7.0×10^{-11} Pa. Suppose that a chamber contains helium at that pressure and at room temperature (300 K). Estimate the mean free path for helium in the chamber. Assume the diameter of a helium atom is 1.0×10^{-10} m.

68. • The mean free path for O_2 molecules at a temperature of 300 K and at 1.00 atm pressure is 7.10×10^{-8} m. Use these data to estimate the size of an O_2 molecule. SSM

14-4 Most substances expand when the temperature increases

69. • Calculate the temperature change needed for a cylinder of gold to increase in length by 0.1%. Example 14-4

70. • Calculate the coefficient of linear expansion for a 10.0-m-long metal bar that shortens by 0.500 cm when the temperature drops from 25.0°C to 10.0°C. Example 14-4

71. • By how much would a 1.00-m-long aluminum rod increase in length if its temperature were raised 8.00°C? Example 14-4

72. • A silver pin is exactly 5.00 cm long when its temperature is 180.0°C. How long is the pin when it cools to 28.0°C? SSM Example 14-4

73. •• A sheet of lead has an 8.00-cm-diameter hole drilled through it while at a temperature of 8.00°C. What will be the diameter of the hole if the sheet is heated to 208.00°C? Example 14-4

74. •• A thin sheet of copper 80.00 cm by 100.00 cm at 28.0°C is heated to 228.0°C. What will be the new area of the sheet? Example 14-4

75. •• A sheet of copper at a temperature of 0.00°C has dimensions of 20.0 cm by 30.0 cm. (a) Calculate the dimensions of the sheet when the temperature rises to 45.0°C. (b) By what percent does the area of the sheet of copper change? Example 14-4

76. •• A 5.00-m-long cylinder of solid aluminum has a radius of 2.00 cm. (a) If the cylinder is initially at a temperature of 5.00°C, how much will the length change when the temperature rises to 30.00°C? (b) Due to the temperature increase, by how much would the density of the aluminum cylinder change? (c) By what percentage does the volume of the cylinder increase? SSM Example 14-4

77. •• A cube of pure iron is heated uniformly to 100.0°C. At that temperature, the volume of the iron is 20.0 cm^3. Find the dimensions of the cube (a) at 100.0°C and (b) at 20.0°C. Example 14-4

78. •• A sphere of gold has an initial radius of 1.00 cm when the temperature is 20.0°C. (a) If the temperature is raised to 80.0°C, calculate the new radius of the sphere. (b) What is the percent change in the volume of the sphere? Example 14-4

79. •A 20.0 m × 25.0 m pool is filled to a depth of 1.50 m at a temperature of 20.0°C. It warms to 35.0°C by the end of a hot summer day. Assuming no evaporation, by how much does the depth of the water change? Example 14-4

80. •• In high-altitude desert, it is not uncommon for the temperature to change by 20.0°C over the course of a summer day. If a vehicle's 45.0-L gas tank is half full when the temperature is 35.0°C, how full will it be the next morning if the temperature is 15.0°C, assuming the vehicle was not driven? Ignore any expansion of the tank and express your answer as a percentage. Example 14-4

81. •• A small copper sphere with a radius of 1.00 cm at a room temperature of 20.0°C has a mass density of 8.96 g/cm^3. Calculate the new mass density of the copper sphere after it has been heated to a temperature of 400.0°C. Example 14-4

14-5 Heat is energy that flows due to a temperature difference

82. • What is the specific heat of a 0.500-kg metal sample that rises 4.80°C when 307 J of heat is added to it? Example 14-5

83. • You wish to heat 0.250 kg of water to make a hot cup of coffee. If the water starts at 20.0°C and you want your coffee to be 95.0°C, calculate the minimum amount of heat required. SSM Example 14-5

84. • In a thermodynamically sealed container, 20.0 g of 15.0°C water is mixed with 40.0 g of 60.0°C water. Calculate the final equilibrium temperature of the water. Example 14-6

85. • How much heat is transferred to the environment when the temperature of 2.00 kg of water drops from 88.0°C to 42.0°C? Example 14-5

86. • A copper pot has a mass of 1.00 kg and is at 100.00°C. How much heat must be removed from it to decrease its temperature to precisely 0.00°C? Example 14-5

87. • A lake has a specific heat of 4186 J/(kg · K). If we transferred 1.70×10^{14} J of heat to the lake and warmed the water from 10.0°C to 15.0°C, what is the mass of the water in the lake? Neglect heat released to the surroundings. SSM Example 14-5

88. •• Calculate the temperature increase in a 1.00-kg sample of water that results from the conversion of gravitational potential energy directly to heat energy in the world's tallest waterfall, the 807-m-tall *Salto Angel* in Canaima National Park, Venezuela. Example 14-5

89. • A superheated iron bar that has a mass of 5.00 kg absorbs 2.50×10^6 J of heat from a blacksmith's fire. Calculate the temperature increase of the iron. Example 14-5

90. • A 0.200-kg block of ice is at −10.0°C. How much heat must be removed to lower its temperature to −40.0°C? Example 14-5

91. •• A 0.250-kg sample of copper is heated to 100.0°C and placed into a cup containing 0.300-kg of water initially at 30.0°C. Ignoring the container holding the water, find the final equilibrium temperature of the copper and water, assuming no heat is lost or gained to the environment. Example 14-6

92. •• A 50.0-g calorimeter cup made from aluminum contains 0.100 kg of water. Both the aluminum and the water are at 25.0°C. A 0.300-kg cube of some unknown metal is heated to 150.0°C and placed into the calorimeter; the final equilibrium temperature for the water, aluminum, and metal sample is 41.0°C. Calculate the specific heat of the unknown metal and make a guess as to its composition. Example 14-6

93. • A hacksaw is used to cut a 20.0-g steel bolt. Each stroke of the saw supplies 30.0 J of energy. How many strokes of the saw will it take to raise the temperature of the bolt from 20.0°C to 80.0°C? Assume none of the energy goes into heating the surroundings. Of course, it will take significantly more than this to also cut the metal! SSM Example 14-5

14-6 Energy must enter or leave an object in order for it to change phase

94. • Calculate the amount of heat required to change 25.0 g of ice at 0°C to 25.0 g of water at 0°C. Example 14-7

95. • How much heat is required to change 25.0 g of ice at −40.0°C to 25.0 g of steam at 140°C? Example 14-7

96. • A sealed container (with negligible heat capacity) holds 30.0 g of 120°C steam. Describe the final state if 100,000 J of heat is removed from the steam. Example 14-7

97. • How much heat is required to melt a 0.400-kg sample of copper that starts at 20.0°C? SSM Example 14-7

98. •• Suppose 20.0 g of ice at −10.0°C is placed into 0.300 kg of water in a 0.200-kg copper calorimeter. If the final temperature of the water and copper calorimeter is 18.0°C, what was the initial common temperature of the water and copper? Example 14-8

99. •• Anindya has 0.700-kg of scorching hot 90.0°C coffee in an insulated travel container. He wants to cool his coffee down to a more manageable 55.0°C. What mass of ice must he place in the container to do this? Assume that no heat is transferred to the container or the environment. Example 14-8

100. •• A 2.5-kg block of −12.0°C ice is placed in a thermally isolated chamber with 3.5 kg of 120.0°C steam. When the system in the chamber reaches equilibrium, what are the final state and temperature of the H_2O? Example 14-8

101. •• What mass of ice at −20.0°C must be added to 50.0 g of steam at 120.0°C to end up with water at 40.0°C? Example 14-8

102. • A 60.0-kg ice hockey player is moving at 8.00 m/s when he skids to a stop. If 40% of his kinetic energy goes into melting ice, how much water is created as he comes to a stop? Assume that the surface layer of the ice in the hockey rink has a temperature of 0°C. Example 14-7

103. • You have 50.0 g of iron at 120°C, 60.0 g of copper at 150°C, and 30.0 g of water at 40.0°C. Which of those would melt the most ice, starting at −5.00°C? How much ice does each melt? SSM Example 14-8

14-7 Heat can be transferred by radiation, convection, or conduction

104. • A heated bar of gold radiates at a temperature of 300°C. (a) By what factor does the radiated power increase if the temperature is increased to 600°C? (b) What if it's increased to 900°C? (c) If the surface area of the gold bar is also doubled, how will the answers be affected?

105. • **Astronomy** An astrophysicist determines the surface temperature of a distant star is 12,000 K. The surface temperature of the Sun is about 5800 K. If the surface temperature of the Sun were to suddenly increase to 12,000 K, by how much would the radiated power increase?

106. • **Astronomy** A star radiates 3.75 times less power than our own Sun. What is the ratio of the temperature of the star to the temperature of our Sun? Assume the star and our Sun have the same radius.

107. • **Astronomy** A distant star radiates 1000 times more power than our own Sun, even though the temperature of the star is only 70% of the Sun's. If both stars have an emissivity

of 1, estimate the radius of the distant star. Recall, the radius of the Sun is 6.96×10^8 m. SSM

108. •• **Biology** The skin temperature of a nude person is 34.0°C, and the surroundings are at 20.0°C. The emissivity of skin is 0.900, and the surface area of the person is 1.50 m^2. (a) What is the rate at which energy radiates from the person? (b) What is the net energy loss from the body in 1 min by radiation?

109. • Calculate the heat through a glass window that is 30.0 cm × 150 cm in area and 1.20 mm thick. Assume the temperature on the inside of the window is 25.0°C while the outside temperature is 8.00°C. SSM Example 14-9

110. • What is the rate of heat transfer due to conduction through a single-pane glass window that is 3.0 m^2 and 3.175 mm thick, when the exterior temperature is 37°C and the interior temperature is 24°C? Example 14-9

111. • The surface area of the human body is about 1.8 m^2. If an average 37.0°C human is surfing in 10.0°C seawater in a 3.0-mm thick neoprene wetsuit with a thermal conductivity of 0.050 W/(m · K), how much heat does the surfer lose to the ocean in 1 hour? Example 14-9

112. • A 37°C chef needs to handle a 165°C cast iron skillet. With a cotton hot pad, the skillet can be handled safely, but if the hot pad is wet, the chef can get burned. If the chef's hot pad is 1.0 mm thick, what is the ratio of heat transfer through 1.0 cm^2 of the pad if it is dry versus if it is wet? Use the thermal conductivity of water for the wet hot pad. Example 14-9

General Problems

113. • What is the average kinetic energy for an ideal gas made up of diatomic molecules that can move in two dimensions when the temperature is 75.0°F? Example 14-3

114. ••• **Astronomy** (a) Calculate the escape speed for hydrogen atoms from the surface (the *photosphere*) of our Sun (see Section 10-3). The Sun has mass 1.99×10^{30} kg and radius 6.96×10^8 m. (b) If the root-mean-square speed of the hydrogen atoms were equal to the speed you found in part (a), what would be the temperature of the Sun's photosphere? The mass of a hydrogen atom is 1.68×10^{-27} kg. (c) Given that the actual temperature of the photosphere is 5800 K, is the Sun likely to lose its atomic hydrogen? Example 14-3

115. •• **Astronomy** Titan, a satellite of Saturn, has a nitrogen atmosphere with a surface temperature of −179°C and pressure of 1.5 atm. The mass of a nitrogen molecule is 4.7×10^{-26} kg, and we can model it as a sphere of diameter 2.4×10^{-10} m. The average temperature of Earth's atmosphere is about 10°C. (a) What is the density of particles in the atmosphere of Titan? (b) Which has a denser atmosphere, Titan or Earth? Justify your answer by calculating the ratio of the particle density on Titan to the particle density on Earth. (c) What is the average distance that a nitrogen molecule travels between collisions on Titan? How does this result compare with the distance for oxygen calculated in Section 14-3? Is your result reasonable? Example 14-3

116. • Derive an approximate formula for the area expansion (ΔA) that a sheet of material undergoes as the temperature changes from T_i to T_f. Assume the linear coefficient of expansion is α for the material. Example 14-4

117. • A 1.00-cm-diameter sphere of copper is placed concentrically over a 0.99-cm-diameter hole in a sheet of aluminum (**Figure 14-22**). Both the copper and the aluminum start at a temperature of 15.0°C. Describe one set of conditions (if any) in which the copper sphere will pass through the hole in the aluminum. Justify your conclusion with calculations. Example 14-4

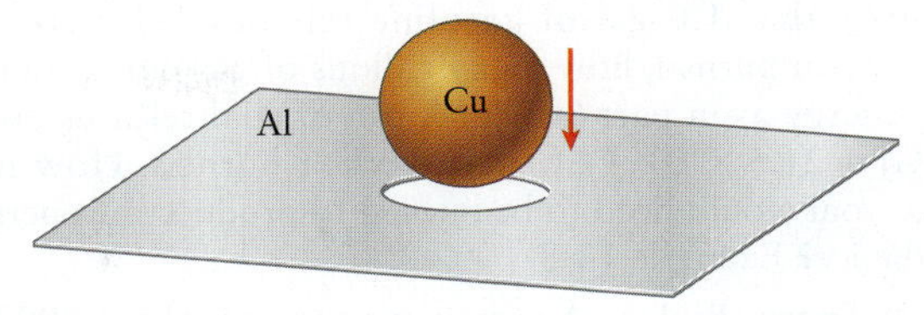

Figure 14-22 Problem 117

118. • A 20.0-m-long bar of steel expands due to a temperature increase. A 10.0-m-long bar of copper also gets longer due to the same temperature rise. If the two bars were originally separated by a gap of 1.5 cm, calculate the change in temperature and the distances that the steel and copper stretch if the gap is exactly "closed" by the expanding bars. Assume the steel and copper bars are fixed on the ends, as shown in **Figure 14-23**. Example 14-4

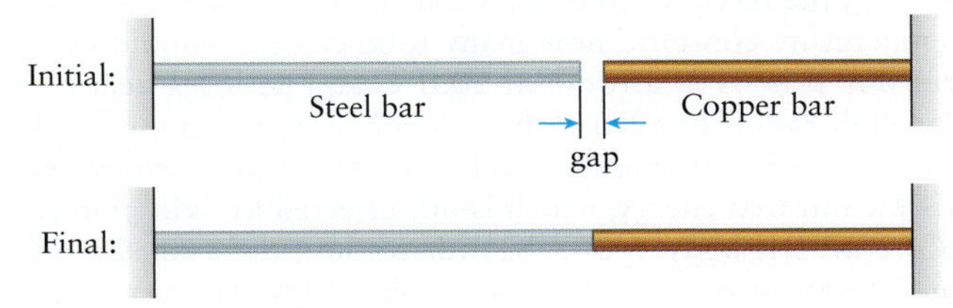

Figure 14-23 Problem 118

119. • Suppose a person who lives in a house next to a busy urban freeway attempts to "harness" the sound energy from the nonstop traffic to heat the water in his home. He places a transducer on his roof, "catches" the sound waves, and converts the sound waves into an electrical signal that warms a cistern of water. However, after running the system for 7 days, the 5 kg of water increases in temperature by only 0.01°C! Assuming 100% transfer efficiency, calculate the acoustic power "caught" by the transducer. SSM Example 14-5

120. • **Biology** Susan ordinarily eats 2000 kcal of food per day. If her mass is 60 kg and her height is 1.7 m, her surface area is probably close to 1.7 m^2. Although the body is only about two-thirds water, we will model it as being all water. (a) Typically 80% of the calories we consume are converted to heat. If Susan's body has no way of getting rid of the heat produced, by how many degrees Celsius would her body temperature rise in a day? (b) Would this be a noticeable increase? (c) How does her body prevent the increase from happening? (See Table 14.3 as needed.) Example 14-5

121. • Helium condenses at −268.93°C and has a latent heat of vaporization of 21,000 J/kg. If you start with 5.00 g of helium gas at 30.0°C, calculate the amount of heat required to change the sample to liquid helium. The specific heat of helium is 5193 J/(kg · K) at 300 K. SSM Example 14-7

122. •• **Biology** Jane's surface area is approximately 1.5 m^2. At what rate is heat released from her body when the temperature difference across the skin is 1.00°C? Assume the average thickness of the skin is 1.00 mm. Example 14-9

123. •• The Arctic perennial sea ice does not melt during the summer and thus lasts all year. NASA found that the perennial sea ice decreased by 14% between 2004 and 2005. The melted ice covered an area of 720,000 km^2 (the size of Texas!) and was 3.00 m thick on average. The ice is pure water (not

salt water) and is 92% as dense as liquid water. Assume that the ice was initially at $-10.0°C$ and see Tables 14-3 and 14-4 as needed. (a) How much heat was required to melt the ice? (b) Given that 1.0 gal of gasoline releases 1.3×10^8 J of energy when burned, how many gallons of gasoline contain as much energy as in part (a)? (c) A ton of coal releases around 21.5 GJ = 21.5×10^9 J of energy when burned. How many tons of coal would need to be burned to produce the energy to melt the ice? Example 14-8

124. •• **Sports, Biology** A person can generate about 300 W of power on a treadmill. If the treadmill is inclined at 3.00° and a 70.0-kg man runs at 3.00 m/s for 45.0 min, (a) calculate the percentage of the power output that goes into heating up his body and the percentage that keeps him moving on the treadmill. (b) How much water would that heat evaporate? Example 14-5

125. •• **Biology** A 1.88-m (6 ft 2 in.) man has a mass of 80.0 kg, a body surface area of 2.1 m^2, and a skin temperature of 30.0°C. Normally 80% of the food calories he consumes go to heat; the rest goes to mechanical energy. To keep his body's temperature constant, how many food calories should he eat per day if he is in a room at 20.0°C and he loses heat only through radiation? Does the answer seem reasonable? His emissivity e is 1 because his body radiates almost entirely non-visible infrared energy, which is not affected by skin pigment. (*Careful*! His body at 30.0°C radiates into the air at 20.0°C, but the air also radiates back into his body. The *net* rate of radiation is $P_{net} = P_{body} - P_{air}$.) SSM

126. • **Astronomy** Calculate the total power radiated by our Sun. Assume it is a perfect emitter of radiation ($e = 1$) with a radius of 6.96×10^8 m and a temperature of 5800 K.

127. ••• **Biology** You may have noticed that small mammals (such as mice) seem to be constantly eating, whereas some large mammals (such as lions) eat much less frequently. Let us investigate the phenomenon. For simplicity we can model an animal as a sphere. (a) Show that the heat energy stored by an animal is proportional to the cube of its radius, but the rate at which the animal radiates energy away is proportional to the square of its radius. (b) Show that the fraction of the animal's stored energy that it radiates away per second is inversely proportional to the animal's radius. (c) Use the result in part (b) to explain why small animals must eat much more per gram of body weight than very large animals.

128. •• **Astronomy** About 65 million years ago an asteroid struck Earth in the area of the Yucatán Peninsula and wiped out the dinosaurs and many other life forms. Judging from the size of the crater and the effects on Earth, the asteroid was about 10.0 km in diameter (assumed spherical) and probably had a density of 2.0 g/cm^3 (typical of asteroids). Its speed was at least 11 km/s. (a) What is the maximum amount of ocean water (originally at 20.0°C) that the asteroid could have evaporated if all of its kinetic energy were transferred to the water? Express your answer in kilograms and treat the ocean as though it were freshwater. (b) If the water were formed into a cube, how high would it be? (See Tables 14-3 and 14-4 as needed.) Example 14-8

129. •• A spherical container is constructed from steel and has a radius of 2.00 m at 15.0°C. The container sits in the Sun all day, and its temperature rises to 38.0°C. The container is initially filled completely with water, but it is not sealed. Describe what will happen to the water after the temperature increase. Example 14-4

James P. Blair/Getty Images

What do you think?

This African bombardier beetle (*Stenaptinus insignis*) defends itself by expelling a foul-smelling mixture of hot liquid and gas at attackers like ants and spiders. To initiate its defense mechanism, the beetle releases certain chemicals into a chamber in its abdomen, where they combine with enzymes secreted by the cells that line the chamber's walls. The resulting reaction releases enough energy to vaporize some of the liquid. The pressure inside the chamber increases rapidly, expelling the hot chemical mixture through an opening at the tip of the abdomen. As the gaseous part of the expelled mixture escapes from this opening, does its temperature (a) stay the same, (b) increase, or (c) decrease?

In this chapter, your goals are to:

- (15-1) Explain the general ideas of the laws of thermodynamics.
- (15-2) Define and be able to apply the first law of thermodynamics.
- (15-3) Describe the nature of isobaric, isothermal, adiabatic, and isochoric processes.
- (15-4) Explain why different amounts of heat are required to change the temperature of a gas depending on whether the gas is held at constant volume or constant pressure.
- (15-5) Define the second law of thermodynamics and its application to heat engines and refrigerators.
- (15-6) Explain the concept of entropy and the circumstances under which entropy changes.

To master this chapter, you should review:

- (6-2) The significance of positive and negative work.
- (14-3) The properties of ideal gases.
- (14-5) The relationship between the quantity of heat added to an object and the resulting temperature change.

15-1 The laws of thermodynamics involve energy and entropy

In Chapter 14 we learned about the zeroth law of thermodynamics, which basically says that the concept of temperature is a useful one. But what are the other laws of thermodynamics, and what do they tell us?

In this chapter we'll begin by learning about the *first law of thermodynamics*, which is a generalized statement about the conservation of energy. This law tells us about the interplay between the internal energy of an object or system, the work that the system does on its surroundings, and the heat that flows into or out of the system (**Figure 15-1a**). We'll use the first law to analyze a variety of *thermodynamic processes* in which the state of a system—as measured by its pressure, volume, and temperature—changes due to external influences.

According to the first law, some very remarkable things could happen: Room-temperature water could spontaneously lower its temperature and freeze at the same time that the room-temperature glass holding the water spontaneously raises its temperature and melts. The *second law of thermodynamics* describes why such remarkable

A cat demonstrates the first law of thermodynamics: The energy it takes in as food is either stored (as fat), converted into work, or released to its surroundings as heat.
(a)
Trevor Hirst/ iStockphoto

An aircraft engine demonstrates the second law of thermodynamics: The burning of fuel releases energy in the form of heat, but it is impossible to convert 100% of this heat into useful work.
(b)
Roger A. Freedman

Figure 15-1 The laws of thermodynamics The laws of thermodynamics are universal: They apply to (a) natural systems such as living organisms and (b) manufactured devices such as engines.

things are never observed in the real world. It also places a firm limit on the efficiency of *heat engines,* devices that convert heat into work: It tells us that no matter how carefully the engine is designed or built, only part of that heat can be converted to work. Most vehicles in our technological society use a heat engine, so this aspect of the second law is of tremendous importance (**Figure 15-1b**).

We'll finish the chapter with a discussion of *entropy*, a physical quantity that measures the amount of disorder in a system. We'll find that as a result of the second law of thermodynamics, entropy is not conserved: The total amount of entropy in the universe is increasing.

TAKE-HOME MESSAGE FOR Section 15-1

✔ The first law of thermodynamics relates changes in a system's internal energy to the work that the system does and the heat that flows into the system.

✔ The second law of thermodynamics describes what thermodynamic processes are possible. It says that heat cannot of itself flow from one object to a hotter object and that the universe tends toward ever-greater disorder.

15-2 The first law of thermodynamics relates heat flow, work done, and internal energy change

Suppose you take some unpopped kernels of popcorn at room temperature, put them in a pot with cooking oil, and put a lid on the pot. You then put the pot on the stove and warm it. In a few minutes the popcorn has popped and expanded so much that it has lifted the lid off the pot (**Figure 15-2**). By popping the popcorn you've changed its volume (each kernel has expanded), its temperature (which has increased), and its

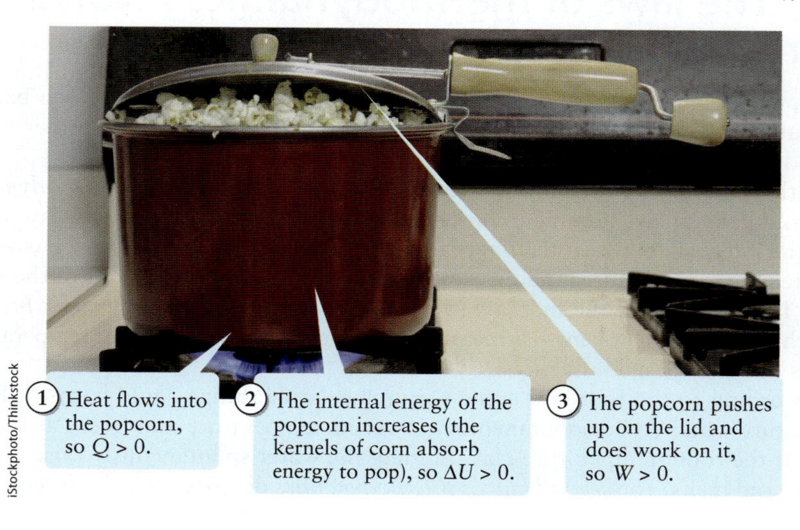

iStockphoto/Thinkstock
(1) Heat flows into the popcorn, so $Q > 0$.
(2) The internal energy of the popcorn increases (the kernels of corn absorb energy to pop), so $\Delta U > 0$.
(3) The popcorn pushes up on the lid and does work on it, so $W > 0$.

Figure 15-2 A thermodynamic process In a general thermodynamic process the internal energy of a system changes, heat enters or leaves the system, and work is done by or on the system.

internal energy, which is the sum of all of the kinetic and potential energies of the molecules that make up a system such as the popcorn (the chemical reaction involved in popping the kernels requires that each kernel absorb energy, so the internal energy has increased). Quantities such as volume, temperature, pressure, and internal energy are called **state variables** because they depend on the state or condition of the popcorn. Any process that changes the state of a system such as the popcorn is called a **thermodynamic process**.

Note that the values of the state variables of a system do *not* depend on the system's history—that is, the details of how the system got to that state. For example, you can pop popcorn on the stove as in Figure 15-2 or in a microwave oven, and in either case the results are the same. The state of a glass of water at room temperature is the same whether the water came out of the tap at room temperature, was heated in a pot and allowed to cool, or was originally a collection of ice cubes that gradually warmed and melted. In other words the state variables of a system in a given state do not depend on the particular thermodynamic processes that led to that state.

Figure 15-2 shows that in popping the popcorn, heat flows into the popcorn from the stove; the internal energy of the popcorn increases; and the popcorn does work on its surroundings as it lifts the lid of the pot. All three of these processes involve *energy* in one way or another, so let's look at each of them more closely. We'll then see how they're related.

We'll use the same convention as in Chapter 14 that heat Q is positive if it flows *into* a system and negative if it flows *out of* a system (see Section 14-5). Since heat flows into the popcorn in Figure 15-2 from the stove, $Q > 0$ for the popcorn.

We use the symbol "U" for the *internal* energy of the popcorn. Since the internal energy of the popcorn increases, its change is positive and $\Delta U > 0$. Generally speaking, if the temperature of a system increases in a thermodynamic process, the internal energy also increases and so $\Delta U > 0$. If the system temperature decreases, the internal energy generally decreases and so $\Delta U < 0$.

In thermodynamics we'll use the symbol "W" to represent the work done *by* a system on its surroundings. For the process shown in Figure 15-2, the popcorn does positive work on the lid: The popcorn exerts an upward force on the lid, and the lid's displacement is also upward. Hence $W > 0$ for the popcorn. In general $W > 0$ for a system if that system expands and pushes against its surroundings, as in Figure 15-2. If the system is compressed, like air being compressed in a bicycle pump, positive work is done *on* the system, so $W < 0$. Remember that if an object does negative work on a second object, the second object does positive work on the first one (see Section 6-2). You do positive work on the air in the bicycle pump to compress it, so the air does negative work on you. (Note that in most discussions of thermodynamics in chemistry textbooks, W is defined to be the work done *on* a system, which is the exact opposite of how it's defined in physics. Keep this in mind if you're taking physics and chemistry at the same time!) **Figure 15-3** shows the conventions that we use for the sign of Q and the sign of W.

For the popcorn shown in Figure 15-2, heat flows into the popcorn ($Q > 0$), the internal energy of the popcorn increases ($\Delta U > 0$), and the popcorn does work on its surroundings ($W > 0$). Careful measurement shows that the heat flow into the system is *exactly* equal to the sum of the change in internal energy plus the work done. In other words, the energy that enters the popcorn in the form of heat goes either into changing the popcorn's internal energy or into doing work. None of the energy is lost. We can write this statement as an equation:

$$Q = \Delta U + W \tag{15-1}$$

It's conventional to rearrange this equation with ΔU on one side and Q and W on the other side:

The first law of thermodynamics (15-2)

$$\Delta U = Q - W \tag{15-2}$$

During a thermodynamic process the **change in the internal energy of a system...**

...equals the **heat that flows into the system** during the process...

...minus the **work that the system does** during the process.

WATCH OUT! Internal energy isn't the same as potential energy.

When we studied mechanical energy, we used the symbol "U" to represent the *potential energy* associated with a system. So it can be confusing that the same symbol "U" is used in thermodynamics to represent the *internal energy* of a system because these two kinds of energy are *not* the same. In thermodynamics the value of U includes the kinetic energy of its molecules and the potential energy of their interactions with each other, but U does *not* include the gravitational potential energy of the system as a whole. For example, if you carry the popcorn in Figure 15-2 to your neighbor's apartment above you, the gravitational potential energy associated with the popcorn increases (the popcorn is at a greater height) but the popcorn's internal energy stays the same. In this chapter and throughout your study of thermodynamics, keep the new definition of the quantity U in mind.

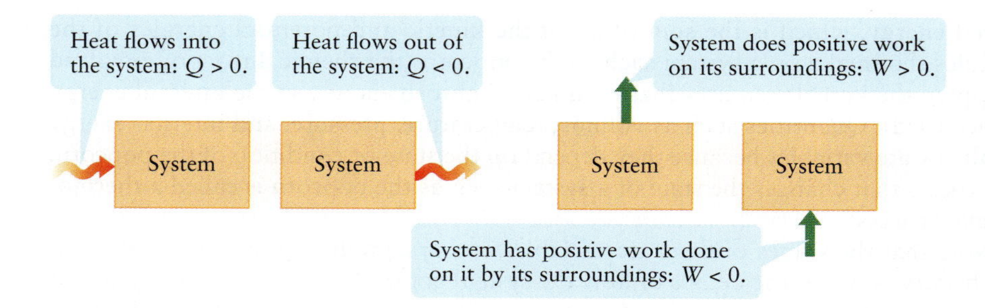
Heat flows into the system: $Q > 0$.
Heat flows out of the system: $Q < 0$.
System does positive work on its surroundings: $W > 0$.
System
System
System
System
System has positive work done on it by its surroundings: $W < 0$.

Figure 15-3 The signs of work and heat A quantity of heat Q is considered positive if it enters a system, negative if it leaves. The sign of the work W for a system depends on which object does positive work.

Equation 15-2 is a mathematical statement of the **first law of thermodynamics.** It's a generalization of the law of conservation of energy. It says that the internal energy of a system can change *only* if the system gains energy from its surroundings (if heat flows in from the surroundings, or its surroundings do positive work on the system) or if the system loses energy to its surroundings (if heat flows out to the surroundings, or the system does positive work on its surroundings). No exception to this rule has ever been found.

BIO-Medical For the system shown in Figure 15-2, the quantity of heat that flows into the popcorn is greater than the amount of work that the popcorn does to lift the lid. So while Q and W are both positive, Q is greater than W and $\Delta U = Q - W$ is positive. In other situations the quantities ΔU, Q, and W have different signs. For example, think of your body as a thermodynamic system that you are exercising on a stationary bicycle in the gym. You do work on the pedals of the bicycle ($W > 0$), and heat flows out of your body ($Q < 0$). Since Q is negative and W is positive, $\Delta U = Q - W$ is negative and you lose internal energy. That's the thermodynamic explanation of why exercise helps you lose weight: The internal energy you lose is extracted from energy stored in your body, part of which is in the form of fat (see Example 15-1).

Internal Energy Change and Thermodynamic Paths

As we mentioned above, the internal energy U is a state function that depends only on the current state of the system, not on how it got into that state. To see what this implies let's consider two *different* thermodynamic processes that take a system from the *same* initial state to the same final state. To be specific, let's consider two ways to change the state of a system made up of a quantity of gas.

In **Figure 15-4a** a cylinder with a moveable piston encloses a quantity of gas. The gas has an initial pressure, volume, and temperature. We use a candle to let heat flow slowly into the gas (so $Q > 0$) and allow the piston to move so that the gas expands. The expanding gas does positive work on the piston (the gas pushes the piston to the right as the piston moves to the right), so $W > 0$. If we regulate how fast the piston moves, we can arrange it so that the rate at which the gas does work on the piston is exactly the same as the rate at which heat flows into the gas. Then $Q = W$, and from the first law of thermodynamics (Equation 15-2), the internal energy of the gas will remain unchanged: $\Delta U = Q - W = 0$. For an ideal gas the average energy per molecule is proportional to the Kelvin temperature T of the gas (see Section 14-3), so the internal energy U—the average energy per molecule, multiplied by the number of molecules—is also proportional to T. Since U remains constant for the process shown in Figure 15-4a, the temperature of the gas also remains constant.

Now imagine a second thermodynamic process that starts with the gas in the same initial state. In **Figure 15-4b** we use the same quantity of gas as in Figure 15-4a, and the gas starts at the same pressure, volume, and temperature. The difference is that now the cylinder is equipped with a thin barrier of lightweight material instead of a piston, and there is a vacuum between the barrier and the other end of the cylinder. If we puncture the barrier, the gas expands freely and fills the entire cylinder. Since the gas doesn't push against anything as it expands (it expands into a vacuum), the gas does no work so $W = 0$. Furthermore, if we insulate the walls of the cylinder so that no heat can flow into or out of the gas, $Q = 0$ for this process. So the first law of thermodynamics tells us that in this process as well the internal energy of the gas will not change: $\Delta U = Q - W = 0 - 0 = 0$. And since U is proportional to Kelvin temperature T for an ideal gas, it follows that the temperature of the gas remains constant for this process. In other words *both* the processes shown in Figure 15-4 take the same quantity of gas from the

(a)
When energy is added slowly to gas trapped in a cylinder with a moveable piston...
Gas
Air
...the gas expands, pushing the piston back.
Gas
The gas has done work on the piston.

(b)
A thin barrier of lightweight material separates a gas from a vacuum.

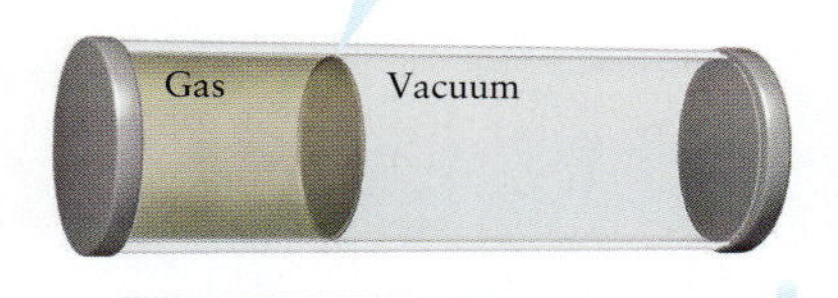
Gas
Vacuum

When the barrier ruptures, the gas expands to fill the entire cylinder. No energy was added to the gas in this case, but the final pressure and volume are the same as in part (a).

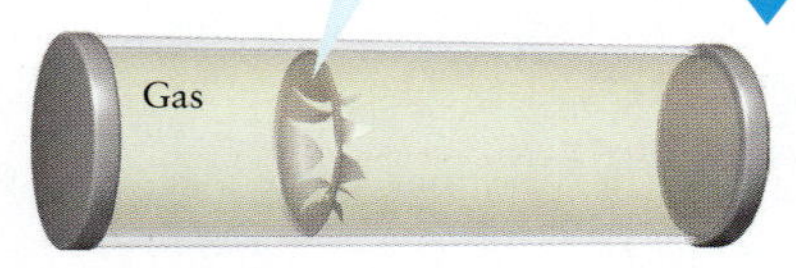
Gas

Figure 15-4 Two thermodynamic paths (a) If heat is added to an expanding gas at just the right rate, the gas temperature remains constant. (b) The gas temperature is also unchanged in a free expansion. So both of these processes have the same initial and final thermodynamic states but follow different paths between these states.

same initial volume to the same final volume while leaving the temperature unchanged. (The ideal gas law $pV = nRT$, Equation 14-7, tells us that the final pressure p will also be the same. That's because the final volume V, final temperature T, and number of moles n are the same in both processes.)

The two very different processes shown in Figure 15-4 take the system from the same initial state to the same final state but follow different thermodynamic *paths*. The first path required heat to be added to the system and the system to do positive work on its surroundings. The second path required neither of these things. This is why we *cannot* say a particular state of a system *contains* work or heat. Rather, the amount of heat or work required depends on the path the system takes. For any thermodynamic process that takes a system between the same two states, the change in internal energy is the *same*, independent of the path.

BIO-Medical EXAMPLE 15-1 Cycling It Off

At the end of a 15-min session on a stationary bicycle at the gym, the bike's display indicates that you have delivered 165 kJ of energy to the bike in the form of work. Your heart rate monitor shows that you have "burned off" 165 food calories of energy. How much energy has your body lost to your surroundings in the form of heat?

Set Up

In this problem your body is the thermodynamic system. You do work on your surroundings (the bicycle), so W is positive and equal to 165 kJ. The 165 food calories (165 kcal) that you have "burned off" represent a decrease in your body's internal energy, so ΔU is negative and equal to -165 kcal. We'll use the first law of thermodynamics to determine the heat Q that flows out of your body.

First law of thermodynamics:

$$\Delta U = Q - W \qquad (15\text{-}2)$$

Your body does work on the bicycle: $W > 0$.
system = your body
Heat flows out of your body: $Q < 0$.

Solve

First convert ΔU from kilocalories to kilojoules.

1 kcal = 4186 J = 4.186 kJ, so

$$\Delta U = -165 \text{ kcal}\left(\frac{4.186 \text{ kJ}}{1 \text{ kcal}}\right) = -691 \text{ kJ}$$

Rearrange Equation 15-2 to solve for the quantity of heat Q that flows *into* your body. A negative value of Q means that heat flows *out of* your body.

From Equation 15-2,

$$\begin{aligned} Q &= \Delta U + W \\ &= -691 \text{ kJ} + 165 \text{ kJ} \\ &= -526 \text{ kJ} \end{aligned}$$

This is negative, so 526 kJ of energy flows out of your body in the form of heat as you exercise.

Reflect

During this exercise session your body expended 691 kJ of internal energy, of which 165 kJ went into doing work and 526 kJ was lost in the form of heat. Only (165 kJ)/(691 kJ) = 0.239, or 23.9%, of the energy that you expended went into doing work, so your body has an *efficiency* of 23.9% in this situation.

Note that 3500 kcal is the energy content of one pound (0.45 kg) of fat. If all of the 165 kcal that you expended came from stored fat, the net result of your exercise is that you will have lost an amount of fat equal to (165 kcal)/(3500 kcal/lb) = 0.047 lb (0.021 kg, or 0.75 ounce). Losing weight requires a *lot* of exercise!

GOT THE CONCEPT? 15-1 An Expanding Ideal Gas

? An ideal gas is enclosed in a thermally isolated cylinder so that no heat can flow into or out of the gas. One end of the cylinder is sealed by a moveable piston like that shown in Figure 15-4a. If you allow the gas to expand slowly by pulling back the piston, what will happen to the temperature T of the gas? (a) T increases; (b) T decreases; (c) T stays the same; (d) T increases to begin with, then decreases; (e) T decreases to begin with, then increases.

TAKE-HOME MESSAGE FOR Section 15-2

✔ The internal energy of a system (the sum of the kinetic energy and potential energy of every atom and molecule in the system) describes the state of the system in a way that does not depend on the specific thermodynamic processes that led to it.

✔ By itself, heat flow into a system increases the internal energy of a system.

✔ Energy leaves a system when that system does positive work.

✔ The first law of thermodynamics relates the change in internal energy of a thermodynamic system to the heat that enters the system and the work done by that system.

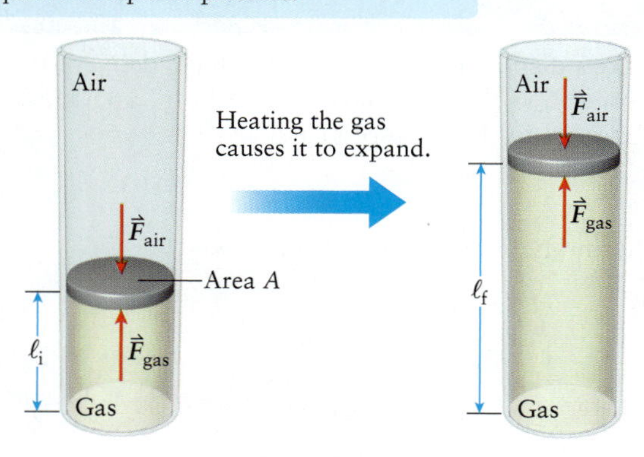
The piston reaches its equilibrium position when the pressure of the gas in the cylinder equals atmospheric pressure.
Air
Heating the gas causes it to expand.
$\vec{F}_{air}$
Area A
ℓ_i
$\vec{F}_{gas}$
Gas
Air
$\vec{F}_{air}$
$\vec{F}_{gas}$
ℓ_f
Gas
Even after the gas expands, the gas pressure must equal atmospheric pressure. Because neither gas pressure nor atmospheric pressure has changed, the expansion is isobaric.

Figure 15-5 An isobaric expansion The piston exerts a constant pressure on the gas as it is heated, so the pressure of the gas itself remains constant.

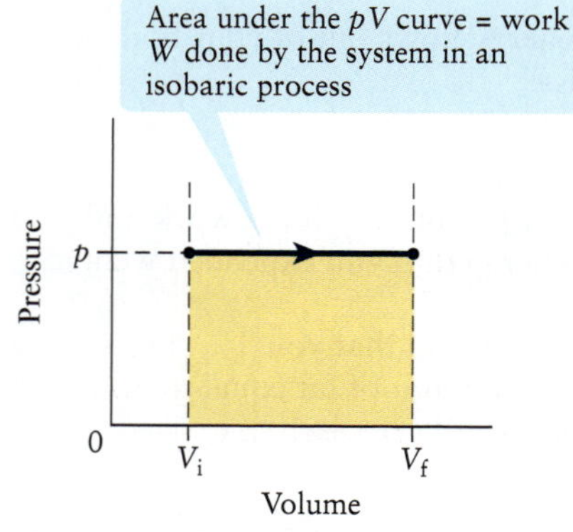
Area under the pV curve = work W done by the system in an isobaric process
Pressure
p
0
V_i
V_f
Volume

Figure 15-6 Work done in an isobaric expansion On a pV diagram (a graph of pressure versus volume) for a thermodynamic process, the area under the curve represents the work done by the system.

15-3 A graph of pressure versus volume helps to describe what happens in a thermodynamic process

A thermodynamic process involves a change in the state of a system. In most cases this involves a change in the pressure of the system, the volume that the system occupies, or both. That's the case for the popping popcorn in Figure 15-2 and the gas that expands against a piston in Figure 15-4a. In this section we'll use the first law of thermodynamics to help us analyze four special but important kinds of thermodynamic processes:

- **Isobaric process:** The pressure of the system remains constant.
- **Isothermal process:** The temperature of the system remains constant.
- **Adiabatic process:** There are no heat transfers into or out of the system.
- **Isochoric process:** The volume of the system remains constant.

Many thermodynamic processes do not fit into any of these simple categories. But by analyzing these four special cases, we'll learn general rules that can be applied to *any* thermodynamic process.

Isobaric Processes

Isobaric processes occur with no change in pressure. An example is when you cook food in a frying pan without a lid: Heat flows into the food and its temperature increases, but the pressure that the atmosphere exerts on the food remains constant.

Let's consider an isobaric process in which work is done. Suppose our system is a quantity of gas inside a cylinder that's sealed with a moveable piston of negligible mass, as in **Figure 15-5**. At equilibrium the upward force on the piston due to the pressure of the gas equals the downward force due to atmospheric pressure. These forces also balance if we heat the gas gradually and allow it to expand slowly so that the piston moves up at constant velocity: The net force on the piston remains zero. Since atmospheric pressure doesn't change, the pressure in the gas must remain constant as well. The process shown in Figure 15-5 is therefore isobaric: Although the volume of the gas changes, its pressure remains constant.

To make it easier to visualize and understand what happens in this process, it's useful to make a pV **diagram**. This is a graph that plots the pressure of a system on the vertical axis versus the volume of the system on the horizontal axis. On a pV diagram an isobaric process is represented by a horizontal line (**Figure 15-6**).

In Figure 15-5 the expanding gas does work W on the piston to lift it. To calculate how much work is done, let the constant pressure of the gas be p and the surface area of the piston be A. Figure 15-5 shows that the original length of the cylinder containing the gas is ℓ_i, and the final length of the cylinder after the gas has expanded is ℓ_f. Recall that pressure is force per unit area, so force is pressure multiplied by area. This means that the force that the gas exerts on the piston has magnitude $F_{gas} = pA$. This force acts upward, the same direction that the piston moves, and the piston moves a distance $\ell_f - \ell_i$. So the total work done by the gas as it expands is

$$W = F(\ell_f - \ell_i) = pA(\ell_f - \ell_i) = p(A\ell_f - A\ell_i) \quad (15\text{-}3)$$

The volume of a cylinder is equal to the area of its base multiplied by its height. So $A\ell_i$ and $A\ell_f$ are, respectively, the initial volume V_i and final volume V_f occupied by the gas, and we can rewrite Equation 15-3 as

Go to Picture It 15-1 for more practice dealing with pV diagrams.

Work that a system does in an isobaric process

Volume of the system at the end of the process

$$W = p(V_f - V_i)$$

Constant pressure of the system

Volume of the system at the beginning of the process

Work done in an isobaric (constant-pressure) process (15-4)

The work done by the gas in this isobaric process is the pressure p multiplied by the change in volume $(V_f - V_i)$. Figure 15-6 shows that this is the *area* of the shaded region under the straight line that represents the process on the pV diagram. This region is a rectangle with height p and length $V_f - V_i$, so its area equals its height multiplied by its length or $p(V_f - V_i)$.

Equation 15-4 is valid for an isobaric process whether the volume increases or decreases. If $V_f > V_i$ so that the system expands like the gas in Figure 15-5, then $V_f - V_i$ is positive and the system does positive work on its surroundings. Such a process is called an *isobaric expansion*. If instead $V_f < V_i$, which would be like the process in Figure 15-5 in reverse, then $V_f - V_i$ is negative and the system does *negative* work on its surroundings—that is, the surroundings (in Figure 15-5, the piston) do positive work on the system. This is called an *isobaric compression*.

Equation 15-4 is valid *only* for the case of constant pressure. But it turns out to be true in general that in *any* process the work done is the area under the curve that represents the process on a pV diagram. We'll use this idea later in this section.

EXAMPLE 15-2 Boiling Water Isobarically

Boiling 1.00 g (1.00 cm^3) of water at 1.00 atm results in 1671 cm^3 of water vapor. A cylinder like that shown in Figure 15-5 contains 1.00 g of water at 100°C. The pressure on the outside of the piston is 1.00 atm. Determine the change in internal energy of the water when just enough heat is added to the water to boil all of the liquid.

Set Up

In this process heat flows into the water to change its phase from liquid to vapor, and the vapor does work on the piston as it expands. The pressure on the outside of the piston is constant, so the pressure that the piston exerts on the water (liquid plus vapor) is also constant. So this is an isobaric process. We'll apply the first law of thermodynamics to determine the change in internal energy.

First law of thermodynamics:

$$\Delta U = Q - W \quad (15\text{-}2)$$

Work done in an isobaric (constant-pressure) process:

$$W = p(V_f - V_i) \quad (15\text{-}4)$$

Heat required for a phase change:

$$Q = \pm mL \quad (14\text{-}21)$$

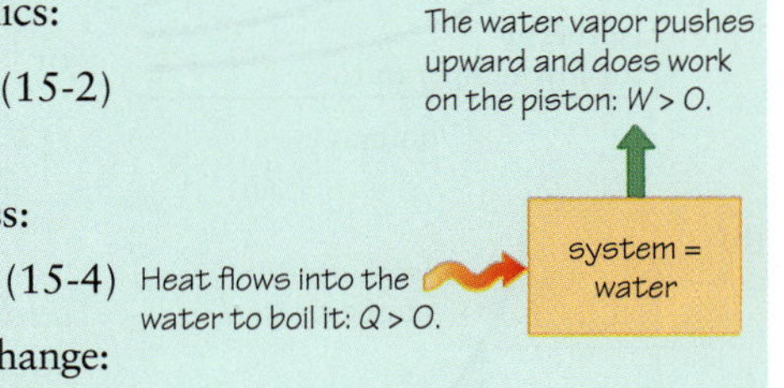
The water vapor pushes upward and does work on the piston: W > 0.
system = water
Heat flows into the water to boil it: Q > 0.

Solve

Use Equation 14-21 to find the heat Q that flows into the water to vaporize it.

Heat flows into the water, so Q in Equation 14-21 is positive. The mass of water is $m = 1.00$ g, and the latent heat of vaporization is $L_V = 2260 \times 10^3$ J/kg $= 2.26 \times 10^6$ J/kg (from Table 14-4). So

$$Q = mL_V$$
$$= (1.00 \text{ g})\left(\frac{1 \text{ kg}}{10^3 \text{ g}}\right)(2.26 \times 10^6 \text{ J/kg})$$
$$= 2.26 \times 10^3 \text{ J}$$

Use Equation 15-4 to find the work W that the expanding water vapor does on the piston.

The pressure of the water vapor is $p = 1.00$ atm $= 1.01 \times 10^5$ Pa $= 1.01 \times 10^5$ N/m^2 and remains constant in this isobaric expansion. The volume occupied by the water increases from $V_i = 1.00$ cm^3 in the liquid state to $V_f = 1671$ cm^3 in the vapor state. In calculating the work done by the expanding vapor, we must convert cm^3 to m^3 using the relationship $1 \text{ m} = 10^2 \text{ cm}$:

$$W = p(V_f - V_i)$$
$$= (1.01 \times 10^5 \text{ N/m}^2)(1671 \text{ cm}^3 - 1.00 \text{ cm}^3)\left(\frac{1 \text{ m}}{10^2 \text{ cm}}\right)^3$$
$$= (1.01 \times 10^5 \text{ N/m}^2)(1670 \times 10^{-6} \text{ m}^3)$$
$$= 169 \text{ N} \cdot \text{m} = 169 \text{ J}$$

Substitute the values for Q and W into Equation 15-2, the first law of thermodynamics. This tells us the change in internal energy of the water in this process.

$$\Delta U = Q - W$$
$$= 2.26 \times 10^3 \text{ J} - 169 \text{ J}$$
$$= 2.09 \times 10^3 \text{ J}$$

Reflect

We have to add 2.26×10^3 J of energy to boil the water. Of this energy, 2.09×10^3 J goes into increasing the internal energy of the system. The rest of the energy (169 J) leaves the system in the form of work done on the piston.

Isothermal Processes

In an isothermal process temperature remains constant. Although the process in Example 15-2 is isobaric (constant pressure), it is also isothermal: When a substance freezes or boils, the phase transition occurs with no change in temperature.

Gases can also expand or be compressed in such a way as to maintain a constant temperature, even though no phase transition occurs. As we discussed in Section 15-2, the internal energy U of an ideal gas is directly proportional to the Kelvin temperature of the gas. So if the temperature of an ideal gas remains constant during a thermodynamic process, the internal energy U remains constant as well and there is zero change in the internal energy: $\Delta U = 0$. The first law of thermodynamics, Equation 15-2, tells us that $\Delta U = Q - W$, so for an isothermal process that involves an ideal gas, we have

$$0 = Q - W$$

or

$$Q = W \quad (15\text{-}5)$$

(isothermal process for an ideal gas)

So *when an ideal gas undergoes an isothermal process, the quantity of heat Q added to the gas is equal to the work W done by the gas*. Put another way, whatever energy is added to the gas in the form of heat appears as work done by the gas. If heat flows into the gas so that $Q > 0$, it must be that $W > 0$: The gas does positive work on its surroundings, which means it must expand. This is called an *isothermal expansion*. If heat flows out of the gas so that $Q < 0$, then $W < 0$ and the gas does negative work on its surroundings. This means that the surroundings do positive work on the gas, so the volume of the gas decreases. This is called an *isothermal compression*.

How much work is done by a gas that undergoes an isothermal expansion or compression? To get insight into the answer, let's use a pV diagram to depict such a process. The ideal gas law (which we studied in Section 14-3) says that the pressure p, volume V, and temperature T of n moles of an ideal gas are related by

$$pV = nRT \quad (14\text{-}7)$$

In an isothermal process T is constant, so the product pV is likewise constant. If the gas expands and the volume V increases, the pressure p must decrease; if the gas compresses and the volume V decreases, the pressure p must increase. **Figure 15-7a** shows three **isotherms**, or curves of constant temperature, on a graph of pressure versus volume.

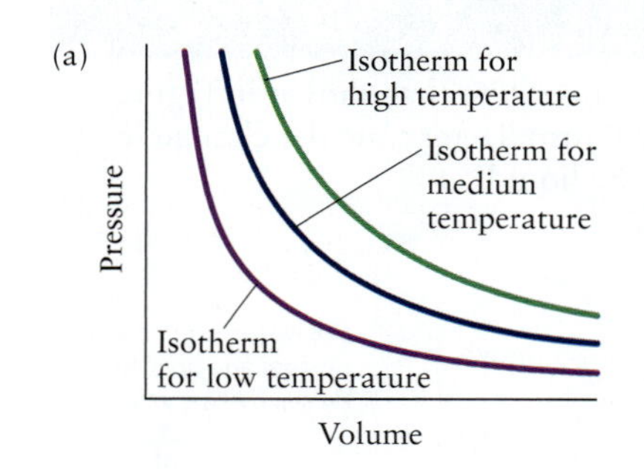

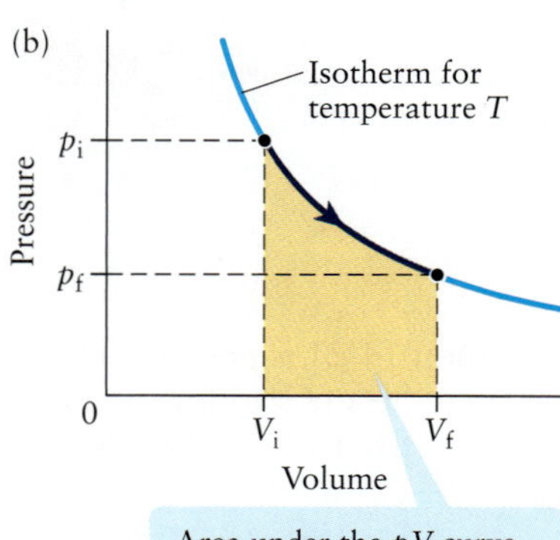

Figure 15-7 Isothermal processes in an ideal gas (a) This pV diagram shows isotherms (curves of constant temperature) for an ideal gas. (b) The work done by an ideal gas in an isothermal expansion.

On a pV diagram, the area under the curve represents the work W done by the gas; **Figure 15-7b** shows this area for an isothermal expansion. Unlike the isobaric case shown in Figure 15-6, the area under the curve is *not* a rectangle, so we cannot find W simply by multiplying the pressure and the change in volume. Using calculus it can be shown that the work that the gas does as the volume changes is

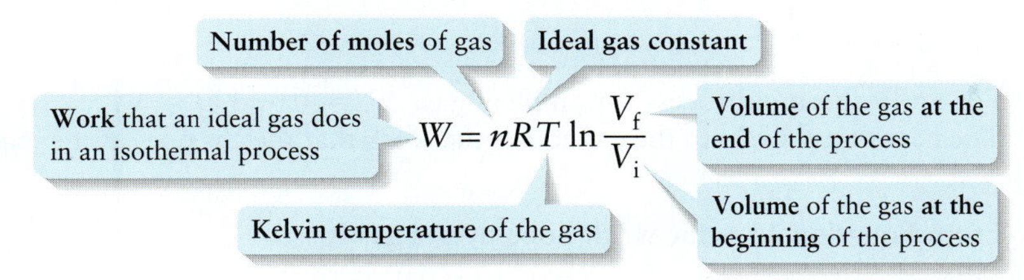

Work done by an ideal gas in an isothermal (constant-temperature) process (15-6)

In Equation 15-6 "ln" is the natural logarithm function. It has the properties that $\ln x$ is positive if x is greater than 1 and $\ln x$ is negative if x is less than 1; if $x = 1$, then $\ln x = \ln 1 = 0$. So Equation 15-6 says that the gas does positive work ($W > 0$) if the gas expands, so V_f is greater than V_i and V_f/V_i is greater than 1. The gas does negative work (that is, positive work is done on the gas by its surroundings) if the gas is compressed, so V_f is less than V_i and V_f/V_i is less than 1. An alternative version of Equation 15-6 in terms of the number of molecules N present in the gas is

$$W = NkT \ln \frac{V_f}{V_i} \qquad (15\text{-}7)$$

$\sqrt{x}$ *See the Math Tutorial for more information on logarithms.*

Go to Interactive Exercise 15-1 for more practice dealing with internal energy.

EXAMPLE 15-3 An Isothermal Expansion

A cylinder sealed with a moveable piston as in Figure 15-5 contains 0.100 mol of an ideal gas at a temperature of 295 K. Heat is transferred slowly into the gas, and it is allowed to expand isothermally from an initial volume of $1.00 \times 10^{-3}\ \text{m}^3$ to $3.00 \times 10^{-3}\ \text{m}^3$ (that is, from 1.00 to 3.00 L). (a) What are the initial and final pressures of the gas? (b) How much heat is transferred into the gas during this process?

Set Up

Since this is an ideal gas, we can use the ideal gas equation to determine the pressure p from the volume V, number of moles n, and temperature T. The internal energy of an ideal gas does not change if the temperature remains constant, so $\Delta U = 0$ and Q (the quantity of heat that flows into the gas, which is what we're trying to find) equals the work W done by the gas. We'll find W and hence Q with Equation 15-6.

Ideal gas law:

$$pV = nRT \qquad (14\text{-}7)$$

Isothermal process for an ideal gas:

$$Q = W \qquad (15\text{-}5)$$

Work done by an ideal gas in an isothermal (constant-temperature) process:

$$W = nRT \ln \frac{V_f}{V_i} \qquad (15\text{-}6)$$

The gas pushes upward and does work on the piston: $W > 0$.

system = gas

Heat flows into the gas to make it expand: $Q > 0$.

Solve

(a) Use Equation 14-7 to find the initial and final pressures.

Rewrite the ideal gas law $pV = nRT$ to solve for the pressure:

$$p = \frac{nRT}{V}$$

The pressure p_i when the gas occupies the initial volume $V_i = 1.00 \times 10^{-3}\ \text{m}^3$ is

$$p_i = \frac{nRT}{V_i} = \frac{(0.100\ \text{mol})[8.314\ \text{J/(mol}\cdot\text{K)}](295\ \text{K})}{1.00 \times 10^{-3}\ \text{m}^3}$$

$$= 2.45 \times 10^5\ \frac{\text{J}}{\text{m}^3}\left(\frac{1\ \text{N}\cdot\text{m}}{1\ \text{J}}\right)$$

$$= 2.45 \times 10^5\ \text{N/m}^2 = 2.45 \times 10^5\ \text{Pa}$$

(We used the relationship that 1 J = 1 N · m.) The pressure when the gas occupies the final volume $V_f = 3.00 \times 10^{-3}$ m^3 is

$$p_f = \frac{nRT}{V_f} = \frac{(0.100 \text{ mol})[8.314 \text{ J/(mol} \cdot \text{K)}](295 \text{ K})}{3.00 \times 10^{-3} \text{ m}^3}$$

$$= 8.18 \times 10^4 \text{ Pa}$$

(b) From the first law of thermodynamics, the quantity of heat that enters the gas equals the work done by the gas.

For an isothermal process for an ideal gas, $\Delta U = 0$ and $Q = W$. The heat that flows into the gas equals the work that the gas does as it expands.

Use Equation 15-6 to calculate the work done by the gas and so the quantity of heat that is added to the gas.

From Equation 15-6 the work done by the gas is

$$W = nRT \ln \frac{V_f}{V_i}$$

$$= (0.100 \text{ mol})[8.314 \text{ J/(mol} \cdot \text{K)}](295 \text{ K}) \times \ln\left(\frac{3.00 \times 10^{-3} \text{ m}^3}{1.00 \times 10^{-3} \text{ m}^3}\right)$$

$$= (245 \text{ J}) \ln 3.00 = (245 \text{ J})(1.10)$$

$$= 269 \text{ J}$$

Since $Q = W$, the quantity of heat added to the gas is $Q = 269$ J.

Reflect

For an ideal gas that expands isothermally, all of the heat that is transferred into the gas (in this case 269 J) is converted to work done by the gas on its surroundings.

Note that to keep the temperature constant the gas pressure must decrease during the expansion. To make this happen it's not enough to add heat slowly to the gas: The pressure on the other side of the piston must also decrease as the volume increases to match the desired pressure of the gas (so that the piston moves slowly and doesn't accelerate). This tells us that an isothermal expansion or compression of an ideal gas requires rather special circumstances.

Adiabatic Processes

In an adiabatic process there is no heat transfer into or out of a system. An important system that undergoes adiabatic processes is our atmosphere. Air is a very poor conductor of heat (see Section 14-7), so there is little heat flow into or out of a mass of air as its pressure and volume change. This helps explain why many mountain ranges have a relatively wet climate on one side and a dry climate on the other (**Figure 15-8**).

When winds carry air up and over mountains, the lower pressure at higher altitudes causes the air mass to expand and do work on the surrounding air ($W > 0$). This expansion occurs nearly adiabatically thanks to the low thermal conductivity of air, so Q is effectively zero. From the first law of thermodynamics, Equation 15-2, the change in the internal energy of the air is

$$\Delta U = Q - W = 0 - W \quad \text{or} \quad \Delta U = -W \text{(adiabatic process)} \qquad (15\text{-}8)$$

Figure 15-8 The adiabatic winds of Moloka'i The prevailing winds push air up and over the mountains that form the backbone of the Hawaiian island of Moloka'i. The temperature of the uplifted air drops and airborne moisture falls as rain, creating the lush rainforest visible as bright green in this satellite photograph. The air that descends on the southwestern side of the mountains is dry, resulting in an arid plain on the southern and western parts of the island.

Since $W > 0$, it follows that $\Delta U < 0$: The internal energy U of the air decreases as it is pushed up the flanks of the mountains. For an ideal gas U is proportional to the Kelvin temperature of the gas, so the air temperature drops by as much as 10°C for every 1 km of elevation gained. As the temperature decreases, water vapor in the air condenses into clouds and falls as rain, which helps to promote lush vegetation on the windward side of the mountains. By the time the air reaches the other side of the mountains, it has been depleted of moisture, so little rain falls on that side.

The air we have described expands as it is pushed up the slopes of the mountains, so this is an *adiabatic expansion*. In an *adiabatic compression* the volume of the gas decreases, so $W < 0$ (the surroundings do work on the gas) and $\Delta U > 0$ (the internal energy and temperature of the gas increase).

WATCH OUT! Remember that heat and temperature are not the same thing.

Even though no heat flows into or out of air pushed up a range of mountains, the temperature of the air changes. If this statement seems contradictory, it's probably because you're still thinking of *heat* as meaning approximately the same thing as *temperature*. Remember that these two concepts are actually quite different: Heat is energy that flows into or out of an object due to a temperature difference, while temperature is a measure of the kinetic energy of an object's molecules. The temperature can change even if there is *no* heat flow, provided work is done by (or done on) the object.

Isochoric Processes

An isochoric process is one for which the volume remains constant. As an example, consider a quantity of gas sealed inside a rigid container. If heat flows into the gas, the pressure and temperature of the gas will increase but the volume of the gas will remain the same because the container cannot expand.

The pressure of the gas exerts forces on the walls of its container, but since the walls do not move (there is no moveable piston), there is zero displacement. Work is force multiplied by displacement, so the gas does *zero* work in an isochoric process: $W = 0$. From the first law of thermodynamics, Equation 15-2,

$$\Delta U = Q - W = Q - 0 \quad \text{or} \quad \Delta U = Q \text{(isochoric process)} \qquad \textbf{(15-9)}$$

This says that for an isochoric process any heat flow into or out of a system goes entirely into changing the internal energy of the system. If Q is positive so that heat flows into the system, the process is called *isochoric heating*; if Q is negative and heat flows out of the system, the process is called *isochoric cooling*.

EXAMPLE 15-4 Two Thermodynamic Processes: Isochoric and Isobaric

The pV diagram in **Figure 15-9** shows two thermodynamic processes that occur in an ideal diatomic gas, for which the internal energy is $U = (5/2)nRT$. The gas is in a cylinder with a moveable piston (see Figure 15-5) and is initially in the state labeled A on the diagram, at pressure $p_i = 3.03 \times 10^5$ Pa and volume $V_i = 1.20 \times 10^{-3}$ m^3. The piston is first locked in place so that the volume of the gas cannot change, and the cylinder is cooled so that the temperature and pressure of the gas both decrease. When the gas is in state B, the pressure of the gas is $p_f = 1.01 \times 10^5$ Pa $= 1.00$ atm, the same as the air pressure outside the cylinder. The piston is then unlocked so that it is free to move, and the cylinder is slowly heated so that the gas expands at constant pressure. The temperature of the gas increases until in the final state (C in Figure 15-9), when the gas is at the same temperature as in the initial state A. (a) What is the final volume of the gas? (b) Find the values of W, ΔU, and Q for the isochoric process $A \rightarrow B$. (c) Find the values of W, ΔU, and Q for the isobaric process $B \rightarrow C$. (d) Find the values of W, ΔU, and Q for the net process $A \rightarrow B \rightarrow C$.

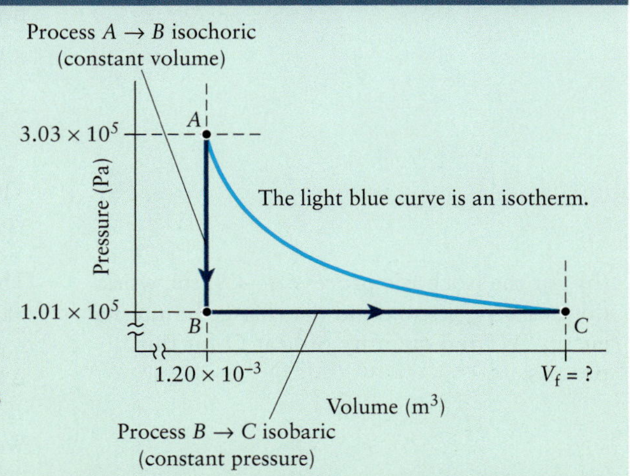

Figure 15-9 Isochoric cooling and isobaric expansion In this two-step process for an ideal gas, how much work is done, how much heat flows into the gas, and by how much does the internal energy change?

Set Up

We'll use the ideal gas law to determine the final volume of the gas. For the isochoric (constant-volume) process $A \rightarrow B$, the gas does no work, so the quantity of heat Q that enters the gas is equal to the internal energy change ΔU of the gas. (The gas is *cooled* in this process, so $Q < 0$ and $\Delta U < 0$.) We'll determine ΔU from the temperature change. For the isobaric (constant-pressure) process $B \rightarrow C$, the work W that the gas does is given by Equation 15-4; this is positive because the gas expands. The internal energy change ΔU is again given by the temperature change. We'll then determine Q for this process from the first law of thermodynamics.

Ideal diatomic gas:

$$U = \frac{5}{2}nRT$$

Ideal gas law:

$$pV = nRT \qquad (14\text{-}7)$$

First law of thermodynamics:

$$\Delta U = Q - W \qquad (15\text{-}2)$$

Work done in an isobaric (constant-pressure) process:

$$W = p(V_f - V_i) \qquad (15\text{-}4)$$

Isochoric (constant-volume) process:

$$\Delta U = Q \qquad (15\text{-}9)$$

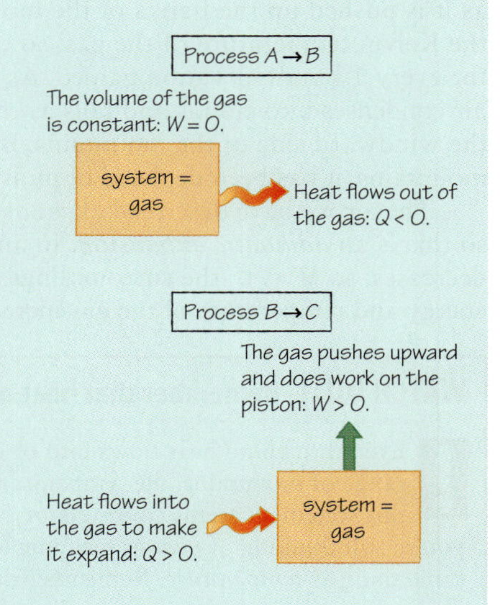

Process A→B
The volume of the gas is constant: W = 0.
system = gas
Heat flows out of the gas: Q < 0.
Process B→C
The gas pushes upward and does work on the piston: W > 0.
Heat flows into the gas to make it expand: Q > 0.
system = gas

Solve

(a) Use the ideal gas law to find the final volume V_f.

The initial pressure and volume of the gas are

$$p_i = 3.03 \times 10^5 \text{ Pa}$$
$$V_i = 1.20 \times 10^{-3} \text{ m}^3$$

From the ideal gas law, Equation 14-7,

$$p_i V_i = nRT_i$$

where T_i is the initial temperature of the gas in state A.

The final pressure of the gas is

$$p_f = 1.01 \times 10^5 \text{ Pa}$$

The ideal gas law tells us that

$$p_f V_f = nRT_f$$

where T_f is the final temperature of the gas in state C. The initial and final temperatures are the same ($T_i = T_f$), so

$$p_i V_i = p_f V_f$$

$$V_f = \frac{p_i V_i}{p_f} = \frac{(3.03 \times 10^5 \text{ Pa})(1.20 \times 10^{-3} \text{ m}^3)}{1.01 \times 10^5 \text{ Pa}} = 3.60 \times 10^{-3} \text{ m}^3$$

The final pressure is one-third of the initial pressure, and the final volume is three times the initial volume.

(b) For the isochoric process $A \rightarrow B$, the work done is $W_{A \rightarrow B} = 0$. Find the change in internal energy ΔU and quantity of heat Q for this process.

The change in internal energy U is the difference between the value of U in state B and the value of U in state A:

$$\Delta U_{A \rightarrow B} = U_B - U_A = \frac{5}{2}nRT_B - \frac{5}{2}nRT_i$$

We don't know the number of moles of gas n, nor do we know the temperatures T_i (in the initial state A, for which the pressure is p_i and the volume is V_i) and T_B (in state B, for which the pressure is p_f and the volume is V_i). But from the ideal gas law $pV = nRT$, so

$$p_i V_i = nRT_i \text{ and } p_f V_i = nRT_B$$

So the change in internal energy is

$$\Delta U_{A\rightarrow B} = \frac{5}{2}p_f V_i - \frac{5}{2}p_i V_i = \frac{5}{2}(p_f - p_i)V_i$$
$$= \frac{5}{2}(1.01 \times 10^5 \text{ Pa} - 3.03 \times 10^5 \text{ Pa})(1.20 \times 10^{-3} \text{ m}^3)$$
$$= -606 \text{ J}$$

(Remember that 1 Pa = 1 N/m^2 and 1 N · m = 1 J.) From Equation 15-9 for an isochoric process,

$$Q_{A\rightarrow B} = \Delta U_{A\rightarrow B} = -606 \text{ J}$$

Heat flows out of the gas and the internal energy of the gas decreases.

(c) Find W, ΔU, and Q for the isobaric process $B \rightarrow C$.

The work done in an isobaric process is given by Equation 15-4. The constant pressure is $p_f = 1.01 \times 10^5$ Pa, and the volume increases from $V_i = 1.20 \times 10^{-3}$ m^3 to $V_f = 3.60 \times 10^{-3}$ m^3:

$$W_{B\rightarrow C} = p_f(V_f - V_i)$$
$$= (1.01 \times 10^5 \text{ Pa})(3.60 \times 10^{-3} \text{ m}^3 - 1.20 \times 10^{-3} \text{ m}^3)$$
$$= +242 \text{ J}$$

The gas does positive work as it expands. As in part (b) we calculate the internal energy change with the aid of the ideal gas law:

$$\Delta U_{B\rightarrow C} = U_C - U_B = \frac{5}{2}nRT_f - \frac{5}{2}nRT_B$$
$$= \frac{5}{2}p_f V_f - \frac{5}{2}p_f V_i = \frac{5}{2}p_f(V_f - V_i)$$

This is just 5/2 times the above expression for $W_{B\rightarrow C}$, so

$$\Delta U_{B\rightarrow C} = \frac{5}{2}W_{B\rightarrow C} = \frac{5}{2}(+242 \text{ J}) = +606 \text{ J}$$

From the first law of thermodynamics, the heat that flows into the gas in this process is

$$Q_{B\rightarrow C} = \Delta U_{B\rightarrow C} + W_{B\rightarrow C} = (+606 \text{ J}) + (+242 \text{ J})$$
$$= +848 \text{ J}$$

(d) Find the values of W, ΔU, and Q for the combined process $A \rightarrow B \rightarrow C$.

The total amount of work done by the gas is

$$W_{A\rightarrow B} + W_{B\rightarrow C} = 0 + (+242 \text{ J}) = +242 \text{ J}$$

The total change in internal energy of the gas is

$$\Delta U_{A\rightarrow B} + \Delta U_{B\rightarrow C} = (-606 \text{ J}) + (+606 \text{ J}) = 0$$

The total quantity of heat that flows into the gas is

$$Q_{A\rightarrow B} + Q_{B\rightarrow C} = -606 \text{ J} + (+848 \text{ J}) = +242 \text{ J}$$

Reflect

For the combined process, there is *zero* net change in the internal energy U of the ideal gas. That's because U depends only on the number of moles of gas and the temperature, which have the same values in the final state C as in the initial state A. In the combined process 242 J of heat flows into the gas, and the gas uses this energy to do 242 J of work.

GOT THE CONCEPT? 15-2 Ranking Thermodynamic Processes

The pV diagram in **Figure 15-10** shows four thermodynamic processes for a certain quantity of an ideal gas: (i) $a \to b$, (ii) $a \to c$, (iii) $a \to d$, and (iv) $a \to e$. All four processes start in the same initial state a. The volume is the same for states b, c, and d. Process $a \to c$ lies along an isotherm. (a) Rank these four processes according to the internal energy change ΔU of the gas, from most positive to most negative. If two processes have the same value of ΔU, say so. (b) Rank these four processes according to the work W done by the gas, from most positive to most negative. If two processes have the same value of W, say so.

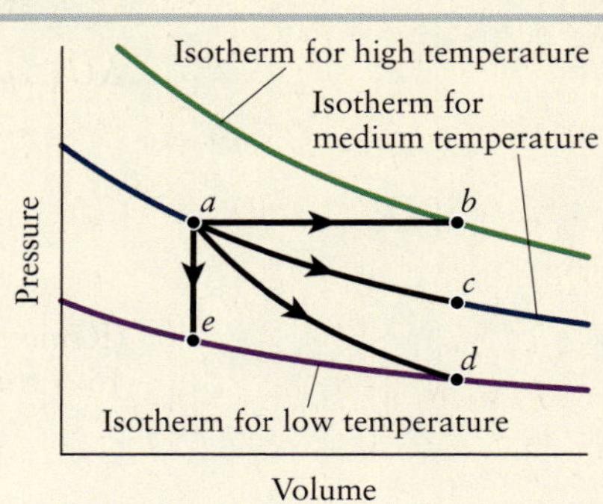

Figure 15-10 Four thermodynamic processes What are the properties of the processes $a \to b$, $a \to c$, $a \to d$, and $a \to e$?

GOT THE CONCEPT? 15-3 Which Could Be Adiabatic?

The pV diagram in Figure 15-10 shows four thermodynamic processes for a certain quantity of an ideal gas: $a \to b$, $a \to c$, $a \to d$, and $a \to e$. All four processes start in the same initial state a. The volume is the same for states b, c, and d. Process $a \to c$ lies along an isotherm. One of these processes could be adiabatic. Which one is it? (a) $a \to b$; (b) $a \to c$; (c) $a \to d$; (d) $a \to e$; (e) not enough information given to decide.

TAKE-HOME MESSAGE FOR Section 15-3

✔ A pV diagram can be used to describe thermodynamic processes that take a system from one state to another.

✔ Isobaric processes occur with no change in pressure.

✔ Temperature remains constant in isothermal processes.

✔ There is no heat flow into or out of a system during an adiabatic process.

✔ Isochoric processes occur with no change in volume.

15-4 More heat is required to change the temperature of an ideal gas isobarically than isochorically

Our discussion in Section 15-3 of various types of thermodynamic processes was fairly general. Let's now look more closely at some of these processes for the special case in which the thermodynamic system is an *ideal gas*. This special case has a tremendous number of applications: It will help us understand why a bicycle pump gets warm when you use it to inflate a tire and how a diesel engine can make a fuel–air mixture ignite even though there are no spark plugs in such an engine.

Specific Heats of an Ideal Gas

Suppose you have a quantity of ideal gas. How much heat does it take to raise the temperature of the gas to a new, higher temperature? The best answer is "It depends on how you do it." The pV diagram in **Figure 15-11** helps us understand why this is so.

This figure shows two possible ways that an ideal gas in a cylinder with a moveable piston can be slowly heated from a lower temperature to a higher one. Process $A \to B$ is *isochoric:* It takes place at constant volume (the moveable piston is locked in position), so the pressure of the gas increases as you heat the gas. Process $A \to C$ is *isobaric:* It takes place at constant pressure. The piston is allowed to move so that the gas expands as it is heated, and the pressure inside the cylinder remains equal to the pressure on the other side of the piston. The final temperature is the same for both processes because both state B and state C are on the same isotherm in Figure 15-11.

The internal energy U of an ideal gas depends only on the number of moles and the temperature, so the internal energy change ΔU is the same for both processes. But the amount of work W done by the gas is *different* for the two processes: $W = 0$ for process $A \to B$ (the gas does no work if the volume it occupies doesn't change), while

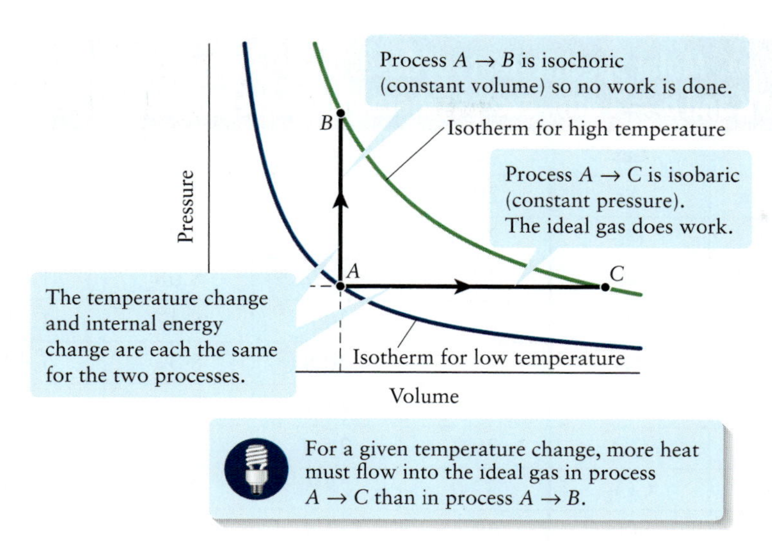
Process $A \rightarrow B$ is isochoric (constant volume) so no work is done.
Isotherm for high temperature
Process $A \rightarrow C$ is isobaric (constant pressure). The ideal gas does work.
The temperature change and internal energy change are each the same for the two processes.
Isotherm for low temperature
Pressure
Volume
A
B
C
For a given temperature change, more heat must flow into the ideal gas in process $A \rightarrow C$ than in process $A \rightarrow B$.

Figure 15-11 Two ways to increase temperature Processes $A \rightarrow B$ and $A \rightarrow C$ for an ideal gas lead to different final states but the same final temperature.

$W > 0$ for process $A \rightarrow C$ (the expanding gas does work on the piston to make it move). From the first law of thermodynamics, Equation 15-2,

$$\Delta U = Q - W \quad \text{so} \quad Q = \Delta U + W$$

Because ΔU is the same for both processes, *more* heat is required for process $A \rightarrow C$ (in which positive work is done) than for process $A \rightarrow B$ (for which $W = 0$). So the heat required to increase the temperature of an ideal gas is *greater* if the temperature is increased at constant *pressure* than if it is increased at constant *volume*.

Let's be more quantitative about the quantity of heat required to change the temperature of a certain quantity of ideal gas by a given amount. In Section 14-5 we wrote an equation for the heat required to change the temperature of a mass m of substance by an amount ΔT:

$$Q = mc\Delta T \qquad (14\text{-}20)$$

In Equation 14-20 the quantity c is the *specific heat* of the substance. For an ideal gas we usually write the internal energy U in terms of the number of *moles* of gas present rather than the amount of mass. We follow the same approach for the quantity of heat Q, so we write

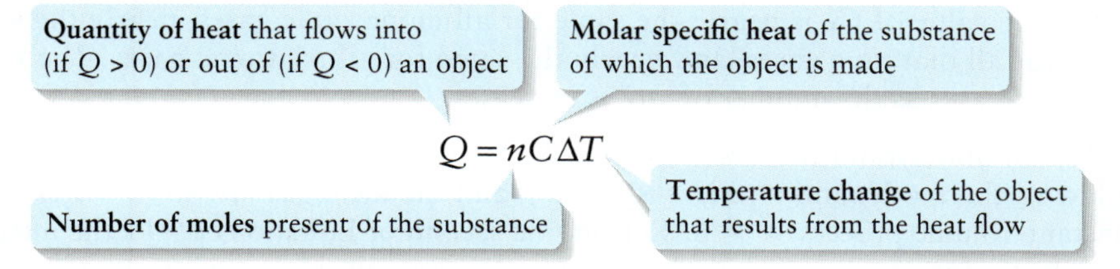
Quantity of heat that flows into (if $Q > 0$) or out of (if $Q < 0$) an object
Molar specific heat of the substance of which the object is made
$Q = nC\Delta T$
Number of moles present of the substance
Temperature change of the object that results from the heat flow

Quantity of heat and the resulting temperature change in terms of molar specific heat (15-10)

The quantity C in Equation 15-10 is called the **molar specific heat** of the substance. It has units of joules per mole per kelvin, or $\text{J}/(\text{mol} \cdot \text{K})$.

WATCH OUT! Don't confuse specific heat and molar specific heat.

It's important to recognize the difference between the specific heat c that appears in Equation 14-20 and the molar specific heat C in Equation 15-10. Specific heat c (lowercase) tells you the number of joules of heat required to raise the temperature of one *kilogram* of a substance by one kelvin. Molar specific heat C (uppercase) tells you the number of joules of heat required to raise the temperature of one *mole* of the substance by one kelvin. If the substance has a molar mass (mass per mole) of M, the two quantities are related by $C = Mc$.

We've seen that the quantity of heat Q required to make an ideal gas undergo a temperature change ΔT is different if the pressure p is held constant than if the volume V is held constant. To keep track of this difference, we use the symbol C_p for the

TABLE 15-1 Molar Specific Heats of Gases (in J/[mol • K] for a gas at $T = 300$ K and $p = 1$ atm, except where otherwise indicated)

	Gas	C_p	C_V	$C_p - C_V$	$\gamma = C_p/C_V$
Monatomic	Ar	20.8	12.5	8.28	1.66
	He	20.8	12.5	8.28	1.66
	Ne	20.8	12.7	8.08	1.63
Diatomic	H_2	28.8	20.4	8.37	1.41
	N_2	29.1	20.8	8.33	1.40
	O_2	29.4	21.1	8.33	1.39
	CO	29.3	21.0	8.28	1.39
Triatomic	CO_2	37.0	28.5	8.49	1.30
	SO_2	40.4	31.4	9.00	1.29
	H_2O (373 K)	34.3	25.9	8.37	1.32

molar specific heat at constant pressure and the symbol C_V for the **molar specific heat at constant volume.** Then we can write Equation 15-10 for these two special cases as

$$Q = nC_p\Delta T \quad \text{(constant pressure)}$$
$$Q = nC_V\Delta T \quad \text{(constant volume)} \tag{15-11}$$

Our discussion of Figure 15-11 showed that for a given temperature increase ΔT, the value of Q was greater for a constant-pressure process than for a constant-volume process. So we conclude that C_p, the molar specific heat at constant pressure, must be greater than the molar specific heat at constant volume, C_V. **Table 15-1**, which lists experimental values of C_p and C_V for various gases, shows that this is indeed the case.

C_p, C_V, and Degrees of Freedom

Notice the following from Table 15-1:

(a) The *difference* $C_p - C_V$ between the two different molar specific heats has almost the same value for *all* gases.
(b) The value of C_V is nearly the same for all monatomic gases, nearly the same for all diatomic gases, and nearly the same for all triatomic gases. The value increases as the number of atoms increases.

We can understand the first of these observations by using Equation 15-10, the first law of thermodynamics ($\Delta U = Q - W$), and the ideal gas law ($pV = nRT$). For a constant-volume process $W = 0$, so from the second of Equations 15-11 the internal energy change for n moles of an ideal gas that undergoes a given temperature change ΔT is

$$\Delta U = Q - W = nC_V\Delta T - 0 = nC_V\Delta T \tag{15-12}$$
(ideal gas, constant volume)

For a constant-*pressure* process that starts at the same temperature, the gas changes its volume by ΔV and does an amount of work

$$W = p\Delta V = p(V_f - V_i) = pV_f - pV_i$$

where V_f and V_i are the final and initial volumes of the gas. But from the ideal gas law,

$$pV_f = nRT_f \quad \text{and} \quad pV_i = nRT_i$$

where T_f and T_i are the final and initial temperatures of the gas, respectively. The temperature change of the gas is $\Delta T = T_f - T_i$, so we can write the work done by the gas in a constant-pressure process as

$$W = pV_f - pV_i = nRT_f - nRT_i = nR(T_f - T_i) = nR\Delta T \tag{15-13}$$
(ideal gas, constant pressure)

Using Equation 15-13 and the first of Equations 15-11, we can now write the internal energy change for this constant-pressure process as

$$\Delta U = Q - W = nC_p\Delta T - nR\,\Delta T = n(C_p - R)\Delta T \tag{15-14}$$

(ideal gas, constant pressure)

Equation 15-12 for a constant-volume process and Equation 15-14 for a constant-pressure process look rather different from each other. But we know that the internal energy change ΔU for an ideal gas is the *same* in both cases, as long as the number of moles n and the temperature change ΔT are the same for both processes. So ΔU in Equation 15-12 is equal to ΔU in Equation 15-14:

$$nC_V\Delta T = n(C_p - R)\Delta T$$

For this to be true it must be that $C_V = C_p - R$, or

Molar specific heat of an ideal gas at **constant pressure**

$$C_p - C_V = R \tag{15-15}$$

Ideal gas constant

Molar specific heat of an ideal gas at **constant volume**

Molar specific heats of an ideal gas (15-15)

Equation 15-15 says that for an ideal gas C_p is greater than C_V by $R = 8.314$ J/(mol · K), the ideal gas constant. The experimentally determined values of $C_p - C_V$ given in Table 15-1 for *real* gases are very close to this theoretical prediction.

How can we understand our second observation from Table 15-1, that the value of C_V is essentially the same for all gases with the same number of atoms? To see the answer, note from Equation 15-12 that for an ideal gas with constant volume $\Delta U = nC_V\Delta T$. In other words C_V tells us about the change in internal energy of a gas that results from a temperature change. We learned something important about the internal energy of an ideal gas in Section 14-3: On average each molecule in an ideal gas at temperature T has energy $(1/2)kT$ for each of its *degrees of freedom*. (Recall that k is the Boltzmann constant, equal to the ideal gas constant R divided by Avogadro's number N_A.) A monatomic gas has three degrees of freedom, corresponding to its ability to move in the x, y, and z directions. A diatomic gas at room temperature has two additional degrees of freedom that correspond to its ability to rotate around either of two axes perpendicular to the long axis of the molecule, and a triatomic gas (which at room temperature can both rotate and vibrate) has even more degrees of freedom.

We'll use the symbol D to represent the number of degrees of freedom. Then the average energy per gas molecule is $(D/2)kT$. One mole of the gas contains Avogadro's number of molecules, so the energy per mole is $N_A(D/2)kT = (D/2)(N_Ak)T = (D/2)RT$. (Again recall that $k = R/N_A$, so $R = N_Ak$.) Then the internal energy of n moles of ideal gas is

$$U = n\left(\frac{D}{2}R\right)T \tag{15-16}$$

(ideal gas, D degrees of freedom)

If the temperature of the gas changes by ΔT, it follows from Equation 15-16 that the internal energy change is

$$\Delta U = n\left(\frac{D}{2}R\right)\Delta T \tag{15-17}$$

(ideal gas, D degrees of freedom)

But we saw above that for an ideal gas $\Delta U = nC_V\,\Delta T$. Comparing this to Equation 15-17 we see that

Number of degrees of freedom for a molecule of the gas

Molar specific heat of an ideal gas at **constant volume**

$$C_V = \left(\frac{D}{2}\right)R \tag{15-18}$$

Ideal gas constant

Molar specific heat at constant volume for an ideal gas (15-18)

Let's see how well Equation 15-18 works for the real gases listed in Table 15-1. For a monatomic gas there are three degrees of freedom per molecule, so $D = 3$ and Equation 15-18 becomes

$$C_V = \frac{3}{2}R = \frac{3}{2}(8.314 \text{ J/[mol}\cdot\text{K]}) = 12.47 \text{ J/(mol}\cdot\text{K)} \quad \text{(monatomic gas)}$$

Table 15-1 shows that this is a very close fit to the experimental values of C_V for argon, helium, and neon. For a diatomic gas with five degrees of freedom ($D = 5$), we expect

$$C_V = \frac{5}{2}R = \frac{5}{2}(8.314 \text{ J/[mol}\cdot\text{K]}) = 20.79 \text{ J/(mol}\cdot\text{K)} \quad \text{(diatomic gas)}$$

This is also a very close fit to the experimental values for the diatomic atoms listed in Table 15-1. Indeed, this is how we know that such molecules have five degrees of freedom! Experiments show that at temperatures well above room temperature, the value of C_V for diatomic molecules increases and becomes closer to $7R/2 = 29.10$ J/(mol · K). What's happening is that at higher temperatures it becomes possible for these molecules to vibrate, giving each molecule two additional degrees of freedom (one for the kinetic energy of vibration, the other for the potential energy of vibration).

We encourage you to use Equation 15-18 to determine the number of degrees of freedom for the triatomic gases listed in Table 15-1.

The molar specific heat at constant pressure C_p also depends on D, the number of degrees of freedom. Since $C_p - C_V = R$ (Equation 15-15) and $C_V = (D/2)R$ (Equation 15-18), you can see that

$$C_p = C_V + R = \left(\frac{D}{2}\right)R + R = \left(\frac{D+2}{2}\right)R \tag{15-19}$$

We can get an expression for the number of degrees of freedom D by taking the *ratio* of C_p to C_V. This **ratio of specific heats** is denoted by the Greek letter γ ("gamma"):

$$\gamma = \frac{C_p}{C_V} \quad \text{(ratio of specific heats)} \tag{15-20}$$

If we substitute Equation 15-18 for C_V and Equation 15-19 for C_p into Equation 15-20, we get

$$\gamma = \frac{C_p}{C_V} = \frac{\left(\frac{D+2}{2}\right)R}{\left(\frac{D}{2}\right)R} = \frac{D+2}{D} \quad \text{(ideal gas)} \tag{15-21}$$

You can check Equation 15-21 against the experimental values for γ given in Table 15-1. For example, for a monatomic ideal gas with three degrees of freedom ($D = 3$), Equation 15-21 predicts

$$\gamma = \frac{3+2}{3} = 1.67$$

which is a very good match to the experimental values for argon, helium, and neon. Try Equation 15-21 for a diatomic gas with $D = 5$; how well does the predicted value of γ compare to the experimental values in Table 15-1?

Adiabatic Processes for an Ideal Gas

The ratio of specific heats $\gamma = C_p/C_V$ for an ideal gas plays an important role in *adiabatic* processes in which there is zero heat flow into or out of the gas. If we have a quantity of gas in a cylinder with a moveable piston and then allow the piston to move in or out, the gas will compress or expand adiabatically if the cylinder and piston are made of an insulating material that does not allow heat to flow through it. It can be shown that if an ideal gas expands or contracts adiabatically, the pressure p and volume V of the gas are related by

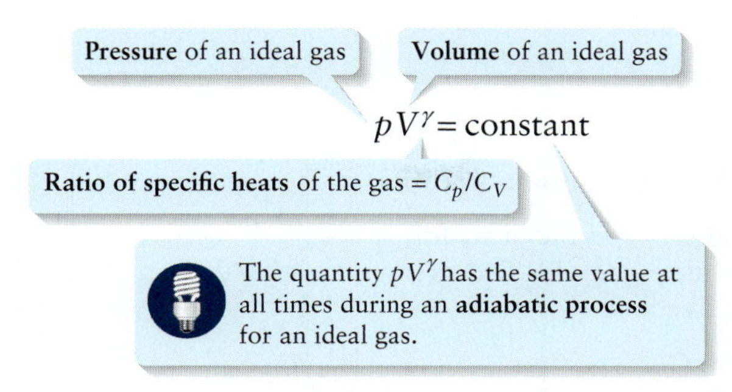
Pressure of an ideal gas
Volume of an ideal gas
$pV^\gamma = \text{constant}$
Ratio of specific heats of the gas = C_p/C_V
The quantity pV^γ has the same value at all times during an adiabatic process for an ideal gas.

Pressure and volume for ideal gas, adiabatic process (15-22)

As long as the thermodynamic changes that take place in an ideal gas are adiabatic, the product of pressure and volume to the power γ remains the same. As volume increases in an adiabatic expansion, pressure decreases; as volume decreases in an adiabatic compression, pressure increases.

The ideal gas law $pV = nRT$ also applies to this gas, so $p = nRT/V$. If we substitute this expression for the pressure p into Equation 15-22, we get

$$\frac{nRT}{V}V^\gamma = nRTV^{\gamma-1} = \text{constant}$$

Since the number of moles n does not change (no gas enters or leaves the system) and R is itself a constant, we can rewrite this expression as

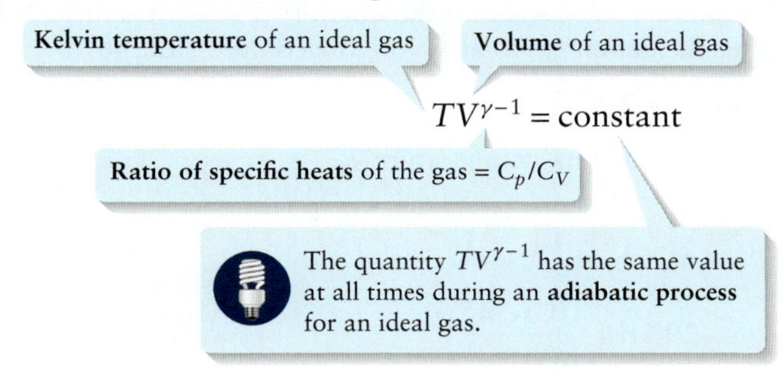
Kelvin temperature of an ideal gas
Volume of an ideal gas
$TV^{\gamma-1} = \text{constant}$
Ratio of specific heats of the gas = C_p/C_V
The quantity $TV^{\gamma-1}$ has the same value at all times during an adiabatic process for an ideal gas.

Temperature and volume for ideal gas, adiabatic process (15-23)

The value of γ is greater than 1 for all gases (see Table 15-1), so $\gamma - 1$ is greater than zero. So as volume increases in an adiabatic expansion, temperature decreases; as volume decreases in an adiabatic compression, temperature increases.

Equation 15-23 explains what happens when you use a bicycle pump to inflate a tire: The pump rapidly becomes warm to the touch. Although the pump is not made of an insulating material, the air in the pump cylinder is compressed so rapidly that there's no time for heat to flow into or out of the air during the compression. As a result, the compression is adiabatic, so the temperature of the gas (and the pump that holds it) increases as the air is compressed.

You can demonstrate an adiabatic expansion using your own breath. If you open your mouth and breathe on the back of your hand, you can feel that the expelled breath is warm. But the same breath feels cold if you form your lips into a small "O" and then breathe on the back of your hand. In the latter case the air expands rapidly as it exits through the small aperture through your lips, and this expansion is nearly adiabatic. Hence the expelled air drops in temperature and feels cold on your hand.

EXAMPLE 15-5 Burning Paper Without Heat

In a classroom physics demonstration a piece of paper is placed in the bottom of a test tube with a sealed plunger. When the plunger is quickly depressed, compressing the air inside, the paper combusts. How is this possible? Let the initial volume of the gas in the test tube be 10.0 cm^3 and the final volume be 1.00 cm^3, and let the initial temperature be room temperature (20°C). Calculate the final temperature of the gas in the test tube. Assume that air is an ideal, diatomic gas.

Set Up

Because the compression happens quickly, there is no time for any heat transfer into or out of the air in the tube. So we can consider this to be an adiabatic process. We use Equation 15-23 to relate volume and temperature for this process, and we use Equation 15-21 to find the ratio of specific heats γ for air. For diatomic gases, there are five degrees of freedom.

Ideal gas, adiabatic process:

$$TV^{\gamma-1} = \text{constant} \quad (15\text{-}23)$$

Ratio of specific heats for an ideal gas with D degrees of freedom:

$$\gamma = \frac{C_p}{C_V} = \frac{D+2}{D} \quad (15\text{-}21)$$

$V_i = 10.0\ \text{cm}^3$
20°C
paper
$V_f = 1.00\ \text{cm}^3$
temperature = ?

Solve

First find the value of γ for the gas and convert the temperature from Celsius to Kelvin.

From Equation 15-21 with $D = 5$,

$$\gamma = \frac{C_p}{C_V} = \frac{5+2}{5} = \frac{7}{5} = 1.40$$

The initial temperature of the air is 20.0°C or

$$T_i = (20 + 273.15)\ \text{K} = 293\ \text{K}$$

We know the initial values of the temperature ($T_i = 293$ K) and volume ($V_i = 10.0\ \text{cm}^3$) and the final value of the volume ($V_f = 1.00\ \text{cm}^3$). Use these in Equation 15-23 to solve for the final value of temperature.

Equation 15-23 says that the quantity $TV^{\gamma-1}$ has the same value after the compression as before the compression, so

$$T_f V_f^{\gamma-1} = T_i V_i^{\gamma-1}$$

Solve for the final temperature T_f:

$$T_f = \frac{T_i V_i^{\gamma-1}}{V_f^{\gamma-1}} = T_i\left(\frac{V_i}{V_f}\right)^{\gamma-1}$$

$$= (293\ \text{K})\left(\frac{10.0\ \text{cm}^3}{1.00\ \text{cm}^3}\right)^{\left(\frac{7}{5}-1\right)}$$

$$= (293\ \text{K})(10.0)^{\frac{2}{5}}$$

$$= 736\ \text{K or } 463°\text{C}$$

Reflect

The final temperature is above the flash point of paper (the temperature at which paper catches fire—around 450°C), so the paper will burn until the supply of paper or oxygen is used up. This same process is used in diesel engines: A mixture of fuel and air is rapidly compressed in the cylinders of the engine, raising the temperature to such a high value that the mixture spontaneously ignites without a spark plug (as is used in a conventional gasoline engine).

GOT THE CONCEPT? 15-4 Heating Helium

? A quantity of helium gas at room temperature is placed in a cylinder with a moveable piston. If the cylinder is locked in place so that the volume cannot change and a quantity of heat Q_0 is added to the gas, the temperature of the gas increases by ΔT_0. Now the gas is cooled to its original temperature, and the piston is unlocked so that the gas is free to expand. If the pressure exerted on the gas by the piston is held constant and the same quantity of heat Q_0 is added to the gas, what will be the temperature change of the gas? (a) ΔT_0, the same as before; (b) less than ΔT_0; (c) more than ΔT_0.

GOT THE CONCEPT? 15-5 Helium versus Nitrogen

? Two cylinders, each equipped with a moveable piston, contain equal numbers of moles of a gas. One cylinder contains helium (He), while the other contains nitrogen (N_2). If the same quantity of heat Q is added to each gas and each is allowed to expand at constant pressure, what can you say about the temperature change ΔT and amount of work done W for the two gases? (a) ΔT and W are the same for both gases; (b) ΔT and W are both less for He than for N_2; (c) ΔT and W are both greater for He than for N_2; (d) ΔT is less for He than for N_2, but W is greater for He than for N_2; (e) ΔT is greater for He than for N_2, but W is less for He than for N_2; (f) none of these.

TAKE-HOME MESSAGE FOR Section 15-4

✔ More heat must be added to a gas at constant pressure than at constant volume in order to produce the same change in temperature.

✔ Thermodynamic processes that are carried out quickly are often adiabatic because heat transfer is relatively slow.

15-5 The second law of thermodynamics describes why some processes are impossible

Figure 15-12 shows two frames from a video in which someone inflates a balloon until it pops. How do you know that the left frame happens before the right one? None of the physical laws that we have studied so far preclude a set of balloon fragments from flying toward each other and reassembling as an intact balloon.

For example, neither conservation of energy nor momentum would be violated were the balloon fragments to reassemble. What about the pictures of an ice cube melting on a piece of granite in **Figure 15-13**? Each image was taken before the one below it, but would it be possible for this sequence to run in reverse? Here again the reverse process would be permitted by the fundamental laws we know. The total energy of the system would be conserved if the granite drew enough energy from the water to cause it to freeze: The amount of energy required to melt a chunk of ice is exactly the same as the energy needed to freeze the resulting water. Yet we have declared that heat transfer is always from a hotter to a cooler object. Why?

For both the balloon in Figure 15-12 and the ice cube in Figure 15-13, events happen in a sequence that tends to decrease the level of *organization* or *order*. In Figure 15-12 the air that was confined within the balloon in an orderly way spreads out across the room when the balloon is popped, and the balloon itself breaks into a collection of random fragments. In Figure 15-13 the water molecules begin in an ordered state in which they are locked into the crystalline structure of the ice cube, and they end up free to move in a disordered way within the puddle of melted water. The reverse processes, in which air and balloon fragments spontaneously coalesce into a filled balloon or water spontaneously freezes, never occur in nature. That is, natural events always happen in a direction from ordered to disordered, or *random*. We can generalize this observation to the **second law of thermodynamics**:

The amount of disorder in an isolated system either always increases or, if the system is in equilibrium, stays the same.

We emphasize that a system described by the second law of thermodynamics must be *isolated*. This is to ensure that nothing outside the system can cause its state to change in an unnatural way. For example, we could manually sort out the fragments of the balloon in Figure 15-12, carefully reassemble them, and then inflate the reassembled balloon. The balloon system would experience increasing order, but only as a result of our external intervention; we could not treat the balloon as isolated.

Let's see how to use the second law of thermodynamics to analyze systems that convert heat to work or work to heat. We'll look in particular at mechanical systems such as the internal combustion engine in a car (which takes the heat released by burning fuel and converts it to work) and a kitchen refrigerator (which takes work done by the refrigerator

Ted Kinsman/Science Source

Figure 15-12 Bursting your balloon Both images were taken from a video of a person blowing up a balloon until it pops. How do you know that the left frame happens before the right one?

David Tauck

Figure 15-13 Melting ice An ice cube melts atop a piece of granite. Each image in the sequence was taken after the one above it. Could this sequence run in reverse?

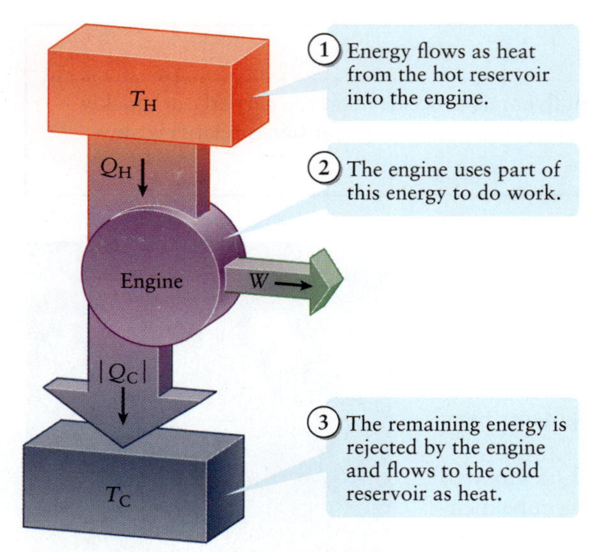

T_H
Q_H
Engine
W
$|Q_C|$
T_C
1 Energy flows as heat from the hot reservoir into the engine.
2 The engine uses part of this energy to do work.
3 The remaining energy is rejected by the engine and flows to the cold reservoir as heat.

Figure 15-14 What a heat engine does A generic heat engine, represented by a circle, is thermally connected to a reservoir of higher temperature (T_H) and to a reservoir of lower temperature (T_C). Energy Q_H flows from the hot reservoir into the engine in the form of heat. The engine uses some of the energy to do work W; the remaining energy is rejected as heat to the cold reservoir. (Note that Q_C is negative because it is heat that *leaves* the engine, so the energy that enters the cold reservoir is $|Q_C|$.)

motor and uses it to transfer heat out of the refrigerator's contents). We'll see that the second law imposes strict limits on how efficient an engine can be and on how much performance you can get from a refrigerator.

In the following section we'll extend the second law to include the concept of *entropy*, which is a measure of the amount of randomness in a system. We'll use the entropy concept to gain further insight into the second law and to see what this law tells us about living systems.

Heat Engines and the Second Law

A system or device that converts heat to work is a **heat engine** or simply an *engine*. Heat engines are cyclic: Some part of the system absorbs energy, work is done, and the system returns to its original state in order for the cycle to begin again. Although the term "engine" might call to mind the complex device that powers an automobile, in thermodynamics a heat engine can be as simple as gas in a piston that expands as heat is added and contracts as the gas cools.

BIO-Medical There are also tiny molecular engines in living cells. Though much smaller than a piston or the engine in a car, these engines perform the same function, transforming energy into motion. In the mitochondria of all human cells, hydrogen ions flow through ATP synthase, a curious enzyme that has a structure not unlike a waterwheel. The hydrogen ions cause the stalk of the ATP synthase to spin, which in turn pushes phosphate ions and adenosine diphosphate molecules together to form adenosine triphosphate (ATP). ATP synthase is in many ways analogous to an engine that converts the kinetic energy of hydrogen ions into a high-energy chemical bond that stores energy in a form that can be used by all cells. Enzymes such as ATP synthase follow a cycle, using energy to do work and then returning to their original state so that the cycle can start again.

Figure 15-14 shows a simple model for any heat engine. The engine is thermally connected to a *reservoir* of higher (hotter) temperature (T_H) and to a reservoir of lower (cooler) temperature (T_C). A **reservoir** is a part of a system large enough to either absorb or supply heat without a change in temperature. In an old-time steam engine, for example, the furnace serves as the hot reservoir and the surrounding atmosphere acts as the cold reservoir. Energy Q_H flows from the hot reservoir into the engine; during this process some of the energy goes into work W, and the remainder $|Q_C|$ flows out of the engine and into the cold reservoir in the form of heat.

WATCH OUT! The quantity Q_C is negative.

Remember our convention that Q is positive if it flows into the object in question and negative if it flows out of the object. So for our engine Q_H is positive (heat flows into the engine from the hot reservoir) and Q_C is negative (heat flows out of the engine into the cold reservoir). However, in analyzing the engine in Figure 15-14 we'll be interested in how much energy flows *into* the *cold reservoir*. So we've labeled Figure 15-14 with $|Q_C|$ (the absolute value of Q_C, which is positive) to indicate the positive quantity of energy that flows into the cold reservoir in the form of heat.

In a perfect engine *all* of the energy Q_H taken in from the hot reservoir would be converted to work, with no heat at all going into the cold reservoir (that is, $Q_C = 0$ in Figure 15-14). This would be a 100% efficient engine: None of the energy from the hot reservoir would be "thrown away" to the cold reservoir. Unfortunately the second law of thermodynamics says that such a perfect engine is *impossible*. To see why, note that there is an increase in order of the hot reservoir: Because energy flows out of this reservoir, some of the molecules of the reservoir must lose their random motion and so become more orderly. Over one cycle of operation the engine itself returns to its original state, so it has zero net change in order, and there is no change at all in the cold reservoir (because no energy flows into it). So the net result would be that over one cycle the system of engine plus reservoirs undergoes an increase in order. But this violates the second law of thermodynamics, which says that the order of an isolated system can decrease or stay the same but cannot increase! We conclude that you simply cannot build a perfect engine that's 100% efficient. The *Kelvin–Planck statement* is a rewording of the second law of thermodynamics that describes the inherent inefficiency of heat engines:

No process is possible in which heat is absorbed from a reservoir and converted completely into work.

In order for a heat engine to satisfy the second law of thermodynamics, some heat *must* flow into the cold reservoir in order to make the cold reservoir less ordered. (This happens because the energy added to the cold reservoir increases the random motion of the reservoir's molecules.) The decrease in order of the cold reservoir can then compensate for the increase in order of the hot reservoir. So not all of the energy that flows from the hot reservoir can be converted to work; some of this energy must flow into the cold reservoir and cannot be recovered. That's what happens in a car's gasoline engine: Only about a quarter of the energy released by burning gasoline gets converted into work to propel the car. The remainder goes into heating up the engine and the exhaust gases. Similarly, in the human body only about a quarter of the energy in food is actually used by your cells to do work. The rest of the energy appears as body heat.

Let's apply the *first* law of thermodynamics, Equation 15-2, to the generic engine shown in Figure 15-14. In one cycle the engine takes in energy Q_H from the hot reservoir in the form of heat and sends energy $|Q_C|$ to the cold reservoir in the form of heat, so the *net* quantity of heat that goes into the engine is $Q = Q_H - |Q_C|$. In that same cycle the engine does work W. Because the cycle returns the engine to its initial state at the beginning of the cycle, there is zero net change in its internal energy: $\Delta U = 0$. So the first law of thermodynamics, $\Delta U = Q - W$, becomes

$$0 = Q_H - |Q_C| - W$$

or

Heat that flows from the hot reservoir into the engine in one cycle

Work done by a heat engine in one cycle

$$W = Q_H - |Q_C|$$

Heat that flows from the engine to the cold reservoir in one cycle

Work done by a heat engine (15-24)

The work done equals the difference between the energy taken in from the hot reservoir and the energy discarded into the cold reservoir. The **efficiency** of the engine is defined as the work W divided by the energy Q_H taken in to do the work:

$$e = \frac{W}{Q_H} \quad (15\text{-}25)$$

Efficiency is essentially what you get out of an engine divided by what you put in. If all of the input energy were converted to work, then W would equal Q_H and the efficiency would be 100%, or $e = 1$. But the Kelvin–Planck form of the second law of thermodynamics tells us this is impossible. The best combustion engines, such as the ones found in automobiles, operate at an efficiency of $e \approx 0.30$ (about 30%). Similarly, when human muscles contract only about 25% of the input energy does work, with the rest going into body heat, so $e \approx 0.25$. Using Equation 15-24 we can express the efficiency of a heat engine (Equation 15-25) in terms of the heat taken in from the hot reservoir and the heat rejected to the cold reservoir:

Go to Picture It 15-2 for more practice dealing with efficiency.

$$e = \frac{W}{Q_H} = \frac{Q_H - |Q_C|}{Q_H} = 1 - \frac{|Q_C|}{Q_H} \text{ (efficiency of a heat engine)} \quad (15\text{-}26)$$

EXAMPLE 15-6 Efficiency of an Engine

The combustion of gasoline in a lawnmower engine releases 44.0 J of energy per cycle. Of that, 31.4 J are lost to warming the body of the engine and the surrounding air. (a) How much work does the lawnmower engine do per cycle? (b) What is the efficiency of the engine?

Set Up

We use Equation 15-24 to determine the work done by the engine and Equation 15-26 to find the efficiency.

Work done by a heat engine:

$$W = Q_H - |Q_C| \quad (15\text{-}24)$$

Efficiency of a heat engine:

$$e = \frac{W}{Q_H} = \frac{Q_H - |Q_C|}{Q_H} = 1 - \frac{|Q_C|}{Q_H} \quad (15\text{-}26)$$

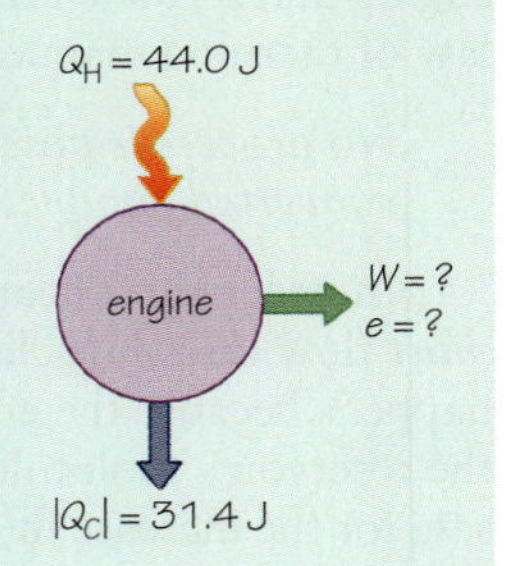

Solve

(a) The work done by the engine equals the difference between the heat Q_H taken in by burning gasoline and the heat $|Q_C|$ that is lost by the engine.

We are given that in one cycle

$$Q_H = 44.0 \text{ J}$$
$$|Q_C| = 31.4 \text{ J}$$

From Equation 15-24 the work that the engine does in one cycle is

$$W = Q_H - |Q_C|$$
$$= 44.0 \text{ J} - 31.4 \text{ J} = 12.6 \text{ J}$$

(b) The efficiency of the engine equals the work done divided by the heat taken in. It's also equal to 1 minus the ratio of heat thrown away to heat taken in.

From Equation 15-26 the efficiency is

$$e = \frac{W}{Q_H} = \frac{12.6 \text{ J}}{44.0 \text{ J}} = 0.286$$

Alternatively,

$$e = 1 - \frac{|Q_C|}{Q_H} = 1 - \frac{31.4 \text{ J}}{44.0 \text{ J}}$$
$$= 1 - 0.714 = 0.286$$

Reflect

An efficiency of around 0.25 to 0.30 (25% to 30%) is typical for a gasoline engine.

Figure 15-15 shows the pV diagram for the cyclic process (called the *Otto cycle*) used in a typical automobile engine. (The process is somewhat simplified in this figure.) Beginning at the state marked 1, a mixture of fuel and air is rapidly compressed in a cylinder with a moveable piston. Because the compression is rapid, step $1 \rightarrow 2$ is adiabatic and there is no heat flow into the fuel–air mixture. However, work is done *on* the mixture by the piston, so for this step the work done *by* the fuel–air mixture is negative: $W_{1\rightarrow2} < 0$. At state 2 a spark ignites the fuel, resulting in heat Q_H flowing into

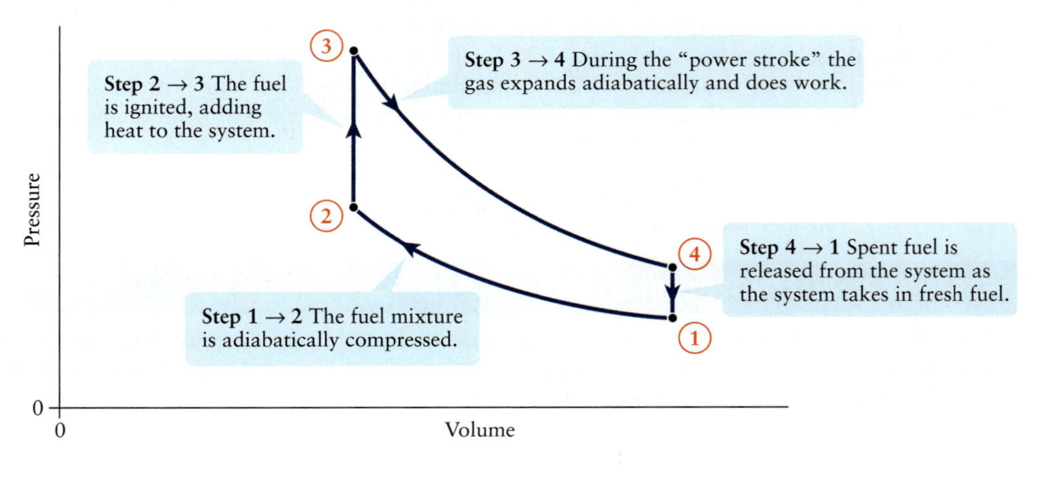

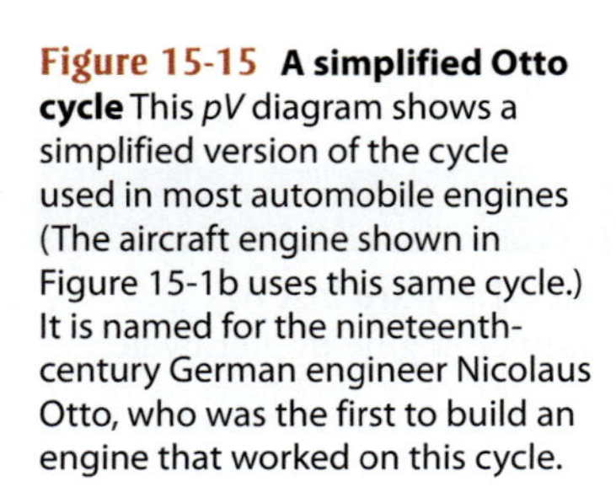

Figure 15-15 A simplified Otto cycle This pV diagram shows a simplified version of the cycle used in most automobile engines (The aircraft engine shown in Figure 15-1b uses this same cycle.) It is named for the nineteenth-century German engineer Nicolaus Otto, who was the first to build an engine that worked on this cycle.

the system. The associated increase in pressure and temperature from state 2 to state 3 occurs so rapidly that the volume of the gas in the piston remains nearly constant, so no work is done. At state 3, the high pressure causes the gas to expand rapidly and so adiabatically. Because $Q = 0$ in an adiabatic process, the first law of thermodynamics ($\Delta U = Q - W$) tells us that all of the change in internal energy in step $3 \rightarrow 4$ results in positive work done by the gas on the piston: $W_{3 \rightarrow 4} > 0$. As the system returns to its initial state (step $4 \rightarrow 1$, for which the volume again remains constant so no work is done) exhaust is expelled and new fuel is injected into the cylinder.

The net work output in the cycle is the sum of the positive work in step $3 \rightarrow 4$ (in which the gas expands) and the negative work in step $1 \rightarrow 2$ (in which the gas compresses). Remember that the work done equals the area under the curve in a pV diagram; step $3 \rightarrow 4$ happens at higher pressure than step $1 \rightarrow 2$, so there is more area under the curve for step $3 \rightarrow 4$ than for step $1 \rightarrow 2$ and there is more positive work than negative work. So the net work output of the engine over one cycle is positive. This work is used to turn the driveshaft of the automobile and make the wheels turn.

The Carnot Cycle

The cycle shown in Figure 15-15 is not the only cycle used in heat engines, nor does it have the highest efficiency. It turns out that the most efficient of all possible cyclic processes in a heat engine is the **Carnot cycle** (pronounced "car-noe"), first proposed in the early 1800s by French engineer Sadi Carnot. Carnot based his cycle on a requirement that all of the processes that comprise the cycle be **reversible processes**—that is, the reverse process is physically possible. Let's explore what this means.

Most thermodynamic processes are **irreversible processes**: They can proceed in only one direction. An example is what happens when ice at 0°C is placed in a large metal pot at 20°C. Heat flows from the pot into the ice, so the ice begins to melt, and the temperature of the pot begins to drop. When the ice and pot come to equilibrium, the ice has completely melted, and the pot and melted water both reach the same final temperature. The reverse of this process—in which liquid water in a pot spontaneously freezes, and the pot warms up—never happens in nature.

An example of a reversible process is ice at 0°C placed in a large metal pot that is also at 0°C. If a small amount of heat is made to flow from the pot into the ice, the ice will partially melt and the pot will contain ice and liquid water at 0°C. Because the amount of heat is so small, the temperature of the pot changes hardly at all and is still 0°C. If the same small amount of heat is now made to flow from the water–ice mixture back into the pot, the water will freeze back to ice and the ice and pot will both be at 0°C. In general a process that involves heat flow will be reversible *only* if the objects between which heat flows are at the same temperature. If they are at different temperatures, the heat flow will always be from the higher-temperature object to the lower-temperature object and the process will be irreversible.

A process can also be reversible if it involves *no* heat flow ($Q = 0$). An example is a *slow* adiabatic expansion or compression, such as would happen with gas in an insulated cylinder. If the gas is allowed to expand slightly, it does a small amount of positive work on the piston ($W > 0$) and the internal energy of the gas decreases slightly (from the first law of thermodynamics, $\Delta U = Q - W$ is negative because Q is zero, and W is positive). In the reverse process the gas is compressed slightly and does negative work on the piston ($W < 0$, meaning that the piston does work on the gas), and the internal energy of the gas increases. The system then returns to its initial state. By contrast, a free expansion of a gas, in which a gas under pressure suddenly expands to a larger volume (as in Figure 15-4b), is *irreversible*. It is not possible to return the system to the initial pressure and volume without adding energy by doing work.

Carnot understood that energy is conserved in both reversible and irreversible processes. But he also understood that in an irreversible process much of the energy becomes *unavailable*. For example, when energy flows in the form of heat from the 20°C pot into the 0°C ice to melt it, the energy that went into the ice cannot be returned to the pot. In order not to "lose" energy in this way, Carnot realized that the most efficient heat engine cycle possible must therefore involve only reversible processes. As we've seen, this means either slow processes that involve heat flow with zero temperature difference or slow adiabatic processes.

The Carnot cycle consists of:
- Two adiabatic processes, for which no heat is transferred in or out of the system.
- Two isothermal processes, one at the temperature of the hot reservoir and one at the temperature of the cold reservoir.

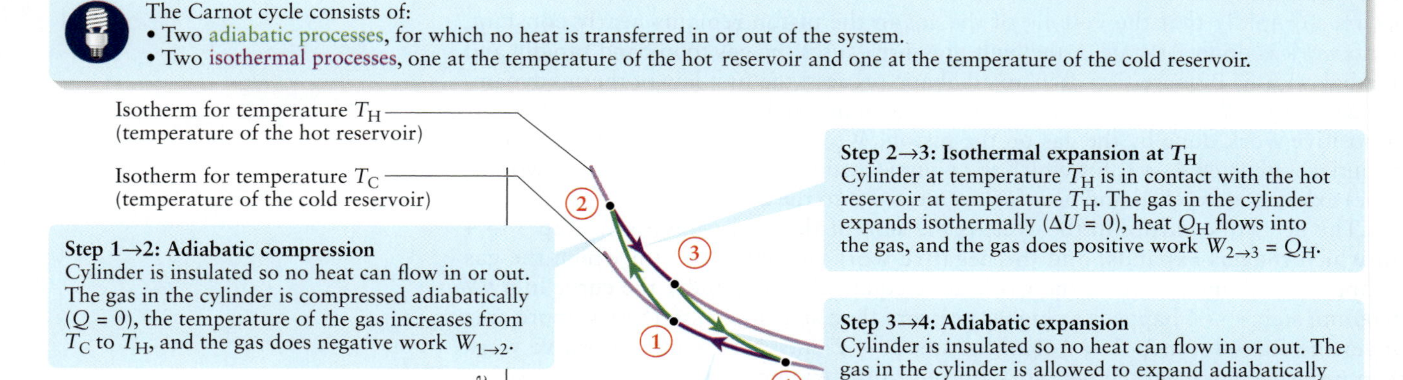

Isotherm for temperature T_H
(temperature of the hot reservoir)
Isotherm for temperature T_C
(temperature of the cold reservoir)
Step 1→2: Adiabatic compression
Cylinder is insulated so no heat can flow in or out. The gas in the cylinder is compressed adiabatically ($Q = 0$), the temperature of the gas increases from T_C to T_H, and the gas does negative work $W_{1\to2}$.
Step 2→3: Isothermal expansion at T_H
Cylinder at temperature T_H is in contact with the hot reservoir at temperature T_H. The gas in the cylinder expands isothermally ($\Delta U = 0$), heat Q_H flows into the gas, and the gas does positive work $W_{2\to3} = Q_H$.
Step 3→4: Adiabatic expansion
Cylinder is insulated so no heat can flow in or out. The gas in the cylinder is allowed to expand adiabatically ($Q = 0$), the temperature of the gas drops from T_H to T_C, and the gas does work $W_{3\to4}$.
1
2
3
4
Pressure
0
0
Volume
Step 4→1: Isothermal compression at T_C
Cylinder at temperature T_C is in contact with the cold reservoir at temperature T_C. The gas in the cylinder is compressed isothermally ($\Delta U = 0$), heat $|Q_C|$ flows into the gas, and the gas does negative work $W_{4\to1} = -|Q_C|$.

Figure 15-16 The Carnot cycle The most efficient cycle possible for a heat engine is made up of four reversible steps.

Figure 15-16 shows the pV diagram for the Carnot cycle. The system is an ideal gas in a cylinder with a moveable piston. In state 1 the cylinder is in contact and in equilibrium with the cold reservoir at temperature T_C, so the gas is also at temperature T_C. Here are the four *reversible* steps that make up the cycle:

- **Step 1 → 2:** Adiabatic compression. You take the cylinder filled with gas and wrap it with a perfect insulator so there can be no heat flow into or out of the gas, and you do work on the moveable piston to slowly push it inward (so the gas does negative work). As the gas is compressed, its temperature increases from T_C to T_H.
- **Step 2 → 3:** Isothermal expansion. You remove the insulation and put the cylinder in contact with the hot reservoir at temperature T_H (which is the same as the current temperature of the gas). You now allow the gas to expand so that it does positive work on the piston. The temperature remains equal to T_H because the gas and hot reservoir are in equilibrium.
- **Step 3 → 4:** Adiabatic expansion. You again wrap the cylinder with a perfect insulator and allow the gas to expand. The expanding gas does more positive work on the piston, and the temperature of the gas decreases from T_H back down to T_C.
- **Step 4 → 1:** Isothermal compression. You again remove the insulation, but now you put the cylinder in contact with the cold reservoir at temperature T_C (which is the same as the current temperature of the gas). You do work on the piston to slowly push it inward, so the gas does negative work. The temperature remains equal to T_C because the gas and cold reservoir are in equilibrium. You stop when the gas is once again in state 1, with the same values of pressure, volume, and temperature as initially.

You can see from the description that the Carnot cycle is an idealization (like a massless rope or a frictionless ramp) and is *not* a practical cycle for a real engine. The Carnot cycle is nonetheless tremendously important. Because all four of its processes are reversible, the Carnot cycle is the *most efficient* heat engine cycle possible. It therefore represents the standard against which all other cycles should be compared. Let's see how to determine the efficiency of a Carnot engine.

From Equation 15-26 the efficiency of any heat engine is $e = 1 - |Q_C|/Q_H$. For the Carnot cycle depicted in Figure 15-16, Q_H is the heat that flows into the ideal gas during step 2 → 3, when the gas cylinder is in contact with the hot reservoir at temperature T_H, and Q_C is the (negative) heat that flows into the gas during step 4 → 1, when the gas cylinder is in contact with the cold reservoir at temperature T_C. Both of these processes are isothermal, so $Q = W$ (Equation 15-5): The quantity of heat Q that enters the gas equals the work W done by the gas. Furthermore, we know from Equation 15-6 that the work done by n moles of an ideal gas in an isothermal process at temperature T is $W = nRT \ln (V_f/V_i)$, where V_i and V_f are the volume of the gas at the beginning and end of the process, respectively. So for the two isothermal steps in the Carnot cycle, we have

$$Q_H = W_{2\to 3} = nRT_H \ln\left(\frac{V_3}{V_2}\right)$$

$$Q_C = W_{4\to 1} = nRT_C \ln\left(\frac{V_1}{V_4}\right)$$

Note that Q_C is negative for step $4 \to 1$. To see this, notice that V_1, the volume in state 1, is less than V_4, the volume in state 4 (see Figure 15-16). So the ratio V_1/V_4 is less than 1, and the natural logarithm of a number less than 1 is negative. Since $\ln(1/x) = -\ln x$, we can write the absolute value of Q_C as

$$|Q_C| = -Q_C = nRT_C \ln\left(\frac{V_4}{V_1}\right)$$

Then we can write the efficiency of the Carnot cycle using Equation 15-26 as

$$e_{\text{Carnot}} = 1 - \frac{|Q_C|}{Q_H} = 1 - \frac{nRT_C \ln\left(\frac{V_4}{V_1}\right)}{nRT_H \ln\left(\frac{V_3}{V_2}\right)} \qquad (15\text{-}27)$$

Equation 15-27 looks rather messy, but we can simplify it by using Equation 15-23, which says that $TV^{\gamma-1}$ = constant for any adiabatic process. If we apply this to the adiabatic steps $1 \to 2$ and $3 \to 4$, we get

$$\text{Step } 1 \to 2: T_C V_1^{\gamma-1} = T_H V_2^{\gamma-1} \quad \text{or} \quad \frac{T_C}{T_H} = \frac{V_2^{\gamma-1}}{V_1^{\gamma-1}}$$

$$\text{Step } 3 \to 4: T_C V_4^{\gamma-1} = T_H V_3^{\gamma-1} \quad \text{or} \quad \frac{T_C}{T_H} = \frac{V_3^{\gamma-1}}{V_4^{\gamma-1}}$$

If we set the two expressions for T_C/T_H equal to each other and rearrange, we get

$$\frac{V_2^{\gamma-1}}{V_1^{\gamma-1}} = \frac{V_3^{\gamma-1}}{V_4^{\gamma-1}} \quad \text{so} \quad \frac{V_2}{V_1} = \frac{V_3}{V_4} \quad \text{and} \quad \frac{V_4}{V_1} = \frac{V_3}{V_2}$$

The last of these tells us that the quantities $\ln(V_4/V_1)$ and $\ln(V_3/V_2)$ in Equation 15-27 are equal and so cancel out, as do the factors of nR. So we're left with a very simple expression for the efficiency of a Carnot engine:

Efficiency of a Carnot engine

Kelvin temperature of the cold reservoir

$$e_{\text{Carnot}} = 1 - \frac{T_C}{T_H} \qquad (15\text{-}28)$$

Kelvin temperature of the hot reservoir

Efficiency of a Carnot engine (15-28)

The efficiency of an ideal Carnot heat engine depends *only* on the hot and cold temperatures between which the engine operates. Equation 15-28 is the maximum theoretical efficiency of *any* heat engine operating between temperatures T_H and T_C. It is impossible for any engine to do better than this.

The efficiency of an ideal Carnot engine can be made closer and closer to e_{Carnot} = 1 (100% efficiency) by decreasing T_C and increasing T_H. In order for the efficiency to be exactly one, T_C must equal 0 K. It is not possible, however, to attain a temperature of absolute zero. This statement is referred to as the **third law of thermodynamics:**

> *It is possible for the temperature of a system to be arbitrarily close to absolute zero, but it can never reach absolute zero.*

A consequence of the third law of thermodynamics is that it is impossible to make a heat engine that is 100% efficient.

Go to Interactive Exercises 15-2 and 15-3 for more practice dealing with efficiency.

EXAMPLE 15-7 Carnot Efficiency and Actual Efficiency

An automotive engine is designed to operate between 290 and 450 K. The engine produces 1.50×10^2 J of mechanical energy for every 6.00×10^2 J of heat absorbed from the combustion of fuel. Calculate (a) the actual efficiency of this engine and (b) the theoretical maximum efficiency of a Carnot engine operating between these temperatures.

Set Up

We'll use Equation 15-26 to determine the actual efficiency and Equation 15-28 to find the efficiency of a Carnot engine operating between 290 and 450 K.

Efficiency of a heat engine:

$$e = \frac{W}{Q_H} = \frac{Q_H - |Q_C|}{Q_H} = 1 - \frac{|Q_C|}{Q_H} \quad (15\text{-}26)$$

Efficiency of a Carnot engine:

$$e_{Carnot} = 1 - \frac{T_C}{T_H} \quad (15\text{-}28)$$

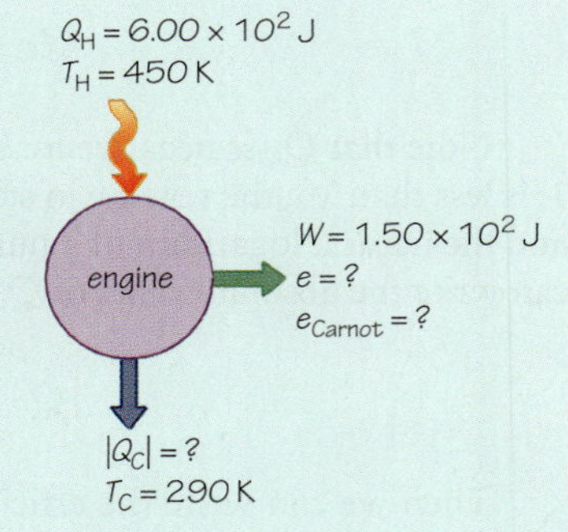

Solve

(a) We are given $W = 1.50 \times 10^2$ J and $Q_H = 6.00 \times 10^2$ J. Use this and the first expression in Equation 15-26 to find the actual efficiency of the engine.

Actual efficiency:

$$e = \frac{W}{Q_H} = \frac{1.50 \times 10^2 \text{ J}}{6.00 \times 10^2 \text{ J}} = 0.250$$

(b) The efficiency of the engine in (a) is less than the theoretical maximum efficiency of a Carnot engine with $T_C = 290$ K and $T_H = 450$ K.

Efficiency for a theoretical Carnot engine operating between these temperatures:

$$e_{Carnot} = 1 - \frac{T_C}{T_H} = 1 - \left(\frac{290 \text{ K}}{450 \text{ K}}\right) = 0.356$$

Reflect

Note that even a Carnot engine—the most efficient heat engine that could possibly operate between these temperatures—is not terribly efficient: $e_{Carnot} = 0.356$ means that only 35.6% of the energy released by combustion can be used to do work. The other 64.4% is thrown away in the form of heat. This is one reason why electric cars, which are powered by electric motors that are not subject to the limitations of heat engines, are much more energy efficient than gasoline-powered cars.

Refrigerators

A **refrigerator** is a device that takes in energy and uses it to transfer heat from an object at low temperature to an object at high temperature. An ordinary kitchen refrigerator works this way: Energy is supplied to the refrigerator in the form of electricity, and the refrigerator extracts heat from the contents of the refrigerator (which you want to keep cold) and exhausts heat into the kitchen (normally through the back of the refrigerator, which is why it's warm back there). In many ways a refrigerator is simply an engine running backward. Like heat engines, refrigerators are cyclic, and like heat engines, refrigerators are constrained by the second law of thermodynamics. The *Clausius form* of the second law of thermodynamics, named after the nineteenth-century German physicist Robert Clausius, says that:

> *No process is possible in which heat is absorbed from a cold reservoir and transferred completely to a hot reservoir.*

In other words it isn't possible to make a perfect refrigerator.

The diagram of a generic refrigerator shown in **Figure 15-17** looks similar to the generic heat engine shown in Figure 15-14, but the heat and work arrows are in the opposite direction. Work is an *input* to this device; as a result, heat Q_C is made to go from the colder region to the hotter one. This transfer is, of course, exactly what you'd want in order to cool down a container in which to store food (a refrigerator) or a building (an air conditioner). A heat pump, a device used to warm up a building, is identical to a refrigerator; the difference is that a heat pump transfers heat from the cold exterior of the building into the warm interior to keep it warm.

Figure 15-18 shows the cycle used in a typical kitchen refrigerator, which uses the inert compound R134a (chemical formula CH_2FCF_3). R134a in the liquid state is forced through a tube (the evaporator coil) that passes through the freezer and refrigerator compartments. The R134a absorbs energy from the air in these compartments to cool them, which raises the temperature of the R134a and turns it into a gas. A compressor pump then pressurizes the R134a gas, which drives the gas around the system and also increases its temperature even more. As the hot gas passes through the long, narrow condenser coil tube (usually in the back of the refrigerator), heat is transferred from the gaseous R134a to the cooler air in the kitchen. This part of the process cools the gas enough to cause it to liquefy, and the liquid R134a is further cooled by letting it expand rapidly as it passes through a valve from a narrow tube to a much wider one. The cold liquid then makes its way back to the evaporator coil to start the process again. In this way energy is removed from the interior of the refrigerator and delivered to the surroundings.

The less work required to extract heat Q_C from the cold reservoir, the more efficient the refrigerator. A perfect refrigerator would require no work to be done in order to cause heat to be transferred from the cold to the hot reservoir, but this process isn't possible according to the second law of thermodynamics. The efficiency of a refrigerator is given by the **coefficient of performance**, for which we use symbol CP:

$$CP = \frac{|Q_C|}{|W|} \qquad (15\text{-}29)$$

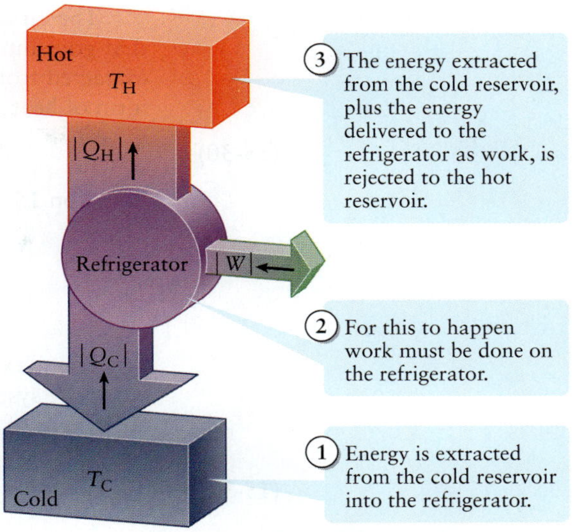

Hot
T_H
$|Q_H|$
Refrigerator
$|W|$
$|Q_C|$
Cold
T_C
(3) The energy extracted from the cold reservoir, plus the energy delivered to the refrigerator as work, is rejected to the hot reservoir.
(2) For this to happen work must be done on the refrigerator.
(1) Energy is extracted from the cold reservoir into the refrigerator.

Figure 15-17 What a refrigerator does A generic refrigerator, represented by a circle, extracts heat Q_C from a cold reservoir at temperature T_C. To do this, work must be done on the refrigerator (so W, the work done *by* the refrigerator, is negative, and the work done *on* the refrigerator is $|W|$). All of the energy added to the engine as heat and by work is rejected to the hot reservoir at temperature T_H. Here Q_H is negative (it represents heat that *leaves* the refrigerator), so the heat that is rejected *into* the hot reservoir is $|Q_H|$.

As with efficiency, CP is essentially what you get out divided by what you put in. The smaller the amount of work W for a given quantity of heat Q_C extracted from the cold reservoir, the larger the coefficient of performance CP and the better the performance of the refrigerator. A typical value for CP for a kitchen refrigerator or a home air conditioner is 3: This means that for every 1 J of work input to the refrigerator, 3 J of heat is extracted from the cold interior of the refrigerator or home and 1 J + 3 J = 4 J is rejected to the warmer surroundings.

A typical kitchen refrigerator uses the inert compound R134a (chemical formula CH_2FCF_3). The R134a is used to remove energy from the interior of the refrigerator and deliver it to the surroundings.

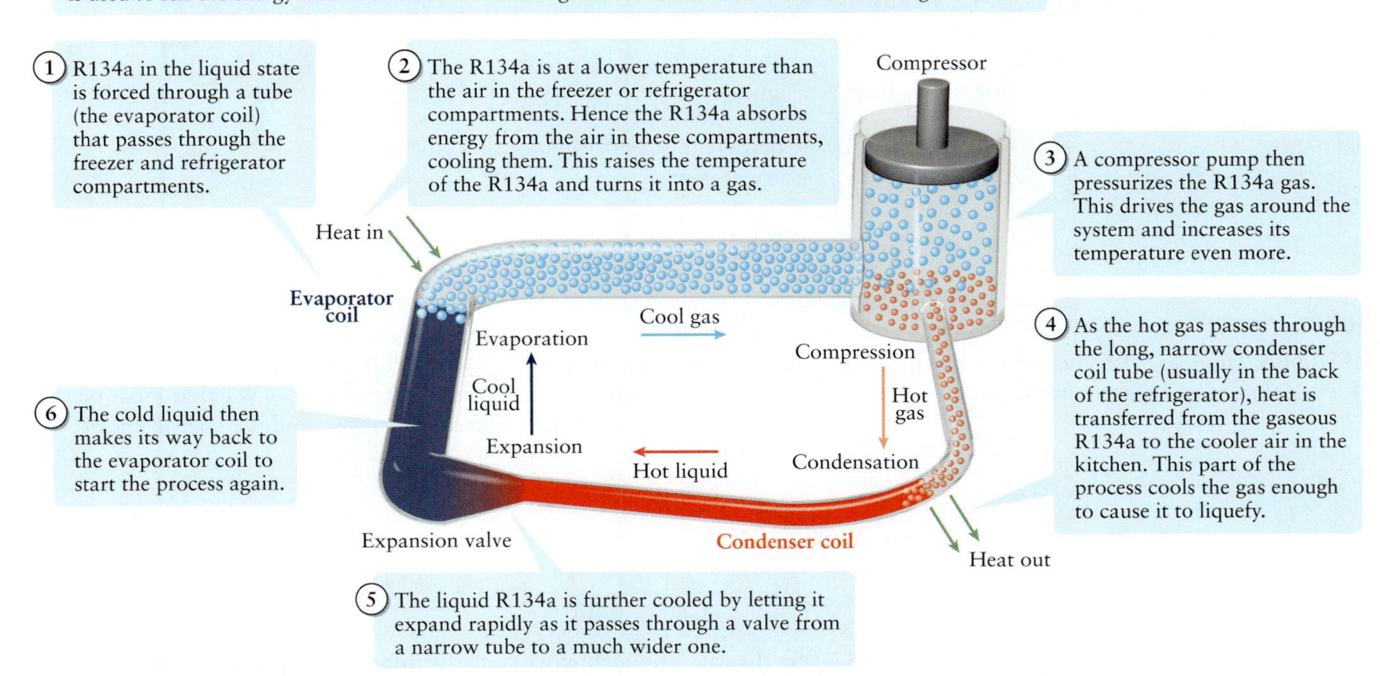

(1) R134a in the liquid state is forced through a tube (the evaporator coil) that passes through the freezer and refrigerator compartments.
(2) The R134a is at a lower temperature than the air in the freezer or refrigerator compartments. Hence the R134a absorbs energy from the air in these compartments, cooling them. This raises the temperature of the R134a and turns it into a gas.
Compressor
(3) A compressor pump then pressurizes the R134a gas. This drives the gas around the system and increases its temperature even more.
Heat in
Evaporator coil
Evaporation
Cool gas
Compression
Cool liquid
Hot gas
Expansion
Hot liquid
Condensation
(4) As the hot gas passes through the long, narrow condenser coil tube (usually in the back of the refrigerator), heat is transferred from the gaseous R134a to the cooler air in the kitchen. This part of the process cools the gas enough to cause it to liquefy.
(6) The cold liquid then makes its way back to the evaporator coil to start the process again.
Expansion valve
Condenser coil
Heat out
(5) The liquid R134a is further cooled by letting it expand rapidly as it passes through a valve from a narrow tube to a much wider one.

Figure 15-18 Inside a refrigerator This simplified diagram shows the cyclic process that cools a kitchen refrigerator. This process involves compressing and expanding a refrigerant that transitions between being a liquid and a gas.

The first law of thermodynamics requires that in one cycle, the total energy that goes into the refrigerator (in the form of work $|W|$ done on it and heat $|Q_C|$ extracted from the cold reservoir) equals the amount of energy that comes out (in the form of heat $|Q_H|$ delivered to the hot reservoir). So

$$|Q_C| + |W| = |Q_H| \quad \text{or} \quad |W| = |Q_H| - |Q_C| \text{ (refrigerator)} \tag{15-30}$$

Equation 15-29 then becomes

$$CP = \frac{|Q_C|}{|W|} = \frac{|Q_C|}{|Q_H| - |Q_C|} \tag{15-31}$$

(coefficient of performance of a refrigerator)

The maximum possible coefficient of performance of a refrigerator is obtained when the system is based on the Carnot cycle run in reverse. It can be shown that the coefficient of performance of a Carnot refrigerator is

$$CP_{\text{Carnot}} = \frac{T_C}{T_H - T_C} \tag{15-32}$$

(coefficient of performance of a Carnot refrigerator)

Like the Carnot cycle for engines, a Carnot refrigerator is not a practical device. The significance of Equation 15-32 is that it tells us the theoretical maximum coefficient of performance of a refrigerator operating between cold temperature T_C and hot temperature T_H.

EXAMPLE 15-8 Kitchen Refrigerator

A kitchen refrigerator has a coefficient of performance of 3.30. When the refrigerator pump is running, it removes energy from the interior of the refrigerator at a rate of 760 J/s. (a) How much work must be done per second to extract this energy?(b) How much energy is rejected into the kitchen per second in this process? (c) If the refrigerator operates between 2°C (the interior of the refrigerator) and 25°C (the kitchen), what is the theoretical maximum coefficient of performance?

Set Up

The energy removed from the cold interior of the refrigerator is represented by $|Q_C|$, and the work done to remove that energy is represented by $|W|$. We can relate these to the coefficient of performance by using Equation 15-31, and to the energy $|Q_H|$ rejected into the warm kitchen by using Equation 15-30. The theoretical maximum value of CP is the Carnot value, given by Equation 15-32.

Coefficient of performance of a refrigerator:

$$CP = \frac{|Q_C|}{|W|} = \frac{|Q_C|}{|Q_H| - |Q_C|} \tag{15-31}$$

Energy relationships for a refrigerator:

$$|Q_C| + |W| = |Q_H| \text{ or}$$
$$|W| = |Q_H| - |Q_C| \tag{15-30}$$

Coefficient of performance of a Carnot refrigerator:

$$CP_{\text{Carnot}} = \frac{T_C}{T_H - T_C} \tag{15-32}$$

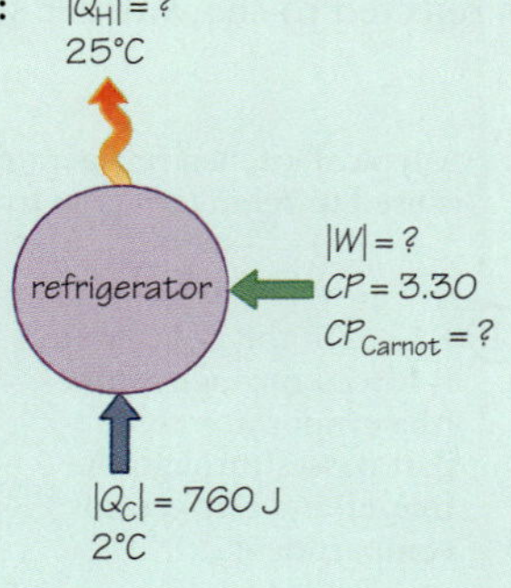

Solve

(a) Use Equation 15-31 to determine the amount of work that must be done per second.

In 1 s an amount of energy $|Q_C| = 760$ J is removed from the interior of the refrigerator. The amount of work $|W|$ that must be done to make this happen depends on $|Q_C|$ and the coefficient of performance:

$$CP = \frac{|Q_C|}{|W|}$$

$$|W| = \frac{|Q_C|}{CP} = \frac{760\ \text{J}}{3.30} = 230\ \text{J}$$

So the rate at which work must be done is 230 J/s = 230 W.

(b) Find the rate at which energy is rejected into the kitchen.

From Equation 15-30 the amount of heat $|Q_H|$ rejected into the kitchen in 1 s is

$$|Q_H| = |Q_C| + |W| = 760 \text{ J} + 230 \text{ J} = 990 \text{ J}$$

So the rate at which the refrigerator rejects heat is 990 J/s = 990 W.

(c) Calculate the coefficient of performance of a Carnot refrigerator operating between the same hot and cold temperatures.

The temperatures between which the refrigerator operates are

2°C or $T_C = (273.15 + 2)$ K = 275 K and

25°C or $T_H = (273.15 + 25)$ K = 298 K

The coefficient of performance of a Carnot refrigerator operating between these temperatures is

$$CP_{\text{Carnot}} = \frac{T_C}{T_H - T_C} = \frac{275 \text{ K}}{298 \text{K} - 275 \text{ K}} = \frac{275 \text{ K}}{23 \text{ K}}$$
$$= 12$$

Reflect

The coefficient of performance CP is substantially less than what could be achieved by an ideal Carnot refrigerator. Notice that the higher the actual value of CP, the less work must be done to achieve the same amount of refrigeration. Less work means less waste, and as a result, less heating of the kitchen. This actual refrigerator rejects heat into the kitchen at a rate of 990 J/s or 990 W, about the same as the power output of an electric space heater.

GOT THE CONCEPT? 15-6 A Super Engine

A fellow student announces that he has invented a gasoline engine that is 90% efficient. He claims that he gets such efficiency by running the engine at very high temperatures. The cylinders of his engine are made of titanium, which has a melting temperature of 1941 K. If he asks you to invest in this engine, should you agree?

TAKE-HOME MESSAGE FOR Section 15-5

✔ One statement of the second law of thermodynamics is that randomness in systems tends to increase over time.

✔ A system or device that converts heat to work is a heat engine. It is not possible to create an engine or process in which heat is absorbed from a reservoir and converted completely into work.

✔ A refrigerator is a heat engine in reverse. It uses work to move heat from a cold reservoir to a hot reservoir.

✔ The third law of thermodynamics states that the temperature of a system can never attain absolute zero.

15-6 The entropy of a system is a measure of its disorder

The second law of thermodynamics uses the concept of the *order* of a system. How can we quantify just how ordered or disordered a system is? Searching for the answer to this question will lead us to introduce a new physical quantity called *entropy* that plays an essential role in thermodynamics.

To get the best understanding of how ordered or disordered a thermodynamic system is, you would have to note the position and behavior of every one of its molecules. That's a challenging task for a system even as small as a single drop of water, which contains about 10^{22} H_2O molecules. (That's roughly 1000 times more molecules than there are grains of sand on all of Earth's beaches.) Instead let's look at a system with a much smaller number of constituents. Imagine that your sock drawer is an unorganized jumble of an equal number of blue socks and red socks. If you grab four socks without looking at their colors, what distribution of colors within the four-sock "thermodynamic system" are you most likely to have?

TABLE 15-2 Possible Combinations of Red and Blue Socks

Blue	Red	Combinations	Number
0	4	(••••)	1
1	3	(••••), (••••), (••••), (••••)	4
2	2	(••••), (••••), (••••), (••••), (••••), (••••)	6
3	1	(••••), (••••), (••••), (••••)	4
4	0	(••••)	1

There is only one way to end up with four red socks: Each individual selection must be a red sock. There are four ways to have three red socks and one blue one, however. One way is to get a red sock during each of the first three selections and then get one blue one. Another way is for the first sock chosen to be blue and the last three red. We can summarize the possible ways to get three red and one blue sock this way:

(••••), (••••), (••••), and (••••)

Table 15-2 summarizes all 16 possible four-sock combinations. The most probable grouping is two red socks and two blue socks (6 of the 16 possible combinations). This is also the most randomly distributed, least ordered four-sock state. The most ordered state would be one that was made of either four red socks or four blue socks. Table 15-2 shows that such states are also the least likely ones.

These conclusions about socks taken from a drawer turn out to be true in general: *The most probable final or equilibrium state of a system is the one that is the least ordered.* For example, a hot object and a cold object put into thermal contact come to the same final temperature not because no other possibility is allowed but because this state is the least ordered and therefore the most probable. Energy would still be conserved if, say, the hot object got hotter while the cold one got colder. But this would be highly improbable because this is a highly ordered state (the energy of the system ends up concentrated in the hot object rather than being spread between the two objects).

To represent order, we must take the list of variables that describe the state of a system—its temperature T, pressure p, volume V, and internal energy U—and add to it a new variable that we call **entropy**. Entropy, which for historical reasons we give the rather unexpected symbol S, is defined so that the larger its value the *less order* and the *more disorder* is present in a system. So we can think of entropy S as a measure of *disorder*. The second law of thermodynamics therefore states that *the entropy of an isolated system must either remain constant or increase.*

Entropy Change in a Reversible Process

The second law of thermodynamics suggests that what matters most is not the *amount* of order in a system but rather how much *change* in order there is in a thermodynamic process. The first law of thermodynamics says the same thing about internal energy. The first law tells us that the change in internal energy ΔU must be equal to $Q - W$ (the quantity of heat that enters the system minus the work that the system does). Let's see how to express the entropy change for the special case of a system that undergoes a process that happens at a constant temperature T. To be specific we'll look at the reversible isothermal expansion of an ideal gas.

Figure 15-19 shows a quantity of ideal gas at temperature T confined within a cylinder with a moveable piston. We put the cylinder in thermal contact with a large reservoir at the same temperature T then allow the gas to expand. Because the gas is in thermal contact with the reservoir, its temperature remains constant and the

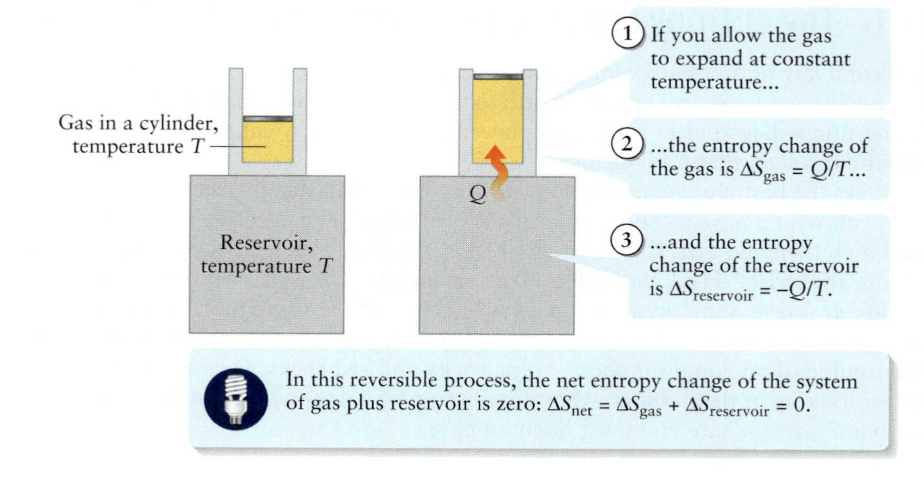

Figure 15-19 Entropy changes in an isothermal expansion In a reversible isothermal expansion of an ideal gas, the entropy of the gas increases; the entropy of the reservoir with which the gas is in contact decreases by an equal amount.

internal energy of the gas does not change: $\Delta U = 0$. The gas does work on the piston as it expands, so $W > 0$. From the first law of thermodynamics, $\Delta U = Q - W$ and $Q = \Delta U + W$, or in this case (since ΔU is zero) $Q = W$. All of the energy that flows into the gas from the reservoir in the form of heat Q is used to do work on the piston.

As the gas expands, the volume V that the gas occupies increases, and the molecules of the gas are now more spread out. This makes the gas more disordered, just as taking socks from a drawer and strewing them around the larger volume of your bedroom makes the socks more disordered. The greater the quantity of heat that enters the gas, the more the gas expands, and the greater the increase in disorder. So in this isothermal expansion the entropy change ΔS of the gas is *proportional* to the quantity of heat Q that flows into the gas.

The entropy change ΔS in an isothermal expansion also depends on the temperature T. For a given quantity of heat Q, the increase in disorder is greater if the system is at a low temperature (so that the molecules are moving slowly and are relatively well ordered to start with) than if the system is at a high temperature (so that the molecules are moving rapidly in a relatively disordered state). We conclude that the entropy change ΔS in this reversible isothermal process is *inversely proportional* to the Kelvin temperature T.

Putting these ideas together, we *define* the entropy change ΔS in a reversible isothermal process as

Heat that flows into the system ($Q > 0$ if heat flows in, $Q < 0$ if heat flows out)

Entropy change of a system in **a reversible isothermal process**

$$\Delta S = \frac{Q}{T}$$

Kelvin temperature of the system

Entropy change in a reversible isothermal process (15-33)

Equation 15-33 shows that the units of entropy are the units of heat (that is, energy) divided by the units of temperature, or joules per kelvin (J/K).

We can also apply Equation 15-33 to the *reservoir* with which the gas cylinder is in contact. If a positive quantity of heat $Q > 0$ flows *into* the gas *from* the reservoir, the heat that flows into the reservoir from the gas is $-Q$. (This says that the quantity of heat that enters the gas equals the quantity of heat that leaves the reservoir.) The gas and reservoir are both at the same temperature T, so the *net* entropy change of the system of gas plus reservoir is

$$\Delta S_{\text{net}} = \Delta S_{\text{gas}} + \Delta S_{\text{reservoir}} = \frac{Q}{T} + \frac{(-Q)}{T} = 0$$

In a reversible isothermal expansion the entropy of the gas increases (it becomes more disordered), but the entropy of the reservoir decreases (it becomes *less* disordered because it has transferred some of its random thermal energy to the gas). The net change in entropy is zero.

EXAMPLE 15-9 Calculating Entropy Change

A cylinder containing n moles of an ideal gas at temperature T changes its volume isothermally from an initial volume V_i to a final volume V_f. Find expressions for (a) the entropy change of the gas and (b) the entropy change of the thermal reservoir with which the gas is in contact. (c) Evaluate these for the special case in which 0.050 mol of gas at 20.0°C has an initial volume of 1.00×10^{-3} m^3 and expands to a final volume of 2.00×10^{-3} m^3.

Set Up

Equation 15-33 tells us the entropy change of the gas in terms of its temperature T and the heat Q that flows into the gas during the volume change. From Equation 15-5 the quantity of heat Q equals the work W that the gas does. Equation 15-6 tells us the work done in this process. We'll put all of these pieces together to find the entropy changes of the gas and the thermal reservoir.

Entropy change in a reversible isothermal process:

$$\Delta S = \frac{Q}{T} \quad (15\text{-}33)$$

Isothermal process for an ideal gas:

$$Q = W \quad (15\text{-}5)$$

Work done by an ideal gas in an isothermal process:

$$W = nRT \ln \frac{V_f}{V_i} \quad (15\text{-}6)$$

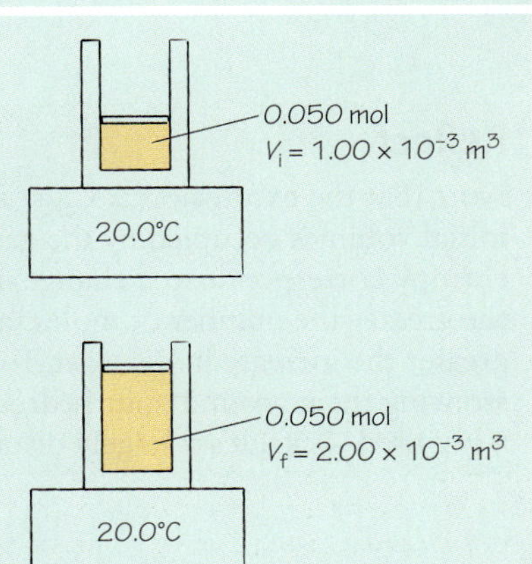

Solve

(a) Find an expression for the entropy change of the gas.

Combine Equation 15-5 and Equation 15-6 to get an expression for the heat that enters the gas:

$$Q_{gas} = W = nRT \ln \frac{V_f}{V_i}$$

From Equation 15-33 the entropy change of the gas in this process is

$$\Delta S_{gas} = \frac{Q_{gas}}{T} = \frac{nRT}{T} \ln \frac{V_f}{V_i} = nR \ln \frac{V_f}{V_i}$$

If the gas expands so that the final volume is greater than the initial volume, the ratio V_f/V_i is greater than 1 and ΔS_{gas} is positive (the entropy of the gas increases). If the gas is compressed so that the final volume is less than the initial volume, the ratio V_f/V_i is less than 1 and ΔS_{gas} is negative (the entropy of the gas decreases).

(b) Find an expression for the entropy change of the reservoir.

The heat that enters the gas comes from the reservoir, so the value of Q for the reservoir equals the negative of the value of Q for the gas:

$$Q_{reservoir} = -Q_{gas} = -nRT \ln \frac{V_f}{V_i}$$

The entropy change of the reservoir is then

$$\Delta S_{reservoir} = \frac{Q_{reservoir}}{T} = -\frac{nRT}{T} \ln \frac{V_f}{V_i} = -nR \ln \frac{V_f}{V_i}$$

If we compare this to the result from part (a), we see that $\Delta S_{reservoir} = -\Delta S_{gas}$

(c) Calculate the numerical values of ΔS_{gas} and $\Delta S_{reservoir}$.

With $n = 0.050$ mol, $V_i = 1.00 \times 10^{-3}\ \text{m}^3$, and $V_f = 2.00 \times 10^{-3}\ \text{m}^3$, we get

$$\begin{aligned}\Delta S_{gas} &= nR \ln \frac{V_f}{V_i} \\ &= (0.050\ \text{mol})\left(8.314\ \frac{\text{J}}{\text{mol} \cdot \text{K}}\right) \ln \left(\frac{2.00 \times 10^{-3}\ \text{m}^3}{1.00 \times 10^{-3}\ \text{m}^3}\right) \\ &= (0.416\ \text{J/K}) \ln 2.00 \\ &= 0.288\ \text{J/K}\end{aligned}$$

$$\Delta S_{reservoir} = -\Delta S_{gas} = -0.288\ \text{J/K}$$

The net entropy change of the system of gas and reservoir is

$$\begin{aligned}\Delta S_{net} &= \Delta S_{gas} + \Delta S_{reservoir} \\ &= 0.288\ \text{J/K} + (-0.288\ \text{J/K}) = 0\end{aligned}$$

Reflect

Note that the expression $\Delta S_{gas} = nR \ln (V_f/V_i)$ just depends on the number of moles n and the ratio of the final and initial volumes occupied by the gas. The temperature doesn't matter. This is consistent with the idea that a change in entropy corresponds to a change in disorder: The greater the increase in volume, the more disordered the system becomes; the greater the number of moles (and hence the greater the number of molecules) that expand into the new volume, the greater the increase in disorder. To see why the number of moles matter, think of taking socks from your sock drawer and strewing them around your bedroom. If there are only two socks involved, the amount of disorder isn't great; if there are a hundred socks, it's a hugely disordered mess.

In Example 15-9 the gas and reservoir together make up an *isolated* system: The gas and reservoir are in thermal contact with each other but are not in thermal contact with anything else. In this case, in which the thermodynamic process is *reversible*, the entropy of that isolated system remains constant.

Entropy Change in an Irreversible Process

Let's now consider a different isothermal process that's *irreversible*. In **Figure 15-20** a metal bar is used to put a hot reservoir at temperature T_H in thermal contact with a cold reservoir at temperature T_C. Heat flows from the hot reservoir to the cold one, but the reservoirs are so large that their temperatures remain essentially unchanged. Just as we did for the reservoir in Example 15-9, we can calculate the entropy change of each reservoir using Equation 15-33, $\Delta S = Q/T$. If a quantity of heat Q is transferred between the reservoirs, then $Q_C = +Q$ (heat flows *into* the cold reservoir from the hot one) and $Q_H = -Q$ (an equal quantity of heat flows *out* of the hot reservoir into the cold one). The entropy of the metal bar doesn't change: As much heat enters one end of the bar as leaves the other. The entropy changes of the two reservoirs are then

$$\text{Cold reservoir: } \Delta S_C = \frac{(+Q)}{T_C} \text{ (positive)}$$

$$\text{Hot reservoir: } \Delta S_H = \frac{(-Q)}{T_H} \text{ (negative)}$$

$$\text{Net entropy change: } \Delta S_{net} = \Delta S_C + \Delta S_H = \frac{(+Q)}{T_C} + \frac{(-Q)}{T_H}$$

As in Example 15-9 the entropy of one object (the cold reservoir) increases while the entropy of the other object (the hot reservoir) decreases. The difference is that the process shown in Figure 15-20 is *irreversible* because the two objects are at *different* temperatures T_C and T_H: By itself heat will flow only from the high-temperature object at T_H to the low-temperature one at T_C, never the other way, so the process cannot run in reverse. And because the two temperatures are different, the entropy changes ΔS_C and ΔS_H in this irreversible process do *not* cancel. To be specific suppose a quantity of heat $Q = 1.20 \times 10^3$ J flows from a hot reservoir at $T_H = 400$ K to a cold reservoir at $T_C = 300$ K, as shown in Figure 15-20. In this case

$$\text{Cold reservoir: } \Delta S_C = \frac{(+Q)}{T_C} = \frac{1.20 \times 10^3 \text{ J}}{300 \text{ K}} = +4 \text{ J/K}$$

$$\text{Hot reservoir: } \Delta S_H = \frac{(-Q)}{T_H} = \frac{-1.20 \times 10^3 \text{ J}}{400 \text{ K}} = -3 \text{ J/K}$$

$$\text{Net entropy change: } \Delta S_{net} = \Delta S_C + \Delta S_H = 4 \text{ J/K} + (-3 \text{ J/K}) = +1 \text{ J/K}$$

(1) A quantity of heat $Q = 1.20 \times 10^3$ J flows through the metal rod from the hot reservoir to the cold reservoir.

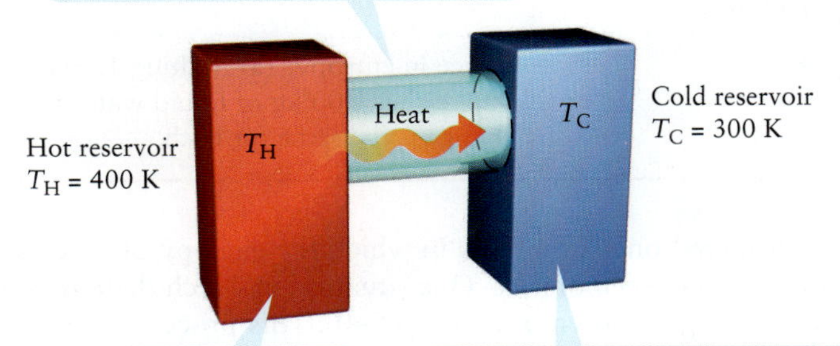

(2) Entropy change of the hot reservoir:

$$\Delta S_H = \frac{(-Q)}{T_H} = \frac{(-1.20 \times 10^3 \text{ J})}{400 \text{ K}} = -3 \text{ J/K}$$

(3) Entropy change of the cold reservoir:

$$\Delta S_C = \frac{(+Q)}{T_C} = \frac{(+1.20 \times 10^3 \text{ J})}{300 \text{ K}} = +4 \text{ J/K}$$

The net entropy change in this irreversible process is positive: $\Delta S_{net} = \Delta S_H + \Delta S_C = (-3 \text{ J/K}) + (+4 \text{ J/K}) = +1$ J/K. The net entropy always increases in an irreversible process.

Figure 15-20 Entropy changes in irreversible heat flow In irreversible heat flow from a hot object to a cold one, the *net* entropy of the two objects together increases.

For this irreversible process the cold reservoir gains more entropy than the hot reservoir loses, so there is a *net increase* in the entropy of the system of two reservoirs. So the system becomes increasingly disordered. While energy is conserved, the energy becomes less available to do useful work (for instance, to run a heat engine between the temperatures of the high and low reservoirs).

Entropy and the Second Law of Thermodynamics

We can summarize our observations about reversible and irreversible processes as follows:

In a reversible process there is no net change in entropy. In an irreversible process the net entropy increases.

This is the *entropy statement* of the second law of thermodynamics.

WATCH OUT! The entropy of a system can go down but only through a process that raises the entropy of another system.

For both the reversible isothermal expansion of an ideal gas (Figure 15-19) and the irreversible flow of heat from a hot reservoir to a cold one (Figure 15-20), one of the objects undergoes a *decrease* in entropy. This may seem to contradict the statement that entropy stays the same in a reversible process and increases in an irreversible process. In fact there's no contradiction, because this statement is about *net* entropy. Whenever an object undergoes a decrease in entropy, it's because there's a second object that undergoes an *increase* in entropy (the gas in the example of the reversible isothermal expansion and the cold reservoir in the example of irreversible heat flow). When all of the interacting objects are taken into account, the *net* entropy never decreases.

The entropy statement of the second law of thermodynamics gives us additional insight into the Carnot cycle, which is the most efficient heat engine theoretically possible. Each step in the Carnot cycle—an isothermal compression, an adiabatic compression, an isothermal expansion, and an adiabatic expansion—is *reversible*. So there is no net entropy change of the system of a Carnot engine, a hot reservoir, and a cold reservoir. (In Example 15-9 we saw that $\Delta S_{\text{net}} = 0$ for an isothermal expansion; the same is true for an isothermal compression, in which the gas loses entropy and the reservoir with which it is in contact gains an equal amount of entropy. There is no heat flow at all in a reversible adiabatic compression or expansion, so the entropy does not change in either of these steps.) At the end of each cycle, the system is just as ordered as it was at the beginning of the cycle. By contrast, if we use a heat engine that utilizes irreversible processes (like the idealized automotive engine cycle shown in Figure 15-15), there is a net increase in entropy—and hence in disorder—per cycle. So the *most efficient* engine is also the one that gives rise to the *least* increase in net entropy.

GOT THE CONCEPT? 15-7 Melting versus Vaporization

For water the latent heat of fusion (melting) is $L_F = 3.34 \times 10^5$ J/kg, and the latent heat of vaporization is $L_V = 2.26 \times 10^6$ J/kg. Which involves a greater change in entropy? (a) Melting 1.00 kg of ice at 0°C; (b) vaporizing 1.00 kg of liquid water at 100°C; (c) the entropy change is the same for both.

We've considered only situations in which the entropy of objects changes due to the transfer of heat between objects. One situation in which there is *no* heat flow is the free expansion of a gas shown in Figure 15-4b. This process is irreversible: After the barrier has ruptured and the expansion has taken place, it would be impossible for all of the gas molecules to spontaneously reassemble on the left-hand part of the cylinder in Figure 15-4b. So we would expect the entropy of the gas to increase in the irreversible free expansion. However, we can't use Equation 15-33 to calculate the entropy change; there is no heat flow into the gas (the cylinder that encloses the gas is insulated), so the equation $\Delta S = Q/T$ predicts incorrectly that there would be *zero* entropy change. The reason Equation 15-33 doesn't work is that it assumes that the process occurs slowly enough that the system is never far from equilibrium, which is definitely not the case for the sudden rush of gas in a free expansion. Instead we find the entropy change by noting that the final state—pressure, volume, and temperature—of the gas after a free expansion is the same as if it had undergone a reversible isothermal expansion to the same final volume (compare parts (a) and (b) of Figure 15-4). We saw in Example 15-9 that the entropy of the gas increases in such a process, and we saw how to calculate ΔS_{gas}. Because entropy is a state variable whose value does not depend on the history of the system, the change in entropy ΔS_{gas} must be the same for a free expansion. In a reversible isothermal expansion the gas is in thermal contact with a reservoir at the same temperature, and the entropy of the reservoir decreases by as much as the entropy of the gas increases so that the net entropy change is zero. But for a free expansion the cylinder is isolated, so the only entropy change is the increase in entropy ΔS_{gas} of the gas. So there is indeed a net entropy increase, just as we expect for an irreversible process.

It's instructive to ask what the entropy statement of the second law of thermodynamics says about the existence and evolution of life. Living organisms such as the cat shown in Figure 15-1a undergo a decrease in entropy as they develop. For example, humans start out as a single cell and grow into a highly ordered network of cells by rearranging raw materials (food) from the environment. Complex organisms are more efficient, require less energy, and are thereby better able to survive in an environment of limited resources. The processes, like evolution, by which complex organisms arise from simpler ones necessarily introduce more order and therefore lower entropy. Does this mean that they violate the second law of thermodynamics?

The answer to this question is no: Life most definitely does *not* violate the second law of thermodynamics. While the entropy of a living organism decreases as it develops, the entropy of its surroundings *increases*. That's because when an organism consumes and metabolizes food, it uses some of the energy from the food to do work (for example, to grow) but also wastes some of that energy, too. It is this inefficiency that causes the *net* effect to be an increase in entropy. Another way to say this is that the entropy of an organism decreases, but the entropy of its waste products—for example, the carbon dioxide that you exhale, the perspiration that you release from your skin, and the liquid and solid material that you excrete—increases more than the organism's entropy decreases.

TAKE-HOME MESSAGE FOR Section 15-6

✔ The entropy of a system is a measure of the disorder of the system. Systems with a low value of entropy are more ordered.

✔ The most probable final state of a system is the one that is the most disordered and has the highest entropy.

✔ In a process in which two objects interact, it's possible for the entropy of one object to decrease. However, the *net* entropy remains constant in a reversible process and increases in an irreversible process.

Key Terms

adiabatic process
Carnot cycle
coefficient of performance
efficiency
entropy
first law of thermodynamics
heat engine
internal energy
irreversible process
isobaric process
isochoric process
isotherm
isothermal process
molar specific heat
molar specific heat at constant pressure
molar specific heat at constant volume
pV diagram
ratio of specific heats
refrigerator
reservoir
reversible process
second law of thermodynamics
state variable
thermodynamic process
third law of thermodynamics

Chapter Summary

Topic	Equation or Figure
The first law of thermodynamics: In any thermodynamic process the change in internal energy of the system is related to the heat Q that enters the system and the work W that the system does during the process. There are simple rules for the signs of Q and W.	During a thermodynamic process the **change in the internal energy of a system**... $\Delta U = Q - W$ (15-2) ...equals the **heat that flows into the system** during the process... ...minus the **work that the system does** during the process.
	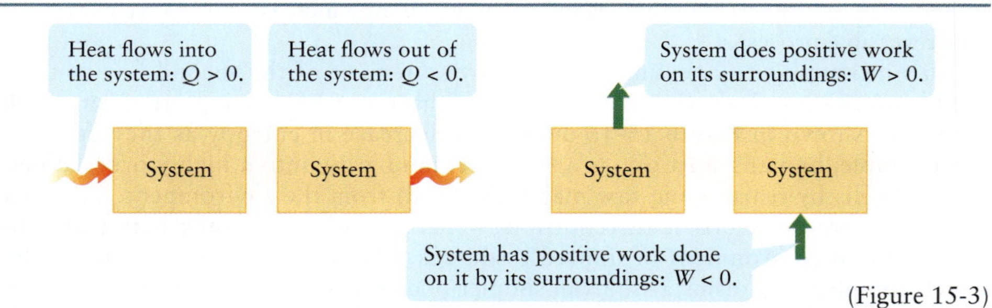 (Figure 15-3)
Thermodynamic processes: Four important types of thermodynamic processes are isobaric (constant pressure), isothermal (constant temperature), adiabatic (no heat flow), and isochoric (constant volume). In *any* process, the work done is represented by the area under the curve of the pV diagram for the process. There are simple expressions for the work done in an isobaric process and for the work done by an ideal gas in an isothermal process. In an isochoric process zero work is done.	**Work** that a system does in an isobaric process; **Volume** of the system **at the end** of the process. $W = p(V_f - V_i)$ (15-4) **Constant pressure** of the system; **Volume** of the system **at the beginning** of the process
	Number of moles of gas; **Ideal gas constant**; **Work** that an ideal gas does in an isothermal process. $W = nRT \ln \frac{V_f}{V_i}$ (15-6) **Volume** of the gas **at the end** of the process; **Volume** of the gas **at the beginning** of the process; **Kelvin temperature** of the gas
Molar specific heats of a gas: The molar specific heat is the energy required to raise the temperature of one mole of substance by one kelvin. For an ideal gas, the molar specific heats at constant pressure (C_p) and constant volume (C_V) differ by a fixed amount; the value of C_V depends on the number of degrees of freedom of a gas molecule.	**Quantity of heat** that flows into (if $Q > 0$) or out of (if $Q < 0$) an object; **Molar specific heat** of the substance of which the object is made. $Q = nC\Delta T$ (15-10) **Number of moles** present of the substance; **Temperature change** of the object that results from the heat flow
	Molar specific heat of an ideal gas at **constant pressure**. $C_p - C_V = R$ (15-15) **Ideal gas constant**; **Molar specific heat** of an ideal gas at **constant volume**
	Number of degrees of freedom for a molecule of the gas; **Molar specific heat** of an ideal gas at **constant volume**. $C_V = \left(\frac{D}{2}\right)R$ (15-18) **Ideal gas constant**

Adiabatic processes for an ideal gas: How pressure, volume, and temperature change in an adiabatic process for an ideal gas depends on the value of the ratio of specific heats for the gas, $\gamma = C_p/C_V$.

Pressure of an ideal gas

Volume of an ideal gas

$$pV^{\gamma} = \text{constant} \tag{15-22}$$

Ratio of specific heats of the gas = C_p/C_V

The quantity pV^{γ} has the same value at all times during an **adiabatic process** for an ideal gas.

Kelvin temperature of an ideal gas

Volume of an ideal gas

$$TV^{\gamma-1} = \text{constant} \tag{15-23}$$

Ratio of specific heats of the gas = C_p/C_V

The quantity $TV^{\gamma-1}$ has the same value at all times during an **adiabatic process** for an ideal gas.

The second law of thermodynamics: The second law says that the amount of order in an isolated system cannot increase. This implies that a heat engine—a device that converts heat into work—cannot be 100% efficient but can convert only part of its heat intake Q_H (from a high-temperature source) into work. The rest is rejected as heat $|Q_C|$ to the engine's low-temperature surroundings. The efficiency is $e = W/Q_H$.

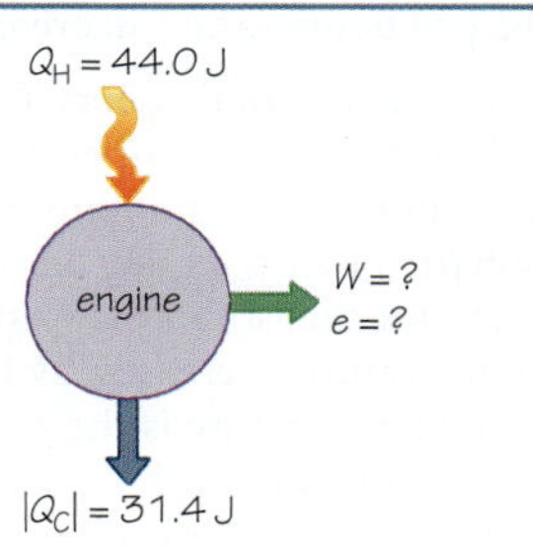

(Example 15-6)

The Carnot cycle: The most efficient heat engine theoretically possible is a Carnot engine. It uses the Carnot cycle, which is based on two reversible adiabatic processes and two reversible isothermal processes. No other engine can have a greater efficiency. A Carnot engine run in reverse is also the best-performing refrigerator theoretically possible.

Efficiency of a **Carnot engine**

Kelvin temperature of the **cold reservoir**

$$e_{\text{Carnot}} = 1 - \frac{T_C}{T_H} \tag{15-28}$$

Kelvin temperature of the **hot reservoir**

Entropy: The entropy of a system is a measure of the amount of disorder present in the system. In a reversible process in which two objects interact with each other, the entropy of one object increases while the entropy of the other object decreases by the same amount. In an irreversible process, the entropy changes do not cancel and there is a net increase in entropy.

Heat that flows into the system ($Q > 0$ if heat flows in, $Q < 0$ if heat flows out)

Entropy change of a system in a **reversible isothermal process**

$$\Delta S = \frac{Q}{T} \tag{15-33}$$

Kelvin temperature of the system

Answer to What do you think? Question

(c) The gas escapes very rapidly through the opening in the beetle's abdomen, moving from the high-pressure reaction chamber into the low-pressure air. The gas expands when it encounters lower pressure, but due to its high speed it has no time to exchange heat with the surrounding air. So the expansion of this gas is adiabatic, which means that its temperature drops even though there is no heat flow (see Section 15-3). The droplets of hot liquid do not expand, however, so they maintain their high temperature and deliver a painful (or even fatal) shock to any insect that dares attack the bombardier beetle.

Answers to Got the Concept? Questions

15-1 (b) We can answer this question using the first law of thermodynamics, $\Delta U = Q - W$. In this situation $Q = 0$ because there is no heat flow into or out of the gas. The gas pushes on the moving piston in the same direction that the piston moves, so the gas does positive work on the piston and W is positive. So $\Delta U = Q - W$ is equal to zero minus a positive number, and ΔU is negative—that is, the internal energy U of the gas decreases. For an ideal gas the internal energy U is directly proportional to the Kelvin temperature T, so a decrease in internal energy means that the gas temperature decreases.

15-2 (a) (i), (ii), (iii) and (iv) (tie) The internal energy U of an ideal gas is directly proportional to its Kelvin temperature. So the internal energy change ΔU is positive if the temperature increases, as is the case for the isobaric process $a \rightarrow b$; ΔU is zero if the temperature stays the same, as for the isothermal process $a \rightarrow c$; and ΔU is negative if the temperature decreases, as for the processes $a \rightarrow d$ and $a \rightarrow e$. The final temperature is the same for processes $a \rightarrow d$ and $a \rightarrow e$ (states d and e are both on the same isotherm), so ΔU has the same negative value for both processes $a \rightarrow d$ and $a \rightarrow e$. (b) (i), (ii), (iii), (iv) The work done in each process is indicated by the area under the curve for that process in the pV diagram. This is greatest for $a \rightarrow b$, smaller for $a \rightarrow c$, and even smaller for $a \rightarrow d$. For all three of these processes, the volume increases, and $W > 0$. For the isochoric process $a \rightarrow e$ there is no volume change, so the gas does no work at $W = 0$.

15-3 (c) The first law of thermodynamics, Equation 15-2, tells us that $\Delta U = Q - W$ or $Q = \Delta U + W$. For an adiabatic process $Q = 0$ and so $\Delta U = -W$. For process $a \rightarrow b$ we have $\Delta U > 0$ (the internal energy increases because the temperature increases) and $W > 0$ (the gas expands), so $Q = \Delta U + W$ is positive and the process cannot be adiabatic. For process $a \rightarrow c$ we have $\Delta U = 0$ (the internal energy remains the same because the temperature is constant) and $W > 0$ (the gas expands), so again $Q = \Delta U + W$ is positive and the process cannot be adiabatic. For process $a \rightarrow e$ $\Delta U < 0$ (the internal energy decreases because the temperature decreases) and $W = 0$ (the volume does not change and so the gas does zero work), so $Q = \Delta U + W$ is negative and the process cannot be adiabatic. The only process that could be adiabatic is $a \rightarrow d$, for which $\Delta U < 0$ (the internal energy decreases because the temperature decreases) and $W > 0$ (the gas expands). For the process the quantity $Q = \Delta U + W$ could be adiabatic.

15-4 (b) For the first, constant-volume process the relationship between heat and temperature change is $Q = nC_V \Delta T$, while for the second, constant-pressure process the relationship is $Q = nC_p \Delta T$ (see Equations 15-11). The relationship $C_p - C_V = R$ (Equation 15-15) tells us that C_p is greater than C_V, so for a given quantity of heat Q the temperature change ΔT will be less for a constant-pressure process. You can get the same result from the first law of thermodynamics, $\Delta U = Q - W$ or $Q = \Delta U + W$. For the constant-volume process $W = 0$ and $Q = \Delta U$: All of the heat goes into increasing the internal energy of the gas and so into raising its temperature. But for the constant-pressure process, the expanding gas must do positive work W on the piston. Some of the heat goes into doing this work, so only a fraction of Q is available to increase the internal energy of the gas.

15-5 (c) From Equations 15-11 the relationship between heat and temperature change for these constant-pressure processes is $Q = nC_p \Delta T$. Table 15-1 shows that the molar specific heat at constant pressure C_p has a smaller value for He than for N_2, so for a given quantity of heat Q the temperature change ΔT is greater for He than for N_2. Equation 15-13 tells us that the work done by an ideal gas in a constant-pressure process is $W = nR\,\Delta T$; the number of moles n is the same for both gases and R is the ideal gas constant, so more work is done by the gas that undergoes the greater temperature change ΔT (in this example, helium).

15-6 No. A gasoline engine uses ambient temperature (about 20°C or 293 K) as the cold temperature. The engine your fellow student has designed uses a hot temperature that must be less than 1941 K, the melting temperature of titanium, or else the cylinders would melt. The efficiency of a Carnot cycle operating between $T_C = 293$ K and $T_H = 1941$ K is $e_{Carnot} = 1 - (T_C/T_H) = 1 - [(293\text{ K})/(1941\text{ K})] = 1 - 0.151 = 0.849$, or 84.9%. Your fellow student is claiming an even higher efficiency, which is impossible because no heat engine can have greater efficiency than a Carnot engine. This would be a poor investment!

15-7 (b) Both melting and vaporization take place at a constant temperature, so we can use Equation 15-33, $\Delta S = Q/T$, to calculate the entropy change in each process. For melting at 0°C or $T = 273.15$ K, $Q = mL_F = (1.00\text{ kg})(3.34 \times 10^5\text{ J/kg}) = 3.34 \times 10^5$ J and $\Delta S_{melting} = Q/T = (3.34 \times 10^5\text{ J})/(273.15\text{ K}) = 1.22 \times 10^3$ J/K. For vaporization at 100°C or $T = (100 + 273.15)$ K $= 373.15$ K, $Q = mL_V = (1.00\text{ kg})(2.26 \times 10^6\text{ J/kg}) = 2.26 \times 10^6$ J and $\Delta S_{vaporization} = Q/T = (2.26 \times 10^6\text{ J})/(373.15\text{ K}) = 6.06 \times 10^3$ J/K. The entropy change is about five times greater for vaporization than for melting, which agrees with the idea that entropy is a measure of disorder. The molecules of liquid water are able to move while those of ice cannot, so the entropy of liquid water is greater than that of ice. But the molecules of water vapor are free to move over a much greater volume, so the entropy of water vapor is *very* much greater than that of liquid water.

Questions and Problems

In a few problems you are given more data than you actually need; in a few other problems you are required to supply data from your general knowledge, outside sources, or informed estimate.

Interpret as significant all digits in numerical values that have trailing zeros and no decimal points. For all problems use $g = 9.80 \text{ m/s}^2$ for the free-fall acceleration due to gravity. Neglect friction and air resistance unless instructed to do otherwise.

- • Basic, single-concept problem
- •• Intermediate-level problem; may require synthesis of concepts and multiple steps
- ••• Challenging problem
- SSM *Solution is in Student Solutions Manual*
- Example *See worked example for a similar problem*

Conceptual Questions

1. • Can a system absorb heat without increasing its internal energy? Explain.

2. • Why is it possible for the temperature of a system to remain constant even though heat is released or absorbed by the system?

3. • In a slow, steady isothermal expansion of an ideal gas against a piston, the work done is equal to the heat input. Is this consistent with the first law of thermodynamics?

4. • Clearly define and give an example of each of the following thermodynamic processes: (a) isothermal, (b) adiabatic, (c) isobaric, and (d) isochoric. SSM

5. • Why does the temperature of a gas increase when it is quickly compressed?

6. • When we say "engine," we think of something mechanical with moving parts. In such an engine friction always reduces the engine's efficiency. Why? SSM

7. • Why do engineers designing a steam-electric generating plant always try to design for as high a feed-steam temperature as possible?

8. • The frictional drag of the atmosphere causes an orbiting satellite to move closer to Earth and to gain kinetic energy. In what way does energy become unavailable for doing work in this irreversible process?

9. • Is the operation of an automobile engine reversible?

10. • Is a process necessarily reversible if there is no exchange of heat between the system in which the process takes place and its surroundings?

11. • Conduction across a temperature difference is an irreversible process, but the object that lost heat can always be rewarmed, and the one that gained heat can be recooled. An object sliding across a rough table slows down and warms up as mechanical energy dissipates. This process is irreversible, but the object can be cooled and set moving again at its original speed. So in just what sense are these processes "irreversible"? SSM

12. • There are people who try to keep cool on a hot summer day by leaving the refrigerator door open, but you can't cool your kitchen this way! Why not?

13. • If the coefficient of performance is greater than 1, do we get more energy out than we put in, violating conservation of energy? Why or why not?

14. • How does the time required to freeze water vary with each of the following parameters: mass of water, power of the refrigerator, and temperature of the outside air?

15. • How is the entropy of the universe changed when heat is released from a hotter object to a colder one? In what sense does this correspond to energy becoming unavailable for doing work? SSM

16. • If you drop a glass cup on the floor, it will shatter into fragments. If you then drop the fragments on the floor, why will they not become a glass cup?

17. • Why is the entropy of 1 kg of liquid iron greater than that of 1 kg of solid iron? Explain your answer.

18. • If a gas expands freely into a larger volume in an insulated container so that no heat is added to the gas, its entropy increases. Explain this using the idea that this process is irreversible.

19. • In discussing the Carnot cycle, we say that extracting heat from a reservoir isothermally does not change the entropy of the universe. In a real process, this is a limiting situation that can never quite be reached. Why not? What is the effect on the entropy of the universe? SSM

20. • A pot full of hot water is placed in a cold room, and the pot gradually cools. How does the entropy of the water change?

Multiple-Choice Questions

21. • An ideal gas trapped inside a thermally isolated cylinder expands slowly by pushing back against a piston. The temperature of the gas
    A. increases.
    B. decreases.
    C. remains the same.
    D. increases if the process occurs quickly.
    E. remains the same if the process occurs quickly. SSM

22. • A gas is compressed adiabatically by a force of 800 N acting over a distance of 5.0 cm. The net change in the internal energy of the gas is
    A. +800 J.
    B. +40 J.
    C. −800 J.
    D. −40 J.
    E. 0.

23. • An ideal gas is contained in a closed cylinder of fixed length and diameter. Eighty joules of heat is added to the gas. The work done by the gas on the walls of the cylinder is
    A. 80 J.
    B. 0 J.
    C. less than 80 J.
    D. more than 80 J.
    E. not specified by the information given.

24. • In an isothermal process there is no change in
    A. pressure.
    B. temperature.
    C. volume.
    D. heat.
    E. internal energy *or* pressure.

25. • In an isobaric process there is no change in
    A. pressure.
    B. temperature.
    C. volume.
    D. internal energy.
    E. internal energy *or* pressure. SSM

26. • In an isochoric process there is no change in
   A. pressure.
   B. temperature.
   C. volume.
   D. internal energy.
   E. internal energy *or* pressure.

27. • A gas quickly expands in an isolated environment. During the process the gas exchanges no heat with its surroundings. The process is
   A. isothermal.
   B. isobaric.
   C. isochoric.
   D. adiabatic.
   E. isotonic.

28. • The statement that no process is possible in which heat is absorbed from a cold reservoir and transferred completely to a hot reservoir is
   A. not always true.
   B. only true for isothermal processes.
   C. the first law of thermodynamics.
   D. the second law of thermodynamics.
   E. the zeroth law of thermodynamics.

29. • Carnot's heat engine employs
   A. two adiabatic processes and two isothermal processes.
   B. two adiabatic processes and two isobaric processes.
   C. two adiabatic processes and two isochoric processes.
   D. two isothermal processes and two isochoric processes.
   E. two isothermal processes and two isobaric processes. SSM

30. • Compare two methods to improve the theoretical efficiency of a heat engine: lower T_C by 10 K or raise T_H by 10 K. Which one is better?
   A. Lower T_C by 10 K.
   B. Raise T_H by 10 K.
   C. Both changes would give the same result.
   D. The better method would depend on the difference between T_C and T_H.
   E. There is nothing you can do to improve the theoretical efficiency of a heat engine.

Estimation/Numerical Analysis

31. • (a) Estimate the work (in J) done in raising a book from the floor to the table. (b) Estimate the temperature rise in a glass of water if that amount of energy were added to it.

32. • (a) Estimate the work (in J) done in driving a car across America. (b) If the energy required to do that work were added to an Olympic-sized swimming pool, how much would the temperature of the water rise?

33. • Estimate the internal energy increase in a 1-L sample of oxygen that increases in temperature from 20°C to 100°C. Assume that the volume of this ideal gas is constant in the process. SSM

34. • Estimate the pressure acting on a 2-L sample of nitrogen, an ideal gas, when it is held at 300 K.

35. • Estimate how much energy is expended by all the runners in the New York City Marathon.

36. • Estimate the efficiency of an average internal combustion engine in Canada.

37. • Estimate the distance a car can be driven on a tank of gas. Assume that the gas releases 125,000 BTU of energy per gallon. SSM

38. • Estimate the amount of energy that is wasted each day in Europe due to an additional inefficiency of 5% (on top of the thermodynamic efficiency) from cars that are not lubricated properly. You will have to make some assumptions about how many cars are poorly maintained, how far Europeans drive each day, and so on.

39. • **Biology** Estimate the efficiency of the human body acting as a heat engine.

Problems

15-1 The laws of thermodynamics involve energy and entropy

15-2 The first law of thermodynamics relates heat flow, work done, and internal energy change

40. • If 800 J of heat is added to a system that does no external work, how much does the internal energy of the system increase? Example 15-1

41. • Five hundred joules of heat is absorbed by a system that does 200 J of work on its surroundings. What is the change in the internal energy of the system? Example 15-1

15-3 A graph of pressure versus volume helps to describe what happens in a thermodynamic process

42. • Calculate the amount of work done *on* a gas that undergoes a change of state described by the pV diagram shown in **Figure 15-21**.

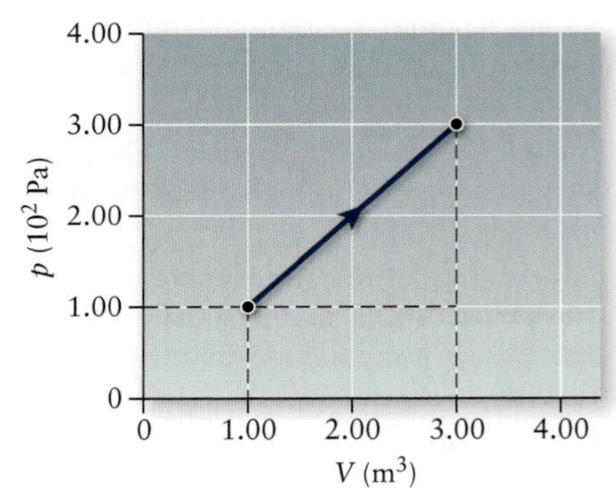

Figure 15-21 Problem 42

43. • Calculate the amount of work done *on* a gas that undergoes a change of state described by the pV diagram shown in **Figure 15-22**.

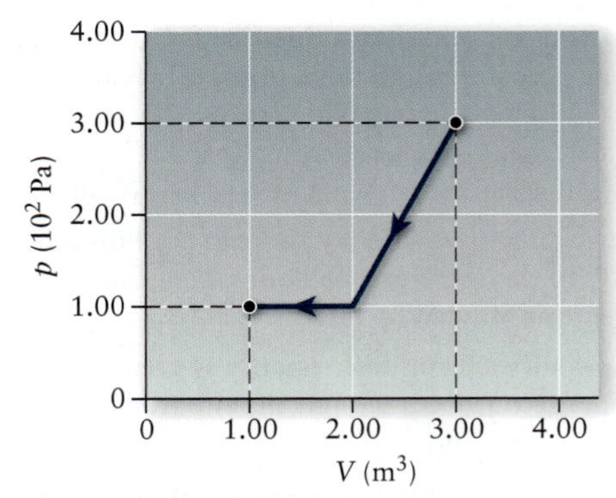

Figure 15-22 Problem 43

44. • A gas is heated and is allowed to expand such that it follows a horizontal line path on a pV diagram from its initial state (1.0×10^5 Pa, 1.0 m^3) to its final state (1.0×10^5 Pa, 2.0 m^3). Calculate the work done by the gas on its surroundings. Example 15-4

45. • A gas is heated such that it follows a vertical line path on a pV diagram from its initial state (1.0×10^5 Pa, 3.0 m^3) to its final state (2.0×10^5 Pa, 3.0 m^3). Calculate the work done by the gas on its surroundings. SSM Example 15-4

46. • A sealed cylinder has a piston and contains 8.00×10^3 cm^3 of an ideal gas at a pressure of 8.00 atm. Heat is slowly introduced, and the gas isothermally expands to 1.60×10^4 cm^3. How much work does the gas do on the piston? Example 15-3

47. • An ideal gas expands isothermally, performing 8.80 kJ of work in the process. Calculate the heat absorbed during the expansion. SSM Example 15-3

48. • Heat is added to 8.00 m^3 of helium gas in an expandable chamber that increases its volume by 2.00 m^3. If in the isothermal expansion process 2.00 kJ of work is done by the gas, what was its original pressure? Example 15-3

49. • A cylinder that has a piston contains 2.00 mol of an ideal gas and undergoes a reversible isothermal expansion at 400 K from an initial pressure of 12.0 atm down to 3.00 atm. Determine the amount of work done by the gas. Example 15-3

50. • A gas contained in a cylinder that has a piston is kept at a constant pressure of 2.80×10^5 Pa. The gas expands from 0.500 m^3 to 1.50 m^3 when 300 kJ of heat is added to the cylinder. What is the change in internal energy of the gas? Example 15-2

51. • The pressure in an ideal gas is slowly reduced to $\frac{1}{4}$ its initial value, while being kept in a container with rigid walls. In the process 800 kJ of heat leaves the gas. What is the change in internal energy of the gas during this process? SSM Example 15-4

52. • An ideal gas is compressed adiabatically to half its volume. In doing so 1888 J of work is done on the gas. What is the change in internal energy of the gas? Example 15-3

15-4 More heat is required to change the temperature of an ideal gas isobarically than isochorically

53. • Two moles of an ideal monatomic gas expand adiabatically, performing 8.00 kJ of work in the process. What is the change in temperature of the gas during the expansion? Example 15-5

54. • One mole of an ideal monatomic gas ($\gamma = 1.66$), initially at a temperature of 0.00°C, undergoes an adiabatic expansion from a pressure of 10.0 atm to a pressure of 2.00 atm. How much work is done on the gas? Example 15-5

55. • A container holds 32.0 g of oxygen gas at a pressure of 8.00 atm. How much heat is required to increase the temperature by 100°C at constant pressure?

56. • A container holds 32.0 g of oxygen gas at a pressure of 8.00 atm. How much heat is required to increase the temperature by 100°C at constant volume?

57. • The temperature of 4.00 g of helium gas is increased at constant volume by 1.00°C. Using the same amount of heat, the temperature of what mass of oxygen gas will increase at constant volume by 1.00°C? SSM

58. • Heat is added to 1.00 mol of air at constant pressure, resulting in a temperature increase of 100°C. If the same amount of heat is instead added at constant volume, what is the temperature increase? The molar specific heat ratio $\gamma = C_p/C_V$ for the air is 1.4.

59. • The volume of a gas is halved during an adiabatic compression that increases the pressure by a factor of 2.6. What is the molar specific heat ratio $\gamma = C_p/C_V$?

60. •• The volume of a gas is halved during an adiabatic compression that increases the pressure by a factor of 2.5. By what factor does the temperature increase?

61. • What ratio of initial volume to final volume V_i/V_f will raise the temperature of air from 27.0°C to 857°C in an adiabatic process? The molar specific heat ratio $\gamma = C_p/C_V$ for air is 1.4. SSM Example 15-5

62. • A monatomic ideal gas at a pressure of 1.00 atm expands adiabatically from an initial volume of 1.50 m^3 to a final volume of 3.00 m^3. What is the new pressure?

15-5 The second law of thermodynamics describes why some processes are impossible

63. • An engine doing work takes in 10.0 kJ and exhausts 6.00 kJ. What is the efficiency of the engine? SSM Example 15-6

64. • What is the theoretical maximum efficiency of a heat engine operating between 100°C and 500°C? Example 15-7

65. • A heat engine operating between 473 K and 373 K runs at 70.0% of its theoretical maximum efficiency. What is its efficiency? Example 15-7

66. • An engine operates between 10.0°C and 200°C. At the very best how much heat should we be prepared to supply in order to output 1.00×10^3 J of work? Example 15-7

67. • A furnace supplies 28.0 kW of thermal power at 300°C to an engine that exhausts waste energy at 20.0°C. At the very best how much work could we expect to get out of the system per second? SSM Example 15-7

68. • A kitchen refrigerator extracts 75.0 kJ per second of energy from a cool chamber while exhausting 1.00×10^2 kJ per second to the room. What is its coefficient of performance? Example 15-8

69. • What is the coefficient of performance of a Carnot refrigerator operating between 0.00°C and 80.0°C? Example 15-8

70. •• An electric refrigerator removes 13.0 MJ of heat from its interior for each kilowatt-hour of electric energy used. What is its coefficient of performance? Example 15-8

71. • A certain refrigerator requires 35.0 J of work to remove 190 J of heat from its interior. (a) What is its coefficient of performance? (b) How much heat is ejected to the surroundings at 22.0°C? (c) If the refrigerator cycle is reversible, what is the temperature inside the refrigerator? SSM Example 15-8

15-6 The entropy of a system is a measure of its disorder

72. • A reservoir at a temperature of 400 K gains 100 J of heat from another reservoir. What is its entropy change? Example 15-9

73. • What is the minimum change of entropy that occurs in 0.200 kg of ice at 273 K when 6.68×10^4 J of heat is added so that it melts to water? SSM Example 15-9

74. • If, in a reversible process, enough heat is added to change a 500-g block of ice to water at a temperature of 273 K, what is the change in the entropy of the ice/water system? Example 15-9

75. • A room is at a constant 295 K maintained by an air conditioner that pumps heat out. What is the entropy change of the room for each 5.00 kJ of heat removed? Example 15-9

76. •• One mole of ideal gas expands isothermally from 1.00 m^3 to 2.00 m^3. What is the entropy change for the gas?

77. •• A 1.80×10^3-kg car traveling at 80.0 km/h crashes into a concrete wall. If the temperature of the air is 27.0°C, what is the entropy change of the universe as a result of the crash? Assume all of the car's kinetic energy is converted into heat. SSM Example 15-9

78. •• A 1.00×10^3-kg rock at 20.0°C falls 1.00×10^2 m into a large lake, also at 20.0°C. Assuming that all of the rock's kinetic energy on entering the lake converts to thermal energy absorbed by the lake, what is the change in entropy of the lake? Example 15-9

79. • **Astronomy** The surface of the Sun is about 5700 K, and the temperature of Earth's surface is about 293 K. What entropy change occurs when 8000 J of energy is transferred by heat from the Sun to Earth? Example 15-9

80. • A 0.750-L cup of coffee at 70°C is left outside where the temperature is 4°C. When the coffee reaches thermal equilibrium with the atmosphere, by how much has the entropy of the atmosphere changed? Assume the properties of coffee are identical to those of water. Example 15-9

81. • A balloon containing 0.50 mol of helium in a 20.0°C chamber undergoes an isothermal expansion as the pressure in the chamber is slowly reduced. If the pressure drops to half its initial value during this process, what is the entropy change of the helium? Example 15-9

General Problems

82. ••• A vertical, insulated cylinder contains an ideal gas. The top of the cylinder is closed off by a piston of mass m that is free to move up and down with no appreciable friction. The piston is a height h above the bottom of the cylinder when the gas alone supports it. Sand is now very slowly poured onto the piston until the weight of the sand is equal to the weight of the piston. Find the new height of the piston (in terms of h) if the ideal gas in the cylinder is (a) oxygen, O_2; (b) helium, He; or (c) hydrogen, H_2. Example 15-5

83. • A Carnot engine on a ship extracts heat from seawater at 18.0°C and exhausts the heat to evaporating dry ice at −78.0°C. If the ship's engines are to run at 8.00×10^3 horsepower, what is the minimum amount of dry ice the ship must carry for the ship to run for a single day? SSM Example 15-7

84. • A Carnot engine removes 1.20×10^3 J of heat from a high-temperature source and dumps 6.00×10^2 J to the atmosphere at 20.0°C. (a) What is the efficiency of the engine? (b) What is the temperature of the hot reservoir? Example 15-7

85. •• A certain engine has a second-law efficiency of 85.0%. During each cycle it absorbs 4.80×10^2 J of heat from a reservoir at 300°C and dumps 3.00×10^2 J of heat to a cold temperature reservoir. (a) What is the temperature of the cold reservoir? (b) How much more work could be done by a Carnot engine working between the same two reservoirs and extracting the same 4.80×10^2 J of heat in each cycle? Example 15-7

86. • A refrigerator is rated at 370 W. Its interior is at 0°C, and its surroundings are at 20°C. If the second law efficiency of its cycle is 66%, how much heat can it remove from its interior in 1 min? Example 15-8

87. •• **Medical** During a high fever a 60.0-kg-patient's normal metabolism is increased by 10.0%. This results in an increase of 10.0% in the heat given off by the person. When the person slowly walks up five flights of stairs (20.0 m), she normally releases 1.00×10^5 J of heat. Compare her efficiency when she has a fever to when her temperature is normal. SSM Example 15-6

88. •• A rigid 5.50-L pressure cooker contains steam initially at 100°C under a pressure of 1.00 atm. Consult Table 15-1 as needed and assume that the values given there remain constant. The mass of a water molecule is 2.99×10^{-26} kg. (a) To what temperature (in °C) would you have to heat the steam so that its pressure was 1.25 atm? (b) How much heat would you need in part (a)? (c) Calculate the specific heat of the steam in part (a) in units of J/(kg • K).

89. • Liquid nitrogen, with a latent heat of fusion of $L_F = 25.3 \times 10^3$ J/kg, solidifies at a temperature of 63 K. What is the change in entropy of a 1.5-kg sample of liquid nitrogen as it transitions from a liquid to a solid? Example 15-9

90. •• A certain electric generating plant produces electricity by using steam that enters its turbine at a temperature of 320°C and leaves it at 40°C. Over the course of a year, the plant consumes 4.40×10^{16} J of heat and produces an average electric power output of 600 MW. What is its second law efficiency? Example 15-7

91. •• As we drill down into the rocks of Earth's crust, the temperature typically increases by 3.0°C for every 100 m of depth. Oil wells can be drilled to depths of 1830 m. If water is pumped into the shaft of the well, it will be heated by the hot rock at the bottom and the resulting steam can be used as a heat engine. Assume that the surface temperature is 20°C. (a) Using such a 1830-m well as a heat engine, what is the maximum efficiency possible? (b) If a combination of such wells is to produce a 2.5-MW power plant, how much energy will it absorb from the interior of Earth each day? SSM Example 15-7

92. ••• The energy efficiency ratio (or rating)—the EER—for air conditioners, refrigerators, and freezers is defined as the ratio of the input rate of heat ($|Q_C|/t$, in BTU/h) to the output rate of work (W/t, in W): $\text{EER} = \dfrac{Q_C/t(\text{BTU/hr})}{W/t(\text{W})}$. (a) Show that the EER can be expressed as $\dfrac{Q_C(\text{BTU})}{W(\text{W} \cdot \text{h})}$ and is therefore nothing more than the coefficient of performance CP expressed in mixed units. (b) Show that the EER is related to the coefficient of performance CP by the equation $CP = \text{EER}/3.412$. (c) Typical home freezers have EER ratings of about 5.1 and operate between an interior freezer temperature of 0.00°F and an outside kitchen temperature of about 70.0°F. What is the coefficient of performance for such a freezer, and how does it compare to the coefficient of performance of the best possible freezer operating between those temperatures? (d) What is the EER of the best possible freezer in part (c)? Example 15-8

93. •• Your energy-efficient home freezer has an EER of 6.50 (see Problem 15-92). In preparation for a picnic you put 1.50 L of water at 20.0°C into the freezer to make ice at 0.00°C for your ice chest. (See Tables 11-1, 14-3, and 14-4 as needed.) (a) How much electrical energy (which runs the freezer) is required to make the ice? Express your answer in J and kWh. (b) How much heat is ejected into your kitchen, which is at

22.0°C during the process? (c) How much does making the ice change the entropy of your kitchen? Example 15-8

94. •• **Biology** The volume of air taken in during a typical breath is 0.5 L. The inhaled air is heated to 37°C (the internal body temperature) as it enters the lungs. Because air is about 80% nitrogen N_2, we can model it as an ideal gas. Suppose that the outside air is at room temperature (20°C) and that you take two breaths every 3.0 s. Assume that the pressure does not change during the process. (a) How many joules of heat does it take to warm the air in a single breath? (b) How many food calories (kcal) are used up per day in heating the air you breathe? Is this a significant amount of typical daily caloric intake?

95. •• A heat engine works in a cycle between reservoirs at 273 K and 490 K. In each cycle, the engine absorbs 1250 J of heat from the high-temperature reservoir and does 475 J of work. (a) What is its efficiency? (b) By how much is the entropy of the universe changed when the engine goes through one full cycle? (c) How much energy becomes unavailable for doing work when the engine goes through one full cycle? SSM Example 15-7

96. ••• You have a cabin on the plains of central Saskatchewan. It is built on a 8.50-m by 12.5-m rectangular foundation with walls 3.00 m tall. The wooden walls and flat roof are made of white pine that is 9.00 cm thick. To conserve heat the windows are negligibly small. The floor is well insulated, so you lose negligible heat through it. The cabin is heated by an electrically powered heat pump operating on the Carnot cycle between the inside and outside air. When the outside temperature is a frigid −10.0°F, how much electrical energy does the heat pump consume per second to keep the interior temperature a steady and toasty 70.0°F? Assume that the surfaces of the walls and roof are at the same temperature as the air with which they are in contact and neglect radiation. (Consult Table 14-5 as needed.) Example 15-7

97. • **Sports** In an international diving competition divers fall from a platform 10.0 m above the surface of the water into a very large pool. The diver leaves the platform with negligible initial speed. What is the maximum change in the entropy of the water in the pool at 25.0°C when a 75.0-kg diver executes his dive? Does the pool's entropy increase or decrease? Example 15-9

98. •• A heat engine works in a cycle between reservoirs at 273 and 490 K. In each cycle the engine absorbs 1250 J of heat from the high-temperature reservoir and does 475 J of work. (a) What is its efficiency? (b) What is the change in entropy of the universe when the engine goes through one complete cycle? (c) How much energy becomes unavailable for doing work when the engine goes through one complete cycle? Example 15-9

99. •• **Biology** A 68.0-kg person typically eats about 2250 kcal per day, 20.0% of which goes to mechanical energy and the rest to heat. If she spends most of her time in her apartment at 22.0°C, how much does the entropy of her apartment change in one day? Does the entropy of the apartment increase or decrease? SSM Example 15-9

100. ••• Consider an engine in which the working substance is 1.23 mol of an ideal gas for which $\gamma = 1.41$. The engine runs reversibly in the cycle shown on the pV diagram (**Figure 15-23**). The cycle consists of an isobaric (constant-pressure) expansion a at a pressure of 15.0 atm, during which the temperature of the gas increases from 300 to 600 K, followed by an isothermal expansion b until its pressure becomes 3.00 atm. Next is an isobaric compression c at a pressure of 3.00 atm, during which the temperature decreases from 600 to 300 K, followed by an isothermal compression d until its pressure returns to 15 atm. Find the work done by the gas, the heat absorbed by the gas, the internal energy change, and the entropy change of the gas, first for each part of the cycle and then for the complete cycle. Example 15-9

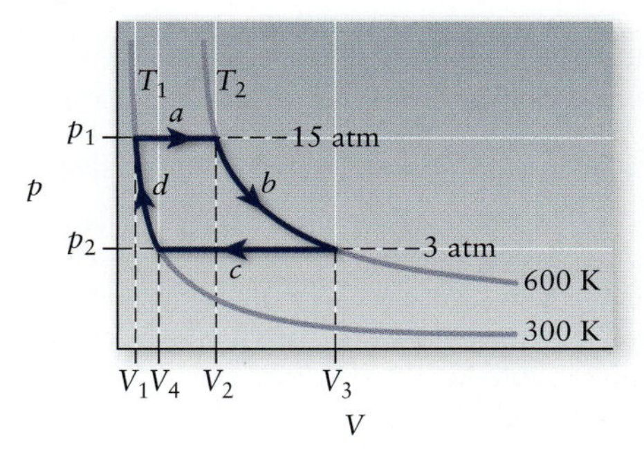

Figure 15-23 Problem 100

APPENDIX A

SI Units and Conversion Factors

Base Units*

Length	The *meter* (m) is the distance traveled by light in a vacuum in 1/299,792,458 s.
Time	The *second* (s) is the duration of 9,192,631,770 periods of the radiation corresponding to the transition between the two hyperfine levels of the ground state of the ^{133}Cs atom.
Mass	The *kilogram* (kg) is the mass of the international standard body preserved at Sèvres, France.
Mole	The *mole* (mol) is the amount of substance of a system which contains as many elementary entities as there are atoms in 0.012 kg of carbon-12.
Current	The *ampere* (A) is that constant current which, if maintained in two straight parallel conductors of infinite length, of negligible circular cross section, and placed 1 m apart in vacuum, would produce between the conductors a force equal to 2×10^{-7} N/m of length.
Temperature	The *kelvin* (K) is 1/273.16 of the thermodynamic temperature of the triple point of water.
Luminous intensity	The *candela* (cd) is the luminous intensity in a given direction, of a source that emits monochromatic radiation of frequency 540×10^{12} Hz and that has a radiant intensity, in that direction of 1/683 W/steradian.

*These definitions are found on the Internet at http://physics.nist.gov/cuu/Units/current.html.

Derived Units

Force	newton (N)	$1\ \text{N} = 1\ \text{kg}\cdot\text{m/s}^2$
Work, energy	joule (J)	$1\ \text{J} = 1\ \text{N}\cdot\text{m}$
Power	watt (W)	$1\ \text{W} = 1\ \text{J/s}$
Frequency	hertz (Hz)	$1\ \text{Hz} = \text{cy/s}$
Charge	coulomb (C)	$1\ \text{C} = 1\ \text{A}\cdot\text{s}$
Potential	volt (V)	$1\ \text{V} = 1\ \text{J/C}$
Resistance	ohm (Ω)	$1\ \Omega = 1\ \text{V/A}$
Capacitance	farad (F)	$1\ \text{F} = 1\ \text{C/V}$
Magnetic field	tesla (T)	$1\ \text{T} = 1\ \text{N/(A}\cdot\text{m)}$
Magnetic flux	weber (Wb)	$1\ \text{Wb} = 1\ \text{T}\cdot\text{m}^2$
Inductance	henry (H)	$1\ \text{H} = 1\ \text{J/A}^2$

Conversion Factors

Conversion factors are written as equations for simplicity; relations marked with an asterisk are exact.

Length

1 km = 0.6214 mi

1 mi = 1.609 km

1 m = 1.0936 yard = 3.281 ft = 39.37 in.

*1 in. = 2.54 cm

*1 ft = 12 in. = 30.48 cm

*1 yard = 3 ft = 91.44 cm

1 light-year = 1 $c \cdot$ y = 9.461×10^{15} m

*1 Å = 0.1 nm

Area

*1 $m^2 = 10^4\ cm^2$

1 km^2 = 0.3861 mi^2 = 247.1 acres

*1 $in.^2$ = 6.4516 cm^2

1 $ft^2 = 9.29 \times 10^{-2}\ m^2$

1 m^2 = 10.76 ft^2

*1 acre = 43 560 ft^2

1 mi^2 = 640 acres = 2.590 km^2

Volume

*1 $m^3 = 10^6\ cm^3$

*1 L = 1000 $cm^3 = 10^{-3}\ m^3$

1 gal = 3.785 L

1 gal = 4 qt = 8 pt = 128 oz = 231 $in.^3$

1 $in.^3$ = 16.39 cm^3

1 ft^3 = 1728 $in.^3$ = 28.32 L
= $2.832 \times 10^4\ cm^3$

Time

*1 h = 60 min = 3.6 ks

*1 d = 24 h = 1440 min = 86.4 ks

1 y = 365.25 day = 3.156×10^7 s

Speed

*1 m/s = 3.6 km/h

1 km/h = 0.2778 m/s = 0.6214 mi/h

1 mi/h = 0.4470 m/s = 1.609 km/h

1 mi/h = 1.467 ft/s

Angle and Angular Speed

*π rad = 180°

1 rad = 57.30°

1° = 1.745×10^{-2} rad

1 rev/min = 0.1047 rad/s

1 rad/s = 9.549 rev/min

Mass

*1 kg = 1000 g

*1 tonne = 1000 kg = 1 Mg

1 u = 1.6605×10^{-27} kg
931.49 MeV/c^2

1 kg = 6.022×10^{26} u

1 slug = 14.59 kg

1 kg = 6.852×10^{-2} slug

Density

*1 g/cm^3 = 1000 kg/m^3 = 1 kg/L

(1 g/cm^3)g = 62.4 lb/ft^3

Force

1 N = 0.2248 lb = 10^5 dyn

*1 lb = 4.448222 N

(1 kg)g = 2.2046 lb

Pressure

*1 Pa = 1 N/m^2

*1 atm = 101.325 kPa = 1.01325 bar

1 atm = 14.7 $lb/in.^2$ = 760 mmHg
= 29.9 in.Hg = 33.9 ftH_2O

1 $lb/in.^2$ = 6.895 kPa

1 torr = 1 mmHg = 133.32 Pa

1 bar = 100 kPa

Energy

*1 kW · h = 3.6 MJ

*1 cal = 4.186 J

1 ft · lb = 1.356 J = 1.286×10^{-3} BTU

*1 L · atm = 101.325 J

1 L · atm = 24.217 cal

1 BTU = 778 ft · lb = 252 cal = 1054.35 J

1 eV = 1.602×10^{-19} J

1 u · c^2 = 931.49 MeV

*1 erg = 10^{-7} J

Power

1 horsepower = 550 ft · lb/s = 745.7 W

1 BTU/h = 2.931×10^{-4} kW

1 W = 1.341×10^{-3} horsepower
= 0.7376 ft · lb/s

Magnetic Field

*1 T = 10^4 G

Thermal Conductivity

1 W/(m · K) = 6.938 BTU · in./(h · ft^2 · °F)

1 BTU · in./(h · ft^2 · °F) = 0.1441 W/(m · K)

APPENDIX B
Numerical Data

Terrestrial Data

Free-fall acceleration g	
Standard value (at sea level at 45° latitude)*	$9.806\ 65\ m/s^2$; $32.1740\ ft/s^2$
At equator*	$9.7804\ m/s^2$
At poles*	$9.8322\ m/s^2$
Mass of Earth M_E	5.98×10^{24} kg
Radius of Earth R_E, mean	6.38×10^6 m; 3960 mi
Escape speed	1.12×10^4 m/s; 6.96 mi/s
Solar constant†	$1.37\ kW/m^2$
Standard temperature and pressure (STP):	
Temperature	273.15 K
Pressure	101.3 kPa (1.00 atm)
Molar mass of air	28.97 g/mol
Density of air (273.15 K, 101.3 kPa), ρ_{air}	$1.29\ kg/m^3$
Speed of sound (273.15 K, 101.3 kPa)	331 m/s
Latent heat of fusion of H_2O (0°C, 1 atm)	334 kJ/kg
Latent heat of vaporization of H_2O (100°C, 1 atm)	2.26 MJ/kg

* Measured relative to Earth's surface.

† Average power incident normally on 1 m^2 outside Earth's atmosphere at the mean distance from Earth to the Sun.

Astronomical Data*

Earth	
Distance to the Moon, mean†	3.844×10^8 m; 2.389×10^5 mi
Distance to the Sun, mean†	1.496×10^{11} m; 9.32×10^7 mi; 1.00 AU
Orbital speed, mean	2.98×10^4 m/s
Moon	
Mass	7.35×10^{22} kg
Radius	1.737×10^6 m
Period	27.32 day
Acceleration of gravity at surface	$1.62\ m/s^2$
Sun	
Mass	1.99×10^{30} kg
Radius	6.96×10^8 m

* Additional solar system data are available from NASA at http://nssdc.gsfc.nasa.gov/planetary/planetfact.html.

† Center to center.

Physical Constants*

Universal constant of gravitation	G	$6.674\,08(31) \times 10^{-11}\ \mathrm{N \cdot m^2/kg^2}$
Speed of light	c	$2.997\,924\,58 \times 10^{8}\ \mathrm{m/s}$
Fundamental charge	e	$1.602\,176\,6208(98) \times 10^{-19}\ \mathrm{C}$
Avogadro's constant	N_A	$6.022\,140\,857(74) \times 10^{23}$ particles/mol
Gas constant	R	$8.314\,4598(48)\ \mathrm{J/(mol \cdot K)}$
		$1.987\,2036(11)\ \mathrm{cal/(mol \cdot K)}$
		$8.205\,7338(47) \times 10^{-2}\ \mathrm{L \cdot atm/(mol \cdot K)}$
Boltzmann constant	$k = R/N_A$	$1.380\,648\,52(79) \times 10^{-23}\ \mathrm{J/K}$
		$8.617\,3303(50) \times 10^{-5}\ \mathrm{eV/K}$
Stefan-Boltzmann constant	$\sigma = (\pi^2/60)k^4/(\hbar^3c^2)$	$5.670\,367(13) \times 10^{-8}\ \mathrm{W/(m^2 \cdot K^4)}$
Atomic mass constant	$m_u = (1/12)m(^{12}\mathrm{C})$	$1.660\,539\,040(20) \times 10^{-27}\ \mathrm{kg} = 1\ \mathrm{u}$
Permeability of free space	μ_0	$4\pi \times 10^{-7}\ \mathrm{N/A^2}$
		$1.256\,637 \ldots \times 10^{-6}\ \mathrm{N/A^2}$
Permittivity of free space	$\epsilon_0 = 1/(\mu_0 c^2)$	$8.854\,187\,817 \ldots \times 10^{-12}\ \mathrm{C^2/(N \cdot m^2)}$
Coulomb constant	$k = 1/(4\pi\epsilon_0)$	$8.987\,551\,787 \ldots \times 10^{9}\ \mathrm{N \cdot m^2/C^2}$
Planck's constant	h	$6.626\,070\,040(81) \times 10^{-34}\ \mathrm{J \cdot s}$
		$4.135\,667\,662(25) \times 10^{-15}\ \mathrm{eV \cdot s}$
	$\hbar = h/(2\pi)$	$1.054\,571\,800(13) \times 10^{-34}\ \mathrm{J \cdot s}$
		$6.582\,119\,514(40) \times 10^{-16}\ \mathrm{eV \cdot s}$
Mass of electron	m_e	$9.109\,383\,56(11) \times 10^{-31}\ \mathrm{kg}$
		$0.510\,998\,9461(31)\ \mathrm{MeV}/c^2$
Mass of proton	m_p	$1.672\,621\,898(21) \times 10^{-27}\ \mathrm{kg}$
		$938.272\,0813(58)\ \mathrm{MeV}/c^2$
Mass of neutron	m_n	$1.674\,927\,471(21) \times 10^{-27}\ \mathrm{kg}$
		$939.565\,4133(58)\ \mathrm{MeV}/c^2$
Bohr magneton	$m_B = e\hbar/(2m_e)$	$9.274\,009\,994(57) \times 10^{-24}\ \mathrm{J/T}$
		$5.788\,381\,8012(26) \times 10^{-5}\ \mathrm{eV/T}$
Nuclear magneton	$m_n = e\hbar/(2m_p)$	$5.050\,783\,699(31) \times 10^{-27}\ \mathrm{J/T}$
		$3.152\,451\,2550(15) \times 10^{-8}\ \mathrm{eV/T}$
Magnetic flux quantum	$\phi_0 = h/(2e)$	$2.067\,833\,831(13) \times 10^{-15}\ \mathrm{T \cdot m^2}$
Quantized Hall resistance	$R_K = h/e^2$	$2.581\,280\,745\,55(59) \times 10^{4}\ \Omega$
Rydberg constant	R_H	$1.097\,373\,156\,8508(65) \times 10^{7}\ \mathrm{m^{-1}}$
Josephson frequency–voltage quotient	$K_J = 2e/h$	$4.835\,978\,525(30) \times 10^{14}\ \mathrm{Hz/V}$
Compton wavelength	$\lambda_C = h/(m_e c)$	$2.426\,310\,2367(11) \times 10^{-12}\ \mathrm{m}$

* Updated values for these and other constants may be found on the Internet at http://physics.nist.gov/cuu/Constants/index.html. The numbers in parentheses represent the uncertainties in the last two digits. (For example, 2.044 43(13) stands for 2.044 43 ± 0.000 13.) Values without uncertainties are exact, including those values with ellipses (such as the value of π, which is exactly 3.1415. . .).

APPENDIX C

Periodic Table of Elements*

1	2	3	4	5	6	7	8	9	10	11	12	13	14	15	16	17	18
1 H																	2 He
3 Li	4 Be											5 B	6 C	7 N	8 O	9 F	10 Ne
11 Na	12 Mg											13 Al	14 Si	15 P	16 S	17 Cl	18 Ar
19 K	20 Ca	21 Sc	22 Ti	23 V	24 Cr	25 Mn	26 Fe	27 Co	28 Ni	29 Cu	30 Zn	31 Ga	32 Ge	33 As	34 Se	35 Br	36 Kr
37 Rb	38 Sr	39 Y	40 Zr	41 Nb	42 Mo	43 Tc	44 Ru	45 Rh	46 Pd	47 Ag	48 Cd	49 In	50 Sn	51 Sb	52 Te	53 I	54 Xe
55 Cs	56 Ba	57–71 Lanthanoids	72 Hf	73 Ta	74 W	75 Re	76 Os	77 Ir	78 Pt	79 Au	80 Hg	81 Tl	82 Pb	83 Bi	84 Po	85 At	86 Rn
87 Fr	88 Ra	89–103 Actinoids	104 Rf	105 Db	106 Sg	107 Bh	108 Hs	109 Mt	110 Ds	111 Rg	112 Cn	113 Nh	114 Fl	115 Mc	116 Lv	117 Ts	118 Og

Lanthanoids	57 La	58 Ce	59 Pr	60 Nd	61 Pm	62 Sm	63 Eu	64 Gd	65 Tb	66 Dy	67 Ho	68 Er	69 Tm	70 Yb	71 Lu
Actinoids	89 Ac	90 Th	91 Pa	92 U	93 Np	94 Pu	95 Am	96 Cm	97 Bk	98 Cf	99 Es	100 Fm	101 Md	102 No	103 Lr

* From https://iupac.org/what-we-do/periodic-table-of-elements/.

Atomic Numbers and Atomic Weights*

Atomic Number	Name	Symbol	Weight
1	Hydrogen	H	[1.007 84; 1.008 11]
2	Helium	He	4.002602(2)
3	Lithium	Li	[6.938; 6.997]
4	Beryllium	Be	9.0121831(5)
5	Boron	B	[10.806; 10.821]
6	Carbon	C	[12.009 6; 12.011 6]
7	Nitrogen	N	[14.006 43; 14.007 28]
8	Oxygen	O	[15.999 03; 15.999 77]
9	Fluorine	F	18.998403163(6)
10	Neon	Ne	20.1797(6)
11	Sodium	Na	22.98976928(2)
12	Magnesium	Mg	[24.304, 24.307]
13	Aluminum	Al	26.9815385(7)
14	Silicon	Si	[28.084; 28.086]
15	Phosphorus	P	30.973761998(5)
16	Sulfur	S	[32.059; 32.076]
17	Chlorine	Cl	[35.446; 35.457]
18	Argon	Ar	39.948(1)
19	Potassium	K	39.0983(1)
20	Calcium	Ca	40.078(4)
21	Scandium	Sc	44.955908(5)
22	Titanium	Ti	47.867(1)
23	Vanadium	V	50.9415(1)
24	Chromium	Cr	51.9961(6)
25	Manganese	Mn	54.938044(3)
26	Iron	Fe	55.845(2)
27	Cobalt	Co	58.933194(4)
28	Nickel	Ni	58.6934(4)
29	Copper	Cu	63.546(3)
30	Zinc	Zn	65.38 (2)
31	Gallium	Ga	69.723(1)
32	Germanium	Ge	72.630(8)
33	Arsenic	As	74.921595(6)
34	Selenium	Se	78.971(8)
35	Bromine	Br	[79.901, 79.907]
36	Krypton	Kr	83.798(2)
37	Rubidium	Rb	85.4678(3)
38	Strontium	Sr	87.62(1)
39	Yttrium	Y	88.90584(2)
40	Zirconium	Zr	91.224(2)
41	Niobium	Nb	92.90637(2)
42	Molybdenum	Mo	95.95(1)
43	Technetium	Tc	
44	Ruthenium	Ru	101.07(2)
45	Rhodium	Rh	102.90550(2)
46	Palladium	Pd	106.42(1)
47	Silver	Ag	107.8682(2)
48	Cadmium	Cd	112.414(4)
49	Indium	In	114.818(1)
50	Tin	Sn	118.710(7)
51	Antimony	Sb	121.760(1)
52	Tellurium	Te	127.60(3)
53	Iodine	I	126.90447(3)
54	Xenon	Xe	131.293(6)
55	Cesium	Cs	132.90545196(6)
56	Barium	Ba	137.327(7)
57	Lanthanum	La	138.90547(7)
58	Cerium	Ce	140.116(1)
59	Praseodymium	Pr	140.90766(2)
60	Neodymium	Nd	144.242(3)
61	Promethium	Pm	
62	Samarium	Sm	150.36(2)
63	Europium	Eu	151.964(1)
64	Gadolinium	Gd	157.25(3)
65	Terbium	Tb	158.92535(2)
66	Dysprosium	Dy	162.500(1)
67	Holmium	Ho	164.93033(2)
68	Erbium	Er	167.259(3)
69	Thulium	Tm	168.93422(2)
70	Ytterbium	Yb	173.045(10)
71	Lutetium	Lu	174.9668(1)
72	Hafnium	Hf	178.49(2)
73	Tantalum	Ta	180.94788(2)
74	Tungsten	W	183.84(1)
75	Rhenium	Re	186.207(1)
76	Osmium	Os	190.23(3)
77	Iridium	Ir	192.217(3)
78	Platinum	Pt	195.084(9)
79	Gold	Au	196.966569(5)
80	Mercury	Hg	200.592(3)
81	Thallium	Tl	[204.382; 204.385]
82	Lead	Pb	207.2(1)
83	Bismuth	Bi	208.98040(1)
84	Polonium	Po	
85	Astatine	At	
86	Radon	Rn	
87	Francium	Fr	
88	Radium	Ra	
89	Actinium	Ac	
90	Thorium	Th	232.0377(4)
91	Protactinium	Pa	231.03588(2)
92	Uranium	U	238.02891(3)
93	Neptunium	Np	
94	Plutonium	Pu	
95	Americium	Am	
96	Curium	Cm	
97	Berkelium	Bk	
98	Californium	Cf	
99	Einsteinium	Es	
100	Fermium	Fm	
101	Mendelevium	Md	
102	Nobelium	No	
103	Lawrencium	Lr	
104	Rutherfordium	Rf	
105	Dubnium	Db	
106	Seaborgium	Sg	
107	Bohrium	Bh	
108	Hassium	Hs	
109	Meitnerium	Mt	
110	Darmstadtium	Ds	
111	Roentgenium	Rg	
112	Copernicium	Cn	
113	Nihonium	Nh	
114	Flerovium	Fl	
115	Moscovium	Mc	
116	Livermorium	Lv	
117	Tennessine	Ts	
118	Oganesson	Og	

*Some weights are listed as intervals ([*a*; *b*]; $a \le$ atomic weight $\le b$) because these weights are not constant but depend on the physical, chemical, and nuclear histories of the samples used. Atomic weights are not listed for some elements because these elements do not have stable isotopes. Exceptions are thorium, protactinium, and uranium. From http://www.ciaaw.org/atomic-weights.htm.

Table of Atomic Masses

Element	Symbol	Mass number (*indicates radioactive)	Atomic mass	Percent abundance	Half-life and decay mode (if unstable)	
(Neutron)	*n*	1*	1.008665		10.4 m	β^-
Hydrogen	H	1	1.007825	99.985		
Deuterium	D	2	2.014102	0.015		
Tritium	T	3*	3.016049		12.33 y	β^-
Helium	He	3	3.016029	0.00014		
		4	4.002602	99.99986		
		6*	6.018886		0.81 s	β^-
		8*	8.033922		0.12 s	β^-
Lithium	Li	6	6.015121	7.5		
		7	7.016003	92.5		
		8*	8.022486		0.84 s	β^-
		9*	9.026789		0.18 s	β^-
		11*	11.043897		8.7 ms	β^-
Beryllium	Be	7*	7.016928		53.3 d	ec
		9	9.012174	100		
		10*	10.013534		1.5×10^6 y	β^-
		11*	11.021657		13.8 s	β^-
		12*	12.026921		23.6 ms	β^-
		14*	14.042866		4.3 ms	β^-
Boron	B	8*	8.024605		0.77 s	β^+
		10	10.012936	19.9		
		11	11.009305	80.1		
		12*	12.014352		0.0202 s	β^-
		13*	13.017780		17.4 ms	β^-
		14*	14.025404		13.8 ms	β^-
		15*	15.031100		10.3 ms	β^-
Carbon	C	9*	9.031030		0.13 s	β^+
		10*	10.016854		19.3 s	β^+
		11*	11.011433		20.4 m	β^+
		12	12.000000	98.90		
		13	13.003355	1.10		
		14*	14.003242		5730 y	β^-
		15*	15.010599		2.45 s	β^-
		16*	16.014701		0.75 s	β^-
		17*	17.022582		0.20 s	β^-

(*Continued*)

Element	Symbol	Mass number (*indicates radioactive)	Atomic mass	Percent abundance	Half-life and decay mode (if unstable)	
Nitrogen	N	12*	12.018613		0.0110 s	β^+
		13*	13.005738		9.96 m	β^+
		14	14.003074	99.63		
		15	15.000108	0.37		
		16*	16.006100		7.13 s	β^-
		17*	17.008450		4.17 s	β^-
		18*	18.014082		0.62 s	β^-
		19*	19.017038		0.24 s	β^-
Oxygen	O	13*	13.024813		8.6 ms	β^+
		14*	14.008595		70.6 s	β^+
		15*	15.003065		122 s	β^+
		16	15.994915	99.71		
		17	16.999132	0.039		
		18	17.999160	0.20		
		19*	19.003577		26.9 s	β^-
		20*	20.004076		13.6 s	β^-
		21*	21.008595		3.4 s	β^-
Fluorine	F	17*	17.002094		64.5 s	β^+
		18*	18.000937		109.8 m	β^+
		19	18.998404	100		
		20*	19.999982		11.0 s	β^-
		21*	20.999950		4.2 s	β^-
		22*	22.003036		4.2 s	β^-
		23*	23.003564		2.2 s	β^-
Neon	Ne	18*	18.005710		1.67 s	β^+
		19*	19.001880		17.2 s	β^+
		20	19.992435	90.48		
		21	20.993841	0.27		
		22	21.991383	9.25		
		23*	22.994465		37.2 s	β^-
		24*	23.993999		3.38 m	β^-
		25*	24.997789		0.60 s	β^-
Sodium	Na	21*	20.997650		22.5 s	β^+
		22*	21.994434		2.61 y	β^+
		23	22.989767	100		
		24*	23.990961		14.96 h	β^-
		25*	24.989951		59.1 s	β^-
		26*	25.992588		1.07 s	β^-
Magnesium	Mg	23*	22.994124		11.3 s	β^+
		24	23.985042	78.99		
		25	24.985838	10.00		
		26	25.982594	11.01		
		27*	26.984341		9.46 m	β^-
		28*	27.983876		20.9 h	β^-
		29*	28.375346		1.30 s	β^-

Element	Symbol	Mass number (*indicates radioactive)	Atomic mass	Percent abundance	Half-life and decay mode (if unstable)	
Aluminum	Al	25*	24.990429		7.18 s	β^+
		26*	25.986892		7.4×10^5 y	β^+
		27	26.981538	100		
		28*	27.981910		2.24 m	β^-
		29*	28.980445		6.56 m	β^-
		30*	29.982965		3.60 s	β^-
Silicon	Si	27*	26.986704		4.16 s	β^+
		28	27.976927	92.23		
		29	28.976495	4.67		
		30	28.973770	3.10		
		31*	30.975362		2.62 h	β^-
		32*	31.974148		172 y	β^-
		33*	32.977928		6.13 s	β^-
Phosphorus	P	30*	29.978307		2.50 m	β^+
		31	30.973762	100		
		32*	31.973762		14.26 d	β^-
		33*	32.971725		25.3 d	β^-
		34*	33.973636		12.43 s	β^-
Sulfur	S	31*	30.979554		2.57 s	β^+
		32	31.972071	95.02		
		33	32.971459	0.75		
		34	33.967867	4.21		
		35*	34.969033		87.5 d	β^-
		36	35.967081	0.02		
Chlorine	Cl	34*	33.973763		32.2 m	β^+
		35	34.968853	75.77		
		36*	35.968307		3.0×10^5 y	β^-
		37	36.965903	24.23		
		38*	37.968010		37.3 m	β^-
Argon	Ar	36	35.967547	0.337		
		37*	36.966776		35.04 d	ec
		38	37.962732	0.063		
		39*	38.964314		269 y	β^-
		40	39.962384	99.600		
		42*	41.963049		33 y	β^-
Potassium	K	39	38.963708	93.2581		
		40*	39.964000	0.0117	1.28×10^9 y	β^+, ec, β^-
		41	40.961827	6.7302		
		42*	41.962404		12.4 h	β^-
		43*	42.960716		22.3 h	β^-

(*Continued*)

Element	Symbol	Mass number (*indicates radioactive)	Atomic mass	Percent abundance	Half-life and decay mode (if unstable)
Calcium	Ca	40	39.962591	96.941	
		41*	40.962279		1.0×10^5 y ec
		42	41.958618	0.647	
		43	42.958767	0.135	
		44	43.955481	2.086	
		46	45.953687	0.004	
		48	47.952534	0.187	
Scandium	Sc	41*	40.969250		0.596 s β^+
		43*	42.961151		3.89 h β^+
		45	44.955911	100	
		46*	45.955170		83.8 d β^-
Titanium	Ti	44*	43.959691		49 y ec
		46	45.952630	8.0	
		47	46.951765	7.3	
		48	47.947947	73.8	
		49	48.947871	5.5	
		50	49.944792	5.4	
Vanadium	V	48*	47.952255		
		50*	49.947161	0.25	15.97 d β^+
		51	50.943962	99.75	1.5×10^{17} y β^+
Chromium	Cr	48*	47.954033		21.6 h ec
		50	49.946047	4.345	
		52	51.940511	83.79	
		53	52.940652	9.50	
		54	53.938883	2.365	
Manganese	Mn	53*	52.941292		3.74×10^6 y ec
		54*	53.940361		312.1 d ec
		55	54.938048	100	
		56*	55.938908		2.58 h β^-
Iron	Fe	54	53.939613	5.9	
		55*	54.938297		2.7 y ec
		56	55.934940	91.72	
		57	56.935396	2.1	
		58	57.933278	0.28	
		60*	59.934078		1.5×10^6 y β^-
Cobalt	Co	57*	56.936294		271.8 d ec
		58*	57.935755		70.9 h ec, β^+
		59	58.933198	100	
		60*	59.933820		5.27 y β^-
		61*	60.932478		1.65 h β^-
Nickel	Ni	58	57.935346	68.077	
		59*	58.934350		7.5×10^4 y ec, β^+
		60	59.930789	26.223	
		61	60.931058	1.140	
		62	61.928346	3.634	
		63*	62.929670		100 y β^-
		64	63.927967	0.926	

Element	Symbol	Mass number (*indicates radioactive)	Atomic mass	Percent abundance	Half-life and decay mode (if unstable)
Copper	Cu	63	62.929599	69.17	
		64*	63.929765		12.7 h ec
		65	64.927791	30.83	
		66*	65.928871		5.1 m β^-
Zinc	Zn	64	63.929144	48.6	
		66	65.926035	27.9	
		67	66.927129	4.1	
		68	67.924845	18.8	
		70	69.925323	0.6	
Gallium	Ga	69	68.925580	60.108	
		70*	69.926027		21.1 m β^-
		71	70.924703	39.892	
		72*	71.926367		14.1 h β^-
Germanium	Ge	69*	68.927969		39.1 h ec, β^+
		70	69.924250	21.23	
		72	71.922079	27.66	
		73	72.923462	7.73	
		74	73.921177	35.94	
		76	75.921402	7.44	
		77*	76.923547		11.3 h β^-
Arsenic	As	73*	72.923827		80.3 d ec
		74*	73.923928		17.8 d ec, β^+
		75	74.921594	100	
		76*	75.922393		1.1 d β^-
		77*	76.920645		38.8 h β^-
Selenium	Se	74	73.922474	0.89	
		76	75.919212	9.36	
		77	76.919913	7.63	
		78	77.917307	23.78	
		79*	78.918497		$\leq 6.5 \times 10^4$ y β^-
		80	79.916519	49.61	
		82*	81.916697	8.73	1.4×10^{20} y $2\beta^-$
Bromine	Br	79	78.918336	50.69	
		80*	79.918528		17.7 m β^+
		81	80.916287	49.31	
		82*	81.916802		35.3 h β^-
Krypton	Kr	78	77.920400	0.35	
		80	79.916377	2.25	
		81*	80.916589		2.11×10^5 y ec
		82	81.913481	11.6	
		83	82.914136	11.5	
		84	83.911508	57.0	
		85*	84.912531		10.76 y β^-
		86	85.910615	17.3	

(Continued)

Element	Symbol	Mass number (*indicates radioactive)	Atomic mass	Percent abundance	Half-life and decay mode (if unstable)
Rubidium	Rb	85	84.911793	72.17	
		86*	85.911171		18.6 d β^-
		87*	86.909186	27.83	4.75×10^{10} y β^-
		88*	87.911325		17.8 m β^-
Strontium	Sr	84	83.913428	0.56	
		86	85.909266	9.86	
		87	86.908883	7.00	
		88	87.905618	82.58	
		90*	89.907737		29.1 y β^-
Yttrium	Y	88*	87.909507		106.6 d ec, β^+
		89	88.905847	100	
		90*	89.914811		2.67 d β^-
Zirconium	Zr	90	89.904702	51.45	
		91	90.905643	11.22	
		92	91.905038	17.15	
		93*	92.906473		1.5×10^6 y β^-
		94	93.906314	17.38	
		96	95.908274	2.80	
Niobium	Nb	91*	90.906988		6.8×10^2 y ec
		92*	91.907191		3.5×10^7 y ec
		93	92.906376	100	
		94*	93.907280		2×10^4 y β^-
Molybdenum	Mo	92	91.906807	14.84	
		93*	92.906811		3.5×10^3 y ec
		94	93.905085	9.25	
		95	94.905841	15.92	
		96	95.904678	16.68	
		97	96.906020	9.55	
		98	97.905407	24.13	
		100	99.907476	9.63	
Technetium	Tc	97*	96.906363		2.6×10^6 y ec
		98*	97.907215		4.2×10^6 y β^-
		99*	98.906254		2.1×10^5 y β^-
Ruthenium	Ru	96	95.907597	5.54	
		98	97.905287	1.86	
		99	98.905939	12.7	
		100	99.904219	12.6	
		101	100.905558	17.1	
		102	101.904348	31.6	
		104	103.905428	18.6	
Rhodium	Rh	102*	101.906794		207 d ec
		103	102.905502	100	
		104*	103.906654		42 s β^-

Element	Symbol	Mass number (*indicates radioactive)	Atomic mass	Percent abundance	Half-life and decay mode (if unstable)
Palladium	Pd	102	101.905616	1.02	
		104	103.904033	11.14	
		105	104.905082	22.33	
		106	105.903481	27.33	
		107*	106.905126		6.5×10^6 y β^-
		108	107.903893	26.46	
		110	109.905158	11.72	
Silver	Ag	107	106.905091	51.84	
		108*	107.905953		2.39 m ec, β^+, β^-
		109	108.904754	48.16	
		110*	109.906110		24.6 s β^-
Cadmium	Cd	106	105.906457	1.25	
		108	107.904183	0.89	
		109*	108.904984		462 d ec
		110	109.903004	12.49	
		111	110.904182	12.80	
		112	111.902760	24.13	
		113*	112.904401	12.22	9.3×10^{15} y β^-
		114	113.903359	28.73	
		116	115.904755	7.49	
Indium	In	113	112.904060	4.3	
		114*	113.904916		1.2 m β^-
		115*	114.903876	95.7	4.4×10^{14} y β^-
		116*	115.905258		54.4 m β^-
Tin	Sn	112	111.904822	0.97	
		114	113.902780	0.65	
		115	114.903345	0.36	
		116	115.901743	14.53	
		117	116.902953	7.68	
		118	117.901605	24.22	
		119	118.903308	8.58	
		120	119.902197	32.59	
		121*	120.904237		55 y β^-
		122	121.903439	4.63	
		124	123.905274	5.79	
Antimony	Sb	121	120.903820	57.36	
		123	122.904215	42.64	
		125*	124.905251		2.7 y β^-
Tellurium	Te	120	119.904040	0.095	
		122	121.903052	2.59	
		123*	122.904271	0.905	1.3×10^{13} y ec
		124	123.902817	4.79	
		125	124.904429	7.12	
		126	125.903309	18.93	
		128*	127.904463	31.70	$> 8 \times 10^{24}$ y $2\beta^-$
		130*	129.906228	33.87	1.2×10^{21} y $2\beta^-$

(Continued)

Element	Symbol	Mass number (*indicates radioactive)	Atomic mass	Percent abundance	Half-life and decay mode (if unstable)
Iodine	I	126*	125.905619		13 d ec, β^+, β^-
		127	126.904474	100	
		128*	127.905812		25 m β^-, ec, β^+, β^-
		129*	128.904984		1.6×10^7 y
Xenon	Xe	124	123.905894	0.10	
		126	125.904268	0.09	
		128	127.903531	1.91	
		129	128.904779	26.4	
		130	129.903509	4.1	
		131	130.905069	21.2	
		132	131.904141	26.9	
		134	133.905394	10.4	
		136	135.907215	8.9	
Cesium	Cs	133	132.905436	100	
		134*	133.906703		2.1 y β^-
		135*	134.905891		2×10^6 y β^-
		137*	136.907078		30 y β^-
Barium	Ba	130	129.906289	0.106	
		132	131.905048	0.101	
		133*	132.905990		10.5 y ec
		134	133.904492	2.42	
		135	134.905671	6.593	
		136	135.904559	7.85	
		137	136.905816	11.23	
		138	137.905236	71.70	
Lanthanum	La	137*	136.906462		6×10^4 y ec
		138*	137.907105	0.0902	1.05×10^{11} y ec, β^+
		139	138.906346	99.9098	
Cerium	Ce	136	135.907139	0.19	
		138	137.905986	0.25	
		140	139.905434	88.43	
		142	141.909241	11.13	
Praseodymium	Pr	140*	139.909071		3.39 m ec, β^+
		141	140.907647	100	
		142*	141.910040		25.0 m β^-
Neodymium	Nd	142	141.907718	27.13	
		143	142.909809	12.18	
		144*	143.910082	23.80	2.3×10^{15} y α
		145	144.912568	8.30	
		146	145.913113	17.19	
		148	147.916888	5.76	
		150	149.920887	5.64	
Promethium	Pm	143*	142.910928		265 d ec
		145*	144.912745		17.7 y ec
		146*	145.914698		5.5 y ec
		147*	146.915134		2.623 y β^-

Element	Symbol	Mass number (*indicates radioactive)	Atomic mass	Percent abundance	Half-life and decay mode (if unstable)
Samarium	Sm	144	143.911996	3.1	
		146*	145.913043		1.0×10^8 y α
		147*	146.914894	15.0	1.06×10^{11} y α
		148*	147.914819	11.3	7×10^{15} y α
		149	148.917180	13.8	
		150	149.917273	7.4	
		151*	150.919928		90 y β^-
		152	151.919728	26.7	
		154	153.922206	22.7	
Europium	Eu	151	150.919846	47.8	
		152*	151.921740		13.5 y ec, β^+
		153	152.921226	52.2	
		154*	153.922975		8.59 y β^-
		155*	154.922888		4.7 y β^-
Gadolinium	Gd	148*	147.918112		75 y α
		150*	149.918657		1.8×10^6 y α
		152*	151.919787	0.20	1.1×10^{14} y α
		154	153.920862	2.18	
		155	154.922618	14.80	
		156	155.922119	20.47	
		157	156.923957	15.65	
		158	157.924099	24.84	
		160	159.927050	21.86	
Terbium	Tb	158*	157.925411		180 y ec, β^+, β^-
		159	158.925345	100	
		160*	159.927551		72.3 d β^-
Dysprosium	Dy	156	155.924277	0.06	
		158	157.924403	0.10	
		160	159.925193	2.34	
		161	160.926930	18.9	
		162	161.926796	25.5	
		163	162.928729	24.9	
		164	163.929172	28.2	
Holmium	Ho	165	164.930316	100	
		166*	165.932282		1.2×10^3 y β^-
Erbium	Er	162	161.928775	0.14	
		164	163.929198	1.61	
		166	165.930292	33.6	
		167	166.932047	22.95	
		168	167.932369	27.8	
		170	169.935462	14.9	
Thulium	Tm	169	168.934213	100	
		171*	170.936428		1.92 y β^-

(Continued)

Element	Symbol	Mass number (*indicates radioactive)	Atomic mass	Percent abundance	Half-life and decay mode (if unstable)
Ytterbium	Yb	168	167.933897	0.13	
		170	169.934761	3.05	
		171	170.936324	14.3	
		172	171.936380	21.9	
		173	172.938209	16.12	
		174	173.938861	31.8	
		176	175.942564	12.7	
Lutetium	Lu	173*	172.938930		1.37 y ec
		175	174.940772	97.41	
		176*	175.942679	2.59	3.8×10^{10} y β^-
Hafnium	Hf	174*	173.940042	0.162	2.0×10^{15} y α
		176	175.941404	5.206	
		177	176.943218	18.606	
		178	177.943697	27.297	
		179	178.945813	13.629	
		180	179.946547	35.100	
Tantalum	Ta	180	179.947542	0.012	
		181	180.947993	99.988	
Tungsten (Wolfram)	W	180	179.946702	0.12	
		182	181.948202	26.3	
		183	182.950221	14.28	
		184	183.950929	30.7	
		186	185.954358	28.6	
Rhenium	Re	185	184.952951	37.40	
		187*	186.955746	62.60	4.4×10^{10} y β^-
Osmium	Os	184	183.952486	0.02	
		186*	185.953834	1.58	2.0×10^{15} y α
		187	186.955744	1.6	
		188	187.955744	13.3	
		189	188.958139	16.1	
		190	189.958439	26.4	
		192	191.961468	41.0	
		194*	193.965172		6.0 y β^-
Iridium	Ir	191	190.960585	37.3	
		193	192.962916	62.7	
Platinum	Pt	190*	189.959926	0.01	6.5×10^{11} y α
		192	191.961027	0.79	
		194	193.962655	32.9	
		195	194.964765	33.8	
		196	195.964926	25.3	
		198	197.967867	7.2	
Gold	Au	197	196.966543	100	
		198*	197.968217		2.70 d β^-
		199*	198.968740		3.14 d β^-

Element	Symbol	Mass number (*indicates radioactive)	Atomic mass	Percent abundance	Half-life and decay mode (if unstable)
Mercury	Hg	196	195.965806	0.15	
		198	197.966743	9.97	
		199	198.968253	16.87	
		200	199.968299	23.10	
		201	200.970276	13.10	
		202	201.970617	29.86	
		204	203.973466	6.87	
Thallium	Tl	203	202.972320	29.524	
		204*	203.973839		3.78 y β^-
		205	204.974400	70.476	
	(Ra E″)	206*	205.976084		4.2 m β^-
	(Ac C″)	207*	206.977403		4.77 m β^-
	(Th C″)	208*	207.981992		3.053 m β^-
	(Ra C″)	210*	209.990057		1.30 m β^-
Lead	Pb	202*	201.972134		5×10^4 y ec
		204	203.973020	1.4	
		205*	204.974457		1.5×10^7 y ec
		206	205.974440	24.1	
		207	206.975871	22.1	
		208	207.976627	52.4	
	(Ra D)	210*	209.984163		22.3 y β^-
	(Ac B)	211*	210.988734		36.1 m β^-
	(Th B)	212*	211.991872		10.64 h β^-
	(Ra B)	214*	213.999798		26.8 m β^-
Bismuth	Bi	207*	206.978444		32.2 y ec, β^+
		208*	207.979717		3.7×10^5 y ec
		209	208.980374	100	
	(Ra E)	210*	209.984096		5.01 d α, β^-
	(Th C)	211*	210.987254		2.14 m α
	(Ra C)	212*	211.991259		60.6 m α, β^-
		214*	213.998692		19.9 m β^-
		215*	215.001836		7.4 m β^-
Polonium	Po	209*	208.982405		102 y α
	(Ra F)	210*	209.982848		138.38 d α
	(Ac C′)	211*	210.986627		0.52 s α
	(Th C′)	212*	211.988842		0.30 μs α
	(Ra C′)	214*	213.995177		164 μs α
	(Ac A)	215*	214.999418		0.0018 s α
	(Th A)	216*	216.001889		0.145 s α
	(Ra A)	218*	218.008965		3.10 m α
Astatine	At	215*	214.998638		$\approx$100 μs α
		218*	218.008685		1.6 s α
		219*	219.011297		0.9 m α
Radon	Rn				
	(An)	219*	219.009477		3.96 s α
	(Tn)	220*	220.011369		55.6 s α
	(Rn)	222*	222.017571		3.823 d α

(Continued)

Element	Symbol	Mass number (*indicates radioactive)	Atomic mass	Percent abundance	Half-life and decay mode (if unstable)
Francium		221*	221.01425		4.18 m α
	Fr	222*	222.017585		14.2 m β^-
	(Ac K)	223*	223.019733		22 m β^-
Radium	Ra	221*	221.01391		29 s α
	(Ac X)	223*	223.018499		11.43 d α
	(Th X)	224*	224.020187		3.66 d α
		225*			14.9 d β^-
	(Ra)	226*	226.025402		1600 y α
	($MsTh_1$)	228*	228.031064		5.75 y β^-
Actinium	Ac	225*			10 d α
	($Ms Th_2$)	227*	227.027749		21.77 y β^-
		228*	228.031015		6.15 h β^-
		229*			1.04 h β^-
Thorium	Th				
	(Rd Ac)	227*	227.027701		18.72 d α
	(Rd Th)	228*	228.028716		1.913 y α
		229*	229.031757		7300 y α
	(Io)	230*	230.033127		75,000 y α, sf
	(UY)	231*	231.036299	100	25.52 h β^-
	(Th)	232*	232.038051		1.40×10^{10} y α
	(UX_1)	234*	234.043593		24.1 d β^-
Protactinium	Pa	231*	231.035880		32,760 y α
	(UZ)	234*	234.043300		6.7 h β^-
Uranium	U	231*	231.036264		4.2 d β^+
		232*	232.037131		69 y α
		233*	233.039630		1.59×10^5 y α
	(UII)	234*	234.040946	0.0055	2.45×10^5 y α
	(Ac U)	235*	235.043924	0.720	7.04×10^8 y α
	(UI)	236*	236.045562		2.34×10^7 y α
		238*	238.050784	99.2745	4.47×10^9 y α
		239*	239.054290		23.5 m β^-
Neptunium	Np	235*	235.044057		396 d α
		236*	236.046559		1.54×10^5 y ec
		237*	237.048168		2.14×10^6 y α
Plutonium	Pu	236*	236.046033		2.87 y α, sf
		238*	238.049555		87.7 y α, sf
		239*	239.052157		24,120 y α, sf
		240*	240.053808		6560 y α, sf
		241*	241.056846		14.4 y β^-
		242*	242.058737		3.7×10^5 y α, sf
		244*	244.064200		8.1×10^7 y α, sf
Americium	Am	240*	240.055285		2.12 d ec
		241*	241.056824		432 y α, sf
Curium	Cm	247*	247.070347		1.56×10^7 y α
		248*	248.072344		3.4×10^5 y α, sf

Element	Symbol	Mass number (*indicates radioactive)	Atomic mass	Percent abundance	Half-life and decay mode (if unstable)
Berkelium	Bk	247*	247.070300		1380 y α
		249*	249.074979		327 d β^-
Californium	Cm	250*	250.076400		13.1 y α, sf
		251*	251.079580		898 y α
Einsteinium	Es	252*	252.082974		1.29 y α
		253*	253.084817		2.02 d α, sf
Fermium	Fm	253*	253.085173		3.00 d ec
		254*	254.086849		3.24 h α, sf
Mendelevium	Md	256*	256.093988		75.6 m ec, β^+
		258*	258.098594		55 d α
Nobelium	No	257*	257.096855		25 s α
		259*	259.100932		58 m α, sf
Lawrencium	Lr	259*	259.102888		6.14 s α, sf
		260*	260.105346		3.0 m α, sf
Rutherfordium	Rf	260*	260.160302		24 ms sf
		261*	261.108588		65 s α, sf
Dubnium	Db	261*	261.111830		1.8 s α
		262*	262.113763		35 s α
Seaborgium	Sg	263*	263.118310		0.78 s α, sf
Bohrium	Bh	262*	262.123081		0.10 s α, sf
Hassium	Hs	265*	265.129984		1.8 ms α
		267*	267.131770		60 ms α
Meitnerium	Mt	266*	266.137789		3.4 ms α, sf
		268*	268.138820		70 ms α
Darmstadtium	Ds	269*	269.145140		0.17 ms α
		271*	271.146080		1.1 ms α
		273*	272.153480		8.6 ms α
Roentgenium	Rg	272*	272.153480		1.5 ms α
Copernicium	Cn	277*	?		0.2 ms α
Nihonium	Nh	284*	?		? α
Flerovium	Fl	289*	?		? α
Moscovium	Mc	288*	?		? α
Livermorium	Lv	292*	?		? α
Tennessine	Ts	293*	?		? α
Oganesson	Og	294*	?		? α

GLOSSARY

absolute pressure Total pressure at a point in a fluid, equal to the sum of the gauge and the atmospheric pressures.

absolute zero The lowest temperature that is theoretically possible, at which the motion of particles is at a minimum; 0 on the Kelvin scale; −273.15°C or −459.67°F.

absorption lines Dark lines in an otherwise continuous spectrum. These indicate certain wavelengths of light that are absorbed by the atoms of the intervening medium.

absorption spectrum A continuous spectrum, broken by a specific pattern of dark lines or bands, observed when light traverses a particular absorbing medium.

ac (alternating current) An electric current that reverses its direction many times a second at regular intervals.

acceleration due to gravity Acceleration of a body due to the pull of gravity; an object in free fall near Earth's surface has an acceleration of approximately 9.8 m/s^2.

acceleration vector (*or* instantaneous acceleration vector) The change in a velocity vector per unit over time.

acceleration The rate of change of velocity, due to changes in its direction or magnitude.

ac generator Alternating current generator; as its coil rotates in a magnetic field, an oscillating emf is generated in the turns of wire that make up the coil.

adiabatic process In thermodynamics, a process that occurs without transfer of heat or matter in or out of a system.

alpha decay Type of radioactive decay in which an atomic nucleus emits an alpha particle (helium nucleus) and thereby transforms into a nucleus with a mass number that is reduced by four and an atomic number that is reduced by two.

alpha particle A positively charged particle, indistinguishable from a helium nucleus and consisting of two protons and two neutrons.

alternating current (ac) *See* ac (alternating current).

ampere The SI unit of electric current, equal to a flow of one coulomb per second.

Ampère's law For any closed loop path, the circulation of a magnetic field created by an electric current is equal to the size of that electric current times the permeability of free space; discovered by the French physicist André-Marie Ampère.

Amperian loop An imaginary closed path in space around a current-carrying conductor.

amplitude (of a wave) The maximum displacement from equilibrium that occurs as a wave moves through its medium.

angular acceleration The rate of change of angular velocity.

angular displacement The angle through which an object has been rotated.

angular frequency Frequency of a periodic process (as electric oscillation or sound vibration) expressed in radians per second, equivalent to the frequency in cycles multiplied by 2π.

angular position On a rotating object, the angle of a line on the object from its original position.

angular resolution The angle that separates two point objects that are just barely resolved through a circular aperture.

angular speed The magnitude or absolute value of angular velocity.

angular velocity The rate of change of angular displacement.

angular wave number The reciprocal of wavelength multipled by 2π.

antimatter Particles with the same properties as ordinary particles, but with the opposite electric charge.

antineutrino The antiparticle of a neutrino.

antinode Positions along a standing wave at which the oscillation is maximal.

antiquark The antiparticle of a quark.

apparent weight Weight of an object submerged in a fluid. This is less than its true weight due to buoyant forces.

apparent weightlessness The perceived state of an object that is accelerating along with its surroundings.

Archimedes' principle The buoyant force on an object immersed in a fluid is equal to the weight of the fluid that the object displaces.

atmosphere (unit of pressure) The average value of atmospheric pressure at sea level; equal to 1.01325×10^5 Pa or about 14.7 pounds per square inch.

atomic number The number of protons in an atomic nucleus; determines the chemical properties of an element and its place in the periodic table.

average acceleration The change in velocity divided by the elapsed time.

average angular acceleration The change in angular velocity divided by the elapsed time.

average angular velocity The change in angular displacement divided by the elapsed time.

average speed The total distance an object travels divided by the time it takes for the object to travel that distance.

average velocity The total displacement of an object divided by the elapsed time.

baryon Any hadron that can be made up of three quarks.

battery An electrochemical cell that can set charges into motion.

beat frequency Rate of the periodic variations of amplitude when two waves of different frequencies interfere.

beats Periodic variations in amplitude when two waves of different frequency interfere.

becquerel The SI unit of radioactive decay rate, equal to one disintegration per second.

Bernoulli's equation A relationship among pressure, speed, and height in an ideal fluid in motion.

Bernoulli's principle In a moving fluid, the pressure is low where the fluid is moving rapidly.

beta decay A type of radioactive decay in which a beta particle (an electron or a positron) is emitted from an atomic nucleus.

beta-minus decay When a neutron (charge zero) changes into a proton, an electron, and an electron antineutrino.

beta-plus decay When a proton changes into a neutron (charge zero), a positron, and an electron neutrino.

Big Bang The rapid expansion of matter from a state of extremely high density and temperature that marked the origin of the universe.

binding energy The energy required to disassemble an atomic nucleus into its component protons and neutrons.

blackbody An object that does not reflect any light at all but absorbs all radiation falling on it.

blackbody radiation Light emitted by a perfect blackbody of emissivity = 1.

Bohr model Theory of atomic structure in which a small, positively charged nucleus is surrounded by electrons that travel in circular orbits around the nucleus.

Bohr orbit In the Bohr model, one of the orbits in which electrons in an atom travel around the nucleus.

Boltzmann constant Physical constant in the ideal gas law which has the same value for all gases and relates the average kinetic energy of the particles in a gas to the temperature of the gas; equal to 1.38065×10^{-23} J/K.

boundary layer In a fluid, the layer next to a solid surface within which the fluid speed increases from zero at the surface to full speed at the edge of the layer.

Brewster's angle An angle of incidence at which light with a particular polarization is perfectly transmitted through a transparent dielectric surface, with no reflection.

British thermal unit The quantity of heat required to increase the temperature of 1 lb of pure water from 63°F to 64°F.

bulk modulus A measure of how resistant to compression a substance is, defined as the ratio of the pressure increase to the resulting relative decrease of the volume.

buoyant force The upward force exerted by a fluid on an object placed in it.

calorie The quantity of heat required to increase the temperature of one gram (1 g) of pure water from 14.5°C to 15.5°C.

capacitance The ability of a system to store an electric charge.

capacitor A system or device that can store positive and negative charge, consisting of one or more pairs of conductors that may be separated by an insulator.

Carnot cycle An ideal, reversible thermodynamic cycle consisting of two isothermal processes and two adiabatic processes; the most efficient cycle in a heat engine; first proposed in the early 1800s by Sadi Carnot.

Cavendish experiment Experiment used to determine the value of the gravitational constant.

Celsius scale The most common temperature scale; based on the work of the eighteenth-century Swedish astronomer Anders Celsius. In this scale the freezing point of water is approximately 0°C, and the boiling point is approximately 100°C.

center of curvature (of a mirror) The center of the sphere defined by the surface of a concave mirror.

center of mass A point representing the average position of the matter in a body or system; moves as though all of the body's mass were concentrated at that point and all external forces act on it.

centripetal acceleration The rate of change of tangential velocity of an object in circular motion; points toward the center of the circle defined by the object's trajectory.

centripetal force The force that points toward the inside of an object's curving trajectory and produces the centripetal acceleration.

charged object A body with net electric charge.

circuit A complete loop in which a charge flows through a wire continuously from one terminal of a battery to the other.

circuit element Any component of a circuit, such as a battery, resistor, inductor, or capacitor.

circulation (of a magnetic field) For an Amperian loop, the sum of products (of the component of magnetic field parallel to each loop segment multiplied by the segment length) along the loop.

closed pipe A pipe which is open at one end and blocked at the other.

coefficient of kinetic friction The ratio of the force of friction acting on a sliding object to the object's normal force; depends on the properties of the surfaces in contact.

coefficient of linear expansion The fractional change in length of an object per unit change in temperature.

coefficient of performance (of a heat pump or refrigerator) The ratio of useful heating or cooling provided to work required.

coefficient of rolling friction The ratio of the force of friction acting on the point of a rolling object that is in contact with the surface to the object's normal force; depends on the properties of the surfaces in contact.

coefficient of static friction The ratio of the force of friction acting on an unmoving object to the object's normal force; depends on the properties of the surfaces in contact.

coefficient of volume expansion The fractional change in the volume of an object per unit change in temperature.

completely inelastic collision Encounter in which two objects stick together after they collide and the most mechanical energy is lost.

component (of a vector) The projection of a vector onto a coordinate axis.

compressible fluid A fluid that can be easily compacted by squeezing.

compression Pressure applied to all sides of a body, which may result in a reduction in volume.

compressive strain The length by which an object shrinks expressed as a fraction of its relaxed length.

compressive stress The force applied to an object being squeezed divided by the object's cross-sectional area.

Compton scattering Elastic scattering of a photon by a free charged particle, usually an electron.

concave Curved inward, such as a mirror.

condensation The phase change from gas to liquid.

conduction Transfer of heat by the direct collision of particles, with no net displacement of the particles.

conductor A substance in which charges can move freely.

conservation of angular momentum If there is no net external torque on a system, the angular momentum of the system is conserved.

conservative force A force that can be associated with a potential energy, such as the gravitational force or the force exerted by an ideal spring.

constant acceleration A situation in which velocity changes at a steady rate.

constant velocity Movement at a steady speed in the same direction.

constructive interference The mutual reinforcement of waves such that the amplitude of the total wave is the sum of the amplitudes of the individual waves.

contact force A force that arises only when two objects come in contact with one another.

contact time The amount of time colliding objects are in contact.

convection Energy transfer by the motion of a liquid or gas (such as air) caused by the tendency of hotter, less dense material to rise and colder, denser material to sink under the influence of gravity.

converging lens Lens that takes incoming parallel light rays and brings them to a focus on the principal axis.

convex Curved outward, such as a mirror.

coordinates Quantities indicating the position of an object in reference to the origin on a coordinate system.

coordinate system A system that can be used to denote the position of an object at a given time.

cosmic background radiation (cosmic microwave background) The thermal radiation left over from the Big Bang.

cosmological redshift A redshift caused by the expansion of the universe.

cosmology The science of the origin and evolution of the universe.

coulomb The unit of electric charge, equal to the amount of electricity conveyed in one second by a current of one ampere; named after the eighteenth-century French physicist Charles-Augustin de Coulomb, who uncovered the fundamental law that governs the interaction of charges.

Coulomb's constant A proportionality constant used to determine the magnitude of the electric forces between two point charges; equal to $8.99 \times 10^9\ \text{N} \cdot \text{m}^2/\text{C}^2$.

Coulomb's law The magnitude of the force of electrostatic attraction or repulsion acting between two electric charges is directly proportional to the product of the charges and inversely proportional to the square of the distance between them.

critical angle The angle of incidence beyond which total internal reflection occurs.

critically damped oscillations Case in which the minimum amount of damping is applied to result in a nonoscillatory response; when displaced from equilibrium, the system returns smoothly to equilibrium with no overshoot and hence no oscillation.

critical point A point on a phase diagram at which both the liquid and gas phases of a substance have the same density and are therefore indistinguishable.

cross product (vector product) The product of two vectors in three dimensions that is itself a vector at right angles to both the original vectors, with a direction given by the right-hand rule and a magnitude equal to the product of the magnitudes of the original vectors and the sine of the angle between their directions.

current The rate at which charge flows past any point in a circuit.

current loop A single loop that carries a current provided by a source of emf.

damped oscillations Oscillations that are diminished by a frictional force.

damping coefficient Proportionality constant related to the physical characteristics of a particular system and determining the degree to which oscillations are diminished.

dark energy A repulsive force that counteracts gravity and causes the universe to expand at an accelerating rate.

dark matter Nonluminous material that is postulated to exist in space and that is the dominant form of matter in the universe.

dc (direct current) An electric current that does not change direction in a circuit.

de Broglie wavelength The wavelength of a particle, given by Planck's constant divided by the momentum of the particle.

decay constant Proportionality between the size of a population of radioactive nuclei and the rate at which the population decreases because of radioactive decay.

decibel The units of sound intensity level.

degree of freedom In thermodynamics, a possible form of motion of an object.

density The mass of a substance divided by the volume that it occupies.

deposition The phase change from gas to solid.

destructive interference When two waves cancel each other out so that the amplitude of the total wave is zero.

diamagnetic Tending to become magnetized in a direction opposite to that of the applied magnetic field.

dielectric A material that is both an insulator and polarizable.

dielectric constant The greater the dielectric constant of a material, the more the material is polarized when it is placed in the electric field between the plates of a charged capacitor.

diffraction The bending of light around an obstacle or aperture.

diffraction maxima Locations in a diffraction pattern where wavelets interfere constructively, producing a bright fringe.

diffraction minima Locations in a diffraction pattern where wavelets interfere destructively, producing a dark fringe.

diffraction pattern The distinctive pattern of bright and dark fringes caused when light is diffracted through a slit or aperture.

diffuse light Light that reflects from an object's surface in many random directions.

dimensional analysis A method of checking the relations of physical quantities by identifying their dimensions.

diode A single-junction semiconductor device.

diopter A unit of power for a lens or mirror that is equal to the reciprocal of the focal length (in meters).

direct current (dc) *See* dc (direct current).

dispersion The separation of light according to wavelength due to differing propagation speeds of different wavelengths of light; a prism separates white light into its component colors because light of each color travels at a different speed through the glass.

displacement The difference between the positions of an object at two separate times.

displacement current A quantity appearing in Maxwell's equations that is defined in terms of the rate of change of the electric displacement field.

displacement vector Vector drawn from the starting point of an object's motion to its endpoint.

diverging lens Lens that takes incoming parallel light rays and causes them to spread away from the principal axis.

doping Adding small amounts of a different kind of atom to a substance.

Doppler effect The change in frequency of a wave caused by an observer moving relative to its source or its source moving relative to the observer.

drag force The force that resists the motion of an object through a liquid or a gas.

drift speed The average speed at which charges move through a conductor.

driving angular frequency The angular frequency of a driving force in an oscillation.

eccentricity Parameter determining the circularity of an ellipse; a perfect circle has zero eccentricity.

efficiency The useful work divided by the amount of energy taken in to do the work.

elapsed time The duration of a time interval.

elastic An elastic object returns to its original shape after being squeezed or stretched.

elastic collision A collision in which the forces between the colliding objects are conservative; both total momentum and total mechanical energy are conserved.

electric charge The physical property of matter that causes it to experience a force near other charged material.

electric dipole A combination of two point charges of the same magnitude but opposite signs.

electric energy density The energy per unit volume stored in an electric field, such as in a capacitor.

electric field lines Lines showing the direction of an electric field.

electric flux The area of a surface multiplied by the component of the electric field that's perpendicular to that surface.

electric force The force between electric charges.

electric potential The electric potential energy for a charge at a given position divided by the value of that charge.

electric potential difference The difference in electric potential between two locations.

electric potential energy Potential energy that results from conservative Coulomb forces; associated with the configuration of a particular set of point charges.

electromagnetic induction The process whereby a changing magnetic field induces an electric field.

electromagnetic spectrum The range of electromagnetic waves according to wavelength.

electromagnetic wave Waves that are propagated by simultaneous periodic variations of electric and magnetic fields. These include radio waves, infrared, visible light, ultraviolet, x rays, and gamma rays.

electromagnetism An umbrella term to cover both electricity and magnetism, since both involve interactions between charges.

electron volt A unit of energy equal to the work done on an electron in accelerating it through a potential difference of one volt.

emf (electromotive force) The voltage developed by any source of electrical energy such as a battery.

emission lines Bright lines in the emission spectrum of a gas.

emission spectrum A spectrum that consists only of specific emitted wavelengths.

emissivity How well or how poorly a surface radiates.

energy The capacity to do work.

energy level Quantized value of energy, for example of electrons in an atom.

entropy A measure of the amount of disorder present in a system.

equation of continuity In fluid dynamics, $A_1 v_1 = A_2 v_2$: the product of a pipe's cross-sectional area A and the flow speed v is conserved; it has the same value at point 1 as at point 2.

equation of hydrostatic equilibrium The equation $p = p_0 + \rho g d$, which must be satisfied for a fluid to remain at rest.

equation of state A relationship among the quantities of pressure, volume, and temperature.

equilibrium State in which the net external force on an object is zero.

equipartition theorem Principle stating that the energy of a molecule is shared equally among each degree of freedom.

equipotential A curve along which the electric potential has the same value at all points.

equipotential surface A surface on which the electric potential has the same value at all points.

equipotential volume Space inside a conductor in which the electric potential has the same value everywhere.

equivalent capacitance Effective capacitance of an arrangement of two or more connected capacitors.

equivalent resistance Effective resistance of an arrangement of two or more connected resistance.

escape speed The minimum speed at which an object must be launched from Earth's surface to escape to infinity.

event Something that happens at a certain point in time.

exchange particle A virtual particle that interacts with ordinary particles to mediate forces, producing the effects of attraction and repulsion.

exponent The superscript following the number 10 that denotes the number of zeros needed to write the long form of a number in scientific notation.

exponential function The irrational number *e* raised to a power.

external forces Forces exerted on an object by other objects.

Fahrenheit scale The official temperature scale used in the United States, the Cayman Islands, and Belize; originated by the German scientist Daniel Fahrenheit. On this scale water freezes at 32°F and boils at 212°F.

failure The point at which the structure of a material starts to lose its integrity, which eventually leads to the object breaking apart.

farad The SI unit of electrical capacitance, equal to the capacitance of a capacitor in which one coulomb of charge causes a potential difference of one volt; named after English physicist Michael Faraday.

Faraday's law (of induction) An emf is induced in a loop if the magnetic flux through that loop changes.

ferromagnetic A ferromagnetic material has a high susceptibility to magnetization, the strength of which depends on that of the applied magnetizing field and that may persist after removal of the applied field.

first law of thermodynamics In a thermodynamic process, the change in the internal energy of a system equals the heat that flows into the system during the process minus the work that the system does during the process.

fluid A substance (a gas or a liquid) that can flow because its molecules can move freely with respect to each other.

fluid resistance The resistance experienced by an object as it moves through a fluid.

focal length The distance from the focal point to the center of a mirror or lens.

focal point The point along the principal axis of a mirror or lens at which incident rays parallel to the principal axis converge and come to a common focus.

force A push or a pull.

forced oscillations Case in which a periodic driving force causes a system to oscillate at the frequency of that driving force.

force pair In an interaction between objects A and B, the forces of A on B and of B on A.

frame of reference (*or* reference frame) A coordinate system with respect to which we can make observations or measurements.

free-body diagram A graphical representation of all external forces acting on a body.

free fall The state of an object falling toward Earth without any effect from air resistance.

freezing The phase change from liquid to solid.

frequency The number of cycles of an oscillation per unit of time.

friction force Force resisting the sliding of an object across a surface, acting parallel to the surface and opposite to the motion of an object.

fringe The series of bright and dark patches in an interference or diffraction pattern.

front (of a wave) Wave crest.

fundamental frequency The lowest natural frequency of an oscillating object.

fundamental mode For an oscillating object, the standing wave mode corresponding to the fundamental frequency.

fusion The change from solid to liquid; melting.

galaxy A collection of a tremendous number of stars held together by gravity.

Galilean transformation Set of equations used to translate between the coordinates of two reference frames which differ only by constant relative motion.

Galilean velocity transformation Set of equations used to translate between the velocity of an object in two reference frames which differ only by constant relative motion.

gamma decay A radioactive process in which an atomic nucleus loses energy by emitting a gamma ray without a change in its atomic number or mass number.

gas A fluid that expands to fill whatever volume is available to it.

gauge pressure The amount by which the pressure exceeds atmospheric pressure.

Gaussian surface A closed surface used to enclose charge in order to apply Gauss's law.

Gauss's law (for the electric field) The net electric flux through a closed surface (called a Gaussian surface) equals the net charge enclosed by that surface divided by the permittivity. Charges outside the surface have no effect on the net electric flux through the surface; named after nineteenth-century German mathematician and physicist Carl Friedrich Gauss.

Gauss's law for the magnetic field For any closed Gaussian surface, there is zero net flux of the magnetic field through that surface.

general theory of relativity Albert Einstein's theory which provides a unified description of gravity as a geometric property of space and time, or spacetime.

geometrical optics The science of mirrors and lenses.

global warming An increase in the global average surface temperature caused by the greenhouse effect.

gluon A subatomic particle of a class that is thought to bind quarks together.

gravitational bending of light Effect in which a light beam passing near a massive object will curve under the influence of the object's gravity.

gravitational constant The constant involved in the calculation of gravitational force between two objects; equal to $6.67 \times 10^{-11}\ \text{N} \cdot \text{m}^2/\text{kg}^2$.

gravitational force The force of attraction between all masses in the universe; especially the attraction of Earth's mass for bodies near its surface.

gravitational potential energy The ability to do work related to an object's vertical position in the presence of gravity.

gravitational slowing of time Effect in which gravity influences the rate of a ticking clock; clocks on the ground floor of a building tick more slowly than clocks on the top floor, which are farther from Earth's center.

gravitational waves Small variations in the curvature of spacetime that spread away from moving massive objects.

greenhouse effect Warming effect in which the atmosphere prevents some of the radiation emitted by Earth's surface from escaping into space.

greenhouse gas One of several gases in the atmosphere that are transparent to visible light but not to infrared radiation.

hadron A particle that can experience the strong force.

Hagen-Poiseuille equation Relates the pressure difference between the two ends of a pipe to the resulting flow rate of a viscous fluid.

half-life The time taken for the radioactivity of a specified isotope to fall to half its original value.

harmonic property Characteristic of simple harmonic motion in which the angular frequency, period, and frequency of an oscillation are independent of the amplitude if the restoring force obeys Hooke's law.

heat A form of energy arising from the random motion of the molecules of bodies, which may be transferred by conduction, convection, or radiation.

heat engine A system or device that converts heat to work.

Heisenberg uncertainty principle The shorter the duration of a phenomenon, the greater the uncertainty in the energy of that phenomenon.

hertz The SI unit of frequency, equal to one cycle per second.

Higgs field The theoretical field that gives fundamental particles their mass.

Higgs particle Subatomic particle associated with the Higgs field.

hole The lack of an electron at a position where one could exist in an atom.

Hooke's law The force needed to extend a spring by a certain distance is proportional to that distance.

Hubble constant The ratio of the speed of recession of a galaxy (due to the expansion of the universe) to its distance from the observer.

Hubble law The observation that the speed of recession of distant galaxies is proportional to their distance from the observer.

Huygens' principle The wave front at a later time is the superposition of all of the wavelets emitted at the starting time and is tangent to the leading edges of the wavelets.

hydrostatic equilibrium State in which a fluid is at rest.

ideal gas A theoretical gas composed of a set of randomly moving, noninteracting point particles.

ideal gas constant A constant in the ideal gas law, expressed in units of energy per temperature increment per mole; equal to $8.314 \text{ J/(mol} \cdot \text{K)}$.

ideal gas law Equation of the state of a hypothetical ideal gas.

image An appearance of an object formed by light rays reflected by a mirror or focused by a lens.

image distance The distance from a mirror to an object's reflected image, or the distance from a lens to the image.

image height The height of an object's image reflected in a mirror or focused by a lens.

impedance The measure of the opposition that a circuit presents to an alternating current when an alternating voltage is applied.

impulse The product of the force acting on an object and the duration of the time interval over which that force acts.

incident light Incoming light that strikes a surface.

incompressible fluid A fluid whose volume and density change very little when squeezed.

index of refraction A measure of the speed of light traveling in a medium; the speed of light in a vacuum divided by the speed of light in the medium.

induced emf An emf induced around a loop by a changing magnetic field.

induced magnetic field A magnetic field produced by the current in a loop that is caused by an induced emf in that loop.

inductance For a current-carrying conducting coil, the ratio of the magnetic flux through the coil divided by the current that produces the flux.

inelastic collision A collision in which mechanical energy is not conserved.

inertia The tendency of an object to resist change in motion.

inertial frame of reference A frame of reference attached to an object that does not accelerate.

instantaneous acceleration Acceleration of an object at a specific instant.

instantaneous speed Speed of an object at a specific instant.

instantaneous velocity Velocity of an object at a specific instant.

insulator Substance in which charges are not able to move freely.

intensity Average wave power per unit area.

interference The combination of two or more electromagnetic waves to form a resultant wave in which the displacement is either reinforced or canceled.

interference maxima Locations in an interference pattern where wavelets interfere constructively, producing a bright fringe.

interference minima Locations in an interference pattern where wavelets interfere destructively, producing a dark fringe.

interference pattern Alternating bright and dark fringes produced when two or more light waves combine or cancel each other out.

internal energy The energy within an object due to the kinetic and potential energies associated with the individual molecules that comprise the object.

internal forces Forces exerted by one part of an object or system on another part.

internal resistance The resistance that mobile charges encounter as they pass through a battery.

inverse Lorentz velocity transformation Set of equations relating the velocity of an object in two reference frames moving relative to one another, consistent with special relativity.

inverse-square law for waves Intensity is inversely proportional to the square of the distance from the source.

inverted image Image (produced by a mirror or lens) that is flipped upside down relative to the object.

inviscid flow Flow of a fluid that is assumed to have no viscosity.

ionizing radiation Photons with enough energy to dislodge an electron from an atom.

irreversible process A process that cannot return both the system and the surroundings to their original conditions.

irrotational flow Flow in which the speed varies gradually from one part of the fluid to another, with no abrupt jumps.

isobaric process Process in which the pressure of the system remains constant.

isochoric process Process in which the volume of the system remains constant.

isotherm Contours of constant temperature.

isothermal process Process in which the temperature of the system remains constant.

isotope Variants of a particular chemical element that share the same number of protons in the nucleus of each atom but differ in neutron numbers.

joule The SI unit of work or energy, equal to the work done by a force of one newton when its point of application moves one meter in the direction of action of the force.

junction Points in a circuit where either the current breaks into two currents or two currents come together into one.

kelvin The SI base unit of thermodynamic temperature, equal in magnitude to the degree Celsius.

Kelvin scale Temperature scale based on the relationship between pressure and temperature of low-density gases; first proposed by the nineteenth-century Scottish physicist William Thomson (1st Baron Kelvin).

kilogram The SI unit of mass.

kinetic energy The energy that an object possesses by virtue of being in motion.

kinetic friction Force acting on a sliding object that opposes the object's motion.

Kirchhoff's junction rule The sum of the currents flowing into a junction equals the sum of the currents flowing out of it.

Kirchhoff's loop rule The sum of the changes in electric potential around a closed loop in a circuit must equal zero.

laminar flow Smooth fluid flow in which each object follows the object directly in front of it.

latent heat (of fusion or vaporization) The amount of heat per unit mass that must flow into or out of the substance to cause the phase change.

lateral magnification The ratio of the height of an image to the height of the corresponding object.

law of areas A line joining the Sun and a planet sweeps out equal areas in equal intervals of time, regardless of the position of the planet in the orbit.

law of conservation of angular momentum *See* conservation of angular momentum

law of conservation of energy One kind of energy can transform into another, but the total amount of energy of all forms remains the same.

law of conservation of momentum If the net external force on a system of objects is zero, then the total momentum of the system does not change.

law of orbits The orbit of each planet is an ellipse with the Sun located at one focus of the ellipse.

law of periods The square of the period of a planet's orbit is proportional to the cube of the semimajor axis of the orbit.

law of reflection The angle of the reflected light is the same as the angle of the incident light.

***LC* circuit** A circuit made up of an inductor of inductance L and a capacitor of capacitance C connected by ideal, zero-resistance wires.

length contraction The shortening of an object or distance moving at nearly the speed of light along the direction of motion; predicted by the theory of special relativity.

lens A piece of glass or other transparent material with a curved surface for concentrating or dispersing light rays.

lens equation Expression relating object distance and lens distance to focal length.

lensmaker's equation Expression relating the focal length of a lens in air to its index of refraction and curvature.

Lenz's law The direction of the magnetic field induced within a conducting loop opposes the change in magnetic flux that created it; named after Russian physicist Heinrich Lenz.

lepton A particle that is not affected by the strong force.

lever arm The perpendicular distance from the rotation axis of a body to the line of action of a force applied to that body.

light-emitting diode (LED) A semiconductor device that emits visible light when an electric current passes through it; produces much more light for a given power input than do incandescent light bulbs or fluorescent lamps.

linearly polarized light Light for which the electric field is oriented completely along one direction.

linear mass density Mass per unit length.

linear momentum (momentum) The product of an object's mass and its velocity vector.

linear motion Motion in a straight line.

line of action For a force applied to a body, an extension of the force vector through the point of application.

liquid A fluid that maintains the same volume regardless of the shape and size of its container.

longitudinal wave (pressure wave) A traveling disturbance or vibration in which the individual parts of the wave medium move in the direction parallel to the direction of wave propagation.

Lorentz transformation Set of equations used to translate between the coordinates of two reference frames moving relative to one another, consistent with special relativity.

Lorentz velocity transformation Set of equations relating the velocity of an object in two reference frames moving relative to one another, consistent with special relativity.

Mach angle For supersonic flow, the angle a shock wave makes with the direction of motion, determined by the velocity of the object and the velocity of shock propagation.

Mach cone The conical pressure wave front produced by a body moving at a speed greater than that of sound.

Mach number The ratio of an object's speed to the speed of sound.

magnet A material or object that produces a magnetic field.

magnetic dipole A pair of equal and opposite magnetic poles separated by a small distance; the magnetic field points away from the magnet's north pole and toward the magnet's south pole.

magnetic energy The energy required to set up a magnetic field in and around an inductor.

magnetic energy density The energy per unit volume stored in a magnetic field, such as in an inductor.

magnetic field A region around a magnetic material or a moving electric charge within which the force of magnetism acts.

magnetic flux The area of a surface multiplied by the component of the magnetic field that's perpendicular to that surface.

magnetic force Force of attraction or repulsion that arises between magnets, or between electrically charged particles because of their motion.

magnetic poles The two ends of a magnetic field.

magnetic resonance imaging (MRI) A form of medical imaging that measures the response of atomic nuclei in body tissues to radio waves when placed in a strong magnetic field, producing detailed images of internal organs.

magnetism The interaction between magnets or between electrically charged particles due to their motion.

magnitude (of a vector) The straight-line distance from the starting point of a vector to its endpoint.

mass The measure of the amount of material in an object.

mass number The total number of protons and neutrons in an atomic nucleus.

mass spectrometer A device used to determine the masses of individual atoms and molecules.

Maxwell–Ampère law Magnetic fields can be generated in two ways: by electrical current and by changing electric fields.

Maxwell's equations Four basic equations that describe all electromagnetic phenomena.

mean free path The average distance that a molecule travels from the time at which it collides with one molecule to when it collides with another molecule.

mechanical wave A propagating oscillation of matter that transfers energy through a medium.

medium The substance through which a mechanical wave propagates.

meson A subatomic particle that is intermediate in mass between an electron and a proton and that transmits the strong interaction that binds nucleons together in the atomic nucleus.

meter The SI unit of measurement for length.

Michelson–Morley experiment An experiment performed in 1887 attempting to detect the velocity of Earth with respect to the hypothetical luminiferous ether; discovered no evidence for its existence.

millimeters of mercury Unit of measurement for pressure; 760 mmHg equals 1 atm.

mirror equation Expression relating object distance and image distance to the focal length of a mirror.

mobile charge Charged particles that are free to move throughout a conducting material, as in a circuit.

molar specific heat The quantity of heat required to raise the temperature of one mole of the substance by one kelvin.

molar specific heat at constant pressure The quantity of heat required to make one mole of a substance undergo a temperature change of one kelvin if the pressure is held constant.

molar specific heat at constant volume The quantity of heat required to make one mole of a substance undergo a temperature change of one kelvin if the volume is held constant.

mole Unit measuring the quantity of a substance; one mole equals the number of atoms in exactly 12 grams of carbon-12, given by Avogadro's number: 6.022×10^{22}.

moment of inertia A property of a body that defines its resistance to a change in angular velocity about an axis of rotation.

momentum (linear momentum) *See* linear momentum.

monochromatic light Light that has a single definite wavelength.

motional emf A changing emf due to the motion of a conductor in a magnetic field.

motion diagram A diagram that visualizes the motion of an object using distance gained between equal time intervals.

motion in a plane Two-dimensional motion.

multiloop circuit A circuit with more than one pathway that a moving charge can take from the positive terminal of the battery through the circuit to the negative terminal.

mutual inductance An effect in a transformer in which a change in the current in one coil induces an emf and current in a second coil.

natural angular frequency The angular frequency at which a system oscillates when not subjected to an external force.

negative displacement The distance, in the negative direction along a defined coordinate axis, between the position of an object at one time and its position at an earlier time.

negative work Work done whenever the angle between the displacement of an object and the force acting on that object is greater than 90°.

net external force (net force) The vector sum of all external forces acting on an object.

neutral matter Matter that contains equal amounts of positive and negative charge.

neutrino A nearly massless, neutral particle.

neutron-induced fission The radioactive decay of an atomic nucleus initiated by the collision of a neutron.

neutron number The number of neutrons in the nucleus of an atom.

newton The SI unit of force; equal to the force that would give a mass of one kilogram an acceleration of one meter per second per second.

Newton's first law An object at rest tends to stay at rest, and an object in uniform motion tends to stay in motion with the same speed and in the same direction, unless acted upon by a net force.

Newton's law of universal gravitation Any two objects exert a gravitational force of attraction on each other in a direction along the line joining the objects, with a magnitude proportional to the product of the masses of the objects and inversely proportional to the square of the distance between them.

Newton's laws of motion Isaac Newton's three fundamental relationships between force and motion.

Newton's second law If a net external force acts on an object, the object accelerates. The net external force is equal to the product of the object's mass and the object's acceleration.

Newton's third law If object A exerts a force on object B, object B exerts a force on object A that has the same magnitude but is in the opposite direction. These two forces act on different objects.

node (of a wave) Any point where the displacement of a wave is always zero.

nonconservative force A dissipative force that does not have a defined potential energy, such as friction.

noninertial frame A frame of reference attached to an accelerated object.

normal A line that is perpendicular to a surface.

normal force The support force exerted upon an object that is in contact with another stable object, acting perpendicular to the surface of contact.

no-slip condition Requirement that the velocity of a fluid be zero next to a solid surface.

***n*-type semiconductor** A type of semiconductor doping provides extra electrons to the host material, creating an excess of negative electron charge carriers.

nuclear fission A nuclear reaction in which a heavy nucleus splits spontaneously or on impact with another particle, releasing energy.

nuclear fusion A nuclear reaction in which atomic nuclei of low atomic number fuse to form a heavier nucleus, releasing energy.

nuclear radiation The emission by a nucleus of either energy (in the form of a photon) or particles (such as an alpha particle).

nucleon A proton or neutron.

nuclide An atomic species characterized by the specific constitution of its nucleus, that is, by its number of protons and its number of neutrons.

object Anything that acts as a source of light rays for an optical device.

object distance An object's distance from a mirror or lens.

object height The vertical extent of an object that is reflected in a mirror or refracted by a lens.

ohm The SI unit of electrical resistance; the resistance in a circuit transmitting a current of one ampere when subjected to a potential difference of one volt.

one-dimensional wave A wave that propagates along a single dimension of space.

open pipe A pipe that is unblocked on both ends.

optical device An instrument that changes the direction of light rays in a regular way.

orbital period The time required to complete an orbit.

origin The location from which the points on a coordinate system are measured. Also called reference position.

oscillation The regular movement of an object back and forth around a point of equilibrium.

overdamped oscillations Case in which high damping is applied, resulting in a nonoscillatory response; when displaced from equilibrium, the system returns to equilibrium with no overshoot and hence no oscillation.

parabola A particular u-shaped curve; the shape of the path of a projectile under the influence of gravity.

parallel (capacitors) An arrangement of capacitors connected along multiple paths (not in series), resulting in a multiloop circuit; the total capacitance is equal to the sum of all the individual capacitances.

parallel (resistors) An arrangement of resistors connected along multiple paths (not in series), resulting in a multiloop circuit; the reciprocal of the total resistance is equal to the sum of the reciprocals of all the individual resistances.

parallel-axis theorem Expression determining the moment of inertia of an object about a given axis in terms of the moment of inertia about a parallel axis running through its center of mass and the distance between the two axes.

parallel-plate capacitor A capacitor formed using two parallel metal plates.

paramagnetic Tendency to be weakly attracted by the poles of a magnet but not retaining any permanent magnetism; if the magnetic field is turned off, random thermal motion will cause the atomic current loops to return to their original, nonaligned orientations.

partially polarized light Light in which the orientation of the electric field changes randomly but is more likely to be in one orientation than in other orientations.

particle physics The branch of physics that concerns the fundamental constituents of matter and how they interact.

pascal Unit of measurement for pressure; equal to one newton per square meter.

Pascal's principle Pressure applied to a confined, static fluid is transmitted undiminished to every part of the fluid as well as to the walls of the container.

path length difference The difference in the distance traversed by two waves traveling to the same point from different locations.

Pauli exclusion principle The quantum mechanical principle that no two electrons may occupy the same quantum state simultaneously.

pendulum A system that oscillates back and forth due to the restoring force of gravity.

period The time for one complete cycle of an oscillation.

permeability of free space A constant involved in the relationship between an electric current and the magnetic field that it produces in a vacuum.

permittivity of free space A constant involved in the relationship between an electric charge and the electric field that it produces in a vacuum.

phase A physically distinctive form of matter, such as a solid, liquid, gas, or plasma.

phase angle The amount that a wave is shifted, indicating where in the oscillation cycle the object is at $t = 0$.

phase change The transformation from one state of matter to another.

phase diagram A graph of pressure p versus temperature T for a substance, showing the values of p and T for each phase of the substance.

phase difference The mathematical difference between two phase angles, such as the phase angles of two different waves.

photoelectric effect The emission, or ejection, of electrons from the surface of a material in response to incident light.

photoelectrons Electrons emitted through the photoelectric effect.

photon The quantum of electromagnetic energy, regarded as a discrete particle having zero mass, no electric charge, and an indefinitely long lifetime.

photovoltaic solar cell An electrical device that converts the energy of light directly into electricity by the photovoltaic effect.

physical pendulum A pendulum whose mass is distributed throughout its volume.

Planck's constant A physical constant relating the ratio of the energy of a photon to its frequency; equal to $6.626 \times 10^{-34}\ \text{J}\cdot\text{s}$.

plane mirror A flat, reflecting surface.

plastic The state of an object when the tensile stress exceeds the yield strength and an object deforms permanently.

plates Two pieces of metal used in a capacitor to store charge.

***pn* junction** The region inside a semiconductor where p-type and n-type semiconductors meet.

point charges Very small charged objects whose size is much smaller than the separation between the charges.

polarization The orientation of the electric field in a light wave.

polarizing filter A transparent sheet which contains long-chain molecules that are all oriented in the same direction, used to polarize the light passing through it.

position The location of an object on a coordinate system.

position vector A vector that extends from an origin to a point where an object is located at a specific moment in time.

positive displacement The distance, in the positive direction along a defined coordinate axis, between the position of an object at one time and its position at an earlier time.

positron A subatomic particle with the same mass as an electron and a numerically equal but positive charge.

potential energy An ability to do work based on an object's position.

pound The English unit of force.

power The rate at which work is done or energy is transferred.

power (of a lens or mirror) The reciprocal of the focal length (in meters).

power of ten An alternative name for the exponent given in scientific notation, referring to how many tens must be multiplied together to give the desired number.

pressure The magnitude of the force per unit area on the surface of an object.

pressure amplitude In a sound wave, the maximum pressure variation above or below the pressure of the undisturbed air.

pressure wave (longitudinal wave) *See* longitudinal wave.

primary coil The winding of a transformer connected to the input voltage.

principal axis A line passing through the center of the surface of a lens or spherical mirror and through the centers of curvature of all segments of the lens or mirror.

principle of equivalence A gravitational field is equivalent to an accelerated frame of reference in the absence of gravity.

principle of Newtonian relativity The laws of motion are the same in all inertial frames of reference.

projectile Object undergoing free-fall motion under the influence of gravity.

projectile motion Free-fall motion under the influence of gravity, involving both vertical and horizontal motion.

propagation speed The rate at which a wave travels in a medium.

proper time The time interval between two events in a frame of reference in which the events occur at the same place.

***p*-type semiconductor** A type of semiconductor with an absence of negative charge and hence an abundance of positive charge carriers or holes.

***pV* diagram** A graph that plots the pressure of a system on the vertical axis versus the volume of the system on the horizontal axis.

quantized Restricted to only certain values, as energy.

quantum mechanics Branch of physics that deals with the motions and interactions of atoms and subatomic particles, incorporating the concepts of quantization of energy, wave-particle duality, and the uncertainty principle.

quantum number Number describing the value of a physical quantity in a quantum mechanical system, such as an atom.

quark Any of a number of subatomic particles carrying a fractional electric charge, postulated as building blocks of the hadrons.

quark confinement The phenomenon wherein quarks can never be removed from the hadrons they compose.

radiation Energy transfer by the emission (or absorption) of electromagnetic waves.

radioactive A radioactive nuclide is one that decays into another nuclide by emitting ionizing radiation or particles.

radius of curvature The distance from a curved surface (such as a mirror) to its center of curvature.

ratio of specific heats For a given substance, the ratio of molar specific heat at constant pressure to the molar specific heat at constant volume.

ray An arrow that points in the direction of light propagation.

ray diagram Drawing used to determine the position of an image made by a mirror or lens.

reaction force The force exerted by object A on object B in reaction to having a force exerted by object B on object A.

real image An image formed by light rays coming together.

redshift The displacement of spectral lines toward longer wavelengths (the red end of the spectrum) in radiation from distant galaxies and celestial objects.

reference frame *See* frame of reference.

reference position *See* origin.

refraction The change in direction of a beam of light that travels from one medium into another; the angle that the refracted light makes to the normal is not equal to the angle of the incident light to the normal.

refrigerator A device that takes in energy and uses it to transfer heat from an object at low temperature to an object at high temperature.

relativistic gamma A dimensionless quantity that is equal to 1 when an object is at rest and becomes infinitely large as the speed of the object approaches the speed of light.

relativistic speed A speed that is a significant proportion of the speed of light.

reservoir A part of a system large enough either to absorb or supply heat without a change in temperature.

resistance For an electrical conductor, the resistivity of the material of which the conductor is made multiplied by the length of the conductor and divided by its cross-sectional area.

resistivity A measure of how well or poorly a material inhibits the flow of electric charge.

resistor A circuit component intended to add resistance to the flow of current.

resolve To optically distinguish; as in telling two closely spaced objects apart.

resonance The condition in which an object or system is subjected to an oscillating force having a frequency close to its own natural frequency.

rest energy Energy of a particle that is not in motion.

restoring force A force that tends to bring an object back toward equilibrium.

reversible process An ideal process that can return both the system and the surroundings to their original conditions without increasing entropy; throughout the entire reversible process the system is in thermodynamic equilibrium with its surroundings.

Reynolds number The ratio of the forces due to pressure differences acting on a small piece of a fluid to the viscous forces acting on the same piece of fluid.

right-hand rule A rule that uses the right hand to determine the orientation of vector quantities normal to a plane; for example, used to find the direction of the angular momentum vector around an axis of rotation.

rigid object An object with a fixed shape; the distance between any two points in the object remains constant in time regardless of external forces exerted on it.

rolling friction Force acting on the point of contact between a rolling object and the surface opposite to the direction of motion.

rolling without slipping The state in which an object rolls uniformly across a surface without skidding.

root-mean-square speed (rms speed) (of molecules) A measure of how fast gas molecules move; the square root of the average value of the square of individual speeds.

root mean square value (rms value) (of an ac circuit) The square root of the average value of the ac voltage squared.

rotation Motion in which an object spins around an axis.

rotational kinematics The study of rotational motion, including angular velocities and angular acceleration, in the absence of forces.

rotational kinetic energy Energy possessed by an object by virtue of its rotational motion.

scalar A physical quantity that has only magnitude, not direction.

scientific notation A standard shorthand system used by physicists for extremely large or small numbers.

second The SI unit of measurement for time.

secondary coil The winding of a transformer that is the source of the output voltage.

second law of thermodynamics The amount of disorder in an isolated system either always increases or, if the system is in equilibrium, stays the same.

self-inductance The induction of a voltage in a current-carrying coil when the current in the coil itself is changing. The induced emf will oppose any change in the current.

semiconductors Substances with electrical properties that are intermediate between those of insulators and conductors.

semimajor axis Half of the distance of the longest diameter of an ellipse (the major axis).

series (*LRC* circuit) A circuit containing an inductor, resistor, and capacitor in series.

series (*RC* circuit) A circuit that contains both a resistor and capacitor in series.

series (capacitors) An arrangement of capacitors connected along a single path (not in parallel); the reciprocal of the total capacitance is equal to the sum of the reciprocals of the individual capacitances.

series (resistors) An arrangement of resistors connected along a single path (not in parallel); the total resistance is equal to the sum of all the individual resistances.

shear modulus A measure of the rigidity and resistance to deformation of a material.

shear strain The change in an object's shape caused by shear stress.

shear stress Forces applied parallel to the plane in which an object lies, deforming the object without making it expand or contract.

shock wave A sudden pressure increase in a narrow region of a medium (for example, air), such as that caused by a body moving faster than the speed of sound.

significant figures The number of digits in a figure that can be known with some degree of confidence.

simple harmonic motion (SHM) Oscillatory motion under a Hooke's law restoring force (which is proportional to the displacement from the equilibrium position).

simple pendulum A pendulum in which all of the mass is concentrated at a single point.

single-loop circuit A circuit with only a single path that moving charges can follow.

single-slit diffraction Experiment in which a wave passes through a narrow opening, producing a pattern of bright and dark fringes.

sinusoidal function A mathematical curve that describes a smooth repetitive oscillation, such as the sine and cosine functions.

sinusoidal wave A wave in which the wave pattern at any instant is a sinusoidal function.

Snell's law of refraction Expression describing the relationship between the angles of incidence and refraction when referring to light or other waves passing through a boundary between two different media, such as water, glass, and air.

solenoid A straight helical coil of wire.

solid A substance whose individual molecules cannot move freely but remain in essentially fixed positions relative to one another.

sonic boom A loud, explosive noise caused by the shock wave from an aircraft traveling faster than the speed of sound.

sound intensity level The power carried by sound waves per unit area.

sound wave A wave in air consisting of periodic variations in air pressure.

source of emf A device that originates voltage in a circuit, such as a battery.

special theory of relativity Theory developed by Albert Einstein that states: (1) All laws of physics are the same in all inertial frames, and (2) the speed of light in a vacuum is the same in all inertial frames, independent of both the speed of the source of the light and the speed of the observer.

specific gravity The density of a substance divided by the density of 4°C liquid water.

specific heat The amount of heat per unit mass required to raise the temperature by one kelvin.

specular reflection Type of surface reflection in which light rays moving in a single direction reflect from a smooth surface in a single outgoing direction.

speed The rate at which an object is moving; equal to the magnitude of the object's velocity.

speed of light The speed at which all electromagnetic waves—including radio waves, x rays, and others—travel in a vacuum.

spin An intrinsic characteristic of electrons, protons, and neutrons akin to the angular momentum of a rotating sphere.

spring constant Measure of the stiffness of a spring.

spring potential energy The energy stored in a spring, based on whether it is relaxed, stretched, or compressed.

Standard Model A mathematical description of the elementary particles of matter and the electromagnetic, weak, and strong forces by which they interact.

standing wave A wave in which each point in the medium has a constant amplitude, giving it the appearance of being stationary.

standing wave mode The conditions under which a standing wave is possible; for example, a whole number of half-wavelengths must fit onto a string.

state variables Quantities such as volume, temperature, pressure, and internal energy that depend on the state or condition of a substance.

static friction Force acting on a stationary object that opposes the object's sliding motion.

steady flow Type of fluid flow in which the flow pattern does not change with time.

Stefan–Boltzmann constant The constant of proportionality in the Stefan–Boltzmann law: The total energy radiated per unit surface area of a blackbody in unit time is proportional to the fourth power of the thermodynamic temperature; equal to $5.670 \times 10^{-8}\ \text{W}/(\text{m}^2 \cdot \text{K}^4)$.

step-down transformer A transformer in which the number of windings in the secondary coil is less than the number of windings in the primary coil, so the output voltage is less than the input voltage.

step-up transformer A transformer in which the number of windings in the secondary coil is greater than the number of windings in the primary coil, so the output voltage is greater than the input voltage.

strain The amount of deformation that results from applied stress.

streamlines The paths followed by bits of fluid in laminar flow.

stress Force per area exerted on an object tending to cause the object to change in size or shape.

strong nuclear force An attractive force between protons and neutrons that is stronger than the repulsive electric force between protons.

sublimation The phase change from solid directly to gas.

surface charge density The amount of charge per unit area on a surface.

surface tension The attractive force exerted on molecules at the surface of a liquid by the molecules beneath, causing the liquid to assume the shape having the least surface area.

surface wave A wave that propagates along the interface between two media (for example, a seismic wave that travels along the surface of the Earth).

Système International (SI) The standard system of units based on the fundamental quantities.

temperature A measure of the kinetic energy associated with molecular motion.

tensile strain The distance an object stretches when pulled, expressed as a fraction of its relaxed length.

tensile stress Stretching force applied to an object divided by the object's cross-sectional area.

tension Stretching force applied at each end of an object; for example, the force exerted by a rope on an object it tows.

tesla The SI units of magnetic field strength.

test charge A point charge with such a small magnitude that it negligibly affects the field in which it is placed.

thermal conductivity A measure of how easily heat passes through a specified material.

thermal contact A state in which two or more systems can exchange thermal energy.

thermal energy Energy associated with the random motion of atoms and molecules.

thermal equilibrium The condition in which two objects in physical contact exchange no heat energy; in thermal equilibrium the objects are said to be at the same temperature.

thermal expansion The tendency of matter to increase in length, area, or volume in response to an increase in temperature.

thermodynamic process Any process that changes the state of a system.

thermodynamics The branch of physics that deals with relationships among properties of substances such as temperature, pressure, and volume, as well as the energy and flow of energy associated with these properties.

thermometer Instrument used to measure temperature changes.

thin film A very fine layer of a substance on a supporting material; for example, the thin lining behind the retina of the eyes of some animals.

thin lens A lens with a thickness that is negligible compared to the radii of curvature of the lens surfaces.

third law of thermodynamics It is possible for the temperature of a system to be arbitrarily close to absolute zero, but it can never reach absolute zero.

time constant The product of resistance multiplied by capacitance.

time dilation A difference of elapsed time between two events as measured by observers either moving relative to each other or located at different distances from large gravitational masses.

time interval A set length of time.

torque A force that causes rotation.

torr A unit measuring pressure; used especially in measuring partial vacuums; equal to 1/760 atm or 133.32 pascals.

total internal reflection The phenomenon occurring when a wave strikes a medium boundary at an angle larger than the critical angle with respect to the normal; 100% of the incident light is reflected back into the first medium.

total mechanical energy The sum of kinetic and potential energy.

total momentum The momentum of a system of objects.

trajectory The path followed by a projectile or an object.

transformer A device that can raise or lower an ac voltage to a desired value.

translation Motion in which an object as a whole moves through space.

translational kinetic energy The energy possessed by an object by virtue of its motion as a whole through space.

transverse wave A traveling disturbance or vibration in which the individual parts of the wave medium move in a direction perpendicular to the direction of wave propagation.

traveling wave A wave of the form $y(x,t) = A \cos (kx - vt)$ in which a disturbance propagates from one location to another.

triple point The particular combination of pressure and temperature of a material at which the solid, liquid, and vapor phases all coexist.

turbulent flow Type of fluid motion in which the velocity of the flow at any point is continuously undergoing changes in both magnitude and direction.

two-dimensional motion Vertical and horizontal movement of an object; motion confined to a plane.

ultimate strength The maximum tensile stress that a material can withstand before failure.

underdamped oscillation Case in which a small damping coefficient applied to an oscillating system causes the system to oscillate with ever decreasing amplitude.

uniform circular motion The motion of an object going around a circular path at a constant speed.

uniform density Constant mass density throughout a volume.

units The standard measurement for a specific quantity; for example, the second is a unit of measurement for time.

unpolarized light Natural light such as that emitted by the Sun or an ordinary light bulb, in which the orientation of the electric field changes randomly from one moment to the next.

unsteady flow Fluid motion in which the velocity changes with time.

upright image Image (produced by a lens or mirror) that has the same orientation as the object.

vaporization The phase change from solid to liquid.

vector A physical quantity that has both a magnitude and a direction.

vector addition The combination of two or more displacements to find the vector sum.

vector difference The result of subtracting one vector from another.

vector multiplication by a scalar An operation in which the product of a scalar and a vector pointing in a given direction is a new vector that has a magnitude equal to the product of the absolute value of the scalar and the magnitude of the original vector. The direction of the new vector is either the same as that of the original vector (if the scalar is positive) or opposite to it (if the scalar is negative).

vector product *See* cross product.

vector subtraction The process of taking a vector difference; the inverse of vector addition.

vector sum The result of vector addition.

velocity The magnitude and direction of the rate of change of an object's position.

velocity vector (or instantaneous velocity vector) The rate of change of the position of an object at a given point in time, expressed as a magnitude (speed) and direction.

virtual image An image from which rays of reflected or refracted light appear to diverge; for example, the image seen in a plane mirror.

virtual particle A particle whose existence is allowed by the uncertainty principle and that exhibits many of the characteristics of an ordinary particle, but that exists for a limited time.

viscosity A measure of the resistance to flow of a fluid.

visible light The range of wavelengths visible to the human eye.

volt The SI unit of electromotive force or electric potential; the emf required to drive one ampere of current against one ohm resistance.

voltage Electric potential difference.

volume flow rate Volume of fluid per unit time passing a given point.

volume strain The change in volume of an object under stress divided by the original volume.

volume stress Force applied perpendicularly to all faces of an object; pressure change.

v_x–t graph Chart depicting an object's velocity along the x axis versus time.

watt The SI unit of power; equal to one joule per second.

wave A disturbance or vibration that travels through space.

wave function A mathematical description of the properties of a wave, expressing the displacement of the wave medium at every position and at every time.

wavelength The distance between successive crests of a wave.

wavelet A tiny segment of a larger wave.

wave-particle duality The dual character of both light and matter, which have both wave and particle characteristics.

weak force An interaction between elementary particles, often involving neutrinos or antineutrinos, that is responsible for certain kinds of radioactive decay.

weight The magnitude of the gravitational force that acts on an object.

weighted average A mean calculated by giving values in a data set more influence according to some attribute of the data, such as how often a given value appears in the data set.

work The transfer of energy from one object to another.

work-energy theorem When an object undergoes a displacement, the work done on it by the net force equals the object's kinetic energy at the end of the displacement minus its kinetic energy at the beginning of the displacement.

work function The minimum amount of energy required to remove a single electron from a material.

***x* component** The component of a vector parallel to the *x* axis.

***x–t* graph** Chart depicting an object's position along the *x* axis versus time.

***x–y* plane** The coordinate plane formed by the *x* axis and the *y* axis.

***y* component** The component of a vector parallel to the *y* axis.

yield strength The tensile strength at which an object is permanently deformed and can no longer return to its normal strength.

Young's modulus A measure of the stiffness of a given material.

zeroth law of thermodynamics If two objects are each in thermal equilibrium with a third object, they are also in thermal equilibrium with each other.

Math Tutorial

In this tutorial, we review some of the basic results of algebra, geometry, trigonometry, and calculus. In many cases, we merely state results without proof. **Table M-1** lists some mathematical symbols.

M-1 Significant figures

Many numbers we work with in science are the result of measurement and are therefore known only within a degree of uncertainty. This uncertainty should be reflected in the number of digits used. For example, if you have a 1-meter-long rule with scale spacing of 1 cm, you know that you can measure the height of a box to within a fifth of a centimeter or so. Using this rule, you might find that the box height is 27.0 cm. If there is a scale with a spacing of 1 mm on your rule, you might perhaps measure the box height to be 27.03 cm. However, if there is a scale with a spacing of 1 mm on your rule, you might not be able to measure the height more accurately than 27.03 cm because the height might vary by 0.01 cm or so, depending on where you measure the height of the box. When you write down that the height of the box is 27.03 cm, you are stating that your best estimate of the height is 27.03 cm, but you are not claiming that it is exactly 27.030000 . . . cm high. The four digits in 27.03 cm are called **significant figures**. Your measured length, 27.03 cm, has four significant digits. Significant figures are also called significant digits.

The number of significant digits in an answer to a calculation will depend on the number of significant digits in the given data. When you work with numbers that have uncertainties, you should be careful not to include more digits than the certainty of measurement warrants. *Approximate* calculations (order-of-magnitude estimates) always result in answers that have only one significant digit or none. When you multiply, divide, add, or subtract numbers, you must consider the accuracy of the results. Listed below are some rules that will help you determine the number of significant digits of your results.

(1) When multiplying or dividing quantities, the number of significant digits in the final answer is no greater than that in the quantity with the fewest significant digits.
(2) When adding or subtracting quantities, the number of decimal places in the answer should match that of the term with the smallest number of decimal places.
(3) Exact values have an unlimited number of significant digits. For example, a value determined by counting, such as 2 tables, has no uncertainty and is an exact value. In addition, the conversion factor 0.0254000 . . . m/in. is an exact value because 1.000 . . . inches is exactly equal to 0.0254000 . . . meters. (The yard is, by definition, equal to exactly 0.9144 m, and 0.9144 divided by 36 is exactly equal to 0.0254.)
(4) Sometimes zeros are significant and sometimes they are not. If a zero is before a leading nonzero digit, then the zero is not significant. For example, the number 0.00890 has three significant digits. The first three zeroes are not significant digits but are merely markers to locate the decimal point. Note that the zero after the nine is significant.

TABLE M-1
Mathematical Symbols

Symbol	Meaning
$=$	is equal to
$\neq$	is not equal to
$\approx$	is approximately equal to
$\sim$	is of the order of
$\propto$	is proportional to
$>$	is greater than
$\geq$	is greater than or equal to
$\gg$	is much greater than
$<$	is less than
$\leq$	is less than or equal to
$\ll$	is much less than
Δx	change in x
$\lvert x \rvert$	absolute value of x
$n!$	$n(n-1)(n-2)\ldots 1$
Σ	sum

(5) Zeros that are between nonzero digits are significant. For example, 5603 has four significant digits.
(6) The number of significant digits in numbers with trailing zeros and no decimal point is ambiguous. For example, 31,000 could have as many as five significant digits or as few as two significant digits. To prevent ambiguity, you should report numbers by using scientific notation or by using a decimal point.

EXAMPLE M-1 Finding the Average of Three Numbers

Find the average of 19.90, −7.524, and −11.8179.

Set Up

You will be adding 3 numbers and then dividing the result by 3. The first number has four significant digits, the second number has four, and the third number has six.

Solve

Sum the three numbers.

$$19.90 + (-7.524) + (-11.8179) = 0.55\mathit{81}$$

If the problem only asked for the sum of the three numbers, we would round the answer to the least number of decimal places among all the numbers being added—the answer would be 0.56 (0.5581 rounds up to 0.56 to two significant digits). However, we must divide this intermediate result by 3, so we use the intermediate answer with the two extra digits (italicized and red).

$$\frac{0.5581}{3} = 0.18\mathit{60333}\ldots$$

Only two of the digits in the intermediate answer, 0.55*81* . . . , are significant digits, so we must round the final number to get our final answer. The number 3 in the denominator is a whole number and has an unlimited number of significant digits. Thus, the final answer has the same number of significant digits as the numerator, which is 2.

The final answer is 0.19.

Reflect

The sum in step 1 has two significant digits following the decimal point, the same as the number being summed with the least number of significant digits after the decimal point.

M-2 Equations

An **equation** is a statement written using numbers and symbols to indicate that two quantities, written on either side of an equal sign (=), are equal. The quantity on either side of the equal sign may consist of a single term, or of a sum or difference of two or more **terms**. For example, the equation $x = 1 - (ay + b)/(cx - d)$ contains three terms, x, 1, and $(ay + b)/(cx - d)$.

You can perform the following operations on equations:

(1) The same quantity can be added to or subtracted from each side of an equation.
(2) Each side of an equation can be multiplied or divided by the same quantity.
(3) Each side of an equation can be raised to the same power.

These operations are meant to be applied to each *side* of the equation rather than each term in the equation. (Because multiplication is distributive over addition, operation 2—and only operation 2—of the preceding operations also applies term by term.)

Caution: *Division by zero is forbidden at* any *stage in solving an equation; results (if any) would be invalid.*

Adding or Subtracting Equal Amounts
To find x when $x - 3 = 7$, add 3 to both sides of the equation: $(x - 3) + 3 = 7 + 3$; thus, $x = 10$.

Multiplying or Dividing by Equal Amounts
If $3x = 17$, solve for x by dividing both sides of the equation by 3; thus, $x = \frac{17}{3}$, or 5.7.

EXAMPLE M-2 Simplifying Reciprocals in an Equation

Solve the following equation for x:

$$\frac{1}{x} + \frac{1}{4} = \frac{1}{3}$$

Equations containing reciprocals of unknowns occur in many circumstances in physics. Two instances of this are geometric optics and electric circuit analysis.

Set Up

In this equation, the term containing x is on the same side of the equation as a term not containing x. Furthermore, x is found in the denominator of a fraction. We'll start by isolating the $1/x$ term, find common denominators, and then multiply both sides of the equation by appropriate quantities.

Solve

Subtract $\frac{1}{4}$ from each side.

$$\frac{1}{x} = \frac{1}{3} - \frac{1}{4}$$

Simplify the right side of the equation by using the lowest common denominator.

Begin by multiplying both terms on the right-hand side by appropriate forms of 1.

$$\frac{1}{x} = \frac{1}{3}\frac{4}{4} - \frac{1}{4}\frac{3}{3} = \frac{4}{12} - \frac{3}{12}$$

$$= \frac{4-3}{12} = \frac{1}{12} \quad \text{so} \quad \frac{1}{x} = \frac{1}{12}$$

Multiply both sides of the equation by $12x$ to determine the value of x.

$$12\cancel{x}\frac{1}{\cancel{x}} = \cancel{12}x\frac{1}{\cancel{12}}$$

$$12 = x$$

Reflect

To check our answer, substitute 12 for x in the left side of original equation.

$$\frac{1}{x} + \frac{1}{4} = \frac{1}{12} + \frac{3}{12} = \frac{4}{12} = \frac{1}{3}$$

M-3 Direct and inverse proportions

When we say variable quantities x and y are **directly proportional**, we mean that as x and y change, the ratio x/y is constant. To say that two quantities are proportional is to say that they are directly proportional. When we say variable quantities x and y are **inversely proportional**, we mean that as x and y change, the ratio xy is constant.

Relationships of direct and inverse proportion are common in physics. Objects moving at the same velocity have momenta directly proportional to their masses. The ideal gas law ($PV = nRT$) states that pressure P is directly proportional to (absolute) temperature T, when volume V remains constant, and is inversely proportional to volume, when temperature remains constant. Ohm's law ($V = IR$) states that the voltage V across a resistor is directly proportional to the electric current in the resistor when the resistance remains constant.

Constant of Proportionality

When two quantities are directly proportional, the two quantities are related by a *constant of proportionality*. If you are paid for working at a regular rate R in dollars per day, for example, the money m you earn is directly proportional to the time t you work;

the rate R is the constant of proportionality that relates the money earned in dollars to the time worked t in days:

$$\frac{m}{t} = R \quad \text{or} \quad m = Rt$$

If you earn \$400 in 5 days, the value of R is \$400/(5 days) = \$80/day. To find the amount you earn in 8 days, you could perform the calculation

$$m = (\$80/\text{day})(8\text{ days}) = \$640$$

Sometimes the constant of proportionality can be ignored in proportion problems. Because the amount you earn in 8 days is $\frac{8}{5}$ times what you earn in 5 days, this amount is

$$m_{8\text{ days}} = 8\text{ days}\frac{\$400}{5\text{ days}} = \$640$$

EXAMPLE M-3 Painting Cubes

You need 15.4 mL of paint to cover one side of a cube. The area of one side of the cube is 426 cm^2. What is the relation between the volume of paint needed and the area to be covered? How much paint do you need to paint one side of a cube on which the one side has an area of 503 cm^2?

Set Up

To determine the amount of paint for the side whose area is 503 cm^2 we will set up a proportion.

Solve

The volume V of paint needed increases in proportion to the area A to be covered.

V and A are directly proportional.

That is, $\frac{V}{A} = k$ or $V = kA$

where k is the proportionality constant

Determine the value of the proportionality constant using the given values $V_1 = 15.4$ mL and $A_1 = 426$ cm^2.

$$k = \frac{V_1}{A_1} = \frac{15.4\text{ mL}}{426\text{ cm}^2} = 0.0362\text{ mL/cm}^2$$

Determine the volume of paint needed to paint a side of a cube whose area is 503 cm^2 using the proportionality constant in step 1.

$$V_2 = kA_2 = (0.0362\text{ mL/cm}^2)(503\text{ cm}^2) = 18.2\text{ mL}$$

Reflect

Our value for V_2 is greater than the value for V_1, as expected. The amount of paint needed to cover an area equal to 503 cm^2 should be greater than the amount of paint needed to cover an area of 426 cm^2 because 503 cm^2 is larger than 426 cm^2.

M-4 Linear equations

A **linear equation** is an equation of the form $x + 2y - 4z = 3$. That is, an equation is linear if each term either is constant or is the product of a constant and a variable raised to the first power. Such equations are said to be linear because the plots of these equations form straight lines or planes. The equations of direct proportion between two variables are linear equations.

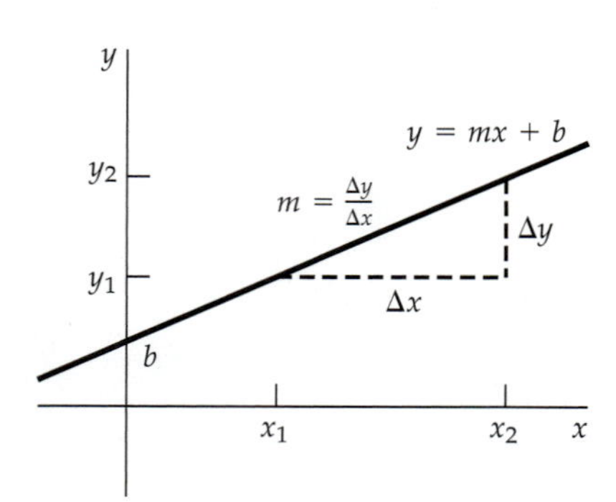

Figure M-1 Graph of the linear equation $y = mx + b$, where b is the y intercept and $m = \Delta y/\Delta x$ is the slope.

Graph of a Straight Line

A linear equation relating y and x can always be put into the standard form

$$y = mx + b \tag{M-1}$$

where m and b are constants that may be either positive or negative. **Figure M-1** shows a graph of the values of x and y that satisfy Equation M-1. The constant b, called the **y intercept**, is the value of y at $x = 0$. The constant m is the **slope** of the line, which equals the ratio of the change in y to the corresponding change in x. In the figure, we have indicated two points on the line, (x_1, y_1) and (x_2, y_2), and the changes $\Delta x = x_2 - x_1$ and $\Delta y = y_2 - y_1$. The slope m is then

$$m = \frac{y_2 - y_1}{x_2 - x_1} = \frac{\Delta y}{\Delta x}$$

If x and y are both unknown in the equation $y = mx + b$, there are no unique values of x and y that are solutions to the equation. Any pair of values (x_1, y_1) on the line in Figure M-1 will satisfy the equation. If we have two equations, each with the same two unknowns x and y, the equations can be solved simultaneously for the unknowns. Example M-4 shows two methods for simultaneously solving two linear equations.

EXAMPLE M-4 Using Two Equations to Solve for Two Unknowns

Find any and all values of x and y that simultaneously satisfy

$$3x - 2y = 8 \qquad \text{(M-2)}$$

and

$$y - x = 2 \qquad \text{(M-3)}$$

Set Up

Graph the two equations. At the point where the lines intersect, the values of x and y satisfy both equations.

We can solve two simultaneous equations by first solving either equation for one variable in terms of the other variable and then substituting the result into the second equation.

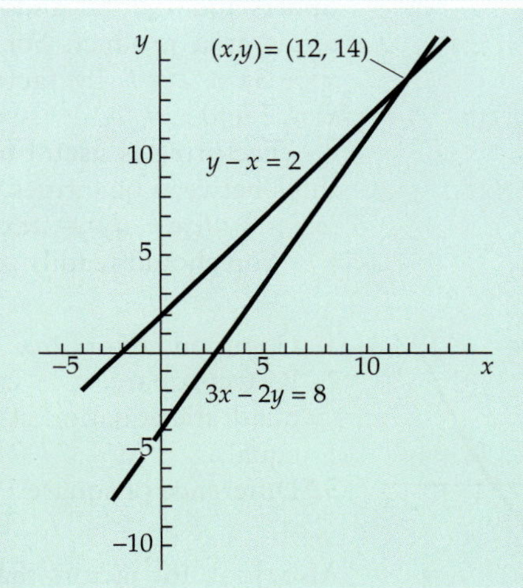

Figure M-2 Graph of Equations M-2 and M-3. At the point where the lines intersect, the values of *x* and *y* satisfy both equations.

Solve

Solve Equation M-3 for y.

$$y = x + 2$$

Substitute this value for y into Equation M-2.

$$3x - 2(x + 2) = 8$$

Simplify the equation and solve for x.

$$3x - 2x - 4 = 8$$
$$x - 4 = 8$$
$$x = 12$$

Use your solution for x and one of the given equations to find the value of y.

Return to Equation M-3 and substitute $x = 12$.

$$y - x = 2, \text{ where } x = 12$$
$$y - 12 = 2$$
$$y = 2 + 12 = 14$$

Reflect

An alternative method is to multiply one equation by a constant such that one of the unknown terms is eliminated when the equations are added or subtracted.

We can multiply through Equation M-3 by 2

$$2(y - x) = 2(2)$$
$$2y - 2x = 4$$

Add the result to Equation M-2 and solve for x:

$$\cancel{2y} - 2x = 4$$
$$3x - \cancel{2y} = 8$$
$$\overline{3x - 2x = 12 \Rightarrow x = 12}$$

Substitute into Equation M-3 and solve for y:

$$y - 12 = 2 \Rightarrow y = 14$$

M-5 Quadratic equations and factoring

A **quadratic equation** is an equation of the form $ax^2 + bxy + cy^2 + ex + fy + g = 0$, where x and y are variables and a, b, c, e, f, and g are constants. In each term of the equation the powers of the variables are integers that sum to 2, 1, or 0. The designation *quadratic equation* usually applies to a much simpler equation of one variable that can be written in the standard form

$$ax^2 + bx + c = 0 \tag{M-4}$$

where a, b, and c are constants. The quadratic equation has two solutions or **roots**—values of x for which the equation is true.

Factoring

We can solve some quadratic equations by **factoring.** Very often terms of an equation can be grouped or organized into other terms. When we factor terms, we look for multipliers and multiplicands—which we now call **factors**—that will yield two or more new terms as a product. For example, we can find the roots of the quadratic equation $x^2 - 3x + 2 = 0$ by factoring the left side to get $(x - 2)(x - 1) = 0$. The roots are $x = 2$ and $x = 1$.

Factoring is useful for simplifying equations and for understanding the relationships between quantities. You should be familiar with the multiplication of the factors $(ax + by)(cx + dy) = acx^2 + (ad + bc)xy + bdy^2$.

You should readily recognize some typical factorable combinations:

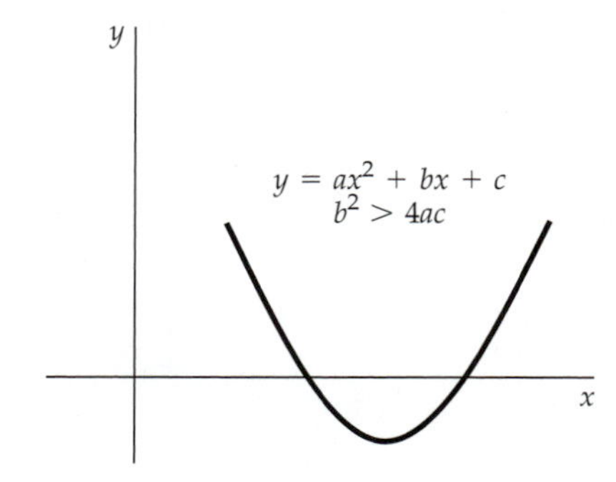

Figure M-3 Graph of y versus x when $y = ax^2 + bx + c$ for the case $b^2 > 4ac$. The two values of x for which $y = 0$ satisfy the quadratic equation (Equation M-4).

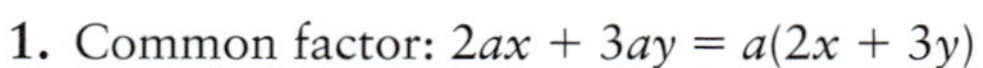

1. Common factor: $2ax + 3ay = a(2x + 3y)$
2. Perfect square: $x^2 - 2xy + y^2 = (x - y)^2$ (If the expression on the left side of a quadratic equation in standard form is a perfect square, the two roots will be equal.)
3. Difference of squares: $x^2 - y^2 = (x + y)(x - y)$

Also, look for factors that are prime numbers (2, 5, 7, etc.) because these factors can help you simplify terms quickly. For example, the equation $98x^2 - 140 = 0$ can be simplified because 98 and 140 share the common factor 2. That is, $98x^2 - 140 = 0$ becomes $2(49x^2 - 70) = 0$, so we have $49x^2 - 70 = 0$.

This result can be further simplified because 49 and 70 share the common factor 7. Thus, $49x^2 - 70 = 0$ becomes $7(7x^2 - 10) = 0$, so we have $7x^2 - 10 = 0$.

The Quadratic Formula

Not all quadratic equations can be solved by factoring. However, *any* quadratic equation in the standard form $ax^2 + bx + c = 0$ can be solved by the **quadratic formula,**

$$x = \frac{-b \pm \sqrt{b^2 - 4ac}}{2a} = -\frac{b}{2a} \pm \frac{1}{2a}\sqrt{b^2 - 4ac} \tag{M-5}$$

When b^2 is greater than $4ac$, there are two solutions corresponding to the + and − signs, respectively. **Figure M-3** shows a graph of y versus x where $y = ax^2 + bx + c$. The curve, a **parabola**, crosses the x axis twice. (The simplest representation of a parabola in (x, y) coordinates is an equation of the form $y = ax^2 + bx + c$.) The two roots of this equation are the values for which $y = 0$; that is, they are the x *intercepts.*

When b^2 is less than $4ac$, the graph of y versus x does not intersect the x axis, as is shown in **Figure M-4**; there are still two roots, but they are not real numbers. When $b^2 = 4ac$, the graph of y versus x is tangent to the x axis at the point $x = -b/2a$; the two roots are each equal to $-b/2a$.

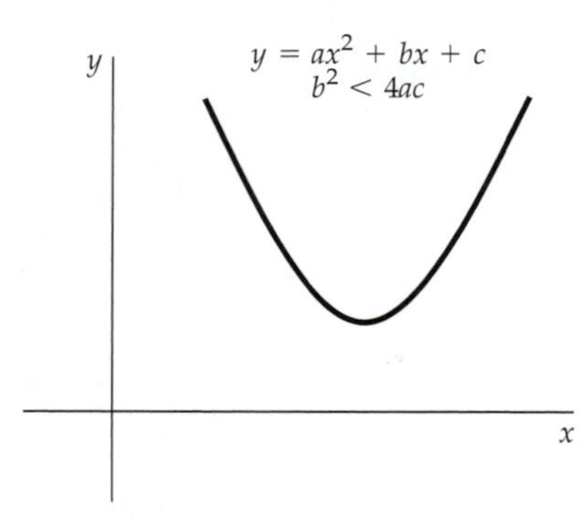

Figure M-4 Graph of y versus x when $y = ax^2 + bx + c$ for the case $b^2 < 4ac$. In this case, there are no real values of x for which $y = 0$.

EXAMPLE M-5 Factoring a Second-Degree Polynomial

Factor the expression $6x^2 + 19xy + 10y^2$.

Set Up

We examine the coefficients of the terms to see whether the expression can be factored without resorting to more advanced methods. Remember that the multiplication $(ax + by)(cx + dy) = acx^2 + (ad + bc)xy + bdy^2$.

Solve

The coefficient of x^2 is 6, which can be factored two ways.

$ac = 6$
$3 \cdot 2 = 6$ or $6 \cdot 1 = 6$

The coefficient of y^2 is 10, which can also be factored two ways.

$bd = 10$
$5 \cdot 2 = 10$ or $10 \cdot 1 = 10$

List the possibilities for a, b, c, and d in a table. Include a column for $ad + bc$.

If $a = 3$, then $c = 2$, and vice versa. In addition, if $a = 6$, then $c = 1$, and vice versa. For each value of a there are four values for b.

a	b	c	d	$ad + bc$
3	5	2	2	16
3	2	2	5	19
3	10	2	1	23
3	1	2	10	32
2	5	3	2	19
2	2	3	5	16
2	10	3	1	32
2	1	3	10	23
6	5	1	2	17
6	2	1	5	32
6	10	1	1	16
6	1	1	10	61
1	5	6	2	32
1	2	6	5	17
1	10	6	1	61
1	1	6	10	16

Find a combination such that $ad + bc = 19$. As you can see from the table there are two such combinations.

$ad + bc = 19$
$3 \cdot 5 + 2 \cdot 2 = 19$ and
$2 \cdot 2 + 5 \cdot 3 = 19$

It doesn't matter which combination we choose. To finish this problem we will use the combination in the second row of the table to factor the expression in question:

$$6x^2 + 19xy + 10y^2 = (3x + 2y)(2x + 5y)$$

Reflect

As a check, expand $(3x + 2y)(2x + 5y)$ to see if we return to the original equation.

$$(3x + 2y)(2x + 5y) = 6x^2 + 15xy + 4xy + 10y^2$$
$$= 6x^2 + 19xy + 10y^2$$

You should be able to show that the combination in the fifth row is also an acceptable factoring.

M-6 Exponents and logarithms

Exponents

The notation x^n stands for the quantity obtained by multiplying x by itself n times. For example, $x^2 = x \cdot x$ and $x^3 = x \cdot x \cdot x$. The quantity n is called the **power**, or the **exponent**, of x (the **base**). Listed below are some rules that will help you simplify terms that have exponents.

(1) When two powers of x are multiplied, the exponents are added:

$$(x^m)(x^n) = x^{m+n} \qquad \text{(M-6)}$$

Example: $x^2x^3 = x^{2+3} = (x \cdot x)(x \cdot x \cdot x) = x^5$.

(2) Any number (except 0) raised to the 0 power is defined to be 1:

$$x^0 = 1 \tag{M-7}$$

(3) Based on rule 2,

$$x^n x^{-n} = x^0 = 1$$

$$x^{-n} = \frac{1}{x^n} \tag{M-8}$$

(4) When two powers are divided, the exponents are subtracted:

$$\frac{x^n}{x^m} = x^n x^{-m} = x^{n-m} \tag{M-9}$$

(5) When a power is raised to another power, the exponents are multiplied:

$$(x^n)^m = x^{nm} \tag{M-10}$$

(6) When exponents are written as fractions, they represent the roots of the base. For example,

$$x^{1/2} \cdot x^{1/2} = x$$

so

$$x^{1/2} = \sqrt{x} \quad (x > 0)$$

EXAMPLE M-6 Simplifying a Quantity That Has Exponents

Simplify $\frac{x^4 x^7}{x^8}$.

Set Up

According to rule 1, when two powers of x are multiplied, the exponents are added.

$$(x^m)(x^n) = x^{m+n} \tag{M-6}$$

Rule 4 states that when two powers are divided, the exponents are subtracted.

$$\frac{x^n}{x^m} = x^n x^{-m} = x^{n-m} \tag{M-9}$$

Solve

Simplify the numerator $x^4 x^7$ using rule 1.

$$x^4 x^7 = x^{4+7} = x^{11}$$

Simplify $\frac{x^{11}}{x^8}$ using rule 4.

$$\frac{x^{11}}{x^8} = x^{11} x^{-8} = x^{11-8} = x^3$$

Reflect

Use the value $x = 2$ to test our answer.

$$\frac{2^4 2^7}{2^8} = 2^3 = 8$$

$$\frac{2^4 2^7}{2^8} = \frac{(16)(128)}{256} = \frac{2048}{256} = 8$$

Logarithms

Any positive number can be expressed as some power of any other positive number except one. If y is related to x by $y = a^x$, then the number x is said to be the **logarithm** of y to the **base** a, and the relation is written

$$x = \log_a y$$

Thus, logarithms are *exponents*, and the rules for working with logarithms correspond to similar laws for exponents. Listed below are some rules that will help you simplify terms that have logarithms.

(1) If $y_1 = a^n$ and $y_2 = a^m$, then

$$y_1 y_2 = a^n a^m = a^{n+m}$$

Correspondingly,

$$\log_a y_1 y_2 = \log_a a^{n+m} = n + m = \log_a a^n + \log_a a^m = \log_a y_1 + \log_a y_2 \tag{M-11}$$

It then follows that

$$\log_a y^n = n \log_a y \tag{M-12}$$

(2) Because $a^1 = a$ and $a^0 = 1$,

$$\log_a a = 1 \tag{M-13}$$

and

$$\log_a 1 = 0 \tag{M-14}$$

There are two bases in common use: logarithms to base 10 are called **common logarithms,** and logarithms to base e (where $e = 2.718\ldots$) are called **natural logarithms.**

In this text, the symbol ln is used for natural logarithms and the symbol log, without a subscript, is used for common logarithms. Thus,

$$\log_e x = \ln x \quad \text{and} \quad \log_{10} x = \log x \tag{M-15}$$

and $y = \ln x$ implies

$$x = e^y \tag{M-16}$$

Logarithms can be changed from one base to another. Suppose that

$$z = \log x \tag{M-17}$$

Then

$$10^z = 10^{\log x} = x \tag{M-18}$$

Taking the natural logarithm of both sides of Equation M-18, we obtain

$$z \ln 10 = \ln x$$

Substituting $\log x$ for z (see Equation M-17) gives

$$\ln x = (\ln 10)\log x \tag{M-19}$$

EXAMPLE M-7 Converting between Common Logarithms and Natural Logarithms

The steps leading to Equation M-19 show that, in general, $\log_b x = (\log_b a)\log_a x$, and thus that conversion of logarithms from one base to another requires only multiplication by a constant. Describe the mathematical relation between the constant for converting common logarithms to natural logarithms and the constant for converting natural logarithms to common logarithms.

Set Up

We have a general mathematical formula for converting logarithms from one base to another. We look for the mathematical relation by exchanging a for b and vice versa in the formula.

Solve

We have a formula for converting logarithms from base a to base b.

$$\log_b x = (\log_b a)\log_a x$$

To convert from base b to base a, exchange all a for b and vice versa.

$$\log_a x = (\log_a b)\log_b x$$

Divide both sides of the equation in step 1 by $\log_a x$.

$$\frac{\log_b x}{\log_a x} = \log_b a$$

Divide both sides of the equation in step 2 by $(\log_a b)\log_a x$.

$$\frac{1}{\log_a b} = \frac{\log_b x}{\log_a x}$$

The results show that the conversion factors $\log_b a$ and $\log_a b$ are reciprocals of one another.

$$\frac{1}{\log_a b} = \log_b a$$

Reflect

For the value of $\log_{10} e$, your calculator will give 0.43429. For ln 10, your calculator will give 2.3026. Multiply 0.43429 by 2.3026; you will get 1.0000.

M-7 Geometry

The properties of the most common **geometric figures**—bounded shapes in two or three dimensions whose lengths, areas, or volumes are governed by specific ratios—are a basic analytical tool in physics. For example, the characteristic ratios within triangles give us the laws of *trigonometry* (see Section M-8), which in turn give us the theory of vectors, essential in analyzing motion in two or more dimensions. Circles and spheres are essential for understanding, among other concepts, angular momentum and the probability densities of quantum mechanics.

Basic Formulas in Geometry

Circle The ratio of the circumference of a circle to its diameter is a number π, which has the approximate value

$$\pi = 3.141\,592$$

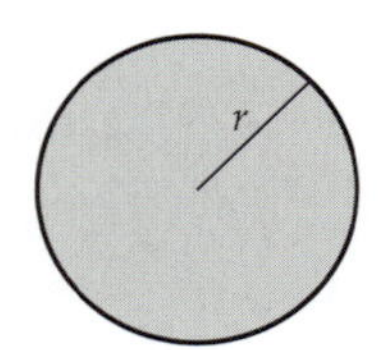

Area of a circle $A = \pi r^2$

Figure M-5 Area of a circle.

The circumference C of a circle is thus related to its diameter d and its radius r by

(M-20) $$C = \pi d = 2\pi r \quad \text{circumference of circle}$$

The area of a circle is (**Figure M-5**)

(M-21) $$A = \pi r^2 \quad \text{area of circle}$$

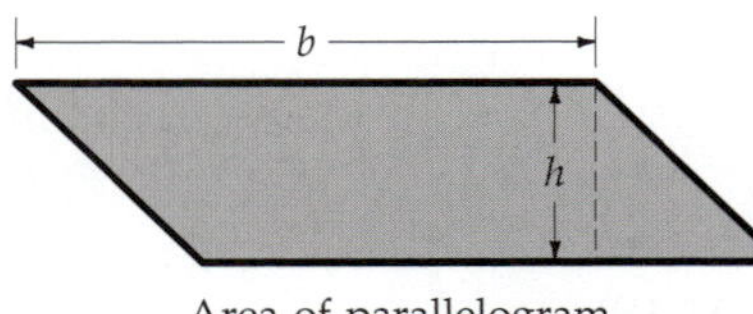

Area of parallelogram
$A = bh$

Figure M-6 Area of a parallelogram.

Parallelogram The area of a parallelogram is the base b multiplied by the height h (**Figure M-6**):

$$A = bh$$

Triangle The area of a triangle is one-half the base multiplied by the height (**Figure M-7**):

$$A = \frac{1}{2}bh$$

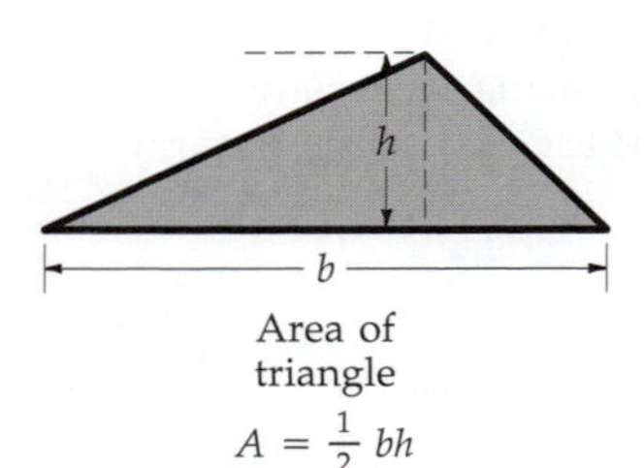

Area of triangle
$A = \frac{1}{2}bh$

Figure M-7 Area of a triangle.

Sphere A sphere of radius r (**Figure M-8**) has a surface area given by

(M-22) $$A = 4\pi r^2 \quad \text{surface area of sphere}$$

and a volume given by

(M-23) $$V = \frac{4}{3}\pi r^3 \quad \text{volume of sphere}$$

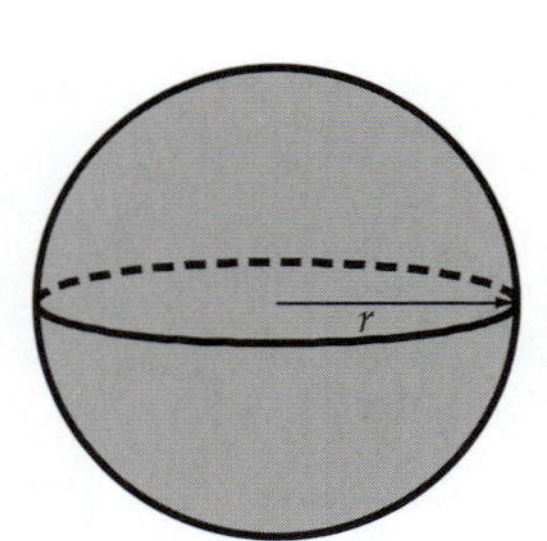

Spherical surface area
$A = 4\pi r^2$
Spherical volume
$V = \frac{4}{3}\pi r^3$

Figure M-8 Surface area and volume of a sphere.

Cylinder A cylinder of radius r and length L (**Figure M-9**) has a surface area (not including the end faces) of

(M-24) $$A = 2\pi rL \quad \text{surface of cylinder}$$

and volume of

(M-24) $$V = \pi r^2 L \quad \text{volume of cylinder}$$

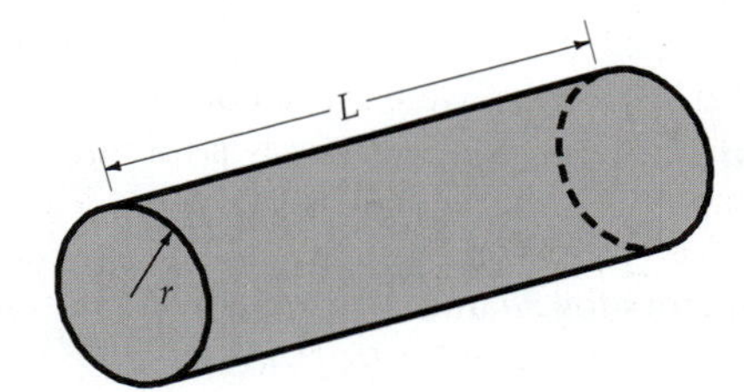

Cylindrical surface area
$A = 2\pi rL$
Cylindrical volume
$V = \pi r^2 L$

Figure M-9 Surface area (not including the end faces) and the volume of a cylinder.

EXAMPLE M-8 Calculating the Volume of a Spherical Shell

An aluminum spherical shell has an outer diameter of 40.0 cm and an inner diameter of 38.0 cm. What is the volume of the aluminum in this shell?

Set Up

The volume of the aluminum in the spherical shell is the volume that remains when we subtract the volume of the inner sphere having $d_i = 2r_i = 38.0$ cm from the volume of the outer sphere having $d_o = 2r_o = 40.0$ cm.

Spherical Volume:

$$V = \frac{4}{3}\pi r^3 \qquad \text{(M-23)}$$

Solve

Subtract the volume of the sphere of radius r_i from the volume of the sphere of radius r_o.

$$V = V_o - V_i = \frac{4}{3}\pi r_o^3 - \frac{4}{3}\pi r_i^3 = \frac{4}{3}\pi(r_o^3 - r_i^3)$$

Substitute 20.0 cm for r_o and 19.0 cm for r_i.

$$V = \frac{4}{3}\pi[(20.0\text{ cm})^3 - (19.0\text{ cm})^3]$$
$$= 4.78 \times 10^3\text{ cm}^3$$

Reflect

The volume calculated is less than the volume of the outer sphere.

$$V_o = \frac{4}{3}\pi r_o^3 = \frac{4}{3}\pi(20.0\text{ cm})^3$$
$$= 3.35 \times 10^4\text{ cm}^3$$

M-8 Trigonometry

Trigonometry, which gets its name from Greek roots meaning "triangle" and "measure," is the study of some important mathematical functions, called **trigonometric functions**. These functions are most simply defined as ratios of the sides of right triangles. However, these right-triangle definitions are of limited use because they are valid only for angles between zero and 90°. However, the validity of the right-triangle definitions can be extended by defining the trigonometric functions in terms of the ratio of the coordinates of points on a circle of unit radius drawn centered at the origin of the *xy* plane.

In physics, we first encounter trigonometric functions when we use vectors to analyze motion in two dimensions. Trigonometric functions are also essential in the analysis of any kind of periodic behavior, such as circular motion, oscillatory motion, and wave mechanics.

Angles and Their Measure: Degrees and Radians

The size of an angle formed by two intersecting straight lines is known as its **measure**. The standard way of finding the measure of an angle is to place the angle so that its **vertex**, or point of intersection of the two lines that form the angle, is at the center of a circle located at the origin of a graph that has Cartesian coordinates and one of the lines extends rightward on the positive *x* axis. The distance traveled *counterclockwise* on the circumference from the positive *x* axis to reach the intersection of the circumference with the other line defines the measure of the angle. (Traveling clockwise to the second line would simply give us a negative measure; to illustrate basic concepts, we position the angle so that the smaller rotation will be in the counterclockwise direction.)

One of the most familiar units for expressing the measure of an angle is the **degree,** which equals 1/360 of the full distance around the circumference of the circle. For greater precision, or for smaller angles, we either show degrees plus minutes (′) and seconds (″), with $1' = 1°/60$ and $1'' = 1'/60 = 1°/3600$; or show degrees as an ordinary decimal number.

For scientific work, a more useful measure of an angle is the **radian** (rad). Again, place the angle with its vertex at the center of a circle and measure counterclockwise rotation around the circumference. The measure of the angle in radians is then defined as the length of the circular arc from one line to the other divided by the radius of the

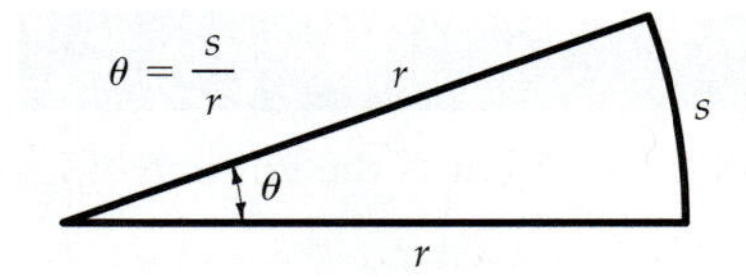

Figure M-10 The angle θ in radians is defined to be the ratio s/r, where s is the arc length intercepted on a circle of radius r.

circle (**Figure M-10**). If s is the arc length and r is the radius of the circle, the angle θ measured in radians is

$$\theta = \frac{s}{r} \tag{M-26}$$

Because the angle measured in radians is the ratio of two lengths, it is dimensionless. The relation between radians and degrees is

$$360° = 2\pi \text{ rad}$$

or

$$1 \text{ rad} = \frac{360°}{2\pi} = 57.3°$$

Figure M-11 shows some useful relations for angles.

The Trigonometric Functions

Figure M-12 shows a right triangle formed by drawing the line segment BC perpendicular to AC. The lengths of the sides are labeled a, b, and c. The right-triangle definitions of the trigonometric functions $\sin\theta$ (the **sine**), $\cos\theta$ (the **cosine**), and $\tan\theta$ (the **tangent**) for an acute angle θ are

$$\sin\theta = \frac{a}{c} = \frac{\text{opposite side}}{\text{hypotenuse}} \tag{M-27}$$

$$\cos\theta = \frac{b}{c} = \frac{\text{adjacent side}}{\text{hypotenuse}} \tag{M-28}$$

$$\tan\theta = \frac{a}{b} = \frac{\text{opposite side}}{\text{adjacent side}} = \frac{\sin\theta}{\cos\theta} \tag{M-29}$$

(**Acute angles** are angles whose positive rotation around the circumference of a circle measures less than 90° or $\pi/2$.) Three other trigonometric functions—the **secant** (sec), the **cosecant** (csc), and the **cotangent** (cot), defined as the reciprocals of these functions—are

$$\csc\theta = \frac{c}{a} = \frac{1}{\sin\theta} \tag{M-30}$$

$$\sec\theta = \frac{c}{b} = \frac{1}{\cos\theta} \tag{M-31}$$

$$\cot\theta = \frac{b}{a} = \frac{1}{\tan\theta} = \frac{\cos\theta}{\sin\theta} \tag{M-32}$$

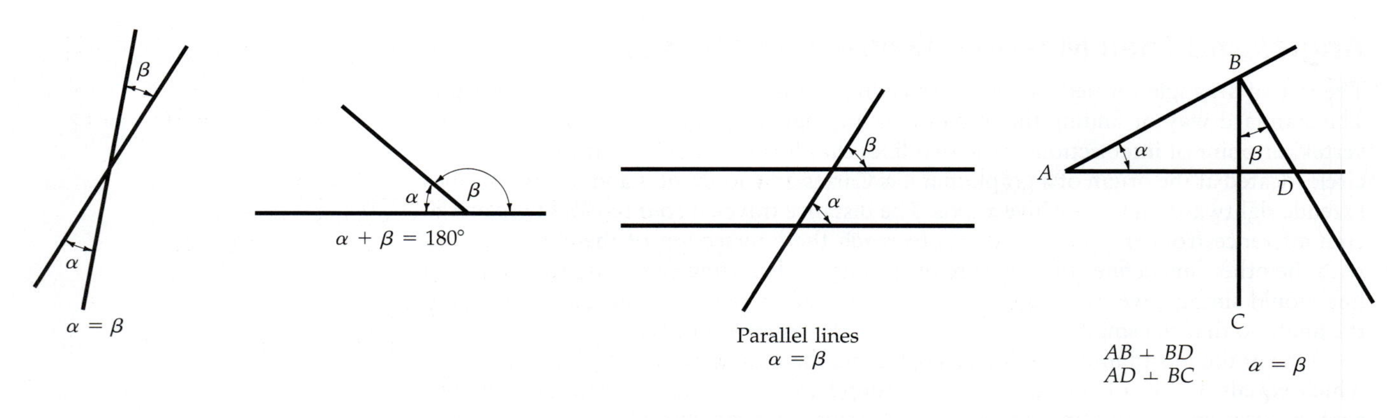

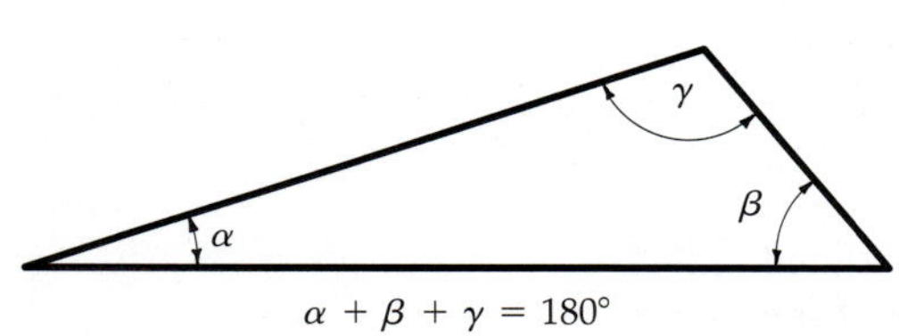

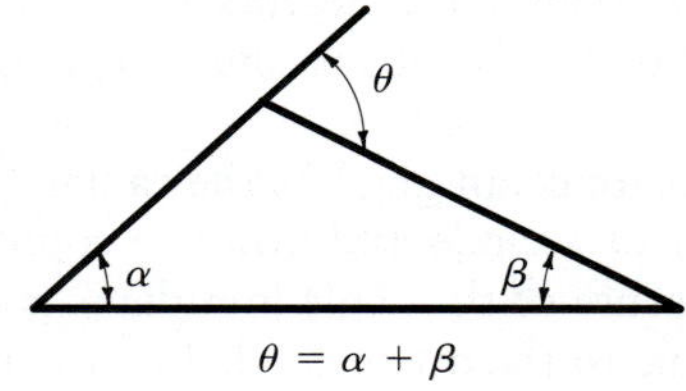

Figure M-11 Some useful relations for angles.

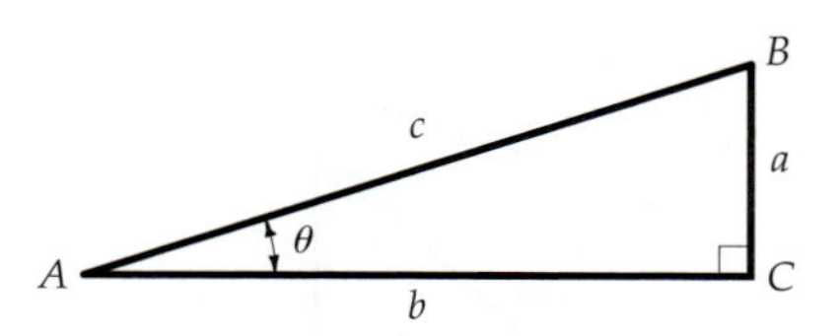

Figure M-12 A right triangle with sides of length a and b and a hypotenuse of length c.

The angle θ, whose sine is x, is called the arcsine of x, and is written $\sin^{-1}x$. That is, if

$$\sin\theta = x$$

then

$$\theta = \arcsin x = \sin^{-1}x \qquad \text{(M-33)}$$

The arcsine is the inverse of the sine. The inverse of the cosine and tangent are defined similarly. The angle whose cosine is y is the arccosine of y. That is, if

$$\cos\theta = y$$

then

$$\theta = \arccos y = \cos^{-1}y \qquad \text{(M-34)}$$

The angle whose tangent is z is the arctangent of z. That is, if

$$\tan\theta = z$$

then

$$\theta = \arctan z = \tan^{-1}z \qquad \text{(M-35)}$$

Trigonometric Identities

We can derive several useful formulas, called **trigonometric identities**, by examining relationships between the trigonometric functions. Equations M-30 through M-32 list three of the most obvious identities, formulas expressing some trigonometric functions as reciprocals of others. Almost as easy to discern are identities derived from the **Pythagorean theorem**,

$$a^2 + b^2 = c^2 \qquad \text{(M-36)}$$

Simple algebraic manipulation of Equation M-36 gives us three more identities. First, if we divide each term in Equation M-36 by c^2, we obtain

$$\frac{a^2}{c^2} + \frac{b^2}{c^2} = 1$$

or, from the definitions of $\sin\theta$ (which is a/c) and $\cos\theta$ (which is b/c),

$$\sin^2\theta + \cos^2\theta = 1 \qquad \text{(M-37)}$$

Similarly, we can divide each term in Equation M-36 by a^2 or b^2 and obtain

$$1 + \cot^2\theta = \csc^2\theta \qquad \text{(M-38)}$$

and

$$1 + \tan^2\theta = \sec^2\theta \qquad \text{(M-39)}$$

Table M-2 lists these last three and many more trigonometric identities. Notice that they fall into four categories: functions of sums or differences of angles, sums or differences of squared functions, functions of double angles (2θ), and functions of half angles ($\frac{1}{2}\theta$). Notice that some of the formulas contain paired alternatives, expressed with the signs $\pm$ and $\mp$; in such formulas, remember to always apply the formula with either all the upper or all the lower alternatives.

TABLE M-2 Trigonometric Identities

$$\sin(A \pm B) = \sin A\cos B \pm \cos A\sin B$$

$$\cos(A \pm B) = \cos A\cos B \mp \sin A\sin B$$

$$\tan(A \pm B) = \frac{\tan A \pm \tan B}{1 \mp \tan A\tan B}$$

$$\sin A \pm \sin B = 2\sin\left[\frac{1}{2}(A \pm B)\right]\cos\left[\frac{1}{2}(A \mp B)\right]$$

$$\cos A + \cos B = 2\cos\left[\frac{1}{2}(A + B)\right]\cos\left[\frac{1}{2}(A - B)\right]$$

$$\cos A - \cos B = 2\sin\left[\frac{1}{2}(A + B)\right]\sin\left[\frac{1}{2}(B - A)\right]$$

$$\tan A \pm \tan B = \frac{\sin(A \pm B)}{\cos A\cos B}$$

$$\sin^2\theta + \cos^2\theta = 1;\ \sec^2\theta - \tan^2\theta = 1;\ \csc^2\theta - \cot^2\theta = 1$$

$$\sin 2\theta = 2\sin\theta\cos\theta$$

$$\cos 2\theta = \cos^2\theta - \sin^2\theta = 2\cos^2\theta - 1 = 1 - 2\sin^2\theta$$

$$\tan 2\theta = \frac{2\tan\theta}{1 - \tan^2\theta}$$

$$\sin\frac{1}{2}\theta = \pm\sqrt{\frac{1 - \cos\theta}{2}};\ \cos\frac{1}{2}\theta = \pm\sqrt{\frac{1 + \cos\theta}{2}};$$

$$\tan\frac{1}{2}\theta = \pm\sqrt{\frac{1 - \cos\theta}{1 + \cos\theta}}$$

Some Important Values of the Functions

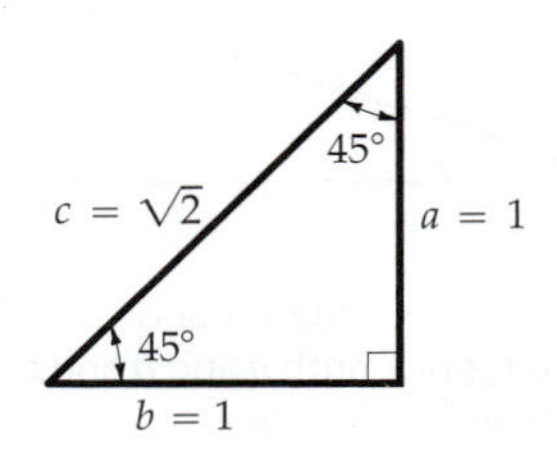

Figure M-13 An isosceles right triangle.

Figure M-13 is a diagram of an *isosceles* right triangle (an isosceles triangle is a triangle with two equal sides), from which we can find the sine, cosine, and tangent of 45°. The two acute angles of this triangle are equal. Because the sum of the three angles in a triangle must equal 180° and the right angle is 90°, each acute angle must be 45°. For convenience, let us assume that the equal sides each have a length of 1 unit. The Pythagorean theorem gives us a value for the hypotenuse of

$$c = \sqrt{a^2 + b^2} = \sqrt{1^2 + 1^2} = \sqrt{2} \text{ units}$$

We calculate the values of the functions as follows:

$$\sin 45° = \frac{a}{c} = \frac{1}{\sqrt{2}} = 0.707 \quad \cos 45° = \frac{b}{c} = \frac{1}{\sqrt{2}} = 0.707 \quad \tan 45° = \frac{a}{b} = \frac{1}{1} = 1$$

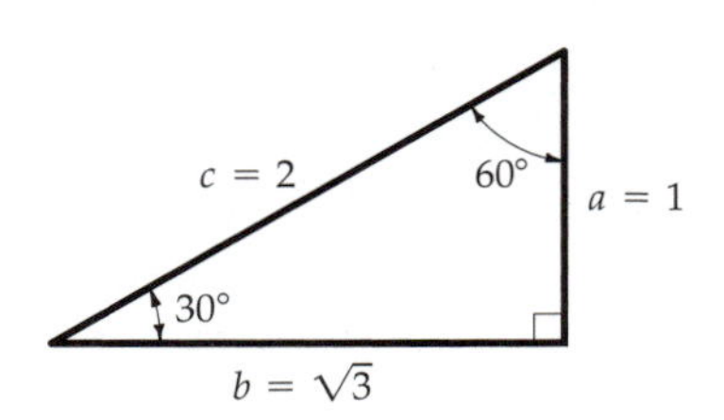

Figure M-14 A 30°–60°–90° right triangle.

Another common triangle, a 30°–60°–90° right triangle, is shown in **Figure M-14**. Because this particular right triangle is in effect half of an *equilateral triangle* (a 60°–60°–60° triangle or a triangle having three equal sides and three equal angles), we can see that the sine of 30° must be exactly 0.5 (**Figure M-15**). The equilateral triangle must have all sides equal to c, the hypotenuse of the 30°–60°–90° right triangle. Thus, side a is one-half the length of the hypotenuse, and so

$$\sin 30° = \frac{1}{2}$$

To find the other ratios within the 30°–60°–90° right triangle, let us assign a value of 1 to the side opposite the 30° angle. Then

$$c = \frac{1}{0.5} = 2 \qquad b = \sqrt{c^2 - a^2} = \sqrt{2^2 - 1^2} = \sqrt{3}$$

$$\cos 30° = \frac{b}{c} = \frac{\sqrt{3}}{2} = 0.866 \qquad \tan 30° = \frac{a}{b} = \frac{1}{\sqrt{3}} = 0.577$$

$$\sin 60° = \frac{b}{c} = \cos 30° = 0.866 \qquad \cos 60° = \frac{a}{c} = \sin 30° = \frac{1}{2}$$

$$\tan 60° = \frac{b}{a} = \frac{\sqrt{3}}{1} = 1.732$$

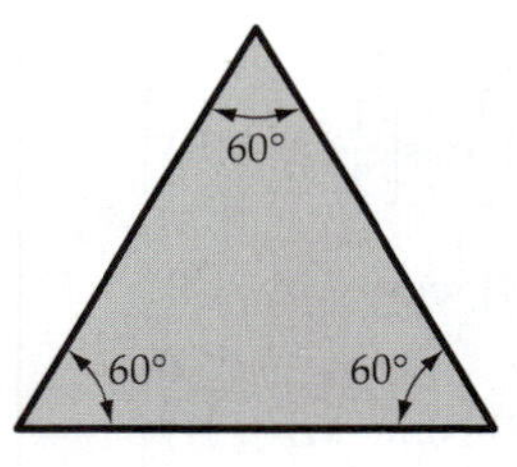

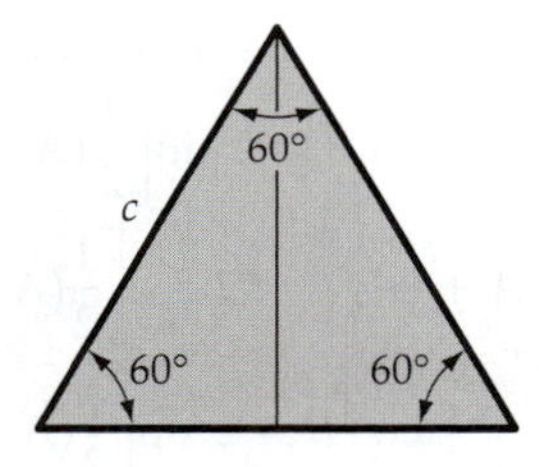

Figure M-15 (a) An equilateral triangle. (b) An equilateral triangle that has been bisected to form two 30°–60°–90° right triangles.

Small-Angle Approximation

For small angles, the length a is nearly equal to the arc length s, as can be seen in **Figure M-16**. The angle $\theta = s/c$ is therefore nearly equal to $\sin\theta = a/c$:

(M-40) $$\sin\theta \approx \theta \quad \text{for small values of } \theta$$

Similarly, the lengths c and b are nearly equal, so $\tan\theta = a/b$ is nearly equal to both θ and $\sin\theta$ for small values of θ:

(M-41) $$\tan\theta \approx \sin\theta \approx \theta \quad \text{for small values of } \theta$$

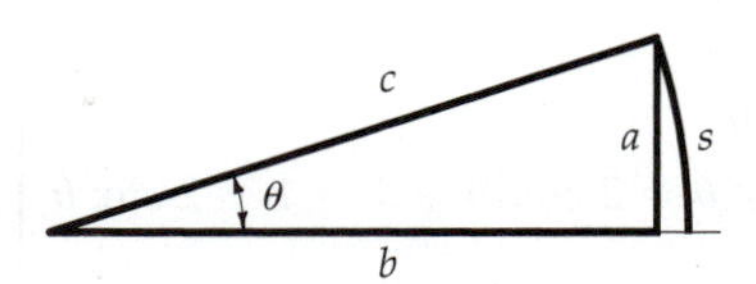

Figure M-16 For small angles, $\sin\theta = a/c$, $\tan\theta = a/b$, and the angle $\theta = s/c$ are all approximately equal.

Equations M-40 and M-41 hold only if θ is measured in radians. Because $\cos\theta = b/c$, and because these lengths are nearly equal for small values of θ, we have

(M-42) $$\cos\theta \approx 1 \quad \text{for small values of } \theta$$

Figure M-17 shows graphs of θ, $\sin\theta$, and $\tan\theta$ versus θ for small values of θ. If accuracy of a few percent is needed, small-angle approximations can be used only for angles of about a quarter of a radian (or about 15°) or less. Below this value, as the angle becomes smaller, the approximation $\theta \approx \sin\theta \approx \tan\theta$ is even more accurate.

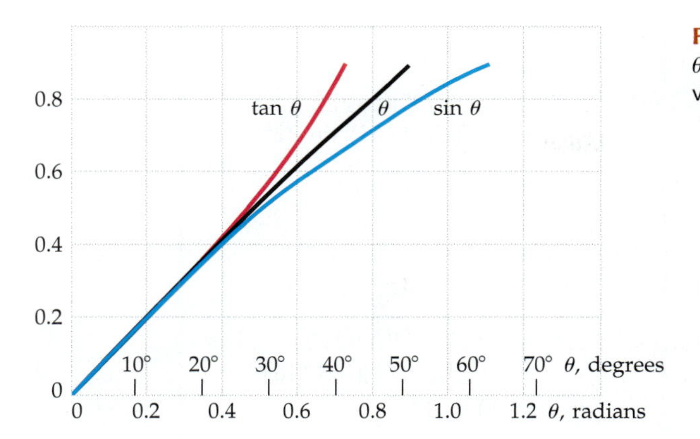

Figure M-17 Graphs of tan θ, θ, and sin θ versus θ for small values of θ.

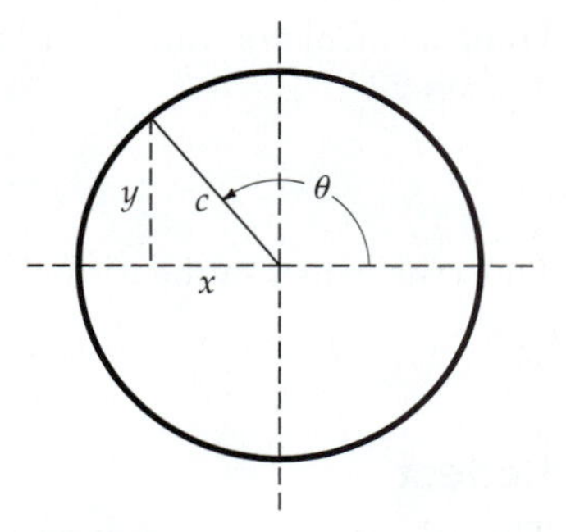

Figure M-18 Diagram for defining the trigonometric functions for an obtuse angle.

Trigonometric Functions as Functions of Real Numbers

So far we have illustrated the trigonometric functions as properties of angles. **Figure M-18** shows an *obtuse* angle with its vertex at the origin and one side along the x axis. The trigonometric functions for a "general" angle such as this are defined by

$$\sin \theta = \frac{y}{c} \tag{M-43}$$

$$\cos \theta = \frac{x}{c} \tag{M-44}$$

$$\tan \theta = \frac{y}{x} \tag{M-45}$$

It is important to remember that values of x to the left of the vertical axis and values of y below the horizontal axis are negative; c in the figure is always regarded as positive. **Figure M-19** shows plots of the general sine, cosine, and tangent functions versus θ. The sine and cosine functions have a period of 2π rad. Thus, for any value of θ, $\sin(\theta + 2\pi) = \sin \theta$, and so forth. That is, when an angle changes by 2π rad, the function returns to its original value. The tangent function has a period of π rad. Thus, $\tan(\theta + \pi) = \tan \theta$, and so forth. Some other useful relations are

$$\sin(\pi - \theta) = \sin \theta \tag{M-46}$$

$$\cos(\pi - \theta) = -\cos \theta \tag{M-47}$$

$$\sin\left(\frac{1}{2}\pi - \theta\right) = \cos \theta \tag{M-48}$$

$$\cos\left(\frac{1}{2}\pi - \theta\right) = \sin \theta \tag{M-49}$$

Because the radian is dimensionless, it is not hard to see from the plots in Figure M-21 that the trigonometric functions are functions of all real numbers.

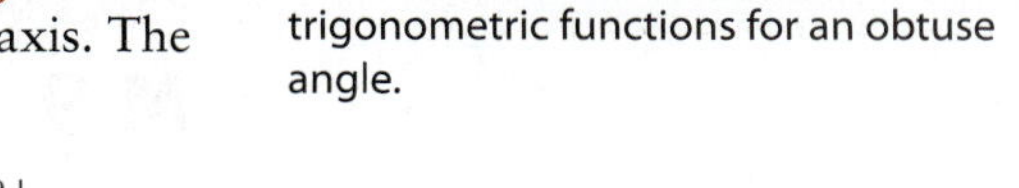
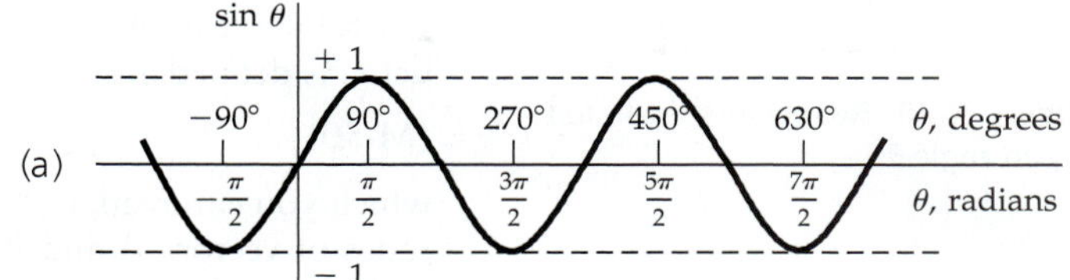

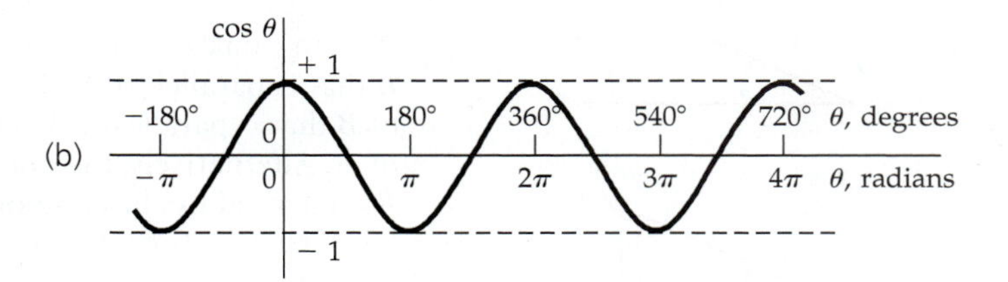

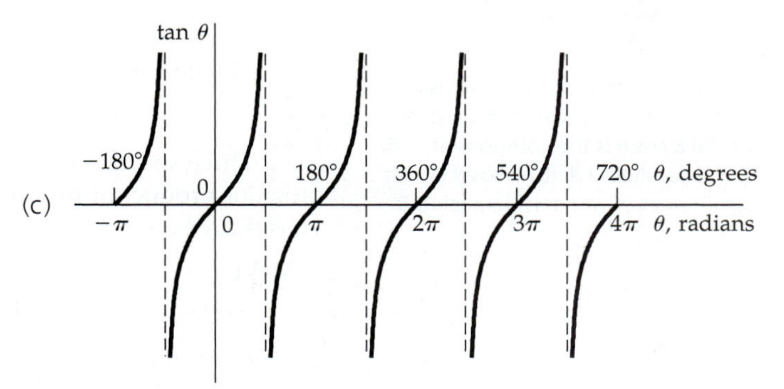

Figure M-19 The trigonometric functions sin θ, cos θ, and tan θ versus θ.

EXAMPLE M-9 Cosine of a Sum

Using the suitable trigonometric identity from Table M-2, find $\cos(135° + 22°)$. Give your answer with four significant figures.

Set Up

As long as all angles are given in degrees, there is no need to convert to radians, because all operations are numerical values of the functions. Be sure, however, that your calculator is in degree mode. The suitable identity is $\cos(A \pm B) = \cos A \cos B \mp \sin A \sin B$, where the upper signs are appropriate.

Solve

Write the trigonometric identity for the cosine of a sum, with $A = 135°$ and $B = 22°$:

$$\cos(135° + 22°) = (\cos 135°)(\cos 22°) - (\sin 135°)(\sin 22°)$$

Using a calculator, find cos 135°, sin 135°, cos 22°, and sin 22°:

$$\cos 135° = -0.7071$$
$$\cos 22° = 0.9272$$
$$\sin 135° = 0.7071$$
$$\sin 22° = 0.3746$$

Enter the values in the formula and calculate the answer:

$$\cos(135° + 22°) = (-0.7071)(0.9272) - (0.7071)(0.3746) = -0.9205$$

Reflect

The calculator shows that $\cos(135° + 22°) = \cos(157°) = -0.9205$.

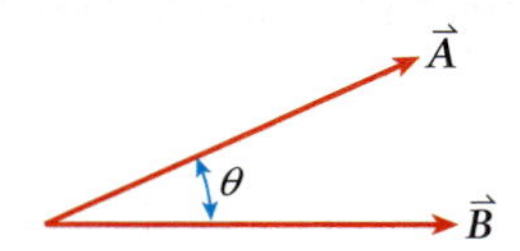

Figure M-20 Two vectors separated by an angle θ.

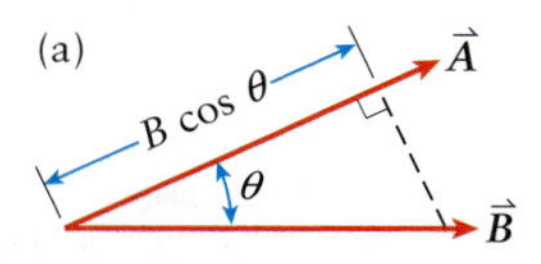

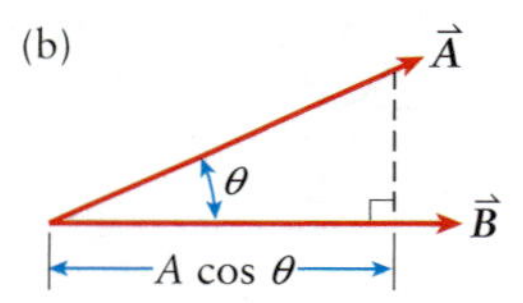

Figure M-21 The dot product is a measure of how parallel two vectors are. (a) $B \cos\theta$ is the component of $\vec{B}$ that is parallel to $\vec{A}$. (b) $A \cos\theta$ is the component of $\vec{A}$ that is parallel to $\vec{B}$.

M-9 The dot product

For two vectors $\vec{A}$ and $\vec{B}$ separated by angle θ, as shown in **Figure M-20**, their dot product C is defined as

$$C = \vec{A} \cdot \vec{B} = AB \cos\theta \tag{M-50}$$

which you can read as, "C equals A dot B." In Equation M-50, A and B are the magnitudes of vectors $\vec{A}$ and $\vec{B}$, respectively. As a result, the dot product of two vectors is a scalar quantity. This is why $\vec{A} \cdot \vec{B}$ is also called the **scalar product** of $\vec{A}$ and $\vec{B}$.

Physically, the dot product $\vec{A} \cdot \vec{B}$ is a measure of how parallel the two vectors are. We can think of it as the magnitude of vector $\vec{A}$ multiplied by the component of vector $\vec{B}$ that is parallel to $\vec{A}$. Referring to **Figure M-21a**, we see that $B \cos\theta$ is the component of $\vec{B}$ that is parallel to $\vec{A}$. That is, $B \cos\theta$ tells us how much of $\vec{B}$ points in the direction of $\vec{A}$. Alternatively, the dot product $\vec{A} \cdot \vec{B}$ can be thought of as the magnitude of vector $\vec{B}$ multiplied by the component of vector $\vec{A}$ parallel to $\vec{B}$ (**Figure M-21b**).

The dot product is commutative; the order of the vectors in a dot product does not affect the result:

$$\vec{A} \cdot \vec{B} = \vec{B} \cdot \vec{A}$$

The dot product is also distributive, which means

$$\vec{A} \cdot (\vec{B} + \vec{C}) = \vec{A} \cdot \vec{B} + \vec{A} \cdot \vec{C}$$

Three special cases of the dot product are particularly important in physics. First, the dot product of two vectors $\vec{A}$ and $\vec{B}$ that point in the same direction (so $\theta = 0$ and $\cos\theta = \cos 0 = 1$) equals the product of their magnitudes:

$$\vec{A} \cdot \vec{B} = AB \cos 0 = AB \quad \text{(if } \vec{A} \text{ and } \vec{B} \text{ point in the same direction)} \tag{M-51}$$

(As an example, the dot product of a vector $\vec{A}$ with itself is equal to the square of its magnitude: $\vec{A} \cdot \vec{A} = AA \cos 0 = A^2$.)

Second, the dot product of two perpendicular vectors $\vec{A}$ and $\vec{B}$ (so $\theta = 90°$ and $\cos\theta = \cos 90° = 0$) is zero:

$$\vec{A} \cdot \vec{B} = AB \cos 90° = 0 \quad \text{(if } \vec{A} \text{ and } \vec{B} \text{ are perpendicular)} \tag{M-52}$$

Third, if two vectors $\vec{A}$ and $\vec{B}$ point in opposite directions (so $\theta = 180°$ and $\cos\theta = \cos 180° = -1$), their dot product equals the *negative* of the product of their magnitudes:

$$\vec{A} \cdot \vec{B} \cos 180° = -AB \quad \text{(if } \vec{A} \text{ and } \vec{B} \text{ point in opposite directions)} \tag{M-53}$$

Finally, it's useful to know how to calculate the dot product of two vectors $\vec{A}$ and $\vec{B}$ that are expressed in terms of their components A_x, A_y, A_z, and B_x, B_y, and B_z:

$$\vec{A} \cdot \vec{B} = A_x B_x + A_y B_y + A_z B_z \tag{M-54}$$

You can verify that Equation M-54 is correct by thinking of $\vec{A}$ as the sum of three vectors: $\vec{A}_1$, which has only an x-component A_x; $\vec{A}_2$, which has only a y-component A_y; and $\vec{A}_3$, which has only a z-component A_z. From the definition of the dot product, $\vec{A}_1 \cdot \vec{B}$ is equal to A_x multiplied by the component of $\vec{B}$ in the direction of $\vec{A}_1$, or $\vec{A}_1 \cdot \vec{B} = A_xB_x$. Similarly, $\vec{A}_2 \cdot \vec{B} = A_yB_y$ and $\vec{A}_3 \cdot \vec{B} = A_zB_z$. Since $\vec{A} = \vec{A}_1 + \vec{A}_2 + \vec{A}_3$ and the dot product is distributive, it follows that

$$\vec{A} \cdot \vec{B} = (\vec{A}_1 + \vec{A}_2 + \vec{A}_3) \cdot \vec{B} = \vec{A}_1 \cdot \vec{B} + \vec{A}_2 \cdot \vec{B} + \vec{A}_3 \cdot \vec{B} = A_xB_x + A_yB_y + A_zB_z$$

That's the same as Equation M-54. If the vectors have only x- and y-components, Equation M-54 simplifies to $\vec{A} \cdot \vec{B} = A_xB_x + A_yB_y$.

EXAMPLE M-10 The Dot Product

(a) Calculate the dot product of vector $\vec{A}$ with magnitude 5.00 pointed in a horizontal direction 36.9° north of east and vector $\vec{B}$ of magnitude 1.50 pointed in a horizontal direction 53.1° south of west. (b) What is the dot product of vector $\vec{C}$ with components $C_x = 4.00$, $C_y = 3.00$ and vector $\vec{D}$ with components $D_x = -0.900$, $D_y = -1.20$?

Set Up

In part (a) we know the magnitude and direction of the vectors, so we'll use Equation M-50. In part (b) the vectors are given in terms of components, so we'll evaluate the dot product using Equation M-54.

$$\vec{A} \cdot \vec{B} = AB\cos\theta \qquad \text{(M-50)}$$

Dot product of two vectors in terms of components:

$$\vec{A} \cdot \vec{B} = A_xB_x + A_yB_y + A_zB_z \qquad \text{(M-54)}$$

Solve

(a) The drawing shows that the angle between $\vec{A}$ and $\vec{B}$ is $\theta = 163.8°$. We use this in Equation M-50 to evaluate the dot product.

$$\begin{aligned}\vec{A} \cdot \vec{B} &= AB\cos\theta \\ &= (5.00)(1.50)\cos 163.8° \\ &= (5.00)(1.50)(-0.960) \\ &= -7.20\end{aligned}$$

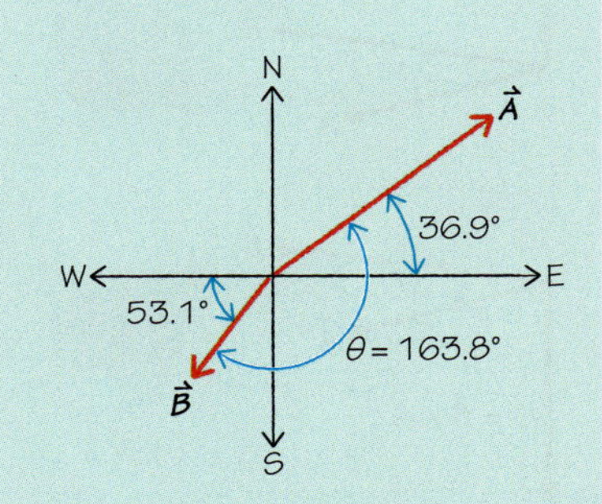

(b) Both $\vec{C}$ and $\vec{D}$ are in the x-y plane and have no z components, so we just need the first two terms in Equation M-54 to calculate their dot product.

$$\begin{aligned}\vec{C} \cdot \vec{D} &= C_xD_x + C_yD_y \\ &= (4.00)(-0.900) + (3.00)(-1.20) \\ &= -7.20\end{aligned}$$

Reflect

It's not a coincidence that we got the same result in part (b) as in part (a): Vectors $\vec{A}$ and $\vec{C}$ are the same, as are vectors $\vec{B}$ and $\vec{D}$. (You can verify this by using the techniques from Chapter 3 to calculate the components of the vectors $\vec{A}$ and $\vec{B}$ in part (a). You'll find that the components are the same as those of $\vec{C}$ and $\vec{D}$ in part (b).) This should give you confidence that the method of calculating the dot product using components gives you the same result as the method that involves the magnitudes and directions of the vectors.

Notice that the angle between vectors $\vec{A}$ and $\vec{B}$ is between 90° and 180°, and the dot product is negative.

M-10 The cross product

The dot product, described in section M-9, is only one way to multiply two vectors. We can also multiply two vectors $\vec{A}$ and $\vec{B}$ using the **cross product**

$$\vec{C} = \vec{A} \times \vec{B} \quad \text{(M-55)}$$

Figure M-22 The cross product is a vector $\vec{C}$ that is perpendicular to both $\vec{A}$ and $\vec{B}$, and has a magnitude $AB \sin \phi$, which equals the area of the parallelogram shown.

The symbol "×" represents the mathematical operation known as the cross product. As you can see from Equation M-55, the result of taking the cross product of two vectors is also a vector. The magnitude of the resulting vector is the product of the magnitudes of the two vectors and the sine of the angle between them. That is, the magnitude of the cross product of $\vec{A}$ and $\vec{B}$ is

$$C = |\vec{A} \times \vec{B}| = AB \sin \phi \quad \text{(M-56)}$$

where according to convention ϕ is defined as the angle that goes from $\vec{A}$ to $\vec{B}$. $\vec{C}$ points in the direction perpendicular to both $\vec{A}$ and $\vec{B}$ as shown in **Figure M-22**.

The magnitude of the cross product $\vec{A} \times \vec{B}$ can be interpreted as the magnitude of vector $\vec{A}$ multiplied by the component of vector $\vec{B}$ perpendicular to $\vec{A}$, or the magnitude of vector $\vec{B}$ multiplied by the component of vector $\vec{A}$ perpendicular to $\vec{B}$.

Note that the order of the two vectors in a cross product makes a difference. The cross product of $\vec{B}$ and $\vec{A}$ is the negative of the cross product of $\vec{A}$ and $\vec{B}$ or

$$\vec{A} \times \vec{B} = -\vec{B} \times \vec{A} \quad \text{(M-57)}$$

This results from the definition of the angle ϕ in Equation M-56. Since ϕ is directed from the first vector to the second vector, if you travel the angle from the second vector to the first—in reverse direction—ϕ becomes negative. And the sine of a negative angle is also negative.

In addition, the cross product obeys the distributive law under addition:

$$\vec{A} \times (\vec{B} + \vec{C}) = \vec{A} \times \vec{B} + \vec{A} \times \vec{C} \quad \text{(M-58)}$$

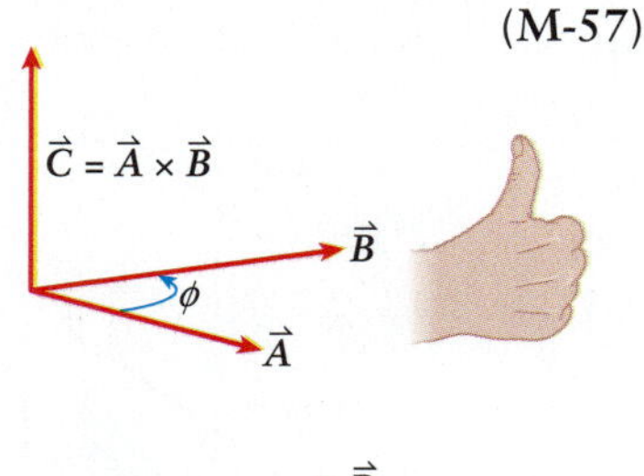

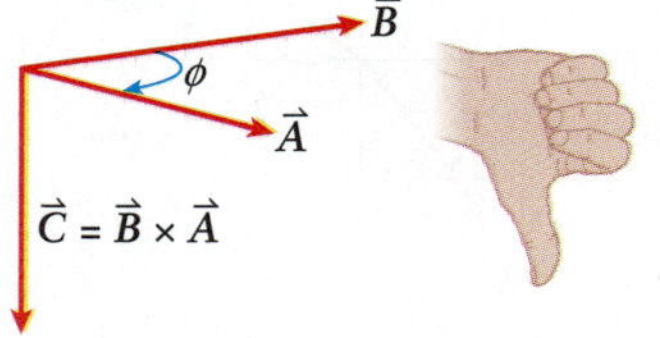

Figure M-23 (a) To find the direction of , point the fingers of your right hand in the direction of vector $\vec{A}$, then curl them toward vector $\vec{B}$. Your thumb points in the direction of the cross product. (b) The direction of $\vec{B} \times \vec{A}$ points in the opposite direction of.

To determine the direction of the cross product $\vec{C} = \vec{A} \times \vec{B}$, you can use the right-hand rule. To apply this rule, point the fingers of your right hand in the direction of the first vector of the cross product (in this case $\vec{A}$). Then curl your fingers toward the second vector, $\vec{B}$. If you stick your thumb straight out, it points in the direction of the cross product, vector $\vec{C}$ (**Figure M-23a**). If you instead want to find the direction of the cross product $\vec{B} \times \vec{A}$, begin by pointing the fingers of your right hand in the direction of vector $\vec{B}$. Then curl them toward vector $\vec{A}$. Your thumb again points in the direction of the cross product (**Figure M-23b**). Note that because you must curl your fingers in the opposite direction as for $\vec{C} = \vec{A} \times \vec{B}$, the cross product of $\vec{B} \times \vec{A}$ points in the opposite direction of $\vec{A} \times \vec{B}$, which is just what we stated in Equation M-58.

There are two special cases of the cross product that are worth pointing out. The first is the cross product for two perpendicular vectors, for which $\phi = 90°$, so $\sin \phi = 1$.

$$|\vec{A} \times \vec{B}| = AB \sin 90° = AB(1) = AB$$

(magnitude of the cross product of two perpendicular vectors)

The second special case is the cross product of two parallel vectors, for which $\phi = 0$, so $\sin \phi = 0$.

$$|\vec{A} \times \vec{B}| = AB \sin 0 = AB(0) = 0$$

(magnitude of the cross product for two parallel vectors)

One example of a cross product of two parallel vectors is the cross product of a vector with itself: $\vec{A} \times \vec{A} = 0$.

EXAMPLE M-11 The Cross Product

Evaluate $\vec{A} \times \vec{B}$, in which the components of vector $\vec{A}$ are $A_x = 5$, $A_y = 0$, and the components of vector $\vec{B}$ are $B_x = 9$, $B_y = 7$.

Set Up

We will use the definition of the magnitude of the cross product, Equation M-56, to find the magnitude of the cross product, and the right-hand rule to determine the direction of the cross product.

We will have to use the components of vector $\vec{B}$ to determine its magnitude and the angle it makes with the x axis and vector $\vec{A}$.

$$C = |\vec{A} \times \vec{B}| = AB \sin \phi \quad \text{(M-56)}$$

Finding vector magnitude and direction from vector components:

$$A = \sqrt{A_x^2 + A_y^2}$$

$$\tan \theta = \frac{A_y}{A_x} \quad \text{(3-2)}$$

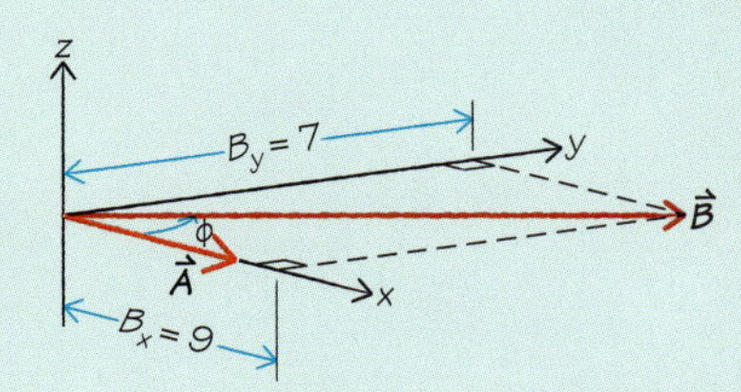

Solve

Begin by determining the magnitude and direction of vector $\vec{B}$ using its components and Equations 3-2.

Determine the magnitude of vector $\vec{B}$ from its components:

$$B = \sqrt{B_x^2 + B_y^2} = \sqrt{9^2 + 7^2} = 11.4$$

Determine the angle $\vec{B}$ makes with the x axis (and vector $\vec{A}$) from its components:

$$\tan \phi = \frac{B_y}{B_x} = \frac{7}{9} = 0.778, \text{ so}$$

$$\phi = \arctan 0.778 = 37.9°$$

Because vector $\vec{A}$ has only an x component, its magnitude is equal to its x component, and the angle it makes with the x axis is 0.

Determine the magnitude of vector $\vec{A}$ from its components:

$$A = \sqrt{A_x^2 + A_y^2} = \sqrt{5^2 + 0^2} = 5$$

Because $\vec{A}$ has only an x component, it makes an angle of zero degrees with the x axis.

Now that we know both the magnitude and direction of the vectors, we can use Equation M-56 to determine the magnitude of the cross product.

Apply Equation M-56 to the two vectors:

$$\begin{aligned} |\vec{A} \times \vec{B}| &= AB \sin \phi \\ &= (5.00)(11.4)\sin 37.9° \\ &= (5.00)(11.4)(0.614) \\ &= 35.0 \end{aligned}$$

Use the right-hand rule to determine the direction of the cross product.

From the figure, if we first point the fingers of our right hand in the direction of $\vec{A}$ (along the x axis), and then curl them toward $\vec{B}$, we see that the thumb points in the positive z direction. So the cross product $\vec{A} \times \vec{B}$ has a magnitude of 35 in the $+z$ direction.

Reflect

The vectors $\vec{A}$ and $\vec{B}$ lie in the xy plane, so the cross product, which must be perpendicular to both vectors, should point along the z axis, which is just what we found.

ANSWERS TO ODD PROBLEMS

Chapter 1

1. The meter (SI unit for length) is defined by the distance light travels in a vacuum in a tiny fraction (1/299,792,458) of a second. The SI unit for mass is the kilogram. It is defined as the mass of a carefully protected prototype block of platinum and iridium that is stored in France. The SI unit of time is the second. It is defined as the time required for a cesium atom to emit 9,192,631,770 complete cycles of radio waves due to transitions between two particular energy levels of the cesium atom.

3. To be a useful standard of measurement, an object, system, or process should be unchanging, replicable, and possible to measure precisely so that errors in its measurement do not carry over into calibration errors in every other measurement.

5. No. The equation 3 meters = 70 meters has consistent units, but it is false. The same goes for 1 = 2, which consistently has no units.

7. (a) The fewest number of significant figures in 61,000 is two—the 6 and the 1. (b) If the period is acting as a decimal point, then the trailing zeros are significant and the quantity 61,000. would have five significant figures. (c) When numbers are written in scientific notation, all the digits before the power of 10 are significant. Therefore, 6.10×10^4 has three significant figures.

9. B

11. C

13. C

15. D

17. C

19. The distance from home plate to the center field fence is about 100 m. A well-hit ball leaves the bat at around 100 mph, or 45 m/s. Assuming the ball comes off the bat horizontal to the ground, this gives an estimate of $100\text{ m} \times \frac{1\text{ s}}{45\text{ m}} = 2\text{ s}$. This is probably a little low but is still reasonable.

21. An average person uses about 100 L of water for showering. A total of 10 L of water is used for cooking, drinking, and washing hands; 24 L of water is used in the toilet; and 20 L is used for a load of laundry in a top-loading washer. So a reasonable estimate of daily water use is about 150 L.

23. There is no one answer to this question. When estimating, keep in mind that one story is about 10 feet.

25. We can estimate the number of cells in the human body by determining the mass of a cell and comparing it to the mass of a human. An average human male has a mass of 80 kg. A person is mainly water, so we can approximate the density of a human body (and its cells. as 1000 kg/m^3. We are told that the volume of a cell is the same as a sphere with a radius of 10^{-5} m, or approximately $4 \times 10^{-15}\text{ m}^3$; the mass of a single cell is $4 \times 10^{-15}\text{ m}^3 \times \frac{1000\text{ kg}}{1\text{ m}^3} = 4 \times 10^{-12}\text{ kg}$.
The number of cells in the body is then
$n_{\text{cell}} = \frac{m_{\text{body}}}{m_{\text{cell}}} = \frac{80\text{ kg}}{4 \times 10^{-12}\text{ kg}} = 2 \times 10^{13}$.

27. A. 2.37×10^2
B. 2.23×10^{-3}
C. 4.51×10^1
D. 1.115×10^3
E. 1.487×10^4
F. 2.1478×10^2
G. 4.42×10^{-6}
H. 1.234578×10^7

29. A. kilo (k)
B. giga (G)
C. mega (M)
D. tera (T)
E. milli (m)
F. pico (p)
G. micro (μ)
H. nano (n)

31. A. 1. 25 m
B. 2.33×10^{-1} kg
C. 7.86×10^{-1} s
D. 4.54×10^8 mg
E. $2.08 \times 10^{-2}\text{ m}^2$
F. $4.44 \times 10^6\text{ cm}^2$
G. $1.25 \times 10^{-5}\text{ m}^3$
H. $1.44 \times 10^8\text{ cm}^3$

33. A. 0.328 L
B. 0.112 m^3
C. $2.2 \times 10^6\text{ m}^2$
D. 4.43 hectares
E. 0.225 m^3
F. 1.72×10^8 L
G. 2.253×10^{-2} hectare · m
H. 2×10^9 mL

35. (a) 3×10^5 m
(b) 3.37×10^{-5} m
(c) 7.75×10^{10} W

37. 13 mpg

39. A. 11
B. 7
C. 6
D. 22.4
E. 14.8
F. 3
G. 199.0
H. 266

41. The first term has dimensions of length; the second term has dimensions of length divided by time squared and multiplied by time squared, which is length; the third has dimensions of length divided by time and multiplied by time, which is length; the fourth is length. Therefore, the equation is consistent.

43. Dimensions of α: SI unit is s^{-1}
Dimensions of k: SI unit is m^{-1}
Dimensions of ω: SI unit is s^{-1}

45. The first term has dimensions of time. The second term has, inside the square root, dimensions of length and divided by length divided by time squared, which is the square root of time squared, which is time. Therefore, the equation is consistent.

47. 10 cm

49. (a) 2.5×10^{-3} kg
(b) $0.15 \frac{kg}{m^3}$
(c) 5×10^{-16} m^3
(d) One tablet: 8.1×10^{-5} kg
One bottle: 8.1×10^{-3} kg
(e) $2.0 \times 10^{-8} \frac{m^3}{s}$
(f) $1400 \frac{kg}{m^3}$

51. $\frac{m^3}{kg \cdot s^2}$

53. $3.00 \times 10^8 \frac{m}{s}$

55. (a) $0.4 \frac{g}{cm^3}$; $400 \frac{kg}{m^3}$
(b) 40 times the density of water
(c) 9×10^{-5} g
(d) 5×10^{-4}%

Chapter 2

1. It's always a good idea to include the units of every quantity throughout your calculation. Eventually, once they become more comfortable, most people will convert all of their values into SI units and stop including them in the intermediate steps of the calculation. Using SI units in all calculations minimizes calculation errors and ensures the answer will be in SI units as well.

3. Average velocity is a vector quantity—the *displacement* over the time interval. Average speed is a scalar quantity—the *distance* over the time interval.

5. The largest speed, 1 m/s, will give the largest displacement in a fixed time.

7. An object will slow down when its acceleration vector points in the opposite direction to its velocity vector. Recall that acceleration is the change in velocity over the change in time.

9. One advantage is that people in countries that use the metric system will have a chance of understanding the distance part of the unit. The disadvantages are that no country lists speed limits in m/s, so it is hard to figure out how long it takes to get to highway speeds. (For comparison, 65.0 mph = 105 km/h = 29.1 m/s.) Further, the raw numbers will be smaller than if another unit system were used, so careless readers not comparing units will get the impression that the car accelerates slowly. It would be better to use km/h/s, as speed limits around the world are generally expressed in km/h. The best plan would be to tailor the units to the individual market.

11. The acceleration due to gravity is constant in both magnitude (g) and direction (down). When a ball is thrown straight up, its acceleration vector points in the opposite direction to its velocity vector, which means it slows down and eventually stops. Assuming the braking acceleration of the car is constant in magnitude (g) and points opposite to the car's velocity vector, the car, too, will slow down and eventually stop. If the ball and the car start at the same initial speed and have the same acceleration, the ball and the car will take the same amount of time to come to rest.

13. The acceleration of a ball thrown straight up in the air is constant because it is under the influence of Earth's gravity.

15. There is no reason why "up" cannot be labeled as "negative" or "left" as "positive."
Usually up and right are chosen as positive, since these correspond to positive x and y in a standard Cartesian coordinate system.

17. D

19. E

21. C

23. The average car takes about 10–15 s to reach highway speeds. Cars can brake faster than this, say, 5–10 s.

25. (a) Stopping from 10 m/s in 2 s on a muddy field will have an acceleration of only 5 m/s^2. Stopping from 10 m/s in 0.5 s on a high-traction surface will require an acceleration of about 20 m/s^2. (b) For the male athlete, the accelerations will be approximately the same.

27. A 5-kg cat would need to reach a velocity of roughly 4.4 m/s to jump up to the counter (about 1 m high). If the cat's legs extend about 10 cm while jumping, this would require an acceleration of 98 m/s^2.

29. Cruise ships accelerate very slowly, reaching a cruising speed of 10 m/s over half an hour. This gives an acceleration of about 0.006 m/s^2.

31. (a) The ball goes from 0 m/s to about 45 m/s over the course of about 0.5 s as the pitcher winds up and throws the ball. This would equate to about 90 m/s^2. (b) The soccer ball accelerates from 0 m/s to about 30 m/s over the course of about 0.1 s as the player's foot makes contact with the ball. This means the acceleration is about 300 m/s^2.

33. With $x(t)$ in m and t in s:
$x(t) = 6t - 12;\ 0 \le t \le 4$
$x(t) = 3(t - 4) + 12;\ 4 \le t \le 5$
$x(t) = 15,\ 5 \le t \le 8$
$x(t) = 1.5t^2 - 22.5t + 99;\ 8 \le t \le 13$
$x(t) = 5(t - 13) + 60;\ 13 \le t \le 21$
$x(t) = 10(21 - t) + 100;\ 21 \le t \le 24$
$x(t) = 70;\ 24 \le t \le 25$

35. A. $108 \frac{km}{h}$
B. $22.5 \frac{km}{h}$
C. $2.01 \times 10^5 \frac{mi}{h}$
D. $60.0 \frac{mi}{h}$
E. $40 \frac{m}{s}$ (to one significant figure)

37. $4.00\frac{\text{km}}{\text{h}}$

39. (a) Displacement = 0 m, so his average velocity = 0

(b) $4.00\frac{\text{km}}{\text{h}}$

(c) $2.70\frac{\text{m}}{\text{s}}$

41. 39.2 km

43. 17.0 m

45. $0.54\frac{\text{m}}{\text{s}}$

47. (a) At the 4-s mark, velocity transitions from 0 to −0.33 m/s over an extremely short time interval, suggesting a very large negative acceleration at this time.
(b) Speed is greatest in the interval between 4 and 5 s. However, since we are limited to one significant figure in the answer, this speed can be said to be roughly equal to the speed in the interval between 0.5 and 2 s.
(c) 0

49. A. velocity = −; acceleration = − (Velocity is negative and increasing during the interval.)
B. velocity = −; acceleration = + (Velocity is negative and decreasing during the interval.)
C. velocity = +; acceleration = + (Velocity is positive and increasing during the interval.)
D. velocity = +; acceleration = − (Velocity is positive and decreasing during the interval.)
E. velocity = −; acceleration = − (Velocity is negative and increasing during the interval.)
F. velocity = −; acceleration = + (Velocity is negative and decreasing during the interval.)

51.

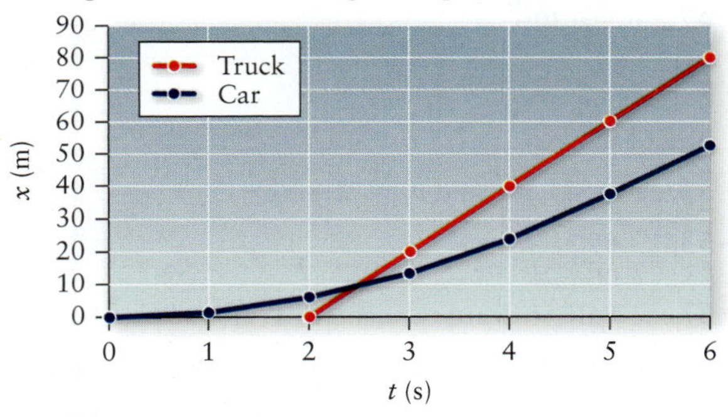

53. $49.1\frac{\text{m}}{\text{s}^2}$

55. (a) $a_{\text{A}x} = 0.556\frac{\text{m}}{\text{s}^2}$

$a_{\text{B}x} = 0.556\frac{\text{m}}{\text{s}^2}$

(b) $x_{\text{A}} = 55.6$ m

$x_{\text{B}} = 97.2$ m

57. Accelerations:

$a_{\text{Bugatti}, x} = 11.2\frac{\text{m}}{\text{s}^2}$

$a_{\text{Caparo}, x} = 10.7\frac{\text{m}}{\text{s}^2}$

$a_{\text{Ultimo}, x} = 10.3\frac{\text{m}}{\text{s}^2}$

$a_{\text{SSC}, x} = 9.92\frac{\text{m}}{\text{s}^2}$

$a_{\text{Saleen}, x} = 9.57\frac{\text{m}}{\text{s}^2}$

59. $-5.63\frac{\text{m}}{\text{s}^2}$

61. 4.08 s

63. (a) $17.1\frac{\text{m}}{\text{s}}$

(b) 2.26 s

65. (a) $4.08\frac{\text{m}}{\text{s}}$

(b) 0.832 s

67. $8.20\frac{\text{m}}{\text{s}}$; $1.60\frac{\text{m}}{\text{s}}$; $31.0\frac{\text{m}}{\text{s}}$; 1.84 s

69. $2.2\frac{\text{m}}{\text{s}}$

71. (a) 2.87 s
(b) 0.413 m

73. Geoff wins the race.

75. 2.83 m

77. 125 m

79. (a) 0.494 s
(b) 2.20 m
(c) 0.730 s

81. 123 m apart

Chapter 3

1. A scalar is a number, whereas a vector has a direction associated with a number. Distance and speed are examples of scalar quantities, whereas displacement and velocity are examples of vectors.

3. A. No, the sum of two vectors that have different magnitudes can never equal zero. The only time two vectors can add together to have zero magnitude is when they point in opposite directions and have the same magnitude.
B. Yes, the sum of three (or more) vectors with different magnitudes can be equal to zero.

5. Gravity on the Moon is about one-sixth that on Earth. If you were playing tennis on the Moon, you would have to hit the ball more softly than on Earth in order for it to stay in bounds. The trajectories should still look parabolic, though.

7. During the motion of a projectile, only v_x, $a_x(=0)$, and a_y $(=-g$, if up is positive) are constant. The position (x, y) of the projectile is constantly changing while it is moving. The acceleration in the y direction is nonzero, which means v_y is changing.

9. The rock's speed will be greater than the speed with which it was thrown. Gravity is accelerating the rock in the y direction, so the magnitude of the y component of its velocity will increase.

11. No. A long jumper should take off at an angle of 45° in order to achieve the maximum range.

13. Yes. The force from the vine is accelerating the ape upward; at the bottom of the swing is when the vertical component of the velocity changes from downward to upward, which requires an acceleration in the upward direction.

15. Yes. You are accelerating because the direction of the velocity vector is changing. The acceleration vector points east.

17. B

19. D

21. D

23. D

25. B

27. $r_x = 16$; $r_y = 9.0$

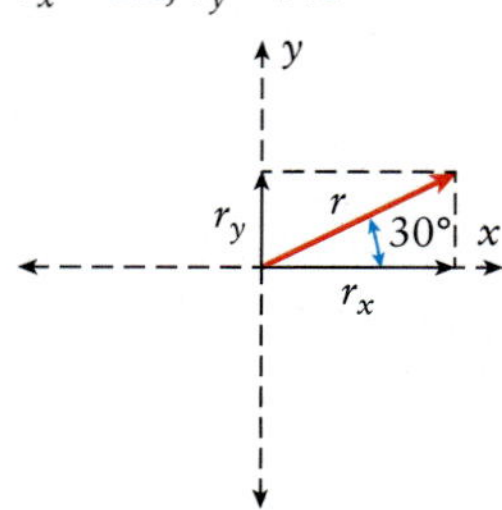

29. The ball peaked at about 110 ft in the vertical direction and about 300 ft in the horizontal direction. The launch angle is approximately 30°. Using an acceleration of 32 ft/s^2, that corresponds to an initial speed of about 53 m/s. Without air resistance, because the ball reached maximum height at 300 ft, we might expect the ball to travel 600 ft. According to the graph, we expect the ball might travel 540 ft, suggesting that the second half of the ball's flight showed a 20-foot decrease in flight distance. Accounting for this loss due to drag, we might estimate an initial speed of up to 65 m/s for the baseball.

31. (a) A plot of v_x versus time is constant and greater than zero for all time.
(b) A plot of v_y versus time is linear, starts at a maximum $+v_{0y}$, crosses zero when the ball is at its highest point, and has a constant slope of -9.80 m/s^2. The ball hits the ground when $v_y = -v_{0y}$.
(c) A plot of a_x versus time is equal to zero for all time.
(d) A plot of a_y versus time is equal to -9.80 m/s^2 for all time.

33. 5.6; 0.46 rad

35. A. 4; 300°
B. 3; 100°
C. 2; 270°

37. $C = 6.40$; 38.7°

39. (a) $C_x = 21\frac{\text{m}}{\text{s}}$ and $C_y = 61\frac{\text{m}}{\text{s}}$
(b) $D_x = 21\frac{\text{m}}{\text{s}}$ and $D_y = -19\frac{\text{m}}{\text{s}}$
(c) $E_x = 42\frac{\text{m}}{\text{s}}$ and $E_y = 82\frac{\text{m}}{\text{s}}$

41. 50 m/s; 53° west of north

43.

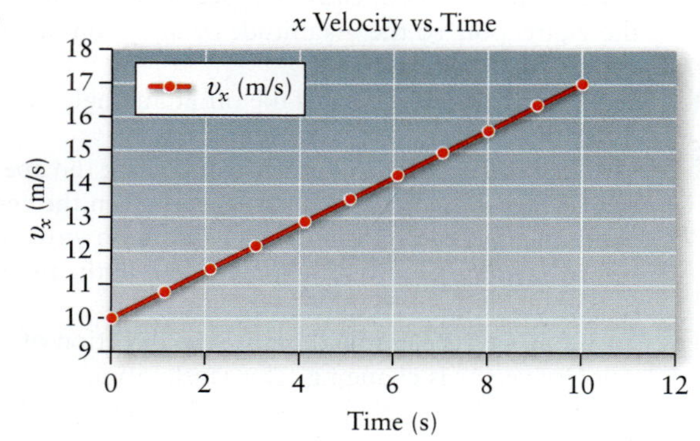

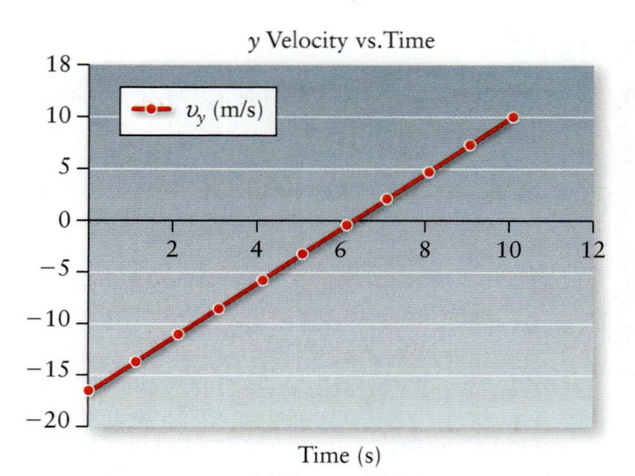

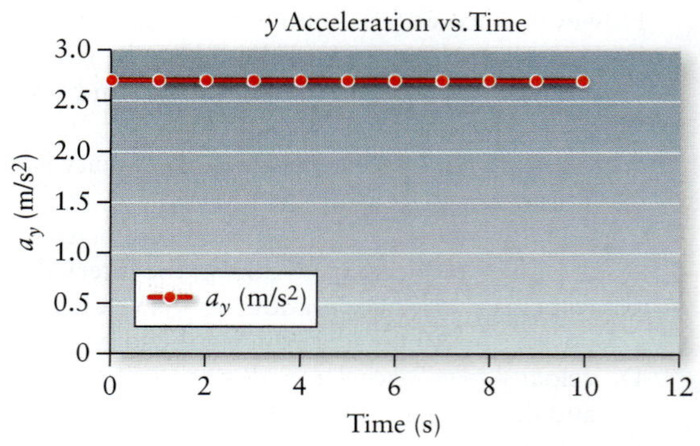

45. (a) $v_x = -4.00\frac{\text{km}}{\text{h}}$ and $v_y = 2.50\frac{\text{km}}{\text{h}}$
(b) Average velocities are the same.

47. $v_{0x} = 30\frac{\text{m}}{\text{s}}$
$v_{0y} = 20\frac{\text{m}}{\text{s}}$

49. 4 m

51. $v_x = 21.7\frac{\text{m}}{\text{s}}$ and $v_y = \pm 8.82\frac{\text{m}}{\text{s}}$

53. 3.49 s

55. 4.00 m/s

57. 17 m/s^2 toward the center of the track

59. (a) $5.9 \times 10^{-3}\frac{\text{m}}{\text{s}^2}$ toward the center of the Sun
(b) 8.00 m/s^2 toward the center of the circle

61. $r = 0.07$ m; $v = 700\frac{\text{m}}{\text{s}}$

63. 0.25 s

65. 6.7 km east and 13 km south

67. (a) 6.86 s
(b) 7.16 s

69. (a) $t = 9.24$ s
(b) 271 m
(c) 282 m

71. (a) 14.9 m/s
(b) 1.43 s
(c) 7.15 m

73. Maximum height = $6.05h$; maximum range = $6.05R$

75. Moon: increased by a factor of 6; Mars: increased by a factor of 2.6

77. (a) 68.8 m/s^2; $7g$
(b) 2.26 s; time for the pilot must be increased by 15%

79. 0.309 m

Chapter 4

1. Yes, the net force vector will determine the direction of the acceleration vector for the mass in question.
3. $[F] = \text{N} = \text{kg} \cdot \text{m/s}^2$
5. Zero. When you stand on a motionless horizontal bathroom scale, it is not accelerating. Therefore the ground must push up on the scale with the same magnitude of force you push down on the scale.
7. If the box moves at a constant speed when a 60-N force is applied, the floor must oppose the motion with a force of 60 N so that the net force and, therefore, the acceleration are zero. When a force of 80 N is applied, the net force on the box is, therefore, 20 N, which results in a nonzero acceleration. If this force is applied to the box for the remainder of its motion across the room, the box will continue to increase in speed until it hits the wall.
9. Lifting a truck requires applying an upward force greater than the weight of the truck. Since the weight of a truck on the Moon is approximately one-sixth of the weight of the same truck on Earth, a smaller lifting force is required on the Moon.
11. If the fisherman reels in the fish horizontally, the maximum 4-lb force the line can exert on the fish contributes to its horizontal acceleration. So as long as the fish pulls on the hook with less than 4 lbs of force in the horizontal direction, the line will not break.
13. The boxer's claim that both his jaw and his opponent's fist experience the same force is correct. However, the fist is relatively hard while the jaw and the surrounding tissue are not. So one would imagine that it's more painful to take a blow to the face than one to the hand.
15. When you walk, you push "backward" on the floor with your shoe. According to Newton's third law, the floor must then push "forward" on your shoe. It is this forward push that gives you a forward acceleration as you start to walk.
17. If you try to hang the 850-N object stationary, the rope will break. However, if the object's downward acceleration is large enough to require at least 50 N of net force, the rope will not break.
19. The scale always reads the force that it exerts on the chair. As the elevator begins to ascend, everything in it accelerates upward. So the upward force on the chair from the scale must be greater than the downward force of the chair's weight. When moving at constant speed either up or down, the chair does not accelerate. So the scale must exert a force equal in magnitude to the chair's weight. Both when the elevator slows at the top, and when the elevator begins to descend, the acceleration of the chair is downward. So the upward force of the scale on the chair must be less than the scale's weight. As the elevator slows at the bottom, it again accelerates upward, and the force of the scale on the chair is greater than the scale's weight.
21. Seatbelts increase the time over which passengers come to rest, which decreases the magnitude of their accelerations and, correspondingly, the magnitude of the net force acting on them.
23. The pain you feel in your hand as it strikes the wall is approximately proportional to the amount of force the wall exerts on your hand in slowing it to a stop. Assuming you strike both the solid wall and a padded wall with the same blow (that is, same hand speed before your hand intersects the wall), both cases will slow your hand from its initial speed to zero. However, the hard wall will achieve this over a shorter distance than the padded wall, requiring a larger acceleration on your hand. According to Newton's second law, this means the solid wall requires a greater force to achieve the larger acceleration of your hand. Since the force exerted on your hand by the solid wall is thus greater than that exerted by the padded wall, your hand will experience more pain from the solid wall than the padded wall, even though your initial hand speed was the same in both cases.
25. B
27. D
29. B
31. C
33. D
35. Apple ≈ 1 N; penny ≈ 0.03 N; textbook ≈ 25 N; calculator ≈ 5 N; 1 L of water ≈ 10 N
37. A baseball throw can be in the range of 40 m/s. A baseball has a mass of around 0.15 kg. The powerful part of the throw lasts about 0.05 s. This gives
$$F = m\frac{\Delta v}{\Delta t} \approx 120 \text{ N}$$
39. Tennis balls weigh around 57 g. A typical ball launcher sends the ball at 12 m/s. This process takes around 0.1 s. The force is about 7 N.
41. 2000 to 50,000 N
43. 1 to 5 N
45.
47. 4000 N
49. 12.00 m/s^2
51. $F_2 = 1.39$ N
$F_3 = 1.63$ N
53. 2450 N
55. 0 N
57.
59.

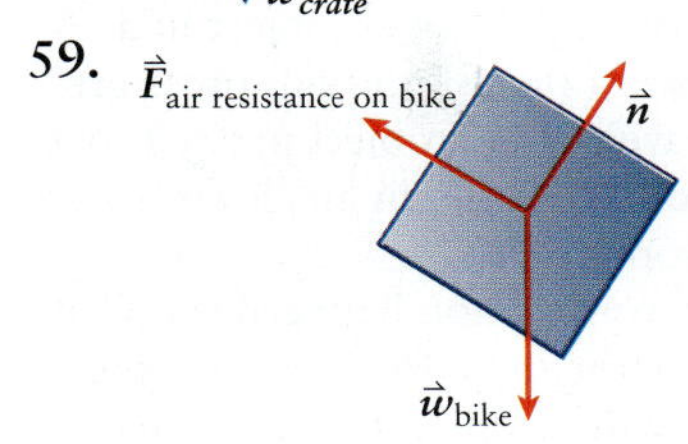

61. (a) 45 N in the vertical, upward direction; (b) 10 N in the vertical, upward direction; (c) 0 N
63. 760 N
65. 410 N
67. 10,000 N

69. 481 N
71. 4.20 s
73. 1.50 m/s^2 to the west
75. 3.31 kg
77. $3.50\frac{\text{m}}{\text{s}}$
79. $157.5\frac{\text{m}}{\text{s}}$
81. (a) 25.6 N
(b) 3.76 kg
83. (a) 2.80 m/s^2
(b) The elevator is moving down.
85. 1.70×10^4 N
87. (a) 189 N; (b) 294 N; (c) 414 N
89. The scale will initially read 700 N while the elevator is at rest. Then it will read 914 N as the elevator accelerates upward. The scale will again read 700 N as the elevator travels at constant speed. The scale will read 557 N as the elevator slows to a stop. It will then read 0 N as the elevator free-falls to the bottom.
91. (a) $a = 2.10 \times 10^3$ m/s^2; time is $t = 1.38$ ms
(b) $F_n = 25.9$ mN
$F_g = 0.12$ mN
(c) $F_{\text{ground}} = 25.9$ mN, which is 215 times the weight of the froghopper.
93. (a) 644 N; (b) 92.9 kg
95. (a) 466 N; (b)537 N
97. $T_1 = 2.49$ N and $T_2 = 7.46$ N
99. $a = 3.36$ m/s^2; $T = 19.3$ N
101. $a = 3.50$ m/s^2; $T_2 = 25.2$ N; $T_1 = 8.40$ N
103. (a) The largest pull force before a string breaks is 73.2 N.
(b) $T_A = 26.2$ N; $T_B = 45.0$ N

Chapter 5

1. Sometimes. If the coefficient of static friction is less than 1, the normal force will be larger than the static friction force. If the coefficient of static friction is more than 1, the static friction force can exceed the normal force or not, depending on how much friction is required to prevent slipping.
3. When a pickup truck accelerates forward, the contents will also accelerate forward due to the force of static friction between the box and the truck's bed.
5. (a) Weight stays the same; (b) normal force decreases; (c) frictional force is the same; (d) maximum static frictional force decreases
7. An object on a slope (Figure 5-11a); an object in a stack of objects (the lower objects' normal forces are greater than their own weight); an object pressed to a wall (Figure 5-13); an object sitting in an elevator as the elevator starts or stops
9. Although the coefficients of friction between synovial fluid and bone should be the same for everyone, the normal forces involved will be different from person to person. This means the magnitudes of the frictional forces will also be different from person to person.
11. His acceleration decreases because the net force on him decreases. Net force is equal to his weight minus his drag force. The faster an object is moving, the larger the drag force it experiences as a function of its speed. Because drag force increases with increasing speed, net force and acceleration decrease. As the drag force approaches his weight, the acceleration approaches zero.
13. Yes. If he lies flat in relation to the wind and she does not, she will have a smaller profile in relation to the wind. Thus she will experience less drag and be able to catch up.
15. If you know the car's instantaneous speed and the radius of curvature, you can use the formula $a_{\text{cent}} = v^2/r$. Another method would be to hang a small object from a light string and measure the angle θ of the string with respect to the vertical. The magnitude of the car's acceleration is the acceleration due to gravity multiplied by the tangent of the angle: $a_{\text{cent}} = g \tan \theta$.
17. If the water stays in the bucket, it is because the whirling is so fast that at the top of the circle, the bucket is accelerating downward faster than the acceleration of gravity. The bottom of the bucket is needed to pull the water down! The sides of the bucket prevent the water from sloshing ahead or behind with the rest of the water that is going in the same circle. The weight is always pointing straight down (whether at the top or at the bottom), but the normal force N from the bottom of the bucket changes direction, pointing down at the top of the swing and up at the lowest point. Thus, $\sum F_{\text{top}} = mg + N$, while $\sum F_{\text{bottom}} = mg - N$.
19. The frictional force between tires and road provides the centripetal force that keeps the car moving in a circle. If the frictional force is not strong enough to keep the car in a circle, the car continues in a straight line and starts to skid.
21. A
23. A
25. A and D are both correct.
27. B
29. 0.7 for standard tires, but as high as 0.9 for racing tires
31. Assume the initial speed of the runner is 7 m/s and the slide covers 3 m. The acceleration would be approximately -8 m/s^2. Then $\mu_k = a/g \approx 0.8$.
33. A hockey puck may leave the stick of a player with a speed of 8.0 m/s and slide about 30 m before coming to a stop. Then $\mu_k \approx 0.1$.
35. For a waterslide inclined 30° from the horizontal, assume it takes about 2 s to slide 4 m down the slide. This corresponds to an acceleration of 2 m/s^2. The coefficient of kinetic friction is then approximately 0.3.
37. We can plot the force versus time.

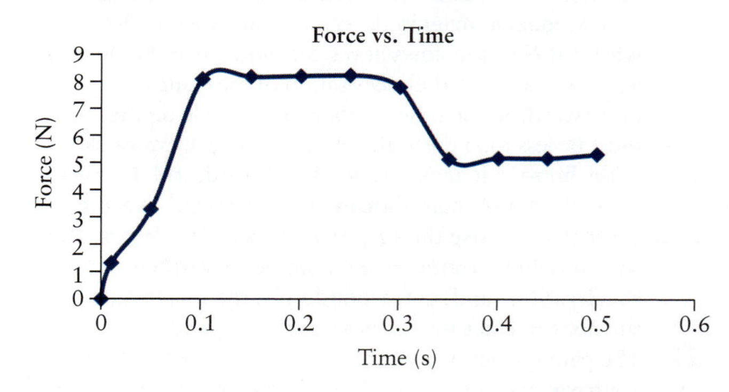

Then we see that the crate is likely stationary up until $t = 0.3$ s. So we use the average of points at $t = 0.1$ s through $t = 0.25$ s to find the coefficient of static friction. We then use the points at $t = 0.35$ s through $t = 0.5$ s for the coefficient of kinetic friction. So $\mu_k = 0.535$.

39. 41.0 N
41. 0.5
43. (a) 245 N; (b) 25.0 N; (c) 0.102
45. 28.9 N
47. 80.2 N
49. 2.20 kg
51. $a = 2.21\ \mathrm{m/s^2}$; $T = 36.1\ \mathrm{N}$
53. 6.98×10^{-9} N
55. (a) 6.17 m/s; (b) 0.274 kg/m
57. 6.06 m/s
59. (a) 1.75 N; (b) 1.26 N
61. 1.18×10^4 N
63. $9.0 \times 10^{13}\ \mathrm{m/s^2}$
65. 6.29×10^3 N
67. 0.885
69. 4.52 kg
71. The maximum available static friction is greater than the horizontal component of the leaning force which opposes it; therefore, the table will not move (but only barely).
73. 62.5 N
75. $F_{\mathrm{net},\,x} = F - 9\mu_{\mathrm{k}}Mg$
77. (a) The crate will not slide. (b) 2.16 s
79. 3.92 N
81. (a) 8.12 s; (b) 8.39 s
83. $3.51\ \mathrm{kg} < m_2 < 52.6\ \mathrm{kg}$
85. 2.19×10^4 N
87. $T = 221$ N; $v = 4.20$ m/s
89. (a) $3.5\ \mathrm{m/s^2}$; (b) 16 N

Chapter 6

1. If the magnitude of the force or the magnitude of the displacement is zero, then the work done can be zero. In addition if the force and displacement are nonzero but perpendicular, then the work done will also be zero.
3. Zero. Your roommate does positive work on the block as she lifts it (she exerts an upward force in the same direction as the displacement). Assuming she sets the block down gently on the floor, she does an equal amount of negative work in setting the block down (she exerts an upward force, but the block moves downward). Similarly your roommate must exert a horizontal force in the direction of motion, doing positive work to begin to move it across the room. But she must do an equal amount of negative work to stop the block's horizontal motion. So she does no net work on the block.
5. (a) Yes, she is expending energy. (b) No, she does no work on the boulder. The energy she is expending is being dissipated inside herself.
7. Yes, every time the kinetic energy decreases from a positive value to a smaller positive value, the change in kinetic energy is negative.
9. The orbital path of the satellite is around a circle. Its instantaneous displacement vector is tangent to the circle. The gravitational force points toward the center of the circle along the radius. The radius is perpendicular to the tangent of a circle, meaning that the displacement vector is perpendicular to the force. Therefore, no work can be done.
11. (a) When it reaches the bottom; (b) When it reaches terminal velocity. This will occur when the force of air resistance is equal and opposite to the force of gravity. This may or may not be before the rock reaches the bottom.
13. The system of you and Earth has more gravitational potential energy when you are at the top of the hill. It is harder to pedal up the hill because you are converting chemical energy in your body to mechanical energy, which is stored as potential energy as you climb the hill.
15. (a) Conservation of mechanical energy would imply that the ball comes back to exactly the same height at which it was released. The small amount of mechanical energy lost as heat through friction implies it can come back no higher than this. (b) You might be hit by the ball if you push it ever so slightly rather than just releasing it.
17. Before each firing of the toy gun, it is "loaded" by a person expending energy to compress its spring, which is held in place by the gun's trigger. The person firing the gun also expends energy in pulling the trigger, which releases the spring.
19. C
21. C
23. E
25. C
27. An 8-kg suitcase being lifted into a trunk that has an elevation of 0.75 m will gain 60 J of potential energy, but that energy is not the minimum required to get the suitcase in. The lip of the trunk is about 1.1 m above the ground, so the suitcase will at one point have an increase of 90 J in potential energy.
29. Since we know by experience that walking on level ground is far easier than climbing a mountain, we expect that most of the energy required in this walk is to achieve the change in gravitational potential energy associated with a 1.5-km increase in height. For a 75-kg person, this energy is roughly 1.1×10^6 J.
31. The cue ball has a mass of roughly 0.17 kg and travels at a speed of about 1.5 m/s, which yields 0.5 J for the kinetic energy.
33. Pumas have been seen to execute a 5.4-m vertical leap. Supposing that pumas are of average weight (62 kg), the maximum kinetic energy required is about 3300 J.
35. The recommended daily caloric intake for adult males is 2500 food calories (2000 for females). This equates to 10.5 million J (8.4 million J for females). If weight is to remain constant, the energy must be metabolized over 24 hours (86,400 s). The rate of metabolism is thus approximately 120 W for males, 100 W for females.
37. (a) 0.812 J; (b) 2.069 J; (c) 1.257 J; (d) 2.37 J
39. 8700 J
41. Moe does 6.00×10^3 J of work, Larry does -3.60×10^3 J of work, and Curly does 0 J of work.
43. (a) 2800 J; (b) -1000 J; (c) -1800 J; (d) 0 J
45. 1.25×10^{-2} J
47. -2.7×10^9 J
49. 13.3 m/s
51. 0.251
53. 4.50 m/s
55. 58.5 N
57. (a) 0.424 m/s; (b) 0.612 m/s; (c) 0.707 m/s

59. 3×10^2 J
61. 5.69×10^5 J
63. -8.79×10^4 J
65. -3.14×10^5 J
67. 9.69×10^{-2} J
69. (a) 103 J; (b) 103 J
71. 227 kJ
73. The kinetic energy has decreased by a factor of 8/9 (approximately 89%). This energy is converted into sound and heat.
75. 15.6 m/s
77. 9.66 m
79. Point B: 6 m/s; Point C: 8 m/s; Point D: 6 m/s; Point E: 4 m/s.
81. 2.05×10^3 N
83. 88.1% was "lost" to air resistance
85. 13.3°
87. Neil: $14.8\frac{\text{m}}{\text{s}}$; Gus: 14.2 m/s
89. 2.64×10^5 W
91. 54 s
93. 0.370 m
95. 7.0 m/s
97. (a) 19.9 m; (b) 9.90 m/s
99. No, there was no friction.
101. 0.882
103. (a) 24 m/s; (b) 710 J/kg is lost to nonconservative forces
105. $v_i = \sqrt{3gR}$
107. 9.2×10^2 W

Chapter 7

1. Yes, if the tennis ball moves 18 times faster
3. If an object, or a system of objects, experiences no net external force, its momentum is conserved. No external forces act on the two-sphere system; its momentum is thus conserved. Each individual sphere does experience a net external force during its collision with the other sphere; as a result, individual sphere momentum will not be conserved.
5. Look for "loss" of energy. Did the collision make a sound? Did the objects deform? If you can accurately gauge their velocities, do the objects have less total kinetic energy than before? Any of these conditions indicates an inelastic collision.
7. When two balls are raised and released on a Newton's cradle, the balls collide with the balls at rest on the cradle and transfer their energy and momentum to the two balls on the other side of the collision. The fact that two balls are always moving is required by the combination of conservation of momentum and conservation of energy. If you only applied conservation of momentum, it would be possible for the three stationary balls to move off with a lower speed than the two incoming balls. This does not happen.
9. An impulse, commonly speaking, is a brief push; or, psychologically, a spur-of-the-moment desire. This fits a common use for the physicist's notion of impulse, which is a collision. The forces in collisions will tend to be brief. Changes in momentum, of course, occur in other situations, and the term makes less sense there.
11. The force of the impact is less on carpet, which can be seen by considering the momentum (the impulse is spread over more time) or by considering energy (the work is done over more distance).
13. Pushing off the boat to step onto the pier will push the boat away. You will make less progress than if you are on solid ground and can easily fall in the gap.
15. (a) Momentum is conserved during the explosion but not before or after, since an external force (namely, gravity) is acting on the system. (b) Mechanical energy is not conserved, since much of the energy will be "lost" to heat, light, and sound. (c) The center of mass of the system will follow the parabolic path both before and after the explosion.
17. D
19. A
21. B
23. D
25. B
27. A typical car has a mass of 1000 kg. A speed of 65 mph is around 30 m/s. Therefore, the magnitude of the car's momentum is 3×10^4 kg·m/s.
29. The mass of a tennis ball is 0.06 kg, and a professional tennis player can serve at around 125 mph (55 m/s). The magnitude of the ball's momentum is around 3 kg·m/s.
31. A bumblebee has a mass of around 0.5 g and can fly at a speed of 4 m/s. The magnitude of its momentum is 0.002 kg·m/s.
33. A softball has a mass of around 0.18 kg and is pitched at around 30 m/s, giving it a momentum of about $5\frac{\text{kg}\cdot\text{m}}{\text{s}}$. A baseball has a mass of around 0.145 kg and is pitched at around 40 m/s, giving it a momentum of about $6\frac{\text{kg}\cdot\text{m}}{\text{s}}$.
35. 7 m
37. $2.0 \times 10^5 \frac{\text{kg}\cdot\text{m}}{\text{s}}$ to the east
39. Initial momentum: $-6250 \frac{\text{kg}\cdot\text{m}}{\text{s}}$; final momentum: $17{,}500 \frac{\text{kg}\cdot\text{m}}{\text{s}}$; change in momentum: $2.38 \times 10^4 \frac{\text{kg}\cdot\text{m}}{\text{s}}$
41. (a) 8 times larger; (b) 16 times larger
43. (a) 6.40 m/s; (b) She moves in the opposite direction.
45. 0.8 m/s to the right
47. $v_{1x} = 1.1v_0$, $v_{1y} = -1.1v_0$, $v_{2x} = 0.95v_0$, $v_{2y} = 0.55v_0$
49. The lighter sheep wins.
51. (a) $6.67\frac{\text{m}}{\text{s}}$ to the east; (b) before $K_i = 2.00 \times 10^6$ J, after $K_f = 6.67 \times 10^5$ J
53. (a) 9.81 m/s; (b) before $K_i = 552$ J, after $K_f = 530$ J
55. 5.99 m/s, 45.2° deflected toward the sidelines from the running back's original path
57. The 2.00-kg ball goes 1.00 m/s to the left, and the 4.00-kg ball goes 2.00 m/s to the right.
59. Ball 1's final velocity is 1.72 m/s to the left, and ball 2's final velocity is 4.28 m/s to the right.
61. 2.00 kg

63. $2.3 \times 10^2 \frac{\text{m}}{\text{s}}$

65. 19 N in the direction of initial bag velocity

67. x component = −3.6 m; y component = 0.0 m

69. $0.038R$

71. $13.5 \frac{\text{kg} \cdot \text{m}}{\text{s}}$ downward

73. (a) 8.63×10^{10} kg·m/s while the people were falling, but there is no change from before the fall to after the fall. (b) While the people were falling, Earth temporarily moved toward them at a velocity of 1.45×10^{-14} m/s.

75. 69.4 mph

77. 421 kg

79. The final velocity of the snowboarder is zero, and the final velocity of the skier is equal to $9.6 \frac{\text{m}}{\text{s}}$, in the snowboarder's original direction.

81. 76.5 km/h at 80.8° north of east

83. 488 g

85. (a) $-3.75 \frac{\text{kg} \cdot \text{m}}{\text{s}}$; (b) $-2.65 \frac{\text{kg} \cdot \text{m}}{\text{s}}$; (c) 264 N in the direction opposite to the ball's initial motion.

Chapter 8

1. One radian is the angle produced by going one radius of length around the edge of a circle. Radians should be included if the resulting quantity is intrinsically angular in nature (for example, one can include it in angular momentum) but not otherwise. For example, rotational kinetic energy is just energy, so the units are joules, not joules multiplied by radians squared.

3. We can use Table 8-1 to look up the moments of inertia for the various shapes listed. Ranking the moments of inertia from greatest to least:

$$(I_{\text{hoop}} = MR^2) > \left(I_{\text{hollow sphere}} = \frac{2}{3}MR^2\right) > \left(I_{\text{solid cylinder}} = \frac{1}{2}MR^2\right) > \left(I_{\text{solid sphere}} = \frac{2}{5}MR^2\right)$$

5. The moment of inertia of an object determines how that object responds to an applied torque.

7.

Translational motion	Rotational motion
$x = x_0 + v_0t + at^2/2$	$\theta = \theta_0 + \omega_0 t + \alpha t^2/2$
$v = v_0 + at$	$\omega - \omega_0 + \alpha t$
$p = mv$	$L = I\omega$
$\sum F = ma = \Delta p/\Delta t$	$\sum \tau = I\alpha = \Delta L/\Delta t$
$K = mv^2/2 = p^2/(2m)$	$k_{\text{rot}} = I\omega^2/2 = L^2/(2I)$

9. The magnitude of the torque exerted on the bolt is equal to $FR \sin(\phi)$, where F is the magnitude of the force, R is the distance from the rotation axis (the bolt, in this case) to the location of the force, and ϕ is the angle between these two vectors. Since F is the same in all four cases, the torque will be the largest when the force is applied to the end farthest from the bolt *and* perpendicular to the wrench (C). The torques in (B) and (D) are the next largest and are equal to one another. Finally, (A) is the smallest, since R is much smaller in this case compared to the other three. To reiterate: (A) < (B) = (D) < (C).

11. (a) iii; (b) vi; (c) i; (d) vii; (e) iv; (f) ix;
(g)

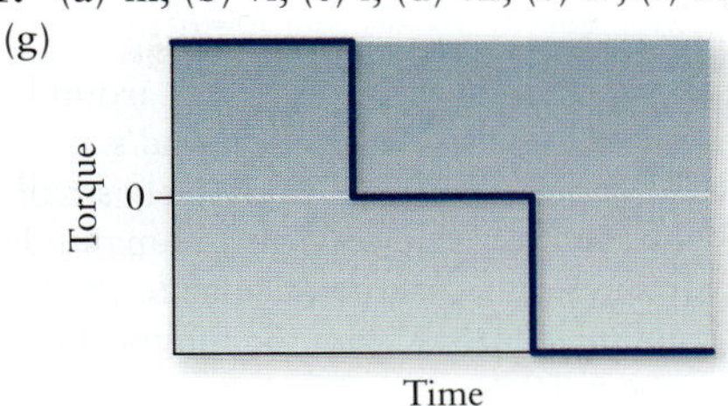

13. N·m is the same unit as energy because 1 Joule = 1 kg·m²·s⁻² = N·m. However, even though the units are the same, we cannot equate torque and energy.

15. An ice skater is changing the way her mass is distributed and, therefore, changing her moment of inertia when she moves her arms in and out. Since there are no external torques acting on the skater once she is in the pirouette, angular momentum is conserved. This is why her angular velocity changes.

17. An object moving in a straight line can have a nonzero angular momentum relative to a rotation axis as long as its velocity vector cannot be extended through the axis.

19. This collision can be considered as the rotational analogue to a completely inelastic linear collision between a moving and stationary object. No net external torque acts on the dough–turntable system; system angular momentum in the direction of the rotation axis of the turntable is thus conserved. If the dough is dropped vertically onto the turntable, its distance from the axis of rotation remains constant, and system moment of inertia is unchanged. However, precollision all angular momentum is carried by the turntable. Postcollision it is carried by both the turntable and the dough, which means that final joint angular velocity must be less than the initial angular velocity of the turntable. Since the final velocity of the dough–turntable is less than initial velocity, angular acceleration slows the turntable during the collision, at the same time speeding up the dough. These accelerations are the result of internal system torques exchanged between the turntable and dough during the collision. Until the rotational speed of the dough matches that of the turntable, kinetic friction occurs between the surfaces in contact. Nonconservative work is done during this time, reducing system rotational kinetic energy.

21. (a) They are leaning to change the torque that the force of gravity acting on them applies to the seesaw. Leaning backward increases the lever arm and hence the torque exerted by gravity due to the person at the top. Simultaneously, leaning forward decreases the lever arm and hence the torque exerted by gravity due to the person at the bottom. These two effects combine to provide a net torque that causes the seesaw to rotate, so the person at the top descends, and the person at the bottom rises. (b) If the riders are going up and down by this method and are situated equally far from the center, their masses must be quite similar.

23. B
25. E
27. C
29. B
31. B
33. The radius is around 25 m and the speed is around 15 m/s, so the angular speed is about 0.6 rad/s.
35. A typical door is about 1 m wide. A relatively small force, say 10 N, will open a door with well-maintained hinges if the force is applied perpendicular to the face of the door along the edge far from the hinges. The torque in this case is 10 N · m.
37. Female figure skaters are often around 50 kg. Model the skater's body as a cylinder of mass M_{body} equal to 40 kg and radius R equal to 0.15 m; model her arms as rods each of mass M_{arm} equal to 5 kg and length L equal to 0.50 m. When her arms are extended the moment of inertia of her body around her central axis is $I = \left(\frac{1}{2}\right) M_{body}R^2 = 0.45\ \text{kg}\cdot\text{m}^2$. Her arms each contribute $I = \left(\frac{1}{12}\right)M_{arm}L^2 + M_{arm}h^2$, where h is the distance from her central axis to the center of mass of her arms, approximately halfway from shoulder to hand. Here h equals $0.15\ \text{m} + \left(\frac{1}{2}\right)0.50\ \text{m}$ or 0.40 m. Each arm therefore contributes $I = 0.90\ \text{kg}\cdot\text{m}^2$; her moment of inertia around her central axis is therefore $0.45\ \text{kg}\cdot\text{m}^2 + 2(0.90\ \text{kg}\cdot\text{m}^2) = 2.3\ \text{kg}\cdot\text{m}^2$. With arms drawn tightly, let M_{body} equal 50 kg and neglect her arms; $I = 0.56\ \text{kg}\cdot\text{m}^2$.
39. One rotation per day yields $\omega = 0.000073$ rad/s.
41. If the ball's mass is 0.25 kg, its diameter is negligibly small and the radius of its motion is 1 m, its moment of inertia is $0.25\ \text{kg}\cdot\text{m}^2$. If it completes one revolution per second, the angular moment is $7.2\ \text{kg}\cdot\text{m}^2/\text{s}$.
43. 1.05 rad/s
45. 4.71 rad/s; 3.49 rad/s; 39.5 rad/s
47. 11.1 J
49. $19.4\ \text{km}\cdot\text{m}^2$
51. $0.72\ \text{kg}\cdot\text{m}^2$
53. The twirler is correct.
55. $(7/5)\ MR^2$
57. $0.391\ \text{kg}\cdot\text{m}^2$
59. $r = \sqrt{\frac{7}{2}}R$
61. $\frac{2}{5}MR^2 + M(L + R)^2$
63. 2.57×10^{29} J
65. 54.6 rpm
67. 3.32 m/s
69. 14 cm above the top of the loop
71. (a) $314.2\ \text{s} = 3 \times 10^2\ \text{s}$; (b) $157\ \text{rev} = 2 \times 10^2\ \text{rev}$
73. (a) $0.0438\ \frac{\text{rad}}{\text{s}^2}$; (b) 40.5 rad; (c) 3.77 m/s
75. (a) 1.3 rev; (b) 1574 rev
77. (a) 0.262 rad; (b) 6.94×10^{-4} rpm; 7.27×10^{-5} rad/s
79. 37.1 rad/s^2
81. 28 N · m
83. 102 kg
85. (a) $21.5\ \text{kg}\cdot\text{m}^2$; (b) 5.63 N · m
87. (a) $1.27\ \frac{\text{rad}}{\text{s}^2}$, clockwise; (b) 1.00 N, directed to the right
89. (a) 6.27 m/s^2 and points to the right for block 1 and downward for block 2. Tensions are 12.5 N and 14.1 N. (b) 0.847 s; (c) 133 rad/s
91. $0.0127\ \frac{\text{kg}\cdot\text{m}^2}{\text{s}}$
93. $2.68 \times 10^{40}\ \frac{\text{kg}\cdot\text{m}^2}{\text{s}}$
95. $2.19 \times 10^6\ \frac{\text{m}}{\text{s}}$
97. 0.530 m
99. $4.4 \times 10^{-3}\ \frac{\text{rad}}{\text{s}}$
101. (a) 1.72 N · m; (b) 0.392 N · m; (c) 1.72 N · m; (d) stick is still balanced
103. 3.00×10^2 N
105. The man must walk 3.50 m to achieve equilibrium. Zero.
107. $0.63\ ML^2$
109. (a) 147,000 rad/s^2; (b) 0.292 rotations
111. 2.0×10^2 rpm
113. 5.13 rad/s
115. (a) $R_{ring} = R\sqrt{\frac{2}{5}}$; (b) same angular momentum
117. (a) 14.0 rad/s; (b) 3.0 rad/s; (c) -4.0×10^2 J
119. (a) This relationship is not unique; it will hold as long as $m_1 = 2m_2$. (b) This relationship is unique.
121. (a) 2.57×10^{29} J; (b) 3.9×10^8 y; (c) 2.9×10^8 y
123. (a) Decrease; (b) increase
125. (a) $0.03\ \frac{\text{rad}}{\text{s}^2}$; (b) $3\ \frac{\text{rad}}{\text{s}}$; (c) $9g$ (d) 2×10^1 s
127. 4.3 m/s

Chapter 9

1. When an object experiences a stress below the yield strength, it should return its original shape unchanged.
3. Yes. Real cables have weight, and that weight can be sufficient to break the cable.
5. The brass wire stretches by $\frac{4\ \text{mgL}}{\pi D^2 Y_{brass}}$, and the steel wire stretches $\frac{4\ \text{mgL}}{\pi D^2 Y_{steel}}$.
7. The skin loses some of its elasticity as people age and the tension in the skin decreases. Since the skin is not held as tightly, it will wrinkle.
9. When determining the volume of an object by completely immersing it in a liquid, we assume that the object does not absorb any of the liquid. In addition we assume a high bulk modulus for the object, so its volume does not appreciably decrease when subjected to the increased pressure associated with immersion in the liquid.
11. (a) Young's modulus accounts for the stretching or compressing of an object in one dimension. The bulk modulus describes the expansion or compression of an entire volume. (b) Both of the variables have the same units (Pa, MPa, GPa, and so on).

13. The shear modulus is a measure of how much an object will deform under a given stress. The larger the shear modulus, the less it will deform; it is more rigid. So the term "rigidity" applies because a material will deform less if it has a larger shear modulus.

15. The compressive stress due to the top of a cone is smaller than the compressive stress due to a cylinder.

17. Yield strength is the tensile stress at which a material becomes permanently deformed.
Ultimate strength is the maximum tensile stress the material can withstand before failure.

19. Leg bones need to hold form in order to fulfill their role in holding the body together; they are cushioned by hooves and cartilage. The small strain of the leg bone means that the bones do not change length by a large amount as weight is first placed on a leg and then taken off the leg. Antlers, on the other hand, need to be able to bend a great deal in order to survive the strong impacts to which they are subjected.

21. B

23. C

25. A

27. A

29. A

31. 10 GPa (approximately the same as for oak wood)

33. Assuming the breaking strength is about 10 MPa and the diameter of your bone is about 1 cm, this leads to about 800 N; a value of 1000 N or more is probably accurate.

35. 3.5×10^7 N

37.

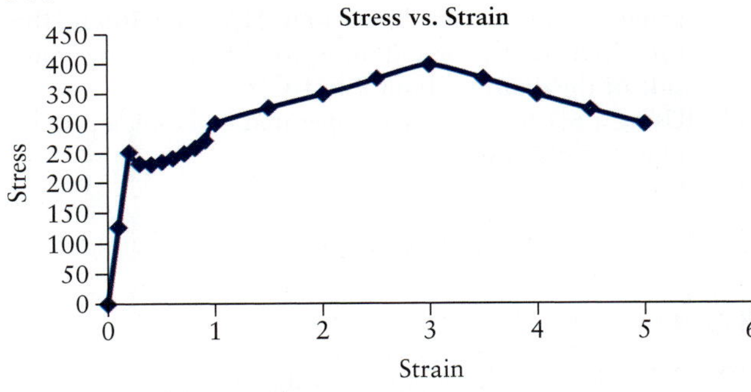

(a) Yield strength = 250 GPa;
(b) ultimate strength = 400 GPa;
(c) Young's modulus = 1250 GPa

39. 3.4×10^3 N

41. (a) 9.5×10^5 N; (b) strain = 0.14

43. Brass

45. (a) The strain on the aluminum bar is 2.9 times the strain on the steel bar. (b) The strain is independent of the relative lengths of the bars.

47. 0.27 MPa

49. (a) 64.2 mm; (b) 6.42 mm

51. 2.0×10^8 Pa

53. 1×10^7 N

55. 2.0

57. 6.63 m

59. 1.6 kN

61. (a) 5.0; (b) No, they cannot use this rope.

63. 0.02 m = 2 cm

65. 1.3 kN; 2.9×10^2 lb

67. (a) 1.2×10^8 strands; (b) 2.2 cm; yes, this seems reasonable.

69. (a) -0.0023 cm^3; (b) -0.082 cm

71. 5.4×10^{-5} m = 54 μm

73. $1.9 \times 10^5 \dfrac{\text{N}}{\text{m}^2}$

75. (a) 0.69 cm; (b) 0.040 cm; (c) $3.5 \times 10^2 \dfrac{\text{N}}{\text{m}}$; (d) A spring constant of $3.5 \times 10^2 \dfrac{\text{N}}{\text{m}}$ is comparable to everyday springs in the lab.

77. (a) 9.8×10^3 N; (b) 0.34 cm^2

79. $3 \times 10^8 \dfrac{\text{N}}{\text{m}^2}$

81. $2.5 \times 10^4 \dfrac{\text{N}}{\text{m}^2}$

Chapter 10

1. Newton reasoned that the Moon, as it orbits, falls back toward the center of Earth due to the pull of gravity. Incrementally, it will move tangent to the orbital path; then it will fall back to Earth. The combined motion leads to the familiar circular path. An apple falls straight toward Earth without the tangential motion. The vertical motion is the same for both the apple and the Moon.

3. (a) When the mass of one object is doubled, the force between two objects doubles. (b) If both masses are halved, the force between two objects is one-fourth the original value.

5. The mass cancels when applying $F = ma$ to get the acceleration.

7. Orbiting the Sun is an incomplete description of the Moon's motion, but the Moon is definitely orbiting the Sun as a part of the Earth–lunar system.

9. The gravitational potential energy is negative because we choose to place the zero point of potential energy at $r = \infty$. With that choice the gravitational potential energy is negative because the gravitational force is attractive. The two objects will gain kinetic energy as they approach one another.

11. Humans had to evolve to support their body against the force of gravity (human legs had to evolve to support the weight of a human). The cardiovascular system had to evolve to be able to move blood back from the legs to the heart against the force of gravity. The way humans run is also evolved to the particular gravity of Earth. There are many other examples.

13. The most noticeable effects would be the comparison between the dynamics at relatively small distances (such as the Earth–Moon system) with the dynamics at extremely large distances (such as galaxies and galaxy clusters). The deviation in the inverse square law would result in a different value for G or M. Also noncircular orbits would not be elliptical, although if the power is close enough to 2, the difference in orbit could be too small to notice without watching for a very long time.

15. From Kepler's law the larger the orbit, the longer the period.

17. B
19. E
21. A
23. B
25. A
27. Assuming it is possible to measure a force of about 10^{-7} N, about 1 cm center-to-center separation
29. 104 N; as a result the gravitational attraction between the two skyscrapers can be neglected during engineering calculations.
31. 0.14 m; the escape speed would be $10^{-4}\ \frac{\text{m}}{\text{s}}$.
33. 4.00×10^{-10} N
35. 9.35×10^{-16} N
37. The force of attraction to Earth is 9.80 N. The force of attraction to the Moon is 3.44×10^{-5} N.
39. 2.37×10^{20} N toward the Sun
41. The point between Earth and the Moon where the net gravitational force due to the two is zero is 3.46×10^{8} m from Earth and 3.83×10^{7} m from the Moon.
43. 1.91×10^{28} J
45. (a) $1.12 \times 10^{4}\ \frac{\text{m}}{\text{s}}$; (b) v_{escape} would be smaller.
47. (a) -9.1×10^{24} J; (b) -9.4×10^{24} J; (c) 29 $\frac{\text{m}}{\text{s}}$
49. (a) 1.51 h; (b) 15.9 $\frac{\text{sunrise}}{\text{day}}$
51. 70.2 h
53. 1.43×10^{12} m
55. 1.99×10^{30} kg
57. 0.759 y
59. (a) Europa: 3.551 d; Ganymede: 1.071×10^{9} m = 1.071×10^{6} km; Callisto: 1.883×10^{9} m = 1.070×10^{6} km; (b) 1.90×10^{27} kg
61. (a) 5.67×10^{5} m = 5.7×10^{2} km; (b) weight in orbit = 6.9×10^{2} N; weight on Earth's surface = 8.2×10^{2} N
63. (a) 5.3 km; (b) 7.5 km
65. (a) 0.923 au, or 1.38×10^{11} m; (b) 324 d; (c) fastest at the perihelion and slowest at the aphelion; (d) 1.47
67. 1.9×10^{41} kg

Chapter 11

1. The dam could be the same thickness.
3. The collection bag should be held below the body so that the blood can flow down into it.
5. While jumping up the person will have very little blood pumped to the brain, as most of the blood will be in the lower trunk and in the legs due to gravity. This may result in a fainting spell. Once the person is horizontal the blood will evenly distribute in the body and will reach the brain, and consciousness will be regained.
7. The pressure in a fluid depends only on the depth. Since the height of the water is the same in both vessels, they will have the same pressure at the bottom. The total weight of the fluid is not directly related to the pressure.
9. No. The pieces could be different sizes.
11. The water level will fall. The boat displaces a volume of water equal to the weight of the boat. If the boat is removed from the water, the water will no longer be displaced and the water level will fall.
13. The equation of continuity applies, assuming no water is lost or gained along the way. The speed will increase when the cross-sectional area decreases. Therefore, the speed of the water in the wide valley will be slower than the speed of the water in the narrow channel.
15. The landing planes should approach from the east (into the wind). This allows pilots to use the wind to provide lift so that they can cut the plane's engines and drift down in a controlled manner. If the wind is at their back, pilots will have to maintain airspeed with the engines to stay airborne. In addition, the speed of the plane at impact will be much greater and require significantly more braking to come to a stop.
17. Horizontal
19. D
21. B
23. A
25. A
27. A wind that is 33 m/s (barely hurricane level) passing over the top of a 1-m^2 prone body that is on top of still air results in a lift of approximately 700 N.
29. The Hoover Dam is approximately 700 ft (213 m) tall and approximately 1200 ft (366 m) long. Pressure increases linearly with water depth, so we can take an average pressure at mid-depth on the water side, which is atmospheric pressure plus roughly 10^{6} Pa. Approximating the dam as a flat surface (admittedly a rough estimate) with dimensions 213 m by 366 m, the total hydrostatic force (that is, total force on the water side of the dam) is about 9×10^{10} N.
31. Using a 6:1 hydraulic jack operated by hand, a 300-kg object could easily be lifted.
33. 62 g
35. (a) $5.9 \times 10^{2}\ \frac{\text{kg}}{\text{m}^3}$; (b) perhaps fresh wood; (c) 0.59
37. 42 L
39. $3.7 \times 10^{2}\ \frac{\text{kg}}{\text{m}^3}$; the container will float.
41. $5.50 \times 10^{3}\ \frac{\text{kg}}{\text{m}^3}$
43. (a) 3.11 MPa; (b) 311 MPa
45. 2.0×10^{5} Pa
47. 1.0×10^{2} m
49. 880 mmHg
51. 1.01 atm
53. 550 kPa
55. 1.0 kN
57. 3.79 MN toward the inside of the drum
59. 2.88×10^{1} N
61. 0.183 m
63. 9.76 cm
65. 7 people
67. 12.8%
69. $1.0 \times 10^{4}\ \frac{\text{m}}{\text{s}}$

71. $6.25v_1$
73. 1.25 h
75. 2.49×10^5 Pa
77. 6.29 m
79. 1.21×10^3 Pa
81. $0.291\frac{\text{L}}{\text{s}}$
83. (a) 5.3 kg; (b) 9.6×10^{12}
85. $1.79 \times 10^1 \frac{\text{mi}}{\text{kg}}$
87. (a) 1.90×10^6 Pa; (b) 1.89×10^2 m
89. 194/154
91. 18.8 m/s
93. 2

Chapter 12

1. Simple harmonic motion also requires that the displacement point be in a direction that is opposite to the force.
3. Oscillatory motion includes simple harmonic motion, but it also includes circular motion, decaying oscillations, and oscillations that have shapes other than pure sine waves.
5. (a) Radians/second; (b) radians
7. It is an offset of where in the oscillation cycle we choose to set the zero time. If, for example, the phase angle is π rather than zero, the solution starts at $-A$ instead of $+A$. In either case, however, the initial velocity is zero, and the motion is the same.
9. In simple harmonic motion, potential energy is at a maximum when kinetic energy is at a minimum. One quarter-cycle later, potential energy is at a minimum, and kinetic energy is at a maximum. The phase angle between kinetic and potential energy is thus 90°, or $\pi/2$ radians.
11. A simple pendulum is made from a long, thin string that is tied to a small mass. We can treat it as a point mass attached to the end of a string. A physical pendulum has a massive connecting structure from the pivot point to the end; that is, there is a continuously varying mass spread over the length of the pendulum.
13. The force of gravity depends on elevation and determines the period of the pendulum. Measuring the period and length of the pendulum will yield enough information to estimate the strength of the gravitational force; doing so very precisely would enable useful comparison to established values.
15. The frequency of the driving force is just how often the applied force repeats. That depends on things outside the oscillator. The natural frequency of the oscillator is the frequency it oscillates at most readily or the frequency it will oscillate at if displaced from equilibrium and released.
17. B
19. E
21. A
23. D
25. B
27. (a)

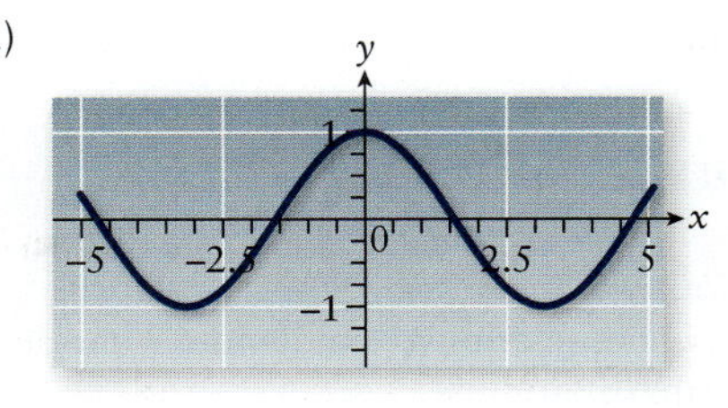

(b) The graph is shifted 30° ($\pi/6$) to the left.
29. Swings are typically around 2 m in length, so we get a period of around 3 s.
31. 50 Hz
33. A yo-yo is about 1 m long when unraveled, so the period would be about 2.0 s if we ignore the mass of the shoelace.
35. The fly landing on the pendulum would reduce the frequency of the pendulum by an almost imperceptible amount ($\Delta f \approx 3 \times 10^{-6}$ Hz $\approx 5 \times 10^{-4}$%), one that would be noticeable only over a significant length of time.
37. A broom that has a 1.5-m-long, 0.5-kg handle with moment of inertia 1/3 (mL^2) about the end of the broom handle as well as a 0.5-kg broom head with moment of inertia mL^2 will have a period of about 2.3 s.
39. About 3 years
41. 80 Hz
43. (a) 0.067 s; (b) 1.8×10^3 cycles
45. (a) $x(t) = A\cos((k/M)^{1/2}t)$; (b) $-\frac{A}{2}\sqrt{\frac{3k}{M}}$; (c) 0
47. 0.379 S
49. (a) 3.16 J; (b) 14 cm; (c) 0.790 J
51. (a, c)

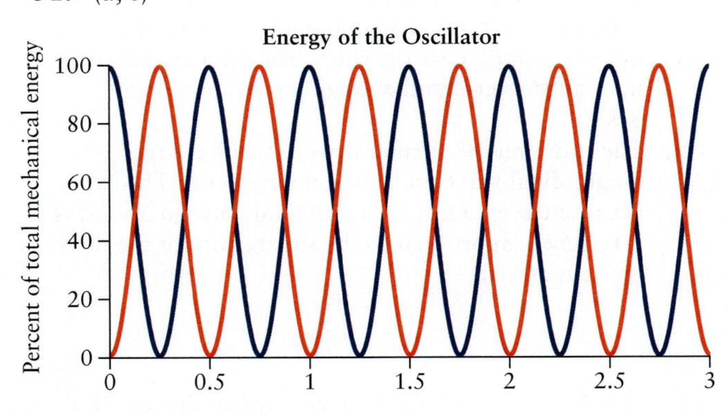

(b) $v(t) = -\omega A\sin(\omega t)$
53. 1.26 m
55. 0.792 m
57. (a) 2.010 s; (b) 1.001 T_0
59. The time between $\pm$ 8.0° is 0.71 s. The time between $\pm$ 4° is 0.24 s.
61. 2.5 s
63. 0.53 s
65. (a) 0.36 Hz; (b) 0.25; (c) 0.35 Hz; (d) 9.1 s
67. 1.0 m
69. $3.2\ \frac{\text{m}}{\text{s}}$
71. (a) 0.41 m/s; (b) 0.74 m/s^2
73. (a) 1.55 m/s; (b) 7.5 cm or -7.5 cm from equilibrium
75. 30.0 loss per cycle

77. (a) 57 cm; (c) 0.90 m
79. 1.63 s
81. $(3.09 \text{ cm}) \sin\left(\left(114\frac{\text{rad}}{\text{s}}\right)t\right)$

83. (a) 35 N/m; (b) 17.6 m
85. 0.85 rad; the oscillations are not simple harmonic.
87. (a) 1.45 cm; (b) no

Chapter 13

1. Longitudinal waves are those motions where the displacements are along the axis of wave propagation; transverse waves are those where the displacements are perpendicular to that axis. Longitudinal waves include sound and waves along a spring produced by extending and contracting it. Transverse waves include the vibrations of string instruments and "the wave" in a stadium.

3.

wavelength
amplitude
trough
crest

5. A P wave moves the rock through which it moves along the direction of propagation much like a sound wave through air. This push and pull can cause tall buildings to sway and perhaps topple. S waves cause up–down displacements, which could cause objects to fall off shelves. In earthquakes the most damage results from waves that propagate along the surface, in a way similar to waves on the surface of the ocean. Surface waves can be transverse undulations or, like ocean waves crashing near the shore, rolling waves that have both a transverse and a longitudinal component.

7. Yes

9. The rod with the smaller density will have the longitudinal wave with the greater speed. This is because the speed of a longitudinal wave in a solid is inversely proportional to the square root of the density of the material:

$$v_p = \sqrt{\frac{Y}{\rho}}$$

11. The phase difference is how two waves line up. If it's 0, then they line up perfectly and add to each other directly, resulting in a wave with double the amplitude of either individual wave. If it's 180°, the two waves cancel perfectly, resulting in no wave. If it's 90° or 270°, then they add to each other, resulting in a wave with $\sqrt{2}$ times the amplitude of a single wave and phase directly in between that of the original two waves. If it's 300°, then again they add differently, but the result is a wave with $\sqrt{3}$ times the original amplitude of either individual wave, and a phase again directly between the two original waves.

13. (a) A transverse standing wave is a pair of transverse waves with opposite wave vectors, with the result that the wave does not appear to travel anywhere. Indeed, its shape does not even change except by growing and shrinking. (b) So long as one particular pair of opposite wave vectors dominates, the result will be a standing wave. This can happen if the string is, say, plucked in the middle. The even modes will not be excited by this motion, and the higher odd modes will decay faster than the first. At that time the result will be a standing wave. Alternately, the string could be shaken periodically at the frequency of the first mode for a time.

15. There is no physical way that an antinode on the open end can have another antinode on the closed end, so there are only odd numbers of $\frac{1}{4}$ waves allowed in the standing wave patterns for closed pipes.

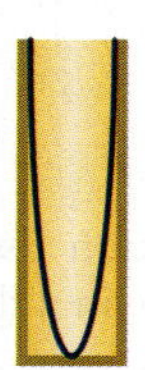

17. (a) Intensity is the power per unit area that is emitted by a source of sound. It has nothing to do with the reception of the sound through the process of "hearing." The units are W/m^2. Sound level is a mathematical relationship that puts the intensity onto a logarithmic scale. The units are decibels (dB). Loudness is the physiological response to sound waves. Different frequencies with the same intensity will be more sensitively heard by humans. Power is the energy per unit time emitted by the source. (b) Intensity depends on the inverse of the distance squared, so intensity increases as the source of sound moves closer. (c) Sound level depends on the log of intensity, so sound level increases as the source of sound moves closer. (d) The total power emitted by the source does not change as it moves closer to the observer.

19. A rider on the train, moving at the same speed as the horn was when it emitted the sound, will not experience a Doppler effect on the whistle because of that shared velocity. There can be slight differences in the two velocities, though, from building up speed, braking, or going around a curve. In such cases there will be a very slight Doppler effect.

21. B
23. D
25. B
27. D
29. C

31. (a) $\Delta x \approx \left(343\frac{\text{m}}{\text{s}}\right)(10 \text{ s}) = 3430 \text{ m}$; (b) The speed of sound depends upon the temperature of air. For example, at 0°C the speed of sound is about 331 m/s, and at 10°C the speed is 337 m/s. So if the temperature starts at 20°C and drops 10°, the distance will decrease by about 60 m (or about 1.75%). If the temperature drops another 10°, the distance will decrease another 60 m for a total of about 3.5%. Note that this does not take into account the changes due to wind.

33. (a) If the perimeter of the arena is around 500 m, it will take about 30 s for the human wave to make it around the entire arena. This corresponds to a speed of about 17 m/s. (b) The wavelength might be related to the size of the group of people who stand or raise their hands at any one time. The frequency would be the number of times people raise their hands per unit time interval. The amplitude might be related to how high their hands are raised.

35. The wavelength is equal to the speed multiplied by the period. In this case it should be on the order of a few centimeters.

37. (a) Typically waves in the open water of the ocean have amplitudes of about 5 m, a frequency of about 0.05 Hz, and wavelengths of about 150 m. (b) $v_p = \lambda f = (150\ \text{m})(0.05\ \text{Hz}) = 8\ \dfrac{\text{m}}{\text{s}}$

39. Frequency: 1.000×10^2 Hz; angular frequency: $6.283 \times 10^2\ \dfrac{\text{rad}}{\text{s}}$

41. (a) $y(x, t) = (0.20\ \text{m}) \sin\left((17.95\ \text{m}^{-1})x - \left(12.57\ \dfrac{\text{rad}}{\text{s}}\right)t\right)$;
(b) $y(x, t) = (0.20\ \text{m}) \cos\left((17.95\ \text{m}^{-1})x - \left(12.57\ \dfrac{\text{rad}}{\text{s}}\right)t\right)$;
(c) $y(x, t) = -(0.20\ \text{m}) \cos\left((17.95\ \text{m}^{-1})x - \left(12.57\ \dfrac{\text{rad}}{\text{s}}\right)t\right)$;
(d) $y(x, t) = (0.20\ \text{m}) \cos\left((17.95\ \text{m}^{-1})x - \left(12.57\ \dfrac{\text{rad}}{\text{s}}\right)t + 0.927\right)$

43. $y(x, t) = (0.10\ \text{m}) \sin\left((\pi \text{m}^{-1})x - \left(20\pi\ \dfrac{\text{rad}}{\text{s}}\right)\text{t}\right)$

45. (a) 1.0 atm; (b) 6.0 m^{-1}; (c) 0.64 Hz; (d) 0.67 m/s

47. $y(x, t) = 2 \sin\left(\pi x - \dfrac{\pi}{2}t + 0.85\right)$

49. 2

51. $2.2\ \dfrac{\text{kg}}{\text{m}}$

53. $1.86 \times 10^3\ \dfrac{\text{m}}{\text{s}}$

55. $1.03 \times 10^3\ \dfrac{\text{kg}}{\text{m}^3}$

57. Two waves of equal amplitude A:

One wave of amplitude 3A, one of the amplitude $-A$:

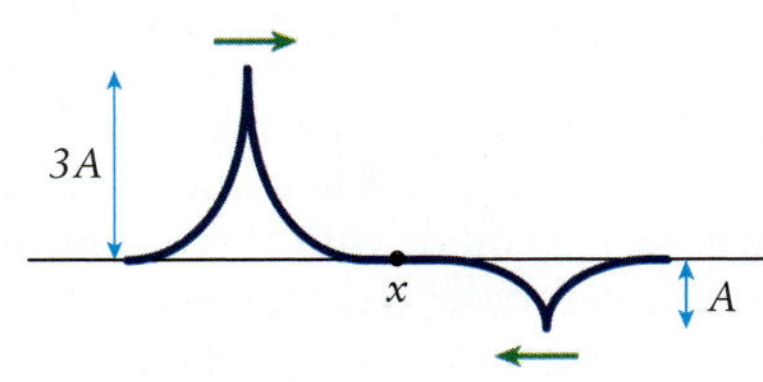

One wave of amplitude 0.5A, one of the amplitude 1.5A:

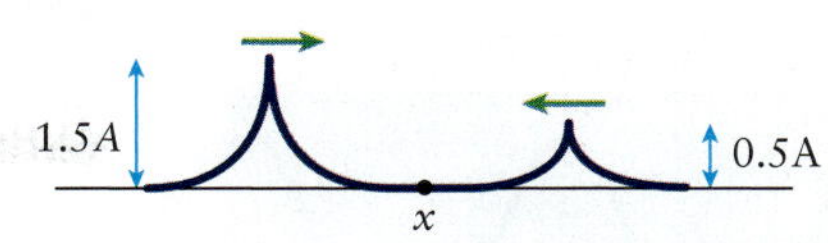

59. (a) The resulting waveform will have the same width and general shape as the incoming waveforms but an amplitude of $A = A_1 + A_2 = 2A$; (b) The resulting waveform will have the same width and general shape as the incoming waveforms but an amplitude of $A = A_1 + A_2 = 1.5A$; (c) The resulting waveform will have an amplitude of $A = A_1 - A_2 = 0$; (d) The resulting waveform will have the same width and general shape as the incoming waveforms but an amplitude of $A = A_1 - A_2 = -A/2$.

61. 2.9 m

63.

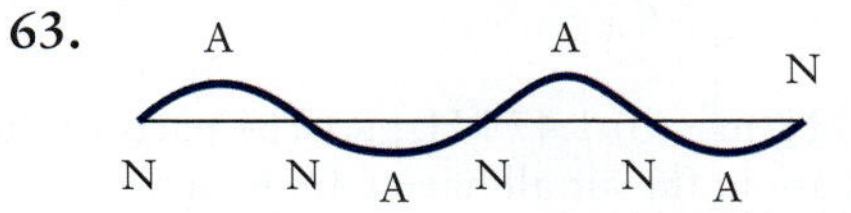

65. (a) $f_1 = 27.0$ Hz; $f_2 = 54.0$ Hz; $f_3 = 81.0$ Hz; $f_4 = 108$ Hz
(b)

67. 11.6 kg

69. 70.5 N

71. Open at both ends

73. The length is 3.38 m, and the pipe is open at both ends.

75. 4.9 m. That's definitely on the large size but not outrageous.

77. 8.79 to 44.49 kHz

79. (a) 5 Hz; (b) Try tightening the string just a little tiny bit. If that makes the beat faster, loosen it a bit.

81. 1.04×10^2 N

83. (a) 3.16; (b) 1.73; (c) 1.41; (d) No difference, except that you may be able to get a larger factor decrease by moving to the side rather than directly away from the source.

85. 43 J per day. She'd do better by completely replacing the plan with something else. Barring that, she could move the collector closer to the highway (above the sound blocking barrier would be best), or make it much larger (much of the cost would be the electronics rather than the collector). Focusing wouldn't do much in the horizontal direction as the source is diffuse, but vertical focusing reflectors could help.

87. 102.8 dB

89. (a) 1.26; (b) 100

91. 50 μW

93. 53 Hz

95. 2.11 m/s

97. 0.33 m/s

99. (a)

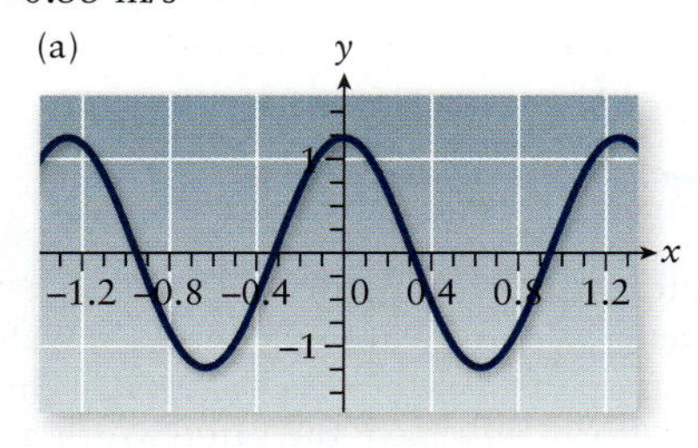

(b)

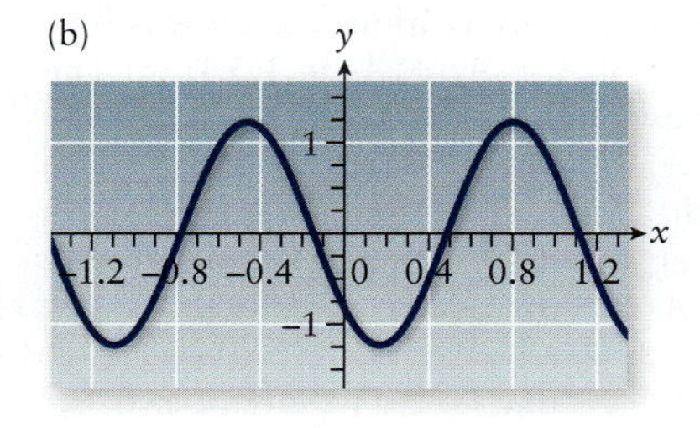

101. (a) $1.69 \times 10^5\ \text{N/m}^2$; (b) 1.47 kHz; (c) The parts of the resonator that are in the air all have a frequency increase from the substitution of some helium into the air; the parts of the resonator that are in the body do not. This produces a difference mix of frequencies and a different general sound.

103. (a) 1760 Hz; (b) $T/4$

105. 32°C

107. 69.0 dB

109. (a) 1.1 mm; (b) 360 J

111. 2.98 m/s

113. $f_0 = 390$ Hz; $f_1 = 1200$ Hz

Chapter 14

1. If A is in thermal equilibrium with B, and B is in thermal equilibrium with C, then A is in thermal equilibrium with C. This is used to establish that temperature is actually a property of objects.

3. The temperature at which the pressure of all gases should become zero is −273.15 K. This value, the same for all gases, sets the zero point of the Kelvin temperature scale.

5. It is true that a 1-degree Celsius (or Kelvin) change is equal to a 1.8-degree Fahrenheit change, but the length of a thermometer is based on the maximum and minimum values it is designed to read, not the temperature scale it employs.

7. Thermometer A (200–270 K) can be used as a freezer gauge. Thermometer B (230–270 K) can be used as a meteorological gauge (for outdoor temperatures). Thermometer C (300–550 K) can be used as an oven gauge. Thermometer D (300–315 K) can be used as a medical thermometer for humans.

9. Place the thermometer in an ice bath and let it come to equilibrium. The level of the mercury corresponds to 0°C. Then boil some water, place the thermometer in it, and let it reach equilibrium. The level of the mercury in this case corresponds to 100°C. Then we can divide the distance between the 0 mark and the 100 mark by 10 to get the marks for 10°C, 20°C, and so on. Dividing these regions into 10 again will give the individual degree markings.

11. Since the two gases are in thermal contact for a long time, they will eventually reach thermal equilibrium, which means $T_A = T_B$. From the ideal gas law, $pV = NkT$, so $p_A V_A = p_B V_B$. We cannot say anything about the individual values of the pressure and volume, only their product.

13. (a) $PV = NkT$ is most appropriate if one is counting molecules (deriving a quantity). $PV = nRT$ is most appropriate if one is counting moles (analyzing lab results). (b) P equals the pressure of the gas, V is the volume the gas occupies, N equals the number of molecules, n is the number of moles of molecules, k is the Boltzmann constant, R is the universal gas constant, and T is the temperature of the gas.

15. In addition to making repairs easier, the gaps between smaller sections allow for more expansion during warmer months.

17. The wall lengthens by 0.281 cm.

19. The reason that so many people live near coastal regions (versus in landlocked locales) is related to this thermodynamic argument. The temperature range in landlocked regions is much larger because the specific heat of dirt is so much lower than that of water. The average temperature in Kansas City is about 12°C, and the average in San Francisco is 13°C. However, maximum and minimum temperatures range from below −18°C to above 38°C in the Midwest, while in the San Francisco area the range is from −1°C to less than 38°C. The water that surrounds coastal cities tends to balance the temperature changes due to the large value for water's specific heat. The average temperature of a coastal city is basically the average temperature of the ocean right off the coast.

21. The word "latent" means "hidden." In the context of latent heat, a system is absorbing or releasing heat, but the temperature is not changing while a phase change is taking place.

23. Heated air expands and becomes more buoyant. Cooler, denser air near the ceiling descends, displacing the warmer air and pushing it upward. Thus the warm air rises and the cooler air falls.

25. Radiation is heat transfer by electromagnetic waves through a medium or through empty space. Convection is heat transfer by the motion of a fluid. A warm, high-energy region of the fluid is less dense than the colder fluid around it. The cold fluid sinks and pushes the warmer fluid upward. As the warm fluid rises, it carries heat energy with it. Conduction is heat transfer by the collision of particles from one material to another or within a single material. When faster-moving particles in a warm region collide with slower-moving particles in a neighboring cooler region, some of the energy of the faster-moving particles is transferred to the slower-moving particles.

27. Fiberglass insulation resists the flow of heat energy because it provides a maximum amount of air pockets in the space between interior and exterior walls. Since air has a very low thermal conductivity, this inhibits the thermal energy from leaving our homes in the cold months and entering during the hot months.

29. B.
31. A
33. D
35. A
37. C
39. 2.03×10^{22} J
41. 4.493×10^{6} J
43. 2 or 3 min
45. 30 kJ
47. The final temperature is 84% of the initial temperature.
49. Conduction carries away about 200 times more heat.
51. (a) \$0.10 per kg; (b) \$0.75 per kg
53. (a) 82°F; (b) 14°C; (c) −229°F; (d) 290 K; (e) 99°F
55. $T_F = \left(\frac{9}{5}\right)T_C + 32$
57. Fahrenheit: $T_{F,\,low} = 98.1°F$; $T_{F,\,high} = 98.6°F$
Kelvin: $T_{K,\,low} = 309.8$ K; $T_{K,\,high} = 310.2$ K
59. (a) $T_B < T_A$; (b) $T_B > T_A$
61. 6×10^{3} J
63. 17.0 L
65. $872\ \frac{m}{s}$
67. 1.3×10^{9} m
69. 70.4 K
71. 1.78×10^{-4} m
73. 8.04 cm
75. (a) Length: 20.015 cm; width: 30.022 cm; (b) increases by 0.15%
77. (a) 2.7144 cm; (b) 2.7118 cm
79. 0.004 m
81. 8.90 g/cm^3
83. 78.5 kJ
85. 3.85×10^{5} J
87. 8.12×10^{9} kg
89. 1110°C
91. 35.0°C
93. 19 strokes
95. 79.4 kJ
97. 248 kJ
99. 0.175 kg
101. 235 g
103. The iron melts 7.86 g of ice, the copper melts 10.1 g of ice, and the water melts 14.6 g of ice. Water melts the most ice.
105. A factor of 18.32 times
107. 65 solar radii or 4.5×10^{10} m
109. 6.12 kW
111. 2.9×10^{6} J
113. 8.20×10^{-21} J
115. (a) $1.2 \times 10^{26}\ \frac{\text{particles}}{m^3}$; (b) Titan; (c) 3.3×10^{-8} m
117. Heat up just the aluminum to 470°C, while leaving the copper at room temperature.
119. 346 μW
121. 7870 J must be removed.
123. (a) 7.1×10^{20} J; (b) 5.4×10^{12} gal; (c) 3.3×10^{8} tons
125. 3250 kcal, not unreasonable for a man of this size
127. (a) The stored energy is proportional to the animal's mass, which is proportional to its volume, which is proportional to the cube of its radius $\left(V = \frac{4}{3}\pi R^3\right)$. The rate at which the animal radiates energy is proportional to its surface area, which is proportional to the square of its radius ($A = 4\pi R^2$). (b) The fraction of its stored energy that it radiates away per second is $\left(\frac{\Delta E}{\Delta t}\right)/E$. The numerator is proportional to R^2, and the denominator is proportional to R^3, so the radiated ratio is $\left(\frac{\left(\frac{\Delta E}{\Delta t}\right)}{E} \propto \left(\frac{R^2}{R^3}\right) \propto \left(\frac{1}{R}\right)\right)$.
(c) For a small animal (small R) the fraction of its energy radiated per second is larger than it is for a large animal because that fraction is inversely proportional to R, as shown in part (b). To replace this energy in order to maintain a constant body temperature, the small animal must eat more frequently than the large animal.
129. 0.129 m^3 of water leaks out of the sphere.

Chapter 15

1. Yes, a system can absorb heat without increasing its internal energy, for example, if the heat absorbed by the system is equal to the work done by the system.
3. Yes, this is consistent with the first law of thermodynamics because there is no change in internal energy for an isothermal process. Therefore, the work done by the gas must equal the heat put into the system.
5. The work done on the gas to compress it results in an increase in temperature because there is insufficient time for heat to flow out of the system. Thus the internal energy of the system increases, and the temperature increases.
7. High efficiency is almost always a primary objective for a steam-electric generating plant. In such a plant the steam is the working substance for the heat engine that drives the electric generator. The temperature of the steam is the temperature of the high-temperature heat source. The temperature of the low-temperature reservoir is usually fixed by circumstances, such as the temperature of a nearby lake. Thus increasing the feed-steam temperature is the only way to increase the Carnot efficiency limit for the generator.
9. No. Although they are pretty close, there is always some irreversible thermodynamic change that accompanies the running of real-world engines: A bit of the metal wears away, there is an increase in the friction between the piston and the cylinder, and so on. We spend lots of effort trying to make sure these losses are minimized, especially with lubrication, but we will never create the "perfect" engine that is not degraded by continued use.

11. The irreversible processes are a bit like one-way streets—you can go around the block, but you cannot take a U-turn. More concretely, general heat loss is undone not by letting the heat back into the hot object but by finding something even hotter or by doing work on the object to heat it; the halting of a ball that rolled to a stop is not undone by letting the grass kick it back into motion but by going and fetching the ball. Note that the reversible adiabatic cooling does not let heat escape to the environment but rather stores the thermal energy as mechanical energy in another part of the system.

13. A coefficient of performance only gauges how efficiently existing energy is shifted from one place to another; it does not indicate violation of conservation of energy.

15. It increases by the amount of the heat divided by the temperature of the cool object $\left(\frac{Q}{T_c}\right)$, minus the amount of heat divided by the temperature of the hot object $\left(\frac{Q}{T_h}\right)$. When there was a temperature difference, that difference could be harnessed to do work. Once the two objects have equilibrated, that way of extracting work is no longer possible because the difference no longer exists.

17. The entropy of 1 kg of liquid iron is larger than that of 1 kg of solid iron because the atoms in liquid iron are less ordered than those in solid iron. Also, heat had to be added to solid iron in order to melt it.

19. As the piston recedes it will sap energy from the gas molecules striking it. This lowers the gas temperature, requiring it to be reheated by the hot reservoir—but that is then not isothermal. The slower the piston recedes, the less energy it saps, but if the engine is to actually run at all, it does have to move. By not acting isothermally, there is heat, so Q/T will be nonzero, and thus there is an entropy increase.

21. B

23. B

25. A

27. D

29. A

31. (a) Raising a 1-kg book 1 m would take about 10 J. (b) This would raise the temperature of a glass of water by 0.02°C.

33. About 40 J.

35. The number of runners in the New York City Marathon is approximately 50,000. Trained runners expend about 75 kcal/km (depending on the mass of the runner), so each runner will expend about $75\ \text{kcal/km} \times 40\ \text{km} = 3000$ kcal per marathon race. Thus the total amount of energy expended by all the runners is about $3000\ \text{kcal/runner} \times 50{,}000$ runners = 150 million kcal $= 6 \times 10^{11}$ J.

37. 500 km

39. $e = \dfrac{T_{body} - T_{environment}}{T_{body}}$; using $T_{body} = 310$ K and $T_{environment} = 293$ K, e is about 5.5. $[310 - 293]/310 = 5.5$. This is far lower than real human body efficiency; the body is not well modeled as a heat engine.

41. 300 J

43. 300 J

45. 0 J

47. 8.80 kJ

49. 9.22 kJ

51. −800 kJ

53. −321 kJ

55. 2.91 kJ

57. 19.2 g

59. 1.4

61. 27.5

63. 40.0%

65. 14.8%

67. 13.7 kJ

69. 3.41

71. (a) 5.43; (b) 225 J; (c) −24°C

73. 245 J/K

75. −16.9 J/K

77. 1.48 kJ/K

79. 25.9 J/K

81. 2.88 J/K

83. 1.83×10^6 kg

85. (a) 47.3°C; (b) 31.7 J

87. The efficiency drops from 10.5% to 9.66%.

89. −603 J/K

91. (a) 15.8%; (b) 1.37 TJ

93. (a) 329 kJ = 0.0914 kWh; (b) 956 kJ; (c) 3.24 kJ/K

95. (a) 38.0%; (b) 0.288 J/K; (c) 78.6 J

97. The water's entropy increases by 24.7 J/K.

99. 25.5 kJ/K; increases

INDEX

Note: Page numbers preceded by A indicate appendices; those preceded by M indicate the Math Tutorial.

A

B

C

D

F

G

H

I

J

K

L

M

N

O

P

S

T

W

X

Y

Z